COLLEGE PHYSICS

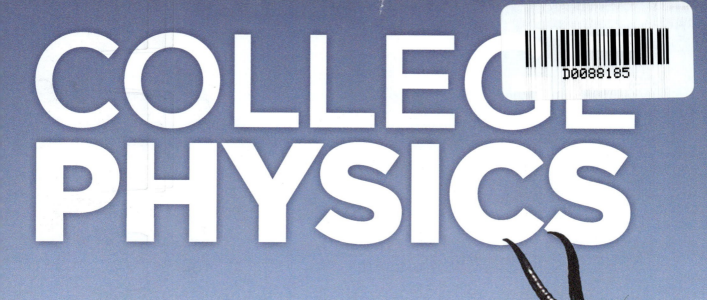

$$y = v_{0y}\, t - \tfrac{1}{2}\, g t^2$$

$$v_y = v_{0y} - g t$$

$$\vec{w} = m\vec{g}$$

THIRD EDITION

Roger A. Freedman

Todd G. Ruskell

Philip R. Kesten

David L. Tauck

macmillan learning

Austin • Boston • New York • Plymouth

Senior Vice President, STEM: Daryl Fox
Program Director, Physical Sciences: Brooke Suchomel, Andrew Dunaway
Program Manager: Lori Stover
Marketing Manager: Leah Christians
Executive Content Development Manager, STEM: Debbie Hardin
Development Editors: Barbara Price, Meg Rosenburg
Executive Project Manager, Content, STEM: Katrina Mangold
Editorial Project Manager: Lucinda Bingham
Director of Content, Physical Sciences: Heather Southerland
Content Team Lead: Josh Hebert
Associate Media Editor: Kevin Davidson
Editorial Assistant: George Hajjar
Marketing Assistant: Morgan Psiuk
Director of Content Management Enhancement: Tracey Kuehn
Senior Managing Editor: Lisa Kinne
Lead Content Project Manager: Kerry O'Shaughnessy
Senior Workflow Project Manager: Paul Rohloff
Director of Design, Content Management: Diana Blume
Design Services Manager: Natasha Wolfe
Cover Design Manager: John Callahan
Text Designer: Lumina Datamatics, Inc.
Art Manager: Matthew McAdams
Illustrations: Lumina Datamatics, Inc.
Director of Digital Production: Keri deManigold
Executive Media Project Manager: Chris Efstratiou
Permissions Manager: Michael McCarty
Photo Permissions Manager: Jennifer MacMillan
Photo Editors: Krystyna Borgen, Cheryl Du Bois
Production Supervisor: Robert Cherry
Composition: Lumina Datamatics, Inc.
Printing and Binding: King Printing Co., Inc.
Cover Image: CarGe/Getty Images

Library of Congress Control Number: 2020942143

ISBN-13: 978-1-319-25534-3
ISBN-10: 1-319-25534-5

In 1946, William Freeman founded W. H. Freeman and Company and published Linus Pauling's *General Chemistry*, which revolutionized the chemistry curriculum and established the prototype for a Freeman text. W. H. Freeman quickly became a publishing house where leading researchers can make significant contributions to mathematics and science. In 1996, W. H. Freeman joined Macmillan and we have since proudly continued the legacy of providing revolutionary, quality educational tools for teaching and learning in STEM.

ROGER:

To the memory of S/Sgt Ann Kazmierczak Freedman, WAC, and Pvt. Richard Freedman, AUS.

TODD:

To Susan and Allison, whose never-ending patience, love, and support made it possible, and Jody Rendall, who first introduced me to the wonderful world of physics.

PHIL:

To my parents for instilling in me a love of learning, to my wife for her unconditional support, and to my children for letting their kooky dad infuse so much of their lives with science.

DAVE:

To my parents, Bill and Jean, for showing me how to lead a wonderful life, and to my sister and friends, teachers and students for helping me do it.

Dear Students and Instructors,

Welcome to *College Physics*! We are excited to bring you this new approach to learning algebra-based physics. Grounded in real-world applications, balanced with rigor and thoroughness, and infused with enthusiasm for the subject, no other college physics text presents material in quite this way. To create this unique learning experience, we have assembled a unique author team that includes a trusted and highly successful textbook author, physicists who have spent years focusing on how students learn physics best, and a biology professor who brings a life scientist's perspective on what makes physics interesting to the students who take this course.

Learning research tells us that students learn best and remember more readily when they can connect new topics to previously learned material. *College Physics* presents a list of topics to review at the beginning of every chapter, and it also artfully integrates this information throughout the narrative, explaining the relevance of previously learned material to the topic at hand. In addition, the authors show students how what they learn in one chapter will be relevant to their studies in future chapters, providing students with everything they need to connect the dots.

Our innovations, coupled with a clear and engaging writing style, help students master tough physics concepts, acquire competency in logical reasoning, and practice the problem-solving skills they need to do well in this course and beyond — throughout their lives. Every chapter highlights biological connections and real-world applications that show students that physics is part of everyday life, with examples from both the natural world and today's technology. In addition, accompanying this book is our ground-breaking media program and assessment platform *Achieve*, which supports student learning every step of the way.

Our aim is to instill in students a deeper appreciation of physics — by showing how physics connects to their lives and their future careers, and by helping them succeed in the course. We hope you enjoy exploring physics and using this text!

Best Regards,

Roger, Todd, Phil, Dave

Transform every student's potential into success

with Freedman's *College Physics* and Achieve

Now available for the first time with Macmillan's new online learning platform Achieve, Freedman's *College Physics* makes it easy for instructors to support students by using best teaching practices in their algebra-based physics courses. With resources for before, during, and after class, students are engaged and supported at every step of the learning process.

Key Features of *College Physics*

- **Set Up, Solve, Reflect** 3-step problem-solving strategy is demonstrated in the many worked Examples in every chapter.
  - *Set Up* drawings help students visualize the problem.
  - *Solve* steps students through the required math and logical reasoning.
  - *Reflect* connects the worked example to a larger concept.
- **Word balloons** annotate figures and equations, highlighting important physics concepts and terms so that students can learn in easy-to-manage pieces.
- Every chapter includes numerous **biological examples** and **real-world physics applications**. End-of-chapter problems are tagged for **biology**, **medical**, **astronomy**, and **sports** applications, and **biological applications** in each chapter are listed in a separate index.

- **Got the Concept?** Questions at strategic points in the narrative allow students to check their conceptual understanding.
- **Watch Out!** boxes draw attention to common student misconceptions and correct them immediately.
- Outcome-based **learning objectives** provided as **chapter goals** guide students through the most important topics to understand.
- Every chapter section ends with a **Take-Home Message**, which summarizes key points from the reading and aligns with the **chapter goals**.
- Students are directed to relevant sections of the accompanying **Math Tutorial** in the Appendix at key junctures in each chapter, for extra help.
- Icons placed strategically in the margin direct students to online content for additional practice with concepts being discussed.

NEW ACHIEVE *COLLEGE PHYSICS* ONLINE HOMEWORK

- Resources to support learning **before** class, or for flipped classrooms, include our popular videos and assessment from FlipIt Physics.
- To support learning **during** class, In-Class Activities written by Roger Freedman have been fully adapted for implementation through our easy-to-use In-Class Activity Guides. iClicker questions and deep integration with Achieve also supports active participation during class time.
- For homework and study **after** class, Achieve features over 3000 unique homework questions, including questions with algorithmically generated values to enhance security. Every question in Achieve offers a hint, answer-specific feedback, and a fully worked solution.
- LearningCurve **adaptive quizzing** keeps students learning, whatever their level. With hints and redirects to remedial material, students are never "stuck."
- HTML5 **PhET simulations** from the University of Colorado at Boulder give students a visual understanding of concepts and illustrate cause-and-effect relationships. Linked tutorial

questions guide further quantitative exploration, while addressing specific problem-solving needs.

- **Interactive word balloon questions** in Achieve help students identify important physics concepts in key figures and equations.
- 250 **P'Cast** whiteboard-style videos, based on worked examples from the text, simulate the experience of watching an instructor walk through the steps of solving a problem.
- All worked examples from *College Physics* are available as a downloadable **Pocket Worked Example** for mobile devices.

END-OF-CHAPTER QUESTIONS

Every end-of-chapter question has been reviewed and authored to the rigorous Macmillan standards and are now are all available for assigning in Achieve. Questions have been evaluated by multiple subject matter experts for accuracy and clarity, ensuring that instructors have the freedom to construct the assignments they desire, with the confidence that students will receive the support they need from our expertly written and detailed feedback, solutions, and hints.

Achieve is the culmination of years of development work put toward creating the most powerful online learning tool for physics students. It houses all of our renowned assessments, multimedia assets, e-books, and instructor resources in a powerful new platform.

Achieve supports educators and students throughout the full range of instruction, including assets suitable for pre-class preparation, in-class active learning, and post-class study and assessment. The pairing of a powerful new platform with outstanding physics content provides an unrivaled learning experience.

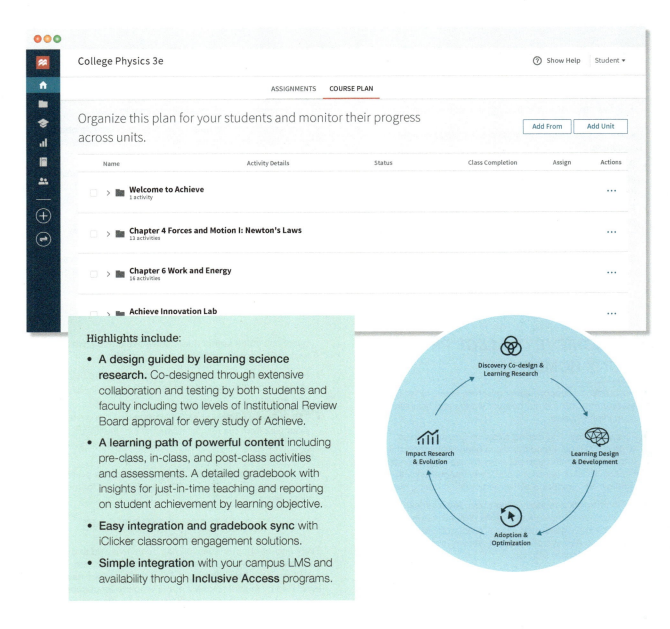

Highlights include:

- **A design guided by learning science research.** Co-designed through extensive collaboration and testing by both students and faculty including two levels of Institutional Review Board approval for every study of Achieve.

- **A learning path of powerful content** including pre-class, in-class, and post-class activities and assessments. A detailed gradebook with insights for just-in-time teaching and reporting on student achievement by learning objective.

- **Easy integration and gradebook sync** with iClicker classroom engagement solutions.

- **Simple integration** with your campus LMS and availability through **Inclusive Access** programs.

For more information or to sign up for a demonstration of Achieve, contact your local Macmillan representative or visit **macmillanlearning.com/achieve**

Engage Every Student

Achieve includes LearningCurve, a full learning path of content for preclass preparation, in-class active learning, and post-class engagement and assessment. It's the perfect solution for engaging all students in an inclusive teaching experience.

LearningCurve

LearningCurve's game-like quizzing motivates each student to engage with the course content, and reporting tools help teachers get a handle on what their class needs.

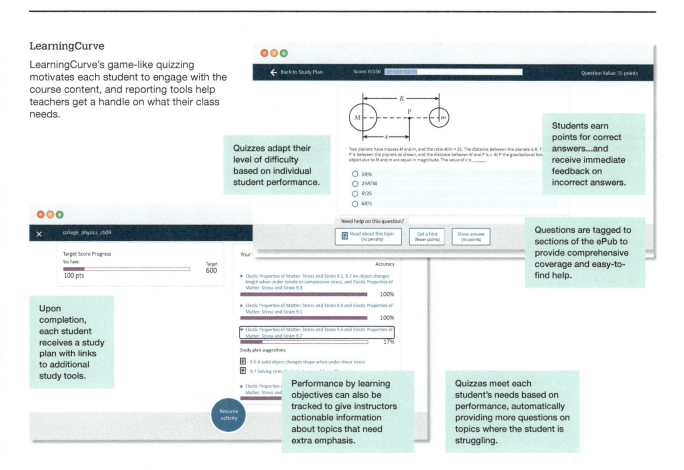

Quizzes adapt their level of difficulty based on individual student performance.

Students earn points for correct answers...and receive immediate feedback on incorrect answers.

Questions are tagged to sections of the ePub to provide comprehensive coverage and easy-to-find help.

Upon completion, each student receives a study plan with links to additional study tools.

Performance by learning objectives can also be tracked to give instructors actionable information about topics that need extra emphasis.

Quizzes meet each student's needs based on performance, automatically providing more questions on topics where the student is struggling.

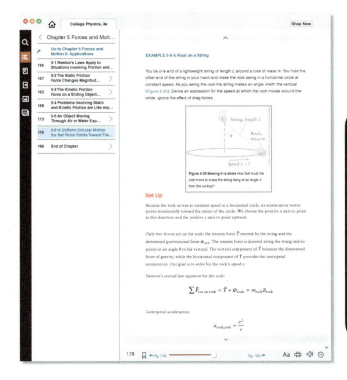

Now available as an ePub from **VitalSource Bookshelf**, the leading eTextbook platform that allows you to access course materials from your browser, download files to your desktop, and use on your mobile device.

- Students can print select pages, have the book read aloud to them, and search for key terms, phrases, and figures. Additionally, students can highlight, take notes, and create flashcards for studying.

- Uniquely suited to *College Physics'* successful pedagogical approach, the ePub allows students click on a reference to previously learned material in one chapter and go to the place in the book where the topic is fully explained, allowing students to review, build long-term memory, and tie it all together.

- The platform seamlessly integrates with Achieve, giving students the single sign-on experience that makes learning about physics all about physics, not about the software.

Making the Most out of Student Study Before Class, During Class, and After Class

Prelecture videos introduce students to core physics concepts before class, and active learning exercises in class provide hands-on exploration. Homework problems support learning by applying core concepts to assessments.

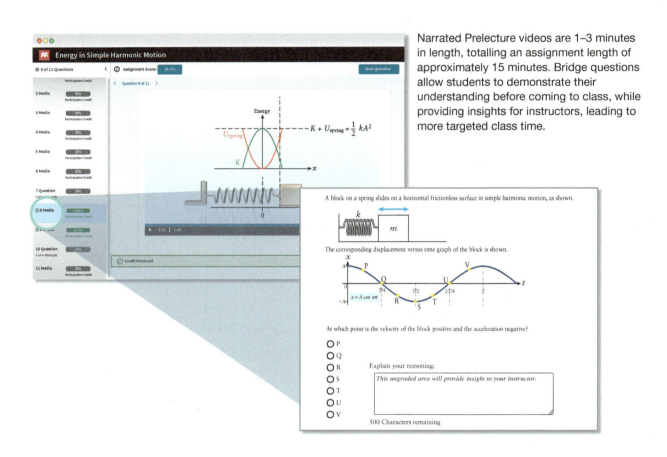

Narrated Prelecture videos are 1–3 minutes in length, totalling an assignment length of approximately 15 minutes. Bridge questions allow students to demonstrate their understanding before coming to class, while providing insights for instructors, leading to more targeted class time.

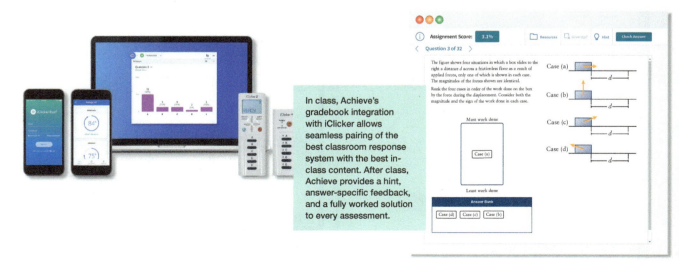

In class, Achieve's gradebook integration with iClicker allows seamless pairing of the best classroom response system with the best in-class content. After class, Achieve provides a hint, answer-specific feedback, and a fully worked solution to every assessment.

Powerful analytics, viewable in an elegant dashboard, offer instructors a window into student progress. Achieve gives you the insight to address students' weaknesses and misconceptions before they struggle on a test.

Reports provide powerful views of student performance and can be viewed in a number of different ways.

At-a-glance assignment level details let you know how far your students have gone so far and how quickly they're grasping the material.

Performance by Learning Objectives and can also be tracked to give instructors actionable information about topics that need extra emphasis.

Unit and subunit reports give quick performance overviews based on the way you've decided to group your assignments.

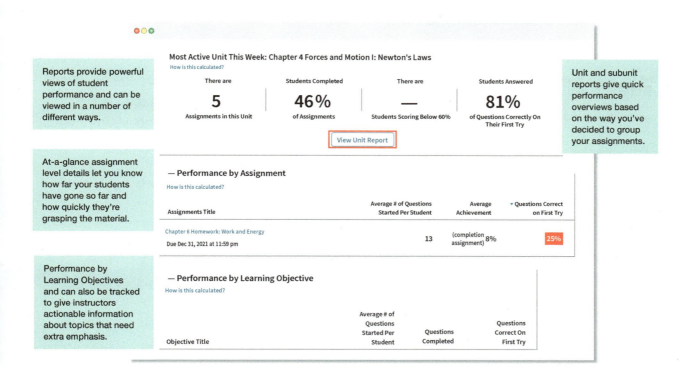

Insights give instructors information regarding student performance versus learning objectives.

Insights highlight the different assignments that address different learning objectives and show visualizations of student performance.

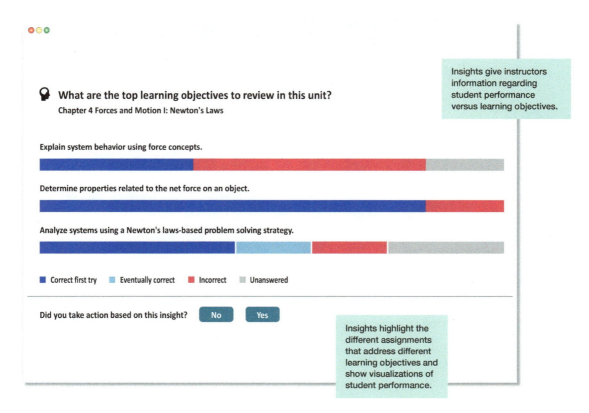

Problem Solving in *College Physics*
Setting a Strong Foundation

BREAKING DOWN CONCEPTS AND PROBLEMS TO MANAGEABLE SIZE

Word Balloons: Comic-style word balloons explain key equations, highlighting and defining each variable.

Set Up, Solve, Reflect: Worked examples contain every step of required math, while providing figures to help students visualize the problem. The Reflect section includes detail and context to connect the worked example to a larger concept.

I really like the annotations on the equations, it reinforces the idea that equations are shorthand for relationships between physical quantities that could be expressed in (a lot more) words.

–Juan Cabanela, *Minnesota State University, Moorhead*

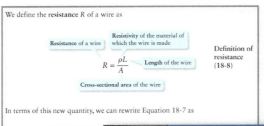

We define the **resistance** *R* of a wire as

Resistance of a wire   **Resistivity of the material of which the wire is made**

$$R = \frac{\rho L}{A}$$ **Length of the wire**

Definition of resistance (18-8)

Cross-sectional area of the wire

In terms of this new quantity, we can rewrite Equation 18-7 as

Potential difference

Current in th

BioMedical EXAMPLE 18-5   Resistance of a Potassium Channel

Using an instrument called a patch clamp, scientists are able to control the electrical potential difference across a tiny patch of cell membrane and measure the current through individual potassium channels. In one experiment a voltage of 0.120 V was applied across a patch of membrane, and as a result K^+ ions carried 6.60 pA of current (1 pA = 1 picoampere = 10^{-12} A). What is the resistance of this K^+ channel?

Set Up

Whether the mobile charges are negative electrons or positive ions, we can use Equation 18-9 to relate potential difference (voltage), current, and resistance.

Relationship among potential difference, current, and resistance:

$$V = iR \qquad (18\text{-}9)$$

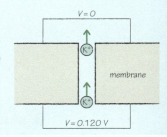

Solve

Rewrite Equation 18-9 to solve for the resistance.

From Equation 18-9,

$$R = \frac{V}{i}$$

The current is $i = 6.60$ pA $= 6.60 \times 10^{-12}$ A, so

$$R = \frac{0.120 \text{ V}}{6.60 \times 10^{-12} \text{ A}} = 1.82 \times 10^{10} \text{ } \Omega$$

Reflect

This is an *immensely* high resistance compared to the values found in circuits like that shown in Figure 18-6. The explanation is that a potassium channel is only about 10^{-9} m wide—so small that K^+ ions must pass through it in single file—and so the resistance to current flow through the channel is very high. There are other types of channels in cell membranes, and all typically have resistances in the range of 10^9 to 10^{11} Ω.

Although the current through a single potassium channel is very small, the electric field required to produce that current is tremendous. The thickness of the cell membrane is only about 7.5 nm = 7.5×10^{-9} m. Can you use Equation 18-5 to show that a potential difference of 0.120 V across this distance corresponds to an electric field magnitude of 1.6×10^7 V/m? (By comparison, the electric field inside the wires of a flashlight is only about 10 V/m.)

EXAMPLES FROM BIOLOGY AND REAL-WORLD APPLICATIONS

College Physics features relevant examples that connect physics to everyday life. Special emphasis has been placed on biological and biomedical applications, and a list of biological examples, tagged to each chapter, is provided in a separate index.

10-4 A solid object changes shape when under shear stress

BioMedical

When you run, play sports, or do any sort of physical activity, blood flow increases to your active tissues. The arteries need to expand or *dilate* to accommodate this increased blood flow. In addition to the dilation caused by increased blood pressure pushing outward against arterial walls, the deformation of endothelial cells that line the inside of the artery also leads to dilation of the vessel (Figure 10-11a, b). The force of blood pushing on the exposed surfaces of the cells deforms them in the direction of blood flow (Figure 10-11c). Because the cells are firmly attached to the arterial wall, the change in shape is neither a stretch nor a compression. Instead, the cells deform in a way similar to a cube of gelatin dessert being pushed parallel to the top face. In response to this deformation, the endothelial cells release nitric oxide gas. This gas diffuses to the muscle cells in the arterial wall and causes them to relax. As a result the artery dilates, thus relieving the stress on the endothelial cells (Figure 10-11d).

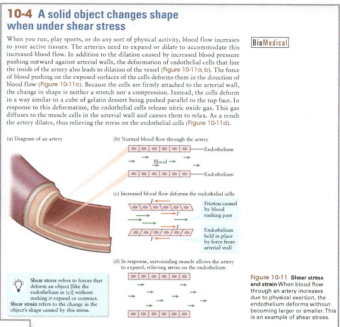

(a) Diagram of an artery

(b) Normal blood flow through the artery

(c) Increased blood flow deforms the endothelial cells

(d) In response, surrounding muscle allows the artery to expand, relieving stress on the endothelium

Shear stress refers to forces that deform an object [like the endothelium in (c)] without making it expand or contract. Shear strain refers to the change in the object's shape caused by this stress.

Figure 10-11 Shear stress and strain When blood flow through an artery increases due to physical exertion, the endothelium deforms without becoming larger or smaller. This is an example of shear stress.

BioMedical GOT THE CONCEPT? 12-12 Mosquitoes Listening for Each Other

The sounds made by male mosquitoes in flight are close to 650 Hz, while those made by females are close to 400 Hz. The antennae of mosquitoes differ from male to female (Figure 12-23): The natural frequency of a male's antennae is about 400 Hz, while that of the female's is about 200 Hz. Biologists suspect that some mosquitoes can detect the presence of others based on nerve signals generated when their antennae vibrate. Would you expect that (a) male mosquitoes can detect other males; (b) females can detect other females; (c) males can detect females; (d) females can detect males; (e) more than one of these; or (f) none of these?

Figure 12-23 Mosquito antennae Mosquitoes may be able to use their antennae to detect oscillations produced by other mosquitoes. What does the physics of resonance tell us about this?

Male mosquito

Female mosquito

Eric V. Grave/Science Source

> [*College Physics'*] biggest strength is its worked examples, which are always illustrated beautifully and always worked through the way an expert **student** would, without skipping steps or assumptions that are trivial to a physicist but not to the learner. It is a text that students will find is actually worth reading before class, rather than just for consulting later during their homework sessions.
>
> —Colin West, *University of Colorado, Boulder*

Contents

Biological Applications

Unique and fully integrated physiological, medical, and biological applications are found throughout the text and are indicated by a **BioMedical** icon. Below is a list of select biological applications organized by chapter section for easy reference.

Acknowledgments

Bringing a textbook into its third edition requires the coordinated effort of an enormous number of talented professionals. We are grateful for the dedicated support of our in-house team at W. H. Freeman; thank you for sustaining our concept and developing this book into a beautiful new edition.

We especially want to thank our development editors, Barbara Price and Meg Rosenburg, for their dedication to this book. We would also like to thank our Program Manager, Lori Stover, for encouraging us and leading our team; Editorial Project Manager Lucinda Bingham for keeping us on schedule; Content Director Heather Southerland, Executive Content Development Manager for STEM Debbie Hardin, Associate Media Editor Kevin Davidson, Executive Media Project Manager Chris Efstratiou, and Physics Content Team Lead Josh Hebert for their thoughtful contributions; and Editorial Assistant George Hajjar, Permissions Manager Jennifer MacMillan, and Senior Content Project Manager Kerry O'Shaughnessy for their patience and attention to detail. Of course, none of this would be possible without the support of Program Director Brooke Suchomel and Senior Vice President for STEM Daryl Fox. Special thanks also go to our talented marketing team, Marketing Manager Leah Christians and Marketing Assistant Morgan Psiuk for their insight and assistance.

Friends and Family

One of us (RAF) thanks his wife, Caroline Robillard, for her patience with the seemingly endless hours that went into preparing this textbook. I also thank my students at the University of California, Santa Barbara, for giving me the opportunity to test and refine new ideas for making physics more accessible. Go Gauchos!

One of us (TGR) thanks his wife, Susan, and daughter, Allison, for their limitless patience and understanding with the countless hours spent working on this book. I also thank my parents who showed me how to live a balanced life.

One of us (PRK) would like to acknowledge valuable and insightful conversations on physics and physics teaching with Richard Barber, John Birmingham, and J. Patrick Dishaw of Santa Clara University, and to offer these colleagues my gratitude. Finally, I offer my gratitude to my wife Kathy and my children Sam and Chloe for their unflagging support during the arduous process that led to the book you hold in your hands.

One of us (DLT) thanks his family and friends for accommodating my tight schedule during the years that writing this book consumed. I especially want to thank my parents, Bill and Jean, for their boundless encouragement and support, and for teaching me everything I've ever really needed to know—they've shown me how to live a good life, be happy, and age gently. I greatly appreciate my sister for encouraging me not to abandon a healthy lifestyle just to write a book. I also want to thank my nonbiological family, Holly and Geoff, for leading me to Sonoma County and for making Sebastopol feel like home.

Accuracy checker:
Jose Lozano, *Bradley University*

Student Workbook authors:
Perry Hilburn, *Gannon University*
Seong-Gon Kim, *Mississippi State University*
Garrett Yoder, *Eastern Kentucky University*
Linda McDonald, *North Park University*
Avishek Kumar, *Arizona State University*
Michael Dunham, *SUNY Fredonia*

Test Bank authors:
Anna Harlick, *University of Calgary*
Debashis Dasgupta, *University of Wisconsin at Milwaukee*
Elizabeth Holden, *University of Wisconsin at Platteville*

Questions and Problems editors:
David Abbott
Dante Gordon
Steven Mielke
Betsy Olson
Michael Scott
Felicia Tam

Third-Edition Reviewers
We would also like to thank the many colleagues who carefully reviewed chapters for us. Their insightful comments significantly improved our book.

Edward Ackad, *Southern Illinois University, Edwardsville*
Fariba Ansari, *El Paso Community College*
Tetyana Antimirova, *Ryerson University*
Vasudeva Aravind, *Clarion University*
Mohammad Arif, *Northern Virginia Community College*
Mircea Atanasiu, *Mount Royal University*
Gilbert Ayuk, *Bakersfield College*
Bernhard Bach, *University Nevada, Reno*
Yiyan Bai, *Houston Community College*
Harinder Singh Bawa, *California State University, Fresno*
L. Y. Beaulieu, *Memorial University*
Christine Berven, *University of Idaho*
Dave Z. Besson, *University of Kansas*
Daniel Bissonnette, *Thompson Rivers University*
Steven Blusk, *Syracuse University*
Ken Bolland, *The Ohio State University*
Derrick Boucher, *Florida Gulf Coast University*
John B. Bulman, *Loyola Marymount University*

John B. Buncher, *North Dakota State University*
Lior M. Burko, *Georgia Gwinnett College*
Temple Burling, *Carthage College*
Juan Cabanela, *Minnesota State University, Moorhead*
Alex Chediak, *California Baptist University*
Orion Ciftja, *Prairie View Texas A&M University*
Leon Cole, *Christopher Newport University*
Joshua Colwell, *University of Central Florida*
Kazuo Cottrell, *Lamoille Union High School*
Jose D'Arruda, *University of North Carolina, Pembroke*
Debashis Dasgupta, *University of Wisconsin, Milwaukee*
Alakabha Datta, *University of Mississippi*
Ethan Deneault, *University of Tampa*
Radovan Dermisek, *Indiana University, Bloomington*
Sharvil Desai, *The Ohio State University*
Ethan Dolle, *Northern Arizona University*
Steven Doty, *Denison University*
James Dove, *Metropolitan State University of Denver*
Diana Driscoll, *Case Western Reserve University*
Donald D. Driscoll, Jr., *Kent State University*
Steve Drum, *Orange Coast College*
Michael M. Dunham, *SUNY Fredonia*
Davene Eyres, *North Seattle College*
Cassandra Fallscheer, *Central Washington University*
Xiaojuan Fan, *Marshall University*
AmirAli Farokhniaee, *University of Connecticut*
Nancy Forde, *Simon Fraser University*
Eduardo Galiano-Riveros, *Laurentian University*
Sambandamurthy Ganapathy, *SUNY Buffalo*
Delena Bell Gatch, *Georgia Southern University*
Bryan Gibbs, *Richland College*
Jennifer Gimmell, *College of DuPage*
Svetlana Gladycheva, *Towson University*
Romulus Godang, *University of South Alabama*
Juan Gomez, *Abraham Baldwin Agricultural College*
Chien-Wei Han, *PIMA Community College*
Michael Harder, *British Columbia Institute of Technology*
Erik Helgren, *California State University, East Bay*
Inge Heyer, *Loyola University Maryland*
Pei-Chun Ho, *California State University, Fresno*
Elizabeth Holden, *University of Wisconsin, Madison*
Benne Willem Holwerda, *University of Louisville*
Maria Hossu, *North Lake College*
Meenu Jain, *Middlesex County College*
Timothy A. Jenkins, *University of Oregon*
Adam G. Jensen, *University of Nebraska at Kearney*
Matthew C. Jewell, *University of Wisconsin, Eau Claire*
Yong Joe, *Ball State University*
Adam Johnston, *Weber State University*
R. H. Kadala, *Hawaii Pacific University*
Preeti Kapoor, *College of DuPage*
Wafaa Khattou, *Valencia College*
Kevin R. Kimberlin, *Bradley University*
David Klinger, *Loyola University, Chicago*
Armen Kocharian, *California State University, Los Angeles*
Elena Kuchina, *Thomas Nelson Community College*
Samuel Levenson, *William Rainey Harper College*
Ran Li, *Kent State University*
Ramon Lopez, *University of Texas, Arlington*
Jose Lozano, *Bradley University*
Kingshuk Majumdar, *Grand Valley State University*
Mark E. Mattson, *James Madison University*

Will McElgin, *Louisiana State University & A&M College*
Rahul Mehta, *University of Central Arkansas*
Vesna Milosevic-Zdjelar, *University of Winnipeg*
Mark Morgan-Tracy, *University of New Mexico*
Raman Nambudripad, *Massachusetts Bay Community College*
Kriton Papavasiliou, *Virginia Tech*
Marina Papenkova, *East Los Angeles College*
Galen Pickett, *California State University, Los Angeles*
Michael Pravica, *University of Nevada, Las Vegas*
Vahid Rezania, *Grant MacEwan University*
Desiree Rose, *California State University, Los Angeles*
Bibhudutta Rout, *University of North Texas*
Ben Rybolt, *Kennesaw State University*
Jeffrey Sabby, *Southern Illinois University, Edwardsville*
Rihab Sawah, *Lincoln Land Community College*
Jeremy Schmit, *Kansas State University*
Brian Schuft, *North Carolina A&T State University*
Jerry Shakov, *Tulane University*
Andreas Shalchi, *University of Manitoba*
Leah B. Sherman, *Appalachian State University*
Christopher Sirola, *University of Southern Mississippi*
Michael T. Smitka, *The Pennsylvania State University*
Gina A. Sorci, *University of Louisiana at Lafayette*
Bryan Spring, *Northeastern University*
David Sturm, *University of Maine*
Maxim Sukharev, *Arizona State University Polytechnic*
Cynthia Swift, *Santiago Canyon College*
Tatsu Takeuchi, *Virginia Tech*
Fiorella Terenzi, *Florida International University*
Dmitri Tsybychev, *SUNY Stony Brook*
Gregory Tucker, *Brown University*
A. R. Umar, *University of South Florida*
Diego Valente, *University of Connecticut*
Matthew D. Vannette, *Saginaw Valley State University*
Melissa Vigil, *Marquette University*
Chuji Wang, *Mississippi State University*
Laura Weinkauf, *Jacksonville State University*
Colin G. West, *University of Colorado, Boulder*
James Wetzel, *Augustana College*
Lee Widmer, *Xavier University*
Luc T. Wille, *Florida Atlantic University*
Bill Williams, *Hudson Valley Community College*
John Wilson, *Sam Houston State University*
Brian Woodahl, *Indiana University–Purdue University, Indianapolis*
Frankie Wood-Black, *Northern Oklahoma College*
Alexander Wurm, *Western New England University*
Garett Yoder, *Eastern Kentucky University*
Todd Young, *Wayne State College*
Sharon T. Zane, *University of Miami*
Ulrich Zurcher, *Cleveland State University*

Second-Edition Reviewers

In addition, we would like to acknowledge the many reviewers who reviewed the previous edition of this text, as their contributions live on in this edition.

Miah Muhammad Adel, *University of Arkansas at Pine Bluff*
Mikhail M. Agrest, *The Citadel*
Vasudeva Rao Aravind, *Clarion University of Pennsylvania*
Yiyan Bai, *Houston Community College*

E. C. Behrman, *Wichita State University*
Antonia Bennie-George, *Green River College*
Ken Bolland, *The Ohio State University*
Matthew Joseph Bradley, *Santa Rosa Junior College*
Matteo Broccio, *University of Pittsburgh*
Daniel J. Costantino, *The Pennsylvania State University*
Adam Davis, *Wayne State College*
Sharvil Desai, *The Ohio State University*
Eric Deyo, *Fort Hays State University*
Diana I. Driscoll, *Case Western Reserve University*
Davene Eyres, *North Seattle College*
William Falls, *Erie Community College*
Sambandamurthy Ganapathy, *SUNY Buffalo*
Vladimir Gasparyan, *California State University, Bakersfield*
Frank Gerlitz, *Washtenaw Community College*
Svetlana Gladycheva, *Towson University*
Romulus Godang, *University of South Alabama*
Javier Gomez, *The Citadel*
Rick Goulding, *Memorial University of Newfoundland*
Ania Harlick, *Memorial University of Newfoundland*
Erik Helgren, *California State University, East Bay*
Perry G. Hillburn, *Gannon University*
Zdeslav Hrepic, *Columbus State University*
Patrick Huth, *Community College of Allegheny County*
Matthew Jewell, *University of Wisconsin, Eau Claire*
Wafaa Khattou, *Valencia College*
Seong-Gon Kim, *Mississippi State University*
Patrick Koehn, *Eastern Michigan University*
Ameya S. Kolarkar, *Auburn University*
Maja Krcmar, *Grand Valley State University*
Elena Kuchina, *Thomas Nelson Community College*
Avishek Kumar, *Arizona State University*
Chunfei Li, *Clarion University of Pennsylvania*
Jose Lozano, *Bradley University*

Dan MacIsaac, *SUNY Buffalo State College*
Linda McDonald, *North Park University*
Francis Mensah, *Virginia Union University*
Victor Migenes, *Texas Southern University*
Krishna Mukherjee, *Slippery Rock University of Pennsylvania*
Rumiana Nikolova-Genov, *College of DuPage*
Moses Ntam, *Tuskegee University*
Martin O. Okafor, *Georgia Perimeter College*
Gabriela Petculescu, *University of Louisiana at Lafayette*
Yuriy Pinelis, *University of Houston, Downtown*
Sulakshana Plumley, *Community College of Allegheny Count*
Lawrence Rees, *Brigham Young University*
David Richardson, *Northwest Missouri State University*
Carlos Roldan, *Central Piedmont Community College*
Jeffrey Sabby, *Southern Illinois University Edwardsville*
Arun Saha, *Albany State University*
Haiduke Sarafian, *The Pennsylvania State University*
Katrin Schenk, *Randolph College*
Surajit Sen, *SUNY Buffalo*
Jerry Shakov, *Tulane University*
Douglas Sharman, *San Jose State University*
Ananda Shastri, *Minnesota State University, Moorhead*
Marllin L. Simon, *Auburn University*
Stanley J. Sobolewski, *Indiana University of Pennsylvania*
Erin C. Sutherland, *Kennesaw State University*
Sarah F. Tebbens, *Wright State University*
Fiorella Terenzi, *Florida International University*
Dmitri Tsybychev, *Stony Brook University*
Laura Whitlock, *Georgia Perimeter College*
Pushpa Wijesinghe, *Arizona State University*
Fengyuan Yang, *The Ohio State University*
Garett Yoder, *Eastern Kentucky University*
Jiang Yu, *Fitchburg State University*
Ulrich Zurcher, *Cleveland State University*

About the Authors

Christine Pang

Roger Freedman This groundbreaking text boasts an exceptionally strong writing team that is uniquely qualified to write a college physics textbook. The *College Physics* author team is led by Roger Freedman, an accomplished textbook author of such best-selling titles as *Universe* (W. H. Freeman), *Investigating Astronomy* (W. H. Freeman), and *University Physics* (Pearson). Dr. Freedman is a teaching professor in physics at the University of California, Santa Barbara. He was an undergraduate at the University of California campuses in San Diego and Los Angeles, and did his doctoral research in theoretical nuclear physics at Stanford University. He came to UCSB in 1981 after 3 years of teaching and doing research at the University of Washington. At UCSB, Dr. Freedman has taught in both the Department of Physics and the College of Creative Studies, a branch of the university intended for highly gifted and motivated undergraduates. He has been a pioneer in the use of classroom response systems and the "flipped" classroom model at UCSB. Roger holds a commercial pilot's license and is academic advisor to UCSB Women's and Men's Rowing. He was an early organizer of the San Diego Comic-Con, now the world's largest popular culture convention, and today is a member of the board of directors of San Diego Comic Fest, an annual convention that recaptures the spirit of Comic-Con in its early years.

Mark Ramirez

Todd Ruskell As a Teaching Professor of Physics at the Colorado School of Mines, Todd Ruskell focuses on teaching at the introductory level and continually develops more effective ways to help students learn. He was instrumental in developing and implementing Studio Physics in the introductory physics courses at Mines, and has recently turned his attention to methods of bringing the "studio mindset" to physics courses taught online. Dr. Ruskell brings his experience in improving students' conceptual understanding to the text, as well as a strong liberal arts perspective. Dr. Ruskell's love of physics began with a BA in physics from Lawrence University in Appleton, Wisconsin. He went on to receive an MS and a PhD in optical sciences from the University of Arizona. He has received awards for teaching excellence, including Colorado School of Mines' Alumni Teaching Award.

Philip Kesten

Philip Kesten, Associate Professor of Physics and Associate Vice Provost for Undergraduate Studies at Santa Clara University, holds a BS in physics from the Massachusetts Institute of Technology and received his PhD in high-energy particle physics from the University of Michigan. Since joining the Santa Clara faculty in 1990, Dr. Kesten has also served as Chair of Physics, Faculty Director of the ATOM and da Vinci Residential Learning Communities, and Director of the Ricard Memorial Observatory. He has received awards for teaching excellence and curriculum innovation, was Santa Clara's Faculty Development Professor for 2004–2005, and was named the California Professor of the Year in 2005 by the Carnegie Foundation for the Advancement of Education. Dr. Kesten is co-founder of Docutek (a SirsiDynix Company), an Internet software company, and has served as the Senior Editor for *Modern Dad*, a newsstand magazine.

Philip Kesten

David L. Tauck Unlike any other physics text on the market, *College Physics* includes a physiologist as a primary author. David Tauck, Associate Professor of Biology, holds both a BA in biology and an MA in Spanish from Middlebury College. He earned his PhD in physiology at Duke University and completed postdoctoral fellowships at Stanford University and Harvard University in anesthesia and neuroscience, respectively. Since joining the Santa Clara University faculty in 1987, he has served as Chair of the Biology Department, the College Committee on Rank and Tenure, and the Institutional Animal Care and Use Committee; he has also served as president of the local chapter of Phi Beta Kappa. Dr. Tauck currently serves as the Faculty Director in Residence of the da Vinci Residential Learning Community.

Introduction to Physics

In this chapter, your goals are to:

- (1-1) Explain the roles that concepts and equations play in physics.
- (1-2) Describe three key steps in solving any physics problem.
- (1-3) Identify the fundamental units used for measuring physical quantities and convert from one set of units to another.
- (1-4) Use significant figures in calculations.
- (1-5) Use dimensional analysis to check algebraic results.

To master this chapter, you should review:

- The principles of algebra and how to solve an equation for a particular variable.

What do you think?

The size of a hummingbird's head is closest to which of these lengths?
(a) 1 micrometer (1 μm);
(b) 1 millimeter (1 mm);
(c) 1 centimeter (1 cm);
(d) 1 meter (1 m).

The answer to the *What do you think?* question can be found at the end of the chapter.

1-1 Physicists use both words and equations to describe the natural world

BioMedical

A hummingbird is a beautiful sight to anyone. As seen through the eyes of a physicist, however, a hummingbird also illustrates several physical principles at work. The hummingbird's beating wings push downward on the air, which causes the air to push upward in response and keep the bird aloft. The flow of blood through the hummingbird's body is governed by the density of the blood and the interplay of pressure and friction in the blood vessels. Light rays bend as they enter the bird's eye, forming an image on the retina and providing the hummingbird with vision. And the interference of light waves reflecting off the feathers at the bird's throat gives them their characteristic iridescence. To truly understand hummingbirds or any other living organism, including the human body, we need to understand the principles of physics. The purpose of this book is to help you gain that understanding.

The Polish-French physicist Marie Skłodowska Curie (the only person ever to win the Nobel Prize in both physics and chemistry) once said that "I was taught that the way of progress was neither swift nor easy." You will find that making progress in your study of physics will take time and will often be challenging. But the time and effort spent will be rewarding if you go about your study properly. For many students, one of the greatest challenges in making progress in physics is that its key concepts are expressed not just in words but also in terms of mathematics.

Figure 1-1 $E = mc^2$ When you burn natural gas in a kitchen range, you are converting some of the rest energy of methane and oxygen into light and heat.

Einstein's equation for rest energy
(1-1)

There's a very good reason physicists use mathematics so extensively: An equation can express in a compact way a concept that might take several sentences to write out in terms of words. But if you're just beginning your study of physics, the sheer number of equations you'll find in this book (or any physics textbook) may be daunting. The key is to recognize that not all of these equations are equally important. A much smaller set of *fundamental* equations expresses the key ideas of physics. By starting with these fundamental equations and combining them using the rules of algebra, you'll be able to solve all of the physics problems that you encounter in this book.

Here's an example of how a very simple equation can be shorthand for a remarkably profound concept. Consider the German-American physicist Albert Einstein's famous equation

$$E = mc^2$$

In this equation E represents energy, m represents mass, and c represents the speed at which light travels in a vacuum. But what the equation *means* is something deeper. In this book we'll often write fundamental equations with explanations attached, such as this one:

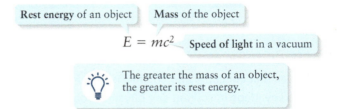

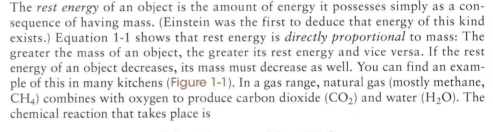

$$E = mc^2$$

The rest energy of an object is the amount of energy it possesses simply as a consequence of having mass. (Einstein was the first to deduce that energy of this kind exists.) Equation 1-1 shows that rest energy is *directly proportional* to mass: The greater the mass of an object, the greater its rest energy and vice versa. If the rest energy of an object decreases, its mass must decrease as well. You can find an example of this in many kitchens (Figure 1-1). In a gas range, natural gas (mostly methane, CH_4) combines with oxygen to produce carbon dioxide (CO_2) and water (H_2O). The chemical reaction that takes place is

$$CH_4 + 2O_2 \longrightarrow CO_2 + 2H_2O$$

This process converts some of the rest energy of the reactants (a CH_4 molecule and two O_2 molecules) into heat and light. So the combined rest energy of the products (a CO_2 molecule and two H_2O molecules) is less than the rest energy of the reactants. Equation 1-1 then makes the remarkable statement that the combined *mass* of the product CO_2 and H_2O molecules must also be slightly less than the combined mass of the CH_4 and O_2 molecules present before the reaction. Sensitive measurements reveal this statement to be true. Equation 1-1 itself isn't the important part of this story: It's the *concept* behind this equation that's truly important. When we come across an important concept like this one, or a situation that requires special care, we'll put it in a *Watch Out!* box like the one on this page.

$\sqrt{x}$ **See the Math Tutorial for more information on equations.**

WATCH OUT! **You need to be careful.**

When we come across a situation that requires special care, we'll put it in a *Watch Out!* box like this one.

TAKE-HOME MESSAGE FOR Section 1-1

✔ Like all other sciences, physics is a collection of interlinked concepts.

✔ The equations of physics help us summarize complex ideas in mathematical shorthand.

✔ It's essential to understand the concepts behind each fundamental equation.

1-2 Success in physics requires well-developed problem-solving skills

The process of learning physics involves learning to solve problems. Physics problems usually can't be done simply by selecting one special equation and plugging in values. Instead, you need to apply a consistent strategy for solving physics problems of all kinds. The strategy that we recommend, and that we'll use for all of the worked examples in this book, involves three steps: *Set Up, Solve,* and *Reflect.*

EXAMPLE 1-1 Strategy for Solving Problems

Set Up

At the beginning of any problem, ask yourself these questions: What is the problem asking for? Which quantities are known, and which are unknowns whose values you need to determine to solve the problem? What concepts will help solve the problem, and which equations should you use to apply these principles? Once you've answered these questions, you'll have a good idea of how to get started with solving the problem. Then do the following:

Draw a picture. Just drawing a simple picture will help you visualize what's going on in the problem. A good problem-solving picture should capture the motion, the process, the geometry, or whatever else defines the problem. You don't need to be an artist; it's perfectly okay to represent most complex objects as squares, circles, or even dots.

Label all quantities. It makes sense to label all quantities such as length, speed, and temperature right on your picture. But be careful! First, many problems have more than one of a given kind of variable. For example, if two objects are moving at different speeds, but you use the same symbol v for the speeds of both objects, confusion is bound to result. Subscripts are useful in such a case, for example, v_A and v_B for the speeds of the two objects. Also, don't hesitate to label lengths, velocities, and other quantities even if you haven't been given a value for them. You might be able to figure out the value of one of these quantities and use it later in the problem, or you might gain an insight that will guide you to a solution.

Look for connections. Often the best strategy for solving a physics problem is to look for connections between quantities. These connections could be relationships between two or more variables or a statement that a certain quantity has the same value for two objects. These connections will often enable you either to eliminate variables or to find a value for an unknown variable. Always be on the lookout for connections such as these when starting a problem.

Solve

Once you've selected the concepts to use and chosen the appropriate equations, finding the answer is often just an exercise in mathematics. In the worked examples, we'll show many intermediate steps as we describe our reasoning and thought processes. We recommend that you do the same in your own work. Trying to do two or more steps of a calculation in your head is an invitation to making careless mistakes!

Reflect

Once you get the answer, it's essential that you *review* the answer to see what it tells you. If the problem calls for a numerical answer, ask yourself, "Do the number and the units make sense?" For example, if you've been asked to calculate the diameter of Earth and you come up with 40 centimeters, or if you come up with an answer in kilograms (the unit of mass), it's time to go back and check your calculations. If your answer is a formula rather than a single number, check that what the formula says is reasonable. As an example, if you've found a formula that relates the weight of a cat to its volume, your answer should show that a heavier cat has a larger volume than a lightweight one. If your answer doesn't, there's an error in your calculations that you need to find and fix.

Learning physics will be easier if you study all of the worked examples in this book, and if you get in the habit of following the same steps of *Set Up, Solve,* and *Reflect* in your own problem solving.

TAKE-HOME MESSAGE FOR Section 1-2

✔ Success in physics requires having a methodical approach to problem solving.

✔ When approaching any worked example, follow these three steps: **Set Up** a problem by selecting the appropriate

concepts and equations, **Solve** by working through the mathematics, and **Reflect** on the results to see what they mean.

1-3 Measurements in physics are based on standard units of time, length, mass, and other quantities

Our discussion of physics concepts and equations may give the impression that physics is just about theories. But at its heart physics is an *experimental* science based on measurements of the natural world. The concepts and equations that physicists use are based on these experimental measurements. Here's how the Scots-Irish physicist Lord Kelvin described the importance of measurement: "If you can measure that of which you speak, and can express it by a number, you know something of your subject; but if you cannot measure it, your knowledge is unsatisfactory."

Any measurement requires a standard of comparison. To compare the heights of two children, for example, you might stand them side by side and use one child's height as a reference for the other. A more reliable method would be to use a vertical rod with equally spaced markings (in other words, a ruler) that acts as a standard of length.

In physics, three fundamental quantities for which it is essential to have standards are *time, length,* and *mass*. The standards, or **units,** for these three quantities are the *second,* the *meter,* and the *kilogram,* respectively, and the system of units based on these quantities is called the **Système International,** or **SI** for short. Table 1-1 lists all of the fundamental SI quantities, including others that we'll encounter in later chapters.

The definition of the **second** (abbreviated s) is based on the properties of the cesium atom. When such atoms are excited, they emit radio waves of a very definite frequency. The second is defined to be the amount of time required for cesium to emit 9,192,631,770 complete cycles of these waves. This may seem like a bizarre way to measure time, but modern radio technology is highly refined, and it's relatively easy to prepare a gas of cesium atoms with the right properties. Hence a physicist anywhere in the world could readily repeat these measurements.

In the past, objects such as metal bars were used as standards of length. These were not very good standards because their length changes slightly but measurably when the temperature changes. Today light waves are used to define the **meter** (abbreviated m), the SI standard of length. The speed of light in a vacuum is defined to be precisely 299,792,458 meters per second, so the meter is the distance that light travels in 1/299,792,458 of a second. This is a robust and repeatable standard because light in a vacuum travels at the same speed anywhere on Earth and, as best we can tell, everywhere in the universe. (We specify a vacuum because light travels more slowly in a material such as glass, water, or even air.)

Mass is a measure of the amount of material in an object. A feather has less mass than a baseball, which in turn has less mass than a cinder block. Prior to 2019, the standard for the SI unit of mass, the **kilogram** (abbreviated kg), was the mass of a special cylinder of platinum-iridium alloy kept in a repository near Paris, France (Figure 1-2). Not only was this standard inconvenient, it was unsatisfactory because a small number of atoms were scraped off the cylinder whenever it was handled, so repeated measurements of its mass gave different values. Since 2019, physicists have defined the kilogram in terms of a fundamental constant of nature known as *Planck's constant.* (We'll discuss the value

TABLE 1-1	**Fundamental Quantities and Their SI Units**	
Quantity	**Unit**	**Abbreviation**
time	second	s
length	meter	m
mass	kilogram	kg
temperature	kelvin	K
electric charge	coulomb	C
amount of substance	mole	mol
luminous intensity	candela	cd

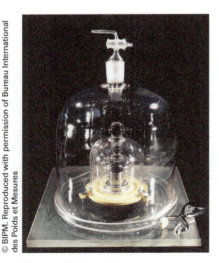

Figure 1-2 An old-fashioned mass standard Prior to 2019, 1 kilogram was defined to be equal to the mass of this special cylinder, stored at the International Bureau of Weights and Measures in France. Today the kilogram is defined in terms of a fundamental constant of nature called Planck's constant.

of Planck's constant later in this section, and see its physical significance in Chapter 22.) Because this constant has the same value everywhere, a physicist will measure its value to be the same no matter where their laboratory is located and can use this to determine the value of the kilogram. (The *gram*, equal to 1/1000 of a kilogram, is not a fundamental SI unit.)

WATCH OUT! Mass and weight are different quantities.

! It's important not to confuse *mass* and *weight*. An object's mass tells you how much material that object contains; its weight tells you how strongly gravity pulls on that object's material. Consider a person who weighs 110 pounds on Earth, corresponding to a mass of 50 kilograms. Gravity is only about one-sixth as strong on the Moon as it is on Earth, so on the Moon this person would weigh only one-sixth of 110 pounds, or about 18 pounds. But that person's *mass* of 50 kilograms is the *same* on the Moon; wherever you go in the universe, you take all of your material along with you. We'll explore the relationship between mass and weight in Chapter 4.

Units for many other physical quantities can be derived from the fundamental units of time, length, and mass. For example, speed is measured in meters per second (m/s), and weight is measured in kilograms times meters per second per second, or kilogram-meters per second squared ($kg \cdot m/s^2$). Some derived quantities are used so frequently that physicists have named a new unit to represent them. For example, the SI unit of weight is the newton (N), equal to $1\ kg \cdot m/s^2$, or about 0.225 pounds.

When we begin our study of heat and temperature in Chapter 14, we will need to introduce another unit called the *kelvin*, which is the fundamental unit of temperature. We'll also need another unit called the *coulomb*, a unit of electric charge, when we begin our study of electricity in Chapter 16. Until then, all of the units we'll work with will be based on seconds, meters, and kilograms.

Unit Conversions

Besides SI units, a variety of other units are used for time, length, and mass. We'll see that it's straightforward to do **unit conversions** between different units for the same quantity.

In addition to the second, some other common units of time include the following:

$$1 \text{ minute (min)} = 60 \text{ s}$$
$$1 \text{ hour (h)} = 3600 \text{ s}$$
$$1 \text{ day (d)} = 86,400 \text{ s}$$

Some other units of length that are handy when discussing sizes and distances include millimeters (mm), centimeters (cm), and kilometers (km). These units of length are related to the meter as follows:

$$1 \text{ mm} = 0.001 \text{ m}$$
$$1 \text{ cm} = 0.01 \text{ m}$$
$$1 \text{ km} = 1000 \text{ m}$$

Although the English system of inches (in), feet (ft), and miles (mi) is much older than the SI, today the English system is actually based on the SI system: The inch is *defined* to be exactly 2.54 cm. A useful set of conversions is as follows:

$$1 \text{ in} = 2.54 \text{ cm}$$
$$1 \text{ ft} = 0.3048 \text{ m}$$
$$1 \text{ mi} = 1.609 \text{ km}$$

You'll find more conversion factors in Appendix A.

We can use conversions such as these to convert any quantity from one set of units to another. For example, the glass skyscraper in London, England, called the Shard is 1016 feet tall (Figure 1-3). How can we convert this height to meters? Figure 1-4 shows the technique. Just like numbers, you can multiply units together or divide one by the other, and units cancel if the same unit appears in the numerator and denominator.

Figure 1-3 From feet to meters
The tallest building in the United Kingdom, the Shard, stands 1016 feet tall. How tall is this in meters?

Figure 1-4 Converting units To convert units, remember that a quantity does not change if you multiply by 1. The only thing to remember is to write 1 in an appropriate way.

(1) To convert units—for example, to convert a distance in feet (ft) to a distance in meters (m)—multiply by 1 in the appropriate form. Multiplying a number by 1 does not change the number.

$$1016 \text{ ft} = 1016 \text{ ft} \times 1$$

(2) To find the appropriate form of 1, begin with the conversion between the two units...

(3) ...and divide one side by the other. (If you divide a number by an equal number, the result is always equal to 1.) This gives two ways to express the conversion between feet and meters.

$$1 \text{ ft} = 0.3048 \text{ m, so} \quad \frac{1 \text{ ft}}{0.3048 \text{ m}} = 1 \text{ and } \frac{0.3048 \text{ m}}{1 \text{ ft}} = 1$$

INCORRECT APPROACH:

(4) If you try multiplying 1016 ft by the first form of 1...

(5) ...the result is not in the desired units.

$$1016 \text{ ft} = 1016 \text{ ft} \times 1 = 1016 \text{ ft} \times \frac{1 \text{ ft}}{0.3048 \text{ m}} = 3333 \frac{\text{ft}^2}{\text{m}} \quad ✗$$

CORRECT APPROACH:

(6) If instead you multiply 1016 ft by the second form of 1...

(7) ...the units of ft cancel and the result is in m, as desired.

$$1016 \text{ ft} = 1016 \text{ ft} \times 1 = 1016 \cancel{\text{ft}} \times \frac{0.3048 \text{ m}}{1 \cancel{\text{ft}}} = 309.7 \text{ m} \quad \text{OK}$$

To convert a quantity from one set of units to another, multiply the quantity by 1 (equal to the ratio of the two units). Choose the ratio that enables the unwanted units to cancel out.

WATCH OUT! Take care with conversions that have multiple steps.

Some conversions cannot be easily carried out in a single step. As an illustration, let's find the number of seconds in a day. Since 1 day (d) equals 24 hours (h), 1 h equals 60 minutes (min), and 1 min equals 60 seconds (s), we convert units three times:

Notice that we were careful to check the cancellation of each pair of like units (d and d, h and h, min and min). We strongly recommend that you carry out conversions such as this one in an equally methodical way.

$$1 \text{ d} = 1 \cancel{\text{d}} \left(\frac{24 \cancel{\text{h}}}{1 \cancel{\text{d}}} \right) \left(\frac{60 \cancel{\text{min}}}{1 \cancel{\text{h}}} \right) \left(\frac{60 \text{ s}}{1 \cancel{\text{min}}} \right) = 86,400 \text{ s}$$

BioMedical EXAMPLE 1-2 Converting Units: Speed

The world's fastest bird, the white-throated needletail (*Hirundapus caudacutus*), can move at speeds up to 47.0 m/s in level flight. (Falcons can go even faster, but they must dive to do so.) What is this speed in km/h? In mi/h?

Set Up

This is a problem in converting units. To find the speed in km/h, we will have to convert meters (m) to kilometers (km), and seconds (s) to hours (h). To then find the speed in mi/h, we will have to convert km to miles (mi). In each stage of the conversion, we'll follow the procedure shown in Figure 1-4.

Conversions:

$$1 \text{ h} = 3600 \text{ s}$$
$$1 \text{ km} = 1000 \text{ m}$$
$$1 \text{ mi} = 1.609 \text{ km}$$

Solve

We write the first two conversions above as a ratio equal to 1, such as (1 km)/(1000 m) = 1. We then multiply each of these by 47.0 m/s to find the speed in km/h. Note that we must set up the ratios so that the units of meters (m) and seconds (s) cancel as shown.

$$47.0 \text{ m/s} = \left(47.0 \frac{\cancel{\text{m}}}{\cancel{\text{s}}} \right) \left(\frac{3600 \cancel{\text{s}}}{1 \text{ h}} \right) \left(\frac{1 \text{ km}}{1000 \cancel{\text{m}}} \right)$$
$$= 169 \text{ km/h}$$

To convert this speed to mi/h, we use the ratio (1 mi)/(1.609 km) = 1.

$$169 \text{ km/h} = \left(169\,\frac{\text{km}}{\text{h}}\right)\left(\frac{1 \text{ mi}}{1.609\,\cancel{\text{km}}}\right)$$
$$= 105 \text{ mi/h}$$

Reflect

Since there are 3600 s in an hour, an object that moves at 1 m/s will travel 3600 m, or 3.6 km, in 1 h. So 1 m/s = 3.6 km/h. This says that *any* speed in km/h will be 3.6 times greater than the speed in m/s. You can check our first result by verifying that 169 is just 3.6 times 47.0.

To check our second result, note that a mile is a greater distance than a kilometer. Hence our bird travels fewer miles than kilometers in a 1-hour time interval, and the speed in mi/h will be less than in km/h. That's just what we found.

WATCH OUT! Use the correct conversion factor.

! You can get into trouble if you are careless in applying the method of taking the number whose units are to be converted and multiplying it by a conversion factor equal to 1. For example, suppose we tried to find the speed of the white-throated needletail (Example 1-2) in mi/h by multiplying 169 km/h by the conversion factor 1 = (1.609 km)/(1 mi).

$$169 \text{ km/h} = \left(169\,\frac{\text{km}}{\text{h}}\right)\left(\frac{1.609 \text{ km}}{1 \text{ mi}}\right) = 272 \text{ km}^2/(\text{h}\cdot\text{mi})$$

We would then get a result with strange units (kilometers squared per hour per mile) because the unwanted units of kilometers didn't cancel. What's more, the speed in mi/h should be less than in km/h, not more. If you *always* keep track of units when doing conversions, you'll avoid mistakes of this kind. (See Figure 1-4 for another example of incorrect versus correct unit conversion.)

Quantities Versus Units

In this book we'll often use symbols to denote the values of physical quantities. We'll always write these symbols using *italics*. Some examples include the symbols v for speed, B for magnetic field, and E for energy. (As these examples show, the symbol is not always the first letter of the name of the quantity.) By contrast, the letters that we use to denote units, such as m (meters), s (seconds), and kg (kilograms), are not italicized. If the unit is named for a person, the letter is capitalized. An example is the unit of power, the watt (abbreviated W), named for the Scottish inventor James Watt.

In some cases the same symbol is used for more than one quantity. One example is the symbol T, which can denote a temperature, a period of time, or the tension in a stretched rope. The letter T (not italicized) is also used as an abbreviation for the unit of magnetic field, the tesla (named for the Serbian-American physicist Nikola Tesla). You'll often need to pay careful attention to the context in which a symbol is used to discern the quantity or unit it represents.

Scientific Notation

Physicists investigate objects that vary in size from the largest structures in the universe, including galaxies and clusters of galaxies, down to atomic nuclei and the particles found within nuclei. The time intervals that they analyze range from the age of the universe to a tiny fraction of a second. To describe such wide ranges of values, we need an equally wide range of both large and small numbers.

To avoid such confusing terms as "a million billion billion," physicists use a standard shorthand system called **scientific notation**. All the cumbersome zeros that accompany a large number are consolidated into one term consisting of 10 followed by an **exponent**, which is written as a superscript. The exponent indicates how many zeros you would need to write out the long form of the number. The exponent also tells you how many tens must be multiplied together to give the desired number, which is why the exponent is also called the **power of ten**. For example, ten thousand can be written as 10^4 ("ten to the fourth" or "ten to the fourth power") because $10^4 = 10 \times 10 \times 10 \times 10 = 10,000$. Some powers of ten are so common that we have created abbreviations for them that appear as a prefix to the actual unit. For example, rather than writing 10^9 watts (10^9 W), we can write 1 gigawatt, or 1 GW. Table 1-2 lists the prefix and its abbreviation for several common powers of ten.

TABLE 1-2 Prefixes

Factor	Prefix	Symbol
10^{-24}	yocto	y
10^{-21}	zepto	z
10^{-18}	atto	a
10^{-15}	femto	f
10^{-12}	pico	p
10^{-9}	nano	n
10^{-6}	micro	μ
10^{-3}	milli	m
10^{-2}	centi	c
10^{-1}	deci	d
10^{1}	deka	da
10^{2}	hecto	h
10^{3}	kilo	k
10^{6}	mega	M
10^{9}	giga	G
10^{12}	tera	T
10^{15}	peta	P
10^{18}	exa	E
10^{21}	zetta	Z
10^{24}	yotta	Y

GOT THE CONCEPT? 1-1
Smallest to Biggest

(?) Each of these numbers represents the size of something. Arrange them from smallest to largest.
(a) 0.1 mm; (b) 7 μm;
(c) 6380 km; (d) 165 cm;
(e) 200 nm.

(Answers to *Got the Concept?* questions can be found at the end of the chapter.)

In scientific notation, numbers are usually written as a figure between 1 and 10 multiplied by the appropriate power of ten. The approximate distance between Earth and the Sun, for example, can be written as 1.5×10^8 km. This is far more convenient than writing 150,000,000 km or one hundred and fifty million kilometers. A power output of 2500 watts (2500 W) can be written in scientific notation as 2.5×10^3 W. (Alternatively, we can write this as 2.5 kW, where 1 kW = 1 kilowatt = 10^3 W.)

Most electronic calculators, including those on mobile devices, use a shorthand for scientific notation. To enter the number 1.5×10^8 you first enter 1.5, then press a key labeled EXP, EE, or 10^x, then enter the exponent 8. (The EXP, EE, or 10^x key takes care of the × 10 part of the expression.) The number will then appear on the display as 1.5 E 8, 1.5 8, or some variation of these; typically the × 10 is not displayed as such. There are some variations from one calculator to another, so you should spend a few minutes to make sure you know the correct procedure for working with numbers in scientific notation. You'll use this notation continually in your study of physics, so this is time well spent.

You can use scientific notation for numbers that are less than one by using a minus sign in front of the exponent. A negative exponent tells you to *divide* by the appropriate number of tens. For example, 10^{-2} ("ten to the minus two") means to divide by 10 twice, so $10^{-2} = (1/10) \times (1/10) = 1/100 = 0.01$. Table 1-2 also includes negative powers of ten.

WATCH OUT! Know how your calculator expresses scientific notation.

(!) Confusion can result from the way calculators display scientific notation. Since 1.5×10^8 is displayed as 1.5 8 or 1.5 E 8, it is not uncommon to think that 1.5×10^8 is the same as 1.5^8. That is not correct, however; 1.5^8 is equal to 1.5 multiplied by itself eight times, or 25.63, which is not even close to $150,000,000 = 1.5 \times 10^8$. Another common mistake is to write 1.5×10^8 as 15^8. If you are inclined to do this, perhaps you are thinking that you can multiply 1.5 by 10, then tack on the exponent later. This also does not work; 15^8 is equal to 15 multiplied by itself eight times, or 2,562,890,625, which again is nowhere near 1.5×10^8. Reading over the manual for your calculator will help you to avoid these common errors.

GOT THE CONCEPT? 1-2
Comparing Sizes

(?) For each pair, determine which quantity is larger and find the ratio of the larger quantity to the smaller quantity.
(a) 1 mg, 1 kg (b) 1 mm, 1 cm
(c) 1 MW, 1 kW (d) 10^{-10} m, 10^{-14} m
(e) 10^{10} m, 10^{14} m

You can also use scientific notation to express a number such as 0.00245: $0.00245 = 2.45 \times 0.001 = 2.45 \times 10^{-3}$. (Again, in scientific notation the first figure is a number between 1 and 10.) This notation is particularly useful when dealing with very small numbers. As an example, the diameter of a hydrogen atom is much more convenient to state in scientific notation (1.1×10^{-10} m) than as a decimal (0.00000000011 m) or a fraction (110 trillionths of a meter).

One of the smallest numbers in physics is Planck's constant (symbol h), which in the SI system is the basis for the definition of the kilogram. Its value is defined to be $h = 6.62607015 \times 10^{-34}$ kg·m²/s. Since the units of h involve the kilogram, the meter, and the second, if a physicist measures h in the laboratory, they can use their result and the definitions of the meter and second to determine the value of the kilogram.

BioMedical EXAMPLE 1-3 Unit Conversion and Scientific Notation: Hair Growth

The hair on a typical person's head grows at an average rate of 1.5 cm per month (Figure 1-5). (a) What is this rate in meters per second? (b) Approximately how far (in meters) will the ends of your hair move during a 50-min physics class? Express the answer using scientific notation.

Figure 1-5 **Hair growth** How much does a person's hair grow during a 50-min class period?

Prostock-studio/Shutterstock

Set Up

Like Example 1-2, part (a) is a problem in converting units. We'll again follow the procedure shown in Figure 1-4. Not all months have the same number of days, so we'll assume an average 30-day month. In part (b) we'll use the idea that since speed equals distance per time, the distance the hair moves equals the growth speed multiplied by the time in one class period.

Conversions:
$$1 \text{ min} = 60 \text{ s}$$
$$1 \text{ d} = 86{,}400 \text{ s}$$
$$1 \text{ month (average)} = 30 \text{ d}$$
$$1 \text{ cm} = 0.01 \text{ m}$$
$$\text{speed} = \frac{\text{distance}}{\text{time}}$$
$$\text{so distance} = \text{speed} \times \text{time}$$

Solve

(a) Let's first express the speed at which hair grows in SI units (m/s). We use the conversion factors relating meters to centimeters, months to days, and days to seconds, and cancel units. We must move the decimal point to the right by nine steps to put the result in scientific notation, so the exponent is −9.

1.5 cm/month

$$= \left(1.5 \frac{\cancel{\text{cm}}}{\cancel{\text{month}}}\right)\left(\frac{0.01 \text{ m}}{1 \cancel{\text{cm}}}\right)\left(\frac{1 \cancel{\text{month}}}{30 \cancel{\text{d}}}\right)\left(\frac{1 \cancel{\text{d}}}{86{,}400 \text{ s}}\right)$$

$$= 0.0000000058 \text{ m/s} = 5.8 \times 10^{-9} \text{ m/s}$$

(b) To find the distance that hair grows in 50 min, we multiply this speed by 50 min and include the conversion factor relating seconds and minutes. We then put the result into scientific notation. Note that the first number in the result is between 1 and 10. Note also that when multiplying together powers of ten, we simply add the exponents.

$$\text{distance} = \left(5.8 \times 10^{-9} \frac{\text{m}}{\cancel{\text{s}}}\right)(50 \cancel{\text{min}})\left(\frac{60 \cancel{\text{s}}}{1 \cancel{\text{min}}}\right)$$

$$= (5.8 \times 10^{-9})(3 \times 10^{3}) \text{ m}$$

$$= 5.8 \times 3 \times 10^{-9} \times 10^{3} \text{ m}$$

$$= 17 \times 10^{-9} \times 10^{3} \text{ m}$$

$$= 1.7 \times 10^{1} \times 10^{-9} \times 10^{3} \text{ m}$$

$$= 1.7 \times 10^{1-9+3} \text{ m}$$

$$= 1.7 \times 10^{-5} \text{ m}$$

Reflect

You can also express our result from part (b) in micrometers (μm). Note that when dividing powers of ten, we subtract the exponents. A distance of 17 μm is about 1/40 mm. Do you think you could notice this change in length with the unaided eye?

$$1.7 \times 10^{-5} \text{ m} = (1.7 \times 10^{-5} \text{ m})\left(\frac{1 \text{ μm}}{10^{-6} \text{ m}}\right)$$

$$= 1.7 \times \frac{10^{-5}}{10^{-6}} \text{ μm}$$

$$= 1.7 \times 10^{-5-(-6)} \text{ μm}$$

$$= 1.7 \times 10^{1} \text{ μm} = 17 \text{ μm}$$

TAKE-HOME MESSAGE FOR Section 1-3

✔ Many physical quantities can be measured in units that are combinations of the SI units seconds, meters, and kilograms.

✔ To convert a physical quantity from one set of units to another, multiply it by an appropriate conversion factor equal to 1.

✔ In scientific notation, quantities are expressed as a number between 1 and 10 multiplied by 10 raised to a power.

✔ Prefixes can be used instead of scientific notation to indicate 10 raised to a power.

1-4 Correct use of significant figures helps keep track of uncertainties in numerical values

Although physicists strive to make their measurements as precise as possible, the unavoidable fact is that there is always some uncertainty in every measurement. This is not a result of a lack of care on the part of the person doing the measurement; rather, it is due to fundamental limitations on how well measurements can be carried out.

 See the Math Tutorial for more information on significant figures.

If you try to measure your thumbnail using a ruler with markings in millimeters, you might find that it's between 15 and 16 mm in length. With keen eyes you might be able to tell that the length is very nearly halfway between those two markings on the ruler, so you might say that the length is approximately 15.5 mm. However, armed only

(a) Earth's Moon

NASA/JPL/USGS

(b) Neptune's moon Nereid

NASA/JPL

Figure 1-6 Significant figures
The diameter of the Moon is
known to much greater precision
than the diameter of Neptune's
small moon Nereid (shown here
in the highest-resolution image
yet made). That's why we express
the diameter of the Moon to four
significant figures (3.476×10^3 km)
but the diameter of Nereid to
only two significant figures
(3.4×10^2 km).

GOT THE CONCEPT? 1-3
Significant Figures

(?) All of these numbers have
the same value but are
written with a different
number of significant figures:
(a) 123, (b) 1.230×10^2,
(c) 123.000, (d) 1.23×10^2,
(e) 0.12300×10^3. Rank them in
order of least to most significant
figures.

with your ruler it would be impossible to say whether the length was 15.47, 15.51, or 15.54 mm. We would say that the fourth digit is *not significant*, that is, it does not contain meaningful information. Hence this measurement has only three **significant figures**. We specify this implicitly by giving the measured value as 15.5 mm, by which we mean that it's between 15.4 and 15.6 mm. More precise measurements have more significant figures; less precise measurements have fewer significant figures.

An alternative way to present the number would be to give the length of your thumbnail as 15.5 ± 0.1 mm, with the number after the $\pm$ sign representing the uncertainty. This kind of presentation is very common in experimental work. In this book, however, we'll use significant figures to represent uncertainty.

Note that some numbers are *exact*, with no uncertainty whatsoever. Three examples are the number of meters in a kilometer (exactly 1000), the speed of light (defined to be exactly 299,792,458 m/s, as we learned in Section 1-3), and the value of π. The value of any measured quantity, however, will have some uncertainty.

Scientific notation is helpful in representing the number of significant figures in a measurement. When we say that the diameter of the Moon (Figure 1-6a) at its equator is 3476 km, we imply that there are four significant figures in the value and that the uncertainty in the diameter is only about 1 km. Since there are 1000 m in 1 km, we could also express the diameter as 3,476,000 m. This is misleading, however, since it gives the impression that there are *seven* significant figures in the value. That isn't correct; we do *not* know the value of the Moon's equatorial radius to within 1 meter. A better way to express the diameter in meters is as 3.476×10^6 m, which makes it explicit that there are only four significant figures in the value. The diameter of Nereid, one of the moons of Neptune (Figure 1-6b), is less well known and is about 340 km; only the first two digits are significant, so it's more accurate to express this diameter as 3.4×10^2 km or 3.4×10^5 m.

There are special rules to help determine whether zeros in a number are significant. *Leading zeros* are not significant, so the number 0.0035 has only two significant figures. *Trailing zeros* to the right of the decimal point are significant, so the number 0.3500 has four significant figures. When we have trailing zeros to the left of the decimal point, as in 350, things get ambiguous. To make it clear that the trailing zero is significant, it is best to write 3.50×10^2. However, there are times when it's more expedient to write exact numbers without scientific notation. For example, there are exactly 60 s in 1 min, so the 60 has as many significant figures as we need, and we won't bother writing 6.0000×10^1 s or 60.000 to indicate that the value should have five significant figures.

Calculations with Significant Figures

In most problems in this book, we'll use numbers with two, three, or four significant figures. For example, you might be asked about the motion of an ostrich that runs 43.9 m (three significant figures) in 2.3 s (two significant figures). If you use your calculator to find the speed of this ostrich (the distance traveled divided by the elapsed time), the result will be

$$\frac{43.9 \text{ m}}{2.3 \text{ s}} = 19.08695652 \text{ m/s} \quad \text{(incorrect)}$$

This answer is incorrect because there are *not* 10 significant figures in the answer. You don't know the result that precisely! The rule for *multiplying or dividing* numbers is that the result has the same number of significant figures as the input number with the *fewest* significant figures. In our example the elapsed time of 2.3 s has fewer significant figures (two) than the distance of 43.9 m, so the result has only two significant figures. Rounding our result, we get

3 significant figures   Result has only 2 significant figures

$$\frac{43.9 \text{ m}}{2.3 \text{ s}} = 19 \text{ m/s} \quad \text{(correct)}$$

2 significant figures

Following the same rule, if a strip of cloth is 35.65 cm long and 2.49 cm wide, the area (length times width) has only three significant figures: (35.65 cm)(2.49 cm) = 88.7685 cm^2, which rounds to 88.8 cm^2.

There's a slightly different rule for *adding or subtracting* numbers. Suppose that you drive 44.3 km (according to the odometer of your car) from Central City to the parking lot at the Metropolis shopping mall, then walk an additional 108 m or 0.108 km (according to your pedometer) to your favorite clothing store. The total distance you have traveled is (44.3 km) + (0.108 km), which according to your calculator is 44.408 km. This isn't quite right, however, because you know your driving distance only to the nearest tenth of a kilometer. As a result, you need to round off the answer to the nearest tenth of a kilometer, or 44.4 km. So

1 digit to the right of the decimal point

3 digits to the right of the decimal point

$$44.3 \text{ km} + 0.108 \text{ km} = 44.4 \text{ km}$$

Result has only 1 digit to the right of the decimal point

Thus the rule for adding or subtracting numbers is that the result can have no less uncertainty than the most uncertain input number. In our example that's the 44.3-km distance, which has an uncertainty of about 0.1 km. Stated another way, the answer may have no more digits to the right of the decimal point than the input number with the fewest digits to the right of the decimal point.

Let's summarize our general rules for working with significant figures:

Rule 1 When multiplying or dividing numbers, the result should have the same number of significant figures as the input number with the fewest significant figures. The same rule applies to squaring, taking the square root, calculating sines and cosines, and so on.

Rule 2 When adding or subtracting numbers, the result should have the same number of digits to the right of the decimal point as the input number with the fewest digits to the right of the decimal point.

We'll use these rules for working with significant figures in all of the worked examples in the book. (If you look back, you'll see that we actually used them in Examples 1-2 and 1-3 in Section 1-3.) Make sure that you follow these same rules in your own problem solving.

WATCH OUT! Be careful with significant figures and scientific notation.

For a number given in scientific notation, the exponent of 10 is known exactly; it doesn't affect the number of significant figures. So 1.45×10^{-2} has three significant figures, and 2.2×10^{-4} has two significant figures. If you *multiply* these together, by Rule 1 their product has only two significant figures: $(1.45 \times 10^{-2})(2.2 \times 10^{-4}) = 3.19 \times 10^{-2+(-4)} = 3.19 \times 10^{-6}$, which rounds to 3.2×10^{-6}. To *add* these two numbers together, first write them using the same exponent of 10:

$1.45 \times 10^{-2} + 2.2 \times 10^{-4} = 1.45 \times 10^{-2} + 0.022 \times 10^{-2} = (1.45 + 0.022) \times 10^{-2}$. In this sum, there are two significant figures after the decimal point for the first number versus three significant figures for the second number. By Rule 2 the sum can have only two significant figures after the decimal point, so we must round $1.45 + 0.022 = 1.472$ to 1.47, and the final answer is 1.47×10^{-2}.

WATCH OUT! Be mindful of significant figures when rounding numbers.

If you're doing a calculation with several steps, it's best to retain *all* of the digits in your calculation until the very end. Then you can round off the result as necessary to give the answer to the correct number of significant figures. For example, suppose you want to calculate $32.1 \times 4.998 \times 4.87$. Rule 1 above tells us that the final answer has only three significant figures. The product of the first two numbers is $32.1 \times 4.998 = 160.4358$, which if you round off at this point

becomes 160. If you multiply this rounded number by 4.87, you get 779.2, which rounds to 779. If, however, you calculate $32.1 \times 4.998 \times 4.87$ directly using your calculator *without* rounding the intermediate answer, you get 781.322346. Rounding this to three significant figures gives the correct answer, which is 781. Moral: Do *not* round until the end of your calculation!

EXAMPLE 1-4 Significant Figures: Combining Volumes

What is the volume (in cubic kilometers) of Earth? What is the volume of the Moon? What is the combined volume of the two worlds together? The radius of Earth is 6378 km, and the radius of the Moon is 1738 km.

Set Up

From the Math Tutorial the volume of a sphere of radius R is $4\pi R^3/3$. Since we multiply the radius by itself, the answer can have no more significant figures than the value of the radius.

Radius of Earth $R_E = 6378$ km
Radius of Moon $R_M = 1738$ km

Solve

For Earth, R_E has four significant figures, so the volume has only four significant figures.

Volume of Earth:

$$V_E = \frac{4\pi(6378 \text{ km})^3}{3}$$
$$= 1.086781293 \times 10^{12} \text{ km}^3$$

Round to four significant figures:

$$V_E = 1.087 \times 10^{12} \text{ km}^3$$

The radius of the Moon also has four significant figures, so its volume does as well.

Volume of Moon:

$$V_M = \frac{4\pi(1738 \text{ km})^3}{3}$$
$$= 2.199064287 \times 10^{10} \text{ km}^3$$

Round to four significant figures:

$$V_M = 2.199 \times 10^{10} \text{ km}^3$$

To find the combined volume we add V_E and V_M. To make sure we retain the correct number of significant figures in our answer, we express both numbers in scientific notation with the same exponent. According to Rule 2 for addition, our answer must have only three significant figures to the right of the decimal point.

$$V_E = 1.087 \times 10^{12} \text{ km}^3$$
$$V_M = 2.199 \times 10^{10} \text{ km}^3$$
$$= 0.02199 \times 10^{12} \text{ km}^3$$

To avoid intermediate rounding errors we use the unrounded values and then round the result to three digits to the right of the decimal point at the end of the calculation.

$$V_E + V_M = (1.086781293 \times 10^{12} \text{ km}^3)$$
$$+ (0.02199064287 \times 10^{12} \text{ km}^3)$$
$$= (1.086781293 + 0.02199064287) \times 10^{12} \text{ km}^3$$
$$= 1.10877193587 \times 10^{12} \text{ km}^3$$

This has too many significant figures, so we round to the final answer of

$$V_E + V_M = 1.109 \times 10^{12} \text{ km}^3$$

Reflect

In many problems you'll have to use both the significant figure rule for multiplication and the significant figure rule for addition, just as we did here. Note that if we had added the rounded values of the individual volumes to calculate the combined volume, we would have gotten $V_E + V_M = 1.087 \times 10^{12} \text{ km}^3 + 0.02199 \times 10^{12} \text{ km}^3 = 1.10899 \times 10^{12} \text{ km}^3$, which still rounds to $1.109 \times 10^{12} \text{ km}^3$. But this won't always be the case, so you should always wait to round until the very end of your calculations.

TAKE-HOME MESSAGE FOR Section 1-4

✔ Significant figures characterize the precision of a measurement or the uncertainty of a numerical value.

✔ When multiplying or dividing numbers the result has the same number of significant figures as the input number with the fewest significant figures.

✔ When adding or subtracting numbers the result has the same number of digits to the right of the decimal point as the input number with the fewest number of digits to the right of the decimal point *when all numbers are expressed in scientific notation with the same exponent.*

✔ When writing a number all trailing zeros to the right of the decimal point are significant figures.

1-5 Dimensional analysis is a powerful way to check the results of a physics calculation

Once you've solved a problem, it's useful to be able to quickly determine how likely it is that you've solved it correctly. One valuable approach is to do a **dimensional analysis**. This approach is particularly useful when you calculate an algebraic result and want to check it before substituting numerical values.

To understand what we mean by dimensional analysis, first note that any physical quantity has a *dimension*. Three physical quantities that we introduced in Section 1-3 are mass, length, and time. (Table 1-1 in Section 1-3 lists additional physical quantities.) We say that the mass of a hydrogen atom has dimensions of mass, the diameter of the atom has dimensions of length, and the amount of time it takes light to travel that diameter has dimensions of time. Mass, length, and time are *fundamental* dimensions because they cannot be expressed in terms of other more fundamental quantities. (Temperature and electric charge are also fundamental dimensions.)

Dimensions *have* units but are not units themselves. Units are used to help express the numerical value of a dimension. A dimension of length may be expressed in units of meters; for example, the length of a typical house cat is 0.65 m. The dimension of time might be expressed in units of nanoseconds, and the dimension of mass might be expressed in units of kilograms.

The dimensions of many quantities are made up of a combination of fundamental dimensions. One example is speed, which has dimensions of distance per time, or (distance)/(time) for short. Another is volume, which has dimensions of distance cubed, or (distance)^3.

Here's the key to using dimensional analysis: *In any equation, the dimensions must be the same on both sides of the equation.* As an example, suppose you hold a ball at height h above the ground and then let it fall (Figure 1-7). The ball takes a time t to reach the ground. Your friend calculates that the height h and the time t are related by an equation that also involves the speed v of the ball just before it hits the ground:

$$h = vt^2 \quad \text{(is this correct?)}$$

Is this equation dimensionally correct? To find out, replace each symbol by the dimensions of the quantity that it represents. The height h has dimensions of length or distance, speed v has dimensions of (distance)/(time), and time t has dimensions of time. Dimensions cancel just like algebraic quantities, so if we simplify this, we get

$$\text{distance} = \frac{\text{distance}}{\text{time}} \times (\text{time})^2$$

$$= \frac{\text{distance}}{\text{time}} \times \text{time} \times \text{time}$$

$$= \text{distance} \times \text{time}$$

This is *inconsistent*: The left side of the equation above (h) has dimensions of distance, while the right side (vt^2) has dimensions of distance multiplied by time. So this equation *cannot* be correct.

WATCH OUT!
Dimensions are not the same as units.

❗ Be careful to make the distinction between *dimensions* and *units*. For example, volume may be expressed in any number of different units such as m^3, cm^3, mm^3, or km^3. But no matter what units are used, volume *always* has dimensions of (distance)^3.

---

① You drop a ball from a height h above the ground. The ball reaches the ground a time t later. What equation correctly relates h and t?

② To decide whether an equation relating h and t could be correct, use dimensional analysis.

Figure 1-7 Dimensional analysis of a falling ball To determine whether an equation in physics—such as that for the motion of a falling ball—could be correct, use the rule of dimensional analysis: The dimensions must be the same on both sides of the equation.

Tom Wang/Alamy

It turns out that the correct relationship between the height h from which the ball is dropped and the time t it takes to reach the ground is

$$h = \frac{1}{2}gt^2 \quad \text{(correct!)}$$

where g is a constant. Let's use dimensional analysis to figure out the dimensions of g. Again h has units of distance and t has units of time. The number 1/2 has no dimensions at all, so it can be ignored (it is a pure number with no units, so it is *dimensionless*). We rewrite the equation in terms of dimensions and solve for the dimensions of g:

$$\text{distance} = (\text{dimensions of } g) \times (\text{time})^2$$
$$(\text{dimensions of } g) = \frac{\text{distance}}{(\text{time})^2}$$

The SI units of g must then be m/s² (meters per second squared, or meters per second per second). We'll learn the significance of the quantity g (called the *acceleration due to gravity*) in Chapter 2.

WATCH OUT! Dimensional analysis can't tell you everything.

Suppose that as a result of an algebraic error, we failed to include the factor of 1/2 in the correct equation above and instead found $h = gt^2$. Because the pure number 1/2 has no dimensions, it wouldn't change the dimensional analysis we did above. So a dimensional analysis would not catch this mistake. Performing dimensional analysis can tell you whether an answer is wrong, but even if the dimensions of your answer are correct, the numerical value might still be wrong.

GOT THE CONCEPT? 1-4 Dimensional Analysis

A block of mass m oscillates back and forth on the end of a spring. (Flip forward in the book and look at Figure 12-2.) You'll find that the time T needed for the block to complete one full back-and-forth cycle depends on a constant k that has dimensions of mass divided by time squared according to

$$T = 2\pi\sqrt{\frac{m}{k}}$$

Use dimensional analysis to determine whether this result could be correct.

TAKE-HOME MESSAGE FOR Section 1-5

✔ Dimensional analysis is the technique of checking the dimensions of your algebraic answer before you substitute values to compute a numeric result.

✔ The dimensions of any quantity can be put together from a few fundamental dimensions, including mass, distance, and time.

✔ The key to using dimensional analysis: *In any equation, the dimensions must be the same on both sides of the equation.*

Key Terms

dimensional analysis
exponent
kilogram
mass

meter
power of ten
scientific notation
second

significant figures
Système International (SI)
unit conversion
units

Chapter Summary

Topic	Equation or Figure
Problem-solving strategy: We'll solve all problems using a strategy that incorporates three steps.	*Set Up* the problem *Solve* for the desired quantities *Reflect* on the answer

Unit conversion: Units can be converted by multiplying any value by an appropriate representation of 1.

$$47.0 \text{ m/s} = \left(47.0\frac{\cancel{m}}{\cancel{s}}\right)\left(\frac{3600\ \cancel{s}}{1\text{ h}}\right)\left(\frac{1\text{ km}}{1000\ \cancel{m}}\right)$$
$$= 169 \text{ km/h}$$

Scientific notation: Any number can be written in scientific notation, which usually consists of a number between 1 and 10 multiplied by 10 raised to some power. Prefixes can be used as abbreviations for many powers of ten.

$$150{,}000{,}000 \text{ km} = 1.5 \times 10^8 \text{ km}$$
$$2500 \text{ watts} = 2.5 \times 10^3 \text{ W}$$
$$= 2.5 \text{ kilowatts} = 2.5 \text{ kW}$$

Significant figures: Any measured quantity has a finite number of significant digits. You need to keep track of the number of significant digits in all your calculations. There are different rules for addition/subtraction and multiplication/division.

Rule 1 When multiplying or dividing numbers, the result should have the same number of significant figures as the input number with the fewest significant figures. The same rule applies to squaring, taking the square root, calculating sines and cosines, and so on.

3 significant figures — Result has only 2 significant figures

$$\frac{43.9 \text{ m}}{2.3 \text{ s}} = 19 \text{ m/s} \quad \text{(correct)}$$

2 significant figures

Rule 2 When adding or subtracting numbers, the result should have the same number of digits to the right of the decimal point as the input number with the fewest digits to the right of the decimal point.

1 digit to the right of the decimal point — 3 digits to the right of the decimal point

$$44.3 \text{ km} + 0.108 \text{ km} = 44.4 \text{ km}$$

Result has only 1 digit to the right of the decimal point

Dimensional analysis: Analyzing the dimensions of an equation can provide a clue as to the correctness of that equation. Dimensions and units are not the same thing.

$$h = \frac{1}{2}gt^2$$
$$\text{distance} = (\text{dimensions of } g) \times (\text{time})^2$$
$$(\text{dimensions of } g) = \frac{\text{distance}}{(\text{time})^2}$$

Answer to What do you think? Question

(c) 1 cm. The length of a typical hummingbird is 7.5 to 13 cm, and the head is 1 to 2 cm in length.

Answers to Got the Concept? Questions

1-1 From smallest to largest, the order is (e), (b), (a), (d), and (c). Some examples of objects of about the size of the five distances include (e) 200 nanometers is the diameter of a typical virus; (b) 7 μm is the diameter of a human red blood cell; (a) 0.1 mm is the diameter of a typical human hair; (d) 165 cm is the average height of a woman in the United States; and (c) 6380 km is the mean radius of Earth.

1-2 (a) One kilogram is bigger than 1 milligram by a factor of 1 million; 1 mg equals 0.001 g equals 0.000001 kg. (b) One centimeter is 10 times bigger than 1 mm; 1 mm equals 0.001 m, and 1 cm equals 0.01 m. (c) One megawatt is 1000 times bigger than 1 kW; 1 MW equals 1,000,000 W, and 1 kW equals 1000 W. (d) A distance of 10^{-10} m is 10^4 (10,000) times bigger than 10^{-14} m: $(10^{-10} \text{ m})/(10^{-14} \text{ m}) = 10^{-10-(-14)} = 10^{-10+14} = 10^4$.

(e) A distance 10^{14} m is 10^4 (10,000) times bigger than 10^{10} m; $(10^{14} \text{ m})/(10^{10} \text{ m}) = 10^{14-10} = 10^4$.

1-3 123 has three significant figures as does 1.23×10^2. 1.230×10^2 is written with four significant figures, 0.12300×10^3 has five significant figures, and 123.000 has six.

1-4 Since m has dimensions of mass, k has dimensions of mass per time squared, and 2π is dimensionless, the dimensions of the right side of the equation $T = 2\pi\sqrt{m/k}$ are

$$\sqrt{\frac{\text{mass}}{\text{mass}/(\text{time})^2}} = \sqrt{\text{mass} \times \frac{(\text{time})^2}{\text{mass}}}$$
$$= \sqrt{(\text{time})^2} = \text{time}$$

The left side of the equation, T, also has dimensions of time. Hence you can have some confidence that this relationship is correct. You cannot say, however, that this equation is *exactly* correct because the pure number 2π has no dimensions and therefore does not play a role in the dimensional analysis. Had the equation been, say, $T = 5\sqrt{m/k}$, the dimensional analysis would be exactly the same.

Questions and Problems

In a few problems you are given more data than you actually need; in a few other problems you are required to supply data from your general knowledge, outside sources, or informed estimate.

•	Basic, single-concept problem
••	Intermediate-level problem; may require synthesis of concepts and multiple steps
•••	Challenging problem
Example	See worked example for a similar problem

Conceptual Questions

1. • Define the basic quantities in physics (length, mass, and time) and list the appropriate SI unit that is used for each quantity.

2. • Is it possible to define a system of units in which length is not one of the fundamental properties?

3. • What properties should an object, system, or process have for it to be a useful standard of measurement of a physical quantity such as length or time?

4. • Why do physicists and other scientists prefer metric units and prefixes (for example, micro, milli, and centi) to English system units, for example, inches and pounds?

5. • All valid equations in physics have consistent units. Are all equations that have consistent units valid? Support your answer with examples.

6. • Mass density is defined as mass per unit volume. To determine the density of a solution in grams per milliliter, the mass and volume of two samples are measured. The results are recorded in the table.

	Sample 1	Sample 2
Mass (g)	9.85	9.96
Volume (mL)	10.1	10.19

For each sample, which measurement limits the number of significant digits in the calculated density?

7. • Consider the number 61,000. (a) What is the least number of significant figures this might have? (b) The greatest number? (c) If the same number is expressed as 6.10×10^4, how many significant figures does it have?

8. • Acceleration has dimensions L/T^2, where L is length and T is time. What are the SI units of acceleration?

Multiple-Choice Questions

9. • Which of the following are fundamental quantities?
   A. density (mass per volume)
   B. length
   C. area

D. mass
E. all of the above

10. • Which length is the largest?
   A. 10 nm
   B. 10 cm
   C. 10^2 mm
   D. 10^{-2} m
   E. 1 m

11. • One nanosecond is
   A. 10^{-15} s
   B. 10^{-6} s
   C. 10^{-9} s
   D. 10^{-3} s
   E. 10^9 s

12. • How many cubic meters are there in a cubic centimeter?
   A. 10^2
   B. 10^6
   C. 10^{-2}
   D. 10^{-3}
   E. 10^{-6}

13. • How many square centimeters are there in a square meter?
   A. 10
   B. 10^2
   C. 10^4
   D. 10^{-2}
   E. 10^{-4}

14. • Calculate $1.4 + 15 + 7.15 + 8.003$ using the proper number of significant figures.
   A. 31.553
   B. 31.550
   C. 31.55
   D. 31.6
   E. 32

15. • Calculate $0.688/0.28$ using the proper number of significant figures.
   A. 2.4571
   B. 2.457
   C. 2.46
   D. 2.5
   E. 2

16. • Which of the following relationships is dimensionally consistent with a value for acceleration that has dimensions of distance per time per time? In these equations x is distance, t is time, and v is speed.
   A. v^2/t
   B. v/t
   C. v/t^2
   D. v/x^2
   E. v^2/x^2

17. • Calculate 25.8×70.0 using the proper number of significant figures.
- A. 1806.0
- B. 1806
- C. 1810
- D. 1800
- E. 2000

Problems

1-3 Measurements in physics are based on standard units of time, length, and mass

18. • Write the following numbers in scientific notation:
- A. 237
- B. 0.00223
- C. 45.1
- D. 1115
- E. 14,870
- F. 214.78
- G. 0.00000442
- H. 12,345,678 Example 1-3

19. • Write the following numbers using decimals:
- A. 4.42×10^{-3}
- B. 7.09×10^{-6}
- C. 8.28×10^{2}
- D. 6.02×10^{6}
- E. 456×10^{3}
- F. 22.4×10^{-3}
- G. 0.375×10^{-4}
- H. 138×10^{-6} Example 1-3

20. • Write the metric prefixes for the following powers of ten:
- A. $10^{3} = $ _____
- B. $10^{9} = $ _____
- C. $10^{6} = $ _____
- D. $10^{12} = $ _____
- E. $10^{-3} = $ _____
- F. $10^{-12} = $ _____
- G. $10^{-6} = $ _____
- H. $10^{-9} = $ _____ Example 1-3

21. • Write the numerical values for the following prefixes as powers of ten:
- A. $p = $ _____
- B. $m = $ _____
- C. $M = $ _____
- D. $\mu = $ _____
- E. $f = $ _____
- F. $G = $ _____
- G. $T = $ _____
- H. $c = $ _____ Example 1-3

22. • Complete the following conversions:
- A. 125 cm = _____ m
- B. 233 g = _____ kg
- C. 786 ms = _____ s
- D. 454 kg = _____ mg
- E. 208 cm^2 = _____ m^2
- F. 444 m^2 = _____ cm^2
- G. 12.5 cm^3 = _____ m^3
- H. 144 m^3 = _____ cm^3 Example 1-2

23. • Complete the following conversions:
- A. 238 ft = _____ m
- B. 772 in = _____ cm
- C. 1220 in^2 = _____ cm^2
- D. 559 oz = _____ L
- E. 973 lb = _____ g
- F. 122 ft^3 = _____ m^3
- G. 1.28 mi^2 = _____ km^2
- H. 442 in^3 = _____ cm^3 Example 1-2

24. • Complete the following conversions:
- A. 328 cm^3 = _____ L
- B. 112 L = _____ m^3
- C. 220 hectares = _____ m^2
- D. 44,300 m^2 = _____ hectares
- E. 225 L = _____ m^3
- F. 17.2 hectares·m = _____ L
- G. 225,300 L = _____ hectares·m
- H. 2000 m^3 = _____ mL Example 1-2

25. • Complete the following conversions:
- A. 33.5 gal = _____ L
- B. 62.8 L = _____ gal
- C. 216 acre·ft = _____ L
- D. 1770 gal = _____ m^3
- E. 22.8 fl oz = _____ cm^3
- F. 54.2 cm^3 = _____ cups
- G. 1.25 hectares = _____ acre
- H. 664 mm^3 = _____ qt Example 1-2

26. • Write the following quantities in scientific notation without prefixes: (a) 300 km, (b) 33.7 μm, and (c) 77.5 GW. Example 1-3

27. • Write the following with prefixes and not using scientific notation: (a) 3.45×10^{-4} s, (b) 2.00×10^{-11} W, (c) 2.337×10^{8} m, and (d) 6.54×10^{4} g. Example 1-3

28. • **Biology** Kinesin is a motor protein that walks along microtubules and transports molecular cargo. The maximum walking speed of kinesin obtained at ATP-saturated condition is about 400 nm/s. Convert this value to miles/hr. Example 1-2

29. • **Biology** Mechanical unfolding experiments of proteins using an atomic force microscope (AFM) involve pulling on the protein molecules with the AFM cantilever. A typical spring constant of the cantilever is $k = 100$ pN/nm. What is this value in N/m? Example 1-2

30. •• The United States is about the only country that still uses the units feet, miles, and gallons. However, you might see some car specifications that give fuel efficiency as 7.6 km per kilogram of fuel. Given that a mile is 1.609 km, a gallon is 3.785 liters, and a liter of gasoline has a mass of 0.729 kg, what is the car's fuel efficiency in miles per gallon? Example 1-2

1-4 Correct use of significant figures helps keep track of uncertainties in numerical values

31. • Give the number of significant figures in each of the following numbers:
- A. 112.4
- B. 10
- C. 3.14159
- D. 700
- E. 1204.0
- F. 0.0030
- G. 9.33×10^{3}
- H. 0.02240 Example 1-4

32. • Complete the following operations using the correct number of significant figures:
- A. $5.36 \times 2.0 = $ _____
- B. $\dfrac{14.2}{2} = $ _____
- C. $2 \times 3.14159 = $ _____
- D. $4.040 \times 5.55 = $ _____
- E. $4.444 \times 3.33 = $ _____
- F. $\dfrac{1000}{333.3} = $ _____
- G. $2.244 \times 88.66 = $ _____
- H. $133 \times 2.000 = $ _____ Example 1-4

33. • Complete the following operations using the correct number of significant figures:

A.	B.	C.	D.
4.55	80.00	71.1	200
+21.6	−112.3	+3.70	+33.7

Example 1-4

1-5 Dimensional analysis is a powerful way to check the results of a physics calculation

34. • In physics lab, you measure three quantities: x, y, and z. Suppose that $x = 2.4$ kg, $y = 0.8$ m, and $z = 4.1$ kg/m. Which of the following mathematical combinations might be meaningful?

   A. $x + y + z$
   B. $x - (y/z)$
   C. $(x/y) - z$
   D. xyz
   E. $(x/z) - y$
   F. $(x - y)/z$
   G. $(xz) + y$
   H. xy/z

35. • One equation that describes motion of an object is $x = vt + x_0$, where x is the position of the object, v is its speed, t is time, and x_0 is the initial position. Show that the dimensions in the equation are consistent.

36. • The motion of a vibrating system is described by $y(x, t) = A_0 e^{-\alpha t} \sin(kx - \omega t)$. Find the SI units for k, ω, and α.

37. • The kinetic energy of a particle is $K = \frac{1}{2} mv^2$, where m is the mass of the particle and v is its speed. Show that 1 joule (J), the SI unit of energy, is equivalent to 1 kg·m^2/s^2.

38. • The period T of a simple pendulum, the time for one complete oscillation, is given by $T = 2\pi\sqrt{(L/g)}$, where L is the length of the pendulum and g is the acceleration due to gravity. Show that the dimensions in the equation are consistent.

39. • Which of the following could be correct based on a dimensional analysis?

   A. The volume flow rate is 64 m^3/s.
   B. The height of the Transamerica Pyramid is 332 m^2.
   C. The time required for a fortnight is 66 m/s.
   D. The speed of a train is 9.8 m/s^2.
   E. The weight of a standard kilogram mass is 2.2 lb.
   F. The density of gold is 19.3 kg/m^2.

General Problems

40. • **Biology** A typical human cell is approximately 10 μm in diameter and enclosed by a membrane that is 5.0 nm thick. (a) What is the volume of the cell? (b) What is the volume of the cell membrane? (c) What percentage of the cell volume does its membrane occupy? Model the cell as a sphere. Example 1-3

41. • **Medical** Express each quantity in the standard SI units requested. (a) An adult should have no more than 2500 mg of sodium per day. What is the limit in kg? (b) A 240-mL cup of whole milk contains 35 mg of cholesterol. Express the cholesterol concentration in the milk in kg/m^3 and in mg/mL. (c) A typical human cell is about 10 μm in diameter, modeled as a sphere. Express its volume in cubic meters. (d) A low-strength aspirin tablet (sometimes called a baby aspirin) contains 81 mg of the active ingredient. How many kg of the active ingredient does a 100-tablet bottle of baby aspirin contain? (e) The average flow rate of urine out of the kidneys is typically 1.2 mL/min. Express the rate in m^3/s. (f) The density of blood proteins is about 1.4 g/cm^3. Express the density in kg/m^3. Example 1-2

42. • **Medical** The concentration of PSA (prostate-specific antigen) in the blood is sometimes used as a screening test for possible prostate cancer in men. The value is normally reported as the number of nanograms of PSA per milliliter of blood. A PSA concentration of 1.7 is considered low. Express that value in (a) g/L, (b) standard SI units of kg/m^3, and (c) μg/L. Example 1-2

43. • The acceleration g (units m/s^2) of a falling object near a planet is given by the equation $g = GM/R^2$. If the planet's mass M is expressed in kg and the distance of the object from the planet's center R is expressed in m, determine the units of the gravitational constant G. Example 1-2

44. ••• **Medical** At a resting pulse rate of 75 beats per minute, the human heart typically pumps about 70 mL of blood per beat. Blood has a density of 1060 kg/m^3. Circulating all of the blood in the body through the heart takes about 1 min in a person at rest. (a) How much blood (in L and m^3) is in the body? (b) On average, what mass of blood (in g and kg) does the heart pump each beat? Example 1-2

45. • During a certain experiment, light is found to take 37.1 μs to traverse a measured distance of 11.12 km. Determine the speed of light from the data. Express your answer in SI units and in scientific notation, using the appropriate number of significant figures. Example 1-3

46. •• **Medical** The body mass index (BMI) estimates the amount of fat in a person's body. It is defined as the person's mass m in kg divided by the square of the person's height h in m. (a) Write the formula for BMI in terms of m and h. (b) In the United States, most people measure weight in pounds and height in feet and inches. Show that with weight W in pounds and height h in inches, the BMI formula is BMI $= 703 W/h^2$. (c) A person with a BMI between 25.0 and 30.0 is considered overweight. If a person is 5′11″ tall, for what range of mass will he be considered overweight? Example 1-2

47. •• **Medical** A typical prostate gland has a mass of about 20 g and is about the size of a walnut. The gland can be modeled as a sphere 4.50 cm in diameter and of uniform density. (a) What is the density (mass per volume) of the prostate? Express your answer in g/cm^3 and in standard SI units. (b) How does the density compare to that of water? (c) During a biopsy of the prostate, a thin needle is used to remove a series of cylindrical tissue samples. If the cylinders have a total length of 28.0 mm and a diameter of 0.100 mm, what is the total mass (in g) of tissue taken? (d) What percentage of the mass of the prostate is removed during the biopsy? Example 1-2

2 Motion in One Dimension

In this chapter, your goals are to:

- (2-1) State and explain the definition of motion in one dimension.
- (2-2) Explain the meanings of and relationships among displacement, average velocity, and constant velocity.
- (2-3) Explain the distinction between velocity and acceleration, and interpret graphs of velocity versus time.
- (2-4) Use and interpret the equations and graphs for motion in one dimension with constant acceleration.
- (2-5) Solve constant-acceleration problems for motion in one dimension.
- (2-6) Solve problems involving objects in free fall.

To master this chapter, you should review:

- (1-2) The three key steps in solving any physics problem.
- (1-4) How to use significant figures in calculations.

What do you think?

When a jumper is at the high point of her trajectory, is she accelerating? If so, is she accelerating upward or downward?

The answer to the *What do you think?* question can be found at the end of the chapter.

2-1 Studying motion in a straight line is the first step in understanding physics

We live in a universe of motion. We are surrounded by speeding cars, scampering animals, and moving currents of air. Our planet is in motion as it spins on its axis and orbits the Sun. There is motion, too, within our bodies, including the pulsations of the heart and the flow of blood through the circulatory system. One of the principal tasks of physics is to describe motion in all of its variety, and it is with the description of motion—a subject called *kinematics*—that we begin our study of physics.

In general, objects can move in all three dimensions: forward and back, left and right, and up and down. In this chapter, however, we'll concentrate on the simpler case of **motion in one dimension** (also called *linear motion*), or motion in a straight line. Examples of motion in one dimension include an airliner accelerating down the runway before it takes off, a blood cell moving along a capillary, and a rocket climbing upward from the launch pad. All of the concepts that we'll study in this chapter apply directly to motion in two or three dimensions, which we'll study in Chapter 3. You'll use the ideas of this chapter and the next throughout your study of physics, so the time you spend understanding them now will be an excellent investment for the future.

TAKE-HOME MESSAGE FOR Section 2-1

✔ Motion in one dimension (also called *linear motion*) is motion in a straight line.

✔ We'll build on the ideas of motion in one dimension in later chapters to understand more complex kinds of motion, such as motion in two or three dimensions.

2-2 Constant velocity means moving at a steady speed in the same direction

Figure 2-1 shows objects moving in one dimension: three swimmers traveling down the lanes of a swimming pool. Their motion is particularly simple because each swimmer has a **constant velocity**: The swimmer's speed stays the same, and the swimmer always moves in the same direction. Let's look at this kind of motion more closely.

Each swimmer moves with a steady speed and always moves in the same direction. Hence each swimmer has a **constant velocity**.

Fuse/Getty Images

Figure 2-1 Constant velocity An object has a constant velocity when its speed remains the same (it neither speeds up nor slows down) and its direction of motion remains the same. These three swimmers all move in the same direction yet have different constant velocities because their speeds are different: The middle swimmer (who reaches the end of the pool first) is the fastest.

Coordinates, Displacement, and Average Velocity

A useful way to keep track of the changing position of a moving object (such as a swimmer) is a **coordinate system** as shown in Figure 2-2a. We use the symbol x to denote the position of the object at a given time t. The value of x can be positive or negative, depending on where the object is relative to a reference point, or **origin**, that we take to be at $x = 0$. We can choose the origin to be anywhere that's convenient. For example, we might choose $x = 0$ to be the point where we're sitting alongside a swimming pool. If our seat is 10.0 meters (10.0 m for short) from one end of a swimming pool with a length of 50.0 m, then one end of the pool is at $x = -10.0$ m and the other end is at $x = +40.0$ m. We are free to choose the positive x direction (the direction in which x increases) as we wish; in Figure 2-2a, we've chosen it to be to the right.

The **coordinates**, or **position**, of an object relative to an origin are what we call **vector** quantities. Vectors are special quantities that contain two kinds of information: size, or **magnitude**, and direction. For instance, from $x = 0$ the magnitude of the position of the right-hand end of the pool in Figure 2-2a is 40.0 m, and to get there from $x = 0$ you must travel in the positive x direction. We summarize these statements by saying that this end is at $x = +40.0$ m. The left-hand end of the pool in Figure 2-2a is 10.0 m away from $x = 0$, and to get there you must travel in the negative x direction (the direction in which x decreases). So the position of this end is $x = -10.0$ m.

Distance, by comparison, is *not* a vector quantity: It does not contain information about direction. (If you say something is 50.0 m away, you're stating its distance; if you say it's 50.0 m away to the east, you're stating its position.) That's why distance, unlike position, is always given as a positive number. If you were asked, "How far is it from New York to Boston?" you would never answer, "Negative 300 km!"

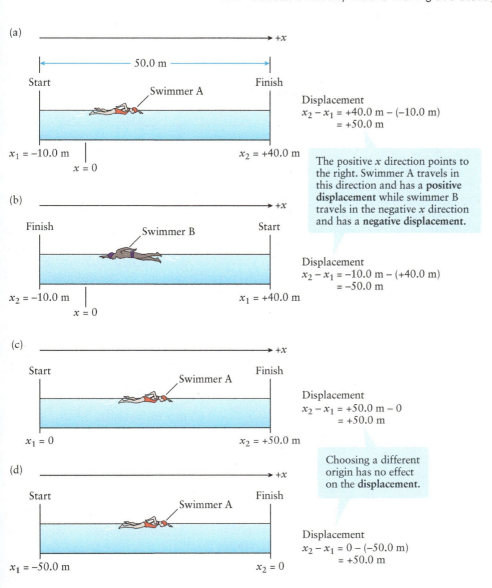

(a)

50.0 m

Start — Finish

Swimmer A

$x_1 = -10.0$ m $x_2 = +40.0$ m

$x = 0$

Displacement
$x_2 - x_1 = +40.0$ m $- (-10.0$ m$)$
$= +50.0$ m

(b)

Finish — Start

Swimmer B

$x_2 = -10.0$ m $x_1 = +40.0$ m

$x = 0$

Displacement
$x_2 - x_1 = -10.0$ m $- (+40.0$ m$)$
$= -50.0$ m

The positive x direction points to the right. Swimmer A travels in this direction and has a **positive displacement** while swimmer B travels in the negative x direction and has a **negative displacement**.

(c)

Start — Finish

Swimmer A

$x_1 = 0$ $x_2 = +50.0$ m

Displacement
$x_2 - x_1 = +50.0$ m $- 0$
$= +50.0$ m

Choosing a different origin has no effect on the **displacement**.

(d)

Start — Finish

Swimmer A

$x_1 = -50.0$ m $x_2 = 0$

Displacement
$x_2 - x_1 = 0 - (-50.0$ m$)$
$= +50.0$ m

Figure 2-2 Coordinate systems and displacement With our choice of coordinate system, as they travel from the starting line to the finish line (a) swimmer A has a positive displacement and (b) swimmer B has a negative displacement. (c), (d) Displacement is independent of the coordinate system.

WATCH OUT! A quantity that can be positive or negative might not be a vector.

Some quantities that can be either positive or negative are *not* vectors. For instance, the temperature might be +35°C at noon in midsummer but –10°C on a cold winter night. But there is no *direction* such as "up" or "left" associated with temperature, so it is not a vector quantity. You need to rely on the context to decide whether the sign of a quantity refers to a direction.

The **displacement** of an object is the difference between its positions at two different times. It tells us how far and in what direction the object moves, so displacement, like position, is a vector. For example, if swimmer A travels the length of the swimming pool from left to right (Figure 2-2a), she starts at $x_1 = -10.0$ m and ends at $x_2 = +40.0$ m. (The subscripts "1" and "2" help us keep track of which position is which.) Her displacement is the *later* position minus the *earlier* position: $x_2 - x_1 = +40.0$ m $- (-10.0$ m$) = +50.0$ m. The positive value of displacement means that her value of x increases and she moves 50.0 m in the positive x direction. By comparison, swimmer B (Figure 2-2b) travels the length of the pool in the opposite direction. For him $x_1 = +40.0$ m and $x_2 = -10.0$ m, and his displacement is $x_2 - x_1 = (-10.0$ m$) - 40.0$ m $= -50.0$ m. Swimmer B moves 50.0 m in the negative x direction, and his displacement is negative. The *sign* (plus or minus) of the displacement tells you the *direction* in which each swimmer moves.

WATCH OUT! Displacement and distance are not the same thing.

Remember that distance has no direction and is always a positive number. But displacement, like position, is a vector quantity that has both a magnitude and a direction. Swimmer A in Figure 2-2a and swimmer B in Figure 2-2b both travel the same distance, 50.0 m, but their displacements are different (+50.0 m for swimmer A, −50.0 m for swimmer B) because the *directions* of their motion are different.

Note that the displacement of either swimmer does *not* depend on our choice of origin. As an example, if we choose $x = 0$ to be at the left-hand end of the pool, the right-hand end is at $x = 50.0$ m, and the displacement of swimmer A is $x_2 - x_1 = 50.0$ m $- 0 = 50.0$ m (Figure 2-2c). If instead we choose $x = 0$ to be at the right-hand end of the pool, the left-hand end is at $x = -50.0$ m, and swimmer A again has displacement $x_2 - x_1 = 0 - (-50.0$ m$) = 50.0$ m (Figure 2-2d).

To determine how *fast* an object moves, we define a new quantity, **average speed**. This is the total distance the object travels divided by the amount of time it takes the object to travel that distance—that is, the **elapsed time** or *duration* of that **time interval**. We use the symbol t to denote a value of the time, and the symbol $v_{average}$ ("*v*-average") for average speed. For example, suppose you start timing swimmer A just as she starts off at $x_1 = -10.0$ m, so the time at this instant is $t_1 = 0$ (Figure 2-2a). If she reaches the other end of the pool at time $t_2 = 25.0$ s, the elapsed time is $t_2 - t_1 = 25.0$ s $- 0 = 25.0$ s. She traveled a distance of 50.0 m in that time, so her average speed is

$$v_{average} = \frac{distance}{time\ interval} = \frac{50.0\ m}{25.0\ s} = 2.00\ m/s\ for\ swimmer\ A$$

Slow-moving swimmer B (Figure 2-2b) also travels a distance of 50.0 m, but it takes him 100.0 s. His average speed is

$$v_{average} = \frac{distance}{time\ interval} = \frac{50.0\ m}{100.0\ s} = 0.500\ m/s\ for\ swimmer\ B$$

Note that average speed, like distance, tells you nothing about the direction of motion. Like distance, average speed is never negative.

To get direction information, we need to calculate the **average velocity** of an object, which is a vector quantity. We use the symbol $v_{average,x}$ ("*v*-average-*x*") for average velocity. This is very close to the symbol for average speed but includes the subscript x to remind us that the object can move in either the positive or negative x direction. The average velocity equals the displacement divided by the elapsed time for that displacement. For swimmer A in Figure 2-2a, who starts at position $x = -10.0$ m and ends at position $x = +40.0$ m after 25.0 s, the average velocity is

$$v_{average,x} = \frac{x_2 - x_1}{t_2 - t_1} = \frac{40.0\ m - (-10.0\ m)}{25.0\ s - 0}$$

$$= \frac{50.0\ m}{25.0\ s} = +2.00\ m/s\ for\ swimmer\ A$$

So swimmer A moves in the positive x direction (as shown by the positive sign of $v_{average,x}$), or from left to right in Figure 2-2a, at an average *speed* of 2.00 meters per second. The standard SI units for average velocity and average speed are meters per second (m/s). Table 2-1 lists some other common units for these quantities.

WATCH OUT! Speed and velocity are not the same thing.

Average speed and average velocity are related but are *not* the same quantity. Average speed $v_{average}$ is equal to the *distance* traveled divided by the elapsed time. Distances are always positive, so average speed is always positive. Average velocity $v_{average,x}$ can be positive or negative, depending on the direction in which the object moves. Average velocity is a vector.

TABLE 2-1	**Units of Velocity and Speed**

The SI unit of velocity and speed is meters per second (m/s). Here are some other common units:

1 kilometer per hour (km/h) = 0.2778 m/s
1 mile per hour (mi/h) = 0.4470 m/s = 1.609 km/h
1 foot per second (ft/s) = 0.3048 m/s = 1.097 km/h = 0.6818 mi/h
1 knot = 1 nautical mile per hour = 0.5148 m/s = 1.853 km/h = 1.152 mi/h = 1.689 ft/s

By comparison, swimmer B in Figure 2-2b starts at $x_1 = +40.0$ m at $t_1 = 0$, and at $t_2 = 100.0$ s he reaches $x_2 = -10.0$ m. Hence the average velocity of Swimmer B is

$$v_{average,x} = \frac{x_2 - x_1}{t_2 - t_1} = \frac{-10.0 \text{ m} - 40.0 \text{ m}}{100.0 \text{ s} - 0}$$

$$= \frac{-50.0 \text{ m}}{100.0 \text{ s}} = -0.500 \text{ m/s for swimmer B}$$

Swimmer B moves at only 0.500 m/s. The minus sign of $v_{average,x}$ says that he travels in the negative x direction, or from right to left in Figure 2-2b. Equation 2-1 summarizes these ideas about average velocity:

Average velocity of an object in motion in one dimension

Displacement (change in position) of the object over a certain time interval: The object moves from x_1 to x_2, so $\Delta x = x_2 - x_1$.

$$v_{average,x} = \frac{x_2 - x_1}{t_2 - t_1} = \frac{\Delta x}{\Delta t}$$

For both the displacement and the elapsed time, subtract the earlier value (x_1 or t_1) from the later value (x_2 or t_2).

Elapsed time for the time interval: The object is at x_1 at time t_1 and x_2 at time t_2, so the elapsed time is $\Delta t = t_2 - t_1$.

Average velocity for motion in one dimension equals displacement divided by elapsed time (2-1)

WATCH OUT! Be sure you understand the meaning of Δ.

! The symbol Δx does *not* mean a quantity Δ multiplied by a quantity x! Instead, it means the *change* in the value of x.

In Equation 2-1, Δx ("delta-x") is an abbreviation for the change $x_2 - x_1$ in the position x, and Δt ("delta-t") is an abbreviation for the change $t_2 - t_1$ in the time t; that is, the elapsed time. Throughout our study of physics we'll use the symbol Δ ("delta") to represent the change in a quantity.

WATCH OUT! Average speed is not always the magnitude of the average velocity.

! The motion of the swimmers may make you think that average speed is simply the magnitude of average velocity. That's true if the object moves in a straight line and doesn't turn around. But consider a swimmer who does a complete lap in the pool, and so starts and ends at the same location. Her total displacement is zero, and her average *velocity* (displacement divided by time interval) is also zero. But was her average *speed* also zero? No! She traveled a nonzero distance equal to two lengths of the pool. So her average speed (distance divided by time interval) is *not* zero, and is not the magnitude of her average velocity.

Constant Velocity, Motion Diagrams, and Graphs of *x* versus *t*

We've calculated the average velocity for each swimmer's entire trip from one end of the pool to the other. But we can also calculate the average velocity for any *segment* of a swimmer's trip between the two ends. A swimmer has **constant velocity** if the average velocity calculated for *any* segment of the trip has the same value as for any other segment. As an example, in Figure 2-3a swimmer A is at $x_1 = 0.0$ m at $t_1 = 5.0$ s and at $x_2 = 30.0$ m at $t_2 = 20.0$ s. Her average velocity for this part of her motion is

$$v_{average,x} = \frac{x_2 - x_1}{t_2 - t_1} = \frac{30.0 \text{ m} - 0.0 \text{ m}}{20.0 \text{ s} - 5.0 \text{ s}}$$

$$= \frac{30.0 \text{ m}}{15.0 \text{ s}} = +2.00 \text{ m/s for swimmer A}$$

That's the same as we calculated above for the entire trip from one end of the pool to the other. If the average velocity has the same value for all segments, we call it simply the *velocity* and give it the symbol v_x ("*v-x*").

Figure 2-3a shows how to depict swimmer A's entire motion in a way that makes it clear that her velocity is constant. We draw a dot to represent her position at equal time intervals (say, $t = 0$, $t = 5.0$ s, $t = 10.0$ s, $t = 15.0$ s, and so on). Taken together, these dots make up a **motion diagram**. The spacing between adjacent dots shows you the distance that the object traveled during the corresponding time interval. The spacing is the same between any two adjacent dots in Figure 2-3a, which tells you that swimmer A travels equal distances in the +*x* direction—that is, has equal displacement—in equal time intervals. That's just what we mean by saying that swimmer A has constant velocity.

Figure 2-3b shows a motion diagram for swimmer B. Since he moves more slowly than swimmer A (speed 0.5 m/s compared to 2.00 m/s), he travels a smaller distance in the same time interval. Hence the dots in swimmer B's motion diagram are more closely spaced than those for swimmer A.

Yet another useful way to depict motion is in terms of a graph of position versus time, also known as an *x–t* **graph**. In Figure 2-4a, we've turned swimmer A's motion diagram on its side so that the *x* axis now runs upward. Then to emphasize that each dot in the diagram corresponds to a specific time *t*, we've moved each dot to the right by an amount that corresponds to the value of *t*. We add a horizontal axis for time to make it easy to read the value of *t* that corresponds to each dot. The result is an *x–t* graph, a graph of the swimmer's coordinate *x* versus the time *t*. To finish the graph, we draw a smooth curve connecting the dots. For this constant-velocity motion, the graph is a straight line.

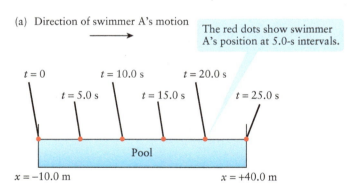

(a) Direction of swimmer A's motion

The red dots show swimmer A's position at 5.0-s intervals.

$t = 0$ $t = 5.0$ s $t = 10.0$ s $t = 15.0$ s $t = 20.0$ s $t = 25.0$ s

Pool

$x = -10.0$ m $x = +40.0$ m

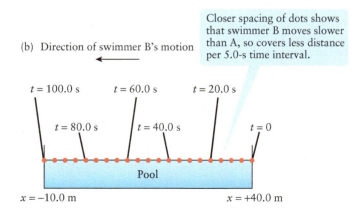

Closer spacing of dots shows that swimmer B moves slower than A, so covers less distance per 5.0-s time interval.

(b) Direction of swimmer B's motion

$t = 100.0$ s $t = 80.0$ s $t = 60.0$ s $t = 40.0$ s $t = 20.0$ s $t = 0$

Pool

$x = -10.0$ m $x = +40.0$ m

Figure 2-3 Motion diagrams for constant velocity A series of dots shows the positions of (a) swimmer A and (b) swimmer B at equal time intervals.

WATCH OUT! Horizontal motion can be graphed on the vertical axis.

! Although the swimmer's motion in the x direction in Figure 2-3a is *horizontal*, we've drawn the x–t graph of Figure 2-4a with the x axis *vertical*. We do this because in graphs that show how a quantity varies with time, we always put time on the horizontal axis. For example, a graph of how a

stock mutual fund performs always has time (that is, different dates) on the horizontal axis and the value of the fund on the vertical axis. That's why we put the coordinate x on the vertical axis, no matter how the object's motion is actually oriented in space.

Figure 2-4b shows the x–t graph for swimmer B. This swimmer also has a constant velocity, and the x–t graph is again a straight line. However, this graph differs from the x–t graph for swimmer A (Figure 2-4a) in two ways: where the line touches the vertical axis, and the *slope* of the line (how steep it is and whether it slopes up or down).

The point on the graph where the line touches the vertical (x) axis shows you the position x of the object at $t = 0$. We give this the symbol x_0 ("x-zero"). If the motion starts at $t = 0$, then x_0 represents the *initial position* of the object. The slope of the line tells you the value of the object's *velocity*. If you consider two points on the line, corresponding to times t_1 and t_2 and positions x_1 and x_2, the slope is the vertical difference $x_2 - x_1$ between those two points (the "rise" of the graph) divided by the horizontal

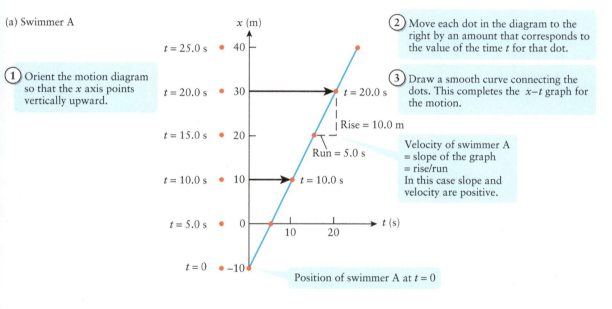

(a) Swimmer A

1 Orient the motion diagram so that the x axis points vertically upward.

2 Move each dot in the diagram to the right by an amount that corresponds to the value of the time t for that dot.

3 Draw a smooth curve connecting the dots. This completes the x–t graph for the motion.

Rise = 10.0 m

Run = 5.0 s

Velocity of swimmer A = slope of the graph = rise/run
In this case slope and velocity are positive.

Position of swimmer A at $t = 0$

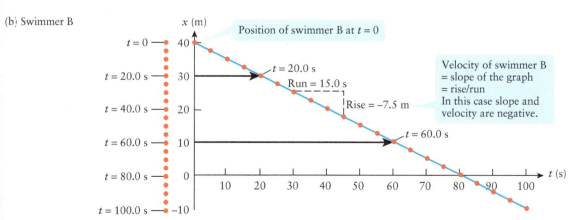

(b) Swimmer B

Position of swimmer B at $t = 0$

Run = 15.0 s

Rise = −7.5 m

Velocity of swimmer B = slope of the graph = rise/run
In this case slope and velocity are negative.

Figure 2-4 From motion diagrams to *x*–*t* graphs: Constant velocity How to convert the motion diagrams from Figure 2-3 for (a) swimmer A and (b) swimmer B into graphs of position x versus time t. If the velocity is constant, the x–t graph is a straight line. The slope (rise divided by run) of the straight line equals the velocity.

difference $t_2 - t_1$ (the "run" of the graph). But that's just our definition of the average velocity, Equation 2-1. The x–t graph in Figure 2-4a has a steep slope (so swimmer A is moving rapidly) and slopes upward (so she is moving in the positive x direction). In Figure 2-4b the x–t graph has a shallow slope (so swimmer B is moving slowly) and slopes downward (so he is moving in the negative x direction).

The Equation for Constant-Velocity Motion in One Dimension

We can write a simple and useful equation for constant-velocity motion by starting with Equation 2-1, $v_{average,x} = (x_2 - x_1)/(t_2 - t_1)$. Since the velocity is constant, we replace $v_{average,x}$ by v_x. We choose the earlier time t_1 in this equation to be $t = 0$ and use x_0 instead of x_1 as the symbol for the object's position at this time. We also use the symbols t and x instead of t_2 and x_2 to represent the later time and the object's position at that time. Then Equation 2-1 becomes

$$v_x = \frac{x - x_0}{t - 0} = \frac{x - x_0}{t}$$

We multiply both sides of this equation by t then add x_0 to both sides. The result is

Position versus time for motion in one dimension with constant velocity (2-2)

Position at time t of an object in motion in one dimension with constant velocity · Position at time $t = 0$ of the object

$$x = x_0 + v_x t$$

Constant velocity of the object · Time at which the object is at position x

This equation gives the position x of the object as a function of the time t. As we'll see below, Equation 2-2 is an important tool for solving problems about motion in one dimension with constant velocity. But even with this equation, an essential part of solving any such problem is drawing x–t graphs like those shown in Figure 2-4.

→ *Go to Interactive Exercise 2-1 for more practice dealing with velocity.*

GOT THE CONCEPT? 2-1 Interpreting x–t Graphs

(?) Figure 2-5 shows x–t graphs for four objects, all drawn on the same axes. Each object moves at a constant velocity from $t = 0$ to $t = 5.0$ s. (a) Which object moves at the greatest speed? (b) Which one travels the greatest distance from $t = 0$ to $t = 5.0$ s? (c) Which one has the most positive displacement from $t = 0$ to $t = 5.0$ s? (d) Which one has the most negative displacement from $t = 0$ to $t = 5.0$ s? (e) Which one ends up at the most positive value of x? (f) Which one ends up at the most negative value of x?

(Answers to *Got the Concept?* questions can be found at the end of the chapter.)

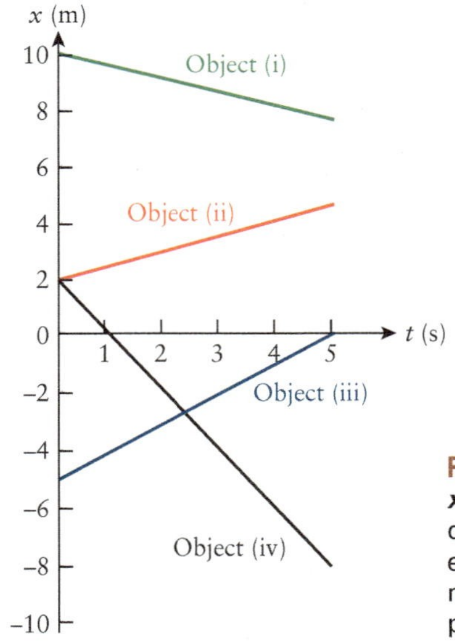

Figure 2-5 Four different x–t graphs Each graph depicts a different example of straight-line motion. What are the properties of each motion?

EXAMPLE 2-1 Average Velocity

You drive from Bismarck, North Dakota, to Fargo, North Dakota, on Interstate Highway 94, an approximately straight road 315 km in length. You leave Bismarck and travel 210 km at constant velocity in 2.50 h, then stop for 0.50 h at a rest area. You then drive the remaining 105 km to Fargo at constant velocity in 1.00 h. You then turn around immediately and drive back to Bismarck nonstop at constant velocity in 2.75 h. (a) Draw an x–t graph for the entire round trip. Then calculate the average velocity for (b) the trip from Bismarck to the rest area, (c) the trip from the rest area to Fargo, (d) the entire outbound trip from Bismarck to Fargo, (e) the return trip from Fargo to Bismarck, and (f) the round trip from Bismarck to Fargo and back to Bismarck.

Set Up

This is an example of motion in one dimension along the straight highway between Bismarck and Fargo. We set up our coordinates as shown, with $x = 0$ at the starting position (Bismarck) and the positive x direction from Bismarck toward Fargo. We also choose $t = 0$ to be when the car leaves Bismarck. Using Equation 2-1 will tell us the average velocity for each portion of the trip.

Average velocity for motion in one dimension:

$$v_{average,x} = \frac{x_2 - x_1}{t_2 - t_1}$$

$$= \frac{\Delta x}{\Delta t} \qquad (2\text{-}1)$$

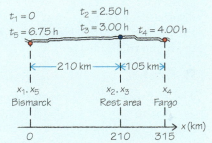

Solve

(a) To draw the x–t graph for the car's motion, we first put a dot on the graph to indicate the car's position and the corresponding time for the beginning and end of each leg of the trip. The velocity is constant on each leg, so we draw straight lines connecting successive points. The slope of the graph is positive (upward) when the car is moving in the positive x direction toward Fargo and negative (downward) when the car is moving in the negative x direction back toward Bismarck. The slope is *zero* (the line is flat) while the car is at the rest area and not moving.

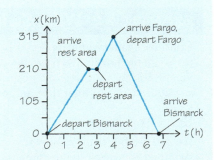

(b) Calculate the average velocity for the trip from Bismarck (position $x_1 = 0$ at time $t_1 = 0$) to the rest area (position $x_2 = 210$ km at time $t_2 = 2.50$ h).

Bismarck to rest area:

$$v_{average,x} = \frac{\Delta x}{\Delta t} = \frac{x_2 - x_1}{t_2 - t_1}$$

$$= \frac{210 \text{ km} - 0}{2.50 \text{ h} - 0} = +84.0 \text{ km/h}$$

(c) Calculate the average velocity for the trip from the rest area (position $x_3 = 210$ km at time $t_3 = 2.50$ h $+ 0.50$ h $= 3.00$ h) to Fargo (position $x_4 = 315$ km at time $t_4 = 3.00$ h $+ 1.00$ h $= 4.00$ h).

Rest area to Fargo:

$$v_{average,x} = \frac{\Delta x}{\Delta t} = \frac{x_4 - x_3}{t_4 - t_3}$$

$$= \frac{315 \text{ km} - 210 \text{ km}}{4.00 \text{ h} - 3.00 \text{ h}} = +105 \text{ km/h}$$

(d) Calculate the average velocity for the entire outbound trip from Bismarck (position $x_1 = 0$ at time $t_1 = 0$) to Fargo (position $x_4 = 315$ km at time $t_4 = 4.00$ h).

Bismarck to Fargo

$$v_{average,x} = \frac{\Delta x}{\Delta t} = \frac{x_4 - x_1}{t_4 - t_1}$$

$$= \frac{315 \text{ km} - 0}{4.00 \text{ h} - 0} = +78.8 \text{ km/h}$$

(e) Calculate the average velocity for the return trip from Fargo (position $x_4 = 315$ km at time $t_4 = 4.00$ h) to Bismarck (position $x_5 = 0$ at time $t_5 = 4.00$ h $+ 2.75$ h $= 6.75$ h).

Fargo back to Bismarck:

$$v_{\text{average},x} = \frac{\Delta x}{\Delta t} = \frac{x_5 - x_4}{t_5 - t_4}$$

$$= \frac{0 - 315 \text{ km}}{6.75 \text{ h} - 4.00 \text{ h}} = -115 \text{ km/h}$$

(f) For the round trip from Bismarck (position $x_1 = 0$ at time $t_1 = 0$) to Fargo and back to Bismarck (position $x_5 = 0$ at time $t_5 = 6.75$ h), the net displacement $\Delta x = x_5 - x_1$ is *zero*: The car ends up back where it started. Hence the average velocity for the round trip is also zero.

Bismarck to Fargo and back to Bismarck:

$$v_{\text{average},x} = \frac{\Delta x}{\Delta t} = \frac{x_5 - x_1}{t_5 - t_1}$$

$$= \frac{0 - 0}{6.75 \text{ h} - 0} = 0$$

Reflect

The average velocity is positive for the trips from Bismarck toward Fargo (since the car moves in the positive x direction) and negative for the trip from Fargo back to Bismarck (since the car moves in the negative x direction). The result $v_{\text{average},x} = 0$ for the round trip means that the net displacement for the round trip was zero. The average *speed* for the trip was definitely not zero, however: The car traveled a total distance of $2(315 \text{ km}) = 630$ km in a total elapsed time of 6.75 h, so the average speed was $(630 \text{ km})/(6.75 \text{ h}) = 93.3$ km/h. Displacement and distance are not the same thing; likewise, average velocity and average speed are not the same thing.

EXAMPLE 2-2 When Swimmers Pass

Consider again swimmers A and B, whose motion is depicted in Figure 2-3. Both of them begin swimming at $t = 0$. (a) Write the position-versus-time equation, Equation 2-2, for each swimmer. Use the coordinate system shown in Figure 2-3. (b) Calculate the position of each swimmer at $t = 10.0$ s. (c) Calculate the time when each swimmer is at $x = 20.0$ m. (d) At what time and position do the two swimmers pass each other?

Set Up

We can use Equation 2-2 because each swimmer is moving in one dimension with constant velocity. Figure 2-3 shows where each swimmer starts and each swimmer's constant velocity, which are just the quantities we need to substitute into Equation 2-2. (We use subscripts "A" and "B" to denote the quantities that pertain to each swimmer.)

$$x = x_0 + v_x t \qquad (2\text{-}2)$$

Swimmer A starts at $x = -10.0$ m at $t = 0$ and has velocity $+2.00$ m/s, so

$$x_{A0} = -10.0 \text{ m}$$
$$v_{Ax} = +2.00 \text{ m/s}$$

Swimmer B starts at $x = +40.0$ m at $t = 0$ and has velocity -0.500 m/s, so

$$x_{B0} = +40.0 \text{ m}$$
$$v_{Bx} = -0.500 \text{ m/s}$$

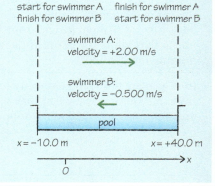

Solve

(a) Use the given information to write Equation 2-2 for each swimmer. It's useful to draw the x–t graphs for both swimmers on the same graph (see Figure 2-4).

Swimmer A:

$$x_A = x_{A0} + v_{Ax} t \quad \text{so}$$
$$x_A = -10.0 \text{ m} + (2.00 \text{ m/s})t$$

Swimmer B:

$$x_B = x_{B0} + v_{Bx} t \quad \text{so}$$
$$x_B = +40.0 \text{ m} + (-0.500 \text{ m/s})t$$

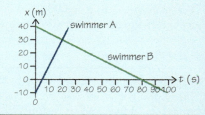

(b) To find the position of each swimmer at $t = 10.0$ s, just substitute this value of t into the equations for x_A and x_B from (a).

Swimmer A:

$$x_A = -10.0 \text{ m} + (2.00 \text{ m/s})(10.0 \text{ s})$$
$$= -10.0 \text{ m} + 20.0 \text{ m}$$
$$= +10.0 \text{ m}$$

Swimmer B:

$$x_B = +40.0 \text{ m} + (-0.500 \text{ m/s})(10.0 \text{ s})$$
$$= +40.0 \text{ m} + (-5.00 \text{ m})$$
$$= +35.0 \text{ m}$$

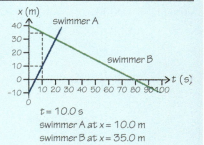

(c) To solve for the times when the two swimmers are at $x = 20.0$ m, we must first rearrange the equations from (a) so that t (the quantity we are trying to find) is by itself on one side of each equation. We then substitute 20.0 m for both x_A and x_B.

Swimmer A:

$x_A = -10.0$ m $+ (2.00$ m/s$)t$ so
$x_A + 10.0$ m $= (2.00$ m/s$)t$

$$t = \frac{x_A + 10.0 \text{ m}}{2.00 \text{ m/s}}$$

Substitute $x_A = 20.0$ m:

$$t = \frac{20.0 \text{ m} + 10.0 \text{ m}}{2.00 \text{ m/s}} = 15.0 \text{ s for A}$$

Swimmer B:

$x_B = +40.0$ m $+ (-0.500$ m/s$)t$ so
$x_B - 40.0$ m $= (-0.500$ m/s$)t$

$$t = \frac{x_B - 40.0 \text{ m}}{(-0.500 \text{ m/s})}$$

Substitute $x_B = 20.0$ m:

$$t = \frac{20.0 \text{ m} - 40.0 \text{ m}}{(-0.500 \text{ m/s})} = 40.0 \text{ s for B}$$

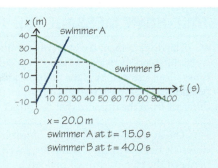

$x = 20.0$ m
swimmer A at $t = 15.0$ s
swimmer B at $t = 40.0$ s

(d) The two swimmers pass when they are at the *same* value of x (so $x_A = x_B$) at the *same* time t. We first find the time *when* this happens by setting the expressions for x_A and x_B from (a) equal to each other, then solving for the corresponding value of t.

Swimmer A:

$x_A = -10.0$ m $+ (2.00$ m/s$)t$

Swimmer B:

$x_B = +40.0$ m $+ (-0.500$ m/s$)t$

When the two swimmers pass, $x_A = x_B$

-10.0 m $+ (2.00$ m/s$)t$
$\quad = 40.0$ m $+ (-0.500$ m/s$)t$
$(2.00$ m/s$)t - (-0.500$ m/s$)t$
$\quad = 40.0$ m $- (-10.0$ m$)$
$(2.50$ m/s$)t = 50.0$ m

$$t = \frac{50.0 \text{ m}}{2.50 \text{ m/s}} = 20.0 \text{ s}$$

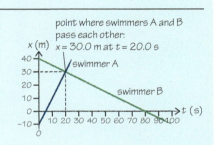

point where swimmers A and B pass each other:
$x = 30.0$ m at $t = 20.0$ s

To find *where* the two swimmers pass, substitute the value of t for when they pass into either the equation for x_A or the equation for x_B.

(Both should give the same answer, since at this time $x_A = x_B$.)

At $t = 20.0$ s swimmer A's position is

$x_A = -10.0$ m $+ (2.00$ m/s$)t$
$\quad = -10.0$ m $+ (2.00$ m/s$)(20.0$ s$)$
$\quad = 30.0$ m

$t = 20.0$ s is the time when the two swimmers pass, so the position where they pass is $x = 30.0$ m.

Reflect

To verify our results, note that swimmer B moves 1/4 as fast as swimmer A. This checks out for part (b): In 10.0 s, swimmer A moves 20.0 m in the positive x direction from her starting point at $x_{A0} = -10.0$ m, and swimmer B moves 5.00 m (that is, 1/4 as far) in the negative x direction from his starting point at $x_{B0} = +40.0$ m. It also checks out for part (c): Swimmer A travels 30.0 m in 15.0 s, while slow-moving swimmer B travels a shorter distance (20.0 m) in a longer time (40.0 s). And in part (d), the two swimmers meet 40.0 m from where swimmer A started but 10.0 m (1/4 as far) from where swimmer B started.

You can also check the answer to part (d) by substituting $t = 20.0$ s into the equation for x_B. You should find $x_B = 30.0$ m, the same as x_A (since $t = 20.0$ s is when the two swimmers pass). Do you?

GOT THE CONCEPT? 2-2 Average Velocity

It takes you 15 min to drive 6.0 mi in a straight line to the local hospital. It takes 10 min to go the last 3.0 mi, 2.0 min to go the last mile, and only 30 s ($= 0.50$ min) to go the last 0.50 mi. What is your average velocity for the trip? Take the positive x direction to be from your starting point toward the hospital.

TAKE-HOME MESSAGE FOR Section 2-2

✔ The displacement Δx of an object moving in one dimension is the difference between the object's position x at two different times (displacement Δx is the *later* position minus the *earlier* position). If the object changes direction, the displacement can be different from the distance that the object travels.

✔ The average velocity of the object between those two different times, $v_{\text{average},x}$, is the displacement Δx (the difference between the two positions) divided by the elapsed time Δt (the difference between the two times).

✔ The average speed of the object between those two different times, v_{average}, is the total distance traveled by the object divided by the elapsed time. In general, the average speed is not the same as the magnitude of the average velocity.

✔ An object has constant velocity if $v_{\text{average},x}$ has the same value for any time interval. If the velocity is constant, the object always moves in the same direction and maintains a steady speed.

✔ Motion diagrams and x–t graphs are important tools for interpreting what happens during motion in one dimension. The slope of an x–t graph equals the object's velocity.

✔ You can solve many problems that involve straight-line motion by using Equation 2-1, which relates average velocity, displacement, and elapsed time.

✔ If the velocity is constant, you can also use Equation 2-2 to relate velocity, position, and time.

2-3 Velocity is the rate of change of position, and acceleration is the rate of change of velocity

In Section 2-2 we considered motion in one dimension with constant velocity: an object that moves in a straight line with unvarying speed and always in the same direction. But in many important situations the speed, the direction of motion, or both can change as an object moves. Some examples include a car speeding up to get through an intersection before the light turns red; the same car slowing down as a police officer signals the driver to pull over; or a dog that runs away from you to pick up a thrown stick, then changes direction and returns the stick to you.

Whenever an object changes its speed or direction of motion—that is, whenever its velocity changes—we say that the object *accelerates* and undergoes an *acceleration*. Before we study acceleration, however, we first need to take a closer look at the idea of velocity.

Instantaneous Velocity

Figure 2-6 is a motion diagram of an object that moves with a changing velocity: a jogger who starts at rest, speeds up, then slows down toward the end of a 100-m run that takes him 38.0 s. The dots show his position at equal time intervals of 2.0 s. Because he speeds up and slows down, these dots are *not* equally spaced: His displacement, and so his average velocity, is *not* the same for all 2.0-s intervals.

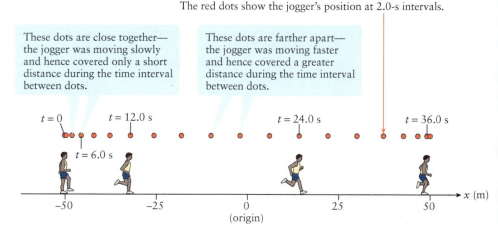

The red dots show the jogger's position at 2.0-s intervals.

These dots are close together—the jogger was moving slowly and hence covered only a short distance during the time interval between dots.

These dots are farther apart—the jogger was moving faster and hence covered a greater distance during the time interval between dots.

$t = 0$ $t = 12.0$ s $t = 24.0$ s $t = 36.0$ s

$t = 6.0$ s

−50   −25   0 (origin)   25   50 x (m)

Figure 2-6 Motion with varying velocity A jogger's velocity changes as he speeds up and slows down. This motion diagram shows these velocity changes.

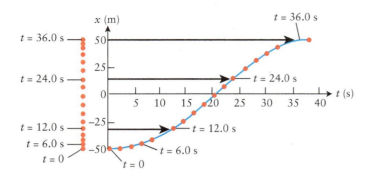

Figure 2-7 From motion diagram to *x*–*t* graph: Varying velocity Converting the jogger's motion diagram from Figure 2-6 to an *x*–*t* graph. Compare to Figure 2-4.

Figure 2-7 shows the *x*–*t* graph for the jogger. The graph is not a straight line because the velocity is not constant. In Section 2-2 we saw that if the velocity *is* constant, as for the two swimmers whose motion is graphed in Figure 2-4, the slope of the straight-line *x*–*t* graph tells us the value of the velocity. We'll now show that the same is true when the velocity is not constant.

In Figure 2-8a, we've picked out two points during the jogger's motion, representing a time interval from $t = 6.0$ s to $t = 24.0$ s. The rise of a straight line connecting these points on the *x*–*t* graph equals the displacement Δx during the interval, and the run of this line equals the duration Δt of the interval. From Equation 2-1, the average velocity for this interval is $v_{average,x} = \Delta x/\Delta t$; this is the rise divided by the run, or the *slope* of this straight line. For the case shown in Figure 2-8a, $\Delta x = +59.5$ m, and $\Delta t = 18.0$ s, so $v_{average,x} = (+59.5$ m$)/(18.0$ s$) = +3.31$ m/s. The plus sign means that the straight line slopes upward, which tells us that *x* increased during the time interval and the jogger moved in the positive *x* direction.

The straight line in Figure 2-8a doesn't exactly match the shape of the *x*–*t* curve between the two points. However, if we decrease the time interval from $t = 6.0$ s to

(a) Average *x* velocity on an *x*–*t* graph

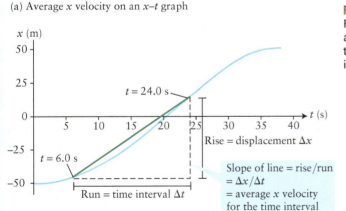

Figure 2-8 Average and instantaneous velocity Finding the average velocity of the jogger over shorter and shorter time intervals. The slope of a line that is tangent to the *x*–*t* graph at a given time *t* gives the instantaneous velocity at time *t*.

Rise = displacement Δx

Slope of line = rise/run
= $\Delta x/\Delta t$
= average *x* velocity for the time interval

(b) Average *x* velocity for a shorter time interval

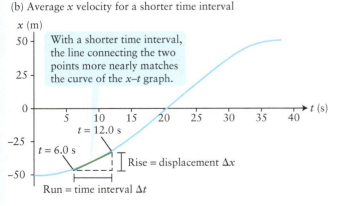

(c) Instantaneous *x* velocity on an *x*–*t* graph

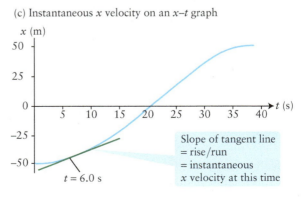

Slope of tangent line
= rise/run
= instantaneous
x velocity at this time

WATCH OUT! 🔆

Instantaneous speed is the magnitude of instantaneous velocity.

⚠ Unlike average speed, which may or may not equal the absolute value of the average velocity, the *instantaneous* speed is always equal to the absolute value of the *instantaneous* velocity.

$t = 12.0$ s, as in Figure 2-8b, the straight line is a better match to the *x–t* curve. And if we make the time interval *very* short—say, from $t = 6.000$ s to 6.001 s—the straight line becomes a nearly perfect match to the *x–t* curve during that interval. Indeed, if we make the time interval infinitesimally small in duration, the straight line has the same slope as a line tangent to the *x–t* curve, as in Figure 2-8c. This slope equals the average velocity during an infinitesimally brief interval around an instant. We call this quantity v_x (with no "average"), or the **instantaneous velocity** at the instant in question. (We used the same symbol v_x in Section 2-2 for objects that move with constant velocity. If the velocity is constant, v_x has the same value at all times.)

The sign of the instantaneous velocity tells us what direction the object is moving at the instant in question: v_x is positive if the object is moving in the positive *x* direction, and v_x is negative if the object is moving in the negative *x* direction. The absolute value or magnitude of v_x is the **instantaneous speed** of the object, which we denote by the symbol v without a subscript. Instantaneous speed is always positive or zero, never negative. A car's speedometer is a familiar device for measuring instantaneous speed.

As an example, consider the jogger's *x–t* graph in Figure 2-8c and the tangent line at the point for $t = 6.0$ s. If you carefully measure the rise and the run of the tangent line, then take their quotient, you'll find that the slope of this line is $+1.50$ m/s. So at $t = 6.0$ s the jogger's instantaneous velocity is $v_x = +1.50$ m/s; at this instant he is moving in the positive *x* direction at speed $v = 1.50$ m/s. Henceforth we'll use the terms **velocity** to refer to instantaneous velocity (which we'll use much more often than average velocity) and **speed** to refer to instantaneous speed.

Mathematically, an object's velocity v_x is the *rate of change* of the object's position as given by the coordinate *x*. The faster the position changes, the greater the magnitude of the velocity v_x and the greater the speed of the object. Positive v_x means that *x* is increasing, and the object is moving in the positive *x* direction; negative v_x means that *x* is decreasing, and the object is moving in the negative *x* direction.

EXAMPLE 2-3 **Decoding an *x–t* Graph**

Figure 2-9 shows an *x–t* graph for the motion of an object. (a) Determine the direction in which the object is moving at $t = 0$, 10 s, 20 s, 30 s, and 40 s. (b) Describe the object's motion in words and draw a motion diagram for the object for the period from $t = 0$ to $t = 40$ s.

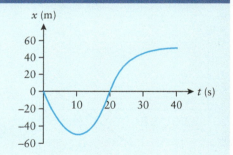

Figure 2-9 An *x–t* graph to interpret How must an object move to produce this *x–t* graph?

Set Up

We'll use the idea that the slope of the *x–t* graph for an object tells us the object's velocity v_x. The algebraic sign of the velocity (plus or minus) tells us whether the object is moving in the positive or negative *x* direction.

v_x = slope of *x–t* graph

Solve

(a) On a copy of Figure 2-9, we've drawn green lines tangent to the blue *x–t* curve at the five times of interest. The slope is negative at $t = 0$ (point 1 in the figure), so $v_x < 0$, and the object is moving in the negative *x* direction. The object is momentarily at rest ($v_x = 0$) at $t = 10$ s (point 2 in the figure), so it is not moving in either direction. At $t = 20$ s and 30 s (points 3 and 4 in the figure) the object is moving in the positive *x* direction ($v_x > 0$); at $t = 40.0$ s (point 5 in the figure) the object is again at rest ($v_x = 0$).

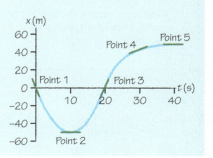

(b) At time $t = 0$ the object is at $x = 0$ and moving in the negative x direction. The x–t graph has a negative slope until $t = 10$ s, so the object continues to move in the negative x direction; because the slope of the x–t graph becomes shallower, the object is slowing down so that the dots are closer together. At $t = 10$ s, the object comes to rest momentarily at $x = -50$ m. After $t = 10$ s, the x–t graph has a positive slope, so the object moves in the positive x direction. From $t = 10$ s to 20 s, the object is speeding up because the slope of the x–t graph is increasing. The object has the fastest speed (because the graph has the greatest slope) at $t = 20$ s when the object again passes through $x = 0$. After $t = 20$ s the slope of the x–t graph is still positive but getting shallower; so the object is still moving in the positive x direction but its speed is decreasing so that the dots again get closer together. The object finally comes to rest at $x = +50$ m at $t = 40$ s.

Reflect

It's a common mistake to look at Figure 2-9 and say that the velocity at $t = 10$ s is negative. For this value of time, the curve is below the horizontal axis and so the *coordinate* x is negative. However, the x–t graph has zero *slope* at $t = 10$ s, so at this instant the coordinate x is neither increasing nor decreasing, and the velocity is *zero*. Remember that the *value* of an x–t graph shows you the object's *position* (in other words, where the object is), while the *slope* of the x–t graph tells you its *velocity* (that is, how fast and in which direction the object is moving).

Average Acceleration and Instantaneous Acceleration

Just as velocity tells you how rapidly position changes, acceleration tells you how rapidly velocity changes. Suppose an object like the jogger in Figure 2-6 has velocity v_{1x} at time t_1 and velocity v_{2x} at time t_2. We define the object's **average acceleration** $a_{\text{average},x}$ over this time interval to be the change in velocity $\Delta v_x = v_{2x} - v_{1x}$ divided by the elapsed time $\Delta t = t_2 - t_1$:

Average acceleration of an object in motion in one dimension

Change in **velocity** of the object over a certain time interval: The velocity changes from v_{1x} to v_{2x}.

$$a_{\text{average},x} = \frac{\Delta v_x}{\Delta t} = \frac{v_{2x} - v_{1x}}{t_2 - t_1}$$

Average acceleration for motion in one dimension

(2-3)

Elapsed time for the time interval: The object has velocity v_{1x} at time t_1 and has velocity v_{2x} at time t_2.

If a race car initially at rest blasts away from the starting line and rapidly comes to its top speed, its velocity changes substantially (Δv_x is large) in a short time (Δt is small), and its average acceleration $a_{\text{average},x}$ has a large magnitude. If, instead, an elderly horse starts from rest and gradually speeds up to a slow amble, the horse's velocity changes only a little (Δv_x is small) and takes a long time to change (Δt is large). In this case the horse's average acceleration $a_{\text{average},x}$ has only a small magnitude.

Velocity has units of meters per second (m/s) and time has units of seconds (s), so the SI units of acceleration are meters per second per second, or meters per second squared (m/s²). Saying that an object has an acceleration of $+2.0$ m/s² means that the object's velocity becomes more positive by 2.0 m/s every second; an acceleration of -4.5 m/s² means that the velocity becomes more negative by 4.5 m/s every second.

Like velocity, average acceleration is a vector and contains direction information, as we explain below.

Interpreting Positive and Negative Acceleration

We saw in Section 2-2 that a positive velocity means that the object is moving in the positive x direction; a negative velocity means the object is moving in the negative x direction. In a similar way, the sign of the average acceleration $a_{\text{average},x}$ tells us the direction in which the object is *accelerating*: in the positive x direction if $a_{\text{average},x}$ is positive and in the negative x direction if $a_{\text{average},x}$ is negative. To see what this means, let's look again at our

Motion diagram

(a)

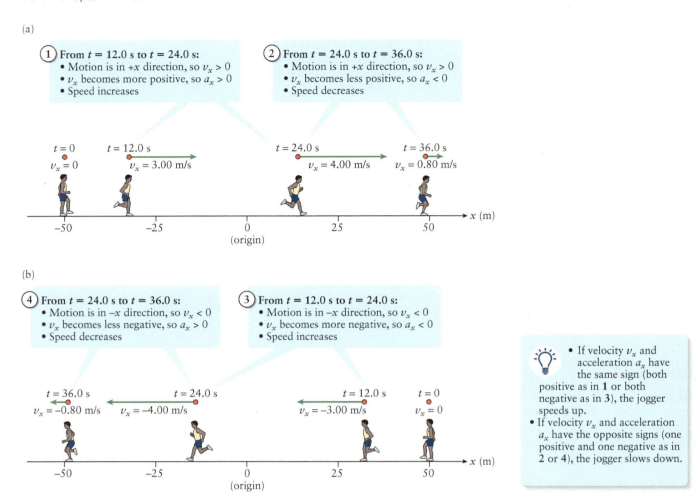

① From $t = 12.0$ s to $t = 24.0$ s:
• Motion is in $+x$ direction, so $v_x > 0$
• v_x becomes more positive, so $a_x > 0$
• Speed increases

② From $t = 24.0$ s to $t = 36.0$ s:
• Motion is in $+x$ direction, so $v_x > 0$
• v_x becomes less positive, so $a_x < 0$
• Speed decreases

$t = 0$
$v_x = 0$

$t = 12.0$ s
$v_x = 3.00$ m/s

$t = 24.0$ s
$v_x = 4.00$ m/s

$t = 36.0$ s
$v_x = 0.80$ m/s

x (m)

-50 -25 0
(origin)   25   50

(b)

④ From $t = 24.0$ s to $t = 36.0$ s:
• Motion is in $-x$ direction, so $v_x < 0$
• v_x becomes less negative, so $a_x > 0$
• Speed decreases

③ From $t = 12.0$ s to $t = 24.0$ s:
• Motion is in $-x$ direction, so $v_x < 0$
• v_x becomes more negative, so $a_x < 0$
• Speed increases

$t = 36.0$ s
$v_x = -0.80$ m/s

$t = 24.0$ s
$v_x = -4.00$ m/s

$t = 12.0$ s
$v_x = -3.00$ m/s

$t = 0$
$v_x = 0$

x (m)

-50 -25 0
(origin)   25   50

• If velocity v_x and acceleration a_x have the same sign (both positive as in **1** or both negative as in **3**), the jogger speeds up.
• If velocity v_x and acceleration a_x have the opposite signs (one positive and one negative as in **2** or **4**), the jogger slows down.

Figure 2-10 **Interpreting positive and negative acceleration** You must consider the algebraic signs (plus or minus) of acceleration *and* velocity to determine whether an object is slowing down or speeding up.

jogger (Figure 2-10a). He moves in the positive x direction at all times. During the time interval from $t = 12.0$ s to $t = 24.0$ s, his velocity increases from $+3.00$ m/s to $+4.00$ m/s, and the change in his velocity is positive: $\Delta v_x = (+4.00$ m/s$) - (+3.00$ m/s$) = +1.00$ m/s. The jogger's average acceleration for this time interval is also positive: $a_{\text{average},x} = \Delta v_x / \Delta t = (+1.00$ m/s$)/(24.0$ s $- 12.0$ s$) = +0.0833$ m/s². We say that the jogger accelerates in the positive x direction. His acceleration is in the *same direction* as his velocity, and he *speeds up*.

Now look at the time interval from $t = 24.0$ s to $t = 36.0$ s in Figure 2-10a. The jogger's velocity decreases from $+4.00$ m/s to $+0.80$ m/s, and the change in his velocity is *negative*: $\Delta v_x = (+0.80$ m/s$) - (+4.00$ m/s$) = -3.20$ m/s. This means that the average acceleration for this time interval is also negative: $a_{\text{average},x} = \Delta v_x / \Delta t = (-3.20$ m/s$)/(36.0$ s $- 24.0$ s$) = -0.267$ m/s². We say that the jogger accelerates in the negative x direction. In this case the jogger's acceleration is *opposite* to his velocity, and he *slows down*.

It's common to think that positive acceleration always corresponds to speeding up and that negative acceleration always corresponds to slowing down. That's *not* always true, however! Figure 2-10b shows an example: The jogger from Figure 2-10a repeats the same motion but in the negative x direction, so his velocity is negative at all times. The figure shows that the jogger has *negative* average acceleration when he is speeding up and *positive* acceleration when he is slowing down. Here's a simple rule to help you understand when an object is speeding up and when it is slowing down:

When an object moving in one dimension speeds up, its velocity and acceleration have the same sign (both plus or both minus). When an object moving in one dimension slows down, its velocity and acceleration have opposite signs (one plus and the other minus).

WATCH OUT! Acceleration doesn't have to mean increasing speed.

In everyday language, "acceleration" is used to mean "speeding up" and "deceleration" is used to mean "slowing down." In physics, however, acceleration refers to *any* change in velocity and so includes *both* speeding up and slowing down (Figure 2-11). Speeding up is "accelerating forward" (that is, in the direction of motion) and slowing down is "accelerating backward" (that is, opposite to the direction of motion).

Figure 2-11 Two pedals that cause acceleration The right-hand pedal in a car is called the accelerator because it's used to make the car go faster. But the left-hand pedal, which controls the brakes, is also an accelerator pedal because it's used to make the car slow down.

SuperStock/AGE Fotostock

EXAMPLE 2-4 Calculating Average Acceleration

The object described in Example 2-3 has velocity $v_x = -10$ m/s at $t = 0$, $v_x = 0$ at $t = 10$ s, $v_x = +10$ m/s at $t = 20$ s, $v_x = +1.4$ m/s at $t = 30$ s, and $v_x = 0$ at $t = 40$ s. Find the object's average acceleration for the time intervals (a) $t = 0$ to 10 s, (b) $t = 10$ s to 20 s, (c) $t = 20$ s to 30 s, and (d) $t = 30$ s to 40 s.

Set Up

The figures in Example 2-3 show the x–t graph and motion diagram for this object. For each time interval we'll use the definition of average acceleration, Equation 2-3.

Average acceleration for motion in one dimension:

$$a_{\text{average},x} = \frac{\Delta v_x}{\Delta t} = \frac{v_{2x} - v_{1x}}{t_2 - t_1} \quad (2\text{-}3)$$

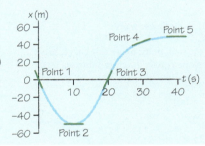

Solve

(a) For the time interval $t_1 = 0$ to $t_2 = 10$ s (from point 1 to point 2 in the x–t graph), we have $v_{x1} = -10$ m/s and $v_{x2} = 0$. Because the velocity becomes more positive (it goes from negative to zero), the average x acceleration is positive.

$$a_{\text{average},x} = \frac{v_{2x} - v_{1x}}{t_2 - t_1}$$

$$= \frac{0 - (-10 \text{ m/s})}{10 \text{ s} - 0} = \frac{+10 \text{ m/s}}{10 \text{ s}}$$

$$= +1.0 \text{ m/s}^2$$

(b) For the time interval $t_2 = 10$ s to $t_3 = 20$ s (from point 2 to point 3 in the x–t graph), we have $v_{x2} = 0$ and $v_{x3} = +10$ m/s. The velocity again becomes more positive (it goes from zero to a positive value), so the average x acceleration is again positive.

$$a_{\text{average},x} = \frac{v_{3x} - v_{2x}}{t_3 - t_2}$$

$$= \frac{(+10 \text{ m/s}) - 0}{20 \text{ s} - 10 \text{ s}} = \frac{+10 \text{ m/s}}{10 \text{ s}}$$

$$= +1.0 \text{ m/s}^2$$

(c) For the time interval $t_3 = 20$ s to $t_4 = 30$ s (from point 3 to point 4 in the x–t graph), we have $v_{x3} = +10$ m/s and $v_{x4} = +1.4$ m/s. The velocity becomes more negative (less positive), so the average x acceleration is negative.

$$a_{\text{average},x} = \frac{v_{4x} - v_{3x}}{t_4 - t_3}$$

$$= \frac{(+1.4 \text{ m/s}) - (+10 \text{ m/s})}{30 \text{ s} - 20 \text{ s}} = \frac{-8.6 \text{ m/s}}{10 \text{ s}}$$

$$= -0.86 \text{ m/s}^2$$

(d) For the time interval $t_4 = 30$ s to $t_5 = 40$ s (from point 4 to point 5 in the x–t graph), we have $v_{x4} = +1.4$ m/s and $v_{x5} = 0$. Again the velocity becomes more negative (less positive) and the average acceleration is negative.

$$a_{average,x} = \frac{v_{5x} - v_{4x}}{t_5 - t_4}$$

$$= \frac{0 - (+1.4 \text{ m/s})}{40 \text{ s} - 30 \text{ s}} = \frac{-1.4 \text{ m/s}}{10 \text{ s}}$$

$$= -0.14 \text{ m/s}^2$$

Reflect

Let's confirm that these results agree with the rule about the algebraic signs of velocity v_x and average acceleration $a_{average,x}$. From $t_1 = 0$ to $t_2 = 10$ s, v_x is negative (the x–t graph slopes downward), but $a_{average,x} = +1.0$ m/s^2 is positive. Because v_x and $a_{average,x}$ have different signs, the speed decreases during this interval (from 10 m/s to zero). From $t_2 = 10$ s to $t_3 = 20$ s, v_x is positive (the x–t graph slopes upward), and $a_{average,x} = +1.0$ m/s^2 is also positive. During this interval, v_x and $a_{average,x}$ have the same sign and the speed increases (from 0 to 10 m/s). During the intervals from $t_3 = 20$ s to $t_4 = 30$ s (for which $a_{average,x} = -0.86$ m/s^2) and from $t_4 = 30$ s to $t_5 = 40$ s (for which $a_{average,x} = -0.14$ m/s^2), v_x and $a_{average,x}$ have opposite signs (v_x is positive and $a_{average,x}$ is negative), and the speed decreases (from 10 m/s to 1.4 m/s to zero). Remember that the sign of the acceleration alone doesn't determine whether the speed increases or decreases—what matters is how the signs of velocity and acceleration compare.

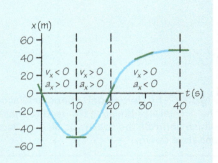

Instantaneous Acceleration and the v_x–t Graph

An object's acceleration can change from one moment to the next. For example, a car can accelerate forward and gain speed, cruise at a steady velocity with zero acceleration, then accelerate backward when the driver steps on the brakes. The **instantaneous acceleration**, or simply **acceleration**, describes how the velocity is changing at a given instant. We use the symbol a_x (without the word "average") to denote acceleration.

Just as an object's velocity v_x is the rate of change of its position as expressed by its coordinate x, the object's acceleration a_x is the rate of change of its velocity v_x. In other words, *acceleration is to velocity as velocity is to position*. The slope of a graph of an object's position versus time (an x–t graph) indicates its velocity v_x. In the same way, the slope of a graph of the object's velocity versus time—that is, a v_x–t **graph**—tells us its acceleration a_x. If the v_x–t graph has an upward (positive) slope, the velocity is becoming more positive, and the acceleration is positive; if the v_x–t graph has a downward (negative) slope, the velocity is becoming more negative, and the acceleration is negative. If the velocity is not changing so that the acceleration is zero, the v_x–t graph is horizontal and has zero slope.

You can see these properties of the v_x–t graph by inspecting Figure 2-12, which shows both the x–t graph and the v_x–t graph for the jogger whose motion we depicted in Figures 2-6, 2-7, and 2-10a. From $t = 0$ to $t = 16$ s his velocity becomes more positive, which we can see in two ways: The x–t graph has an *increasing* slope (which means v_x is becoming more positive), and the v_x–t graph has a *positive* slope (which means a_x is positive). From $t = 16$ s to $t = 28$ s, the jogger has a constant velocity, which is why the x–t graph has a *constant* slope (which means v_x is not changing), and the v_x–t graph is a horizontal line with *zero* slope (which means a_x is zero) during these times. Finally, from $t = 28$ s to $t = 38$ s, the jogger's velocity is becoming more negative, so the x–t graph has a *decreasing* slope (which means v_x is becoming more negative), and the v_x–t graph has a *negative* slope (which means a_x is negative). Table 2-2 summarizes these key features of x–t and v_x–t graphs.

(a) Jogger's x–t graph

At each point on an x–t curve, the slope equals the instantaneous velocity v_x.

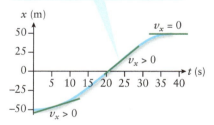

(b) Jogger's v_x–t graph

At each point on a v_x–t curve, the slope equals the instantaneous acceleration a_x.

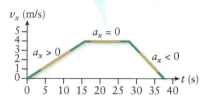

Figure 2-12 From x–t graph to v_x–t graph Graphs of a jogger's (a) position versus time and (b) velocity versus time.

TABLE 2-2 Interpreting x–t Graphs and v_x–t Graphs

A graph of coordinate x versus time t (x–t graph) and a graph of x velocity v_x versus time t (v_x–t graph) are different ways to depict the motion of an object along a straight line.

Type of graph	x–t graph	v_x–t graph
The *value* of the graph tells you...	the coordinate x of the object at a given time t	the velocity v_x of the object at a given time t
The *slope* of the graph tells you...	the velocity v_x of the object at a given time t	the acceleration a_x of the object at a given time t
Changes in the slope of the graph tell you...	the acceleration a_x of the object at a given time t	whether the acceleration is changing

EXAMPLE 2-5 Decoding a v_x–t Graph

Figure 2-13 shows both the x–t graph and the v_x–t graph for the motion of the same object that we examined in Examples 2-3 and 2-4. Using these graphs alone, determine whether the object's acceleration a_x is positive, negative, or zero at (a) $t = 0$, (b) $t = 10$ s, (c) $t = 20$ s, and (d) $t = 30$ s.

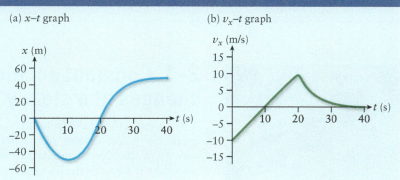

(a) x–t graph

(b) v_x–t graph

Figure 2-13 Two graphs to interpret The x–t graph and v_x–t graph for the object from Examples 2-3 and 2-4.

Set Up

We use two ideas: (i) Acceleration is the rate of change of velocity, and (ii) the value of the acceleration equals the slope of the v_x–t graph (Table 2-2).

Solve

(a) At $t = 0$ the v_x–t graph has a positive (upward) slope, which means that v_x is increasing and hence that a_x is positive. The x–t graph shows the same thing: It has a steep negative slope at $t = 0$, which means v_x is negative, but as time increases the slope becomes shallower as v_x becomes closer to zero. Because v_x is changing from a negative value toward zero, the velocity is increasing and a_x is positive.

(b) At $t = 10$ s the v_x–t graph has the same upward slope as at $t = 0$, which means that the acceleration a_x is again positive. At this instant the velocity is zero and the object is momentarily at rest. It's still accelerating, however; the object has a negative velocity just before $t = 10$ s, and a positive velocity just after $t = 10$ s. This is the time at which the object turns around. You can also see this from the x–t graph, the slope of which is changing from negative to zero to positive around $t = 10$ s.

(c) At $t = 20$ s the v_x–t graph has zero slope, so the acceleration is zero. At this instant the velocity is neither increasing nor decreasing. That's why the x–t graph at this time is nearly a straight line of constant slope, indicating that the velocity isn't changing at this instant.

(d) At $t = 30$ s the v_x–t graph has a negative (downward) slope, so a_x is negative. The x–t graph has a positive slope at this instant (v_x is positive) but is flattening out, so the slope and v_x are both becoming more negative. That's another way of saying that the acceleration is negative.

Reflect

We can check our answers by seeing what they tell us about how the *speed* of the object is changing. At $t = 0$ the object has negative velocity and positive acceleration. Because v_x and a_x have opposite signs, the object is slowing down. At $t = 20$ s the velocity is positive but the acceleration is zero; the velocity is not changing at this instant, so the object is neither speeding up nor slowing down. Finally, at $t = 30$ s, the velocity is positive and the acceleration is negative, so the object is once again slowing down. You can confirm these conclusions by comparing with the motion diagram in Example 2-3. Are they consistent? Can you also use Figure 2-13 to show that at $t = 40$ s the object is at rest (it has zero velocity) and is remaining at rest (its acceleration is zero, so the velocity remains equal to zero)?

GOT THE CONCEPT? 2-3 Interpreting the Sign of Acceleration

 Which statement about the motion depicted in Figure 2-13 is correct? (a) At $t = 5$ s, a_x is positive and the object is speeding up. (b) At $t = 5$ s, a_x is positive and the object is slowing down. (c) At $t = 5$ s, a_x is negative and the object is speeding up. (d) At $t = 5$ s, a_x is negative and the object is slowing down.

TAKE-HOME MESSAGE FOR Section 2-3

✔ Instantaneous velocity v_x (or "velocity" for short) tells you an object's speed and direction of motion at a given instant. It is equal to the rate of change of the object's position and equal to the slope of the object's x–t graph.

✔ The average acceleration of an object, $a_{\text{average},x}$, is the change in its velocity v_x over a given time interval divided by the duration Δt of the interval.

✔ Instantaneous acceleration a_x (or "acceleration" for short) equals the rate of change of an object's velocity at a given instant. It also equals the slope of the object's v_x–t graph.

✔ An object speeds up when its velocity and acceleration have the same sign and slows down when they have opposite signs.

Figure 2-14 An object with constant acceleration As this car brakes to a halt, its velocity is changing at a steady rate. Hence its acceleration—the rate of change of velocity—remains constant.

(2-4)

2-4 Constant acceleration means velocity changes at a steady rate

When a basketball falls from a player's hand or a car comes to a halt with the brakes applied, the acceleration of the object turns out to be nearly *constant*—that is, the velocity of the object changes at a steady rate (Figure 2-14). This situation is so common that it's worth deriving a set of kinematic equations that describe motion in a straight line with **constant acceleration**. These equations turn out to be useful even in situations where the acceleration isn't strictly constant.

If the acceleration is constant, the *instantaneous* acceleration a_x at any instant is the same as the *average* acceleration $a_{\text{average},x}$ for any time interval. Thus we can use a_x in place of $a_{\text{average},x}$ in any equation. Let's consider a time interval from $t = 0$ to some other time t. We use the symbol v_{0x} for the velocity at $t = 0$, which we call the initial velocity, and the symbol v_x (with no zero) for the velocity at time t. From the definition of average acceleration given in Equation 2-3,

$$a_x = \frac{v_x - v_{0x}}{t - 0}$$

We can rewrite Equation 2-4 to get an expression for the velocity v_x at any time t. Multiply both sides of the equation by t, add v_{0x} to both sides, and rearrange:

Velocity at time t of an object in motion in one dimension with constant acceleration

Velocity at time $t = 0$ of the object

Velocity, acceleration, and time for constant acceleration only

(2-5)

$$v_x = v_{0x} + a_x t$$

Constant acceleration of the object

Time at which the object has velocity v_x

Equation 2-5 tells us how the object's velocity v_x varies with time. If $a_x = 0$ so that the object is not accelerating, v_x at any time t is the same as the velocity v_{0x} at $t = 0$: In other words, the velocity is constant. If a_x is not zero and the object is accelerating, the presence of the term $a_x t$ in this equation shows that the velocity changes with time.

Yauhen_D/Shutterstock

WATCH OUT! Motion in one dimension doesn't have to be horizontal motion.

! You may have the impression that our discussion of motion in one dimension is just for horizontal motion. (Our use of x as the symbol for the direction of motion may have helped to give you this impression.) But in fact motion in one dimension can be in *any* direction: horizontal, vertical, or any direction in between. As an example, consider a skier descending a mountain with a uniform slope (Figure 2-15). Although the skier is moving both vertically and horizontally, the skier is traveling in a straight line, and so this is motion in one dimension. All of the equations we've developed in this chapter apply to the skier's motion if we simply choose the x axis to lie along the skier's path.

Cultura Creative/Alamy

Figure 2-15 **Motion in one (tilted) dimension** This skier follows a straight-line path down the mountain, so the skier is moving in one dimension (measured along the path).

We'd also like to have an expression that shows how the object's x coordinate varies with time. To obtain such an expression first note that if the object is at coordinate x_0 at $t = 0$ and at coordinate x at time t, its x displacement during the time interval from 0 to t is $\Delta x = x - x_0$. From Equation 2-1, Δx divided by the duration $\Delta t = t - 0$ of the time interval equals the average velocity for this time interval:

$$v_{\text{average},x} = \frac{\Delta x}{\Delta t} = \frac{x - x_0}{t - 0}$$

We can rewrite this expression using the same steps we used in writing Equation 2-5:

$$x = x_0 + v_{\text{average},x}t \qquad (2\text{-}6)$$

Equation 2-6 tells us that to determine where the object is at time t—that is, what its x coordinate is—we need to know the average velocity $v_{\text{average},x}$ for the time interval from 0 to t.

The average velocity is easy to find if the acceleration is constant. For example, suppose that the initial velocity is $v_{0x} = 3.0$ m/s and the acceleration is $a_x = 2.0$ m/s^2. From Equation 2-5, the velocity is

$$v_x = (3.0 \text{ m/s}) + (2.0 \text{ m/s}^2)(0 \text{ s}) = 3.0 \text{ m/s at } t = 0$$
$$v_x = (3.0 \text{ m/s}) + (2.0 \text{ m/s}^2)(1.0 \text{ s}) = 5.0 \text{ m/s at } t = 1.0 \text{ s}$$
$$v_x = (3.0 \text{ m/s}) + (2.0 \text{ m/s}^2)(2.0 \text{ s}) = 7.0 \text{ m/s at } t = 2.0 \text{ s}$$

To find the average of these three velocities—that is, the average velocity $v_{\text{average},x}$ for the time interval from $t = 0$ to $t = 2.0$ s—we add them together and divide by 3:

$$v_{\text{average},x} = \frac{3.0 \text{ m/s} + 5.0 \text{ m/s} + 7.0 \text{ m/s}}{3} = \frac{15.0 \text{ m/s}}{3} = 5.0 \text{ m/s}$$

Notice that $v_{\text{average},x} = 5.0$ m/s is also the value of the instantaneous velocity v_x at the midpoint of the interval ($t = 1.0$ s). Furthermore, $v_{\text{average},x}$ is equal to the average of the instantaneous velocities at the *beginning and end* of the time interval ($v_x = 3.0$ m/s at $t = 0$ and $v_x = 7.0$ m/s at $t = 2.0$ m/s; the average of these is

(3.0 m/s + 7.0 m/s)/2 = (10.0 m/s)/2 = 5.0 m/s). This is *always* the case if the acceleration is constant. So for constant a_x, the average velocity for the time interval from time 0 to time t is the average of v_{0x} and v_x:

(2-7)
$$v_{\text{average},x} = \frac{v_{0x} + v_x}{2}$$

If we substitute the expression for v_x from Equation 2-5 into Equation 2-7, we get

(2-8)
$$v_{\text{average},x} = \frac{v_{0x} + v_{0x} + a_x t}{2} = \frac{2v_{0x} + a_x t}{2} = v_{0x} + \frac{1}{2} a_x t$$

Equation 2-8 gives the average velocity of an object moving with constant acceleration for the interval from 0 to t.

If we substitute the expression for $v_{\text{average},x}$ from Equation 2-8 into Equation 2-6, we get an equation for the coordinate of the object x at time t:

$$x = x_0 + v_{\text{average},x} t = x_0 + \left(v_{0x} + \frac{1}{2} a_x t \right) t$$

or, simplifying,

Position, acceleration, and time for constant acceleration only (2-9)

| Position at time t of an object in motion in one dimension with constant acceleration | Velocity at time $t = 0$ of the object | Constant acceleration of the object |

$$x = x_0 + v_{0x} t + \frac{1}{2} a_x t^2$$

| Position at time $t = 0$ of the object | Time at which the object has position x |

If we know the object's initial position x_0, its initial velocity v_{0x}, and its constant acceleration a_x, Equation 2-9 tells us its position x at any time t.

How can we interpret the three terms on the right-hand side of Equation 2-9? The first term says that the object starts at $x = x_0$ at time $t = 0$. The second term says that if the object has a nonzero initial velocity v_{0x}, after time t the object has a displacement $v_{0x}t$ due to that initial velocity. And the third term says that if the object also has a constant nonzero acceleration a_x, the object has an additional displacement of $(1/2)a_x t^2$ after time t. Note that if the velocity is constant so that $v_x = v_{0x}$, the acceleration is zero ($a_x = 0$) and Equation 2-9 becomes $x = x_0 + v_{0x}t = x_0 + v_x t$. That's the same as Equation 2-2 for motion in one dimension with constant velocity, which we found in Section 2-2.

WATCH OUT! Instantaneous velocity and average velocity are different quantities.

! Equations 2-5 and 2-8 look quite similar, but they describe two very different things. The first of these, $v_x = v_{0x} + a_x t$, tells us the *instantaneous velocity* at the end of the time interval from 0 to t. The second of these, $v_{\text{average},x} = v_{0x} + (1/2)a_x t$, is a formula for the *average velocity during* this time interval.

Graphing Motion with Constant Acceleration

A good way to interpret Equations 2-5 and 2-9 is to draw the associated motion diagram, x–t graph, and v_x–t graph. Figure 2-16 shows three examples. Because the acceleration is constant and nonzero, in each case the spacing between the dots of the motion diagram changes as the object's velocity changes. In addition, the x–t graph is a special curve called a **parabola**, while the v_x–t graph is a straight line. If the acceleration is positive, the x–t graph curves upward and the v_x–t graph slants upward (its slope, which denotes a_x, is positive); if the acceleration is negative, the x–t graph curves downward and the v_x–t graph slants downward. The value of the initial velocity v_{0x} determines the initial slope of the x–t graph and the initial value of the

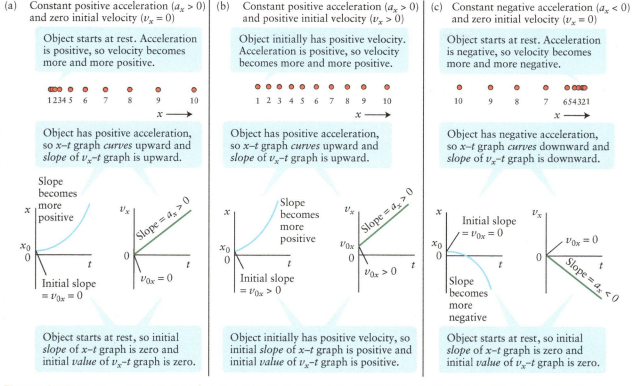

Figure 2-16 **Motion with constant acceleration: Three examples** Each of these three sets of a motion diagram, an x–t graph, and a v_x–t graph depicts a different example of motion in one dimension with constant acceleration. In each of the cases shown, we've assumed that x_0 is positive. You should redraw each of the x–t graphs assuming instead that x_0 is negative. (This change has no effect on the v_x–t graphs. Can you see why not?)

v_x–t graph; both are positive if the initial velocity is positive, and both are negative if the initial velocity is negative.

WATCH OUT! Know your graphs.

 Be careful not to confuse the x–t graph and the v_x–t graph for constant nonzero acceleration. The x–t graph is *curved* because its slope at each point represents the *velocity* at the corresponding time t. The velocity is continually changing, so the slope is different at each point and the graph is curved. (Constant acceleration means the velocity changes at the same rate at all times.) By contrast, the v_x–t graph is a *straight line* because its slope at each point represents the *acceleration* at the corresponding time. The acceleration is constant, so the v_x–t graph has a constant slope (which makes the graph a straight line).

GOT THE CONCEPT? 2-4 Drawing x–t and v_x–t Graphs

(?) Draw the x–t graph and v_x–t graph for motion with constant acceleration for each of the following cases: (a) $a_x > 0$, $v_{0x} < 0$; (b) $a_x < 0$, $v_{0x} > 0$; and (c) $a_x < 0$, $v_{0x} < 0$. (Assume that x_0 is positive in each case.) Use Figure 2-16 as a guide.

The Kinematic Equations for Motion with Constant Acceleration

Equations 2-5 and 2-9 give the object's x coordinate and velocity at a given time t. There is also a third equation that is often useful in analyzing motion in one dimension with constant acceleration. This equation expresses the object's velocity when the object is at a given x coordinate, without reference to time. To create this equation, first solve Equation 2-5 for the time at which the object has velocity v_x:

$$t = \frac{v_x - v_{0x}}{a_x}$$

Then substitute this expression into Equation 2-9, $x = x_0 + v_{0x}t + \frac{1}{2}a_x t^2$, every place you see the quantity t:

(2-10)

$$x = x_0 + v_{0x}\left(\frac{v_x - v_{0x}}{a_x}\right) + \frac{1}{2}a_x\left(\frac{v_x - v_{0x}}{a_x}\right)^2$$

You should fill in the algebraic steps needed to rewrite this equation as

Velocity, acceleration, and position for constant acceleration only
(2-11)

Velocity at position x of an object in motion in one dimension with constant acceleration

Velocity at position x_0 of the object

$$v_x^2 = v_{0x}^2 + 2a_x(x - x_0)$$

Constant acceleration of the object

Two positions of the object

Equations 2-5, 2-9, and 2-11 allow you to solve *any* problem in kinematics that involves motion in one dimension with constant acceleration. Situations with constant or nearly constant acceleration are quite common in nature and technology, so these equations will be useful to us throughout our study of physics. In the following section we'll see how to tackle problems of this kind.

GOT THE CONCEPT? 2-5 Interpreting Motion with Constant Acceleration

At $t = 0$ an object has velocity +2.00 m/s. Its acceleration is constant and equal to −1.00 m/s². State whether the object is (i) speeding up, (ii) slowing down, or (iii) neither at each of the following times: (a) $t = 1.00$ s, (b) $t = 2.00$ s, (c) $t = 3.00$ s. (*Hint*: Use Equation 2-5.)

TAKE-HOME MESSAGE FOR Section 2-4

✔ If an object moving in one dimension has a constant acceleration a_x, its velocity v_x changes at a steady rate.

✔ The v_x–t graph for an object moving in one dimension with constant acceleration a_x is a straight line. The slope of the line tells you the acceleration.

✔ The x–t graph for an object moving in one dimension with constant acceleration a_x is a parabola. The parabola curves upward if a_x is positive and curves downward if a_x is negative. The slope of the parabola where it meets the vertical axis tells you the velocity at $t = 0$.

2-5 Solving one-dimensional motion problems: Constant acceleration

Table 2-3 summarizes the properties of Equations 2-5, 2-9, and 2-11 for motion in one dimension with constant acceleration. The three equations in Table 2-3 involve six physical quantities: the time t, the object's initial position coordinate x_0 at time 0, the object's position x at time t, its velocity v_{0x} at time 0, its velocity v_x at time t, and its constant acceleration a_x. However, not all of these quantities appear in any one of the three constant-acceleration equations. In the following worked examples, we'll show how to decide which equation or equations are the proper ones to use for a given problem.

As we'll see, there are situations where you'll need to use more than one of the equations in Table 2-3 to solve the problem. In these situations you'll need to use your skills at solving simultaneous equations. In addition, Equation 2-9 is a *quadratic* equation that involves the square of the time t. This quadratic equation can have two, one, or no solutions—you'll have to decide which of these is the case for the particular problem at hand. See the Math Tutorial to review how to work with simultaneous equations and with quadratic equations.

 See the Math Tutorial for more information on solving simultaneous equations and quadratic equations.

TABLE 2-3 Equations for Motion in One Dimension with Constant Acceleration

By using one or more of these equations, you can solve any problem involving motion in a straight line for which the acceleration is constant. The check marks indicate which of the quantities t (time), x_0 (the x coordinate at time 0), x (the x coordinate at time t), v_{0x} (the velocity at time 0), v_x (the velocity at time t), and a_x (the constant acceleration) appear in each equation.

Equation		Which quantities the equation includes					
		t	x	x_0	v_{0x}	v_x	a_x
$v_x = v_{0x} + a_x t$	(2-5)	✓			✓	✓	✓
$x = x_0 + v_{0x}t + \frac{1}{2}a_x t^2$	(2-9)	✓	✓	✓	✓		✓
$v_x^2 = v_{0x}^2 + 2a_x(x - x_0)$	(2-11)		✓	✓	✓	✓	✓

EXAMPLE 2-6 Motion with Constant Acceleration I: Cleared for Takeoff!

After winning the lottery you are shopping for a private jet. During a demonstration flight with a salesperson, you notice that a particular type of jet takes off after traveling 1.20 km down the runway and after reaching a speed of 226 km/h (140 mi/h). (a) Sketch a motion diagram, an x–t graph, and a v_x–t graph for the jet's motion as it rolls down the runway. (b) What is the magnitude of the airplane's acceleration (assumed constant) during the takeoff roll? (c) How long does it take the airplane to reach takeoff speed?

Set Up

We choose the positive x axis to point along the runway in the direction that the jet travels, choose the origin at the point where the jet starts to move, and choose $t = 0$ to be when it starts to move. We are told the jet's initial velocity ($v_{0x} = 0$ because the jet starts at rest) as well as its velocity $v_x = 226$ km/h at position $x = 1.20$ km.

To find a_x from x_0, x, v_{0x}, and v_x:

$$v_x^2 = v_{0x}^2 + 2a_x(x - x_0) \quad (2\text{-}11)$$

To find t from v_{0x}, v_x, and a_x:

$$v_x = v_{0x} + a_x t \quad (2\text{-}5)$$

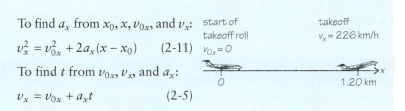

Our goals in parts (b) and (c) are to find the values of *two* unknown quantities: the jet's acceleration a_x and the time t when the jet attains $v_x = 226$ km/h. Hence we need two equations. Table 2-3 tells us that to find a_x from x_0, x, v_x, and v_{0x}, we should use Equation 2-11. Once we know the value of a_x, we can use Equation 2-5 to find the time t at which $v_x = 226$ km/h.

Solve

(a) Because the jet speeds up as it rolls down the runway, successive dots in its motion diagram are spaced increasingly far apart. The x–t graph is a parabola (because a_x is constant) that starts at $x_0 = 0$ at $t = 0$ (the jet is initially at the origin), has zero slope at $t = 0$ (the initial velocity is zero), and curves upward (the slope increases as v_x increases). The v_x–t graph starts at $v_{0x} = 0$ at $t = 0$ (the initial velocity is zero) and is a straight line that slopes upward (the acceleration is constant and positive). (Compare Figure 2-16a.) We draw the graphs only to the point of takeoff because after that point the jet is flying and no longer simply moving along the x axis.

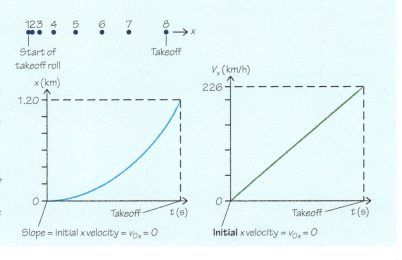

(b) We first convert $v_x = 226$ km/h to m/s and convert $x = 1.20$ km to m. We then solve Equation 2-11 for a_x and substitute in the known values.

$$v_x = (226 \text{ km/h})\left(\frac{1000 \text{ m}}{1 \text{ km}}\right)\left(\frac{1 \text{ h}}{60 \text{ min}}\right)\left(\frac{1 \text{ min}}{60 \text{ s}}\right) = 62.8 \text{ m/s}$$

$$x = (1.20 \text{ km})\left(\frac{1000 \text{ m}}{1 \text{ km}}\right) = 1.20 \times 10^3 \text{ m}$$

Substitute into Equation 2-11:

$$a_x = \frac{v_x^2 - v_{0x}^2}{2(x - x_0)} = \frac{(62.8 \text{ m/s})^2 - (0 \text{ m/s})^2}{2(1.20 \times 10^3 \text{ m} - 0 \text{ m})} = 1.64 \text{ m/s}^2$$

(c) We now know the value of a_x from part (b), so we can solve Equation 2-5 for the time t at which $v_x = 226$ km/h $= 62.8$ m/s. When you enter the values for v_x and a_x into your calculator, retain the extra digits from your calculations above and round the value for the time t only at the end.

Substitute a_x from part (b) into Equation 2-5:

$$t = \frac{v_x - v_{0x}}{a_x} = \frac{(62.8 \text{ m/s}) - (0 \text{ m/s})}{1.64 \text{ m/s}^2} = 38.2 \text{ s}$$

Reflect

We can check our answers to part (b) and part (c) by substituting $x = 0$, $v_{0x} = 0$, $a_x = 1.64$ m/s^2, and $t = 38.2$ s into Equation 2-9, the one constant-acceleration equation that we haven't used yet. We get $x = 1.20$ km, which means that when the jet reaches takeoff speed, it is 1.20 km down the runway—just as we were told in the statement of the problem. That tells us our results are consistent.

Substitute a_x from part (b) and t from part (c) into Equation 2-9:

$$x = x_0 + v_{0x}t + \frac{1}{2}a_xt^2$$

$$= 0 + (0)(38.2 \text{ s}) + \frac{1}{2}(1.64 \text{ m/s}^2)(38.2 \text{ s})^2$$

$$= 1.20 \times 10^3 \text{ m} = 1.20 \text{ km}$$

EXAMPLE 2-7 Motion with Constant Acceleration II: Which Solution Is Correct?

Consider again the jet takeoff described in Example 2-6. There is a marker alongside the runway 328 m from the point where the jet starts to move. How long after beginning its takeoff roll does the jet pass the marker?

Set Up

We use the same x axis and choice of origin as in Example 2-6. Again we have $x_0 = 0$ and $v_{0x} = 0$, and we know the value of a_x from our work in Example 2-6. Our goal is to calculate the time t at which $x = 328$ m. We don't know the jet's velocity at this point, nor are we asked for its value.

Table 2-3 tells us that the equation we need to find t from the given information (x and v_x) is Equation 2-9.

Position, acceleration, and time for constant acceleration only:

$$x = x_0 + v_{0x}t + \frac{1}{2}a_xt^2 \qquad (2\text{-}9)$$

Solve

First let's substitute the known values $x = 328$ m, $x_0 = 0$, $v_{0x} = 0$, and $a_x = 1.64$ m/s^2 into Equation 2-9.

Substitute $x_0 = 0$ and $v_{0x} = 0$, then simplify:

$$x = 0 + (0)t + \frac{1}{2}a_xt^2 = \frac{1}{2}a_xt^2$$

Then substitute $x = 328$ m, $a_x = 1.64$ m/s^2:

$$328 \text{ m} = \frac{1}{2}(1.64 \text{ m/s}^2)t^2$$

Rearrange the equation to solve for t.

$$t^2 = \frac{2(328 \text{ m})}{1.64 \text{ m/s}^2} = 4.00 \times 10^2 \text{ s}^2$$

$$t = \pm\sqrt{4.00 \times 10^2 \text{ s}^2} = \pm 20.0 \text{ s}$$

The ± symbol in front of 20.0 s means that the square of either $t = +20.0$ s or $t = -20.0$ s is equal to $t^2 = 4.00 \times 10^2$ s^2. In this case we clearly want a positive value of time: The jet must reach the marker at $x = 328$ m some time *after* beginning its takeoff roll at $t = 0$. So the correct answer is $t = 20.0$ s.

Choose the value that corresponds to a time after the jet started moving:

$t = 20.0$ s

Reflect

In Example 2-6 we found that the jet takes 38.2 s to travel 1.20 km. Our answer says that it takes less time (20.0 s) to travel a shorter distance (328 m), which is reasonable.

What may not seem reasonable is that we got *two* answers to the problem, only one of which made sense. To see the meaning of the other answer, $t = -20.0$ s, remember that we assumed that the jet has *constant* acceleration $a_x = 1.64$ m/s^2. As far as the equations are concerned, the jet has *always* had this acceleration, even before $t = 0$! According to these equations, before $t = 0$ the jet had the same positive value of a_x but had a *negative* velocity v_x and was moving *backward* along the runway in the negative x direction. The answer $t = -20.0$ s refers to the time during this fictitious motion when the backward-moving jet passed through the point $x = 328$ m.

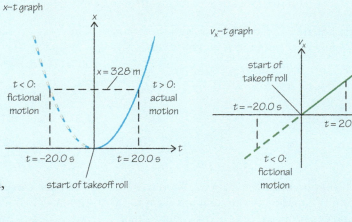

EXAMPLE 2-8 Motion with Constant Acceleration III: Demolition Derby

As a stunt for a movie, two cars are to collide with each other head-on (Figure 2-17). The two cars initially are 125 m apart, car A is heading straight for car B at 30.0 m/s, and car B is at rest. Car A maintains the same velocity, while car B accelerates toward car A at a constant 4.00 m/s^2. (a) When do the cars collide? (b) Where do the cars collide? (c) How fast is car B moving at the moment of collision?

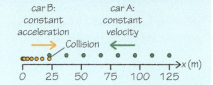

Figure 2-17 A head-on collision When and where will these two cars collide?

Set Up

This situation is similar to the two swimmers in Example 2-2 in Section 2-2. The two cars collide when they are at the same place—in other words, at the time t when they have the same x coordinate, as the motion diagram shows. We choose the x axis so that initially car B is at the origin and car A is at $x = x_{A0} = 125$ m. Then car A has initial velocity $v_{A0x} = -30.0$ m/s (it is moving in the negative x direction, toward the origin) and zero acceleration, while car B has zero initial velocity (it starts at rest) and constant acceleration $a_{Bx} = +4.00$ m/s^2 (it accelerates in the positive x direction).

The two cars collide at the time t when their x–t graphs cross. Note from the v_x–t graph that the two cars never have the same *velocity* because they move in opposite directions.

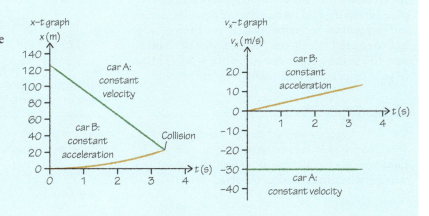

Our goals are to find the value of t when the two cars collide, the common value of x for the two cars at this time, and the velocity of car B at this time. Because time is an essential quantity in this problem, we use the two equations from Table 2-3 that involve t.

Position, acceleration, and time for constant acceleration only:

$$x = x_0 + v_{0x}t + \frac{1}{2}a_xt^2 \qquad (2\text{-}9)$$

Velocity, acceleration, and time for constant acceleration only:

$$v_x = v_{0x} + a_xt \qquad (2\text{-}5)$$

Solve

(a) Write Equation 2-9 twice, once with the values for car A and once with the values for car B.

Car A ($x_{A0} = 125$ m, $v_{A0x} = -30.0$ m/s, $a_{Ax} = 0$):

$$x_A = 125 \text{ m} + (-30.0 \text{ m/s})t + \frac{1}{2}(0)t^2$$

$$= 125 \text{ m} - (30.0 \text{ m/s})t$$

Car B ($x_{B0} = 0$, $v_{B0x} = 0$, $a_{Bx} = +4.00 \text{ m/s}^2$):

$$x_B = 0 + (0)t + \frac{1}{2}(4.00 \text{ m/s}^2)t^2 = (2.00 \text{ m/s}^2)t^2$$

The collision occurs when $x_A = x_B$. We rewrite the resulting equation in the standard form for a quadratic equation, $ax^2 + bx + c = 0$, but using t in place of x.

When the cars collide, $x_A = x_B$, so
$125 \text{ m} - (30.0 \text{ m/s})t = (2.00 \text{ m/s}^2)t^2$ or
$(2.00 \text{ m/s}^2)t^2 + (30.0 \text{ m/s})t - 125 \text{ m} = 0$

In the quadratic equation for t, we have $a = 2.00 \text{ m/s}^2$, $b = 30.0$ m/s, and $c = -125$ m. We substitute these values into the formula for the general solution to a quadratic equation and get two solutions for t. The collision took place after the time we called $t = 0$, so we choose the positive solution for t.

   Note that before rounding, the positive solution is $t = 3.397247$ s. We'll use this unrounded result in part (b).

Solve for time t:

$$t = \frac{-b \pm \sqrt{b^2 - 4ac}}{2a}$$

$$= \frac{-(30.0 \text{ m/s}) \pm \sqrt{(30.0 \text{ m/s})^2 - 4(2.00 \text{ m/s}^2)(-125 \text{ m})}}{2(2.00 \text{ m/s}^2)}$$

$$= \frac{-(30.0 \text{ m/s}) \pm \sqrt{1900 \text{ m}^2/\text{s}^2}}{2(2.00 \text{ m/s}^2)}$$

$$= \frac{-(30.0 \text{ m/s}) \pm (43.6 \text{ m/s})}{4.00 \text{ m/s}^2}$$

$$= 3.40 \text{ s or} -18.4 \text{ s}$$

The answer must be positive, so the desired answer is $t = 3.40$ s.

(b) To find where the collision took place, we can plug t from part (a) into either the equation for x_A or the equation for x_B because at this time x_A and x_B are equal. Let's choose the first of these. Make sure to use the unrounded value of $t = 3.397247$ s in your calculation and round only at the end.

Solve for the position of car A at $t = 3.40$ s:

$$x_A = 125 \text{ m} - (30.0 \text{ m/s})t$$

$$= 125 \text{ m} - (30.0 \text{ m/s})(3.397247 \text{ s})$$

$$= 23.1 \text{ m}$$

So the collision takes place 23.1 m from the origin (the place where car B started).

(c) Now that we know from part (a) the time at which the collision occurred, we can use Equation 2-5 to calculate how fast car B was moving when the collision occurred.

Solve for the velocity of car B at $t = 3.40$ s. Its initial velocity is 0, and its acceleration is 4.00 m/s^2.

$$v_x = v_{0x} + a_xt = 0 + (4.00 \text{ m/s}^2)(3.40 \text{ s})$$

$$= 13.6 \text{ m/s}$$

Reflect

You can check the answer for part (b) by verifying that you get the same result if you substitute t from part (a) into the equation for x_B. Be sure to try this!

Here's another way to look at the answer to part (b). When the two cars collide, car B has moved 23.1 m (from the origin to $x = 23.1$ m) while car A has moved more than four times as far (from $x = 125$ m to $x = 23.1$ m). These results make sense: Car B started at rest and reached a speed of only 13.6 m/s, while car A moved at a faster steady speed of 30.0 m/s and so covered more distance in the same amount of time.

GOT THE CONCEPT? 2-6 Don't Trust Your Calculator Too Much

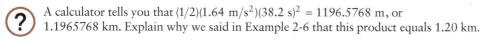

A calculator tells you that $(1/2)(1.64 \text{ m/s}^2)(38.2 \text{ s})^2 = 1196.5768$ m, or 1.1965768 km. Explain why we said in Example 2-6 that this product equals 1.20 km.

GOT THE CONCEPT? 2-7 Distance Traveled While Slowing

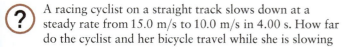

A racing cyclist on a straight track slows down at a steady rate from 15.0 m/s to 10.0 m/s in 4.00 s. How far do the cyclist and her bicycle travel while she is slowing down? (a) 40.0 m; (b) 50.0 m; (c) 60.0 m; (d) not enough information given to decide.

TAKE-HOME MESSAGE FOR Section 2-5

✔ To solve problems that involve motion in one dimension with constant acceleration, you need just three equations (see Table 2-3) that involve six quantities: time t, initial position x_0, position x at time t, velocity v_{0x} at time $t = 0$, velocity v_x at time t, and (constant) acceleration a_x.

✔ Keeping track of known and unknown quantities as part of solving any problem helps you select which equations will help you find the desired quantity of interest.

2-6 Objects falling freely near Earth's surface have constant acceleration

If we can ignore the effects of the air, the players and the ball are all in **free fall**: Their acceleration is constant and downward.

Perhaps the most important case of constant acceleration is the motion of falling objects near the surface of Earth (Figure 2-18). Experiment shows that if air resistance isn't important, an object falling near the surface has a constant, downward acceleration. We will learn later that this acceleration is due to a gravitational force exerted by Earth on the object.

When is air resistance important? To help understand this, drop a sheet of notebook paper. It speeds up for a second or so, then wafts downward at a slow, roughly constant speed. The speed doesn't keep increasing, so this is definitely *not* a case of constant acceleration. But now take a second, identical piece of notebook paper and wad it into a ball. If you drop the first sheet and the second, wadded sheet side by side, you'll find that the wadded one falls more rapidly and *does* keep on gaining speed as it falls. The difference is that by wadding up the second sheet, you've given it a smaller cross-sectional area and thus made it less susceptible to air resistance. (Competition bicyclists use this same trick when they bend low over the handlebars. This reduces their cross section and the retarding effects of air resistance so that they can go faster for the same effort.)

But even the wadded-up paper experiences some air resistance. Imagine, though, that you could somehow crush the paper to an infinitesimally small sphere. There would be essentially no air resistance on this tiny sphere, and only the pull of gravity would affect its fall. In this idealized situation, called **free fall**, the sphere would have a constant downward acceleration.

Free fall is an idealization, but many real-life situations come very close to this ideal. Some examples of nearly free fall include the basketball and players shown in

• An object in free fall has a constant downward acceleration.
• The magnitude g of the acceleration is the same no matter what the size of the object.
• The acceleration is the same whether the object is moving upward, moving downward, or momentarily at rest.

Figure 2-18 Free fall If an airborne object is moving relatively slowly, the effects of the air on the object can be ignored. Then the object moves with a constant acceleration due to gravity.

Goodshoot/Getty Images

Figure 2-18, a high diver descending toward the water, and a leaping frog in midair. In each of these cases, the falling object experiences minimal air resistance because it has a relatively slow speed, a small cross section, or both and so falls with constant downward acceleration.

WATCH OUT! Be careful when falling isn't free.

⚠ If an object is falling at high speed, or has a relatively large cross section, we *cannot* ignore air resistance and so the motion is *not* free fall. Some examples include a sky diver descending under a parachute, a falling leaf, and a hawk with folded wings plummeting at high speed downward onto its prey. In each of these cases, the falling object reaches a maximum speed, or *terminal speed*, at which the retarding effects of air resistance balance the downward pull of gravity. We'll return to situations of this kind in Chapter 5.

Measuring Free-Fall Acceleration

Figure 2-19 shows an experiment to measure the acceleration of a freely falling object near the surface of Earth. In this strobe photograph, a billiard ball falls after being dropped from rest. The image of the ball is recorded at intervals of 0.10 s. It's conventional to call the vertical axis the y axis and to choose the positive y direction to point *upward*. Then increasing values of y correspond to increasing height. After the ball is dropped from rest, it accelerates downward (in the negative y direction) and acquires a negative velocity v_y (the subscript "y" reminds us that we are now discussing motion along the y axis). We choose the initial position of the ball to be $y_0 = 0$ so that as the ball descends its y coordinate becomes a larger and larger negative number.

Careful measurements of the y coordinate and velocity of the ball yield the values shown in Table 2-4. We plot the values of v_y and y versus time in Figure 2-20. We learned in Section 2-3 that the slope of an object's velocity–time graph tells us that object's acceleration. The v_y–t graph in Figure 2-20a is a straight line with constant slope, so the acceleration of the falling ball is constant. We can determine the value of this acceleration by measuring the slope of the v_y–t graph. Table 2-4 says that

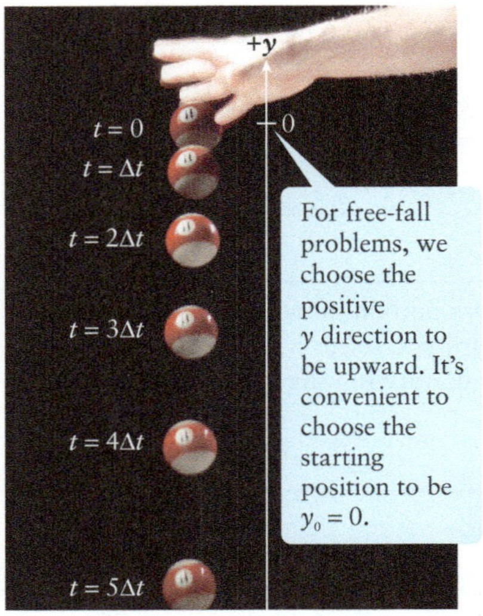

For free-fall problems, we choose the positive y direction to be upward. It's convenient to choose the starting position to be $y_0 = 0$.

When an object is dropped from rest and falls freely, it travels a greater distance in successive time intervals: That is, it accelerates downward.

Figure 2-19 Analyzing a freely falling object Measurements of this falling ball confirm that it has a constant acceleration.

TABLE 2-4 Motion of a Falling Ball

This table lists the y coordinate and y velocity at 0.10-s intervals for the falling ball shown in Figure 2-19. Initially the ball is at rest at $y = 0$. The positive y direction is upward, so as the ball descends it has a negative y velocity and it moves to increasingly negative values of y.

Time t (s)	y coordinate (m)	y velocity (m/s)
0	0.00	0.00
0.10	−0.05	−0.98
0.20	−0.20	−1.96
0.30	−0.44	−2.94
0.40	−0.78	−3.92
0.50	−1.23	−4.91
0.60	−1.76	−5.89

(a) v_y–t graph

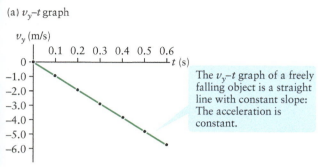

(b) y–t graph

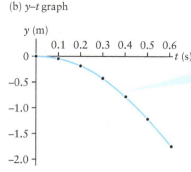

The v_y–t graph of a freely falling object is a straight line with constant slope: The acceleration is constant.

The y–t graph of a freely falling object is a parabola, as it should be for an object that moves with constant acceleration.

Figure 2-20 Graphing free fall Graphs of velocity versus time and position versus time for the freely falling object shown in Figure 2-19.

the velocity is $v_{1y} = -1.96$ m/s at $t_1 = 0.20$ s, and the velocity is $v_{2y} = -3.92$ m/s at $t_2 = 0.40$ s, so the acceleration a_y of the falling ball is

$$a_y = \text{slope of } v_y\text{–}t \text{ graph} = \frac{v_{2y} - v_{1y}}{t_2 - t_1} = \frac{-3.92 \text{ m/s} - (-1.96 \text{ m/s})}{0.40 \text{ s} - 0.20 \text{ s}} = -9.8 \text{ m/s}^2$$

We can confirm this value of acceleration in free fall by comparing the graph of position versus time in Figure 2-20b with the positions we can calculate by using the value $a_y = -9.8$ m/s². To make these calculations, we use Equation 2-9 for motion with constant acceleration, $x = x_0 + v_{0x}t + \frac{1}{2}a_x t^2$, but change every "$x$" in this equation to a "y." Using the initial velocity $v_{0y} = 0$ and initial y coordinate $y_0 = 0$, this equation becomes $y = \frac{1}{2}a_y t^2$. If we substitute $t = 0.30$ s and $a_y = -9.8$ m/s², our predicted position for the falling ball is

$$y = \frac{1}{2}(-9.8 \text{ m/s}^2)(0.30 \text{ s})^2 = -0.44 \text{ m}$$

which agrees with the value in Table 2-4.

Acceleration Due to Gravity

Experiments such as that depicted in Figure 2-19 show that an object in free fall near Earth's surface has a downward acceleration of approximate magnitude 9.80 m/s². Physicists refer to this magnitude as the **acceleration due to gravity** and denote it by the symbol g:

$$g = 9.80 \text{ m/s}^2 \qquad (2\text{-}12)$$

There are slight variations in the value of g from one place to another, related to differences in elevation and in the density of material beneath the surface. Keep in mind that 9.80 m/s² is an *average* value, which we state to three significant figures for convenience in calculation. The value of g also decreases very slightly with altitude: It is about 0.3% less at 10,000 m (32,800 ft) above sea level than at sea level. As we will learn in Chapter 7, a spacecraft journeying far beyond Earth encounters very different values of g. The value of g is also totally different on the surfaces of the Moon and other planets.

The downward acceleration due to gravity also acts on objects that are moving upward or are even momentarily stationary. If a ball is tossed straight up, as it ascends its velocity is positive (upward), and its acceleration is negative (downward). Since velocity and acceleration have opposite signs, the ball slows down (see our discussion in Section 2-3). After the ball has reached the high point of its motion and is descending, its velocity and acceleration are both negative (downward). The signs of velocity and acceleration are the same, so the ball speeds up as it falls. At all times that the ball is in free fall, it has the *same* magnitude and direction of acceleration (Figure 2-21).

WATCH OUT! *g* is always positive, never negative.

! Note that g is a *positive* number because it is the *magnitude* (absolute value) of the acceleration due to gravity. We always take the positive y direction to point upward, as in Figures 2-18 and 2-19, so the acceleration due to gravity is negative: $a_y = -g = -9.80$ m/s². But no matter whether you take the positive direction to be upward or downward, the quantity g is *always* positive.

Figure 2-21 **Free-fall acceleration**
The acceleration due to gravity does not depend on how the object is moving.

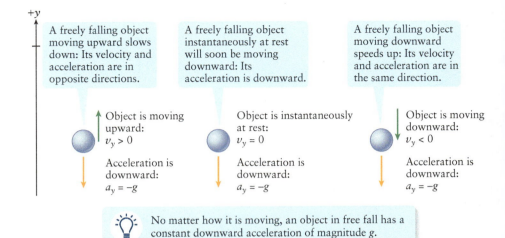

A freely falling object moving upward slows down: Its velocity and acceleration are in opposite directions.

A freely falling object instantaneously at rest will soon be moving downward: Its acceleration is downward.

A freely falling object moving downward speeds up: Its velocity and acceleration are in the same direction.

Object is moving upward:
$v_y > 0$

Object is instantaneously at rest:
$v_y = 0$

Object is moving downward:
$v_y < 0$

Acceleration is downward:
$a_y = -g$

Acceleration is downward:
$a_y = -g$

Acceleration is downward:
$a_y = -g$

No matter how it is moving, an object in free fall has a constant downward acceleration of magnitude g.

WATCH OUT! What happens when an object in free fall reaches its highest point?

It's a common error to think that if a ball is tossed upward, its acceleration at the high point of its motion is zero. It's true that the *velocity* is zero at this point because the ball is momentarily at rest. The ball's velocity is changing, however; its velocity was upward a moment before reaching the high point, and its velocity is downward a moment after. Since its velocity is changing, the ball is accelerating at the high point of the motion, with the same value of acceleration as on the way up and as on the way down (see Figure 2-21). If the acceleration at the high point of the motion actually *were* zero, the tossed ball would reach the high point and simply stop in midair: With no acceleration and therefore no change in velocity, the ball's zero velocity would remain zero forever! That doesn't happen in the real world, so the acceleration can't be zero at the high point.

The Equations of Free Fall: A Special Case of Constant Acceleration

Let's see how to describe free fall that involves motion that is straight up and down—that is, free fall in *one dimension*. As we did in Figures 2-18, 2-19, 2-20, and 2-21, we call the vertical axis the y axis and choose the positive y direction to point upward. Then the acceleration in free fall points in the negative y direction, so the acceleration is $a_y = -g$. (Remember from Equation 2-12 that g is a *positive* number.) We can write the equations for free fall by replacing every "x" in Equations 2-5, 2-9, and 2-11 with a "y" and by using $-g$ for the acceleration a_y:

Velocity, acceleration, and time for free fall
(2-13)

Velocity at time t of an object in free fall   Velocity at time $t = 0$ of the object

$$v_y = v_{0y} - gt$$

Acceleration due to gravity (g is positive)   Time at which the object has position v_y

Position, acceleration, and time for free fall
(2-14)

Position at time t of an object in free fall   Velocity at time $t = 0$ of the object   Acceleration due to gravity (g is positive)

$$y = y_0 + v_{0y}t - \frac{1}{2}gt^2$$

Position at time $t = 0$ of the object   Time at which the object is at position y

Velocity at position y of an object in free fall	Velocity at position y_0 of the object	Position of object when it has velocity v_{0y}

$$v_y^2 = v_{0y}^2 - 2g\,(y - y_0)$$

Acceleration due to gravity (g is positive)   Position of object when it has velocity v_y

Velocity, acceleration, and position for free fall
(2-15)

Note the minus sign in front of each term containing g, which is a reminder that the acceleration is always *downward*, in the negative y direction. If a freely falling object is rising, it slows down; if the object is descending, it speeds up (Figure 2-21). When solving problems that involve free fall, make sure that you always check your answers to see that they agree with these commonsense principles! If they don't—if, say, your answers tell you that a ball thrown straight up goes faster and faster as it climbs—you need to go back and check your work.

WATCH OUT! The free-fall equations are the same as the constant-acceleration equations.

Equations 2-13, 2-14, and 2-15 are simply specific versions of the more general Equations 2-5, 2-9, and 2-11 for motion in one dimension with constant acceleration. Remember that there are only *two* differences between these two sets of equations: (i) In the equations for free fall the axis of motion is called y rather than x, with the positive y direction chosen to be upward, and (ii) in the equations for free fall the constant acceleration is equal to $a_y = -g$, where g is always a *positive* number.

Solving Free-Fall Problems

Table 2-5 summarizes the three equations we can use to analyze the motion of an object experiencing free fall. We use the same approach to solve problems that we used in Section 2-5. After all, free fall is just a special case of motion in one dimension with constant acceleration. As you read through the following examples, keep in mind that the process we use to solve each problem is no less important than the final answer.

GOT THE CONCEPT? 2-8
Interpreting Graphs for Free Fall

Figure 2-16 on page 41 shows three possible combinations of x–t graphs and v_x–t graphs. If we interpret these as y–t graphs and v_y–t graphs, with the positive y direction chosen to be upward, which of these graph combinations are *not* possible for an object in free fall?

 Go to Interactive Exercise 2-2 for more practice dealing with free fall.

TABLE 2-5 Equations for Linear Free Fall

By using one or more of these equations, you can solve any problem involving free-fall motion that is straight up and down. The check marks indicate which of the quantities t (time), y_0 (the y coordinate at time 0), y (the y coordinate at time t), v_{0y} (the y velocity at time 0), v_y (the y velocity at time t), and $a_y = -g$ (the constant, downward y acceleration) appear in each equation. Note that g, the magnitude of the acceleration due to gravity, is a *positive* quantity.

Equation		Which quantities the equation includes					
		t	y	y_0	v_{0y}	v_y	$a_y = -g$
$v_y = v_{0y} - gt$	(2-13)	✓			✓	✓	✓
$y = y_0 + v_{0y}t - \frac{1}{2}gt^2$	(2-14)	✓	✓	✓	✓		✓
$v_y^2 = v_{0y}^2 - 2g(y - y_0)$	(2-15)		✓	✓	✓	✓	✓

WATCH OUT! Up or down, it's still free fall.

In many free-fall problems an object first moves upward then falls back downward (see Example 2-10). Many students make such problems more complicated by writing separate sets of equations for the upward and downward motions. But remember that the acceleration is the same, $a_y = -g$, the whole time the object is in free fall (see Figure 2-21). One set of equations is all you need for both the upward and downward motions, as Example 2-10 shows.

EXAMPLE 2-9 Free Fall I: A Falling Monkey

A monkey drops from a tree limb to grab a piece of fruit on the ground 1.80 m below (Figure 2-22). (a) How long does it take the monkey to reach the ground? (b) How fast is the monkey moving just before it reaches the ground? Neglect the effects of air resistance, so you can treat the monkey as being in free fall.

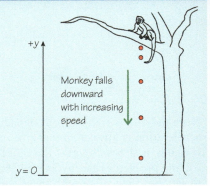

Monkey falls downward with increasing speed

Figure 2-22 A monkey in free fall What are the monkey's position and velocity as functions of time?

Set Up

To describe our freely falling monkey, we choose the positive y direction to be upward (see Figure 2-22). Then the monkey's constant downward acceleration is $a_y = -g = -9.80$ m/s^2 and we can use the equations in Table 2-5 as needed. Our goals are to find the time when he reaches the ground and the magnitude of his velocity just *before* reaching the ground. (Once the monkey is *on* the ground, his speed is zero.)

As Figure 2-22 shows, we choose the origin ($y = 0$) to be on the ground. Then the monkey's initial y coordinate is $y_0 = 1.80$ m, his initial velocity is $v_{0y} = 0$ (he starts at rest), and his y coordinate when he reaches the ground is zero.

We've drawn the y–t graph and v_y–t graph for the monkey's motion using these values of y_0 and v_{0y}. Because the acceleration is negative, the y–t graph curves downward (the slope, which represents the velocity v_y, becomes increasingly negative as time goes by) and the v_y–t graph has a downward (negative) slope.

We know the values of y_0, y, v_{0y}, and a_y, and want to find the values of (a) t and (b) v_y. To solve for t from the known information, we use Equation 2-14. Once we know the value of t, we can substitute it into Equation 2-13 to solve for the monkey's velocity v_y.

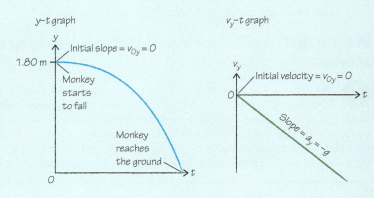

Position, acceleration, and time for free fall:

$$y = y_0 + v_{0y}t - \frac{1}{2}gt^2 \qquad (2\text{-}14)$$

Velocity, acceleration, and time for free fall:

$$v_y = v_{0y} - gt \qquad (2\text{-}13)$$

Solve

(a) Substitute the known values into Equation 2-14. Then solve for t by rearranging the equation so that t^2 is by itself on the left-hand side of the equation and by taking the square root. The equation for t has both a positive solution and a negative solution. The monkey started to fall at $t = 0$, and we are looking for a time after this, so we must choose the positive value of t.

To find the time t when the monkey reaches the ground, substitute values into Equation 2-14:

$$0 = 1.80 \text{ m} + (0)t - \frac{1}{2}(9.80 \text{ m/s}^2)t^2$$

Solve for t:

$$\frac{1}{2}(9.80 \text{ m/s}^2)t^2 = 1.80 \text{ m}$$

$$t^2 = \frac{1.80 \text{ m}}{(1/2)(9.80 \text{ m/s}^2)} = 0.367 \text{ s}^2$$

$$t = \pm\sqrt{0.367 \text{ s}^2} = \pm 0.606 \text{ s}$$

Choose the positive value of t:
$$t = 0.606 \text{ s}$$

(b) To find the monkey's velocity v_y just before it reaches the ground, substitute the known values and $t = 0.606$ s from part (a) into Equation 2-13. We want the monkey's *speed*, which is a positive number and is just the absolute value of v_y.

To find the velocity just before reaching the ground, substitute into Equation 2-13:

$$v_y = 0 - (9.80 \text{ m/s}^2)(0.606 \text{ s})$$
$$= -5.94 \text{ m/s so}$$

Speed = 5.94 m/s

Reflect

The 1.80-m distance that the monkey falls is about the height of a human adult. Try dropping a pencil or a wadded-up piece of paper from this height and estimate the time of its fall. Our result in part (a) says that the fall takes less than a second. Can you verify that this is about right?

As a check on our result in part (b), we can also calculate the speed using Equation 2-15. If you substitute $v_{0y} = 0$, $g = 9.80$ m/s^2, $y_0 = 1.80$ m, and $y = 0$ into this equation, you should again find that the monkey's speed is 5.94 m/s. Do you?

Velocity, acceleration, and position for free fall:

$$v_y^2 = v_{0y}^2 - 2g(y - y_0) \tag{2-15}$$

BioMedical **EXAMPLE 2-10** **Free Fall II: A Pronking Springbok**

When startled, a springbok (*Antidorcas marsupialis*) like the one shown in Figure 2-23 leaps upward several times in succession, reaching a height of several meters. (This behavior, called *pronking*, may inform predators that the springbok knows of their presence.) During one such jump, a springbok leaves the ground at a speed of 6.00 m/s. (a) What maximum height above the ground does the springbok attain? (b) How long does it take the springbok to attain this height? (c) What is the springbok's velocity when it is 1.25 m above the ground? (d) At what times is the springbok 1.25 m above the ground?

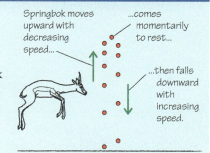

Figure 2-23 Pronking When a springbok jumps into the air, it is in free fall.

Set Up

Like the monkey in Example 2-9, the springbok is in free fall once it leaves the ground. Unlike the monkey, however, the springbok is moving initially and so v_{0y} is *not* zero.

At the moment when the springbok attains its maximum height, its velocity is zero. So parts (a) and (b) are really asking for the springbok's vertical coordinate and the time t when $v_y = 0$. In part (c) our goal is to find the springbok's velocity v_y at a certain vertical coordinate, and in part (d) we wish to find the times t at which the springbok is located at this coordinate.

We again choose the y axis to point upward and again place the origin at ground level. Then the springbok has initial y coordinate $y_0 = 0$ and initial velocity $v_{0y} = +6.00$ m/s (positive because it's moving upward). The drawings show the springbok's y–t graph and v_y–t graph. Because of the initial upward velocity, the y–t graph has an upward slope at $t = 0$, and the v_y–t graph starts with a positive value at $t = 0$.

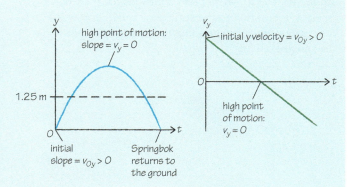

In parts (a) and (b) we want to calculate the value of y and the time t when the springbok reaches its highest point. We know the values of y_0 and v_{0y}, so we can use Equation 2-15 to find the position when the velocity v_y is zero and Equation 2-13 to determine the time t when the velocity is zero. In part (c) we want the value of v_y at a height corresponding to $y = 1.25$ m. We can again use Equation 2-15. In part (d) we'll use Equation 2-13 to find the time t when the springbok has this value of y. We expect the springbok to pass through this height twice, once moving upward and once moving back downward.

For parts (a) and (c):

$$v_y^2 = v_{0y}^2 - 2g(y - y_0) \tag{2-15}$$

For parts (b) and (d):

$$v_y = v_{0y} - gt \tag{2-13}$$

Solve

(a) To determine the value of y corresponding to the high point of the motion, we substitute the values of the initial velocity and g into Equation 2-15. Then we solve this equation for y, the height of the springbok above its launch point ($y_0 = 0$).

Substitute into Equation 2-15:

$$0^2 = (6.00 \text{ m/s})^2 - 2(9.80 \text{ m/s}^2)(y - 0)$$

Solve for y at the high point of the motion:

$$2(9.80 \text{ m/s}^2)y = (6.00 \text{ m/s})^2$$

$$y = \frac{(6.00 \text{ m/s})^2}{2(9.80 \text{ m/s}^2)} = 1.84 \text{ m}$$

(b) To find the time when the springbok is at the high point of its motion, use Equation 2-13 and substitute the values of the initial velocity and g.

Substitute into Equation 2-13:

$$0 = 6.00 \text{ m/s} - (9.80 \text{ m/s}^2)t$$

Solve for t at the high point of the motion:

$$(9.80 \text{ m/s}^2)t = 6.00 \text{ m/s}$$

$$t = \frac{6.00 \text{ m/s}}{9.80 \text{ m/s}^2} = 0.612 \text{ s}$$

(c) Now we want to determine the value of v_y when $y = 1.25$ m. Substitute the values of the initial velocity and g into Equation 2-15 to solve for v_y. When we take the square root of v_y^2, we get two solutions for v_y: a positive one corresponding to the springbok moving upward and a negative one corresponding to the springbok moving downward.

Substitute into Equation 2-15:

$$v_y^2 = (6.00 \text{ m/s})^2 - 2(9.80 \text{ m/s}^2)(1.25 \text{ m} - 0)$$
$$= 36.0 \text{ m}^2/\text{s}^2 - 24.5 \text{ m}^2/\text{s}^2$$
$$= 11.5 \text{ m}^2/\text{s}^2$$

Solve for v_y when $y = 1.25$ m:

$$v_y = \pm\sqrt{11.5 \text{ m}^2/\text{s}^2}$$

Solutions:

$v_y = +3.39$ m/s (springbok moving up)

$v_y = -3.39$ m/s (springbok moving down)

(d) From part (c) we know the two values of v_y for when the springbok passes through $y = 1.25$ m. To find the times t when it passes through this height, we substitute each value of v_y into Equation 2-13, along with $v_{0y} = 6.00$ m/s and $g = 9.80$ m/s^2.

Substitute into Equation 2-13:
Springbok moving up ($v_y = +3.39$ m/s):

$+3.39$ m/s $= +6.00$ m/s $- (9.80$ m/s$^2)t$

Solve for t:

$(9.80$ m/s$^2)t = 6.00$ m/s $- 3.39$ m/s

$$t = \frac{6.00 \text{ m/s} - 3.39 \text{ m/s}}{9.80 \text{ m/s}^2} = \frac{2.61 \text{ m/s}}{9.80 \text{ m/s}^2}$$

$$= 0.266 \text{ s}$$

Springbok moving down ($v_y = -3.39$ m/s):

-3.39 m/s $= +6.00$ m/s $- (9.80$ m/s$^2)t$

Solve for t:

$(9.80$ m/s$^2)t = 6.00$ m/s $- (-3.39$ m/s$)$

$$t = \frac{6.00 \text{ m/s} - (-3.39 \text{ m/s})}{9.80 \text{ m/s}^2} = \frac{9.39 \text{ m/s}}{9.80 \text{ m/s}^2}$$

$$= 0.958 \text{ s}$$

Reflect

A professional basketball player can jump to a height of about 1 m; our result in part (a) says a springbok can jump even higher. This makes sense because a springbok has four powerful legs to propel it upward rather than just two!

We can check our result in part (c) by noticing that 3.39 m/s is less than 6.00 m/s. In other words, when the springbok is rising through $y = 1.25$ m, it is going more slowly than when it left the ground—as it should be because gravity slows the springbok as it ascends.

You can also check the times that we calculated in part (d) by solving for them directly without first calculating the values of v_y. You can do this with Equation 2-14: Substitute $y = 1.25$ m, $y_0 = 0$, $v_{0y} = 6.00$ m/s, and $g = 9.80$ m/s^2 and solve for t in the same way that we did in Example 2-8 in Section 2-5. Try this! You should get the same two answers for t, 0.266 s and 0.958 s.

GOT THE CONCEPT? 2-9 How Many Times Does It Pass?

You throw a ball straight upward at an initial speed of 5.00 m/s. How many times does the ball pass through a point 2.00 m above the point where it left your hand? Ignore the effects of air resistance. (a) Two; (b) one; (c) zero; (d) not enough information given to decide.

TAKE-HOME MESSAGE FOR Section 2-6

✔ An object is in free fall if its motion is influenced by gravity alone (that is, if air resistance is so small that it can be ignored). Free fall includes moving upward, moving downward, and the high point of the motion.

✔ Free-fall motion near the surface of Earth is a special case of constant-acceleration motion in one dimension, with a downward acceleration of magnitude $g = 9.80$ m/s^2. The magnitude g is always a positive number.

✔ When discussing free fall, we always choose the positive y direction to be upward. Then the acceleration of an object in free fall is $a_y = -g$. For an object in free fall, the y–t graph is a parabola with a downward curvature, and the v_y–t graph is a straight line with downward (negative) slope.

✔ You can use the three equations in Table 2-5 to solve any problems involving free fall in one dimension.

✔ The same techniques that we used in Section 2-5 for general constant-acceleration problems can also be used for free-fall problems.

Key Terms

acceleration	displacement	origin
acceleration due to gravity	elapsed time	parabola
average acceleration	free fall	position
average speed	instantaneous acceleration	speed
average velocity	instantaneous speed	time interval
constant acceleration	instantaneous velocity	v_x–t graph
constant velocity	magnitude	x–t graph
coordinates	motion diagram	vector
coordinate system	motion in one dimension	velocity

Chapter Summary

Topic	Equation or Figure	
Linear motion: We describe the motion of an object along a straight line by stating how its position x changes with the passage of time t. The change in the object's position over a time interval is called the displacement.	(a) Direction of swimmer A's motion The red dots show swimmer A's position at 5.0-s intervals. $t = 0$ $t = 10.0$ s $t = 20.0$ s $t = 5.0$ s $t = 15.0$ s $t = 25.0$ s Pool $x = -10.0$ m $x = +40.0$ m	(Figure 2-3a)
Average velocity and instantaneous velocity: The average velocity of an object during a time interval describes how fast the object moved and in what direction. If the time interval is very short, the average velocity becomes the instantaneous velocity v_x ("velocity" for short).	Average velocity of an object in motion in one dimension	Displacement (change in position) of the object over a certain time interval: The object moves from x_1 to x_2, so $\Delta x = x_2 - x_1$. $$v_{\text{average},x} = \frac{x_2 - x_1}{t_2 - t_1} = \frac{\Delta x}{\Delta t}$$ For both the displacement and the elapsed time, subtract the earlier value (x_1 or t_1) from the later value (x_2 or t_2). Elapsed time for the time interval: The object is at x_1 at time t_1 and x_2 at time t_2, so the elapsed time is $\Delta t = t_2 - t_1$.	(2-1)
Constant-velocity linear motion: If an object has the same velocity v_x at all times, it maintains a steady speed and always moves in the same direction. A graph of position x versus time t for constant-velocity motion is a straight line with slope v_x.	Position at time t of an object in motion in one dimension with constant velocity Position at time $t = 0$ of the object $$x = x_0 + v_x t$$ Constant velocity of the object Time at which the object is at position x	(2-2)

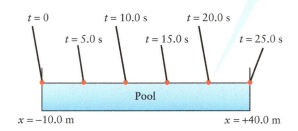

(a) Swimmer A

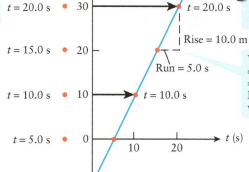

① Orient the motion diagram so that the *x* axis points vertically upward.

② Move each dot in the diagram to the right by an amount that corresponds to the value of the time *t* for that dot.

③ Draw a smooth curve connecting the dots. This completes the *x*–*t* graph for the motion.

Velocity of swimmer A = slope of the graph = rise/run In this case slope and velocity are positive.

Position of swimmer A at *t* = 0

(Figure 2-4a)

Average acceleration and instantaneous acceleration: An object's average acceleration over a time interval equals the average rate of change of the object's velocity during that interval. If the time interval is very short, average acceleration becomes instantaneous acceleration a_x ("acceleration" for short).

Average acceleration of an object in motion in one dimension

Change in **velocity** of the object over a certain time interval: The velocity changes from v_{1x} to v_{2x}.

$$a_{average,x} = \frac{\Delta v_x}{\Delta t} = \frac{v_{2x} - v_{1x}}{t_2 - t_1}$$ (2-3)

Elapsed time for the time interval: The object has velocity v_{1x} at time t_1 and has velocity v_{2x} at time t_2.

Interpreting velocity and acceleration: The slope of an object's *x*–*t* graph (graph of position versus time) is its velocity v_x. The slope of its v_x–*t* graph (graph of velocity versus time) is its acceleration a_x. An object speeds up if v_x and a_x have the same sign and slows down if v_x and a_x have opposite signs.

(a) *x*–*t* graph

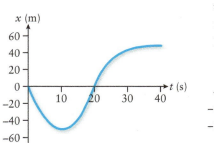

(b) v_x–*t* graph

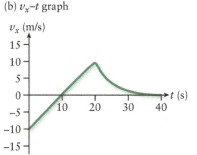

(Figure 2-13)

Motion with constant acceleration: The velocity of an object changes at the same rate at all times. The *x*–*t* graph for such an object is a parabola, and the v_x–*t* graph is a straight line.

Velocity at time *t* of an object in motion in one dimension with constant acceleration

Velocity at time *t* = 0 of the object

$$v_x = v_{0x} + a_x t$$ (2-5)

Constant acceleration of the object

Time at which the object has velocity v_x

Position at time t of an object in motion in one dimension with constant acceleration

Velocity at time $t = 0$ of the object

Constant acceleration of the object

$$x = x_0 + v_{0x}t + \frac{1}{2}a_xt^2 \tag{2-9}$$

Position at time $t = 0$ of the object

Time at which the object has position x

Velocity at position x of an object in motion in one dimension with constant acceleration

Velocity at position x_0 of the object

$$v_x^2 = v_{0x}^2 + 2a_x(x - x_0) \tag{2-11}$$

Constant acceleration of the object

Two positions of the object

Free fall: An object that moves upward or downward under the influence of gravity alone is in free fall. It has a constant downward acceleration of magnitude g. Near the surface of Earth, $g = 9.80 \ \text{m/s}^2$.

Velocity at time t of an object in free fall

Velocity at time $t = 0$ of the object

$$v_y = v_{0y} - gt \tag{2-13}$$

Acceleration due to gravity (g is positive)

Time at which the object has position v_y

Position at time t of an object in free fall

Velocity at time $t = 0$ of the object

Acceleration due to gravity (g is positive)

$$y = y_0 + v_{0y}t - \frac{1}{2}gt^2 \tag{2-14}$$

Position at time $t = 0$ of the object

Time at which the object is at position y

Velocity at position y of an object in free fall

Velocity at position y_0 of the object

Position of object when it has velocity v_{0y}

$$v_y^2 = v_{0y}^2 - 2g(y - y_0) \tag{2-15}$$

Acceleration due to gravity (g is positive)

Position of object when it has velocity v_y

Answer to What do you think? Question

The athlete is accelerating downward at this point and at *all* times from when her feet leave the ground to when they again touch the ground. This is a characteristic of free fall, discussed in Section 2-6.

Answers to Got the Concept? Questions

2-1 (a) (iv), (b) (iv), (c) (iii), (d) (iv), (e) (i), (f) (iv). Object (iv) has the steepest slope, so this object moves fastest *and* travels the greatest distance. The slope is negative, so the object moves in the negative x direction, but speed and distance are both positive quantities. Object (iii) has the most positive (upward) slope, so this object has the most positive velocity and hence the most positive displacement (from $x_1 = -5.0$ m to $x_2 = 0$ m, so $x_2 - x_1 = +5.0$ m). Object (iv) has the most negative (downward) slope, so this object has the most negative velocity and hence the most negative displacement (from $x_1 = +2.0$ m to $x_2 = -8.0$ m, so $x_2 - x_1 = -10.0$ m). The graphs show that at $t = 5.0$ s object (i) ends up at the most positive value of x (7.5 m), and object (iv) ends up at the most negative value of x (–8.0 m).

2-2 24 mi/h. Average velocity equals the net displacement (6.0 mi) divided by the elapsed time (15 min = 0.25 h), or $v_x = (6.0 \text{ mi})/(0.25 \text{ h}) = 24.0$ mi/h. The other information tells you how the velocity varied during the trip, but isn't relevant to the average velocity of the trip as a whole.

2-3 (b) We can determine the sign of the acceleration a_x from the slope of the v_x–t graph in Figure 2-13b. At $t = 5$ s, this graph has a positive slope, so a_x is positive and the velocity is becoming more positive. The v_x–t graph also shows that the value of the velocity v_x at $t = 5$ s is negative. Because v_x and a_x have opposite signs at $t = 5$ s, it follows that the object is slowing down at this time. You can confirm this result from Figure 2-13b: The object has $v_x = -10$ m/s at $t = 0$ and so is moving at 10 m/s in the negative x direction, while the object has $v_x = 0$ at $t = 10$ s and so is at rest. So the speed decreases from $t = 0$ to $t = 10$ s, and the object slows down.

2-4 (a) Because the acceleration is positive, the x–t graph is a parabola that curves upward, and the v_x–t graph slopes upward. The slope of the x–t graph is negative at $t = 0$ because $v_{0x} < 0$. (b) Because the acceleration is negative, the x–t graph is a parabola that curves downward, and the v_x–t graph slopes downward. The slope of the x–t graph is positive at $t = 0$ because $v_{0x} > 0$. (c) Because the acceleration is negative, the x–t graph is a parabola that curves downward, and the v_x–t graph slopes downward. The slope of the x–t graph is negative at $t = 0$ because $v_{0x} < 0$.

2-5 (a) (ii), (b) (iii), (c) (i). To decide how the speed of an accelerating object is changing, we need to compare the signs of its velocity and its acceleration. (a) At $t = 1.00$ s, $v_x = v_{0x} + a_x t = +2.00 \text{ m/s} + (-1.00 \text{ m/s}^2)(1.00 \text{ s}) = +1.00$ m/s. The velocity is positive, but the acceleration $a_x = -1.00 \text{ m/s}^2$ is negative, so the object is slowing down. (b) At $t = 2.00$ s, $v_x = +2.00 \text{ m/s} + (-1.00 \text{ m/s}^2)(2.00 \text{ s}) = 0$. The object is instantaneously at rest. It has just come to a stop (slowing down), but it is also speeding up because of the positive acceleration. (c) At $t = 3.00$ s, $v_x = +2.00 \text{ m/s} + (-1.00 \text{ m/s}^2)(3.00 \text{ s}) = -1.00$ m/s. The velocity and acceleration are both negative, so the object is speeding up.

2-6 We learned in Section 1-4 that when two numbers are multiplied together, the number with the smaller number of significant figures determines the number of significant figures in the product. Because 1.64 m/s^2 and 38.2 s each have only three significant figures, their product can only have

three significant figures. (The factor of 1/2 is an exact number, so it doesn't affect how many significant figures are in the result.)

2-7 (b) "Slowing down at a steady rate" means that the cyclist has a constant acceleration. In this situation, the average velocity over a certain time interval is just the average of the instantaneous velocities at the beginning and end of the interval (see Equation 2-7). Because $v_{0x} = 15.0$ m/s and $v_x = 10.0$ m/s, the average velocity over the 4.00-s time interval is

$$v_{\text{average},x} = \frac{v_{0x} + v_x}{2} = \frac{15.0 \text{ m/s} + 10.0 \text{ m/s}}{2} = 12.5 \text{ m/s}$$

The cyclist's displacement during the time interval equals the average velocity multiplied by the elapsed time, or (12.5 m/s)(4.00 s) = 50.0 m. You can find this same result using the equations in Table 2-3, but it's much easier to use this basic fact about the average velocity when acceleration is constant.

2-8 (a), (b) The acceleration in free fall is $a_y = -g$, which is negative. Hence the impossible graphs are (a) and (b), which assume positive acceleration. The possible case is (c): The y–t graph is a downward-curved parabola, and the v_y–t graph is a straight line with a negative slope.

2-9 (c) The ball *never* reaches a height 2.00 m above the release point. You can see this in more than one way. One is to calculate the maximum vertical height of the ball (where the ball comes momentarily to rest) by using Equation 2-15 and setting $v_y = 0$. Let the point where you release the ball be at $y = 0$.

$$v_y^2 = v_{0y}^2 - 2g(y - y_0)$$
$$0 = (5.00 \text{ m/s})^2 - 2(9.80 \text{ m/s}^2)(y - 0)$$
$$y = \frac{(5.00 \text{ m/s})^2}{2(9.80 \text{ m/s}^2)} = 1.28 \text{ m}$$

The maximum height attained is 1.28 m, so the ball never reaches a height of 2.00 m above the release point. Another way to reach the same conclusion is to use Equation 2-14 to calculate the time t when the ball is at $y = 2.00$ m:

$$y = y_0 + v_{0y}t - \frac{1}{2}gt^2$$

$$2.00 \text{ m} = 0 + (5.00 \text{ m/s})t - \frac{1}{2}(9.80 \text{ m/s}^2)t^2$$

$$(4.90 \text{ m/s}^2)t^2 + (-5.00 \text{ m/s})t + 2.00 \text{ m} = 0$$

$$t = \frac{-(-5.00 \text{ m/s}) \pm \sqrt{(-5.00 \text{ m/s})^2 - 4(4.90 \text{ m/s}^2)(2.00 \text{ m})}}{2(4.90 \text{ m/s}^2)}$$

The quantity inside the square root is

$$(-5.00 \text{ m/s})^2 - 4(4.90 \text{ m/s}^2)(2.00 \text{ m})$$
$$= 25.0 \text{ m}^2/\text{s}^2 - 39.2 \text{ m}^2/\text{s}^2 = -14.2 \text{ m}^2/\text{s}^2$$

The square root of a negative number is *imaginary*: There is no real number that is equal to $\sqrt{-14.2}$. So this problem has no solution, and we conclude that the ball never reaches a height of 2.00 m.

Questions and Problems

In a few problems you are given more data than you actually need; in a few other problems you are required to supply data from your general knowledge, outside sources, or informed estimate.

Interpret as significant all digits in numerical values that have trailing zeros and no decimal points. For all problems use $g = 9.80$ m/s² for the free-fall acceleration due to gravity. Neglect friction and air resistance unless instructed to do otherwise.

- •   Basic, single-concept problem
- ••   Intermediate-level problem; may require synthesis of concepts and multiple steps
- •••   Challenging problem
- Example   See worked example for a similar problem

Conceptual Questions

1. • Under what circumstances is it acceptable to omit the units during a physics calculation? What is the advantage of using SI units in *all* calculations, no matter how trivial?

2. • Compare the concepts of *speed* and *velocity*. Do these two quantities have the same units? When can you interchange these two with no confusion? When would it be problematic?

3. • Explain the difference between average *speed* and average *velocity*.

4. • Under what circumstances will the magnitude of the displacement and the distance traveled be the same? When will they be different?

5. • Which speed gives the largest straight-line displacement in a fixed time: 1 m/s, 1 km/h, or 1 mi/h?

6. • Under what circumstance(s) will the average velocity of a moving object be the same as the instantaneous velocity?

7. • What happens to an object's velocity when the object's acceleration is in the opposite direction to the velocity?

8. • Discuss the direction and magnitude of the velocity and acceleration of a ball that is thrown straight up, from the time it leaves your hand until it returns and you catch it.

9. • What are the units of the slopes of the following graphs: (a) displacement versus time? (b) velocity versus time? (c) distance fallen by a dropped rock versus time?

10. • The upper limit of the braking acceleration for most cars is about the same magnitude as the acceleration due to gravity on Earth. Compare the braking motion of a car with a ball thrown straight upward. Both have the same initial speed. Ignore air resistance.

11. • A video is made of a ball being thrown up into the air and then falling. Is there any way to tell whether the video is being played backward? Explain your answer.

12. • The velocity of a ball thrown straight up decreases as it rises. Does its acceleration increase, decrease, or remain the same as the ball rises? Explain your answer.

13. • A device launches a ball straight up from the edge of a cliff so that the ball falls and hits the ground at the base of the cliff. The device is then turned so that a second, identical ball is launched straight down from the same height. Does the second ball hit the ground with a velocity that is higher than, lower than, or the same as the first ball? Explain your answer.

Multiple-Choice Questions

14. • Figure 2-24 shows a position-versus-time graph for a moving object. At which lettered point is the object moving the slowest?
    - A. A
    - B. B
    - C. C
    - D. D
    - E. E

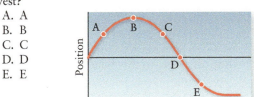

Figure 2-24 Problem 14

15. • Figure 2-25 shows a position-versus-time graph for a moving object. At which lettered point is the object moving the fastest?
    - A. A
    - B. B
    - C. C
    - D. D
    - E. E

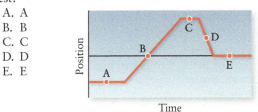

Figure 2-25 Problem 15

16. • A person is driving a car down a straight road. The instantaneous acceleration is constant and in the direction of the car's motion. The speed of the car is
    - A. increasing.
    - B. decreasing.
    - C. constant.
    - D. increasing but will eventually decrease.
    - E. decreasing but will eventually increase.

17. • A person is driving a car down a straight road. The instantaneous acceleration is constant and directed opposite the direction of the car's motion. The speed of the car is
    - A. increasing.
    - B. decreasing.
    - C. constant.
    - D. increasing but will eventually decrease.
    - E. decreasing but will eventually increase.

18. • When throwing a ball straight up, which of the following are correct about the magnitudes of its velocity (v) and its acceleration (a) at the highest point in its path?
    - A. both $v = 0$ and $a = 0$
    - B. $v \neq 0$ but $a = 0$
    - C. $v = 0$ but $a = 9.80$ m/s²
    - D. $v \neq 0$ but $a = 9.80$ m/s²
    - E. not enough information to determine the magnitudes of the velocity (v) and acceleration (a)

Problems

2-2 Constant velocity means moving at a steady speed in the same direction

19. • Convert the following speeds:
    - A. 30.0 m/s = _____ km/h
    - B. 14.0 mi/h = _____ km/h

C. 90.0 km/s = _____ mi/h
D. 88.0 ft/s = _____ mi/h
E. 100 mi/h = _____ m/s

20. • A bowling ball moves down the lane from $x_1 = 3.50$ cm to $x_2 = -4.70$ cm during the time interval from $t_1 = 3.00$ s to $t_2 = 5.50$ s. What is the ball's average velocity? Example 2-1

21. • Lucy took exactly 5 minutes to drive 850 m through her neighborhood to her friend Dongmei's house. Measured along a straight line, Dongmei's house is 775 m east of Lucy's house. What was Lucy's average speed during the trip in meters per second? Example 2-1

22. • **Sports** The Olympic record for the marathon set in 2008 is 2 h, 6 min, 32 s. The marathon distance is 26.2 mi. What was the average speed of the record-setting runner in km/h? Example 2-1

23. •• **Sports** Jack and Jill exercise in a 25-m-long swimming pool. Jack swims nine lengths of the pool in 4 min, 10 s. Jill, the faster swimmer, covers ten lengths in the same time interval. Find the average velocity and average speed of each swimmer. Example 2-1

24. •• A student rides her bicycle home from physics class to get her physics book and then heads back to class. It takes her 21.0 min to make the 12.2-km trip home and 13.0 min to get back to class. (a) Calculate the average velocity of the student for the round trip (from the lecture hall to home and back to the lecture hall). (b) Calculate her average speed for the trip from the lecture hall to her home. (c) Calculate her average speed for the trip from her home back to the lecture hall. (d) Calculate her average speed for the round trip. Example 2-1

25. • On a recent road trip, it took Juan 2.6 h to drive from a suburb of Phoenix to a small town outside of Yuma. According to his dashboard computer, he averaged 77 mi/h on the drive. How far did Juan drive in miles? Example 2-1

26. •• A car traveling at 80.0 km/h is 1500 m behind a truck traveling at 70.0 km/h. How long will it take the car to catch up with the truck? Example 2-2

27. • **Medical** Alcohol consumption slows people's reaction times. In a controlled government test, it takes a certain driver 0.320 s to hit the brakes in a crisis when unimpaired and 1.00 s when drunk. When the car is initially traveling at 90.0 km/h, how much farther does the car travel before coming to a stop when the person is drunk compared to sober? Example 2-1

28. •• A jet takes off from SFO (San Francisco, CA) and flies to ORD (Chicago, IL). The distance between the airports is 3.00×10^3 km. After a 1.00-h layover, the jet returns to San Francisco. The total time for the round trip (including the layover) is 9 h, 52 min. If the westbound trip (from ORD to SFO) takes 24 more minutes than the eastbound portion, calculate the time required for each leg of the trip. What is the average speed for the overall trip? What is the average speed *without* the layover? Example 2-1

29. • A trainer times his racehorse as it completes a workout on a long, straight track. The trainer's graph of his horse's position versus time is shown in Figure 2-26. Use the graph to calculate the average speed of the horse between (a) 0 and 10 s, (b) 10 and 30 s, and (c) 0 and 50 s. Example 2-1

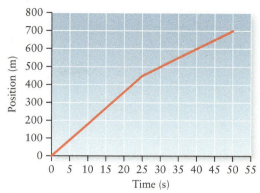

Figure 2-26 Problem 29

2-3 Velocity is the rate of change of position, and acceleration is the rate of change of velocity

30. • **Biology** The position-versus-time graph for a red blood cell leaving the heart is shown in Figure 2-27. Determine the instantaneous speed of the red blood cell when $t = 10$ ms. Recall that 1 ms = 0.001 s, 1 mm = 0.001 m. Example 2-3

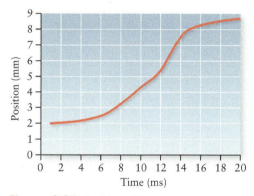

Figure 2-27 Problem 30

31. • Figure 2-28 shows an x–t graph for some object. (a) At $t = 4$ s how far is the object from its position at $t = 0$? (b) At about what time is the object's *speed* greatest? (c) What is the velocity of the object between $t = 2$ s and $t = 4$ s? Example 2-3

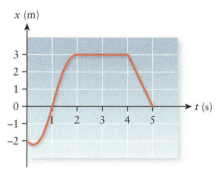

Figure 2-28 Problems 31–32

32. • Figure 2-28 shows an x–t graph for some object. (a) Over what time interval is the acceleration negative? (b) At about what time is the object's *speed* the greatest? (c) What is the acceleration between $t = 4$ s and $t = 5$ s? Example 2-4

33. • Figure 2-29 shows a v_x–t graph for some object. (a) What is the object's maximum velocity? (b) How many

times does the object change direction? (c) What is the acceleration of the object between $t = 2$ s and $t = 3$ s? (d) What is the object's average acceleration between $t = 0$ and $t = 6$ s? Example 2-5

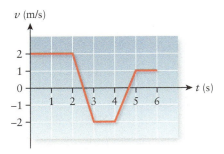

Figure 2-29 Problem 33

34. •• Consider the x–t plot in **Figure 2-30**. For each defined interval identify the *sign* (that is, + or −) of velocity v and acceleration a. Example 2-5

A. $0 \text{ s} \le t \le 2 \text{ s}$
B. $2 \text{ s} \le t \le 3 \text{ s}$
C. $3 \text{ s} \le t \le 4 \text{ s}$
D. $4 \text{ s} \le t \le 5 \text{ s}$
E. $5 \text{ s} \le t \le 5.8 \text{ s}$
F. $t \ge 5.8 \text{ s}$

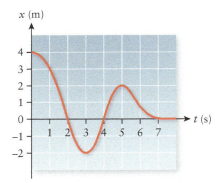

Figure 2-30 Problem 34

35. •• A driver is traveling along a straight road at the speed limit v. After 2 min, she slows at a constant rate to a stop at a traffic light. Two minutes later the light turns green, and she accelerates at a constant rate back up to the speed limit. Five minutes later she again slows at a constant rate to a stop. After 3 min she performs a u-turn then accelerates at a constant rate back up to speed v. Two minutes later she reaches her destination and slows at a constant rate to a stop. Sketch a graph of velocity versus time that represents her trip. (Since you are not given values for acceleration, your graph need only be qualitatively correct.) Example 2-5

36. ••• A car is stopped at a traffic light, defined as position $x = 0$. At $t = 0$ the light turns green, and the car accelerates constantly at 3 m/s² until it reaches 15 m/s at $t = 5$ s, at which time it continues on at that velocity. At $t = 2$ s a speeding truck passes through the traffic light in the same direction traveling at a constant velocity of 20 m/s. On the *same* set of axes, draw the position-versus-time graphs for both the car and the truck out to $t = 6$ s. Example 2-5

2-5 Solving one-dimensional motion problems: Constant acceleration

37. •• Two trains, traveling toward one another on a straight track, are 300 m apart when the engineers on both trains become aware of the impending collision and hit their brakes. The eastbound train, initially moving at 98.0 km/h, slows down at 3.50 m/s². The westbound train, initially moving at 120 km/h, slows down at 4.20 m/s². Will the trains stop before colliding? If so, what is the distance between them once they stop? If not, what initial separation would have been needed to avert a disaster? Example 2-8

38. • Paola can flex her legs from a bent position through a distance of 20.0 cm. Paola leaves the ground when her legs are straight, at a speed of 4.43 m/s. Calculate the magnitude of her acceleration, assuming that it is constant. Example 2-6

39. • A runner starts from rest and achieves a maximum speed of 8.97 m/s. If her acceleration is 9.77 m/s², how long does it take her to reach that speed? Example 2-7

40. • A car traveling at 35.0 km/h speeds up to 45.0 km/h in a time of 5.00 s. The same car later speeds up from 65.0 km/h to 75.0 km/h in a time of 5.00 s. (a) Calculate the magnitude of the constant acceleration for each of these intervals. (b) Determine the distance traveled by the car during each of these time intervals. Example 2-6

41. • A car starts from rest and reaches a maximum speed of 34.0 m/s in a time of 12.0 s. Calculate the magnitude of its average acceleration. Example 2-4

42. • The world's fastest cars are rated by the time required for them to accelerate from 0 to 60.0 mi/h. Convert 60.0 mi/h to km/h and then calculate the acceleration in m/s² for each of the cars on the following list: Example 2-4

1. Bugatti Veyron 16.4   2.4 s
2. Caparo T1   2.5 s
3. Ultima GTR   2.6 s
4. SSC Ultimate Aero TT   2.7 s
5. Saleen S7 Twin Turbo   2.8 s

43. •• Using the information for the Bugatti Veyron in the previous problem, calculate the distance that the car would travel in the time it takes to reach 90.0 km/h, which is the speed limit on many Canadian roads. Compare your answer to the distance that the Saleen S7 Twin Turbo would travel while accelerating to the same final speed. Example 2-6

44. • A driver is traveling at 30.0 m/s when he sees a moose crossing the road 80.0 m ahead. The moose becomes distracted and stops in the middle of the road. If the driver of the car slams on the brakes, what is the minimum constant acceleration he must undergo to stop short of the moose and avert an accident? (In British Columbia, Canada, alone, there are more than 4000 moose–car accidents each year.) Example 2-6

45. • **Biology** A sperm whale can accelerate at about 0.100 m/s² when swimming on the surface. How far will a sperm whale travel if it starts at a speed of 1.00 m/s and accelerates to a speed of 2.25 m/s? Assume the whale travels in a straight line. Example 2-6

46. • At the start of a race, a horse accelerates out of the gate at a rate of 3.00 m/s². How long does it take the horse to cover the first 25.0 m of the race? Example 2-7

2-6 Objects falling freely near Earth's surface have constant acceleration

47. • A ball is dropped from rest at a height of 25.0 m above the ground. (a) How fast is the ball moving when it is 10.0 m above the ground? (b) How much time is required for it to reach the ground? Ignore the effects of air resistance. Example 2-9

48. •• Alex climbs to the top of a tall tree while his friend Gary waits on the ground below. Gary throws a ball up to Alex, who allows the ball to go past him before catching it on its way down. The ball has an initial speed of 10 m/s and is caught 3.5 m above where it was thrown. How long after the ball was thrown does Alex catch it? Ignore the effects of air resistance. Example 2-10

49. •• **Biology** A fox locates its prey, usually a mouse, under the snow by slight sounds the rodent makes. The fox then leaps straight into the air and burrows its nose into the snow to catch its next meal. If a fox jumps to a height of 85.0 cm, calculate (a) the speed at which the fox leaves the snow and (b) how long the fox is in the air. Ignore the effects of air resistance. Example 2-10

50. • **Medical** More people end up in U.S. emergency departments because of fall-related injuries than for any other reason. At what speed would someone who accidentally stepped off the top rung of a 6-ft-tall stepladder hit the ground? (That step is usually embossed with the phrase "Warning! Do not stand on this step.") Ignore the effects of air resistance. Example 2-9

51. •• A ball is thrown straight up at 18.0 m/s. How fast is the ball moving after 1.00 s? after 2.00 s? after 5.00 s? When does the ball reach its maximum height? Ignore the effects of air resistance. Example 2-10

52. •• While standing at the edge of a building's roof, Chad throws an egg upward with an initial speed of 12 m/s. The egg subsequently smashes on the ground, 7 m beneath the height from which Chad threw it. (a) At what speed does the egg pass the point from which it was thrown? (b) How long is the egg in the air? (c) What total distance did the egg travel? Ignore the effects of air resistance. Example 2-10

General Problems

53. •• Katie serves a ping pong ball by tossing the ball straight up and hitting the ball once it returns to the same point where it was tossed. If the ball is in the air for 0.45 s, what is the ball's initial speed after being tossed? Ignore the effects of air resistance. Example 2-10

54. ••• **Biology** Cheetahs can accelerate to a speed of 20.0 m/s in 2.50 s and can continue to accelerate to reach a top speed of 29.0 m/s. Assume the acceleration is constant until the top speed is reached and is zero thereafter. (a) Express the cheetah's top speed in mi/h. (b) Starting from a crouched position, how long does it take a cheetah to reach its top speed, and how far does it travel in that time? (c) If a cheetah sees a rabbit 120 m away, how long will it take to reach the rabbit, assuming the rabbit does not move? Example 2-6

55. ••• **Medical** Very large accelerations can injure the body, especially if they last for a considerable length of time. One model used to gauge the likelihood of injury is the severity index (SI), defined as $SI = a^{5/2}t$, where a is the acceleration in multiples of g, and t is the time the acceleration lasts (in seconds). In one set of studies of rear-end motor vehicle collisions, a person's velocity increases by 15.0 km/h with an acceleration of 35.0 m/s². (a) What is the severity index for the collision? (b) How far does the person travel during the collision if the car was initially moving forward at 5.00 km/h? Example 2-6

56. •• Destiny is eating lunch at her favorite cafe when her friend Jada calls and says she wants to meet her. Jada is calling from a city 45 miles away and wants to meet Destiny somewhere between the two locations. Jada says she will start driving right away, but Destiny needs 10 min to finish her lunch before she can begin driving. Jada plans to drive at 60 mi/h, whereas Destiny usually averages about 75 mi/h. (a) How long will Jada be driving before she meets Destiny? (b) How far will Destiny have traveled when they meet? Ignore acceleration and assume the highway is a straight line. Example 2-8

57. ••• Blythe and Geoff compete in a 1.00-km race. Blythe's strategy is to run the first 600 m of the race at a constant speed of 4.00 m/s, and then accelerate to her maximum speed of 7.50 m/s, which takes her 1.00 min, and then finish the race at that speed. Geoff decides to accelerate to his maximum speed of 8.00 m/s at the start of the race and to maintain that speed throughout the rest of the race. It takes Geoff 3.00 min to reach his maximum speed. Assuming all accelerations are constant, who wins the race and by how many seconds?

58. •• Figure 2-31 is a graph of the velocity of a moving car. What is its instantaneous acceleration at times $t = 2$ s, $t = 4.5$ s, $t = 6$ s, and $t = 8$ s? Example 2-5

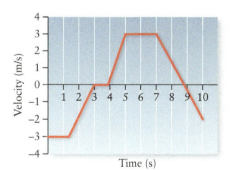

Figure 2-31 Problem 58

59. ••• A ball is dropped from an upper floor, some unknown distance above your apartment. As you look out of your window, which is 1.50 m tall, you observe that it takes the ball 0.180 s to traverse the length of the window. Determine how high above the top of your window the ball was dropped. Ignore the effects of air resistance. Example 2-6

60. •• A ball is thrown straight up at 15.0 m/s. (a) How much time does it take for the ball to be 5.00 m above its release point? (b) How fast is the ball moving when it is 7.00 m above its release point? (c) How much time is required for the ball to reach a point that is 7.00 m above its release point? Why are there two answers to part (c)? Example 2-10

61. •• Ada drops a rock into a deep well and hears the sounds of it hitting the bottom 5.50 s later. If the speed of sound is 340 m/s, determine the depth of the well. Example 2-6

62. ••• A rocket is fired straight up from the ground. It contains a payload and two stages of solid rocket fuel that are designed to burn for 10.0 s and 5.00 s, respectively, with no

time interval between them. In stage 1 the upward acceleration is 15.0 m/s². In stage 2 the acceleration is 12.0 m/s². Neglecting air resistance, (a) how fast is the payload moving at the end of stage 1? (b) How fast is it moving at the end of stage 2? (c) What is the altitude of the rocket at the end of stage 1? (d) What is the altitude of the rocket at the end of stage 2? (e) What is the maximum altitude obtained by the payload? (f) What is the total travel time of the payload from launch to landing? Assume the acceleration due to gravity is constant. Example 2-6

63. •• **Biology** A black mamba snake has a length of 4.30 m and a top speed of 8.90 m/s! Suppose a mongoose and a black mamba find themselves nose to nose. In an effort to escape, the snake accelerates past the mongoose at 18 m/s² from rest. (a) How much time does it take for the snake to reach its top speed? (b) How far does the snake travel in that time? (c) How much time does the mongoose have to react before the black mamba's tail passes the mongoose's nose? Example 2-6

64. ••• **Sports** Kharissia wants to complete a 1000-m race with an average speed of 8.00 m/s. After 750 m, she has averaged 7.20 m/s. What average speed must she maintain for the remainder of the race in order to attain her goal? Example 2-1

65. •• A small object begins a free fall from a height of 24 m. After 2 s, a second small object is launched vertically upward from the ground with an initial velocity of 36 m/s. At what height above the ground will the two objects first meet? Example 2-9

66. •• A small bag of sand is released from an ascending hot-air balloon whose constant, upward velocity is 2.5 m/s. If the balloon is 220 m above the ground when the sandbag is released, how much time does it take for the sandbag to reach the ground? Example 2-6

67. • The ants of the North Creek Ant Colony traverse a straight trail between their colony and a nearby picnic table. One ant, returning after quenching its thirst with a portion of a tasty beverage dripping from the table, is first observed along the trail 3.2 m from the colony. Then, after traveling a total of 1.4 m while staggering back and forth along the trail, the ant falls unconscious 2.9 m from the colony exactly 35 s later. Defining the entrance of the colony as the origin and the positive direction toward the picnic table, determine the ant's average velocity during this time interval. Example 2-1

68. •• **Biology** Estimate the acceleration that a cat undergoes as it jumps from the floor to a typical countertop.

69. ••• Classes are canceled due to snow, so Ulrich decides to conduct a fun physics experiment. He fastens a large toy rocket to the back of a sled and takes it to a large, flat, snowy field. Then he ignites the rocket and observes that the sled accelerates forward from rest at a rate of 12 m/s² for 3.0 s before the engine shuts off. The sled then experiences a constant backward acceleration of 0.6 m/s² due to friction. After the rocket turns off, how much time does it take for the sled to come to a stop?

John Bingham/Alamy

Motion in Two or Three Dimensions

In this chapter, your goals are to:

- (3-1) Recognize the differences among one-, two-, and three-dimensional motion.
- (3-2) Describe the properties of a vector and how to find the sum or difference of two vectors.
- (3-3) Relate the components of a vector to its magnitude and direction, and use components in calculations involving vectors.
- (3-4) Explain how displacement, velocity, and acceleration are described in terms of vectors.
- (3-5) Identify the key features of projectile motion and how to interpret this kind of motion.
- (3-6) Solve problems involving projectile motion.
- (3-7) Describe why an object moving in a circle is always accelerating.
- (3-8) Explain how the same object can have different velocities as measured by different observers.

To master this chapter, you should review:

- (2-2, 2-3) The ideas of displacement, velocity, and acceleration for motion in a straight line.
- (2-4, 2-5) The concepts and equations of straight-line motion with constant acceleration.
- (2-6) The concepts and equations of free fall.

What do you think?

If air resistance isn't a factor, does a high jumper (shown here in a time-lapse photo) take more time to (a) climb from one meter below the bar to her maximum height or (b) descend from her maximum height back to one meter below the bar? Or (c) do the climb and descent take the same amount of time?

3-1 The ideas of linear motion help us understand motion in two or three dimensions

The high jumper shown in the photo introducing this chapter follows a curved path. That means that we can't describe her motion simply by using the ideas of Chapter 2 for motion in a straight line (one-dimensional motion). But we can extend the same concepts of displacement, velocity, and acceleration to motion along curved paths in two or three dimensions—that is, to motion in a plane or to generalized motion in space (Figure 3-1a).

We'll look in detail at an important kind of two-dimensional motion called *projectile motion*. This kind of motion in a plane occurs when the only source of an object's acceleration is the downward pull of gravity (Figure 3-1b). Projectile motion is not a perfect description of the flight of batted baseballs or kicked footballs, for which air resistance can be important, but it can still give useful approximate results.

We'll also examine the important case of motion in a circle (Figure 3-1c). We'll discover that even if an object travels around a circle at a constant speed, it is nonetheless accelerating. That's because the object is *turning*—the direction of its motion is continuously changing. In later chapters we'll use this observation to help explain why birds bank their wings when they turn, why cars sometimes skid when turning on a rain-slicked road, and other aspects of the natural and technological world.

The most general kind of motion is three dimensional. In many cases, however, the motion is in a plane and so is two dimensional.

This aircraft follows a complicated, three-dimensional path.

This motorcycle's path through the air is two dimensional (in a plane): It moves left to right as well as up and down.

Each car on this Ferris wheel moves in the plane of the wheel, so its path is two dimensional.

(a)  (b)  (c)

Design Pics Inc/Alamy

Stephen Barnes/Science/Alamy

Holly Kuchera/Getty Images

Figure 3-1 Motion in three and two dimensions Three examples of motion.

Finally, we'll explore the idea of *relative motion*. When we say that an object like the airplane in Figure 3-1a has a certain velocity, we naturally mean its velocity relative to us watching its flight from the ground. But that airplane's velocity can have a very different value as seen by the pilot of a second airplane. (If the two airplanes are flying side by side in perfect formation, the velocity of each airplane relative to the other is exactly zero.) In the final section of this chapter we'll see how to relate the velocities of the same object as measured by two different observers.

You might worry that motion in two or three dimensions is two or three times harder to analyze than one-dimensional motion along a straight line. Not so! By using the idea of a *vector* (a quantity with both a magnitude and a direction) that we introduced in Section 2-2, we'll be able to use all of the same techniques that we used in Chapter 2 for studying straight-line motion. The only difference is that in two- or three-dimensional motion, the vectors for position, displacement, velocity, and acceleration do *not* always point along the same straight line. In the following section we'll see how to work with vectors in two and three dimensions.

TAKE-HOME MESSAGE FOR Section 3-1

✔ Motion in two or three dimensions uses the same ideas of velocity and acceleration as linear motion.

✔ In two- and three-dimensional motion, velocity and acceleration have to be treated as vector quantities.

3-2 A vector quantity has both a magnitude and a direction

One afternoon you decide to take your textbook to study at your favorite coffee house. You can take different paths to get there, as Figure 3-2a shows, but no matter which path you choose you end up 520 m from your apartment in a direction 60° north of east (Figure 3-2b). This change in your position is your *displacement*, and it has both a *magnitude* (520 m) and a *direction* (60° north of east). This is an extension of the idea of displacement that we introduced in Chapter 2 for motion along a straight line. In that case the sign of the displacement (positive or negative) told us whether the displacement was in the positive or negative *x* direction. Because your motion from apartment to coffee house is in a *plane*, however, we need to expand our definition of displacement.

Vectors and Scalars

Any quantity that has both a magnitude and a direction is called a **vector**. An example is the **displacement vector** in Figure 3-2b, which we depict as an arrow that points from your starting point (your apartment) to your destination (the coffee house). The length of the arrow denotes the **magnitude** of the vector, which in this case is 520 m, the straight-line distance from starting point to destination. The **direction** of the displacement vector in this case is 60° north of east. Two vectors are equal *only* if they have the same magnitude and the same direction (Figure 3-2c). If two vectors differ in their magnitude, direction, or both, they are not equal to each other (Figure 3-2d).

(a)

No matter what path you take from your apartment to the coffee house...

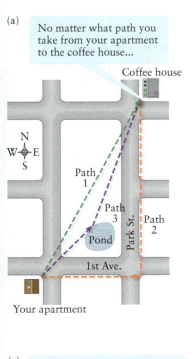

Coffee house

Path 1

Path 3

Path 2

Park St.

Pond

1st Ave.

Your apartment

(b)

...the displacement is the same. The displacement is represented by the vector $\vec{E}$.

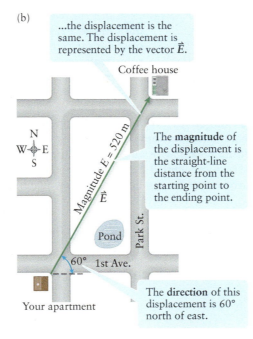

Coffee house

Magnitude E = 520 m

$\vec{E}$

The **magnitude** of the displacement is the straight-line distance from the starting point to the ending point.

Park St.

Pond

60° 1st Ave.

Your apartment

The **direction** of this displacement is 60° north of east.

Figure 3-2 **Vectors** (a) Three paths from your apartment to your favorite coffee house. (b) Each path has the same displacement: from apartment to coffee house. (c) Two equal vectors. (d) Three unequal vectors.

(c)

These two displacements have different starting points and different ending points...

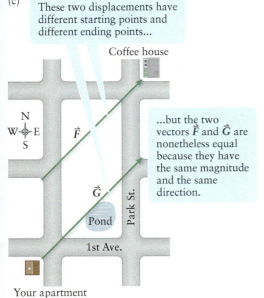

Coffee house

$\vec{F}$

...but the two vectors $\vec{F}$ and $\vec{G}$ are nonetheless equal because they have the same magnitude and the same direction.

$\vec{G}$

Park St.

Pond

1st Ave.

Your apartment

(d)

Vectors $\vec{H}$ and $\vec{J}$ have different magnitudes but the same direction.

Vectors $\vec{J}$ and $\vec{K}$ have the same magnitude but different directions.

$\vec{H}$ $\vec{J}$ $\vec{K}$

Vectors $\vec{H}$ and $\vec{K}$ have different magnitudes and different directions.

A vector has only two attributes: magnitude and direction.

Many important physical quantities have both a magnitude and a direction, and hence are vectors. When we say that a pigeon is flying at 10 m/s due west, we are stating its *velocity* vector (Figure 3-3). The pigeon's *speed* is the magnitude of this vector. Just as for the displacement vector, we use an arrow to denote the vector, and

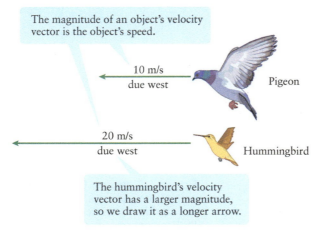

The magnitude of an object's velocity vector is the object's speed.

10 m/s
due west

Pigeon

20 m/s
due west

Hummingbird

The hummingbird's velocity vector has a larger magnitude, so we draw it as a longer arrow.

Figure 3-3 Velocity vectors The velocity of a flying bird is a vector: It has both a magnitude and a direction (the direction of the bird's travel).

we use the length of the vector to denote the magnitude. That's why we draw the velocity vector for a hummingbird flying at 20 m/s as having twice the length of the velocity vector for the 10-m/s pigeon in Figure 3-3: The hummingbird's speed (the magnitude of its velocity) is twice as great. Other vectors that we will encounter in our study of physics include force (a push or pull), acceleration (which describes changes in velocity), and electric and magnetic fields.

Other physical quantities have *no* direction associated with them, so they are *not* vectors. These quantities are called **scalars** and can be described simply by stating a number and a unit. For example, the average temperature inside a home is 20°C, and the duration of a typical university lecture is 50 min. Other examples of scalar quantities are area, volume, mass, and density. Note that some scalar quantities can be negative: For example, the temperature inside a typical home freezer is −18°C or below.

We use different symbols for scalars and vectors to help us distinguish between them. We always use an *italic* letter to denote a scalar quantity, such as T for temperature or t for time. In handwriting you use ordinary letters for these quantities. By contrast, we always denote vectors using *boldface italic* letters with an arrow on top. Thus we write displacement vectors like those shown in Figure 3-2 as $\vec{E}, \vec{F}, \vec{G}$, and so on. In handwriting you should *always* draw an arrow over the symbol for a vector quantity.

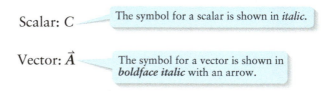

Scalar: C — The symbol for a scalar is shown in *italic*.

Vector: $\vec{A}$ — The symbol for a vector is shown in *boldface italic* with an arrow.

The symbol for the magnitude of a vector is the same symbol as for the vector itself, but *without* the arrow over the symbol and in italic rather than boldface italic. For example, the displacement $\vec{E}$ shown in Figure 3-2b has magnitude $E = 520$ m. We also sometimes use absolute value signs to denote the magnitude of a vector: $|\vec{E}| = E = 520$ m.

WATCH OUT! Use vector notation correctly.

! Note that it is *never* correct to write an equation such as $\vec{E} = 520$ m. A vector is not simply equal to its magnitude; its direction is just as important. You must instead say, "$\vec{E}$ has magnitude 520 m and points 60° north of east." (Later in this chapter we'll see a different, shorthand way to represent a vector mathematically.)

Like a scalar, the magnitude of a vector is given by a number and a unit. The unit is meters (m) for the displacement shown in Figure 3-2b. The magnitude of a vector can *never* be a negative number. If asked the distance from your apartment to the coffee house in Figure 3-2b, you would never say, "It's negative 520 m from here"! In the same way, speed (the magnitude of velocity) is never negative. That's why there are no negative numbers on a speedometer.

Adding Vectors

Scalars add together according to the rules of ordinary arithmetic: If your house has a floor area of 200 m² and you build an extra room with a floor area of 15 m², the net result is a house with a floor area of $200 \text{ m}^2 + 15 \text{ m}^2 = 215 \text{ m}^2$. Vectors behave differently, however. As an example, suppose you walk due east from your apartment to the corner of 1st Avenue and Park Street, a displacement $\vec{P}$, then walk due north to the coffee house, a displacement $\vec{Q}$ (Figure 3-4a). Your net displacement is from your apartment to the coffee house; the combined effect of these two displacements is the same as the single displacement $\vec{E}$. Combining $\vec{P}$ and $\vec{Q}$ to get $\vec{E}$ is an example of **vector addition**, and we say that $\vec{E}$ is the **vector sum** of $\vec{P}$ and $\vec{Q}$. That is, $\vec{E} = \vec{P} + \vec{Q}$. Figure 3-4a shows that another way to get to the coffee house is to travel first from your apartment to the duck pond (displacement $\vec{R}$) and then from the duck pond to the coffee house (displacement $\vec{S}$). So it's also true that $\vec{E} = \vec{R} + \vec{S}$.

Figure 3-4a illustrates that to perform vector addition we draw the two vectors in sequence, with the tail of the second vector up against the tip of the first vector. The sum of these two vectors then points from the tail of the first vector to the tip of the second vector. Even if the two vectors are not originally tip to tip, you can make them that way by moving or *translating* one vector's tail to the other vector's tip while keeping the directions of the vectors the same, as Figure 3-4b shows. This figure also shows that the order in which you add vectors doesn't matter, so $\vec{A} + \vec{B} = \vec{B} + \vec{A}$. Ordinary scalar addition behaves the same way: for example, $3 + 4 = 4 + 3$.

(a)

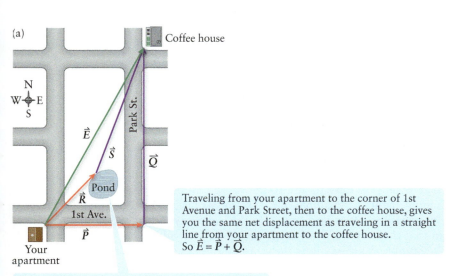

Traveling from your apartment to the corner of 1st Avenue and Park Street, then to the coffee house, gives you the same net displacement as traveling in a straight line from your apartment to the coffee house. So $\vec{E} = \vec{P} + \vec{Q}$.

A trip via the duck pond also gives you the same net displacement as a straight-line trip. So $\vec{E} = \vec{R} + \vec{S}$.

Figure 3-4 Vector addition (a) All three paths from your apartment to the coffee house give the same total displacement vector: $\vec{E} = \vec{P} + \vec{Q} = \vec{R} + \vec{S}$. (b) How to place two vectors $\vec{A}$ and $\vec{B}$ so they can be added.

(b)

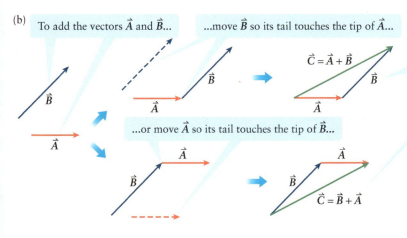

To add the vectors $\vec{A}$ and $\vec{B}$... ...move $\vec{B}$ so its tail touches the tip of $\vec{A}$... ...or move $\vec{A}$ so its tail touches the tip of $\vec{B}$...

...and the sum $\vec{C}$ of the two vectors runs from the tail of the first vector to the tip of the second vector. You get the same result no matter which order you add the vectors: $\vec{C} = \vec{A} + \vec{B} = \vec{B} + \vec{A}$.

When adding two vectors, place the tip of the first vector so it touches the tail of the second vector. The vector sum then extends from the tail of the first vector to the tip of the second vector.

WATCH OUT! Vscator addition is not the same as ordinary addition.

It's a common error to conclude that the magnitude of $\vec{A} + \vec{B}$, the sum of two vectors, is equal to the sum of the magnitude of $\vec{A}$ and the magnitude of $\vec{B}$. This usually gives the *wrong* answer when adding vectors. As an example, in **Figure 3-5a** the vectors $\vec{A}$ and $\vec{B}$ are perpendicular, so these two vectors and their sum $\vec{C} = \vec{A} + \vec{B}$ make up three sides of a right triangle. In this case the Pythagorean theorem tells us that the magnitude of $\vec{C}$ is given by $C^2 = A^2 + B^2$,

so $C = \sqrt{A^2 + B^2}$. This is less than $A + B$. The one situation in which the magnitude $|\vec{A} + \vec{B}|$ *is* equal to the sum of magnitudes $A + B$ is when $\vec{A}$ and $\vec{B}$ point in the same direction (**Figure 3-5b**). **Figure 3-5c** shows that if $\vec{A}$ and $\vec{B}$ point in *opposite* directions, the magnitude $|\vec{A} + \vec{B}|$ is equal to the *difference* of A and B. In Section 3-3 we'll learn a technique for easily calculating the magnitude and direction of a vector sum.

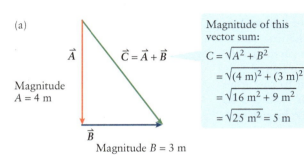

(a)

Magnitude $A = 4$ m
Magnitude $B = 3$ m

Magnitude of this vector sum:

$C = \sqrt{A^2 + B^2}$
$= \sqrt{(4 \text{ m})^2 + (3 \text{ m})^2}$
$= \sqrt{16 \text{ m}^2 + 9 \text{ m}^2}$
$= \sqrt{25 \text{ m}^2} = 5$ m

(b) $\vec{A}$ Magnitude $A = 4$ m $\vec{B}$ Magnitude $B = 3$ m

$\vec{C} = \vec{A} + \vec{B}$

Magnitude of this vector sum:

$C = A + B$
$= 4 \text{ m} + 3 \text{ m}$
$= 7$ m

(c) $\vec{A}$ Magnitude $A = 4$ m

$\vec{C} = \vec{A} + \vec{B}$ $\vec{B}$ Magnitude $B = 3$ m

Magnitude of this vector sum:

$C = A - B$
$= 4 \text{ m} - 3 \text{ m}$
$= 1$ m

The magnitude of the vector sum $\vec{A} + \vec{B}$ is equal to the sum of the magnitudes of $\vec{A}$ and $\vec{B}$ only if $\vec{A}$ and $\vec{B}$ point in the same direction, as in (b).

Figure 3-5 Special cases of vector addition The sum $\vec{A} + \vec{B}$ of two vectors that point (a) perpendicular to each other, (b) in the same direction, and (c) in opposite directions.

Subtracting Vectors

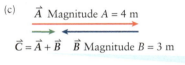
Go to Picture It 3-1 for more practice dealing with adding and subtracting vectors.

A simple problem in arithmetic is "What is 7 minus 4?" What this problem is really asking is, "What do I have to add to the second number (in our example, 4) to get the first number (in our example, 7)?" The answer is $7 - 4 = 3$ because 3 added to 4 gives 7. **Vector subtraction** works in much the same way. When we say that $\vec{D} - \vec{E} = \vec{F}$ we mean that the vector $\vec{F}$ is what we would have to add to the vector $\vec{E}$ to get the vector $\vec{D}$. **Figure 3-6a** shows how the vectors $\vec{D}$, $\vec{E}$, and $\vec{F}$ are related.

(a) Subtracting vectors:

The difference $7 - 4 = 3$ is what you must add to 4 to get to 7.

$\vec{F} = \vec{D} - \vec{E}$

Similarly, the vector difference
$\vec{F} = \vec{D} - \vec{E}$
is what you must add to $\vec{E}$ to get $\vec{D}$.

(b) How to calculate $\vec{F} = \vec{D} - \vec{E}$:

① Form the vector $-\vec{E}$ (with the same magnitude as $\vec{E}$ but the opposite direction).

② Add $\vec{D}$ and $-\vec{E}$ by placing them tip to tail. Their sum is $\vec{D} + (-\vec{E}) = \vec{D} - \vec{E}$.

Figure 3-6 Vector subtraction Subtracting $\vec{E}$ from $\vec{D}$ is the same as adding $-\vec{E}$ to $\vec{D}$. (a) Subtracting one vector from another. (b) How to place the vectors $\vec{D}$ and $\vec{E}$ so they can be subtracted.

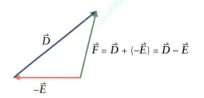

$\vec{F} = \vec{D} + (-\vec{E}) = \vec{D} - \vec{E}$

To see how to carry out vector subtraction, let's first notice something about ordinary subtraction: When we *subtract* 4 from 7, we're really *adding* −4 to 7, so that $7 − 4 = 7 + (−4) = 3$. In the same way, when we *subtract* the vector $\vec{E}$ from the vector $\vec{D}$ to calculate the **vector difference** $\vec{D} − \vec{E}$, we're really *adding* the vector $−\vec{E}$ to the vector $\vec{D}$: that is, $\vec{D} − \vec{E} = \vec{D} + (−\vec{E})$. To take the *negative of a vector,* we keep its magnitude the same and reverse its direction (Figure 3-6b). Thus if $\vec{E}$ is a displacement vector of magnitude 600 m that points due *east,* the vector $−\vec{E}$ also has magnitude 600 m but points due *west.* Figure 3-6b shows how to carry out the subtraction $\vec{D} − \vec{E}$ by adding $\vec{D}$ and $−\vec{E}$. If you know how to add two vectors, you also know how to subtract them!

A number of different animal species use vector addition and subtraction to navigate. As an example, honeybees (Figure 3-7a) fly along a jagged path from flower to flower in search of nectar, but they fly straight back—on a beeline—to their hive. A bee manages this feat by keeping track of the individual distances and directions it travels and then computing the vector needed to go home. That is, if a honeybee's individual displacements from the hive are $\vec{A}$, $\vec{B}$, and $\vec{C}$, it knows that its net displacement from the hive is $\vec{A} + \vec{B} + \vec{C}$ and that it must travel through an additional displacement $−(\vec{A} + \vec{B} + \vec{C})$ to return to the hive (Figure 3-7b). Fiddler crabs, desert ants, and hamsters also navigate using vectors. We'll use vector subtraction in Section 3-4 to help us extend the ideas of velocity and acceleration to motion in two or three dimensions.

Multiplying a Vector by a Scalar

A third useful bit of mathematics involving vectors is **vector multiplication by a scalar.** Figure 3-8 shows the simple rules for this. If c is a *positive* scalar, then the product of c and a vector $\vec{A}$ is a new vector $c\vec{A}$ that points in the same direction as $\vec{A}$ but has a different magnitude, equal to cA (the product of c and the magnitude of $\vec{A}$). If c is a *negative* scalar, then $c\vec{A}$ points in the direction opposite to $\vec{A}$ and has magnitude $|c|A$ (the absolute value of c multiplied by the magnitude of $\vec{A}$). We actually used this idea already when we defined the negative of a vector (Figure 3-6b): The vector $−\vec{E}$ equals the product of the negative scalar $−1$ and the vector $\vec{E}$, so $−\vec{E}$ points in the direction opposite to $\vec{E}$ and has magnitude $|−1| E = (1)E = E$ (that is, the same magnitude as $\vec{E}$).

We can also *divide* a vector by a scalar. This is really the same thing as multiplying by a scalar. As an example, dividing the vector $\vec{A}$ by 2 is the same as multiplying $\vec{A}$ by $1/2$: $\vec{A}/2 = (1/2)\vec{A}$. So $\vec{A}/2$ is a vector that points in the same direction as $\vec{A}$ and has one-half the magnitude (Figure 3-8).

We'll multiply and divide a vector by a scalar frequently in our study of physics. In Section 3-4 we'll use these ideas to help define the velocity and acceleration vectors.

Rob Flynn/USDA ARS

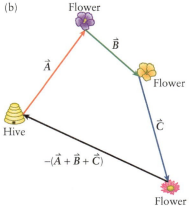

(b)

Flower

$\vec{B}$

$\vec{A}$

Flower

$\vec{C}$

Hive

$−(\vec{A} + \vec{B} + \vec{C})$

Flower

Figure 3-7 Honeybee vectors (a) A honeybee (genus *Apis*) visiting a flower. (b) Honeybees keep track of their vector displacements $\vec{A}, \vec{B}, \vec{C}, \ldots$ and use vector subtraction to calculate the direct route back to the hive.

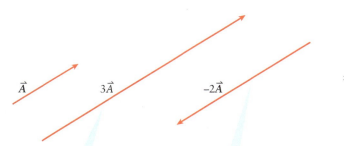

$\vec{A}$ $3\vec{A}$ $−2\vec{A}$ $\dfrac{\vec{A}}{2} = \dfrac{1}{2}\vec{A}$

The number 3 is positive, so the vector $3\vec{A}$ points in the same direction as $\vec{A}$ but has 3 times the magnitude.

The number −2 is negative, so the vector $−2\vec{A}$ points in the direction opposite to $\vec{A}$ and has 2 times the magnitude.

Dividing a vector by 2 is the same as multiplying it by 1/2, so the vector $\vec{A}/2$ points in the same direction as $\vec{A}$ (since 1/2 is positive) and has 1/2 the magnitude.

Figure 3-8 Multiplying a vector by a scalar How to multiply a vector $\vec{A}$ by a positive scalar or a negative scalar, and how to divide a vector by a scalar.

GOT THE CONCEPT? 3-1 Adding and Subtracting Vectors

Vector $\vec{A}$ points due north, and vector $\vec{B}$ points due west. Both vectors have the same magnitude.

Which of the following vectors points southwest? (a) $\vec{A} + \vec{B}$; (b) $\vec{A} − \vec{B} = \vec{A} + (−\vec{B})$; (c) $\vec{B} − \vec{A}$; (d) $−\vec{A} − \vec{B}$; (e) none of these.

TAKE-HOME MESSAGE FOR Section 3-2

✔ Vectors are quantities that have both a magnitude and a direction. Scalars are ordinary numbers.

✔ To find the vector sum $\vec{A} + \vec{B}$ of two vectors, place the tail of the second vector ($\vec{B}$) against the tip of the first vector ($\vec{A}$). The vector sum points from the tail of the first vector to the tip of the second vector.

✔ The vector difference of two vectors, $\vec{A} - \vec{B}$, is the sum of $\vec{A}$ and $-\vec{B}$ (the negative of vector $\vec{B}$, which has the same magnitude as $\vec{B}$ but points in the opposite direction).

✔ The product of a scalar c and a vector $\vec{A}$ is a new vector that has magnitude $|c|A$ (the absolute value of c multiplied by the magnitude of $\vec{A}$). This vector points in the same direction as $\vec{A}$ if c is positive but in the opposite direction if c is negative.

3-3 Vectors can be described in terms of components

We saw in Figure 3-5 that adding two vectors is straightforward if the vectors are perpendicular, point in the same direction, or point in opposite directions. But in many situations vectors are not so conveniently arranged. What can we do then?

A powerful technique that we can use for all kinds of calculations with vectors, no matter how they're oriented, is the method of **components**. An example is the arc of a kicked football in the absence of air resistance. Gravity pulls the football vertically downward but does not affect the football's horizontal motion. Likewise, wind blowing horizontally does not affect the vertical motion of the ball. As a result we can analyze the two-dimensional motion of the football as two separate, one-dimensional motions—vertical motion and horizontal motion. The football's velocity is a vector, so it makes sense to break that vector into horizontal and vertical components.

Figure 3-9a shows a pair of mutually perpendicular coordinate axes labeled x and y. These two axes define a plane, and any point in the plane can be identified by its x and y coordinates. For instance, a point located at $x = 3$ and $y = 5$ has coordinates (3, 5). The

Figure 3-9 Components of a vector (a) Determining the x component A_x and y component A_y of a vector $\vec{A}$. (b), (c), (d) Whether the x and y components are positive or negative depends on the direction of the vector.

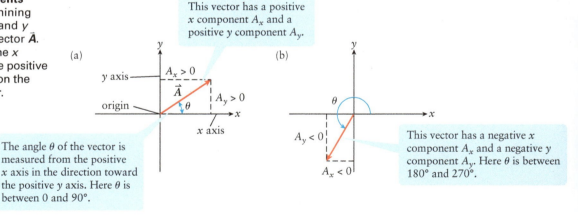

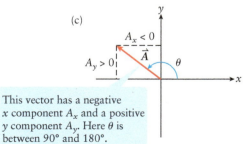

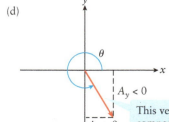

figure also shows a vector $\vec{A}$ whose direction lies in the $x - y$ plane. If we place the tail of this vector at the **origin**—that is, at the point $x = 0$, $y = 0$—the coordinates of the tip of the vector $\vec{A}$ are (A_x, A_y). The quantities A_x (we say "A-sub-x") and A_y (we say "A-sub-y") are called the components of the vector $\vec{A}$: A_x is called the **x component**, and A_y is called the **y component**.

As Figure 3-9a shows, to find the x component A_x we first draw the vector $\vec{A}$ with its tail at the origin, then draw a line perpendicular to the x axis from the tip of $\vec{A}$ to the x axis. Where this line intersects the x axis tells you the value of A_x. In a similar way, to find the value of A_y, we draw a line perpendicular to the y axis from the tip of $\vec{A}$ to the y axis. Depending on the direction of the vector, the components can be both positive (Figure 3-9a), both negative (Figure 3-9b), or of different signs (Figures 3-9c and 3-9d).

Once you have stated the x component and y component of a vector, you have defined the vector completely. So you can describe a vector such as $\vec{A}$ in terms of either (1) its magnitude A and the angle θ or (2) its components A_x and A_y. You can use whichever description is more convenient. To see how to convert between these two descriptions, note that the vector $\vec{A}$ of magnitude A, its x component A_x, and its y component A_y form three sides of a right triangle. So we can use the laws of trigonometry to relate the magnitude and angle to the x and y components, as Figure 3-10 shows.

These relationships in Figure 3-10 give us two equations that tell us how to find the components from the magnitude and direction, and two that tell us how to find the magnitude and direction from the components:

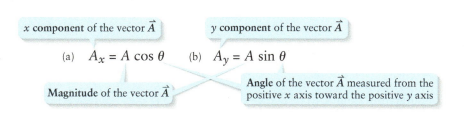

x component of the vector $\vec{A}$ y component of the vector $\vec{A}$

(a) $A_x = A \cos \theta$ (b) $A_y = A \sin \theta$

Magnitude of the vector $\vec{A}$

Angle of the vector $\vec{A}$ measured from the positive x axis toward the positive y axis

Finding vector components from vector magnitude and direction (3-1)

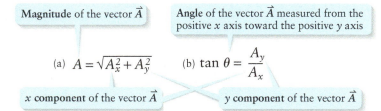

Magnitude of the vector $\vec{A}$

Angle of the vector $\vec{A}$ measured from the positive x axis toward the positive y axis

(a) $A = \sqrt{A_x^2 + A_y^2}$ (b) $\tan \theta = \dfrac{A_y}{A_x}$

x component of the vector $\vec{A}$ y component of the vector $\vec{A}$

Finding vector magnitude and direction from vector components (3-2)

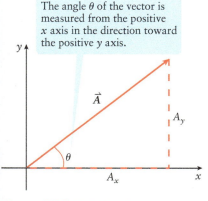

The angle θ of the vector is measured from the positive x axis in the direction toward the positive y axis.

① From the definitions of cosine and sine,

$$\cos \theta = \frac{\text{adjacent side}}{\text{hypotenuse}} = \frac{A_x}{A}$$

$$\sin \theta = \frac{\text{opposite side}}{\text{hypotenuse}} = \frac{A_y}{A}$$

so $A_x = A \cos \theta$ and $A_y = A \sin \theta$
Use these equations to find the components A_x and A_y from the magnitude A and angle θ.

② From the Pythagorean theorem,

$$A^2 = A_x^2 + A_y^2$$

so $A = \sqrt{A_x^2 + A_y^2}$

From the definition of tangent,

$$\tan \theta = \frac{\text{opposite side}}{\text{adjacent side}} = \frac{A_y}{A_x}$$

Use these equations to find the magnitude A and angle θ from the components A_x and A_y.

Figure 3-10 Relating the two ways to describe a vector You can describe a vector $\vec{A}$ in terms of its magnitude A and direction θ or in terms of its components A_x and A_y.

WATCH OUT! Be careful with angles when doing calculations with vector components.

Many students get frustrated when working with Equations 3-1 and 3-2 because they didn't realize their calculator was set to "radians" mode for angles. If you're given an angle θ in degrees, or are trying to calculate an angle in degrees, always check that your calculator is in "degrees" mode. In addition, note that Equations 3-1 and 3-2 are valid *only* if the angle θ is measured from the positive x axis in the direction toward the positive y axis, as in Figures 3-9 and 3-10. We recommend that you *always* measure θ in that way. If you choose to measure the angle θ in some other way, you'll need to apply the trigonometric relations carefully to properly relate the magnitude, x component, and y component to your chosen angle.

Vector Arithmetic with Components

Describing vectors in terms of their components greatly simplifies the vector arithmetic that we described in Section 3-2. Here are the rules:

(1) Figure 3-11 shows that if you add the vectors $\vec{A}$ and $\vec{B}$ to form the vector sum $\vec{C} = \vec{A} + \vec{B}$, each component of $\vec{C}$ is just the sum of the corresponding components of $\vec{A}$ and $\vec{B}$:

x component of the vector $\vec{C} = \vec{A} + \vec{B}$

x component of the vector $\vec{A}$

x component of the vector $\vec{B}$

Rules for vector addition using components (3-3)

(a) $C_x = A_x + B_x$

(b) $C_y = A_y + B_y$

y component of the vector $\vec{C} = \vec{A} + \vec{B}$

y component of the vector $\vec{A}$

y component of the vector $\vec{B}$

(2) If we subtract $\vec{B}$ from $\vec{A}$ to form the vector difference $\vec{D} = \vec{A} - \vec{B} = \vec{A} + (-\vec{B})$, each component of $\vec{D}$ is equal to the sum of the corresponding components of $\vec{A}$ and $-\vec{B}$. The components of $-\vec{B}$ are $-B_x$ and $-B_y$, so the components of $\vec{D}$ are $D_x = A_x + (-B_x)$ and $D_y = A_y + (-B_y)$, or

x component of the vector $\vec{D} = \vec{A} - \vec{B}$

x component of the vector $\vec{A}$

x component of the vector $\vec{B}$

Rules for vector subtraction using components (3-4)

(a) $D_x = A_x - B_x$

(b) $D_y = A_y - B_y$

y component of the vector $\vec{D} = \vec{A} - \vec{B}$

y component of the vector $\vec{A}$

y component of the vector $\vec{B}$

Figure 3-11 Vector addition using components The simplest way to add two vectors is in terms of their components.

The y component of the vector $\vec{C} = \vec{A} + \vec{B}$ equals the sum of the y components of $\vec{A}$ and $\vec{B}$:

$C_y = A_y + B_y$

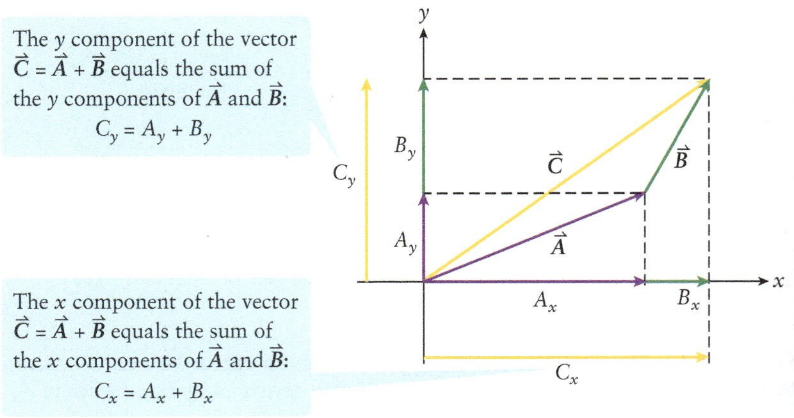

Each component of a vector sum $\vec{C} = \vec{A} + \vec{B}$ equals the sum of the corresponding components of the vectors $\vec{A}$ and $\vec{B}$.

The x component of the vector $\vec{C} = \vec{A} + \vec{B}$ equals the sum of the x components of $\vec{A}$ and $\vec{B}$:

$C_x = A_x + B_x$

That is, each component of $\vec{D} = \vec{A} - \vec{B}$ equals the difference between the corresponding components of $\vec{A}$ and $\vec{B}$.

(3) Finally, multiplying a vector $\vec{A}$ by a scalar c gives a new vector $\vec{E} = c\vec{A}$ whose components are just the components of $\vec{A}$ multiplied by c:

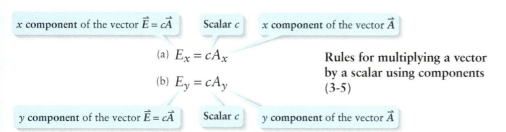

x component of the vector $\vec{E} = c\vec{A}$ Scalar c x component of the vector $\vec{A}$

(a) $E_x = cA_x$

(b) $E_y = cA_y$

Rules for multiplying a vector by a scalar using components (3-5)

y component of the vector $\vec{E} = c\vec{A}$ Scalar c y component of the vector $\vec{A}$

If c is negative, the sign of each vector component is reversed and so the direction of the vector as a whole is reversed. That's just what we saw in Figure 3-8.

To use these rules you need to know the vectors $\vec{A}$ and $\vec{B}$ in terms of their components. If instead you're given $\vec{A}$ and $\vec{B}$ in terms of their magnitudes and directions, you can find their components using Equations 3-1. Once you've found the components of the desired result (such as a vector sum or a vector difference), you can determine its magnitude and direction using Equations 3-2.

GOT THE CONCEPT? 3-2
Vector Components

(?) What are the signs (positive or negative) of the x and y components of the four vectors $\vec{A}$, $\vec{B}$, $\vec{C}$, and $\vec{D}$ shown in the accompanying figure?

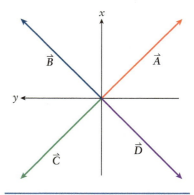

EXAMPLE 3-1 **Different Descriptions of a Vector**

(a) Find the x and y components of a velocity vector $\vec{v}$ that has magnitude 35.0 m/s and points in a direction 36.9° west of north. (b) Find the magnitude and direction of a displacement vector $\vec{r}$ that has x component −24.0 m and y component −11.0 m. Choose the positive x direction to point east and the positive y direction to point north.

Set Up

Our drawing shows the two vectors as well as the x and y axes. We'll use Equations 3-1 to find the x and y components of the vector $\vec{v}$ in part (a) and Equations 3-2 to find the magnitude and direction of the vector $\vec{r}$ in part (b).

Finding vector components from vector magnitude and direction:

$$A_x = A \cos \theta \quad (3\text{-}1a)$$
$$A_y = A \sin \theta \quad (3\text{-}1b)$$

Finding vector magnitude and direction from vector components:

$$A = \sqrt{A_x^2 + A_y^2} \quad (3\text{-}2a)$$
$$\tan \theta = \frac{A_y}{A_x} \quad (3\text{-}2b)$$

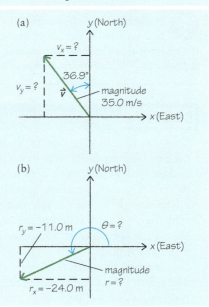

Solve

(a) The angle 36.9° is measured from the positive y axis. To use Equations 3-1, the angle of the vector $\vec{v}$ must be measured from the positive x axis. The figure shows that the angle we need is $90° + 36.9° = 126.9°$. Use this angle in Equations 3-1 along with the magnitude $v = 35.0$ m/s of the vector $\vec{v}$.

$$v_x = (35.0 \text{ m/s}) \cos 126.9°$$
$$= (35.0 \text{ m/s})(-0.600)$$
$$= -21.0 \text{ m/s}$$

$$v_y = (35.0 \text{ m/s}) \sin 126.9°$$
$$= (35.0 \text{ m/s})(0.800)$$
$$= +28.0 \text{ m/s}$$

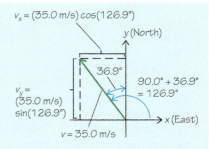

(b) We are given the x and y components of the vector $\vec{r}$, so we can use Equation 3-2a to calculate the magnitude r of this vector.

$$r = \sqrt{r_x^2 + r_y^2}$$
$$= \sqrt{(-24.0 \text{ m})^2 + (-11.0 \text{ m})^2}$$
$$\text{so } r = 26.4 \text{ m}$$

From Equation 3-2b, the tangent of the angle θ shown in the figure equals r_y divided by r_x, or 0.458. So the angle θ is the *inverse* tangent, or arctangent, of 0.458. If you use your calculator to find this, the result you'll get is 24.6°. However, the tangent function has the property that $\tan \phi = \tan (180° + \phi)$. Hence 0.458 is equal to *both* $\tan 24.6°$ and $\tan (180° + 24.6°) = \tan 204.6°$. The figure shows that the angle of the vector $\vec{r}$ measured from the positive x axis is greater than 180°, so the answer we want is $\theta = 204.6°$.

$$\tan \theta = \frac{r_y}{r_x} = \frac{-11.0 \text{ m}}{-24.0 \text{ m}} = 0.458$$
$$\theta = \tan^{-1} 0.458$$
$$= 24.6° \text{ or } 204.6°$$
$$= 205° \text{ to 3 significant figures}$$

Reflect

The drawing of $\vec{v}$ in part (a) shows that this vector points in the negative x direction and the positive y direction. This agrees with our results that v_x is negative and v_y is positive. The drawing of $\vec{r}$ in part (b) helped us decide which value of θ was the correct one. As this example shows, it's important to *always* draw a picture for any problem that involves vectors.

WATCH OUT! Use the inverse tangent function properly.

Students sometimes confuse the inverse tangent function that we used in the above example, $\tan^{-1} 0.458$, with the reciprocal of the tangent function, $1/(\tan 0.458)$. (The superscript "−1" can be the source of this confusion.)

These two quantities have very different numerical values ($\tan^{-1} 0.458 = 24.6°$ or 204.6°, while $1/(\tan 0.458°) = 125$), so don't make this mistake!

BioMedical EXAMPLE 3-2 At What Angle Is Your Heart?

One way to help diagnose health issues such as lung disease, congenital heart disease, and hypertension is to measure the *electrical axis* of the heart. In your heart pacemaker cells generate electrical signals that trigger a contraction of the heart as the signals spread from the top of the heart to the bottom. The electrical axis is the direction in which these signals travel, and reflects the angle at which the heart is positioned in the chest cavity.

The direction of a person's electrical axis is found using a technique called *electrocardiography*. In a simplified version of electrocardiography, electrodes are placed on the person's left wrist, right wrist, and right ankle (Figure 3-12a). The electrical signal between the two wrist electrodes tells us the horizontal, right-to-left component of the heart's electrical axis vector. The signal between the right wrist and right ankle gives us the vertical, top-to-bottom component.

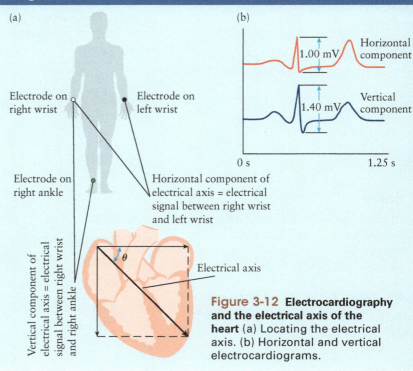

Figure 3-12 **Electrocardiography and the electrical axis of the heart** (a) Locating the electrical axis. (b) Horizontal and vertical electrocardiograms.

The graphs in Figure 3-12b show sample recordings of the electrical signal versus time for these two pairs of electrodes. These graphs are called *electrocardiograms*, abbreviated ECG or EKG (after their German name *Electrokardiogramm*). In this particular example, the maximum ECG signal between the wrists is 1.00 millivolt (mV), and the maximum ECG signal between the right wrist and right ankle is 1.40 mV. At what angle θ does the electrical signal propagate across the heart?

Set Up

Our goal is to find the angle θ shown in the figure, which is the direction of the heart's electrical axis measured relative to the horizontal. We choose the x axis to be horizontal and choose the y axis to be vertically downward. (We can choose the axes to be whatever we want, provided they're mutually perpendicular.) The sketch shows the electrical axis as a green vector; it has an x component of 1.00 mV (shown in red) and a y component of 1.40 mV (shown in blue). As in Example 3-1, we'll use Equation 3-2b to find this angle.

Finding vector direction from vector components:

$$\tan \theta = \frac{A_y}{A_x} \quad (3\text{-}2b)$$

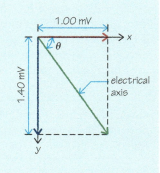

Solve

The angle θ is measured from the positive x axis toward the positive y axis, so we can safely use Equation 3-2b. Using your calculator you'll find that $\tan^{-1} 1.40 = 54.5°$. We saw in Example 3-1 that a second solution is 180° plus the result from your calculator, or 180° + 54.5°, or 234.5°, or 234.5°. However, the figure above shows that θ is between 0 and 90°, so the correct answer is 54.5°.

$$\tan \theta = \frac{y \text{ component}}{x \text{ component}}$$
$$= \frac{1.40 \text{ mV}}{1.00 \text{ mV}} = 1.40$$

$\theta = \tan^{-1} 1.40 = 54.5°$ or $234.5°$

Correct answer: $\theta = 54.5°$

Reflect

If the horizontal and vertical signals were equally strong, the electrical axis would be tilted at an angle of 45° from the horizontal. In this example the vertical component is somewhat larger than the horizontal component, so the angle is a bit more than 45°.

EXAMPLE 3-3 **Adding Vectors Using Components**

You travel 62.0 km in a direction 30.0° north of east then turn to a direction 50.0° south of east and travel an additional 23.0 km (Figure 3-13). How far and in what direction are you from your starting point?

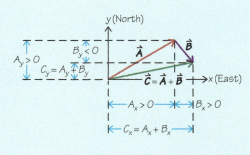

Figure 3-13 A problem in vector addition What is your net displacement?

Set Up

Each leg of the trip constitutes a displacement vector, which we show as $\vec{A}$ and $\vec{B}$ in the sketch. To find the magnitude and direction of the total displacement, which is just the vector sum $\vec{C} = \vec{A} + \vec{B}$, we will first find the x and y components of $\vec{A}$ and $\vec{B}$ using Equations 3-1. We'll then find the components of the vector sum $\vec{C}$ using Equations 3-3. Finally, we'll find the magnitude and direction of $\vec{C}$ using Equations 3-2.

$$A_x = A \cos \theta \quad (3\text{-}1a)$$
$$A_y = A \sin \theta \quad (3\text{-}1b)$$
$$C_x = A_x + B_x \quad (3\text{-}3a)$$
$$C_y = A_y + B_y \quad (3\text{-}3b)$$
$$A = \sqrt{A_x^2 + A_y^2} \quad (3\text{-}2a)$$
$$\tan \theta = \frac{A_y}{A_x} \quad (3\text{-}2b)$$

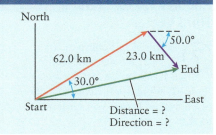

Solve

We choose the positive x axis to point east and the positive y axis to point north, and draw both $\vec{A}$ and $\vec{B}$ with their tails at the origin. The vector $\vec{A}$ points north of east at an angle of 30.0°, and the vector $\vec{B}$ points south of east at an angle of −50.0° (negative since this angle is measured from the positive x axis in the direction *away from* the positive y axis). We use Equations 3-1 to find their components.

$$A_x = (62.0 \text{ km}) \cos 30.0°$$
$$= 53.7 \text{ km}$$
$$A_y = (62.0 \text{ km}) \sin 30.0°$$
$$= 31.0 \text{ km}$$
$$B_x = (23.0 \text{ km}) \cos(-50.0°)$$
$$= 14.8 \text{ km}$$
$$B_y = (23.0 \text{ km}) \sin(-50.0°)$$
$$= -17.6 \text{ km}$$

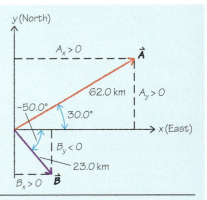

Given the components of $\vec{A}$ and $\vec{B}$, we can now find the components of $\vec{C} = \vec{A} + \vec{B}$ using Equations 3-3.

$$C_x = A_x + B_x = 53.7 \text{ km} + 14.8 \text{ km} = 68.5 \text{ km}$$
$$C_y = A_y + B_y = 31.0 \text{ km} + (-17.6 \text{ km}) = 13.4 \text{ km}$$

Now calculate the magnitude and direction of $\vec{C}$ using Equations 3-2. Both components of $\vec{C}$ are positive, so the vector points north of east, and the angle θ_C is between 0 and 90°. Hence the desired angle is 11.1°, not $180° + 11.1° = 191.1°$.

$$C = \sqrt{C_x^2 + C_y^2}$$
$$= \sqrt{(68.5 \text{ km})^2 + (13.4 \text{ km})^2}$$
$$C = 69.8 \text{ km}$$

$$\tan \theta_C = \frac{C_y}{C_x} = \frac{13.4 \text{ km}}{68.5 \text{ km}} = 0.195$$

$$\theta_C = \tan^{-1} 0.195 = 11.1°$$

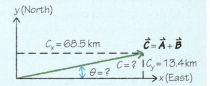

Reflect

The net displacement takes you 69.8 km from the starting point at an angle of 11.1° north of east. These numbers would be very difficult to get simply from the drawing of the two vectors shown above.

We now have all the mathematical tools that we need to study motion in two or three dimensions. In the next section we'll begin this study by seeing how to relate the position, velocity, and acceleration of an object using vectors.

GOT THE CONCEPT? 3-3
Adding One Vector to Another Vector to Get a Third Vector

 Vector $\vec{A}$ has components $A_x = 5$, $A_y = 3$, and vector $\vec{B}$ has components $B_x = 7$, $B_y = -2$. What are the components of the vector $\vec{D}$ that you must add to $\vec{A}$ to get $\vec{B}$?

TAKE-HOME MESSAGE FOR Section 3-3

✔ The easiest way to do arithmetic with vectors (adding, subtracting, and multiplying by a scalar) is to use components.

✔ Equations 3-1 let you find the components of a vector from its magnitude and direction, and Equations 3-2 let you find the magnitude and direction of a vector from its components.

✔ Equations 3-3, 3-4, and 3-5 show how to add vectors, subtract vectors, and multiply a vector by a scalar using components.

✔ For any problem involving vectors, it's essential to draw a sketch of the situation and to measure angles from the positive x direction toward the positive y direction.

✔ Be cautious using Equation 3-2b, $\tan \theta = A_y / A_x$. This equation actually gives two answers for the direction θ of a vector $\vec{A}$; your sketch will help you decide which answer is correct.

3-4 For motion in a plane, velocity and acceleration are vector quantities

A hawk glides over a meadow at a shallow angle then steepens its flight path to dive onto a mouse before sailing upward again (Figure 3-14a). How can an ornithologist describe the hawk's motion using the language of physics?

Just as for the straight-line motion that we discussed in Chapter 2, to have a complete description of the hawk's motion means that we must know its position at every instant of time. This knowledge requires that we have a reference point, or origin, from which to measure positions. Let's choose the origin to be at the point where the ornithologist stands and observes the motion. Then at each instant we can visualize a **position vector** $\vec{r}$ that extends from the origin to the hawk's position at that instant. Figure 3-14b shows one such position vector. If we know what $\vec{r}$ is at each instant of time t, we know the path or **trajectory** that the hawk follows through space as well as what time the hawk passes through each point along that trajectory. In Figure 3-14b we have drawn dots to show the hawk's position at 1-s intervals, so this figure also shows the hawk's motion diagram.

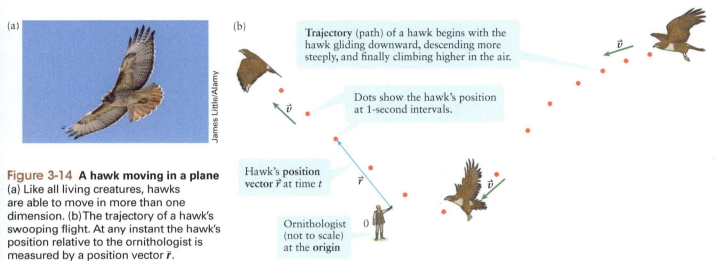

(a)

(b)

Trajectory (path) of a hawk begins with the hawk gliding downward, descending more steeply, and finally climbing higher in the air.

Dots show the hawk's position at 1-second intervals.

Hawk's **position vector** $\vec{r}$ at time t

Ornithologist (not to scale) at the **origin**

$\vec{v}$

$\vec{r}$

0

James Little/Alamy

Figure 3-14 A hawk moving in a plane (a) Like all living creatures, hawks are able to move in more than one dimension. (b) The trajectory of a hawk's swooping flight. At any instant the hawk's position relative to the ornithologist is measured by a position vector $\vec{r}$.

For simplicity let's assume that our hawk is in **two-dimensional motion,** so it moves in only two of the three dimensions of space. This means that the hawk may move up and down as well as forward and back, but doesn't turn left or right. Many real-life motions are two-dimensional—for example, the flight of a thrown baseball and Earth's orbit around the Sun. In these situations we need only two coordinate axes, which we typically call x and y. In mathematics two such axes define a plane, so another name for two-dimensional motion is **motion in a plane.** If we use x and y to define the two perpendicular directions within the plane, we'll refer to this plane as the **x–y plane.**

The Velocity Vector

Just as for the one-dimensional motion that we studied in Chapter 2, we'll use the concepts of displacement and velocity to describe how an object moves along a two- or three-dimensional trajectory. Consider how the hawk shown in Figure 3-14b moves during a time interval between two instants of time that we call t_1 and t_2 (see Figure 3-15). The change of the hawk's position during that interval is its displacement vector $\Delta\vec{r}$ for that time interval. This vector points from the object's position at t_1 to its position at t_2 and is the difference between the object's position vector $\vec{r}_2$ at t_2 and its position vector $\vec{r}_1$ at t_1:

Displacement vector (change in position) of the object over the time interval from time t_1 to a later time t_2

Position vector of the object at later time t_2

Position vector of the object at earlier time t_1

$$\Delta\vec{r} = \vec{r}_2 - \vec{r}_1$$

To calculate the displacement vector, **subtract** the earlier value from the later value (just as for motion in a straight line).

Displacement vector equals the change in position vector over a time interval

(3-6)

Figure 3-15 The displacement vector and velocity vector The displacement and velocity vectors both point along the trajectory.

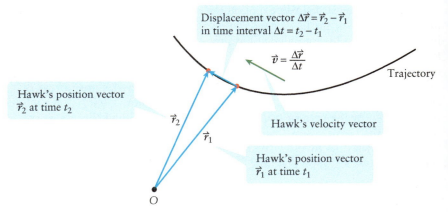

Displacement vector $\Delta\vec{r} = \vec{r}_2 - \vec{r}_1$ in time interval $\Delta t = t_2 - t_1$

$\vec{v} = \dfrac{\Delta\vec{r}}{\Delta t}$

Trajectory

Hawk's position vector $\vec{r}_2$ at time t_2

$\vec{r}_2$

$\vec{r}_1$

Hawk's velocity vector

Hawk's position vector $\vec{r}_1$ at time t_1

O

💡 • An object's displacement during a short time interval is a vector that points from the object's position at the beginning of the interval to its position at the end of the interval.
• The object's velocity vector and displacement vector for that short time interval both point along the object's trajectory.

If the time interval $t_2 - t_1$ between these two points is very small, $\Delta\vec{r}$ in Figure 3-15 is a very small vector and points very nearly along the hawk's trajectory, even if the trajectory is curved. By analogy to what we did for one-dimensional motion in Chapter 2, we define the hawk's **instantaneous velocity vector** $\vec{v}$ (or just **velocity vector** for short) at a given point along the trajectory as follows:

Velocity vector equals displacement vector divided by time interval (3-7)

Velocity vector for the object over a very short time interval from time t_1 to a later time t_2

Displacement vector (change in position) of the object over the short time interval

$$\vec{v} = \frac{\Delta\vec{r}}{\Delta t} = \frac{\vec{r}_2 - \vec{r}_1}{t_2 - t_1}$$

For both the displacement and the elapsed time, **subtract** the earlier value from the later value.

Elapsed time for the time interval

Equation 3-7 tells us to multiply the displacement vector $\Delta\vec{r}$ by $1/\Delta t$ (see Figure 3-8). Since $1/\Delta t$ is positive, the velocity vector $\vec{v}$ defined by Equation 3-7 points in the same direction as $\Delta\vec{r}$. So $\vec{v}$ at each point also points along the trajectory (Figure 3-16).

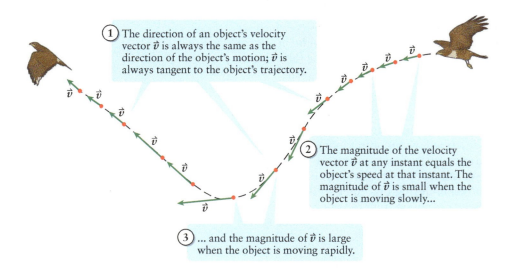

① The direction of an object's velocity vector $\vec{v}$ is always the same as the direction of the object's motion; $\vec{v}$ is always tangent to the object's trajectory.

② The magnitude of the velocity vector $\vec{v}$ at any instant equals the object's speed at that instant. The magnitude of $\vec{v}$ is small when the object is moving slowly...

③ ... and the magnitude of $\vec{v}$ is large when the object is moving rapidly.

Figure 3-16 Velocity vectors on a curved trajectory The varying velocity of the swooping hawk depicted in Figure 3-14.

💡 The direction and magnitude at any instant of an object's velocity vector $\vec{v}$ tells you the direction of the object's motion and the object's speed at that instant.

The *magnitude* v of the velocity vector is the magnitude of $\Delta\vec{r}$ divided by Δt: $v = |\vec{v}| = |\Delta\vec{r}|/\Delta t$. Note that $|\Delta\vec{r}|$ is just the distance that the hawk travels during the very short time interval of duration Δt. So v is the instantaneous speed (distance per time) of the hawk at the instant in question. That's why the velocity vectors drawn in Figure 3-16 have different lengths at different points along the trajectory: $\vec{v}$ has a large magnitude where the hawk is moving rapidly and a small magnitude where the hawk is moving slowly (compare Figure 3-3). Like velocity and speed for motion in a straight line, the magnitude of the velocity vector $\vec{v}$ has units of meters per second (m/s).

Velocity Components

Like any other vector, we can describe an object's velocity vector $\vec{v}$ in terms of its components. The x component of $\vec{v}$, which we call v_x, equals the x component of displacement (the change in x coordinate during the time interval, $\Delta x = x_2 - x_1$) divided by the time interval $\Delta t = t_2 - t_1$. We define v_y, the y component of $\vec{v}$, in a similar way. Thus

x component of the velocity vector $\vec{v}$

x component of the **displacement** during a short time interval

(a) $v_x = \dfrac{\Delta x}{\Delta t} = \dfrac{x_2 - x_1}{t_2 - t_1}$

(b) $v_y = \dfrac{\Delta y}{\Delta t} = \dfrac{y_2 - y_1}{t_2 - t_1}$

y component of the **displacement** during a short time interval

y component of the velocity vector $\vec{v}$

Elapsed time for the time interval

Components of the velocity vector
(3-8)

Equation 3-8a is the same expression that we wrote in Section 2-2 for the average velocity along the x axis in straight-line motion. When the time interval is short, this expression also gives the x component of the instantaneous velocity. Equation 3-8b is the corresponding equation for straight-line motion along the y axis. So our two-dimensional vector definition of velocity is a natural extension of our definition of straight-line motion.

The Acceleration Vector and Its Direction

You are driving a car along a country road. You speed up along a straight part of the road, then slow down when you see a curve up ahead, and finally go around the curve at a constant speed (Figure 3-17). In which of these three cases are you accelerating?

The answer is that you are accelerating in *all three* of these cases: speeding up, slowing down, and turning. Remember from Section 2-4 that an object undergoes an acceleration whenever its velocity changes. Because velocity is a vector, this means that an object accelerates if there is a change in *any* aspect of the velocity vector $\vec{v}$. When your car speeds up (Figure 3-18a) or slows down (Figure 3-18b) while moving in a straight line, the *magnitude* of $\vec{v}$ changes, and there is an acceleration. When your car is turning (Figures 3-18c and 3-18d), the *direction* of $\vec{v}$ changes and so there is also an acceleration, even if the car's speed doesn't change.

Unlike velocity, the direction of a car's acceleration vector may or may not be along its trajectory. To explore this, first recall from Section 2-4 that the acceleration at a given time equals the change in velocity during a very short time interval around that time divided by the duration of the time interval. For an object that moves in a plane, we have to consider the change in its velocity *vector* during a time interval. We again consider an infinitesimally short time interval from time t_1, when the object's velocity is $\vec{v}_1$, to a later time t_2, when the object has velocity $\vec{v}_2$. The object's **acceleration vector** $\vec{a}$ (short for **instantaneous acceleration vector**) at the time of this infinitesimal time interval is equal to the change in velocity divided by the duration of the interval:

Arterra Picture Library/Alamy

Figure 3-17 An accelerating car A car is accelerating whenever its velocity vector $\vec{v}$ is changing in any way. That's the case if the car is speeding up or slowing down, so the magnitude of $\vec{v}$ is changing. It's also the case if a car is traveling at a constant speed but following a winding road: In this situation the magnitude of $\vec{v}$ is constant, but its direction is changing as the car turns.

Acceleration vector for the object over a very short time interval from time t_1 to a later time t_2

Change in velocity of the object over the short time interval

$$\vec{a} = \dfrac{\Delta \vec{v}}{\Delta t} = \dfrac{\vec{v}_2 - \vec{v}_1}{t_2 - t_1}$$

For both the velocity change and the elapsed time, **subtract** the earlier value from the later value.

Elapsed time for the time interval

Acceleration vector equals change in velocity vector divided by time interval
(3-9)

(a) Speeding up in a straight line

The magnitude of the velocity vector is changing, so the object is accelerating.

(b) Slowing down in a straight line

The magnitude of the velocity vector is changing, so the object is accelerating.

(c) Turning right at a steady speed

(d) Turning left at a steady speed

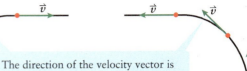

The direction of the velocity vector is changing, so the object is accelerating.

An object accelerates if it speeds up, slows down, or changes direction. That's because in each case the velocity vector changes in either its magnitude or its direction.

Figure 3-18 Four different examples of acceleration An object is accelerating if *any* aspect of its velocity vector $\vec{v}$ is changing.

Figure 3-19 Finding the direction of the acceleration vector An object's acceleration vector $\vec{a}$ for a given brief time interval points in the same direction as the difference $\Delta\vec{v} = \vec{v}_2 - \vec{v}_1$ between the object's velocities at the end and beginning of the interval. The four examples shown here correspond to the four situations in Figure 3-18.

The acceleration vector $\vec{a}$ points in the same direction as $\Delta\vec{v}$ and its magnitude equals the magnitude of $\Delta\vec{v}$ divided by Δt: $|\vec{a}| = |\Delta\vec{v}|/\Delta t$. Just like acceleration for motion in a straight line, the magnitude of the acceleration vector has units of meters per second squared (m/s^2).

Figure 3-19 shows how to find the direction of the acceleration vector $\Delta\vec{v}$ for each situation shown in Figure 3-18. In **Figure 3-19a** a car is speeding up as it moves in a

(a) Speeding up in a straight line.

1. Add $\vec{v}_2$ and $-\vec{v}_1$ to form $\Delta\vec{v} = \vec{v}_2 - \vec{v}_1$:

2. Average acceleration $\vec{a}_{average}$ is in the same direction as $\Delta\vec{v}$—in this case, in the direction of motion.

(b) Slowing down in a straight line.

1. Add $\vec{v}_2$ and $-\vec{v}_1$ to form $\Delta\vec{v} = \vec{v}_2 - \vec{v}_1$:

2. Average acceleration $\vec{a}_{average}$ is in the same direction as $\Delta\vec{v}$—in this case, opposite to the direction of motion.

(c) Turning right at a steady speed.

1. Add $\vec{v}_2$ and $-\vec{v}_1$ to form $\Delta\vec{v} = \vec{v}_2 - \vec{v}_1$:

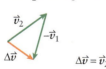

$\Delta\vec{v} = \vec{v}_2 - \vec{v}_1$

2. Average acceleration $\vec{a}_{average}$ is in the same direction as $\Delta\vec{v}$—in this case, toward the inside of the turn.

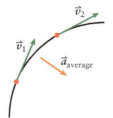

(d) Turning left at a steady speed.

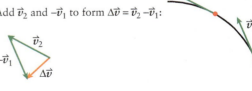

1. Add $\vec{v}_2$ and $-\vec{v}_1$ to form $\Delta\vec{v} = \vec{v}_2 - \vec{v}_1$:

2. Average acceleration $\vec{a}_{average}$ is in the same direction as $\Delta\vec{v}$—in this case, toward the inside of the turn.

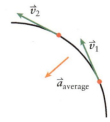

straight line. The velocity change $\Delta\vec{v}$ points in the same direction as the car's motion, and so its acceleration vector $\vec{a}$ is likewise in the direction of motion (the car "accelerates forward"). If instead the car slows down as it moves in a straight line, as shown in Figure 3-19b, $\Delta\vec{v}$ and the acceleration $\vec{a}$ both point opposite to the car's motion (it "accelerates backward"). In Figures 3-19c and 3-19d the car is going around a curve at a constant speed. In this case the acceleration vector $\vec{a}$ is *perpendicular* to the car's trajectory and points toward the *inside* of the curve. If the car is turning to the right, as in Figure 3-19c, the acceleration $\vec{a}$ is to the right; if it is turning to the left, as in Figure 3-19d, the acceleration $\vec{a}$ is to the left. We can say that a car going around a curve "accelerates sideways."

Here's how you can convince yourself that Figures 3-19c and 3-19d correctly show the direction of acceleration while turning: Note that *when you are riding in a vehicle that accelerates, you feel a push in the direction opposite the vehicle's acceleration.* (We'll see why this is so in Chapter 4.) If you're riding in a car that speeds up, you feel pushed *backward* into your seat. This push is opposite to the car's *forward* acceleration vector $\vec{a}$ (see Figure 3-19a). If the car brakes suddenly, you feel thrown *forward* in a direction opposite to the *backward* acceleration vector of the car (see Figure 3-19b). And if the car makes a turn at constant speed, you feel pushed to the *outside* of the turn. The acceleration vector $\vec{a}$ is opposite to this direction and hence must point to the *inside* of the turn, as Figures 3-19c and 3-19d show.

Figure 3-20 shows another useful way to think about the acceleration vector $\vec{a}$. At a given point along an object's trajectory, $\vec{a}$ has two components: one that is perpendicular to the trajectory and one that points along the trajectory in the direction that the object is moving. The acceleration component perpendicular to the trajectory tells us about the change in the object's *direction*, and the component along the trajectory tells us about the change in the object's *speed*.

Figure 3-20 Finding the acceleration of a swooping hawk How to find the acceleration vector of an object from its velocity vectors.

① Find the direction of $\vec{a}$ for (a) the interval from point 1 to point 2 and for (b) the interval from point 2 to point 3.

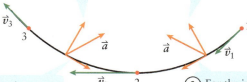

A hawk speeds up as it moves from point 1 to point 2 and slows down as it moves from point 2 to point 3.

For each interval we can break the acceleration $\vec{a}$ into a component along the trajectory and a component perpendicular to the trajectory.

② For both intervals there is a component of acceleration that points perpendicular to the trajectory (toward the inside of the turn). This describes changes in the hawk's direction of motion.

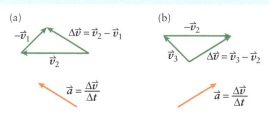

④ For the interval from point 2 to point 3, there is a component of acceleration opposite to the direction of the hawk's motion. This describes the decrease in the hawk's speed.

③ For the interval from point 1 to point 2, there is a component of acceleration in the same direction as the hawk's motion. This describes the increase in the hawk's speed.

 If an object's speed is increasing, $\vec{a}$ has a component in the direction of motion.
If an object's speed is decreasing, $\vec{a}$ has a component opposite to the direction of motion.
If an object's direction of motion is changing (the trajectory is curved), $\vec{a}$ has a component perpendicular to its trajectory that points *toward* the inside of the turn.

It's also often useful to express the acceleration vector $\vec{a}$ of an object in terms of its x and y components. Use the symbols v_{1x} and v_{1y} for the velocity components of the object at time t_1 and the symbols v_{2x} and v_{2y} for the velocity components at time t_2. Then a_x, the x component of $\vec{a}$ equals the x component of velocity change ($\Delta v_x = v_{2x} - v_{1x}$) divided by the time interval $\Delta t = t_2 - t_1$, and the y component of $\vec{a}$ equals the y component of velocity change ($\Delta v_y = v_{2y} - v_{1y}$) divided by Δt:

Components of the acceleration vector in two-dimensional motion (3-10)

x component of the acceleration vector $\vec{a}$

Change in the x component of the velocity during a short time interval

(a) $a_x = \dfrac{\Delta v_x}{\Delta t} = \dfrac{v_{2x} - v_{1x}}{t_2 - t_1}$

Change in the y component of the velocity during a short time interval

(b) $a_y = \dfrac{\Delta v_y}{\Delta t} = \dfrac{v_{2y} - v_{1y}}{t_2 - t_1}$

Elapsed time for the time interval

y component of the acceleration vector $\vec{a}$

In the next section we'll use the component description of velocity and acceleration to help us analyze the motion of a *projectile*—an object moving through the air that is acted on by gravity alone.

GOT THE CONCEPT? 3-4 Acceleration and Changing Speed

(?) Figure 3-21 shows the acceleration vector $\vec{a}$ at several points along an object's curved trajectory. At each of points 1, 2, 3, and 4, is the object (a) speeding up, (b) slowing down, or (c) maintaining a constant speed?

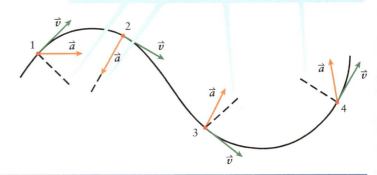

At each point, the dashed line is perpendicular to the trajectory.

Figure 3-21 Acceleration on a curved trajectory At each of the points shown, is the object's speed increasing, decreasing, or staying the same?

TAKE-HOME MESSAGE FOR Section 3-4

✔ If an object moves during a time interval, its displacement $\Delta \vec{r}$ is a vector that points from its position at the beginning of the interval to its position at the end of the interval.

✔ The velocity $\vec{v}$ of an object is a vector. It equals the displacement $\Delta \vec{r}$ for a very short interval divided by the duration of the interval. The velocity vector always points along the object's trajectory.

✔ The acceleration $\vec{a}$ of an object is also a vector. It equals the change in velocity $\Delta \vec{v}$ for a very short interval divided by the duration of the interval. The direction of the acceleration vector depends on how the object's speed is changing and on whether the object is changing direction.

3-5 A projectile moves in a plane and has a constant acceleration

We now have all the tools we need to analyze the kind of motion shown in the photograph that opens this chapter, which shows the arc of a high jumper's motion. The curved path followed by the high jumper is the same as those followed by streams of water from a fountain and by clumps of lava emerging from a volcano (Figure 3-22). The direction of motion of an object that follows such a curved path is constantly changing. Hence the object is *accelerating* at all points during the motion.

(a)

Just as for vertical free fall, the downward acceleration $\vec{a}$ of a projectile has

If the object is moving through the air at a relatively slow speed so that air resistance can be neglected, the object's acceleration is due to gravity alone. Then the object is in *free fall* and has a constant acceleration pointed downward, just as we discussed in Section 2-6. In that section we considered only straight up-and-down free fall; now we want to consider **projectile motion**, which is free fall with both vertical motion *and* horizontal motion. (We often refer to an object undergoing this kind of motion as a **projectile**.) Just as for vertical free fall, the downward acceleration $\vec{a}$ of a projectile has a magnitude g, called the *acceleration due to gravity*. We'll use the value $g = 9.80 \text{ m/s}^2$.

(b)

Projectile motion is always *two*-dimensional, so the motion lies in a plane. To see why, think about tossing a basketball upward and toward the east (Figure 3-23). When the ball leaves your hand, its velocity vector has a horizontal component toward the east, a vertical component that points upward, and zero component along the north–south direction. Because the acceleration vector points straight downward, there is *no* north–south component of acceleration. Therefore, the north–south velocity doesn't change—this component of velocity starts at zero and remains zero. So the basketball always remains in a vertical plane that's oriented east–west, as Figure 3-23 shows.

Figure 3-22 Two examples of projectile motion (a) Water in a fountain. (b) Material ejected from an erupting volcano.

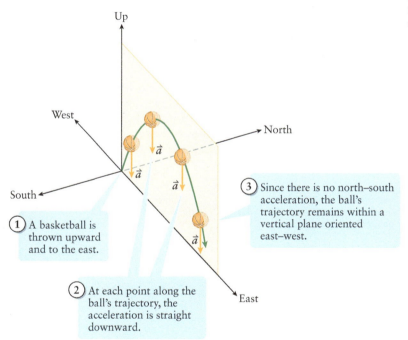

① A basketball is thrown upward and to the east.

② At each point along the ball's trajectory, the acceleration is straight downward.

③ Since there is no north–south acceleration, the ball's trajectory remains within a vertical plane oriented east–west.

Figure 3-23 Why projectile motion is two-dimensional motion The trajectory of a projectile is in a vertical plane because the acceleration due to gravity points downward.

Figure 3-24 shows the acceleration vector $\vec{a}$ at several points along a projectile's trajectory as well as the components of $\vec{a}$ along and perpendicular to the trajectory. There is a perpendicular component of $\vec{a}$ at *all* points, so the projectile's direction of motion is always changing (that is, the projectile's trajectory is curved all along its length). When the projectile is ascending there is also a component of acceleration opposite to its motion, so the projectile is slowing down. When the projectile is descending there is an acceleration component in the direction of motion, so the projectile is speeding up. The only point at which the speed is instantaneously *not* changing is at the highest point of the trajectory, where the acceleration vector is perpendicular to the trajectory. At this point the projectile has its *minimum* speed.

Figure 3-24 Projectile motion
Acceleration components along and perpendicular to the trajectory.

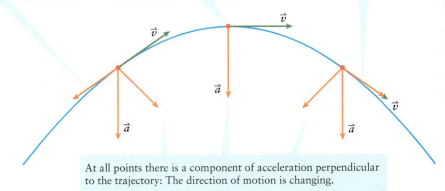

There is a component of acceleration opposite to the direction of motion: The projectile is slowing down.

There is no component of acceleration along the direction of motion: The speed of the projectile has its minimum value.

There is a component of acceleration in the direction of motion: The projectile is speeding up.

At all points there is a component of acceleration perpendicular to the trajectory: The direction of motion is changing.

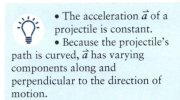

- The acceleration $\vec{a}$ of a projectile is constant.
- Because the projectile's path is curved, $\vec{a}$ has varying components along and perpendicular to the direction of motion.

WATCH OUT! A projectile is a special case.

Not every object in flight can be regarded as a projectile. As an example, a gliding bird is definitely *not* a projectile: The air exerts lift and drag forces on the bird that can't be ignored because these forces are comparable to the force of gravity. A batted baseball also feels a substantial drag force exerted by the air because of its high speed (much greater than the speed of the basketball in Figure 3-23), so this object isn't really a projectile either. Nonetheless, studying projectile motion is an important first step toward understanding these more complex kinds of motion.

The Equations of Projectile Motion

Let's be more quantitative about how a projectile's velocity changes during its flight. We use the x and y axes to define the plane in which the projectile moves, with the positive x axis pointing in a horizontal direction, and the positive y axis pointing vertically upward (Figure 3-25). With this choice of axes, the components of the constant, vertically downward acceleration vector $\vec{a}$ are

Components of the acceleration vector in projectile motion
(3-11)

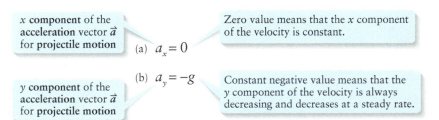

x component of the acceleration vector $\vec{a}$ for projectile motion

(a) $a_x = 0$

Zero value means that the x component of the velocity is constant.

y component of the acceleration vector $\vec{a}$ for projectile motion

(b) $a_y = -g$

Constant negative value means that the y component of the velocity is always decreasing and decreases at a steady rate.

WATCH OUT! Projectile motion means constant vertical acceleration.

Just as for vertical free fall, the vertical (y) component of acceleration is the *same* ($a_y = -g$) at all times during projectile motion. This is true whether the projectile is moving upward, at the high point of its motion, or moving downward.

Because both the x and y components of acceleration are constant, we can use our results from Chapter 2 to easily write the equations that show how the x and y components of the projectile's velocity $\vec{v}$ vary with time. We'll let $t = 0$ represent the instant when the projectile begins its flight. At this moment the projectile has x component of velocity v_{0x} and y component of velocity v_{0y}. (The "0" in the subscripts reminds us that these are the components at $t = 0$.) Equation 2-5 from Section 2-4 tells us that if the

Figure 3-25 Projectile motion: Horizontal and vertical components of velocity and acceleration The acceleration due to gravity has a constant vertical component and no horizontal component. This determines how the velocity components vary with time.

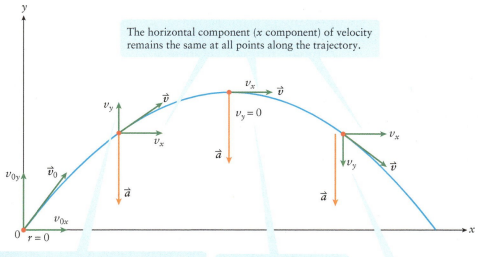

The horizontal component (x component) of velocity remains the same at all points along the trajectory.

$v_y = 0$

The vertical component (y component) of velocity changes from positive (upward)...

... to zero (neither upward nor downward)...

... to negative (downward).

x component of acceleration is constant, the x component of velocity v_x as a function of time is $v_x = v_{0x} + a_x t$. The same equation rewritten for the y direction is $v_y = v_{0y} + a_y t$. If we substitute the values $a_x = 0$ and $a_y = -g$ from Equations 3-11, we get

x **component** of an object's **velocity** vector $\vec{v}$ at time t for **projectile motion**

x component of the object's velocity vector at $t = 0$

(a) $v_x = v_{0x}$

Acceleration due to gravity (g is positive)

(b) $v_y = v_{0y} - gt$

Time at which the object has velocity components v_x and v_y

y **component** of an object's **velocity** vector $\vec{v}$ at time t for **projectile motion**

y component of the object's velocity vector at $t = 0$

Velocity, acceleration, and time for projectile motion
(3-12)

Equation 3-12a tells us that *in projectile motion the horizontal component of velocity doesn't change*. The vertical component of velocity *does* change with time, however; as Equation 3-12b shows, the y component of velocity decreases at a steady rate. If v_{0y} is positive as shown in Figure 3-25, v_y decreases to zero (at the high point of the trajectory) and then continues to decrease or becomes increasingly negative (that is, the projectile moves downward at an ever-faster rate).

We can use another constant-acceleration equation from Chapter 2 to write the equations for the projectile's x and y coordinates at any time t. We let x_0 and y_0 be the projectile's coordinates at time $t = 0$. From Equation 2-9 in Section 2-4, the x coordinate is given by $x = x_0 + v_{0x}t + (1/2)a_x t^2$, and the same equation rewritten for the y direction is $y = y_0 + v_{0y}t + (1/2)a_y t^2$. For projectile motion the constant components of acceleration are $a_x = 0$ and $a_y = -g$, so these equations become

x and y **coordinates** at time t of an object in **projectile motion**

x and y components at $t = 0$ of the object's velocity vector

Time at which the object has coordinates x and y

(a) $x = x_0 + v_{0x}t$

(b) $y = y_0 + v_{0y}t - \dfrac{1}{2}gt^2$

x and y coordinates at $t = 0$ of the object

Acceleration due to gravity (g is positive)

Position, acceleration, and time for projectile motion
(3-13)

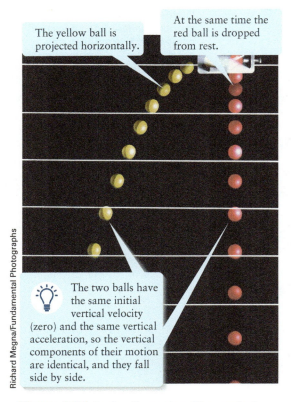

The yellow ball is projected horizontally.

At the same time the red ball is dropped from rest.

The two balls have the same initial vertical velocity (zero) and the same vertical acceleration, so the vertical components of their motion are identical, and they fall side by side.

Richard Megna/Fundamental Photographs

Figure 3-26 Projectile motion: The vertical and horizontal components of motion are independent Although these two balls have different initial horizontal velocities, their initial vertical velocities — and hence their subsequent vertical motions — are identical.

The curved path that the projectile follows in accordance with Equations 3-13 is called a **parabola**. You can see this shape in the photograph that opens this chapter and in Figure 3-22.

You probably recognize Equations 3-12b and 3-13b from our discussion of free fall in Section 2-6; they're exactly the same as Equations 2-13 and 2-14, respectively. So *we can think of projectile motion as a combination of horizontal motion with constant velocity and vertical free fall*. What's more, *the horizontal and vertical motions of a projectile are independent*. To see this, notice that the expressions for v_x and x in Equations 3-12 and 3-13 involve only x quantities, while the expressions for y and v_y involve only y quantities. The multiflash photograph in Figure 3-26 helps to show the independence of the horizontal and vertical motions of a projectile.

Another way to think about projectile motion is to first imagine what would happen if gravity didn't exist, so that $g = 0$. Then a thrown projectile would have zero acceleration, its velocity would remain constant, and its trajectory would be a straight line. In fact, a projectile's trajectory curves downward because gravity *does* exist. Equation 3-13b tells us that after a time t, the projectile has been dragged downward by a distance $(1/2)gt^2$ from the straight-line path it would follow in the absence of gravity (Figure 3-27).

The demonstration depicted in Figure 3-28 illustrates this way of thinking about projectile motion. A projectile is aimed directly at a hanging target. Just as the projectile starts its flight toward the target, the target is released and begins to fall. If gravity didn't exist, the target wouldn't fall, the projectile would follow a straight-line path, and the projectile would score a direct hit on the target. Because gravity does exist, however, the target does fall, and the projectile follows a curved path. But the projectile still scores a direct hit because gravity pulls *both* the projectile and the target downward by the *same* distance $(1/2)gt^2$ in a time t.

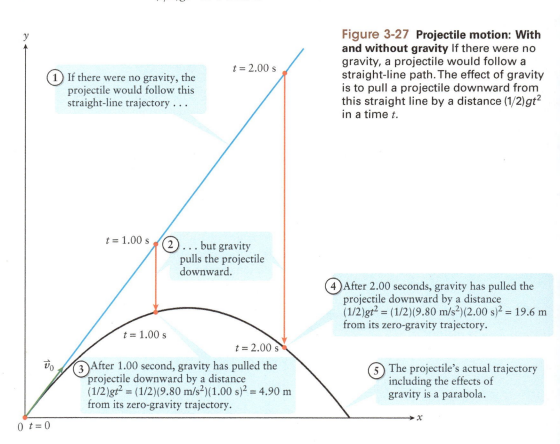

① If there were no gravity, the projectile would follow this straight-line trajectory . . .

$t = 2.00$ s

$t = 1.00$ s

② . . . but gravity pulls the projectile downward.

$t = 1.00$ s

$t = 2.00$ s

$\vec{v}_0$

③ After 1.00 second, gravity has pulled the projectile downward by a distance $(1/2)gt^2 = (1/2)(9.80 \text{ m/s}^2)(1.00 \text{ s})^2 = 4.90$ m from its zero-gravity trajectory.

④ After 2.00 seconds, gravity has pulled the projectile downward by a distance $(1/2)gt^2 = (1/2)(9.80 \text{ m/s}^2)(2.00 \text{ s})^2 = 19.6$ m from its zero-gravity trajectory.

⑤ The projectile's actual trajectory including the effects of gravity is a parabola.

0 $t = 0$

Figure 3-27 Projectile motion: With and without gravity If there were no gravity, a projectile would follow a straight-line path. The effect of gravity is to pull a projectile downward from this straight line by a distance $(1/2)gt^2$ in a time t.

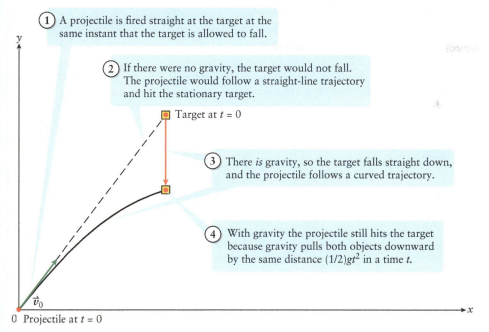

(1) A projectile is fired straight at the target at the same instant that the target is allowed to fall.

(2) If there were no gravity, the target would not fall. The projectile would follow a straight-line trajectory and hit the stationary target.

Target at $t = 0$

(3) There *is* gravity, so the target falls straight down, and the projectile follows a curved trajectory.

(4) With gravity the projectile still hits the target because gravity pulls both objects downward by the same distance $(1/2)gt^2$ in a time t.

$\vec{v}_0$

0 Projectile at $t = 0$

Figure 3-28 Hitting a dropped target This experiment demonstrates that all objects in free fall have the same downward acceleration.

Thus we have two alternative ways of thinking about projectile motion: (1) a combination of horizontal, constant-velocity motion and vertical free fall or (2) straight-line motion in the direction of the initial velocity, but pulled downward by a distance $(1/2)gt^2$ in time t. As we'll see in the next section, both of these interpretations can be useful when solving problems with projectile motion.

GOT THE CONCEPT? 3-5
The Changing Velocity of a Projectile

(?) An object launched at some initial speed and angle follows a parabolic arc. At what point during the trajectory is the magnitude of its velocity the smallest? At what point, if any, will the velocity of the object be zero?

TAKE-HOME MESSAGE FOR Section 3-5

✔   A projectile is an object that is launched and then falls freely (that is, moves without air resistance).

✔   Only gravity acts on a projectile, so the acceleration is downward and has the same magnitude throughout the motion.

✔   Projectile motion can be thought of as a combination of horizontal motion with constant velocity and vertical free fall.

✔   Projectile motion can also be thought of as straight-line motion in the direction of the initial velocity with an additional downward motion due to gravity.

3-6 You can solve projectile motion problems using techniques learned for straight-line motion

Equations 3-12 and 3-13 are the principal equations we'll use for solving problems in projectile motion. These equations are identical to those from Chapter 2 for constant-velocity motion (along the x direction) and free fall (along the y direction). So we'll be able to use many of the same problem-solving techniques learned in that chapter.

In Table 3-1 we've collected Equations 3-12 and 3-13 along with information about which quantities are involved in each equation. In the same way that you used Table 2-3 in Section 2-5, you can use this table to decide which equations to use for a given projectile problem.

 Go to Interactive Exercise 3-1 for more practice dealing with projectile motion.

TABLE 3-1   Equations for Projectile Motion

By using one or more of these equations, you can solve any problem involving projectile motion. The check marks indicate which of the quantities t (time), x_0 (the x position at time 0), x (the x position at time t), v_{0x} (the x velocity at time 0), v_x (the x velocity at time t), y_0 (the y position at time 0), y (the y position at time t), v_{0y} (the y velocity at time 0), and v_y (the y velocity at time t) are present in each equation. Note that g, the magnitude of the acceleration due to gravity, is a *positive* quantity.

Equation		Which quantities the equation includes								
		t	x_0	x	v_{0x}	v_x	y_0	y	v_{0y}	v_y
$v_x = v_{0x}$	(3-12a)				✓	✓				
$v_y = v_{0y} - gt$	(3-12b)	✓							✓	✓
$x = x_0 + v_{0x}t$	(3-13a)	✓	✓	✓	✓					
$y = y_0 + v_{0y}t - \frac{1}{2}gt^2$	(3-13b)	✓					✓	✓	✓	

EXAMPLE 3-4   Cliff Diving I

A diver leaps from a cliff in La Quebrada (Acapulco), Mexico. The cliff is 30.0 m above the surface of the water. He leaves the cliff moving at 2.00 m/s at 40.0° above the horizontal. If air resistance can be neglected, (a) how long does it take him to hit the water, and (b) how far does he travel horizontally before reaching the ocean?

Set Up

We can treat the diver as a projectile because air resistance is negligible. We choose the origin ($x_0 = 0$, $y_0 = 0$) to be where the diver leaves the cliff and starts to move as a projectile. The diver's initial velocity (of magnitude $v_0 = 2.00$ m/s and angle $\theta_0 = 40.0°$) has positive x and y components that we can find using trigonometry. In part (a) we'll find the time t when he reaches the water at $y = -30.0$ m. We'll use this value of t in part (b) to find his x position (that is, how far he has traveled horizontally) at this time. Table 3-1 tells us which equations to use for each part.

$$v_{0x} = v_0 \cos \theta_0$$
$$v_{0y} = v_0 \sin \theta_0$$

For part (a):

$$y = y_0 + v_{0y}t - \frac{1}{2}gt^2 \quad \text{(3-13b)}$$

For part (b):

$$x = x_0 + v_{0x}t \quad \text{(3-13a)}$$

Solve

(a) First find the x and y components of the diver's initial velocity.

$$v_{0x} = v_0 \cos \theta_0$$
$$= (2.00 \text{ m/s}) \cos 40.0°$$
$$= 1.53 \text{ m/s}$$
$$v_{0y} = v_0 \sin \theta_0$$
$$= (2.00 \text{ m/s}) \sin 40.0°$$
$$= 1.29 \text{ m/s}$$

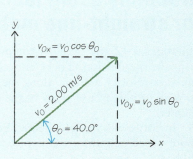

Then solve Equation 3-13b for time t and insert the known values $y_0 = 0$, $y = -30.0$ m, $v_{0y} = 1.29$ m/s, and $g = 9.80$ m/s^2.

$$y = y_0 + v_{0y}t - \frac{1}{2}gt^2 \text{ or } \frac{1}{2}gt^2 - v_{0y}t + (y - y_0) = 0$$

This is a quadratic equation of the form $at^2 + bt + c = 0$, with $a = g/2$, $b = -v_{0y}$, and $c = y - y_0$.

$$t = \frac{-b \pm \sqrt{b^2 - 4ac}}{2a} = \frac{-(-v_{0y}) \pm \sqrt{(-v_{0y})^2 - 4\left(\frac{g}{2}\right)(y - y_0)}}{2\left(\frac{g}{2}\right)}$$

$$= \frac{v_{0y} \pm \sqrt{v_{0y}^2 - 2g(y - y_0)}}{g}$$

$$= \frac{(1.29 \text{ m/s}) \pm \sqrt{(1.29 \text{ m/s})^2 - 2(9.80 \text{ m/s}^2)(-30.0 \text{ m})}}{9.80 \text{ m/s}^2}$$

$$= \frac{(1.29 \text{ m/s}) \pm \sqrt{590 \text{ m}^2/\text{s}^2}}{9.80 \text{ m/s}^2} = \frac{(1.29 \text{ m/s}) \pm (24.3 \text{ m/s})}{9.80 \text{ m/s}^2}$$

$$= 2.61 \text{ s or } -2.35 \text{ s}$$

The diver reaches the water after leaving the cliff at $t = 0$, so the solution for t that we want is the positive one.

$t = 2.61$ s

(b) We substitute the diver's constant x component of velocity $v_{0x} = 1.53$ m/s along with $t = 2.61$ s from part (a) into Equation 3-13a to determine the diver's final horizontal position.

$$x - x_0 = v_{0x}t$$
$$= (1.53 \text{ m/s})(2.61 \text{ s})$$
$$= 4.00 \text{ m}$$

Reflect

We can check our results by using one of our interpretations of projectile motion. If there were no gravity, the diver wouldn't fall but would continue along a straight line at an angle of 40.0° above the horizontal. Then in 2.61 s the diver would rise 3.37 m above the cliff. In fact the diver falls a distance $\frac{1}{2}gt^2 = 33.4$ m below his zero-gravity trajectory in 2.61 s. The net result is that the diver goes through a vertical displacement of −30.0 m—exactly the displacement from the height of the cliff to sea level.

If gravity did not exist, in time $t = 2.61$ s the diver would travel horizontally: $x - x_0 = v_{0x}t$

$$= (1.53 \text{ m/s})(2.61 \text{ s})$$
$$= 4.00 \text{ m}$$

vertically: $y - y_0 = v_{0y}t$

$$= (1.29 \text{ m/s})(2.61 \text{ s})$$
$$= 3.37 \text{ m}$$

When gravity is considered the diver falls a distance

$$\frac{1}{2}gt^2 = \frac{1}{2}(9.80 \text{ m/s}^2)(2.61 \text{ s})^2 = 33.4 \text{ m}$$

so the diver's net vertical displacement in 2.61 s is 3.37 m − 33.4 m = −30.0 m

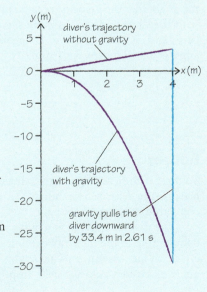

Part (a) of this example illustrates that projectile motion problems can have more than one solution, just like problems in linear motion with constant acceleration (see Example 2-8 in Section 2-5) or free-fall problems (see Example 2-10 in Section 2-6). Be careful!

 Go to Interactive Exercise 3-2 for more practice dealing with projectile motion.

EXAMPLE 3-5 Cliff Diving II

What is the velocity (magnitude and direction) of the diver in Example 3-4 when he enters the water?

Set Up

From Example 3-4 we know the diver's initial ($t = 0$) velocity components, $v_{0x} = 1.53$ m/s and $v_{0y} = 1.29$ m/s, as well as the time $t = 2.61$ s that it takes for him to fall to the surface of the water. We'll use this information and Equations 3-12 to calculate his velocity components v_x and v_y when he reaches the water. We'll then use Equations 3-2 to find the magnitude and direction of his velocity at that point.

Velocity, acceleration, and time for projectile motion:

$$v_x = v_{0x} \quad (3\text{-}12a)$$

$$v_y = v_{0y} - gt \quad (3\text{-}12b)$$

Finding vector magnitude and direction from vector components:

$$A = \sqrt{A_x^2 + A_y^2} \quad (3\text{-}2a)$$

$$\tan \theta = \frac{A_y}{A_x} \quad (3\text{-}2b)$$

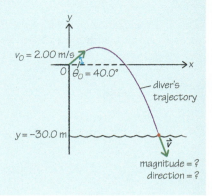

Solve

Equation 3-12a tells us that v_x doesn't change as the diver plummets toward the water, so it has the same value as when he left the cliff at $t = 0$. By contrast, Equation 3-12b tells us that v_y decreases by an amount gt.

$$v_x = v_{0x} = 1.53 \text{ m/s}$$
$$v_y = v_{0y} - gt$$
$$= 1.29 \text{ m/s} - (9.80 \text{ m/s}^2)(2.61 \text{ s})$$
$$= -24.3 \text{ m/s}$$

The diver's x and y components of velocity and his speed v make up the three sides of a right triangle, so we can find v from v_x and v_y using the Pythagorean theorem as in Equation 3-2a.

We can use Equation 3-2b to calculate the angle that the velocity vector makes with the $+x$ axis. Note from the sketch that because v_x is positive and v_y is negative, the angle θ must be between 0 and $-90°$.

Magnitude of $\vec{v}$:

$$v = \sqrt{v_x^2 + v_y^2}$$
$$= \sqrt{(1.53 \text{ m/s})^2 + (-24.3 \text{ m/s})^2}$$
$$= 24.3 \text{ m/s}$$

Direction of $\vec{v}$:

$$\tan \theta = \frac{v_y}{v_x}$$
$$= \frac{-24.3 \text{ m/s}}{1.53 \text{ m/s}} = -15.9$$
$$\theta = \tan^{-1}(-15.9)$$
$$= -86.4°$$

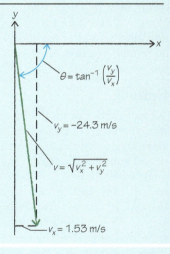

Reflect

To three significant figures, the diver's speed v is the same as the absolute value of his y component of velocity v_y. That's because the diver is descending at a very steep angle of almost $-90°$, so the x component of his velocity is very small compared to the y component.

The diver is moving at a substantial speed when he enters the water. You can show that $v = 87.5$ km/h or 54.4 mi/h, comparable to the highway driving speed of an automobile. (In fact the diver's speed will be somewhat less due to the effects of air resistance.) That's why it's essential that the diver enter the water with his body oriented at just the right angle to avoid a painful "belly flop."

EXAMPLE 3-6 How High Does It Go?

A frog jumps so that it leaves the ground at speed v_0 and at an angle θ_0 above the horizontal. What is the maximum height that the frog reaches during its leap? Ignore air resistance.

Set Up

We considered a similar problem in Example 2-10 (Section 2-6), in which a springbok jumped straight up. For both the springbok and the frog, the y

Velocity, acceleration, and time for projectile motion:

$$v_y = v_{0y} - gt \quad (3\text{-}12b)$$

component of velocity is zero at the peak of the motion (when the animal is moving neither up nor down). Unlike the springbok, the frog also has an x component of velocity, but this has no effect on the up-and-down motion.

We use a coordinate system as shown in the sketch, with the origin at the frog's starting point so that its initial coordinates are $x_0 = 0$, $y_0 = 0$. We'll use Equation 3-12b to find the time t_1 when the frog's y component of velocity is zero so that it is at the highest point of its leap. We'll then use Equation 3-13b to find the frog's y coordinate y_1 at this time.

Position, acceleration, and time for projectile motion:

$$y = y_0 + v_{0y} t - \frac{1}{2} g t^2 \quad (3\text{-}13b)$$

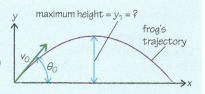

Solve

First express the frog's initial y component of velocity v_{0y} in terms of its initial speed v_0 and launch angle θ_0. Then, substitute this expression into Equation 3-12b. Set $v_y = 0$ (corresponding to the high point of the motion) at time $t = t_1$, then solve for t_1.

$v_{0y} = v_0 \sin \theta_0$ so

$v_y = v_0 \sin \theta_0 - gt$

$0 = v_0 \sin \theta_0 - gt_1$

$gt_1 = v_0 \sin \theta_0$

$t_1 = \dfrac{v_0 \sin \theta_0}{g}$

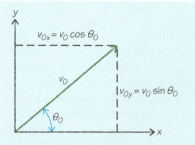

Now use Equation 3-13b to solve for the frog's y coordinate at time t_1. Substitute $y_0 = 0$, $v_{0y} = v_0 \sin \theta_0$, and $t_1 = (v_0 \sin \theta_0)/g$ from above.

$y_1 = 0 + (v_0 \sin \theta_0) \left(\dfrac{v_0 \sin \theta_0}{g} \right)$

$\quad - \dfrac{1}{2} g \left(\dfrac{v_0 \sin \theta_0}{g} \right)^2$

$\quad = \dfrac{v_0^2 \sin^2 \theta_0}{g} - \dfrac{v_0^2 \sin^2 \theta_0}{2g}$ so

$y_1 = \dfrac{v_0^2 \sin^2 \theta_0}{2g}$

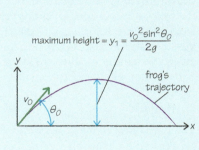

Reflect

In this example our result is a formula rather than a number. Can you verify that our answer for y_1 has the correct units?

To interpret what this formula tells us, let's look at how the maximum height y_1 depends on the values of the frog's initial speed v_0 and launch angle θ_0 as well as on the value of g. Because y_1 is proportional to the square of v_0, doubling the frog's launch speed makes it go *four* times as high. (Different frog species jump with very different launch speeds; v_0 is about 1.6 m/s for the northern leopard frog *Rana pipens*, but can be more than 4 m/s for the American bullfrog *Rana catesbeiana*.) Note also that y_1 is proportional to $\sin^2 \theta_0$. The sine function has its maximum value for $\theta_0 = 90°$, which corresponds to the frog jumping straight up.

Finally, note that y_1 is *inversely* proportional to g (the factor of g is in the denominator). Hence, if the value of g were smaller, you would divide by a smaller number to calculate y_1 and you would get a larger answer. In other words, if gravity were weaker, the frog would reach a greater height.

EXAMPLE 3-7 How Far Does It Go?

When the frog in Example 3-6 returns to the ground, how far from its launch point does it land?

Set Up

The frog started its flight at $x_0 = 0$, $y_0 = 0$. When the frog lands, it returns to the same height $y = 0$ at horizontal coordinate x_2. The frog's distance from its launch point when it lands is the x component of its displacement, $x_2 - x_0$. To determine x_2, we'll first use Equation 3-13b to calculate the time t_2 when the frog lands and returns to $y = 0$. We'll then substitute this time into Equation 3-13a to calculate the frog's x coordinate when it lands, x_2.

Position, acceleration, and time for projectile motion:

$$x = x_0 + v_{0x} t \quad (3\text{-}13a)$$

$$y = y_0 + v_{0y} t + \frac{1}{2} g t^2 \quad (3\text{-}13b)$$

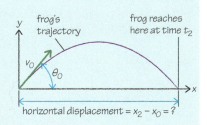

Solve

From Example 3-6 we know that the frog's initial y component of velocity is $v_{0y} = v_0 \sin \theta_0$. When the frog returns to ground level its y coordinate is the same as when it left ground, so $y = y_0 = 0$. We substitute these into Equation 3-13b and set $t = t_2$ (the time when the frog returns to ground level).

One solution to this equation is $t_2 = 0$, which is the time when the frog *leaves* the ground. However, the solution we want for t_2 is greater than zero. To find this solution we divide our expression by t_2, eliminating the quadratic, and solve for t_2.

Note that the time $t_2 = (2v_0 \sin \theta_0)/g$ for the frog to return to its starting height is exactly *twice* the time $t_1 = (v_0 \sin \theta_0)/g$ that we calculated in Example 3-6 for the frog to reach its maximum height. In other words, a projectile takes just as long to ascend from a given height to its maximum height as it does to descend from its maximum height back to that given height.

To find the frog's x component of displacement $x_2 - x_0$ when it returns to the ground, substitute $t = t_2$ into Equation 3-13a. We also need the frog's initial x component of velocity, which from Example 3-6 is $v_{0x} = v_0 \cos \theta_0$. The distance $x_2 - x_0$ that the frog travels horizontally during its projectile motion is called the *horizontal range*.

$v_{0y} = v_0 \sin \theta_0$

Equation 3-13b becomes

$$0 = 0 + (v_0 \sin \theta_0)t_2 - \frac{1}{2} g t_2^2$$

$$0 = (v_0 \sin \theta_0) - \frac{1}{2} g t_2 \text{ or}$$

$$t_2 = \frac{2v_0 \sin \theta_0}{g}$$

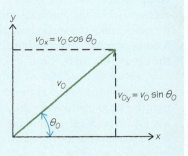

$x_2 - x_0 = v_{0x} t_2$

At time $t_2 = (2v_0 \sin \theta_0)/g$, find the displacement $x_2 - x_0$:

Horizontal range:

$$x_2 - x_0 = (v_0 \cos \theta_0)\left(\frac{2v_0 \sin \theta_0}{g} \right) = \frac{2v_0^2 \sin \theta_0 \cos \theta_0}{g}$$

Reflect

The horizontal range is proportional to v_0^2 (increasing the launch speed makes the frog go farther) and inversely proportional to g (the frog would also go farther if gravity were weaker).

To understand how the horizontal range depends on the launch angle θ_0, it's helpful to rewrite our result (using a formula from trigonometry) as $v_0^2 \sin 2\theta_0 /g$. This expression shows that horizontal range is greatest for $\theta_0 = 45°$, so that $2\theta_0 = 90°$ and the function $\sin 2\theta_0$ has its greatest value ($\sin 90° = 1$). For launch angles less than 45°, the horizontal range is shorter because the projectile is in the air for a relatively short time and so cannot travel as far horizontally. For launch angles greater than 45°, the projectile spends more time in the air yet has a shorter horizontal range because its initial velocity is mostly upward (see the sketch).

Notice that for $\theta_0 = 90°$ the horizontal range is *zero*: The projectile lands right back where it started. However, with this launch angle the projectile reaches the greatest possible height, as we discussed in Example 3-6.

Horizontal range

$$= \frac{2v_0^2 \sin \theta_0 \cos \theta_0}{g}$$

From trigonometry,

$$2 \sin \theta_0 \cos \theta_0 = \sin 2\theta_0$$

so

$$\text{horizontal range} = \frac{v_0^2 \sin 2\theta_0}{g}$$

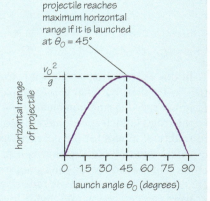

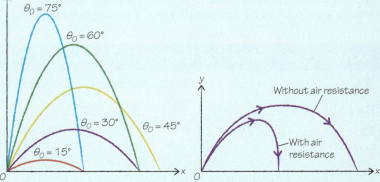

WATCH OUT! Air resistance affects range.

The rule that a launch angle of 45° gives the greatest horizontal range doesn't work for objects such as kicked footballs or batted baseballs. Air resistance plays a significant role in the flight of these objects, so their acceleration is not constant, their trajectories are not parabolic, and the formulas we've developed in this section don't apply. Detailed calculations that account for air resistance show that a batted baseball has the greatest horizontal range for a launch angle of about 35° above the horizontal.

GOT THE CONCEPT? 3-6 How fast is a flying frog?

Suppose the frog in Examples 3-6 and 3-7 leaves the ground at a launch angle $\theta_0 = 45°$. If the frog's speed is v_0 when it leaves the ground, what is its speed when it is at its maximum height? (a) v_0; (b) between v_0 and $v_0/2$; (c) $v_0/2$; (d) between $v_0/2$ and zero; (e) zero; (f) not enough information given to decide.

TAKE-HOME MESSAGE FOR Section 3-6

✔ In solving problems that involve projectile motion, you can use many of the same techniques that we used in Section 2-6 to solve one-dimensional free-fall problems.

✔ Equations 3-12 and 3-13 are essential tools for solving any projectile motion problem.

3-7 An object moving in a circle is accelerating even if its speed is constant

It takes a harder tug on a car's steering wheel to round a tight corner at high speed than it does to take a gentle curve at low speed. We saw in Section 3-4 that a car following a curved path experiences an acceleration directed toward the inside of the curve and that the driver causes this acceleration by turning the steering wheel. So the car's acceleration must be greater for a high-speed, tight turn than for a low-speed, gentle turn. But *how much* greater? What is the relationship between the car's acceleration, its speed, and the radius of its circular path?

To see the answer, let's look at an object in **uniform circular motion**, that is, going around a circular path at a constant speed. We saw in Section 3-4 that if an object follows a curved trajectory with a constant speed, its acceleration vector $\vec{a}$ is perpendicular to its velocity vector $\vec{v}$ and points toward the inside of the turn (see Figures 3-19c and 3-19d). For uniform circular motion, this means that $\vec{a}$ always points toward the *center* of the circle (Figure 3-29). An acceleration that points in this direction is called a **centripetal acceleration**, which comes from a Latin term meaning "seeking the center." Let's see how to calculate the magnitude of the centripetal acceleration vector.

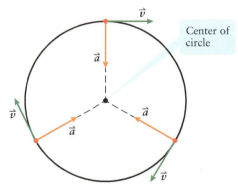

- In uniform circular motion the speed is *constant*, but the object accelerates because the velocity $\vec{v}$ is always changing direction.
- Because the speed is constant, the acceleration $\vec{a}$ is always perpendicular to the velocity $\vec{v}$.
- $\vec{a}$ always points toward the center of the circle.

Figure 3-29 Uniform circular motion An object moving around a circle at a constant speed.

Analyzing Motion in a Circle

Figure 3-30a shows two points along the trajectory of an object in uniform circular motion around a circular path of radius r. From point 1 to point 2, the object travels a distance Δs around the circle, and an imaginary line drawn from the center of the circle to the object rotates through an angle $\Delta\theta$. We can write a simple relationship between Δs and $\Delta\theta$ provided that we express $\Delta\theta$ in *radians*:

$$\Delta\theta = \frac{\Delta s}{r} \quad (\Delta\theta \text{ in radians})$$

$$(3\text{-}14)$$

(a)

An object in uniform circular motion moves a distance Δs from point 1 to point 2.

(b)

The three lengths r, r, and Δs approximate the three sides of a triangle.

(c)

The velocity vector of the object rotates through the same angle $\Delta\theta$.

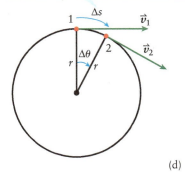

(d)

The vectors $-\vec{v}_1$, $\vec{v}_2$, and $\Delta\vec{v}$ form the three sides of a triangle... $\Delta\vec{v}$ is in the direction of the object's acceleration.

(e)

Both $-v_1$ and v_2 have the same magnitude v. This triangle is similar to that in part (b).

Figure 3-30 Analyzing uniform circular motion (a) An object's circular path between points 1 and 2. (b) A triangle relating the radius r of the circle, the distance traveled Δs, and the angle $\Delta\theta$. (c) Relating the velocity vectors at points 1 and 2. (d) Calculating the velocity change between points 1 and 2. (e) A triangle relating the speed v, the magnitude Δv of the velocity change, and the angle $\Delta\theta$.

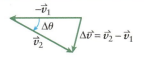

You can check this formula by thinking about the case in which the object goes all the way around the circle so that the distance Δs equals $2\pi r$, the circumference of the circle. Then Equation 3-14 says that $\Delta\theta = (2\pi r)/r = 2\pi$, which is just the number of radians that make up a complete circle.

If the time interval between points 1 and 2 is very short, the angle $\Delta\theta$ is small, and the curved arc of length Δs is nearly a straight line. Then the lines from the center of the circle to points 1 and 2 and the curved arc form a triangle with two sides of length r separated by an angle $\Delta\theta$ and a third side of length Δs given by Equation 3-14 (see Figure 3-30b). Thus Equation 3-14 is a relationship between the length Δs of the short side of the triangle, the length r of the two long sides, and the angle $\Delta\theta$.

Figure 3-30c shows why Equation 3-14 is useful: The velocity vector rotates through the same angle $\Delta\theta$ as the object moves from point 1 (velocity $\vec{v}_1$) to point 2 (velocity $\vec{v}_2$). Because the speed is constant, $\vec{v}_1$ and $\vec{v}_2$ have the same magnitude v. In Figure 3-30d we've converted $\vec{v}_1$ to $-\vec{v}_1$ by reversing its direction and placed the head of $-\vec{v}_1$ against the tail of $\vec{v}_2$ to form the vector difference $\Delta\vec{v} = \vec{v}_2 - \vec{v}_1$. This vector diagram is equivalent to a triangle with two sides of length v separated by an angle $\Delta\theta$ and a third side of length Δv, which is the magnitude of the velocity change between points 1 and 2 (Figure 3-30e).

You can see that the triangle in Figure 3-30e is *similar* to the triangle in Figure 3-30b. Both triangles have two equal sides separated by the same angle, so the relationship between the lengths of their sides is the same. Comparing the two figures shows that Δv and v in Figure 3-30e correspond to Δs and r, respectively, in Figure 3-30b. So we can write an equation involving Δv and v that's the equivalent of Equation 3-14:

(3-15)
$$\Delta\theta = \frac{\Delta v}{v}$$

We'd like to calculate the magnitude of the object's acceleration, which is equal to $a_{\text{cent}} = \Delta v/\Delta t$. (We use the subscript "cent" to remind us that the acceleration is *centripetal*.) To do this calculation we first set our two expressions for $\Delta\theta$, Equations 3-14 and 3-15, equal to each other. This gives a new equation that relates Δv, v, Δs, and r:

$$\frac{\Delta v}{v} = \frac{\Delta s}{r}$$

To get an expression for $a_{\text{cent}} = \Delta v/\Delta t$, multiply both sides of this equation by v and divide both sides by Δt. Then we get

$$a_{cent} = \frac{\Delta v}{\Delta t} = \left(\frac{\Delta s}{r}\right)\left(\frac{v}{\Delta t}\right) = \left(\frac{\Delta s}{\Delta t}\right)\left(\frac{v}{r}\right) \qquad (3\text{-}16)$$

Now the object's speed v is equal to the distance Δs that it travels divided by the time interval Δt required to travel this distance: $v = \Delta s/\Delta t$. Therefore, we can replace $\Delta s/\Delta t$ in Equation 3-16 by v, and our expression for the object's centripetal acceleration becomes $a_{cent} = v(v/r)$, or

Centripetal acceleration: magnitude of the acceleration of an object in uniform circular motion

Speed of an object as it moves around the circle

$$a_{cent} = \frac{v^2}{r}$$

Radius of the circle

Centripetal acceleration for motion in a circle (3-17)

The units of v and r are m/s and m, respectively. Hence v^2/r in Equation 3-17 has units $(m/s)^2/m$ or m/s^2, which are the correct units for acceleration.

Equation 3-17 tells us that the magnitude of the centripetal acceleration is proportional to the square of the speed and inversely proportional to the radius of the circle. So there is greater centripetal acceleration the greater the object's speed and the smaller the radius (and hence the tighter the turn). That's entirely consistent with our discussion at the beginning of this section.

Although we derived Equation 3-17 for an object moving around a circle at constant speed, the same equation applies to motion at constant speed around any *portion* of a circle. As an example, suppose a bird initially flying north makes a 90° right turn along a semicircular arc at constant speed until it is heading east. The part of the bird's flight during which it is turning constitutes uniform circular motion, with an acceleration directed toward the center of the circle of which the arc is part (Figure 3-31).

Equation 3-17 is also valid when the object's speed changes (a case called *nonuniform* circular motion). In this case, as we saw in Section 3-4, the acceleration vector has both a component along the direction of motion (which describes changes in speed) and a component perpendicular to the motion (which describes changes in direction). It turns out that the magnitude of this perpendicular component at each point around the circular path is given by Equation 3-17, $a_{cent} = v^2/r$, provided that we set v equal to the instantaneous value of the speed at that point (Figure 3-32).

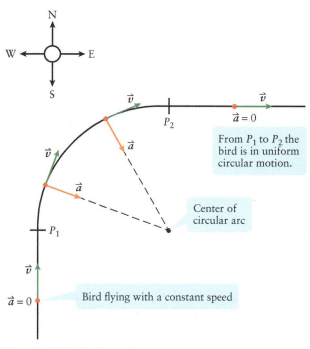

From P_1 to P_2 the bird is in uniform circular motion.

Center of circular arc

Bird flying with a constant speed

Figure 3-31 Uniform circular motion along an arc An object doesn't have to travel all the way around a complete circle to be in uniform circular motion.

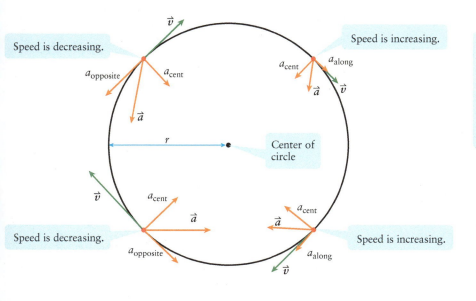

Speed is decreasing.

Speed is increasing.

Center of circle

Speed is decreasing.

Speed is increasing.

 • If an object moves around a circle with varying speed, the acceleration $\vec{a}$ has a component along or opposite the direction of motion.
• Whether the speed is increasing, decreasing, or remaining constant, the acceleration always has a component $a_{cent} = v^2/r$ directed toward the center of the circle.

Figure 3-32 Nonuniform circular motion If the speed varies in circular motion, the acceleration vector is not directed toward the center of the circle.

EXAMPLE 3-8 Around the Bend

A certain curve in a level road is signposted for a maximum recommended speed of 65 km/h. The curve is a circular arc with a radius of 95 m. What is the centripetal acceleration of a car that takes this curve at the maximum recommended speed?

Set Up

We are given the car's speed $v = 65$ km/h and the radius $r = 95$ m of its trajectory, so we can calculate its centripetal acceleration a_{cent} using Equation 3-17.

Centripetal acceleration:

$$a_{cent} = \frac{v^2}{r} \qquad (3\text{-}17)$$

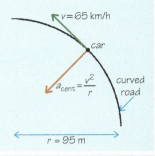

$v = 65$ km/h
car
$a_{cent} = \frac{v^2}{r}$
curved road
$r = 95$ m

Solve

We are given r in m and v in km/h. To calculate the car's acceleration in m/s², we must first convert v to m/s.

$$v = 65 \text{ km/h} \times \left(\frac{1000 \text{ m}}{1 \text{ km}} \right) \times \left(\frac{1 \text{ h}}{3600 \text{ s}} \right)$$

$$= 18 \text{ m/s}$$

Given the value of v we calculate a_{cent} using Equation 3-17.

$$a_{cent} = \frac{v^2}{r} = \frac{(18 \text{ m/s})^2}{95 \text{ m}} = 3.4 \text{ m/s}^2$$

Reflect

The magnitude of the centripetal acceleration is just over one-third of $g = 9.8$ m/s², the magnitude of the acceleration of an object in free fall.

EXAMPLE 3-9 Orbital Speed

The International Space Station (ISS) orbits Earth at an altitude of about 400 km above Earth's surface. The acceleration required to keep the ISS in orbit along a circular path is provided by Earth's gravity, which is directed toward the center of Earth and hence toward the center of the orbit as shown in Figure 3-33. (Earth's gravitational pull decreases gradually with altitude, so the acceleration due to gravity that acts on the ISS is only 8.68 m/s².) (a) What is the orbital speed of the ISS in meters per second, kilometers per hour, and miles per hour? (b) How long (in minutes) does it take the ISS to complete an orbit? Note that the radius of Earth is 6378 km.

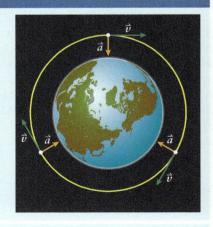

Figure 3-33 An orbiting space station The centripetal acceleration of an orbiting space station or other satellite is provided by Earth's gravity.

Set Up

We are given the ISS's centripetal acceleration a_{cent} and orbital radius r (the sum of Earth's radius and the altitude of the ISS, as the sketch shows). Using this information we can calculate the speed v using Equation 3-17. Once we know v we can calculate the time T for the ISS to complete one orbit (that is, to travel a distance equal to the circumference of the orbit) using the relationship (speed) = (distance)/(time).

Centripetal acceleration:

$$a_{cent} = \frac{v^2}{r} \qquad (3\text{-}17)$$

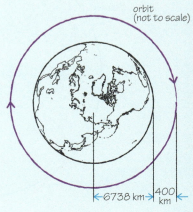

orbit (not to scale)

6738 km   400 km

Solve

(a) We want an expression for the speed v of the space station, so we rearrange Equation 3-17 to put v by itself on the left-hand side.

$$a_{cent} = \frac{v^2}{r}$$
$$v^2 = ra_{cent}$$
$$v = \sqrt{ra_{cent}}$$

To calculate the orbital radius r, add Earth's radius and the altitude of the ISS. Then convert the result to meters.

$$r = 6378 \text{ km} + 400 \text{ km}$$
$$= 6778 \text{ km} \times \left(\frac{10^3 \text{ m}}{1 \text{ km}}\right) = 6.778 \times 10^6 \text{ m}$$

Substitute our calculated value of r into the expression for v.

$$v = \sqrt{(6.778 \times 10^6 \text{ m})(8.68 \text{ m/s}^2)}$$
$$= \sqrt{5.88 \times 10^7 \text{ m}^2/\text{s}^2} = 7.67 \times 10^3 \text{ m/s}$$

Convert the speed to km/h and mi/h.

$$v = 7.67 \times 10^3 \text{ m/s} \times \left(\frac{1 \text{ km}}{10^3 \text{ m}}\right) \times \left(\frac{3600 \text{ s}}{1 \text{ h}}\right)$$
$$= 2.76 \times 10^4 \text{ km/h}$$
$$v = 2.76 \times 10^4 \text{ km/h} \times \left(\frac{1 \text{ mi}}{1.609 \text{ km}}\right)$$
$$= 1.72 \times 10^4 \text{ mi/h}$$

(b) The distance that the ISS travels in one orbit is equal to the orbit circumference $2\pi r$. Because (speed) = (distance)/(time), the time T for one orbit equals the orbit circumference divided by the speed. We convert this answer to minutes.

$$\text{distance} = 2\pi(6.778 \times 10^6 \text{ m})$$
$$= 4.259 \times 10^7 \text{ m}$$
$$\text{time } T = \frac{\text{distance}}{\text{speed}} = \frac{4.259 \times 10^7 \text{ m}}{7.67 \times 10^3 \text{ m/s}}$$
$$= 5.55 \times 10^3 \text{ s} = (5.55 \times 10^3 \text{ s}) \times \left(\frac{1 \text{ min}}{60 \text{ s}}\right)$$
$$= 92.5 \text{ min}$$

Reflect

The sketch shows why there is one particular speed for a circular orbit of a given altitude. If you drop a ball, it falls straight downward. If you throw it horizontally, it follows a curved path before hitting the ground. The faster you throw the ball, the farther it travels prior to impact, and the more gradually curved its trajectory. If you throw the ball with a sufficiently great speed, the curvature of its trajectory precisely matches the curvature of Earth. Then the ball remains at the same height above Earth's surface and never falls to the ground. Putting a satellite such as the International Space Station into orbit works the same way, except that a rocket rather than a strong pitching arm is used to provide the necessary speed. Gravity is making the ISS "fall," but because of its high speed, it falls *around* Earth rather than *toward* Earth. As we calculated in part (b), the ISS makes a complete orbit about every hour and a half.

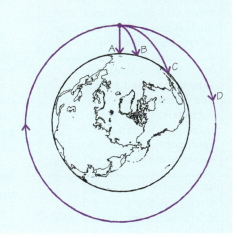

GOT THE CONCEPT? 3-7 Acceleration in Uniform Circular Motion

 If an object is in uniform circular motion, its speed is constant. Is its *acceleration* constant?

GOT THE CONCEPT? 3-8 Finding the Road Radius

 If the car in Example 3-8 went around a curve at twice the speed assumed in that example, how much larger would the radius of the curve have to be for the centripetal acceleration to be the same? (a) $\sqrt{2}$ times as large; (b) twice as large; (c) $2\sqrt{2}$ times as large; (d) 4 times as large; (e) 16 times as large.

How We Sense Acceleration

BioMedical

When you ride in a car or bus, you can sense its acceleration when it suddenly speeds up, stops, or makes a sharp turn. You can also perceive much more subtle accelerations, such as the gentle swaying of a hammock, even with your eyes closed. How do our bodies detect such small accelerations?

The answer lies in the *vestibular system* of the ear, which is a series of tiny, interconnected canals and chambers in the skull (Figure 3-34). In particular, two chambers at the center of the vestibular system detect the horizontal and vertical components of the acceleration vector. A layer of sensory cells is oriented vertically in one chamber and horizontally in the other. A bundle of microscopic, hair-like extensions called *cilia* stick up from each sensory cell into a gelatinous substance coated with a layer of calcium carbonate crystals. The rock-like appearance of these crystals gives them their name, *otoliths* ("ear stones" in Greek).

Figure 3-34 The inner ear The organs that detect acceleration are located inside the fluid-filled chambers of the inner ear.

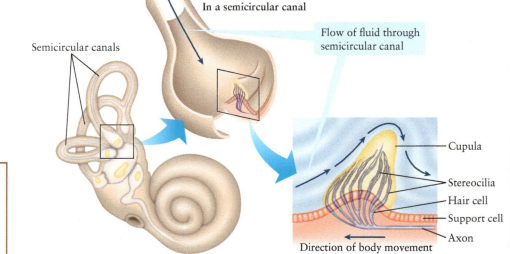

Semicircular canals

In a semicircular canal

Flow of fluid through semicircular canal

Cupula

Stereocilia

Hair cell

Support cell

Axon

Direction of body movement

TAKE-HOME MESSAGE FOR Section 3-7

✔ If an object follows a curved trajectory, it has an acceleration even if its speed remains constant. Motion in a circle is an important example of motion of this kind.

✔ Uniform circular motion is motion around a circle at a constant speed. The acceleration of an object in uniform circular motion is centripetal: It points toward the center of the circle.

✔ Problems in uniform circular motion involve the relationships among centripetal acceleration, speed, and radius expressed by Equation 3-17.

✔ The vestibular system of the ear detects the acceleration vector of your head. It separately measures the horizontal and vertical components of acceleration.

When the head accelerates, the relatively massive otoliths lag behind due to their inertia, which causes the cilia to bend. This bending causes the sensory cells to transmit an electrical signal to the brain. The sensory cells produce a signal even when the tip of a hair-cell bundle moves by less than $0.5\,\mu m$, so these cells are exquisitely sensitive to even gentle accelerations. In addition, each individual hair cell responds best to being bent in one particular direction. Hence any acceleration of the head results in a unique pattern of nerve signals that enables your brain to determine both the magnitude and direction of the head's acceleration.

The other organs of the vestibular system detect *angular* acceleration (that is, the rate of change of your head's rotational motion). This is accomplished by three fluid-filled *semicircular canals* that are oriented perpendicular to one another (see Figure 3-34). Whenever the head rotates around the axis of one of the canals, the fluid inside the canal moves. This bends a structure inside the canal called a *cupula* whose hair cells function like the cilia. The response of the brain to the resulting electrical signals it receives gives you the feeling of angular acceleration. Rotations of the head as small as $0.005°$ can be sensed in this way.

If the vestibular system malfunctions, you feel dizzy and may not be able to walk or even stand upright. In addition, if information from the vestibular system doesn't match the information from the eyes, the result is a sense of disorientation, like the motion sickness some people experience in a small boat on choppy water. Some medications to prevent motion sickness work in part by drying out the fluid in the semicircular canals and essentially shutting off that part of the vestibular system.

3-8 The velocity you measure for an object depends on how you are moving

Up to this point in this chapter we've been measuring velocity and acceleration relative to a *stationary* observer. For example, when we say that one of the race cars in Figure 3-35 is traveling north at 36 m/s (130 km/h, or 81 mi/h), we're stating its speed and direction relative to a stationary spectator. But to the driver of a second race car trying to pass the first one, what matters is the velocity of her car *relative* to the first one. (For example, if the second car is going 38 m/s while the first one is going 36 m/s in the same direction, the velocity of the second car relative to the first one has magnitude 38 m/s − 36 m/s = 2 m/s.) In this section we'll explore this concept of **relative velocity**. We'll use this concept numerous times throughout this book, including our discussions of orbiting spacecraft (Chapter 7), electromagnetic induction (Chapter 20), and Einstein's theory of relativity (Chapter 25).

Let's derive a simple equation for the velocity of one object relative to another. Just as we did in the discussion of velocity in Section 3-4, we'll begin our discussion with position, then look at what happens when position changes. Figure 3-36 shows the two moving cars from Figure 3-35, which we'll call A and B, at a certain time t, as well as a stationary spectator S. There are three position vectors in the figure: the position vector $\vec{r}_{\text{A relative to S}}$ of A relative to S, the position vector $\vec{r}_{\text{B relative to S}}$ of B relative to S, and the position vector $\vec{r}_{\text{B relative to A}}$ of B relative to A. (Note that the direction of the vector $\vec{r}_{\text{A relative to S}}$ is from S to A, and similarly for the other two vectors.) You can see that the tips and tails of these three vectors are arranged just like those in Figure 3-4 in Section 3-2, which illustrates vector addition. So we conclude that the vectors in Figure 3-36 are related according to

$$\vec{r}_{\text{B relative to S}} = \vec{r}_{\text{B relative to A}} + \vec{r}_{\text{A relative to S}} \qquad \text{(positions at time } t\text{)} \qquad (3\text{-}18)$$

The velocity of one race car relative to the other determines who wins the race.

Figure 3-35 When relative velocity matters The velocity of one race car relative to the other determines who wins the race.

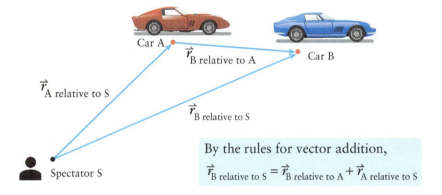

Figure 3-36 Relative position vectors A simple vector addition equation relates the position vectors of car A relative to spectator S, car B relative to S, and B relative to A.

By the rules for vector addition,

$$\vec{r}_{\text{B relative to S}} = \vec{r}_{\text{B relative to A}} + \vec{r}_{\text{A relative to S}}$$

We can write a similar equation for the positions of the two cars relative to S and relative to each other a short time Δt later. We'll add a prime (′) to each position vector in Equation 3-18 to denote the position vector at this later time:

$$\vec{r}'_{\text{B relative to S}} = \vec{r}'_{\text{B relative to A}} + \vec{r}'_{\text{A relative to S}} \qquad \text{(positions at time } t + \Delta t\text{)} \qquad (3\text{-}19)$$

For each relative position, the difference $\vec{r}' - \vec{r}$ equals the change in that position over the time interval of duration Δt — that is, the corresponding displacement $\Delta \vec{r}$. So if we subtract Equation 3-18 from Equation 3-19, we get

$$\Delta\vec{r}_{\text{B relative to S}} = \Delta\vec{r}_{\text{B relative to A}} + \Delta\vec{r}_{\text{A relative to S}} \qquad \text{(displacements during time interval } \Delta t\text{)} \qquad (3\text{-}20)$$

Finally, we divide each displacement in Equation 3-20 by the time interval Δt and recall that the velocity vector $\vec{v}$ equals the displacement $\Delta \vec{r}$ divided by Δt. So our final result is

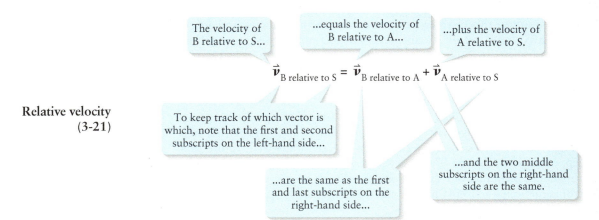

Relative velocity
(3-21)

The velocity of B relative to S...

...equals the velocity of B relative to A...

...plus the velocity of A relative to S.

$$\vec{v}_{\text{B relative to S}} = \vec{v}_{\text{B relative to A}} + \vec{v}_{\text{A relative to S}}$$

To keep track of which vector is which, note that the first and second subscripts on the left-hand side...

...are the same as the first and last subscripts on the right-hand side...

...and the two middle subscripts on the right-hand side are the same.

WATCH OUT! Take care with the subscripts in the relative velocity equation.

If you have trouble deciding which velocity is which in Equation 3-21, use the simple rules given in that equation. You can think of the two middle subscripts on the right-hand side (in Equation 3-21, the subscripts "A") as cancelling each other out, leaving only the first and last subscripts (in Equation 3-21, "B" and "S," respectively).

You can use vector arithmetic to find any one of the three relative velocities in Equation 3-21 in terms of the other two. For example, if the spectator S in Figure 3-36 measures the velocities relative to her of car A ($\vec{v}_{\text{A relative to S}}$) and of car B ($\vec{v}_{\text{B relative to S}}$), she can calculate the velocity of car B relative to car A ($\vec{v}_{\text{B relative to A}}$) by rearranging Equation 3-21:

$$\vec{v}_{\text{B relative to A}} = \vec{v}_{\text{B relative to S}} - \vec{v}_{\text{A relative to S}}$$

The following examples show how to apply Equation 3-21 to solve problems involving relative velocity.

EXAMPLE 3-10   Relative Velocity on the Highway

You are driving your car eastbound at 105 km/h on a long, straight highway. (a) What is the velocity relative to you of a truck driving eastbound at 80 km/h? (b) What is your velocity relative to a motorcycle driving westbound at 90 km/h?

Set Up

In this problem all of the motions are along a line, so we need only one component of Equation 3-21 for relative velocities. If we choose the positive x direction to be to the east, the x velocities of the eastbound car A and truck T relative to a stationary spectator S are positive; for the westbound motorcycle M, the x velocity relative to S is negative. We want to find the velocity of truck T relative to car A in part (a), and the velocity of car A relative to motorcycle M in part (b).

Relative velocity:

$$\vec{v}_{\text{B relative to S}} = \vec{v}_{\text{B relative to A}} + \vec{v}_{\text{A relative to S}} \quad (3\text{-}21)$$

x component of this equation:

$$v_{\text{B relative to S},x} = v_{\text{B relative to A},x} + v_{\text{A relative to S},x}$$

Velocities of car A, truck T, and motorcycle M relative to spectator S:

$$v_{\text{A relative to S},x} = +105 \text{ km/h}$$
$$v_{\text{T relative to S},x} = +80 \text{ km/h}$$
$$v_{\text{M relative to S},x} = -90 \text{ km/h}$$

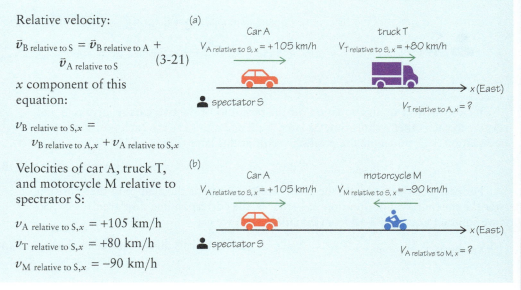

Solve

(a) We're using the symbol T for the second vehicle (the truck) rather than B as in Figure 3-36. So we replace B with T in the x component of Equation 3-22 and solve for the velocity of T relative to car A.

In the relative velocity equation, replace B with T:

$$v_{T \text{ relative to } S,x} = v_{T \text{ relative to } A,x} + v_{A \text{ relative to } S,x}$$

Solve for $v_{T \text{ relative to } A,x}$:

$$v_{T \text{ relative to } A,x} = v_{T \text{ relative to } S,x} - v_{A \text{ relative to } S,x}$$
$$= (+80 \text{ km/h}) - (+105 \text{ km/h})$$
$$= -25 \text{ km/h}$$

(b) We are interested in the velocity of car A relative to motorcycle M. So in the x component of Equation 3-21 we replace A with M and replace B with A, then solve.

In the relative velocity equation, replace A with M and replace B with A:

$$v_{A \text{ relative to } S,x} = v_{A \text{ relative to } M,x} + v_{M \text{ relative to } S,x}$$

Solve for $v_{A \text{ relative to } M,x}$:

$$v_{A \text{ relative to } M,x} = v_{A \text{ relative to } S,x} - v_{M \text{ relative to } S,x}$$
$$= (+105 \text{ km/h}) - (-90 \text{ km/h})$$
$$= +195 \text{ km/h}$$

Reflect

In part (a) we found that the x velocity of truck T relative to car A is negative and equal to –25 km/h. From the perspective of the car, the truck is moving westward (backward relative to the car) at 25 km/h. In part (b) the x velocity of car A relative to motorcycle M is positive and equal to 195 km/h. From the motorcycle's perspective, the car is speeding eastward at 195 km/h!

Can you show that the x velocity of the *motorcycle* relative to the *truck*, $v_{M \text{ relative to } T,x}$, is –170 km/h?

EXAMPLE 3-11   Relative Velocity in Bird Migration

The arctic tern (*Sterna paradisaea*, Figure 3-37) is a migratory bird whose annual travels take it from its breeding grounds in and near the Arctic all the way to the Antarctic coast, then back again. Their flying speed (measured relative to the air) is about 35 km/h. (a) If an arctic tern is flying toward the Arctic across the North Atlantic and encounters a wind from west to east of 15 km/h, in what direction must it point its body so that it will travel due north relative to the surface? (b) What will be the tern's speed relative to a spectator on the ground in this situation?

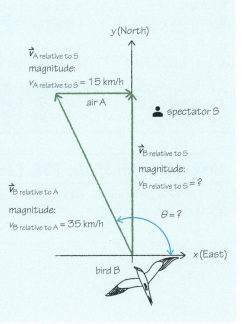
Chris Grady/Alamy

Figure 3-37 **A migratory bird** The arctic tern travels up to 90,000 km per year as it migrates between the Arctic and the Antarctic, the longest migrations known in the animal kingdom.

Set Up

There are three velocities in this problem. First, the wind velocity—that is, the velocity of the air (A) relative to a stationary spectator on the ground (S), $\bar{v}_{A \text{ relative to } S}$ —has magnitude 15 km/h and points toward the east. Second, the velocity of the bird (B) relative to the air A, $\bar{v}_{B \text{ relative to } A}$, has magnitude 35 km/h (the bird's flying airspeed); our task in part (a) is to the find the direction of this vector, which is the direction in which the bird has to point itself to compensate for the wind. Finally, the velocity of the bird B relative to spectator S, $\bar{v}_{B \text{ relative to } S}$, must be pointed northward so that the bird travels due north; in part (b) we are to find the magnitude of this velocity. We'll use the equation for relative velocity, Equation 3-21, for both parts of the problem. We'll express this equation in component form; the figure shows our choice of coordinate axes.

Relative velocity:

$$\bar{v}_{B \text{ relative to } S} = \bar{v}_{B \text{ relative to } A} + \bar{v}_{A \text{ relative to } S} \quad (3\text{-}21)$$

Components of this equation:

$$v_{B \text{ relative to } S,x} = v_{B \text{ relative to } A,x} + v_{A \text{ relative to } S,x}$$

$$v_{B \text{ relative to } S,y} = v_{B \text{ relative to } A,y} + v_{A \text{ relative to } S,y}$$

Solve

(a) We express all three of the velocity vectors in terms of their components. Note that the angle θ of the vector $\vec{v}_{\text{B relative to A}}$ is measured from the positive x axis (east) in the direction toward the positive y axis (north), so we can use Equation 3-1 for the components of that vector.

Components of $\vec{v}_{\text{B relative to S}}$:

$v_{\text{B relative to S},x} = 0$

$v_{\text{B relative to S},y} = v_{\text{B relative to S}} = ?$

Components of $\vec{v}_{\text{B relative to A}}$:

$v_{\text{B relative to A},x} = v_{\text{B relative to A}} \cos\theta = (35 \text{ km/h}) \cos\theta$

$v_{\text{B relative to A},y} = v_{\text{B relative to A}} \sin\theta = (35 \text{ km/h}) \sin\theta$

Components of $\vec{v}_{\text{A relative to S}}$:

$v_{\text{A relative to S},x} = v_{\text{A relative to S}} = 15 \text{ km/h}$

$v_{\text{A relative to S},y} = 0$

Substitute these components into the component form of Equation 3-21. Then solve the x component equation for the angle θ of the bird's velocity relative to the air.

x component of relative velocity equation becomes

$0 = v_{\text{B relative to A}} \cos\theta + v_{\text{A relative to S}}$

y component of relative velocity equation becomes

$v_{\text{B relative to S}} = v_{\text{B relative to A}} \sin\theta - 0$

Solve the x equation for θ:

$v_{\text{B relative to A}} \cos\theta = -v_{\text{A relative to S}}$

$\cos\theta = -\dfrac{v_{\text{A relative to S}}}{v_{\text{B relative to A}}} = -\dfrac{15 \text{ km/h}}{35 \text{ km/h}} = -0.43$

$\theta = \cos^{-1}(-0.43) = 115°$

The bird must point at an angle 115° north of east, or 115° − 90° = 25° west of north.

(b) Substitute the value of θ into the y component equation and solve for the speed $v_{\text{B relative to S}}$ of the bird relative to a spectator on the ground.

Solve the y equation for $v_{\text{B relative to S}}$:

$v_{\text{B relative to S}} = v_{\text{B relative to A}} \sin\theta = (35 \text{ km/h}) \sin 115°$
$\qquad\qquad = 32 \text{ km/h}$

Reflect

The bird must point its body 25° to the west of north to compensate for the wind. (If it did not, the 15-km/h wind would blow it off course by 15 km for every hour flown.) Note that even though the wind is from the west rather than from the north, it still slows the bird's northward motion from 35 km/h to 32 km/h; some northward velocity is lost because of having to fight the crosswind.

GOT THE CONCEPT? 3-9 Relative Velocity

Consider again the bird in Example 3-11, which has a speed relative to the air of 35 km/h and is flying in a wind from west to east of 15 km/h. If the bird flies so that it is pointed due north, what will be its speed relative to the ground? (a) 50 km/h; (b) between 50 and 35 km/h; (c) 35 km/h; (d) between 35 and 20 km/h; (e) 20 km/h.

TAKE-HOME MESSAGE FOR Section 3-8

✔ The velocity of any object is stated relative to some other object or point of reference.

✔ A simple vector equation relates the velocity of two objects relative to each other and the velocities of these objects relative to a spectator (or a third object).

Key Terms

acceleration vector (or instantaneous acceleration vector)
centripetal acceleration
component
direction
displacement vector
magnitude
motion in a plane
origin
parabola

position vector
projectile
projectile motion
relative velocity
scalar
trajectory
two-dimensional motion
uniform circular motion
vector
vector addition

vector difference
vector multiplication by a scalar
vector subtraction
vector sum
velocity vector (or instantaneous velocity vector)
x component
y component
x–y plane

Chapter Summary

Topic	Equation or Figure
Vectors and scalars: A vector quantity such as displacement or velocity has both a magnitude and a direction. By contrast, a scalar quantity such as temperature or time has no direction. We write a vector in boldface italic with an arrow above it. The magnitude of a vector $\vec{A}$ is written as A.	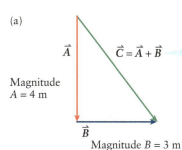 (Figure 3-2d)
Vector arithmetic: To find the sum of two vectors $\vec{A}$ and $\vec{B}$, place them tip to tail. Then $\vec{A} + \vec{B}$ points from the tail of the first vector to the tip of the second vector. To find the difference $\vec{A} - \vec{B}$ of two vectors, add $\vec{A}$ and $-\vec{B}$ (a vector with the same magnitude as $\vec{B}$ that points opposite to $\vec{B}$). Multiplying a vector by a scalar can change the vector's magnitude, direction, or both.	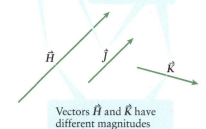 (Figure 3-5a)

(d) Vectors $\vec{H}$ and $\vec{J}$ have different magnitudes but the same direction.

Vectors $\vec{J}$ and $\vec{K}$ have the same magnitude but different directions.

Vectors $\vec{H}$ and $\vec{K}$ have different magnitudes and different directions.

 A vector has only two attributes: magnitude and direction.

(a)

$\vec{A}$ $\vec{C} = \vec{A} + \vec{B}$

Magnitude $A = 4$ m

$\vec{B}$
Magnitude $B = 3$ m

Magnitude of this vector sum:
$$C = \sqrt{A^2 + B^2}$$
$$= \sqrt{(4\text{ m})^2 + (3\text{ m})^2}$$
$$= \sqrt{16\text{ m}^2 + 9\text{ m}^2}$$
$$= \sqrt{25\text{ m}^2} = 5\text{ m}$$

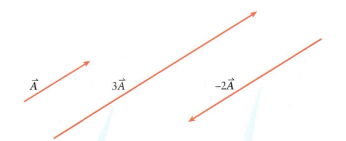

(Figure 3-8)

The number 3 is positive, so the vector $3\vec{A}$ points in the same direction as $\vec{A}$ but has 3 times the magnitude.

The number –2 is negative, so the vector $-2\vec{A}$ points in the direction opposite to $\vec{A}$ and has 2 times the magnitude.

Dividing a vector by 2 is the same as multiplying it by 1/2, so the vector $\vec{A}/2$ points in the same direction as $\vec{A}$ (since 1/2 is positive) and has 1/2 the magnitude.

Vector components: A vector can be expressed in terms of either its magnitude and direction or its x and y components. The components can be calculated from the magnitude and direction, and the magnitude and direction can be calculated from the components.

This vector has a positive x component A_x and a positive y component A_y.

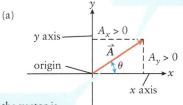

(a)

(Figure 3-9a)

The angle θ of the vector is measured from the positive x axis in the direction toward the positive y axis. Here θ is between 0 and 90°.

x **component** of the vector $\vec{A}$ y **component** of the vector $\vec{A}$

(a) $A_x = A \cos \theta$ (b) $A_y = A \sin \theta$ (3-1)

Magnitude of the vector $\vec{A}$

Angle of the vector $\vec{A}$ measured from the positive x axis toward the positive y axis

Magnitude of the vector $\vec{A}$ **Angle** of the vector $\vec{A}$ measured from the positive x axis toward the positive y axis

(a) $A = \sqrt{A_x^2 + A_y^2}$ (b) $\tan \theta = \dfrac{A_y}{A_x}$ (3-2)

x **component** of the vector $\vec{A}$ y **component** of the vector $\vec{A}$

Vector arithmetic with components: To find the components of a vector sum $\vec{A} + \vec{B}$, add the respective components. To find the components of a vector difference $\vec{A} - \vec{B}$, subtract the respective components. To find the components of a vector $\vec{A}$ multiplied by a scalar c, multiply each component of $\vec{A}$ by c.

x **component** of the vector $\vec{C} = \vec{A} + \vec{B}$

x **component** of the vector $\vec{A}$ x **component** of the vector $\vec{B}$

(a) $C_x = A_x + B_x$

(b) $C_y = A_y + B_y$ (3-3)

y **component** of the vector $\vec{C} = \vec{A} + \vec{B}$

y **component** of the vector $\vec{A}$ y **component** of the vector $\vec{B}$

x component of the vector $\vec{D} = \vec{A} - \vec{B}$

x component of the vector $\vec{A}$

x component of the vector $\vec{B}$

(a) $D_x = A_x - B_x$

(b) $D_y = A_y - B_y$

(3-4)

y component of the vector $\vec{D} = \vec{A} - \vec{B}$

y component of the vector $\vec{A}$

y component of the vector $\vec{B}$

---

x component of the vector $\vec{E} = c\vec{A}$

Scalar c

x component of the vector $\vec{A}$

(a) $E_x = cA_x$

(b) $E_y = cA_y$

(3-5)

y component of the vector $\vec{E} = c\vec{A}$

Scalar c

y component of the vector $\vec{A}$

The displacement and velocity vectors: The displacement $\Delta\vec{r}$ of an object during a time interval is a vector that points from the object's position at the earlier time t_1 to its position at the later time t_2. The velocity vector $\vec{v}$ at a given time points along the object's trajectory; its magnitude is the object's speed. The components of $\vec{v}$ are given by the same expressions we used in Chapter 2 for velocity in linear motion.

Displacement vector (change in position) of the object over the time interval from time t_1 to a later time t_2

Position vector of the object at later time t_2

(3-6)

$$\Delta\vec{r} = \vec{r}_2 - \vec{r}_1$$

Position vector of the object at earlier time t_1

To calculate the displacement vector, **subtract** the earlier value from the later value (just as for motion in a straight line).

Velocity vector for the object over a very short time interval from time t_1 to a later time t_2

Displacement vector (change in position) of the object over the short time interval

(3-7)

$$\vec{v} = \frac{\Delta\vec{r}}{\Delta t} = \frac{\vec{r}_2 - \vec{r}_1}{t_2 - t_1}$$

For both the displacement and the elapsed time, **subtract** the earlier value from the later value.

Elapsed time for the time interval

x component of the velocity vector $\vec{v}$

x component of the **displacement** during a short time interval

(a) $v_x = \dfrac{\Delta x}{\Delta t} = \dfrac{x_2 - x_1}{t_2 - t_1}$

y component of the **displacement** during a short time interval

(3-8)

(b) $v_y = \dfrac{\Delta y}{\Delta t} = \dfrac{y_2 - y_1}{t_2 - t_1}$

y component of the velocity vector $\vec{v}$

Elapsed time for the time interval

The acceleration vector: An object has a nonzero acceleration vector $\vec{a}$ whenever it is speeding up, slowing down, *or* changing direction. The x and y components of $\vec{a}$ are given by the same expressions we used in Chapter 2 for acceleration in linear motion. Alternatively, we can express $\vec{a}$ in terms of its component along the direction of motion (which describes changes in speed) and its component perpendicular to the direction of motion (which describes changes in direction).

Acceleration vector for the object over a very short time interval from time t_1 to a later time t_2

Change in velocity of the object over the short time interval (3-9)

$$\vec{a} = \frac{\Delta \vec{v}}{\Delta t} = \frac{\vec{v}_2 - \vec{v}_1}{t_2 - t_1}$$

For both the velocity change and the elapsed time, **subtract** the earlier value from the later value.

Elapsed time for the time interval

x component of the acceleration vector $\vec{a}$

Change in the x component of the velocity during a short time interval

(a) $a_x = \dfrac{\Delta v_x}{\Delta t} = \dfrac{v_{2x} - v_{1x}}{t_2 - t_1}$

Change in the y component of the velocity during a short time interval (3-10)

(b) $a_y = \dfrac{\Delta v_y}{\Delta t} = \dfrac{v_{2y} - v_{1y}}{t_2 - t_1}$

Elapsed time for the time interval

y component of the acceleration vector $\vec{a}$

Projectile motion: A projectile is a moving object acted on by gravity alone. It has a constant downward acceleration of magnitude g. In the equations for the motion of a projectile, we always choose the x axis to be horizontal and the y axis to point vertically upward.

- The acceleration $\vec{a}$ of a projectile is constant.
- Because the projectile's path is curved, $\vec{a}$ has varying components along and perpendicular to the direction of motion.

There is a component of acceleration opposite to the direction of motion: The projectile is slowing down.

There is no component of acceleration along the direction of motion: The speed of the projectile has its minimum value.

There is a component of acceleration in the direction of motion: The projectile is speeding up.

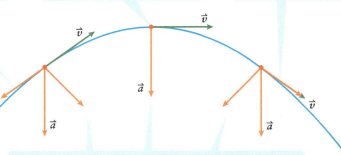

At all points there is a component of acceleration perpendicular to the trajectory: The direction of motion is changing.

(Figure 3-24)

x component of an object's velocity vector $\vec{v}$ at time t for **projectile motion**

x component of the object's velocity vector at $t = 0$ (3-12)

(a) $v_x = v_{0x}$ Acceleration due to gravity (g is positive)

(b) $v_y = v_{0y} - gt$

y component of an object's velocity vector $\vec{v}$ at time t for **projectile motion**

Time at which the object has velocity components v_x and v_y

y component of the object's velocity vector at $t = 0$

x and y coordinates at time t of an object in **projectile motion**

x and y components at $t = 0$ of the object's velocity vector

Time at which the object has coordinates x and y

(a) $x = x_0 + v_{0x}t$ (3-13)

(b) $y = y_0 + v_{0y}t - \dfrac{1}{2}gt^2$

x and y coordinates at $t = 0$ of the object

Acceleration due to gravity (g is positive)

Motion in a circle: For an object moving at constant speed around a circular path, the acceleration is centripetal (directed toward the center of the circle).

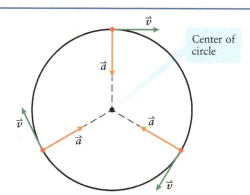

Center of circle

- In uniform circular motion the speed is *constant*, but the object accelerates because the velocity $\vec{v}$ is always changing direction.
- Because the speed is constant, the acceleration $\vec{a}$ is always perpendicular to the velocity $\vec{v}$.
- $\vec{a}$ always points toward the center of the circle.

(Figure 3-29)

Centripetal acceleration: magnitude of the acceleration of an object in uniform circular motion

Speed of an object as it moves around the circle

$$a_{\text{cent}} = \frac{v^2}{r}$$

Radius of the circle

(3-17)

Relative velocity: The velocity you measure for an object depends on how you are moving. The relative velocity equation relates the velocities of two objects A and B measured by a spectator S and the velocity of one of the objects relative to the other.

The velocity of B relative to S...

...equals the velocity of B relative to A...

...plus the velocity of A relative to S.

$$\vec{v}_{\text{B relative to S}} = \vec{v}_{\text{B relative to A}} + \vec{v}_{\text{A relative to S}}$$

To keep track of which vector is which, note that the first and second subscripts on the left-hand side...

...are the same as the first and last subscripts on the right-hand side...

...and the two middle subscripts on the right-hand side are the same.

(3-21)

Answer to What do you think? Question

(c) The high jumper takes the *same* amount of time to climb from a given height to her maximum height as she does to descend from her maximum height back to that same given height. See Example 3-7 in Section 3-6.

Answers to Got the Concept? Questions

3-1   (c) The illustration shows the vectors $\vec{A}$ and $\vec{B}$ and the four combinations. One way to end up with a vector pointing southwest is to add a vector pointing west to a vector pointing south. Vector $\vec{B}$ points west. If we multiply vector $\vec{A}$ (which points north) by -1, the vector $-\vec{A}$ will point south. So $\vec{B} + (-\vec{A}) = \vec{B} - \vec{A}$ points southwest.

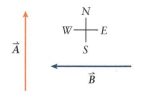

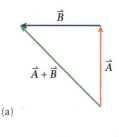

(a)

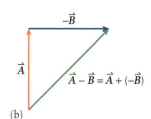

(b)

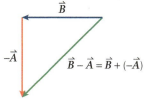

(c)

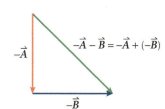

(d)

3-2 A_x is positive, A_y is negative; B_x is positive, B_y is positive; C_x is negative, C_y is positive; D_x is negative, D_y is negative. These answers may surprise you if you assumed that the positive x direction points to the right and the positive y direction points upward. But as the figure shows, for this question we've chosen the positive x direction to point *upward* and the positive y direction to point to the *left*. When thinking about vector components, always be mindful of the directions of the axes!

3-3 $D_x = 2$, $D_y = -5$. Our goal is to find a vector $\vec{D}$ with the property that $\vec{A} + \vec{D} = \vec{B}$, so $\vec{D} = \vec{B} - \vec{A}$. From Equations 3-4, the components of $\vec{D}$ are

$$D_x = B_x - A_x = 7 - 5 = 2$$
$$D_y = B_y - A_y = (-2) - 3 = -5$$

In other words, the x component $\vec{D} = \vec{B} - \vec{A}$ is the x component of $\vec{B}$ minus the x component of $\vec{A}$, and similarly for the y component of $\vec{D} = \vec{B} - \vec{A}$. Our result tells us that the vector $\vec{D}$ has components $D_x = 2$, $D_y = -5$. If you wanted to, you could express $\vec{D}$ in terms of its magnitude and direction. But it's perfectly fine to write $\vec{D}$ in terms of its components as we have done here.

3-4 Point 1, (a); point 2, (c); point 3, (b); point 4, (a). At points 1 and 4 the object is speeding up because its acceleration vector $\vec{a}$ points partially in the direction of motion (that is, partially in the direction of the velocity vector $\vec{v}$). At point 3 the acceleration vector points partially opposite to the direction of $\vec{v}$ so at this point the object is slowing down. At point 2 the object is neither speeding up nor slowing down but is maintaining a constant speed. The acceleration vector is perpendicular to the velocity vector. Note that at all four points on this trajectory, the acceleration vector $\vec{a}$ points in whole or in part toward the inside of the trajectory's curve.

3-5 The magnitude of velocity is smallest at the peak of motion. At the peak the y component of velocity is zero, so regardless of the x component of velocity, the magnitude of $\vec{v}$ must be smallest. If the x component of velocity is zero, then the total velocity, both magnitude and direction, is zero at the peak. Because the x component of velocity v_x does not change, v_x can be zero only if the object initially has no x component of velocity ($v_{0x} = 0$); that is, if it is launched straight up.

3-6 (b) At its maximum height the y component of the frog's velocity is zero, and the x component of its velocity equals the x component of its initial velocity: $v_x = v_{0x} = v_0 \cos \theta_0$. Since $\cos 45° = 1/\sqrt{2} = 0.707$, at the frog's maximum height $v_x = 0.707v_0$ and its speed is $0.707v_0$ — that is, between v_0 and $v_0/2$.

3-7 No. The *magnitude* of the acceleration $a_{cent} = v^2/r$ is constant, since speed v is constant, but the *direction* of the acceleration vector is continually changing so that $\vec{a}$ points toward the center of the circle at all times (see Figure 3-29). Therefore, the acceleration is *not* constant.

3-8 (d) The centripetal acceleration is given by Equation 3-17, $a_{cent} = v^2/r$. If the speed v is doubled, the numerator v^2 in the expression for a_{cent} increases by a factor of $2^2 = 4$. The radius r must also increase by a factor of 4 to maintain the same value of a_{cent}.

3-9 (b) The accompanying figure shows the vector addition diagram for the relative velocity equation $\vec{v}_{B\ relative\ to\ S} = \vec{v}_{B\ relative\ to\ A} + \vec{v}_{A\ relative\ to\ S}$. The vector $\vec{v}_{B\ relative\ to\ A}$ has magnitude 35 km/h and points north (the direction the bird is pointed); the vector $\vec{v}_{A\ relative\ to\ S}$ has magnitude 15 km/h and points east (the wind is from west to east). These two vectors form the two perpendicular sides of a right triangle with $\vec{v}_{B\ relative\ to\ S}$ as the hypotenuse. So the bird's speed relative to the surface is $v_{B\ relative\ to\ S} = \sqrt{(35\ km/h)^2 + (15\ km/h)^2} = 38\ km/h$.

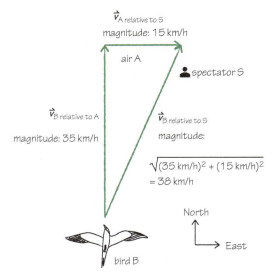

Questions and Problems

In a few problems you are given more data than you actually need; in a few other problems, you are required to supply data from your general knowledge, outside sources, or informed estimate.

   Interpret as significant all digits in numerical values that have trailing zeros and no decimal points. For all problems use $g = 9.80$ m/s^2 for the free-fall acceleration due to gravity.

- • Basic, single-concept problem
- •• Intermediate-level problem; may require synthesis of concepts and multiple steps
- ••• Challenging problem
- Example See worked example for a similar problem

Conceptual Questions

1. • What is the difference between a scalar and a vector? Give an example of a scalar and an example of a vector.

2. • (a) Can the sum of two vectors that have different magnitudes ever be equal to zero? If so, give an example. If not, explain why the sum of two vectors cannot be equal to zero. (b) Can the sum of three vectors that have different magnitudes ever be equal to zero?

3. • Describe a situation in which the average velocity and the instantaneous velocity vectors are identical. Describe a situation in which these two velocity vectors are different.

4. • **Astronomy** If you were playing tennis on the Moon, what adjustments would you need to make for your shots to stay within the boundaries of the court? Would the trajectories of the balls look different on the Moon compared to on Earth?

5. • During the motion of a projectile, which of the following quantities are constant during the flight: x, y, v_x, v_y, a_x, a_y? (Neglect any effects due to air resistance.)

6. • For a given, fixed launch speed, at what angle should you launch a projectile to achieve (a) the longest range, (b) the longest time of flight, and (c) the greatest height? (Neglect any effects due to air resistance.)

7. • A rock is thrown from a bridge at an angle 20° below horizontal. At the instant of impact, is the rock's speed greater than, less than, or equal to the speed with which it was thrown? Explain your answer. (Neglect any effects due to air resistance.)

8. • **Sports** A soccer player kicks a ball at an angle 60° from the ground. The soccer ball hits the ground some distance away. Is there any point at which the velocity and acceleration vectors are perpendicular to each other? Explain your answer. (Neglect any effects due to air resistance.)

9. • **Sports** Suppose you are the coach of a champion long jumper. Would you suggest that she take off at an angle less than 45°? Why or why not?

10. • (a) Explain the difference between an object undergoing uniform circular motion and an object experiencing projectile motion. (b) In what ways are these kinds of motion similar?

11. • An ape swings through the jungle by hanging from a vine. At the lowest point of its motion, is the ape accelerating? If so, what is the direction of its acceleration?

12. • A cyclist rides around a flat, circular track at constant speed. Is his acceleration vector zero? Explain your answer.

13. • You are driving your car in a circular path on flat ground with a constant speed. At the instant you are driving north and turning right, are you accelerating? If so, what is the direction of your acceleration at that moment? If not, why not?

14. • **Sports** A pitching machine is programmed to pitch baseballs horizontally at a speed of 80 mph. The machine is mounted on a truck and aimed backward (Figure 3-38). As the truck drives away from a man at a speed of 40 mph, the pitching machine shoots baseballs toward him. (a) What is the speed of the pitching machine relative to the truck? (b) What is the speed of the baseballs relative to the truck? (c) What is the speed of the pitching machine relative to the man? (d) What is the speed of the baseballs relative to the man?

Figure 3-38 Problem 14

15. • What does the phrase "all motion is relative, there is no absolute motion" mean?

Multiple-Choice Questions

16. • Which of the following is not a vector?
   A. average velocity
   B. instantaneous velocity
   C. distance
   D. displacement
   E. acceleration

17. • Vector $\bar{A}$ has an x component and a y component that are equal in magnitude. Which of the following is the angle that vector $\bar{A}$ makes with respect to the x axis in the same x–y coordinate system?
   A. 0°
   B. 45°
   C. 60°
   D. 90°
   E. 120°

18. • The vector in Figure 3-39 has a length of 4.00 units and makes a 30.0° angle with respect to the y axis as shown. What are the x and y components of the vector?
   A. 3.46, 2.00
   B. –2.00, 3.46
   C. –3.46, 2.00
   D. 2.00, –3.46
   E. –3.46, –2.00

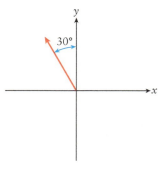

Figure 3-39 Problem 18

19. • The acceleration of a particle in projectile motion
   A. points along the parabolic path of the particle.
   B. is directed horizontally.
   C. vanishes at the particle's highest point.
   D. is vertically downward.
   E. is zero.

20. • Adam drops a ball from rest from the top floor of a building at the same time that Bob throws a ball horizontally from the same location. Which ball hits the level ground below first? (Neglect any effects due to air resistance.)
   A. Adam's ball
   B. Bob's ball
   C. They both hit the ground at the same time.
   D. It depends on how fast Bob throws the ball.
   E. It depends on how fast the ball falls when Adam drops it.

21. • **Sports** Two golf balls are hit from the same point on a flat field. Both are hit at an angle of 30° above the horizontal. Ball 2 has twice the initial speed of ball 1. If ball 1 lands a distance d_1 from the initial point, at what distance d_2 does ball 2 land from the initial point? (Neglect any effects due to air resistance.)
   A. $d_2 = 0.5d_1$
   B. $d_2 = d_1$
   C. $d_2 = 2d_1$
   D. $d_2 = 4d_1$
   E. $d_2 = 8d_1$

22. • A zookeeper is trying to shoot a monkey sitting at the top of a tree with a tranquilizer gun. If the monkey drops from the tree at the same instant that the zookeeper fires, where should the zookeeper aim if he wants to hit the monkey? (Neglect any effects due to air resistance.)
   A. Aim straight at the monkey.
   B. Aim lower than the monkey.
   C. Aim higher than the monkey.
   D. Aim to the right of the monkey.
   E. It's impossible to determine.

23. • The acceleration vector of a particle in uniform circular motion
   A. points along the circular path of the particle and in the direction of motion.
   B. points along the circular path of the particle and opposite the direction of motion.
   C. is zero.

D. points toward the center of the circle.

E. points outward from the center of the circle.

24. • If the speed of an object in uniform circular motion remains constant while the radial distance is doubled, the magnitude of the radial acceleration decreases by what factor?

    A. 2

    B. 3

    C. 4

    D. 6

    E. 1

25. • You toss a ball into the air at an initial angle 40° from the horizontal. At what point in the ball's trajectory does the ball have the smallest speed? (Neglect any effects due to air resistance.)

    A. just after it is tossed

    B. at the highest point in its flight

    C. just before it hits the ground

    D. halfway between the ground and the highest point on the rise portion of the trajectory

    E. halfway between the ground and the highest point on the fall portion of the trajectory

Problems

3-2 A vector quantity has both a magnitude and a direction

3-3 Vectors can be described in terms of components

26. • Vector $\bar{A}$ has components $A_x = 6$ and $A_y = 9$. Vector $\bar{B}$ has components $B_x = 7$, $B_y = -3$, and vector $\bar{C}$ has components $C_x = 0$, $C_y = -6$. Determine the components of the following vectors: (a) $\bar{A} + \bar{B}$, (b) $\bar{A} - 2\bar{C}$, (c) $\bar{A} + \bar{B} - \bar{C}$, and (d) $\bar{A} + \frac{1}{2}\bar{B} - 3\bar{C}$. Example 3-3

27. • Calculate the magnitude and direction of the vector $\bar{r}$ using Figure 3-40. Example 3-1

Figure 3-40 Problem 27

28. • What are the components A_x and A_y of vector $\bar{A}$ in the three coordinate systems shown in Figure 3-41? Example 3-1

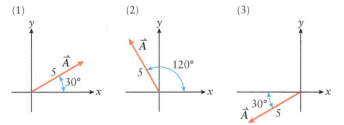

Figure 3-41 Problem 28

29. • Each of the following vectors is given in terms of its x and y components. Find the magnitude of each vector and the angle it makes with respect to the +x axis. Example 3-1

    A. $A_x = 3$, $A_y = -2$

    B. $A_x = -2$, $A_y = 2$

    C. $A_x = 0$, $A_y = -2$

30. •• $\bar{A}$ is 66.0 m long at a 28° angle with respect to the +x axis. $\bar{B}$ is 40.0 m long at a 56° angle above the −x axis. What is the sum of vectors $\bar{A}$ and $\bar{B}$ (magnitude and angle with the +x axis)? Example 3-3

31. •• Given the vector $\bar{A}$ with components $A_x = 2.00$ and $A_y = 6.00$, and the vector $\bar{B}$ with components $B_x = 3.00$ and $B_y = -2.00$, calculate the magnitude and angle with respect to the +x axis of the vector sum $\bar{C} = \bar{A} + \bar{B}$. Example 3-3

32. •• Given the vector $\bar{A}$ with components $A_x = 2.00$, $A_y = 6.00$; the vector $\bar{B}$ with components $B_x = 2.00$, $B_y = -2.00$; and the vector $\bar{D} = \bar{A} - \bar{B}$, calculate the magnitude and angle with the +x axis of the vector $\bar{D}$. Example 3-3

33. •• Two velocity vectors are given as follows: $\bar{A} = 30$ m/s, 45° north of east and $\bar{B} = 40$ m/s, due north. Calculate each of the resultant velocity vectors: (a) $\bar{A} + \bar{B}$, (b) $\bar{A} - \bar{B}$, (c) $2\bar{A} + \bar{B}$. Example 3-3

34. •• Consider the two vectors in Figure 3-42. Nathan says the magnitude of the resultant vector is 7, and the resultant vector points in a direction 37° in the northeasterly direction. What, if anything, is wrong with his statement? If something is wrong, explain the error(s) and how to correct it (them). Example 3-3

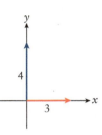

Figure 3-42 Problem 34

3-4 For motion in a plane, velocity and acceleration are vector quantities

35. •• What are the magnitude and direction of the change in velocity if the initial velocity is 30 m/s south and the final velocity is 40 m/s west? Example 3-3

36. •• The two vectors shown in Figure 3-43 represent the initial and final velocities of an object during a trip that took 5 s. Calculate the average acceleration during this trip. Is it possible to determine whether the acceleration was uniform from the information given in the problem? Example 3-3

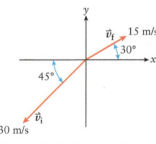

Figure 3-43 Problem 36

37. ••• An object travels with a constant acceleration for 10 s. The vectors in Figure 3-44 represent the final and initial velocities. Carefully graph the x component of the velocity versus time, the y component of the velocity versus time, and the y component of the acceleration versus time. Example 3-3

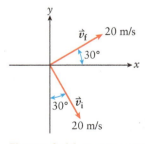

Figure 3-44 Problem 37

38. •• An object experiences a constant acceleration of 2.00 m/s² along the −x axis for 2.70 s, attaining a velocity of 16.0 m/s in a direction 45° from the +x axis. Calculate the initial velocity vector of the object. Example 3-3

39. •• Cody starts at a point 6.00 km to the east and 4.00 km to the south of a location that represents the origin

of a coordinate system for a map. He ends up at a point 10.0 km to the west and 6.00 km to the north of the map origin. (a) What was his average velocity if the trip took him 4.00 h to complete? (b) Cody walks to his destination at a constant rate. His friend Marcus covers the distance with a combination of jogging, walking, running, and resting so that the total trip time is also 4.00 h. How do their average velocities compare? Example 3-3

3-5 A projectile moves in a plane and has a constant acceleration
3-6 You can solve projectile motion problems using techniques learned for straight-line motion

40. • An object is undergoing projectile motion as shown from the side in Figure 3-45. Assume the object starts its motion at ground level. For the five positions shown, draw to scale vectors representing the magnitudes of (a) the x components of the velocity, (b) the y components of the velocity, and (c) the accelerations. (Neglect any effects due to air resistance.) Example 3-5

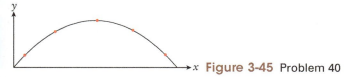

Figure 3-45 Problem 40

41. •• An object undergoing projectile motion travels 1.00×10^2 m in the horizontal direction before returning to its initial height. If the object is thrown initially at a 30.0° angle from the horizontal, determine the x component and the y component of the initial velocity. (Neglect any effects due to air resistance.) Example 3-7

42. •• Five balls are thrown off a cliff at the angles shown in Figure 3-46. Each has the same initial speed. Rank (a) the time required for each to hit the ground, (b) the vertical displacement of each, and (c) the vertical component of final velocity. (Neglect any effects due to air resistance.) Example 3-5

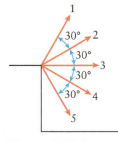

Figure 3-46 Problem 42

43. • **Biology** A Chinook salmon can jump out of water with a speed of 6.30 m/s. How far horizontally can a Chinook salmon travel through the air if it leaves the water with an initial angle of 40°? (Neglect any effects due to air resistance.) Example 3-7

44. • **Biology** A tiger leaps horizontally out of a tree that is 4.00 m high. If he lands 5.00 m from the base of the tree, calculate his initial speed. (Neglect any effects due to air resistance.) Example 3-4

45. • A football is kicked at 25.0 m/s at an angle of 30.0° above the horizon. What is the velocity vector of the ball when it is 5.00 m above ground level? Assume it starts 1.00 m above ground level. (Neglect any effects due to air resistance.) Example 3-5

46. •• A dart is thrown at a dartboard 2.37 m away. When the dart is released at the same height as the center of the dartboard, it hits the center in 0.447 s. At what angle relative to the floor was the dart thrown? (Neglect any effects due to air resistance.) Example 3-5

47. • **Sports** In the 1970 National Basketball Association championship, Jerry West made a 60-ft shot from beyond half court to lead the Los Angeles Lakers to an improbable tie at the buzzer with the New York Knicks. Neglecting air resistance, estimate the initial speed of the ball. (The Knicks won the game in overtime.) Example 3-5

48. • Use a spreadsheet program or a graphing calculator to make (a) a graph of v_x versus time, (b) a graph of v_y versus time, (c) a graph of a_x versus time, and (d) a graph of a_y versus time for an object that undergoes projectile motion. Identify the points where the object reaches its highest point and where it hits the ground at the end of its flight. Example 3-6

3-7 An object moving in a circle is accelerating even if its speed is constant

49. • A ball attached to a string is twirled in a circle of radius 1.25 m. If the constant speed of the ball is 2.25 m/s, what is the time required to travel once around the circle? Example 3-9

50. • A ball is twirled on a 0.870-m-long string with a constant speed of 3.36 m/s. Calculate the acceleration of the ball. Be sure to specify the direction of the acceleration. Example 3-8

51. • A car driving around a circular track has a centripetal acceleration of 80.0 m/s² until it hits a slick spot and slides straight off the track. If the radius of the track is 15.0 m, what is the car's velocity when it leaves the track? Example 3-9

52. •• In 1892, George W. G. Ferris designed a carnival ride in the shape of a large wheel. This Ferris wheel had a diameter of 76 m and rotated one revolution every 20 min. What was the magnitude of the centripetal acceleration that riders experienced? Example 3-8

53. • A car races at a constant speed of 330 km/h around a flat, circular track 1.00 km in diameter. What is the car's centripetal acceleration in m/s²? Example 3-8

54. • **Astronomy** We know that the Moon revolves around Earth during a period of 27.3 days. The average distance from the center of Earth to the center of the Moon is 3.84×10^8 m. What is the acceleration of the Moon due to its motion around Earth? Example 3-9

55. •• Calculate the accelerations of (a) Earth as it orbits the Sun with an average orbital radius of 1.5×10^{11} m and (b) a car traveling along a circular path that has a radius of 50.0 m at a speed of 20.0 m/s. Example 3-9

56. • **Biology** In a vertical dive, a peregrine falcon can accelerate at 0.6 times the free-fall acceleration (that is, at 0.6g) in reaching a speed of about 100 m/s. If a falcon pulls out of a dive into a circular arc at this speed and can sustain a centripetal acceleration of 0.6g, what is the minimum radius of the turn? Example 3-9

57. •• Commercial ultracentrifuges can rotate at rates of 100,000 rpm (revolutions per minute). As a consequence, they can create accelerations on the order of 800,000g. (A "g" represents an acceleration of 9.80 m/s².) Calculate the distance from the rotation axis of the sample chamber in such a device. What is the speed of an object traveling under the given conditions? Example 3-9

3-8 The velocity you measure for an object depends on how you are moving

58. • A motorcyclist drives at 22 m/s in a direction 35° east of north relative to a car, and at 8.0 m/s due north relative to Earth. (a) What is the magnitude of the car's velocity relative to Earth? (b) What is the direction of the car's velocity relative to Earth, measured counterclockwise from due east? Example 3-11

59. •• Daniel takes his two dogs, Pauli and Newton, out to a field and lets them loose. The dogs run in different directions while Daniel stands still. From Daniel's point of view, Newton runs due north at 4.5 m/s. From Pauli's point of view, Newton appears to be moving at 1.5 m/s due east. (a) What is the magnitude of Pauli's velocity relative to Daniel? (b) What is the direction of Pauli's velocity, measured counterclockwise from due east? Example 3-11

60. • A bicyclist rides north at 8.00 m/s. A car moving north at 25.0 m/s is initially behind the cyclist. A truck moving south at 15.0 m/s approaches the cyclist and the car from the north. (a) Make a sketch of the three vehicles and their respective velocity vectors, measured relative to the ground. (b) Calculate the velocity of each vehicle relative to each of the other two (for a total of six velocities). Example 3-10

61. • A sailboat departs from the pier at Fisherman's Wharf in San Francisco at 4.00 m/s, heading directly toward Alcatraz. A kitesurfer heads directly away from Alcatraz and toward the boat at a relative speed of 6.00 m/s according to the skipper of the sailboat. Calculate the velocity of the kitesurfer relative to the pier. Example 3-11

General Problems

62. ••• **Sports** Aaron Rodgers stands on the 20-yard line, poised to throw long. He throws the ball at initial velocity $v_0 = 15.0$ m/s and releases it at an angle $\theta = 45.0°$. Having faked an end around, Jordy Nelson comes racing past Aaron at a constant velocity $v_J = 8.00$ m/s, heading straight down the field. Assuming that Jordy catches the ball at the same height above the ground that Aaron throws it, how long must Aaron wait to throw, after Jordy goes past, so that the ball falls directly into Jordy's hands? Example 3-5

63. ••• You throw a ball from the balcony onto the court in the basketball arena. You release the ball at a height of 7.00 m above the court, with an initial velocity equal to 9.00 m/s at 33.0° above the horizontal. A friend of yours, standing on the court 11.0 m from the point directly beneath you, waits for a period of time after you release the ball and then begins to move directly away from you at an acceleration of 1.80 m/s². (She can only do this for a short period of time!) If you throw the ball in a line with her, how long after you release the ball should she wait to start running directly away from you so that she'll catch the ball exactly 1.00 m above the floor of the court? Example 3-5

64. •• One day, Sofia goes hiking at a nearby nature preserve. At first, she follows the straight, clearly marked trails. From the trailhead, she travels 2.00 miles down the first trail. Then, she turns 30° to her left to follow a second trail for 1.60 miles. Next, she turns 150° to her right to follow a third trail for 2.40 miles. At this point, Sofia is getting tired and would like to get back as quickly as possible, but all of the available trails seem to lead her deeper into the woods. She would like to take a shortcut directly through the woods, ignoring the trails. How far to her right should she turn and how far does she have to walk to return directly to the trailhead? Example 3-3

65. • Nathan walks due east a certain distance and then walks due south twice that distance. He finds himself 15.0 km from his starting position. How far east and how far south does Nathan walk? Example 3-3

66. •• A water balloon is thrown horizontally at a speed of 2.00 m/s from the roof of a building that is 6.00 m above the ground. At the same instant the balloon is released, a second balloon is thrown straight down at 2.00 m/s from the same height. Determine which balloon hits the ground first and how much sooner it hits the ground than the other balloon does. (Neglect any effects due to air resistance.) Example 3-5

67. •• You throw a rock from the upper edge of a 75.0-m vertical dam with a speed of 25.0 m/s at 65.0° above the horizon. How long after throwing the rock will it (a) return to its initial height and (b) hit the water flowing out at the base of the dam? (Neglect any effects due to air resistance.) Example 3-4

68. • An airplane releases a ball as it flies parallel to the ground at a height of 255 m (Figure 3-47). If the ball lands on the ground at a horizontal displacement of exactly 255 m from the release point, calculate the airspeed of the plane. (Neglect any effects due to air resistance.) Example 3-4

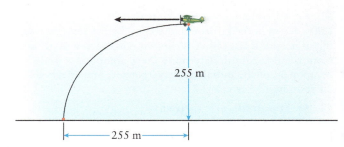

255 m

255 m

Figure 3-47 Problem 68

69. •• An airplane flying upward at 35.3 m/s and an angle of 30.0° relative to the horizontal releases a ball when it is 255 m above the ground. Calculate (a) the time it takes the ball to hit the ground, (b) the maximum height of the ball, and (c) the horizontal distance the ball travels from the release point to the ground. (Neglect any effects due to air resistance.) Example 3-4

70. •• **Sports** In 1993, Javier Sotomayor set a world record of 2.45 m in the men's outdoor high jump. He is 195 cm (6 ft, 5 in) tall. By treating his body as a point located at half his height, and given that he left the ground a horizontal distance from the bar of 1.5 m at a takeoff angle of 65.0°, determine Javier Sotomayor's takeoff speed. (Neglect any effects due to air resistance.) Example 3-6

71. •• **Sports** In the sport of billiards, event organizers sometimes remove the rails on a pool table to allow players to measure the speed of their break shots, as shown in Figure 3-48. A competitor wants to know if he has broken the world speed record for the break shot of 35 mph. If the winner's ball landed a distance of $d = 4.80$ m from the table's edge and the surface of the pool table is $h = 75$ cm above the

floor, (a) calculate the speed v_0 of his break shot. (b) At what speed v_1 did his ball hit the ground? (c) How long was the ball in the air? Example 3-5

Figure 3-48 Problem 71

72. • **Sports** Gabriele Reinsch threw a discus 76.80 m on July 9, 1988, to set the women's world record. Assume that she launched the discus with an elevation angle of 45.0° and that her hand was 2.00 m above the ground at the instant of launch. What was the initial speed of the discus required to achieve that range? (Neglect any effects due to air resistance.) Example 3-5

73. •• **Astronomy** The froghopper, a tiny insect, is a remarkable jumper. Suppose you raised a colony of the little critters on the Moon, where the acceleration due to gravity is only 1.62 m/s^2. If on Earth a froghopper's maximum jump height is h and maximum horizontal range is R, what would its maximum height and range be on the Moon in terms of h and R? Assume a froghopper's takeoff speed is the same on the Moon and on Earth. Example 3-7

74. •• **Sports** In 1998, Jason Elam kicked a record field goal. The football started on the ground a horizontal distance of 63.0 yards from the goal posts and just barely cleared the 10-ft-high bar. If the initial trajectory of the football was 40.0° above the horizontal, (a) what was its initial speed and (b) how long after the ball was struck did it pass through the goal posts? (Neglect any effects due to air resistance.) Example 3-4

75. ••• **Sports** In the hope that the Moon and Mars will one day become tourist attractions, a golf course is built on each. An average golfer on Earth can drive a ball from the tee about 63% of the distance to the hole. If this is to be true on the Moon and on Mars, by what factor should the dimensions of the golf courses on the Moon and Mars be changed relative to a course on Earth? (Neglect any effects due to air resistance.) Example 3-7

76. •• **Biology** Anne is working on a research project that involves the use of a centrifuge. Her samples must first experience an acceleration of 100g, but then the acceleration must increase by a factor of 8. By how much will the rotation speed have to increase? Express your answer as a multiple of the initial rotation rate. Example 3-9

77. • **Medical** In a laboratory test of tolerance for high centripetal acceleration, pilots were swung in a circle 13.4 m in diameter. It was found that they blacked out when they were spun at 30.6 rpm (rev/min). (a) At what acceleration (in SI units and in multiples of g) did the pilots black out? (b) To decrease the acceleration by 25.0% without changing the diameter of the circle, technicians change the time in which a pilot completes one revolution. At what percentage of the original time should the technicians set the apparatus? Example 3-8

78. • **Medical** Aerobatic pilots can survive centripetal accelerations up to 9g (88 m/s^2). What acceleration does a pilot flying in a horizontal 3320-m-radius circle at a constant speed of 515 m/s experience? Example 3-8

79. • **Sports** A girl's fast-pitch softball player does a windmill pitch, moving her hand through a circular arc with her arm straight. She releases the ball at a speed of 24.6 m/s. Just before the ball leaves her hand, the ball's centripetal acceleration is 1960 m/s^2. What is the length of her arm from the pivot point at her shoulder? Example 3-9

80. •• A particular cheetah has a top speed of 120 km/h, which it can only maintain for a distance of 500 m. It is hunting an impala, which has a top speed of 90 km/h. When the cheetah reaches its top speed, the impala is 200 m in front of the cheetah and also running at top speed. (a) What is the velocity of the cheetah relative to the impala? (b) How far can the impala run before the cheetah has to slow down? (c) Can the cheetah catch the impala while running at its top speed? Assume that the cheetah and impala run in a straight line. Example 3-10

81. ••• Robin would like to shoot an orange in a tree with his bow and arrow. The orange is hanging $y_f = 5.00$ m above the ground. On his first try, Robin looses his arrow at $v_0 = 35.0$ m/s at an angle of 30.0° above the horizontal. The initial height of the arrow is $y_0 = 1.50$ m and its tip is $x = 55.0$ m away from the target orange. (a) Treating the arrow as a point projectile and neglecting air resistance, what is the height of the arrow when it reaches the horizontal position of the orange? For his second try, Robin changes his position but looses the arrow with the same speed and angle as his first try. (b) At what two horizontal distances from the orange will Robin hit his target? Example 3-7

4

Forces and Motion I: Newton's Laws

What do you think?

The tug is pulling the airliner with just enough force to maintain a constant velocity. If the tug instead pulls with double this force, what will happen to the airliner? (a) It will move with a faster constant speed. (b) It will move for a while with a constant speed and then its speed will increase. (c) Its speed will increase continuously.

In this chapter, your goals are to:

- (4-1) Describe the importance of forces in determining how an object moves.
- (4-2) Use Newton's second law to relate the net force on an object to the object's acceleration.
- (4-3) Recognize the distinctions among mass, weight, and inertia.
- (4-4) Draw and use free-body diagrams in problems that involve forces.
- (4-5) Describe how Newton's third law relates forces that act on different objects.
- (4-6) Apply the sequence of steps used in solving all problems involving forces, including those that also involve kinematics.

To master this chapter, you should review:

- (2-2, 2-3) The ideas of displacement, velocity, and acceleration for motion in a straight line.
- (2-4, 2-5, and 2-6) The concepts and equations of straight-line motion with constant acceleration, including free fall.
- (3-2, 3-3) How to add vectors and how to do vector calculations using components.
- (3-4) The ideas of displacement, velocity, and acceleration in two dimensions.

4-1 How objects move is determined by the forces that act on them

In the language of physics, the tug in the above photo exerts a *force* on the airliner. Force is a vector: It has both magnitude and direction. For example, the force the tug exerts on the airliner points in the direction of the airliner's motion, while the ground exerts a friction force on the airliner that points backward (opposite to the airliner's motion). The magnitude of each force is a measure of how strong it pushes or pulls on the airliner.

What determines how the airliner moves is not any one single force but rather the *combined* effect of *all* the forces that act on it. If the forward force from the tug has the same magnitude as the backward friction force, these two oppositely directed force vectors cancel and add to zero. In this case the airliner moves with a constant velocity. But if the tug pulls harder than the force of friction, the forces do not cancel, and there is a *net* force in the forward direction. Then the airliner accelerates.

Experiment shows that no matter how hard the tug pulls on the airliner, the airliner pulls back on the tug with an equally strong force. This turns out to be true for forces of all kinds: If one object exerts a force on a second object, the second object necessarily exerts a force on the first one.

These observations are at the heart of *Newton's laws of motion*, a time-tested set of physical principles that have a tremendous range of applicability. We'll use Newton's laws throughout the remainder of this book. We'll devote this chapter and the next to understanding these laws and some of their most important applications.

4-2 If a net external force acts on an object, the object accelerates

If you want to start a soccer ball rolling along the ground, you have to give it a push or kick. If you're a hockey goalie and want to deflect a puck away from the goal that you're defending, you have to hit the puck with your hockey stick. And if your dog runs off in pursuit of the neighbor's cat, you have to pull on the dog's leash to slow it down and bring it to a halt. In each of these cases you're changing the velocity of an object, either making it speed up (for the soccer ball), changing the direction of its motion (for the hockey puck), or making it slow down (for your dog). And in each case to cause the object's velocity to change, you have to exert a **force**—that is, a push or a pull—on that object.

Forces are *vectors* because they have both *magnitude* and *direction*. For example, you might pull a glass of your favorite beverage gently toward you or push a plate of cafeteria food forcefully away from you. When more than one force acts on an object, what determines how the object moves is the vector *sum* of all of the forces acting on that object, also called the **net external force** on the object (or, for short, the **net force**). If the individual forces acting on an object are $\vec{F}_1$, $\vec{F}_2$, $\vec{F}_3$, and so on, we can write the net force on that object as

The **net external force** acting on an object...

...equals the **vector sum** of all of the individual forces that act on the object.

$$\sum \vec{F}_{\text{ext}} = \vec{F}_1 + \vec{F}_2 + \vec{F}_3 + \ldots$$

Net external force on an object
(4-1)

The sum includes only external forces (forces exerted on the object by other objects).

For example, three forces act on a baseball player as he slides to get safely to base (Figure 4-1). One of these forces is the downward pull of Earth's gravity, called the **gravitational force** (denoted by a lowercase $\vec{w}$). The other two forces are exerted by the surface on which the player slides. The **friction force** on the player (which we

TAKE-HOME MESSAGE FOR Section 4-1

✔ What determines how an object moves is the combined effect of all of the forces (pushes or pulls) that act on the object.

✔ If one object exerts a force on a second object, the second object must also exert a force on the first object.

Figure 4-1 Forces on a baseball player (a) A baseball player sliding to get safely on base. (b) The red vectors indicate the directions of each external force that acts on the player.

(a)

Jamie Roach/Shutterstock

(b)

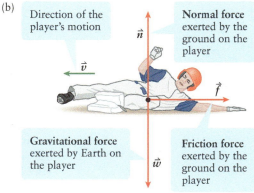

Direction of the player's motion

$\vec{v}$

Normal force exerted by the ground on the player

$\vec{n}$

$\vec{f}$

Gravitational force exerted by Earth on the player

Friction force exerted by the ground on the player

$\vec{w}$

 The only forces that affect the player's motion are *external* forces (forces exerted on the player by other objects). These include Earth's gravitational force (the player's weight) and forces from objects that the player is touching (in this case the normal force and friction force exerted by the ground).

denote by a lowercase $\vec{f}$) acts parallel to the surface and opposite to the player's motion. The friction force arises from chemical bonds that form between atoms on the ground and atoms on the player's uniform. These bonds resist being broken, so the player feels a force slowing him down as he slides. The **normal force** on the player, by contrast, is directed perpendicular to the surface. (In physics and mathematics "normal" is not the opposite of "abnormal"; rather, it's another word for "perpendicular.") Like the friction force, the normal force is a contact force that arises from interactions between atoms: The atoms in the player's uniform resist being squeezed into the atoms on the ground. This force, which we denote by a lowercase $\vec{n}$, prevents the player from falling through the ground below him. If you're sitting in a chair right now, you're feeling an upward normal force exerted by the chair on your rear end; if you're standing, you're feeling an upward normal force exerted on your feet by the ground.

The three forces that act on the baseball player in Figure 4-1 are called **external forces** because they are exerted by other objects outside of the player's body. The gravitational force is exerted by Earth as a whole, while the friction force and normal force are exerted by the ground that's in contact with the player.

Internal forces are also exerted by one part of the player's body on another. One example is the force that the player's shoulder exerts on his left arm to keep it from falling off. While internal forces are important for understanding the interactions of one part of the player's body with another, these forces are *not* included in the sum in Equation 4-1. That's why the left-hand side of that equation has the subscript "ext" for "external." In a moment we'll see why it's appropriate to include only external forces in Equation 4-1.

Newton's Second Law

In the late seventeenth century the English physicist and mathematician Sir Isaac Newton (1642–1727) published a treatise in which he described three fundamental relationships between force and motion. These relationships, called **Newton's laws of motion**, explain nearly all physical phenomena in our everyday experience. One of these relationships, commonly known as **Newton's second law**, is the most important for our discussion of forces. It states:

If a net external force acts on an object, the object accelerates. The net external force is equal to the product of the object's mass and the object's acceleration:

If a net external force acts on an object...

...the object accelerates. The acceleration is in the same direction as the net force.

Newton's second law of motion
(4-2)

$$\sum \vec{F}_{\text{ext}} = m\vec{a}$$

The magnitude of acceleration that the net external force causes depends on the mass m of the object (the quantity of material in the object). The greater the mass, the smaller the acceleration.

Newton's second law is simple to state but can be challenging to fully understand. Here are its three essential features:

(1) *A net force in a certain direction causes acceleration in that direction.* The baseball player shown in Figure 4-1 has a backward acceleration (he slows down) because the net force acting on him points opposite to the direction of his motion (Figure 4-2a). If we ignore air resistance, a falling basketball is acted on by only a single force, the downward gravitational force. So the net force on the ball is downward, and it accelerates downward (Figure 4-2b). This relationship between net force and acceleration is the reason we devoted so much effort in Chapters 2 and 3 to understanding the nature of acceleration.

(2) *The magnitude of acceleration caused by the net force depends on the object's mass.* **Mass** is a measure of how much matter an object has. The SI unit of mass is the **kilogram**, abbreviated kg. A liter of water has a mass of 1 kg. Newton's second law says that the product of an object's mass and the object's acceleration equals the net external force on the object. So if you deliver identically strong kicks to a tennis ball

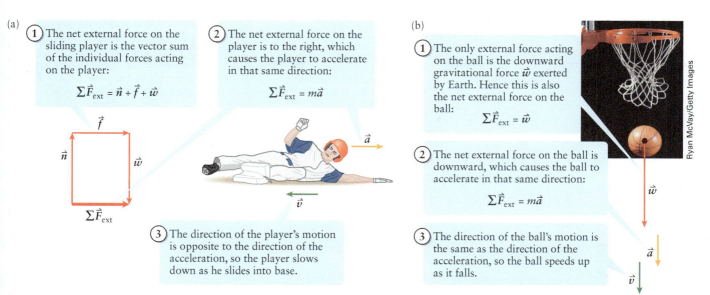

(a)

1 The net external force on the sliding player is the vector sum of the individual forces acting on the player:

$$\sum \vec{F}_{ext} = \vec{n} + \vec{f} + \vec{w}$$

2 The net external force on the player is to the right, which causes the player to accelerate in that same direction:

$$\sum \vec{F}_{ext} = m\vec{a}$$

3 The direction of the player's motion is opposite to the direction of the acceleration, so the player slows down as he slides into base.

(b)

1 The only external force acting on the ball is the downward gravitational force $\vec{w}$ exerted by Earth. Hence this is also the net external force on the ball:

$$\sum \vec{F}_{ext} = \vec{w}$$

2 The net external force on the ball is downward, which causes the ball to accelerate in that same direction:

$$\sum \vec{F}_{ext} = m\vec{a}$$

3 The direction of the ball's motion is the same as the direction of the acceleration, so the ball speeds up as it falls.

Ryan McVay/Getty Images

Figure 4-2 The net external force determines acceleration The net external force on an object is the vector sum $\sum \vec{F}_{ext}$ of the individual forces that act on that object. The acceleration $\vec{a}$ produced by the net external force is in the same direction as $\sum \vec{F}_{ext}$.

(mass $m = 0.058$ kg) and to a soccer ball (mass $m = 0.43$ kg), the more massive soccer ball will experience less acceleration while it's in contact with your foot and will fly off with a slower speed.

(3) *Only external forces acting on an object affect that object's acceleration.* As an example, sit in your chair with your feet off the ground and pull upward on your belt with both hands. No matter how hard you try, you can't lift yourself out of the chair! Your body has to accelerate to rise out of the chair (it has to go from being stationary to being in motion), but the force of your hands on your belt is an *internal* force (one part of your body pulls on another part). Experiment shows that this force can't produce an acceleration. A helpful friend could lift you out of your chair, but that would happen because your friend exerts an *external* force (one that originates outside your body).

To see the significance of mass more clearly, divide both sides of Equation 4-2 by the object's mass m:

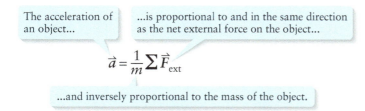

The acceleration of an object...

...is proportional to and in the same direction as the net external force on the object...

$$\vec{a} = \frac{1}{m} \sum \vec{F}_{ext}$$

...and inversely proportional to the mass of the object.

Newton's second law, alternative form (4-3)

Figure 4-3 illustrates the ideas of Equation 4-3.

WATCH OUT! Mass and weight are not the same thing.

! An object's mass is the same no matter where the object is; for example, an astronaut walking on the Moon has the same mass as she has on Earth (the amount of matter in her body is exactly the same on either world).

However, the astronaut has less *weight* on the Moon because gravity is weaker on the Moon than on Earth. We'll discuss the distinction between mass and weight in more detail in Section 4-3.

The SI unit of force is the **newton**, abbreviated N. (An uppercase abbreviation is used for units that bear a person's name.) A net force of one newton applied to an object with a mass of 1 kilogram gives the object an acceleration of 1 m/s². Because $\sum \vec{F}_{ext} = m\vec{a}$ from Equation 4-2, it follows that

$$1 \text{ N} = 1 \text{ kg} \times 1 \text{ m/s}^2 = 1 \text{ kg} \cdot \text{m/s}^2$$

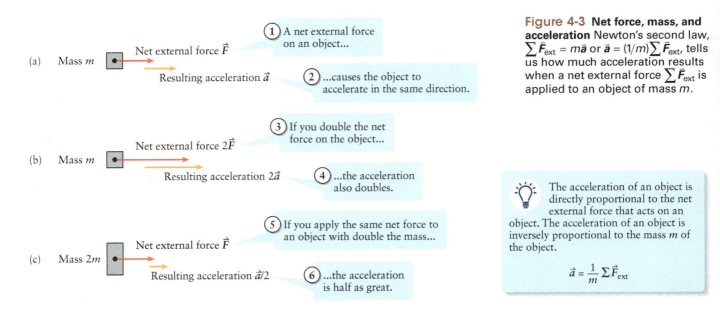

(a) Mass m

Net external force $\vec{F}$

Resulting acceleration $\vec{a}$

① A net external force on an object...

② ...causes the object to accelerate in the same direction.

(b) Mass m

Net external force $2\vec{F}$

Resulting acceleration $2\vec{a}$

③ If you double the net force on the object...

④ ...the acceleration also doubles.

(c) Mass $2m$

Net external force $\vec{F}$

Resulting acceleration $\vec{a}/2$

⑤ If you apply the same net force to an object with double the mass...

⑥ ...the acceleration is half as great.

Figure 4-3 Net force, mass, and acceleration Newton's second law, $\sum \vec{F}_{\text{ext}} = m\vec{a}$ or $\vec{a} = (1/m)\sum \vec{F}_{\text{ext}}$, tells us how much acceleration results when a net external force $\sum \vec{F}_{\text{ext}}$ is applied to an object of mass m.

The acceleration of an object is directly proportional to the net external force that acts on an object. The acceleration of an object is inversely proportional to the mass m of the object.

$$\vec{a} = \frac{1}{m}\sum \vec{F}_{\text{ext}}$$

The English unit of force is the **pound** (abbreviated lb). To three significant figures, 1 lb equals 4.45 N, and 1 N = 0.225 lb. One newton is a bit less than a quarter of a pound.

WATCH OUT! It's the net force that matters.

Newton's second law tells us that a net force is required to make an object accelerate, that is, to *change* its velocity. However, your experience may suggest that a force has to act on an object to make it move even at a *constant* velocity. After all, you might reason, you have to exert a force to make a book slide across a table (Figure 4-4a). In fact there's no contradiction between these two statements! The explanation is that the *net* force on the sliding book is zero. The forward force that you exert on the book just balances the backward force of friction that the table exerts on the book, just as the upward normal force exerted on the book by the table just balances the downward gravitational force. Thus the vector sum of *all* forces on the book is zero or $\sum \vec{F}_{\text{ext}} = 0$. From Newton's second law this means that $\vec{a} = 0$, so the book moves over the table with a constant velocity (Figure 4-4a). If you push harder on the book, the force that you exert is greater in magnitude than the friction force and so the net force $\sum \vec{F}_{\text{ext}}$ points forward; then the book has a forward acceleration and speeds up (Figure 4-4b). If you let go of the sliding book, the net force $\sum \vec{F}_{\text{ext}}$ points backward, so the book has a backward acceleration and slows to a stop (Figure 4-4c). Just remember that it's the *net* force on an object that determines its acceleration, not any one particular force.

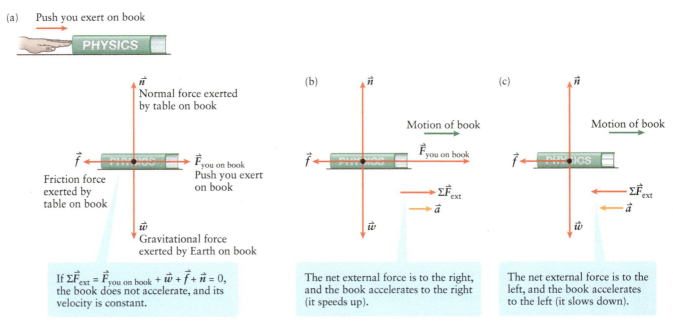

(a) Push you exert on book

$\vec{n}$ Normal force exerted by table on book

$\vec{f}$ Friction force exerted by table on book

$\vec{F}_{\text{you on book}}$ Push you exert on book

$\vec{w}$ Gravitational force exerted by Earth on book

If $\sum \vec{F}_{\text{ext}} = \vec{F}_{\text{you on book}} + \vec{w} + \vec{f} + \vec{n} = 0$, the book does not accelerate, and its velocity is constant.

(b) $\vec{n}$

Motion of book

$\vec{f}$ $\vec{F}_{\text{you on book}}$

$\sum \vec{F}_{\text{ext}}$ $\vec{a}$

$\vec{w}$

The net external force is to the right, and the book accelerates to the right (it speeds up).

(c) $\vec{n}$

Motion of book

$\vec{f}$

$\sum \vec{F}_{\text{ext}}$ $\vec{a}$

$\vec{w}$

The net external force is to the left, and the book accelerates to the left (it slows down).

Figure 4-4 It's the net external force that matters (a) If you consider only the force that *you* apply to this book, you might think that a force is required to maintain constant velocity. If instead you consider *all* of the forces that act on the book, you'll see that a zero *net* external force keeps the book's velocity constant. If the net external force on the book is not zero, the book accelerates in the direction of the net force.

The sum $\sum \vec{F}_{ext}$ in Newton's second law involves vector addition. We saw in Chapter 3 that it's usually easiest to add vectors if we use components, so we'll often use Equation 4-2 in component form:

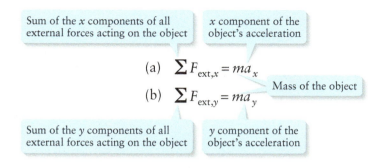

Sum of the x components of all external forces acting on the object

x component of the object's acceleration

(a) $\sum F_{ext,x} = ma_x$

Mass of the object

(b) $\sum F_{ext,y} = ma_y$

Sum of the y components of all external forces acting on the object

y component of the object's acceleration

Newton's second law in component form
(4-4)

The two examples below illustrate how to use Newton's second law to relate net force, mass, and acceleration.

BioMedical EXAMPLE 4-1 Small but Forceful

Microtubules are assembled from protein molecules (Figure 4-5). Microtubules help cells maintain their shape and are responsible for various kinds of movements within cells, such as pulling apart chromosomes during cell division. Measurements show that microtubules can exert forces from a few pN (1 pN = 1 piconewton = 10^{-12} N) up to hundreds of nN (1 nN = 1 nanonewton = 10^{-9} N). A particular bacterial chromosome has a mass of 2.00×10^{-17} kg. If a microtubule applies a force of 1.00 pN to the chromosome, what is the magnitude of the chromosome's acceleration? (Ignore any other forces that might act on the chromosome.)

Figure 4-5 Microtubule Microtubules (shown here in green within a fertilized sea urchin egg undergoing division) are protein molecules found within cells. They help cells hold their shape and are responsible for various kinds of movements within cells.

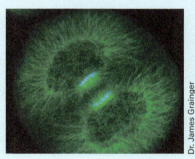

Dr. James Grainger

Set Up

Newton's second law tells us that the acceleration $\vec{a}_{chromosome}$ of the chromosome is determined by the net force acting on the chromosome and the chromosome's mass. We will use this law to determine the magnitude of $\vec{a}_{chromosome}$.

Newton's second law of motion:

$$\sum \vec{F}_{ext\ on\ chromosome} = m_{chromosome} \vec{a}_{chromosome} \quad (4.2)$$

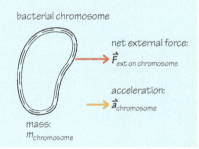

Solve

Take the magnitude of both sides of Equation 4-2. (Note that mass is always positive.) Then solve for the magnitude $a_{chromosome}$ of the acceleration of the chromosome.

Magnitude of both sides of Equation 4-2:

$$\left|\sum \vec{F}_{ext\ on\ chromosome}\right| = m_{chromosome}\left|\vec{a}_{chromosome}\right| \text{ so}$$

$$\left|\sum \vec{F}_{ext\ on\ chromosome}\right| = m_{chromosome} a_{chromosome}$$

$$a_{chromosome} = \frac{\left|\sum \vec{F}_{ext\ on\ chromosome}\right|}{m_{chromosome}}$$

The net force on the chromosome is just the force exerted on it by the microtubule, of magnitude 1.00 pN. Use this value and $m_{chromosome} = 2.00 \times 10^{-17}$ kg to solve for the magnitude $a_{chromosome}$.

$$a_{chromosome} = \frac{1.00 \text{ pN}}{2.00 \times 10^{-17} \text{ kg}} = \frac{1.00 \times 10^{-12} \text{ N}}{2.00 \times 10^{-17} \text{ kg}}$$

$$= 5.00 \times 10^4 \text{ N/kg}$$

Use the definition $1 \text{ N} = 1 \text{ kg} \cdot \text{m/s}^2$:

$$a_{\text{chromosome}} = \left(5.00 \times 10^4 \text{ N/kg}\right) \left(\frac{1 \text{ kg} \cdot \text{m/s}^2}{1 \text{ N}}\right)$$

$$= 5.00 \times 10^4 \text{ m/s}^2$$

Reflect

This acceleration is about 5000 times g, the acceleration due to gravity! It is important to note, however, that the force exerted by the microtubule is opposed by resistive forces that we chose to neglect. The acceleration would not be quite this big in reality. Note also that the forces exerted by microtubules act for only very short periods of time.

EXAMPLE 4-2   A Three-Person Tug-of-War

Jesse, Karim, and Luis, three hungry fraternity brothers, are each pulling on a cafeteria tray laden with desserts. The mass of the tray and its contents is 2.50 kg. The tray sits on a horizontal dining table, and each of the fraternity brothers pulls horizontally on the tray. Jesse pulls due south with a force of magnitude 85.0 N, and Karim pulls with a force of magnitude 90.0 N in a direction 35.0° north of east. If the tray accelerates due north at 20.0 m/s², how hard and in what direction does Luis pull on the tray? Ignore the effects of friction.

Set Up

The tray accelerates due to the net force that acts on it, which is the vector sum of the individual forces exerted by Jesse, Karim, and Luis. We've drawn these individual force vectors with their tails together. (We're told to ignore friction, so no other horizontal forces act on the tray. The tray doesn't accelerate vertically, so there's no net vertical force—the upward normal force exerted by the tabletop balances the downward gravitational force.) Because we know the mass and acceleration of the tray, we can use Newton's second law to calculate the net force on the tray. We'll then use vector addition and subtraction to determine the unknown force exerted by Luis.

$$\sum \vec{F}_{\text{ext on tray}} = m_{\text{tray}} \vec{a}_{\text{tray}} \qquad (4\text{-}2)$$

In component form:

$$\sum F_{\text{ext on tray},x} = m_{\text{tray}} a_{\text{tray},x} \qquad (4\text{-}4a)$$

$$\sum F_{\text{ext on tray},y} = m a_{\text{tray},y} \qquad (4\text{-}4b)$$

Net force on tray = sum of forces exerted by Jesse, Karim, and Luis:

$$\sum \vec{F}_{\text{ext on tray}}$$

$$= \vec{F}_{\text{Jesse on tray}} + \vec{F}_{\text{Karim on tray}} + \vec{F}_{\text{Luis on tray}}$$

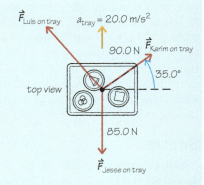

Solve

Vector arithmetic is easiest if we use vector components. The forces the three brothers exert are all in the horizontal plane, so we choose x and y axes that lie in this plane with $+x$ pointing east and $+y$ pointing north. We've drawn all three force vectors with their tails at the origin. The tray's acceleration vector $\vec{a}_{\text{tray}}$ points in the positive y direction.

$$a_{\text{tray},x} = 0$$

$$a_{\text{tray},y} = 20.0 \text{ m/s}^2$$

Use Newton's second law in component form, Equations 4-4, to calculate the x and y components of the net force on the tray. Recall that $1 \text{ kg} \cdot \text{m/s}^2 = 1 \text{ N}$.

$$\sum F_{\text{ext on tray},x} = m_{\text{tray}} a_{\text{tray},x} = m_{\text{tray}}(0) = 0$$

$$\sum F_{\text{ext on tray},y} = m_{\text{tray}} a_{\text{tray},y} = (2.50 \text{ kg})(20.0 \text{ m/s}^2)$$

$$= 50.0 \text{ kg} \cdot \text{m/s}^2 = 50.0 \text{ N}$$

The x component of the net force on the tray is the sum of the x components of the individual forces on the tray, and similarly for the y component.

$$\sum F_{\text{ext on tray},x} = 0$$
$$= F_{\text{Jesse on tray},x} + F_{\text{Karim on tray},x} + F_{\text{Luis on tray},x}$$
$$\sum F_{\text{ext on tray},y} = 50.0 \text{ N}$$
$$= F_{\text{Jesse on tray},y} + F_{\text{Karim on tray},y} + F_{\text{Luis on tray},y}$$

Solve for the components of the force that Luis exerts on the tray:

$$F_{\text{Luis on tray},x} = -F_{\text{Jesse on tray},x} - F_{\text{Karim on tray},x}$$
$$F_{\text{Luis on tray},y} = 50.0 \text{ N} - F_{\text{Jesse on tray},y} - F_{\text{Karim on tray},y}$$

To proceed we need to know the x and y components of the forces that Jesse and Karim exert on the tray.

$\vec{F}_{\text{Jesse on tray}}$:

magnitude 85.0 N, points due south (in the negative y direction), so

$$F_{\text{Jesse on tray},x} = 0$$
$$F_{\text{Jesse on tray},y} = -85.0 \text{ N}$$

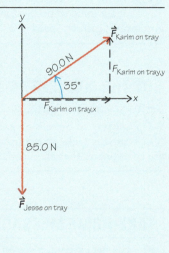

$\vec{F}_{\text{Karim on tray}}$:

magnitude 90.0 N, points 35.0° north of east, so $F_{\text{Karim on tray},x}$ and $F_{\text{Karim on tray},y}$ are both positive.

$$F_{\text{Karim on tray},x} = (90.0 \text{ N}) \cos 35.0°$$
$$= 73.7 \text{ N}$$
$$F_{\text{Karim on tray},y} = (90.0 \text{ N}) \sin 35.0°$$
$$= 51.6 \text{ N}$$

Substitute these values into the expressions for the components $F_{\text{Luis on tray},x}$ and $F_{\text{Luis on tray},y}$ and solve.

$$F_{\text{Luis on tray},x} = -(0) - (73.7 \text{ N})$$
$$= -73.7 \text{ N}$$
$$F_{\text{Luis on tray},y} = 50.0 \text{ N} - (-85.0 \text{ N}) - (51.6 \text{ N})$$
$$= 83.4 \text{ N}$$

The force that Luis exerts, $\vec{F}_{\text{Luis on tray}}$, has a negative x component and positive y component, so it points west of north. Calculate the magnitude of $\vec{F}_{\text{Luis on tray}}$ using the Pythagorean theorem and the direction of $\vec{F}_{\text{Luis on tray}}$ using trigonometry.

Magnitude of $\vec{F}_{\text{Luis on tray}}$:

$$F_{\text{Luis on tray}} = \sqrt{\left(F_{\text{Luis on tray},x}\right)^2 + \left(F_{\text{Luis on tray},y}\right)^2}$$
$$= \sqrt{(-73.7 \text{ N})^2 + (83.4 \text{ N})^2}$$
$$= 111 \text{ N}$$

Angle θ of $\vec{F}_{\text{Luis on tray}}$ measured west of north:

$$\tan \theta = \frac{73.7 \text{ N}}{83.4 \text{ N}} = 0.884$$
$$\theta = \tan^{-1} 0.884 = 41.5°$$

Reflect

Luis pulls on the tray with a force of 111 N at an angle of 41.5° west of north. In terms of components, Luis pulls west (in the negative x direction) with a force of 73.7 N, thus canceling Karim's 73.7-N pull to the east. Luis also pulls north (in the positive y direction) with a force of 83.4 N; combined with Karim's 51.6-N northward pull, Luis's pull overwhelms Jesse's 85.0-N pull toward the south. Hence the net force on the tray and its delicious contents is northward, and its acceleration is northward as well.

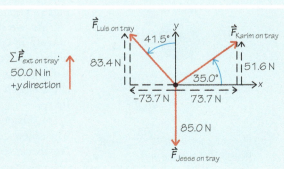

Example 4-2 illustrates an important point: *It's always wise to choose coordinate axes that align with one or more of the vectors.* In this example we chose the axes to be east–west and north–south so that the tray's acceleration $\vec{a}_{\text{tray}}$ and the force $\vec{F}_{\text{Jesse on tray}}$ exerted by Jesse were both along one of the coordinate axes. This choice made it easy to express these vectors in terms of components and so simplified the calculation. Using a different choice of coordinate axes (say, one in which the positive x axis pointed 30° north of west and the positive y axis pointed 30° east of north), you would have ended up with the same result, but with a good deal more effort.

GOT THE CONCEPT? 4-1 Newton's Second Law

(?) Rank the following objects in order of the magnitude of the net force that acts on the object, from greatest magnitude to smallest magnitude. (a) A 1250-kg automobile gaining speed at 2.00 m/s²; (b) a 200,000-kg airliner flying in a straight line at a constant 280 m/s; (c) a 4500-kg truck slowing down at 0.600 m/s²; (d) an 80.0-kg hiker climbing up a 4.00° slope at a constant 1.20 m/s.

GOT THE CONCEPT? 4-2 Measuring Mass in Zero Gravity

(?) Imagine you are aboard a spacecraft far from any planet or star, so there is no gravitational force on you or anything else in the spacecraft. Floating in front of you are two spheres that are identical in size and appearance. One is made of lead, and the other is made of plastic; the lead sphere is considerably more massive than the plastic one. Can you devise a simple experiment to determine which sphere is lead and which is plastic?

TAKE-HOME MESSAGE FOR Section 4-2

✔ The net external force on an object is the vector sum of all of the individual forces that act on it from other objects.

✔ Newton's second law states that an object accelerates if the net external force on the object is not zero.

✔ The acceleration is in the same direction as the net force, is directly proportional to the magnitude of the net force, and is inversely proportional to the mass of the object (the quantity of material in the object).

---

① If an object falls without air resistance, the net external force on the object equals the downward gravitational force on that object, which has magnitude w.

② An object falling without air resistance accelerates downward. The magnitude of the acceleration is g.

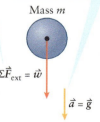

Mass m

$\Sigma \vec{F}_{ext} = \vec{w}$

$\vec{a} = \vec{g}$

③ Newton's second law, $\Sigma \vec{F}_{ext} = m\vec{a}$, tells us that $\vec{w} = m\vec{g}$ and so $w = mg$.

Figure 4-6 Gravitational force
Applying Newton's second law to a freely falling object shows that the magnitude of the gravitational force is $w = mg$.

4-3 Mass, weight, and inertia are distinct but related concepts

How much do you weigh? If you grew up in the United States, chances are your answer will be in pounds. If not, you'll probably use kilograms. We've already declared that the SI units of mass are kilograms. Does that mean that pounds and kilograms are both units of the same quantity? Are weight and mass the same?

The answer to both questions is "no!" Although people often use the terms "mass" and "weight" interchangeably in everyday conversation, they have very different meanings in science.

As we discussed in Section 4-2, the mass of an object describes how much matter is contained in the object. By contrast the **weight** of that object is the magnitude of the gravitational force that acts on the object. Because it's a force, weight is measured in newtons (*not* kilograms!) in the SI system and pounds in the English system. We'll use the symbols w for weight and $\vec{w}$ for gravitational force.

Consider an object that has mass m and is acted on by only the gravitational force. The object could be a ball falling in a vacuum, so there is no air resistance (Figure 4-6). Newton's second law, Equation 4-2, tells us that the gravitational force $\vec{w}$ equals the object's mass multiplied by its acceleration $\vec{a}$:

$$\vec{w} = m\vec{a}$$

The magnitude of the gravitational force is the weight w, so

$$w = ma$$

where a is the magnitude of the object's acceleration. We know from Section 2-6 that if only gravity acts on an object, the object's acceleration has magnitude g. So we get the following expression for the weight of an object of mass m:

Weight of an object (equal to the magnitude of the gravitational force on that object)

$$w = mg$$

Mass of the object   Magnitude of the **acceleration due to gravity**

Weight of an object of mass m
(4-5)

That is, *the weight of an object is equal to the object's mass multiplied by g, the acceleration due to gravity*. This statement makes it clear that the units of weight cannot be the same as the units of mass. In Chapters 2 and 3 we used an average value of $g = 9.80$ m/s^2 but recognized that the exact value of g varies slightly from place to place on Earth. Therefore, the weight of an object must change as it is moved from place to place. *Hence weight is not an intrinsic property of an object.* Not only does the weight of an object change depending on where you are on Earth, but as we will see in Chapter 7, it can be significantly different on other worlds such as the Moon and Mars.

WATCH OUT! Objects have weight even if they are not accelerating.

! We arrived at Equation 4-5 for the weight of an object by assuming that the object was falling freely. However, Equation 4-5 is true even if the object is not accelerating.

A 10.0-kg object near Earth's surface has a weight of 98.0 N whether it's sitting on a table, falling off the table, or being pulled across the floor.

GOT THE CONCEPT? 4-3 Most Massive to Least Massive

? Rank the following objects according to their mass, from largest to smallest. The acceleration due to gravity at the equator of Mars is 3.69 m/s^2. (a) A rock on the Martian equator that weighs 10.0 N; (b) a rock at Earth's

equator that weighs 10.0 N; (c) a rock on the Martian equator that has a mass of 10.0 kg; (d) a rock at Earth's equator that has a mass of 10.0 kg.

Consider dropping two different objects side by side in a vacuum so that the only force acting on each object is the gravitational force (Figure 4-7). The heavier object weighs twice as much as the other, so it is subjected to twice as much gravitational force and hence twice as much net force as the lighter object.

So you might expect that the heavier object would have a greater downward acceleration $\vec{a}$. But the heavier object also has a greater mass, and Equation 4-5 tells us that weight and mass are directly proportional. If the heavier object has twice the weight of the lighter one, it also has twice the mass. An object's acceleration is equal to the net force that acts on it divided by its mass (recall the alternative form of Newton's second law, $\vec{a} = \dfrac{1}{m}\sum \vec{F}_{ext}$, from Equation 4-3). Hence the quotient $\dfrac{1}{m}\sum \vec{F}_{ext}$ has the *same* value for both objects, and they both fall with the same acceleration in a vacuum.

Can you lift an object that weighs 10 N? A 1000-N object? In Section 4-2, we mentioned the following conversions, which are valid to three significant figures:

$$1\text{ N} = 0.225\text{ lb}$$
$$1\text{ lb} = 4.45\text{ N}$$

(You can find more precise conversion factors in Appendix A.) According to Equation 4-5, the mass of an object that weighs 10.0 N is $m = w/g = (10.0\text{ N})/(9.80\text{ m/s}^2) = 1.02$ kg. A 1-L bottle of water has a mass of 1 kg; that is, it weighs about 10 N. In

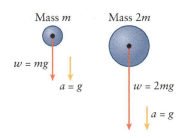

The gravitational force on an object is proportional to its mass. The more massive object experiences a greater gravitational force than the less massive object but has the same acceleration.

Figure 4-7 **Greater mass means greater gravitational force** Both of these objects are falling in a vacuum, so only the gravitational force acts on each object.

pounds, that's 10 multiplied by 0.225 lb or about 2.2 lb. You'd have no problem lifting that amount. Many people can lift an object that weighs 1000 N (225 lb); that's a mass of about 100 kg. Remembering that 1 liter of water has a mass of 1 kg and a weight of 10 N can help you get an intuitive feel for the numerical values of masses and forces.

Mass, Inertia, and Newton's First Law

The concept of mass is intertwined with the observation that all objects *resist changes* in their state of motion. We use the term **inertia** for the tendency of an object to resist a change in motion. For example, a *stationary* object (such as a rock lying on the ground or a roommate sleeping on the couch) tends to remain stationary. An object *in motion*, like a fast-moving hockey puck sliding on the ice, tends to keep moving in the same direction with the same speed.

WATCH OUT! Don't confuse mass, weight, and inertia.

! Note the differences among mass, weight, and inertia. Mass is the *quantity of material* in an object; no matter where in the universe you take the object, its mass remains the same. Weight is the *magnitude of the gravitational force* on an object; it's proportional to the mass but depends on the value of g at the object's location. Inertia is the *tendency* of *all* objects to maintain the same motion. Unlike mass or weight, we do not give numerical values to the inertia of an object.

Colette Sutton

Figure 4-8 An object at rest Kjeragbolten is a boulder suspended above an abyss in Norway that is 984 m (3228 ft) deep. It remains at rest because the vector sum of the external forces acting on it — the downward force exerted on it by Earth's gravity, the additional downward force exerted on it by a brave tourist, and the upward forces exerted on it by the surrounding rock — is zero.

Newton described inertia as an intrinsic, unchanging property of mass and of objects. This property is defined in **Newton's first law**:

An object at rest tends to stay at rest, and an object in uniform motion tends to stay in motion with the same speed and in the same direction unless acted upon by a net force.

Newton's first law contradicts the older theories of motion developed by the ancient Greeks, principally by the philosopher Aristotle (384 B.C.–322 B.C.). Aristotle believed that every object had a natural place in the world. For example, heavy objects, such as rocks, are naturally at rest on Earth, and light objects, such as clouds, are naturally at rest in the sky. In Aristotle's view, a moving object tended to stop when it found its natural rest position. To keep an object moving, reasoned Aristotle, a force had to be applied to it; to make it move faster, a greater force was required. Newton saw more deeply. He identified the concept of the *net* force on an object and realized that the net force on an object determines its acceleration rather than its speed. (This realization is Newton's second law.) If the net force on an object is zero, the object has zero acceleration and moves with a constant velocity—that is, with the same speed and in the same direction. (You should review the discussion of Figure 4-4 in Section 4-2.)

At this point you can think of Newton's first law as a special case of the *second* law that applies when the net force on an object is zero. We'll nonetheless call these laws "first" and "second" in the same manner in which Newton numbered them. In equation form we can write the first law as follows:

Newton's first law of motion (4-6)

If the net external force on an object is zero...   ...the object does not accelerate...

$$\text{If } \sum \vec{F}_{\text{ext}} = 0, \text{ then } \vec{a} = 0 \text{ and } \vec{v} = \text{constant}$$

...and the velocity of the object remains constant. If the object is at rest, it remains at rest; if it is in motion, it continues in motion in a straight line at a constant speed.

We say that an object is in **equilibrium** if the net external force on it is zero (Figure 4-8). A chandelier hanging from the ceiling is in equilibrium: The chain from which the chandelier is suspended exerts an upward force that exactly balances the downward gravitational force on the chandelier. The chandelier's velocity is zero and remains zero. But a moving airliner is also in equilibrium if it flies in a straight line at a constant speed. The forward thrust provided by the airliner's engines balances the backward drag force that the air exerts on the airliner, and the upward lift provided by air flowing around the wings balances the downward gravitational force. The airliner has a nonzero velocity, but its *acceleration* is zero and so the airliner is in equilibrium.

The following example illustrates how to use Newton's first law to analyze an object at rest.

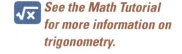

See the Math Tutorial for more information on trigonometry.

EXAMPLE 4-3   Let Sleeping Cats Lie

A 40.0-N cat is asleep on a ramp that is tilted by an angle of 15.0° from the horizontal. Three forces act on the cat: the downward gravitational force, a normal force perpendicular to the ramp, and a friction force directed uphill parallel to the ramp. The cat remains at rest while sleeping. Determine the magnitude of each force that acts on the cat.

Set Up

As in Example 4-2, we've drawn all of the external force vectors on the cat with their tails touching. The cat's weight is $w = 40.0$ N; this is just the magnitude of the gravitational force on the cat ($\vec{w}_{cat}$). We know the directions of the normal force $\vec{n}$ (perpendicular to the ramp) and the friction force $\vec{f}$ (uphill parallel to the ramp). Our task is to find the magnitudes of $\vec{n}$ and $\vec{f}$. We'll do this using Newton's first law: Because the cat remains at rest, the sum of $\vec{w}_{cat}$, $\vec{n}$, and $\vec{f}$ must be zero. So we can solve this problem by using vector addition.

Newton's first law of motion:

$$\sum \vec{F}_{\text{ext on cat}} = \vec{w}_{cat} + \vec{n} + \vec{f} = 0 \tag{4-6}$$

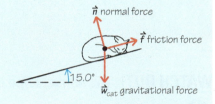

Solve

Like most vector addition problems, this one is most easily solved using components. If $\sum \vec{F}_{\text{ext on cat}} = 0$, then each of the components of $\sum \vec{F}_{\text{ext on cat}}$ must also be equal to zero. It's convenient to choose the x axis to be along the tilted ramp and the y axis to be perpendicular to the ramp. (The x and y axes must be perpendicular to each other but don't have to be horizontal or vertical.) Using this choice, two of the vectors, $\vec{n}$ and $\vec{f}$, lie either directly along or directly opposite to one of the axes.

Newton's first law in component form:

$$\sum F_{\text{ext on cat},x} = w_{cat,x} + n_x + f_x = 0$$
$$\sum F_{\text{ext on cat},y} = w_{cat,y} + n_y + f_y = 0$$

Write the x and y components of each of the external force vectors in terms of their magnitudes w, n, and f and the angle $\theta = 15.0°$ of the ramp. Note that $\vec{w}_{cat}$ has a positive x component (down the ramp) and a negative y component (into the ramp), $\vec{n}$ has only a positive y component (perpendicular to the ramp), and $\vec{f}$ has only a negative x component (up the ramp).

Components of gravitational force $\vec{w}_{cat}$:

$$w_{cat,x} = +w_{cat} \sin \theta$$
$$w_{cat,y} = -w_{cat} \cos \theta$$

Components of normal force $\vec{n}$:

$$n_x = 0$$
$$n_y = +n$$

Components of $\vec{f}$:

$$f_x = -f$$
$$f_y = 0$$

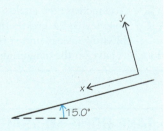

Substitute the expressions for $w_{cat,x}$, $w_{cat,y}$, n_x, n_y, f_x, and f_y into Newton's first law and solve for the magnitudes n and f.

Substitute into Newton's first law in component form:

$$\sum F_{ext\ on\ cat,x} = w_{cat} \sin\theta + 0 + (-f) = 0$$
$$\sum F_{ext\ on\ cat,y} = -w_{cat} \cos\theta + n + 0 = 0$$

From the y equation:

$$n = w_{cat} \cos\theta = (40.0\ N)\cos 15.0° = 38.6\ N$$

From the x equation:

$$f = w_{cat} \sin\theta = (40.0\ N)\sin 15.0° = 10.4\ N$$

Reflect

The downward gravitational force has magnitude 40.0 N, the normal force acting perpendicular to the ramp has magnitude 38.6 N, and the friction force acting uphill has magnitude 10.4 N. Note that no single force "balances" any of the other forces: All three forces are needed to mutually balance each other and keep the cat at rest.

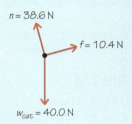

$n = 38.6\ N$

$f = 10.4\ N$

$w_{cat} = 40.0\ N$

WATCH OUT! The normal force isn't always equal to the weight.

! It's a common mistake to assume that the normal force on an object (in this case, the sleeping cat) has the same magnitude as the object's weight. That's clearly *not* true in this case: The normal-force magnitude $n = 38.6$ N is *less* than the cat's 40.0-N weight. As a general rule it's *always* safest to treat the magnitude of the normal force as an unknown, as we have done in this example.

GOT THE CONCEPT? 4-4 Tilt the Cat, but Don't Wake It Up

? You increase the angle θ of the ramp in Example 4-3 to a new value greater than 15.0°. The cat remains blissfully asleep and at rest at the new angle. What effect does this have on the magnitude n of the normal force and the magnitude f of the friction force? (a) n and f both increase. (b) n and f both decrease. (c) n increases and f decreases. (d) n decreases and f increases. (e) None of these. (f) The answer depends on the new value of θ.

TAKE-HOME MESSAGE FOR Section 4-3

✔ The weight of an object is the magnitude of the gravitational force that acts on it.

✔ Newton's first law states that if the external forces on an object sum to zero, the object does not accelerate. If the object is stationary, it remains at rest, and if it is in motion, it continues to move in a straight line with constant velocity.

4-4 Making a free-body diagram is essential in solving any problem involving forces

You need the right key to open a locked door. And you need the proper map to navigate in a foreign city. The key that unlocks physics problems involving forces and the map that allows you to navigate these problems is the *free-body diagram*.

A **free-body diagram** is a graphical representation of all of the external forces acting on an object. It's useful because Newton's second law and Newton's first law both involve the sum of all external forces on an object, $\sum \vec{F}_{ext}$. The term "free body" means that we draw only the object on which the forces act, *not* the other objects that exert those forces. We drew free-body diagrams for the tray in Example 4-2 and the cat in Example 4-3 (see the Set Up step in each example). As another example, Figure 4-9a shows a block on a horizontal table. A wind is blowing from left to right, and the wind pushes on the block and makes it slide. Figure 4-9b lists the steps involved in constructing this block's free-body diagram.

(a) A block on a horizontal table.

(b) Drawing the free-body diagram for the block.

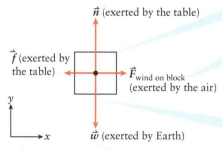

$\vec{n}$ (exerted by the table)

$\vec{f}$ (exerted by the table)

$\vec{F}_{\text{wind on block}}$ (exerted by the air)

$\vec{w}$ (exerted by Earth)

1. Sketch the object on which the forces act.

2. Identify what other objects exert forces on this object. This includes Earth (which exerts a gravitational force) and anything that touches the object.

3. For each force that acts on the object, draw the force vector with its tail at the center of the object. (It helps to draw a dot at the object's center.) Make sure each vector points in the correct direction. Do not include forces that the object exerts on the other objects.

4. Label each force with its symbol. Be certain that you can identify what other object exerts each force. If you're not sure what exerts a force, it's probably not a real force!

5. Choose, draw, and label the directions of the positive x and y axes. Choose the axes so that as many force vectors as possible lie along one of the axes.

Figure 4-9 Constructing a free-body diagram (a) The wind pushes to the right on this block, which slides on a horizontal table. (b) How to create a free-body diagram that depicts all of the external forces that act on the block. (Note that we followed these same steps in Example 4-2 and Example 4-3.)

Often we'll label a force to indicate what object is exerting the force and what object the force acts on. For example, if a girl pushes on a book, we will label that force $\vec{F}_{\text{girl on book}}$. (We used a label of this kind in Figure 4-9b for the force of the wind blowing on the block.) However, we'll label some commonly encountered forces in a different way. We'll use $\vec{w}$ for the gravitational force that Earth exerts on an object. If the object is in contact with a surface, we'll use $\vec{f}$ for the friction force on the object that acts parallel to the surface and $\vec{n}$ for the normal force on the object that acts perpendicular to the surface (see Figure 4-9b). The friction force and normal force are **contact forces**: They arise only when two objects are in contact with each other. By contrast the gravitational force is *not* a contact force: When a basketball is in midair, the gravitational force pulls it downward even though the ball isn't in contact with the ground.

Examining Free-Body Diagrams

Let's look at some examples of free-body diagrams and what we can learn from them. In Figure 4-10a, a stationary box rests on a table with a horizontal top. The free-body diagram in Figure 4-10b shows the two forces that act on the box: the downward gravitational force $\vec{w}$ and the upward normal force $\vec{n}$ exerted by the table.

We choose the y axis to be vertical; then both forces lie along this axis. The sum of $\vec{w}$ and $\vec{n}$ must be zero for the box to remain at rest, so the free-body diagram tells us that $\vec{w}$ and $\vec{n}$ must be opposite in direction and have the same magnitude. Note that the *box* exerts a downward force on the surface of the *table*, but we don't include that force in the free-body diagram for the box. A diagram for the box includes only those forces that act *on* the box.

In Figure 4-11a a box is placed on an inclined ramp with a rough surface. There is friction between the box and the ramp, so the free-body diagram in Figure 4-11b includes a friction force $\vec{f}$ as well as the gravitational force $\vec{w}$ and the normal force $\vec{n}$. Because the ramp is tilted by an angle θ from the horizontal, the normal force (which is perpendicular to the ramp) is tilted from the vertical by the same angle θ. We've chosen the x axis to be parallel to the ramp and the y axis to be perpendicular to the ramp; then $\vec{f}$ has only an x component, and $\vec{n}$ has only a y component.

(a)

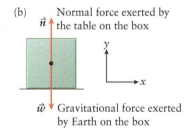

(b)

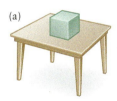

$\vec{n}$ Normal force exerted by the table on the box

$\vec{w}$ Gravitational force exerted by Earth on the box

💡 The box exerts a downward force on the surface of the table, but we don't include this force in the free-body diagram of the box.

Figure 4-10 A free-body diagram: A stationary box atop a table (a) A box rests on a horizontal tabletop. (b) The free-body diagram for the box.

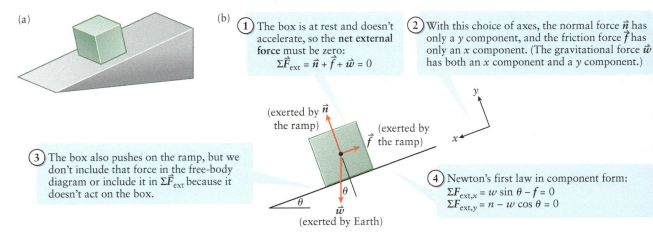

(a)

(b)

① The box is at rest and doesn't accelerate, so the **net external force** must be zero:
$\Sigma \vec{F}_{\text{ext}} = \vec{n} + \vec{f} + \vec{w} = 0$

② With this choice of axes, the normal force $\vec{n}$ has only a y component, and the friction force $\vec{f}$ has only an x component. (The gravitational force $\vec{w}$ has both an x component and a y component.)

③ The box also pushes on the ramp, but we don't include that force in the free-body diagram or include it in $\Sigma \vec{F}_{\text{ext}}$ because it doesn't act on the box.

(exerted by $\vec{n}$ the ramp)

(exerted by $\vec{f}$ the ramp)

y

x

θ

θ

$\vec{w}$
(exerted by Earth)

④ Newton's first law in component form:
$\Sigma F_{\text{ext},x} = w \sin \theta - f = 0$
$\Sigma F_{\text{ext},y} = n - w \cos \theta = 0$

Figure 4-11 A free-body diagram: A stationary box on a ramp (a) Friction keeps the box from sliding down the ramp. (b) The free-body diagram for the box.

WATCH OUT! Draw the force vectors in a free-body diagram with their tails touching.

⚠ In Figures 4-9, 4-10, 4-11, and 4-12, we've drawn the individual force vectors that act on an object so that their tails all touch within the object. (We also did this in Examples 4-2 and 4-3.) This makes it easier to calculate the x and y components of these vectors.

Figure 4-12 also includes vectors for the net external force $\sum \vec{F}_{\text{ext}}$ and the acceleration $\vec{a}$. If you add vectors like this to your free-body diagram, draw them *alongside* (not touching) the object, as in Figure 4-12. In your diagram only the individual forces that act on the object should be touching it.

By assumption the block remains at rest, so the net external force $\sum \vec{F}_{\text{ext}}$ on it— that is, the vector sum of $\vec{w}$, $\vec{n}$, and $\vec{f}$ —must be zero. So the friction force $\vec{f}$ must point in the negative x direction to cancel the (positive) x component of the gravitational force $\vec{w}$; the normal force $\vec{n}$ points in the positive y direction and must cancel the (negative) y component of $\vec{w}$. (This is the same situation as in Example 4-3 in Section 4-3.) Note that the x and y components of the gravitational force $\vec{w}$ are less than the magnitude of the force, which is the box's weight w. The free-body diagram therefore tells us something useful: The magnitudes of the friction force and normal force are both *less* than the weight of the box. Compare this situation to the box on a horizontal surface shown in Figure 4-9, for which the magnitude of the normal force is just equal to the weight.

What happens if we grease the surface of the ramp so that there is no more friction? In this case the friction force $\vec{f}$ goes to zero, and the free-body diagram is as shown in Figure 4-12. Now the vector sum $\sum \vec{F}_{\text{ext}}$ of the external forces on the box cannot be zero: For the normal force and gravitational force to cancel each other, they would have to point in opposite directions, which the free-body diagram in Figure 4-12 shows they do not. Instead the net external force points down the ramp in the positive x direction. According to Newton's second law, $\sum \vec{F}_{\text{ext}} = m\vec{a}$ (Equation 4-2), the box must therefore have a nonzero acceleration $\vec{a}$ down the ramp. We learn this simply by examining the free-body diagram for the box.

Neither the normal force nor the gravitational force in Figure 4-12 depends on how the box is moving. Whether the box is sliding up the ramp, momentarily at rest, or sliding down the ramp, these forces are the same. This means the acceleration is also the same in each of these cases. If the box is moving up the ramp, it slows

① With the friction force absent, the remaining normal force $\vec{n}$ and gravitational force $\vec{w}$ do not cancel. So the net external force $\Sigma \vec{F}_{\text{ext}} = \vec{n} + \vec{w}$ is not zero.

② The box neither leaps off the ramp nor sinks down into it, so it doesn't accelerate in the y direction. So the normal force must cancel the (negative) y component of the gravitational force.

Figure 4-12 A free-body diagram: A box sliding on a ramp The free-body diagram for a box sliding on a frictionless ramp.

③ Nothing cancels the positive x component of the gravitational force, so the net force is in the positive x direction...

④ ...and the acceleration of the block is in the positive x direction, the same as the direction of the net force.

(exerted by $\vec{n}$ the ramp)

y

x

$\Sigma \vec{F}_{\text{ext}} = \vec{n} + \vec{w}$

θ

$\vec{a}$

θ

$\vec{w}$
(exerted by Earth)

⑤ Newton's second law in component form:
$\Sigma F_{\text{ext},x} = w \sin \theta = ma_x$
$\Sigma F_{\text{ext},y} = n - w \cos \theta = 0$

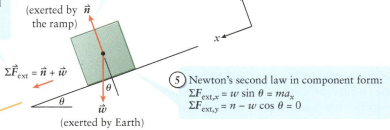

down; if it is momentarily at rest, the box starts moving down the ramp; and if it is moving down the ramp, it speeds up. So the free-body diagram also tells us that the box shown in Figure 4-12 moves in a straight line with a constant acceleration—which means we can analyze its motion using the techniques that we learned in Chapter 2.

WATCH OUT! Don't add nonexistent forces to a free-body diagram.

Students are sometimes tempted to add a force in a free-body diagram that is somehow associated with the motion. Common names for this force are "the force of velocity" or "the force of acceleration." If you feel this temptation, resist it! Remember that neither velocity nor acceleration is a force. Keep in mind also that acceleration is the *result* of force: For example, the acceleration of the box in Figure 4-12 is a result of the nonzero force in the x direction. (It's true that Newton's second law says that $\sum \vec{F}_{ext} = m\vec{a}$, which may mislead you into thinking that the quantity $m\vec{a}$ is itself a force. But remember the real meaning of Newton's second law: A net external force $\sum \vec{F}_{ext}$ on an object makes

the object accelerate. So $m\vec{a}$ is *not* a force, and equation $\sum \vec{F}_{ext} = m\vec{a}$ just tells you the magnitude and direction of that resulting acceleration $\vec{a}$.) If you're still not convinced, ask yourself what external object could exert a "force of velocity" or a "force of acceleration" on the object you're analyzing. (Remember only *external* forces can affect an object's motion. You may want to review the discussion in Section 4-2.) In Figure 4-12 we've already accounted for Earth's gravity, and we've already accounted for the one and only thing that touches the box (the surface of the ramp). So there can be no external forces on the box other than the two we've already drawn in Figure 4-12.

We'll use free-body diagrams throughout our study of forces and motion, and you should as well. Before we delve more deeply into how to solve problems using Newton's first and second laws, we'll introduce Newton's *third* law. This will give us deeper insight into the way in which objects exert forces on each other.

GOT THE CONCEPT? 4-5 Choose the Correct Free-Body Diagram

When a box is placed on an inclined ramp and released, it moves down the ramp with increasing speed. There is friction between the ramp and the box. Which of the free-body diagrams shown in Figure 4-13 correctly depicts the forces on the box as it slides down the ramp?

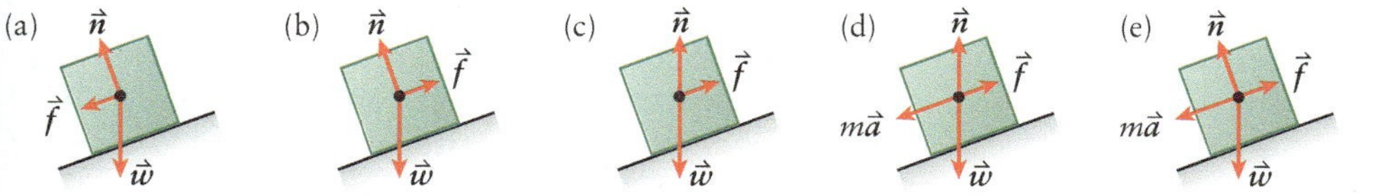

(a)

(b)

(c)

(d)

(e)

Figure 4-13 A box sliding down a ramp with friction Which of these free-body diagrams is correct?

TAKE-HOME MESSAGE FOR Section 4-5

✔   To solve a problem about how an object moves and the forces that act on it, you must draw a free-body diagram for the object. This diagram depicts all of the external forces that act on the object.

✔   Choose the x and y axes in a free-body diagram so that most of the forces point along one or the other of the axes.

4-5 Newton's third law relates the forces that two objects exert on each other

If you hold a book in your palm or dribble a basketball, you're exerting a force on an object. However, you know from experience that the *object* exerts a force on *you* as well. You can feel the book pushing back against your palm, and you can feel the slap of the basketball against your hand as you push down on it. What is the connection between the force that you exert on an object and the force that the object exerts back on you?

Here's an experiment that offers an answer to this question. Take two spring scales (such as you might use for weighing a fish) and connect their upper ends together (Figure 4-14a). You hold the free end of one scale, and you attach the free end of the other scale to a hook mounted to the wall (Figure 4-14b). When you pull on the free end of the first scale, the wall pulls on the free end of the other scale. Remarkably, no matter how hard you pull, the reading on the scale that you hold (which measures the amount of force you exert) is *identical* to the reading on the scale attached to the wall (which measures the amount of force that the wall exerts). In other words, the force that you exert on the wall has the *same* magnitude as the force that the wall exerts back on you (Figure 4-14c).

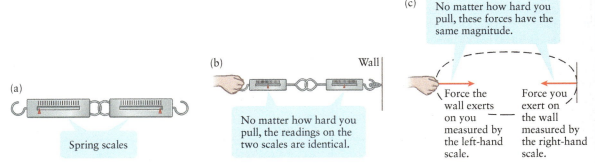

Figure 4-14 Comparing the forces that objects exert on each other (a) Two spring scales connected together. Each scale measures the force exerted on it. (b) One scale is connected to a wall, and you pull on the other end. (c) The force you exert on the wall with the scales is equal in magnitude to the force that the wall exerts back on you.

The car and trailer are linked together by two spring scales (like those shown in Figure 4-14) connected end to end.

💡 Each spring scale has the same reading: The force that the car exerts on the trailer always has the same magnitude as the force that the trailer exerts on the car. This is true no matter how the car and trailer move (and is also true if they aren't moving at all).

Figure 4-15 Comparing the forces that moving objects exert on each other A pair of spring scales like those shown in Figure 4-14a are used to connect a car and trailer.

In Figure 4-14 both you and the wall are stationary. We can try the same experiment with two objects in motion—say, a car and a trailer connected together by the same arrangement of two spring scales (Figure 4-15). The readings on each scale depend on whether the car and trailer are moving at a constant speed, speeding up, or slowing down. In all of these cases, however, experiment shows that the spring scale attached to the car reads *exactly* the same as the spring scale attached to the trailer. So the force that the car exerts on the trailer always has the same magnitude as the force that the trailer exerts on the car. The only difference between the two forces is that they are in *opposite* directions: The car pulls forward on the trailer, while the trailer pulls backward on the car.

We can summarize these observations in **Newton's third law:**

If object A exerts a force on object B, object B exerts a force on object A that has the same magnitude but is in the opposite direction. These two forces act on different objects.

In declaring this law Newton added: "If you press a stone with your finger, the finger is also pressed by the stone." This quote reminds us that Newton's third law refers to two forces that act on *different* objects. In Newton's example these two forces are your finger pressing on the stone and the stone pressing on your finger. Physicists speak of the *interaction* between two objects and refer to the two forces involved in an interaction as a **force pair.** (These two forces are sometimes called the *action* and the *reaction.*)

We can express Newton's third law as an equation relating the force that an object A exerts on another object B (written as $\vec{F}_{A \text{ on } B}$) to the **reaction force** that object B exerts on object A (written as $\vec{F}_{B \text{ on } A}$):

Force that object A exerts on object B   Force that object B exerts on object A

Newton's third law of motion
(4-7)

$$\vec{F}_{A \text{ on } B} = -\vec{F}_{B \text{ on } A}$$

 If object A exerts a force on object B, object B must exert a force on object A with the same magnitude but opposite direction.

The minus sign means that $\vec{F}_{B\,on\,A}$ has the same magnitude as $\vec{F}_{A\,on\,B}$ but is in the opposite direction (Figure 4-16). The two forces in a force pair are sometimes called "equal and opposite," which means that they are equal *in magnitude* and opposite *in direction*.

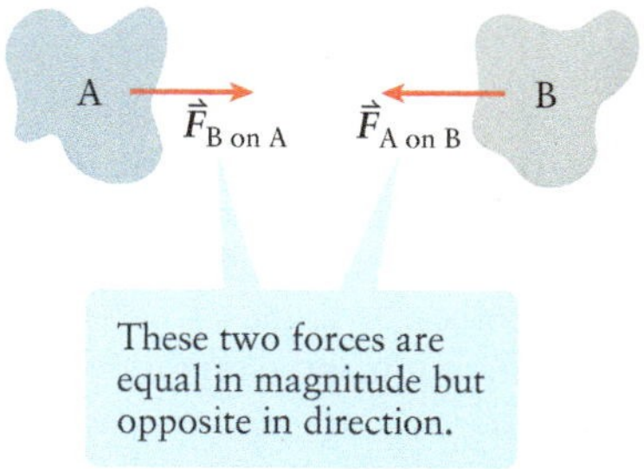

Comparing Newton's Three Laws

We've emphasized that Newton's third law is fundamentally different from the first and second laws because it tells us about forces that act on *two different* objects. By contrast, Newton's first and second laws tell us about the effect of the net external force acting on a *single* object. Here's an example that illustrates this important distinction. Figure 4-17a shows a ball held at rest in a person's hand. The drawing shows three forces: the gravitational force of Earth pulling down on the ball (the ball's weight), the force of the person's hand on the ball, and the force of the ball on the person's hand. All three forces have equal magnitudes but for different reasons!

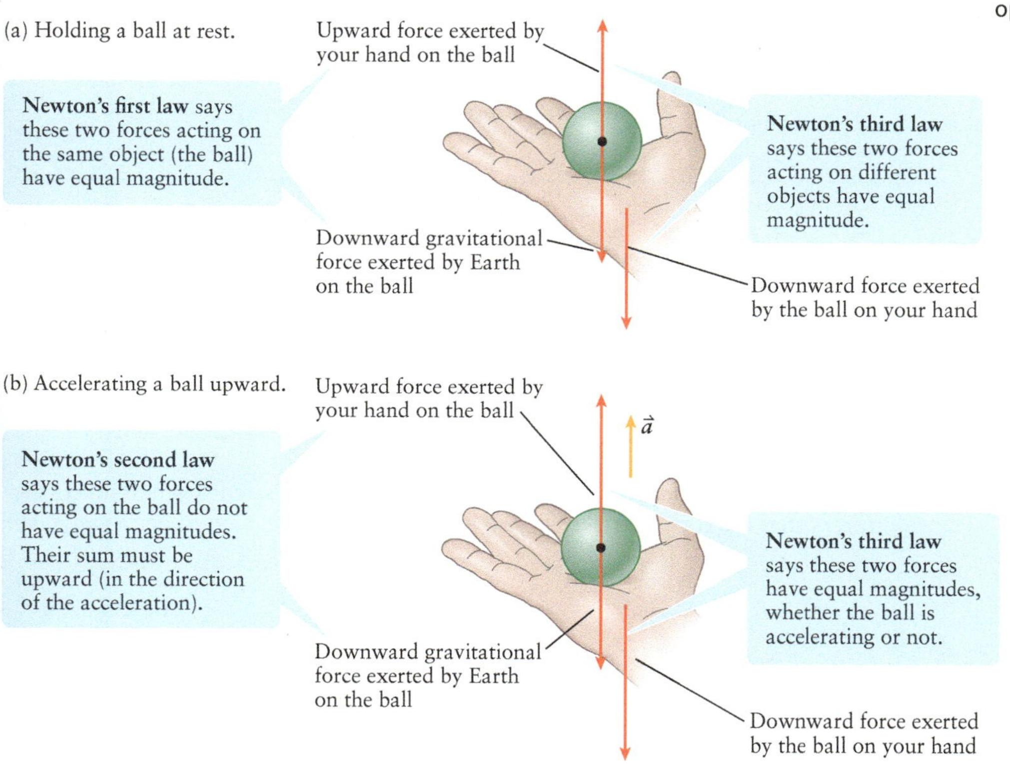

These two forces are equal in magnitude but opposite in direction.

Figure 4-16 Newton's third law If object A exerts a force $\vec{F}_{A\,on\,B}$ on object B, then object B necessarily exerts a force $\vec{F}_{B\,on\,A}$ on object A with the same magnitude but the opposite direction.

(a) Holding a ball at rest.

Upward force exerted by your hand on the ball

Newton's first law says these two forces acting on the same object (the ball) have equal magnitude.

Newton's third law says these two forces acting on different objects have equal magnitude.

Downward gravitational force exerted by Earth on the ball

Downward force exerted by the ball on your hand

(b) Accelerating a ball upward.

Upward force exerted by your hand on the ball

$\vec{a}$

Newton's second law says these two forces acting on the ball do not have equal magnitudes. Their sum must be upward (in the direction of the acceleration).

Newton's third law says these two forces have equal magnitudes, whether the ball is accelerating or not.

Downward gravitational force exerted by Earth on the ball

Downward force exerted by the ball on your hand

Figure 4-17 Comparing Newton's three laws (a) Newton's first law and third law applied to a ball at rest in your palm. (b) Newton's second law and third law applied to a ball that you accelerate upward.

- The force of Earth on the ball and the force of the person's hand on the ball have equal magnitudes and opposite directions according to Newton's *first* law. These are the only two external forces acting *on* the ball, so their vector sum $\sum \vec{F}_{ext}$ must be zero (that is, the two forces must balance each other) in order that the ball remain at rest and not accelerate.

- The force of the person's hand on the ball and the force of the ball on the person's hand have equal magnitudes and opposite directions according to Newton's *third* law. These two forces are a force pair: If A represents the hand and B represents the ball, these are the forces of A on B and of B on A.

Figure 4-17b shows the situation when the person's hand is accelerating the ball upward. Now the vector sum $\sum \vec{F}_{ext}$ of external forces on the ball is not zero, so the force of Earth on the ball does *not* have the same magnitude as the force of the person's hand on the ball. However, the force of the person's hand on the ball still *does* have the same magnitude of the force of the ball on the hand because Newton's third law works at all times, even when the objects are accelerating.

Here's a general rule to keep in mind:

Use Newton's first law or second law when dealing with the forces that act on a given object. Use Newton's third law when relating the forces that act between two objects.

WATCH OUT! Newton's third law involves only two objects.

! Two forces can be a force pair only if they are exerted between the *same* two objects. Consider the situation in which you are holding a rock in your hand. The weight force exerted by Earth on the rock cannot ever be a force pair to the normal force of your hand on the same rock. These two forces involve three objects: Earth, rock, and hand. So these two forces cannot be a force pair.

GOT THE CONCEPT? 4-6 Is Motion Impossible?

(?) You ask a friend to push a large crate across a floor (Figure 4-18). Your friend declares that this feat isn't possible because whatever force he applies to the box will be met with an equal but opposite force. The vector sum of these forces, he argues, will be zero, so the crate will have zero acceleration and won't move. What's wrong with this argument?

Figure 4-18 Pushing a crate Do Newton's laws make it impossible to push this crate across the floor?

GOT THE CONCEPT? 4-7 Which Forces Act?

(?) Two crates are at rest, one touching the other, on a horizontal surface. You push horizontally on crate 1 as shown by the red arrow in Figure 4-19. Which of the following is a force that acts on crate 2? (a) A force exerted by crate 1; (b) a force exerted by you; (c) a reaction force exerted by crate 2; (d) a force due to the acceleration of crate 2; (e) more than one of these.

Figure 4-19 Pushing two crates If you push on crate 1 as shown by the red arrow, what forces act on crate 2?

Newton's Third Law and Tension

We can use Newton's third law to help us understand another important type of contact force. **Tension** is the force exerted by a rope (or thread, string, cable, or wire) on an object to which the rope is connected. As an example, the rope shown in Figure 4-20a exerts a tension force on the sailor who holds on to the left-hand end of the rope and exerts a tension force on the boat tied to its right-hand end. We'll use the symbol $\vec{T}$ for a tension force.

Here's an important fact about tension: If a rope has sufficiently small mass, the tension forces that the rope exerts at its two ends have the *same* magnitude. To see why

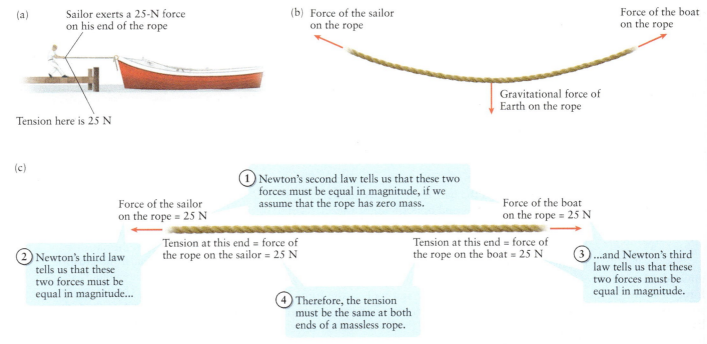

(a) Sailor exerts a 25-N force on his end of the rope

Tension here is 25 N

(b) Force of the sailor on the rope

Force of the boat on the rope

Gravitational force of Earth on the rope

(c)

(1) Newton's second law tells us that these two forces must be equal in magnitude, if we assume that the rope has zero mass.

Force of the sailor on the rope = 25 N

Force of the boat on the rope = 25 N

(2) Newton's third law tells us that these two forces must be equal in magnitude...

Tension at this end = force of the rope on the sailor = 25 N

Tension at this end = force of the rope on the boat = 25 N

(3) ...and Newton's third law tells us that these two forces must be equal in magnitude.

(4) Therefore, the tension must be the same at both ends of a massless rope.

Figure 4-20 Comparing the tensions at the two ends of a rope (a) A sailor docking a boat pulls on one end of a rope. (b) The forces that act on the rope. (c) The forces that act on a massless rope. In this case the tension force must have the same magnitude at either end of the rope.

these forces have the same magnitude, consider a sailor who docks a boat by pulling on one end of a rope whose other end is tied to the boat (Figure 4-20a). If he pulls with a given force—say, 25 N—on his end of the rope, Newton's third law tells us that the rope pulls back on him with a force of 25 N. How much force does the *other* end of the rope exert on the boat?

To see the answer, consider Figure 4-20b, which shows the forces that act *on* the rope. The weight of the rope causes the rope to sag so that the angle of the rope is different at the two ends. However, in many situations the weight of the rope is very small compared to the pulling forces exerted on the rope by the objects attached to its ends. That's the case in Figure 4-20c. So we can safely approximate the rope as having *zero* mass and hence *zero* weight. Using this approximation the rope won't sag at all under its own weight (because it has none) and will be perfectly straight. Then the only forces acting on the rope are those at either end. Because the rope is straight, these two forces point in opposite directions (Figure 4-20c).

Newton's *second* law says that the sum of external forces on the rope, $\sum \vec{F}_{ext}$, is equal to $m\vec{a}$, the product of the rope's mass m and its acceleration $\vec{a}$. However, if the rope has zero mass ($m = 0$), the product $m\vec{a}$ is always equal to zero, even if the rope is accelerating. Therefore, the sum of external forces on the rope must be zero $\left(\sum \vec{F}_{ext} = 0\right)$ under all circumstances. For this to be the case, the oppositely directed forces acting on either end of the rope (Figure 4-20c) must have the same magnitude whether the rope is accelerating or not.

Newton's *third* law tells us that the force exerted *on* each end of the rope has the same magnitude as the tension forces exerted *by* each end of the rope. Because equal-magnitude forces act on each end of a massless rope, it follows that the tension is also the same at *both* ends: If the tension is 25 N at the left-hand end, it is 25 N at the right-hand end (Figure 4-20c). So we are left with the following general rule:

> *The tension is the same at both ends of a rope or string, provided the weight of the rope or string is much less than the forces exerted on the ends of the rope or string.*

Newton's Third Law and Propulsion

Newton's third law plays an essential role in biology: All living systems (including you) use it for *propulsion.* As an example, suppose you start running from a standing start. It's common to say that you propel *yourself* forward, but this isn't strictly correct: It takes an external force to accelerate you forward from rest, and an external force has to come from outside your body. What happens is that to start running, you use your feet to exert a *backward* force on the surface of the running track. By Newton's third law, the track exerts a *forward* force of the same magnitude on your feet. It's this force exerted by the track that's responsible for propelling you forward.

Bird flight also depends on Newton's third law. As they flap, the wings of a bird in flight push air both backward and downward. Newton's third law tells us that the air therefore exerts forward and upward components of force on the bird's wings. The forward component of force exerted on the bird's wings by the air balances the backward force of air resistance that acts on the bird's body, so the net horizontal external force on the bird is zero. The upward component of force that the air exerts on the bird's wings balances the downward gravitational force on the bird's body, so the net vertical external force on the bird is also zero. Since there is no net external force on the bird, it continues flying with zero acceleration (that is, in a straight line with constant speed).

Fish typically have a number of fins of various shapes and sizes (Figure 4-21a) that they use to take full advantage of Newton's third law. In almost all species of fish, the primary means of forward propulsion is the side-to-side motion of the caudal fin and the rear part of the fish's body. Figure 4-21b shows the fish from above as it sweeps its caudal fin to the right. In doing so the fish exerts a force $\vec{F}_{fish\,on\,water}$ on the surrounding water that acts perpendicular to the fish's body, in a manner similar to a normal force. In accordance with Newton's third law, the water exerts a force on the fish, $\vec{F}_{water\,on\,fish}$, of equal magnitude in the opposite direction. The forward component of this force on the fish, labeled $\vec{F}_{thrust}$ in Figure 4-21b, propels the fish forward against

BioMedical

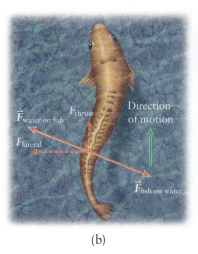

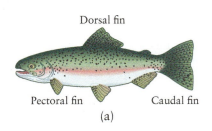

Dorsal fin

Pectoral fin   Caudal fin

(a)

(b)

**TAKE-HOME MESSAGE
FOR Section 4-5**

✔ Newton's third law tells us
that if object A exerts a force
on object B, object B has to
exert a force on object A of
the same magnitude but in the
opposite direction.

✔ A rope exerts the same
tension at both ends, provided
the mass of the rope can be
ignored.

---

the backward resistance of water on the front of the fish. In some species the motion of
the dorsal fin (Figure 4-21a) also contributes to forward propulsion.

Figure 4-21b shows that there is also a force component F_{lateral} on the fish that
acts perpendicular to the net motion of the fish. These lateral forces contribute to the
stability of the fish and help it turn and maneuver. Most fish can also make subtle
adjustments to the orientation of the pectoral fins on the sides of the fish, enabling it to
stop, start, and change direction quickly.

4-6 All problems involving forces can be solved using the same series of steps

In this section we'll use Newton's laws to solve a variety of problems that involve
forces and acceleration. We can set up *any* force problem using just two steps.

(1) Make a free-body diagram for the object or objects of interest using the rules that
we outlined in Section 4-4. Choose the coordinate axes and label the positive directions. (As we've mentioned before, it's best to choose these axes so that one or more
of the vectors in the problem—the acceleration and the individual forces—lie along
one of the axes.)

(2) Write Newton's second law for each object and in each direction separately. Use
the free-body diagram to make sure that for each object, all of the forces that
act on that object are included in the sum of external forces. Then solve for the
unknowns.

Problems Involving a Single Object

Our first few examples involve just a single object. In some problems the goal
is to find the object's acceleration; in other cases you need to determine one or
more of the forces acting on the object. Sometimes determining the acceleration or
unknown force is not the goal itself but just a necessary step to arrive at the final
answer.

*See the Math Tutorial
for more information on
trigonometry.*

One thing to think about in these problems is the choice of *x* and *y* axes for your
force diagram. If the object is on an incline, choose one of the axes to point along the
incline. If the object is moving in a known direction, it's best to choose one of the axes
to point in that direction. If you're not sure which way the object will accelerate (say,
whether to the left or to the right), don't worry about it—choose an axis to point in
one direction or the other. If it turns out that the acceleration is in the direction opposite to the one you chose, you'll know because the acceleration will turn out to be
negative.

EXAMPLE 4-4   What's the Angle?

A pair of fuzzy dice hangs on a lightweight thread from the rearview mirror of your high-performance sports car. At what angle from the vertical do the dice hang when the car is accelerating forward at a constant 6.50 m/s²?

Set Up

The dice move along with the car and so have the same forward acceleration of 6.50 m/s². So there must be a net forward force on the dice. The individual forces on the dice are the gravitational force $\vec{w}_{dice}$ exerted by Earth and the tension force $\vec{T}$ exerted by the thread. Because the thread is lightweight, the gravitational force on it is negligible, and the thread will be straight when taut. So the angle of the thread from the vertical—which is the quantity we're trying to find—is the same as the angle θ of the tension force $\vec{T}$. We'll find this angle by applying Newton's second law to the dice.

Newton's second law in component form:

$$\sum F_{\text{ext on dice},x} = m_{dice}\, a_{\text{dice},x} \quad \text{(4-4a)}$$

$$\sum F_{\text{ext on dice},y} = m_{dice}\, a_{\text{dice},y} \quad \text{(4-4b)}$$

Net force on dice = sum of tension force and gravitational force:

$$\sum \vec{F}_{\text{ext on dice}} = \vec{T} + \vec{w}_{dice}$$

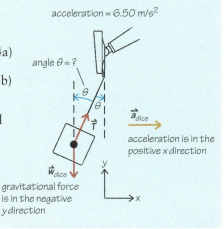

acceleration = 6.50 m/s²

angle θ = ?

$\vec{a}_{dice}$

acceleration is in the positive x direction

$\vec{T}$

$\vec{w}_{dice}$

gravitational force is in the negative y direction

Solve

Write the x and y components of the external forces that act on the dice and the acceleration of the dice.

Components of tension force $\vec{T}$:

$$T_x = T \sin\theta$$
$$T_y = T \cos\theta$$

Components of gravitational force $\vec{w}_{dice}$ with magnitude $w_{dice} = m_{dice}g$:

$$w_{dice,x} = 0$$
$$w_{dice,y} = -w_{dice} = -m_{dice}g$$

Components of acceleration $\vec{a}_{dice}$:

$$a_{dice,x} = a_{dice} = 6.50 \text{ m/s}^2$$
$$a_{dice,y} = 0$$

$T_x = T \sin\theta$

$T_y = T \cos\theta$

$\vec{T}$

$\vec{w}_{dice}$

Substitute the expressions for the components of $\vec{w}, \vec{T}$, and $\vec{a}_{dice}$ into Newton's second law. This gives two equations, one from the x components and one from the y components.

Newton's second law in component form:

x: $\sum F_{\text{ext on dice},x} = T_x + w_{dice,x} = m_{dice}\, a_{dice,x}$, so

$$T \sin\theta + 0 = m_{dice}\, a_{dice}$$

y: $\sum F_{\text{ext on dice},y} = T_y + w_{dice,y} = m_{dice}\, a_{dice,y}$, so

$$T \cos\theta + (-m_{dice}g) = 0$$

We don't know either the magnitude T or the angle θ of the tension force. All we are asked to solve for is the value of θ, so we eliminate T from the two equations above. We also use the definition of the tangent function, $\tan\theta = (\sin\theta)/(\cos\theta)$.

From the y equation:

$T \cos\theta = m_{dice}g$, so

$$T = \frac{m_{dice}\, g}{\cos\theta}$$

Substitute this into the x equation:

$T \sin\theta = m_{dice}\, a_{dice}$, so

$$\frac{m_{dice}\, g}{\cos\theta}\sin\theta = m_{dice}\, a_{dice}$$

Divide through by m_{dice} and g and apply the definition of tangent.

$$\frac{\sin\theta}{\cos\theta} = \tan\theta = \frac{a_{dice}}{g}$$

Solve for θ:

$$\theta = \tan^{-1}\left(\frac{a_{\text{dice}}}{g}\right) = \tan^{-1}\left(\frac{6.50 \text{ m/s}^2}{9.80 \text{ m/s}^2}\right) = \tan^{-1}(0.663)$$
$$= 33.6°$$

Reflect

Note that our answer doesn't depend on the mass m_{dice}. The angle θ would be the same whether the dice were made of lightweight plastic or solid gold.

   We can check our result by using a range of values of acceleration. When the car is not accelerating (as would be the case if its velocity is constant), $a_{\text{dice}} = 0$ and $\theta = \tan^{-1} 0 = 0$. The dice hang straight down, which is expected. The greater the acceleration a_{dice}, the greater the value of θ—which is also what we would expect.

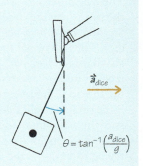

WATCH OUT! Newton's laws apply only in inertial frames of reference.

You might argue that the dice in the preceding example accelerate *backward* rather than forward. After all, from your perspective as the driver, the dice swing backward when you step on the accelerator pedal. The driver's seat isn't a good place from which to apply Newton's laws, however, because you are yourself accelerating! It turns out that Newton's laws work *only* if you observe from a vantage point, or *frame of reference*, that is *not* accelerating. Such a vantage point is called an **inertial frame of reference**. (A more detailed study of Newton's first law than we can present in this book usually considers Newton's first law as a *definition* of inertial frames of reference. That's really why Newton's first law gets its own number.) One such frame of reference would be a person standing by the road as you accelerate away. The dice initially will lag behind the car due to their inertia, which is why the dice swing backward relative to the driver. Once the dice have stabilized, however, they move with the same acceleration as the car. The same inertial effect explains why you feel pushed back in your seat when the car accelerates forward, and you feel thrown forward against your safety belt if the car brakes rapidly. No real force acts on you in either case: It's just your inertia. (If you think inertia *is* a real force, ask yourself what could be exerting it. If you think it's your own body, remember that only external forces affect motion; if you think it's "the force of acceleration" or "the force of velocity," remember from Section 4-4 that these forces don't exist.)

EXAMPLE 4-5 Slipping and Sliding I

One wintry day you accidentally drop your physics book (mass 2.50 kg) on an icy, essentially frictionless sidewalk that tilts at an angle of 10.0° with respect to the horizontal. What is the book's acceleration as it slides downhill?

Set Up

We'll find the book's acceleration using Newton's second law. Because there is no friction in this situation, the only forces acting on the book are the gravitational force $\vec{w}_{\text{book}}$ (which points straight down) and the normal force $\vec{n}$ (which points perpendicular to the surface of the inclined sidewalk). This situation is similar to the block shown in Figure 4-12 (Section 4-4). Because the motion is down the ramp, we choose the x axis to be in that direction.

Newton's second law in component form applied to the book:

$$\sum F_{\text{ext on book},x} = m_{\text{book}} a_{\text{book},x} \quad \text{(4-4a)}$$
$$\sum F_{\text{ext on book},y} = m_{\text{book}} a_{\text{book},y} \quad \text{(4-4b)}$$

Net force on book = sum of normal force and gravitational force:

$$\sum \vec{F}_{\text{ext on book}} = \vec{n} + \vec{w}_{\text{book}}$$

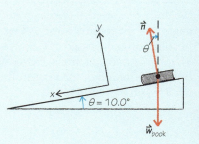

Solve

Write the forces and the acceleration in component form. The book doesn't move perpendicular to the ramp, so the y component of its acceleration is zero.

Components of normal force $\vec{n}$:

$n_x = 0$

$n_y = n$

Components of gravitational force $\vec{w}_{book}$:

magnitude $w_{book} = m_{book}g$

$w_{book,x} = w_{book}\ \sin\theta$
$\qquad = m_{book}g\ \sin\theta$

$w_{book,y} = -w_{book}\ \cos\theta$
$\qquad = -m_{book}g\ \cos\theta$

Components of acceleration $\vec{a}_{book}$:
$a_{book,x}$: to be determined

$a_{book,y} = 0$

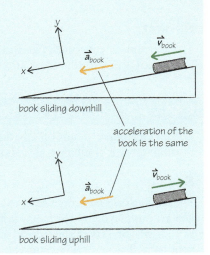

$n_y = n$

$w_{book,y} = -w_{book}\cos\theta$

$w_{book,x} = +w_{book}\sin\theta$

Use the components to write Newton's second law for the book in component form.

Newton's second law in component form:

x: $\sum F_{ext\ on\ book,x} = n_x + w_x = 0 + m_{book}g\sin\theta$
$\qquad\qquad\quad = m_{book}\,a_{book,x}$, so

$m_{book}g\sin\theta = m_{book}\,a_{book,x}$

y: $\sum F_{ext\ on\ book,y} = n_y + w_y = n - m_{book}g\cos\theta = m_{book}\,a_{book,y}$, so

$n - m_{book}g\cos\theta = 0$

All we are asked to find is the acceleration of the book, so we need only the x equation.

Dividing the x equation by m_{book}:

$a_{book,x} = g\sin\theta = (9.80\ \text{m/s}^2)\sin 10.0°$
$\qquad = 1.70\ \text{m/s}^2$

Reflect

Our result tells us that the book accelerates in the positive x direction (downhill). This result makes sense: When no friction opposes its motion, the book should gain speed as it slides downhill. If the slope were steeper, θ and $\sin\theta$ would be greater, and the acceleration $a_{book,x} = g\sin\theta$ would be greater as well. Just as for an object in free fall, the acceleration doesn't depend on the book's mass.

Note also that our analysis would be exactly the same if you put the book near the bottom of the incline and gave it an upward push to start it moving. The gravitational force and normal force would be unchanged, so the magnitude and direction of the acceleration would likewise be unchanged. Because the acceleration is downhill and the velocity uphill, the book would slow down as it moved up the incline—again, just as we would expect.

book sliding downhill

acceleration of the book is the same

book sliding uphill

EXAMPLE 4-6 **Slipping and Sliding II**

If there is friction between the book and the incline in the previous example, the book's downhill acceleration will be less than we calculated above. If the book gains speed at only 0.200 m/s² as it slides downhill, what is the magnitude of the friction force on the 2.50-kg book?

Set Up

The situation is much the same as in Example 4-5 but with an additional friction force $\vec{f}$ which points up the incline, opposing the book's motion. We are given the book's acceleration, so we'll use Newton's second law to determine the magnitude of the friction force.

Newton's second law in component form applied to the book:

$$\sum F_{\text{ext on book},x} = m_{\text{book}} a_{\text{book},x} \quad (4\text{-}4a)$$

$$\sum F_{\text{ext on book},y} = m_{\text{book}} a_{\text{book},y} \quad (4\text{-}4b)$$

Net force on book = sum of normal force, gravitational force, and friction force:

$$\sum \vec{F}_{\text{ext on book}} = \vec{n} + \vec{w}_{\text{book}} + \vec{f}$$

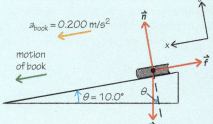

Solve

The vectors $\vec{n}$, $\vec{w}_{\text{book}}$, and $\vec{a}_{\text{book}}$ break into components the same way that they did in Example 4-5. We complete the list with the vector components of $\vec{f}$.

Components of friction force $\vec{f}$:

$$f_x = -f$$
$$f_y = 0$$

Use the components to write Newton's second law for the book in component form.

Newton's second law in component form:

$x:$ $\sum F_{\text{ext on book},x} = n_x + w_x + f_x = 0 + m_{\text{book}} g \sin\theta - f = m_{\text{book}} a_{\text{book},x}$, so

$$m_{\text{book}} g \sin\theta - f = m_{\text{book}} a_{\text{book},x}$$

$y:$ $\sum F_{\text{ext on book},y} = n_y + w_y + f_y = n - m_{\text{book}} g \cos\theta + 0 = m_{\text{book}} a_{\text{book},y}$, so

$$n - m_{\text{book}} g \cos\theta = 0$$

Solve the x equation for the magnitude f of the friction force.

$m_{\text{book}} g \sin\theta - f = m_{\text{book}} a_{\text{book},x}$, so

$$f = m_{\text{book}} g \sin\theta - m_{\text{book}} a_{\text{book},x}$$
$$= m_{\text{book}} (g \sin\theta - a_{\text{book},x})$$
$$= (2.50 \text{ kg})\left[(9.80 \text{ m/s}^2) \sin 10.0° - 0.200 \text{ m/s}^2 \right]$$
$$= 3.75 \text{ kg} \cdot \text{m/s}^2 = 3.75 \text{ N}$$

Reflect

This example is very much like Example 4-3 in Section 4-3. The only difference is that the net force on the cat in Example 4-3 is zero (the cat is at rest), while the net force on the sliding book in this example is *not* zero.

EXAMPLE 4-7 How to "Lose Weight" the Easy Way

A 70.0-kg student stands on a bathroom scale in an elevator. What is the reading on the scale (a) when the elevator is moving upward at a constant 0.500 m/s? (b) When the elevator is accelerating downward at 1.00 m/s²?

Set Up

The reading on a bathroom scale tells you how hard you are pushing on it, which may not be the same as your weight. (Try pushing on a bathroom scale with your hand. The reading measures how strong you are, not how much you weigh.) Newton's third law tells us that the downward normal force of the student on the scale is equal in magnitude to the upward normal force exerted by the scale on the student, $\vec{n}_{\text{scale on student}}$. So what we want to know is the magnitude of $\vec{n}_{\text{scale on student}}$ in each situation, which we'll find by applying Newton's second law to the student. The only other force acting on the student is the gravitational force $\vec{w}_{\text{student}}$. Both forces and the acceleration in part (b) are in the vertical direction, so we only need the y component.

y component of Newton's second law applied to the student:

$$\sum F_{\text{ext on student},y} = m_{\text{student}} a_{\text{student},y} \quad (4\text{-}4b)$$

Net force on student = sum of normal force and gravitational force:

$$\sum \vec{F}_{\text{ext on student}} = \vec{n}_{\text{scale on student}} + \vec{w}_{\text{student}}$$

Solve

(a) If the elevator is moving at a constant velocity, the student's acceleration is zero. So the sum of the external forces on the student must be zero. We then solve for the magnitude of the normal force that the scale exerts on the student.

Student's y velocity is constant, so

$$a_{\text{student},y} = 0$$

Newton's second law in component form:

$$y: \sum F_{\text{ext on student},y}$$

$$= n_{\text{scale on student}} + (-w_{\text{student}})$$

$$= m_{\text{student}} \, a_{\text{student},y} = 0, \text{ so}$$

$$n_{\text{scale on student}}$$

$$= w_{\text{student}} = m_{\text{student}} \, g$$

$$= (70.0 \text{ kg})(9.80 \text{ m/s}^2)$$

$$= 686 \text{ N}$$

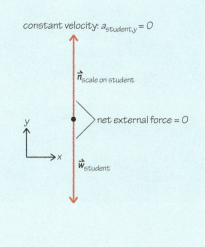

constant velocity: $a_{\text{student},y} = 0$

$\vec{n}_{\text{scale on student}}$

net external force = 0

$\vec{w}_{\text{student}}$

(b) If the elevator, and so also the student, is accelerating downward (in the negative y direction) at 1.00 m/s², the net force on the student must be downward. The net force can be downward only if the upward normal force exerted by the scale is smaller than the downward gravitational force. Again we solve for the magnitude $n_{\text{scale on student}}$ using Newton's second law.

The student's acceleration has a downward (negative) y component:

$$a_{\text{student},y} = -1.00 \text{ m/s}^2$$

Newton's second law in component form:

$$y: \sum F_{\text{ext on student},y}$$

$$= n_{\text{scale on student}} + (-w_{\text{student}})$$

$$= m_{\text{student}} \, a_{\text{student},y}, \text{ so}$$

$$n_{\text{scale on student}}$$

$$= w_{\text{student}} + m_{\text{student}} \, a_{\text{student},y}$$

$$= 686 \text{ N} + (70.0 \text{ kg})(-1.00 \text{ m/s}^2) = 616 \text{ N}$$

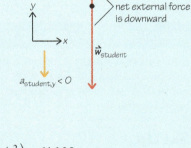

downward acceleration

$\vec{n}_{\text{scale on student}}$

net external force is downward

$a_{\text{student},y} < 0$

$\vec{w}_{\text{student}}$

Reflect

According to the scale the student "weighs" 70 N less when the elevator accelerates downward. You've probably experienced this phenomenon in an elevator when it starts moving downward from rest or when it comes to a halt when moving upward. A similar phenomenon occurs in an elevator that accelerates *upward*. Then $a_{\text{student},y}$ is positive, $n_{\text{scale on student}}$ is greater than the student's weight w_{student}, and the student feels heavier than normal.

If the elevator is in free fall,

$$a_{\text{student},y} = -g$$

$$n_{\text{scale on student}}$$

$$= w_{\text{student}} + m_{\text{student}} \, a_{\text{student},y}$$

$$= m_{\text{student}} \, g + m_{\text{student}} (-g)$$

$$= 0$$

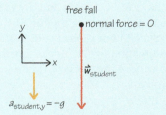

free fall

normal force = 0

$a_{\text{student},y} = -g$

$\vec{w}_{\text{student}}$

What if the elevator is in free fall? Then $a_{\text{student},y} = -g = -9.80 \text{ m/s}^2$ (negative because the acceleration is downward). Substituting this value into our result gives $n_{\text{scale on student}} = 0$. The scale reads zero, and the student will feel weightless! This situation is called *apparent weightlessness*. (The student isn't truly weightless because the gravitational force still acts on her.) We'll see in Chapter 7 that astronauts in Earth orbit are also in free fall, so they *feel* weightless just like the student in the free-falling elevator.

Problems Involving Ropes and Tension

What object is pulling the water skier shown in Figure 4-22? You might be tempted to say that it's the boat—but that can't be correct because the boat doesn't touch the skier. Instead the boat exerts a force on the tow rope, and it's the tow rope that pulls on the skier. The force that pulls the skier is therefore a *tension* force.

As we learned in Section 4-5, if the weight of the tow rope is small compared to the forces exerted by the objects to which it's attached (in this case, the boat and the water skier), we can ignore its mass altogether. Then the magnitude of the tension force

 Go to Picture It 4-1 for more practice dealing with weight

 Go to Interactive Exercises 4-1 and 4-2 for more practice dealing with tension

David Holbrook/Alamy

Figure 4-22 Towing a water skier The boat exerts a force on the tow rope, and the tow rope pulls on the skier.

is the *same* at both ends of the rope no matter how the boat, skier, and rope are moving. We'll use this principle in the next two examples, both of which involve moving objects that are connected by a lightweight string.

EXAMPLE 4-8 **Tension in a String**

You pull two blocks connected by a lightweight string across a horizontal table. Block 1 has a mass of 1.00 kg, and block 2 has a mass of 2.00 kg. There is negligible friction between the blocks and the table. The string will break if the tension is greater than 6.00 N. What is the maximum pull that you can apply to block 1 without breaking the string?

Set Up

As in the previous examples we start by drawing a free-body diagram. Note that here there are *two* bodies (block 1 and block 2), so we have to be careful and draw *two* free-body diagrams to include the forces that act on each block. We choose the positive *x* axis to be in the direction of the blocks' horizontal motion and choose the positive *y* axis to be vertically upward.

$$\sum \vec{F}_{\text{ext on 1}} = m_1 \vec{a}_1$$
$$\sum \vec{F}_{\text{ext on 2}} = m_2 \vec{a}_2 \qquad (4\text{-}2)$$

Net force on block 1 = sum of force you exert, normal force, gravitational force, and tension:

$$\sum \vec{F}_{\text{ext on 1}} = \vec{F}_{\text{you on 1}} + \vec{n}_1 + \vec{w}_1 + \vec{T}_{\text{string on 1}}$$

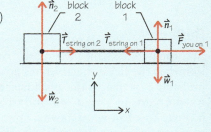

We are told that the string connecting the blocks is lightweight, so the tension at either end of the string has the same magnitude: $T_{\text{string on 1}} = T_{\text{string on 2}}$. We'll call this magnitude T for short. Our goal is to find the maximum value of $F_{\text{you on 1}}$ (the magnitude of the force that you exert on block 1) so that T is no greater than 6.00 N.

Net force on block 2 = sum of normal force, gravitational force, and tension:

$$\sum \vec{F}_{\text{ext on 2}} = \vec{n}_2 + \vec{w}_2 + \vec{T}_{\text{string on 2}}$$

Tension is the same at both ends of the string:

$$T_{\text{string on 1}} = T_{\text{string on 2}} = T$$

Solve

We then write Newton's second law for each block in component form. If we assume that the string doesn't stretch, the two blocks move together and have the same *x* component of acceleration, which we call a_x.

Both blocks have the same *x* acceleration:

$$a_{1x} = a_{2x} = a_x$$

Both blocks have zero *y* acceleration.

Newton's second law in component form for block 1:

$$x: \sum F_{\text{ext on 1},x} = F_{\text{you on 1}} + (-T)$$
$$= m_1 a_x$$
$$y: \sum F_{\text{ext on 1},y} = n_1 + (-w_1) = 0$$

Newton's second law in component form for block 2:

$$x: \sum F_{\text{ext on 2},x} = T = m_2 a_x$$
$$y: \sum F_{\text{ext on 2},y} = n_2 + (-w_2) = 0$$

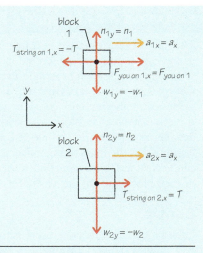

We want to find the value of $F_{\text{you on 1}}$ when $T = 6.00$ N (the maximum tension that will prevent the string from breaking). We use the x equation for block 2 to find the x acceleration a_x in this situation and then we use this value of a_x in the x equation for block 1 to solve for $F_{\text{you on 1}}$.

First solve for the value of a_x when $T = 6.00$ N. Use the x equation for block 2:

$$T = m_2 a_x$$
$$a_x = \frac{T}{m_2} = \frac{6.00 \text{ N}}{2.00 \text{ kg}} = 3.00 \text{ N/kg} = 3.00 \text{ m/s}^2$$

Now substitute $a_x = 3.00$ m/s^2 into the x equation for block 1 and solve for $F_{\text{you on 1}}$:

$$F_{\text{you on 1}} - T = m_1 a_x$$
$$F_{\text{you on 1}} = T + m_1 a_x = 6.00 \text{ N} + (1.00 \text{ kg})(3.00 \text{ m/s}^2)$$
$$= 6.00 \text{ N} + 3.00 \text{ kg} \cdot \text{m/s}^2 = 6.00 \text{ N} + 3.00 \text{ N} = 9.00 \text{ N}$$

Reflect

We can check our answers by referring back to Newton's second law in the x direction for each block. You exert a forward force $F_{\text{you on 1}} = 9.00$ N on block 1, and the string exerts a backward force of 6.00 N. So the net forward force on block 1 is 9.00 N – 6.00 N = 3.00 N. This block has a mass of 1.00 kg, so the 3.00-N net force gives block 1 an acceleration of 3.00 m/s^2. The only horizontal force on block 2 is the forward tension force of 6.00 N; because block 2 has a mass of 2.00 kg, the forward acceleration produced is (6.00 N)/(2.00 kg) = 3.00 m/s^2. This acceleration is the same as for block 1, as it must be because the two blocks move together.

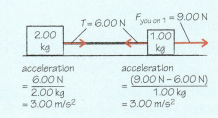

Because the two blocks move together, we could have treated the two blocks together as a single unit with a mass equal to $m_1 + m_2 = 3.00$ kg. That wouldn't have helped us solve this problem, however. The reason is that the tension forces are *internal* forces (they involve one part of the combined unit pulling on another part) in this approach. As a result there would be no way to apply the condition that the tension cannot exceed 6.00 N.

Here's the lesson: When solving problems that involve forces and accelerations for more than one object, it's safest to *always* apply Newton's laws to each object separately.

However, the combined system of two blocks can be used as another way to find the acceleration of the two blocks moving together. Can you solve the equations for the two blocks as a unit to show again that this acceleration is 3.00 m/s^2?

Newton's second law in component form for the two blocks as a unit:

$$x: \sum F_{\text{ext on unit},x}$$
$$= F_{\text{you on 1}}$$
$$= (m_1 + m_2) a_{\text{on unit},x}$$
$$y: \sum F_{\text{ext on unit},y}$$
$$= n_{\text{unit}} + (-w_{\text{unit}}) = 0$$

We can't solve for $F_{\text{you on 1}}$ because the tension T does not appear in these equations.

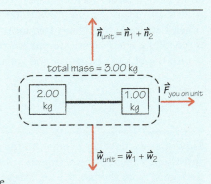

EXAMPLE 4-9 Atwood's Machine

Two blocks are connected by a lightweight string that passes over a pulley as shown in Figure 4-23. (This arrangement is known as *Atwood's machine* after Rev. George Atwood, who devised it in 1784 as a way to demonstrate motion with constant acceleration.) Block 1 is more massive than block 2, so when released block 1 accelerates downward, and block 2 accelerates upward. Assume that the mass of the pulley is so small that we can neglect it. Derive expressions for (a) the acceleration of each block and (b) the tension in the string.

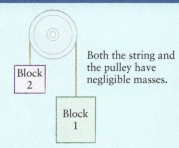

Figure 4-23 Atwood's machine When the blocks are released from rest, block 1 falls, and block 2 rises. What are their accelerations, and what is the tension in the string?

Set Up

Each block has only two forces acting on it: a downward gravitational force and an upward tension force. We choose the positive y direction for block 1 to be downward and the positive y direction for block 2 to be upward. Using this choice, both blocks accelerate in the positive y direction and both blocks have the same y component of acceleration (because we assume that the string doesn't stretch). One of our goals is to calculate this acceleration, which we call a_y.

Because we can neglect the mass of the pulley, the magnitude of the tension force has the same value T at both ends of the lightweight string. Our other goal is to calculate the value of T.

Newton's second law in the y direction applied to each block:

$$\sum F_{\text{ext on }1,y} = m_1 a_{1y}$$
$$\sum F_{\text{ext on }2,y} = m_2 a_{2y} \qquad (4\text{-}4b)$$

Net force on block 1 = sum of gravitational force and tension:
$$\sum \vec{F}_{\text{ext on }1} = \vec{w}_1 + \vec{T}_{\text{string on }1}$$

Net force on block 2 = sum of gravitational force and tension:
$$\sum \vec{F}_{\text{ext on }2} = \vec{w}_2 + \vec{T}_{\text{string on }2}$$

Both blocks have the same y acceleration:
$$a_{1y} = a_{2y} = a_y$$

Magnitude of tension is the same at both ends of the string:
$$T_{\text{string on }1} = T_{\text{string on }2} = T$$

Solve

Write the y component of Newton's second law for each block.

Newton's second law in component form for block 1:
$$y\text{:} \sum F_{\text{ext on }1,y} = w_1 + (-T) = m_1 a_y \text{ or } m_1 g - T = m_1 a_y$$
Newton's second law in component form for block 2:
$$y\text{:} \sum F_{\text{ext on }2,y} = T + (-w_2) = m_2 a_y \text{ or } T - m_2 g = m_2 a_y$$

We have two equations (one for block 1 and one for block 2) and two quantities that we want to find (a_y and T). One way to proceed is to first solve for one unknown in terms of the other. Let's solve the block 1 equation for T in terms of a_y. Then substitute this expression for T into the block 2 equation and solve for a_y.

Solve the block 1 equation for T:
$$m_1 g - T = m_1 a_y \text{ so } T = m_1 g - m_1 a_y$$
Substitute this into the block 2 equation:
$$(m_1 g - m_1 a_y) - m_2 g = m_2 a_y$$
Solve for a_y:
$$m_1 g - m_2 g = m_1 a_y + m_2 a_y$$
$$(m_1 - m_2)g = (m_1 + m_2)a_y \text{ or}$$
$$a_y = \left(\frac{m_1 - m_2}{m_1 + m_2}\right)g$$

Now that we have an expression for a_y, substitute it back into the equation for T that we found from the block 1 equation. This resulting expression gives us our final answer for T.

From the block 1 equation for T:
$$T = m_1 g - m_1 a_y$$
Substitute our expression for a_y and simplify:
$$T = m_1 g - m_1 \left(\frac{m_1 - m_2}{m_1 + m_2}\right)g = m_1 g \left(1 - \frac{m_1 - m_2}{m_1 + m_2}\right)$$
$$= m_1 g \left(\frac{m_1 + m_2}{m_1 + m_2} - \frac{m_1 - m_2}{m_1 + m_2}\right) = m_1 g \left(\frac{m_1 + m_2 - m_1 + m_2}{m_1 + m_2}\right) \text{ or}$$
$$T = \frac{2 m_1 m_2 g}{m_1 + m_2}$$

Reflect

Do these results make sense? As a start you should confirm that our expression for a_y (an acceleration) has units of m/s², and our expression for T (a force magnitude) has units of newtons. Do they?

To check our expression for acceleration a_y in more detail, notice that it depends on the difference $m_1 - m_2$ between the two masses. If $m_1 > m_2$ as we assumed here, a_y is positive; then both blocks accelerate in the positive y direction, so block 1 (the more massive and heavier one) accelerates downward, and block 2 accelerates upward.

If instead $m_1 < m_2$ so that block 2 is more massive and heavier, the acceleration a_y is negative. Then block 1 (which is now the lighter one) accelerates upward and block 2 downward. Finally, if the two blocks have equal mass so that $m_1 = m_2$, our expression says that $a_y = 0$, or the blocks don't accelerate at all. If the two equal-mass blocks begin at rest, they will remain at rest and in balance, just as we would expect.

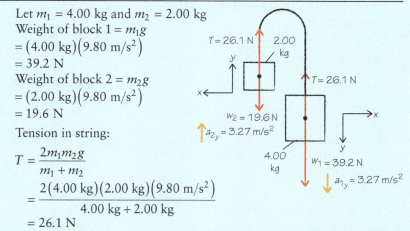

What about our expression for the tension T? To check this out let's choose some specific values of m_1 and m_2. With $m_1 = 4.00$ kg and $m_2 = 2.00$ kg, the tension is $T = 26.1$ N. That's less than the 39.2-N weight of block 1 but greater than the 19.6-N weight of block 2. This result makes sense because for block 1 the downward gravitational force overwhelms the upward tension force and the block accelerates downward, while for block 2 the upward tension force overwhelms gravity, and the block accelerates upward. Note that because the tension force "holds back" the downward acceleration of block 1, the value of a_y is less than g.

Let $m_1 = 4.00$ kg and $m_2 = 2.00$ kg

Weight of block 1 $= m_1 g$
$= (4.00 \text{ kg})(9.80 \text{ m/s}^2)$
$= 39.2$ N

Weight of block 2 $= m_2 g$
$= (2.00 \text{ kg})(9.80 \text{ m/s}^2)$
$= 19.6$ N

Tension in string:

$$T = \frac{2m_1 m_2 g}{m_1 + m_2}$$

$$= \frac{2(4.00 \text{ kg})(2.00 \text{ kg})(9.80 \text{ m/s}^2)}{4.00 \text{ kg} + 2.00 \text{ kg}}$$

$$= 26.1 \text{ N}$$

You should repeat these calculations for different values of the two masses. Try $m_1 = 1.00$ kg and $m_2 = 3.00$ kg: How does the tension compare to the weights of the two blocks in this case? Is the acceleration positive or negative in this case?

Acceleration of either block:

$$a_y = \left(\frac{m_1 - m_2}{m_1 + m_2}\right) g$$

$$= \left(\frac{4.00 \text{ kg} - 2.00 \text{ kg}}{4.00 \text{ kg} + 2.00 \text{ kg}}\right)(9.80 \text{ m/s}^2)$$

$$= 3.27 \text{ m/s}^2$$

Kinematics and Newton's Laws

All of the objects that we considered in Examples 4-4 through 4-9 move in straight lines. What's more, the forces acting on each object are constant. So the *accelerations* that these forces produce are constant as well. That means we can apply all of the kinematic formulas for straight-line motion with constant acceleration that we learned in Chapter 2. If you can solve problems that use these formulas, and can also solve problems that involve Newton's laws, it's straightforward to solve problems that involve both of these. Here's an example.

EXAMPLE 4-10 Down the Slopes

A 60.0-kg skier starts from rest at the top of a hill with a 30.0° slope. She reaches the bottom of the slope 4.00 s later. If there is a constant 72.0-N friction force that resists her motion, how long is the hill?

Set Up

All three of the forces that act on the skier—the normal force $\vec{n}$, the gravitational force $\vec{w}_{\text{skier}}$, and the friction force $\vec{f}$—are constant. So the net force on the skier is constant, as is her acceleration. We can therefore use a constant-acceleration formula to find the length of the hill. We'll find the skier's downhill acceleration using Newton's second law.

$$x = x_0 + v_{0x}t + \frac{1}{2}a_x t^2 \quad (2\text{-}9)$$

Newton's second law in component form for the skier:

x: $\sum F_{\text{ext on skier},x} = m_{\text{skier}} a_x$

y: $\sum F_{\text{ext on skier},y} = m_{\text{skier}} a_y$

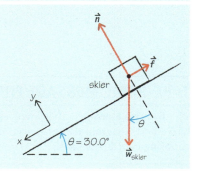

We choose the positive x axis to point down the slope and the positive y axis to point perpendicular to the slope as shown. (Compare Example 4-6.) Then the skier's acceleration points along the x axis only.

Net force on skier = sum of normal force, gravitational force, and friction force:

$$\sum \vec{F}_{\text{ext on skier}} = \vec{n} + \vec{w}_{\text{skier}} + \vec{f}$$

Solve

The normal force $\vec{n}$ has only a positive y component. The gravitational force $\vec{w}_{\text{skier}}$ has a positive (downhill) x component and a negative y component, while the friction force has only a negative (uphill) x component. Use these components and Newton's second law to solve for the skier's x component of acceleration a_x.

Find the skier's x component of acceleration from the x component of Newton's second law:

$$\sum F_{\text{ext on skier},x} = w_{\text{skier}} \sin \theta + (-f)$$
$$= m_{\text{skier}} a_x$$

Weight of skier:

$w_{\text{skier}} = m_{\text{skier}} g$, so

$$m_{\text{skier}} a_x = m_{\text{skier}} g \sin \theta - f$$

$$a_x = g \sin \theta - \frac{f}{m_{\text{skier}}}$$

$$= \left(9.80 \text{ m/s}^2\right) \sin 30.0° - \frac{72.0 \text{ N}}{60.0 \text{ kg}}$$

$$= 4.90 \text{ m/s}^2 - 1.20 \text{ m/s}^2$$

$$= 3.70 \text{ m/s}^2$$

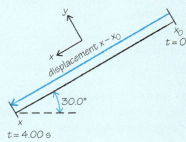

Now that we know the skier's acceleration in the x direction, we can use the kinematic equation to solve for her final position, which will be equal to the length of the hill as long as she starts at $x_0 = 0$. Because the skier starts at rest, her initial velocity is $v_{0x} = 0$.

Find the skier's final position after 4.00 s, assuming she starts at $x_0 = 0$.

$$x = x_0 + v_{0x} t + \frac{1}{2} a_x t^2$$

$$= 0 + 0 + \frac{1}{2}\left(3.70 \text{ m/s}^2\right)\left(4.00 \text{ s}\right)^2$$

$$= 29.6 \text{ m}$$

Reflect

Is our answer consistent? With an acceleration of 3.70 m/s², our skier reaches a velocity of 14.8 m/s after 4.00 s. Because she started at rest, her average x component of velocity for the trip is half the final velocity, or 7.40 m/s. If you travel at 7.40 m/s for 4.00 s, you cover a distance of 29.6 m—which is just the answer that we obtained.

Skier's velocity at $t = 4.00$ s:

$$v_x = v_{0x} + a_x t \tag{2-5}$$

The skier's initial velocity is $v_{0x} = 0$, so

$$v_x = 0 + \left(3.70 \text{ m/s}^2\right)\left(4.00 \text{ s}\right) = 14.8 \text{ m/s}$$

Average velocity for the skier's motion with constant acceleration:

$$v_{\text{average},x} = \frac{v_{0x} + v_x}{2} = \frac{0 + 14.8 \text{ m/s}}{2} = 7.40 \text{ m/s} \tag{2-7}$$

Distance traveled in 4.00 s:

$$x - x_0 = v_{\text{average},x} t = 0 + \left(7.40 \text{ m/s}\right)\left(4.00 \text{ s}\right) = 29.6 \text{ m}$$

By now you should have a pretty good sense of how to attack problems that involve force and acceleration. In the next chapter we'll use the same techniques to tackle problems involving a more detailed description of friction as well as motion in a circle.

TAKE-HOME MESSAGE FOR Section 4-6

✔ To solve a problem that involves forces, begin by making a free-body diagram for the object (or objects) of interest and choosing coordinate axes.

✔ Once you have drawn the free-body diagram, write Newton's second law for each coordinate axis and solve for the unknowns.

Key Terms

contact force	inertia	Newton's laws of motion
equilibrium	inertial frame of reference	Newton's second law
external force	internal force	Newton's third law
force	kilogram	normal force
force pair	mass	pound
free-body diagram	net external force (or net force)	reaction force
friction force	newton	tension
gravitational force	Newton's first law	weight

Chapter Summary

Topic	Equation or Figure	
Forces and Newton's second law of motion: The net external force on an object is the vector sum of all forces exerted on that object by other objects. Common forces include the gravitational force, the normal force, friction, and the tension force. Newton's second law states that if the net external force on an object is nonzero, the object accelerates.	The **net external force** acting on an object... ...equals the **vector sum** of all of the individual forces that act on the object. $$\sum \vec{F}_{\text{ext}} = \vec{F}_1 + \vec{F}_2 + \vec{F}_3 + \ldots$$ The sum includes only external forces (forces exerted on the object by other objects).	(4-1)
	If a net external force acts on an object... ...the object accelerates. The acceleration is in the same direction as the net force. $$\sum \vec{F}_{\text{ext}} = m\vec{a}$$ The magnitude of acceleration that the net external force causes depends on the mass m of the object (the quantity of material in the object). The greater the mass, the smaller the acceleration.	(4-2)
	The acceleration of an object... ...is proportional to and in the same direction as the net external force on the object... $$\vec{a} = \frac{1}{m} \sum \vec{F}_{\text{ext}}$$ ...and inversely proportional to the mass of the object.	(4-3)
Mass, weight, inertia, and Newton's first law of motion: The mass of an object is the quantity of material that it contains. Its weight is the magnitude of the gravitational force on the object. Inertia is the tendency of an object to maintain the same motion. Newton's first law states that if the net external force on an object is zero, it maintains a constant velocity.	**Weight** of an object (equal to the magnitude of the gravitational force on that object) $$w = mg$$ **Mass** of the object **Magnitude** of the **acceleration due to gravity**	(4-5)
	If the net external force on an object is zero... ...the object does not accelerate... $$\text{If } \sum \vec{F}_{\text{ext}} = 0, \text{ then } \vec{a} = 0 \text{ and } \vec{v} = \text{constant}$$ ...and the velocity of the object remains constant. If the object is at rest, it remains at rest; if it is in motion, it continues in motion in a straight line at a constant speed.	(4-6)

Free-body diagrams:
A free-body diagram depicts all of the external forces that act on an object. The coordinate axes are chosen so that as many of the vectors as possible (force and acceleration) point along one of the axes.

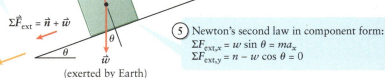

① With the friction force absent, the remaining normal force $\vec{n}$ and gravitational force $\vec{w}$ do not cancel. So the net external force $\Sigma\vec{F}_{ext} = \vec{n} + \vec{w}$ is not zero.

② The box neither leaps off the ramp nor sinks down into it, so it doesn't accelerate in the y direction. So the normal force must cancel the (negative) y component of the gravitational force.

③ Nothing cancels the positive x component of the gravitational force, so the net force is in the positive x direction...

④ ...and the acceleration of the block is in the positive x direction, the same as the direction of the net force.

(exerted by $\vec{n}$ the ramp)

$\Sigma\vec{F}_{ext} = \vec{n} + \vec{w}$

$\vec{a}$

θ

$\vec{w}$
(exerted by Earth)

(Figure 4-12)

⑤ Newton's second law in component form:
$\Sigma F_{ext,x} = w\sin\theta = ma_x$
$\Sigma F_{ext,y} = n - w\cos\theta = 0$

Newton's third law of motion:
The third law relates the forces that two objects exert on each other. This law is valid no matter how the objects move.

Force that object A exerts on object B

Force that object B exerts on object A

$$\vec{F}_{A\ on\ B} = -\vec{F}_{B\ on\ A}$$

(4-7)

💡 If object A exerts a force on object B, object B must exert a force on object A with the same magnitude but opposite direction.

Solving force problems: All problems that involve forces can be solved by (1) first drawing a free-body diagram for the object in question and (2) then writing out Newton's second law in component form and solving for the unknowns.

Sum of the x components of all external forces acting on the object

x component of the object's acceleration

(a) $\sum F_{ext,x} = ma_x$

Mass of the object

(b) $\sum F_{ext,y} = ma_y$

(4-4)

Sum of the y components of all external forces acting on the object

y component of the object's acceleration

Answer to What do you think? Question

(c) The forward force exerted on the airliner from the tug now exceeds the backward force of friction on the airliner. Hence there is a net forward force on the airliner, and it accelerates forward—that is, it speeds up.

Answers to Got the Concept? Questions

4-1 (c), (a), (b) and (d) [tie]. From Newton's second law, $\sum\vec{F}_{ext} = m\vec{a}$, the magnitude of the net force $\sum\vec{F}_{ext}$ equals the object's mass m multiplied by the magnitude of the object's acceleration $\vec{a}$. For case (a) this product is $(1250\text{ kg})(2.00\text{ m/s}^2) = 2500\text{ kg}\cdot\text{m/s}^2 = 2500\text{ N}$; for case (c) it is $(4500\text{ kg})(0.600\text{ m/s}^2) = 2700\text{ N}$. (The truck is slowing down rather than speeding up, but all that counts is the *magnitude* of the acceleration.) So there is a greater net force on the truck in (c) than on the automobile in (a). In cases (b) and (d) the objects are not accelerating (the velocities are constant), so $\vec{a} = 0$ and the net force on the object is zero.

4-2 Newton's second law holds everywhere in the universe, even in the absence of gravity. The second law says that if you apply the same force to each sphere, the massive lead sphere will have a smaller acceleration than the less massive plastic sphere. You can use this same principle here on Earth: Hold this textbook in one hand and a pencil in the other; then try to shake both hands side to side. The hand holding this textbook will move far less than the other hand because this book is quite a bit more massive than a pencil. So the acceleration of the book (and the hand holding it) will be much smaller than the acceleration of the pencil, assuming that each hand exerts roughly the same force.

4-3 (c) and (d) [tie], (a), (b). From Equation 4-5 an object's mass m is related to its weight w and the acceleration due to gravity g by $m = w/g$. So the mass of rock (a) is $m_a = (10.0 \text{ N})/(3.69 \text{ m/s}^2) = 2.71$ kg, and the mass of rock (b) is $m_b = (10.0 \text{ N})/(9.80 \text{ m/s}^2) = 1.02$ kg. Both rock (c) and rock (d) have a mass of 10.0 kg.

4-4 (d) Example 4-3 shows that $n = w_{cat} \cos \theta$ and $f = w_{cat} \sin \theta$, where w_{cat} is the weight of the cat. As the value of θ increases, the values of both $\cos \theta$ and n decrease, and the value of $\sin \theta$ increases. This says that the steeper the angle, the greater the friction force that the ramp must exert to keep the cat from sliding. As we'll learn in Chapter 5, there is a limit to how much friction force a surface can exert. So if you increase the angle too much, the ramp will no longer be able to provide the needed amount of friction, and the unhappy cat will start to slide.

4-5 (b) Three external forces act on the box: the gravitational force $\vec{w}$ (exerted by Earth), the normal force $\vec{n}$ (exerted by the surface of the ramp), and the friction force $\vec{f}$ (also exerted by the surface of the ramp). Acceleration is not a force, so the quantity $m\vec{a}$ doesn't belong in the free-body diagram. This rules out choices (d) and (e). The normal force $\vec{n}$ is not vertical in this situation but points perpendicular to the inclined surface of the ramp; this rules out choices (c) and (d). Finally, the friction force $\vec{f}$ opposes sliding, so if the box is sliding down the ramp it must be that $\vec{f}$ points up the ramp. This rules out choice (a). Hence only choice (b) is correct.

Note that this is the same free-body diagram as shown in Figure 4-11b for a block at rest on the ramp. The difference here is that friction is weaker than in Figure 4-11b, so the block slides down the ramp rather than remaining stationary.

4-6 It is true that when your friend pushes on the crate, the crate pushes back on him according to Newton's third law. However, these two forces act on two *different* objects. The force exerted by the crate acts on your friend, *not* on the crate. So we don't include it in the sum of external forces acting on the crate. (Remember that one of the keys to applying Newton's second law is considering *only* those external forces that act *on* the object you're considering—in this case, the crate.) The only horizontal forces that act on the crate are your friend's forward push and the backward push of friction exerted by the floor. If your friend can overcome that friction force, the crate will accelerate and start moving!

4-7 (a) Crate 1 is in contact with crate 2 and so exerts a force on it. Answer (b) is incorrect: You do *not* exert a force on crate 2 because you are not in direct contact with it. (It's true that crate 1 exerts a force on crate 2 only because you push on crate 1, but crate 2 doesn't "know" that: All it "feels" is the push from crate 1.) Crate 2 does exert a reaction force to the push it receives from crate 1. But Newton's third law tells us that this reaction acts on crate 1, *not* on crate 2, so answer (c) is incorrect. As we discussed in Section 4-4, acceleration is caused by force, not the other way around, and there is no such thing as "the force of acceleration." So answer (d), too, is incorrect.

Questions and Problems

In a few problems you are given more data than you actually need; in a few other problems you are required to supply data from your general knowledge, outside sources, or informed estimate.

  Interpret as significant all digits in numerical values that have trailing zeros and no decimal points.

  For all problems use $g = 9.80 \text{ m/s}^2$ for the free-fall acceleration due to gravity.

* •   Basic, single-concept problem
* ••   Intermediate-level problem; may require synthesis of concepts and multiple steps
* •••   Challenging problem
* Example   See worked example for a similar problem

Conceptual Questions

1. • According to Newton's second law, does the direction of the net force always equal the direction of the acceleration?

2. • If the sum of the forces acting on an object equals zero, does this imply that the object is at rest?

3. • What are the basic SI units (kg, m, s) for force according to Newton's second law $\sum \vec{F}_{ext} = m\vec{a}$?

4. • Explain why the force that a horizontal surface exerts on an object at rest on the surface is called the normal force.

5. • What is the net force on a bathroom scale when a 75-kg person stands on it?

6. • Two forces of 30 N and 70 N act on an object. What are the minimum and maximum magnitudes for the sum of these two forces?

7. • A definition of the inertia of an object is that it is a measure of the quantity of matter. How does this definition compare with the definition discussed in the chapter?

8. • **Astronomy** Why would it be easier to lift a truck on the Moon's surface than it is on Earth?

9. • When constructing a free-body diagram, why is it a good idea to choose your coordinate system so that the motion of an object is along one of the axes?

10. • **Sports** How can a fisherman land a 5-lb fish using fishing line that is rated at 4 lb?

11. • List all the forces acting on a bottle of water if it were sitting on your desk.

12. • **Astronomy** We know that the Sun pulls on Earth. Does Earth also pull on the Sun? Why or why not?

13. • **Medical** Use Newton's third law to explain the forces involved in walking.

14. • **Biology** Use Newton's third law to explain how birds are able to fly forward.

15. • A certain rope will break under any tension greater than 800 N. How can it be used to lower an object weighing 850 N over the edge of a cliff without the rope breaking?

16. • Tension is a very common force in day-to-day life. Identify five ordinary situations that involve the force of tension.

17. • A chair is mounted on a scale inside an elevator in the physics building. Describe the variation in the scale reading as the elevator begins to ascend, goes up at constant speed,

stops, begins to descend, goes down at a constant speed, and stops again.

18. • **Medical** Why does the American Academy of Pediatrics recommend that all infants sit in rear-facing car seats starting with their first ride home from the hospital? Explain your answer.

19. • **Medical** Why should the driver and passengers in a car wear seat belts? Explain your answer.

Multiple-Choice Questions

20. • The net force on a moving object suddenly becomes zero and remains zero. The object will
   A. stop abruptly.
   B. reduce speed gradually.
   C. continue at constant velocity.
   D. increase speed gradually.
   E. reduce speed abruptly.

21. • Which has greater monetary value, a newton of gold on Earth or a newton of gold on the Moon?
   A. a newton of gold on Earth
   B. a newton of gold on the Moon
   C. The value is the same, regardless of location.
   D. One cannot say without checking the weight on the Moon.
   E. a newton of gold on the Moon but only when inside a spaceship

22. • According to Newton's second law of motion, when a net force acts on an object, the acceleration is
   A. zero.
   B. inversely proportional to the object's mass.
   C. independent of mass.
   D. inversely proportional to the net force.
   E. directly proportional to the object's mass.

23. • In the absence of a net force, an object can be
   A. at rest.
   B. in motion with a constant velocity.
   C. accelerating.
   D. at rest or in motion with a constant velocity.
   E. It's not possible to know without more information.

24. • **Medical** A car stops suddenly during a head-on collision, causing the driver's brain to slam into the skull. The resulting injury would most likely be to which part of the brain?
   A. frontal portion of the brain
   B. rear portion of the brain
   C. middle portion of the brain
   D. left side of the brain
   E. right side of the brain

25. • **Medical** During the sudden impact of a car accident, a person's neck can experience abnormal forces, resulting in an injury commonly known as whiplash. If a victim's head and neck move in the manner shown in **Figure 4-24**, the car was hit from the
   A. front.
   B. rear.
   C. right side.
   D. left side.
   E. top during a rollover.

Time

Figure 4-24 Problem 25

26. • When a net force acts on an object, the object
   A. is at rest.
   B. is in motion with a constant velocity.
   C. has zero speed.
   D. is accelerating.
   E. is at rest or in motion with a constant velocity.

27. • In the absence of a net force, an object cannot be
   A. at rest.
   B. in motion with a constant velocity.
   C. accelerating.
   D. moving with an acceleration of zero.
   E. experiencing opposite but equal forces.

28. • How does the magnitude of the normal force exerted by the ramp on the block in **Figure 4-25** compare to the weight of the block? The normal force is

   Figure 4-25 Problem 28
   A. equal to the weight of the block.
   B. greater than the weight of the block.
   C. less than the weight of the block.
   D. possibly equal to or less than the weight of the block, depending on whether the ramp surface is smooth.
   E. possibly greater than or equal to the weight of the block, depending on whether the ramp surface is smooth.

29. • While an elevator traveling upward slows down to stop, the normal force on the feet of a passenger is _____ her weight. While an elevator traveling downward slows down to stop, the normal force on the feet of a passenger is _____ his weight.
   A. larger than; smaller than
   B. larger than; larger than
   C. smaller than; smaller than
   D. smaller than; larger than
   E. equal to; equal to

30. • Case (a) in **Figure 4-26** shows block A accelerated across a frictionless table by a hanging 10-N block (1.02 kg). In case (b) the same block A is accelerated by a steady 10-N tension in the string. Treat the masses of the strings, as well as the masses and friction of the pulleys, as negligible. The acceleration of block A in case (b) is
   A. greater than its acceleration in case (a).
   B. less than its acceleration in case (a).
   C. equal to its acceleration in case (a).
   D. twice its acceleration in case (a).
   E. half its acceleration in case (a).

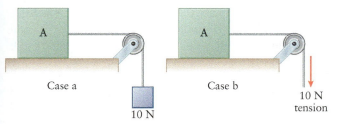

Figure 4-26 Problem 30

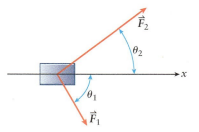

Figure 4-27 Problem 36

31. • A 1-kg ball and a 10-kg ball are dropped from a height of 10 m at the same time. In the absence of air resistance
    A. the 1-kg ball hits the ground first.
    B. the 10-kg ball hits the ground first.
    C. the two balls hit the ground at the same time.
    D. the 10-kg ball will take 10 times the amount of time to reach the ground.
    E. There is not enough information to determine which ball hits the ground first.

Problems

4-2 If a net external force acts on an object, the object accelerates

32. • A hockey player uses her stick to exert a force of 2.00 N on a stationary puck, which is resting on a nearly frictionless ice surface. The hockey puck has a mass of 160 g. What is the acceleration of the hockey puck? Example 4-1

33. • What net force is needed to accelerate a 2.00×10^3-kg car at 2.00 m/s^2? Example 4-1

34. •• Applying a constant net force to an object causes it to accelerate at 10 m/s^2. What will the acceleration of the object be if (a) the force is doubled, (b) the mass is halved, (c) the force is doubled and the mass is doubled, (d) the force is doubled and the mass is halved, (e) the force is halved, (f) the mass is doubled, (g) the force is halved and the mass is halved, and (h) the force is halved and the mass is doubled? Example 4-1

35. • Suppose that the maximum power delivered by a car's engine results in a force of 15,000 N being applied to the car by the road. In the absence of any other forces, what is the maximum acceleration this engine can produce in a 1250-kg car? Example 4-1

36. •• Three rugby players are pulling horizontally on ropes attached to a box, which remains stationary. Player 1 exerts a force F_1 equal to 1.00×10^2 N at an angle θ_1 equal to 60.0° with respect to the +x direction (Figure 4-27). Player 2 exerts a force F_2 equal to 2.00×10^2 N at an angle θ_2 equal to 37.0° with respect to the +x direction. The view in the figure is from above. Ignore friction and note that gravity can be ignored in this problem. (a) Determine the force $\vec{F}_3$ exerted by player 3. State your answer by giving the components of $\vec{F}_3$ in the directions perpendicular to and parallel to the positive x direction. (b) Redraw the diagram and add the force $\vec{F}_3$ as carefully as you can. (c) Player 3's rope breaks, and player 2 adjusts by pulling with a force of magnitude F_2' equal to 1.50×10^2 N at the same angle as before. In which direction is the acceleration of the box relative to the +x direction shown? (d) In part (c) the magnitude of the acceleration is measured to be 10.0 m/s^2. What is the mass of the box? Example 4-2

37. •• Three forces act on a 2.00-kg object at angles $\theta_1 = 40.0°, \theta_2 = 60.0°$, and $\theta_3 = 20.0°$, as shown in Figure 4-28. Find the magnitude of $\vec{F}_2$ and $\vec{F}_3$ if the magnitude of $\vec{F}_1$ is 1.00 N and the acceleration of the object is 1.50 m/s^2 toward the +x direction. Example 4-2

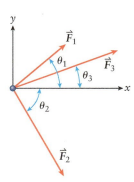

Figure 4-28 Problem 37

4-3 Mass, weight, and inertia are distinct but related concepts

38. • What is the weight on Earth of a wrestler who has a mass of 120 kg? Example 4-3

39. • A bluefin tuna has a mass of 250 kg. What is its weight? Example 4-3

40. •• **Astronomy** An astronaut has a mass of 80.0 kg. How much would the astronaut weigh on Mars, where surface gravity is 38.0% of that on Earth? Example 4-3

41. • What is the net force on an apple that weighs 3.5 N when you hold it at rest in your hand? Example 4-3

42. •• A 1300-kg car is capable of a maximum acceleration of 5.0 m/s^2. If this car is required to push a stalled car of mass 1700 kg, what is the maximum magnitude of acceleration of the two-car system? Example 4-3

4-4 Making a free-body diagram is essential in solving any problem involving forces

43. • Draw a free-body diagram for a heavy crate being lowered by a steel cable straight down at a constant speed. Example 4-3

44. • Draw a free-body diagram for a box being pushed horizontally by a person across a smooth, frictionless floor at a steadily increasing speed. Example 4-3

45. • Draw a free-body diagram for a bicycle rolling down a hill. Ignore the friction between the bicycle wheels and the hill, but consider any air resistance. Example 4-3

46. •• A tugboat uses its winch to pull up a sinking sailboat with an upward force of 4500 N. The mass of the boat is 200 kg and the water acts on the sailboat with a drag force of 2000 N. Draw a free-body diagram for the sailboat and describe the motion of the sailboat. Example 4-3

4-6 All problems involving forces can be solved using the same series of steps

47. •• Box A weighs 80 N and rests on a table (Figure 4-29). A rope that connects boxes A and B drapes over a pulley so that box B hangs above the table, as shown in the figure. The pulley and rope are massless, and the pulley is frictionless. What force does the table exert on box A if box B weighs (a) 35 N, (b) 70 N, (c) 90 N? Example 4-9

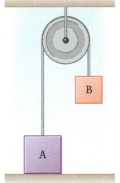

Figure 4-29 Problem 47

48. •• Two forces act on an object of mass $M = 3.00$ kg as shown in Figure 4-30. Because of these forces, the object experiences an acceleration of $a = 7.00$ m/s² in the $+x$ direction. If $\theta_1 = 30.0°$, (a) calculate the magnitude of $\vec{F}_2$ and (b) make a careful drawing to show its direction, given that $F_1 = 20.0$ N. Example 4-2

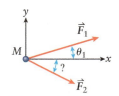

Figure 4-30 Problem 48

49. •• A 1.00×10^2-kg streetlight is supported equally by two ropes as shown in Figure 4-31. One rope pulls up and to the right, 40.0° above the horizontal; the other rope pulls up and to the left, 40.0° above the horizontal. What is the magnitude of the tension in each rope? Example 4-3

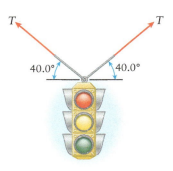

Figure 4-31 Problem 49

50. ••• A 2.00×10^2-N sign is supported by two ropes as shown in Figure 4-32. If $\theta_L = 45.0°$ and $\theta_R = 30.0°$, what is the magnitude of the tension in each rope? Example 4-3

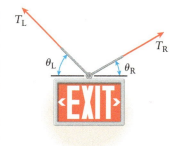

Figure 4-32 Problem 50

51. •• The distance between two telephone poles is 50.0 m. When a 0.500-kg bird lands on the telephone wire midway between the poles, the wire sags 0.15 m. How much tension does the bird produce in the wire? Ignore the weight of the wire. Example 4-3

52. •• A locomotive pulls 10 identical freight cars with an acceleration of 2.00 m/s². (a) What is the magnitude of the force between the third and fourth cars? (b) What is the magnitude of the force between the seventh and eighth cars? Example 4-8

53. • A locomotive pulls 10 identical freight cars. The force between the locomotive and the first car is 1.00×10^5 N, and the acceleration of the train is 2.00 m/s². There is no friction to consider. Find the force between the ninth and tenth cars. Example 4-8

54. • A 0.0100-kg block and a 2.00-kg block are attached to the ends of a rope. A student holds the 2.00-kg block and lets the 0.0100-kg block hang below it; then he lets go. What is the magnitude of the tension in the rope while the blocks are falling, before either hits the ground? Air resistance can be neglected. Example 4-8

55. • While parachuting, a 66.0-kg person experiences a downward acceleration of 2.50 m/s². What is the downward force on the parachute from the person? Example 4-4

56. • A bicycle and 50.0-kg rider accelerate at 1.00 m/s² up an incline of 10.0° above the horizontal. What is the magnitude of the force that the bicycle exerts on the rider? Example 4-6

57. •• A car uniformly accelerates from 0 to 28.0 m/s. A 60.0-kg passenger experiences a horizontal force of 4.00×10^2 N. How much time does it take for the car to reach 28.0 m/s? Example 4-10

58. •• A car accelerates from 0 to 1.00×10^2 km/h in 4.50 s. What force does a 65.0-kg passenger experience during this acceleration? Example 4-10

59. • Adam and Ben pull hand over hand on opposite ends of a rope while standing on a frictionless frozen pond. Adam's mass is 75.0 kg, and Ben's mass is 50.0 kg. If Adam's acceleration is 1.00 m/s² to the east, what are the magnitude and direction of Ben's acceleration? Example 4-8

60. •• Two blocks of masses M_1 and M_2 are connected by a massless string that passes over a massless pulley (Figure 4-33). M_2, which has a mass of 20.0 kg, rests on a long ramp of angle $\theta = 30.0°$. Friction can be ignored in this problem. (a) What is the value of M_1 for which the two blocks are in equilibrium (no acceleration)? (b) If the actual mass of M_1 is 5.00 kg and the system is allowed to move, what is the magnitude of the acceleration of the two blocks? (c) In part (b) does M_2 move up or down the ramp? (d) In part (b) how far does block M_2 move in 2.00 s? Example 4-9

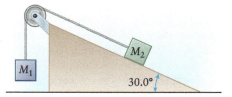

Figure 4-33 Problem 60

61. •• In Figure 4-34, the block on the left incline is 6.00 kg. If $\theta_1 = 60.0°$ and $\theta_2 = 25.0°$, find the mass of the block on the right incline so that the system is in equilibrium (no acceleration). All surfaces are frictionless. Example 4-9

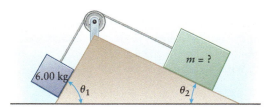

Figure 4-34 Problem 61

General Problems

62. • **Medical** During an arthroscopic surgery repair of a patient's knee, a portion of healthy tendon is grafted in place of a damaged anterior cruciate ligament (ACL). To test the results the surgeon applies a known force to the ligament and increases it at 1-s intervals for 10 s. Plot the data listed in the table as a function of time in a spreadsheet or on a graphing calculator and extrapolate the best-fit line to estimate the force that would correspond to a time of 15 s.

Time (s)	Force (mN)
0	0
1	2.28
2	5.73
3	11.28
4	12.27
5	12.54
6	12.38
7	14.11
8	16.79
9	21.08
10	28.51

63. •• Use a spreadsheet or graphing calculator to plot the velocity versus height of an express elevator. The elevator starts from rest on the ground floor, has constant acceleration until reaching its maximum speed of 10 m/s at the 4th floor (the distance between each floor is 3.5 m), remains at this speed until the 20th floor, and then has constant acceleration until it stops at the 30th floor. Identify the point(s) on the graph when the weight of a rider as measured by a spring scale will reach its maximum value.

64. • A reckless coyote engineer straps on a pair of ice skates and attaches a rocket capable of producing 5.0×10^3 N of thrust to his back. Together, the coyote and the rocket have a mass of 120 kg. If the coyote starts at rest on level, frictionless ice and bends over such that the rocket thrust is directed parallel to the ice, what is his final speed if the rocket burns for 5.0 s? Example 4-5

65. • You stand at the base of a 4.00-m long frictionless ramp that is inclined at an angle of 9.00°. You want to slide a 2.00-kg object up the ramp so that it stops just as it reaches the top. What initial velocity must you give the object? Example 4-5

66. • You and a friend are ice skating. Standing in the middle of the rink, you give your 85-kg friend a push with a force of 3.0×10^2 N. Assuming that the ice surface is frictionless, what magnitude of acceleration does your friend experience? Example 4-5

67. • A skier starts at rest atop a ski slope that has a slope of exactly 40.0°. Her total mass is 72.0 kg. Approximating the slope to be frictionless and assuming that the skier skis straight down the slope, what will her speed be after 25.0 s? Example 4-5

68. •• A 5.0-kg block slides in a straight line. The velocity of the block as a function of time is displayed in the v_x–t graph in Figure 4-35. Calculate the net force on the block for the time intervals $t = 0$ to 1 s, 1 to 3 s, 3 to 5 s, 5 to 6 s, and 6 to 7 s.

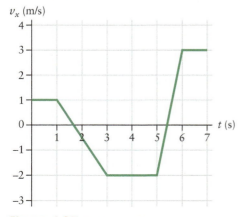

Figure 4-35 Problem 68

69. • The Atwood machine in Figure 4-36 consists of two masses hanging from the ends of a rope that passes over a pulley. The masses have the values $m_1 = 20.0$ kg and $m_2 = 12.0$ kg. (a) Determine the magnitude of the acceleration of the masses. (b) Calculate the tension in the rope. Assume that the rope and pulley are massless and that the pulley is frictionless. Example 4-9

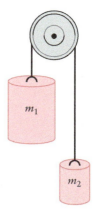

Figure 4-36 Problem 69

70. • **Medical, Astronomy** A spaceship takes off vertically from rest with an acceleration of 29.0 m/s². What force is exerted on a 75.0-kg astronaut during takeoff? Express your answer in newtons and also as a multiple of the astronaut's weight on Earth. Example 4-1

71. • A girl of mass 50.0 kg stands on a weight scale in an elevator that is initially at rest. As the elevator begins to move, the scale displays a value of 350 N. What acceleration does the girl experience? Assume that the positive direction is upward. Example 4-7

72. •• A child on a sled starts from rest at the top of a 20.0° slope. Assuming that there are no forces resisting the sled's motion, how long will the child take to reach the bottom of the slope, 210 m from the top? Example 4-10

73. •• A person weighing W newtons stands on a scale in an elevator that is initially at rest. The elevator accelerates upward to a constant speed, travels at that constant speed for some amount of time, and then slows to a stop. Which graph in Figure 4-37 best represents the reading on the scale as a function of time? Example 4-7

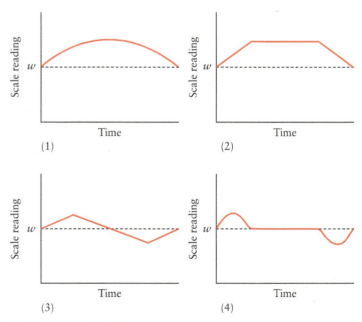

Figure 4-37 Problem 73

74. •• Your friend's car runs out of fuel, and you volunteer to push it to the nearest gas station. You carefully drive your car so that the bumpers of the two cars are in contact and then slowly accelerate to a speed of 2.00 m/s over the course of 1.00 min. If the mass of your friend's car is 1200 kg, what is the contact force between the two bumpers? Example 4-10

75. •• An 80.0-kg person stands on a scale in an elevator. Calculate the reading on the scale when the elevator (a) accelerates at 3.50 m/s² downward, (b) when the elevator cruises down at a steady speed, and (c) when the elevator accelerates at 4.00 m/s² upward. Example 4-7

76. •• A person weighs 588 N. If she stands on a scale while riding on the Inclinator (the lift at the Luxor Hotel in Las Vegas), what will be the reading on the scale? Assume the Inclinator moves at a constant acceleration of 1.25 m/s², in a direction 39.0° above the horizontal, and that the rider stands vertically in the elevator car. Example 4-7

77. •• A delivery person carries a stack of three boxes labeled 1, 2, and 3. Box 3 is on the bottom of the stack and box 1 is on the top. The masses of boxes 1, 2, and 3 are $m_1 = 1.5$ kg, $m_2 = 2.5$ kg, and $m_3 = 6.0$ kg, respectively. The delivery person places the stack of boxes on an elevator floor, which then accelerates upward with a magnitude of $a = 0.90$ m/s². (a) Calculate the contact force $\vec{F}_{1 \text{ on } 2}$ that box 1 exerts on box 2 during the acceleration. (b) Calculate the contact force $\vec{F}_{3 \text{ on } 2}$ that box 3 exerts on box 2 during the acceleration. Assume that the positive direction is upward. Example 4-7

78. •• A fuzzy die that has a weight of 1.80 N hangs from the ceiling of a car by a massless string. The car travels with a forward acceleration of 2.70 m/s² on a horizontal road. The string makes an angle θ with respect to the vertical, shown in Figure 4-38. What is the angle θ? Example 4-4

Figure 4-38 Problem 78

79. •• **Biology** On average, froghopper insects have a mass of 12.3 mg and jump to a height of 428 mm. The takeoff velocity is achieved as the little critter flexes its legs over a distance of approximately 2.00 mm. Assume a vertical jump with constant acceleration. (a) How long does the jump last (the jump itself, not the time in the air), and what is the froghopper's acceleration during that time? (b) Make a free-body diagram of the froghopper during its leap (but before it leaves the ground). (c) What force did the ground exert on the froghopper during the jump? Express your answer in millinewtons and as a multiple of the insect's weight. Example 4-10

80. •• **Medical** A car traveling at 28.0 m/s hits a bridge abutment. A passenger in the car, who has a mass of 45.0 kg, moves forward a distance of 55.0 cm while being brought to rest by an inflated air bag. Assuming that the force that stops the passenger is constant, what is the magnitude of this force? Example 4-10

81. •• Two mountain climbers are working their way up a glacier when one falls into a crevasse (Figure 4-39). The icy slope, which makes an angle of $\theta = 45.0°$ with the horizontal, can be

considered frictionless. Sue's weight is pulling Paul up the 45.0° slope. If Sue's mass is 66.0 kg and she falls 2.00 m in 10.0 s starting from rest, calculate (a) the tension in the rope joining them and (b) Paul's mass. Example 4-9

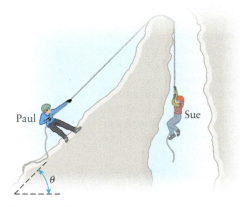

Figure 4-39 Problem 81

82. •• **Medical** During a front-end car collision, the acceleration limit for the chest is 60g. If a car was initially traveling at 48.0 km/h, (a) how much time does it take for the car to come to rest, assuming a constant acceleration equal to the threshold acceleration for damage to the chest? (b) Draw a free-body diagram of a person during the crash. (c) What force in newtons does the air bag exert on the chest of a 72.0-kg person? The trunk of the body comprises about 43.0% of body weight. (d) Why doesn't this force injure the person? Example 4-10

83. •• A window washer sits in a bosun's chair that dangles from a massless rope that runs over a massless, frictionless pulley and back down to the man's hand. The combined mass of man and chair is 95.0 kg. With how much force must he pull downward to raise himself (a) at constant speed and (b) with an upward acceleration of 1.50 m/s²? Example 4-9

84. •• A 2.00×10^2-kg block is hoisted by pulleys that are massless and frictionless, as shown in Figure 4-40. If a force of 1.50×10^3 N is applied to the massless rope, what is the acceleration of the suspended mass? Example 4-9

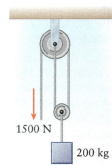

1500 N

200 kg

Figure 4-40 Problem 84

85. ••• Two blocks connected by a light string are being pulled across a frictionless horizontal tabletop by a hanging 10.0-N weight (block C) (Figure 4-41). Block A has a mass of 2.00 kg. The mass of block B is only 1.00 kg. The strings remain taut at all times. Assuming the pulley is massless and frictionless, what are the values of the tensions T_1 and T_2? Example 4-8

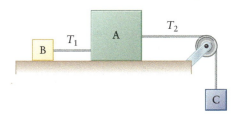

Figure 4-41 Problem 85

86. ••• Three boxes are lined up so that they are touching each other on a nearly frictionless plane, as shown in Figure 4-42. Box A has a mass of 20.0 kg, box B has a mass of 30.0 kg, and box C has a mass of 50.0 kg. If an external force (*F*) pushes on box A toward the right, and the force that box B exerts on box C is 2.00×10^2 N, what is the acceleration of the boxes and the magnitude of the external force *F*? Ignore friction. Example 4-8

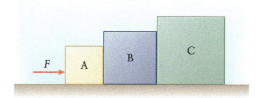

Figure 4-42 Problem 86

87. •• A 2.00-kg object A is connected with a massless string across a massless, frictionless pulley to a 3.00-kg object B (Figure 4-43). The smaller object rests on a nearly frictionless plane, which is tilted at an angle of $\theta = 40.0°$ as shown. What are the acceleration of the system and the tension in the string? Example 4-9

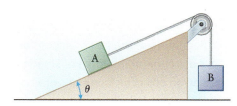

Figure 4-43 Problem 87

88. ••• A 1.00-kg object A is connected with a string to a 2.00-kg object B, which is connected with a second string over a massless, frictionless pulley to a 4.00-kg object C (Figure 4-44). Calculate the acceleration of the system and the tension in both strings. The strings have negligible mass and do not stretch, and the level tabletop is frictionless. Example 4-8

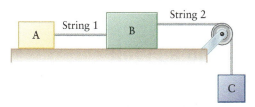

Figure 4-44 Problem 88

89. ••• A 1.00-kg object A is connected with a string to a 2.00-kg object B, which is connected with a second string over a massless, frictionless pulley to a 4.00-kg object C (Figure 4-45). The first two objects are placed onto a frictionless inclined plane that makes an angle $\theta = 30.0°$ as shown. Calculate the acceleration of the masses and the tensions in both strings. Example 4-8

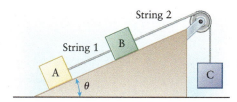

Figure 4-45 Problem 89

90. ••• **Sports** An athlete drops from rest from a platform 10.0 m above the surface of a 5.00-m-deep pool. Assuming that the athlete enters the water vertically and moves through the water with constant acceleration, what is the minimum average force the water must exert on a 62.0-kg diver to prevent her from hitting the bottom of the pool? Express your answer in newtons and also as a multiple of the diver's weight. Air resistance during the athlete's dive can be ignored in this problem. Example 4-10

91. ••• A person pulls three crates over a smooth horizontal floor as shown in Figure 4-46. The crates are connected to each other by identical horizontal strings A and B, each of which can support a maximum tension of 45.0 N before breaking. (a) What is the largest pulling force that can be exerted without breaking either of the strings? (b) What are the tensions in A and B just before one of the strings breaks? Example 4-8

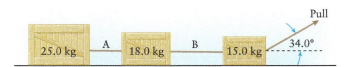

Figure 4-46 Problem 91

5

Forces and Motion II: Applications

What do you think?

When an airplane or a bird flies around a circular path at a constant speed, is there a nonzero *net* force on the airplane or bird? If so, does the net force point (a) inward toward the center of the circle, (b) outward away from the center of the circle, or (c) in some other direction?

In this chapter, your goals are to:

- (5-1) Appreciate that Newton's laws apply to all forces and in all situations.
- (5-2) Recognize what determines the magnitude of the static friction force.
- (5-3) Be able to find the magnitude and direction of the force of kinetic friction.
- (5-4) Solve problems in which the forces on an object include static or kinetic friction.
- (5-5) Analyze situations in which fluid resistance is important.
- (5-6) Apply Newton's laws to objects in uniform circular motion.

To master this chapter, you should review:

- (3-4)   The concepts and equations of motion in two or three dimensions.
- (3-7)   The concepts and equations of uniform circular motion.
- (4-2 to 4-6) How to solve problems using Newton's laws of motion.

5-1 Newton's laws apply to situations involving friction and drag as well as to motion in a circle

In Chapter 4 we learned the basic laws that describe the motions of objects, and we saw how external forces acting on an object determine the magnitude and direction of the object's acceleration. In this chapter we'll see how to apply these laws to a variety of very common situations. In particular, we'll take a much closer look at situations involving friction forces and drag forces—two forces that resist the motion of one object relative to another. We'll also analyze the direction of the forces acting on an object moving in a circular path and describe the resulting acceleration of that object.

The force of *friction* plays an important role in many physical situations. Friction acts in almost every case when two surfaces are in contact, and it can be present whether or not the two surfaces are sliding past each other. In Chapter 4 we used the friction force in some examples; in this chapter we'll learn more about the nature of friction and see how to solve more complex problems that involve friction. Another important situation arises when objects move through a fluid (a gas or liquid), such as a football flying through the air or a submarine cruising beneath the surface of the ocean.

The fluid exerts a *drag force* on each of these objects. The drag force makes it more difficult for an object to move through a fluid and is greater the faster an object travels.

The forces of friction and drag are special *kinds* of forces. Another important situation is one that can involve forces of *any* kind. Consider an object moving in a circular path at constant speed, like the airplanes shown in the photograph that opens this chapter. In this situation, no matter what forces act on the object, the net force must give the object an acceleration toward the center of the circle.

5-2 The static friction force changes magnitude to offset other applied forces

Friction is a force that resists the sliding of one object past another. If you push or pull on an object to make it slide but it doesn't move, we call the force that is preventing sliding a **static friction** force (Figure 5-1a). If the object is sliding, we call the force that opposes this motion a **kinetic friction** force (Figure 5-1b). Both static and kinetic friction forces result from interactions between the microscopic irregularities (bumps) of the two surfaces in contact. Even surfaces that appear or feel smooth can be rough at the microscopic level. Irregularities on one surface can catch on or be impeded by irregularities on the other surface.

We'll concentrate on static friction in this section and analyze kinetic friction in Section 5-3.

Properties of Static Friction

To demonstrate some properties of static friction, gently rest your open palm and fingertips against a wall and then try to drag your hand downward while maintaining contact with the wall. You'll find that it doesn't take much effort to make your hand slide. If you push a bit harder against the wall, you'll notice that it takes a bit more effort to get your hand to move. If you push as hard as you can on the wall, you'll find it very challenging to slide your hand down the wall. If the wall has a smooth surface, however, you'll find that your hands slip more easily than on a rough-surfaced wall.

Here's what we learn from this simple experiment:

(1) In each case the force that tries to keep your stationary hand from sliding is the force of static friction. We'll use the symbol $\vec{f}_s$ ("f-sub-s") for this force. (The subscript "s" reminds us that we're talking about *static* friction.) The static friction force $\vec{f}_s$ exerted on your hand by the wall acts parallel to your hand and parallel to the wall, that is, parallel to the surfaces in contact.

(2) You can overcome static friction by pushing against it with sufficient force. This says that in a given situation, the static friction force has a maximum magnitude $f_{s,max}$. You overcome the static friction force, and so start your hand sliding, if you apply a force in a direction parallel to the surfaces, whose magnitude is greater than $f_{s,max}$.

(3) The value of $f_{s,max}$ depends on how hard the two surfaces are pushed together—that is, on the *normal force* that acts perpendicular to the two surfaces. The greater the magnitude n of the normal force, the greater the maximum static friction force and the more difficult it is to make the surfaces slide past each other.

(4) The force of static friction depends on the properties of the surfaces in contact.

We can combine these observations into a simple mathematical relationship:

(a)

The ground exerts a static friction force on the stationary children's feet. This force prevents their feet from sliding.

(b)

The water slide exerts a kinetic friction force on the moving child that slows her down. The water reduces the friction but doesn't eliminate it.

Figure 5-1 Static friction versus kinetic friction Both (a) static friction and (b) kinetic friction arise from the interaction between two surfaces in contact.

① The magnitude of the **force of static friction** on an object...

② ...is less than or equal to a certain maximum value.

$$f_s \leq f_{s,max}$$

③ The **maximum force of static friction** depends on...

$$f_{s,max} = \mu_s n$$

④ ...the **coefficient of static friction** (which depends on the properties of the two surfaces in contact)...

⑤ ...and the **normal force** pressing the object against a surface.

Magnitude of the static friction force
(5-1a)

(5-1b)

The quantity μ_s in Equation 5-1b is called the **coefficient of static friction**. Its value depends on the properties of both surfaces in contact; for example, μ_s has different values for steel on steel, for steel on glass, and for glass on glass. Because both f_s and n have units of force, Equation 5-1b tells us that μ_s must be dimensionless.

Typical values of the coefficient of static friction are between about 0.1 and 0.8. Some approximate values of μ_s are 0.1 for a ski on snow, 0.5 for two pieces of wood in contact, and 0.7 for two pieces of metal in contact.

BioMedical

Lubrication reduces the frictional force between surfaces and hence the value of μ_s. In human joints, such as the elbow and knee, the presence of lubricating fluid results in a coefficient of static friction of about 0.01. This relatively low value of μ_s minimizes sticking and reduces wear and tear of the cartilage that supports these joints.

WATCH OUT! Static friction is described by an inequality, not an equality.

All of the mathematical relationships we've seen so far in our study of physics have been equalities, with an equals sign relating two quantities. But Equation 5-1a is an *inequality*. Figure 5-2 helps show why this is so. Imagine that you apply a horizontal force to a stationary block on a horizontal table. You find that if you apply a 20.0-N force, the block remains at rest. Therefore, the static friction force must have the same 20.0-N magnitude to balance the force that you apply (Figure 5-2a).

What happens if you apply a lesser force, say only 10.0 N? The block would still remain at rest, so the force of static friction is now only 10.0 N in magnitude to exactly cancel the applied force (Figure 5-2b). The block also remains at rest if you don't push on it at all, so in this case the static friction force is zero (Figure 5-2c). If you push enough to overcome the maximum force of static friction given by Equation 5-1b, the object will slide. In Section 5-3 we'll discuss what happens then.

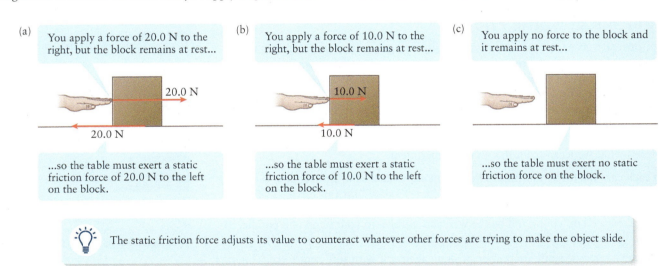

(a) You apply a force of 20.0 N to the right, but the block remains at rest...

20.0 N

20.0 N

...so the table must exert a static friction force of 20.0 N to the left on the block.

(b) You apply a force of 10.0 N to the right, but the block remains at rest...

10.0 N

10.0 N

...so the table must exert a static friction force of 10.0 N to the left on the block.

(c) You apply no force to the block and it remains at rest...

...so the table must exert no static friction force on the block.

The static friction force adjusts its value to counteract whatever other forces are trying to make the object slide.

Figure 5-2 **The magnitude of the static friction force** For a given object at rest on a given surface, the magnitude of the static friction force is *not* always the same: It depends on the other forces acting on the object.

WATCH OUT! The direction of the static friction force is parallel to the contact surface.

Equations 5-1 relate only the *magnitudes* of the static friction force and the normal force. These forces are *not* in the same direction! The friction force always acts parallel to the surface of contact, while the normal force always acts perpendicular to the surface. (Remember that "normal" is another word for "perpendicular.")

Static friction explains why a nail does not slide out of the wood into which it is hammered, even if the nail is upside down (Figure 5-3a). The wood fits tightly around the nail and so exerts strong normal forces on the sides of the nail (Figure 5-3b). Equation 5-1b tells us that the maximum static friction force is therefore quite large, more than enough to keep the nail in place. If you live in a wood-frame house, static friction deserves much of the credit for holding it together! The same effect explains why it is difficult to take off a tight-fitting pair of boots or to pull an overstuffed pillow out of its pillowcase.

(a)

(b)

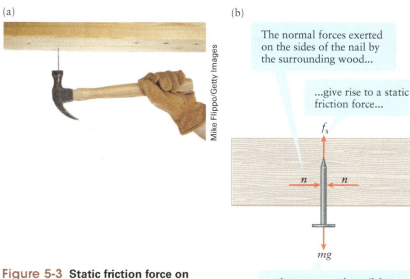

The normal forces exerted on the sides of the nail by the surrounding wood...

...give rise to a static friction force...

f_s

n n

mg

Figure 5-3 **Static friction force on a nail** (a) Hammering a nail upward into a piece of wood. (b) Forces on the nail.

...that prevents the nail from sliding out under the influence of the gravitational force.

Figure 5-4 **Walking on a glass wall** A gecko uses static friction to support its weight while walking on a vertical surface. It can do this thanks to the specialized architecture of its feet.

BioMedical

Geckos (tropical lizards of the family *Gekkonidae*) use static friction to walk on vertical surfaces, even on a smooth plate of glass (Figure 5-4). At first glance this seems impossible: On a vertical surface there is nothing pushing the gecko into the surface, so the normal force would be zero, and there would be zero static friction force (see Equation 5-1b). However, at the end of each of a gecko's toes are hundreds of thousands of tiny hairs or *setae*, each of which splits into hundreds of even smaller, flat pads. Each pad or *spatula* is only about 0.2 μm (2×10^{-7} m) across—so small that it can nestle in one of the tiny crevices on the surface of a pane of glass. Like what happens to the nail in Figure 5-3, the normal force between a spatula and the walls of its crevice give rise to a static friction force. Multiply that tiny force by the millions of spatulae on a gecko's four feet, and the result is a maximum force of friction as great as 20 N—more than adequate to support the weight of an average gecko, which is less than 0.5 N.

Measuring the Coefficient of Static Friction

Here's a simple experiment for measuring the coefficient of static friction. Place a block of weight w on a ramp whose angle θ can be varied (Figure 5-5a). If the angle θ is small enough, the block will remain at rest. However, the block will start to slide if the angle is greater than a certain critical value. Let's see how to determine the value of μ_s from this critical angle, which we'll call θ_{slip}.

We'll follow the two-step method for solving force problems that we used in Section 4-6. (You should review that strategy because we'll be using it a lot in this chapter.) Figure 5-5b shows the free-body diagram for the block when $\theta = \theta_{slip}$. (See Figure 4-9 in Section 4-4 for a reminder of how to draw free-body diagrams.) Three forces act on the block: the normal force of magnitude n, which is perpendicular to the surface of the ramp; the gravitational force of magnitude w, which points downward; and the static friction force of magnitude f_s, which points opposite to the direction the block would move if it could. At this angle the block is just about to slip, which means the friction force must be at its maximum value. So from Equations 5-1, $f_s = f_{s,max} = \mu_s n$.

We set the sum of the external forces on the block in each direction equal to its mass multiplied by its acceleration in that direction (Newton's second law). We take the positive x direction to point down the ramp and the positive y direction to point perpendicular to the ramp, as Figure 5-5b shows. The block remains at rest, so $a_x = a_y = 0$. Then the equations of Newton's second law are

(a) $\quad x: \sum F_{ext,x} = n_x + w_x + f_{s,max,x} = ma_x = 0$

(b) $\quad y: \sum F_{ext,y} = n_y + w_y + f_{s,max,y} = ma_y = 0$

(5-2)

(a) A block at rest on an inclined ramp

Block, weight w

Ramp whose angle θ from the horizontal can be varied

θ

(b) Free-body diagram for the block at the angle where it is just about to slip

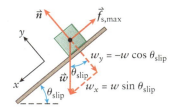

$\vec{n}$ $\vec{f}_{s,max}$

y

$w_y = -w \cos\theta_{slip}$

$\vec{w}$ θ_{slip}

x $w_x = w \sin\theta_{slip}$

θ_{slip}

Figure 5-5 **Measuring the coefficient of static friction** The angle at which a block on an incline just begins to slip tells you the coefficient of static friction μ_s.

Figure 5-5b shows that the gravitational force has a positive (downhill) x component $w_x = w \sin \theta_{slip}$ and a negative y component $w_y = -w \cos \theta_{slip}$. The normal force points in the positive y direction (so $n_x = 0$ and $n_y = n$), and the static friction force points in the negative x direction (so $f_{s,max,x} = -f_{s,max}$ and $f_{s,max,y} = 0$). So Equations 5-2 become

(5-3) $\quad$ (a) $\quad x$: $\sum F_{ext,x} = 0 + w \sin \theta_{slip} + (-f_{s,max}) = 0$

$\quad\quad\quad$ (b) $\quad y$: $\sum F_{ext,y} = n + (-w \cos \theta_{slip}) + 0 = 0$

Our goal is to find the coefficient of static friction μ_s and so we need to incorporate μ_s into our equations. We do this by substituting Equation 5-1b, $f_{s,max} = \mu_s n$, into Equation 5-3a. Then, by rearranging Equations 5-3, we get the following expressions:

(5-4) $\quad$ (a) $\quad \mu_s n = w \sin \theta_{slip}$

$\quad\quad\quad$ (b) $\quad n = w \cos \theta_{slip}$

Equation 5-4b tells us the value of n, the magnitude of the normal force, so we can substitute this quantity into Equation 5-4a and solve for μ_s:

$$\mu_s(w \cos \theta_{slip}) = w \sin \theta_{slip}$$

$$\mu_s \cos \theta_{slip} = \sin \theta_{slip}$$

Or, using $\tan \theta = (\sin \theta)/(\cos \theta)$,

Angle at which an object slips
on an incline
(5-5)

Coefficient of static friction for an object at rest on a surface $\quad$ Angle of the surface at which the object **begins to slip**

$$\mu_s = \tan \theta_{slip}$$

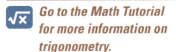 **Go to the Math Tutorial for more information on trigonometry.**

The greater the angle θ_{slip} beyond which the box begins to slide, the greater the value of $\tan \theta_{slip}$ and hence the greater the coefficient of static friction must be. Note that Equation 5-5 does *not* involve the weight of the block. A plastic block on a wooden ramp will begin to slide at the same angle θ regardless of its weight.

WATCH OUT! The coefficient of static friction can be greater than one.

! Because typical values of the coefficient of static friction are less than one, it is tempting to assume that μ_s is *always* less than one. Not so! Figure 5-6 shows a wooden block that rests without slipping on a ramp covered with sandpaper. The ramp is set at a 50° angle; if this is the angle beyond which the block will slip, then from Equation 5-5 we find that $\mu_s = \tan 50° = 1.2$. Whenever θ_{slip} is greater than 45°, the coefficient of static friction is greater than $\tan 45° = 1$.

Figure 5-6 **A block on a sandpaper surface** This block can remain at rest at an angle of up to 50°, so the coefficient of static friction for this block on sandpaper is $\tan 50° = 1.2$ (see Equation 5-5).

David Tauck

WATCH OUT! The normal force exerted on an object is not always equal to the object's weight.

! Since the maximum force of static friction is related to the normal force, it's important to remember that the normal force for an object on a surface is *not* always equal to the object's weight. (We cautioned you about this in Section 4-3.) The situation shown in Figure 5-5 is one example: Equation 5-4b shows that the normal force on this object is the object's weight multiplied by the cosine of the angle of the incline. Resist the temptation to always set the normal force equal to the weight!

BioMedical EXAMPLE 5-1 Friction in Joints

The wrist is made up of eight small bones called *carpals* that glide back and forth as you wave your hand from side to side. A thin layer of cartilage covers the surfaces of the carpals, making them smooth and slippery. In addition, the spaces between the bones contain synovial fluid, which provides lubrication. During a laboratory experiment, a physiologist applies a compression force to squeeze the bones together along their nearly planar bone surfaces. She then measures the force that must be applied parallel to the surface of contact to make them move. (Figure 5-7 shows the contact region between these two carpal bone surfaces.) When the compression force is 11.2 N, the minimum force required to move the bones is 0.135 N. What is the coefficient of static friction in the joint?

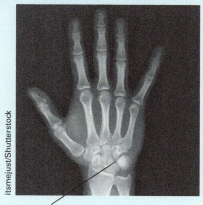

itsmejust/Shutterstock

Figure 5-7 Static friction in the wrist This x-ray image shows the carpal bones of the wrist.

Interface between two nearly planar carpal bone surfaces

Set Up

We'll consider the forces that act on one of the two carpals in question, which we call carpal A, when it is just about to slip relative to carpal B. (For simplicity we draw each carpal as a rectangle.) The physiologist exerts two of the four forces acting on carpal A: the compression force $\vec{F}_{compression}$ (of magnitude 11.2 N) and the force $\vec{F}_{slide}$ applied parallel to the contact surface (of magnitude 0.135 N) to make bone A slide relative to bone B. The other two forces acting on carpal A are exerted by carpal B: the normal force $\vec{n}$ and the static friction force $\vec{f}_{s,max}$. (This is the *maximum* force of static friction because the carpals are just about to slip.) Our goal is to find the value of μ_s, so we'll also use the relationship between $f_{s,max}$, μ_s, and n given by Equation 5-1b.

Newton's second law for carpal A:

$$\sum \vec{F}_{ext\,on\,A}$$
$$= \vec{F}_{compression} + \vec{F}_{slide} + \vec{n} + \vec{f}_{s,max}$$
$$= m_A \vec{a}_A$$

Magnitude of the static friction force:

$$f_{s,max} = \mu_s n \qquad (5\text{-}1b)$$

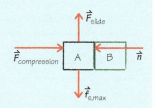

Solve

To draw a free-body diagram for carpal A, we move the force vectors so their tails are at the same point (which we indicate by a black dot). Let the positive x axis point up (in the direction in which we're trying to make carpal A slide) and the positive y axis point to the left. Since we are considering the situation in which the carpals have not yet begun to slide, carpal A is at rest and $a_{A,x} = a_{A,y} = 0$. We use this information to write Newton's second law for carpal A in component form.

Newton's second law for carpal A in component form:

$x:\ \sum F_{ext\,on\,A,x}$
$$= 0 + F_{slide} + 0 + (-f_{s,max})$$
$$= m_A a_{A,x} = 0$$

$y:\ \sum F_{ext\,on\,A,y}$
$$= (-F_{compression}) + 0 + n + 0$$
$$= m_A a_{A,y} = 0$$

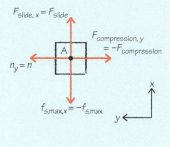

To find the value of μ_s using Equation 5-1b, we need to know the values of n and $f_{s,max}$. We find these quantities by using Newton's second law equations and the known values $F_{compression} = 11.2$ N and $F_{slide} = 0.135$ N.

Solve Equation 5-1b for μ_s:

$$\mu_s = \frac{f_{s,max}}{n}$$

Solve Newton's second law equations for $f_{s,max}$ and n. From the x component equation:

$$f_{s,max} = F_{slide} = 0.135 \text{ N}$$

From the y component equation:

$$n = F_{compression} = 11.2 \text{ N so}$$

$$\mu_s = \frac{0.135 \text{ N}}{11.2 \text{ N}} = 1.21 \times 10^{-2} = 0.0121$$

Reflect

This coefficient of static friction is very small, which means that very little effort is required to make your wrist move. The lubricating fluid between the bones is the key to keeping the friction so small. The fluid also helps reduce wear on the joints. If the bones or even the cartilage surfaces were in direct contact, our joints would wear down relatively quickly.

GOT THE CONCEPT? 5-1 Face to Face with Friction

(?) A rectangular block made of steel has dimensions 1.00 cm × 2.00 cm × 3.00 cm. You place it on a wooden board, and then you tilt the board until the block begins to slide down the board. Which face of the block should be in contact with the board so that it begins to slide at the *shallowest* angle? (a) The 1.00-cm × 2.00-cm face; (b) the 1.00-cm × 3.00-cm face; (c) the 2.00-cm × 3.00-cm face. (d) The block will begin to slide at the same angle in each case. (e) The answer depends on the type of steel and the type of wood.

TAKE-HOME MESSAGE FOR Section 5-2

✔ If two objects are at rest with respect to each other and an external force is applied to try to make them slide past each other, static friction acts to oppose that external force.

✔ The magnitude of the static friction force on an object adjusts itself to balance the other forces acting on the object.

However, there is a maximum possible magnitude of the static friction force, given by Equations 5-1.

✔ If an object is placed on a tilted ramp, it will slide if the angle of the ramp is greater than the slip angle given by Equation 5-5.

5-3 The kinetic friction force on a sliding object has a constant magnitude

In the previous section, we saw that if a block is on a tilted ramp (Figure 5-5) and the ramp is inclined by more than a certain critical angle, not even the maximum force of static friction will prevent the block from sliding. The block still experiences a resistive force as it slides. However, this force is no longer static friction but *kinetic friction*, the frictional force between two surfaces that move relative to each other.

The Magnitude of the Kinetic Friction Force

Like static friction, the force of kinetic friction $\vec{f}_k$ ("f-sub-k") between two objects is proportional to the normal force that one object exerts on the other perpendicular to their surface of contact—in other words, it is proportional to how hard the two surfaces are pushed together. However, the kinetic friction force is typically less than the maximum force of static friction.

We'll model the kinetic friction force as being independent of how fast the two objects slide past each other, as well as independent of the area of contact between the objects. (This model works well in many situations.) Then we can write an expression for the magnitude f_k of the kinetic friction force:

Magnitude of the kinetic friction force
(5-6)

The **force of kinetic friction** depends on...

$$f_k = \mu_k n$$

...the **coefficient of kinetic friction** (which depends on the properties of the two surfaces in contact)...

...and the **normal force** exerted on the object by the surface on which it slides.

 **→ Go to Interactive Exercises 5-1 and 5-2 for more practice dealing with kinetic friction.**

Like the coefficient of static friction μ_s, the value of the **coefficient of kinetic friction** μ_k depends on the kinds of surfaces that are in contact. For typical surfaces μ_k is close to but less than μ_s. That's why it takes less force to keep a block sliding at constant speed over a horizontal surface than it does to start it moving in the first place (Figure 5-8). Values of the coefficient of kinetic friction for typical materials range between 0.1 and 0.8.

(a)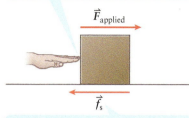
You push a stationary block with a horizontal force less than the maximum force of static friction.

$\vec{F}_{applied}$

$\vec{f}_{s}$

Static friction balances applied force: block remains at rest.

(b)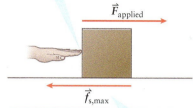
You push a stationary block with a horizontal force equal to the maximum force of static friction.

$\vec{F}_{applied}$

$\vec{f}_{s,max}$

Static friction just balances applied force: block is at rest but just about to slip.

(c)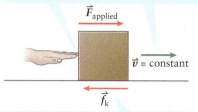
Once the block is in motion, you can keep it moving at constant velocity by exerting a horizontal force equal to the force of kinetic friction.

$\vec{F}_{applied}$

$\vec{v}$ = constant

$\vec{f}_{k}$

Force of kinetic friction has a smaller magnitude than the maximum force of static friction.

If you push hard enough on an object at rest, you can overcome the force of static friction.

Figure 5-8 From static friction to kinetic friction If you push hard enough on an object to overcome static friction and start the object sliding, the force of friction becomes the kinetic friction force.

In Figure 5-9a the block moves at a constant velocity. That's because the horizontal force that you apply just balances the force of kinetic friction of magnitude f_k opposing the motion. Hence the net force exerted on the block is zero and its acceleration is zero as well. If the horizontal force that you apply is greater in magnitude than f_k, the net force points forward and the block speeds up (Figure 5-9b). If you apply a horizontal force with a magnitude less than f_k to the sliding block, the net force points backward and the block slows down (Figure 5-9c).

(a)
You can keep a block sliding with constant velocity over a horizontal surface by exerting a horizontal force equal to the force of kinetic friction.

$\vec{F}_{applied}$

$\vec{v}$ = constant

$\vec{f}_{k}$

(b)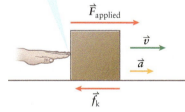
If you apply a horizontal force to the moving block greater than the force of kinetic friction, it speeds up.

$\vec{F}_{applied}$

$\vec{v}$

$\vec{a}$

$\vec{f}_{k}$

(c)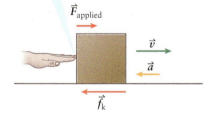
If you apply a horizontal force to the moving block less than the force of kinetic friction, it slows down.

$\vec{F}_{applied}$

$\vec{v}$

$\vec{a}$

$\vec{f}_{k}$

The kinetic friction force on a sliding object does not depend on how hard you push the object, nor does it depend on the speed of the object.

Figure 5-9 Kinetic friction: Three examples Depending on how hard you push on an object sliding with kinetic friction, it will (a) maintain a constant velocity, (b) speed up, or (c) slow down.

BioMedical EXAMPLE 5-2 Sliding with Kinetic Friction

Antarctic penguins travel long distances from the sea to their nesting area. Rather than doing a slow, waddling walk, penguins often lie on their stomach and slide or *toboggan* over the snowy surface (Figure 5-10), especially if the surface has a downward incline. Suppose the coefficient of kinetic friction between the incline and a sliding 32.0-kg emperor penguin (the largest of all penguin species) is 0.350. (a) What is the downhill acceleration of the penguin if the incline is at a 30.0° angle from the horizontal? (b) What is the acceleration if the angle is 10.0°?

Figure 5-10 A tobogganing penguin What will be the acceleration of this emperor penguin (*Aptenodytes forsteri*) as it slides down a snowy incline?

Frank Pey/Alamy

Set Up

The penguin moves down the incline of angle θ, so the kinetic friction force (which always opposes sliding) must point up the incline. To find the net external force on the penguin and hence the penguin's acceleration, we need to determine the magnitude of the friction force by using Equation 5-6. Once we've found a general expression for the penguin's acceleration in terms of θ, we'll substitute the values $\theta = 30.0°$ and $\theta = 10.0°$.

Newton's second law applied to the penguin:

$$\sum \vec{F}_{\text{ext on penguin}} = \vec{n} + \vec{w}_{\text{penguin}} + \vec{f}_k$$
$$= m_{\text{penguin}} \vec{a}_{\text{penguin}}$$

Magnitude of the kinetic friction force:

$$f_k = \mu_k n \tag{5-6}$$

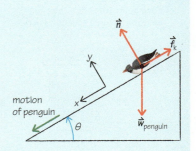

Solve

We draw the free-body diagram for the penguin and use it to help us write Newton's second law for the penguin in component form. Because the penguin's acceleration is along the incline (that is, along the x direction), it follows that $a_{\text{penguin},y} = 0$.

Newton's second law for the penguin in component form:

x: $\sum \vec{F}_{\text{ext on penguin},x}$
$= 0 + w_{\text{penguin}} \sin \theta + (-f_k)$
$= m_{\text{penguin}} a_{\text{penguin},x}$

y: $\sum \vec{F}_{\text{ext on penguin},y}$
$= n - w_{\text{penguin}} \cos \theta + 0$
$= m_{\text{penguin}} a_{\text{penguin},y} = 0$

To find the magnitude of the kinetic friction force, we solve the y component equation for the magnitude of the normal force n and then substitute this expression into the equation for f_k. Next we substitute our result for f_k into the x component equation.

From the y component equation,

$n = w_{\text{penguin}} \cos \theta$

Substitute this into Equation 5-6:

$f_k = \mu_k n = \mu_k (w_{\text{penguin}} \cos \theta)$

Substitute this expression for f_k into the x equation:

$w_{\text{penguin}} \sin \theta - \mu_k w_{\text{penguin}} \cos \theta = m_{\text{penguin}} a_{\text{penguin},x}$

Now we can complete our solution for the penguin's acceleration $a_{\text{penguin},x}$. Use the relationship between gravitational force and mass.

Relationship between the mass m_{penguin} of the penguin and the gravitational force w_{penguin} on the penguin:

$w_{\text{penguin}} = m_{\text{penguin}} g$

Substitute into the x equation and solve for $a_{\text{penguin},x}$:

$m_{\text{penguin}} g \sin \theta - \mu_k m_{\text{penguin}} g \cos \theta = m_{\text{penguin}} a_{\text{penguin},x}$

The factor of m_{penguin} cancels, so

$a_{\text{penguin},x} = g \sin \theta - \mu_k g \cos \theta$ or

$a_{\text{penguin},x} = g(\sin \theta - \mu_k \cos \theta)$

Use this general formula for $a_{\text{penguin},x}$ for the given value of μ_k and the two given values of θ.

(a) Substitute $\theta = 30.0°$:

$a_{\text{penguin},x} = g(\sin \theta - \mu_k \cos \theta)$
$= (9.80 \text{ m/s}^2)[\sin 30.0° - (0.350)(\cos 30.0°)]$
$= (9.80 \text{ m/s}^2)[0.500 - (0.350)(0.866)]$
$= 1.93 \text{ m/s}^2$

(b) Substitute $\theta = 10.0°$:

$a_{\text{penguin},x} = g(\sin \theta - \mu_k \cos \theta)$
$= (9.80 \text{ m/s}^2)[\sin 10.0° - (0.350)(\cos 10.0°)]$
$= (9.80 \text{ m/s}^2)[0.174 - (0.350)(0.985)]$
$= -1.68 \text{ m/s}^2$

Reflect

The penguin has a positive (downhill) acceleration for $\theta = 30.0°$, so it speeds up as it slides downhill. However, when the incline is at a shallower angle of $\theta = 10.0°$, the penguin

Downhill component of gravitational force:

$w_{\text{penguin},x} = w_{\text{penguin}} \sin \theta = m_{\text{penguin}} g \sin \theta$

Uphill component of kinetic friction force:

$f_k = \mu_k n = \mu_k m_{\text{penguin}} g \cos \theta_k$

has a *negative* (uphill) acceleration; its speed decreases as it moves down the incline. (In such a case a penguin would have to use its flippers to sustain its downhill motion.)

To see why the speed decreases, let's calculate the downhill (x) component of the gravitational force and the uphill component of the kinetic friction force for each angle. The shallower the angle θ, the smaller the downhill gravitational force and the greater the uphill force of kinetic friction. If the angle is small enough, the friction force will dominate over the gravitational force—in which case the net force and acceleration point up the incline.

Can you show that the acceleration of the penguin will be zero (that is, it will slide down the incline with constant velocity) if $\theta = 19.3°$?

If $\theta = 30.0°$,

$$w_{\text{penguin},x} = (32.0 \text{ kg})(9.80 \text{ m/s}^2)\sin 30.0°$$
$$= 157 \text{ N}$$

$$f_k = (0.350)(32.0 \text{ kg})(9.80 \text{ m/s}^2)\cos 30.0°$$
$$= 95.1 \text{ N}$$

Gravity dominates over kinetic friction and so the net force is downhill.

If $\theta = 10.0°$,

$$w_{\text{penguin},x} = (32.0 \text{ kg})(9.80 \text{ m/s}^2)\sin 10.0°$$
$$= 54.5 \text{ N}$$

$$f_k = (0.350)(32.0 \text{ kg})(9.80 \text{ m/s}^2)\cos 10.0°$$
$$= 108 \text{ N}$$

Kinetic friction dominates over gravity and so the net force is uphill.

EXAMPLE 5-3 How Far Up the Incline?

Suppose that after tobogganing downhill, the 32.0-kg penguin in Example 5-2 has a speed of 2.00 m/s. It then starts moving uphill on an incline that tilts 30.0° above the horizontal. The coefficient of kinetic friction between penguin and incline is again 0.350. How far up the incline does the penguin slide before coming to a stop?

Set Up

As in Example 5-2, the three forces acting on the sliding penguin are the normal force, gravitational force, and kinetic friction force. But now the friction force points downhill rather than uphill because the penguin is now sliding uphill. (Remember that the kinetic friction force always opposes sliding.)

This problem asks for a distance, so it's really a kinematics problem. Because all three forces that act on the penguin are constant, the net external force and acceleration are constant as well. Therefore, we can use the formulas from Section 2-4 for linear motion with constant acceleration. We choose an equation that relates the penguin's initial velocity at the bottom of the incline v_{0x}, its final velocity v_x, and its acceleration $a_{\text{penguin},x}$ to its displacement $x - x_0$, which is what we are asked to find.

Newton's second law applied to the penguin:

$$\sum \vec{F}_{\text{ext on penguin}} = \vec{n} + \vec{w}_{\text{penguin}} + \vec{f}_k$$
$$= m_{\text{penguin}}\vec{a}_{\text{penguin}}$$

Magnitude of the kinetic friction force:

$$f_k = \mu_k n \qquad (5\text{-}6)$$

Velocity, acceleration, and position for constant acceleration only:

$$v_x^2 = v_{0x}^2 + 2a_{\text{penguin},x}(x - x_0) \quad (2\text{-}11)$$

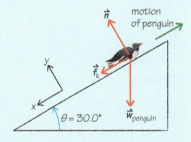

Solve

Use the free-body diagram to help us write Newton's second law for the penguin in component form. The equations are the same as in Example 5-2 except that the friction force now has a positive (downhill) x component. The penguin's acceleration is along the x axis only, so $a_{\text{penguin},y} = 0$.

Newton's second law for the penguin in component form:

$$x: \sum F_{\text{ext on penguin},x}$$
$$= 0 + w_{\text{penguin}} \sin \theta + f_k$$
$$= m_{\text{penguin}} a_{\text{penguin},x}$$

$$y: \sum F_{\text{ext on penguin},y}$$
$$= n - w_{\text{penguin}} \cos \theta + 0$$
$$= m_{\text{penguin}} a_{\text{penguin},y} = 0$$

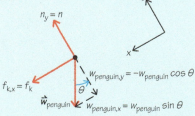

Use these equations and Equation 5-6 to solve for the acceleration of the penguin.

From the y component equation and $w_{penguin} = m_{penguin}g$,

$$n = w_{penguin} \cos \theta = m_{penguin} g \cos \theta$$

Substitute into Equation 5-6 and find the magnitude of the kinetic friction force:

$$f_k = \mu_k n = \mu_k m_{penguin} g \cos \theta$$

Substitute into the x component equation and solve for $a_{penguin,x}$:

$$m_{penguin} g \sin \theta + \mu_k m_{penguin} g \cos \theta = m_{penguin} a_{penguin,x}$$
$$g \sin \theta + \mu_k g \cos \theta = a_{penguin,x}$$
$$\begin{aligned} a_{penguin,x} &= g(\sin \theta + \mu_k \cos \theta) \\ &= (9.80 \text{ m/s}^2)[\sin 30.0° + (0.350)(\cos 30.0°)] \\ &= (9.80 \text{ m/s}^2)[0.500 + (0.350)(0.866)] \\ &= 7.87 \text{ m/s}^2 \end{aligned}$$

(Acceleration is positive because it points downhill.)

Now we can solve for the displacement of the penguin from the bottom of the incline to where it stops. Its initial velocity v_{0x} is negative because the penguin starts off moving uphill. Its final velocity v_x is zero because it is at rest after traveling the maximum distance up the incline. Use Equation 2-11 to find $x - x_0$.

Equation 2-11:

$$v_x^2 = v_{0x}^2 + 2a_{penguin,x}(x - x_0)$$

with $v_{0x} = -2.00$ m/s, $v_x = 0$, and $a_{penguin,x} = 7.87$ m/s^2

Solve for $x - x_0$:

$$\begin{aligned} x - x_0 &= \frac{v_x^2 - v_{0x}^2}{2a_{penguin,x}} \\ &= \frac{0 - (-2.00 \text{ m/s})^2}{2(7.87 \text{ m/s}^2)} \\ &= -0.254 \text{ m} = -25.4 \text{ cm} \end{aligned}$$

(Displacement is negative since the penguin moves uphill.)
So the penguin travels only 25.4 cm uphill.

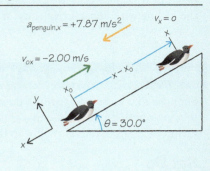

Reflect

The downhill acceleration of the penguin (7.87 m/s^2) is much greater than in Example 5-2 (1.93 m/s^2) because the force of kinetic friction now points downhill and so augments the downhill component of the gravitational force.

We can check our result by seeing how far the penguin would travel if the surface were horizontal, so $\theta = 0$. In this case the penguin travels 58.3 cm, which is more than twice the distance we found for the penguin on the 30.0° incline. It travels a greater distance if the surface is horizontal because the force of gravity is no longer helping to slow the penguin down. This result is physically reasonable and gives us confidence that our answer is correct.

If the surface is horizontal ($\theta = 0$):

$$\begin{aligned} a_{penguin,x} &= g(\sin 0 + \mu_k \cos 0) \\ &= g[0 + \mu_k(1)] \\ &= \mu_k g \\ &= (0.350)(9.80 \text{ m/s}^2) \\ &= 3.43 \text{ m/s}^2 \end{aligned}$$

Then the displacement of the penguin from where it starts to where it stops is

$$x - x_0 = \frac{v_x^2 - v_{0x}^2}{2a_{penguin,x}} = \frac{0 - (-2.00 \text{ m/s})^2}{2(3.43 \text{ m/s}^2)}$$
$$= -0.583 \text{ m} = -58.3 \text{ cm}$$

So the penguin travels 58.3 cm.

Rolling Friction

In addition to static friction and kinetic friction, a third important type of friction is **rolling friction**. If you start a billiard ball rolling across a horizontal surface, it will roll for quite a distance but will eventually come to a stop. A force must therefore act opposite to its motion, and it's this force that we call rolling friction (**Figure 5-11**). This force always acts opposite to the motion of a rolling object, whether it's rolling uphill, downhill, or on a horizontal surface.

Just as for static and kinetic friction, the force of rolling friction is proportional to the normal force that acts between the rolling object and the surface over which it rolls. So we can express the magnitude f_r of the rolling friction force as

$$f_r = \mu_r n \quad \text{(force of rolling friction)} \qquad (5\text{-}7)$$

The **coefficient of rolling friction** μ_r can be a very small number—about 0.02 for rubber automobile tires on concrete and about 0.002 for the steel wheels of a railroad car on steel rails. As a result, rolling friction is generally much smaller than kinetic friction, so a round object will roll much farther than a brick-shaped object of the same mass will slide. In general, the more rigid the rolling object and the more rigid the surface over which it rolls, the smaller the value of μ_r. The tires of your car are an example: If you don't keep them properly inflated, the coefficient of rolling friction for your tires will be greater than it should be, and your fuel consumption will be greater as well.

Figure 5-12 summarizes the three varieties of friction force.

The ball moves to the left...

...but rolls instead of sliding.

Normal force n

$\vec{v}$

Rolling friction force
$f_r = \mu_r n$

Gravitational force mg

Figure 5-11 A rolling billiard ball A ball can roll much farther on a horizontal surface than a block can slide on the same surface. The reason is that rolling friction is a much weaker force than kinetic friction.

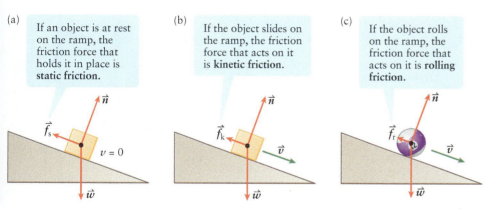

(a) If an object is at rest on the ramp, the friction force that holds it in place is **static friction**.

$\vec{n}$

$\vec{f_s}$

$v = 0$

$\vec{w}$

(b) If the object slides on the ramp, the friction force that acts on it is **kinetic friction**.

$\vec{n}$

$\vec{f_k}$

$\vec{v}$

$\vec{w}$

(c) If the object rolls on the ramp, the friction force that acts on it is **rolling friction**.

$\vec{n}$

$\vec{f_r}$

$\vec{v}$

$\vec{w}$

Figure 5-12 Three kinds of friction Comparing (a) static friction, (b) kinetic friction, and (c) rolling friction.

GOT THE CONCEPT? 5-2 A Question of Mass

Suppose the penguin in Examples 5-2 and 5-3 had a mass of 64.0 kg rather than 32.0 kg. How would this change the magnitude of the acceleration of the penguin as it moves along the ramp? (a) The magnitude would be greater going both downhill and uphill. (b) The magnitude would be less going both downhill and uphill. (c) The magnitude would be greater going downhill but less going uphill. (d) The magnitude would be less going downhill but greater going uphill. (e) None of these.

GOT THE CONCEPT? 5-3 Sliding Uphill to a Stop

Consider again the penguin sliding up the 30.0° incline in Example 5-3. For which of the following values of the coefficient of *static* friction will the penguin remain at rest after it stops? (a) 0.350 (the same as μ_k); (b) 0.600; (c) 0.750; (d) both (b) and (c); (e) all of (a), (b), and (c).

TAKE-HOME MESSAGE FOR Section 5-3

✔ If an object is sliding over a surface, the kinetic friction force on the object acts in the direction that opposes the sliding.

✔ The magnitude of the kinetic friction force is proportional to the normal force exerted on the object by the surface on which it is sliding (Equation 5-6).

✔ Like kinetic friction, rolling friction opposes motion and is proportional to the normal force, but the coefficient of friction for rolling is much smaller than that for sliding.

5-4 Problems involving static and kinetic friction are like any other problem with forces

In this section we'll look at a number of problems that involve static friction, kinetic friction, or both. We'll solve these examples using the same two-step approach that we introduced in Section 4-6. First, draw a free-body diagram for each object of interest; second, use the diagram to write down a Newton's second law equation for each object and solve for the unknowns. The only new features for problems that involve friction are the expressions for the magnitudes of the static friction force ($f_s \leq f_{s,max} = \mu_s n$) and the kinetic friction force ($f_k = \mu_k n$).

EXAMPLE 5-4 Two Boxes, a String, and Friction

Two boxes are connected by a lightweight string that passes over a massless, frictionless pulley (Figure 5-13). Box 1 (mass $m_1 = 2.00$ kg) sits on a ramp inclined at 30.0° from the horizontal, while box 2 (mass $m_2 = 4.00$ kg) hangs from the other end of the string. When the boxes are released, the string remains taut, box 1 accelerates up the ramp, and box 2 accelerates downward. The coefficient of kinetic friction for box 1 sliding on the ramp is $\mu_k = 0.250$. What is the acceleration of each box?

Figure 5-13 A pulley problem with friction If there is friction between box 1 and the ramp, what is the acceleration of the boxes?

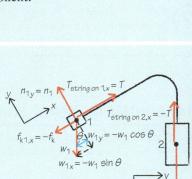

Set Up

Because box 2 has more mass than box 1, we expect that box 2 will accelerate downward and pull box 1 up the ramp. So we choose to have the positive x axis point up the ramp for box 1 and straight down for box 2. Box 1 has four forces acting on it: a normal force $\vec{n}_1$ from the surface of the ramp, its weight $\vec{w}_1$, the tension force from the string $\vec{T}_{\text{string on 1}}$, and the kinetic friction force from the ramp $\vec{f}_{k1}$. Two forces act on box 2: its weight $\vec{w}_2$ and the tension force from the string $\vec{T}_{\text{string on 2}}$. Because the string is lightweight and the pulley is massless and frictionless, the tensions $\vec{T}_{\text{string on 1}}$ and $\vec{T}_{\text{string on 2}}$ at either end have the same magnitude T. If the string doesn't stretch, the two boxes move together and their accelerations $\vec{a}_1$ and $\vec{a}_2$ have the same x component a_x. (Note that the positive x direction for each box is in the direction that the box moves.) Our goal is to calculate a_x.

Newton's second law applied to box 1:

$$\sum \vec{F}_{\text{ext on 1}}$$
$$= \vec{n}_1 + \vec{w}_1 + \vec{T}_{\text{string on 1}} + \vec{f}_{k1}$$
$$= m_1 \vec{a}_1$$

Magnitude of the kinetic friction force on box 1:

$$f_{k1} = \mu_k n_1$$

Newton's second law applied to box 2:

$$\sum \vec{F}_{\text{ext on 2}}$$
$$= \vec{w}_2 + \vec{T}_{\text{string on 2}} = m_2 \vec{a}_2$$

Tensions have the same magnitude:

$$T_{\text{string on 1}} = T_{\text{string on 2}} = T$$

Accelerations have the same x component:

$$a_{1x} = a_{2x} = a_x$$

Solve

Use the free-body diagrams to help write Newton's second law for each box in component form. Use T for the magnitude of each tension and a_x for the x component of each acceleration. Box 1 doesn't move in the y direction, so its y component of acceleration is zero and the net external force on box 1 has zero y component.

Newton's second law in component form for box 1:

x: $0 + (-w_1 \sin \theta) + T + (-f_k) = m_1 a_x$
y: $n_1 + (-w_1 \cos \theta) + 0 + 0 = 0$

Newton's second law in component form for box 2:

x: $w_2 + (-T) = m_2 g - T = m_2 a_x$

Calculate the normal force n_1 that acts on box 1 from the y component equation for that box. Use this expression to find the magnitude of the kinetic friction force on box 1 and substitute the result into the x equation for that box.

From the y component equation for box 1,

$$n_1 = w_1 \cos \theta = m_1 g \cos \theta$$

So the kinetic friction force on box 1 has magnitude

$$f_{k1} = \mu_k n_1 = \mu_k m_1 g \cos \theta$$

Substitute this expression and $w_1 = m_1 g$ into the x component equation for box 1:

$$-m_1 g \sin \theta + T - \mu_k m_1 g \cos \theta = m_1 a_x$$

We now have two equations that involve the unknown quantities T and a_x. We're only asked to find the value of a_x, so we eliminate T between the two equations.

x component equation for box 1:

$$-m_1 g \sin \theta + T - \mu_k m_1 g \cos \theta = m_1 a_x$$

x component equation for box 2:

$$m_2 g - T = m_2 a_x$$

Solve for T using the x component equation for box 1:

$$T = m_1 g \sin \theta + \mu_k m_1 g \cos \theta + m_1 a_x$$

Substitute this into the x component equation for box 2:

$$m_2 g - m_1 g \sin \theta - \mu_k m_1 g \cos \theta - m_1 a_x = m_2 a_x$$

Rearrange and solve for a_x:

$$m_2 g - m_1 g \sin \theta - \mu_k m_1 g \cos \theta = m_1 a_x + m_2 a_x$$

$$(m_2 - m_1 \sin \theta - \mu_k m_1 \cos \theta)g = (m_1 + m_2)a_x$$

$$a_x = \frac{(m_2 - m_1 \sin \theta - \mu_k m_1 \cos \theta)}{(m_1 + m_2)} g$$

$$= \frac{[4.00 \text{ kg} - (2.00 \text{ kg}) \sin 30.0° - (0.250)(2.00 \text{ kg}) \cos 30.0°]}{(2.00 \text{ kg} + 4.00 \text{ kg})}$$

$$\times (9.80 \text{ m/s}^2)$$

$$= 4.19 \text{ m/s}^2$$

Reflect

If there were no friction to slow the boxes down, we would expect the acceleration to be greater. We can check this conclusion by looking at the case $\mu_k = 0$ (which implies zero kinetic friction). Sure enough, a_x is greater in this case.

Returning to the case where friction is present, you can imagine that if box 1 were sufficiently more massive than box 2, the kinetic friction force would be so great that the boxes would slow down rather than speed up as they move. In this case a_x would be negative. Using $m_2 = 4.00$ kg and $\mu_k = 0.250$, can you show that a_x would be negative if m_1 were greater than 5.58 kg?

If there were no friction between box 1 and the ramp:

$\mu_k = 0$ so

$$a_x = \frac{(m_2 - m_1 \sin \theta)}{(m_1 + m_2)} g$$

$$= \frac{[4.00 \text{ kg} - (2.00 \text{ kg}) \sin 30.0°]}{(2.00 \text{ kg} + 4.00 \text{ kg})}(9.80 \text{ m/s}^2)$$

$$= 4.90 \text{ m/s}^2$$

EXAMPLE 5-5 Pinned Against a Wall

You place a block of plastic that weighs 33.0 N against a vertical wall and push it toward the wall with a force of 55.0 N (Figure 5-14). The coefficients of static and kinetic friction between the plastic and the wall are $\mu_s = 0.420$ and $\mu_k = 0.400$, respectively. Does the block remain at rest? If not, with what acceleration does it slip down the wall? Neglect any friction between you and the block.

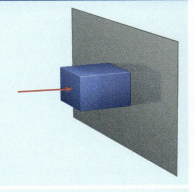

Figure 5-14 Pushing a block against a wall If you push a block against a wall with a force of a given magnitude, will the block remain at rest or slide down the wall?

Set Up

Four forces act on the block: the downward gravitational force $\vec{w}_{block}$, the force $\vec{F}_{you\ on\ block}$ you exert to push it against the wall, the normal force $\vec{n}_{block}$ that the wall exerts, and the upward friction force $\vec{f}$. The block will remain at rest only if the force of static friction is large enough to balance the 33.0-N gravitational force.

We'll use Equation 5-1b to determine the maximum force of static friction, which will help us decide whether the block will slip. If the block slips, we'll use Equation 5-6 for the force of kinetic friction to determine its acceleration.

Newton's second law applied to the block:

$$\sum \vec{F}_{ext\ on\ block}$$
$$= \vec{w}_{block} + \vec{F}_{you\ on\ block} + \vec{n}_{block} + \vec{f}$$
$$= m_{block}\vec{a}_{block}$$

Magnitude of the static friction force:

$$f_{s,max} = \mu_s n \qquad (5\text{-}1b)$$

Magnitude of the kinetic friction force:

$$f_k = \mu_k n \qquad (5\text{-}6)$$

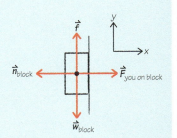

Solve

We use Newton's second law to determine the normal force that the wall exerts on the block and Equation 5-1b to find the maximum static friction force available. Only 23.1 N of static friction is available, which is less than the gravitational force on the block, so we conclude that the block will slip.

Newton's second law in component form for the block, assuming the block remains at rest (so the friction force is static friction):

x: $F_{you\ on\ block} + (-n_{block}) = 0$

y: $f_s + (-w_{block}) = 0$

From the x component equation,

$$n_{block} = F_{you\ on\ block} = 55.0\ \text{N}$$

From Equation 5-1b, the maximum static friction force available is

$$f_{s,max} = \mu_s n_{block} = (0.420)(55.0\ \text{N}) = 23.1\ \text{N}$$

From the y component equation, the required static friction force is

$$f_s = w_{block} = 33.0\ \text{N}$$

This force is more than the maximum static friction force available, so the block can't remain at rest.

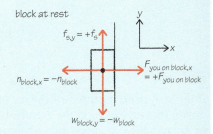

To find the acceleration of the block as it slides down the wall, we again apply Newton's second law. Again we find the normal force that the wall exerts on the block, and we use this value to find the kinetic friction force. Note that we are given the block's weight, not its mass, so we have to calculate its mass m_{block}.

Newton's second law in component form for the block, assuming the block is sliding downward (so the friction force is kinetic friction):

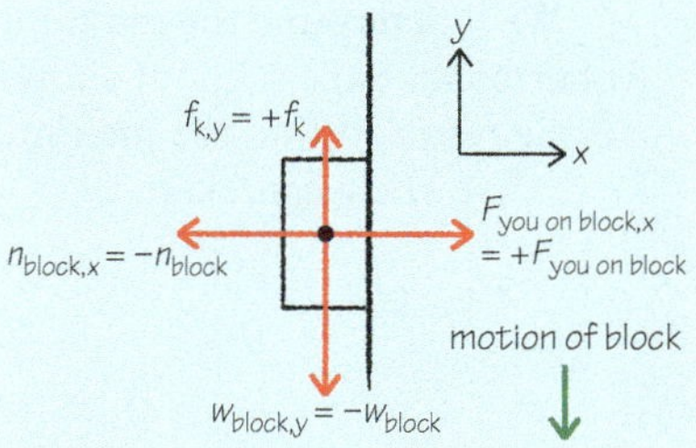

x: $F_{you\ on\ block} + (-n_{block}) = 0$

y: $f_k + (-w_{block}) = m_{block} a_{block,y}$

From the x component equation,

$n_{block} = F_{you\ on\ block} = 55.0\ \text{N}$

From Equation 5-6,

$f_k = \mu_k n_{block} = (0.400)(55.0\ \text{N}) = 22.0\ \text{N}$

Find the mass of the block:

$w_{block} = m_{block} g$

$m_{block} = \dfrac{w_{block}}{g} = \dfrac{33.0\ \text{N}}{9.80\ \text{m/s}^2} = 3.37\ \text{kg}$

Substitute into the y equation and solve for $a_{block,y}$:

$a_{block,y} = \dfrac{f_k - w_{block}}{m_{block}} = \dfrac{22.0\ \text{N} - 33.0\ \text{N}}{3.37\ \text{kg}} = -3.27\ \text{m/s}^2$

(figure labels, top right:)
$f_{k,y} = +f_k$
$n_{block,x} = -n_{block}$
$F_{you\ on\ block,x} = +F_{you\ on\ block}$
motion of block
$w_{block,y} = -w_{block}$

Reflect

We chose the positive y direction to be upward, so the negative value of $a_{block,y}$ means that the block accelerates downward. The magnitude of the acceleration is somewhat less than $g = 9.80\ \text{m/s}^2$ because kinetic friction prevents the block from attaining free fall.

Can you show that the block would remain at rest if you pushed with a force of at least 78.6 N?

EXAMPLE 5-6 A Sled Ride

Two children, Roberto (mass 35.0 kg) and Mary (mass 30.0 kg), go out sledding one winter day. Roberto sits on the sled of mass 5.00 kg, and Mary gives the sled a forward push (Figure 5-15). The coefficients of friction between Roberto and the upper surface of the sled are $\mu_s = 0.300$ and $\mu_k = 0.200$; the friction force between the sled and the icy ground is so small that we can ignore it. Mary finds that if she pushes too hard on the sled, Roberto slides toward the back of the sled. What is the maximum pushing force she can exert without this happening?

Mary, mass 30.0 kg
Roberto, mass 35.0 kg
Sled, mass 5.00 kg

Figure 5-15 Pushing a sled How hard can Mary push without making Roberto slide toward the back of the sled?

Set Up

If Roberto doesn't slip, the friction force that acts between him and the sled is *static* friction. (The sled and Roberto are sliding over the ice, but they're not sliding relative to each other.) The maximum force of static friction sets a limit on how great the forward acceleration of Roberto and the sled together can be.

This suggests that we should consider two objects: (a) Roberto by himself and (b) Roberto and the sled considered as a unit. Roberto is pushed forward by the friction force exerted on him by the sled, while Roberto and the sled together are pushed forward by Mary's push $\vec{F}_{Mary\ on\ unit}$. (Note that Mary doesn't touch Roberto, so her push doesn't act on him directly.)

Newton's second law equation for Roberto:

$\sum \vec{F}_{ext\ on\ Roberto}$
$\quad = \vec{n}_{sled\ on\ Roberto} + \vec{w}_{Roberto}$
$\quad\quad + \vec{f}_{s,sled\ on\ Roberto}$
$\quad = m_{Roberto} \vec{a}_{Roberto}$

Newton's second law equation for Roberto and the sled considered as a unit:

$\sum \vec{F}_{ext\ on\ unit}$
$\quad = \vec{n}_{ground\ on\ unit} + \vec{w}_{unit} + \vec{F}_{Mary\ on\ unit}$
$\quad = m_{unit} \vec{a}_{unit}$

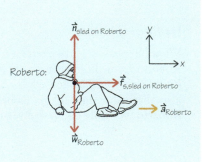
$\vec{n}_{sled\ on\ Roberto}$
Roberto:
$\vec{f}_{s,sled\ on\ Roberto}$
$\vec{a}_{Roberto}$
$\vec{w}_{Roberto}$

We'll determine the maximum acceleration Roberto can have without slipping and then use the result to find the maximum forward force that Mary can exert.

Magnitude of the static friction force:

$$f_{s,max} = \mu_s n \qquad (5\text{-}1b)$$

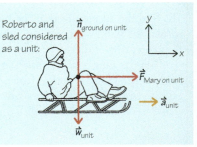

Roberto and sled considered as a unit:

Solve

First let's consider just Roberto and the forces that act on him. He is pushed forward by the static friction force that the sled exerts on him. If Roberto is just about to slip, the static friction force has its maximum value. To determine this value we first find the normal force that the sled exerts on Roberto.

Newton's second law in component form for Roberto, assuming that the static friction force has its maximum value:

x: $f_{s,max\ of\ sled\ on\ Roberto} = m_{Roberto} a_{Roberto,x}$

y: $n_{sled\ on\ Roberto} + (-w_{Roberto}) = 0$

From the y component equation,

$$n_{sled\ on\ Roberto} = w_{Roberto} = m_{Roberto}g$$

So the maximum force of static friction that the sled exerts on Roberto is

$$f_{s,max\ of\ sled\ on\ Roberto} = \mu_s n_{sled\ on\ Roberto}$$
$$= \mu_s m_{Roberto}g$$

So from the x component equation, Roberto's maximum acceleration is

$$a_{Roberto,x} = \frac{f_{s,max\ of\ sled\ on\ Roberto}}{m_{Roberto}}$$
$$= \frac{\mu_s m_{Roberto}g}{m_{Roberto}} = \mu_s g$$

Because Roberto does not slip, he and the sled have the same acceleration. So we can treat Roberto and the sled as a single unit that accelerates forward due to Mary's push. (For this combined unit we don't have to consider the forces that Roberto and the sled exert on each other. That's because these are *internal* forces.) Because we know Roberto's (and the sled's) maximum acceleration that will prevent slipping, we can calculate the maximum force that Mary can exert.

Newton's second law in component form applied to Roberto and the sled considered as a unit:

x: $F_{Mary\ on\ unit} = m_{unit} a_{unit,x}$

y: $n_{ground\ on\ unit} + (-w_{unit}) = 0$

If Roberto has his maximum acceleration and does not slip on the sled,

$$a_{unit,x} = a_{Roberto,x} = \mu_s g$$

So the maximum force that Mary can exert without causing Roberto to slip is

$$F_{Mary\ on\ unit} = m_{unit}\mu_s g$$
$$= (m_{Roberto} + m_{sled})\mu_s g$$
$$= [(35.0\ kg) + (5.00\ kg)](0.300)(9.80\ m/s^2)$$
$$= 118\ kg \cdot m/s^2 = 118\ N$$

Reflect

What happens if Mary exerts a force greater than 118 N on the sled? There will not be enough static friction to prevent Roberto from sliding across the top of the sled. The net force on him will still be forward, but now it will be due to *kinetic* friction. This force has a smaller magnitude than the maximum force of static friction (we are told that μ_k has a smaller value than μ_s). So Roberto's forward acceleration will be less than that of the sled. From Mary's perspective, Roberto will slide backward relative to the sled.

GOT THE CONCEPT? 5-4 Speeding Up with Friction?

 Can the force of friction cause an object to *speed up*? (a) Yes, but only the force of kinetic friction; (b) yes, but only the force of static friction; (c) yes, both kinetic and static friction can cause an object to speed up; (d) no.

TAKE-HOME MESSAGE FOR Section 5-4

✔ If friction has to be included in a problem that involves forces, approach the problem using the same steps as described in Section 4-6: Draw a free-body diagram for each object of interest; then use the diagram to write a Newton's second law equation for each object.

✔ If an object is not sliding, use Equations 5-1 to describe the static friction force. If the problem asks for a limiting case such as "when the object just begins to slip," assume the static friction force is maximum.

✔ If an object is sliding, use Equation 5-6 to describe the kinetic friction force.

5-5 An object moving through air or water experiences a drag force

When a solid object slides over another solid object, the force of kinetic friction provides resistance to its motion. Another important kind of resistance is the **fluid resistance** experienced by an object when it moves through a *fluid*—that is, a liquid or gas (Figure 5-16). The force that resists the motion of an object through a liquid such as water or a gas such as air is called a **drag force**.

(a)

This sea turtle feels a backward force of fluid resistance (drag) as it moves through the water.

To maintain a constant forward velocity, the turtle pushes backward on the water. The water responds by exerting a forward force (thrust) on the flippers (Newton's third law).

(b)

These skydivers feel an upward force of fluid resistance (drag) as they move downward through the air.

Fluid resistance opposes gravity. This slows their fall, giving them more time to enjoy the descent.

Figure 5-16 Fluid resistance When an object moves through either (a) a liquid such as water or (b) a gas such as air, it experiences fluid resistance (a drag force).

Drag Force on Microscopic Objects

Unlike the force of kinetic friction, the drag force depends on the object's *speed v* as it moves through the fluid. The faster the object's speed, the greater the magnitude of the drag force. For a very slow-moving object that is very small, such as a living cell or virus, experiment shows that the magnitude of the drag force is directly proportional to the speed v:

Magnitude of the **drag force** on a **small object** moving at a **low speed**

$$F_{\text{drag}} = bv$$

Constant that depends on the properties of the object and of the fluid

Speed of the object relative to the fluid

Drag force for small objects at low speeds
(5-8)

The value of b in Equation 5-8 depends on the size, shape, and surface characteristics of the object and the properties of the liquid or gas through which the object is moving.

As an example, for an algal spore (which has a spherical shape) of radius 0.020 mm moving in water, the coefficient b is 3.8×10^{-7} N·s/m. At a speed of 5.0×10^{-5} m/s, the drag force on this spore has magnitude

$$F_{\text{drag}} = bv = (3.8 \times 10^{-7} \text{ N·s/m})(5.0 \times 10^{-5} \text{ m/s}) = 1.9 \times 10^{-11} \text{ N}$$

This force is so small that you might think it would have no effect on the spore at all. However, the mass of a spore with this size is only 3.6×10^{-11} kg. If the only force acting on the spore is the drag force (Figure 5-17), the resulting acceleration is

$$a_{\text{spore},x} = \frac{F_{\text{drag},x}}{m_{\text{spore}}} = \frac{-1.9 \times 10^{-11} \text{ N}}{3.6 \times 10^{-11} \text{ kg}} = -0.53 \text{ m/s}^2$$

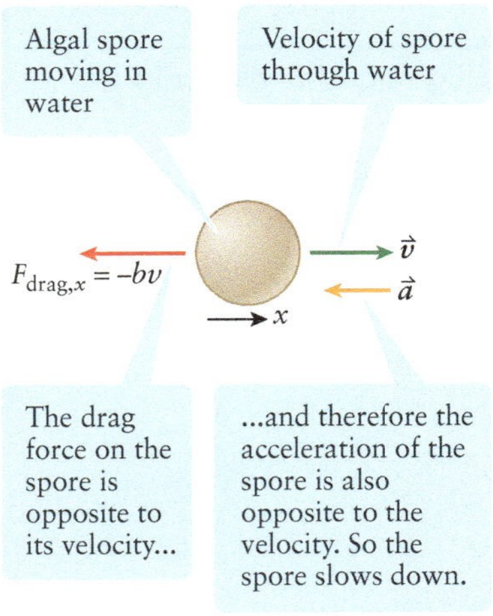

The negative values of $F_{\text{drag},x}$ and $a_{\text{spore},x}$ mean that the drag force acts opposite to the spore's motion, as in Figure 5-17. This substantial acceleration brings the slow-moving spore nearly to a halt relative to the water in a fraction of a second. Since such microscopic spores move very little *relative* to the water around them, they move readily from place to place only if the water is itself in motion; the drag force exerted by the moving water carries the spores along with it. Thus, thanks to fluid resistance, the location where an algal spore finally germinates and produces a new organism depends crucially on water currents.

Algal spore moving in water

Velocity of spore through water

$F_{\text{drag},x} = -bv$ $\vec{v}$ $\vec{a}$ x

The drag force on the spore is opposite to its velocity...

...and therefore the acceleration of the spore is also opposite to the velocity. So the spore slows down.

Figure 5-17 Drag on a microscopic spore A spherical algal spore moves through water in the positive *x* direction.

Drag Force on Larger Objects

For a larger object moving at relatively high speeds, such as the sea turtle or skydivers in Figure 5-16, experiment shows that the drag force has a different dependence on the speed v: Its magnitude is approximately proportional to the *square* of v. In this case we can write

Drag force for larger objects at faster speeds (5-9)

Magnitude of the **drag force** on a **larger object** moving at a **faster speed**

$$F_{\text{drag}} = cv^2$$

Constant that depends on the properties of the object and of the fluid

Speed of the object relative to the fluid

Like the coefficient b in Equation 5-8, the quantity c in Equation 5-9 depends on the fluid through which the object moves and the size, shape, and surface properties of the object. For a baseball flying through the air, $c = 1.3 \times 10^{-3}$ N·s^2/m^2. If the baseball is traveling at 25 m/s (a relatively slow pitching speed for a professional player), the magnitude of the drag force on the ball is

$$F_{\text{drag}} = (1.3 \times 10^{-3} \text{ N·s}^2/\text{m}^2)(25 \text{ m/s})^2 = 0.81 \text{ N}$$

This drag force is substantial: F_{drag} is a bit more than half the magnitude of the gravitational force on a regulation baseball, which weighs 1.4 N or 5 ounces. You can see that the drag force plays an important role in baseball and other ball sports.

Giving an object a streamlined shape can lower its value of c. For example, a dolphin (which has a streamlined body) moving in water has a value of c that is only about 1% as large as that of a sphere of the same cross-sectional area. Furthermore, because air is less dense than water, the value of c for a dolphin moving through air is only about 10^{-3} as great as for a dolphin moving through water. That explains why dolphins jump out of the water, or *porpoise*, when moving at high speeds: The dramatic reduction in drag force that they experience while airborne more than compensates for the effort required to leap clear of the water (Figure 5-18a).

Another species that minimizes its value of the quantity c in Equation 5-9 to achieve high speed is the peregrine falcon (Figure 5-18b). This species of predatory bird preys on other, smaller birds, and attacks its prey by diving on it from high altitude. To achieve maximum speed in a dive, which maximizes the impact force it can impart when it strikes its prey, a peregrine falcon streamlines its shape by folding back its wings and tail and tucking in its feet. As a result, a diving falcon can attain remarkable speeds, as the following example shows.

(a)

(b)

Neil Walker/Alamy

Figure 5-18 Two ways to minimize drag (a) By leaping into the air, these bottlenose dolphins (*Tursiops truncatus*) experience less drag than they do swimming in water. (b) As it dives on its prey, a peregrine falcon (*Falco peregrinus*) makes itself more streamlined to maximize its diving speed.

BioMedical

BioMedical **EXAMPLE 5-7 Terminal Speed**

When a peregrine falcon dives straight down toward its prey (Figure 5-18b), two forces act on it: a downward gravitational force and a drag force of magnitude F_{drag} given by Equation 5-9 directed vertically upward (opposite to the direction of the falcon's motion through the air). As the falcon falls and its speed v increases, the value of F_{drag} also increases. When the drag force becomes equal in magnitude to the gravitational force, the net force on the falcon is zero and the falcon ceases to accelerate. It has reached its *terminal speed*, so it no longer speeds up nor does it slow down. Find the terminal speed of a female peregrine falcon of mass 1.2 kg, for which the value of c is 1.6×10^{-3} N·s²/m².

Set Up

The free-body diagram shows the two forces acting on the falcon. We use Equation 5-9 to find the value of the speed v at which the sum of these forces is zero, so that the acceleration is zero and the downward velocity is constant.

$\sum \vec{F}_{\text{ext on falcon}}$
$= \vec{F}_{\text{drag on falcon}} + \vec{w}_{\text{falcon}}$
$= m\vec{a}_{\text{falcon}} = 0$

Drag force for larger objects at faster speeds:

$$F_{\text{drag}} = cv^2 \qquad (5\text{-}9)$$

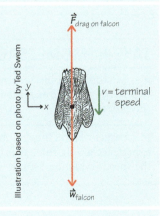

Illustration based on photo by Ted Swem

$\vec{F}_{\text{drag on falcon}}$

v = terminal speed

$\vec{w}_{\text{falcon}}$

Solve

Write Newton's second law in component form and solve for the terminal speed v_{term}.

Newton's second law in component form applied to the falcon:

y: $F_{\text{drag on falcon}} + (-w_{\text{falcon}}) = 0$

At the terminal speed v_{term},

$F_{\text{drag on falcon}} = cv_{\text{term}}^2$ so

$cv_{\text{term}}^2 - w_{\text{falcon}} = 0$

$cv_{\text{term}}^2 = w_{\text{falcon}} = mg$

$v_{\text{term}}^2 = \dfrac{mg}{c}$

$v_{\text{term}} = \sqrt{\dfrac{mg}{c}}$

Substitute the numerical values of m and c for the falcon.

Using $m = 1.2$ kg and $c = 1.6 \times 10^{-3}$ N·s²/m²,

$$v_{\text{term}} = \sqrt{\dfrac{(1.2 \text{ kg})(9.80 \text{ m/s}^2)}{1.6 \times 10^{-3} \text{ N·s}^2/\text{m}^2}}$$

$= 86$ m/s $= 310$ km/h $= 190$ mi/h

Reflect

The high diving speeds attained by a peregrine falcon make it the fastest member of the animal kingdom.

The relationship $v_{\text{term}} = \sqrt{\dfrac{mg}{c}}$ explains the common notion that "heavier objects fall faster." A baseball and an iron ball of the same radius falling side by side in air have the same value of c (because they have the same shape and size), but the iron ball will have a greater terminal speed because its mass m is greater. So a heavier object *does* fall faster if we take the drag force into account. If the baseball and iron ball were dropped side by side in a vacuum, however, they would have the same acceleration of magnitude g and so would always have the same speed.

WATCH OUT! The drag force is not a constant force.

! Unlike other forces we've considered, the magnitude of the drag force on an object changes as the object's speed changes. So if a drag force acts on an object, the net force on the object and hence the object's acceleration are *not* constant. As an example, a skydiver who falls without opening her parachute reaches half of her terminal speed ($v = v_{\text{term}}/2$) in 3 to 5 s. To achieve nearly terminal speed takes more than twice as long, about 15 to 20 s after beginning the fall. So her acceleration from rest to $v = v_{\text{term}}/2$ is much greater than that from $v = v_{\text{term}}/2$ to $v = v_{\text{term}}$. It would therefore be quite incorrect to try to apply the constant-acceleration formulas from Chapters 2 and 3.

GOT THE CONCEPT? 5-5 Comparing Drag Forces

? A baseball and a second object of the same mass are thrown into the air. When the second object is moving at one-half the speed of the baseball, the drag force on the second object is 4 times greater than the drag force on the baseball. Compared to the value of the quantity c for the baseball, the value of c for the second object is (a) 2 times greater, (b) 4 times greater, (c) 8 times greater, (d) 16 times greater, (e) none of these.

TAKE-HOME MESSAGE FOR Section 5-5

✔ Microscopic objects moving through a fluid at low speeds experience a drag force that is proportional to the speed (Equation 5-8).

✔ Larger objects moving through a fluid at faster speeds experience a drag force that is proportional to the square of the speed (Equation 5-9).

✔ A falling object reaches its terminal speed when the upward drag force just balances the downward gravitational force.

5-6 In uniform circular motion the net force points toward the center of the circle

Think of thrill seekers on a fast-moving amusement park ride or a motorcycle making a sharp turn (Figure 5-19). In both cases an object is following a curved path, so the direction of its velocity vector is changing. Because the direction of the velocity vector is changing, the object is *accelerating* (the object's speed can be constant or changing). In Section 3-5 we learned that when an object follows a curved path, the acceleration vector $\vec{a}$ must have a component directed toward the inside of the curve. We went on to discover in Section 3-8 that the magnitude of this component, called the *centripetal acceleration*, depends on the radius of the curve and on the object's speed. From Section 4-2 we know that for an object to accelerate there must be a *net external force* that points in the direction of the acceleration vector $\vec{a}$. In this section we'll put all of these ideas together and complete our analysis of objects that move along curved paths.

What kinds of force can provide the acceleration needed to make an object follow a curved path? The answer is simple: *any kind of force*, provided that it has a component perpendicular to the object's trajectory. In Figure 5-19a, for example, the force responsible is the normal force exerted on each person by the seat on which they sit. Because the seat is tilted or *banked*, the normal force $\vec{n}$ on the person is not

Each person on this amusement park ride moves in a circle at constant speed...

This motorcycle rider also moves along a circular path at constant speed...

(a)

(b)

F_{net}

F_{net}

F_{net}

F_{net}

...so the acceleration of each person and the net force on each person both point toward the center of the person's circular path.

...so his acceleration and the net force on him both point toward the center of the circular path.

Figure 5-19 Forces in uniform circular motion An object moving on a circular path at a constant speed is in uniform circular motion. Since the object's direction of motion is continuously changing, the object is accelerating and a net external force must act on it.

vertical, but also has a horizontal component; it's this component of $\bar{n}$ that provides the required acceleration. For the motorcycle in Figure 5-19b, the acceleration is provided by the horizontal friction force that the ground exerts on the motorcycle's tires.

One of the simplest situations in which an object follows a curved trajectory is *uniform circular motion*, or motion around a circular path at a constant speed. We saw in Section 3-8 that the acceleration at any point in uniform circular motion is directed toward the center of the circle (that is, it is *centripetal*) and has a magnitude a_{cent} that depends on the object's speed v and the radius r of the circle:

Centripetal acceleration: Magnitude of the acceleration of an object in uniform circular motion

Speed of the object as it moves around the circle

$$a_{cent} = \frac{v^2}{r}$$

Radius of the circle

Centripetal acceleration for motion in a circle
(3-17)

Because the acceleration is always toward the center of the circle, it follows from Newton's second law that at any instant the net external force acting on the object of mass m must likewise be directed toward the center of the circle. Furthermore, the magnitude of the net external force must be equal to ma_{cent}. In equation form,

Net force on an object in uniform circular motion

Mass of the object

$$\sum F_{cent} = \frac{mv^2}{r}$$

Speed of the object as it moves around the circle

The direction of the net force is toward the center of the object's circular path.

Radius of the circle

Newton's second law equation for uniform circular motion
(5-10)

This equation says that in uniform circular motion the magnitude of the net external force on the object is equal to the object's mass m multiplied by the magnitude of the centripetal acceleration $a_{cent} = v^2/r$.

Let's look at several examples of how to apply these ideas to problems in which an object moves around a circle at constant speed.

→ *Go to Interactive Exercise 5-3 for more practice dealing with centripetal acceleration.*

WATCH OUT! "Centripetal force" is not a special *kind* of force—it describes the *direction* of the net force.

The force that points toward the inside of an object's curving trajectory and produces the centripetal acceleration is sometimes called the **centripetal force**. Keep in mind, however, that this is not a separate kind of force that you need to include in a free-body diagram. The word "centripetal" simply describes the direction of the force. Just as we might write F_x to denote the x component of a force, we write F_{cent} to denote the component of force directed toward the center of an object's circular trajectory. As we'll see in the following examples, in different circumstances different forces play the role of the centripetal force.

EXAMPLE 5-8 A Rock on a String

You tie one end of a lightweight string of length L around a rock of mass m. You hold the other end of the string in your hand and make the rock swing in a horizontal circle at constant speed. As you swing the rock the string makes an angle θ with the vertical (Figure 5-20). Derive an expression for the speed at which the rock moves around the circle. Ignore the effect of drag forces.

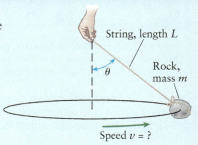

Figure 5-20 Moving in a circle How fast must the rock move to make the string hang at an angle θ from the vertical?

Set Up

Because the rock moves at constant speed in a horizontal circle, its acceleration vector points horizontally toward the center of the circle. We choose the positive x axis to point in this direction and the positive y axis to point upward.

Only two forces act on the rock: the tension force $\vec{T}$ exerted by the string and the downward gravitational force $\vec{w}_{\text{rock}}$. The tension force is directed along the string and so points at an angle θ to the vertical. The vertical component of $\vec{T}$ balances the downward force of gravity, while the horizontal component of $\vec{T}$ provides the centripetal acceleration. Our goal is to solve for the rock's speed v.

Newton's second law equation for the rock:

$$\sum \vec{F}_{\text{ext on rock}} = \vec{T} + \vec{w}_{\text{rock}} = m_{\text{rock}}\vec{a}_{\text{rock}}$$

Centripetal acceleration:

$$a_{\text{rock,cent}} = \frac{v^2}{r} \qquad (3\text{-}17)$$

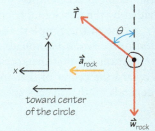

toward center of the circle

Solve

Write Newton's second law for the rock in component form. Note that the quantity v, which is what we're trying to find, doesn't appear in these equations. We'll introduce it in the next step through the expression for the rock's centripetal acceleration.

Newton's second law in component form applied to the rock:

x: $T \sin \theta = m_{\text{rock}} a_{\text{rock,cent}}$

y: $T \cos \theta + (-w)$
   $= T \cos \theta - m_{\text{rock}} g = 0$

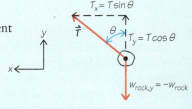

Use Equation 3-17 to express the acceleration of the rock. Note that because the string is at an angle, the radius of the rock's circular path is *not* equal to the length of the string.

Radius of the rock's circular path:

$r = L \sin \theta$

so the rock's centripetal acceleration is

$$a_{\text{rock,cent}} = \frac{v^2}{r} = \frac{v^2}{L \sin \theta}$$

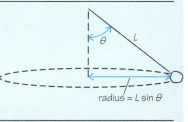

Substitute the expression for centripetal acceleration into the Newton's second law equation for the x direction and solve for v.

Newton's second law equations for the rock become

x: $T \sin \theta = m_{\text{rock}} a_{\text{rock,cent}} = \dfrac{m_{\text{rock}} v^2}{L \sin \theta}$

y: $T \cos \theta = m_{\text{rock}} g$

Solve the y component equation for T:

$$T = \frac{m_{\text{rock}} g}{\cos \theta}$$

Substitute this expression into the x component equation and solve for v:

$$\frac{m_{rock}g}{\cos\theta}\sin\theta = \frac{m_{rock}v^2}{L\sin\theta}$$

$$\frac{g\sin\theta}{\cos\theta} = \frac{v^2}{L\sin\theta}$$

$$v^2 = \frac{gL\sin^2\theta}{\cos\theta}$$

$$v = \sqrt{\frac{gL\sin^2\theta}{\cos\theta}}$$

Reflect

You can experiment with this motion yourself by tying a piece of string around an eraser or other small object and whirling it in a horizontal circle. You'll find that the faster the object moves, the greater the angle θ. Let's check that our answer agrees with this conclusion.

If θ is a small angle close to zero so that the string hangs down almost vertically, $\sin\theta$ is small and $\cos\theta$ is nearly equal to 1. (Remember that $\sin 0 = 0$ and $\cos 0 = 1$.) So the ratio $(\sin^2\theta)/(\cos\theta)$ is a small number, and the speed v will be small. If θ is a large angle close to 90° so that the string is nearly horizontal, $\sin\theta$ is a little less than 1 and $\cos\theta$ is small (recall $\sin 90° = 1$ and $\cos 90° = 0$). In this case the ratio $(\sin^2\theta)/(\cos\theta)$ is a large number, and the speed v will be large. So our formula agrees with the experiment!

WATCH OUT! No "centrifugal force," please.

In Example 5-8 you may have been tempted to add a force that points toward the *outside* of the rock's curved trajectory (Figure 5-21a). Don't do it! If you felt this temptation, it's because you're thinking of the so-called centrifugal force you feel when riding in a car that makes a sharp turn. This "centrifugal force" seems to push you toward the outside of the turn ("centrifugal" means "fleeing the center"). This force is *fictitious*, however; nothing is exerting an outward force on you. Rather, you're just feeling your body's inertia, which is the tendency of all objects to want to continue in a straight line. Inertia explains what would happen if you let go of the string attached to the rock: The rock would fly off in a straight line as seen from above (Figure 5-21b) and so it would move away from the center of its circular path. That's not because of a mysterious "centrifugal force" but rather because of the *absence* of the inward tension force. The lesson here is that you should never include a centrifugal force in any of your free-body diagrams. (If you still feel that you have to add an outward force to "keep the object in balance," remember that an object going around a curve is *not* in balance: The object is accelerating, so the net external force on it is *not* zero!)

(a) *Incorrect* free-body diagram of the rock

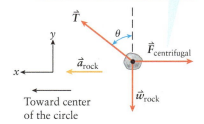

There is no object that exerts $F_{centrifugal}$. So this force does not exist and should not be included in the free-body diagram.

(b) A top view of the rock's trajectory

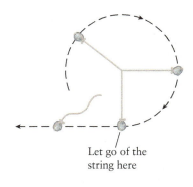

Let go of the string here

Figure 5-21 How *not* to draw a free-body diagram for uniform circular motion (a) A free-body diagram should never include a "centrifugal force." (b) If the string is let go, the rock flies off in a straight line due to its inertia — *not* a "centrifugal force."

EXAMPLE 5-9 Making an Airplane Turn I: Lift and Apparent Weight

An airplane banks to one side in order to turn in that direction. By banking the plane, the *lift force*—a force that acts perpendicular to the direction of flight, caused by the motion of air over the airplane's wings—ends up with both a vertical component and a horizontal component (Figure 5-22). The horizontal component of this force provides the centripetal acceleration needed to make the airplane move around a circle. The airplane of mass $m_{airplane}$ and weight $w_{airplane} = m_{airplane}g$ is traveling at speed v and is banked by an angle θ; the pilot of the airplane has mass m_{pilot} and weight $w_{pilot} = m_{pilot}g$. (a) What is the magnitude of the lift force? (b) If the pilot is sitting on a scale as the airplane makes its banked turn, what does the scale read?

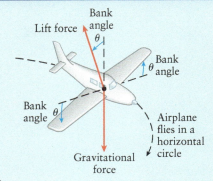

Figure 5-22 Forces on an airplane in a banked turn In addition to the lift force and gravitational force, there is a forward thrust force and a backward drag force. The thrust and drag forces cancel, which is why they're not shown here.

Set Up

We draw two free-body diagrams, one for the airplane of mass $m_{airplane}$ and one for the pilot of mass m_{pilot}. (In both cases the airplane and pilot are coming at us head on.) Both diagrams are almost identical to the free-body diagram for the rock in Example 5-8: A force that acts at an angle has a vertical component that balances the gravitational force on the object and a horizontal component that gives the object a centripetal acceleration.

In part (a) we want to find the magnitude L of the lift force; in part (b) our goal is to determine the magnitude n of the normal force. This is the magnitude of the force that the scale exerts on the pilot, which is the same as the reading on the scale. The value of n is the pilot's *apparent weight*, or how much she *feels* that she weighs during the turn.

Newton's second law equation for the airplane:

$$\sum \vec{F}_{ext\ on\ airplane}$$
$$= \vec{L} + \vec{w}_{airplane} = m_{airplane}\vec{a}_{airplane}$$

Newton's second law equation for the pilot:

$$\sum \vec{F}_{ext\ on\ pilot}$$
$$= \vec{n} + \vec{w}_{pilot} = m_{pilot}\vec{a}_{pilot}$$

Centripetal acceleration:

$$a_{airplane} = a_{pilot} = a_{cent} = \frac{v^2}{r} \quad (3\text{-}17)$$

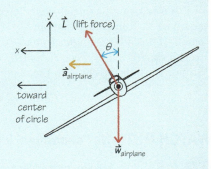

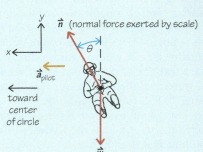

Solve

(a) Once again we begin by writing Newton's second law in component form. The airplane's acceleration is in the x direction toward the center of the turn, so it is a centripetal acceleration and equal to v^2/r. We don't know the radius r of the turn, so we can't use the x equation to solve for the lift force L. Instead we find this force from the y component equation.

Newton's second law in component form applied to the airplane:

x: $L \sin \theta = m_{airplane}a_{cent} = m_{airplane}\dfrac{v^2}{r}$

y: $L \cos \theta + (-w_{airplane}) = L \cos \theta - m_{airplane}g = 0$

From the y equation,

$L \cos \theta = m_{airplane}g$

$$L = \frac{m_{airplane}g}{\cos \theta} = \frac{w_{airplane}}{\cos \theta}$$

(b) The component form of Newton's second law for the pilot is very similar to that for the airplane. The main difference is that the lift force L is replaced by the normal force n. Again we use only the y component equation to solve for n.

Newton's second law in component form applied to the pilot:

$$x: n \sin \theta = m_{\text{pilot}} a_{\text{cent}} = m_{\text{pilot}} \frac{v^2}{r}$$

$$y: n \cos \theta + (-w_{\text{pilot}}) = n \cos \theta - m_{\text{pilot}} g = 0$$

Solve the y component equation for n:

$$n \cos \theta = m_{\text{pilot}} g$$

$$n = \frac{m_{\text{pilot}} g}{\cos \theta} = \frac{w_{\text{pilot}}}{\cos \theta}$$

Reflect

Our result in part (a) says that the steeper the bank angle θ, the greater the lift force required. That's because the vertical component of lift $L \cos \theta$ must always equal the airplane's weight. Since $\cos \theta$ decreases as θ increases, the magnitude L must increase as the bank angle increases. (You should recall that $\cos 0 = 1$, $\cos 45° = 1/\sqrt{2} = 0.707$, and $\cos 90° = 0$.) Note that the airplane's speed doesn't appear in our result, so our conclusions are the same whether the airplane is flying slowly or rapidly. The strength of the wings determines the maximum safe value of L and hence maximum safe bank angle. The same result applies to a bird in flight, which also banks its wings to turn.

BioMedical Our result in part (b) shows that the pilot's apparent weight is greater than her true weight. She *feels* heavier than normal, and her apparent weight increases with increasing bank angle in the same way that the lift does. The same is true for every part of the pilot's body, including her blood. That's the physiological risk of too steep a bank angle: If θ is too great, the pilot's heart can't pump this "heavy" blood up to her brain. A typical pilot will lose consciousness after 5 to 10 s at an apparent weight of 4 to 5 times $m_{\text{pilot}} g$.

To counter these effects, aerobatic pilots (who often make very sharp turns at steep bank angles, such as in the photo that opens this chapter) flex the muscles in their abdomen and legs. This displaces blood away from these contracted muscles and upward toward the brain.

Birds also bank steeply while turning (Figure 5-23), but don't have to worry about passing out. That's because their heads are at the same level as their hearts.

The greater the bank angle θ, the greater the lift force L on the airplane, the greater the pilot's scale reading, and the heavier the pilot feels:

Lift force: $L = \dfrac{w_{\text{airplane}}}{\cos \theta}$

Pilot's actual weight: w_{pilot}

Pilot's apparent weight: $n = \dfrac{w_{\text{pilot}}}{\cos \theta}$

For $\theta = 0$: $\cos 0 = 1$

$L = w_{\text{airplane}}$ and $n = w_{\text{pilot}}$

For $\theta = 45°$: $\cos 45° = 0.707$

$$L = \frac{w_{\text{airplane}}}{0.707} = 1.41 w_{\text{airplane}} \text{ and } n = 1.41 w_{\text{pilot}}$$

(The pilot feels "1.41 g.")

For $\theta = 60°$: $\cos 60° = 0.500$

$$L = \frac{w_{\text{airplane}}}{0.500} = 2.00 w_{\text{airplane}} \text{ and } n = 2.00 w_{\text{pilot}}$$

(The pilot feels "2.00 g.")

For $\theta = 80°$: $\cos 80° = 0.174$

$$L = \frac{w_{\text{airplane}}}{0.174} = 5.76 w_{\text{airplane}} \text{ and } n = 5.76 w_{\text{pilot}}$$

(The pilot feels "5.76 g" and might pass out.)

Figure 5-23 A tightly turning bird The northern fulmar (*Fulmaris glacialis*) can make very tight turns by flying slowly and using a very steep bank angle. Its wingspan is about 1.1 m.

Derren Fox/Alamy

EXAMPLE 5-10   Making an Airplane Turn II: Turning Radius

Find the radius of the banked turn that the airplane makes in Example 5-9.

Set Up

The free-body diagram for the airplane and the corresponding Newton's second law equations are the same as in Example 5-9. In that example we used the y equation to find the magnitude L of the lift force. Here our goal is to find the radius r of the airplane's turn, so we'll use the x equation (which includes r in the expression v^2/r for centripetal acceleration).

Newton's second law in component form applied to the airplane:

x: $L \sin \theta = m_{\text{airplane}} a_{\text{cent}}$

$$= m_{\text{airplane}} \frac{v^2}{r}$$

y: $L \cos \theta + (-w_{\text{airplane}})$

$$= L \cos \theta - m_{\text{airplane}} g = 0$$

From the y equation,

$$L = \frac{m_{\text{airplane}} g}{\cos \theta} = \frac{w_{\text{airplane}}}{\cos \theta}$$

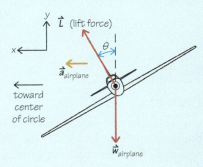

Solve

Since we know the value of the lift force L from Example 5-9, we substitute it into the x equation and solve for r.

Substitute $L = m_{\text{airplane}} g / \cos \theta$ into the x equation:

$$\left(\frac{m_{\text{airplane}} g}{\cos \theta} \right) \sin \theta = m_{\text{airplane}} \frac{v^2}{r}$$

The mass of the airplane cancels, and $(\sin \theta)/(\cos \theta) = \tan \theta$. Solve for r:

$$g \tan \theta = \frac{v^2}{r}$$

$$r = \frac{v^2}{g \tan \theta}$$

Reflect

The value of r depends on the airplane's speed v and its bank angle θ. The slower the speed, the smaller the turning radius; the greater the bank angle, the greater the value of $\tan \theta$ and again the smaller the turning radius. (Recall that $\tan 0 = 0$, $\tan 45° = 1$, and $\tan 90° =$ infinity.)
BioMedical Our result for the turning radius r doesn't involve the mass of the airplane, so it applies to airplanes of any size *and* to birds of any size. One bird species that's adept at making tight turns is the northern fulmar (*Fulmaris glacialis*), which can make very steeply banked turns while flying slowly (Figure 5-23).

Gliding speed of a northern fulmar:

$v = 11$ m/s (40 km/h, or 25 mi/h)

Bank angle in Figure 5-23:

$\theta = 70°$

Turning radius of this northern fulmar:

$$r = \frac{v^2}{g \tan \theta}$$

$$= \frac{(11 \text{ m/s})^2}{(9.80 \text{ m/s}^2) \tan 70°}$$

$$= \frac{(11 \text{ m/s})^2}{(9.80 \text{ m/s}^2)(2.75)}$$

$$= 4.5 \text{ m}$$

(This is about the same as the turning radius of a small car driving at low speed.)

EXAMPLE 5-11   A Turn on a Level Road

A car of mass m_{car} exits the freeway and travels around a circular off-ramp at a constant speed of 15.7 m/s (56.5 km/h, or 35.1 mi/h). The ramp is level (it is not banked), so the centripetal force has to be provided by the force of friction. If the coefficient of static friction between the car's tires and the road is 0.550 (a typical value when the road surface is wet), what radius of curvature is required for the off-ramp so that the car does not skid?

Set Up

This example is very similar to Example 5-6 in Section 5-4. The only difference is that in Example 5-6 the static friction force gave Roberto a *forward* acceleration, while in this example the static friction force causes a *centripetal* acceleration.

The free-body diagram shows the car head-on. Note that even though the car is rolling on its tires, the force that provides the centripetal acceleration is *static* friction rather than rolling friction. That's because the force of interest is the one that prevents the car from sliding sideways away from the center of its circular path. (By contrast, rolling friction acts opposite to the car's forward motion and so doesn't contribute to the centripetal acceleration.) The minimum value of the radius r corresponds to the maximum centripetal acceleration $a_{cent} = v^2/r$, which in turn corresponds to the maximum value $f_{s,max}$ of the static friction force.

Newton's second law equation for the car:

$$\sum \vec{F}_{\text{ext on car}}$$
$$= \vec{n} + \vec{w}_{car} + \vec{f}_s$$
$$= m_{car}\vec{a}_{car}$$

Centripetal acceleration:

$$a_{car} = a_{cent} = \frac{v^2}{r} \qquad (3\text{-}17)$$

Magnitude of the static friction force:

$$f_{s,max} = \mu_s n \qquad (5\text{-}1b)$$

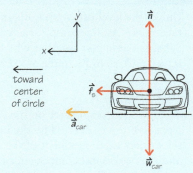

toward
center
of circle

Solve

We write Newton's second law for the car in component form, with the static friction force set equal to its maximum value $f_{s,max}$. The horizontal (x) acceleration equals the centripetal acceleration; there is zero vertical (y) acceleration. We then solve for the radius r of the car's circular path.

Newton's second law in component form for the car, assuming that the static friction force has its maximum value:

$$x: f_{s,max} = m_{car}a_{cent}$$
$$y: n + (-w_{car}) = 0$$

From the y component equation,

$$n = w_{car} = m_{car}g$$

So the maximum force of static friction of the tires is

$$f_{s,max} = \mu_s n = \mu_s m_{car}g$$

Substitute this expression and $a_{cent} = v^2/r$ into the x component equation, then solve for r:

$$\mu_s m_{car}g = m_{car}\frac{v^2}{r}$$
$$r = \frac{v^2}{\mu_s g}$$

Using $\mu_s = 0.550$ and $v = 15.7$ m/s,

$$r = \frac{(15.7 \text{ m/s})^2}{(0.550)(9.80 \text{ m/s}^2)} = 45.7 \text{ m}$$

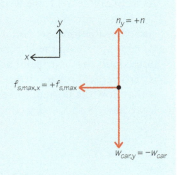

Reflect

Our answer is the *minimum* radius of the off-ramp. If the radius is larger, the car will have a smaller centripetal acceleration $a_{cent} = v^2/r$ and the static friction force required will be less than the maximum force available.

Note that the minimum radius r increases if the car is going faster (v is greater). Note also that for a given speed, r will be smaller if the coefficient of static friction μ_s is greater (so that the tires have a better "grip" on the road). Hence the car would be able to make a tighter turn of smaller radius if the road were dry, in which case μ_s is about 1.0.

EXAMPLE 5-12 A Turn on a Banked Road

Unlike the level off-ramp in Example 5-11, many freeway off-ramps are banked. The same is true for corners at racetracks. As an example, each corner at the Indianapolis Motor Speedway is banked by 9.20° and has radius 256 m. What is the fastest speed at which a race car could take one of these corners if the coefficient of static friction is $\mu_s = 0.550$, as in Example 5-11?

Set Up

The net centripetal force on the car now comes from *two* forces exerted by the banked road: the horizontal component of the normal force and the horizontal component of the static friction force. Because we want the car to move as fast as possible and hence have the maximum possible centripetal acceleration, it has to have the maximum available centripetal force exerted on it. In this extreme case the static friction force will be at its maximum value. Our goal is to find the value of v in this situation.

Newton's second law equation for the car:

$$\sum \vec{F}_{\text{ext on car}}$$
$$= \vec{n} + \vec{w}_{\text{car}} + \vec{f}_s$$
$$= m_{\text{car}} \vec{a}_{\text{car}}$$

Centripetal acceleration:

$$a_{\text{car}} = a_{\text{cent}} = \frac{v^2}{r} \qquad (3\text{-}17)$$

Magnitude of the static friction force:

$$f_{s,\text{max}} = \mu_s n \qquad (5\text{-}1b)$$

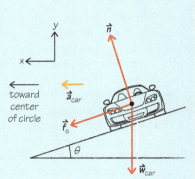

Solve

As in Example 5-11, we write Newton's second law for the car in component form, with the static friction force set equal to its maximum value $f_{s,\text{max}}$. Again the horizontal (x) acceleration equals the centripetal acceleration; there is zero vertical (y) acceleration. The difference is that now the normal force *and* the friction force have both x and y components.

Newton's second law in component form for the car, assuming that the static friction force has its maximum value:

x: $n \sin \theta + f_{s,\text{max}} \cos \theta = m_{\text{car}} a_{\text{cent}}$

y: $n \cos \theta + (-w_{\text{car}})$
$\qquad + (-f_{s,\text{max}} \sin \theta) = 0$

Substitute the expressions for the maximum force of static friction, the centripetal acceleration, and the weight of the car:

$$f_{s,\text{max}} = \mu_s n \quad ; \quad a_{\text{cent}} = \frac{v^2}{r} \quad ; \quad w_{\text{car}} = m_{\text{car}} g$$

Then the Newton's law equations become

x: $n \sin \theta + \mu_s n \cos \theta = m_{\text{car}} \dfrac{v^2}{r}$

y: $n \cos \theta - m_{\text{car}} g - \mu_s n \sin \theta = 0$

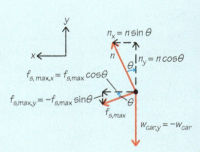

Solve the y component equation for the normal force n.

Rearrange the y component equation to find n:

$n \cos \theta - \mu_s n \sin \theta = m_{\text{car}} g$

$n(\cos \theta - \mu_s \sin \theta) = m_{\text{car}} g$

$$n = \frac{m_{\text{car}} g}{\cos \theta - \mu_s \sin \theta}$$

Substitute this expression for n into the x component equation and solve for the speed v.

Rearrange the x component equation:

$n \sin \theta + \mu_s n \cos \theta = m_{\text{car}} \dfrac{v^2}{r}$

$n \left(\sin \theta + \mu_s \cos \theta \right) = m_{\text{car}} \dfrac{v^2}{r}$

Substitute the expression for n and solve for v:

$$\frac{m_{car}\,g}{\cos\theta - \mu_s \sin\theta}(\sin\theta + \mu_s \cos\theta) = m_{car}\frac{v^2}{r}$$

$$g\left(\frac{\sin\theta + \mu_s \cos\theta}{\cos\theta - \mu_s \sin\theta}\right) = \frac{v^2}{r}$$

$$v^2 = gr\left(\frac{\sin\theta + \mu_s \cos\theta}{\cos\theta - \mu_s \sin\theta}\right)$$

$$v = \sqrt{gr\left(\frac{\sin\theta + \mu_s \cos\theta}{\cos\theta - \mu_s \sin\theta}\right)}$$

Use the values of r, θ, and μ_s given for the Indianapolis Motor Speedway.

Using $r = 256$ m, $\theta = 9.20°$, and $\mu_s = 0.550$,

$\sin\theta = \sin 9.20° = 0.160$

$\cos\theta = \cos 9.20° = 0.987$

$$v = \sqrt{(9.80\text{ m/s}^2)(256\text{ m})\left[\frac{0.160 + (0.550)(0.987)}{0.987 - (0.550)(0.160)}\right]}$$

$$= 44.3\text{ m/s}$$

$$= (44.3\text{ m/s})\left(\frac{1\text{ km}}{1000\text{ m}}\right)\left(\frac{3600\text{ s}}{1\text{ h}}\right) = 159\text{ km/h}$$

$$= (159\text{ km/h})\left(\frac{1\text{ mi}}{1.61\text{ km}}\right) = 99.1\text{ mi/h}$$

Reflect

This speed is quite a bit less than that reached by race cars on the straightaway (200 to 300 km/h), so it's necessary to slow down before turning this corner.

We can check our result by looking at two special cases: (a) If the curve is *not* banked, then $\theta = 0$, $\sin\theta = 0$, and $\cos\theta = 1$. In this case our result reduces to what we found for the flat road in Example 5-10. (b) If the curve is banked but there is no friction, then $\mu_s = 0$ (because then the maximum force of friction is $f_{s,max} = \mu_s n = 0$). In this situation our result simplifies to what we found for the airplane making a level turn in Example 5-10. (The airplane isn't on a road, but the lift force $\vec{L}$ on a turning airplane plays exactly the same role as the normal force $\vec{n}$ on a car on a frictionless banked curve.)

You can show that in case (a) the car's maximum speed would be only 37.1 m/s, while in case (b) it would be only 20.2 m/s. A banked curve with friction allows the highest speeds!

(a) If the curve is not banked (so $\theta = 0$):

$$v = \sqrt{gr\left(\frac{\sin 0 + \mu_s \cos 0}{\cos 0 - \mu_s \sin 0}\right)}$$

$$= \sqrt{gr\left(\frac{0 + \mu_s(1)}{1 - \mu_s(0)}\right)}$$

$$= \sqrt{\mu_s gr}$$

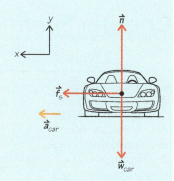

This equation is equivalent to what we found in Example 5-10 for a flat road:

$$r = \frac{v^2}{\mu_s g}, \text{ so } v^2 = \mu_s gr \text{ and } v = \sqrt{\mu_s gr}$$

(b) If the curve is banked but there is no friction (so $\mu_s = 0$):

$$v = \sqrt{gr\left(\frac{\sin\theta + (0)\cos\theta}{\cos\theta - (0)\sin\theta}\right)}$$

$$= \sqrt{gr\left(\frac{\sin\theta}{\cos\theta}\right)}$$

$$= \sqrt{gr\tan\theta}$$

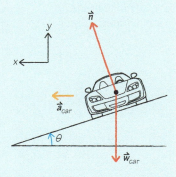

This equation is equivalent to what we found in Example 5-10 for an airplane making a banked turn:

$$r = \frac{v^2}{g\tan\theta}, \text{ so } v^2 = gr\tan\theta \text{ and }$$

$$v = \sqrt{gr\tan\theta}$$

GOT THE CONCEPT? 5-6 Comparing Circular Motions

(?) Five objects all move in uniform circular motion. The objects have different masses, move at different speeds, and travel in circular paths of different radii. Rank these objects in order of the magnitude of the centripetal force that acts on them, from greatest to least: (a) a 2.00-kg object moving at 10.0 m/s in a circle of radius 8.00 m; (b) a 2.00-kg object moving at 5.00 m/s in a circle of radius 4.00 m; (c) a 4.00-kg object moving at 5.00 m/s in a circle of radius 2.00 m; (d) a 4.00-kg object moving at 5.00 m/s in a circle of radius 8.00 m; (e) an 8.00-kg object moving at 5.00 m/s in a circle of radius 1.00 m.

TAKE-HOME MESSAGE FOR Section 5-6

✔ If an object is in uniform circular motion, its acceleration is directed toward the center of the circle and given by Equation 3-17.

✔ The net force on an object in uniform circular motion is also directed toward the center of the circle

(Equation 5-10). The net force in this case is called the centripetal force.

✔ Any kind of force (or combination of forces) can provide the centripetal force on an object in uniform circular motion.

Key Terms

centripetal force
coefficient of kinetic friction
coefficient of rolling friction
coefficient of static friction

drag force
fluid resistance
friction
kinetic friction

rolling friction
static friction

Chapter Summary

Topic	Equation or Figure

Static friction: If an object is at rest on a surface, a static friction force $\vec{f}_s$ arises if other forces are trying to make the object move. The static friction force can have any magnitude up to a maximum value.

(a) You apply a force of 20.0 N to the right, but the block remains at rest...

20.0 N

20.0 N

...so the table must exert a static friction force of 20.0 N to the left on the block.

(b) You apply a force of 10.0 N to the right, but the block remains at rest...

10.0 N

10.0 N

...so the table must exert a static friction force of 10.0 N to the left on the block.

(c) You apply no force to the block and it remains at rest...

...so the table must exert no static friction force on the block.

The static friction force adjusts its value to counteract whatever other forces are trying to make the object slide.

(Figure 5-2)

 ① The magnitude of the **force of static friction** on an object... 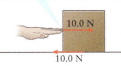 ② ...is less than or equal to a certain maximum value.

$$f_s \le f_{s,\text{max}}$$ (5-1a)

 ③ The **maximum force of static friction** depends on...

$$f_{s,\text{max}} = \mu_s n$$ (5-1b)

④ ...the **coefficient of static friction** (which depends on the properties of the two surfaces in contact)...  ⑤ ...and the **normal force** pressing the object against a surface.

An object at rest on an incline: Static friction will keep an object at rest on an inclined surface provided the angle θ of the incline is less than a critical value θ_{slip}.

(a) A block at rest on an inclined ramp

Block, weight w

Ramp whose angle θ from the horizontal can be varied

θ

(Figure 5-5a)

Coefficient of static friction for an object at rest on a surface

Angle of the surface at which the object **begins to slip**

$$\mu_s = \tan \theta_{\text{slip}}$$

(5-5)

Kinetic friction: If an object is sliding over a surface, that surface exerts a kinetic friction force $\vec{f}_k$ that opposes sliding. We model this force as being independent of the object's speed. There is a similar expression for the force of rolling friction; the difference is that the coefficient for rolling has a smaller value than that for sliding.

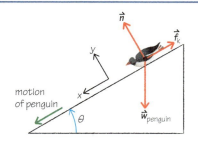

(Example 5-2)

The **force of kinetic friction** depends on...

$$f_k = \mu_k n$$

(5-6)

...the **coefficient of kinetic friction** (which depends on the properties of the two surfaces in contact)...

...and the **normal force** exerted on the object by the surface on which it slides.

Drag force: An object moving through a fluid (gas or liquid) experiences a resistive force that opposes its motion. If the object is small and moving at a low speed v relative to the fluid, the magnitude of the drag force is proportional to v. For larger objects moving at faster speeds, the drag force is proportional to the square of v.

Magnitude of the **drag force** on a **small object** moving at a **low speed**

$$F_{\text{drag}} = bv$$

(5-8)

Constant that depends on the properties of the object and of the fluid

Speed of the object relative to the fluid

Magnitude of the **drag force** on a **larger object** moving at a **faster speed**

$$F_{\text{drag}} = cv^2$$

(5-9)

Constant that depends on the properties of the object and of the fluid

Speed of the object relative to the fluid

Uniform circular motion: If an object moves at constant speed around a circular path, the net force on the object points toward the center of the path. Any kind of force or combination of forces can contribute to this net force.

Centripetal acceleration: Magnitude of the acceleration of an object in uniform circular motion

Speed of the object as it moves around the circle

$$a_{\text{cent}} = \frac{v^2}{r}$$

(3-17)

Radius of the circle

Net force on an object in uniform circular motion

Mass of the object

$$\sum F_{\text{cent}} = \frac{mv^2}{r}$$

(5-10)

Speed of the object as it moves around the circle

The direction of the net force is toward the center of the object's circular path.

Radius of the circle

Answer to What do you think? Question

(a) There is a net force, and it points inward toward the center of the circle. Newton's second law tells us that the net force must be in the same direction as the airplane's acceleration, and in uniform circular motion the acceleration is inward. If you're riding in the airplane it may *feel* as though a force is pushing you outward. This outward "force" isn't real, however; it's just a consequence of your body's inertia (see Section 5-6).

Answers to Got the Concept? Questions

5-1 (d) As shown in Equation 5-1b, the maximum static friction force between the block and board depends only on the normal force n and the coefficient of static friction μ_s. It does *not* depend on the area of contact between the block and board. Therefore, the angle at which the block begins to slide will be the same no matter which face is in contact with the board.

5-2 (e) The acceleration of the penguin doesn't depend on its mass at all. The acceleration is caused by the kinetic friction force and the downhill component of the gravitational force. Both of these are proportional to the mass of the penguin. (The kinetic friction force has magnitude $f_k = \mu_k n$, and the normal force n is proportional to the penguin's weight and hence its mass.) The mass cancels out when we calculate the acceleration of the penguin, which is equal to the net external force on the penguin divided by its mass.

5-3 (d) We saw in Section 5-2 that an object at rest on an incline with angle θ will just barely remain at rest (that is, the maximum force of static friction is just enough to balance the downhill component of the gravitational force) if $\mu_s = \tan \theta$. For an incline with $\theta = 30.0°$, $\mu_s = \tan 30.0° = 0.577$. If μ_s has a greater value than 0.577, the maximum static friction force available is greater than the downhill component of the gravitational force, and the object will still remain at rest. So both $\mu_s = 0.600$ and $\mu_s = 0.750$ will keep the object at rest. If

$\mu_s = 0.350$, there isn't enough static friction and so the penguin will start sliding back downhill.

5-4 (c) Example 5-6 illustrates that both kinetic friction and static friction can cause an object to speed up. If Roberto doesn't slip on the upper surface of the sled, the force that accelerates him forward is the force of static friction. If he does slip, he nonetheless accelerates forward due to the force of kinetic friction. (In the latter case it's true that he slips backward relative to the sled. However, both he and the sled are accelerating forward; it's just that Roberto's acceleration is less than the acceleration of the sled.)

5-5 (d) The drag force for both objects is given by $F_{\text{drag}} = cv^2$, so $c = F_{\text{drag}}/v^2$. Because the drag force on the second object is 4 times greater and 1/2 the speed, the quantity c is $(4)/(1/2)^2 = 16$ times greater for the second object than for the baseball.

5-6 (e), (c), (a), (b), and (d) [tie] The centripetal force on an object of mass m moving at speed v in a circle of radius r is equal to mv^2/r—that is, to m multiplied by the object's centripetal acceleration $a_{\text{cent}} = v^2/r$. For the five cases this quantity is (a) $(2.00\,\text{kg})(10.0\,\text{m/s})^2/(8.00\,\text{m}) = 25.0\,\text{N}$; (b) $(2.00\,\text{kg})(5.00\,\text{m/s})^2/(4.00\,\text{m}) = 12.5\,\text{N}$; (c) $(4.00\,\text{kg})(5.00\,\text{m/s})^2/(2.00\,\text{m}) = 50.0\,\text{N}$; (d) $(4.00\,\text{kg})(5.00\,\text{m/s})^2/(8.00\,\text{m}) = 12.5\,\text{N}$; (e) $(8.00\,\text{kg})(5.00\,\text{m/s})^2/(1.00\,\text{m}) = 2.00 \times 10^2\,\text{N}$.

Questions and Problems

In a few problems you are given more data than you actually need; in a few other problems you are required to supply data from your general knowledge, outside sources, or informed estimate.

Interpret as significant all digits in numerical values that have trailing zeros and no decimal points. For all problems use $g = 9.80 \text{ m/s}^2$ for the free-fall acceleration due to gravity.

- •   Basic, single-concept problem
- ••   Intermediate-level problem; may require synthesis of concepts and multiple steps
- •••   Challenging problem
- Example See worked example for a similar problem

Conceptual Questions

1. • Complete the sentence: The static frictional force between two surfaces is (never/sometimes/always) less than the normal force. Explain your answer.

2. • You want to push a heavy box across a rough floor. You know that the coefficient of static friction (μ_s) is larger than the coefficient of kinetic friction (μ_k). Should you push the box for a short distance, rest, push the box another short distance, and then repeat the process until the box is where you want it, or will it be easier to keep pushing the box across the floor once you get it moving?

3. • If the force of friction always opposes the sliding of an object, how then can a frictional force cause an object to increase in speed?

4. • A solid rectangular block has sides of three different areas. You can choose to rest any of the sides on the floor as you apply a horizontal force to the block. Does the choice of side on the floor affect how hard it is to push the block? Explain your answer.

5. • You press a book against the wall with your hand. As you get tired you exert less force, but the book remains in the same spot on the wall. Do each of the following forces increase, decrease, or not change in magnitude when you reduce the force you are applying to the book: (a) weight of the book, (b) normal force of the wall on the book, (c) frictional force of the wall on the book, and (d) maximum static frictional force of the wall on the book?

6. • An antilock braking system (ABS) prevents wheels from skidding while drivers stomp on the brakes in emergencies. How far will a car with an ABS move before finally stopping, as compared to a car without an ABS?

7. • Describe two situations in which the normal force acting on an object is not equal to the object's weight.

8. •• You are sliding a piece of furniture across the floor at constant velocity when a friend jumps on top of it. Assume $\mu_k < \mu_s$ and use the ideas discussed in this chapter to describe what happens and why.

9. • *Synovial fluid* lubricates the surfaces where bones meet in joints, making the coefficient of friction between bones very small. Why is the minimum force required for moving bones in a typical knee joint different for different people, in spite of the fact that their joints have the same coefficient of friction?

For simplicity assume the surfaces in the knee are flat and horizontal.

10. • At low speeds the drag force on an object moving through a fluid is proportional to its velocity. According to Newton's second law, force is proportional to acceleration. As acceleration and velocity aren't the same quantity, is there a contradiction here?

11. • As a skydiver falls faster and faster through the air, does the magnitude of his acceleration increase, decrease, or remain the same? Explain your answer.

12. • Why do raindrops fall from the sky at different speeds? Explain your answer.

13. • James Bond leaps without a parachute from a burning airplane flying at 15,000 ft. Ten seconds later his assistant, who was following behind in another plane, dives after him, wearing her parachute and clinging to one for her hero. Is it possible for her to catch up with Bond and save him?

14. • What distinguishes the forces that act on a car driving over the top of a hill from those acting on a car driving through a dip in the road? Explain how the forces relate to the sensations passengers in the car experience during each situation.

15. • Explain how you might measure the centripetal acceleration of a car rounding a curve.

16. • For an object moving in a circle, which of the following quantities are zero over one revolution: (a) displacement, (b) average velocity, (c) average acceleration, (d) instantaneous velocity, and (e) instantaneous centripetal acceleration?

17. • Why does water stay in a bucket that is whirled around in a vertical circle? Contrast the forces acting on the water when the bucket is at the lowest point on the circle to when the bucket is at the highest point on the circle.

18. • Explain why curves in roads and cycling velodromes are banked.

19. • Why might your car start to skid if you drive too fast around a curve?

20. •• Consider the effects of air resistance on a projectile. Describe qualitatively how the projectile's velocities and accelerations in the vertical and horizontal directions differ when the effects of air resistance are ignored and when the effects are considered.

Multiple-Choice Questions

21. • If a sport utility vehicle (SUV) drives up a slope of 45°, what must be the minimum coefficient of static friction between the SUV's tires and the road?

    A. 1.0   D. 0.9
    B. 0.5   E. 0.05
    C. 0.7

22. • A block of mass m slides down a rough incline with constant speed. If a similar block that has a mass of $4m$ were placed on the same incline, it would

    A. slide down at the same constant speed.
    B. accelerate down the incline.
    C. slowly slide down the incline and then stop.

D. slide down with a faster constant speed.

E. not move.

23. • A 10-kg crate sits on a horizontal conveyor belt moving with a constant speed. The crate does not slip. If the coefficients of friction between the crate and the belt are $\mu_s = 0.50$ and $\mu_k = 0.30$, what is the frictional force exerted on the crate?

   A. 98 N   D. 9.8 N

   B. 49 N   E. 0

   C. 29 N

24. • **Biology** The *Escherichia coli* (*E. coli*) bacterium propels itself through water by means of long, thin structures called flagella. If the force exerted by the flagella doubles, the velocity of the bacterium

   A. doubles.

   B. decreases by half.

   C. does not change.

   D. increases by a factor of 4.

   E. cannot be determined without more information.

25. • A 1-kg wood ball and a 5-kg lead ball have identical sizes, shapes, and surface characteristics. They are dropped simultaneously from a tall tower. Air resistance is present. How do their accelerations compare?

   A. The 1-kg wood ball has the larger acceleration, but it cannot be calculated without more information.

   B. The 5-kg lead ball has the larger acceleration, but it cannot be calculated without more information.

   C. The accelerations are the same.

   D. The 5-kg ball accelerates at 5 times the acceleration of the 1-kg ball.

   E. The 1-kg ball accelerates at 5 times the acceleration of the 5-kg ball.

26. • A skydiver is falling at his terminal speed. Immediately after he opens his parachute

   A. his speed will be larger than his terminal speed.

   B. the magnitude of the drag force on the skydiver will decrease.

   C. the net force on the skydiver is in the downward direction.

   D. the magnitude of the drag force is larger than the skydiver's weight.

   E. the net force on the skydiver is zero.

27. • Two identical rocks are tied to massless strings and whirled in nearly horizontal circles at the same speed. One string is twice as long as the other. What is the tension T_1 in the shorter string compared to the tension T_2 in the longer one?

   A. $T_1 = T_2/4$

   B. $T_1 = T_2/2$

   C. $T_1 = T_2$

   D. $T_1 = 2T_2$

   E. $T_1 = 4T_2$

28. • You are on a Ferris wheel moving in a vertical circle. When you are at the bottom of the circle, how does the magnitude of the normal force n exerted by your seat compare to your weight mg?

   A. $n = mg$

   B. $n > mg$, but cannot be exactly calculated without more information

   C. $n < mg$, but cannot be exactly calculated without more information

   D. $n = mg/2$

   E. $n = 2mg$

29. • Two rocks are tied to massless strings and whirled in nearly horizontal circles so that the time to travel around the circle once is the same for both. One string is twice as long as the other. The tension in the longer string is twice the tension in the shorter one. What is the mass m_1 of the rock at the end of the shorter string compared to the mass m_2 of the rock at the end of the longer one?

   A. $m_1 = m_2/4$

   B. $m_1 = m_2/2$

   C. $m_1 = m_2$

   D. $m_1 = 2m_2$

   E. $m_1 = 4m_2$

Problems

5-2 The static friction force changes magnitude to offset other applied forces

30. • What is the minimum horizontal force that will cause a 5.00-kg box to begin to slide on a horizontal surface when the coefficient of static friction is 0.670? Example 5-1

31. • A 7.60-kg object rests on a level floor with a coefficient of static friction of 0.550. What minimum horizontal force will cause the object to start sliding? Example 5-1

32. • Draw a free-body diagram for the situation shown in **Figure 5-24**. An object of mass M rests on a ramp; there is friction between the object and the ramp. Example 5-1

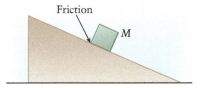

Friction

M

Figure 5-24 Problem 32

5-3 The kinetic friction force on a sliding object has a constant magnitude

33. • A book is pushed across a horizontal table at a constant speed. If the horizontal force applied to the book is equal to one-half of the book's weight, calculate the coefficient of kinetic friction between the book and the tabletop. Example 5-2

34. • An object on a level surface experiences a horizontal force of 12.7 N due to kinetic friction. If the coefficient of kinetic friction is 0.37, what is the mass of the object? Example 5-2

35. • A 25.0-kg crate rests on a level floor. A horizontal force of 50.0 N accelerates the crate at 1.00 m/s². Calculate (a) the normal force of the floor on the crate, (b) the frictional force of the floor on the crate, and (c) the coefficient of kinetic friction between the crate and the floor. Example 5-2

36. • A mop is pushed across the floor with a force F of 50.0 N at an angle of $\theta = 50.0°$ (**Figure 5-25**). The mass of the mop head is 3.75 kg. Calculate the acceleration of the mop head if the coefficient of kinetic friction between the head and the floor is 0.400. Example 5-2

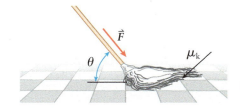

$\vec{F}$

θ

μ_k

Figure 5-25
Problem 36

37. • If the coefficient of kinetic friction between an object with mass $M = 3.00$ kg and a flat surface is 0.400, what magnitude of force applied at an angle of $\theta = 30.0°$ will cause the object to accelerate at 2.50 m/s²? (Figure 5-26). Example 5-2

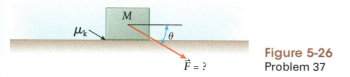

Figure 5-26
Problem 37

5-4 Problems involving static and kinetic friction are like any other problem with forces

38. •• A horizontal force $\vec{F}$ with magnitude 10.0 N is applied to a stationary block with a mass M of 2.00 kg as shown in Figure 5-27. The coefficient of static friction between the block and the floor is 0.750; the coefficient of kinetic friction is 0.450. Find the acceleration of the box. Example 5-5

Figure 5-27 Problem 38

39. •• A taut string connects a crate with mass $M_1 = 5.00$ kg to a crate with mass $M_2 = 12.0$ kg (Figure 5-28). The coefficient of static friction between the smaller crate and the floor is 0.573; the coefficient of static friction between the larger crate and the floor is 0.443. What is the minimum magnitude of force $\vec{F}$ required to start the crates in motion? Example 5-4

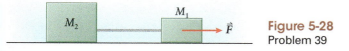

Figure 5-28
Problem 39

40. ••• A box of mass $M_{box} = 2.00$ kg rests on top of a crate with mass $M_{crate} = 5.00$ kg (Figure 5-29). The coefficient of static friction between the box and the crate is 0.667. The coefficient of static friction between the crate and the floor is 0.400. Calculate the minimum magnitude of force $\vec{F}$ that is required to move the crate to the right and the corresponding magnitude of the tension T in the rope that connects the box to the wall when the crate is moved. Example 5-6

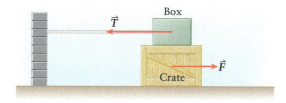

Figure 5-29 Problem 40

41. • Two blocks are connected over a massless, frictionless pulley (Figure 5-30). The mass of block 2 is 8.00 kg, and the coefficient of kinetic friction between block 2 and the incline is 0.220. The angle θ of the incline is 28.0°. Block 2 slides down the incline at constant speed. What is the mass of block 1? Example 5-4

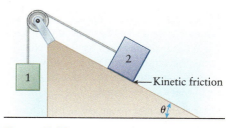

Figure 5-30 Problems 41 and 42

42. • Two blocks are connected over a massless, frictionless pulley (Figure 5-30). The mass of block 2 is 10.0 kg, and the coefficient of kinetic friction between block 2 and the incline is 0.200. The angle θ of the incline is 30.0°. If block 2 moves up the incline at constant speed, what is the mass of block 1? Example 5-4

43. •• An object with mass M_1 of 2.85 kg is held in place on an inclined plane that makes an angle θ of 40.0° with the horizontal (Figure 5-31). The coefficient of kinetic friction between the plane and the object is 0.552. A second object that has a mass M_2 of 4.75 kg is connected to the first object with a massless string over a massless, frictionless pulley. Calculate the magnitude and direction of the initial acceleration of the system and the tension in the string once the objects are released. Example 5-4

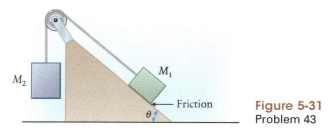

Figure 5-31
Problem 43

44. • Draw free-body diagrams for the situation shown in Figure 5-32. An object of mass M_2 rests on a frictionless table, and an object of mass M_1 sits on it; there is friction between the objects. A horizontal force $\vec{F}$ is applied to the lower object as shown. Example 5-6

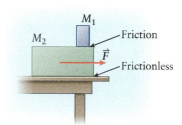

Figure 5-32 Problem 44

5-5 An object moving through air or water experiences a drag force

45. • **Biology** A single-celled animal called a *paramecium* propels itself quite rapidly through water by using its hairlike *cilia*. A certain paramecium experiences a drag force of magnitude $F_{drag} = cv^2$ in water, where the drag coefficient c is approximately 0.310. What propulsion force does this paramecium generate when moving at a constant (terminal) speed v of 0.150×10^{-3} m/s? Example 5-7

46. • **Biology** The bacterium *Escherichia coli* (*E. coli*) propels itself with long, thin structures called flagella. When its flagella exert a force of 1.50×10^{-13} N, the bacterium swims through water at a speed of 20.0 μm/s. Find the speed of the bacterium in water when the force exerted by its flagella is 3.00×10^{-13} N. Example 5-7

47. • We model the drag force of the atmosphere as $F_{drag} = cv^2$, where the value of c for a 70.0-kg person with a parachute is 18.0 kg/m. (a) What is the person's terminal velocity? (b) Without a parachute, the same person's terminal velocity would be about 50.0 m/s. What would be the value of the proportionality constant c in that case? Example 5-7

48. •• A girl rides her scooter on a hill that is inclined at 10.0° with the horizontal. The combined mass of the girl and scooter is 50.0 kg. On the way down, she coasts at a constant speed of 12.0 m/s, while experiencing a drag force that is proportional to the square of her velocity. What force, parallel to the surface of the hill, is required to increase her speed to 20.0 m/s? Neglect any other resistive forces, including the friction between the scooter and the hill. Example 5-7

5-6 In uniform circular motion, the net force points toward the center of the circle

49. • A hockey puck that has a mass of 0.170 kg is tied to a light string and spun in a circle of radius 1.25 m on frictionless ice. If the string breaks under a tension that exceeds 5.00 N, what is the maximum speed of the puck without breaking the string? Example 5-10

50. • A 1.50×10^3-kg truck rounds an unbanked curve on the highway at a speed of 20.0 m/s. If the maximum frictional force between the surface of the road and all four of the tires is 8.00×10^3 N, calculate the minimum radius of curvature for the curve to prevent the truck from skidding off the road. Example 5-11

51. • A 25.0-g metal washer is tied to a 60.0-cm-long string and whirled around in a vertical circle at a constant speed of 6.00 m/s. Calculate the tension in the string (a) when the washer is at the bottom of the circular path and (b) when it is at the top of the path. Example 5-8

52. • A centrifuge spins small tubes in a circle of radius 10.0 cm at a rate of 1.20×10^3 rev/min. What is the centripetal force on a sample that has a mass of 1.00 g? Example 5-8

53. • **Biology** Very high speed ultracentrifuges are useful devices to sediment materials quickly or to separate materials. An ultracentrifuge spins a small tube in a circle of radius 10.0 cm at 6.00×10^4 rev/min. What is the centripetal force experienced by a sample that has a mass of 3.00 g? Example 5-8

54. • **Astro** What centripetal force is exerted on the Moon as it orbits about Earth at a center-to-center distance of 3.84×10^8 m with a period of 27.4 days? What is the source of the force? The mass of the Moon is equal to 7.35×10^{22} kg. Example 5-8

55. • At the Fermi National Accelerator Laboratory (Fermilab), a large particle accelerator makes protons travel in a circular orbit 6.3 km in circumference at a speed of nearly 3.0×10^8 m/s.

What is the centripetal acceleration of one of the protons? Example 5-10

56. • In the game of tetherball a 1.25-m rope connects a 0.750-kg ball to the top of a vertical pole so that the ball can spin around the pole as shown in **Figure 5-33**. What is the speed of the ball as it rotates around the pole when the angle θ of the rope is 35.0° with the vertical? Example 5-8

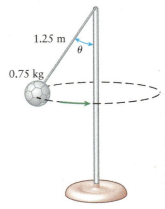

Figure 5-33 Problem 56

57. • What is the magnitude of the force that a jet pilot feels against his seat as he completes a vertical loop that is 5.00×10^2 m in radius at a speed of 2.00×10^2 m/s? Assume his mass is 70.0 kg and that he is located at the bottom of the loop. Example 5-9

58. • The radius of Earth is 6.38×10^6 m, and it completes one revolution in one day. What is the centripetal acceleration of an object located (a) on the equator and (b) at latitude 40.0° north? Example 5-10

59. •• A coin that has a mass of 25.0 g rests on a phonograph turntable that rotates at 78.0 rev/min. The center of the coin is 13.0 cm from the turntable axis. If the coin does not slip, what is the minimum value of the coefficient of static friction between the coin and the turntable surface? Example 5-11

60. • **Sports** In executing a windmill pitch, a fast-pitch softball player moves her hand through a circular arc of radius 0.310 m. The 0.190-kg ball leaves her hand at 24.0 m/s. What is the magnitude of the force exerted on the ball by her hand immediately before she releases it? Example 5-8

General Problems

61. • The following table lists the forces applied to a 1.00-kg crate at particular times. The crate begins at rest on a rough, horizontal surface until it begins to move at a constant speed in the horizontal direction. Using the data, estimate the coefficient of static friction and the coefficient of kinetic friction between the crate and the surface. Example 5-6

t (s)	F (N)	t (s)	F (N)
0	0	0.25	8.26
0.01	1.33	0.30	7.84
0.05	3.28	0.35	5.17
0.10	8.11	0.40	5.21
0.15	8.20	0.45	5.22
0.20	8.24	0.50	5.37

62. ••• A block of mass M rests on a block of mass $M_1 = 5.00$ kg which is on a tabletop (**Figure 5-34**). A light string passes over a massless, frictionless pulley and connects the blocks. The coefficient of kinetic friction μ_k at both surfaces equals 0.330. A force of magnitude 60.0 N pulls the upper block to the left and the lower block to the right. The blocks are moving at a constant speed. Determine the mass of the upper block. Example 5-4

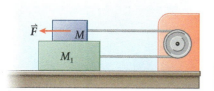

Figure 5-34
Problem 62

63. •• A wedding cake sits atop a table. A mischievous ring-bearer wants to show off his strength by lifting one end of the table. If the cake weighs 53.0 N and the coefficient of static friction between the cake base and the table is 0.400, to what angle can the table tilt before the cake starts to slide? Example 5-1

64. • You lean against a table such that your weight exerts a force on the edge of the table that is directed at an angle of 11.0° below a line drawn parallel to the table's surface (Figure 5-35). The table has a mass of 25.0 kg and the coefficient of static friction between its feet and the ground is 0.550. What is the maximum force with which you can lean against the table before it slides? Example 5-10

11°

Figure 5-35
Problem 64

65. • You find yourself pushing a 42.0-kg box up a 6.00° ramp into a moving truck. The coefficient of kinetic friction between the box and the metal ramp is 0.320. To save energy, you only push hard enough to move the box up the ramp at constant velocity. Assuming that your pushing force is directed parallel to the ramp, with what force do you push? Example 5-3

66. •• You are a fitness trainer developing an exercise that involves pulling a heavy crate across a rough surface. You attach a rope to a crate of mass $M = 45$ kg and place it on a carpeted track. The coefficient of kinetic friction between the crate and the carpet is $\mu_k = 0.40$. To test the apparatus, you have one of your employees pull on the rope as hard as she can. If she pulls with a force of $F = 300$ N along the direction of the rope while the rope makes an angle of $\theta = 25°$ with the floor (Figure 5-36), what is the acceleration a of the box? Example 5-2

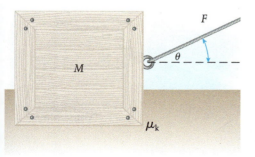

F

M

θ

μ_k

Figure 5-36
Problem 66

67. • Luis absentmindedly leaves his physics textbook on the horizontal top of his 1825-kg car while getting in. The coefficients of friction between the 1.84-kg textbook and the top of the car are $\mu_s = 0.180$ and $\mu_k = 0.113$. If the car is on level ground, what is the magnitude of the maximum acceleration the car can have before Luis's textbook slides off the car? Example 5-6

68. • A librarian moonlighting as a magician wants to pull one book out of a stack of books. The stack is seven books high and all the books are of roughly equal mass M and of comparable size, and the coefficient of kinetic friction between the books is μ_k. The librarian needs the third book in the stack, counting from the bottom. The librarian gives the third book a hard, straight yank with a force of magnitude F perpendicular to the stack. In terms of M, g, and μ_k, what is the net horizontal force on the yanked book? Example 5-5

69. • A large block is being pushed against a smaller block such that the smaller block remains elevated while being pushed (Figure 5-37). The mass of the smaller block is $m = 0.75$ kg. The blocks need to have a minimum acceleration of $a = 15$ m/s^2 for the smaller block to remain elevated and not slide down. What is the coefficient of static friction between the two blocks? Example 5-5

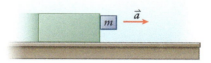

m $\vec{a}$

Figure 5-37
Problem 69

70. •• A truck driver hauls a large log in the back of a flatbed truck. The log has a mass of $m = 800$ kg and the truck has a mass of $M = 9500$ kg. The truck can accelerate from 0 to 55 mi/h in 25 s. Calculate the minimum coefficient of static friction μ_s needed to keep the log from sliding off the back of the truck. Example 5-6

71. •• The coefficient of static friction between a rubber tire and dry pavement is about 0.800. Assume that a car's engine only turns the two rear wheels and that the weight of the car is uniformly distributed over all four wheels. (a) What limit does the coefficient of static friction place on the time required for a car to accelerate from rest to 60 mi/h (26.8 m/s)? (b) How can friction accelerate a car *forward* when friction *opposes* motion? Example 5-2

72. •• Two blocks are connected over a massless, frictionless pulley (Figure 5-38). Block 1 has a mass $m_1 = 1.00$ kg and block 2 has a mass $m_2 = 0.400$ kg. The angle θ of the incline is 30.0°. The coefficients of static friction and kinetic friction between block 1 and the incline are $\mu_s = 0.500$ and $\mu_k = 0.400$, respectively. What is the magnitude of the tension in the string? Example 5-5

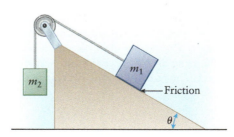

m_2 m_1 —Friction

θ

Figure 5-38
Problems 72 and 73

73. •• Two blocks are connected over a massless, frictionless pulley (Figure 5-38). Block 1 has a mass m_1 of 1.00 kg and block 2 has a mass m_2 of 2.00 kg. The angle θ of the incline is 30.0°. The coefficients of static friction and kinetic friction between block 1 and the incline are $\mu_s = 0.500$ and $\mu_k = 0.400$. What is the acceleration of block 1? Example 5-4

74. • A runaway ski slides down a 250-m-long slope inclined at 37.0° with the horizontal. If the initial speed is 10.0 m/s, how long does it take the ski to reach the bottom of the incline if the coefficient of kinetic friction between the ski and snow is (a) 0.100 and (b) 0.150? Example 5-2

75. • A 2.50-kg package slides down a 12.0-m-long inclined plane that makes an angle of 20.0° with the horizontal. The package has an initial speed of 2.00 m/s at the top of the incline. What must the coefficient of kinetic friction between the package and the inclined plane be so that the package reaches the bottom with no speed? Example 5-2

76. ••• In Figure 5-39, two blocks are connected to each other by a massless string over a massless and frictionless pulley. The mass of block 1 is $m_1 = 6.00$ kg. Assuming the coefficient of static friction $\mu_s = 0.542$ for all surfaces, find the range of values of the mass m_2 so that the system is in equilibrium. Example 5-5

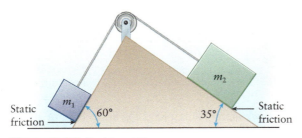

Figure 5-39 Problem 76

77. • Biomedical laboratories routinely use ultracentrifuges, some of which are able to spin at 1.00×10^5 rev/min about the central axis. The turning rotor in certain models is about 20.0 cm in diameter. At its top spin speed, what force does the rotor exert on a 2.00-g sample that is positioned at the greatest distance from the spin axis? Would the force be appreciably different if the sample were spun in a vertical or a horizontal circle? Why or why not? Example 5-11

78. •• An amusement park ride called the Rotor debuted in 1955 in Germany. Passengers stand in the cylindrical drum of the Rotor as it rotates around its axis. Once the Rotor reaches its operating speed, the floor drops but the riders remain pinned against the wall of the cylinder. Suppose the cylinder makes 25.0 rev/min and has a radius of 3.50 m. What is the minimum coefficient of static friction between the wall of the cylinder and the backs of the riders? Example 5-11

79. ••• An object of mass $m_1 = 0.125$ kg undergoes uniform circular motion and is connected by a massless string through a hole in a frictionless table to a larger object of mass $m_2 = 0.225$ kg (Figure 5-40). If the larger object is stationary, calculate the tension in the string and the speed of the circular motion of the smaller object. The radius R of the circular path is equal to 1.00 m. Example 5-10

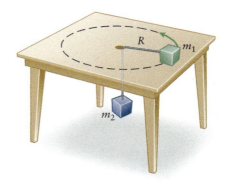

Figure 5-40
Problem 79

80. •• An object that has a mass M hangs from a support by a massless string of length L (Figure 5-41). The support is rotated so that the object follows a circular path at an angle θ from the vertical as shown. The object makes N revolutions per second. Derive an expression for the angle θ in terms of M, L, N, and any necessary physical constants. Example 5-8

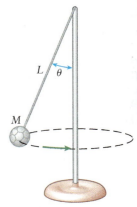

Figure 5-41 Problem 80

81. • **Medical** Occupants of cars hit from behind often suffer serious neck injury from whiplash. During a low-speed rear-end collision, a person's head suddenly pivots about the base of the neck through a 60.0° angle, a motion that lasts 250 ms. The distance from the base of the neck to the center of the head is typically about 0.20 m, and the head normally comprises about 6.0% of body weight. We can model the motion of the head as having uniform speed over the course of its pivot. (a) What is the acceleration of the head during the collision? (b) What force (in newtons and in pounds) does the neck exert on the head of a 75-kg person in the collision? (As a first approximation, neglect the force of gravity on the head.) (c) Would headrests mounted to the backs of the car seats help protect against whiplash? Why or why not? Example 5-10

82. • A curve that has a radius of 1.00×10^2 m is banked at an angle of 10.0° (Figure 5-42). If a 1.00×10^3-kg car navigates the curve at 65.0 km/h without skidding, what is the minimum coefficient of static friction between the pavement and the tires? Example 5-12

Figure 5-42
Problem 82

blickwinkel/Alamy

Work and Energy

6

In this chapter, your goals are to:

- (6-1) Explain the relationship between work and energy.
- (6-2) Calculate the work done by a constant force on an object moving in a straight line.
- (6-3) Describe what kinetic energy is and understand the work-energy theorem.
- (6-4) Apply the work-energy theorem to solve problems.
- (6-5) Recognize why the work-energy theorem applies even for curved paths and varying forces like the spring force.
- (6-6) Explain the meaning of potential energy and how conservative forces such as the gravitational force and the spring force give rise to gravitational potential energy and spring potential energy.
- (6-7) Explain the differences between, and when you can apply, the generalized law of conservation of energy and the conservation of total mechanical energy.
- (6-8) Identify which kinds of problems are best solved with energy conservation and the steps to follow in solving these problems.
- (6-9) Describe what power is and its relationship to work and energy.

To master this chapter, you should review:

- (2-4) The equations for constant-acceleration motion in a straight line.
- (4-2, 4-5) Newton's second and third laws of motion.

What do you think?

The ostrich (*Struthio camelus*) is the world's largest bird, with twice the mass of an adult human. Despite its size, an ostrich can run at a steady 50 km/h (14 m/s, or 31 mi/h) for extended periods. Which anatomical structure is most important for sustaining a running ostrich's motion? (a) The ostrich's long neck; (b) its feathers; (c) the tendons in its legs; (d) its clawed feet.

6-1 The ideas of work and energy are intimately related

Energy is used in nearly every living process—moving, breathing, circulating blood, digesting food, and absorbing nutrients. You've no doubt heard that there are different kinds of energy, including kinetic, potential, and internal, and that energy can be transformed from one of these kinds into another. But what *is* energy?

In this chapter we'll see that if an object or system has **energy**, then that object or system may be able to do *work*. Work has many different meanings in everyday life. In the language of physics, however, work has a very specific definition that describes what happens when a force is exerted on an object as it moves. One example of doing work is lifting a book from your desk to a high bookshelf; another is a football player

The football player applies a force F to the sled while it moves a distance d. Hence he does an amount of work on the sled equal to Fd.

(a)

AP Photo/Kevin Wolf

The girl does work on the basketball: She exerts a force on the ball as she pushes it away from her. As a result the ball acquires kinetic energy (energy of motion).

(b)

(c)

Jens Karlsson/Getty Images

As the ball hits its target and slows down, work is done on the target. Some kinetic energy also goes into deforming the ball and giving it elastic potential energy, which converts again to kinetic energy when the ball springs back.

Figure 6-1 Work and energy In this chapter we'll explore the ideas of (a) work, (b) kinetic energy, and (c) potential energy.

pushing a blocking sled across a field (Figure 6-1a). By combining the definition of work with our knowledge of Newton's second law, we'll be led to the idea of *kinetic energy*, which is the energy that an object has due to its motion (Figure 6-1b). An object with kinetic energy has the ability to do work: For example, a moving ball has the ability to displace objects in its path (Figure 6-1c).

Sometimes kinetic energy is lost: It gets converted to a form of energy that can't be used to do work. One example is when an egg thrown at high speed splatters against a wall. But if that egg is replaced with a rubber ball, the ball bounces off with nearly the same speed at which it was thrown and so we recover most of the ball's kinetic energy. We say that as the ball compresses against the wall, its kinetic energy is converted into *potential energy*— energy associated not with the ball's motion but with its shape and position. This potential energy is again converted into kinetic energy as the ball bounces back (Figure 6-1c).

In this chapter we'll look at the ways in which kinetic energy, potential energy, and *internal energy*—energy stored within an object, sometimes in a way that can't easily be extracted—can transform into each other. (As an example, the energy you need to read this chapter is extracted from the internal energy of food that you consumed earlier.) In this way we'll discover the *law of conservation of energy*, which proves to be one of the great unifying principles of the natural sciences in general.

6-2 The work that a constant force does on a moving object depends on the magnitude and direction of the force

The man depicted in Figure 6-2 is doing *work* as he pushes a crate up a ramp. The amount of work that he does depends not only on how hard he pushes on the crate (that is, on the magnitude of the force that he exerts) but also on the distance over which he moves the crate (that is, the displacement of the crate). In a similar way, the football players shown in Figure 6-1a will be more exhausted if the coach asks them to push the blocking sled all the way down the field rather than a short distance.

These examples suggest how we should define the **work** done by a force on an object. Let's begin with the case of an object moving in a straight line through a displacement $\vec{d}$. As the object moves, a constant force $\vec{F}$ acts on it that points in the same

The man exerts a constant force $\vec{F}$ on the crate. The direction of the force is parallel to the ramp.

As he exerts the force, the crate moves through a displacement $\vec{d}$ up the ramp.

stevecoleimages/E+/Getty Images

 If the force $\vec{F}$ the man exerts on the crate is in the same direction as the displacement $\vec{d}$ of the crate, the work W that he does on the crate is the product of force and displacement: $W = Fd$.

Figure 6-2 Work is force times distance Calculating how much work the man does to push the crate up the ramp.

direction as $\vec{d}$. Then the work done by the force equals the product of the force magnitude F and the distance d over which the object moves:

> Work done on an object by a constant force $\vec{F}$ that points in the **same direction** as the object's displacement $\vec{d}$

> Magnitude of the constant force $\vec{F}$

$$W = Fd$$

> Magnitude of the displacement $\vec{d}$

Work done by a constant force that points in the same direction as the straight-line displacement (6-1)

Note that Equation 6-1 refers only to situations in which the motion is in a straight line and the force acts in the *same* direction as the displacement. You've already seen two situations of this sort: The football players in Figure 6-1a push the sled backward as it moves backward, and the man in Figure 6-2 pushes the crate uphill as it moves uphill. Later we'll consider cases in which force and displacement are *not* in the same direction and the motion isn't along a straight line.

We saw in Chapters 4 and 5 that it's important to keep track of what object *exerts* a given force and on what object that force is exerted. It's equally important to keep track of both the object that exerts a force and the object on which the force does work. For example, in Figure 6-2, the object exerting a force is the man, and the object on which the force does work is the crate.

The unit of work, the **joule** (J), is named after the nineteenth-century English physicist James Joule, who did fundamental research on the relationship between motion and work. From Equation 6-1,

$$1\,J = 1\,N \times 1\,m \text{ or } 1\,J = 1\,N \cdot m$$

You do 1 J of work when you exert a 1-N push on an object as it moves through a distance of 1 m.

WATCH OUT! Work and weight have similar symbols.

! Since *work* and *weight* begin with the same letter, it's important to use different symbols to represent them in equations. We'll use an uppercase W for work and a lowercase w for weight (the magnitude of the gravitational force), and we recommend that you do the same to prevent confusion.

EXAMPLE 6-1 Lifting a Book

How much work must you do to lift a textbook with a mass of 2.00 kg—roughly the mass of the printed version of this book—by 5.00 cm? (You lift the book at a constant speed.)

Set Up

Newton's first law tells us that the net force on the book must be zero if it is to move upward at a constant speed. Hence the upward force you apply must be constant and equal in magnitude to the gravitational force on the book. Since the book moves in a straight line, the applied force is constant, and this force is in the direction in which the book moves, we can use Equation 6-1 to calculate the work done.

Work done on an object, force in the same direction as displacement:

$$W = Fd \qquad (6\text{-}1)$$

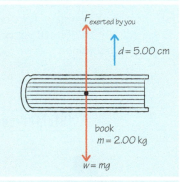

Solve

Calculate the magnitude F of the force that you exert (equal to the gravitational force on the book). Then substitute this and the displacement $d = 5.00$ cm into Equation 6-1 to determine the work that you do.

$$F = mg = (2.00 \text{ kg})(9.80 \text{ m/s}^2)$$
$$= 19.6 \text{ N}$$
$$W = Fd = (19.6 \text{ N})(5.00 \text{ cm})$$
$$= (19.6 \text{ N})(0.0500 \text{ m})$$
$$= 0.980 \text{ N} \cdot \text{m}$$
$$= 0.980 \text{ J}$$

Reflect

The work that you do is almost exactly 1 joule. The actual amount of energy that you would need to *expend* is several times more than 0.980 J, however. That's because your body isn't 100% efficient at converting energy into work. Some of the energy goes into heating your muscles. In fact, your muscles can consume energy even when they do *no* work, as we describe below. The value of 0.980 J that we calculated is just the amount of work that you do *on the book*.

BioMedical EXAMPLE 6-2 Work Done by Actin

In order to fertilize eggs, the sperm of the horseshoe crab (Figure 6-3) must penetrate two protective layers of the egg with a combined thickness of about 40 μm = 40×10^{-6} m. To achieve this, a bundle of the protein actin on the outer surface of the sperm pushes through the egg's protective layers with a constant force of 1.9×10^{-9} N. How much work does the actin bundle do in this process?

J Hindman/Shutterstock

Figure 6-3 A horseshoe crab Horseshoe crabs (family Limulidae) have existed on Earth for 450 million years. Despite their name, they are more closely related to spiders and scorpions than to true crabs. Like the cells of other animals and plants, horseshoe crab cells contain an important protein called actin.

Set Up

The force exerted is both constant and in the same direction as the straight-line motion of the end of the bundle. Hence we can once again use Equation 6-1.

$$W = Fd \qquad (6-1)$$

Solve

The work done by the actin bundle equals the force exerted by the actin bundle multiplied by the distance that it pushes through the outer layers of the egg.

$$\begin{aligned} W_{\text{done by bundle}} &= Fd \\ &= (1.9 \times 10^{-9} \text{ N})(40 \times 10^{-6} \text{ m}) \\ &= 7.6 \times 10^{-14} \text{ J} \end{aligned}$$

Reflect

The amount of work done by the actin bundle seems ridiculously small. But such a bundle is microscopic, with a length of only 60 μm (slightly greater than the thickness of the layers of the egg that it must penetrate) and a mass $m_{\text{bundle}} \approx 10^{-16}$ kg. (The symbol ≈ means "approximately equal to.") Hence the amount of work that the actin bundle does per kilogram of mass is $W_{\text{done by bundle}}/m_{\text{bundle}} \approx 10^3$ J/kg. If a 60-kg person could do that much work on a per-kilogram basis on the book in Example 6-1, he or she could lift the book 3 km (nearly 10,000 ft)! An actin bundle is small, but it packs a big punch.

$$\frac{\text{Work done by actin bundle}}{\text{Mass of actin bundle}} = \frac{7.6 \times 10^{-14} \text{ J}}{10^{-16} \text{ kg}} \approx 10^3 \text{ J/kg}$$

If a 60-kg person could do that much work on a per-kilogram basis,

$$\text{Work done} = W \approx (10^3 \text{ J/kg})(60 \text{ kg}) = 6 \times 10^4 \text{ J}$$

Height d to which this amount of work could raise a book of mass 2.00 kg and weight 19.6 N:

$$W = mgd, \text{ so } d = \frac{W}{mg} = \frac{6 \times 10^4 \text{ J}}{19.6 \text{ N}} \approx 3 \times 10^3 \text{ m} = 3 \text{ km}$$

Muscles and Doing Work

BioMedical

Figure 6-4 Getting tired while doing zero work Weights that you hold stationary in your outstretched arms undergo no displacement, so you do zero work on them. Why, then, do your arms get tired?

Pick up a heavy object and hold it in your hand at arm's length (Figure 6-4). After a while you'll notice your arm getting tired: It feels like you're doing work to hold the object in midair. But Equation 6-1 says that you're doing *no* work on the object because it isn't moving (its displacement d is zero). So why does your arm feel tired?

To see the explanation for this seeming paradox, notice that the muscles in your arm (and elsewhere in your body) exert forces on their ends by contracting. Skeletal muscles (those that control the motions of your arms, legs, fingers, toes, and other structures) are made up of bundles of muscle cells (Figure 6-5a). Muscle cells consist of bundles of myofibrils that are segmented into thousands of tiny structures called sarcomeres. Connected to each other, end to end, sarcomeres contain interdigitated filaments of the contractile proteins actin and myosin (Figure 6-5b). As the filaments slide past each other the sarcomeres can shorten, resulting in the muscle contracting; when the muscle relaxes, sarcomeres lengthen.

The reason your arm tires while holding a heavy object is that shortening the sarcomeres and maintaining the contraction against the load of the object requires energy. Metabolic pathways convert energy stored in the chemical bonds of fat and sugar to a form that your muscle cells can use, but these reactions generate waste products that change the conditions in the muscle. This leads to the feeling that you've been doing work—even though you're doing no work on the *object* you're holding.

(a)

(b)

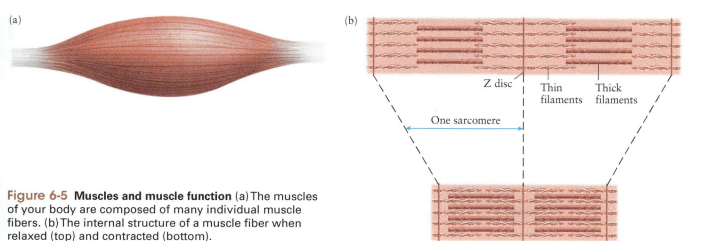

Z disc

Thin filaments

Thick filaments

One sarcomere

Figure 6-5 Muscles and muscle function (a) The muscles of your body are composed of many individual muscle fibers. (b) The internal structure of a muscle fiber when relaxed (top) and contracted (bottom).

If you again hold the object in one hand, but now rest your elbow on a table, you'll be able to hold the object for a longer time without fatigue. That's because in this case you're using only the muscles of your hand and forearm, not those of your upper arm. Fewer muscle fibers have to stay contracted, fewer crossbridges between filaments have to be reattached to maintain the contraction, and you expend less energy.

Work by Forces Not Parallel to Displacement and Negative Work

For an object that moves in a straight line, how can we calculate the work done by a constant force that is *not* in the direction of the object's motion? As an example, in Figure 6-6a a groundskeeper is using a rope to pull a screen across a baseball diamond to smooth out the dirt. The net tension force $\vec{F}$ that the rope exerts on the screen is at an angle θ with respect to the straight-line displacement $\vec{d}$ of the screen, so only the *component* of the force along the displacement does work on the screen (Figure 6-6b). This component is $F \cos \theta$, so the amount of work done by the force is

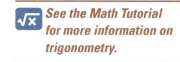 **See the Math Tutorial for more information on trigonometry.**

> Work done on an object by a constant force $\vec{F}$ that points **at an angle** θ to the object's displacement $\vec{d}$
>
> Magnitude of the constant force $\vec{F}$
>
> $$W = (F \cos \theta)d$$
>
> **Angle** between the directions of $\vec{F}$ and $\vec{d}$ Magnitude of the displacement $\vec{d}$

Calculating the work done by a constant force at an angle θ to the straight-line displacement (6-2)

So *the work done by a constant force equals the component of force in the direction of the displacement, multiplied by the displacement.* Note that we can also write Equation 6-2 as $W = F(d \cos \theta)$, where $d \cos \theta$ is the component of the displacement along the force (Figure 6-6c). So, equivalently, *the work done by a constant force equals the force multiplied by the component of displacement in the direction of the force.*

To better understand Equation 6-2, let's look at some special cases.

- If $\vec{F}$ is in the same direction as the motion, $\theta = 0$ and $\cos \theta = \cos 0 = 1$. This is the same situation as shown in Figure 6-2, and in this case Equation 6-2 gives the same result as Equation 6-1: $W = Fd$.

- If the angle θ between the force and displacement is more than zero but less than 90°, as in Figure 6-6b, then $\cos \theta$ is less than one but still positive. The work W done by the force $\vec{F}$ is positive but less than Fd.

- If $\vec{F}$ is perpendicular to the direction of motion, $\theta = 90°$ and $\cos 90° = 0$. In this case Equation 6-2 tells us that force $\vec{F}$ does *zero* work. An example

(a)

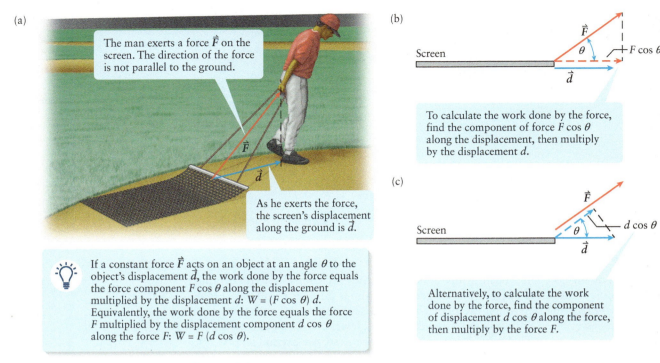

The man exerts a force $\vec{F}$ on the screen. The direction of the force is not parallel to the ground.

$\vec{F}$

As he exerts the force, the screen's displacement along the ground is $\vec{d}$.

$\vec{d}$

(b)

Screen

$\vec{F}$

θ

$F \cos \theta$

$\vec{d}$

To calculate the work done by the force, find the component of force $F \cos \theta$ along the displacement, then multiply by the displacement d.

(c)

Screen

$\vec{F}$

θ

$d \cos \theta$

$\vec{d}$

Alternatively, to calculate the work done by the force, find the component of displacement $d \cos \theta$ along the force, then multiply by the force F.

If a constant force $\vec{F}$ acts on an object at an angle θ to the object's displacement $\vec{d}$, the work done by the force equals the force component $F \cos \theta$ along the displacement multiplied by the displacement d: $W = (F \cos \theta)\,d$. Equivalently, the work done by the force equals the force F multiplied by the displacement component $d \cos \theta$ along the force F: $W = F\,(d \cos \theta)$.

Figure 6-6 When force is not aligned with displacement (a) A man exerts a force on a screen to pull it across a baseball field. (b), (c) Finding the work that the man does on the screen (seen from the side).

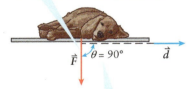

A lazy dog exerts a downward force on the screen.

$\vec{F}$ $\theta = 90°$ $\vec{d}$

The angle between the force and the displacement $\vec{d}$ of the screen is 90° (the force and displacement are perpendicular). Hence the force has zero component in the direction of $\vec{d}$ and does zero work.

Figure 6-7 Doing zero work A dog resting atop the screen exerts a force on the screen as it slides but does zero work on the screen.

is the force exerted by a lazy dog lying on top of the screen from Figure 6-6a (Figure 6-7). The dog exerts a downward force on the screen as the screen moves horizontally, so $\theta = 90°$ and the lazy dog does no work at all.

- If the angle θ in Equation 6-2 is greater than 90°, the value of $\cos \theta$ is negative. In this case force $\vec{F}$ does **negative work**: $W < 0$. As an example, Figure 6-8 shows a cart rolling along the floor as a person tries to slow it down. The force $\vec{F}$ that the person exerts on the cart is directed opposite to the cart's displacement $\vec{d}$, so the angle θ in Equation 6-2 is 180°. Since $\cos 180° = -1$, this means that the work the person does on the cart is $W_{\text{person on cart}} = -Fd$, which is negative.

What does it mean to do negative work on an object? Remember that when the force on an object (and hence its acceleration) points in the opposite direction of the object's velocity and displacement, the object slows down (see Figure 2-10). So if you want to slow an object down, you must do negative work on it. Conversely, if you do positive work on an object, you can speed it up.

Here's another way to understand what it means to do negative work. The person in Figure 6-8 does negative work $W_{\text{person on cart}} = -Fd$ on the cart because she exerts a force $\vec{F}$ on the cart opposite to the cart's displacement $\vec{d}$. But by Newton's third law, the cart exerts a force $-\vec{F}$ on her (the same magnitude of force but in the opposite direction). The person has the same displacement $\vec{d}$ as the cart, and the force the cart exerts on her is in the *same* direction as her displacement ($\theta = 0$ and $\cos \theta = +1$ in Equation 6-2). So the work that the *cart* does on the *person* is positive and equal to $W_{\text{cart on person}} = +Fd$. *If object A does negative work on object B, then object B does an equal amount of positive work on object A.* Another example of this is when a moving cue ball hits a stationary 8-ball on a pool table. The cue ball does *positive* work on the 8-ball: It pushes the 8-ball forward (in the direction that the 8-ball moves) and makes the 8-ball speed up. The 8-ball does *negative* work on the cue ball: It pushes back on the cue ball (in the direction opposite to the cue ball's motion) and makes the cue ball slow down.

Table 6-1 summarizes the relationship between the angle at which a force acts on an object and the work the force does on the object.

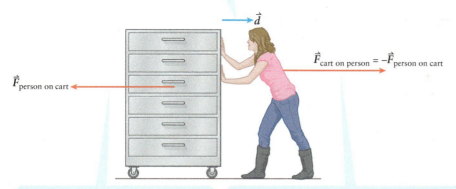

① As the person tries to make the cart slow down, the cart and the person's hands move together to the right (they have the same displacement).

$\vec{d}$

$\vec{F}_{\text{cart on person}} = -\vec{F}_{\text{person on cart}}$

$\vec{F}_{\text{person on cart}}$

② The force of the person on the cart is opposite to the cart's displacement. Hence $\theta = 180°$, $\cos \theta = -1$, and the person does negative work on the cart.

③ By Newton's third law, the cart exerts an equally strong force on the person, but in the opposite direction—that is, in the same direction of the displacement of the person's hands. So the cart does positive work on the person.

If one object (like the person) does negative work on a second object (like the cart), the second object does an equal amount of positive work on the first object.

Figure 6-8 Doing negative work As the cart rolls to the right, the person pushes on the cart to the left to make it slow down. As a result, she does negative work on the cart.

TABLE 6-1 Work Done by a Constant Force

If the angle between a force $\vec{F}$ on an object and the displacement $\vec{d}$ of the object is . . .	. . . then the work done by the force is . . .
Less than 90°	positive
90°	zero
More than 90°	negative

Calculating Work Done by Multiple Forces

What if more than one force acts on an object as it moves? Then the *total* work done on the object is the *sum* of the work done on the object by each force individually. For example, four forces act on the screen in Figure 6-6a: the tension in the rope, the gravitational force, the normal force, and the force of friction. The free-body diagram in Figure 6-9 shows all of these forces, as well as the displacement vector $\vec{d}$ (which points in the direction of motion).

We use Equation 6-2 to determine the work that each force does on the screen:

Tension force: This force $\vec{T}$ (which we labeled $\vec{F}$ in Figure 6-6a) acts at an angle θ with respect to the direction of motion, so $W_{\text{tension}} = Td \cos \theta$.

Gravitational force: The gravitational force $\vec{w}$ is perpendicular to the direction of motion, so the angle θ in Equation 6-2 is 90° for this force. Since $\cos 90° = 0$, the work done by the gravitational force is $W_{\text{gravity}} = 0$.

Normal force: The normal force $\vec{n}$ is also perpendicular to the direction of motion. So like the gravitational force, the normal force does no work at all: $W_{\text{normal}} = 0$.

Kinetic friction force: Because friction opposes sliding, the kinetic friction force vector $\vec{f}_k$ points in the direction opposite to the motion. The angle to use in Equation 6-2 is therefore 180°, and $W_{\text{friction}} = f_k d \cos 180° = -f_k d$. (The friction force does negative work on the screen because it opposes the screen's motion.)

The total work done on the screen by all four forces is the sum of these:

$$W_{\text{total}} = W_{\text{tension}} + W_{\text{gravity}} + W_{\text{normal}} + W_{\text{friction}} = Td \cos \theta + 0 + 0 + (-f_k d)$$
$$= Td \cos \theta - f_k d = (T \cos \theta - f_k)d$$

If the horizontal component $T \cos \theta$ of the tension force (which does positive work) is greater in magnitude than the kinetic friction force f_k (which does negative work), the total work done on the screen is *positive* and the screen *speeds up*. But if $T \cos \theta$ is less than f_k, the total work done on the screen is *negative* and the screen *slows down*.

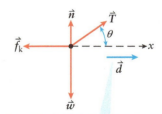

$\vec{n}$ $\vec{T}$

θ

$\vec{f}_k$ ← --------- → x

$\vec{d}$

$\vec{w}$

Always draw the displacement vector $\vec{d}$ to one side so you don't confuse it with the force vectors.

Figure 6-9 Calculating the work done by multiple forces The free-body diagram for the screen shown in Figure 6-6a. The motion of the screen is in the positive x direction. We use Equation 6-2 to calculate the work done on the screen by each force, then sum these to find the total work done on the screen.

EXAMPLE 6-3 Up the Hill

You need to push a box of supplies (weight 225 N) from your car to your campsite, a distance of 6.00 m up a 5.00° incline. You exert a force of 85.0 N parallel to the incline, and a 56.0-N kinetic friction force acts on the box. Calculate (a) how much work you do on the box, (b) how much work the force of gravity does on the box, (c) how much work the friction force does on the box, and (d) the net work done on the box by all forces.

Set Up

Since this problem involves forces, we draw a free-body diagram for the box. The forces do not all point along the direction of the box's displacement $\vec{d}$, so we'll have to use Equation 6-2 to calculate the work done by each force. Once we've calculated these, we'll sum them to find the net work done on the box by all forces.

Work done by a constant force, straight-line displacement:

$$W = Fd \cos \theta \qquad (6\text{-}2)$$

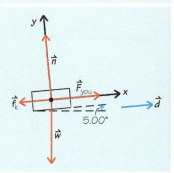

Solve

(a) The force you exert on the box is in the same direction as the box's displacement, so $\theta = 0°$ in Equation 6-2.

$$\begin{aligned} W_{you} &= F_{you}d \cos 0° \\ &= (85.0 \text{ N})(6.00 \text{ m})(1) \\ &= 5.10 \times 10^2 \text{ N} \cdot \text{m} \\ &= 5.10 \times 10^2 \text{ J} \end{aligned}$$

(b) The angle between the gravitational force and the displacement is $\theta = 90.0° + 5.00° = 95.0°$. Since this is more than 90°, $\cos \theta$ is negative and the gravitational force does negative work on the box.

$$\begin{aligned} W_{grav} &= wd \cos 95.0° \\ &= (225 \text{ N})(6.00 \text{ m})(-0.0872) \\ &= -1.18 \times 10^2 \text{ J} \end{aligned}$$

(c) The friction force points opposite to the displacement, so for this force $\theta = 180°$ and $\cos \theta = -1$.

$$\begin{aligned} W_{friction} &= f_k d \cos 180° \\ &= (56.0 \text{ N})(6.00 \text{ m})(-1) \\ &= -3.36 \times 10^2 \text{ J} \end{aligned}$$

(d) The net work is the sum of the work done by all four forces that act on the box. The normal force points perpendicular to the displacement, so it does zero work (for this force, $\theta = 90°$ and $\cos \theta = 0°$).

$$\begin{aligned} W_{total} &= W_{you} + W_{grav} + W_{friction} + W_n \\ &= (5.10 \times 10^2 \text{ J}) + (-1.18 \times 10^2 \text{ J}) \\ &\quad + (-3.36 \times 10^2 \text{ J}) + 0 \\ &= 0.56 \times 10^2 \text{ J} = 56 \text{ J} \end{aligned}$$

Reflect

To check our result, let's calculate how much work is done by the *net* force that acts on the box. Our discussion of Equation 6-2 tells us that we only need the component of the net force in the direction of the displacement, which in our figure is the positive *x* component. So the work done by the net force is just $\sum F_x$ multiplied by *d*.

We learn two things from this. One, the work done by the net force has the *same* value as the sum of the work done by the individual forces. Two, in this case the net force is in the direction of the displacement, which means that the box picks up speed as it moves *and* the net work done on the box is positive. We'll use both of these observations in the next section to relate the net work done on an object to the change in its speed.

Net force up the incline:

$$\begin{aligned} \sum F_x &= F_{you} - f_k - w \sin 5.00° \\ &= 85.0 \text{ N} - 56.0 \text{ N} - (225 \text{ N})(0.0872) \\ &= 9.4 \text{ N} \end{aligned}$$

(positive, so the net force is uphill)

Work done by net force:

$$\begin{aligned} W_{net} &= \left(\sum F_x\right)d \\ &= (9.4 \text{ N})(6.00 \text{ m}) \\ &= 56 \text{ J} \end{aligned}$$

This equals the sum of the work done by the four forces individually.

GOT THE CONCEPT? 6-1 Work: Positive, Negative, or Zero?

For each of the following cases, state whether the work you do on the baseball is positive, negative, or zero. (a) You catch a baseball and your hand moves backward as you bring the ball to a halt. (b) You throw a baseball. (c) You carry a baseball in your hand as you ride your bicycle in a straight, level line at constant speed. (d) You carry a baseball in your hand as you ride your bicycle in a straight, level line at increasing speed.

TAKE-HOME MESSAGE FOR Section 6-2

✔ If a force acts on an object that undergoes a displacement, the force can do work on that object.

✔ For a constant force and straight-line displacement, the amount of work done equals the displacement multiplied by the component of the force parallel to that displacement.

✔ Whether the work done is positive, negative, or zero depends on the angle between the direction of the force and the direction of the displacement.

✔ If one object does negative work on a second object, the second object must do an equal amount of positive work on the first object.

6-3 Kinetic energy and the work-energy theorem give us an alternative way to express Newton's second law

In Example 6-3 we considered a box being pushed uphill that gained speed as it moved. We found that the total amount of work being done on this box was positive. Let's now show that there's a *general* relationship between the total amount of work done on an object and the change in that object's speed. To find this important relationship, we'll combine our definition of work from Section 6-2 with what we learned about one-dimensional motion in Chapter 2 and our knowledge of Newton's laws from Chapter 4.

Let's begin by considering an object that moves along a straight line that we'll call the x axis, traveling from initial coordinate x_i to final coordinate x_f with a constant acceleration a_x (Figure 6-10). Equation 2-11 gives us a relation between the object's velocity v_{ix} at x_i and its velocity v_{fx} at x_f:

$$v_{fx}^2 = v_{ix}^2 + 2a_x(x_f - x_i) \qquad (6\text{-}3)$$

In Equation 6-3 v_{fx}^2 is the square of the velocity at x_f, but it also equals the square of the object's *speed* v_f at x_f. That's because v_{fx} is equal to $+v_f$ if the object is moving in the positive x direction and equal to $-v_f$ if moving in the negative x direction. In either case, $v_{fx}^2 = v_f^2$. For the same reason $v_{ix}^2 = v_i^2$, where v_i is the object's speed at x_i. So we can rewrite Equation 6-3 as

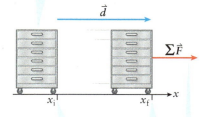

① A cart of mass m rolls along a straight line that we'll call the x axis.

② The cart slides from initial position x_i to final position x_f.

③ As the cart moves, a constant net force in the x direction acts on the cart.

Figure 6-10 Deriving the work-energy theorem A cart moves along a straight line, traveling from initial coordinate x_i to final coordinate x_f. The net force is constant, so the cart has a constant acceleration a_x.

| Speed at position x_f of an object in linear motion with constant acceleration | Speed at position x_i of the object |

$$v_f^2 = v_i^2 + 2a_x\,(x_f - x_i)$$

| Constant acceleration of the object | Two positions of the object |

Relating speed, acceleration, and position for straight-line motion with constant acceleration (6-4)

In Equation 6-4 $x_f - x_i$ is the displacement of the object along the x axis. From Equation 6-2, if we multiply this by the x component of the net force on the object, ΣF_x—that is, by the net force component in the direction of the displacement—we get W_{net}, the work done by the net force on the object. (This assumes that the net force on the object is constant, which is consistent with our assumption that the object's acceleration is constant.) That is,

$$W_{net} = \left(\Sigma F_x\right)(x_f - x_i) \qquad (6\text{-}5)$$

Newton's second law tells us that $\sum F_x = ma_x$, where m is the mass of the object. So we can rewrite Equation 6-5 as

Calculating the work done by a constant net force, straight-line motion

(6-6)

Work done on an object by the net force on that object

Mass of the object

$$W_{net} = ma_x(x_f - x_i)$$

Constant acceleration of the object

Displacement of the object

We can get another expression for the quantity $ma_x(x_f - x_i)$ by multiplying Equation 6-4 by $m/2$ and rearranging:

$$\frac{1}{2}mv_f^2 = \frac{1}{2}mv_i^2 + ma_x(x_f - x_i)$$

$$ma_x(x_f - x_i) = \frac{1}{2}mv_f^2 - \frac{1}{2}mv_i^2$$

The left-hand side of this equation is the work done on the object by the net force (see Equation 6-6). So

(6-7)

$$W_{net} = \frac{1}{2}mv_f^2 - \frac{1}{2}mv_i^2$$

The right-hand side of Equation 6-7 is the *change* in the quantity $\frac{1}{2}mv^2$ over the course of the displacement (the value at the end of the displacement, where the speed is v_f, minus the value at the beginning, where the speed is v_i). We call this quantity the **kinetic energy** K of the object:

Kinetic energy of a moving object

(6-8)

Kinetic energy of an object

Mass of the object

$$K = \frac{1}{2}mv^2$$

Speed of the object

The units of kinetic energy are $kg \cdot m^2/s^2$. Since $1\ J = 1\ N \cdot m$ and $1\ N = 1\ kg \cdot m/s^2$, you can see that kinetic energy is measured in joules, the same as work. Using the definition of kinetic energy given in Equation 6-8, we can rewrite Equation 6-7 as

The work-energy theorem

(6-9)

Work done on an object by the net force on that object

$$W_{net} = K_f - K_i$$

Kinetic energy of the object after the work is done on it

Kinetic energy of the object before the work is done on it

When an object undergoes a displacement, the work done on it by the net force equals the object's kinetic energy at the end of the displacement minus its kinetic energy at the beginning of the displacement. This statement is called the **work-energy theorem**. It is valid as long as the object acted on is rigid—that is, it doesn't deform like a rubber ball might. Although we have derived this theorem for the special case of straight-line motion with constant forces, it turns out to be valid even if the object follows a curved path and is acted on by varying forces. (We'll justify this claim later in Section 6-5.)

The Meaning of the Work-Energy Theorem

What is kinetic energy, and how does the work-energy theorem help us solve physics problems? To answer these questions let's first return to the cart from Figure 6-8 and imagine that it starts at rest on a horizontal frictionless surface. If you give the cart a push as in Figure 6-11a, the net force on the cart equals the force that you exert (the upward normal force on the cart cancels the downward gravitational force), so the work that you do is the work W_{net} done by the net force. The cart starts at rest, so $v_i = 0$ and the cart's initial kinetic energy $K_i = \frac{1}{2}mv_i^2$ is zero. After you've finished the push, the cart has final speed $v_f = v$ and kinetic energy $K_f = \frac{1}{2}mv^2$. So the work-energy theorem states that

$$W_{you\ on\ cart} = K_f - K_i = \frac{1}{2}mv^2 - 0 = \frac{1}{2}mv^2$$

This gives us our first interpretation of kinetic energy: *An object's kinetic energy equals the work that was done to accelerate it from rest to its present speed.*

Now suppose your friend stands in front of the moving cart and brings it to a halt (Figure 6-11b). If we apply the work-energy theorem to the part of the motion where she exerts a force on the cart, the cart's initial kinetic energy is $K_i = \frac{1}{2}mv^2$, and its final kinetic energy is $K_f = 0$ (the cart ends up at rest). The net force on the cart is the force exerted by your friend, so the work she does equals the net work:

$$W_{friend\ on\ cart} = K_f - K_i = 0 - \frac{1}{2}mv^2 = -\frac{1}{2}mv^2$$

Our discussion in Section 6-2 tells us that the *cart* does an amount of work on your *friend* that's just the negative of the work that your friend does on the cart. So we can write

$$W_{cart\ on\ friend} = -W_{friend\ on\ cart} = \frac{1}{2}mv^2$$

This gives us a second interpretation of kinetic energy: *An object's kinetic energy equals the amount of work it can do in the process of coming to a halt from its present speed.*

(a) Making the cart speed up

① The cart rolls without friction and the normal force balances the gravitational force.

(b) Making the cart slow down

① The cart rolls without friction and the normal force balances the gravitational force.

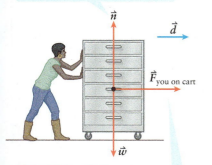

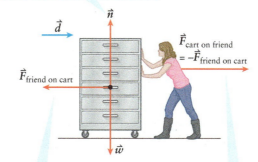

② The force you exert therefore equals the net force on the cart. You do positive work on the cart, and the cart gains speed and kinetic energy.

② The force your friend exerts therefore equals the net force on the cart. She does negative work on the cart, and the cart loses speed and kinetic energy.

③ As the cart loses kinetic energy, it pushes on your friend and does positive work on her.

 An object's kinetic energy equals the work that was done to accelerate it from rest to its present speed.

 An object's kinetic energy equals the work it can do in the process of coming to a halt from its present speed.

Figure 6-11 The meaning of kinetic energy We can interpret kinetic energy in terms of (a) what it takes to accelerate an object from rest or (b) what the object can do in slowing to a halt.

This should remind you of the general definition of energy as the ability to do work, which we introduced in Section 6-1.

You already have a good understanding of this interpretation of kinetic energy. If you see a thrown basketball coming toward your head, you know intuitively that due to its mass and speed it has a pretty good amount of kinetic energy $K = \frac{1}{2}mv^2$, which means that it can do a pretty good amount of work on you (it can exert a force on your nose that pushes it inward a painful distance). You don't want this to happen, which is why you duck!

WATCH OUT! Kinetic energy is a scalar.

We've become comfortable breaking an object's motion into components, so you might be tempted to resolve kinetic energy into components as well. However, kinetic energy depends on *speed*, not velocity. Because speed is not a vector, kinetic energy is a *scalar* quantity and *not* a vector. It wouldn't be meaningful to set up components of kinetic energy in different directions. Note also that we've only discussed the kinetic energy associated with the motion of an object as a whole, which we call **translational kinetic energy**. Kinetic energy is also associated with the rotation of an object around its own axis. We'll return to this *rotational kinetic energy* in a later chapter.

Net Work and Net Force

In Equation 6-9 we interpret W_{net} as the work done by the net force. But we saw in Example 6-3 in Section 6-2 that the work done by the net force equals the *sum* of the amount of work done by each individual force acting on the object. So we can also think of W_{net} in Equation 6-9 as the *net work* done by all of the individual forces. It turns out that this statement is true not just for the situation in Example 6-3; it applies to all situations.

This gives us a simplified statement of the work-energy theorem: *The net work done by all forces on an object, W_{net}, equals the difference between its final kinetic energy K_f and its initial kinetic energy K_i.* If the net work done is positive, then $K_f - K_i$ is positive, the final kinetic energy is greater than the initial kinetic energy, and the object gains speed. If the net work done is negative, then $K_f - K_i$ is negative, the final kinetic energy is less than the initial kinetic energy, and the object loses speed. If zero net work is done, the kinetic energy does not change, and the object maintains the same speed (Table 6-2). This agrees with the observations we made in Section 6-1.

Example 6-4 illustrates how to use the work-energy theorem.

TABLE 6-2 The Work-Energy Theorem

If the net work done on an object is ...	... then the change in the object's kinetic energy $K_f - K_i$ is ...	... and the speed of the object ...
$W_{net} > 0$ (positive net work)	$K_f - K_i > 0$ (kinetic energy increases)	increases (object speeds up)
$W_{net} < 0$ (negative net work)	$K_f - K_i < 0$ (kinetic energy decreases)	decreases (object slows down)
$W_{net} = 0$ (zero net work)	$K_f - K_i = 0$ (kinetic energy stays the same)	is unchanged (object maintains the same speed)

EXAMPLE 6-4 Up the Hill Revisited

Consider again the box of supplies that you pushed up the incline in Example 6-3 (Section 6-2). Use the work-energy theorem to find how fast the box is moving when it reaches your campsite, assuming it is moving uphill at 0.75 m/s at the bottom of the incline.

Set Up

Our goal is to find the final speed v_f of the box. In Example 6-3 we were given the weight (and hence the mass m) of the box, and we found that the net work done on the box is $W_{net} = 56$ J. Since we're given $v_i = 0.75$ m/s, we can use Equations 6-8 and 6-9 to solve for v_f.

Work-energy theorem:

$$W_{net} = K_f - K_i \qquad (6-9)$$

Kinetic energy:

$$K = \frac{1}{2}mv^2 \qquad (6-8)$$

Solve

Use the work-energy theorem to solve for v_f in terms of the initial speed v_i, the net work done W_{net}, and the mass m. To do this, isolate the $\frac{1}{2}mv_f^2$ term on one side of the work-energy theorem equation, then solve for v_f by multiplying through by $2/m$ and taking the square root.

Combine Equations 6-8 and 6-9:

$$W_{net} = \frac{1}{2}mv_f^2 - \frac{1}{2}mv_i^2$$

$$\frac{1}{2}mv_f^2 = \frac{1}{2}mv_i^2 + W_{net}$$

$$v_f^2 = v_i^2 + \frac{2W_{net}}{m}$$

$$v_f = \sqrt{v_i^2 + \frac{2W_{net}}{m}}$$

From Example 6-3 we know that $W_{net} = 56$ J. We find the mass m of the box from its known weight $w = 225$ N and the relationship $w = mg$. Use these to find the value of v_f.

$$m = \frac{w}{g} = \frac{225 \text{ N}}{9.80 \text{ m/s}^2} = 23.0 \text{ kg}$$

$$v_f = \sqrt{(0.75 \text{ m/s})^2 + \frac{2(56 \text{ J})}{23.0 \text{ kg}}}$$

$$= 2.3 \text{ m/s}$$

Reflect

We *could* have solved this problem by first finding the net force on the box, then calculating its acceleration from Newton's second law, and finally solving for v_f by using one of the kinematic equations from Chapter 2. But using the work-energy theorem is easier.

As a check on our result, note that we found in Example 6-3 that the net force on the box is 9.4 N uphill. Can you follow the procedure we just described to find the final velocity after moving 6.00 m up the incline?

Example 6-4 illustrates how using the work-energy theorem can simplify problem solving. In the following section we'll see more examples of how to apply this powerful theorem to various cases of straight-line motion. In Section 6-5 we'll see how the work-energy theorem can be applied to problems in which the motion is along a curved path and in which the forces are not constant.

GOT THE CONCEPT? 6-2 Slap Shot

 A hockey player does work on a hockey puck to propel it from rest across the ice. When a constant force is applied over a certain distance, the puck leaves her hockey stick at speed v. If instead she wants the puck to leave at speed $2v$, by what factor must she increase the distance over which she applies the same force? (a) $\sqrt{2}$; (b) 2; (c) $2\sqrt{2}$; (d) 4; (e) 8.

TAKE-HOME MESSAGE FOR Section 6-3

✔ The net work done on an object (the sum of the work done on it by all forces) as it undergoes a displacement is equal to the change in the object's kinetic energy during that displacement.

✔ The formula for the kinetic energy of an object of mass m and speed v is $K = \frac{1}{2}mv^2$.

✔ The kinetic energy of an object is equal to the amount of work that was done to accelerate the object from rest to its present speed.

✔ The kinetic energy of an object is also equal to the amount of work the object can do in the process of coming to a halt from its present speed.

6-4 The work-energy theorem can simplify many physics problems

In this section we'll explore the relationships among work, force, and speed by applying the definitions of work and kinetic energy (Equations 6-2 and 6-8) and the work-energy theorem (Equation 6-9) to a variety of physical situations. Even if the problem could be solved using Newton's second law, the work-energy theorem often makes the solution easier as well as giving additional insight.

Strategy: Problems with the Work-Energy Theorem

The work-energy theorem is a useful tool for problems that involve an object that moves a distance along a straight line while being acted on by constant forces. You can use this theorem to relate the forces, the distance traveled (the displacement), and the speed of the object at the beginning and end of the displacement.

Note that the work-energy theorem makes no reference to the *time* it takes the object to move through this displacement. It also makes no reference to the *acceleration* of the object. If the problem requires you to use or find either the time or the acceleration, you should use a different approach.

Set Up: Always draw a picture of the situation that shows the object's displacement. Include a free-body diagram that shows all of the forces that act on the object. Draw the direction of each force carefully, since the direction is crucial for determining how much work each force does on the object. Decide what unknown quantity the problem is asking you to determine (for example, the object's final speed or the magnitude of one of the forces).

Solve: Use Equation 6-2 to find expressions for the work done by each force. The sum of the work done by each force is the total or net work done on the object, W_{net}. Then use Equations 6-8 and 6-9 to relate this to the object's initial and final kinetic energies. Solve the resulting equations for the desired unknown.

Reflect: Always check whether the numbers have reasonable values and that each quantity has the correct units.

EXAMPLE 6-5 How Far to Stop?

The U.S. National Highway Traffic Safety Administration lists the minimum braking distance for a car traveling at 40.0 mi/h (64.4 km/h) to be 101 ft (30.8 m). If the braking force is the same at all speeds, what is the minimum braking distance for a car traveling at 65.0 mi/h (105 km/h)?

Set Up

The gravitational force, the normal force exerted by the road, and the braking force all act on the car. But if the road is level, only the braking force does work on the car. (The other two forces are perpendicular to the displacement, so they do no work because $Fd \cos 90° = 0$.) We'll use Equations 6-2, 6-8, and 6-9 to relate the initial speed v_i, the braking force, and the distance the car must travel to have the final speed be $v_f = 0$.

Work done by a constant force, straight-line displacement:

$$W = Fd \cos \theta \qquad (6\text{-}2)$$

Kinetic energy:

$$K = \frac{1}{2}mv^2 \qquad (6\text{-}8)$$

Work-energy theorem:

$$W_{net} = K_f - K_i \qquad (6\text{-}9)$$

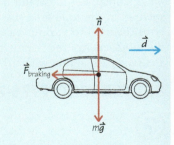

Solve

Since the car ends up at rest, its final kinetic energy is zero. So the work-energy theorem tells us that the net work done on the car (equal to the work done on it by the braking force) equals the negative of the initial kinetic energy.

Work-energy theorem with final kinetic energy equal to zero:

$$W_{net} = 0 - K_i = -\frac{1}{2}mv_i^2$$

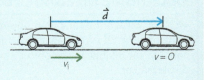

The braking force of magnitude $F_{braking}$ is directed opposite to the displacement, so $\theta = 180°$ in the expression for the work $W_{braking}$ done by this force.

Net work = work done by the braking force:

$$W_{net} = W_{braking}$$
$$= F_{braking}d \cos 180° = -F_{braking}d$$

Substitute $W_{net} = -F_{braking}d$ into the work-energy theorem and solve for the braking distance d. The result tells us that d is proportional to the square of the initial speed v_i.

Work-energy theorem becomes

$$-F_{braking}d = -\frac{1}{2}mv_i^2$$

Solve for d:

$$d = \frac{mv_i^2}{2F_{braking}}$$

We aren't given the mass m of the car or the magnitude $F_{braking}$ of the braking force, but we can set up a ratio between the values of distance for $v_{i1} = 40.0$ mi/h and $v_{i2} = 65.0$ mi/h.

Use this expression to set up a ratio for the value of d at two different initial speeds. The mass of the car and the braking force are the same for both initial speeds:

$$\frac{d_2}{d_1} = \frac{mv_{i2}^2/2F_{braking}}{mv_{i1}^2/2F_{braking}} = \frac{v_{i2}^2}{v_{i1}^2}$$

Solve for the stopping distance d_2 corresponding to $v_{i2} = 65.0$ mi/h:

$$d_2 = d_1\frac{v_{i2}^2}{v_{i1}^2} = (101 \text{ ft})\left(\frac{65.0 \text{ mi/h}}{40.0 \text{ mi/h}}\right)^2$$

$$= 267 \text{ ft} = 81.3 \text{ m}$$

Reflect

This example demonstrates what every driver training course stresses: The faster you travel, the more distance you should leave between you and the car ahead in case an emergency stop is needed. Increasing your speed by 62.5% (from 40.0 to 65.0 mi/h) increases your stopping distance by 164% (from 101 to 267 ft). The reason is that stopping distance is proportional to kinetic energy, which is proportional to the *square* of the speed.

EXAMPLE 6-6 Work and Kinetic Energy: Force at an Angle

At the start of a race, a four-man bobsleigh crew pushes their sleigh as fast as they can down the 50.0-m straight, relatively flat starting stretch (Figure 6-12). The net force that the four men together apply to the 325-kg sleigh has magnitude 285 N and is directed at an angle of 20.0° below the horizontal. As they push, a 60.0-N kinetic friction force also acts on the sleigh. What is the speed of the sleigh right before the crew jumps in at the end of the starting stretch?

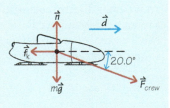

sampics/Corbis/Getty Images

Figure 6-12 Bobsleigh start The success of a bobsleigh team depends on the team members giving the sleigh a competitive starting speed.

Set Up

Again we'll use the work-energy theorem, this time to determine the final speed v_f of the bobsleigh (which starts at rest, so its initial speed is $v_i = 0$). The normal force and gravitational force do no work on the sleigh since they act perpendicular to the displacement. The four men of the crew do positive work, while the friction force does negative work. The new wrinkle is that the force exerted by the crew points at an angle to the displacement. We'll deal with this using Equation 6-2.

The sleigh gains speed and kinetic energy during the motion, so the net work done by friction and by the crew must be positive.

Work done by a constant force, straight-line displacement:

$$W = Fd \cos \theta \qquad (6\text{-}2)$$

Kinetic energy:

$$K = \frac{1}{2}mv^2 \qquad (6\text{-}8)$$

Work-energy theorem:

$$W_{net} = K_f - K_i \qquad (6\text{-}9)$$

Solve

The sleigh starts at rest, so its initial kinetic energy is zero. From Equation 6-9 the net work done on the sleigh is therefore equal to its final kinetic energy, which is related to the final speed v_f.

Work-energy theorem with initial kinetic energy equal to zero:

$$W_{net} = K_f - 0 = \frac{1}{2}mv_f^2$$

The force of the crew is at $\theta = 20.0°$ to the displacement, and the friction force is at $\theta = 180°$. Use these in Equation 6-2 to find the net work done on the sleigh. (Note that the net work is positive, as we predicted.)

Net work = work done by the crew plus work done by friction

$$W_{net} = W_{crew} + W_{friction}$$
$$= F_{crew}d \cos 20.0° + f_k d \cos 180°$$
$$= (285 \text{ N})(50.0 \text{ m})(0.940)$$
$$\quad + (60.0 \text{ N})(50.0 \text{ m})(-1)$$
$$= 1.04 \times 10^4 \text{ J}$$

Substitute W_{net} into the work-energy theorem and solve for the final speed v_f.

Work-energy theorem says $W_{net} = \dfrac{1}{2}mv_f^2$

Solve for v_f: $\quad v_f = \sqrt{\dfrac{2W_{net}}{m}} = \sqrt{\dfrac{2(1.04 \times 10^4 \text{ J})}{325 \text{ kg}}}$

$$= 8.00 \text{ m/s}$$

Reflect

This answer for v_f is relatively close to the speed of world-class sprinters, so it seems reasonable. In reality the start of an Olympic four-man bobsleigh race is slightly downhill, so the sleigh's speed at the end of the starting stretch is typically even faster (11 or 12 m/s).

EXAMPLE 6-7 Find the Work Done by an Unknown Force

An adventurous parachutist of mass 70.0 kg drops from the top of a tall waterfall. The parachutist deploys his chute after falling 295 m, at which point her speed is 54.0 m/s. During the 295-m drop, (a) what was the net work done on her and (b) what was the work done on her by the force of air resistance?

Set Up

We are given the parachutist's mass and the distance that she falls, so we can find the gravitational force that acts on her and the work done by that force using Equation 6-2. But the force of air resistance also acts on the parachutist. We don't know its magnitude, so we can't directly calculate the work done by air resistance. Instead, we'll use the work-energy theorem. We'll assume that the parachutist starts at rest, so her initial kinetic energy is zero.

Work done by a constant force, straight-line displacement:

$$W = Fd \cos \theta \qquad (6\text{-}2)$$

Work-energy theorem:

$$W_{net} = K_f - K_i \qquad (6\text{-}9)$$

Kinetic energy:

$$K = \frac{1}{2}mv^2 \qquad (6\text{-}8)$$

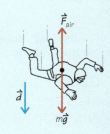

Solve

(a) Since $K_i = 0$, Equation 6-9 says that the net work done on the parachutist is just equal to her final kinetic energy.

$$W_{net} = K_f = \frac{1}{2}mv_f^2$$
$$= \frac{1}{2}(70.0 \text{ kg})(54.0 \text{ m/s})^2$$
$$= 1.02 \times 10^5 \text{ J}$$

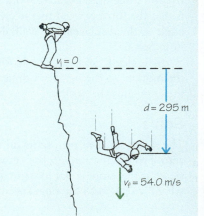

(b) The net work is the sum of the work done by the gravitational force, W_{grav}, and the work done by air resistance, W_{air}. Hence W_{air} is the difference between W_{net} and W_{grav}.

Work done by the gravitational force: Displacement is in the same direction (downward) as the force, so

$$W_{grav} = mgd$$
$$= (70.0 \text{ kg})(9.80 \text{ m/s}^2)(295 \text{ m})$$
$$= 2.02 \times 10^5 \text{ J}$$

Hence the work done by air resistance is

$$W_{air} = W_{net} - W_{grav}$$
$$= 1.02 \times 10^5 \text{ J} - 2.02 \times 10^5 \text{ J}$$
$$= -1.00 \times 10^5 \text{ J}$$

Reflect

The work done by the force of air resistance is negative because the force is directed upward, opposite to the downward displacement. We can use the calculated value of W_{air} to find an average value of the air resistance force, considered to be a constant. This is only an *average* value because this force is *not* constant: The faster the parachutist falls, the greater the force of air resistance.

Average upward force of air resistance:

$$W_{air} = F_{air}d \cos 180° = -F_{air}d$$

Hence

$$F_{air} = -\frac{W_{air}}{d} = -\frac{(-1.00 \times 10^5 \text{ J})}{295 \text{ m}}$$
$$= 339 \text{ J/m} = 339 \text{ N}$$

GOT THE CONCEPT? 6-3 Pushing Boxes

(?) You push two boxes from one side of a room to the other. Each box begins and ends at rest. The contact surfaces and areas between box and floor are the same for both, but one box is heavy while the other is light. There is friction between each box and the floor. (a) How does the work *you* do on each box compare? (i) You do more work on the heavy box. (ii) You do more work on the light box. (iii) You do the same amount of work on both boxes. (b) How does the *net* work done on each box compare? (i) There is more net work on the heavy box. (ii) There is more net work on the light box. (iii) There is the same net work on both boxes.

TAKE-HOME MESSAGE FOR Section 6-4

✔ The work-energy theorem allows us to explore the relationship between force, distance, and speed in a variety of physical situations.

✔ To solve problems using the work-energy theorem, begin by drawing a free-body diagram that shows all the forces that act on the object in question. Then write expressions for the kinetic energies at the beginning and end of the displacement and for the total work done on the object (the sum of the work done by each force). Relate these using the work-energy theorem and solve for the unknown quantity.

6-5 The work-energy theorem is also valid for curved paths and varying forces

We've derived the work-energy theorem only for the special case of straight-line motion with constant forces. Since we already know how to solve problems for that kind of motion, you may wonder why this theorem is important. Here's the answer: The work-energy theorem also works for motion along a *curved* path and in cases where the forces are *not* constant (Figure 6-13a).

To prove this, consider Figure 6-13b, which shows an object's curved path from an initial point i to a final point f. We mark a large number N of equally spaced intermediate points 1, 2, 3, . . . , N along the path and imagine breaking the path into short segments (from i to 1, from 1 to 2, from 2 to 3, and so on). If each segment is sufficiently short, we can treat it as a straight line. Furthermore, the forces on the object don't change very much during the brief time the object traverses one of these segments, so we can treat the forces as constant over that segment. (The forces may be different from

Figure 6-13 Motion along a curved path
(a) The weight travels in a curved path during the bicep curl. (b) Analyzing motion along a general curved path.

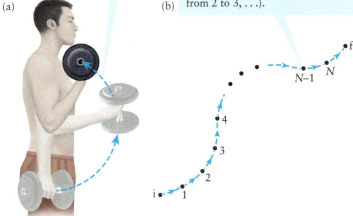

A weight lifter doing bicep curls makes the weight move through a curved path. The force that he exerts on the weight varies during its travel.

An object's curved path from an initial point i to a final point f. We mark a large number N of equally spaced intermediate points 1, 2, 3, . . . , N along the path and imagine breaking the path into short segments (from i to 1, from 1 to 2, from 2 to 3, . . .).

one segment to the next.) Since each segment is a straight line with constant forces, we can safely apply the work-energy theorem in Equation 6-9 to every segment:

$$W_{\text{net, i to 1}} = K_1 - K_i$$
$$W_{\text{net, 1 to 2}} = K_2 - K_1$$
$$W_{\text{net, 2 to 3}} = K_3 - K_2$$
$$\cdots$$
$$W_{\text{net, N to f}} = K_f - K_N$$

Now add all of these equations together:

$$W_{\text{net, i to 1}} + W_{\text{net, 1 to 2}} + W_{\text{net, 2 to 3}} + \cdots + W_{\text{net, N to f}}$$

(6-10)
$$= (K_1 - K_i) + (K_2 - K_1) + (K_3 - K_2) + \cdots + (K_f - K_N)$$

The left-hand side of Equation 6-10 is the sum of the amounts of work done by the net force on each segment. This sum equals the *net* work done by the net force along the entire path from i to f, which we call simply W_{net}. On the right-hand side of the equation, $-K_1$ in the second term cancels K_1 in the first term, $-K_2$ in the third term cancels K_2 in the second term, and so on. The only quantities that survive on the right-hand side are $-K_i$ and K_f, so we are left with

$$W_{\text{net}} = K_f - K_i$$

This is *exactly* the same statement of the work-energy theorem as Equation 6-9. So the work-energy theorem is valid for *any* path and for *any* forces, whether constant or not. This is why the work-energy theorem is so important: You can apply it to situations where using forces and Newton's laws would be difficult or impossible.

Work Done by the Gravitational Force

One of the forces acting on the weight in Figure 6-13a as it follows a curved path is the gravitational force. Since gravity plays a role in so many physics problems, it's important to answer the question, "How much work does the gravitational force do on an object that follows a curved path?" As we'll see, the answer turns out to be quite simple.

In Figure 6-14a an object of mass m travels along a curved path from an initial position $x = x_i, y = y_i$ to a final position $x = x_f, y = y_f$. (We take the x axis to be horizontal and choose the positive y direction to be upward as shown.) The gravitational force $\vec{w} = m\vec{g}$ that Earth exerts on the object remains constant as the object moves:

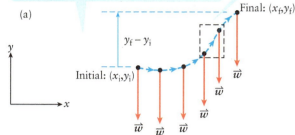

1 As an object moves along a curved path, the gravitational force $\vec{w} = m\vec{g}$ that acts on it has the same magnitude $w = mg$ and the same downward direction at all points along the path. The net vertical displacement on this path is $y_f - y_i$.

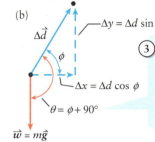

2 Look at a closeup of one segment of the curved path that's short enough to regard as a straight line. The displacement $\Delta\vec{d}$ on this segment is at an angle ϕ with respect to the x axis, has x component $\Delta x = \Delta d \cos \phi$, and has y component $\Delta y = \Delta d \sin \phi$.

3 The angle between the gravitational force $\vec{w}$ and the displacement $\Delta\vec{d}$ is $\theta = \phi + 90°$, so the work done by the gravitational force during this segment is $\Delta W_{grav} = mg\Delta d \cos \theta = mg\Delta d \cos (\phi + 90°)$. Trigonometry shows this is equal to $\Delta W_{grav} = -mg\Delta y$.

The work done on the object by the gravitational force during a segment of the path depends on the vertical component Δy of the displacement, not the horizontal component. The same is true for the work done by the gravitational force over the entire curved path: It depends only on the net vertical displacement $y_f - y_i$, not on the horizontal displacement or the shape of the path.

Figure 6-14 Calculating the work done by the gravitational force No matter what the shape of the curved path that an object follows, the work done on the object by the gravitational force just depends on the object's net vertical displacement.

It always points downward in the negative y direction and always has the same magnitude $w = mg$. As we did for the path of the object in Figure 6-13b, we break this object's path into many small segments, each of which is short enough that we can regard it as a straight line. Figure 6-14b shows a closeup of one such segment, during which the object moves through a small displacement $\Delta\vec{d}$. (The displacement vector $\Delta\vec{d}$ is different for each individual segment.) As Figure 6-14b shows, the displacement vector is at an angle ϕ with respect to the horizontal x axis, so the angle between the downward gravitational force $\vec{w}$ and the displacement $\Delta\vec{d}$ is $\theta = \phi + 90°$. Using Equation 6-2, the small amount of work that the force $\vec{w} = m\vec{g}$ does during the small displacement $\Delta\vec{d}$ is

$$\Delta W_{grav} = w\Delta d \cos \theta = mg\Delta d \cos(\phi + 90°) \tag{6-11}$$

We know from trigonometry that $\cos(\phi + 90°) = \cos \phi \cos 90° - \sin \phi \sin 90°$ (see Table M-2 in the Math Tutorial at the back of this book). Since $\cos 90° = 0$ and $\sin 90° = 1$, this simplifies to $\cos(\phi + 90°) = -\sin \phi$. Hence Equation 6-11 becomes $\Delta W_{grav} = -mg\Delta d \sin \phi$. But as Figure 6-14b shows, $\Delta d \sin \phi$ is equal to Δy, the y component of the displacement $\Delta\vec{d}$ over this short segment. So we can rewrite Equation 6-11 for the small work done by the gravitational force over this short segment as

$$\Delta W_{grav} = -mg\Delta y \tag{6-12}$$

Equation 6-12 agrees with our discussion of work in Section 6-2: The work done by the constant gravitational force equals the force w multiplied by the component of displacement $-\Delta y$ in the direction of the force. (The minus sign is there because the force is in the negative y direction.) If the object moves upward over the segment, as in Figure 6-14b, Δy in Equation 6-12 is positive and the gravitational force does negative work ΔW_{grav} (the force is downward but the displacement is upward); if the object moves downward over the segment, Δy is negative and the gravitational force does positive work ΔW_{grav} (the force and displacement are both downward).

We can now find how much work the gravitational force does on the object over the entire curved path shown in Figure 6-14a. If there are N segments of this path numbered $t = 1, 2, 3, \ldots, N$, over each of these segments the y component of displacement is $\Delta y_1, \Delta y_2, \Delta y_3, \ldots, \Delta y_N$. From Equation 6-12 the work done by the gravitational force over each of these segments is $-mg\Delta y_1, -mg\Delta y_2, -mg\Delta y_3, \ldots, -mg\Delta y_N$.

The net amount of work W_{grav} done over the entire path is the sum of the work done over all of the segments:

(6-13)

$$W_{grav} = (-mg\Delta y_1) + (-mg\Delta y_2) + (-mg\Delta y_3) + \cdots + (-mg\Delta y_N)$$
$$= -mg(\Delta y_1 + \Delta y_2 + \Delta y_3 + \cdots + \Delta y_N)$$

In Equation 6-13 the sum $\Delta y_1 + \Delta y_2 + \Delta y_3 + \cdots + \Delta y_N$ is the *total* vertical displacement of the object over the curved path. From Figure 6-14a this equals $y_f - y_i$, the difference between the final and initial y coordinates of the object. So we can rewrite Equation 6-13 as

Work done by the gravitational force

(6-14)

Work done by the gravitational force on an object as the object follows a curved path

Mass of the object

Acceleration due to gravity

$$W_{grav} = -mg(y_f - y_i)$$

Final y coordinate of the object at the end of the path

Initial y coordinate of the object at the beginning of the path

Equation 6-14 says something quite remarkable: The work done on an object by the gravitational force depends only on the object's vertical displacement $y_f - y_i$, not on the shape of the path that the object follows. This observation helps makes it possible to apply the work-energy theorem to situations that would be very difficult to analyze using Newton's laws alone, as the following example shows.

EXAMPLE 6-8 A Swinging Spider

Figure 6-15 shows a South African kite spider (*Gasteracantha*) swinging on a strand of spider silk. Suppose a momentary gust of wind blows on a spider of mass 1.00×10^{-4} kg (0.100 g) that's initially hanging straight down on such a strand. As a result, the spider acquires a horizontal velocity of 1.3 m/s. How high will the spider swing? Ignore air resistance.

Emil von Maltitz/Getty Images

Figure 6-15 A spider swinging on silk If we know the spider's speed at the low point of its arc, how do we determine how high it swings?

Set Up

Because the spider follows a curved path, this would be a very difficult problem to solve using Newton's laws directly. Instead, we'll use the work-energy theorem. We are given the spider's mass m and initial speed $v_i = 1.3$ m/s (which means we know its initial kinetic energy). We want to find its maximum height h at the point where the spider comes momentarily to rest (so its speed and kinetic energy are zero) before swinging back downward.

The free-body diagram shows that only two forces act on the swinging spider: the gravitational force and the tension force exerted by the silk. We'll use Equation 6-14 for the work done by the gravitational force. To calculate the work done on the spider by the tension force, we'll break the curved path into a large number of segments as in Figures 6-13 and 6-14.

Work-energy theorem:

$$W_{net} = K_f - K_i \qquad (6-9)$$

Kinetic energy:

$$K = \frac{1}{2}mv^2 \qquad (6-8)$$

Work done by a constant force, straight-line displacement:

$$W = Fd\cos\theta \qquad (6-2)$$

Work done by the gravitational force:

$$W_{grav} = -mg(y_f - y_i) \qquad (6-14)$$

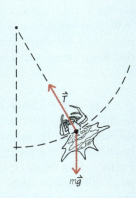

Solve

Rewrite the work-energy theorem in terms of the initial and final speeds of the spider and the work done by each force. Use Equation 6-14 to find the work done by the gravitational force.

Combine Equations 6-9 and 6-8:

$$W_{net} = \frac{1}{2}mv_f^2 - \frac{1}{2}mv_i^2$$

Net work is the sum of the work done by the tension force T and the work done by the gravitational force:

$$W_{net} = W_{grav} + W_T$$

The final speed of the spider at the high point of its motion is $v_f = 0$, so the work-energy theorem becomes

$$W_{grav} + W_T = -\frac{1}{2}mv_i^2$$

The initial and final y coordinates of the spider are $y_i = 0$ and $y_f = h$, so the work done by gravity is

$$W_{grav} = -mg(h - 0) = -mgh$$

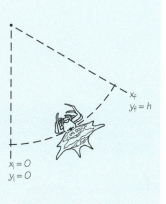

To calculate the work done by the tension force along the spider's curved path, break the path up into segments so short that each one can be considered as a straight line. On each segment of the path the tension force $\vec{T}$ points radially inward, perpendicular to the path and hence perpendicular to the displacement $\Delta\vec{d}$. Hence $\theta = 90°$ in Equation 6-2, which means that the tension force does *no* work during this displacement.

Work done by the tension force as the spider moves through a short segment of its path:

$$\Delta W_T = T\Delta d \cos 90° = 0$$

This is true for each short segment of the path. So the net work done by the tension force over the entire path is

$$W_T = 0$$

Angle between $\vec{T}$ and $\Delta\vec{d}$: $\theta = 90°$

Substitute the expressions for W_{grav} and W_T into the work-energy theorem and solve for h.

Work-energy theorem:

$$W_{grav} + W_T = -mgh + 0 = -\frac{1}{2}mv_i^2$$

$$h = \frac{v_i^2}{2g} = \frac{(1.3 \text{ m/s})^2}{2(9.80 \text{ m/s}^2)} = 0.086 \text{ m} = 8.6 \text{ cm}$$

Reflect

The maximum height reached is the *same* as if the spider had initially been moving *straight up* at 1.3 m/s without being attached to the silk. In both cases the gravitational force does the same amount of (negative) work to reduce the spider's kinetic energy to zero.

The tension force in this problem is complicated because its magnitude and direction change as the spider moves through its swing. But in the work-energy approach we don't have to worry about the tension force at all because it does no work on the spider.

If the spider is initially moving straight up:

$$W_{net} = \frac{1}{2}mv_f^2 - \frac{1}{2}mv_i^2$$

$$W_{grav} = -mgh = 0 - \frac{1}{2}mv_i^2$$

$$h = \frac{v_i^2}{2g} = 8.6 \text{ cm}$$

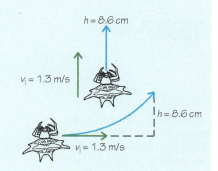

Example 6-8 illustrates an important point: *If a force always acts perpendicular to an object's curved path, it does zero work on the object.* This enabled us to ignore the effects of the tension force. We'll use this same idea again in Section 6-6.

Work Done by a Varying Force

In Example 6-8 the tension force acting on the spider varied in magnitude but did no work (because it acts perpendicular to the spider's displacement). In many situations, however, a force of variable magnitude *does* do work on an object. As an example, you must do work to stretch a spring. The force you exert to do this is not constant: The farther you stretch the spring, the greater the magnitude of the force you must exert. How can we calculate the amount of work that you do while stretching the spring?

To see the answer let's first consider a *constant* force F that acts on an object in the direction of its straight-line motion. Figure 6-16 shows a graph of this force versus position as the object undergoes a displacement d. The *area* shown as a colored rectangle under the graph of force versus position equals the height of the rectangle (the force F) multiplied by its width (the displacement d). But the area Fd is just equal to the work done by the constant force. So *on a graph of force versus position, the work done by the force equals the area under the graph.*

Let's apply this "area rule" to the work that you do in stretching a spring. Experiment shows that if you stretch a spring by a relatively small amount, the force that the *spring* exerts on you is directly proportional to the amount of stretch (Figure 6-17). We can write this relationship, known as **Hooke's law**, as

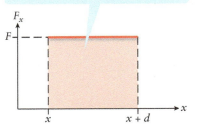

Work done by a constant force along direction of motion = Fd = area under graph of force versus position.

Figure 6-16 The "area rule" for work Finding the work done by a constant force using a graph of force versus position.

Hooke's law for the force exerted by an ideal spring (6-15)

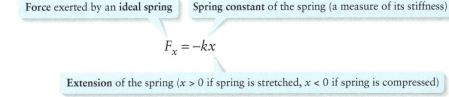

Force exerted by an **ideal spring**   Spring constant of the spring (a measure of its stiffness)

$$F_x = -kx$$

Extension of the spring ($x > 0$ if spring is stretched, $x < 0$ if spring is compressed)

The minus sign in Equation 6-15 means that the force that the spring exerts on you is in the direction *opposite* to the stretch. If you pull one end of the spring in the positive x direction so $x > 0$, the spring pulls back on you in the negative x direction. (As we will see below, Hooke's law also describes situations in which the spring is compressed rather than stretched.) The quantity k, called the **spring constant**, measures the stiffness of the spring: The greater the value of k, the stiffer the spring. If the force is measured in newtons and the extension in meters, k has units of N/m.

Equation 6-15 gives the force that the *spring* exerts on *you* as you stretch it. By Newton's third law, the force that *you* exert on the *spring* has the same magnitude but the opposite direction. So the x component of the force you exert has the opposite sign to the force in Equation 6-15:

(6-16) $F_{\text{you on spring},x} = +kx$ (force that you exert on the spring)

Stretching the spring means $x > 0$, and to do this you must exert a force in the same direction. So $F_{\text{you on spring},x} > 0$ if $x > 0$, which is just what Equation 6-16 tells us.

Figure 6-18 graphs the force that you exert as a function of the distance x that the spring has been stretched. If the spring is initially stretched a distance x_1 and you

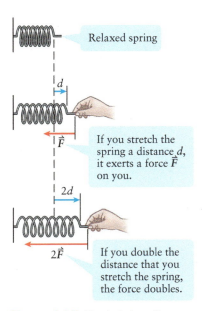

Relaxed spring

If you stretch the spring a distance d, it exerts a force $\vec{F}$ on you.

If you double the distance that you stretch the spring, the force doubles.

Figure 6-17 Hooke's law If you stretch an ideal spring, the force that it exerts on you is directly proportional to its extension.

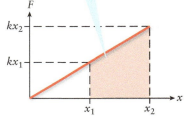

Area under curve = work you do to stretch spring from x_1 to x_2.

Figure 6-18 Applying the "area rule" to an ideal spring We can use the same technique as in Figure 6-16 to find the work required to stretch a spring.

stretch it further to x_2, the work W that you do is equal to the area of the colored trapezoid in Figure 6-18. From geometry, this area is equal to the *average* height of the graph multiplied by the width $x_2 - x_1$. Using Equation 6-16, we find

$$W = \left[\frac{(\text{force you exert at } x_1) + (\text{force you exert at } x_2)}{2} \right](x_2 - x_1)$$

$$= \frac{(kx_1 + kx_2)}{2}(x_2 - x_1) = \frac{1}{2}k(x_1 + x_2)(x_2 - x_1)$$

We can simplify this to

Work that must be done on a spring to stretch it from $x = x_1$ to $x = x_2$

Spring constant of the spring (a measure of its stiffness)

$$W = \frac{1}{2}kx_2^2 - \frac{1}{2}kx_1^2$$

x_1 = initial stretch of the spring
x_2 = final stretch of the spring

Go to Picture It 6-1 for more practice dealing with springs.

Work done to stretch a spring
(6-17)

WATCH OUT! Who's doing the work?

Note that Equation 6-17 tells us the work that *you* do on the *spring*. The work that the *spring* does on *you* is equal to the negative of Equation 6-17.

Example 6-9 shows how to use Equation 6-17 to attack a problem that would have been impossible to solve with the force techniques from Chapters 4 and 5.

EXAMPLE 6-9   Work Those Muscles!

An athlete stretches a set of exercise cords 47 cm from their unstretched length. The cords behave like a spring with spring constant 860 N/m. (a) How much force does the athlete exert to hold the cords in this stretched position? (b) How much work did he do to stretch them? (c) The athlete loses his grip on the cords. If the mass of the handle is 0.25 kg, how fast is it moving when it hits the wall to which the other end of the cords is attached? (You can ignore gravity and assume that the cords themselves have only a small mass.)

Set Up

In part (a) we'll use Equation 6-16 to find the force that the athlete exerts, and in part (b) Equation 6-17 will tell us the work that he does. In part (c) we'll see how much work the *cords* do as they go from being stretched to relaxed. This work goes into the kinetic energy of the handle. We're ignoring the mass of the cords and therefore assuming that they have no kinetic energy of their own.

$F_{\text{athlete on cords},x} = +kx$ (6-16)

$W = \frac{1}{2}kx_2^2 - \frac{1}{2}kx_1^2$ (6-17)

$W_{\text{net}} = K_f - K_i$ (6-9)

$K = \frac{1}{2}mv^2$ (6-8)

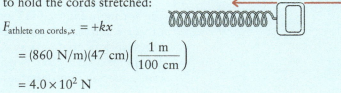

Solve

(a) The force that the athlete exerts is proportional to the distance $x = 47$ cm that the cords are stretched.

Force exerted by the athlete to hold the cords stretched:

$F_{\text{athlete on cords},x} = +kx$

$= (860 \text{ N/m})(47 \text{ cm})\left(\frac{1 \text{ m}}{100 \text{ cm}}\right)$

$= 4.0 \times 10^2 \text{ N}$

(b) The cords are not stretched at all to start (so $x_1 = 0$) and end up stretched by 47 cm (so $x_2 = 47$ cm $= 0.47$ m).

Work done by the athlete to stretch the cords:

$$W_{\text{athlete on cords}} = \frac{1}{2}kx_2^2 - \frac{1}{2}kx_1^2$$

$$= \frac{1}{2}(860 \text{ N/m})(0.47 \text{ m})^2 - \frac{1}{2}(860 \text{ N/m})(0 \text{ m})^2$$

$$= 95 \text{ N} \cdot \text{m} = 95 \text{ J}$$

(c) When the athlete releases the handle, the cords relax from their new starting position ($x_1 = 0.47$ m) to their final, unstretched position ($x_2 = 0$). The work that the cords do on the handle is given by the negative of Equation 6-17.

Work done by the cords on the handle as they relax:

$$W_{\text{cords on handle}}$$

$$= -\left(\frac{1}{2}kx_2^2 - \frac{1}{2}kx_1^2 \right)$$

$$= -\left[\frac{1}{2}(860 \text{ N/m})(0 \text{ m})^2 - \frac{1}{2}(860 \text{ N/m})(0.47 \text{ m})^2 \right]$$

$$= -(-95 \text{ N} \cdot \text{m}) = 95 \text{ J}$$

We'll ignore the force of gravity (that is, we assume the handle flies back horizontally and doesn't fall). Then the net work done on the handle equals the work done by the cords. This is equal to the change in the handle's kinetic energy. Use this to find the handle's speed when the cords are fully relaxed.

Net work done on handle:

$$W_{\text{net}} = W_{\text{cords on handle}} = 95 \text{ J}$$

Work-energy theorem applied to handle:

$$W_{\text{net}} = K_f - K_i = \frac{1}{2}mv_f^2 - \frac{1}{2}mv_i^2$$

Handle is initially at rest, so $v_i = 0$. Solve for final speed of the handle:

$$\frac{1}{2}mv_f^2 = W_{\text{net}} = 95 \text{ J}$$

$$v_f^2 = \frac{2W_{\text{net}}}{m}$$

$$v_f = \sqrt{\frac{2W_{\text{net}}}{m}} = \sqrt{\frac{2(95 \text{ J})}{0.25 \text{ kg}}} = 28 \text{ m/s}$$

Reflect

A common *incorrect* way to approach part (b) is to use the formula for the work done by a *constant* force: Multiply the force needed to hold the cords fully stretched by the cords' displacement. This gives the wrong answer because the force needed to pull the cords is *not* constant.

Notice that the amount of work that the athlete does to stretch the cords (95 J) is the *same* as the amount of work that the cords do on the handle when they relax. This suggests that the athlete stores energy in the cords by stretching them. We'll explore this idea in Section 6-6 as we introduce a new kind of energy, called *potential energy*.

Incorrect way to calculate work that the athlete does on the cords:

$$W_{\text{athlete on cords}} = F_x d = (4.0 \times 10^2 \text{ N})(0.47 \text{ m}) = 190 \text{ J}$$

(actual value: $W_{\text{athlete on cords}} = 95$ J)

More on Hooke's Law and Its Limitations

The spring in Figure 6-17 exerts a force when it is stretched. A spring also exerts a force when it is *compressed* (Figure 6-19a). One example is a car's suspension, whose springs compress as you load passengers and luggage into the car. The force that an ideal spring exerts when compressed is given by Equation 6-15, $F_x = -kx$, the *same* equation that describes the force exerted by a *stretched* spring (Figure 6-19b).

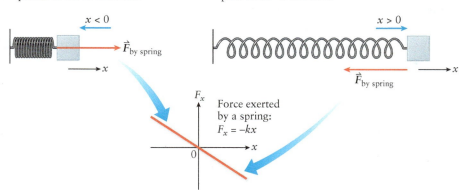

(a) A compressed spring ($x < 0$) pushes in the $+x$ direction.

(b) A stretched spring ($x > 0$) pulls in the $-x$ direction.

$x < 0$

$\vec{F}_{\text{by spring}}$

x

$x > 0$

$\vec{F}_{\text{by spring}}$

x

F_x

Force exerted by a spring: $F_x = -kx$

x

0

Figure 6-19 Compressing and stretching an ideal spring Hooke's law, $F_x = -kx$, applies equally well to an ideal spring whether it is (a) compressed or (b) stretched.

The only difference is that x is negative if the spring is compressed. Hence if you compress the spring (push its end in the negative x direction), the spring pushes back on you in the positive x direction.

Like Equation 6-15, Equation 6-17 is also valid when a spring is compressed. If a spring with spring constant $k = 1000$ N/m is initially relaxed (so $x_1 = 0$) and you stretch it by 20 cm (so $x_2 = +20$ cm $= +0.20$ m), the work that you do is

$$W = \frac{1}{2}kx_2^2 - \frac{1}{2}k(0)^2 = \frac{1}{2}(1000 \text{ N/m})(0.20 \text{ m})^2 = 20 \text{ J}$$

The work that you do to *compress* the same spring by 20 cm (so $x_2 = -20$ cm $= -0.20$ m) is

$$W = \frac{1}{2}kx_2^2 - \frac{1}{2}k(0)^2 = \frac{1}{2}(1000 \text{ N/m})(-0.20 \text{ m})^2 = 20 \text{ J}$$

So you have to do positive work to compress a spring as well as to stretch it.

Hooke's law and Equations 6-15, 6-16, and 6-17 are only *approximate* descriptions of how real springs, elastic cords, and tendons behave. As an example, Figure 6-20 is a graph of the force needed to stretch a human patellar tendon (which connects the kneecap to the shin). The curve isn't a straight line, which means that the force isn't directly proportional to the amount of stretch. The force you have to apply to the tendon is also greater when you stretch than when you let it relax. What's more, the

BioMedical

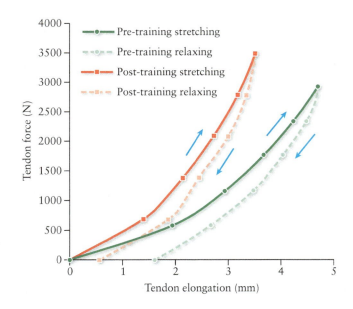

Pre-training stretching
Pre-training relaxing
Post-training stretching
Post-training relaxing

Tendon force (N)

Tendon elongation (mm)

Figure 6-20 Tendons are not ideal springs This graph of force versus extension for a human patellar tendon is very different from that for an ideal spring (compare to Figure 6-19). The graph is not a straight line; there is a different graph for relaxing than for stretching the tendon; and the graph for tendons that have undergone exercise (square, red data points) is different from that for tendons that have not (circular, green data points).

tendon can change its properties: The two sets of curves in Figure 6-20 are for males in their 70s before and after a 14-week course of physical training, which caused the patellar tendon to become much stronger and stiffer.

It's nonetheless true that Hooke's law is a useful approximation for the behavior of many materials when stretched or compressed, provided the amount of stretch or compression is small. We'll use this law in the next section.

GOT THE CONCEPT? 6-4 Double the Work

 In Example 6-9 the athlete did 95 J of work to stretch the exercise cords by 47 cm. How far would he have to stretch the cords to do double the amount of work (2×95 J = 190 J)? (a) $\sqrt{2} \times 47$ cm = 66 cm; (b) 2×47 cm = 94 cm; (c) 4×47 cm = 188 cm.

TAKE-HOME MESSAGE FOR Section 6-5

✔ The work-energy theorem applies even when the object follows a curved path or the forces that act on the object are not constant.

✔ A force that always acts perpendicular to an object's curved path does zero work on the object.

✔ An ideal spring exerts a force proportional to the distance that it is stretched or compressed (Hooke's law). The work required to stretch or compress a spring by a given distance is proportional to the square of that distance.

6-6 Potential energy is energy related to an object's position

We've seen that an object in motion has the ability to do work, as measured by its kinetic energy. But a *stationary* object can also have the ability to do work. An example is the barbell in Figure 6-21. When held at rest above the weight lifter's head, the barbell has no kinetic energy. If the weight lifter should drop the barbell, however, it will fall to the ground with a resounding crash and leave a dent in the floor. In other words, the barbell will exert a downward force on the floor over a distance and so will do work. Thus the barbell has the *potential* to move and to do work simply because it's held so high. We use the term **potential energy** to refer to an ability to do work that's related to an object's *position*.

There are many examples of potential energy in your environment. The spring in a mousetrap has the potential to do very destructive work on any mouse foolish enough to release the trap. And the positively charged protons in a uranium nucleus, which want to push away from each other but are prevented from doing so by the strong force between the constituents of the nucleus, have the potential to do work if the nucleus is broken apart—a process that provides the energy released in a nuclear reactor.

Figure 6-21 Potential energy
The barbell in this photo is at rest and so has zero kinetic energy. But it has the *potential* to acquire kinetic energy if the weight lifter should drop it. So there is potential energy in this situation, a kind we call *gravitational potential energy.*

Paul Bradbury/Getty Images

Gravitational Potential Energy

Let's quantify the amount of gravitational potential energy associated with the barbell in Figure 6-21. If the barbell has mass m and is initially a height h above the floor, as it falls its vertical displacement is $y_f - y_i = -h$ (its y coordinate decreases by h). From Equation 6-14, the work done on the barbell as it falls by the gravitational force $W_{grav} = -mg(y_f - y_i) = -mg(-h) = mgh$. (This is positive because the gravitational force and the displacement are both in the same downward direction.) From the work-energy theorem this is equal to the kinetic energy that the dropped barbell has just before it hits the floor. (We're ignoring any effects of air resistance.) So we say that the gravitational potential energy (symbol U_{grav}) of the Earth–barbell system before the barbell was dropped was $U_{grav} = mgh$, and that this potential energy was converted to kinetic energy as the barbell fell. Note that potential energy has units of joules, the same as work and kinetic energy.

Where did the potential energy come from? To see the answer, consider what happens as the weight lifter *raises* the barbell from the floor to a height h (Figure 6-22). During the lifting, the barbell begins with zero kinetic energy (sitting on the floor) and ends up with zero kinetic energy (at rest above the weight lifter's head). The net *change* in its kinetic energy is zero, so the net work done on the barbell is also zero. Hence the positive work that the weight lifter did to raise the barbell must just balance the negative work done by the gravitational force of magnitude mg. The vertical displacement of the barbell during the lifting is $y_f - y_i = +h$ (its y coordinate increases by h), so from Equation 6-14 the work done by gravity is $W_{grav} = -mg(y_f - y_i) = -mg(+h) = -mgh$. (This is negative since the downward gravitational force and the upward displacement are in opposite directions. From our discussion in Section 6-5, the work done by the gravitational force is the same whether the barbell moves straight upward or along the curved path shown in Figure 6-22.) Hence the work done by the weight lifter to raise the barbell to height h is $W_{weight\ lifter} = +mgh$, which is exactly equal to the gravitational potential energy mgh associated with the barbell when it is at height h. So the weight lifter is the source of the gravitational potential energy. We call $U = mgh$ the **gravitational potential energy** since it arises from the weight lifter doing work against the gravitational force.

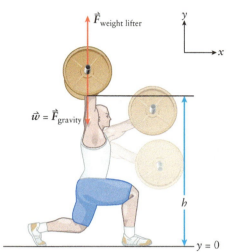

Figure 6-22 Adding gravitational potential energy The net change in the kinetic energy of the barbell as it is raised by the weight lifter is zero. The work that he does to lift the barbell a height h goes into increasing the gravitational potential energy.

WATCH OUT! To what object does gravitational potential energy "belong"?

We've been careful *not* to say, "The barbell *has* gravitational potential energy." The reason is the gravitational potential energy is really a property of the *system* made up of the barbell and Earth (which exerts the gravitational force on the barbell). The weight lifter adds potential energy to this system by moving the barbell away from Earth. The gravitational potential energy mgh would be the same if the barbell remained at rest but Earth as a whole were pushed down a distance h. You should always be careful to define your system when using work and energy to solve problems.

In general, if an object of mass m is at a vertical coordinate y, the gravitational potential energy is

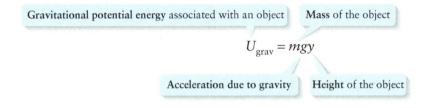

Gravitational potential energy associated with an object | Mass of the object

$$U_{grav} = mgy$$

Acceleration due to gravity | Height of the object

Gravitational potential energy associated with an object (6-18)

When the weight lifter in Figure 6-22 raises the barbell (so y increases), the gravitational potential energy increases; when he lowers or drops the barbell, y decreases, and the gravitational potential energy decreases.

Interpreting Potential Energy

Here's how the work-energy theorem describes what happens to a dropped object in the absence of air resistance: As the object falls, the gravitational force does work on it, and this work goes into changing the object's kinetic energy. In equation form,

$$W_{grav} = \Delta K = K_f - K_i \text{ (only the gravitational force does work)} \quad \textbf{(6-19)}$$

Now let's *reinterpret* this statement in terms of gravitational potential energy and generalize to the case in which the object follows a curved path (as would be the case if the object were thrown rather than dropped) rather than falling straight down. The

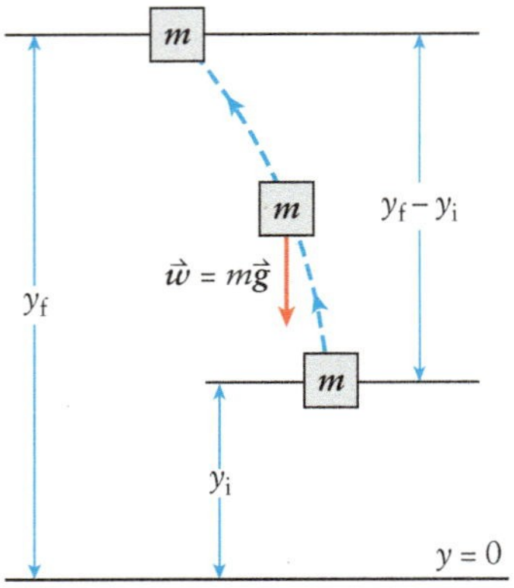

Figure 6-23 Raising an object along a curved path The work done on the object by the gravitational force $\vec{w}$ as it follows this curved path between height y_i and height y_f is $W_{grav} = -mg(y_f - y_i)$. This equals the negative of the change in gravitational potential energy: $W_{grav} = -\Delta U_{grav}$. In the case shown here W_{grav} is negative (the force is downward but the displacement is upward), so ΔU_{grav} is positive and the gravitational potential energy increases.

object in **Figure 6-23** moves along a curve from an initial y coordinate y_i to a final coordinate y_f. As we learned in Section 6-5, as the object moves, the gravitational force does an amount of work on it equal to $W_{grav} = -mg(y_f - y_i)$ (Equation 6-14). We can rewrite this in terms of the gravitational potential energy as given by Equation 6-18, $U_{grav} = mgy$:

$$\begin{aligned} W_{grav} &= -mg(y_f - y_i) = -mgy_f + mgy_i = -(mgy_f - mgy_i) \\ &= -(U_{grav,f} - U_{grav,i}) = -\Delta U_{grav} \end{aligned}$$

(6-20)

In other words, *the work done by gravity on an object equals the negative of the change in gravitational potential energy of the Earth–object system.* If an object rises as in Figure 6-23, the downward gravitational force does negative work on it and the gravitational potential energy of the system increases (its change is positive). If an object descends, the downward gravitational force does positive work on it and the gravitational potential energy of the system decreases (its change is negative). If an object begins and ends its motion at the same height, the gravitational force does zero net work on it and there is zero net change in the system's gravitational potential energy.

We can now restate the work-energy theorem for any object on which the only force that does work is the gravitational force (for example, a falling barbell without air resistance). If we substitute W_{grav} from Equation 6-20 into Equation 6-19, we get

Change in kinetic energy if only the gravitational force does work

(6-21)

If the **only force that does work** on a moving object is the **gravitational force...**

$$\Delta K = -\Delta U_{grav}$$

...then the **change in the object's kinetic energy** K... | ...is equal to the **negative of the change in gravitational potential energy** U_{grav}.

If K increases, U_{grav} decreases and vice versa.

Go to Interactive Exercise 6-1 for more practice dealing with work, kinetic energy, and potential energy.

If the object rises, gravitational potential energy increases and kinetic energy decreases (the object slows down). If the object descends, gravitational potential energy decreases and kinetic energy increases (the object speeds up). From this perspective we no longer need to talk about the *work* done by the gravitational force: That's accounted for completely by the change in the system's gravitational potential energy.

WATCH OUT! The choice of $y = 0$ for gravitational potential energy doesn't matter.

The value of $U_{grav} = mgy$ depends on what height you choose to be $y = 0$. But Equation 6-21 shows that what matters is the *change* in gravitational potential energy, and that does *not* depend on your choice of $y = 0$. That's because the change in gravitational potential energy depends only on the *difference* between the initial and final heights, not the heights themselves: $\Delta U_{grav} = U_f - U_i = mgy_f - mgy_i = mg(y_f - y_i)$. Example 6-10 illustrates this important point.

EXAMPLE 6-10 A Ski Jump

A skier of mass m starts at rest at the top of a ski jump ramp (Figure 6-24). The vertical distance from the top of the ramp to the lowest point is a distance H, and the vertical distance from the lowest point to where the skier leaves the ramp is D. The first part of the ramp is at an angle θ from the horizontal, and the second part is at an angle ϕ. Derive an expression for the speed of the skier when she leaves the ramp. Assume that her skis are well waxed, so that there is negligible friction between the skis and the ramp, and ignore air resistance.

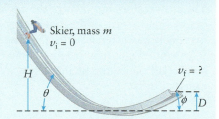

Figure 6-24 Flying off the ramp
What is the skier's speed when she leaves the frictionless ramp?

Set Up

The only forces that act on the skier are the gravitational force and the normal force exerted by the ramp. The normal force does *no* work on her because it always acts perpendicular to the ramp and hence perpendicular to her direction of motion. (That's just like the tension force on the swinging spider in Example 6-8 in Section 6-5. That force, too, did no work.) This means that only the gravitational force does work, so we can use Equation 6-21 for the Earth–skier system to calculate the change in the skier's kinetic energy and hence her final speed v_f.

Change in kinetic energy if only the gravitational force does work:

$$\Delta K = -\Delta U_{grav} \qquad (6\text{-}21)$$

Gravitational potential energy:

$$U_{grav} = mgy \qquad (6\text{-}18)$$

Kinetic energy:

$$K = \frac{1}{2}mv^2 \qquad (6\text{-}8)$$

Solve

Let's take $y = 0$ to be at the low point of the ramp. The skier then begins at rest ($v_i = 0$) at $y_i = H$ and is at $y_f = D$ when she leaves the ramp. Use Equations 6-8, 6-18, and 6-21 to solve for v_f.

At starting point:

$$K_i = \frac{1}{2}mv_i^2 = 0$$

$$U_{grav,i} = mgy_i = mgH$$

At the point where skier leaves the ramp:

$$K_f = \frac{1}{2}mv_f^2$$

$$U_{grav,f} = mgy_f = mgD$$

Use Equation 6-21:

$$-\Delta U_{grav} = \Delta K, \text{ where}$$

$$\Delta U_{grav} = U_{grav,f} - U_{grav,i} = mgD - mgH$$
$$= mg(D - H) = -mg(H - D)$$

(this is negative since $D < H$) and

$$\Delta K = K_f - K_i$$

$$= \frac{1}{2}mv_f^2 - 0 = \frac{1}{2}mv_f^2$$

(this is positive since the skier is moving faster at the end of the ski jump than at the beginning). So

$$+mg(H - D) = \frac{1}{2}mv_f^2$$

$$v_f^2 = 2g(H - D)$$
$$v_f = \sqrt{2g(H - D)}$$

Reflect

The answer does *not* involve the angles θ or ϕ, or any other aspect of the ramp's shape. All that matters is the difference $H - D$ between the skier's initial and final heights. If the ramp had a different shape, the final speed v_f would be exactly the same.

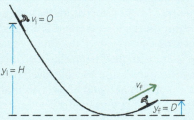 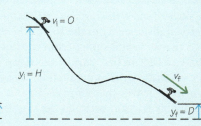

As we mentioned in the **Watch Out!** feature just before this example, our answer also shouldn't depend on our having chosen $y = 0$ to be at the low point of the ramp. For example, if we instead chose $y = 0$ to be where the skier leaves the ramp, the result for v_f would be the same. Try this yourself: Take $y = 0$ to be at the skier's starting point, so the end of the ramp is at $y = -(H - D)$. Do you get the same ΔU_{grav} with this choice?

With $y = 0$ at the point where the skier leaves the ramp:

At starting point:

$$U_{grav,i} = mgy_i$$
$$= mg(H - D)$$

At the point where skier leaves the ramp:

$$U_{grav,f} = mgy_f = 0, \text{ So}$$
$$\Delta U_{grav} = U_{grav,f} - U_{grav,i}$$
$$= 0 - mg(H - D)$$
$$= -mg(H - D)$$

This is the same as with our previous choice of $y = 0$, so we'll find the same value of v_f.

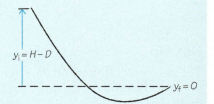

GOT THE CONCEPT? 6-5 Choosing a Point of Reference

A ball is dropped from the top of an 11-story building, 45 m above the ground, to a balcony on the back of the ninth floor, 36 m above the ground. In which case is the change in the potential energy associated with the ball greatest? (a) If we choose the ground to be at $y = 0$; (b) if we choose the balcony to be at $y = 0$; (c) if we choose the top of the building to be at $y = 0$; (d) The change is the same in all three cases.

Spring Potential Energy

In Example 6-10 the work done by the gravitational force on the skier is $W_{grav} = -\Delta U_{grav}$ (see Equation 6-20). We saw that this work depends only on the skier's final and initial positions, *not* on the path that she took between those two points. That's why we can express the work done by the gravitational force in terms of the difference ΔU_{grav} between the gravitational potential energy values at the two points.

The same is true for the work done by an ideal spring. We saw in Section 6-5 that the work you must do to stretch a spring from an extension x_i to an extension x_f is $\frac{1}{2}kx_f^2 - \frac{1}{2}kx_i^2$ (this is Equation 6-17, with 1 replaced by i and 2 replaced by f). The work that the *spring* does is the negative of this: $W_{spring} = -\left(\frac{1}{2}kx_f^2 - \frac{1}{2}kx_i^2\right)$. Note that this also depends on only the initial and final extensions of the spring, not on the details of how you got from one to the other. Hence just as we did for the work done by the gravitational force, we can write the work done by an ideal spring in terms of a change in potential energy:

$$W_{spring} = -\left(\frac{1}{2}kx_f^2 - \frac{1}{2}kx_i^2\right)$$
$$= -(U_{spring,f} - U_{spring,i}) = -\Delta U_{spring}$$

(6-22)

The quantity U_{spring} in Equation 6-22 is called the **spring potential energy**:

Spring potential energy of a stretched or compressed spring | Spring constant of the spring

$$U_{\text{spring}} = \frac{1}{2}kx^2$$

Extension of the spring ($x > 0$ if spring is stretched, $x < 0$ if spring is compressed)

Spring potential energy
(6-23)

The spring potential energy is zero if the spring is relaxed ($x = 0$) and positive if the spring is stretched ($x > 0$) or compressed ($x < 0$) (Figure 6-25). This says that we have to do work to either stretch the spring or compress it, and the work that we do goes into the spring potential energy.

While a human tendon is not an ideal spring, we can think of it as storing spring potential energy when it is stretched. When you are running and one of your feet is in contact with the ground, the Achilles tendon at the back of that leg is stretched. The spring potential energy stored in that tendon literally "springs" you back in the air, helping you to sustain your running pace.

Conservative and Nonconservative Forces

A force that can be associated with a potential energy, like the gravitational force or the force exerted by an ideal spring, is called a **conservative force**. (In Section 6-7 we'll see the reason for this term.) By contrast, the friction force is an example of a **nonconservative force** for which we *cannot* use the concept of potential energy. The reason is that unlike the gravitational force or the force exerted by an ideal spring, the work done by the friction force *does* depend on the path taken from the initial point to the final point. In Figure 6-26a we slide a book across a tabletop from an initial point to a final point along two different paths. Along either path the kinetic friction force has the same magnitude and points opposite to the direction of motion. Hence the friction force does more (negative) work along the curved path than along the straight path. Since the work done by friction depends on more than just the initial and final positions, we can't write it in terms of a change in potential energy. That's why there's no such thing as "friction potential energy."

Here's an equivalent way to decide whether a certain kind of force is conservative: If the work done by the force on a *round trip* (that is, one where the initial and final positions are the same) is *zero*, the force is conservative and we can use the idea of potential energy. This is the case for the gravitational force: If $y_f = y_i$, then from Equation 6-14 $W_{\text{grav}} = -mg(y_f - y_i) = 0$. If you toss a ball straight up, the gravitational force does negative work on it as it rises and an equal amount of positive work as it falls back to your hand. The same is true for the spring force: If $x_f = x_i$, then $W_{\text{spring}} = \frac{1}{2}kx_f^2 - \frac{1}{2}kx_i^2 = 0$. But if you slide a book on a round trip on a tabletop, the total amount of work that the friction force does on the book is negative and *not* zero (Figure 6-26b). Hence the friction force is nonconservative. (To keep the book moving, you have to do an equal amount of positive work on the book as you push it.)

WATCH OUT! For spring potential energy, $x = 0$ means a relaxed spring.

⚠ For gravitational potential energy we're free to choose $y = 0$ to be anywhere we like (see Example 6-10). We don't have that kind of freedom for spring potential energy: In Equation 6-23, we *must* choose $x = 0$ to be where the spring is neither compressed nor stretched.

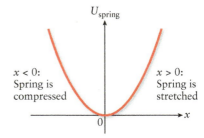

U_{spring}

$x < 0$:
Spring is compressed

$x > 0$:
Spring is stretched

0 x

Figure 6-25 Spring potential energy The potential energy in a spring is proportional to the square of its extension x. (See Equation 6-23.)

(a) Friction does more negative work along path 2 than along path 1. Since the work done by friction depends on the path, we conclude that friction is **not** a conservative force.

(b) The book makes a round trip that begins and ends at this point. Friction does a nonzero, negative amount of work on the book for this round trip. So we again conclude that friction is **not** a conservative force.

Path 1   Path 2

Path 1   Path 2

Figure 6-26 Kinetic friction is a nonconservative force Because the work done by kinetic friction depends on the path, and it does nonzero net work on a round trip, it is a nonconservative force.

BioMedical

Figure 6-20 shows that the force exerted by a human tendon is also nonconservative. The tendon exerts more force on its end while being stretched than while relaxing to its original length, so it does more negative work on the muscle attached to its end as it stretches than it does positive work as it relaxes. Hence during this "round trip" the tendon does a nonzero (and negative) amount of work, which means the muscle has to do an equal amount of *positive* work on the tendon to make it go through a complete cycle (just like the positive work you must do to push the book around the path in Figure 6-26b).

What happens to the work that you do in these cases? Before we can answer this question, we'll combine the ideas of kinetic energy and potential energy into yet another kind of energy, called *total mechanical energy*. This will lead us to one of the central ideas of physics, the idea of *conservation of energy*.

TAKE-HOME MESSAGE FOR Section 6-6

✔ The work done by a conservative force depends only on the initial and final positions of the object, not on the path the object followed from one position to the other. The gravitational force and the force exerted by an ideal spring are examples of conservative forces.

✔ A kind of energy, called potential energy, is associated with each kind of conservative force.

✔ The work done by a conservative force is equal to the negative of the change in the associated potential energy. If the force does positive work, the potential energy decreases; if the force does negative work, the potential energy increases.

6-7 If only conservative forces do work, total mechanical energy is conserved

Take a pencil in your hand and toss it upward. As the pencil ascends, it loses speed and its kinetic energy $K = \frac{1}{2}mv^2$ decreases. At the same time the pencil gains height so that the gravitational potential energy $U_{\text{grav}} = mgy$ increases. After the pencil reaches its maximum height and falls downward, its kinetic energy increases as it gains speed and the gravitational potential energy decreases as it loses height. This way of thinking about the pencil's up-and-down motion suggests that energy is *transformed* from one form (kinetic) into a different form (gravitational potential) as the pencil rises and is transformed back as the pencil descends.

BioMedical

Another example of energy transformation is happening inside your body right now. When your heart contracts it pushes blood into the arteries, stretching their walls to accommodate the increased volume. The stretched arterial walls behave like a stretched spring and so possess spring potential energy. In between heart contractions the arterial walls relax back to their equilibrium size and lose their potential energy, just as a stretched spring does when it relaxes to its unstretched length. As a result blood gains kinetic energy as the arterial walls push on it between heartbeats. Thus spring potential energy of the arteries is transformed into kinetic energy of the blood (Figure 6-27).

In both of these situations the *sum* $K + U$ of kinetic energy and potential energy—a sum that we call the *total mechanical energy* E—keeps the same value and is *conserved*. When the system's kinetic energy K increases, its potential energy U decreases, and when K decreases, U increases. It's like having two bank accounts: You can transfer money from one account to the other, but the total amount of money in the two accounts remains the same. In this section we'll learn the special circumstances in which total mechanical energy is conserved, and we'll see how to express these ideas in equation form.

Suppose that as an object moves from an initial position to a final position, the only force that does work on the object is the gravitational force. [An example is the skier on a ski jump in Example 6-10 (Section 6-6). In addition to the gravitational force on the skier, there is a normal force exerted by the ramp, but it does no work on the skier because this force is always perpendicular to the skier's path.] According to Equation 6-21 in this situation the work-energy theorem tells us that the change in the object's kinetic energy is equal to the negative of the change in its potential energy:

$$\Delta K = -\Delta U_{\text{grav}} \text{ (if only the gravitational force does work)} \qquad (6\text{-}24)$$

The change in kinetic energy equals the final value minus the initial value, and likewise for the gravitational potential energy. So we can rewrite Equation 6-24 as

$$K_{\text{f}} - K_{\text{i}} = -(U_{\text{grav,f}} - U_{\text{grav,i}}) = -U_{\text{grav,f}} + U_{\text{grav,i}}$$
$$\text{(if only the gravitational force does work)}$$

Let's rearrange this equation so that all the terms involving the initial situation are on the left-hand side of the equals sign and all the terms involving the final situation are on the right-hand side. We get

$$K_{\text{i}} + U_{\text{grav,i}} = K_{\text{f}} + U_{\text{grav,f}} \text{ (if only the gravitational force does work)} \qquad (6\text{-}25)$$

If only the gravitational force does work on the object as it moves, its speed v and kinetic energy $K = \frac{1}{2}mv^2$ can change, and its height y and gravitational potential energy $U_{\text{grav}} = mgy$ can change. But the *sum* of K and U_{grav} has the same value at the end of the motion as at the beginning.

We can generalize these ideas and Equation 6-25 even more. Our results from Section 6-6 show that the work done by *any* conservative force can be written as the negative of the change in the associated potential energy. (Equations 6-22 and 6-23 show this for the case of the work done by an ideal spring: $W_{\text{spring}} = -\Delta U_{\text{spring}}$, where the spring potential energy is $U_{\text{spring}} = \frac{1}{2}kx^2$ and x is the extension of the spring.) If a number of conservative forces $\vec{F}_A, \vec{F}_B, \vec{F}_C, \dots$ do work on an object as it moves, and the potential energies associated with each of these forces are $U_A, U_B, U_C, \dots$, then the total work done by these forces is

$$\begin{aligned}
W_{\text{conservative}} &= W_A + W_B + W_C + \dots \\
&= (-\Delta U_A) + (-\Delta U_B) + (-\Delta U_C) + \dots \\
&= (-U_{A,\text{f}} + U_{A,\text{i}}) + (-U_{B,\text{f}} + U_{B,\text{i}}) + (-U_{C,\text{f}} + U_{C,\text{i}}) + \dots \\
&= -(U_{A,\text{f}} + U_{B,\text{f}} + U_{C,\text{f}} + \dots) + (U_{A,\text{i}} + U_{B,\text{i}} + U_{C,\text{i}} + \dots) \\
&= -U_{\text{f}} + U_{\text{i}} = -\Delta U
\end{aligned} \qquad (6\text{-}26)$$

In Equation 6-26 the quantity U is the *total* potential energy, which is just the sum of the individual potential energies $U_A, U_B, U_C, \dots$ If these conservative forces are the only forces that do work on the object, then $W_{\text{conservative}}$ in Equation 6-26 equals the net work done on the object, which from the work-energy theorem equals $\Delta K = K_{\text{f}} - K_{\text{i}}$. So if only conservative forces do work, we have

$$K_{\text{f}} - K_{\text{i}} = -U_{\text{f}} + U_{\text{i}}$$

or, rearranging,

Values of the **kinetic energy** K of an object at two points (i and f) during its motion

$$K_{\text{i}} + U_{\text{i}} = K_{\text{f}} + U_{\text{f}}$$

Values of the **potential energy** U at the same two points

Mechanical energy is conserved if only conservative forces do work
(6-27)

> If only conservative forces do work on an object, the sum $K + U$ maintains the same value throughout the motion. This sum is called the **total mechanical energy** E.

Figure 6-27 Kinetic and potential energy in the arteries When you feel the pulse in your radial artery, you are actually feeling spring potential energy of the arterial walls being converted to kinetic energy of the blood. This energy transformation, which occurs about once per second, is essential for life.

Philippe Garo/Science Source

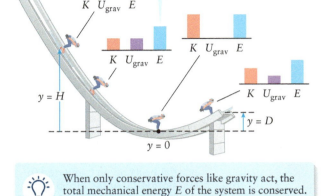

The kinetic energy K and the gravitational potential energy U_{grav} both change, but their sum E remains the same.

K U_{grav} E

K U_{grav} E

K U_{grav} E

K U_{grav} E

$y = H$

$y = D$

$y = 0$

When only conservative forces like gravity act, the total mechanical energy E of the system is conserved.

Figure 6-28 **Total mechanical energy is conserved** As the skier moves on this frictionless ramp, the kinetic energy K and the gravitational potential energy U_{grav} both change, but their sum $E = K + U_{grav}$—the total mechanical energy—always has the same value.

You can see that this is a more general form of Equation 6-25, which refers to the case in which only one conservative force—the gravitational force—does work. Note that we can take the initial point i and the final point f to be *any* two points during the motion, as long as only conservative forces do work.

The sum of kinetic energy K and potential energy U is called the **total mechanical energy** E. So Equation 6-27 tells us that if only conservative forces do work on an object, the total mechanical energy is *conserved*; that is, it maintains the same value during the motion (see Figure 6-28). For this energy conservation to take place, your system must include the sources of the conservative forces—for example, Earth and springs. You can now see the origin of the term "conservative" for forces like the gravitational force and the spring force: If only forces of this kind do work, the total mechanical energy of the system is conserved.

Another way to write Equation 6-27 is to move all of the terms in that equation to the same side of the equals sign:

$$K_f - K_i + U_f - U_i = 0$$

Since $K_f - K_i = \Delta K$ (the change in kinetic energy) and $U_f - U_i = \Delta U$ (the change in potential energy), we can rewrite this equation as

Change in the kinetic energy K of an object during its motion

Change in the potential energy U associated with an object during its motion

Conservation of mechanical energy—alternative version (6-28)

$$\Delta K + \Delta U = 0$$

If only conservative forces do work on the object, the sum of these two changes is zero: Total mechanical energy can transform between kinetic and potential forms, but there is no net change in the amount of total mechanical energy.

 Go to Picture It 6-2 for more practice dealing with mechanical energy.

Equation 6-28 refers to the change in kinetic energy ΔK and the change in potential energy ΔU during the motion. If the only forces that do work on the object are conservative forces, the sum of these is zero: Any increase in kinetic energy comes with an equal decrease in potential energy, and vice versa.

WATCH OUT! Total mechanical energy is a property of a system, not a single object.

 We saw in Section 6-6 that gravitational potential energy for an object like a baseball is really a shared property of a *system* of two objects, the baseball and Earth. Thus the total mechanical energy for a baseball in flight, which incorporates gravitational potential energy, is likewise a property of that system. Likewise, if a weight is attached to a horizontal spring and allowed to oscillate back and forth, the total mechanical energy is a shared property of the weight (which has kinetic energy) and the spring (which has spring potential energy). Whenever you think about total mechanical energy, you should always be able to state to what system that mechanical energy belongs.

In Section 6-8 we'll use the idea of conservation of mechanical energy to solve a number of physics problems. First, however, let's see how this idea has to be modified if there are also *nonconservative* forces that do work.

Generalizing the Idea of Conservation of Energy: Nonconservative Forces

There are many situations in which total mechanical energy is *not* conserved. As an example, at the beginning of this section we asked you to toss a pencil into the air. During the toss you had to do *positive* work on the pencil: You exerted an upward force on the pencil as you pushed it upward. As a result the total mechanical energy of the pencil increased (Figure 6-29). If you subsequently catch the pencil as it falls, you do *negative* work on the pencil because you exert an upward force on it as the pencil moves downward. As Figure 6-29 shows, the total mechanical energy decreases during the catch.

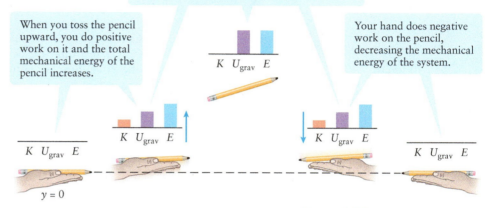

 Because you do work on the pencil, the total mechanical energy E—the sum of kinetic energy K and gravitational energy U_{grav}—is not conserved throughout this process.

When you toss the pencil upward, you do positive work on it and the total mechanical energy of the pencil increases.

While the pencil is in free fall, the value of the total mechanical energy stays the same. The energy shifts from kinetic energy to potential energy, then back to kinetic energy.

Your hand does negative work on the pencil, decreasing the mechanical energy of the system.

K U_{grav} E

K U_{grav} E

K U_{grav} E

K U_{grav} E

K U_{grav} E

$y = 0$

Figure 6-29 Tossing and catching a pencil The total mechanical energy associated with the pencil increases as you toss it upward, remains constant while the pencil is in flight, then decreases as you catch it.

The total mechanical energy changes in these situations because the force exerted by your hand is not conservative. We can write a modified version of Equation 6-27 that can accommodate situations like these in which the total mechanical energy is *not* conserved. To do this let's go back to the work-energy theorem, $W_{net} = K_f - K_i$. If both conservative and nonconservative forces do work on an object, the net work is the sum of the work done by the two kinds of forces:

$$W_{net} = W_{conservative} + W_{nonconservative}$$ (6-29)

Equation 6-26 says that the work done by conservative forces is $W_{conservative} = -U_f + U_i = -\Delta U$. Substituting this and Equation 6-29 into the work-energy theorem, we get

$$W_{net} = (-U_f + U_i) + W_{nonconservative} = K_f - K_i$$

We can rearrange this to

Values of the **kinetic energy** K of an object at two points (i and f) during its motion

$$K_i + U_i + W_{nonconservative} = K_f + U_f$$

Values of the **potential energy** U at the same two points

Change in mechanical energy when nonconservative forces act (6-30)

 If nonconservative forces do work on an object, the total mechanical energy $E = K + U$ changes its value.
If $W_{nonconservative} > 0$, $E_f > E_i$ and the total mechanical energy increases.
If $W_{nonconservative} < 0$, $E_f < E_i$ and the total mechanical energy decreases.

Equation 6-30 gives us more insight into the difference between conservative and nonconservative forces. In Figure 6-29 only the conservative force of gravity acts on the pencil from the time when it leaves your hand to when it returns to your hand. So the mechanical energy $E = K + U$ remains constant during this part of the motion. Kinetic energy decreases and changes into potential energy as the pencil ascends, but the potential energy is turned back into kinetic energy as the pencil falls. If the force of your hand were also a conservative force, all of the energy you expended to toss the pencil would be returned to your body when you caught the pencil; you could toss the pencil up and down all day and you would never get tired. This is *not* the case, however, and you *do* get tired. (If you don't believe this, try it with a book rather than a pencil and try tossing the book up and catching it a few dozen times.) Because the energy you expend in the toss is *not* returned to you in the catch, the force of your hand is nonconservative.

BioMedical

Here's an alternative way to think of the nonconservative forces depicted in Figure 6-29. To toss the pencil upward you actually use energy stored in the chemical bonds of adenosine triphosphate (ATP), carbohydrates, and fat in the muscles of your arm. We don't include this kind of energy in the quantity U in Equation 6-30, which refers only to potential energy associated with the pencil. Instead, we'll use the symbol E_{other} for **internal energy**, a catch-all kind of energy that is not included in K or U in Equation 6-30. Not all of the released chemical energy goes into moving your muscles; some goes into warming your muscles (which is why you get warm when you exercise) and hence into **thermal energy**, which is energy associated with the random motion of atoms and molecules (in this case inside your arm). Thermal energy is also considered part of E_{other}. So during the toss, some chemical energy is converted into thermal energy (and so remains within E_{other}), and the rest is used to do work on the pencil (**Figure 6-30**). Hence during the toss for which $W_{nonconservative} > 0$, E_{other} of your body decreases so $\Delta E_{other} < 0$.

When you catch the pencil you again use your muscles and so you use more of the energy stored in chemical bonds. In the catch, however, both your arm and the pencil warm by a slight but measurable amount, corresponding to an increase in thermal energy, and this increase is greater than the loss of chemical energy (Figure 6-30). So during the catch, for which $W_{nonconservative} < 0$, E_{other} of you and the pencil increases so $\Delta E_{other} > 0$.

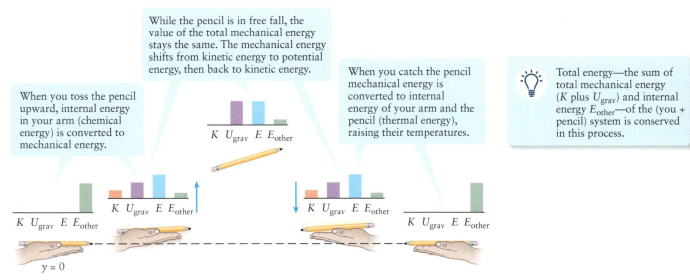

While the pencil is in free fall, the value of the total mechanical energy stays the same. The mechanical energy shifts from kinetic energy to potential energy, then back to kinetic energy.

When you catch the pencil mechanical energy is converted to internal energy of your arm and the pencil (thermal energy), raising their temperatures.

Total energy—the sum of total mechanical energy (K plus U_{grav}) and internal energy E_{other}—of the (you + pencil) system is conserved in this process.

When you toss the pencil upward, internal energy in your arm (chemical energy) is converted to mechanical energy.

K U_{grav} E E_{other}

$y = 0$

Figure 6-30 Tossing and catching a pencil, revisited When you toss the pencil some internal energy of your body in chemical form (E_{other}) is converted to mechanical energy. When you catch the pencil this added mechanical energy is converted back into internal energy.

These observations suggest that the amount of nonconservative work done is related to the change in internal energy. Indeed, many careful measurements show that in *all* situations, $W_{nonconservative}$ is *exactly* equal to the negative of ΔE_{other}:

$$W_{nonconservative} = -\Delta E_{other} \qquad (6\text{-}31)$$

With Equation 6-31 in mind we can rewrite Equation 6-30 as

$$K_i + U_i - \Delta E_{other} = K_f + U_f$$

If we move all of the terms in this equation to the same side of the equals sign, we get

$$K_f - K_i + U_f - U_i + \Delta E_{other} = 0$$

Now $K_f - K_i$ is the change in kinetic energy ΔK and $U_f - U_i$ is the change in potential energy ΔU. So we can rewrite this equation as

> **Change in the kinetic energy K** of an object during its motion

> **Change in the potential energy U associated** with an object during its motion

$$\Delta K + \Delta U + \Delta E_{other} = 0$$

> **Change in the internal energy E_{other}** during the object's motion

 Energy can transform between kinetic, potential, and internal forms, but there is no net change in the amount of energy.

The law of conservation of energy
(6-32)

Equation 6-32 suggests that we broaden our definition of energy to include both total mechanical energy ($E = K + U$) and internal energy E_{other}. In this case we have to expand the system to include not only the object and the sources of the conservative forces that produce ΔU but also the sources of the nonconservative forces that result in ΔE_{other} (like your hand in the case of the thrown pencil). As the object moves, K, U, and E_{other} can all change values, but the sum of these changes is zero: One kind of energy can transform into another, but the total amount of energy of all forms remains the same (Figure 6-31). This is the most general statement of the **law of conservation of energy**.

Unlike the similar-appearing Equation 6-28, which applies only if nonconservative forces do no work, Equation 6-32 is *always* true as long as the system contains all objects that are exerting or feeling forces. Scientists have made an exhaustive search for situations in which the law of conservation of energy does not hold. No such situation has ever been found. So we conclude that conservation of energy is an absolute law of nature. In the following section we'll see how to apply this law to solve physics problems.

BioMedical

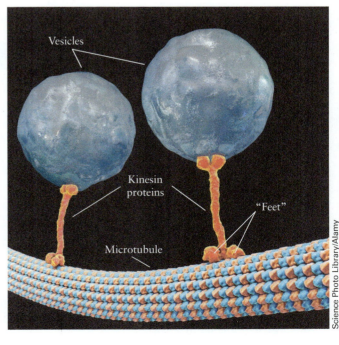

Figure 6-31 Energy transformation at the molecular level
A class of proteins called *kinesins* found within living cells behave like workers carrying packages: One end of the kinesin is attached to a "package" in the form of a vesicle (a ball of fluid held together by a surface layer of lipid molecules), and the other end has footlike structures that "walk" along the microtubules that make up the skeleton of a cell. Taking one microscopic "step" along a microtubule means giving the kinesin kinetic energy ($\Delta K > 0$ in Equation 6-32); this comes from the energy released from the hydrolysis of a single ATP molecule ($\Delta E_{other} < 0$ in Equation 6-32).

Science Photo Library/Alamy

GOT THE CONCEPT? 6-6 How High?

(?) A block is released from rest on the left-hand side of an asymmetric skateboard ramp at a height H (Figure 6-32). The left-hand side of the ramp is twice as steep as the right-hand side of the ramp. If the ramp is frictionless, to what height does the block rise on the right-hand side before stopping and sliding back down? (a) $2H$; (b) H; (c) $H/2$; (d) not enough information given to decide.

Figure 6-32 An asymmetric skateboard ramp If the block is released at height H on the left-hand side, to what height will it rise on the more gently sloped right-hand side?

TAKE-HOME MESSAGE FOR Section 6-7

✔ If the only forces that do work on an object are conservative forces, then the total mechanical energy E (the sum of the kinetic energy K and the potential energy U) is conserved. The values of K and U may change as the object moves, but the value of E remains the same.

✔ If nonconservative forces also do work on the object, the total mechanical energy changes.

The value of E increases if positive nonconservative work is done and decreases if negative nonconservative work is done.

✔ Work done by a nonconservative force is associated with a change in internal energy. The total energy—the sum of kinetic, potential, and internal energies—is always conserved.

6-8 Energy conservation is an important tool for solving a wide variety of problems

→ Go to Interactive Exercise 6-2 for more practice dealing with energy conservation.

Let's now see how to use the energy conservation equations that are valid when only conservative forces do work (Equations 6-27 and 6-28) and those that are valid when nonconservative forces also do work (Equations 6-30 and 6-32). As a rule, if a problem involves a conserved quantity (that is, a quantity whose value remains unchanged), it simplifies the problem tremendously. Often the conservation equation, which relates the value of the conserved quantity at one point in an object's motion to the value at a different point, is all you need to solve for the desired unknown. We'll see several problems of that kind in this section.

EXAMPLE 6-11 A Ski Jump, Revisited

As in Example 6-10 (Section 6-6), a skier of mass m starts at rest at the top of a ski jump ramp (Figure 6-33). (a) Use an energy conservation equation to find an expression for the skier's speed as she flies off the ramp. There is negligible friction between the skis and the ramp, and you can ignore air resistance. (b) After the skier reaches the ground at point P in Figure 6-33, she begins braking and comes to a halt at point Q. Find an expression for the magnitude of the constant friction force that acts on her between points P and Q.

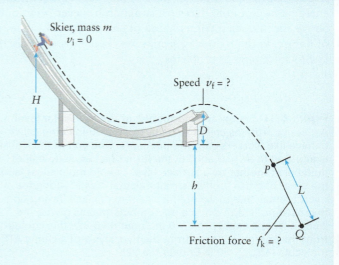

Figure 6-33 Flying off the ramp and coming to a halt What is the skier's speed as she leaves the ramp? How much friction is required to bring her to a halt?

Set Up

In part (a) of this problem the skier travels from an initial point i at the top of the ramp to a final point f at the end of the ramp. During this motion only the conservative force of gravity does work on her. (The ramp also exerts a normal force on her, but this force does no work since it is always perpendicular to her motion.) So we can use Equation 6-27. (There are no springs, so there is no spring potential energy.) Our goal for this part is to find the skier's speed v_f as she leaves the ramp.

For part (b) we consider the skier's entire motion from the top of the ramp (which we again call point i) to where she finally comes to rest at point Q. A normal force also acts on her when she reaches the ground, but again this force does no work on her. Between points P and Q a nonconservative *friction* force also acts and does negative work on her (the force is opposite to her motion). So for the motion as a whole, we must use Equation 6-30. Our goal for part (b) is to find the magnitude f_k of the friction force.

Conservation of mechanical energy:

$$K_i + U_i = K_f + U_f \tag{6-27}$$

Gravitational potential energy:

$$U_{grav} = mgy \tag{6-18}$$

Change in mechanical energy when nonconservative forces act:

$$K_i + U_i + W_{nonconservative} = K_f + U_f \tag{6-30}$$

Kinetic energy:

$$K = \frac{1}{2}mv^2 \tag{6-8}$$

$m\vec{g}$
on ramp

$m\vec{g}$
in midair

$m\vec{g}$
between
P and Q

Solve

(a) Write expressions for the skier's kinetic energy K and the gravitational potential energy U_{grav} at the top of the ramp (at $y_i = H$, where the speed is $v_i = 0$) and the end of the ramp (at $y_f = D$, where the speed is v_f).

At the top of the ramp:

$$K_i = \frac{1}{2}mv_i^2 = 0$$

$$U_{grav,i} = mgy_i = mgH$$

At the end of the ramp:

$$K_f = \frac{1}{2}mv_f^2$$

$$U_{grav,f} = mgy_f = mgD$$

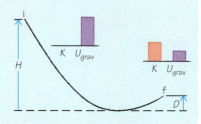

Now substitute the energies into Equation 6-27 and solve for v_f.

Equation 6-27:

$K_i + U_i = K_f + U_f$, so

$$0 + mgH = \frac{1}{2}mv_f^2 + mgD$$

$$\frac{1}{2}mv_f^2 = mgH - mgD = mg(H - D)$$

$$v_f^2 = 2g(H - D)$$

$$v_f = \sqrt{2g(H - D)}$$

(b) Now the final position is at point Q, a distance h below the low point of the ramp so $y_Q = -h$. At this point the skier is again at rest, so $v_Q = 0$. The friction force f_k acts opposite to her motion as she moves a distance L, so this force does work $-f_k L$ on her.

At the top of the ramp:

$$K_i = \frac{1}{2}mv_i^2 = 0$$

$$U_{grav,i} = mgy_i = mgH$$

At point Q:

$$K_Q = \frac{1}{2}mv_Q^2 = 0$$

$$U_{grav,Q} = mgy_Q$$
$$= mg(-h) = -mgh$$

Work done by the friction force, which points opposite to the skier's displacement:

$$W_{nonconservative} = f_k L \cos 180° = f_k L(-1)$$
$$= -f_k L$$

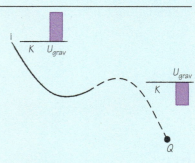

Now use Equation 6-30 to solve for f_k.

Equation 6-30:

$K_i + U_i + W_{nonconservative} = K_Q + U_Q$, so

$0 + mgH + (-f_k L) = 0 + (-mgh)$

$f_k L = mgH + mgh = mg(H + h)$

$f_k = mg \left(\dfrac{H + h}{L} \right)$

Reflect

Our answer for part (a) is the same one that we found in Example 6-10, where we used the expression $W_{grav} = -\Delta U_{grav}$ for the work done by gravity. That's as it should be, since we used that expression in Section 6-7 to help derive Equation 6-27.

   Our result for f_k in part (b) has the correct dimensions of force, the same as mg, since the dimensions of $H + h$ (the total vertical distance that the skier descends) cancel those of L (the skier's stopping distance). Note that the smaller the stopping distance L compared to $H + h$, the greater the friction force required to bring the skier to a halt.

   An alternative way to solve part (b) would be to treat the second part of the skier's motion separately, with the initial point at the end of the ramp [at height D, where the skier's speed is $\sqrt{2g(H - D)}$ as found in part (a)] and the final point is at point Q (at height $-h$, where the skier's speed is zero). You should try solving the problem this way using Equation 6-30 and $W_{nonconservative} = -f_k L$; you should get the same answer for f_k. Do you?

EXAMPLE 6-12   Warming Skis and Snow

When the skier in Example 6-11 brakes to a halt, her skis and the snow over which she slides both warm up. What is the increase in the internal energy of the skis and snow in this process?

Set Up

This problem involves internal energy, so we'll use the most general statement of energy conservation. This says that during the skier's motion (from the top of the ramp to when she finally comes to rest at point Q) the sum of the changes in kinetic energy K, potential energy U, and internal energy E_{other} must be zero. We'll use this to find how much internal energy is added to the ski and snow during the braking.

The law of conservation of energy:

$\Delta K + \Delta U + \Delta E_{other} = 0$ (6-32)

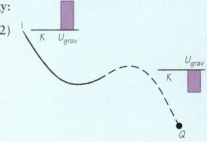

Solve

Substitute the change in kinetic energy ΔK and the change in potential energy ΔU into Equation 6-32 and solve for the increase in internal energy. We'll refer to Example 6-11 for the initial and final values of y and v.

At the top of the ramp: $K_i = \dfrac{1}{2} mv_i^2 = 0$

$U_{grav,i} = mgy_i = mgH$

At point Q: $K_Q = \dfrac{1}{2} mv_Q^2 = 0$

$U_{grav,Q} = mgy_Q = mg(-h) = -mgh$

Change in kinetic energy: $\Delta K = K_Q - K_i = 0 - 0 = 0$

Change in potential energy: $\Delta U = U_{grav,Q} - U_{grav,i}$

$= -mgh - mgH = -mg(H + h)$

From Equation 6-32, $\Delta K + \Delta U + \Delta E_{other} = 0$

$0 + [-mg(H + h)] + \Delta E_{other} = 0$

$\Delta E_{other} = mg(H + h)$

Reflect

The overall motion of the skier begins and ends with zero kinetic energy. The net result is that the (gravitational) potential energy decreases by an amount $mg(H + h)$ and the internal energy increases by the same amount. This represents

$W_{nonconservative} = -\Delta E_{other}$ (6-31)

From Example 6-11, $W_{nonconservative} = -f_k L$, and

$f_k = mg \left(\dfrac{H + h}{L} \right)$, so

the amount of thermal energy that is added to the skier's skis and the snow over which she slides as she brakes to a halt.

We can check our result by using Equation 6-31, which relates the change in internal energy to the amount of nonconservative work done. We get the same answer with this approach, as we must.

$$W_{\text{nonconservative}} = -mg\left(\frac{H+h}{L}\right)L = -mg(H+h)$$

$$\Delta E_{\text{other}} = -W_{\text{nonconservative}} = mg(H+h)$$

EXAMPLE 6-13   A Ramp and a Spring

A child's toy uses a spring with spring constant $k = 36$ N/m to shoot a block up a ramp inclined at $\theta = 30°$ from the horizontal (Figure 6-34). The mass of the block is $m = 8.0$ g. When the spring is compressed 4.2 cm and released, the block slides up the ramp, loses contact with the spring, and comes to rest a distance d along the ramp from where it started. Find the value of d. Neglect friction between the block and the surface of the ramp.

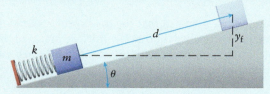

Figure 6-34 Spring potential energy and gravitational potential energy If the spring is compressed a certain distance, how far up the frictionless ramp will the block go?

Set Up

The forces on the block are a normal force, the gravitational force, and a spring force that acts while the block is in contact with the spring. The normal force is always perpendicular to the block's motion and so does no work. So only conservative forces (gravity and the spring force) do work, and total mechanical energy is conserved. We'll use Equation 6-27 to solve for the distance d. The initial point i is where the block is released, and the final point f is a distance d up the ramp from the initial point.

$$K_i + U_i = K_f + U_f \qquad (6\text{-}27)$$

$$U_{\text{grav}} = mgy \qquad (6\text{-}18)$$

$$U_{\text{spring}} = \frac{1}{2}kx^2 \qquad (6\text{-}23)$$

$$K = \frac{1}{2}mv^2 \qquad (6\text{-}8)$$

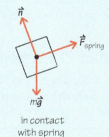

in contact with spring   after losing contact with spring

Solve

The block is at rest with zero kinetic energy at both the initial point and the final point. If we take $y = 0$ to be the height of the block where it is released, then $y_i = 0$; the final height y_f is related to the distance d by trigonometry. We use Equation 6-18 to write the initial and final gravitational potential energies. The initial spring potential energy is given by Equation 6-23 with $k = 36$ N/m and $x_i = -4.2$ cm; the final spring potential energy is zero because the spring is relaxed.

Before block is released:

$$K_i = \frac{1}{2}mv_i^2 = 0$$

$$U_{\text{grav},i} = mgy_i = 0$$

$$U_{\text{spring},i} = \frac{1}{2}kx_i^2$$

When block has traveled a distance d up the ramp:

$$K_f = \frac{1}{2}mv_f^2 = 0$$

From Figure 6-34,

$$y_f = d\sin\theta,\text{ so}$$

$$U_{\text{grav},f} = mgy_f = mgd\sin\theta$$

$$U_{\text{spring},f} = \frac{1}{2}kx_f^2 = 0$$

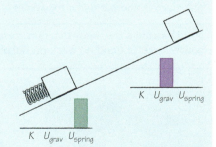

K U_{grav} U_{spring}

K U_{grav} U_{spring}

Substitute the energies into Equation 6-27. (Remember that the *total* potential energy U is the sum of the gravitational and spring potential energies.) Then solve for the distance d.

Equation 6-27:

$$K_i + U_i = K_f + U_f, \text{ or}$$

$$K_i + U_{\text{grav},i} + U_{\text{spring},i} = K_f + U_{\text{grav},f} + U_{\text{spring},f}, \text{ so}$$

$$0 + 0 + \frac{1}{2}kx_i^2 = 0 + mgd\sin\theta + 0$$

$$d = \frac{kx_i^2}{2mg\sin\theta}$$

Insert numerical values (being careful to express distances in meters and masses in kilograms) and find the value of d. Recall that $1\text{ N} = 1\text{ kg·m/s}^2$.

$$x_i = (-4.2\text{ cm})\left(\frac{1\text{ m}}{100\text{ m}}\right) = -0.042\text{ m}$$

$$m = (8.0\text{ g})\left(\frac{1\text{ kg}}{1000\text{ g}}\right) = 0.0080\text{ kg}$$

$$d = \frac{(36\text{ N/m})(-0.042\text{ m})^2}{2(0.0080\text{ kg})(9.80\text{ m/s}^2)\sin 30°}$$

$$= 0.81\frac{\text{N·s}^2}{\text{kg}} = 0.81\frac{\text{kg·m}}{\text{s}^2}\frac{\text{s}^2}{\text{kg}}$$

$$= 0.81\text{ m} = 81\text{ cm}$$

Reflect

In our solution we assumed that the final spring potential energy is zero, which means that the spring ends up neither compressed nor stretched. In other words, we assumed that the block moves so far up the ramp that it loses contact with the spring. That's consistent with our result: The block moves 81 cm up the ramp, much greater than the 4.2-cm distance by which the spring was originally compressed.

Our solution shows that the net effect of the block's motion is to convert spring potential energy at the initial position to gravitational potential energy at the final position. The block begins and ends with zero kinetic energy, just as the skier does in Examples 6-11 and 6-12. Note that the block *does* have kinetic energy at intermediate points during the motion, when it's moving up the ramp. However, we didn't need to worry about these intermediate stages of the motion.

This problem would have been impossible to solve using the techniques of Chapters 4 and 5 because the net force on the block isn't constant: The magnitude of the spring force starts off large, then decreases to zero as the block moves uphill and the spring relaxes. The energy approach is the *only* method that will lead us to the answer to this problem.

GOT THE CONCEPT? 6-7 Where Is It Zero?

? In Example 6-13, which of the following are you free to choose as you see fit? (a) The point on the ramp where gravitational potential energy is zero; (b) the point on the ramp where spring potential energy is zero; (c) both of these; (d) neither of these.

TAKE-HOME MESSAGE FOR Section 6-8

✔   The kinds of problems that are best solved using the law of conservation of energy are those in which an object moves from one place to another under the action of well-defined forces.

✔   Drawing a free-body diagram is essential for deciding which forces act and do work on the object.

✔   If only conservative forces do work, total mechanical energy is conserved. If nonconservative forces also do work, you must use the more general statement of energy conservation.

Figure 6-35 Running versus walking Both running and walking involve doing work; the difference is the rate at which work is done.

6-9 Power is the rate at which energy is transferred

If you walk for a kilometer, your heart rate will increase above its resting value. But if you run for a kilometer, your heart rate will increase to an even higher value (Figure 6-35). Now that we've learned about energy, this may seem paradoxical: You covered the same distance and did roughly the same amount of work in both cases, so why is a higher heart rate needed for running? The answer, as we will see, is related to the *rate* of transferring energy and the rate of doing work.

The rate at which energy is transferred from one place to another, or from one form to another, is called **power**. The unit of power is the joule per second, or **watt** (abbreviated W): $1\text{ W} = 1\text{ J/s}$. As an example, a 50-W light bulb is designed so that when it is in operation in an electric circuit, 50 J of energy is delivered to it every second by that circuit. Other common units of power are the *kilowatt* ($1\text{ kW} = 1000\text{ W}$) and the *horsepower* ($1\text{ hp} = 746\text{ W}$ is a typical rate at which a horse does work by pulling on a plow). We'll usually denote power by the uppercase symbol P.

Let's apply the concept of power to the rate of *doing work*. If you are using a stationary exercise bike at the gym, a reading of 200 W on the bike's display means that you are doing 200 J of work on the bike every second. So the power you are applying to the pedals equals the amount of work that you do divided by the time it takes for you to do that work:

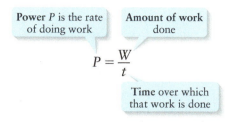

$$P = \frac{W}{t}$$

Power, work done, and time
(6-33)

The quantity P in this equation is sometimes called the power *delivered* to the object on which work is being done.

To help interpret this equation, let's think again about the groundskeeper pulling a screen across a baseball diamond that we described in Section 6-2 (Figure 6-6). The groundskeeper exerted a force of magnitude F on the screen at an angle θ to the screen's displacement. So to move the screen a distance d over the ground at a steady speed, the groundskeeper did an amount of work $W = (F \cos \theta)d$, as described by Equation 6-2. If the groundskeeper took a time t to pull the screen this distance, the speed of the screen was $v = d/t$. So Equation 6-33 tells us that the power output that the groundskeeper delivered to the screen was

$$P = \frac{W}{t} = \frac{(F \cos \theta)d}{t} = (F \cos \theta)v \qquad (6\text{-}34)$$

This equation says that the greater the force F that the groundskeeper exerts and the faster the speed v at which he makes the screen move, the more power he delivers to the screen.

The same equation also applies to the power that you deliver to the pedals of a bicycle or of a stationary exercise bike. When your foot pushes straight down on one of the pedals, as in Figure 6-36a, the angle θ in Equation 6-34 is zero (the force is in the direction that the pedal is moving). Since $\cos 0 = 1$, the power you deliver is then $P = Fv$. But if the force is at an angle of $\theta = 45°$, as in Figure 6-36b, the power you deliver is less: $P = (F \cos 45°)v = 0.707Fv$. So the power you deliver to the pedal is maximum when you push straight down on a downward-moving pedal. (It's also maximum when you pull straight up on an upward-moving pedal, so again $\theta = 0$.) You can notice this most easily when you stand up on the pedals; you can feel greater power being delivered when your feet move vertically.

We can now use the concept of power to explain the difference between walking and running that we mentioned at the beginning of this section. Whether you walk or run, you expend some of the internal energy E_{other} stored in your body and use it to do nonconservative work $W_{\text{nonconservative}}$ to propel yourself. (This work is nonconservative because you don't get the energy back after you stop.) From Equation 6-31 in Section 6-7,

$$W_{\text{nonconservative}} = -\Delta E_{\text{other}}$$

Here $W_{\text{nonconservative}}$ is positive (you do positive work) and ΔE_{other} is negative (your internal energy decreases). If we divide both sides of this equation by the time t that you walk or run, we get

$$\frac{W_{\text{nonconservative}}}{t} = \frac{-\Delta E_{\text{other}}}{t} \qquad (6\text{-}35)$$

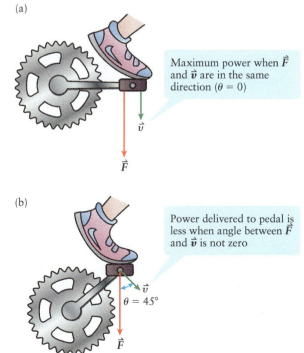

Figure 6-36 Power in cycling The power you deliver to a bicycle pedal depends on the angle between the force you exert and the direction of the pedal's motion.

BioMedical

If you compare Equation 6-35 to Equation 6-34, you'll see that $W_{\text{nonconservative}}/t$ is just the power P that you deliver to make yourself move. The term $\Delta E_{\text{other}}/t$ is the rate at which you lose internal energy as you move. When you run rather than walk, you cover the same distance in a shorter time t, $P = W_{\text{nonconservative}}/t$ has a greater value, and you do work at a faster rate. To do work more rapidly your muscles consume internal energy more rapidly and also require more oxygen per second. Hence your respiration rate goes up so that you will inhale more oxygen per second, and your heart rate goes up in order to more rapidly supply your muscles with oxygenated blood. This elevated heart rate is why running is superior to walking as cardiovascular exercise.

In walking, running, and all other forms of exercise, only part of the chemical energy that you expend goes into doing work. The rest goes into increasing the thermal energy of your body and your surroundings. The faster the rate at which you do work—that is, the greater your power output—the more rapidly you use up your chemical energy and the more rapidly the thermal energy increases (and so the more rapidly you warm up). Example 6-14 explores this for one particular type of vigorous exercise.

EXAMPLE 6-14 Power in a Rowing Race

Rowing demands one of the highest average power outputs of all competitive sports (Figure 6-37). In a typical rowing race a racing shell travels 2000 m in 6.00 min, during which each rower has an average power output of 4.00×10^2 W and expends 7.20×10^5 J = 720 kJ of chemical energy. (a) How much work does each rower do to propel the shell? (b) By how much does the thermal energy of each rower increase?

Figure 6-37 **Power and energy in rowing** These athletes are converting chemical energy into work and into thermal energy.

Sherene DuBois

Set Up

We are given the rower's power output P and the time t for which that power is delivered, so we can use Equation 6-33 to find the total (nonconservative) work that the rower does. We can then use Equation 6-35 to find the change ΔE_{other} in internal energy. This is the sum of the change in chemical energy (which we are given) and the change in thermal energy (which we want to find).

$$P = \frac{W_{\text{nonconservative}}}{t} \tag{6-33}$$

$$\frac{W_{\text{nonconservative}}}{t} = -\frac{\Delta E_{\text{other}}}{t} \tag{6-35}$$

Solve

(a) Solve Equation 6-33 for the work done by the rower. Remember that $1\ \text{W} = 1\ \text{J/s}$, and be careful to express the time t in seconds.

Equation 6-33:

$P = \dfrac{W_{\text{nonconservative}}}{t}$, so

$W_{\text{nonconservative}} = Pt$

$P = 4.00 \times 10^2\ \text{W} = 4.00 \times 10^2\ \text{J/s}$

$t = (6.00\ \text{min})\left(\dfrac{60\ \text{s}}{1\ \text{min}}\right) = 3.60 \times 10^2\ \text{s}$

$W_{\text{nonconservative}} = (4.00 \times 10^2\ \text{W})(3.60 \times 10^2\ \text{s})$
$= 1.44 \times 10^5\ \text{J} = 144\ \text{kJ}$

(b) We can now find the total internal change ΔE_{other}, which is the negative of the work that the rower does. We know the change in the rower's chemical energy is $\Delta E_{\text{chemical}} = -7.20 \times 10^5$ J (this is negative because the chemical energy decreases). So we can find the change in thermal energy $\Delta E_{\text{thermal}}$.

Multiply Equation 6-35 by time t:

$W_{\text{nonconservative}} = -\Delta E_{\text{other}}$, so

$\Delta E_{\text{other}} = -W_{\text{nonconservative}} = -1.44 \times 10^5\ \text{J}$

This is the sum of $\Delta E_{\text{chemical}}$ and $\Delta E_{\text{thermal}}$:

$\Delta E_{\text{other}} = \Delta E_{\text{chemical}} + \Delta E_{\text{thermal}}$, so

$\Delta E_{\text{thermal}} = \Delta E_{\text{other}} - \Delta E_{\text{chemical}}$
$= (-1.44 \times 10^5\ \text{J}) - (-7.20 \times 10^5\ \text{J})$
$= +5.76 \times 10^5\ \text{J} = +576\ \text{kJ}$

Reflect

In most of the world chemical energy or food energy is expressed in kilojoules, or kJ. In the United States it is expressed in food calories (symbol C), where 1 C = 4.186 kJ. So our rower "burned" 720 kJ or 172 food calories during the race

In accordance with the law of conservation of energy, we can account for 100% of the chemical energy used: 0.200 or 20.0% went into doing work, and 0.800 or 80.0% went into increasing the thermal energy.

Chemical energy used:

$$(720 \text{ kJ})\left(\frac{1 \text{ C}}{4.186 \text{ kJ}}\right) = 172 \text{ C}$$

Fraction of chemical energy that went into doing work:

$$\frac{144 \text{ kJ}}{720 \text{ kJ}} = 0.200$$

Fraction of chemical energy that went into thermal energy:

$$\frac{576 \text{ kJ}}{720 \text{ kJ}} = 0.800$$

GOT THE CONCEPT? 6-8 Automotive Power

 The power that a gasoline engine delivers to a car's wheels is roughly the same at all speeds. Is the forward force on the car greater at (a) low speeds or (b) high speeds? (*Hint*: Consider Equation 6-34.)

TAKE-HOME MESSAGE FOR Section 6-9

✔ Power is the rate of transferring energy from one place or form to another.

✔ The power delivered to an object is related to the speed of the object and the magnitude and direction of the applied force.

Key Terms

conservative force	law of conservation of energy	thermal energy
energy	negative work	total mechanical energy
gravitational potential energy	nonconservative force	translational kinetic energy
Hooke's law	potential energy	watt
internal energy	power	work
joule	spring constant	work-energy theorem
kinetic energy	spring potential energy	

Chapter Summary

Topic	Equation or Figure
Work done by a force in the direction of displacement: If an object moves in a straight line while a constant force is applied in the same direction as the displacement, the work is just equal to the force times the displacement.	**Work done on an object** by a constant force $\vec{F}$ that points in the **same direction** as the object's displacement $\vec{d}$ **Magnitude of the constant force** $\vec{F}$ $$W = Fd \qquad (6\text{-}1)$$ **Magnitude of the displacement** $\vec{d}$

Work done by a constant force at an angle θ to the straight-line displacement: If an object moves in a straight line while a constant force is applied at some angle θ to the displacement, the work is equal to the force times the displacement multiplied by the cosine of the angle between the force and displacement.

Work done on an object by a constant force $\vec{F}$ that points at an angle θ to the object's displacement $\vec{d}$ Magnitude of the constant force $\vec{F}$

$$W = (F \cos \theta)d \qquad (6\text{-}2)$$

Angle between the directions of $\vec{F}$ and $\vec{d}$ Magnitude of the displacement $\vec{d}$

Kinetic energy and the work-energy theorem: The work-energy theorem states that the net work done in a displacement—the sum of the work done on the object by individual forces—is equal to the change in the object's kinetic energy, or energy of motion, during that displacement. This theorem is valid whether the path is curved or straight and whether the forces are constant or varying. For a path of any shape, the work done by the gravitational force depends only on the vertical displacement.

Work done on an object by the **net force** on that object

$$W_{\text{net}} = K_f - K_i \qquad (6\text{-}9)$$

Kinetic energy of the object after the work is done on it   Kinetic energy of the object before the work is done on it

Kinetic energy of an object   Mass of the object

$$K = \frac{1}{2}mv^2 \qquad (6\text{-}8)$$

Speed of the object

Work done by the gravitational force on an object as the object follows a curved path   Mass of the object   Acceleration due to gravity

$$W_{\text{grav}} = -mg(y_f - y_i) \qquad (6\text{-}14)$$

Final y coordinate of the object at the end of the path   Initial y coordinate of the object at the beginning of the path

The spring force and Hooke's law: An ideal spring exerts a force that is proportional to how far it is stretched or compressed, and it is always opposite to the stretch or compression. The spring force is not constant, so the work needed to stretch or compress a spring is not simply the force magnitude multiplied by the displacement.

Force exerted by an **ideal spring**   Spring constant of the spring (a measure of its stiffness)

$$F_x = -kx \qquad (6\text{-}15)$$

Extension of the spring ($x > 0$ if spring is stretched, $x < 0$ if spring is compressed)

Work that must be done on a spring to stretch it from $x = x_1$ to $x = x_2$ Spring constant of the spring (a measure of its stiffness)

$$W = \frac{1}{2}kx_2^2 - \frac{1}{2}kx_1^2 \qquad (6\text{-}17)$$

x_1 = initial stretch of the spring
x_2 = final stretch of the spring

Potential energy: Unlike kinetic energy, potential energy is associated with the position of an object. Gravitational potential energy increases with height. The potential energy of a spring increases with the stretch or compression of the spring. Only conservative forces, for which the work done does not depend on the path taken, are associated with a potential energy.

Gravitational potential energy associated with an object   Mass of the object

$$U_{\text{grav}} = mgy \qquad (6\text{-}18)$$

Acceleration due to gravity   Height of the object

Spring potential energy of a stretched or compressed spring   Spring constant of the spring

$$U_{\text{spring}} = \frac{1}{2}kx^2 \qquad (6\text{-}23)$$

Extension of the spring ($x > 0$ if spring is stretched, $x < 0$ if spring is compressed)

Total mechanical energy and its conservation: If only conservative forces do work on an object, the total mechanical energy is conserved. (Other forces can act on the object, but they cannot do any work.) The kinetic and potential energies can change, but their sum remains constant.

Values of the **kinetic energy** K of an object at two points (i and f) during its motion

$$K_i + U_i = K_f + U_f$$

Values of the **potential energy** U at the same two points

$$(6\text{-}27)$$

 If only conservative forces do work on an object, the sum $K + U$ maintains the same value throughout the motion. This sum is called the **total mechanical energy** E.

Change in the kinetic energy K of an object during its motion   Change in the potential energy U associated with an object during its motion

$$\Delta K + \Delta U = 0$$

$$(6\text{-}28)$$

 If only conservative forces do work on the object, the sum of these two changes is zero: Total mechanical energy can transform between kinetic and potential forms, but there is no net change in the amount of total mechanical energy.

Generalized law of conservation of energy: If nonconservative forces do work, total mechanical energy is not conserved. If we broaden our definition of energy to include the internal energy of the interacting objects, then total energy is conserved in *all* cases.

Values of the **kinetic energy** K of an object at two points (i and f) during its motion

$$K_i + U_i + W_{\text{nonconservative}} = K_f + U_f$$

Values of the **potential energy** U at the same two points

$$(6\text{-}30)$$

 If nonconservative forces do work on an object, the total mechanical energy $E = K + U$ changes its value.
If $W_{\text{nonconservative}} > 0$, $E_f > E_i$ and the total mechanical energy increases.
If $W_{\text{nonconservative}} < 0$, $E_f < E_i$ and the total mechanical energy decreases.

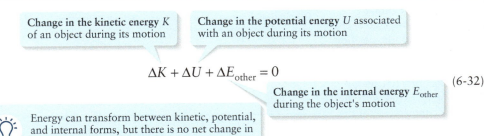

$$\Delta K + \Delta U + \Delta E_{\text{other}} = 0 \qquad \text{(6-32)}$$

Energy can transform between kinetic, potential, and internal forms, but there is no net change in the amount of energy.

Power: Power is the rate at which energy is transferred from one place or form to another place or form. If work is being done on an object, the power delivered equals the rate at which work is done.

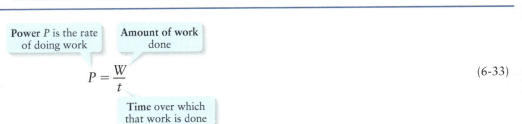

$$P = \frac{W}{t} \qquad \text{(6-33)}$$

Answer to What do you think? Question

(c) The long, thick Achilles tendon on the back of an ostrich's leg acts like a spring. When a running ostrich lands on one foot, this tendon stretches and converts kinetic energy into spring potential energy. When this foot subsequently leaves the ground, the tendon relaxes, and this stored spring energy is turned back into kinetic energy. Were it not for this form of energy storage, the ostrich's leg muscles would have to push much harder with each step to maintain the animal's motion.

Answers to Got the Concept? Questions

6-1 (a) negative, (b) positive, (c) zero, (d) positive. In (a) you exert a force opposite to the ball's displacement to slow it down, so the work you do on the ball is negative. In (b) the force you exert on the ball as you throw it is in the same direction as the ball's displacement, so you do positive work on the ball. In (c) you exert an upward force on the ball to keep it from falling, but there is no forward or backward force on the ball. Its displacement is forward, so the force you exert is perpendicular to the ball's displacement and no work is done. In (d) you must push the ball forward (in the direction of its displacement) to make it speed up, so you do positive work on the ball.

6-2 (d) The work-energy theorem equates the work done on an object to the change in its kinetic energy. The work done is proportional to the distance over which the force is applied, so this distance is directly proportional to the change in kinetic energy. When an object starts from rest, the change in kinetic energy (from Equation 6-9) equals the final kinetic energy, which is proportional to the square of the final speed. So the distance is proportional to the square of the final speed, and doubling the speed therefore requires that the force be applied over $2^2 = 4$ times the distance.

6-3 (a) (i), (b) (iii). The floor exerts a greater normal force n on the heavy box, so the kinetic friction force $f_k = \mu_k n$ is greater for the heavy box. Hence you must push harder on the heavy box to balance the greater kinetic friction force.

You therefore do more work on the heavy box than on the light box because you exert a greater force over the same distance. The work-energy theorem tells us that *net* work done on *both* boxes is the same and equal to zero. That's because both boxes have initial kinetic energy $K_i = 0$ (they start at rest) and final kinetic energy $K_f = 0$ (they end at rest), so $W_{\text{net}} = K_f - K_i = 0 - 0 = 0$. You do positive work on each box, but the floor does an equal amount of negative work through the kinetic friction force.

6-4 (a) From Example 6-10 the work that the athlete must do to stretch the cords by a distance x_2 from their relaxed state ($x_1 = 0$) is $W_{\text{athlete on cords}} = \frac{1}{2}kx_2^2 - \frac{1}{2}kx_1^2 = \frac{1}{2}kx_2^2 - 0 = \frac{1}{2}kx_2^2$. This says that the work done is proportional to the square of the stretch. Equivalently, this says that the stretch is proportional to the square root of the work done. So to increase the work done by a factor of 2, the stretch must increase by a factor of $\sqrt{2}$.

6-5 (d) Equation 6-20 tells us that the change in gravitational potential energy as the ball moves from height y_i (the top of the building) to height y_f (the height of the balcony) is $\Delta U_{\text{grav}} = U_f - U_i = mgy_f - mgy_i = mg(y_f - y_i)$. The difference $y_f - y_i$ is 36 m − 45 m = −9 m; that is, the balcony is 9 m below the top of the building. This distance doesn't depend at all on where you choose $y = 0$! (If $y = 0$ is at the balcony, $y_i = 9$ m and $y_f = 0$, so $y_f - y_i = 0 - 9$ m = −9 m; if $y = 0$ is at the top of the building, $y_i = 0$ and $y_f = -9$ m, and

$y_f - y_i = -9$ m $- 0 = -9$ m.) As far as *changes* in gravitational potential energy are concerned, you can choose the level $y = 0$ to be wherever you like.

6-6 (b) There is no friction, and the normal force does no work on the block (it always acts perpendicular to the path of the block). Hence only the conservative gravitational force does work and so total mechanical energy is conserved. Let point i be where the block is released and point f be where it stops momentarily on the right-hand side of the ramp. The block is not moving at either point, so $K_i = K_f = 0$. Equation 6-27 then says that $U_i = U_f$, which means that the gravitational potential energy is the same at both points. Since $U_{grav} = mgy$, this means that the block is at the same height y at both points. Hence the block reaches the same height $y = H$ on the right-hand side of the ramp as on the left-hand side.

6-7 (a) We learned in Section 6-6 that we can choose $y = 0$ to be wherever we like when calculating gravitational potential energy $U_{grav} = mgy$. However, we do not have the same freedom when calculating spring potential energy $U_{spring} = \frac{1}{2}kx^2$; we *must* choose $x = 0$ to be where the spring is relaxed (neither stretched nor compressed).

6-8 (a) The forward force is in the same direction as the car's motion, so the power delivered is given by Equation 6-34 with $\theta = 0$: $P = (F \cos 0)v = Fv$. The power P, and hence the product Fv, is the same at low speed and high speed, so the force F has a large value when the speed v has a small value. As a result a car powered by a gasoline engine has a greater forward force (and greater acceleration) when it is first starting from rest than it does when it's already moving fast.

Questions and Problems

In a few problems you are given more data than you actually need; in a few other problems you are required to supply data from your general knowledge, outside sources, or informed estimate.

Interpret as significant all digits in numerical values that have trailing zeros and no decimal points. For all problems use $g = 9.80$ m/s^2 for the free-fall acceleration due to gravity.

- •   Basic, single-concept problem
- ••   Intermediate-level problem; may require synthesis of concepts and multiple steps
- •••   Challenging problem
- Example   See worked example for a similar problem

Conceptual Questions

1. • Using Equation 6-2, explain how the work done on an object by a force can be equal to zero.

2. • Why do seasoned hikers step *over* logs that have fallen in their path rather than stepping *onto* them?

3. • Your roommate lifts a cement block, carries it across the room, and sets it back down on the floor. Is the net work she did on the block positive, negative, or zero? Explain your answer.

4. • Can the normal force ever do work on an object? Explain your answer.

5. • One of your classmates in physics reasons, "If there is no displacement, then a force will perform no work." Suppose the student pushes with all of her might against a massive boulder, but the boulder doesn't move. (a) Does she expend energy pushing against the boulder? (b) Does she do work on the boulder? Explain your answer.

6. • Can kinetic energy ever have a negative value? Explain your answer.

7. • Can a change in kinetic energy ever have a negative value? Explain your answer.

8. • Define the concept of "nonconservative force" in your own words. Give three examples.

9. • A satellite orbits Earth in a circular path at a high altitude. Explain why the gravitational force does zero work on the satellite.

10. • Model rockets are propelled by an engine containing a combustible propellant. Two rockets, one twice as heavy as the other, are launched using engines that contain the same amount of propellant. (a) Is the maximum kinetic energy of the heavier rocket less than, equal to, or more than the maximum kinetic energy of the lighter one? (b) Is the speed of the heavier rocket just after the propellant is used up less than, equal to, or greater than the launch speed of the lighter one?

11. • When does the kinetic energy of a rock that is dropped from the edge of a high cliff reach its maximum value? Answer the question (a) when the air resistance is negligible and (b) when there is significant air resistance.

12. • Can gravitational potential energy have a negative value? Explain your answer.

13. • Bicycling to the top of a hill is much harder than coasting back down to the bottom. Do you have more gravitational potential energy at the top of the hill or at the bottom? Explain your answer.

14. • Analyze the types of energy that are associated with the "circuit" that a snowboarder follows from the bottom of a mountain, to its peak, and back down again. Be sure to explain all changes in energy.

15. • A common classroom demonstration involves holding a bowling ball attached by a rope to the ceiling close to your face and releasing it from rest. In theory, you should not have to worry about being hit by the ball as it returns to its starting point. However, some of the most "exciting" demonstrations *have* involved the professor being hit by the ball! (a) Why should you expect not to be hit by the ball? (b) Why *might* you be hit if you actually perform the demonstration?

16. • Why are the ramps for people with disabilities quite long instead of short and steep?

17. • A rubber dart can be launched over and over again by the spring in a toy gun. In terms of work and energy, describe the entire process responsible for delivering kinetic energy to the dart.

18. • The energy provided by your electric company is *not* sold by the joule. Instead, you are charged by the kilowatt-hour (typically $0.25 per kWh including taxes). Explain why this makes sense for the average consumer. It might be helpful to read the electrical specifications of an appliance (such as a hair dryer or a blender).

Multiple-Choice Questions

19. • Box 2 is pulled up a rough incline by box 1 in an arrangement as shown in Figure 6-38. How many forces are doing work on box 2?
    A. one
    B. two
    C. three
    D. four
    E. five

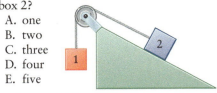

Figure 6-38
Problem 19

20. • Figure 6-39 shows four situations in which a box slides to the right a distance d across a frictionless floor as a result of applied forces, *one* of which is shown. The magnitudes of the forces shown are identical. Rank the four cases in order of increasing work done on the box by the force shown during the displacement.

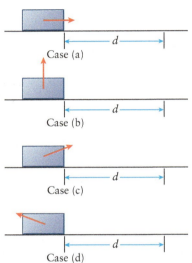

Case (a)

Case (b)

Case (c)

Case (d)

Figure 6-39 Problem 20

21. • A box is dragged a distance d across a floor by a force $\vec{F}$ which makes an angle θ with the horizontal (Figure 6-40). If the magnitude of the force is held constant but the angle θ is increased up to 90°, the work done by the force in dragging the box
    A. remains the same.
    B. increases.
    C. decreases.
    D. first increases, then decreases to its initial value.
    E. first decreases, then increases to its initial value.

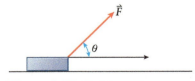

Figure 6-40 Problem 21

22. • A car moves along a straight, level road. If the car's velocity changes from 60 mi/h due east to 60 mi/h due west, the kinetic energy of the car
    A. remains the same.
    B. increases.
    C. decreases.
    D. first increases, then decreases.
    E. first decreases, then increases.

23. • A boy swings a ball on a string at constant speed in a horizontal circle that has a circumference equal to 6 m. What is the work done on the ball by the 10-N tension force in the string during one revolution of the ball?
    A. 190 J   D. 15 J
    B. 60 J   E. 0
    C. 30 J

24. • As part of a lab experiment, Allison uses an air-track cart of mass m to compress a spring of constant k by an amount x from its equilibrium length. The air track has negligible friction. When Allison lets go, the spring launches the cart. What cart velocity should she expect after it is launched by the spring?
    A. $\sqrt{2kx/m}$ B. $\sqrt{kx/m}$ C. $(\sqrt{k/m})x$ D. $(\sqrt{2k/m})x$

25. • Adam and Bobby are twins who have the same weight. Adam drops to the ground from a tree at the same time that Bobby begins his descent down a frictionless slide. If they both start at the same height above the ground, how do their kinetic energies compare when they hit the ground?
    A. Adam has greater kinetic energy than Bobby.
    B. Bobby has greater kinetic energy than Adam.
    C. They have the same kinetic energy.
    D. Bobby has twice the kinetic energy as Adam.
    E. More information is required to compare their kinetic energies.

26. • Three balls are thrown off a tall building with the same speed but in different directions. Ball A is thrown in the horizontal direction; ball B starts out at 45° above the horizontal; ball C begins its flight at 45° below the horizontal. Which ball has the greatest speed just before it hits the ground? Ignore any effects due to air resistance.
    A. ball A
    B. ball B
    C. ball C
    D. All balls have the same speed.
    E. Balls B and C have the same speed, which is greater than the speed of ball A.

Problems

6-2 The work that a constant force does on a moving object depends on the magnitude and direction of the force

27. • A crane lifts a 2.00×10^2-kg crate at a constant rate through a vertical distance of 15.0 m. How much work does the crane do on the crate? How much work does gravity do on the crate? Example 6-1

28. • **Sports** In the 2016 Olympic games, Deng Wei set a women's record of 147 kg in the clean and jerk. How much work did she do on the weights in lifting them above her head, a change in height of approximately 1.60 m? Example 6-1

29. • A 350-kg box is pulled 7.00 m up a 30.0° inclined plane by an external force of 5.00×10^3 N that acts parallel to the frictionless plane. Calculate the work done by (a) the external force, (b) gravity, and (c) the normal force. Example 6-3

30. • Three clowns try to move a 3.00×10^2-kg crate 12.0 m to the right across a smooth, low-friction floor. Moe pushes to the right with a force of 5.00×10^2 N, Larry pushes to the left with 3.00×10^2 N, and Curly pushes straight down with 6.00×10^2 N. Calculate the work done by each of the clowns. Example 6-3

31. • An assistant for the football team carries a 30.0-kg cooler of water from the top row of the stadium, which is 20.0 m above the field level, down to the bench area on the field. (a) If the speed of the cooler is constant throughout the trip, calculate the work done by the assistant on the cooler of water. (b) How much work is done by the force of gravity on the cooler of water? Example 6-1

32. •• A statue is crated and moved for cleaning. The mass of the statue and the crate is 150 kg. As the statue slides down a ramp inclined at 40.0°, the curator pushes up, parallel to the ramp's surface, so that the crate does not accelerate (Figure 6-41). If the statue slides 3.0 m down the ramp, and the coefficient of kinetic friction between the crate and the ramp is 0.54, calculate the work done on the crate by each of the following: (a) the gravitational force, (b) the curator, (c) the friction force, and (d) the normal force between the ramp and the crate. Example 6-3

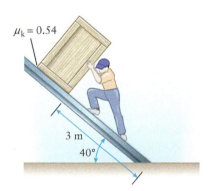

$\mu_k = 0.54$

3 m

40°

Figure 6-41 Problem 32

6-3 Kinetic energy and the work-energy theorem give us an alternative way to express Newton's second law

33. • A 1250-kg car moves at 20.0 m/s. How much work must be done on the car to increase its speed to 30.0 m/s? Example 6-4

34. • A bumblebee has a mass of about 0.25 g. If its speed is 10.0 m/s, calculate its kinetic energy. Example 6-4

35. • A small truck has a mass of 2100 kg. How much work is required to decrease the speed of the vehicle from 22.0 m/s to 12.0 m/s on a level road? Example 6-4

36. • An 8500-metric-ton freight train is out of control and moving at 90 km/h on level track. How much work must a superhero do on the train to bring it to a halt? Example 6-4

37. • A box with mass $m = 6.00$ kg is pushed $x = 15.0$ m across a level floor by a constant applied force $F = 12.0$ N. The coefficient of kinetic friction between the box and the floor is 0.150 (a) Assuming the box starts from rest, what is the final velocity of the box at the 15.0-m point? (b) If there were no friction between the box and the floor, what applied force would be required to give the box the same final velocity? Example 6-4

6-4 The work-energy theorem can simplify many physics problems

38. • A force of 1200 N pushes a man on a bicycle forward. Air resistance pushes against him with a force of 800 N. If he starts from rest and is on a level road, how fast will he be moving after 20.0 m? The mass of the bicyclist and his bicycle is 90.0 kg. Example 6-5

39. •• A book slides across a level, carpeted floor at an initial speed of 4.00 m/s and comes to rest after 3.25 m. Calculate the coefficient of kinetic friction between the book and the carpet. Assume the only forces acting on the book are friction, weight, and the normal force. Example 6-7

40. •• Calculate the final speed of the 2.00-kg object that is pushed for 22.0 m by a 40.0-N force directed 20.0° below the horizontal on a level, frictionless floor (Figure 6-42). Assume the object starts from rest. Example 6-6

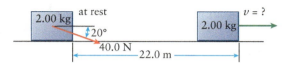

Figure 6-42 Problem 40

41. • A 325-g model boat facing east floats on a pond. The wind in its sail provides a force of 1.85 N that points 25° north of east. The force on its keel is 0.782 N pointing south. The drag force of the water on the boat is 0.750 N toward the west. If the boat starts from rest and heads east, how fast is it moving after it travels for a distance of 3.55 m? Example 6-6

42. • **Sports** A catcher in a baseball game stops a pitched ball originally moving at 44.0 m/s at the moment it first comes into contact with his glove. After contacting the glove, the ball traveled an additional 12.5 cm before coming to a complete stop. The mass of the ball is 0.145 kg. What is the magnitude of the average force that the glove imparts to the ball during the catch? Comment on the force that the catcher's hand experiences during the catch. Example 6-5

6-5 The work-energy theorem is also valid for curved paths and varying forces

43. • A sled of mass $m = 22.0$ kg is accelerated from rest on a frictionless surface to a final velocity of $v = 12.5$ m/s. Figure 6-43 displays a graph of force versus distance for the sled in terms of F_{max}. Use the graph to help you solve for F_{max} in terms of m, v, and numerical factors.

Figure 6-43 Problem 43

44. • An object attached to the free end of a horizontal spring of constant 450 N/m is pulled from a position 12 cm beyond equilibrium to a position 18 cm beyond equilibrium. Calculate the work the spring does on the object. Example 6-9

45. • A 5.00-kg object is attached to one end of a horizontal spring that has a negligible mass and a spring constant of 250 N/m. The other end of the spring is fixed to a wall. The spring is compressed by 10.0 cm from its equilibrium position and released from rest. (a) What is the speed of the object when it is 8.00 cm from equilibrium? (b) What is the speed when the object is 5.00 cm from

equilibrium? (c) What is the speed when the object is at the equilibrium position? Example 6-9

46. • A pendulum is constructed by attaching a small metal ball to one end of a 1.25-m-long string that hangs from the ceiling (Figure 6-44). The ball is released when it is raised high enough for the string to make an angle of 30.0° with the vertical. How fast is it moving at the bottom of its swing? Does the mass of the ball affect the answer? Example 6-8

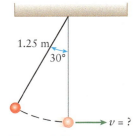

Figure 6-44 Problem 46

47. • Logan is standing on a dock holding on to a rope swing that is 4 m long and suspended from a tree branch above (Figure 6-45). The rope is taut and makes a 30° angle with the vertical direction. Logan swings in a circular arc until he releases the rope when it makes an angle of 12° from vertical, but on the other side. If Logan's mass is 75 kg, how much work does gravity do on him up to the point where he releases the rope? Example 6-8

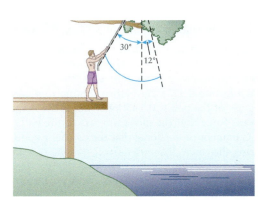

Figure 6-45 Problem 47

48. • Jack and Jill, whose masses are identical, go up a 300-m hill to fetch a pail of water. Jack climbs a sheer rock face to reach the top, while Jill follows a meandering path 835 m long. What is the magnitude of the difference between the work Jack and Jill do against gravity to get to the top? Example 6-8

49. • Wei drags a heavy piece of driftwood for 910 m along an irregular path. If Wei ends 750 m from where he started and exerted a constant force of 625 N, parallel to his path, the entire time, how much work did he do? Example 6-8

50. • Earth orbits the Sun at a radius of about 1.5×10^8 km. At this distance the force of gravity on Earth due to the Sun is 3.6×10^{22} N. Assuming Earth's orbit to be perfectly circular, how much work does the Sun's gravity do on Earth in one year? Example 6-8

51. •• Mikaela is out for a bike ride on a breezy day. The wind blows out of the west such that it exerts a constant drag force of 115 N pointing east. Initially riding north on flat roads, Mikaela traverses a 1.2-km-long *circular arc* at a constant speed that ends with her heading directly into the wind; the arc is a quarter circle that starts pointing north and ends pointing west (Figure 6-46). How much work does the wind

do on her as she rounds this curve from point A to point B? Example 6-8

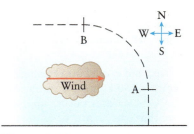

Figure 6-46 Problem 51

6-6 Potential energy is energy related to an object's position

52. • What is the gravitational potential energy relative to the ground associated with a 1.00-N Gravenstein apple hanging from a limb 2.50 m above the ground? Example 6-10

53. • Pilings are driven into the ground at a building site by dropping a 2000-kg object onto them. What change in gravitational potential energy does the object undergo if it is released from rest 18.0 m above the ground and ends up 2.00 m above the ground? Example 6-10

54. • A 40.0-kg boy steps on a skateboard and pushes off from the top of a hill. What change in gravitational potential energy takes place as the boy glides down to the bottom of the hill, 4.35 m below the starting level? Example 6-10

55. • How much additional potential energy is stored in a spring that has a spring constant of 15.5 N/m if the spring starts 10.0 cm from its unstretched length and ends up 15.0 cm from its unstretched length?

56. • A spring that has a spring constant of 2.00×10^2 N/m is oriented vertically with one end on the ground. (a) What distance must the spring compress for a 2.00-kg object placed on its upper end to reach equilibrium? (b) By how much does the potential energy stored in the spring increase during the compression?

57. • An external force moves a 3.50-kg box at a constant speed up a frictionless ramp (Figure 6-47). The force acts in a direction parallel to the ramp. (a) Calculate the work done on the box by this force as it is pushed up the 5.00-m ramp to a height of 3.00 m. (b) Compare the value with the change in gravitational potential energy that the box undergoes as it rises to its final height. Example 6-10

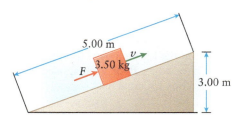

Figure 6-47 Problem 57

58. • Over 630 m in height, the Burj Khalifa is the world's tallest skyscraper. What is the change in gravitational potential energy of a $20 gold coin (33.5 g) when it is carried

from ground level up to the top of the Burj Khalifa? Neglect any slight variations in the acceleration due to gravity. Example 6-10

59. • For a great view and a thrill, check out EdgeWalk at the CN Tower in Toronto, where you can walk on the roof of the tower's main pod, 356 m above the ground. What is the gravitational potential energy relative to the surface of Earth of a 65.0-kg sightseer on EdgeWalk? Neglect any slight variations in the acceleration due to gravity. Example 6-10

60. • A spring that is compressed by 12.5 cm stores 3.33 J of potential energy. Determine the spring constant.

6-7 If only conservative forces do work, total mechanical energy is conserved

6-8 Energy conservation is an important tool for solving a wide variety of problems

61. • **Sports** A baseball player throws a ball at a nearby boulder. The ball hits the boulder with a speed of 40.0 m/s and then bounces straight up to a height of 12.0 m. Assuming that the rotational motion of the ball is negligible, what fraction f of the ball's initial mechanical energy was dissipated in the collision with the boulder? Example 6-12

62. • A ball is thrown straight up with an initial speed of 15.0 m/s. At what height will the ball have one-half of its initial speed? Example 6-11

63. • A water balloon is thrown straight down with an initial speed of 12.0 m/s from a second-floor window, 5.00 m above ground level. How fast is the balloon moving when it hits the ground? Example 6-11

64. • A gold coin (33.5 g) is dropped from the top of the Burj Khalifa building, 630 m above ground level. In the absence of air resistance, how fast would it be moving when it hit the ground? Example 6-11

65. •• Starting from rest, a 30.0-kg child rides a 9.00-kg sled down a frictionless ski slope. At the bottom of the hill, her speed is 7.00 m/s. If the slope makes an angle of 15.0° with the horizontal, how far did she slide on the sled? Example 6-11

66. •• **Sports** During a long jump Olympic champion Carl Lewis's center of mass rose 1.2 m from the launch point to the top of the arc. What minimum speed did he need at launch if he was traveling at 6.6 m/s at the top of the arc? Example 6-11

67. • An ice cube starts at rest at point A, and slides down a frictionless track as shown in Figure 6-48. Calculate the speed of the cube at points B, C, D, and E. Example 6-11

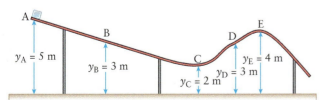

Figure 6-48 Problem 67

68. • A driver slams on the brakes, leaving 88.0-m-long skid marks on the level road as the car comes to a stop. The coefficient of kinetic friction is estimated to be 0.480. How fast was the car moving when the driver hit the brakes? Example 6-11

69. •• A 65.0-kg woman steps off a 10.0-m diving platform and drops straight down into the water. If she reaches a depth of 4.50 m, what is the average resistance force exerted on her by the water? Ignore air resistance. Example 6-11

70. •• A skier leaves the starting gate at the top of a ski jump with an initial speed of 4.00 m/s (Figure 6-49). The starting position is 120 m higher than the end of the ramp, which is 3.00 m above the snow. Find the final speed of the skier if he lands 145 m down the 20.0° slope. Assume there is no friction on the ramp, but air resistance causes a 50% loss in the final kinetic energy. Example 6-11

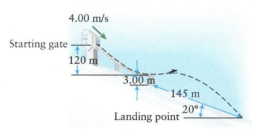

Figure 6-49 Problem 70

71. • An 18.0-kg suitcase falls from a hot-air balloon that is at rest at a height of 385 m above the surface of Earth. The suitcase reaches a speed of 30.0 m/s just before it hits the ground. Calculate the percentage of the initial energy that is "lost" to air resistance. Example 6-12

72. •• An ideal spring is used to stop blocks as they slide along a table without friction (Figure 6-50). A 0.85-kg block traveling at a speed of 2.1 m/s can be stopped over a distance of 0.15 m, once it makes contact with the spring. What distance would a 1.3-kg block travel after making contact with the spring, if the block were traveling at a speed of 3.3 m/s? Example 6-13

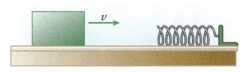

Figure 6-50 Problem 72

73. • A child is ready to slide down a snow-covered slope on a sled. At the top of the hill, her mother gives her a push to start her off with a speed of 1.00 m/s. The frictional force acting on the sled is one-fifth of the combined weight of the child and the sled. If she travels for a distance of 25.0 m and her speed at the bottom is 4.00 m/s, calculate the angle that the hill makes with the horizontal. Example 6-12

74. • Neil and Gus are having a competition to see who can launch a marble higher in the air using his own spring. Neil has a firm spring ($k_{Neil} = 50.8$ N/m), but it can be compressed only a maximum of $y_{Neil,max} = 14.0$ cm. Gus's spring is less firm than Neil's ($k_{Gus} = 12.7$ N/m), but its maximum compression is greater ($y_{Gus,max} = 27.0$ cm). Given that the mass of the marble is 4.5 g and ignoring the effects of air resistance,

who can launch the marble higher and what is the winning height? Assume the marble starts from the same height in both cases when the springs are at maximum compression. Example 6-13

75. • Continuing with the previous problem, what is the speed of the marble as it leaves each spring? Example 6-13

6-9 Power is the rate at which energy is transferred

76. • The power output of professional cyclists averages about 350 W when climbing mountains. How much energy does a typical 70-kg pro cyclist expend climbing a 12.3-km-long mountain road with a 3.9° average slope at an average speed of 22.5 km/h? Example 6-14

77. • One of the fastest elevators in the world is found in the Taipei 101 building. It ascends at 16.83 m/s and rises 382.2 m. Each elevator has a maximum load capacity of 1600 kg. What power is required to raise a maximum load from the lowest level to the top floor? Example 6-14

78. • Neglecting extraneous factors like wind resistance, and given that the human body is only about 25% efficient, how much power must an 80.0-kg runner produce to run at a constant speed of 3.75 m/s up an 8.0° incline? Example 6-14

79. • A tower crane has a hoist motor rated at 167 hp. Assuming the crane is limited to using 70% of its maximum hoisting power for safety reasons, what is the shortest time in which the crane can lift a 5700-kg load over a distance of 85 m? Example 6-14

General Problems

80. • The data below represent a changing force that acts on an object in the x direction (the force is parallel to the displacement). Graph F versus x using a graphing calculator or a spreadsheet. Calculate the work done by the force (a) in the first 0.100 m, (b) in the first 0.200 m, (c) from 0.100 to 0.200 m, and (d) for the entire motion (for $0 < x < 0.250$ m).

x (m)	F (N)	x (m)	F (N)	x (m)	F (N)
0	0.00	0.0900	12.00	0.180	12.50
0.0100	2.00	0.100	12.48	0.190	12.50
0.0200	4.00	0.110	12.48	0.200	12.50
0.0300	6.00	0.120	12.48	0.210	12.48
0.0400	8.00	0.130	12.60	0.220	9.36
0.0500	10.00	0.140	12.60	0.230	6.24
0.0600	10.50	0.150	12.70	0.240	3.12
0.0700	11.00	0.160	12.70	0.250	0.00
0.0800	11.50	0.170	12.60		

81. • Starting from rest a 75.0-kg skier skis down a slope 8.00×10^2 m long that has an average incline of 40.0°. The speed of the skier at the bottom of the slope is 20.2 m/s. How much work was done by nonconservative forces? Example 6-5

82. • You push a 20.0-kg crate at constant velocity up a ramp inclined at an angle of 33.0° to the horizontal. The coefficient of kinetic friction between the ramp and the crate μ_k is equal to 0.200. How much work must you do to push the crate a distance of 2.00 m? Example 6-7

83. •• A 12.0-kg block (M) is released from rest on a frictionless incline that makes an angle of 28.0°, as shown in Figure 6-51. Below the block is a spring that has a spring constant of 13,500 N/m. The block momentarily stops when it compresses the spring by 5.50 cm. How far does the block move down the incline from its release point to the stopping point? Example 6-13

Figure 6-51 Problem 83

84. •• **Sports** A man on his luge (total mass of 88.0 kg) emerges onto a horizontal straight track at the bottom of a hill with a speed of 28.0 m/s. If the luge and rider slow at a constant rate of 2.80 m/s², what is the total work done on them by the force that slows them to a stop? Example 6-5

85. • **Biology** An adult dolphin is about 5.00 m long and weighs about 1600 N. How fast must he be moving as he leaves the water in order to jump to a height of 2.50 m? Ignore any effects due to air resistance. Example 6-10

86. • A 3.00-kg block is placed at the top of a track consisting of two frictionless quarter circles of radius $R = 2.00$ m connected by a 7.00-m-long, straight, horizontal surface (Figure 6-52). The coefficient of kinetic friction between the block and the horizontal surface is $\mu_k = 0.100$. The block is released from rest. What maximum vertical height does the block reach on the right-hand section of the track? Example 6-11

Figure 6-52 Problem 86

87. • An object is released from rest on a frictionless ramp of angle $\theta_1 = 60.0°$, at a (vertical) height $H_1 = 12.0$ m above the base of the ramp (Figure 6-53). The bottom end of the ramp *merges smoothly* with a second frictionless ramp that rises at angle $\theta_2 = 37.0°$. (a) How far along the second ramp does the object slide before coming to a momentary stop? (b) When the object is on its way back down the second ramp, what is its speed at the moment that it is a (vertical) height $H_2 = 7.00$ m above the base of the ramp? Example 6-10

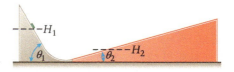

Figure 6-53 Problem 87

88. •• **Biology** An average froghopper insect has a mass of 12.3 mg and reaches a maximum height of 290 mm when its takeoff angle is 58.0° above the horizontal. What is the takeoff speed of the froghopper? Example 6-10

89. ••• A 20.0-g object is placed against the free end of a spring (k equal to 25.0 N/m) that is compressed 10.0 cm (**Figure 6-54**). Once released, the object slides 1.25 m across the tabletop and eventually lands 1.60 m from the edge of the table on the floor, as shown. Is there friction between the object and the tabletop? If there is, what is the coefficient of kinetic friction? The sliding distance on the tabletop includes the 10.0-cm compression of the spring, and the tabletop is 1.00 m above the floor level. Example 6-13

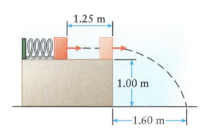

Figure 6-54 Problem 89

90. •• A 1.00-kg object is attached by a thread of negligible mass, which passes over a pulley of negligible mass, to a 2.00-kg object. The objects are positioned so that they are the same height from the floor and then released from rest. What are the speeds of the objects when they are separated vertically by 1.00 m? Example 6-10

91. •• Gravel-filled runaway truck lanes are designed to stop trucks that have lost their brakes on mountain grades. Typically such a lane is horizontal (if possible) and about 35.0 m long. We can think of the ground as exerting a frictional drag force on the truck. If a truck enters the gravel lane with a speed of 55.0 mi/h (24.6 m/s), use the work-energy theorem to find the minimum coefficient of kinetic friction between the truck and the lane to be able to stop the truck. Example 6-5

92. ••• In 2006, the United States produced 282×10^9 kilowatt-hours (kWh) of electrical energy from 4138 hydroelectric plants ($1.00 \text{ kWh} = 3.60 \times 10^6 \text{ J}$). On average each plant is 90% efficient at converting mechanical energy to electrical energy, and the average dam height is 50.0 m. (a) At 282×10^9 kWh of electrical energy produced in one year, what is the average power output per hydroelectric plant? (b) What total mass of water flowed over the dams during 2006?

(c) What was the average mass of water per dam and the average volume of water per dam that provided the mechanical energy to generate the electricity? (The density of water is 1000 kg/m^3.) (d) A gallon of gasoline contains $45.0 \times 10^6 \text{ J}$ of energy. How many gallons of gasoline did the 4138 dams save? Example 6-14

93. ••• **Astronomy** Violent gas eruptions have been observed on Mars, where the acceleration due to gravity is 3.7 m/s². The jets throw sand and dust about 75.0 m above the surface. (a) What is the speed of the material just as it leaves the surface? (b) Scientists estimate that the jets originate as high-pressure gas speeds through vents just underground at about 160 km/h. How much energy per kilogram of material is lost due to nonconservative forces as the high-speed matter forces its way to the surface and into the air? Example 6-12

94. • A 3.00-kg block is sent up a ramp of angle θ equal to 37.0° with an initial speed $v_0 = 20.0$ m/s. Between the block and the ramp, the coefficient of kinetic friction is $\mu_k = 0.50$, and the coefficient of static friction is $\mu_s = 0.80$. How far up the ramp (measured along the ramp) does the block go before it comes to a stop? Example 6-5

95. ••• A small block of mass M is placed halfway up on the inside of a vertical, frictionless, circular loop of radius R (**Figure 6-55**). The size of the block is very small compared to the radius of the loop. Determine an expression for the minimum downward speed with which the block must be released to guarantee that it will make a full circle. Example 6-10

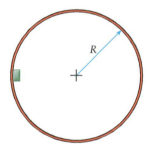

Figure 6-55 Problem 95

96. •• Niagara Falls in Canada has the highest flow rate of any waterfall in the world with a height of more than 50 m. The total average flow rate of the falls is 2400 m³/s and its average height is 51 m. Given that the density of water is 1000 kg/m³, calculate the average power output of Niagara Falls.

97. •• The Carmatech SAR12 pneumatic air rifle can accelerate a 3.0-g paintball from zero to 280 ft/s over a barrel length of 20 in. Calculate the average amount of power delivered to the paintball when fired.

7

Gravitation

What do you think?

The International Space Station (ISS) orbits Earth about 400 km (about 250 mi) above the surface of our planet, which has a radius of about 6370 km (3960 mi). Astronauts on the ISS feel as though they are weightless. Compared to the gravitational force that Earth exerts on an astronaut standing on our planet's surface, how great is the gravitational force that Earth exerts on the same astronaut when on the ISS? (a) The same; (b) about 10% less; (c) about 50% less; (d) about 75% less; (e) essentially zero.

In this chapter, your goals are to:

- (7-1) Identify what it means to say that gravitation is universal.
- (7-2) Explain how Newton's law of universal gravitation describes the attractive gravitational force between any two objects.
- (7-3) Describe the general expression for gravitational potential energy and how to relate it to the expression used near Earth's surface.
- (7-4) Apply the law of universal gravitation and the expression for gravitational potential energy to analyze the orbits of satellites and planets.
- (7-5) Explain the origin of ocean tides and apparent weightlessness.

To master this chapter, you should review:

- (3-2, 3-3) How to add vectors using components.
- (3-7) The properties of uniform circular motion and how the inner ear senses acceleration.
- (4-3) The nature of the gravitational force near Earth's surface.
- (5-6) Newton's laws applied to uniform circular motion.
- (6-3, 6-6, and 6-7) The ideas of kinetic energy, potential energy, and the conservation of total mechanical energy.

7-1 Gravitation is a force of universal importance

Eighty years ago the idea of humans orbiting Earth or sending spacecraft to other worlds was regarded as science fiction. Today science fiction has become common-place reality. Humans live and work in Earth orbit aboard the International Space Station (see photograph above), have ventured as far as the Moon, and have sent robotic spacecraft to explore all the planets of the solar system.

While we think of spaceflight as an innovation of the twentieth century, we can trace its origins to Isaac Newton's revolutionary seventeenth-century notion of *universal gravitation*—the idea that all objects in the universe attract each other through the force of gravity. In Chapters 3 and 4 we learned that Earth's gravitational force acts on all objects near our planet's surface. In this chapter we'll extend

(a) Earth

(b) Saturn, its rings, and three of its moons

(c) Galaxy NGC 6744

Figure 7-1 Gravitation is universal The force of gravitation is responsible for (a) keeping our Earth from flying apart, (b) keeping moons and rings in orbit around their planets, and (c) holding galaxies — among the largest structures in the universe — together.

Earth is held together by the mutual gravitational attraction of all its parts.

Saturn's rings are made of countless small objects that, like Saturn's moons, are held in orbit by the planet's gravitation.

This galaxy (a near-twin of our Milky Way) contains more than 10^{11} stars. The gravitational attraction of all parts of the galaxy for each other holds the galaxy together.

our discussion to consider the gravitational forces exerted by any massive object on any other massive object. Gravitational forces between the components of our planet are partly responsible for holding Earth together (Figure 7-1a), gravitational forces exerted by Saturn keep its moons and other small bodies in orbit (Figure 7-1b), and gravitational forces between stars and the material between the stars hold entire galaxies of stars together (Figure 7-1c).

In this chapter we will learn about the properties of the gravitational force, including how Newton deduced that this force must grow weaker as objects move farther apart. We will see how to treat the gravitational potential energy of a system of two objects interacting with each other, such as a satellite and Earth. We'll then use these ideas about gravitational force and gravitational potential energy to understand the nature of orbits. We'll find that the same ideas that apply to a satellite orbiting Earth also apply to planets orbiting the Sun. Indeed, we will see how Newton's idea of universal gravitation allowed him to explain properties of planetary motion that had been discovered, but not understood, decades before. We'll also use the idea of universal gravitation to explain the origin of ocean tides and to understand why astronauts in orbit feel weightless.

TAKE-HOME MESSAGE FOR Section 7-1

✔ Isaac Newton deduced that gravitation is universal — it is a force that acts between any two massive objects in the universe.

✔ The ideas of universal gravitation will help us understand the motions of satellites around Earth and of planets around the Sun.

7-2 Newton's law of universal gravitation explains the orbit of the Moon

Newton deduced much of what we now know about gravitation by considering the motion of the Moon (Figure 7-2). It had been understood for centuries that the Moon orbits around Earth. Newton knew from his laws of motion that there must therefore be a force acting on the Moon to keep it from flying off into space, and that Earth must exert this force on the Moon. This is just the force of Earth's gravitation. But how strong is that force? Is this force the same, stronger, or weaker than it would be if the Moon were closer? Let's see how Newton answered these questions by using the ideas of circular motion.

Newton knew that the Moon orbits around Earth at a nearly constant speed in a nearly circular path. The evidence for this is that the Moon changes its position against the background stars. It moves against this background at a nearly constant rate (which tells us that its speed is nearly constant), and as it moves its apparent size changes very little (which tells us that the Moon maintains a roughly constant distance from Earth, and so its orbit is nearly circular).

Figure 7-2 The Moon Isaac Newton used the known distance to the Moon and the Moon's speed to deduce its acceleration, which he concluded was caused by Earth's gravitational attraction. This gave him an important clue about how the gravitational force depends on distance.

As we learned in Section 3-7, an object that travels at constant speed v around a circular path of radius r—that is, an object in *uniform circular motion*—has an acceleration toward the center of the circle of magnitude.

Centripetal acceleration for motion in a circle (3-17)

Centripetal acceleration: magnitude of the acceleration of an object in uniform circular motion

Speed of an object as it moves around the circle

$$a_{\text{cent}} = \frac{v^2}{r}$$

Radius of the circle

Let's apply Equation 3-17 to the Moon. The Moon's acceleration is due to the gravitational pull of Earth on the Moon, so a_{cent} is equal to the value of g—the acceleration due to gravity—at the position of the Moon. Equation 3-17 shows that to determine the Moon's acceleration Newton had to know both the values of the radius r of the Moon's orbit—that is, the distance from the center of Earth to the Moon—and the speed v of the Moon in its orbit.

The value of the radius r of the Moon's orbit was well known in Newton's time, even though travel to the Moon was centuries in the future. This was possible because as seen from different locations on Earth, the Moon appears to be in slightly different positions relative to the background of stars (Figure 7-3). By measuring the differences in apparent position and using trigonometry, it's possible to determine the distance D in Figure 7-3. From these measurements, the distance from the center of Earth to the Moon (equal to Earth's radius $R = 6370$ km, which was also known in Newton's time, plus the distance D in Figure 7-3 from the surface of Earth to the Moon) turns out to be $384{,}000$ km $= 3.84 \times 10^8$ m. This is the radius r of the Moon's orbit.

Newton also knew the value of the Moon's speed v in its orbit. From observations of the Moon's apparent motion relative to the stars, it was known in Newton's time that the Moon makes one complete orbit around Earth in a time $T = 27.3$ days, or 2.36×10^6 s. The speed v of the Moon in its orbit is therefore the circumference of its orbit, equal to 2π times the radius $r = 3.84 \times 10^8$ m of its orbit, divided by the time $T = 2.36 \times 10^6$ s to complete an orbit. The result is $v = 1.02 \times 10^3$ m/s (about 3670 km/h or 2280 mi/h).

If you substitute these values for v and r into Equation 3-17, you'll find that the Moon's acceleration is, to two significant digits,

$$a_{\text{cent}} = \frac{v^2}{r} = \frac{(1.02 \times 10^3 \text{ m/s})^2}{3.84 \times 10^8 \text{ m}} = 2.7 \times 10^{-3} \text{ m/s}^2$$

Note that a_{cent} is *very* much less than the value $g = 9.80$ m/s^2 at the surface of Earth. This implies that Earth's gravitational attraction—the force that acts on the Moon to give it its centripetal acceleration—gets weaker with increasing distance. Newton

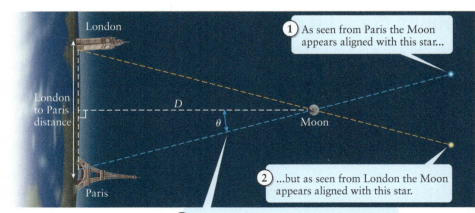

Figure 7-3 Measuring the distance to the Moon The Moon is close enough that it appears to be in slightly different positions against the backdrop of distant stars when viewed from different locations on Earth. This made it possible to calculate the distance to the Moon many centuries before humans were able to send a spacecraft there.

London

1 As seen from Paris the Moon appears aligned with this star...

London to Paris distance

D

θ

Moon

Paris

2 ...but as seen from London the Moon appears aligned with this star.

3 Given the distance from London to Paris and the angle θ, the distance D from Earth's surface to the Moon can be calculated.

observed that the Moon's acceleration is less than the value of g at Earth's surface by a factor of

$$\frac{a_{\text{cent,Moon}}}{g_{\text{surface of Earth}}} = \frac{2.7 \times 10^{-3} \text{ m/s}^2}{9.80 \text{ m/s}^2} = 2.8 \times 10^{-4} = \frac{1}{3600} = \frac{1}{60^2}$$

He also noted that the distance from the center of Earth to Earth's surface is smaller than the distance from the center of Earth to the Moon by a factor of

$$\frac{r_{\text{Earth center to surface}}}{r_{\text{Earth center to Moon}}} = \frac{6370 \text{ km}}{384,000 \text{ km}} = \frac{1}{60}$$

From this Newton deduced that

the acceleration due to Earth's gravity decreases in proportion to the square of the distance from Earth's center.

If you could stand atop a tower 6370 km tall—that is, a tower whose height equals the radius of Earth—you would be twice as far from Earth's center as a person at sea level, and the acceleration due to gravity would be $1/(2)^2 = 1/4$ as great as at sea level. By contrast if you were to stand atop Earth's tallest mountain, Mount Everest, you would be a little less than 9 km above sea level. The distance from Earth's center to the peak of Mount Everest is 6379 km as compared to 6370 km from Earth's center to sea level, which is greater by a factor of $(6379 \text{ km})/(6370 \text{ km}) = 1.0014$. So the value of g atop Mount Everest is about $1/(1.0014)^2 = 0.997$ of the value at sea level, or about 0.3% less than the sea-level value. This shows why there is only a small difference in the value of g at different elevations on Earth.

The Law of Universal Gravitation

Let's see how Newton deduced a general equation for the gravitational force that one object exerts on another. We'll begin with Newton's observation that the value of the acceleration g due to Earth's gravitation decreases in proportion to the square of the distance r from the center of Earth. In equation form, this says

$$g = \frac{(\text{a constant})}{r^2} \tag{7-1}$$

The value of the constant in this equation is chosen so that at a point on Earth's surface, where r equals Earth's radius of 6370 km, the value of g equals 9.80 m/s². We learned in Section 4-3 that the magnitude of Earth's gravitational force on an object of mass m is mg. So it follows from Equation 7-1 that if an object of mass m is located a distance r from Earth's center, the gravitational force of Earth on this object is

$$F_{\text{Earth on object}} = \frac{(\text{a constant})m}{r^2}$$

That is, Earth's gravitational force on an object of mass m is directly proportional to the mass m of the object that experiences the force.

Newton generalized this to say that *any* object of mass m_1 exerts a gravitational force on a second object of mass m_2, that this force is directly proportional to the mass m_2 of the object that experiences the force, and that this force is inversely proportional to the square of the distance r between the centers of the objects (Figure 7-4a):

$$F_{1 \text{ on } 2} = \frac{(\text{a constant})m_2}{r^2} \tag{7-2}$$

For the force of Earth on the Moon, m_2 is the Moon's mass, and r is the Earth–Moon distance. If Equation 7-2 is true for *any* two objects, then it must also be true that the object of mass m_2 exerts a gravitational force on the object of mass m_1, and this force is directly proportional to the mass m_1. But Newton's third law (Section 4-5) tells us that the forces that the two objects exert on each other have opposite directions and the same magnitude: $F_{2 \text{ on } 1} = F_{1 \text{ on } 2}$. Hence the gravitational force that each object

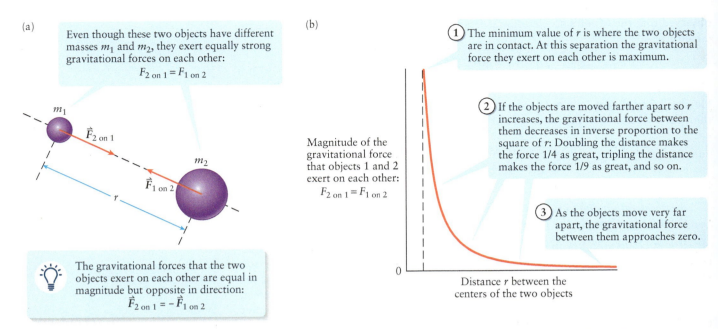

Figure 7-4 The law of universal gravitation (a) Any two objects exert gravitational forces on each other. These forces are equal in magnitude but opposite in direction: $\vec{F}_{2 \text{ on } 1} = -\vec{F}_{1 \text{ on } 2}$. (b) The magnitude of the gravitational force is inversely proportional to the square of the center-to-center distance r.

exerts on the other must be directly proportional to *both* m_1 and m_2. This chain of reasoning leads us to **Newton's law of universal gravitation:**

Newton's law of universal gravitation

(7-3)

Gravitational constant (same for any two objects)   Masses of the two objects

Any two objects (1 and 2) exert equally strong gravitational forces on each other.

$$F_{1 \text{ on } 2} = F_{2 \text{ on } 1} = \frac{G m_1 m_2}{r^2}$$

Center-to-center distance between the two objects

The gravitational forces are attractive: $\vec{F}_{1 \text{ on } 2}$ pulls object 2 toward object 1 and $\vec{F}_{2 \text{ on } 1}$ pulls object 1 toward object 2.

As Figure 7-4a shows, the forces that the two objects exert on each other are attractive and are directed along the line that connects the centers of the two objects.

The proportionality constant G in Equation 7-3 is called the **gravitational constant.** We'll see later in this section how the value of G (which Newton did not know) is determined. Its currently accepted value, to three significant figures, is

$$G = 6.67 \times 10^{-11} \text{ N} \cdot \text{m}^2/\text{kg}^2$$

Figure 7-4b graphs the magnitude of the gravitational force between the two objects as a function of the distance r between their centers.

WATCH OUT! Be sure to understand the limitations of Equation 7-3.

Strictly speaking, Equation 7-3 applies to two infinitesimally small objects with masses m_1 and m_2. It's much more challenging to precisely calculate the gravitational force between two extended objects with complicated shapes, such as between Earth and the Moon (neither of which is a perfect sphere). However, in practice we can safely use Equation 7-3 if the *distance* between the two objects is large compared to the *size* of either object. That's the case for Earth and the Moon, which are separated by a distance of 384,000 km—much larger than the radius of either Earth (6370 km) or the Moon (1740 km).

It's also safe to use Equation 7-3 if one object is very much smaller than the other, as is the case for a satellite orbiting Earth, as long as we can approximate the large object (in this case Earth) as being roughly spherically symmetric.

Note that Equation 7-3 is *not* valid if one object is *inside* the other. For example, we can't use this equation to find the gravitational force on you when you're in a tunnel deep inside Earth's interior.

WATCH OUT! The distance *r* in Equation 7-3 is the center-to-center distance.

Always remember that *r* in Equation 7-3 is the distance between the *centers* of the two objects. If you're calculating the gravitational force that Earth exerts on a person standing on Earth's surface, the distance *r* equals the radius of Earth—that is, the distance from the center of Earth to that person.

Equation 7-3 shows that the gravitational force that one object exerts on another is proportional to $1/r^2$, which is the reciprocal of the square of the distance *r* or "the inverse square"

of *r* for short. That's why the law of universal gravitation is sometimes called *the inverse-square law for gravitation*. We'll learn in later chapters that the *electric* force that one charged particle exerts on another (for instance, the force between an electron and a proton) also obeys an inverse-square law.

Because the gravitational constant has such a small value, gravitational forces between two objects also tend to be pretty small unless at least one of the objects is relatively massive (that is, planet-sized). The following examples illustrate this.

EXAMPLE 7-1 You and Your Backpack

While you sit studying, what is the magnitude of the gravitational force that your 10.0-kg backpack sitting on your desk exerts on you? Assume that your mass is 70.0 kg and that the center of your body is 0.60 m from the center of your backpack.

Set Up

We'll use Newton's law of universal gravitation, Equation 7-3, to find the magnitude of the gravitational force.

Newton's law of universal gravitation:

$$F_{\text{bag on you}} = \frac{Gm_{\text{bag}}m_{\text{you}}}{r^2} \qquad (7\text{-}3)$$

Solve

Substitute the given values of G, $m_{\text{bag}} = 10.0$ kg, $m_{\text{you}} = 70.0$ kg, and $r = 0.60$ m into the expression for gravitational force. The distance is given to only two significant figures, so our result has only two significant figures.

$$
\begin{aligned}
F_{\text{bag on you}} &= \frac{Gm_{\text{bag}}m_{\text{you}}}{r^2} \\
&= \frac{(6.67 \times 10^{-11}\ \text{N} \cdot \text{m}^2/\text{kg}^2)(10.0\ \text{kg})(70.0\ \text{kg})}{(0.60\ \text{m})^2} \\
&= 1.3 \times 10^{-7}\ \text{N}
\end{aligned}
$$

Reflect

The gravitational force your bag exerts on you is *very* small, about one ten-millionth of a newton. This is equivalent to the weight of a few specks of dust. This force is far smaller than the gravitational force Earth exerts on you (see the next example). It's also far smaller than the force of friction between you and your chair should you start to slide. That's why we generally neglect the gravitational force between everyday objects.

EXAMPLE 7-2 Earth's Gravitational Force on You

Calculate the magnitude of the gravitational force that Earth exerts on you, again assuming you have a mass of 70.0 kg. Earth has mass 5.97×10^{24} kg and radius 6370 km.

Set Up

Again we'll find the magnitude of the gravitational force using Newton's law of universal gravitation, Equation 7-3. The distance *r* between the centers of the two objects (Earth and you) equals Earth's radius, 6370 km. The extra meter or so distance from Earth's surface to the center of your body is insignificant compared to the size of Earth's radius, so we ignore it.

Newton's law of universal gravitation:

$$F_{\text{Earth on you}} = \frac{Gm_{\text{Earth}}m_{\text{you}}}{r^2} \qquad (7\text{-}3)$$

r = distance from center of Earth to you
  $= R_{\text{Earth}}$ = radius of Earth

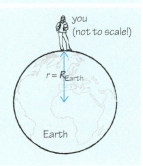

Solve

To use the distance $r = R_{Earth} = 6370$ km in Equation 7-3, we first convert it to meters.

$$r = R_{Earth}$$

$$= (6370 \text{ km}) \times \frac{1000 \text{ m}}{1 \text{ km}} = 6.37 \times 10^6 \text{ m}$$

Substitute the values of G, m_{Earth}, m_{you}, and r into Equation 7-3.

$$F_{Earth \text{ on you}} = \frac{Gm_{Earth}m_{you}}{r^2} = \frac{Gm_{Earth}m_{you}}{R_{Earth}^2}$$

$$= \frac{(6.67 \times 10^{-11} \text{N} \cdot \text{m}^2/\text{kg}^2)(5.97 \times 10^{24} \text{ kg})(70.0 \text{ kg})}{(6.37 \times 10^6 \text{ m})^2}$$

$$= 687 \text{ N} = 690 \text{ N to two significant figures}$$

Reflect

We can check our result by using the expression $w = mg$ for gravitational force that we introduced in Section 4-3. We get the same result as above to two significant figures, which shows that the values of G and m_{Earth} are compatible with the value of g at Earth's surface.

Using the expression $w = mg$ for gravitational force,

$$F_{Earth \text{ on you}} = m_{you}g$$
$$= (70.0 \text{ kg})(9.80 \text{ m/s}^2)$$
$$= 686 \text{ N}$$
$$= 690 \text{ N to two significant figures}$$

Finding the Value of *G* and the Mass of Earth

Example 7-2 shows that there is a connection between the value of g at Earth's surface and the values of G and m_{Earth}, the gravitational constant and Earth's mass. To see this more clearly, take the two expressions for the gravitational force on you and set them equal to each other:

$$F_{Earth \text{ on you}} = m_{you}g = \frac{Gm_{Earth}m_{you}}{R_{Earth}^2}$$

Then divide both sides of the equation by your mass m_{you} to get an expression for g:

Value of *g* at Earth's surface
(7-4)

Gravitational constant   Mass of Earth

Acceleration due to gravity at Earth's surface

$$g = \frac{Gm_{Earth}}{R_{Earth}^2}$$

Radius of Earth

This equation tells us that whenever we calculate a weight or an acceleration using g, we are simply using a shorthand form of the law of universal gravitation applied near Earth's surface. So the familiar expression $w = mg$ is just a special form of the general law of universal gravitation, Equation 7-3.

Equation 7-4 relates the value of g (which we can measure by observing a freely falling object) to the values of Earth's radius R_{Earth}, Earth's mass m_{Earth}, and the gravitational constant G. The value of R_{Earth} has been known since the third century B.C., when Eratosthenes, a scholar from Cyrene in modern-day Libya, first determined its value. (Figure 7-5 shows how Eratosthenes was able to do this.) But that leaves two unknown values in Equation 7-4, G and m_{Earth}. With only one equation to relate them, it seems impossible to determine the values of both of these unknowns.

The solution to this problem was provided by the British scientist Henry Cavendish in 1798, more than a century after Newton published the law of universal gravitation. Cavendish conducted the elegant experiment shown schematically in Figure 7-6, now known simply as the **Cavendish experiment**. The gravitational force of each large sphere on the nearby small sphere makes the wooden rod rotate, which twists the wire from which the rod is suspended. By measuring the wire before including the large spheres, Cavendish knew how much force on each small sphere was required to twist the wire through a given rotation angle. By measuring the amount of twist in the experiment shown in Figure 7-6, he could then determine the gravitational force

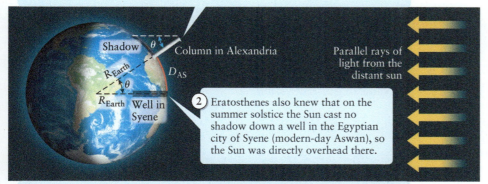

① As the story is told, on the day of the summer solstice (when the Sun is highest in the sky as seen from the northern hemisphere), Eratosthenes observed the shadow cast by a column in the city of Alexandria.

② Eratosthenes also knew that on the summer solstice the Sun cast no shadow down a well in the Egyptian city of Syene (modern-day Aswan), so the Sun was directly overhead there.

③ Given the known distance D_{AS} from Alexandria to Syene and his measurement of the angle θ, Eratosthenes was able to calculate the radius R_{Earth}.

Figure 7-5 How Eratosthenes measured the radius of Earth Around 240 B.C. the scholar Eratosthenes measured Earth's radius by analyzing the shadows cast by the Sun at two different locations.

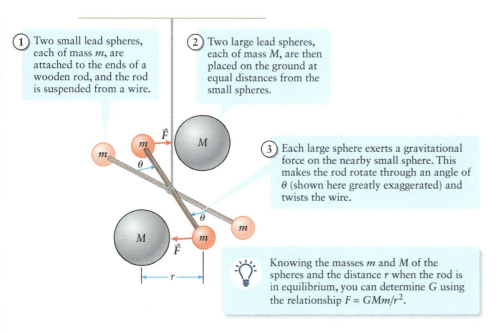

① Two small lead spheres, each of mass m, are attached to the ends of a wooden rod, and the rod is suspended from a wire.

② Two large lead spheres, each of mass M, are then placed on the ground at equal distances from the small spheres.

③ Each large sphere exerts a gravitational force on the nearby small sphere. This makes the rod rotate through an angle of θ (shown here greatly exaggerated) and twists the wire.

Knowing the masses m and M of the spheres and the distance r when the rod is in equilibrium, you can determine G using the relationship $F = GMm/r^2$.

Figure 7-6 The Cavendish experiment This experiment is used to determine the value of the gravitational constant G. In the original (1798) version of the experiment, the wooden rod was 1.8 m in length, each small lead sphere had a mass of 0.73 kg, each large lead sphere had a mass of 158 kg, and the distance r was about 23 cm.

of each large sphere on the neighboring small sphere. With this technique Cavendish could measure extremely small forces on the order of 10^{-7} N, which made his experiment possible.

Given the magnitude of the force F, the masses m and M of the small and large spheres, respectively, and the distance r between neighboring large and small spheres, we can use Newton's law of universal gravitation to solve for the value of G:

$$F = \frac{GMm}{r^2} \quad \text{so} \quad G = \frac{Fr^2}{Mm}$$

Modern experiments to measure G to great precision still use variations on Cavendish's original apparatus. Once the value of G is known, we can use Equation 7-4 to determine the value of Earth's mass m_{Earth} (now known to be 5.97×10^{24} kg).

Armed with knowledge of the value of G, we have all the tools we need to apply the law of universal gravitation to the orbits of satellites and planets. To analyze these orbits it's important to first consider the potential energy associated with the gravitational force, which is the topic of the next section. First, however, here are two examples that make use of the law of universal gravitation.

Go to Picture It 7-1 for more practice dealing with gravitational force.

EXAMPLE 7-3 How Does *g* Vary with Altitude?

(a) For a satellite at an altitude *h* above Earth's surface, how does the value of *g* compare to the value at the surface given by Equation 7-4? (b) Find the value of *g* for a satellite at *h* = 400 km (the altitude of the International Space Station shown in the photograph that opens this chapter) and at *h* = 2.02×10^4 km (the altitude of the satellites used in the Global Positioning System, or GPS).

Set Up

Equation 7-4 tells us the value of *g* at Earth's surface, which we'll call $g_{surface}$. Our goal is to find the value of *g* at a height *h* above the surface. We'll use the law of universal gravitation and the relationship between the gravitational force on an object and the value of *g*. Remember that the distance *r* in the law of universal gravitation is *not* the height above the surface but rather the distance to Earth's center. This equals the radius of Earth plus the height of the object.

Value of *g* at Earth's surface:

$$g_{surface} = \frac{Gm_{Earth}}{R_{Earth}^2} \tag{7-4}$$

Newton's law of universal gravitation:

$$F_{Earth\ on\ object} = \frac{Gm_{Earth}m_{object}}{r^2} \tag{7-3}$$

Gravitational force on an object in terms of *g*:

$$F_{Earth\ on\ object} = m_{object}g$$

Distance from Earth's center to a height *h* above the surface:

$$r = R_{Earth} + h$$

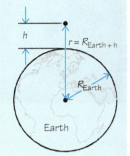

Solve

(a) Use the two expressions for the gravitational force on an object to solve for *g* at a height *h* above the surface. Our result shows that the greater the value of *h*, the smaller the value of *g*.

Set the two expressions for $F_{Earth\ on\ object}$ equal to each other:

$$F_{Earth\ on\ object} = m_{object}g = \frac{Gm_{Earth}m_{object}}{r^2}$$

To solve for *g* divide through by m_{object}. Then substitute $r = R_{Earth} + h$ in the expression for *g*:

$$g = \frac{Gm_{Earth}}{r^2} = \frac{Gm_{Earth}}{(R_{Earth} + h)^2}$$

To compare this to $g_{surface}$, calculate the ratio of *g* to $g_{surface}$:

$$\frac{g}{g_{surface}} = \frac{Gm_{Earth}/(R_{Earth} + h)^2}{Gm_{Earth}/R_{Earth}^2} = \frac{R_{Earth}^2}{(R_{Earth} + h)^2}$$

(Note that Gm_{Earth} cancels.) We can rewrite this as

$$g = g_{surface} \frac{R_{Earth}^2}{(R_{Earth} + h)^2}$$

(b) Solve for the values of *g* at *h* = 400 km and *h* = 2.02×10^4 km.

Use $g_{surface} = 9.80$ m/s² and $R_{Earth} = 6370$ km:

At altitude *h* = 400 km,

$$g = g_{surface} \frac{R_{Earth}^2}{(R_{Earth} + h)^2} = g_{surface} \frac{(6370\ km)^2}{(6370\ km + 400\ km)^2}$$

$$= 0.885 g_{surface} = 8.68\ m/s^2$$

At altitude *h* = 2.02×10^4 km,

$$g = g_{surface} \frac{R_{Earth}^2}{(R_{Earth} + h)^2} = g_{surface} \frac{(6370\ km)^2}{(6370\ km + 2.02 \times 10^4\ km)^2}$$

$$= 0.0575 g_{surface} = 0.563\ m/s^2$$

Reflect

Our calculation shows that $g = Gm_{Earth}/r^2$. In words this says that g, like the gravitational force, is inversely proportional to the square of the distance r from Earth's center. At the relatively low 400-km altitude of the International Space Station, $r = 6370$ km $+ 400$ km $= 6770$ km isn't much larger than Earth's radius $R_{Earth} = 6370$ km, so g is still 88.5% as great as at Earth's surface. But at the much higher 2.02×10^4-km altitude of a GPS satellite, $r = 6370$ km $+ 2.02 \times 10^4$ km $= 2.66 \times 10^4$ km, which is 4.17 times $R_{Earth} = 6370$ km. That's why at this altitude, g is only 5.75% as great as at Earth's surface.

EXAMPLE 7-4   Finding *g* on Other Worlds

Figure 7-7 shows Earth, Mars, and the Moon to scale. Find the value of the acceleration due to gravity g at the surface of Mars and at the surface of the Moon, and compare to the value of g at Earth's surface. The table below lists the masses and radii of these three worlds.

	Mass (kg)	Radius (m)
Earth	5.97×10^{24}	6.37×10^6
Mars	6.42×10^{23}	3.40×10^6
Moon	7.35×10^{22}	1.74×10^6

Figure 7-7 **Three worlds to scale** How does the value of g differ on the surfaces of Earth, Mars, and the Moon?

Set Up

Equation 7-4 tells us the value of g at the surface of Earth. But this same equation also tells us the value of g at the surface of *any* world (planet or moon) if we simply replace the mass and radius of Earth with that world's mass and radius. We'll use this along with the given data for Mars and the Moon to calculate g on these worlds' surfaces. We'll then compare these values to $g_{surface of Earth} = 9.80$ m/s².

Value of g at Earth's surface:

$$g_{surface\ of\ Earth} = \frac{Gm_{Earth}}{R_{Earth}^2} \qquad (7\text{-}4)$$

So the value of g at the surface of any world of mass m_{world} and radius R_{world} is

$$g_{surface\ of\ world} = \frac{Gm_{world}}{R_{world}^2}$$

Solve

Calculate the value of g on each world.

For Mars,

$$g_{surface\ of\ Mars} = \frac{Gm_{Mars}}{R_{Mars}^2}$$
$$= \frac{(6.67 \times 10^{-11}\ N\cdot m^2/kg^2)(6.42 \times 10^{23}\ kg)}{(3.40 \times 10^6\ m)^2}$$
$$= 3.70\ N/kg = 3.70\ m/s^2$$

(Recall that $1\ N = 1\ kg\cdot m/s^2$, so $1\ N/kg = 1\ m/s^2$.)
For the Moon,

$$g_{surface\ of\ Moon} = \frac{Gm_{Moon}}{R_{Moon}^2}$$
$$= \frac{(6.67 \times 10^{-11}\ N\cdot m^2/kg^2)(7.35 \times 10^{22}\ kg)}{(1.74 \times 10^6\ m)^2}$$
$$= 1.62\ N/kg = 1.62\ m/s^2$$

To compare each value of g to the value on Earth's surface, calculate the ratio $g_{\text{surface of world}}/g_{\text{surface of Earth}}$.

Compare the values of g on Mars and on the Moon to the value on Earth:

$$\frac{g_{\text{surface of Mars}}}{g_{\text{surface of Earth}}} = \frac{3.70 \text{ m/s}^2}{9.80 \text{ m/s}^2} = 0.378$$

$$\frac{g_{\text{surface of Moon}}}{g_{\text{surface of Earth}}} = \frac{1.62 \text{ m/s}^2}{9.80 \text{ m/s}^2} = 0.165$$

Reflect

The equation $g_{\text{surface of world}} = Gm_{\text{world}}/R_{\text{world}}^2$ shows that the value of g at the surface of a world (planet or moon) depends on both its mass and its radius. Decreasing the mass by itself makes g smaller (since m_{world} is in the numerator), while decreasing the radius by itself makes g larger (since R_{world} is in the denominator). From Earth to Mars to the Moon, both m_{world} and R_{world} decrease, so it might not be obvious which factor is more important. But our results show that g decreases as we go from Earth to Mars (where g is only 0.378 as great as on Earth) to the Moon (where g is only 0.165 as great as on Earth). That's because as we go from larger to smaller worlds, the mass decreases much more rapidly than the radius: For example, the Moon has only 0.0123 as much mass as Earth, but has 0.273 the radius. Generally speaking, the smaller the size of a planet or moon, the smaller the value of g at its surface.

Your weight w when standing on the surface of any world is the product of your mass m and the value of g on that world's surface: $w_{\text{surface of world}} = mg_{\text{surface of world}}$. So our results show that compared to your weight on Earth, your weight on Mars would be 0.378 times as great and your weight on the Moon would be just 0.165 times as great. Your *mass* would be the same, however, since your body would still contain the same amount of material.

GOT THE CONCEPT? 7-1 Gravitational Force

Compared to the gravitational force that Earth exerts on a 50-kg object located three Earth radii above the north pole, the gravitational force that Earth exerts on a 50-kg object located at the north pole is (a) the same; (b) 3 times greater; (c) 4 times greater; (d) 9 times greater; (e) 16 times greater.

GOT THE CONCEPT? 7-2 Gravitational Acceleration

Compared to the acceleration due to Earth's gravity at a point three Earth radii above the north pole, the acceleration due to Earth's gravity at the north pole is (a) the same; (b) 3 times greater; (c) 4 times greater; (d) 9 times greater; (e) 16 times greater.

TAKE-HOME MESSAGE FOR Section 7-2

✔ Newton's law of universal gravitation describes the gravitational force that any object exerts on another object. It states that the force is directly proportional to the product of the masses of two objects and inversely proportional to the square of the distance between their centers.

✔ The relationship we use for weight near Earth's surface, $w = mg$, is simply an application of the law of universal gravitation.

7-3 The gravitational potential energy of two objects is negative and increases toward zero as the objects are moved farther apart

In Chapter 6 we introduced the ideas of kinetic energy (Section 6-3) and potential energy (Section 6-6). We found that if a force is *conservative*—so that the work done by that force on an object as it moves depends only on the object's starting point and ending point, not on the path the object takes between them—then we can associate a potential energy with that force. If only conservative forces do work on an object as it moves, the total mechanical energy—the sum of the kinetic energy and potential energy of the system—always keeps the same value and is *conserved* (Section 6-7).

One particularly important conservative force is the gravitational force on an object of mass m near Earth's surface, $w = mg$. We found that the potential energy associated with this force is $U_{grav} = mgy$, where y is the height of the object above some level that we choose to call $y = 0$. But because this simple expression for U_{grav} is based on the expression $w = mg$ for gravitational force, it *cannot* be correct for the more general form of the gravitational force between two objects $F = Gm_1m_2/r^2$. What, then, is the correct expression for the gravitational potential energy of two objects of mass m_1 and m_2 separated by a distance r?

Deriving this expression from the law of universal gravitation is a task that requires calculus, which is beyond our scope. Instead we'll just present the answer, then examine it to see why it makes sense:

Gravitational constant (same for any two objects) | Masses of the two objects

Gravitational potential energy of a system of two objects (1 and 2)

$$U_{grav} = -\frac{Gm_1m_2}{r}$$

Center-to-center distance between the two objects

Gravitational potential energy
(7-5)

The gravitational potential energy is zero when the two objects are infinitely far apart. If the objects are brought closer together (so r is made smaller), U_{grav} decreases (it becomes more negative).

The expression in Equation 7-5 looks nothing like the familiar formula $U_{grav} = mgy$. To understand the difference let's consider two different cases in which gravitational potential energy changes (Figure 7-8).

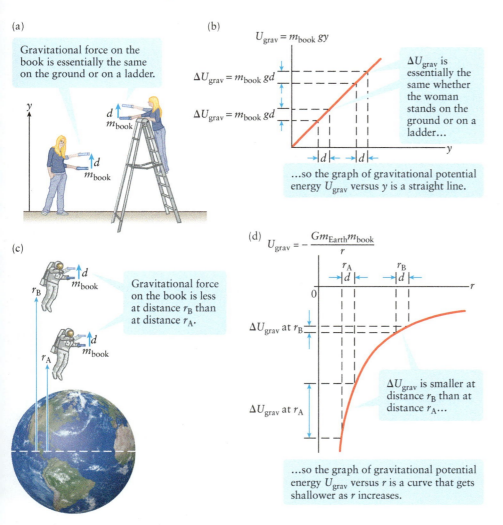

(a) Gravitational force on the book is essentially the same on the ground or on a ladder.

(b) $U_{grav} = m_{book}gy$

$\Delta U_{grav} = m_{book}gd$

$\Delta U_{grav} = m_{book}gd$

ΔU_{grav} is essentially the same whether the woman stands on the ground or on a ladder...

...so the graph of gravitational potential energy U_{grav} versus y is a straight line.

(c) Gravitational force on the book is less at distance r_B than at distance r_A.

(d) $U_{grav} = -\dfrac{Gm_{Earth}m_{book}}{r}$

ΔU_{grav} at r_B

ΔU_{grav} at r_A

ΔU_{grav} is smaller at distance r_B than at distance r_A...

...so the graph of gravitational potential energy U_{grav} versus r is a curve that gets shallower as r increases.

Figure 7-8 Gravitational potential energy near Earth's surface and far away (a) A woman lifts a book of mass m_{book} a distance d while standing on the ground. She then repeats the process while standing on a ladder. (b) The gravitational force and the change in gravitational potential energy are basically the same in both cases. (c) The woman again raises the book a distance d, first at distance r_A from Earth's center, then at distance r_B. (d) The gravitational force and the change in gravitational potential energy are less at distance r_B than at distance r_A.

In Figure 7-8a, a woman on Earth's surface raises a book of mass m_{book} a distance d. The downward gravitational force on the book has magnitude $m_{book}g$, so she must exert an upward force of the same magnitude. She does an amount of work $m_{book}gd$ in the process. The book begins and ends at rest, so the work $m_{book}gd$ that she does goes into increasing the gravitational potential energy associated with the book: $\Delta U_{grav} = m_{book}gd$. Earth's gravitational force changes hardly at all near the surface, so if the woman stands on a ladder and once again lifts the book by a distance d, the gravitational force she works against is still $m_{book}g$. She again does work $m_{book}gd$ to lift the book a distance d, and the gravitational potential energy increases by the same amount $m_{book}gd$ as before. It follows that the gravitational potential energy increases in direct proportion to the object's height y, so the graph of U_{grav} versus height is a straight line: $U_{grav} = m_{book}gy$ (Figure 7-8b).

Now the same woman dons a spacesuit and brings the book with her to a point in space a distance r_A from the center of Earth (Figure 7-8c). At this point the gravitational force that Earth exerts on the book has magnitude $F_A = Gm_{Earth}m_{book}/r_A^2$. If she moves the book an additional distance d away from Earth, she has to exert a force of the same magnitude on the book and so does an amount of work $F_A d = (Gm_{Earth}m_{book}/r_A^2)d$. Just as on Earth, this work goes into increasing the gravitational potential energy. If she now moves to a greater distance r_B from the center of Earth and repeats the process, Earth's gravitational force has a smaller magnitude $F_B = Gm_{Earth}m_{book}/r_B^2$. So at this greater distance she does a smaller amount of work $F_B d = (Gm_{Earth}m_{book}/r_B^2)d$ and so adds a smaller amount of gravitational potential energy. This means that at great distances the gravitational potential energy does *not* increase in direct proportion to the distance r. The graph of U_{grav} versus distance is not a straight line but a curve that becomes shallower with increasing values of r (Figure 7-8d). This is just what Equation 7-5 tells us, which helps to justify this expression for gravitational potential energy.

More on Gravitational Potential Energy

Here are three important attributes of Equation 7-5 for gravitational potential energy.

(1) *The gravitational potential energy of two objects interacting is a shared property of the two objects.* We learned in Section 6-6 that the gravitational potential energy $U_{grav} = mgy$ for an object of mass m near Earth's surface actually "belongs" to both Earth and the object, since it's associated with their mutual gravitational attraction. In the same way $U_{grav} = -Gm_1m_2/r$ as given by Equation 7-5 "belongs" to the *system* made up of an object of mass m_1 and an object of mass m_2.

(2) *The gravitational potential energy as given by Equation 7-5 is never positive.* The minus sign in Equation 7-5 means that the value of U_{grav} given by this expression is always negative, except when the two objects are infinitely far apart so $r \rightarrow \infty$ and $U_{grav} = 0$ (see Figure 7-8d). Don't be concerned by this! We learned in Section 6-6 that what's really physically meaningful are not the values of potential energy, but rather the *differences* between the values of potential energy at two different points. That means we can choose the point where potential energy equals zero to be wherever it's convenient. With our choice that $U_{grav} = 0$ when an object of mass m_1 and an object of mass m_2 are infinitely far apart, U_{grav} as given by Equation 7-5 decreases and becomes more and more negative as the objects move toward each other, just as gravitational potential energy decreases as a ball falls near Earth's surface (Figure 7-9).

(3) *The gravitational potential energy as given by Equation 7-5 is consistent with the formula $U_{grav} = mgy$ from Section 6-6.* To see how this can be, imagine an object of mass m a height y_i above Earth's surface—that is, at a distance $r = R_{Earth} + y_i$ from Earth's center. If you lift this object to a new height y_f above the surface, its new distance from Earth's center is $r = R_{Earth} + y_f$. The change in the gravitational potential energy in this process is

$$\Delta U_{grav} = U_{grav,f} - U_{grav,i} = \left(-\frac{Gm_{Earth}m}{R_{Earth} + y_f}\right) - \left(-\frac{Gm_{Earth}m}{R_{Earth} + y_i}\right)$$

$$= Gm_{Earth}m\left(\frac{1}{R_{Earth} + y_i} - \frac{1}{R_{Earth} + y_f}\right)$$

(a) Near Earth's surface

(1) If we choose $y = 0$ to be at ground level, the gravitational potential energy for this golf ball of mass m is $U_{grav} = mgy = 0$.

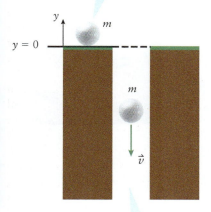

$y = 0$

(2) If the ball falls into a hole, y is less than zero and gravitational potential energy $U_{grav} = mgy$ is less than zero. $E = K + U_{grav}$ is conserved; potential energy U_{grav} decreases, so kinetic energy K increases.

(b) Far from Earth

(1) Infinitely far from Earth at $r = \infty$, the gravitational potential energy for this rock of mass m is $U_{grav} = -Gm_{Earth}m/r = 0$.

$r = \infty$

(2) If the rock falls toward Earth, r is less than infinity and gravitational potential energy $U_{grav} = -Gm_{Earth}m/r$ is less than zero. $E = K + U_{grav}$ is conserved; potential energy U_{grav} decreases, so kinetic energy K increases.

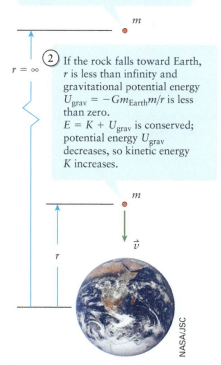

NASA/JSC

Figure 7-9 Understanding negative potential energy A negative value of gravitational potential energy U_{grav} just means that the value is less than at the point where you chose U_{grav} to be zero. This is true whether the object is (a) near Earth's surface, so you use $U_{grav} = mgy$, or (b) at a great distance from Earth, so you use $U_{grav} = -Gm_{Earth}m/r$.

If we express this in terms of a common denominator, you should be able to show that we get

$$\Delta U_{grav} = m\left[\frac{Gm_{Earth}}{(R_{Earth} + y_i)(R_{Earth} + y_f)}\right](y_f - y_i) \tag{7-6}$$

This looks like a horrible mess! But if the heights y_i and y_f are both small compared to Earth's radius R_{Earth} — that is, if the object of mass m remains close to Earth's surface — then the quantities $R_{Earth} + y_i$ and $R_{Earth} + y_f$ are both approximately equal to R_{Earth}. In that case the quantity in square brackets in Equation 7-6 becomes Gm_{Earth}/R_{Earth}^2, which we found in the previous section is equal to the value of g at Earth's surface (see Equation 7-4). Then Equation 7-6 becomes

$$\begin{aligned}\Delta U_{grav} &= U_{grav,f} - U_{grav,i} \\ &= mg(y_f - y_i) = mgy_f - mgy_i\end{aligned}$$

That's precisely the result we would have obtained using $U_{grav} = mgy$. In other words, if we restrict ourselves to motion of the object near the surface of Earth, we can safely use the expression from Section 6-6 for gravitational potential energy.

Gravitational Potential Energy and Conservation of Total Mechanical Energy

If only the gravitational force does work, the total mechanical energy E — the sum of the kinetic energy K and the gravitational potential energy U_{grav} given by Equation 7-5 — is *conserved*. An example is a rocket that's launched from Earth's surface with a certain initial speed. Earth is so massive compared to the rocket that we can assume that there is no effect on our planet. Then, if we ignore the effects of air resistance during the rocket's initial climb through the atmosphere, only gravity does work on the rocket, and the total mechanical energy of the Earth–rocket system is conserved. The following example demonstrates how to use this idea.

EXAMPLE 7-5 How High Does It Go?

You launch a rocket of mass 2.40×10^4 kg straight upward from Earth's surface. The rocket's engines burn for a short time, giving the rocket an initial speed of 9.00 km/s (32,400 km/h or 20,100 mi/h), then shut off (Figure 7-10). To what maximum height above Earth's surface will this rocket rise? Earth's mass is 5.97×10^{24} kg, and its radius is 6370 km $= 6.37 \times 10^6$ m.

Figure 7-10 A rocket launch What is the relationship between the speed imparted to the rocket at launch and the maximum height that it reaches?

Set Up

We'll ignore the force of air resistance on the rocket as it ascends through the atmosphere (this force, like kinetic friction, is nonconservative). Then Earth's gravity is the only force that does work on the rocket, and total mechanical energy is conserved.

With such a tremendous launch speed, we expect the rocket to reach a very high altitude. So the rocket will not remain close to Earth's surface, and we must use the more general form for gravitational potential energy given by Equation 7-5.

Initially the rocket has speed $v_i = 9.00$ km/s $= 9.00 \times 10^3$ m/s and is a distance r_i from Earth's center equal to the radius of Earth. At its maximum height the rocket is at rest ($v_f = 0$) and is a distance h above the surface. Our goal is to determine h.

Total mechanical energy is conserved:

$$K_i + U_{\text{grav,i}} = K_f + U_{\text{grav,f}} \qquad (6\text{-}27)$$

Gravitational potential energy:

$$U_{\text{grav}} = -\frac{Gm_{\text{Earth}}m_{\text{rocket}}}{r} \qquad (7\text{-}5)$$

Kinetic energy of rocket:

$$K = \frac{1}{2}m_{\text{rocket}}v^2 \qquad (6\text{-}8)$$

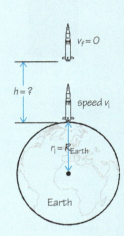

Solve

We first determine the total mechanical energy of the system using the given values for the rocket's initial speed and position. (Recall that $1\text{ J} = 1\text{ kg}\cdot\text{m}^2/\text{s}^2 = 1\text{ N}\cdot\text{m}$.)

When the rocket is initially launched at the surface, its kinetic energy is

$$K_i = \frac{1}{2}m_{\text{rocket}}v_i^2 = \frac{1}{2}(2.40 \times 10^4 \text{ kg})(9.00 \times 10^3 \text{ m/s})^2$$
$$= 9.72 \times 10^{11} \text{ kg}\cdot\text{m}^2/\text{s}^2 = 9.72 \times 10^{11} \text{ J}$$

The gravitational potential energy is

$$U_{\text{grav,i}} = -\frac{Gm_{\text{Earth}}m_{\text{rocket}}}{r_i} = -\frac{Gm_{\text{Earth}}m_{\text{rocket}}}{R_{\text{Earth}}}$$
$$= -\frac{(6.67 \times 10^{-11}\text{N}\cdot\text{m}^2/\text{kg}^2)(5.97 \times 10^{24} \text{ kg})(2.40 \times 10^4 \text{ kg})}{6.37 \times 10^6 \text{ m}}$$
$$= -1.50 \times 10^{12} \text{ N}\cdot\text{m} = -1.50 \times 10^{12} \text{ J}$$

The total mechanical energy is

$$E_i = K_i + U_{\text{grav,i}} = (9.72 \times 10^{11} \text{ J}) + (-1.50 \times 10^{12} \text{ J}) = -5.28 \times 10^{11} \text{ J}$$

When the rocket is at the high point of its trajectory, it is momentarily at rest, and its kinetic energy is zero. At this point the (conserved) total mechanical energy is equal to the gravitational potential energy. Use this to solve for the rocket's distance r_f from Earth's center at its high point.

At the high point of the trajectory, rocket speed is $v_f = 0$, and its kinetic energy is

$$K_f = \frac{1}{2} m_{rocket} v_f^2 = 0$$

The total mechanical energy has the same value as when the rocket was launched:

$$E_f = K_f + U_{grav,f} = E_i = -5.28 \times 10^{11} \text{ J}$$

Since $K_f = 0$, this means that

$$U_{grav,f} = -\frac{Gm_{Earth}m_{rocket}}{r_f} = E_i = -5.28 \times 10^{11} \text{ J}$$

Solve for r_f: $r_f = -\dfrac{Gm_{Earth}m_{rocket}}{E_i}$

$$= -\frac{(6.67 \times 10^{-11} \text{N} \cdot \text{m}^2/\text{kg}^2)(5.97 \times 10^{24} \text{ kg})(2.40 \times 10^4 \text{ kg})}{(-5.28 \times 10^{11} \text{ J})}$$

$$= 1.81 \times 10^7 \text{ m} = 1.81 \times 10^4 \text{ km}$$

Subtract Earth's radius from r_f to find the rocket's final height above the surface.

The rocket's final distance from Earth's center, r_f, equals the radius of Earth (R_{Earth}) plus the rocket's final height h above Earth's surface:

$$r_f = R_{Earth} + h$$

Solve for h:

$$h = r_f - R_{Earth} = 1.81 \times 10^4 \text{ km} - 6.37 \times 10^3 \text{ km} = 1.17 \times 10^4 \text{ km}$$

Reflect

The final height is nearly twice the radius of Earth! This justifies our decision to use Equation 7-5 for gravitational potential energy.

Had we used the expression from Chapter 6, $U_{grav} = m_{rocket}gy$, we would have gotten an incorrect answer because that expression assumes that g has the same value at all heights. In fact, g decreases substantially in value at greater distances from Earth, which means that the rocket climbs much higher than the expression $U_{grav} = m_{rocket}gy$ would predict.

In Example 7-3 (Section 7-2) we found that the value of g at a distance r from Earth's center is $g = \dfrac{Gm_{Earth}}{r^2}$; can you use this to show that at the rocket's maximum height, $g = 1.22 \text{ m/s}^2$? (This is *much* less than the value at Earth's surface, $g = 9.80 \text{ m/s}^2$.)

A common *incorrect* way to calculate h would be to use $U_{grav} = m_{rocket}gy$, which is based on the (false!) assumption that $g = 9.80 \text{ m/s}^2$ at all altitudes. Then conservation of total mechanical energy would have told us that

$$\frac{1}{2} m_{rocket} v_i^2 + m_{rocket}gy_i = \frac{1}{2} m_{rocket} v_f^2 + m_{rocket}gy_f$$

With $y_i = 0$ (rocket at surface), $y_f = h$ (rocket at maximum height), and $v_f = 0$ (rocket momentarily at rest at maximum height), this becomes

$$\frac{1}{2} m_{rocket} v_i^2 + 0 = 0 + m_{rocket}gh$$

$$h = \frac{v_i^2}{2g} = \frac{(9.00 \times 10^3 \text{ m/s})^2}{2(9.80 \text{ m/s}^2)} = 4.13 \times 10^6 \text{ m} = 4.13 \times 10^3 \text{ km}$$

(Incorrect answer!)

The actual height that the rocket reaches is almost three times higher, 1.17×10^4 km, because in fact the value of g decreases with altitude.

Escape Speed

Like a ball thrown vertically upward, the rocket in Example 7-5 reaches a peak height and eventually returns to the surface. If the rocket is launched with a great enough speed, however, it will *never* return. That's because the gravitational force that Earth exerts on the rocket decreases with increasing distance from Earth and approaches zero as the rocket moves to infinitely great distances (see Equation 7-3 and Figure 7-4b). As in Example 7-5, the rocket engines give it a certain initial speed, then shut off. The rocket then coasts upward and loses speed due to Earth's gravity. As the rocket climbs, the gravitational force on the rocket decreases, so the rocket loses an ever-smaller amount of speed per second. If the initial speed of the rocket is great enough, there will still be some speed remaining when the rocket is very far from Earth, and it will keep on going forever. We say that the rocket has *escaped* from Earth.

WATCH OUT! You can never be "beyond the pull of Earth's gravity."

! There's a common misconception that if a rocket travels far enough from Earth, it reaches a point where it's "beyond" the pull of our planet's gravity. But Equation 7-3 shows that in fact the gravitational force that Earth exerts on an object of mass m, $F = Gm_{Earth}m/r^2$, becomes zero only when the distance r from Earth to the object is *infinity*. The force gets progressively weaker as the object moves farther away from Earth, but it never entirely disappears. A rocket can escape from Earth if it has enough speed to overcome the attraction of Earth's gravity and travel to infinity. But the rocket still feels that attraction at all points along its infinite voyage.

We can find the minimum speed for a projectile to escape from Earth by using Equation 7-5 for gravitational potential energy. As in Example 7-5, we'll ignore the effects of air resistance during the time that the projectile is passing through Earth's atmosphere. The projectile (mass m) is launched from Earth's surface (that is, from a distance $r_i = R_{Earth}$ from Earth's center) with a speed v_i. The total mechanical energy just as the projectile is launched is then

$$(7\text{-}7) \qquad E_i = K_i + U_{grav,i} = \frac{1}{2}mv_i^2 + \left(-\frac{Gm_{Earth}m}{R_{Earth}} \right)$$

When the projectile is a distance r from Earth's center and moving at a speed v, the total mechanical energy is

$$(7\text{-}8) \qquad E = K + U_{grav} = \frac{1}{2}mv^2 + \left(-\frac{Gm_{Earth}m}{r} \right)$$

Since total mechanical energy is conserved, E_i as given by Equation 7-7 is equal to E as given by Equation 7-8:

$$(7\text{-}9) \qquad \frac{1}{2}mv_i^2 + \left(-\frac{Gm_{Earth}m}{R_{Earth}} \right) = \frac{1}{2}mv^2 + \left(-\frac{Gm_{Earth}m}{r} \right)$$

Now let's suppose that the projectile just barely escapes Earth, so its speed is zero when it is infinitely far away. That corresponds to replacing v with zero and r with infinity in Equation 7-9. The reciprocal of infinity is zero, so in this case both terms on the right-hand side of Equation 7-9 are zero: The projectile ends up with zero kinetic energy (it is at rest), and the gravitational potential energy ends up with a zero value (the projectile is infinitely far from Earth; see Figure 7-8d). Then Equation 7-9 becomes

$$(7\text{-}10) \qquad \frac{1}{2}mv_i^2 + \left(-\frac{Gm_{Earth}m}{R_{Earth}} \right) = 0$$

The initial speed v_i in Equation 7-10 is called the **escape speed**, which we denote by the symbol v_{escape}. This is the minimum speed at which an object must be launched from Earth's surface to escape to infinity. To find its value replace v_i in Equation 7-10 with v_{escape}, add $Gm_{Earth}m/R_{Earth}$ to both sides, and multiply both sides by $2/m$.

$$\frac{1}{2}mv_{escape}^2 = \frac{Gm_{Earth}m}{R_{Earth}}$$

$$v_{escape}^2 = \frac{2Gm_{Earth}}{R_{Earth}}$$

Finally, take the square root of both sides:

Gravitational constant | Mass of Earth

Speed that a projectile must have at Earth's surface in order to escape Earth

$$v_{escape} = \sqrt{\frac{2Gm_{Earth}}{R_{Earth}}}$$

Radius of Earth

Escape speed
(7-11)

If you substitute $G = 6.67 \times 10^{-11}$ N·m²/kg², $m_{Earth} = 5.97 \times 10^{24}$ kg, and $R_{Earth} = 6.37 \times 10^6$ m into Equation 7-11, you'll find that $v_{escape} = 1.12 \times 10^4$ m/s = 11.2 km/s (about 40,300 km/h or 25,000 mi/h). Rockets designed to send spacecraft to other planets must have powerful engines (see Figure 7-10) to accelerate their payload to such high speeds.

Notice that the launch speed of the rocket in Example 7-5 was only 9.00 km/s, which is less than escape speed. That's why the rocket in that example did *not* escape but reached a maximum distance from Earth before falling back. To have the rocket escape it would have to be launched at 11.2 km/s or faster.

We can also use Equation 7-11 to find the escape speed from *any* planet or moon: Just replace m_{Earth} and R_{Earth} with the mass and radius of the object from which the projectile is launched.

EXAMPLE 7-6 Escaping from a Martian Moon

You have landed your spacecraft on Phobos, the larger of the two airless moons of Mars (Figure 7-11). Phobos is roughly spherical with an average radius of 11.1 km = 1.11×10^4 m. By dropping a baseball while standing on Phobos, you find that g has a very small value at the moon's surface, only 0.00580 m/s². At what speed would you have to throw the baseball to have it escape Phobos?

Figure 7-11 Phobos What is the escape speed from this miniature moon of Mars, just 11.1 km in radius?

Set Up

To find the escape speed v_{escape} using Equation 7-11, we need the radius of Phobos, which we are given, and the mass of Phobos, which we are *not* given. However, we recall from Equation 7-4 in Section 7-2 that the value of g on Earth's surface is related to Earth's mass and radius. We can write this same equation for Phobos, which will let us use the measured value of g on Phobos to determine its mass.

Escape speed from Phobos:

$$v_{escape} = \sqrt{\frac{2Gm_{Phobos}}{R_{Phobos}}} \quad (7-11)$$

Value of g at the surface of Phobos:

$$g = \frac{Gm_{Phobos}}{R_{Phobos}^2} \quad (7-4)$$

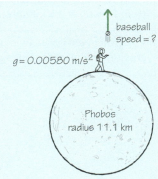

baseball speed = ?

$g = 0.00580$ m/s²

Phobos radius 11.1 km

Solve

First use Equation 7-4 to determine the mass of Phobos. (Recall that 1 N = 1 kg·m/s².)

From Equation 7-4 $\quad g = \frac{Gm_{Phobos}}{R_{Phobos}^2}$

Solve this for the mass of Phobos:

$$m_{Phobos} = \frac{gR_{Phobos}^2}{G} = \frac{(0.00580 \text{ m/s}^2)(1.11 \times 10^4 \text{ m})^2}{6.67 \times 10^{-11} \text{ N·m}^2/\text{kg}^2}$$

$$= 1.07 \times 10^{16} \frac{\text{kg}^2 \cdot \text{m/s}^2}{\text{N}} = 1.07 \times 10^{16} \text{ kg}$$

Then use Equation 7-11 to find the escape speed from Phobos.

Escape speed:

$$v_{\text{escape}} = \sqrt{\frac{2Gm_{\text{Phobos}}}{R_{\text{Phobos}}}}$$

$$= \sqrt{\frac{2(6.67 \times 10^{-11}\ \text{N} \cdot \text{m}^2/\text{kg}^2)(1.07 \times 10^{16}\ \text{kg})}{1.11 \times 10^4\ \text{m}}}$$

$$= \sqrt{129\ \frac{\text{N} \cdot \text{m}}{\text{kg}}} = \sqrt{129\ \frac{\text{kg} \cdot \text{m}}{\text{s}^2}\frac{\text{m}}{\text{kg}}} = \sqrt{129\ \frac{\text{m}^2}{\text{s}^2}}$$

$$= 11.3\ \text{m/s} = 40.9\ \text{km/h} = 25.4\ \text{mi/h}$$

Reflect

The escape speed on Phobos is almost exactly 1/1000 the escape speed on Earth! In general the smaller the planet or moon, the lower the escape speed.

A typical high school baseball pitcher can easily throw a baseball at more than 25 m/s (90 km/h or 56 mi/h), so a well-trained astronaut should have no problem throwing a baseball at the 11.3-m/s escape speed on Phobos.

GOT THE CONCEPT? 7-3 Which Direction for Escape?

? In Example 7-6 suppose you throw the baseball with a speed faster than 11.3 m/s. Figure 7-12 shows four possible directions in which you could throw the ball at that speed. In which of these directions would the ball escape from Phobos? (a) Direction A only; (b) direction A or B; (c) direction A, B, or C; (d) direction A, B, C, or D; (e) none of these.

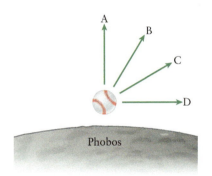

Figure 7-12 Four directions for escaping Phobos If a ball is thrown at escape speed, in what direction(s) can it be thrown so that it *will* escape?

TAKE-HOME MESSAGE FOR Section 7-3

✔ When analyzing situations in which an object does not remain close to the surface of Earth, you must use the expression for gravitational potential energy that is derived from Newton's law of universal gravitation.

✔ Using Newton's law of universal gravitation, the gravitational potential energy of two objects is always negative or zero. It is zero when the two objects are infinitely far apart, and it becomes more and more negative as the objects move closer together.

✔ Escape speed is the speed at which a projectile must be launched from a planet or moon so that it never falls back to the world from which it was launched.

7-4 Newton's law of universal gravitation explains Kepler's laws for the orbits of planets and satellites

We saw in Section 7-2 that Newton deduced the law of universal gravitation from the properties of the Moon's orbit around Earth. Because gravitation is universal, the same principles apply to the orbit of *any* celestial object around another.

Circular Orbits: Orbital Speed and Orbital Period

The simplest type of orbit to analyze is a circular orbit. Many Earth satellites, including the International Space Station (see the photograph that opens this chapter) and the satellites that provide signals for Global Positioning System (GPS) navigation, are in circular or nearly circular orbits.

Although the first Earth satellite was put in orbit in 1957, Newton understood what was required to put a satellite in a circular orbit almost three centuries earlier

(a)

Page 6.

Photo by Christina Micek

(b)

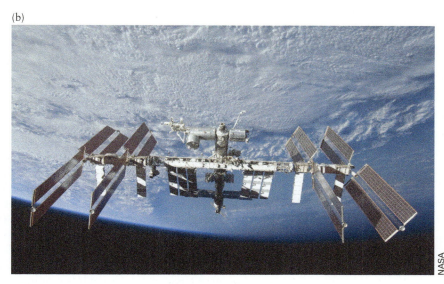

NASA

Figure 7-13 Newton's recipe for an Earth satellite (a) Isaac Newton imagined that if a cannonball were fired horizontally from the top of a mountain with sufficient speed, the rate at which it fell could be made to match the rate at which Earth's surface fell away. The cannonball would therefore end up orbiting Earth. Newton created this illustration for the same 1687 book in which he presented the law of universal gravitation. (b) The International Space Station is in almost perfectly circular orbit, just like the orbit that Newton described.

(Figure 7-13). If a cannonball is dropped from a great height such as the top of a mountain, it will fall straight down. If it is thrown horizontally at a moderate speed, the cannonball will follow a curved arc before hitting the ground. But if it is thrown horizontally with just the right speed, the surface of Earth will fall away below the cannonball so that the cannonball always remains at the same height above the surface. Put another way, Earth's gravitational attraction will cause the cannonball to accelerate toward Earth's center. If the cannonball's speed is just right, the result will be uniform circular motion of the sort that we studied in Section 3-7, in which the acceleration is always directed toward the center of the cannonball's circular path (Figure 7-14).

We can find the speed v required for a circular orbit of radius r from Equation 3-17 for the acceleration in uniform circular motion: $a_{\text{cent}} = v^2/r$. For a satellite of mass m orbiting Earth, the acceleration is provided by Earth's gravitational force. From Newton's second law, this says

$$F_{\text{Earth on satellite}} = ma_{\text{cent}}$$

or, from Equation 3-17 for uniform circular motion and Equation 7-3 for the law of universal gravitation,

$$\frac{Gm_{\text{Earth}}m}{r^2} = m\frac{v^2}{r}$$

To solve for v, multiply both sides of this equation by r/m:

$$\frac{Gm_{\text{Earth}}}{r} = v^2$$

and take the square root:

Gravitational constant   Mass of Earth

Speed of an Earth satellite in a circular orbit

$$v = \sqrt{\frac{Gm_{\text{Earth}}}{r}}$$

Radius of the satellite's orbit

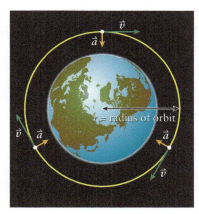

Figure 7-14 A circular orbit A satellite in a circular orbit is in uniform circular motion: The acceleration (due to Earth's gravitational force) has a constant magnitude and is always directed toward Earth's center, and the velocity has a constant magnitude.

Speed of an Earth satellite in a circular orbit
(7-12)

Many satellites, including the International Space Station (Figure 7-13b), are in *low Earth orbit*: Their height above the surface is a few hundred kilometers, which is a short distance compared to Earth's radius of 6370 km. So we can find the approximate speed of a satellite in low Earth orbit by replacing r in Equation 7-12 with the radius of

Earth and substituting the values $G = 6.67 \times 10^{-11}$ N·m²/kg², $m_{\text{Earth}} = 5.97 \times 10^{24}$ kg, and $R_{\text{Earth}} = 6.37 \times 10^6$ m:

$$v = \sqrt{\frac{Gm_{\text{Earth}}}{R_{\text{Earth}}}} = 7.91 \times 10^3 \text{ m/s} = 7.91 \text{ km/s}$$

(orbital speed for low Earth orbit)

This speed (about 28,500 km/h or 17,700 mi/h) is slower than the escape speed of 11.2 km/s (which we discussed in Section 7-3) by a factor of $1/\sqrt{2} = 0.707$.

Equation 7-12 shows that increasing the orbital radius r decreases the speed for a circular orbit. We saw an example of this in Section 7-2: The Moon orbits Earth at an average distance of 3.84×10^8 m, about 60 times Earth's radius, and its orbital speed is only 1.02×10^3 m/s = 1.02 km/s. You can understand why the speed decreases with increasing orbital radius by considering a ball on the end of a string. Imagine whirling the ball on a string around your hand so that the ball makes a circular orbit around your hand. To make the ball move at high speed around a small circle, you have to exert a substantial pull on the string. But if you lengthen the string and make the same ball move at low speed around a large circle, much less pull is required. Similarly, a satellite that orbits close to Earth experiences a substantial gravitational pull and so moves at high speed, while a planet in a larger orbit experiences less gravitational force and moves at a lower speed.

Another useful way to describe how rapidly a satellite moves around its circular orbit is in terms of the **orbital period**, which is the time required to complete one orbit. In uniform circular motion the speed v is constant, so the orbital period T is just the circumference $2\pi r$ of the orbit (the distance around the circular orbit of radius r) divided by the speed: $T = 2\pi r/v$. Using Equation 7-12 for v this becomes

$$T = \frac{2\pi r}{v} = 2\pi r \sqrt{\frac{r}{Gm_{\text{Earth}}}}$$

It's convenient to square both sides of this equation so as to eliminate the square root. Note that the square of r is r^2 and the square of $\sqrt{r}$ is r, so we end up with a factor of $r^2 \times r = r^3$ on the right-hand side of the equation:

Relationship between orbital period and radius for a circular orbit
(7-13)

Period of an Earth satellite in a circular orbit   Radius of the satellite's orbit

Gravitational constant $$T^2 = \frac{4\pi^2}{Gm_{\text{Earth}}} r^3$$ Mass of Earth

WATCH OUT! When a satellite orbits Earth, Earth moves too.

One aspect of orbital motion that we've ignored in this discussion is Newton's third law: If Earth exerts a gravitational force on a satellite, the satellite exerts an equally strong force on Earth. So Earth also moves in a small orbit. However, even the largest satellite humans have ever placed in orbit has a tiny mass compared to the mass of Earth, so the motion of Earth is negligible. But the Moon's mass is a reasonable fraction of Earth's mass (about 1.2%), so for detailed calculations of the Moon's motion this effect must be taken into account (Figure 7-15).

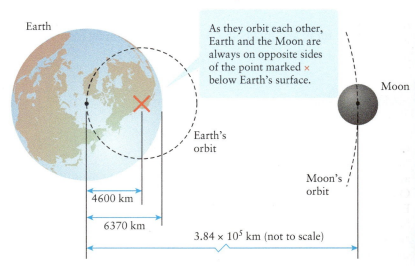

Figure 7-15 The motions of Earth and Moon
Because Earth and the Moon both exert gravitational forces on each other, it's not correct to say that the Moon orbits Earth. Rather, both objects orbit around a point between their centers. This point is inside Earth, about 4600 km from our planet's center. (In Chapter 8 we'll see the significance of this point, called the *center of mass* of the Earth–Moon system.)

Circular Orbits: Energy

Before placing a satellite into a circular orbit, an important question to ask is "How much energy will it take to put the satellite in the desired orbit?" The answer determines how powerful a rocket will be needed for the task. We have all the tools we need to answer this question: Equation 7-12 tells us the speed the satellite must have in its circular orbit, which allows us to determine the required kinetic energy, and Equation 7-5 tells us the gravitational potential energy. So the kinetic energy for a satellite of mass m in a circular orbit of radius r is

$$K = \frac{1}{2}mv^2 = \frac{1}{2}m\left(\sqrt{\frac{Gm_{Earth}}{r}}\right)^2 = \frac{Gm_{Earth}m}{2r}$$

Making the orbital radius r larger means the satellite moves more slowly, and the kinetic energy decreases. The gravitational potential energy is

$$U_{grav} = -\frac{Gm_{Earth}m}{r}$$

As we discussed in Section 7-3, the gravitational potential energy is negative. Making the orbital radius r larger makes the gravitational potential energy less negative— that is, closer to zero—which means that the gravitational potential energy increases. The total mechanical energy is the sum of K and U_{grav}:

$$E = K + U_{grav} = \frac{1}{2}mv^2 + \left(-\frac{Gm_{Earth}m}{r}\right) = \frac{Gm_{Earth}m}{2r} + \left(-\frac{Gm_{Earth}m}{r}\right)$$

or

Gravitational constant   Mass of Earth

Total mechanical energy for a satellite in a **circular orbit** around Earth

$$E = -\frac{Gm_{Earth}m}{2r}$$

Mass of satellite

Radius of the satellite's orbit

Total mechanical energy for a circular orbit
(7-14)

The greater the radius r of the circular orbit, the greater (less negative) the total mechanical energy.

The total mechanical energy E is always negative because of the way we defined gravitational potential energy. A zero value of E corresponds to an orbit with an infinitely large radius r. In this limiting case the gravitational potential energy is zero, and the satellite would have zero speed and zero kinetic energy.

As an example, for a satellite in low Earth orbit (with $r = R_{Earth}$) the total mechanical energy is $E = -Gm_{Earth}m/2R_{Earth}$. Before the spacecraft was launched and was sitting on Earth's surface, its kinetic energy was zero, and E was equal to the gravitational potential energy: $E = U_{grav} = -Gm_{Earth}m/R_{Earth}$. So the amount of energy that had to be imparted to the satellite to put it into low Earth orbit was

$$\Delta E = E_{in\ orbit} - E_{on\ surface} = \left(-\frac{Gm_{Earth}m}{2R_{Earth}}\right) - \left(-\frac{Gm_{Earth}m}{R_{Earth}}\right) = +\frac{Gm_{Earth}m}{2R_{Earth}}$$

For a 1-kg satellite ($m = 1.00$ kg) $\Delta E = 3.13 \times 10^7$ J. So 3.13×10^7 J of energy has to be imparted to each kilogram of a satellite placed into low Earth orbit. In practice the energy requirements are much greater because the rocket and its fuel (which are much more massive than the satellite itself) also have to be given kinetic energy. One way to *reduce* the energy requirements is to take advantage of Earth's rotation. Every point on Earth's surface is rotating to the east (which is why we see the Sun rise in the east), and the speed in meters per second is greatest near the equator. That's why NASA and the European Space Agency launch satellites into orbit from locations relatively close to the equator (Florida and French Guiana, respectively) and launch them toward the east (Figure 7-16).

East →

Cultura Creative (RF)/Alamy

Figure 7-16 Launching toward the east This time-exposure photograph shows a night launch of a satellite from Florida. The launch vehicle climbs upward and to the east (to the right in the photograph). In this way the spacecraft takes advantage of Earth's eastward rotation, which gives the spacecraft 406 m/s of speed even before it lifts off.

WATCH OUT! When a satellite falls to Earth, it's not just gravity's fault.

Orbiting satellites do sometimes fall out of orbit and crash back to Earth. When this happens, however, the real culprit is not gravity but air resistance. A satellite in a relatively low orbit is actually flying through the tenuous outer wisps of Earth's atmosphere. Air resistance does negative work on the spacecraft, just like a kinetic friction force. This causes the total mechanical energy E to decrease and become more negative. From Equation 7-14 this means that the orbital radius r becomes smaller, and the satellite sinks to a lower altitude, where it encounters more air resistance and sinks even lower. Eventually the satellite either strikes Earth or burns up in flight due to air friction. By contrast the Moon and planets orbit in the near-vacuum of interplanetary space. Hence they are unaffected by this kind of air resistance, and they have remained in orbit around the Sun since the solar system formed 4.56 billion years ago.

EXAMPLE 7-7 A Satellite for Satellite Television

The broadcasting satellites used in a satellite television system orbit Earth's equator with a period of exactly 24 hours, the same as Earth's rotation period. As a result, these satellites are *geostationary*: They always remain over the same spot on the equator, so they always appear to be in the same position in the sky as seen from anywhere on Earth's surface. (A satellite TV receiver "dish" is aimed to receive the signal broadcast from one of these satellites.) (a) At what distance from Earth's center must a television satellite orbit? (b) What must be its orbital speed? (c) If a television satellite has a mass of 3.50×10^3 kg, how much energy must it be given to place it in orbit?

Set Up

A geostationary satellite has $T = 24$ h; we'll use this information and Equation 7-13 to find the radius r of the orbit. Once we find the value of r, we'll use Equation 7-12 to find the orbital speed of the satellite.

Equation 7-14 tells us the total mechanical energy of the spacecraft in orbit, and Equation 7-5 will tell us the total mechanical energy when the spacecraft is at rest on Earth's surface before being launched. The difference between these two values is the energy that must be given to the satellite.

Relationship between period and radius for a circular orbit:

$$T^2 = \frac{4\pi^2}{Gm_{Earth}}r^3 \tag{7-13}$$

Speed of an Earth satellite in a circular orbit:

$$v = \sqrt{\frac{Gm_{Earth}}{r}} \tag{7-12}$$

Total mechanical energy for a circular orbit:

$$E = -\frac{Gm_{Earth}m}{2r} \tag{7-14}$$

Gravitational potential energy:

$$U_{grav} = -\frac{Gm_1m_2}{r} \tag{7-5}$$

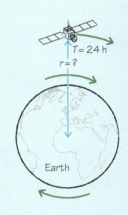

$T = 24$ h
$r = ?$

Earth

Solve

(a) To use Equation 7-13 to find the orbital radius r, first convert the orbital period T to seconds.

Period of the satellite orbit:

$$T = 24 \text{ h}\left(\frac{60 \text{ min}}{1 \text{ h}}\right)\left(\frac{60 \text{ s}}{1 \text{ min}}\right) = 8.64 \times 10^4 \text{ s}$$

Solve Equation 7-13 for r^3:

$$r^3 = \frac{Gm_{Earth}T^2}{4\pi^2}$$

$$= \frac{(6.67 \times 10^{-11} \text{ N} \cdot \text{m}^2/\text{kg}^2)(5.97 \times 10^{24} \text{ kg})(8.64 \times 10^4 \text{ s})^2}{4\pi^2}$$

$$= 7.53 \times 10^{22} \frac{\text{N} \cdot \text{m}^2 \cdot \text{s}^2}{\text{kg}} = 7.53 \times 10^{22} \text{ m}^3$$

(We used 1 N = 1 kg·m/s².)

Take the cube root to find r:

$$r = \sqrt[3]{r^3} = \sqrt[3]{7.53 \times 10^{22} \text{ m}^3}$$

$$= 4.22 \times 10^7 \text{ m} = 4.22 \times 10^4 \text{ km}$$

This is 6.63 times Earth's radius ($R_{Earth} = 6.37 \times 10^3$ km).

(b) Then use Equation 7-12 to find the orbital speed of the satellite.

Orbital speed:

$$v = \sqrt{\frac{Gm_{\text{Earth}}}{r}}$$

$$= \sqrt{\frac{(6.67 \times 10^{-11}\ \text{N} \cdot \text{m}^2/\text{kg}^2)(5.97 \times 10^{24}\ \text{kg})}{4.22 \times 10^7\ \text{m}}}$$

$$= \sqrt{9.43 \times 10^6\ \frac{\text{N} \cdot \text{m}}{\text{kg}}} = \sqrt{9.43 \times 10^6\ \frac{\text{kg} \cdot \text{m}}{\text{s}^2}\frac{\text{m}}{\text{kg}}}$$

$$= \sqrt{9.43 \times 10^6\ \frac{\text{m}^2}{\text{s}^2}}$$

$$= 3.07 \times 10^3\ \text{m/s} = 1.11 \times 10^4\ \text{km/h} = 6.87 \times 10^3\ \text{mi/h}$$

(Again we used $1\ \text{N} = 1\ \text{kg} \cdot \text{m/s}^2$.)

(c) Find the total mechanical energy in orbit, the total mechanical energy on Earth's surface, and the difference between these values for the satellite of mass $m = 3.50 \times 10^3$ kg.

When the satellite is in orbit, the total mechanical energy is

$$E_{\text{in orbit}} = -\frac{Gm_{\text{Earth}}m}{2r}$$

$$= -\frac{(6.67 \times 10^{-11}\ \text{N} \cdot \text{m}^2/\text{kg}^2)(5.97 \times 10^{24}\ \text{kg})(3.50 \times 10^3\ \text{kg})}{2(4.22 \times 10^7\ \text{m})}$$

$$= -1.65 \times 10^{10}\ \text{N} \cdot \text{m}$$

$$= -1.65 \times 10^{10}\ \text{J}$$

With the spacecraft at rest on Earth's surface, the total mechanical energy is just the gravitational potential energy:

$$E_{\text{on surface}} = U_{\text{grav, on surface}} = -\frac{Gm_{\text{Earth}}m}{r}$$

$$= -\frac{(6.67 \times 10^{-11}\ \text{N} \cdot \text{m}^2/\text{kg}^2)(5.97 \times 10^{24}\ \text{kg})(3.50 \times 10^3\ \text{kg})}{6.37 \times 10^6\ \text{m}}$$

$$= -2.19 \times 10^{11}\ \text{N} \cdot \text{m} = -2.19 \times 10^{11}\ \text{J}$$

The amount of energy that must be given to the satellite to put it into orbit is

$$E_{\text{in orbit}} - E_{\text{on surface}} = (-1.65 \times 10^{10}\ \text{J}) - (-2.19 \times 10^{11}\ \text{J})$$

$$= 2.02 \times 10^{11}\ \text{J}$$

$E_{\text{in orbit}} = -1.65 \times 10^{10}\ \text{J}$

$E_{\text{on surface}} = -2.19 \times 10^{11}\ \text{J}$

Earth

Reflect

Our answer to part (a) shows that geostationary broadcast satellites orbit at a tremendous distance from Earth. Part (c) shows that it takes an enormous amount of energy to put them there.

You can check the result for part (b) by confirming that the satellite's orbital speed v equals $2\pi r$ (the circumference of the circular orbit of radius r) divided by T (the time to complete one orbit). Does it?

Kepler's Laws of Planetary Motion

Decades before Newton published his law of universal gravitation, astronomers had carefully observed and recorded the positions of the planets as they traced out their orbits. By analyzing these observations, the German mathematician and astronomer Johannes Kepler discovered three laws that summarize the motions of all of the planets.

Kepler's laws describe the shape of planetary orbits, the speed at which a planet moves along its orbit, and the time it takes a planet to complete an orbit. One of Newton's great accomplishments was to show that his law of universal gravitation, in conjunction with his laws of motion, explained *why* the planets move according to Kepler's laws. Because gravitation is universal, Kepler's laws also describe the orbits of a moon around a planet and a satellite around Earth. Let's look at Kepler's three laws in turn.

The first of Kepler's laws is the **law of orbits**:

The orbit of each planet is an ellipse with the Sun located at one focus of the ellipse.

Figure 7-17 **Ellipses** (a) As a planet orbits the Sun, its orbit is an ellipse with the Sun at one focus. (b) An ellipse can have any eccentricity from $e = 0$ (a circle) to just under $e = 1$ (virtually a straight line).

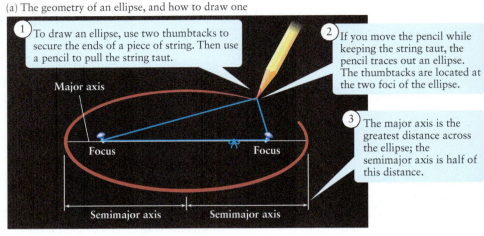

(a) The geometry of an ellipse, and how to draw one

1. To draw an ellipse, use two thumbtacks to secure the ends of a piece of string. Then use a pencil to pull the string taut.

2. If you move the pencil while keeping the string taut, the pencil traces out an ellipse. The thumbtacks are located at the two foci of the ellipse.

3. The major axis is the greatest distance across the ellipse; the semimajor axis is half of this distance.

Major axis

Focus   Focus

Semimajor axis   Semimajor axis

(b) Ellipses with the same major axis but different eccentricities

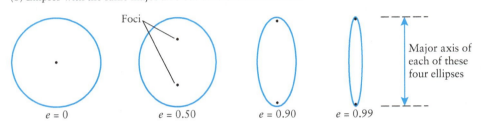

Foci

$e = 0$ $e = 0.50$ $e = 0.90$ $e = 0.99$

Major axis of each of these four ellipses

You can draw an ellipse by using a loop of string, two thumbtacks, and a pencil (**Figure 7-17a**). Each thumbtack in the figure is at a *focus* (plural foci) of the ellipse; an ellipse has two foci. The longest diameter of an ellipse, called the *major axis*, passes through both foci. Half of that distance is called the **semimajor axis** and is usually designated by the symbol a. (Unfortunately a is also the symbol for acceleration. We promise never to use semimajor axis and acceleration in the same equation!) A circle is a special case of an ellipse in which the two foci are at the same point; this corresponds to using only a single thumbtack in Figure 7-17a. The semimajor axis of a circle is equal to its radius. So a circular orbit is a special case of an elliptical orbit.

The **eccentricity** (symbol e) of an ellipse describes how elongated the ellipse is. The value of e can range from 0 (a circle) to just under 1 (nearly a straight line). The greater the eccentricity, the more elongated the ellipse (**Figure 7-17b**). All of the objects that orbit the Sun have orbits that are at least slightly elliptical, with e greater than zero. The most circular of any planetary orbit is that of Venus, with an eccentricity e of just 0.007; Mercury's orbit has $e = 0.206$. The minor planet Pluto has an orbit with $e = 0.249$, and a number of small bodies called comets move in very elongated orbits with eccentricities just less than 1 (**Figure 7-18**). For any elliptical orbit, the Sun is at one focus; there is nothing at the other focus.

One of Newton's triumphs was to show that the planets can have elliptical orbits only if the force F attracting them toward the Sun is in inverse proportion to the square of the distance r from Sun to planet. In other words, the elliptical shape of the planets is a verification of the law of universal gravitation, which states that $F = Gm_{sun}m_{planet}/r^2$. (The proof is mathematically complex and beyond our scope.)

Kepler's second law states that unlike the case of a circular orbit, a planet does *not* move at a constant speed on an elliptical orbit. His **law of areas** describes how the speed changes around the orbit:

A line joining the Sun and a planet sweeps out equal areas in equal intervals of time, regardless of the position of the planet in the orbit.

Figure 7-19 illustrates this law. Suppose that it takes 30 days for a planet to go from point A to point B. During that time an imaginary line joining the Sun and the planet sweeps out a nearly triangular area. Kepler discovered that a line joining the Sun and the planet also sweeps out exactly the same area during any other 30-day interval.

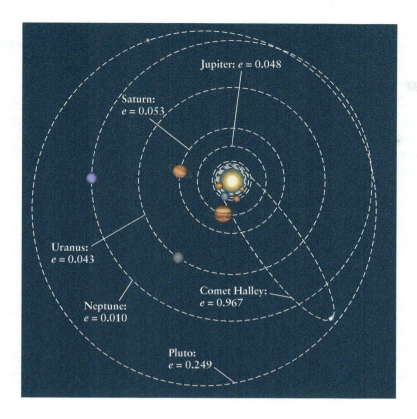

Figure 7-18 Orbits in the solar system The elliptical orbits of most of the planets have relatively small eccentricities, so these orbits are nearly circular. By contrast, the orbit of minor planet Pluto is clearly noncircular (eccentricity $e = 0.249$); at times during its orbit it is closer to the Sun than Neptune is. Comet Halley, which completes one orbit every 76 years, has a very elongated orbit with eccentricity $e = 0.967$.

In other words, if the planet also takes 30 days to go from point C to point D, then the two shaded segments in Figure 7-19 are equal in area.

The law of areas tells us that the planet moves fastest at *perihelion* (the point in its orbit closest to the Sun) and slowest at *aphelion* (the point when the planet is farthest from the Sun). Note that this is consistent with the law of universal gravitation: A planet speeds up as it moves toward the Sun and approaches perihelion, and slows down as it moves away from the Sun and toward aphelion. But why does the speed vary in the particular manner described by the law of areas? Newton showed that this is a direct consequence of the *direction* of the gravitational force that the Sun exerts on a planet: The law of areas is obeyed if and only if this gravitational force points directly toward the Sun, not at an angle. (The proof of this is, again, beyond our scope. But in Chapter 9 we'll return to the law of areas and motivate why it's true.)

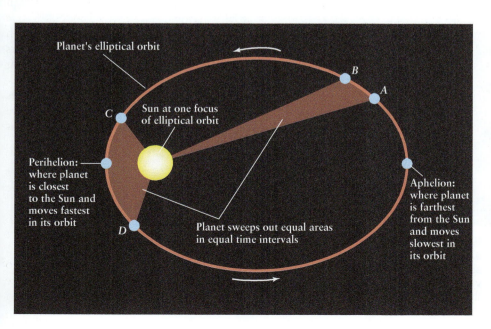

Figure 7-19 Kepler's law of orbits and law of areas According to Kepler's law of orbits, a planet travels around the Sun along an elliptical orbit with the Sun at one focus. (There is nothing at the other focus.) According to his law of areas, as the planet moves, an imaginary line joining the planet and the Sun sweeps out equal areas in equal intervals of time (from A to B or from C to D). By using these laws in his calculations, Kepler found a perfect fit to the apparent motions of the planets.

The first two of Kepler's laws describe how a given planet moves in its orbit. His third law, the **law of periods**, compares the orbital periods of orbits of different sizes:

The square of the period of a planet's orbit is proportional to the cube of the semimajor axis of the orbit.

This is precisely the relationship that we found for *circular* orbits in Equation 7-13. Newton was able to show that the law of periods follows from the law of universal gravitation even for elliptical orbits, if we replace the orbital radius *r* in Equation 7-13 with the semimajor axis *a*. For objects orbiting the Sun we must also replace Earth's mass m_{Earth} with the mass of the Sun:

Newton's form of Kepler's law of periods (7-17)

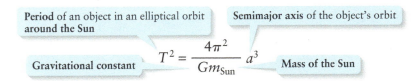

Period of an object in an elliptical orbit around the Sun   Semimajor axis of the object's orbit

Gravitational constant $$T^2 = \frac{4\pi^2}{Gm_{Sun}} a^3$$ Mass of the Sun

We've described Kepler's laws, and Newton's explanations of these laws, for the orbits of the planets around the Sun. But the same laws apply to the orbits of satellites around Earth, to moons orbiting Saturn (see Figure 7-1b), and in general to any system where one object orbits another. One exciting recent application of Kepler's and Newton's ideas is to the study of *extrasolar planets*, which are planets orbiting other stars. Such planets are too faint to be seen with even the most powerful telescopes. Nonetheless thousands of extrasolar planets have been discovered, most by detecting the faint, periodic dimming of a star that occurs when a planet of that star passes in front of it. (A dimming occurs once per orbit of the planet, so the time interval between dimming events equals the orbital period.) Figure 7-20 shows an artist's impression of a system of seven extrasolar planets discovered in this way. Observations confirm that the orbits

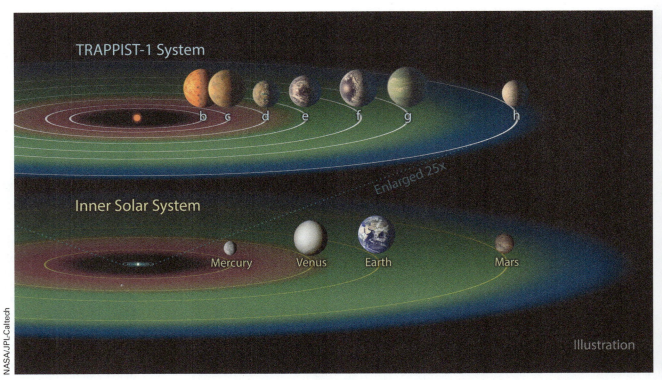

NASA/JPL-Caltech

Figure 7-20 Planets orbiting another star The upper illustration shows the seven extrasolar planets known to orbit TRAPPIST-1, a faint star that lies 39.6 light years from our solar system. (One light year, equal to 3.09×10^{13} km, is the distance that light travels in a vacuum in one year.) As the lower illustration of our solar system shows, the orbits of all of these planets around their star are much smaller than Mercury's orbit around the Sun. The region shaded green in both planetary systems is where conditions are suitable (neither too hot nor too cold) for liquid water to exist on a planet's surface. This raises the tantalizing question of whether life might have evolved on the TRAPPIST-1 planets e, f, or g.

of these planets obey Equation 7-17: The larger the semimajor axis of a planet's orbit around their star, the longer that planet's orbital period.

Another technique for searching for extrasolar planets is by using an effect that we ignored in our discussion of planets orbiting the Sun: The planets exert gravitational forces on the Sun, so the Sun does not remain at rest as we have assumed. In fact these forces cause the Sun to "wobble" in the same way the Moon's gravitational force causes Earth to move (Figure 7-15). Although this wobble is larger in size than the radius of the Sun (6.96×10^5 km), we've neglected it because the wobble is small compared to the semimajor axes of planetary orbits. But astronomers can measure the equally small wobble of other stars, and in this way have detected planets orbiting those stars.

Newton's explanation of Kepler's three laws was perhaps his ultimate achievement. By extrapolating from the laws of nature that he saw here on Earth, he became the first human to understand the behavior of objects in the heavens. The following example demonstrates one application of Newton's form of Kepler's law of periods, Equation 7-17.

EXAMPLE 7-8 A Comet's Orbit

Distances in the solar system are typically measured not in meters or kilometers but in *astronomical units* (au), where 1 au is the semimajor axis of Earth's orbit around the Sun (1 au = 1.50×10^8 km). Earth's orbital period around the Sun is one year. Find the orbital period of a comet that is 0.50 au from the Sun at perihelion and 17.50 au from the Sun at a phelion.

Set Up

We'll use the data about the comet to find its semimajor axis a. From this we'll use Newton's form of Kepler's law of periods to determine the period T.

Newton's form of Kepler's law of periods:

$$T^2 = \frac{4\pi^2}{Gm_{\text{Sun}}} a^3 \qquad (7\text{-}17)$$

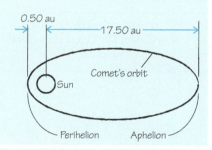

Solve

The drawing shows that the length of the major axis of the comet's orbit is the sum of the perihelion and aphelion distances. The semimajor axis a is one-half the length of the major axis.

Length of the major axis: (distance from Sun to comet at perihelion) + (distance from Sun to comet at aphelion)

= 0.50 au + 17.50 au = 18.0 au

Semimajor axis:

a_{comet}

$= \dfrac{1}{2} \times$ (length of the major axis)

= 9.0 au

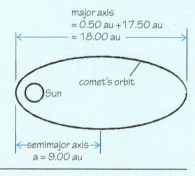

We could calculate T directly from Equation 7-17 by substituting the values of G, m_{Sun}, and the semimajor axis of the comet's orbit (which we would have to convert to meters). But a much simpler approach is to compare the comet's orbit to that of Earth, for which the semimajor axis is 1 au and the period is one year.

For Earth: $T_{\text{Earth}}^2 = \dfrac{4\pi^2}{Gm_{\text{Sun}}} a_{\text{Earth}}^3$

For the comet: $T_{\text{comet}}^2 = \dfrac{4\pi^2}{Gm_{\text{Sun}}} a_{\text{comet}}^3$

If we divide the second equation by the first one, the factor of $4\pi^2/Gm_{\text{Sun}}$ cancels out, leaving

$$\frac{T_{\text{comet}}^2}{T_{\text{Earth}}^2} = \frac{a_{\text{comet}}^3}{a_{\text{Earth}}^3}$$

Substitute values:

$$\frac{T_{\text{comet}}^2}{(1 \text{ y})^2} = \frac{(9.0 \text{ au})^3}{(1 \text{ au})^3} = 729$$

$$T_{\text{comet}} = \sqrt{729} \times 1 \text{ y} = 27 \text{ y}$$

Reflect

Our technique for solving this problem really used Kepler's original form of the law of periods: The square of the period is directly proportional to the cube of the semimajor axis. The semimajor axis for the comet's orbit is 9.0 times larger than that of Earth's orbit, so the square of the orbital period for the comet must be $(9.0)^3 = 729$ times larger than the square of the one-year orbital period of Earth. Therefore the comet's orbital period is $\sqrt{729}$ years, or 27 years.

GOT THE CONCEPT? 7-4 Comparing Orbits

(?) Rank the following five objects in order of their orbital period around the Sun, from shortest period to longest period. (a) An asteroid in a circular orbit with radius 3.00×10^8 km; (b) an asteroid in a circular orbit with radius 4.00×10^8 km; (c) an asteroid in an elliptical orbit that is 3.00×10^8 km from the Sun at perihelion and 4.00×10^8 km from the Sun at aphelion; (d) an asteroid in an elliptical orbit that is 2.00×10^8 km from the Sun at perihelion and 5.00×10^8 km from the Sun at aphelion; and (e) an asteroid in an elliptical orbit that is 2.00×10^8 km from the Sun at perihelion and 4.00×10^8 km from the Sun at aphelion.

TAKE-HOME MESSAGE FOR Section 7-4

✔ For an object in a circular orbit, the orbital speed is inversely proportional to the square root of the orbital radius. The square of the orbital period is directly proportional to the cube of the orbital radius.

✔ The total mechanical energy E for an object in a circular orbit is negative. The larger the orbital radius, the greater (less negative) the value of E.

✔ Kepler's three laws of planetary motion—the law of orbits, the law of areas, and the law of periods—were deduced from observations of how the planets move, but can all be explained by Newton's law of universal gravitation.

7-5 The properties of the gravitational force explain Earth's tides and space travelers' apparent weightlessness

BioMedical

Our discussion of Newton's law of universal gravitation may have given you the impression that this law is important only for describing the motions through space of planets, moons, and spacecraft. But the properties of the gravitational force also have great importance for many living creatures (Figure 7-21). In this section we'll explain

(a) Ocean tides

(b) Apparent weightlessness

Figure 7-21 Gravitational forces and biology (a) The sea star *Pisaster ochraceus* has evolved to thrive on Pacific shorelines, where tides alternately expose them and cover them with water. These tides are a consequence of gravitational forces exerted by the Moon and Sun. (b) Due to the nature of gravitational forces, astronauts on an orbiting spacecraft — and their meals — seem to be weightless. Humans have not evolved to handle apparent weightlessness, which can have major physiological effects.

how gravitation gives rise to *tides* in Earth's oceans as well as the *apparent weightlessness* experienced by astronauts on board an orbiting spacecraft.

Tides

Newton's law of universal gravitation (Equation 7-3) states that the gravitational force of one object on another decreases in proportion to the square of the distance between their centers. Ocean tides on Earth are a direct consequence of this relationship. To see how this arises, imagine three billiard balls of the same mass that are initially at rest some distance from the Moon (Figure 7-22). When released, the three balls experience different gravitational forces because of their slightly different distances from the Moon, and so spread apart. The net result is the same as if *tidal forces* pushed the leading and trailing billiard balls away from the center one, causing the system of three balls to elongate along a line toward the center of the Moon. In fact the tidal force on the 1 ball is not a new kind of force, but is just the *difference* between the Moon's gravitational forces on the 1 ball and on the 2 ball. Similarly, the tidal force on the 3 ball is just the difference between the gravitational forces on the 2 ball and 3 ball.

Unlike the billiard balls in Figure 7-22, Earth does not fall toward the Moon. But just as for the billiard balls, the gravitational force that the Moon exerts on a portion of Earth is stronger if that portion is on the side closest to the Moon, weaker if it's at Earth's center, and weakest if it's on the side farthest from the Moon. As for the billiard balls, this causes our planet to elongate along a line toward the center of the Moon. The solid Earth itself deforms less than a meter in response. But liquid water in the oceans deforms quite a bit more (Figure 7-23a). As Earth rotates under the deformed oceans, an observer on Earth experiences **tides**: The ocean level is highest (high tide) twice a day and lowest (low tide) twice a day. (In this simplified picture we've assumed that Earth is completely covered with water, and so have ignored the complicating effects of the continents.)

The Sun also causes tides on Earth's oceans. The Sun is much more massive than the Moon, so its gravitational force on Earth is much greater than the Moon's. But because the Sun is much farther from Earth than the Moon is, the difference in the Sun's gravitational force from one side of Earth to the other is smaller than the difference in the Moon's gravitational force. The result is that the tidal effect of the Sun is

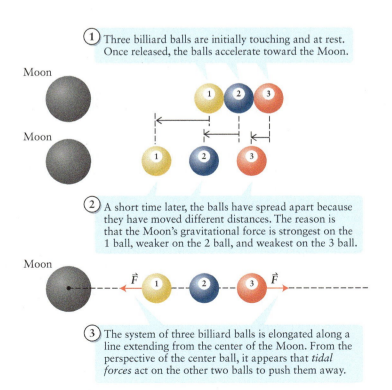

① Three billiard balls are initially touching and at rest. Once released, the balls accelerate toward the Moon.

Moon

② A short time later, the balls have spread apart because they have moved different distances. The reason is that the Moon's gravitational force is strongest on the 1 ball, weaker on the 2 ball, and weakest on the 3 ball.

Moon

③ The system of three billiard balls is elongated along a line extending from the center of the Moon. From the perspective of the center ball, it appears that *tidal forces* act on the other two balls to push them away.

Figure 7-22 The origin of tidal forces Because the gravitational force decreases in magnitude with increasing distance, this system of three billiard balls in space gets stretched out along a line extending from the center of the Moon.

Figure 7-23 Tides in Earth's oceans (a) The Moon's gravitational forces cause our planet's oceans to elongate along a line toward the Moon, just as for the billiard balls in Figure 7-22. The Sun's gravitational forces also cause an elongation of the oceans; this produces (b) strong tides when the Sun, the Moon, and Earth are along a line and (c) weak tides when the Sun, the Moon, and Earth form a right angle.

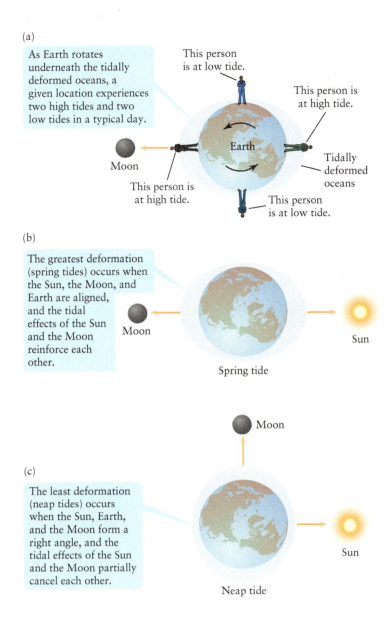

(a)

As Earth rotates underneath the tidally deformed oceans, a given location experiences two high tides and two low tides in a typical day.

This person is at low tide.

This person is at high tide.

Moon

Earth

Tidally deformed oceans

This person is at high tide.

This person is at low tide.

(b)

The greatest deformation (spring tides) occurs when the Sun, the Moon, and Earth are aligned, and the tidal effects of the Sun and the Moon reinforce each other.

Moon

Sun

Spring tide

(c)

Moon

The least deformation (neap tides) occurs when the Sun, Earth, and the Moon form a right angle, and the tidal effects of the Sun and the Moon partially cancel each other.

Sun

Neap tide

only about 40% as great as the tidal effect of the Moon. When the Sun, the Moon, and Earth lie along a line (Figure 7-23b), the tidal effects of the Sun and Moon act together to maximize the deformation of the oceans. The result is particularly strong tides (a large difference between high and low tide) called *spring tides*. But when the line from Earth to the Moon is perpendicular to the line from Earth to the Sun (Figure 7-23c), the two tidal deformations partially cancel each other. The result is *neap tides*. The tides are particularly weak, with a reduced difference between high and low tide.

Earth also exerts tidal forces on the Moon, and these are much greater than the tidal forces of the Moon on Earth. (To see why, imagine replacing the Moon with Earth in the situation shown in Figure 7-22. Because Earth's mass is 81 times greater than the Moon's mass, the gravitational force on each billiard ball will be 81 times greater, as will be the difference in accelerations between neighboring balls. So there will be a much greater tendency to spread the billiard balls apart.) Since the Moon has no oceans, you might think the tidal forces from Earth would be unimportant. But the Moon is not a symmetrical body; it is longer along one particular axis than along any other. The effect of Earth's tidal forces is to keep this longest axis continually pointed toward Earth, just as the longest axis of Earth's deformed oceans continually points toward the Moon (see Figure 7-23a). This explains why only one side of the Moon

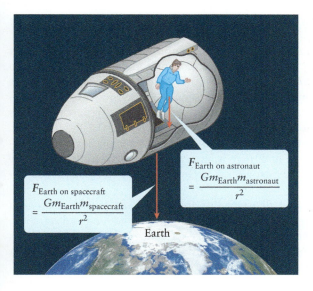

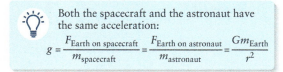

Figure 7-24 The origin of apparent weightlessness If the only external force acting on a spacecraft is the gravitational force, then the spacecraft and its occupant have the same acceleration $g = F/m$. As a result the spacecraft and the astronaut fall freely along the same trajectory, and the astronaut feels "weightless."

Both the spacecraft and the astronaut have the same acceleration:
$$g = \frac{F_{\text{Earth on spacecraft}}}{m_{\text{spacecraft}}} = \frac{F_{\text{Earth on astronaut}}}{m_{\text{astronaut}}} = \frac{Gm_{\text{Earth}}}{r^2}$$

is visible from Earth. Most of the moons of the solar system experience similar tidal effects, and so always keep their longest axis directed toward their parent planet. (One such moon is Phobos, the moon of Mars shown in Figure 7-11.)

Apparent Weightlessness

The tides in Earth's oceans arise because the gravitational force depends on distance, so objects at different distances from the Moon have different accelerations as in Figure 7-22. The **apparent weightlessness** of the astronaut and fruit in Figure 7-21b, by contrast, arises because two objects at the same position have the *same* acceleration g due to gravity, even if their masses are different. Since astronauts are the same distance from Earth as their spacecraft, they have the same acceleration as the spacecraft: They are independent satellites of Earth in the same orbit as the spacecraft, and so "fall together" with the spacecraft (Figure 7-24). The term "apparent weightlessness" means that an astronaut or other object orbiting with the spacecraft has no tendency to move toward the floor, ceiling, or any other part of the spacecraft, but merely "floats."

WATCH OUT! What causes apparent weightlessness—and what doesn't.

It's a common misconception that an astronaut in orbit feels weightless due to being "beyond the pull of Earth's gravity." In fact, if gravity didn't act on an orbiting spacecraft or its occupants, the spacecraft would have *zero* acceleration and hence couldn't stay in orbit: It would maintain a constant velocity and simply fly off into space in a straight line. In fact an orbiting spacecraft remains in orbit precisely because it *does* feel the pull of gravity, which provides the acceleration necessary to make the spacecraft follow a circular trajectory around Earth. The astronaut and fruit in Figure 7-21b *seem* to be weightless because they are following the same trajectory as their spacecraft.

Note that it's not necessary for a spacecraft to be in orbit around a planet for the occupants to experience apparent weightlessness. All that's needed is for the spacecraft to be falling freely, so the only external force on the spacecraft is the gravitational force. Astronauts will feel weightless whether they are orbiting Earth aboard the International Space Station (as in Figure 7-21b), on a mission from Earth to the Moon, or on an interplanetary voyage to Mars. (For a mission to Mars or any other planet, the spacecraft follows an elliptical orbit with the Sun at one focus, just as in Figure 7-19. This orbit is chosen so that it intersects the orbit of the destination planet.) They also feel weightless when they are outside the spacecraft, as in the photo that opens this chapter. The only times during the flight when the astronauts do not feel weightless are when the rockets are firing, either to put the spacecraft into the desired orbit or to take it out of orbit for landing.

BioMedical

At first glance it might seem that being in a state of apparent weightlessness would be relaxing. In fact there are many very negative effects on an astronaut's body. One common problem is *space adaptation syndrome* (also called *space sickness*). About half of all astronauts suffer from this condition during their first few days of spaceflight. The symptoms include motion sickness and disorientation. Research suggests that space adaptation syndrome is related to the vestibular system of the inner ear, which senses gravity and acceleration (see Section 3-7). If an astronaut has a more sensitive vestibular system than normal and there is an asymmetry between the vestibular systems in the left and right ears, the astronaut is likely to be plagued by space adaptation syndrome.

BioMedical

Two other problems that bedevil all astronauts are *loss of muscle mass* and *blood loss*. You exercise the muscles in your calves and spine simply by walking and standing. But in apparent weightlessness these muscles are not exercised and can lose as much as 5% of their mass per week. During a very long duration spaceflight, such as a mission to Mars (which could take 10 months each way), astronauts might lose up to 40% of their capacity to do physical work, equivalent to a 40-year-old astronaut's muscles deteriorating to those of an 80-year-old. For this reason astronauts on board the International Space Station exercise for 2 hours a day using special equipment designed to mimic gravity. (Building muscle mass before going into space does not appear to help: Astronauts who begin their missions with the greatest muscle mass also show the greatest decline while in space.)

The cause of blood loss is more subtle. When you stand upright on Earth, blood pools in your feet due to gravity, so your blood pressure is higher in your feet than in your brain. (That's why blood pressure is always measured at the same vertical position in the body, typically at your upper arm at the same level as your heart.) In apparent weightlessness, however, blood distributes itself more evenly through the body: Blood moves out of the legs and into the large veins in the upper body where stretch receptors respond to the apparent increase in blood volume. (This gives astronauts in space a characteristic puffy-faced appearance.) Activation of these stretch receptors initiates a reflex to reduce blood volume. The reflex involves the release of a hormone from the heart, inhibition of the release of a hormone from the brain, partial inhibition of the neural signals that cause arteries to contract, and the production of copious amounts of dilute urine. As a result astronauts can lose up to 22% of their blood volume within 3 days of apparent weightlessness. The weightless environment reduces demand on the heart because it does not have to overcome the effects of gravity to provide the brain with blood, so the heart muscle begins to atrophy as well. This is another reason for astronauts to do cardiovascular exercise while in space.

Abnormalities of vision are other possible consequences of apparent weightlessness. Astronauts' eyesight after long periods in orbit is so often impaired that the International Space Station carries a supply of drugstore eyeglasses with different magnifications. One suspected explanation is that as blood distributes itself more evenly throughout the body, the pressure inside the skull increases, and this can cause flattening of the back of the eyeball as well as other effects.

Astronauts recover muscle strength relatively quickly after returning from space (at a rate of about one day of recovery per day in space), blood volume can be restored within a few days by drinking fluids, and most vision problems resolve themselves within a few months. More problematic is recovery from the *bone loss* that occurs in space. Astronauts' bones atrophy at a rate of about 1% per month in apparent weightlessness, and recovery from a 6-month mission may require 2 or 3 years back on Earth coupled with a program of strenuous exercise.

Another threat to astronauts' well-being is that while human muscle and bone tend to atrophy in apparent weightlessness, microbes such as *E. coli* and *Staphylococcus* actually reproduce more rapidly under these conditions. This leads to increased risk for contamination and serious infection during a long-duration spaceflight. Such are the biological challenges that confront the future of human exploration of space.

TAKE-HOME MESSAGE FOR Section 7-5

✔ Tides in Earth's oceans result primarily from the Moon's gravitational pull, which depends on distance from the Moon and so causes the oceans to deform.

✔ Astronauts in space feel apparent weightlessness because their acceleration is the same as that of the spacecraft.

✔ The physiological challenges of apparent weightlessness include space sickness; loss of muscle mass, blood, and bone; and increased risk of microbial infection.

Key Terms

apparent weightlessness	law of areas	orbital period
Cavendish experiment	law of orbits	semimajor axis
eccentricity	law of periods	tides
escape speed	Newton's law of universal	
gravitational constant	gravitation	

Chapter Summary

Topic	Equation or Figure
Newton's law of universal gravitation: Any two objects attract each other with a gravitational force. This force is proportional to the product of the masses of the objects and inversely proportional to the square of the distance between the centers of the two objects.	Gravitational constant (same for any two objects) Masses of the two objects Any two objects (1 and 2) exert equally strong gravitational forces on each other. $$F_{1 \text{ on } 2} = F_{2 \text{ on } 1} = \frac{Gm_1 m_2}{r^2}$$ Center-to-center distance between the two objects (7-3) The gravitational forces are attractive: $\vec{F}_{1 \text{ on } 2}$ pulls object 2 toward object 1 and $\vec{F}_{2 \text{ on } 1}$ pulls object 1 toward object 2.
Value of g at Earth's surface: The acceleration due to gravity at our planet's surface is related to Earth's mass and radius. At greater distances r from Earth's center, the value of g decreases in inverse proportion to the square of r.	Gravitational constant Mass of Earth Acceleration due to gravity at Earth's surface $$g = \frac{Gm_{\text{Earth}}}{R^2_{\text{Earth}}}$$ Radius of Earth (7-4)
The Cavendish experiment: The value of the gravitational constant G in the law of universal gravitation can be determined by measuring the gravitational attraction between objects of known mass.	① Two small lead spheres, each of mass m, are attached to the ends of a wooden rod, and the rod is suspended from a wire. ② Two large lead spheres, each of mass M, are then placed on the ground at equal distances from the small spheres. 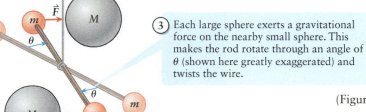 ③ Each large sphere exerts a gravitational force on the nearby small sphere. This makes the rod rotate through an angle of θ (shown here greatly exaggerated) and twists the wire. (Figure 7-6) Knowing the masses m and M of the spheres and the distance r when the rod is in equilibrium, you can determine G using the relationship $F = GMm/r^2$.

Gravitational potential energy: The general expression for the gravitational potential energy of two objects follows from the law of universal gravitation. In this expression the potential energy is zero when the two objects are infinitely far apart; for any finite separation, U_{grav} is negative.

Gravitational constant (same for any two objects)   Masses of the two objects

Gravitational potential energy of a system of two objects (1 and 2)

$$U_{grav} = -\frac{Gm_1m_2}{r}$$

Center-to-center distance between the two objects

(7-5)

The gravitational potential energy is zero when the two objects are infinitely far apart. If the objects are brought closer together (so r is made smaller), U_{grav} decreases (it becomes more negative).

Escape speed: If an object is launched from Earth's surface at the escape speed or faster, it will never fall back to Earth but will escape to infinity.

Gravitational constant   Mass of Earth

Speed that a projectile must have at Earth's surface in order to escape Earth

$$v_{escape} = \sqrt{\frac{2Gm_{Earth}}{R_{Earth}}}$$

Radius of Earth

(7-11)

Circular orbits: A satellite in a circular orbit around Earth moves with a constant speed v. The orbital period T and the total mechanical energy E in a circular orbit both increase with increasing orbital radius r.

Gravitational constant   Mass of Earth

Speed of an Earth satellite in a circular orbit

$$v = \sqrt{\frac{Gm_{Earth}}{r}}$$

Radius of the satellite's orbit

(7-12)

Period of an Earth satellite in a circular orbit   Radius of the satellite's orbit

$$T^2 = \frac{4\pi^2}{Gm_{Earth}}r^3$$

Gravitational constant   Mass of Earth

(7-13)

Gravitational constant   Mass of Earth

Total mechanical energy for a satellite in a circular orbit around Earth

$$E = -\frac{Gm_{Earth}m}{2r}$$

Mass of satellite

Radius of the satellite's orbit

(7-14)

The greater the radius r of the circular orbit, the greater (less negative) the total mechanical energy.

Elliptical orbits: A circular orbit is actually a special case of an elliptical orbit. The planets move in elliptical orbits with the Sun at one focus. The speed varies in accordance with Kepler's law of areas. The orbital period is given by the same expression as for circular orbits around Earth, with Earth's mass replaced by the mass of the Sun and the orbital radius replaced by the semimajor axis (one-half the length of the long axis of the ellipse).

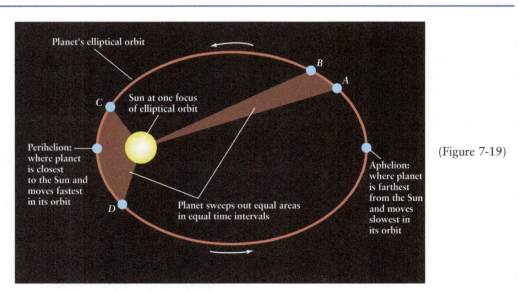

Planet's elliptical orbit

B

A

C

Sun at one focus of elliptical orbit

Perihelion: where planet is closest to the Sun and moves fastest in its orbit

D

Planet sweeps out equal areas in equal time intervals

Aphelion: where planet is farthest from the Sun and moves slowest in its orbit

(Figure 7-19)

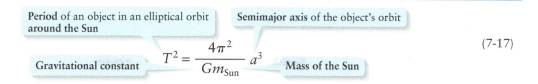

Period of an object in an elliptical orbit around the Sun

Semimajor axis of the object's orbit

Gravitational constant

$$T^2 = \frac{4\pi^2}{Gm_{\text{Sun}}} a^3$$

Mass of the Sun

(7-17)

Tides: Tides in Earth's oceans are due primarily to the Moon's gravitation. Because the Moon exerts a stronger force on the near side of Earth than the far side, the oceans are deformed and a person on Earth experiences high and low tides as our planet rotates. The Sun also contributes to ocean tides on Earth.

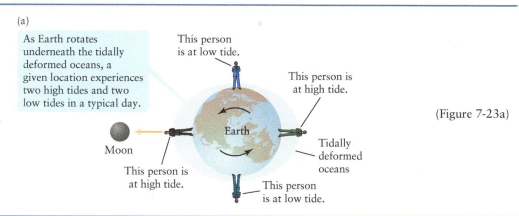

(a)

As Earth rotates underneath the tidally deformed oceans, a given location experiences two high tides and two low tides in a typical day.

This person is at low tide.

This person is at high tide.

Moon

Earth

This person is at high tide.

Tidally deformed oceans

This person is at low tide.

(Figure 7-23a)

Apparent weightlessness: Because gravitation is universal, astronauts riding in a spacecraft (with the rockets off) have the same acceleration as the spacecraft. As a result they "fall" along with the spacecraft and feel as though they are weightless.

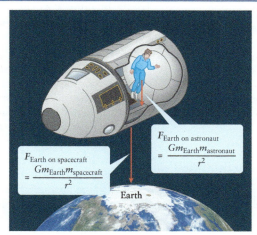

$F_{\text{Earth on astronaut}}$
$= \dfrac{Gm_{\text{Earth}}m_{\text{astronaut}}}{r^2}$

$F_{\text{Earth on spacecraft}}$
$= \dfrac{Gm_{\text{Earth}}m_{\text{spacecraft}}}{r^2}$

Earth

(Figure 7-24)

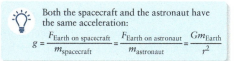

Both the spacecraft and the astronaut have the same acceleration:

$$g = \frac{F_{\text{Earth on spacecraft}}}{m_{\text{spacecraft}}} = \frac{F_{\text{Earth on astronaut}}}{m_{\text{astronaut}}} = \frac{Gm_{\text{Earth}}}{r^2}$$

Answer to What do you think? Question

(b) Newton's law of universal gravitation states that Earth's gravitational force is inversely proportional to the square of the distance from Earth's center. At the altitude of the ISS, the distance from Earth's center is 6370 km + 400 km = 6770 km, which is (6770 km)/(6370 km) = 1.063 times Earth's radius. Hence the gravitational force at the altitude of the ISS is $1/(1.063)^2 = 0.89$ as great as at Earth's surface, or about 10%

less than at Earth's surface. It's true that an astronaut aboard the ISS *feels* weightless, but that's *not* because there is no gravitational force acting. Rather, there are gravitational forces on both the astronaut and the ISS, which give both objects the same acceleration. Hence both objects fall together around Earth in the same orbit. (See Sections 7-2 and 7-5.)

Answers to Got the Concept? Questions

7-1 (e) The gravitational force is inversely proportional to the square of the distance from Earth's center. A point at the north pole is one Earth radius from the center, while the other point is four Earth radii from the center (three Earth radii from the surface). So the point at the north pole is 1/4 the distance from Earth's center, and the gravitational force there is greater by a factor of $1/(1/4)^2 = 4^2 = 16$.

7-2 (e) The gravitational acceleration g is also inversely proportional to the square of the distance from Earth's center. By the same line of reasoning as in the previous question, the value of g is 16 times greater at the north pole (one Earth radius from the planet's center) than at a point three Earth radii above the surface (four Earth radii from the center).

7-3 (d) If the ball has escape speed or greater, the total mechanical energy is enough for the ball to travel to infinity. The total mechanical energy is a scalar, not a vector, so the direction in which the ball is thrown doesn't matter. Changing the direction in which the ball is thrown simply affects the path that the ball follows to infinity.

7-4 (a) and (e) [tie], (c) and (d) [tie], (b). The square of the orbital period T is proportional to the cube of the semimajor axis a. So a ranking in order of the orbital period is the same as a ranking in order of the semimajor axis. For a circular orbit the semimajor axis is the orbital radius; for an elliptical orbit the semimajor axis is half the major (long) axis, and the long axis is the sum of the distances at perihelion and aphelion. So the semimajor axes of the orbits are (a) 3.00×10^8 km, (b) 4.00×10^8 km, (c) $\frac{1}{2} \times (3.00 \times 10^8$ km $+ 4.00 \times 10^8$ km $= 3.50 \times 10^8$ km), (d) $\frac{1}{2} \times (2.00 \times 10^8$ km $+ 5.00 \times 10^8$ km) $= 3.50 \times 10^8$ km, and (e) $\frac{1}{2} \times (2.00 \times 10^8$ km $+ 4.00 \times 10^8$ km) $= 3.00 \times 10^8$ km.

Questions and Problems

In a few problems you are given more data than you actually need; in a few other problems you are required to supply data from your general knowledge, outside sources, or informed estimate.

Interpret as significant all digits in numerical values that have trailing zeros and no decimal points. For all problems use $g = 9.80$ m/s^2 for the free-fall acceleration due to gravity. Neglect friction and air resistance unless instructed to do otherwise. Refer to Appendix B for any required astronomical data like the masses or orbital radii of the planets.

- • Basic, single-concept problem
- •• Intermediate-level problem; may require synthesis of concepts and multiple steps
- ••• Challenging problem
- Example See worked example for a similar problem

Conceptual Questions

1. • Isaac Newton probably gained no insight by being hit on the head with a falling apple, but he did believe that "the Moon is falling." Explain this statement.

2. • Would the magnitude of the acceleration due to gravity near Earth's surface increase more if Earth's mass were doubled or if Earth's radius were cut in half? Justify your answer.

3. • According to Equation 7-2 what happens to the gravitational force between two objects if (a) the mass of one object is doubled and (b) if both masses are halved?

4. • Every object in the universe with mass experiences an attractive gravitational force due to every other object with mass. Why don't you feel a pull from objects close to you?

5. • The gravitational force acts on all objects in proportion to their mass. Why don't heavy objects fall faster than light ones, if you neglect air resistance?

6. • A spacecraft traveling from Earth to the Moon doesn't need to fire its thrusters for the entire trip. Identify the physical reasons that explain why the thrusters only need to fire for part of the trip.

7. • The Sun gravitationally attracts the Moon. Does the Moon orbit the Sun?

8. • The fabrication of precision ball bearings may be better performed on the International Space Station than on the surface of Earth. Explain why.

9. • Why is the gravitational potential energy of two objects negative?

10. • When a frog is at the bottom of a well, the frog–Earth system has a negative amount of gravitational potential energy with respect to the ground level, where we choose to set potential energy equal to zero. Can the frog escape from the well if it jumps upward with a positive amount of kinetic energy that is larger than the negative potential energy when the frog is at the bottom of the well? Explain.

11. • Describe some ways that Newton's law of gravity may have affected human evolution.

12. •• Imagine a world where the force of gravity is proportional to the inverse *cube* of the distance between two objects ($F_{gravity} \propto 1/r^3$). How would Kepler's law of periods be changed?

13. •• Much attention has been devoted to the exact numerical power in Newton's law of universal gravitation ($F_{gravity} \propto 1/r^2$). Some theorists have investigated whether the dependence might be slightly larger or smaller than 2. What would be the significance (what impact would there be) if the power was not exactly 2?

14. • Earth moves faster in its orbit around the Sun during the winter in the northern hemisphere than it does during the summer in the northern hemisphere. Is Earth closer to the Sun during the northern hemisphere's winter or during the northern hemisphere's summer? Explain your answer.

15. • A satellite is to be raised from one circular orbit to one farther from Earth's surface. What will happen to its period?

16. • Geostationary satellites remain stationary over one point on Earth. How is this accomplished?

Multiple-Choice Questions

17. • Planet A orbits a star with a period T. Planet B circles the same star at four times the distance of planet A. Planet B is four times as massive as planet A. Planet B orbits the star with a period of
   A. $4T$.
   B. $8T$.
   C. $16T$.
   D. T.
   E. $T/4$.

18. • According to Newton's universal law of gravitation, $F = \dfrac{Gm_1 m_2}{r^2}$, if both masses are doubled, the force is
   A. four times as much as the original value.
   B. twice as much as the original value.
   C. the same as the original value.
   D. one-half of the original value.
   E. one-fourth of the original value.

19. • According to Newton's universal law of gravitation, $F = \dfrac{Gm_1 m_2}{r^2}$, if the distance r is doubled, the force is

    A. four times as much as the original value.
    B. twice as much as the original value.
    C. the same as the original value.
    D. one-half of the original value.
    E. one-fourth of the original value.

20. • Which is larger, the Sun's pull on Earth or Earth's pull on the Sun?

    A. The Sun's pull on Earth is larger.
    B. Earth's pull on the Sun is larger.
    C. They pull on each other equally.
    D. The Sun's pull on Earth is twice as large as Earth's pull on the much larger Sun.
    E. There is no pull or force between Earth and the Sun.

21. • Compare the weight of a mountain climber when she is at the bottom of a mountain with her weight when she is at the top of the mountain. In which case is her weight larger?

    A. She weighs more at the bottom.
    B. She weighs more at the top.
    C. Both are the same.
    D. She weighs twice as much at the top.
    E. She weighs four times as much at the top.

22. • A satellite and the International Space Station have the same mass and are going around Earth in concentric orbits. The distance of the satellite from Earth's center is twice that of the International Space Station's distance. What is the ratio of the centripetal force acting on the satellite compared to that acting on the International Space Station?

    A. 1/4
    B. 1/2
    C. 1
    D. 2
    E. 4

23. • A satellite in a low-altitude circular orbit just above a planet's surface has speed v. To escape the planet, its speed must be

    A. v.
    B. $\sqrt{2}v$.
    C. $2v$.
    D. $v/2$.
    E. $v/\sqrt{2}$.

24. • The escape speed from planet X is v. Planet Y has the same radius as planet X but is twice as dense. The escape speed from planet Y is

    A. v.
    B. $\sqrt{2}v$.
    C. $2v$.
    D. $v/2$.
    E. $v/\sqrt{2}$.

25. • Two satellites having equal masses are in circular orbits around Earth. Satellite A has a smaller orbital radius than satellite B. Which statement is true?

    A. Satellite A has more kinetic energy, less potential energy, and less mechanical energy (kinetic energy plus potential energy) than satellite B.
    B. Satellite A has less kinetic energy, less potential energy, and less mechanical energy than satellite B.
    C. Satellite A has more kinetic energy, more potential energy, and less mechanical energy than satellite B.

    D. Satellite A and satellite B have the same amount of mechanical energy.
    E. Satellite A and satellite B have the same amount of kinetic energy and no potential energy because they are in motion.

26. •• The eight planets, as well as Pluto and Comet Halley, orbit the Sun according to Newton's law of universal gravitation. The table has data for the periods (T) and semimajor axes (a) for each orbital ellipse. (a) Derive a best-fit constant, similar to that referred to in Kepler's law of periods. (b) Predict the value of a for the elliptical orbit of Comet Halley.

Object	T (days)	a (au)
Mercury	87.97	0.3871
Venus	224.7	0.7233
Earth	365.2	1.000
Mars	687.0	1.5234
Jupiter	4332	5.204
Saturn	10,832	9.582
Uranus	30,799	19.23
Neptune	60,190	30.10
Pluto	90,613.3	39.48
Halley	27,507	?

Problems

7-1 Gravitation is a force of universal importance

7-2 Newton's law of universal gravitation explains the orbit of the Moon

27. • A 5.00×10^2-kg tree stump is located 1.00×10^3 m from a 12,000-kg boulder. Determine the magnitude and direction of the gravitational force exerted by the tree stump on the boulder. Example 7-1

28. • Kramer goes bowling and decides to employ the force of gravity to "pick up a spare." He rolls the 7.0-kg bowling ball very slowly so that it comes to rest a center-to-center distance of 0.20 m from the one remaining 1.5-kg bowling pin. Determine the force of gravity between the ball and the pin at this point and comment on the efficacy of the technique. Treat the ball and pin as point objects for this problem. Example 7-1

29. • A 150-g baseball is 1.00×10^2 m from a 935-g bat. What is the force of gravitational attraction between the two objects? Ignore the size of the bat and ball (treat them as point objects) for this problem. Example 7-1

30. • In 1994 the performer Rod Stewart drew over 3 million people to a concert in Rio de Janeiro, Brazil. (a) If the people in the group had an average mass of 80.0 kg, what collective gravitational force would the group have on a 4.50-kg eagle soaring 3.00×10^2 m above the throng? If you treat the group as a point object, you will get an upper limit for the gravitational force. (b) What is the ratio of that force of attraction to the force between Earth and the eagle? Example 7-2

31. • Compare Earth's gravitational force on a 1-kg apple that is on the surface of Earth versus the gravitational force due to the Moon on the same apple in the same location on the surface of Earth. Assume that Earth and the Moon are spherical and that both have their masses concentrated at their respective centers. Example 7-2

32. • Determine the average magnitude of the force of gravity between the Sun and Earth. Example 7-4

33. • During a solar eclipse, the Moon is positioned directly between Earth and the Sun. (a) Find the magnitude of the net gravitational force acting on the Moon during the solar eclipse due to both Earth and the Sun. (b) What is the direction of this force? Example 7-4

34. • A star that has a mass equal to the mass of our Sun is located 7.50×10^9 km from another star that has a mass that is one-half of the Sun's mass. At what point(s) will the gravitational force from the two stars on a 1.00×10^5-kg space probe be equal to zero? Example 7-4

35. • At what point between Earth and the Moon will a 5.00×10^4-kg space probe experience no net gravitational force? Example 7-4

36. • Compare the weight of a 5.00-kg object on Earth's surface to the gravitational force between a 5.00-kg object that is one Earth radius from another object of mass equal to 5.98×10^{24} kg. Use Newton's universal law of gravitation for the second part of the question. Example 7-2

37. • **Biology** The tibia, the long vertical bone in the lower leg, is the major weight-bearing bone in the human body. Maintaining a healthy bone density requires a constant load on the tibia and other bones. This load is greatly reduced in a low-gravity environment, which may contribute to bone loss in future astronauts exploring the solar system. During normal walking, the typical force (mechanical load) on the tibia of each leg is about three times the body weight. (This is greater than the body weight since the tibia must also accelerate the body upward during a stride.) Calculate the mechanical load on each tibia of a 70.0-kg astronaut walking on the surface of each of the following worlds: (a) Earth; (b) the Moon; (c) Mimas, a small moon of Saturn with mass 3.75×10^{19} kg and radius 198 km. Example 7-4

7-3 The gravitational potential energy of two objects is negative and increases toward zero as the objects are moved farther apart

38. • How much energy would be required to move the Moon from its present orbit around Earth to a location that is twice as far away? Assume the Moon's orbit around Earth is nearly circular and has a radius of 3.84×10^8 m. Example 7-5

39. • How much work is done by the force of gravity as a 10.0-kg object moves from a point that is 6.00×10^3 m above sea level to a point that is 1.00×10^3 m above sea level? Example 7-5

40. • (a) What is the escape speed of a space probe that is launched from the surface of Earth? (b) Would the answer change if the launch occurs on top of a very tall mountain? Explain your answer. Example 7-6

41. • A small asteroid that has a mass of 1.00×10^2 kg is moving at 2.00×10^2 m/s when it is 1.00×10^3 km above the Moon. (a) How fast will the asteroid be traveling when it impacts the lunar surface if it is heading straight toward the center of the Moon? (b) How much work does the Moon do in stopping the asteroid if neither the Moon nor the asteroid heats up in the process? The radius of the Moon is 1.737×10^6 m. Example 7-5

42. •• The volume of water in the Pacific Ocean is about 7.0×10^8 km^3. The density of seawater is about 1030 kg/m^3. (a) Determine the gravitational potential energy of the Moon–Pacific Ocean system when the Pacific is facing away from the Moon. (b) Repeat the calculation when Earth has rotated so that the Pacific Ocean faces toward the Moon. (c) Estimate the maximum speed of the water in the Pacific Ocean due to the tidal influence of the Moon. For the sake of the calculations, treat the Pacific Ocean as a pointlike object (obviously a very rough approximation). Example 7-5

7-4 Newton's law of universal gravitation explains Kepler's laws for the orbits of planets and satellites

43. • The orbit of Mars around the Sun has a radius that is 1.524 times greater than the radius of Earth's orbit. Determine the time required for Mars to complete one revolution. Example 7-8

44. • The space shuttle usually orbited Earth at altitudes of around 3.00×10^5 m. (a) Determine the time for one orbit of the shuttle about Earth. (b) How many sunrises per day did the astronauts witness? Example 7-7

45. • What was the speed of a space shuttle that orbited Earth at an altitude of 3.00×10^5 m? Example 7-7

46. • A satellite orbits Earth at an altitude of 8.00×10^4 km. Determine the time for the satellite to orbit Earth once. Example 7-7

47. • Astronomers discover an exoplanet, a planet orbiting a star other than our Sun, that has an orbital period of 3.5 Earth years in a circular orbit around its star, which has a measured mass of 3.60×10^{30} kg. Determine the radius of the exoplanet's orbit. Example 7-8

48. • The orbital period of Saturn is 29.46 years. Determine the semimajor axis of its orbit. Example 7-8

49. •• Suppose that humans have created a colony outside of our solar system on a planet called Cheops5. Cheops5 has a mass of 2.25×10^{25} kg and a day that lasts 27.5 h (which defines the rotational period of the planet). The colony is located on the planet's equator. The colonists set up a communications satellite that orbits Cheops5. The satellite has a circular orbit that keeps it positioned directly above the colony. Calculate the radius of the satellite's orbit in kilometers. Example 7-7

50. •• An artificial satellite is in a circular orbit 6.50×10^2 km from the surface of a planet of radius 5.50×10^3 km. The period of revolution of the satellite around the planet is 3.00 h. What is the average density of the planet? Example 7-7

51. • The Moon orbits Earth in a nearly circular orbit that lasts 27.32 days. Determine the distance from the surface of the Moon to the surface of Earth. Example 7-7

52. • A planet orbits a star with an orbital radius of 1.00 au. If the star has a mass that is 1.75 times our own Sun's mass, determine the time for one revolution of the planet around the star. Example 7-8

53. • Tidal forces arise on Earth because the Moon exerts different gravitational forces on different parts of Earth. The Moon has a radius of 1.74×10^6 m. Consider a 1.00-kg object on the side of Earth closest to the Moon and an identical 1.00-kg object on the side of Earth farthest from the Moon. (a) Determine the difference between the magnitudes of the forces that the Moon exerts on the two objects; this difference is a measure of the strength of the tidal forces that the Moon causes on Earth. To measure the strength of the tidal forces

that Earth causes on the Moon, consider a 1.00-kg object on the side of the Moon closest to Earth and an identical 1.00-kg object on the side of the Moon farthest from Earth. (b) Determine the difference between the magnitudes of the forces that Earth exerts on the two objects. (c) Which do you conclude is stronger: the tidal forces that the Moon causes on Earth or the tidal forces that Earth causes on the Moon?

General Problems

54. •• A 425-kg satellite is launched into a circular orbit that has a period of 702 min and a radius of 20,100 km around Earth. (a) Determine the gravitational potential energy of the satellite's orbit. (b) Estimate the energy required to place the satellite in orbit around Earth. Example 7-7

55. •• **Astronomy** The four largest of Jupiter's moons are listed in the table below. Using these data, Kepler's three laws, and the law of universal gravitation, (a) complete the table and (b) determine the mass of Jupiter. Example 7-8

Moon	Semimajor Axis (km)	Orbital Period (days)
Io	421,700	1.769
Europa	671,034	?
Ganymede	?	7.155
Callisto	?	16.689

56. •• (a) What speed is needed to launch a rocket due east from near the equator into an orbit in which the rocket skims along near the surface of Earth with an orbital radius $r \approx R_E$? (b) Repeat the calculation for a rocket fired due west. Example 7-7

57. •• The former Soviet Union launched the first artificial Earth satellite, *Sputnik*, in 1957. Its mass was 84 kg, and it made one orbit every 96 min. (a) Determine the altitude of *Sputnik*'s orbit above Earth's surface, assuming circular orbits. (b) What was *Sputnik*'s weight in orbit and at Earth's surface? Example 7-7

58. • **Astronomy** The 2004 landings of the Mars rovers *Spirit* and *Opportunity* involved many stages, resulting in each probe having zero vertical velocity about 12 m above the surface of Mars. Determine (a) the time required for the final free-fall descent of the probes and (b) the vertical velocity at impact. The mass of Mars is 6.419×10^{23} kg, and its radius is 3.397×10^6 m. Example 7-2

59. ••• **Biology, Astronomy** On Earth, froghoppers can jump upward with a takeoff speed of 2.8 m/s. Suppose you took some of the insects to an asteroid. If it is small enough, they can jump free of it and escape into space. (a) What is the diameter (in kilometers) of the largest spherical asteroid from which they could jump free? Assume a typical asteroid density of 2.0 g/cm³. (b) Suppose that one of the froghoppers jumped

horizontally from a small hill on an asteroid. What would the diameter (in kilometers) of the asteroid need to be so that the insect could go into a circular orbit just above the surface? Example 7-6

60. • **Astronomy** The International Space Station (ISS) orbits Earth in a nearly circular orbit that is 400 km above Earth's surface. (a) How many hours does it take for the ISS to make each orbit? (b) Some of the experiments performed by astronauts in the ISS involve the effects of "weightlessness" on objects. What gravitational force does Earth exert on a 10.0-kg object in the ISS? Express your answer in newtons and as a fraction of the force that Earth would exert on the object at Earth's surface. (c) Considering your answer in part (b), how can an object be considered *weightless* in the ISS? Example 7-7

61. •• **Astronomy** Measurements on the asteroid Apophis have shown that its aphelion (farthest distance from the Sun) is 1.099 au, its perihelion (closest distance from the Sun) is 0.746 au, and its mass is 2.7×10^{10} kg. (a) Determine the semimajor axis of Apophis in astronomical units and in meters. (b) How many days does it take Apophis to orbit the Sun? (c) At what point in its orbit is Apophis traveling fastest, and at what point is it traveling slowest? (d) Determine the ratio of its maximum speed to its minimum speed. Example 7-8

62. •• A rack of seven spherical bowling balls (each 8.00 kg, radius of 11.0 cm) is positioned along a line 1.00 m from a point *P*, as shown in **Figure 7-25**. Determine the gravitational force the bowling balls exert on a ping-pong ball of mass 2.70 g centered at point *P*. Example 7-4

Figure 7-25
Problem 62

63. • **Astronomy** The Sun and solar system actually are not at rest in our Milky Way galaxy. We orbit around the center of the Milky Way galaxy once every 225,000,000 years, at a distance of 27,000 light years. (One light year is the distance that light travels in one year: 1 ly = 9.46×10^{12} km = 9.46×10^{15} m.) If the mass of the Milky Way responsible for keeping us in this orbit were concentrated at the center of the galaxy, what would be the mass of the galaxy? Example 7-7

64. •• Locate the point(s) along the line $\overline{AB}$ where a small, 1.00-kg object could rest such that the net gravitational force on it due to the two objects shown is exactly zero (**Figure 7-26**). Example 7-4

Figure 7-26 **Problem 64**

Jim Zipp/Science Source

8

Momentum, Collisions, and the Center of Mass

What do you think?

Cooper's hawk (*Accipiter cooperi*) is a species of raptor that hunts and captures other birds in flight. Suppose a fast-moving Cooper's hawk attacks a slow-moving pigeon from behind and captures the pigeon in its talons. Compared to the total kinetic energy of the two separate birds immediately before the capture, the kinetic energy of the system of two birds moving together immediately after the capture is (a) greater; (b) the same; (c) less.

In this chapter, your goals are to:

- (8-1) Comprehend the significance of momentum and the center of mass.
- (8-2) Define the linear momentum of an object and explain how it differs from kinetic energy.
- (8-3) Explain the conditions under which the total momentum of a system is conserved and why total momentum is conserved in a collision.
- (8-4) Identify the differences and similarities between elastic, inelastic, and completely inelastic collisions.
- (8-5) Apply momentum conservation and energy conservation to problems about elastic collisions.
- (8-6) Relate the momentum change of an object, the force that causes the change, and the time over which the force acts.
- (8-7) Explain the physical significance of the center of mass and describe how the net force on a system affects the motion of the system's center of mass.

To master this chapter, you should review:

- (3-2)   Multiplying a vector by a scalar.
- (3-4)   The definitions of the velocity and acceleration vectors.
- (4-2, 4-5) Newton's second and third laws of motion.
- (6-2, 6-3) The ideas of work, kinetic energy, and the work-energy theorem.

8-1 Newton's third law helps lead us to the idea of momentum

The ideas of kinetic energy and the work-energy theorem that we introduced in Chapter 6 gave us new ways to think about motion. These ideas derive fundamentally from Newton's second law, which states that the net external force on an object determines the object's acceleration. In this chapter we'll learn even more new physics

(a)

(b)

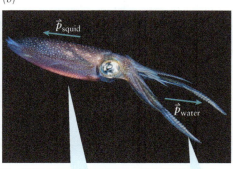

$\vec{p}_{squid}$

$\vec{p}_{water}$

(c)

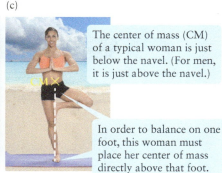

The center of mass (CM) of a typical woman is just below the navel. (For men, it is just above the navel.)

CM

In order to balance on one foot, this woman must place her center of mass directly above that foot.

When a baseball bat collides with a baseball, it transfers some of its momentum $\vec{p} = m\vec{v}$ to the ball.

A squid propels itself by ejecting water at high speed. The water acquires momentum in one direction, and the squid acquires momentum of the same magnitude in the opposite direction.

Figure 8-1 Momentum and center of mass In this chapter we'll explore (a) momentum, (b) conservation of momentum, and (c) the center of mass of a system.

by reconsidering Newton's *third* law: When two objects interact with each other, they exert forces of equal magnitude but opposite direction on each other.

An important way to express this concept is in terms of the *momentum* of an object (written as $\vec{p}$), which is the product of an object's mass and its velocity. Since velocity is a vector, momentum is as well. We'll see that we can describe the behavior of two interacting objects, such as a baseball bat and a ball (Figure 8-1a) or a squid and the water it ejects to propel itself (Figure 8-1b), in terms of momentum: Each object in the pair undergoes a change in velocity and hence in momentum, and the momentum changes are equal in magnitude but opposite in direction. This important observation will lead us to a physical principle called the *law of conservation of momentum*. This law will turn out to be essential for analyzing what happens when two objects interact with each other.

Intimately related to the idea of momentum is the notion of *center of mass*. This is a special point associated with a system of objects that moves as though all of the mass of the system were concentrated there and as though all of the forces on the system act at that point. You've used the idea of the center of mass if you've ever balanced on one foot (Figure 8-1c). If all of the mass of your body were concentrated into a small blob placed on the ground, that blob would remain at rest because it would be supported from below. In the same way, to have your body remain at rest while standing on the ground on one foot, you lean so that your center of mass (located near your navel) is above your foot and so is supported from below. We'll develop the idea of the center of mass by expanding on what we'll have learned about momentum.

We'll begin our exploration of the momentum concept by considering a simple system: a person standing on a skateboard who decides to jump off.

BioMedical

TAKE-HOME MESSAGE FOR Section 8-1

✔ The idea of momentum is useful in situations where two or more objects interact with each other.

8-2 Momentum is a vector that depends on an object's mass, speed, and direction of motion

You're standing atop a stationary skateboard (Figure 8-2a). You then jump straight to the left off the skateboard, and the skateboard rolls away to the right (Figure 8-2b). If you try this, you'll find that the skateboard rolls to the right much faster than you fly through the air to the left. Why is this? What determines how much faster the skateboard moves than you do?

Like any question about motion, the best way to answer this question is by using Newton's laws. Figure 8-2c shows the free-body diagrams for you and your skateboard while your foot is in contact with the skateboard and you're pushing off. The vertical forces on the skateboard cancel, as do the vertical forces on you, so the net force on the skateboard and the net force on you are both horizontal. There is negligible friction

(c) Gino's Premium Images/Alamy

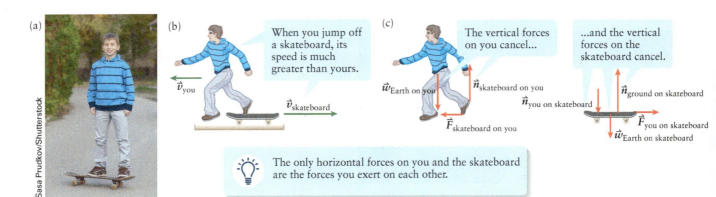

Figure 8-2 **Jumping off a skateboard** You are (a) initially at rest atop a skateboard then (b) jump off horizontally. (c) Free-body diagrams for you and the skateboard during the push-off.

between the ground and the wheels of the skateboard, so the only horizontal forces that act are the forces that you and the skateboard exert on each other. If we apply Newton's second law to the skateboard and to you, we get the equations

$$\sum \vec{F}_{\text{ext on skateboard}} = \vec{F}_{\text{you on skateboard}} = m_{\text{skateboard}}\vec{a}_{\text{skateboard}}$$
$$\sum \vec{F}_{\text{ext on you}} = \vec{F}_{\text{skateboard on you}} = m_{\text{you}}\vec{a}_{\text{you}}$$

(8-1)

The vectors $\vec{a}_{\text{skateboard}}$ and $\vec{a}_{\text{you}}$ are the accelerations of the skateboard and you, respectively, during the push-off.

Equations 8-1 involve the *accelerations* of the skateboard and you, but we're interested in the *velocities* of the skateboard and you after the push-off. To see how to get these, let's go back to the definition of the acceleration vector in Section 3-4:

Acceleration vector equals change in velocity vector divided by time interval
(3-9)

Acceleration vector for the object over a very short time interval from time t_1 to a later time t_2	Change in velocity of the object over the short time interval

$$\vec{a} = \frac{\Delta\vec{v}}{\Delta t} = \frac{\vec{v}_2 - \vec{v}_1}{t_2 - t_1}$$

For both the velocity change and the elapsed time, subtract the earlier value from the later value.

Elapsed time for the time interval

In Equation 3-9 the elapsed time Δt is how long the push-off lasts. The initial velocity before the push-off is $\vec{v}_1 = 0$ for both the skateboard and you, since both begin at rest. We'll use the symbols $\vec{v}_{\text{skateboard}}$ and $\vec{v}_{\text{you}}$ for the velocities of the skateboard and you just after the push-off (corresponding to $\vec{v}_2$ in Equation 3-9). When we substitute these and Equation 3-9 into Equations 8-1, then multiply both sides of these equations by Δt (the duration of the push-off), we get

$$\vec{F}_{\text{you on skateboard}}\Delta t = m_{\text{skateboard}}\vec{v}_{\text{skateboard}}$$
$$\vec{F}_{\text{skateboard on you}}\Delta t = m_{\text{you}}\vec{v}_{\text{you}}$$

(8-2)

Now remember that from Newton's third law, the force that the skateboard exerts on you has the same magnitude as the force that you exert on the skateboard but points in the opposite direction (Figure 8-2c): $\vec{F}_{\text{skateboard on you}} = -\vec{F}_{\text{you on skateboard}}$. If we substitute this into Equations 8-2, you can see that the quantities $m_{\text{skateboard}}\vec{v}_{\text{skateboard}}$ and $m_{\text{you}}\vec{v}_{\text{you}}$ likewise have the same magnitude but point in opposite directions:

(8-3)
$$m_{\text{skateboard}}\vec{v}_{\text{skateboard}} = -m_{\text{you}}\vec{v}_{\text{you}}$$

This relationship is a direct consequence of Newton's second and third laws of motion. Here's how to interpret Equation 8-3:

- The minus sign in Equation 8-3 tells us that $\vec{v}_{\text{skateboard}}$ and $\vec{v}_{\text{you}}$ are in opposite directions. As Figure 8-2b shows, if you push the skateboard to the right, you must move to the left.

- If we take the magnitude of both sides of this equation, we get

$$m_{\text{skateboard}}v_{\text{skateboard}} = m_{\text{you}}v_{\text{you}} \quad \text{or} \quad v_{\text{skateboard}} = \left(\frac{m_{\text{you}}}{m_{\text{skateboard}}}\right)v_{\text{you}} \qquad (8\text{-}4)$$

The speed of the skateboard is equal to your speed multiplied by the ratio $m_{\text{you}}/m_{\text{skateboard}}$. You are much more massive than the skateboard, so this ratio is a large number and the skateboard moves much faster than you do.

- If you push off from the skateboard with greater force, both the skateboard and you will fly off at faster speeds. But this force doesn't appear in either Equation 8-3 or 8-4. That means the ratio of the skateboard's speed to yours will be the same no matter how hard you push.

Our analysis of the skateboard problem, and especially Equation 8-3, suggests that we think about a quantity that's equal to the product of an object's mass m and its velocity vector $\vec{v}$. This quantity is called the object's **linear momentum**:

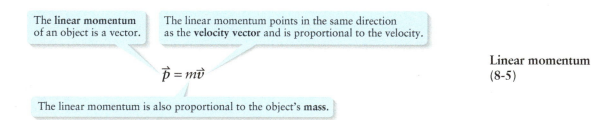

The **linear momentum** of an object is a vector.

The linear momentum points in the same direction as the **velocity vector** and is proportional to the velocity.

$$\vec{p} = m\vec{v}$$

The linear momentum is also proportional to the object's **mass**.

Linear momentum
(8-5)

We'll usually call $\vec{p}$ simply the **momentum** of the object, which is how most physicists refer to it. Note that since the mass m is a positive scalar quantity, the momentum $\vec{p}$ points in the same direction as the velocity $\vec{v}$ (see Section 3-2). Figure 8-3 compares the momentum vectors for three different objects.

Equation 8-5 shows that the units of momentum are the units of mass multiplied by the units of velocity, or kg·m/s. This combination of units doesn't have a special name in the SI system, so we simply say that the momentum of a 1000-kg car driving at 20 m/s has magnitude $p = mv = (1000 \text{ kg})(20 \text{ m/s}) = 20{,}000 \text{ kg·m/s}$, or "twenty thousand kilogram-meters per second."

We can use the definition of momentum, Equation 8-5, to rewrite Equation 8-3 for the skateboarder and you: $m_{\text{skateboard}}\vec{v}_{\text{skateboard}} = -m_{\text{you}}\vec{v}_{\text{you}}$ becomes

$$\vec{p}_{\text{skateboard}} = -\vec{p}_{\text{you}} \qquad (8\text{-}6)$$

Just after the push-off, you and the skateboard each have the same magnitude of momentum, but your momentum is directly opposite to the skateboard's momentum. We'll use this observation in Example 8-1.

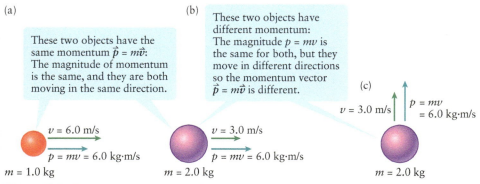

(a)

These two objects have the same momentum $\vec{p} = m\vec{v}$: The magnitude of momentum is the same, and they are both moving in the same direction.

$v = 6.0$ m/s

$p = mv = 6.0$ kg·m/s

$m = 1.0$ kg

(b)

These two objects have different momentum: The magnitude $p = mv$ is the same for both, but they move in different directions so the momentum vector $\vec{p} = m\vec{v}$ is different.

$v = 3.0$ m/s

$p = mv = 6.0$ kg·m/s

$m = 2.0$ kg

(c)

$v = 3.0$ m/s

$p = mv = 6.0$ kg·m/s

$m = 2.0$ kg

Figure 8-3 The momentum vector Comparing the momentum vectors for three different objects.

WATCH OUT! Don't confuse momentum and kinetic energy.

! Like momentum, the kinetic energy K of an object (introduced in Chapter 6) depends on its mass m and its speed v: $K = \frac{1}{2}mv^2$. But kinetic energy is a scalar quantity (it has no direction), while momentum $\vec{p} = m\vec{v}$ is a vector quantity. Furthermore, two objects with the same magnitude of momentum can have very different kinetic energies. Example 8-1 explores this further.

EXAMPLE 8-1 Momentum and Kinetic Energy

Suppose that you have a mass of 50.0 kg. Your skateboard has a mass of 2.50 kg. Just after you push off as in Figure 8-2, you are moving to the left at 0.600 m/s. (a) What is the magnitude of your momentum and the magnitude of the skateboard's momentum just after you push off? (b) What is the speed of the skateboard just after you push off? (c) What is your kinetic energy and the skateboard's kinetic energy just after you push off?

Set Up

We use the definitions of linear momentum and kinetic energy. Our discussion above and Equation 8-6 tell us how to find the speed of the skateboard from your speed.

$$\vec{p} = m\vec{v} \tag{8-5}$$

$$K = \frac{1}{2}mv^2 \tag{6-8}$$

$$\vec{p}_{\text{skateboard}} = -\vec{p}_{\text{you}} \tag{8-6}$$

$m_{\text{you}} = 50.0\ \text{kg}$
$v_{\text{you}} = 0.600\ \text{m/s}$
$v_{\text{skateboard}} = ?$
$m_{\text{skateboard}} = 2.50\ \text{kg}$

Solve

(a) Equation 8-5 gives us the magnitude of your momentum. Equation 8-6 tells us that the skateboard has as much momentum to the right as you have to the left.

Your momentum vector:

$$\vec{p}_{\text{you}} = m_{\text{you}}\vec{v}_{\text{you}}$$

Magnitude of this vector:

$$p_{\text{you}} = m_{\text{you}}v_{\text{you}}$$
$$= (50.0\ \text{kg})(0.600\ \text{m/s})$$
$$= 30.0\ \text{kg}\cdot\text{m/s}$$

Since $\vec{p}_{\text{skateboard}} = -\vec{p}_{\text{you}}$, you and the skateboard have the same magnitude of momentum:

$\vec{p}_{\text{you}} = -\vec{p}_{\text{skateboard}}$: same magnitude, opposite directions

$$p_{\text{skateboard}} = m_{\text{skateboard}}v_{\text{skateboard}} = p_{\text{you}} = m_{\text{you}}v_{\text{you}} = 30.0\ \text{kg}\cdot\text{m/s}$$

(b) We know the skateboard's mass and momentum, so we can solve for its speed.

$p_{\text{skateboard}} = m_{\text{skateboard}}v_{\text{skateboard}}$ so

$$v_{\text{skateboard}} = \frac{p_{\text{skateboard}}}{m_{\text{skateboard}}}$$
$$= \frac{30.0\ \text{kg}\cdot\text{m/s}}{2.50\ \text{kg}}$$
$$= 12.0\ \text{m/s}$$

$v_{\text{you}} = 0.600\ \text{m/s}$
$v_{\text{skateboard}} = 12.0\ \text{m/s}$
(20.0 times faster than you)

(c) Using Equation 6-8, we find the kinetic energies from the masses and speeds of the two objects.

Your kinetic energy:

$$K_{\text{you}} = \frac{1}{2}m_{\text{you}}v_{\text{you}}^2$$
$$= \frac{1}{2}(50.0\ \text{kg})(0.600\ \text{m/s})^2$$
$$= 9.00\ \text{kg}\cdot\text{m}^2/\text{s}^2 = 9.00\ \text{J}$$

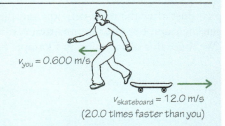
$K_{\text{you}} = 9.00\ \text{J}$
$K_{\text{skateboard}} = 180\ \text{J}$
(20.0 times more than you)

$$K_{\text{skateboard}} = \frac{1}{2}m_{\text{skateboard}}v_{\text{skateboard}}^2 = \frac{1}{2}(2.50\ \text{kg})(12.0\ \text{m/s})^2 = 180\ \text{J}$$

Reflect

The skateboard's speed is 12.0 m/s, about the same as the speed limit for cars in a residential neighborhood. It's not difficult to shove a skateboard hard enough to give it that speed.

As we discussed above, the skateboard moves much faster than you do after the push-off because it has much less mass. The ratio of your mass to that of the skateboard is 20.0 to 1: This same factor of 20.0 appears in the ratio of speeds (the skateboard is faster). It also appears in the ratio of kinetic energies: The skateboard has 20.0 times more kinetic energy than you do. If two objects have the same magnitude of momentum, the less massive one always has more kinetic energy.

Skateboard speed: $v_{\text{skateboard}} = \left(12.0\ \dfrac{\text{m}}{\text{s}}\right)\left(\dfrac{1\ \text{km}}{1000\ \text{m}}\right)\left(\dfrac{3600\ \text{s}}{1\ \text{h}}\right)$
$= 43.2\ \text{km/h} = 26.8\ \text{mi/h}$

Ratio of masses: $\dfrac{m_{\text{you}}}{m_{\text{skateboard}}} = \dfrac{50.0\ \text{kg}}{2.50\ \text{kg}} = 20.0$

Ratio of speeds: $\dfrac{v_{\text{you}}}{v_{\text{skateboard}}} = \dfrac{0.600\ \text{m/s}}{12.0\ \text{m/s}} = 0.0500 = \dfrac{1}{20.0}$

Ratio of kinetic energies: $\dfrac{K_{\text{you}}}{K_{\text{skateboard}}} = \dfrac{9.00\ \text{J}}{180\ \text{J}} = 0.0500 = \dfrac{1}{20.0}$

We can get more insight into why the two objects in Example 8-1 have the same magnitude of momentum but different kinetic energies. Let's combine Equations 8-2 and 8-3 with the definition $\vec{p} = m\vec{v}$ for momentum:

$$\vec{F}_{\text{you on skateboard}}\Delta t = m_{\text{skateboard}}\vec{v}_{\text{skateboard}} = \vec{p}_{\text{skateboard}}$$
$$\vec{F}_{\text{skateboard on you}}\Delta t = m_{\text{you}}\vec{v}_{\text{you}} = \vec{p}_{\text{you}}$$

(8-7)

Equations 8-7 say that the *momentum* that each object acquires during the push-off is equal to the net force that acts on it multiplied by the *time* over which the force acts. Newton's third law tells us that you and the skateboard exert forces of equal magnitude on each other and that when one force is present the other one must be as well. Hence both forces act for the same time Δt, and the skateboard and you end up with equal magnitudes of momentum (although in opposite directions). The *kinetic energy* that each object acquires, however, is equal to the work done on that object during the push-off, and work equals force multiplied by the *distance* over which the force acts (see Section 6-3). The skateboard moves faster than you do during the push-off, so it travels a greater distance than you do and so has more work done on it by the same magnitude of force. Therefore, the skateboard ends up with more kinetic energy (Figure 8-4).

We've seen how the idea of momentum is useful for the special case of a person pushing off a skateboard. In the next section we'll see how to apply this idea to the general case in which two objects exert forces of any kind on each other.

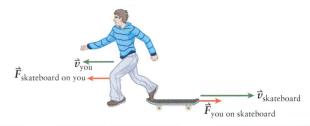

① $\vec{F}_{\text{skateboard on you}}$ and $\vec{F}_{\text{you on skateboard}}$ have the same magnitude.

② These forces act on you and the skateboard for the same amount of time, so you and the skateboard acquire the same magnitude of momentum.

③ The skateboard moves faster and covers a greater distance while the forces act, so more work is done on the skateboard and it acquires more kinetic energy.

Figure 8-4 Same momentum but different kinetic energy During the push-off you and the skateboard acquire the same magnitude of momentum, but the skateboard acquires more kinetic energy.

GOT THE CONCEPT? 8-1 Robin Hood on Rollerblades

You are challenged to an archery competition while wearing rollerblades. You release the arrow, and you recoil backward. Compared to the kinetic energy of the arrow just after it leaves the bow, the kinetic energy of you and the bow together is (a) greater; (b) the same; (c) less; (d) the answer depends on how powerful the bow is.

TAKE-HOME MESSAGE FOR Section 8-2

✔ The linear momentum of an object is a vector. It is equal to the product of the object's mass and its velocity.

✔ If two objects at rest push away from each other, each object acquires an equal amount of momentum in opposite directions (if no other net forces act on the objects).

✔ If two objects of different masses have the same magnitude of momentum, the object with the smaller mass has more kinetic energy than the other object.

8-3 The total momentum of a system of objects is conserved under certain conditions

In Section 8-2 we found that just after you pushed off a skateboard, your momentum and the momentum of the skateboard had the same magnitude but were in opposite directions. In terms of vectors (Equation 8-6), we wrote this as

$$\vec{p}_{\text{skateboard}} = -\vec{p}_{\text{you}}$$

Let's rearrange this equation so that both momentum vectors are on the same side of the equals sign:

(8-8)
$$\vec{p}_{\text{you}} + \vec{p}_{\text{skateboard}} = 0$$

The left-hand side of Equation 8-8 is the **total momentum** of you and the skateboard together just after the push-off. Since momentum is a vector, we have to use the rules of vector arithmetic to add them together: $\vec{p}_{\text{you}}$ and $\vec{p}_{\text{skateboard}}$ have equal magnitudes but opposite directions, so they add to zero. Hence the system of you and the skateboard has zero total momentum just after the push-off.

The total momentum just *before* the push-off is also zero. Initially neither you nor the skateboard is moving, so both objects have zero momentum and the sum of these is likewise zero. So for this special case the total momentum of the system of you and the skateboard is *conserved*: It has the same value (in this case zero) before and after the push-off (Figure 8-5).

We saw in Chapter 6 that the *total mechanical energy E* of a system is conserved only under special circumstances (if no work is done by nonconservative forces). Is something similar true for the total momentum of a system? And if so, what are the special circumstances under which total momentum is conserved? Let's find the answers to these questions, using our discussion from Section 8-2 as a guide.

① Before the push-off: No motion, so total momentum of the system of you and skateboard is zero:
$$\vec{p}_{\text{you}} + \vec{p}_{\text{skateboard}} = 0$$

② After the push-off: Both objects have momentum, but the total momentum is still zero:
$$\vec{p}_{\text{you}} + \vec{p}_{\text{skateboard}} = 0$$

 In this situation the total momentum of the system is conserved: The push-off does not affect its value.

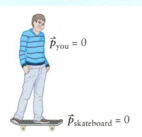

$$\vec{p}_{\text{you}} = 0$$

$$\vec{p}_{\text{skateboard}} = 0$$

$$\vec{p}_{\text{you}}$$

$$\vec{p}_{\text{skateboard}}$$

Figure 8-5 Total momentum When you push off from the skateboard, your momentum and the momentum of the skateboard both change. But the *total* momentum of the system of you and the skateboard is unchanged.

A System of Several Objects: Internal and External Forces

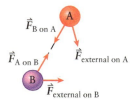

Figure 8-6 shows a system of two objects. These objects could be billiard balls on a billiard table, oxygen molecules in the air around you, or a planet and its moon. Two kinds of forces can act on each object: forces exerted by other members of the system, which we call **internal forces**, and forces exerted by objects outside the system, which we call **external forces**. For two billiard balls the internal forces are the forces that one ball exerts on another when they collide; the external forces are the normal and friction forces exerted by the billiard table, the gravitational force exerted by Earth, and the force you exert on the cue ball with the cue. For the two objects A and B shown in Figure 8-6, we can write Newton's second law as

$$\sum \vec{F}_{\text{on A}} = \sum \vec{F}_{\text{external on A}} + \vec{F}_{\text{B on A}} = m_A \vec{a}_A$$
$$\sum \vec{F}_{\text{on B}} = \sum \vec{F}_{\text{external on B}} + \vec{F}_{\text{A on B}} = m_B \vec{a}_B \qquad (8\text{-}9)$$

We can reduce Equations 8-9 to a single equation by adding them together. This has several advantages. First, it leaves us with a single equation to analyze. Second, from Newton's third law, we know that $\vec{F}_{\text{B on A}} = -\vec{F}_{\text{A on B}}$. So when we add Equations 8-9, these two terms cancel each other and we end up with

$$\sum \vec{F}_{\text{external on A}} + \sum \vec{F}_{\text{external on B}} = m_A \vec{a}_A + m_B \vec{a}_B \qquad (8\text{-}10)$$

The internal forces that the two objects exert on each other have disappeared from Equation 8-10. This will prove to be incredibly important for analyzing collisions: We'll be able to calculate the effects of a collision without knowing the details of those internal forces.

Over a time interval Δt the velocity of A changes from $\vec{v}_{Ai}$ to $\vec{v}_{Af}$ ("i" for initial, "f" for final), and the velocity of B changes from $\vec{v}_{Bi}$ to $\vec{v}_{Bf}$. Using Equation 3-9, $\vec{a} = \Delta \vec{v}/\Delta t$, we can rewrite each of the accelerations in Equation 8-10 in terms of the changes in velocity:

$$\sum \vec{F}_{\text{external on A}} + \sum \vec{F}_{\text{external on B}} = \frac{m_A(\vec{v}_{Af} - \vec{v}_{Ai})}{\Delta t} + \frac{m_B(\vec{v}_{Bf} - \vec{v}_{Bi})}{\Delta t} \qquad (8\text{-}11)$$

We can further rewrite Equation 8-11 by multiplying both sides of each equation by Δt, using the definition of momentum $\vec{p} = m\vec{v}$, and rearranging terms on the right-hand side:

$$\left(\sum \vec{F}_{\text{external on A}} + \sum \vec{F}_{\text{external on B}}\right)\Delta t = m_A \vec{v}_{Af} - m_A \vec{v}_{Ai} + m_B \vec{v}_{Bf} - m_B \vec{v}_{Bi}$$
$$\left(\sum \vec{F}_{\text{external on A}} + \sum \vec{F}_{\text{external on B}}\right)\Delta t = (\vec{p}_{Af} + \vec{p}_{Bf}) - (\vec{p}_{Ai} + \vec{p}_{Bi}) \qquad (8\text{-}12)$$

In Equation 8-12 $\vec{p}_{Ai}$ and $\vec{p}_{Bi}$ are the momenta of A and B, respectively, at the beginning of the time interval Δt, and $\vec{p}_{Af}$ and $\vec{p}_{Bf}$ are the momenta of A and B, respectively, at the end of the time interval.

The quantity in parentheses on the left-hand side of Equation 8-12 is the sum of all of the external forces that act on the system of objects A and B. We'll call this $\sum \vec{F}_{\text{external on system}}$ for short. The right-hand side of the equation is the difference between the *total* momentum of the system after the time interval Δt, $\vec{P}_f = \vec{p}_{Af} + \vec{p}_{Bf}$, and the *total* momentum of the system before the time interval, $\vec{P}_i = \vec{p}_{Ai} + \vec{p}_{Bi}$. So we can rewrite Equation 8-12 as

> The **sum of all external forces** acting on a system of objects

> **Duration** of a time interval over which the external forces act

$$\left(\sum \vec{F}_{\text{external on system}}\right)\Delta t = \vec{P}_f - \vec{P}_i = \Delta\vec{P}$$

> **Change** during that time interval **in the total momentum** of the objects that make up the system

External force and total momentum change for a system of objects (8-13)

Equation 8-13 says something quite remarkable: *Only the external forces acting on a system can affect the system's total momentum.* The internal forces of one object on another allow momentum to be transferred between the objects (for example, when a moving cue ball hits an eight-ball at rest and sends the eight-ball flying), but they don't affect the value of the *total* momentum.

We now have the answer to the question "When is momentum conserved?" Equation 8-13 says that the total momentum does not change over the time interval Δt if the net external force on the system is zero. Then the left-hand side of Equation 8-13 is zero and so there is zero difference between the final total momentum $\vec{P}_f = \vec{p}_{Af} + \vec{p}_{Bf}$ and the initial total momentum $\vec{P}_i = \vec{p}_{Ai} + \vec{p}_{Bi}$:

If the net external force on a system of objects is zero, **the total momentum of the system** is conserved.

Law of conservation of momentum (8-14)

$$\vec{P}_f = \vec{P}_i$$

Then the total momentum of the system at the end of a time interval...

...is equal to the total momentum of the system at the beginning of that time interval.

Equation 8-14 is the **law of conservation of momentum.** It explains why the total momentum was conserved for the system of you and the skateboard in Section 8-2: There were external forces acting on you and on the skateboard, but the vector sum of these external forces was zero. Hence the total momentum of you and the skateboard had the same value (zero) both before and after the push-off.

WATCH OUT! Remember that momentum is a vector.

Although we have said that total momentum is conserved, it may seem like momentum appears from "out of nowhere" in Figure 8-5: There is zero momentum before the push-off, but after the push-off both you and the skateboard have acquired momentum. Isn't this a contradiction? The answer is no, and the reason is that momentum is a *vector*. Your momentum vector $\vec{p}_{you}$ points to the left, while the skateboard's momentum vector $\vec{p}_{skateboard}$ points to the right and has the same magnitude. Hence the *vector sum* of these two vectors is zero, and the *total* momentum has the same value before and after the push-off. If you always keep in mind that momentum is a vector, you'll avoid a lot of confusion.

Momentum Conservation and Collisions

The law of conservation of momentum turns out to be useful even when the net external force on a system is *not* zero. One example is a collision between two automobiles (Figure 8-7). During the collision the net vertical force on each car is zero: The upward normal force exerted by the ground balances the downward gravitational force. But the ground also exerts a horizontal friction force on each car, and there is nothing to balance the friction forces. So there is a net external force on the system of two cars. Nonetheless, it turns out that we can ignore this net external force! To see why this is so, let's rewrite Equation 8-12 the way it would appear if we didn't immediately cancel the internal forces when we added Equations 8-9 (the internal forces still appear on the left-hand side of the equation):

(8-15)

$$\left(\sum \vec{F}_{\text{external on A}} + \sum \vec{F}_{\text{external on B}}\right)\Delta t + (\vec{F}_{\text{B on A}} + \vec{F}_{\text{A on B}})\Delta t$$
$$= \vec{p}_{Af} + \vec{p}_{Bf} - \vec{p}_{Ai} - \vec{p}_{Bi}$$

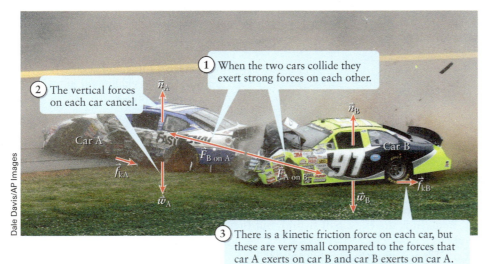

Figure 8-7 Forces in a collision
During a collision between two cars, the internal forces (the forces of one car on another) are so great that we can ignore the external forces on the system.

1 When the two cars collide they exert strong forces on each other.

2 The vertical forces on each car cancel.

$\vec{n}_A$
$\vec{n}_B$

Car A
Car B

$\vec{F}_{\text{B on A}}$
$\vec{F}_{\text{A on B}}$

$\vec{f}_{kA}$
$\vec{f}_{kB}$

$\vec{w}_A$
$\vec{w}_B$

Dale Davis/AP Images

3 There is a kinetic friction force on each car, but these are very small compared to the forces that car A exerts on car B and car B exerts on car A.

So we can say that:
• The net force on car A is $\vec{F}_{\text{B on A}}$ and the net force on car B is $\vec{F}_{\text{A on B}}$.
• These forces are internal to the system of two cars.

The forces $\vec{F}_{\text{B on A}}$ and $\vec{F}_{\text{A on B}}$ that the cars exert on each other are *very* large: They can deform metal and shatter both plastic and glass. By comparison, the external friction forces are much smaller in magnitude. So it's a very good approximation to ignore the external friction forces during the brief duration Δt of the collision and set the terms $\sum \vec{F}_{\text{external on A}}$ and $\sum \vec{F}_{\text{external on B}}$ in Equation 8-15 equal to zero:

$$(\vec{F}_{\text{B on A}} + \vec{F}_{\text{A on B}})\Delta t = \vec{p}_{\text{Af}} + \vec{p}_{\text{Bf}} - \vec{p}_{\text{Ai}} - \vec{p}_{\text{Bi}} \qquad (8\text{-}16)$$

But by Newton's third law $\vec{F}_{\text{B on A}} = -\vec{F}_{\text{A on B}}$, so the left-hand side of Equation 8-16 for the two colliding cars is zero. Thus the change in total momentum during the collision is still zero, or

$$\vec{p}_{\text{Af}} + \vec{p}_{\text{Bf}} = \vec{p}_{\text{Ai}} + \vec{p}_{\text{Bi}} \qquad (8\text{-}17)$$

This is the same as Equation 8-14, the law of conservation of momentum. So we conclude that

If the internal forces during a collision are much greater in magnitude than the external forces, the total momentum of the colliding objects has essentially the same value just before and just after the collision.

In *most* collisions the internal forces are much larger in magnitude than the external forces. For example, when two hockey players collide on the ice, the internal forces of one player on the other are strong enough to knock the wind out of the players or even cause injury. These internal forces are much greater than the friction forces that the ice exerts on the players. Similarly, the forces that act between a tennis ball and tennis racquet in a serve are so large that the ball distorts noticeably (Figure 8-8). The external forces that act on the ball and racquet during the time they are in contact—the gravitational forces and the force of the player's hand on the racquet—are very feeble by comparison. So it's almost always safe to say that the total momentum of a system of colliding objects just *after* the collision is the same as just *before* the collision.

Ted Foxx/Alamy

Figure 8-8 Strong internal forces, weak external forces During a serve, the internal force of a tennis racquet on the ball is great enough to cause tremendous distortion. The external forces on the system of racquet and ball are so much weaker by comparison that they can be neglected.

 Go to Interactive Exercise 8-1 for more practice dealing with conservation of momentum.

WATCH OUT! In a collision momentum is conserved only *during* the collision.

The general rule that we have discovered says that the total momentum of colliding objects is conserved only during the very brief period during which the collision takes place. Once the collision is over the strong internal forces no longer act, the external forces become dominant, and the total momentum is no longer conserved. For example, the two colliding cars shown in Figure 8-7 continued to move after the collision (with a total momentum just after the collision that equals their total momentum just before the collision). Once the collision has ended, however, friction with the track and grass soon caused the cars to come to a halt and lose all of their momentum.

WATCH OUT! Momentum conservation means each component of momentum is separately conserved.

It's important to note that Equation 8-14, the law of conservation of momentum, is a vector equation. If the vectors $\vec{P}_{\text{f}}$ and $\vec{P}_{\text{i}}$ are equal, it must be that the x components of the two vectors are equal *and* the y components of the two vectors are equal. Examples 8-2, 8-3, and 8-4 illustrate how to solve problems that involve the vector nature of momentum.

EXAMPLE 8-2 Conservation of Momentum: A Collision on the Ice

Gordie, a 100-kg hockey player, is initially moving to the right at 5.00 m/s directly toward Mario, a stationary 80.0-kg player. After the two players collide head-on, Mario is moving to the right at 3.75 m/s. (a) In what direction and at what speed is Gordie moving after the collision? (b) What was the change in Gordie's momentum in the collision? What was the change in Mario's momentum?

Set Up

The system that we're considering is made up of the two players. The vertical forces on each player (the normal force exerted by the ice and the gravitational force) cancel each other, so there is no net external force in the vertical direction. The friction forces between the players and the ice are small compared to the forces that Gordie and Mario exert on each other. So we can treat the total momentum of the system as conserved during the collision. We'll use this to find Gordie's final velocity and the changes in momentum of each player. We choose the positive x direction to be to the right, as shown.

Momentum conservation

$$\vec{P}_f = \vec{P}_i \qquad (8\text{-}14)$$

So $\vec{P} = \vec{p}_G + \vec{p}_M$ ("G" for Gordie, "M" for Mario) has the same value just before and just after the collision.

$$\vec{p} = m\vec{v} \qquad (8\text{-}5)$$

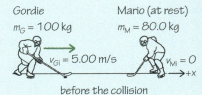

Gordie $m_G = 100$ kg
Mario (at rest) $m_M = 80.0$ kg
$v_{Gi} = 5.00$ m/s
$v_{Mi} = 0$
$\rightarrow +x$
before the collision

Solve

(a) Write the equation of momentum conservation, using the subscript "i" (for initial) for values just before the collision and subscript "f" (for final) for values just after. Note that $\vec{v}_{Mi} = 0$ (Mario's velocity just before the collision) since Mario is originally at rest. We need to find $\vec{v}_{Gf}$ (Gordie's velocity just after the collision).

Just before the collision the total momentum is

$$\vec{P}_i = \vec{p}_{Gi} + \vec{p}_{Mi} = m_G\vec{v}_{Gi} + m_M\vec{v}_{Mi}$$

Just after the collision the total momentum is

$$\vec{P}_f = \vec{p}_{Gf} + \vec{p}_{Mf} = m_G\vec{v}_{Gf} + m_M\vec{v}_{Mf}$$

Momentum is conserved:

$$\vec{P}_f = \vec{P}_i$$
$$m_G\vec{v}_{Gf} + m_M\vec{v}_{Mf} = m_G\vec{v}_{Gi} + m_M\vec{v}_{Mi}$$

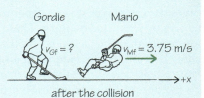

Gordie
$v_{Gf} = ?$
Mario
$v_{Mf} = 3.75$ m/s
$\rightarrow +x$
after the collision

Since the motion is entirely along the x axis, we only need the x component of the momentum conservation equation. Solve for Gordie's final x velocity, v_{Gfx}; then substitute the values of the players' masses, Gordie's velocity before the collision, and Mario's velocity after the collision.

$$m_G v_{Gfx} + m_M v_{Mfx} = m_G v_{Gix} + m_M v_{Mix}$$

Mario is originally at rest, so $v_{Mix} = 0$ and

$$m_G v_{Gfx} + m_M v_{Mfx} = m_G v_{Gix}$$
$$m_G v_{Gfx} = m_G v_{Gix} - m_M v_{Mfx}$$

$$v_{Gfx} = v_{Gix} - \frac{m_M v_{Mfx}}{m_G}$$

Players' masses: $m_G = 100$ kg, $m_M = 80.0$ kg

Gordie's initial x velocity: $v_{Gix} = +5.00$ m/s

Mario's final x velocity: $v_{Mfx} = +3.75$ m/s

$$v_{Gfx} = (+5.00 \text{ m/s}) - \frac{(80.0 \text{ kg})(+3.75 \text{ m/s})}{100 \text{ kg}}$$

$$= +5.00 \text{ m/s} - 3.00 \text{ m/s}$$

$$= +2.00 \text{ m/s}$$

Gordie ends up moving at 2.00 m/s to the right (in the positive x direction).

(b) The change in each player's momentum equals his momentum after the collision minus his momentum before the collision.

Gordie: $\Delta p_{Gx} = m_G v_{Gfx} - m_G v_{Gix}$
$= (100 \text{ kg})(+2.00 \text{ m/s}) - (100 \text{ kg})(+5.00 \text{ m/s})$
$= 200 \text{ kg} \cdot \text{m/s} - 500 \text{ kg} \cdot \text{m/s}$
$= -300 \text{ kg} \cdot \text{m/s}$

Mario: $\Delta p_{Mx} = m_M v_{Mfx} - m_M v_{Mix}$
$= (80.0 \text{ kg})(+3.75 \text{ m/s}) - (80.0 \text{ kg})(0 \text{ m/s})$
$= +300 \text{ kg} \cdot \text{m/s}$

Reflect

Our answer to part (a) tells us that after the collision Gordie is still moving in the positive x direction but with reduced speed: He has lost x momentum, while Mario (who was originally at rest) has gained x momentum. In fact, as part (b) shows, the amount of x momentum that Gordie loses (300 kg·m/s) is exactly the same as the amount of x momentum that Mario gains. Thus we can think of the collision between the two players as a transfer of momentum between Gordie and Mario. No momentum is lost in the collision; it simply changes hands from one player to the other.

Notice that in this problem we needed to know Mario's final x velocity v_{Mfx} to find Gordie's final x velocity v_{Gfx}. That's because the statement that momentum is conserved in the x direction gave us only one equation which relates v_{Mfx} and v_{Gfx}. Hence we were able to solve for only one unknown quantity. If we didn't know either of the players' final velocities, we would have needed additional information—in particular, the duration of the collision and the magnitude of the force that one player exerts on the other during the collision—which we unfortunately do not have. The law of conservation of momentum is a great tool, but by itself it can't tell you everything!

Although we used conservation of momentum in this example, note that neither player's individual momentum is conserved: Gordie's momentum after the collision is different than before the collision and likewise for Mario's momentum. It's only the total momentum of Gordie and Mario's system that is conserved. *In general the momentum of any particular object within a system will not be conserved.*

EXAMPLE 8-3 A Collision at the Bowl-a-Rama

The sequence of images in Figure 8-9 shows a bowling ball striking a stationary pin. The second image shows that when the collision occurs the ball strikes a bit to the left of the horizontal center of the pin. The third image shows what happens after the collision: The ball moves on a path to the left of its original direction, and the pin shoots off to the right. Just after one such collision, the ball is moving off horizontally at a 10.0° angle and the pin is moving off horizontally at a 60.0° angle (measured relative to the original direction of motion of the ball). The mass of the ball is 3.50 times greater than the mass of the pin. Just after the collision is the pin moving faster or slower than the ball, and by what factor?

Figure 8-9 Hitting a bowling pin off-center After the bowling ball strikes the left-hand side of the pin, the ball deflects to the left and the pin moves off to the right. How do their speeds compare?

Set Up

The forces that the ball and pin exert on each other are very strong (imagine what it would feel like if your finger were between the ball and pin when they hit!). Compared to these we can ignore the external forces exerted on the system of ball and pin, so the total momentum of this system is conserved. The collision is two dimensional, so we must account for both the x and y components of momentum.

Momentum conservation
$$\vec{P}_f = \vec{P}_i \qquad (8\text{-}14)$$

So $\vec{P} = \vec{p}_B + \vec{p}_P$ ("B" for ball, "P" for pin) has the same value just before and just after the collision.

Linear momentum:
$$\vec{p} = m\vec{v} \qquad (8\text{-}5)$$

We are given the directions of motion of the ball and pin before and after the collision, and we want to find the ratio of the pin's final speed v_{Pf} to the ball's final speed v_{Bf}. We don't know the mass of either the ball or the pin, but we do know the ratio of their masses.

Ratio of the mass of the ball to the mass of the pin:

$$\frac{m_B}{m_P} = 3.50$$

Solve

The total momentum before the collision (subscript "i") equals the total momentum after the collision (subscript "f"). Write this for both the x component and the y component of momentum.

Momentum conservation:

$\vec{p}_{Bf} + \vec{p}_{Pf} = \vec{p}_{Bi} + \vec{p}_{Pi}$, or

$m_B\vec{v}_{Bf} + m_P\vec{v}_{Pf} = m_B\vec{v}_{Bi} + m_P\vec{v}_{Pi}$, or

$m_B v_{Bfx} + m_P v_{Pfx} = m_B v_{Bix} + m_P v_{Pix}$
(total x momentum is conserved)

$m_B v_{Bfy} + m_P v_{Pfy} = m_B v_{Biy} + m_P v_{Piy}$
(total y momentum is conserved)

Just prior to the collision, the ball has no component of velocity in the y direction (so $v_{Biy} = 0$) and the pin is at rest (so $v_{Pix} = v_{Piy} = 0$). Just after the collision the ball is moving at a 10.0° angle with positive x and y components of velocity, while the pin is moving at a 60.0° angle with a positive x velocity and a negative y velocity. Use these facts to rewrite the equations for the conservation of x and y momentum.

x equation:

$m_B v_{Bf} \cos 10.0°$
$\quad + m_P v_{Pf} \cos 60.0° = m_B v_{Bi}$

y equation:

$m_B v_{Bf} \sin 10.0°$
$\quad + (-m_P v_{Pf} \sin 60.0°) = 0$

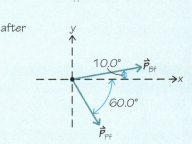

Our goal is to find the ratio of the speed of the pin just after the collision (v_{Pf}) to the speed of the ball just after the collision (v_{Bf}). The equation for x momentum isn't useful for this because we aren't given the value of the ball's speed v_{Bi} before the collision. Instead we use the equation for y momentum to solve for v_{Pf}/v_{Bf}, and then substitute the ratio of the masses of the two objects.

$m_B v_{Bf} \sin 10.0° = m_P v_{Pf} \sin 60.0°$

$$\frac{v_{Pf}}{v_{Bf}} = \frac{m_B}{m_P}\frac{\sin 10.0°}{\sin 60.0°} = (3.50)\frac{0.174}{0.866} = 0.702$$

After the collision the speed of the pin is 0.702 times that of the ball.

Reflect

The pin moves relatively slowly because it suffered only a glancing blow from the ball. With a more head-on impact the pin can end up moving away from the collision at a much faster speed than the ball. We'll explore this in Section 8-5.

EXAMPLE 8-4   A Tricky Billiards Shot

On a billiards table the 7-ball and the 8-ball are initially at rest and touching each other. You hit the cue ball in such a way that it acquires a speed of 1.7 m/s before hitting the 7-ball and 8-ball simultaneously. After the collision the cue ball is at rest, and the other two balls are each moving at 45° from the direction that the cue ball was moving. What are the speeds of the 7-ball and 8-ball immediately after the collision? Each billiard ball has a mass of 0.16 kg.

Set Up

This is a more complicated collision than in the previous two examples, but the fundamental principle is the same: Total momentum is conserved.

Initially all of the momentum is in the cue ball and points along the direction of its motion. After the collision the 7-ball and 8-ball must travel on opposite sides of the cue ball's initial path as shown. (If they were on the same side, there would be a nonzero total momentum perpendicular to the cue ball's initial path, and momentum would not be conserved.)

As in Example 8-2 we know the directions of motion of the balls before and after the collision. We know the initial speed $v_{cue,i} = 1.7$ m/s and final speed $v_{cue,f} = 0$ of the cue ball, and we want to find the final speeds v_{7f} and v_{8f} of the 7-ball and 8-ball.

Momentum conservation

$$\vec{P}_f = \vec{P}_i \qquad (8\text{-}14)$$

So $\vec{P} = \vec{p}_{cue} + \vec{p}_7 + \vec{p}_8$ (cue ball, 7-ball, and 8-ball) has the same value just before and just after the collision

Linear momentum:

$$\vec{p} = m\vec{v} \qquad (8\text{-}5)$$

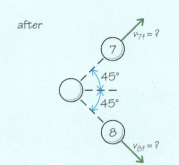

Solve

Write in component form the momentum conservation equation, which says that the total momentum before the collision (subscript "i") equals the total momentum after the collision (subscript "f"). Take the positive x axis to be in the direction that the cue ball was moving before the collision. Note that all three balls have the same mass $m = 0.16$ kg.

Momentum conservation:

$$\vec{p}_{cue,f} + \vec{p}_{7f} + \vec{p}_{8f} = \vec{p}_{cue,i} + \vec{p}_{7i} + \vec{p}_{8i}$$

The 7-ball and 8-ball are not moving before the collision, so $\vec{p}_{7i} = \vec{p}_{8i} = 0$, and the cue ball is not moving after the collision, so $\vec{p}_{cue,f} = 0$. So the momentum conservation equation becomes

$$\vec{p}_{7f} + \vec{p}_{8f} = \vec{p}_{cue,i}$$
$$m\vec{v}_{7f} + m\vec{v}_{8f} = m\vec{v}_{cue,i}$$

or in component form

$$mv_{7fx} + mv_{8fx} = mv_{cue,ix} \text{ (total } x \text{ momentum is conserved)}$$
$$mv_{7fy} + mv_{8fy} = mv_{cue,iy} \text{ (total } y \text{ momentum is conserved)}$$

The cue ball's initial velocity had zero y component, so $v_{cue,iy} = 0$. Of the 7-ball and 8-ball, one has positive final y velocity and the other negative final y velocity to satisfy the condition that total y momentum is conserved. Use these observations to simplify the momentum equations.

x equation:

$$mv_{7f} \cos 45°$$
$$+ mv_{8f} \cos 45° = mv_{cue,i}$$

y equation:

$$mv_{7f} \sin 45°$$
$$+ (-mv_{8f} \sin 45°) = 0$$

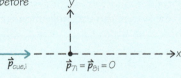

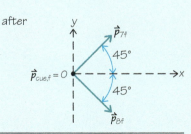

Solve the momentum conservation equations for the final speeds of the 7-ball and 8-ball.

From the y equation

$$mv_{7f} \sin 45° = mv_{8f} \sin 45° \text{ so}$$
$$v_{7f} = v_{8f}$$

After the collision the 7-ball and the 8-ball move with the same speed. To find this common speed, replace v_{8f} in the x equation with v_{7f} and then solve for v_{7f}:

$$mv_{7f}\cos 45° + mv_{7f}\cos 45° = mv_{\text{cue,i}}$$

$$2mv_{7f}\cos 45° = mv_{\text{cue,i}}$$

$$v_{7f} = \frac{v_{\text{cue,i}}}{2\cos 45°} = \frac{1.7\ \text{m/s}}{2\cos 45°} = 1.2\ \text{m/s}$$

After the collision both the 7-ball and the 8-ball are moving at 1.2 m/s.

Reflect

Note that we didn't need the value of the billiard ball mass. That's because all three balls have the same mass m, so m canceled out in the equations. That didn't happen in Example 8-3 because the bowling ball and pin had *different* masses.

There were three objects in this collision rather than two as in Examples 8-2 and 8-3. But the same ideas of momentum conservation apply no matter how many objects are involved in the collision.

GOT THE CONCEPT? 8-2 Which Way Did the Other One Go?

 An object moving due west collides with a second object that is initially at rest. Just after the collision the first object is moving toward the southeast. Just after the collision the second object must be moving toward the (a) northeast; (b) southeast; (c) southwest; (d) northwest; (e) west.

TAKE-HOME MESSAGE FOR Section 8-3

✔ In a system of objects internal forces are those that one object in the system exerts on another object in the system. External forces are those exerted on objects in the system by other, outside objects.

✔ If the net external force on a system of objects is zero, the total momentum of the system is conserved. Objects within the system can exchange momentum with each other, but the vector sum of the momentum of all members of the system remains constant.

✔ During a collision the internal forces are typically much larger in magnitude than any external forces. Hence the external forces can be ignored, and the total momentum just after the collision equals the total momentum just before the collision.

8-4 In an inelastic collision some of the mechanical energy is lost

We've learned that momentum is conserved in collisions of all kinds. We also learned in Chapter 6 that *mechanical energy* is conserved in certain circumstances. Are *both* mechanical energy and momentum conserved in collisions?

The answer to this question is "sometimes." In Section 6-7 we learned that mechanical energy is conserved if *only* conservative forces do work. An example of this is two objects that behave like ideal springs when they collide: They compress, converting some of the kinetic energy of the colliding objects into spring potential energy; then they relax and convert the potential energy back into kinetic energy (Figure 8-10a). A collision of this kind, in which the forces between the colliding objects are conservative, is called an **elastic collision**. In an elastic collision both total momentum *and* total mechanical energy are conserved.

Figure 8-10 Collision variations
(a) The ideal springs between these colliding objects exert conservative forces. Mechanical energy is conserved in this elastic collision.
(b) The forces between colliding billiard balls are very nearly conservative, and the collision is very nearly elastic. (c) The forces between colliding cars are not conservative. Mechanical energy is lost, and the collision is inelastic.

(a)

Because the ideal springs between these colliding objects exert conservative forces, mechanical energy is conserved in this **elastic collision**.

Before collision   During collision   After collision

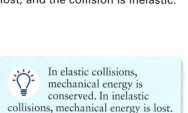

In elastic collisions, mechanical energy is conserved. In inelastic collisions, mechanical energy is lost.

(b)

The collision of the billiard balls is elastic: It does not cause any permanent deformation.

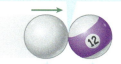

The forces between colliding ideal billiard balls are conservative. No mechanical energy is lost in this **elastic collision**.

(c)

The collision of the two cars is inelastic: It causes a permanent deformation.

The forces between colliding cars are not conservative. Mechanical energy is lost in this **inelastic collision**.

Elastic collisions are happening all around you: When oxygen and nitrogen molecules in the air collide with each other, the collisions are almost always elastic. That's because the forces between molecules (which are fundamentally electric in nature) are conservative forces. On a larger scale a collision between billiard balls is very nearly elastic (Figure 8-10b). A billiard ball deforms slightly when hit but quickly springs back to its original shape with very little loss of energy.

Something very different usually happens when two automobiles collide: The bodies of the automobiles deform and do not spring back (Figure 8-10c). The forces involved in this deformation are *nonconservative* so mechanical energy is lost and is converted to internal energy. (This is by design. By absorbing mechanical energy as it deforms, the structure of an automobile prevents that energy from being used to do potentially harmful work on the automobile's occupants.) A collision in which mechanical energy is *not* conserved is called an **inelastic collision**.

If we know the masses of the colliding objects and their velocities before and after the collision, it's straightforward to determine whether the collision is elastic or inelastic. Here's the idea: By analogy to a collision between ideal springs (Figure 8-10a), there is zero potential energy just before and just after the collision (corresponding to the springs being relaxed). So just before and just after the collision, the total mechanical energy of the system is equal to the sum of the *kinetic energies* of the colliding objects. If the total kinetic energy has the same value before and after the collision, the collision is elastic; if there is less total kinetic energy after the collision, the collision is inelastic. Example 8-5 illustrates this technique for analyzing collisions.

EXAMPLE 8-5   Elastic or Inelastic?

Determine whether the following collisions are elastic or inelastic: (a) The collision of two hockey players in Example 8-2 (Section 8-3); (b) the collision of three billiard balls in Example 8-4 (Section 8-3).

Set Up

For each collision we calculate the total kinetic energy (the sum of the individual kinetic energies of the colliding objects) just before and just after the collision. If these are equal, total mechanical energy is conserved and the collision is elastic; if they are not equal, the collision is inelastic.

Kinetic energy:

$$K = \frac{1}{2}mv^2 \qquad (6\text{-}8)$$

Solve

(a) The collision between the two hockey players in Example 8-2 is one dimensional (that is, along a straight line). So the square of each player's speed v is the same as the square of each player's x velocity v_x. The figure shows the values (from Example 8-2) of the masses of each player and their speeds before and after the collision.

Players' masses:

$m_G = 100$ kg, $m_M = 80.0$ kg

Before the collision Gordie has x velocity $v_{Gix} = +5.00$ m/s and Mario is at rest. So the *initial* total kinetic energy is

$$K_{Gi} + K_{Mi} = \frac{1}{2}m_G v_{Gix}^2 + \frac{1}{2}m_M v_{Mix}^2$$

$$= \frac{1}{2}(100 \text{ kg})(5.00 \text{ m/s})^2$$

$$+ \frac{1}{2}(80.0 \text{ kg})(0 \text{ m/s})^2$$

$$= 1.25 \times 10^3 \text{ J}$$

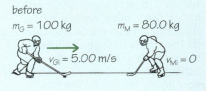

before

$m_G = 100$ kg $m_M = 80.0$ kg

$v_{GI} = 5.00$ m/s $v_{MI} = 0$

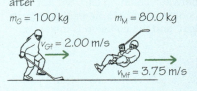

after

$m_G = 100$ kg $m_M = 80.0$ kg

$v_{Gf} = 2.00$ m/s

$v_{Mf} = 3.75$ m/s

After the collision, Gordie has x velocity $v_{Gfx} = +2.00$ m/s and Mario has x velocity $v_{Mfx} = +3.75$ m/s. So the *final* total kinetic energy is

$$K_{Gf} + K_{Mf} = \frac{1}{2}m_G v_{Gfx}^2 + \frac{1}{2}m_M v_{Mfx}^2$$

$$= \frac{1}{2}(100 \text{ kg})(2.00 \text{ m/s})^2 + \frac{1}{2}(80.0 \text{ kg})(3.75 \text{ m/s})^2$$

$$= 763 \text{ J}$$

The total kinetic energy is *less* after the collision than before, so this collision is *inelastic*.

(b) Repeat the calculation for the three billiard balls in Example 8-4. The figure shows the values from that example of the masses of the billiard balls, their speeds before the collision, and their speeds after the collision.

Each ball has mass $m = 0.16$ kg. Before the collision the cue ball is moving at speed $v_{cue,i} = 1.7$ m/s, and both the 7-ball and 8-ball are at rest. So the *initial* total kinetic energy is

$$K_{cue,i} + K_{7i} + K_{8i}$$

$$= \frac{1}{2}mv_{cue,i}^2 + \frac{1}{2}mv_{7i}^2 + \frac{1}{2}mv_{8i}^2$$

$$= \frac{1}{2}(0.16 \text{ kg})(1.7 \text{ m/s})^2$$

$$+ \frac{1}{2}(0.16 \text{ kg})(0 \text{ m/s})^2$$

$$+ \frac{1}{2}(0.16 \text{ kg})(0 \text{ m/s})^2 = 0.23 \text{ J}$$

before

$m = 0.16$ kg for each ball

$v_{cue,i} = 1.7$ m/s

7 $v_{7i} = 0$

8 $v_{8i} = 0$

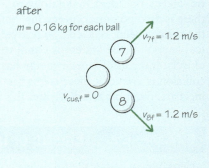

after

$m = 0.16$ kg for each ball

$v_{7f} = 1.2$ m/s

7

$v_{cue,f} = 0$

8 $v_{8f} = 1.2$ m/s

After the collision the cue ball is at rest, the 7-ball is moving at 1.2 m/s, and the 8-ball is also moving at 1.2 m/s. So the *final* total kinetic energy is

$$K_{cue,f} + K_{7f} + K_{8f} = \frac{1}{2}mv_{cue,f}^2 + \frac{1}{2}mv_{7f}^2 + \frac{1}{2}mv_{8f}^2$$

$$= \frac{1}{2}(0.16 \text{ kg})(0)^2 + \frac{1}{2}(0.16 \text{ kg})(1.2 \text{ m/s})^2$$

$$+ \frac{1}{2}(0.16 \text{ kg})(1.2 \text{ m/s})^2$$

$$= 0.23 \text{ J}$$

The total kinetic energy is *unchanged* by the collision, so this collision is *elastic*.

Reflect

The human body is not as "springy" as a billiard ball, so the forces between Gordie and Mario when they collide are nonconservative. The lost mechanical energy went into increasing the two players' internal energies, which means that their temperatures increased slightly as a result of the collision.

Completely Inelastic Collisions

The type of collision in which the *most* mechanical energy is lost is a **completely inelastic collision**, in which two objects stick together after they collide. A collision between two cars is completely inelastic if the cars lock together and don't separate. (If the cars deform and then bounce apart, the collision is inelastic but not *completely* inelastic.) You are the result of a completely inelastic collision in which a sperm cell fused with an ovum, which 9 months later led to your birth (Figure 8-11).

When an object of mass m_A and velocity $\vec{v}_{Ai}$ undergoes a completely inelastic collision with a second object of mass m_B and velocity $\vec{v}_{Bi}$, we can regard what remains after the collision as a single object of mass $m_A + m_B$. Momentum conservation then gives us an equation for the velocity $\vec{v}_f$ of this combined object just after the collision:

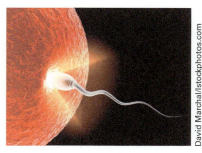

Figure 8-11 A completely inelastic collision A sperm cell and an ovum move together after they collide.

Velocities of objects A and B **before** they undergo a completely inelastic collision (in which they stick together)

$$m_A\vec{v}_{Ai} + m_B\vec{v}_{Bi} = (m_A + m_B)\vec{v}_f$$

Masses of objects A and B

Velocity of the two objects moving together **after** the collision

Momentum conservation in a completely inelastic collision (8-18)

→ *Go to Interactive Exercise 8-2 for more practice dealing with inelastic collisions.*

Examples 8-6 and 8-7 illustrate how to use this equation.

EXAMPLE 8-6 A Head-On, Completely Inelastic Collision

In a scene from an action movie, a 1.50×10^3-kg car moving north at 35.0 m/s collides head-on with a 7.50×10^3-kg truck moving south at 25.0 m/s. The car and truck stick together after the collision. (a) How fast and in what direction is the wreckage traveling just after the collision? (b) How much mechanical energy is lost in the collision?

Set Up

Since the two vehicles stick together, this is a completely inelastic collision. We'll use Equation 8-18 to find the final velocity $\vec{v}_f$ of the wreckage. We'll then compare the final kinetic energy of the wreckage to the combined kinetic energies of the car and truck before the collision; the difference is the amount of mechanical energy that's lost in the collision.

Momentum conservation in a completely inelastic collision:

$$m_{car}\vec{v}_{car,i} + m_{truck}\vec{v}_{truck,i} = (m_{car} + m_{truck})\vec{v}_f \quad (8\text{-}18)$$

Kinetic energy:

$$K = \frac{1}{2}mv^2 \quad (6\text{-}8)$$

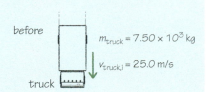
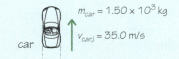

Solve

(a) The collision is along a straight line, which we call the x axis. We take the positive x axis to be to the north, in the direction of the car's motion before the collision. We use the x component of Equation 8-18 to solve for the final velocity.

Equation for conservation of x momentum in a completely inelastic collision:

$$m_{car}v_{car,ix} + m_{truck}v_{truck,ix} = (m_{car} + m_{truck})v_{fx}$$

Solve for final velocity of wreckage:

$$v_{fx} = \frac{m_{car}v_{car,ix} + m_{truck}v_{truck,ix}}{m_{car} + m_{truck}}$$

With our choice of x axis, we have

$$v_{car,ix} = +35.0 \text{ m/s}$$
$$v_{truck,ix} = -25.0 \text{ m/s}$$

Substitute values:

$$v_{fx} = \frac{\left[\begin{array}{c}(1.50 \times 10^3 \text{ kg})(+35.0 \text{ m/s}) \\ - (7.50 \times 10^3 \text{ kg})(-25.0 \text{ m/s})\end{array}\right]}{1.50 \times 10^3 \text{ kg} + 7.50 \times 10^3 \text{ kg}}$$
$$= -15.0 \text{ m/s}$$

The wreckage moves at 15.0 m/s to the south (in the negative x direction).

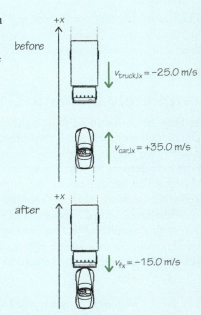

(b) Calculate the total kinetic energies before and after the collision, and find the difference between them.

The change in kinetic energy is -2.25×10^6 J, so the mechanical energy lost is the absolute value of that change, or $+2.25 \times 10^6$ J.

Total kinetic energy before the collision:

$$K_{car,i} + K_{truck,i} = \frac{1}{2}m_{car}v_{car,i}^2 + \frac{1}{2}m_{truck}v_{truck,i}^2$$
$$= \frac{1}{2}(1.50 \times 10^3 \text{ kg})(35.0 \text{ m/s})^2$$
$$+ \frac{1}{2}(7.50 \times 10^3 \text{ kg})(25.0 \text{ m/s})^2$$
$$= 9.19 \times 10^5 \text{ J} + 2.34 \times 10^6 \text{ J}$$
$$= 3.26 \times 10^6 \text{ J}$$

Total kinetic energy after the collision:

$$K_{car,f} + K_{truck,f} = \frac{1}{2}(m_{car} + m_{truck})v_f^2$$
$$= \frac{1}{2}(1.50 \times 10^3 \text{ kg} + 7.50 \times 10^3 \text{ kg})(15.0 \text{ m/s})^2$$
$$= 1.01 \times 10^6 \text{ J}$$

To find the change in mechanical energy during the collision, calculate the difference between the final and initial energies:

$$(K_{car,f} + K_{truck,f}) - (K_{car,i} + K_{truck,i})$$
$$= 1.01 \times 10^6 \text{ J} - 3.26 \times 10^6 \text{ J}$$
$$= -2.25 \times 10^6 \text{ J}$$

Reflect

It makes sense that the wreckage moves in the direction of the truck's initial motion. Before the collision the magnitude of the southbound truck's momentum was much greater than that of the northbound car, so the total momentum was to the south.

Our result in part (b) shows that 2.25×10^6 J of mechanical energy is lost in the collision, which is more than two-thirds of the original mechanical

Momentum of car before the collision:

$$p_{car,ix} = m_{car}v_{car,ix}$$
$$= (1.50 \times 10^3 \text{ kg})(+35.0 \text{ m/s})$$
$$= +5.25 \times 10^4 \text{ kg} \cdot \text{m/s}$$

Momentum of truck before the collision:

$$p_{truck,ix} = m_{truck}v_{truck,ix}$$
$$= (7.50 \times 10^3 \text{ kg})(-25.0 \text{ m/s})$$
$$= -1.88 \times 10^5 \text{ kg} \cdot \text{m/s}$$

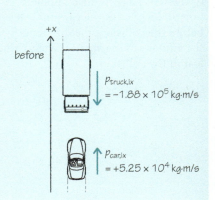

energy. This lost energy goes into doing work to deform the two vehicles.

Total momentum before the collision:

$$P_{ix} = p_{car,ix} + p_{truck,ix}$$
$$= +5.25 \times 10^4 \text{ kg} \cdot \text{m/s}$$
$$+ (-1.88 \times 10^5 \text{ kg} \cdot \text{m/s})$$
$$= -1.35 \times 10^5 \text{ kg} \cdot \text{m/s}$$

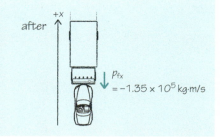

EXAMPLE 8-7 A Completely Inelastic Collision in the Skies

The photograph that opens this chapter shows a Cooper's hawk attacking a pigeon from behind. Each bird has a mass of 0.600 kg. The pigeon is gliding due west at 8.00 m/s at a constant altitude, and the Cooper's hawk is diving on the pigeon at a speed of 12.0 m/s at an angle of 30.0° below the horizontal. Just after the Cooper's hawk grabs the pigeon in its talons, how fast and in what direction are the two birds moving?

Set Up

This is a completely inelastic collision because the two birds move together afterward. Unlike Example 8-6, this collision takes place in two dimensions, so we will need to consider more than one component of Equation 8-18 to find the final velocity $\bar{v}_f$ of the two birds after the collision.

Momentum conservation in a completely inelastic collision:

$$m_{pigeon}\bar{v}_{pigeon,i} + m_{hawk}\bar{v}_{hawk,i}$$
$$= (m_{pigeon} + m_{hawk})\bar{v}_f \quad (8\text{-}18)$$

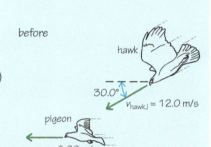

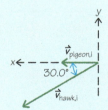

Solve

We choose the positive x axis to be horizontal and to the west, and we choose the positive y axis to be upward. We write the components of the initial velocities of the hawk and pigeon.

Pigeon is initially moving in the positive x direction, so

$$v_{pigeon,ix} = v_{pigeon,i} = +8.00 \text{ m/s}$$
$$v_{pigeon,iy} = 0$$

Hawk is initially moving in the positive x direction and the negative y direction, so

$$v_{hawk,ix} = v_{hawk,i} \cos 30.0° = (12.0 \text{ m/s}) \cos 30.0°$$
$$= +10.4 \text{ m/s}$$
$$v_{hawk,iy} = -v_{hawk,i} \sin 30.0° = -(12.0 \text{ m/s}) \sin 30.0°$$
$$= -6.00 \text{ m/s}$$

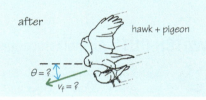

Use conservation of momentum in component form to find the components of the final velocity of the two birds.

Equation for conservation of x momentum in the collision:

$$m_{pigeon}v_{pigeon,ix} + m_{hawk}v_{hawk,ix} = (m_{pigeon} + m_{hawk})v_{fx}$$

Solve for the final x velocity of the two birds:

$$v_{fx} = \frac{m_{pigeon}v_{pigeon,ix} + m_{hawk}v_{hawk,ix}}{m_{pigeon} + m_{hawk}}$$
$$= \frac{(0.600 \text{ kg})(+8.00 \text{ m/s}) + (0.600 \text{ kg})(+10.4 \text{ m/s})}{0.600 \text{ kg} + 0.600 \text{ kg}}$$
$$= +9.20 \text{ m/s}$$

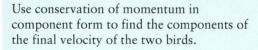

Equation for conservation of y momentum in the collision:

$$m_{\text{p.geon}} v_{\text{pigeon,iy}} + m_{\text{hawk}} v_{\text{hawk,iy}} = (m_{\text{pigeon}} + m_{\text{hawk}}) v_{\text{fy}}$$

Solve for the final y velocity of the two birds:

$$
\begin{aligned}
v_{\text{fy}} &= \frac{m_{\text{pigeon}} v_{\text{pigeon,iy}} + m_{\text{hawk}} v_{\text{hawk,iy}}}{m_{\text{pigeon}} + m_{\text{hawk}}} \\
&= \frac{(0.600 \text{ kg})(0 \text{ m/s}) + (0.600 \text{ kg})(-6.00 \text{ m/s})}{0.600 \text{ kg} + 0.600 \text{ kg}} \\
&= -3.00 \text{ m/s}
\end{aligned}
$$

Given the components of the final velocity vector $\vec{v}_{\text{f}}$, use trigonometry to find the magnitude and direction of $\vec{v}_{\text{f}}$.	Speed after collision = magnitude of final velocity vector: $$\begin{aligned} v_{\text{f}} &= \sqrt{v_{\text{fx}}^2 + v_{\text{fy}}^2} \\ &= \sqrt{(+9.20 \text{ m/s})^2 + (-3.00 \text{ m/s})^2} \\ &= 9.67 \text{ m/s} \end{aligned}$$ Direction of final velocity vector: $$\tan\theta = \frac{v_{\text{fy}}}{v_{\text{fx}}} = \frac{-3.00 \text{ m/s}}{+9.20 \text{ m/s}} = -0.326$$ $$\theta = \arctan(-0.326) = -18.1°$$ After the collision the two birds move at 9.67 m/s in a direction 18.1° below the horizontal.

Reflect

The impact causes the hawk to slow down from 12.0 m/s to 9.67 m/s and causes the pigeon to speed up from 8.00 m/s to 9.67 m/s. This is just what we would expect when a slow-moving pigeon is hit from behind by a fast-moving hawk.

WATCH OUT! An inelastic collision doesn't have to be *completely* inelastic.

 A common misconception is that a collision is inelastic if two objects come together and stick, and it is elastic if they bounce off each other. Only the first part of this statement is true! Part (a) of Example 8-5 describes a collision between two hockey players that is inelastic even though the players bounce off each other. For a collision to be inelastic, some mechanical energy has to be lost in the collision, and this can happen even if the colliding objects don't stick together.

Three Special Cases of Completely Inelastic Collisions

Figure 8-12a shows an object of mass m_A moving with velocity $\vec{v}_{Ai}$ and about to collide with a second object of mass m_B that is initially at rest, so $\vec{v}_{Bi} = 0$. If the collision

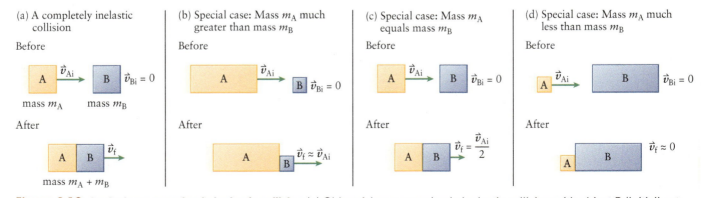

Figure 8-12 Analyzing a completely inelastic collision (a) Object A has a completely inelastic collision with object B (initially at rest). (b) If A is much more massive than B, the final velocity is almost the same as the initial velocity of A. (c) If A and B have the same mass, the final velocity is one-half the initial velocity of A. (d) If A is much less massive than B, the final velocity is nearly zero.

is completely inelastic, we can use Equation 8-18 to find the velocity $\bar{v}_f$ after the collision. Since $\bar{v}_{Bi} = 0$,

$$m_A \bar{v}_{Ai} = (m_A + m_B)\bar{v}_f$$

and so

$$\bar{v}_f = \frac{m_A \bar{v}_{Ai}}{m_A + m_B} \qquad (8\text{-}19)$$

(completely inelastic collision, object B initially at rest)

Let's consider three important special cases of this equation.

(1) *The moving object has much greater mass than the object at rest* (Figure 8-12b). Whenever one quantity in an expression is significantly larger than another, it's a good approximation to ignore the smaller quantity if the two quantities are added or subtracted. So when the mass of object A is much larger than the mass of object B, we can safely replace $m_A + m_B$ in Equation 8-19 with m_A:

$$\bar{v}_f \approx \frac{m_A \bar{v}_{Ai}}{m_A} = \bar{v}_{Ai}$$

In this case the stationary object is so small that the moving object is essentially unaffected by the collision, and its velocity remains the same. If a fast-moving car runs into a hovering fly, the car slows down only imperceptibly.

(2) *The two objects have the same mass* (Figure 8-12c). If $m_A = m_B$, Equation 8-19 becomes

$$\bar{v}_f = \frac{\bar{v}_{Ai}}{2}$$

This says that when a moving object collides with and sticks to a stationary one of the same mass, the combined system moves at half the initial speed.

(3) *The moving object has much less mass than the object at rest* (Figure 8-12d). In this case the mass of object A is much less than the mass of object B, so the quantity m_A in the numerator of Equation 8-19 is very much less than the sum $m_A + m_B$ in the denominator. Therefore $m_A/(m_A + m_B)$ is close to zero, and in this approximation Equation 8-19 becomes

$$\bar{v}_f \approx 0$$

This says that because object A has so little mass, object B hardly moves at all when it is struck. An example is using an axe to split logs for the fireplace (Figure 8-13). Each impact is a completely inelastic collision between the axe (object A) and a large, massive, stationary object B made up of the log, the stump on which the log rests, and the ground below the stump. The final velocity of the axe and log is zero, which means *all* of the kinetic energy of the axe is lost. This lost energy goes into heating the axe and heating and ripping apart the log.

Figure 8-13 Firewood physics In splitting this log, why does the axe blade get hot as it repeatedly strikes the log?

GOT THE CONCEPT? 8-3 **How to Stop a Truck**

In Example 8-6, a 1.50×10^3-kg car moving north at 35.0 m/s collides head-on with a 7.50×10^3-kg truck moving south. If the collision is completely inelastic, how fast must the truck be moving so that the wreckage is at rest immediately after the collision? (a) 20.0 m/s; (b) 14.0 m/s; (c) 10.0 m/s; (d) 7.0 m/s; (e) 5.0 m/s.

TAKE-HOME MESSAGE FOR **Section 8-4**

✔ In an elastic collision the forces between the colliding objects are conservative. Both momentum and mechanical energy are conserved in an elastic collision.

✔ An inelastic collision is one in which there are nonconservative forces between the colliding objects. In this

case momentum is conserved in the collision, but mechanical energy is not conserved.

✔ In a completely inelastic collision the colliding objects stick together. Momentum is conserved, but mechanical energy is not.

8-5 In an elastic collision both momentum and mechanical energy are conserved

In Section 8-4 we introduced the idea of an *elastic* collision in which the forces between colliding objects are conservative and no mechanical energy is lost in the collision. Hence *both* total momentum (a vector) and total mechanical energy (a scalar) are conserved in an elastic collision. We call the two colliding objects A and B, with masses m_A and m_B. We then have two conservation equations that relate the velocities $\vec{v}_{Af}$ and $\vec{v}_{Bf}$ of the objects after the collision to the velocities $\vec{v}_{Ai}$ and $\vec{v}_{Bi}$ before the collision:

(8-20)

(a)
$$m_A\vec{v}_{Af} + m_B\vec{v}_{Bf} = m_A\vec{v}_{Ai} + m_B\vec{v}_{Bi}$$

(momentum is conserved in an elastic collision)

(b)
$$\frac{1}{2}m_A v_{Af}^2 + \frac{1}{2}m_B v_{Bf}^2 = \frac{1}{2}m_A v_{Ai}^2 + \frac{1}{2}m_B v_{Bi}^2$$

(mechanical energy is conserved in an elastic collision)

Equations 8-20 require some effort to solve, and even after that they often don't tell the whole story of an elastic collision. As an example, consider a collision on a pool table between a moving cue ball (A) with known initial velocity $\vec{v}_{Ai}$ and an 8-ball (B) that is initially at rest so $\vec{v}_{Bi} = 0$. Such a collision is almost perfectly elastic. If you're a pool player, you would like to know how fast and in what direction each ball will go after the collision, which means that you want to know the x and y components of their final velocities. So there are four unknowns in this problem: v_{Afx}, v_{Afy}, v_{Bfx}, and v_{Bfy}. But Equations 8-20 give us only *three* equations for these four unknowns: two from Equation 8-20a for the x and y components of momentum and one from Equation 8-20b. So these equations by themselves aren't enough to solve the problem. (For a collision on a pool table, you'd also need to know how far off-center the two balls collide, which makes the problem quite complicated.)

Although we don't have enough information to solve Equations 8-20 for a two-dimensional collision, we can solve these equations completely for a *one-dimensional* elastic collision, so all of the motions are along a straight line. To make the problem even simpler, we'll consider the case in which object B is initially at rest (Figure 8-14). An example is a moving cue ball that hits a stationary 8-ball head-on. We call that axis along which the motion takes place the x axis. Then $v_{Bix} = 0$ (the initial velocity of object B is zero) and Equations 8-20 become

(8-21)

(a)
$$m_A v_{Afx} + m_B v_{Bfx} = m_A v_{Aix}$$

(momentum is conserved, head-on elastic collision, object B initially at rest)

(b)
$$\frac{1}{2}m_A v_{Afx}^2 + \frac{1}{2}m_B v_{Bfx}^2 = \frac{1}{2}m_A v_{Aix}^2$$

(mechanical energy is conserved, head-on elastic collision, object B initially at rest)

Since each object moves along the x axis only, in Equation 8-21b we've replaced the square of each object's speed with the square of its x velocity.

The two unknowns in Equations 8-21 are the final velocities v_{Afx} and v_{Bfx} of the two objects. Solving these equations takes quite a few steps, so we'll just present the results:

(8-22)

$$v_{Afx} = v_{Aix}\left(\frac{m_A - m_B}{m_A + m_B}\right)$$

$$v_{Bfx} = v_{Aix}\left(\frac{2m_A}{m_A + m_B}\right)$$

(final velocities in a head-on elastic collision, object B initially at rest)

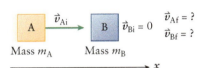

Because this elastic collision is head-on, the final velocities of the two objects will be along the same x axis as the initial velocity of object A.

Figure 8-14 A head-on, one-dimensional elastic collision If momentum and mechanical energy are both conserved in this head-on collision, what are the final velocities of the two objects?

You can verify Equations 8-22 by substituting them into Equations 8-21 and simplifying.

Equations 8-22 show that the final velocity of each object is proportional to the initial velocity v_{Aix} of object A, and also depends on a factor in parentheses that involves the masses of the two objects. The *direction* of each object's motion is given by the sign of its final velocity. If the sign of an object's final velocity is the same as the sign of v_{Aix}, the object leaves the collision in the same direction as object A was moving initially. If the final velocity has the opposite sign from v_{Aix}, the object is moving in the direction opposite to the initial motion. For example, if v_{Afx} is negative while v_{Aix} is positive, object A has bounced backward after colliding with initially stationary object B.

WATCH OUT! Some equations are less important than others.

Equations 8-22 are useful for solving a certain, very special kind of collision problem, but they are not *essential* equations that you should worry about trying to memorize or even remember in detail. What is worth remembering is the key physics of an elastic collision: Both momentum and kinetic energy are conserved.

EXAMPLE 8-8   When Molecules Collide Elastically

In air, at room temperature, a typical molecule of oxygen (O_2) moves at about 500 m/s. (In Chapter 14 we'll explore the relationship between the temperature of a gas and the speeds of molecules in that gas.) The mass of an O_2 molecule is 32.0 u, where 1 u = 1 atomic mass unit = 1.66×10^{-27} kg. Suppose an O_2 molecule is initially moving at 5.00×10^2 m/s in the positive x direction and has a head-on elastic collision with a molecule of hydrogen (H_2, of mass 2.02 u) that is at rest. Determine the final velocity of each molecule, the initial and final momentum of each molecule, and the initial and final kinetic energy of each molecule.

Set Up

Since this is a one-dimensional elastic collision in which one object is initially at rest, we can find the final velocities using Equations 8-22. Here A represents the moving oxygen molecule and B represents the hydrogen molecule. We can then use the definitions of momentum and kinetic energy to find the initial and final values of these quantities.

Final velocities in a head-on elastic collision, object B initially at rest:

$$v_{Afx} = v_{Aix}\left(\frac{m_A - m_B}{m_A + m_B}\right)$$

$$v_{Bfx} = v_{Aix}\left(\frac{2m_A}{m_A + m_B}\right) \quad (8\text{-}22)$$

$$\bar{p} = m\bar{v} \quad (8\text{-}5)$$

$$K = \frac{1}{2}mv^2 \quad (6\text{-}8)$$

before

after

Solve

Calculate the final velocities using Equations 8-22. Since these equations involve the ratios of the masses, we can use the mass values in atomic mass units without needing to convert to kilograms.

Oxygen molecule A is initially moving in the positive x direction, so

$v_{Aix} = +5.00 \times 10^2$ m/s

Mass of molecule A (O_2) is $m_A = 32.0$ u
Mass of molecule B (H_2) is $m_B = 2.02$ u

Final velocity of O_2 molecule:

$$v_{Afx} = v_{Aix}\left(\frac{m_A - m_B}{m_A + m_B}\right)$$

$$= (+5.00 \times 10^2 \text{ m/s})\left(\frac{32.0 \text{ u} - 2.02 \text{ u}}{32.0 \text{ u} + 2.02 \text{ u}}\right)$$

$$= +4.41 \times 10^2 \text{ m/s}$$

Final velocity of H_2 molecule:

$$v_{Bfx} = v_{Aix}\left(\frac{2m_A}{m_A + m_B}\right)$$

$$= (+5.00 \times 10^2 \text{ m/s})\left[\frac{2(32.0 \text{ u})}{32.0 \text{ u} + 2.02 \text{ u}}\right]$$

$$= +9.41 \times 10^2 \text{ m/s}$$

Both final velocities are positive, so after the collision both molecules are moving in the same direction as the initial motion of the O_2 molecule. The H_2 molecule is moving almost twice as fast as the initial velocity of the O_2 molecule.

We now know both the initial and final velocities, so we can calculate the initial and final values of momentum and kinetic energy for both molecules. For these calculations we need the masses in kilograms.

Masses of the molecules in kilograms:

m_A = mass of O_2 molecule

$\qquad = (32.0 \text{ u})(1.66 \times 10^{-27} \text{ kg/u}) = 5.31 \times 10^{-26} \text{ kg}$

m_B = mass of H_2 molecule

$\qquad = (2.02 \text{ u})(1.66 \times 10^{-27} \text{ kg/u}) = 3.35 \times 10^{-27} \text{ kg}$

Calculate the initial and final values of x momentum:

For molecule A (O_2):

$$p_{Aix} = m_A v_{Aix} = (5.31 \times 10^{-26} \text{ kg})(+5.00 \times 10^2 \text{ m/s})$$

$$= +2.66 \times 10^{-23} \text{ kg} \cdot \text{m/s}$$

$$p_{Afx} = m_A v_{Afx} = (5.31 \times 10^{-26} \text{ kg})(+4.41 \times 10^2 \text{ m/s})$$

$$= +2.34 \times 10^{-23} \text{ kg} \cdot \text{m/s} = 0.881 p_{Aix}$$

For molecule B (H_2):

$$p_{Bix} = m_B v_{Bix} = 0 \text{ (molecule B initially at rest)}$$

$$p_{Bfx} = m_B v_{Bfx} = (3.35 \times 10^{-27} \text{ kg})(+9.41 \times 10^2 \text{ m/s})$$

$$= +3.15 \times 10^{-24} \text{ kg} \cdot \text{m/s} = 0.119 p_{Aix}$$

In the collision the O_2 molecule transfers 0.119 of its initial momentum to the H_2 molecule, leaving 0.881 for itself.

Calculate the initial and final values of kinetic energy:

For molecule A (O_2):

$$K_{Ai} = \frac{1}{2}m_A v_{Aix}^2 = \frac{1}{2}(5.31 \times 10^{-26} \text{ kg})(+5.00 \times 10^2 \text{ m/s})^2$$

$$= 6.64 \times 10^{-21} \text{ J}$$

$$K_{Af} = \frac{1}{2}m_A v_{Afx}^2 = \frac{1}{2}\left(5.31 \times 10^{-26} \text{ kg}\right)\left(+4.41 \times 10^2 \text{ m/s}\right)^2$$

$$= 5.16 \times 10^{-21} \text{ J} = 0.777 K_{Ai}$$

For molecule B (H_2):

$$K_{Bi} = \frac{1}{2}m_B v_{Bix}^2 = 0 \text{ (molecule B initially at rest)}$$

$$K_{Bf} = \frac{1}{2}m_B v_{Bfx}^2 = \frac{1}{2}(3.35 \times 10^{-27} \text{ kg})(9.41 \times 10^2 \text{ m/s})^2$$

$$= 1.48 \times 10^{-21} \text{ J} = 0.223 K_{Ai}$$

In the collision the O_2 molecule transfers 0.223 of its initial kinetic energy to the H_2 molecule, leaving 0.777 for itself.

Reflect

Both the *total* momentum and the *total* kinetic energy of the system of two molecules are conserved in this collision: Each of these quantities has the same numerical value after the collision as before the collision.

In the case of an O_2 molecule (object A) hitting a stationary H_2 molecule (object B), moving object A is much more massive than stationary target object B. We found that object A slows down only slightly and loses only a little momentum and kinetic energy. Object B flies off moving faster than the original speed of object A.

We invite you to repeat these calculations for the case in which the moving O_2 molecule collides with a second O_2 molecule that is at rest. You'll find that the moving O_2 molecule ends up at rest and the stationary O_2 molecule ends up moving with the original velocity of the first molecule, 5.00×10^2 m/s. In this case the first O_2 molecule transfers all of its momentum and all of its kinetic energy to the second O_2 molecule.

You should also repeat these calculations for the case in which the moving O_2 molecule collides with a protein molecule of mass 3.20×10^4 u that is at rest. In this case you'll find that the O_2 molecule has a final velocity of -4.99×10^2 m/s. The negative sign means that the O_2 molecule *bounces back*, and the magnitude means that the O_2 molecule ends up with nearly the same speed as it had originally (5.00×10^2 m/s). You'll also find that the massive protein molecule ends up with a final velocity of just $+0.999$ m/s, so it moves very slowly in the initial direction of motion of the O_2 molecule. Be sure to also calculate the final momentum and kinetic energy of the protein molecule; you'll find that it ends up with about twice the initial momentum of the O_2 molecule but has little kinetic energy.

Below we'll see that these results are just what we would expect for elastic collisions.

Example 8-8 suggests that just as we did in our exploration of completely inelastic collisions (Section 8-4), we can gain insight into the physics of elastic collisions by considering three special cases involving the relative masses of the two objects. Again we consider a moving object A that has a head-on elastic collision with a stationary object B as is shown in Figure 8-15, so we can use Equations 8-22.

(1) *The moving object has much greater mass than the object at rest* (Figure 8-15a). If the mass of object A is much larger than the mass of object B, we can safely replace both $m_A - m_B$ and $m_A + m_B$ in Equations 8-22 with m_A:

$$v_{Afx} = v_{Aix}\left(\frac{m_A - m_B}{m_A + m_B}\right) \approx v_{Aix}\left(\frac{m_A}{m_A}\right) = v_{Aix}$$

$$v_{Bfx} = v_{Aix}\left(\frac{2m_A}{m_A + m_B}\right) \approx v_{Aix}\left(\frac{2m_A}{m_A}\right) = 2v_{Aix}$$

In this extreme case the motion of massive object A is unaffected by the elastic collision, but the much lighter object B flies off at twice the initial speed of object A. An example is the nearly elastic collision between a bowling ball and a stationary bowling pin, which has only about one-quarter the mass of the ball: The ball continues with nearly the same speed, while the pin flies off at a faster speed.

Go to Picture It 8-1 for more practice dealing with elastic collisions.

(2) *The two objects have the same mass* (Figure 8-15b). If $m_A = m_B$, Equations 8-22 become

$$v_{Afx} = v_{Aix}\left(\frac{m_A - m_B}{m_A + m_B}\right) = 0$$

$$v_{Bfx} = v_{Aix}\left(\frac{2m_A}{m_A + m_B}\right) = v_{Aix}\left(\frac{2m_A}{2m_A}\right) = v_{Aix}$$

In this case object A comes to a stop during the elastic collision and object B leaves the collision with the initial velocity of A. If a cue ball hits an 8-ball head-on, the result of this nearly elastic collision is that the cue ball stops and the 8-ball rolls away with

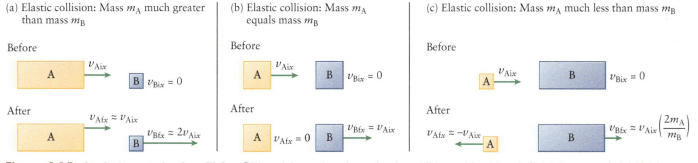

Figure 8-15 Analyzing an elastic collision Object A has a head-on elastic collision with object B (initially at rest). (a) If A is much more massive than B, A is almost unaffected and B flies off with twice the initial velocity of A. (b) If A and B have the same mass, A comes to rest and B flies off with the initial velocity of A. (c) If A is much less massive than B, A bounces back with nearly its initial speed and B moves forward very slowly.

(a) Before collision

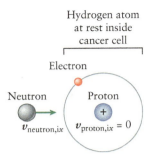

Hydrogen atom
at rest inside
cancer cell

Electron

Neutron   Proton

$v_{neutron,ix}$ $v_{proton,ix} = 0$

(b) After collision: Neutron has transferred
almost all of its momentum and kinetic energy
to the proton

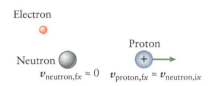

Electron

Proton

Neutron

$v_{neutron,fx} \approx 0$ $v_{proton,fx} \approx v_{neutron,ix}$

BioMedical **Figure 8-16 Fast neutron radiotherapy** In one type of radiation therapy for cancer, a beam of fast-moving neutrons (with speeds of about 10^8 m/s, or one-third the speed of light) is directed at a site with cancer cells. A neutron has almost exactly the same mass as a proton ($m_{neutron} = 1.0014\ m_{proton}$), so if a neutron collides elastically with a proton in a hydrogen atom inside a cancer cell, the result is like that shown in Figure 8-15b: The neutron essentially stops and the proton recoils with nearly the same speed as the neutron had. Because this fast-moving proton has an electric charge, it causes tremendous amounts of ionization as it travels through the cancer cell, and this ionization can be great enough to sever both strands of the cell's DNA molecules. Hence the cancer cell cannot reproduce. Fast neutron therapy is used for certain types of cancer, including salivary gland tumors, that do not respond well to other forms of radiation therapy.

the speed that the cue ball had before the collision. This same effect is used in one type of cancer radiotherapy (Figure 8-16).

(3) *The moving object has much less mass than the object at rest* (Figure 8-15c). In this case the mass of object A is small compared to the mass of object B, so $m_A - m_B$ is approximately equal to $-m_B$, and $m_A + m_B$ is approximately equal to m_B. Then Equations 8-22 become

$$v_{Afx} = v_{Aix}\left(\frac{m_A - m_B}{m_A + m_B}\right) \approx v_{Aix}\left(\frac{-m_B}{m_B}\right) = -v_{Aix}$$

$$v_{Bfx} = v_{Aix}\left(\frac{2m_A}{m_A + m_B}\right) \approx v_{Aix}\left(\frac{2m_A}{m_B}\right)$$

In this case object A is reflected back from the elastic collision, moving with the same speed it had initially but in the opposite direction. Hence object A has lost almost no kinetic energy, but its x momentum has changed from positive ($p_{Aix} = +m_A v_{Aix}$) to negative ($p_{Afx} = -m_A v_{Aix}$). Object B is moving very slowly (because m_A is much less than m_B, the ratio $2m_A/m_B$ is much less than 1). But because its mass is so great, object B has *twice* as much positive x momentum as object A had before the collision:

$$p_{Bfx} = m_B v_{Bfx} = m_B v_{Aix}\left(\frac{2m_A}{m_B}\right) = 2m_A v_{Aix}$$

Then the total momentum of the two objects after the collision is

$$p_{Afx} + p_{Bfx} = -m_A v_{Aix} + 2m_A v_{Aix} = m_A v_{Aix}$$

which is the same as the momentum of object A before the collision. So total momentum is conserved in this case, as it must be. An example is a pebble hitting a boulder: The pebble bounces back, while the recoil of the boulder is so little as to be imperceptible. You should review Example 8-8, which describes elastic collisions of a moving O_2 molecule with a stationary molecule of smaller mass, the same mass, or greater mass.

Figure 8-17 summarizes the differences between elastic collisions, inelastic collisions, and completely inelastic collisions.

(a) Elastic collision: Momentum conserved, mechanical energy conserved

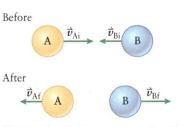

(b) Inelastic collision: Momentum conserved, mechanical energy not conserved

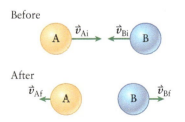

(c) Completely inelastic collision: Momentum conserved, mechanical energy not conserved, colliding objects stick together

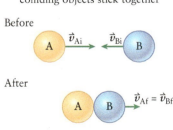

Figure 8-17 Three types of collision Comparing (a) elastic, (b) inelastic, and (c) completely inelastic collisions.

GOT THE CONCEPT? 8-4 Classifying Collisions

(?) For each of the following collisions, state whether it is elastic, inelastic, or completely inelastic. (a) After a ball with 200 J of kinetic energy hits a stationary ball on a horizontal surface, the first ball has 125 J of kinetic energy, and the second has 75 J of kinetic energy. (b) A dog leaps in the air and catches a ball in its teeth. (c) A ball is dropped from rest and hits the ground. It rebounds to half of its original height.

8-6 What happens in a collision is related to the time the colliding objects are in contact

During a collision that brings a car to a sudden stop, the air bags installed in most cars inflate (Figure 8-18). Their purpose is to prevent serious injury by cushioning the impact between the car's occupants and hard surfaces in the vehicle.

The principle of the air bag is to minimize the *force* on the occupants by maximizing the *time* that it takes to bring the occupants to rest. To see how this works, let's look at the relationship between the net force on an object, the time that the net force acts, and the change in the object's momentum caused by the net force.

Figure 8-18 A sudden stop with air bags How does a deployed air bag help prevent serious injury?

Collision Force, Contact Time, and Momentum Change

We learned in Section 8-3 that if external forces act on a *system* of objects for a time Δt, the result is a change in the momentum of the system from an initial value $\vec{P}_i$ to a final value $\vec{P}_f$. This change is given by Equation 8-13:

$$\left(\sum \vec{F}_{\text{external on system}}\right)\Delta t = \vec{P}_f - \vec{P}_i$$

Now suppose the system is made up of only a single object. Then we can replace the symbol $\vec{P}$ (for the momentum of a system) by $\vec{p}$ (for the momentum of a single object), and Equation 8-13 becomes

The **sum of all external forces** acting on an object

Duration of the time interval over which the external forces act

$$\left(\sum \vec{F}_{\text{external on object}}\right)\Delta t = \vec{p}_f - \vec{p}_i = \Delta\vec{p}$$

Change in the momentum $\vec{p}$ of the object during that time interval

Impulse-momentum theorem: External force and momentum change for an object (8-23)

Equation 8-23 is called the **impulse-momentum theorem**. (It's a *theorem* because it's a consequence of a more fundamental principle, Newton's second law.) The quantity $\left(\sum \vec{F}_{\text{external on object}}\right)\Delta t$ is called the **impulse** that acts on the object. This equation tells us that to change the momentum of an object from $\vec{p}_i$ to $\vec{p}_f$, a certain amount of impulse is required. This impulse can be the result of a large external force $\sum \vec{F}_{\text{external on system}}$ that acts for a short time Δt or a smaller external force that acts for a longer time.

If the momentum change occurs as a result of a collision, the net external force on an object is predominantly due to the other object with which it is colliding; any other forces are very weak by comparison (see Section 8-3). So for the special case of a collision, Equation 8-23 becomes

Force exerted on an object during a collision

Duration of the collision = **contact time**

$$\vec{F}_{\text{collision}}\,\Delta t = \vec{p}_f - \vec{p}_i = \Delta\vec{p}$$

Change in the momentum $\vec{p}$ of the object during the collision

Collision force, contact time, and momentum change (8-24)

The **contact time** in Equation 8-24 is the amount of time that the colliding objects are in contact and hence the amount of time that the objects exert forces on each other.

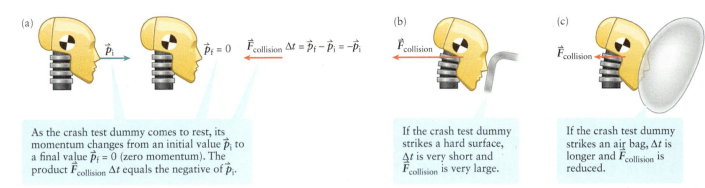

(a) As the crash test dummy comes to rest, its momentum changes from an initial value $\vec{p}_i$ to a final value $\vec{p}_f = 0$ (zero momentum). The product $\vec{F}_{collision} \Delta t$ equals the negative of $\vec{p}_i$.

$$\vec{F}_{collision} \Delta t = \vec{p}_f - \vec{p}_i = -\vec{p}_i$$

(b) If the crash test dummy strikes a hard surface, Δt is very short and $\vec{F}_{collision}$ is very large.

(c) If the crash test dummy strikes an air bag, Δt is longer and $\vec{F}_{collision}$ is reduced.

Figure 8-19 Contact time (a) To bring a crash test dummy to rest requires that $\vec{F}_{collision} \Delta t$ have a certain large value. (b) A short contact time Δt means that the force $\vec{F}_{collision}$ will be large. (c) A longer contact time Δt reduces the magnitude of $\vec{F}_{collision}$.

Equation 8-24 explains the principle of the air bag shown in Figure 8-19. If a car comes to a sudden stop, the momentum of one of its occupants changes from a large initial value $\vec{p}_i$ to a final value of zero ($\vec{p}_f = 0$). So the momentum change $\vec{p}_f - \vec{p}_i$ is large, which means the product $\vec{F}_{collision} \Delta t$ also has a large value (Figure 8-19a). If there are no air bags, the occupant has a "hard" collision with the structure of the car and also comes to a sudden stop. In this case the contact time Δt is very short and so $\vec{F}_{collision}$ must have a very large value: The car's structure exerts a tremendous force on the occupant that is likely to cause injury, as Figure 8-19b shows. But if the car is equipped with air bags, the occupant covers a greater distance as it comes to rest and so is in contact with the air bag for a longer time Δt (Figure 8-19c). Hence $\vec{F}_{collision}$ (which represents the force exerted on the occupant by the air bag) is greatly reduced, and injury is avoided. For the same reason, if you jump down from a table it's less painful if you bend your knees when landing: This maximizes the contact time Δt that it takes for you to come to a stop.

In deriving Equations 8-23 and 8-24, we have assumed that the forces acting on the object are constant forces that do not change during the time Δt. This is usually *not* the case: The force on a colliding object usually varies with time during the collision (Figure 8-20). We can nonetheless still use these equations: Just treat $\sum \vec{F}_{external\ on\ system}$ in Equation 8-23 and $\vec{F}_{collision}$ in Equation 8-24 as the *average* values of these forces over the time interval Δt.

Example 8-9 shows how to apply the idea of contact time to a collision in which an object speeds up rather than slowing down.

Figure 8-20 Collision force
During the time Δt over which a collision takes place, the force that acts on a colliding object increases to a maximum and then decreases.

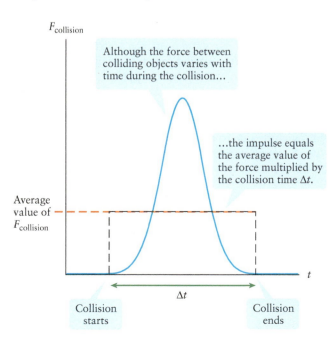

$F_{collision}$

Although the force between colliding objects varies with time during the collision...

...the impulse equals the average value of the force multiplied by the collision time Δt.

Average value of $F_{collision}$

Collision starts

Δt

Collision ends

t

EXAMPLE 8-9 Follow-Through

Tennis players are taught to *follow through* as they serve the ball—that is, to continue to swing the racket after first striking the ball. The rationale is that by increasing the time that the ball and racket are in contact, the ball might have a larger speed as it leaves the racket. Using her racket, a player can exert an average force of 560 N on a 57-g tennis ball during an overhand serve (Figure 8-21). At what speed does the ball leave her racket if the contact time between them is 0.0050 s? What would the speed of the ball be if she improved her follow-through and increased the contact time by a factor of 1.3 to 0.0065 s? During an overhand serve, the tennis ball is struck when it is essentially motionless.

PCN Photography/Alamy

Figure 8-21 Tennis physics How does an improved follow-through affect the speed of a tennis ball?

Set Up

We are given the force that the racket exerts on the ball and the contact time (the time that the ball is touching the racket). We also know that the ball starts at rest and so with zero momentum. We can use Equation 8-24 to calculate the final momentum of the ball, and then use Equation 8-5 to determine the final speed.

Collision force, contact time, and momentum change:

$$\vec{F}_{collision}\Delta t = \vec{p}_f - \vec{p}_i \tag{8-24}$$

$$\vec{p} = m\vec{v} \tag{8-5}$$

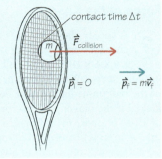

Solve

The racket follows a curved path during the serve. But since the collision between racket and ball lasts such a short time, the ball is in contact with the racket for just a short segment of that path, which we can treat as a straight line. We choose the positive x axis to be in the direction of the motion of the ball.

The collision is one dimensional, so we just need the x component of Equation 8-24:

$$F_{collision,x}\Delta t = p_{fx} - p_{ix}$$

The ball is initially at rest, so $v_{ix} = 0$ and

$$p_{ix} = 0$$

After the collision with the racket, the ball has (unknown) velocity v_{fx}, so

$$p_{fx} = mv_{fx}$$

Then the x component of Equation 8-24 becomes

$$F_{collision,x}\Delta t = mv_{fx} - 0 = mv_{fx}$$

Solve for the final velocity of the ball in each case.

Rewrite the x component of Equation 8-24 to solve for v_{fx}:

$$v_{fx} = \frac{F_{collision,x}\Delta t}{m}$$

In both cases $F_{collision,x} = 560$ N and

$$m = 57\ g\left(\frac{1\ kg}{1000\ g}\right) = 0.057\ kg$$

Calculate v_{fx} if the contact time is 0.0050 s. Recall that $1\ N = 1\ kg\cdot m/s^2$:

$$v_{fx} = \frac{F_{collision,x}\Delta t}{m} = \frac{(560\ N)(0.0050\ s)}{0.057\ kg}$$

$$= 49\frac{N\cdot s}{kg}\left(\frac{1\ kg\cdot m/s^2}{1\ N}\right) = 49\ m/s$$

With improved follow-through and a contact time of 0.0065 s,

$$v_{fx} = \frac{F_{collision,x}\Delta t}{m} = \frac{(560\ N)(0.0065\ s)}{0.057\ kg} = 64\ m/s$$

Reflect

For a given force the change in momentum and therefore the change in velocity during a collision are directly proportional to the contact time. Increasing the contact time Δt from 0.0050 s to (1.3)(0.0050 s) = 0.0065 s results in an increase in the final speed of the tennis ball from 49 m/s to (1.3)(49 m/s) = 64 m/s.

The Impulse-Momentum Theorem Versus the Work-Energy Theorem

The impulse-momentum theorem, Equation 8-23, says that the product of the net force on an object and the *time* that the force acts equals the change in *momentum* of the object. This should remind you of the work-energy theorem from Section 6-3, which we can write as

$$W_{net} = Fd \cos \theta = K_f - K_i$$

(8-25)

This says that the net work on an object—which is the product of the net force on an object parallel to its displacement and the *displacement* of the object during the time that the force acts—equals the change in *kinetic energy* of the object. What is the fundamental difference between these two theorems?

To see the answer, consider a 1-kg object and a 2-kg object that both start at rest. You apply the same net force to both objects for the same time Δt, causing both objects to accelerate. Since both objects are subjected to the same *impulse* (net force multiplied by time), both end up with the same final *momentum* $\bar{p} = m\bar{v}$. But since the 2-kg object has a greater mass m than the 1-kg object, it has a slower final speed v. The 2-kg object also had less acceleration due to its greater mass, and so traveled a shorter distance than the 1-kg object during the contact time Δt. That means the net force did less *work* (force times distance) on the 2-kg object than on the 1-kg object, and so the 2-kg object ends up with less *kinetic energy* $K = \frac{1}{2}mv^2$ than the 1-kg object. (We came to a similar conclusion about momentum and kinetic energy in Section 8-2 when we discussed the results of Example 8-1.)

Here's a shorthand way to remember the difference between the impulse-momentum and work-energy theorems: Force multiplied by time (impulse) affects momentum; force multiplied by distance (work) affects kinetic energy.

GOT THE CONCEPT? 8-5
Force of the Kick

(?) A soccer player's foot is in contact with a 0.43-kg ball for 8.6×10^{-3} s. The ball is initially at rest. What is the speed of the ball immediately after being kicked if the player exerts a force of 1700 N on it? (a) 1.5 m/s; (b) 3.4 m/s; (c) 15 m/s; (d) 34 m/s; (e) not enough information given to decide.

TAKE-HOME MESSAGE FOR Section 8-6

✔ The change in momentum of an object during a collision equals the force exerted on the object during the collision multiplied by the contact time—the length of time that the object touches another during the collision.

✔ The same change in momentum can be produced by a large force acting for a short time or by a small force acting for a longer time.

8-7 The center of mass of a system moves as though all of the system's mass were concentrated there

The concept of momentum can help us answer a question that may have been lurking in the back of your mind since Chapter 4. We learned in that chapter that the acceleration of an object is determined by the external forces on the object as described by Newton's second law: $\sum \vec{F}_{ext} = m\vec{a}$. But if the object is rotating, like a gymnast tumbling in midair or a barrel rolling downhill, different parts of the object have different accelerations. So the quantity $\vec{a}$ in Newton's second law must refer to the acceleration of a specific point on the object. But what point is that?

The point we are looking for is called the **center of mass** of the object. As we will see, the center of mass moves as though all of the object's mass were squeezed into a tiny blob at that point and all of the external forces acted on that blob. We'll begin by showing how to calculate the position of the center of mass; then we'll use ideas about momentum to justify why the center of mass behaves as it does.

Averages and Weighted Averages

The position of an object's center of mass is a kind of *average* that takes into account the masses and positions of all of the pieces that make up that object. If the object is a car, you can think of these pieces as the various components of a car, or the individual atoms that comprise the car. To see how this average is defined, let's first remind ourselves how to calculate an ordinary average.

To find the average of a set of N numbers, add them together and divide by N. For example, the average of the three numbers ($N = 3$) 4, 8, and 18 is

$(4 + 8 + 18)/3 = 30/3 = 10$. The average of the set of six numbers $n_1 = 4$, $n_2 = 4$, $n_3 = 8$, $n_4 = 8$, $n_5 = 8$, $n_6 = 10$ is

$$\bar{n} = \frac{1}{6}\sum_{i=1}^{6} n_i = \frac{4 + 4 + 8 + 8 + 8 + 10}{6} = \frac{42}{6} = 7$$

Here n_i represents the ith value in the set, and the symbol $\bar{n}$ (n with a bar on top) denotes the average value of the numbers n_i. Notice that because some of the values in the set occur more than once, we could write this average as

$$\bar{n} = \frac{2(4) + 3(8) + 10}{6} = \frac{2}{6}(4) + \frac{3}{6}(8) + \frac{1}{6}(10) = 7 \quad (8\text{-}26)$$

Each term in this sum includes the fraction of the entire set comprised of the value that follows it: 2/6 of the set has value 4, 3/6 of the set has value 8, and 1/6 of the set has value 10. In writing the sum this way we have defined a **weighted average**, in which each value affects the result to a greater or lesser extent depending on how often it appears in the set. Here *weight* does not refer to a force due to gravity but rather to the amount that each value contributes to the result.

Let's see how to use a weighted average to define the center of mass of a complex object. Imagine a waiter who uses one hand to balance a tray that has a bowl of soup on one side and a small mint on the other (Figure 8-22). To do this the waiter must support the tray at a position much closer to the soup than the mint. This position is the center of mass of the three objects: If the tray, soup bowl, and mint were all squeezed into a single blob, the waiter would put his hand directly under the blob to support it. The position of the center of mass isn't the ordinary average of the positions of tray, soup bowl, and mint, which would be a point at the geometrical center of the tray. (If the waiter tried to balance the tray there, the result would be soup everywhere.) Instead, the center of mass is a *weighted* average that puts more emphasis on the position of the high-mass soup bowl than on the position of the low-mass mint.

Equation 8-26 provides some guidance as to how to write such a weighted average. Suppose we have a system of N objects with different masses m_1, m_2, m_3, ..., m_N. The total mass M_{tot} of the system is the sum of these N individual masses:

$$M_{tot} = m_1 + m_2 + m_3 + \ldots + m_N = \sum_{i=1}^{N} m_i \quad (8\text{-}27)$$

The mass of each individual object represents a fraction m_i/M of the total mass. Now suppose the N objects are at different positions along the x axis: x_1, x_2, x_3, ..., x_N. (Figure 8-23). By analogy to Equation 8-26 we write the position of the center of mass as a weighted average of these positions:

Position of the center of mass of a system of N objects

m_i = **masses** of the individual objects

$$x_{CM} = \frac{m_1}{M_{tot}}x_1 + \frac{m_2}{M_{tot}}x_2 + \frac{m_3}{M_{tot}}x_3 + \ldots + \frac{m_N}{M_{tot}}x_N = \frac{1}{M_{tot}}\sum_{i=1}^{N} m_i x_i$$

Position of the center of mass of a system (8-28)

Total mass of all N objects together x_i = **positions** of the individual objects

The more massive a given object, the greater the ratio m_i/M_{tot} and the greater the importance of that object's position x_i in the sum given by Equation 8-28. Example 8-10 illustrates this property of the position of the center of mass.

Lightweight mint   Heavier soup bowl

The system behaves as though all of its mass were concentrated at the center of mass. That's where the waiter supports the tray.

Figure 8-22 Center of mass at the restaurant The tray balances if it is supported at the center of mass of the system of tray, soup bowl, and mint.

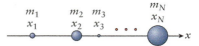

m_1 m_2 m_3 m_N
x_1 x_2 x_3 ... x_N
→x

Figure 8-23 A system of objects The position of the center of mass of this collection of objects depends on the masses and positions of the individual objects (Equation 8-28).

 Go to Picture It 8-2 for more practice dealing with center of mass.

EXAMPLE 8-10 Locating the Center of Mass

Two objects are located on the x axis at the positions $x_1 = 2.0$ m and $x_2 = 8.0$ m. For the system made up of these two objects, find the position of the center of mass (a) if m_1 and m_2 are both equal to 3.0 kg and (b) if $m_1 = 3.0$ kg and $m_2 = 33$ kg.

Set Up

Since there are just two objects, the total mass of the system is just the sum of their two masses. We use Equation 8-28 in each case to locate the center of mass.

Position of the center of mass of a system:

$$x_{CM} = \frac{m_1}{M_{tot}} x_1 + \frac{m_2}{M_{tot}} x_2 + \frac{m_3}{M_{tot}} x_3$$

$$+ \ldots + \frac{m_N}{M_{tot}} x_N$$

$$= \frac{1}{M_{tot}} \sum_{i=1}^{N} m_i x_i \qquad (8\text{-}28)$$

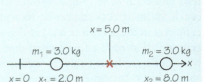

Total mass of the system of two objects:

$$M_{tot} = m_1 + m_2 \qquad (8\text{-}27)$$

Solve

(a) Calculate the position of the center of mass for the case in which both objects have the same mass.

The two objects have equal masses:

$$m_1 = m_2 = 3.0 \text{ kg}$$

Total mass of the system:

$$M_{tot} = m_1 + m_2$$
$$= 3.0 \text{ kg} + 3.0 \text{ kg} = 6.0 \text{ kg}$$

Position of the center of mass:

$$x_{CM} = \frac{m_1}{M_{tot}} x_1 + \frac{m_2}{M_{tot}} x_2$$

$$= \frac{3.0 \text{ kg}}{6.0 \text{ kg}} (2.0 \text{ m}) + \frac{3.0 \text{ kg}}{6.0 \text{ kg}} (8.0 \text{ m}) = 5.0 \text{ m}$$

(b) Repeat part (a) for the case in which m_2 is much larger than m_1.

The two objects have different masses:

$$m_1 = 3.0 \text{ kg}, m_2 = 33 \text{ kg}$$

Total mass of the system:

$$M_{tot} = m_1 + m_2$$
$$= 3.0 \text{ kg} + 33 \text{ kg} = 36 \text{ kg}$$

Position of the center of mass:

$$x_{CM} = \frac{m_1}{M_{tot}} x_1 + \frac{m_2}{M_{tot}} x_2$$

$$= \frac{3.0 \text{ kg}}{36 \text{ kg}} (2.0 \text{ m}) + \frac{33 \text{ kg}}{36 \text{ kg}} (8.0 \text{ m}) = 7.5 \text{ m}$$

Reflect

Part (a) shows that if two objects have the same mass, their center of mass will be exactly halfway between them. In part (b), by contrast, the center of mass is 5.5 m from the less massive object 1 ($m_1 = 3.0$ kg) and only 0.5 m from the more massive object 2 ($m_2 = 33$ kg). If one object in the system is much more massive than the others, the center of mass is closest to that most massive object. We saw this idea in Section 7-4; Figure 7-15 shows the center of mass of the Earth–Moon system, which is much closer to the center of Earth than to the much less massive Moon.

It's important to note that the relative position of the center of mass does *not* depend on the choice of origin. As an example, suppose we choose $x = 0$ to be at the position of object 1 so that object 2 is at position $x_2 = 6.0$ m. In this case we find the center of mass in case (b) to be at $x_{CM} = 5.5$ m. That looks different from the value of 7.5 m that we found above but is still 5.5 m from object 1 and 0.5 m from object 2—exactly as we found before.

Find the center of mass in case (b), with two objects of different mass, but now with the origin chosen to be at the position of object 1:

$$m_1 = 3.0 \text{ kg}, m_2 = 33 \text{ kg so } M_{tot} = 36 \text{ kg}$$

$$x_1 = 0, x_2 = 6.0 \text{ m}$$

Position of the center of mass:

$$x_{CM} = \frac{m_1}{M_{tot}} x_1 + \frac{m_2}{M_{tot}} x_2$$

$$= \frac{3.0 \text{ kg}}{36 \text{ kg}} (0) + \frac{33 \text{ kg}}{36 \text{ kg}} (6.0 \text{ m}) = 5.5 \text{ m}$$

More on the Position of the Center of Mass

In writing Equation 8-28 we assumed that all of the objects are arranged along a line that we call the x axis. If the objects are arranged in a plane defined by the x and y axes, we need to specify the x and y coordinates of the position of the center of mass. Equation 8-28 gives the x coordinate, and a very similar equation gives us the y coordinate:

$$y_{CM} = \frac{m_1}{M_{tot}} y_1 + \frac{m_2}{M_{tot}} y_2 + \frac{m_3}{M_{tot}} y_3 + \ldots + \frac{m_N}{M_{tot}} y_N = \frac{1}{M_{tot}} \sum_{i=1}^{N} m_i y_i \qquad (8\text{-}29)$$

(y coordinate of the position of the center of mass)

If the objects are arranged in three dimensions, you would also need an equation similar to Equations 8-28 and 8-29 for the z coordinate z_{CM} of the center of mass.

The center of mass of the system in part (a) of Example 8-10 is midway between the two objects, at the *geometrical* center of the system. This is true whenever the system is *symmetrical* so that the mass is distributed in the same way on both sides of the center of mass. For example, the center of mass of a billiard ball is at its geometrical center, as is the center of mass of a uniform metal rod. In part (b) of Example 8-10 the center of mass is *not* at the geometrical center because the system is not symmetrical: Masses m_1 and m_2 have very different values.

WATCH OUT! There doesn't have to be any mass at the center of mass.

Note that in both parts (a) and (b) of Example 8-10, there's nothing physically located at the center of mass. The same is true for any symmetrical object with a hole at its center, like a Blu-ray disc, a doughnut, or a ping-pong ball: In each case the center of mass is located in the center of the empty hole.

Momentum, Force, and the Motion of the Center of Mass

It's now time to justify the statement that the center of mass moves as though all of the mass of the system were concentrated there. To begin, let's combine Equations 8-28 and 8-29 into a single *vector* equation for the position of the center of mass:

$$\vec{r}_{CM} = \frac{m_1}{M_{tot}} \vec{r}_1 + \frac{m_2}{M_{tot}} \vec{r}_2 + \frac{m_3}{M_{tot}} \vec{r}_3 + \ldots + \frac{m_N}{M_{tot}} \vec{r}_N = \frac{1}{M_{tot}} \sum_{i=1}^{N} m_i \vec{r}_i \qquad (8\text{-}30)$$

(vector position of the center of mass)

In Equation 8-30 $\vec{r}_{CM}$ is the position vector of the center of mass, and the vectors $\vec{r}_1, \vec{r}_2, \vec{r}_3, \ldots, \vec{r}_N$ represent the positions of the N individual objects that make up the system. Each vector $\vec{r}_i$ has components x_i and y_i (and, if a third dimension is required, z_i), and $\vec{r}_{CM}$ has components x_{CM} and y_{CM} (and z_{CM} if required).

Let's now allow the objects that make up the system to move. (The objects need not be connected, so they may or may not move together.) During a time interval Δt each of the vectors $\vec{r}_i$ in Equation 8-30 changes by an amount $\Delta \vec{r}_i$, which can have a different value for each object. From Equation 8-30 the change $\Delta \vec{r}_{CM}$ in the position vector of the center of mass during Δt is

$$\Delta \vec{r}_{CM} = \frac{m_1}{M_{tot}} \Delta \vec{r}_1 + \frac{m_2}{M_{tot}} \Delta \vec{r}_2 + \frac{m_3}{M_{tot}} \Delta \vec{r}_3 + \ldots + \frac{m_N}{M_{tot}} \Delta \vec{r}_N = \frac{1}{M_{tot}} \sum_{i=1}^{N} m_i \Delta \vec{r}_i \qquad (8\text{-}31)$$

(change in vector position of the center of mass)

Equation 3-7 in Section 3-4 tells us that if an object changes its position by $\Delta \vec{r}$ during a time Δt, the velocity of the object is $\vec{v} = \Delta \vec{r} / \Delta t$. If we divide Equation 8-31 through by Δt, we get an expression for the velocity of the center of mass:

$$\begin{aligned} \vec{v}_{CM} &= \frac{\Delta \vec{r}_{CM}}{\Delta t} \\ &= \frac{1}{\Delta t} \left(\frac{m_1}{M_{tot}} \Delta \vec{r}_1 + \frac{m_2}{M_{tot}} \Delta \vec{r}_2 + \frac{m_3}{M_{tot}} \Delta \vec{r}_3 + \ldots + \frac{m_N}{M_{tot}} \Delta \vec{r}_N \right) \\ &= \frac{m_1}{M_{tot}} \vec{v}_1 + \frac{m_2}{M_{tot}} \vec{v}_2 + \frac{m_3}{M_{tot}} \vec{v}_3 + \ldots + \frac{m_N}{M_{tot}} \vec{v}_N = \frac{1}{M_{tot}} \sum_{i=1}^{N} m_i \vec{v}_i \end{aligned} \qquad (8\text{-}32)$$

(vector velocity of the center of mass)

Just as the position of the center of mass is a weighted average of the positions of the objects that make up the system, the *velocity* of the center of mass is a weighted average of the *velocities* of the objects.

The quantity $m_i\vec{v}_i$ on the right-hand side of Equation 8-32 is just the *momentum* of the ith object in the system. So $\sum_{i=1}^{N} m_i\vec{v}_i$ is the vector sum of the momentum of all objects that make up the system. This is just the total momentum of the system, which we denote as $\vec{P}$. If we multiply Equation 8-32 by the total mass of the system M_{tot}, we get

> The **total momentum** of a system...   ...equals the **vector sum** of the momentum of all objects in the system...

$$\vec{P} = \sum_{i=1}^{N} m_i\vec{v}_i = M_{\text{tot}}\vec{v}_{\text{CM}}$$

> ...and also equals the **total mass of the system** multiplied by the **velocity of the center of mass.**

In other words, the total momentum of a system of objects of total mass M_{tot} is the *same* as if all of the objects were squeezed into a single blob moving at the velocity of the center of mass (Figure 8-24). This means that the total linear momentum of a football doesn't depend on how fast the football is spinning but only on the velocity of the center of mass of the football (which is at the football's geometrical center because the football is symmetrical).

We can now use Equation 8-33 to see what affects the motion of the center of mass. This equation shows that for the velocity of the center of mass to change by an amount $\Delta\vec{v}_{\text{CM}}$, there must be a change in the total momentum of the system:

$$\Delta\vec{P} = M_{\text{tot}}\Delta\vec{v}_{\text{CM}}$$

(The mass M_{tot} doesn't change since the system always includes the same set of objects.) But Equation 8-13 in Section 8-3 tells us that the total momentum can change during a time Δt only if a net external force acts on the system during that time:

$$\left(\sum \vec{F}_{\text{external on system}}\right)\Delta t = \Delta\vec{P}$$

Combining these two equations, we get

$$\left(\sum \vec{F}_{\text{external on system}}\right)\Delta t = M_{\text{tot}}\Delta\vec{v}_{\text{CM}}$$

Divide both sides of this equation by Δt, and recall that the acceleration $\vec{a}$ of an object equals the change in its velocity divided by the time over which the velocity changes: $\vec{a} = \Delta\vec{v}/\Delta t$ (Equation 3-9 in Section 3-4). The result is

> The **net external force** on a system...   ...causes the **center of mass** of the system to **accelerate.**

$$\sum \vec{F}_{\text{external on system}} = M_{\text{tot}}\frac{\Delta\vec{v}_{\text{CM}}}{\Delta t} = M_{\text{tot}}\vec{a}_{\text{CM}}$$

Equation 8-34 looks almost exactly like Newton's second law for a single object of mass m:

$$\sum \vec{F}_{\text{ext}} = m\vec{a}$$

So what Equation 8-34 tells us is that the center of mass of a system of objects moves exactly as if the entire mass M_{tot} of the system were concentrated at the center of mass, and all of the external forces on the system acted on that concentrated mass. This justifies the statements we made at the beginning of this section about the significance of the center of mass.

Equation 8-34 also says that only *external forces* affect the motion of the center of mass. If various parts of the system exert forces on each other, these forces can affect how those parts move relative to each other, but they have *no* effect on the motion of the center of mass. As an example, consider the snowboarder shown in Figure 8-25.

Total momentum and the velocity of the center of mass (8-33)

The total momentum of the system is $\vec{P} = m_1\vec{v}_1 + m_2\vec{v}_2 + m_3\vec{v}_3...$

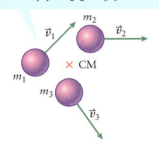

$M_{\text{tot}} = m_1 + m_2 + m_3 +...$

$\vec{v}_{\text{CM}}$

...which is the same as the momentum of all the mass moving together at the velocity of the center of mass: $\vec{P} = M_{\text{tot}}\vec{v}_{\text{CM}}$.

Figure 8-24 Total momentum of a system Two ways to represent the total momentum of a system of objects.

How the net external force on a system affects the center of mass (8-34)

BioMedical

Figure 8-25 Center of mass of a snowboarder The motion of the snowboarder's body is very complex, but the motion of his center of mass is very simple.

The snowboarder's center of mass (shown as a red ×) follows a parabolic trajectory, just as if all of the snowboarder's mass were concentrated there.

Cultura RM/Alamy

During his flight, his arms, legs, head, and torso move in complicated ways relative to each other. But his center of mass moves in a simple parabolic trajectory. This is the same trajectory that would be followed by a blob with the same mass as the snowboarder that was acted on by only the external force of gravity.

If the net external force on a system is zero, Equation 8-34 says that the center of mass cannot accelerate even if the various parts of the system move relative to each other. As an example, if the snowboarder in Figure 8-25 were in gravity-free outer space, his center of mass would continue moving in a straight line at a constant speed no matter how he moved the various parts of his body.

The center of mass will play an important role when we discuss rotational motion in Chapter 9. We'll see that if an object is rotating as it moves through space, like a spinning football in flight or a spinning tire on a moving car, its motion can naturally be thought of as being in two pieces: the motion of the object's center of mass, plus rotation of the object around its center of mass.

GOT THE CONCEPT? 8-6 An Exploding Cannon Shell

(?) An army artillery unit fires a cannon shell over level ground at a target 200 m away. The cannon is perfectly aimed to hit the target, and air resistance can be neglected. At the high point of the shell's trajectory, the shell explodes into two identical halves, both of which hit the ground at the same time. One half falls vertically downward from the point of the explosion. Where does the other half land? (a) On the target; (b) 50 m short of the target; (c) 50 m beyond the target; (d) 100 m beyond the target; (e) 150 m beyond the target.

TAKE-HOME MESSAGE FOR Section 8-7

✔ The center of mass of a system of objects is a weighted average of the positions of the individual objects.

✔ The total momentum of the system equals the system's total mass multiplied by the velocity of the center of mass.

✔ If there is a net external force on the system, the center of mass accelerates as though all of the mass of the system were concentrated into a blob at the center of mass. If there is zero net external force on the system, the center of mass does not accelerate.

Key Terms

center of mass	impulse	linear momentum (momentum)
completely inelastic collision	impulse-momentum theorem	momentum (linear momentum)
contact time	inelastic collision	total momentum
elastic collision	internal forces	weighted average
external forces	law of conservation of momentum	

Chapter Summary

Topic	Equation or Figure
Linear momentum: The momentum of an object is a vector that points in the same direction as its velocity. It depends on both the mass and velocity of the object. Do not confuse momentum with kinetic energy, which is a scalar quantity.	The **linear momentum** of an object is a vector. The linear momentum points in the same direction as the **velocity vector** and is proportional to the velocity. $$\vec{p} = m\vec{v} \qquad (8\text{-}5)$$ The linear momentum is also proportional to the object's **mass**.
External force and total momentum change for a system of objects: The total momentum of a system of objects is the vector sum of the individual momentum of each object in the system. Only external forces (which originate from objects outside the system) affect the total momentum; internal forces (exerted by one member of the system on another) do not.	The **sum of all external forces** acting on a system of objects **Duration** of a time interval over which the external forces act $$\left(\sum \vec{F}_{\text{external on system}} \right) \Delta t = \vec{P}_{\text{f}} - \vec{P}_{\text{i}} = \Delta\vec{P} \qquad (8\text{-}13)$$ **Change** during that time interval **in the total momentum** of the objects that make up the system
Collisions and the law of conservation of momentum: The total momentum of a system is conserved (maintains the same value) if there is zero net external force on the system. If the internal forces during a collision are much greater in magnitude than the external forces, the total momentum of the colliding objects is conserved.	If the net external force on a system of objects is zero, **the total momentum of the system** is conserved. $$\vec{P}_{\text{f}} = \vec{P}_{\text{i}} \qquad (8\text{-}14)$$ Then the total momentum of the system at the end of a time interval... ...is equal to the total momentum of the system at the beginning of that time interval.

Kinds of collisions: Momentum is conserved in collisions of all types. In an elastic collision mechanical energy is also conserved. Mechanical energy is not conserved in an inelastic collision. In a completely inelastic collision the colliding objects stick together after the collision.

(a) Elastic collision: Momentum conserved, mechanical energy conserved

(b) Inelastic collision: Momentum conserved, mechanical energy not conserved

(c) Completely inelastic collision: Momentum conserved, mechanical energy not conserved, colliding objects stick together

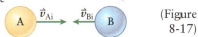
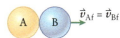

(Figure 8-17)

Topic	Equation or Figure
Completely inelastic collisions: The common final velocity in a completely inelastic collision depends on the masses and initial velocities of the colliding objects.	**Velocities** of objects A and B **before** they undergo a **completely inelastic collision** (in which they stick together) $$m_A\vec{v}_{\text{Ai}} + m_B\vec{v}_{\text{Bi}} = (m_A + m_B)\vec{v}_{\text{f}} \qquad (8\text{-}18)$$ **Masses** of objects A and B **Velocity** of the two objects moving together **after the collision**

Impulse and external force: To cause a certain change in an object's momentum requires a certain impulse (the product of the net external force on an object and the time that those forces act).

The **sum of all external forces** acting on an object

Duration of the time interval over which the external forces act

$$\left(\sum \vec{F}_{\text{external on object}} \right) \Delta t = \vec{p}_{\text{f}} - \vec{p}_{\text{i}} = \Delta \vec{p} \tag{8-23}$$

Change in the momentum $\vec{p}$ of the object during that time interval

---

Contact time: The impulse in a collision is determined by the force in the collision and the contact time (the time that the colliding objects interact with each other). The same momentum change can be caused by a strong force with a short contact time or a weak force with a long contact time.

Force exerted on an object during a collision

Duration of the collision = **contact time**

$$\vec{F}_{\text{collision}} \, \Delta t = \vec{p}_{\text{f}} - \vec{p}_{\text{i}} = \Delta \vec{p} \tag{8-24}$$

Change in the momentum $\vec{p}$ of the object during the collision

---

Center of mass: The center of mass of a system of objects is a weighted average of the positions of the individual objects.

Position of the center of mass of a system of N objects

m_i = **masses** of the individual objects

$$x_{\text{CM}} = \frac{m_1}{M_{\text{tot}}} x_1 + \frac{m_2}{M_{\text{tot}}} x_2 + \frac{m_3}{M_{\text{tot}}} x_3 + \dots + \frac{m_N}{M_{\text{tot}}} x_N = \frac{1}{M_{\text{tot}}} \sum_{i=1}^{N} m_i x_i$$

$$\tag{8-28}$$

Total mass of all N objects together

x_i = **positions** of the individual objects

Total momentum and the motion of the center of mass: The total momentum of the system is the same as if all of the mass were concentrated at the center of mass and moving with the center-of-mass velocity.

The **total momentum** of a system...

...equals the **vector sum** of the momentum of all objects in the system...

$$\vec{P} = \sum_{i=1}^{N} m_i \vec{v}_i = M_{\text{tot}} \vec{v}_{\text{CM}} \tag{8-33}$$

...and also equals the **total mass of the system** multiplied by the **velocity of the center of mass.**

Motion of the center of mass: The center of mass of a system behaves as though all of the mass of the system were concentrated there and all of the external forces acted at that point. If there is no net external force on the system, the center of mass does not accelerate.

The **net external force** on a system...

...causes the **center of mass** of the system to **accelerate.**

$$\sum \vec{F}_{\text{external on system}} = M_{\text{tot}} \frac{\Delta \vec{v}_{\text{CM}}}{\Delta t} = M_{\text{tot}} \vec{a}_{\text{CM}} \tag{8-34}$$

Answer to What do you think? Question

(c) If two objects moving at different velocities collide and stick together, kinetic energy is always lost. That's because some portion of one or both of the objects is compressed in the collision. Hence negative work is done on the object or objects, and kinetic energy is lost. We discuss collisions of this kind (called *completely inelastic* collisions) in Section 8-4.

Answers to Got the Concept? Questions

8-1 (c) This situation is exactly like the case of you pushing off the skateboard. The bow and the arrow exert forces of equal magnitude on each other, and these forces act for the same amount of *time*. So just after the shot, the bow (with you and your rollerblades attached) has the same magnitude of *momentum* as does the arrow. Because the arrow has such a small mass, however, it ends up with much more kinetic energy than the combination of the bow, you, and your rollerblades. (Another way to see this is that the arrow travels a much greater distance while it is being released than the bow, you, and your rollerblades do. So the product of force times *distance* is much greater for the arrow.) Since kinetic energy is a measure of how much work an object can do in coming to rest (see Section 6-3), it follows that the arrow will do much more damage to whatever it hits than will you recoiling on your rollerblades. (That's an important reason an archer holds the bow and lets the arrow fly rather than the other way around!)

8-2 (d) Momentum is conserved in this collision. The collision is two dimensional, so we need both x and y axes; we'll take the positive x axis to point west (the direction in which the first object is moving initially) and the positive y axis to point north. Then before the collision the total momentum (which is equal to the momentum of the first object) has a positive x component and a zero y component. Because momentum is conserved, the total momentum after the collision must also point in the positive x direction. After the collision the first object has negative x momentum and negative y momentum, so the second object must have a large positive x momentum (so that the total x momentum can be positive) and a positive y momentum (so that the total y momentum can be zero). Therefore, the second object is moving west and north.

8-3 (d) For the system of car and truck to be at rest after this completely inelastic collision, the total momentum of the system must be zero. The car has a northward momentum of magnitude $p_{car,i} = m_{car}v_{car,i} = (1.50 \times 10^3 \text{ kg}) \times (35.0 \text{ m/s}) = 5.25 \times 10^4 \text{ kg·m/s}$, so the truck must have a southward momentum of the same magnitude for the total momentum to be zero. Therefore, $p_{truck,i} = m_{truck}v_{truck,i} = (7.50 \times$

$10^3 \text{ kg})v_{truck,i} = 5.25 \times 10^4 \text{ kg·m/s}$, and $v_{truck,i} = (5.25 \times 10^4 \text{ kg·m/s})/(7.50 \times 10^3 \text{ kg}) = 7.00 \text{ m/s}$.

8-4 (a) elastic, (b) completely inelastic, (c) inelastic. In (a) the total kinetic energy just after the collision is $125 \text{ J} + 75 \text{ J} = 200 \text{ J}$, the same as just before the collision. So mechanical energy is conserved, and the collision is elastic. In (b) the colliding objects (the dog and the ball) move together after the collision, so this must be a completely inelastic collision as described in Section 8-4. In (c) the ball collides with Earth, which is very much more massive than the ball. If the collision were elastic, the ball would rebound with nearly the same kinetic energy as it had just before hitting the ground and so it would bounce back to its original height. Since this does not happen, mechanical energy must have been lost and the collision must have been inelastic. (The collision would be *completely* inelastic only if the ball didn't rebound at all, in which case the colliding ball and Earth would move together after the collision.)

8-5 (d) This is the same situation as Example 8-9, so we can use the same equation for the final velocity of the ball: $v_{fx} = F_{collision,x}\Delta t/m = (1700 \text{ N})(8.6 \times 10^{-3} \text{ s})/(0.43 \text{ kg}) = 34 \text{ m/s}$ (about 120 km/h or 76 mi/h).

8-6 (d) If the shell did not explode, its center of mass would hit the target. The force that blows the cannon shell apart is *internal* to the shell, so it does not affect the motion of the center of mass. So the center of mass still hits the target. The explosion happens when the shell is halfway along its trajectory, so the half that falls vertically lands 100 m short of the target (half the horizontal distance from cannon to target). We saw in Example 8-10 that the center of mass of a system of two equal masses is halfway between the two masses. So the other mass must land 100 m on the other side of the target.

Questions and Problems

In a few problems you are given more data than you actually need; in a few other problems you are required to supply data from your general knowledge, outside sources, or informed estimate. Interpret as significant all digits in numerical values that have trailing zeros and no decimal points.

For all problems use $g = 9.80 \text{ m/s}^2$ for the free-fall acceleration due to gravity.

- •   Basic, single-concept problem
- ••   Intermediate-level problem; may require synthesis of concepts and multiple steps
- •••   Challenging problem
- Example   See worked example for a similar problem

Conceptual Questions

1. • If the mass of a basketball is 18 times that of a tennis ball, can they ever have the same momentum? Explain your answer.

2. • Starting from Newton's second law explain how a collision that is free from external forces conserves momentum. In other words, explain how the momentum of the system remains constant with time.

3. • The classic collision problem involves two hard spheres colliding. When you define the system as *both* spheres, momentum is conserved, but when you define the system as only *one*

of the spheres, momentum is *not* conserved. Explain why momentum is conserved in the first case but not the second.

4. • Two objects have equal kinetic energies. Are the magnitudes of their momenta equal? Explain your answer.

5. • How would you determine if a collision is elastic or inelastic?

6. • Cite two examples of totally inelastic collisions that occur in your daily life.

7. •• Why is conservation of energy alone not sufficient to explain the motion of a Newton's cradle toy, shown in Figure 8-26? Consider the case when two balls are raised and released together.

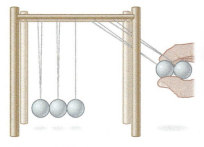

Figure 8-26 Problem 7

8. • An arrow shot into a straw target penetrates a distance that depends on the speed with which it strikes the target. How does the penetration distance change if the arrow's speed is doubled? Be sure to list all the assumptions you make while arriving at your answer.

9. • Using the common definition of the word "impulse," comment on physicists' choice to define impulse as the change in momentum.

10. • A glass will break if it falls onto a hardwood floor but not if it falls from the same height onto a padded, carpeted floor. Describe the different outcomes in the collision between a glass and the floor in terms of fundamental physical quantities.

11. • A child stands on one end of a long wooden plank that rests on a frictionless icy surface. (a) Describe the motion of the plank when she runs to the other end of the plank. (b) Describe the motion of the center of mass of the system. (c) How would your answers change if she had walked the plank rather than run down it?

12. • Based on what you know about center of mass, why is it potentially dangerous to step off of a small boat before it is secured to the dock?

13. • A man and his large dog sit at opposite ends of a rowboat, floating on a still pond. You notice from shore that the boat moves as the dog walks toward his owner. Describe the motion of the boat from your perspective.

14. • After being thrown into the air, a lit firecracker explodes at the apex of its parabolic flight. (a) Is momentum conserved during the flight before the explosion? (b) Is momentum conserved during the explosion? (c) Is momentum conserved during the flight after the explosion? (d) Is the mechanical energy conserved during the flight before the explosion? (e) Is the mechanical energy conserved during the explosion? (f) Is the mechanical energy conserved during the flight after the explosion? (g) What is the path of the center of mass? Neglect the effects of air resistance and explain your answers.

Multiple-Choice Questions

15. • A large semitrailer truck and a small car have equal momentum. How do their speeds compare?
   A. The truck has a much higher speed than the car.
   B. The truck has only a slightly higher speed than the car.
   C. Both have the same speed.
   D. The truck has only a slightly lower speed than the car.
   E. The truck has a much lower speed than the car.

16. • A tennis player smashes a ball of mass m horizontally at a vertical wall. The ball rebounds at the same speed v with which it struck the wall. Has the momentum of the ball changed, and if so, what is the magnitude of the change?
   A. mv
   B. 0
   C. $mv/2$
   D. $2mv$
   E. $4mv$

17. • You throw a bouncy rubber ball and a wet lump of clay, both of mass m, at a wall. Both strike the wall at speed v, but while the ball bounces off with no loss of speed, the clay sticks. What is the change in momentum of the clay and ball, respectively, assuming that toward the wall is the positive direction?
   A. 0; mv
   B. mv; 0
   C. 0; $-2mv$
   D. $-mv$; $-mv$
   E. $-mv$; $-2mv$

18. • Consider a completely inelastic, head-on collision between two particles that have equal masses and equal speeds. Describe the velocities of the particles after the collision.
   A. The velocities of both particles are zero.
   B. Both of their velocities are reversed.
   C. One of the particles continues with the same velocity, and the other comes to rest.
   D. One of the particles continues with the same velocity, and the other reverses direction at twice the speed.
   E. More information is required to determine the final velocities.

19. • An object is traveling in the positive x direction with speed v. A second object that has half the mass of the first is traveling in the opposite direction with the same speed. The two experience a completely inelastic collision. The final x component of the velocity is
   A. 0.
   B. $v/2$.
   C. $v/3$.
   D. $2v/3$.
   E. v.

20. • Consider a completely elastic head-on collision between two particles that have the same mass and the same speed. What are the velocities after the collision?
   A. Both are zero.
   B. The magnitudes of the velocities are the same, but the directions are reversed.
   C. One of the particles continues with the same velocity, and the other comes to rest.
   D. One of the particles continues with the same velocity, and the other reverses direction at twice the speed.
   E. More information is required to determine the final velocities.

21. • Two small, identical steel balls collide completely elastically. Initially, ball 1 is moving along the x axis with velocity v_{1x}, and ball 2 is stationary. After the collision, the final velocities of ball 1 and ball 2, respectively, are
   A. $\frac{1}{2}v_{1x}$; $\frac{1}{2}v_{1x}$.
   B. v_{1x}; $2v_{1x}$.
   C. 0; v_{1x}.
   D. $-v_{1x}$; 0.
   E. $-v_{1x}$; $2v_{1x}$.

22. • Two ice skaters, Lilly and John, face each other while stationary and push against each other's hands. John's mass is twice that of Lilly. How do their speeds compare after the push-off?
   A. Lilly's speed is one-fourth of John's speed.
   B. Lilly's speed is one-half of John's speed.
   C. Lilly's speed is the same as John's speed.
   D. Lilly's speed is twice John's speed.
   E. Lilly's speed is four times John's speed.

23. • A friend throws a heavy ball toward you while you are standing on smooth ice. You can either catch the ball or deflect it back toward your friend. Which of the following options will maximize your speed right after your interaction with the ball?
   A. You should catch the ball.
   B. You should deflect the ball back toward your friend at the same speed with which it hit your hand.
   C. You should let the ball go past you without touching it.
   D. It doesn't matter—your speed is the same regardless of what you do.
   E. You should deflect the ball back toward your friend at half the speed with which it hit your hand.

24. ••• Two blocks are released from rest on either side of a friction-less half-pipe (Figure 8-27). Block B has twice the mass of block A. The height H_B from

Figure 8-27 Problem 24

which block B is released is twice the height H_A from which block A is released. The blocks collide elastically on the flat section. After the collision, which is correct?
- A. Block A rises to a height greater than H_A, and block B rises to a height less than H_B.
- B. Block A rises to a height less than H_A, and block B rises to a height greater than H_B.
- C. Block A rises to height H_A, and block B rises to height H_B.
- D. Block A rises to height H_B, and block B rises to height H_A.
- E. The heights to which the blocks rise depend on where along the flat section they collide.

Problems

8-1 Newton's third law helps lead us to the idea of momentum

8-2 Momentum is a vector that depends on an object's mass, speed, and direction of motion

25. • A 1.00×10^4-kg train car moves east at 15 m/s. Determine the momentum of the train car. Example 8-1

26. • **Sports** The magnitude of the instantaneous momentum of a 57-g tennis ball is 2.6 kg·m/s. What is its speed? Example 8-1

27. • Determine the initial momentum, final momentum, and change in momentum of a 1250-kg car initially backing up at 5.00 m/s, then moving forward at 14.0 m/s. Example 8-1

28. • **Sports** What is the magnitude of momentum of a 135-kg defensive lineman running at 7.00 m/s? Example 8-1

29. • One ball has four times the mass and twice the speed of another. (a) How does the magnitude of momentum of the more massive ball compare to the magnitude of momentum of the less massive one? (b) How does the kinetic energy of the more massive ball compare to the kinetic energy of the less massive one? Example 8-1

30. • A girl who has a mass of 55.0 kg rides a skateboard to class at a speed of 6.00 m/s. (a) What is the magnitude of her momentum? (b) If the magnitude of momentum of the skateboard itself is 30.0 kg·m/s, what is its mass? Example 8-1

31. • Blythe and Geoff are ice skating together. Blythe has a mass of 50.0 kg, and Geoff has a mass of 80.0 kg. Blythe pushes Geoff in the chest when both are at rest, causing him to move away at a speed of 4.00 m/s. (a) Determine Blythe's speed after she pushes Geoff. (b) In what direction does she move? Example 8-1

8-3 The total momentum of a system of objects is conserved under certain conditions

32. • A 2.00-kg object is moving east at 4.00 m/s when it collides with a 6.00-kg object that is initially at rest. After the collision the larger object moves east at 1.00 m/s. What is the final velocity of the smaller object after the collision? Assume no external forces act on the objects. Example 8-2

33. • A 3.00-kg object is moving toward the right at 6.00 m/s. A 5.00-kg object moves to the left at 4.00 m/s. After the two objects collide the 3.00-kg object moves toward the left at 2.00 m/s. What is the final velocity of the 5.00-kg object? Assume no external forces act on the objects. Example 8-2

34. •• An object of mass $3M$, moving in the +x direction at speed v_0, breaks into two pieces of mass M and $2M$ as shown in Figure 8-28. If $\theta_1 = 45°$ and $\theta_2 = 30°$, determine the final velocities of the resulting pieces in terms of v_0. Example 8-4

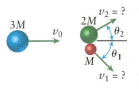

Figure 8-28 Problem 34

35. •• In a game of pool the cue ball is rolling at 2.00 m/s in a direction 30.0° north of east when it collides with the 8-ball (initially at rest). The mass of the cue ball is 170 g, but the mass of the 8-ball is only 156 g. After the collision the cue ball heads off at 10.0° north of east, and the 8-ball moves off due north. What are the final speeds of each ball after the collision? Example 8-3

36. • **Biology** During mating season, male bighorn sheep establish dominance with head-butting contests which can be heard up to a mile away. When two males butt heads, the "winner" is the one that knocks the other backward. In one contest a sheep of mass 95.0 kg moving at 10.0 m/s runs directly into a sheep of mass 80.0 kg moving at 12.0 m/s. Which ram wins the head-butting contest? Example 8-2

8-4 In an inelastic collision some of the mechanical energy is lost

37. • Two hockey players collide on the ice and go down together in a tangled heap. Player 1 has a mass of 105 kg, and player 2 has a mass of 92 kg. Before the collision player 1 had a velocity of $v_1 = 6.3$ m/s in the +x direction, and player 2 had a velocity of $v_2 = 5.6$ m/s at an angle of 72° with respect to the +x axis as shown in Figure 8-29. At what speed and in what direction do they slide together on the ice after the collision? Example 8-7

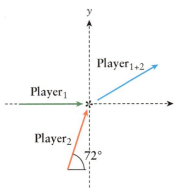

Figure 8-29 Problem 37

38. • A superhero and a supervillain are having a battle. The hero is flying parallel to the ground at a speed of 51.8 m/s when the villain on the ground picks up and hurls a car at him. The 1860-kg car is moving parallel to the ground at 19.1 m/s when it strikes the 102-kg hero at an angle of −46° with respect to the hero's initial direction of motion, as shown in Figure 8-30. The hero and the car are stuck together after the collision. What is their

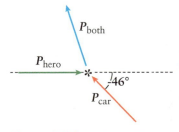

Figure 8-30 Problem 38

speed, and in what direction relative to the hero's initial velocity do they travel after the collision? Example 8-7

39. • A 1.00×10^4-kg train car moving due east at 20.0 m/s collides with and couples to a 2.00×10^4-kg train car that is initially at rest. (a) What is the common velocity of the two-car train after the collision? (b) What is the total kinetic energy of the two train cars before and after the collision? Example 8-6

40. • A large fish has a mass of 25.0 kg and swims at 1.00 m/s toward and then swallows a smaller fish that is not moving. If the smaller fish has a mass of 1.00 kg, what is the speed of the larger fish immediately after it finishes lunch? Example 8-6

41. • A 5.00-kg howler monkey is swinging due east on a vine. It overtakes and grabs onto a 6.00-kg monkey also moving east on a second vine. The first monkey is moving at 12.0 m/s at the instant it grabs the second, which is moving at 8.00 m/s. (a) After they join on the same vine, what is their common speed? (b) What is the total kinetic energy of the two monkeys before and after they join the same vine? Example 8-6

42. • A 1200-kg car is moving at 20.0 m/s due north. A 1500-kg car is moving at 18.0 m/s due east. The two cars simultaneously approach an icy intersection where, with no brakes or steering, they collide and stick together. Determine the speed and direction of the combined two-car wreck immediately after the collision. Example 8-7

43. • **Sports** An 85.0-kg linebacker is running at 8.00 m/s directly toward the sideline of a football field. He tackles a 75.0-kg running back moving at 9.00 m/s straight toward the goal line (perpendicular to the original direction of the linebacker). Determine their common speed and direction immediately after they collide. Example 8-7

8-5 In an elastic collision both momentum and mechanical energy are conserved

44. •• One way that scientists measure the mass of an unknown particle is to bounce a known particle, such as a proton or an electron, off the unknown particle in a bubble chamber. The initial and rebound velocities of the known particle are measured from photographs of the bubbles it creates as it moves; the information is used to determine the mass of the unknown particle. (a) If a known particle of mass m and initial speed v_0 collides elastically, head-on with a stationary unknown particle and then rebounds with speed v, find an expression for the mass of the unknown particle in terms of m, v, and v_0. (b) If the known particle is a proton and the unknown particle is a neutron, what will be the recoil speed of the proton and the final speed of the neutron? Example 8-8

45. • A 2.00-kg ball moves at 3.00 m/s toward the right. It collides elastically with a 4.00-kg ball that is initially at rest. Determine the velocities of the balls after the collision. Example 8-8

46. • A 10.0-kg block of ice is sliding due east at 8.00 m/s when it collides elastically with a 6.00-kg block of ice that is sliding in the same direction at 4.00 m/s. Determine the velocities of the blocks of ice after the collision. Example 8-8

47. • A 0.170-kg ball moves at 4.00 m/s toward the right. It collides elastically with a 0.155-kg ball moving at 2.00 m/s toward the left. Determine the final velocities of the balls after the collision. Example 8-8

48. • A neutron traveling at 2.00×10^5 m/s collides elastically with a deuteron that is initially at rest. Determine the final speeds of the two particles after the collision. The mass of a neutron is 1.67×10^{-27} kg, and the mass of a deuteron is 3.34×10^{-27} kg. Example 8-8

8-6 What happens in a collision is related to the time the colliding objects are in contact

49. •• A sudden gust of wind exerts a force of 20.0 N for 1.20 s on a bird that had been flying at 5.00 m/s. As a result the bird ends up moving in the opposite direction at 7.00 m/s. What is the mass of the bird? Example 8-9

50. • Determine the average force exerted on your hand as you catch a 0.200-kg ball moving at 20.0 m/s. Assume the time of contact is 0.0250 s. Example 8-9

51. • An expert boxer delivers a 3.20×10^3-N punch to the head of his opponent. If the punch contacts the unfortunate boxer's 4.82-kg head for 0.35 s, what is the speed of the head when the punch loses contact? Assume no other forces act on the opponent's head. (In reality, the force of the neck on the head plays an important role.) Example 8-9

52. • **Sports** A baseball of mass 0.145 kg is thrown at a speed of 40.0 m/s. The batter strikes the ball with a force of 25,000 N; the bat and ball are in contact for 0.500 ms. Assuming that the force is exactly opposite to the original direction of the ball, determine the final speed of the ball. Example 8-9

53. • During a neighborhood baseball game in a vacant lot, a particularly wild hit sends a 145-g baseball crashing through the pane of a second-floor window in a nearby building. The ball strikes the glass at 15.0 m/s, shatters the glass as it passes through, and leaves the window at 12.5 m/s with no change of direction. (a) Calculate the magnitude of the impulse that the glass imparts to the baseball. (b) If the baseball is in contact with the glass for 8.50 ms, what is the magnitude of the average force of the glass on the ball? Example 8-9

54. •• **Sports** A baseball bat strikes a ball when both are moving at 31.3 m/s (relative to the ground) toward each other. The bat and ball are in contact for 1.20 ms, after which the ball is traveling opposite its initial direction at a speed of 42.5 m/s. The mass of the bat and the ball are 850 and 145 g, respectively. Calculate the magnitude and direction of the impulse given to (a) the ball by the bat and (b) the bat by the ball. (c) What average force does the bat exert on the ball? (d) Why doesn't the force shatter the bat? Example 8-9

8-7 The center of mass of a system moves as though all of the system's mass were concentrated there

55. • Find the coordinates of the center of mass of the three objects shown in Figure 8-31 if $m_1 = 4.00$ kg, $m_2 = 2.00$ kg, and $m_3 = 3.00$ kg. Distances are in meters. Example 8-10

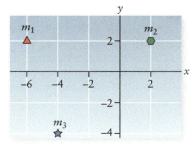

Figure 8-31 Problem 55

56. • What are the coordinates of the center of mass for the combination of the three objects shown in **Figure 8-32**? The uniform rod has a mass of 10.0 kg, has a length of 30.0 cm, and is located at $x = 50.0$ cm. The oval football has a mass of 2.00 kg, a semimajor axis of 15.0 cm, and a semiminor axis of 8.00 cm, and it is centered at $x = -50.0$ cm. The spherical volleyball has a mass of 1.00 kg, has a radius of 10.0 cm, and is centered at $y = -30$ cm. Assume both balls are of uniform mass density. Example 8-10

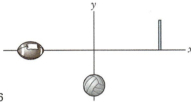

Figure 8-32 Problem 56

57. •• Four beads each of mass M are attached at various locations to a hoop of mass M and radius R (**Figure 8-33**). Find the center of mass of the hoop and beads. Example 8-10

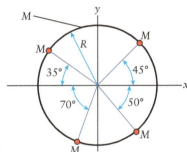

Figure 8-33 Problem 57

58. ••• A 3.0-kg block (A) is attached to a 1.0-kg block (B) by a massless spring that is compressed and locked in place, as shown in **Figure 8-34**. The blocks slide without friction along the x direction at an initial constant speed of 2.0 m/s. At time $t = 0$, the positions of blocks A and B are $x = 1.0$ m and $x = 1.2$ m, respectively, at which point a mechanism releases the spring, and the blocks begin to oscillate as they slide. If 2.0 s later block B is located at $x = 6.0$ m, where will block A be located? Example 8-10

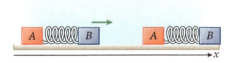

Figure 8-34 Problem 58

General Problems

59. •• A 3.0-kg object constrained to move along the x axis is subjected to a time-varying force, as shown in **Figure 8-35**. (a) What is the object's change in velocity between 0 s and 6 s? (b) What is the object's change in velocity between 5 s and 10 s? (c) What is the object's change in velocity between 10 s and 17 s?

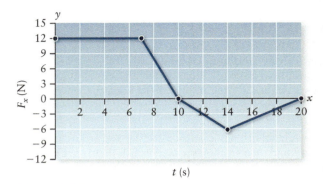

Figure 8-35 Problem 59

60. •• **Sports** A major league baseball has a mass of 0.145 kg. Neglecting the effects of air resistance, determine the momentum of the ball when it hits the ground if it falls from rest on the roof of the Willis Tower in Chicago, Illinois, a height of 442.0 m. Example 8-1

61. ••• Forensic scientists can determine the speed at which a rifle fires a bullet by shooting into a heavy block hanging by a wire. As the bullet embeds itself in the block, the block and embedded bullet swing up; the impact speed is determined from the maximum angle of the swing. (a) Which would make the block swing higher, a 0.204 Ruger bullet of mass 2.14 g and muzzle speed 1290 m/s or a 7-mm Remington Magnum bullet of mass 9.71 g and muzzle speed 948 m/s? Assume the bullets enter the block right after leaving the muzzle of the rifle. (b) Using your answer in part (a), determine the mass of the block so that when hit by the bullet it will swing through a 60.0° angle. The block hangs from wire of length 1.25 m and negligible mass. (c) What is the speed of an 8.41-g bullet that causes the block to swing upward through a 30.0° angle? Example 8-6

62. •• A friend suggests that if all the people in the United States dropped down from a 1-m-high table at the same time, Earth would move in a noticeable way. To test the credibility of this proposal, (a) determine the momentum imparted to Earth by 300 million people, of an average mass of 65.0 kg, dropping from 1.00 m above the surface. Assume no one bounces. (b) What change in Earth's speed would result? Example 8-6

63. ••• Sally finds herself stranded on a frozen pond so slippery that she can't stand up or walk on it. To save herself, she throws one of her heavy boots horizontally, directly away from the closest shore. Sally's mass is 60 kg, the boot's mass is 5 kg, and Sally throws the boot with speed equal to 30.0 m/s. (a) What is Sally's speed immediately after throwing the boot? (b) Where is the center of mass of the Sally–boot system, relative to where she threw the boot, after 10.0 s? (c) How long does it take Sally to reach the shore, a distance of 30.0 m away from where she threw the boot? For all parts, assume the ice is frictionless. Example 8-1

64. ••• You have been called to testify as an expert witness in a trial involving a head-on collision. Car A weighs 680 kg and was traveling eastward. Car B weighs 500 kg and was traveling westward at 72.0 km/h. The cars locked bumpers and slid eastward with their wheels locked for 6.00 m before stopping. You have measured the coefficient of kinetic friction between

the tires and the pavement to be 0.750. How fast (in miles per hour) was car A traveling just before the collision? Example 8-6

65. • A 5000-kg open train car is rolling at a speed of 20.0 m/s when it begins to rain heavily, and 200 kg of water collects quickly in the car. If only the total mass has changed, what is the speed of the flooded train car? For simplicity, assume that all of the water collects at one instant and that the train tracks are frictionless. Example 8-2

66. • An 8000-kg open train car is rolling at a speed of 20.0 m/s when it begins to rain heavily. After water has collected in the car, it slows to 19.0 m/s. What mass of water has collected in the car? For simplicity, assume that all of the water collects at one instant and that the train tracks are frictionless. Example 8-2

67. •• An open rail car of initial mass 10,000 kg is moving at 5.00 m/s when rocks begin to fall into it from a conveyor belt. The rate at which the mass of rocks increases is 500 kg/s. Find the speed of the train car after rocks have fallen into the car for a total of 3.00 s. Example 8-2

68. • A 65-kg novice skier stops to rest partway down a slope. An inattentive snowboarder with the same 65-kg mass is barreling down the same hill at 9.6 m/s and crashes right into the back of the skier. Miraculously, the collision is perfectly elastic and nobody falls over. What are the velocities of the skier and snowboarder after the collision? Example 8-8

69. •• **Sports** The sport of curling is quite popular in Canada. A curler slides a 19.1-kg stone so that it strikes a competitor's stationary stone at 6.40 m/s before moving at an angle of 60° from its initial direction. The competitor's stone moves off at 5.60 m/s. Determine the final speed of the first stone and the final direction of the second one. Example 8-3

70. •• **Biology** Lions can run at speeds up to approximately 80.0 km/h. A hungry 135-kg lion running northward at top speed attacks and holds onto a 29.0-kg Thomson's gazelle running eastward at 60.0 km/h. Find the speed and direction of travel of the lion–gazelle system just after the lion attacks. Example 8-7

71. •• **Biology** The mass of a pigeon hawk is twice that of the pigeons it hunts. Suppose a pigeon is gliding north at a speed of $v_P = 23.0$ m/s when a hawk swoops down, grabs the pigeon, and flies off (Figure 8-36). The hawk was flying north at speed of $v_H = 35.0$ m/s, at an angle $\theta = 45°$ below the

horizontal, at the instant of the attack. Find the final velocity vector of the birds just after the attack. Example 8-7

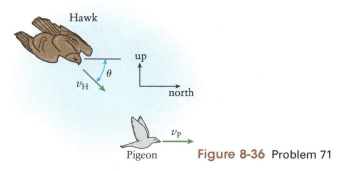

Figure 8-36 Problem 71

72. ••• A 12.0-g bullet is fired into a block of wood with speed $v = 250$ m/s (Figure 8-37). The block is attached to a spring that has a spring constant of 200 N/m. The block with the embedded bullet compresses the spring a distance $d = 30.0$ cm to the right, before momentarily coming to a stop. Determine the mass of the wooden block. Example 8-6

Figure 8-37 Problem 72

73. ••• In a ballistic pendulum experiment, a small marble is fired into a cup attached to the end of a pendulum. If the mass of the marble is 0.00750 kg and the mass of the pendulum is 0.250 kg, how high will the pendulum swing if the marble has an initial speed of 6.00 m/s? Assume that the mass of the pendulum is concentrated at its end. Example 8-6

74. ••• A 0.0750-kg ball is thrown horizontally at 25.0 m/s toward a brick wall. (a) Determine the impulse that the wall imparts to the ball when it hits and rebounds at 25.0 m/s in the opposite direction. (b) Determine the impulse that the wall imparts to the ball when it hits and rebounds at 25.0 m/s at an angle of 45.0°. (c) If the ball thrown in part (b) contacts the wall for 0.0100 s, determine the magnitude and direction of the average force that the wall exerts on the ball. Example 8-9

[left] Kevin Lee—ISU/Getty Images; [right] Rick Bowmer/AP Images

Rotational Motion

What do you think?

When a figure skater pulls her arms in, her rotational speed increases and she spins faster. Her arms stay at the same height and there is almost no friction between her and the ice, so external forces do no work on her. As she pulls her arms in, does the kinetic energy of her spinning body (a) increase, (b) decrease, (c) increase and then decrease, (d) decrease and then increase, or (e) remain the same because no work is done by external forces?

In this chapter, your goals are to:

- (9-1) Define translation and rotation.
- (9-2) Identify the equations of rotational kinematics and know how to use them to solve problems.
- (9-3) Define the concept of lever arm and know how to use it to calculate the torque generated by a force.
- (9-4) Describe the techniques for finding the moment of inertia of an object, including use of the parallel-axis theorem.
- (9-5) Apply Newton's second law for rotational motion to problems involving rotating objects, including rolling without slipping.
- (9-6) Calculate the rotational kinetic energy of a rotating object in terms of its moment of inertia, and use it to apply the conservation of mechanical energy to rotating objects.
- (9-7) Define what is meant by the angular momentum of a rotating object and of a moving particle, and explain the circumstances under which angular momentum is conserved.
- (9-8) Explain how to find the direction of the angular momentum, angular velocity, and torque vectors.

To master this chapter, you should review:

- (2-2, 2-4) The definitions of average and instantaneous velocity and acceleration.
- (3-7) The equation for the length of an arc in terms of its radius and angle.
- (4-2) Newton's second law of motion.
- (6-3, 6-7) The definition of kinetic energy and the principle of conservation of energy.
- (8-2, 8-3) The concept of linear momentum and when linear momentum is conserved.
- (8-7) The concept of the center of mass of an object.

9-1 Rotation is an important and ubiquitous kind of motion

Up to this point we've considered the motions of objects such as a thrown baseball, a speeding car, and a soaring bird. We use the word "**translation**" to refer to these kinds of motion, in which an object as a whole moves through space. However, we

A Ferris wheel rotates around a fixed (stationary) axis.

This mother-of-pearl moth caterpillar (*Pleurotya ruralis*) wraps itself into a circle for self-defense: It can roll away from predators 40 times faster than it can walk. As it rolls, the caterpillar moves as a whole (it translates) as well as rotates.

Our planet rotates to the east once per day around an axis that extends from the north pole to the south pole. Earth also translates: It takes one year to move in an orbit around the Sun.

(a)

(b)

(c)

Holly Kuchera/Getty Images

Dr. John Brackenbury/Science Source

NASA/NOAA/GSFC/Suomi NPP/VIIRS/ Norman Kuring

Figure 9-1   Rotational motion Three examples of rotating objects.

haven't worried much about the kind of motion called **rotation**, in which an object spins around an axis. (One example is the figure skater in the photographs on the previous page, who rotates around an axis that extends upward from the skate that touches the ground. Figure 9-1 shows three more examples.) In this chapter we'll investigate rotational motion.

In some situations, an object undergoes rotation but not translation. An example is the Ferris wheel shown in Figure 9-1a: The Ferris wheel as a whole remains at the same location in the amusement park, but the wheel rotates continuously to entertain its riders. Other objects that rotate without translation are a spinning ceiling fan and the rotating platter in a microwave oven. In other situations, like a rolling caterpillar (Figure 9-1b) or the wheels of a fast-moving bicycle, an object undergoes rotation and translation at the same time. You've spent your entire life on a planet that experiences both rotation and translation (Figure 9-1c). In this chapter we'll look at rotation both without and with translation.

Although rotation and translation are different kinds of motion, many of the same concepts that we developed for translational motion apply as well to rotational motion. We'll begin by looking at the rotational analogs of velocity and acceleration, which we call *angular velocity* (which measures how fast and in what direction the object is rotating) and *angular acceleration* (which measures the rate of change of the angular velocity). We'll then introduce the concept of *torque*, which measures the extent to which a force acting on an object tends to change its rotation. We'll see that just as a net external force acting on an object gives that object an acceleration, a net external torque on an object causes an angular *acceleration*. We'll also learn about the rotational equivalents of kinetic energy and momentum, called *rotational kinetic energy* and *angular momentum*, and see how to use these to solve many problems involving rotational motion. Finally, we'll apply the idea of torque to help us understand systems that are intended not to rotate at all but to be in *equilibrium*, such as a gymnast balancing on one foot.

> **TAKE-HOME MESSAGE FOR Section 9-1**
>
> ✔   Translation refers to the motion through space of an object as a whole.
>
> ✔   Rotation refers to the spinning motion of an object.
>
> ✔   An object can rotate around a fixed axis (rotation without translation), or its axis can move through space as the object rotates (rotation with translation).

9-2   The equations for rotational kinematics are almost identical to those for linear motion

To keep things simple, we'll begin our study of rotational motion by considering rotation without translation. To be specific, we'll examine a **rigid object**—that is, one whose shape doesn't change as it rotates. The spinning wind turbines shown in Figure 9-2 are examples of rotating rigid objects. (By contrast, the figure skater in the photographs that open this chapter can change her shape as she spins.) We'll also assume that the rotation axis is *fixed*—that is, it keeps the same orientation in space. The wind turbines in Figure 9-2 rotate around fixed axes.

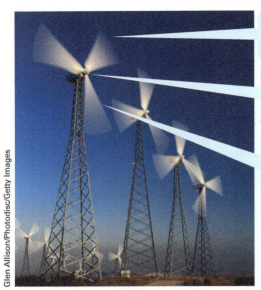

Each blade of this rotating wind turbine is rigid: It maintains the same shape at all times.

The wind turbine rotates around a fixed axis: The axis remains in the same place and keeps the same orientation.

While all parts of the blade rotate together, different parts move at different speeds: The blurring shows that the outer parts (at a greater distance r from the axis) move faster than the inner parts of the blade.

Glen Allison/Photodisc/Getty Images

Figure 9-2 **A rotating rigid object** The blades of a wind turbine undergo rotation but do not undergo translation (the wind turbine as a whole doesn't go anywhere).

Angular Velocity and Angular Speed

The language we use to describe the rotation of a rigid object is called **rotational kinematics**. It's the analog of the linear kinematics that we used in Chapter 2 to describe straight-line motion in terms of displacement, velocity, and acceleration. To get started, let's consider the motion of one of the wind turbine blades in Figure 9-2. As **Figure 9-3** shows, during a short time interval Δt, the entire blade rotates through a small angle $\Delta \theta$ (pronounced "delta-theta"), which we call the **angular displacement** of the blade. By analogy to how we defined average velocity for straight-line motion in Section 2-2, we define the **average angular velocity** of the blade to be

<img class="math-tutorial"> **See the Math Tutorial for more information on trigonometry.**

Average angular velocity of a rotating object over a time interval (can be + or −)

Angular displacement of the object during the time interval (can be + or −)

$$\omega_{\text{average},z} = \frac{\Delta \theta}{\Delta t}$$

Duration of the time interval

Average angular velocity
(9-1)

Here ω is the Greek letter omega, and the z in the subscript tells us that the blade is rotating around an axis that we call the z axis (Figure 9-3). For any rotating rigid object, the value of $\omega_{\text{average},z}$ is the same for all pieces of the object. The average angular velocity can be positive or negative, depending on the direction in which the object rotates. A common choice is to take counterclockwise rotation to be positive and clockwise rotation to be negative. With this choice, since the blade in Figure 9-3 is rotating counterclockwise, its angular displacement $\Delta \theta$ and average angular velocity $\omega_{\text{average},z}$ are both positive.

Just as we did for ordinary velocity in Section 2-4, we'll define the *instantaneous* angular velocity ω_z, or just **angular velocity** for short, as the rate at which the object is rotating at a given instant. That is, the instantaneous angular velocity is the average angular velocity for a very short time interval.

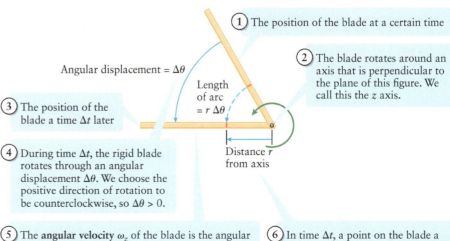

① The position of the blade at a certain time

② The blade rotates around an axis that is perpendicular to the plane of this figure. We call this the z axis.

Angular displacement = $\Delta \theta$

Length of arc = $r \Delta \theta$

③ The position of the blade a time Δt later

④ During time Δt, the rigid blade rotates through an angular displacement $\Delta \theta$. We choose the positive direction of rotation to be counterclockwise, so $\Delta \theta > 0$.

Distance r from axis

⑤ The **angular velocity** ω_z of the blade is the angular displacement $\Delta \theta$ divided by the elapsed time Δt:

$$\omega_z = \frac{\Delta \theta}{\Delta t}$$

The **angular speed** ω is the absolute value of ω_z.

⑥ In time Δt, a point on the blade a distance r from the axis moves through an arc of length $r \Delta \theta$. The speed of this point is $v = r\omega$.

Figure 9-3 **Angular velocity** We define angular velocity by analogy to the way that we defined velocity for straight-line motion in Chapter 2.

The preferred units of angular velocity are radians per second, or rad/s. As we discussed in Section 3-8, radians are a measure of angle; there are 2π radians in a circle. We can also describe angle in terms of the number of revolutions (rev) that an object makes, so another common set of units for angular velocity is revolutions per second (rev/s). Since one revolution equals 2π radians, 1 rev/s = 2π rad/s.

We also use rad/s and rev/s as the units for **angular speed**, which is the magnitude or absolute value of angular velocity. An object that rotates 5.0 radians in 1 s has angular velocity $\omega_z = +5.0$ rad/s if it rotates in the positive direction and angular velocity $\omega_z = -5.0$ rad/s if it rotates in the negative direction, but in either case its angular speed is 5.0 rad/s. We'll use the symbol ω (with no subscript) for angular speed.

Speed of a Point on a Rotating Object

All points on the wind turbine blade in Figure 9-3 have the same angular displacement $\Delta\theta$ during time interval Δt. But during this time interval a point near the rotation axis travels only a short distance, while a point farther from the axis travels a greater distance. Hence a point farther out along the blade moves faster and has a greater speed than a point closer to the axis. You can see this in Figure 9-2: The blade is more blurred the farther it is from the axis. So the speed v in meters per second (m/s) is different for different parts of the blade. To find the speed of a given point on the rotating blade, consider the point shown in red in Figure 9-3 located a distance r from the rotation axis. In time Δt this point moves through a circular arc of radius r and angle $\Delta\theta$, which we've taken to be positive. We saw in Section 3-7 that if the angle $\Delta\theta$ is measured in radians and r is measured in meters, the length of such an arc in meters is $r\Delta\theta$ (see Figure 3-30); this is the distance that this point on the blade travels in a time Δt. The speed v of this point is the distance traveled divided by the time interval, or

$$v = \frac{r\Delta\theta}{\Delta t} \tag{9-2}$$

But $\Delta\theta/\Delta t$ is just the angular speed ω of the rotating blade. (Since the blade is rotating in the positive direction, its angular velocity $\omega_z = \Delta\theta/\Delta t$ is positive and so the angular speed ω—which is always positive—is the same as ω_z.) So we can rewrite Equation 9-2 for the speed of a point on the blade as

Speed (in m/s) of a point on a rotating rigid object   Distance (in m) from the axis of rotation to the point

$$v = r\omega$$

Angular speed (in rad/s) of the rotating rigid object

Speed of a point on a rotating rigid object
(9-3)

On an object rotating with angular speed ω, such as the wind turbine blade in Figures 9-2 and 9-3, points that are farther from the axis are at a greater distance r and so are moving at a greater speed $v = r\omega$.

WATCH OUT! Use correct units when relating speed and angular speed.

! We derived Equation 9-3 from the statement that the length of the arc in Figure 9-3 is $r\Delta\theta$. This is true *only* if the angle $\Delta\theta$ is measured in radians (see Section 3-7). Hence you can safely use Equation 9-3 only if the angular speed ω is measured in radians per second (rad/s). If you are given the angular speed in other units, such as revolutions per second (rev/s), revolutions per minute (rev/min), or degrees per second, you must convert it to rad/s before you can use Equation 9-3.

We derived Equation 9-3 assuming that the object is rotating in the positive direction. But it's equally valid if the object rotates in the negative direction, since the speed v and angular speed ω are always positive no matter what the direction of motion.

EXAMPLE 9-1   Speed versus Angular Speed

A turbine blade is rotating at 25.0 rev/min. How fast (in m/s) is a point on the blade moving that is (a) 0.500 m from the rotation axis and (b) 1.00 m from the rotation axis?

Set Up

For each point we are given the angular speed ω of the blade and the distance r from the rotation axis. We'll use Equation 9-3 to find the speed v of that piece.

Speed of a point on a rotating object:

$$v = r\omega \qquad (9\text{-}3)$$

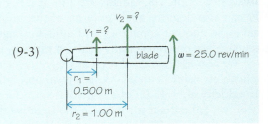

Solve

We are given the angular speed in revolutions per minute. We first need to convert this to radians per second so we can safely use Equation 9-3.

$\omega = 25.0$ rev/min

Convert revolutions to radians and minutes to seconds:

$$\omega = \left(25.0\ \frac{\text{rev}}{\text{min}}\right)\left(\frac{2\pi\ \text{rad}}{1\ \text{rev}}\right)\left(\frac{1\ \text{min}}{60\ \text{s}}\right)$$

$$= \frac{(25.0)(2\pi)}{60}\ \frac{\text{rad}}{\text{s}} = 2.62\ \text{rad/s}$$

For each case substitute $\omega = 2.62$ rad/s and the value of the distance r of the point from the axis.

(a) $v_1 = r_1\omega = (0.500\ \text{m})(2.62\ \text{rad/s}) = 1.31\ \text{m/s}$

(b) $v_2 = r_2\omega = (1.00\ \text{m})(2.62\ \text{rad/s}) = 2.62\ \text{m/s}$

Reflect

Note that piece (b) is twice as far from the rotation axis as is piece (a) and has twice the speed. That's because in the same amount of time, piece (b) travels in a circle that has twice the radius and twice the circumference of the circle traveled by piece (a); so piece (b) must cover twice as much distance as piece (a).

Note that the units of radians disappeared when we calculated the speed of each piece. We did this because a radian isn't truly a unit but simply a way of counting the number of revolutions or fractions of a revolution through which an object has rotated.

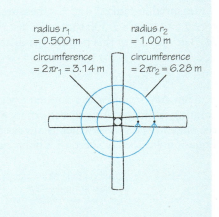

Angular Acceleration

At time t_1 an object rotates with angular velocity ω_{1z}.

At a later time t_2 the object rotates with a different angular velocity ω_{2z}.

 The object's average angular acceleration equals the change in angular velocity divided by the elapsed time:

$$\alpha_{\text{average},z} = \frac{\omega_{2z} - \omega_{1z}}{t_2 - t_1} = \frac{\Delta\omega_z}{\Delta t}$$

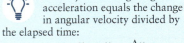

Figure 9-4 Defining angular acceleration The greater the magnitude of the average angular acceleration $\alpha_{\text{average},z}$ of a rotating object, the more rapidly the angular velocity changes.

In many situations we need to know not just how rapidly an object is rotating, but also how rapidly its angular velocity is changing—that is, its *angular acceleration*. Angular acceleration is an important factor in the design of rotating machinery. When you turn on an electric fan on a hot day, you want the fan to come up to speed right away. But if you're riding on a carnival merry-go-round, you want it to gain speed gradually at the beginning of the ride (and lose speed in the same way at the end of the ride) rather than in a sudden jerk.

We'll define angular acceleration in much the same way that we defined the acceleration of an object moving in a straight line in Section 2-4. In that section we described the average acceleration $a_{\text{average},x}$ of an object moving along the x axis as the change Δv_x in its x velocity during a time interval divided by the duration Δt of the time interval:

$$a_{\text{average},x} = \frac{\Delta v_x}{\Delta t} \quad \text{(average } x \text{ acceleration)}$$

If the duration Δt of the time interval becomes very short, the ratio $\Delta v_x/\Delta t$ becomes the *instantaneous* acceleration a_x or, for short, simply the acceleration.

Let's follow the same steps for rotational motion. Suppose a rigid object is rotating around the z axis with angular velocity ω_{1z} at time t_1 and with angular velocity ω_{2z} at a later time t_2 (Figure 9-4). The **average angular acceleration** $\alpha_{\text{average},z}$ (the Greek letter α, or alpha) for the time interval between t_1 and t_2 is the change in angular velocity, $\Delta\omega_z = \omega_{2z} - \omega_{1z}$, divided by the duration of the time interval, $\Delta t = t_2 - t_1$:

Average angular acceleration of a rotating object	Change in angular velocity of the object over a certain time interval: The angular velocity changes from ω_{1z} to ω_{2z}.

$$\alpha_{\text{average},z} = \frac{\omega_{2z} - \omega_{1z}}{t_2 - t_1} = \frac{\Delta\omega_z}{\Delta t}$$

Average angular acceleration
(9-4)

Elapsed time for the time interval: The object has angular velocity ω_{1z} at time t_1, and has angular velocity ω_{2z} at time t_2.

If the two times t_1 and t_2 are very close to each other, so that Δt is very short, this becomes the *instantaneous* angular acceleration α_z, or just **angular acceleration** for short. At any instant angular acceleration is equal to the rate of change of angular velocity at that instant. Since angular velocity is measured in rad/s, angular acceleration is measured in rad/s^2.

Just as all pieces of a rigid object rotate with the same angular velocity ω_z at any time t, all pieces have the same angular acceleration. The value of α_z is positive if the angular velocity ω_z is becoming more positive and negative if ω_z is becoming more negative.

WATCH OUT! The sign of angular acceleration can be misleading.

Just as for the acceleration of an object moving along a straight line, a positive value of angular acceleration α_z does *not* necessarily correspond to speeding up and a negative value of z does not necessarily correspond to slowing down. An object's rotation speeds up if ω_z and α_z have the same algebraic sign—if both are positive, the object is rotating in the positive direction and speeding up, while if both are negative, the object is rotating in the negative direction and speeding up. If ω_z and α_z have opposite signs (one positive and one negative), the object's rotation is slowing down (Figure 9-5).

(a) $\omega_z > 0$
$\alpha_z > 0$:
Angular speed increasing

(b) $\omega_z < 0$
$\alpha_z < 0$:
Angular speed increasing

(c) $\omega_z > 0$
$\alpha_z < 0$:
Angular speed decreasing

(d) $\omega_z < 0$
$\alpha_z > 0$:
Angular speed decreasing

Figure 9-5 **The sign of angular acceleration** A rotating object is speeding up if its angular velocity ω_z and angular acceleration α_z are either (a) both positive or (b) both negative. It is slowing down if (c) ω_z is positive and α_z is negative or (d) ω_z is negative and α_z is positive.

Motion with Constant Angular Acceleration

We saw in Chapter 2 that an important case of motion in a straight line is when the acceleration a_x is constant so that the velocity v_x changes at a steady rate. Let's look at the analogous situation of rotational motion with constant *angular* acceleration. A ball or wheel rolling downhill from rest moves with nearly constant angular acceleration, as do many types of rotating machinery as they start up or slow down.

It's straightforward to write down the equations for rotational motion with constant angular acceleration. To see how, take a look at the equations for linear velocity v_x and angular velocity ω_z, and the equations for linear acceleration a_x and angular acceleration α_z:

 Go to Interactive Exercise 9-1 for more practice dealing with rotational kinematics.

Linear motion	Rotational motion
$v_x = \dfrac{\Delta x}{\Delta t}$ (linear velocity)	$\omega_z = \dfrac{\Delta\theta}{\Delta t}$ (angular velocity)
$a_x = \dfrac{\Delta v_x}{\Delta t}$ (linear acceleration)	$\alpha_z = \dfrac{\Delta\omega_z}{\Delta t}$ (angular acceleration)

Comparing these equations shows that the rotational quantities θ, ω_z, and α_z are related to each other in exactly the same way that x, v_x, and a_x are related to each other. So we can take the equations for constant linear acceleration and convert them to the equations for constant angular acceleration by replacing x with θ, v_x with ω_z, and a_x with α_z. The equations for linear motion are

(2-5)
$$v_x = v_{0x} + a_x t \text{ (constant } x \text{ acceleration only)}$$

(2-9)
$$x = x_0 + v_{0x}t + \frac{1}{2}a_x t^2 \text{ (constant } x \text{ acceleration only)}$$

(2-11)
$$v_x^2 = v_{0x}^2 + 2a_x(x - x_0) \text{ (constant } x \text{ acceleration only)}$$

Hence the equations for constant angular acceleration are

Angular velocity, angular acceleration, and time for constant angular acceleration only (9-5)

Angular velocity at time t of a rotating object with constant angular acceleration

Angular velocity at time $t = 0$ of the object

$$\omega_z = \omega_{0z} + \alpha_z t$$

Constant angular acceleration of the object   Time at which the object has angular velocity ω_z

Angular position, angular acceleration, and time for constant angular acceleration only (9-6)

Angular position at time t of a rotating object with constant angular acceleration

Angular velocity at time $t = 0$ of the object

Constant angular acceleration of the object

$$\theta = \theta_0 + \omega_{0z}t + \frac{1}{2}\alpha_z t^2$$

Angular position at time $t = 0$ of the object   Time at which the object is at angular position θ

Angular velocity, angular acceleration, and angular position for constant angular acceleration only (9-7)

Angular velocity at angular position θ of a rotating object with constant angular acceleration

Angular velocity at angular position θ_0 of the object

$$\omega_z^2 = \omega_{0z}^2 + 2\alpha_z(\theta - \theta_0)$$

Constant angular acceleration of the object   Two angular positions of the object

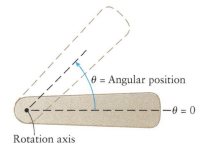

θ = Angular position

$\theta = 0$

Rotation axis

Figure 9-6 Angular position Imagine a line drawn on this rotating object from its rotation axis outward. The angle of this line at a given time is the object's angular position θ at that time.

To see what we mean by *angular position*, imagine a line drawn on the object outward from the rotation axis (Figure 9-6). The angle of this line from a reference direction changes as the object rotates, and it's this angle that we call the **angular position θ**. The angular displacement from time 0 to time t is $\theta - \theta_0$.

We can use Equations 9-5, 9-6, and 9-7 to solve a wide range of problems in rotational motion, just as we used Equations 2-5, 2-9, and 2-11 for linear motion problems in Chapter 2. Here's a representative example.

WATCH OUT! Equations 9-5, 9-6, and 9-7 are for rigid objects only.

The three equations we've just presented can be used only if *all parts* of a rotating object have the same angular velocity and angular acceleration at any given time—in other words, only if the object is *rigid*. They *cannot* be used for a rotating object that isn't rigid, such as water in a bathtub as it swirls down the drain or a cake mix being stirred in a bowl. In Example 9-2, and in the problems at the end of the chapter, we'll always assume that the rotating objects are rigid.

EXAMPLE 9-2 A Stopping Top

A top spinning at 4.00 rev/s comes to a complete stop in 64.0 s. Assuming the top slows down at a constant rate, how many revolutions does it make before coming to a stop?

Set Up

The statement that the top slows down at a constant rate means that its angular acceleration is constant, so we can use Equations 9-5 and 9-6. (We won't use Equation 9-7 because that equation doesn't involve time, which is one of the quantities we're given.) These two equations involve six quantities (t, θ_0, θ, ω_{0z}, ω_z, and α_z).

In this case we are given the initial angular velocity $\omega_{0z} = 4.00$ rev/s, the final angular velocity $\omega_z = 0$, and the elapsed time $t = 64.0$ s. If we let the top start at an angular position of $\theta_0 = 0$, we want to find the top's final angular position θ during this time but don't know the value of the angular acceleration α_z. We'll use Equation 9-5 to determine α_z from the known information then substitute this value into Equation 9-6 to find the value of θ.

To find constant angular acceleration α_z from ω_{0z}, ω_z, and t:

$$\omega_z = \omega_{0z} + \alpha_z t \qquad (9\text{-}5)$$

To find the final angular position θ from θ_0, ω_{0z}, α_z, and t:

$$\theta = \theta_0 + \omega_{0z} t + \frac{1}{2}\alpha_z t^2 \qquad (9\text{-}6)$$

Solve

Rewrite Equation 9-5 to solve for α_z. Then substitute the values of ω_z, ω_{0z}, and t. (Note that we don't need to convert revolutions to radians: We just have to be consistent with units throughout our solution.)

Rewrite Equation 9-5: $\omega_z - \omega_{0z} = \alpha_z t$

$$\alpha_z = \frac{\omega_z - \omega_{0z}}{t}$$

The resulting value of α_z is negative while ω_{0z} is positive, which correctly says that the rotation is slowing down.

Substitute values: $\alpha_z = \dfrac{0 - 4.00 \text{ rev/s}}{64.0 \text{ s}} = -0.0625 \text{ rev/s}^2$

Solve for θ by plugging this value of α_z into Equation 9-6 along with the given values of ω_{0z} and t.

Equation 9-6: $\theta = \theta_0 + \omega_{0z}t + \dfrac{1}{2}\alpha_z t^2$

$= 0 \text{ rev} + (4.00 \text{ rev/s})(64.0 \text{ s}) + \dfrac{1}{2}(-0.0625 \text{ rev/s}^2)(64.0 \text{ s})^2$

$= 256 \text{ rev} - 128 \text{ rev}$

$= 128 \text{ rev}$

Reflect

We can check our answer by substituting the values for ω_z, ω_{0z}, and α_z into Equation 9-7 (which we did not use above) and solving for θ. This gives us the same answer for θ as above.

Note that by writing $\omega_{0z} = 4.00$ rev/s, we made the assumption that the top is initially rotating in the positive direction. Can you show that if you instead make $\omega_{0z} = -4.00$ rev/s, so that the top is initially rotating in the negative direction, you get $\theta - \theta_0 = -128$ rev? (This means that the top rotates 128 rev but in the negative direction.)

Equation 9-7: $\omega_z^2 = \omega_{0z}^2 + 2\alpha_z(\theta - \theta_0)$

Rewrite to solve for θ:

$\omega_z^2 - \omega_{0z}^2 = 2\alpha_z(\theta - \theta_0)$

$\theta = \dfrac{\omega_z^2 - \omega_{0z}^2}{2\alpha_z} + \theta_0 = \dfrac{(0 \text{ rev/s})^2 - (4.00 \text{ rev/s})^2}{2(-0.0625 \text{ rev/s}^2)} + 0 \text{ rev}$

$= 128 \text{ rev}$

Figure 9-7 **Tangential acceleration** If the angular speed of a rotating rigid object is increasing, a point on that object (shown as a red dot) has a tangential component of acceleration, $a_{\text{tangential}}$. (a) If the angular speed is increasing, $a_{\text{tangential}}$ is in the same direction as the velocity $\vec{v}$ of the point. (b) If the angular speed is decreasing, $a_{\text{tangential}}$ is in the direction opposite to $\vec{v}$. (Compare with Figure 3-32.)

(a) Angular speed is increasing: $a_{\text{tangential}}$ is in the direction of the point's motion

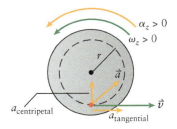

(b) Angular speed is decreasing: $a_{\text{tangential}}$ is opposite to the point's motion

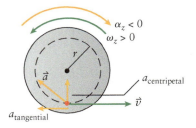

Red dot: A point on the rotating object
Dashed circle: Circular path of radius r followed by this point

Red dot: A point on the rotating object
Dashed circle: Circular path of radius r followed by this point

Tangential Acceleration of a Point on a Rotating Object

A point on a rotating object is moving in a circle. From what we learned in Section 3-7, this means that such a point has a component of acceleration directed toward the center of the circle (that is, a *centripetal* acceleration). But if the angular speed of the rotating object is changing, a point on the object will also have a component of acceleration directed either in the direction of its velocity $\vec{v}$ (if the angular speed is increasing) or opposite the direction of $\vec{v}$ (if the angular speed is decreasing). This component is tangent to the circular path that the point follows as the object rotates, so we call this component its **tangential acceleration** (Figure 9-7).

The centripetal acceleration tells us that the direction of the velocity $\vec{v}$ of the point in Figure 9-7 changes as it moves around the circular path. The *tangential* acceleration tells us how the *magnitude* of $\vec{v}$—that is, the speed v of the point—changes: v is increasing if the angular speed ω is increasing (Figure 9-7a), and v is decreasing if the angular speed ω is decreasing (Figure 9-7b). So the value of the tangential acceleration $a_{\text{tangential}}$ is equal to the rate of change of the speed v. From Equation 9-5 the speed of a point on the rotating object a distance r from the rotation axis is $v = r\omega$, so if the angular speed ω changes from ω_1 at time t_1 to ω_2 at time t_2, the tangential acceleration during this time interval is

$$a_{\text{tangential}} = \frac{r\omega_2 - r\omega_1}{t_2 - t_1} = r\left(\frac{\omega_2 - \omega_1}{t_2 - t_1}\right) = r\frac{\Delta\omega}{\Delta t} \tag{9-8}$$

(Because the object is rigid, the value of r doesn't change during the time interval: The point shown as a red dot in Figure 9-7 stays the same distance from the rotation axis.) If the object is rotating in the positive (counterclockwise) direction as in Figure 9-7, then the angular velocity ω_z is positive and so is equal to the angular speed ω (which is always positive). Then $\Delta\omega/\Delta t$ in Equation 9-8 is equal to $\Delta\omega_z/\Delta t$, which from Equation 9-4 is just the angular acceleration α_z. (We're assuming that the time interval Δt is very short, so $\Delta\omega_z/\Delta t$ is the instantaneous angular acceleration.) So our expression for tangential acceleration becomes

$$a_{\text{tangential}} = r\alpha_z \tag{9-9}$$

(tangential acceleration of a point on an object rotating in the positive direction)

Recall that in Equation 9-5, $v = r\omega$, we must always express the angular speed ω in rad/s, not rev/s. For the same reason, in Equation 9-9 we must always express the angular acceleration α_z in rad/s^2, not rev/s^2.

As an application, consider the top in Example 9-2, for which we found $\alpha_z = -0.0625$ rad/s^2. We can use Equation 9-9 to find the tangential acceleration of a point on this top 2.00 cm = 0.0200 m from the rotation axis, being careful to convert α_z from rev/s^2 to rad/s^2:

$$a_{\text{tangential}} = r\alpha_z = (0.0200 \text{ m})(-0.0625 \text{ rev/s}^2)\left(\frac{2\pi \text{ rad}}{1 \text{ rev}}\right) = -7.85 \times 10^{-3} \text{ m/s}^2$$

The minus sign tells us that the tangential acceleration points opposite to the velocity as the top slows down, as in Figure 9-7b. The value of the tangential acceleration is quite small because the top is slowing down very gradually.

The idea of tangential acceleration turns out to be an important one. In the following section we'll see how to use this idea to find the rotational equivalent of Newton's second law. Just as the second law that we introduced in Chapter 4 tells us how external forces can make an object accelerate, the rotational equivalent of the second law tells us how suitably applied forces can give an object an *angular* acceleration.

GOT THE CONCEPT? 9-1 Rate of Change of Angular Speed

(?) Rank the following rotating objects in order of the rate at which the speed of their rotation is changing, from speeding up at the fastest rate to slowing down at the fastest rate. (a) A disk with $\omega_z = 2.00$ rad/s and $\alpha_z = -1.00$ rad/s^2; (b) a wheel with $\omega_z = 3.00$ rad/s and $\alpha_z = -2.00$ rad/s^2; (c) a ceiling fan with $\omega_z = -2.00$ rad/s and $\alpha_z = -1.00$ rad/s^2; (d) a flywheel with $\omega_z = 0$ and $\alpha_z = -0.750$ rad/s^2; (e) a circular saw with $\omega_z = 3.00$ rad/s and $\alpha_z = 0.750$ rad/s^2.

TAKE-HOME MESSAGE FOR Section 9-2

✔ Angular velocity is a measure of how rapidly and in what direction an object rotates. Angular speed is the magnitude of angular velocity.

✔ Just as acceleration in a straight line is the rate of change of velocity, angular acceleration is the rate of change of angular velocity.

✔ If the angular acceleration is constant for a rigid object, there are three simple equations that relate time, angular position, angular velocity, and angular acceleration.

✔ The speed of a point on a rotating object depends on the point's distance r from the rotation axis and the object's angular speed; similarly, the tangential acceleration of such a point depends on r and the object's angular acceleration.

9-3 Torque is to rotation as force is to translation

Where do you push or pull on a door to open it most easily? In what direction should you exert that push or pull? These may seem like odd questions to ask in a chapter about rotational motion. But keep in mind that when you open a door you are giving it an *angular acceleration*: The door starts at rest and begins rotating around its hinges, so you're changing the door's angular velocity. So we can rephrase our questions like this: Where, and in what direction, should we push or pull on a door to give it an angular acceleration around its hinges?

Experience tells you that it's nearly impossible to open a door by pushing or pulling near its hinges, as with force $\vec{F}_1$ in Figure 9-8. A much better place to apply a force is on the opposite end of a door from its hinges, which is why the doorknob is located there. Even then, it's ineffective to push or pull on the doorknob in a direction parallel to the plane of the door (as with force $\vec{F}_2$ in Figure 9-8). The easiest way to open the

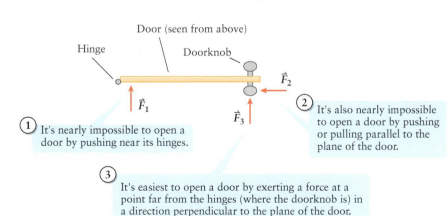

Door (seen from above)

Hinge

Doorknob

$\vec{F}_2$

(1) It's nearly impossible to open a door by pushing near its hinges.

$\vec{F}_1$

$\vec{F}_3$

(2) It's also nearly impossible to open a door by pushing or pulling parallel to the plane of the door.

(3) It's easiest to open a door by exerting a force at a point far from the hinges (where the doorknob is) in a direction perpendicular to the plane of the door.

Figure 9-8 Opening a door When you push or pull on a door to open it, it matters where and in what direction the force is applied.

door is to exert a force far from the hinges (that is, at the doorknob) in a direction perpendicular to the plane of the door, as with force $\vec{F}_3$ in Figure 9-8.

This example shows that to give an object an angular acceleration, what matters is not just how hard you push or pull on the object, but also *where* and *in what direction* that push or pull is applied. A physical quantity that relates all of these aspects of an applied force is the *torque* associated with that force. In this section we'll find that just as the net force acting on an object determines its translational or linear acceleration (Newton's second law), the net *torque* acting on an object determines its *angular* acceleration.

Defining the Magnitude and Direction of Torque

Suppose that you exert a force $\vec{F}$ on an object as shown in Figure 9-9a. We use the symbol $\vec{r}$ to denote the vector from the rotation axis to the point where the force is applied, and we use the symbol ϕ (the Greek letter phi) for the angle between the directions of $\vec{r}$ and $\vec{F}$. The component of $\vec{F}$ that points straight out from the rotation axis, $F \cos \phi$, doesn't have any tendency to make the object rotate. But the perpendicular component of $\vec{F}$, $F \sin \phi$, *does* tend to make the object rotate—in this case in a clockwise direction. We describe the rotational effect of the force $\vec{F}$ by a quantity called the **torque** τ (the Greek letter tau) associated with the force. This is given by

**Magnitude of torque
(9-10)**

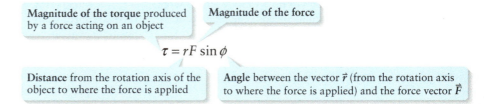

Magnitude of the torque produced by a force acting on an object

Magnitude of the force

$$\tau = rF \sin \phi$$

Distance from the rotation axis of the object to where the force is applied

Angle between the vector $\vec{r}$ (from the rotation axis to where the force is applied) and the force vector $\vec{F}$

Equation 9-10 tells us the *magnitude* of the torque, but torque also has *direction*. For the force shown in Figure 9-9a, we say that the associated torque is clockwise because it tends to make the object rotate clockwise. By contrast, the force depicted in Figure 9-9b tends to make the object rotate counterclockwise, so we say that the torque associated with this force is counterclockwise. (Later in this chapter we'll see that torque itself can be regarded as a vector. But for rotation around an axis with fixed direction, like a door rotating around its hinges, all we need is the idea that torque can be clockwise or counterclockwise.)

Views of object from along its rotation axis

(a) A force that tends to cause clockwise rotation

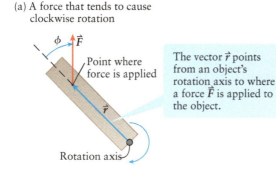

Point where force is applied

ϕ $\vec{F}$

$\vec{r}$

Rotation axis

The vector $\vec{r}$ points from an object's rotation axis to where a force $\vec{F}$ is applied to the object.

(b) A force that tends to cause counterclockwise rotation

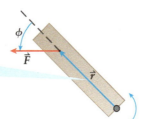

ϕ

$\vec{F}$

$\vec{r}$

(c) Determining the line of action of a force and its lever arm

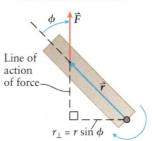

ϕ $\vec{F}$

Line of action of force

$\vec{r}$

$r_\perp = r \sin \phi$

The lever arm $r_\perp$ is perpendicular to the line of action of the force, and extends from the rotation axis to the line of action of the force.

Figure 9-9 **Torque** (a) The vector $\vec{r}$ points from an object's rotation axis to where a force $\vec{F}$ is applied to the object. The torque produced by the force has magnitude $\tau = rF \sin \phi$. In this case the torque is clockwise because it tends to make the object rotate clockwise. (b) This torque is counterclockwise because it tends to make the object rotate counterclockwise. (c) Finding the torque in terms of the lever arm $r_\perp$.

In Section 9-2 we denoted the rotation axis as the z axis and used the symbol ω_z to denote angular velocity, a quantity that can be positive or negative depending on the direction of rotation. We did this to distinguish ω_z from the angular speed ω, which is the magnitude of angular velocity. In the same way we'll use τ_z as the symbol for torque and τ (with no subscript) to denote the magnitude of torque, as in Equation 9-10. If we choose the positive rotation direction to be counterclockwise, then in Figure 9-9a the torque is negative (it tends to cause clockwise rotation) and $\tau_z < 0$. In Figure 9-9b the torque has the same magnitude τ (the distance r, force magnitude F, and angle ϕ are the same as in Figure 9-9a), but the torque tends to cause counterclockwise rotation and so $\tau_z > 0$.

Figure 9-9c shows another way to calculate the magnitude of the torque for the situation in Figure 9-9a. As this figure shows, the **line of action** of the force is just an extension of the force vector $\vec{F}$ through the point where the force is applied. The **lever arm** of the force (also called the *moment arm*) is the perpendicular distance from the rotation axis to the line of action of the force, which is why we denote it by the symbol $r_\perp$ ("$\perp$" is mathematical shorthand for "perpendicular"). Trigonometry shows that the lever arm $r_\perp$ equals $r \sin \phi$, the same quantity that appears in Equation 9-10. Hence we can rewrite that equation as

 Go to Interactive Exercise 9-2 for more practice dealing with torque.

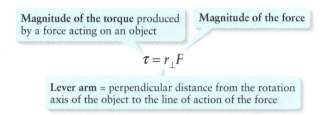

Magnitude of the torque produced by a force acting on an object   Magnitude of the force

$$\tau = r_\perp F$$

Lever arm = perpendicular distance from the rotation axis of the object to the line of action of the force

Magnitude of torque in terms of lever arm
(9-11)

Equation 9-11 is mathematically equivalent to Equation 9-10 but defines torque in a slightly different way: It says that the magnitude of the torque produced by a force equals the lever arm of the force multiplied by the force magnitude. The longer the lever arm for a given force magnitude, the greater the torque.

Figure 9-10 illustrates these ideas about torque using the door that we discussed previously (see Figure 9-8). For a given force magnitude F, the lever arm $r_\perp$ and hence the torque magnitude τ will be as large as possible if r is as large as possible and if $\sin \phi$ has its maximum value of 1, which occurs if $\phi = 90°$. In other words, a force produces the maximum torque if it is applied as far from the rotation axis as possible and in a direction perpendicular to a line from the rotation axis to the point of application (see Figure 9-10a). That's just what we said in our discussion of the door. By contrast, if the force is applied close to the rotation axis so the distance r is small—like pushing on a door at a point close to the hinges—the lever arm $r_\perp = r \sin \phi$ is short and the torque magnitude is small (Figure 9-10b). The torque is also small if the force is applied in nearly the same direction as the vector $\vec{r}$, as in Figure 9-10c. In that case the angle ϕ is nearly zero, $\sin \phi$ has a small value, and the lever arm $r_\perp = r \sin \phi$ is again short. This is like pulling on a door in a direction nearly parallel to the door's plane. (The same is true if the force is directed nearly opposite to $\vec{r}$. Then ϕ is close to 180° and again $\sin \phi$ is close to zero.)

From Equation 9-11, the SI units of torque are newtons multiplied by meters, or newton-meters (abbreviated N·m).

(a) Pulling on a door far from the hinge (view as seen looking down on the door from above): Large lever arm $r_\perp$.

(b) Pulling on a door close to the hinge: Small lever arm $r_\perp$.

(c) Pushing on a door directly toward the hinge: Zero lever arm $r_\perp$.

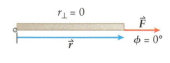

Figure 9-10 **Torques on a door** (a) For a force $\vec{F}$ of a given magnitude, the torque is greatest when r is large and $\phi = 90°$. The torque is small if (b) r is small or (c) $\sin \phi$ is small.

WATCH OUT! Torque and work have the same units but are very different quantities.

! Both torque and work have units of newton-meters. But you can't set one equal to the other. Work involves the product of the component of a force parallel to a displacement.

Torque is the product of the component of a force perpendicular to a lever arm. Because physically they represent two very different things, it makes no sense to set one equal to the other.

GOT THE CONCEPT? 9-2 Using Torque to Loosen a Bolt

? A socket wrench, like the one in Figure 9-11, can be used to loosen a bolt. If a bolt has become frozen in place and is especially difficult to loosen, a common trick is to slide a section of pipe over the handle of the wrench and turn the bolt while gripping the end of the pipe. Why does this work? Why do many experienced craftspeople tend to avoid using this trick?

Joe Belanger/Shutterstock

Figure 9-11 Wrench torque How does lengthening the handle of a wrench make it easier to loosen a bolt?

(a)

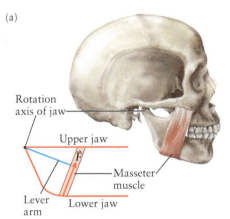

(b)

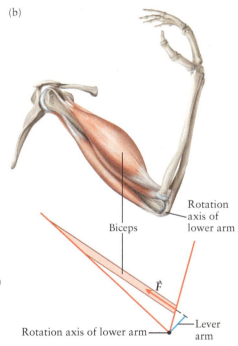

Biceps

Rotation axis of lower arm

$\vec{F}$

Rotation axis of lower arm —— Lever arm

Figure 9-12 Lever arms in the human jaw and arm (a) The masseter muscle exerts a force on the lower jaw with a relatively long lever arm, so the torque on the lower jaw is quite large. (b) The lever arm for the force of the biceps muscle on the forearm is relatively short, so the torque is relatively small.

Lever Arms in Anatomy

BioMedical Equation 9-11 shows that even a small force F can generate a large torque τ if the lever arm $r_\perp$ is long enough. Humans and other animals take advantage of this principle through the arrangement of their muscles and bones. As an example, Figure 9-12a shows that the point at which the masseter muscle is attached to the lower jawbone is far from the joint around which the jaw rotates. Hence the force exerted by the masseter muscle has a long lever arm, and so produces a large torque on the lower jawbone. This large torque enables you to crack a nut between your back teeth or use your front teeth to tear into a raw carrot.

A very different arrangement is used in your forearm. One end of the biceps muscle attaches to the bone of the upper arm and the other to the lower arm just below the elbow (Figure 9-12b). Even though the biceps muscle can exert a large force, this muscle–joint arrangement doesn't generate a large torque because the lever arm is relatively short. In chimpanzees the bicep is attached farther from the elbow joint, giving it the ability to generate more forearm torque than a human despite the chimpanzee's smaller size and smaller muscles.

Torque, Angular Acceleration, and Newton's Second Law for Rotation

We have suggested that a torque acting on an object (like the torque produced by pushing or pulling on a door) affects its rotation. What is the precise relationship between torque and the rotational effect that it causes? To find out, let's recall Newton's second law for *linear* motion along the *x* axis:

(4-4a)
$$\sum F_{\text{ext},x} = ma_x$$

This equation states that an object accelerates in response to the sum of all the external forces that act on the object: This sum $\sum F_{\text{ext},x}$ equals the object's mass m multiplied by its acceleration a_x.

To find the rotational equivalent of this equation, consider a rigid object like one blade of a wind turbine (see Figures 9-2 and 9-3). Figure 9-13 shows this blade, which

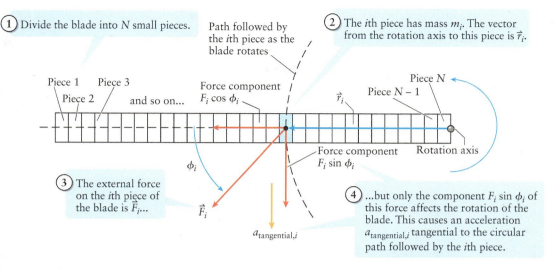

① Divide the blade into N small pieces.

Path followed by the ith piece as the blade rotates

② The ith piece has mass m_i. The vector from the rotation axis to this piece is $\vec{r}_i$.

Piece 1   Piece 3

Piece 2   and so on...

Force component $F_i \cos \phi_i$

$\vec{r}_i$

Piece $N-1$

Piece N

③ The external force on the ith piece of the blade is $\vec{F}_i$...

ϕ_i

$\vec{F}_i$

Force component $F_i \sin \phi_i$

Rotation axis

$a_{\text{tangential},i}$

④ ...but only the component $F_i \sin \phi_i$ of this force affects the rotation of the blade. This causes an acceleration $a_{\text{tangential},i}$ tangential to the circular path followed by the ith piece.

Figure 9-13 Forces on a rigid object To find how the forces on a rigid object affect its rotation, we imagine dividing the object into many small pieces.

is free to rotate around the z axis (perpendicular to the plane of the figure). The wind can exert a different force on each portion of the blade, so we imagine that the blade is divided into many small pieces. There are N such small pieces, where N is a large number. We label each piece using the subscript "i," where i can equal any integer from 1 to N: $i = 1, 2, 3,..., N$. The ith piece has mass m_i and is a distance r_i from the rotation axis. The force $\vec{F}_i$ that the wind exerts on the ith piece of the blade has magnitude F_i and is directed at an angle ϕ_i to the long axis of the blade. This force has a component $F_i \cos \phi_i$ directed along the long axis of the blade, which does nothing to affect the blade's rotation. But the component of $\vec{F}_i$ perpendicular to the long axis of the blade, $F_i \sin \phi_i$, is in just the right direction to affect the blade's rotation. This force component acting alone gives the ith piece of the blade of mass m_i a *tangential* acceleration $a_{\text{tangential},i}$ given by Equation 4-4a:

$$F_i \sin \phi_i = m_i a_{\text{tangential},i}$$

If we multiply both sides of this equation by r_i and use Equation 9-9 for tangential acceleration, $a_{\text{tangential},i} = r_i \alpha_z$, we get

$$r_i F_i \sin \phi_i = m_i r_i a_{\text{tangential},i} = m_i r_i (r_i \alpha_z) = m_i r_i^2 \alpha_z \qquad (9\text{-}12)$$

Comparing this with Equation 9-10 shows that the left-hand side of Equation 9-12, $r_i F_i \sin \phi_i$, is just the *torque* that the wind exerts on the ith piece of the wind turbine blade. If we denote this by the symbol τ_{iz}, Equation 9-12 becomes

$$\tau_{iz} = m_i r_i^2 \alpha_z \qquad (9\text{-}13)$$

We can write an equation like Equation 9-13 for each of the N pieces of the wind turbine blade. Because the blade rotates as a rigid object, at any instant all of the pieces will have the same angular velocity ω_z and the same angular acceleration α_z. If we add all N of these equations together, we get

$$\tau_{1z} + \tau_{2z} + \tau_{3z} + ... + \tau_{Nz} = m_1 r_1^2 \alpha_z + m_2 r_2^2 \alpha_z + m_3 r_3^2 \alpha_z + ... \, m_N r_N^2 \alpha_z$$
$$= (m_1 r_1^2 + m_2 r_2^2 + m_3 r_3^2 + ... \, m_N r_N^2) \alpha_z$$

which we can write as

$$\sum_{i=1}^{N} \tau_{iz} = \left(\sum_{i=1}^{N} m_i r_i^2 \right) \alpha_z \qquad (9\text{-}14)$$

The left-hand side of Equation 9-14 is the sum of all of the external torques—that is, the *net* external force—on the wind turbine blade. So this equation looks very much like Newton's second law, Equation 4-4a, with net force replaced by net torque and linear acceleration a_x replaced by angular acceleration α_z. What's new is the quantity $\sum_{i=1}^{N} m_i r_i^2$ on the right-hand side of Equation 9-14. Although it involves a sum over all the little pieces into which we've divided the turbine blade, this is *not* simply the total mass M of the blade. That sum would be $M = \sum_{i=1}^{N} m_i$, without the factor of r_i^2. Instead the sum $\sum_{i=1}^{N} m_i r_i^2$ is a new quantity that tells us how the mass of the blade is *distributed*: It depends on both the mass of each small piece (m_i) and how far away from the rotation axis that piece is (r_i). This quantity is called the **moment of inertia** (sometimes called *rotational inertia*) of the turbine blade. We represent it by the symbol I:

Moment of inertia of an object (9-15)

To find the **moment of inertia** I of an object... ...we imagine dividing the object into N **small pieces,** and calculate the sum...

$$I = \sum_{i=1}^{N} m_i r_i^2$$

...of the product of the **mass m_i of the ith piece**... ...and the square of the **distance r_i** from the ith piece to the rotation axis.

The SI units of moment of inertia are kilograms multiplied by meters squared ($\text{kg} \cdot \text{m}^2$). Note that the value of I depends on where the rotation axis is: If we change the position of the rotation axis for a given object, the distances r_i to the various pieces of the object in Figure 9-13 will be different, so the value of I will be different as well.

WATCH OUT! Take a moment to understand the meaning of "moment."

Despite its name, the term "*moment* of inertia" has nothing to do with a particular moment in time. Rather, "moment" is a mathematical term for a quantity that tells you how things are distributed. For example, when your class takes an exam, the average score on that exam (which tells you something about how the scores were distributed between zero and 100%) is an example of a moment.

Using the definition of moment of inertia from Equation 9-15, we can rewrite Equation 9-14 as

Newton's second law for rotational motion (9-16)

The **net torque,** or sum of all external torques acting on an object for the axis about which it can rotate

The **angular acceleration** produced by the net torque

$$\sum \tau_{\text{ext}, z} = I \alpha_z$$

The object's **moment of inertia** for the axis about which it can rotate

An object acquires an *angular* acceleration α_z in response to the sum $\sum \tau_{\text{ext}, z}$ of all of the external *torques* that act on the object, and this angular acceleration is directly proportional to the net torque. The magnitude of the angular acceleration also depends on the object's moment of inertia I around the rotation axis. The basic physics is the same as the original form of Newton's second law, Equation 4-4a. Note that Equation 9-16 for the *rotational* form of Newton's second law is the same as the formula for the *translational* form of this law, $\sum F_{\text{ext}, x} = ma_x$, with forces replaced by torques, mass m replaced by moment of inertia I, and linear acceleration replaced by angular acceleration.

Note also that in deriving Equation 9-16 we used Equation 9-9, $a_{\text{tangential}, i} = r_i \alpha_z$, which is only correct if the angular acceleration α is expressed in rad/s^2. So in Equation 9-16, too, angular acceleration is always expressed in rad/s^2.

To use Equation 9-16 to solve problems about the rotational motion of a rigid object, we need to be able to determine the moment of inertia I of the object. We'll devote the following section of this chapter to learning some techniques for this. For now, let's see how to apply Equation 9-16 to a situation in which the moment of inertia is particularly easy to calculate.

EXAMPLE 9-3 Can He Swing from a Thread? Torque and Angular Acceleration

In Example 6-8 (Section 6-5) we considered a South African kite spider (*Gasteracantha*) swinging on a thread of spider silk. We can consider the spider and thread together to be a rigid object that can rotate around the upper end of the thread, which is attached to a tree. The spider has mass 1.0×10^{-4} kg and is about 10 mm in size; the length of the lightweight thread is 0.50 m. (a) Find the moment of inertia of the spider around the upper end of the thread (its rotation axis). (b) Find the angular acceleration of the spider around the upper end of the thread if the thread is deflected to the right by 25° from the vertical and if the thread is vertical.

Set Up

The spider has a complicated shape, so it may seem that calculating its moment of inertia I would be very difficult. But because the thread is lightweight and the spider is so small compared to the length of the thread, it's a good approximation to treat the spider as a particle of mass 1.0×10^{-4} kg at the end of a massless thread of length 0.50 m. Then the sum in Equation 9-15 has just one term in it (for the spider).

The free-body diagram shows the forces on the spider. (For clarity we've drawn the vector $\vec{r}_{spider}$, which shows the position of the spider relative to the rotation axis, to one side.) We'll calculate the torque on the spider due to each of these forces using Equation 9-10, then use Equation 9-16 to find the spider's angular acceleration.

Moment of inertia:

$$I = \sum_{i=1}^{N} m_i r_i^2 \qquad (9\text{-}15)$$

Definition of torque:

$$\tau = rF \sin \phi \qquad (9\text{-}10)$$

Newton's second law for rotational motion:

$$\sum \tau_{ext,z} = I\alpha_z \qquad (9\text{-}16)$$

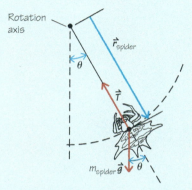

Positive rotation direction: counterclockwise

Solve

(a) Find the moment of inertia using Equation 9-15, treating the spider as a single particle located at the end of the thread.

From Equation 9-15:

$$I = \sum_{i=1}^{N} m_i r_i^2 = m_{spider} r_{spider}^2$$

$$= (1.0 \times 10^{-4} \text{ kg})(0.50 \text{ m})^2 = 2.5 \times 10^{-5} \text{ kg} \cdot \text{m}^2$$

(b) Two forces act on the spider, the gravitational force $m_{spider}\vec{g}$ and the tension force $\vec{T}$ exerted by the thread. From Equation 9-10, if the vector from the rotation axis to where a force is applied is $\vec{r}$, and the force $\vec{F}$ is directed at an angle ϕ from $\vec{r}$, the torque due to that force has magnitude $\tau = rF \sin \phi$. Use this to find the torque due to each force.

Both forces on the spider (considered as a particle) act at the same point, so for both $r = r_{spider}$.

For $m_{spider}\vec{g}$, the angle ϕ equals the angle θ by which the thread is deflected from the vertical, so

$$\tau_{gravitational} = r_{spider} m_{spider} g \sin \phi = r_{spider} m_{spider} g \sin \theta$$

If the thread is deflected to the right as in the figure above, the gravitational torque tends to make it swing to the left in the clockwise direction. If we make the usual choice that the counterclockwise direction is positive, this means that the z component of the torque is negative. So

$$\tau_{gravitational,z} = -r_{spider} m_{spider} g \sin \theta$$

Tension force $\vec{T}$ points directly toward the rotation axis, opposite to the vector $\vec{r}_{spider}$, so for this force $\phi = 180°$ and

$$\tau_{tension} = r_{spider} T \sin 180° = 0$$

Find the net torque on the spider and solve for its angular acceleration using Equation 9-16.

Equation 9-16: $\sum \tau_{\text{ext},z} = I\alpha_z$

Net torque:

$$\sum \tau_{\text{ext},z} = \tau_{\text{gravitational},z} + \tau_{\text{tension},z}$$
$$= -r_{\text{spider}} m_{\text{spider}} g \sin\theta + 0$$
$$= -r_{\text{spider}} m_{\text{spider}} g \sin\theta$$

Solve Equation 9-16 for α_z:

$$\alpha_z = \frac{\sum \tau_{\text{ext},z}}{I} = \frac{-r_{\text{spider}} m_{\text{spider}} g \sin\theta}{m_{\text{spider}} r_{\text{spider}}^2}$$
$$= -\frac{g \sin\theta}{r_{\text{spider}}}$$

If $\phi = 25°$, the angular acceleration of the spider is

$$\alpha_z = -\frac{g \sin 25°}{r_{\text{spider}}} = -\frac{(9.8 \text{ m/s}^2)(0.423)}{0.50 \text{ m}} = -8.3 \text{ rad/s}^2$$

If the string is vertical so $\phi = 0$, the angular acceleration is zero:

$$\alpha_z = -\frac{g \sin 0}{r_{\text{spider}}} = 0 \text{ (since } \sin 0 = 0)$$

Reflect

Note that you can also find the torque due to each force in terms of lever arm by using Equation 9-11.

Magnitude of torque in terms of lever arm:

$$\tau = r_\perp F \qquad (9\text{-}11)$$

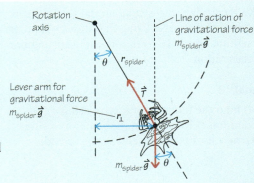

Rotation axis

Line of action of gravitational force $m_{\text{spider}}\vec{g}$

θ

r_{spider}

$\vec{\tau}$

Lever arm for gravitational force $m_{\text{spider}}\vec{g}$

$r_\perp$

$m_{\text{spider}}\vec{g}$

θ

For the gravitational force $m_{\text{spider}}\vec{g}$ the lever arm is the opposite side of a right triangle of hypotenuse r_{spider} and angle θ:

$$r_\perp = r_{\text{spider}} \sin\theta, \text{ so}$$
$$\tau_{\text{gravitational}} = (r_{\text{spider}} \sin\theta)(m_{\text{spider}} g) = r_{\text{spider}} m_{\text{spider}} g \sin\theta$$

For the tension force $\vec{T}$ the lever arm is zero because the line of action of the force passes through the rotation axis. Hence this force produces zero torque.

Our results show that if the thread is deflected to the right at $\theta = 25°$, the angular acceleration α_z is negative (clockwise). If instead the thread is deflected to the *left* at 25°, then $\theta = -25°$ and α_z is *positive* (counterclockwise). You can see that as the spider swings back and forth its angular acceleration changes from positive to negative, and is zero when the thread is vertical. We'll explore this kind of *oscillating* motion in Chapter 12.

If $\theta = -25°$, the angular acceleration of the spider is positive:

$$\alpha_z = -\frac{g \sin(-25°)}{r_{\text{spider}}} = -\frac{(9.8 \text{ m/s}^2)(-0.423)}{0.50 \text{ m}} = +8.3 \text{ rad/s}^2$$

BioMedical **GOT THE CONCEPT? 9-3** **Rotating the Human Jaw**

When you chew, your jaw rotates primarily around the axis labeled y in Figure 9-14, while moving your jaw from side to side involves rotating it around the axis labeled z. (You are probably not able to rotate your jaw around the axis labeled x.) Measurements show that for a lower jaw with a mass of approximately 0.4 kg, typical values for the moments of inertia are $I_y = 3 \times 10^{-4}$ kg·m^2 and $I_z = 9 \times 10^{-4}$ kg·m^2. Is more torque required (a) to open your jaw or (b) to rotate it from side to side?

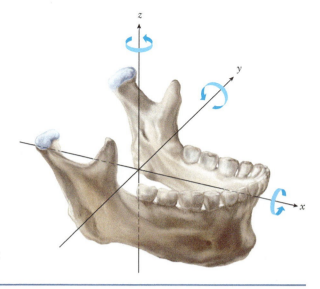

Figure 9-14 **The human jaw** The human jaw has a different moment of inertia for each of the three axes x, y, and z.

TAKE-HOME MESSAGE FOR **Section 9-3**

✔ If a force acting on an object tends to change that object's rotation, the force generates a torque. The magnitude of the torque depends on where the force is applied relative to the rotation axis and on the orientation of the force.

✔ The magnitude of a torque is equal to the magnitude of the force that causes the torque multiplied by the lever arm of the force (the perpendicular distance from the rotation axis to the line of action of the force).

✔ The moment of inertia of an object depends on the object's mass and how that mass is distributed relative to the object's rotation axis.

✔ Newton's second law for rotation says that if a net external torque $\sum \tau_{ext,z}$ acts on an object, the object acquires an angular acceleration α_z. These are related via the moment of inertia I: $\sum \tau_{ext,z} = I\alpha_z$.

9-4 An object's moment of inertia depends on its mass distribution and the choice of rotation axis

We saw in Section 9-3 that the moment of inertia I of an object plays the same role for rotational motion that the object's mass m does for translational motion. The moment of inertia of an object depends on how its mass is *distributed* throughout the object. An example is the door of a bank vault (Figure 9-15). The door takes effort to move in part because it's so massive. But an additional factor is that a substantial fraction of the door's mass is in its locking mechanism, and this mechanism is located a good distance away from the hinges around which the door rotates. The greater the fraction of the door's mass that's located far from the hinges, the more difficult it is to start the door rotating if it's at rest or to stop it moving if it's already in motion.

Because the moment of inertia of an object is such an important property, we'll devote this section to some examples of calculating its value. We'll see that the moment of inertia depends both on how the object's mass is distributed and on the particular axis around which the object rotates.

Figure 9-15 **Moving a bank vault door** How easy or difficult it is to start this door moving depends on its moment of inertia for rotation around the hinges. The moment of inertia is determined by the mass of the door and how that mass is distributed.

Moment of Inertia of a Collection of Small Pieces

From Section 9-3, if we imagine dividing an object into a large number of small objects of masses $m_1, m_2, m_3, \ldots m_N$, the moment of inertia I of the object is

$$I = \sum_{i=1}^{N} m_i r_i^2 \qquad (9\text{-}15)$$

(a)

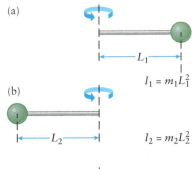

$I_1 = m_1 L_1^2$

(b)

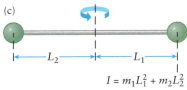

$I_2 = m_2 L_2^2$

(c)

$I = m_1 L_1^2 + m_2 L_2^2$

Figure 9-16 Calculating moment of inertia The sum of the moments of inertia of the objects in (a) and (b) equals the moment of inertia of the composite object in (c).

→ *Go to Picture It 9-1 for more practice dealing with moment of inertia.*

In this expression r_i is the distance from the axis around which the object rotates to the position of the ith small piece.

Equation 9-15 reveals three important properties of the moment of inertia:

(1) *The moment of inertia is additive.* Each term $m_i r_i^2$ in Equation 9-15 represents the moment of inertia of the ith piece of the object, and the total moment of inertia is the sum of the terms for all pieces. So if you know the moments of inertia of two objects around some rotation axis and you attach them to form a single object, the moment of inertia of the new object—around the same axis—is the sum of the moments of inertia of the two separate objects (Figure 9-16).

(2) *The farther away from the rotation axis an object's mass lies, the greater the object's moment of inertia.* In Equation 9-15, the farther the small pieces of the object are from the rotation axis, the greater the values of r_i and the greater the moment of inertia. A bit of mass far from the rotation axis has a larger effect on the value of the moment of inertia than the same amount of mass close to the axis. Example 9-4 demonstrates this idea.

(3) *The moment of inertia depends on the rotation axis.* If a different rotation axis is used, the quantities r_i in Equation 9-15 change, so the moment of inertia changes as well. In general, the moment of inertia for a given object is different for each different rotation axis. We'll see how this works in Example 9-5.

Let's illustrate these statements by looking at objects that we can describe as being composed of just a few pieces.

EXAMPLE 9-4 **Moment of Inertia for Dumbbells I**

A dumbbell is made up of two small, massive spheres connected by a lightweight rod. For a dumbbell made with two identical 50.0-kg spheres, find the moment of inertia for rotation around the midpoint of the dumbbell if the spheres are separated by (a) 1.20 m or (b) 2.40 m.

Set Up

The rod connecting the spheres is "lightweight," so we can ignore its mass. The two spheres are also described as "small," so we can treat each sphere as an infinitesimally small particle of mass (like we did for the spider in Example 9-3). Then the dumbbell is made up of only two small pieces, and the sum in Equation 9-15 has only two terms. That is, the moment of inertia of the dumbbell is the sum of the moments of inertia of the two spheres.

Moment of inertia:

$$I = \sum_{i=1}^{N} m_i r_i^2 = m_1 r_1^2 + m_2 r_2^2$$

$$(9\text{-}15)$$

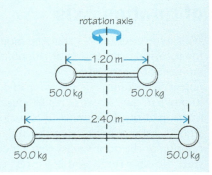

Solve

Each sphere has mass 50.0 kg, so $m_1 = m_2 = 50.0$ kg in each case. For the first dumbbell each sphere is one-half of 1.20 m from the rotation axis, so $r_1 = r_2 = 0.600$ m; similarly, in the second case each sphere is one-half of 2.40 m from the rotation axis, so $r_1 = r_2 = 1.20$ m. Substitute these values into the formula for moment of inertia I.

For both dumbbells,

$$I = m_1 r_1^2 + m_2 r_2^2 = (50.0 \text{ kg}) r_1^2 + (50.0 \text{ kg}) r_2^2$$

(a) For the first dumbbell:

$$I = (50.0 \text{ kg})(0.600 \text{ m})^2 + (50.0 \text{ kg})(0.600 \text{ m})^2 = 36.0 \text{ kg} \cdot \text{m}^2$$

(b) For the second dumbbell:

$$I = (50.0 \text{ kg})(1.20 \text{ m})^2 + (50.0 \text{ kg})(1.20 \text{ m})^2 = 144 \text{ kg} \cdot \text{m}^2$$

Reflect

Although both dumbbells have the same total mass $m_1 + m_2 = 50.0\text{ kg} + 50.0\text{ kg} = 100.0\text{ kg}$, the mass is distributed in different ways. Hence, they have very different values of the moment of inertia. The farther the massive spheres are from the rotation axis, the greater the value of the moment of inertia.

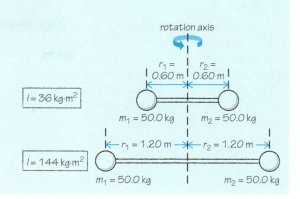

EXAMPLE 9-5   Moment of Inertia for Dumbbells II

Consider again the second dumbbell from Example 9-4, which has two small 50.0-kg spheres separated by 2.40 m. (a) Calculate the moment of inertia of this same dumbbell if it rotates around the center of one of the spheres. (b) Calculate the angular acceleration of this dumbbell if a torque of 576 N·m is applied to it if its rotation axis is at its midpoint and if its rotation axis is at the center of one of its spheres.

Set Up

As in Example 9-4 the moment of inertia I of the dumbbell is the sum of two terms, one for each sphere. We expect a different answer for I in part (a) because the values of r_1 and r_2 are different from those in Example 9-4. Once we know the value of I for each choice of axis, we can calculate the angular acceleration using Equation 9-16.

Moment of inertia:

$$I = \sum_{i=1}^{N} m_i r_i^2 = m_1 r_1^2 + m_2 r_2^2 \qquad (9\text{-}15)$$

Newton's second law for rotational motion:

$$\sum \tau_{\text{ext},z} = I\alpha_z \qquad (9\text{-}16)$$

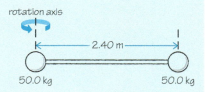

Solve

(a) Again each sphere has mass 50.0 kg, so $m_1 = m_2 = 50.0$ kg in each case. Now one sphere is *on* the rotation axis, so $r_1 = 0$; the other sphere is 2.40 m from the rotation axis, so $r_2 = 2.40$ m. Substitute these values into the formula for I.

$$I = m_1 r_1^2 + m_2 r_2^2 = (50.0\text{ kg})(0)^2 + (50.0\text{ kg})(2.40\text{ m})^2$$
$$= 288\text{ kg}\cdot\text{m}^2$$

(b) In Example 9-4 we found that the moment of inertia of this dumbbell for rotation around its midpoint was $I = 144$ kg·m². Use this in Equation 9-16 to solve for the angular acceleration for this choice of rotation axis.

For rotation around the midpoint of the dumbbell,

$$\sum \tau_{\text{ext},z} = I\alpha_z$$

$$576\text{ N}\cdot\text{m} = (144\text{ kg}\cdot\text{m}^2)\alpha_z$$

$$\alpha_z = \frac{576\text{ N}\cdot\text{m}}{144\text{ kg}\cdot\text{m}^2} = 4.00\ \frac{\text{N}}{\text{kg}\cdot\text{m}} = 4.00\text{ rad/s}^2$$

(Recall that 1 N = 1 kg·m/s² and that α_z in Equation 9-16 must be in rad/s².)

Repeat the calculation for rotation around one of the spheres. We found above that for this axis, $I = 288$ kg·m².

For rotation around one of the spheres of the dumbbell,

$$\sum \tau_{\text{ext},z} = I\alpha_z$$

$$576\text{ N}\cdot\text{m} = (288\text{ kg}\cdot\text{m}^2)\alpha_z$$

$$\alpha_z = \frac{576\text{ N}\cdot\text{m}}{288\text{ kg}\cdot\text{m}^2} = 2.00\text{ rad/s}^2$$

Reflect

The dumbbell in this example is exactly the same object as the second dumbbell in Example 9-4, yet its moment of inertia is different. This shows that the moment of inertia of an object depends not only on its mass and how that mass is distributed, but also on the particular axis around which the object rotates.

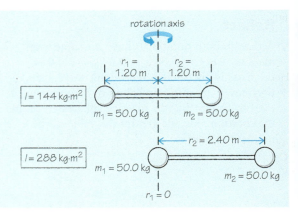

WATCH OUT! An object has more than one moment of inertia.

If your friend points to an object and asks, "What is the moment of inertia of that object?" it could be a trick question! Example 9-5 shows that objects do *not* have a single moment of inertia; rather, the value of the moment of inertia depends on the specific rotation axis. *Always* make sure that you identify the axis of rotation before determining the moment of inertia of an object!

The Parallel-Axis Theorem

Finding the moment of inertia of an object can be challenging for certain choices of rotation axis. An example is the toy hoop of mass M and radius R in Figure 9-17a. The child rotates the hoop around an axis that is perpendicular to the plane of the hoop and is near the hoop's rim. If we imagine dividing the hoop into a large number of small pieces labeled $i = 1, 2, 3, \ldots, N$ (Figure 9-17b), every piece is a different distance r_i from the rotation axis, so it's very difficult to calculate the moment of inertia using Equation 9-15.

However, a remarkable relationship exists between the moment of inertia of an object for an axis through its *center of mass*, and the moment of inertia when the object rotates around any other parallel axis. (We introduced the idea of center of mass in Section 8-7. This would be a good time to review that section.) This relationship, called the **parallel-axis theorem**, is useful because it's often relatively easy to determine an object's moment of inertia around its center of mass.

Here's the statement of the parallel-axis theorem: Suppose the moment of inertia of an object for a certain rotation axis is I. The moment of inertia of the same object for a second axis that's parallel to the first one but passes through the object's center of mass is I_{CM} (Figure 9-18). If the distance between the two axes is h and the mass of the object is M, the relationship between the two values of moment of inertia is

Moment of inertia of an object for a **certain rotation axis**

Moment of inertia of the same object for a second, **parallel axis through its center of mass**

The parallel-axis theorem
(9-17)

$$I = I_{CM} + Mh^2$$

Mass of the object

Distance between the two parallel axes

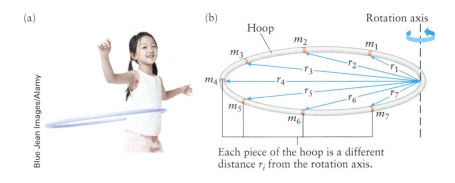

(a)

(b)

Figure 9-17 A spinning hoop
(a) A hoop rotating around an axis that passes through its rim. (b) Calculating the hoop's moment of inertia for this axis.

Each piece of the hoop is a different distance r_i from the rotation axis.

(a) Hoop rotating around an axis through its center

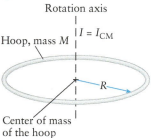

Rotation axis

$I = I_{CM}$

Hoop, mass M

R

Center of mass of the hoop

(b) Hoop rotating around an axis through its rim

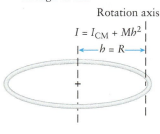

Rotation axis

$I = I_{CM} + Mh^2$

$\leftarrow h = R \rightarrow$

(c) Top view of the hoop

Rotation axis perpendicular to plane of figure

R R

R R

If the rotation axis passes through the center of the hoop, all parts of the hoop are a distance R from the axis.

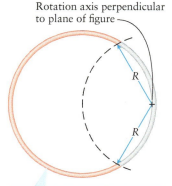

Rotation axis perpendicular to plane of figure

R

R

With this choice of axis, most of the hoop is farther than R from the axis.

Figure 9-18 Two moments of inertia for a hoop and the parallel-axis theorem The parallel-axis theorem relates the hoop's moment of inertia for two different axes.

To see the parallel-axis theorem in action, let's return to the hoop of mass M and radius R shown in Figure 9-17b. The theorem tells us that to determine the hoop's moment of inertia for an axis at its rim, we should first calculate its moment of inertia for a parallel axis that passes through its center of mass (Figure 9-18a). The center of mass of a symmetrical object such as a hoop is at the geometrical center of the object. (Recall from Section 8-7 that there doesn't actually have to be any mass *at* an object's center of mass.) If we divide the hoop into many small pieces, each of those pieces is the *same* distance R from this rotation axis. So in Equation 9-15 for the moment of inertia we can replace r_i for each value of i by R. This equation then becomes

$$I_{CM} = \sum_{i=1}^{N} m_i r_i^2 = \sum_{i=1}^{N} m_i R^2 = \left(\sum_{i=1}^{N} m_i \right) R^2 \qquad (9\text{-}18)$$

The quantity $\sum_{i=1}^{N} m_i$ in Equation 9-18 is just the sum of the masses of all of the individual pieces that make up the hoop. This sum is just the mass M of the hoop as a whole, so we can rewrite Equation 9-18 as

$$I_{CM} = MR^2 \qquad (9\text{-}19)$$

(moment of inertia of a hoop of mass M and radius R, axis through its center of mass and perpendicular to its plane)

Now we can use the parallel-axis theorem, Equation 9-17, to find the moment of inertia of this hoop for a parallel axis that passes through the rim of the hoop (Figure 9-18b). The distance between the two axes is just the radius R of the hoop, so we let $h = R$ in Equation 9-17. Using $I_{CM} = MR^2$ from Equation 9-19, we find

$$I = I_{CM} + MR^2 = MR^2 + MR^2$$

or

$$I = 2MR^2 \qquad (9\text{-}20)$$

(moment of inertia of a hoop of mass M and radius R, axis through its rim and perpendicular to its plane)

This result would have been extremely difficult to obtain without using the parallel-axis theorem.

Why is the moment of inertia of the hoop of radius R for an axis through its rim (Equation 9-20) greater than that for an axis through its center (Equation 9-19)? The explanation is that with the rotation axis at the rim, the average distance from the axis to the pieces that make up the hoop is greater than R (Figure 9-18c). (A portion of the hoop is closer than R, but a larger portion is farther away than R.) As we have seen before, the farther an object's mass lies from the rotation axis, the greater its moment of inertia.

EXAMPLE 9-6 Using the Parallel-Axis Theorem I

Use the parallel-axis theorem and the results of Example 9-4 to find the moment of inertia of the dumbbell in Example 9-5 rotated about one end of the dumbbell.

Set Up

The two spheres at the ends of the dumbbell have the same mass, so the center of mass of the dumbbell is at its midpoint. In Example 9-4 we found the moment of inertia of this dumbbell for an axis that passes through the midpoint, so this expression is I_{CM}. The dumbbell in Example 9-5 is identical but has a different, parallel rotation axis, so we can find its moment of inertia using the parallel-axis theorem, Equation 9-17.

Parallel-axis theorem:

$$I = I_{CM} + Mh^2 \qquad (9\text{-}17)$$

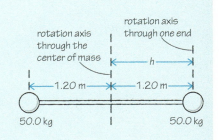

Solve

From Example 9-4, the moment of inertia of the dumbbell through its center of mass is $I_{CM} = 144 \text{ kg} \cdot \text{m}^2$. The axis of rotation for the dumbbell in Example 9-5 is a distance $h = 1.20$ m from the center of mass, and the total mass M of the dumbbell is the sum of the masses of the two 50.0-kg spheres. Substitute these values into the parallel-axis theorem.

$$I_{CM} = 144 \text{ kg} \cdot \text{m}^2$$
$$h = 1.20 \text{ m}$$
$$M = 50.0 \text{ kg} + 50.0 \text{ kg} = 100.0 \text{ kg}$$
$$I = I_{CM} + Mh^2 = 144 \text{ kg} \cdot \text{m}^2 + (100.0 \text{ kg})(1.20 \text{ m})^2$$
$$= 288 \text{ kg} \cdot \text{m}^2$$

Reflect

We get the same answer as in Example 9-5, which is a nice check that the parallel-axis theorem works.

The Moment of Inertia for Common Shapes

It was straightforward to calculate the moment of inertia for the hoop shown in Figure 9-18 because all parts of the hoop are the same distance from its rotation axis. For solid objects with other shapes, calculating the moment of inertia is more difficult and requires the use of calculus, which is beyond our scope. Instead, we'll just present the results for some common shapes that show up in various situations involving rotation (Table 9-1). We assume that these objects are *uniform*; that is, each object has the same composition throughout its volume.

You can see from Table 9-1 that the moment of inertia of a thin cylindrical shell of mass M and radius R around its central axis ($I = MR^2$) is twice that of a solid cylinder of the same mass and radius ($I = \frac{1}{2}MR^2$). You should expect the cylindrical shell to have a larger moment of inertia because the moment of inertia of an object is strongly influenced by how far the mass is from the rotation axis. In the case of the thin cylindrical shell, all of the mass is located a distance R from the axis, while for the solid cylinder only a fraction of the mass is that far from the axis. Hence the moment of inertia of the shell must be larger than the moment of inertia of the solid cylinder.

We can use the parallel-axis theorem to verify some of the results shown in Table 9-1. For example, consider a thin rod that has mass M and length L. The center of mass of this rod is at its geometric center. From Table 9-1, the moment of inertia of such a rod rotating around an axis perpendicular to the rod and through its center is

$$I_{CM} = \frac{ML^2}{12}$$

(moment of inertia of a thin uniform rod, rotation axis through its center)

What is the moment of inertia of this same rod for an axis through one end and perpendicular to the rod? This axis is parallel to an axis through the center of mass of

TABLE 9-1 Moments of Inertia of Uniform Objects of Various Shapes

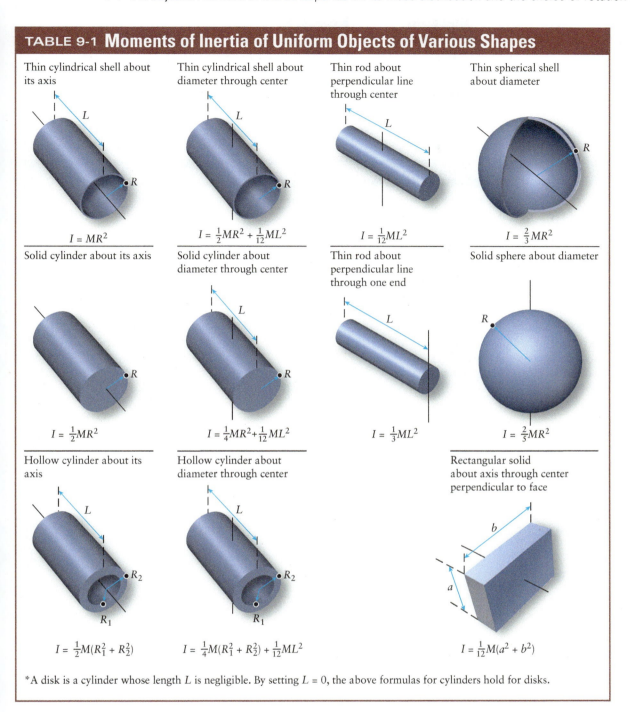

Thin cylindrical shell about its axis

$I = MR^2$

Thin cylindrical shell about diameter through center

$I = \frac{1}{2}MR^2 + \frac{1}{12}ML^2$

Thin rod about perpendicular line through center

$I = \frac{1}{12}ML^2$

Thin spherical shell about diameter

$I = \frac{2}{3}MR^2$

Solid cylinder about its axis

$I = \frac{1}{2}MR^2$

Solid cylinder about diameter through center

$I = \frac{1}{4}MR^2 + \frac{1}{12}ML^2$

Thin rod about perpendicular line through one end

$I = \frac{1}{3}ML^2$

Solid sphere about diameter

$I = \frac{2}{5}MR^2$

Hollow cylinder about its axis

$I = \frac{1}{2}M(R_1^2 + R_2^2)$

Hollow cylinder about diameter through center

$I = \frac{1}{4}M(R_1^2 + R_2^2) + \frac{1}{12}ML^2$

Rectangular solid about axis through center perpendicular to face

$I = \frac{1}{12}M(a^2 + b^2)$

*A disk is a cylinder whose length L is negligible. By setting $L = 0$, the above formulas for cylinders hold for disks.

the rod and a distance $h = L/2$ from that axis. From Equation 9-17, the moment of inertia for an axis through the end is

$$I = I_{CM} + Mh^2 = \frac{ML^2}{12} + M\left(\frac{L}{2}\right)^2$$

$$= \left(\frac{1}{12} + \frac{1}{4}\right)ML^2 = \left(\frac{1}{12} + \frac{3}{12}\right)ML^2 = \frac{4ML^2}{12}$$

or, simplifying,

$$I = \frac{ML^2}{3}$$

(moment of inertia of a thin uniform rod, axis through one end, perpendicular to rod)

This expression is just the result shown in Table 9-1 for the rod rotating around its end.

If one of the shapes shown in Table 9-1 rotates around an axis that is different than but parallel to the axis shown in the table, we can use the parallel-axis theorem to find the new moment of inertia. Because the moment of inertia is additive, you can also use the results shown in Table 9-1 to find the moment of inertia of a more complex object made up of two or more of the simple objects shown in the table.

EXAMPLE 9-7 Using the Parallel-Axis Theorem II: Earring Moment of Inertia

An earring is a thin, uniform disk that has a mass M and a radius R. The earring hangs from the earring post by a small hole near the edge of the disk and is free to rotate. Find the moment of inertia of the disk around this rotation axis.

Set Up

A solid disk is an example of a solid cylinder like that shown in Table 9-1. This table shows that the moment of inertia of such a disk for an axis perpendicular to the plane of the disk and passing through its center is $I = MR^2/2$. (Unlike the cylinder shown in the table, the length L of the earring is much less than its radius R. But the length has no effect on the value of I.)

Because the disk is uniform, its geometrical center is its center of mass, so $MR^2/2$ equals I_{CM}. The axis we want is parallel to the axis through the disk's center, so we can use the parallel-axis theorem, Equation 9-17, to find the moment of inertia around the axis at the rim.

Parallel-axis theorem:

$$I = I_{CM} + Mh^2 \qquad (9\text{-}17)$$

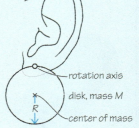
rotation axis
disk, mass M
R
center of mass

Solve

The edge of the disk is a distance $h = R$ from the center of mass of the disk. Substitute this value and $I_{CM} = MR^2/2$ into the parallel-axis theorem.

$$I = \frac{MR^2}{2} + MR^2$$

or

$$I = \frac{3MR^2}{2}$$

Reflect

If the wearer of these earrings nods her head up and down while dancing at a club, she can make the disks rotate back and forth around the posts in an eye-catching way. The more massive the disks and the larger their radius, the greater their moment of inertia and the more torque will be required to get them moving (recall the rotational form of Newton's second law, $\sum \tau_{ext,z} = I\alpha_z$). If the earrings are small and lightweight, less effort will be required from the wearer!

GOT THE CONCEPT? 9-4 "X" Marks the Moment of Inertia, Part I

 Two thin, uniform rods, each of which has mass M and length L, are attached at their centers to form an "X" or a "+" shape. What is the moment of inertia when this configuration is rotated around an axis that passes through their centers, perpendicular to the plane of the two rods? (a) $ML^2/24$; (b) $ML^2/12$; (c) $ML^2/6$; (d) $ML^2/4$; (e) $ML^2/3$.

GOT THE CONCEPT? 9-5 "X" Marks the Moment of Inertia, Part II

 Four thin, uniform rods, each of which has mass $M/2$ and length $L/2$, are attached at their ends to form an "X" or a "+" shape. What is the moment of inertia when this configuration is rotated around the axis that passes through the center of the "X" perpendicular to the plane of the rods? (a) $ML^2/24$; (b) $ML^2/12$; (c) $ML^2/6$; (d) $ML^2/4$; (e) $ML^2/3$.

TAKE-HOME MESSAGE FOR Section 9-4

✔ The moment of inertia of an object is additive and increases as the object's mass is moved farther from the rotation axis.

✔ An object's moment of inertia depends on the axis around which the object rotates.

✔ The parallel-axis theorem relates the moment of inertia for an axis through an object's center of mass to the moment of inertia for a second, parallel axis.

9-5 The techniques used for solving problems with Newton's second law also apply to rotation problems

Now that we have learned more about how to determine the moment of inertia for a rigid object around a rotation axis, it's time to return to the rotational form of Newton's second law. The examples in this section show how to solve problems using this law. Just as for the translational form of this law, in each example we'll begin our solution by drawing a free-body diagram.

In many problems it's necessary to use *both* the translational and rotational forms of Newton's second law. Example 9-9 shows how to tackle a problem like this.

WATCH OUT! In rotational problems, it matters *where* the force acts.

! In problems that involve Newton's second law for translational motion, we draw all of the forces on an object as acting at a single point. But in problems that involve rotational motion, it's crucial to indicate in the free-body diagram *where* on the object each force acts. That's because the torque generated by a force depends on where the force is applied relative to the rotation axis. In the examples below, study closely how forces are depicted in free-body diagrams and how the corresponding torques are calculated.

EXAMPLE 9-8 A Simple Pulley I

A pulley is a solid uniform cylinder of mass M_{pulley} and radius R that is free to rotate around an axis through its center. A lightweight rope is wound around the pulley. You exert a constant force of magnitude F on the rope, which makes the rope unwind and rotates the pulley (Figure 9-19). The rope does not stretch and does not slip on the pulley. What is the angular acceleration of the pulley about its axis?

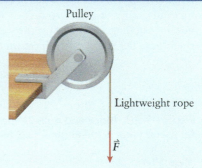

Pulley

Lightweight rope

$\vec{F}$

Figure 9-19 Exerting a torque on a pulley What is the angular acceleration of the pulley as the force $\vec{F}$ makes the rope unwind?

Set Up

We begin by drawing a free-body diagram for the pulley, taking care to draw each force at the point where it acts. We are told that the rope is lightweight (that is, it has much less mass than the pulley), so the force that you exert on the free end of the rope has the same magnitude F as the force that the rope exerts on the pulley. This force exerts a torque on the pulley and causes an angular acceleration. We'll use the rotational form of Newton's second law to determine this angular acceleration.

Newton's second law for rotational motion:

$$\sum \tau_{ext,z} = I_{pulley}\alpha_{pulley,z} \qquad (9\text{-}16)$$

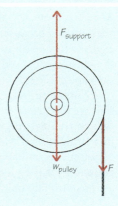

$F_{support}$

w_{pulley} F

Solve

The free-body diagram shows that only the tension force F exerts a torque on the pulley, causing it to rotate around its axis. (The support force F_{support} and the weight of the pulley W_{pulley} both act at its rotation axis, so the lever arm is zero for both of these forces.) The lever arm for the tension force is R, so the tension torque is R multiplied by F.

From Table 9-1 in Section 9-4, the moment of inertia of a solid cylinder around its central axis is $I = MR^2/2$. Insert this into Newton's second law for rotational motion and solve for $\alpha_{\text{pulley},z}$.

Torque due to tension force:

$$\tau_z = r_\perp F = RF$$

(This torque makes the pulley rotate in the clockwise direction, so we take clockwise to be the positive rotation direction.)
This is the only torque acting on the pulley, so

$$\sum \tau_{\text{ext},z} = RF$$

$$\sum \tau_{\text{ext},z} = I_{\text{pulley}} \alpha_{\text{pulley},z}$$

Substitute expressions for $\sum \tau_{\text{ext},z}$ and $I_{\text{pulley}} = M_{\text{pulley}} R^2/2$:

$$RF = \frac{1}{2} M_{\text{pulley}} R^2 \alpha_{\text{pulley},z}$$

Solve for angular acceleration $\alpha_{\text{pulley},z}$: $\alpha_{\text{pulley},z} = \dfrac{2RF}{M_{\text{pulley}} R^2} = \dfrac{2F}{M_{\text{pulley}} R}$

Reflect

Our result for the angular acceleration $\alpha_{\text{pulley},z}$ depends on the pulling force F, the pulley mass M_{pulley}, and the pulley radius R. It makes sense that our result is proportional to the ratio F/M_{pulley}: A greater force F means a stronger pull and a greater angular acceleration, while a greater mass M_{pulley} means the pulley is more difficult to rotate and gives a smaller angular acceleration.

Our result also shows that the larger the pulley radius R, the smaller the angular acceleration $\alpha_{\text{pulley},z}$. This may seem backward, since a larger radius means that the force F causes a larger torque $\tau = RF$. However, increasing the radius increases the moment of inertia $I = M_{\text{pulley}} R^2/2$ by a larger factor than it increases the torque. (Doubling the radius doubles the torque but quadruples the moment of inertia.) So the moment of inertia plays a more important role, which is why $\alpha_{\text{pulley},z}$ decreases with increasing pulley radius.

EXAMPLE 9-9 A Simple Pulley II

You attach a block of mass M_{block} to the free end of the rope in Example 9-8 (Figure 9-20). When the block is released from rest and falls downward, the rope unwinds, and the pulley rotates around its central axis. As in Example 9-8, the rope neither stretches nor slips on the pulley. Derive expressions for the acceleration of the block, the angular acceleration of the pulley, and the tension in the string.

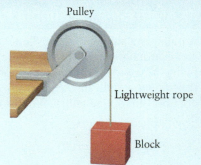

Figure 9-20 Block and pulley What is the downward acceleration of the block as it falls, and what is the angular acceleration of the pulley as the rope unwinds?

Set Up

As in Example 9-8, the pulley has an angular acceleration because the tension of the rope exerts a torque on it. The difference here is that the tension is caused by the block attached to the rope's free end.

Another difference from Example 9-8 is that we have *two* moving objects: the pulley (which rotates) and the block (which moves in a straight line). Hence we have to write two Newton's second law equations: a rotational equation for the pulley, as in Example 9-8, and a translational equation for the block.

Newton's second law for rotational motion applies to the pulley:

$$\sum \tau_{\text{ext},z} = I_{\text{pulley}} \alpha_{\text{pulley},z} \qquad (9\text{-}16)$$

Newton's second law for translational motion applies to the block:

$$\sum F_{\text{ext},x} = M_{\text{block}} a_{\text{block},x} \qquad (4\text{-}4a)$$

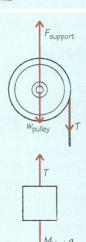

These two equations won't be enough to solve the problem, since we're trying to find *three* quantities: the block's acceleration $a_{block,x}$, the pulley's angular acceleration $\alpha_{pulley,z}$, and the string tension T. Happily, we can get a third equation because the string doesn't stretch. This tells us that the speed of the block equals the speed of a point on the pulley rim. We'll use this to find a relationship between $a_{block,x}$ and $\alpha_{pulley,z}$.

Statement that rope doesn't stretch:

$$v_{block} = v_{rim\ of\ pulley}$$

Solve

As in Example 9-8, the only force that exerts a torque on the pulley is the rope tension, which we call T. The lever arm of this force is R, so the net torque is $\sum \tau_{ext,z} = RT$. (As in Example 9-8 this torque makes the pulley rotate in the clockwise direction. So we take clockwise to be the positive rotation direction.) Insert this and $I_{pulley} = M_{pulley}R^2/2$ into the rotational Newton's second law equation and solve for the tension.

For the pulley,

$$\sum \tau_{ext,z} = I_{pulley}\alpha_{pulley,z}$$

Substitute values of $\sum \tau_{ext,z}$ and I_{pulley}:

$$RT = \frac{1}{2}M_{pulley}R^2\alpha_{pulley,z}$$

Divide through by R:

$$T = \frac{1}{2}M_{pulley}R\alpha_{pulley,z}$$

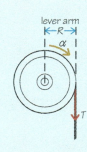

We take the positive x direction to point downward, which is the direction in which the block will move when the pulley rotates clockwise. With this choice the acceleration $a_{block,x}$ of the block and the angular acceleration $\alpha_{pulley,z}$ of the pulley are both positive, the gravitational force $M_{block}g$ is in the positive x direction, and the tension force T on the block is in the negative x direction. (The rope has negligible mass, so the tension is the same throughout its length.) Use this to write the Newton's second law equation for the block.

For the block,

$$\sum F_{ext,x} = M_{block}a_{block,x}$$

Substitute the individual forces:

$$M_{block}g - T = M_{block}a_{block,x}$$

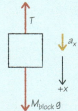

Since the rope doesn't stretch, at any instant the speed of the block v_{block} equals the speed of a point on the pulley rim, which is equal to $R\omega$ (the product of the pulley radius and the pulley angular speed given by Equation 9-3). As the block falls it gains speed and the pulley gains angular speed, but the speed of the block remains equal to the speed of the pulley's rim. Use this to relate the linear acceleration of the block to the angular acceleration of the pulley.

At any time during the motion,

$$v_{block} = v_{rim\ of\ pulley}\ \text{or}$$
$$v_{block,x} = R\omega_{pulley} = R\omega_{pulley,z}$$

(The block is moving downward in the positive x direction, so its x velocity $v_{block,x}$ is positive and equal to its speed v_{block}. The pulley is rotating in the positive direction, so its angular *velocity* $\omega_{pulley,z}$ is positive and equal to the pulley's angular *speed* ω_{pulley}.)

Hence at time t_1 we have

$$v_{block,x1} = R\omega_{pulley,z1}$$

and at a slightly later time t_2 we have

$$v_{block,x2} = R\omega_{pulley,z2}$$

Subtract the first equation from the second and divide by $t_2 - t_1$:

$$\frac{v_{block,x2} - v_{block,x1}}{t_2 - t_1} = \frac{R\omega_{pulley,z2} - R\omega_{pulley,z1}}{t_2 - t_1} = R\left(\frac{\omega_{pulley,z2} - \omega_{pulley,z1}}{t_2 - t_1}\right)$$

The ratios on the two sides are just the acceleration of the block and the angular acceleration of the pulley. So

$$a_{block,x} = R\alpha_{pulley,z}\ \text{or}\ \alpha_{pulley,z} = \frac{a_{block,x}}{R}$$

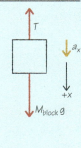

Combine the three equations to solve for the three unknowns.

Substitute $\alpha_{\text{pulley},z} = a_{\text{block},x}/R$ into the pulley equation:

$$T = \frac{1}{2} M_{\text{pulley}} R \alpha_{\text{pulley},z} = \frac{1}{2} M_{\text{pulley}} R \left(\frac{a_{\text{block},x}}{R} \right) = \frac{1}{2} M_{\text{pulley}} a_{\text{block},x}$$

Substitute this expression for T into the block equation:

$$M_{\text{block}} g - T = M_{\text{block}} a_{\text{block},x} \text{ so}$$

$$M_{\text{block}} g - \frac{1}{2} M_{\text{pulley}} a_{\text{block},x} = M_{\text{block}} a_{\text{block},x}$$

Solve for $a_{\text{block},x}$:

$$M_{\text{block}} g = M_{\text{block}} a_{\text{block},x} + \frac{1}{2} M_{\text{pulley}} a_{\text{block},x}$$

$$= \left(M_{\text{block}} + \frac{1}{2} M_{\text{pulley}} \right) a_{\text{block},x}$$

$$a_{\text{block},x} = \left(\frac{M_{\text{block}}}{M_{\text{block}} + M_{\text{pulley}}/2} \right) g$$

Now find $\alpha_{\text{pulley},z}$ by substituting this into $\alpha_{\text{pulley},z} = a_{\text{block},x}/R$:

$$\alpha_{\text{pulley},z} = \frac{a_{\text{block},x}}{R} = \left(\frac{M_{\text{block}}}{M_{\text{block}} + M_{\text{pulley}}/2} \right) \frac{g}{R}$$

Finally, solve for T from the pulley equation:

$$T = \frac{1}{2} M_{\text{pulley}} R \alpha_{\text{pulley},z} = \frac{1}{2} M_{\text{pulley}} R \left(\frac{M_{\text{block}}}{M_{\text{block}} + M_{\text{pulley}}/2} \right) \frac{g}{R}$$

$$= M_{\text{block}} g \left(\frac{M_{\text{pulley}}}{2M_{\text{block}} + M_{\text{pulley}}} \right)$$

Reflect

The downward acceleration of the block is less than g because the tension of the rope exerts an upward pull on the block that partially cancels the downward pull of gravity. The expression for tension T confirms this: T is less than $M_{\text{block}} g$, which says that the tension force doesn't completely balance out the gravitational force. If it did, the block would just hang there when released and wouldn't move at all!

Here's a way to check our results: If the pulley is very much lighter than the block, so M_{pulley} is nearly zero, it's as though the block is connected to nothing at all. In this case the block's downward acceleration would be g (it would be in free fall), and there would be no tension in the rope. Can you use our results for $a_{\text{block},x}$ and T to show that this is the case?

$$a_{\text{block},x} = \left(\frac{M_{\text{block}}}{M_{\text{block}} + M_{\text{pulley}}/2} \right) g$$

Inside the parentheses the numerator M_{block} is less than the denominator $M_{\text{block}} + M_{\text{pulley}}/2$, so the ratio inside the parentheses is less than 1 and $a_{\text{block},x} < g$.

$$T = M_{\text{block}} g \left(\frac{M_{\text{pulley}}}{2M_{\text{block}} + M_{\text{pulley}}} \right)$$

Inside the parentheses the numerator M_{pulley} is less than the denominator $2M_{\text{block}} + M_{\text{pulley}}$, so the ratio inside the parentheses is less than 1 and $T < M_{\text{block}} g$.

Combined Translation and Rotation and Rolling Without Slipping

In Example 9-9 we had one object that had rotational motion but no translational motion (the pulley) and one object that had translational motion but no rotational motion (the block). But many situations' objects have *both* translational *and* rotational motion simultaneously. Examples include a ship's propeller that rotates as it moves forward with the ship, a maple seed that rotates as it falls, and the wrench shown in Figure 9-21 that rotates as it slides across a table. How can we analyze objects that have both translation and rotation?

1. The center of mass of this wrench (shown as a red cross) moves in a straight line...

2. ...and the wrench rotates around the center of mass.

Berenice Abbott/Science Source

Figure 9-21 **Combined translation and rotation** In this time-lapse image, you are looking down on a spinning wrench as it slides across a table. The wrench naturally rotates around its center of mass (shown by the red "cross"), so we can think of its motion as the translation of the center of mass plus the rotation of the wrench around the center of mass.

It turns out that combined translational and rotational motion naturally separates into two parts: (i) the translational motion of the center of mass of the object, and (ii) the rotational motion of the object around its center of mass. We learned in Section 8-7 that the center of mass of an object moves as though all of the object's mass were concentrated there. If there is a net external force on the object, the center of mass accelerates in response; if there is no net external force on the object, the velocity of the center of mass is constant. For example, the wrench shown in Figure 9-21 slides on a nearly frictionless table, so the net external force on the wrench is essentially zero. Hence the center of mass has zero *acceleration* and moves in a straight line with constant speed. There is also essentially zero net external torque on the wrench, so it has zero angular acceleration and rotates with constant angular velocity.

To describe combined translation and rotation of an object thus takes two equations. The first is Newton's second law for translation of the center of mass, Equation 8-34. This involves the mass m of the object:

$$\sum \vec{F}_{\text{ext}} = m\vec{a}_{\text{CM}} \qquad (8\text{-}34)$$

The second equation is Newton's second law for rotation around the center of mass, Equation 9-16, which we write as

$$\sum \tau_{\text{ext},z} = I_{\text{CM}}\alpha_z \qquad (9\text{-}21)$$

The subscript "CM" on the symbol for moment of inertia reminds us that we must use a rotation axis that passes through the center of mass of the object in our calculations. If the object is symmetrical, like a bicycle tire or bowling ball, the center of mass will be at the object's geometrical center. In such a case Table 9-1 in Section 9-4 will help you determine the value of I_{CM} for that object. (The wrench shown in Figure 9-21 is not symmetric: It has more mass at one end than the other, so its center of mass is *not* at its geometrical center.)

Note that in Equation 9-21 we've assumed that the rotation axis, or z axis, always has the same orientation in space. There are situations in which this isn't the case, but we won't consider them in this chapter.

One of the most important situations that involves combined translation and rotation is **rolling without slipping**. The bicycle tires shown in Figure 9-22a are rolling without slipping—that is, they do not slip or slide over the road. The same is true for the tumbleweed shown in Figure 9-22b. An important feature of rolling without slipping is that there is a specific relationship between the speed of the center of the wheel, v_{CM}, and the angular speed ω of the wheel's rotation (Figure 9-23a). To see what this relationship is, consider what happens when the wheel rolls through a small angle $\Delta\theta$ over a time interval Δt (Figure 9-23b). Imagine that a thread has been wrapped around the circumference of the wheel and unwinds as the wheel rolls, marking the distance traveled. During the interval Δt a point on the rim of the wheel of radius R rotates a distance $\Delta s = R\Delta\theta$. Since the wheel doesn't slip on the road, $\Delta s = R\Delta\theta$ is also the length of thread that the wheel unwinds onto the road and hence the distance that the wheel as a whole moves down the road. Hence v_{CM}, the speed of the wheel as a whole, is equal to the distance Δs divided by the duration Δt of the time interval:

$$v_{\text{CM}} = \frac{\Delta s}{\Delta t} = \frac{R\Delta\theta}{\Delta t} = R\frac{\Delta\theta}{\Delta t}$$

(a) Bicycle tires roll without slipping for transportation.

Purestock/Alamy

(b) Tumbleweeds roll without slipping for reproduction.

mikeledray/Shutterstock

BioMedical **Figure 9-22** **Rolling without slipping** (a) The tires of this bicycle roll without slipping on the road. This greatly minimizes friction, which is why bicycling is such an efficient means of transportation. (b) The prickly Russian thistle (*Kali tragus*) is a flowering plant that uses rolling without slipping for reproduction. When the plant begins to die, it dries out and becomes brittle and the base of the stem breaks off. The plant then rolls as a tumbleweed in the wind, dispersing its seeds as it travels.

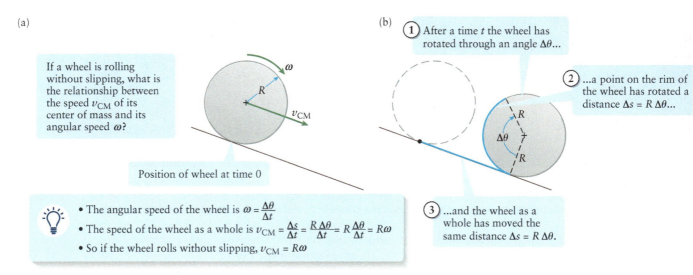

(a)

If a wheel is rolling without slipping, what is the relationship between the speed v_{CM} of its center of mass and its angular speed ω?

Position of wheel at time 0

- The angular speed of the wheel is $\omega = \dfrac{\Delta\theta}{\Delta t}$
- The speed of the wheel as a whole is $v_{CM} = \dfrac{\Delta s}{\Delta t} = \dfrac{R\,\Delta\theta}{\Delta t} = R\dfrac{\Delta\theta}{\Delta t} = R\omega$
- So if the wheel rolls without slipping, $v_{CM} = R\omega$

(b)

① After a time t the wheel has rotated through an angle $\Delta\theta$...

② ...a point on the rim of the wheel has rotated a distance $\Delta s = R\,\Delta\theta$...

③ ...and the wheel as a whole has moved the same distance $\Delta s = R\,\Delta\theta$.

Figure 9-23 A wheel rolling downhill If a wheel rolls without slipping, there is a definite relationship between its angular speed and the speed of its center of mass.

But $\Delta\theta/\Delta t$ is just equal to ω, the angular speed of the wheel, so

Condition for rolling without slipping (9-22)

Speed of the center of mass of a rolling object

$$v_{CM} = R\omega$$

Radius of the object   Angular speed of the object's rotation

Equation 9-22 says that there is a direct proportionality between the linear speed v_{CM} and angular speed ω of a circular object that rolls without slipping. As the wheel in Figure 9-23 rolls downhill it gains speed so v_{CM} increases, and it rotates faster so ω increases, but these increases are always proportional to each other so that v_{CM} and ω are always related by Equation 9-22.

If an object is rolling but does not obey Equation 9-22, it is slipping. For example, if you are driving on an icy road without chains, the tires are likely to slip: They will spin too fast for the speed the car is traveling, and $R\omega$ will be greater than v_{CM}. If you are driving on a dry road but apply the brakes too forcefully, the wheels may "lock" and spin only slowly. In this case $R\omega$ will be less than v_{CM}. To prevent this, most modern cars are equipped with antilock brakes, which keep $R\omega$ equal to v_{CM} at all times so that the tires always roll without slipping.

For an object that rolls without slipping there is also a relationship between the *acceleration* of the object's center of mass and the *angular acceleration* of the object. In Example 9-9 we had the equation $v_{block,x} = R\omega_{pulley,z}$, which is very similar to Equation 9-22, and this led to the relationship $a_{block,x} = R\alpha_{pulley,z}$. The same mathematical steps (which we won't repeat here) lead us to the same relationship between the linear acceleration $a_{CM,x}$ of the center of mass and the angular acceleration α_z for an object that rolls without slipping:

Acceleration and angular acceleration for rolling without slipping (9-23)

Acceleration of the center of mass of a rolling object

$$a_{CM,x} = R\alpha_z$$

Radius of the object   Angular acceleration of the object's rotation

In this equation the center of mass of the object is moving in the x direction, and the object is rotating around a z axis that passes through the center of mass. The positive directions are chosen so that $a_{CM,x}$ and α_z have the same sign. For example, if in Figure 9-23 we take the positive x direction to be downhill along the incline (the direction in which the wheel's center of mass is accelerating), then the positive rotation direction must be *clockwise* (since the wheel is spinning clockwise with increasing angular speed).

We now have all the tools that we need to analyze an object that is rolling without slipping: Equation 8-34 (Newton's second law for translation of the center of mass), Equation 9-21 (Newton's second law for rotation around the center of mass), and Equation 9-23 (the relationship between $a_{CM,x}$ and α_z). The following example shows how these equations work together.

EXAMPLE 9-10   Rolling Without Slipping

A uniform disk of mass M and radius R is initially at rest on a ramp inclined at an angle θ to the horizontal. When released, the disk rolls without slipping down the ramp. Determine the downhill acceleration of the disk's center of mass.

Set Up

Unlike Example 9-9, in which we had one object (the pulley) undergoing purely rotational motion and another object (the block) undergoing purely translational motion, here we have *one* object that undergoes *both* rotational and translational motion. The principles are the same, however. We'll use the equations for both the rotational and translational forms of Newton's second law for the disk. We'll also use Equation 9-23 to relate the disk's center of mass acceleration and its angular acceleration. This equation holds because the disk rolls without slipping.

Newton's second law for motion of the center of mass of the disk—note that we include both x and y equations:

$$\sum F_{ext,x} = Ma_{CM,x}$$

$$\sum F_{ext,y} = Ma_{CM,y} \qquad (8\text{-}34)$$

Newton's second law for rotation around the center of mass:

$$\sum \tau_{ext,z} = I_{CM}\alpha_z \qquad (9\text{-}21)$$

Disk rolls without slipping:

$$a_{CM,x} = R\alpha_z \qquad (9\text{-}23)$$

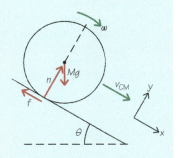

Solve

Three forces act on the disk as it rolls. Besides the downward force of gravity Mg and the normal force n that acts perpendicular to the surface of the ramp, there *must* be an uphill force of friction f. If this force weren't present, the disk wouldn't roll; instead, it would move downhill without rotating at all.

To see why this is so, note that according to Equation 9-21 there must be a torque around its center of mass for the disk to start rotating. The force of gravity exerts no torque around the center of mass because it acts at the disk's center of mass, so its lever arm is $r_\perp = 0$. The normal force also exerts no torque: It acts along a line that passes through the center of mass, so its lever arm is also zero. Only the friction force exerts a torque around the center of mass: For this force, $r_\perp = R$ and so $\tau_{friction,z} = Rf$.

From Table 9-1 in Section 9-4, the moment of inertia for a solid disk about its axis is $I_{CM} = MR^2/2$. We substitute all of these into Newton's second law for rotation.

Newton's second law for rotation:

$$\sum \tau_{ext,z} = I_{CM}\alpha_z$$

Substitute $\sum \tau_{ext,z} = \tau_{friction,z} = Rf$ and

$I_{CM} = MR^2/2$:

$$Rf = \frac{1}{2}MR^2\alpha_z$$

Divide both sides of this equation by R:

$$f = \frac{1}{2}MR\alpha_z$$

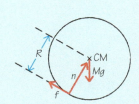

For translational motion, write the equations for the net force in the x and y directions. There is no acceleration in the y direction (perpendicular to the ramp) so $a_{CM,y} = 0$. The acceleration downhill in the x direction, $a_{CM,x}$, results from the forces of gravity and friction.

Net force in each direction:

$$\sum F_x = Mg \sin \theta - f$$

$$\sum F_y = n - Mg \cos \theta$$

Substitute into Newton's second law for translational motion:

$$Mg \sin \theta - f = Ma_{CM,x}$$

$$n - Mg \cos \theta = 0$$

(since $a_{CM,y} = 0$)

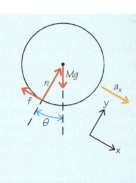

The y equation for linear motion isn't useful (it would be of interest only if we wanted to calculate the normal force n). The other two Newton's second law equations—one for rotational motion and one for linear motion along the x axis—involve three unknowns: f, α_z, and $a_{CM,x}$. So we need a third equation. This comes from the condition for rolling without slipping, $a_{CM,x} = R\alpha_z$.

Rotation: $f = \frac{1}{2}MR\alpha_z$

Translation along x axis: $Mg \sin \theta - f = Ma_{CM,x}$

Rolling without slipping: $a_{CM,x} = R\alpha_z$ or $\alpha_z = a_{CM,x}/R$

Combine the three equations to solve for $a_{CM,x}$.

Substitute $\alpha_z = a_{CM,x}/R$ into the rotational equation:

$$f = \frac{1}{2}MR\alpha_z = \frac{1}{2}MR\left(\frac{a_{CM,x}}{R}\right) = \frac{1}{2}Ma_{CM,x}$$

Substitute this expression for f into the translation equation and solve for $a_{CM,x}$:

$$Mg \sin \theta - f = Ma_{CM,x} \text{ so } Mg \sin \theta - \frac{1}{2}Ma_{CM,x} = Ma_{CM,x}$$

$$Mg \sin \theta = Ma_{CM,x} + \frac{1}{2}Ma_{CM,x} = \frac{3}{2}Ma_{CM,x}$$

$$g \sin \theta = \frac{3}{2}a_{CM,x}$$

$$a_{CM,x} = \frac{2}{3}g \sin \theta$$

Reflect

We can also solve for the magnitude of the friction force. We find that for this case where there is rolling without slipping, the friction force must have a very definite value, equal to $\frac{1}{3}Mg \sin \theta$. That provides exactly enough angular acceleration to satisfy the condition for rolling without slipping, $a_{CM,x} = R\alpha_z$.

If there were *no* friction, the disk would slide downhill rather than roll and its acceleration would be $g \sin \theta$ (the component along the ramp of the acceleration due to gravity), as in Example 4-5. Friction provides the torque to give the disk its angular acceleration, but also slows the disk's downhill translation. That's why the acceleration of the disk is only two-thirds of $g \sin \theta$.

To solve for the friction force, substitute our result for $a_{CM,x}$ into the equation for f:

$$f = \frac{1}{2}Ma_{CM,x} = \frac{1}{2}M\left(\frac{2}{3}g \sin \theta\right)$$

$$= \frac{1}{3}Mg \sin \theta$$

Rolling Without Slipping: What Kind of Friction Acts?

You may wonder what kind of friction acts on the disk rolling downhill in Example 9-10. Although the disk is moving, this is *not* kinetic friction! The reason is that kinetic friction is present only when an object slides over a surface (see Section 5-3). Since the disk rolls without slipping, it is not sliding. In fact the kind of friction present is *static* friction. That's because the point where the disk touches the ramp is at rest during the instant that this point and the ramp are in contact. As we learned in Section 5-2, the force of static friction can adjust to have any value up to a maximum that depends on the normal force and on the nature of the surfaces in contact. That's how the friction force in Example 9-10 was able to have the particular value $f = \frac{1}{3} Mg \sin \theta$ required to have rolling without slipping. (Note that a steeper ramp angle θ means a greater value of $\sin \theta$ and so a greater required value of the friction force f. If θ is too great, the required friction force can exceed the maximum static friction force available. In that case the disk will slip as it moves downhill.)

Note that although the disk in Example 9-10 is rolling, it does *not* experience the *rolling friction* that we discussed in Section 5-3. That's because we've assumed that the disk and the ramp on which it rolls are perfectly rigid. Rolling friction will be present only if either the rolling object or the surface on which it rolls is *not* perfectly rigid—that is, if it can *deform*. An example is a bicycle tire or automobile tire, which flexes and flattens on its bottom where it contacts the road. As the tire rolls, different parts of the tire successively get compressed as they touch the ground and relax as they lose contact with the ground. This flexing causes the tire to get warmer, which expends mechanical energy. The energy comes from the kinetic energy of the rotating tire, which means that left to itself on a horizontal surface the tire will lose kinetic energy and slow down. (In the following section we'll see how to express the kinetic energy of a rotating object.) The net effect is the same as if a force were acting to oppose the rolling tire's motion. Rolling friction is also present if the surface on which the object rolls is deformable, like a wrestling mat or the green baize that covers a billiards table.

You can minimize rolling friction on your bicycle or automobile by keeping tires properly inflated. Underinflated tires flex more, which means that there is more rolling friction. That's also why transporting goods via railroad is more efficient than using trucks: Railroad cars have very rigid steel wheels, which flex much less than even properly inflated truck tires. The tumbleweed shown in Figure 9-22b is *very* deformable, so it has a lot of rolling friction. That's why a tumbleweed slows down to a stop very quickly unless there is a wind to push it along.

In this chapter we'll assume that any object that rolls without slipping is perfectly rigid, as is the surface on which it rolls. With that assumption there is no rolling friction, and the only friction force present will be static friction. In the following section we'll look again at rolling without slipping, not from the perspective of forces and torques but rather in terms of *energy*.

GOT THE CONCEPT? 9-6 A Uniform Sphere Rolling Downhill

 Suppose we replace the disk in Example 9-10 with a uniform solid sphere like a billiard ball. Compared to the linear acceleration of the disk, what will be the linear acceleration of the sphere? (a) Greater; (b) the same; (c) less.

TAKE-HOME MESSAGE FOR Section 9-5

✔ Always draw a free-body diagram as part of the solution of any problem that involves forces and/or torques. To be able to calculate torques correctly, draw each force at the point where it is applied.

✔ In problems of this kind apply Newton's second law for translation to any object that changes position and Newton's second law for rotation to any object that rotates. Both laws must be applied to any object that undergoes both translation and rotation.

✔ In problems that involve both translation and rotation you will need to relate the translational acceleration to the rotational acceleration.

✔ If an object rolls without slipping on a surface, the speed of its center of mass is directly proportional to its angular speed, and the acceleration of its center of mass is directly proportional to its angular acceleration. If the rolling object and the surface are perfectly rigid, the friction force present is static friction, not kinetic friction or rolling friction.

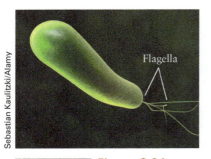

BioMedical **Figure 9-24**
Rotational motion on the microscopic scale *Helicobacter pylori* bacteria, which can cause diseases of the human gastrointestinal tract, are propelled by a set of flagella at one end. These flagella rotate at several hundred revolutions per minute. A typical *H. pylori* bacterium is about 3 µm (3×10^{-6} m) in length.

9-6 An object's rotational kinetic energy is related to its angular speed and its moment of inertia

We've seen how to analyze rotational motion using the rotational version of Newton's second law. But it can often be useful to describe rotation in terms of *energy*, just as we did for translational motion in Chapter 6. As an example, thinking about the kinetic energy that an object has due to its rotation—that is, its **rotational kinetic energy**—helps us understand the motion of bacteria that propel themselves using flagella (Figure 9-24). The flagella on bacteria of this kind are attached to a microscopic motor powered by chemical energy. The motor converts this into rotational kinetic energy of the flagella, which act like spinning propellers. This kinetic energy is in turn used to push on the surrounding fluid, which responds by pushing the bacterium forward and giving it translational kinetic energy.

Rotational Kinetic Energy

Since different parts of a rotating rigid object move at different speeds v, we cannot simply use the formula $K = \frac{1}{2}mv^2$ to determine its kinetic energy. Instead, we'll divide the object into pieces, each small enough that the linear speed is the same for the entire piece. This is what we've done in Figure 9-25, which shows the same rotating wind turbine blade that we analyzed in Figure 9-13 (Section 9-3) to develop the rotational form of Newton's second law. The blade is rotating around a fixed axis, so its kinetic energy is purely due to rotation.

All of the pieces of the blade, which we label $i = 1, 2, 3, \ldots, N$, rotate at the same angular speed ω, but the speed v is different for each piece. The ith piece has mass m_i and is a distance r_i from the rotation axis, so from Equation 9-3 its speed is $v_i = r_i\omega$. If we measure ω in radians per second and r_i in meters, v_i will be in meters per second (m/s). The kinetic energy of the ith piece of the blade is therefore

$$K_i = \frac{1}{2}m_i v_i^2 = \frac{1}{2}m_i(r_i\omega)^2 = \frac{1}{2}m_i r_i^2 \omega^2$$

The total kinetic energy of the entire rotating turbine blade—that is, its rotational kinetic energy—is the sum of the kinetic energies of all pieces of the blade:

(9-24)
$$K_{\text{rotational}} = K_1 + K_2 + K_3 + \ldots + K_N = \sum_{i=1}^{N} K_i = \sum_{i=1}^{N} \frac{1}{2}m_i r_i^2 \omega^2 = \frac{1}{2}\left(\sum_{i=1}^{N} m_i r_i^2\right)\omega^2$$

The subscript "rotational" in Equation 9-24 reminds us that this is the kinetic energy of the turbine blade due to its rotation. The quantities $\frac{1}{2}$ and ω^2 have the same values

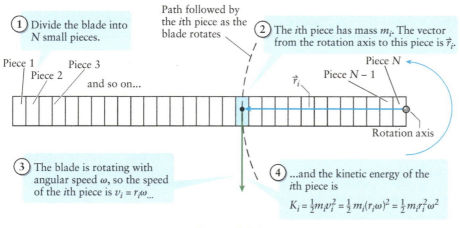

① Divide the blade into N small pieces.

Path followed by the ith piece as the blade rotates

② The ith piece has mass m_i. The vector from the rotation axis to this piece is $\vec{r}_i$.

Piece 1 Piece 3
Piece 2 and so on...

Piece N
Piece $N-1$
$\vec{r}_i$
Rotation axis

③ The blade is rotating with angular speed ω, so the speed of the ith piece is $v_i = r_i\omega$...

④ ...and the kinetic energy of the ith piece is
$K_i = \frac{1}{2}m_i v_i^2 = \frac{1}{2}m_i(r_i\omega)^2 = \frac{1}{2}m_i r_i^2 \omega^2$

- The total kinetic energy of the rotating blade is the sum of the kinetic energies of the N pieces:
$K_{\text{rotational}} = K_1 + K_2 + K_3 + \ldots K_N$
$= \sum_{i=1}^{N} K_i = \sum_{i=1}^{N} \frac{1}{2}m_i r_i^2 \omega^2$

- We can write this in terms of the moment of inertia of the blade:
$K_{\text{rotational}} = \frac{1}{2}I\omega^2$ where $I = \sum_{i=1}^{N} m_i r_i^2$

Figure 9-25 Calculating rotational kinetic energy and moment of inertia The kinetic energy of a rotating object is the sum of the kinetic energies of all of its component pieces.

for each term in the sum, which is why we could factor them out of the sum in Equation 9-24. (If you're not sure whether it's correct to factor constant terms out of a sum, try it with a few numbers; for example, $2(3) + 2(4) + 2(5) = 2(3 + 4 + 5)$.) You'll recognize the term in parentheses in Equation 9-24 as the moment of inertia of the blade for the rotation axis shown in Figure 9-25, $I = \sum_{i=1}^{N} m_i r_i^2$ (see Equation 9-15). So Equation 9-24 for rotational kinetic energy becomes

Rotational kinetic energy of a rigid object spinning around an axis

Moment of inertia of the object for that rotation axis

$$K_{\text{rotational}} = \frac{1}{2} I\omega^2$$

Angular speed of the object

Rotational kinetic energy of a rigid object
(9-25)

In deriving Equation 9-25 we used Equation 9-3, $v = r\omega$, which is valid only if angles are measured in radians. To have the units of rotational kinetic energy work out properly with Equation 9-25, you *must* express angular speed ω in radians per second (rad/s).

We've derived Equation 9-25 for a rotating turbine blade, but it applies to *any* rotating rigid object. Note that this expression for *rotational* kinetic energy is the same as the formula for *translational* kinetic energy, $K_{\text{translational}} = \frac{1}{2} mv^2$, with mass m replaced by moment of inertia I and speed v replaced by angular speed ω.

WATCH OUT!   When calculating rotational kinetic energy, use the appropriate moment of inertia.

!   Furthermore, as we learned in Section 9-4, the moment of inertia I for a given rigid object depends on where the rotation axis is located. So when calculating rotational kinetic energy with Equation 9-25, you must take care to use the appropriate value of I. The following example illustrates this point.

EXAMPLE 9-11   Rotational Kinetic Energy for Dumbbells

In Examples 9-4 and 9-5 (Section 9-4) we considered a rotating dumbbell made up of two small, identical 50.0-kg spheres connected by a lightweight rod 2.40 m in length. You give the dumbbell an angular speed of 2.00 rad/s. (a) What is the angular speed in revolutions per minute? (b) What is the rotational kinetic energy of the dumbbell if it rotates around the midpoint of the dumbbell? If it rotates around the center of one of the spheres?

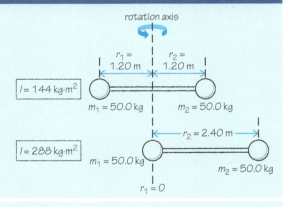

Set Up

We are given the value of the angular speed ω in rad/s; we are asked to convert this to rev/min. When we calculate the rotational kinetic energy using Equation 9-25, however, we must use the value of ω in rad/s. The value of the moment of inertia I to use in this equation is different for the two rotation axes. We found these values in Examples 9-4 and 9-5: $I = 144$ kg·m² for an axis through the midpoint and $I = 288$ kg·m² for an axis through one of the spheres.

Rotational kinetic energy:

$$K_{\text{rotational}} = \frac{1}{2} I\omega^2 \qquad (9\text{-}25)$$

Solve

(a) Calculate the angular speed in revolutions per minute.	We are given $\omega = 2.00$ rad/s. Since 1 rev = 2π rad and 1 min = 60 s, $$\omega = \left(2.00 \ \frac{\text{rad}}{\text{s}}\right)\left(\frac{1 \ \text{rev}}{2\pi \ \text{rad}}\right)\left(\frac{60 \ \text{s}}{1 \ \text{min}}\right) = 19.1 \ \text{rev/min}$$ The dumbbell makes 19.1 rotations around its axis every minute.

(b) Calculate the rotational kinetic energy for each of the two rotation axes. Substitute this value of ω and the given value of I into the expression for rotational kinetic energy. We can drop the "rad" from the units of the final answer because radians aren't a true unit.	For rotation around the midpoint, $I = 144 \ \text{kg} \cdot \text{m}^2$. The rotational kinetic energy is $$K_{\text{rotational}} = \frac{1}{2}I\omega^2 = \frac{1}{2}(144 \ \text{kg} \cdot \text{m}^2)(2.00 \ \text{rad/s})^2$$ $$= 288 \ \text{kg} \cdot \text{m}^2/\text{s}^2 = 288 \ \text{J}$$ For rotation around the center of one of the spheres, $I = 288 \ \text{kg} \cdot \text{m}^2$. The rotational kinetic energy is $$K_{\text{rotational}} = \frac{1}{2}I\omega^2 = \frac{1}{2}(288 \ \text{kg} \cdot \text{m}^2)(2.00 \ \text{rad/s})^2$$ $$= 576 \ \text{kg} \cdot \text{m}^2/\text{s}^2 = 576 \ \text{J}$$

Reflect

For the same angular speed ω, you must supply twice as much kinetic energy—that is, you have to do twice as much work—to make the dumbbell rotate around the center of one of its spheres compared to making it rotate around its midpoint. That's because the moment of inertia is twice as great for the axis through one of the spheres.

We can check our results for rotational energy in the two cases by finding the *translational* kinetic energies of the individual spheres and adding them together. (Remember from Example 9-4 that we can treat each sphere as a point object, so we don't have to worry about the kinetic energy of a sphere spinning on its axis.) Our calculations show that changing the rotation axis while keeping the angular speed ω the same causes the speed v of the two spheres to change and hence changes the total kinetic energy—with numerical values that exactly match the ones we calculated above.

For rotation around the midpoint of the dumbbell at angular speed $\omega = 2.00$ rad/s: Each sphere is a distance $r = 1.20$ m from the rotation axis, so the speed of each sphere is

$$v = r\omega = (1.20 \ \text{m})(2.00 \ \text{rad/s}) = 2.40 \ \text{m/s}$$

Each 50.0-kg sphere has translational kinetic energy

$$K_{\text{sphere}} = \frac{1}{2}mv^2 = \frac{1}{2}(50.0 \ \text{kg})(2.40 \ \text{m/s})^2 = 144 \ \text{J}$$

So the total kinetic energy of the system of two spheres rotating around its midpoint is 144 J + 144 J = 288 J.

For rotation around one of the spheres at angular speed $\omega = 2.00$ rad/s: One sphere is at the rotation axis and is essentially at rest, so it has zero kinetic energy. The other sphere is a distance $r = 2.40$ m from the rotation axis, so the speed of this sphere is

$$v = r\omega = (2.40 \ \text{m})(2.00 \ \text{rad/s}) = 4.80 \ \text{m/s}$$

The translational kinetic energy of this 50.0-kg sphere is

$$K_{\text{sphere}} = \frac{1}{2}mv^2 = \frac{1}{2}(50.0 \ \text{kg})(4.80 \ \text{m/s})^2 = 576 \ \text{J}$$

So the total kinetic energy of the system of two spheres rotating around one of the spheres is 0 J + 576 J = 576 J.

Conservation of Mechanical Energy in Rotational Motion

We learned in Section 6-7 that the mechanical energy of a system—the sum of kinetic energy and potential energy—is *conserved* if no nonconservative forces do work on that system. Does conservation of energy still hold true if we include rotational kinetic energy? And in particular does it hold true for objects that are *both* translating (moving through space) *and* rotating, like a ball rolling downhill or the spinning wheels of a car?

    Happily, the answer to both questions is "yes" because rotational kinetic energy isn't really a new kind of energy. As we saw in Example 9-11, the rotational kinetic energy of a rigid body is just the combined translational kinetic energy of all of its component pieces due to the motion of those pieces around the rotation axis. So the principle of conservation of energy also holds if we include rotational kinetic energy.

We can also use the principle of conservation of energy for a rigid object that's both moving through space as a whole and rotating. In such a situation it turns out that we can write the object's *total* kinetic energy as the sum of two terms: the *translational* kinetic energy associated with the motion of the object's center of mass (see Section 8-7) and the *rotational* kinetic energy associated with the object's rotation around its center of mass. (This is analogous to what we learned in Section 9-5: to use Newton's second law to analyze an object that undergoes both translation and rotation, apply the translational form of the second law to the motion *of* the center of mass and apply the rotational form of the second law to the motion *around* the center of mass.)

If the object has mass M and its center of mass is moving with speed v_{CM}, its translational kinetic energy is $K_{translational} = \frac{1}{2}Mv_{CM}^2$; if the object's moment of inertia for an axis through its center of mass is I_{CM} and it rotates with angular speed ω, its rotational kinetic energy is $K_{rotational} = \frac{1}{2}I_{CM}\omega^2$. The total kinetic energy (translational plus rotational) of the object is then

(1) **Total kinetic energy** of a system that is both translating and rotating

(2) The **translational kinetic energy** is the same as if all of the mass of the system were moving at the speed of the center of mass.

$$K = K_{translational} + K_{rotational} = \frac{1}{2}Mv_{CM}^2 + \frac{1}{2}I_{CM}\omega^2$$

Total kinetic energy for a rigid object undergoing both translation and rotation
(9-26)

(3) The **rotational kinetic energy** is the same as if the system were not translating, and all of the mass were rotating around the center of mass.

As we learned in Section 9-3, the center of mass is at the object's geometrical center if the object is symmetrical, like a uniform sphere.

Whenever we have a situation involving a rigid object that undergoes both translation and rotation, we can make use of the energy relationships from Section 6-7 provided that we use Equation 9-26 as the expression for kinetic energy. If only conservative forces do work, mechanical energy (the sum of kinetic energy K and potential energy U) is conserved, so the value of $K + U$ is the same at any two times during the motion:

$$K_i + U_i = K_f + U_f \qquad (6\text{-}27)$$

(if only conservative forces do work)

If nonconservative forces also do work, the final mechanical energy $K_f + U_f$ is equal to the initial mechanical energy $K_i + U_i$ plus $W_{nonconservative}$, the amount of nonconservative work that is done during the motion:

$$K_i + U_i + W_{nonconservative} = K_f + U_f \qquad (6\text{-}30)$$

(if nonconservative forces do work)

In both Equations 6-27 and 6-30, the kinetic energy K is now given by Equation 9-26 and the potential energy U includes a term for each conservative force that acts (for example, the gravitational force or a spring force).

Example 9-12 illustrates how to use these ideas to solve a simple problem that involves both translational and rotational motion.

EXAMPLE 9-12 Flying Disc Energy

A flying disc with mass 0.175 kg and diameter 0.266 m is used in the team sport called Ultimate. A player takes a disc at rest and does 1.00 J of work on it, causing the disc to fly off in a horizontal direction. When the disc leaves the player's hand, nine-tenths of its kinetic energy is translational and one-tenth is rotational. Find (a) the speed of the flying disc's center of mass and (b) the angular speed of the rotating disc. Treat the disc as uniform.

Set Up

The work done by the player counts as nonconservative work, so we use Equation 6-30 to describe how the disc's energy changes. We use Equation 9-26 to describe the kinetic energy of the disc, which involves its moment of inertia I_{CM} for an axis through its center of mass (the same as its geometrical center). As in Example 9-7 in Section 9-4, this moment of inertia is $I_{CM} = MR^2/2$, where $M = 0.175$ kg and R is one-half of the disc's diameter (see the first entry in the second row of Table 9-1). Our goal is to find the values of v_{CM} and ω just after the disc leaves the player's hand.

If nonconservative forces do work:

$$K_i + U_i + W_{nonconservative} = K_f + U_f \qquad (6\text{-}30)$$

Total kinetic energy for a rigid object undergoing both translation and rotation:

$$K = K_{translational} + K_{rotational}$$
$$= \frac{1}{2}Mv_{CM}^2 + \frac{1}{2}I_{CM}\omega^2 \qquad (9\text{-}26)$$

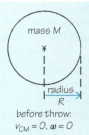

mass M
radius R

before throw:
$v_{CM} = 0,\ \omega = 0$

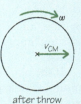

after throw

Solve

During the throw the disc moves horizontally, so its height stays the same and its gravitational potential energy is constant; that is, $U_f = U_i$. The disc starts at rest, so $K_i = 0$ (it has zero initial kinetic energy). So the work done on the disc by the player is transformed into the disc's final kinetic energy K_f.

$$K_i + U_i + W_{nonconservative} = K_f + U_f$$

Since $U_f = U_i$, $K_i = 0$, and $W_{nonconservative} = 1.00$ J, this becomes

$$1.00\ \text{J} = K_f$$

The disc's 1.00 J of kinetic energy is divided so that 9/10 is translational and 1/10 is rotational.

$$K_{translational} = \frac{1}{2}Mv_{CM}^2 = (9/10)(1.00\ \text{J}) = 0.900\ \text{J}$$

$$K_{rotational} = \frac{1}{2}I_{CM}\omega^2 = (1/10)(1.00\ \text{J}) = 0.100\ \text{J}$$

(a) Find the speed of the disc's center of mass.

From the expression for translational kinetic energy,

$$v_{CM}^2 = \frac{2K_{translational}}{M}$$

$$v_{CM} = \sqrt{\frac{2K_{translational}}{M}} = \sqrt{\frac{2(0.900\ \text{J})}{0.175\ \text{kg}}} = \sqrt{\frac{2(0.900\ \text{kg} \cdot \text{m}^2/\text{s}^2)}{0.175\ \text{kg}}}$$

$$= 3.21\ \text{m/s}$$

(b) Calculate the disc's moment of inertia for the axis of rotation through its center, then solve for its angular speed ω. Note that if $K_{rotational}$ is in joules and I_{CM} is in kg·m², ω is in radians per second.

Calculate the disc's moment of inertia through its center of mass:

$$I_{CM} = \frac{1}{2}MR^2 = \frac{1}{2}(0.175\ \text{kg})\left(\frac{0.266\ \text{m}}{2}\right)^2$$

$$= 1.55 \times 10^{-3}\ \text{kg} \cdot \text{m}^2$$

From the expression for rotational kinetic energy,

$$\omega^2 = \frac{2K_{rotational}}{I_{CM}}$$

$$\omega = \sqrt{\frac{2K_{rotational}}{I_{CM}}} = \sqrt{\frac{2(0.100\ \text{J})}{1.55 \times 10^{-3}\ \text{kg} \cdot \text{m}^2}}$$

$$= \sqrt{\frac{2(0.100\ \text{kg} \cdot \text{m}^2/\text{s}^2)}{1.55 \times 10^{-3}\ \text{kg} \cdot \text{m}^2}} = 11.4\ \text{rad/s}$$

Reflect

This is a problem for which the conservation of energy equation is essential but not sufficient. To solve for the two unknowns, v_{CM} and ω, we needed a second relationship telling us how the kinetic energy was distributed—the statement that the disc's kinetic energy was 9/10 translational and 1/10 rotational.

Note that v_{CM} is greater than a typical walking speed (1 to 2 m/s) but less than the speed of a world-class sprinter (about 10 m/s). Since 2π (just over 6) radians represents a full revolution, the value of ω tells us that the disc spins just under two revolutions per second. Are these values what you might expect for a hand-thrown flying disc?

(a) A symmetrical rigid object rolling downhill without slipping

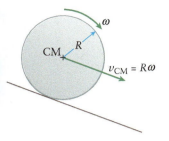

(b) Forces acting on the object

Neither the gravitational force $m\vec{g}$ nor the normal force $\vec{n}$ exerts any torque around the center of mass (CM).

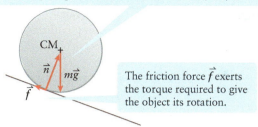

The friction force $\vec{f}$ exerts the torque required to give the object its rotation.

 Because the object does not slide but rolls, the friction force does no work on the object. Hence mechanical energy is conserved.

Figure 9-26 Rolling without slipping: Is mechanical energy conserved? (a) An example of rolling without slipping. (b) Although a friction force acts on the object, it does no work. Hence only the conservative gravitational force does work on the object as it rolls downhill, and mechanical energy is conserved.

Rolling Without Slipping Revisited

A special and important case of combined translation and rotation is *rolling without slipping*, which we introduced in Section 9-5. An example is a symmetrical rigid object of radius R, such as a uniform disk or uniform sphere, rolling down a ramp (Figure 9-26a). In rolling without slipping there is a definite relationship between the speed v_{CM} of the center of mass and the angular speed ω, given by Equation 9-22: $v_{CM} = R\omega$. As the object in Figure 9-26a rolls down the ramp the gravitational potential energy U_{grav} decreases; in addition, both v_{CM} and ω increase as the object rolls, so the translational kinetic energy $K_{translational}$ and rotational kinetic energy $K_{rotational}$ both increase. Is the total mechanical energy $E = K_{translational} + K_{rotational} + U_{grav}$ conserved as the object rolls downhill?

We saw in Example 9-10 (Section 9-5) that there must be a friction force acting on the object as it rolls downhill, as Figure 9-26b shows. (If there were no friction force, there would be no torque on the object around its center of mass. In this case the object would not start to rotate if it were released from rest, and so would slide rather than roll.) So you might be inclined to think that mechanical energy is lost due to friction as the object rolls. It's true that if an object *slides* over a surface, the kinetic friction force does work on the object and changes its mechanical energy (see Section 6-7). But there is *no* sliding if the object rolls without slipping, and the force of friction in Figure 9-26b is *static* friction. As the object rolls, each part of its rim sets down on the ramp and then lifts off again. Since there is no sliding, the friction force does not act over any distance and the work done by friction (force times distance) is zero. Hence mechanical energy *is* conserved in rolling without slipping. The friction force still plays an important role: It's what makes the object roll. So as the object in Figure 9-26 rolls downhill and gravitational potential energy is converted to kinetic energy, the friction force ensures that a portion of the kinetic energy goes into the rotational form.

As we discussed in Section 9-5, rolling friction will also be present if either the object in Figure 9-26 or the ramp on which it rolls can deform. In this case mechanical energy will be lost as the object rolls. In the following example we'll assume that nothing deforms as an object rolls without slipping, so mechanical energy is conserved.

EXAMPLE 9-13 Downhill Race: Disk versus Hoop

A uniform disk and a hoop are both allowed to roll without slipping down a ramp of height H (Figure 9-27). Both objects have the same radius R and the same mass M. If both objects start from rest at the same time, which one reaches the bottom of the ramp first?

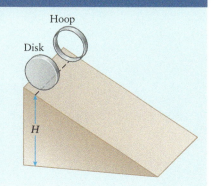

Figure 9-27 Who wins the race? A competition between two objects rolling down a ramp.

Set Up

Both objects roll without slipping, so mechanical energy is conserved as they move. We can use Equations 6-27, 9-22, and 9-26 to describe their motion. The only real difference between the two objects is their moment of inertia: From Table 9-1, $I_{CM} = MR^2/2$ for the uniform disk while $I_{CM} = MR^2$ for the hoop. To decide which one wins the race, we'll calculate $v_{CM,f}$ (the speed at the bottom of the ramp) for a general object that rolls down the ramp starting from rest. We'll then plug in the value of I_{CM} for the disk and hoop. The one with the faster speed will get to the bottom first and be declared the winner.

Conservation of mechanical energy:

$$K_i + U_i = K_f + U_f \qquad (6\text{-}27)$$

Condition for rolling without slipping:

$$v_{CM} = R\omega \qquad (9\text{-}22)$$

Kinetic energy for an object that undergoes both translation and rotation:

$$K = \frac{1}{2}Mv_{CM}^2 + \frac{1}{2}I_{CM}\omega^2 \qquad (9\text{-}26)$$

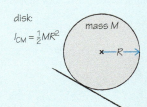

disk:
$I_{CM} = \frac{1}{2}MR^2$

mass M

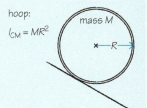

hoop:
$I_{CM} = MR^2$

mass M

Solve

Initially the object is at rest with zero kinetic energy, so $K_i = 0$. The initial gravitational potential energy is $U_i = Mgy_i$, and the final gravitational potential energy is $U_f = Mgy_f$. The figure shows that $y_i - y_f = H - 0 = H$, so we can solve for the object's final kinetic energy K_f at the bottom of the ramp.

Conservation of energy equation:

$$0 + Mgy_i = K_f + Mgy_f, \text{ or}$$

$$\begin{aligned} K_f &= Mgy_i - Mgy_f \\ &= Mg(y_i - y_f) \end{aligned}$$

Since $y_i = H$ and $y_f = 0$,

$$K_f = MgH$$

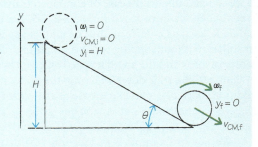

The kinetic energy is part translational and part rotational. We can use Equation 9-22 to write ω_f (the final angular velocity at the bottom of the ramp) in terms of $v_{CM,f}$ (the final linear speed at the bottom of the ramp) and so can express the final kinetic energy in terms of the linear speed $v_{CM,f}$ only.

Kinetic energy at the bottom of the ramp:

$$K_f = \frac{1}{2}Mv_{CM,f}^2 + \frac{1}{2}I_{CM}\omega_f^2$$

Rolling without slipping:

$$v_{CM,f} = R\omega_f, \text{ so}$$

$$\omega_f = \frac{v_{CM,f}}{R}$$

Substitute into kinetic energy equation:

$$K_f = \frac{1}{2}Mv_{CM,f}^2 + \frac{1}{2}I_{CM}\left(\frac{v_{CM,f}}{R}\right)^2 = \frac{1}{2}Mv_{CM,f}^2 + \frac{1}{2}\left(\frac{I_{CM}}{R^2}\right)v_{CM,f}^2$$

$$= \frac{1}{2}\left(M + \frac{I_{CM}}{R^2}\right)v_{CM,f}^2$$

We now have two expressions for K_f. Set them equal to each other and solve for $v_{CM,f}$.

$$\frac{1}{2}\left(M + \frac{I_{CM}}{R^2}\right)v_{CM,f}^2 = MgH$$

$$v_{CM,f}^2 = \frac{2MgH}{M + \dfrac{I_{CM}}{R^2}}$$

$$v_{CM,f} = \sqrt{\frac{2MgH}{M + \dfrac{I_{CM}}{R^2}}}$$

Finally, substitute $I_{CM} = MR^2/2$ for the disk and $I_{CM} = MR^2$ for the hoop and find the final speed for each object.

The speed of the disk at the bottom of the ramp ($\sqrt{(4/3)gH}$) is faster than that of the hoop ($\sqrt{gH}$). Both objects will accelerate down the ramp, but at any given position down the ramp, the disk will be moving faster than the hoop. In a race the disk would win.

For the uniform disk:

$$I_{CM} = \frac{MR^2}{2}$$

$$M + \frac{I_{CM}}{R^2} = M + \frac{M}{2} = \frac{3M}{2}$$

$$v_{CM,f} = \sqrt{\frac{2MgH}{3M/2}} = \sqrt{\left(\frac{2}{3M}\right)(2MgH)} = \sqrt{\frac{4gH}{3}}$$

For the hoop:

$$I_{CM} = MR^2$$

$$M + \frac{I_{CM}}{R^2} = M + M = 2M$$

$$v_{CM,f} = \sqrt{\frac{2MgH}{2M}} = \sqrt{gH}$$

Reflect

Why does the disk win? Both objects have the same mass and descend the same distance, so they both convert the same amount of gravitational potential energy into kinetic energy. But since the uniform disk has a smaller moment of inertia ($MR^2/2$) than does the hoop (MR^2), a smaller fraction of its kinetic energy is rotational and a larger fraction is translational. This means that the disk travels faster down the ramp and wins the race.

We can verify this by calculating the ratio of rotational kinetic energy $K_{rotational}$ to translational kinetic energy $K_{translational}$ for a rolling object. This ratio is I_{CM}/MR^2, so the smaller the moment of inertia I_{CM} for a given mass M and radius R—like the uniform disk compared to the hoop—the smaller the rotational kinetic energy and the more energy is available for translation.

Ratio of rotational kinetic energy to translational kinetic energy:

$$\frac{K_{rotational}}{K_{translational}} = \frac{\frac{1}{2}I_{CM}\omega^2}{\frac{1}{2}Mv_{CM}^2} = \frac{I_{CM}\omega^2}{Mv_{CM}^2}$$

If the object is rolling without slipping,

$$v_{CM} = R\omega \text{ and } \omega = \frac{v_{CM}}{R}, \text{ so}$$

$$\frac{K_{rotational}}{K_{translational}} = \frac{I_{CM}\omega^2}{Mv_{CM}^2} = \frac{I_{CM}}{Mv_{CM}^2}\left(\frac{v_{CM}}{R}\right)^2 = \frac{I_{CM}}{MR^2}$$

For the uniform disk:

$$\frac{K_{rotational}}{K_{translational}} = \frac{I_{CM}}{MR^2} = \frac{MR^2/2}{MR^2} = \frac{1}{2}$$

For the hoop:

$$\frac{K_{rotational}}{K_{translational}} = \frac{I_{CM}}{MR^2} = \frac{MR^2}{MR^2} = 1$$

Recall that we've already studied a uniform disk rolling down a ramp. In Example 9-10 we used Newton's second law to find the downhill acceleration $a_{CM,x}$ of the disk's center of mass. To check that our result in this example is consistent with Example 9-10, we can calculate $a_{CM,x}$ for the disk using our result for v_{CM} along with one of the equations for constant acceleration. Happily, we get the same answer $a_{CM,x} = \frac{2}{3}g\sin\theta$ as in Example 9-10.

The forces on the uniform disk (shown in Figure 9-26) are constant as it rolls downhill, so its acceleration must be constant. Take the positive x direction to be downhill. For constant acceleration,

$$v_{CM,x}^2 = v_{CM,x0}^2 + 2a_{CM,x}(x - x_0)$$
(2-11)

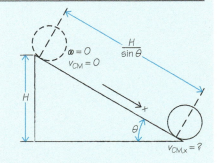

The disk starts from rest, so its initial velocity is $v_{CM,x0} = 0$; we calculated that at the bottom of the ramp, its velocity is $v_{CM,x} = \sqrt{4gH/3}$. From trigonometry, as the disk descends a distance H vertically its displacement along the ramp is $x - x_0 = H/\sin\theta$. Substitute into Equation 2-11 and solve for $a_{CM,x}$:

$$(\sqrt{4gH/3})^2 = 0 + 2a_{CM,x}(H/\sin\theta)$$

$$a_{CM,x} = \left(\frac{4gH}{3}\right)\left(\frac{\sin\theta}{2H}\right) = \frac{2}{3}g\sin\theta$$

GOT THE CONCEPT? 9-7 Does Size Matter?

In Example 9-13 suppose the uniform disk had half the radius and half the mass of the hoop. In a downhill race between these two, would the disk still win? Why or why not?

GOT THE CONCEPT? 9-8 A Three-Way Downhill Race

A uniform disk, a uniform sphere, and a block are released from rest down adjacent ramps of identical angle. The disk and sphere roll without slipping, while the block slides without friction. All three objects start from the same height. Rank the three objects in the order in which they reach the bottom of the ramp, starting with the first to arrive.

TAKE-HOME MESSAGE FOR Section 9-6

✔ The rotational kinetic energy of an object rotating around an axis depends on its angular speed and its moment of inertia for that axis.

✔ The total kinetic energy K of an object of mass M is the sum of the translational kinetic energy of its center of mass, $K_{\text{translational}} = \frac{1}{2}Mv_{\text{CM}}^2$, and the rotational kinetic energy of the object's rotation around its center of mass, $K_{\text{rotational}} = \frac{1}{2}I_{\text{CM}}\omega^2$.

✔ The same energy conservation equations we learned in Chapter 6 also apply when there is rotational motion. The only difference is that the kinetic energy K now includes both translational and rotational kinetic energies.

✔ If a perfectly rigid object rolls without slipping on a perfectly rigid surface, a static friction force is present but does no work on the object as it rolls.

9-7 Angular momentum is conserved when there is zero net torque on a system

When a spinning figure skater pulls her arms and legs in close to her body, the rate at which she spins automatically increases (see the photographs that open this chapter). You can demonstrate this same effect by sitting in an office chair that's free to rotate. Sit with your arms outstretched and hold a weight, like a brick or a full water bottle, in each hand. Now use your feet to start your body and the chair rotating, lift your feet off the ground, and then pull your arms inward. Your rotation will speed up quite noticeably!

Why does this happen? For both cases, the figure skater and you in the office chair, there's no torque acting to make the object rotate faster. Instead, both of these situations are examples of the *conservation of angular momentum*. This principle is the rotational analog of the conservation of *linear* momentum that we introduced in Chapter 8.

Angular Momentum and Angular Momentum Conservation

To see what we mean by angular momentum, consider an object rotating around a fixed z axis that doesn't change its orientation. The object's moment of inertia around this axis is I and its angular velocity is ω_z. We've seen that these quantities are analogous to the mass m and velocity v_x of an object in straight-line motion along the x axis. From Section 8-2, the linear momentum for such straight-line motion is

(9-27)
$$p_x = mv_x$$

By analogy to Equation 9-27 we'll define a new quantity called the **angular momentum** of an object rotating around the z axis. We'll give this the symbol L_z:

The **angular momentum** of a rigid object rotating around a fixed axis...

...is proportional to the **angular velocity** at which the object rotates around that axis.

Angular momentum for rotation around a fixed axis
(9-28)

$$L_z = I\omega_z$$

The angular momentum is also proportional to the object's **moment of inertia** for that rotation axis.

The greater the moment of inertia and the faster the angular velocity, the greater the value of angular momentum. Note that like angular velocity ω_z, angular momentum L_z can be positive or negative depending on which way the object is rotating and which direction of rotation you choose to be positive.

WATCH OUT! Angular momentum and linear momentum are different quantities.

Although linear momentum p_x and angular momentum L_z are analogous to each other, they are *not* the same quantity and do not have the same units. Linear momentum has units of mass times velocity, or kg·m/s. The units of angular momentum are the units of moment of inertia (kg·m^2) multiplied by the units of angular velocity (rad/s), which we can write as kg·m^2/s. (We can eliminate the "rad" since radians are not a true unit.)

Why is angular momentum important? We saw in Section 8-3 that the total *linear* momentum of a system of two or more objects is *conserved* if there is no net external force on the system. The analogous statement about angular momentum is

If there is no net external torque on a system, the angular momentum of the system is conserved.

This is the principle of **conservation of angular momentum**. It explains what happens to a spinning figure skater when she pulls her limbs in. As she does so her moment of inertia I decreases because more of her body's mass is closer to her rotation axis. There are no external torques on her—the force of gravity and the normal force don't have any effect on her rotation, and the friction force on her skates is very small. Hence her angular momentum L_z keeps the same value, and as I decreases her angular velocity ω_z has to increase so that the product $L_z = I\omega_z$ doesn't change. The same thing happens in the office chair experiment we described at the beginning of this section.

GOT THE CONCEPT? 9-9 Tetherball Physics

In the child's game of tetherball, a rope attached to the top of a tall pole is tied to a ball. Players hit the ball in opposite directions in an attempt to wrap the rope around the pole. As the ball wraps around the pole, does the angular speed of the ball (a) decrease, (b) stay the same, or (c) increase? Explain in terms of angular momentum. Treat the rope as having negligible mass and neglect any resistive forces (such as air resistance and friction).

EXAMPLE 9-14   Spinning Figure Skater

A figure skater executing a "scratch spin" gradually pulls her arms in toward her body while spinning on one skate. During the spin her angular speed increases from 1.50 rad/s (approximately one revolution every 4 seconds) to 15.0 rad/s (approximately 2.5 rev/s). (a) By what factor does her moment of inertia around her central axis change as she pulls in her arms? (b) By what factor does her rotational kinetic energy change?

Set Up

There is very little friction between the skater and the ice, so the net torque on her is essentially zero and her angular momentum can be considered constant. We'll use this to determine her moment of inertia after she pulls her arms in, I_{after}, in terms of her moment of inertia before pulling them in, I_{before}. (Note that we are given the values $\omega_{before} = 1.50$ rad/s and $\omega_{after} = 15.0$ rad/s.) Then we'll compare her rotational kinetic energy before and after pulling her arms in by using Equation 9-25.

Angular momentum of the skater:

$$L_z = I\omega_z \qquad (9\text{-}28)$$

Rotational kinetic energy of the skater:

$$K_{rotational} = \frac{1}{2}I\omega^2 \qquad (9\text{-}25)$$

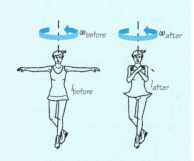

Solve

(a) Since angular momentum is conserved, we write an equation saying that the angular momentum after she pulls her arms in equals the angular momentum before she pulls them in. Then we solve for I_{after} in terms of I_{before}.

Conservation of angular momentum:

$$L_{after,z} = L_{before,z}$$

From Equation 9-28, this says

$$I_{after}\omega_{after,z} = I_{before}\omega_{before,z}$$

Let the direction of the skater's rotation be positive. Then her angular velocity ω_z at any instant is positive and equal to her angular speed ω at that instant. So the angular momentum conservation equation says

$$I_{after}\omega_{after} = I_{before}\omega_{before}$$

Rearrange to find the ratio of I_{after} to I_{before}:

$$\frac{I_{after}}{I_{before}} = \frac{\omega_{before}}{\omega_{after}}$$

$$= \frac{1.50 \text{ rad/s}}{15.0 \text{ rad/s}} = 0.100$$

The moment of inertia after she pulls her arms in is only 0.100 times (one-tenth) as large as the value before she pulls her arms in.

(b) Now we can use Equation 9-25 to compare the skater's rotational kinetic energy before and after she pulls her arms in.

$$K_{rotational,before} = \frac{1}{2}I_{before}\omega_{before}^2$$

$$K_{rotational,after} = \frac{1}{2}I_{after}\omega_{after}^2$$

Calculate the ratio of her rotational kinetic energy before and after:

$$\frac{K_{rotational,after}}{K_{rotational,before}} = \frac{(1/2)I_{after}\omega_{after}^2}{(1/2)I_{before}\omega_{before}^2} \quad \text{so}$$

$$\frac{K_{rotational,after}}{K_{rotational,before}} = \left(\frac{I_{after}}{I_{before}}\right)\left(\frac{\omega_{after}}{\omega_{before}}\right)^2 = (0.100)\left(\frac{15.0 \text{ rad/s}}{1.50 \text{ rad/s}}\right)^2$$

$$= (0.100)(10.0)^2 = 10.0$$

Her rotational kinetic energy after she pulls her arms in is 10.0 times as great as the value before she pulls her arms in.

Reflect

The skater's moment of inertia changes by a factor of $0.100 = 1/10.0$. This might seem large, because her arms are a relatively small fraction of her total mass. However, the moment of inertia depends on the square of distance (see Equation 9-15), so holding her arms close to her body rather than extended has a significant effect on the skater's moment of inertia.

Although the skater's angular momentum maintains the same value as she pulls her arms in, her rotational kinetic energy does not. You can see why by rewriting the expression for rotational kinetic energy in terms of the angular momentum magnitude $L = I\omega$ and the angular speed ω. Since L is unchanged and ω increases when the skater pulls her arms in, her rotational kinetic energy must increase. The extra kinetic energy comes from the skater herself: Chemical energy from food she ate is used to flex muscles and bring her arms in.

Moment of inertia of the skater:

$$I = \sum_{i=1}^{N} m_i r_i^2 \qquad (9\text{-}15)$$

Rotational kinetic energy:

$$K_{rotational} = \frac{1}{2}I\omega^2 = \frac{1}{2}(I\omega)\omega = \frac{1}{2}L\omega$$

The skater's angular speed ω increases while the magnitude L of her angular momentum stays the same, so $K_{rotational}$ increases.

Angular Momentum of a Particle

Example 9-14 demonstrates how the angular momentum concept helps us solve problems about rotating objects. It can also be useful to think about the angular momentum of a single *particle*. We've already seen the utility of this concept for the example of a tetherball moving around a pole (see Got the Concept? 9-9). Let's look at the general case of a particle of mass m moving with velocity $\vec{v}$ so that its linear momentum is $\vec{p} = m\vec{v}$. As an example, let's consider a girl (who we regard as a particle) running at constant speed toward a playground merry-go-round (Figure 9-28). We take the rotation axis to be the axis of the merry-go-round. At a given instant the vector $\vec{r}$ points directly away from the rotation axis to the position of the girl, and there is an angle ϕ between the vectors $\vec{r}$ and $\vec{p}$. What is the girl's angular momentum around this rotation axis?

In Equation 9-11 we defined the magnitude of the torque due to a force $\vec{F}$ as $\tau = r_\perp F$, where $r_\perp$ is the perpendicular distance from the rotation axis to the point where the force is applied. We define the angular momentum of a particle with linear momentum $\vec{p}$ in an analogous way:

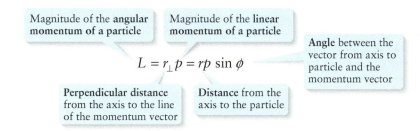

Magnitude of the **angular momentum of a particle**

Magnitude of the **linear momentum of a particle**

Angle between the vector from axis to particle and the momentum vector

$$L = r_\perp p = rp \sin \phi$$

Perpendicular distance from the axis to the line of the momentum vector

Distance from the axis to the particle

Magnitude of the angular momentum of a particle (9-29)

For example, in Figure 9-28 the perpendicular distance $r_\perp$ equals the radius R of the merry-go-round.

Does Equation 9-29 agree with our earlier definition $L_z = I\omega_z$ (Equation 9-28)? To find out, note that $p \sin \phi = mv \sin \phi = mv_\perp$, where $v_\perp$ is the component of the particle's velocity that's perpendicular to $\vec{r}$ in Figure 9-28. We can also write $v_\perp = r\omega$,

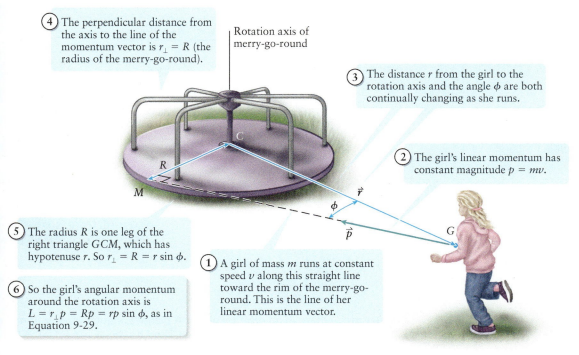

④ The perpendicular distance from the axis to the line of the momentum vector is $r_\perp = R$ (the radius of the merry-go-round).

Rotation axis of merry-go-round

③ The distance r from the girl to the rotation axis and the angle ϕ are both continually changing as she runs.

② The girl's linear momentum has constant magnitude $p = mv$.

⑤ The radius R is one leg of the right triangle GCM, which has hypotenuse r. So $r_\perp = R = r \sin \phi$.

① A girl of mass m runs at constant speed v along this straight line toward the rim of the merry-go-round. This is the line of her linear momentum vector.

⑥ So the girl's angular momentum around the rotation axis is $L = r_\perp p = Rp = rp \sin \phi$, as in Equation 9-29.

Figure 9-28 A girl running with linear momentum and angular momentum The angular momentum of the girl around the rotation axis is $L = r_\perp p = rp \sin \phi$ (see Equation 9-29).

where ω is the angular speed of the particle around the rotation axis. If we substitute these into Equation 9-29 for the magnitude of the particle's angular momentum, we get

(9-30)
$$L = rp \sin \phi = rmv \sin \phi = rmv_{\perp} = rm(r\omega) = mr^2\omega$$

Equation 9-15 tells us that if there's only a single particle of mass m a distance r from the rotation axis, its moment of inertia is $I = mr^2$. Then Equation 9-30 for the magnitude of the particle's angular momentum becomes $L = I\omega$. But this is just the magnitude of the angular momentum $L_z = I\omega_z$ as given by Equation 9-28. So Equation 9-29 for the angular momentum of a particle is completely consistent with our earlier work.

The girl in Figure 9-28 has a constant linear momentum because the net external force acting on her is zero. She also has a constant *angular* momentum: Figure 9-28 shows that her angular momentum is $L = r_{\perp}p = Rp$, where R is the radius of the merry-go-round. Since R and her linear momentum p are both constant, her angular momentum L is constant. That happens because there is also zero net *torque* acting on her.

If there *is* a net force that acts on a particle, but if that force is always directed toward a fixed point that acts as a rotation axis, that force exerts zero torque (there is a 180° angle between the vector $\vec{r}$ from axis to particle and the force vector $\vec{F}$ that points from particle to axis, so the torque is $\tau = rF \sin \phi = rF \sin 180° = 0$). In that case the angular momentum is again conserved. One important example of this is the motion of a planet in an elliptical orbit around the Sun (Figure 9-29). As the planet moves around its orbit, the distance r from the Sun to the planet and the planet's linear momentum $p = mv$ both change, but the planet's angular momentum $L = rp \sin \phi$ remains the same. Using more mathematics than we have space for here, we can show that this means that a line from the Sun to the planet sweeps out equal areas in equal intervals of time. That's just the law of equal areas, or Kepler's second law, that we encountered in Section 7-4. (Kepler deduced this law from observations of the planets. Isaac Newton used this law to conclude that the force on a planet must point directly toward the Sun.)

Another everyday example of conservation of angular momentum is the motion of water as it circles the drain of a bathtub. The water is pulled directly toward the drain, so the force on each parcel of water exerts no torque and the angular momentum of the water is conserved. As the water approaches the drain, $r_{\perp}$ decreases and so $p = mv$ increases to keep the angular momentum $L = r_{\perp}p$ constant. The same thing happens on a much more dramatic scale as air is drawn into the low pressure at the center of a hurricane. The circulating air gains tremendous speed, which gives the hurricane great destructive power.

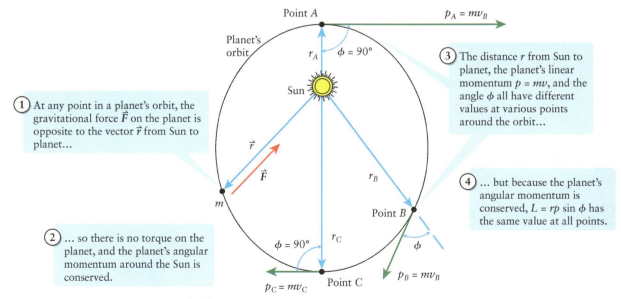

① At any point in a planet's orbit, the gravitational force $\vec{F}$ on the planet is opposite to the vector $\vec{r}$ from Sun to planet...

② ... so there is no torque on the planet, and the planet's angular momentum around the Sun is conserved.

③ The distance r from Sun to planet, the planet's linear momentum $p = mv$, and the angle ϕ all have different values at various points around the orbit...

④ ... but because the planet's angular momentum is conserved, $L = rp \sin \phi$ has the same value at all points.

Point A

$p_A = mv_B$

r_A $\phi = 90°$

Planet's orbit

Sun

$\vec{r}$

$\vec{F}$

m

$\phi = 90°$

r_C

$p_C = mv_C$

Point C

r_B

Point B

ϕ

$p_B = mv_B$

Figure 9-29 Angular momentum of an orbiting planet A planet's angular momentum is conserved as it orbits the Sun. At point A (closest to the Sun) $\phi = 90°$, $r_{\perp} = r \sin 90° = r$ has its minimum value, and the linear momentum $p = mv$ has its maximum value; at point C (farthest from the Sun) $\phi = 90°$, $r_{\perp} = r \sin 90° = r$ has its maximum value, and $p = mv$ has its minimum value.

Angular Momentum and a Rotational "Collision"

Here's another application of the angular momentum of a particle. The girl shown in Figure 9-28 runs across the playground and finally jumps onto the merry-go-round so that both end up rotating together. This seems very much like the inelastic collisions we discussed in Section 8-4, in which a moving object collided with an initially stationary one and the two stuck together afterward. We approached those inelastic collisions by demanding that linear momentum is conserved. In Figure 9-28, however, the girl has linear momentum before the collision, and the system of girl and merry-go-round has *zero* total linear momentum after the collision (the system rotates but does not move bodily from one place to another). So *linear* momentum is not conserved in this collision.

However, the *angular* momentum of the system of girl and merry-go-round *is* conserved because there are no external torques acting on the system around the merry-go-round's rotation axis. (The vertical force of gravity and normal force don't affect the rotation, and there is very little friction in the bearings of the merry-go-round.) During the collision the girl and the merry-go-round exert torques on each other, but these torques are *internal* to the system of girl plus merry-go-round, so they don't affect the *total* angular momentum. All that happens in the collision is that angular momentum is transferred between the girl and the merry-go-round. Example 9-15 shows how to analyze what happens.

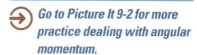

→ **Go to Picture It 9-2 for more practice dealing with angular momentum.**

EXAMPLE 9-15   A Girl on a Merry-Go-Round

The girl in Figure 9-28 has a mass of 30.0 kg. She runs toward the merry-go-round at 3.0 m/s, then jumps on. The merry-go-round is initially at rest, has a mass of 100 kg and a radius of 2.0 m, and can be treated as a uniform disk. Find the rotation speed of the merry-go-round after the girl jumps on.

Set Up

As we described above, angular momentum is conserved in this process. Before she jumps on, the girl has all of the angular momentum, with a magnitude given by Equation 9-29. After she jumps on, she and the merry-go-round rotate together with the same angular speed ω. Together they behave like a single object made up of a rotating disk with an extra mass (the girl) at its rim.

Angular momentum of the girl before jumping on:

$$L = rp_\perp = rp \sin \phi \qquad (9\text{-}29)$$

Angular momentum of the merry-go-round with the girl riding on it:

$$L_z = I\omega_z \qquad (9\text{-}28)$$

Moment of inertia of the merry-go-round plus girl:

$$I = I_{\text{merry-go-round}} + I_{\text{girl}} \qquad (9\text{-}15)$$

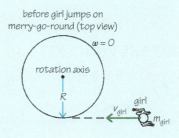

before girl jumps on
merry-go-round (top view)

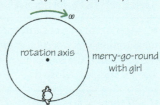

after girl jumps on (top view)

Solve

Figure 9-28 shows that the distance $r \sin \phi$ is just equal to the radius of the merry-go-round, $R = 2.0$ m. Use this to find the magnitude of the angular momentum before she jumps on.

Before the girl jumps on, her linear momentum has magnitude

$$p = m_{\text{girl}} v_{\text{girl}} = (30 \text{ kg})(3.0 \text{ m/s}) = 90 \text{ kg} \cdot \text{m/s}$$

Her angular momentum is

$$L = rp \sin \phi = (r \sin \phi)p$$
$$= Rp = (2.0 \text{ m})(90 \text{ kg} \cdot \text{m/s}) = 180 \text{ kg} \cdot \text{m}^2/\text{s}$$

After she jumps on, the magnitude of the angular momentum is $L = I\omega$, where ω is what we are trying to find. Calculate the value of I for the system of merry-go-round plus girl.

Moment of inertia of the merry-go-round alone, considered as a uniform cylinder: From Table 9-1,

$$I_{\text{merry-go-round}} = \frac{1}{2}M_{\text{merry-go-round}}R^2 = \frac{1}{2}(100 \text{ kg})(2.0 \text{ m})^2 = 200 \text{ kg}\cdot\text{m}^2$$

Moment of inertia of the girl alone, considered as an object of mass m_{girl} rotating a distance R from the axis:

$$I_{\text{girl}} = m_{\text{girl}}R^2 = (30 \text{ kg})(2.0 \text{ m})^2 = 120 \text{ kg}\cdot\text{m}^2$$

Moment of inertia of the merry-go-round plus girl:

$$I = I_{\text{merry-go-round}} + I_{\text{girl}} = 200 \text{ kg}\cdot\text{m}^2 + 120 \text{ kg}\cdot\text{m}^2 = 320 \text{ kg}\cdot\text{m}^2$$

Angular momentum is conserved, so set the magnitude of angular momentum before the girl jumps on equal to the magnitude after she jumps on. Then solve for the final angular speed ω. Note that if L is in kg·m²/s and I is in kg·m², then ω is in rad/s.

Before the girl jumps on, $L = 180 \text{ kg}\cdot\text{m}^2/\text{s}$.

After she jumps on, angular momentum is $L_z = I\omega_z$ with magnitude $L = I\omega = (320 \text{ kg}\cdot\text{m}^2)\omega$.

Angular momentum is conserved, so the magnitude L is the same before and after:

$$180 \text{ kg}\cdot\text{m}^2/\text{s} = (320 \text{ kg}\cdot\text{m}^2)\omega$$

$$\omega = \frac{180 \text{ kg}\cdot\text{m}^2/\text{s}}{320 \text{ kg}\cdot\text{m}^2} = 0.56 \text{ rad/s}$$

Reflect

To get a better sense of our result, we convert it to rev/min. Our answer says that the merry-go-round spins 5.4 times per minute. In other words, it makes one revolution every (1/5.4) minute, or about once every 11 s—a reasonable pace for a merry-go-round.

Convert ω to revolutions per minute:

$$\omega = (0.56 \text{ rad/s})\left(\frac{1 \text{ rev}}{2\pi \text{ rad}}\right)\left(\frac{60 \text{ s}}{1 \text{ min}}\right) = \frac{(0.56)(60)}{2\pi}\frac{\text{rev}}{\text{min}}$$

$$= 5.4 \text{ rev/min}$$

GOT THE CONCEPT? 9-10 Conserving angular momentum

 For which of the following objects is angular momentum conserved as it rotates? (a) The pulley in Example 9-8 (Section 9-5); (b) the disk in Example 9-10 (Section 9-5); (c) both the pulley and the disk; or (d) neither the pulley nor the disk.

TAKE-HOME MESSAGE FOR Section 9-7

✔ Just as an object's linear momentum is the product of its mass and velocity, a rotating object's angular momentum is the product of its moment of inertia around its rotation axis and its angular velocity.

✔ If there is no net external torque on a system, the angular momentum of the system is conserved. This holds true even if the system includes an object whose shape and moment of inertia change.

✔ A particle moving past a point has angular momentum relative to that point, even if the particle is not following a curved path.

9-8 Rotational quantities such as angular momentum and torque are actually vectors

Figure 9-30 shows an athlete doing the long jump. Although his body is essentially in an upright, running position at the beginning of the jump, while in midair his legs swing forward. By the time he lands, his legs are out in front of his torso. This significantly increases the distance of his jump. How does the athlete manage this in midair, with nothing to push against? What is the physics behind his motion?

The answer is that there was no net torque acting on the athlete during his leap, so his angular momentum was conserved. (Only the force of gravity acted on him. This acted at his center of mass and so exerted no torque around the center of mass.) As the athlete left the ground, the net angular momentum of his body around its center of mass was zero, or nearly so. Once in the air he rotated his legs up and forward (counterclockwise in Figure 9-30), giving them a nonzero angular momentum. For angular momentum to be conserved, some other part of his body had to rotate in such a way that the two contributions to his *net* angular momentum canceled. He rotated his arms rapidly clockwise while in the air, which helped rotate his upper torso clockwise in Figure 9-30. The angular momentum of the arms in one direction thus canceled the angular momentum of his legs in the opposite direction.

Angular Velocity and Angular Momentum as Vectors

The long jumper example uses the idea that angular momentum has a direction. This suggests that we should consider angular momentum as a *vector* $\vec{L}$. In what direction does $\vec{L}$ point? We need to account for the direction of rotation, but there is no way to align a single vector in the plane of the rotation to indicate this direction (Figure 9-31a). Hence we take the angular momentum vector to point along the *axis of rotation*, which we call the *z* axis.

By convention the specific direction of $\vec{L}$ is given by a **right-hand rule**. Curl the fingers of your right hand in the direction of motion and stick your thumb straight out. By the right-hand rule, your thumb points in the direction of the angular momentum vector (Figure 9-31b). We use this same rule to determine the direction of the angular *velocity* vector $\vec{\omega}$: This also points along the rotation axis. If the object rotates in the direction shown in Figure 9-31b, $\vec{\omega}$ and $\vec{L}$ point in the positive *z* direction and their components ω_z and L_z are positive; if the object rotates in the opposite direction, the vectors $\vec{\omega}$ and $\vec{L}$ point in the negative *z* direction and their components ω_z and L_z are negative. That's the origin of the positive and negative signs for ω_z and L_z that we introduced in Sections 9-2 and 9-7, respectively. (Strictly speaking, $\vec{L}$ points along the rotation axis only if the object's mass is arranged symmetrically around the rotation axis. We'll consider only situations of this kind, however, so you needn't worry about this point.)

How does this apply to the long jumper? If more than one element of a system is rotating, the net angular momentum $\vec{L}$ of the system is the vector sum of the angular momenta of each individual element. The angular momentum vector for the jumper's arms and upper torso has the same magnitude but the opposite direction to that of his legs, so the total $\vec{L}$ of his body is zero. The result is that both arms and legs are forward at the end of the jump, resulting in a longer flight.

Figure 9-30   Angular momentum during a long jump This long jumper rotates his legs counterclockwise (upward) by swinging his arms clockwise. In this way he keeps the total angular momentum of his body equal to zero.

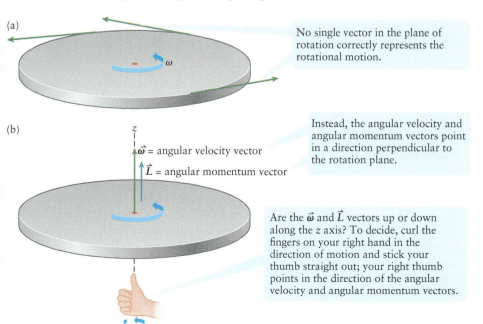

(a)

No single vector in the plane of rotation correctly represents the rotational motion.

(b)

z

$\vec{\omega}$ = angular velocity vector

$\vec{L}$ = angular momentum vector

Instead, the angular velocity and angular momentum vectors point in a direction perpendicular to the rotation plane.

Are the $\vec{\omega}$ and $\vec{L}$ vectors up or down along the *z* axis? To decide, curl the fingers on your right hand in the direction of motion and stick your thumb straight out; your right thumb points in the direction of the angular velocity and angular momentum vectors.

Figure 9-31   The angular velocity and angular momentum vectors Rather than lying in (a) the plane of rotation, the angular velocity and angular momentum vectors of a rotating object are (b) oriented along the rotation axis. The directions of $\vec{\omega}$ and $\vec{L}$ are given by a right-hand rule.

Figure 9-32 A falling cat (a) A falling cat usually lands on its feet, which means rotating its body in midair. (b) To accomplish this rotation, this cat uses the vector nature of angular momentum. (To see which direction each part of the cat is rotating, point your right thumb along the direction of the associated angular momentum vector. The fingers of your right hand will then curl in the direction of the rotation.)

(a) Stage in the cat's fall   (b) How the cat changes its orientation while falling

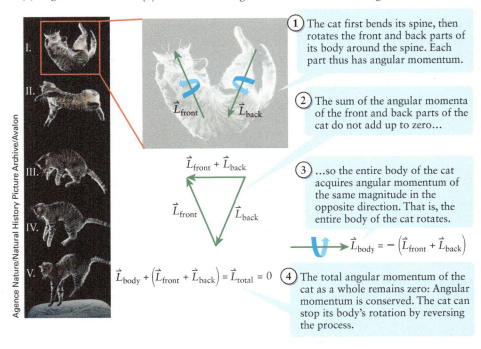

Agence Nature/Natural History Picture Archive/Avalon

1. The cat first bends its spine, then rotates the front and back parts of its body around the spine. Each part thus has angular momentum.

$\vec{L}_{front}$ $\vec{L}_{back}$

2. The sum of the angular momenta of the front and back parts of the cat do not add up to zero...

$\vec{L}_{front} + \vec{L}_{back}$

$\vec{L}_{front}$ $\vec{L}_{back}$

3. ...so the entire body of the cat acquires angular momentum of the same magnitude in the opposite direction. That is, the entire body of the cat rotates.

$\vec{L}_{body} = - \left(\vec{L}_{front} + \vec{L}_{back} \right)$

$\vec{L}_{body} + \left(\vec{L}_{front} + \vec{L}_{back} \right) = \vec{L}_{total} = 0$

4. The total angular momentum of the cat as a whole remains zero: Angular momentum is conserved. The cat can stop its body's rotation by reversing the process.

BioMedical

A falling cat (Figure 9-32a) also uses conservation of angular momentum to land on its feet. As for the long jumper in Figure 9-30, the only external force acting on the cat as it falls is the force of gravity acting at its center of mass. So just as for the long jumper, there is zero net external torque around the center of mass and the total angular momentum $\vec{L}$ of the cat is conserved. If the cat is initially not rotating, $\vec{L}$ for the cat as a whole is zero and so must remain zero. But unlike the long jumper, the cat has an extraordinarily flexible spine. By bending in the middle (stage I in Figure 9-32a) the cat divides its body into two parts, each of which can rotate around a different axis. The cat uses its muscles to rotate its front part and rear part around the spine. To cancel the angular momentum of these rotations (Figure 9-32b), the entire body of the cat rotates in the opposite direction (stage II in Figure 9-32a). When the legs are pointed downward (stage III in Figure 9-32a), the front and rear parts rotate the opposite direction from before to stop the rotation. The cat is now ready to make a feet-first landing (stages IV and V in Figure 9-32a).

Torque as a Vector

Like angular momentum $\vec{L}$, torque can be expressed as a vector $\vec{\tau}$ that has both a magnitude and a direction. Equation 9-10 from Section 9-3 tells us the magnitude of the torque vector:

(9-10)
$$\tau = rF \sin \phi$$

In Equation 9-10, r is the position from the rotation axis to the point where a force of magnitude F is applied. Both the position $\vec{r}$ and the force $\vec{F}$ are vectors, and ϕ is the angle between $\vec{r}$ and $\vec{F}$. The direction of $\vec{\tau}$ is given by a right-hand rule that's similar to the one we described to find the direction of the angular momentum or angular velocity vectors: Curl the fingers of your right hand in the direction that the force $\vec{F}$ would tend to make the object rotate, and your right thumb will point in the direction of $\vec{\tau}$ (Figure 9-33). Alternatively, point the fingers of your right hand along the direction of $\vec{r}$ so that your palm faces the vector $\vec{F}$ then curl your fingers from the direction of $\vec{r}$ to the direction of $\vec{F}$ along the shortest path. Your extended right thumb then points in the direction of $\vec{\tau}$. (Practice this with the two situations shown in Figure 9-33.)

Note that $\vec{\tau}$ is perpendicular to the plane defined by the vectors $\vec{r}$ and $\vec{F}$. Mathematically, a combination of two vectors $\vec{r}$ and $\vec{F}$ that has these properties is called the **cross product** or **vector product** of $\vec{r}$ and $\vec{F}$. You can learn more about the properties of the cross product in the Math Tutorial. In this text, though, we will rely on Equation 9-10 to calculate the magnitude of the torque and on the right-hand rule to determine its direction.

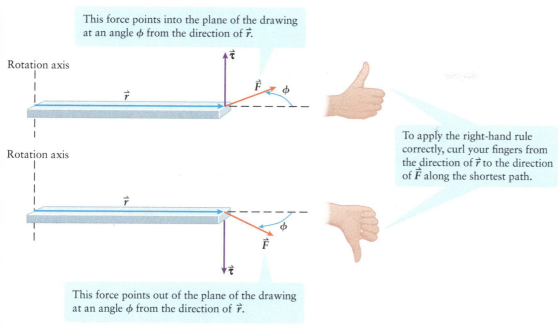

This force points into the plane of the drawing at an angle ϕ from the direction of $\vec{r}$.

To apply the right-hand rule correctly, curl your fingers from the direction of $\vec{r}$ to the direction of $\vec{F}$ along the shortest path.

This force points out of the plane of the drawing at an angle ϕ from the direction of $\vec{r}$.

Figure 9-33 **The torque vector** A force $\vec{F}$ acts on an object at a point that is separated from the rotation axis by a vector $\vec{r}$. The torque $\vec{\tau}$ produced by this force is a vector that is perpendicular to both $\vec{r}$ and $\vec{F}$. The magnitude of $\vec{\tau}$ is $\tau = rF\sin\phi$, and the direction of $\vec{\tau}$ is given by a right-hand rule.

WATCH OUT! The direction of torque is not the same as the direction of motion.

The torques shown in Figure 9-33 are directed either straight up or straight down, but they don't cause the bar on which the torque acts to *move* either up or down.

Instead, the bar rotates around the rotation axis. In fact, there is *no* up or down motion at all in Figure 9-33! The direction of the torque vector is simply a mathematical tool.

If the net torque on an object is zero—that is, if the *vector* sum $\sum \vec{\tau}$ of all external torques acting on the object is zero—the angular momentum of the object is conserved and its angular acceleration is zero. That's the case for the falling cat in Figure 9-32. (It's true that there are *internal* torques exerted by one part of the cat on another to cause the twisting motions shown in Figure 9-32b. But thanks to Newton's third law, if one part of the cat exerts a torque $\vec{\tau}$ on a second part, the second part will exert a torque $-\vec{\tau}$ on the first part with the same magnitude but the opposite direction. Hence these internal torques cancel out.)

Torques, Forces, and Equilibrium

An important application of the idea of torque is to an object in **equilibrium**. This means that (i) the object's center of mass has zero acceleration and (ii) the object is not rotating. Figure 9-34a shows a colorful example: a flamingo balancing on one leg. For the flamingo's center of mass to remain at rest, the sum of the external *forces* must be zero. Hence in Figure 9-34b the upward normal force $\vec{n}$ exerted by the ground must have the same magnitude as the downward gravitational force $\vec{w}$. In addition, for the bird not to start rotating and topple over, the sum of the net external *torques* must also be zero. Since $\vec{w}$ acts at the center of mass, it exerts no torque around that point; the same is true for the normal force $\vec{n}$ if the flamingo positions itself with its center of mass directly over its foot, since then the line of action of $\vec{n}$ passes directly through the center of mass. Hence the net external torque on the flamingo is zero, and it remains in equilibrium. (Try balancing on one foot; you'll find that like the flamingo, you must position your center of mass over your foot to keep from falling over.)

In *any* problem where an object is in equilibrium, you must satisfy the equilibrium conditions that the sum of the external forces and the sum of the external torques are both zero. The following examples illustrate how to use these conditions.

Figure 9-34
**A flamingo
in equilibrium** (a) Flamingos
like this one (*Phoenicopterus*)
typically stand on just one leg.
(One proposed explanation for
this behavior is that given its
anatomy, a flamingo expends less
energy to maintain a one-legged
stance than a two-legged one.)
(b) To remain in equilibrium, both
the net external force and the net
external torque on the flamingo
must be zero.

(a) A flamingo in equilibrium   (b) Forces on the flamingo

The flamingo
positions its
center of mass
(CM) directly
above its
foot...

Line of action
of $\vec{n}$

CM

$\vec{w}$

...so the line of
action of the
normal force $\vec{n}$
passes through
the CM.

$\vec{n}$

robertharding/Alamy

EXAMPLE 9-16   Balancing on a Seesaw

A seesaw is a uniform plank supported at its midpoint. Akeelah and her little sister Bree sit on opposite sides of the
seesaw. Akeelah weighs 1.50 times as much as Bree, and Akeelah sits 80.0 cm from the midpoint of the seesaw. If the
seesaw is tilted by an angle θ from the horizontal, where should Bree sit so that they just balance?

Set Up

When the system made up of Akeelah, Bree,
and the seesaw is balanced, the system is in
equilibrium: It has no tendency to start rotating
one way or the other. Hence the net torque on
the system must be zero.

We are told that Akeelah's distance from
the pivot is $r_A = 80.0$ cm $= 0.800$ m and that
Akeelah's weight is 1.50 times that of Bree:
that is, $w_A = 1.50 w_B$. Our goal is to find the
distance r_B from the pivot to Bree that will
satisfy the condition of zero torque. We'll
write this condition in terms of the torque
vector, using Equation 9-10 to calculate
the magnitudes of the torques, and the
right-hand rule to determine their
directions.

Condition that the net
external force is zero:

$$\sum \vec{F} = 0$$

Condition that the net
torque is zero:

$$\sum \vec{\tau} = 0$$

Magnitude of the torque:

$$\tau = rF \sin \phi \qquad (9\text{-}10)$$

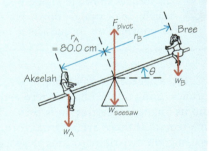

Solve

Neither the weight of the seesaw nor the
support force exerts any torque on the system
around the pivot point, since both of these
forces act at the pivot and have $\vec{r} = 0$.
The only torques are due to the weight of
Akeelah, $\vec{w}_A$, and the weight of Bree, $\vec{w}_B$. The
right-hand rule tells us that the corresponding
torques $\vec{\tau}_A$ and $\vec{\tau}_B$ are in opposite directions,
out of and into the page, respectively, so their
vector sum can be zero (as it must be to keep
the seesaw in balance).

Net torque on the system of
Akeelah, Bree, and the seesaw:

$$\sum \vec{\tau} = \vec{\tau}_A + \vec{\tau}_B = 0$$

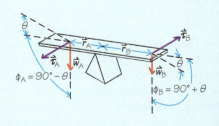

For the net torque to be zero, the opposite torques $\vec{\tau}_A$ and $\vec{\tau}_B$ must have the same magnitude. The angle between $\vec{r}_A$ and $\vec{w}_A$ is $\phi_A = 90° - \theta$, and the angle between $\vec{r}_B$ and $\vec{w}_B$ is $\phi_B = 90° + \theta$. From trigonometry, $\sin(90° - \theta) = \sin(90° + \theta) = \cos\theta$.

Opposite torques $\vec{\tau}_A$ and $\vec{\tau}_B$ have the same magnitude:

$$\tau_A = \tau_B$$

$$r_A w_A \sin\phi_A = r_B w_B \sin\phi_B$$

Substitute $\phi_A = 90° - \theta$, $\phi_B = 90° + \theta$:

$$r_A w_A \sin(90° - \theta) = r_B w_B \sin(90° + \theta)$$

$$r_A w_A \cos\theta = r_B w_B \cos\theta$$

$$r_A w_A = r_B w_B$$

Solve for Bree's distance from the pivot, r_B.

From $r_A w_A = r_B w_B$, $r_B = r_A \dfrac{w_A}{w_B}$

Substitute $r_A = 0.800$ m, $w_A = 1.50 w_B$:

$$r_B = r_A \frac{w_A}{w_B} = (0.800 \text{ m})\left(\frac{1.50 w_B}{w_B}\right) = (0.800 \text{ m})(1.50)$$

$$= 1.20 \text{ m}$$

Bree must sit 1.20 m from the pivot to balance Akeelah.

Reflect

Notice the significance of the lever arm in this problem. Bree, who is lighter than Akeelah, can create a balance by sitting farther from the center. A small force can give rise to a large torque when the lever arm is large. Bree has $1/1.50$ the weight of Akeelah, so Bree's lever arm must be 1.50 times greater than Akeelah's and she must sit 1.50 times as far from the pivot as Akeelah.

Note also that the answer doesn't depend on the angle θ. The greater the value of θ, the shorter each girl's lever arm, but the *ratio* of the two lever arms is the same for any value of θ. So the seesaw will balance no matter what the angle θ.

Finally, note that the center of mass of the system of Akeelah, Bree, and the seesaw is located at the pivot point. (Can you show this?) That means that the pivot force $\vec{F}_{pivot}$ acts at the center of mass, and so exerts no torque around that point. Hence just like the flamingo in Figure 9-34, the seesaw is in equilibrium.

In this example we didn't have to use the condition that the net *force* on the system is zero. Can you use this to find the magnitude of the pivot force in terms of the weights of Akeelah, Bree, and the seesaw?

Lever arm for Akeelah:

$$r_{A\perp} = r_A \cos\theta$$

Lever arm for Bree:

$$r_{B\perp} = r_B \cos\theta$$

Ratio of the two lever arms:

$$\frac{r_{B\perp}}{r_{A\perp}} = \frac{r_B \cos\theta}{r_A \cos\theta} = \frac{r_B}{r_A} = 1.50$$

This is the same for any value of θ.

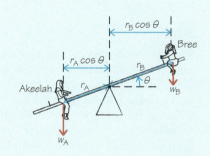

EXAMPLE 9-17 Leaning on a Wall

You lean a ladder of length L and weight w against a vertical wall so that it is at an angle θ from the horizontal. The vertical wall has very little friction. To keep the ladder from sliding, what must be the magnitude of the friction force that the floor exerts on the ladder?

Set Up

The diagram shows the forces acting on the ladder. We take the positive x direction to be the right, the positive y direction to be upward, and the positive z direction to be out of the plane of the drawing so that counterclockwise rotation is positive. There are three unknown force magnitudes: the normal force n_{wall} exerted by the frictionless vertical wall, the normal force n_{floor} exerted by the floor, and the static friction force f_s exerted by the floor. (This must be *static* friction because the ladder doesn't slide.) Our goal is find f_s.

Since we have three unknowns, we need three equations. These will be the x and y components of the force equation $\sum \vec{F} = 0$ and the z component of the torque equation $\sum \vec{\tau} = 0$. It's convenient to write the magnitude of each torque in terms of its lever arm around the center of mass; we'll do this using Equation 9-11.

Condition that the net external force is zero:

$$\sum \vec{F} = 0$$

Condition that the net torque is zero:

$$\sum \vec{\tau} = 0$$

Magnitude of the torque in terms of lever arm:

$$\tau = r_\perp F \qquad (9\text{-}11)$$

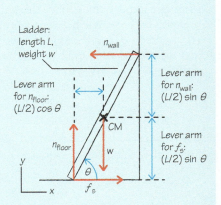

Ladder: length L, weight w

n_{wall}

Lever arm for n_{floor}: $(L/2)\cos\theta$

Lever arm for n_{wall}: $(L/2)\sin\theta$

CM

n_{floor}

w

Lever arm for f_s: $(L/2)\sin\theta$

y

θ

x

f_s

Solve

Write the equations for the net force and the net torque around the center of mass in component form. Note that the normal force exerted by the wall and the friction force both produce positive (counterclockwise) torques around the center of mass, while the normal force exerted by the floor produces a negative (clockwise) torque. The gravitational force acts at the center of mass and so produces no torque.

Components of the net force on the ladder:

$$\sum F_x = f_s - n_{\text{wall}} = 0$$
$$\sum F_y = n_{\text{floor}} - w = 0$$

z component of the net torque on the ladder around its center of mass:

$$\sum \tau_z = \left(\frac{L}{2}\sin\theta\right)n_{\text{wall}} + \left(\frac{L}{2}\sin\theta\right)f_s - \left(\frac{L}{2}\cos\theta\right)n_{\text{floor}} = 0$$

Solve the force equations for n_{wall} in terms of f_s and n_{floor} in terms of w.

From the x force equation,

$$n_{\text{wall}} = f_s$$

From the y force equation,

$$n_{\text{floor}} = w$$

Substitute our results for n_{wall} and n_{floor} into the torque equation and solve for f_s.

The torque equation becomes

$$\left(\frac{L}{2}\sin\theta\right)f_s + \left(\frac{L}{2}\sin\theta\right)f_s - \left(\frac{L}{2}\cos\theta\right)w = 0$$

Divide through by the common factor $L/2$, combine the first two terms, and solve for f_s:

$$2f_s\sin\theta - w\cos\theta = 0$$
$$2f_s\sin\theta = w\cos\theta$$
$$f_s = \frac{w\cos\theta}{2\sin\theta} = \frac{w}{2\tan\theta}$$

(Recall that $\dfrac{\sin\theta}{\cos\theta} = \tan\theta$.)

Reflect

Our result shows that the required friction force f_s is proportional to the weight w of the ladder: For a given angle θ, a heavier ladder requires more friction to keep it from sliding. The friction force is also *inversely* proportional to $\tan \theta$. Since the tangent function increases as θ increases from $0°$ to $90°$, less friction is needed for large values of θ (that is, when the ladder is more nearly vertical).

We can check our result by repeating our calculation using a different choice of rotation axis. If the ladder is in equilibrium, it cannot rotate around *any* axis, so we are free to choose any axis we wish. (We don't have that freedom if the object is rotating around a fixed axis or simultaneously translating and rotating.) We'll take our new axis to be perpendicular to the plane of the drawing and passing through the point where the ladder touches the floor. With this choice neither n_{floor} nor f_s produces any torque, while n_{wall} produces a counterclockwise (positive) torque and w produces a clockwise (negative) torque.

z component of the net torque on the ladder around the lower end of the ladder:

$$\sum \tau_z = (L \sin \theta)n_{wall}$$
$$- \left(\frac{L}{2} \cos \theta\right)w = 0$$

Substitute $n_{wall} = f_s$ from the x component of the force:

$$(L \sin \theta)f_s - \left(\frac{L}{2} \cos \theta\right)w = 0$$

Divide through by $L/2$ and solve for f_s:

$$2f_s \sin \theta - w \cos \theta = 0$$
$$f_s = \frac{w \cos \theta}{2 \sin \theta} = \frac{w}{2 \tan \theta}$$

This is the same answer as we found by calculating torques around an axis through the center of mass.

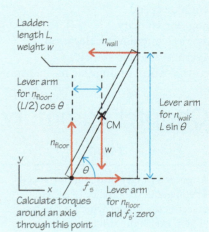

Ladder: length L, weight w

n_{wall}

Lever arm for n_{floor}: $(L/2) \cos \theta$

Lever arm for n_{wall}: $L \sin \theta$

CM

n_{floor}

w

f_s

Lever arm for n_{floor} and f_s: zero

Calculate torques around an axis through this point

TAKE-HOME MESSAGE FOR Section 9-8

✔ The angular velocity and angular momentum of a rotating object are vectors that lie along the object's rotation axis. The direction of these vectors is determined by a right-hand rule.

✔ Torque is also a vector whose direction is determined by a right-hand rule. If the object is free to rotate around an axis, the torque around that axis points along the axis.

✔ If an object is in equilibrium, the net external force and the net external torque on it must both be zero.

GOT THE CONCEPT? 9-11 Torque and angular momentum vectors

? In each of the four cases shown in Figure 9-35, a uniform disk is rotating around an axis through its center and perpendicular to the plane of the drawing, with a force of magnitude F applied on its rim. In each case the force is in the plane of the drawing. (a) In which case(s) is the angular momentum $\vec{L}$ of the disk out of the plane of the picture? In which case(s) is $\vec{L}$ into the plane? (b) In which case(s) is the torque $\vec{\tau}$ due to the force disk out of the plane of the picture? In which case(s) is $\vec{\tau}$ into the plane? In which case(s) is $\vec{\tau}$ zero? (c) In which case(s) does the torque cause the rotation to speed up? In which case(s) does it cause the rotation to slow down? In which case(s) does the rotation speed stay the same?

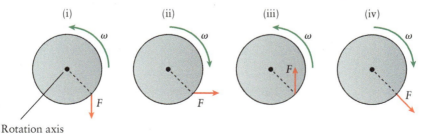

Figure 9-35 **Forces on a rotating disk** What are the directions of the angular momentum and torque vectors in each case?

Key Terms

angular acceleration
angular displacement
angular momentum
angular position
angular speed
angular velocity
average angular acceleration
average angular velocity
conservation of angular momentum

cross product (vector product)
equilibrium
lever arm
line of action
moment of inertia
parallel-axis theorem
right-hand rule
rigid object
rolling without slipping

rotation
rotational kinematics
rotational kinetic energy
tangential acceleration
torque
translation
vector product (cross product)

Chapter Summary

Topic	Equation or Figure
Angular velocity and angular speed: The angular velocity ω_z of a rigid object is the rate at which it rotates. (The z axis is the rotation axis of the object.) Angular velocity can be positive or negative, depending on which direction of rotation you choose to be positive. Angular speed ω is the magnitude of angular velocity. A point a distance r from the rotation axis has speed $v = r\omega$.	(1) The position of the blade at a certain time (2) The blade rotates around an axis that is perpendicular to the plane of this figure. We call this the z axis. Angular displacement = $\Delta\theta$ Length of arc = $r\,\Delta\theta$ (3) The position of the blade a time Δt later Distance r from axis (Figure 9-3) (4) During time Δt, the rigid blade rotates through an angular displacement $\Delta\theta$. We choose the positive direction of rotation to be counterclockwise, so $\Delta\theta > 0$. (5) The **angular velocity** ω_z of the blade is the angular displacement $\Delta\theta$ divided by the elapsed time Δt: $$\omega_z = \frac{\Delta\theta}{\Delta t}$$ The **angular speed** ω is the absolute value of ω_z. (6) In time Δt, a point on the blade a distance r from the axis moves through an arc of length $r\,\Delta\theta$. The speed of this point is $v = r\omega$.
Rotational kinematics: Angular acceleration is to angular velocity as ordinary acceleration is to velocity. If the angular acceleration of a rigid rotating object is constant, three basic equations relate the time, angular position, angular velocity, and angular acceleration of the object.	Angular velocity at time t of a rotating object with constant angular acceleration Angular velocity at time $t = 0$ of the object $$\omega_z = \omega_{0z} + \alpha_z t \qquad (9\text{-}5)$$ Constant angular acceleration of the object Time at which the object has angular velocity ω_z Angular position at time t of a rotating object with constant angular acceleration Angular velocity at time $t = 0$ of the object Constant angular acceleration of the object $$\theta = \theta_0 + \omega_{0z} t + \frac{1}{2}\alpha_z t^2 \qquad (9\text{-}6)$$ Angular position at time $t = 0$ of the object Time at which the object is at angular position θ Angular velocity at angular position θ of a rotating object with constant angular acceleration Angular velocity at angular position θ_0 of the object $$\omega_z^2 = \omega_{0z}^2 + 2\alpha_z(\theta - \theta_0) \qquad (9\text{-}7)$$ Constant angular acceleration of the object Two angular positions of the object

Views of object from along its rotation axis

(a) A force that tends to cause clockwise rotation

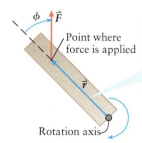

Point where force is applied

Rotation axis

The vector $\vec{r}$ points from an object's rotation axis to where a force $\vec{F}$ is applied to the object.

(b) A force that tends to cause counterclockwise rotation

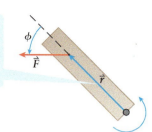

(c) Determining the line of action of a force and its lever arm

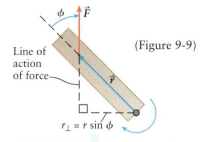

(Figure 9-9)

Line of action of force

$r_\perp = r \sin \phi$

Torque: If a force $\vec{F}$ acts at a position $\vec{r}$ from an object's rotation axis, it can change the rotation of the object. The torque associated with the force has magnitude $\tau = rF \sin \phi = r_\perp F$, where $r_\perp$ is the lever arm of the force. The torque τ_z can be positive or negative.

The lever arm $r_\perp$ is perpendicular to the line of action of the force, and extends from the rotation axis to the line of action of the force.

Moment of inertia: The quantity that plays the role of mass for rotational motion is an object's moment of inertia. This depends on the mass of the object but also on how the components of the object are positioned relative to the rotation axis. (Table 9-1 lists the moments of inertia for various common shapes.)

To find the **moment of inertia** I of an object...

...we imagine dividing the object into N **small pieces**, and calculate the sum...

$$I = \sum_{i=1}^{N} m_i r_i^2 \qquad (9\text{-}15)$$

...of the product of the **mass** m_i of the ith piece...

...and the square of the **distance** r_i from the ith piece to the rotation axis.

Parallel-axis theorem: If an object of mass M has a moment of inertia I_{CM} for an axis through its center of mass, its moment of inertia for a second axis parallel to the center-of-mass axis and separated by a distance h is $I = I_{CM} + Mh^2$.

(a) Hoop rotating around an axis through its center

Rotation axis

$I = I_{CM}$

Hoop, mass M

R

Center of mass of the hoop

(b) Hoop rotating around an axis through its rim

Rotation axis

$I = I_{CM} + Mh^2$

$h = R$

(Figure 9-18 a and b)

Rotational form of Newton's second law: A net external torque on an object gives the object an angular acceleration. To analyze an object that is able to translate as well as rotate (such as one that rolls without slipping), both Newton's second law in its original form $\left(\sum \vec{F} = m\vec{a} \right)$ and this rotational form must be used.

The **net torque**, or sum of all external torques acting on an object for the axis about which it can rotate

The **angular acceleration** produced by the net torque

$$\sum \tau_{\text{ext}, z} = I\alpha_z \qquad (9\text{-}16)$$

The object's **moment of inertia** for the axis about which it can rotate

Rolling without slipping: If an object rolls on a surface without slipping, the speed of the center of mass of the object must be related to the angular speed of the object's rotation. And the acceleration of its center of mass must be related to its angular acceleration. A friction force acts on this object to maintain this relationship. If the object is completely rigid, however, this friction force does no work on the object.

Speed of the center of mass of a rolling object

$$v_{CM} = R\omega \tag{9-22}$$

Radius of the object   Angular speed of the object's rotation

Acceleration of the center of mass of a rolling object

$$a_{CM,x} = R\alpha_z \tag{9-23}$$

Radius of the object   Angular acceleration of the object's rotation

Rotational kinetic energy, total kinetic energy, and conservation of mechanical energy: The rotational kinetic energy of an object is related to its moment of inertia and angular speed. The total kinetic energy K of an object is the sum of its translational kinetic energy and rotational kinetic energy. With this expression for K, the same equations for conservation of energy that we learned in Chapter 6 apply for an object that is rotating, translating, or both.

Rotational kinetic energy of a rigid object spinning around an axis   Moment of inertia of the object for that rotation axis

$$K_{rotational} = \frac{1}{2} I\omega^2 \tag{9-25}$$

Angular speed of the object

① Total kinetic energy of a system that is both translating and rotating   ② The **translational kinetic energy** is the same as if all of the mass of the system were moving at the speed of the center of mass.

$$K = K_{translational} + K_{rotational} = \frac{1}{2} Mv_{CM}^2 + \frac{1}{2} I_{CM}\omega^2 \tag{9-26}$$

③ The **rotational kinetic energy** is the same as if the system were not translating, and all of the mass were rotating around the center of mass.

Angular momentum and the conservation of angular momentum: A rotating rigid object has an angular momentum L_z whose sign depends on the direction of rotation. A particle moving relative to an axis also has angular momentum around that axis. If there is no net external torque on an object or system of objects, its angular momentum remains constant.

The **angular momentum** of a rigid object rotating around a fixed axis...   ...is proportional to the **angular velocity** at which the object rotates around that axis.

$$L_z = I\omega_z \tag{9-28}$$

The angular momentum is also proportional to the object's **moment of inertia** for that rotation axis.

Magnitude of the **angular momentum of a particle**   Magnitude of the **linear momentum of a particle**

Angle between the vector from axis to particle and the momentum vector

$$L = r_\perp p = rp \sin \phi \tag{9-29}$$

Perpendicular distance from the axis to the line of the momentum vector   Distance from the axis to the particle

Rotational quantities as vectors:
The angular velocity $\vec{\omega}$, angular momentum $\vec{L}$, and torque $\vec{\tau}$ are actually all vector quantities that lie along the rotation axis. Their directions are given by a right-hand rule.

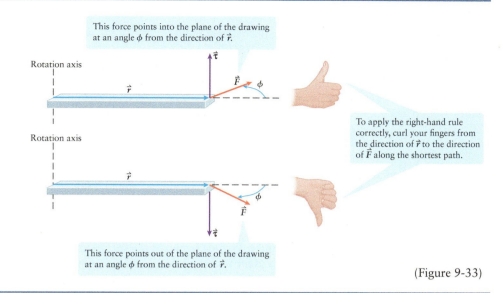

This force points into the plane of the drawing at an angle ϕ from the direction of $\vec{r}$.

Rotation axis

Rotation axis

To apply the right-hand rule correctly, curl your fingers from the direction of $\vec{r}$ to the direction of $\vec{F}$ along the shortest path.

This force points out of the plane of the drawing at an angle ϕ from the direction of $\vec{r}$.

(Figure 9-33)

Equilibrium: An object in equilibrium is one that does not rotate and whose center of mass does not accelerate. For an object to be in equilibrium, the net external force and net external torque on the object must both be zero.

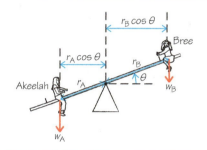

Answer to What do you think? Question

(a) The skater's rotational kinetic energy increases. There are no external torques acting on the figure skater, so her angular momentum $L_z = I\omega_z$ —the product of her moment of inertia I and her angular velocity ω_z —remains the same as she pulls her arms in. Since her moment of inertia I decreases, her angular velocity ω_z increases. Her rotational kinetic energy is

$K_{\text{rotational}} = \frac{1}{2}I\omega_z^2 = \frac{1}{2}(I\omega_z)\omega_z = \frac{1}{2}L_z\omega_z$; since L_z remains constant and ω_z increases, it follows that $K_{\text{rotational}}$ increases. The work done to provide the extra rotational kinetic energy comes from the skater herself, who must do work to pull her arms in (see Example 9-14).

Answers to Got the Concept? Questions

9-1 (c), then (d) and (e) are tied for second place, then (a), then (b). The angular velocity ω_z and angular acceleration α_z have the same sign for objects (c) and (e), so the rotation is speeding up for both objects; (c) is speeding up at a greater rate because the absolute value of α_z is greater than for (e). Object (d) is at rest and has a nonzero α_z, so it must also be speeding up; its α_z has the same absolute value as object (e), so these two objects are speeding up at the same rate. Objects (a) and (b) are slowing down because ω_z and α_z have opposite signs; object (b) is slowing down at a faster rate because it has a larger absolute value of α_z.

9-2 Extending the handle of the wrench with the pipe lengthens the lever arm, because it increases the distance between the rotation axis and the point where the force is applied.

Equation 9-11 shows that torque increases as the lever arm increases, even when the force remains the same, making it more likely that the bolt will rotate. The downside is that with this arrangement, applying even a small force may result in a torque large enough to break the bolt or the wrench!

9-3 (b) The torque required to create a certain angular acceleration is proportional to the moment of inertia (Equation 9-16). Because I_y is only about one-third as great as I_z, up-and-down motions (rotation around the y axis) are easier than side-to-side motions (rotation around the z axis). About three times as much torque is required to rotate the jaw from side to side (rotation around the z axis) as compared to the torque required to cause the jaw to open and close (rotation around the y axis).

9-4 (c) Table 9-1 shows that the moment of inertia of a thin, uniform rod rotating around its center is $I_{\text{rod}} = ML^2/12$. Because the moment of inertia is additive, the moment of inertia of the two rods together is $I = 2I_{\text{rod}} = 2(ML^2/12) = ML^2/6$.

9-5 (c) The moment of inertia of a thin, uniform rod that has a mass m and a length l rotating around its end is $ml^2/3$. Each of the four rods in this problem (with $m = M/2$, $l = L/2$) rotates around its end, so each rod has a moment of inertia

$$I_{\text{rod}} = \frac{(M/2)(L/2)^2}{3} = \frac{ML^2}{24}$$

when rotating around its end. Because the moment of inertia is additive, the moment of inertia of the four rods together is

$$I = 4I_{\text{rod}} = 4\left(\frac{ML^2}{24}\right) = \frac{ML^2}{6}$$

Notice that the final configuration of the four rods in this problem is identical to that of the two rods in the previous one. So it's not surprising that the result is the same.

9-6 (a) The situation is the same as in Example 9-10 but with a moment of inertia $I_{\text{CM}} = (2/5)MR^2$ rather than $I_{\text{CM}} = (1/2)MR^2$ (see Table 9-1). Since the sphere has a moment of inertia smaller than that of the disk, it requires a smaller torque to give it an angular acceleration. Hence the uphill friction force f acting on the sphere will be less than that on the disk. As a result, the net downhill force on the sphere, and thus its downhill acceleration, will be greater for the sphere than for the disk.

You can also get this result by redoing the calculation in Example 9-10 using $I_{\text{CM}} = (2/5)MR^2$. The rotational form of Newton's second law becomes

$$Rf = \frac{2}{5}MR^2\alpha_z \text{ or } f = \frac{2}{5}MR\alpha_z$$

With $\alpha_z = a_{\text{CM},x}/R$, this becomes

$$f = \frac{2}{5}MR\left(\frac{a_{\text{CM},x}}{R}\right) = \frac{2}{5}Ma_{\text{CM},x}$$

Substitute this into the Newton's second law equation for the sphere's motion along the x axis:

$$Mg\sin\theta - f = Ma_{\text{CM},x}$$

$$Mg\sin\theta - \frac{2}{5}Ma_{\text{CM},x} = Ma_{\text{CM},x}$$

$$Mg\sin\theta = Ma_{\text{CM},x} + \frac{2}{5}Ma_{\text{CM},x} = \frac{7}{5}Ma_{\text{CM},x}$$

$$a_{\text{CM},x} = \frac{5}{7}g\sin\theta$$

Since 5/7 (about 0.714) is greater than 2/3 (about 0.667), the acceleration of the solid sphere is in fact greater than the acceleration of the disk, $a_{\text{CM},x} = (2/3)g\sin\theta$.

9-7 The uniform disk would still win the race. As Example 9-13 shows, the speeds after moving a vertical distance H down the ramp are $v_{\text{CM}} = \sqrt{4gH/3}$ for the solid disk and $v_{\text{CM}} = \sqrt{gH}$ for the hoop. These don't depend at all on either the mass or the radius of either object. We saw a similar result

in Example 6-10 (Section 6-6) for the purely translational motion of a skier on a ski jump: The final speed didn't depend on the skier's mass.

9-8 Block, uniform sphere, uniform disk. Because it doesn't rotate and has no work done on it by friction, *all* of the block's gravitational potential energy is transformed into translational kinetic energy. By contrast, as both the disk and the sphere roll downhill, part of the gravitational potential energy is transformed into rotational kinetic energy. That means that at any location, the hoop and sphere will be moving more slowly than the block. Hence the block reaches the bottom first. The sphere arrives second, since from Table 9-1 it has a smaller moment of inertia than the disk ($I_{\text{CM}} = 2MR^2/5$ for the sphere versus $I_{\text{CM}} = MR^2/2$ for the disk), so less of the sphere's kinetic energy goes into rotation (see Example 9-13). The disk, with its large moment of inertia, comes in last.

9-9 (c) The angular speed of the ball increases. Once the ball has been hit and set in motion it has angular momentum around the pole, and because we neglect resistive forces there is no external torque on it. (Yes, there is a force on the ball due to tension in the rope. This force is straight inward toward the pole, however, so the angle ϕ between $\vec{r}$ and $\vec{F}$ in Equation 9-16, the definition of torque, is 180°. Hence the magnitude of the torque is zero: $\tau = rF\sin\phi = rF\sin 180° = 0$.) Angular momentum L_z is therefore conserved. As the rope loops around the pole, the ball gets closer to the rotational axis and its moment of inertia I decreases. Hence the angular speed (which is the magnitude of the angular velocity ω_z) must increase for the magnitude of $L_z = I\omega_z$ to remain constant.

9-10 (d) The angular momentum L_z of a system is conserved only if no net torque acts on the system. This isn't the case for *either* the pulley or the disk: A net torque due to the tension force acts on the pulley as it rotates, and a net torque due to the force of friction acts on the disk as it rolls downhill. For both objects the moment of inertia I remains unchanged (the objects don't change shape) and the angular velocity ω_z increases, so $L_z = I\omega_z$ increases. Conservation laws are great when they apply, but remember that they do *not* apply in all situations!

9-11 (a) $\vec{L}$ is out of the plane in cases (i) and (iii), into the plane in cases (ii) and (iv). In each case curl the fingers of your right hand around the disk in the direction of rotation. Your right thumb will then point in the direction of $\vec{L}$. (b) $\vec{\tau}$ is out of the plane in cases (ii) and (iii), into the plane in case (i), and zero in case (iv). In cases (i), (ii), and (iii) curl the fingers of your right hand in the direction that the force would tend to make the object rotate if it started at rest. Your right thumb will then point in the direction of $\vec{\tau}$. In case (iv) the line of action of the force passes through the rotation axis, so there is zero torque. (c) The disk speeds up in case (iii), slows down in cases (i) and (ii), and maintains the same angular speed in case (iv). The angular speed increases if $\vec{\tau}$ and $\vec{L}$ are in the same direction, decreases if $\vec{\tau}$ and $\vec{L}$ are in opposite directions, and stays the same if $\vec{\tau} = 0$.

Questions and Problems

In a few problems you are given more data than you actually need; in a few other problems you are required to supply data from your general knowledge, outside sources, or informed estimate. Interpret as significant all digits in numerical values that have trailing zeros and no decimal points.

For all problems use $g = 9.80$ m/s^2 for the free-fall acceleration due to gravity.

- •   Basic, single-concept problem
- ••   Intermediate-level problem; may require synthesis of concepts and multiple steps
- •••   Challenging problem
- Example See worked example for a similar problem

Conceptual Questions

1. • Define the SI unit radian. The unit appears in some physical quantities (for example, the angular velocity of a turntable is 3.5 rad/s) and it is omitted in others (for example, the translational velocity at the rim of a turntable is 0.35 m/s). Because the formula relating rotational and translational quantities involves multiplying by a radian ($v = r\omega$), discuss when it is appropriate to include radians and when the unit should be dropped.

2. • What are the units of angular velocity (ω_z)? Why are factors of 2π present in many equations describing rotational motion?

3. • Consider a situation in which a merry-go-round, starting from rest, speeds up in the counterclockwise direction. It eventually reaches and maintains a maximum angular velocity. After a short time the merry-go-round then starts to slow down and eventually stops. Assume the accelerations experienced by the merry-go-round have constant magnitudes and the counterclockwise direction is positive. (a) Which graph in Figure 9-36 describes the angular velocity as the merry-go-round speeds up? (b) Which graph describes the angular position as the merry-go-round speeds up? (c) Which graph describes the angular velocity as the merry-go-round travels at its maximum velocity? (d) Which graph describes the angular position as the merry-go-round travels at its maximum velocity? (e) Which graph describes the angular velocity as the merry-go-round slows down? (f) Which graph describes the angular position as the merry-go-round slows down? (g) Draw a graph of the torque experienced by the merry-go-round as a function of time during the scenario described in the problem.

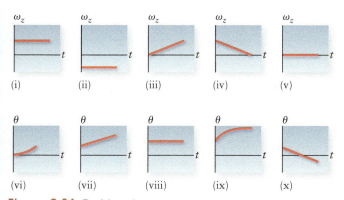

Figure 9-36 Problem 3

4. • Describe what a "torque wrench" is (look up the definition, if not known) and discuss any difficulties that a Canadian auto or bicycle mechanic might have working with an American mechanic's tools (and vice versa).

5. • Rank the torques exerted on the bolts in a–d (Figure 9-37) from least to greatest. Note that the forces in b and d make an angle of 45° with the wrench. Assume the wrenches and the magnitude of the force F are identical.

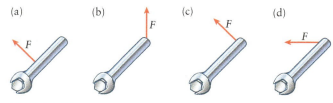

Figure 9-37 Problem 5

6. • Describe any inconsistencies in the following statement: "The units of torque are N·m, but that's not the same as the units of energy."

7. • In Chapter 4 you learned that the mass of an object determines how that object responds to an applied force. Write a rotational analog to that idea based on the concepts of this chapter.

8. • A student cannot open a door at her school. She pushes with ever-greater force, and still the door will not budge! Knowing that the door does push open, it is not locked, and a minimum torque is required to open the door, give a few reasons why this might be occurring.

9. • Why is it critical to define the axis of rotation when you set out to find the moment of inertia of an object?

10. • The four solids shown in Figure 9-38 have equal heights, widths, and masses. The axes of rotation are located at the center of each object and are perpendicular to the plane of the paper. Rank the moments of inertia from greatest to least.

Hoop   Solid cylinder   Solid sphere   Hollow sphere

Figure 9-38 Problem 10

11. • A hollow cylinder rolls without slipping up an incline, stops, then rolls back down. Which of the graphs in Figure 9-39 shows the (a) angular acceleration and (b) angular velocity for the motion? Assume that up the ramp is the positive direction.

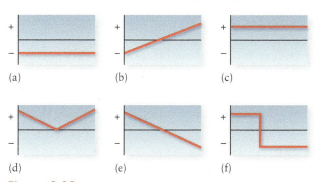

Figure 9-39 Problem 11

12. • What is the ratio of rotational kinetic energy for two balls, each tied to a light string and spinning in a circle with a radius equal to the length of the string? The first ball has a mass m and a string of length L, and rotates at a rate of ω. The second ball has a mass $2m$ and a string of length $2L$, and rotates at a rate of 2ω.

13. • Using the rotational concepts of this chapter, explain why a uniform solid sphere beats a uniform solid cylinder which beats a ring when the three objects "race" down an inclined plane while rolling without slipping.

14. • Explain which physical quantities change when an ice skater moves her arms in and out as she rotates in a pirouette. What causes her angular velocity to change, if it changes at all?

15. • Explain how an object moving in a straight line can have a nonzero angular momentum.

16. • Which quantity is larger: the angular momentum of Earth rotating on its axis each day or the angular momentum of Earth revolving about the Sun each year? Try to determine the answer without using a calculator.

17. • Analyze the following statement and determine if there are any physical inconsistencies: While rotating a ball on the end of a string of length L, the rotational kinetic energy remains constant as long as the length and angular speed are fixed. When the ball is pulled inward and the length of the string is shortened, the rotational kinetic energy will remain constant due to conservation of energy, but the angular momentum will not because there is an external force acting on the ball to pull it inward. The moment of inertia and angular speed will, of course, remain the same throughout the process because the ball is rotating in the same plane throughout the motion.

18. • A freely rotating turntable moves at a steady angular velocity. A glob of cookie dough falls straight down and attaches to the very edge of the turntable. Describe which quantities (angular velocity, angular acceleration, torque, rotational kinetic energy, moment of inertia, or angular momentum) are conserved during the process and describe qualitatively what happens to the motion of the turntable.

19. • What are the units of the following quantities: (a) rotational kinetic energy, (b) moment of inertia, and (c) angular momentum?

20. • While watching two people on a seesaw, you notice that the person at the top always leans backward, while the person at the bottom always leans forward. (a) Why do the riders do this? (b) Assuming they are sitting equidistant from the pivot point of the seesaw, what, if anything, can you say about the relative masses of the two riders?

21. • Referring to the time-lapse photograph of a falling cat in Figure 9-32a, do you think that a cat will fall on her feet if she does not have a tail? Explain your answer using the concepts of this chapter.

22. • In describing rotational motion it is often useful to develop an analogy with translational motion. First, write a set of equations describing translational motion. Then write the rotational analogs (for example, $\theta = \theta_0 \ldots$) of the translational equations (for example, $x = x_0 + v_{0x}t + \frac{1}{2}a_x t^2$) using the following legend:

$$x \Leftrightarrow \theta \quad v_x \Leftrightarrow \omega_z \quad a_x \Leftrightarrow \alpha_z \quad F_x \Leftrightarrow \tau_z \quad m \Leftrightarrow I$$
$$p_x \Leftrightarrow L_z \quad K_{\text{translational}} \Leftrightarrow K_{\text{rotational}}$$

Multiple-Choice Questions

23. • Todd and Susan are riding on a merry-go-round. Todd rides on a horse toward the outside of the circular platform, and Susan rides on a horse toward the center of the circular platform. When the merry-go-round is rotating at a constant angular speed, Todd's angular speed is
    A. exactly half as much as Susan's.
    B. larger than Susan's.
    C. smaller than Susan's.
    D. the same as Susan's.
    E. exactly twice as much as Susan's.

24. • Todd and Susan are riding on a merry-go-round. Todd rides on a horse toward the outer edge of a circular platform, and Susan rides on a horse toward the center of the circular platform. When the merry-go-round is rotating at a constant angular speed ω, Todd's speed v is
    A. exactly half as much as Susan's.
    B. larger than Susan's.
    C. smaller than Susan's
    D. the same as Susan's.
    E. exactly twice as much as Susan's.

25. • You give a quick push to a ball at the end of a massless, rigid rod, causing the ball to rotate clockwise in a horizontal circle (Figure 9-40). The rod's pivot is frictionless. After the push has ended, the ball's angular velocity

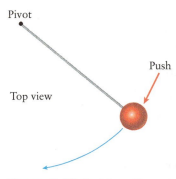

Figure 9-40 Problem 25

    A. steadily increases.
    B. increases for a while, then remains constant.
    C. decreases for a while, then remains constant.
    D. remains constant.
    E. steadily decreases.

26. • The moment of inertia of a thin ring about its symmetry axis is $I_{\text{CM}} = MR^2$. What is the moment of inertia if you twirl a large ring around your finger, so that in essence it rotates about a point on the ring, about an axis parallel to the symmetry axis?
    A. $5MR^2$
    B. $2MR^2$
    C. MR^2
    D. $1.5MR^2$
    E. $0.5MR^2$

27. •• You have two steel spheres; sphere 2 has twice the radius of sphere 1. What is the ratio of the moment of inertia $I_2{:}I_1$ measured about an axis through the center of the spheres?
    A. 2
    B. 4
    C. 8
    D. 16
    E. 32

28. • A solid sphere of radius R, a solid cylinder of radius R, and a rod of length R all have the same mass, and all three are rotating with the same angular velocity. The sphere is rotating around an axis through its center. The cylinder is rotating around its long axis, and the rod is rotating around an axis

through its center but perpendicular to the rod. Which one has the greatest rotational kinetic energy?
- A. the sphere
- B. the cylinder
- C. the rod
- D. The rod and cylinder have the same rotational kinetic energy.
- E. They all have the same kinetic energy.

29. •• A solid ball, a solid disk, and a hoop, all with the same mass and the same radius, are set rolling without slipping up an incline, all with the same initial energy. Which goes farthest up the incline?
- A. the ball
- B. the disk
- C. the hoop
- D. The hoop and the disk roll to the same height, farther than the ball.
- E. They all roll to the same height.

30. •• A solid ball, a solid disk, and a hoop, all with the same mass and the same radius, are set rolling without slipping up an incline, all with the same initial linear speed. Which goes farthest up the incline?
- A. the ball
- B. the disk
- C. the hoop
- D. The hoop and the disk roll to the same height, farther than the ball.
- E. They all roll to the same height.

31. • How would a flywheel's (spinning disk's) kinetic energy change if its moment of inertia were five times larger but its angular speed were five times smaller?
- A. 0.1 times as large as before
- B. 0.2 times as large as before
- C. same as before
- D. 5 times as large as before
- E. 10 times as large as before

32. • While a gymnast is in the air during a leap, which of the following quantities must remain constant for her?
- A. position
- B. velocity
- C. momentum
- D. angular velocity
- E. angular momentum about her center of mass

Problems

9-1 Rotation is an important and ubiquitous kind of motion

9-2 The equations for rotational kinematics are almost identical to those for linear motion

33. • What is the angular speed, in rad/s, of an object that completes 2.00 rev every 12.0 s? Example 9-1

34. • A car rounds a curve with a translational speed of 12.0 m/s. If the radius of the curve is 7.00 m, calculate the angular speed in rad/s. Example 9-1

35. • Convert the following: Example 9-1

$$45.0 \text{ rev/min} = \underline{\hspace{2cm}} \text{ rad/s}$$
$$33\tfrac{1}{3} \text{ rpm} = \underline{\hspace{2cm}} \text{ rad/s}$$
$$2\pi \text{ rev/s} = \underline{\hspace{2cm}} \text{ rad/s}$$

36. • Calculate the angular speed of the Moon as it orbits Earth. The Moon completes one orbit about Earth in 27.4 days and the Earth–Moon distance is 3.84×10^8 m. Example 9-1

37. • Suppose a roulette wheel is spinning at 1 rev/s. (a) How long will it take for the wheel to come to rest if it experiences an angular acceleration of −0.02 rad/s²? (b) How many rotations will it complete in that time? Example 9-2

38. • A spinning top completes 6.00×10^3 rotations before it starts to topple over. The average speed of the rotations is 8.00×10^2 rpm. Calculate how long the top spins before it begins to topple. Example 9-2

39. • A child pushes a merry-go-round that has a diameter of 4.00 m and goes from rest to an angular speed of 18.0 rpm in a time of 43.0 s. (a) Calculate the average angular acceleration (in rad/s²) of the merry-go-round. (b) Calculate the angular displacement (in rad) of the merry-go-round during this time interval. (c) What is the maximum tangential speed of the child if she rides on the edge of the platform? Example 9-2

40. • Allison twirls an umbrella around its central axis so that it completes 24.0 rotations in 30.0 s. (a) If the umbrella starts from rest, calculate the average angular acceleration (in rad/s²) of a point on the outer edge. (b) What is the maximum tangential speed of a point on the edge if the umbrella has a radius of 55.0 cm? Example 9-2

41. • Prior to the music CD, stereo systems had a phonographic turntable on which vinyl disk recordings were played. A particular phonographic turntable starts from rest and achieves a final constant angular speed of $33\tfrac{1}{3}$ rpm in a time of 4.5 s. (a) How many rotations did the turntable undergo during that time? (b) The classic Beatles album *Abbey Road* is 47 min and 7 s in duration. If the turntable requires 8 s to come to rest once the album is over, calculate the total number of rotations for the complete start-up, playing, and slow-down of the album. Example 9-2

42. • A CD player varies its speed as it changes circular tracks on the CD. A CD player is rotating at 300 rpm. To read another track the angular speed is increased to 450 rpm in a time of 0.75 s. Calculate the average angular acceleration in rad/s² during the change. Example 9-2

43. • **Astronomy** A communication satellite circles Earth in a geosynchronous orbit such that the satellite remains directly above the same point on the surface of Earth. (a) What angular displacement (in radians) does the satellite undergo in 1 h of its orbit? (b) Calculate the angular speed of the satellite in rev/min and rad/s. Example 9-2

9-3 Torque is to rotation as force is to translation

44. • What is the torque about your shoulder axis if you hold a 10.0-kg barbell in one hand straight out and at shoulder height? Assume your hand is 75 cm from your shoulder and neglect the torque due to the weight of your arm. Example 9-3

45. • **Medical** When the palmaris longus muscle in the forearm is flexed, the wrist moves back and forth (Figure 9-41). If the muscle generates a force of 45.0 N and it is acting with an effective lever arm of 22.0 cm, what is the torque that the muscle produces on the wrist? Curiously, many people lack this muscle. Some studies correlate the absence of the muscle with carpal tunnel syndrome. Example 9-3

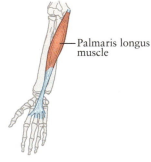

Figure 9-41 Problem 45

Palmaris longus muscle

46. • A torque wrench is used to tighten a nut on a bolt. The wrench is 25 cm long, and a force of 120 N is applied at the end of the wrench as shown in Figure 9-42. Calculate the torque about the axis that passes through the bolt. Example 9-3

Figure 9-42 Problem 46

47. • An 85.0-cm-wide door is pushed open with a force of $F = 75.0$ N. Calculate the torque about an axis that passes through the hinges in each of the cases in Figure 9-43. Example 9-3

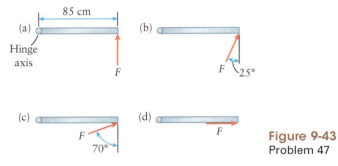

Figure 9-43
Problem 47

48. • A robotic arm lifts a barrel of radioactive waste (Figure 9-44). If the maximum torque delivered by the arm about the axis O is 3.00×10^3 N·m and the distance r in the diagram is 3.00 m, what is the maximum mass of the barrel? Example 9-3

Figure 9-44 Problem 48

49. • A typical adult can deliver about 10 N·m of torque when attempting to open a twist-off cap on a bottle. What is the maximum force that the average person can exert with his fingers if most bottle caps are about 2 cm in diameter? Example 9-3

50. • A driver applies a horizontal force of 20.0 N (to the right) to the top of a steering wheel, as shown in Figure 9-45. The steering wheel has a radius of 18.0 cm and a moment of inertia of 0.0970 kg·m². Calculate the angular acceleration of the steering wheel about the central axis due to this force. Example 9-3

Figure 9-45 Problem 50

51. •• A potter's wheel is initially at rest. A constant external torque of 75.0 N·m is applied to the wheel for 15.0 s, giving the wheel an angular speed of 5.00×10^2 rev/min. (a) What is the moment of inertia of the wheel? (b) The external torque is then removed, and a brake is applied. If it takes the wheel 2.00×10^2 s to come to rest after the brake is applied, what is the magnitude of the torque exerted by the brake? Example 9-3

9-4 An object's moment of inertia depends on its mass distribution and the choice of rotation axis

52. • What is the combined moment of inertia for the three point objects about the axis O in Figure 9-46? Example 9-4

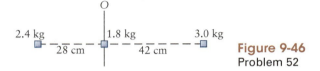

Figure 9-46
Problem 52

53. • What is the combined moment of inertia of three point objects ($m_1 = 1.00$ kg, $m_2 = 1.50$ kg, $m_3 = 2.00$ kg) tied together with massless strings and rotating about the axis O as shown in Figure 9-47? Example 9-4

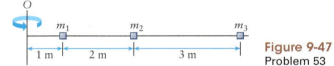

Figure 9-47
Problem 53

54. •• A baton twirler in a marching band complains that her baton is defective (Figure 9-48). The manufacturer specifies that the baton should have an overall length of $L = 60.0$ cm and a total mass between 940 and 950 g (there is one 350-g object on each end). Also according to the manufacturer, the moment of inertia about the central axis passing through the baton should fall between 0.0750 and 0.0800 kg·m². The twirler (who has completed a class in physics) claims this is impossible. Who's right? Explain your answer. Example 9-4

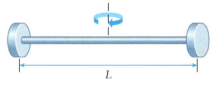

Figure 9-48
Problem 54

55. • What is the moment of inertia of a steering wheel about the axis that passes through its center? Assume the rim of the wheel has a radius R and a mass M. Assume that there are five radial spokes that connect in the center as shown in Figure 9-49. The spokes are thin rods of uniform mass density with length R and mass $\frac{1}{2}M$, evenly spaced around the wheel. Example 9-4

Figure 9-49 Problem 55

56. • Using the parallel-axis theorem, calculate the moment of inertia for a solid, uniform sphere about an axis that is tangent to its surface (Figure 9-50). Example 9-7

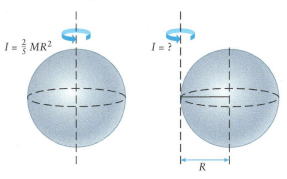

$I = \frac{2}{5}MR^2$ $I = ?$

Figure 9-50 Problem 56

57. • Calculate the moment of inertia for a uniform, solid cylinder (mass M, radius R) if the axis of rotation is tangent to the side of the cylinder as shown in Figure 9-51. Example 9-7

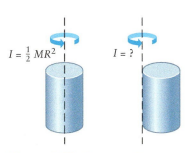

$I = \frac{1}{2}MR^2$ $I = ?$

Figure 9-51 Problem 57

58. • Calculate the moment of inertia for a thin uniform rod that is 1.25 m long and has mass 2.25 kg. The axis of rotation passes through the rod at a point one-third of the way from the left end (Figure 9-52). Example 9-6

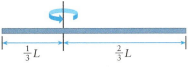

$\frac{1}{3}L$ $\frac{2}{3}L$

Figure 9-52 Problem 58

59. •• Calculate the moment of inertia of a thin plate that is 5.00 cm × 7.00 cm in area and has a uniform mass density of 1.50 g/cm². The axis of rotation is located at the left side, as shown in Figure 9-53. Example 9-7

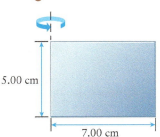

5.00 cm

7.00 cm

Figure 9-53 Problem 59

60. •• Calculate the radius of a solid uniform sphere of mass M that has the same moment of inertia about an axis through its center of mass as a second solid uniform sphere of radius R and mass M which has the axis of rotation passing tangent to the surface and parallel to the center of mass axis (Figure 9-54). Example 9-7

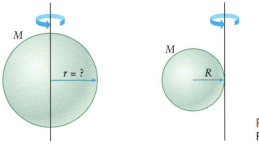

M $r = ?$ M R

Figure 9-54
Problem 60

61. •• Two uniform, solid spheres, one with mass M and radius R and the other with mass M and radius $2R$, are connected by a thin, uniform rod of mass M and length $3R$ (Figure 9-55). Find the moment of inertia about the axis through the center of the rod. Example 9-7

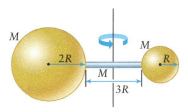

Figure 9-55 Problem 61

62. ••• Two thin disks are rigidly attached as shown in Figure 9-56. The angle $\theta = 45°$. The mass and radius of disk A are 0.80 kg and 20 cm, respectively. The mass and radius of disk B are 0.50 kg and 10 cm, respectively What is the moment of inertia of the composite object about an axis perpendicular to the page and passing through point P? Example 9-7

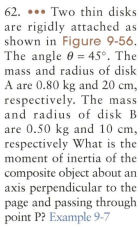

Figure 9-56 Problem 62

9-5 The techniques used for solving problems with Newton's second law also apply to rotation problems

63. •• A 250-g mass hangs from a string that is wrapped around a pulley, as shown in Figure 9-57. The pulley is suspended in such a way that it can rotate freely. When the mass is released, it accelerates toward the floor as the string unwinds. Model the pulley as a uniform solid cylinder of mass 1.00 kg and radius 5.00 cm. Assume that the thread has negligible mass and does not slip or stretch as it unwinds.

Figure 9-57 Problem 63

(a) Determine the magnitude of the pulley's angular acceleration. (b) Determine the magnitude of the acceleration of the descending weight. (c) Calculate the tension in the string. Example 9-9

64. •• Figure 9-58 shows a solid, uniform cylinder of mass 7.00 kg and radius 0.450 m with a light string wrapped around it. A 3.00-N tension force is applied to the string, causing the cylinder to roll without slipping across a level surface as shown. Assume that to the right and into the page are the positive directions. (a) What is the angular acceleration of the cylinder? (b) Determine the frictional force that acts on the cylinder. Example 9-10

3.00 N

Figure 9-58
Problem 64

65. • A crane winch is lifting a 2300-kg mass. The winch can be modeled as a cable of negligible mass wound around a solid uniform cylinder with a 12-cm radius and a mass of 320 kg that rotates around its central axis. How much torque is required to accelerate the 2300-kg mass straight upward at 0.35 m/s²? Assume there is no friction in the system and that the winch radius remains constant. Example 9-10

66. ••• A block with mass m_1 = 2.00 kg rests on a frictionless table. It is connected with a light string over a pulley to a hanging block of mass m_2 = 4.00 kg. The pulley is a uniform disk with a radius of 4.00 cm and a mass of 0.500 kg

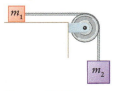

Figure 9-59 Problem 66

(Figure 9-59). (a) Calculate the acceleration of each block and the tension in each segment of the string. (b) How long does it take the blocks to move a distance of 2.25 m? (c) What is the angular speed of the pulley at this time? Example 9-9

67. •• A yo-yo with a mass of 0.0750 kg and a rolling radius of r = 2.50 cm rolls down a string with a linear acceleration of 6.50 m/s² (Figure 9-60). (a) Calculate the tension in the string and the angular acceleration of the yo-yo. (b) What is the moment of inertia of this yo-yo? Example 9-10

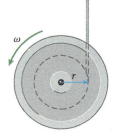

Figure 9-60
Problem 67

9-6 An object's rotational kinetic energy is related to its angular speed and its moment of inertia

68. • If a 0.250-kg point object rotates at 3.00 rev/s about an axis that is 0.500 m away, what is the kinetic energy of the object? Example 9-11

69. • What is the rotational kinetic energy of an object that has a moment of inertia of 0.280 kg·m² about the axis of rotation when its angular speed is 4.00 rad/s? Example 9-11

70. • What is the moment of inertia of an object that rotates at 13.0 rev/min about an axis and has a rotational kinetic energy of 18.0 J? Example 9-11

71. • What is the angular speed of a rotating wheel that has a moment of inertia of 0.330 kg·m² and a rotational kinetic energy of 2.75 J? Give your answer in both rad/s and rev/min. Example 9-11

72. • **Sports** A bowling ball that has a radius of 11.0 cm and a mass of 5.00 kg rolls without slipping on a level lane at 2.00 rad/s. Calculate the ratio of the translational kinetic energy to the rotational kinetic energy of the bowling ball. Example 9-13

73. • **Astronomy** Earth is approximately a solid sphere, has a mass of 5.98×10^{24} kg, a radius of 6.38×10^6 m, completes one rotation about its central axis each day, and orbits the Sun once each year (the Earth–Sun distance is 1.50×10^{11} m). (a) Calculate the rotational kinetic energy of Earth as it spins on its axis. (b) Calculate the rotational kinetic energy of Earth as it orbits the Sun. Example 9-11

74. • A potter's flywheel is made of a 5.00-cm-thick, round slab of concrete that has a mass of 60.0 kg and a diameter of 35.0 cm. This disk rotates about an axis that passes through

its center, perpendicular to its round area. Calculate the angular speed of the slab about its center if the rotational kinetic energy is 15.0 J. Express your answer in both rad/s and rev/min. Example 9-11

75. •• **Sports** A flying disk (160 g, 25.0 cm in diameter) spins at a rate of 3.00×10^2 rpm with its center balanced on a fingertip. What is the rotational kinetic energy of the disk if it has 70.0% of its mass on the outer edge (basically a thin ring 25.0 cm in diameter) and the remaining 30.0% is a nearly flat disk 25.0 cm in diameter? Example 9-11

76. • A uniform, solid cylinder of radius 5.00 cm and mass 3.00 kg starts from rest at the top of an inclined plane that is 2.00 m long and tilted at an angle of 25.0° with the horizontal and rolls without slipping down the ramp. What is the cylinder's speed at the bottom of the ramp? Example 9-13

77. • A uniform, solid sphere of radius 5.00 cm and mass 3.00 kg starts with a translational speed of 2.00 m/s at the top of an inclined plane that is 2.00 m long and tilted at an angle of 25.0° with the horizontal and rolls without slipping down the ramp. What is the sphere's speed at the bottom of the ramp? Example 9-13

78. ••• A 25-mm-diameter glass marble with a mass of 20 g rolls without slipping down a track toward a vertical loop of radius R = 10 cm, as shown in Figure 9-61. Approximate the minimum translational speed the marble must have when it is H = 25 cm above the bottom of the loop in order to complete the loop without falling off the track. Example 9-13

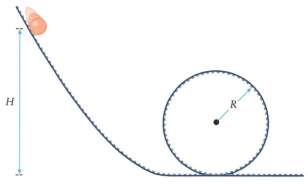

Figure is not to scale.

Figure 9-61 Problem 78

79. ••• A billiard ball of mass 160 g and radius 2.50 cm starts with a translational speed of 2.00 m/s at point A on the track as shown in Figure 9-62. If point B is at the top of a hill that has a radius of curvature of 60 cm, what is the normal force acting on the ball at point B? Assume the billiard ball rolls without slipping on the track. Example 9-13

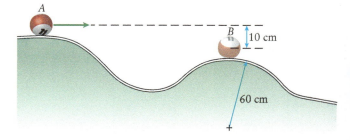

Figure 9-62 Problem 79

9-7 Angular momentum is conserved when there is zero net torque on a system

80. • What is the angular momentum about the central axis of a thin disk that is 18.0 cm in diameter, has a mass of 2.50 kg, and rotates at a constant 1.25 rad/s? Example 9-14

81. • What is the angular momentum of a 0.300-kg tetherball when it whirls around the central pole at 60.0 rpm at a radius of 125 cm? Example 9-14

82. • **Astronomy** Calculate the angular momentum of Earth as it orbits the Sun. Recall that the mass of Earth is 5.98×10^{24} kg, the distance between Earth and the Sun is 1.50×10^{11} m, and the time for one orbit is 365.3 days. Example 9-14

83. • **Astronomy** Calculate the angular momentum of Earth as it spins on its central axis once each day. Assume Earth is approximately a uniform, solid sphere that has a mass of 5.98×10^{24} kg and a radius of 6.38×10^{6} m. Example 9-14

84. • What is the speed of an electron in the lowest energy orbital of hydrogen, of radius equal to 5.29×10^{-11} m? The mass of an electron is 9.11×10^{-31} kg, and its angular momentum in this orbital is 1.055×10^{-34} J·s. Example 9-14

85. • What is the angular momentum of a 70.0-kg person riding on a Ferris wheel that has a diameter of 35.0 m and rotates once every 25.0 s? Example 9-14

86. • A professor sits on a rotating stool that spins at 10.0 rpm while she holds a 1.00-kg weight in each of her hands. Her outstretched arms are 0.750 m from the axis of rotation, which passes through her head into the center of the stool. When she draws the weights in toward her body, her angular speed increases to 20.0 rpm. Neglecting the mass of her arms, how far are the weights from the rotational axis at the increased speed? Example 9-14

87. •• A brave child decides to grab onto an already spinning merry-go-round. The child is initially at rest and has a mass of 30 kg. The child grabs and clings to a bar that is 1.5 m from the center of the merry-go-round, causing the angular velocity of the merry-go-round to abruptly drop from 45 to 20 rpm. What is the moment of inertia of the merry-go-round with respect to its central axis? Example 9-15

88. • A giant toroidal space station is being evacuated. Model the 1.0×10^{10}-kg station as a cylindrical hoop of radius 1.8 km rotating at a rate of 2π rad/min. If each wave of evacuees consists of one thousand 4.0×10^{5}-kg escape pods whose launchers are pointed radially outward from the station rim, how much does the angular speed of the station *change* with the first wave of evacuees? Express your answer in rad/sec. Example 9-15

89. • A chef is tossing 0.500 kg of pizza crust dough. With each toss the dough has an initial angular speed of 5.20 rad/s. During one particular toss, the dough starts out uniformly distributed throughout a 20.0-cm-diameter disk and expands to 22.0 cm in diameter. Assuming the mass remains uniformly distributed, what is the angular speed of the dough when the chef catches it? Example 9-14

9-8 Rotational quantities such as angular momentum and torque are actually vectors

90. • A 50.0-g meter stick is balanced at its midpoint (50.0 cm). Then a 0.100-kg and a 0.200-kg mass are hung with light string from the 10.0-cm and 70.0-cm points, respectively.

(Figure 9-63). Defining the counterclockwise direction to be positive, calculate the torques acting on the board due to the four forces shown about an axis pointing out of the page at the following points: (a) the 0-cm point, (b) the 50-cm point, and (c) the 100-cm point. (d) Is the meter stick still balanced after the two masses have been added? Example 9-16

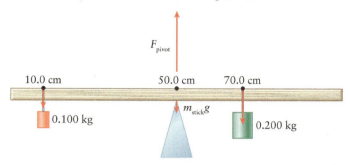

Figure 9-63 Problem 90

91. •• A tightrope walker is walking between two buildings using a 15.0-m-long, 18.0-kg pole for balance (Figure 9-64). The daredevil grips the pole with each hand 0.600 m from the center of the pole. A 0.540-kg bird lands on the very end of the left-hand side of the pole. Assuming the daredevil applies forces with each hand in a direction perpendicular to the pole, how much force must *each* hand exert to counteract the torque of the bird? The +x axis points in the walking direction, through the center of the pole. What are the directions of the torque vectors due to the bird, the left hand, and the right hand? Example 9-16

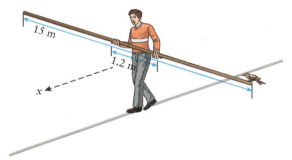

Figure 9-64 Problem 91

92. • A rigid, uniform, 4.00-m-long, 40.0-kg beam rests on a pivot placed 3.00 m from one end (Figure 9-65). An 80.0-kg man walks up the beam from the end resting on the ground. How far along the beam must the man walk before the low end lifts off the ground? What is the direction of the net torque vector on the beam until the man reaches that critical point? Example 9-16

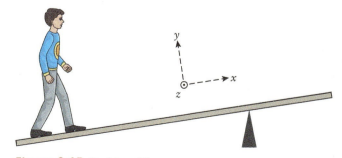

Figure 9-65 Problem 92

93. •• A yo-yo with an inner cylinder radius of $r = 2.5$ cm and outer disk radius of $R = 7.5$ cm is at rest on a 30.0° incline.

It is held in place by friction and the tension in the string that is wrapped around its inner cylinder, as shown in **Figure 9-66**. If the mass of the yo-yo is 420 g, what is the friction force? Assume that the positive direction points down the incline. Example 9-17

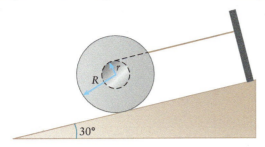

Figure 9-66 Problem 93

General Problems

94. • A 325-kg merry-go-round with a radius of 1.40 m is spinning counterclockwise as viewed from above at 4.70 rad/s. A 36.0-kg child is hanging on tightly 1.25 m from the rotation axis. Her father applies friction to the outer rim to slow the merry-go-round to a stop in 5 s. How much torque must he apply? Model the merry-go-round as a solid disk and assume that counterclockwise rotation is positive. Example 9-5

95. • A circular revolving door has a radius of $R = 1.00$ m, total moment of inertia of $I = 119$ kg·m², and is free to rotate on frictionless bearings as shown in **Figure 9-67**. Each of the door "arms" makes a tight seal with the frame, so that each arm provides $f_k = 45.0$ N of frictional force when the door rotates. If an 85.0-kg adult pushes perpendicularly to one of the doors at a distance $r = 0.600$ m from the rotation axis, what force must the person exert to make the door spin at a constant rate, assuming all four arms are in contact with the frame? Example 9-5

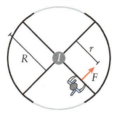

Figure 9-67 Problem 95

96. • Using a spreadsheet and the following data, calculate the average angular speed of the rotating object over the first 10 s. Calculate the average angular acceleration from 15 to 25 s. If the object has a moment of inertia of 0.25 kg·m² about the axis of rotation, calculate the average torque during the following time intervals: $0 < t < 10$ s, 10 s $< t < 15$ s, and 15 s $< t < 25$ s. Example 9-2

t (s)	θ (rad)	t (s)	θ (rad)	t (s)	θ (rad)
0	0	9	3.14	18	6.48
1	0.349	10	3.50	19	8.53
2	0.700	11	3.50	20	11.0
3	1.05	12	3.49	21	14.1
4	1.40	13	3.50	22	17.6
5	1.75	14	3.51	23	21.6
6	2.10	15	3.51	24	26.2
7	2.44	16	3.98	25	31.0
8	2.80	17	5.01		

97. •• A baton is constructed by attaching two small objects that each have a mass M to the ends of a uniform rod that has a length L and a mass M. Find an expression for the moment of inertia of the baton when it is rotated around a point $(3/8) L$ from one end. Example 9-6

98. • The outside diameter of the playing area of an optical Blu-ray disc is 11.75 cm, and the inside diameter is 4.50 cm. When movies are being viewed, the disc rotates so that a laser maintains a constant linear speed relative to the disc of 7.50 m/s as it tracks over the playing area. (a) What are the maximum and minimum angular speeds (in rad/s and rpm) of the disc? (b) At which location of the laser on the playing area do these speeds occur? (c) What is the average angular acceleration of a Blu-ray disc as it plays an 8.0-h set of movies? Assume that the player starts at the inner track and ends at the outer track. Example 9-2

99. • A table saw has a 25.0-cm-diameter blade that rotates at a rate of 7000 rpm. It is equipped with a safety mechanism that can stop the blade within 5.00 ms if something like a finger is accidentally placed in contact with the blade. (a) What average angular acceleration occurs if the saw starts at 7000 rpm and comes to rest in this time? (b) How many rotations does the blade complete during the stopping period? Example 9-2

100. •• In 1932 Albert Dremel of Racine, Wisconsin, created a rotary tool that has come to be known as a dremel. (a) Suppose a dremel starts from rest and achieves an operating speed of 35,000 rev/min. If it requires 1.20 s to reach operating speed and is held at that speed for 45.0 s, how many rotations has the bit made? (b) Suppose it requires another 8.50 s for the tool to return to rest. What are the average angular accelerations for the start-up and the slow-down periods? (c) How many rotations does the tool complete from start to finish? Example 9-2

101. •• **Medical, Sports** On average, both arms and hands together account for 13% of a person's mass, while the head is 7.0% and the trunk and legs account for 80%. We can model a spinning skater with his arms outstretched as a vertical cylinder (head + trunk + legs) with two solid uniform rods (arms + hands) extended horizontally. Suppose a 62.0-kg skater is 1.80 m tall, has arms that are each 65.0 cm long (including the hands), and a trunk that can be modeled as being 35.0 cm in diameter. If the skater is initially spinning at 70.0 rpm with his arms outstretched, what will his angular velocity be (in rpm) when he pulls in his arms until they are at his sides parallel to his trunk? Example 9-14

102. •• Because of your success in physics class you are selected for an internship at a prestigious bicycle company in its research and development division. Your first task involves designing a wheel made of a hoop that has a mass of 1.00 kg and a radius of 50.0 cm, and spokes with a mass of 10.0 g each. The wheel should have a total moment of inertia 0.280 kg·m². (a) How many spokes are necessary to construct the wheel? (b) What is the mass of the wheel? Example 9-5

103. •• Two beads that each have a mass M are attached to a thin rod that has a length $2L$ and a mass $M/8$ (**Figure 9-68**). Each bead is initially a distance $L/4$ from the center of the rod. The whole system is set into uniform rotation about the center of the rod, with initial angular frequency $\omega_i = 20\pi$ rad/s. If the beads are then allowed to slide to the ends of the rod, what will the angular frequency become? Example 9-14

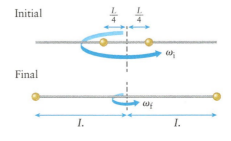

Figure 9-68 Problem 103

104. •• A uniform disk that has a mass $M = 0.300$ kg and a radius $R = 0.270$ m rolls up a ramp of angle $\theta = 55°$ with initial speed $v = 4.8$ m/s. If the disk rolls without slipping, how far up the ramp does it go? Example 9-10

105. •• In a new model of a machine, a spinning solid spherical part of radius R must be replaced by a ring of the same mass that is to have the same kinetic energy. Both parts need to spin at the same rate, the sphere about an axis through its center and the ring about an axis perpendicular to its plane at its center. (a) What should the radius of the ring be in terms of R? (b) Will both parts have the same angular momentum? If not, which one will have more? Example 9-14

106. • Many 2.5-in.-diameter (6.35-cm) computer hard disks spin at a constant 7200 rpm operating speed. The disks have a mass of about 7.50 g and are essentially uniform throughout with a very small hole at the center. If they reach their operating speed 2.50 s after being turned on, what average torque does the disk drive supply to the disk during the acceleration? Example 9-5

107. •• **Sports** At the 1984 Olympics, the great diver Greg Louganis won one of his 10 gold medals for the reverse $3\frac{1}{2}$ somersault tuck dive. In the dive, Louganis began his $3\frac{1}{2}$ turns with his body tucked in at a maximum height of approximately 2.0 m above the platform, which itself was 10.0 m above the water. He spun uniformly $3\frac{1}{2}$ times and straightened out his body just as he reached the water. A reasonable approximation is to model the diver as a thin uniform rod 2.0 m long when he is stretched out and as a uniform solid cylinder of diameter 0.75 m when he is tucked in. (a) What was Louganis's average angular speed as he fell toward the water with his body tucked in? *Hint*: How long did it take him to reach the water from his highest point? (b) What was his angular speed just after he stretched out? (c) How much did Louganis's rotational kinetic energy change while extending his body if his mass was 75 kg? Example 9-14

108. ••• **Medical** The bones of the forearm (radius and ulna) are hinged to the humerus at the elbow (Figure 9-69). The biceps muscle connects to the bones of the forearm about 2 cm beyond the joint. Assume the forearm has a mass of 2 kg and a length of 0.4 m. When the humerus and the biceps muscle are nearly vertical and the forearm is horizontal, if a person wishes to hold an object of mass M so that her forearm remains motionless, what is the relationship between the force exerted by the biceps muscle and the mass of the object? Example 9-16

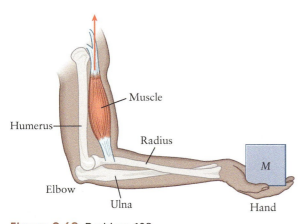

Figure 9-69 Problem 108

109. ••• **Medical** The femur of a human leg (mass 10 kg, length 0.9 m) is in traction (Figure 9-70). The center of gravity of the leg is one-third of the distance from the pelvis to the bottom of the foot. Two objects, with masses m_1 and m_2, are hung at both ends of the leg using pulleys to provide upward support. A third object of 8 kg is hung to provide tension along the leg. The body provides tension as well. (a) What is the mathematical relationship between m_1 and m_2? Is this relationship unique in the sense that there is only one combination of m_1 and m_2 that maintains the leg in static equilibrium? (b) How does the relationship change if the tension force due to m_1 is applied at the leg's center of mass? Example 9-16

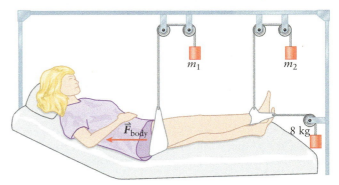

Figure 9-70 Problem 109

110. • **Astronomy** It is estimated that 60,000 tons of meteors and other space debris accumulate on Earth each year. Assume the debris is accumulated uniformly across the surface of Earth. (a) How much does Earth's rotation rate change per year as a result of this accumulation? (That is, find the change in angular velocity.) (b) How long would it take the accumulation of debris to change the rotation period by 1 s? Example 9-14

111. •• **Astronomy** Suppose we decided to use the rotation of Earth as a source of energy. (a) What is the maximum amount of energy we could obtain from this source? (b) By the year 2025 the projected rate at which the world uses energy is expected to be 6.6×10^{20} J/y. If energy use continues at that rate, for how many years would the spin of Earth supply our energy needs? Does this seem long enough to justify the effort and expense involved? (c) How long would it take before our day was extended to 48 h instead of 24 h? Assume that Earth is uniform throughout. Example 9-11

112. •• **Astronomy** In a little more than 5 billion years, our Sun will collapse to a white dwarf approximately 16,000 km in diameter. (Ignore the fact that the Sun will lose mass as it ages.) (a) What will our Sun's angular momentum and rotation rate be as a white dwarf? (Express your answers as multiples of its present-day values.) (b) Compared to its present value, will the Sun's rotational kinetic energy increase, decrease, or stay the same when it becomes a white dwarf? If it does change, by what factor will it change? The radius of the Sun is presently 6.96×10^8 m. Example 9-14

113. • **Astronomy** (a) If all the people in the world (~7 billion) lined up along the equator, would Earth's rotation rate increase or decrease? Justify your answer. (b) How would the rotation rate change if all people were no longer on Earth? Assume the average mass of a human is 70.0 kg. Example 9-14

114. • A 1.00×10^3-kg merry-go-round (a flat, solid cylinder) supports 10 children, each with a mass of 50.0 kg, located at the axis of rotation (thus you may assume the children have no angular momentum at that location). Describe a plan to move the children such that the angular velocity of the merry-go-round decreases to one-half its initial value. Example 9-14

115. •• One way for pilots to train for the physical demands of flying at high speeds is with a device called the human centrifuge. It involves having the pilots travel in circles at high speeds so that they can experience forces greater than their own weight. The diameter of the centrifuge used by NASA is 17.8 m. (a) Suppose a pilot starts at rest and accelerates at a constant rate so that he undergoes 30 rev in 2 min. What is his angular acceleration (in rad/s^2)? (b) What is his angular velocity (in rad/s) at the end of that time? (c) After the 2-min period, the centrifuge moves at a constant speed. The g-force experienced is the centripetal force keeping the pilot moving along a circular path. What is the g-force experienced by the pilot (1 g = mass × 9.80 m/s^2)? (d) The pilot can tolerate 12 g in the horizontal direction. How long would it take the centrifuge to reach that state if it starts at the angular speed found in part (c) and accelerates at the rate found in part (a)? Example 9-2

116. •• A flywheel of mass 35.0 kg and diameter 60.0 cm is spinning at 400 rpm when it experiences a sudden power loss. The flywheel slows due to friction in its bearings during the 20.0 s the power is off. If the flywheel makes 200 complete revolutions during the power failure, (a) At what rate is the flywheel spinning when the power comes back on? (b) How long would it have taken for the flywheel to come to a complete stop? Example 9-2

117. • A uniform cylinder of radius 2.50 cm and a uniform solid sphere of radius 8.00 cm are rolling without slipping along the same floor. The two objects have the same mass. If they are to have the same total kinetic energy, what should the ratio $\omega_{cylinder}/\omega_{sphere}$ of their angular speeds be? Example 9-13

118. ••• A fan is designed to last for a certain amount of time before it will have to be replaced (that is, planned obsolescence). The fan has one speed (at a maximum of 750 rpm), which it reaches in 2 s starting from rest. It takes the fan 10 s to stop rotating once it is turned off. If the manufacturer specifies that the fan will operate up to 1 billion rotations, estimate how many days you will be able to use the fan if you run it for 4 hours per day, on average. Example 9-2

119. •• Model a wooden pencil as a 20-cm-long uniform rod with a mass of 5.5 g. What is its moment of inertia as it is spun about its center by a nervous student during an exam? Example 9-4

Vladimir Pcholkin/Exactostock-1672/Superstock

10
Elastic Properties of Matter: Stress and Strain

In this chapter, your goals are to:

- (10-1) Explain the meaning of stress and strain in physics.
- (10-2) Define Hooke's law for an object under tension or compression.
- (10-3) Explain the importance of the minus sign in the relationship between volume stress and volume strain.
- (10-4) Define what is meant by shear, and explain what shear stress and shear strain are.
- (10-5) Explain what must happen to a material for it to go from elastic to plastic to failure, and why biological materials do not become plastic.

To master this chapter, you should review:

- (1-4) The rules for working with significant figures.
- (6-5) Hooke's law for the force required to stretch or compress an ideal spring.

What do you think?

This athlete's arms actually shorten slightly as they support her weight. If the diameter of her arms were only half as much, would the amount that her arms shorten be (a) the same, (b) greater, or (c) less?

Figure 10-1 Stress and strain
When an object such as an Achilles tendon is put under stress, it undergoes strain.

10-1 When an object is under stress, it deforms

BioMedical So far in our study of physics we've treated objects as perfectly rigid. That is, an object's shape and size don't change no matter what forces we apply to it. But what about real objects that *do* change shape when forces are applied? As an example, look at the Achilles tendon that connects the heel bone to the muscles of the calf (Figure 10-1). When you bend your ankle to point your toes upward, the calf muscles pull upward on the top of the Achilles tendon just as hard as the heel bone pulls downward on the bottom of the tendon. Hence the net force on the tendon is zero, and the tendon as a whole doesn't

This Achilles tendon is under **stress**: The muscles at its upper end exert an upward force *F* on the tendon, and the heel bone at its lower end exerts an equally strong downward force on the tendon. The tendon responds by undergoing **strain** (it stretches).

This Achilles tendon is relaxed.

Rachel Torres/Alamy

💡 **Stress** refers to forces that act to deform an object (change its size or shape). **Strain** refers to the deformation caused by these forces.

Figure 10-2 Three kinds of stress You apply (a) tensile stress when you stretch an object along its length, (b) volume stress when you squeeze it on all sides, and (c) shear stress when you apply offset forces (as with a pair of shears).

(a)

This rope is under **tensile stress:** Forces are applied along its length to stretch it.

Tetra Images, LLC/Alamy

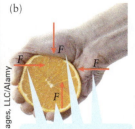

(b)

This orange is under **volume stress:** Forces are applied on all sides to squeeze it.

pbombaert/Shutterstock

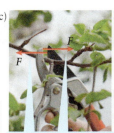

(c)

This branch is under **shear stress:** Offset forces are applied to deform it.

accelerate. However, individual *parts* of the tendon move as the tendon stretches and deforms. In this chapter, we'll examine the physics of stretching, squeezing, compressing, and twisting materials.

We say that the applied forces in Figure 10-1 exert a **stress** on the tendon that makes it deform, and we call the resulting deformation the **strain**. We'll see that there are three distinct kinds of stress, which differ in the manner in which they are applied (Figure 10-2). One of our principal goals will be to see how much an object deforms when we apply a particular amount of stress—for example, how much a tendon stretches when you flex a muscle, or how much a skin diver's body compresses due to water pressure as she dives to the bottom of a lake.

Some materials, such as your Achilles tendon, your earlobes, or the tip of your nose, are *elastic*: If you apply a stress to stretch or squeeze them, they snap back to their original shape after the stress is removed. Other materials, like a well-chewed piece of bubble gum, are *plastic*: They exhibit a permanent change of shape when stretched. Even an elastic material like rubber can remain irreversibly deformed if large enough forces are applied. The material might even break apart and *fracture*, like a plastic fork that's bent too far. We'll look at the differences among these kinds of behavior later in this chapter.

WATCH OUT! **Stress and strain are different quantities.**

In everyday language the words "stress" and "strain" can mean pretty much the same thing. You might say, "I'm under a lot of stress" or "I'm under a lot of strain." In physics, however, stress refers to forces that make a body deform, and strain refers to how much deformation those forces cause.

TAKE-HOME MESSAGE FOR Section 10-1

✔ An object is said to be under stress when forces act on the object to deform it (change its shape or size). The amount of deformation that results is called the strain.

✔ Elastic materials return to their original shape and size when the stress is removed, while plastic materials exhibit a permanent change in shape.

10-2 An object changes length when under tensile or compressive stress

(a)

This gymnast's upper leg is under **tension:** The ring pulls up on her leg and her hip pulls down on her leg.

Muriel de Seze/Getty Images

Figure 10-3 Tension and compression (a) An object under tension tends to stretch along its length; (b) an object under compression tends to shorten along its length.

(b)

This gymnast's arm is under **compression:** The shoulder pushes down on her arm and the ring in her hand pushes up on her arm.

Christoph Wilhelm/Getty Images

The Achilles tendon shown in Figure 10-1 and the rope shown in Figure 10-2a are under a kind of stress called **tension** or *tensile stress*: A force is applied at each end that makes the tendon or rope stretch. Figure 10-3a shows a gymnast whose upper legs are under tension. (Don't confuse this with the emotion of feeling "under tension," such as may happen before taking a physics exam!) A closely related kind of stress is **compression** or *compressive stress*, which occurs when a force is applied to each end of an object that tends to squeeze the object. The gymnast's arms in Figure 10-3b are under compression. If you grab each end of a pencil and try to pull the ends apart, you're putting the pencil under tension; if you push the ends toward each other, you're putting the pencil under compression.

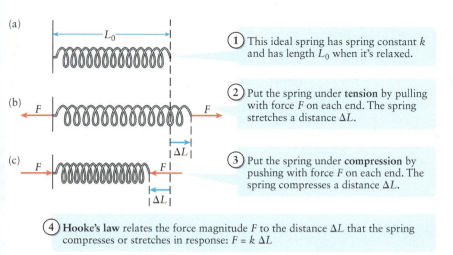

(a)

L_0

① This ideal spring has spring constant k and has length L_0 when it's relaxed.

(b)

F F

② Put the spring under **tension** by pulling with force F on each end. The spring stretches a distance ΔL.

$|\Delta L|$

(c)

F F

③ Put the spring under **compression** by pushing with force F on each end. The spring compresses a distance ΔL.

$|\Delta L|$

④ **Hooke's law** relates the force magnitude F to the distance ΔL that the spring compresses or stretches in response: $F = k\,\Delta L$

Figure 10-4 Hooke's law for an ideal spring When an ideal spring is under tension or compression, the change in the spring's length is directly proportional to the forces applied to its ends.

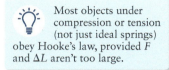
Most objects under compression or tension (not just ideal springs) obey Hooke's law, provided F and ΔL aren't too large.

Hooke's Law

How do objects behave when they are under tension or compression? We learned the answer to this question in Chapter 6 for one kind of object: an ideal spring (Figure 10-4a). When we stretch a spring along its length, we apply forces of the same magnitude F to both ends (Figure 10-4b). (This ensures that the net external force on the spring is zero, so the spring as a whole doesn't accelerate from rest.) According to **Hooke's law**, which we first encountered in Section 6-5, the force magnitude F is proportional to the distance ΔL that the spring stretches, provided that ΔL is small compared to the relaxed length of the spring, L_0:

Magnitude of the **force you apply to each end** of the object

Spring constant k tells you how stiff the object is: Larger k means stiffer.

$$F = k\,\Delta L$$

Magnitude of the **change in the object's length** as a result of the force you apply

Hooke's law
(10-1)

(In Chapter 6 we used the symbol x for the amount that the spring stretches. We've changed this to ΔL to emphasize that this distance is the *change* in the overall length of the spring.)

The quantity k in Equation 10-1 is called the **spring constant**: It's a measure of the stiffness of the spring. To see this, note that we can rewrite Equation 10-1 as $k = F/\Delta L$. The stiffer the spring, the shorter the amount of stretch ΔL for a given force magnitude F and so the greater the value of $k = F/\Delta L$. If the spring is loose and floppy, there is more stretch for a given force magnitude and $k = F/\Delta L$ has a smaller value. The SI units of k are newtons per meter (N/m).

Hooke's law also applies to an ideal spring under compression (Figure 10-4c): When compressed by forces of magnitude F, an ideal spring decreases in length by a distance ΔL given by Equation 10-1. The value of k for an ideal spring is the same for compression as for tension. Hence if stretching an ideal spring with forces of magnitude F makes its length increase by 1.00 mm, squeezing that spring with forces of the same magnitude F will make its length decrease by 1.00 mm.

Hooke's law is important because it doesn't apply to just ideal springs. Experiment shows that it holds true for most objects when they are stretched, provided that the amount of stretch is small. An example is a rubber strip made by cutting apart a rubber band (Figure 10-5a). When put under tension, the strip stretches by a distance ΔL that's directly proportional to the force magnitude F (Figure 10-5b). Likewise, when the strip is compressed, it shrinks by a distance ΔL that's directly proportional to F.

Experiment shows that the spring constant k for a given object depends on three things: its relaxed length L_0, its cross-sectional area A, and the material of which it is made (for example, the particular kind of rubber). The last of these is given by

Figure 10-5 Hooke's law for tension and compression An object under tension obeys Hooke's law if the stress is not too great. The same is true for an object under compression.

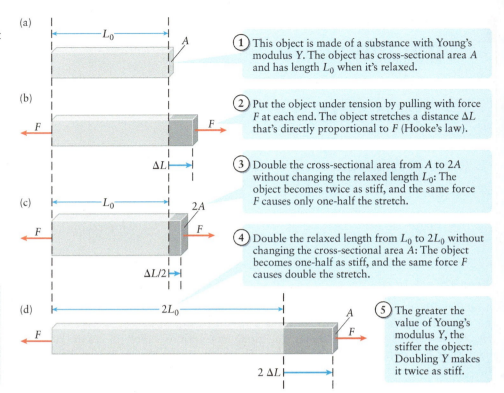

(a)

① This object is made of a substance with Young's modulus Y. The object has cross-sectional area A and has length L_0 when it's relaxed.

(b)

② Put the object under tension by pulling with force F at each end. The object stretches a distance ΔL that's directly proportional to F (Hooke's law).

③ Double the cross-sectional area from A to 2A without changing the relaxed length L_0: The object becomes twice as stiff, and the same force F causes only one-half the stretch.

(c)

④ Double the relaxed length from L_0 to $2L_0$ without changing the cross-sectional area A: The object becomes one-half as stiff, and the same force F causes double the stretch.

The object's spring constant is proportional to its Young's modulus and its cross-sectional area but inversely proportional to its length: $k = Y\dfrac{A}{L_0}$.

(d)

⑤ The greater the value of Young's modulus Y, the stiffer the object: Doubling Y makes it twice as stiff.

a quantity called **Young's modulus** (or elastic modulus) that measures how stiff the material is. The greater the value of Young's modulus, denoted by the symbol Y, the greater the material's intrinsic stiffness. Experiment shows that k is directly proportional to Y and A but inversely proportional to L_0 (**Figures 10-5c** and **10-5d**). The mathematical relationship that describes these observations is

Spring constant for an object under tension or compression

(10-2)

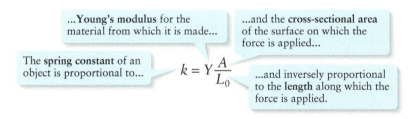

...**Young's modulus** for the material from which it is made...

...and the **cross-sectional area** of the surface on which the force is applied...

The **spring constant** of an object is proportional to...

$$k = Y\frac{A}{L_0}$$

...and inversely proportional to the **length** along which the force is applied.

This says that a short, thick object (small L_0, large A) is stiffer than a long, thin object (large L_0, small A) made of the same material and hence with the same value of Y. That's why a rubber eraser is harder to stretch than a rubber band. Equation 10-2 also says that the stiffer the material used to make an object—that is, the greater the value of Y—the stiffer the object will be.

WATCH OUT! Young's modulus is not the same as the spring constant.

Don't confuse Young's modulus Y with the spring constant k. Young's modulus tells you how stiff a given *material* is; the spring constant refers to a particular *object* made of that material. If two objects have the same shape and size (and so the same values of A and L_0) but are made of different materials, Equation 10-2 says the object made of the material with greater Y will be harder to stretch or compress and will have a greater value of k. Equation 10-2 also says that if two objects are made of the same material (so Y is the same) but have different dimensions (so A, L_0, or both are different), their spring constants k can be different.

The first column of Table 10-1 gives specific values of Young's modulus Y for various materials. The SI units of Y are newtons per square meter, or N/m^2. This unit

TABLE 10-1 Elastic Properties of Selected Materials

Material	Young's Modulus (10^9 Pa)	Bulk Modulus (10^9 Pa)	Shear Modulus (10^9 Pa)
aluminum	70	76	26
brass	100	80	40
concrete	30	13	15
iron	211	170	82
nylon	3		4.1
rubber band	0.005		0.003
steel	200	140	78
air		1.41×10^{-4}	
ethyl alcohol		1.0	
water		2.2	
human ACL	0.1		
human lung		$1.5 – 9.8 \times 10^{-6}$	
pig endothelial cell			2×10^{-5}

is also called the **pascal**, abbreviated Pa: 1 Pa $= 1$ N/m^2. As an example, $Y = 5 \times 10^6$ Pa for the rubber in a typical rubber band, while $Y = 2 \times 10^{11}$ Pa for steel. So compared to the force required to stretch a rubber band, the force required to stretch by the same amount a steel rod of the same cross-sectional area and length would be $(2 \times 10^{11})/(5 \times 10^6) = 4 \times 10^4$ times greater. Materials found in the human body have a wide range of values of Y (Figure 10-6).

GOT THE CONCEPT? 10-1 Compare the Spring Constants

(?) Rank the following four rods in order of how easy they are to stretch, from easiest to most difficult. All four rods are made of the same material. If any two rods are equally easy to stretch, say so. (a) A rod 5.00 cm long with cross-sectional area 2.00 mm^2; (b) a rod 2.50 cm long with cross-sectional area 2.00 mm^2; (c) a rod 5.00 cm long with cross-sectional area 1.00 mm^2; (d) a rod 2.50 cm long with cross-sectional area 1.00 mm^2.

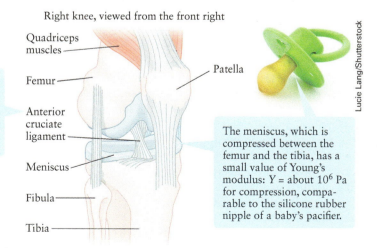

Right knee, viewed from the front right

Quadriceps muscles

Patella

Femur

The anterior cruciate ligament (ACL) has a moderate value of Young's modulus: Y = about 10^8 Pa for tension.

Anterior cruciate ligament

Meniscus

The meniscus, which is compressed between the femur and the tibia, has a small value of Young's modulus: Y = about 10^6 Pa for compression, comparable to the silicone rubber nipple of a baby's pacifier.

Fibula

Tibia

Lucie Lang/Shutterstock

Figure 10-6 Young's modulus in the knee How do the values of Young's modulus for common materials compare to the values for two components of the human knee?

 An ACL is about 1/2000 as stiff as a steel rod ($Y = 2 \times 10^{11}$ Pa) of the same dimensions, but 20 times stiffer than a rubber band ($Y = 5 \times 10^6$ Pa). Hence your knee can bend freely (which an ACL of steel would not allow) without letting your tibia flop around loosely (as would happen with an ACL of rubber).

Relating Stress and Strain

Here's a useful way to rewrite Hooke's law for an object under tension or compression. This will also help us define *stress* and *strain* more precisely. First we substitute k from Equation 10-2 into Equation 10-1:

$$F = \left(Y \frac{A}{L_0} \right) \Delta L$$

Then we divide both sides by the cross-sectional area A:

> **Tensile or compressive stress** on the object: applied force F divided by the object's cross-sectional area A

> **Young's modulus** Y tells you how stiff the material is of which the object is made: Larger Y means stiffer.

Hooke's law for an object under tension or compression (10-3)

$$\frac{F}{A} = Y \frac{\Delta L}{L_0}$$

> **Tensile or compressive strain** of the object: resulting change in length ΔL divided by the object's relaxed length L_0

Figure 10-7 depicts the quantities included in Equation 10-3. The ratio F/A, equal to the force *per area* exerted on each end of the object, is called the **tensile stress** (if the object is being stretched) or **compressive stress** (if the object is being squeezed). Like Young's modulus, stress has units of newtons per square meter (N/m^2), or pascals (Pa).

The ratio $\Delta L/L_0$ that appears on the right-hand side of Equation 10-3 is called the **tensile strain** or **compressive strain**: It equals the distance that the object stretches or shrinks expressed as a fraction of its relaxed length. Tensile or compressive strain is a length divided by another length, so it's a dimensionless quantity.

Equation 10-3 tells us that we can express Hooke's law for tension or compression as follows:

(10-4)
$$\text{(tensile or compressive stress)} = \text{(Young's modulus)} \times \text{(tensile or compressive strain)}$$

That is, *stress and strain are directly proportional*, and the constant of proportionality is Young's modulus. This gives us an alternative way to interpret Figure 10-5. Compared to the object under tension in Figure 10-5b, the object in Figure 10-5c is under one-half the tensile stress (the force is the same but the cross-sectional area is twice as great). Hence the tensile strain is also reduced by one-half: Both objects have the same relaxed length L_0, so the object in Figure 10-5c stretches only half as far as the object in Figure 10-5b. The objects in Figures 10-5b and 10-5d have the same force at each end and the same cross-sectional area, so both objects are under the same tensile stress and the tensile strain is the same for both. Since the object in Figure 10-5d is twice as long, it stretches twice as far as the object in Figure 10-5b so that the ratio of the stretch distance to the relaxed length is the same.

Here's an analogy that illustrates the importance of the ratio of applied force (F) to the area over which it's applied (A). Consider how it feels to stand barefoot on a floor compared to how it might feel to stand barefoot on the head of a nail. In both cases the force F of the supporting surface on you is equal to your weight. When that force is applied over the relatively large surface area of your feet, so F/A is small and there is little compressive stress on your foot, you will be comfortable. But if you stand on the small surface area of the nail head so F/A is large, it will definitely hurt!

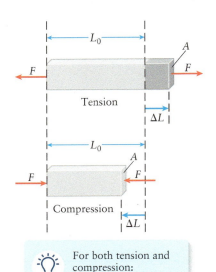

For both tension and compression:
stress = F/A
strain = $\Delta L/L_0$
Hooke's law: stress = $Y \times$ strain

Figure 10-7 Stress and strain for tension or compression Hooke's law relates the stress on an object to the resulting strain of the object.

GOT THE CONCEPT? 10-2 Stress, Strain, and Young's Modulus

You pull with equal force on each end of two rods of identical cross-sectional area. Both rods obey Hooke's law. Rod 1, made of material P, stretches 1.00 mm. Rod 2, made of material Q, stretches 2.00 mm. Which material has the greater value of Young's modulus? (a) Material P; (b) material Q; (c) they both have the same Young's modulus; (d) not enough information given to decide. *Hint:* Read the details of this question carefully.

WATCH OUT! Hooke's law has important limitations.

❗ Hooke's law in Equation 10-3 applies *only* when the tensile strain is small, that is, when the change in length is small relative to the overall initial length. In Section 10-5 we'll investigate what can happen if the strain is *not* small. Hooke's law also applies only to materials that are of a uniform composition, such as a solid bar of iron. Furthermore, for metals the value of Young's modulus is the same for compression as for tension; for other materials the values for compression and tension may be very different. In the following example and in Section 10-5 we'll also look at biological materials like ligaments and tendons, which are not of uniform composition and so don't obey Hooke's law very faithfully. For many purposes, however, it's a reasonable approximation to apply Hooke's law even to biological materials like these.

BioMedical EXAMPLE 10-1 Tensile Stress and Strain

You are doing a laboratory experiment with a specimen cut from the sternoclavicular ligament in the shoulder of a cadaver (Figure 10-8). The segment is 2.00 by 2.00 mm in cross section and 30.0 mm long. When you apply a force of 1.00 N to each end of the specimen, the specimen stretches by 1.20 mm. Determine (a) the tensile stress on the specimen, (b) the tensile strain of the specimen, and (c) Young's modulus for the specimen. (d) If you repeated the experiment applying the same forces to a second specimen of ligament that has the same length but is 4.00 by 4.00 mm in cross section, how would the answers to (a), (b), and (c) change? How far would this specimen stretch?

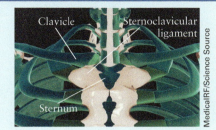

Figure 10-8 Sternoclavicular ligament Ligaments connect one bone to another in the body; the sternoclavicular ligaments connect the sternum (breastbone) to the clavicles.

Set Up

We are given the dimensions of the specimen (shown in the figure) and the force $F = 1.00$ N applied to each end. We'll find the tensile stress and tensile strain using the definitions of these quantities, and we'll find Young's modulus Y using Hooke's law for tension.

$$\text{tensile stress} = \frac{F}{A}$$

$$\text{tensile strain} = \frac{\Delta L}{L_0}$$

$$(\text{tensile stress}) = (\text{Young's modulus}) \times (\text{tensile strain}) \qquad (10\text{-}4)$$

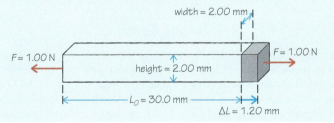

Solve

(a) The tensile stress is the applied force $F = 1.00$ N divided by the cross-sectional area. We convert millimeters to meters.

$$\text{area } A = (2.00 \text{ mm}) \times (2.00 \text{ mm})$$
$$= (2.00 \times 10^{-3} \text{ m}) \times (2.00 \times 10^{-3} \text{ m}) = 4.00 \times 10^{-6} \text{ m}^2$$

$$\text{tensile stress} = \frac{F}{A} = \frac{1.00 \text{ N}}{4.00 \times 10^{-6} \text{ m}^2}$$

$$= 2.50 \times 10^5 \text{ N/m}^2 = 2.50 \times 10^5 \text{ Pa}$$

(b) The tensile strain is the change in length $\Delta L = 1.20$ mm divided by the relaxed length $L_0 = 30.0$ mm.

$$\text{tensile strain} = \frac{\Delta L}{L_0} = \frac{1.20 \text{ mm}}{30.0 \text{ mm}}$$

$$= 0.0400 = 4.00 \times 10^{-2}$$

(c) Rearranging Hooke's law tells us that Young's modulus for this specimen equals the tensile stress divided by the tensile strain.

$$\text{Young's modulus } Y = \frac{\text{tensile stress}}{\text{tensile strain}} = \frac{2.50 \times 10^5 \text{ Pa}}{4.00 \times 10^{-2}}$$

$$= 6.25 \times 10^6 \text{ Pa}$$

(d) If each of the cross-sectional dimensions doubles from 2.00 to 4.00 mm, the cross-sectional area A increases by a factor of $2 \times 2 = 4$, so the tensile stress is $\frac{1}{4}$ as much as the previous value of 2.50×10^5 Pa. The value of Young's modulus is *unaffected*, since this depends on the type of material and not on its dimensions. From Hooke's law, the new tensile strain equals the new tensile stress divided by Y. Since Y is the same but the tensile stress is $\frac{1}{4}$ the previous value, the tensile strain is likewise $\frac{1}{4}$ of the previous value of 4.00×10^{-2}. The new specimen therefore stretches by $\frac{1}{4}$ as much as the first one.

New cross-sectional area A

$$= (4.00 \text{ mm}) \times (4.00 \text{ mm})$$
$$= (4.00 \times 10^{-3} \text{ m}) \times (4.00 \times 10^{-3} \text{ m})$$
$$= 1.60 \times 10^{-5} \text{ m}$$

New tensile stress

$$= \frac{F}{A} = \frac{1.00 \text{ N}}{1.60 \times 10^{-5} \text{ m}^2} = 6.25 \times 10^4 \text{ N/m}^2 = 6.25 \times 10^4 \text{ Pa}$$

Young's modulus $Y = 6.25 \times 10^6$ Pa

New tensile strain

$$= \frac{\text{new tensile stress}}{\text{Young's modulus}} = \frac{6.25 \times 10^4 \text{ Pa}}{6.25 \times 10^6 \text{ Pa}} = 1.00 \times 10^{-2}$$

$$\Delta L = (\text{new tensile strain}) \times L_0 = (1.00 \times 10^{-2})(30.0 \text{ mm})$$
$$= 0.300 \text{ mm}$$

Reflect

This example emphasizes the key message of Hooke's law: For relatively small stresses the stress is proportional to the strain. Note that the value of Young's modulus for the sternoclavicular ligament is quite a bit less than that for the ACL ($Y = 1 \times 10^8$ Pa). This is because different ligaments in different parts of the body are made of different materials.

For this 2.00 mm × 2.00 mm × 30.0 mm segment of sternoclavicular ligament, can you show that a stretching force of 0.025 N would be required to increase the 30.0-mm length of the segment by 0.10%?

GOT THE CONCEPT? 10-3 Compressive Stress and Strain

(?) A brass rod ($Y = 1 \times 10^{11}$ Pa) and a steel rod ($Y = 2 \times 10^{11}$ Pa) of the same dimensions are subjected to identical squeezing forces on their ends. (a) Which experiences the greater compressive stress, the brass rod or the steel rod? (b) Which undergoes the greater compressive strain, the brass rod or the steel rod? (c) Which undergoes the greatest change in length, the brass rod or the steel rod?

GOT THE CONCEPT? 10-4 Tensile Stress

(?) Four rods that obey Hooke's law are each put under tension. Rank them according to the tensile stress on each rod, from smallest to largest value. (a) A rod 50.0 cm long with cross-sectional area 1.00 mm² and with a 200-N force applied on each end; (b) a rod 25.0 cm long with cross-sectional area 1.00 mm² and with a 200-N force applied on each end; (c) a rod with cross-sectional area 2.00 mm² with a 100-N force applied on each end; (d) a rod with Young's modulus 2.00×10^{10} Pa and relaxed length 50.0 cm that stretches 0.025 mm due to forces on its ends.

TAKE-HOME MESSAGE FOR Section 10-2

✔ Young's modulus Y is a measure of how easy it is to stretch or compress an object.

✔ Tensile stress equals the stretching force exerted on each end of an object divided by the object's cross-sectional area. The resulting change in length relative to the relaxed length, $\Delta L/L_0$, is called the tensile strain.

✔ We use the terms "compressive stress" and "compressive strain" for the case in which an object is squeezed along a single direction rather than stretched.

✔ If an object obeys Hooke's law, stress and strain are directly proportional.

✔ In problems involving tension or compression, use Hooke's law, Equation 10-3, to relate the stress (force divided by cross-sectional area) to the strain (change in length divided by relaxed length) and Young's modulus for the material. Then solve for the unknown quantity.

10-3 An object expands or shrinks when under volume stress

We saw in Section 10-2 that applied forces can cause an object to stretch or compress along its length. Applied forces can also cause changes in an object's *volume*. You can see this effect when a fire extinguisher is discharged (Figure 10-9a). Inside the fire extinguisher, the gas (which is the object in question) is under high pressure, and this pressure keeps the gas confined within the small volume of the extinguisher cylinder. When you trigger the fire extinguisher, the released gas moves into the low-pressure outside air and expands to a much larger volume. The *volume stress* on the extinguisher gas is the *change* in the pressure that pushes on the gas, and the *volume strain* is a measure of how much the volume of this gas changes in response to that pressure change.

BioMedical For the gas that escapes from the fire extinguisher in Figure 10-9a, the volume stress is negative (the pressure on the gas decreases), and the volume strain is positive (the volume of the gas increases). The reverse is true for a descending scuba diver (Figure 10-9b). The deeper in the ocean she dives, the greater the pressure on her body, so the volume stress is positive (the pressure on her body increases). This causes gases within her body to compress, so their volume strain is negative (the volume of the gas decreases). These compressed gases dissolve in her body fluids. If a diver ascends too quickly, causing a rapid decrease in pressure, the gases may pop out of solution and form bubbles that can cause excruciating pain (a condition called *decompression sickness*). To avoid this, scuba divers are trained to return to the surface at a gradual pace so that the pressure on their bodies decreases gradually. The dissolved gases then expand slowly and can safely be exhaled.

Relating Volume Stress and Volume Strain: Hooke's Law Revisited

Let's be more quantitative about volume stress and volume strain. First notice that every object around you has forces exerted on it by the surrounding material (Figure 10-10a). Your body has forces exerted on it by the air around you; a submerged fish has forces exerted on it by the water in which it swims; and a rock buried below Earth's surface has forces exerted on it by the surrounding dirt. The magnitude of the force per unit area on the surface of the object is called the **pressure** p. (Unfortunately, we used this same symbol in Chapter 7 for momentum. We promise never to use these two quantities in the same equation!) The units of pressure are newtons per square meter (N/m^2) or pascals (Pa). This is the same unit we used for the tensile or compressive stress (force per unit area), which can also be thought of as a pressure.

As Figure 10-10 illustrates, an object's volume changes when the pressure p changes. We call the pressure change Δp the **volume stress**, and we call the resulting change in volume ΔV divided by the original volume V_0—that is, the quantity $\Delta V/V_0$—the **volume strain**. If the pressure increases so that the volume stress Δp is positive, then the object shrinks, and the volume strain is negative (Figure 10-10b). If the pressure decreases so that the volume stress is negative, then the object expands, and the volume strain is positive (Figure 10-10c). Note that volume stress, like pressure, has units of N/m^2 or Pa, while volume strain is dimensionless (it's a volume divided by another volume). These are the same units as the quantities tensile stress and tensile strain that we studied in Sections 10-2 and 10-3.

Experiment shows that if the volume stress on an object is not too large, the resulting volume strain is *directly proportional* to the volume stress. That is, Hooke's law relates volume stress and strain in the same manner as it does for tensile stress and strain:

$$\text{(volume stress)} = -(\text{bulk modulus}) \times (\text{volume strain})$$ (10-5)

That is, *volume stress and volume strain are directly proportional.* The proportionality constant in Equation 10-5 is called the **bulk modulus** of the material of which the object is made. Like Young's modulus, the bulk modulus B has units of N/m^2 or Pa. The second column of Table 10-1 lists the values of the bulk modulus

(a) Triggering a fire extinguisher releases the high-pressure gases inside the tank. The pressure is much lower outside the tank, so the gases expand.

Michael Blann/Getty Images

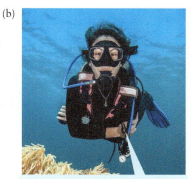

(b) As this scuba diver descends, the pressure on her increases. This compresses gases inside her body and makes them dissolve in her body fluids.

Image Source/Alamy

 Volume stress refers to the change in pressure on an object.
Volume strain refers to the change in the object's volume caused by this pressure change.

Figure 10-9 Volume stress and strain Volume stress involves a change in pressure. (a) A pressure decrease makes the volume increase. (b) A pressure increase makes the volume decrease.

Figure 10-10 Hooke's law for volume stress An object under volume stress obeys Hooke's law if the stress is not too great.

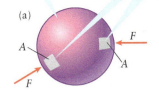

① Initially this object has volume V_0. It is made of a material with bulk modulus B.

② The object's surroundings exert a force of magnitude F (shown by the red vectors) on each area A of the object's surface. The pressure is $p = F/A$.

(a)

③ If the pressure on the object changes by Δp, the object's volume changes by ΔV as indicated by the blue arrows.

(b)

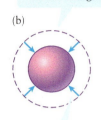

(c)

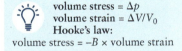

volume stress = Δp
volume strain = $\Delta V/V_0$
Hooke's law:
volume stress = $-B \times$ volume strain

④ If the pressure increases ($\Delta p > 0$), the volume decreases ($\Delta V < 0$).

⑤ If the pressure decreases ($\Delta p < 0$), the volume increases ($\Delta V > 0$).

of various substances. The minus sign in Equation 10-5 tells us that volume stress and volume strain have opposite signs; if one is positive, the other is negative (see parts (b) and (c) of Figure 10-10).

In terms of the quantities Δp, V_0, and ΔV depicted in Figure 10-10, we can rewrite Equation 10-5 as

Volume stress on the object: change in the pressure p on all sides of the object

Bulk modulus B: how difficult it is to compress the material of which the object is made; larger B means harder to compress

Hooke's law for an object under volume stress
(10-6)

$$\Delta p = -B\frac{\Delta V}{V_0}$$

Important minus sign!
If Δp is positive, then ΔV is negative; if Δp is negative, then ΔV is positive.

Volume strain of the object: resulting change in volume ΔV divided by the object's initial volume V_0

To get more insight into the meaning of Equation 10-6, multiply both sides by $-V_0/B$ to rewrite it so that ΔV is by itself on one side of the equation:

(10-7)

$$\Delta V = -\frac{\Delta p}{B}V_0$$

Equation 10-7 says that for a given pressure change Δp and initial volume V_0, making the bulk modulus B larger makes the change in volume ΔV smaller. A material that has a relatively large bulk modulus is relatively *incompressible*, which means that its volume doesn't change much even when it experiences a large volume stress. From the entries in Table 10-1 you can see that materials such as iron and concrete have very large bulk moduli; they are hard to compress, which makes them good materials for constructing buildings. By contrast, air has a bulk modulus more than a million times smaller than that of steel and is relatively easy to compress. The human lung is made of even more compressible material: Under physiologic conditions, the bulk modulus of the human lung varies between about 1500 and 10,000 Pa in young adults and increases with age. These low values of B minimize the effort needed to make the lungs expand and contract during respiration.

Another interpretation of the bulk modulus can be gained by examining Equation 10-7 in the case when Δp is equal to B. When this happens we see that ΔV is equal to $-V_0$. That is, the object has shrunk to zero volume! So we can think of B as the *increase* in pressure required to completely shrink an object made of that material. It's pretty challenging to shrink most objects to a point that has zero volume. That's one reason B is so large. If Δp is equal to $-B$, then ΔV is equal to V_0. That is, the volume of the object has doubled. So bulk modulus can also be interpreted as the *decrease* in pressure required for an object to double its volume.

In the following examples we'll look at two problems that involve volume stress and strain. Note that the ideas of volume stress and strain apply equally well to objects that are gases, liquids, and solids.

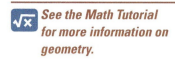
See the Math Tutorial for more information on geometry.

EXAMPLE 10-2   A Submerged Cannonball

A solid iron cannonball falls into the Pacific near the Galapagos Islands, where the ocean is 2400 m deep and the pressure is 2.40×10^7 Pa at the bottom. (By contrast, air pressure at the surface is 1.01×10^5 Pa.) By what percentage does the cannonball shrink in volume as it sinks to the bottom?

Set Up

We use Hooke's law, Equation 10-6, to find the volume strain $\Delta V/V_0$ of the cannonball due to the additional pressure exerted on it (the volume stress Δp). Since $\Delta V/V_0$ expresses the change in volume as a fraction, we can convert it to a percentage by multiplying it by 100.

Hooke's law for an object under volume stress:

$$\Delta p = -B\frac{\Delta V}{V_0} \qquad (10\text{-}6)$$

cannonball at surface
$p = 1.01 \times 10^5$ Pa,
volume V_0

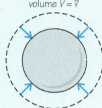
submerged cannonball
$p = 2.40 \times 10^7$ Pa,
volume $V = ?$

Solve

The volume stress is the difference between the pressure at the bottom of the ocean and the pressure at the surface.

Volume stress Δp:

= (pressure at bottom of ocean) − (pressure at surface)

= 2.40×10^7 Pa − 1.01×10^5 Pa = 2.39×10^7 Pa

Use the value of B for iron from Table 10-1 and solve for $\Delta V/V_0$. Then convert this to a percentage.

$$\frac{\Delta V}{V_0} = -\frac{\Delta p}{B} = -\frac{2.39 \times 10^7 \text{ Pa}}{1.7 \times 10^{11} \text{ Pa}}$$

$$= -1.4 \times 10^{-4}$$

Percentage change in volume:

$$\frac{\Delta V}{V_0} \times 100\% = -1.4 \times 10^{-4} \times 100\%$$

$$= -0.014\%$$

Reflect

The minus sign tells us that $\Delta V < 0$ and so the volume of the cannonball decreases. Because iron has such a large bulk modulus B, even this immense pressure increase causes only a miniscule change in the cannonball's volume, just 14/1000 of 1%.

Note that while the volume stress Δp is given to three significant figures, the answer has only two significant figures. Can you see why? (See the discussion of significant figures in Section 1-4 if you're not sure.)

EXAMPLE 10-3   Bubbles Rising

A scuba diver 10.0 m below the surface of a lake releases a spherical air bubble 2.00 cm in radius. As it rises to the surface, the pressure exerted by the surrounding water on the bubble decreases by 9.80×10^4 Pa due to the decreasing amount of water above the bubble. This pressure decrease causes the bubble to expand. Use Hooke's law to find the radius of the bubble as it breaks the surface.

Set Up

As in Example 10-2, we use the Hooke's law relationship between volume stress and volume strain, Equation 10-6. This gives us the change in volume ΔV versus the initial volume V_0. We assume that the amount of air in the bubble doesn't change as it rises. Thus the change in volume of the bubble is due only to the volume stress, which equals the change in water pressure pushing on the outside of the bubble.

$$\Delta p = -B \frac{\Delta V}{V_0} \quad (10\text{-}6)$$

volume stress = Δp
$= -9.80 \times 10^4$ Pa

(negative because the pressure on the outside of the bubble decreases)

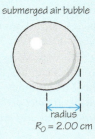
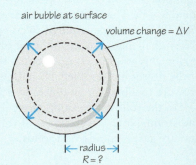

submerged air bubble

air bubble at surface

volume change = ΔV

radius
$R_0 = 2.00$ cm

radius
$R = ?$

Solve

Use the bulk modulus of air from Table 10-1 and the value of Δp to solve for the volume strain $\Delta V/V_0$. The volume strain is positive because the bubble expands.

Bulk modulus of air from Table 10-1:
$B = 1.41 \times 10^5$ Pa, so

$$\frac{\Delta V}{V_0} = -\frac{\Delta p}{B} = -\frac{(-9.80 \times 10^4 \text{ Pa})}{1.41 \times 10^5 \text{ Pa}} = 0.695$$

Next we find the initial volume V_0 of the bubble using the formula for a sphere of radius $R_0 = 2.00$ cm (see Math Tutorial). The final volume of the bubble is the initial volume V_0 plus the volume change ΔV.

Initial volume of a spherical bubble:

$$V_0 = \frac{4\pi}{3} R_0^3 = \frac{4\pi}{3} (2.00 \text{ cm})^3$$

$$= 33.5 \text{ cm}^3$$

Final volume:

$$V = V_0 + \Delta V = V_0 \left(1 + \frac{\Delta V}{V_0} \right)$$

$$= (33.5 \text{ cm}^3)(1 + 0.695) = (33.5 \text{ cm}^3)(1.695)$$

$$= 56.8 \text{ cm}^3$$

Solve for the final radius R of the bubble using the formula $V = (4\pi/3)R^3$ for the volume of a sphere of radius R.

$$V = \frac{4\pi}{3} R^3$$

$$R = \sqrt[3]{\frac{3V}{4\pi}} = \sqrt[3]{\frac{3(56.8 \text{ cm}^3)}{4\pi}} = 2.38 \text{ cm}$$

Reflect

Our calculation shows that as the bubble rises to the surface, its volume increases by a factor of 1.695, and its radius increases from 2.00 to 2.38 cm. Consider the implications of this result on decompression sickness (see the discussion above of Figure 10-9b). When a diver ascends too quickly, not only do bubbles form, but they also get bigger as the diver rises to the surface.

It turns out that the answers we calculated here aren't exactly correct. The reason is that the air pressure *inside* the bubble also decreases as it expands, and as a result the bubble expands a bit less than we have calculated. A more detailed calculation shows that the volume actually increases by a factor of 1.623, and the final radius is 2.35 cm. Our calculations with Hooke's law are pretty close, however, which shows that this law can give quite useful results.

GOT THE CONCEPT? 10-5 Volume Stress and Volume Strain

Two identical balloons are filled to the same size, one with water and one with air. When the two balloons are subjected to the same volume stress (the same force per unit surface area pushing in on the balloon), which of the two will exhibit the larger change in volume: (a) the water-filled balloon or (b) the air-filled balloon?

GOT THE CONCEPT? 10-6 Pressure and Force

You apply a pressure of 5.00×10^6 N/m² on all sides of a sphere of radius 2.00 m. What is the net force that is exerted on the sphere as a result? (a) 6.28×10^7 N; (b) 1.68×10^8 N; (c) 2.51×10^8 N; (d) zero.

TAKE-HOME MESSAGE FOR Section 10-3

✔ Volume stress equals the *change* in pressure on an object.

✔ An object under volume stress either expands (if the stress is negative) or contracts (if the stress is positive); the volume strain is the change in the object's volume divided by its initial volume.

✔ If an object obeys Hooke's law, volume stress and volume strain are directly proportional; the constant of

proportionality is the bulk modulus B for the material of which the object is made.

✔ In problems involving volume stress and strain, use Hooke's law, Equation 10-6, to relate the stress (the change in pressure on an object) to the strain (change in the object's volume divided by its original volume) and the bulk modulus of the material. Then solve for the unknown quantity.

10-4 A solid object changes shape when under shear stress

When you run, play sports, or do any sort of physical activity, blood flow increases to your active tissues. The arteries need to expand or *dilate* to accommodate this increased blood flow. In addition to the dilation caused by increased blood pressure pushing outward against arterial walls, the deformation of endothelial cells that line the inside of the artery also leads to dilation of the vessel (Figure 10-11a, b). The force of blood pushing on the exposed surfaces of the cells deforms them in the direction of blood flow (Figure 10-11c). Because the cells are firmly attached to the arterial wall, the change in shape is neither a stretch nor a compression. Instead, the cells deform in a way similar to a cube of gelatin dessert being pushed parallel to the top face. In response to this deformation, the endothelial cells release nitric oxide gas. This gas diffuses to the muscle cells in the arterial wall and causes them to relax. As a result the artery dilates, thus relieving the stress on the endothelial cells (Figure 10-11d).

BioMedical

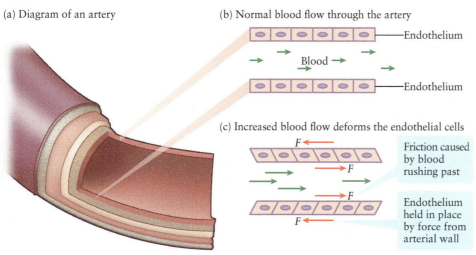

(a) Diagram of an artery

(b) Normal blood flow through the artery
— Endothelium
Blood →
— Endothelium

(c) Increased blood flow deforms the endothelial cells
F
F
F
F
Friction caused by blood rushing past
Endothelium held in place by force from arterial wall

(d) In response, surrounding muscle allows the artery to expand, relieving stress on the endothelium

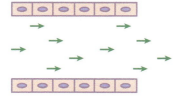

Shear stress refers to forces that deform an object [like the endothelium in (c)] without making it expand or contract. **Shear strain** refers to the change in the object's shape caused by this stress.

Figure 10-11 Shear stress and strain When blood flow through an artery increases due to physical exertion, the endothelium deforms without becoming larger or smaller. This is an example of shear stress.

The kind of stress shown in Figure 10-11c is called **shear**. Unlike tensile stress and compressive stress (which are applied along a line within the object) or volume stress (which is applied over the entire surface of the object), shear stress is caused by forces applied *parallel* (tangential) to the faces of an object. As Figure 10-11c shows, these forces are *offset* from each other: They do not act along the same line. An object is said to experience shear when one face is made to move or slide relative to the opposite face.

Like the other types of stress we've encountered, **shear stress** is defined as a force per unit area. As an example, consider the object shown in Figure 10-12a. In Figure 10-12b opposite forces of the same magnitude $F_\parallel$ are applied to the upper and lower surfaces of this object. The subscript "$\parallel$" means "parallel," and reminds us that these forces act parallel to the surfaces as in Figure 10-12b. (For the endothelial cells depicted in Figure 10-11c, one force is exerted by the blood that flows past the cells. The other force is exerted by the arterial wall to which the cells are attached.) The shear stress associated with these forces is the force magnitude $F_\parallel$ divided by the area A over which this force is applied, or $F_\parallel/A$. Just like tensile and volume stress, the units of shear stress are N/m^2 or Pa.

Figure 10-12 Shear stress and strain Shear stress involves oppositely directed forces that act on different surfaces of an object. These offset forces cause a twisting deformation (shear strain).

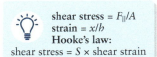

shear stress = $F_\parallel/A$
strain = x/h
Hooke's law:
shear stress = $S \times$ shear strain

① This object has height h. Its upper and lower surfaces each have area A. It is made of a material with shear modulus S.

(a)

② Apply a force of magnitude $F_\parallel$ parallel to the upper surface...

(b)

③ ...and apply an opposite force of the same magnitude $F_\parallel$ parallel to the lower surface. Note that the two forces are offset (they do not act along the same line).

④ As a result of these forces, the upper and lower surfaces move a distance x relative to each other.

Figure 10-12 shows that this shear stress causes the object to deform. We measure the amount of deformation by the **shear strain**, which is the displacement of one surface relative to the other (shown as x in Figure 10-12b) divided by the distance between the surfaces (in Figure 10-12b, the distance h). Thus the shear strain equals x/h.

Experiment shows that for most materials the shear strain is proportional to the shear stress that produces it, provided that the stress is not too great. This is yet another version of Hooke's law, this time for shear:

(10-8)   shear stress = shear modulus × shear strain

The **shear modulus** S is a measure of the rigidity and resistance to deformation of a material. The more rigid a material is, the larger its shear modulus. For example, the shear modulus for steel is about 7.8×10^{10} Pa, while that of rubber (which is much more deformable than steel) is only about 10^6 Pa.

We can use the above definitions of shear stress and shear strain (shown in Figure 10-12) to rewrite Equation 10-8 as

Shear stress on the object: force $F_\parallel$ applied to each of two opposite surfaces divided by the area A of each surface

Shear modulus S: how difficult it is to deform the material of which the object is made: larger S means more difficult

Hooke's law for an object under shear stress
(10-9)

$$\frac{F_\parallel}{A} = S\frac{x}{h}$$

Shear strain of the object: distance x that the two surfaces move relative to each other divided by the distance h between the surfaces

WATCH OUT! There is no shear modulus for a liquid or gas.

(!) While liquids and gases can sustain a *volume* stress that causes a change in volume, they cannot sustain a *shear* stress due to forces applied tangentially to their surfaces. Instead of deforming like the object shown in Figure 10-12, a force tangential to the surface of a liquid or gas causes it to flow rather than deform. This means that the shear modulus is defined *only* for solids.

Here are two examples of how to solve problems involving shear stress and shear strain.

BioMedical EXAMPLE 10-4 Endothelial Cells and Shear

Endothelial cells in an artery experience shear stress as a result of blood flow. In an *in vitro* study of arterial endothelial cells from pigs, the shear strain was observed to be proportional to the shear stress when this stress was 8.6×10^3 N/m². The shear modulus of a typical endothelial cell is 2.0×10^4 N/m². Figure 10-13 shows normal endothelial cells and cells after undergoing shear stress. At what angle ϕ does one surface of a cell move relative to the opposite surface? (The angle ϕ is labeled in Figure 10-13b.)

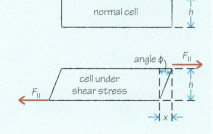

(a) Normal endothelial cells (b) Endothelial cells after shear stress

Republished with permission of Elsevier Science and Technology Journals, from Noria, S., et al., "Assembly and Reorientation of Stress Fibers Drives Morphological Changes to Endothelia Cells Exposed to Shear Stress," *American Journal of Pathology,* 164(4):1211–1223, Fig. 2, 2004; permission conveyed through the Copyright Clearance Center, Inc.

Figure 10-13 Shear stress on endothelial cells (a) Normal endothelial cells. The brighter areas are where neighboring cells either overlap (upper arrow) or touch (lower arrowhead). (b) A different but similar group of cells after experiencing shear stress. Before the stress was imposed, the center cell was roughly rectangular, with one side as shown by the dashed yellow line of height *h*. The stress pushed one side of the cell by a distance *x* relative to the opposite side, so that the cell is distorted by an angle ϕ. The arrow shows where the stress has pulled projections out of this cell's membrane.

Set Up

The observation that shear stress and shear strain are proportional means that Hooke's law applies in this situation, so we can use Equation 10-9. We are given the values of the shear stress and the shear modulus, so we can find the ratio x/h (the shear strain). Once we know this ratio we'll use trigonometry to determine the angle ϕ shown in Figure 10-13.

Hooke's law for shear:

$$\frac{F_{\parallel}}{A} = S\frac{x}{h} \qquad (10\text{-}9)$$

Solve

Rearrange Equation 10-9 to calculate the shear strain. Then substitute $F_{\parallel}/A = 8.6 \times 10^3$ Pa for the shear stress and $S = 2.0 \times 10^4$ Pa for the shear modulus.

Divide both sides of Equation 10-9 by S:

$$\frac{x}{h} = \frac{(F_{\parallel}/A)}{S} = \frac{8.6 \times 10^3 \text{ Pa}}{2.0 \times 10^4 \text{ Pa}} = 0.43$$

The lengths x and h make up two sides of a right triangle with included angle ϕ. The tangent of ϕ equals the opposite side (x) divided by the adjacent side (h). Use this to solve for ϕ.

$$\tan \phi = \frac{x}{h} = 0.43$$

So ϕ is the angle whose tangent is 0.43, which means that ϕ is the inverse tangent of 0.43:

$$\phi = \tan^{-1} 0.43 = 23°$$

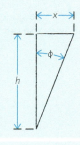

Reflect

The value of $\phi = 23°$ is quite large, which means that these endothelial cells undergo a dramatic change in shape (see Figure 10-13). This illustrates that endothelial cells in arteries, and the vascular system in general, are *very* sensitive to shear stress due to fluid flow. As depicted in Figure 10-11, endothelial cells release nitric oxide gas when under shear stress; this gas release causes the underlying muscle to relax, thus expanding the artery. This increases blood flow while reducing the flow speed and relieving the shear stress on the endothelial cells—a classic example of negative feedback in physiology.

EXAMPLE 10-5 Earthquake Damage

The dedication plate mounted to the base of a building was deformed during an earthquake. The plate, made from a metal alloy of shear modulus 4.00×10^{10} Pa, was originally 80.0 cm high, 50.0 cm long, and 5.00 mm thick. The earthquake shifted the top surface of the plate 0.100 mm relative to the bottom surface. What shear force did the plate experience during the earthquake?

Set Up

The displacement of the two surfaces of the plate ($x = 0.100$ mm) is small compared to the distance between them ($h = 80.0$ cm), so the shear strain x/h is very small. This suggests that we can use Hooke's law for shear, Equation 10-9. We are given the value of the shear modulus ($S = 4.00 \times 10^{10}$ Pa) and the dimensions of the plate, so we can solve for the shear force $F_\parallel$:

Hooke's law for shear:

$$\frac{F_\parallel}{A} = S\frac{x}{h} \qquad (10\text{-}9)$$

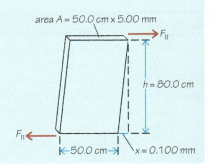

area $A = 50.0$ cm x 5.00 mm
$F_\parallel$
$h = 80.0$ cm
$F_\parallel$
$\leftarrow 50.0\text{ cm} \rightarrow$ $x = 0.100$ mm

Solve

First calculate the shear stress $F_\parallel/A$ using Equation 10-9, making sure to convert to SI units.

$$\frac{F_\parallel}{A} = S\frac{x}{h} = (4.00 \times 10^{10}\text{ Pa})\left(\frac{0.100\text{ mm}}{80.0\text{ cm}}\right)$$

$$= (4.00 \times 10^{10}\text{ N/m}^2)\left(\frac{1.00 \times 10^{-4}\text{ m}}{0.800\text{ m}}\right)$$

$$= (4.00 \times 10^{10}\text{ N/m}^2)(1.25 \times 10^{-4})$$

$$= 5.00 \times 10^6\text{ N/m}^2$$

The area A over which the force is applied is the area of the top or bottom surface of the plate. Multiply the shear stress by A to calculate the force magnitude $F_\parallel$.

$$A = (50.0\text{ cm})(5.0\text{ mm}) = (0.500\text{ m})(5.00 \times 10^{-3}\text{ m})$$

$$= 2.50 \times 10^{-3}\text{ m}$$

$$F_\parallel = \frac{F_\parallel}{A}A = (5.00 \times 10^6\text{ N/m}^2)(2.50 \times 10^{-3}\text{ m})$$

$$= 1.25 \times 10^4\text{ N}$$

Reflect

The shear force was 12,500 N, or about 1.25 tons! This force is about the same as the weight of a small car. It takes a *lot* of force to deform such a large piece of metal.

GOT THE CONCEPT? 10-7 Shear Stress and Shear Strain

? Figure 10-14 shows three objects that are all made of the same material. The forces applied to the three objects all have the same magnitude. (a) Rank the objects in order of the shear stress acting on them, from greatest to least. (b) Rank the objects in order of the shear strain that they will undergo, from greatest to least.

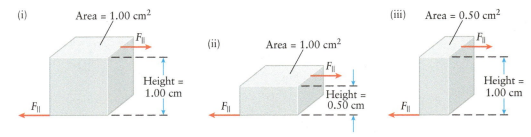

Figure 10-14 Three objects under shear stress The same forces act on all three objects, all of which are made of material with the same shear modulus. Which one experiences the greatest strain?

10-5   Objects deform permanently or fail when placed under too much stress

(a) Elastic: stretched skin

Your arteries, ears, nose, and ligaments, as well as many other structures in your body, can be stretched, compressed, or bent without deforming them permanently. Somewhat like a rubber band, after you stretch your earlobe or squeeze the skin on your forearm by applying a modest force, it quickly returns to its original shape (Figure 10-15a). Structures with this property are called **elastic**. (We used the term "elastic" in Chapter 8 to refer to collisions in which mechanical energy is conserved. Our new definition is really the same. If two elastic objects collide with each other, they may deform during the collision but will spring back to their original shapes afterward. Hence no energy is used up during the collision, and the total mechanical energy is conserved.) But if a structure is deformed too much, it will not spring back to its original size and shape (Figure 10-15b) and may even tear or fracture (Figure 10-15c). This is what can happen when too much force is applied to a human knee, making the anterior cruciate ligament (ACL) stretch too far and creating a tear in the ACL. How can we describe what happens when a structure or material is deformed so much that it is no longer elastic?

(b) Plastic: overextended spring

From Elastic to Plastic to Failure

If the tensile stress on an object is increased so that the change in the object's length ΔL becomes large relative to its initial length L_0, the deformation of the object becomes permanent. This is what happens if you stretch a metal spring too far: The spring is permanently deformed and can no longer return to its original shape (Figure 10-15b). The tensile stress at which this occurs is called the **yield strength**. If the tensile stress exceeds the yield strength and the object deforms permanently, we say that the object has become **plastic**.

(c) Failure: broken bread

WATCH OUT!   When is a plastic object plastic?

The term "plastic" may seem like a strange one to use to describe objects that don't spring back to their original shape. After all, most objects that we think of as "plastic"—such as a plastic fork or plastic bottle—are actually quite springy.

But to make these objects, the plastic material is heated so that it can flow into a mold, then cooled so that it solidifies into the desired shape. Strictly speaking, the material is "plastic" only when it is flowing into its new shape.

Figure 10-15 **Three responses to stress: Elastic, plastic, and failure** (a) This child's skin is elastic: When she stops applying stress with her fingers, her mouth will return to its original shape. (b) This spring has been stretched so far that it no longer returns to its original coiled shape: The metal of which it is made became plastic during the stretch. (c) The baker applied so much stress to this bread that it underwent failure and broke into two pieces.

Figure 10-16 shows elastic and plastic behavior in an idealized graph of tensile stress versus tensile strain. In the elastic regime the tensile strain is *reversible*: If you increase the amount of stress by a small amount, then release that extra stress (say, by pulling a little harder on the ends of a spring, then easing off on the pull), the object will relax to the length it had before you increased the stress. By contrast, in the plastic region the tensile strain is *irreversible*. If you pull a little harder on the ends of the object, it stretches a bit more and stays stretched even after you stop pulling.

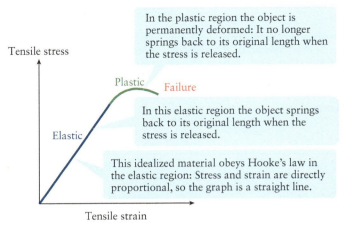

In the plastic region the object is permanently deformed: It no longer springs back to its original length when the stress is released.

In this elastic region the object springs back to its original length when the stress is released.

This idealized material obeys Hooke's law in the elastic region: Stress and strain are directly proportional, so the graph is a straight line.

Tensile stress

Plastic

Failure

Elastic

Tensile strain

Figure 10-16 Elastic behavior, plastic behavior, and failure This graph shows tensile stress versus tensile strain for an idealized material. The curve bends downward for large strain: Once the material has been permanently deformed and stretched beyond a certain point, it gets easier to stretch it further.

When the strain on an object is large enough, it undergoes **failure**: The structure of the material starts to lose its integrity, which eventually leads to the object breaking apart. The maximum tensile stress that the material can withstand before failure is called the **ultimate strength**.

Biological Tissue: From Elastic to Failure

BioMedical Most biological tissue, including your skin and ligaments—such as the anterior cruciate ligament, or ACL, in your knee (Figure 10-6); the sternoclavicular ligament in your shoulder (Figure 10-8); and the skin on your face (Figure 10-15a)—responds to increased tensile stress in a different way than the idealized material graphed in Figure 10-16. This tissue does *not* have a plastic regime: Once a stress is applied that exceeds the ability of biological tissue to spring back to its initial shape, it undergoes partial or complete failure. You can demonstrate this by holding each end of a string bean—a simple biological structure—with your fingertips and trying to pull the two ends apart. If you pull too hard, the string bean will break in two rather than acquire a permanent stretch.

The reason biological tissue is different from metals has to do with its composition. As we mentioned in Section 10-2, stretchable biological materials are *not* uniform but are composed of a combination of tiny fibrils of collagen and elastin (see Figure 10-17a). As a result these materials do *not* obey Hooke's law if the tensile stress is small. If the tensile stress is too great, the fibrils do not acquire a permanent stretch but start breaking apart (Figure 10-17b).

Figure 10-18 shows an idealized curve of stress versus strain for biological tissue. The curve is not a straight line for small stresses (region I in Figure 10-18), which means that stress and strain are not directly proportional, and Hooke's law is not strictly obeyed. Hooke's law *is* more valid for moderate stresses (region II in Figure 10-18). If the stress is too great and the elastin fibers begin to tear, the tissue fails (region III in Figure 10-18). This may be referred to as *fracture*, *rupture*, or *tearing*. That's what happens when an athlete suffers a knee injury and ends up with a partially or completely torn ACL.

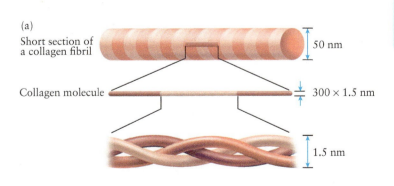

(a)
Short section of a collagen fibril

50 nm

Collagen molecule

300×1.5 nm

1.5 nm

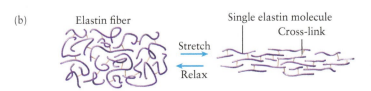

(b)
Elastin fiber

Single elastin molecule
Cross-link

Stretch

Relax

Figure 10-17 Microscopic structure of biological tissue Connective tissues get their rigidity from the protein collagen and their elastic properties from the protein elastin.

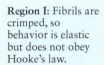

Region I: Fibrils are crimped, so behavior is elastic but does not obey Hooke's law.

Region II: Fibrils are straight. The behavior is elastic and obeys Hooke's law.

Region III: Fibrils begin to tear.

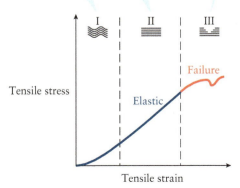

I

II

III

Tensile stress

Elastic

Failure

Tensile strain

Figure 10-18 Elastic behavior and failure for biological materials This graph shows tensile stress versus tensile strain for typical biological materials. Because such materials are made of elastin and collagen fibrils, this behavior is very different from that of the idealized material shown in Figure 10-16.

WATCH OUT! Materials can be elastic yet not obey Hooke's law.

> ! While biological tissue does not obey Hooke's law in the region of low tensile stress (region I in Figure 10-18), it is nonetheless elastic in this region. That's because it springs back to its original length when the tensile stress is removed.

The same is true for the larger stresses within region II in Figure 10-18, in which Hooke's law is more nearly valid. An object that obeys Hooke's law is necessarily elastic, but an elastic object does *not* necessarily obey Hooke's law!

In this section we've considered only the effects of *tensile* stress. Failure can also occur when an object is under compression (think of what would happen if you were to push the two ends of a peeled banana together). It can also happen under shear (which is what happens when you hold a piece of paper in your hands and tear it in two).

In the following three examples we'll examine the stress–strain behavior of a typical biological material, the anterior cruciate ligament (ACL) found in the human knee.

BioMedical EXAMPLE 10-6 **Human ACL I: Maximum Force**

A study of the properties of human ACLs revealed that the ultimate strength for an ACL in a younger person is, on average, 3.8×10^7 Pa. A typical cross-sectional area of the ACL is 4.4×10^{-5} m^2. What is the force exerted on the ACL when the tensile stress is at the maximum value that the ACL can withstand?

Set Up

The ultimate strength of an object is the maximum tensile stress, or maximum force per area, that the object can withstand. Use F_{max} to denote this maximum force, which is what we want to find. We use the definition of tensile stress from Equation 10-3 to relate the ultimate strength to the cross-sectional area A and the force F_{max}.

$$\text{tensile stress} = \frac{F}{A}$$

so ultimate strength

$$= \text{maximum tensile stress} = \frac{F_{max}}{A}$$

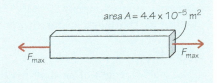

Solve

Rearrange the equation for ultimate strength to solve for F_{max}. Then substitute the known values and use 1 Pa = 1 N/m^2.

$$F_{max} = (\text{ultimate strength}) \times A$$
$$= (3.8 \times 10^7 \text{ Pa}) \times (4.4 \times 10^{-5} \text{ m}^2)$$
$$= (3.8 \times 10^7 \text{ N/m}^2) \times (4.4 \times 10^{-5} \text{ m}^2)$$
$$= 1.7 \times 10^3 \text{ N}$$

Reflect

Our result shows that a single knee ligament can withstand a sizeable force of 1.7×10^3 N. That's greater than the weight of the heaviest professional players in American football, who have a mass of about 150 kg. In normal use (walking, running, and so on) the forces that act on the ACL are only a fraction of F_{max}, the maximum force that the ACL can withstand.

Offensive tackle in American football:

mass $m = 150$ kg

weight $mg = (150 \text{ kg})(9.80 \text{ m/s}^2)$
$$= 1.5 \times 10^3 \text{ N}$$

BioMedical EXAMPLE 10-7 **Human ACL II: Breaking Strain**

A study of the properties of human anterior cruciate ligaments found that at the point of ACL failure in a younger person, the tensile strain of the ACL is approximately 0.60. The typical length of the ACLs studied was 2.7 cm. How far can the ACL stretch before it breaks?

Set Up

When an object is under tensile stress, its tensile strain equals how far it stretches ΔL divided by its total unstretched length L_0 (Equation 10-3). We know the tensile strain at the point at which the ACL fails (when the ACL is stretched as far as possible) as well as its unstretched length, so we can solve for its maximum stretch ΔL_{max}.

$$\text{tensile strain} = \frac{\Delta L}{L_0}$$

so tensile strain at failure

$$= \text{maximum tensile strain} = \frac{\Delta L_{max}}{L_0}$$

area $A = 4.4 \times 10^{-5}$ m^2

$L_0 = 2.7$ cm

Solve

Rearrange the equation for the tensile strain at failure and solve for ΔL_{max}.

$$\Delta L_{max} = (\text{tensile strain at failure}) \times L_0$$
$$= (0.60)(2.7 \text{ cm}) = 1.6 \text{ cm}$$

Reflect

If 1.6 cm (almost two-thirds of an inch) seems like a large stretch, it is! A maximum tensile strain of 0.60 means that the ACL can stretch by 60% before failing.

BioMedical **EXAMPLE 10-8** **Human ACL III: The Point of No Return**

A study of the properties of human anterior cruciate ligaments found that the yield strength of the ACL of a younger person is approximately 3.3×10^7 Pa. By what percentage of its initial length can the ACL stretch before it will no longer return to its original length intact? Use $Y_{ACL} = 1.1 \times 10^8$ Pa as the value for Young's modulus for an ACL.

Set Up

The yield strength of an object is the tensile stress beyond which the object is no longer elastic. In Figure 10-17, which shows a representative stress–strain curve for biological tissue, this corresponds to the boundary between region II and region III. While Hooke's law (Equation 10-4) is not strictly valid throughout the elastic regions I and II, we can use it to estimate the tensile strain at the point where the tensile stress equals the yield strength. This and the definition of tensile strain will tell us the percentage increase in the length of the ACL.

Hooke's law for tension expressed in terms of stress and strain:

tensile stress

$$= \text{Young's modulus} \times \text{tensile strain} \quad (10\text{-}4)$$

$$\text{tensile strain} = \frac{\Delta L}{L_0}$$

Solve

We first solve Hooke's law, Equation 10-4, for the value of the tensile strain of the ACL when the tensile stress equals the yield strength.

$$\text{tensile strain} = \frac{\text{tensile stress}}{\text{Young's modulus}}$$

tensile stress = yield strength = 3.3×10^7 Pa

Young's modulus = $Y_{ACL} = 1.1 \times 10^8$ Pa

Hence when the tensile stress equals the yield strength,

$$\text{tensile strain} = \frac{3.3 \times 10^7 \text{ Pa}}{1.1 \times 10^8 \text{ Pa}} = 0.30$$

Convert the tensile strain to a percentage increase in the length of the ACL.

$$\text{tensile strain} = \frac{\Delta L}{L_0} = \text{fractional change in the length of the ACL}$$

To convert a fraction to a percentage, multiply by 100%:

tensile strain = $(0.30)(100\%) = 30\%$

Reflect

We have found that the human ACL is elastic enough to withstand being stretched by a factor of 30% while still being able to return undamaged to its normal length. In Example 10-7 we discovered that a human ACL tears completely when stretched by about 60%. For strains between 30% and 60% some damage—for example, a partial tear—can occur.

GOT THE CONCEPT? 10-8 Hooke's Law and Elastic Behavior

 To stretch a certain object by 1.00 cm takes a force of 25.0 N applied at each end. To stretch the same object by 2.00 cm takes a force of 60.0 N applied at each end. In both cases the object returns to its initial length when the force is released. (a) Does the object obey Hooke's law? (b) Does the object display elastic behavior when the 60.0-N forces are applied?

TAKE-HOME MESSAGE FOR Section 10-5

✔ When an object is subjected to a stress that results in a relatively small strain, it deforms but returns to its initial shape after the stress is removed (the elastic regime).

✔ If the stress on an object is increased so that the change in length becomes large relative to the object's initial length, the deformation of the object becomes irreversible (the plastic regime) and represents the onset of failure, which eventually leads to the object breaking apart.

✔ Biological materials do not have a plastic regime: As the stress is increased, they go from the elastic regime directly to failure.

Key Terms

bulk modulus
compression
compressive strain
compressive stress
elastic
failure
Hooke's law
pascal
plastic

pressure
shear
shear modulus
shear strain
shear stress
spring constant
strain
stress
tensile strain

tensile stress
tension
ultimate strength
volume strain
volume stress
yield strength
Young's modulus

Chapter Summary

Topic	Equation or Figure
Stress and strain: An object is under stress when forces act to deform it (change its size or shape). Stress has units of force per area and is measured in pascals ($1 \text{ Pa} = 1 \text{ N/m}^2$). The amount of deformation that results is the strain, which is a dimensionless quantity.	This Achilles tendon is relaxed. This Achilles tendon is under **stress**: The muscles at its upper end exert an upward force F on the tendon, and the heel bone at its lower end exerts an equally strong downward force on the tendon. The tendon responds by undergoing **strain** (it stretches).  (Figure 10-1)

 Stress refers to forces that act to deform an object (change its size or shape). **Strain** refers to the deformation caused by these forces.

Tension, compression, and Hooke's law: An object is under tension if forces act to lengthen it along one dimension; it is under compression if forces act to shorten it along one dimension. If the object obeys Hooke's law, the strain is directly proportional to the stress; the constant of proportionality, called Young's modulus, depends on what the object is made of but not on its dimensions. The spring constant k for an object like that shown here equals YA/L_0.

For both tension and compression:
stress $= F/A$
strain $= \Delta L/L_0$
Hooke's law: stress $= Y \times$ strain

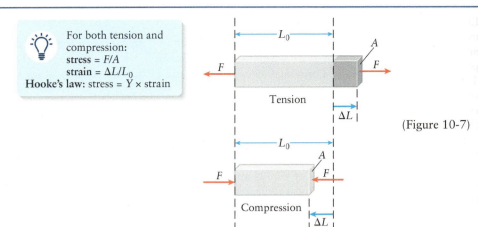

Tension

Compression

(Figure 10-7)

Volume stress and strain: An object is under volume stress if there is a *change* in the pressure p on the object. The stress is equal to the pressure change Δp. If Δp is small, Hooke's law applies and the strain (the fractional change in volume, $\Delta V/V_0$) is proportional to the stress. The proportionality constant is the bulk modulus of the material. The signs of stress and strain are opposite: Positive volume stress ($\Delta p > 0$) causes a negative volume strain ($\Delta V/V_0 < 0$) and vice versa.

① Initially this object has volume V_0. It is made of a material with bulk modulus B.

② The object's surroundings exert a force of magnitude F (shown by the red vectors) on each area A of the object's surface. The pressure is $p = F/A$.

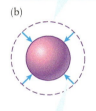

(a)

(Figure 10-10)

③ If the pressure on the object changes by Δp, the object's volume changes by ΔV as indicated by the blue arrows.

(b)

(c)

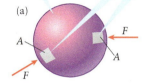

volume stress $= \Delta p$
volume strain $= \Delta V/V_0$
Hooke's law:
volume stress $= -B \times$ volume strain

④ If the pressure increases ($\Delta p > 0$), the volume decreases ($\Delta V < 0$).

⑤ If the pressure decreases ($\Delta p < 0$), the volume increases ($\Delta V > 0$).

Shear stress and strain: An object is under shear stress if offset forces $F_\parallel$ act to change the object's shape. If the shear stress is not too great, Hooke's law applies and the shear strain is proportional to the shear stress.

① This object has height h. Its upper and lower surfaces each have area A. It is made of a material with shear modulus S.

② Apply a force of magnitude $F_\parallel$ parallel to the upper surface...

③ ...and apply an opposite force of the same magnitude $F_\parallel$ parallel to the lower surface. Note that the two forces are offset (they do not act along the same line).

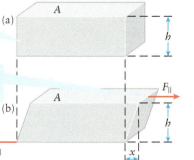

(a)

(b)

(Figure 10-12)

shear stress $= F_\parallel/A$
strain $= x/h$
Hooke's law:
shear stress $= S \times$ shear strain

④ As a result of these forces, the upper and lower surfaces move a distance x relative to each other.

Elasticity, plasticity, and failure: Most materials return to their original size and shape when the stress is removed, provided the stress is not too great. This is called elastic behavior. If the stress is greater than a certain value, the material becomes plastic: It retains the deformation even after the stress is removed. Beyond a certain even greater value, the material undergoes failure (it ruptures or fractures). Biological materials do not become plastic, but go from elastic behavior to failure if the stress is made too great.

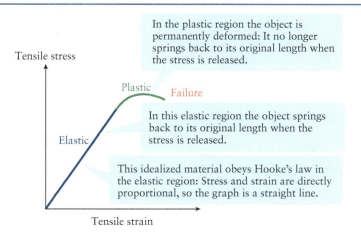

In the plastic region the object is permanently deformed: It no longer springs back to its original length when the stress is released.

In this elastic region the object springs back to its original length when the stress is released.

This idealized material obeys Hooke's law in the elastic region: Stress and strain are directly proportional, so the graph is a straight line.

(Figure 10-16)

Answer to What do you think? Question

(b) In Section 10-2 we learn that the amount that an object's length decreases due to compression is proportional to the compressive stress F/A, which is the force F applied (in this case to the athlete's arms) divided by the object's cross-sectional area A. If the arms had one-half the diameter, they would have one-quarter the area A and the compressive stress F/A would be four times as much (assuming that F is unchanged). Hence her arms would shorten by four times the distance.

Answers to Got the Concept? Questions

10-1 (c), (a) and (d) (tie), (b). The easier a rod is to stretch, the smaller its spring constant k. Equation 10-2 tells us that $k = YA/L_0$. All four rods have the same Young's modulus Y (since they are all made of the same material). So the rod with the smallest value of k is the one with the smallest ratio A/L_0 — that is, the one with the smallest cross-sectional area and the longest length. The values of this ratio for each of the four rods are (a) $(2.00 \text{ mm}^2)/(5.00 \text{ cm}) = 0.400 \text{ mm}^2/\text{cm}$, (b) $(2.00 \text{ mm}^2)/(2.50 \text{ cm}) = 0.800 \text{ mm}^2/\text{cm}$, (c) $(1.00 \text{ mm}^2)/(5.00 \text{ cm}) = 0.200 \text{ mm}^2/\text{cm}$, (d) $(1.00 \text{ mm}^2)/(2.50 \text{ cm}) = 0.400 \text{ mm}^2/\text{cm}$.

10-2 (d) Not enough information is given to decide. To see why not, rewrite Equation 10-3 as

$$Y = \frac{F/A}{\Delta L/L_0} = \frac{\text{tensile stress}}{\text{tensile strain}}$$

To compare the value of Y for the two materials, we need to compare the ratio of tensile stress to tensile strain for the two rods. The force F and cross-sectional area A are the same for both rods, so both are under the same tensile stress. We also know that ΔL is twice as great for rod 2 as for rod 1. But we don't know how the tensile *strain* compares for the two rods because we aren't told anything about the relaxed lengths L_0 of the two rods. (That was the detail we asked you to look for.) So we don't have enough information to decide which material has the greater value of Y.

10-3 (a) Same for both, (b) brass rod, (c) brass rod. Each rod has the same cross-sectional area A and is subjected to forces of the same magnitude F, so the compressive *stress* F/A is the same for both. From Hooke's law the compressive strain equals the compressive stress divided by Young's modulus Y: $\Delta L/L = (F/A)(1/Y)$. The brass rod has a smaller value of Y, so it will experience the greater compressive strain. The distance

ΔL that each rod contracts equals the original length (which is the same for both rods) multiplied by the compressive strain, so the brass rod (which experiences greater strain) contracts more than the steel rod.

10-4 (c), then (d), then (a) and (b) (tie). For the first three cases use the definition of tensile stress as force divided by cross-sectional area:

(a) tensile stress $= \dfrac{F}{A} = \dfrac{200 \text{ N}}{(1.00 \text{ mm}^2)}\left(\dfrac{100 \text{ mm}}{1 \text{ m}}\right)^2$
$= 2.0 \times 10^6 \text{ N/m}^2$

(b) Same as (a); the length is different, but this has no effect on calculating the tensile stress

(c) tensile stress $= \dfrac{F}{A} = \dfrac{100 \text{ N}}{(2.00 \text{ mm}^2)}\left(\dfrac{100 \text{ mm}}{1 \text{ m}}\right)^2$
$= 5.0 \times 10^5 \text{ N/m}^2$

(d) For the fourth case we use Hooke's law, which states that tensile stress = (Young's modulus) × (tensile strain) $= Y\dfrac{\Delta L}{L_0} =$

$(2.00 \times 10^{10} \text{ Pa})\left(\dfrac{0.025 \text{ mm}}{50.0 \text{ cm}}\right)\left(\dfrac{1 \text{ cm}}{10 \text{ mm}}\right) = 1.0 \times 10^6 \text{ N/m}^2$

10-5 (b) The volume of the balloon filled with air will change the most. From Table 10-1 you can see that the bulk modulus of water is about 2×10^4 times larger than that of air. Water is therefore about 2×10^4 times *less* compressible than air.

10-6 (d) Because the pressure is the same on all sides, the forces pushing on opposite sides of the sphere are of equal magnitude but opposite direction. Hence the *net* force is zero!

10-7 (a) (iii), (i) and (ii) (tie); (b) (iii), (i) and (ii) (tie). In part (a) the shear stress equals $F_{\parallel}/A$. The force $F_{\parallel}$ is the same in each case. Hence object (iii) has the greatest shear stress since it has the smallest value of A. Objects (i) and (ii) have the same area and so have the same shear stress.

Part (b) uses Hooke's law: Shear stress and shear strain are proportional. All three objects are made of the same material with shear modulus S, so the constant or proportionality is the same in each case. Therefore, the ranking in order of shear strain is the same as the ranking in order of shear stress.

10-8 (a) No; (b) yes. If an object obeys Hooke's law, doubling the stretch (and hence doubling the tensile strain) would require double the force (and hence double the tensile stress). This object does *not* obey Hooke's law since doubling the stretch from 1.00 to 2.00 cm requires more than double the force (from 25.0 to 60.0 N). The object *is* nonetheless elastic because it returns to its unstretched length when the tensile forces go away.

Questions and Problems

In a few problems you are given more data than you actually need; in a few other problems you are required to supply data from your general knowledge, outside sources, or informed estimate.

Interpret as significant all digits in numerical values that have trailing zeros and no decimal points. For all problems use $g = 9.80$ m/s^2 for the free-fall acceleration due to gravity. Neglect friction and air resistance unless instructed to do otherwise.

- •   Basic, single-concept problem
- ••   Intermediate-level problem; may require synthesis of concepts and multiple steps
- •••   Challenging problem
- Example See worked example for a similar problem

Conceptual Questions

1. • Describe the small-stretch limit of Hooke's law for a spring.

2. • Give a few reasons why Hooke's law is intuitively obvious and a few reasons why it is counterintuitive.

3. • Is it possible for a long cable hung vertically to break under its own weight? Explain your answer.

4. • A horizontally oriented 2"×4" pine stud is securely clamped at one end to an immovable object. A heavy weight hangs from the free end of the wood, causing it to bend. (a) Which part of the plank is under compression? (b) Which part of the plank is under tension? (c) Is there any part that is neither stretched nor compressed?

5. • A steel wire and a brass wire, each of length L and diameter D, are joined together to form a wire of length $2L$. If this wire is then used to hang an object of mass m, describe the amount of stretch in the two segments of wire.

6. • What can cause nylon tennis racket strings to break when they are hit by the ball?

7. • Human skin is under tension like a rubber glove that has had air blown into it. Why does skin acquire wrinkles as people get older?

8. • Balloons expand as they rise through the atmosphere. Why does this happen?

9. • One way of determining the volume of an irregular solid is to submerge it in a liquid to measure the volume change of the liquid in its container. What assumptions are being made about the object and the object's bulk modulus when performing such experiments?

10. • Marshmallows and iron have vastly different bulk moduli. One has a much larger value than the other. Suppose a marshmallow cube and an iron cube of approximately the same size were placed in a vacuum chamber. Explain what would happen to each of the cubes as the air was slowly removed from the chamber.

11. • (a) What is the difference between Young's modulus and bulk modulus? (b) What are the units of these two physical quantities?

12. • Is it possible, when tightening the lug nuts on the wheel of your car, to use too much torque and break off one of the bolts? Explain your answer.

13. • Shear modulus (S) is sometimes known as *rigidity*. Can you explain why rigidity is an appropriate synonym?

14. • Devise a simple way of determining which modulus (Young's, bulk, or shear) is appropriate for any given stress–strain problem.

15. • Why are tall mountains typically shaped like cones rather than a straight vertical column-like structure?

16. • (a) Describe some common features of strain that were defined in this chapter. (b) We encountered three types of strain (tensile, volume, and shear). What are some distinguishing features of these quantities?

17. • Define the terms "yield strength" and "ultimate strength."

18. • **Biology** In some recent studies it has been shown that women are more susceptible to torn ACLs than men when competing in similar sports (most notably in soccer and basketball). What are some reasons this disparity might exist?

19. • **Biology** The leg bone of a cow has an ultimate strength of about 150×10^6 N/m^2 and a maximum strain of about 1.5%. The antler of a deer has an ultimate strength of about 160×10^6 N/m^2 and a maximum strain of about 12%. Explain the relationship between structure and function in these data.

Multiple-Choice Questions

20. • The units for strain are
    A. N/m.
    B. N/m^2.
    C. N.
    D. N·m^2.
    E. none of the above.

21. • The units for stress are
    A. N/m.
    B. N/m^2.
    C. N.
    D. N·m.
    E. N·m^2.

22. • When tension is applied to a metal wire of length L, it stretches by ΔL. If the same tension is applied to a wire of the

same material with the same cross-sectional area, but of length 2L, by how much will it stretch?

   A. ΔL
   B. $2\Delta L$
   C. $0.5\Delta L$
   D. $3\Delta L$
   E. $4\Delta L$

23. • A steel cable lifting a heavy box stretches by ΔL. If you want the cable to stretch by only half of ΔL, by what factor must you increase its diameter?

   A. 2
   B. 4
   C. $\sqrt{2}$
   D. 1/2
   E. 1/4

24. • A wire is stretched just to its breaking point by a force F. A longer wire made of the same material has the same diameter. The force that will stretch it to its breaking point is

   A. larger than F.
   B. smaller than F.
   C. equal to F.
   D. much smaller than F.
   E. much larger than F.

25. • Two solid rods have the same length and are made of the same material with circular cross sections. Rod 1 has a radius r, and rod 2 has a radius $r/2$. If a compressive force F is applied to both rods, their lengths are reduced by ΔL_1 and ΔL_2, respectively. The ratio $\Delta L_1/\Delta L_2$ is

   A. 1/4.
   B. 1/2.
   C. 1.
   D. 2.
   E. 4.

26. • A wall mount for a television consists in part of a mounting plate screwed or bolted flush to the wall. Which kinds of stress play a role in keeping the mount securely attached to the wall?

   A. compression stress
   B. tension stress
   C. shear stress
   D. bulk stress
   E. A, B, and C

27. • When choosing building construction materials, what kinds of materials would you choose, all other things being equal?

   A. materials with a relatively large bulk modulus
   B. materials with a relatively small bulk modulus
   C. materials with either a large or a small bulk modulus
   D. any kind (it doesn't matter) as long as the building is not too tall
   E. materials with a relatively small shear modulus

28. • A book is pushed sideways, deforming it as shown in Figure 10-19. To describe the relationship between stress and strain for the book in this situation, you would use

   A. Young's modulus.
   B. bulk modulus.
   C. shear modulus.
   D. both Young's modulus and bulk modulus.
   E. both shear modulus and bulk modulus.

Figure 10-19 Problem 28

29. • A steel cable supports an actor as he swings onto the stage. The weight of the actor stretches the steel cable. To describe the relationship between stress and strain for the steel cable, you would use

   A. Young's modulus.
   B. bulk modulus.
   C. shear modulus.
   D. both Young's modulus and bulk modulus.
   E. both shear modulus and bulk modulus.

Problems

10-1 When an object is under stress, it deforms

10-2 An object changes length when under tensile or compressive stress

30. • A cylindrical steel rod is originally 250 cm long and has a diameter of 0.254 cm. A force is applied longitudinally and the rod stretches 0.85 cm. What is the magnitude of the force? Example 10-1

31. • A solid band of rubber, which has a circular cross section of radius 0.25 cm, is stretched a distance of 3.0 cm by a force of 87 N. Calculate the original length of the cylinder of rubber. Example 10-1

32. • A bar of aluminum has a cross section of 1.0 cm × 1.0 cm and a length of 88 cm. (a) What force would be needed to stretch the bar to 1.00 m? (b) What is the tensile strain of the aluminum bar at that point? Example 10-1

33. • The tensile stress on a concrete block is 0.52×10^9 N/m². What is its tensile strain under this force? Example 10-1

34. • A physicist examines a metal sample and measures the ratio of the tensile stress to the tensile strain to be 95×10^9 N/m². From what material might the sample be made? Example 10-1

35. • A 10.0-m-long copper wire is pulled with a force of 1200 N and it stretches 10.0 cm. Calculate the radius of the wire if the value of Young's modulus is 110×10^9 N/m². Example 10-1

36. • (a) Calculate the ratio of the tensile strain on an aluminum bar to that on a steel bar if both bars have the same cross-sectional area and the same force is applied to each bar. (b) Does the original length of each bar affect your answer to part (a)? Example 10-1

37. • **Biology** (a) What is the compressive stress on your feet if your weight is spread out evenly over both soles of your shoes? (b) By what factor does the compressive stress change if only 10.0 cm² of each shoe is in contact with the floor (as might be the case for high-heeled shoes)? Assume your mass is 55.0 kg and each shoe has an area of 200.0 cm². Example 10-1

38. • **Medical** The anterior cruciate ligament in a woman's knee is 2.5 cm long and has a cross-sectional area of 0.54 cm². If a force of 3.0×10^3 N is applied longitudinally, how much will the ligament stretch? Example 10-1

39. •• **Medical** One model for the length of a person's ACL (L_{ACL}, in millimeters) relates it to the person's height (h, in centimeters) with the linear function $L_{ACL} = 0.4606h - 41.29$. Age, gender, and weight do not significantly influence the relationship. If a basketball player has a height of 2.29 m, (a) approximately how long is his ACL? (b) If a pressure of 10.0×10^6 N/m² is applied longitudinally to his ligament, how much will it stretch? Example 10-1

10-3 An object expands or shrinks when under volume stress

40. • A cube of lead (each side is 5.0 cm) is pressed equally on all six sides with forces of 1.0×10^5 N. What will the new dimensions of the cube be after the forces are applied ($B = 46 \times 10^9$ N/m^2 for lead)? Example 10-2

41. • A rigid cube (each side is 0.10 m) is filled with water and frozen solid. When water freezes its volume expands about 9%. How much pressure is exerted on the sides of the cube? *Hint:* Imagine trying to squeeze the block of ice back into the original cube. Example 10-2

42. • A sphere of copper that has a radius of 5.00 cm is compressed uniformly by a force of 2.00×10^8 N. Calculate (a) the change in volume of the sphere and (b) the sphere's final radius. The bulk modulus for copper is 140×10^9 N/m^2. Example 10-2

10-4 A solid object changes shape when under shear stress

43. • Shear forces act on a steel door during an earthquake (Figure 10-20). If the shear strain is 0.005, calculate the force acting on the door with dimensions 0.044 m × 0.81 m × 2.03 m. Example 10-5

Figure 10-20 Problem 43

44. • A brass nameplate is 2.00 cm × 10.0 cm × 20.0 cm in size. If a force of 2.00×10^5 N acts on the upper left side and the bottom right side (Figure 10-21), find the shear strain (x/h) and the angle ϕ. Example 10-4

2.00 × 10^5 N

2.00 × 10^5 N

Figure 10-21 Problem 44

45. • **Medical** In patients with asthma, an increased thickness of the airways causes a local reduction in stress through the airway walls. The effect can be as much as a 50% reduction in the local shear modulus of the airways of an asthmatic patient compared to those of a healthy person. Calculate the ratio of the shear strain in an asthmatic airway to that of a healthy airway. Example 10-4

46. •• A force of 5.0×10^6 N is applied tangentially at the center of one side of a brass cube. The angle of shear ϕ is measured to be 0.65°. Calculate the volume of the original cube. Example 10-4

47. • An enormous piece of granite 2.00×10^2 m thick with a shear modulus of 5.00×10^{10} N/m^2 is sheared from its geologic formation by an earthquake force of 275×10^9 N. The area on which the force acts is a square of side x as shown in Figure 10-22. Find the value of x if the shear force produces a shear strain of 0.125. Example 10-5

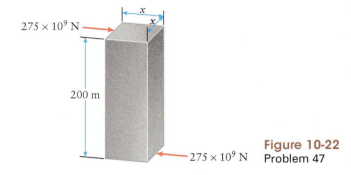

275 × 10^9 N

200 m

275 × 10^9 N

Figure 10-22 Problem 47

48. • **Medical** In regions of the cardiovascular system where there is steady laminar blood flow, the shear stress on cells lining the walls of the blood vessels is about 20 dyne/cm^2. If the shear strain is about 0.008, estimate the shear modulus for the affected cells. Note 1 dyne = $1 \, \text{g} \cdot \text{cm/s}^2$ and 1 N = 10^5 dyne. Example 10-5

10-5 Objects deform permanently or fail when placed under too much stress

49. • A piece of steel piano wire is 1.60 m long and has a diameter of 0.20 cm. What is the magnitude of the tension required to break it? The ultimate strength of steel is 5.0×10^8 N/m^2. Example 10-6

50. • Steel will ultimately fail if the shear stress exceeds 4.00×10^8 N/m^2. Determine the force required to shear a steel bolt that is 0.50 cm in diameter. Example 10-6

51. • A theater rigging company uses a safety factor of 10 for all its ropes, which means that all ultimate breaking strengths will be underestimated by a factor of 10 just to be safe. Suppose a rope with an ultimate breaking strength of 1.0×10^4 N is tied with a knot that decreases rope strength by 50%. (a) If the rope is used to support a load of 1.0×10^3 N, what is the safety factor? (b) Will the rigging company be able to use the rope with the knot? Example 10-6

52. • Standard 12-gauge copper wiring (commonly used in home electrical wiring) is 2.053 mm in diameter. Copper has a yield strength of 70 MPa (1 MPa = 10^6 Pa). (a) What is the maximum amount of weight that a 12-gauge copper wire can hold before it can no longer return to its original length intact? (b) The yield strength of silk produced by a silkworm is 500 MPa. What minimum diameter of thread of a silkworm's silk would be needed to hold the same weight as the 12-gauge copper wire, provided that the elasticity of the thread is preserved? Example 10-8

General Problems

53. • The following data are associated with an alloy of steel and are plotted in Figure 10-23. What are (a) the yield strength, (b) the ultimate strength, and (c) Young's modulus for the material?

Strain (%)	Stress (10^9 N/m^2)	Strain (%)	Stress (10^9 N/m^2)
0	0	1.0	300
0.1	125	1.5	325
0.2	250	2.0	350
0.3	230	2.5	375
0.4	230	3.0	400
0.5	235	3.5	375
0.6	240	4.0	350
0.7	250	4.5	325
0.8	260	5.0	300
0.9	270		

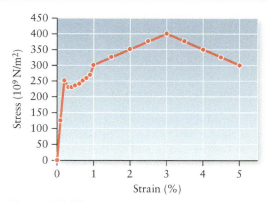

Figure 10-23 Problem 53

54. • **Biology** A galloping horse experiences the following stresses and corresponding strains on its front leg bone. Plot a graph of the stress versus strain for the bone and identify (a) the elastic region, (b) the yield strength, and (c) Young's modulus for this type of bone.

Stress (10^6 N/m^2)	Strain (%)
35	0.2
70	0.4
105	0.6
140	1.0
175	1.5

55. •• The elastic limit of an alloy is 0.6×10^9 N/m^2. What is the minimum radius of a 4-m-long wire made from the alloy if a single strand is designed to support a commercial sign that has a weight of 8000 N and hangs from a fixed point? To stay within safety codes, the wire cannot stretch more than 5 cm. Example 10-8

56. • A 50.0-kg air-conditioning unit slips from its window mount, but the end of the electrical cord gets caught in the mounting bracket. In the process the cord (which is 0.50 cm in diameter) stretches from 3.0 to 4.5 m. What is Young's modulus for the cord? Example 10-1

57. • **Medical** The largest tendon in the body, the Achilles tendon connects the calf muscle to the heel bone of the foot. This tendon is typically 25.0 cm long, is 5.0 mm in diameter, and has a Young's modulus of 1.47×10^9 N/m^2. If an athlete has stretched the tendon to a length of 26.1 cm, what is the tension (in newtons and pounds) in the tendon? Example 10-1

58. • The strongest human-made fiber is a type of carbon fiber that has an ultimate strength of 6.37 GPa (1 GPa = 10^9 Pa). Those fibers, however, have an approximate diameter of only 5.0×10^{-6} m. If the fibers could be made thicker while maintaining the same strength, what minimum diameter of fiber would be needed to lift an adult human (m = 75 kg)? Example 10-6

59. •• **Biology** Many caterpillars construct cocoons from silk, one of the strongest naturally occurring materials known. Each thread is typically 2.0 μm in diameter and the silk has a Young's modulus of 4.0×10^9 N/m^2. (a) How many strands would be needed to make a rope 9.0 m long that would stretch only 1.00 cm when supporting a pair of 85-kg mountain climbers? (b) Assuming that there is no appreciable space between the parallel strands, what would be the diameter of the rope? Does the diameter seem reasonable for a rope that mountain climbers might carry? Example 10-1

60. •• A 2.8-carat diamond is grown under a high pressure of 58×10^9 N/m^2. (a) By how much does the volume of a spherical 2.8-carat diamond expand once it is removed from the chamber and exposed to atmospheric pressure? (b) What is the increase in the diamond's radius? One carat is 0.200 g, and you can use 3.52 g/cm^3 for the density of diamond, and 4.43×10^{11} N/m^2 for the bulk modulus of diamond. Example 10-3

61. • A glass marble that has a diameter of 1.00 cm is dropped into a graduated cylinder that contains 20.0 cm of mercury. (a) By how much does the volume of the marble shrink while at the bottom of the mercury, where the pressure is 2.7×10^4 N/m^2 greater than at the surface? (b) What is the corresponding change in radius associated with the compression? The bulk modulus of glass is 50×10^6 N/m^2. Example 10-2

62. • **Sports** During the 2004 Olympic clean-and-jerk weight-lifting competition, Hossein Rezazadeh lifted 263.5 kg. Mr. Rezazadeh himself had a mass of 163 kg. Ultimately the weight is all supported by the tibia (shin bone) of the lifter's legs. The average length of a tibia is 385 mm, and its diameter (modeling it as having a round cross section) is about 3.0 cm. Young's modulus for bone is typically about 2.0×10^{10} N/m^2. (a) By how much did the lift compress the athlete's tibia, assuming that the bone is solid? (b) Is it necessary to include the lifter's weight in your calculations? Why or why not? Example 10-1

63. • When a house is moved it is gradually raised and supported on wooden blocks. A typical house averages about 54,000 kg. The house is supported uniformly on six stacks of blocks of Douglas fir wood (which has a Young's modulus of 13×10^9 N/m^2). Each block is 25 by 75 cm. If the wood is stacked 1.5 m high, by how much will the house compress the supporting stack of blocks? Example 10-1

64. •• (a) What diameter steel cable is needed to support a large diesel engine with a mass of 4.0×10^3 kg? (b) By how much will the 10.0-m-long cable stretch once the engine is raised up off the ground? Assume the ultimate breaking strength of steel is 4.00×10^8 N/m^2. Example 10-6

65. • **Sports** A runner's foot pushes backward on the ground as shown in Figure 10-24. This results in a 25-N shearing force exerted in the forward direction by the ground, and in the backward direction by the foot, distributed over an area

Figure 10-24 Problem 65

of 15 cm² and a 1.0-cm-thick sole. If the shear angle θ is 5.0°, what is the shear modulus of the sole? Example 10-4

66. • The spherical bubbles near the surface of a glass of water are 2.5 mm in diameter at sea level, where the atmosphere exerts a pressure of 1.01×10^5 N/m² over the surface of each bubble. If the glass of water is taken to the elevation of Golden, Colorado, where the atmosphere exerts a pressure of 8.22×10^4 N/m² over the bubble surface, what will be the diameter of the bubbles? Example 10-3

67. •• **Biology** A particular human hair has a Young's modulus of 4.0×10^9 N/m² and a diameter of 150 μm. (a) If a 250-g object is suspended by the single strand of hair that is originally 20.0 cm long, by how much will the hair stretch? (b) If the same object were hung by an aluminum wire of the same dimensions as the hair, by how much would the aluminum stretch? (Try to do this part without repeating the previous calculation but use proportional reasoning instead.) (c) If we think of the strand of hair as a spring, what is its spring constant? (d) How does the hair's spring constant compare with that of ordinary springs in your physics laboratory? Example 10-1

68. • **Medical** The femur bone in the human leg has a minimum effective cross section of 3.0 cm². How much compressive force can it withstand before breaking? Assume the ultimate strength of the bone to be 1.7×10^8 N/m². Example 10-6

69. ••• A beam is attached to a vertical wall with a hinge. The mass of the beam is 1000 kg, and it is 4 m long. A steel support wire is tied from the end of the beam to the wall, making an angle of 30° with the beam (Figure 10-25). (a) By summing the torque about the axis passing through the hinge, calculate the tension in the support wire. Assume the beam is uniform so that the weight acts at its exact center. (b) What is the minimum cross-sectional area of the steel wire so that it is not permanently stretched? The yield strength (elastic limit) for steel is 290×10^6 N/m², and the ultimate breaking strength is 400×10^6 N/m². Example 10-8

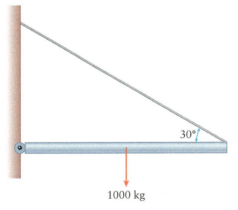

30°

1000 kg

Figure 10-25
Problem 69

70. •• **Biology** The typical compressive ultimate stress in the transverse direction is 133×10^6 N/m² for human bones and 178×10^6 N/m² for cow bones. (a) Is a human or a cow more likely to suffer a transverse break in a bone? Why? (b) The compressive longitudinal yield stress of human bones is approximately 182×10^6 N/m², and the compressive longitudinal ultimate stress is about 195×10^6 N/m². For cows, the compressive longitudinal yield stress of bone is about 196×10^6 N/m², and the compressive longitudinal ultimate stress is 237×10^6 N/m². Explain how a cow's bones are much

more capable of supporting their extreme weight in comparison to a human's bones. (c) A bone in a woman's leg has an effective cross-sectional area of 3.00 cm². If the bone is 35 cm long, how much compressive force can it withstand before breaking? How much will her bone compress if it is subjected to a force one-tenth the magnitude of the force that breaks it? The longitudinal elastic modulus of human bone is about 9.6×10^9 N/m². Example 10-6

71. •• An 85-kg mass falls 15 m from rest before it is jerked to a stop by a safety cable in 0.07 s. The cable is attached to a 9-mm-diameter eyebolt sunk into a rigid wooden beam (Figure 10-26). Assuming the beam does not deform at all, what is the average shear stress on the bolt? Example 10-5

Figure 10-26 Problem 71

72. ••• **Sports** Bicycle caliper brakes consist of rubber pads mounted in rigid metal "shoes." The normal force applied by a bicycle caliper brake pad to the wheel rim is about 4.00×10^2 N. The contact area of the brake pad is approximately 4.0 cm × 0.70 cm, with the long dimension parallel to the rim's rotation. The pad is 1.0 cm thick. The coefficient of kinetic friction between the rubber pad and the aluminum rim is 0.50. Given that the shear modulus of the rubber is 0.30 MPa, calculate the deformation angle φ of the brake pad. Example 10-4

73. ••• A warehouse worker accidentally slides into a shelving unit, stopping himself with his hands by exerting a force of 3700 N on the top shelf. The shelving unit is 1.5 m tall, 2.5 m long, and 0.67 m deep. During the collision the shelving unit is knocked into a parallelogram shape with the top overhanging the bottom by 13 cm along the long axis before it springs back into shape. Assuming all the force of the collision was translated into shear forces acting parallel to the top surface, calculate the shear modulus of the shelving unit. Example 10-5

74. • A construction crane is lifting a 3000-kg load of drywall with a 0.75-cm diameter steel cable that is 40.0 m long. Once the cable is taut, how much cable will the crane reel in before the load is actually lifted off the ground? Example 10-1

75. •• The maximum shear stress of steel is 400×10^6 N/m². A 200-kg steel beam is to be supported by standard M3 steel bolts (diameter 3 mm). If specifications require a minimum safety margin of a factor of 10, how many bolts are required to support the beam to avoid shearing (breaking) of the bolts? Example 10-7

76. ••• You are building a backyard tire swing. The tire has a mass of 9.0 kg and is supported by a single rope that is 3.0 m long. The rope hangs off a horizontal steel bolt screwed into a support. You decide that if the tire swing can support your 130-kg uncle, it will be safe enough for your 3-year-old daughter. You guess your uncle can reach a maximum speed of 10 m/s at the bottom of the swing. (a) Assuming that the ultimate shear strength of steel is about 210 MPa, what is the smallest diameter steel bolt that won't shear off under the load of your swinging uncle? (b) If the nylon rope is 1 cm in diameter, what is the maximum distance your uncle will make it stretch? Example 10-7

Jane Gould/Alamy

Fluids

11

What do you think?

Each of these yellowback fusilier fish (*Caesio xanthonota*) experiences an upward buoyant force exerted by the water surrounding it. If you replaced one of these fish with a life-size fish sculpture made of iron (which is denser than a fish's body), would the magnitude of the buoyant force (a) increase, (b) decrease, or (c) remain the same?

In this chapter, your goals are to:

- (11-1) Describe the similarities and differences between liquids and gases.
- (11-2) Recognize how to apply the definition of density.
- (11-3) Explain the origin of fluid pressure in terms of molecular motion.
- (11-4) Calculate the pressure at a given depth in a fluid in hydrostatic equilibrium.
- (11-5) Explain the difference between absolute pressure and gauge pressure.
- (11-6) Calculate the force on an object due to a difference in pressure on its sides.
- (11-7) Explain how to apply Pascal's principle to a fluid at rest.
- (11-8) Apply Archimedes' principle to find the buoyant force on an object in a fluid.
- (11-9) Use the equation of continuity to analyze the flow of an incompressible fluid.
- (11-10) Apply Bernoulli's principle to relate fluid pressure and flow speed in an incompressible fluid.
- (11-11) Explain what happens in flows where viscosity is important.
- (11-12) Describe the role of surface tension in the behavior of liquids.

To master this chapter, you should review:

- (6-7) What happens to the total mechanical energy of a system when nonconservative forces do work.
- (10-3) How pressure is defined.

11-1 Liquids and gases are both examples of fluids

Liquids and gases—collectively known as *fluids* for their ability to flow—are the most common states of matter in the universe. Indeed, part of our planet's interior, the entirety of our Sun, and most of your body are composed of fluids (Figure 11-1). Solid objects in our daily environment such as rocks and trees are exceptional cases in a universe dominated by fluids.

Solids are substances with individual molecules that cannot move freely but remain in essentially fixed positions relative to one another. That's why a solid object maintains its shape, though it may bend or deform if you apply forces to it. In contrast, **fluids** are substances that can flow because their molecules can move with respect to each other and are not tied to fixed locations.

433

Figure 11-1 **Fluids** Materials that can flow make up (a) part of our Earth, (b) all of our Sun (as well as all of the stars visible in the night sky, which are objects like the Sun), and (c) most of our bodies. Note that the photo in (a) shows at least three different fluids: molten lava flowing down the rocks, liquid water in the ocean, and the air in our atmosphere.

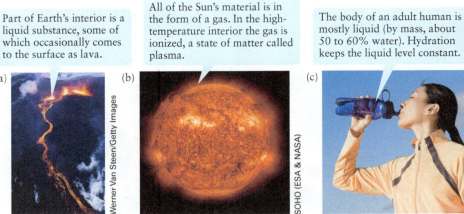

Part of Earth's interior is a liquid substance, some of which occasionally comes to the surface as lava.

All of the Sun's material is in the form of a gas. In the high-temperature interior the gas is ionized, a state of matter called plasma.

The body of an adult human is mostly liquid (by mass, about 50 to 60% water). Hydration keeps the liquid level constant.

(a) Werner Van Steen/Getty Images

(b) SOHO (ESA & NASA)

(c) Image Source Plus/Alamy

Fluids fall into two broad categories: liquids and gases. **Liquids** tend to maintain the same volume regardless of the shape and size of their container. For example, 1 cubic meter (1 m^3) of water taken from a storage tank will exactly fill a 1-m^3 container into which it is poured (**Figure 11-2a**). **Gases**, however, expand to fill whatever volume is available to them. If you place a cylinder containing 1 m^3 of compressed oxygen gas in a large room and open the valve, the oxygen gas spreads out over the entire room so that its volume is much greater than 1 m^3 (**Figure 11-2b**). It can also be compressed further to fill an even smaller container. (The gas in the interior of the Sun, shown in Figure 11-2b, is ionized: The gas atoms have lost one or more electrons. Such an ionized gas is called a *plasma*. Despite having a similar name, blood plasma is a liquid and is not ionized.)

What explains the difference between these two types of fluids? In a liquid, molecules are close to one another (almost touching). At close range the attractive forces between molecules in a liquid are strong, keeping the molecules together so that the volume of the liquid stays the same. In a gas, however, the molecules are much farther apart; in the air you're breathing now, the average distance between molecules is about 10 times the size of a single molecule. At these greater separations the molecules exert little or no attractive forces on each other, so nothing prevents the molecules from spreading out to occupy all of the available volume.

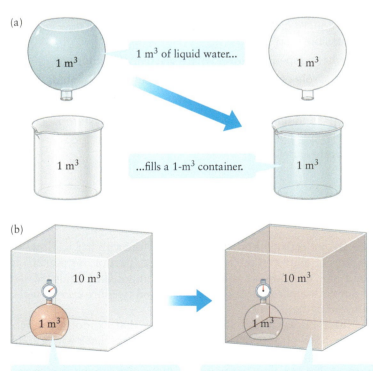

Figure 11-2 **Liquids versus gases** (a) A liquid maintains the same volume when you pour it from one container to another. (b) A gas expands to fill whatever volume is available.

(a)

1 m^3 of liquid water...

1 m^3

1 m^3

1 m^3

...fills a 1-m^3 container.

1 m^3

(b)

10 m^3

1 m^3

10 m^3

1 m^3

1 m^3 of compressed oxygen gas...

...expands to fill a much larger volume.

In this chapter we will touch on many of the key aspects of fluids. The central concepts of *density* and *pressure*, coupled with the notion of what it means for a fluid to be in *equilibrium*, will help us understand why water pressure increases with increasing depth. They also explain the phenomenon of *buoyancy* that enables marine animals to float under water without sinking or rising. Fluids in *motion* are influenced not only by pressure and density but also by *viscous* forces, which are the frictional forces of one part of the fluid rubbing past another. By examining some of the myriad ways in which fluids flow, we will gain insight into the flight of airplanes and birds, the behavior of blood in the circulatory system, and the peculiar challenges that face very small swimming organisms. Finally, we will see how the mutual attraction of molecules in a liquid gives rise to *surface tension*, a phenomenon that impacts the function of your lungs.

TAKE-HOME MESSAGE FOR Section 11-1

✔ Fluids (which include liquids and gases) do not have a definite shape. They can flow to take on the shape of their container because their molecules are in constant motion.

✔ Liquids maintain essentially the same volume as they flow, but gases will expand or contract as necessary to fill the available space.

✔ The attractive forces between molecules are strong in liquids but are weak in gases because the molecules are farther apart.

11-2 Density measures the amount of mass per unit volume

One quantity that helps describe how tightly packed the molecules are in a gas, liquid, or solid is the *density*. The **density** of a substance, denoted by the Greek letter ρ ("rho"), is its *mass per volume*: the mass of the substance divided by the volume that it occupies. The greater the density of a substance, the more kilograms of that substance are packed into a given number of cubic meters of volume:

The **density** ρ ("rho") of a certain substance...

...equals the **mass** m of a given quantity of that substance...

$$\rho = \frac{m}{V}$$

...divided by the **volume** V of that quantity of the substance.

Definition of density
(11-1)

A given number of water molecules of combined mass m occupy a smaller volume V in liquid water than they do in water vapor (a gas). Hence the density $\rho = m/V$ of liquid water is greater than that of water vapor.

WATCH OUT! Density alone doesn't tell you how closely packed the molecules are.

! A certain number of oxygen gas molecules have 14% greater density than the same number of nitrogen gas molecules with the same spacing between molecules. That's because density is *mass* per unit volume, and the mass of an oxygen molecule is 14% greater than that of a nitrogen molecule. To find the spacing between molecules in a given substance, you need to know both the density of the substance *and* the mass per molecule.

Density is usually stated in kilograms per cubic meter (kg/m^3) or grams per cubic centimeter (g/cm^3; $1\ g/cm^3 = 1000\ kg/m^3$). As an example, one cubic meter of liquid water at a temperature of 4°C has a mass of 1000 kg and therefore a density of $\rho = (1000\ kg)/(1\ m^3) = 1000\ kg/m^3 = 1\ g/cm^3$. Twice the mass of liquid water (2000 kg) occupies twice the volume (2 m^3) and so its density is the same: $\rho = (2000\ kg)/(2\ m^3) = 1000\ kg/m^3$. This example shows that density depends only on the nature of the substance, *not* on how much of the substance is present. When we say that gold has a density of $19{,}300\ kg/m^3 = 1.93 \times 10^4\ kg/m^3$, we mean that *any* quantity of gold has this same density. Table 11-1 lists the densities of some common substances.

TABLE 11-1 Densities of Various Substances

Substance	Density (kg/m³)
air at sea level at 15°C	1.23
dry timber (white pine)	370
fresh wood (red oak)	740
gasoline	740
ethanol (ethyl alcohol) at 20°C	789
lipids (fats and oils)	915–945
ice at 0°C	917
fresh water at 4°C	1.000×10^3
fresh water at 20°C	998
seawater at 4°C	1.025×10^3
muscle	1.06×10^3
deer antler (low-density bone)	1.76×10^3
cow femur (typical bone)	2.06×10^3
tympanic bulla of a fin whale (densest bone)	2.47×10^3
aluminum (Al)	2.70×10^3
mollusk shell	2.7×10^3
calcite (mineral of shell)	2.71×10^3
tooth enamel, human	2.9×10^3
apatite (mineral of bone, tooth)	3.2×10^3
planet Earth (average density)	5.52×10^3
iron (Fe)	7.8×10^3
mercury (Hg)	13.595×10^3
gold (Au)	19.3×10^3

An alternative way to describe the density of a substance is to compare it to another substance whose density serves as a standard. One common standard is liquid water at 4°C ($\rho = 1000$ kg/m³), and the **specific gravity** of a substance is equal to its density divided by that of 4°C liquid water. Human blood, for instance, has a specific gravity of 1.06, which means it is 1.06 times as dense as water: $\rho_{blood} = (1.06)(1000$ kg/m³$) = 1.06 \times 10^3$ kg/m³. The term "specific gravity" in this case means that a certain volume of blood has 1.06 times the mass and hence 1.06 times the weight as the same volume of water. On average, red blood cells make up about 38% of the volume of blood in women and 42% in men; these cells are only slightly denser than water, which is why the specific gravity of blood is just a little greater than 1.

In general, the density of a liquid depends on temperature. If you take water from the refrigerator (temperature about 4°C) and let it warm to room temperature (about 20°C), the water expands slightly. Because the mass m of the water is the same but its volume V has increased, the density $\rho = m/V$ of 20°C water is slightly less (by about 0.2%) than that of 4°C water. Most liquids expand and become less dense when the temperature increases. Between 0°C and 4°C, water is a notable exception to this general rule: Within this temperature range, water *contracts* and becomes *more* dense as the temperature increases. That's why we give the densities of water and salt water in Table 11-1 at 4°C, the temperature at which these liquids are densest.

The density of gases is sensitive to both temperature and pressure. Heated air expands, which lowers its density; applying pressure makes the air occupy a smaller volume, which raises its density. Because gases can easily be compressed by squeezing, they are **compressible fluids**. By contrast, for most practical purposes liquids are **incompressible fluids**: The volume and the density of a liquid change very little when it is squeezed because the molecules of a liquid are almost touching and resist being squeezed much closer together. You can easily compress air to a higher density with a hand pump used for inflating bicycle tires, but to compress water appreciably takes tremendous pressures. (At the deepest point in the Pacific Ocean, at a depth of 11 km, the pressure is about a thousand times greater than at the surface. Yet the density of seawater there is only about 5% greater than at the surface. In Section 11-3 we'll discuss why pressure increases with depth.)

While the main focus of this chapter is on fluids, the idea of density applies to solids as well. The following examples show some applications.

EXAMPLE 11-1 Density of Chicken

A typical whole chicken on sale at a supermarket has a mass of 2.3 kg and is approximately spherical in shape, with a radius of about 8.0 cm. What is the approximate density of such a chicken in kg/m³? Compare the result to the densities of the biological materials listed in Table 11-1.

Set Up

We are given the chicken's mass and its radius, and our goal is to calculate its density ρ. We use the definition of density in Equation 11-1, $\rho = m/V$.

The chicken is very nearly spherical, so to determine its volume we use the expression for the volume of a sphere of radius R (see the Math Tutorial).

Definition of density:

$$\rho = \frac{m}{V} \qquad (11\text{-}1)$$

Volume of a sphere of radius R:

$$V = \frac{4\pi R^3}{3}$$

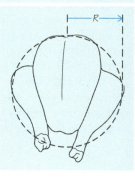

Solve

To calculate the density in the desired units, we need to find the chicken's volume in cubic meters (m^3). We first express its radius in meters.	$R = (8.0 \text{ cm})\left(\dfrac{1 \text{ m}}{100 \text{ cm}}\right) = 8.0 \times 10^{-2} \text{ m}$
We then calculate the volume of the chicken in cubic meters.	$V = \dfrac{4\pi R^3}{3} = \dfrac{4\pi(8.0 \times 10^{-2} \text{ m})^3}{3} = 2.1 \times 10^{-3} \text{ m}^3$
The density is the mass of the chicken ($m = 2.3$ kg) divided by this volume.	$\rho = \dfrac{m}{V} = \dfrac{2.3 \text{ kg}}{2.1 \times 10^{-3} \text{ m}^3} = 1.1 \times 10^3 \text{ kg/m}^3$

Reflect

Our answer is close to some of the values for biological materials given in Table 11-1, but isn't a precise match for any of them. That's because a chicken's body is not a pure substance but a mixture of substances of different densities: fats, water, muscle, and bone. Our calculated density of 1.1×10^3 kg/m^3 is actually the *average* density of the chicken. We explore the idea of average density further in Example 11-2.

You may be wondering if it's reasonable to approximate a chicken as being spherical. If you look at the sketch, some parts of the chicken extend beyond the sphere, and some of the sphere is empty. But the volume of the chicken outside the sphere is nearly equal to the volume of the "empty" part of the sphere, so the volume of the chicken is actually pretty close to the volume of the sphere.

EXAMPLE 11-2 Average Density

A submersible vessel used for deep-sea research has a metal hull with a large cavity in its interior for the crew and their equipment. Suppose that the hull is made of 1.130×10^3 m^3 of a particular steel alloy of density 7.910×10^3 kg/m^3 and that the cavity in the hull has volume 7.600×10^3 m^3 and is filled with air of density 1.200 kg/m^3. Find the *average* density of the submersible, that is, the density of the submersible as a whole.

Set Up

The submersible contains two materials—the steel of the hull and the air in the interior spaces—which have different densities and occupy different volumes. We want to find a third density, the *average* density of the entire submersible, equal to its total mass m_{sub} divided by its total volume V_{sub}. We again use the definition of density, Equation 11-1. We'll apply this definition three times, once for each density.

Definition of density:

$$\rho = \frac{m}{V} \qquad (11\text{-}1)$$

steel hull
density $\rho_{steel} = 7.910 \times 10^3$ kg/m^3
volume $V_{hull} = 1.130 \times 10^3$ m^3

interior spaces filled with air
density $\rho_{air} = 1.200$ kg/m^3
volume $V_{cavity} = 7.600 \times 10^3$ m^3

Solve

The submersible's average density $\rho_{average}$ is its total mass m_{sub} divided by its total volume V_{sub}, so we need to calculate both m_{sub} and V_{sub}. The volume of the submersible (V_{sub}) is the sum of the volume of the hull (V_{hull}) and the volume of the cavity (V_{cavity}).

Average density of the sub as a whole:

$$\rho_{average} = \frac{m_{sub}}{V_{sub}}$$

Total volume of the sub:

$$V_{sub} = V_{hull} + V_{cavity} = 1.130 \times 10^3 \text{ m}^3 + 7.600 \times 10^3 \text{ m}^3$$
$$= 8.730 \times 10^3 \text{ m}^3$$

From Equation 11-1 the mass of steel in the hull (m_{hull}) and the mass of air in the cavity (m_{cavity})

Rearrange Equation 11-1:

$$m = \rho V$$

are given by their respective densities multiplied by their respective volumes.

Solve for m_{hull} and m_{cavity}:

$$m_{hull} = \rho_{steel}V_{hull} = (7.910 \times 10^3 \text{ kg/m}^3)(1.130 \times 10^3 \text{ m}^3)$$
$$= 8.938 \times 10^6 \text{ kg}$$

$$m_{cavity} = \rho_{air}V_{cavity} = (1.200 \text{ kg/m}^3)(7.600 \times 10^3 \text{ m}^3)$$
$$= 9.120 \times 10^3 \text{ kg}$$

The total mass of the submersible (m_{sub}) is the sum of the mass of steel in the hull and the mass of air in the cavity.

$$m_{sub} = m_{hull} + m_{cavity} = 8.938 \times 10^6 \text{ kg} + 9.120 \times 10^3 \text{ kg}$$
$$= 8.947 \times 10^6 \text{ kg}$$

To get the average density, substitute the total volume V_{sub} and the total mass m_{sub} into the expression for $\rho_{average}$.

$$\rho_{average} = \frac{m_{sub}}{V_{sub}} = \frac{8.947 \times 10^6 \text{ kg}}{8.730 \times 10^3 \text{ m}^3}$$
$$= 1.025 \times 10^3 \text{ kg/m}^3$$

Reflect

No one component of the submersible has a density of 1.025×10^3 kg/m^3; instead, this represents an average of the densities of the two components. Note that the average density is intermediate between the density of air (1.200 kg/m^3) and the density of steel (7.910×10^3 kg/m^3).

   This particular value of average density is greater than the density of fresh water (1.000×10^3 kg/m^3) but is the *same* as that of seawater (1.025×10^3 kg/m^3). As we will see in Section 11-8, this means that the submersible will sink in fresh water but will float when submerged in seawater—which is a useful thing for a deep-sea research vessel to do.

EXAMPLE 11-3   How Much Space for a Molecule?

A single molecule of water (H$_2$O) has a mass of 2.99×10^{-26} kg. Find the average volume per water molecule (a) in liquid water at 4°C and (b) in water vapor (that is, water in gas form) at 120°C, which has a density of 0.559 kg/m^3. In each case, what is the length of each side of a cube with this volume? Compare to the size of a water molecule, which is about 2.0×10^{-10} m across.

Set Up

The density ρ of a substance tells us the mass per volume. We want to find the volume per *molecule* in the substance. We'll do this calculation by combining the value of ρ for each substance (liquid water at 4°C and water vapor at 120°C) with the mass per water molecule. We'll imagine that each molecule is at the center of a cube of side L, and we'll determine L from the formula for the volume of a cube.

Definition of density:

$$\rho = \frac{m}{V} \qquad (11\text{-}1)$$

Volume of a cube of side L:

$$V = L \times L \times L = L^3$$

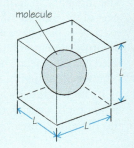

molecule

Solve

Rewrite Equation 11-1 to solve for the volume occupied by a quantity m of the substance.

From Equation 11-1,

$$V = \frac{m}{\rho}, \text{ or}$$

$$\text{volume occupied by a molecule} = \frac{\text{(mass of a molecule)}}{\text{(density)}}$$

(a) Table 11-1 tells us that water at 4°C has a density of 1.000×10^3 kg/m^3. Use this to find the volume per molecule and the length L of a cube with this volume.

For water at 4°C,

volume occupied by a molecule = V

$$= \frac{2.99 \times 10^{-26} \text{ kg}}{1.00 \times 10^3 \text{ kg/m}^3}$$
$$= 2.99 \times 10^{-29} \text{ m}^3$$

The volume V is equal to L^3; that is, V is the cube of L. So L is the cube *root* of V:

$$L = \sqrt[3]{V} = \sqrt[3]{2.99 \times 10^{-29} \ m^3}$$
$$= 3.10 \times 10^{-10} \ m$$

This is about 50% larger than the size of a water molecule (roughly 2.0×10^{-10} m).

(b) Repeat the calculation for water vapor at 120°C. The molecule is the same and so has the same mass, but the density is far less than that for the liquid at 4°C.

For water vapor at 120°C,

volume occupied by a molecule = V

$$= \frac{2.99 \times 10^{-26} \ kg}{0.559 \ kg/m^3}$$
$$= 5.35 \times 10^{-26} \ m^3$$

The length L of a cube with this volume is

$$L = \sqrt[3]{V} = \sqrt[3]{5.35 \times 10^{-26} \ m^3}$$
$$= 3.77 \times 10^{-9} \ m$$

This is larger than the size of a water molecule by a factor of $(3.77 \times 10^{-9} \ m)/(2.0 \times 10^{-10} \ m) = 19$.

Reflect

In both liquids and gases, the molecules are in constant motion. That's why the volumes we calculated in this example are the *average* volume per molecule.

Our results show that in the liquid state a water molecule almost fills the volume that it occupies on average. So there is very little extra room to move the molecules closer together, which means that it is very difficult to compress a liquid. In the gaseous state, however, on average each molecule is surrounded by lots of empty space. In this state it's much easier to push the molecules closer together, so a gas is relatively easy to compress.

GOT THE CONCEPT? 11-1 Weight and Density

 Which is heavier, a block of iron or a block of ice? (a) The block of iron; (b) the block of ice; (c) they have the same weight; (d) not enough information given to decide.

TAKE-HOME MESSAGE FOR Section 11-2

✔ The density of a substance describes how much mass of a particular substance occupies a given volume.

✔ The specific gravity of a substance equals the density of that substance relative to the density of fresh water at 4°C.

✔ An object's average density equals its total mass divided by its overall volume. It is intermediate in value between the densities of the least dense and most dense constituents of the object.

11-3 Pressure in a fluid is caused by the impact of molecules

Pressure is a commonplace idea. Everyone has had their blood pressure measured during a visit to the physician. An air-filled balloon bursts if the pressure within it is too great. And weather reports refer to low-pressure and high-pressure weather systems in the atmosphere. But what *is* pressure?

Figure 11-3 shows how pressure is defined. If you place a small, thin, flat plate in a fluid—a gas or a liquid—you'll find that the fluid exerts forces that act perpendicular to either face of the plate. You can feel these forces if you stick your hand into a swimming pool or a sink full of water. The same kind of forces act on your hand when it's surrounded by air, but are very feeble compared to the forces exerted by water. The net force that the fluid exerts on either face of the plate acts perpendicular to the face. The **pressure** of the fluid acting on the face is the magnitude of this perpendicular force divided by the area of the face:

Definition of pressure
(11-2)

The **pressure** p exerted by a fluid on an object in the fluid...

...equals the **force** $F_\perp$ that the fluid exerts **perpendicular to the face of the object...**

$$p = \frac{F_\perp}{A}$$

...divided by the **area** A of that face.

WATCH OUT! The symbols for pressure and density are not the same.

! We use a lowercase p to represent pressure and the lowercase Greek letter ρ ("rho") to represent density. Unfortunately these two symbols look a lot alike. Be careful, especially when you write them, that you properly distinguish between the two.

To understand why a fluid exerts forces like those shown in Figure 11-3, keep in mind that the molecules of a liquid or gas are in constant motion. Any object placed in a fluid is struck repeatedly by these moving molecules. The force of each individual impact is miniscule, but there are so many molecules and so many impacts that their combined effect is appreciable. It's that combined force that gives rise to the forces labeled $F_\perp$ in Figure 11-3. As an example, more than 10^{23} air molecules collide with a square centimeter of your skin every second, exerting a combined force of about 10 N on that square centimeter. In a fluid at rest (such as water in the kitchen sink that is not flowing or draining out, or the air in a room with no wind currents), the molecules of the fluid move in entirely random directions. At any point in the fluid there are as many molecules moving to the left as to the right, as many moving upward as downward, and so on. Hence a small, thin, flat plate placed in the fluid (Figure 11-3) experiences the same number of impacts, and hence the same magnitude of force, on both of its sides no matter which way the plate is oriented.

If we double the surface area A exposed to the fluid, the force magnitude $F_\perp$ also doubles because there is twice as much area to be hit by the molecules of the fluid. However, the pressure $p = F_\perp/A$ remains the same. Pressure is a property of the fluid, not of the object that's immersed in the fluid. Furthermore, because the pressure is the same no matter how the disk in Figure 11-3 is oriented, pressure has no direction: Unlike force, pressure is *not* a vector.

The SI units of pressure are newtons per square meter (N/m^2). As we discussed in Section 10-2, one newton per square meter is also called a **pascal** (abbreviated Pa): $1\ Pa = 1\ N/m^2$. We'll discuss other common units of pressure in Section 11-5.

A small, thin plate of area A is immersed in a fluid.

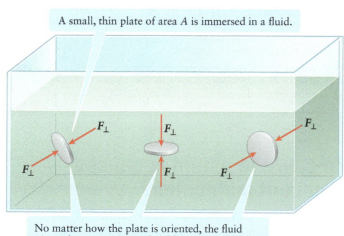

Figure 11-3 **Pressure on a thin plate in a fluid** The fluid pressure on this plate does not depend on its orientation.

No matter how the plate is oriented, the fluid exerts the same magnitude of force $F_\perp$ and the same pressure $p = F_\perp/A$ on both sides of the plate.

WATCH OUT! Don't confuse force and pressure.

! Although force and pressure are related, they are *not* the same quantity. Here's an example. If you take off your shoes and stand up, your mass is distributed evenly over the surface area of your two feet. The pressure on the soles of your feet is your mass divided by the surface area of the bottom of your two feet and is probably not uncomfortable. But if you stood on a bed of nails, it would hurt! Your weight hasn't changed, so the force you exert on the surface supporting you hasn't changed. However, because the supporting surface—the tips of a few nails—has a much smaller surface area, the pressure on your feet is much greater (and much more painful) in this case. However, people can comfortably *lie* on a suitably prepared bed of nails. By lying flat, a person's weight is distributed over the area of the tips of many nails, resulting in a tolerable pressure (Figure 11-4).

Figure 11-4 **A bed of nails** It doesn't hurt to lie on this bed of nails. That's because the force that the nails exert is distributed over the area of many nails, resulting in a tolerable pressure on the person's back.

At sea level on Earth, the pressure of the atmosphere is about 1.01×10^5 N/m², or 1.01×10^5 Pa. By contrast, the pressure at an altitude of 12,500 m (41,000 ft, about the highest altitude at which airliners fly) is only 1.78×10^4 Pa, or 17.6% of the pressure at sea level. One reason the pressure is lower at high altitude is that the air is less dense there (0.289 kg/m³ at 12,500 m versus 1.23 kg/m³ at sea level). Hence there are fewer air molecules to produce pressure through collisions. A second reason is that the molecules at high altitude move more slowly than those near sea level, so when collisions do take place they are relatively gentle. We'll learn in Chapter 14 that gas molecules move more slowly in a cold gas than a warm gas. The air at 12,500 m is indeed quite cold: The average temperature there is −56°C (−69°F) compared to 15°C (59°F) at sea level. The air pressure at 12,500 m is too low to sustain human life, which is why airliners have *pressurized* cabins: Air from the jet engine intakes is pumped into the cabin, increasing the pressure inside the cabin to what it would be at an altitude of 1800–2400 m (6000–8000 ft) above sea level.

The variation of pressure with elevation isn't unique to our atmosphere. In oceans, in lakes, and even in a glass of water there is greater pressure as you go deeper and lower pressure as you ascend. In the next section we'll learn more about how and why pressure increases with depth in a fluid.

EXAMPLE 11-4 Pressures and Forces on an Airliner Door

A typical passenger door on an airliner (Figure 11-5) is 183 cm high and 84 cm wide, and has an area (height times width) of 1.54 m². When the airliner is sitting on the ground at sea level, the air pressure on the outside of the door (in the surrounding air) and on the inside of the door (inside the cabin) are both equal to 1.01×10^5 Pa. When the airliner is cruising at an altitude of 12,500 m above sea level, the air pressure outside the cabin is 1.78×10^4 Pa and is 7.53×10^4 Pa inside the pressurized cabin. Find the forces that the air exerts on the inside and outside of the door, and the net force that the air exerts on the door, (a) on the ground at sea level and (b) at an altitude of 12,500 m.

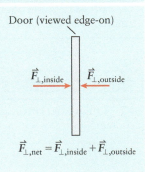

Passenger door

Figure 11-5 **An airliner door** How do the forces on this airliner door due to air pressure vary as the airliner climbs to its cruising altitude?

Set Up

Most airliner doors are curved to match the shape of the fuselage, but for simplicity we'll assume that the door is flat. We're given the pressure p on each side of the door as well as its area A, so we can use Equation 11-2 to solve for the force $F_\perp$ that the air exerts perpendicular to each side of the door. The forces on the two sides of the door are in opposite directions: Their vector sum is the *net* force that the air exerts on the door.

Definition of pressure:

$$p = \frac{F_\perp}{A} \qquad (11\text{-}2)$$

Door (viewed edge-on)

$\vec{F}_{\perp,\text{inside}}$ $\vec{F}_{\perp,\text{outside}}$

$\vec{F}_{\perp,\text{net}} = \vec{F}_{\perp,\text{inside}} + \vec{F}_{\perp,\text{outside}}$

Solve

Rewrite Equation 11-2 to solve for the force exerted perpendicular to each side of the door.

From Equation 11-2,

$$F_\perp = pA$$

(a) At sea level the pressure is the same on both sides of the door, so the air exerts forces of equal magnitude on both sides of the door.

At sea level,

$$p_{\text{inside}} = p_{\text{outside}} = 1.01 \times 10^5 \text{ Pa} = 1.01 \times 10^5 \text{ N/m}^2$$

$$\begin{aligned} F_{\perp,\text{inside}} = p_{\text{inside}} A &= (1.01 \times 10^5 \text{ N/m}^2)(1.54 \text{ m}^2) \\ &= 1.56 \times 10^5 \text{ N} \end{aligned}$$

$$F_{\perp,\text{outside}} = p_{\text{outside}} A = 1.56 \times 10^5 \text{ N}$$

Since the forces on the inside and outside have the same magnitude but are in opposite directions, the *net* force exerted by the air on the door is zero:

$$F_{\perp,\text{net}} = 0$$

(b) Repeat the calculation for an altitude of 12,500 m.

At an altitude of 12,500 m,

$$p_{\text{inside}} = 7.53 \times 10^4 \text{ Pa} = 7.53 \times 10^4 \text{ N/m}^2$$

$$\begin{aligned} F_{\perp,\text{inside}} = p_{\text{inside}} A &= (7.53 \times 10^4 \text{ N/m}^2)(1.54 \text{ m}^2) \\ &= 1.16 \times 10^5 \text{ N} \end{aligned}$$

$$p_{\text{outside}} = 1.78 \times 10^4 \text{ Pa} = 1.78 \times 10^3 \text{ N/m}^2$$

$$\begin{aligned} F_{\perp,\text{outside}} = p_{\text{outside}} A &= (1.78 \times 10^3 \text{ N/m}^2)(1.54 \text{ m}^2) \\ &= 2.74 \times 10^4 \text{ N} \end{aligned}$$

The outward force on the inside of the door is greater in magnitude than the inward force on the outside of the door. Hence the net force exerted by the air on the door is outward. The magnitude of the net force is

$$\begin{aligned} F_{\perp,\text{net}} &= F_{\perp,\text{inside}} - F_{\perp,\text{outside}} \\ &= (1.16 \times 10^5 \text{ N}) - (2.74 \times 10^4 \text{ N}) \\ &= 8.86 \times 10^4 \text{ N} \end{aligned}$$

Reflect

The net force on the airliner door while cruising at 12,500 m is tremendous, equivalent to the weight of an object of mass 9040 kg! The door must be carefully engineered to withstand such a force.

Mass m of an object with weight $w = mg = 8.86 \times 10^4$ N:

$$\begin{aligned} m = \frac{w}{g} &= \frac{8.86 \times 10^4 \text{ N}}{9.80 \text{ m/s}^2} = \frac{8.86 \times 10^4 \text{ kg} \cdot \text{m/s}^2}{9.80 \text{ m/s}^2} \\ &= 9.04 \times 10^3 \text{ kg} \end{aligned}$$

GOT THE CONCEPT? 11-2 Something's Fishy

? A tropical fish is at rest in the middle of an aquarium. Compared to the pressure on the left side of the fish, the pressure on the right side of the fish has (a) the same magnitude and the same direction; (b) the same magnitude and opposite direction; (c) a different magnitude and the same direction; (d) a different magnitude and opposite direction; (e) none of these.

TAKE-HOME MESSAGE FOR Section 11-3

✔ A fluid exerts pressure on any object with which it is in contact.

✔ The pressure equals the force that the fluid exerts perpendicular to the object's face divided by the area of that face.

11-4 In a fluid at rest pressure increases with increasing depth

Figure 11-6 shows the results of an experiment with air in our atmosphere. An empty plastic bottle is opened at an altitude of 9000 ft (2740 m) above sea level so that the bottle fills with air at that altitude. The bottle is then tightly capped (Figure 11-6a). As the bottle is brought back to sea level, it compresses as though it had been squeezed by an invisible hand (see Figure 11-6b).

Why does this happen? The answer is that the pressure in a fluid *increases* as you go *deeper* into the fluid. For example, air pressure in the atmosphere is about 40% greater at sea level than it is at an altitude of 2740 m. The air inside the bottle in Figure 11-6 was at the pressure found at 2740 m, so it compressed when it was brought to sea level and was surrounded by air at higher pressure. In this section we'll explore the relationship between pressure and depth in a fluid.

(a)

Roger Freedman

(b)

Roger Freedman

Hydrostatic Equilibrium

Let's consider a large tank filled with a fluid at rest (Figure 11-7). When a fluid is at rest (that is, not flowing), we say it is in **hydrostatic equilibrium**. The term "hydrostatic" specifically refers to water at rest, but it's common to use this term to refer to *any* kind of fluid in equilibrium.

Now imagine a box-shaped volume of fluid within the tank, as in Figure 11-7. The area of the top and bottom of the box is A, and the height of the box is d. We won't put an actual box in the fluid; instead we simply imagine a boundary that separates the box-shaped volume from the rest of the fluid.

The weight of the fluid above the box exerts a downward force of magnitude F_{down} on the box. Because fluids exert a force in all directions, the fluid below the box pushes upward on the box with a force of magnitude F_{up}. (We'll see in a moment that F_{up} is *not* equal to F_{down}.) The fluid also exerts forces on the sides of the box, but the *net* force on the sides is zero: There is as much force on the left side as on the right side, and these forces cancel. Finally, there is a downward force of gravity on the fluid inside the box—that is, the weight $\vec{w}$ of the fluid. Figure 11-7 shows these three forces on the box, $\vec{F}_{up}$, $\vec{F}_{down}$, and $\vec{w}$.

Figure 11-6 Air pressure varies with elevation (a) This empty plastic bottle was opened to the air in a nonpressurized airplane at an altitude of 9000 ft (2740 m) then tightly capped. (b) Outside air pressure made the bottle collapse when it was returned to sea level.

Consider a box-shaped volume of fluid that is part of a larger quantity of fluid at rest.

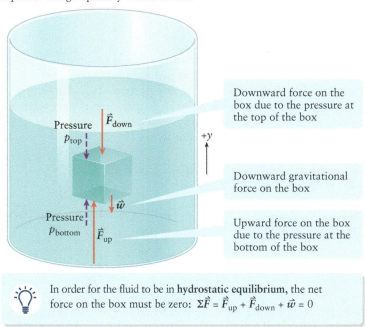

Pressure p_{top} | $\vec{F}_{down}$

Downward force on the box due to the pressure at the top of the box

$+y$

$\vec{w}$

Downward gravitational force on the box

Pressure p_{bottom} | $\vec{F}_{up}$

Upward force on the box due to the pressure at the bottom of the box

💡 In order for the fluid to be in **hydrostatic equilibrium**, the net force on the box must be zero: $\Sigma\vec{F} = \vec{F}_{up} + \vec{F}_{down} + \vec{w} = 0$

Figure 11-7 Hydrostatic equilibrium We can derive the equation of hydrostatic equilibrium by considering a box-shaped portion of a fluid at rest. For the box to be in equilibrium, the pressure at its bottom must be greater than the pressure at its top.

We started with the assumption that the fluid in the tank is at rest. This requires that there be no *net* force on the fluid volume; that is, the vector sum of the three forces acting on the box must be zero:

$$\Sigma \vec{F} = \vec{F}_{up} + \vec{F}_{down} + \vec{w} = 0 \tag{11-3}$$

All of these forces are vertically upward or downward, so we need only the y components of these forces. Taking the positive y direction to be upward, Equation 11-3 becomes

$$\Sigma F_y = F_{up} - F_{down} - w = 0 \tag{11-4}$$

We can see that because the fluid in the box has a nonzero weight w, the forces F_{up} and F_{down} *cannot* be equal. From Equation 11-2, $p = F_\perp/A$, we can express the magnitude F_{up} of the upward force (which acts perpendicular to the underside of the box) as the pressure p_{bottom} at the bottom of the box multiplied by the area A of the bottom of the box. Similarly, the magnitude F_{down} of the downward force equals the pressure p_{top} at the top of the box multiplied by the area A of the top of the box. Furthermore, the weight w of the fluid in the box equals the product of the mass m of fluid in the box and the acceleration due to gravity g. So Equation 11-4 becomes

$$p_{bottom} A - p_{top} A - mg = 0 \quad \text{or} \quad p_{bottom} A = p_{top} A + mg \tag{11-5}$$

Equation 11-5 tells us that because of the weight mg of the fluid in the box, the pressure p_{bottom} has to be greater than the pressure p_{top}. In other words the pressure of the fluid is greater at a point that's deep in the fluid (the bottom of the box) than at a point that's less deep (the top of the box) and so pressure increases with increasing depth in the fluid.

A Uniform-Density Fluid: The Equation of Hydrostatic Equilibrium

We can simplify Equation 11-5 even further if we assume that the fluid has **uniform density**; that is, the density of the fluid has the same value throughout the fluid. The mass m of the volume of fluid in the box is the product of the density of the fluid, ρ, and the volume of the box. The height of the rectangular box is d, so its volume is $V = Ad$, and the mass of the fluid in the box is $m = \rho V = \rho Ad$. So Equation 11-5, the statement that the fluid in the box remains at rest, becomes

$$p_{bottom} A = p_{top} A + (\rho A d)g \tag{11-6}$$

Now divide each term in Equation 11-6 by A:

$$p_{bottom} = p_{top} + \rho g d \tag{11-7}$$

This says that the pressure at the bottom of the box is greater than the pressure at the top of the box. Furthermore, the box is just an imaginary construct, so the real meaning of Equation 11-7 is as follows: At *any* two points in the fluid with vertical separation d, the pressure at the lower point is greater than the pressure at the upper point by an amount $\rho g d$. If we let p_0 be the pressure at the upper point, the pressure p at a point a distance d below the upper point is

Variation of pressure with depth in a fluid with uniform density (equation of hydrostatic equilibrium)
(11-8)

| Pressure at a certain point in a fluid at rest | Density of the fluid (same at all points in the fluid) |

| Pressure at a second, lower point in the fluid | Acceleration due to gravity |

$$p = p_0 + \rho g d$$

Depth of the second point where the pressure is p below the point where the pressure is p_0

Equation 11-8 must be satisfied for a fluid to remain at rest, so this is also called the **equation of hydrostatic equilibrium**.

WATCH OUT! Understand what Equation 11-8 tells you.

⚠ Be careful with the sign of the $\rho g d$ term in Equation 11-8: Remember that the symbol d means *depth*. If the second point is *below* the point where the pressure is p_0, d is positive and p is greater than p_0. If the second point is *above* the point where the pressure is p_0, d is *negative* and p is less than p_0. In other words, pressure increases as you go deeper in a fluid such as the ocean but decreases as you ascend. Note also that Equation 11-8 is based on the assumption that the fluid has the same density ρ at all points. If the density is noticeably different at different depths (as is the case for the air in our atmosphere), Equation 11-8 gives only approximate results.

EXAMPLE 11-5 Air Pressure Versus Water Pressure

You dig a swimming pool in your backyard that is 2.00 m deep. What is the pressure at the bottom of the pool (a) before you fill it with water and (b) after it is filled with water? At the time that you make the measurements, the air pressure at ground level is 1.0100×10^5 Pa.

Set Up

We are given the pressure $p_0 = 1.0100 \times 10^5$ Pa at the top of the pool and want to find the pressure p at a depth $d = 2.00$ m below the top of the pool. We'll use Equation 11-8 for this purpose. The only difference between parts (a) and (b) is the density ρ of the fluid: In part (a) we use the density of air, while in part (b) we use the density of water.

Variation of pressure with depth:

$$p = p_0 + \rho g d \qquad (11\text{-}8)$$

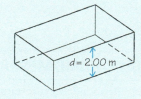

Solve

(a) From Table 11-1 the density of air at sea level is $\rho = 1.23$ kg/m^3. Substitute this into Equation 11-8.

Pressure at the bottom of the pool filled with air:

$$p = (1.0100 \times 10^5 \text{ Pa}) + (1.23 \text{ kg/m}^3)(9.80 \text{ m/s}^2)(2.00 \text{ m})$$
$$= 1.0100 \times 10^5 \text{ Pa} + 24.1 \text{ kg/(m} \cdot \text{s}^2)$$

Convert units: Recall that

$1 \text{ N} = 1 \text{ kg} \cdot \text{m/s}^2$ so

$1 \text{ Pa} = 1 \text{ N/m}^2 = 1 \text{ kg/(m} \cdot \text{s}^2)$

So the pressure at the bottom of the air-filled pool is

$$p = 1.0100 \times 10^5 \text{ Pa} + 24.1 \text{ Pa}$$
$$= 1.0100 \times 10^5 \text{ Pa} + 0.000241 \times 10^5 \text{ Pa}$$
$$= 1.0102 \times 10^5 \text{ Pa}$$

(b) When the pool is filled with water, the calculation is the same as in part (a) except that we use the density of water from Table 11-1, $\rho = 1.000 \times 10^3$ kg/m^3.

Pressure at the bottom of the pool filled with water:

$$p = (1.0100 \times 10^5 \text{ Pa}) + (1.000 \times 10^3 \text{ kg/m}^3)(9.80 \text{ m/s}^2)(2.00 \text{ m})$$
$$= 1.0100 \times 10^5 \text{ Pa} + 1.96 \times 10^4 \text{ kg/(m} \cdot \text{s}^2)$$
$$= 1.0100 \times 10^5 \text{ Pa} + 0.196 \times 10^5 \text{ Pa}$$
$$= 1.206 \times 10^5 \text{ Pa}$$

Reflect

The pressure at the bottom of the air-filled pool is only 0.02% greater than at the top of the pool. That's such a tiny difference that we can ignore it. It takes a much greater variation in altitude than 2.00 m for the pressure difference in the air to be noticeable. (Your ears—which are sensitive pressure sensors—may "pop" when you drive up into the mountains, but won't "pop" when you climb a ladder or a single flight of stairs.)

There *is* a noticeable pressure difference between the top and bottom of the *water*-filled pool: The pressure is about 20% greater at the bottom than at the surface. The difference is that water is almost a thousand times denser than air. A swimming pool full of water weighs almost a thousand times more than a swimming pool full of air, so it produces an additional pressure almost a thousand times greater.

GOT THE CONCEPT? 11-3 Watching a U-tube

(?) You pour water into a U-shaped tube, as shown in Figure 11-8. The left-hand leg of the tube is 2.00 cm in radius, while the right-hand leg is 1.00 cm in radius. When the water is in equilibrium, how will the height of the water in the left-hand leg and the pressure at the bottom of the left-hand leg compare to the height of water in the right-hand leg and the pressure at the bottom of the right-hand leg? (a) Greater height, greater pressure; (b) same height, same pressure; (c) lower height, lower pressure; (d) same height, greater pressure; (e) none of these.

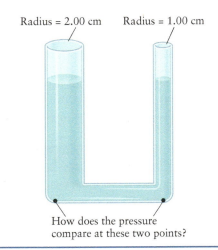

Radius = 2.00 cm   Radius = 1.00 cm

How does the pressure compare at these two points?

Figure 11-8 A hollow U-tube If water is poured into this hollow tube, how will the water level in each leg and the pressure at the bottom of each leg compare?

TAKE-HOME MESSAGE FOR Section 11-4

✔ In a fluid in hydrostatic equilibrium (at rest), pressure increases as you go deeper into the fluid.

✔ The pressure difference between any two points within a uniform fluid in hydrostatic equilibrium depends only on the difference in the vertical height of the two points and the density of the fluid.

11-5 Scientists and medical professionals use various units for measuring fluid pressure

The air around you pushes in on you, exerting a considerable pressure from every direction. The air pressure at sea level varies somewhat due to the passage of weather systems, but on average is equal to 1.01325×10^5 Pa in SI units or about 14.7 pounds per square inch (lb/in^2) in English units. This average value of atmospheric pressure at sea level is defined to be one **atmosphere**, abbreviated atm:

$$1 \text{ atm} = 1.01325 \times 10^5 \text{ Pa} = 14.7 \text{ lb/in}^2$$

To three significant figures, 1 atm = 1.01×10^5 Pa. Since 1 Pa = 1 N/m^2, this means that if you paint a square 1 m on a side onto the ground at sea level, the weight of the column of air that sits atop that square and extends to the upper limit of Earth's atmosphere is 1.01×10^5 N. A metric ton has a mass $m = 1000$ kg and a weight $mg = (1000 \text{ kg})(9.80 \text{ m/s}^2) = 9.80 \times 10^3$ N; the weight of air above that 1-m square is 10.3 times greater (10.3 metric tons). If you prefer to think in English units, paint a 1-in. square onto the ground at sea level: The weight of air above that square is 14.7 lb.

The atmosphere (atm) is a convenient unit for dealing with relatively large pressures. For example, the pressures found at the bottom of the ocean can exceed 10^3 atm, and high-pressure gas systems can operate at pressures of hundreds or even thousands of atmospheres. Very *low* pressures are often measured in pascals rather than atmospheres; for example, a high-quality vacuum pump can reduce the pressure inside a container to 10^{-5} Pa or lower.

EXAMPLE 11-6 Diver's Rule of Thumb

To what depth in a freshwater lake would a diver have to descend for the pressure to be 1.00 atm greater than at the surface?

Set Up

Equation 11-8 from Section 11-4 tells us how the pressure p at a depth d below the lake's surface compares to the pressure p_0 at the surface. We can use this information to find the value of d such that p is greater than p_0 by 1.00 atm.

Variation of pressure with depth:

$$p = p_0 + \rho g d \qquad (11\text{-}8)$$

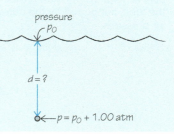

Solve

Rearrange Equation 11-8 to solve for the depth d at which the pressure has a certain value p.

From Equation 11-8,

$$p - p_0 = \rho g d$$

$$d = \frac{p - p_0}{\rho g}$$

We want the pressure p at depth d to be greater than the pressure p_0 at the surface by 1.00 atm, so $p - p_0 = 1.00$ atm. Substitute this value as well as the value of density ρ for fresh water from Table 11-1. Then use the conversion factors $1 \text{ atm} = 1.01 \times 10^5 \text{ Pa} = 1.01 \times 10^5 \text{ N/m}^2$ and $1 \text{ N} = 1 \text{ kg} \cdot \text{m/s}^2$.

Pressure difference:

$$p - p_0 = 1.00 \text{ atm}$$

Density of fresh water from Table 11-1:

$\rho = 1.000 \times 10^3 \text{ kg/m}^3$ so

$$d = \frac{p - p_0}{\rho g}$$

$$= \frac{1.00 \text{ atm}}{(1.000 \times 10^3 \text{ kg/m}^3)(9.80 \text{ m/s}^2)} \times \left(\frac{1.01 \times 10^5 \text{ N/m}^2}{1 \text{ atm}} \right)$$

$$= 10.3 \frac{\text{m}^3}{\text{kg}} \frac{\text{s}^2}{\text{m}} \frac{\text{N}}{\text{m}^2} \times \left(\frac{1 \text{ kg} \cdot \text{m/s}^2}{1 \text{ N}} \right)$$

$$= 10.3 \text{ m}$$

Reflect

Our answer is a little bit more than 10 m, so it's a good approximation (a "rule of thumb" for divers) to say that our diver experiences a pressure change of 1 atm going from the surface to a depth of 10 m. You can see that descending an additional 10 m would increase the pressure by another atmosphere, and so on.

Measuring Pressure

Blood pressure readings and atmospheric pressure values in weather reports are usually not given in units of atmospheres, Pa, or lb/in^2. Instead, the units used are commonly either inches or millimeters of mercury. To understand these units of pressure, we'll examine how to construct a simple device called a *barometer* that's used to measure air pressure.

Fill a tube closed at one end with mercury (Hg), which has density ρ_{Hg}. While keeping the open end sealed, turn the tube upside down into a pan partially filled with mercury. (**Warning:** Mercury is poisonous, so do *not* try this experiment at home!) Now release the seal on the end of the tube, which is beneath the surface of the mercury in the pan. No air can get into the tube, so an equal volume of vacuum must remain in the place of whatever volume of fluid drains from the tube. Figure 11-9 shows this process.

Let's consider what happens once the system has come to hydrostatic equilibrium and a column of mercury of height H (measured from the surface of the mercury of the pan) remains in the tube. The pressure at the top of this column is $p_0 = 0$ because the pressure in a vacuum is zero. The pressure at the exposed surface of the mercury

Figure 11-9 Constructing a simple barometer A simple barometer to measure atmospheric pressure can be made by inverting a tube of fluid such as mercury in a pan. A height H of liquid remains in the tube, supported by the outside air pressure.

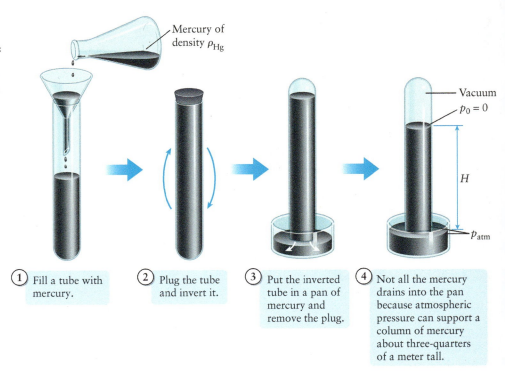

Mercury of density ρ_{Hg}

Vacuum
$p_0 = 0$

H

p_{atm}

(1) Fill a tube with mercury.

(2) Plug the tube and invert it.

(3) Put the inverted tube in a pan of mercury and remove the plug.

(4) Not all the mercury drains into the pan because atmospheric pressure can support a column of mercury about three-quarters of a meter tall.

in the pan equals atmospheric pressure p_{atm}. Since the mercury is in hydrostatic equilibrium, the pressure at a point inside the mercury column at the same elevation as the exposed mercury surface is *also* equal to p_{atm}. From the equation of hydrostatic equilibrium, Equation 11-8, the pressure p_{atm} is greater than p_0 by $\rho_{Hg}gH$:

$$p_{atm} = p_0 + \rho_{Hg}gH$$

Since $p_0 = 0$ in the vacuum at the top of the mercury column, this equation becomes

(11-9)
$$p_{atm} = \rho_{Hg}gH$$

(atmospheric pressure measured with a mercury barometer)

As the atmospheric pressure p_{atm} rises and falls, the height H of the mercury column rises and falls along with it. Hence the height of the mercury column is a direct measure of atmospheric pressure.

A mercury barometer like that shown in Figure 11-9 is not a perfect instrument for measuring pressure: The density of mercury ρ_{Hg} changes with temperature, and the value of the acceleration due to gravity g depends on location. But at a temperature of 0°C so that the density of mercury is $\rho_{Hg} = 1.35951 \times 10^3$ kg/m^3 and at a location where $g = 9.80665$ m/s^2 (a "standard" value of g), the height of a column of mercury supported by exactly 1 atm of pressure is found to be 760 mm. For that reason pressures are sometimes given in **millimeters of mercury**, or mmHg for short. Thus 1 atm = 760 mmHg. An equivalent (and preferred) unit is the **torr**, defined as 1/760 atm. This unit is named for the Italian scientist and mathematician Evangelista Torricelli, who invented the mercury barometer in the 1640s. The torr is a more appropriate unit than mmHg since most modern barometers work on different principles that don't involve columns of mercury.

If mercury is poisonous, why was it used in barometers? It's a case of convenience triumphing over safety. Equation 11-9 shows that the greater the density of the fluid, the lower the height H of the fluid column in the barometer. Because mercury is so dense—13.6 times denser than water—a mercury barometer is relatively compact and convenient to use. If you wanted to make a barometer like that shown in Figure 11-9 but using water rather than mercury, it would have to be 13.6 times taller than a mercury barometer. Can you show that if $p_{atm} = 1$ atm, the height of the liquid column in a water barometer would be 10.3 m—far too tall to be convenient?

Gauge Pressure

What is the air pressure inside a flat automobile tire? It's not zero: Rather, it's equal to atmospheric pressure (about 1 atm, 1.01×10^5 Pa, or 14.7 lb/in^2 at sea level). The same is true for a balloon that hasn't been inflated and for an empty, open bottle, because air at atmospheric pressure is inside these vessels. But if you use a tire pressure gauge to measure the air pressure in the deflated tire, it *will* read zero. This **gauge pressure** shows how much the pressure *exceeds* atmospheric pressure. The **absolute pressure**—that is to say, the true value of pressure—inside a deflated tire is equal to atmospheric pressure. If a tire pressure gauge shows that the air inside an inflated tire is at a pressure of 30.0 lb/in^2, that's the gauge pressure. *The absolute pressure equals the gauge pressure plus atmospheric pressure.* So if the gauge pressure inside the tire is 30.0 lb/in^2 and atmospheric pressure is 14.7 lb/in^2, the absolute pressure inside the tire is (30.0 lb/in^2) + (14.7 lb/in^2) = 44.7 lb/in^2.

GOT THE CONCEPT? 11-4 Rank the Pressures

 You have two balloons at sea level. One is deflated, and the other is inflated to 0.50 atm according to a pressure gauge connected to the mouth of the balloon. Rank the following from highest to lowest: (a) the gauge pressure inside the deflated balloon; (b) the absolute pressure inside the deflated balloon; (c) the gauge pressure inside the inflated balloon; (d) the absolute pressure inside the inflated balloon.

Blood Pressure

Most of us encounter a pressure gauge when we visit the doctor and have our blood pressure measured (Figure 11-10). Because the heart is a pump, the pressure in the system changes over the period of each heartbeat. Blood pressure is usually given as two numbers, for example, "120 over 80." These values are measured in units of mmHg and represent the high and low values of the *gauge* pressure in the arteries at the level of the heart.

Blood pressure is usually measured using a cuff around your upper arm, at the same elevation as your heart. We can understand why this is so by using the relationship between pressure and depth given by Equation 11-8. The blood within the body behaves rather like the fluid in the container shown in Figure 11-7: The farther down you go within the fluid, the greater the pressure. Since the heart is the principal organ of the circulatory system, let's have p_0 be the pressure at the position of the heart. Then Equation 11-8 tells us that the pressure of blood in the feet will be higher than the pressure in the heart, and the pressure in the head will be lower. To measure the actual pressure produced by the heart's pumping action, it's important to measure that pressure at the same elevation as the heart itself—which explains the placement of a blood pressure cuff.

The average blood pressure (gauge pressure) in an upright person's feet is about 100 mmHg higher than when it leaves the heart, or about twice as great as the gauge pressure measured in a person's arm. In contrast, the pressure in the head is about 50 mmHg less than that at the heart. If you could somehow stretch your neck upward by a meter or so, the blood pressure in your head would drop to zero and no blood would reach your brain. So it's all for the best that you can't stretch your neck that far! (Some animals do have necks that are even longer: The head of an adult giraffe is about 2 m higher than his heart. To get blood to the brain, the giraffe heart has to generate much higher pressure than a human heart does—about 260 mmHg.)

These differences in pressure between different regions of the body affect the flow of blood in the veins returning to the heart from the head and neck. Blood pressure is always lower in veins than in arteries; if you can feel a pulse in your neck, you're feeling the carotid *artery* with its relatively high pressure. The gauge pressure of blood entering the heart is very close to zero. Since the pressure above the heart is even lower, the veins of the head and neck tend to collapse! As blood continues to enter the veins, pressure builds up, and the vessels reopen. But as the blood flows through them, the pressure will again drop, and the veins will collapse again. The result is intermittent venous blood flow in the head and neck.

You may be wondering why the gauge pressure of the blood entering the heart is so much lower than the gauge pressure of the blood leaving the heart, when the blood

BioMedical

Figure 11-10 Measuring blood pressure A blood pressure cuff is applied around the upper arm, at the same elevation as the patient's heart. If the pressure were measured at a different elevation, the values would be either lower or higher than the actual pressures produced by the heart.

Panther Media GmbH/Alamy

is at essentially the same height. Equation 11-8 seems to indicate that these two pressures should be the same. However, you need to remember that Equation 11-8 is valid only under the conditions of *hydrostatic equilibrium*. Because blood is pumped by the heart and flows through the circulatory system, the blood is certainly *not* in hydrostatic equilibrium and we cannot use Equation 11-8. In Section 11-11 we'll discuss viscous flow, which helps to explain the difference in arterial and venous pressure.

TAKE-HOME MESSAGE FOR Section 11-5

✔ The pressure exerted by air at sea level is approximately equal to 1 atm. This pressure is due to the weight of a column of air that extends to the top of our atmosphere.

✔ Gauge pressure is measured relative to some reference pressure. Usually this reference pressure is atmospheric pressure.

✔ Blood pressure is measured in mmHg, which is the height of a column of mercury the pressure could support. An equivalent unit to mmHg which physicists prefer is the torr.

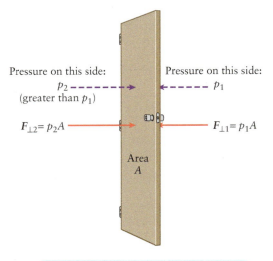

Pressure on this side:
p_2
(greater than p_1)

Pressure on this side: p_1

$F_{\perp 2} = p_2 A$

$F_{\perp 1} = p_1 A$

Area A

Net force on the door due to the pressure difference:
$F_{\text{net}} = F_{\perp 2} - F_{\perp 1} = (p_2 - p_1)A$

Figure 11-11 Net force due to a pressure difference Because there are different pressures p_1 and p_2 on the two sides of this door, there is a net force on the door. (Compare Example 11-4 in Section 11-3.)

11-6 A difference in pressure on opposite sides of an object produces a net force on the object

What makes a cork fly out of a champagne bottle? What makes the pistons move in the cylinders of an automobile engine? And what makes your beverage move up the straw into your mouth as you sip? In each case the cause is not strictly pressure but rather pressure *difference*. The cork pops out because carbon dioxide gas inside the champagne bottle is at higher pressure than the air outside the bottle. Burning gasoline in the cylinder produces hot gas on one side of the piston that's at much higher pressure than the air on the other side, forcing the piston to move. And when you suck on a straw, you reduce the pressure in your mouth to a lower value than in the outside air, and the pressure difference drives liquid up the straw.

As a simple example of the force produced by a pressure difference, consider an object like a door with fluid on each side. (This would be a good time to review Example 11-4 in Section 11-3.) The object has area A, there is pressure p_1 on one side of the object, and there is a greater pressure p_2 on the other side (Figure 11-11). By rearranging Equation 11-2, $p = F_{\perp}/A$, we can calculate the magnitude of the force that each fluid exerts perpendicular to the object's surface:

Force on side 1: magnitude $F_{\perp 1} = p_1 A$

Force on side 2: magnitude $F_{\perp 2} = p_2 A$

Since $p_2 > p_1$, the force $F_{\perp 2}$ on side 2 is greater than the force $F_{\perp 1}$ on side 1. These forces are in opposite directions, so the *net* force on the object has magnitude

$$F_{\text{net}} = F_{\perp 2} - F_{\perp 1} = p_2 A - p_1 A$$

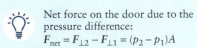

A **pressure difference** on two opposite sides of an object **produces a net force.**

Area of each side of the object

$$F_{\text{net}} = (p_2 - p_1)\, A$$

High pressure on one side of the object

Lower pressure on the other side of the object

Net force on an object due to a pressure difference
(11-10)

That is, the net force on the object due to the pressures acting on it is proportional to the difference in pressure between the object's two sides. Note that Equation 11-10 refers *only* to the net force on an object due to a difference in pressure on its two sides. It doesn't include other forces such as gravity that may act on the object.

EXAMPLE 11-7   Submarine Hatch

The air in the crew compartment of a research submarine is maintained at a pressure of 1.00 atm so that the crew can breathe normally. The sub operates at a depth of 85.0 m below the surface of the ocean. At this depth what force would be required to push open a hatch of area 2.00 m²?

Set Up

There is 1.00 atm of air pressure on the inside of the hatch (call this p_{air}) pushing the hatch outward, but this is much less than the outside water pressure (call this p_{water}) pushing the hatch inward. The net inward force F_{net} due to this pressure difference is given by Equation 11-10; $A = 2.00$ m² is the surface area of the hatch. To open the hatch the crew would have to exert an outward force of the same magnitude F_{net}.

    To calculate F_{net} we need to know the value of the water pressure p_{water}. We'll find this value using Equation 11-8 for the pressure at depth $d = 85.0$ m. In this equation p_0 equals the pressure at the surface of the ocean (that is, at zero depth), which we also take to be 1.00 atm.

Net force on the hatch due to the pressure difference:

$$F_{net} = (p_{water} - p_{air})A \qquad (11\text{-}10)$$

Variation of pressure with depth:

$$p_{water} = p_0 + \rho g d \qquad (11\text{-}8)$$

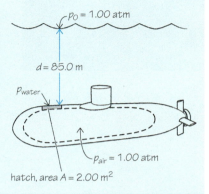

Solve

First find the pressure of the water at depth $d = 85$ m. Use the density of seawater from Table 11-1, $\rho = 1.025 \times 10^3$ kg/m³; convert atmospheres to newtons per square meter; and note that 1 kg/(m·s²) = 1 N/m².

Water pressure on the outside of the hatch:

$$p_{water} = p_0 + \rho g d$$

In this, $p_0 = 1.00$ atm $= 1.01 \times 10^5$ N/m², so

$$
\begin{aligned}
p_{water} &= 1.01 \times 10^5 \text{ N/m}^2 \\
&\quad + (1.025 \times 10^3 \text{ kg/m}^3)(9.80 \text{ m/s}^2)(85.0 \text{ m}) \\
&= 1.01 \times 10^5 \text{ N/m}^2 + 8.54 \times 10^5 \text{ kg/(m·s}^2) \\
&= 1.01 \times 10^5 \text{ N/m}^2 + 8.54 \times 10^5 \text{ N/m}^2 \\
&= 9.55 \times 10^5 \text{ N/m}^2
\end{aligned}
$$

Now we can calculate the magnitude of the net force on the hatch due to the pressure difference. This is the same as the force that the crew must exert to open the hatch.

Air pressure on the inside of the hatch:

$p_{air} = 1.00$ atm $= 1.01 \times 10^5$ N/m², so

$$
\begin{aligned}
F_{net} &= (p_{water} - p_{air})A \\
&= (9.55 \times 10^5 \text{ N/m}^2 - 1.01 \times 10^5 \text{ N/m}^2)(2.00 \text{ m}^2) \\
&= (8.54 \times 10^5 \text{ N/m}^2)(2.00 \text{ m}^2) \\
&= 1.71 \times 10^6 \text{ N}
\end{aligned}
$$

Reflect

This is an immense amount of force, equivalent to nearly 200 U.S. tons! (It's also more than 10 times the force on the airliner door in Example 11-4.) Clearly the crew would not be able to open the hatch. For a crew member in diving gear to exit the submerged vessel, a portion of the crew compartment would have to be flooded with water from the outside. This action will equalize the inside and outside pressures and make it possible to open the hatch.

The Lungs

BioMedical

Pressure differences drive air into and out of the lungs. We humans suck air into our lungs by contracting the diaphragm and the muscles of the chest wall to increase the volume of our chest cavity. This pulls the lungs open, increasing their volume and therefore decreasing the pressure inside them. As this pressure becomes less than atmospheric pressure, the pressure difference pushes air through the airways and into the expanding lungs. Other animals create a pressure difference in different ways (Figure 11-12).

Michael Willis/Alamy

BioMedical **Figure 11-12** **How frogs breathe** Frogs do not have an expandable chest cavity like humans and so cannot breathe in the same way. Instead a frog closes its mouth and lowers the floor of its mouth, causing the mouth cavity to expand, then takes air into its mouth through its nostrils. It then closes its nostrils and raises the floor of its mouth, increasing the pressure there and forcing the trapped air into the lungs. This process is relatively inefficient, so frogs take in most of their oxygen by diffusion into blood vessels that lie just below the frog's permeable skin.

It takes energy for us to contract the muscles required for *inhalation*. However, normal *exhalation* does not require our muscles to do additional work on our chest wall. Relaxing the diaphragm and chest muscles allows the volume of the chest cavity to decrease. This process causes the pressure in the chest to rise above atmospheric pressure and forces air out of the lungs.

Breathing becomes more difficult if the chest is under pressure, such as if you're at the bottom of a football pile-on. The outside pressure on the chest makes it more difficult to expand the chest to inhale. Measurements show that if the pressure on a person's chest exceeds atmospheric pressure by 0.05 atm, the chest can't expand and inhalation is no longer possible. Example 11-8 shows that this places a limit on how deep underwater a person can be and still breathe through a hollow tube that opens above the surface.

EXAMPLE 11-8 Breathing Underwater

Secret agent Cassian Andor dives into a shallow freshwater pond to avoid capture by his nemesis. He intends to lie flat on his back on the bottom of the pond and breathe through a hollow reed. What is the length of the longest reed for which this could work?

Set Up

Cassian needs to ensure that the difference between the water pressure p on his chest and the pressure p_0 of the atmosphere at the surface (where the open end of the reed is) does not exceed 0.05 atm. Equation 11-8 tells us how the water pressure varies with depth d, so our goal is to find the depth such that $p - p_0$ equals 0.05 atm.

Variation of pressure with depth:

$$p = p_0 + \rho g d \qquad (11\text{-}8)$$

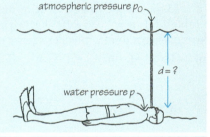

atmospheric pressure p_0

$d = ?$

water pressure p

Solve

Solve Equation 11-8 to find the depth d in terms of the water pressure on Cassian's chest (p) and atmospheric pressure (p_0). Then find the depth d for which $p - p_0 = 0.05$ atm. Use the density of fresh water given in Table 11-1 as well as the definitions 1 atm = 1.01×10^5 N/m² and 1 N = 1 kg·m/s².

Rearrange Equation 11-8:

$$p - p_0 = \rho g d$$

Solve for d:

$$d = \frac{p - p_0}{\rho g}$$

$$= \frac{0.05 \text{ atm}}{(1.000 \times 10^3 \text{ kg/m}^3)(9.80 \text{ m/s}^2)} \times \frac{1.01 \times 10^5 \text{ N/m}^2}{1 \text{ atm}}$$

$$= 0.5 \frac{\text{N} \cdot \text{s}^2}{\text{kg}} \times \frac{1 \text{ kg} \cdot \text{m/s}^2}{1 \text{ N}} = 0.5 \text{ m}$$

Reflect

The longest hollow reed through which Cassian can breathe underwater is about half a meter—not that long, really. To check this result let's calculate the weight of water half a meter deep on the area of a person's chest, about 30 cm × 60 cm. The answer is about 900 N or 200 lb. Certainly if a 200-lb person were sitting on your chest, you'd find it hard to breathe!

Area of chest:

$$A = 30 \text{ cm} \times 60 \text{ cm}$$
$$= 0.30 \text{ m} \times 0.60 \text{ m}$$
$$= 0.18 \text{ m}^2$$

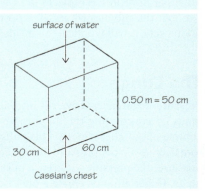

surface of water

0.50 m = 50 cm

30 cm   60 cm

Cassian's chest

Scuba divers can descend to depths much greater than 0.5 m, but to do so they carry tanks of pressurized air. The regulator on a scuba tank delivers this air to the diver's mouth at the same pressure as the surrounding water, so $p - p_0 = 0$. Hence it's just as easy for a scuba diver to breathe underwater as on dry land.

Mass of water in a rectangular volume with base 0.18 m^2 and height 0.50 m:

$$m = \rho V$$
$$= (1.000 \times 10^3 \text{ kg/m}^3) \times (0.180 \text{ m}^2)(0.50 \text{ m})$$
$$= 90 \text{ kg}$$

Weight of that water:

$$mg = (90 \text{ kg})(9.8 \text{ m/s}^2)$$
$$= 900 \text{ N} = 200 \text{ lb}$$

GOT THE CONCEPT? 11-5 Paper and Pressure

Hold a piece of paper in your two hands so that the plane of the paper is horizontal. Considering the air pressure on the top of the paper and the air pressure on the bottom of the paper, is the force F_{net} on the paper described by Equation 11-10 (a) upward, (b) downward, or (c) zero?

TAKE-HOME MESSAGE FOR Section 11-6

✔   If the fluid pressure is different on two sides of an object, the result is a net force on the object. This net force is proportional to the pressure difference and to the area of the object.

11-7 A pressure increase at one point in a fluid causes a pressure increase throughout the fluid

When you squeeze one end of a tube of toothpaste, the pressure you apply is transmitted throughout the tube, and toothpaste comes out the other end. This is a simple example of a more general principle first proposed by the seventeenth-century French philosopher, mathematician, and scientist Blaise Pascal, who did pioneering investigations of the nature of pressure in fluids:

> *Pascal's principle: Pressure applied to a confined, static fluid is transmitted undiminished to every part of the fluid as well as to the walls of the container.*

We've already seen **Pascal's principle** in action in Section 11-4. There we learned that if a fluid is at rest—that is, in hydrostatic equilibrium—and the pressure at a certain point in the fluid is p_0, the pressure at a second point a distance d deeper in the fluid is $p = p_0 + \rho g d$ (Equation 11-8). If we were to increase the value of the pressure p_0 at the first point by, say, 50 Pa, the value of the pressure p at the second point would also have to increase by 50 Pa to maintain hydrostatic equilibrium.

A common practical application of Pascal's principle is a *hydraulic jack*. This device makes it possible to lift heavy objects using a force much less than the object's weight. You'll find these devices in operation wherever cars are being worked on: A hydraulic lift raises the car so that the mechanic can work on the car's underside. Smaller versions are used to raise and lower the chairs you sit in at the dentist and the barber. Figure 11-13 shows the construction of a simplified hydraulic lift. Start with a tube bent into a "U" shape. On the side labeled "1" the tube is much narrower than on the side labeled "2"; we'll call the cross-sectional areas of the two sides of the tube A_1 and A_2, respectively. We partially fill the tube with an incompressible liquid such as oil so that its density does not change when pressure is applied. The liquid rises to the same height on both sides of the tube. (If you're not sure why, see the "Got the Concept?" question at the end of Section 11-4.) We'll put a moveable cap at each end of the U-tube to keep the fluid from leaking out.

Now apply a downward force of magnitude F_1 to the cap on side 1. This causes the pressure in the fluid under the cap to increase by an amount $\Delta p = F_1/A_1$. According to Pascal's principle, this change in pressure is transmitted throughout the tube. We will therefore see this same pressure increase Δp below the cap on side 2, and that

Figure 11-13 A hydraulic jack
Hydraulic jacks are used in workshops to elevate cars under repair and in dental offices to elevate the patient's chair. They operate using Pascal's principle.

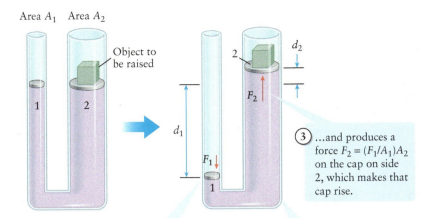

① When a downward force F_1 is applied to the cap on side 1, which has area A_1...

② ...the increase in pressure $\Delta p = F_1/A_1$ is transmitted throughout the fluid...

③ ...and produces a force $F_2 = (F_1/A_1)A_2$ on the cap on side 2, which makes that cap rise.

will cause an upward force $F_2 = \Delta p \times A_2$ on that cap. Substituting $\Delta p = F_1/A_1$ into this expression for F_2 gives

$$F_2 = \Delta p \times A_2 = \left(\frac{F_1}{A_1}\right)A_2$$

or, rearranging,

(11-11)
$$F_2 = F_1\left(\frac{A_2}{A_1}\right)$$

Since A_2 is greater than A_1, F_2 is greater than F_1, which means that a small force applied to side 1 results in a larger force applied to the cap on side 2. This is why a hydraulic jack is useful to auto mechanics: A relatively small force applied to side 1 can produce enough force on side 2 to raise a heavy car off the ground.

Equation 11-11 may make it seem like we get extra force "for free." However, there is a trade-off: The cap on side 1 must be pushed down a large distance to make the cap on side 2 move up a short distance. Here's why: When force F_1 is applied to the cap on side 1, that cap moves down a distance d_1 and displaces a volume of liquid $V_1 = d_1A_1$. The increase in pressure throughout the tube causes liquid to rise on side 2, pushing up the cap on that side by a distance d_2. The increase in liquid volume under cap 2 is $V_2 = d_2A_2$. Since the liquid used in the U-tube is assumed to be incompressible, these two volumes of liquid must be equal. Hence

$$d_2 A_2 = d_1 A_1$$

or

(11-12)
$$d_2 = d_1\left(\frac{A_1}{A_2}\right)$$

√x̄ *See the Math Tutorial for more information on direct and inverse proportions.*

The factor A_1/A_2 in Equation 11-12 is the reciprocal of the factor that relates F_2 to F_1 in Equation 11-11. Although the *force* that is applied on side 2 is larger than the force on side 1, the *distance* the cap on side 2 moves is smaller than the distance the cap on side 1 was moved.

GOT THE CONCEPT? 11-6 Work and a Hydraulic Lift

? Consider a hydraulic lift like that shown in Figure 11-13. The cap on side 1 has a radius of 1.00 cm, and the cap on side 2 has a radius of 5.00 cm. As you push down on the cap on side 1, you do a certain amount of work W_1. How much work is done on the cap on side 2 as it rises up? (a) $25W_1$; (b) $5W_1$; (c) W_1; (d) $W_1/5$; (e) $W_1/25$.

TAKE-HOME MESSAGE FOR Section 11-7

✔ Pascal's principle states that pressure applied to a confined, static fluid is transmitted undiminished to every part of the fluid as well as to the walls of the container.

✔ Pascal's principle explains the hydraulic lift, which allows a small force exerted on a fluid at one location to translate into a larger force exerted by the fluid at another location.

11-8 Archimedes' principle helps us understand buoyancy

Drop a plastic toy into a pond, and you're not surprised when it pops back up to the surface. A boat anchor won't come back up—which is also not surprising. But why does a boat, which weighs much more than its anchor, float on the surface rather than sink?

To answer these questions, think of a fluid (gas or liquid) at rest. To maintain hydrostatic equilibrium of any portion of the fluid, like the shaded blob shown in Figure 11-14a, the net force on that portion of fluid must be zero. Hence the combined force due to the pressure of the surrounding fluid—greater pressure on the bottom, less pressure on top—must be an upward force that exactly balances the weight of the shaded blob of fluid. We call this combined upward force the **buoyant force**; it "buoys up" the blob of fluid just enough to keep it from sinking.

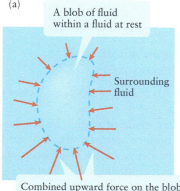

(a) A blob of fluid within a fluid at rest

Surrounding fluid

Combined upward force on the blob due to the surrounding fluid just balances the weight of the blob.

(b) Replace the blob with an object of the same volume.

Surrounding fluid

Combined upward force on the object is the same as on the fluid blob in (a).

Figure 11-14 Archimedes' principle (a) The blob is a portion of a fluid at rest. The pressure from the surrounding fluid is greater at the bottom than at the top. (b) If the blob is replaced by some other object of the same volume, that object feels the same upward (buoyant) force as did the blob.

Now we replace the blob of fluid in Figure 11-14a with an object of exactly the same dimensions (Figure 11-14b). We say that this object *displaces* a volume of fluid just equal to the volume of the object that's immersed in the fluid. For a boat floating in water, the displaced volume equals that portion of the boat's volume that's underwater. For a submarine under water, a balloon in air, or any other object that's totally immersed in a fluid, the displaced volume equals the total volume of the object.

Here's the critical observation: At each point on the object's surface, the *pressure* of the surrounding fluid (caused by fluid molecules colliding with the object) is *exactly* the same as it was before we swapped the object for the blob of fluid. So the submerged object feels the *same* buoyant force as the blob of fluid did, with a magnitude just equal to the weight of that fluid blob. This statement about buoyant force is called **Archimedes' principle**:

The buoyant force on an object immersed in a fluid is equal to the weight of the fluid that the object displaces.

To be specific, suppose the object displaces a volume $V_{displaced}$ of fluid. If the fluid has density ρ_{fluid}, the fluid that the object displaces has mass $m_{fluid} = \rho_{fluid}V_{displaced}$ and weight $m_{fluid}g = \rho_{fluid}V_{displaced}g$. Hence the magnitude F_b of the buoyant force that acts on the object is

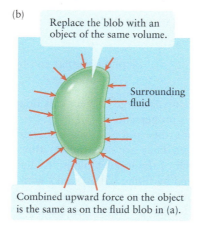

Surrounding fluid

(a) F_b

w_{object}

(b) F_b

w_{object}

(c) F_b

w_{object}

$F_b = W_{object}$
Object floats

$F_b < W_{object}$
Object sinks

$F_b > W_{object}$
Object rises

Figure 11-15 Buoyancy: Floating, sinking, and rising All three of these objects are immersed in the same fluid. All three have the same shape and size and so experience the same buoyant force F_b. Whether the object floats, sinks, or rises depends on its weight.

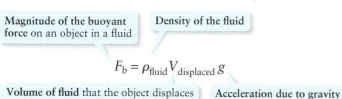

Magnitude of the buoyant force on an object in a fluid

Density of the fluid

$$F_b = \rho_{fluid} V_{displaced} g$$

Volume of fluid that the object displaces

Acceleration due to gravity

Buoyant force
(11-13)

If the object's weight is exactly the same as the magnitude of the buoyant force F_b—that is, exactly the same as the weight of the displaced fluid—the net force on the object is zero and the object neither sinks nor rises; instead, it floats, remaining at the same height within the fluid (Figure 11-15a). If the object weighs more than

(a)

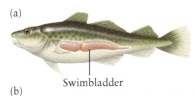

Swimbladder

(b)

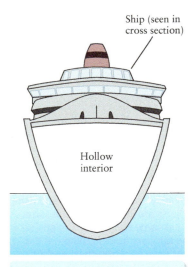

Image Source/Alamy

Figure 11-16 Marine animals with and without swimbladders (a) The gas-filled swimbladder gives this fish the same average density as the surrounding water so that it floats while submerged. Many fish species have a swimbladder, including salmon, herring, mackerel, cod, and the fish shown in the photo that opens this chapter. (b) Fish that live near the ocean bottom, like flounder and dogfish, lack swimbladders. They are denser than water and hence cannot float; they must swim continuously to avoid sinking to the bottom. The same is true for sharks like the great hammerhead (*Sphyrna mokarran*) shown here.

Ship (seen in cross section)

Hollow interior

Steel is denser than water, but ships made of steel are hollow inside; this makes their overall density less than that of water and so allows them to remain afloat while only partially submerged.

Figure 11-17 How steel ships float What determines whether an object floats is how its overall density compares to the density of water.

the displaced fluid, the object's weight overwhelms the buoyant force; hence the net force is downward and the object sinks (Figure 11-15b). If the object weighs less than the displaced fluid, the buoyant force overwhelms the object's weight so that the net force is upward and the object rises (Figure 11-15c).

Floating: Submarines, Fish, Ships, and Balloons

Many types of fish, as well as submarines, are able to float while submerged. In this case the volume $V_{displaced}$ of displaced fluid in Equation 11-13 is the same as the volume of the object. If the object has density ρ, its mass is $m = \rho V_{displaced}$ and its weight is $\rho V_{displaced} g$. Then the condition that the buoyant force on the object has the same magnitude as the object's weight is

$$\rho_{fluid} V_{displaced} g = \rho V_{displaced} g$$

or, dividing through by $V_{displaced} g$,

$$\rho_{fluid} = \rho$$

(condition that an object of density ρ floats while submerged in a fluid of density ρ_{fluid})

In words, a submerged object can float only if its density is the same as the fluid in which it is immersed.

Submarines and deep-sea research vessels are made of steel, which is far denser than water. They are nonetheless able to float underwater because they are hollow, with much of the internal volume filled with low-density air. The amount of air can be adjusted by filling ballast tanks with seawater or emptying the ballast tanks so that the submarine's *average* density (the mass of the submarine divided by its volume) is equal to that of water (see Example 11-2 in Section 11-2). For the same reason, **BioMedical** many species of fish have a flexible gas-filled sac called a *swimbladder* (Figure 11-16). These fish are able to float while submerged because they can regulate the amount of gas in the swimbladder to make their average density equal to that of water. (In cod and other species this regulation is done by exchanging gas between the swimbladder and the blood.)

Common aquarium fish have swimbladders and are able to float while submerged. But they move their fins constantly even if they are standing still. The reason is that the equilibrium provided by a swimbladder is *unstable*. If a water current displaces the fish slightly upward, the pressure at its new position is slightly less than before, and Pascal's principle (Section 11-7) tells us that this reduced pressure will be transmitted throughout the fish to the swimbladder. The reduced pressure makes the gas in the swimbladder expand, which decreases the average density of the fish as a whole. As a result the fish is now less dense than the surrounding water and will continue to move upward. Similarly, if the fish should be displaced below its equilibrium position, the swimbladder will shrink in response to the increased pressure, thus increasing the density of the fish and making it sink even farther. The continuous motions of the fish's fins are an effort to maintain a constant depth.

An object that is *less* dense than water floats on the surface with only part of its volume submerged; the lower the object's density relative to that of water, the higher it floats (see Example 11-8). Steel is denser than water, but ships made of steel are hollow inside; this makes their overall density less than that of water and so allows them to remain afloat while only partially submerged (Figure 11-17).

We've mostly been discussing objects in liquids such as water, but there is also a buoyant force on an object immersed in a gas such as the atmosphere. The buoyant force on a balloon filled with helium (which is less dense than air) is greater than the balloon's weight, which is why it rises if released. By contrast, a balloon filled with room-temperature air falls if released: The air inside the balloon is at higher pressure and has a higher density than the outside air, so the balloon's weight is more than the buoyant force acting on it. Hot air is less dense than cold air, so an air-filled balloon can be made to float if it's equipped with a burner to warm the balloon's contents. That's the principle of a hot-air balloon.

EXAMPLE 11-9 What Lies Beneath the Surface?

A solid block of density ρ_{block} is placed in a liquid of density ρ_{liquid}. If the block is less dense than the liquid (that is, $\rho_{block} < \rho_{liquid}$), derive an expression for the fraction of the block's volume that is submerged.

Set Up

Archimedes' principle tells us that the buoyant force exerted on the block by the liquid equals the weight of the displaced liquid. In order for the block to float, the upward buoyant force must balance the weight of the block. We'll use this principle to compare the volume of displaced liquid $V_{displaced}$—equal to the volume of the block that's submerged in the liquid—to the total volume of the block.

Buoyant force:

$$F_b = \rho_{liquid} V_{displaced} g \qquad (11\text{-}13)$$

block (seen from the side), volume V_{block}
$V_{displaced}$

Solve

First write expressions for the block's mass m_{block} and weight w_{block}. When the block is in equilibrium, the upward buoyant force on the block equals the block's weight. Use this and the expression for w_{block} to solve for the ratio of $V_{displaced}$ to V_{block}.

volume of block = V_{block}
mass of block = (density of block) × (volume of block):

$$m_{block} = \rho_{block} V_{block}$$

weight of block = (mass of block) × g:

$$w_{block} = m_{block} g = \rho_{block} V_{block} g$$

Net force on the block in equilibrium is zero:

$$\sum F_y = F_b - w_{block} = 0 \text{ so}$$
$$F_b = w_{block}$$

From Equation 11-13,

$$F_b = \rho_{liquid} V_{displaced} g, \text{ so}$$
$$\rho_{liquid} V_{displaced} g = \rho_{block} V_{block} g$$

Divide through by g:

$$\rho_{liquid} V_{displaced} = \rho_{block} V_{block}$$

Solve for ratio of $V_{displaced}$ to V_{block}:

$$\text{fraction of block's volume that is submerged} = \frac{V_{displaced}}{V_{block}} = \frac{\rho_{block}}{\rho_{liquid}}$$

F_b
w_{block}

Reflect

The fraction of the block's volume that is submerged, $V_{displaced}/V_{block}$, is equal to the ratio of the block's density ρ_{block} to the density of the liquid, ρ_{liquid}. Let's try this for a couple of real-life examples: ice floating in fresh water and ice floating in salt water. We find that a cube of pure ice in a glass of water floats with 91.7% of its volume submerged, while an iceberg in salt water floats a bit higher with 89.5% of its volume below the surface.

Block of ice floating in fresh water:

ρ_{block} = density of ice = 917 kg/m^3
ρ_{liquid} = density of fresh water = 1.000×10^3 kg/m^3

$$\text{fraction of ice that's submerged} = \frac{V_{displaced}}{V_{block}} = \frac{\rho_{block}}{\rho_{liquid}}$$
$$= \frac{917 \text{ kg/m}^3}{1.000 \times 10^3 \text{ kg/m}^3} = 0.917$$

So 91.7% of the ice is submerged.

Block of ice floating in salt water:

ρ_{block} = density of ice = 917 kg/m^3
ρ_{liquid} = density of salt water = 1.025×10^3 kg/m^3

$$\text{fraction of ice that's submerged} = \frac{V_{displaced}}{V_{block}} = \frac{\rho_{block}}{\rho_{liquid}}$$
$$= \frac{917 \text{ kg/m}^3}{1.025 \times 10^3 \text{ kg/m}^3} = 0.895$$

So 89.5% of the ice is submerged.

EXAMPLE 11-10 Underwater Float

A solid plastic ball of density 6.00×10^2 kg/m³ and radius 2.00 cm is attached by a lightweight string to the bottom of an aquarium filled with fresh water. What is the tension in the string?

Set Up

Three forces act on the ball: the downward force of gravity, the upward buoyant force exerted by the water, and the downward tension force exerted by the string (which is what we want to find). We'll first use Equation 11-13 to determine the buoyant force on the ball and then use Newton's first law to solve for the tension force.

Buoyant force:

$$F_b = \rho_{water} V_{displaced} g \qquad (11\text{-}13)$$

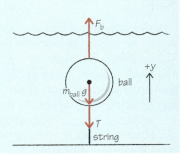

Solve

Newton's first law tells us that since the ball is at rest, the net force on the ball must be zero. Solve for the tension T in the string in terms of the buoyant force F_b and the weight of the ball $m_{ball} g$.

Newton's first law:

$$\Sigma F_y = F_b - T - m_{ball} g = 0 \text{ so}$$
$$T = F_b - m_{ball} g$$

The entire ball is submerged, so the volume of water that it displaces is equal to the volume of the ball of radius $R_{ball} = 2.00$ cm $= 2.00 \times 10^{-2}$ m. Use this to calculate the buoyant force.

Buoyant force:

$$F_b = \rho_{water} V_{displaced} g$$
$$V_{displaced} = V_{ball}$$
$$= \frac{4}{3} \pi R_{ball}^3 = \frac{4}{3} \pi (2.00 \times 10^{-2} \text{ m})^3$$
$$= 3.35 \times 10^{-5} \text{ m}^3 \text{ so}$$
$$F_b = (1.000 \times 10^3 \text{ kg/m}^3)(3.35 \times 10^{-5} \text{ m}^3)(9.80 \text{ m/s}^2)$$
$$= 0.328 \text{ kg} \cdot \text{m/s}^2 = 0.328 \text{ N}$$

Use the density of the ball to calculate its weight.

weight of ball $= m_{ball} g = \rho_{ball} V_{ball} g$
$$= (6.00 \times 10^2 \text{ kg/m}^3)(3.35 \times 10^{-5} \text{ m}^3)(9.80 \text{ m/s}^2)$$
$$= 0.197 \text{ kg} \cdot \text{m/s}^2 = 0.197 \text{ N}$$

Finally, use the relationship that we derived from Newton's first law to solve for the tension T.

$$T = F_b - m_{ball} g = 0.328 \text{ N} - 0.197 \text{ N}$$
$$= 0.131 \text{ N}$$

Reflect

The ball is less dense than water, so the buoyant force ($F_b = 0.328$ N) is greater than the weight of the ball ($m_{ball} g = 0.197$ N). Hence the string must exert a downward tension force to keep the ball from rising. If the string were to break, the ball would rise to the surface of the water. Can you see from Example 11-9 that the ball would end up floating with 60.0% of its volume submerged?

Apparent Weight

In Example 11-10 the submerged plastic ball is less dense than water and so has to be tethered to keep it from floating upward. Let's now think about a submerged object that's denser than water. The problem now is how to keep the object from sinking, which we solve by suspending it from above. Figure 11-18a shows such a submerged object hanging from a spring balance, and Figure 11-18b shows an identical object that is surrounded by vacuum. Although both objects have the same mass M and the same volume V, the weights registered on the spring balances are *not* the same. Why? The difference is that for the submerged object in Figure 11-18a, the buoyant force due to the displaced water opposes the force of gravity and makes the object seem to weigh less. Thus the scale reads the object's **apparent weight** rather than its true weight. (In Section 7-5 we saw that an object in orbit has an apparent weight of zero due to its acceleration. The kind of apparent weight we're discussing here is for an object that's *not* accelerating but is acted on by a buoyant force.)

You can experience the difference between apparent weight and true weight in a swimming pool. If you swim below the surface of the water, you feel as though you weigh less than normal. Just as for the plastic ball in Example 11-10, the upward buoyant force on you partially cancels the force of gravity. Astronauts use this effect to train for working in low gravity: They practice while submerged in an immense pool of water.

How is the apparent weight related to the true weight? Figure 11-18c shows the free-body diagram for the submerged object on the left in Figure 11-18a. The three forces acting on the object are the downward force of gravity (whose magnitude is the object's true weight w), the upward buoyant force F_b exerted by the water, and the upward force exerted by the spring balance. The scale on the spring balance measures how much force the balance exerts, so this force is just equal to the apparent weight $w_{apparent}$. The net force on the hanging object is zero (because the object is at rest), so

$$\Sigma F_y = w_{apparent} + F_b - w = 0$$

and

$$w_{apparent} = w - F_b \qquad \textbf{(11-14)}$$

(apparent weight of a submerged object)

The apparent weight of a submerged object equals its true weight minus the buoyant force exerted on the object by the surrounding fluid.

If the object is completely submerged, the volume of fluid that it displaces equals the volume V of the object. Then we can replace $V_{displaced}$ in Equation 11-13 (the statement of Archimedes' principle) by V, and Equation 11-14 becomes

$$w_{apparent} = w - \rho_{fluid}Vg \qquad \textbf{(11-15)}$$

(apparent weight of an object of weight w
and volume V completely submerged in a fluid of density ρ_{fluid})

So by measuring the apparent weight of an object while submerged ($w_{apparent}$) and its true weight when not submerged (w), you can determine the object's volume. This method can be quite handy when the object is irregularly shaped!

We can also write the apparent weight in terms of the average density ρ of the object. Since the object has volume V, its mass is $m = \rho V$ and its true weight is $w = mg = \rho Vg$. Then Equation 11-15 becomes

$$w_{apparent} = \rho Vg - \rho_{fluid}Vg = (\rho - \rho_{fluid})Vg \qquad \textbf{(11-16)}$$

(apparent weight of an object of average density ρ
and volume V completely submerged in a fluid of density ρ_{fluid})

Equation 11-16 says that if we know an object's volume V, we can determine its average density ρ by measuring its apparent weight $w_{apparent}$ when submerged in a fluid of density ρ_{fluid}. According to legend, Archimedes came to this realization while he was himself submerged in his bathtub—a discovery that led him to his understanding of buoyancy.

A useful application of apparent weight is to find the average density ρ of an object composed of two materials of known density. This makes it possible to determine what fraction of the object is composed of each material. Example 11-11 applies this idea to the human body, which is largely composed of lean muscle and fat.

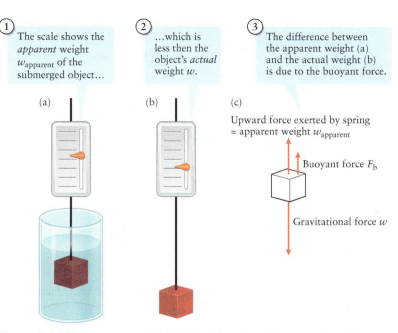

① The scale shows the *apparent* weight $w_{apparent}$ of the submerged object…

② …which is less then the object's *actual* weight w.

③ The difference between the apparent weight (a) and the actual weight (b) is due to the buoyant force.

(a)　(b)　(c)

Upward force exerted by spring = apparent weight $w_{apparent}$

Buoyant force F_b

Gravitational force w

Figure 11-18 Apparent weight (a) An object hung from a spring balance has an apparent weight that depends on the fluid in which it is submerged. The reading on the scale indicates the force exerted by the spring, which is equal in magnitude to the object's apparent weight. (b) The reading on the scale shows the object's actual weight only if the object is surrounded by vacuum, so no buoyant force acts. (c) A free-body diagram showing the three forces acting on the submerged object in (a).

Go to Picture It 11-1 for more practice dealing with the buoyant force.

Go to Interactive Exercises 11-1 and 11-2 for more practice dealing with the buoyant force.

BioMedical EXAMPLE 11-11 Measuring Body Fat

In the human body the density of lean muscle is about 1.06×10^3 kg/m³, and the density of fat tissue is about 9.30×10^2 kg/m³. Two-thirds of Americans have a percent body fat of 25% or more and are therefore overweight or obese. Adult men with between 18 and 24% body fat are considered healthy. An adult male patient with a mass of 85.0 kg comes to your clinic with the claim that he has 20.0% body fat. To test this claim, you measure his apparent weight when submerged in water. If his claim is correct, what are (a) the volume of his body, (b) his average density, and (c) the apparent weight you will measure? (d) What do you conclude if the apparent weight you measure is 19.7 N? Assume that he is made of muscle and fat only, and ignore the bones (which typically make up about 15% of body mass).

Set Up

We're given the densities of fat and of muscle, the fluid density $\rho_{\text{fluid}} = 1000$ kg/m³, and the person's mass $m = 85.0$ kg (from which we can find his actual weight $w = mg$). We're also given his claimed percentage of body fat. We'll use Equation 11-1 to determine the volume of his body that is fat and the volume that is muscle according to his claim, then use these together to calculate what his claim says about his overall volume V and his average density ρ. We can then use either Equation 11-15 or Equation 11-16 to calculate what his apparent weight should be if his claim is correct.

Definition of density:

$$\rho = \frac{m}{V} \qquad (11\text{-}1)$$

Two equations for the apparent weight w_{apparent} of an object of weight w, volume V, and average density ρ submerged in fluid of density ρ_{fluid}:

$$w_{\text{apparent}} = w - \rho_{\text{fluid}}Vg \qquad (11\text{-}15)$$
$$w_{\text{apparent}} = (\rho - \rho_{\text{fluid}})Vg \qquad (11\text{-}16)$$

fat mass: m_{fat}
muscle mass: m_{muscle}
total mass: $m = m_{\text{fat}} + m_{\text{muscle}}$

Solve

(a) Use the person's claim about his percentage of body fat to determine the mass of fat and mass of muscle in his body. Then use Equation 11-1 to determine the volume of the fat, V_{fat}, and the volume of the muscle, V_{muscle}. The total volume V of the person is the sum of V_{fat} and V_{muscle}.

The person claims that 20.0% = 0.200 of their body is fat, so 80.0% = 0.800 is muscle:

$m_{\text{fat}} = 0.200m = (0.200)(85.0 \text{ kg}) = 17.0$ kg
$m_{\text{muscle}} = 0.800m = (0.800)(85.0 \text{ kg}) = 68.0$ kg

Rearrange Equation 11-1 to solve for the volume of fat and volume of muscle in his body:

$$\rho = \frac{m}{V}, \text{ so } V = \frac{m}{\rho}$$

$$V_{\text{fat}} = \frac{m_{\text{fat}}}{\rho_{\text{fat}}} = \frac{17.0 \text{ kg}}{9.30 \times 10^2 \text{ kg/m}^3} = 0.0182 \text{ m}^3$$

$$V_{\text{muscle}} = \frac{m_{\text{muscle}}}{\rho_{\text{muscle}}} = \frac{68.0 \text{ kg}}{1.06 \times 10^3 \text{ kg/m}^3} = 0.0642 \text{ m}^3$$

The person's total volume V is

$$V = V_{\text{fat}} + V_{\text{muscle}} = 0.0182 \text{ m}^3 + 0.0642 \text{ m}^3$$
$$= 0.0824 \text{ m}^3$$

(b) If the person's body fat claim is correct, his average density from Equation 11-1 will be his mass divided by the volume V that we found in part (a).

Average density of person from Equation 11-1:

$$\rho = \frac{m}{V} = \frac{85.0 \text{ kg}}{0.0824 \text{ m}^3}$$
$$= 1.031 \times 10^3 \text{ kg/m}^3$$

or, rounded to 3 significant figures,

$$\rho = 1.03 \times 10^3 \text{ kg/m}^3$$

Since the person is part fat and part muscle, the value of ρ is intermediate between the values of ρ_{fat} and ρ_{muscle}.

(c) Use Equation 11-15 to calculate the apparent weight of the person submerged in water, provided his body fat claim is correct.

From Equation 11-15,

$$w_{\text{apparent}} = w - \rho_{\text{fluid}}Vg$$

The person's true weight is

$$w = mg = (85.0 \text{ kg})(9.80 \text{ m/s}^2)$$
$$= 833 \text{ kg} \cdot \text{m/s}^2 = 833 \text{ N}$$

The fluid density $\rho_{\text{fluid}} = 1.00 \times 10^3$ kg/m³ is only a little less than the person's average density $\rho = 1.03 \times 10^3$ kg/m³, so the buoyant force of magnitude $\rho_{\text{fluid}}Vg$ is only a little less than the person's weight w. Hence the apparent weight (true weight minus buoyant force) is only a small fraction of the true weight.

The fluid in which he is immersed is water with density $\rho_{\text{fluid}} = 1.00 \times 10^3$ kg/m³. Using his body volume V from part (a),

$$w_{\text{apparent}} = 833 \text{ N} - (1.00 \times 10^3 \text{ kg/m}^3)(0.0824 \text{ m}^3)(9.80 \text{ m/s}^2)$$
$$= 833 \text{ N} - 808 \text{ N}$$
$$= 25 \text{ N}$$

(d) The person's measured apparent weight is 19.7 N, which is less than the 25 N that we calculated in part (c) based on the person's claim of 20.0% body fat. The *smaller* measured value of w_{apparent} means that the person has a *greater* body fat percentage than his claim suggests.

From Equation 11-15, the apparent weight is

$$w_{\text{apparent}} = w - \rho_{\text{fluid}}Vg$$

The measured value of w_{apparent} is 19.7 N, versus the 25-N value we calculated based on the person's claim. Since the person's weight $w = 833$ N is known, the only way to get a lower value of w_{apparent} is for the buoyant force $\rho_{\text{fluid}}Vg$ to have a greater value. This means that the volume V of the person must be *greater* than we calculated in part (a), and so the person's average density $\rho = m/V$ must be *less* than we calculated in part (b). For this to be so, the person's body must have a greater percentage of low-density fat ($\rho_{\text{fat}} = 9.30 \times 10^2$ kg/m³) and a smaller percentage of high-density muscle ($\rho_{\text{muscle}} = 1.06 \times 10^3$ kg/m³) than he claims.

Reflect

We can check our calculation for the person's apparent weight (based on his optimistic body fat claim) by using Equation 11-16. Happily we get the same result as we did using Equation 11-15.

Can you show that the person's actual body fat percentage is 25.0%? (See Problem 11-63.)

Use Equation 11-16 for an alternative calculation of the person's apparent weight based on his body fat claim:

$$w_{\text{apparent}} = (\rho - \rho_{\text{fluid}})Vg$$

Use the value of V from part (a) and the unrounded value $\rho = 1.031 \times 10^3$ kg/m³ from part (b):

$$w_{\text{apparent}} = (1.031 \times 10^3 \text{ kg/m}^3 - 1.00 \times 10^3 \text{ kg/m}^3)(0.0824 \text{ m}^3)(9.80 \text{ m/s}^2)$$
$$= (31 \text{ kg/m}^3)(0.0824 \text{ m}^3)(9.80 \text{ m/s}^2)$$
$$= 25 \text{ N}$$

GOT THE CONCEPT? 11-7 Buoyancy I

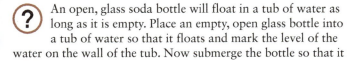

A plastic cube with a coin taped to its top surface is floating partially submerged in water. Mark the level of the water on the cube then remove the coin and tape it to the *bottom* of the cube. Will the cube sit (a) higher in the water, (b) lower in the water, or (c) the same as it was when the coin was on the top of the cube? *Hint:* The buoyant force on an object is the weight of the fluid that it displaces, and in this problem water is displaced by whatever is below the surface.

GOT THE CONCEPT? 11-8 Buoyancy II

An open, glass soda bottle will float in a tub of water as long as it is empty. Place an empty, open glass bottle into a tub of water so that it floats and mark the level of the water on the wall of the tub. Now submerge the bottle so that it fills with water and sinks to the bottom. Is the level of the water in the tub (a) higher, (b) lower, or (c) the same as it was when the bottle was floating?

TAKE-HOME MESSAGE FOR Section 11-8

✔ An object surrounded by a fluid experiences a buoyant force equal to the weight of the fluid displaced by the object.

✔ An object immersed in a fluid has an apparent weight equal to its true weight minus the magnitude of the buoyant force on the object.

11-9 Fluids in motion behave differently depending on the flow speed and the fluid viscosity

In the preceding two sections we have considered fluids at rest. All around us, however, we find fluids in *motion*. Masses of air shift through the atmosphere to bring today's weather, river water courses downhill, and the oceans move in and out with the tides. Within our bodies moving fluids—blood, lymph, and the air used in respiration—play essential roles in sustaining life.

In a wind tunnel smoke trails display the smooth, steady pattern of laminar flow around this automobile.

The smoke rising from this incense stick changes from laminar flow to chaotic, turbulent flow.

Although water rotates around the center of a whirlpool, the flow is largely irrotational (a physicist's way of saying that the flow velocity changes in a special and gradual way from one point to another).

(a)
culture-images GmbH/Alamy

(b)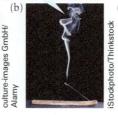
iStockphoto/Thinkstock

(c)
Nick Stephens/Alamy

Figure 11-19 **Some examples of fluid flow** Three very different situations in which a fluid is in motion.

Fluid flows in both nature and technology are breathtaking in their diversity (Figure 11-19). We'll begin our study of fluids in motion by examining how physicists classify different types of fluid flow.

Steady Flow and Unsteady Flow

The simplest type of fluid flow is one in which the flow pattern doesn't change with time, like a stream in which water moves very smoothly with no variations. Such fluid motion is called **steady flow**. The direction and speed of the flow can be different from one point in the fluid to another (Figure 11-20). At any one point, however, the flow velocity remains constant from one moment to the next.

In **unsteady flow** the velocity at a given point can change with time. You experience the unsteady flow of air when you stand outside on a gusty day: The direction and speed of the wind at your position change erratically. A less erratic example of unsteady flow is the pulsing motion of blood as it exits your heart through the aorta. Ocean waves, too, involve an unsteady pulsing motion that carries water (and anyone bobbing in the water) alternately up and down.

As a particle of fluid moves through this pipe from (a) to (b) to (c), its velocity changes in direction and magnitude...

Flow (a) v (b) v

...but at any individual point in the fluid, the velocity of the fluid at that point is always the same. We call this steady flow.

(c) v

Flow

Figure 11-20 **Steady flow of fluid in a pipe** In steady fluid flow the velocity can be different at different points in the flow, but the velocity at any one point maintains the same value at all times.

WATCH OUT! Steady flow doesn't mean that the fluid velocity is the same everywhere.

! Note that even though the fluid velocity at any one position in steady flow remains constant, the velocity of any given quantity of fluid *can* change as it moves. For example, water flowing through a curved pipe such as the one shown in Figure 11-20 changes its direction of motion as it follows the bends of the pipe. When we say that the flow is steady, we mean that when the next quantity of fluid passes through the same point in the pipe, its velocity at that given point will be the same as the previous quantity of fluid that passed through that same point.

Laminar Flow and Turbulent Flow

Imagine a multilane highway filled with moving cars. If none of the cars ever changed lanes, you'd have an efficient, smooth-running transportation system in which each car would follow exactly the same trajectory as the one in front of it and the one behind it (Figure 11-21a). If you now replace the cars with bits of fluid, you would have a type of fluid motion called **laminar flow**. Each bit of fluid follows a path called a **streamline** that is the direct equivalent of one of the lanes of our idealized highway (Figure 11-21b). The smoke trails in Figure 11-19a show the streamlines of laminar flow in a wind tunnel.

Real highways are not as well organized as those in our imaginary example. Some cars change lanes, some cars move faster than others in the same lane, and occasionally there are collisions (Figure 11-22a). A fluid that behaves in this way is undergoing **turbulent flow** (Figure 11-22b). There are no streamlines in this case, since adjacent bits of fluid can follow very different paths. Figure 11-19b shows the turbulent flow of smoke.

Generally speaking, a given type of flow changes from laminar to turbulent if the flow speed exceeds some critical value (which depends on the particular type of fluid). That's why airliners approaching Denver, which is downwind of the Rocky Mountains, can have a smooth ride when the wind is light and the airflow is laminar but a much bumpier ride when the wind is howling and the airflow turbulent.

BioMedical   Turbulence is much noisier than laminar flow. As an example, when you make a hissing or "S" sound, you blow air past your teeth in a way that produces turbulence. By contrast, if you form your lips into an "o" and blow with equal force, your teeth are out of the way, the airflow is more laminar, and the sound is much softer. Your diastolic blood pressure—the second of the two numbers in a blood pressure report such as 120/80 (see Section 11-5)—is measured by putting a high-pressure cuff around your upper arm, gradually lowering the pressure of the cuff, and listening for a change in sound. At pressures just above the diastolic value, the artery is partially compressed, making the flow turbulent and noisy; at the diastolic pressure and below, the noise disappears because the artery is fully expanded and the flow is once again in its normal laminar state.

Viscous Flow and Inviscid Flow

When adjacent parts of a fluid move at different velocities, the parts rub and exert frictional forces on each other. Just as kinetic friction opposes the sliding motion of a block on a ramp, this rubbing opposes the sliding of one bit of fluid past another. This intrinsic resistance to flow is called **viscosity**. Motor oil is more *viscous*—that is, it has a greater viscosity—than water, which in turn is more viscous than air. Many liquids are less viscous at high temperatures; for example, warm maple syrup flows more easily than cold. Most automobiles with gasoline engines use a *multiviscosity* oil designed to flow (and hence lubricate the engine's moving parts) equally well over a broad range of temperatures.

Fluids also experience friction when they flow past a solid surface. This friction is so great that it leads to what is called the **no-slip condition**: Right next to the solid surface, the velocity of the fluid is *zero* so the fluid does not "slip" over the surface. Instead, if fluid is flowing past the surface with a speed *v*, there is a **boundary layer** next to the surface within which the fluid speed increases from zero at the solid surface to the full speed *v* at the edge of the layer (Figure 11-23). A boundary layer of this kind develops around an automobile in motion, which is why driving even at freeway speeds doesn't blow dirt off the car. Any dirt particles lie well within the boundary layer, where the air is hardly moving at all. Even running water over the car won't dislodge all of the dirt particles, since there is also a boundary layer for water flow; only scrubbing with a sponge will complete the job.

Figure 11-23 shows that at different depths within the boundary layer, the fluid flows at different speeds. Hence frictional (viscous) forces act in the boundary layer, and these forces oppose the flow past the solid surface. This effect is called *viscous*

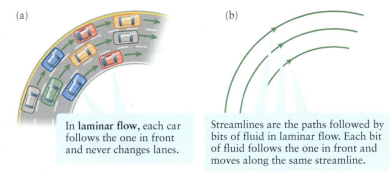

(a)

In **laminar flow**, each car follows the one in front and never changes lanes.

(b)

Streamlines are the paths followed by bits of fluid in laminar flow. Each bit of fluid follows the one in front and moves along the same streamline.

Figure 11-21 Laminar flow of cars and a fluid (a) An idealized highway in which cars move in laminar flow. (b) Streamlines in a fluid with laminar flow.

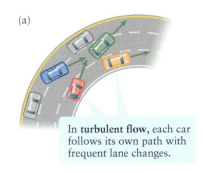

(a)

In **turbulent flow**, each car follows its own path with frequent lane changes.

(b)

There are no streamlines for a fluid in turbulent flow. While each bit of fluid follows its own path, the one behind it can follow a completely different path.

Figure 11-22 Turbulent flow of cars and a fluid (a) A less-than-ideal highway in which cars move in turbulent flow. (b) In a fluid with turbulent flow, bits of fluid move in a haphazard and seemingly unpredictable way.

Figure 11-23 A boundary layer in a viscous fluid A boundary layer in a fluid moving past a flat surface.

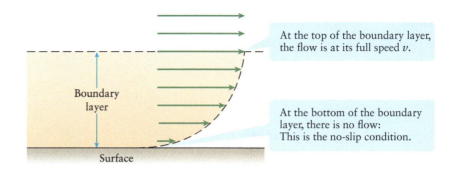

At the top of the boundary layer, the flow is at its full speed v.

At the bottom of the boundary layer, there is no flow: This is the no-slip condition.

drag, and it makes an important contribution to the force of air resistance felt by a moving car. If the flow in the boundary layer is turbulent, there is a greater difference in flow speeds between bits of fluid close to the car's skin, so the viscous forces are greater and the vehicle experiences more drag. Giving a car a "streamlined" shape is a way to make the flow of air around the car less turbulent and more laminar so that the air follows streamlines, as in Figure 11-19a. This reduces the drag so that less power has to be provided by the engine, and the car gets better mileage.

While every fluid has some viscosity, in many physical situations the viscosity is relatively unimportant. (An example is the airflow around a bird in flight. The viscous forces are much less important than the forces due to the pressure of air on the bird.) In these situations it's reasonable to simplify the problem by imagining that the moving fluid has *zero* viscosity. Such **inviscid flow** is an idealization, just like the frictionless ramps and massless strings that we considered in Chapter 4. We'll make use of this idealization in Section 11-10.

Finally, as we did in our earlier consideration of static fluids, we'll examine only the special case of incompressible fluids—that is, fluids in which changes in pressure do not affect the density of the fluid. Flowing water, blood, and air usually behave as incompressible fluids. All of these fluids *can* be compressed by exerting pressure on them, but in many practical situations the pressures are low enough that the compression can be ignored.

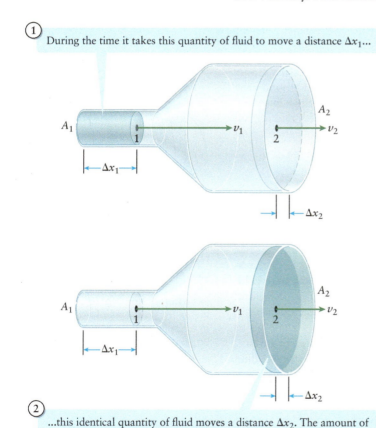

① During the time it takes this quantity of fluid to move a distance Δx_1...

② ...this identical quantity of fluid moves a distance Δx_2. The amount of fluid between points 1 and 2 remains constant.

Figure 11-24 The equation of continuity This illustration shows the flow of an incompressible fluid through a pipe of varying diameter. A slug of fluid of volume $A_1\Delta x_1$ moving at speed v_1 enters the region between points 1 and 2, which causes a slug of fluid of volume $A_2\Delta x_2$ moving at speed v_2 to exit the region. The equation of continuity says that fluid enters the region at the same volume flow rate that it leaves the region.

The Equation of Continuity

No matter what other properties a moving fluid may have, it must obey the following restriction: Mass can be neither created nor destroyed as the fluid flows. Therefore, in a *steady* flow, if a certain mass of fluid flows *into* a given region of space (say, the interior of a certain segment of a pipe) in a given time interval, the same amount of mass must flow *out of* that region during that same time interval. This is called the *principle of continuity*. For example, if 1 kg of fluid flows into one end of a pipe each second, then 1 kg/s must flow out the other end. (If the principle of continuity were violated, the amount of fluid within the pipe would either increase or decrease. But in this case the flow wouldn't be steady as we originally assumed it to be.)

We can make a simple mathematical statement of the principle of continuity for the case of an *incompressible* fluid. The principle of continuity tells us that in a given time interval the same *volume* of incompressible fluid must flow into a certain region of space as flows out of that region. Figure 11-24 illustrates this principle for a pipe with a cross-sectional area that varies along its length. Let 1 and 2 be two different points along the pipe, and consider the segment of the pipe between these two points. While a quantity of fluid of volume $A_1\Delta x_1$ enters the pipe segment at point 1, a quantity of volume $A_2\Delta x_2$ exits the segment at point 2. The principle of continuity then tells us that

$$A_1\Delta x_1 = A_2\Delta x_2 \qquad (11\text{-}17)$$

It takes some time interval Δt for the quantity of fluid to enter the region at point 1. And since the volume of fluid between the points must remain constant, the quantity of fluid at point 2 must exit the region in the same time interval. If we divide Equation 11-17 by the time interval Δt it takes for the quantities of fluid to enter or exit the region, we arrive at

$$A_1 \frac{\Delta x_1}{\Delta t} = A_2 \frac{\Delta x_2}{\Delta t} \qquad (11\text{-}18)$$

To see why we divided through by Δt, note that the fluid at point 1 moves a distance Δx_1 during the time interval Δt and so has speed $v_1 = \Delta x_1/\Delta t$. Similarly, the fluid at point 2 (which moves a distance Δx_2 during the same time interval) has speed $v_2 = \Delta x_2/\Delta t$. So we can rewrite Equation 11-18 as the following relationship, called the **equation of continuity**:

Cross-sectional area of the flow at point 1   Flow speed at point 1

$$A_1v_1 = A_2v_2$$

Cross-sectional area of the flow at point 2   Flow speed at point 2

In other words, the product of the pipe's cross-sectional area A and the flow speed v has the same value at point 1 as at point 2. The choice of these points is quite arbitrary; they could be any two points along the pipe's length. So Equation 11-19 tells us that the product Av has the same value *everywhere* along the pipe. *Where a pipe is narrow, an incompressible fluid moves rapidly; where the pipe is broad, the fluid moves more slowly.* If the flow is laminar, the "pipe" doesn't have to be a solid object with walls; it can be a volume enclosed by a set of streamlines (Figure 11-25). It follows that *where streamlines are close together, an incompressible fluid moves rapidly; where streamlines are far apart, the fluid moves more slowly.* The airflow around the car shown in Figure 11-19a is mostly incompressible, so the flow is rapid over the roof of the car where streamlines are close together.

The equation of continuity explains what happens when you put your thumb over the end of a garden hose. This action reduces the area through which water can flow out of the hose and so the water emerges at a faster speed. The same principle explains why there are often strong winds through mountain passes, where the air is forced into a "pipe" of narrow cross section.

Note that the product Av in Equation 11-19 has units of $(\text{m}^2)(\text{m/s}) = \text{m}^3/\text{s}$, or volume per unit time. This quantity, called the **volume flow rate**, tells you the number of cubic meters of fluid that pass a given point each second. For example, the average volume flow rate of blood through the aorta of a resting human is about 10^{-4} m^3/s, or 6 L/min. (This is an *average* volume flow rate since the flow is in pulses rather than steady.) So the equation of continuity says that *the volume flow rate of an incompressible fluid moving through a pipe is the same at all points.*

BioMedical   This principle needs a little restatement if the pipe branches into a number of small pipes. An example is the human circulatory system, where the flow of oxygenated blood from the aorta is spread out into an enormous number of narrow capillaries (Figure 11-26). In this case the volume flow rate in the aorta is the same as the *combined* volume flow rate through all of the tiny capillaries. If the combined cross-sectional area of the capillaries, A_c, were the same as the cross-sectional area A_a of the aorta, blood would flow at the same speed throughout the system. In fact, in the human circulatory system A_c for the capillaries is greater than A_a for

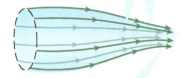

In laminar flow, fluid enclosed by these streamlines remains in the enclosed volume.

Fluid moves fastest where streamlines are closest together.

Fluid moves slowest where streamlines are far apart.

Figure 11-25 Streamline spacing and flow speed In laminar flow of an incompressible fluid, the spacing between streamlines tells you about the flow speed.

Equation of continuity for steady flow of an incompressible fluid (11-19)

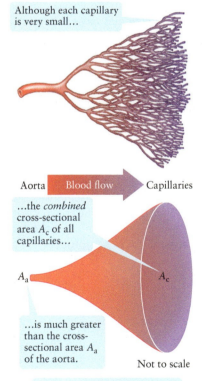

Although each capillary is very small…

Aorta   Blood flow   Capillaries

…the *combined* cross-sectional area A_c of all capillaries…

A_a A_c

…is much greater than the cross-sectional area A_a of the aorta.

Not to scale

Hence the *speed* at which blood flows is much slower in the capillaries then in the aorta.

Figure 11-26 The human circulatory system Blood leaving the heart flows first through a single vessel, the aorta. Branches off the aorta eventually lead to billions of very small capillaries.

the aorta, so the flow speed in the capillaries is *slower* than in the aorta. This slower speed gives the blood in the capillaries more time for exchange of nutrients, gases, and waste products between the blood and the surrounding tissues.

WATCH OUT! The equation of continuity is for incompressible flow only.

Note that Equation 11-19 applies only if the flow is *incompressible*, so its density is the same in all circumstances. In certain cases of fluid motion, however, the fluid does change density, and we say that the flow is *compressible*. Air flowing faster than sound behaves like a compressible fluid: When such fast-moving air enters a constriction such as a narrow pipe, it slows down and the molecules of the fluid get squeezed together so that the density of the fluid increases. In such a case

Equation 11-19 doesn't apply. An analogy for a compressible fluid is how automobiles behave on a highway. You can think of the cars as "molecules" that make up a "fluid" that "flows" along the highway. In light traffic the cars in this "fluid" are far apart, but when this "fluid" passes through a narrower "pipe"—for example, a section of a two-lane highway where one lane is closed due to construction—the flow slows down, the cars get squeezed together, and a traffic jam results.

EXAMPLE 11-12 Flow in a Constriction

An incompressible fluid in a pipe of cross-sectional area 4.0×10^{-4} m^2 flows at a speed of 3.0 m/s. What is the flow speed in a part of the pipe that is constricted and has a cross-sectional area of 2.0×10^{-4} m^2?

Set Up

We are given the cross-sectional area and flow speed at one point in the pipe, and we wish to determine the flow speed at another point in the same pipe that has a different cross-sectional area. Since the fluid is incompressible, we can use the equation of continuity, Equation 11-19.

Equation of continuity for steady flow of an incompressible fluid:

$$A_1 v_1 = A_2 v_2 \qquad (11\text{-}19)$$

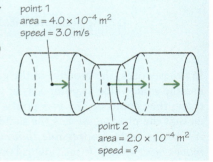

point 1
area = 4.0×10^{-4} m^2
speed = 3.0 m/s

point 2
area = 2.0×10^{-4} m^2
speed = ?

Solve

Let point 1 be a location where the pipe has its full cross-sectional area. Then $A_1 = 4.0 \times 10^{-4}$ m^2 and $v_1 = 3.0$ m/s. Choose point 2 to be in the constriction, where $A_2 = 2.0 \times 10^{-4}$ m^2. Use the equation of continuity to solve for v_2.

Rearrange Equation 11-19 to solve for v_2:

$$v_2 = \frac{A_1}{A_2} v_1$$

Substitute values:

$$v_2 = \frac{(4.0 \times 10^{-4} \text{ m}^2)}{(2.0 \times 10^{-4} \text{ m}^2)} (3.0 \text{ m/s})$$
$$= 6.0 \text{ m/s}$$

Reflect

Our result shows that the speed in the narrow constriction is faster than in the broad part of the pipe, just as the equation of continuity tells us it must be.

BioMedical EXAMPLE 11-13 How Many Capillaries?

The inner diameter of the human aorta is about 2.50 cm, while that of a typical capillary is about 6.00 μm = 6.00×10^{-6} m (see Figure 11-26). In a person at rest, the average flow speed of blood is about 20.0 cm/s in the aorta and about 1.00 mm/s in a capillary. Calculate (a) the volume flow rate (in m^3/s) of blood in the aorta, (b) the volume flow rate in a single capillary, and (c) the total number of open capillaries into which blood from the aorta is distributed at any one time.

Set Up

Figure 11-26 shows the situation. We are given the dimensions of the aorta and each capillary as well as the flow speed in each of these pipes. Our goal is to determine the volume flow rate in the aorta and in a capillary as well as the number of capillaries into which the aorta empties. The volume flow rate in a pipe equals its cross-sectional area times the speed of the fluid in the pipe. Like water, blood acts like an incompressible fluid. (It will compress appreciably only under pressures much higher than those found in the body.) So we can use the equation of continuity: The volume flow rate through the aorta must be equal to the flow rate through all of the open capillaries combined.

Equation of continuity for steady flow of an incompressible fluid:

$$A_1 v_1 = A_2 v_2 \qquad (11\text{-}19)$$

1 = aorta
2 = all open capillaries combined

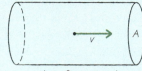

volume flow rate = Av

Solve

(a) The volume flow rate in the aorta is equal to the product of its cross-sectional area and the flow speed of aortal blood ($v_{\text{aorta}} = 20.0$ cm/s $= 0.200$ m/s).

Radius of aorta:

$$r_{\text{aorta}} = (1/2) \times (\text{diameter of aorta}) = (1/2) \times 2.50 \text{ cm}$$
$$= 1.25 \text{ cm} = 1.25 \times 10^{-2} \text{ m}$$

Cross-sectional area of aorta:

$$A_{\text{aorta}} = \pi r_{\text{aorta}}^2 = \pi (1.25 \times 10^{-2} \text{ m})^2$$
$$= 4.91 \times 10^{-4} \text{ m}^2$$

Volume flow rate in aorta:

$$A_{\text{aorta}} v_{\text{aorta}} = (4.91 \times 10^{-4} \text{ m}^2)(0.200 \text{ m/s})$$
$$= 9.82 \times 10^{-5} \text{ m}^3/\text{s}$$

(b) Do the same calculations for a single capillary, in which the flow speed is $v_{\text{capillary}} = 1.00$ mm/s $= 1.00 \times 10^{-3}$ m/s.

Radius of a capillary:

$$r_{\text{capillary}} = (1/2) \times (\text{diameter of capillary}) = (1/2) \times 6.00 \times 10^{-6} \text{ m}$$
$$= 3.00 \times 10^{-6} \text{ m}$$

Cross-sectional area of a capillary:

$$A_{\text{capillary}} = \pi r_{\text{capillary}}^2 = \pi (3.00 \times 10^{-6} \text{ m})^2$$
$$= 2.83 \times 10^{-11} \text{ m}^2$$

Volume flow rate in a capillary:

$$A_{\text{capillary}} v_{\text{capillary}} = (2.83 \times 10^{-11} \text{ m}^2)(1.00 \times 10^{-3} \text{ m/s})$$
$$= 2.83 \times 10^{-14} \text{ m}^3/\text{s}$$

(c) Our results from (a) and (b) show that compared to the volume flow rate through a *single* capillary, the volume flow rate through the aorta is 3.47×10^9 times greater. The idea of continuity tells us that the combined volume flow rate through *all* the open capillaries must be equal to the volume flow rate through the aorta. We therefore learn the total number of open capillaries.

$$\frac{\text{volume flow rate in aorta}}{\text{volume flow rate in a capillary}} = \frac{9.82 \times 10^{-5} \text{ m}^3/\text{s}}{2.83 \times 10^{-14} \text{ m}^3/\text{s}}$$
$$= 3.47 \times 10^9$$

(volume flow rate in aorta) = (total volume flow rate in all open capillaries combined) . . . so there must be 3.47×10^9 open capillaries.

Reflect

Our results show that the human circulatory system is truly extensive!

   As a check on our results, note that the *combined* cross-sectional areas of all capillaries is 9.82×10^{-2} m^2, which is 200 times greater

Total cross-sectional area of all open capillaries combined:

$$A_{\text{all open capillaries}} = (3.47 \times 10^9) A_{\text{capillary}}$$
$$= (3.47 \times 10^9)(2.83 \times 10^{-11} \text{ m}^2)$$
$$= 9.82 \times 10^{-2} \text{ m}^2$$

than the cross-sectional area of the aorta. By the equation of continuity, the flow speed in the capillaries should therefore be *slower* than in the aorta by a factor of $1/(2.00 \times 10^2)$; that is, $v_{capillary} = v_{aorta}/(2.00 \times 10^2) = (0.200 \text{ m/s})/(2.00 \times 10^2) = 1.00 \times 10^{-3}$ m/s. This gives us back one of the numbers we started with, so our calculation is consistent.

$$\frac{\text{area of all open capillaries combined}}{\text{area of aorta}} = \frac{9.82 \times 10^{-2} \text{ m}^2}{4.91 \times 10^{-4} \text{ m}^2}$$
$$= 2.00 \times 10^2$$

BioMedical EXAMPLE 11-14 From Capillaries to the Vena Cavae

Blood returns to the heart from the capillaries through two veins known as the *vena cavae*. The combined cross-sectional area of the vena cavae is 10.0 cm^2. At what average speed does blood move through these veins?

Set Up

We know the net volume flow rate of blood in the aorta from Example 11-13, and we're given the cross-sectional area $A_{vc} = 10.0$ cm^2 for the vena cavae. We'll use the equation of continuity to find the flow speed of blood in the vena cavae, v_{vc}.

Equation of continuity for steady flow of an incompressible fluid:

$$A_1 v_1 = A_2 v_2 \qquad (11\text{-}19)$$

1 = aorta
2 = vena cavae

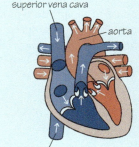

superior vena cava
aorta
inferior vena cava

Solve

If we assume no blood volume is lost as it circulates through the body, the volume flow rate in the aorta is the same as in the vena cavae.

From Example 11-13,

volume flow rate in aorta = $A_{aorta} v_{aorta}$
$$= (4.91 \times 10^{-4} \text{ m}^2)(0.200 \text{ m/s})$$
$$= 9.82 \times 10^{-5} \text{ m}^3/\text{s}$$

volume flow rate in venae cavae = $A_{vc} v_{vc}$, where

$$A_{vc} = (10.0 \text{ cm}^2)\left(\frac{1 \text{ m}}{100 \text{ cm}}\right)^2$$
$$= 1.00 \times 10^{-3} \text{ m}^2$$

The two volume flow rates are the same, so

$$A_{aorta} v_{aorta} = A_{vc} v_{vc}$$
$$v_{vc} = \frac{A_{aorta} v_{aorta}}{A_{vc}}$$
$$= \frac{9.82 \times 10^{-5} \text{ m}^3/\text{s}}{1.00 \times 10^{-3} \text{ m}^2}$$
$$= 0.0982 \text{ m/s} = 9.82 \text{ cm/s} = 98.2 \text{ mm/s}$$

Reflect

It's a common misconception that the slow-moving blood in the capillaries continues to move slowly as it returns to the heart. This example shows that this isn't the case! Compared to an average speed of 1.00 mm/s in the capillaries, blood moves 98.2 times faster

Speed of blood in vena cavae compared to speed of blood in capillaries is

$$\frac{v_{vc}}{v_{\text{all open capillaries}}} = \frac{98.2 \text{ mm/s}}{1.00 \text{ mm/s}} = 98.2$$

in the vena cavae. The reason is that the cross-sectional area of the vena cavae is 1/98.2 as great as the combined cross-sectional area of the open capillaries (see Example 11-13). So the blood has to speed up as it returns to the heart to compensate for the decreased cross-sectional area.

Note that blood does flow more slowly in the vena cavae (0.0982 m/s) than in the aorta (0.200 m/s). That's because the venae cavae have a larger cross-sectional area than the aorta.

Explanation: Cross-sectional area of vena cavae compared to cross-sectional area of all open capillaries combined is

$$\frac{A_{vc}}{A_{\text{all open capillaries}}} = \frac{1.00 \times 10^{-3} \text{ m}^2}{9.82 \times 10^{-2} \text{ m}^2} = \frac{1}{98.2}$$

GOT THE CONCEPT? 11-9 How Diameter Goes with the Flow

? While walking past a construction site, you notice a pipe sticking out of a second-floor window, with water rushing out. As the water flows to the ground, it must speed up due to the effect of gravity. How does the diameter of the flowing stream of water change as it descends? If we assume that the flow remains laminar, the diameter (a) increases; (b) stays the same; (c) decreases.

TAKE-HOME MESSAGE FOR Section 11-9

✔ In steady flow, the flow velocity at any one point remains constant from one moment to the next.

✔ In laminar flow, each bit of fluid follows a streamline. In turbulent flow, adjacent bits of fluid can follow very different paths.

✔ The viscosity of a fluid is a measure of its resistance to flow. It results in the formation of a boundary layer next to a solid surface. Fluid speed increases from zero at the solid surface to the full speed v of the flow at the edge of the boundary layer.

✔ In steady, incompressible flow, the product of the cross-sectional area of the flow and the speed of the flow is constant: If the cross-sectional area increases, the speed of the flow decreases, and vice versa.

11-10 Bernoulli's equation helps us relate pressure and speed in fluid motion

Hold a piece of notebook paper by two corners so that the paper droops downward. Then blow on the top of the paper as shown in Figure 11-27. You might expect that the force of the air expelled from your mouth would push the paper downward. But remarkably, the paper actually lifts *up* where it is struck by the moving air.

What's happened is that by making air move over the top of the paper, you've lowered the pressure of that air. Hence there is greater air pressure on the underside of the paper than on the top, and the paper lifts up. This simple experiment illustrates that certain kinds of fluid flow have a property described by **Bernoulli's principle**: In a moving fluid, the pressure is low where the fluid is moving rapidly. This principle was first identified by the eighteenth-century mathematician Daniel Bernoulli.

Bernoulli's principle explains why an open door may swing closed on a windy day. The pressure in the moving air outside the house is lower than the pressure of the still air inside the house. The difference in air pressure on the two sides of the door pulls the door toward the outside, slamming it shut. An umbrella bulges upward in the wind for the same reason; there is low-pressure, fast-moving air on top of the umbrella, but high-pressure still air in the space underneath.

Blowing on the top of a sheet of paper lowers the pressure there.

Higher pressure below pushes the paper up.

Figure 11-27 Bernoulli's principle A piece of paper lifts up if you blow over the top of the paper.

Figure 11-28 illustrates the origins of Bernoulli's principle. A pipe of varying diameter carries an incompressible fluid from left to right. According to the equation of continuity that we discussed in Section 11-9, the fluid moves fastest in the narrow part of the pipe (point 2). Hence a parcel of fluid must speed up as it enters the narrow part. A net force must act on the fluid to change its velocity (that is, cause it to accelerate). This net force must be to the right as a parcel of fluid enters the narrow part. If we assume that there is negligible viscosity (see Section 11-9), the only forces that could be acting are those due to differences in pressure on the left and right sides of the parcel. As Figure 11-28 shows, to produce the required acceleration, the pressure must be lower in the narrow part of the pipe (where the fluid moves rapidly) than in the wide part (where the fluid moves slowly). This is just what Bernoulli's principle says: The pressure is lowest where the fluid moves the fastest.

WATCH OUT! Pressure differences cause velocity changes, not the other way around.

! It's *not* correct to say that the pressure differences in Figure 11-28 are *caused* by the changing velocity of the fluid. In fact, just the reverse is true: The changes in fluid velocity are caused by the pressure differences! Recall the meaning of Newton's second law: An object accelerates (that is, changes its velocity) in response to a net force acting on it. In other words, a change in velocity is a result of a net force, not the other way around. In the same way, the meaning of Bernoulli's principle is that a fluid undergoes a change in velocity as a result of pressure differences, not the reverse.

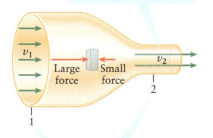

Fluid must move faster at point 2 than at point 1 to satisfy the equation of continuity.

To make the fluid speed up between point 1 and point 2, there must be a higher pressure at 1 than at 2.

Figure 11-28 **Interpreting Bernoulli's principle** Pressure differences drive a fluid through a constriction. The larger pressure on the left means that a parcel of fluid entering the constriction has a stronger force pushing it to the right than to the left, so the parcel speeds up as it approaches the constriction.

Bernoulli's Equation

Let's expand on these ideas and see how to express Bernoulli's principle as a rather simple and useful equation. By seeing where this equation comes from, we'll be able to understand why Bernoulli's principle holds only under certain special conditions. Many kinds of flow satisfy these conditions, but many others—including, for example, the flow of blood in the circulatory system—do not.

The shaded volume in Figure 11-29a shows a quantity of a fluid in motion. If we assume that the flow is *laminar* (see Section 11-9), we can think of this quantity of fluid as being enclosed within streamlines just as though it were flowing inside a pipe. As time goes by this quantity of fluid moves along its "pipe," displacing other fluid at its front end and being displaced by fluid at its back end. Figure 11-29b shows our quantity of fluid a brief time interval Δt after the instant shown in Figure 11-29a. The fluid has vacated a volume $A_1\Delta x_1$ and has moved into a volume $A_2\Delta x_2$.

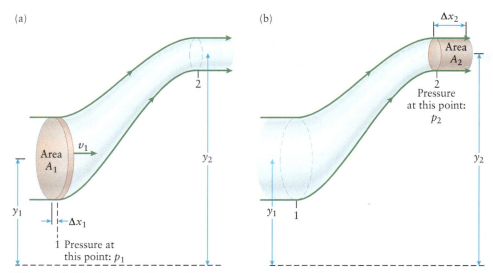

Figure 11-29 **Deriving Bernoulli's equation** (a) A slug of incompressible fluid of volume $A_1\Delta x_1$ moving at speed v_1 takes a time Δt to enter the region between points 1 and 2. (b) During this same time Δt, a slug of fluid of volume $A_2\Delta x_2$ moving at speed v_2 exits the region. The equation of continuity says that fluid enters the region at the same rate that it exits. (Compare with Figure 11-28.)

Let's further assume that the flow is *steady* so that the fluid velocity at any fixed position in the fluid remains the same. In this case the motion of the shaded volume of fluid from the situation in Figure 11-29a to that in Figure 11-29b is the same as if the fluid had "lost" the volume $A_1\Delta x_1$ at its back end and "gained" the volume $A_2\Delta x_2$ at its front end, with no other changes. If we make an additional assumption that the fluid is *incompressible*—so that its density ρ is the same at all points in the fluid—then the volume "lost" at the back end must be the same as the volume "gained" at the front end:

$$\text{volume "lost"} = A_1\Delta x_1 = A_2\Delta x_2 = \text{volume "gained"} \tag{11-20}$$

While the volume of the moving incompressible fluid remains the same, its *energy* can change. Over the time interval Δt the shaded volume of fluid has lost the kinetic energy and gravitational potential energy associated with the "lost" volume at point 1 but has gained the kinetic and potential energies associated with the "gained" volume at point 2. The total energy change associated with the shaded volume of fluid is

$$\Delta E = \Delta K + \Delta U \tag{11-21}$$

We can write the mass of this volume of fluid as $\rho A_1\Delta x_1$ (when it is at point 1) or $\rho A_2\Delta x_2$ (when it is at point 2). We'll use the symbols v_1 and v_2 to denote the fluid speed at points 1 and 2, respectively, and use the symbols y_1 and y_2 for the heights at points 1 and 2. Then, using the familiar formulas $K = \frac{1}{2}mv^2$ for kinetic energy and $U = mgy$ for gravitational potential energy of the fluid–Earth system, we can write ΔK and ΔU in Equation 11-21 as

$$\Delta K = \text{change in kinetic energy} = +\frac{1}{2}(\rho A_2\Delta x_2)v_2^2 - \frac{1}{2}(\rho A_1\Delta x_1)v_1^2 \tag{11-22}$$

$$\Delta U = \text{change in gravitational potential energy} = +(\rho A_2\Delta x_2)gy_2 - (\rho A_1\Delta x_1)gy_1 \tag{11-23}$$

We learned in Section 6-7 that the total mechanical energy of a system (in this case Earth and the shaded volume of fluid) changes only when work is done on it by nonconservative forces. For simplicity we'll assume that there's no friction of any kind so that the fluid is *inviscid* (that is, it has zero viscosity). Then the only nonconservative forces that could act on the shaded volume of fluid are forces from the pressure of the surrounding fluid. There is pressure on all sides of the shaded volume of fluid, but work is done only by those forces that act on the *moving* parts of the fluid. From Figure 11-29 we can determine that the forces (pressure times area) on the back and front ends of the shaded volume of fluid are p_1A_1 and p_2A_2, respectively. The force on the back end pushes in the same direction as the displacement Δx_1 and so does positive work; the force on the front end does negative work because it pushes opposite to the displacement Δx_2. Hence the total nonconservative work done on the shaded volume of fluid in Figure 11-29 is

$$W_{\text{nonconservative}} = +(p_1A_1)\Delta x_1 - (p_2A_2)\Delta x_2 \tag{11-24}$$

Now we can put all the pieces together. From Section 6-7 the total change in mechanical energy given by Equations 11-21, 11-22, and 11-23 is equal to the nonconservative work done on the fluid given by Equation 11-24. That is,

$$\Delta K + \Delta U = W_{\text{nonconservative}}$$

or

$$+\frac{1}{2}(\rho A_2\Delta x_2)v_2^2 - \frac{1}{2}(\rho A_1\Delta x_1)v_1^2 + (\rho A_2\Delta x_2)gy_2 - (\rho A_1\Delta x_1)gy_1$$
$$= +(p_1A_1)\Delta x_1 - (p_2A_2)\Delta x_2 \tag{11-25}$$

Equation 11-25 looks like a horrible mess. But because we assumed the fluid is incompressible, $A_1\Delta x_1$ is equal to $A_2\Delta x_2$ (see Equation 11-20). Hence we can divide out all the factors of $A_1\Delta x_1$ and $A_2\Delta x_2$, leaving the simpler expression

$$+\frac{1}{2}\rho v_2^2 - \frac{1}{2}\rho v_1^2 + \rho gy_2 - \rho gy_1 = p_1 - p_2$$

We can rearrange this expression to read

$$(11\text{-}26) \qquad p_1 + \frac{1}{2}\rho v_1^2 + \rho g y_1 = p_2 + \frac{1}{2}\rho v_2^2 + \rho g y_2$$

Equation 11-26 says that the quantity $p + \frac{1}{2}\rho v^2 + \rho g y$ has the same value at point 1 (the back end of the shaded volume of fluid) as at point 2 (the front end). But we could have chosen the back and front ends to be *anywhere* along the length of the shaded volume of fluid in Figure 11-29. So the quantity $p + \frac{1}{2}\rho v^2 + \rho g y$ must have the same value at *any* point within the streamlines that enclose this shaded volume.

What about a volume of fluid immediately adjacent to the shaded one we've been discussing (Figure 11-30)? In principle this fluid could be moving at a very different speed than the fluid in the shaded volume. But in real fluids in which the flow is laminar, there is enough viscosity—that is, enough frictional force within the fluid—that adjoining bits of fluid move at nearly the same speed. Then the speed varies gradually from one part of the fluid to another, with no abrupt jumps. Physicists say that such a flow is **irrotational**. This does *not* mean that the fluid can't rotate—it certainly can, like water in a whirlpool (Figure 11-19c) or air in a hurricane. Rather, the curious term "irrotational" refers to what would happen to a little paddlewheel that was put in the fluid and allowed to move with it. If the fluid speed on one side of the paddlewheel were sharply different from that on the other side, the paddlewheel would start turning. In the idealized case we're discussing, there are no such differences, and the paddlewheel wouldn't rotate—which is why we call the flow irrotational.

Let's now make a final assumption that the fluid flow shown in Figure 11-30 is irrotational. This assumption is actually in contradiction to our earlier approximation that the flow is inviscid; *some* viscosity must be present to make the flow irrotational. So our approximation is really that the fluid has a little viscosity but not too much! With this additional assumption the quantity $p + \frac{1}{2}\rho v^2 + \rho g y$ has the same value in the shaded volume in Figure 11-29 as it does in any adjacent volume. It also has the same value in the next volume over, and so on. So our idealized fluid has the same value of $p + \frac{1}{2}\rho v^2 + \rho g y$ at *all* points in the fluid. Put another way, points 1 and 2 in Equation 11-26 could be any two points in the fluid.

Our final result is **Bernoulli's equation:**

For this "tube" of fluid $p + \frac{1}{2}\rho v^2 + \rho g y =$ a constant.

For this "tube" of fluid $p + \frac{1}{2}\rho v^2 + \rho g y$ = another constant.

If the flow is irrotational, the constant is the same for all tubes of fluid.

Figure 11-30 Irrotational flow and Bernoulli's equation Two adjacent tubes of fluid. If the flow is irrotational, the quantity $p + \frac{1}{2}\rho v^2 + \rho g y$ has the same value for both tubes and for *all* parts of the fluid.

Bernoulli's equation (11-27)

Fluid pressure at a given point in the fluid

Speed of the fluid at that point

Vertical coordinate of that point

$$p + \frac{1}{2}\rho v^2 + \rho g y = \text{a constant with the same value throughout the fluid}$$

Density of the fluid (uniform throughout the fluid)

Acceleration due to gravity

Note that if the fluid is at rest, so that $v = 0$ at all points, Equation 11-27 becomes $p + \rho g y =$ constant; that is, as y decreases and you go deeper in a static fluid, the quantity $\rho g y$ decreases and the pressure p increases. This is just the relationship for pressure at various depths in a static fluid that we found in Section 11-4. Note also that if we compare two points at the same height y, Equation 11-27 tells us that the pressure p is high where the fluid speed v is low—which is just the statement of Bernoulli's principle that we made at the beginning of this section.

To derive Bernoulli's equation, we had to make several approximations: The fluid flow had to be *laminar, steady, incompressible, inviscid,* and *irrotational.* In many real-life situations these assumptions aren't valid. For example, the flow of water down a waterfall is turbulent, not laminar; viscosity is important for blood flow in capillaries; and the air flowing around a supersonic airplane undergoes substantial compression. But there are a number of situations where Bernoulli's equation gives reasonably good results. In the examples that follow we'll examine how to use Bernoulli's equation in some of these situations.

EXAMPLE 11-15 Lift on a Wing

Figure 11-31 shows a computer simulation of air flowing around an airplane wing (seen end-on). As the figure shows, air flows faster over the top of the wing so that there is lower pressure on the top of the wing than on the bottom. Suppose the air (density 1.20 kg/m³) moves at 75 m/s past the lower surface of a small airplane's wing and at 85 m/s past the upper surface. If the area of the wing as seen from above is 10.0 m² and the top-to-bottom thickness of the wing is 7.0 cm, what is the overall upward force (the *lift*) that the air exerts on the wing?

Figure 11-31 **A computer simulation of airflow around a wing** A vertical column of parcels of air (shown by colored dots) starts at the left and moves to the right, flowing around a wing. The parcels passing over the top of the wing go faster than those passing below the bottom of the wing (they're spaced farther apart horizontally), so there must be lower pressure on the top. (The dots are colored red in regions of low pressure.)

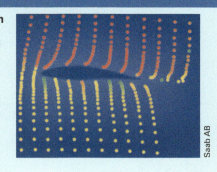

Saab AB

Set Up

The pressure on the bottom of the wing exerts an upward force, and the pressure on the top exerts a downward force. Our goal is to calculate the *combined* vertical force that these pressures exert on the wing. We use Bernoulli's equation to relate the pressure on the two surfaces of the wing. Pressure is force per area, so the force on either surface of the wing is the pressure multiplied by the wing area.

Bernoulli's equation:

$$p + \frac{1}{2}\rho v^2 + \rho g y = \text{a constant} \quad (11\text{-}27)$$

Definition of pressure:

$$p = \frac{F_\perp}{A} \quad (11\text{-}2)$$

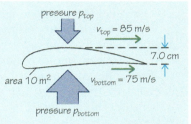

Solve

We want to calculate the net upward force on the wing, or the lift L. Let p_{bottom} and p_{top} be the pressures on the bottom and top of the wing, respectively. Express L in terms of these pressures and the wing area A.

From Equation 11-2 the upward force on the bottom of the wing is

$$F_{\text{bottom}} = p_{\text{bottom}}A$$

and the downward force on the top of the wing is

$$F_{\text{top}} = p_{\text{top}}A$$

Combined upward force on the wing:

$$L = F_{\text{bottom}} - F_{\text{top}}$$
$$= p_{\text{bottom}}A - p_{\text{top}}A$$
$$= (p_{\text{bottom}} - p_{\text{top}})A$$

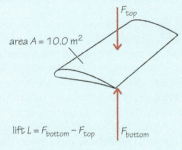

Use Bernoulli's equation to write an expression for the pressure difference $p_{\text{bottom}} - p_{\text{top}}$.

$p + \frac{1}{2}\rho v^2 + \rho g y$ has the same value on the top and bottom of the wing, so

$$p_{\text{bottom}} + \frac{1}{2}\rho v_{\text{bottom}}^2 + \rho g y_{\text{bottom}} = p_{\text{top}} + \frac{1}{2}\rho v_{\text{top}}^2 + \rho g y_{\text{top}}$$

Solve for $p_{\text{bottom}} - p_{\text{top}}$:

$$p_{\text{bottom}} - p_{\text{top}} = \frac{1}{2}\rho v_{\text{top}}^2 - \frac{1}{2}\rho v_{\text{bottom}}^2 + \rho g y_{\text{top}} - \rho g y_{\text{bottom}}$$

Calculate the pressure difference using $\rho = 1.20$ kg/m³, $v_{\text{top}} = 85$ m/s, $v_{\text{bottom}} = 75$ m/s, and $y_{\text{top}} - y_{\text{bottom}} = 7.0$ cm $= 7.0 \times 10^{-2}$ m (the top-to-bottom thickness of the wing).

$$\frac{1}{2}\rho v_{\text{top}}^2 - \frac{1}{2}\rho v_{\text{bottom}}^2 = \frac{1}{2}\rho(v_{\text{top}}^2 - v_{\text{bottom}}^2)$$

$$= \frac{1}{2}(1.20 \text{ kg/m}^3)[(85 \text{ m/s})^2 - (75 \text{ m/s})^2]$$

$$= 9.6 \times 10^2 \text{ kg/(m} \cdot \text{s}^2) = 9.6 \times 10^2 \text{ N/m}^2$$

$$\rho g y_{\text{top}} - \rho g y_{\text{bottom}} = \rho g(y_{\text{top}} - y_{\text{bottom}})$$
$$= (1.20 \text{ kg/m}^3)(9.80 \text{ m/s}^2)(7.0 \times 10^{-2} \text{ m})$$
$$= 0.82 \text{ kg/(m} \cdot \text{s}^2) = 0.82 \text{ N/m}^2$$

so

$$p_{\text{bottom}} - p_{\text{top}} = 9.6 \times 10^2 \text{ N/m}^2 + 0.82 \text{ N/m}^2$$
$$= 9.6 \times 10^2 \text{ N/m}^2 \text{ to two significant figures}$$

Calculate the lift L by multiplying the pressure difference by the wing area.

$$L = (p_{\text{bottom}} - p_{\text{top}}) A = (9.6 \times 10^2 \text{ N/m}^2)(10.0 \text{ m}^2)$$
$$= 9.6 \times 10^3 \text{ N}$$

Reflect

In straight-and-level flight the airplane is not accelerating vertically, so the upward force of lift must exactly balance the weight of the airplane. So our wing can keep an airplane of weight 9.6×10^3 N (about 2200 lb) in the air.

Our result shows that for given speeds of airflow along the top and bottom of the wing, the lift L is proportional to the wing area A. The heavier the airplane, the larger the wing required to maintain flight. The same principle applies to gliding birds: An eagle or vulture weighs more than a hawk and so has a larger wing.

Even if the airplane were not moving, there would still be a small pressure difference of $\rho g y_{\text{top}} - \rho g y_{\text{bottom}} = 0.82$ N/m^2 between the upper and lower surfaces of the wing. This pressure difference means that the air exerts a small buoyant force on the wing. Our calculations show that this small pressure difference is totally negligible compared to the pressure difference $\frac{1}{2}\rho v_{\text{top}}^2 - \frac{1}{2}\rho v_{\text{bottom}}^2 = 9.6 \times 10^3$ N/m^2 due to air traveling faster past the top of the wing than past the bottom of the wing.

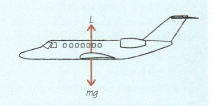

WATCH OUT! Air molecules don't meet at the trailing edge of a wing.

Why does air travel at different speeds over the two surfaces of the wing? A common misconception is that air traveling along the top of the wing takes just as much time to go from the wing's leading edge to its trailing edge as does the air that travels along the bottom. According to this misconception, air has to move faster along the top of the wing because the upper surface is curved more than the lower, and molecules that parted company at the leading edge must somehow meet up at the trailing edge. But the computer simulation in Figure 11-31 shows that this isn't the case at all. Air travels over the upper surface of the wing *much* faster than the common misconception would have us believe, and the molecules *don't* meet up at the trailing edge. A better explanation is that air follows the curvature of the wing due to its (slight) viscosity. This curvature is such that the wing pushes air downward. By Newton's third law the air must push the wing *upward* equally hard. To produce this upward force, or lift, there has to be a pressure difference between the two surfaces of the wing, and this pressure difference is what causes the air to flow at different speeds over the top and bottom of the wing.

EXAMPLE 11-16 A Venturi Meter

Figure 11-32 shows a simple device called a *Venturi meter* for measuring fluid velocity in a gas such as air. When gas passes from left to right through the horizontal pipe, it speeds up as it passes through the constriction at point 2. Bernoulli's principle tells us that the gas pressure must be lower at point 2 than at point 1, and the pressure difference causes the liquid in the U-tube to drop on the left-hand side and rise on the right-hand side. Suppose the gas is air (density 1.20 kg/m^3) that enters the left-hand side of the Venturi meter at 25.0 m/s. The horizontal tube has cross-sectional area 2.00 cm^2 at point 1 and cross-sectional area 1.00 cm^2 at point 2. If the liquid in the U-tube is water, what is the difference in height between the water columns on the left-hand and right-hand sides of the tube?

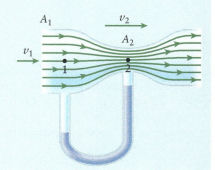

Figure 11-32 A Venturi meter A Venturi meter, or flow meter, can be used to measure the flow speed of a gas.

Set Up

In this problem there are *two* fluids, the air that flows through the horizontal pipe and the water in the U-tube. Hence we'll use Bernoulli's equation twice: once to relate the moving air at point 1 to the moving air at point 2, and once to relate the heights of the water on the two sides of the U-tube (which is what we're trying to find). We'll also use the equation of continuity to relate the speeds of the air at points 1 and 2.

Bernoulli's equation:

$$p + \frac{1}{2}\rho v^2 + \rho g y = \text{a constant}$$

(11-27)

Equation of continuity for steady flow of an incompressible fluid:

$$A_1 v_1 = A_2 v_2 \qquad (11\text{-}19)$$

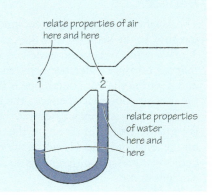

relate properties of air here and here

relate properties of water here and here

Solve

We need to find the difference in air pressure at points 1 and 2, since this is what causes the difference in height of the water on the two sides of the U-tube. Find this difference using Bernoulli's equation and the equation of continuity applied to the air in the horizontal pipe, keeping in mind that we know the values of v_1, A_1, and A_2.

Bernoulli's equation for the air at points 1 and 2:

$$p_1 + \frac{1}{2}\rho_{\text{air}}v_1^2 + \rho_{\text{air}}g y_1 = p_2 + \frac{1}{2}\rho_{\text{air}}v_2^2 + \rho_{\text{air}}g y_2$$

At the center of the horizontal pipe, $y_1 = y_2$, so

$$p_1 + \frac{1}{2}\rho_{\text{air}}v_1^2 = p_2 + \frac{1}{2}\rho_{\text{air}}v_2^2$$

$$p_1 - p_2 = \frac{1}{2}\rho_{\text{air}}v_2^2 - \frac{1}{2}\rho_{\text{air}}v_1^2$$

From the equation of continuity,

$$v_2 = \frac{A_1}{A_2}v_1 \text{ so}$$

$$p_1 - p_2 = \frac{1}{2}\rho_{\text{air}}(v_2^2 - v_1^2) = \frac{1}{2}\rho_{\text{air}}\left[\left(\frac{A_1}{A_2}\right)^2 v_1^2 - v_1^2\right]$$

$$= \frac{1}{2}\rho_{\text{air}}v_1^2\left[\left(\frac{A_1}{A_2}\right)^2 - 1\right]$$

$$= \frac{1}{2}(1.20 \text{ kg/m}^3)(25.0 \text{ m/s})^2\left[\left(\frac{2.00 \text{ cm}^2}{1.00 \text{ cm}^2}\right)^2 - 1\right]$$

$$= 1.13 \times 10^3 \text{ kg/(m·s}^2) = 1.13 \times 10^3 \text{ N/m}^2$$

The pressure difference between points 1 and 2 is also the pressure difference between the water at two points: the top of the water column on the left-hand side of the U-tube and the water column on the right-hand side. The water is at rest ($v = 0$) at both points. Use this technique to find the height difference between the two water columns.

Bernoulli's principle for the water at the tops of the two columns:

$$p_1 + \frac{1}{2}\rho_{\text{water}}v_{\text{water},1}^2 + \rho_{\text{water}}g y_{\text{water},1}$$

$$= p_2 + \frac{1}{2}\rho_{\text{water}}v_{\text{water},2}^2 + \rho_{\text{water}}g y_{\text{water},2}$$

Water is at rest: $v_{\text{water},1} = v_{\text{water},2} = 0$

Solve for the height difference of the two water columns:

$$p_1 + \rho_{\text{water}}g y_{\text{water},1} = p_2 + \rho_{\text{water}}g y_{\text{water},2}$$

$$\rho_{\text{water}}g y_{\text{water},2} - \rho_{\text{water}}g y_{\text{water},1} = p_1 - p_2$$

$$y_{\text{water},2} - y_{\text{water},1} = \frac{p_1 - p_2}{\rho_{\text{water}}g} = \frac{1.13 \times 10^3 \text{ kg/(m·s}^2)}{(1.000 \times 10^3 \text{ kg/m}^3)(9.80 \text{ m/s}^2)}$$

$$= 0.115 \text{ m} = 11.5 \text{ cm}$$

Reflect

This height difference is large enough to easily measure, so the Venturi meter is a practical device.

We made an approximation in our solution that the pressure difference between points 1 and 2 within the horizontal pipe is the same as the pressure difference between the two water columns. This isn't exactly correct because there is a greater weight of air above the column on the left-hand side. However, the resulting additional pressure difference is so small that we can ignore it.

Additional pressure difference between the tops of the two water columns due to the extra weight of air on the left-hand side:

$$\rho_{air}g(y_{water,2} - y_{water,1}) = (1.20 \text{ kg/m}^3)(9.80 \text{ m/s}^2)(0.115 \text{ m})$$
$$= 1.35 \text{ kg/(m}\cdot\text{s}^2) = 1.35 \text{ N/m}^2$$

This is 0.12% of the pressure difference calculated above ($p_1 - p_2 = 1.13 \times 10^3 \text{ N/m}^2$), so we can neglect it.

Applications of Bernoulli's Principle

Can a thrown baseball be a "curveball"—that is, can it be made to follow a path that curves left or right? In the early days of baseball, most people thought a curveball was just an optical illusion. It wasn't until 1941, when *Life* magazine published photographs of a curveball taken with a strobe light, that baseball fans (and everyone else) had proof that a properly thrown baseball can be made to curve. Bernoulli's principle helps to explain why.

Imagine a baseball thrown without any spin, what ball players call a knuckleball. As the ball flies, the air rushes past the left and right sides of the ball at the same speed. Since the speed of the air is the same on both sides, according to Bernoulli's equation the air pressure in the air is also the same on both sides. Hence there is no tendency for the ball's trajectory to curve either right or left.

Now imagine the ball spinning as it moves (Figure 11-33). In this figure we're looking at the ball from above, and it is rotating counterclockwise as seen from this vantage point. The ball is moving in the direction shown by the arrow, and air rushes past it in the opposite direction. Because the surface of the ball is rough, and because the baseball has raised seams that hold the leather cover of the ball together, viscosity drags a layer of air around in the same direction as the spin of the ball (compare Figure 11-23). As a result, the net speed of air relative to the ball is slower on the right-hand side of the ball than on the left-hand side. Bernoulli's principle tells us that a speed difference corresponds to a pressure difference, and the slower speed to the right of the ball means that there is higher pressure there. This pressure difference means that as viewed from above, this baseball feels a force to the left and will curve to the left!

BioMedical Bernoulli's principle can help baseball pitchers win games; it also helps fish such as mackerel stay alive. A mackerel swims with its mouth open, allowing water to enter through the mouth, pass through the gills where oxygen is extracted, and exit through an aperture called the *operculum*. The streamlines of water flow around the mackerel to follow the contours of its body. Figure 11-34 shows that these streamlines are close together near the operculum. As we learned in Section 11-9, flow in an incompressible fluid such as water is rapid where streamlines are close together but slow where the streamlines are far apart. Hence the water flow is *faster* at the operculum than at the mouth, and so according to Bernoulli's principle the pressure at the operculum is *lower* than at the mouth. This pressure difference helps to force water from mouth to operculum by way of the mackerel's gills. This effect, called *ram ventilation*, is essential to the mackerel's ability to extract sufficient oxygen from the water. Other species of fish use ram ventilation only at high swimming speeds; at lower speeds, where ram ventilation is less effective, they use muscular action to pump water through their gills.

As the ball rotates counterclockwise (in this picture) a layer of air close to the ball is also dragged around counterclockwise. This boundary layer moves in the same direction as the air rushing past the ball on the left but opposes the airflow on the right.

Motion of ball from the pitcher toward the batter

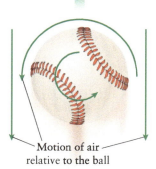

Motion of air relative to the ball

 According to Bernoulli's equation, higher net air speed on the left results in lower pressure. The pressure difference between the right and left sides of the ball results in a net force, so this ball will curve to the left!

Figure 11-33 What makes a curveball curve If a baseball rotates, viscosity drags a layer of air around the ball in the same direction as the ball's rotation.

GOT THE CONCEPT? 11-10 Bernoulli in Action

 Imagine holding two pieces of paper vertically, with a small gap between them, and then blowing gently into the gap. Would you expect the pieces of paper to (a) be drawn together, (b) be pushed apart, or (c) be unaffected? (Try it!)

TAKE-HOME MESSAGE FOR Section 11-10

✔ Bernoulli's equation relates the pressure, speed, and height of a fluid whose flow is *laminar, steady, incompressible, inviscid,* and *irrotational.*

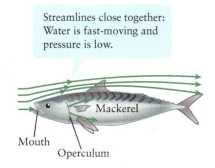

Streamlines close together: Water is fast-moving and pressure is low.

Mouth   Operculum   Mackerel

Figure 11-34 Bernoulli's principle and fish respiration Mackerel breathe by letting water flow from the mouth through the gills (where oxygen is extracted from the water) and out the operculum. Bernoulli's principle tells us that the water pressure at the operculum is lower when the fish is in motion, thus facilitating flow through the gills.

11-11 Viscosity is important in many types of fluid flow

We saw in Section 11-10 that Bernoulli's equation holds true only with certain very special assumptions. In particular we saw that the fluid had to have a little bit of viscosity to make the laminar flow be irrotational. But we couldn't allow the fluid to have too much viscosity, since that would mean that *viscous forces*—that is, forces on fluid due to friction—would play an important role in determining the acceleration of a bit of moving fluid.

Water exiting hose

Faucet end of hose

Open end of hose: Pressure is actually lower here than at the faucet.

Figure 11-35 When Bernoulli's equation doesn't work A horizontal garden hose attached to a faucet. If Bernoulli's equation held true, the water pressure would be the same everywhere along the length of the hose. Bernoulli's equation does not apply here because viscous forces are important in this situation.

For many situations involving fluid flow, there is "too much" viscosity, and Bernoulli's equation is *not* valid. One example is water flowing in a garden hose (Figure 11-35). The hose has the same cross-sectional area A throughout its length, so to satisfy the equation of continuity (which says that the product Av is a constant), the flow speed v must be the same at all points along the hose's length. Furthermore, if the hose is horizontal, then all points along the hose are at the same height y. If Bernoulli's equation, Equation 11-27 (which says that $p + \frac{1}{2}\rho v^2 + \rho gy$ is a constant), applied to this situation, then the water pressure p would also have to have the same value everywhere inside the hose. But in fact the pressure at the open end of the hose is substantially *lower* than at the end attached to the faucet. The reason is that frictional forces oppose the flow of water through the hose, so there has to be additional pressure at the faucet to sustain the flow.

The human circulatory system is another situation where Bernoulli's equation doesn't hold. The speed of blood flow in capillaries (about 10^{-3} m/s) is hundreds of times slower than in the aorta (about 0.2 m/s), so the term $\frac{1}{2}\rho v^2$ in Equation 11-27 has a much smaller value in the capillaries. Bernoulli's equation would therefore predict that blood pressure is much higher in the capillaries than in the aorta. In fact, the pressure is higher in the *aorta*; this is necessary to push against the frictional forces that the blood encounters on its way to the capillaries.

Let's be more quantitative about viscosity and its consequences. Figure 11-36 shows one way that physicists measure the viscosity of a fluid. Two glass plates, each of area A, are separated by a distance d and have a quantity of fluid filling the space between them. The lower plate is held stationary while the upper plate is pulled sideways at a constant speed v. Experiment shows that if the speed v is not too great, the fluid at the top remains attached to the upper

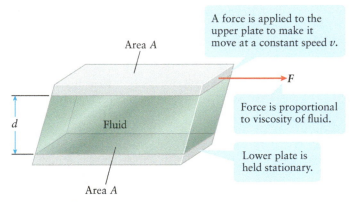

A force is applied to the upper plate to make it move at a constant speed v.

Area A

F

d Fluid

Area A

Force is proportional to viscosity of fluid.

Lower plate is held stationary.

Figure 11-36 Measuring viscosity An experiment to measure the viscosity of a fluid.

plate and the fluid at the bottom remains attached to the lower plate, while the fluid in between slides sideways in sheets like a deck of cards. The force (magnitude F) that must be applied to the upper plate to make the fluid move in this way is relatively large for a viscous fluid like oil but relatively low for a fluid like water. The force magnitude F is also proportional to the speed v and the area A but inversely proportional to the plate spacing d. We can summarize these observations in the following expression for F, which defines what we mean by the *viscosity* η (the Greek letter "eta") of a fluid:

Definition of viscosity
(11-28)

Force required to pull the upper plate in Figure 11-36

Viscosity of the fluid

Area of either plate in Figure 11-36

$$F = \frac{\eta A v}{d}$$

Spacing between the plates

Speed of one plate relative to the other

Equation 11-28 says that the greater the viscosity η, the greater the force that must be applied to move one plate relative to the other. The units of viscosity are pascals times seconds, or Pa·s for short. Viscosities are also sometimes given in *poise*, named for the nineteenth-century French scientist Jean Poiseuille; 1 poise equals 0.1 Pa·s.

Table 11-2 lists values of viscosity for several common fluids. Note that the viscosity of a fluid generally depends on temperature; liquids such as maple syrup and motor oil become less viscous and so flow more easily at higher temperatures.

TABLE 11-2 Viscosities of Various Substances

Substance	Viscosity (Pa·s)
air at 0°C	1.709×10^{-5}
air at 20°C	1.808×10^{-5}
air at 40°C	1.904×10^{-5}
fresh water at 0°C	1.787×10^{-3}
fresh water at 20°C	1.002×10^{-3}
fresh water at 40°C	6.53×10^{-4}
seawater at 20°C	1.072×10^{-3}
mercury (Hg) at 20°C	1.554×10^{-3}
blood at 37°C	0.003–0.004
typical motor oil at 20°C	0.1
glycerin at 20°C	1.490
chocolate syrup at 20°C	1.5
ketchup at 20°C	5.1
honey at 20°C	10
peanut butter at 20°C	25.5

Reynolds Number: Comparing Viscous Forces and Forces Due to Pressure Differences

The experiment shown in Figure 11-36 is useful for measuring viscosity because the only forces making the fluid move are the frictional (viscous) forces that act between parts of the fluid and between the fluid and plates. In most real-life situations, however, there are also forces due to pressure differences acting on the fluid (for example, the pressure difference between the two ends of a garden hose as you water the lawn). To describe the relative importance of forces due to pressure difference and viscous forces in fluid flow, physicists use a dimensionless quantity—that is, a number with no units—called the **Reynolds number**:

Reynolds number
(11-29)

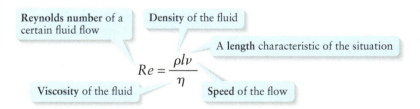

Reynolds number of a certain fluid flow

Density of the fluid

A length characteristic of the situation

$$Re = \frac{\rho l v}{\eta}$$

Viscosity of the fluid

Speed of the flow

Roughly speaking, the Reynolds number Re (*not* R_e) is the ratio of the forces due to pressure differences acting on a small piece of the fluid to the viscous forces acting on the same piece of fluid. Because the viscosity η is in the denominator in this expression, the Reynolds number increases as the viscosity *decreases*, and vice versa. Thus Re has a large value if viscosity is relatively unimportant, but it has a small value if viscosity plays a dominant role.

Definition of the length l in Equation 11-29 depends on the situation. For a fluid flowing in a tube, such as water in a pipe or air through your nostrils, l could be the diameter of the tube; the narrower the pipe, the smaller the Reynolds number, the more important the role of viscosity, and the less likely it is that Bernoulli's equation will be valid.

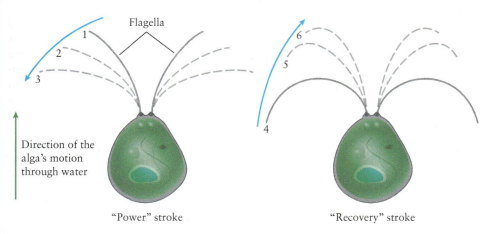

Flagella

1
2
3

Direction of the
alga's motion
through water

"Power" stroke

6
5
4

"Recovery" stroke

**Figure 11-37 Viscous forces
on swimming algae** (a) The tiny
flagellate *Chlamydomonas*, which
is just a few micrometers in
diameter, moves forward through
water during a power stroke.
(b) Due to the viscosity of water,
Chlamydomonas actually moves
backward during the recovery
stroke.

For a fluid flowing *around* an object, such as air around a bird's wing or water around a fish, the length *l* could be the front-to-back dimension of the wing or fish. Hence the Reynolds number *Re* is smaller, and viscosity more important, for a small organism moving through water than for a larger organism moving through water at the same speed. If the small organism moves at a slow speed v, the Reynolds number is even smaller and viscosity can become extremely important. As an example, the viscosity of water is relatively unimportant for a human doing the breaststroke underwater. The swimmer accelerates forward when she pulls her arms back in a "power stroke," pushing against the water, and continues forward even when she pulls her arms forward in the "recovery stroke." By contrast, viscosity is a huge effect for the single-celled alga *Chlamydomonas*, which propels itself slowly through water in a breaststroke-like way using two flagella (Figure 11-37). When the cell pulls its flagella back in the "recovery stroke," the cell actually *backs up* by about a third of the distance that it moves forward during a "power stroke." You can better empathize with the plight of *Chlamydomonas* by imagining yourself swimming in a tub of molasses!

Bio Medical

Low Reynolds Number: Laminar Flow

If the viscosity of a flowing fluid is sufficiently high, and hence *Re* is sufficiently low, the flow will be laminar but will not obey Bernoulli's equation. An important example of this is the flow of fluids within pipes, such as the flow of vaccine in a hypodermic needle or of blood in your circulatory system. (Blood vessels can curve and can vary in diameter along their length, but for simplicity we'll consider straight pipes with circular cross sections.) To sustain a steady flow there must be a pressure difference between the ends of the pipe; fluid flows from the high-pressure end to the low-pressure end. That's why a hypodermic syringe is equipped with a plunger; the force of a thumb on the plunger provides the necessary high pressure on one end of the needle.

Figure 11-38 shows the laminar flow of fluid within a circular pipe seen in cross section. Due to friction the fluid that's right up against the inside walls of the pipe doesn't move at all. This is the *no-slip condition* that we described in Section 11-9. The fluid a small distance from the walls moves a little, the fluid a bit farther away moves a bit faster, and so on. The fastest flow is at the very center of the pipe. This flow pattern is called *parabolic* (since the dashed curve in Figure 11-36 is a parabola lying on its side). A derivation using calculus gives the following expression, called the **Hagen–Poiseuille equation**, relating the pressure difference between the two ends of the pipe to the resulting flow rate. (Non-French speakers can pronounce Poiseuille as "pwa-zoo-yah.")

Cross section
of a pipe

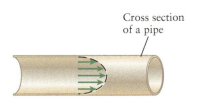

**Figure 11-38 Laminar flow of
a viscous liquid in a circular pipe**
The viscous fluid moves fastest
at the center of the pipe and does
not move at all at the pipe walls.

Volume flow rate of
fluid through a pipe

Radius of the pipe

Pressure difference between
the two ends of the pipe

$$Q = \frac{\pi R^4}{8\eta L}\,\Delta p$$

Viscosity of the fluid

Length of the pipe

Hagen–Poiseuille equation for
laminar flow of a viscous fluid
(11-30)

The longer the pipe (greater L) and the smaller its radius R, the less the volume flow rate Q caused by a given pressure difference Δp. You can understand this by comparing a volume of fluid that fills the inside of a short, wide pipe to the same volume of fluid that fills a long, narrow pipe. More of the fluid in the long, narrow pipe is in contact with the pipe walls and hence is not in motion, so the flow rate is less.

Equation 11-30 states that for the same pressure difference Δp, using a slightly wider pipe gives a much larger flow rate. Doubling the radius increases the flow rate by a factor of $2^4 = 16$! This equation also tells us that if we want to get the same volume flow rate Q through a narrower pipe (smaller R), we have to apply a greater pressure difference Δp between the two ends of the pipe. This helps explain how a diet high in cholesterol can lead to *hypertension*, or high blood pressure; cholesteric plaque building up within arteries decreases their inner radius R. Also, because the plaque buildup makes the arteries less elastic, hardened arteries expand less than healthy ones to accommodate blood pumped into them by the heart. Therefore, the heart must supply higher pressure to sustain the same blood flow.

High Reynolds Number: Turbulent Flow

Suppose fluid is moving through a circular pipe with the flow pattern shown in Figure 11-38. If there's a small imperfection in the pipe walls (biological pipes are never perfectly circular) or if something should disturb the flow, what happens? If the Reynolds number is low enough (that is, if the flow is sufficiently slow or the viscosity sufficiently high), the answer is "not much": The flow pattern is deformed a little, but thanks to viscous forces the pattern recovers its parabolic shape a short distance downstream of the disturbance. If the Reynolds number is greater than a certain value, however, the flow loses its regularity and becomes *turbulent* like the smoke trails in Figure 11-19b. For fluid flow in circular pipes, the critical value of the Reynolds number is about $Re = 2000$, where l in Equation 11-29 is the diameter (not the radius) of the pipe.

Unlike laminar flow, turbulent flow is messy as well as difficult to describe mathematically. It is also very common in nature. The flow of air around the wing of a gliding bird is laminar near the wing's leading edge but becomes turbulent as the air moves along the wing's upper surface. The effect is more pronounced if the bird glides at a slower speed, in which case its wings are inclined at a greater angle to the direction of flight to provide enough lift. The turbulent air moving over the wing's upper surface can cause individual wing feathers to deflect upward (Figure 11-39). If the bird flies too slowly and tilts the wing up too sharply, the air on the upper surface becomes so turbulent that it no longer follows the contours of the wing. Since the wing produces lift by forcing air downward, this onset of turbulence means that the lifting force of the wing decreases dramatically (a phenomenon called an *aerodynamic stall*) and the bird drops suddenly. Birds recover from this drop by dipping their wings slightly to restore a more laminar flow. Airplane wings are subject to the same effect, which pilots deal with in the same way.

Blood flow in the human circulatory system is often approximately laminar. But the Reynolds number in the larger blood vessels is close to the critical value that separates laminar and turbulent flows. Hence a relatively small obstruction in the aorta—the large artery that carries blood from the heart to the rest of the body—can trigger turbulence. Turbulent flow in the aorta generates noise, which is one thing that a physician listens for (and hopes not to hear) when using a stethoscope to check your heart. A totally benign noise caused by turbulence is the sound you hear on a windy day as turbulent air blows past your ears.

BioMedical

Turbulent flow over the top of the bird's wing makes these feathers deflect upward.

iStockphoto/Thinkstock

Figure 11-39 Turbulent airflow over a buzzard's wings The airflow over the top of this buzzard's wings is largely turbulent, causing some of the small feathers near mid-span to be pushed up.

BioMedical

EXAMPLE 11-17   Laminar or Turbulent?

Is the flow of water at 20°C laminar or turbulent if (a) the pipe has radius 0.500 mm $= 5.00 \times 10^{-4}$ m and the average flow speed is 0.800 m/s; (b) the radius is 0.125 mm $= 1.25 \times 10^{-4}$ m and the average flow speed is 12.8 m/s? At this temperature the density of water is 998 kg/m^3.

Set Up

The key to solving this problem is the Reynolds number, which for flow in a pipe of circular cross section is given by Equation 11-29. In this equation l is the diameter of the pipe. The flow is laminar if Re is less than 2000; otherwise it is turbulent.

Reynolds number:

$$Re = \frac{\rho l v}{\eta} \qquad (11\text{-}29)$$

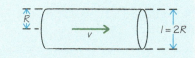

Solve

(a) Use the diameter of the pipe (twice the radius) and the flow speed to calculate the Reynolds number. We use Table 11-2 to find the viscosity of water.

$l = 2(5.00 \times 10^{-4} \text{ m}) = 1.00 \times 10^{-3} \text{ m}$

Reynolds number of the flow:

$$Re = \frac{(9.98 \times 10^2 \text{ kg/m}^3)(1.00 \times 10^{-3} \text{ m})(0.800 \text{ m/s})}{1.002 \times 10^{-3} \text{ Pa}\cdot\text{s}}$$

$$= 797$$

Since the Reynolds number is less than 2000, flow is laminar.

(b) Repeat the calculation for the new pipe diameter and flow speed.

$l = 2(1.25 \times 10^{-4} \text{ m}) = 2.50 \times 10^{-4} \text{ m}$

Reynolds number of the flow:

$$Re = \frac{(9.98 \times 10^2 \text{ kg/m}^3)(2.50 \times 10^{-4} \text{ m})(12.8 \text{ m/s})}{1.002 \times 10^{-3} \text{ Pa}\cdot\text{s}}$$

$$= 3.19 \times 10^3$$

In this case the Reynolds number is greater than 2000, so the flow is turbulent.

Reflect

Notice that compared to the pipe in (a), the pipe in (b) has 1/4 the radius and hence 1/16 the cross-sectional area (which is proportional to the square of the radius). Notice also that the average flow speed in (b) is 16 times that in (a). So the volume flow rate (the product of cross-sectional area A and flow speed v) is the *same* in both (a) and (b), but the flow is turbulent in the narrower pipe of (b). In a similar way an obstruction in a blood vessel (which reduces the vessel's radius) doesn't change the flow rate but does increase the Reynolds number and so can induce turbulence.

Volume flow rate in first pipe:

$$\begin{aligned}A_1 v_1 &= \pi R_1^2 v_1 \\ &= \pi (5.00 \times 10^{-4} \text{ m})^2 \\ &\quad \times (0.800 \text{ m/s}) \\ &= 6.28 \times 10^{-7} \text{ m}^3/\text{s}\end{aligned}$$

Volume flow rate in second pipe:

$$\begin{aligned}A_2 v_2 &= \pi R_2^2 v_2 \\ &= \pi (1.25 \times 10^{-4} \text{ m})^2 (12.8 \text{ m/s}) \\ &= 6.28 \times 10^{-7} \text{ m}^3/\text{s}\end{aligned}$$

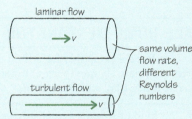

EXAMPLE 11-18 Viscous Laminar Flow in a Long, Narrow Pipe

Water at 20°C moves in laminar flow at an average flow speed of 0.800 m/s through a pipe of radius 0.500 mm = 5.00×10^{-4} m. The pipe is 2.00 m long. Determine what the pressure difference must be between the two ends of the pipe.

Set Up

We saw in part (a) of Example 11-17 that the flow is laminar for water at this temperature moving through such a pipe at this speed. Because the pipe is long and narrow, viscosity is important and laminar flow is an example of Hagen–Poiseuille flow. (The density and viscosity are given in Example 11-17.) Our goal is to find the pressure difference Δp between the ends of the pipe, and we use the Hagen–Poiseuille equation, Equation 11-30.

Hagen–Poiseuille equation for laminar flow of a viscous fluid:

$$Q = \frac{\pi R^4}{8\eta L} \Delta p \qquad (11\text{-}30)$$

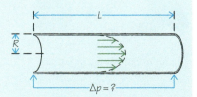

Solve

In Example 11-17 we calculated the volume flow rate.

Volume flow rate:

$$Q = 6.28 \times 10^{-7} \ \text{m}^3/\text{s}$$

Solve the Hagen–Poiseuille equation for the pressure difference Δp and substitute values.

$$\Delta p = \frac{8\eta L Q}{\pi R^4}$$

$$= \frac{8(1.002 \times 10^{-3} \ \text{Pa} \cdot \text{s})(2.00 \ \text{m})(6.28 \times 10^{-7} \ \text{m}^3/\text{s})}{\pi (5.00 \times 10^{-4} \ \text{m})^4}$$

$$= 5.13 \times 10^4 \ \text{Pa}$$

Reflect

Even though the viscosity of water is low, the pressure required to sustain the flow is substantial (about half an atmosphere) because the pipe is so narrow and long.

$$\Delta p = (5.13 \times 10^4 \ \text{Pa})\left(\frac{1 \ \text{atm}}{1.01325 \times 10^5 \ \text{Pa}}\right)$$

$$= 0.506 \ \text{atm}$$

GOT THE CONCEPT? 11-11 When Is Viscosity More Important?

 Two identical tubes have fluid flowing through them at the same speed. In one tube the fluid is air, and in the other tube the fluid is fresh water. In which tube is viscosity more important? (a) the tube with water; (b) the tube with air; (c) viscosity is equally important in each tube; (d) not enough information given to decide.

TAKE-HOME MESSAGE FOR Section 11-11

✔ The Reynolds number for a fluid flow is large if viscous forces are relatively unimportant in the flow and small if viscous forces play a dominant role.

✔ If a fluid flow has a sufficiently low Reynolds number, the flow is laminar. A sufficiently high Reynolds number means that the flow is turbulent.

✔ The Hagen–Poiseuille equation describes the laminar flow of a viscous fluid. This equation relates the pressure difference between the two ends of a pipe to the resulting flow rate.

11-12 Surface tension explains the shape of raindrops and how respiration is possible

Why does rain fall in droplets? Why does water roll off a duck's back? And why is it useful to use detergent when washing clothes? We can answer all of these questions by examining an important property of liquids called *surface tension*.

A liquid maintains an essentially constant volume because its molecules exert strong attractive forces on each other. As a result the molecules try to huddle close to each other. The potential energy associated with these forces is lowest if as many of the molecules as possible are completely surrounded by other molecules, and if as few as possible are on the surface (where they are surrounded by other molecules on one side only). Therefore, to minimize this potential energy, the molecules arrange themselves so as to minimize the surface area of the liquid (Figure 11-40). A sphere has the smallest surface area for a given volume, which is why raindrops are spherical. The drop behaves as though its surface were a membrane that resists being stretched, which is why this behavior is called **surface tension**.

An irregular blob of liquid tends to become spherical due to surface tension.

Figure 11-40 Surface tension The effect of surface tension on a liquid blob or drop.

WATCH OUT! Raindrops aren't shaped like "teardrops."

! It's a common belief that raindrops are teardrop-shaped, with a rounded lower half and a pointed upper half. But the aerodynamic forces that would force a raindrop into this shape are quite weak compared to the attractive forces between molecules. In fact, due to surface tension, raindrops are nearly spherical in shape.

A spherical drop of radius R has volume $(4\pi/3)R^3$ and surface area $4\pi R^2$. The importance of surface tension is proportional to the ratio of the drop's surface area (which is the size of the "membrane" enclosing the drop) to its volume (which tells you the total amount of fluid in the drop). This ratio is $4\pi R^2/[(4\pi/3)R^3] = 3/R$, which *increases* as the radius R *decreases*. Hence surface tension is more important for small drops than for large ones. If you pour enough liquid into a container, surface tension is overwhelmed by gravitational forces and the liquid assumes a shape that fills the container and has a horizontal upper surface. In zero gravity even large quantities of liquid form spherical drops (Figure 11-41).

The surface tension of water enables ducks to remain warm and dry even after submerging themselves in search of a meal. A duck's feather has a crisscross grid of tiny barbules that form a fine mesh (Figure 11-42a). To push through this mesh the water surface must stretch as shown in Figure 11-42b. Surface tension resists such stretching of the surface "membrane," so water does not penetrate the mesh, and the duck's skin remains dry. The same effect makes it difficult for water to penetrate between the closely spaced fibers of clothing. That's why clothing is best washed using a detergent, which is a substance that reduces the surface tension of water.

Figure 11-41 Surface tension on a weightless water drop In the apparent weightlessness of the orbiting International Space Station, surface tension allows this astronaut to squeeze water out of her beverage container and form a spherical drop about 3 cm in diameter.

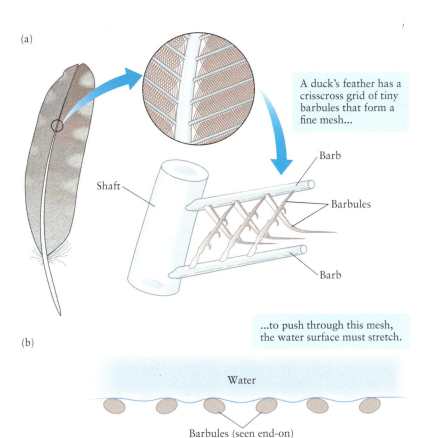

(a)

A duck's feather has a crisscross grid of tiny barbules that form a fine mesh...

Barb

Shaft

Barbules

Barb

(b)

...to push through this mesh, the water surface must stretch.

Water

Barbules (seen end-on)

Figure 11-42 Surface tension and bird feathers (a) This drawing of a bird feather shows the tiny interlocking barbules. (b) Due to surface tension, water cannot easily pass through the tiny spaces between barbules.

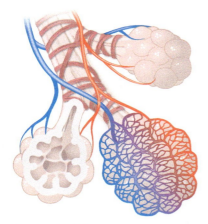

Figure 11-43 **Alveoli** This illustration shows a cluster of the tiny balloon-like sacs called alveoli. A human lung contains about 3×10^8 alveoli.

BioMedical Surface tension also plays an important role in human respiration. In the lungs oxygen exchange takes place across the walls of tiny, balloon-like structures called *alveoli* (Figure 11-43). The inside surface of the alveoli is wet, and the surface tension of this fluid generates a pressure that tends to pull the walls of the alveoli inward. The magnitude of this pressure is proportional to the surface tension and inversely proportional to the radius of the alveolus:

pressure = (2 × surface tension)/(alveolar radius)

The problem is that the alveoli do not all have the same radius, so smaller alveoli experience a greater pressure than larger ones. As a result, small alveoli would tend to collapse into larger ones if not for a substance called a *surfactant* that coats the inside surface of each alveolus. This remarkable substance decreases surface tension as a function of its concentration in the fluid lining the inside of the alveoli. As the radius of the alveolus decreases during exhalation, the concentration of surfactant increases on the surface of the fluid lining the alveolus and therefore decreases surface tension and the resulting pressure in the alveolus. This prevents smaller alveoli from collapsing into larger ones. Also, during inhalation the reduced surface tension makes it easier to inflate the lungs with the relatively small pressure difference between inhaled air and the low-pressure air in the thoracic cavity—about 130 Pa, or 0.13% of atmospheric pressure. A condition called respiratory distress syndrome occurs in premature infants whose lungs are deficient in surfactant; the increased surface tension pressure can make their alveoli collapse. Administering a replacement surfactant helps these newborn babies survive.

TAKE-HOME MESSAGE FOR **Section 11-12**

✔ The molecules of a liquid minimize their energy by surrounding themselves with their fellow molecules. As a result, liquids tend to form spherical drops.

✔ The surface of a liquid drop resists being stretched. This behavior is called surface tension.

Key Terms

absolute pressure	gas	pressure
apparent weight	gauge pressure	Reynolds number
Archimedes' principle	Hagen–Poiseuille equation	solid
atmosphere (unit of pressure)	hydrostatic equilibrium	specific gravity
Bernoulli's equation	incompressible fluid	steady flow
Bernoulli's principle	inviscid flow	streamline
boundary layer	irrotational flow	surface tension
buoyant force	laminar flow	torr
compressible fluid	liquid	uniform density
density	millimeters of mercury	unsteady flow
equation of continuity	no-slip condition	turbulent flow
equation of hydrostatic equilibrium	pascal	viscosity
fluid	Pascal's principle	volume flow rate

Chapter Summary

Topic	Equation or Figure

Fluids: The molecules in a fluid are not arranged in any organized structure. A fluid does not have any shape of its own, and flows to conform to the shape of the container in which it is placed. A quantity of liquid maintains nearly the same volume, while a quantity of gas expands to fill the volume available.

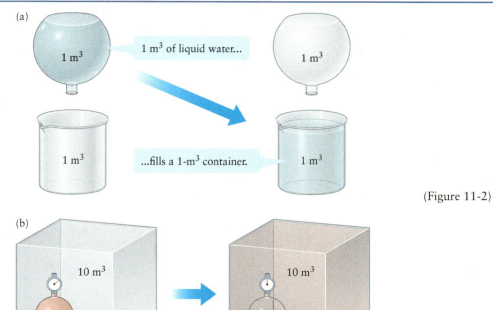

(a) 1 m³ of liquid water... ...fills a 1-m³ container.

(b) 1 m³ of compressed oxygen gas... ...expands to fill a much larger volume.

(Figure 11-2)

Density: The density ρ of a substance is its mass per volume. Density depends on how closely packed the molecules of the substance are, as well as on the mass of the individual molecules.

The **density** ρ ("rho") of a certain substance... ...equals the **mass** m of a given quantity of that substance...

$$\rho = \frac{m}{V}$$

...divided by the **volume** V of that quantity of the substance.

(11-1)

Pressure: The pressure p exerted by a fluid is the perpendicular force it exerts on a given area in the fluid divided by the area. Unlike force, pressure is not a vector. Pressure measured relative to atmospheric pressure is called gauge pressure.

The **pressure** p exerted by a fluid on an object in the fluid... ...equals the **force** $F_\perp$ that the fluid exerts **perpendicular to the face of the object**...

$$p = \frac{F_\perp}{A}$$

...divided by the **area** A of that face.

(11-2)

Fluids at rest: A stationary fluid must be in hydrostatic equilibrium so that the sum of the forces on any parcel of fluid is zero. In a fluid of uniform density, the pressure at a given point increases with depth to support the weight of the fluid above that point.

Pressure at a certain point in a fluid at rest | Density of the fluid (same at all points in the fluid)

Pressure at a second, lower point in the fluid | Acceleration due to gravity

$$p = p_0 + \rho g d$$

(11-8)

Depth of the second point where the pressure is p below the point where the pressure is p_0

Pressure difference: If there are different pressures on the two sides of an object, the different pressures produce a net force on the object.

A pressure difference on two opposite sides of an object **produces a net force.**

Area of each side of the object

$$F_{net} = (p_2 - p_1)\,A \tag{11-10}$$

High pressure on one side of the object

Lower pressure on the other side of the object

Pascal's principle: Pressure applied to a confined, static fluid is transmitted undiminished to every part of the fluid as well as to the walls of the container. A hydraulic jack, which converts a small force acting over a large distance into a large force acting over a small distance, relies on Pascal's principle.

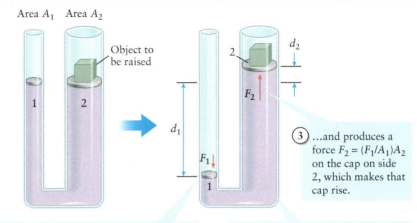

Area A_1 Area A_2

Object to be raised

③ ...and produces a force $F_2 = (F_1/A_1)A_2$ on the cap on side 2, which makes that cap rise.

① When a downward force F_1 is applied to the cap on side 1, which has area A_1...

② ...the increase in pressure $\Delta p = F_1/A_1$ is transmitted throughout the fluid...

(Figure 11-13)

Buoyancy: An object in a fluid feels an upward buoyant force F_b equal to the weight of fluid that the object displaces. A submerged object will neither sink nor rise if its average density equals that of the fluid.

Magnitude of the buoyant force on an object in a fluid

Density of the fluid

$$F_b = \rho_{fluid}\,V_{displaced}\,g \tag{11-13}$$

Volume of fluid that the object displaces

Acceleration due to gravity

Fluid flow: The motion of a fluid is classified as being steady or unsteady; as laminar or turbulent; and as viscous (with internal friction) or inviscid (without internal friction).

(b)

(b)

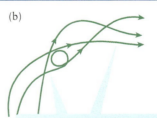

Streamlines are the paths followed by bits of fluid in laminar flow. Each bit of fluid follows the one in front and moves along the same streamline.

There are no streamlines for a fluid in turbulent flow. While each bit of fluid follows its own path, the one behind it can follow a completely different path.

(Figure 11-21b)

(Figure 11-22b)

Equation of continuity: If an incompressible (constant-density) fluid flows through a pipe, the size of the pipe and the flow speed may change, but the volume flow rate (the cross-sectional area of the flow multiplied by the flow speed) remains the same.

Cross-sectional area of the flow at point 1

Flow speed at point 1

$$A_1 v_1 = A_2 v_2 \tag{11-19}$$

Cross-sectional area of the flow at point 2

Flow speed at point 2

Bernoulli's principle and Bernoulli's equation: The pressure in a moving fluid is reduced at locations where the fluid moves rapidly. The mathematical statement of this principle, called Bernoulli's equation, is accurate only for idealized laminar, inviscid flows.

Fluid pressure at a given point in the fluid

Speed of the fluid at that point

Vertical coordinate of that point

$$p + \frac{1}{2}\rho v^2 + \rho g y = \text{a constant with the same value throughout the fluid}$$ (11-27)

Density of the fluid (uniform throughout the fluid)

Acceleration due to gravity

Viscosity: The viscosity η of a fluid is a measure of how much friction is present when different parts of the fluid slide past one another. One way to measure viscosity is to see how much force is required to slide one plate relative to another, parallel plate if there is fluid between the plates.

Force required to pull the upper plate in Figure 11-36

Viscosity of the fluid

Area of either plate in Figure 11-36

$$F = \frac{\eta A v}{d}$$ (11-28)

Spacing between the plates

Speed of one plate relative to the other

Turbulent viscous flow: The flow of a viscous liquid becomes turbulent if the Reynolds number Re of the flow is too large. This happens if the flow speed is too great, the characteristic dimension of the flow (for example, the diameter of the pipe through which the fluid moves) is too large, or both.

Reynolds number of a certain fluid flow

Density of the fluid

A length characteristic of the situation

$$Re = \frac{\rho l v}{\eta}$$ (11-29)

Viscosity of the fluid

Speed of the flow

Laminar viscous flow: To make a viscous fluid flow through a pipe, there must be a pressure difference between the two ends of the pipe. A greater pressure difference is required for long, thin pipes than for short, broad ones.

Volume flow rate of fluid through a pipe

Radius of the pipe

Pressure difference between the two ends of the pipe

$$Q = \frac{\pi R^4}{8 \eta L} \Delta p$$ (11-30)

Viscosity of the fluid

Length of the pipe

Surface tension: Attractive forces between the molecules of a liquid make it favorable for the liquid to minimize its surface area. This effect causes water to form droplets and makes it difficult to force water through small apertures.

An irregular blob of liquid tends to become spherical due to surface tension.

(Figure 11-40)

Answer to What do you think? Question

(c) The upward buoyant force on an object immersed in a fluid, such as a fish immersed in water, is equal in magnitude to the weight of the fluid that the object displaces. A life-size sculpture of a fish displaces the same volume of water, and hence the same weight of water, as an actual fish. So the buoyant force would not change if you replaced the fish with a sculpture of the same size and shape, even if the sculpture is made of iron. The difference is that a fish's weight is about the same as the upward buoyant force exerted by the surrounding water, so a fish can float with little effort. By contrast, the weight of an iron fish sculpture is much greater than the buoyant force, so the sculpture would sink. (See Section 11-8.)

Answers to Got the Concept? Questions

11-1 (d) If you chose the block of iron, be careful! Table 11-1 tells us that iron has a much greater density ρ (7.8×10^3 kg/m^3) than does ice (0.917×10^3 kg/m^3), but the mass m (and hence the weight) of each block depends on its volume V. From Equation 11-1, $m = \rho V$. If the two blocks have the same volume, the block of iron will weigh more than the block of ice. But if the block of ice is the size of a house while the block of iron is the size of your fingernail, the ice will definitely outweigh the iron.

11-2 (e) Remember that, unlike force, pressure is *not* a vector and so has no direction. Therefore any statement that refers to the direction of pressure is an incorrect and misleading one! The pressure on each side of the fish is the same.

11-3 (b) In hydrostatic equilibrium the pressure in the water depends only on the vertical position. To convince yourself of this assertion, let p_0 be the pressure at the bottom of the left-hand leg, and p be the pressure at the bottom of the right-hand leg. These two pressures are related by Equation 11-8, $p = p_0 + \rho g d$. Since both points are at the same height, $d = 0$, and $p = p_0$. That is, the pressure p at the bottom of the left-hand tube must be the *same* as the pressure at the bottom of the right-hand tube, since these points are at the same level. (If the pressure were higher on one side or the other, water would flow from the region of high pressure to the region of low pressure until the pressures equalized.) To see that the height of each column is the same, now let the pressure at the top of each column of water, which is equal to air pressure, be written as p_0. Since the pressure p_0 at the top of the water column is the same on both sides and the pressure p at the bottom of the water column is the same on both sides, it follows from $p = p_0 + \rho g d$ that the depth of water d must be the same on both sides as well.

11-4 (d), (b), (c), (a). The deflated balloon has air at atmospheric pressure inside, so its gauge pressure is zero and its absolute pressure is about 1.0 atm. The inflated balloon has a gauge pressure of 0.50 atm, and its absolute pressure is about (0.50 atm) + (1.0 atm) = 1.5 atm.

11-5 (a) We know from Section 11-4 that pressure increases with increasing depth in a fluid, including the atmosphere. So the air pressure on the bottom of the paper is slightly greater than the pressure on the top, and the net force due to these pressures is upward (although very small). It's true that the sum of *all* forces acting on the paper—including the force of gravity and the force exerted by your hands—is zero, because the paper is at rest. But that's not what the question is asking.

11-6 (c) The cap on side 2 has 5 times the radius and so 25 times the area of the cap on side 1 (that is, $A_2/A_1 = 25$). So from Equation 11-11 the force on side 2 (F_2) is 25 times greater than the force on side 1 (F_1). But from Equation 11-12, the displacement of the cap on side 2 (d_2) is only 1/25 as much as the displacement of the cap on side 1 (d_1). Since work equals the product of force and displacement, the work done on side 2 is the *same* as the work done on side 1, because $25 \times (1/25) = 1$. The hydraulic jack doesn't increase the amount of work that you can do: It just makes it possible for you to do that work by applying a relatively small force F_1.

11-7 (a) Regardless of whether the coin is on the top or bottom of the cube, since the cube and coin are floating, the buoyant force of the water displaced must equal the weight of the cube plus the weight of the coin. (If that weren't true, the cube and coin would sink.) So the amount of water displaced is the same in both cases. However, when the coin is on the top, only the cube is displacing water, but when the coin is on the bottom, both the cube and the coin displace water. In this second case, then, the volume of the coin is contributing to the buoyant force, which means that the cube itself won't displace as much water as when the buoyant force was due totally to the amount of the cube that was submerged. For that reason, more of the cube will be above the surface of the water when the coin is taped to the bottom, and the cube will sit higher in the water.

11-8 (b) When the bottle was floating it displaced a volume of water equal to its weight. That's because to float, the buoyant force must equal the bottle's weight, and according to Archimedes' principle, the buoyant force is also equal to the weight of the displaced water. However, when the bottle has sunk to the bottom of the tub, the volume of water it displaces is equal only to the volume of the glass that forms the bottle. Since the bottle sank, we know that the glass must be denser than water, which means the glass itself takes up less volume than the equivalent weight of water. The net result is that when the bottle sinks, the water level goes down.

11-9 (c) According to the equation of continuity, the product of the cross-sectional area of a flow of fluid and the speed of flow is constant. For that reason, as the speed of the flow increases, the cross-sectional area must decrease. This "neck-down" effect is visible even for height differences of about a meter and is quite noticeable in waterfalls before the flow turns turbulent.

11-10 (a) According to Bernoulli's principle when two regions in a flowing fluid have different flow speed, the pressure must be lower in the region in which the flow is faster. In the experiment in which you blow between two papers, because air flows faster between the papers than in the region surrounding the papers, the pressure between them is lower than in the surrounding region, causing the papers to be drawn together—not pushed apart.

11-11 (b) The Reynolds number $Re = \rho l v/\eta$ is a measure of the relative importance of pressure forces and viscous forces in a flowing fluid. The smaller the value of Re, the more important viscosity is. The two tubes have the same dimensions (so the length l is the same), and the fluid flows at the same speed (so v is the same). Hence what's different between the two flows is the value of ρ/η, the ratio of density to viscosity. From Tables 11-1 and 11-2, the values of this ratio for the two fluids are as follows.

Water at 20°C:

$$\rho/\eta = (998 \text{ kg/m}^3)/(1.002 \times 10^{-3} \text{ Pa} \cdot \text{s})$$
$$= 9.96 \times 10^5 \text{ kg}/(\text{m}^3 \cdot \text{Pa} \cdot \text{s})$$

Air at around 15 to 20°C:

$$\rho/\eta = (1.23 \text{ kg/m}^3)/(1.808 \times 10^{-5} \text{ kg/m}^3)$$
$$= 6.80 \times 10^4 \text{ kg}/(\text{m}^3 \cdot \text{Pa} \cdot \text{s})$$

The value of ρ/η for air is less than one-tenth of the value for water, so viscous effects are actually more important for air than for water.

Questions and Problems

In a few problems you are given more data than you actually need; in a few other problems you are required to supply data from your general knowledge, outside sources, or informed estimate.

Interpret as significant all digits in numerical values that have trailing zeros and no decimal points.

For all problems use $g = 9.80$ m/s^2 for the free-fall acceleration due to gravity. Neglect friction and air resistance unless instructed to do otherwise.

•	Basic, single-concept problem
••	Intermediate-level problem; may require synthesis of concepts and multiple steps
•••	Challenging problem
Example	See worked example for a similar problem

Conceptual Questions

1. • A dam holds back a very long, deep lake. If the lake were twice as long, but still just as deep, how much thicker would the dam need to be? Why?

2. • **Medical** When you cut your finger badly, why might it be wise to hold it high above your head?

3. • **Medical** When you donate blood, is the collection bag held below or above your body? Why?

4. • **Medical** Usually blood pressure is measured on the arm at the same level as the heart. How would the results differ if the measurement were made on the leg instead?

5. • **Medical** After sitting for many hours during a trans-Pacific flight, a passenger jumps up quickly as soon as the "fasten seat belt" light is turned off and immediately falls to the floor, unconscious! Within seconds of being in the horizontal position, however, consciousness is restored. What happened?

6. • A dam will be built across a river to create a reservoir. Does the pressure in the reservoir at the base of the dam depend on the shape of the dam? Explain your answer.

7. • Two identically shaped containers in the shape of a truncated cone are placed on a table, but one is inverted such that the small end is resting on the table. The containers are filled with the same height of water. The pressure at the bottom of each container is the same. However, the weight of the water in each container is different. Explain why this statement is correct.

8. • An ice cube floats in a glass of water so that the water level is exactly at the rim. After the ice cube melts will all the water still be in the glass? Explain your answer.

9. • Aluminum is more dense than plastic. You are given two identical, closed, and opaque boxes. One contains a piece of aluminum, and the other contains a piece of plastic. Without opening the boxes is it possible to tell which box contains the plastic and which box contains the aluminum? Explain your answer.

10. • The salinity of the Great Salt Lake varies from place to place ranging from about two to as much as eight times the salinity of ocean water. Is it easier or harder for a person to float in the Great Salt Lake compared to floating in ocean water? Why?

11. • A wooden boat floats in a small pond, and the level of the water at the edge of the pond is marked. Will the level of the water rise, fall, or stay the same when the boat is removed from the pond?

12. • You are given two objects of identical size, one made of aluminum and the other of lead. You hang each object separately from a spring balance. Because lead is denser than aluminum, the lead object weighs more. Now you weigh each object while it is submerged in water. Will the difference between the measured weights of the aluminum object and the lead object be greater than, less than, or the same as it was when the objects were weighed in air? Explain your answer.

13. • A river runs through a wide valley and then through a narrow channel. How do the velocities of the flows of water compare between the wide valley and the narrow channel?

14. • Why does the stream of water from a faucet become narrower as it falls?

15. • The wind is blowing from west to east. Should landing airplanes approach the runway from the west or the east? Explain your answer.

16. • Use Bernoulli's equation to explain why a house roof is easily blown off during a tornado or hurricane.

17. • A cylindrical container is filled with water. If a hole is cut on the side of the container so that the water shoots out, what is the direction of the water flow the instant it leaves the container?

Multiple-Choice Questions

18. • Object A has density ρ_1. Object B has the same shape and dimensions as object A, but it is three times as massive. Object B has density ρ_2 such that
    A. $\rho_2 = 3\rho_1$.
    B. $\rho_2 = \dfrac{\rho_1}{3}$.
    C. $\rho_2 = \rho_1$.
    D. $\rho_2 = 2\rho_1$.
    E. $\rho_2 = \dfrac{\rho_1}{2}$.

19. • If the gauge pressure is doubled, the absolute pressure will
    A. be halved.
    B. be doubled.
    C. be unchanged.
    D. be increased, but not necessarily doubled.
    E. be decreased, but not necessarily halved.

20. • A toy floats in a swimming pool. The buoyant force exerted on the toy depends on the volume of
    A. water in the pool.
    B. the pool.
    C. the toy under water.
    D. the toy above water.
    E. none of the above choices.

21. • An object floats in water with 5/8 of its volume submerged. The ratio of the density of the object to that of water is
    A. 8/5.
    B. 5/8.
    C. 1/2.
    D. 2/1.
    E. 3/8.

22. • An ice cube floats in a glass of water. As the ice melts what happens to the water level?

   A. It rises.

   B. It remains the same.

   C. It falls by an amount that cannot be determined from the information given.

   D. It falls by an amount proportional to the volume of the ice cube.

   E. It falls by an amount proportional to the volume of the ice cube that was initially above the water line.

23. • Water flows through a 0.5-cm-diameter pipe connected to a 1-cm-diameter pipe. Compared to the speed of the water in the 0.5-cm pipe, the speed in the 1-cm pipe is

   A. one-quarter the speed in the 0.5-cm pipe.

   B. one-half the speed in the 0.5-cm pipe.

   C. the same as the speed in the 0.5-cm pipe.

   D. double the speed in the 0.5-cm pipe.

   E. quadruple the speed in the 0.5-cm pipe.

24. • **Medical** Blood flows through an artery that is partially blocked. As the blood moves from the wider region into the narrow region, the blood speed

   A. increases.

   B. decreases.

   C. stays the same.

   D. drops to zero.

   E. alternately increases and then decreases.

25. • Two wooden boxes of equal mass but different density are held beneath the surface of a large container of water. Box A has smaller average density than box B. When the boxes are released, they accelerate upward to the surface. Which box has the greater acceleration?

   A. box A

   B. box B

   C. They are the same.

   D. We need to know the actual densities of the boxes to answer the question.

   E. It depends on the contents of the boxes.

Problems

11-1 Liquids and gases are both examples of fluids

11-2 Density measures the amount of mass per unit volume

26. • Determine the mass of a cube of iron that is 2 cm × 2 cm × 2 cm in size. Example 11-1

27. • What is the radius of a sphere made of aluminum, if its mass is 24 kg? Example 11-1

28. • (a) Determine the average density of a cylinder that is 20 cm long and 1.0 cm in radius, and has a mass of 37 g. (b) Referring to Table 11-1, of what material might the cylinder be made? (c) What is the specific gravity of the material? Example 11-1

29. • How long are the sides of an ice cube if its mass is 0.35 kg? Example 11-3

30. • **Biology** Approximately 65% of a person's body weight is water. What is the volume of water in a 65-kg man? Example 11-3

31. • A regulation men's basketball has an inflated circumference of 75 cm and an uninflated mass of 623.69 g. What is the average density of the basketball when it is inflated with air at an absolute pressure of 1.544 atm? Air at this pressure has a density of 1.89 kg/m³. You may assume the vinyl shell has negligible thickness. Example 11-2

32. • A 20 ft × 8.0 ft × 8.5 ft shipping container has a mass of 2350 kg when empty and has about 33 m³ of cargo space. The container is 2/3 full of bags of cat litter that have a density of 0.54 g/cm³. What is the average density of the container? Will it float if it falls off a ship? Example 11-2

33. • **Astronomy** When a massive star reaches the end of its life, it is possible for a supernova to occur. This may result in the formation of a very small, but very dense, neutron star, the density of which is about the same as a neutron. Determine the radius of a neutron star that has the mass of our Sun and the same density as a neutron. A neutron has a mass of 1.7×10^{-27} kg and an approximate radius of 1.2×10^{-15} m. Example 11-3

34. • Earth has radius 6380 km and mass 5.98×10^{24} kg. Determine the average density of our planet. Example 11-1

11-3 Pressure in a fluid is caused by the impact of molecules

35. • **Biology** An elephant that has a mass of 3000 kg evenly distributes her weight on all four feet. (a) If her feet are approximately circular and each has a diameter of 50 cm, estimate the pressure on each foot. (b) Compare the answer in part (a) with the pressure on each of your feet when you are standing up. Make some rough but reasonable assumptions about the area of your feet. Example 11-4

36. • The head of a nail is 0.32 cm in diameter. You hit it with a hammer with a force of 25 N. (a) What is the pressure on the head of the nail? (b) If the pointed end of the nail, opposite to the head, is 0.032 cm in diameter, what is the pressure on that end? Example 11-4

11-4 In a fluid at rest pressure increases with increasing depth

37. • Calculate the absolute pressure at the bottom of a 0.25-m-tall graduated cylinder that is half full of mercury and half full of water. Example 11-5

38. • A 5.0 m × 10 m swimming pool is filled to a depth of 10 m. What is the pressure on the bottom of the pool? Example 11-5

39. • **Medical** What is the difference in blood pressure (mmHg) between the top of the head and bottom of the feet of a 1.75-m-tall person standing vertically? The density of blood is 1.06×10^3 kg/m³. Example 11-5

40. • At 25°C the density of ether is 713 kg/m³ and the density of ethanol is 789 kg/m³. A cylinder is filled with ethanol to a depth of 1.5 m. How tall would a cylinder filled with ether need to be so that the pressure at the bottom is the same as the pressure at the bottom of the cylinder filled with ethanol? Example 11-5

41. • Estimate the total hydrostatic *force* on the Hoover Dam. Assume that the top 50 m of the dam is above the surface of Lake Mead. Example 11-5

11-5 Scientists and medical professionals use various units for measuring fluid pressure

42. • A diver is 10.0 m below the surface of the ocean. The surface pressure is 1 atm. What are the absolute pressure and gauge pressure experienced by the diver? Example 11-6

43. • **Medical** Suppose that your pressure gauge for determining the blood pressure of a patient measured absolute pressure instead of gauge pressure. How would you write the normal value of systolic blood pressure, 120 mmHg, in such a case?

44. • Convert the following pressures to the SI unit of pascals (Pa), where $1\ Pa = 1\ N/m^2$: (a) 1500 kPa; (b) 35 psi; (c) 2.85 atm; (d) 883 torr.

45. • Assuming the atmospheric pressure is exactly 1 atm at sea level, determine the atmospheric pressure in Death Valley, California, 85 m below sea level. Example 11-6

46. • Elaine wears her wide-brimmed hat at the beach. If the atmospheric pressure at the beach is exactly 1 atm, determine the weight of the imaginary column of air that "rests" on her hat if its diameter is 45 cm. Example 11-6

47. • What is the *absolute* pressure in pascals (Pa) of the air inside a bicycle tire that is inflated to 65 psi?

11-6 A difference in pressure on opposite sides of an object produces a net force on the object

48. •• A rectangular swimming pool is 8.0 m × 35 m in area. The depth varies uniformly from 1.0 m in the shallow end to 2.0 m in the deep end. (a) Determine the pressure at the bottom of the deep end of the pool and at the shallow end. (b) What is the net force on the bottom of the pool due to the water in the pool? (Ignore the effects of the air above the pool for this part.) Example 11-7

49. • What is the net force on an airplane window of area $1000\ cm^2$ if the pressure inside the cabin is 0.95 atm and the pressure outside is 0.85 atm? Example 11-7

50. • Suppose that the hatch on the side of a Mars lander is built and tested on Earth so that the internal pressure just balances the external pressure. The hatch is a disk 50.0 cm in diameter. When the lander goes to Mars, where the external pressure is $650\ N/m^2$, what will be the net force (in newtons and pounds) on the hatch, assuming that the internal pressure is the same in both cases? Will it be an inward or an outward force? Example 11-7

51. • What is the net force on the walls of a 55-gal drum that is on the bottom of the ocean at a depth of 250 m? Assume the drum is a cylinder with a height of $34\frac{1}{2}$ in. and a diameter of $21\frac{5}{8}$ in. (Remember to convert inches to meters!) The interior pressure of the drum is exactly 1 atm. Neglect the variation in pressure over the height of the drum. Example 11-7

52. **Biology** An Andean condor with a wingspan of 300 cm and mass 12 kg soars along a horizontal path. Modeling its wings as a rectangle with a width of 30 cm, what must be the difference in pressure between the top and bottom surfaces of its wings, $p_{top} - p_{bottom}$? Example 11-8

11-7 A pressure increase at one point in a fluid causes a pressure increase throughout the fluid

53. • What is the maximum weight that can be raised by the hydraulic lift shown in Figure 11-44?

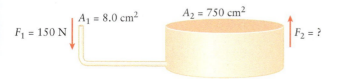

Figure 11-44 Problem 53

54. • A hydraulic lift is designed to raise a 9.00×10^2-kg car. If the "large" piston has a radius of 35.0 cm and the "small" piston has a radius of 2.00 cm, determine the minimum force exerted on the small piston to accomplish the task.

55. • (a) What force will the large piston provide if the small piston in a hydraulic lift is moved down as shown in Figure 11-45? (b) If the small piston is depressed a distance of Δy_1, by how much will the large piston rise? (c) How much work is done in pushing down the small piston compared to the work done in raising the large piston if $\Delta y_1 = 0.20$ m?

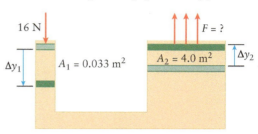

Figure 11-45 Problem 55

56. • A hydraulic lift has a leak so that it is only 75.0% efficient in raising its load. If the large piston exerts a force of 150 N when the small piston is depressed with a force of 15.0 N and the radius of the small piston is 5.00×10^{-2} m, what is the radius of the large piston?

11-8 Archimedes' principle helps us understand buoyancy

57. • A rectangular block of wood, 10.0 cm × 15.0 cm × 40.0 cm, has a specific gravity of 0.600. (a) Determine the buoyant force that acts on the block when it is placed in a pool of fresh water. *Hint:* Draw a free-body diagram labeling all of the forces on the block. (b) What fraction of the block is submerged? (c) Determine the weight of the water that is displaced by the block. Example 11-9

58. • A rectangular block of wood floats in fresh water with its lower 10.0 cm submerged. What distance will be submerged when it floats in seawater (specific gravity 1.025)? Example 11-9

59. •• A cube of side s is completely submerged in a pool of fresh water. (a) Derive an expression for the pressure difference between the bottom and top of the cube. (b) After drawing a free-body diagram, derive an algebraic expression for the net force on the cube. (c) What is the weight of the displaced water when the cube is submerged? Your expressions may include some or all of the following quantities: atmospheric pressure, the density of fresh water, the length of the side of the cube, the mass of the cube, and the acceleration due to gravity. Example 11-10

60. • A log raft is 3.00 m × 4.00 m × 0.150 m and is made from trees that have an average density of 7.00×10^2 kg/m³. How many people can stand on the raft and keep their feet dry, assuming an average person has a mass of 70.0 kg? Example 11-9

61. • A crown that is supposed to be made of solid gold is under suspicion. When the crown is weighed in air, it has a weight of 5.15 N. When it is suspended from a digital balance and lowered into water, its apparent weight is measured to be 4.88 N. Given that the specific gravity of gold is 19.3, (a) what should the apparent weight of the crown be? (b) Identify the material from which the crown is made. Example 11-11

62. • A woman floats in a region of the Great Salt Lake where the water is about four times saltier than the ocean and has a density of about 1130 kg/m³. The woman has a mass of 55 kg, and her density is 985 kg/m³ after exhaling as much air as possible from her lungs. Determine the percentage of her volume that will be above the waterline. Example 11-9

63. The person in Example 11-11 has a weight of 833 N. When submerged in water (density 1.00×10^3 kg/m³), his apparent weight is 19.7 N. (a) What is his average density? (b) What percent of his body mass is fat? (Assume that the person is made of muscle and fat only.) Example 11-11

11-9 Fluids in motion behave differently depending on the flow speed and the fluid viscosity

64. • (a) A hose is connected to a faucet and used to fill a 5.0-L container in a time of 45 s. Determine the volume flow rate in m³/s. (b) Determine the velocity of the water in the hose in part (a) if it has a radius of 1.0 cm. Example 11-12

65. • At what speed is the water leaving the 7.5-mm-diameter nozzle of a hose with a volume flow rate of 0.45 m³/s? Example 11-12

66. • Determine the time required for a 50.0-L container to be filled with water when the speed of the incoming water is 25.0 cm/s and the cross-sectional area of the hose carrying the water is 3.00 cm². Example 11-12

67. • Medical A cylindrical blood vessel is partially blocked by the buildup of plaque. At one point, the plaque decreases the diameter of the vessel by 60.0%. The blood approaching the blocked portion has speed v_0. Just as the blood enters the blocked portion of the vessel, what will be its speed in terms of v_0? Example 11-12

68. • Medical You inject your patient with 2.50 mL of medicine. If the inside diameter of the 31-gauge needle is 0.114 mm and the injection lasts 0.650 s, determine the average speed of the fluid as it leaves the needle. Example 11-12

69. • In July 1995 a spillway gate broke at the Folsom Dam in California. During the uncontrolled release the flow rate through the gate peaked at 40,000 ft³/s, and about 1.35 billion gallons of water were lost (nearly 40% of the reservoir). Estimate the time that the gate was open. Example 11-12

70. • The return-air ventilation duct in a home has a cross-sectional area of 9.0×10^2 cm². The air in a room with dimensions 7.0 m × 10.0 m × 2.4 m is to be completely circulated in a 30-min cycle. What is the speed of the air in the duct? Example 11-12

11-10 Bernoulli's equation helps us relate pressure and speed in fluid motion

71. •• Water flows from a fire truck through a hose that is 11.7 cm in diameter and has a nozzle that is 2.00 cm in diameter. The firefighters stand on a hill 5.00 m above the level of the truck. When the water leaves the nozzle, it has a speed of 20.0 m/s. Determine the minimum gauge pressure in the truck's water tank. Example 11-15

72. • At one point, Hurricane Katrina had maximum sustained winds of 175 mi/h (282 km/h) and a low pressure in the eye of 676.52 mmHg (0.890 atm). (a) Using the given air speed, determine the ratio of the pressure predicted by

Bernoulli's equation to the measured value. (b) Explain why the value does not equal 1. Assume the pressure of the air is normally 1.00 atm. Example 11-16

73. • When the atmospheric pressure is 1.00 atm, a water fountain ejects a stream of water that rises to a height of 5.00 m. There is a 1.00-cm-radius pipe that leads from a pressurized tank to the opening that ejects the water. Calculate the height of the fountain if it were operational while the eye of a hurricane passed by. Assume that the atmospheric pressure in the eye is 0.877 atm and the tank's pressure remains the same. Example 11-15

74. •• A cylinder that is 20 cm tall and open on top is filled with water (Figure 11-46). If a hole is made in the side of the cylinder, 5 cm below the top level, how far from the base of the cylinder will the stream land? Assume that the cylinder is large enough so that the level of the water in the cylinder does not drop significantly. Example 11-16

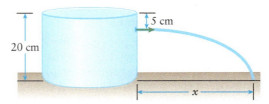

Figure 11-46 Problem 74

11-11 Viscosity is important in many types of fluid flow

75. • Water flows at 0.500 mL/s through a horizontal tube that is 30.0 cm long and has an inside diameter of 1.50 mm. Determine the pressure difference required to drive this flow if the viscosity of water is 1.00 mPa·s. Assume laminar flow. Example 11-18

76. • Biology Blood takes about 1.50 s to pass through a 2.00-mm-long capillary. If the diameter of the capillary is 5.00 μm and the pressure drop is 2.60 kPa, calculate the viscosity of blood. Assume laminar flow. Example 11-18

77. •• A very large tank is filled to a depth of 250 cm with oil that has a density of 860 kg/m³ and a viscosity of 180 mPa·s. If the container walls are 5.00 cm thick and a cylindrical hole of radius 0.750 cm has been bored through the base of the container, what is the initial volume flow rate (in L/s) of the oil through the hole? Example 11-18

78. ••• Water flows to an outlet from a pumping station 5.00 km away. From the pumping station to the outlet there is a net vertical rise of 19 m. Take the viscosity of water to be 0.0100 poise. The pipe from the pumping station to the outlet is 1.00 cm in diameter, and the gauge pressure in the pipe where it exits the pumping station is 520 kPa. At what volume flow rate does water flow from the outlet? Example 11-18

General Problems

79. •• Medical The human body contains about 5.0 L of blood that has a density of 1060 kg/m³. Approximately 45% (by mass) of the blood is cells and the rest is plasma. The density of blood cells is approximately 1125 kg/m³, and about 1% of the cells are white blood cells, with the rest being red blood cells. The red blood cells are about 7.5 μm across (modeled as

spheres). (a) What is the mass of the blood in a typical human? (b) Approximately how many blood cells (of both types) are in the blood? Example 11-3

80. • In 2009 a person in Italy won a lottery prize of 146.9 million euros, which at the time was worth 211.8 million U.S. dollars. Gold at the time was worth $953 per troy ounce, and silver was worth $14.16 per troy ounce. A troy ounce is 31.1035 g, the density of gold is 19.3 g/cm^3, and the density of silver is 10.5 g/cm^3. (a) If the lucky lottery winner opted to be paid in a single cube of pure gold, how high would the cube be? (b) What would be the height of a silver cube of the same value? Example 11-3

81. • A hybrid car travels about 50.0 mi/gal of gasoline. The density of gasoline is 737 kg/m^3 and 1 gal equals 3.788 L. Express the car's mileage in miles per kilogram (mi/kg) of gas. Example 11-3

82. • **Medical** Blood pressure is normally expressed as the ratio of the *systolic* pressure (when the heart just ejects blood) to the *diastolic* pressure (when the heart is relaxed). The measurement is made at the level of the heart (usually at the middle of the upper arm), and the pressures are given in millimeters of mercury, although the units are not usually written. Normal blood pressure is typically 120/80. How would you write normal blood pressure in (a) pascals, (b) atmospheres, and (c) pounds per square inch (lb/in.^{2}, psi)? (d) Is the blood pressure, as typically stated, the absolute pressure or the gauge pressure? Explain your answer.

83. • **Astronomy** Oceans as deep as 0.50 km once may have existed on Mars. The acceleration due to gravity on Mars is 0.379g. (a) If there were any organisms in the Martian ocean in the distant past, what pressure (absolute and gauge) would they have experienced at the bottom, assuming the surface pressure was the same as it is on present-day Earth? Assume that the salinity of Martian oceans was the same as oceans on Earth. (b) If the bottom-dwelling organisms in part (a) were brought from Mars to Earth, how deep could they go in our ocean without exceeding the maximum pressure they experienced on Mars? Example 11-5

84. • **Medical** Blood pressure is normally taken on the upper arm at the level of the heart. Suppose, however, that a patient has both arms in casts so that you cannot take his blood pressure in the usual way. If you have him stand up and take the blood pressure at his calf, which is 95.0 cm below his heart, what would normal blood pressure be? The density of blood is 1060 kg/m^3. Example 11-5

85. • **Medical** A syringe which has an inner diameter of 0.60 mm is attached to a needle with an inner diameter of 0.25 mm. A nurse uses the syringe to inject an ideal fluid into a patient's artery where blood pressure is 140/100. What minimum force must the nurse apply to the syringe? Example 11-16

86. •• A large water tank is being filled by a 1.0-cm-diameter hose. The water in the hose has a uniform speed of 15 cm/s. Meanwhile, the tank springs a leak at the bottom. The hole has a diameter of 0.50 cm. Determine the equilibrium level of the water in the tank if water continues flowing into the tank at the same rate. Example 11-16

87. • A large water tank is 18 m above the ground (Figure 11-47). Suppose a pipe with a diameter of 8.0 cm is connected to the base of the tank and leads down to the ground. How fast does the water rush out of the pipe at ground level? Assume that the tank is open to the atmosphere. Example 11-6

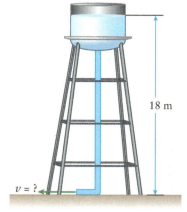

Figure 11-47 Problem 87

88. • **Medical** The aorta is approximately 25 mm in diameter. The mean pressure there is about 100 mmHg, and the blood flows through the aorta at approximately 60 cm/s. Suppose that at a certain point a portion of the aorta is blocked so that the cross-sectional area is reduced to 3/4 of its original area. The density of blood is 1060 kg/m^3. (a) How fast is the blood moving just as it enters the blocked portion of the aorta? (b) What is the gauge pressure (in mmHg) of the blood just as it enters the blocked portion? Example 11-16

89. ••• An equilateral prism of wood with specific gravity 0.6 is placed into a freshwater pool (Figure 11-48). Determine the ratio y_d/y_u of the depth of submersion when the prism is pointed down to the depth of submersion when the prism is pointed up. The prism has a side of s. Example 11-19

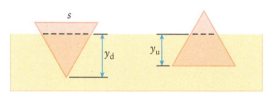

Figure 11-48 Problem 89

90. •• **Sports** The air around the spinning baseball in Figure 11-49 (top view) experiences a faster speed on the left than the right side and hence a smaller pressure on the left than the right. Suppose the ball is traveling in the x direction at 38.0 m/s and it breaks 15.0 cm to the left in the $-y$ direction. Determine the pressure difference between the right and left sides of the ball. Assume the ball has a mass of 142 g and a radius of 3.55 cm. In addition, assume it travels at a constant speed of 38 m/s in the x direction from the pitcher's mound to home plate (60.5 ft = 18.44 m). *Hint:* The pressure difference equals the leftward force divided by the cross-sectional area of the ball, πr^2. Example 11-15

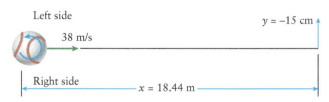

Figure 11-49 Problem 90

12

Oscillations

leonello calvetti/Alamy

What do you think?

In a typical adult at rest, the heart beats 60 times per minute. During moderate cardiovascular exercise the heart rate can increase to 120 beats per minute. If the walls of the heart move in and out by the same distance for both heart rates, how does the maximum acceleration of the heart walls during moderate exercise compare to that when resting? (a) $\frac{1}{4}$ as great; (b) $\frac{1}{2}$ as great; (c) the same; (d) twice as great; (e) 4 times as great.

In this chapter, your goals are to:

- (12-1) Define oscillation.
- (12-2) Describe the key properties of oscillations, including what makes oscillation happen.
- (12-3) Explain the connection between Hooke's law and the special kind of oscillation called simple harmonic motion.
- (12-4) Discuss how kinetic energy and potential energy vary during an oscillation.
- (12-5) Explain what determines the period, frequency, and angular frequency of a simple pendulum.
- (12-6) Explain what determines the period, frequency, and angular frequency of a physical pendulum.
- (12-7) Describe what happens in underdamped, critically damped, and overdamped oscillations.
- (12-8) Identify the circumstances under which resonance occurs in a forced oscillation.

To master this chapter, you should review:

- (2-4)   How to interpret x–t, v_x–t, and a_x–t graphs.
- (6-5, 6-6)   The force exerted by an ideal spring that obeys Hooke's law and the potential energy stored in such a spring.
- (3-7, 9-2)   The relationships among speed, angular speed, and centripetal acceleration in circular motion.
- (9-3)   Newton's second law for rotational motion and how to calculate the torque exerted by a force.

12-1 We live in a world of oscillations

Try sitting by yourself in your room and keeping absolutely still. If no one else is in the room, it may seem as though there is no motion anywhere around you. Not so! Your heart is in continuous motion, pulsing at a rate of about one cycle per second. Your eardrums are also vibrating softly in response to the sound of your breathing. And in the electrical wires within the walls of your room, electrons move back and

The wings of common hummingbirds oscillate up and down about 50 times per second.

The timekeeper at the heart of most wristwatches is a tiny piece of quartz that oscillates 32,768 times per second.

This MRI image is made by mapping how protons oscillate within molecules in the brain in response to radio waves.

Figure 12-1 Oscillations The kind of motion that we call oscillation is commonplace (a) in nature, (b) in technology, and (c) in medicine.

(a)

Paul Piebinga/Vetta/Getty Images

(b)

Photodisc/Thinkstock/Getty Images

(c)

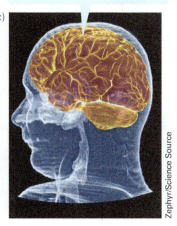

Zephyr/Science Source

forth about 60 times per second as part of the processes that supply energy to the room lights, your computer, and other appliances. These are just three examples of a kind of motion called **oscillation**, in which an object moves back and forth around a position of *equilibrium*—that is, a point at which it experiences zero net force. (We introduced the idea of equilibrium in Section 4-3.) As an example, your eardrum is in equilibrium when it is relaxed and displaced neither inward nor outward. Figure 12-1 shows three examples of oscillations in nature and technology.

We'll begin this chapter by studying the causes of oscillation. We'll find that oscillation is possible only when an object experiences a *restoring force*—one that always pushes or pulls the object toward the equilibrium position. Oscillations are particularly simple in nature if the restoring force is directly proportional to how far the object is displaced from equilibrium. This proves to be an important special case that's a very good approximation to many kinds of oscillation.

Just as for other kinds of motion, we'll find that it's useful to describe oscillations in terms of kinetic and potential energy. The energy approach will also help us to understand the role of *damping* in oscillations—for example, what happens to make the oscillations of a swinging pendulum die out so that the pendulum eventually ends up at rest. Finally, we'll examine *forced* oscillations, in which a rhythmic force is applied to an object to sustain its oscillations even when damping is present. The ideas of forced oscillation will help us understand how the right kind of sound can break a wine glass, how some insects fly, and why architects must design buildings to avoid catastrophic vibrations.

12-2 Oscillations are caused by the interplay between a restoring force and inertia

Figure 12-2 shows a simple way to make an object oscillate. A block is attached to a horizontal spring and is free to slide back and forth. (For simplicity we'll assume that there's no friction between the block and the surface over which it slides.) We'll call the horizontal direction along which the block can move the x axis, and let $x = 0$ be the point where the spring is relaxed and the block is in equilibrium (Figure 12-2a). Then the quantity x represents the *position* of the block as well as the *displacement* of the block from equilibrium.

When the spring is stretched, corresponding to $x > 0$, it exerts a force on the block that points in the negative x direction so $F_x < 0$ (Figure 12-2b). If instead the spring is compressed, corresponding to $x < 0$, it exerts a force on the block that points in the positive x direction so $F_x > 0$ (Figure 12-2c). In other words, no matter which way the block is displaced from equilibrium, the spring exerts a force that tends to pull or push the block back toward the equilibrium position. A force of this kind is called a **restoring force**.

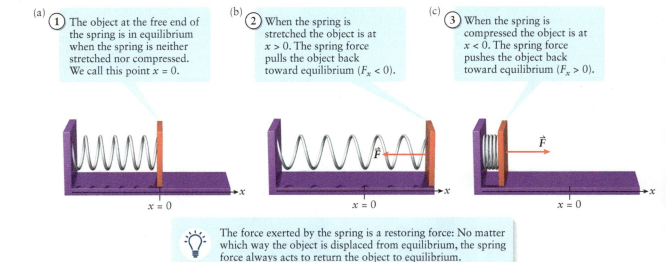

(a) ① The object at the free end of the spring is in equilibrium when the spring is neither stretched nor compressed. We call this point $x = 0$.

(b) ② When the spring is stretched the object is at $x > 0$. The spring force pulls the object back toward equilibrium ($F_x < 0$).

(c) ③ When the spring is compressed the object is at $x < 0$. The spring force pushes the object back toward equilibrium ($F_x > 0$).

$x = 0$

$x = 0$

$x = 0$

The force exerted by the spring is a restoring force: No matter which way the object is displaced from equilibrium, the spring force always acts to return the object to equilibrium.

Figure 12-2 A restoring force For oscillations to take place, there must be a force that always acts to return the oscillating object to equilibrium.

To start the block in Figure 12-2 oscillating, move the block so as to stretch the spring by a distance A and then release it (Figure 12-3a). The restoring force of the spring pulls the block back toward the equilibrium position at $x = 0$, so the block accelerates and continues to gain speed as it moves back toward equilibrium (Figure 12-3b). The block reaches its maximum speed as it passes through equilibrium (Figure 12-3c). Even though there is zero net force on the block at the equilibrium position, the block's inertia causes it to overshoot this point. As a result the block compresses the spring, so the restoring force is now directed opposite to the block's motion as it slows down (Figure 12-3d). Thus the block eventually comes momentarily to rest with the spring compressed (Figure 12-3e). The restoring force of the spring still acts, so the block again starts moving and gains speed as it heads back toward the equilibrium position (Figure 12-3f). The object reaches equilibrium a second time and again overshoots the equilibrium position (Figure 12-3g). The block stretches the spring as it overshoots, and the restoring force causes the block to slow down (Figure 12-3h). The block ends up once again at its initial position with zero speed (Figure 12-3i). The cycle then repeats. If there is no friction or air drag, the oscillation will go on forever!

Not every oscillation involves a spring, but *every* oscillation follows the same general sequence of steps shown in Figure 12-3. This figure shows that the two key ingredients for oscillation are a restoring force that always draws the oscillating object back toward the equilibrium position and inertia that causes the object to overshoot equilibrium.

Oscillation Period and Frequency

The time for one complete cycle of an oscillation is called the **period** of the oscillation. We'll use the symbol T for period. Since T is a time, its units are seconds (s). But it's useful to think of these units as "seconds per cycle."

Intuition should tell you that a stiffer spring (which will exert a stronger restoring force) will push and pull the block more rapidly through the steps shown in Figure 12-3 and so will cause the period to be shorter. That same intuition will tell you that a more massive block, which has more inertia, will go through the steps more slowly and hence have a longer period. We'll verify that these ideas are correct later in this section.

A quantity related to the oscillation period is the **frequency** f of the oscillation. The frequency is equal to the number of cycles per unit time. The unit of frequency is the **hertz** (abbreviated Hz): 1 Hz = 1 cycle per second. A pendulum that oscillates once per second has a frequency $f = 1$ Hz. If you play the A4 key on a piano (A above middle C, also called "concert A"), the piano string vibrates 440 times per second, so the frequency is $f = 440$ Hz. The vibrating string compresses the air at this same frequency, which is the frequency of sound that you hear.

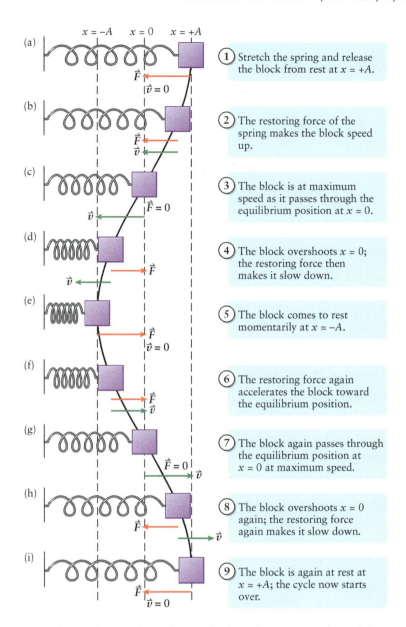

Figure 12-3 A cycle of oscillation
Nine stages in the oscillation of a block attached to a spring.

(a) ① Stretch the spring and release the block from rest at $x = +A$.

(b) ② The restoring force of the spring makes the block speed up.

(c) ③ The block is at maximum speed as it passes through the equilibrium position at $x = 0$.

(d) ④ The block overshoots $x = 0$; the restoring force then makes it slow down.

(e) ⑤ The block comes to rest momentarily at $x = -A$.

(f) ⑥ The restoring force again accelerates the block toward the equilibrium position.

(g) ⑦ The block again passes through the equilibrium position at $x = 0$ at maximum speed.

(h) ⑧ The block overshoots $x = 0$ again; the restoring force again makes it slow down.

(i) ⑨ The block is again at rest at $x = +A$; the cycle now starts over.

Since period T is the number of seconds that elapse per cycle and frequency f is the number of cycles that happen per second, it follows that these quantities are the reciprocals of each other:

Period T of an oscillation

$$f = \frac{1}{T} \quad \text{and} \quad T = \frac{1}{f}$$

Frequency f of the oscillation

Frequency and period
(12-1)

EXAMPLE 12-1 Frequency and Period

(a) What is the oscillation period of your eardrum when you are listening to the A4 note on a piano (frequency 440 Hz)?
(b) A bottle floating in the ocean bobs up and down once every 2.00 min. What is the frequency of this oscillation?

Set Up

We'll use Equation 12-1 to find the period T from the frequency f and vice versa.

Frequency and period:

$$f = \frac{1}{T} \quad \text{and} \quad T = \frac{1}{f}$$

(12-1)

Solve

(a) The period of this oscillation is the reciprocal of the frequency. Note that a cycle is not a true unit but simply a way of counting, so we can remove it from the final answer as needed.

$$T = \frac{1}{f} = \frac{1}{440 \text{ Hz}} = \frac{1}{440 \text{ cycles/s}} = \frac{1}{440} \frac{\text{s}}{\text{cycle}}$$
$$= 2.3 \times 10^{-3} \text{ s}$$

(b) The frequency of this oscillation is the reciprocal of the period. To get a result in Hz, we must first convert the period to seconds. Note that the period is the number of seconds per cycle, so we can add "cycle" to the calculation as needed.

$$T = 2.00 \text{ min} \times \frac{60 \text{ s}}{1 \text{ min}} = 120 \text{ s}$$
$$f = \frac{1}{T} = \frac{1}{120 \text{ s}} = \frac{1}{120 \text{ s/cycle}} = \frac{1}{120} \frac{\text{cycle}}{\text{s}}$$
$$= 8.33 \times 10^{-3} \text{ Hz}$$

Reflect

Since frequency and period are the reciprocals of each other, a large value of f (high frequency) corresponds to a small value of T (short period) as in (a) and a small value of f (low frequency) corresponds to a large value of T (long period) as in (b).

Oscillation Amplitude

If the spring in Figures 12-2 and 12-3 exerts the same magnitude of force when it is stretched a certain distance as when it is compressed the same distance, the block in Figure 12-3 will move as far to the left of equilibrium ($x < 0$) as it does to the right of equilibrium ($x > 0$). In this chapter we'll consider only restoring forces that have this symmetric property. In this case the block will oscillate between $x = +A$ and $x = -A$, as shown in Figure 12-3. The distance A is called the **amplitude** of the oscillation. It's equal to the maximum displacement of the object from equilibrium. The amplitude is *always* a positive number, never a negative one.

WATCH OUT! **The amplitude of oscillation is the maximum displacement from equilibrium.**

! It's tempting to think of amplitude as the distance from where the spring is stretched the most (Figure 12-3a) to where the spring is compressed the most (Figure 12-3e). That's not correct: The amplitude A equals the distance from the *equilibrium* position to the point of maximum stretch, and it also equals the distance from the equilibrium position to the point of maximum compression. Keep this principle in mind to avoid giving answers that are wrong by a factor of 2.

 GOT THE CONCEPT? 12-1 **Heartbeats**

(?) Jess has a resting pulse of 50 heartbeats per minute. When Jess sprints down the track, her pulse increases to 150 beats per minute. Compared to when she is at rest, the oscillation period of her heart when she is sprinting is (a) 9 times as great; (b) 3 times as great; (c) the same; (d) 1/3 as great; (e) 1/9 as great.

TAKE-HOME MESSAGE FOR Section 12-2

✔ Oscillation is caused by the interplay between (i) a restoring force that returns an object to its equilibrium position (where it experiences zero net force) and (ii) inertia that makes the object overshoot equilibrium.

✔ The period of an oscillation is the time for one complete cycle. The frequency, which is the reciprocal of the period, is the number of cycles per second.

✔ The amplitude of an oscillation equals the maximum displacement from equilibrium.

12-3 The simplest form of oscillation occurs when the restoring force obeys Hooke's law

A complete description of an oscillation means knowing more than just the period, frequency, and amplitude. We also need to know the position, velocity, and acceleration of the oscillating object at all times during its motion. We can actually find these for the case in which the restoring force is *directly proportional* to the distance that the object is displaced from equilibrium. This relationship is called **Hooke's law** for the force exerted by an ideal spring (Section 6-5). Hooke's law is important because experiment shows that if you stretch a spring by a relatively small amount, the force that the spring exerts on you is directly proportional to the amount of stretch:

Force exerted by an **ideal spring**   Spring constant of the spring (a measure of its stiffness)

$$F_x = -kx$$

Hooke's law
(6-15)

Extension of the spring ($x > 0$ if spring is stretched, $x < 0$ if spring is compressed)

The minus sign in Equation 6-15 tells us that if the spring is stretched, $x > 0$, it exerts a force in the negative x direction so $F_x < 0$. It also tells us that if the spring is compressed, $x < 0$, it exerts a force in the positive x direction so $F_x > 0$. That's exactly the relationship between displacement x and spring force F_x shown in Figure 12-2.

As we did in Section 6-5, we'll call a spring that obeys Equation 6-15 an *ideal* spring. Real springs deviate from Hooke's law a little. But experiment shows that if the value of x is kept small enough, so that the oscillations have small amplitude, almost every spring obeys Hooke's law quite faithfully.

If the spring in Figures 12-2 and 12-3 is an ideal one, then the net force on the block in Figures 12-2 and 12-3 is equal to the spring force given by Equation 6-15. (There's also a downward gravitational force on the block plus an upward normal force exerted by the surface over which it slides. But these forces cancel each other so that the net force on the block has zero vertical component.) Newton's second law for the x motion of the block tells us that this net force is equal to the mass m of the block multiplied by the block's acceleration a_x:

$$-kx = ma_x$$

If we divide both sides of this equation by the mass m of the block, we get

$$a_x = -\left(\frac{k}{m}\right)x$$

(12-2)

Equation 12-2 tells us that if a block oscillates because it is attached to an ideal spring that obeys Hooke's law, the acceleration of the block is proportional to the negative of the displacement. The acceleration is negative when the displacement x is positive, positive when the displacement is negative, and zero when the displacement is zero (that is, when the block is at equilibrium).

Because the acceleration of the block is not constant, we *cannot* describe the motion of the block in Figure 12-3 using the equations from Chapter 2 for straight-line motion with constant acceleration. (Those are nice, simple equations, but they simply don't apply here.) Instead we'll have to solve Equation 12-2 to find x, v_x, and a_x as functions of time t. This sounds like a complicated mathematical problem, but it turns out that we can solve it by using what we know about another, seemingly unrelated, kind of motion: an object moving around a circle at constant speed.

Uniform Circular Motion and Hooke's Law

Figure 12-4a shows an object of mass m moving at constant speed v around a circular path of radius A. (We'll see shortly why we chose this symbol for radius.) The object travels once around the circle, an angle of 2π radians, in a time T, so the object's constant *angular* speed is $\omega = 2\pi/T$.

Figure 12-4 Oscillation and uniform circular motion If you look at one component of uniform circular motion, what you see is an oscillation.

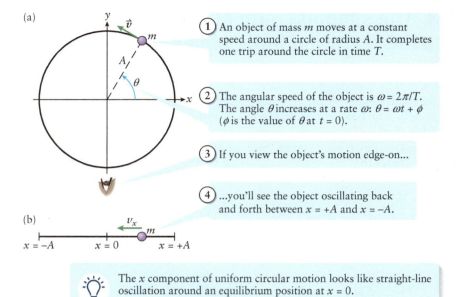

(a)

(1) An object of mass m moves at a constant speed around a circle of radius A. It completes one trip around the circle in time T.

(2) The angular speed of the object is $\omega = 2\pi/T$. The angle θ increases at a rate ω: $\theta = \omega t + \phi$ (ϕ is the value of θ at $t = 0$).

(3) If you view the object's motion edge-on...

(4) ...you'll see the object oscillating back and forth between $x = +A$ and $x = -A$.

(b)

$x = -A$ $\qquad$ $x = 0$ $\qquad$ $x = +A$

The x component of uniform circular motion looks like straight-line oscillation around an equilibrium position at $x = 0$.

How does this uniform circular motion relate to oscillation? Imagine that you view the circular path edge-on so that all you can see is the x component of the object's motion (Figure 12-4b). From this vantage point you'll see the object oscillating back and forth along a straight-line path, moving a distance A to the left of center and the same distance A to the right of center (Figure 12-4b). The object will take a time T for one back-and-forth cycle. So what you'll see is the object moving in the same way as the block on a spring in Figure 12-3, with period T and amplitude A.

Let's show that the x component of the uniform circular motion obeys Equation 12-2 so that the x component of the object's acceleration is proportional to the object's x coordinate. To do this we'll use three important facts about uniform circular motion. From Section 9-2 the *linear* speed v of an object in uniform circular motion is equal to the *angular* speed ω multiplied by the radius of the circle (see Equation 9-3). The radius of the circle in Figure 12-4a is A, so the speed of the object's circular motion is

$$v = \omega A \tag{12-3}$$

We also know from Section 3-7 that an object in uniform circular motion has a centripetal acceleration (that is, an acceleration directed toward the center of the circle). This acceleration has a constant magnitude a_{cent} equal to the square of the speed v divided by the radius of the circle, which in this case is equal to A. Using Equation 12-3 we can write the magnitude of the centripetal acceleration as

$$a_{cent} = \frac{v^2}{A} = \frac{(\omega A)^2}{A} = \frac{\omega^2 A^2}{A} = \omega^2 A \tag{12-4}$$

Finally, from Equation 9-6 we have the following equation for the angle θ from the positive x axis for the angular position of an object at time t, assuming that the object has constant angular acceleration:

$$\theta = \theta_0 + \omega_{0z} t + \frac{1}{2} \alpha_z t^2 \tag{9-6}$$

In Equation 9-6 α_z is the angular acceleration of the object, which is zero in this case because the motion is uniform and the angular velocity is constant. The quantity ω_{0z} is the initial z component of the angular velocity of the object, which is equal to the object's constant angular speed ω. (Figure 12-4a shows that the object moves around the circle in the positive, counterclockwise direction, so the angular velocity is positive and equal to the angular speed.) The angle θ_0 is the value of θ at time $t = 0$; we'll use the symbol ϕ (the Greek letter "phi") for this quantity. Then we can write Equation 9-6 as $\theta = \phi + \omega t$, or

$$\theta = \omega t + \phi \tag{12-5}$$

(a)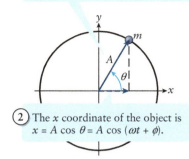

① The vector from the center of the circle to the object has length A and is at an angle θ from the +x axis.

② The x coordinate of the object is $x = A \cos \theta = A \cos (\omega t + \phi)$.

(b)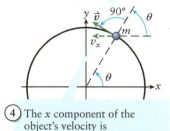

③ The object's velocity vector has magnitude $v = \omega A$ and is at an angle $90° + \theta$ from the +x axis.

④ The x component of the object's velocity is
$v_x = v \cos (90° + \theta) = -v \sin \theta$
or $v_x = -\omega A \sin (\omega t + \phi)$.

(c)

⑤ The object's acceleration vector has magnitude $a_{\text{cent}} = \omega^2 A$ and is at an angle $180° + \theta$ from the +x axis.

⑥ The x component of the object's acceleration is
$a_x = a_{\text{cent}} \cos (180° + \theta) = -a_{\text{cent}} \cos \theta$
or $a_x = -\omega^2 A \cos (\omega t + \phi)$.

 Compare the equations for x and a_x: You'll see that $a_x = -\omega^2 x$. That's the same relationship as for Hooke's law, $a_x = -(k/m)x$, with $\omega^2 = k/m$. So the x component of uniform circular motion is the same as the simple harmonic motion of an object oscillating on the end of an ideal, Hooke's law spring.

Figure 12-5 Oscillation, uniform circular motion, and Hooke's law Finding the x components of the position, velocity, and acceleration in uniform circular motion reveals an important connection to simple harmonic motion.

Now let's put all of these pieces together. Figure 12-5a shows that the x coordinate of the object as it moves around the circle is $x = A \cos \theta$. From Equation 12-5 we can write this as

$$x = A \cos (\omega t + \phi) \tag{12-6}$$

Figure 12-5b shows that the x velocity of the object is $v_x = -v \sin \theta$. From Equation 12-3 and Equation 12-5, we can write this as

$$v_x = -\omega A \sin (\omega t + \phi) \tag{12-7}$$

From Figure 12-5c the x acceleration of the object is $a_x = -a_{\text{cent}} \cos \theta$. Equation 12-4 and Equation 12-5 together tell us that we can write the x acceleration as

$$a_x = -\omega^2 A \cos (\omega t + \phi) \tag{12-8}$$

Here's the payoff for all of this mathematics: If we compare Equation 12-6 and Equation 12-8, we see that the x acceleration of an object in uniform circular motion is directly proportional to the object's x coordinate:

$$a_x = -\omega^2 x \tag{12-9}$$

That's *precisely* the relationship between acceleration a_x and displacement x that we found for an object of mass m attached to a spring with spring constant k that obeys Hooke's law: $a_x = -(k/m)x$ (Equation 12-2). The two equations are identical if we set the square of the angular speed ω in Figure 12-4 and Figure 12-5 equal to the ratio of the quantities k/m:

$$\omega^2 = \frac{k}{m} \quad \text{or} \quad \omega = \sqrt{\frac{k}{m}} \tag{12-10}$$

In other words, Equation 12-9 and Equation 12-10 tell us something quite profound and unexpected: A block of mass m attached to an ideal, Hooke's law spring of spring constant k oscillates in precisely the same way as the x coordinate of an object in uniform circular motion with angular speed ω, provided that $\omega = \sqrt{k/m}$. We made the detour into analyzing uniform circular motion so that we could arrive at this insight.

With this observation in mind, we can now go back and write a complete description of how an object oscillates when acted on by a restoring force that obeys Hooke's law.

√x **See the Math Tutorial for more information on trigonometry.**

Simple Harmonic Motion: Angular Frequency, Period, and Frequency

We've seen that the period T of the oscillation in Figure 12-4b is the same as the time T that it takes the object in uniform circular motion in Figure 12-4a to travel once around the circle. The angular speed ω equals $2\pi/T$, so T equals $2\pi/\omega$. Using Equation 12-10 for the value of ω along with Equation 12-1 for the relationship between period T and frequency f, we get the following results for the oscillations of a block of mass m attached to an ideal, Hooke's law spring of spring constant k:

Angular frequency, period, and frequency for a block attached to an ideal spring
(12-11)

$$\text{Angular frequency} \quad \omega = \sqrt{\frac{k}{m}}$$

k = **spring constant** of the spring

m = **mass** of the object connected to the spring

$$\text{Period} \quad T = \frac{2\pi}{\omega} = 2\pi\sqrt{\frac{m}{k}}$$

$$\text{Frequency} \quad f = \frac{1}{T} = \frac{1}{2\pi}\sqrt{\frac{k}{m}} = \frac{\omega}{2\pi}$$

In the first of Equations 12-11, we've introduced a new quantity called the **angular frequency**. This has the same symbol and the same value as the angular *speed* of the circular motion in Figure 12-4. That's because our analysis of Figure 12-4 showed that the back-and-forth motion of a block attached to an ideal spring is equivalent to the x component of uniform circular motion. If you find this confusing to remember, just keep in mind that the ordinary frequency f equals $\omega/2\pi$, or equivalently $\omega = 2\pi f$ (see the third of Equations 12-11). As we'll see, the quantity ω appears naturally in the equations for oscillation, which is why we use it. Like angular speed, angular frequency has units of radians per second (rad/s).

The second of Equations 12-11 tells us that the period of oscillation T is proportional to the square root of the mass m and inversely proportional to the square root of the spring constant k. This means that increasing the mass m makes the period longer, while increasing the spring constant k makes the period shorter. That's exactly what we predicted in Section 12-2 for the oscillations of a block attached to a spring: a more massive block (greater m) has more inertia and so requires more time to complete one cycle (greater T), while a stiffer spring (greater k) exerts more restoring force and so helps the block complete one cycle in less time (smaller T). The last of Equations 12-11 is just the statement that the frequency f is the reciprocal of the period T. This equation says that increasing the mass causes the frequency to decrease, while increasing the spring constant makes the frequency increase.

Note that the amplitude A does *not* appear in Equations 12-11, which means that the angular frequency, period, and frequency of an oscillation are *independent* of the amplitude if the restoring force obeys Hooke's law. This is called the **harmonic property**. This may come as a surprise, since doubling the amplitude means the oscillating object has to cover twice as much distance during an oscillation and so should take a longer time to complete an oscillation. But doubling the amplitude also means that the object reaches larger displacements x and so experiences a stronger Hooke's law restoring force $F_x = -kx$. This effect makes the object move faster during the oscillation. The net result is that the period T, the time for one oscillation, is unaffected by changing the amplitude. That means the frequency $f = 1/T$ and the angular frequency $\omega = 2\pi f = 2\pi/T$ are unaffected as well. This is the case *only* if the restoring force obeys Hooke's law. If the restoring force depends on displacement in a different way, the oscillations do not have the harmonic property.

The word "harmonic" sounds like a musical term, and indeed the harmonic property is important in musical instruments. We'll learn in Chapter 13 that the pitch of a musical sound is determined primarily by the frequency of oscillation of the musical instrument that makes the sound, while the loudness of the sound is determined primarily by the amplitude. If changing the amplitude caused a change in frequency, playing the same key on a piano would make a different pitch if you pressed the key softly or pushed hard on the key. That would render a piano almost completely useless. But because the strings of a piano have the harmonic property, they vibrate at the same frequency and produce sounds of the same pitch whether they are played softly or loudly.

As we've described, the harmonic property is a direct result of the restoring force being directly proportional to the displacement from equilibrium (Hooke's law, Equation 6-15). For this reason the kind of oscillation that results from a Hooke's law restoring force is called **simple harmonic motion (SHM)**. Later in this chapter we'll encounter kinds of oscillations that are not simple harmonic motion; the oscillations of a pendulum are one example. We'll see that in many such cases, however, these kinds of oscillations are *approximately* SHM if the amplitude of oscillation is sufficiently small.

WATCH OUT! Simple harmonic motion is not the same as circular motion.

Even though we describe simple harmonic motion in terms of an *angular* frequency, the oscillating object is *not* actually moving in a circle! As we've seen we can *visualize* simple harmonic motion as one component of the motion of an object moving in a circle with an angular speed ω, but there doesn't need to be an object that's actually moving in a circle.

EXAMPLE 12-2 SHM I: Angular Frequency, Period, and Frequency

An object of mass 0.80 kg is attached to an ideal horizontal spring of spring constant 1.8×10^2 N/m and set into oscillation as in Figure 12-3. (a) Calculate the angular frequency, period, and frequency of the object. (b) Calculate the angular frequency, period, and frequency if the mass of the object is quadrupled to 4×0.80 kg = 3.2 kg.

Set Up

An ideal spring obeys Hooke's law, so the oscillations are simple harmonic motion. This means that in part (a) we can use Equations 12-11 to calculate the angular frequency ω, period T, and frequency f from the spring constant $k = 1.8 \times 10^2$ N/m and the mass $m = 0.80$ kg. In part (b) we'll use these equations to determine how the values of ω, T, and f change when we change the value of the mass m.

Angular frequency, period, and frequency:

$$\omega = \sqrt{\frac{k}{m}}$$

$$T = \frac{2\pi}{\omega} = 2\pi\sqrt{\frac{m}{k}}$$

$$f = \frac{1}{T} = \frac{1}{2\pi}\sqrt{\frac{k}{m}} = \frac{\omega}{2\pi} \qquad (12\text{-}11)$$

Solve

(a) Substitute the given values for the mass m and spring constant k into the first of Equations 12-11 to calculate the angular frequency.

$$\omega = \sqrt{\frac{k}{m}} = \sqrt{\frac{1.8 \times 10^2 \text{ N/m}}{0.80 \text{ kg}}}$$

Recall that 1 N = 1 kg·m/s², so

$$\frac{1 \text{ N/m}}{\text{kg}} = \frac{1 \text{ kg/s}^2}{\text{kg}} = \frac{1}{\text{s}^2} = \frac{1 \text{ rad}^2}{\text{s}^2}$$

(A radian is a way of counting, not a true unit, so we can insert it or remove it as needed.) Then

$$\omega = \sqrt{\frac{k}{m}} = \sqrt{\frac{1.8 \times 10^2}{0.80} \frac{\text{rad}^2}{\text{s}^2}} = 15 \text{ rad/s}$$

Then use the second and third of Equations 12-11 to find the period and frequency.

If the angular frequency ω is in rad/s, the period T will be in s:

$$T = \frac{2\pi}{\omega} = \frac{2\pi}{15 \text{ rad/s}} = 0.42 \text{ s}$$

The reciprocal of the period in s is the frequency in Hz:

$$f = \frac{1}{T} = \frac{1}{0.42 \text{ s}} = 2.4 \text{ Hz}$$

(b) We could substitute the new value of the mass m into Equations 12-11 to find the new values of ω, T, and f. Instead we use proportional reasoning to find the new values from the old ones.

The relationships $\omega = \sqrt{\dfrac{k}{m}}$ and $f = \dfrac{1}{2\pi}\sqrt{\dfrac{k}{m}}$ tell us that the angular frequency and the frequency are inversely proportional to the square root of the mass. The new mass is four times greater than the old mass, so the new values of ω and f are $1/\sqrt{4} = 1/2$ the old values:

$$\omega_{\text{new}} = \frac{\omega_{\text{old}}}{\sqrt{4}} = \frac{\omega_{\text{old}}}{2} = \frac{15 \text{ rad/s}}{2} = 7.5 \text{ rad/s}$$

$$f_{\text{new}} = \frac{f_{\text{old}}}{2} = \frac{2.4 \text{ Hz}}{2} = 1.2 \text{ Hz}$$

In the same way, the relationship $T = 2\pi\sqrt{\dfrac{m}{k}}$ tells us that the period T is directly proportional to the square root of the mass. Again the mass has increased by a factor of 4, so the new period is

$$T_{\text{new}} = \sqrt{4}\,T_{\text{old}} = 2T_{\text{old}} = 2(0.42 \text{ s}) = 0.84 \text{ s}$$

Reflect

Our results show that when the mass of the oscillating object increases, the frequency decreases and the period increases. To check our results, can you confirm that f_{new} is equal to $\omega_{\text{new}}/2\pi$ and that T_{new} is equal to $1/f_{\text{new}}$?

Simple Harmonic Motion: Position, Velocity, and Acceleration

Let's now look in more detail at the position x, velocity v_x, and acceleration a_x of an object moving in simple harmonic motion. From Equation 12-6, the position is given by

Position as a function of time for simple harmonic motion (12-12)

> **Position** of an object in SHM at time t
>
> **Amplitude** of the oscillation = maximum displacement from the equilibrium position

$$x = A \cos(\omega t + \phi)$$

> **Angular frequency** of the oscillation ($\omega = 2\pi f = 2\pi/T$)
>
> **Phase angle** ϕ (in radians)

The cosine function has values from $+1$ to -1. So Equation 12-12 says that the value of x varies from $x = A$ (when the oscillating object is at its most positive displacement from equilibrium) through $x = 0$ (the equilibrium position) to $x = -A$ (when the object is at its most negative displacement from equilibrium). The cosine is one of the simplest of all oscillating functions, which helps justify the name *simple* harmonic motion. The cosine and sine functions are called **sinusoidal functions** (from the Latin word *sinus*, meaning "bent" or "curved") because of their shape.

The angular frequency ω is measured in rad/s and time t is measured in s, so the product ωt in Equation 12-12 is in radians—just as the argument of the cosine function should be. One cycle of oscillation lasts a time T, over which time the product ωt varies from 0 at $t = 0$ to ωT at $t = T$. But since $T = 2\pi/\omega$ from the second of Equations 12-11, it follows that $\omega T = \omega(2\pi/\omega) = 2\pi$. So over the course of one cycle the value of ωt varies from 0 to $\omega T = 2\pi$. The cosine function goes through one complete cycle when its value increases by 2π, so this tells us that the position x goes through one complete cycle when the time increases by T. That's just what we mean by saying that T is the period of the oscillation.

The one quantity in Equation 12-12 that seems a bit mysterious is the **phase angle** ϕ, which is measured in radians. To explain its significance recall from our comparison between uniform circular motion and oscillation that ϕ represents the angular position at $t = 0$ of an object in uniform circular motion. For oscillation the value of ϕ tells us where in the oscillation cycle the object is at $t = 0$. As an example, if $\phi = 0$ Equation 12-12 becomes

(12-13)
$$x = A \cos \omega t \quad \text{(simple harmonic motion, phase angle } \phi = 0\text{)}$$

Figure 12-6a is the x–t graph that corresponds to Equation 12-13. This graph of x versus t is an ordinary cosine function, which has its most positive value ($x = +A$) when the argument of the function is zero (at $t = 0$) and its most negative value ($x = -A$) one-half cycle later (at $t = T/2$). You can see that the slope of this x–t graph is zero at $t = 0$, which means that the object has zero velocity at $t = 0$. In other words, the value $\phi = 0$ corresponds to starting the object at rest at $t = 0$ at the position $x = A$. That's just the motion that we depicted in Figure 12-3.

By comparison Figure 12-6b is the graph that corresponds to Equation 12-12, $x = A \cos(\omega t + \phi)$, for a case in which ϕ is *not* zero but has a positive value. As the figure shows, you can think of the phase angle as shifting the axes to the right if ϕ is positive, as in Figure 12-6b, or shifting the axes to the left when ϕ is negative. Note that in this x–t graph the value of x at $t = 0$ is *not* +A; rather, from Equation 12-12 it is equal to

$$x(0) = A \cos(0 + \phi) = A \cos \phi$$

So for an object in simple harmonic motion the value of the phase angle ϕ tells you the point in the oscillation cycle that corresponds to $t = 0$. Note that Figures 12-6a and 12-6b really depict the *same* oscillation, with the same amplitude A and period T; the only difference is the point in the cycle that we call $t = 0$.

Just as Figure 12-6 shows how the graph of Equation 12-12 changes if you vary the value of the phase angle ϕ, Figure 12-7 shows how the graph changes if you vary the value of the amplitude A or the angular frequency ω. Decreasing the value of A makes the graph smaller in its vertical dimensions while leaving the horizontal dimensions unchanged (Figure 12-7a). Decreasing the value of ω means decreasing the value of the frequency $f = \omega/2\pi$ and so increasing the value of the period $T = 2\pi/\omega = 1/f$; this stretches the graph out horizontally while leaving the vertical dimensions unchanged (Figure 12-7b).

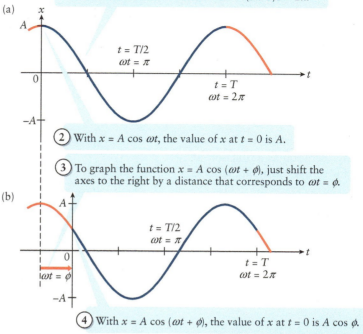

(a)

1. The blue curve graphs the first full cycle of the function $x = A \cos \omega t$, from $t = 0$ to $t = T$. This corresponds to values of ωt from $\omega t = 0$ to $\omega t = \omega T = (2\pi/T)T = 2\pi$.

2. With $x = A \cos \omega t$, the value of x at $t = 0$ is A.

3. To graph the function $x = A \cos(\omega t + \phi)$, just shift the axes to the right by a distance that corresponds to $\omega t = \phi$.

(b)

4. With $x = A \cos(\omega t + \phi)$, the value of x at $t = 0$ is $A \cos \phi$.

Figure 12-6 **Phase angle** In these x–t graphs of Equation 12-12, the two cycles shown in blue differ by a phase angle ϕ.

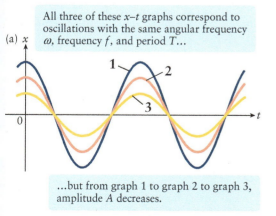

(a) x

All three of these x–t graphs correspond to oscillations with the same angular frequency ω, frequency f, and period T...

...but from graph 1 to graph 2 to graph 3, amplitude A decreases.

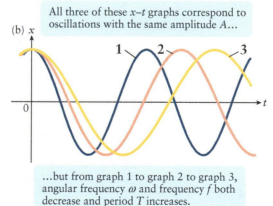

(b) x

All three of these x–t graphs correspond to oscillations with the same amplitude A...

...but from graph 1 to graph 2 to graph 3, angular frequency ω and frequency f both decrease and period T increases.

Figure 12-7 **Amplitude and angular frequency** In the x–t graph of Equation 12-12, the dimensions of the graph change if (a) you decrease the amplitude while keeping the angular frequency the same or (b) you decrease the angular frequency while keeping the amplitude the same. (In each of these graphs the phase angle is $\phi = 0$.)

The phase angle ϕ, amplitude A, and angular frequency ω also appear in the expressions for velocity and acceleration in simple harmonic motion (Equations 12-7 and 12-8):

Velocity as a function of time for simple harmonic motion
(12-14)

Velocity of an object in SHM at time t **Amplitude** of the oscillation = maximum displacement from the equilibrium position

$$v_x = -\omega A \sin(\omega t + \phi)$$

Angular frequency of the oscillation ($\omega = 2\pi f = 2\pi/T$)   **Phase angle** (in radians)

Acceleration as a function of time for simple harmonic motion
(12-15)

Acceleration of an object in SHM at time t **Amplitude** of the oscillation = maximum displacement from the equilibrium position

$$a_x = -\omega^2 A \cos(\omega t + \phi)$$

Angular frequency of the oscillation ($\omega = 2\pi f = 2\pi/T$)   **Phase angle** (in radians)

In the case $\phi = 0$, Equations 12-14 and 12-15 become

(12-16)

$$v_x = -\omega A \sin \omega t$$
$$a_x = -\omega^2 A \cos \omega t$$

(simple harmonic motion, phase angle $\phi = 0$)

Figure 12-8 graphs x versus t, v_x versus t, and a_x versus t for the case $\phi = 0$. In this case the x–t graph is a cosine function, the v_x–t graph is a negative sine function, and

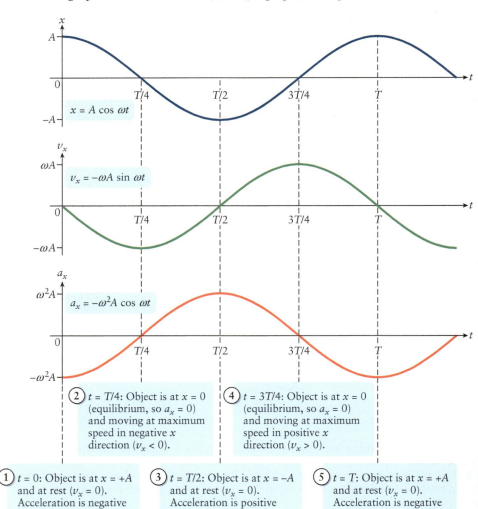

Figure 12-8 Displacement, velocity, and acceleration in SHM These graphs show displacement x (Equation 12-13) as well as velocity v_x and acceleration a_x (Equations 12-16) for simple harmonic motion for the case $\phi = 0$. Compare to Figure 12-3, which shows the same motion that these graphs describe.

② $t = T/4$: Object is at $x = 0$ (equilibrium, so $a_x = 0$) and moving at maximum speed in negative x direction ($v_x < 0$).

④ $t = 3T/4$: Object is at $x = 0$ (equilibrium, so $a_x = 0$) and moving at maximum speed in positive x direction ($v_x > 0$).

① $t = 0$: Object is at $x = +A$ and at rest ($v_x = 0$). Acceleration is negative (toward $x = 0$).

③ $t = T/2$: Object is at $x = -A$ and at rest ($v_x = 0$). Acceleration is positive (toward $x = 0$).

⑤ $t = T$: Object is at $x = +A$ and at rest ($v_x = 0$). Acceleration is negative (toward $x = 0$).

the a_x–t graph is a negative cosine function. You can confirm that these graphs are consistent with each other by recalling from Section 2-4 that the slope of the x–t graph equals the velocity v_x and the slope of the v_x–t graph equals the acceleration a_x. Note also that the a_x–t graph looks like the negative of the x–t graph. That's what we expect from Hooke's law, $a_x = -(k/m)x$ (Equation 12-2), which tells us that the acceleration in simple harmonic motion is proportional to the negative of the displacement x.

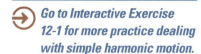

Go to Interactive Exercise 12-1 for more practice dealing with simple harmonic motion.

If the phase angle ϕ is not equal to zero, the vertical axes of all of the graphs in Figure 12-8 are shifted to the left or right as in Figure 12-6. In this case the velocity at $t = 0$ is equal to

$$v_x(0) = -\omega A \sin(0 + \phi) = -\omega A \sin \phi$$

If $\phi = 0$, the velocity at $t = 0$ is zero and the oscillating object is at rest at that instant (see Figure 12-8). If ϕ is not zero, at $t = 0$ the velocity may not be zero. Figure 12-6b illustrates this: In this case the x–t graph has a negative slope at $t = 0$, which shows that the velocity at that instant is not zero.

EXAMPLE 12-3 SHM II: Position, Velocity, and Acceleration

As in Example 12-2, an object of mass 0.80 kg is attached to an ideal horizontal spring of spring constant 1.8×10^2 N/m and set into oscillation (see Figure 12-3). The amplitude of the motion is 2.0×10^{-2} m, and the phase angle is $\phi = \pi/2$. (a) Find the maximum speed of the object and the maximum magnitude of its acceleration during the oscillation. (b) Find the position, velocity, and acceleration of the object at $t = 0$. (c) Sketch graphs of the position, velocity, and acceleration of the object as functions of time.

Set Up

We can use the simple harmonic motion equations because the spring is ideal (it obeys Hooke's law). From Example 12-2 we know that the angular frequency is $\omega = \sqrt{k/m} = 15$ rad/s, and we are given the values of the amplitude $A = 2.0 \times 10^{-2}$ m and the phase angle $\phi = \pi/2$. Notice that a_x from Equation 12-15 is equal to $-\omega^2$ times x from Equation 12-12.

Position as a function of time in SHM:
$$x = A \cos(\omega t + \phi) \quad (12\text{-}12)$$
Velocity as a function of time in SHM:
$$v_x = -\omega A \sin(\omega t + \phi) \quad (12\text{-}14)$$
Acceleration as a function of time in SHM:
$$a_x = -\omega^2 A \cos(\omega t + \phi) \quad (12\text{-}15)$$

Solve

(a) Equation 12-14 tells us the velocity v_x as a function of time. The sine function has values from +1 to −1, so v_x has values from $+\omega A$ to $-\omega A$. The maximum speed is therefore ωA.

Since $v_x = -\omega A \sin(\omega t + \phi)$, the maximum value of the speed is
$$v_{max} = \omega A = (15 \text{ rad/s})(2.0 \times 10^{-2} \text{ m}) = 0.30 \text{ m/s}$$
(A radian is not a true unit, so we can insert it or remove it as needed.) From Figures 12-3 and 12-7, the object has maximum speed when passing through the equilibrium position $x = 0$.

Equation 12-15 tells us the acceleration a_x as a function of time. The cosine varies from +1 to −1, so a_x has values from $+\omega^2 A$ to $-\omega^2 A$ and the maximum magnitude of a_x is $\omega^2 A$.

Since $a_x = -\omega^2 A \cos(\omega t + \phi)$, the maximum value of the acceleration is
$$a_{max} = \omega^2 A = (15 \text{ rad/s})^2 (2.0 \times 10^{-2} \text{ m}) = 4.5 \text{ m/s}^2$$
From Figures 12-3 and 12-7, the object has maximum acceleration when it is farthest from the equilibrium position (at $x = A$ or $x = -A$).

(b) To find the values of x, v_x, and a_x at $t = 0$, substitute the values of A, ω, and ϕ as well as $t = 0$ into Equations 12-12, 12-14, and 12-15. Note that $\cos(\pi/2) = 0$ and $\sin(\pi/2) = 1$. We find that the object is at $x = 0$ at $t = 0$, so it has zero acceleration and has the maximum speed that we found in part (a). Since $v_x(0)$ is negative, at $t = 0$ the object is moving in the negative x direction.

Position at $t = 0$:
$$x(0) = A \cos(0 + \phi) = (2.0 \times 10^{-2} \text{ m}) \cos(\pi/2)$$
$$= (2.0 \times 10^{-2} \text{ m})(0) = 0$$
Velocity at $t = 0$:
$$v_x(0) = -\omega A \sin(0 + \phi) = -(15 \text{ rad/s})(2.0 \times 10^{-2} \text{ m}) \sin(\pi/2)$$
$$= -(15 \text{ rad/s})(2.0 \times 10^{-2} \text{ m})(1) = -0.30 \text{ m/s}$$
Acceleration at $t = 0$:
$$a_x(0) = -\omega^2 A \cos(0 + \phi) = -(15 \text{ rad/s})^2 (2.0 \times 10^{-2} \text{ m}) \cos(\pi/2)$$
$$= -(15 \text{ rad/s})^2 (2.0 \times 10^{-2} \text{ m})(0) = 0$$

(c) To draw the x–t graph, recognize that this graph must be sinusoidal and that the slope of the x–t graph is the velocity v_x. The values of x and v_x at $t = 0$ that we found in part (b) tell us that at $t = 0$ the x–t graph has value 0 and a negative slope. The value of x oscillates between $+A = +2.0 \times 10^{-2}$ m and $-A = -2.0 \times 10^{-2}$ m. The graph repeats itself after a time t equal to the period $T = 0.42$ s.

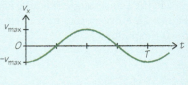

The v_x–t graph is also sinusoidal, and its slope is the acceleration a_x. The values of v_x and a_x at $t = 0$ tell us that at $t = 0$ the v_x–t graph has a negative value and zero slope. From part (a) the value of v_x oscillates between $+v_{max} = +0.30$ m/s and $-v_{max} = -0.30$ m/s. The graph repeats itself after a time t equal to the period $T = 0.42$ s.

Finally, comparing Equations 12-12 and 12-15 shows that $a_x = -\omega^2 x$, so the a_x–t graph looks like the negative of the x–t graph (compare Figure 12-8). From part (a) the value of a_x oscillates between $+a_{max} = +4.5$ m/s^2 and $-a_{max} = -4.5$ m/s^2. The graph repeats itself after a time t equal to the period $T = 0.42$ s.

Reflect

To start an oscillation going so that the graphs look like what we have drawn, begin with the block at rest at equilibrium ($x = 0$) and give it a sharp hit with your hand in the negative x direction so that it starts with a negative x velocity v_x. Note that the x–t, v_x–t, and a_x–t graphs look like those in Figure 12-8 but with the vertical axis shifted to the right by a quarter-cycle. That's what we would expect: Figure 12-6 shows that the effect of a phase angle ϕ is to shift the vertical axis horizontally by ϕ, and $\pi/2$ is one-quarter of 2π (the number of radians in a complete cycle).

BioMedical

Our discussion of simple harmonic motion has concentrated on blocks attached to springs. But the same basic equations, and the sinusoidal graph of the oscillation as a function of time, apply to many different oscillating systems. Figure 12-9a shows an example from cell biology. In some cells the intracellular concentration of calcium ions (Ca^{2+}) oscillates sinusoidally. (These oscillations can regulate enzyme activity, mitochondrial metabolism, or gene expression.) The red sinusoidal curve drawn on top of the data matches these oscillations well. Figure 12-9b shows another biological example, the volume of air in a person's lungs as a function of time. Again we've drawn a red sinusoidal curve on top of the data; the close match between the curves shows that the oscillations are essentially simple harmonic motion.

(a)

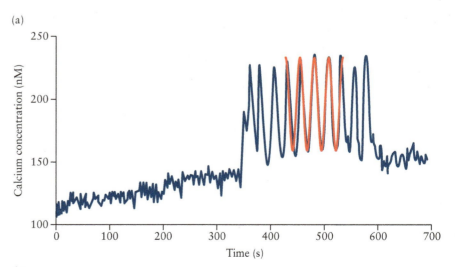

Figure 12-9 Simple harmonic motion in physiology (a) The concentration of calcium ions (Ca^{2+}) within a cell can oscillate under certain conditions. A sinusoidal curve (red) is a good match to these oscillations. (b) This graph shows the volume of air in a person's lungs as a function of time. The oscillations in lung volume are also well matched to a sinusoidal curve (red).

(b)

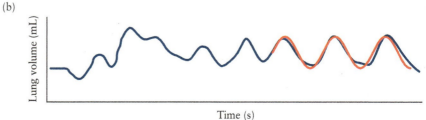

GOT THE CONCEPT? 12-2 Changing the Amplitude

(?) Suppose you were to increase the amplitude of the oscillation in Example 12-3 by a factor of 4, from 2.0×10^{-2} to 8.0×10^{-2} m. The spring constant and mass remain the same as in Example 12-3. This would cause the frequency of the oscillation to (a) increase by a factor of 4; (b) increase by a factor of 2; (c) decrease by a factor of 1/4; (d) decrease by a factor of 1/2; (e) remain unchanged.

GOT THE CONCEPT? 12-3 Changing the Spring Constant

(?) Suppose you were to increase the spring constant of the spring in Example 12-3 by a factor of 4, from 1.8×10^2 to 7.2×10^2 N/m. The mass and amplitude remain the same as in Example 12-3. This would cause the maximum speed of the block during the oscillation to (a) increase by a factor of 4; (b) increase by a factor of 2; (c) decrease by a factor of 1/4; (d) decrease by a factor of 1/2; (e) remain unchanged.

GOT THE CONCEPT? 12-4 Changing the Phase Angle

(?) Suppose you were to change the phase angle in Example 12-3 to $\phi = \pi$. The spring constant, mass, and amplitude remain the same as in Example 12-3. This would cause the displacement x of the block at $t = 0$ to be (a) $A = 2.0 \times 10^{-2}$ m; (b) positive, but less than A; (c) zero; (d) $-A = -2.0 \times 10^{-2}$ m; (e) negative, but between zero and $-A$.

TAKE-HOME MESSAGE FOR Section 12-3

✔ The amplitude of an oscillation is the maximum positive displacement from equilibrium.

✔ An object oscillating under the influence of a Hooke's law restoring force is called simple harmonic motion (SHM). The angular frequency, period, and frequency in SHM are unaffected by changes in the oscillation amplitude.

✔ SHM is identical to the projection onto the x axis of the motion of an object in uniform circular motion.

✔ Graphs of the position, velocity, and acceleration in SHM are sinusoidal curves. In each of these graphs the phase angle shifts the location of the vertical axis (where time equals zero) to the right or left.

12-4 Mechanical energy is conserved in simple harmonic motion

We've been discussing how a spring can provide the restoring force necessary for oscillation to take place. But as we learned in Section 6-6, a spring can also be used to store potential energy. One creature that applies this principle to oscillation is the kangaroo (Figure 12-10). A kangaroo's tendons act like springs to store spring potential energy, which is transformed into kinetic energy as the animal hops; this increases the efficiency of movement. Dolphins use a similar mechanism in

BioMedical

The tendons on the backs of this kangaroo's legs are fully stretched. As such they store spring potential energy...

...which is released during a hop, as the kangaroo straightens its legs and lets its tendons relax...

...and the released spring potential energy is converted into kinetic energy of the kangaroo.

David Tauck

Figure 12-10 Spring potential energy When stretched, tendons in a kangaroo's legs store potential energy in much the same way as a spring.

swimming. Even when a dolphin increases its speed by beating its tail faster, there is hardly any change in the rate at which it consumes oxygen (an indication of the dolphin's power output). The explanation is that the tissue in a dolphin's tail acts much like a spring, storing potential energy as the tail flips either up or down and then transforming the potential energy into kinetic energy. As a result, the dolphin can swim efficiently at high speeds.

These observations suggest that it's worthwhile to look at oscillations again from an energy perspective. To do this we'll study the interplay between kinetic energy and potential energy for a block connected to the free end of an ideal, Hooke's law spring, oriented horizontally as in Figures 12-2 and 12-3.

Our starting point is the expression from Section 6-6 for the potential energy stored in an ideal spring:

> **Spring potential energy** of a stretched or compressed spring   **Spring constant** of the spring

Spring potential energy
(6-23)

$$U_{\text{spring}} = \frac{1}{2} k x^2$$

> **Extension** of the spring ($x > 0$ if spring is stretched, $x < 0$ if spring is compressed)

This equation tells us that the potential energy stored in a spring varies during a cycle of oscillation. When the object is at equilibrium ($x = 0$), the spring is neither stretched nor compressed, and the spring potential energy is zero: $U_{\text{spring}} = 0$. Whenever the spring is stretched ($x > 0$) or compressed ($x < 0$), the spring potential energy is greater than zero. The potential energy is greatest when the spring is either at its maximum extension (when $x = A$) or maximum compression ($x = -A$). At either of those points $U_{\text{spring}} = \frac{1}{2} k A^2$.

The *kinetic* energy of the oscillating block, $K = \frac{1}{2} m v^2$, also varies during an oscillation cycle. The block has its maximum speed v_{max} when passing through the equilibrium position, so the kinetic energy has its maximum value $K = \frac{1}{2} m v_{\text{max}}^2$ there as well. At any other point in the oscillation the block is moving more slowly than when it passes through equilibrium, so the kinetic energy is less than its maximum value. The kinetic energy has its minimum value $K = 0$ when the oscillating block is momentarily at rest; this happens when the spring is at maximum extension ($x = A$) or maximum compression ($x = -A$).

This analysis shows that for a block oscillating on an ideal spring, the kinetic energy is maximum where the spring potential energy is minimum (at $x = 0$), and the kinetic energy is minimum where the spring potential energy is maximum (at $x = A$ and $x = -A$). This is just what we would expect! If there are no nonconservative forces such as friction acting, the total mechanical energy E (the sum of the kinetic energy of the block and the potential energy of the spring) should remain constant, and energy will be transformed back and forth between its kinetic and potential forms as the block oscillates.

To analyze this in more detail, let's look at the expression for the total mechanical energy E of the system of block and spring:

(12-17)
$$E = K + U_{\text{spring}} = \frac{1}{2} m v^2 + \frac{1}{2} k x^2$$

Let's use our results from Section 12-3 to see how E, K, and U_{spring} change during the course of an oscillation. From Equation 12-12 the position as a function of time is $x = A \cos (\omega t + \phi)$, so the spring potential energy as a function of time is

(12-18)
$$U_{\text{spring}} = \frac{1}{2} k x^2 = \frac{1}{2} k A^2 \cos^2 (\omega t + \phi)$$

Since the value of $\cos (\omega t + \phi)$ ranges from $+1$ through zero to -1, you can see that the value of $\cos^2 (\omega t + \phi)$ ranges from 0 to 1. Hence the value of U_{spring} ranges from 0 to a maximum value $\frac{1}{2} k A^2$, just as we mentioned above.

Equation 12-14 tells us that the velocity as a function of time is $v_x = -\omega A \sin(\omega t + \phi)$. The velocity can be negative, but the square of the velocity is positive and equal to the square of the speed. Hence the kinetic energy as a function of time is

$$K = \frac{1}{2}mv^2 = \frac{1}{2}m\omega^2 A^2 \sin^2(\omega t + \phi)$$

From the first of Equations 12-11, we know that the angular frequency for a block oscillating on the end of an ideal spring is $\omega = \sqrt{k/m}$. So $m\omega^2 = m(k/m) = k$, and we can rewrite our expression for the kinetic energy as a function of time as

$$K = \frac{1}{2}mv^2 = \frac{1}{2}kA^2 \sin^2(\omega t + \phi) \tag{12-19}$$

Like the cosine function the value of $\sin(\omega t + \phi)$ ranges from +1 through 0 to –1, and the value of $\sin^2(\omega t + \phi)$ ranges from 0 to 1. So just like the spring potential energy, the kinetic energy of the block ranges in value from 0 to a maximum value $\frac{1}{2}kA^2$.

> $\sqrt{x}$ **See the Math Tutorial for more information on trigonometry.**

We can now insert our expressions for K from Equation 12-19 and U_{spring} from Equation 12-18 into Equation 12-17 for the total mechanical energy E. We get

$$E = K + U_{spring} = \frac{1}{2}kA^2 \sin^2(\omega t + \phi) + \frac{1}{2}kA^2 \cos^2(\omega t + \phi)$$

$$= \frac{1}{2}kA^2[\sin^2(\omega t + \phi) + \cos^2(\omega t + \phi)] \tag{12-20}$$

Equation 12-20 looks like a complicated function of time. But we can simplify it thanks to an important result from trigonometry:

$$\sin^2 \theta + \cos^2 \theta = 1 \qquad \text{for any value of } \theta$$

Then Equation 12-20 becomes

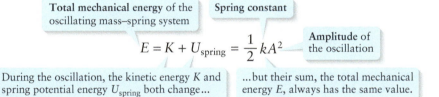

Total mechanical energy of the oscillating mass–spring system	Spring constant

$$E = K + U_{spring} = \frac{1}{2}kA^2$$

Amplitude of the oscillation

During the oscillation, the kinetic energy K and spring potential energy U_{spring} both change... ...but their sum, the total mechanical energy E, always has the same value.

Total mechanical energy of a mass oscillating on an ideal spring
(12-21)

Because both k and A are constant for a specific motion of a given oscillator, the total energy of the oscillating system is constant. This is a statement of the conservation of energy: The energy of the system transforms between the kinetic and potential forms, but the total mechanical energy remains constant (Figure 12-11).

Equations 12-18 and 12-19 give the spring potential energy and kinetic energy as functions of time. It's also helpful to write these relationships as functions of the displacement x. We know from Equation 6-23 that the spring potential energy is $U_{spring} = (1/2)kx^2$; the graph of this as a function of x is a parabola with its minimum value at $x = 0$ (Figure 12-12). From Equation 12-21 the kinetic energy K is equal to the total mechanical energy $E = (1/2)kA^2$ (a constant) minus the potential energy U_{spring}:

$$K = E - U_{spring} = \frac{1}{2}kA^2 - \frac{1}{2}kx^2 \tag{12-22}$$

This is an "upside-down" parabola that has its *maximum* value at $x = 0$, as Figure 12-12 shows.

Note that U_{spring} is greatest at the two extremes of displacement ($x = +A$ and $x = -A$) and zero at equilibrium ($x = 0$). The kinetic energy K is zero at the two extremes, where the block momentarily comes to a stop, and greatest as the object passes through $x = 0$. So at $x = 0$ the energy is 100% kinetic and 0% potential, while at $x = +A$ and $x = -A$ the energy is 0% kinetic and 100% potential.

Figure 12-11 Energy during a cycle of oscillation The bar graphs show the kinetic energy K, spring potential energy U_{spring}, and total mechanical energy $E = K + U_{spring}$ at nine stages in the oscillation of a block attached to a spring (compare with Figure 12-3).

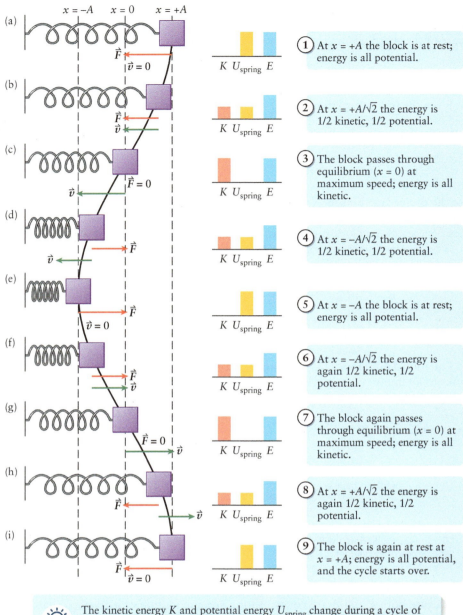

(1) At $x = +A$ the block is at rest; energy is all potential.

(2) At $x = +A/\sqrt{2}$ the energy is 1/2 kinetic, 1/2 potential.

(3) The block passes through equilibrium ($x = 0$) at maximum speed; energy is all kinetic.

(4) At $x = -A/\sqrt{2}$ the energy is 1/2 kinetic, 1/2 potential.

(5) At $x = -A$ the block is at rest; energy is all potential.

(6) At $x = -A/\sqrt{2}$ the energy is again 1/2 kinetic, 1/2 potential.

(7) The block again passes through equilibrium ($x = 0$) at maximum speed; energy is all kinetic.

(8) At $x = +A/\sqrt{2}$ the energy is again 1/2 kinetic, 1/2 potential.

(9) The block is again at rest at $x = +A$; energy is all potential, and the cycle starts over.

The kinetic energy K and potential energy U_{spring} change during a cycle of oscillation, but the total mechanical energy $E = K + U_{spring}$ remains constant.

Figure 12-12 Kinetic energy and potential energy versus displacement The total mechanical energy — the sum of the kinetic energy and spring potential energy — associated with an object in simple harmonic motion is constant. The percentage of the total mechanical energy that is in the kinetic and potential forms depends on the object's displacement from equilibrium.

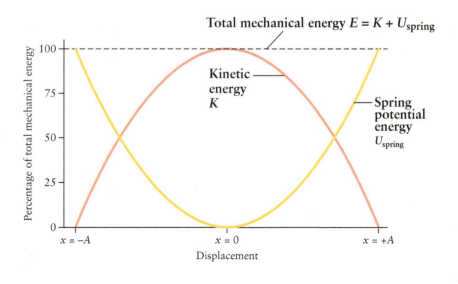

EXAMPLE 12-4   SHM III: Kinetic and Potential Energy

As in Example 12-3 an object of mass 0.80 kg is attached to an ideal horizontal spring of spring constant 1.8×10^2 N/m. The object is initially at rest at equilibrium. You then start the object oscillating with amplitude 2.0×10^{-2} m. (a) How much work do you have to do on the object to set it into oscillation? (b) What is the speed of the object when the spring is compressed by 1.0×10^{-2} m? (c) How far is the object from equilibrium when the kinetic energy of the object equals the potential energy in the spring?

Set Up

We'll use energy ideas to answer these questions. Equation 12-21 will let us find the (constant) total mechanical energy E of the system from the given spring constant and amplitude. Equation 6-23 will let us find the spring potential energy for any displacement x; we can then find the kinetic energy K using Equation 12-21, and from that find the speed for that value of x.

Total mechanical energy of a mass oscillating on an ideal spring:

$$E = K + U_{spring} = \frac{1}{2}kA^2 \quad (12\text{-}21)$$

Spring potential energy:

$$U_{spring} = \frac{1}{2}kx^2 \quad (6\text{-}23)$$

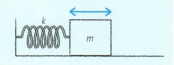

Solve

(a) Initially the system has zero kinetic energy and zero potential energy. You can start the oscillation by pulling the object so that you stretch the spring by a distance $A = 2.0 \times 10^{-2}$ m from equilibrium. The work you do goes into the potential energy of the spring.

(work you do) = (change of potential energy of the spring as you stretch it from $x = 0$ to $x = A$)

Initial potential energy $= \frac{1}{2}k(0)^2 = 0$

Final potential energy $= \frac{1}{2}kA^2$

So the work you do is

$$W = \frac{1}{2}kA^2 - 0$$

$$= \frac{1}{2}(1.8 \times 10^2 \text{ N/m})(2.0 \times 10^{-2} \text{ m})^2$$

$$= 3.6 \times 10^{-2} \text{ N·m} = 3.6 \times 10^{-2} \text{ J}$$

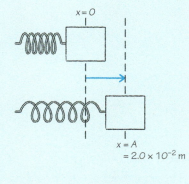

(b) When the spring is compressed by 1.0×10^{-2} m, the displacement is $x = -1.0 \times 10^{-2}$ m. The first step in determining the block's speed at this value of x is to determine its kinetic energy.

From (a) the total mechanical energy is

$$E = \frac{1}{2}kA^2 = 3.6 \times 10^{-2} \text{ J}$$

From Equation 6-19 the spring potential energy for $x = -1.0 \times 10^{-2}$ m is

$$U_{spring} = \frac{1}{2}kx^2$$

$$= \frac{1}{2}(1.8 \times 10^2 \text{ N/m})(-1.0 \times 10^{-2} \text{ m})^2$$

$$= 9.0 \times 10^{-3} \text{ J}$$

So from Equation 12-21 the kinetic energy at this value of x is

$$K = E - U_{spring} = 3.6 \times 10^{-2} \text{ J} - 9.0 \times 10^{-3} \text{ J}$$

$$= 2.7 \times 10^{-2} \text{ J}$$

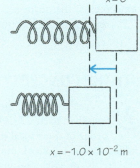

Given the kinetic energy K of the block and its mass $m = 0.80$ kg, calculate the speed of the block.

Solve for the speed v:

Kinetic energy is $K = \frac{1}{2}mv^2$, so

$$v^2 = \frac{2K}{m}$$

$$v = \sqrt{\frac{2K}{m}} = \sqrt{\frac{2(2.7 \times 10^{-2} \text{ J})}{0.80 \text{ kg}}} = 0.26 \text{ m/s}$$

(Recall from Chapter 6 that if the kinetic energy is in joules and the mass is in kilograms, the speed is in meters per second.)

The *velocity* of the block could be +0.26 m/s or −0.26 m/s, depending on what direction it is moving as it passes through this point.

(c) Use the same ideas as in part (b) to solve for the value of x at which the kinetic energy K equals the spring potential energy U_{spring}.

From above the total mechanical energy is

$$E = \frac{1}{2}kA^2 = 3.6 \times 10^{-2} \text{ J}$$

The spring potential energy is

$$U_{spring} = \frac{1}{2}kx^2$$

and the kinetic energy is

$$K = E - U_{spring}$$

At the value of x for which $K = U_{spring}$,

$$\frac{1}{2}kA^2 - \frac{1}{2}kx^2 = \frac{1}{2}kx^2$$

Multiply both sides by $2/k$ and solve for x:

$$A^2 - x^2 = x^2 \quad \text{so} \quad 2x^2 = A^2 \quad \text{and} \quad x^2 = \frac{A^2}{2}$$

$$x = \pm\sqrt{\frac{A^2}{2}} = \pm\frac{A}{\sqrt{2}}$$

$$= \pm\frac{2.0 \times 10^{-2} \text{ m}}{\sqrt{2}} = \pm 1.4 \times 10^{-2} \text{ m}$$

Note that x can be either positive (the spring is stretched by 1.4×10^{-2} m) or negative (the spring is compressed by 1.4×10^{-2} m).

Reflect

The result for part (b) would have been very difficult to find without using the energy approach. You would have needed to use the equations from Section 12-3 to solve for the time t at which the block passes through $x = -1.0 \times 10^{-2}$ m, then use this value of t to find the velocity of the block at that time. The energy approach makes this much easier.

Note that the points $x = \pm A/\sqrt{2} = \pm 0.71A$ where the kinetic and potential energies are equal are shown in Figure 12-11 (see parts (b), (d), (f), and (h) of that figure). These points are *not* halfway between the equilibrium position ($x = 0$) and the extremes of the motion ($x = \pm A$); they are actually closer to the extremes. Can you show that at $x = \pm A/2$, the energy is 75% kinetic and 25% potential? (*Hint:* See the solution to part (b).)

GOT THE CONCEPT? 12-5 Rank the Energies

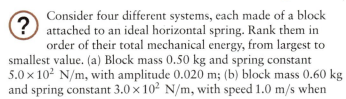

Consider four different systems, each made of a block attached to an ideal horizontal spring. Rank them in order of their total mechanical energy, from largest to smallest value. (a) Block mass 0.50 kg and spring constant 5.0×10^2 N/m, with amplitude 0.020 m; (b) block mass 0.60 kg and spring constant 3.0×10^2 N/m, with speed 1.0 m/s when passing through equilibrium; (c) block mass 1.2 kg and spring constant 4.0×10^2 N/m, with speed 0.50 m/s when passing through $x = -0.010$ m; (d) block mass 2.0 kg and spring constant 2.0×10^2 N/m, with speed 0.20 m/s when passing through $x = 0.050$ m.

TAKE-HOME MESSAGE FOR Section 12-4

✔ Energy is transformed back and forth from kinetic energy to potential energy in systems which contain mechanical or biological springs.

✔ When an object attached to a spring is at its maximum displacement, the energy of the system of spring and object is entirely potential energy.

✔ When the object passes through equilibrium, the energy of the system of spring and object is entirely kinetic energy.

✔ The total energy of the system, the maximum potential energy, and the maximum kinetic energy are all equal to $(1/2)kA^2$.

12-5 The motion of a pendulum is approximately simple harmonic

BioMedical You've probably had a physician strike the patellar tendon just below your kneecap with a hammer then watched your lower leg swing upward and back (Figure 12-13a). For a physician this is a test of your patellar reflex: Hitting that tendon sends a signal to your spinal cord, which in turn sends a signal to the quadriceps muscle on your upper thigh. This makes that muscle flex and makes the lower leg move. But to a physicist this is the same kind of motion as a mass swinging at the end of a string (Figure 12-13b). When allowed to move freely, both your leg and the mass hang straight down in their equilibrium positions thanks to the influence of gravity. When either your leg or the mass is displaced from equilibrium, released, and allowed to move freely, the force of gravity pulls it back toward equilibrium—that is, gravity serves as a restoring force. And in either case inertia causes the object (leg or mass) to overshoot the equilibrium position, resulting in an oscillation. Both a swinging leg and a mass on the end of a string are examples of **pendulums**—systems that oscillate back and forth due to the restoring force of gravity.

Since pendulums actually *rotate* around a point, we'll describe them using the language of rotational motion that we developed in Chapter 9. So instead of relating a restoring force to the acceleration that it produces, as we did in Section 12-3 for a block attached to an ideal spring, we'll describe pendulum motion in terms of a restoring *torque* that produces an *angular* acceleration.

In this section we'll explore oscillations of a **simple pendulum**. This is one in which all of the mass is concentrated at a single point. It's an idealized version of the mass-and-string arrangement shown in Figure 12-13b in which the string of length L has zero mass and all of the mass m is compressed to a point. (In Section 12-6 we'll look at other, less idealized pendulums.)

Figure 12-14a shows a simple pendulum. In equilibrium the string hangs straight down; the figure shows the pendulum displaced by an angle θ from the vertical. The blue dashed curve shows the path that the object of mass m at the end of the pendulum will take to return to equilibrium. Figure 12-14b shows the free-body diagram for this object. The two forces that act on this object are a tension force of magnitude T exerted by the string and a gravitational force of magnitude mg. The tension force points along the string, so the line of action of the tension force passes through the pivot point.

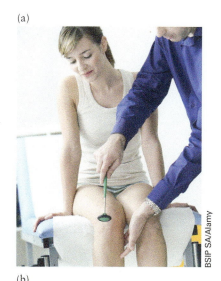
(a)

BSIP SA/Alamy

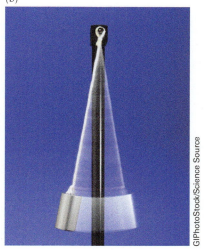
(b)

GIPhotoStock/Science Source

Figure 12-13 Swinging leg, swinging pendulum The motions of both (a) your lower leg in a physician's office and (b) a mass hanging from a string have the ingredients for oscillation: a restoring force (gravity) that tends to pull the object toward equilibrium (hanging straight down), and inertia that causes the object to overshoot equilibrium.

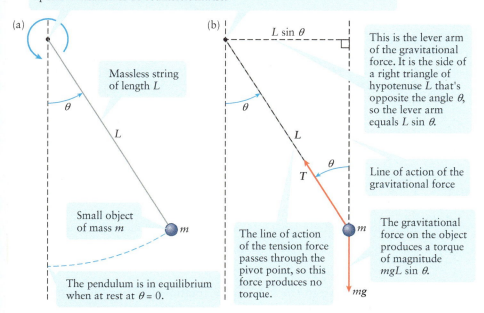

Pivot point: The positive z axis (the rotation axis of the pendulum) points outward from this point, perpendicular to the plane of the drawing. We take positive rotation to be counterclockwise.

(a)

Massless string of length L

θ

L

Small object of mass m

m

The pendulum is in equilibrium when at rest at $\theta = 0$.

(b)

$L \sin \theta$

This is the lever arm of the gravitational force. It is the side of a right triangle of hypotenuse L that's opposite the angle θ, so the lever arm equals $L \sin \theta$.

θ

L

θ

T

Line of action of the gravitational force

m

The line of action of the tension force passes through the pivot point, so this force produces no torque.

The gravitational force on the object produces a torque of magnitude $mgL \sin \theta$.

mg

Figure 12-14 A simple pendulum (a) Layout of a simple pendulum. (b) The forces on a simple pendulum.

Hence the tension force exerts no torque around the pivot (it does nothing to make the pendulum rotate around the pivot). The gravitational force does exert a torque, however. From Section 9-3 its magnitude is the magnitude mg of the gravitational force multiplied by the lever arm of the force (the perpendicular distance from the line of action of the gravitational force to the pivot). Figure 12-14b shows that if the pendulum is at an angle θ from the vertical, this lever arm is $L \sin \theta$. Hence we can write the torque as:

(12-23)
$$\tau_z = -mgL \sin \theta$$

The subscript z indicates that the pendulum tends to rotate around the z axis shown in Figure 12-14a. The minus sign in Equation 12-23 indicates that this is a *restoring* torque. If the pendulum swings to the right of vertical (counterclockwise) so that θ and $\sin \theta$ are positive, as in Figure 12-14a, the torque will be negative (clockwise): That means it acts to decrease θ and so return the pendulum to the equilibrium position $\theta = 0$. If the pendulum swings instead to the left of vertical (clockwise) so that θ and $\sin \theta$ are negative, the torque will be positive (counterclockwise) and so will again act to restore the pendulum to equilibrium.

The torque in Equation 12-23 is the only torque that acts on the pendulum, so this is also the *net* torque. Newton's second law for rotational motion (Section 9-3) says that the net torque on an object is equal to the moment of inertia multiplied by the angular acceleration. For an object made up of a single point of mass m a distance L from the rotation axis, the moment of inertia is $I = mL^2$. Using the torque from Equation 12-23, we have

$$-mgL \sin \theta = mL^2 \alpha_z$$

If we divide both sides of this equation by mL^2 and rearrange, we get

(12-24)
$$\alpha_z = -\frac{g}{L} \sin \theta$$

Let's see whether Equation 12-24 is equivalent to Hooke's law. If it is, that means the oscillations of simple pendulums are simple harmonic motion and we can use all of our results from Section 12-3. We saw in Section 12-3 that we can write Hooke's law for straight-line oscillation as

$$a_x = -\omega^2 x$$

That is, the acceleration is directly proportional to the displacement, and the proportionality constant is the negative of the square of the angular frequency ω. From Section 9-2 the quantities x and a_x for straight-line motion correspond to the quantities θ and α_z, respectively, for rotational motion. So the rotational version of Hooke's law is

(12-25)
$$\alpha_z = -\omega^2 \theta$$

The difference between Equation 12-24 for the pendulum and Equation 12-25, the rotational version of Hooke's law, is that Equation 12-24 involves $\sin \theta$ rather than θ. So in general the oscillations of a simple pendulum do *not* obey Hooke's law, and so the motion of the pendulum is *not* simple harmonic motion.

If, however, the angle θ is relatively small and we measure θ in radians, it turns out that $\sin \theta$ is *approximately* equal to θ (Figure 12-15):

$$\sin \theta \approx \theta \quad \text{when } \theta \text{ is small}$$

(You can confirm this with your calculator. Try calculating $\sin \theta$ for $\theta = 1$ rad, 0.5 rad, 0.2 rad, 0.1 rad, and 0.01 rad. You'll find that as you try smaller values of θ, the value of $\sin \theta$ gets closer and closer to θ.) With this approximation, Equation 12-24 for the simple pendulum becomes

(12-26)
$$\alpha_z \approx -\frac{g}{L} \theta \quad \text{if } \theta \text{ is small}$$

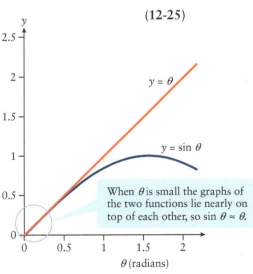

Figure 12-15 **Approximating sin θ** If the value of θ is small, $\sin \theta$ is approximately equal to θ (measured in radians).

When θ is small the graphs of the two functions lie nearly on top of each other, so $\sin \theta \approx \theta$.

Compare this to Hooke's law for rotational motion, Equation 12-25, and you'll see that the oscillations of a pendulum *with small amplitude* (so that θ is always small) obey Hooke's law and that the angular frequency of the simple pendulum's oscillations is given by

$$\omega^2 = \frac{g}{L} \quad \text{or} \quad \omega = \sqrt{\frac{g}{L}}$$

As in Section 12-3 the period T is equal to $2\pi/\omega$, and the frequency f is equal to $1/T$ or $\omega/2\pi$. So we can write the following results for the small-amplitude oscillations of a simple pendulum:

Angular frequency $\quad \omega = \sqrt{\dfrac{g}{L}}$

g = acceleration due to gravity

L = length of the simple pendulum

Period $\quad T = \dfrac{2\pi}{\omega} = 2\pi\sqrt{\dfrac{L}{g}}$

Frequency $\quad f = \dfrac{1}{T} = \dfrac{1}{2\pi}\sqrt{\dfrac{g}{L}} = \dfrac{\omega}{2\pi}$

Angular frequency, period, and frequency for a simple pendulum (small amplitude)
(12-27)

Equations 12-27 show that the angular frequency, period, and frequency for a simple pendulum do *not* depend on the mass of the object at the end of the pendulum. That's because the restoring torque is provided by the gravitational force. Doubling the mass doubles the moment of inertia, which by itself would make the oscillations happen more slowly, but it also doubles the restoring torque, which by itself would make the oscillations happen more rapidly. The two effects cancel each other out so that ω, T, and f are unaffected by changes in the pendulum mass.

Because the small-amplitude oscillations of a simple pendulum obey Hooke's law, it also follows that the angular frequency, period, and frequency do not depend on the *amplitude* of the oscillations. Here the amplitude is the maximum angle from the vertical that the pendulum attains. If you were to pull the simple pendulum in Figure 12-14a to an angle $\theta = 0.1$ rad—about 6°—and let it go, the pendulum would oscillate between $\theta = +0.1$ rad (to the right of vertical) and $\theta = -0.1$ rad (to the left of vertical), and the amplitude of the oscillation would be 0.1 rad.

Go to Picture It 12-1 for more practice dealing with simple pendulums.

WATCH OUT! Large-amplitude oscillations of a simple pendulum are *not* simple harmonic motion.

Equations 12-27 apply only if the angle θ reached by the pendulum is always small enough that $\sin \theta$ is approximately equal to θ. Figure 12-15 shows that this is not a good approximation if θ is greater than about 0.5 rad (about 30°). If the amplitude is larger than this, the motion is *not* simple harmonic and the values of ω, T, and f do depend on the amplitude. As an example, if the amplitude is $\pi/2$ (90°), the period of oscillation is about 18% greater than the value given by the second of Equations 12-27.

Just as we did for the oscillations of a block on a spring in Section 12-4, we can also interpret the oscillations of a pendulum in terms of kinetic energy, potential energy, and total mechanical energy. The only difference is that for the pendulum there is *gravitational* potential energy instead of spring potential energy. As the pendulum rises, the gravitational potential energy U increases and the kinetic energy K decreases; as it descends, U decreases and K increases. At the lowest point of the pendulum's motion, U is minimum and the mass has its maximum speed and maximum kinetic energy K. At the highest points of the motion, U is maximum, the mass is momentarily at rest, and K is zero. At all points during the motion, the total mechanical energy $E = K + U$ remains the same. (The tension force exerted by the string does no work because it always acts perpendicular to the direction of motion of the mass. Hence only the conservative force of gravity does work, and mechanical energy is conserved.)

EXAMPLE 12-5 **Changing a Pendulum**

You make a simple pendulum by hanging a small 60-g marble from an elastic thread of negligible mass. With the marble attached the thread is 0.40 m long. (a) Find the period when the marble is pulled a small angle to one side and released. (b) You replace the marble by another small one of mass 260 g. When you do this the thread stretches an additional 0.10 m. Find the new period.

Set Up

We'll use the second of Equations 12-27 to find the period of this simple pendulum.

Period of a simple pendulum:

$$T = 2\pi\sqrt{\frac{L}{g}} \qquad (12\text{-}27)$$

Solve

(a) Find the period of the pendulum with the initial length $L = 0.40$ m.

Then the period is

$$T = 2\pi\sqrt{\frac{L}{g}} = 2\pi\sqrt{\frac{0.40\text{ m}}{9.80\text{ m/s}^2}} = 2\pi\sqrt{\frac{0.40\text{ s}^2}{9.80}} = 1.3\text{ s}$$

(b) Repeat the calculation with the new value of the length.

The new length of the pendulum is

$$L_{new} = 0.40\text{ m} + 0.10\text{ m} = 0.50\text{ m}$$

So the new period is

$$T_{new} = 2\pi\sqrt{\frac{L_{new}}{g}} = 2\pi\sqrt{\frac{0.50\text{ m}}{9.80\text{ m/s}^2}} = 1.4\text{ s}$$

Reflect

Although the mass of the marble changed, we did *not* have to use its mass in the calculation. As we've seen, the period of a simple pendulum depends only on its length, not on its mass.

Note that we know the value of g to three significant figures, but our answers for the period are given to only two significant figures. Can you see why? (*Hint:* Review Section 1-4 if you're not sure.)

GOT THE CONCEPT? 12-6 **Simple Pendulum I**

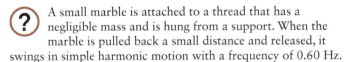

A small marble is attached to a thread that has a negligible mass and is hung from a support. When the marble is pulled back a small distance and released, it swings in simple harmonic motion with a frequency of 0.60 Hz.

What is the frequency of the pendulum after the length of the thread is increased by a factor of 4 but the marble is released in the same way? (a) 2.4 Hz; (b) 1.2 Hz; (c) 0.60 Hz; (d) 0.30 Hz; (e) 0.15 Hz.

GOT THE CONCEPT? 12-7 **Simple Pendulum II**

A small marble is attached to a thread that has a negligible mass and is hung from a support. When the marble is pulled back a small distance and released, it swings in simple harmonic motion with a frequency of 0.60 Hz. If the marble is replaced by one that has four times the mass

of the original one, and the new marble is pulled back by half the distance of the original marble before being released, what is the new frequency? Assume that the length of the thread remains the same. (a) 2.4 Hz; (b) 1.2 Hz; (c) 0.60 Hz; (d) 0.30 Hz; (e) 0.15 Hz.

GOT THE CONCEPT? 12-8 **Pendulum on an Asteroid**

A small marble is attached to a thread that has a negligible mass and is hung from a support. When the marble is pulled back a small distance and released, it swings in simple harmonic motion with a frequency of 0.60 Hz.

What is the frequency of the pendulum if it is transported to the surface of a small asteroid, where the acceleration due to gravity is only 1/100 that on Earth's surface? (a) 60 Hz; (b) 6.0 Hz; (c) 0.60 Hz; (d) 0.060 Hz; (e) 0.0060 Hz.

12-6 A physical pendulum has its mass distributed over its volume

We began the previous section by comparing two examples of pendulum motion: the swing of your lower leg when a physician tests your patellar reflex (Figure 12-13a) and the motion of a mass on the end of a string (Figure 12-13b). The mass on a string is nearly an ideal simple pendulum, since all of the mass is concentrated into a very small volume. But your lower leg is *not* a simple pendulum because its mass is distributed along the entire distance from the knee to the foot. A pendulum like your lower leg whose mass is distributed throughout its volume is called a **physical pendulum**. Other examples of physical pendulums include a swinging church bell, a chandelier swaying back and forth after an earth tremor, and the pendulum of an old-fashioned grandfather clock. Let's see how to find the angular frequency, period, and frequency for the oscillations of a physical pendulum.

Figure 12-16 shows an example of a physical pendulum of mass m. Two forces act on the pendulum, a gravitational force of magnitude mg that acts at the center of mass of the pendulum, a distance h from the pivot point, and a support force that acts at the pivot. We haven't drawn the support force because it acts at the pivot point and so exerts no torque around that point. The torque exerted by the gravitational force is equal to the magnitude mg of the gravitational force multiplied by the lever arm of the gravitational force, which is the perpendicular distance from the line of action of the gravitational force to the pivot. Figure 12-16 shows that if the pendulum is displaced from the vertical by an angle θ, the lever arm equals $h \sin \theta$. Hence the torque on the pendulum is

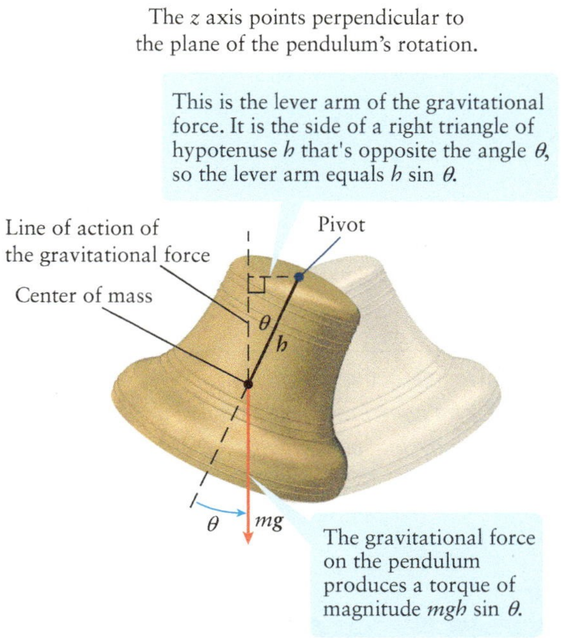

$$\tau_z = -mgh \sin \theta \tag{12-28}$$

Just as for the torque on a simple pendulum, the minus sign in Equation 12-28 means that this is a restoring torque that always acts to pull the pendulum back toward its equilibrium position $\theta = 0$, with the center of mass directly below the pivot.

The torque in Equation 12-28 is the only torque on the pendulum and so is equal to the net torque $\sum \tau_z$. From Newton's second law for rotation (Section 9-3), $\sum \tau_z = I\alpha_z$, we have

$$-mgh \sin \theta = I\alpha_z \tag{12-29}$$

In Equation 12-29 the quantity I is the moment of inertia of the pendulum around the pivot point. Just as for the simple pendulum, if the angle θ is small then $\sin \theta$ is approximately equal to θ (measured in radians). With this approximation we can rewrite Equation 12-29 as

$$\alpha_z \approx -\left(\frac{mgh}{I}\right)\theta \quad \text{if } \theta \text{ is small} \tag{12-30}$$

Compare Equation 12-30 to the corresponding equation for the simple pendulum that we derived in Section 12-5, $\alpha_z \approx -(g/L)\theta$ (Equation 12-26): The equation is identical except that g/L has been replaced by mgh/I. So we conclude that just as for the simple pendulum, the oscillations of a physical pendulum are simple harmonic motion provided that the amplitude is relatively small. To find the angular frequency, period,

The z axis points perpendicular to the plane of the pendulum's rotation.

This is the lever arm of the gravitational force. It is the side of a right triangle of hypotenuse h that's opposite the angle θ, so the lever arm equals $h \sin \theta$.

Line of action of the gravitational force

Pivot

Center of mass

θ

h

θ mg

The gravitational force on the pendulum produces a torque of magnitude $mgh \sin \theta$.

Figure 12-16 **A physical pendulum** The forces on a physical pendulum (one whose mass is not all concentrated in a single small blob, unlike the simple pendulum shown in Figure 12-14).

and frequency of a physical pendulum, we simply take Equations 12-27 for a simple pendulum and replace g/L with mgh/I:

Angular frequency, period, and frequency for a physical pendulum (small amplitude) (12-31)

Angular frequency ▸ $\omega = \sqrt{\dfrac{mgh}{I}}$

Period ▸ $T = \dfrac{2\pi}{\omega} = 2\pi\sqrt{\dfrac{I}{mgh}}$

Frequency ▸ $f = \dfrac{1}{T} = \dfrac{1}{2\pi}\sqrt{\dfrac{mgh}{I}} = \dfrac{\omega}{2\pi}$

m = mass of physical pendulum

g = acceleration due to gravity

h = distance from pivot point to center of mass of physical pendulum

I = moment of inertia of physical pendulum about pivot

Here's a check on our results for a physical pendulum. If all of the mass of the physical pendulum is concentrated at the center of mass, then the moment of inertia of the physical pendulum around the pivot is $I = mh^2$. Then the quantity mgh/I in Equations 12-31 becomes

$$(12\text{-}32) \qquad \frac{mgh}{I} = \frac{mgh}{mh^2} = \frac{g}{h}$$

 Go to Interactive Exercise 12-2 for more practice dealing with physical pendulums.

If we change the symbol for the distance from the pivot to the center of mass from h to L, the quantity in Equation 12-32 becomes g/L. Then, from the first of Equations 12-31, the angular frequency for a physical pendulum with all of its mass concentrated at the center of mass becomes $\omega = \sqrt{g/L}$. That's exactly the result we found in Section 12-5 for a simple pendulum, which is just a physical pendulum with all of its mass concentrated a distance L from the pivot (see Figure 12-14 and the first of Equations 12-27). So our results for a *physical* pendulum give us the correct answers for the special case of a *simple* pendulum, just as they should.

EXAMPLE 12-6 An Oscillating Rod

A uniform rod has a length L and a mass m, and is supported so that it can swing freely from one end. Derive an expression for the period of oscillation when the rod is pulled slightly from the vertical and released.

Set Up

Equations 12-31 tell us the period of a physical pendulum. The center of mass of a uniform rod is at its center, and its moment of inertia around one end is given in Table 9-1 (Section 9-4).

Period of a physical pendulum:

$T = 2\pi\sqrt{\dfrac{I}{mgh}}$ (from Equations 12-31)

Moment of inertia for a uniform rod of mass m and length L rotating about one end:

$I = \dfrac{1}{3}mL^2$

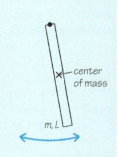

center of mass

m, L

Solve

Given the moment of inertia I and the distance h, solve for the period.

The center of mass is a distance $L/2$ from the pivot, so

$h = \dfrac{L}{2}$

Substitute I and h into the expression for the period T from Equations 12-31:

$T = 2\pi\sqrt{\dfrac{(1/3)mL^2}{mg(L/2)}} = 2\pi\sqrt{\dfrac{2mL^2}{3mgL}}$

$= 2\pi\sqrt{\dfrac{2L}{3g}}$

Reflect

Compared to the period of a simple pendulum of length L, $T_{\text{simple pendulum}} = 2\pi\sqrt{L/g}$, the period of a uniform rod free to rotate about its end is smaller by a factor of $\sqrt{2/3} = 0.816$. In other words, the physical pendulum oscillates more quickly. That may seem surprising if you just compare the torques on the two objects. The center of mass of the rod is a distance $h = L/2$ from the pivot, so from Equation 12-28 the torque on the rod is $\tau_z = -mgh \sin\theta = -(mgL/2)\sin\theta$. By contrast, for a simple pendulum all of the mass m is a distance L from the pivot, so $h = L$ and the torque on the pendulum is $\tau_z = -mgh \sin\theta = -mgL \sin\theta$. So the rod experiences half as much torque pulling it back toward equilibrium as does the simple pendulum. Why, then, does the rod oscillate more rapidly?

To see the explanation we also need to compare the moment of inertia of the rod ($I = mL^2/3$) to the moment of inertia of the simple pendulum ($I = mL^2$). So while the rod experiences half as much torque as the simple pendulum, the rod has only one-third as much inertia as does the simple pendulum. Hence the rod ends up oscillating faster.

BioMedical EXAMPLE 12-7   Moment of Inertia of a Human Leg

How you walk is affected in part by how your legs swing around the rotation axis created by your hip joints. Because your leg has a complex shape and contains bone, muscle, skin, and other materials, it would be very difficult to calculate the moment of inertia I of a leg around the hip. Instead I can be found experimentally by measuring the period of the leg when allowed to swing freely. In a clinical study a person's leg of length 0.88 m is estimated to have a mass of 6.5 kg and a center of mass 0.37 m from the rotation axis through the hip. When allowed to swing freely it oscillates with a period of 1.2 s. Determine the moment of inertia of the leg in rotation around the axis through the hip. Compare your answer to a uniform rod that has the same mass and length (see Example 12-6).

Set Up

The expression for the period T of a physical pendulum depends on the moment of inertia I for rotation around the pivot, the mass m, and the distance h from pivot to center of mass. We're given the values of $T = 1.2$ s, $m = 6.5$ kg, and $h = 0.37$ m, so we can solve for I. We'll then compare our result to a uniform rod with the same mass ($m = 6.5$ kg) and length ($L = 0.88$ m) as a leg.

Period of a physical pendulum:

$$T = 2\pi\sqrt{\frac{I}{mgh}} \quad \text{(from Equations 12-31)}$$

Moment of inertia for a uniform rod of mass m and length L rotating about one end:

$$I = \frac{1}{3}mL^2$$

Solve

Rearrange the expression for period T to find a formula for the moment of inertia I, then substitute the known values.

Begin with the expression

$$T = 2\pi\sqrt{\frac{I}{mgh}}$$

Square both sides to get rid of the square root, then solve for I:

$$T^2 = \frac{4\pi^2 I}{mgh} \quad \text{so} \quad I = \frac{mghT^2}{4\pi^2}$$

Substitute numerical values:

$$I = \frac{(6.5\text{ kg})(9.80\text{ m/s}^2)(0.37\text{ m})(1.2\text{ s})^2}{4\pi^2} = 0.86\text{ kg}\cdot\text{m}^2$$

As a comparison, find the moment of inertia of a uniform rod with the same mass and length of the leg rotating about one end.

For a rod rotating around one end,

$$I = \frac{1}{3}mL^2 = \frac{1}{3}(6.5\text{ kg})(0.88\text{ m})^2 = 1.7\text{ kg}\cdot\text{m}^2$$

Reflect

The moment of inertia of the uniform rod is *twice* as large as the moment of inertia of that leg. Is that reasonable? Consider that your upper leg (above the knee) is more massive than your lower leg, as suggested by the sketch above, so more of the leg's mass is above the knee than below the knee. By contrast, the mass of a uniform rod is distributed uniformly along its length. We learned in Section 9-4 that the farther an object's mass lies from the rotation axis, the greater the moment of inertia for that axis. So it's not surprising that the uniform rod, which has more of its mass farther from the rotation axis, has a greater moment of inertia than the leg.

GOT THE CONCEPT? 12-9 Period of a Rod

 A uniform rod has the same length and mass as the leg in the previous example. When supported at one end and allowed to rotate, would the period of this rod be (a) greater than the period of the leg; (b) the same as the period of the leg; (c) less than the period of the leg; or (d) not enough information given to decide?

GOT THE CONCEPT? 12-10 Frequency of a Physical Pendulum

 You measure the oscillation frequency of a physical pendulum when it is pivoted at one end. You then rearrange the experiment so that the physical pendulum is pivoted around its center of mass. Compared to the frequency when it is pivoted at one end, the frequency when it is pivoted at the center of mass is (a) greater; (b) the same; (c) less, but not zero; (d) zero; (e) not enough information given to decide. (*Hint:* Think about how the pendulum will move when supported at its center of mass.)

TAKE-HOME MESSAGE FOR Section 12-6

✔   A physical pendulum has its mass distributed throughout its volume. Unlike a simple pendulum, it cannot be treated as a point object suspended at the end of a string of negligible mass.

✔   The gravitational torque that arises when the pendulum is displaced from equilibrium acts as a restoring torque that causes the pendulum to return to its equilibrium position.

✔   The angular frequency, period, and frequency of a physical pendulum depend on its mass, the distance from the pivot point to the pendulum's center of mass, and the pendulum's moment of inertia around that point.

12-7 When damping is present, the amplitude of an oscillating system decreases over time

Stand up and let one arm hang limp by your side. With the other hand pull the limp arm forward, then let it go. The limp arm will swing back and forth like a pendulum, but within a few swings the oscillations die out, and your arm is once again hanging by your side. In the same way the vibrations of your eardrum, in response to a source of sound, quickly die away when the source is cut off, so do the vibrations of a cymbal after it has been struck with a drumstick. In all of these cases the oscillations are diminished or *damped* by some kind of friction force. Let's examine the properties of these **damped oscillations**.

In Section 5-3 we learned about the kinetic friction force that arises when one object slides over another. In that section we assumed that the friction force did not depend on the speed of the sliding. For many oscillating systems, however, it turns out to be a better description to say that the force that causes damping *does* depend on speed. The simplest such damping force is one that is proportional to the speed of the oscillating object:

$$\vec{F}_{\text{damping}} = -b\vec{v} \tag{12-33}$$

The minus sign says that the damping force always opposes the motion of the object: If the object moves to the left, the damping force is to the right, and so on. The quantity b is called the **damping coefficient**. Its value depends on various physical characteristics of a particular system. For example, think of an oscillating object that's immersed in a fluid such as air, water, or oil. The value of b in this case depends on the viscosity of the fluid as well as on the size of the oscillating object. The units of force are newtons, where $1\ \text{N} = 1\ \text{kg} \cdot \text{m/s}^2$, and the units of velocity are m/s, so it follows that the units of the damping coefficient b are $\text{N/(m/s)} = (\text{kg} \cdot \text{m/s}^2)(\text{s/m}) = \text{kg/s}$.

Let's consider an object of mass m attached to an ideal, Hooke's law spring of spring constant k that oscillates along the x axis as in Figure 12-2. The force exerted by the spring has only an x component, equal to $-kx$. If the object is also subject to a damping force given by Equation 12-33, that force also has only an x component (because the velocity has only an x component): $F_{\text{damping},x} = -bv_x$. The net force on the object is the sum of the spring and damping forces, so Newton's second law for the object is

$$ma_x = -kx - bv_x \tag{12-34}$$

Equation 12-34 is a difficult equation to solve, so we'll just present the solution for x as a function of time. It turns out that the character of the solution depends on the size of the damping coefficient b and that there are three distinct solutions that depend on whether the damping coefficient is relatively small, equal to a certain critical value, or greater than that critical value.

Underdamped Oscillations

If b is relatively small (less than $2\sqrt{km}$), the system still oscillates but with ever-decreasing amplitude. (This is what you observed in the experiment we described at the beginning of this section about letting your arm swing.) This is called **underdamped oscillation**. If the amplitude when the oscillation begins at $t = 0$ is A, the amplitude at a later time t is

The **amplitude** of an underdamped oscillator at time t after it is set in motion...

...equals the **initial amplitude** A at $t = 0$...

$$A(t) = Ae^{-(b/2m)t}$$

...multiplied by an **exponential function** that decreases with time.

Damping coefficient

Mass of the oscillating object

Amplitude of an underdamped oscillation
(12-35)

Figure 12-17 is a graph of this function. The decreasing amplitude means that the oscillating object makes smaller and smaller excursions around the equilibrium position as time goes by.

The presence of the damping force also changes the oscillation angular frequency from the value $\omega = \sqrt{k/m}$ without damping. With damping present the angular frequency is

Angular frequency of an underdamped mass–spring combination

Spring constant of the spring

$$\omega_{\text{damped}} = \sqrt{\frac{k}{m} - \left(\frac{b}{2m}\right)^2}$$

Damping coefficient

Mass of the oscillating object

Angular frequency of an underdamped oscillation
(12-36)

The presence of the term $(b/2m)^2$ inside the square root in Equation 12-36 means that the angular frequency is less with damping present than if there is no damping. The presence of the damping force acts to slow down the motion, so it makes sense that damping should decrease the angular frequency and so increase the oscillation period.

If we combine the information in Equations 12-35 and 12-36, we get the following expression for the displacement x as a function of time for a mass–spring combination that is underdamped, that is, that has a relatively small damping force:

$$x = Ae^{-(b/2m)t} \cos (\omega_{\text{damped}} t + \phi) \tag{12-37}$$

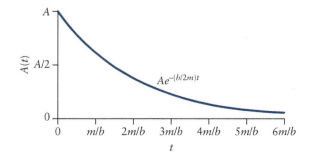

Figure 12-17 **Amplitude of an underdamped oscillation** If an oscillating system has relatively little damping, its oscillations are like simple harmonic motion but with an amplitude $A(t)$ that decreases with time.

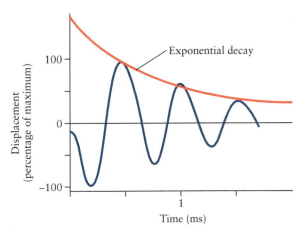

Figure 12-18 **Damped oscillations of a frog's eardrum** When stimulated by a burst of sound, the eardrum of a frog oscillates with decreasing amplitude (blue curve). The red curve is a decreasing exponential function like that shown in Figure 12-17.

The cosine function in Equation 12-37 tells us that an underdamped mass–spring combination oscillates with angular frequency ω_{damped}; the exponential function tells us that the amplitude of those oscillations decreases with time. Equation 12-35 tells us that the greater the value of the damping coefficient b, the more rapidly the amplitude decreases; Equation 12-36 tells us that a greater value of b makes the value of the angular frequency ω_{damped} smaller and so makes the oscillations happen more slowly (smaller frequency and greater period).

BioMedical Figure 12-18 shows a real-life underdamped oscillation. The blue curve shows measurements of the displacement of a frog's eardrum after it is set into oscillation by a burst of sound. The red curve is an exponential function like that in Equations 12-35 and 12-37. There is a close match between the theoretical red curve and the peaks of the experimental blue curve, which shows that the amplitude decays just as we would expect in an underdamped oscillation. Indeed, Equation 12-37 turns out to be an accurate description of the shape of the blue curve.

BioMedical EXAMPLE 12-8 **Oscillations in the Inner Ear**

As we discussed in Section 3-7, special sensory cells in the human inner ear play a role in detecting acceleration and in enabling you to balance. In response to a prolonged stimulus, these cells send out an oscillating signal that can resemble the blue curve in Figure 12-19. The red dashed line follows Equation 12-35; for the data shown $b/2m = 42 \text{ s}^{-1}$ and the period of the oscillations is about 3.2×10^{-3} s. Calculate the time it takes for the amplitude of the oscillations to fall to 20% of the initial amplitude. Express the answer in seconds and as a multiple of the oscillation period.

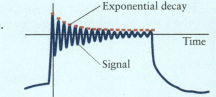

Figure 12-19 **Damped oscillations in the inner ear** The blue curve shows a signal sent out by sensory cells in the human inner ear. The red dashed line shows the decay in the amplitude of these signals.

Set Up

The signal oscillates with a decaying amplitude, so this is an underdamped oscillation. Our goal is to find the time t when $A(t)$ equals 0.20 times the initial amplitude A. We'll use Equation 12-35 for amplitude as a function of time.

Amplitude of an underdamped oscillation:

$$A(t) = Ae^{-(b/2m)t} \qquad (12\text{-}35)$$

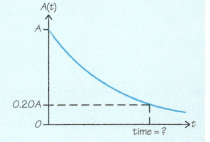

Solve

Set the time-dependent amplitude equal to 0.20 of the original amplitude and solve for t. To eliminate the exponential, use the properties of the natural logarithm function.

We want the value of t that satisfies the equation

$$A(t) = 0.20A = Ae^{-(b/2m)t}$$

Divide through by the initial amplitude A:

$$0.20 = e^{-(b/2m)t}$$

To get an expression for the time t, we need to "undo" the exponential function. We can do this using the natural logarithm function (ln), which has the property that

$$\ln e^x = x \text{ for any } x$$

With this in mind take the natural logarithm of both sides of the above equation:

$$\ln 0.20 = \ln e^{-(b/2m)t}, \text{ so}$$
$$\ln 0.20 = -(b/2m)t$$

Then solve for t:

$$t = -\frac{\ln 0.20}{(b/2m)} = -\frac{(-1.6)}{(42 \text{ s}^{-1})} = \frac{1.6}{42} \text{ s} = 0.038 \text{ s}$$

Express the answer in terms of the period of the oscillation.	The period of the oscillation is 3.2×10^{-3} s, so we can write the time t as

$$t = (0.038 \text{ s}) \times \left(\frac{1 \text{ period}}{3.2 \times 10^{-3} \text{ s}}\right) = 12 \text{ periods}$$

Reflect

Our results show that the time for the signal amplitude to decay to 0.20 of its initial value corresponds to about 12 complete oscillations. You can use a ruler to make measurements on Figure 12-18 to verify the height of the peak (that is, the amplitude) of the 12th cycle in the data is close to 20% of the height of the first peak.

Note that the signal shown in Figure 12-18 is an *electrical* signal, not a mechanical one, yet it obeys the same rules as mechanical oscillations. This suggests that the key physics behind oscillation also applies to nonmechanical oscillations. In later chapters we will discover the tremendous importance of electrical oscillations for understanding how power is transmitted over wires and how radio communication works.

Critically Damped and Overdamped Oscillations

Equation 12-36 shows that the angular frequency of oscillation gets smaller and smaller as the damping coefficient b is increased. You can see that the angular frequency becomes *zero* if the quantity under the square root in Equation 12-36 becomes zero. Zero angular frequency means that there are *no* cycles of oscillation per second, which means that the system does not oscillate at all! In this situation we say that the oscillations are **critically damped**: When displaced from equilibrium, the system returns smoothly to equilibrium with *no* overshoot, so there is no oscillation. The suspension of an automobile is designed to give nearly critical damping. (The damping is provided by the shock absorbers.) If the suspension were underdamped, as can happen if the shock absorbers are worn out, an automobile that drove over a bump would oscillate up and down on its suspension several times before settling down, which would make for a very uncomfortable ride.

The value of b that corresponds to critical damping is the value that makes $\omega_{\text{damping}} = 0$ in Equation 12-36. This will happen if

$$\frac{k}{m} - \left(\frac{b}{2m}\right)^2 = 0 \quad \text{or} \quad \left(\frac{b}{2m}\right)^2 = \frac{k}{m}$$

To solve this equation for the value of b for which $\omega_{\text{damped}} = 0$, take the square root of both sides and multiply both sides by $2m$:

$$\frac{b}{2m} = \sqrt{\frac{k}{m}} \quad \text{and so} \quad b = 2m\sqrt{\frac{k}{m}}$$

Simplifying, we get

$$b = 2\sqrt{km} \text{ (condition for critical damping)} \tag{12-38}$$

As an example, in Examples 12-2, 12-3, and 12-4 (Section 12-3) we considered an object of mass 0.80 kg oscillating on a spring of spring constant 1.8×10^2 N/m. From Equation 12-38, this system will be critically damped if we add a damping force for which the damping coefficient b is

$$b = 2\sqrt{km} = 2\sqrt{(1.8 \times 10^2 \text{ N/m}) (0.80 \text{ kg})}$$

$$= 2\sqrt{(1.8 \times 10^2 \text{ kg/s}^2) (0.80 \text{ kg})}$$

$$= 24 \text{ kg/s}$$

(We used the definition of the newton: $1 \text{ N} = 1 \text{ kg} \cdot \text{m/s}^2$, so $1 \text{ N/m} = 1 \text{ kg/s}^2$.)

Figure 12-20 Underdamped, critically damped, and overdamped oscillations The motion of a damped oscillating system depends on whether it is underdamped ($b < 2\sqrt{km}$), critically damped ($b = 2\sqrt{km}$), or overdamped ($b > 2\sqrt{km}$). In each case the system is displaced from equilibrium by a distance A and then released.

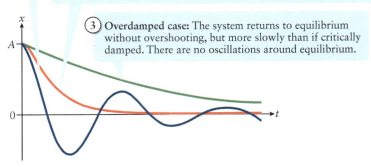

① **Underdamped case:** The system oscillates around the equilibrium position, and the amplitude continually decreases.

② **Critically damped case:** The system returns to equilibrium without overshooting. There are no oscillations around equilibrium.

③ **Overdamped case:** The system returns to equilibrium without overshooting, but more slowly than if critically damped. There are no oscillations around equilibrium.

→ *Go to Picture It 12-2 for more practice dealing with damped oscillations.*

If b has a value greater than $2\sqrt{km}$, the oscillations are **overdamped**. If an overdamped system is displaced from equilibrium, it again returns to equilibrium without overshooting, so again there are no oscillations. But now the damping is so great that the system takes even longer to return to equilibrium than in the critically damped case. If a system is critically damped, it returns to equilibrium in the shortest possible time. Figure 12-20 illustrates underdamped, critically damped, and overdamped oscillations.

GOT THE CONCEPT? 12-11 Underdamped, Critically Damped, or Overdamped?

 A block is attached to an ideal spring. The system of block and spring is critically damped: When the block is displaced from equilibrium and released, the block returns smoothly to equilibrium in the minimum time possible and does not overshoot equilibrium. If you replace the block with a new one of twice the mass, but the damping coefficient and spring constant remain the same, what kind of oscillations will result? (a) Underdamped; (b) critically damped; (c) overdamped; (d) not enough information given to decide.

TAKE-HOME MESSAGE FOR Section 12-7

✔   A damping force is proportional to the velocity of an oscillating object and opposes its motion.

✔   If the damping is light so that the system is underdamped, the object oscillates in periodic motion with an amplitude that decreases with time. The frequency is less than if damping were absent.

✔   Heavier damping can cause critical damping or overdamping, in which case the system returns to equilibrium without overshooting.

12-8 Forcing a system to oscillate at the right frequency can cause resonance

As you did at the beginning of Section 12-7, stand up and let one arm hang limply by your side. With the other hand, again pull the limp arm forward and let it go, and note the frequency at which the arm tends to oscillate on its own. Now put the other hand in your pocket and swing the limp arm back and forth as you might do while walking. Try making it swing at different frequencies. You'll find that it's easiest to make the arm swing at a frequency close to that at which it oscillates on its own, and that it's harder to make it swing at a slower frequency or higher frequency.

What you've just demonstrated are some of the properties of **forced oscillations**. In a forced oscillation a periodic driving force causes a system to oscillate at the frequency of that driving force. The experiment with your arm shows that it's easiest to force the system to oscillate if the *driving* frequency is near the system's *natural* frequency, which is the frequency at which it would oscillate on its own.

A periodic driving force can take many forms. In some cases it's a single push per cycle, such as what happens when a parent pushes a child on a swing. It turns out that the simplest case to analyze is a *sinusoidal* driving force. Let's see what happens when such a driving force acts on a damped mass–spring combination.

We saw in Section 12-7 that if an underdamped mass–spring combination is displaced from equilibrium, it oscillates with an ever-decreasing amplitude and an angular frequency

$$\omega_{\text{damped}} = \sqrt{\frac{k}{m} - \left(\frac{b}{2m}\right)^2} = \sqrt{\omega_0^2 - \left(\frac{b}{2m}\right)^2} \qquad \text{(12-39)}$$

In Equation 12-39 we use the symbol ω_0 for the quantity $\sqrt{k/m}$. We recognize this from Section 12-3 as the angular frequency that the mass–spring combination would have if there were *no* damping (see the first of Equations 12-11). We'll call ω_0 the **natural angular frequency** for the combination of mass and spring. (The natural *frequency* measured in Hz is ω_0 divided by 2π.) Now let's apply a sinusoidal driving force to the mass:

$$F(t) = F_0 \cos \omega t \qquad \text{(12-40)}$$

The angular frequency ω of the driving force, called the **driving angular frequency**, can have any value. As an example, consider the human eardrum, which behaves like a damped mass–spring system. Sound entering the ear acts as a driving force that makes the eardrum oscillate; the sound can have any of a wide range of frequencies.

With the driving force $F(t)$ from Equation 12-40 acting along with the force of the spring and the damping force, Newton's second law for the object of mass m attached to the spring is

$$ma_x = -kx - bv_x + F_0 \cos \omega t \qquad \text{(12-41)}$$

As was the case for Equation 12-34, the equation for damped oscillations without a driving force that we presented in Section 12-7, solving Equation 12-41 is beyond our scope. But here's what the solutions are like: When the driving force is first applied, the motion is a complicated combination of an oscillation at the angular frequency ω_{damped} given by Equation 12-39 and an oscillation at the driving angular frequency ω. The oscillation at ω_{damped} quickly dies away (it's a damped oscillation), and the system is left oscillating at the driving angular frequency ω. The amplitude of these oscillations is

Amplitude of a mass-spring system driven by a force $F(t) = F_0 \cos \omega t$

Maximum magnitude of the driving force

Damping coefficient

$$A = \frac{F_0}{\sqrt{m^2(\omega_0^2 - \omega^2)^2 + b^2 \omega^2}}$$

Mass of the oscillating object

Natural angular frequency

Driving angular frequency

Amplitude of a damped, driven oscillation

(12-42)

Although we've discussed the case of an underdamped system only, Equation 12-42 turns out to be valid no matter how large the value of b.

Equation 12-42 shows that for a given driving force magnitude F_0, the oscillation amplitude A depends on the driving angular frequency ω. The amplitude is largest when the denominator in Equation 12-42 is smallest. That happens when b is small (so there is very little damping) and when ω is close to the natural angular frequency ω_0 so $\omega_0^2 - \omega^2$ is close to zero. In other words, a system oscillates with large amplitude if you drive at close to its natural angular frequency, and the amplitude is particularly large if the system is only lightly damped. This phenomenon is called **resonance**. Figure 12-21 shows resonance in the eardrum of a frog.

Why does resonance happen? It turns out that if the driving angular frequency ω is equal to the natural angular frequency ω_0, the velocity v_x of the oscillating object is *in phase* with the driving force given by Equation 12-40. That is, the object is passing through equilibrium and has its maximum positive velocity v_x at the time when the force $F(t) = F_0 \cos \omega t$ has its maximum positive value; the object is at its maximum displacement and momentarily at rest ($v_x = 0$) when $F(t)$ equals zero; and so on. The net result is that when ω and ω_0 are equal, the driving force $F_0 \cos \omega t$ (which is

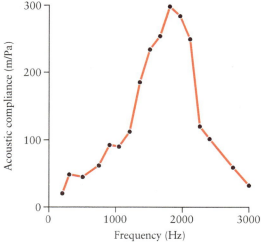

Figure 12-21 Resonance This graph shows a quantity related to the oscillation amplitude of a frog's eardrum (on the vertical axis) as a function of the frequency of sound waves causing the oscillation. The peak around 1800 Hz corresponds to the eardrum being forced at the eardrum's natural frequency of 1800 Hz.

Joe Kirby/Alamy

Figure 12-22 **Destructive resonance** When forced to oscillate at its natural angular frequency, a crystal wine glass can vibrate with such a large amplitude that it shatters.

in the same direction of the velocity) is the *negative* of the damping force $-bv_x$ (which is always opposite to the velocity). So at this driving angular frequency the driving force *cancels* the damping force! In this case Equation 12-41 becomes

$$ma_x = -kx$$

That's the same equation obeyed by a mass attached to an ideal spring with *no* damping and *no* driving force. Such a system oscillates freely at angular frequency $\omega_0 = \sqrt{k/m}$.

Figure 12-22 shows a dramatic example of resonance. A crystal wine glass is placed directly in front of a loudspeaker, and the tone emitted by the speaker is tuned—that is, its angular frequency ω is adjusted—to match the natural angular frequency ω_0 of the glass. The oscillations of the glass have little damping, as you can verify by holding such a glass by the stem and tapping the glass with a fingernail: The glass rings for quite a long time, showing that the oscillations are only lightly damped by the internal friction of the glass and so the damping coefficient b is small. As a result the amplitude of the oscillations when driven at the natural angular frequency is very large—so large that the glass literally tears itself apart. (This doesn't work if you use an ordinary drinking glass. Even if the loudspeaker is tuned to the natural angular frequency of the glass, the larger internal friction makes b greater than for a crystal wine glass, and the amplitude of oscillation of the driven drinking glass isn't large enough to break the glass.)

Bio **Medical** One example of resonance in nature is the flight of flies, wasps, and bees. These insects use resonance to help them beat their wings. Their bodies are equipped with special flight muscle cells that can apply cyclic forces at frequencies as high as 1000 Hz or more. These high frequencies correspond to the natural frequencies of the system of an insect's thorax and wings. The special flight muscle cells force the wings of the insect to oscillate at these natural frequencies. The result is that the wings oscillate with large amplitude A even though the force exerted by the flight muscle cells has a small magnitude F_0 (see Equation 12-42).

Bio **Medical** **GOT THE CONCEPT? 12-12** **Mosquitoes Listening for Each Other**

? The sounds made by male mosquitoes in flight are close to 650 Hz, while those made by females are close to 400 Hz. The antennae of mosquitoes differ from male to female (Figure 12-23): The natural frequency of a male's antennae is about 400 Hz, while that of the female's is about 200 Hz. Biologists suspect that some mosquitoes can detect the presence of others based on nerve signals generated when their antennae vibrate. Would you expect that (a) male mosquitoes can detect other males; (b) females can detect other females; (c) males can detect females; (d) females can detect males; (e) more than one of these; or (f) none of these?

Figure 12-23 **Mosquito antennae** Mosquitoes may be able to use their antennae to detect oscillations produced by other mosquitoes. What does the physics of resonance tell us about this?

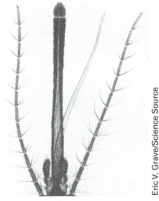

Eric V. Grave/Science Source

Male mosquito   Female mosquito

TAKE-HOME MESSAGE FOR **Section 12-8**

✔ An oscillating system can be forced to oscillate at any frequency.

✔ When the driving angular frequency is the same as the natural angular frequency of the system, the amplitude of oscillations becomes large, and especially large if the damping force is small. This is called resonance.

Key Terms

amplitude	damping coefficient	harmonic property
angular frequency	driving angular frequency	hertz
critically damped oscillations	forced oscillations	Hooke's law
damped oscillations	frequency	natural angular frequency

oscillation
overdamped oscillations
pendulum
period

phase angle
physical pendulum
resonance
restoring force

simple harmonic motion (SHM)
simple pendulum
sinusoidal function
underdamped oscillation

Chapter Summary

Topic	Equation or Figure

Oscillations: A system will oscillate if there is (a) a restoring force that always pulls the system back toward equilibrium and (b) inertia that causes the system to overshoot equilibrium. The period T is the time for one oscillation cycle; the frequency $f = 1/T$ is the number of cycles per second (measured in Hz); the angular frequency ω equals $2\pi f$; and the amplitude A is the maximum displacement from equilibrium.

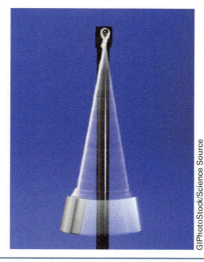

(Figure 12-13b)

GIPhotoStock/Science Source

Simple harmonic motion: Oscillations are particularly simple if the restoring force obeys Hooke's law, $F_x = -kx$. In this case, called simple harmonic motion, the angular frequency, period, and frequency are independent of the oscillation amplitude.

Angular frequency $\quad \omega = \sqrt{\dfrac{k}{m}}$

k = **spring constant** of the spring

m = **mass** of the object connected to the spring

Period $\quad T = \dfrac{2\pi}{\omega} = 2\pi\sqrt{\dfrac{m}{k}}$

Frequency $\quad f = \dfrac{1}{T} = \dfrac{1}{2\pi}\sqrt{\dfrac{k}{m}} = \dfrac{\omega}{2\pi}$

(12-11)

Equations of simple harmonic motion: In simple harmonic motion the position, velocity, and acceleration are all sinusoidal functions of time. The phase angle ϕ describes where in the oscillation cycle the system is at $t = 0$. We can understand this by comparing uniform circular motion to the motion of a block attached to an ideal spring.

Position of an object in SHM at time t

Amplitude of the oscillation = maximum displacement from the equilibrium position

$$x = A\cos(\omega t + \phi)$$

(12-12)

Angular frequency of the oscillation $(\omega = 2\pi f = 2\pi/T)$

Phase angle ϕ (in radians)

Velocity of an object in SHM at time t

Amplitude of the oscillation = maximum displacement from the equilibrium position

$$v_x = -\omega A\sin(\omega t + \phi)$$

(12-14)

Angular frequency of the oscillation $(\omega = 2\pi f = 2\pi/T)$

Phase angle (in radians)

Acceleration of an object in SHM at time t

Amplitude of the oscillation = maximum displacement from the equilibrium position

$$a_x = -\omega^2 A\cos(\omega t + \phi)$$

(12-15)

Angular frequency of the oscillation $(\omega = 2\pi f = 2\pi/T)$

Phase angle (in radians)

Energy in simple harmonic motion: If there is no friction, the total mechanical energy of a system in simple harmonic motion is conserved. At equilibrium the kinetic energy is maximum, and the spring potential energy is zero; at the extremes of the motion, the kinetic energy is zero, and the spring potential energy is maximum.

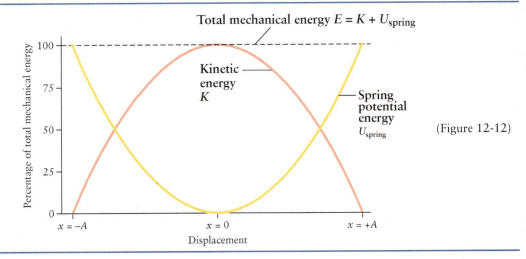

(Figure 12-12)

The simple pendulum: The oscillations of a simple pendulum are simple harmonic motion if the amplitude of oscillation is small. In this case the angular frequency of oscillation of a simple pendulum of length L is $\omega = \sqrt{g/L}$; this does not depend on the mass m.

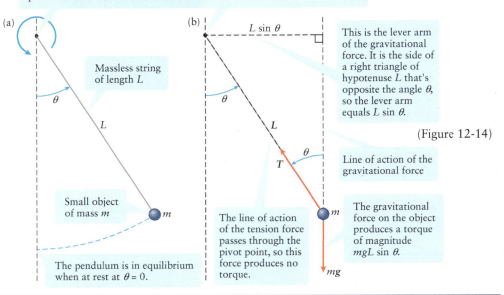

Pivot point: The positive z axis (the rotation axis of the pendulum) points outward from this point, perpendicular to the plane of the drawing. We take positive rotation to be counterclockwise.

(a) Massless string of length L

Small object of mass m

The pendulum is in equilibrium when at rest at $\theta = 0$.

(b) $L \sin \theta$

This is the lever arm of the gravitational force. It is the side of a right triangle of hypotenuse L that's opposite the angle θ, so the lever arm equals $L \sin \theta$.

Line of action of the gravitational force

The gravitational force on the object produces a torque of magnitude $mgL \sin \theta$.

The line of action of the tension force passes through the pivot point, so this force produces no torque.

(Figure 12-14)

The physical pendulum: A pendulum of arbitrary shape also oscillates in simple harmonic motion if the amplitude is small. The angular frequency of oscillation in this case is $\omega = \sqrt{mgh/I}$ for a physical pendulum with mass m whose center of mass is a distance h from the pivot. The pendulum's moment of inertia around the pivot is I.

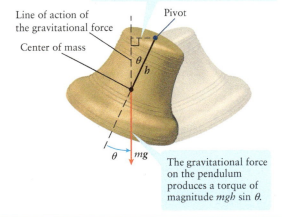

The z axis points perpendicular to the plane of the pendulum's rotation.

This is the lever arm of the gravitational force. It is the side of a right triangle of hypotenuse h that's opposite the angle θ, so the lever arm equals $h \sin \theta$.

Line of action of the gravitational force

Center of mass

Pivot

The gravitational force on the pendulum produces a torque of magnitude $mgh \sin \theta$.

(Figure 12-16)

Damped oscillations: In most cases a damping force acts on an oscillating system. This force can be approximated as being proportional to the oscillating object's speed. The character of the motion depends on the magnitude of the damping coefficient b.

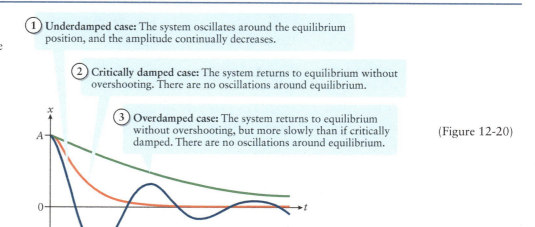

1 **Underdamped case:** The system oscillates around the equilibrium position, and the amplitude continually decreases.

2 **Critically damped case:** The system returns to equilibrium without overshooting. There are no oscillations around equilibrium.

3 **Overdamped case:** The system returns to equilibrium without overshooting, but more slowly than if critically damped. There are no oscillations around equilibrium.

(Figure 12-20)

Driven oscillations: Even if damping is present, a system can be driven to oscillate at any angular frequency ω. The amplitude of the resulting oscillation is largest if b is small and the driving angular frequency is close to ω_0, the angular frequency the system would have if there were no damping. This is called resonance.

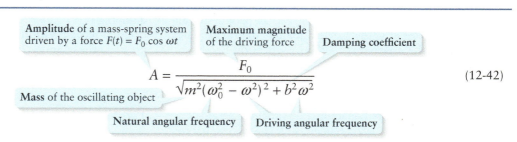

Amplitude of a mass-spring system driven by a force $F(t) = F_0 \cos \omega t$

Maximum magnitude of the driving force

Damping coefficient

Mass of the oscillating object

$$A = \frac{F_0}{\sqrt{m^2(\omega_0^2 - \omega^2)^2 + b^2\omega^2}}$$

(12-42)

Natural angular frequency   **Driving angular frequency**

Answer to What do you think? Question

(e) The maximum acceleration of an object oscillating in simple harmonic motion is $a_{max} = \omega^2 A$, where A is the amplitude of the oscillation and ω is the angular frequency (see Example 12-3 in Section 12-3). The angular frequency is in turn proportional to the frequency f (see Equations 12-11). Both oscillations in the problem have the same amplitude A (the maximum distance that the heart walls move in or out relative to equilibrium), but during exercise the oscillations have twice the frequency (f = (120 beats/min)(1 min/60 s) = 2 Hz versus f = 60 beats/

min = 1 Hz). Because ω is proportional to f, the oscillations during exercise have twice the angular frequency of the oscillations while resting. Since a_{max} is proportional to the square of ω, it follows that the oscillations have 2^2 = 4 times the maximum acceleration of the oscillations while resting. This means that the heart muscles must exert 4 times as much force on the heart walls during exercise than while resting, which helps to develop these muscles—just one of the many benefits of cardiovascular exercise.

Answers to Got the Concept? Questions

12-1 (d) The number of beats per unit time is just the oscillation frequency of the heart. The oscillation period is the reciprocal of the frequency, so when the frequency increases by a factor of 3 (from 50 beats/min to 150 beats/min) the period decreases to 1/3 of its original value.

12-2 (e) This oscillation is an example of simple harmonic motion, which means that the frequency is unaffected by changes in the amplitude (the harmonic property).

12-3 (b) From Example 12-3 the maximum speed of the block is $v_{max} = \omega A$. The angular frequency is $\omega = \sqrt{k/m}$ (see the first of Equations 12-11). So increasing the spring constant k by a factor of 4 increases ω and $v_{max} = \omega A$ by a factor of $\sqrt{4} = 2$.

12-4 (d) From Equation 12-12, the displacement at $t = 0$ is $x(0) = A \cos (0 + \phi) = A \cos \pi$. Since $\cos \pi = -1$, it follows that $x(0) = -A$.

12-5 (b), (d), (c), (a). The total mechanical energy of a block oscillating on an ideal spring is $E = K + U_{spring} = \frac{1}{2}mv^2 + \frac{1}{2}kx^2 = \frac{1}{2}kA^2$. In case (a) the spring constant is $k = 5.0 \times 10^2$ N/m, and the amplitude is $A = 0.020$ m, so $E = \frac{1}{2}kA^2 = \frac{1}{2}(5.0 \times 10^2$ N/m$)$ $(0.020$ m$)^2 = 0.10$ J. In case (b) the mass is $m = 0.60$ kg, $k = 3.0 \times 10^2$ N/m, and $v = 1.0$ m/s at $x = 0$. At $x = 0$ the spring potential energy is zero, so the energy is purely kinetic: $E = K = \frac{1}{2}mv^2 = \frac{1}{2}(0.60$ kg$)(1.0$ m/s$)^2 = 0.30$ J. In case (c) there is both kinetic energy ($m = 1.2$ kg and $v = 0.50$ m/s) and spring potential energy ($k = 4.0 \times 10^2$ N/m and $x = -0.010$ m). So

$E = K + U_{\text{spring}} = \frac{1}{2}mv^2 + \frac{1}{2}kx^2 = \frac{1}{2}(1.2 \text{ kg})(0.50 \text{ m/s})^2 + \frac{1}{2}(4.0 \times 10^2 \text{ N/m})(-0.010 \text{ m})^2 = 0.17$ J. Similarly, in case (d) there is both kinetic energy ($m = 2.0$ kg and $v = 0.20$ m/s) and potential energy ($k = 2.0 \times 10^2$ N/m and $x = 0.050$ m), so $E = K + U_{\text{spring}} = \frac{1}{2}mv^2 + \frac{1}{2}kx^2 = \frac{1}{2}(2.0 \text{ kg})(0.20 \text{ m/s})^2 + \frac{1}{2}(2.0 \times 10^2 \text{ N/m})(0.050 \text{ m})^2 = 0.29$ J.

12-6 (d) The last of Equations 12-27 tells us that $f = (1/2\pi)\sqrt{g/L}$, so the frequency is inversely proportional to the square root of the length L of the pendulum. If L is increased by a factor of 4, the frequency changes by a factor of $1/\sqrt{4} = 1/2$. So the frequency is halved from 0.60 to 0.30 Hz when the length of the pendulum increases by a factor of 4.

12-7 (c) The frequency $f = (1/2\pi)\sqrt{g/L}$ is independent of the mass of the pendulum and independent of the amplitude. So the frequency of the new pendulum is the same as before: $f = 0.60$ Hz.

12-8 (d) The frequency $f = (1/2\pi)\sqrt{g/L}$ is proportional to the square root of g, the acceleration due to gravity. The value of g on the asteroid is 1/100 as great as on Earth, so the frequency of oscillation there is different by a factor of $\sqrt{1/100} = 1/10$. So the new frequency on the surface of the asteroid is $(1/10)(0.60 \text{ Hz}) = 0.060$ Hz.

12-9 (a) We know from Equations 12-31 that the period of a physical pendulum is $T = 2\pi\sqrt{I/mgh}$, so it is proportional to the square root of the ratio I/h. For a uniform rod of length 0.88 m (the same as the length of the leg), the center of mass is at the center of the rod. So the distance from the pivot point at one end of the rod to the center of mass is one-half of 0.88 m, or 0.44 m. For the leg the corresponding distance is 0.37 m, so

h is greater for the rod by a factor of $(0.44 \text{ m})/(0.37 \text{ m}) = 1.2$. In Example 12-7 we found that the moment of inertia I of the rod is twice as large as that for the leg when both rotate around one end. So the period of the rod is greater than that of the leg by a factor of $\sqrt{(2/1.2)} = 1.3$. The effect of the greater moment of inertia overwhelms the effect of the greater distance h, so the rod has a greater period of oscillation and takes longer to complete one back-and-forth swing.

12-10 (d) A physical pendulum supported at its center of mass is in equilibrium no matter what its orientation. That's because the gravitational force acts at the support point, so there is no net torque around that point. (In terms of Equations 12-31, that means $h = 0$: There is zero distance from the pivot point to the center of mass.) Hence the pendulum does not oscillate at all. It goes through *zero* cycles per second, and the frequency is zero.

12-11 (a) From Equation 12-38 the condition for critical damping is $b = 2\sqrt{km}$. This condition was satisfied for the original block. With the new block the mass m is twice as great, so the quantity $2\sqrt{km}$ has increased by a factor of $\sqrt{2}$. However, the damping coefficient still has the same value, so now b is less than $2\sqrt{km}$ and the damping is less than the critical value. So the oscillations are now underdamped.

12-12 (c) The frequency of sound made by the female in flight is well tuned to the natural frequency of a male's antennae, so his antennae will exhibit large-amplitude vibrations (resonance) in the presence of a female mosquito. This matching of natural frequency of antennae and frequency of flight sounds does not occur between pairs of male or pairs of female mosquitoes.

Questions and Problems

In a few problems you are given more data than you actually need; in a few other problems you are required to supply data from your general knowledge, outside sources, or informed estimate.

Interpret as significant all digits in numerical values that have trailing zeros and no decimal points. For all problems use $g = 9.80 \text{ m/s}^2$ for the free-fall acceleration due to gravity. Neglect friction and air resistance unless instructed to do otherwise.

- •   Basic, single-concept problem
- ••   Intermediate-level problem, may require synthesis of concepts and multiple steps
- •••   Challenging problem
- Example   See worked example for a similar problem

Conceptual Questions

1. • One fundamental premise of simple harmonic motion is that a force must be proportional to an object's displacement. Is anything else required?

2. • List several examples of simple harmonic motion that you have observed in everyday life.

3. • Not all oscillatory motion is simple harmonic, but simple harmonic motion is always oscillatory. Explain this statement and give an example to support your explanation.

4. • If the rise and fall of your lungs is considered to be simple harmonic motion, how would you relate the period of the motion to your breathing rate (breaths per minute)?

5. • (a) What are the units of ω? (b) What are the units of ωt?

6. • Explain how *either* a cosine *or* a sine function can describe simple harmonic motion.

7. • Compare $x(t) = A \cos(\omega t)$ to $x(t) = A \cos(\omega t + \phi)$. (a) What is the phase angle ϕ, and what effect does changing it have on a simple harmonic oscillator? (b) Make three sketches of $A \cos(\omega t + \phi)$: one for $\phi = 0$, one for $0 < \phi < 90°$, and one for $-90° < \phi < 0$.

8. • Why is mechanical energy always conserved in an ideal simple harmonic oscillator?

9. • In simple harmonic motion what phase angle separates the kinetic and potential energy?

10. • Galileo was one of the first scientists to observe that the period of a simple harmonic oscillator is independent of its amplitude. (a) Explain what it means that the period is independent of the amplitude. Be sure to mention how the requirement that simple harmonic motion undergo small oscillations is affected by this supposition. (b) Draw two position-versus-time graphs for the same simple harmonic oscillator, each with a different amplitude.

11. • Explain the difference between a simple pendulum and a physical pendulum.

12. • What are three factors that can help you distinguish between a simple pendulum and a physical pendulum?

13. • Explain how you could do an experiment to measure the elevation of your location through the use of a simple pendulum.

14. • In the case of the damped harmonic oscillator, what are the units of the damping constant, b?

15. • Explain the difference between the frequency of the driving force and the natural frequency of an oscillator.

16. • The application of an external force other than gravity on a simple pendulum can create many different outcomes, depending on how frequently the force is applied. Explain what will happen to the amplitude of the motion if an external force is applied to a simple pendulum at the same frequency as the natural frequency of the pendulum. Assume that the pendulum is real, not ideal.

Multiple-Choice Questions

17. • A block is attached to a horizontal spring, the spring is stretched so the block is located at $x = A$, and the spring is released. At what point in the resulting simple harmonic motion is the speed of the block at its maximum?
  A. $x = A$ and $x = -A$
  B. $x = 0$
  C. $x = 0$ and $x = A$
  D. $x = 0$ and $x = -A$
  E. $x = 0$, $x = -A$, and $x = A$

18. • A block is attached to a horizontal spring, the spring is stretched so the block is located at $x = A$, and the spring is released. At what point in the resulting simple harmonic motion is the magnitude of the acceleration of the block at its maximum?
  A. $x = A$ and $x = -A$
  B. $x = 0$
  C. $x = 0$ and $x = A$
  D. $x = 0$ and $x = -A$
  E. $x = 0$, $x = -A$, and $x = A$

19. • A small object is attached to a horizontal spring, pushed to position $x = -A$, and released. In one full cycle of its motion, the total distance traveled by the object is
  A. $A/2$.
  B. $A/4$.
  C. A.
  D. $2A$.
  E. $4A$.

20. • A small object is attached to a horizontal spring and set in simple harmonic motion with amplitude A and period T. How long does it take for the object to travel a total distance of $6A$?
  A. $T/2$
  B. $3T/4$
  C. T
  D. $3T/2$
  E. $2T$

21. • An object–spring system undergoes simple harmonic motion. If the mass of the object is doubled, what will happen to the period of the motion?
  A. The period will increase.
  B. The period will decrease by an unknown amount.
  C. The period will not change.
  D. The period will decrease by a factor of 2.
  E. The period will decrease by a factor of 4.

22. • An object–spring system undergoes simple harmonic motion. If the amplitude increases but the mass of the object is not changed, the total energy of the system
  A. increases.
  B. decreases.
  C. doesn't change.

  D. undergoes a sinusoidal change.
  E. decreases exponentially.

23. • You can double the maximum speed of a simple harmonic oscillator by
  A. doubling the amplitude.
  B. reducing the mass to one-fourth its original value.
  C. increasing the spring constant to four times its original value.
  D. all of the above.
  E. none of the above.

24. • Replacing an object on a spring with an object having one-quarter the original mass will have the result of changing the frequency of the vibrating spring by a factor of
  A. $\frac{1}{4}$.
  B. $\frac{1}{2}$.
  C. 1 (no change).
  D. 2.
  E. 4.

25. • A uniform rod of length L hangs from one end and oscillates with a small amplitude. The moment of inertia for a rod rotating about one end is $I = \frac{1}{3}ML^2$. What is the period of the rod's oscillation?
  A. $2\pi\sqrt{\dfrac{L}{g}}$
  B. $2\pi\sqrt{\dfrac{2L}{3g}}$
  C. $2\pi\sqrt{\dfrac{L}{2g}}$
  D. $2\pi\sqrt{\dfrac{L}{3g}}$
  E. $2\pi\sqrt{\dfrac{L}{6g}}$

26. • If the period of a simple pendulum is T and we increase its length so that it's four times longer, what will the new period be?
  A. $T/2$
  B. T
  C. $2T$
  D. $4T$
  E. It is unchanged.

Problems

12-1 We live in a world of oscillations

12-2 Oscillations are caused by the interplay between a restoring force and inertia

27. • The period of a simple harmonic oscillator is 0.0125 s. What is its frequency? Example 12-1

28. • A mass attached to the end of a spring is set in motion. The mass is observed to oscillate up and down, completing 12 cycles every 3.00 s. (a) What is the period of the oscillation? (b) What is the frequency of the oscillation? Example 12-1

29. • An object on the end of a spring oscillates with a frequency of 15 Hz. Calculate (a) the period of the motion and (b) the number of oscillations that the object undergoes in 120 s. Example 12-1

12-3 The simplest form of oscillation occurs when the restoring force obeys Hooke's law

30. •• High tide occurs at 8:00 A.M. and is 1 m above sea level. Six hours later, low tide is 1 m below sea level. After another

6 h high tide occurs (again 1 m above sea level) then finally one last low tide (6 h later, 1 m below sea level). (a) Modeling the tides' oscillation as simple harmonic, write a mathematical expression that would predict the level of the ocean at this beach at any time of day. (b) Find the times in the day when the ocean level is exactly at sea level. Example 12-3

31. • A spring of unstretched length L and spring constant k is attached to a wall and an object of mass M resting on a frictionless surface (Figure 12-24). The object is pulled such that the spring is stretched a distance A then released. Let the $+x$ direction be to the right. (a) What is the position function $x(t)$ for the object? Take $x = 0$ to be the position of the object when the spring is relaxed. (b) What is the *velocity* of the object at $t = (7/6)T$, where T is the period? (c) What is the acceleration of the object at $t = T/4$? Example 12-3

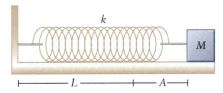

Figure 12-24 Problem 31

32. • What is the mass of an object that is attached to a spring with a force constant of 2.00×10^2 N/m if 14 oscillations occur each 16 s? Example 12-2

33. • A 0.200-kg object is attached to the end of a 55.0 N/m spring. It is displaced 10.0 cm to the right of equilibrium and released on a horizontal, frictionless surface. Calculate the period of the motion. Example 12-2

34. • (a) Plot a graph of the cosine function, cos (x), using a graphing calculator or software program. (b) How does the plot change when a phase angle of 30° is introduced, that is, cos ($x + 30°$)?

35. •• A giant wall clock with diameter d rests vertically on the floor (Figure 12-25). The minute hand sticks out from the face of the clock, and its length is half the clock diameter. A light directly above the clock casts a shadow of the minute hand on the floor. The initial time on the clock is 15 minutes after noon. Let the positive x axis point from the center of the clock toward the 3 o'clock position and the positive y axis point from the center of the clock toward the 12 o'clock position. (a) Construct the algebraic expression for the position $x(t)$ of the tip of the minute hand's shadow as a function of time. (b) Draw the graph for the position of the tip of the shadow as a function of time. Example 12-3

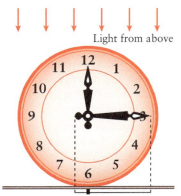

Light from above

Figure 12-25 Problem 35

36. • A force is measured with a force sensor at the times listed in the accompanying table. (a) Make a plot of force versus time

and determine if the force can be modeled using a single sine or cosine function. (b) Can a period be identified for this force? If so, what is it?

t (s)	F (N)	t (s)	F (N)
0	−20	1.3	+10
0.1	−10	1.4	0
0.2	0	1.5	−10
0.3	+10	1.6	−20
0.4	+20	1.7	−10
0.5	+10	1.8	0
0.6	0	1.9	+10
0.7	−10	2.0	+20
0.8	−20	2.1	+10
0.9	−10	2.2	0
1.0	0	2.3	−10
1.1	+10	2.4	−20
1.2	+20	2.5	−10

12-4 Mechanical energy is conserved in simple harmonic motion

37. • An object attached to the end of a spring slides on a horizontal frictionless surface with simple harmonic motion. Determine the location at which the object's kinetic energy and potential energy are the same. Assume the maximum displacement of the object is A so that it oscillates between $+A$ and $-A$. Example 12-4

38. •• A 0.250-kg object attached to a spring oscillates on a frictionless horizontal table with a frequency of 4.00 Hz and an amplitude of 20.0 cm. Calculate (a) the maximum potential energy of the system, (b) the displacement of the object when the potential energy is one-half of the maximum, and (c) the potential energy when the displacement is 10.0 cm. Example 12-4

39. •• The potential energy of an object on a spring is 2.4 J at a location where the kinetic energy is 1.6 J. If the amplitude of the simple harmonic motion is 20.0 cm, (a) calculate the spring constant and (b) find the largest force that the object experiences. Example 12-4

40. •• The potential energy of a simple harmonic oscillator is given by $U = \frac{1}{2}kx^2$. (a) If $x(t) = A \cos(\omega t)$, plot the potential energy versus time for three full periods of motion. (b) Write the expression for the velocity $v(t)$ of the oscillator. (c) Add the plot of the kinetic energy, $K = \frac{1}{2}mv^2$, to your graph. Example 12-4

12-5 The motion of a pendulum is approximately simple harmonic

41. • A simple pendulum on the surface of Earth is 1.24 m long. What is the period of its oscillation? Example 12-5

42. • The period of a simple pendulum on the surface of Earth is 2.25 s. Determine its length. Example 12-5

43. • In 1851 Jean Bernard Léon Foucault suspended a pendulum (later named the Foucault pendulum) from the dome of the Panthéon in Paris. The mass of the pendulum was 28.00 kg and the length of the rope was 67.00 m. The acceleration due to gravity in Paris is 9.809 m/s². Calculate the period of the pendulum. Example 12-5

44. • Geoff counts the number of oscillations of a simple pendulum at a location where the acceleration due to gravity is 9.80 m/s² and finds that it takes 25.0 s for 14 complete cycles. Calculate the length of the pendulum. Example 12-5

45. • **Astronomy** What is the period of a 1.00-m-long simple pendulum on each of the planets in our solar system? You will need to look up the acceleration due to gravity on each planet. Example 12-5

46. • (a) What is the period of a simple pendulum of length 1.000 m at the top of Mount Everest, 8848 m above sea level? (b) Express your answer as a number times T_0, the period at sea level where h equals 0. The acceleration due to gravity in terms of elevation is $g = g_0 \left(\dfrac{R_E}{R_E + h} \right)^2$, where g_0 is the average acceleration due to gravity at sea level, R_E is Earth's radius, and h is the elevation above sea level. Take g_0 to be 9.800 m/s² and Earth's radius R_E to be 6.380×10^6 m. Example 12-5

47. • A simple pendulum of length 0.350 m starts from rest at a maximum displacement of 10.0° from the equilibrium position on the surface of Earth. (a) At what time will the pendulum be located at an angle of displacement of 8.00°? (b) What about 5.00°? (c) When will the pendulum return to its starting position? Example 12-5

48. •• A simple pendulum oscillates between ±8.0° (as measured from the vertical) on the surface of Earth. The length of the pendulum is 0.50 m. Compare the time intervals between ±8.0° and ±4.0°. Example 12-5

12-6 A physical pendulum has its mass distributed over its volume

49. • A pendulum made of a uniform rod with a length of 30.0 cm is set into harmonic motion about one end, on the surface of Earth. Calculate the period of its motion. Example 12-6

50. • A physical pendulum on the surface of Earth consists of a uniform spherical bob that has a mass M of 1.0 kg and a radius R of 0.50 m suspended from a massless string that has a length L of 1.0 m. What is the period T of small oscillations of the pendulum? Example 12-6

51. ••• A thin, round disk made of acrylic plastic with a density of 1.1 g/cm³ is 20.0 cm in diameter and 1.0 cm thick (**Figure 12-26**). A very small hole is drilled through the disk at a point $d = 8.0$ cm from the center. The disk is hung from the hole on a nail and set into simple harmonic motion on the surface of Earth with a maximum angular displacement (measured from vertical) of $\theta = 7.0°$. Calculate the period of the motion. Example 12-6

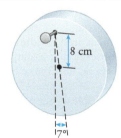
Figure 12-26 Problem 51

52. • A solid sphere, made of acrylic plastic with a density of 1.1 g/cm³, has a radius of 5.0 cm (**Figure 12-27**). A very small "eyelet" is screwed into the surface of the sphere and a horizontal support rod is passed through the eyelet, allowing the sphere to pivot around this fixed axis. If the sphere is displaced slightly from equilibrium on the surface of Earth, find the period of its harmonic motion when it is released. Example 12-6

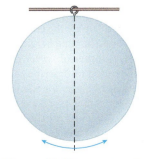
Figure 12-27 Problem 52

12-7 When damping is present, the amplitude of an oscillating system decreases over time

53. • A spring oscillator is designed with a mass of 0.250 kg. It operates while immersed in a damping fluid, selected so that the oscillation amplitude decreases to 1.00% of its initial value in 7.50 s. Determine the damping coefficient b of the system. Example 12-8

54. •• A 0.500-kg object is attached to a spring with a force constant of 2.50 N/m. The object rests on a horizontal surface that has a viscous, oily substance spread evenly on it. The object is pulled 15.0 cm to the right of the equilibrium position and set into harmonic motion. After 3.00 s the amplitude has fallen to 7.00 cm due to frictional losses in the oil. (a) Calculate the natural frequency of the system, (b) the damping constant for the oil, and (c) the frequency of oscillation that will be observed for the motion. (d) How much time will it take before the oscillations have died down to one-tenth of the original amplitude (1.50 cm)? Example 12-8

12-8 Forcing a system to oscillate at the right frequency can cause resonance

55. • A forced oscillator is driven at a frequency of 30.0 Hz with a peak force of 16.5 N. The natural frequency of the physical system is 28.0 Hz. If the damping constant is 1.25 kg/s and the mass of the oscillating object is 0.750 kg, calculate the amplitude of the motion.

56. •• An oscillating system has a natural frequency of 50.0 rad/s. The damping coefficient is 2.0 kg/s. The system is driven by a force $F(t) = (100 \text{ N}) \cos ((50 \text{ rad/s})t)$. What is the amplitude of the oscillations?

57. • A 5.0-kg object oscillates on a spring with a force constant of 180 N/m. The damping coefficient is 0.20 kg/s. The system is driven by a sinusoidal force of maximum value 50.0 N, and an angular frequency of 20.0 rad/s. (a) What is the amplitude of the oscillations? (b) If the driving frequency is varied, at what frequency will resonance occur?

General Problems

58. • A simple pendulum that has a length of 1.25 m and a bob with a mass of 4.0 kg is pulled an angle θ from its natural equilibrium position (hanging straight down). Use a spreadsheet or programmable calculator to calculate the maximum potential energy of the Earth–pendulum system and complete the table. Let the zero point of the potential energy be when the pendulum hangs straight down.

Maximum Angle from Vertical	Maximum Potential Energy of Bob (J)	Maximum Speed of Bob (m/s)
5°		
10°		
15°		
20°		
25°		

59. •• The acceleration of an object that has a mass of 0.025 kg and exhibits simple harmonic motion is given by $a(t) = (10 \text{ m/s}^2) \cos (\pi t + \pi/2)$. Calculate its velocity at $t = 2.0$ s, assuming the velocity of the object is $v = 3.18$ m/s at $t = 0$. Example 12-3

60. • Show that the formulas for the period of an object on a spring ($T = 2\pi\sqrt{m/k}$) and a simple pendulum ($T = 2\pi\sqrt{L/g}$) are dimensionally correct. Example 12-1

61. • The position for a particular simple harmonic oscillator is given by $x(t) = (0.15\text{ m})\cos(\pi t + \pi/3)$. What are (a) the velocity of the oscillator at $t = 1.0$ s and (b) its acceleration at $t = 2.0$ s? Example 12-3

62. • A simple harmonic oscillator is observed to start its oscillations at the maximum amplitude when $t = 0$. Devise a function for the position that is consistent with this initial condition. Repeat when the oscillations start at the equilibrium position when $t = 0$. Example 12-3

63. •• A 0.200-kg object is attached to a spring that has a force constant of 75.0 N/m. The object is pulled 8.00 cm to the right of equilibrium and released from rest to slide on a horizontal, frictionless table. (a) Calculate the maximum speed of the object. (b) Find the location of the object when it has one-third of the maximum speed, is moving to the right, and is speeding up. Example 12-3

64. •• A 0.100-kg object is fixed to the end of a spring that has a spring constant of 15.0 N/m. The object is displaced 15.0 cm to the right and released from rest at $t = 0$ to slide on a horizontal, frictionless table. (a) Calculate the first three times the object is at the equilibrium position. (b) What are the first three times when the object is 10.0 cm to the left of equilibrium? (c) What is the first time that the object is 5.00 cm to the right of equilibrium, moving toward the left? Example 12-3

65. ••• A damped oscillator with a period of 30.0 s shows a reduction of 30.0% in amplitude after 1.0 min. Calculate the percent loss in mechanical energy per cycle. Example 12-8

66. •• A system consisting of a small 1.20-kg object attached to a light spring oscillates on a smooth, horizontal surface. A graph of the position x of the object as a function of time is shown in Figure 12-28. Use the graph to answer the following questions. (a) What are the amplitude, period, frequency, and angular frequency of the motion? (b) What is the spring constant of the spring? (c) What is the maximum speed of the object? (d) What is the maximum acceleration of the object? Example 12-3

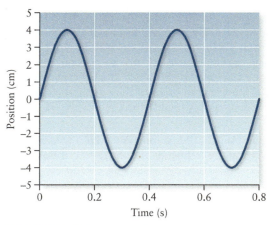

Figure 12-28 Problem 66

67. •• A pendulum made of a uniform rod with length $L_0 = 85$ cm hangs from one end and is allowed to oscillate (Figure 12-29). (a) What length of a simple pendulum L_1 will have the same period of simple harmonic motion? (b) A second pendulum made of a uniform rod hangs from a point that is 5.0 cm from its end. What total length L_2 should it have to give it the same period as the first pendulum? Example 12-6

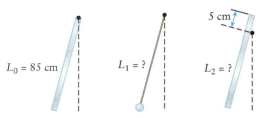

Figure 12-29 Problem 67

68. ••• An object of mass M attached to a spring of force constant k oscillates with amplitude A on a frictionless, horizontal surface. (a) Use conservation of energy to derive an expression for the object's velocity $v(x)$ as a function of position x. (b) If the object has mass $M = 250$ g, the spring constant is 85 N/m, and the amplitude is 10.0 cm, use the formula to calculate the speed of the object at $x = 0$ cm, 2.0 cm, 5.0 cm, 8.0 cm, and 10.0 cm. Example 12-4

69. •• When a small ball swings at the end of a very light, uniform bar, the period of the pendulum is 2.00 s. What will be the period of the pendulum if the bar has the same mass as the ball and the ball is removed and only the bar swings? Example 12-6

70. ••• Derive an algebraic expression for the period of a simple pendulum of mass m and length L that is valid at any altitude h above Earth's surface. Let R_E and M_E denote Earth's radius and mass, respectively. Example 12-5

71. ••• A block with a mass of 0.750 kg resting on a frictionless surface is attached to an unstretched spring with a length of 15.0 cm and a spring constant of $k = 9.80 \times 10^3$ N/m. The spring is attached to a wall at its other end. A 7.50-g, 9-mm-diameter bullet is fired into the block at a speed of 355 m/s and comes to a halt inside the block. Letting the initial position of the block be $x = 0$ and the positive direction be toward the wall, what is the position function $x(t)$ for the bullet–block system after the collision? Example 12-3

72. ••• **Astronomy** Your spaceship lands on a moon of a planet around a distant star. As you initially circled the moon, you measured its diameter to be 5480 km. After landing you observe that a simple pendulum that had a frequency of 3.50 Hz on Earth now has a frequency of 1.82 Hz. (a) What is the mass of the moon? Express your answer in kilograms and as a multiple of our Moon's mass. (b) Could you have used the vibrations of a spring–object system to determine the moon's mass? Explain your reasoning. Example 12-5

73. •• A bungee jumper who has a mass of 80.0 kg leaps off a very high platform. A crowd excitedly watches as the jumper free falls, reaches the end of the bungee cord, then gets "yanked" up by the elastic cord, again and again. One observer measures the time between the low points for the jumper to be 9.5 s. Another observer realizes that simple harmonic motion can be used to describe the process because several of the subsequent bounces for the jumper require 9.5 s also. Finally, the jumper comes to rest a distance of 40.0 m below the jump point. Calculate (a) the effective spring constant for the elastic bungee cord and (b) its unstretched length. Example 12-2

74. • **Biology** Spiderwebs are quite elastic, so when an insect gets caught in a web, its struggles cause the web to vibrate. This alerts the spider to a potential meal. The frequency of vibration

of the web gives the spider an indication of the mass of the insect. (a) Would a rapidly vibrating web indicate a large (massive) or a small insect? Explain your reasoning. (b) Suppose that a 15-mg insect lands on a horizontal web and depresses it 4.5 mm. If we model the web as a spring, what would be its effective spring constant? (c) At what rate would the web in part (b) vibrate, assuming that its mass is negligible compared to that of the insect? Example 12-2

75. •• Andrea is transporting a pendulum to a physics demonstration. The pendulum base is strapped down in the bed of her truck. The pendulum itself is a small 450-g object suspended from a 1.6-m-long, lightweight inflexible wire. While driving at 13.5 m/s Andrea is forced to come to an abrupt stop in 1.2 s. Assuming her acceleration is constant and that the pendulum bob was stationary before she slammed on the brakes, what is the amplitude of the resulting oscillation of the pendulum in radians? Will the oscillations be *simple* harmonic? Example 12-5

76. •• A 2.0-kg object is attached to a spring and undergoes simple harmonic motion. At $t = 0$ the object starts from rest, 10.0 cm from the equilibrium position. If the force constant of the spring k is equal to 75 N/m, calculate (a) the maximum speed and (b) the maximum acceleration of the object. (c) What is the velocity of the object at $t = 5.0$ s? Example 12-3

77. ••• A 475-kg piece of delicate electronic equipment is to be hung by a 2.80-m-long steel cable. If the equipment is disturbed (such as by being bumped), the vertical vibrational frequency must not exceed 25.0 Hz. What is the maximum diameter the cable can have? Example 12-3

78. •• A 0.200-kg particle undergoes simple harmonic motion along the horizontal x axis between the points $x_1 = -0.275$ m and $x_2 = 0.425$ m. The period of oscillation is 0.550 s. Determine (a) the frequency f, (b) the equilibrium position x_{eq}, (c) the amplitude A, (d) the maximum speed v_{max}, (e) the maximum magnitude of acceleration a_{max}, (f) the force constant k, and (g) the total mechanical energy E. Example 12-4

79. ••• A block–spring system hangs vertically and oscillates with simple harmonic motion along the y axis about the $y = 0$ equilibrium position. The block oscillates with amplitude A and angular frequency ω. Figure 12-30 models the motion of the block using the projection of the block's position on the y axis onto a circle of radius A in the x–y plane. The tip of the position vector $\bar{r}$ representing this projection travels about the circle in uniform circular motion with angular frequency ω. The vector shown represents the configuration of the system at time $t = 0$. The labeled points represent positions of the arrowhead at later times. (a) Of the labeled points, which correspond to the system's maximum potential energy? (b) Of the labeled points, which correspond to the system's maximum kinetic energy? (c) Of the labeled points, which correspond to when the system's kinetic energy equals its potential energy? Example 12-4

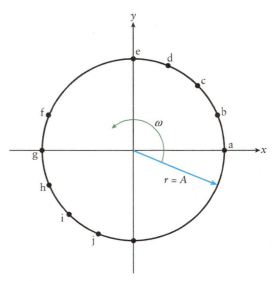

Figure 12-30 Problem 79

13

Waves

Republished with permission of John Wiley & Sons, from J. Cosson. "A Moving Image of Flagella: News and Views on the Mechanisms Involved in Axonemal Beating." *Cell Biology International*, 1996, 20(2), 83–94; permission conveyed through Copyright Clearance Center, Inc.

What do you think?

A spermatozoon moves by beating its tail-like flagellum (typically about 40 µm in length) against the surrounding fluid. Because the flagellum is flexible, waves travel along its length. If a spermatozoon were to beat its flagellum at a faster rate, would the distance along the flagellum from one wave crest to the next be (a) longer, (b) shorter, or (c) unaffected?

In this chapter, your goals are to:

- (13-1) Describe what a mechanical wave is.
- (13-2) Explain the key properties of transverse, longitudinal, and surface waves.
- (13-3) Define the relationship between simple harmonic motion and what happens in a sinusoidal wave.
- (13-4) Explain what determines the propagation speed of a mechanical wave.
- (13-5) Describe what happens when two sinusoidal waves from different sources interfere with each other.
- (13-6) Define the properties of standing waves on a string that is fixed at both ends.
- (13-7) Explain the nature of standing sound waves in open and closed pipes.
- (13-8) Describe how beats arise from combining two sound waves of slightly different frequencies.
- (13-9) Explain what is meant by the intensity and sound intensity level of a sound wave.
- (13-10) Recognize why the frequency of a sound changes if the source and listener are moving relative to each other.

To master this chapter, you should review:

- (6-2, 6-9) The relationship between work, force, and displacement, and the definition of power.
- (10-2) The nature of the Young's modulus and bulk modulus of a substance.
- (11-2) The definition of density.
- (12-2, 12-3, 12-4, and 12-8) The properties of simple harmonic motion and forced oscillations.

13-1 Waves are disturbances that travel from place to place

If you point your index finger downward and flex your wrist repeatedly, your finger oscillates up and down. That's the kind of motion we discussed in Chapter 12: Your finger has an equilibrium position, and it oscillates on either side of that position. But if you dip that finger in a pool of water and repeat this motion, something new happens: A series of ripples spreads away from your finger. The disturbance that you create in this way is called a **wave**. It's actually made up of countless miniature oscillations, since each part of the water's surface oscillates as the wave passes by (Figure 13-1a).

What's remarkable about water waves is that the *disturbance* travels a substantial distance across the water's surface, but the individual water molecules stay in pretty much the same place. The same thing is true for *sound* waves. When you sit in an hour-long lecture, sound waves travel from the lecturer to the back of the lecture hall, but there's no overall flow of air from the front to the back of the hall (Figure 13-1b). If there were, by the end of the hour the air would be denser at the back of the lecture hall than at the front!

In this chapter our subject is disturbances of this kind, called **mechanical waves**. In a mechanical wave a disturbance propagates through a material substance that we call the **medium** for the wave. The medium could be a liquid like water (Figure 13-1a), a gas like air (Figure 13-1b), or a solid. (If you live in a residence hall or an apartment, you certainly know that sound waves can travel through solid walls.) Not all waves are mechanical: As we'll learn in Chapter 22, *electromagnetic* waves such as radio signals, light rays, and x-ray beams don't require a medium to propagate and actually propagate fastest in a vacuum.

We'll begin this chapter by looking at the basic properties of mechanical waves. We'll see how to describe waves using mathematics, and we'll discover what determines the speed at which mechanical waves propagate. We'll go on to investigate what happens when two waves overlap and interfere with each other. In some cases the result is a *standing* wave in which the disturbance no longer propagates; in other cases we get a wave that propagates but whose frequency varies with time in a curious way. We'll study the manner in which we perceive sound waves, and we'll see how the properties of sound change when the source, the listener, or both are in motion.

(a)
A disturbance at one place in the water, such as a falling drop...

...causes a wave to spread outward over the water's surface.

Don Farrall/Getty Images

(b)
Sound waves from the lecturer travel through the classroom, but the air doesn't move along with them.

moodboard/Fotolia.com

Figure 13-1 **Mechanical waves** Two examples of mechanical waves: (a) one in a liquid and (b) one in a gas. Mechanical waves can also travel through solids.

13-2 Mechanical waves can be transverse, longitudinal, or a combination of these

Most mechanical waves are of one of the two types shown in Figure 13-2. The wave shown in Figure 13-2a is called a **transverse wave** because the individual parts of the wave medium (the rope) move in a direction *perpendicular* to the direction in which the wave propagates. By contrast, in the **longitudinal wave** shown in Figure 13-2b, the individual parts of the wave medium (the spring) move in the direction *parallel* to the direction of wave propagation.

You've participated in a kind of transverse wave if you've been to a crowded sports stadium and "done the wave," in which fans jump up section by section and then sit back down (Figure 13-3). The disturbance moves horizontally through the crowd, but individual parts of the wave medium—that is, the individual fans—move up and down, perpendicular to the direction that the "wave" propagates. (Note that this "wave" has the key property of waves that we described in Section 13-1: The disturbance moves through the crowd, but the members of the crowd don't move along with it.)

You can easily make a longitudinal wave just by clapping your hands (Figure 13-4). If you clap your hands together sharply at the opening of a tube, your hands squeeze on the air between them. As a result, the air pressure between your hands increases. When you release your hands this higher-pressure air pushes on the air around it, causing the surrounding air to compress and undergo a pressure increase. The air behind the newly

Figure 13-2 Transverse and longitudinal waves These two types of waves differ in how the parts of the wave medium (the material carrying the wave) move relative to the direction that the wave propagates.

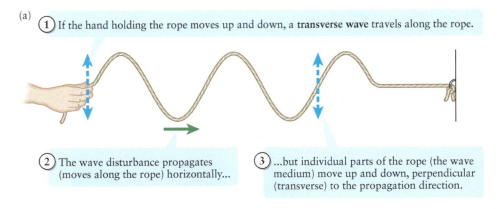

(a)
① If the hand holding the rope moves up and down, a **transverse wave** travels along the rope.

② The wave disturbance propagates (moves along the rope) horizontally...

③ ...but individual parts of the rope (the wave medium) move up and down, perpendicular (transverse) to the propagation direction.

(b)
① If the hand holding the spring moves back and forth, a **longitudinal wave** travels along the spring.

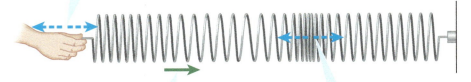

② The wave disturbance propagates (moves along the spring) horizontally...

③ ...but individual parts of the spring (the wave medium) move back and forth, parallel (longitudinal) to the propagation direction.

The "wave" propagates horizontally through the stadium...

...but individual members of the crowd (the wave medium) move up and down, transverse to the propagation direction.

Figure 13-3 A transverse "wave" Sports fans call this "doing the wave," but a more descriptive name would be "doing the transverse wave."

compressed air then relaxes back to its normal pressure. The result is that a pulse of increased pressure travels down the tube. (There's no net flow of air down the tube, just a disturbance in pressure.) If you clap your hands rhythmically, a series of pressure pulses propagates down the tube. We call this propagating disturbance a **sound wave**. Sound waves, and longitudinal waves in general, are also called **pressure waves** because they consist of regions of higher and lower pressure. (You've probably noticed that your ears, which you use to detect sound waves, are sensitive to changes in air pressure.)

Here's how you can see that a sound wave is a longitudinal wave. Within each pressure pulse of the sound wave, air molecules are squeezed together; as a pulse passes, the molecules move apart again. So within the tube air molecules slosh back and forth parallel to the axis of the tube and so parallel to the direction of wave propagation. That's just our definition of a longitudinal wave (see Figure 13-2b).

An earthquake produces *both* transverse *and* longitudinal waves that propagate through the body of Earth. The longitudinal waves travel at about twice the speed of the transverse waves, so a seismic monitoring station some distance away from the earthquake site (the *epicenter*) will receive the longitudinal waves before the transverse waves. The more distant the location from the epicenter, the greater the time delay, so by measuring the time delay at a given location scientists can determine how far from that location the earthquake took place. By correlating such measurements made at three or more locations, they can triangulate the position of the epicenter.

A common type of wave that is actually a *combination* of a transverse wave and a longitudinal wave is a **surface wave**, which is a wave that propagates on the surface of a medium. Two common examples are waves on the surface of the ocean and the waves that spread away from a disturbance in a pond (Figure 13-1a). As a surface wave propagates parallel to the (horizontal) surface of the water, particles near the surface move both vertically and horizontally—that is, both transverse and longitudinal to the direction of wave propagation. You can see this behavior by watching a buoy floating in a harbor: As waves pass by the buoy, the buoy bobs up and down (transverse motion) and moves back and forth (longitudinal motion). Surface waves can occur on the surface of solids as well as liquids. As an example, earthquakes produce not only transverse and longitudinal waves that travel through the body of Earth but also waves that travel along Earth's surface (Figure 13-5).

All of these types of waves have this characteristic in common: The wave propagates through its medium thanks to *restoring forces* that tend to return the medium to its undisturbed state. As an example, for the transverse wave on a rope shown in Figure 13-2a, the restoring force is due to the tension in the rope. To see the importance of tension, imagine detaching the rope from the post on the right-hand side of Figure 13-2a and placing the rope on top of a table. Since the rope is now slack, there is zero tension. If you now wiggle one end of the rope, that end will move in response but *no* wave will travel along the length of the rope. With no restoring force due to tension, the medium (the rope) doesn't resist being disturbed and there is no wave propagation.

The restoring force is different for different types of waves. For the longitudinal waves on a spring shown in Figure 13-2b, it's the Hooke's law force that opposes the spring being either stretched or compressed. In the case of sound waves in air (Figure 13-4), the pressure of the air itself provides the restoring force. A region of higher pressure pushes against the neighboring regions, thus expanding and lowering its pressure back to the equilibrium value; a region of lower pressure contracts due to pressure from its neighbors, thus raising its pressure toward equilibrium. And for surface waves in a pond, lake, or ocean the restoring force is the gravitational force, which tends to make the water surface smooth and level and so opposes the kind of disturbance shown in Figure 13-1a.

We saw in Section 12-2 that the presence of a restoring force was also an essential ingredient for *oscillation*. This suggests that there are deep connections between the physics of oscillation that we studied in Chapter 12 and the physics of waves. We'll explore these connections in the next several sections.

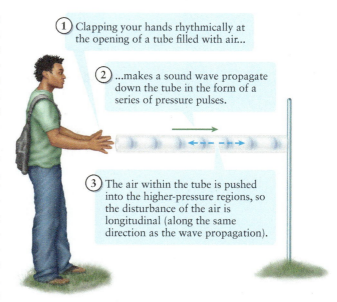

1 Clapping your hands rhythmically at the opening of a tube filled with air...

2 ...makes a sound wave propagate down the tube in the form of a series of pressure pulses.

3 The air within the tube is pushed into the higher-pressure regions, so the disturbance of the air is longitudinal (along the same direction as the wave propagation).

Figure 13-4 A longitudinal wave A sound wave is a longitudinal wave.

GOT THE CONCEPT? 13-1 Restoring Forces in Waves

? Choose the selection that correctly fills in the blanks in the following sentence: In the wave shown in Figure 13-2a the restoring force on a piece of the rope acts in the _____ direction, and in the wave shown in Figure 13-2b the restoring force on a piece of the spring acts in the _____ direction. (a) Vertical, vertical; (b) horizontal, vertical; (c) vertical, horizontal; (d) horizontal, horizontal; (e) any of these, depending on circumstances.

TAKE-HOME MESSAGE FOR Section 13-2

✔ In a transverse wave the elements of the medium are disturbed in the direction perpendicular to the direction in which the wave propagates.

✔ In a longitudinal wave the elements of the medium are disturbed along the direction in which the wave propagates.

✔ Surface waves propagate on the surface of a medium and are a combination of a transverse and longitudinal wave.

✔ In mechanical waves of all types, a restoring force must be present to make the wave propagate.

13-3 Sinusoidal waves are related to simple harmonic motion

In general waves can be very complicated. For example, when you're having a conversation with a friend, the sound waves coming from your mouth spread out in all directions, and the character of the wave changes from one moment to another as you vary the pitch and volume of your voice and pronounce different vowels and consonants. Instead of beginning our mathematical description of waves with complicated cases such as these, we'll start by considering a simple but important kind of wave.

Our first simplification is to restrict ourselves to **sinusoidal waves,** in which the wave pattern at any instant is a sinusoidal function (a sine or cosine). This restriction may seem very limiting, but in fact, *any* wave pattern can be formed by combining sinusoidal waves. That's the principle behind some types of musical *synthesizers*. By combining sinusoidal sound waves with different characteristics, a synthesizer can simulate the sound of other musical instruments or create wholly new sounds.

Figure 13-5 Consequences of a surface wave The surface waves produced by an earthquake cause both vertical and horizontal motion of the surface. The combined rolling motion is often responsible for much of the damage to buildings caused by earthquakes.

We'll also restrict ourselves to waves that travel along a straight line, like the transverse and longitudinal waves shown in Figure 13-2 and the sound wave in a tube in Figure 13-4. We call these **one-dimensional waves** since they propagate along a single dimension of space.

To be specific let's consider a sinusoidal, one-dimensional wave on a rope. This is actually what's drawn in Figure 13-2a. If the hand holding the left-hand end of the rope in Figure 13-2a moves up and down in simple harmonic motion with frequency f and period T, the position of the hand as a function of time is a sinusoidal function (see Section 12-3). The wave that propagates down the rope will then also be a sinusoidal function. Let's look at the properties of such a sinusoidal, one-dimensional wave.

Sinusoidal Waves: Wavelength, Amplitude, Period, Frequency, and Propagation Speed

There are two ways to visualize a wave. One is to imagine taking a "snapshot" or "freeze-frame" of the wave to see what the wave as a whole looks like at a given time. The other is to concentrate on a given piece of the medium and see how that piece moves as a function of time. (For the wave on a rope shown in Figure 13-2a, that would mean focusing on a single small piece of the rope and watching its motion.)

Figure 13-6a shows a "snapshot" of a sinusoidal wave on a rope at a given instant of time. This is a graph of the displacement y of the medium (the rope) from equilibrium versus position x along the length of the medium. For a transverse wave on the rope, the displacement y is indeed perpendicular to x as shown in Figure 13-6a. But Figure 13-6a could also represent a "snapshot" of a longitudinal wave like that in Figure 13-2b; in that case y represents the displacement *in the x direction* of pieces of the medium from their undisturbed, equilibrium positions.

Figure 13-6a shows that the **wavelength** λ (the Greek letter "lambda") is the distance the disturbance travels over one full cycle of the wave. You can measure λ from one *crest* (high point) of the wave to the next crest (Figure 13-6b), one *trough* (low point) to the next trough, or any specific point in the cycle to the analogous point in the next cycle. A wavelength is a distance, so its units in the SI system are meters.

Figure 13-7, the same "snapshot" of a sinusoidal wave as in Figure 13-6a, shows the **amplitude** A of the wave. Just like the amplitude of an oscillation, A is the

(a) "Snapshot" of a wave on a rope

① The "snapshot" of the wave at a given instant is a graph at time t of the displacement y of the medium versus the coordinate x along the length of the medium.

② This is a sinusoidal wave: The graph of displacement y versus coordinate x is a sinusoidal function (sine or cosine).

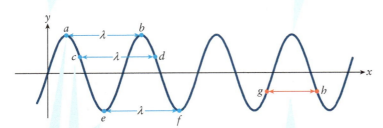

③ Points a and b are one wavelength λ apart: The distance between them corresponds to one complete cycle of the wave. The same is true for points c and d and for points e and f.

④ The displacement is the same at points g and h, but these points are not one wavelength apart: The distance between them does not correspond to a complete cycle.

(b) "Snapshot" of an ocean wave

Kenna Love/Alamy

The wavelength of this ocean wave is the distance between successive crests, just like the distance between points a and b in part (a).

Figure 13-6 A wave "snapshot": Wavelength The wavelength of a wave is the distance over which the wave first repeats itself. This is true for (a) waves on a rope and (b) ocean waves.

maximum displacement from equilibrium that occurs as the wave moves through its medium. It is *not* the difference between the maximum positive and maximum negative displacement.

To help us understand how a wave propagates over time, imagine a series of "snapshots" of a sinusoidal wave equally spaced in time like the frames of a movie. Figure 13-8 shows such a series. As the wave moves through the medium, an individual piece of the medium (like the blue dot in Figure 13-8) oscillates up and down. Just as in Section 12-2, we use the term **period** and the symbol T for the time required for each piece of the medium to go through a complete oscillation cycle. Furthermore, just as in Sections 12-2 and 12-3 the **frequency** f (in hertz, or Hz) is the number of cycles that a piece of the medium goes through per second, and the **angular frequency** ω (in rad/s) equals the frequency multiplied by 2π:

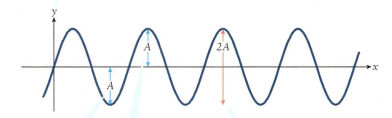

① The "snapshot" of the wave at a given instant is a graph at time t of the displacement y of the medium versus the coordinate x along the length of the medium.

② The amplitude A of the wave is the maximum magnitude of the wave disturbance.

③ The length of the red arrow (from crest to trough) is *not* equal to the amplitude A: It is equal to $2A$.

Figure 13-7 A wave "snapshot": Amplitude The amplitude of a wave is the maximum displacement from equilibrium.

Period of an oscillation

$$f = \frac{1}{T} \quad \text{and} \quad T = \frac{1}{f}$$

Frequency of the oscillation

Frequency and period of a wave
(12-1)

Angular frequency of a wave

$$\omega = 2\pi f = \frac{2\pi}{T}$$

Angular frequency of a wave
(13-1)

The values of T, f, and ω are the same *everywhere* in the medium where the wave is present, so these quantities are properties of the wave as a whole. The frequency of a sound wave helps determine its pitch: The higher the frequency, the higher the pitch.

Figure 13-8 also shows something remarkable about waves: During the time that it takes an individual piece of the medium to complete one cycle of oscillation, the wave travels a distance of one wavelength through the medium. We can use this observation to relate the wavelength and frequency to the speed at which the wave pattern moves through the medium, called the **propagation speed** v_p. Since the wave travels a distance λ in a time T, the propagation speed is

$$v_p = \frac{\lambda}{T}$$

From Equations 12-1, the reciprocal $1/T$ of the period is just the frequency f. So we can rewrite the propagation speed as

Propagation speed of a wave

$$v_p = f\lambda$$

Frequency   Wavelength

Propagation speed, frequency, and wavelength of a wave
(13-2)

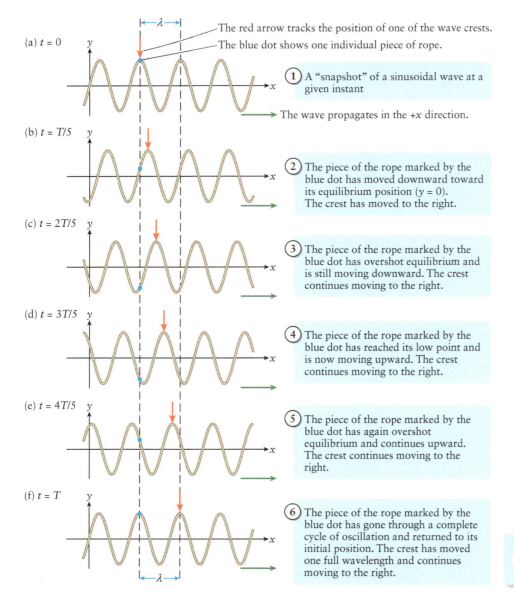

(a) $t = 0$

The red arrow tracks the position of one of the wave crests.
The blue dot shows one individual piece of rope.

(1) A "snapshot" of a sinusoidal wave at a given instant

The wave propagates in the +x direction.

(b) $t = T/5$

(2) The piece of the rope marked by the blue dot has moved downward toward its equilibrium position ($y = 0$). The crest has moved to the right.

(c) $t = 2T/5$

(3) The piece of the rope marked by the blue dot has overshot equilibrium and is still moving downward. The crest continues moving to the right.

(d) $t = 3T/5$

(4) The piece of the rope marked by the blue dot has reached its low point and is now moving upward. The crest continues moving to the right.

(e) $t = 4T/5$

(5) The piece of the rope marked by the blue dot has again overshot equilibrium and continues upward. The crest continues moving to the right.

(f) $t = T$

(6) The piece of the rope marked by the blue dot has gone through a complete cycle of oscillation and returned to its initial position. The crest has moved one full wavelength and continues moving to the right.

Figure 13-8 Scenes from a wave "movi These successive "snapshots" of a wave in motion reveal the connection between the period of a wave and the wavelength. The elapsed time from (a) to (f) is one period T.

 During one period of oscillation T of the wave medium, the wave travels a distance of one wavelength λ.

WATCH OUT! Be careful relating propagation speed, frequency, and wavelength.

Equation 13-2 may give you the impression that by changing the frequency or wavelength of a wave, you can change the propagation speed. In general this is *not* true! For many common types of waves, including sound waves in air and waves on a rope, the propagation speed v_p is the *same* no matter what the frequency (Figure 13-9). For such waves, increasing the frequency f decreases the wavelength λ and vice versa so that the product $f\lambda = v_p$ keeps the same value.

Figure 13-9 Sound waves of different frequencies have the same propagation speed The 88 keys of a piano play notes with frequencies from 27.5 Hz (the lowest pitch) to 4186 Hz (the highest pitch). But no matter what the frequency, a sound wave travels from the piano to the listener at the same speed. If this weren't the case, a listener at the back of the concert hall would hear a time delay between different notes!

Weyo/Alamy

WATCH OUT! Propagation speed is not the same as the speed of pieces of the medium.

! Be careful not to confuse v_p, the propagation speed of the *wave*, with the speed of individual pieces of the medium. As Figure 13-8 shows, such pieces oscillate up and down in a transverse wave (or back and forth in a longitudinal wave), so their speed is continuously changing in magnitude. By contrast, the speed of a wave through a uniform medium (one that has the same properties throughout) stays the same at all times.

The magnitude of the propagation speed can also be very different from the speed of pieces of the wave medium. As an example, the propagation speed of a sound *wave* through dry air at temperature 20°C is $v_p = 343$ m/s (about 1230 km/h, or 767 mi/h). But for a typical sound wave produced by a person speaking, the maximum speed of the *air* due to the wave passing through is less than 10^{-4} m/s (one-tenth of a millimeter per second).

BioMedical EXAMPLE 13-1 **Wave Speed on a Sperm's Flagellum**

The photograph that opens this chapter shows a sea urchin spermatozoon (sperm cell) in motion. These cells move by beating a long, tail-like flagellum against the surrounding fluid. Although some aspects of the motor that drives the flagellum are distributed along the length of the flagellum, we can approximate the flagellum as behaving like a uniform rope with a transverse wave moving along it. The flagellum is 40 μm long, the period of oscillation of the flagellum is 0.030 s, and there are approximately two complete cycles of the wave in the length of the flagellum. Estimate the speed of the transverse wave as it propagates along the flagellum.

Set Up

We are given the oscillation period $T = 0.030$ s and information about the wavelength λ. We use Equations 12-1 to determine the frequency of oscillation of the flagellum. Given this, we use Equation 13-2 to find the wave speed.

Period and frequency:

$$f = \frac{1}{T} \qquad (12\text{-}1)$$

Propagation speed, frequency, and wavelength of a wave:

$$v_p = f\lambda \qquad (13\text{-}2)$$

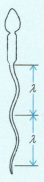

Solve

Find the frequency from the period.

From $T = 0.030$ s, the frequency is

$$f = \frac{1}{T} = \frac{1}{0.030 \text{ s}} = 33 \text{ Hz}$$

Two wavelengths fit into the length of the flagellum, so the wavelength is one-half of the 40-μm length. Use this to determine v_p.

Wavelength of the wave:

$$\lambda = \frac{1}{2} \times 40 \text{ μm} \times \frac{10^{-6} \text{ m}}{1 \text{ μm}} = 2.0 \times 10^{-5} \text{ m}$$

Then the wave speed is

$$v_p = f\lambda = (33 \text{ Hz})(2.0 \times 10^{-5} \text{ m}) = 7 \times 10^{-4} \text{ m/s}$$

to one significant figure.

Reflect

A wave moving at this propagation speed would travel about 2 m per hour. To put this in perspective, the world record holder in the World Snail Racing Championships (held annually in Congham, England) covered the 13-in. (33-cm) course in 2 min, which is just under 10 m/h. The propagation speed on the flagellum is only a factor of 5 slower—not bad for something 1000 times smaller than a snail.

Convert v_p to meters per hour:

$$v_p = 7 \times 10^{-4} \text{ m/s} \times \frac{3600 \text{ s}}{1 \text{ h}} = 2 \text{ m/h}$$

Sinusoidal Waves: Displacement as a Function of Position and Time

For a complete description of a wave, we need to know the displacement y of the medium at every position x and every time t. A function that provides this description is called the **wave function** $y(x,t)$. As its symbol suggests, y is a function of *both* x and t.

To obtain this function for a sinusoidal wave, let's start with a "snapshot" of such a wave with amplitude A and wavelength λ. Figure 13-10a shows such a "snapshot" at a given instant of time, which we'll call $t = 0$. Since the wave is sinusoidal and repeats itself over the distance from $x = 0$ to $x = \lambda$, we can write the wave function at $t = 0$ as

(13-3)
$$y(x,0) = A \cos\left(\frac{2\pi x}{\lambda} + \phi\right)$$

$\sqrt{x}$ **See the Math Tutorial for more information on trigonometry.**

The factor $2\pi x/\lambda$ varies from 0 at $x = 0$ to 2π at $x = \lambda$, so the cosine function goes through a complete cycle over a distance of one wavelength. The **phase angle** ϕ tells us what point in the cycle corresponds to $x = 0$. (We used a phase angle ϕ in a similar way in our description of simple harmonic motion in Section 12-3.) In Figure 13-10a $\phi = 0$, so the function is a cosine that has its maximum value at $x = 0$. In Figure 13-8a, which shows a different sinusoidal wave at $t = 0$, $\phi = -\pi/2$ (the function is shifted by $\pi/2$ radians, or one-quarter cycle, compared to the wave function in Figure 13-10a).

How can we get the wave function for *any* time t from Equation 13-3? Let's suppose that the wave is propagating in the positive x direction at speed v_p. As Figure 13-10b shows we can get the wave function at time t by replacing x in the wave function at $t = 0$ with $x - v_p t$:

$$y(x,t) = A \cos\left[\frac{2\pi}{\lambda}(x - v_p t) + \phi\right]$$
$$= A \cos\left(\frac{2\pi x}{\lambda} - \frac{2\pi v_p t}{\lambda} + \phi\right)$$

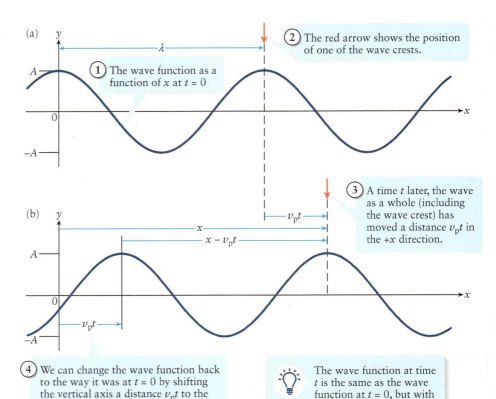

(a) ① The wave function as a function of x at $t = 0$

② The red arrow shows the position of one of the wave crests.

(b) ③ A time t later, the wave as a whole (including the wave crest) has moved a distance $v_p t$ in the $+x$ direction.

Figure 13-10 A sinusoidal wave function The wave function describes the displacement y as a function of position x and time t. If all parts of the wave move at the same propagation speed v_p, we can relate (a) the wave function at time $t = 0$ and (b) the wave function at a later time t.

④ We can change the wave function back to the way it was at $t = 0$ by shifting the vertical axis a distance $v_p t$ to the right, so coordinate x becomes $x - v_p t$.

The wave function at time t is the same as the wave function at $t = 0$, but with x replaced by $x - v_p t$.

We can simplify this expression by using Equation 13-2 for the propagation speed, $v_p = f\lambda$. Dividing both sides of Equation 13-2 by λ tells us that $v_p/\lambda = f$. Then the wave function becomes

$$y(x,t) = A \cos\left(\frac{2\pi x}{\lambda} - 2\pi ft + \phi\right)$$

(13-4)

The quantity $2\pi f$ is the angular frequency ω of the wave (see Equations 13-1), and the combination $2\pi/\lambda$ in Equation 13-4 is called the **angular wave number**. We use the symbol k for this quantity:

Angular wave number

$$k = \frac{2\pi}{\lambda}$$ Wavelength

Angular wave number
(13-5)

Since 2π is the number of radians in one cycle and wavelength λ is in meters, the angular wave number k is measured in radians per meter (rad/m). We use the adjective "angular" since the term "wave number" is typically used for $1/\lambda$, the reciprocal of the wavelength. This quantity multiplied by 2π is the angular wave number $k = 2\pi/\lambda$, in the same fashion that frequency f multiplied by 2π is the angular frequency $\omega = 2\pi f$.

In terms of angular wave number and angular frequency, we can rewrite Equation 13-4 as

Wave function for a sinusoidal wave propagating in the +x direction   Angular wave number of the wave = $2\pi/\lambda$

$$y(x,t) = A \cos (kx - \omega t + \phi)$$ Phase angle

Amplitude of the wave   Angular frequency of the wave = $2\pi f$

Wave function for a sinusoidal wave propagating in the +x direction
(13-6)

As an aid to using Equation 13-6, note that the angular frequency $\omega = 2\pi f$ and the angular wave number $k = 2\pi/\lambda$ are related by the propagation speed. To see this relationship, note that $f = \omega/2\pi$ and $\lambda = 2\pi/k$. Then, from Equation 13-2,

$$v_p = f\lambda = \left(\frac{\omega}{2\pi}\right)\left(\frac{2\pi}{k}\right)$$

or

Propagation speed of a wave   Angular frequency of the wave

$$v_p = \frac{\omega}{k}$$ Angular wave number of the wave

Propagation speed, angular frequency, and angular wave number
(13-7)

For the waves we are considering, the propagation speed v_p is the same no matter what the angular frequency: Increasing ω causes the angular wave number k to increase (that is, causes the wavelength $\lambda = 2\pi/k$ to decrease) so that the ratio $v_p = \omega/k$ stays the same.

We can use Equation 13-6 to verify a statement that we made earlier: If the wave is sinusoidal, each part of the medium oscillates in simple harmonic motion. To see that this is true, think of x as a constant (so that we are considering a single piece of the wave medium). Then Equation 13-6 becomes

$$y(x,t) = A \cos (kx - \omega t + \phi) = A \cos [-\omega t - (-kx - \phi)]$$
$$= A \cos [\omega t + (-kx - \phi)]$$

(13-8)

In the last step we used the trigonometric identity $\cos(-\theta) = \cos\theta$. Equation 13-8 looks *exactly* like Equation 12-12 for simple harmonic motion, $x = A \cos(\omega t + \phi)$, if we interpret the quantity $-kx - \phi$ as the phase angle of the simple harmonic oscillation of the wave medium at position x. Equation 13-8 shows that all parts of the medium oscillate with the same amplitude A and the same angular frequency ω, but with a phase angle that depends on the position x within the medium. Two pieces of the medium oscillate together—that is, they are *in phase*—if they are a whole number (1, 2, 3,...) of wavelengths apart and oscillate opposite to each other—that is, they are *out of phase*—if they are $\frac{1}{2}, 1\frac{1}{2}, 2\frac{1}{2},...$ wavelengths apart.

We developed Equation 13-6 for a wave propagating in the positive x direction. If the wave propagates instead in the *negative x* direction, in time t it moves a distance $v_p t$ to the *left* in Figure 13-10, and the wave function at time t is the same as the wave function at time $t = 0$, with x replaced by $x + v_p t$. The result is that the wave function for a sinusoidal wave propagating in the negative x direction is the same as Equation 13-6, but with $kx - \omega t$ replaced by $kx + \omega t$.

EXAMPLE 13-2 A Wave on a Rope

A wave travels down a stretched rope. The wave has amplitude 2.00 cm = 0.0200 m and wavelength 15.0 cm = 0.150 m, and each part of the rope goes through a complete cycle once every 0.400 s. (a) Find the frequency, angular frequency, angular wave number, and propagation speed for this wave. (b) If the wave propagates in the +x direction and has phase angle $\phi = 0$, find the displacement of the rope at $x = 0.450$ m at time $t = 0.100$ s. (c) If instead the wave propagates in the −x direction and has phase angle $\phi = 0$, find the displacement of the rope at $x = 0.450$ m at time $t = 0.200$ s.

Set Up

We are given the amplitude $A = 2.00$ cm, the oscillation period $T = 0.400$ s, and the wavelength $\lambda = 15.0$ cm. We use Equations 12-1 and 13-1 to determine the frequency and angular frequency, Equation 13-5 to find the angular wave number, and either Equation 13-2 or Equation 13-7 to find the propagation speed. For a wave propagating in the +x direction, we can find the displacement at any position x and any time t using Equation 13-6. If the wave is propagating in the −x direction, we use Equation 13-6 with $kx - \omega t$ replaced by $kx + \omega t$.

Period, frequency, and angular frequency:

$$f = \frac{1}{T} \text{ and } \omega = 2\pi f \quad \text{(12-1 and 13-1)}$$

Angular wave number:

$$k = \frac{2\pi}{\lambda} \quad \text{(13-5)}$$

Propagation speed of a wave:

$$v_p = f\lambda \text{ or } v_p = \frac{\omega}{k} \quad \text{(13-2), (13-7)}$$

Wave function of a sinusoidal wave:

$$y(x,t) = A \cos(kx - \omega t + \phi) \text{ (propagating in +x direction)} \quad \text{(13-6)}$$
$$y(x,t) = A \cos(kx + \omega t + \phi) \text{ (propagating in −x direction)}$$

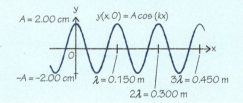

Solve

(a) Calculate the frequency f, angular frequency ω, angular wave number k, and propagation speed v_p.

We know $T = 0.400$ s, so the frequency and angular frequency are

$$f = \frac{1}{T} = \frac{1}{0.400 \text{ s}} = 2.50 \text{ Hz}$$

$$\omega = 2\pi f = (2\pi \text{ rad})(2.50 \text{ Hz}) = 15.7 \text{ rad/s}$$

The angular wave number is

$$k = \frac{2\pi}{\lambda} = \frac{2\pi \text{ rad}}{0.150 \text{ m}} = 41.9 \text{ rad/m}$$

From Equation 13-2, the propagation speed is

$$v_p = f\lambda = (2.50 \text{ Hz})(0.150 \text{ m}) = 0.375 \text{ m/s}$$

Alternatively, from Equation 13-7

$$v_p = \frac{\omega}{k} = \frac{15.7 \text{ rad/s}}{41.9 \text{ rad/m}} = 0.375 \text{ m/s}$$

(b) Find the displacement y at $x = 0.450$ m at time $t = 0.100$ s if the wave propagates in the $+x$ direction.

Since $\phi = 0$, Equation 13-6 becomes $y(x,t) = A \cos (kx - \omega t)$. We have

$$kx = \left(\frac{2\pi}{\lambda}\right) x = \left(\frac{2\pi \text{ rad}}{0.150 \text{ m}}\right)(0.450 \text{ m}) = 6.00\pi \text{ rad}$$

$$\omega t = 2\pi ft = (2\pi \text{ rad})(2.50 \text{ Hz})(0.100 \text{ s}) = 0.500\pi \text{ rad}$$

So

$$
\begin{aligned}
y(x,t) &= (2.00 \text{ cm}) \cos (6.00\pi \text{ rad} - 0.500\pi \text{ rad}) \\
&= (2.00 \text{ cm}) \cos (5.50\pi \text{ rad}) = (2.00 \text{ cm})(0.00) \\
&= 0.00 \text{ cm}
\end{aligned}
$$

(c) Find the displacement y at $x = 0.450$ m at time $t = 0.200$ s if the wave propagates in the $-x$ direction.

Again $\phi = 0$. In Equation 13-6 we replace $kx - \omega t$ by $kx + \omega t$, so $y(x,t) = A \cos (kx + \omega t)$. Now we have

$$kx = \left(\frac{2\pi}{\lambda}\right) x = \left(\frac{2\pi \text{ rad}}{0.150 \text{ m}}\right)(0.450 \text{ m}) = 6.00\pi \text{ rad}$$

$$\omega t = 2\pi ft = (2\pi \text{ rad})(2.50 \text{ Hz})(0.200 \text{ s}) = 1.00\pi \text{ rad}$$

So

$$
\begin{aligned}
y(x,t) &= (2.00 \text{ cm}) \cos (6.00\pi \text{ rad} + 1.00\pi \text{ rad}) \\
&= (2.00 \text{ cm}) \cos (7.00\pi \text{ rad}) = (2.00 \text{ cm})(-1.00) \\
&= -2.00 \text{ cm}
\end{aligned}
$$

Reflect

Note that the calculations of $y(x,t)$ are easier if we express kx and ωt as multiples of π. To understand our answers, note that the point $x = 0.450$ m is 3 wavelengths away from $x = 0$, so there are 3 complete wave cycles between this point and the end of the rope. So at any time the displacement at $x = 0.450$ m is the same as at $x = 0$ (these points oscillate in phase). Furthermore, the time $t = 0.100$ s is equal to one-quarter of a period of oscillation T, and $t = 0.200$ s is equal to one-half of T. In part (b) the wave function has moved by one-quarter of a cycle in the $+x$ direction from $t = 0$ to $t = T/4 = 0.100$ s, and in part (c) it has moved by one-half of a cycle in the $-x$ direction from $t = 0$ to $t = T/2 = 0.200$ s. The red dots in the figures show the point on the rope at $x = 0.450$ m; their displacements match our calculations.

Compared to the wavelength $\lambda = 15.0$ cm $= 0.150$ m, the distance from the end of the rope ($x = 0$) to the point $x = 0.450$ m is

$$\frac{x}{\lambda} = \frac{0.450 \text{ m}}{0.150 \text{ m/cycle}} = 3.00 \text{ cycles}$$

Part (b): $y(x,t)$ at $t = 0.100$ s, wave propagates in $+x$ direction

Since there are a whole number of wavelengths (3) between $x = 0$ and $x = 0.450$ m, these two points oscillate in phase.

Compared to the period $T = 0.400$ s, the times $t = 0.100$ s and $t = 0.200$ s are

$$\frac{t}{T} = \frac{0.100 \text{ s}}{0.400 \text{ s/cycle}} = 0.250 \text{ cycle}$$

$$\frac{t}{T} = \frac{0.200 \text{ s}}{0.400 \text{ s/cycle}} = 0.500 \text{ cycle}$$

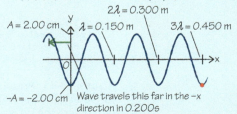

Part (c): $y(x,t)$ at $t = 0.200$ s, wave propagates in $-x$ direction

Sinusoidal Sound Waves

BioMedical

We've described sinusoidal waves in terms of their displacement y as a function of position x and time t. For sound waves in air it's usually more convenient to describe them in terms of *pressure variations* as a function of x and t. Humans hear by detecting very small variations of pressure on the eardrum. The ear can detect pressure changes as small as 2×10^{-5} Pa, equivalent to a change in atmospheric pressure of 1 part in 5 *billion* (5×10^9). If you've ever had difficulty getting your ears to "pop" when driving from the mountains down to sea level or on an airliner descending to land, you know how sensitive to pressure differences your ears can be.

Since a sound wave is a longitudinal wave, a positive value of $y(x,t)$ means that the medium is displaced in the positive x direction, while a negative value of $y(x,t)$ means that the medium is displaced in the negative x direction (**Figure 13-11**). As a result, as a sound wave propagates it squeezes together air molecules at some places along the wave, resulting in regions of air that have higher density and higher pressure. Adjacent regions of air are slightly depleted of air molecules and therefore have lower density and lower pressure. As Figure 13-11 shows, the pressure is greatest or least at points where the displacement is *zero*. So while the pressure is also described by a sinusoidal function, it is shifted by one quarter-cycle from the wave function for displacement $y(x,t)$. That's why we've drawn the displacement graph in Figure 13-11 as a *cosine* curve but the pressure variation graph as a *sine* curve. In particular, if the displacement is described by Equation 13-6,

$$y(x,t) = A \cos (kx - \omega t + \phi)$$

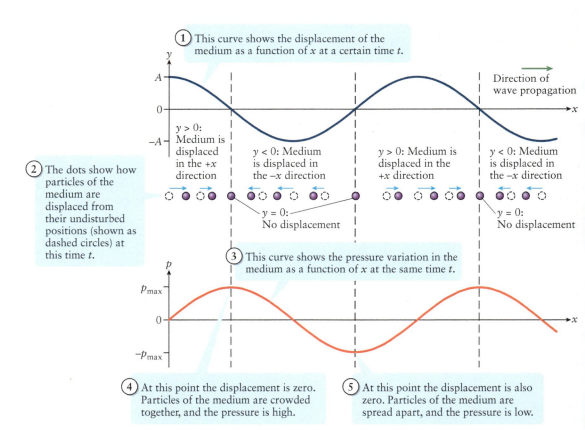

① This curve shows the displacement of the medium as a function of x at a certain time t.

→ Direction of wave propagation

$y > 0$: Medium is displaced in the $+x$ direction

$y < 0$: Medium is displaced in the $-x$ direction

$y > 0$: Medium is displaced in the $+x$ direction

$y < 0$: Medium is displaced in the $-x$ direction

② The dots show how particles of the medium are displaced from their undisturbed positions (shown as dashed circles) at this time t.

$y = 0$: No displacement

$y = 0$: No displacement

③ This curve shows the pressure variation in the medium as a function of x at the same time t.

④ At this point the displacement is zero. Particles of the medium are crowded together, and the pressure is high.

⑤ At this point the displacement is also zero. Particles of the medium are spread apart, and the pressure is low.

Figure 13-11 Pressure and displacement in a sinusoidal sound wave The pressure variation from normal in a sound wave is greatest where the displacement is zero and is zero where the displacement is greatest.

the pressure variation is given by

> **Pressure variation for a sinusoidal sound wave** propagating in the +x direction
>
> **Angular wave number** of the wave = $2\pi/\lambda$
>
> $$p(x,t) = p_{max} \sin(kx - \omega t + \phi)$$
>
> **Phase angle**
>
> **Pressure amplitude** of the wave
>
> **Angular frequency** of the wave = $2\pi f$

Pressure variation for a sinusoidal sound wave propagating in the +x direction (13-9)

The quantity p_{max} is the **pressure amplitude**. It represents the maximum pressure variation above or below the pressure of the undisturbed air. A positive value of $p(x,t)$ corresponds to regions of compression (higher density and pressure), while negative values indicate regions of expansion (lower density and pressure).

WATCH OUT! Pressure variation is not the same as pressure.

Note that $p(x,t)$ in Equation 13-9 is not the total air pressure but rather represents the pressure *variation* in the sound wave relative to the pressure of the undisturbed air. At any point in the air, the pressure is equal to the pressure of the undisturbed air (normally around 10^5 Pa) plus $p(x,t)$.

BioMedical GOT THE CONCEPT? 13-2 Human Hearing

? A person with normal hearing can hear sound waves that range in frequency from approximately 20 Hz to 20 kHz (1 kHz = 1 kilohertz = 10^3 Hz). Compared to a 20-Hz sound wave, a 20-kHz sound wave has (a) a shorter wavelength and a faster propagation speed; (b) a longer wavelength and a faster propagation speed; (c) a shorter wavelength and a slower propagation speed; (d) a longer wavelength and a faster propagation speed; (e) none of these.

TAKE-HOME MESSAGE FOR Section 13-3

✔ In a sinusoidal wave each part of the wave medium goes through simple harmonic motion.

✔ The wavelength λ is the distance the disturbance travels over one full cycle of the wave. In one period T the wave travels (propagates) a distance of one wavelength.

✔ For many common types of wave, the propagation speed is the same no matter what the frequency. Increasing the wave frequency decreases the wavelength and vice versa.

✔ The wave function $y(x,t)$ describes the displacement of the wave medium at any position x and at any time t. For a sinusoidal wave $y(x,t)$ is a sinusoidal function.

13-4 The propagation speed of a wave depends on the properties of the wave medium

As we discussed in the previous section, for many common kinds of mechanical waves the propagation speed does not depend on the frequency or wavelength. So what *does* it depend on? We can see the answer by remembering that for a mechanical wave to propagate through a medium, there must be a *restoring force* that tends to return the medium to equilibrium when it's disturbed. The greater the magnitude of this restoring force, the more rapidly the medium goes back to equilibrium and the faster the wave will propagate. At the same time the *inertia* of the medium acts to slow down the wave propagation. If the same restoring force is applied to a more sluggish medium with more inertia (think of water compared to maple syrup), the medium will return to equilibrium more slowly and the wave will propagate at a lower speed.

Let's look at how the competition between restoring force and inertia determines the propagation speed of transverse waves on a rope and of longitudinal waves in a fluid or solid.

Speed of a Transverse Wave on a String

For a transverse wave on a string or rope, the restoring force is provided by the *tension* in the rope. We'll denote this by the symbol *F*. (We used *T* as the symbol for tension in earlier chapters, but in this chapter we're using *T* to denote the period of a wave.) If you pluck a rope, the wave that you create will move more rapidly if you pull on the end of the rope to increase the tension.

The inertia of the rope depends on its *mass per unit length* or **linear mass density** (SI units kg/m), which we denote by the Greek letter μ ("mu"). If the rope is uniform, μ is just equal to the mass of the rope divided by its length. A thick rope has more mass per unit length than does a piece of ordinary string and so has more inertia. As the mass per unit length of a rope increases, the wave propagation speed decreases.

If we take both restoring force and inertia into account, we find that the propagation speed v_p of a transverse wave on a rope is

Go to Interactive Exercise 13-1 for more practice dealing with tension.

Speed of a transverse wave
on a rope
(13-10)

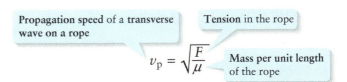

Propagation speed of a transverse wave on a rope   Tension in the rope

$$v_p = \sqrt{\frac{F}{\mu}}$$

Mass per unit length of the rope

Equation 13-10 can be derived by applying Newton's second law to a segment of a rope; however, the derivation is beyond our scope.

BioMedical EXAMPLE 13-3 **Tension in a Sperm's Flagellum**

We found in Example 13-1 (Section 13-3) that transverse waves propagate along the flagellum of a spermatozoon at a speed of about 7×10^{-4} m/s. A flagellum is about 40 μm in length and 0.25 μm in radius, and the material of which the flagellum is made has a density of 1100 kg/m³ (slightly greater than the density of water). Estimate the tension in the flagellum.

Set Up

We can determine the tension *F* in the flagellum from the propagation speed v_p and the linear mass density μ. To determine μ (in kg/m), we'll use the definition of *ordinary* density ρ (mass per unit volume introduced in Section 11-2) to find the total mass of the flagellum, then divide it by the flagellum length *L* = 40 μm.

Speed of a transverse wave on a rope (or flagellum):

$$v_p = \sqrt{\frac{F}{\mu}} \qquad (13\text{-}10)$$

Definition of density:

$$\rho = \frac{m}{V} \qquad (11\text{-}1)$$

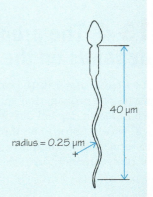

40 μm

radius = 0.25 μm

Solve

The flagellum is like a circular cylinder of radius $r = 0.25\ \mu m$ and length $L = 40\ \mu m$. Its volume is the cross-sectional area *multiplied* by the length. From the definition of density the mass of the flagellum is its density multiplied by its volume. The linear mass density equals the mass *divided* by the flagellum's length. So the length drops out of the calculation.

Cross-sectional area of a cylinder of radius $r = 0.25\ \mu m$:

$$A = \pi r^2 = \pi \left[0.25\ \mu m \times \left(\frac{10^{-6}\ m}{1\ \mu m} \right) \right]^2 = 2.0 \times 10^{-13}\ m^2$$

The volume of the flagellum (a cylinder with this cross-sectional area and length $L = 40\ \mu m$) is $V = AL$.

Mass of the flagellum:

$$m = \rho V = \rho A L$$

The linear mass density of the flagellum is its mass divided by its length L:

$$\mu = \frac{m}{L} = \frac{\rho A L}{L} = \rho A$$
$$= (1.1 \times 10^3\ kg/m^3)(2.0 \times 10^{-13}\ m^2)$$
$$= 2.2 \times 10^{-10}\ kg/m$$

Given the propagation speed v_p and the linear mass density μ, we can calculate the tension F.

The speed of transverse waves on the flagellum is

$$v_p = \sqrt{\frac{F}{\mu}}$$

To solve for F, square both sides:

$$v_p^2 = \frac{F}{\mu}$$
$$F = \mu v_p^2 = (2.2 \times 10^{-10}\ kg/m)(7 \times 10^{-4}\ m/s)^2$$
$$= 1 \times 10^{-16}\ kg \cdot m/s^2 = 1 \times 10^{-16}\ N$$

to one significant figure.

Reflect

The tension in the flagellum is truly miniscule, about 1/10 of a millionth of a billionth of a newton. Trying to measure such an infinitesimal force directly is almost impossible. But as this example shows, we can determine such forces indirectly by using the properties of waves.

The Speed of Longitudinal Waves

Just as for a transverse wave on a rope, the speed of a *longitudinal* wave in a medium depends on the restoring force and the inertia of the medium. For a longitudinal wave in a fluid (a gas or a liquid), we call a longitudinal wave a *sound* wave. The disturbance associated with the wave corresponds to changes in the pressure of the fluid. We saw in Section 10-3 that the *bulk modulus B* of a material is a measure of how it responds to pressure changes. A larger bulk modulus means the material is more difficult to compress, which means that it has a greater tendency to return to its original volume when the pressure is released. In other words, the value of B tells us about the restoring force that acts within a fluid when disturbed by the pressure changes associated with a longitudinal wave. A measure of the inertia of the material is its density ρ, equal to its mass per volume (see Example 13-3 above). A detailed analysis using Newton's laws shows that the speed of sound waves in a fluid is

Propagation speed of a longitudinal wave in a fluid

Bulk modulus of the fluid

$$v_p = \sqrt{\frac{B}{\rho}}$$

Density of the fluid

Speed of a longitudinal wave in a fluid
(13-11)

The greater the bulk modulus B, the faster the propagation speed; the greater the density ρ, the slower the propagation speed.

Figure 13-12 Temperature and the speed of sound in air At the altitudes where most jet airliners fly (about 11,000 m, or 36,000 ft, above sea level), the average temperature is about −55°C. At this low temperature, the speed of sound in air is only about 295 m/s compared to 343 m/s at 20°C. Current airliners fly at speeds equal to about 80% to 85% of the speed of sound.

Sound waves in *air* are particularly important. The bulk modulus B and density ρ of air both depend on temperature and humidity, so Equation 13-11 tells us that the speed of sound in air depends on both of these quantities. For most calculations we will use the speed of sound in dry air at 20°C:

(13-12)
$$v_{\text{sound}} = 343 \text{ m/s}$$

The speed of sound is slower in colder air (Figure 13-12).

The speed of longitudinal waves in a *solid* is given by a different expression than Equation 13-11. If we just consider one-dimensional waves such as those that might travel the length of a solid rod, what matters is not the *bulk* modulus B but rather *Young's modulus* Y. We learned in Section 10-2 that Young's modulus tells us how difficult it is to stretch or compress a piece of material along its length; those are just the kind of stresses experienced in a rod with a longitudinal wave propagating down its long axis. The propagation speed of such a wave is

Speed of a longitudinal wave in a solid rod
(13-13)

Propagation speed of a longitudinal wave along a solid rod · Young's modulus of the solid

$$v_{\text{p}} = \sqrt{\frac{Y}{\rho}}$$ · Density of the solid

We've stated that the propagation speed of waves is independent of frequency or wavelength in many important cases. One notable exception to this general rule is the speed of *surface* waves in water: For such waves the propagation speed turns out to be greater for waves with longer wavelength λ. As an example, waves with $\lambda = 1.0$ m travel at about 1.2 m/s, while longer-wavelength waves with $\lambda = 10$ m move at about 3.9 m/s.

EXAMPLE 13-4 Propagation Speeds

Calculate the speed of propagation of two longitudinal waves: (a) A sound wave in room-temperature water, which has bulk modulus 2.2×10^9 N/m² and density 1.0×10^3 kg/m³; (b) a wave traveling along a steel rail of a railroad track, with Young's modulus 2.1×10^{11} N/m² and density 8.0×10^3 kg/m³.

Set Up

We use Equation 13-11 to calculate the wave speed in water (a fluid) and Equation 13-13 for the wave speed along the steel rod (an example of a solid rod).

Speed of a longitudinal wave in water (a fluid):

$$v_{\text{p}} = \sqrt{\frac{B_{\text{water}}}{\rho_{\text{water}}}} \qquad (13\text{-}11)$$

Speed of a longitudinal wave along a solid steel rod:

$$v_{\text{p}} = \sqrt{\frac{Y_{\text{steel}}}{\rho_{\text{steel}}}} \qquad (13\text{-}13)$$

Solve

(a) For water use $B_{\text{water}} = 2.2 \times 10^9$ N/m² and $\rho_{\text{water}} = 1.0 \times 10^3$ kg/m³.

Speed of sound in water:

$$v_{\text{p}} = \sqrt{\frac{B_{\text{water}}}{\rho_{\text{water}}}} = \sqrt{\frac{2.2 \times 10^9 \text{ N/m}^2}{1.0 \times 10^3 \text{ kg/m}^3}}$$

$$= \sqrt{2.2 \times 10^6 \frac{\text{N} \cdot \text{m}^3}{\text{kg} \cdot \text{m}^2}} = 1.5 \times 10^3 \sqrt{\frac{\text{N} \cdot \text{m}}{\text{kg}}}$$

Since $1 \text{ N} = 1 \text{ kg} \cdot \text{m/s}^2$,

$$v_{\text{p}} = 1.5 \times 10^3 \sqrt{\frac{\text{kg} \cdot \text{m}}{\text{s}^2} \frac{\text{m}}{\text{kg}}} = 1.5 \times 10^3 \sqrt{\frac{\text{m}^2}{\text{s}^2}}$$

$$= 1.5 \times 10^3 \text{ m/s}$$

For the steel rod use $Y_{steel} = 2.1 \times 10^{11}$ N/m^2 and $\rho_{steel} = 8.0 \times 10^3$ kg/m^3.

Speed of longitudinal waves on the rail:

$$v_p = \sqrt{\frac{Y_{steel}}{\rho_{steel}}} = \sqrt{\frac{2.1 \times 10^{11} \text{ N/m}^2}{8.0 \times 10^3 \text{ kg/m}^3}}$$

$$= \sqrt{2.6 \times 10^7 \frac{\text{m}^2}{\text{s}^2}} = 5.1 \times 10^3 \text{ m/s}$$

Reflect

Even though water is almost a thousand times denser than air ($\rho_{water} = 1.0 \times 10^3$ kg/m^3 versus $\rho_{air} = 1.2$ kg/m^3), sound travels over four times faster in water (1.5×10^3 m/s versus 343 m/s in air). The explanation is that the bulk modulus B of water is over *ten thousand* times greater than that for air, 2.2×10^9 N/m^2 versus 1.4×10^5 N/m^2 (which says that water is much more difficult to compress than air). The bulk modulus factor thus overwhelms the density factor so that the wave speed $v_p = \sqrt{B/\rho}$ is greater in water than in air.

In the same way, Young's modulus for steel (2.1×10^{11} N/m^2) is about 100 times greater than the bulk modulus for water (2.2×10^9 N/m^2), which says that steel is *very* difficult to compress, and there is a tremendous restoring force when it is disturbed by a wave. This more than makes up for steel being eight times denser (and so having eight times more inertia) than water. So longitudinal waves travel even faster along a steel rail than they do in water.

GOT THE CONCEPT? 13-3 Crack of a Bat

The speed of sound in air is about 343 m/s, while the speed of *light* in air is 3.0×10^8 m/s. While sitting in the stands at a baseball game, you hear the crack of the bat hitting the ball 0.20 s after you see the batter swing. About how far is your seat from the batter? (a) About 70 m; (b) about 140 m; (c) about 340 m; (d) about 680 m; (e) about 1700 m.

TAKE-HOME MESSAGE FOR Section 13-4

✔ The propagation speed of a mechanical wave depends on the strength of the restoring force in the medium (which makes the wave propagate faster) and on the inertia of the medium (which makes the wave propagate more slowly).

✔ For transverse waves on a rope, the propagation speed depends on the ratio of the tension in the rope to its linear mass density (mass per unit length).

✔ The propagation speed of longitudinal waves (sound waves) in a fluid depends on the ratio of the bulk modulus of the fluid to the density of the fluid. For waves traveling along a solid rod, the speed is determined by the ratio of Young's modulus to the density.

13-5 When two waves are present simultaneously, the total disturbance is the sum of the individual waves

So far in our study of waves, we've considered only single waves in isolation. But in the real world we are bombarded by multiple waves simultaneously. If you are reading this book outside, you may hear sound waves from the conversations of others, from the wind blowing through the leaves, and from vehicles such as buses or airplanes. How can we describe what happens when more than one wave is present in a given medium?

In most cases what happens when more than one wave is present in a medium is this: Nothing unusual happens! If you hear two sounds at the same time, each sound wave is unaffected by the other (Figure 13-13). What you hear is simply the *sum* of the two sound waves. That observation is at the heart of an important physical principle called the *principle of superposition*:

> When two waves are present simultaneously, the total wave is the sum of the two individual waves.

Where the principle of superposition can lead to surprising results is if the two waves have the *same frequency*. As we discussed in Section 13-3, we can write the

Figure 13-13 The superposition of waves If sound waves are present from two sources, such as a musician's voice and his guitar, the result is a total sound wave that is just the sum of the individual waves.

√x̄ **See the Math Tutorial for more information on trigonometry.**

wave function of a one-dimensional sinusoidal wave as $y(x,t) = A \cos (kx - \omega t + \phi)$, where A is the amplitude, $k = 2\pi/\lambda$ is the angular wave number, $\omega = 2\pi f$ is the angular frequency, and ϕ is the phase angle (Equation 13-6). Since the propagation speed is $v_p = \omega/k$ (Equation 13-7), two waves with the same frequency f and hence the same angular frequency ω will have the same angular wave number k and wavelength λ. Suppose we add two waves of the same amplitude and angular frequency but different phase angles. If wave 1 has phase angle zero and wave 2 has phase angle ϕ, we can think of ϕ as the **phase difference** for the two waves (that is, the difference between their phase angles). The wave function of the total wave is

$$
\begin{aligned}
y_{\text{total}}(x,t) &= y_1(x,t) + y_2(x,t) \\
&= A \cos (kx - \omega t) + A \cos (kx - \omega t + \phi) \\
&= A [\cos (kx - \omega t) + \cos (kx - \omega t + \phi)]
\end{aligned}
$$

To add the two cosine functions in this equation, we use the trigonometric identity for the sum of two cosines found in the Math Tutorial at the back of the book. After a little simplification we find

(13-14) $$y_{\text{total}}(x,t) = 2A \cos\left(\frac{\phi}{2}\right) \cos\left(kx - \omega t + \frac{\phi}{2}\right)$$

In other words, the total wave has the *same* angular frequency ω and angular wave number k as each individual wave but has amplitude $2A \cos (\phi/2)$. Let's look at three examples.

(1) *Two waves in phase* ($\phi = 0$). If the phase difference is zero, then the two waves are in phase with each other: Their crests, or high points, occur at the same time t as do their troughs, or low points (Figure 13-14a). Since $\cos 0 = 1$, Equation 13-14 tells us that the total wave has amplitude $2A \cos 0 = 2A$. In other words, the waves reinforce each other so that the amplitude of the total wave is twice that of either individual wave. We call this **constructive interference**. Since each wave repeats itself once per cycle, constructive interference also happens if the phase difference corresponds to a whole number of cycles: $\phi = 2\pi, 4\pi, 6\pi,...$

(2) *Two waves π out of phase* ($\phi = \pi$). Since 2π radians corresponds to one complete cycle, a phase difference of π radians means that the two individual waves are one-half cycle out of step with each other (Figure 13-14b). The waves are *out of phase*: A crest of the first wave occurs at the same time as a trough of the second

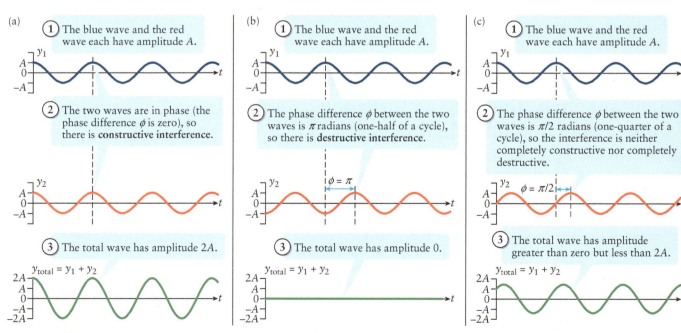

Figure 13-14 Superposition: Combining two waves with the same frequency When two waves arrive at the same location, the total wave is the sum of the two individual waves. The properties of the total wave depend on whether the individual waves are (a) in phase, (b) out of phase, or (c) something in between.

(1) This photograph shows a tank of water viewed from above.

Figure 13-15 Combining water waves Circular surface waves in water spread outward from two sources that oscillate in unison with the same frequency.

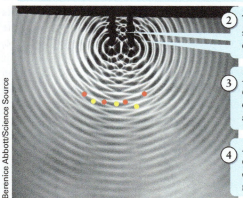

(2) Two "fingers" oscillate up and down and dip into the surface of the water. This produces two sets of waves, each one spreading out in circles from each "finger."

(3) At some positions in the water (shown by the yellow dots ●), a crest of one wave arrives at the same time as a crest of the other wave. There is constructive interference, and the result is a total wave of double the amplitude.

(4) At other positions (shown by the red dots ●), a crest of one wave arrives at the same time as a trough from the other wave. There is destructive interference, and the result is a total wave of nearly zero amplitude.

wave, and vice versa. In this case the two waves cancel each other, so from Equation 13-14 the amplitude of the total wave is $2A \cos \pi/2 = 0$ (recall that $\cos \pi/2 = 0$). This situation is called **destructive interference**. Destructive interference also occurs if the phase difference corresponds to a half cycle plus any whole number of cycles, or π plus an integer multiple of 2π: $\phi = 3\pi, 5\pi, 7\pi,...$

(3) *Two waves with an intermediate phase difference.* If the two waves are neither in phase nor out of phase (Figure 13-14c), we simply say that there is **interference** between the two waves: It is neither completely constructive nor completely destructive.

Figure 13-15 shows a real-life example of such interference. Two oscillating fingers dip in unison into a tank of water, producing two sets of surface waves like those in Figure 13-1a with the same amplitude, same frequency, and same wavelength. In certain locations, some of which are shown by the yellow dots in Figure 13-15, the two waves arrive in phase: A crest of the first wave arrives at the same time as a crest of the second wave. There is constructive interference at these locations, and the total wave has a large amplitude. At other locations, some of which are shown by the red dots, a crest of one wave arrives at the same time as a trough of the other wave. The two waves are out of phase, and there is destructive interference. At these locations the two waves essentially cancel each other, so that the total wave is nearly zero. Because interference of the surface waves occurs at all points in Figure 13-15, we call the overall pattern in that figure an **interference pattern**.

The interference effect shown in Figure 13-15 may seem surprising, since the two wave *sources* are oscillating in phase. It turns out that in this case what matters are the *distances* from each source to the point where the waves meet and interfere. The same principle applies to waves of all kinds, so in Figure 13-16 we've drawn two loudspeakers that emit sinusoidal sound waves of the same amplitude A, the same wavelength λ, and the same frequency f. We've drawn the sound waves as sinusoidal functions that

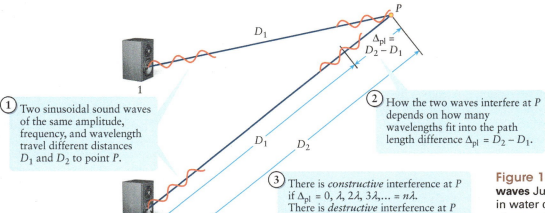

(1) Two sinusoidal sound waves of the same amplitude, frequency, and wavelength travel different distances D_1 and D_2 to point P.

(2) How the two waves interfere at P depends on how many wavelengths fit into the path length difference $\Delta_{pl} = D_2 - D_1$.

(3) There is *constructive* interference at P if $\Delta_{pl} = 0, \lambda, 2\lambda, 3\lambda,... = n\lambda$.
There is *destructive* interference at P if $\Delta_{pl} = \lambda/2, 3\lambda/2, 5\lambda/2,... = (n + \frac{1}{2})\lambda$.

Figure 13-16 Combining sound waves Just as two surface waves in water combine and interfere at any point in Figure 13-15, two sinusoidal sound waves combine at point P.

Berenice Abbott/Science Source

indicate the displacement of air along the direction of propagation. The two loudspeakers oscillate *in phase*, so they both emit the exact same waves at any time t. A listener is at point P, a distance D_1 from speaker 1 and a distance D_2 from speaker 2. We call these distances the *path lengths* from either speaker to point P. The **path length difference** Δ_{pl} ("delta-sub-p-l") for the two waves is the difference between these two distances: $\Delta_{pl} = D_2 - D_1$, so $D_2 = D_1 + \Delta_{pl}$.

At the instant shown in Figure 13-16, wave 1 happens to reach a peak of a cycle at point P. Wave 2 travels farther to reach P, and the additional distance is the path length difference Δ_{pl}. Because the two waves are emitted in phase, each wave hits a peak a distance D_1 from its source. So to determine what happens when the two waves overlap at P, we need only to ask how many cycles of wave 2 fit into the extra distance Δ_{pl}.

If exactly one wavelength λ fits into Δ_{pl}, then wave 2 will be at a peak at point P, just as if it were at distance D_1 from the source. The same is true if 2λ, 3λ, 4λ, or *any* integer number of full wavelengths fit into Δ_{pl}. (It's also true if $\Delta_{pl} = 0$, which is the case if point P is equal distances from the two speakers.) And because we've already established that wave 1 is at the same point in its cycle at point P as wave 2 is at D_1 from source 2, we can make this comparison regardless of where in the cycles the waves are at P. If an integer number of wavelengths fit into Δ_{pl}, the two waves will rise and fall together as they arrive at point P. The two waves are therefore *in phase* at point P, and constructive interference takes place there (see Figure 13-14a).

Suppose instead that exactly one-half of a wavelength, or an odd number of half wavelengths ($\lambda/2$, $3\lambda/2$, $5\lambda/2$,...), fits into the path length difference Δ_{pl} shown in Figure 13-16. In this case when wave 2 is at a peak at distance D_1 from the source, it will be at a trough in its cycle at point P. So waves 1 and 2 will be *out of phase* at point P, resulting in destructive interference. If the amplitudes of the two waves at P are equal, the total wave will be zero as in Figure 13-14b. (The cancellation isn't complete for the situation shown in Figure 13-15. That's because the amplitude of each surface wave decreases with distance from the source. So at points where D_1 and D_2 are different, the two waves have slightly different amplitudes.)

If the path length difference Δ_{pl} is neither a whole number of wavelengths nor an odd number of half-wavelengths, the interference is neither completely constructive nor completely destructive (Figure 13-14c).

Here's a summary of the conditions that must be met for constructive and destructive interference:

Conditions for constructive and destructive interference of waves from two sources (13-15)

> Path length difference from the two wave sources to the point where interference occurs   Wavelength
>
> Constructive interference: $\Delta_{pl} = n\lambda$ where $n = 0, 1, 2, 3,...$
>
> Destructive interference: $\Delta_{pl} = \left[n + \dfrac{1}{2}\right]\lambda$ where $n = 0, 1, 2, 3,...$

Note that for constructive interference, $n = 0$ refers to the case where there is *zero* path length difference, so the two waves naturally arrive in phase. This is the situation for the center yellow dot in Figure 13-15.

WATCH OUT! Interference effects require special conditions.

A sound system with two speakers resembles the setup in Figure 13-16, so you might expect that such a system would have "dead spots" where there is destructive interference between the sound waves coming from the two speakers (like the red dots in Figure 13-15) and "loud spots" where there is constructive interference (like the yellow dots in Figure 13-15). But even if you have such a system, you've probably never noticed any such "dead spots" or "loud spots." One reason is that the positions where destructive interference occurs depend on the wavelength (see Equations 13-15). Music contains sounds of many different frequencies and hence many different wavelengths, and places that are "dead" or "loud" spots for one wavelength will not be "dead" or "loud" for other wavelengths. Another reason is that in a stereo system the signals coming from the left-hand and right-hand speakers are not identical. A final reason is that sound waves also come from the speakers and bounce off the walls or ceiling before reaching you. These have different path lengths than the waves that reach you directly from the speakers, and so they can "smooth out" any interference between the waves that reach you directly.

EXAMPLE 13-5   Stereo Interference

Your sound system consists of two speakers 2.50 m apart. You sit 2.50 m from one of the speakers so that the two speakers are at the corners of a right triangle. As a test you have both speakers emit the same *pure tone* (that is, a sinusoidal sine wave). The speakers emit in phase. At your location what is the lowest frequency for which you will get (a) *destructive* interference? (b) *constructive* interference?

Set Up

There is a path length difference for the two waves that reach your ear: One travels a distance $D_1 = 2.50$ m; the other travels a longer distance D_2 equal to the hypotenuse of the right triangle. Since the product of frequency and wavelength equals the constant speed of sound, the *lowest* frequency for each kind of interference corresponds to the *longest* wavelength.

Constructive interference:

$\Delta_{pl} = n\lambda$ where $n = 0, 1, 2, 3,...$

Destructive interference:

$\Delta_{pl} = \left(n + \dfrac{1}{2}\right)\lambda$

where $n = 0, 1, 2, 3,...$ (13-15)

Propagation speed of a sound wave:

$v_{sound} = f\lambda$ (13-2)

Speed of sound in dry air at 20°C:

$v_{sound} = 343$ m/s   (13-12)

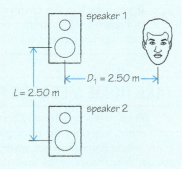

speaker 1

$D_1 = 2.50$ m

$L = 2.50$ m

speaker 2

Solve

(a) The path length difference Δ_{pl} is equal to the difference between the two path lengths D_2 and D_1.

Path length from speaker 2 to the listener:

$D_2 = \sqrt{D_1^2 + L^2} = \sqrt{(2.50 \text{ m})^2 + (2.50 \text{ m})^2}$
$ = 3.54$ m

Path length difference:

$\Delta_{pl} = 3.54$ m $- 2.50$ m $= 1.04$ m

The longest wavelength for which there is destructive interference is the one for which one half-wavelength fits into the distance Δ_{pl}. This corresponds to $n = 0$ in the destructive-interference relation in Equations 13-15.

If one half-wavelength fits into the distance Δ_{pl},

$\Delta_{pl} = \dfrac{1}{2}\lambda$

$\lambda = 2\Delta_{pl} = 2(1.04 \text{ m}) = 2.08$ m

The frequency of a sound wave with this wavelength is

$f = \dfrac{v_{sound}}{\lambda} = \dfrac{343 \text{ m/s}}{2.08 \text{ m}}$
$ = 165 \text{ s}^{-1} = 165$ Hz

If the sound has this frequency, there will be destructive interference at the listener's position and the sound level will be diminished.

(b) The longest wavelength for which there is constructive interference is the one for which one full wavelength fits into the distance Δ_{pl}. This corresponds to $n = 1$ in the constructive-interference relation from Equations 13-15. Note that $n = 0$ is possible only if the path length difference is zero, which in this case it is not.

If one wavelength fits into the distance Δ_{pl},

$\Delta_{pl} = \lambda = 1.04$ m

The frequency of a sound wave with this wavelength is

$f = \dfrac{v_{sound}}{\lambda} = \dfrac{343 \text{ m/s}}{1.04 \text{ m}}$
$ = 330 \text{ s}^{-1} = 330$ Hz

If the sound has this frequency, there will be constructive interference at the listener's position and the sound level will be elevated.

Reflect

The frequencies that we found are of musical importance: 165 Hz is E below middle C (E3 in musical nomenclature), and 330 Hz is E above middle C (or E4). Note that the higher frequency is exactly double that of the lower frequency. In music two such tones are said to be an *octave* apart.

Notice that we didn't consider the number of wavelengths that fit into the distance between you and either speaker. All that matters is the *difference* between the two paths. Can you show that in the destructive case the listener is 1.20 wavelengths away from speaker 1 and 1.70 wavelengths away from speaker 2? Can you also show that in the constructive case the numbers are 2.40 wavelengths and 3.40 wavelengths, respectively?

GOT THE CONCEPT? 13-4 Interference Pattern

 If you increase the frequency at which the fingers in Figure 13-15 dip into the water, would the interference pattern (a) move outward away from the fingers; (b) move inward toward the fingers; (c) remain unchanged; or (d) any of these, depending on the value of the frequency?

TAKE-HOME MESSAGE FOR Section 13-5

✔ When two waves from different sources interfere with each other, they obey the principle of superposition. At every point where more than one wave passes simultaneously, the net disturbance of the medium equals the sum of the displacements that each wave would have caused individually.

✔ If two waves are in phase at some location, the two waves reinforce each other and there is constructive interference. If the two waves are out of phase at some location, they cancel each other and there is destructive interference.

(a) As the pulse arrives at the pole, the string exerts an upward force on the pole...

(b) ...so the pole exerts a downward force on the string. This causes the pulse to become inverted as it is reflected.

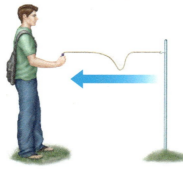

Figure 13-17 Reflecting a wave
(a) A single pulse moves down a stretched string tied to a pole. (b) Since the right-hand end of the string is fixed, the pulse reflected back from the pole is inverted.

13-6 A standing wave is caused by interference between waves traveling in opposite directions

Each of the six strings of a guitar is the same length, and yet each string produces a sound of different pitch when plucked. Why is this? When you turn one of the tuning keys to tighten a string, its pitch goes up. Why? And when one of the strings is pinched against the fingerboard, the pitch also goes up. Again, why?

The answers to these questions will help explain the operation of stringed musical instruments of all kinds, including guitars, violins, harps, and pianos. The basic physics is interference, which we introduced in the last section. But instead of looking at interference between waves coming from two different sources, we'll consider the interference between waves traveling in *opposite directions* along a stretched string.

Standing Waves on a String: Modes

To see how such a situation could arise, consider what happens when you use your hand to send a single pulse down a string tied to a pole (Figure 13-17a). When the pulse reaches the fixed end, the string exerts an upward force on the pole. The pole must then exert a downward force on the string, as required by Newton's third law. This action reflects an inverted pulse back along the string (Figure 13-17b).

If you were to wiggle the end of the string up and down periodically, instead of a pulse you would create a sinusoidal wave on the string. The reflection would be an inverted sinusoidal wave traveling back toward you with the same amplitude, wavelength, and frequency as the incoming wave.

Let's see what happens when we add together two sinusoidal waves traveling in opposite directions. From Equation 13-6 the wave function for such a wave traveling in the positive x direction is $y(x,t) = A \cos(kx - \omega t + \phi)$. For simplicity we'll choose the phase angle ϕ to be zero. As we discussed in Section 13-3, a wave traveling in the *negative x* direction has the same wave function but with $kx - \omega t$ replaced by $kx + \omega t$. We'll also use $\phi = 0$ for this wave but add a minus sign in front of the amplitude to indicate that it is inverted. From the principle of superposition the total wave is the sum of these two sinusoidal waves with the same amplitude A, angular wave number $k = 2\pi/\lambda$, and angular frequency $\omega = 2\pi f$:

$$y_{\text{total}}(x,t) = A \cos(kx - \omega t) + [-A \cos(kx + \omega t)]$$
$$= A[\cos(kx - \omega t) - \cos(kx + \omega t)]$$

We can again use a trigonometric identity to simplify the difference of the two cosine functions (see the Math Tutorial), arriving at

$$y_{\text{total}}(x,t) = 2A(\sin kx)(\sin \omega t) \tag{13-16}$$

The wave function in Equation 13-16 describes a wave pattern unlike any we've seen yet. Note that the x and t terms do not appear in the same sinusoidal function. As a result the crests of the wave do *not* move from place to place: The wave is maximum at points where $\sin kx$ equals either $+1$ or -1, and at those points the value of y_{total} oscillates between $+2A$ and $-2A$ as the function $\sin \omega t$ varies between $+1$ and -1. At the points where $\sin kx = 0$, the wave is *always* zero! Because the wave pattern does not move but stays in the same place, we call it a **standing wave**. By contrast, a wave of the form $y(x,t) = A \cos (kx - \omega t)$ is referred to as a **traveling wave**. Note that we formed the standing wave by combining two traveling waves that propagate in opposite directions. The two traveling waves each carry a disturbance; when we add them together the result is that the disturbance stays in the same place. Pieces of the string still move, but the *pattern* created by the interference between two traveling sinusoidal waves—the standing wave—remains stationary.

Now think again about the string in Figure 13-17. Let L be the length of the string, take $x = 0$ to be at the end of the string you hold in your hand, and let $x = L$ be at the end of the string attached to the pole. Note from Equation 13-16 that at $x = 0$, $y_{\text{total}} = 0$ because $\sin k(0) = \sin 0 = 0$. The end that you hold in your hand moves very little compared to the amplitude of the wave on the string, so it's quite accurate to say that the amplitude of the wave at your hand is zero. So waves will reflect from *both* ends of the string, and a steady standing wave can be set up. At the other end of the string at $x = L$, the transverse displacement y must also be zero at all times because that end is rigidly attached to the pole. So $y_{\text{total}}(L,t) = 0$ for all values of t. Equation 13-16 tells us that this will be true only if $\sin kL = 0$. Now $\sin \theta$ is equal to zero if $\theta = 0, \pi, 2\pi, 3\pi,...$ So it must be true that

$$kL = n\pi, \text{ where } n = 1, 2, 3,... \tag{13-17}$$

(We can't have $kL = 0$, since that would mean either that the string has zero length or the angular wave number $k = 2\pi/\lambda$ is zero, corresponding to the wave having an infinite wavelength.) Equation 13-17 is the condition that the displacement of the string is zero at both ends. Since $k = 2\pi/\lambda$, we can rewrite Equation 13-17 as

$$\frac{2\pi L}{\lambda} = n\pi$$

or, simplifying,

Length of a string held at both ends	Wavelength of a standing wave on the string

$$L = \frac{n\lambda}{2} \quad \text{where } n = 1, 2, 3,...$$

Wavelengths for a standing wave on a string (13-18)

Here λ represents the wavelength of both the standing wave and also the original wave we created on the string. So according to Equation 13-18 a whole number n of half-wavelengths must fit onto the string to generate the standing wave pattern described by Equation 13-16. Figure 13-18 shows the patterns for $n = 1$ (for which half of a wavelength fits), $n = 2$ (for which two half-wavelengths, or one full wavelength, fits), and $n = 3$ (for which three half-wavelengths, or one and a half full wavelengths, fits). Each of these patterns is called a **standing wave mode**.

Figure 13-19 shows photographs of the $n = 1, 2,$ and 3 standing wave modes on a real string. In these photographs the movement of the end of the string that creates the standing waves is extremely slight. This justifies treating the end of this string as fixed, as we described above.

Figures 13-18 and 13-19 show that for all modes except $n = 1$, there are other positions besides the ends at which the displacement of the string is zero for all times. These are the points where $\sin kx = 0$ in Equation 13-16, so the wave function $y_{\text{total}}(x,t)$ equals zero. Any point where the displacement is always zero is called a **node**

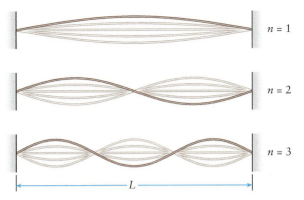

Figure 13-18 Standing waves on a string I These illustrations show the first three standing wave modes of a string connected to two supports. The mode number n counts the number of half-wavelengths that fit into the length of the string.

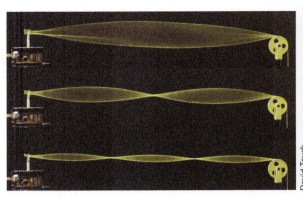

Figure 13-19 Standing waves on a string II These photographs show the first three standing wave modes on a string. The left-hand end of the string moves up and down (slightly) to create each standing wave.

of the standing wave. For the $n = 1$ standing wave mode, there are nodes at each end of the string; for the $n = 2$ mode there is an additional node at the center of the string, $x = L/2$; and for the $n = 3$ mode there are two nodes at $x = L/3$ and $x = 2L/3$.

Positions along the standing wave at which the oscillation of the string is maximal are called **antinodes**. These lie halfway between adjacent nodes. For the $n = 1$ mode in Figures 13-18 and 13-19 there is one antinode at the center of the string ($x = L/2$); for the $n = 2$ mode there are two antinodes, at $x = L/4$ and $x = 3L/4$; and for the $n = 3$ mode there are three antinodes, at $x = L/6$, $x = L/2$, and $x = 5L/6$.

Standing Waves on a String: Frequencies and Musical Sound

We learned in Section 13-4 that the propagation speed of a *traveling* wave on a string under tension F and with linear mass density μ is $v_p = \sqrt{F/\mu}$ (Equation 13-10). Furthermore, the frequency f and wavelength λ of a traveling wave are related by $v_p = f\lambda$ (Equation 13-2). Because a standing wave on a string is a superposition of two traveling waves, we can use these two relationships to find the *frequencies* associated with each of the standing wave modes on a string fixed at both ends.

By combining Equations 13-2 and 13-10, we get the following relationship between the frequency f and wavelength λ of a wave on a string:

$$f\lambda = \sqrt{\frac{F}{\mu}} \quad \text{or} \quad f = \frac{1}{\lambda}\sqrt{\frac{F}{\mu}}$$

To make this specific to *standing* waves on a string, we use the relation $L = n\lambda/2$ from Equation 13-18, which we can rewrite as $(1/\lambda) = n/(2L)$. Substituting this into the above equation, we get

Frequencies for a standing wave on a string (13-19)

nth possible frequency of a standing wave on a string held at both ends

Tension in the string

$$f_n = \frac{n}{2L}\sqrt{\frac{F}{\mu}} \quad \text{where } n = 1, 2, 3, \ldots$$

Length of the string

Linear mass density of the string

The subscript n on the symbol f_n in Equation 13-19 denotes the frequency of the nth standing wave mode (see Figure 13-18 or Figure 13-19). Each of these frequencies represents a *natural* frequency of the string, at which it will oscillate if displaced from equilibrium and released. To make the string oscillate at the $n = 1$ frequency, displace it into the $n = 1$ shape shown in Figure 13-18 and let it go; to make it oscillate at the $n = 2$ frequency, displace it into the $n = 2$ shape; and so on. They are also the *resonant* frequencies of the string. If one end of the string is set into oscillation at one of the frequencies given by Equation 13-19, the string will oscillate with large amplitude (see Figure 13-19). If the end of the string is forced to oscillate at a frequency other than those given by Equation 13-19, the result will be a jumble of small-amplitude wiggles rather than a well-behaved standing wave.

The frequency f_1, which is the lowest natural frequency of the string, is called the **fundamental frequency**. The corresponding ($n = 1$) standing wave mode is called the **fundamental mode**. Equation 13-19 shows that the fundamental frequency is equal to

$$f_1 = \frac{1}{2L}\sqrt{\frac{F}{\mu}}$$

(13-20)

The fundamental frequency is proportional to the square root of the string tension F, inversely proportional to the square root of the string's linear mass density μ, and inversely proportional to the length L of the string.

Equation 13-20 answers the questions we posed at the beginning of this section about the strings of a guitar. For stringed instruments like a guitar, plucking a string normally causes it to vibrate primarily at its fundamental frequency. If it's an acoustic guitar, the oscillation of the string acts as a driving force that causes the body of the guitar to oscillate at the same frequency (see Section 12-8). The oscillations of the body in turn produce a sound wave of the same frequency in the surrounding air, which is what reaches the listener's ear. (If it's an electric guitar, the pickups on the guitar detect the oscillation of the string and generate an electrical signal of the same frequency. This is then amplified in a loudspeaker to produce sound.) So Equation 13-20 is really an equation for the sound frequencies produced by a stringed musical instrument. It shows that for strings of a given length L, the frequency and hence the pitch of the sound is greater for strings with less mass per unit length μ. That's why the first string on a guitar, which has a fundamental frequency of 329.6 Hz, is much thinner than the sixth string, which has a fundamental frequency of 82.4 Hz (Figure 13-20). It also shows that if you increase the tension F on a string, the frequency and pitch increase. And it explains what happens when you pinch one of the strings against the fingerboard: This shortens the length of string that's free to vibrate, thereby causing the frequency and pitch to go up.

Equation 13-19 shows that all of the other natural frequencies of a string held at both ends are integer multiples of f_1:

$$f_n = \frac{n}{2L}\sqrt{\frac{F}{\mu}} = n\left(\frac{1}{2L}\sqrt{\frac{F}{\mu}}\right) = nf_1$$

The $n = 2$ frequency f_2, called the *second harmonic* (or *first overtone*), is equal to $2f_1$; the $n = 3$ frequency f_3, called the *third harmonic* (or *second overtone*), is equal to $3f_1$; and so on.

The word "harmonic" should make you think of music, and indeed stringed musical instruments are an important application of the standing wave modes of a string. When you pluck a guitar string, the string actually vibrates in a *superposition* of standing waves with $n = 1$, $n = 2$, $n = 3$, and so on. (Just as two sinusoidal traveling waves can be present simultaneously on a string and combine to make a sinusoidal standing wave, more than one standing wave can be present on a string at the same time.) If it's an acoustic guitar (Figure 13-20), the guitar body is forced to oscillate simultaneously at the frequencies f_1, f_2, f_3,... and so produces a sound wave that is a superposition of sinusoidal waves at these frequencies. You might think that this would produce a hopeless jumble of sound. In fact, the result is a *nonsinusoidal* sound wave with the same frequency f_1 as the fundamental frequency of the string (Figure 13-21). The shape of the wave function of this nonsinusoidal wave—which depends on the relative amplitudes and phases of the sinusoidal waves at frequencies f_1, f_2, f_3,...—determines the *tone quality* or *timbre* of the sound. Different musical instruments playing the same note all generate sound waves of the same frequency, but with different amounts of the various harmonics. As a result, different instruments produce sound waves with differently shaped wave functions and hence different tone qualities, which is why they sound different.

 Go to Picture It 13-1 for more practice dealing with overtones.

Sixth string: greatest μ, lowest f_1 (82.4 Hz)

First string: smallest μ, highest f_1 (329.6 Hz)

Valery Voennyy/Alamy

Figure 13-20 Guitar strings The six strings on this acoustic guitar all have the same length L and are all under about the same tension F, but have different values of the mass per unit length μ. Hence their fundamental frequencies given by Equation 13-20 are different.

Figure 13-21 Adding harmonics
(a) A sound wave is produced by a string oscillating simultaneously at its fundamental frequency f_1, its second harmonic f_2, and its third harmonic f_3. (b) The total sound wave has frequency f_1 but is not a sinusoidal wave.

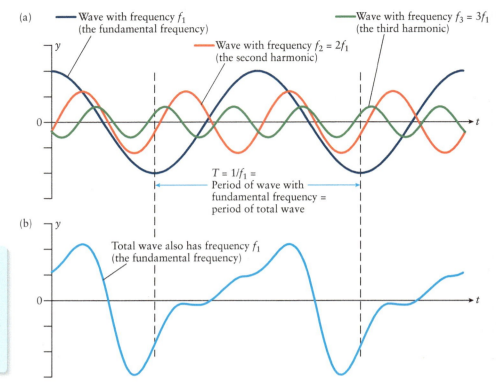

(a) — Wave with frequency f_1 (the fundamental frequency)
— Wave with frequency $f_2 = 2f_1$ (the second harmonic)
— Wave with frequency $f_3 = 3f_1$ (the third harmonic)

$T = 1/f_1 =$ Period of wave with fundamental frequency = period of total wave

(b) Total wave also has frequency f_1 (the fundamental frequency)

• Adding waves of frequencies $f_1, f_2, f_3,...$ (the harmonics of f_1) produces a wave of frequency f_1 with a nonsinusoidal wave function.
• The shape of the sound wave produced by a musical instrument determines the tone quality of the sound.

EXAMPLE 13-6 Guitar String Physics

The third string on a guitar is 0.640 m long and has a linear mass density of 1.14×10^{-3} kg/m. This string is to play a note of frequency 196 Hz (to a musician, this note is G_3 or G below middle C). (a) What must be the tension in the string? (b) What is the wavelength of the fundamental standing wave on the string? (c) What is the speed of transverse waves on this string? (d) What is the wavelength of the sound wave produced by the wave?

Set Up

We are given the string length $L = 0.640$ m and the mass per length of the string $\mu = 1.14 \times 10^{-3}$ kg/m. We'll use Equation 13-19 to find the tension F that gives a fundamental frequency $f_1 = 196$ Hz. We'll use Equation 13-18 to find the wavelength λ of the standing wave and use this along with Equation 13-2 to find the speed of waves on the string. The sound wave produced by the guitar has the same *frequency* as the standing wave on the string. We'll find the *wavelength* of the sound wave using Equations 13-2 and 13-12.

Frequencies for a standing wave on a string:

$$f_n = \frac{n}{2L}\sqrt{\frac{F}{\mu}} \text{ where } n = 1, 2, 3, ...$$

(13-19)

Wavelengths for a standing wave on a string:

$$L = \frac{n\lambda}{2} \text{ where } n = 1, 2, 3, ...$$

(13-18)

Propagation speed, frequency, and wavelength of a wave:

$$v_p = f\lambda$$

(13-2)

Speed of sound in dry air at 20°C:

$$v_{sound} = 343 \text{ m/s}$$

(13-12)

G string

Solve

(a) The fundamental frequency is given by Equation 13-19 with $n = 1$. Solve this equation for the string tension F.

The fundamental frequency ($n = 1$) is

$$f_1 = \frac{1}{2L}\sqrt{\frac{F}{\mu}}$$

To solve for the string tension F, first square both sides to get rid of the square root:

$$f_1^2 = \frac{1}{4L^2}\frac{F}{\mu}$$

Multiply both sides by $4L^2\mu$:

$$F = 4L^2\mu f_1^2$$
$$= 4(0.640 \text{ m})^2(1.14 \times 10^{-3} \text{ kg/m})(196 \text{ Hz})^2$$
$$= 71.8 \text{ kg}\cdot\text{m}\cdot\text{Hz}^2 = 71.8 \text{ kg}\cdot\text{m/s}^2$$
$$= 71.8 \text{ N}$$

(Recall that $1 \text{ Hz} = 1 \text{ s}^{-1}$.)

(b) For the fundamental mode (the $n = 1$ standing wave) of a string held at both ends, the length of the string equals one half-wavelength.

From Equation 13-18,

$$L = \frac{\lambda}{2} \text{ for } n = 1$$

$$\lambda = 2L = 2(0.640 \text{ m}) = 1.28 \text{ m}$$
(wavelength of the standing wave on the string)

(c) Find the wave speed on the string using Equation 13-2.

Speed of transverse waves on this string:

$$v_p = f\lambda$$
$$= (196 \text{ Hz})(1.28 \text{ m})$$
$$= 251 \text{ m/s}$$

(d) Use Equations 13-2 and 13-12 to find the wavelength of the sound wave produced by the guitar.

For the sound wave,

$$v_{\text{sound}} = f\lambda$$

The frequency of the sound wave is the same as the frequency of the standing wave on the string, but the wavelength is different because the wave speed is different:

$$\lambda = \frac{v_{\text{sound}}}{f} = \frac{343 \text{ m/s}}{196 \text{ Hz}} = \frac{343 \text{ m/s}}{196 \text{ s}^{-1}}$$
$$= 1.75 \text{ m}$$

Reflect

There are six strings on the guitar, each of which is under about the same tension. So the total force that acts on the guitar at the points where the strings are attached is about $6 \times 71.8 \text{ N} = 431 \text{ N}$ (about 97 lb). The guitar must be of sturdy construction to withstand these forces.

You can check our result for the speed of transverse waves on the string by using Equation 13-10, $v_p = \sqrt{F/\mu}$, and the value of F that we found in part (a). Do you get the same answer this way?

Note that the sound wave has a greater wavelength λ (1.75 m) than does the standing wave on the string (1.28 m). This makes sense: Although the frequency f is the same for both waves, the propagation speed v_p is greater for the sound wave. Since $v_p = f\lambda$, the wavelength $\lambda = v_p/f$ must be greater for the sound wave.

GOT THE CONCEPT? 13-5 Changing the String Tension I

If the tension in a guitar string is increased by a factor of 2, the wavelength of the $n = 1$ (fundamental) standing wave mode will (a) increase to twice its previous value; (b) increase, but by a factor different than 2; (c) decrease to 1/2 of its previous value; (d) decrease but by a factor different than 1/2; (e) none of these.

GOT THE CONCEPT? 13-6 Changing the String Tension II

If the tension in a guitar string is increased by a factor of 2, the frequency of the $n = 1$ (fundamental) standing wave mode will (a) increase to twice its previous value; (b) increase, but by a factor different than 2; (c) decrease to 1/2 of its previous value; (d) decrease, but by a factor different than 1/2; (e) none of these.

13-7 Wind instruments, the human voice, and the human ear use standing sound waves

You've probably noticed that your singing voice sounds much better in the shower than in the open. You may have also noticed that a flute (a musical instrument based on a long tube) produces lower notes than a piccolo (which uses a shorter tube). And you know that the sound of fingernails on a chalkboard is a particularly unpleasant one—so much so that just thinking about that sound may make you cringe.

We can help to explain all of these effects by again invoking the idea of *standing waves*, which we introduced in the last section. The standing waves we need to consider are not waves on a string, but rather standing *sound* waves that result from traveling sound waves that reflect from the ends of an enclosure or tube.

Figure 13-22 shows an example of such a standing sound wave. Sound waves travel horizontally through the transparent tube and are reflected when they strike the ends, just as transverse waves on a string are reflected at the fixed ends of the string. Sound is a longitudinal wave, so the motion of the air caused by the wave moving horizontally through the tube is also in the horizontal direction. The solid ends of the tube prevent this motion from taking place there, so there is zero displacement at the ends. This is just like the situation for a string with both ends fixed (see Figure 13-18). So just as for a standing wave on a string, a standing *sound* wave will be set up if a whole number of half-wavelengths of the wave fit into the length of the tube (Equation 13-18). In Figure 13-22 the frequency of sound waves provided by the loudspeaker has been adjusted so that two half-wavelengths (one full wavelength) fit into the tube's length, corresponding to $n = 2$ in Figure 13-18 and Equation 13-18.

Note that we can use Figure 13-18 for the displacement in a standing wave on a string to depict the displacement in a standing *sound* wave in a tube like the one shown in Figure 13-22. The difference is that because sound is a longitudinal wave, the displacement is now measured *along* the direction of wave propagation (see Figure 13-11).

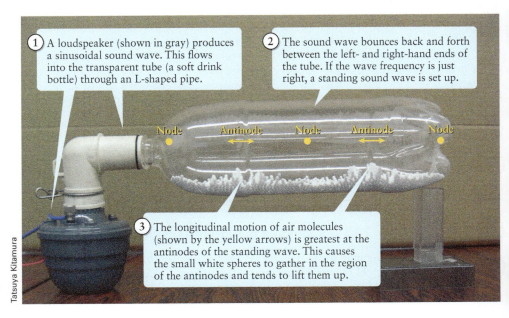

① A loudspeaker (shown in gray) produces a sinusoidal sound wave. This flows into the transparent tube (a soft drink bottle) through an L-shaped pipe.

② The sound wave bounces back and forth between the left- and right-hand ends of the tube. If the wave frequency is just right, a standing sound wave is set up.

Node   Antinode   Node   Antinode   Node

③ The longitudinal motion of air molecules (shown by the yellow arrows) is greatest at the antinodes of the standing wave. This causes the small white spheres to gather in the region of the antinodes and tends to lift them up.

Figure 13-22 Visualizing a standing sound wave The small white spheres inside this apparatus (known as *Kundt's tube* after the German physicist who devised it in the nineteenth century) help to identify the locations of the antinodes of a standing sound wave in air, which would otherwise be invisible.

Tatsuya Kitamura

Note also that the *pressure variation* in a traveling sound wave is a quarter-cycle out of phase with the displacement. Figure 13-11 shows this for a traveling sound wave, but the same is true for a standing sound wave. So where there is zero displacement in a standing sound wave (a displacement node) is where the pressure variation is maximum (a pressure antinode); where the displacement oscillates with maximum amplitude (a displacement antinode) is where the pressure variation is zero, so the pressure is always equal to the undisturbed air pressure (a pressure node).

Each standing wave sound frequency corresponds to a natural frequency of the tube. That means that if you produce a sound in the tube at one of those natural frequencies, *resonance* will take place and a strong standing wave will be set up at that frequency. The net result will be that the sound wave you produce will be enhanced, just like the oscillation of a mass–spring combination is enhanced if you force it to oscillate at its natural frequency (see Section 12-8). That explains what happens when you sing in the shower. In the $n = 1$ standing wave mode, half a wavelength will fit between the walls of a shower stall. A typical shower stall is about 1 m wide, so the wavelength of the $n = 1$ mode is about 2 m. The frequency f_1 of this mode is related to the wavelength λ and the propagation speed v_{sound} for sound waves by $v_{sound} = f_1\lambda$ (Equation 13-2), so

$$f_1 = \frac{v_{sound}}{\lambda} = \frac{343 \text{ m/s}}{2 \text{ m}} \approx 170 \text{ Hz}$$

So if you sing at a frequency near 170 Hz (E or F below middle C), the shower stall will resonate and your voice will sound much fuller than it would outside the shower. The same will happen at frequencies that are multiples of 170 Hz (340 Hz, 510 Hz, 680 Hz, etc.).

Standing Waves and Wind Instruments: Closed Pipes

Just as a guitar or other stringed instrument produces tones by using standing waves on a string, a *wind* instrument makes use of standing waves in an air-filled tube. These instruments are played by setting the air within the tube into oscillation. In a brass instrument like a trumpet or trombone, the musician does this by making a buzzing sound with her lips; in a woodwind like a clarinet or saxophone, the musician blows into a reed that vibrates the air in the instrument. For the sound to get out of the instrument, it has to be *open* at one or both ends. (When a sound wave traveling down the tube reaches the open end, part of it is reflected back. So the ingredients for making a standing wave are still present.) This means that a wind instrument has a different set of standing wave modes than those shown in Figure 13-18.

Most wind instruments behave like a **closed pipe**, in which one end is closed and the other is open to the air. The end that the musician blows into behaves like a closed end: The displacement is essentially zero, and the pressure variation has its maximum amplitude. (This is analogous to what happens at the left-hand end of the string in Figure 13-19. It's at that end that a force is applied to make the string oscillate, but the displacement there is very small. In the same way, there is very little displacement of the air at the musician's end of a brass instrument or woodwind, but there is a large pressure variation due to the musician either buzzing her lips or making a reed vibrate.) The other end of the tube is open to the air. As a result the pressure there is essentially equal to the pressure of the surrounding air, so the pressure variation is close to zero and the displacement amplitude is large. In other words, a closed tube has a displacement node at the closed end but a displacement antinode at the open end.

Figures 13-23 and 13-24 show the first three standing wave modes in a closed pipe. We represent the magnitude of the displacement of the air from equilibrium by the separation of the blue curves from the dashed centerline. Figures 13-23a and 13-24a show that in

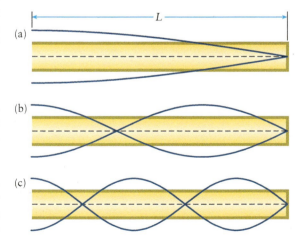

Figure 13-23 Standing sound waves in a closed pipe I (a), (b), and (c) show the displacement patterns for the first three standing sound wave modes of a pipe closed at its right-hand end and open at its left-hand end. (Since sound waves are longitudinal, the displacement of air in these patterns is horizontal.) There is a displacement node at the closed end and a displacement antinode at the open end.

the fundamental mode of this closed pipe, only *one-quarter* of a wavelength λ fits in the length L of the pipe:

$$L = \frac{1}{4}\lambda$$

You can see this most clearly in Figure 13-24a, in which we've extended the sinusoidal representation of the fundamental mode out beyond the open end to show one full wavelength.

WATCH OUT! The standing sound wave in a closed pipe does not extend far beyond the pipe.

! This extension of the wave pattern that we've drawn in Figure 13-24a does *not* represent the standing wave; the standing wave exists inside the pipe only. We've drawn this only to allow you to compare the wavelength of the sound with the length of the pipe.

Figures 13-23b and 13-23c show the next two standing wave modes of a closed pipe, and Figures 13-24b and 13-24c show the displacement patterns extended outside the pipe. You can see that three quarters (3/4) and five quarters (5/4) of a wavelength, respectively, fit into the pipe for these modes. You can also see that the general relationship between L and λ for a closed pipe is

Wavelengths for a standing sound wave in a closed pipe (13-21)

> **Length of a closed pipe** (open at one end, closed at the other)

> **Wavelength of a standing wave** in the pipe

$$L = \frac{n\lambda}{4} \quad \text{where } n = 1, 3, 5, \ldots$$

Notice that only the *odd* harmonics ($n = 1, 3, 5, \ldots$) can exist in a pipe closed at one end. From $v_{\text{sound}} = f\lambda$ (Equation 13-2 for sound) the corresponding natural frequencies for a closed pipe are

(13-22)

$$f_n = \frac{v_{\text{sound}}}{\lambda} = \frac{v_{\text{sound}}}{(4L/n)} = n\left(\frac{v_{\text{sound}}}{4L}\right) \quad \text{where } n = 1, 3, 5, \ldots$$

You can see from Equation 13-22 that the fundamental frequency of a closed pipe of length L is $f_1 = v_{\text{sound}}/4L$. The next harmonic is the *third* harmonic, $f_3 = 3(v_{\text{sound}}/4L) = 3f_1$; the next harmonic after that is the *fifth* harmonic, $f_5 = 5f_1$; and so on. *For a closed pipe the even harmonics are absent.*

Equation 13-22 tells us that to change the fundamental frequency of a closed pipe, you must change the pipe length L. For a brass instrument this is done either with a slide (as in a trombone) or with valves that open and close air passages (as in a trumpet or tuba). For a woodwind like a clarinet or oboe, this is done by opening holes in the pipe that effectively shorten the length of the vibrating air column.

Just like what happens when you pluck a guitar string, blowing into a closed-pipe wind instrument produces standing sound waves at more than one of the natural

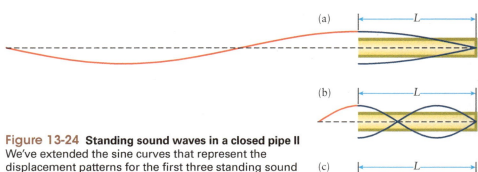

Figure 13-24 Standing sound waves in a closed pipe II We've extended the sine curves that represent the displacement patterns for the first three standing sound wave modes in a closed pipe (compare Figure 13-23). This makes it easier to see what fractional number of wavelengths fit into the pipe in each mode.

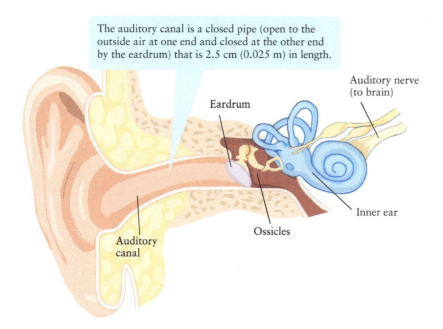

The auditory canal is a closed pipe (open to the outside air at one end and closed at the other end by the eardrum) that is 2.5 cm (0.025 m) in length.

Eardrum

Auditory nerve (to brain)

Inner ear

Ossicles

Auditory canal

Figure 13-25 The human ear Sound waves entering the human ear via the auditory canal cause the eardrum to vibrate, which in turn causes vibrations of the three ossicles (the smallest bones in the human body). These induce vibrations in the fluid that fills the inner ear. These vibrations are detected and converted to an electrical signal that is carried by the auditory nerve to the brain. The auditory canal acts like a closed pipe, which makes hearing especially sensitive for frequencies around 3430 Hz.

frequencies given by Equation 13-22. As a result the sound that emanates from the instrument will be a combination of sinusoidal sound waves at frequencies $f_1, f_3, f_5,...$, and the net result will be a nonsinusoidal wave with frequency f_1 (see Figure 13-21). The particular combination of harmonics is different for different wind instruments playing the same note, which gives each instrument its own distinctive tone quality.

You actually have closed pipes on either side of your head. These are the *auditory canals* of your left and right ears (Figure 13-25), which extend a distance of about 2.5 cm = 0.025 m from the opening (an open end) to the eardrum (a closed end). From Equation 13-22 the fundamental frequency ($n = 1$) of this short closed pipe is

BioMedical

$$f_1 = \frac{v_{\text{sound}}}{4L} = \frac{343 \text{ m/s}}{4(0.025 \text{ m})} = 3430 \text{ Hz}$$

This suggests that if a sound wave with a frequency of about 3430 Hz enters your ear, it will cause resonance to take place and the sound wave will be enhanced. In fact, a person with normal hearing is most sensitive to sounds around this frequency. And an important reason the sound of fingernails scraping across a chalkboard is so unpleasant is that this is a sound in the range from 2000 to 4000 Hz—precisely where your hearing is the most sensitive.

Your vocal tract, which extends from the vocal folds in your throat to your lips, is also a closed pipe (closed at the vocal folds, which behave rather like the reed in a clarinet or oboe, and open at the lips). Unlike other musical instruments, you can reshape your vocal tract by the way that you hold your tongue and your lips. This changes the natural frequencies of the vocal tract (so that they no longer obey Equation 13-22) and makes each spoken sound distinct. To demonstrate this try saying "a, e, i, o, u" and notice how your tongue and lips change position. Then use your fingers to hold your lips together and repeat the experiment; without the ability to reshape your vocal tract, all five vowels will now sound almost the same.

BioMedical

BioMedical EXAMPLE 13-7 An Amorous Frog

The male tree-hole frog of Borneo (*Metaphrynella sundana*) attracts females by croaking out a simple call dominated by a tone of a single frequency. To amplify his call and enhance his attractiveness to potential mates, he finds a tree with a cavity in its trunk, sits inside the cavity, and adjusts the frequency of his call to match the fundamental frequency of the cavity. If the cavity is cylindrical and 11 cm deep, at what frequency should the frog make his mating call?

Set Up

The cavity in the tree trunk acts like a closed pipe (open on the outside of the tree, closed on the inside). Equation 13-22 gives the frequency of the fundamental ($n = 1$) mode of this cavity.

Frequencies for a standing sound wave in a closed pipe:

$$f_n = n\left(\frac{v_{sound}}{4L}\right) \text{ where } n = 1, 3, 5, \dots$$

$$(13\text{-}22)$$

Solve

Calculate the fundamental frequency using $n = 1$ and $v_{sound} = 343$ m/s.

The fundamental frequency ($n = 1$) is

$$f_1 = \frac{v_{sound}}{4L} = \frac{343 \text{ m/s}}{4(11 \text{ cm})}\left(\frac{100 \text{ cm}}{1 \text{ m}}\right)$$
$$= 780 \text{ s}^{-1} = 780 \text{ Hz}$$

Reflect

Our calculated result is close to the measured value of the frequency of a male tree-hole frog's mating call in a cavity 11 cm deep. These frogs have been observed to match the natural frequency of tree cavities from 10 to 15 cm in depth.

Standing Waves and Wind Instruments: Open Pipes

Unlike most other wind instruments, a *flute* is a tube that is open at *both* ends. This is called an **open pipe**. A flautist plays a flute by blowing across the top of a hole in the pipe; this causes a pressure variation that makes the air inside the pipe oscillate.

Because this pipe is open at each end, there is a displacement antinode (pressure node) at each end. Figure 13-26 shows the displacement patterns for the first three standing wave modes of such a pipe. (Just as at the open end of a closed pipe, a sound wave traveling through an open pipe is partially reflected at each open end. The traveling waves moving in opposite directions through the pipe give rise to a standing wave.) You can see that in Figure 13-26a one half-wavelength fits in the length L of the pipe. This represents the fundamental mode of the pipe, so the wavelength λ of the fundamental mode is given by

$$L = \frac{1}{2}\lambda$$

Similarly, the standing wave modes in Figures 13-26b and 13-26c have two half-wavelengths (one full wavelength) and three half-wavelengths, respectively, in

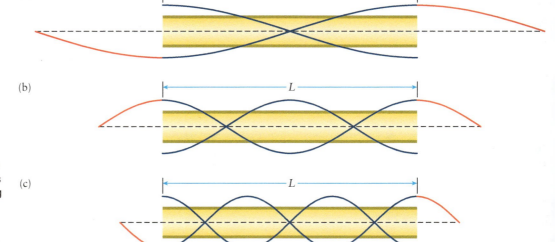

(a)

(b)

Figure 13-26 Standing sound waves in an open pipe (a), (b), and (c) show the displacement patterns for the first three standing sound wave modes of a pipe open at both ends. There is a displacement antinode at each open end.

(c)

the length of the pipe. In general, the relationship between L and λ for an open pipe is

Length of an open pipe (open at both ends)	Wavelength of a standing wave in the pipe

$$L = \frac{n\lambda}{2} \quad \text{where } n = 1, 2, 3, \dots$$

Wavelengths for a standing sound wave in an open pipe
(13-23)

This is the same relationship as for standing waves on a string held at both ends (Equation 13-18) or for a pipe *closed* at both ends (see Figure 13-22). Although the wavelengths are the same as in those two cases, an open pipe has antinodes, not nodes, at its two ends.

The corresponding natural frequencies are

$$f_n = \frac{v_{\text{sound}}}{\lambda} = \frac{v_{\text{sound}}}{(2L/n)} = n\left(\frac{v_{\text{sound}}}{2L}\right) \text{ where } n = 1, 2, 3, \dots$$

(13-24)

Unlike a closed pipe (Equation 13-22), *an open pipe has all harmonics of its fundamental* ($n = 1$) *frequency*. Note that for a given length L, the fundamental frequency of a closed pipe (from Equation 13-22, $f_1 = v_{\text{sound}}/4L$) is one-half of the fundamental frequency of an open pipe (from Equation 13-24, $f_1 = v_{\text{sound}}/2L$). If you cover one end of an open pipe while blowing into it, you change it into a closed pipe and the sound it produces will be lower in frequency and pitch.

GOT THE CONCEPT? 13-7 Filling a Water Bottle

 Imagine that you gently tap the side of a water bottle as you fill it. As the water level rises, the pitch of the sound made by the bottle as you tap it (a) increases; (b) decreases; (c) first increases, then decreases; (d) first decreases, then increases; (e) stays the same.

TAKE-HOME MESSAGE FOR Section 13-7

✔ Traveling sound waves in a pipe can interfere to form a standing sound wave.

✔ A displacement node always appears at a closed end of the pipe, and a displacement antinode always appears at an open end.

✔ For a closed pipe (open at one end, closed at the other) the standing wave frequencies are odd multiples of the fundamental frequency. For an open pipe (open at both ends) the standing wave frequencies include both even and odd multiples of the fundamental frequency.

13-8 Two sound waves of slightly different frequencies produce beats

You might hear a sound something like "wah wah wah" when a guitar player tunes her instrument by plucking two strings at the same time. As she adjusts the tension on one string, the time between the "wahs" gets so long that they can no longer be heard. This phenomenon, known as **beats**, arises from the interference between sound waves. Beats are most pronounced when two waves of nearly identical frequencies interfere.

To see how beats arise let's see what happens when we combine two sinusoidal waves with the same amplitude A but with slightly different frequencies f_1 and f_2. Figure 13-27 shows the result: The total wave is also a sinusoidal wave whose frequency f is the average of f_1 and f_2, but with an amplitude that varies between 0 and $2A$. We use the term "beats" for this up-and-down variation in amplitude. The frequency of the beats, also called the **beat frequency**, is equal to the *absolute value* of the difference between the two frequencies:

Beat frequency heard when sound waves of two similar frequencies interfere	Frequencies of the individual sound waves

$$f_{\text{beats}} = |f_2 - f_1|$$

Beat frequency
(13-25)

Figure 13-27 **Beats** When two sinusoidal waves of nearly the same frequency are added together, the total wave has a rising and falling amplitude, or beats.

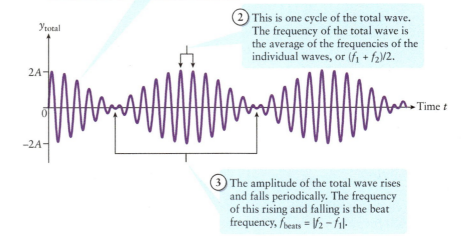

① This is the total wave that results when two individual sine waves of similar frequencies f_1 and f_2 are combined.

② This is one cycle of the total wave. The frequency of the total wave is the average of the frequencies of the individual waves, or $(f_1 + f_2)/2$.

③ The amplitude of the total wave rises and falls periodically. The frequency of this rising and falling is the beat frequency, $f_{beats} = |f_2 - f_1|$.

(The absolute value ensures that we get a positive value for f_{beats}. We do this because a negative frequency has no meaning.) For example, if the two individual sound waves have frequencies $f_1 = 200$ Hz and $f_2 = 202$ Hz, what you will hear is a tone at a frequency of 201 Hz (the average of 200 Hz and 202 Hz) with an amplitude that rises and falls at frequency $f_{beats} = |f_2 - f_1| = |202 \text{ Hz} - 200 \text{ Hz}| = 2$ Hz. That is, the amplitude will rise and fall twice per second, or once every 1/2 s.

The greater the difference between the frequencies f_1 and f_2, the greater the beat frequency and the more rapid the beats. If the frequency difference is large enough, the beats are no longer perceptible and you will hear two distinct frequencies.

WATCH OUT! Don't confuse the beat frequency with the frequency of the sound.

⚠ When you hear two waves with slightly different frequencies f_1 and f_2, the beat frequency given by Equation 13-25 is the frequency at which the amplitude of the sound rises and falls. The frequency of the sound itself is different: It's just the average of f_1 and f_2, or $(f_1 + f_2)/2$.

EXAMPLE 13-8 Guitar Tuning

You have just replaced the first string on your guitar and wish to tune it to the correct frequency of 329.6 Hz (the note E_4). To do this you pluck both this string and the sixth (E_2) string, which you know is properly tuned to 82.4 Hz. You hear one beat every 2.50 s. How far out of tune (in Hz) is the first string?

Set Up

The frequencies 82.4 and 329.6 Hz are far apart, so it may seem surprising that you would get beats when both are sounded. But remember that a plucked string oscillates not only at its fundamental frequency but also at the higher *harmonics* of the string. So what you hear is beats between the sound made by the first string and one of the harmonics of the sixth string.

Beat frequency:

$$f_{beats} = |f_2 - f_1| \qquad (13\text{-}25)$$

Frequencies for a standing wave on a string:

$$f_n = \frac{n}{2L}\sqrt{\frac{F}{\mu}} \text{ where } n = 1, 2, 3, \ldots$$

$$(13\text{-}19)$$

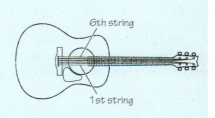

6th string

1st string

Solve

The harmonics of the sixth string are at integer multiples of its fundamental frequency.	For the sixth string, $f_1 = 82.4$ Hz. The harmonics are $f_2 = 2f_1 = 164.8$ Hz $f_3 = 3f_1 = 247.2$ Hz $f_4 = 4f_1 = 329.6$ Hz So if the guitar is properly tuned, the fundamental frequency of the first string is the same as the fourth harmonic frequency of the sixth string. If the first string is mistuned, the frequencies will be different and you will hear beats.
Find the beat frequency and how far out of tune the first string is.	There is one beat per 2.50 s, so the beat frequency (the number of beats per second) is $$f_{beats} = \frac{1}{2.50 \text{ s}} = 0.400 \text{ s}^{-1} = 0.400 \text{ Hz}$$ This is equal to the difference between the fundamental frequency of the first string and the fourth harmonic frequency of the sixth string: $$f_{beats} = 0.400 \text{ Hz} = \left| f_{1 \text{ for first string}} - f_{4 \text{ for sixth string}} \right|$$ $$= \left| f_{1 \text{ for first string}} - 329.6 \text{ Hz} \right|$$ So the first string is out of tune by 0.400 Hz.

Reflect

With just the information given, we don't know whether the first string is tuned 0.400 Hz too *high* (in which case its frequency is 330.0 Hz) or too *low* (in which case its frequency is 329.2 Hz). To find out try increasing the tension on the first string to raise its fundamental frequency. If the beats slow down, the beat frequency is getting closer to zero as you bring the first string into tune, so the first string must have been tuned too low. Keep increasing the tension until the beats stop altogether, at which point the first string is properly tuned. If instead the beats become more rapid, the beat frequency is increasing, so the tuning of the first string is getting worse. This means the first string was tuned too high, so you should decrease the tension on the first string to lower its fundamental frequency and bring it into tune.

GOT THE CONCEPT? 13-8 Tuning Forks

When you strike two tuning forks simultaneously, you hear 3 beats per second. The frequency of the first tuning fork is 440 Hz. What is the frequency of the second tuning fork? (a) 446 Hz; (b) 443 Hz; (c) 437 Hz; (d) 434 Hz; (e) not enough information is given to decide.

TAKE-HOME MESSAGE FOR Section 13-8

✔ When two waves of nearly identical frequencies and wavelengths interfere, the result is a total wave whose amplitude rises and falls periodically. This phenomenon is called beats.

✔ The frequency of the beats equals the absolute value of the difference in frequency between the two waves.

13-9 The intensity of a wave equals the power that it delivers per square meter

Humans hear sounds over a wide range of volumes, from very faint sounds (such as leaves rustling in a gentle breeze) to sounds so loud that they can cause pain (such as the sound of the siren on an emergency vehicle driving nearby). Let's analyze the physical differences between loud sounds and faint sounds.

Wave Energy, Power, and Intensity

An important factor in determining the loudness of a sound is the energy carried by the sound wave. To see why this is so, note that for the ear to detect a sound, the sound wave must exert a force on the eardrum that makes the eardrum move (be displaced). This means that the sound wave must do *work* on the eardrum (recall from Section 6-2 that work is done when a force acts over a distance). Doing work involves the transfer of energy, and in this case the energy comes from the wave itself. So we can think of the ear as a device for detecting the energy in a sound wave.

The ear is particularly sensitive to the *rate* at which sound wave energy is delivered to it. As an analogy, imagine having someone pour a pitcher of water over your outstretched hand. You'll feel very little effect if the water falls on you very slowly, a drop at a time. But if the pitcher is turned over so that all of the water hits your hand in a short time, your hand will feel pushed down. In the same way, your ear has a greater response if sound wave energy is delivered to your ear at a rapid rate.

As we learned in Section 6-9, the rate at which energy is transferred is called **power**. The unit of power is the joule per second, or **watt** (abbreviated W): 1 W = 1 J/s. As an example, if you are using a stationary exercise bicycle at the gym, a reading of 200 W on the bike's display means that you are doing 200 joules of work on the pedals every second. We'll usually denote power by the symbol P.

Let's apply these ideas about energy and power to a sinusoidal sound wave. We've seen that in a sinusoidal wave, pieces of the wave medium undergo simple harmonic motion. In Section 12-4 we learned that if an object of mass m is attached to an ideal spring of spring constant k and oscillates in simple harmonic motion with amplitude A, the total mechanical energy of the mass–spring system is

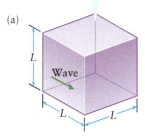

This volume of air has mass $m = \rho L^3$. A sound wave of angular frequency ω and amplitude A is present throughout the volume.

(a)

(13-21)
$$E = \frac{1}{2}kA^2$$

We also learned in Section 12-3 that the angular frequency for an object oscillating on an ideal spring is $\omega = \sqrt{k/m}$, so $\omega^2 = k/m$ and $k = m\omega^2$. If we insert this into Equation 12-21, we get

(13-26)
$$E = \frac{1}{2}m\omega^2 A^2$$

Let's see what this says about a sinusoidal sound wave in air with angular frequency ω and displacement amplitude A. (Remember that for a sound wave A represents the maximum longitudinal displacement of the air as the wave passes through it.) According to Equation 13-26 if this wave is present in a chunk of air of mass m, the *wave energy* present in that chunk is $E = (1/2)m\omega^2 A^2$. The wave carries that energy along with it as it propagates. To get a measure of the *power* associated with the wave, let's imagine a cubical volume in the air with sides of length L (Figure 13-28). If the air has density (mass per volume) ρ, the mass of air in this volume equals the density ρ multiplied by the volume L^3 of the cube: $m = \rho L^3$. From Equation 13-26 the wave energy contained in this volume is

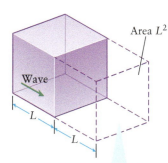

Area L^2

(b)

The volume L^3 of wave energy travels a distance L in time $t = L/v_{\text{sound}}$. The power equals the wave energy E in the volume divided by t. The intensity is the power divided by L^2.

Figure 13-28 Power and intensity in a sound wave (a) This cubical volume encloses a quantity of air with a sound wave propagating through it. (b) If we imagine the volume moving along with the wave, we can find the power and intensity associated with the wave.

(13-27)
$$E = \frac{1}{2}\rho L^3 \omega^2 A^2$$

The energy in this volume moves along with the wave at propagation speed v_{sound}, so the time required for this volume of energy to move through its own length L is $t = L/v_{\text{sound}}$. So the rate at which this energy crosses through an imaginary square of side L (Figure 13-28) — that is, the *power P* delivered across this surface — equals the energy E given by Equation 13-27 divided by the time $t = L/v_{\text{sound}}$:

(13-28)
$$P = \frac{E}{t} = \frac{(1/2)\rho L^3 \omega^2 A^2}{L/v_{\text{sound}}} = \frac{1}{2}\rho L^2 v_{\text{sound}} \omega^2 A^2$$

WATCH OUT! In a wave there is a flow of energy, not of material.

Note that in Figure 13-28 what moves a distance L in time t is the *wave energy* contained within the cubical volume of side L. The wave and its associated energy move together, but the wave *medium* doesn't go anywhere: Each piece of the medium just oscillates around its equilibrium position.

Equation 13-28 says that the power associated with the wave is proportional to the square of the wave amplitude A. This turns out to be a general attribute of waves of all kinds, not just sound waves. Note also that the power is proportional to L^2, which is the cross-sectional area through which the wave energy passes. The power *divided* by the cross-sectional area L^2 is called the **intensity** I of the wave, measured in watts per square meter (W/m^2). From Equation 13-28 this is

Intensity of a sound wave　　Density of the wave medium (usually air)

$$I = \frac{1}{2}\rho v_{sound}\,\omega^2\, A^2$$

Displacement amplitude of the wave

Speed of sound waves in the medium　　Angular frequency of the wave

Sound wave intensity in terms of displacement amplitude A (13-29)

If you increase the displacement amplitude A of a sound wave of a given frequency, the intensity of the sound will increase. For an ear or other sound detector of a given size, the amount of sound wave power that is received equals the intensity I multiplied by the area of the detector. So increasing the intensity means more sound power enters your ear, and you will perceive the sound to be louder.

Note that the larger the area of the ear, the greater the total power (intensity times area) reaching your eardrum. That's why many nocturnal animals such as bats and coyotes have large external ears: This maximizes the cross-sectional area over which they can collect low-amplitude sound waves coming from potential prey or predators. The larger collected power is then directed to their eardrums, improving their sensitivity. Intensity is a useful measure of the strength of a sound wave because the area factor has been divided out: A coyote may have more sensitive ears than yours (because of the greater collection area of its outer ears), but if you both are exposed to the same sound wave, both your external ears experience the same intensity.

BioMedical

Equation 13-29 is expressed in terms of the displacement amplitude A. However, our ears are more directly sensitive to the *pressure* variations in a sound wave. The displacement amplitude A in a sound wave is related to the pressure amplitude p_{max}, or maximum pressure variation, by

$$A = \frac{p_{max}}{\rho v_{sound}\omega} \tag{13-30}$$

If we substitute Equation 13-30 for A into Equation 13-29, we get an alternative expression for the intensity of a sound wave:

$$I = \frac{1}{2}\rho v_{sound}\omega^2\left(\frac{p_{max}}{\rho v_{sound}\omega}\right)^2 = \frac{p_{max}^2}{2}\frac{\rho v_{sound}\omega^2}{\rho^2 v_{sound}^2\omega^2}$$

or, simplifying,

Intensity of a sound wave　　Pressure amplitude of the wave

$$I = \frac{p_{max}^2}{2\rho v_{sound}}$$

Sound wave intensity in terms of pressure amplitude p_{max} (13-31)

Density of the wave medium (usually air)　　Speed of sound waves in the medium

Equation 13-31 says that for sound waves in a medium with density ρ for which the speed of sound is v_{sound}, the intensity is proportional to the square of the pressure amplitude.

A typical human ear can detect sounds with frequencies from 20 to 20,000 Hz. If they were equally sensitive to all frequencies in this range, greater intensity would imply greater loudness. However, our ears are *not* equally sensitive across the audible range: As we learned in Section 13-7, a person with normal hearing is most sensitive to frequencies around 3400 Hz and less sensitive at higher and lower frequencies. At 1000 Hz the lowest intensity that a person with normal hearing can sense (the *threshold of hearing*) is about 10^{-12} W/m^2. The threshold of hearing is substantially higher at frequencies to which the ear is less sensitive, about 10^{-11} W/m^2 at 10,000 Hz and about 10^{-8} W/m^2 at 100 Hz. At most frequencies the greatest intensity that a person can tolerate without pain (the *threshold of pain*) is about 1 W/m^2.

BioMedical EXAMPLE 13-9 Loud and Soft

The amplitude of the motion of your eardrum in response to a sound wave more or less matches the amplitude of the motion of nearby air molecules. (a) What is the oscillation amplitude of your eardrum when you hear a tone that has a frequency of 1000 Hz and an intensity of 1.00 W/m^2, near the threshold of pain? What is the pressure amplitude of this sound? (b) What about a 1000-Hz tone with an intensity of only 1.00×10^{-12} W/m^2, at the threshold of hearing? Use $\rho = 1.20$ kg/m^3 for the density of air and $v_{sound} = 343$ m/s for the speed of sound in air.

Set Up

We'll use Equations 13-29 and 13-31 to relate the intensity of a sound wave to its displacement amplitude A and pressure amplitude p_{max}. The problem statement tells us that the displacement amplitude of the eardrum is about the same as that of the air.

Sound wave intensity in terms of displacement amplitude A:

$$I = \frac{1}{2}\rho v_{sound}\omega^2 A^2 \qquad (13\text{-}29)$$

Sound wave intensity in terms of pressure amplitude p_{max}:

$$I = \frac{p_{max}^2}{2\rho v_{sound}} \qquad (13\text{-}31)$$

Solve

(a) Rearrange Equation 13-29 to find an expression for the displacement amplitude.

From Equation 13-29,

$$A^2 = \frac{2I}{\rho v_{sound}\omega^2}$$

We are given $f = 1000$ Hz, so substitute $\omega = 2\pi f$:

$$A^2 = \frac{2I}{\rho v_{sound}(2\pi f)^2} = \frac{2I}{4\pi^2 \rho v_{sound}f^2} = \frac{I}{2\pi^2 \rho v_{sound}f^2}$$

Take the square root of both sides and substitute $I = 1.00$ W/m^2, $\rho = 1.20$ kg/m^3, $v_{sound} = 343$ m/s, and $f = 1000$ Hz:

$$A = \sqrt{\frac{I}{2\pi^2 \rho v_{sound}f^2}} = \sqrt{\frac{1.00 \text{ W/m}^2}{2\pi^2(1.20 \text{ kg/m}^3)(343 \text{ m/s})(1000 \text{ Hz})^2}}$$

$$= 1.11 \times 10^{-5}\sqrt{\frac{\text{W}\cdot\text{s}}{\text{kg}\cdot\text{Hz}^2}}$$

Note that $1\text{ W} = 1\text{ J/s}, 1\text{ J} = 1\text{ N}\cdot\text{m} = 1\text{ kg}\cdot\text{m}^2/\text{s}^2$, and $1\text{ Hz} = 1/\text{s} = 1\text{ s}^{-1}$. So

$$1\sqrt{\frac{\text{W}\cdot\text{s}}{\text{kg}\cdot\text{Hz}^2}} = 1\sqrt{\frac{\text{J}\cdot\text{s}^2}{\text{kg}}} = 1\sqrt{\frac{\text{kg}\cdot\text{m}^2}{\text{kg}}} = 1\text{ m}$$

The units of the displacement amplitude A are therefore meters, as they should be:

$$A = 1.11\times10^{-5}\text{ m}$$

Use the same approach with Equation 13-31 to find the pressure amplitude p_{max}.

From Equation 13-31,

$$I = \frac{p_{max}^2}{2\rho v_{sound}}$$

Solve for p_{max}:

$$p_{max}^2 = 2\rho v_{sound} I$$

$$p_{max} = \sqrt{2\rho v_{sound} I} = \sqrt{2(1.20\text{ kg/m}^3)(343\text{ m/s})(1.00\text{ W/m}^2)}$$

$$= 28.7\sqrt{\frac{\text{kg}\cdot\text{W}}{\text{m}^4\cdot\text{s}}}$$

Again note that $1\text{ W} = 1\text{ J/s}, 1\text{ J} = 1\text{ N}\cdot\text{m}$, and $1\text{ N} = 1\text{ kg}\cdot\text{m/s}^2$. So $1\text{ W} = 1\text{ N}\cdot\text{m/s}$ and $1\text{ kg} = 1\text{ N}\cdot\text{s}^2/\text{m}$, and the units of our answer are

$$1\sqrt{\frac{\text{kg}\cdot\text{W}}{\text{m}^4\cdot\text{s}}} = 1\sqrt{\frac{1}{\text{m}^4\cdot\text{s}}\frac{\text{N}\cdot\text{s}^2}{\text{m}}\frac{\text{N}\cdot\text{m}}{\text{s}}} = 1\sqrt{\frac{\text{N}^2}{\text{m}^4}} = 1\text{ N/m}^2$$

The units of p_{max} are newtons per square meter or pascals ($1\text{ N/m}^2 = 1\text{ Pa}$), which is correct for pressure:

$$p_{max} = 28.7\text{ N/m}^2 = 28.7\text{ Pa}$$

(b) To find the displacement amplitude A and pressure amplitude p_{max} for the new intensity, we don't have to redo the above calculations. Instead, we notice from our calculations that A and p_{max} are both proportional to the square root of the intensity I.

The new intensity is $1.00\times10^{-12}\text{ W/m}^2$, which is 10^{-12} times the previous intensity. Since A and p_{max} are both proportional to $\sqrt{I}$, the new amplitudes are equal to the previous values multiplied by $\sqrt{10^{-12}} = 10^{-6}$:

$$A = (1.11\times10^{-5}\text{ m})\times10^{-6} = 1.11\times10^{-11}\text{ m}$$

$$p_{max} = (28.7\text{ N/m}^2)\times10^{-6} = 2.87\times10^{-5}\text{ N/m}^2 = 2.87\times10^{-5}\text{ Pa}$$

Reflect

For the painfully loud sound at 1000 Hz, the eardrum oscillates with an amplitude slightly larger than 10 μm, or about 50% larger than the diameter of a human red blood cell. At the threshold of hearing the oscillation amplitude of the eardrum to a 1000-Hz tone is about 10^{-11} m, about one-tenth the diameter of a hydrogen atom. The pressure amplitudes are also very small: 28.9 Pa is about the difference in atmospheric pressure between the floor and ceiling of a bedroom, and 2.89×10^{-5} Pa is about the same pressure that would be exerted on your eardrum by the weight of a single wing of a fly. These remarkable numbers illustrate the exquisite sensitivity of human hearing.

Sound and the Inverse-Square Law

If you've ever convinced a teacher to hold class outside on a warm spring day, you may have noticed that the teacher's voice is much more difficult to hear outdoors than indoors. That's because when a source of sound (like the teacher's voice)

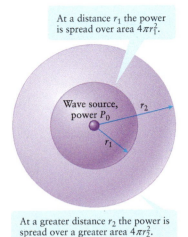

At a distance r_1 the power is spread over area $4\pi r_1^2$.

Wave source, power P_0 r_2

r_1

At a greater distance r_2 the power is spread over a greater area $4\pi r_2^2$.

Figure 13-29 The inverse-square law for sound waves A source of sound waves has power P_0. At greater distances from the source, this power is spread over a greater area, so the intensity is less.

Inverse-square law for
sound waves
(13-32)

radiates sound waves in all directions without anything getting in the way of the waves, the *intensity* of the waves decreases with increasing distance. In contrast, inside the lecture hall, sound bounces off the walls and back toward the audience, so the intensity does not decrease as quickly with increasing distance from the source.

Figure 13-29 shows how this comes about. Since waves carry energy, a source of sound waves is a source of energy. The rate at which the source emits energy in the form of waves is the *power* of the source. None of the power is lost as the waves spread away from the source, but at greater distances the power is distributed over a greater area. Hence the wave intensity (power divided by area) and the perceived loudness of the sound both decrease with increasing distance. If the source has power p_0 and emits equally in all directions as shown in Figure 13-29, at a distance r from the source the power is spread uniformly over a spherical surface of radius r. The area of such a surface is $4\pi r^2$, so the intensity equals

Intensity of a sound wave at a distance r from a source that emits in all directions

Power output of the sound source

$$I = \frac{P_0}{4\pi r^2}$$

Distance from the source to the point where the intensity is measured

Equation 13-32 says that the intensity is inversely proportional to the square of the distance from the source, so this is called the **inverse-square law** for waves.

The inverse-square law explains what happens when class is held outdoors: Compared to the intensity of the teacher's voice at a distance of 1 m, the intensity at 2 m is $(1/2)^2 = 1/4$ as great, and the intensity at 3 m is $(1/3)^2 = 1/9$ as great. So with increasing distance the teacher becomes increasingly difficult to hear. The inverse-square law doesn't apply indoors because sound reflects off the walls and ceiling, so sound energy that would be lost outdoors is reflected back toward you and the other members of the audience.

BioMedical **EXAMPLE 13-10** **Delivering Energy to an Eardrum**

You are at an outdoor concert, standing 20 m from a speaker tower that generates 2500 W of sound power. How much sound energy from the speaker tower strikes one of your eardrums per second? A typical human eardrum has a surface area of 55 mm². Assume that the speaker tower emits sound uniformly in all directions.

Set Up

The inverse-square law for sound waves tells us the intensity of the sound that reaches you. To find the power (energy per time) delivered to your eardrum, we use the idea that intensity equals power per area, so power equals intensity times area.

Inverse-square law for sound waves:

$$I = \frac{P_0}{4\pi r^2} \qquad (13\text{-}32)$$

2500 W

20 m

area = 55 mm²

sound power into ear
= (intensity of sound) × (area of eardrum)

Solve

Use Equation 13-32 to find the sound intensity at your position.

At a distance $r = 20$ m from a source with power $P_0 = 2500$ W,

$$I = \frac{P_0}{4\pi r^2} = \frac{2500 \text{ W}}{4\pi(20 \text{ m})^2} = 0.50 \text{ W/m}^2$$

The sound power that strikes one of your eardrums equals the sound intensity multiplied by the area of the eardrum.

Power (energy per time) on your eardrum:

$P = I \times$ (area of eardrum)

$$= (0.50 \text{ W/m}^2)(55 \text{ mm}^2)\left(\frac{1 \text{ m}}{10^3 \text{ mm}}\right)^2$$

$$= 2.7 \times 10^{-5} \text{ W} = 2.7 \times 10^{-5} \text{ J/s}$$

Reflect

The intensity of the sound is close to the threshold of pain. You would be advised to stand farther away from the sound tower! But notice that even at this very high intensity, the power that reaches the ear is very small (about one *millionth* the amount of power needed to light a 25-watt light bulb). This is another testament to the sensitivity of the human ear.

Sound Intensity Level

The range of sound intensities that we can hear spans the range from about 10^{-12} W/m² to about 1 W/m². It's convenient to compress this broad range of numbers by using *logarithms*. This leads to an alternative way to describe the intensity of a sound wave in terms of *sound intensity level*.

The idea behind logarithms is that any positive number x can be written as $x = b^y$, where b is called the *base*. In this language we say that y is "the logarithm of x to the base b." As an example, 3 is the logarithm of 8 to the base 2: $8 = 2^3$. We write this relationship as follows:

$$\text{If } x = b^y, \text{ then } y = \log_b x.$$

So if x equals b raised to the power y, then y is the logarithm to the base b of x.

A common choice for the base b is 10. With this choice the above relationship becomes

$$\text{If } x = 10^y, \text{ then } y = \log_{10} x.$$

For example, the logarithm to the base 10 of $1000 = 10^3$ is $\log_{10} 10^3 = 3$, and the logarithm to the base 10 of $0.01 = 10^{-2}$ is $\log_{10} 10^{-2} = -2$. (Any number raised to power 0 equals 1, so the logarithm of 1 is 0 in *any* base.) On your calculator the base-10 logarithm is probably denoted by a function key named log.

Note that the logarithm isn't just for numbers that are multiples of 10. As an example, a number between 1 and 10 has a base-10 logarithm that is between 0 (the logarithm of 1) and 1 (the logarithm of $10 = 10^1$). So $\log_{10} 2 = 0.30$, $\log_{10} 3 = 0.48$, and so on.

Let's now see how the logarithm is used for describing the intensity of a sound. The argument of the logarithm must be a pure number without dimensions or units. We satisfy this requirement by taking the ratio of the intensity I of a sound to 10^{-12} W/m², the intensity of a 1000-Hz tone at the threshold of hearing. The **sound intensity level** of a sound with intensity I equals the base-10 logarithm of this ratio multiplied by 10:

Sound intensity level of a sound wave   Intensity of the sound

$$\beta = (10 \text{ dB}) \log_{10}\left[\frac{I}{10^{-12} \text{ W/m}^2}\right]$$

Reference intensity

Sound intensity level
(13-33)

The factor of 10 that multiplies the logarithm in Equation 13-33 is there for convenience. The units of sound intensity level are the **decibel** (dB), named after the

Chapter 13 Waves

<table>
<tr><td colspan="2">TABLE 13-1
Sound Intensity Levels</td></tr>
<tr><td>Sound</td><td>Sound level (dB)</td></tr>
<tr><td>jet engine at 25 m</td><td>150</td></tr>
<tr><td>live amplified music</td><td>120</td></tr>
<tr><td>car horn at 1 m</td><td>110</td></tr>
<tr><td>jackhammer</td><td>100</td></tr>
<tr><td>city street</td><td>90</td></tr>
<tr><td>vacuum cleaner</td><td>70</td></tr>
<tr><td>quiet conversation</td><td>50</td></tr>
<tr><td>rustling leaves</td><td>20</td></tr>
<tr><td>breathing</td><td>10</td></tr>
</table>

Scottish-born scientist and inventor Alexander Graham Bell. The sound intensity level for a 1000-Hz tone at the threshold of hearing is then

$$\beta = (10\ \text{dB})\log_{10}\left(\frac{10^{-12}\ \text{W/m}^2}{10^{-12}\ \text{W/m}^2}\right) = (10\ \text{dB})\log_{10}(1) = 0\ \text{dB}$$

The sound intensity level for a tone at the threshold of pain is

$$\beta = (10\ \text{dB})\log_{10}\left(\frac{1\ \text{W/m}^2}{10^{-12}\ \text{W/m}^2}\right) = (10\ \text{dB})\log_{10}(10^{12}) = 120\ \text{dB}$$

So by using sound intensity level we express the intensities of sound in human hearing by the range from 0 to 120 dB rather than from 10^{-12} to 1 W/m². Table 13-1 lists the sound intensity level of a range of sounds.

WATCH OUT! **A sound intensity level of 0 dB is not the absence of sound.**

We are used to thinking of a quantity with a value of zero as being nothing at all. But a sound with sound intensity level 0 dB, while faint, is still a sound. The zero value just means that this is the faintest sound that someone with normal hearing can hear at a frequency of 1000 Hz.

EXAMPLE 13-11 Double the Intensity

When the intensity of a sound doubles, how does the sound intensity level change?

Set Up

Equation 13-33 relates the sound intensity I to the sound intensity level β. We'll use this to compare two sounds, one of intensity I_1 and the other of intensity $I_2 = 2I_1$.

Sound intensity level:

$$\beta = (10\ \text{dB})\log_{10}\left(\frac{I}{10^{-12}\ \text{W/m}^2}\right) \qquad (13\text{-}33)$$

Solve

Write expressions for the two sound intensity levels.

For sound 1:

$$\beta_1 = (10\ \text{dB})\log_{10}\left(\frac{I_1}{10^{-12}\ \text{W/m}^2}\right)$$

For sound 2:

$$\beta_2 = (10\ \text{dB})\log_{10}\left(\frac{I_2}{10^{-12}\ \text{W/m}^2}\right) = (10\ \text{dB})\log_{10}\left(\frac{2I_1}{10^{-12}\ \text{W/m}^2}\right)$$

Use an important property of logarithms to compare β_1 and β_2.

The logarithm of the *ratio* of two numbers A and B is equal to the *difference* of their logarithms:

$$\log_{10}\frac{A}{B} = \log_{10} A - \log_{10} B$$

Using this property we can express the *difference* between β_2 and β_1 as

$$\beta_2 - \beta_1 = (10\ \text{dB})\log_{10}\left(\frac{2I_1}{10^{-12}\ \text{W/m}^2}\right) - (10\ \text{dB})\log_{10}\left(\frac{I_1}{10^{-12}\ \text{W/m}^2}\right)$$

$$= (10\ \text{dB})\left[\log_{10}\left(\frac{2I_1}{10^{-12}\ \text{W/m}^2}\right) - \log_{10}\left(\frac{I_1}{10^{-12}\ \text{W/m}^2}\right)\right]$$

$$= (10 \text{ dB}) \log_{10} \left(\frac{2I_1}{10^{-12} \text{ W/m}^2} \frac{10^{-12} \text{ W/m}^2}{I_1} \right)$$
$$= (10 \text{ dB}) \log_{10} 2$$
$$= (10 \text{ dB})(0.30) = 3.0 \text{ dB}$$

So a *ratio* of 2 in the intensity corresponds to a *difference* of 3.0 dB in the sound intensity level.

Reflect

A factor of 2 increase in intensity corresponds to adding 3.0 dB to the sound intensity level. So a 54.0-dB sound has twice the intensity of a 51.0-dB sound, a 25.5-dB sound has twice the intensity of a 22.5-dB sound, and so on. Increasing the intensity by a factor of $4 = 2 \times 2$ means adding 3.0 dB + 3.0 dB = 6.0 dB to the sound intensity level; increasing the intensity by $8 = 2 \times 2 \times 2$ means adding 3.0 dB + 3.0 dB + 3.0 dB = 9.0 dB to the sound intensity level; and so on.

GOT THE CONCEPT? 13-9 From One Fan to Two

You measure the sound intensity level of an electric fan to be 60 dB. What is the sound intensity level when a second identical fan is turned on, at the same distance from you as the first one? (a) 120 dB; (b) 90 dB; (c) 66 dB; (d) 63 dB; (e) 62 dB.

TAKE-HOME MESSAGE FOR Section 13-9

✔ The intensity of a sound wave is the power (energy per time) that the wave delivers per unit area. The intensity does not depend on the ear or device used to detect the sound.

✔ The power delivered by a sound wave is equal to the intensity of the wave multiplied by the collecting area.

✔ In the absence of obstructions or reflections, the intensity of sound from a source decreases in inverse proportion to the square of the distance from the source.

✔ The sound intensity level involves the logarithm of a sound wave's intensity divided by a reference value of intensity.

13-10 The frequency of a sound depends on the motion of the source and the listener

Until now we have considered only waves generated by stationary emitters and observed by stationary listeners. Everyday experience, however, suggests that something curious happens when a moving object creates a sound. As a police car with sirens blaring zooms by you, for example, the frequency of the sound is higher as the car approaches and lower after it passes. Even without a siren, a fast-moving car generates a characteristic high-to-low frequency sound (something like "neee-urrrr") as it approaches and passes you. And although we don't often get the chance to experience this phenomenon in reverse, if you were to move at high speed toward a stationary police car with its sirens blaring, the sound of the siren would follow a similar high-to-low frequency shift as you approached and then passed it. This effect is known as the **Doppler effect**, named for the Austrian physicist Christian Doppler who first proposed this phenomenon in 1842.

The Doppler Effect and Frequency Shift

In Figure 13-30a, a police siren emits a periodic sound wave that has a fixed frequency and wavelength. Because the car is stationary, each wave crest—that is, each region of highest pressure along the wave, shown as arcs of circles—is centered on the siren. In Figure 13-30b the car moves from left to right. Wave crest #1 originated

Figure 13-30 The Doppler effect: A moving source I A police car siren emits a periodic sound wave that has a fixed frequency and wavelength. (a) The police car is stationary, so each wave crest, shown as arcs of circles, is centered on the siren. (b) The car moves from left to right. Because the source of the sound waves moves as the waves propagate, the wave crests are closer together as the car approaches a listener's ear and farther apart as the car recedes from the listener.

(a)

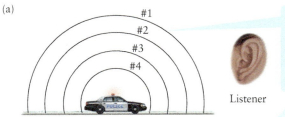

Wave crests spread outward from the siren. The numbers on the crests show the sequence in which they were emitted: First #1, then #2, then #3, then #4, so the most recently emitted crest is closest to the siren.

When the siren is not moving, a listener will hear a constant frequency.

(b)

As the siren approaches the listener, the sound frequency is higher.

The frequency is lower as the siren moves away from the listener.

GOT THE CONCEPT? 13-10
A Car Alarm

? As you run toward your parked car to turn off the blaring alarm system, will the frequency of the tone you hear be (a) higher than, (b) the same as, or (c) lower than the frequency you hear when standing still?

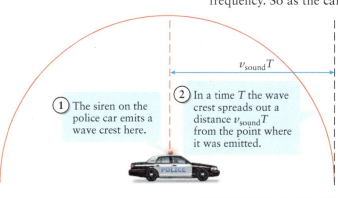

① The siren on the police car emits a wave crest here.

② In a time T the wave crest spreads out a distance $v_{sound}T$ from the point where it was emitted.

$v_{sound}T$

③ In the same time T the police car travels a distance $v_{car}T$.

Figure 13-31 The Doppler effect: A moving source II The wavelength of the sound heard by a stationary listener as a source of sound approaches depends on the distance the wave travels in a certain time and the distance the source travels in that same time.

$v_{car}T$ $\lambda_{listener}$

💡 Because of the car's motion, a listener in front of the police car hears a shortened wavelength.

first, when the car was somewhere to the left, and wave crest #4 is the most recently generated one. Because the car moves as the sound wave propagates, successive wave crests are closer together in front of the car and farther apart behind the car. The distance between wave crests determines the wavelength heard by a stationary listener. So if the car is *moving toward* a listener, as on the left-hand side of Figure 13-30b, the listener detects a *shorter* wavelength than the siren generates. If the car is *moving away* from the listener, as shown the right-hand side of Figure 13-30b, the listener detects a *longer* wavelength than the siren generates.

The relationship $f = v_{sound}/\lambda$ (Equation 13-2 for sound) tells us that a shorter wavelength results in a higher frequency and a longer wavelength results in a lower frequency. So as the car approaches, the listener in Figure 13-30b hears a sound with a shorter wavelength and higher frequency than the siren generates. As the car recedes, the listener hears a sound with a longer wavelength and a lower frequency.

How large are these changes in frequency? To find out, in Figure 13-31 we've shown the moving police car at two instants in time separated by the period T of the tone created by the siren. A crest of the sound wave was created when the car was at the location marked by the red dashed line; in time T the wave crest has propagated a distance $v_{sound}T$, as indicated by the red semicircle. The distance the car moves in that time is $v_{car}T$, where v_{car} is the speed of the car. This new location is where the car will emit the next wave crest. A stationary listener in front of the car detects the distance between two successive wave crests as the wavelength $\lambda_{listener}$ of the siren's tone. As Figure 13-31 shows, these three distances—the distance the sound moves, the distance the car moves, and the distance between two successive wave crests at the place where the listener sits—are related by

$$v_{car}T + \lambda_{listener} = v_{sound}T$$

or

(13-34) $$\lambda_{listener} = v_{sound}T - v_{car}T$$

It's convenient to rewrite Equation 13-34 as a relationship between the *frequency* $f_{listener}$ that the listener hears and the frequency f of the tone emitted by the siren. To do this we use $T = 1/f$ and Equation 13-2 for sound waves:

$$\lambda_{\text{listener}} = \frac{v_{\text{sound}}}{f_{\text{listener}}}$$

Combining these equations with Equation 13-34 gives

$$\frac{v_{\text{sound}}}{f_{\text{listener}}} = \frac{v_{\text{sound}}}{f} - \frac{v_{\text{car}}}{f}$$

or, rearranging,

$$f_{\text{listener}} = \left(\frac{v_{\text{sound}}}{v_{\text{sound}} - v_{\text{car}}} \right) f$$

Because v_{car} is a positive number, the fraction in parentheses must be greater than 1. So when the car is moving toward a stationary listener, $f_{\text{listener}} > f$ (the listener hears a higher frequency).

We can apply a similar approach to determine the observed frequency as the source moves away from a stationary listener, as well as for the cases in which the listener moves relative to a stationary source. Using "source" instead of "car" as a more general way to indicate the source of the sound, here's a single equation for the frequency heard by the listener no matter how the source and listener are moving:

Frequency of sound detected by the listener   Frequency of sound emitted by the source

$$f_{\text{listener}} = \left[\frac{v_{\text{sound}} \pm v_{\text{listener}}}{v_{\text{sound}} \mp v_{\text{source}}} \right] f$$

v_{sound} = speed of sound
v_{listener} = speed of the listener
v_{source} = speed of the source

In both numerator and denominator:
• Use upper sign if source and listener are approaching.
• Use lower sign if source and listener are moving apart.

The Doppler effect
(13-35)

Notice that the plus sign is above the minus sign in the numerator but below it in the denominator. We use the upper sign when the source or listener approaches the other and the lower sign when it retreats from the other. Equation 13-35 can be used for more than one case simultaneously; for example, if the source and listener are both approaching each other, the observed frequency is

$$f_{\text{listener}} = \left(\frac{v_{\text{sound}} + v_{\text{listener}}}{v_{\text{sound}} - v_{\text{source}}} \right) f$$

In this case the numerator is larger than the speed of sound v_{sound}, and the denominator is smaller than v_{sound}; both factors result in the listener hearing a higher frequency than the one emitted by the source.

The speed of a moving object can be determined by measuring the shift in frequency associated with the Doppler effect. A common way to do this is to create a wave of known frequency and then bounce it off the moving object to be studied. A *radar speed gun* uses this principle: It sends out a *radio* wave of a known frequency and compares it to the frequency of the wave after it is reflected by a moving car. (Because this device uses radio waves rather than sound waves, v_{sound} in Equation 13-35 is replaced by the speed of light c.) Bats and dolphins use the same technique with sound waves to track the motion of prey.

Measuring velocity using a Doppler shift has become important in medical applications, too—for example, to measure blood flow and to record real-time images of a moving fetus. Because the frequencies of sound used for this kind of imaging,

BioMedical

between 2 and 10 MHz, are well above the range of human hearing, this technique is called *ultrasonic imaging*. (Frequencies below the range of human hearing are called *infrasonic*.)

Consider the process by which ultrasound is used to determine the speed of blood flow through a heart valve. (An obstructed valve is often characterized by an increase in the speed of the blood flow.) A probe that emits a low-intensity sound wave of (high) frequency f_{emitted} is placed over the chest and focused on a region localized around one heart valve. Because the blood is moving, it experiences a wave of frequency f' that has been shifted according to the Doppler effect for a moving listener. Using Equation 13-35 we write

$$f' = \left(\frac{v_{\text{sound}} \pm v}{v_{\text{sound}}} \right) f_{\text{emitted}}$$

In this expression v is the speed of the blood, and the upper sign applies to motion toward the probe and the lower sign to motion away from the probe. The speed of sound in human tissue is about $v_{\text{sound}} = 1540$ m/s.

The sound wave reflects off the blood and returns to the probe. This reflected wave leaves the blood with frequency f', so it's equivalent to the blood emitting sound at frequency f'. So we again apply Equation 13-35, treating the blood as a moving source. The frequency f'' of the reflected ultrasound detected by the probe is then

$$
\begin{aligned}
f'' &= \left(\frac{v_{\text{sound}}}{v_{\text{sound}} \mp v} \right) f' \\
&= \left(\frac{v_{\text{sound}}}{v_{\text{sound}} \mp v} \right) \left(\frac{v_{\text{sound}} \pm v}{v_{\text{sound}}} \right) f_{\text{emitted}} \\
&= \left(\frac{v_{\text{sound}} \pm v}{v_{\text{sound}} \mp v} \right) f_{\text{emitted}}
\end{aligned}
$$

(13-36)

Equation 13-36 says that when a wave of frequency f_{emitted} from a probe hits blood or another object moving with speed v, the wave that is reflected back to the probe is received with frequency f''. As for Equation 13-35 the upper signs apply to motion toward the probe, and the lower signs to motion away from the probe. If we measure the value of f'' as blood flows through the valve, we can use Equation 13-36 to determine the speed v of blood flow. This same equation is also used in ultrasonic measurements of fetal heartbeats, as the following example shows.

BioMedical EXAMPLE 13-12 Diagnostic Ultrasound

A certain ultrasound device can measure a fetal heart rate as low as 50 beats per minute. This corresponds to the surface of the heart moving at about 4.0×10^{-4} m/s. If the probe generates ultrasound that has a frequency of 2.0 MHz (1 MHz = 1 megahertz = 10^6 Hz), what frequency shift must the machine be able to detect? Use $v_{\text{sound}} = 1540$ m/s for the speed of sound in human tissue.

Set Up

We are given the frequency $f_{\text{emitted}} = 2.0$ MHz of the sound waves that the probe emits. The frequency *shift* is the difference between f_{emitted} and the frequency f'' of the waves that the probe receives after they reflect from the heart.

Frequency received after reflecting from a moving object:

$$f'' = \left(\frac{v_{\text{sound}} \pm v}{v_{\text{sound}} \mp v} \right) f_{\text{emitted}} \quad (13\text{-}36)$$

 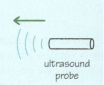

heart   ultrasound probe

Solve

Let's begin by considering a region of the heart that is moving toward the probe. Then we use the upper sign in the numerator and denominator of Equation 13-36.

If the heart wall is approaching the probe,

$$f'' = \left(\frac{v_{\text{sound}} + v}{v_{\text{sound}} - v} \right) f_{\text{emitted}}$$

The numerator $v_{\text{sound}} + v$ is greater than the denominator $v_{\text{sound}} - v$, so the quantity in parentheses is greater than 1 and f'' will be greater than f_{emitted}.

Write an expression for the difference between f'' and $f_{emitted}$.

The frequency shift is

$$f'' - f_{emitted} = \left(\frac{v_{sound} + v}{v_{sound} - v}\right) f_{emitted} - f_{emitted}$$

$$= \left(\frac{v_{sound} + v}{v_{sound} - v} - 1\right) f_{emitted}$$

Substitute $v = 4.0 \times 10^{-4}$ m/s, $v_{sound} = 1540$ m/s, and $f_{emitted} = 2.0 \times 10^6$ Hz:

$$f'' - f_{emitted} = \left(\frac{1540 \text{ m/s} + 4.0 \times 10^{-4} \text{ m/s}}{1540 \text{ m/s} - 4.0 \times 10^{-4} \text{ m/s}} - 1\right)(2.0 \times 10^6 \text{ Hz})$$

$$= 1.0 \text{ s}^{-1} = 1.0 \text{ Hz}$$

The wave received by the probe is 1.0 Hz higher in frequency than the wave emitted by the probe.

Reflect

You should repeat the above calculation for a region of the heart that is moving *away* from the probe. Then you'll use the *lower* sign in the numerator and denominator of Equation 13-36. You should find that the frequency shift in this case is $f'' - f_{emitted} = -1.0$ Hz (that is, the wave received by the probe is 1.0 Hz *lower* in frequency than the wave emitted by the probe). So an ultrasound device must be able to measure a frequency shift that is rather small compared to the emitted frequency $f_{emitted} = 2.0 \times 10^6$ Hz.

Our calculation shows that the frequency shift is proportional to the frequency emitted by the probe. Larger frequency shifts make it easier to detect smaller velocities, so an ultrasound device that operates at a higher frequency has a greater sensitivity. One constraint on this is that absorption of ultrasound waves by human tissue increases with frequency, so higher frequencies cannot penetrate as deeply into the body.

Sound from a Supersonic Source

Something curious occurs when a source of sound moves faster than the speed of sound (that is, the source is *supersonic*). In Figure 13-32a a jet airplane sits on the ground before taking off; the concentric circles centered on the airplane represent the spherical wave crests of the sound it generates. The three drawings in parts (b), (c), and (d) of Figure 13-32 show the airplane flying at increasingly faster speeds. As the airplane increases its speed, it catches up to the sound it generated earlier in the motion, squeezing the wave crests closer together ahead of the airplane. But notice that if the airplane moves at the same speed as the sound, as in Figure 13-32c, the airplane and the sound it generates travel the same distance in any time interval. In this case the crests of all of the sound waves bunch together. If the airplane is supersonic, so that its speed exceeds the speed of sound as in Figure 13-32d, the airplane moves farther in any time interval than the sound that it emits.

Figure 13-33a shows an expanded view of what happens when the airplane is supersonic. Each circle represents the spherical wave crest of a sound generated when the airplane was at the center of the circle. Notice that all of the sound wave crests generated over the time Δt that the airplane has moved from its initial position (the lighter-colored image to the left) to its final position (the image on the right) interfere constructively. This constructive interference takes place on the surface of a cone, known as the **Mach cone** after the Austrian physicist Ernst Mach, who first explained this phenomenon in 1877. In Figure 13-33a (which is only two dimensional), we represent the cone by the two black lines. Note that there is *no* sound to the right of the Mach cone in Figure 13-33a. An observer to the right of the cone will be able to see the airplane but will not be able to hear it: No sound from the airplane will have reached her yet.

Because all of the wave crests pile up at the Mach cone, this is a region of substantially higher pressure than the surrounding air. If you are standing on the ground when a supersonic airplane flies over, you experience this as a **shock wave** (so called because there is no advance warning that it is coming) or as a **sonic boom**. The pressure increase in a shock wave from an airplane is in the range of 5 to 50 Pa.

Crests of sound waves are expanding spheres, drawn as circles.

(a)

The airplane is stationary on the ground.

(b)   (c)   (d)

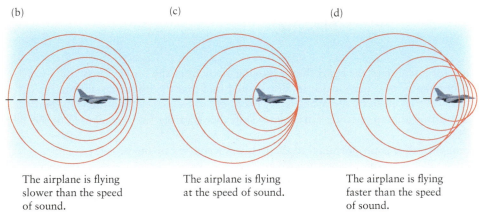

The airplane is flying slower than the speed of sound.

The airplane is flying at the speed of sound.

The airplane is flying faster than the speed of sound.

Figure 13-32 To the speed of sound — and beyond (a) Concentric circles centered on a stationary jet airplane represent the spherical wave fronts of the sound it generates. (b) If the airplane is in flight and moving slower than the speed of the sound waves, the wave pattern is as shown in Figure 13-30b. (c) When the plane moves at the speed of the sound waves, it just exactly catches up to the wave fronts. (d) When the speed of the plane exceeds the speed of sound, the plane moves farther to the right in any time interval than any wave front.

(a)

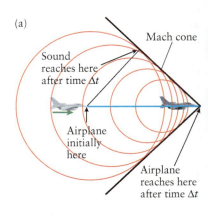

(b)

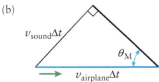

Figure 13-33 Faster than sound (a) In an expanded view of an airplane flying faster than the speed of sound, each circle represents the spherical wave crest of a sound generated when the plane was at the center of the circle. The crests of all of the sound waves generated by the airplane interfere constructively on the surface of the Mach cone. (b) Relating the Mach angle to the speed of sound and the speed of the airplane.

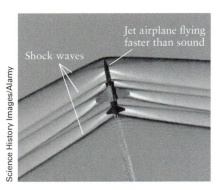

Figure 13-34 Shock waves In this photograph a T-38 jet aircraft is flying at 1.04 times the speed of sound, producing shock waves that originate from various discontinuities on its surface. This photograph was made using a specialized camera in another airplane above the T-38, which recorded the variations in air density caused by the shock waves.

Figure 13-34 is a photograph of an aircraft flying faster than the speed of sound, and shows a number of separate shock waves coming from various discontinuities on the airplane's surface. The region of increased air density and air pressure associated with each shock wave is quite narrow, only a fraction of the length of the airplane. For this reason you would hear a sharp, explosive sound from each shock wave as this aircraft passed overhead.

Note that the sonic boom is *not* the amplified sound of the aircraft engine. As Figure 13-34 shows, shock waves spread out from several parts of the airplane and are not associated with the engine at all. If an airplane is still flying supersonically after it runs out of fuel, it still produces a sonic boom.

The angle of the Mach cone depends on the airplane's speed, which we call v_{airplane}. To see how to calculate this angle, inspect parts (a) and (b) of Figure 13-33. In a time Δt the airplane travels a distance $v_{\text{airplane}}\Delta t$; this is shown by a blue line in Figure 13-33a and Figure 13-33b. The distance $v_{\text{airplane}}\Delta t$ forms the hypotenuse of a right triangle, one leg of which is the distance $v_{\text{sound}}\Delta t$ that the sound travels in time Δt. The angle θ_M in Figure 13-33b is known as the **Mach angle**; the sine of θ_M equals the side opposite this angle ($v_{\text{sound}}\Delta t$) divided by the hypotenuse ($v_{\text{airplane}}\Delta t$). The factors of Δt cancel, leaving

Mach angle
(13-37)

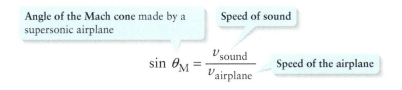

Angle of the Mach cone made by a supersonic airplane

Speed of sound

$$\sin \theta_M = \frac{v_{\text{sound}}}{v_{\text{airplane}}}$$

Speed of the airplane

If an airplane is flying at the speed of sound, the ratio $v_{sound}/v_{airplane}$ equals 1. Since $\sin 90° = 1$, the Mach angle at this speed is $\theta_M = 90°$. As an object's speed increases above v_{sound}, θ_M decreases and the Mach cone lies more tightly around the path of the airplane.

For the case shown in Figure 13-33b, the Mach angle is 45°; in this case Equation 13-37 tells us that $v_{sound}/v_{airplane} = \sin 45° = 0.71$, so $v_{airplane} = v_{sound}/0.71 = 1.4 v_{sound}$ (the airplane is flying at 1.4 times the speed of sound). The ratio of an airplane's speed to the speed of sound is called the **Mach number**: $M = v_{airplane}/v_{sound}$. So you can see that $M = 1.4$ for the airplane shown in Figure 13-33b (it is said to be flying "at Mach 1.4").

In terms of the Mach number, we can rewrite Equation 13-37 as $\sin \theta_M = 1/M$. The airplane in the photograph in Figure 13-34 is flying at $M = 1.04$, so $\sin \theta_M = 1/1.04 = 0.96$ and $\theta_M = \sin^{-1} 0.96 = 74°$. The greater the supersonic speed of an airplane, the smaller its Mach angle θ_M.

WATCH OUT! The Mach cone moves along with the object that creates it.

! Although the individual sound waves that go into making a sonic boom travel at the speed of sound, the Mach cone on which the crests pile up moves along with the supersonic airplane. For example, the sonic boom associated with a jet airplane flying horizontally at Mach 1.4 also moves horizontally at 1.4 times the speed of sound.

EXAMPLE 13-13 Sonic Boom

A supersonic airplane flies directly over you at an altitude of 1.20 km and a Mach number of 1.10. How long after the airplane was directly overhead do you hear the sonic boom? Use 343 m/s for the speed of sound in air.

Set Up

Mach 1.10 means that the speed of the airplane is 1.10 times the speed of sound. You hear the sonic boom when the Mach cone passes over you, a time Δt after the airplane is overhead. As the sketch shows, when this happens the airplane flying at altitude $H = 1.20$ km has flown past you a distance $v_{airplane}\Delta t$, where $v_{airplane} = 1.10v_{sound}$. We'll use Equation 13-37 and trigonometry to solve this problem.

Mach angle:

$$\sin \theta_M = \frac{v_{sound}}{v_{airplane}} \qquad (13\text{-}37)$$

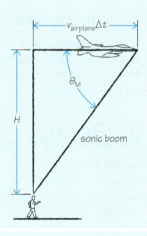

Solve

Since we know the speed of the airplane, we can use Equation 13-37 to find the Mach angle θ_M.

The speed of the airplane is $v_{airplane} = 1.10v_{sound}$, so Equation 13-37 becomes

$$\sin \theta_M = \frac{v_{sound}}{v_{airplane}} = \frac{v_{sound}}{1.10v_{sound}} = \frac{1}{1.10}$$

So θ_M is the angle whose sine is equal to $1/1.10$:

$$\theta_M = \sin^{-1}\left(\frac{1}{1.10}\right) = \sin^{-1} 0.909 = 65.4°$$

The figure above shows that the distances H and $v_{\text{airplane}}\Delta t$ are two sides of a right triangle. Since H is opposite to the angle θ_M and $v_{\text{airplane}}\Delta t$ is adjacent to θ_M, their ratio is the tangent of θ_M.

From the figure,

$$\tan \theta_M = \frac{H}{v_{\text{airplane}}\Delta t}$$

We know the values of $\theta_M = 65.4°$, $H = 1.20$ km, $v_{\text{airplane}} = 1.10 v_{\text{sound}}$, and $v_{\text{sound}} = 343$ m/s, so we can solve for Δt:

$$\Delta t = \frac{H}{v_{\text{airplane}} \tan \theta_M} = \frac{H}{1.10 v_{\text{sound}} \tan \theta_M}$$

$$= \frac{1.20 \text{ km}}{1.10(343 \text{ m/s}) \tan 65.4°} \left(\frac{10^3 \text{ m}}{1 \text{ km}} \right)$$

$$= 1.46 \text{ s}$$

The sonic boom reaches you 1.46 s after the airplane flies overhead.

Reflect

Note that if the airplane emits a sound wave when it is directly above you at altitude $H = 1.20$ km = 1200 m, that sound would need a time $H/v_{\text{sound}} = (1200 \text{ m})/(343 \text{ m/s}) = 3.49$ s to reach you. The sonic boom reaches you *before* that, just 1.46 s after the airplane is directly above you. That's because the shock wave is made up of wave crests that were emitted by the airplane *before* the airplane was directly above you (see Figure 13-33a), so the wave crests have had more time to reach you.

GOT THE CONCEPT? 13-11 Sound from a Supersonic Airplane

 When an airplane is traveling at Mach 2.0 (twice the speed of sound), at what speed do sound waves spread away from the point where the airplane emits them?

(a) 2.0 times the speed of sound; (b) faster than the speed of sound but slower than 2.0 times the speed of sound; (c) the speed of sound; (d) slower than the speed of sound.

TAKE-HOME MESSAGE FOR Section 13-10

✔ The frequency of a sound wave heard by a listener is higher if the listener and source are moving toward each other and lower if they are moving apart. This is called the Doppler effect.

✔ If a source of sound waves moves through the air faster than the speed of sound, a shock wave (sonic boom) is produced.

Key Terms

amplitude	inverse-square law	pressure wave (longitudinal wave)
angular frequency	linear mass density	propagation speed
angular wave number	longitudinal wave (pressure wave)	shock wave
antinode	Mach angle	sinusoidal wave
beat frequency	Mach cone	sonic boom
beats	Mach number	sound intensity level
closed pipe	mechanical wave	sound wave
constructive interference	medium	standing wave
decibel	node	standing wave mode
destructive interference	one-dimensional wave	surface wave
Doppler effect	open pipe	transverse wave
frequency	path length difference	traveling wave
fundamental frequency	period	watt
fundamental mode	phase angle	wave
intensity	phase difference	wave function
interference	power	wavelength
interference pattern	pressure amplitude	

Chapter Summary

Topic	Equation or Figure

Mechanical waves: A mechanical wave is a disturbance that propagates through a material medium. There is no net flow of material through the medium. Mechanical waves can be transverse, longitudinal, or a combination of the two.

(a)

① If the hand holding the rope moves up and down, a **transverse wave** travels along the rope.

② The wave disturbance propagates (moves along the rope) horizontally...

③ ...but individual parts of the rope (the wave medium) move up and down, perpendicular (transverse) to the propagation direction.

(Figure 13-2)

(b)

① If the hand holding the spring moves back and forth, a **longitudinal wave** travels along the spring.

② The wave disturbance propagates (moves along the spring) horizontally...

③ ...but individual parts of the spring (the wave medium) move back and forth, parallel (longitudinal) to the propagation direction.

Sinusoidal waves: In a sinusoidal wave, each piece of the medium undergoes simple harmonic motion with the same amplitude and frequency. At any instant the wave pattern repeats itself over a distance λ known as the wavelength. The product of the frequency and wavelength equals the propagation speed of the wave.

Propagation speed of a wave

$$v_p = f\lambda \tag{13-2}$$

Frequency   Wavelength

Wave function for a sinusoidal wave propagating in the +x direction

Angular wave number of the wave = $2\pi/\lambda$

$$y(x,t) = A\cos(kx - \omega t + \phi) \tag{13-6}$$

Phase angle

Amplitude of the wave   **Angular frequency** of the wave = $2\pi f$

Propagation speed of a wave: The speed at which a wave travels through a medium depends on the type of wave and the properties of the medium.

Propagation speed of a **transverse wave on a rope**

Tension in the rope

$$v_p = \sqrt{\frac{F}{\mu}} \tag{13-10}$$

Mass per unit length of the rope

Propagation speed of a **longitudinal wave in a fluid**

Bulk modulus of the fluid

$$v_p = \sqrt{\frac{B}{\rho}} \tag{13-11}$$

Density of the fluid

Propagation speed of a **longitudinal wave along a solid rod**

Young's modulus of the solid

$$v_p = \sqrt{\frac{Y}{\rho}} \tag{13-13}$$

Density of the solid

Superposition and interference: When two waves are present simultaneously in a medium, the total wave is the sum of the two individual waves. If the two waves have the same frequency but emanate from different places, there will be positions where constructive and destructive interference occur.

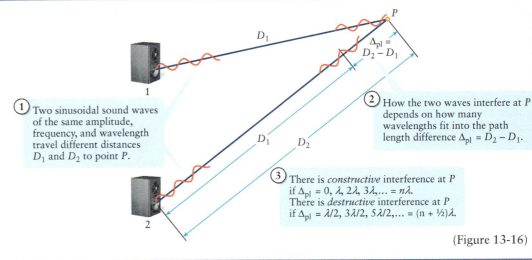

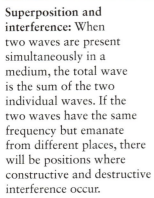

① Two sinusoidal sound waves of the same amplitude, frequency, and wavelength travel different distances D_1 and D_2 to point P.

② How the two waves interfere at P depends on how many wavelengths fit into the path length difference $\Delta_{pl} = D_2 - D_1$.

③ There is *constructive* interference at P if $\Delta_{pl} = 0, \lambda, 2\lambda, 3\lambda,\ldots = n\lambda$.
There is *destructive* interference at P if $\Delta_{pl} = \lambda/2, 3\lambda/2, 5\lambda/2,\ldots = (n + \frac{1}{2})\lambda$.

(Figure 13-16)

Standing waves on a string: When sinusoidal waves reflect back and forth from the ends of a string, the result can be a standing wave. Standing waves are possible only if a whole number of half-wavelengths fit into the length of the string. Each standing wave mode has its own natural frequency.

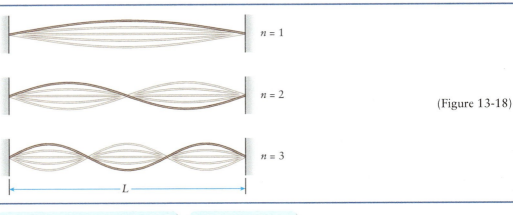

$n = 1$

$n = 2$

$n = 3$

L

(Figure 13-18)

nth possible frequency of a standing wave on a string held at both ends

Tension in the string

$$f_n = \frac{n}{2L} \sqrt{\frac{F}{\mu}} \quad \text{where } n = 1, 2, 3,\ldots$$

(13-19)

Length of the string

Linear mass density of the string

Standing sound waves in a pipe: Standing waves can also occur when sinusoidal sound waves reflect back and forth from the ends of a pipe. The allowed wavelengths and frequencies depend on whether the pipe is closed (closed at one end and open at the other) or open (open at both ends).

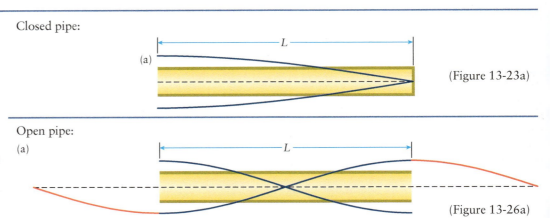

Closed pipe:

(a)

L

(Figure 13-23a)

Open pipe:

(a)

L

(Figure 13-26a)

Beats: When two sound waves of similar frequencies f_1 and f_2 interfere, the result is a total wave with a frequency $(f_1 + f_2)/2$ and an amplitude that rises and falls at the beat frequency $|f_2 - f_1|$.

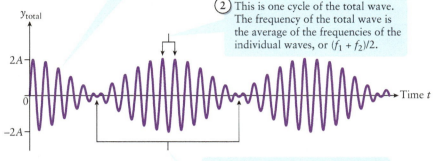

① This is the total wave that results when two individual sine waves of similar frequencies f_1 and f_2 are combined.

② This is one cycle of the total wave. The frequency of the total wave is the average of the frequencies of the individual waves, or $(f_1 + f_2)/2$.

(Figure 13-27)

③ The amplitude of the total wave rises and falls periodically. The frequency of this rising and falling is the beat frequency, $f_{beats} = |f_2 - f_1|$.

Beat frequency heard when sound waves of two similar frequencies interfere

Frequencies of the individual sound waves

$$f_{beats} = |f_2 - f_1| \qquad (13\text{-}25)$$

Wave intensity: The intensity of a sound wave is the amount of power (energy per unit time) that the wave delivers per unit area. It can be expressed in terms of the displacement amplitude or pressure amplitude of the wave. If the wave source emits equally in all directions, the intensity decreases with distance according to an inverse-square law. Sound intensity level is an alternative way to express intensity using a logarithmic scale.

Intensity of a sound wave

Pressure amplitude of the wave

$$I = \frac{p_{max}^2}{2\rho v_{sound}} \qquad (13\text{-}31)$$

Density of the wave medium (usually air)

Speed of sound waves in the medium

Intensity of a sound wave at a distance r from a source that emits in all directions

Power output of the sound source

$$I = \frac{P_0}{4\pi r^2} \qquad (13\text{-}32)$$

Distance from the source to the point where the intensity is measured

The Doppler effect: If a source of sound emits waves with frequency f, a listener will hear a different frequency if the source and listener are moving relative to each other.

Frequency of sound detected by the listener

Frequency of sound emitted by the source

$$f_{listener} = \left[\frac{v_{sound} \pm v_{listener}}{v_{sound} \mp v_{source}} \right] f \qquad (13\text{-}35)$$

v_{sound} = speed of sound
$v_{listener}$ = speed of the listener
v_{source} = speed of the source

In both numerator and denominator:
• Use upper sign if source and listener are approaching.
• Use lower sign if source and listener are moving apart.

Sonic booms: If an object moves through the air faster than the speed of sound, the sound waves from the object combine to form a conical shock wave. The angle of this cone depends on the speed of the object.

(a)

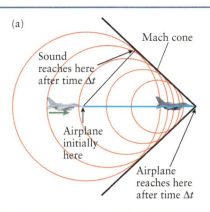

(Figure 13-33a)

Answer to What do you think? Question

(b) We learn in Section 13-3 that the speed of waves on a string (or a flagellum) does not depend on the frequency of the wave.

If the frequency increases, the speed remains the same but the wavelength (the distance from one crest to the next) decreases.

Answers to Got the Concept? Questions

13-1 (c) In the transverse wave shown in Figure 13-2a, the individual pieces of rope (the wave medium) move vertically. So the restoring force that influences their motion must also act vertically. In the same way, since the pieces of the spring in Figure 13-2b move horizontally, the restoring force that acts on them must be in the horizontal direction.

13-2 (e) The propagation speed v_p of sound waves does not depend on the frequency f. The frequency and the wavelength λ are related by Equation 13-2, $v_p = f\lambda$, so increasing the frequency makes the wavelength shorter but has *no* effect on the propagation speed.

13-3 (a) The speed of light is so great that you *see* the batter swing a tiny fraction of a second after the swing took place. The speed of sound is relatively slow by comparison. From the relationship $d = v_{sound}t$ with distance d, propagation speed v_{sound}, and time t, the distance that sound travels in $t = 0.20$ s is $d = (343 \text{ m/s})(0.20 \text{ s}) = 69$ m. Note that the time for light to travel this same distance is $t_{light} = d/v_{light} = (69 \text{ m})/(3.0 \times 10^8 \text{ m/s}) = 2.3 \times 10^{-7}$ s, or about 1/4 of a microsecond (1 microsecond = 1 μs = 10^{-6} s).

13-4 (b) Increasing the frequency decreases the wavelength. The wavelength sets the scale for the entire interference pattern (the positions of destructive and constructive interference depend on the wavelength), so decreasing the wavelength will make all of the distances smaller, and the interference pattern will shrink inward toward the two fingers.

13-5 (e) From Equation 13-18 the wavelength of the $n = 1$ standing wave mode on a string of length L is given by $L = \lambda/2$, so $\lambda = 2L$. This does not depend on the tension in the string, so the wavelength remains the same.

13-6 (b) From Equation 13-19 the frequency of each standing wave mode on the string is proportional to the square root of the tension F. So doubling the tension will cause the

frequency to increase not by a factor of 2 but by a factor of $\sqrt{2} = 1.41$.

13-7 (a) The bottle is like an air-filled pipe that is open at the top and closed on the bottom (at the surface of the water). The fundamental frequency of this pipe is inversely proportional to the length of the pipe (see Equation 13-22). As you fill the bottle the length of the air-filled pipe decreases so that the fundamental frequency goes up. The pitch of the sound from the bottle depends on the fundamental frequency, so the pitch goes up as well.

13-8 (e) The beat frequency equals the difference between the frequencies of the two waves that interfere. The beat frequency here is 3 Hz, so the second tuning fork has a frequency of either 440 Hz + 3 Hz = 443 Hz or 440 Hz − 3 Hz = 437 Hz. Without more information we can't tell which of these two is correct.

13-9 (d) The sound waves from each fan carry the same power, so at any distance from the fans the intensity from each fan is the same. Thus the total intensity doubles when both fans are running. We saw in Example 13-11 that doubling the intensity corresponds to adding 3.0 dB to the sound intensity level. So the sound intensity level due to both fans is 60 dB + 3.0 dB = 63 dB.

13-10 (a) The frequency you hear is higher while you're running compared to when you're standing still. Because you are moving toward the source of the sound, you will intercept more wave crests in a given amount of time than you would were you standing still. The qualitative approach we took using Figure 13-30 for a moving source of sound and a stationary listener applies equally well for a stationary source and a moving listener.

13-11 (c) No matter how a source of sound waves is moving, the waves spread outward at the speed of sound from the point where they were emitted. We used this idea in drawing Figure 13-32.

Questions and Problems

In a few problems you are given more data than you actually need; in a few other problems you are required to supply data from your general knowledge, outside sources, or informed estimate. Interpret as significant all digits in numerical values that have trailing zeros and no decimal points.

For all problems use $g = 9.80$ m/s^2 for the free-fall acceleration due to gravity. Neglect friction and air resistance unless instructed to do otherwise.

- •   Basic, single-concept problem
- ••   Intermediate-level problem; may require synthesis of concepts and multiple steps
- ••• Challenging problem
- Example   See worked example for a similar problem

Conceptual Questions

1. • Explain the difference between longitudinal waves and transverse waves and give two examples of each.

2. • When you talk to your friend, are the air molecules that reach his ear the same ones that were in your lungs? Explain your answer.

3. • Draw a sketch of a transverse wave and label the amplitude, wavelength, a crest, and a trough.

4. • Are water waves longitudinal or transverse? Explain your answer.

5. • Explain the differences and similarities between the concepts of frequency f and angular frequency ω.

6. • If a tree falls in the forest and no humans are there to hear it, was any sound produced?

7. • A sound wave passes from air into water. Give a qualitative explanation of how these properties change: (a) wave speed, (b) frequency, and (c) wavelength of the wave.

8. • Two solid rods have the same Young's modulus, but one has a greater density than the other. In which rod will the speed of longitudinal waves be greater? Explain your answer, making reference to the variables that affect the wave speed.

9. • Search the Internet for the words "rarefaction," "phonon," and "compression" in the context of longitudinal waves. Describe how a longitudinal wave is made up of phonons and explain the connection between rarefaction and compression in such a wave.

10. • Explain the concept of phase difference, ϕ, and predict the outcome of two identical waves interfering when $\phi = 0°$, $\phi = 90°$, $\phi = 180°$, $\phi = 270°$, and $\phi = 300°$.

11. • A car radio is tuned to receive a signal from a particular radio station. While the car slows to a stop at a traffic signal, the reception of the radio seems to fade in and out. Use the concept of interference to explain the phenomenon. *Hint*: In broadcast technology, the phenomenon is known as multipathing.

12. • (a) What is a transverse standing wave? (b) For a string stretched between two fixed points, describe how a disturbance on the string might lead to a standing wave.

13. • How does the length of an organ pipe determine the fundamental frequency?

14. • Why do the sounds emitted by organ pipes that are closed on one end and open on the other *not* have even harmonics? Include in your explanation a sketch of the resonant waves that are formed in the pipes.

15. • Two pianists sit down to play two identical pianos. However, a string is out of tune on Elizabeth's piano. The G$_3^\#$ key (208 Hz) appears to be the problem. When Greg plays the note on his piano and Elizabeth plays hers, a beat frequency of 6 Hz is heard. Luckily a piano tuner is present, and she is ready to correct the problem. However, in all the confusion she inadvertently increases the tension in Greg's G$_3^\#$ string by a factor of 1.058. Now both pianos are out of tune! Yet oddly, when Elizabeth plays her G$_3^\#$ note and Greg plays his G$_3^\#$, there is no beat frequency. Explain what happened.

16. • (a) Explain the differences and similarities among the concepts of intensity, sound level, and power. (b) What happens to intensity as the source of sound moves closer to the observer? (c) What happens to sound level? (d) What happens to power?

17. • (a) Describe in words the nature of a sonic boom. (b) Now, referring to the formula for the Doppler shift, explain the phenomenon.

18. • If you stand beside a railroad track as a train sounding its whistle moves past, you will experience the Doppler effect. Describe any changes in the perceived sound that a person riding on the train will hear.

19. • Discuss several ways that the human body creates or responds to waves.

Multiple-Choice Questions

20. • A visible disturbance propagates around a crowded soccer stadium when fans, section by section, jump up and then sit back down. What type of wave is this?
- A. longitudinal wave
- B. transverse wave
- C. surface wave
- D. spherical wave
- E. pressure wave

21. • Two point sources produce waves of the same wavelength that are in phase. At a point midway between the sources, you would expect to observe
- A. constructive interference.
- B. destructive interference.
- C. alternating constructive and destructive interference.
- D. constructive or destructive interference depending on the wavelength.
- E. no interference.

22. • Standing waves are set up on a string that is fixed at both ends so that the ends are nodes. How many nodes are there in the fourth mode?
- A. 2
- B. 3
- C. 4
- D. 5
- E. 6

23. • Which of the following frequencies are higher harmonics of a string with fundamental frequency of 80 Hz?
- A. 80 Hz
- B. 120 Hz

C. 160 Hz
D. 200 Hz
E. 220 Hz

24. • A trombone has a variable length. When a musician blows air into the mouthpiece and causes air in the tube of the horn to vibrate, the waves set up by the vibrations reflect back and forth in the horn to create standing waves. As the length of the horn is made shorter, what happens to the frequency?
   A. The frequency remains the same.
   B. The frequency will increase.
   C. The frequency will decrease.
   D. The frequency will increase or decrease depending on how hard the horn player blows.
   E. The frequency will increase or decrease depending on the diameter of the horn.

25. • Two tuning forks of frequency 480 and 484 Hz are struck simultaneously. What is the beat frequency resulting from the two sound waves?
   A. 964 Hz
   B. 482 Hz
   C. 4 Hz
   D. 2 Hz
   E. 0 Hz

26. • If a source radiates sound uniformly in all directions, and you triple your distance from the sound source, what happens to the sound intensity at your new position?
   A. The sound intensity increases to three times its original value.
   B. The sound intensity does not change.
   C. The sound intensity drops to 1/3 its original value.
   D. The sound intensity drops to 1/9 its original value.
   E. The sound intensity drops to 1/27 its original value.

27. • If the amplitude of a sound wave is tripled, the intensity will
   A. decrease by a factor of 3.
   B. increase by a factor of 3.
   C. remain the same.
   D. decrease by a factor of 9.
   E. increase by a factor of 9.

28. • A certain sound level is increased by an additional 20 dB. By how much does the intensity increase?
   A. The intensity increases by a factor of 2.
   B. The intensity increases by a factor of 20.
   C. The intensity increases by a factor of 100.
   D. The intensity increases by a factor of 200.
   E. The intensity does not increase.

29. • A person sitting in a parked car hears an approaching ambulance siren at a frequency f_1, and as it passes him and moves away he hears a frequency f_2. The actual frequency f of the source is
   A. $f > f_1$.
   B. $f < f_2$.
   C. $f_2 < f < f_1$.
   D. $f = f_2 + f_1$.
   E. $f = f_2 - f_1$.

Problems

13-1 Waves are disturbances that travel from place to place

13-2 Mechanical waves can be transverse, longitudinal, or a combination of these

13-3 Sinusoidal waves are related to simple harmonic motion

30. • The period of a sound wave is 1.00 ms. Calculate the frequency f and the angular frequency ω. Example 13-2

31. • A wave on a string propagates at 22 m/s. If the frequency is 24 Hz, calculate the wavelength and angular wave number. Example 13-2

32. • (a) Estimate the speed of a human wave like those seen at a large sports venue. (b) How would you define the corresponding concepts of wavelength, frequency, and amplitude for the human wave?

33. • **Biology** Peristalsis is the rhythmic, wavelike contraction of smooth muscles to propel material through the digestive tract. A typical peristaltic wave will only last for a few seconds in the small intestine, traveling at only a few centimeters per second. Estimate the wavelength of the digestive wave.

34. •• A transverse wave on a string has an amplitude of 20 cm, a wavelength of 35 cm, and a frequency of 2.0 Hz. Write the mathematical description of the displacement from equilibrium for the wave if (a) at $t = 0$, $x = 0$ and $y = 0$; (b) at $t = 0$, $x = 0$ and $y = +20$ cm; (c) at $t = 0$, $x = 0$ and $y = -20$ cm; and (d) at $t = 0$, $x = 0$ and $y = 12$ cm. Example 13-2

35. • Show that the dimensions of speed (distance/time) are consistent with both versions of the expression for the propagation speed of a wave: $v_p = \dfrac{\omega}{k}$ and $v_p = \lambda f$. Example 13-1

36. •• Write the wave equation for a periodic transverse wave traveling in the positive x direction at a speed of 20 m/s if it has a frequency of 10 Hz and an amplitude in the y direction of 0.10 m. Example 13-2

37. • A wave on a string is described by the equation $y(x,t) = (0.5$ m$)\cos[(1.0$ rad/m$)x - (10$ rad/s$)t]$. What are (a) the frequency, (b) wavelength, and (c) speed of the wave? Example 13-2

38. •• A pressure wave traveling through air is described by the function $p(x,t) = (1.0$ atm$)\sin[(6.0$ rad/m$)x - (4.0$ rad/s$)t]$. (a) What is the pressure amplitude of the wave? (b) What is the wave number of the wave? (c) What is the frequency of the wave? (d) What is the speed of the wave? Example 13-2

39. •• The equation for a particular wave is $y(x,t) = (0.10$ m$)\cos(kx - \omega t)$. If the frequency of the wave is 2.0 Hz, what is the value of y at $x = 0$ when $t = 4.0$ s? Example 13-2

40. • Using the graph in Figure 13-35, write the mathematical description of the wave if the period of the motion is 4 s and the wave moves to the right (toward the positive x direction). At $x = 0$, $y(0) = 0.6y_{max}$. Example 13-2

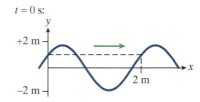

$t = 0$ s:

Figure 13-35 Problem 40

41. • Write a mathematical description of the wave represented by the graphs in Figure 13-36. Example 13-2

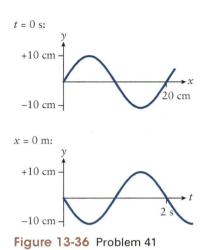

Figure 13-36 Problem 41

13-4 The propagation speed of a wave depends on the properties of the wave medium

42. • If the bulk modulus for liquid A is twice that of liquid B, and the density of liquid A is one-half of the density of liquid B, what does the ratio of the speeds of sound in the two liquids (v_A/v_B) equal? Example 13-4

43. • A string that has a mass of 5.0 g and a length of 2.2 m is pulled taut with a tension of 78 N. Calculate the speed of transverse waves on the string. Example 13-3

44. • A long rope is shaken up and down by a rodeo contestant. The transverse waves travel 12.8 m in a time of 2.1 s. If the tension in the rope is 80.0 N, calculate the mass per unit length for the rope. Example 13-3

45. •• The violin is a four-stringed instrument tuned so that the ratio of the frequencies of adjacent strings is 3 to 2. (This is the ratio when taken as high frequency to lower frequency.) If the diameter of the E string (the highest frequency) on a violin is 0.25 mm, find the diameters of the remaining strings (A, D, and G), assuming they are tuned as indicated (what musicians call intervals of a perfect fifth), they are made of the same material, and they all have the same tension. Example 13-3

46. • At room temperature the bulk modulus of glycerine is about 4.35×10^9 N/m², and the density of glycerine is about 1260 kg/m³. Calculate the speed of sound in glycerin. Example 13-4

47. • The bulk modulus of water is 2.2×10^9 N/m². The density of water is 1000 kg/m³. Calculate the speed of sound in water. Example 13-4

48. • When sound travels through the ocean, where the bulk modulus is 2.34×10^9 N/m², the wavelength associated with 1000-Hz waves is 1.51 m. Calculate the density of seawater. Example 13-4

49. • What is the speed of sound in gasoline? The bulk modulus for gasoline is 1.3×10^9 N/m². The density of gasoline is 0.74 kg/L. Example 13-4

13-5 When two waves are present simultaneously, the total disturbance is the sum of the individual waves

50. • Two waves interfere at the point x in Figure 13-37. The resultant wave is shown. Draw the three possible shapes for the two waves that interfere to produce this outcome. In each case one wave should head toward the right, and one wave should head toward the left.

Figure 13-37 Problem 50

51. • In Figure 13-38, rectangular waveforms approach each other. For each case use the ideas of interference to predict the superposed wave that results when the two waves are coincident. Describe the superposed wave graphically and with text.

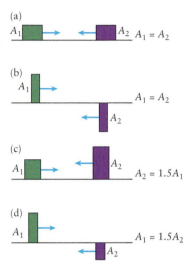

Figure 13-38 Problem 51

52. • Construct the resultant wave that is formed when the two waves shown in each case occupy the same space and interfere (Figure 13-39).

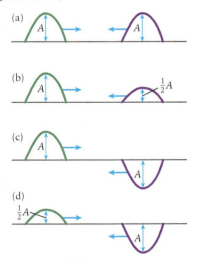

Figure 13-39 Problem 52

53. • Two wave pulses approach each other, as shown in Figure 13-40. Sketch the waveform that results at the instant the centers of the two pulses coincide.

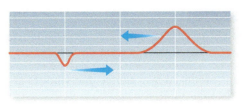

Figure 13-40 Problem 53

54. •• Two identical speakers (1 and 2) are playing a tone with a frequency of 171.5 Hz, in phase (Figure 13-41). The speakers are located 6.00 m apart. Determine what points (A, B, C, D, or E, all separated by 1.00 m) will experience constructive interference along the line that is 6.00 m in front of the speakers. Point A is directly in front of speaker 1. The speed of sound is 343 m/s. Example 13-5

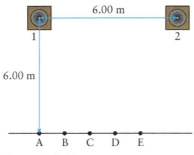

Figure 13-41 Problem 54

55. • Two 60-Hz tone generators are set up such that they interfere constructively at a point midway between them along a line drawn through both speakers. How far must one speaker be moved along the line such that *destructive* interference occurs at the exact same point? Example 13-5

13-6 A standing wave is caused by interference between waves traveling in opposite directions

56. • A string is tied at both ends, and a standing wave is established. The length of the string is 2.00 m, and it vibrates in the fundamental mode ($n = 1$). If the speed of waves on the string is 60.0 m/s, calculate the frequency and wavelength of the waves. Example 13-6

57. • A string is fixed on both ends with a standing wave vibrating in the fourth harmonic. Draw the shape of the wave and label the location of antinodes (A) and nodes (N). Example 13-6

58. • A 2.35-m-long string is tied at both ends, and it vibrates with a fundamental frequency of 24.0 Hz. Find the frequencies and make a sketch of the next four harmonics. What is the speed of waves on the string? Example 13-6

59. •• A string that is 1.25 m long has a mass of 0.0548 kg and a tension of 2.00×10^2 N. The string is tied at both ends and vibrated at various frequencies. (a) What frequencies would you need to apply to excite the first four harmonics? (b) Make a sketch of the first four harmonics for these standing waves. Example 13-6

60. •• A string of length L is tied at both ends, and a harmonic mode is created with a frequency of 40.0 Hz. If the next successive harmonic is at 48.0 Hz and the speed of transverse waves on the string is 56.0 m/s, find the length of the string. Note that the fundamental frequency is not necessarily 40.0 Hz. Example 13-6

61. •• An object of mass M is used to provide tension in a 4.50-m-long string that has a mass of 0.252 kg, as shown in Figure 13-42. A standing wave that has a wavelength equal to 1.50 m is produced by a source that vibrates at 30.0 Hz. Calculate the value of M. Example 13-6

Figure 13-42 Problem 61

62. •• A guitar string has a mass per unit length of 2.35 g/m. If the string is vibrating between points that are 60.0 cm apart, find the tension when the string is designed to play a note of 440 Hz (A4). Example 13-6

63. •• An E string on a violin has a diameter of 0.25 mm and is made from steel (density of 7800 kg/m³). The string is designed to sound a fundamental note of 660 Hz, and its unstretched length is 32.5 cm. Calculate the tension in the string. Example 13-6

13-7 Wind instruments, the human voice, and the human ear use standing sound waves

64. • An organ pipe of length L sounds its fundamental tone at 40.0 Hz. The pipe is open on both ends, and the speed of sound in air is 343 m/s. Calculate (a) the length of the pipe and (b) the frequency of the first four harmonics. Example 13-7

65. • An organ pipe sounds two successive tones at 228.6 and 274.3 Hz. Determine whether the pipe is open at both ends or open at one end and closed at the other. Example 13-7

66. • A narrow glass tube is 0.40 m long and sealed on the bottom end. It is held under a loudspeaker that sounds a tone at 220 Hz, causing the tube to radically resonate in its first harmonic. Find the speed of sound in the room. Example 13-7

67. •• The third harmonic of an organ pipe that is open at both ends and is 2.25 m long excites the fourth harmonic in another organ pipe. Determine the length of the other pipe and whether it is open at both ends or open at one end and closed at the other. Assume the speed of sound is 343 m/s in air. Example 13-7

68. • **Biology** If the human ear canal with a typical length of about 2.8 cm is regarded as a tube open at one end and closed at the eardrum, what is the fundamental frequency that we should hear best? Assume that the speed of sound is 343 m/s in air. Example 13-7

69. • **Biology** A male alligator emits a subsonic mating call that has a frequency of 18 Hz by taking a large breath into his chest cavity and then releasing it. If the hollow chest cavity of an alligator behaves approximately as a pipe open at only one end, estimate its length. Is your answer consistent with the typical size of an alligator? Example 13-7

70. • **Biology** The trunk of a very large elephant may extend up to 3 m! It acts much like an organ pipe open only at one end when the elephant blows air through it. (a) Calculate the fundamental frequency of the sound the elephant can create with its trunk. (b) If the elephant blows even harder, the next harmonic may be sounded. What is the frequency of this first overtone? Example 13-7

71. • The longest pipe in the Mormon Tabernacle Organ in Salt Lake City has a "speaking length" of 9.75 m; the smallest pipe is 1.91 cm long. Assuming that the speed of sound is 343 m/s, determine the range of fundamental frequencies that the organ can produce if both open and closed pipes are used with these lengths. Example 13-7

13-8 Two sound waves of slightly different frequencies produce beats

72. • The sound from a tuning fork of 440 Hz produces beats against the unknown emissions from a vibrating string. If beats are heard at a frequency of 4 Hz, what is the vibrational frequency of the string? Example 13-8

73. • A guitar string is "in tune" at 440 Hz. When a standardized tuning fork rated at 440 Hz is simultaneously sounded with the guitar string, a beat frequency of 5 Hz is heard. (a) How far out of tune is the string? (b) What should you do to correct this? Example 13-8

74. • Two violinists are trying to play in tune. However, whenever they play their A string at the same time, they hear a beat frequency of 7.0 Hz. One of the violinists measures her A string with an electronic tuner and finds that it is tuned to 442 Hz. What are the possible frequencies for the A string of the other violinist? Example 13-8

75. •• A guitar string has a tension of 1.00×10^2 N and is supposed to have a frequency of 1.10×10^2 Hz. When a standard tone of that value is sounded while the string is plucked, a beat frequency of 2.00 Hz is heard. The peg holding the string is loosened (decreasing the tension), and the beat frequency increases. What should the tension in the string be to achieve perfect pitch? Example 13-8

76. •• Two identical guitar strings under 2.00×10^2 N of tension produce sound with frequencies of 290 Hz. If the tension in one string drops to 196 N, what will be the frequency of beats when the two strings are plucked at the same time? Example 13-8

13-9 The intensity of a wave equals the power that it delivers per square meter

77. • By what factor should you move away from a point source of sound waves to lower the intensity by (a) a factor of 10? (b) a factor of 3? (c) a factor of 2? (d) Discuss how this problem would change if it were a point source that was radiating into a hemisphere instead of a sphere. Example 13-10

78. • Calculate the ratio of the acoustic power generated by a blue whale (190 dB) compared to the sound generated by a jackhammer (105 dB). Assume that the receiver of the sound is the same distance from the two sources of sound. Example 13-11

79. •• The steady drone of rush hour traffic persists on a stretch of freeway for 4 h each day. A nearby resident undertakes a plan to harness the wasted sound energy and use it for her home. She mounts a 1-m² "microphone" that absorbs 30% of the sound that hits it. If the ambient sound level is 100 dB at the microphone, calculate the amount of sound energy that is collected. Do you think her plan is "sound"? How could you improve it? Explain, with supporting calculations. Example 13-11

80. • A longitudinal wave has a measured sound level of 85 dB at a distance of 3.0 m from the speaker that created it.

(a) Calculate the intensity of the sound at that point. (b) If the speaker were a point source of sound energy, find its power output. Example 13-11

81. • **Biology** A single goose sounds a loud warning when an intruder enters the farmyard. Some distance from the goose, you measure the sound level of the warning to be 88 dB. If a gaggle of 30 identical geese is present, and they are all approximately the same distance from you, what will the collective sound level be if they all sound off simultaneously? Neglect any interference effects. Example 13-11

82. •• Two sound levels, β_1 and β_2, can be compared relative to each other (rather than to some standardized intensity threshold). Find a formula for this by calculating $\Delta\beta = \beta_2 - \beta_1$. You will need to employ your knowledge of basic logarithm operations to derive the formula. Example 13-11

83. • By what factor must you increase the intensity of a sound to hear (a) a 1-dB rise in the sound level? (b) What about a 20-dB rise? Example 13-11

84. • **Biology** The intensity of a certain sound at your eardrum is 0.0030 W/m². (a) Calculate the rate at which sound energy hits your eardrum. Assume that the area of your eardrum is about 55 mm². (b) What power output is required from a point source that is 2.0 m away to create that intensity? Example 13-10

85. • **Biology** The area of a typical eardrum is 5.0×10^{-5} m². Find the sound power incident on an eardrum at the threshold of pain. Example 13-9

13-10 The frequency of a sound depends on the motion of the source and the listener

86. • A fire engine's siren is 1600 Hz when at rest. What frequency do you detect if you move with a speed of 28 m/s (a) toward the fire engine, and (b) away from it? Assume that the speed of sound is 343 m/s in air. Example 13-12

87. • **Medical** An ultrasound machine can measure blood flow speeds. Assume the machine emits acoustic energy at a frequency of 2.0 MHz, and the speed of the wave in human tissue is taken to be 1500 m/s. What is the beat frequency between the transmitted and reflected waves if blood is flowing in large arteries at 0.020 m/s directly away from the sound source? Example 13-12

88. • A car sounding its horn (rated by the manufacturer at 600 Hz) is moving at 20 m/s toward the east. A stationary observer is standing due east of the oncoming car. (a) What frequency will he hear, assuming that the speed of sound is 343 m/s? (b) What if the observer is standing due west of the car as it drives away? Example 13-12

89. •• **Medical** An ultrasonic scan uses the echo waves coming from something moving (such as the beating heart of a fetus) inside the body and the waves that are directly received from the transmitter to form a measurable beat frequency. This allows the speed of the internal structure to be isolated and analyzed. What is the beat frequency detected when waves with a frequency of 5.00 MHz are used to scan a fetal heartbeat (moving at a speed of ±10.0 cm/s)? The speed of the ultrasound waves in tissue is about 1540 m/s. Example 13-12

90. •• **Biology** A bat emits a high-pitched squeal at 50.0 kHz as it approaches an insect at 10.0 m/s. The insect flies away from the bat, and the reflected wave that echoes off the insect returns to the bat at a frequency of 50.05 kHz. Calculate the speed of the insect as it tries to avoid being the bat's next meal. (A bat can eat more than 3000 mosquitoes in one night!) Example 13-12

General Problems

91. • (a) Using a graphing calculator, spreadsheet, or graphing program, graph the wave function, $y(x)$ described by the data shown. Assume the data were taken at time $t = 0$ s. (b) Write a mathematical function that describes the traveling wave if the period of the motion is 2 s. Example 13-2

x (m)	y (m)	x (m)	y (m)
0	0	7	−4.5
1	4.5	8	−7.8
2	7.8	9	−9
3	9	10	−7.8
4	7.8	11	−4.5
5	4.5	12	0
6	0		

92. •• A large volcanic eruption triggers a tsunami. At a seismic station 250 km away, the instruments record that the time *difference* between the arrival of the tidal wave and the arrival of the sound of the explosion is 9.25 min. Tsunamis typically travel at approximately 800 km/h. (a) Which wave arrives first, the sound in the air or in the tsunami? Prove your answer numerically. (b) How long after the explosion does it take for the first wave to reach the seismic station? (c) How long after the explosion does it take for the second wave to reach the seismic station? Example 13-4

93. • A transverse wave is propagating according to the following wave function:

$y(x,t) = (1.25$ m$) \cos [(5.00$ rad/m$)x − (4.00$ rad/s$)t]$ (a) Plot a graph of y versus x when $t = 0$ s. (b) Repeat for when $t = 1.00$ s. Example 13-2

94. • **Medical** A diagnostic *sonogram* produces a picture of internal organs by passing ultrasound through the tissue. In one application it is used to find the size, location, and shape of the prostate in preparation for surgery or other treatment. The speed of sound in the prostate is 1540 m/s, and a diagnostic sonogram uses ultrasound of frequency 2.00 MHz. The density of the prostate is 1060 kg/m³. (a) What is the wavelength of the sonogram ultrasound? (b) What is the bulk modulus for the prostate gland? Example 13-4

95. •• **Biology** You may have seen a demonstration in which a person inhales some helium and suddenly speaks in a high-pitched voice. Let's investigate the reason for the change in pitch. At 0°C the density of air is 1.40 kg/m³, the density of helium is 0.1786 kg/m³, the speed of sound in helium is 972 m/s, and the speed of sound in air is 331 m/s. (a) What is the bulk modulus of helium at 0°C? (b) If a person produces a sound of frequency 0.500 kHz in the fundamental mode while speaking with his lungs full of air, what frequency sound will that person produce if his respiratory tract is filled with helium instead of air? (c) Use your result to explain why the person sounds strange when he breathes in helium. Example 13-7

96. • **Biology** When an insect ventures onto a spiderweb, a slight vibration is set up, alerting the spider. The density of spider silk is approximately 1.3 g/cm³, and its diameter varies considerably depending on the type of spider, but 3.0 mm is typical. If the web is under a tension of 0.50 N when a small beetle crawls onto it 25 cm from the spider, how long will it take for the spider to receive the first waves from the beetle? Example 13-3

97. •• If two musical notes are an *octave* apart, the frequency of the higher note is twice that of the lower note. The note concert A usually has a frequency of 440 Hz (although there is some variation). (a) What is the frequency of a note that is two octaves above (higher than) concert A in pitch? (b) If a certain string on a viola is tuned to middle C by adjusting its tension to T, what should be the tension (in terms of T) of the string so that it plays a note one octave below middle C? Example 13-6

98. ••• In Western music the octave is divided into 12 notes as follows: C, C#/D^b, D, D#/E^b, E, F, F#/G^b, G, G#/A^b, A, A#/B^b, B, C′. Note that some of the notes are the same, such as C# and D^b. Each of the 12 notes is called a *semitone*. In the ideal *tempered* scale, the ratio of the frequency of any semitone to the frequency of the note below it is the same for all pairs of adjacent notes. So, for example, $f_D/f_{C\#}$ is the same as $f_{A\#}/f_A$. (a) Show that the ratio of the frequency of any semitone to the frequency of the note just below it is $2^{1/12}$. (b) If A is 440 Hz, what is the frequency of F# in the tempered scale? (c) If you want to tune a string from B^b to B by changing only its tension, by what ratio should you change the tension? Should you increase or decrease the tension? Example 13-6

99. •• Find the temperature in an organ loft in Vancouver, British Columbia, if the 5th *overtone* associated with the pipe that is resonating corresponds to 1500 Hz. The pipe is 0.70 m long, and its type (open–open or open–closed) is not specified. The speed of sound in air depends on the centigrade temperature (T) according to the following: Example 13-7

$$v(T) = \sqrt{109{,}700 + 402T}$$

100. • **Biology** An adult female ring-necked duck is typically 16 in. long, and the length of her bill plus neck is about 5.0 cm. (a) Calculate the expected fundamental frequency of the quack of the duck. For a rough but reasonable approximation, assume that the sound is produced only in the neck and bill. (b) An adult male ring-necked duck is typically 18 in. long. If its other linear dimensions are scaled up in the same ratio from those of the female, what would be the fundamental frequency of its quack? (c) Which would produce a higher-pitch quack, the male or female? Example 13-7

101. • When a worker puts on earplugs, the sound level of a jackhammer decreases from 105 to 75 dB. Then the worker moves twice as far from the sound. Determine the sound level with the earplugs at the new location. Example 13-11

102. ••• **Medical** High-intensity focused ultrasound (HIFU) is one treatment for certain types of cancer. During the procedure a narrow beam of high-intensity ultrasound is focused on the tumor, raising its temperature to nearly 90°C to kill it. A range of frequencies and intensities can be used, but in one treatment a beam of frequency 4.0 MHz produced an intensity of 1500 W/cm². The energy was delivered in short pulses for a total time of 2.5 s over an area measuring 1.4 mm by 5.6 mm. The speed of sound in the soft tissue was 1540 m/s, and the density of that tissue was 1058 kg/m³. (a) What was the wavelength of the ultrasound beam? (b) How much energy was delivered to the tissue during the 2.5-s treatment? (c) What was the maximum displacement of the molecules in the tissue as the beam passed through? Example 13-10

103. •• Many natural phenomena produce very high-energy, but inaudible, sound waves at frequencies below 20 Hz (*infrasound*). During the 2003 eruption of the Fuego volcano in Guatemala, sound waves of frequency 10 Hz (and even less) with a sound level of 120 dB were recorded. (a) What was the maximum displacement of the air molecules produced by the waves? (b) How much energy would such a wave deliver to a 2.0-m by 3.0-m wall in 1.0 min? Assume the density of air is 1.2 kg/m³. Example 13-11

104. ••• Two identical 375-g speakers are mounted on parallel springs, each having a spring constant of 50.0 N/cm. Both speakers face in the same direction and produce a steady tone of 1.00 kHz. Both speakers oscillate with an amplitude of 35.0 cm, but they oscillate 180° out of phase with each other. What is the highest beat frequency that a person will hear if she stands in front of the speakers? Example 13-12

105. •• A bicyclist is moving toward a wall while holding a tuning fork rated at 484 Hz. If the bicyclist detects a beat frequency of 6 Hz (between the waves coming directly from the tuning fork and the echo waves coming from the wall), calculate the speed of the bicycle. Assume the speed of sound is 343 m/s. Example 13-12

106. • A jogger hears a car alarm and decides to investigate. While running toward the car, she hears an alarm frequency of 869.5 Hz. After passing the car, she hears the alarm at a frequency of 854.5 Hz. If the speed of sound is 343 m/s, calculate the speed of the jogger. Example 13-12

107. ••• A rescuer in an all-terrain vehicle (ATV) is tracking two injured hikers in the desert, each of whom has an emergency locator transmitter (ELT) stored in his backpack (**Figure 13-43**). The beacons give off radio signals at 121.5 MHz, in phase, and there is a receiver in the ATV that is tuned to that frequency. The speed of the radio waves is 3.00×10^8 m/s.

The ATV is traveling due east, 2.00×10^2 m north of the hikers, and the hikers are 1.00×10^2 m apart. What is the spacing between the points at which the driver detects constructive interference between the two signals? Example 13-5

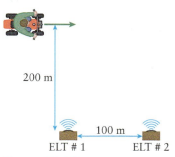

Figure 13-43 Problem 107

108. •• Two identical speakers that face each other and that are separated by a distance of 2.00 m emit a constant tone in phase. You stand in line with the speakers such that your right ear is exactly halfway between the two speakers, resulting in your left ear being closer to one speaker than the other. You notice that what you hear in your right ear is loud, but in your left ear you hear almost nothing. Given that your hearing is fine in both ears and the distance between your ears is 16.0 cm, what are the two lowest frequencies that the speakers could possibly be emitting? Example 13-5

109. •• A weather emergency siren is mounted on a tower, 100 m above the ground. Safety regulations prohibit the siren's sound intensity level from exceeding 120 dB for workers standing on the ground directly beneath the siren. (a) Assuming that the sound is emitted uniformly, what is the siren's maximum permitted power output? (b) How far away from the base of the tower can a person be and still be able to hear the siren? Example 13-10

14

Thermodynamics I: Temperature and Heat

What do you think?

Fire beetles (*Melanophila acuminata*) converge on burning forests from distances of up to 10 km to lay their eggs in trees weakened by fires. These insects can detect a fire from so far away because they are equipped with special chambers that serve as sensitive thermometers. When heated even by a small amount, the fluid inside the chambers expands and causes the pressure within to increase; sensory neurons respond to the pressure change. Compared to how much the volume of the fluid increases in response to the temperature increase, by how much does the volume of the chambers increase? (a) A greater amount; (b) the same amount; (c) by a lesser but nonzero amount; (d) volume actually remains the same; (e) the volume actually decreases.

In this chapter, your goals are to:

- (14-1) Define what thermodynamics is.
- (14-2) Explain the meaning of temperature and thermal equilibrium.
- (14-3) Describe the origin of pressure in an ideal gas and explain the relationship between molecular kinetic energy and gas temperature.
- (14-4) Explain how objects change in size when their temperature changes.
- (14-5) Examine the relationship between the quantity of heat that flows into or out of an object and the temperature change of that object.
- (14-6) Explain how heat must flow in order to cause a substance to change between the solid, liquid, and gas phases.
- (14-7) Describe the key properties of heat transfer by radiation, convection, and conduction.

To master this chapter, you should review:

- (8-4, 8-6)   Elastic collisions and the relationship between force, collision time, and momentum change in a collision.
- (10-2)   The meaning of tensile and compressive stress.
- (11-8)   Archimedes' principle for buoyancy.

14-1 A knowledge of thermodynamics is essential for understanding almost everything around you—including your own body

Thermodynamics, broadly defined, is the branch of physics that deals with relationships among properties of substances such as temperature, pressure, and volume, as well as the energy and flow of energy associated with these properties. If that sounds pretty abstract, that's because thermodynamics is such a general subject that it applies to almost *everything* in your surroundings. Thermodynamics explains how the ice cubes slowly melting in your beverage keep the beverage cold and how the clothing you wear keeps you warm. Your body is an example of a complex thermodynamic system: Some of the

A liquid-in-glass thermometer works on the principle that substances like the liquid expand as the temperature increases.

(a)

iStockphoto/Thinkstock

This ice cube melts because heat flows into it, but its temperature remains at a constant 0°C until it has completely melted.

(b)

Sebastian Duda/Shutterstock

Heat flows into the sleeping dog by radiation from the Sun; heat flows out of the dog due to convection of the air and by conduction into the cool pillow below.

(c)

Juniors Bildarchiv GmbH/Alamy

Figure 14-1 Thermodynamics
The subject matter of thermo-dynamics includes (a) the meaning of temperature and how it is measured, (b) what heat is and what happens when heat flows into or out of an object, and (c) the ways in which heat is transferred from one object to another.

energy released by digesting food goes into maintaining the temperature of your body at a healthy value, and some of the energy is transferred to your surroundings and warms the air around you. Indeed, to understand any aspect of nature or technology that involves the idea of *temperature*, we have to invoke thermodynamic concepts.

Thermodynamics is such a broad subject that it will take us this chapter and the next to introduce and apply its key concepts. We'll begin this chapter by defining what we mean by the temperature of an object, and we'll see that the temperature of a gas has a particularly simple interpretation. We'll go on to examine how objects change their dimensions when the temperature changes (Figure 14-1a). We'll define *heat* as energy that flows from one place to another due to a temperature difference, and we'll see how to analyze the energy flow that takes place when a substance changes phase, such as from solid to liquid (Figure 14-1b). We'll conclude this chapter by looking at the ways in which heat can flow from one object to another (Figure 14-1c).

14-2 Temperature is a measure of the energy within a substance

One of the most fundamental quantities in thermodynamics is the *temperature* of an object. We are used to the idea that we can raise the temperature of an object like a piece of chicken or a slice of eggplant by heating it, say in a frying pan. We also know that we can lower the temperature of an object by cooling it, as in a refrigerator or freezer. But these everyday experiences fail to answer an important question: What *is* temperature?

One way to define what we mean by the temperature of a substance is in terms of what happens on the *microscopic* level—that is, how the individual molecules that make up the substance behave. (In *monatomic* substances, such as helium, the mole-cules are actually individual atoms. In *polyatomic* substances the molecules are made up of combinations of atoms; for example, in water each molecule is made up of two hydrogen atoms and an oxygen atom. We'll use the term *molecule* to refer to the con-stituents of both monatomic and polyatomic substances.)

Molecules in any substance are always in motion. In a gas or liquid, molecules move in every direction with a range of speeds. A molecule in a solid has an equilib-rium position, but it constantly wiggles back and forth around that equilibrium posi-tion. We can define **temperature** as a measure of the kinetic energy associated with molecular motion. As the temperature of a given substance increases, so does the aver-age kinetic energy of a molecule in that substance.

This microscopic definition of temperature isn't very helpful, however. For one thing, it isn't very precise: The relationship between the average kinetic energy of a molecule and the temperature is different for gases, liquids, and solids, and it often depends on what type of molecule makes up the substance. What's more, this defini-tion doesn't tell us how to *measure* temperature, since it's not very practical to analyze individual molecules in a substance to determine the temperature of that substance.

TAKE-HOME MESSAGE FOR Section 14-1

✔   Thermodynamics is the study of the relationships among properties of substances, including temperature, pressure, and volume, as well as the energy and flow of energy associated with these properties.

Thermometer, temperature T_1

Potato, temperature T_2

Thermometer and potato are both at the same intermediate temperature T_{final}; they are in thermal equilibrium.

Figure 14-2 Using a thermometer When a thermometer is placed in contact with an object (such as a baking potato) and allowed to come to thermal equilibrium, the thermometer and the object end up at the same temperature.

A practical approach to measuring temperature is to say that *temperature is the property that you measure with a* **thermometer**. There are many kinds of thermometers that work in different ways, but they're all based on the principle that certain substances change their properties when the temperature changes. For example, nearly all liquids increase in volume as temperature increases. A liquid thermometer uses a column of alcohol or mercury inside a glass tube, and the height of that column goes up as the temperature increases and the liquid expands (see Figure 14-1a).

Strictly speaking, a thermometer measures its *own* temperature. The reason you can use a thermometer to measure the temperature of an object is the phenomenon of *thermal equilibrium*. Experiment shows that when two objects that have different temperatures are in **thermal contact**, so that energy can flow from one to the other, energy flows until both objects reach the same temperature. This final temperature lies between the two original temperatures of the objects. This kind of energy flow that happens due to a temperature difference is what we call *heat transfer*. (We'll present a careful definition of what we mean by *heat* in Section 14-5.) For example, when you pour cold water into a room-temperature glass, energy flows from the glass into the water. As a result, the temperature of the glass decreases and the temperature of the water rises. The energy flow stops when the temperatures of the two objects are the *same*. At this point we say the two objects are in **thermal equilibrium**, and we say that *two objects in thermal equilibrium are at the same temperature*. If one of those objects is a thermometer, the reading on the thermometer indicates the temperature of both the thermometer *and* the object with which it is in contact (Figure 14-2).

A good measurement is one that doesn't change the property being measured. Unfortunately, any thermometer causes a change in the temperature of the object being measured. That's because some heat transfer has to take place between the object and the thermometer, and this will cause the object's temperature to either increase or decrease. A good thermometer minimizes this heat transfer so that the temperature of the object hardly changes before the thermometer and object are in thermal equilibrium. Then the thermometer reading is very close to the temperature that the object had before you made it interact with the thermometer.

Now imagine we put a thermometer C in thermal contact with object A (Figure 14-3a). Heat flows until they reach thermal equilibrium and the reading on the thermometer tells us the temperature of object A. We now put the thermometer C in thermal contact with object B, allowing heat transfer between the thermometer and object B until they come to thermal equilibrium (Figure 14-3b). The reading on the thermometer then tells us the temperature of *both* A and B, so at this point the two objects have the same temperature. Finally, we put A and B into thermal contact with each other (Figure 14-3c). When we do this last step, we find that nothing changes: There is no heat flow between A and B, which means that these two objects at the same temperature must have been in thermal equilibrium *before* we put them in direct contact with each other. In other

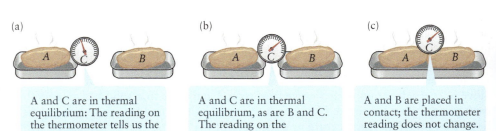

(a)

A and C are in thermal equilibrium: The reading on the thermometer tells us the temperature of A.

(b)

A and C are in thermal equilibrium, as are B and C. The reading on the thermometer tells us the temperature of both A and B.

(c)

A and B are placed in contact; the thermometer reading does not change. Hence A and B are in thermal equilibrium.

Figure 14-3 The zeroth law of thermodynamics If objects A and B are both at the same temperature, they are in thermal equilibrium.

 The **zeroth law of thermodynamics**: If two objects A and B are in thermal equilibrium with a third object C, then A and B are also in thermal equilibrium with each other.

words, *two objects at the same temperature must be in thermal equilibrium, even if they do not interact*. If we combine this observation with our previous observations about thermal equilibrium, we end up with a single statement called the **zeroth law of thermodynamics**:

If two objects are each in thermal equilibrium with a third object, they are also in thermal equilibrium with each other.

From the zeroth law of thermodynamics, we can draw the important conclusion that *temperature is a property of objects in general*, not just of special devices called thermometers. (There are also a *first* and a *second* law of thermodynamics; we'll study these in Chapter 15.)

To measure temperature, we need a scale such as the numbers marked on the tube of a liquid thermometer (Figure 14-1a). The numerical values associated with the markings are arbitrary, and throughout history a variety of schemes have been used to determine what the numbers on a thermometer should be.

The most common temperature scale in everyday life is the **Celsius scale**, based on the work of the eighteenth-century Swedish astronomer Anders Celsius. In this scale the freezing point of water is approximately 0°C and the boiling point is approximately 100°C (both values are at a standard pressure of 1 atm). In a handful of countries (including the United States) the official temperature scale is the **Fahrenheit scale**, originated by the German scientist Daniel Fahrenheit (who also invented the alcohol thermometer and mercury thermometer). On this scale water freezes at 32°F and boils at 212°F at a pressure of 1 atm.

The range of Fahrenheit temperatures between freezing and boiling (212°F − 32°F = 180°F) corresponds almost exactly to a difference of 100°C on the Celsius scale. So a change of 1°C is nearly equivalent to (100/180)°F, or (5/9)°F. If we combine this with the observation that 0°C corresponds to 32°F, we get an approximate conversion between Fahrenheit and Celsius scales:

$$T_C = \frac{5}{9}(T_F - 32) \quad \text{and} \quad T_F = \frac{9}{5}T_C + 32 \tag{14-1}$$

You should use Equations 14-1 to verify that 0°C corresponds to 32°F and 100°C corresponds to 212°F.

The Celsius and Fahrenheit scales were based on the properties of a particular substance (water) under particular circumstances (normal atmospheric pressure on Earth). A scale that has its basis in much more fundamental physics is the **Kelvin scale**, first proposed by the nineteenth-century Scottish physicist William Thomson (1st Baron Kelvin). The Kelvin scale is based on the relationship between the pressure and temperature of low-density gases. (In Section 14-3 we'll see why it's important that the density be low.) Experiment shows that the pressure in a sealed volume of gas decreases as temperature decreases and that a graph of pressure versus temperature is a straight line (Figure 14-5). For a given volume both the slope of this line and the pressure at a given temperature depend on the quantity of gas in the volume. But no matter what kind of gas or what quantity of gas is used, if we extrapolate the lines in Figure 14-5 to low temperatures and pressures, we find that the pressure goes to zero at the *same* temperature. (We have to *extrapolate* each line because at sufficiently low temperatures any gas becomes a liquid.) This temperature is called **absolute zero** because a lower temperature is not physically possible. Zero temperature in the Kelvin scale is absolute zero, so temperatures in the Kelvin scale are never negative.

The unit of temperature used in the Kelvin scale is the **kelvin**, abbreviated K. The value of the kelvin is based on an easily reproduced temperature that can be precisely measured, which is the temperature of the **triple point** of water. The triple point of a substance is the pressure and temperature of that substance at which the solid, liquid, and vapor phases of the substance all coexist. (We'll learn more about the phases of matter and the triple point in Section 14-6.) For water the triple point pressure is 0.00603 atm, and the triple point temperature of water is defined to be 273.16 K.

WATCH OUT! Living organisms avoid thermal equilibrium.

! Although we've emphasized the idea of thermal equilibrium, note that we humans are in general *not* in thermal equilibrium with our surroundings. Instead, our bodies regulate our internal temperature to keep it very constant (Figure 14-4). This is a characteristic of living organisms in general. Our bodies come to thermal equilibrium with their surroundings only after our life functions cease.

(a)

Jaren Wicklund/Alamy

(b)

Michael Honegger/Alamy

Figure 14-4 Maintaining a constant body temperature Whether in (a) a frigid snowstorm or (b) a sweltering desert, your internal body temperature remains nearly constant at about 37°C.

Figure 14-5 Ideal gases and absolute zero The rate at which pressure in a sealed volume of gas decreases as temperature decreases depends on the amount of gas present. However, extrapolating the graphs shows that pressure would become zero for *all* gases at the same temperature. That temperature, –273.15°C, is termed absolute zero.

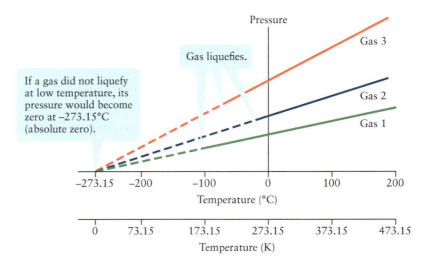

With this choice of the value of the kelvin, a change of 1 degree Celsius is *exactly* equal to a change of 1 kelvin. The triple point of water turns out to be at 0.01°C, so 0°C corresponds to 273.15 K, and 0 K (absolute zero) corresponds to –273.15°C. To convert from a temperature in kelvins to degrees Celsius, you simply *subtract* 273.15, and to convert a Celsius temperature to a Kelvin temperature, you *add* 273.15:

$$(14\text{-}2) \qquad T_C = T_K - 273.15 \text{ and } T_K = T_C + 273.15$$

Water freezes at approximately 0°C = 273.15 K and boils at approximately 100°C = 373.15 K. (The precise values of these temperatures depend on the atmospheric pressure.)

WATCH OUT! Temperature is in kelvins, not "degrees kelvin."

 It's correct to say that an object has a temperature of a certain number of degrees Celsius or of degrees Fahrenheit, as in "20 degrees Celsius." But in the Kelvin scale we say an object has a temperature of a certain number of kelvins, as in "293 kelvins." We do *not* say, "293 degrees kelvin." Note also that the name of the scale (Kelvin) is capitalized, but the unit (kelvin) is not.

Go to Picture It 14-1 for more practice dealing with temperature scales.

As we will see in the following section, the Kelvin scale is a natural one to use for many purposes in physics. Here's an example of how to convert among temperature values in Fahrenheit, Celsius, and Kelvin.

EXAMPLE 14-1 Cold, Hot, and In Between

On the Fahrenheit scale, a cold day might be 5°F. A comfortable temperature is 68°F, which is widely accepted as "room temperature," and a hot day might be 95°F. Express all three temperatures in degrees Celsius and kelvins.

Set Up

We can use the first of Equations 14-1 to convert from degrees Fahrenheit to degrees Celsius and the second of Equations 14-2 to convert from degrees Celsius to temperatures on the Kelvin scale.

Conversion between Celsius and Fahrenheit:

$$T_C = \frac{5}{9}(T_F - 32) \text{ and } T_F = \frac{9}{5}T_C + 32 \qquad (14\text{-}1)$$

Conversion between Celsius and Kelvin:

$$T_C = T_K - 273.15 \text{ and } T_K = T_C + 273.15 \qquad (14\text{-}2)$$

Solve

Convert the cold temperature of 5°F to degrees Celsius and to kelvins.

$$T_{C,\text{cold}} = \frac{5}{9}(T_{F,\text{cold}} - 32) = \frac{5}{9}(5 - 32)$$
$$= -15°C$$
$$T_{K,\text{cold}} = -15 + 273.15$$
$$= 258 \text{ K}$$

Do the conversions for the comfortable room temperature of 68°F.	$T_{C,room} = \dfrac{5}{9}(T_{F,room} - 32) = \dfrac{5}{9}(68 - 32)$ $= 20°C$ $T_{K,room} = 20 + 273.15$ $= 293\ K$
Finally, convert the hot temperature of 95°F to the Celsius and Kelvin scales.	$T_{C,hot} = \dfrac{5}{9}(T_{F,hot} - 32) = \dfrac{5}{9}(95 - 32)$ $= 35°C$ $T_{K,hot} = 35 + 273.15$ $= 308\ K$

Reflect

Our calculations show that typical temperatures on Earth's surface are close to 300 K. That's a good reference number to keep in mind. Another more precise equivalence to remember is that a typical room temperature is 68°F = 20°C = 293 K.

GOT THE CONCEPT? 14-1 Ranking Temperatures

  Rank these temperatures from coldest to warmest: (a) 280 K; (b) –13°C; (c) –13°F.

TAKE-HOME MESSAGE FOR Section 14-2

✔   The temperature of an object is a measure of the kinetic energy of the molecules of which that object is made.

✔   If two objects with different temperatures are in thermal contact so that energy can flow from one to the other, energy will flow until the two objects are at the same temperature. This final state is called thermal equilibrium.

✔   Two objects are in thermal equilibrium if, and only if, they are at the same temperature.

14-3 In a gas, temperature and molecular kinetic energy are directly related

We said in Section 14-2 that the temperature of an object is related to the kinetic energy of its individual molecules. Let's take a look at this relationship in more detail for the case of *gases*, which are in many ways the simplest form of matter. In a liquid or solid, adjacent molecules nearly touch each other and so the forces between molecules are strong. But in a low-density gas, the average distance between molecules is large compared to the size of a molecule (see Example 11-3 in Section 11-2). Because gas molecules are very far apart on average, the forces they exert on each other are so weak that we can ignore them. As we'll see, that leads to tremendous simplifications in the relationship between the temperature of a gas and the kinetic energy of its molecules.

The Ideal Gas Law

Figure 14-5 shows that for a low-density gas at a constant volume, the graph of pressure versus temperature is a straight line, and the pressure p is zero when the Kelvin temperature T is zero. This means that the pressure is directly proportional to the Kelvin temperature. We can write this in equation form as

$$p = (\text{constant}) \times T \qquad (14\text{-}3)$$

(low-density gas at constant volume)

The pressure of a given quantity of gas also depends on the volume V that the gas occupies. For example, think of a quantity of gas in a flexible container whose volume we can change. Experiment shows that if the temperature of the gas is held constant and the volume is decreased, the pressure increases. In fact,

 See the Math Tutorial for more information on direct and inverse proportionality.

the pressure turns out to be *inversely proportional* to the volume, or in equation form

(14-4)
$$p = \frac{(\text{constant})}{V}$$

(low-density gas at constant temperature)

We also find that for a given kind of gas, if the volume and temperature are held fixed but the *mass* of gas in a container is increased, the pressure increases in direct proportion to the mass. Increasing the mass means increasing the number N of molecules present in the gas, so we can write this relationship as

(14-5)
$$p = (\text{constant}) \times N$$

(low-density gas at constant volume and temperature)

We can combine all three equations, 14-3, 14-4, and 14-5, together into a single relationship:

$$p = (\text{constant}) \times \frac{NT}{V}$$

This relationship is most commonly written in the form

Ideal gas law in terms of number of molecules (14-6)

Volume occupied by the gas   Number of molecules present in the gas

Pressure of an ideal gas $pV = NkT$ Temperature of the gas on the Kelvin scale

Boltzmann constant

The constant k in Equation 14-6, called the **Boltzmann constant**, turns out to have the same value for *all* gases. To four significant figures,

$$k = 1.381 \times 10^{-23} \text{ J/K}$$

Equation 14-6 is called the **ideal gas law**, and a gas that would obey this equation exactly is called an **ideal gas**. Real gases are *not* ideal: They do not obey this equation exactly, especially at high pressures or at temperatures close to the point at which the gases become liquids. At relatively low gas densities, however, the ideal gas law does a good job of representing the relationship between pressure, volume, and temperature for real gases. That's why an ideal gas is a useful idealization.

Given the very large number of molecules in most samples of gas, it's often more convenient to write Equation 14-6 in terms of the number of *moles* of a gas. One **mole** of a substance (abbreviated mol) is defined to contain exactly $6.02214076 \times 10^{23}$ molecules. This value is called Avogadro's number N_A. To four significant figures,

$$N_A = 6.022 \times 10^{23} \text{ molecules/mol}$$

The number N of molecules in a substance is therefore equal to the number of moles n multiplied by Avogadro's number: $N = nN_A$. Then Equation 14-6 becomes

$$pV = nN_A kT$$

The product $N_A k$ has a special name: We call it the **ideal gas constant** and give it the symbol R. To four significant figures,

$$R = N_A k = 8.314 \text{ J/(mol·K)}$$

With this definition the ideal gas law (Equation 14-6) becomes

Ideal gas law in terms of number of moles (14-7)

Volume occupied by the gas   Number of moles present in the gas

Pressure of an ideal gas $pV = nRT$ Temperature of the gas on the Kelvin scale

Ideal gas constant

The pressure p, volume V, and temperature T characterize the physical state of a system. A relationship among these quantities, like Equation 14-6 or Equation 14-7 for an ideal gas, is called an **equation of state**. The equation of state for a real gas, a liquid, or a solid can be substantially more complicated than that for an ideal gas.

EXAMPLE 14-2 An Ideal Gas at Room Temperature

At ordinary room temperature (20°C) and a pressure of 1 atm (1.013×10^5 Pa $= 1.013 \times 10^5$ N/m²), how many molecules are there in 1.00 m³ of an ideal gas? How many moles are there?

Set Up

We can use the two forms of the ideal gas law to determine the number of molecules N and number of moles n. Note that the temperature in the ideal gas law is the *Kelvin* temperature, but we're given the value of the *Celsius* temperature. We'll have to convert the temperature using the second of Equations 14-2.

Ideal gas law in terms of number of molecules:

$$pV = NkT \qquad (14\text{-}6)$$

Ideal gas law in terms of number of moles:

$$pV = nRT \qquad (14\text{-}7)$$

Conversion between Celsius and Kelvin:

$$T_C = T_K - 273.15 \text{ and}$$

$$T_K = T_C + 273.15 \qquad (14\text{-}2)$$

$p = 1$ atm
$V = 1.00$ m³
$T = 20°C$
$N = ?$

Solve

First convert the temperature to the Kelvin scale.

From the second of Equations 14-2, a temperature of 20°C corresponds to a Kelvin temperature of

$$T = 20 + 273.15 = 293 \text{ K}$$

Solve Equation 14-6 for the number of molecules N and substitute numerical values.

From Equation 14-6

$pV = NkT$, so

$$N = \frac{pV}{kT} = \frac{(1.013 \times 10^5 \text{ N/m}^2)(1.00 \text{ m}^3)}{(1.381 \times 10^{-23} \text{ J/K})(293 \text{ K})}$$

$$= 2.50 \times 10^{25} \frac{\text{N} \cdot \text{m}}{\text{J}}$$

Since 1 N·m = 1 J, N has *no* units: It is a pure number (in this case, the number of molecules in 1.00 m³). So

$$N = 2.50 \times 10^{25} \text{ molecules}$$

In a similar way, find the number of moles using Equation 14-7.

From Equation 14-7

$pV = nRT$, so

$$n = \frac{pV}{RT} = \frac{(1.013 \times 10^5 \text{ N/m}^2)(1.00 \text{ m}^3)}{(8.314 \text{ J/(mol} \cdot \text{K)})(293 \text{ K})}$$

$$= 41.6 \frac{\text{N} \cdot \text{m} \cdot \text{mol}}{\text{J}} = 41.6 \text{ mol}$$

Reflect

We can check our results by confirming that the number of molecules equals the number of moles multiplied by Avogadro's number (the number of molecules per mole). To put our result into perspective, the number of *stars* in the observable universe is estimated to be about 10^{24}. The number of molecules in a cubic meter of gas is truly astronomical!

We can also calculate N from the number of moles n and Avogadro's number N_A:

$$N = nN_A = (41.6 \text{ mol})(6.022 \times 10^{23} \text{ molecules/mol})$$

$$= 2.50 \times 10^{25} \text{ molecules}$$

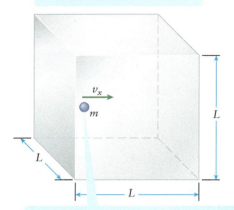

The gas molecule shown has just collided with and been reflected from the wall on the left.

The molecule, moving in the positive x direction, travels a distance L, collides with the right wall, and travels a distance L again before returning to this same position. This round trip takes a time Δt.

Figure 14-6 **The origin of gas pressure** A cubical container contains a gas. This figure shows one representative gas molecule that has just bounced off the left-hand wall. The combined effect of all molecules bouncing off the walls gives rise to the pressure of the gas.

Temperature and Translational Kinetic Energy

With the ideal gas law in hand, let's now see the relationship between pressure, volume, and the *average translational kinetic energy* of molecules in a gas in a container. We'll see that the pressure arises from the forces that individual molecules exert when they collide with the walls of the container.

Figure 14-6 shows a cubical box that has sides of length L and contains N molecules of an ideal gas. Each molecule has mass m. When the gas is in equilibrium, the gas as a whole exhibits no net motion. Although every molecule is moving and has momentum, the *total* momentum of all molecules in the gas averages to zero. In Figure 14-6 one representative molecule has just collided with and bounced off the left wall. We'll follow this molecule for the time Δt that it takes for the single molecule to return to this same position after colliding with the wall on the right side of the container. We'll assume that the molecule doesn't collide with any other molecules during this round trip. (Later in this section we'll see why this is a reasonable assumption to make.) For simplicity we'll initially consider the molecule to be moving in the x direction only. We'll also make the simplifying assumptions that the walls of the container are rigid and the gas molecules behave like hard spheres. Then the collisions are *elastic*, and no energy is lost when a molecule collides with a wall (see Section 8-4).

When the molecule in Figure 14-6 hits the right-hand wall of the box, it reverses direction but loses no kinetic energy. Its velocity changes from $+v_x$ (to the right) to $-v_x$ (to the left) and its momentum changes from $+mv_x$ to $-mv_x$. So the *change* in the molecule's momentum is $(-mv_x) - (+mv_x) = -2mv_x$. This momentum change occurs thanks to the force that the wall exerts on the molecule during the hit. The hit lasts just a short time, and there is one such hit by the right-hand wall on the molecule per time Δt (the time needed for the molecule to make a round-trip and return to the right-hand wall). If we call $F_{\text{wall on molecule},x}$ the *average* force that the right-hand wall exerts on the molecule over the time Δt, Equation 8-24 (Section 8-6) says that this force multiplied by Δt equals the change in the molecule's momentum due to its interaction with the right-hand wall:

$$F_{\text{wall on molecule},x}\Delta t = -2mv_x \quad \text{so} \quad F_{\text{wall on molecule},x} = -\frac{2mv_x}{\Delta t}$$

This is negative because the right-hand wall pushes the molecule to the left, in the negative x direction. By Newton's third law, the average force that the *molecule* exerts on the *wall* is the opposite of this and is *positive* (the molecule exerts a force on the wall in the positive x direction):

$$F_{\text{molecule on wall},x} = -F_{\text{wall on molecule},x} = \frac{2mv_x}{\Delta t}$$

We emphasize that this is the *average* force that this molecule exerts on the wall: During an actual hit the force is large, and the rest of the time (when the molecule is not in contact with the wall) the force is zero. Since Δt is the time it takes the molecule moving at speed v_x to move a distance $2L$ (back and forth across the length L of the box), and time equals distance divided by speed, it follows that $\Delta t = 2L/v_x$. Substituting this into our expression for $F_{\text{molecule on wall},x}$ we get

(14-8)
$$F_{\text{molecule on wall},x} = \frac{2mv_x}{(2L/v_x)} = \frac{mv_x^2}{L}$$

Equation 14-8 tells us the average force on the right-hand wall from *one* molecule. There are N molecules of mass m in the box shown in Figure 14-6, and each of them collides periodically with the right-hand wall. If every molecule had the same value of v_x, the total force from all gas molecules combined would just be N times the force in Equation 14-8. This is extremely unlikely to be the case, however. So we get the total force that the gas as a whole exerts on the wall by replacing v_x^2 in Equation 14-8 with the *average value* of v_x^2 (that is, averaged over its values for all of the molecules in the box) and then multiplying by N:

(14-9)
$$F_{\text{gas on wall},x} = \frac{Nm(v_x^2)_{\text{average}}}{L}$$

WATCH OUT! The average of the square is not the square of the average.

The quantity $(v_x^2)_{average}$ in Equation 14-9 is the *average value* of the *square* of the quantity v_x. Since the square of a quantity is always positive, it follows that $(v_x^2)_{average}$ has a positive value. Suppose instead that you calculated the *square* of the *average value* of v_x. You might think this would give you the same answer, but it doesn't: On average there are as many molecules moving to the left in Figure 14-6 as there are moving to the right, so there are as many with a positive value of v_x as with a negative value. So the average value of v_x is zero, and the square of that is also zero. When we say to take the average value of the square, we mean square first, *then* average!

What we actually measure is the *pressure* that the gas exerts on the walls of the box. The right-hand wall is a square of side L and area L^2, so the pressure (force divided by area) on the right-hand wall is

$$p = \frac{F_{\text{gas on wall},x}}{L^2} = \frac{Nm(v_x^2)_{\text{average}}}{L^3} = \frac{Nm(v_x^2)_{\text{average}}}{V} \tag{14-10}$$

In Equation 14-10 we've used the symbol V for the volume L^3 of the cubical container of side L.

In general the gas molecules are moving in the y and z directions as well as the x direction. Since there's no preferred direction inside the box, we expect that the average values of v_y^2 and v_z^2 are the same as the average value of v_x^2: $(v_x^2)_{\text{average}} = (v_y^2)_{\text{average}} = (v_z^2)_{\text{average}}$. Furthermore, the square of the *speed* v of a given molecule is the sum of the squares of its velocity components: $v^2 = v_x^2 + v_y^2 + v_z^2$. If we put these observations together, we can write an expression for the average value of the square of the speed of gas molecules:

$$(v^2)_{\text{average}} = (v_x^2)_{\text{average}} + (v_y^2)_{\text{average}} + (v_z^2)_{\text{average}} = 3(v_x^2)_{\text{average}} \tag{14-11}$$

Equation 14-11 tells us that the quantity $(v_x^2)_{\text{average}}$ in Equation 14-10 is equal to one-third of the average value of the square of the speed: $(v_x^2)_{\text{average}} = (v^2)_{\text{average}}/3$. Then we can rewrite Equation 14-10 for the gas pressure as

$$p = \frac{Nm(v^2)_{\text{average}}}{3V} \tag{14-12}$$

Now $K_{\text{translational}} = (1/2)mv^2$ is the translational kinetic energy of a single molecule. (We emphasize *translational* since the molecule could also be rotating around its axis and have rotational kinetic energy. Its constituent atoms could also be vibrating within the molecule, so there could be vibrational kinetic energy as well.) So $K_{\text{translational,average}} = (1/2)m(v^2)_{\text{average}}$ is the *average* translational kinetic energy of a single molecule in the gas. In terms of this we can rewrite Equation 14-12 as

$$p = \frac{Nm(v^2)_{\text{average}}}{3V} = \frac{2N}{3V}\left[\frac{1}{2}m(v^2)_{\text{average}}\right] = \frac{2N}{3V}K_{\text{translational,average}} \tag{14-12}$$

The pressure of a given quantity of ideal gas is directly proportional to the average translational kinetic energy of a molecule of the gas. The more kinetic energy that the molecules have on average, the greater the pressure.

We can learn something equally important by comparing Equation 14-12 with the ideal gas law in terms of number of molecules, Equation 14-6: $pV = NkT$, or $p = (N/V)kT$. For both Equations 14-6 and 14-12 to be correct, it must be true that $(2/3)K_{\text{translational,average}} = kT$, or

Average translational kinetic energy of a molecule in an ideal gas

Average value of the square of a gas molecule's speed

Temperature of the gas on the Kelvin scale

Temperature and average translational kinetic energy of an ideal gas molecule (14-13)

$$K_{\text{translational,average}} = \frac{1}{2}m(v^2)_{\text{average}} = \frac{3}{2}kT$$

Mass of a single gas molecule

Boltzmann constant

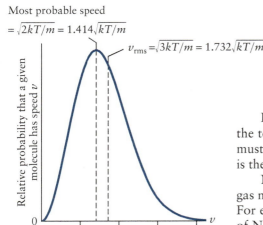

Most probable speed
$= \sqrt{2kT/m} = 1.414\sqrt{kT/m}$

$v_{rms} = \sqrt{3kT/m} = 1.732\sqrt{kT/m}$

Figure 14-7 Molecular speeds in a gas The molecules in an ideal gas have a distribution of speeds. This graph shows the relative probability that any one molecule has a particular speed v. The maximum of the curve is at the most probable speed, which turns out to be $\sqrt{2kT/m} = 1.414\sqrt{kT/m}$. The rms speed $v_{rms} = \sqrt{3kT/m} = 1.732\sqrt{kT/m}$ is a bit faster than this.

→ **Go to Interactive Exercise 14-1 for more practice dealing with rms speed.**

Root-mean-square speed of molecules in an ideal gas
(14-15)

In words, Equation 14-13 says that *the average translational kinetic energy of a molecule in an ideal gas is directly proportional to the Kelvin temperature of the gas.* This justifies the statement we made in Section 14-2 about the physical meaning of temperature. Equation 14-13 relates a *microscopic* property of an ideal gas (the average kinetic energy of an individual gas molecule) to a *macroscopic* property (the temperature of the gas that you might measure with a thermometer).

Equation 14-13 also gives us a microscopic interpretation of absolute zero. If the temperature is 0 K, the average translational kinetic energy of a gas molecule must also be zero and the speed of the molecule must be zero, too. So absolute zero is the temperature at which all molecular motion would cease.

Note from Equation 14-13 that the average translational kinetic energy of a gas molecule does *not* depend on the *mass* of the molecule, just on the temperature. For example, the air around you is mostly nitrogen (N_2) and oxygen (O_2). A mole of N_2 molecules has a mass of 28.0 g, and a mole of O_2 molecules has a mass of 32.0 g, so the mass m of an individual N_2 molecule is $(28.0/32.0) = 0.875$ as much as the mass of an individual O_2 molecule. Nonetheless, on average each kind of molecule in the air has the *same* translational kinetic energy: An average nitrogen molecule must therefore be moving faster than an average oxygen molecule.

A measure of how fast gas molecules move on average is the **root-mean-square speed** or **rms speed**. To see how this is defined, first imagine finding the average value of v^2, the square of the speed v, for all the molecules in a gas that have a given mass m. This is the quantity that we've called $(v^2)_{average}$. Another word for average is *mean*, which is why $(v^2)_{average}$ is called the *mean-square* of the speed. The *root*-mean-square speed is the square root of the mean-square:

$$(14\text{-}14) \qquad v_{rms} = \sqrt{(v^2)_{average}}$$

Note that v_{rms} has units of m/s. From Equation 14-13, $(1/2)m(v^2)_{average} = (3/2)kT$, so $(v^2)_{average} = 3kT/m$. If we substitute this into Equation 14-14, we get

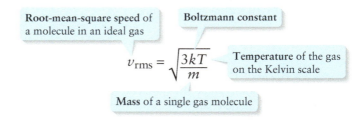

Note that while v_{rms} is a typical speed of molecules in a gas, there is in fact a distribution of speeds. Many molecules move more slowly than v_{rms}, and many move more rapidly (Figure 14-7).

WATCH OUT! The root-mean-square value is not the same as the average value.

We use the root-mean-square speed v_{rms} as a measure of the typical molecular speed in a gas because it follows naturally from Equation 14-13, which relates temperature to translational kinetic energy. It is *not*, however, the same as the *average* molecular speed. To see the difference consider a gas with just five molecules that have speeds 1.00, 2.00, 3.00, 4.00, and 5.00 m/s. The average of these speeds is

$$v_{average} = \frac{\left[\begin{array}{c}1.00\text{ m/s} + 2.00\text{ m/s} + 3.00\text{ m/s} \\ + 4.00\text{ m/s} + 5.00\text{ m/s}\end{array}\right]}{5}$$
$$= 3.00\text{ m/s}$$

but the rms speed is

$$v_{rms} = \sqrt{(v^2)_{average}}$$
$$= \sqrt{\frac{\left[\begin{array}{c}(1.00\text{ m/s})^2 + (2.00\text{ m/s})^2 + (3.00\text{ m/s})^2 \\ + (4.00\text{ m/s})^2 + (5.00\text{ m/s})^2\end{array}\right]}{5}}$$
$$= 3.32\text{ m/s}$$

It turns out that the average value of any collection of numbers is *always* less than the rms value. For the special case of an ideal gas, a detailed analysis shows that the average speed is $\sqrt{8kT/\pi m} = 1.596\sqrt{kT/m}$, which is less than $v_{rms} = \sqrt{3kT/m} = 1.732\sqrt{kT/m}$ but greater than the most probable speed shown in Figure 14-7.

EXAMPLE 14-3 Oxygen at Room Temperature

Calculate (a) the average translational kinetic energy and (b) the root-mean-square speed of an oxygen molecule in air at room temperature (20°C). One mole of oxygen molecules has a mass of 32.0 g = 32.0×10^{-3} kg.

Set Up

Equation 14-13 tells us the average translational kinetic energy, and Equation 14-15 tells us the rms speed of molecules. Note that we're given the mass per mole of O_2, but we'll need to convert this to m, the mass per molecule. We'll also need to convert the temperature from Celsius to Kelvin.

Temperature and average translational kinetic energy of an ideal gas molecule:

$$K_{translational,average} = \frac{1}{2}m(v^2)_{average} = \frac{3}{2}kT \quad (14\text{-}13)$$

Root-mean-square speed of molecules in an ideal gas:

$$v_{rms} = \sqrt{\frac{3kT}{m}} \quad (14\text{-}15)$$

Conversion between Celsius and Kelvin:

$$T_C = T_K - 273.15 \text{ and } T_K = T_C + 273.15 \quad (14\text{-}2)$$

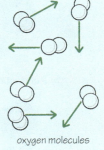

oxygen molecules

Solve

(a) First find the mass per molecule m and the Kelvin temperature T.

The number of molecules per mole is Avogadro's number, $N_A = 6.022 \times 10^{23}$ molecules/mol. So the mass per O_2 molecule is

$$m = \frac{(32.0 \times 10^{-3} \text{ kg/mol})}{(6.022 \times 10^{23} \text{ molecules/mol})}$$
$$= 5.31 \times 10^{-26} \text{ kg/molecule}$$

From the second of Equations 14-2, the Kelvin temperature that corresponds to 20°C is

$$T = 20 + 273.15 = 293 \text{ K}$$

Use Equation 14-13 to find the average kinetic energy per molecule.

From Equation 14-13

$$K_{translational,average} = \frac{3}{2}kT = \frac{3}{2}(1.381 \times 10^{-23} \text{ J/K})(293 \text{ K})$$
$$= 6.07 \times 10^{-21} \text{ J}$$

(b) Use Equation 14-15 to find the rms speed of an O_2 molecule.

From Equation 14-15

$$v_{rms} = \sqrt{\frac{3kT}{m}} = \sqrt{\frac{3(1.381 \times 10^{-23} \text{ J/K})(293 \text{ K})}{5.31 \times 10^{-26} \text{ kg}}}$$
$$= 478 \text{ m/s}$$

Reflect

The rms speed of O_2 molecules in air is tremendous, 478 m/s = 1720 km/h = 1070 mi/h. Very few O_2 molecules travel at precisely that speed: Their speeds range from nearly zero to many times faster than 478 m/s (see Figure 14-7).

Note that *nitrogen* (N_2) molecules in air at 20°C have the *same* translational kinetic energy (which does not depend on the mass m of the molecules) but a *different* rms speed (which is proportional to the reciprocal of $\sqrt{m}$). As we discussed above, an N_2 molecule has 0.875 the mass of an O_2 molecule, so the rms speed for N_2 is $1/\sqrt{0.875} = 1.07$ times faster than the rms speed for O_2.

Degrees of Freedom

The factor of three in Equation 14-13 for the average translational kinetic energy of a molecule, $K_{translational,average} = (1/2)m(v^2)_{average} = (3/2)kT$, is actually rather significant. It arises because $(v^2)_{average}$ is the sum of three terms, one for each component of molecular motion (Equation 14-11): $(v^2)_{average} = (v_x^2)_{average} + (v_y^2)_{average} + (v_z^2)_{average}$. As we

(a) NH_3 at rest

(b) Rotation of NH_3

(c) Bond length oscillations of NH_3

(d) Bond angle oscillations of NH_3

Figure 14-8 Motions of an ammonia molecule If a gas is made of molecules with two or more atoms, a molecule can have energy of motion even if the translational velocity of the molecule is zero. For an ammonia (NH_3) molecule, the molecule can rotate, and both the bond length and the bond angle can oscillate. This means that NH_3 has more degrees of freedom than just the three translational degrees of freedom.

discussed above, each of these terms has the same value. That means we can write the average kinetic energy of a molecule as the sum of three equal terms:

$$\frac{1}{2}m(v^2)_{\text{average}} = \frac{1}{2}m(v_x^2)_{\text{average}} + \frac{1}{2}m(v_y^2)_{\text{average}} + \frac{1}{2}m(v_z^2)_{\text{average}} = \frac{3}{2}kT$$

In other words, on average the translational kinetic energy is shared equally between energies associated with the motions in each of the three possible directions. Each direction of motion available to the molecule contributes an average translational kinetic energy of one-third of $(3/2)kT$, or $(1/2)kT$. We refer to each of these possible directions of motion as a **degree of freedom** of the system. (An object that could move only along a straight line would have one degree of freedom, and one that could move only in a plane would have two degrees of freedom. An object that can move in all three dimensions of space has three degrees of freedom.)

The number of degrees of freedom of a gas can be more than simply the number of directions in which gas molecules can translate. If the molecules contain more than one atom, the molecule can have energy of motion even if its translational velocity is zero. An example is the ammonia molecule, NH_3 (Figure 14-8a). The entire ammonia molecule can rotate (Figure 14-8b), and it can vibrate in two ways: The nitrogen (N) and hydrogen (H) atoms can oscillate along the length of the bond (Figure 14-8c), and the hydrogen atoms can be disturbed so that the angle of the bonds oscillates (Figure 14-8d). Each of these also represents a possible degree of freedom of the gas molecule, and each can contribute an additional $(1/2)kT$ to the average energy per molecule. Thus the average energy per molecule is likely to be more than $(3/2)kT$ for a gas that is composed of polyatomic molecules.

Two other examples in the air around you are O_2 and N_2 at room temperature. For both of these, the total energy per molecule is actually close to $(5/2)kT$. That's because the molecules have five degrees of freedom: They can translate, for which there are three degrees of freedom, and they can rotate, for which there are two degrees of freedom (if the long axis of the O_2 or N_2 molecule is the z axis, one degree of freedom corresponds to rotation around the x axis and the other to rotation around the y axis). Unlike the ammonia molecule depicted in Figure 14-8, however, O_2 and N_2 molecules in the air around you do not oscillate. Quantum mechanics tells us that there's not enough energy at room temperature for bond length oscillations to contribute significantly to the energy of these diatomic molecules.

The fact that the energy of a molecule is shared equally among each degree of freedom is called the **equipartition theorem**.

Mean Free Path

In our analysis of what happens in an ideal gas at the microscopic level, we considered what happens when a gas molecule hits one of the container walls, but we ignored the effects of collisions between gas molecules. How frequently do such collisions between molecules occur? One way to answer this question is in terms of the **mean free path** of a gas molecule. This is the average distance that a molecule travels from the time at which it collides with one molecule to when it collides with another molecule. If this distance is large compared to the size of the container that holds the gas, we can conclude that a molecule is unlikely to have any collisions with other molecules as it bounces back and forth between the container walls. But if the mean free path is short, a molecule may undergo many collisions with other molecules during a single trip across the container.

The mean free path depends on how densely packed the gas molecules are. If there are very few gas molecules in a container, collisions between molecules are unlikely and the mean free path is long. If, however, there are many molecules in the container, the likelihood of a collision is greater and the mean free path is short. The mean free path also depends on the *size* of the molecules: The larger the gas molecules are, the bigger "targets" they are for other molecules, the more likely they are to undergo collisions, and the shorter the mean free path.

If there are N gas molecules in a volume V, and if we model the molecules as spheres of radius r, the mean free path (to which we give the symbol λ, Greek letter "lambda") turns out to be

(14-16)
$$\lambda = \frac{1}{4\sqrt{2}\pi r^2 (N/V)}$$

Just as we expected, the mean free path is short if r is large (the molecules are large in size) or if N/V is large (there are many molecules present per volume, so the molecules are close together). Equation 14-6 tells us that $pV = NkT$, which we can rearrange to read $N/V = p/kT$, so Equation 14-16 becomes

$$\lambda = \frac{kT}{4\sqrt{2}\pi r^2 p}$$

(14-17)

We invite you to use Equation 14-17 to calculate the mean free path for oxygen (O_2) molecules, which are about 2.0×10^{-10} m in radius, at ordinary room temperature (20°C or 293 K) and a pressure of 1 atm = 1.013×10^5 N/m². You'll find that $\lambda = 5.6 \times 10^{-8}$ m, a distance that is large compared to the size of a molecule but hundreds of millions of times smaller than the size of a typical container for air. An O_2 molecule trying to travel from one side to another of a room 2 m wide would undergo tens of millions of collisions along the way!

How, then, can we justify the assumption we made earlier in this section that we can *ignore* collisions between molecules? The explanation is that a given molecule collides with other molecules that are moving in *all* directions. Some collisions make the molecule speed up, and others make it slow down; some deflect it to the left, and others deflect it to the right. The result is that the effects of all these collisions largely cancel out, so on average we can treat molecules as if they were unaffected by collisions. Note that collisions also tend to *randomize* the velocities of the molecules that make up the gas, which justifies our assumption that the average values of v_x^2, v_y^2, and v_z^2 are all the same.

BioMedical The concept of mean free path plays an important role in botany. The leaves of plants "breathe" through their pores or *stomata*, tiny apertures through which they exchange gas molecules with the surrounding air (Figure 14-9). In photosynthesis carbon dioxide (CO_2) is taken in through the pores and oxygen (O_2) is released, in respiration O_2 is taken in and CO_2 is released, and in transpiration water vapor (H_2O) is released. Each pore is like a narrow tube, and gas molecules passing through the tube experience collisions with the walls of the tube as well as with each other. When the diameter of the tube is large compared to the mean free path of the gas molecules, gas molecules collide mostly with each other and only infrequently with the tube walls. As a result, gas can flow relatively easily through the tube. But if the tube diameter is about the same as the mean free path, collisions with the walls become more frequent and the rate of gas flow slows dramatically. Stomata take advantage of this: They are able to expand when conditions require increased gas flow (for example, when intense sunlight shines on the leaf and the rate of photosynthesis goes up).

Figure 14-9 **Stomata on the epidermis of a leaf** This photomicrograph shows fine details on the leaf of an elder tree (*Sambucus nigra*). The size of the stomata, through which the leaf exchanges gases with the atmosphere, is related to the mean free path of gas molecules.

GOT THE CONCEPT? 14-2 Comparing Gases

? A small fraction of the molecules in air are atoms of helium. A mole of helium has a mass of 4.00 g, and a helium atom has a radius of about 3×10^{-11} m. State whether each of the following quantities will be greater, less, or the same for helium than for oxygen (O_2, with a mass of 32.0 g/mol and a radius of about 2.0×10^{-10} m) in a given quantity of air: (a) the average translational kinetic energy per molecule; (b) the total energy of all kinds per molecule; (c) the rms speed; (d) the mean free path.

TAKE-HOME MESSAGE FOR Section 14-3

✔ The ideal gas law relates the pressure, volume, temperature, and number of molecules (or number of moles) for a low-density gas.

✔ The pressure that a gas exerts on the walls of its container is due to collisions that the gas molecules make with the walls.

✔ The temperature of a gas is a measure of the average translational kinetic energy per molecule. In a gas at Kelvin temperature T, this energy is $(3/2)kT$. This does not depend on the mass of the molecule. Molecules with more than one atom can have additional energy associated with rotation or vibration.

✔ The average distance that a molecule travels between collisions with other molecules is called the mean free path. The larger the molecule and the more densely molecules are packed, the shorter the mean free path.

14-4 Most substances expand when the temperature increases

Nearly all objects expand when heated and contract when cooled. For example, the lid on a jar of pickles may be too tight to unscrew when you first take the jar from the refrigerator, but when the jar warms up the lid expands and is easier to remove. This is known as **thermal expansion**. Thermal expansion is the basis of many thermometers, including those that use alcohol or mercury (see Section 14-2 and Figure 14-1a).

In a solid, thermal expansion happens because with increasing temperature the molecules that comprise the solid not only oscillate with greater amplitude around their equilibrium positions but also shift their equilibrium positions so that they are farther away from their neighbors. In a liquid or gas, increasing temperature means that molecules collide with each other with greater momentum; this pushes the molecules farther apart and increases the volume of the liquid or gas.

Linear Expansion

Experiment shows that if the temperature change ΔT of a solid object is not too great, the change in each dimension of the object is proportional to ΔT. If a solid object initially has length L_0, the change ΔL in its length when the temperature changes by ΔT is

Change in length due to a
temperature change
(14-18)

Change in length of an object   Length of the object before the temperature change

$$\Delta L = \alpha L_0 \Delta T$$

Coefficient of linear expansion of the substance of which the object is made   Temperature change of the object that causes the length change

The quantity α (Greek letter "alpha") in Equation 14-18 is called the **coefficient of linear expansion**. It depends on what the object is made of but not on the shape or size of the object. Table 14-1 lists the coefficient of linear expansion for a number of

TABLE 14-1 Coefficients of Linear Expansion			
Substance	Coefficient of linear expansion (α) $K^{-1} \times 10^{-6}$	Substance	Coefficient of linear expansion (α) $K^{-1} \times 10^{-6}$
aluminum	22.2	iron, cast	10.4
antimony	10.4	iron, forged	11.3
beryllium	11.5	iron, pure	12.0
brass	18.7	lead	28.0
brick	5.5	marble	12
bronze	18.0	plaster	25
carbon–diamond	1.2	platinum	9.0
cement	10.0	porcelain	3.0
concrete	14.5	quartz, fused	0.59
copper	16.5	rubber	77
glass, hard	5.9	silver	19.5
glass, plate	9.0	solder	24.0
glass, Pyrex	4.0	steel	13.0
gold	14.2	wood, oak parallel to grain	4.9
graphite	7.9		

substances. Notice in Equation 14-18 that α has dimensions of inverse temperature (often K^{-1}), so the dimensions of $\alpha L_0 \Delta T$ on the right side of the equation match those of ΔL on the left side.

WATCH OUT! The coefficient of linear expansion can have different units.

The units of α can be $(°C)^{-1}$ as well as K^{-1}. That's because the sizes of 1°C and 1 K are defined to be the same, so the temperature *change* ΔT has the same value whether the temperature is measured on the Celsius scale or the Kelvin scale.

Equation 14-18 and Table 14-1 explain why you can loosen a tight-fitting steel lid on a glass jar by running hot water over the lid. Initially the steel lid and the glass mouth of the jar have the same diameter L_0. The hot water makes both the steel and the glass undergo the same temperature change ΔT, but the steel lid expands more (ΔL is greater) because the coefficient of thermal expansion is greater for steel ($\alpha = 13.0 \times 10^{-6}$ K^{-1}) than for glass ($\alpha = 4 \times 10^{-6}$ to 9×10^{-6} K^{-1}). That makes the fit less snug and makes it easier to unscrew the lid.

Equation 14-18 should remind you of Equation 10-3 (Section 10-2), which says that the change ΔL in length of an object due to a tensile or compressive *stress* (force per unit area) is proportional to the initial length L_0 and the stress F/A:

$$\frac{F}{A} = Y \frac{\Delta L}{L_0} \quad \text{or} \quad \Delta L = \left(\frac{1}{Y}\right) L_0 \frac{F}{A}$$

 Go to Picture It 14-2 for more practice dealing with linear expansion.

We saw in Section 10-2 that this direct proportionality between stress and length change is valid if the stress is not too great. In the same fashion, the direct proportionality between ΔL and ΔT given by Equation 14-18 is valid if ΔT is not too great. (If the temperature change is too great, the direct proportionality between ΔL and ΔT breaks down, and we have to treat α as a function of temperature. The character of this function is different for different substances.)

EXAMPLE 14-4 An Expanding Bridge

Bridges are constructed with expansion joints (Figure 14-10) to allow the bridge to expand on hot days without buckling. Suppose the bridge is made in two halves, each of which is made of steel and is 10.0 m long when the temperature is 18.0°C. The left-hand end of the left half is held in place and cannot move, as is the right-hand end of the right half. How large a gap should there be between the two halves at 18.0°C so that the structure does not buckle on an exceptionally hot day when the temperature is 45.0°C?

Figure 14-10 A thermal expansion joint Bridges are built with thermal expansion joints such as this one to account for expansion and contraction of the bridge in hot and cold conditions.

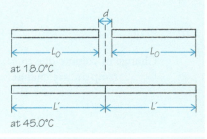

Set Up

At 18.0°C each half of the bridge has length $L_0 = 10.0$ m. We want to choose the gap d between them so that at 45.0°C the gap just closes. For this to happen each half must increase in length by a distance $d/2$ as the temperature increases. We use Equation 14-18 for the length increase of each half of the bridge. Table 14-1 gives the coefficient of thermal expansion α for steel.

Change in length due to a temperature change:

$$\Delta L = \alpha L_0 \Delta T \quad (14\text{-}18)$$

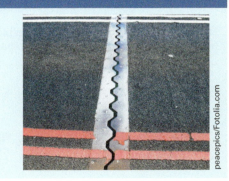

Solve

Write an expression for the length change of each bridge half when the temperature increases from 18.0°C to 45.0°C.

From Table 14-1

$\alpha = 13.0 \times 10^{-6}$ K^{-1} for steel

The temperature change is

$$\Delta T = 45.0°C - 18.0°C = 27.0 \text{ K}$$

(Recall that a temperature change of 1°C is the same as a temperature change of 1 K.)

Then from Equation 14-18 the length change of each bridge half is

$$\begin{aligned}\Delta L &= \alpha L_0 \Delta T \\ &= (13.0 \times 10^{-6} \text{ K}^{-1})(10.0 \text{ m})(27.0 \text{ K}) \\ &= 3.51 \times 10^{-3} \text{ m} = 3.51 \text{ mm}\end{aligned}$$

The length change ΔL of each bridge half is equal to $d/2$ (half the gap at 18.0°C), so d is equal to $2\Delta L$.

The necessary gap between the two bridge halves at 18.0°C is

$$\begin{aligned}d &= 2\Delta L = 2(3.51 \text{ mm}) \\ &= 7.02 \text{ mm} = 0.276 \text{ in.}\end{aligned}$$

Reflect

This gap is small enough that it offers no resistance to a car's tire or a pedestrian's foot. Without a sufficiently large expansion joint, the bridge halves in this problem would push against each other on a hot day and eventually warp or buckle as shown.

WATCH OUT! Holes in an object expand when heated.

Example 14-4 may lead you to conclude that when the temperature of an object increases, the substance of which it is made expands to fill any holes in the object. Not so! *Every* dimension of the object increases, which means the size of the hole also increases (Figure 14-11). That's what happens when you warm the metal lid on a jar to loosen it: The inner diameter of the lid (the size of the "hole" on the inside of the lid) increases so that it no longer fits so tightly on the mouth of the glass jar.

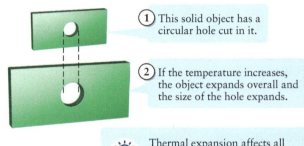

1. This solid object has a circular hole cut in it.

2. If the temperature increases, the object expands overall and the size of the hole expands.

Thermal expansion affects all dimensions of an object equally.

Figure 14-11 Heating a hole When a block with a hole is heated, the hole and the block both expand.

Volume Expansion

When the temperature of a solid object changes, its length, width, and height change according to the linear thermal expansion relationship of Equation 14-18. As a result, its *volume* expands as well. A fluid (a liquid or gas) has no fixed dimensions, however, so Equation 14-18 doesn't apply. But like solids, fluids change their volume when the temperature changes. An equation that describes the volume change with temperature for solids and fluids alike is

Change in volume due to a temperature change (14-19)

Change in volume of a quantity of substance

Volume of the substance before the temperature change

$$\Delta V = \beta V_0 \Delta T$$

Coefficient of volume expansion of the substance

Temperature change of the substance that causes the volume change

The **coefficient of volume expansion** β (Greek letter "beta") also has units of K^{-1} or $(°C)^{-1}$.

For solids it turns out that the coefficient of volume expansion is just three times the coefficient of linear expansion: $\beta = 3\alpha$. (The factor of 3 arises because volume expansion involves all three dimensions of space, while linear expansion involves just one dimension.) So you can find the value of β for solids by multiplying the values of α in Table 14-1 by 3.

Table 14-2 lists values of β for selected liquids. Note that for ethanol, the liquid used in most alcohol thermometers, $\beta = 750 \times 10^{-6}$ K^{-1}. Contrast this to the volume expansion coefficient for glass from Table 14-1: α is between 4.0×10^{-6} K^{-1} and 9.0×10^{-6} K^{-1} for glass, so $\beta = 3\alpha$ is between 12×10^{-6} K^{-1} and 27×10^{-6} K^{-1}. That explains how an alcohol thermometer works: When the temperature goes up the glass tube of the thermometer and the ethanol inside the tube both expand, but the ethanol expands much more (because it has such a large value of β) and so the ethanol rises within the tube. The same is true for a mercury thermometer, since mercury also has a large value of β (see Table 14-2).

We don't list the value of β for gases in Table 14-2. That's because it's most convenient to use Equation 14-7 (the ideal gas law, $pV = nRT$) rather than Equation 14-19 to keep track of the changes in the volume of a gas with temperature.

For nearly all substances β is positive, so the volume increases (the substance expands, and $\Delta V > 0$) when the temperature increases ($\Delta T > 0$), and the volume decreases (the substance contracts, and $\Delta V < 0$) when the temperature decreases ($\Delta T < 0$). But liquid water is a conspicuous exception to this rule at temperatures below 4°C. Below that temperature liquid water actually *expands* as the temperature decreases further (Figure 14-12). So water is less dense at 0°C (the freezing point) than at slightly warmer temperatures. As a result, as the water in rivers and lakes gets cold during the winter, the coldest water (closest to the freezing point) floats to the surface, while denser, warmer water sinks to the bottom. This means that the surface water freezes first and that rivers and lakes freeze from the surface down. Were water like most substances, the coldest water would be denser than warmer water and would sink to the bottom, and bodies of water would freeze from the bottom up. In this case rivers and lakes could freeze throughout their volume during extended periods of cold temperatures, possibly destroying aquatic life. Life on Earth might not have been able to survive if water did not exhibit the unusual property seen in Figure 14-12.

TABLE 14-2 Coefficients of Volume Expansion for Liquids	
Substance	Coefficient of volume expansion (β) K$^{-1} \times 10^{-6}$
gasoline	905
ethanol	750
water	207
mercury	182

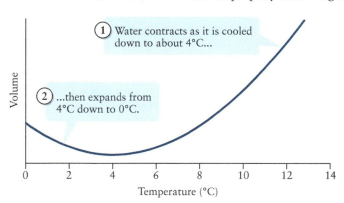

① Water contracts as it is cooled down to about 4°C...

② ...then expands from 4°C down to 0°C.

Volume

Temperature (°C)

Figure 14-12 The strange case of cold water Although most substances expand when heated and contract when cooled, water is different. As water is cooled below 4°C, it stops contracting and actually expands as the temperature decreases from 4°C to the freezing point.

Figure 14-12 shows another aspect of water that affects life on Earth: Above 4°C an increase in temperature causes liquid water to expand. The average global temperature on Earth has been increasing for decades due to the burning of fossil fuels such as coal and gasoline, which releases carbon dioxide (CO_2) in the atmosphere and causes a temperature increase. (We'll discuss the physics of this process in Section 14-7.) As a result, the water in the oceans is expanding, which causes the sea level to rise. At present the global average rate of rise is about 3 mm/y (millimeters per year), about half of which is due to thermal expansion of seawater; the rest is caused by the melting of land-based ice such as glaciers, causing water to run off into the ocean.

BioMedical

GOT THE CONCEPT? 14-3 Warming a Cube

A piece of solid aluminum is in the shape of a cube. If the temperature of the cube is increased so that the length of each side increases by 0.0010 (that is, by 0.10%), by how much does the volume of the cube increase? (a) 0.0010; (b) $2 \times 0.0010 = 0.0020$; (c) $3 \times 0.0010 = 0.0030$; (d) $(0.0010)^2 = 1.0 \times 10^{-6}$; (e) $(0.0010)^3 = 1.0 \times 10^{-9}$.

14-5 Heat is energy that flows due to a temperature difference

We've seen that the temperature of an object is a measure of the average kinetic energy of its molecules. We've also seen that if two objects at different temperatures are placed in contact, their temperatures eventually come to the same value. So there must be a *flow* of energy between objects at different temperatures. We use the term **heat** and the symbol Q for the energy that flows from one object to another as a result of a temperature difference. Since heat is a form of energy, it has units of joules (J). If heat flows into an object, the value of Q for that object is positive; if heat flows out of an object, the value of Q for that object is negative (Figure 14-13).

WATCH OUT! Objects do not contain heat.

! We emphasize that the term *heat* applies only to energy that is in the process of flowing between objects at different temperatures. We use a different term, **internal energy**, for the energy within an object due to the kinetic *and* potential energies associated with the individual molecules that comprise the object. Note also that *temperature* (a measure of the kinetic energy per molecule) and *heat* (energy flowing due to a temperature difference) are *not* the same quantity. The word "hot" doesn't mean "high heat" but rather "high temperature."

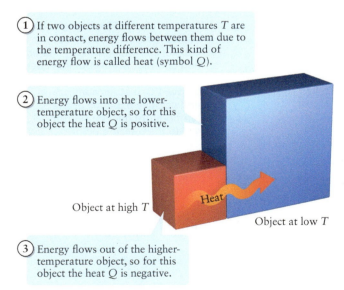

① If two objects at different temperatures T are in contact, energy flows between them due to the temperature difference. This kind of energy flow is called heat (symbol Q).

② Energy flows into the lower-temperature object, so for this object the heat Q is positive.

Object at high T Heat   Object at low T

③ Energy flows out of the higher-temperature object, so for this object the heat Q is negative.

Figure 14-13 Heat flow in thermal contact Energy in the form of heat (Q) flows from the higher-temperature object to the lower-temperature one until the two objects are in thermal equilibrium at the same temperature.

You can understand how heat flows between two objects in contact by considering what happens on the molecular level. Where the two objects touch, molecules on the surfaces of the two objects can collide with each other. The molecules in the higher-temperature object have more kinetic energy than those in the lower-temperature object. As a result, collisions between surface molecules in the two objects end up transferring kinetic energy from the molecules in the higher-temperature object to those in the lower-temperature object. Collisions between neighboring molecules in each object share these energy transfers among all of the object's molecules. The net result is that energy is transferred from the higher-temperature object to the lower-temperature one.

Experiment shows that in most circumstances the flow of heat into or out of an object changes the object's temperature. (The exception is when the object undergoes a *phase change* such as from solid to liquid, liquid to solid, liquid to gas, or gas to liquid. In the next section we'll discuss what happens in a phase change.) If the quantity of heat Q that flows into an object is relatively small, the resulting temperature change ΔT turns out to be *directly* proportional to Q and *inversely* proportional to the mass m of the object. In other words, the greater the quantity of heat Q that flows into an object, the more its temperature changes; the more massive the object and so the more material that makes up the object, the smaller the temperature change for a given quantity of heat Q. (The same quantity of heat that will cook a single meatball will cause hardly any temperature change in a pot roast.) We can write this relationship as

$$\Delta T = (\text{constant}) \times \frac{Q}{m}$$

The constant in this equation depends on the substance of which the object is made. It's conventional to express this equation as

> **Quantity of heat** that flows into (if $Q > 0$) or out of ($Q < 0$) an object

> **Specific heat** of the substance of which the object is made

$$Q = mc\,\Delta T$$

> **Temperature change** of the object that results from the heat flow

> **Mass** of the object

Quantity of heat and the resulting temperature change (14-20)

The quantity c is called the **specific heat** of the material that makes up the object. Its units are joules per kilogram per Kelvin (J/(kg·K) or J·kg^{-1}·K^{-1}). For example, the value of c for aluminum is 910 J/(kg·K); this means that 910 J of heat must flow into a 1-kg block of aluminum to raise its temperature by 1 K (or, equivalently, 1°C). Table 14-3 lists the values of specific heats for a range of substances. The value of the specific heat for any substance varies somewhat with the temperature; however, over the range of temperatures we typically experience, these variations are small enough that we'll ignore them.

Equation 14-20 says that if heat flows *into* an object, so $Q > 0$, it will cause the temperature to increase so $\Delta T > 0$. That's what happens when you warm a saucepan of water on a hot stove: Heat flows from the stove into the saucepan (so $Q > 0$ for the saucepan), and the saucepan's temperature increases (so $\Delta T > 0$ for the saucepan). If heat flows *out* of an object, so $Q < 0$, it will cause the temperature to decrease, so $\Delta T < 0$. This happens to the saucepan when you take it off the stove and allow it to cool: Heat flows from the saucepan to its surroundings (so $Q < 0$ for the saucepan), and the saucepan's temperature decreases (so $\Delta T < 0$ for the saucepan). Example 14-5 shows how to use Equation 14-20 to find the temperature change of an object due to a certain amount of heat flow.

A slightly more complicated application of Equation 14-20 is to find the final temperature of two objects that begin at different temperatures and are placed in contact (for example, hot coffee poured into a cold container). Example 14-6 shows how to solve this sort of problem. In problems of this kind we'll make the simplifying assumption that the two objects are *thermally isolated*; that is, they can exchange energy with each other but don't exchange energy with anything in their environment. So the amount of energy that flows out of the higher-temperature object is equal to the amount of energy that flows into the lower-temperature object.

> **Go to Picture It 14-3 for more practice dealing with specific heat.**

TABLE 14-3 Specific Heats

Substance	Specific heat (J·kg^{-1}·K^{-1})
air (50°C)	1046
aluminum	910
benzene	1750
copper	387
glass	840
gold	130
ice	(−10°C to 0°C) 2093
iron/steel	452
lead	128
marble	858
mercury	138
methyl alcohol	2549
silver	236
steam (100°C)	2009
water (0°C to 100°C)	4186
wood	1700

EXAMPLE 14-5 Camping Thermodynamics

A certain camping stove releases 5.00×10^4 J of energy per minute from burning propane. (This requires that it uses up the propane at a rate of about 1 g/min.) If half of the released energy is transferred to 2.00 kg of water in a pot above the flame, how much does the temperature of the water change in 1.00 min? (The other half of the energy released by the stove goes into warming the surrounding air and the material of the pot.)

Set Up

Energy flows from the stove into the water because of the temperature difference between them, so this is the kind of energy that we call heat. In 1 min heat Q equal to one-half of 5.00×10^4 J flows into the water. Our goal is to find the resulting temperature change ΔT of the water.

Quantity of heat and the resulting temperature change:

$$Q = mc\,\Delta T \qquad (14\text{-}20)$$

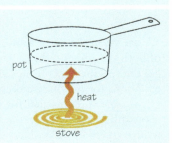

pot

heat

stove

Solve

We are given the values of Q and m, and Table 14-3 lists the value of c for water. Use this and Equation 14-20 to find the temperature change of water in 1.00 min.

Solve Equation 14-20 for the temperature change ΔT:

$$\Delta T = \frac{1}{mc} Q$$

The mass is $m = 2.00$ kg, $c = 4186$ J·kg^{-1}·K^{-1} for water from Table 14-3, and $Q = (1/2) \times (5.00 \times 10^4$ J). So

$$\Delta T = \frac{(1/2)(5.00 \times 10^4 \text{ J})}{(2.00 \text{ kg})(4186 \text{ J·kg}^{-1} \cdot \text{K}^{-1})} = 2.99 \text{ K} = 2.99°\text{C}$$

If the water starts at a temperature of 20.0°C, after 1.00 min its temperature will be 20.0°C + 2.99°C = 23.0°C. After another 1.00 min its temperature will be 23.0°C + 2.99°C = 26.0°C, and so on.

Reflect

If you've ever used a camping stove, you know that it takes quite a while to bring water to a boil. Using the numbers in this problem, we see it would take roughly 27 min to increase the temperature of 2.00 kg of water (corresponding to a volume of 2.00 L) from 20°C to 100°C. That's because water has a very large specific heat, so it takes a lot of heat to raise its temperature by a given amount.

EXAMPLE 14-6 Cooling Coffee

Your expensive coffee maker produces coffee at a temperature of 95.0°C (203°F), which you find is a bit too warm to drink comfortably. To cool the coffee you pour 0.350 kg of brewed coffee (mostly water) at 95.0°C into a 0.250-kg aluminum cup that is initially at room temperature (20.0°C). What is the final temperature of the coffee and cup? Assume that the coffee and cup are thermally isolated.

Set Up

There are three unknown quantities in this problem: the quantity of heat Q_{cup} that flows into the cup, the quantity of heat Q_{coffee} that flows into the coffee (which will be negative since heat flows *out* of the coffee), and the final temperature T_f of the two objects (which is what we want to find). So we need three equations that relate these quantities.

We can get two equations by writing Equation 14-20 twice, once for the cup and once for the coffee. To get a third equation we'll use the idea that the two objects are thermally isolated, so the energy that flows *out* of the hot coffee must equal the energy that flows *into* the cool aluminum cup. So Q_{coffee} and Q_{cup} both involve the same number of joules. However, Q is negative for the coffee (heat flows out of it) and positive for the cup (heat flows into it), so Q_{coffee} is equal to the negative of Q_{cup}.

Quantity of heat and the resulting temperature change:

$$Q = mc\,\Delta T \qquad (14\text{-}20)$$

Energy is conserved:

$$Q_{coffee} = -Q_{cup}$$

coffee
95.0°C
0.350 kg

cup
20.0°C
0.250 kg

coffee in cup;
temperature
$T_f = ?$

Solve

We are given the masses and initial temperatures of both the cup and the coffee, and we can get the specific heats from Table 14-3. Write equations for the three unknowns Q_{cup}, Q_{coffee}, and T_f. Note that for each object ΔT is the difference between the final and initial temperatures for that object.

The cup has mass $m_{cup} = 0.250$ kg and initial temperature $T_{cup,i} = 20.0°C$, is made of aluminum with $c_{Al} = 910$ J·kg^{-1}·K^{-1}, and ends up at temperature T_f. Equation 14-20 for the cup says

$$Q_{cup} = m_{cup}c_{Al}\,\Delta T_{cup} = m_{cup}c_{Al}(T_f - T_{cup,i})$$

The coffee has mass $m_{coffee} = 0.350$ kg and initial temperature $T_{coffee,i} = 95.0°C$, is made almost completely of water with

$c_{\text{water}} = 4186 \text{ J} \cdot \text{kg}^{-1} \cdot \text{K}^{-1}$, and ends up at the same final temperature T_f as the cup. Equation 14-20 for the coffee says

$$Q_{\text{coffee}} = m_{\text{coffee}} c_{\text{water}} \Delta T_{\text{coffee}} = m_{\text{coffee}} c_{\text{water}} (T_f - T_{\text{coffee,i}})$$

The equation of energy conservation is

$$Q_{\text{coffee}} = -Q_{\text{cup}}$$

Combine these three equations to get a single equation for T_f, the quantity we are trying to find.

Substitute Q_{cup} and Q_{coffee} from the first two equations into the third equation for energy conservation:

$$Q_{\text{coffee}} = -Q_{\text{cup}}$$

$$m_{\text{coffee}} c_{\text{water}} (T_f - T_{\text{coffee,i}}) = -m_{\text{cup}} c_{\text{Al}} (T_f - T_{\text{cup,i}})$$

The only unknown quantity in this equation is the final temperature T_f.

Solve for the final temperature T_f. Recall that a temperature change of 1°C is equivalent to a temperature change of 1 K, so we can use kelvins and degrees Celsius interchangeably in our calculation.

Multiply out both sides of the equation:

$$m_{\text{coffee}} c_{\text{water}} T_f - m_{\text{coffee}} c_{\text{water}} T_{\text{coffee,i}} = -m_{\text{cup}} c_{\text{Al}} T_f + m_{\text{cup}} c_{\text{Al}} T_{\text{cup,i}}$$

Rearrange so that all the terms with T_f are on the same side of the equation:

$$m_{\text{coffee}} c_{\text{water}} T_f + m_{\text{cup}} c_{\text{Al}} T_f = m_{\text{coffee}} c_{\text{water}} T_{\text{coffee,i}} + m_{\text{cup}} c_{\text{Al}} T_{\text{cup,i}}$$

or

$$(m_{\text{coffee}} c_{\text{water}} + m_{\text{cup}} c_{\text{Al}}) T_f = m_{\text{coffee}} c_{\text{water}} T_{\text{coffee,i}} + m_{\text{cup}} c_{\text{Al}} T_{\text{cup,i}}$$

Solve for T_f:

$$T_f = \frac{m_{\text{coffee}} c_{\text{water}} T_{\text{coffee,i}} + m_{\text{cup}} c_{\text{Al}} T_{\text{cup,i}}}{m_{\text{coffee}} c_{\text{water}} + m_{\text{cup}} c_{\text{Al}}}$$

$$= \frac{\left[\begin{array}{l} (0.350 \text{ kg})(4186 \text{ J} \cdot \text{kg}^{-1} \cdot \text{K}^{-1})(95.0°\text{C}) \\ \qquad + (0.250 \text{ kg})(910 \text{ J} \cdot \text{kg}^{-1} \cdot \text{K}^{-1})(20.0°\text{C}) \end{array} \right]}{(0.350 \text{ kg})(4186 \text{ J} \cdot \text{kg}^{-1} \cdot \text{K}^{-1}) + (0.250 \text{ kg})(910 \text{ J} \cdot \text{kg}^{-1} \cdot \text{K}^{-1})}$$

$$= 84.9°\text{C}$$

The coffee cools from 95.0°C to 84.9°C, and the aluminum cup warms from 20.0°C to 84.9°C.

Reflect

The temperature of the coffee decreases and the temperature of the cup increases, just as we expected. We can check our results by calculating Q_{cup} and Q_{coffee}. This calculation shows that $Q_{\text{coffee}} = -Q_{\text{cup}}$, which must be true for energy to be conserved.

Note that the final temperature of 84.9°C is closer to the initial temperature of the coffee (95.0°C) than to the initial temperature of the cup (20.0°C). That's because the mass and the specific heat are both greater for the coffee than for the cup, so a given quantity of heat produces a smaller temperature change in the coffee than in the cup.

The temperature change for the cup is

$$\Delta T_{\text{cup}} = T_f - T_{\text{cup,i}} = 84.9°\text{C} - 20.0°\text{C} = +64.9°\text{C} = +64.9 \text{ K}$$

(Recall that 1°C and 1 K represent the same temperature change.) The heat that flows into the cup is

$$\begin{aligned} Q_{\text{cup}} &= m_{\text{cup}} c_{\text{Al}} (T_f - T_{\text{cup,i}}) \\ &= (0.250 \text{ kg})(910 \text{ J} \cdot \text{kg}^{-1} \cdot \text{K}^{-1})(+64.9 \text{ K}) \\ &= 1.48 \times 10^4 \text{ J} \end{aligned}$$

The temperature change for the coffee is

$$\Delta T_{\text{coffee}} = T_f - T_{\text{coffee,i}} = 84.9°\text{C} - 95.0°\text{C} = -10.1°\text{C} = -10.1 \text{ K}$$

The heat that flows into the coffee is

$$\begin{aligned} Q_{\text{coffee}} &= m_{\text{coffee}} c_{\text{water}} \Delta T_{\text{coffee}} \\ &= (0.350 \text{ kg})(4186 \text{ J} \cdot \text{kg}^{-1} \cdot \text{K}^{-1})(-10.1 \text{ K}) \\ &= -1.48 \times 10^4 \text{ J} \end{aligned}$$

This is negative because heat flows *out* of the coffee.

David Tauck

Figure 14-14 Counting calories and kilojoules The label on a package of flavored sugar from Finland lists the energy content in both kilojoules (kJ) and food calories (1 food calorie = 1 kilocalorie = 1 kcal).

In Examples 14-5 and 14-6 we used the SI unit of heat, the joule. However, other units are also commonly used. The **calorie** (cal), equal to 4.186 J, is defined as the quantity of heat required to increase the temperature of 1 gram (1 g) of pure water from 14.5°C to 15.5°C. So in terms of calories, the specific heat of water at 14.5°C is $c = 1 \text{ cal} \cdot \text{g}^{-1} \cdot \text{K}^{-1}$. The energy content of foods is given in *food* calories, often denoted by a capital "C," equal to 1000 calories or 1 kilocalorie: 1 C = 1 kcal = 4186 J = 4.186 kJ. In countries other than the United States, the energy content on food labels is given in units of kilojoules (Figure 14-14).

The unit of heat in the English system is the **British thermal unit** (BTU), defined as the quantity of heat required to increase the temperature of 1 lb of pure water from 63°F to 64°F. The energy flow of air conditioners and heaters is often given in BTUs. One BTU is equal to 1055 J, 252 cal, or 0.252 kcal.

GOT THE CONCEPT? 14-4 Final Temperature I

(?) Two objects that are made of the same substance are placed in thermal contact. Object A is far more massive than object B. Will the final temperature of the two objects be (a) close to the initial temperature of object A, (b) close to the initial temperature of object B, or (c) about midway between the two initial temperatures?

GOT THE CONCEPT? 14-5 Final Temperature II

(?) Two objects that have the same mass are placed in thermal contact. Object A is made of a substance that has a much larger specific heat than the substance from which object B is made. Will the final temperature of the two objects be (a) close to the initial temperature of object A, (b) close to the initial temperature of object B, or (c) about midway between the two initial temperatures?

TAKE-HOME MESSAGE FOR Section 14-5

✔ Heat is energy that flows from one object to another as a result of a temperature difference. Heat Q is positive if energy flows into an object and negative if it flows out.

✔ The specific heat c of a substance is the amount of heat (in J) required to raise the temperature of 1 kg of the substance by 1 K.

14-6 Energy must enter or leave an object for it to change phase

If you add energy to an object, will its temperature increase? The best answer is "Maybe." Suppose you take a cup of ice at –20°C (Figure 14-15a) and let heat flow into it at a constant rate. The temperature of the ice will indeed increase up to its melting point of 0°C (Figure 14-15b). But as you continue to let heat flow into it, the temperature of the ice will *remain* at 0°C as the ice melts (Figure 14-15c). Once all of the ice has melted so that only liquid water remains (Figure 14-15d), the inflow of heat again causes the temperature to increase (Figure 14-15e).

The solid, liquid, and gaseous states of water are called its **phases**, and when ice melts (as happens between the stages in Figure 14-15b and Figure 14-15d) the water undergoes a **phase change**. Energy must be either absorbed or released for a substance to change from one phase to another, but the temperature of the substance stays the same during a phase change. For example, in boiling water at 100°C the energy that the water absorbs goes entirely into rearranging the organization of the water molecules to effect the phase change, with none left over to raise the temperature. You can verify this by putting a thermometer in a pot of water and heating the pot on a stove. The water temperature increases until it reaches 100°C and the water begins to boil. As the water boils away the temperature will remain at 100°C; at this temperature all of the heat flowing into the water goes into changing the phase of the water.

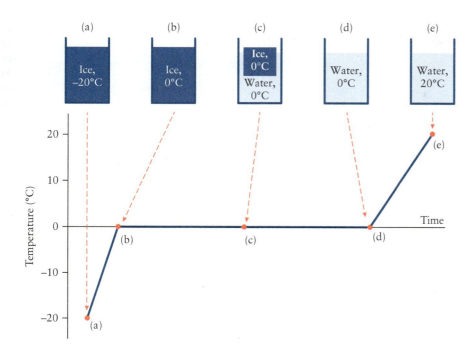

Figure 14-15 A change in phase We allow heat to flow at a constant rate into ice initially at a temperature below 0°C. The temperature increases until the melting point of 0°C, then remains constant until the solid ice has all undergone a phase change to liquid water. Only then will the temperature increase above 0°C. If you reverse the process and allow heat to flow out of liquid water at a constant rate, it will go through the same stages in reverse order.

You're familiar with several kinds of phase change. When a liquid becomes a solid, the process is called **freezing**; when a solid becomes a liquid, the process is called melting or **fusion**. Other common phase changes include from liquid to gas, or **vaporization**, and from gas to liquid, or **condensation**. Some substances can change from solid to gas without an intermediate liquid phase. A common example is carbon dioxide (CO_2); at room temperature atmospheric pressure CO_2 goes from a solid form, commonly known as *dry ice*, directly to CO_2 gas. A phase change from solid directly to gas is called **sublimation**; the reverse process, from gas directly to solid, is called **deposition**.

For a given substance at a given pressure, there is a specific temperature at which a phase change occurs. For example, at 1 atm of pressure water can change between its solid and liquid forms at 0°C. This is the temperature at which these two phases can *coexist*. The stages between Figure 14-15b and Figure 14-15d show solid ice and liquid water coexisting at 0°C as **heat** flows in.

Latent Heat

Suppose a quantity of substance is at the temperature at which a phase change can occur. The amount of heat per unit mass that must flow into or out of the substance to cause the phase change is called the **latent heat**. For solid ice at 0°C the latent heat is the amount of heat per kilogram that must flow into the ice to melt it; for liquid water at 0°C this same latent heat is the amount of heat per kilogram that must flow *out* of the water to freeze it. The units of latent heat are joules per kilogram (J/kg).

The value of the latent heat depends on the substance as well as on the kind of phase change. For example, 2.47×10^5 J of heat must flow into a 1-kg block of iron to melt it at its melting temperature of 1811 K, but even more heat (3.34×10^5 J) must flow into a 1-kg block of ice to melt it at its melting temperature of 273 K. We say that the **latent heat of fusion** (that is, melting) is $L_F = 2.47 \times 10^5$ J/kg for iron and $L_F = 3.34 \times 10^5$ J/kg for water. The **latent heat of vaporization** for water is $L_V = 2.26 \times 10^6$ J/kg, so 2.26×10^6 J of heat must flow into 1 kg of liquid water at its vaporization temperature of 100°C to vaporize it. The values of L_F and L_V for water show that it takes about seven times more energy to vaporize a kilogram of water at 100°C than it does to melt a kilogram of ice at 0°C.

Note that the latent heat of fusion is also the energy *released* per unit mass when a substance changes from a liquid to a solid, and the latent heat of vaporization is the energy *released* per unit mass when a substance changes from gas to liquid. Table 14-4 lists the latent heat of fusion and the latent heat of vaporization for various substances.

TABLE 14-4 **Melting and Boiling Temperatures (T_F and T_V) and Latent Heats of Fusion and Vaporization (L_F and L_V) for Various Substances at 1 atm of Pressure**				
Substance	**T_F (K)**	**T_V (K)**	**L_F(J/kg) $\times 10^3$**	**L_V(J/kg) $\times 10^3$**
alcohol (ethyl)	159	351	104	830
aluminum	933	2792	397	10,900
copper	1357	2835	209	4730
gold	1337	3129	63.7	1645
hydrogen (H_2)	14	20	59.5	445
iron	1811	3134	247	6090
lead	600	2022	24.5	866
nitrogen (N_2)	63	77	25.3	199
oxygen (O_2)	54	90	13.7	213
water	273	373	334	2260

The amount of energy needed to cause a phase change to occur is proportional to the mass m of the substance. The more massive a block of ice or iron, for example, the more energy is required to melt it. We can summarize these relationships in a single equation:

Heat required for a phase change
(14-21)

Quantity of heat that flows into or out of a quantity of substance to cause a phase change

Use + sign if heat flows into the substance; use – sign if heat flows out of the substance.

$$Q = \pm mL$$

Latent heat for the phase change

Mass of the substance

Equation 14-21 applies to all phase changes. For example, if the phase change is fusion (a solid becomes a liquid), then L is the latent heat of fusion L_F and we use the plus sign so Q is positive (heat flows into the substance to make it melt). If the phase change is condensation (a gas becomes a liquid), L is the latent heat of vaporization L_V and we use the minus sign so Q is negative (heat flows out of the substance as it condenses).

WATCH OUT! Equation 14-21 describes what happens in the phase change, not what it takes to get there.

Remember that a phase change can happen only if the substance is at the appropriate temperature (for example, it must be at the vaporization temperature for vaporization or condensation to take place). Equation 14-21 does *not* include the heat that must flow into or out of a substance to get the substance to that temperature. To calculate that, you must use Equation 14-20, $Q = mc\Delta T$. So if you want to know how much heat would be needed to take a quantity of ice at $-20°C$ and melt it to liquid water at $20°C$ (Figure 14-15), you would (1) use Equation 14-20 to find the amount of heat needed to raise the ice temperature from $-20°C$ to $20°C$, (2) use Equation 14-21 to find the amount of heat needed to melt that ice at $0°C$, and (3) use Equation 14-20 again to find the amount of heat to raise the water temperature from $0°C$ to $20°C$.

BioMedical

Equation 14-21 says that energy in the form of heat must flow into a substance to cause a phase change from solid to liquid or from liquid to gas. Often the source of that energy is a second object in thermal contact with the first. For example, as the sweat on our skin evaporates on a hot day, energy flows from our bodies to the beads of perspiration. There is heat flow *into* the sweat, so $Q_{sweat} > 0$, and heat flow *out of* the body, so $Q_{body} < 0$. Energy conservation says that the number of joules of heat that flows out of the body equals the number of joules that flows into the sweat, so $Q_{body} = -Q_{sweat}$. The energy lost from the body to the sweat causes a decrease in the temperature of the body (which, unlike the sweat, does not undergo a phase change). In other words, perspiration helps us to cool off. A typical person perspires about 0.60 kg of sweat per day, and sweat has a latent heat of vaporization of about 2.43×10^6 J/kg. From Equation 14-21 the heat that flows into one day's worth of sweat to make it evaporate is

$$Q_{sweat} = mL_V = (0.60 \text{ kg})(2.43 \times 10^6 \text{ J/kg}) = 1.5 \times 10^6 \text{ J}$$

The heat flow into the body is $Q_{body} = -Q_{sweat} = -1.5 \times 10^6$ J to two significant figures, so the body *loses* 1.5×10^6 J of energy by sweating. Recall from Section 14-5 that one food calorie (1 kcal) equals 4186 J, so the amount of energy that the body loses by sweating is $(1.5 \times 10^6 \text{ J})(1 \text{ kcal}/4186 \text{ J}) = 350$ kcal. So you must consume at least 350 kcal daily to compensate for the energy lost by sweating. (You also need to drink at least 0.60 kg of water, with a volume of 0.60 L, to compensate for the water lost.)

We can apply the same approach to what happens when an ice cube is dropped into a glass of water. Heat is transferred from the water (so $Q_{water} < 0$) into the ice (so $Q_{ice} > 0$) in order for the ice to melt: The number of joules of energy that leaves the water equals the number of joules that enters the ice, so $Q_{water} = -Q_{ice}$. The temperature of the water decreases, and if the water loses enough energy, some or all of it could even freeze.

The following examples illustrate how to solve some typical problems that involve phase changes.

 Go to Interactive Exercise 14-2 for more practice dealing with phase changes.

EXAMPLE 14-7 Melting Ice I

A 1.00-kg block of ice initially at a temperature of 0°C is placed inside an experimental apparatus that allows 250 kJ of heat to flow into the ice. What mass of ice melts as a result? What is the final temperature of the melted ice?

Set Up

The ice starts at the melting temperature, so no heat has to be added to the ice to get it to that temperature. Heat must be added to melt the ice, so $Q_{ice} > 0$. From Table 14-4 the latent heat of fusion for water is $L_F = 334$ kJ/kg $= 334 \times 10^3$ J/kg, so it would take 334 kJ of heat to completely melt 1.00 kg of ice. The heat supplied is less than that, so only part of the ice will melt. We'll use Equation 14-21 to determine the mass that melts.

Heat required to melt a mass m of ice:

$$Q_{ice} = +mL_F \qquad (14\text{-}21)$$

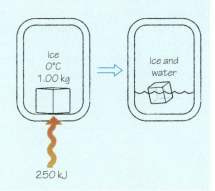

Solve

Find the mass of ice that melts.

From Equation 14-21 the mass that melts is

$$m = \frac{Q_{ice}}{L_F} = \frac{250 \text{ kJ}}{334 \text{ kJ/kg}} = 0.75 \text{ kg}$$

Three-quarters of the 1.00 kg of ice melts. Because all of the heat goes into melting the ice, we end up with solid ice and liquid water that are both at the initial temperature of 0°C.

Reflect

If 334 kJ of heat were transferred to the block of ice, it would completely melt. Any more heat added would go into raising the temperature of the resulting liquid water to a value greater than 0°C.

EXAMPLE 14-8 **Melting Ice II**

A chunk of ice at −10.0°C is placed in an insulated container that holds 1.00 kg of water at 20.0°C. When the system comes to thermal equilibrium, all of the ice has melted and the entire system is at a temperature of 0.00°C. What was the initial mass of the ice?

Set Up

Three processes occur as the system reaches thermal equilibrium: (a) The temperature of the water decreases from 20.0°C to 0.0°C, so heat flows out of the water; (b) the temperature of the ice increases from −10.0°C to 0.0°C, so heat flows into the ice; and (c) the ice melts at 0°C, so additional heat flows into the ice. Equation 14-20 describes the heat flows associated with the temperature changes, and Equation 14-21 describes the heat flow associated with melting the ice. Energy conservation says that the heat that flows into the ice equals the heat that flows out of the water. Our goal is to find the initial mass of ice, m_{ice}.

Quantity of heat and the resulting temperature change:

$$Q = mc \, \Delta T \qquad (14\text{-}20)$$

Heat required to melt a mass m of ice:

$$Q_{ice} = +mL_F \qquad (14\text{-}21)$$

Energy is conserved:

$$Q_{ice} = -Q_{water}$$

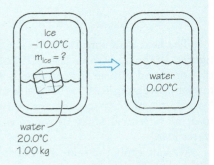

ice
−10.0°C
$m_{ice} = ?$

water

water
20.0°C
1.00 kg

water
0.00°C

Solve

Write Equation 14-20 for the water and for the ice, and write Equation 14-21 for the ice. The specific heats for water and ice are given in Table 14-3, and the latent heat of fusion is given in Table 14-4.

Q_{water} is negative: Heat flows *out* of the water to make its temperature decrease.

Q_{ice} is positive: Both processes require that heat flows *into* the ice. Note that the term that involves melting the ice is about 16 times larger than the term that involves raising the temperature of the ice from −10.0°C to 0.0°C.

(We again use the idea that a temperature change in degrees Celsius is equivalent to a temperature change in kelvins.)

The heat flow into the water that is required to lower its temperature from 20.0°C to 0.0°C:

$$
\begin{aligned}
Q_{water} &= m_{water} c_{water} \Delta T_{water} \\
&= (1.00 \text{ kg})(4186 \text{ J} \cdot \text{kg}^{-1} \cdot \text{K}^{-1})(0.0°C - 20.0°C) \\
&= (1.00 \text{ kg})(4186 \text{ J} \cdot \text{kg}^{-1} \cdot \text{K}^{-1})(-20.0 \text{ K}) \\
&= -8.37 \times 10^4 \text{ J}
\end{aligned}
$$

The heat flow into the ice is the sum of the heat required to raise its temperature by −10.0°C to 0.0°C and the heat required to melt it at 0.0°C:

$$
\begin{aligned}
Q_{ice} &= m_{ice} c_{ice} \Delta T_{ice} + m_{ice} L_F = m_{ice}(c_{ice} \Delta T_{ice} + L_F) \\
&= m_{ice}[(2093 \text{ J} \cdot \text{kg}^{-1} \cdot \text{K}^{-1})[0.0°C - (-10.0°C)] \\
&\quad + 3.34 \times 10^5 \text{ J/kg}] \\
&= m_{ice}(2.093 \times 10^4 \text{ J/kg} + 3.34 \times 10^5 \text{ J/kg}) \\
&= m_{ice}(3.55 \times 10^5 \text{ J/kg})
\end{aligned}
$$

Use energy conservation to relate Q_{water} and Q_{ice}, then solve for m_{ice}.

The energy flow into the ice equals the negative of the (negative) energy flow into the water:

$$Q_{ice} = -Q_{water}$$
$$m_{ice}(3.55 \times 10^5 \text{ J/kg}) = -(-8.37 \times 10^4 \text{ J})$$

Solve for m_{ice}:

$$m_{ice} = \frac{8.37 \times 10^4 \text{ J}}{3.55 \times 10^5 \text{ J/kg}} = 0.236 \text{ kg}$$

Reflect

The amount of ice required has a mass of 0.236 kg. In more familiar terms, 1 kg is the mass of 1 L of water, and an ice cube from a typical ice cube tray has a mass of about 0.04 kg (40 g). So it would take about six ice cubes to reproduce the process described in this problem, which doesn't seem unreasonable.

Note that if the container that holds the ice and water were *not* insulated, we would also have to worry about heat flow between the water–ice mixture and the container. Using an insulated container allows us to avoid this complication.

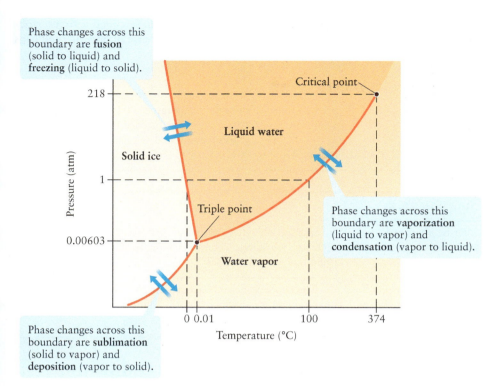

Phase changes across this boundary are **fusion** (solid to liquid) and **freezing** (liquid to solid).

Critical point

218

Liquid water

Solid ice

Pressure (atm)

1

Triple point

0.00603

Phase changes across this boundary are **vaporization** (liquid to vapor) and **condensation** (vapor to liquid).

Water vapor

0 0.01   100   374

Temperature (°C)

Phase changes across this boundary are **sublimation** (solid to vapor) and **deposition** (vapor to solid).

Figure 14-16 **The phase diagram for water** A phase diagram shows the relationship between pressure, temperature, and the phase of the substance. Each of the three red curves in the figure marks a boundary between two phases and shows the temperature at which a phase change is possible as a function of pressure.

Phase Diagrams

A useful tool for visualizing the possible phase changes for a particular substance is a **phase diagram**. This is a graph of pressure versus temperature for the substance. For example, Figure 14-16 is the phase diagram for water. The three red curves in the figure show the boundaries in terms of pressure and temperature between phases. Each point on one of the red curves represents a combination of pressure and temperature at which two phases can coexist and a phase change can take place. For example, at a pressure of 1 atm solid ice and liquid water can coexist at 0°C; this is the melting temperature of ice at that pressure. Similarly, at 1 atm pressure liquid water and water vapor can coexist at 100°C, the temperature at which water boils and vaporizes. The phase diagram shows that at different pressures the phase changes occur at different temperatures. For example, if the pressure is less than 1 atm, the boundary between liquid water and water vapor is at a temperature less than 100°C. That's why water boils at only 95°C at the elevation of Denver or Albuquerque (1610 m or 5280 ft), where air pressure is only about 0.82 atm. Due to the reduced boiling temperature, foods prepared in boiling water at such elevations have to be cooked longer than they would be at sea level.

In Section 14-2 we introduced the idea of the *triple point* of a substance, which is the particular combination of pressure and temperature at which the solid, liquid, and gas phases can all coexist. In Figure 14-16 the triple point is where the three red curves meet. If the pressure is greater than the triple-point pressure, the substance can exist in the solid, liquid, or gas (vapor) phase depending on the temperature. That's the case for water at sea level on Earth: The triple point of water is 0.00603 atm and 0.01°C, so at normal atmospheric pressure (1 atm) water can exist in any of the three phases. By contrast, the triple point of carbon dioxide (CO_2) is 5.10 atm and −56.6°C. Since atmospheric pressure is less than its triple-point pressure, CO_2 can exist in only the solid and gas phases. That's why solid CO_2 sublimates rather than melting at atmospheric pressure and room temperature (Figure 14-17).

Figure 14-16 shows that the red curve that forms the boundary between the liquid and gas (vapor) phases comes to an end at a certain point, which for water occurs at 218 atm and 374°C. This special combination of pressure and temperature is called the **critical point**. If the pressure is greater than the critical-point pressure, the temperature is greater than the critical-point temperature, or both, there is no sharp dividing line

Figure 14-17 **When air pressure is less than the triple-point pressure** Carbon dioxide (CO_2) is a solid at atmospheric pressure (1 atm) and temperatures below −78.5°C. The solid is also called dry ice because it does not melt at room temperature; instead, because atmospheric pressure is less than its triple-point pressure of 5.10 atm, CO_2 sublimates and goes directly to the vapor phase. The escaping cold CO_2 vapor is invisible, but due to its low temperature it causes a fog of water droplets to form in the air.

between gas and liquid: The substance exists in a single phase called a *supercritical fluid*. Above the critical temperature, unlike below it, no increase in pressure will cause a gas to condense into liquid form.

GOT THE CONCEPT? 14-6 Ranking Energies for Water

(?) Use the information in Tables 14-3 and 14-4 to rank the following from largest to smallest: (a) The energy needed to raise the temperature of 1 kg of ice from −5°C to −4°C; (b) the energy needed to melt 1 kg of ice at 0°C; (c) the energy needed to raise the temperature of 1 kg of liquid water from 20°C to 21°C; (d) the energy needed to boil 1 kg of water at 100°C; and (e) the energy needed to raise the temperature of 1 kg of steam from 110°C to 111°C.

TAKE-HOME MESSAGE FOR Section 14-6

✔ Energy in the form of heat must either be absorbed or released for a substance to change from one phase to another.

✔ The latent heat of fusion is the amount of heat per unit mass required to cause a substance to undergo a phase change between the solid and liquid phases. Similarly, the latent heat

of vaporization is the amount of heat per unit mass required to cause a substance to undergo a phase change between the liquid and gaseous phases.

✔ The triple point is the particular combination of pressure and temperature at which the solid, liquid, and gas phases of a substance can coexist.

(a)

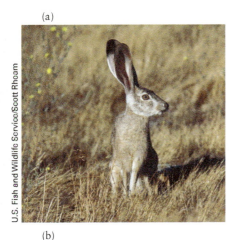

(b)

Figure 14-18 Controlling heat transfer Both (a) jackrabbits and (b) arctic hares belong to the same genus (*Lepus*). However, jackrabbits evolved to radiate heat in the hot desert environment, while arctic hares evolved to conserve heat in the cold conditions of the Arctic.

14-7 Heat can be transferred by radiation, convection, or conduction

BioMedical Animals have evolved to live in a broad range of climates. The jackrabbit in Figure 14-18a has a scrawny body, long skinny legs, enormous ears, and thin fur, all of which promote heat loss and enable the animal to survive in the desert. In contrast, the arctic hare's compact body, stubby legs, relatively small ears, and thick insulating fur help prevent heat loss (Figure 14-18b).

The iguanas shown in Figure 14-19 use different strategies for controlling their body temperature: After a cold swim in the Pacific Ocean, these reptiles warm themselves by lying in the tropical sun atop rocks that have already been heated by the Sun. In this section we'll look at how heat is transferred from one place to another, such as from a jackrabbit to its environment or from a hot rock to a cold iguana.

Heat can be transferred from one object to another through three processes: *radiation, convection,* and *conduction.*

Radiation is energy transfer by the emission (or absorption) of electromagnetic waves. Some examples of electromagnetic waves are visible light, infrared radiation, and microwaves. (We'll see in Chapter 23 that all of these are different manifestations of the same physical phenomenon.) You use radiation when you lie out on a sunny day or warm yourself under a heat lamp. The iguanas in Figure 14-19 are using radiation to gather energy directly from the sun.

Convection is energy transfer by the motion of a liquid or gas (such as air). Convection is what happens in a pot of water boiling on the stove: Warm water from the bottom of the pot rises, carrying energy to the cooler water at the top of the pot. Convection carries energy away from the arctic hare in Figure 14-18b and the iguanas in Figure 14-19, as the air warmed by the animal's body rises away from it.

Conduction is energy transfer by the collision of particles in one object with the particles in another. It requires physical contact between the two objects. If you use an electric blanket to keep warm on a cold winter's night, the energy flows from the blanket into your body by conduction. The iguanas in Figure 14-19 also use conduction to take in energy from the hot rock on which they are lying.

Let's look at each of these processes of heat transfer individually.

Radiation

Both convection and conduction rely on molecules bumping against one another. But because radiation carries energy in the form of electromagnetic waves, it can travel through a vacuum. The energy of radiation emitted by the Sun reaches Earth after passing through 150 million kilometers of almost completely empty space.

Experiment shows that *any* object emits energy in the form of radiation. The rate at which radiation is emitted by an object—that is, the radiated *power P* in joules per second or watts—is proportional to the object's surface area *A* and to the fourth power of the Kelvin temperature *T* of the object:

Rate at which an object emits energy in the form of radiation

Emissivity of the object (a number between 0 and 1)

$$P = e\sigma AT^4$$

Temperature of the object on the Kelvin scale

Stefan–Boltzmann constant = 5.6704×10^{-8} W·m^{-2}·K^{-4}

Surface area of the object

Rate of energy flow in radiation (14-22)

The **Stefan–Boltzmann constant** σ (Greek letter "sigma") has the same value for all objects. The quantity *e* is the **emissivity** of the surface; its value indicates how well or how poorly a surface radiates. A surface with a value of *e* close to 1 is a good radiator of thermal energy.

The factor of T^4 in Equation 14-22 means that the radiated power can change substantially with a change in temperature. For example, a kiln for making ceramic pots can easily reach temperatures of 600 K; a pot at this temperature will radiate energy at $2^4 = 16$ times the rate as when the pot is at room temperature of about 300 K. The red glow from the heating element in an electric toaster is an example of radiation of this kind. The factor *A* in Equation 14-22 shows that for a given surface temperature *T*, an object will radiate energy at a greater rate if it has a larger surface area. That's why it's useful for the jackrabbit in Figure 14-18a to have a lanky frame and big ears; this maximizes its surface area and makes it easier for the jackrabbit to get rid of excess energy in the hot desert. The more compact arctic hare in Figure 14-18b has much less surface area for its volume, so it radiates away less of its internal energy to its frigid surroundings.

Just as for jackrabbits and arctic hares, radiation plays a significant role in heat loss from the human body. Under typical conditions approximately half of the energy transferred from the body to the environment is in the form of radiation. This radiation is emitted almost entirely in the form of infrared light to which your eye is not sensitive, so you can't see yourself glowing in the dark. But specialized devices can detect this radiation (Figure 14-20). One such device now in common use in medicine is the *temporal artery thermometer*. The nurse runs the thermometer over your head in the region of the temporal artery, and the circuitry in the thermometer detects the power radiated by the blood in the artery. This gives a very accurate measurement of body temperature while being noninvasive.

Figure 14-19 Three forms of heat transfer These cold-blooded iguanas lose heat by convection but take heat in by radiation from the Sun and by conduction from the hot rock beneath them.

BioMedical

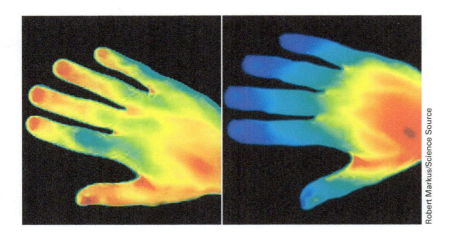

BioMedical **Figure 14-20 Radiation from a human hand** A special infrared camera was used to record these images of a person's right hand before (left) and after (right) smoking a cigarette. The intensity of the radiation from different parts of the hand, and hence the temperature of those parts, is indicated by colors from red (warmest) through yellow, green, and blue (coldest). The temperature of the fingers drops after smoking because the nicotine in the tobacco causes blood vessels to contract and reduces blood circulation to the extremities.

Radiation and Climate

BioMedical

Radiation is also the mechanism that keeps Earth from freezing. Our planet's interior releases very little heat, so what keeps the surface warm is energy that reaches us from the Sun in the form of electromagnetic radiation. In equilibrium the rate at which Earth *absorbs* solar energy (about 1.21×10^{17} W) must be equal to the rate at which it *emits* energy into space as given by Equation 14-22. Since the emission rate depends on the surface temperature, this is what determines our planet's average surface temperature T. If Earth had no atmosphere, T would be about 254 K (= −19°C = −2°F), so cold that oceans and lakes around the world should be frozen over. In fact, Earth's actual average surface temperature is a much more livable 287 K (= 14°C = 57°F). The explanation for this discrepancy is called the **greenhouse effect**: Our atmosphere prevents some of the radiation emitted by Earth's surface from escaping into space. Certain gases in our atmosphere called **greenhouse gases**, among them water vapor and carbon dioxide, are transparent to visible light but not to infrared radiation. Consequently, visible sunlight has no trouble entering our atmosphere and warming the surface. But the infrared radiation coming from the heated surface is partially trapped by the atmosphere and cannot escape into space. This lowers Earth's net emissivity e in Equation 14-22. To have the power radiated into space equal to the power received from the Sun, the factor of T^4 in Equation 14-22 must be greater to compensate for the reduced value of e. The net effect is that Earth's surface is some 33°C (59°F) warmer than it would be without the greenhouse effect, and water remains unfrozen over most of the planet.

The warming caused by the greenhouse effect gives our planet the moderate temperatures needed for the existence of life. For more than a century, however, our technological civilization has been adding greenhouse gases to the atmosphere at an unprecedented rate. Figure 14-21 shows how the concentration of carbon dioxide (CO_2) has varied in our atmosphere over the past 400,000 years. (Data from past centuries come from analyzing bubbles of air trapped in ancient ice in the Antarctic.) While there is natural variation in the CO_2 concentration, the value has skyrocketed since the beginning of the Industrial Revolution thanks to our burning fossil fuels such as coal and petroleum. The result has been an amplification of the greenhouse effect and an increase in the average surface temperature, an effect known as **global warming** (Figure 14-22). Other explanations for

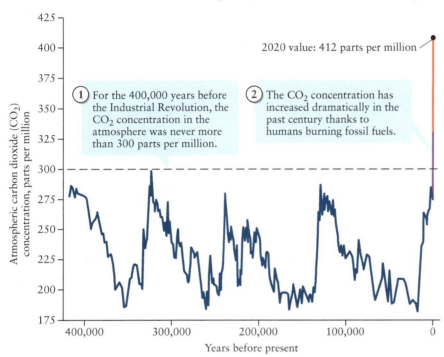

① For the 400,000 years before the Industrial Revolution, the CO_2 concentration in the atmosphere was never more than 300 parts per million.

② The CO_2 concentration has increased dramatically in the past century thanks to humans burning fossil fuels.

2020 value: 412 parts per million

Figure 14-21 Carbon dioxide in the atmosphere Because carbon dioxide (CO_2) absorbs infrared radiation emitted by Earth's surface, its presence in our atmosphere decreases Earth's emissivity and causes the average surface temperature to increase. Thanks to the burning of fossil fuels, the level of CO_2 in our atmosphere is now greater than it has been at any time in the past 400,000 years.

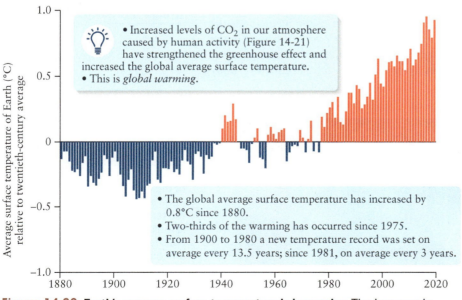

• Increased levels of CO_2 in our atmosphere caused by human activity (Figure 14-21) have strengthened the greenhouse effect and increased the global average surface temperature.
• This is *global warming*.

• The global average surface temperature has increased by 0.8°C since 1880.
• Two-thirds of the warming has occurred since 1975.
• From 1900 to 1980 a new temperature record was set on average every 13.5 years; since 1981, on average every 3 years.

Figure 14-22 Earth's average surface temperature is increasing The increase in greenhouse gases shown in Figure 14-21 has led to elevated temperatures averaged over our planet's surface. The rate of increase has accelerated over recent decades.

global warming have been proposed, such as changes in the sun's brightness, but these do not stand up to close scrutiny. Only greenhouse gases produced by human activity can explain the steep temperature increase shown in Figure 14-22.

The effects of global warming can be seen around the world. Nine out of 10 of the warmest years on record have occurred since 2005, and we have seen increasing numbers of droughts, water shortages, and unprecedented heat waves. Glaciers worldwide are receding, Arctic sea ice is decreasing by 13% per decade, and a portion of the Antarctic ice shelf has broken off. Unfortunately, global warming is predicted to intensify in the decades to come. The UN Intergovernmental Panel on Climate Change predicts that if nothing is done to decrease the rate at which we add greenhouse gases to our atmosphere, the average surface temperature will continue to rise to values 2.5°C to 7.8°C above the preindustrial value by the year 2100. What is worse, temperatures will rise in some regions and decline in others and the patterns of rainfall will be substantially altered. Agriculture depends on rainfall, so these changes in rainfall patterns can cause major disruptions in the world food supply. Studies suggest that the climate changes caused by a 3°C increase in the average surface temperature would cause a worldwide drop in cereal crops of 20 to 400 million tons, putting 400 million more people at risk of hunger.

The solution to global warming will require concerted and thoughtful action. Global warming cannot be stopped completely: Even if we were to immediately halt all production of greenhouse gases, the average surface temperature would increase an additional 2°C by 2100 thanks to the natural inertia of Earth's climate system. Instead, our goal must be to minimize the effects of global warming by changing how we produce energy, making choices about how to decrease our requirements for energy, and searching for ways to remove CO_2 from the atmosphere and trap it in the oceans or beneath our planet's surface. Confronting global warming is perhaps the greatest challenge to face our civilization in the twenty-first century.

Convection

You probably learned long ago that "hot air rises." It does, and the rising air carries energy with it from one region to another in a process called convection. Figure 14-23 shows a *convection current* in a room: a continuously circulating flow that forms when rising air forces the air above it to move out of the way and then downward somewhere else.

Convection is possible because fluids (gases and liquids) expand when heated. As a quantity of fluid expands, it becomes less dense than its surroundings and so is buoyed up according to Archimedes' principle (Section 11-8). The hot air "floats" in cooler air, and the cooler air sinks.

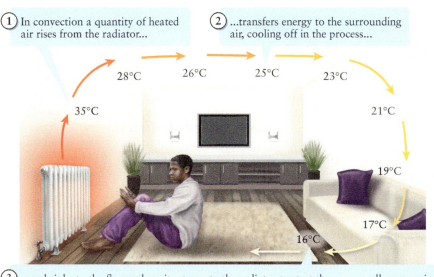

(1) In convection a quantity of heated air rises from the radiator...

(2) ...transfers energy to the surrounding air, cooling off in the process...

28°C   26°C   25°C   23°C

35°C   21°C

19°C

17°C

16°C

(3) ...and sinks to the floor, where it returns to the radiator to start the process all over again.

Figure 14-23 A convection current A continuous circulating flow forms when warm air rises, forcing the air above it to move out of the way and then downward somewhere else.

Figure 14-24 Seaside convection
Because the specific heat of land is much lower than the specific heat of water, the land heats up and cools down more quickly than the water. This creates onshore sea breezes in late afternoon and offshore land breezes during early morning.

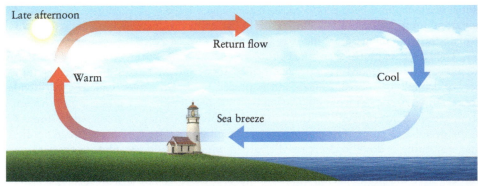

Convection plays an important role in driving motions in the atmosphere. People who live near the coasts of large bodies of water often experience a breeze that blows toward the shore in late afternoon and toward the water in early morning (Figure 14-24). Because the specific heat of land is much lower than the specific heat of water, the land heats up and cools down more quickly than the water. Especially when it is hot and the sky is clear, the temperature of the land becomes higher than the temperature of the water as the day progresses; the air above the land gets warmer and less dense. As this warm air rises, it is replaced by cooler air from above the water, resulting in an onshore, or sea, breeze. In the early morning, after the land and the air above it have cooled to a temperature below the temperature of the water, air rises above the relatively warmer water, and cooler air from above the land flows to replace it. The result is an offshore, or land, breeze.

Conduction

Perhaps you've made the mistake of touching the handle of a metal spoon that has been resting in a pot of soup on the stove. Even though the handle is touching neither the stove nor the soup, it can get hot enough to hurt your fingers if you touch it. The explanation is *conduction*: Energy absorbed from the soup is transmitted along the length of the spoon by collisions between adjacent atoms within the spoon. Through this process, the end of the spoon farthest from the soup eventually will also be hot.

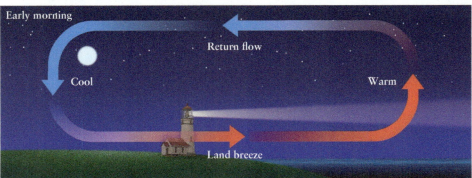

(1) The rate H of heat transfer through the cylinder is proportional to the temperature difference $T_H - T_C$.

(2) The rate of heat transfer through the cylinder is greater when its length L is small and its cross-sectional area A is large.

Figure 14-25 Conduction
The rate of heat exchange by conduction between two objects depends on the temperature difference as well as on the cross-sectional area and length of the contact region between them.

Figure 14-25 shows an idealized situation in which conduction takes place. The cylinder of length L and cross-sectional area A is in thermal contact at one end with a hot object at temperature T_H and in thermal contact at the other end with a cold object at temperature T_C. Experiment shows that the rate of heat transfer is proportional to the temperature difference $T_H - T_C$ (the greater the temperature difference, the more rapidly heat flows). The rate of heat transfer is also greater if the cylinder is short (L is small) and wide (A is large). We can put these observations together into a single equation:

> **Rate of heat transfer H in conduction** = quantity of heat Q that flows divided by the time Δt it takes to flow

> **Temperature difference** of the two places between which heat flows

$$H = \frac{Q}{\Delta t} = k\frac{A}{L}\left(T_{\mathrm{H}} - T_{\mathrm{C}}\right)$$

> **Thermal conductivity** of the material through which the heat flows

> **Cross-sectional area and length** of the material through which the heat flows

Rate of energy flow in conduction
(14-23)

The quantity k, called the **thermal conductivity**, is a constant that depends on the material of which the cylinder is made. The units of $H = Q/\Delta t$ are watts (1 W = 1 J/s), so the units of k are watts per meter per kelvin ($\mathrm{W \cdot m^{-1} \cdot K^{-1}}$). Table 14-5 lists the thermal conductivity of various substances. A good thermal *conductor* has a high value of k; the thermal conductivity of aluminum, for example, is 235 $\mathrm{W \cdot m^{-1} \cdot K^{-1}}$. Materials that are poor thermal conductors—those that make good thermal *insulators*—have k values less than 1 $\mathrm{W \cdot m^{-1} \cdot K^{-1}}$. You can actually feel the difference between thermal conductors and thermal insulators. If you pick up a room-temperature aluminum rod ($k = 235$ $\mathrm{W \cdot m^{-1} \cdot K^{-1}}$) with one hand and a room-temperature wooden stick ($k = 0.12$ $\mathrm{W \cdot m^{-1} \cdot K^{-1}}$) of the same size with the other hand, the aluminum rod will feel colder. Both objects are at the same temperature (which is lower than the temperature of your hand), but the much higher thermal conductivity of the aluminum means that heat can flow into it from your hand at a faster rate.

Table 14-5 shows that the thermal conductivity of air is relatively low. This small value makes air, in particular air trapped between other materials, a good insulator. Most animals rely on air trapped between fur or feathers to slow the rate of heat loss in cold conditions. This phenomenon is even more developed in polar bears. The hair that makes up their fur is hollow; the air trapped within the hair shafts makes it a good thermal insulator. Clothing, particularly that made from cotton, wool, and other woven cloth, keeps us warm primarily because of the air trapped between the fibers. Animal fat is also a good insulator. A typical value of thermal conductivity for human fat is 0.2 $\mathrm{W \cdot m^{-1} \cdot K^{-1}}$, which is about the same as whale blubber and nearly a third smaller than the value of water ($k = 0.58$ $\mathrm{W \cdot m^{-1} \cdot K^{-1}}$). That's one reason blubber is so important to marine mammals in frigid polar waters.

 Go to Picture It 14-4 for more practice dealing with thermal conductivity.

BioMedical

TABLE 14-5 Thermal Conductivities

Material	Thermal conductivity $k(\mathrm{W \cdot m^{-1} \cdot K^{-1}})$	Material	Thermal conductivity $k(\mathrm{W \cdot m^{-1} \cdot K^{-1}})$
air	0.024	hydrogen	0.168
aluminum	235	nitrogen	0.024
brick	0.9	plywood	0.13
copper	401	sand (dry)	0.35
cotton	0.03	silver	429
earth (dry)	1.5	steel	46
human fat	0.2	Styrofoam	0.033
fiberglass	0.04	water	0.58
glass (window)	0.96	wood (white pine)	0.12
granite	1.7–4.0	wool	0.04
gypsum (plaster) board	0.17		

EXAMPLE 14-9 Heat Loss Through a Window

Windows are a major source of heat loss from a house. That's because of the materials typically used in construction, glass has one of the highest thermal conductivities. Determine the rate of heat loss for a house on an evening when the temperatures of the outer and inner surfaces of the windows are 14.0°C and 15.0°C, respectively. Take the total window area of the house to be 28.0 m² and the thickness of the windows to be 3.80 mm.

Set Up

The rate of heat flow H through all of the windows combined is the same as if there were a single window with area $A = 28.0$ m² and thickness $L = 3.80$ mm $= 3.80 \times 10^{-3}$ m. We use Equation 14-23 to calculate the value of H.

Rate of energy flow in conduction:

$$H = k\frac{A}{L}(T_H - T_C) \qquad (14\text{-}23)$$

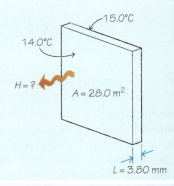

15.0°C
14.0°C
$H = ?$
$A = 28.0$ m²
$L = 3.80$ mm

Solve

Substitute values into Equation 14-23.

From Table 14-5 the thermal conductivity of window glass is $k = 0.96$ W·m⁻¹·K⁻¹. The rate of energy flow through the windows is

$$H = k\frac{A}{L}(T_H - T_C)$$

$$= (0.96 \text{ W·m}^{-1}\text{·K}^{-1})\left(\frac{28.0 \text{ m}^2}{3.80 \times 10^{-3} \text{ m}}\right)(15.0°C - 14.0°C)$$

$$= 7.1 \times 10^3 \text{ W} = 7.1 \text{ kW}$$

(Recall that a temperature difference of 1.0°C is the same as a temperature difference of 1.0 K.)

Reflect

To maintain the interior of the house at the same temperature, energy has to be provided to the interior by the heating system at a rate of 7.1×10^3 W or 7.1 kW. To put that into perspective, a typical microwave oven or portable hair dryer requires about 1 kW of power to operate. So keeping this house warm with electric heat will require the same amount of electric power as seven microwave ovens or seven hair dryers running continuously.

One way to keep heating costs down is to replace the windows in this house with *dual-pane* or *triple-pane* windows. These windows have two or three sheets of glass separated by a gap of 1 to 2 cm, with gas filling the gap. The greater thickness of glass increases the value of L in Equation 14-23, and the gas between the panes reduces the overall thermal conductivity k.

GOT THE CONCEPT? 14-7 Which Plate Is Which?

(?) The photograph in Figure 14-26 was taken 30 s after identical ice cubes were placed on each of two black plates. The plates are the same size and were initially at the same temperature. Which of the following is the most likely composition of the two plates? (a) Left-hand plate is wood, right-hand plate is aluminum; (b) left-hand plate is aluminum, right-hand plate is wood; (c) left-hand plate is brick, right-hand plate is glass; (d) left-hand plate is glass, right-hand plate is brick; (e) not enough information to decide.

University of Illinois at Urbana-Champaign Physics Department

Figure 14-26 Melting ice cubes An ice cube placed on the left-hand plate melts more rapidly than one placed on the right-hand plate. What does this tell you about the compositions of the two plates?

TAKE-HOME MESSAGE FOR Section 14-7

✔ Heat transfer from one place or object to another can occur through three processes: radiation, convection, and conduction.

✔ In radiation, energy flows from one object to a cooler one in the form of electromagnetic waves. The rate at which an object emits energy in the form of radiation is proportional to its surface area and to the fourth power of its Kelvin temperature.

✔ In convection, energy is carried through a liquid or gas by the motion of that liquid or gas.

✔ In conduction, energy flows within an object through collisions between the particles that make up that object. The rate of heat flow through an object by conduction depends on the dimensions of the object and on the temperature difference between its ends.

Key Terms

absolute zero	fusion	radiation
Boltzmann constant	global warming	root-mean-square speed
British thermal unit	greenhouse effect	(rms speed)
calorie	greenhouse gas	specific heat
Celsius scale	heat	Stefan–Boltzmann
coefficient of linear expansion	ideal gas	constant
coefficient of volume expansion	ideal gas constant	sublimation
condensation	ideal gas law	temperature
conduction	internal energy	thermal conductivity
convection	kelvin	thermal contact
critical point	Kelvin scale	thermal equilibrium
degree of freedom	latent heat (of fusion or	thermal expansion
deposition	vaporization)	thermodynamics
emissivity	mean free path	thermometer
equation of state	mole	triple point
equipartition theorem	phase	vaporization
Fahrenheit scale	phase change	zeroth law of
freezing	phase diagram	thermodynamics

Chapter Summary

Topic	Equation or Figure
Temperature and thermal equilibrium: The temperature of an object is a measure of the kinetic energy of its molecules. Two objects are in thermal equilibrium (so no energy flows from one to the other) if and only if they are at the same temperature. We use both the Celsius and Kelvin temperature scales; on the Kelvin scale zero temperature is absolute zero.	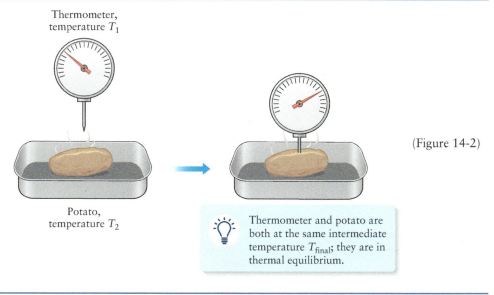 (Figure 14-2)

Ideal gases: In an ideal gas there are no interactions between molecules of the gas. The ideal gas law relates the pressure, volume, and Kelvin temperature of the gas; it can be expressed in terms of either the number of molecules or the number of moles.

Volume occupied by the gas   Number of molecules present in the gas

Pressure of an ideal gas

$$pV = NkT$$

Temperature of the gas on the Kelvin scale   (14-6)

Boltzmann constant

Volume occupied by the gas   Number of moles present in the gas

Pressure of an ideal gas

$$pV = nRT$$

Temperature of the gas on the Kelvin scale   (14-7)

Ideal gas constant

Molecular motion in an ideal gas: The average translational kinetic energy of a molecule of the gas is directly proportional to the Kelvin temperature T. The root-mean-square molecular speed is proportional to the square root of T. If the gas molecules contain more than one atom, there can also be energy in the vibration or rotation of the molecules.

Average translational kinetic energy of a molecule in an ideal gas   Average value of the square of a gas molecule's speed   Temperature of the gas on the Kelvin scale

$$K_{\text{translational,average}} = \frac{1}{2} m(v^2)_{\text{average}} = \frac{3}{2}kT$$

Mass of a single gas molecule   Boltzmann constant   (14-13)

Root-mean-square speed of a molecule in an ideal gas   Boltzmann constant

$$v_{\text{rms}} = \sqrt{\frac{3kT}{m}}$$

Temperature of the gas on the Kelvin scale   (14-15)

Mass of a single gas molecule

Thermal expansion: The length of a solid object changes with temperature, as does the volume of a solid or liquid. In most cases the dimensions increase with increasing temperature; liquid water between 0°C and 4°C is an exception.

Change in length of an object   Length of the object before the temperature change

$$\Delta L = \alpha L_0 \Delta T$$ (14-18)

Coefficient of linear expansion of the substance of which the object is made   Temperature change of the object that causes the length change

Change in volume of a quantity of substance   Volume of the substance before the temperature change

$$\Delta V = \beta V_0 \Delta T$$ (14-19)

Coefficient of volume expansion of the substance   Temperature change of the substance that causes the volume change

Heat and temperature change: Heat is energy that flows from one place to another due to a temperature difference. If there is no change in the phase of an object, heat flow into or out of the object causes a temperature change ΔT.

Quantity of heat that flows into (if $Q > 0$) or out of ($Q < 0$) an object   Specific heat of the substance of which the object is made

$$Q = mc\Delta T$$

Temperature change of the object that results from the heat flow   (14-20)

Mass of the object

Heat and phase change: Solid, liquid, and gas are the different possible phases of a substance. At a given pressure there is a specific temperature at which a substance can change from one phase to another. Each phase change requires heat to flow into or out of the substance. For one particular combination of pressure and temperature, called the triple point, the three phases can coexist.

Quantity of heat that flows into or out of a quantity of substance to cause a phase change

Use + sign if heat flows into the substance; use − sign if heat flows out of the substance.

$$Q = \pm mL$$

Latent heat for the phase change

Mass of the substance

(14-21)

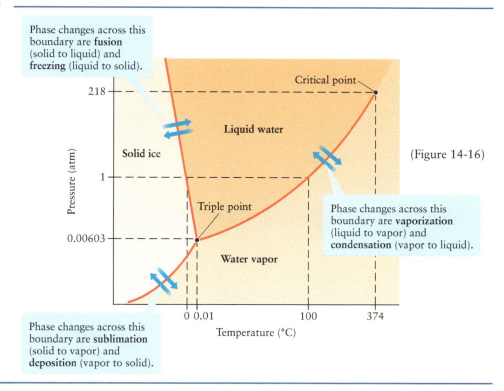

Phase changes across this boundary are **fusion** (solid to liquid) and **freezing** (liquid to solid).

Critical point

Liquid water

Solid ice

Triple point

Water vapor

Phase changes across this boundary are **vaporization** (liquid to vapor) and **condensation** (vapor to liquid).

Phase changes across this boundary are **sublimation** (solid to vapor) and **deposition** (vapor to solid).

(Figure 14-16)

Radiation, convection, and conduction: Energy can flow from place to place by radiation (electromagnetic waves), convection (circulation of a liquid or gas), or conduction (direct contact between solid objects). The rate at which an object emits energy by radiation is proportional to the fourth power of the object's Kelvin temperature. The rate at which energy flows through an object by conduction depends on the temperature difference between the ends of the object, the dimensions of the object, and the object's thermal conductivity.

Rate at which an object emits energy in the form of radiation

Emissivity of the object (a number between 0 and 1)

$$P = e\sigma A T^4$$

Temperature of the object on the Kelvin scale

Stefan–Boltzmann constant = 5.6704×10^{-8} W·m^{-2}·K^{-4}

Surface area of the object

(14-22)

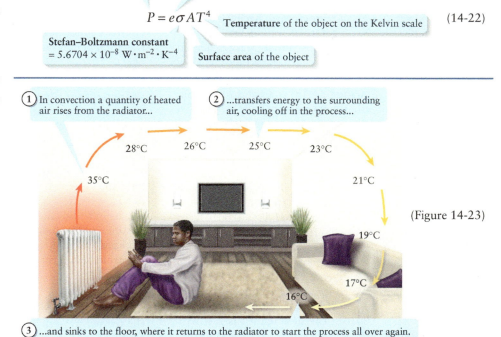

① In convection a quantity of heated air rises from the radiator...

② ...transfers energy to the surrounding air, cooling off in the process...

③ ...and sinks to the floor, where it returns to the radiator to start the process all over again.

(Figure 14-23)

Rate of heat transfer H in conduction = quantity of heat Q that flows divided by the time Δt it takes to flow

Temperature difference of the two places between which heat flows

$$H = \frac{Q}{\Delta t} = k\frac{A}{L}\,(T_{\mathrm{H}} - T_{\mathrm{C}}) \tag{14-23}$$

Thermal conductivity of the material through which the heat flows

Cross-sectional area and length of the material through which the heat flows

Answer to What do you think? Question

(c) Almost every substance expands in response to a temperature increase, so we expect that the chamber walls will expand. If they expanded as much or more than the fluid inside the chambers, we would expect that the pressure of the fluid against the chamber walls would stay the same or decrease.

Because the pressure increases, it must be that the chamber walls expand by a smaller amount than the fluid. A liquid-in-glass thermometer uses the same basic principle: A temperature increase makes the liquid expand more than the glass tube that contains the liquid, so the liquid level rises in the tube.

Answers to Got the Concept? Questions

14-1 (c), (b), (a). To compare these three temperatures, convert them all to the Kelvin scale. Temperature (b) is –13°C or, with the second of Equations 14-2, $T_{\mathrm{K}} = T_{\mathrm{C}} + 273.15 = (-13) + 273.15 = 260$ K. Temperature (c) is –13°F, which from the first of Equations 14-1 is $T_{\mathrm{C}} = \frac{5}{9}(T_{\mathrm{F}} - 32) = \frac{5}{9}(-13 - 32) = \frac{5}{9}(-45) = -25$°C and from the second of Equations 14-2 is $T_{\mathrm{K}} = T_{\mathrm{C}} + 273.15 = (-25) + 273.15 = 248$ K. So temperature (c) (248 K) is coldest, temperature (b) (260 K) is second coldest, and temperature (a) (280 K) is warmest.

14-2 (a) the same, (b) less, (c) greater, (d) greater. The average translational kinetic energy per molecule $K_{\text{translational,average}} = (3/2)kT$ (Equation 14-13) depends on the temperature but not on the kind of molecule, so this is the same for helium as for oxygen in air at a given temperature. Because oxygen is a diatomic molecule, it can rotate as well as translate, so the average energy per O_2 molecule is about $(5/2)kT$; helium is monatomic, so it has only translational kinetic energy with an average value of $(3/2)kT$ per molecule. The rms speed $v_{\text{rms}} = \sqrt{3kT/m}$ (Equation 14-15) is greater for helium because helium has a smaller mass per molecule m. Finally, the mean free path λ (Equation 14-16) is inversely proportional to the square of the molecular radius r, so λ is greater for helium than for oxygen.

14-3 (c) The statement of the problem tells us that if the initial length of one side of the cube is L, the change in length due to the temperature increase is $\Delta L = 0.0010L$. From Equation 14-18, $\Delta L = \alpha L\Delta T$; this says that $\alpha L\Delta T = 0.0010L$, so $\alpha\Delta T = 0.0010$. For solids the coefficient of *volume* expansion β equals 3α, so $\beta\Delta T = 3\alpha\Delta T = 0.0030$. So if the initial volume of the cube is V, the change in its volume due to the temperature increase is given by Equation 14-19: $\Delta V = \beta V\Delta T = (\beta\Delta T)V = 0.0030V$; that is, the volume increases by 0.0030. To check this, note that if the length of each side changes by $\Delta L = 0.0010L$, the new length is $L + \Delta L = L + 0.0010L = 1.0010L$. The volume of the cube then changes from L^3 to $(1.0010L)^3 = (1.0010)^3 L^3$. You can use your calculator to show that $(1.0010)^3 = 1.0030$, so the new volume is

$1.0030L^3$—that is, the volume has increased by 0.0030 of its initial volume L^3.

14-4 (a) When two objects are made of the same substance, the final temperature is closer to the initial temperature of the more massive object. Let's check this conclusion using Equation 14-20. For the two objects this equation says $Q_A = m_A c_A \Delta T_A$ and $Q_B = m_B c_B \Delta T_B$. If the two objects exchange energy with each other only, $Q_A = -Q_B$ (see Example 14-5), so $m_A c_A \Delta T_A = -m_B c_B \Delta T_B$. The two objects are made of the same substance and so have the same specific heat ($c_A = c_B$); thus our equation simplifies to $m_A \Delta T_A = -m_B \Delta T_B$. Since m_A is much greater than m_B, for this equation to be true it must be that ΔT_A is much smaller in absolute value than ΔT_B. In other words, the temperature of A changes much less than does the temperature of B, and the final temperature is closer to the initial temperature of A.

14-5 (a) When two objects have the same mass, the final temperature is closer to the initial temperature of the object with the greater specific heat. Let's check this conclusion using Equation 14-20. For the two objects this equation says $Q_A = m_A c_A \Delta T_A$ and $Q_B = m_B c_B \Delta T_B$. If the two objects exchange energy with each other only, $Q_A = -Q_B$ (see Example 14-5), so $m_A c_A \Delta T_A = -m_B c_B \Delta T_B$. The two objects have the same mass ($m_A = m_B$), so our equation simplifies to $c_A \Delta T_A = -c_B \Delta T_B$. Since c_A is much greater than c_B, for this equation to be true it must be that ΔT_A is much smaller in absolute value than ΔT_B. In other words, the temperature of A changes much less than does the temperature of B, and the final temperature is closer to the initial temperature of A.

14-6 (d), (b), (c), (a), (e). The specific heat of a substance (listed in Table 14-3) is the amount of energy needed to raise the temperature of 1 kg of substance by 1°C (or 1 K). This is 2093 J·kg⁻¹·K⁻¹ for ice, 4186 J·kg⁻¹·K⁻¹ for water, and 2009 J·kg⁻¹·K⁻¹ for steam. The latent heat of fusion and the latent heat of vaporization are, respectively, the amounts of energy needed to melt and vaporize 1 kg of substance at its melting and boiling temperatures. From Table 14-4 the values of these for water are $L_F = 3.34 \times 10^5$ J/kg and $L_V = 2.26 \times 10^6$ J/kg. So the rank of the energies from largest to smallest is (d) energy to boil 1 kg (2.26×10^6 J), (b) energy to melt 1 kg (3.34×10^5 J), (c) energy to raise the temperature of water by 1°C (4186 J), (a) energy to raise the temperature of ice by 1°C (2093 J), and (e) energy to raise the temperature of steam by 1°C (2009 J).

14-7   (b) The ice cube on the left melts much more rapidly because it is placed on a plate with a high thermal conductivity. Table 4-5 shows that the thermal conductivity of aluminum ($k = 235$ W·m^{-1}·K^{-1}) is much greater than that of wood ($k = 0.12$ W·m^{-1}·K^{-1}), so this is a possibility. By contrast, the thermal conductivities of brick ($k = 0.9$ W·m^{-1}·K^{-1}) and glass ($k = 0.96$ W·m^{-1}·K^{-1}) are almost identical, so we wouldn't expect much difference between the melting rate of ice cubes placed on a brick plate and a glass plate.

Questions and Problems

In a few problems you are given more data than you actually need; in a few other problems you are required to supply data from your general knowledge, outside sources, or informed estimate. Interpret as significant all digits in numerical values that have trailing zeros and no decimal points. For all problems use $g = 9.80$ m/s^2 for the free-fall acceleration due to gravity. Neglect friction and air resistance unless instructed to do otherwise.

•	Basic, single-concept problem
••	Intermediate-level problem; may require synthesis of concepts and multiple steps
•••	Challenging problem
Example	See worked example for a similar problem

Conceptual Questions

1. • State the zeroth law of thermodynamics and explain how it is used in physics.

2. • Search the Internet for *temperature scales*. How many different temperature scales can you find, and what are they?

3. • Explain the physical significance of the value −273.15°C.

4. • A physics student has decided that reading the textbook is very time-consuming and has further decided that she will simply attempt the problems in the book without any prior background reading. During the completion of one problem that involves a temperature conversion, the student concludes that the answer is −508°F. Discuss the validity of the answer.

5. • A careful physics student is reading about the concept of temperature and has what she thinks is a bright idea. While reading the text, she discovers that the Celsius temperature scale and the Kelvin temperature scale both have the *same* increments for equal temperature differences. However, the Fahrenheit scale was described as having *larger* increments for the same temperature differences. From this information the student determines that Celsius (or Kelvin) thermometers will always be *shorter* than Fahrenheit thermometers. Explain the parts of her idea that are valid and the parts that are not.

6. • The average normal human body temperature falls within the range of 37 ± 0.5°C or 98.6 ± 0.9°F. Discuss (a) the difficulties in determining an exact value for normal body temperature and (b) why there is a range of values.

7. • Describe some possible uses for the following thermometers whose temperature ranges are provided in kelvins.

Thermometer A: 200–270 K
Thermometer B: 230–370 K
Thermometer C: 300–550 K
Thermometer D: 300–315 K

8. • "Temperature is the physical quantity that is measured with a thermometer." Discuss the limitations of this working definition.

9. • A very old mercury thermometer is discovered in a physics lab. All the markings on the glass have worn away. How could you recalibrate the thermometer?

10. • The phrase "temperature measures kinetic energy" is somewhat difficult to understand and can be seen as contradictory. Give a few reasons why the explanation actually makes sense and a few reasons why it seems contradictory.

11. • If ideal gas A is thermally in contact with ideal gas B for a significant time, what can you say (if anything) about the following variables: P_A versus P_B, V_A versus V_B, and T_A versus T_B? (P, V, and T stand for pressure, volume, and temperature, respectively.)

12. • A skunk is threatened by a great horned owl and emits its foul-smelling fluid to get away. Will the odor be more easily detected on a day when it is cooler or when it is warmer? Explain your answer.

13. • The ideal gas law can be written as $PV = NkT$ or $PV = nRT$. (a) Explain the different contexts in which you might use one or the other and (b) define all the variables and constants.

14. • Why are some sidewalks often formed in small segments, separated by a small gap between the segments?

15. • Older mechanical thermostats use a bimetallic strip to open and close a mercury switch that turns the heat on or off. A bimetallic strip is made of two different metals fastened together. One side is made of one metal, and the other side is made of a different metal. Describe how a bimetallic strip functions in such a thermostat.

16. • A brick wall is composed of 19.0-cm-long bricks ($\alpha = 5.5 \times 10^{-6}$ K^{-1}) and 1.00-cm-long sections of mortar ($\alpha = 8.0 \times 10^{-6}$ K^{-1}) between the bricks. Describe the expansion effects on a 20.0-m-long section of wall that undergoes a temperature change of 25.0°C.

17. •• In Section 14-3 we show that $K_{\text{translational,average}} = (3/2)kT$. The result is valid for a three-dimensional collection of atoms. (a) Discuss any changes in the formula that would correspond to a one-dimensional system. (b) What change would be necessary for a ten-dimensional space?

18. • Using the concepts of heat, temperature changes, and heat capacity, give a simplistic explanation of the global demographic factoid that 90% of the world's population lives within 100 km of a coastline.

19. • (a) Which is hotter: a kilogram of boiling water or a kilogram of steam? (b) Which one can cause a more severe burn? Why? Assume the steam is at the lowest temperature possible for the corresponding pressure.

20. • Explain the term "latent" as it applies to phase changes.

21. • Describe the sequence of thermodynamic steps that a very cold block of ice (below 0°C) will undergo as it transforms into steam at a temperature above 100°C.

22. • **Biology** You have probably heard the old saying "It ain't the heat, it's the humidity that's so unbearable." Discuss what this *really* means and try to include some reference to the definitions of heat and temperature that we have established in this chapter.

23. • There are several different units besides the joule used to measure thermal energy. (a) List the various units associated with thermal energy and (b) indicate in which area of science or technology they are used.

24. • Write a brief definition of the three different modes of heat transfer: radiation, convection, and conduction.

25. • Describe how convection causes hot air to rise in a room.

26. • Explain how a light material, such as fiberglass insulation, makes an effective barrier to keep a home warm in the winter and cool in the summer.

Multiple-Choice Questions

27. • Two objects that have different sizes, masses, and temperatures come into close contact with each other. Thermal energy is transferred
    A. from the larger object to the smaller object.
    B. from the object that has more mass to the one that has less mass.
    C. from the object that has the higher temperature to the object that has the lower temperature.
    D. from the object that has the lower temperature to the object that has the higher temperature.
    E. back and forth between the two objects until they come to equilibrium.

28. • If you halve the value of the root-mean-square speed or v_{rms} of an ideal gas, the absolute temperature must be
    A. reduced to one-half its original value.
    B. reduced to one-quarter its original value.
    C. unchanged.
    D. increased to twice its original value.
    E. increased to four times its original value.

29. • Two gases each have the same number of molecules, same volume, and same atomic radius, but the atomic mass of gas B is twice that of gas A. Compare the mean free paths.
    A. The mean free path of gas A is four times larger than that of gas B.
    B. The mean free path of gas A is four times smaller than that of gas B.
    C. The mean free path of gas A is the same as that of gas B.
    D. The mean free path of gas A is two times larger than that of gas B.
    E. The mean free path of gas A is two times smaller than that of gas B.

30. • If you heat a thin, circular ring so that its temperature increases to twice what it was originally, the ring's hole
    A. becomes larger.
    B. becomes smaller by an unknown amount.

C. remains the same size.
D. becomes four times smaller.
E. becomes two times smaller.

31. • If you add heat to water at 0°C, the water will decrease in volume until it reaches
    A. 1°C.
    B. 2°C.
    C. 3°C.
    D. 4°C.
    E. 100°C.

32. • Two objects that are not initially in thermal equilibrium are placed in close contact. After a while,
    A. the specific heats of both objects will be equal.
    B. the thermal conductivity of each object will be the same.
    C. the increase in temperature of the cooler object will be as great as the temperature decrease of the hotter one.
    D. the temperature of each object will be the same.
    E. the increase in temperature of the cooler object will be twice as great as the temperature decrease of the hotter one.

33. • When a substance goes directly from a solid state to a gaseous form, the process is known as
    A. vaporization.
    B. fusion.
    C. melting.
    D. condensation.
    E. sublimation.

34. • Which heat transfer process(es) is/are important in the transfer of energy from the Sun to Earth?
    A. radiation
    B. convection
    C. conduction
    D. conduction and radiation
    E. conduction, radiation, and convection

35. • A clay pot at room temperature is placed in a kiln, and the pot's temperature doubles. How much more heat per second is the pot radiating when hot compared to when cool?
    A. 2 times
    B. 4 times
    C. 8 times
    D. 16 times
    E. 32 times

36. • If the thickness of a uniform wall is doubled, the rate of heat transfer through the wall is
    A. quadrupled.
    B. doubled.
    C. halved.
    D. unchanged.
    E. one-fourth as much.

Problems

14-1 A knowledge of thermodynamics is essential for understanding almost everything around you—including your own body

14-2 Temperature is a measure of the energy within a substance

37. • Convert the following temperatures: Example 14-1
    A. 28°C = _____ °F
    B. 58°F = _____ °C

C. 128 K = _____ °F

D. 78°F = _____ K

E. 37°C = _____ °F

38. • Convert the following temperatures and comment on any physical significance of each: Example 14-1

    A. 0°C = _____ °F

    B. 212°F = _____ °C

    C. 273 K = _____ °F

    D. 68°F = _____ K

39. • Starting from $T_C = 5/9(T_F - 32)$ (the first of Equations 14-1), derive a formula for converting from °C to °F (the second of Equations 14-1).

40. • The highest temperature ever recorded on Earth is 56.7°C, in Death Valley, California, in 1913. The lowest temperature on record is −89.2°C, measured at Vostok Research Station in Antarctica in 1983. Convert these extreme temperatures to °F and kelvin. Example 14-1

41. • In adults the normal range for oral (under the tongue) temperature is approximately 36.7°C to 37.0°C. (a) Calculate this temperature range in °F and in kelvins. (b) For which temperature scale is the difference between the temperatures the same as the difference in °C? Example 14-1

42. • (a) Write an equation for the temperature T_{eq} at which the temperature in the Celsius and Fahrenheit scales are the same. (b) What is T_{eq}?

43. • A thermally isolated system has a temperature of T_A. The temperature of a second isolated system is T_B. When the two systems are placed in thermal contact with each other, they come to an equilibrium temperature of T_C. Describe all the possibilities regarding the relative magnitudes of T_A and T_B when (a) $T_C < T_A$, and (b) $T_C > T_A$.

14-3 In a gas, temperature and molecular kinetic energy are directly related

44. • One mole of an ideal gas is at a pressure of 1.00 atm and occupies a volume of 1.00 L. (a) What is the temperature of the gas? (b) Convert the temperature to °C and °F. Example 14-2

45. • Calculate the energy of a sample of 1.00 mol of ideal oxygen (O_2) gas molecules at a temperature of 300 K. Assume that the molecules are free to rotate and move in three dimensions. Example 14-3

46. • A 55.0-g sample of a certain gas occupies 4.13 L at 20.0°C and 10.0 atm pressure. What is the gas? Example 14-2

47. • The boiling point of water at 1.00 atm is 373 K. What is the volume occupied by water vapor due to evaporation of 10.0 g of liquid water at 1.00 atm and 373 K? Example 14-2

48. • An ideal monatomic gas is confined to a container at a temperature of 300 K. What is the average translational kinetic energy of an atom of the gas? Example 14-3

49. • Calculate v_{rms} for a helium atom if 1.00 mol of the gas is confined to a 1.00-L container at a pressure of 10.0 atm. Example 14-3

50. • Two gases present in the atmosphere are water vapor (H_2O) and oxygen (O_2). What is the ratio of their rms speeds? Example 14-3

51. • A vacuum chamber can attain pressures as low as 7.0×10^{-11} Pa. Suppose that a chamber contains helium at that pressure and at room temperature (300 K). Estimate the mean free path for helium in the chamber. Assume the diameter of a helium atom is 1.0×10^{-10} m.

52. • The mean free path for O_2 molecules at a temperature of 300 K and at 1.00 atm pressure is 7.10×10^{-8} m. Use these data to estimate the size of an O_2 molecule.

14-4 Most substances expand when the temperature increases

53. • Calculate the temperature change needed for a cylinder of gold to increase in length by 0.1%. Example 14-4

54. • Calculate the coefficient of linear expansion for a 10.0-m-long metal bar that shortens by 0.500 cm when the temperature drops from 25.0°C to 10.0°C. Example 14-4

55. • By how much would a 1.00-m-long aluminum rod increase in length if its temperature were raised 8.00°C? Example 14-4

56. • A silver pin is exactly 5.00 cm long when its temperature is 180.0°C. How long is the pin when it cools to 28.0°C? Example 14-4

57. • A sheet of lead has an 8.00-cm-diameter hole drilled through it while at a temperature of 8.00°C. What will be the diameter of the hole if the sheet is heated to 208.00°C? Example 14-4

58. •• A thin sheet of copper 80.00 cm by 100.00 cm at 28.0°C is heated to 228.0°C. What will be the new area of the sheet? Example 14-4

59. •• A sheet of copper at a temperature of 0.00°C has dimensions of 20.0 cm by 30.0 cm. (a) Calculate the dimensions of the sheet when the temperature rises to 45.0°C. (b) By what percent does the area of the sheet of copper change? Example 14-4

60. •• A 5.00-m-long cylinder of solid aluminum has a radius of 2.00 cm. (a) If the cylinder is initially at a temperature of 5.00°C, how much will the length change when the temperature rises to 30.00°C? (b) Due to the temperature increase, by what percentage does the density of the cylinder change? (c) By what percentage does the volume of the cylinder increase? Example 14-4

61. •• A cube of pure iron is heated uniformly to 100.0°C. At that temperature, the volume of the iron is 20.0 cm³. Find the dimensions of the cube (a) at 100.0°C and (b) at 20.0°C. Example 14-4

62. •• A sphere of gold has an initial radius of 1.00 cm when the temperature is 20.0°C. (a) If the temperature is raised to 80.0°C, calculate the new radius of the sphere. (b) What is the percent change in the volume of the sphere? Example 14-4

63. • A 20.0 m × 25.0 m pool is filled to a depth of 1.50 m at a temperature of 20.0°C. It warms to 35.0°C by the end of a hot summer day. Assuming no evaporation, by how much does the depth of the water change? Example 14-4

64. •• In high-altitude desert, it is not uncommon for the temperature to change by 20.0°C over the course of a summer day. If a vehicle's 45.0-L gas tank is half full when the temperature is 35.0°C, how full will it be the next morning if the temperature is 15.0°C, assuming the vehicle was not driven? Ignore any expansion of the tank and express your answer as a percentage. Example 14-4

65. •• A small copper sphere with a radius of 1.00 cm at a room temperature of 20.0°C has a mass density of 8.96 g/cm³. Calculate the new mass density of the copper sphere after it has been heated to a temperature of 400.0°C. Example 14-4

14-5 Heat is energy that flows due to a temperature difference

66. • What is the specific heat of a 0.500-kg metal sample that rises 4.80°C when 307 J of heat is added to it? Example 14-5

67. • You wish to heat 0.250 kg of water to make a hot cup of coffee. If the water starts at 20.0°C and you want your coffee to be 95.0°C, calculate the minimum amount of heat required. Example 14-5

68. • In a thermodynamically sealed container, 20.0 g of 15.0°C water is mixed with 40.0 g of 60.0°C water. Calculate the final equilibrium temperature of the water. Example 14-6

69. • How much heat is transferred to the environment when the temperature of 2.00 kg of water drops from 88.0°C to 42.0°C? Example 14-5

70. • A copper pot has a mass of 1.00 kg and is at 100.00°C. How much heat must be removed from it to decrease its temperature to precisely 0.00°C? Example 14-5

71. • A lake has a specific heat of 4186 J/(kg·K). If we transferred 1.70×10^{14} J of heat to the lake and warmed the water from 10.0°C to 15.0°C, what is the mass of the water in the lake? Neglect heat released to the surroundings. Example 14-5

72. •• Calculate the temperature increase in a 1.00-kg sample of water that results from the conversion of gravitational potential energy directly to heat energy in the world's tallest waterfall, the 807-m-tall *Salto Angel* in Canaima National Park, Venezuela. Example 14-5

73. • A superheated iron bar that has a mass of 5.00 kg absorbs 2.50×10^6 J of heat from a blacksmith's fire. Calculate the temperature increase of the iron. Example 14-5

74. • A 0.200-kg block of ice is at −10.0°C. How much heat must be removed to lower its temperature to −40.0°C? Example 14-5

75. •• A 0.250-kg sample of copper is heated to 100.0°C and placed into a cup containing 0.300-kg of water initially at 30.0°C. Ignoring the container holding the water, find the final equilibrium temperature of the copper and water, assuming no heat is lost or gained to the environment. Example 14-6

76. •• A 50.0-g calorimeter cup made from aluminum contains 0.100 kg of water. Both the aluminum and the water are at 25.0°C. A 0.300-kg cube of some unknown metal is heated to 150.0°C and placed into the calorimeter; the final equilibrium temperature for the water, aluminum, and metal sample is 41.0°C. Calculate the specific heat of the unknown metal and make a guess as to its composition. Example 14-6

77. • A hacksaw is used to cut a 20.0-g steel bolt. Each stroke of the saw supplies 30.0 J of energy. How many strokes of the saw will it take to raise the temperature of the bolt from 20.0°C to 80.0°C? Assume none of the energy goes into heating the surroundings. Of course, it will take significantly more than this to also cut the metal! Example 14-5

14-6 Energy must enter or leave an object in order for it to change phase

78. • Calculate the amount of heat required to change 25.0 g of ice at 0°C to 25.0 g of water at 0°C. Example 14-7

79. • How much heat is required to change 25.0 g of ice at −40.0°C to 25.0 g of steam at 140°C? Example 14-7

80. • A sealed container (with negligible heat capacity) holds 30.0 g of 120°C steam. Describe the final state if 100,000 J of heat is removed from the steam. Example 14-7

81. • How much heat is required to melt a 0.400-kg sample of copper that starts at 20.0°C? Example 14-7

82. •• Suppose 20.0 g of ice at −10.0°C is placed into 0.300 kg of water in a 0.200-kg copper calorimeter. If the final temperature of the water and copper calorimeter is 18.0°C, what was the initial common temperature of the water and copper? Example 14-8

83. •• Rina is a physics student who enjoys hot tea. She wants to determine how much ice is needed to cool 0.225 kg of tea to an optimal drinking temperature of 57.8°C. The method Rina uses to prepare her tea results in an initial temperature of 80.0°C. The average mass of an ice cube from the ice tray she uses is 20.0 g and has a starting temperature of −5.0°C. (a) What mass of ice in grams is needed to reach the optimal temperature of 57.8°C? (b) How many whole ice cubes should Rina use to cool her tea as close as possible to 57.8°C? Assume that the heat capacity of tea is the same as pure water, and that no heat is lost to the surroundings. Example 14-8

84. •• A 2.5-kg block of −12.0°C ice is placed in a thermally isolated chamber with 3.5 kg of 120.0°C steam. When the system in the chamber reaches equilibrium, what are the final state and temperature of the H_2O? Example 14-8

85. •• What mass of ice at −20.0°C must be added to 50.0 g of steam at 120.0°C to end up with water at 40.0°C? Example 14-8

86. • A 60.0-kg ice hockey player is moving at 8.00 m/s when he skids to a stop. If 40% of his kinetic energy goes into melting ice, how much water is created as he comes to a stop? Assume that the surface layer of the ice in the hockey rink has a temperature of 0°C. Example 14-7

87. • You have 50.0 g of iron at 120°C, 60.0 g of copper at 150°C, and 30.0 g of water at 40.0°C. Which of those would melt the most ice, starting at −5.00°C? How much ice does each melt? Example 14-8

14-7 Heat can be transferred by radiation, convection, or conduction

88. • A heated bar of gold radiates at a temperature of 300°C. (a) By what factor does the radiated power increase if the temperature is increased to 600°C? (b) What if it's increased to 900°C? (c) If the surface area of the gold bar is also doubled, how will the answers be affected?

89. • **Astronomy** An astrophysicist determines the surface temperature of a distant star is 12,000 K. The surface temperature of the Sun is about 5800 K. If the surface temperature of the Sun were to suddenly increase to 12,000 K, by how much would the radiated power increase?

90. • **Astronomy** A star radiates 3.75 times less power than our own Sun. What is the ratio of the temperature of the star to the temperature of our Sun? Assume the star and our Sun have the same radius.

91. • **Astronomy** A distant star radiates 1000 times more power than our own Sun, even though the temperature of the star is only 70% of the Sun's. If both stars have an emissivity of 1, estimate the radius of the distant star. Recall, the radius of the Sun is 6.96×10^8 m.

92. •• **Biology** The skin temperature of a nude person is 34.0°C, and the surroundings are at 20.0°C. The emissivity of

skin is 0.900, and the surface area of the person is 1.50 m². (a) What is the rate at which energy radiates from the person? (b) What is the net energy loss from the body in 1 min by radiation?

93. • Calculate the rate of heat transfer through a glass window that is 30.0 cm × 150 cm in area and 1.20 mm thick. Assume the temperature on the inside of the window is 25.0°C while the outside temperature is 8.00°C. Example 14-9

94. • What is the rate of heat transfer due to conduction through a single-pane glass window that is 3.0 m² and 3.175 mm thick, when the exterior temperature is 37°C and the interior temperature is 24°C? Example 14-9

95. • **Sports, Biology** The surface area of the human body is about 1.8 m². If an average 37.0°C human is surfing in 10.0°C seawater in a 3.0-mm thick neoprene wetsuit with a thermal conductivity of 0.050 W/(m·K), how much heat does the surfer lose to the ocean in 1 hour? Example 14-9

96. • A 37°C chef needs to handle a 165°C cast iron skillet. With a cotton hot pad, the skillet can be handled safely, but if the hot pad is wet, the chef can get burned. If the chef's hot pad is 1.0 mm thick, what is the ratio of heat transfer through 1.0 cm² of the pad if it is dry versus if it is wet? Use the thermal conductivity of water for the wet hot pad. Example 14-9

General Problems

97. • A 1.0-g sample of metal is slowly heated. The temperature increase for each increment of heat added is tabulated in the table. Plot a graph for the data and predict the specific heat for this common metal. If the measurements are ±10%, make a guess as to the type of metal it might be. Example 14-5

Heat (J)	Temperature change (°C)
0	0
3.9	10
7.9	20
20	50
40	100

98. • (a) Plot water density as a function of temperature for the data listed in the table. (b) Use the graph to make a conclusion about the way that lakes freeze.

Density (g/cm³)	Temperature (°C)
0.99990	8.0
0.99996	6.0
1.00000	4.0
0.99996	2.0
0.99988	0.0

99. • What is the average kinetic energy per molecule for an ideal gas made up of diatomic molecules that can move in two dimensions when the temperature is 75.0°F? Example 14-3

100. •• Estimate the amount of energy required to vaporize Lake Superior.

101. ••• **Astronomy** (a) Calculate the escape speed for hydrogen atoms from the surface (the *photosphere*) of our Sun (see Section 10-3). The Sun has mass 1.99×10^{30} kg and radius 6.96×10^8 m. (b) If the root-mean-square speed of the hydrogen atoms were equal to the speed you found in part (a), what would be the temperature of the Sun's photosphere? The mass of a hydrogen atom is 1.68×10^{-27} kg. (c) Given that the actual temperature of the photosphere is 5800 K, is the Sun likely to lose its atomic hydrogen? Example 14-3

102. •• **Astronomy** Titan, a satellite of Saturn, has a nitrogen atmosphere with a surface temperature of −179°C and pressure of 1.5 atm. The mass of a nitrogen molecule is 4.7×10^{-26} kg, and we can model it as a sphere of diameter 2.4×10^{-10} m. The

average surface temperature of Earth's atmosphere is about 10°C. (a) What is the density of atmospheric particles near the surface of Titan? (b) Which has a denser atmosphere, Titan or Earth? Justify your answer by calculating the ratio of the particle density on Titan to the particle density on Earth. (c) What is the average distance that a nitrogen molecule travels between collisions on Titan? How does this result compare with the distance for oxygen calculated in Section 14-3? Is your result reasonable? Example 14-3

103. • Derive an approximate formula for the area expansion (ΔA) of a sheet of material whose temperature changes from T_i to T_f. Assume that the linear coefficient of expansion for the material is α. Example 14-4

104. • A 1.00-cm-diameter sphere of copper is placed concentrically over a 0.99-cm-diameter hole in a sheet of aluminum (Figure 14-27). Both the copper and the aluminum start at a temperature of 15.0°C. Select a single set of conditions (if any) under which the copper sphere will just pass through the hole in the aluminum. Justify your conclusion with calculations. Example 14-4

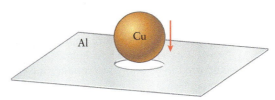

Figure 14-27 Problem 104

105. • A 20.0-m-long bar of steel expands due to a temperature increase. A 10.0-m-long bar of copper also gets longer due to the same temperature rise. If the two bars were originally separated by a gap of 1.5 cm, calculate the change in temperature and the distances that the steel and copper stretch if the gap is exactly "closed" by the expanding bars. Assume the steel and copper bars are fixed on the ends, as shown in Figure 14-28. Example 14-4

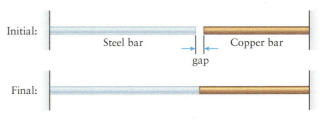

Figure 14-28 Problem 105

106. • Suppose a person who lives in a house next to a busy urban freeway attempts to "harness" the sound energy from the nonstop traffic to heat the water in his home. He places a transducer on his roof, "catches" the sound waves, and converts the sound waves into an electrical signal that warms a cistern of water. However, after running the system for 7 days, the 5 kg of water increases in temperature by only 0.01°C! Assuming 100% transfer efficiency, calculate the acoustic power "caught" by the transducer. Example 14-5

107. • **Biology** Susan ordinarily eats 2000 kcal of food per day. If her mass is 60 kg and her height is 1.7 m, her surface area is probably close to 1.7 m². Although the body is only about two-thirds water, we will model it as being all water. (a) Typically 80% of the calories we consume are converted to heat. If Susan's body has no way of getting rid of the heat produced, by how many degrees Celsius would her body temperature rise in a day? (b) Would this be a noticeable increase?

(c) How does her body prevent the increase from happening? (See Table 14-3 as needed.) Example 14-5

108. • Helium condenses at −268.93°C and has a latent heat of vaporization of 21,000 J/kg. If you start with 5.00 g of helium gas at 30.0°C, calculate the amount of heat that the gas must lose to change the sample to liquid helium. The specific heat of helium is 5193 J/(kg·K) at 300 K. Example 14-7

109. •• **Biology** Jane's surface area is approximately 1.5 m². At what rate is heat released from her body when the temperature difference across the skin is 1.00°C? Assume the average thickness of the skin is 1.00 mm. Example 14-9

110. •• The Arctic perennial sea ice does not melt during the summer and thus lasts all year. NASA found that the perennial sea ice decreased by 14% between 2004 and 2005. The melted ice covered an area of 720,000 km² (the size of Texas!) and was 3.00 m thick on average. The ice is pure water (not salt water) and is 92% as dense as liquid water. Assume that the ice was initially at −10.0°C and see Tables 14-3 and 14-4 as needed. (a) How much heat was required to melt the ice? (b) Given that 1.0 gal of gasoline releases 1.3×10^8 J of energy when burned, how many gallons of gasoline contain as much energy as in part (a)? (c) A ton of coal releases around 21.5 GJ = 21.5×10^9 J of energy when burned. How many tons of coal would need to be burned to produce the energy to melt the ice? Example 14-8

111. •• **Sports, Biology** A person can generate about 300 W of power on a treadmill. If the treadmill is inclined at 3.00° and a 70.0-kg man runs at 3.00 m/s for 45.0 min, (a) calculate the percentage of the power output that goes into heating up his body and the percentage that keeps him moving on the treadmill. (b) How much water would that heat evaporate? Example 14-5

112. •• **Biology** A 1.88-m (6 ft 2 in.) man has a mass of 80.0 kg, a body surface area of 2.1 m², and a skin temperature of 30.0°C. Normally 80% of the food calories he consumes go to heat; the rest goes to mechanical energy. To keep his body's temperature constant, how many food calories should he eat per day if he is in a room at 20.0°C and he loses heat only through radiation? Does the answer seem reasonable? His emissivity e is 1 because his body radiates almost entirely non-visible infrared energy, which is not affected by skin pigment. (*Careful*! His body at 30.0°C radiates into the air at 20.0°C, but the air also radiates back into his body. The *net* rate of radiation is $P_{net} = P_{body} − P_{air}$.)

113. • **Astronomy** Calculate the total power radiated by our Sun. Assume it is a perfect emitter of radiation (e = 1) with a radius of 6.96×10^8 m and a surface temperature of 5800 K.

114. ••• **Biology** You may have noticed that small mammals (such as mice) seem to be constantly eating, whereas some large mammals (such as lions) eat much less frequently. Let us investigate the phenomenon. For simplicity, we can model an animal as a sphere. (a) Show that the heat energy stored by an animal is proportional to the cube of its radius, but the rate at which the animal radiates energy away is proportional to the square of its radius. (b) Show that the fraction of the animal's stored energy that it radiates away per second is inversely proportional to the animal's radius. (c) Use the result in part (b) to explain why small animals must eat much more per gram of body weight than very large animals.

115. •• **Astronomy** About 65 million years ago an asteroid struck Earth in the area of the Yucatán Peninsula and wiped out the dinosaurs and many other life forms. Judging from the size of the crater and the effects on Earth, the asteroid was about 10.0 km in diameter (assumed spherical) and probably had a density of 2.0 g/cm³ (typical of asteroids). Its speed was at least 11 km/s. (a) What is the maximum amount of ocean water (originally at 20.0°C) that the asteroid could have evaporated if all of its kinetic energy were transferred to the water? Express your answer in kilograms and treat the ocean as though it were freshwater. (b) If this amount of water were formed into a cube, how high would it be? (See Tables 14-3 and 14-4 as needed.) Example 14-8

116. •• A spherical container is constructed from steel and has a radius of 2.00 m at 15.0°C. The container sits in the Sun all day, and its temperature rises to 38.0°C. The container is initially filled completely with water, but it is not sealed. Describe what will happen to the water after the temperature increase. Example 14-4

David J. Mitchell/Alamy

15 Thermodynamics II: Laws of Thermodynamics

In this chapter, your goals are to:

- (15-1) Explain the general ideas of the laws of thermodynamics.
- (15-2) Define and be able to apply the first law of thermodynamics.
- (15-3) Describe the nature of isobaric, isothermal, adiabatic, and isochoric processes.
- (15-4) Explain why different amounts of heat are required to change the temperature of a gas depending on whether the gas is held at constant volume or constant pressure.
- (15-5) Define the second law of thermodynamics and its application to heat engines and refrigerators.
- (15-6) Explain the concept of entropy and the circumstances under which entropy changes.

To master this chapter, you should review:

- (6-2) The significance of positive and negative work.
- (14-3) The properties of ideal gases.
- (14-5) The relationship between the quantity of heat added to an object and the resulting temperature change.

15-1 The laws of thermodynamics involve energy and entropy

In Chapter 14 we learned about the zeroth law of thermodynamics, which basically says that the concept of temperature is a useful one. (You should review Section 14-2 about the meaning of temperature and the zeroth law.) But what are the other laws of thermodynamics, and what do they tell us?

In this chapter we'll begin by learning about the *first law of thermodynamics*, which is a generalized statement about the conservation of energy. This law tells us about the interplay between the internal energy of an object or system, the work that the system does on its surroundings, and the heat that flows into or out of the system (Figure 15-1a). We'll use the first law to analyze a variety of *thermodynamic*

What do you think?

Thunderstorm clouds form when moist air rises rapidly to high altitudes. The temperature of the rising air decreases so much that the moisture condenses into droplets of liquid water or pieces of solid ice, which fall to the ground as rain or hail, respectively. Why does the temperature of the rising air decrease? (a) It loses heat to match the low temperature of the atmosphere at high altitude. (b) Its temperature naturally drops as its gravitational potential energy increases. (c) The air expands rapidly as it moves into the low pressure of the upper atmosphere.

Figure 15-1 The laws of thermodynamics The laws of thermodynamics are universal: They apply to (a) natural systems such as living organisms and (b) manufactured devices such as engines.

A cat demonstrates the first law of thermodynamics: The energy it takes in as food is either stored (as fat), converted into work, or released to its surroundings as heat.

(a)

An aircraft engine demonstrates the second law of thermodynamics: The burning of fuel releases energy in the form of heat, but it is impossible to convert 100% of this heat into useful work.

(b)

TAKE-HOME MESSAGE FOR Section 15-1

✔ The first law of thermodynamics relates changes in a system's internal energy to the work that the system does and the heat that flows into the system.

✔ The second law of thermodynamics describes what thermodynamic processes are possible. It also limits how efficient a heat engine can be.

---

processes in which the state of a system—as measured by its pressure, volume, and temperature—changes due to external influences.

According to the first law, some very remarkable things could happen: Room-temperature water could spontaneously lower its temperature and freeze at the same time that the room-temperature glass holding the water spontaneously raises its temperature and melts. The *second law of thermodynamics* describes why such remarkable things are never observed in the real world. It also places a firm limit on the efficiency of *heat engines*, devices that convert heat into work: It tells us that no matter how carefully the engine is designed or built, only part of that heat can be converted to work. Most vehicles in our technological society use a heat engine, so this aspect of the second law is of tremendous importance (Figure 15-1b).

We'll finish the chapter with a discussion of *entropy*, a physical quantity that measures the amount of disorder in a system. We'll find that as a result of the second law of thermodynamics, entropy is not conserved: The total amount of entropy in the universe is increasing.

15-2 The first law of thermodynamics relates heat flow, work done, and internal energy change

Suppose you take some unpopped kernels of popcorn at room temperature, put them in a pot with cooking oil, and put a lid on the pot. You then put the pot on the stove and warm it. In a few minutes the popcorn has popped and expanded so much that it has lifted the lid off the pot (Figure 15-2). By popping the popcorn you've changed its volume (each kernel has expanded) and made its temperature increase. You've also increased its **internal energy**, which is the sum of all of the kinetic and potential energies of the molecules that make up the popcorn. That's because the chemical reaction involved in popping the kernels requires that each kernel absorb energy. (You should review our discussion of internal energy in Section 14-5.) Quantities such as volume,

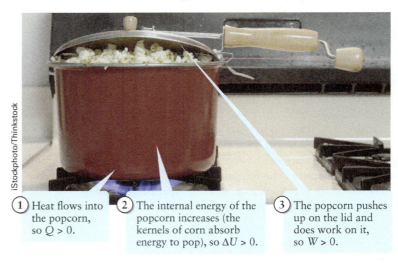

Figure 15-2 A thermodynamic process In a general thermodynamic process the internal energy of a system changes, heat enters or leaves the system, and work is done by or on the system.

① Heat flows into the popcorn, so $Q > 0$.

② The internal energy of the popcorn increases (the kernels of corn absorb energy to pop), so $\Delta U > 0$.

③ The popcorn pushes up on the lid and does work on it, so $W > 0$.

temperature, pressure, and internal energy are called **state variables** because they depend on the state or condition of the popcorn. Any process that changes the state of a system such as the popcorn is called a **thermodynamic process**.

Note that the values of the state variables of a system do *not* depend on the system's history—that is, the details of how the system got to that state. For example, you can pop popcorn on the stove as in Figure 15-2 or in a microwave oven, and in either case the results are the same. The state of a glass of water at room temperature is the same whether the water came out of the tap at room temperature, was heated in a pot and allowed to cool, or was originally a collection of ice cubes that gradually warmed and melted. In other words the state variables of a system in a given state do not depend on the particular thermodynamic processes that led to that state.

Figure 15-2 shows that in popping the popcorn, heat flows into the popcorn from the stove; the internal energy of the popcorn increases; and the popcorn does work on its surroundings as it lifts the lid of the pot. All three of these processes involve *energy* in one way or another, so let's look at each of them more closely. We'll then see how they're related.

We'll use the same convention as in Chapter 14 that heat Q is positive if it flows *into* a system and negative if it flows *out of* a system (see Section 14-5). Since heat flows into the popcorn in Figure 15-2 from the stove, $Q > 0$ for the popcorn.

We use the symbol U for the *internal* energy of the popcorn. Since the internal energy of the popcorn increases, its change is positive and $\Delta U > 0$. Generally speaking, if the temperature of a system increases in a thermodynamic process, the internal energy also increases and so $\Delta U > 0$. If the system temperature decreases, the internal energy generally decreases and so $\Delta U < 0$.

In thermodynamics we'll use the symbol W to represent the work done *by* a system on its surroundings. For the process shown in Figure 15-2, the popcorn does positive work on the lid: The popcorn exerts an upward force on the lid, and the lid's displacement is also upward. If a system increases in volume and pushes against its surroundings, as the popcorn does in Figure 15-2, then $W > 0$. If the system decreases in volume, like air being compressed in a bicycle pump, positive work is done *on* the system, so $W < 0$. Remember that if an object does negative work on a second object, the second object does positive work on the first one (see Section 6-2). You do positive work on the air in the bicycle pump to compress it, so the air does negative work on you. Figure 15-3 shows the conventions that we use for the sign of Q and the sign of W.

For the popcorn shown in Figure 15-2, heat flows into the popcorn ($Q > 0$), the internal energy of the popcorn increases ($\Delta U > 0$), and the popcorn does work on its surroundings ($W > 0$). Careful measurement shows that the heat flow into the system is *exactly* equal to the sum of the change in internal energy plus the work done. In other words, the energy that enters the popcorn in the form of heat goes either into changing the popcorn's internal energy or into doing work. None of the energy is lost. We can write this statement as an equation:

(15-1) $$Q = \Delta U + W$$

It's conventional to rearrange this equation with ΔU on one side and Q and W on the other side:

During a thermodynamic process the **change in the internal energy of a system**...

$$\Delta U = Q - W$$

...equals the **heat that flows into the system** during the process...

...minus the **work that the system does** during the process.

Heat flows into the system: $Q > 0$.

Heat flows out of the system: $Q < 0$.

System does positive work on its surroundings: $W > 0$.

| System | System | System | System |

System has positive work done on it by its surroundings: $W < 0$.

The first law of thermodynamics (15-2)

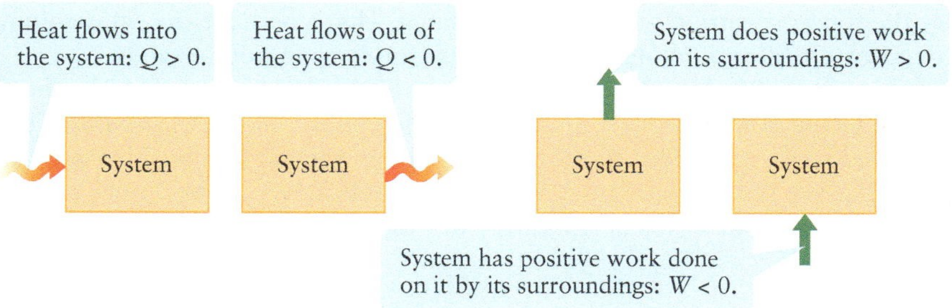

Figure 15-3 The signs of work and heat A quantity of heat Q is considered positive if it enters a system, negative if it leaves. The sign of the work W for a system depends on which object does positive work.

Equation 15-2 is a mathematical statement of the **first law of thermodynamics**. It's a generalization of the law of conservation of energy. It says that the internal energy of a system can change *only* if the system gains energy from its surroundings (if heat flows in from the surroundings, or its surroundings do positive work on the system) or if the system loses energy to its surroundings (if heat flows out to the surroundings, or the system does positive work on its surroundings). No exception to this rule has ever been found.

For the system shown in Figure 15-2, the quantity of heat that flows into the popcorn is greater than the amount of work that the popcorn does to lift the lid. So while Q and W are both positive, Q is greater than W and $\Delta U = Q - W$ is positive. In other situations the quantities ΔU, Q, and W have different signs. For example, think of your body as a thermodynamic system. If you are exercising on a stationary bicycle at the gym, your body does work on the pedals of the bicycle ($W > 0$), and heat flows out of your body ($Q < 0$). Since Q is negative and W is positive, $\Delta U = Q - W$ is negative and you lose internal energy. That's the thermodynamic explanation of why exercise helps you lose weight: The internal energy you lose is extracted from energy stored in your body, part of which is in the form of fat (see Example 15-1).

Internal Energy Change and Thermodynamic Paths

As we mentioned above, the internal energy U is a state function that depends only on the current state of the system, not on how it got into that state. To see what this implies let's consider two *different* thermodynamic processes that take a system from the *same* initial state to the *same* final state.

In Figure 15-4a a cylinder with a moveable piston encloses a thermodynamic system: a quantity of gas. The gas has an initial pressure, volume, and temperature. We use a candle to let heat flow slowly into the gas (so $Q > 0$) and allow the piston to move so that the gas expands. The expanding gas does positive work on the piston (the gas pushes the piston to the right as the piston moves to the right), so $W > 0$. If we regulate how fast the piston moves, we can arrange it so that the rate at which the gas does work on the piston is exactly the same as the rate at which heat flows into the gas. Then $Q = W$, and from the first law of thermodynamics (Equation 15-2), the internal energy of the gas will remain unchanged: $\Delta U = Q - W = 0$. For an ideal gas the average energy per molecule is proportional to the Kelvin temperature T of the gas (see Section 14-3), so the internal energy U—the average energy per molecule, multiplied by the number of molecules—is also proportional to T. Since U remains constant for the process shown in Figure 15-4a, the temperature of the gas also remains constant.

Now imagine a second thermodynamic process that starts with the system (the gas) in the same initial state. In Figure 15-4b we use the same quantity of gas as in Figure 15-4a, and the gas starts at the same pressure, volume, and temperature. The difference is that now the cylinder is equipped with a thin barrier of lightweight material instead of a piston, and there is a vacuum between the barrier and the other end of the cylinder. If we puncture the barrier, the gas expands freely and fills the entire cylinder. Since the gas doesn't push against anything as it expands (it expands into a vacuum), the gas does no work so $W = 0$. Furthermore, if we insulate the walls of the cylinder so that no heat can flow into or out of the gas, $Q = 0$ for this process. So the first law of thermodynamics tells us that in this process as well the internal energy of the gas will not change: $\Delta U = Q - W = 0 - 0 = 0$. And since U is proportional to Kelvin temperature T for an ideal gas, it follows that the temperature of the gas remains constant for this process. In other words *both* the processes shown in Figure 15-4 take the same quantity of gas from the same initial volume to the same final volume while leaving the temperature unchanged. (The ideal gas law $pV = nRT$, Equation 14-7, tells us that the final pressure p will also be the same. That's because the final volume V, final temperature T, and number of moles n are the same in both processes.)

The two very different processes in Figure 15-4 take the system from the same initial state to the same final state, but follow different thermodynamic *paths*. The first path required heat to be added to the system and the system to do positive work on its surroundings. The second path required neither of these things. This is why we *cannot* say a particular state of a system *contains* work or heat. Rather, the amount of heat or

BioMedical

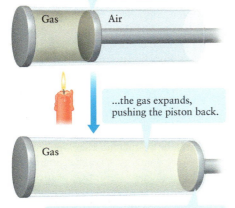

(a) When energy is added slowly to gas trapped in a cylinder with a moveable piston...

...the gas expands, pushing the piston back.

The gas has done work on the piston.

(b) A thin barrier of lightweight material separates a gas from a vacuum.

When the barrier ruptures, the gas expands to fill the entire cylinder. No energy was added to the gas in this case, but the final pressure and volume are the same as in part (a).

Figure 15-4 Two thermodynamic paths (a) If heat is added to an expanding gas at just the right rate, the gas temperature remains constant. (b) The gas temperature is also unchanged in a free expansion. So both of these processes have the same initial and final thermodynamic states but follow different paths between these states.

work required depends on the path the system takes. For any thermodynamic process that takes a system between the same two states, the change in internal energy is the *same*, independent of the path.

BioMedical EXAMPLE 15-1 Cycling It Off

At the end of a 15-min session on a stationary bicycle at the gym, the bike's display indicates that you have delivered 165 kJ of energy to the bike in the form of work. Your heart rate monitor shows that you have "burned off" 155 food calories of energy. How much energy has your body lost to your surroundings in the form of heat?

Set Up

In this problem your body is the thermodynamic system. You do work on your surroundings (the bicycle), so W is positive and equal to 165 kJ. The 155 food calories (155 kcal) that you have "burned off" represent a decrease in your body's internal energy, so ΔU is negative and equal to −155 kcal. We'll use the first law of thermodynamics to determine the heat Q that flows out of your body.

First law of thermodynamics:

$$\Delta U = Q - W \qquad (15\text{-}2)$$

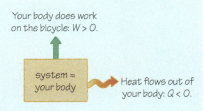

Your body does work on the bicycle: $W > 0$.

system = your body

Heat flows out of your body: $Q < 0$.

Solve

First convert ΔU from kilocalories to kilojoules.

1 kcal = 4186 J = 4.186 kJ, so

$$\Delta U = -155 \text{ kcal}\left(\frac{4.186 \text{ kJ}}{1 \text{ kcal}}\right) = -649 \text{ kJ}$$

Rearrange Equation 15-2 to solve for the quantity of heat Q that flows *into* your body. A negative value of Q means that heat flows *out of* your body.

From Equation 15-2,

$$Q = \Delta U + W$$
$$= -649 \text{ kJ} + 165 \text{ kJ}$$
$$= -484 \text{ kJ}$$

This is negative, so 484 kJ of energy flows out of your body in the form of heat as you exercise.

Reflect

During this exercise session your body expended 649 kJ of internal energy, of which 165 kJ went into doing work and 484 kJ was lost in the form of heat. Only (165 kJ)/(649 kJ) = 0.254, or 25.4%, of the energy that you expended went into doing work. So your body has an *efficiency* of 25.4% in this situation.

Note that 3500 kcal is the energy content of one pound (0.45 kg) of fat. If all of the 155 kcal that you expended came from stored fat, the net result of your exercise is that you will have lost an amount of fat equal to (155 kcal)/(3500 kcal/lb) = 0.044 lb (0.020 kg, or 0.71 ounce). Losing weight requires a *lot* of exercise!

GOT THE CONCEPT? 15-1 An Expanding Ideal Gas

? An ideal gas is enclosed in a thermally isolated cylinder so that no heat can flow into or out of the gas. One end of the cylinder is sealed by a moveable piston like that shown in Figure 15-4a. If you allow the gas to expand slowly by pulling back the piston, what will happen to the temperature T of the gas? (a) T increases. (b) T decreases. (c) T stays the same. (d) T increases to begin with, then decreases. (e) T decreases to begin with, then increases.

TAKE-HOME MESSAGE FOR Section 15-2

✔ The internal energy of a system (the sum of the kinetic energy and potential energy of every atom and molecule in the system) describes the state of the system in a way that does not depend on the specific thermodynamic processes that led to it.

✔ The internal energy of a system changes if heat flows into or out of the system. It also changes if the system does positive or negative work.

✔ The first law of thermodynamics relates the change in internal energy of a thermodynamic system to the heat that enters the system and the work done by that system.

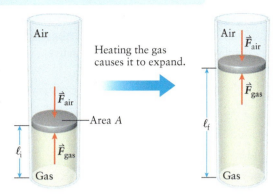

The piston reaches its equilibrium position when the pressure of the gas in the cylinder equals atmospheric pressure.

Heating the gas causes it to expand.

Even after the gas expands, the gas pressure must equal atmospheric pressure. Because neither gas pressure nor atmospheric pressure has changed, the expansion is isobaric.

Figure 15-5 An isobaric expansion The piston exerts a constant pressure on the gas as it is heated, so the pressure of the gas itself remains constant.

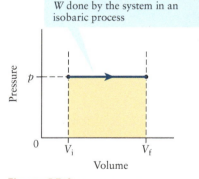

Area under the pV curve = work W done by the system in an isobaric process

Figure 15-6 Work done in an isobaric expansion On a pV diagram (a graph of pressure versus volume) for a thermodynamic process, the area under the curve represents the work done by the system. (If the volume decreases, instead of increasing as shown here, the work done is negative.)

(15-3)

Work done in an isobaric (constant-pressure) process
(15-4)

15-3 A graph of pressure versus volume helps to describe what happens in a thermodynamic process

A thermodynamic process involves a change in the state of a system. In most cases this involves a change in the pressure of the system, the volume that the system occupies, or both. That's the case for the popping popcorn in Figure 15-2 and the gas that expands against a piston in Figure 15-4a. In this section we'll use the first law of thermodynamics to help us analyze four special but important kinds of thermodynamic processes:

- **Isobaric process:** The pressure of the system remains constant.
- **Isothermal process:** The temperature of the system remains constant.
- **Adiabatic process:** There are no heat transfers into or out of the system.
- **Isochoric process:** The volume of the system remains constant.

Many thermodynamic processes do not fit into any of these simple categories. But by analyzing these four special cases, we'll learn general rules that can be applied to *any* thermodynamic process.

Isobaric Processes

Isobaric processes occur with no change in pressure. An example is when you cook food in a frying pan without a lid: Heat flows into the food and its temperature increases, but the pressure that the atmosphere exerts on the food remains constant.

Let's consider an isobaric process in which work is done. Suppose our system is a quantity of gas inside a cylinder that's sealed with a moveable piston of negligible mass, as in Figure 15-5. At equilibrium the upward force on the piston due to the pressure of the gas equals the downward force due to atmospheric pressure. These forces also balance if we heat the gas gradually and allow it to expand slowly so that the piston moves up at constant velocity: The net force on the piston remains zero. Since atmospheric pressure doesn't change, the pressure in the gas must remain constant as well and so the process is isobaric.

To make it easier to visualize and understand what happens in this isobaric process, it's useful to make a *pV* **diagram**. This is a graph that plots the pressure of a system on the vertical axis versus the volume of the system on the horizontal axis. On a *pV* diagram an isobaric process is represented by a horizontal line (Figure 15-6).

In Figure 15-5 the expanding gas does work W on the piston to lift it. To calculate how much work is done, let the constant pressure of the gas be p and the surface area of the piston be A. Figure 15-5 shows that the original length of the cylinder containing the gas is ℓ_i, and the final length of the cylinder after the gas has expanded is ℓ_f. Recall that pressure is force per unit area, so force is pressure multiplied by area. This means that the force that the gas exerts on the piston has magnitude $F_{gas} = pA$. This upward force is in the same direction that the piston moves, and the piston moves a distance $\ell_f - \ell_i$. So the total work done by the gas as it expands is

$$W = F(\ell_f - \ell_i) = pA(\ell_f - \ell_i) = p(A\ell_f - A\ell_i)$$

The volume of a cylinder is equal to the area of its base multiplied by its height. So $A\ell_i$ and $A\ell_f$ are, respectively, the initial volume V_i and final volume V_f occupied by the gas, and we can rewrite Equation 15-3 as

Work that a system does in an isobaric process | Volume of the system **at the end** of the process

$$W = p(V_f - V_i)$$

Constant pressure of the system | Volume of the system **at the beginning** of the process

The work done by the gas in this isobaric process is the pressure p multiplied by the change in volume $(V_f - V_i)$. Figure 15-6 shows that this is the *area* of the shaded region under the straight line that represents the process on the pV diagram. This region is a rectangle with height p and length $V_f - V_i$, so its area equals its height multiplied by its length or $p(V_f - V_i)$.

Equation 15-4 is valid for an isobaric process whether the volume increases or decreases. If $V_f > V_i$ so that the system expands like the gas in Figure 15-5, then $V_f - V_i$ is positive and the system does positive work on its surroundings. Such a process is called an *isobaric expansion*. If instead $V_f < V_i$, which would be like the process in Figure 15-5 in reverse, then $V_f - V_i$ is negative and the system does *negative* work on its surroundings—that is, the surroundings (in Figure 15-5, the piston) do positive work on the system. This is called an *isobaric compression*.

Equation 15-4 is valid *only* for the case of constant pressure. But it turns out to be true in general that in *any* process the work done is the area under the curve that represents the process on a pV diagram. We'll use this idea later in this section.

Go to Picture It 15-1 for more practice dealing with pV diagrams.

EXAMPLE 15-2 Boiling Water Isobarically

Boiling 1.00 g (1.00 cm³) of water at 1.00 atm results in 1671 cm³ of water vapor. A cylinder like that shown in Figure 15-5 contains 1.00 g of water at 100°C. The pressure on the outside of the piston is 1.00 atm. Determine the change in internal energy of the water when just enough heat is added to the water to boil all of the liquid.

Set Up

In this process heat flows into the water to change its phase from liquid to vapor, and the vapor does work on the piston as it expands. The pressure on the outside of the piston is constant, so the pressure that the piston exerts on the water (liquid plus vapor) is also constant. So this is an isobaric process. We'll apply the first law of thermodynamics to determine the change in internal energy.

First law of thermodynamics:

$$\Delta U = Q - W \quad (15\text{-}2)$$

Work done in an isobaric (constant-pressure) process:

$$W = p(V_f - V_i) \quad (15\text{-}4)$$

Heat required for a phase change:

$$Q = \pm mL \quad (14\text{-}21)$$

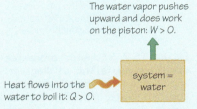

The water vapor pushes upward and does work on the piston: $W > 0$.

Heat flows into the water to boil it: $Q > 0$.

system = water

Solve

Use Equation 14-21 to find the heat Q that flows into the water to vaporize it.

Heat flows into the water, so Q in Equation 14-21 is positive. The mass of water is $m = 1.00$ g, and the latent heat of vaporization is $L_V = 2260 \times 10^3$ J/kg $= 2.26 \times 10^6$ J/kg (from Table 14-4). So

$$Q = mL_V$$

$$= (1.00 \text{ g})\left(\frac{1 \text{ kg}}{10^3 \text{ g}}\right)(2.26 \times 10^6 \text{ J/kg})$$

$$= 2.26 \times 10^3 \text{ J}$$

Use Equation 15-4 to find the work W that the expanding water vapor does on the piston.

The pressure of the water vapor is $p = 1.00$ atm $= 1.01 \times 10^5$ Pa $= 1.01 \times 10^5$ N/m² and remains constant in this isobaric expansion. The volume occupied by the water increases from $V_i = 1.00$ cm³ in the liquid state to $V_f = 1671$ cm³ in the vapor state. In calculating the work done by the expanding vapor, we must convert cm³ to m³ using the relationship 1 m $= 10^2$ cm:

$$W = p(V_f - V_i)$$

$$= (1.01 \times 10^5 \text{ N/m}^2)(1671 \text{ cm}^3 - 1.00 \text{ cm}^3)\left(\frac{1 \text{ m}}{10^2 \text{ cm}}\right)^3$$

$$= (1.01 \times 10^5 \text{ N/m}^2)(1670 \times 10^{-6} \text{ m}^3)$$

$$= 169 \text{ N} \cdot \text{m} = 169 \text{ J}$$

Substitute the values for Q and W into Equation 15-2, the first law of thermodynamics. This tells us the change in internal energy of the water in this process.

$$\Delta U = Q - W$$

$$= 2.26 \times 10^3 \text{ J} - 169 \text{ J}$$

$$= 2.09 \times 10^3 \text{ J}$$

Reflect

We have to add 2.26×10^3 J of energy to boil the water. Of this energy, 2.09×10^3 J goes into increasing the internal energy of the system. The rest of the added energy (169 J) leaves the system in the form of work done on the piston.

(a)

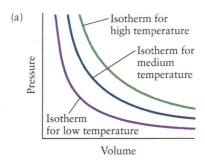

(b)

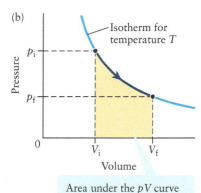

Area under the pV curve
= work W done by the gas
in an isothermal process

Figure 15-7 **Isothermal processes in an ideal gas** (a) This pV diagram shows isotherms (curves of constant temperature) for an ideal gas. (b) The work done by an ideal gas in an isothermal expansion. (If the volume decreases, instead of increasing as shown here, the work done is negative.)

Isothermal Processes

In an isothermal process temperature remains constant. Although the process in **Example 15-2** is isobaric (constant pressure), it is also isothermal: When a substance freezes or boils, the phase transition occurs with no change in temperature.

Gases can also expand or be compressed in such a way as to maintain a constant temperature, even though no phase transition occurs. As we discussed in Section 15-2, the internal energy U of an ideal gas is directly proportional to the Kelvin temperature of the gas. So if the temperature of an ideal gas remains constant during a thermodynamic process, the internal energy U remains constant as well and there is zero change in the internal energy: $\Delta U = 0$. The first law of thermodynamics, Equation 15-2, tells us that $\Delta U = Q - W$, so for an isothermal process that involves an ideal gas, we have

$$0 = Q - W$$

or

$$(15\text{-}5) \qquad Q = W$$

(isothermal process for an ideal gas)

So *when an ideal gas undergoes an isothermal process, the quantity of heat Q added to the gas is equal to the work W done by the gas*. Put another way, whatever energy is added to the gas in the form of heat appears as work done by the gas. If heat flows into the gas so that $Q > 0$, it must be that $W > 0$: The gas does positive work on its surroundings, which means it must expand. This is called an *isothermal expansion*. If heat flows out of the gas so that $Q < 0$, then $W < 0$ and the gas does negative work on its surroundings. This means that the surroundings do positive work *on* the gas, so the volume of the gas decreases. This is called an *isothermal compression*.

How much work is done by a gas that undergoes an isothermal expansion or compression? To get insight into the answer, let's use a pV diagram to depict such a process. The ideal gas law (which we studied in Section 14-3) says that the pressure p, volume V, and temperature T of n moles of an ideal gas are related by

$$(14\text{-}7) \qquad pV = nRT$$

In an isothermal process T is constant, so the product pV is likewise constant. If the gas expands and the volume V increases, the pressure p must decrease; if the gas compresses and the volume V decreases, the pressure p must increase. Figure 15-7a shows three **isotherms**, or curves of constant temperature, on a graph of pressure versus volume.

On a pV diagram, the area under the curve represents the work W done by the gas; Figure 15-7b shows this area for an isothermal expansion. Unlike the isobaric case shown in Figure 15-6, the area under the curve is *not* a rectangle, so we cannot find W simply by multiplying the pressure and the change in volume. An analysis using calculus (which is beyond our scope) shows that the work that the gas does as the volume changes is

Work done by an ideal gas in an isothermal (constant-temperature) process
(15-6)

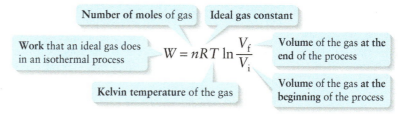

In Equation 15-6 "ln" is the natural logarithm function. It has the properties that $\ln x$ is positive if x is greater than 1 and $\ln x$ is negative if x is less than 1; if $x = 1$, then $\ln x = \ln 1 = 0$. So Equation 15-6 says that the gas does positive work ($W > 0$) if the gas expands, so V_f is greater than V_i and V_f/V_i is greater than 1. The gas does negative work (that is, positive work is done *on* the gas by its surroundings) if the gas is compressed, so V_f less than V_i and V_f/V_i is less than 1. An alternative version of Equation 15-6 in terms of the number of molecules N present in the gas is

$$(15\text{-}7) \qquad W = NkT \ln \frac{V_f}{V_i}$$

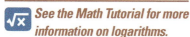

See the Math Tutorial for more information on logarithms.

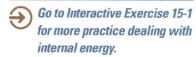

Go to Interactive Exercise 15-1 for more practice dealing with internal energy.

EXAMPLE 15-3 An Isothermal Expansion

A cylinder sealed with a moveable piston as in Figure 15-5 contains 0.100 mol of an ideal gas at a temperature of 295 K. Heat is transferred slowly into the gas, and it is allowed to expand isothermally from an initial volume of 1.00×10^{-3} m^3 to 3.00×10^{-3} m^3 (that is, from 1.00 to 3.00 L). (a) What are the initial and final pressures of the gas? (b) How much heat is transferred into the gas during this process?

Set Up

Since this is an ideal gas, we can use the ideal gas equation to determine the pressure p from the volume V, number of moles n, and temperature T. The internal energy of an ideal gas does not change if the temperature remains constant, so $\Delta U = 0$ and Q (the quantity of heat that flows into the gas, which is what we're trying to find) equals the work W done by the gas. We'll find W and hence Q with Equation 15-6.

Ideal gas law:

$$pV = nRT \qquad (14\text{-}7)$$

Isothermal process for an ideal gas:

$$Q = W \qquad (15\text{-}5)$$

Work done by an ideal gas in an isothermal (constant-temperature) process:

$$W = nRT \ln \frac{V_f}{V_i} \qquad (15\text{-}6)$$

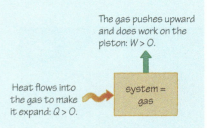

The gas pushes upward and does work on the piston: $W > 0$.

Heat flows into the gas to make it expand: $Q > 0$.

system = gas

Solve

(a) Use Equation 14-7 to find the initial and final pressures.

Rewrite the ideal gas law $pV = nRT$ to solve for the pressure:

$$p = \frac{nRT}{V}$$

The pressure p_i when the gas occupies the initial volume $V_i = 1.00 \times 10^{-3}$ m^3 is

$$p_i = \frac{nRT}{V_i} = \frac{(0.100 \text{ mol})[8.314 \text{ J/(mol·K)}](295 \text{ K})}{1.00 \times 10^{-3} \text{ m}^3}$$

$$= 2.45 \times 10^5 \frac{\text{J}}{\text{m}^3} \left(\frac{1 \text{ N·m}}{1 \text{ J}} \right)$$

$$= 2.45 \times 10^5 \text{ N/m}^2 = 2.45 \times 10^5 \text{ Pa}$$

(We used the relationship that 1 J = 1 N·m.) The pressure when the gas occupies the final volume $V_f = 3.00 \times 10^{-3}$ m^3 is

$$p_f = \frac{nRT}{V_f} = \frac{(0.100 \text{ mol})[8.314 \text{ J/(mol·K)}](295 \text{ K})}{3.00 \times 10^{-3} \text{ m}^3}$$

$$= 8.18 \times 10^4 \text{ Pa}$$

(b) From the first law of thermodynamics, the quantity of heat that enters the gas in this process equals the work done by the gas.

For an isothermal process for an ideal gas, $\Delta U = Q - W = 0$ and $Q = W$. The heat that flows into the gas equals the work that the gas does as it expands.

Use Equation 15-6 to calculate the work done by the gas and so the quantity of heat that is added to the gas.

From Equation 15-6 the work done by the gas is

$$W = nRT \ln \frac{V_f}{V_i}$$

$$= (0.100 \text{ mol})[8.314 \text{ J/(mol·K)}](295 \text{ K}) \times \ln \left(\frac{3.00 \times 10^{-3} \text{ m}^3}{1.00 \times 10^{-3} \text{ m}^3} \right)$$

$$= (245 \text{ J}) \ln 3.00 = (245 \text{ J})(1.10)$$

$$= 269 \text{ J}$$

Since $Q = W$, the quantity of heat added to the gas is $Q = 269$ J.

Adiabatic Processes

In an adiabatic process there is no heat transfer into or out of a system. An important system that undergoes adiabatic processes is our atmosphere. Air is a very poor conductor of heat (see Section 14-7), so there is little heat flow into or out of a mass of air as its pressure and volume change. This helps explain why clouds often appear near mountain summits (Figure 15-8).

When winds carry air up and over mountains, the lower pressure at higher altitudes causes the air mass to expand and do work on the surrounding air ($W > 0$). This expansion occurs nearly adiabatically thanks to the low thermal conductivity of air, so Q is effectively zero. From the first law of thermodynamics, Equation 15-2, the change in the internal energy of the air is

(15-8) $\quad \Delta U = Q - W = 0 - W \quad \text{or} \quad \Delta U = -W \quad$ (adiabatic process)

Since $W > 0$, it follows that $\Delta U < 0$: The internal energy U of the air decreases as it is pushed up the flanks of the mountains. For an ideal gas U is proportional to the Kelvin temperature of the gas, so the air temperature drops by as much as 10°C for every 1 km of elevation gained. As the temperature decreases, water vapor in the air condenses into clouds like the one shown in Figure 15-8.

The air we have described expands as it is pushed up the slopes of the mountains, so this is an *adiabatic expansion*. In an *adiabatic compression* the volume of the gas decreases, so $W < 0$ (the surroundings do work on the gas) and $\Delta U > 0$ (the internal energy and temperature of the gas increase).

Figure 15-8 Adiabatic cooling and mountain clouds The prevailing winds push moist air up and over the mountains that form the Bavarian Alps. The temperature of the uplifted air drops and airborne moisture condenses into droplets, forming cumulus clouds like this one.

imageBROKER/Alamy

WATCH OUT! Remember that heat and temperature are not the same thing.

Even though no heat flows into or out of air pushed up a range of mountains, the temperature of the air changes. If this statement seems contradictory, it's probably because you're still thinking of *heat* as meaning approximately the same thing as *temperature*. Remember that these two concepts are actually quite different: Heat is energy that flows into or out of an object due to a temperature difference, while temperature is a measure of the kinetic energy of an object's molecules. The temperature can change even if there is *no* heat flow, provided work is done by (or done on) the object.

Isochoric Processes

An isochoric process is one for which the volume remains constant. As an example, consider a quantity of gas sealed inside a rigid container. If heat flows into the gas, the pressure and temperature of the gas will increase but the volume of the gas will remain the same because the container cannot expand.

The pressure of the gas exerts forces on the walls of its container, but since the walls do not move (there is no moveable piston), there is zero displacement. Work is force multiplied by displacement, so the gas does *zero* work in an isochoric process: $W = 0$. From the first law of thermodynamics, Equation 15-2,

(15-9) $\quad\quad\quad\quad \Delta U = Q - W = Q - 0 \quad \text{or} \quad \Delta U = Q \quad$ (isochoric process)

This says that for an isochoric process any heat flow into or out of a system goes entirely into changing the internal energy of the system. If Q is positive so that heat flows into the system, the process is called *isochoric heating*; if Q is negative and heat flows out of the system, the process is called *isochoric cooling*.

EXAMPLE 15-4 Two Thermodynamic Processes: Isochoric and Isobaric

The pV diagram in Figure 15-9 shows two thermodynamic processes that occur in an ideal diatomic gas, for which the internal energy is $U = (5/2)nRT$. The gas is in a cylinder with a moveable piston (see Figure 15-5) and is initially in the state labeled A on the diagram, at pressure $p_i = 3.03 \times 10^5$ Pa and volume $V_i = 1.20 \times 10^{-3}$ m^3. The piston is first locked in place so that the volume of the gas cannot change, and the cylinder is cooled so that the temperature and pressure of the gas both decrease. When the gas is in state B, the pressure of the gas is $p_f = 1.01 \times 10^5$ Pa = 1.00 atm, the same as the air pressure outside the cylinder. The piston is then unlocked so that it is free to move, and the cylinder is slowly heated so that the gas expands at constant pressure. The temperature of the gas increases until in the final state (C in Figure 15-9), when the gas is at the same temperature as in the initial state A. (a) What is the final volume of the gas? (b) Find the values of W, ΔU, and Q for the isochoric process $A \to B$. (c) Find the values of W, ΔU, and Q for the isobaric process $B \to C$. (d) Find the values of W, ΔU, and Q for the net process $A \to B \to C$.

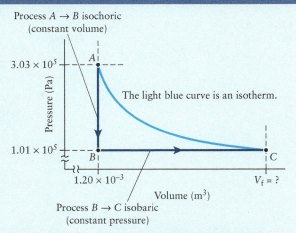

Figure 15-9 Isochoric cooling and isobaric expansion In this two-step process for an ideal gas, how much work is done, how much heat flows into the gas, and by how much does the internal energy change?

Set Up

We'll use the ideal gas law to determine the final volume of the gas. For the isochoric (constant-volume) process $A \to B$, the gas does no work, so the quantity of heat Q that enters the gas is equal to the internal energy change ΔU of the gas. (The gas is *cooled* in this process, so $Q < 0$ and $\Delta U < 0$.) We'll determine ΔU from the temperature change. For the isobaric (constant-pressure) process $B \to C$, the work W that the gas does is given by Equation 15-4; this is positive because the gas expands. The internal energy change ΔU is again given by the temperature change. We'll then determine Q for this process from the first law of thermodynamics.

Ideal diatomic gas:

$$U = \frac{5}{2}nRT$$

Ideal gas law:

$$pV = nRT \qquad (14\text{-}7)$$

First law of thermodynamics:

$$\Delta U = Q - W \qquad (15\text{-}2)$$

Work done in an isobaric (constant-pressure) process:

$$W = p(V_f - V_i) \qquad (15\text{-}4)$$

Isochoric (constant-volume) process:

$$\Delta U = Q \qquad (15\text{-}9)$$

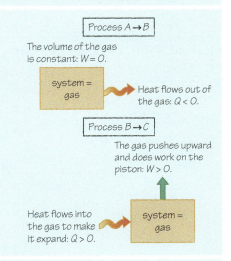

Solve

(a) Use the ideal gas law to find the final volume V_f.

The initial pressure and volume of the gas are

$p_i = 3.03 \times 10^5$ Pa
$V_i = 1.20 \times 10^{-3}$ m^3

From the ideal gas law, Equation 14-7,

$p_i V_i = nRT_i$

where T_i is the initial temperature of the gas in state A.

The final pressure of the gas is

$p_f = 1.01 \times 10^5$ Pa

The ideal gas law tells us that

$p_f V_f = nRT_f$

where T_f is the final temperature of the gas in state C. The initial and final temperatures are the same ($T_i = T_f$), so

$p_i V_i = p_f V_f$

$$V_f = \frac{p_i V_i}{p_f} = \frac{(3.03 \times 10^5 \text{ Pa})(1.20 \times 10^{-3} \text{ m}^3)}{1.01 \times 10^5 \text{ Pa}} = 3.60 \times 10^{-3} \text{ m}^3$$

The final pressure is one-third of the initial pressure, and the final volume is three times the initial volume.

(b) For the isochoric process $A \to B$, the work done is $W_{A \to B} = 0$. Find the change in internal energy ΔU and quantity of heat Q for this process.

The change in internal energy U is the difference between the value of U in state B and the value of U in state A:

$$\Delta U_{A \to B} = U_B - U_A = \frac{5}{2} nRT_B - \frac{5}{2} nRT_i$$

We don't know the number of moles of gas n, nor do we know the temperatures T_i (in the initial state A, for which the pressure is p_i and the volume is V_i) and T_B (in state B, for which the pressure is p_f and the volume is V_i). But from the ideal gas law $pV = nRT$, so

$$p_i V_i = nRT_i \text{ and } p_f V_i = nRT_B$$

So the change in internal energy is

$$\Delta U_{A \to B} = \frac{5}{2} p_f V_i - \frac{5}{2} p_i V_i = \frac{5}{2} (p_f - p_i) V_i$$
$$= \frac{5}{2} (1.01 \times 10^5 \text{ Pa} - 3.03 \times 10^5 \text{ Pa})(1.20 \times 10^{-3} \text{ m}^3)$$
$$= -606 \text{ J}$$

(Remember that $1 \text{ Pa} = 1 \text{ N/m}^2$ and $1 \text{ N} \cdot \text{m} = 1 \text{ J}$.) From Equation 15-9 for an isochoric process,

$$Q_{A \to B} = \Delta U_{A \to B} = -606 \text{ J}$$

Heat flows out of the gas and the internal energy of the gas decreases.

(c) Find W, ΔU, and Q for the isobaric process $B \to C$.

The work done in an isobaric process is given by Equation 15-4. The constant pressure is $p_f = 1.01 \times 10^5$ Pa, and the volume increases from $V_i = 1.20 \times 10^{-3}$ m^3 to $V_f = 3.60 \times 10^{-3}$ m^3:

$$W_{B \to C} = p_f(V_f - V_i)$$
$$= (1.01 \times 10^5 \text{ Pa}) (3.60 \times 10^{-3} \text{ m}^3 - 1.20 \times 10^{-3} \text{ m}^3)$$
$$= +242 \text{ J}$$

The gas does positive work as it expands. As in part (b) we calculate the internal energy change with the aid of the ideal gas law:

$$\Delta U_{B \to C} = U_C - U_B = \frac{5}{2} nRT_f - \frac{5}{2} nRT_B$$
$$= \frac{5}{2} p_f V_f - \frac{5}{2} p_f V_i = \frac{5}{2} p_f (V_f - V_i)$$

This is just $5/2$ times the above expression for $W_{B \to C}$, so

$$\Delta U_{B \to C} = \frac{5}{2} W_{B \to C} = \frac{5}{2} (+242 \text{ J}) = +606 \text{ J}$$

From the first law of thermodynamics, the heat that flows into the gas in this process is

$$Q_{B \to C} = \Delta U_{B \to C} + W_{B \to C} = (+606 \text{ J}) + (+242 \text{ J})$$
$$= +848 \text{ J}$$

(d) Find the values of W, ΔU, and Q for the combined process $A \to B \to C$.

The total amount of work done by the gas is

$$W_{A \to B} + W_{B \to C} = 0 + (+242 \text{ J}) = +242 \text{ J}$$

The total change in internal energy of the gas is

$$\Delta U_{A \to B} + \Delta U_{B \to C} = (-606 \text{ J}) + (+606 \text{ J}) = 0$$

The total quantity of heat that flows into the gas is

$$Q_{A \to B} + Q_{B \to C} = -606 \text{ J} + (+848 \text{ J}) = +242 \text{ J}$$

Reflect

For the combined process $A \to B \to C$, there is *zero* net change in the internal energy U of the ideal gas. That's because U depends only on the number of moles of gas and the temperature, which have the same values in the final state C as in the initial state A. In the combined process 242 J of heat flows into the gas, and the gas uses this energy to do 242 J of work.

GOT THE CONCEPT? 15-2 Ranking Thermodynamic Processes

The pV diagram in Figure 15-10 shows four thermodynamic processes for a certain quantity of an ideal gas: (i) $a \rightarrow b$, (ii) $a \rightarrow c$, (iii) $a \rightarrow d$, and (iv) $a \rightarrow e$. All four processes start in the same initial state a. The volume is the same for states b, c, and d. Process $a \rightarrow c$ lies along an isotherm. (a) Rank these four processes according to the internal energy change ΔU of the gas, from most positive to most negative. If two processes have the same value of ΔU, say so. (b) Rank these four processes according to the work W done by the gas, from most positive to most negative. If two processes have the same value of W, say so.

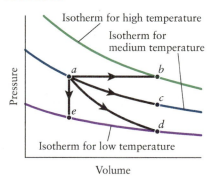

Figure 15-10 Four thermodynamic processes What are the properties of the processes $a \rightarrow b$, $a \rightarrow c$, $a \rightarrow d$, and $a \rightarrow e$?

GOT THE CONCEPT? 15-3 Which Could Be Adiabatic?

The pV diagram in Figure 15-10 shows four thermodynamic processes for a certain quantity of an ideal gas: $a \rightarrow b$, $a \rightarrow c$, $a \rightarrow d$, and $a \rightarrow e$. All four processes start in the same initial state a. The volume is the same for states b, c, and d. Process $a \rightarrow c$ lies along an isotherm. One of these processes could be adiabatic. Which one is it? (a) $a \rightarrow b$; (b) $a \rightarrow c$; (c) $a \rightarrow d$; (d) $a \rightarrow e$; (e) not enough information given to decide.

TAKE-HOME MESSAGE FOR Section 15-3

✔ A pV diagram can be used to describe thermodynamic processes that take a system from one state to another.

✔ Isobaric processes occur with no change in pressure.

✔ Temperature remains constant in isothermal processes.

✔ There is no heat flow into or out of a system during an adiabatic process.

✔ Isochoric processes occur with no change in volume.

15-4 The concept of molar specific heat helps us understand isobaric, isochoric, and adiabatic processes for ideal gases

Our discussion in Section 15-3 of various types of thermodynamic processes was fairly general. Let's now look more closely at some of these processes for the special case in which the thermodynamic system is an *ideal gas*. This special case has a tremendous number of applications: It will help us understand why a bicycle pump gets warm when you use it to inflate a tire and how a diesel engine can make a fuel–air mixture ignite even though there are no spark plugs in such an engine.

Specific Heats of an Ideal Gas

Suppose you have a quantity of ideal gas. How much heat does it take to raise the temperature of the gas to a new, higher temperature? The best answer is "It depends on how you do it." The pV diagram in Figure 15-11 helps us understand why this is so.

This figure shows two possible ways that an ideal gas in a cylinder with a moveable piston can be slowly heated from a lower temperature to a higher one. Process $A \rightarrow B$ is *isochoric*: It takes place

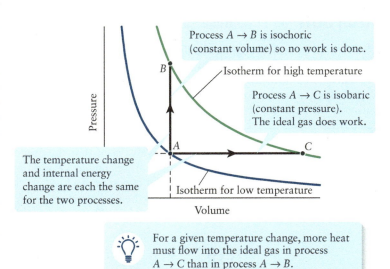

Process $A \rightarrow B$ is isochoric (constant volume) so no work is done.

Isotherm for high temperature

Process $A \rightarrow C$ is isobaric (constant pressure). The ideal gas does work.

The temperature change and internal energy change are each the same for the two processes.

Isotherm for low temperature

For a given temperature change, more heat must flow into the ideal gas in process $A \rightarrow C$ than in process $A \rightarrow B$.

Figure 15-11 Two ways to increase temperature Processes $A \rightarrow B$ and $A \rightarrow C$ for an ideal gas lead to different final states but the same final temperature.

at constant volume (the moveable piston is locked in position), so the pressure of the gas increases as you heat the gas. Process $A \to C$ is *isobaric*: It takes place at constant pressure. The piston is allowed to move so that the gas expands as it is heated, and the pressure inside the cylinder remains equal to the pressure on the other side of the piston. The final temperature is the same for both processes because both state B and state C are on the same isotherm in Figure 15-11.

The internal energy U of an ideal gas depends only on the number of moles and the temperature, so the internal energy change ΔU is the same for both processes. But the amount of work W done by the gas is *different* for the two processes: $W = 0$ for process $A \to B$ (the gas does no work if the volume it occupies doesn't change), while $W > 0$ for process $A \to C$ (the expanding gas does work on the piston to make it move). From the first law of thermodynamics, Equation 15-2,

$$\Delta U = Q - W \quad \text{so} \quad Q = \Delta U + W$$

Because ΔU is the same for both processes, *more* heat is required for process $A \to C$ (in which positive work is done) than for process $A \to B$ (for which $W = 0$). So the heat required to increase the temperature of an ideal gas is *greater* if the temperature is increased at constant *pressure* than if it is increased at constant *volume*.

Let's be more quantitative about the quantity of heat required to change the temperature of a certain quantity of ideal gas by a given amount. In Section 14-5 we wrote an equation for the heat required to change the temperature of a mass m of substance by an amount ΔT:

(14-20)
$$Q = mc\,\Delta T$$

In Equation 14-20 the quantity c is the *specific heat* of the substance. For an ideal gas we usually write the internal energy U in terms of the number of *moles* of gas present rather than the amount of mass. We follow the same approach for the quantity of heat Q, so we write

Quantity of heat and the resulting temperature change in terms of molar specific heat
(15-10)

Quantity of heat that flows into (if $Q > 0$) or out of (if $Q < 0$) an object

Molar specific heat of the substance of which the object is made

$$Q = nC\,\Delta T$$

Number of moles present of the substance

Temperature change of the object that results from the heat flow

The quantity C in Equation 15-10 is called the **molar specific heat** of the substance. It has units of joules per mole per kelvin, or $J/(mol \cdot K)$.

WATCH OUT! Don't confuse specific heat and molar specific heat.

It's important to recognize the difference between the specific heat c that appears in Equation 14-20 and the molar specific heat C in Equation 15-10. Specific heat c (lowercase) tells you the number of joules of heat required to raise the temperature of one *kilogram* of a substance by

one kelvin. Molar specific heat C (uppercase) tells you the number of joules of heat required to raise the temperature of one *mole* of the substance by one kelvin. If the substance has a molar mass (mass per mole) of M, the two quantities are related by $C = Mc$.

We've seen that the quantity of heat Q required to make an ideal gas undergo a temperature change ΔT is different if the pressure p is held constant than if the volume V is held constant. To keep track of this difference, we use the symbol C_p for the **molar specific heat at constant pressure** and the symbol C_V for the **molar specific heat at constant volume**. Then we can write Equation 15-10 for these two special cases as

(15-11)
$$Q = nC_p\,\Delta T \quad \text{(constant pressure)}$$
$$Q = nC_V\,\Delta T \quad \text{(constant volume)}$$

Our discussion of Figure 15-11 showed that for a given temperature increase ΔT, the value of Q was greater for a constant-pressure process than for a constant-volume process. So we conclude that C_p, the molar specific heat at constant pressure, must be greater than the molar specific heat at constant volume C_V. Table 15-1, which lists experimental values of C_p and C_V for various gases, shows that this is indeed the case.

TABLE 15-1 Molar Specific Heats of Gases (in J/[mol·K] for a gas at $T = 300$ K and $p = 1$ atm, except where otherwise indicated)

	Gas	C_p	C_V	$C_p - C_V$	$\gamma = C_p/C_V$
Monatomic	Ar	20.8	12.5	8.28	1.66
	He	20.8	12.5	8.28	1.66
	Ne	20.8	12.7	8.08	1.63
Diatomic	H_2	28.8	20.4	8.37	1.41
	N_2	29.1	20.8	8.33	1.40
	O_2	29.4	21.1	8.33	1.39
	CO	29.3	21.0	8.28	1.39
Triatomic	CO_2	37.0	28.5	8.49	1.30
	SO_2	40.4	31.4	9.00	1.29
	H_2O (373 K)	34.3	25.9	8.37	1.32

C_p, C_V, and Degrees of Freedom

Notice the following from Table 15-1:

(a) The *difference* $C_p - C_V$ between the two different molar specific heats has almost the same value for *all* gases.

(b) The value of C_V is nearly the same for all monatomic gases, nearly the same for all diatomic gases, and nearly the same for all triatomic gases. The value increases as the number of atoms increases.

We can understand the first of these observations by using Equation 15-10, the first law of thermodynamics ($\Delta U = Q - W$), and the ideal gas law ($pV = nRT$). For a constant-volume process $W = 0$, so from the second of Equations 15-11 the internal energy change for n moles of an ideal gas that undergoes a given temperature change ΔT is

$$\Delta U = Q - W = nC_V \Delta T - 0 = nC_V \Delta T \tag{15-12}$$

(ideal gas, constant volume)

For a constant-*pressure* process that starts at the same temperature, the gas changes its volume by ΔV and does an amount of work

$$W = p\Delta V = p(V_f - V_i) = pV_f - pV_i$$

where V_f and V_i are the final and initial volumes of the gas. But from the ideal gas law,

$$pV_f = nRT_f \quad \text{and} \quad pV_i = nRT_i$$

where T_f and T_i are the final and initial temperatures of the gas, respectively. The temperature change of the gas is $\Delta T = T_f - T_i$, so we can write the work done by the gas in a constant-pressure process as

$$W = pV_f - pV_i = nRT_f - nRT_i = nR(T_f - T_i) = nR\Delta T \tag{15-13}$$

(ideal gas, constant pressure)

Using Equation 15-13 and the first of Equations 15-11, we can now write the internal energy change for this constant-pressure process as

$$\Delta U = Q - W = nC_p \Delta T - nR\Delta T = n(C_p - R)\Delta T \tag{15-14}$$

(ideal gas, constant pressure)

Equation 15-12 for a constant-volume process and Equation 15-14 for a constant-pressure process look rather different from each other. But we know that the internal energy change ΔU for an ideal gas is the *same* in both cases, as long as the number of moles n and the temperature change ΔT are the same for both processes. So ΔU in Equation 15-12 is equal to ΔU in Equation 15-14:

$$nC_V\Delta T = n(C_p - R)\Delta T$$

For this to be true it must be that $C_V = C_p - R$, or

Molar specific heats
of an ideal gas
(15-15)

Molar specific heat of an ideal gas at **constant pressure**

$$C_p - C_V = R \quad \text{Ideal gas constant}$$

Molar specific heat of an ideal gas at **constant volume**

Equation 15-15 says that for an ideal gas C_p is greater than C_V by $R = 8.314$ J/(mol·K), the ideal gas constant. The experimentally determined values of $C_p - C_V$ given in Table 15-1 for *real* gases are very close to this theoretical prediction.

How can we understand our second observation from Table 15-1, that the value of C_V is essentially the same for all gases with the same number of atoms? To see the answer, note from Equation 15-12 that for an ideal gas with constant volume $\Delta U = nC_V\Delta T$. In other words, C_V tells us about the change in internal energy of a gas that results from a temperature change. We learned something important about the internal energy of an ideal gas in Section 14-3: On average each molecule in an ideal gas at temperature T has energy $(1/2)kT$ for each of its *degrees of freedom*. (You'll find it helpful to review the discussion of degrees of freedom in Section 14-3. Recall that k is the Boltzmann constant, equal to the ideal gas constant R divided by Avogadro's number N_A.) A monatomic gas has three degrees of freedom, corresponding to its ability to move in the x, y, and z directions. A diatomic gas at room temperature has two additional degrees of freedom that correspond to its ability to rotate around either of two axes perpendicular to the long axis of the molecule, and a triatomic gas (which at room temperature can both rotate and vibrate) has even more degrees of freedom.

We'll use the symbol D to represent the number of degrees of freedom. Then the average energy per gas molecule is $(D/2)kT$. One mole of the gas contains Avogadro's number of molecules, so the energy per mole is $N_A(D/2)kT = (D/2)(N_Ak)T = (D/2)RT$. (Again recall that $k = R/N_A$, so $R = N_Ak$.) Then the internal energy of n moles of ideal gas is

(15-16)

$$U = n\left(\frac{D}{2}R\right)T$$

(ideal gas, D degrees of freedom)

If the temperature of the gas changes by ΔT, it follows from Equation 15-16 that the internal energy change is

(15-17)

$$\Delta U = n\left(\frac{D}{2}R\right)\Delta T$$

(ideal gas, D degrees of freedom)

But we saw above that for an ideal gas $\Delta U = nC_V\Delta T$. Comparing this to Equation 15-17 we see that

Molar specific heat at constant
volume for an ideal gas
(15-18)

Number of degrees of freedom for a molecule of the gas

Molar specific heat of an
ideal gas at **constant volume**
$$C_V = \left(\frac{D}{2}\right)R \quad \text{Ideal gas constant}$$

Let's see how well Equation 15-18 works for the real gases listed in Table 15-1. For a monatomic gas there are three degrees of freedom per molecule, so $D = 3$ and Equation 15-18 becomes

$$C_V = \frac{3}{2}R = \frac{3}{2}(8.314 \text{ J/[mol·K]}) = 12.47 \text{ J/(mol·K)} \quad \text{(monatomic gas)}$$

Table 15-1 shows that this is a very close fit to the experimental values of C_V for argon, helium, and neon. For a diatomic gas with five degrees of freedom ($D = 5$), we expect

$$C_V = \frac{5}{2}R = \frac{5}{2}(8.314 \text{ J/[mol·K]}) = 20.79 \text{ J/(mol·K)} \quad \text{(diatomic gas)}$$

This is also a very close fit to the experimental values for the diatomic atoms listed in Table 15-1. Indeed, this is how we know that such molecules have five degrees of freedom! Experiments show that at temperatures well above room temperature, the value of C_V for diatomic molecules increases and becomes closer to $7R/2 = 29.10$ J/(mol·K). What's happening is that at higher temperatures it becomes possible for these molecules to vibrate, giving each molecule two additional degrees of freedom (one for the kinetic energy of vibration, the other for the potential energy of vibration).

We encourage you to use Equation 15-18 to determine the number of degrees of freedom for the triatomic gases listed in Table 15-1.

The molar specific heat at constant pressure C_p also depends on D, the number of degrees of freedom. Since $C_p - C_V = R$ (Equation 15-15) and $C_V = (D/2)R$ (Equation 15-18), you can see that

$$C_p = C_V + R = \left(\frac{D}{2}\right)R + R = \left(\frac{D+2}{2}\right)R \tag{15-19}$$

We can get an expression for the number of degrees of freedom D by taking the *ratio* of C_p to C_V. This **ratio of specific heats** is denoted by the Greek letter γ ("gamma"):

$$\gamma = \frac{C_p}{C_V} \quad \text{(ratio of specific heats)} \tag{15-20}$$

If we substitute Equation 15-18 for C_V and Equation 15-19 for C_p into Equation 15-20, we get

$$\gamma = \frac{C_p}{C_V} = \frac{\left(\dfrac{D+2}{2}\right)R}{\left(\dfrac{D}{2}\right)R} = \frac{D+2}{D} \quad \text{(ideal gas)} \tag{15-21}$$

You can check Equation 15-21 against the experimental values for γ given in Table 15-1. For example, for a monatomic ideal gas with three degrees of freedom ($D = 3$), Equation 15-21 predicts

$$\gamma = \frac{3+2}{3} = 1.67$$

which is a very good match to the experimental values for argon, helium, and neon. Try Equation 15-21 for a diatomic gas with $D = 5$; how well does the predicted value of γ compare to the experimental values in Table 15-1?

Adiabatic Processes for an Ideal Gas

The ratio of specific heats $\gamma = C_p/C_V$ for an ideal gas plays an important role in *adiabatic* processes in which there is zero heat flow into or out of the gas. If we have a quantity of gas in a cylinder with a moveable piston and then allow the piston to move in or out, the gas will compress or expand adiabatically if the cylinder and piston are made of an insulating material that does not allow heat to flow through it. It can be shown using calculus (which is beyond our scope) that if an ideal gas expands or contracts adiabatically, the pressure p and volume V of the gas are related by

Pressure of an ideal gas   Volume of an ideal gas

Pressure and volume for ideal gas, adiabatic process
(15-22)

$$pV^\gamma = \text{constant}$$

Ratio of specific heats of the gas = C_p/C_V

The quantity pV^γ has the same value at all times during an **adiabatic process** for an ideal gas.

Figure 15-12 Adiabatic expansion of an ideal gas When an ideal gas expands adiabatically from an initial volume to a greater final volume, its pressure drops more than in an isothermal expansion between the same two volumes. Hence the gas temperature decreases. In the reverse process, an adiabatic compression, the gas temperature increases and the pressure increases more than in an isothermal compression between the same two volumes.

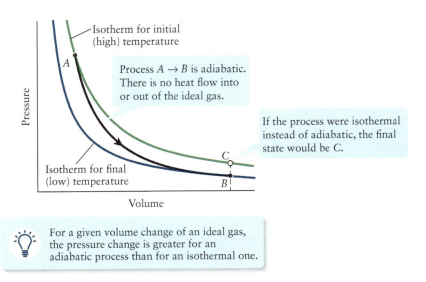

As long as the thermodynamic changes that take place in an ideal gas are adiabatic, the product of pressure and volume to the power γ remains the same. As volume increases in an adiabatic expansion, pressure decreases; as volume decreases in an adiabatic compression, pressure increases. Note that in an adiabatic process, the pressure change is *greater* than in an isothermal expansion or compression between the same initial and final volumes. You can see this in **Figure 15-12**. The path in black is that followed in an adiabatic expansion $A \rightarrow B$ of an ideal gas. If the expansion were isothermal instead, but the initial and final volumes were the same, the path followed would be along the green isotherm in Figure 15-12, which leads to the point labeled C. The pressure at B is less than at C, so there is a greater pressure decrease in an adiabatic expansion than in an isothermal one.

Figure 15-12 also shows that the final state of the ideal gas after an adiabatic expansion lies on an isotherm for a lower temperature than for the initial state. In other words, the temperature of an ideal gas decreases in an adiabatic expansion. This makes sense based on what we learned about adiabatic processes in Section 15-3: In an adiabatic expansion $Q = 0$ and $W > 0$ (the gas does work on its surroundings as it expands), so $\Delta U < 0$: The internal energy of the gas decreases. The internal energy of an ideal gas is proportional to its absolute temperature T, so it follows that T decreases in an adiabatic expansion of an ideal gas. In the same way, the temperature of an ideal gas increases in an adiabatic compression.

We can find a relationship between temperature and volume in an adiabatic process for an ideal gas. To do this we'll use the ideal gas law $pV = nRT$, which we can rewrite as $p = nRT/V$. If we substitute this expression for the pressure p into Equation 15-22, we get

$$\frac{nRT}{V}V^{\gamma} = nRTV^{\gamma-1} = \text{constant}$$

Since the number of moles n does not change (no gas enters or leaves the system) and R is itself a constant, we can rewrite this expression as

Temperature and volume for ideal gas, adiabatic process (15-23)

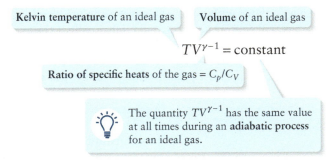

Note that the value of γ is greater than 1 for all gases (see Table 15-1), so $\gamma - 1$ is greater than zero. Hence as the volume V of an ideal gas increases in an adiabatic expansion, the quantity $V^{\gamma-1}$ increases as well. For the quantity $TV^{\gamma-1}$ to maintain the same value, the temperature T must decrease in an adiabatic expansion, just as we described above. In the same way, as volume decreases in an adiabatic compression of an ideal gas, temperature increases.

Equation 15-23 explains what happens when you use a bicycle pump to inflate a tire: The pump rapidly becomes warm to the touch. Although the pump is not made of an insulating material, the air in the pump cylinder is compressed so rapidly that there's no time for heat to flow into or out of the air during the compression. As a result, the compression is adiabatic, so the temperature of the gas (and the pump that holds it) increases as the air is compressed.

You can demonstrate an adiabatic expansion using your own breath. If you open your mouth and breathe on the back of your hand, you can feel that the expelled breath is warm. But the same breath feels cold if you form your lips into a small "O" and then breathe on the back of your hand. In the latter case the air expands rapidly as it exits through the small aperture through your lips, and this expansion is nearly adiabatic. Hence the expelled air drops in temperature and feels cold on your hand.

EXAMPLE 15-5   Burning Paper Without Heat

In a classroom physics demonstration a piece of paper is placed in the bottom of a test tube with a sealed plunger. When the plunger is quickly depressed, compressing the air inside, the paper combusts. How is this possible? Let the initial volume of the gas in the test tube be 10.0 cm^3 and the final volume be 1.00 cm^3, and let the initial temperature be room temperature (20°C). Calculate the final temperature of the gas in the test tube. Assume that air is an ideal, diatomic gas.

Set Up

Because the compression happens quickly, there is no time for any heat transfer into or out of the air in the tube. So we can consider this to be an adiabatic process. We use Equation 15-23 to relate volume and temperature for this process, and we use Equation 15-21 to find the ratio of specific heats γ for air. For diatomic gases, there are five degrees of freedom.

Ideal gas, adiabatic process:

$$TV^{\gamma-1} = \text{constant} \quad (15\text{-}23)$$

Ratio of specific heats for an ideal gas with D degrees of freedom:

$$\gamma = \frac{C_p}{C_V} = \frac{D+2}{D} \quad (15\text{-}21)$$

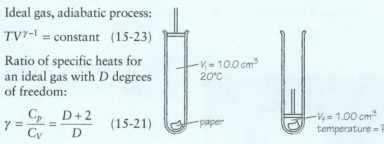

$V_i = 10.0\ \text{cm}^3$
$20°C$
paper

$V_f = 1.00\ \text{cm}^3$
temperature = ?

Solve

First find the value of γ for the gas and convert the temperature from Celsius to Kelvin.

From Equation 15-21 with $D = 5$,

$$\gamma = \frac{C_p}{C_V} = \frac{5+2}{5} = \frac{7}{5} = 1.40$$

The initial temperature of the air is 20.0°C or

$$T_i = (20 + 273.15)\ \text{K} = 293\ \text{K}$$

We know the initial values of the temperature ($T_i = 293$ K) and volume ($V_i = 10.0$ cm^3) and the final value of the volume ($V_f = 1.00$ cm^3). Use these in Equation 15-23 to solve for the final value of temperature.

Equation 15-23 says that the quantity $TV^{\gamma-1}$ has the same value after the compression as before the compression, so

$$T_f V_f^{\gamma-1} = T_i V_i^{\gamma-1}$$

Solve for the final temperature T_f:

$$T_f = \frac{T_i V_i^{\gamma-1}}{V_f^{\gamma-1}} = T_i \left(\frac{V_i}{V_f}\right)^{\gamma-1}$$

$$= (293\ \text{K})\left(\frac{10.0\ \text{cm}^3}{1.00\ \text{cm}^3}\right)^{\frac{7}{5}-1}$$

$$= (293\ \text{K})(10.0)^{\frac{2}{5}}$$

$$= 736\ \text{K or } 463°C$$

Reflect

The final temperature is above the flash point of paper (the temperature at which paper catches fire—around 230°C), so the paper will burn until the supply of paper or oxygen is used up. This same process is used in diesel engines: A mixture of fuel and air is rapidly compressed in the cylinders of the engine, raising the temperature to such a high value that the mixture spontaneously ignites without a spark plug (as is used in a conventional gasoline engine).

GOT THE CONCEPT? 15-4 Heating Helium

(?) A quantity of helium gas at room temperature is placed in a cylinder with a moveable piston. If the cylinder is locked in place so that the volume cannot change and a quantity of heat Q_0 is added to the gas, the temperature of the gas increases by ΔT_0. Now the gas is cooled to its original temperature, and the piston is unlocked so that the gas is free to expand. If the pressure exerted on the gas by the piston is held constant and the same quantity of heat Q_0 is added to the gas, what will be the temperature change of the gas? (a) ΔT_0, the same as before; (b) less than ΔT_0; (c) more than ΔT_0.

GOT THE CONCEPT? 15-5 Helium Versus Nitrogen

(?) Two cylinders, each equipped with a moveable piston, contain equal numbers of moles of a gas. One cylinder contains helium (He), while the other contains nitrogen (N_2). If the same quantity of heat Q is added to each gas and each is allowed to expand at constant pressure, what can you say about the temperature change ΔT and amount of work done W for the two gases? (a) ΔT and W are the same for both gases; (b) ΔT and W are both less for He than for N_2. (c) ΔT and W are both greater for He than for N_2. (d) ΔT is less for He than for N_2, but W is greater for He than for N_2. (e) ΔT is greater for He than for N_2, but W is less for He than for N_2. (f) None of these.

TAKE-HOME MESSAGE FOR Section 15-4

✔ More heat must be added to a gas at constant pressure than at constant volume to produce the same change in temperature.

✔ Thermodynamic processes that are carried out quickly are often adiabatic because heat transfer is relatively slow.

15-5 The second law of thermodynamics describes why some processes are impossible

Figure 15-13 shows two frames from a video in which a person inflates a balloon until it pops. How do you know that the left frame happens before the right one? None of the physical laws that we have studied so far preclude a set of balloon fragments from flying toward each other and reassembling as an intact balloon.

For example, neither conservation of energy nor momentum would be violated were the balloon fragments to reassemble. What about the pictures of a dropped popsicle melting on a carpet in Figure 15-14? Each image was taken after the one to its left, but would it be possible for this sequence to run in reverse? Here again the reverse process would be permitted by the fundamental laws we know. The total energy of the system would be conserved if the carpet drew enough energy from the melted popsicle (mostly water) to cause it to freeze: The amount of energy required to melt a chunk of ice is exactly the same as the energy needed to freeze the resulting water. Yet we have declared that heat transfer is always from a hotter to a cooler object. Why?

For both the balloon in Figure 15-13 and the popsicle in Figure 15-14, events happen in a sequence that tends to decrease the level of *organization* or *order*. In Figure 15-13 the air that was confined within the balloon in an orderly way spreads out across the room when the balloon is popped, and the balloon itself breaks into a collection of random fragments. In Figure 15-14 the water molecules begin in an

Figure 15-13 Bursting your balloon Both images were taken from a video of a person blowing up a balloon until it pops. How do you know that the top frame happens before the bottom one?

Figure 15-14 Melting a popsicle A dropped popsicle melts atop a piece of carpet. Each image in the sequence was taken after the one to its left. Could this sequence run in reverse?

ordered state in which they are locked into the crystalline structure of the popsicle, and they end up free to move in a disordered way into the fibers of the carpet. The reverse processes, in which air and balloon fragments spontaneously coalesce into a filled balloon or water spontaneously freezes, never occur in nature. That is, natural events always happen in a direction from ordered to disordered, or toward increasing *randomness*. We can generalize this observation to the **second law of thermodynamics**:

> *The amount of disorder in an isolated system either always increases or, if the system is in equilibrium, stays the same.*

We emphasize that a system described by the second law of thermodynamics must be *isolated*. This is to ensure that nothing outside the system can cause its state to change in an unnatural way. For example, we could manually sort out the fragments of the balloon in Figure 15-13, carefully reassemble them, and then inflate the reassembled balloon. The balloon system would experience increasing order, but only as a result of our external intervention; we could not treat the balloon as isolated.

Let's see how to use the second law of thermodynamics to analyze systems that convert heat to work or work to heat. We'll look in particular at mechanical systems such as the internal combustion engine in a car (which takes the heat released by burning fuel and converts it to work) and a kitchen refrigerator (which takes work done by the refrigerator motor and uses it to transfer heat out of the refrigerator's contents). We'll see that the second law imposes strict limits on how efficient an engine can be and on how much performance you can get from a refrigerator.

In the following section we'll extend the second law to include the concept of entropy, which is a measure of the amount of randomness in a system. We'll use the entropy concept to gain further insight into the second law and to see what this law tells us about living systems.

Heat Engines and the Second Law

A system or device that converts heat to work is a **heat engine** or simply an *engine*. Heat engines are cyclic: Some part of the system absorbs energy, work is done, and the system returns to its original state for the cycle to begin again. Although the term "engine" might call to mind the complex device that powers an automobile, in thermodynamics a heat engine can be as simple as gas in a piston that expands as heat is added and contracts as the gas cools.

There are actually tiny molecular engines in living cells. In the mitochondria of all human cells, hydrogen ions flow through ATP synthase, an enzyme that has a structure not unlike a waterwheel (Figure 15-15). ATP synthase converts the kinetic energy of hydrogen ions into rotational kinetic energy of the enzyme, which in turn is used to convert adenosine diphosphate (ADP) into adenosine triphosphate (ATP). The chemical bond energy in the resulting ATP is in a form that can be used by all cells.

Figure 15-16 shows a simple model for any heat engine. The engine is thermally connected to a *reservoir* of higher (hotter) temperature (T_H) and to a reservoir of lower (cooler) temperature (T_C). A **reservoir** is a part of a system large enough to either absorb or supply heat without a change in temperature. In an old-time steam engine, for example, the furnace serves as the hot reservoir and the surrounding atmosphere acts as the cold reservoir. Energy Q_H flows from the hot reservoir into the engine; during this process some of the energy goes into work W, and the remainder $|Q_C|$ flows out of the engine and into the cold reservoir in the form of heat.

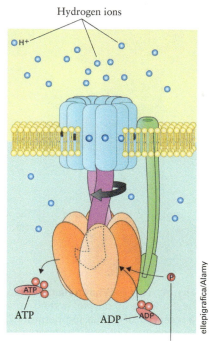

Hydrogen ions

H+

Phosphate ion

ellepigrafica/Alamy

Figure 15-15 A molecular heat engine This illustration shows an ATP synthase enzyme attached to the inner membrane (yellow) of a mitochondrion. The enzyme extracts energy from hydrogen ions flowing through its stationary part (blue) and uses this energy to do work on the stalk of the enzyme (purple and orange) and make it rotate. This rotation pushes phosphate ions and adenosine diphosphate molecules (ADP) together to form adenosine triphosphate (ATP). An ATP synthase molecule is about 10 nm in diameter and 20 nm tall.

BioMedical

WATCH OUT! The quantity Q_C is negative.

Remember our convention that Q is positive if it flows into the object in question and negative if it flows out of the object. So for our engine Q_H is positive (heat flows into the engine from the hot reservoir) and Q_C is negative (heat flows out of the engine into the cold reservoir). However, in analyzing the engine in Figure 15-16 we'll be interested in how much energy flows *into* the *cold reservoir*. So we've labeled Figure 15-16 with $|Q_C|$ (the absolute value of Q_C, which is positive) to indicate the positive quantity of energy that flows into the cold reservoir in the form of heat.

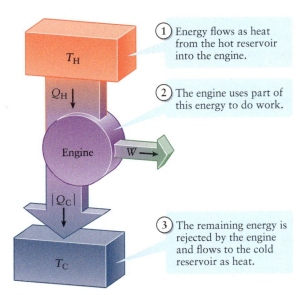

① Energy flows as heat from the hot reservoir into the engine.

② The engine uses part of this energy to do work.

③ The remaining energy is rejected by the engine and flows to the cold reservoir as heat.

Figure 15-16 What a heat engine does A generic heat engine (in purple) is thermally connected to a reservoir of higher temperature T_H (in red) and to a reservoir of lower temperature T_C (in blue). Energy Q_H flows from the hot reservoir into the engine in the form of heat. The engine uses some of the energy to do work W; the remaining energy is rejected as heat to the cold reservoir. (Note that Q_C is negative because it is heat that *leaves* the engine, so the energy that enters the cold reservoir is $|Q_C|$.)

In a perfect engine *all* of the energy Q_H taken in from the hot reservoir would be converted to work, with no heat at all going into the cold reservoir (that is, $Q_C = 0$ in Figure 15-16). This would be a 100% efficient engine: None of the energy from the hot reservoir would be "thrown away" to the cold reservoir. Unfortunately the second law of thermodynamics says that such a perfect engine is *impossible*. To see why, note that there is an increase in order of the hot reservoir: Because energy flows out of this reservoir, some of the molecules of the reservoir must lose their random motion and so become more orderly. Over one cycle of operation the engine itself returns to its original state, so it has zero net change in order, and there is no change at all in the cold reservoir (because no energy flows into it). So the net result would be that over one cycle the system of engine plus reservoirs undergoes an increase in order. But this violates the second law of thermodynamics, which says that the order of an isolated system can decrease or stay the same but cannot increase! We conclude that you simply cannot build a perfect engine that's 100% efficient. The *Kelvin–Planck statement* is a rewording of the second law of thermodynamics that describes the inherent inefficiency of heat engines:

No process is possible in which heat is absorbed from a reservoir and converted completely into work.

For a heat engine to satisfy the second law of thermodynamics, some heat *must* flow into the cold reservoir to make the cold reservoir less ordered. (This happens because the energy added to the cold reservoir increases the random motion of the reservoir's molecules.) The decrease in order of the cold reservoir can then compensate for the increase in order of the hot reservoir. So not all of the energy that flows from the hot reservoir can be converted to work; some of this energy must flow into the cold reservoir and cannot be recovered. That's what happens in a car's gasoline engine: Only about a quarter of the energy released by burning gasoline gets converted into work to propel the car. The remainder goes into heating up the engine and the exhaust gases. Similarly, in the human body only about a quarter of the energy in food is actually used by your cells to do work. The rest of the energy appears as body heat.

Let's apply the *first* law of thermodynamics, Equation 15-2, to the generic engine shown in Figure 15-16. In one cycle the engine takes in energy Q_H from the hot reservoir in the form of heat and sends energy $|Q_C|$ to the cold reservoir in the form of heat, so the *net* quantity of heat that goes into the engine is $Q = Q_H - |Q_C|$. In that same cycle the engine does work W. Because the cycle returns the engine to its initial state at the beginning of the cycle, there is zero net change in its internal energy: $\Delta U = 0$. So the first law of thermodynamics, $\Delta U = Q - W$, becomes

$$0 = Q_H - |Q_C| - W$$

or

Work done by a heat engine (15-24)

Heat that flows from the hot reservoir into the engine in one cycle

Work done by a heat engine in one cycle

$$W = Q_H - |Q_C|$$

Heat that flows from the engine to the cold reservoir in one cycle

The work done equals the difference between the energy taken in from the hot reservoir and the energy discarded into the cold reservoir. The **efficiency** of the engine is defined as the work W divided by the energy Q_H taken in to do the work:

(15-25)

$$e = \frac{W}{Q_H}$$

Efficiency is essentially what you get out of an engine divided by what you put in. If all of the input energy were converted to work, then W would equal Q_H and the efficiency would be 100%, or $e = 1$. But the Kelvin–Planck form of the second law of thermodynamics tells us this is impossible. The best combustion engines, such as the ones found in automobiles, operate at an efficiency of $e \approx 0.30$ (about 30%). Similarly, when human muscles contract only about 25% of the input energy does work, with the rest going into body heat, so $e \approx 0.25$. Using Equation 15-24 we can express the efficiency of a heat engine (Equation 15-25) in terms of the heat taken in from the hot reservoir and the heat rejected to the cold reservoir:

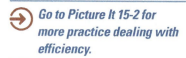

Go to *Picture It 15-2* for more practice dealing with efficiency.

$$e = \frac{W}{Q_H} = \frac{Q_H - |Q_C|}{Q_H} = 1 - \frac{|Q_C|}{Q_H} \qquad (15\text{-}26)$$

(efficiency of a heat engine)

EXAMPLE 15-6   Efficiency of an Engine

The combustion of gasoline in a lawnmower engine releases 44.0 J of energy per cycle. Of that, 31.4 J are lost to warming the body of the engine and the surrounding air. (a) How much work does the lawnmower engine do per cycle? (b) What is the efficiency of the engine?

Set Up

We use Equation 15-24 to determine the work done by the engine and Equation 15-26 to find the efficiency.

Work done by a heat engine:

$$W = Q_H - |Q_C| \qquad (15\text{-}24)$$

Efficiency of a heat engine:

$$e = \frac{W}{Q_H} = \frac{Q_H - |Q_C|}{Q_H} = 1 - \frac{|Q_C|}{Q_H} \qquad (15\text{-}26)$$

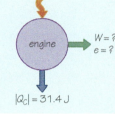

Solve

(a) The work done by the engine equals the difference between the heat Q_H taken in by burning gasoline and the heat $|Q_C|$ that is lost by the engine.

We are given that in one cycle

$$Q_H = 44.0 \text{ J}$$
$$|Q_C| = 31.4 \text{ J}$$

From Equation 15-24 the work that the engine does in one cycle is

$$W = Q_H - |Q_C|$$
$$= 44.0 \text{ J} - 31.4 \text{ J} = 12.6 \text{ J}$$

(b) The efficiency of the engine equals the work done divided by the heat taken in. It's also equal to 1 minus the ratio of heat thrown away to heat taken in.

From Equation 15-26 the efficiency is

$$e = \frac{W}{Q_H} = \frac{12.6 \text{ J}}{44.0 \text{ J}} = 0.286$$

Alternatively,

$$e = 1 - \frac{|Q_C|}{Q_H} = 1 - \frac{31.4 \text{ J}}{44.0 \text{ J}}$$
$$= 1 - 0.714 = 0.286$$

Reflect

An efficiency of around 0.25 to 0.30 (25% to 30%) is typical for a gasoline engine.

Figure 15-17 shows the pV diagram for the cyclic process (called the *Otto cycle*) used in a typical automobile engine. (The process is somewhat simplified in this figure.) Beginning at the state marked 1, a mixture of fuel and air is rapidly compressed in a cylinder with a moveable piston. Because the compression is rapid, step $1 \rightarrow 2$ is adiabatic and there is no heat flow into the fuel–air mixture. However, work is done *on* the

Figure 15-17 A simplified Otto cycle This *pV* diagram shows a simplified version of the cycle used in most automobile engines. (The aircraft engine shown in Figure 15-1b uses this same cycle.) It is named for the nineteenth-century German engineer Nicolaus Otto, who was the first to build an engine that worked on this cycle.

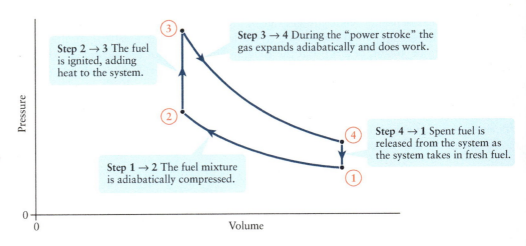

Step 2 → 3 The fuel is ignited, adding heat to the system.

Step 3 → 4 During the "power stroke" the gas expands adiabatically and does work.

Step 4 → 1 Spent fuel is released from the system as the system takes in fresh fuel.

Step 1 → 2 The fuel mixture is adiabatically compressed.

mixture by the piston, so for this step the work done *by* the fuel–air mixture is negative: $W_{1 \to 2} < 0$. At state 2 a spark ignites the fuel, resulting in heat Q_H flowing into the system. The associated increase in pressure and temperature from state 2 to state 3 occurs so rapidly that the volume of the gas in the piston remains nearly constant, so no work is done. At state 3, the high pressure causes the gas to expand rapidly and so adiabatically. Because $Q = 0$ in an adiabatic process, the first law of thermodynamics ($\Delta U = Q - W$) tells us that all of the change in internal energy in step 3 → 4 results in positive work done by the gas on the piston: $W_{3 \to 4} > 0$. As the system returns to its initial state (step 4 → 1, for which the volume again remains constant so no work is done) exhaust is expelled and new fuel is injected into the cylinder.

The net work output in the cycle is the sum of the positive work in step 3 → 4 (in which the gas expands) and the negative work in step 1 → 2 (in which the gas compresses). Remember that the work done equals the area under the curve in a *pV* diagram; step 3 → 4 happens at higher pressure than step 1 → 2, so there is more area under the curve for step 3 → 4 than for step 1 → 2 and there is more positive work than negative work. So the net work output of the engine over one cycle is positive. This work is used to turn the driveshaft of the automobile and make the wheels turn.

The Carnot Cycle

The cycle shown in Figure 15-17 is not the only cycle used in heat engines, nor does it have the highest efficiency. It turns out that the most efficient of all possible cyclic processes in a heat engine is the **Carnot cycle** (pronounced "car-noe"), first proposed in the early 1800s by French engineer Sadi Carnot. Carnot based his cycle on a requirement that all of the processes that comprise the cycle be **reversible processes**—that is, the reverse process is physically possible. Let's explore what this means.

Most thermodynamic processes are **irreversible processes**: They can proceed in only one direction. An example is what happens when ice at 0°C is placed in a large metal pot at 20°C. Heat flows from the pot into the ice, so the ice begins to melt, and the temperature of the pot begins to drop. When the ice and pot come to equilibrium, the ice has completely melted, and the pot and melted water both reach the same final temperature. The reverse of this process—in which liquid water in a pot spontaneously freezes, and the pot warms up—never happens in nature.

An example of a reversible process is ice at 0°C placed in a large metal pot that is also at 0°C. If a small amount of heat is made to flow from the pot into the ice, the ice will partially melt and the pot will contain ice and liquid water at 0°C. Because the amount of heat is so small, the temperature of the pot changes hardly at all and is still 0°C. If the same small amount of heat is now made to flow from the water–ice mixture back into the pot, the water will freeze back to ice and the ice and pot will both be at 0°C. In general a process that involves heat flow will be reversible *only* if the objects between which heat flows are at the same temperature. If they are at different temperatures, the heat flow will always be from the higher-temperature object to the lower-temperature object and the process will be irreversible.

A process can also be reversible if it involves *no* heat flow ($Q = 0$). An example is a *slow* adiabatic expansion or compression, such as would happen with gas in an

The Carnot cycle consists of:
• Two adiabatic processes, for which no heat is transferred in or out of the system.
• Two isothermal processes, one at the temperature of the hot reservoir and one at the temperature of the cold reservoir.

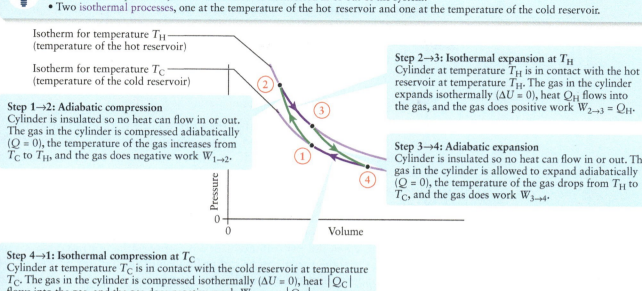

Isotherm for temperature T_H
(temperature of the hot reservoir)

Isotherm for temperature T_C
(temperature of the cold reservoir)

Step 1→2: Adiabatic compression
Cylinder is insulated so no heat can flow in or out.
The gas in the cylinder is compressed adiabatically
($Q = 0$), the temperature of the gas increases from
T_C to T_H, and the gas does negative work $W_{1→2}$.

Step 2→3: Isothermal expansion at T_H
Cylinder at temperature T_H is in contact with the hot
reservoir at temperature T_H. The gas in the cylinder
expands isothermally ($\Delta U = 0$), heat Q_H flows into
the gas, and the gas does positive work $W_{2→3} = Q_H$.

Step 3→4: Adiabatic expansion
Cylinder is insulated so no heat can flow in or out. The
gas in the cylinder is allowed to expand adiabatically
($Q = 0$), the temperature of the gas drops from T_H to
T_C, and the gas does work $W_{3→4}$.

Pressure

0

0   Volume

Step 4→1: Isothermal compression at T_C
Cylinder at temperature T_C is in contact with the cold reservoir at temperature
T_C. The gas in the cylinder is compressed isothermally ($\Delta U = 0$), heat $|Q_C|$
flows into the gas, and the gas does negative work $W_{4→1} = -|Q_C|$.

Figure 15-18 The Carnot cycle The most efficient cycle possible for a heat engine is
made up of four reversible steps.

insulated cylinder. If the gas is allowed to expand slightly, it does a small amount
of positive work on the piston ($W > 0$) and the internal energy of the gas decreases
slightly (from the first law of thermodynamics, $\Delta U = Q - W$ is negative because
Q is zero, and W is positive). In the reverse process the gas is compressed slightly and
does negative work on the piston ($W < 0$, meaning that the piston does work on the gas),
and the internal energy of the gas increases. The system then returns to its initial state. By
contrast, a free expansion of a gas, in which a gas under pressure suddenly expands to a
larger volume (as in Figure 15-4b), is *irreversible*. It is not possible to return the system to
the initial pressure and volume without adding energy by doing work.

Carnot understood that energy is conserved in both reversible and irreversible
processes. But he also understood that in an irreversible process much of the energy
becomes *unavailable*. For example, when energy flows in the form of heat from the
20°C pot into the 0°C ice to melt it, the energy that went into the ice cannot be returned
to the pot. In order not to "lose" energy in this way, Carnot realized that the most efficient heat engine cycle possible must therefore involve only reversible processes. As
we've seen, this means either slow processes that involve heat flow with zero temperature difference or slow adiabatic processes.

Figure 15-18 shows the pV diagram for the Carnot cycle. The system is an ideal
gas in a cylinder with a moveable piston. In state 1 the cylinder is in contact and in
equilibrium with the cold reservoir at temperature T_C, so the gas is also at temperature
T_C. Here are the four *reversible* steps that make up the cycle:

• **Step 1 → 2: Adiabatic compression.** You take the cylinder filled with gas and
wrap it with a perfect insulator so that there can be no heat flow into or out
of the gas, and you do work on the moveable piston to slowly push it inward
(so that the gas does negative work). As the gas is compressed, its temperature
increases from T_C to T_H.

• **Step 2 → 3: Isothermal expansion.** You remove the insulation and put the
cylinder in contact with the hot reservoir at temperature T_H (which is the same
as the current temperature of the gas). You now allow the gas to expand so
that it does positive work on the piston. The temperature remains equal to T_H
because the gas and hot reservoir are in equilibrium.

• **Step 3 → 4: Adiabatic expansion.** You again wrap the cylinder with a perfect
insulator and allow the gas to expand. The expanding gas does more positive

work on the piston, and the temperature of the gas decreases from T_H back down to T_C.

- **Step 4 → 1:** Isothermal compression. You again remove the insulation, but now you put the cylinder in contact with the cold reservoir at temperature T_C (which is the same as the current temperature of the gas). You do work on the piston to slowly push it inward, so the gas does negative work. The temperature remains equal to T_C because the gas and cold reservoir are in equilibrium. You stop when the gas is once again in state 1, with the same values of pressure, volume, and temperature as initially.

You can see from the description that the Carnot cycle is an idealization (like a massless rope or a frictionless ramp) and is *not* a practical cycle for a real engine. The Carnot cycle is nonetheless tremendously important. Because all four of its processes are reversible, the Carnot cycle is the *most efficient* heat engine cycle possible. It therefore represents the standard against which all other cycles should be compared. Let's see how to determine the efficiency of a Carnot engine.

From Equation 15-26 the efficiency of any heat engine is $e = 1 - |Q_C|/Q_H$. For the Carnot cycle depicted in Figure 15-18, Q_H is the heat that flows into the ideal gas during step 2 → 3, when the gas cylinder is in contact with the hot reservoir at temperature T_H, and Q_C is the (negative) heat that flows into the gas during step 4 → 1, when the gas cylinder is in contact with the cold reservoir at temperature T_C. Both of these processes are isothermal, so $Q = W$ (Equation 15-5): The quantity of heat Q that enters the gas equals the work W done by the gas. Furthermore, we know from Equation 15-6 that the work done by n moles of an ideal gas in an isothermal process at temperature T is $W = nRT \ln (V_f/V_i)$, where V_i and V_f are the volume of the gas at the beginning and end of the process, respectively. So for the two isothermal steps in the Carnot cycle, we have

$$Q_H = W_{2\to3} = nRT_H \ln \left(\frac{V_3}{V_2} \right)$$

$$Q_C = W_{4\to1} = nRT_C \ln \left(\frac{V_1}{V_4} \right)$$

Note that Q_C is negative for step 4 → 1. To see this, notice that V_1, the volume in state 1, is less than V_4, the volume in state 4 (see Figure 15-18). So the ratio V_1/V_4 is less than 1, and the natural logarithm of a number less than 1 is negative. Since $\ln (1/x) = -\ln x$, we can write the absolute value of Q_C as

$$|Q_C| = -Q_C = nRT_C \ln \left(\frac{V_4}{V_1} \right)$$

Then we can write the efficiency of the Carnot cycle using Equation 15-26 as

$$(15\text{-}27) \qquad e_{Carnot} = 1 - \frac{|Q_C|}{Q_H} = 1 - \frac{nRT_C \ln \left(\dfrac{V_4}{V_1} \right)}{nRT_H \ln \left(\dfrac{V_3}{V_2} \right)}$$

Equation 15-27 looks rather messy, but we can simplify it by using Equation 15-23, which says that $TV^{\gamma-1} = $ constant for any adiabatic process. If we apply this to the adiabatic steps 1 → 2 and 3 → 4, we get

$$\text{Step 1} \to 2: T_C V_1^{\gamma-1} = T_H V_2^{\gamma-1} \quad \text{or} \quad \frac{T_C}{T_H} = \frac{V_2^{\gamma-1}}{V_1^{\gamma-1}}$$

$$\text{Step 3} \to 4: T_C V_4^{\gamma-1} = T_H V_3^{\gamma-1} \quad \text{or} \quad \frac{T_C}{T_H} = \frac{V_3^{\gamma-1}}{V_4^{\gamma-1}}$$

If we set the two expressions for T_C/T_H equal to each other and rearrange, we get

$$\frac{V_2^{\gamma-1}}{V_1^{\gamma-1}} = \frac{V_3^{\gamma-1}}{V_4^{\gamma-1}} \quad \text{so} \quad \frac{V_2}{V_1} = \frac{V_3}{V_4} \quad \text{and} \quad \frac{V_4}{V_1} = \frac{V_3}{V_2}$$

The last of these tells us that the quantities $\ln(V_4/V_1)$ and $\ln(V_3/V_2)$ in Equation 15-27 are equal and so cancel out, as do the factors of nR. So we're left with a very simple expression for the efficiency of a Carnot engine:

Efficiency of a Carnot engine

Kelvin temperature of the cold reservoir

$$e_{\text{Carnot}} = 1 - \frac{T_C}{T_H}$$

Kelvin temperature of the hot reservoir

Efficiency of a Carnot engine
(15-28)

The efficiency of an ideal Carnot heat engine depends *only* on the hot and cold temperatures between which the engine operates. Equation 15-28 is the maximum theoretical efficiency of *any* heat engine operating between temperatures T_H and T_C. It is impossible for any engine to do better than this.

The efficiency of an ideal Carnot engine can be made closer and closer to $e_{\text{Carnot}} = 1$ (100% efficiency) by decreasing T_C and increasing T_H. For the efficiency to be exactly one, T_C must equal 0 K. It is not possible, however, to attain a temperature of absolute zero. This statement is referred to as the **third law of thermodynamics**:

> *It is possible for the temperature of a system to be arbitrarily close to absolute zero, but it can never reach absolute zero.*

A consequence of the third law of thermodynamics is that it is impossible to make a heat engine that is 100% efficient.

Go to Interactive Exercises 15-2 and 15-3 for more practice dealing with efficiency.

EXAMPLE 15-7   Carnot Efficiency and Actual Efficiency

An automotive engine is designed to operate between 290 and 450 K. The engine produces 1.50×10^2 J of mechanical energy for every 6.00×10^2 J of heat absorbed from the combustion of fuel. Calculate (a) the actual efficiency of this engine and (b) the theoretical maximum efficiency of a Carnot engine operating between these temperatures.

Set Up

We'll use Equation 15-26 to determine the actual efficiency and Equation 15-28 to find the efficiency of a Carnot engine operating between 290 and 450 K.

Efficiency of a heat engine:

$$e = \frac{W}{Q_H} = \frac{Q_H - |Q_C|}{Q_H} = 1 - \frac{|Q_C|}{Q_H}$$

(15-26)

Efficiency of a Carnot engine:

$$e_{\text{Carnot}} = 1 - \frac{T_C}{T_H} \qquad (15\text{-}28)$$

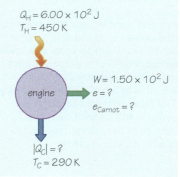

Solve

(a) We are given $W = 1.50 \times 10^2$ J and $Q_H = 6.00 \times 10^2$ J. Use this and the first expression in Equation 15-26 to find the actual efficiency of the engine.

Actual efficiency:

$$e = \frac{W}{Q_H} = \frac{1.50 \times 10^2 \text{ J}}{6.00 \times 10^2 \text{ J}} = 0.250$$

(b) The efficiency of the engine in (a) is less than the theoretical maximum efficiency of a Carnot engine with $T_C = 290$ K and $T_H = 450$ K.

Efficiency for a theoretical Carnot engine operating between these temperatures:

$$e_{\text{Carnot}} = 1 - \frac{T_C}{T_H} = 1 - \left(\frac{290 \text{ K}}{450 \text{ K}}\right) = 0.356$$

Reflect

Note that even a Carnot engine—the most efficient heat engine that could possibly operate between these temperatures—is not terribly efficient: $e_{\text{Carnot}} = 0.356$ means that only 35.6% of the energy released by combustion can be used to do work. The other 64.4% is thrown away in the form of heat. This is one reason electric cars, which are powered by electric motors that are not subject to the limitations of heat engines, are much more energy efficient than are gasoline-powered cars.

For a kitchen refrigerator, the cold reservoir is the interior of the refrigerator ...

... and the hot reservoir is the kitchen outside the refrigerator.

PhotoAlto/Alamy

Figure 15-19 **A refrigerator** Energy is supplied to the refrigerator in the form of electricity, and the refrigerator extracts heat from a cold reservoir (the contents of the refrigerator, which you want to keep at a low temperature) and exhausts heat into a hot reservoir (the kitchen, which is at a higher temperature). The heat exhausts through air vents in the back or near the bottom of the refrigerator, which is why it's warm there.

Refrigerators

A **refrigerator** is a device that takes in energy and uses it to transfer heat from an object at low temperature to an object at high temperature. An ordinary kitchen refrigerator (Figure 15-19) works this way. In many ways a refrigerator is simply an engine running backward.

Like heat engines, refrigerators are cyclic, and like heat engines, refrigerators are constrained by the second law of thermodynamics. The *Clausius form* of the second law of thermodynamics, named after the nineteenth-century German physicist Robert Clausius, says that:

No process is possible in which heat is absorbed from a cold reservoir and transferred completely to a hot reservoir.

In other words it isn't possible to make a perfect refrigerator.

Figure 15-20 is a diagram of a generic refrigerator. This looks similar to the generic heat engine shown in Figure 15-16, but the heat and work arrows are in the opposite direction. Work is an *input* to this device; as a result, heat Q_C is made to go from the colder region to the hotter one. This transfer is, of course, exactly what you'd want to cool down a container in which to store food (a refrigerator) or a building (an air conditioner). A heat pump, a device used to warm up a building, is identical to a refrigerator; the difference is that a heat pump transfers heat from the cold exterior of the building into the warm interior to keep it warm.

Figure 15-21 shows the cycle used in a typical kitchen refrigerator like that in Figure 15-19. This cycle uses the inert compound R134a (chemical formula CH_2FCF_3). R134a in the liquid state is forced through a tube (the evaporator coil) that passes through the freezer and refrigerator compartments. The R134a absorbs energy from the air in these compartments to cool them, which raises the temperature of the R134a and turns it into a gas. A compressor pump then pressurizes the R134a gas, which drives the gas around the system and also increases its temperature even more. As the hot gas passes through the long, narrow condenser coil tube (usually in the back of the refrigerator), heat is transferred from the gaseous R134a to the cooler air in the kitchen. This part of the process cools the gas enough to cause it to liquefy, and the liquid R134a is further cooled by letting it expand rapidly as it passes through a valve from a narrow tube to a much wider one. The cold liquid then makes its way back to the evaporator coil to start the process again. In this way energy is removed from the interior of the refrigerator and delivered to the surroundings.

The less work required to extract heat Q_C from the cold reservoir, the more efficient the refrigerator. A perfect refrigerator would require no work to be done in order to cause heat to be transferred from the cold to the hot reservoir, but this process isn't possible according to the second law of thermodynamics. The efficiency of a refrigerator is given by the **coefficient of performance**, for which we use the symbol CP:

(15-29)
$$CP = \frac{|Q_C|}{|W|}$$

As with efficiency, CP is essentially what you get out divided by what you put in. The smaller the amount of work W for a given quantity of heat Q_C extracted from the cold reservoir, the larger the coefficient of performance CP and the better the performance of the refrigerator. A typical value for CP for a kitchen refrigerator or a home air conditioner is 3: This means that for every 1 J of work input to the refrigerator, 3 J of heat is extracted from the cold interior of the refrigerator or home and 1 J + 3 J = 4 J is rejected to the warmer surroundings.

The first law of thermodynamics requires that in one cycle, the total energy that goes into the refrigerator (in the form of work |W| done on it

Hot T_H

$|Q_H|$

Refrigerator $|W|$

$|Q_C|$

Cold T_C

③ The energy extracted from the cold reservoir, plus the energy delivered to the refrigerator as work, is rejected to the hot reservoir.

② For this to happen work must be done on the refrigerator.

① Energy is extracted from the cold reservoir into the refrigerator.

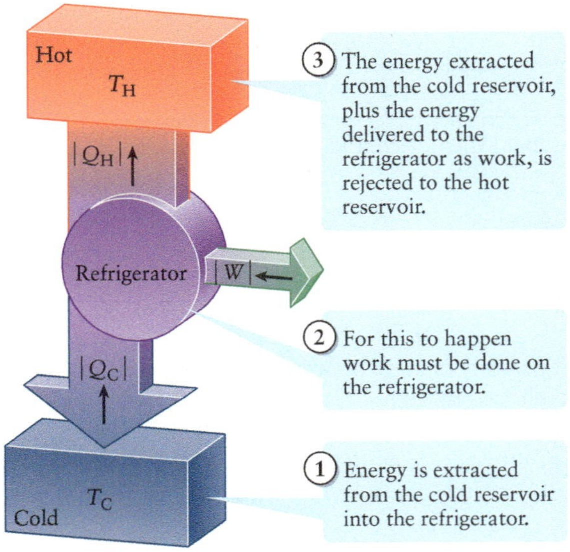

Figure 15-20 **What a refrigerator does** A generic refrigerator (in purple) extracts heat Q_C from a cold reservoir (in blue) at temperature T_C. To do this, work must be done on the refrigerator (so W, the work done *by* the refrigerator, is negative, and the work done *on* the refrigerator is |W|). All of the energy added to the engine as heat and by work is rejected to the hot reservoir (in red) at temperature T_H. Here Q_H is negative (it represents heat that *leaves* the refrigerator), so the heat that is rejected *into* the hot reservoir is $|Q_H|$.

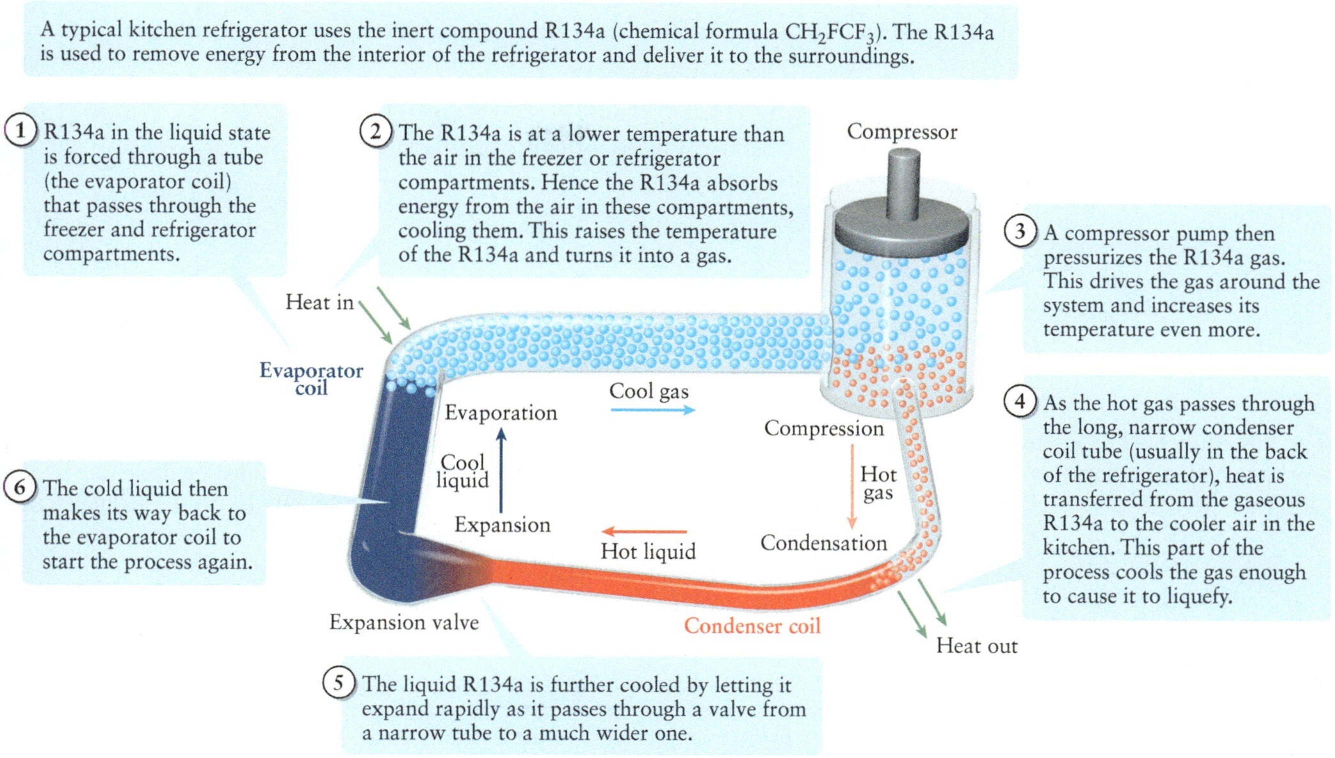

A typical kitchen refrigerator uses the inert compound R134a (chemical formula CH_2FCF_3). The R134a is used to remove energy from the interior of the refrigerator and deliver it to the surroundings.

(1) R134a in the liquid state is forced through a tube (the evaporator coil) that passes through the freezer and refrigerator compartments.

(2) The R134a is at a lower temperature than the air in the freezer or refrigerator compartments. Hence the R134a absorbs energy from the air in these compartments, cooling them. This raises the temperature of the R134a and turns it into a gas.

Compressor

(3) A compressor pump then pressurizes the R134a gas. This drives the gas around the system and increases its temperature even more.

Heat in

Evaporator coil

Evaporation

Cool gas

Cool liquid

Compression

Expansion

Hot liquid

Condensation

Hot gas

(4) As the hot gas passes through the long, narrow condenser coil tube (usually in the back of the refrigerator), heat is transferred from the gaseous R134a to the cooler air in the kitchen. This part of the process cools the gas enough to cause it to liquefy.

(6) The cold liquid then makes its way back to the evaporator coil to start the process again.

Expansion valve

Condenser coil

Heat out

(5) The liquid R134a is further cooled by letting it expand rapidly as it passes through a valve from a narrow tube to a much wider one.

Figure 15-21 Inside a refrigerator This simplified diagram shows the cyclic process that cools a kitchen refrigerator. This process involves compressing and expanding a refrigerant that transitions between being a liquid and a gas.

and heat $|Q_C|$ extracted from the cold reservoir) equals the amount of energy that comes out (in the form of heat $|Q_H|$ delivered to the hot reservoir). So

$$|Q_C| + |W| = |Q_H| \quad \text{or} \quad |W| = |Q_H| - |Q_C| \quad \text{(refrigerator)} \qquad (15\text{-}30)$$

Equation 15-29 then becomes

$$CP = \frac{|Q_C|}{|W|} = \frac{|Q_C|}{|Q_H| - |Q_C|} \qquad (15\text{-}31)$$

(coefficient of performance of a refrigerator)

The maximum possible coefficient of performance of a refrigerator is obtained when the system is based on the Carnot cycle run in reverse. It can be shown that the coefficient of performance of a Carnot refrigerator is

$$CP_{\text{Carnot}} = \frac{T_C}{T_H - T_C} \qquad (15\text{-}32)$$

(coefficient of performance of a Carnot refrigerator)

Like an ideal Carnot engine, an ideal Carnot refrigerator is not a practical device. The significance of Equation 15-32 is that it tells us the theoretical maximum coefficient of performance of a refrigerator operating between cold temperature T_C and hot temperature T_H.

EXAMPLE 15-8 Kitchen Refrigerator

A kitchen refrigerator has a coefficient of performance of 3.30. When the refrigerator pump is running, it removes energy from the interior of the refrigerator at a rate of 760 J/s. (a) How much work must be done per second to extract this energy? (b) How much energy is rejected into the kitchen per second in this process? (c) If the refrigerator operates between 2°C (the interior of the refrigerator) and 25°C (the kitchen), what is the theoretical maximum coefficient of performance?

Set Up

The energy removed from the cold interior of the refrigerator is represented by $|Q_C|$, and the work done to remove that energy is represented by $|W|$. We can relate these to the coefficient of

Coefficient of performance of a refrigerator:

$$CP = \frac{|Q_C|}{|W|} = \frac{|Q_C|}{|Q_H| - |Q_C|} \qquad (15\text{-}31)$$

performance by using Equation 15-31, and to the energy $|Q_H|$ rejected into the warm kitchen by using Equation 15-30. The theoretical maximum value of CP is the Carnot value, given by Equation 15-32.

Energy relationships for a refrigerator:

$|Q_C| + |W| = |Q_H|$ or

$|W| = |Q_H| - |Q_C|$ (15-30)

Coefficient of performance of a Carnot refrigerator:

$$CP_{Carnot} = \frac{T_C}{T_H - T_C} \quad (15\text{-}32)$$

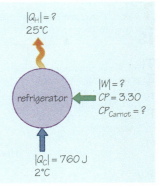

$|Q_H| = ?$
25°C

$|W| = ?$
$CP = 3.30$
$CP_{Carnot} = ?$

refrigerator

$|Q_C| = 760$ J
2°C

Solve

(a) Use Equation 15-31 to determine the amount of work that must be done per second.

In 1 s an amount of energy $|Q_C| = 760$ J is removed from the interior of the refrigerator. The amount of work $|W|$ that must be done to make this happen depends on $|Q_C|$ and the coefficient of performance:

$$CP = \frac{|Q_C|}{|W|}$$

$$|W| = \frac{|Q_C|}{CP} = \frac{760 \text{ J}}{3.30} = 230 \text{ J}$$

So the rate at which work must be done is 230 J/s = 230 W.

(b) Find the rate at which energy is rejected into the kitchen.

From Equation 15-30 the amount of heat $|Q_H|$ rejected into the kitchen in 1 s is

$$|Q_H| = |Q_C| + |W| = 760 \text{ J} + 230 \text{ J} = 990 \text{ J}$$

So the rate at which the refrigerator rejects heat is 990 J/s = 990 W.

(c) Calculate the coefficient of performance of a Carnot refrigerator operating between the same hot and cold temperatures.

The temperatures between which the refrigerator operates are 2°C or $T_C = (273.15 + 2)$ K = 275 K and 25°C or $T_H = (273.15 + 25)$ K = 298 K

The coefficient of performance of a Carnot refrigerator operating between these temperatures is

$$CP_{Carnot} = \frac{T_C}{T_H - T_C} = \frac{275 \text{ K}}{298 \text{ K} - 275 \text{ K}} = \frac{275 \text{ K}}{23 \text{ K}}$$
$$= 12$$

Reflect

The coefficient of performance CP is substantially less than what could be achieved by an ideal Carnot refrigerator. Notice that the higher the actual value of CP, the less work must be done to achieve the same amount of refrigeration. Less work means less waste, and as a result, less heating of the kitchen. This actual refrigerator rejects heat into the kitchen at a rate of 990 J/s or 990 W, about the same as the power output of an electric space heater.

GOT THE CONCEPT? 15-6 A Super Engine

A fellow student announces that he has invented a gasoline engine that is 90% efficient. He claims that he gets such efficiency by running the engine at very high temperatures. The cylinders of his engine are made of titanium, which has a melting temperature of 1941 K. If he asks you to invest in this engine, should you agree?

TAKE-HOME MESSAGE FOR Section 15-5

✔ One statement of the second law of thermodynamics is that randomness in systems tends to increase over time.

✔ A system or device that converts heat to work is a heat engine. It is not possible to create an engine or a process in which heat is absorbed from a reservoir and converted completely into work.

✔ A refrigerator is a heat engine in reverse. It uses work to move heat from a cold reservoir to a hot reservoir.

✔ The third law of thermodynamics states that the temperature of a system can never attain absolute zero.

15-6 The entropy of a system is a measure of its disorder

The second law of thermodynamics uses the concept of the *order* of a system. How can we quantify just how ordered or disordered a system is? Searching for the answer to this question will lead us to introduce a new physical quantity called *entropy* that plays an essential role in thermodynamics.

To get the best understanding of how ordered or disordered a thermodynamic system is, you would have to note the position and behavior of every one of its molecules. That's a challenging task for a system even as small as a single drop of water, which contains about 10^{22} H_2O molecules. (That's roughly 1000 times more molecules than there are grains of sand on all of Earth's beaches.) Instead let's look at a system with a much smaller number of constituents. Imagine that your sock drawer is an unorganized jumble of an equal number of blue socks and red socks. If you grab four socks without looking at their colors, what distribution of colors within the four-sock "thermodynamic system" are you most likely to have?

There is only one way to end up with four red socks: Each individual selection must be a red sock. There are four ways to have three red socks and one blue one, however. One way is to get a red sock during each of the first three selections and then get one blue one. Another way is for the first sock chosen to be blue and the last three red. We can summarize the possible ways to get three red and one blue sock this way:

$$(\bullet\ \bullet\ \bullet\ \bullet),\ (\bullet\ \bullet\ \bullet\ \bullet),\ (\bullet\ \bullet\ \bullet\ \bullet),\ \text{and}\ (\bullet\ \bullet\ \bullet\ \bullet)$$

Table 15-2 summarizes all 16 possible four-sock combinations. The most probable grouping is two red socks and two blue socks (6 of the 16 possible combinations). This is also the most randomly distributed, least ordered four-sock state. The most ordered state would be one that was made of either four red socks or four blue socks. Table 15-2 shows that such states are also the least likely ones.

These conclusions about socks taken from a drawer turn out to be true in general: *The most probable final or equilibrium state of a system is the one that is the least ordered.* For example, a hot object and a cold object put into thermal contact come to the same final temperature not because no other possibility is allowed but because this state is the least ordered and therefore the most probable. Energy would still be conserved if, say, the hot object got hotter while the cold one got colder. But this would be highly improbable because this is a highly ordered state (the energy of the system ends up concentrated in the hot object rather than being spread between the two objects).

To represent order, we must take the list of variables that describe the state of a system—its temperature T, pressure p, volume V, and internal energy U—and add to it a new variable that we call **entropy**. Entropy, which for historical reasons we give the rather unexpected symbol S, is defined so that the larger its value the *less order* and the *more disorder* is present in a system. So we can think of entropy S as a measure of *disorder*. The second law of thermodynamics therefore states that *the entropy of an isolated system must either remain constant or increase.*

Entropy Change in a Reversible Process

The second law of thermodynamics suggests that what matters most is not the *amount* of order in a system but rather how much *change* in order there is in a thermodynamic process. The first law of thermodynamics says the same thing about internal energy. The first law tells us that the change in internal energy ΔU must be equal to $Q - W$ (the quantity of heat that enters the system minus the work that the system does). Let's see how to express the entropy change for the special case of a system that undergoes a process that happens at a constant temperature T. To be specific we'll look at the reversible isothermal expansion of an ideal gas.

Figure 15-22 shows a quantity of ideal gas at temperature T confined within a cylinder with a moveable piston. We put the cylinder in thermal contact with a large reservoir at the same temperature T then allow the gas to expand. Because the gas is

TABLE 15-2 Possible Combinations of Red and Blue Socks

Blue	Red	Combinations	Number
0	4	$(\bullet\ \bullet\ \bullet\ \bullet)$	1
1	3	$(\bullet\ \bullet\ \bullet\ \bullet),\ (\bullet\ \bullet\ \bullet\ \bullet),\ (\bullet\ \bullet\ \bullet\ \bullet),\ (\bullet\ \bullet\ \bullet\ \bullet)$	4
2	2	$(\bullet\ \bullet\ \bullet\ \bullet),\ (\bullet\ \bullet\ \bullet\ \bullet),\ (\bullet\ \bullet\ \bullet\ \bullet),\ (\bullet\ \bullet\ \bullet\ \bullet),$ $(\bullet\ \bullet\ \bullet\ \bullet),\ (\bullet\ \bullet\ \bullet\ \bullet)$	6
3	1	$(\bullet\ \bullet\ \bullet\ \bullet),\ (\bullet\ \bullet\ \bullet\ \bullet),\ (\bullet\ \bullet\ \bullet\ \bullet),\ (\bullet\ \bullet\ \bullet\ \bullet)$	4
4	0	$(\bullet\ \bullet\ \bullet\ \bullet)$	1

Figure 15-22 Entropy changes in an isothermal expansion In a reversible isothermal expansion of an ideal gas, the entropy of the gas increases; the entropy of the reservoir with which the gas is in contact decreases by an equal amount.

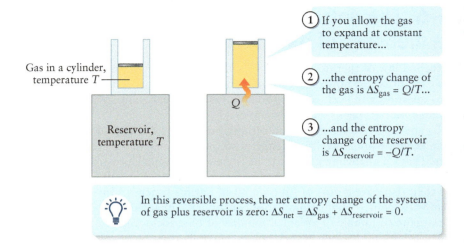

Gas in a cylinder, temperature T

Reservoir, temperature T

Q

(1) If you allow the gas to expand at constant temperature...

(2) ...the entropy change of the gas is $\Delta S_{gas} = Q/T$...

(3) ...and the entropy change of the reservoir is $\Delta S_{reservoir} = -Q/T$.

In this reversible process, the net entropy change of the system of gas plus reservoir is zero: $\Delta S_{net} = \Delta S_{gas} + \Delta S_{reservoir} = 0$.

in thermal contact with the reservoir, its temperature remains constant and the internal energy of the gas does not change: $\Delta U = 0$. The gas does work on the piston as it expands, so $W > 0$. From the first law of thermodynamics, $\Delta U = Q - W$ and $Q = \Delta U + W$, or in this case (since ΔU is zero) $Q = W$. All of the energy that flows into the gas from the reservoir in the form of heat Q is used to do work on the piston.

As the gas expands, the volume V that the gas occupies increases, and the molecules of the gas are now more spread out. This makes the gas more disordered, just as taking socks from a drawer and strewing them around the larger volume of your bedroom makes the socks more disordered. The greater the quantity of heat that enters the gas, the more the gas expands, and the greater the increase in disorder. So in this isothermal expansion the entropy change ΔS of the gas is *proportional* to the quantity of heat Q that flows into the gas.

The entropy change ΔS in an isothermal expansion also depends on the temperature T. For a given quantity of heat Q, the increase in disorder is greater if the system is at a low temperature (so that the molecules are moving slowly and are relatively well ordered to start with) than if the system is at a high temperature (so that the molecules are moving rapidly in a relatively disordered state). We conclude that the entropy change ΔS in this reversible isothermal process is *inversely proportional* to the Kelvin temperature T.

Putting these ideas together, we *define* the entropy change ΔS in a reversible isothermal process as

Entropy change in a reversible isothermal process (15-33)

Heat that flows into the system ($Q > 0$ if heat flows in, $Q < 0$ if heat flows out)

Entropy change of a system in a reversible isothermal process

$$\Delta S = \frac{Q}{T}$$

Kelvin temperature of the system

Equation 15-33 shows that the units of entropy are the units of heat (that is, energy) divided by the units of temperature, or joules per kelvin (J/K).

We can also apply Equation 15-33 to the *reservoir* with which the gas cylinder is in contact. If a positive quantity of heat $Q > 0$ flows *into* the gas *from* the reservoir, the heat that flows into the reservoir from the gas is $-Q$. (This says that the quantity of heat that enters the gas equals the quantity of heat that leaves the reservoir.) The gas and reservoir are both at the same temperature T, so the *net* entropy change of the system of gas plus reservoir is

$$\Delta S_{net} = \Delta S_{gas} + \Delta S_{reservoir} = \frac{Q}{T} + \frac{(-Q)}{T} = 0$$

In a reversible isothermal expansion the entropy of the gas increases (it becomes more disordered), but the entropy of the reservoir decreases (it becomes *less* disordered because it has transferred some of its random thermal energy to the gas). The net change in entropy is zero.

EXAMPLE 15-9 Calculating Entropy Change

A cylinder containing n moles of an ideal gas at temperature T changes its volume isothermally from an initial volume V_i to a final volume V_f. Find expressions for (a) the entropy change of the gas and (b) the entropy change of the thermal reservoir with which the gas is in contact. (c) Evaluate these for the special case in which 0.050 mol of gas at 20.0°C has an initial volume of 1.00×10^{-3} m^3 and expands to a final volume of 2.00×10^{-3} m^3.

Set Up

Equation 15-33 tells us the entropy change of the gas in terms of its temperature T and the heat Q that flows into the gas during the volume change. From Equation 15-5 the quantity of heat Q equals the work W that the gas does. Equation 15-6 tells us the work done in this process. We'll put all of these pieces together to find the entropy changes of the gas and the thermal reservoir.

Entropy change in a reversible isothermal process:

$$\Delta S = \frac{Q}{T} \qquad (15\text{-}33)$$

Isothermal process for an ideal gas:

$$Q = W \qquad (15\text{-}5)$$

Work done by an ideal gas in an isothermal process:

$$W = nRT \ln \frac{V_f}{V_i} \qquad (15\text{-}6)$$

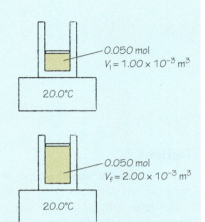

0.050 mol
$V_i = 1.00 \times 10^{-3}$ m^3

20.0°C

0.050 mol
$V_f = 2.00 \times 10^{-3}$ m^3

20.0°C

Solve

(a) Find an expression for the entropy change of the gas.

Combine Equation 15-5 and Equation 15-6 to get an expression for the heat that enters the gas:

$$Q_{gas} = W = nRT \ln \frac{V_f}{V_i}$$

From Equation 15-33 the entropy change of the gas in this process is

$$\Delta S_{gas} = \frac{Q_{gas}}{T} = \frac{nRT}{T} \ln \frac{V_f}{V_i} = nR \ln \frac{V_f}{V_i}$$

If the gas expands so that the final volume is greater than the initial volume, the ratio V_f/V_i is greater than 1, $\ln (V_f/V_i)$ is positive, and ΔS_{gas} is positive (the entropy of the gas increases). If the gas is compressed so that the final volume is less than the initial volume, the ratio V_f/V_i is less than 1, $\ln (V_f/V_i)$ is negative, and ΔS_{gas} is negative (the entropy of the gas decreases).

(b) Find an expression for the entropy change of the reservoir.

The heat that enters the gas comes from the reservoir, so the value of Q for the reservoir equals the negative of the value of Q for the gas:

$$Q_{reservoir} = -Q_{gas} = -nRT \ln \frac{V_f}{V_i}$$

The entropy change of the reservoir is then

$$\Delta S_{reservoir} = \frac{Q_{reservoir}}{T} = -\frac{nRT}{T} \ln \frac{V_f}{V_i} = -nR \ln \frac{V_f}{V_i}$$

If we compare this to the result from part (a), we see that

$$\Delta S_{reservoir} = -\Delta S_{gas}$$

(c) Calculate the numerical values of ΔS_{gas} and $\Delta S_{reservoir}$.

With $n = 0.050$ mol, $V_i = 1.00 \times 10^{-3}$ m³, and $V_f = 2.00 \times 10^{-3}$ m³, we get

$$\Delta S_{gas} = nR \ln \frac{V_f}{V_i}$$

$$= (0.050 \text{ mol})\left(8.314 \frac{\text{J}}{\text{mol} \cdot \text{K}}\right) \ln \left(\frac{2.00 \times 10^{-3} \text{ m}^3}{1.00 \times 10^{-3} \text{ m}^3}\right)$$

$$= (0.416 \text{ J/K}) \ln 2.00$$

$$= 0.288 \text{ J/K}$$

$$\Delta S_{reservoir} = -\Delta S_{gas} = -0.288 \text{ J/K}$$

The net entropy change of the system of gas and reservoir is

$$\Delta S_{net} = \Delta S_{gas} + \Delta S_{reservoir}$$

$$= 0.288 \text{ J/K} + (-0.288 \text{ J/K}) = 0$$

Reflect

Note that the entropy change of the gas, $\Delta S_{gas} = nR \ln (V_f/V_i)$, depends only on the number of moles n and the ratio of the final and initial volumes occupied by the gas. The temperature doesn't matter. This is consistent with the idea that a change in entropy corresponds to a change in disorder. The greater the increase in volume, the more disordered the system becomes; the greater the number of moles (and hence the greater the number of molecules) that expand into the new volume, the greater the increase in disorder. To see why the number of moles matters, think about taking socks from your sock drawer and strewing them around your bedroom. If there are only two socks involved, the amount of disorder isn't great; if there are a hundred socks, it's a hugely disordered mess.

In Example 15-9 the gas and reservoir together make up an *isolated* system: The gas and reservoir are in thermal contact with each other but are not in thermal contact with anything else. In this case, in which the thermodynamic process is *reversible*, the entropy of that isolated system remains constant.

Entropy Change in an Irreversible Process

Let's now consider a different isothermal process that's *irreversible*. In Figure 15-23 we use a metal bar to put a hot reservoir at temperature T_H in thermal contact with a cold reservoir at temperature T_C. Heat flows from the hot reservoir to the cold one, but the

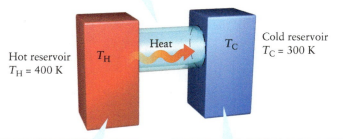

(1) A quantity of heat $Q = 1.20 \times 10^3$ J flows through the metal rod from the hot reservoir to the cold reservoir.

Hot reservoir $T_H = 400$ K

T_H

Heat

T_C

Cold reservoir $T_C = 300$ K

(2) Entropy change of the hot reservoir:

$$\Delta S_H = \frac{(-Q)}{T_H} = \frac{(-1.20 \times 10^3 \text{ J})}{400 \text{ K}} = -3 \text{ J/K}$$

(3) Entropy change of the cold reservoir:

$$\Delta S_C = \frac{(+Q)}{T_C} = \frac{(+1.20 \times 10^3 \text{ J})}{300 \text{ K}} = +4 \text{ J/K}$$

Figure 15-23 Entropy changes in irreversible heat flow In irreversible heat flow from a hot object to a cold one, the *net* entropy of the two objects together increases.

 The net entropy change in this irreversible process is positive: $\Delta S_{net} = \Delta S_H + \Delta S_C = (-3 \text{ J/K}) + (+4 \text{ J/K}) = +1 \text{ J/K}$. The net entropy always increases in an irreversible process.

reservoirs are so large that their temperatures remain essentially unchanged. Just as we did for the reservoir in **Example 15-9**, we can calculate the entropy change of each reservoir using Equation 15-33, $\Delta S = Q/T$. If a quantity of heat Q is transferred between the reservoirs, then $Q_C = +Q$ (heat flows *into* the cold reservoir from the hot one) and $Q_H = -Q$ (an equal quantity of heat flows *out* of the hot reservoir into the cold one). The entropy of the metal bar doesn't change: As much heat enters one end of the bar as leaves the other. The entropy changes of the two reservoirs are then

$$\text{Cold reservoir: } \Delta S_C = \frac{(+Q)}{T_C} \text{ (positive)}$$

$$\text{Hot reservoir: } \Delta S_H = \frac{(-Q)}{T_H} \text{ (negative)}$$

$$\text{Net entropy change: } \Delta S_{net} = \Delta S_C + \Delta S_H = \frac{(+Q)}{T_C} + \frac{(-Q)}{T_H}$$

As in Example 15-9 the entropy of one object (the cold reservoir) increases while the entropy of the other object (the hot reservoir) decreases. The difference is that the process shown in Figure 15-23 is *irreversible* because the two objects are at *different* temperatures T_C and T_H: By itself heat will flow only from the high-temperature object at T_H to the low-temperature one at T_C, never the other way, so the process cannot run in reverse. And because the two temperatures are different, the entropy changes ΔS_C and ΔS_H in this irreversible process do *not* cancel. To be specific, suppose a quantity of heat $Q = 1.20 \times 10^3$ J flows from a hot reservoir at $T_H = 400$ K to a cold reservoir at $T_C = 300$ K, as shown in Figure 15-23. In this case

$$\text{Cold reservoir: } \Delta S_C = \frac{(+Q)}{T_C} = \frac{1.20 \times 10^3 \text{ J}}{300 \text{ K}} = +4 \text{ J/K}$$

$$\text{Hot reservoir: } \Delta S_H = \frac{(-Q)}{T_H} = \frac{-1.20 \times 10^3 \text{ J}}{400 \text{ K}} = -3 \text{ J/K}$$

$$\text{Net entropy change: } \Delta S_{net} = \Delta S_C + \Delta S_H = 4 \text{ J/K} + (-3 \text{ J/K}) = +1 \text{ J/K}$$

For this irreversible process the cold reservoir gains more entropy than the hot reservoir loses, so there is a *net increase* in the entropy of the system of two reservoirs. So the system becomes increasingly disordered. While energy is conserved, the energy becomes less available to do useful work (for instance, to run a heat engine between the temperatures of the high and low reservoirs).

Entropy and the Second Law of Thermodynamics

We can summarize our observations about reversible and irreversible processes as follows:

In a reversible process there is no net change in entropy. In an irreversible process the net entropy increases.

This is the *entropy statement* of the second law of thermodynamics.

WATCH OUT! The entropy of a system can go down, but only through a process that raises the entropy of another system.

For both the reversible isothermal expansion of an ideal gas (Figure 15-22) and the irreversible flow of heat from a hot reservoir to a cold one (Figure 15-23), one of the objects undergoes a *decrease* in entropy. This may seem to contradict the statement that entropy stays the same in a reversible process and increases in an irreversible process. In fact there's no contradiction, because this statement is about *net* entropy. Whenever an object undergoes a decrease in entropy, it's because there's a second object that undergoes an *increase* in entropy (the gas in the example of the reversible isothermal expansion and the cold reservoir in the example of irreversible heat flow). When all of the interacting objects are taken into account, the *net* entropy never decreases.

The entropy statement of the second law of thermodynamics gives us additional insight into the Carnot cycle, which is the most efficient heat engine theoretically possible. Each step in the Carnot cycle—an isothermal compression, an adiabatic compression, an isothermal expansion, and an adiabatic expansion—is *reversible*. So there is no net entropy change of the system of a Carnot engine, a hot reservoir, and a cold reservoir. (In Example 15-9 we saw that $\Delta S_{net} = 0$ for an isothermal expansion; the same is true for an isothermal compression, in which the gas loses entropy and the reservoir with which it is in contact gains an equal amount of entropy. There is no heat flow at all in a reversible adiabatic compression or expansion, so the entropy does not change in either of these steps.) At the end of each cycle, the system is just as ordered as it was at the beginning of the cycle. By contrast, if we use a heat engine that utilizes irreversible processes (like the idealized automotive engine cycle shown in Figure 15-17), there is a net increase in entropy—and hence in disorder—per cycle. So the *most efficient* engine is also the one that gives rise to the *least* increase in net entropy.

GOT THE CONCEPT? 15-7 Melting Versus Vaporization

(?) For water the latent heat of fusion (melting) is $L_F = 3.34 \times 10^5$ J/kg, and the latent heat of vaporization is $L_V = 2.26 \times 10^6$ J/kg. Which involves a greater change in entropy? (a) Melting 1.00 kg of ice at 0°C; (b) vaporizing 1.00 kg of liquid water at 100°C; (c) the entropy change is the same for both.

We've considered only situations in which the entropy of objects changes due to the transfer of heat between objects. One situation in which there is *no* heat flow is the free expansion of a gas shown in Figure 15-4b. This process is irreversible: After the barrier has ruptured and the expansion has taken place, it would be impossible for all of the gas molecules to spontaneously reassemble on the left-hand part of the cylinder in Figure 15-4b. So we would expect the entropy of the gas to increase in the irreversible free expansion. However, we *cannot* use Equation 15-33 to calculate the entropy change; there is no heat flow into the gas (the cylinder that encloses the gas is insulated), so the equation $\Delta S = Q/T$ predicts *incorrectly* that there would be *zero* entropy change.

The reason Equation 15-33 doesn't work is that it assumes that the process occurs slowly enough that the system is never far from equilibrium, which is definitely not the case for the sudden rush of gas in a free expansion. Instead we find the entropy change by noting that the final state—pressure, volume, and temperature—of the gas after a free expansion is the same as if it had undergone a reversible isothermal expansion to the same final volume (compare parts (a) and (b) of Figure 15-4). We saw in Example 15-9 that the entropy of the gas increases in such a process, and we saw how to calculate ΔS_{gas}. Because entropy is a state variable whose value does not depend on the history of the system, the change in entropy ΔS_{gas} must be the same for a free expansion. In a reversible isothermal expansion the gas is in thermal contact with a reservoir at the same temperature, and the entropy of the reservoir decreases by as much as the entropy of the gas increases so that the net entropy change is zero. But for a free expansion the cylinder is isolated, so the only entropy change is the increase in entropy ΔS_{gas} of the gas. So there is indeed a net entropy increase, just as we expect for an irreversible process.

It's instructive to ask what the entropy statement of the second law of thermodynamics says about the existence and evolution of life. Living organisms such as the cat shown in Figure 15-1a undergo a decrease in entropy as they develop. For example, humans start out as a single cell and grow into a highly ordered network of cells by rearranging raw materials (food) from the environment. Complex organisms are more efficient, require less energy, and are thereby better able to survive in an environment of limited resources. The processes, like evolution, by which complex organisms arise from simpler ones necessarily introduce more order and therefore lower entropy. Does this mean that they violate the second law of thermodynamics?

The answer to this question is no: Life most definitely does *not* violate the second law of thermodynamics. While the entropy of a living organism decreases as it develops, the entropy of its surroundings *increases*. That's because when an organism consumes and metabolizes food, it uses some of the energy from the food to do work (for example, to grow) but also wastes some of that energy. It is this inefficiency that causes the *net* effect to be an increase in entropy. Another way to say this is that the entropy of an organism decreases, but the entropy of its waste products—for example, the carbon dioxide that you exhale, the perspiration that you release from your skin, and the liquid and solid material that you excrete—increases more than the organism's entropy decreases.

TAKE-HOME MESSAGE FOR Section 15-6

✔ The entropy of a system is a measure of the disorder of the system. Systems with a low value of entropy are more ordered.

✔ The most probable final state of a system is the one that is the most disordered and has the highest entropy.

✔ In a process in which two objects interact, it's possible for the entropy of one object to decrease. However, the *net* entropy remains constant in a reversible process and increases in an irreversible process.

Key Terms

adiabatic process
Carnot cycle
coefficient of performance
efficiency
entropy
first law of thermodynamics
heat engine
internal energy
irreversible process

isobaric process
isochoric process
isotherm
isothermal process
molar specific heat
molar specific heat at constant
    pressure
molar specific heat at constant
    volume

pV diagram
ratio of specific heats
refrigerator
reservoir
reversible process
second law of thermodynamics
state variable
thermodynamic process
third law of thermodynamics

Chapter Summary

Topic	Equation or Figure
The first law of thermodynamics: In any thermodynamic process the change in internal energy of the system is related to the heat Q that enters the system and the work W that the system does during the process. There are simple rules for the signs of Q and W.	During a thermodynamic process the **change in the internal energy of a system**... $$\Delta U = Q - W \qquad (15\text{-}2)$$ ...equals the **heat that flows into the system** during the process... ...minus the **work that the system does** during the process.

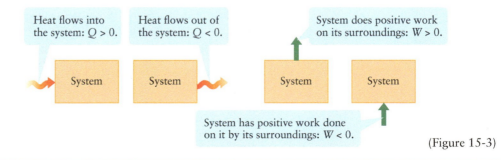

Heat flows into the system: $Q > 0$.

Heat flows out of the system: $Q < 0$.

System does positive work on its surroundings: $W > 0$.

System has positive work done on it by its surroundings: $W < 0$.

(Figure 15-3)

Thermodynamic processes: Four important types of thermodynamic processes are isobaric (constant pressure), isothermal (constant temperature), adiabatic (no heat flow), and isochoric (constant volume). In *any* process, the work done is represented by the area under the curve of the pV diagram for the process. There are simple expressions for the work done in an isobaric process and for the work done by an ideal gas in an isothermal process. In an isochoric process zero work is done.

Work that a system does in an isobaric process | Volume of the system **at the end** of the process

$$W = p(V_f - V_i) \tag{15-4}$$

Constant pressure of the system | Volume of the system **at the beginning** of the process

Number of moles of gas | Ideal gas constant

Work that an ideal gas does in an isothermal process

$$W = nRT \ln \frac{V_f}{V_i} \tag{15-6}$$

Volume of the gas **at the end** of the process

Kelvin temperature of the gas | Volume of the gas **at the beginning** of the process

Molar specific heats of a gas: The molar specific heat is the energy required to raise the temperature of one mole of substance by one kelvin. For an ideal gas, the molar specific heats at constant pressure (C_p) and constant volume (C_V) differ by a fixed amount; the value of C_V depends on the number of degrees of freedom of a gas molecule.

Quantity of heat that flows into (if $Q > 0$) or out of (if $Q < 0$) an object | Molar specific heat of the substance of which the object is made

$$Q = nC\Delta T \tag{15-10}$$

Number of moles present of the substance | Temperature change of the object that results from the heat flow

Molar specific heat of an ideal gas at **constant pressure**

$$C_p - C_V = R \tag{15-15}$$

Ideal gas constant

Molar specific heat of an ideal gas at **constant volume**

Number of degrees of freedom for a molecule of the gas

Molar specific heat of an ideal gas at **constant volume**

$$C_V = \left(\frac{D}{2}\right)R \tag{15-18}$$

Ideal gas constant

Adiabatic processes for an ideal gas: How pressure, volume, and temperature change in an adiabatic process for an ideal gas depends on the value of the ratio of specific heats for the gas $\gamma = C_p/C_V$.

Pressure of an ideal gas | Volume of an ideal gas

$$pV^\gamma = \text{constant} \tag{15-22}$$

Ratio of specific heats of the gas = C_p/C_V

 The quantity pV^γ has the same value at all times during an **adiabatic process** for an ideal gas.

Kelvin temperature of an ideal gas | Volume of an ideal gas

$$TV^{\gamma-1} = \text{constant} \tag{15-23}$$

Ratio of specific heats of the gas = C_p/C_V

 The quantity $TV^{\gamma-1}$ has the same value at all times during an **adiabatic process** for an ideal gas.

The second law of thermodynamics: The second law says that the amount of order in an isolated system cannot increase. This implies that a heat engine—a device that converts heat into work—cannot be 100% efficient but can convert only part of its heat intake Q_H (from a high-temperature source) into work. The rest is rejected as heat $|Q_C|$ to the engine's low-temperature surroundings. The efficiency is $e = W/Q_H$.

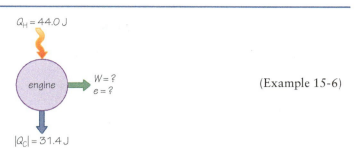

(Example 15-6)

The Carnot cycle: The most efficient heat engine theoretically possible is a Carnot engine. It uses the Carnot cycle, which is based on two reversible adiabatic processes and two reversible isothermal processes. No other engine can have a greater efficiency. A Carnot engine run in reverse is also the best-performing refrigerator theoretically possible.

Efficiency of a Carnot engine   Kelvin temperature of the cold reservoir

$$e_{\text{Carnot}} = 1 - \frac{T_C}{T_H}$$

Kelvin temperature of the hot reservoir

(15-28)

Entropy: The entropy of a system is a measure of the amount of disorder present in the system. In a reversible process in which two objects interact with each other, the entropy of one object increases while the entropy of the other object decreases by the same amount. In an irreversible process, the entropy changes do not cancel and there is a net increase in entropy.

Heat that flows into the system ($Q > 0$ if heat flows in, $Q < 0$ if heat flows out)

Entropy change of a system in a reversible isothermal process

$$\Delta S = \frac{Q}{T}$$

Kelvin temperature of the system

(15-33)

Answer to What do you think? Question

(c) The rising moist air expands when it encounters the lower pressure of the atmosphere at high altitude. Due to its rapid upward motion, the rising air has no time to exchange heat with the surrounding atmosphere. But the rising air loses some of its internal energy, because it must do work to push against the surrounding atmosphere as it expands. This decrease in internal energy causes a decrease in temperature, even though there is no heat flow. A process without heat flow is called *adiabatic* (see Section 15-3).

Answers to Got the Concept? Questions

15-1 (b) We can answer this question using the first law of thermodynamics $\Delta U = Q - W$. In this situation $Q = 0$ because there is no heat flow into or out of the gas. The gas pushes on the moving piston in the same direction that the piston moves, so the gas does positive work on the piston and W is positive. So $\Delta U = Q - W$ is equal to zero minus a positive number, and ΔU is negative—that is, the internal energy U of the gas

decreases. For an ideal gas the internal energy U is directly proportional to the Kelvin temperature T, so a decrease in internal energy means that the gas temperature decreases.

15-2 (a) (i), (ii), (iii) and (iv) (tie). The internal energy U of an ideal gas is directly proportional to its Kelvin temperature. So the internal energy change ΔU is positive if the temperature increases, as is the case for the isobaric process $a \rightarrow b$; ΔU is zero if the temperature stays the same, as for the isothermal process $a \rightarrow c$; and ΔU is negative if the temperature decreases, as for the processes $a \rightarrow d$ and $a \rightarrow e$. The final temperature is the same for processes $a \rightarrow d$ and $a \rightarrow e$ (states d and e are both on the same isotherm), so ΔU has the same negative value for both processes $a \rightarrow d$ and $a \rightarrow e$. (b) (i), (ii), (iii), (iv). The work done in each process is indicated by the area under the curve for that process in the pV diagram. This is greatest for $a \rightarrow b$, smaller for $a \rightarrow c$, and even smaller for $a \rightarrow d$. For all three of these processes, the volume increases, and $W > 0$. For the isochoric process $a \rightarrow e$ there is no volume change, so the gas does no work at $W = 0$.

15-3 (c) The first law of thermodynamics, Equation 15-2, tells us that $\Delta U = Q - W$ or $Q = \Delta U + W$. For an adiabatic process $Q = 0$ and so $\Delta U = -W$. For process $a \rightarrow b$ we have $\Delta U > 0$ (the internal energy increases because the temperature increases) and $W > 0$ (the gas expands), so $Q = \Delta U + W$ is positive and the process cannot be adiabatic. For process $a \rightarrow c$ we have $\Delta U = 0$ (the internal energy remains the same because the temperature is constant) and $W > 0$ (the gas expands), so again $Q = \Delta U + W$ is positive and the process cannot be adiabatic. For process $a \rightarrow e$ $\Delta U < 0$ (the internal energy decreases because the temperature decreases) and $W = 0$ (the volume does not change and so the gas does zero work), so $Q = \Delta U + W$ is negative and the process cannot be adiabatic. The only process that could be adiabatic is $a \rightarrow d$, for which $\Delta U < 0$ (the internal energy decreases because the temperature decreases) and $W > 0$ (the gas expands). For the process the quantity $Q = \Delta U + W$ could be adiabatic.

15-4 (b) For the first, constant-volume process the relationship between heat and temperature change is $Q = nC_V \Delta T$, while for the second, constant-pressure process the relationship is $Q = nC_p \Delta T$ (see Equations 15-11). The relationship $C_p - C_V = R$ (Equation 15-15) tells us that C_p is greater than C_V, so for a given quantity of heat Q the temperature change ΔT will be less for a constant-pressure process. You can get the same result from the first law of thermodynamics $\Delta U = Q - W$ or $Q = \Delta U + W$. For the constant-volume process $W = 0$ and $Q = \Delta U$: All of the heat goes into increasing the internal energy of the gas and so into raising its temperature. But for the constant-pressure process, the expanding gas must do positive work W on the piston. Some of the heat goes into doing this work, so only a fraction of Q is available to increase the internal energy of the gas.

15-5 (c) From Equations 15-11 the relationship between heat and temperature change for these constant-pressure processes is $Q = nC_p \Delta T$. Table 15-1 shows that the molar specific heat at constant pressure C_p has a smaller value for He than for N_2, so for a given quantity of heat Q the temperature change ΔT is greater for He than for N_2. Equation 15-13 tells us that the work done by an ideal gas in a constant-pressure process is $W = nR\Delta T$; the number of moles n is the same for both gases and R is the ideal gas constant, so more work is done by the gas that undergoes the greater temperature change ΔT (in this example, helium).

15-6 No. A gasoline engine uses ambient temperature (about 20°C or 293 K) as the cold temperature. The engine your fellow student has designed uses a hot temperature that must be less than 1941 K, the melting temperature of titanium, or else the cylinders would melt. The efficiency of a Carnot cycle operating between $T_C = 293$ K and $T_H = 1941$ K is $e_{Carnot} = 1 - (T_C/T_H) = 1 - [(293 \text{ K})/(1941 \text{ K})] = 1 - 0.151 = 0.849$, or 84.9%. Your fellow student is claiming an even higher efficiency, which is impossible because no heat engine can have greater efficiency than a Carnot engine. This would be a poor investment!

15-7 (b) Both melting and vaporization take place at a constant temperature, so we can use Equation 15-33, $\Delta S = Q/T$, to calculate the entropy change in each process. For melting at 0°C or $T = 273.15$ K, $Q = mL_F = (1.00 \text{ kg})(3.34 \times 10^5 \text{ J/kg}) = 3.34 \times 10^5$ J and $\Delta S_{melting} = Q/T = (3.34 \times 10^5 \text{ J})/(273.15 \text{ K}) = 1.22 \times 10^3$ J/K. For vaporization at 100°C or $T = (100 + 273.15)$ K $= 373.15$ K, $Q = mL_V = (1.00 \text{ kg})(2.26 \times 10^6 \text{ J/kg}) = 2.26 \times 10^6$ J and $\Delta S_{vaporization} = Q/T = (2.26 \times 10^6 \text{ J})/(373.15 \text{ K}) = 6.06 \times 10^3$ J/K. The entropy change is about five times greater for vaporization than for melting, which agrees with the idea that entropy is a measure of disorder. The molecules of liquid water are able to move while those of ice cannot, so the entropy of liquid water is greater than that of ice. But the molecules of water vapor are free to move over a much greater volume, so the entropy of water vapor is *very* much greater than that of liquid water.

Questions and Problems

In a few problems you are given more data than you actually need; in a few other problems you are required to supply data from your general knowledge, outside sources, or informed estimate.

Interpret as significant all digits in numerical values that have trailing zeros and no decimal points. For all problems use $g = 9.80 \text{ m/s}^2$ for the free-fall acceleration due to gravity. Neglect friction and air resistance unless instructed to do otherwise.

- •   Basic, single-concept problem
- ••   Intermediate-level problem; may require synthesis of concepts and multiple steps
- •••   Challenging problem
- Example See worked example for a similar problem

Conceptual Questions

1. • Can a system absorb heat without increasing its internal energy? Explain.

2. • Why is it possible for the temperature of a system to remain constant even though heat is released or absorbed by the system?

3. • In a slow, steady isothermal expansion of an ideal gas against a piston, the work done is equal to the heat input. Is this consistent with the first law of thermodynamics?

4. • Clearly define and give an example of each of the following thermodynamic processes: (a) isothermal, (b) adiabatic, (c) isobaric, and (d) isochoric.

5. • Why does the temperature of a gas increase when it is quickly compressed?

6. • When we say "engine," we think of something mechanical with moving parts. In such an engine friction always reduces the engine's efficiency. Why?

7. • Why do engineers designing a steam-electric generating plant always try to design for as high a feed-steam temperature as possible?

8. • Is the operation of an automobile engine reversible?

9. • Is a process necessarily reversible if there is no exchange of heat between the system in which the process takes place and its surroundings?

10. • Conduction across a temperature difference is an irreversible process, but the object that lost heat can always be rewarmed, and the one that gained heat can be recooled. An object sliding across a rough table slows down and warms up as mechanical energy dissipates. This process is irreversible, but the object can be cooled and set moving again at its original speed. So in just what sense are these processes "irreversible"?

11. • There are people who try to keep cool on a hot summer day by leaving the refrigerator door open, but you can't cool your kitchen this way! Why not?

12. • If the coefficient of performance is greater than 1, do we get more energy out than we put in, violating conservation of energy? Why or why not?

13. • How does the time required to freeze water vary with each of the following parameters: mass of water, power of the refrigerator, and temperature of the outside air?

14. • How is the entropy of the universe changed when heat is released from a hotter object to a colder one? In what sense does this correspond to energy becoming unavailable for doing work?

15. • If you drop a glass cup on the floor, it will shatter into fragments. If you then drop the fragments on the floor, why will they not become a glass cup?

16. • Why is the entropy of 1 kg of liquid iron greater than that of 1 kg of solid iron? Explain your answer.

17. • If a gas expands freely into a larger volume in an insulated container so that no heat is added to the gas, its entropy increases. Explain this using the idea that this process is irreversible.

18. • In discussing the Carnot cycle, we say that extracting heat from a reservoir isothermally does not change the entropy of the universe. In a real process, this is a limiting situation that can never quite be reached. Why not? What is the effect on the entropy of the universe?

19. • A pot full of hot water is placed in a cold room, and the pot gradually cools. How does the entropy of the water change?

Multiple-Choice Questions

20. • An ideal gas trapped inside a thermally isolated cylinder expands slowly by pushing back against a piston. The temperature of the gas
   A. increases.
   B. decreases.
   C. remains the same.
   D. increases if the process occurs quickly.
   E. remains the same if the process occurs quickly.

21. • A gas is compressed adiabatically by a force of 800 N acting over a distance of 5.0 cm. The net change in the internal energy of the gas is
   A. +800 J.
   B. +40 J.
   C. –800 J.
   D. –40 J.
   E. 0.

22. • An ideal gas is contained in a closed cylinder of fixed length and diameter. Eighty joules of heat is added to the gas. The work done by the gas on the walls of the cylinder is
   A. 80 J.
   B. 0 J.
   C. less than 80 J.
   D. more than 80 J.
   E. not specified by the information given.

23. • In an isothermal process there is no change in
   A. pressure.
   B. temperature.
   C. volume.
   D. heat.
   E. internal energy *or* pressure.

24. • In an isobaric process there is no change in
   A. pressure.
   B. temperature.
   C. volume.
   D. internal energy.
   E. internal energy *or* pressure.

25. • In an isochoric process there is no change in
   A. pressure.
   B. temperature.
   C. volume.
   D. internal energy.
   E. internal energy *or* pressure.

26. • A gas quickly expands in an isolated environment. During the process the gas exchanges no heat with its surroundings. The process is
   A. isothermal.
   B. isobaric.
   C. isochoric.
   D. adiabatic.
   E. isotonic.

27. • The statement that no process is possible in which heat is absorbed from a cold reservoir and transferred completely to a hot reservoir is
   A. not always true.
   B. only true for isothermal processes.
   C. the first law of thermodynamics.
   D. the second law of thermodynamics.
   E. the zeroth law of thermodynamics.

28. • Carnot's heat engine employs
   A. two adiabatic processes and two isothermal processes.
   B. two adiabatic processes and two isobaric processes.
   C. two adiabatic processes and two isochoric processes.
   D. two isothermal processes and two isochoric processes.
   E. two isothermal processes and two isobaric processes.

29. • Compare two methods to improve the theoretical efficiency of a heat engine: lower T_C by 10 K or raise T_H by 10 K. Which one is better?

   A. lower T_C by 10 K

   B. raise T_H by 10 K

   C. Both changes would give the same result.

   D. The better method would depend on the difference between T_C and T_H.

   E. There is nothing you can do to improve the theoretical efficiency of a heat engine.

Problems

15-1 The laws of thermodynamics involve energy and entropy

15-2 The first law of thermodynamics relates heat flow, work done, and internal energy change

30. • If 800 J of heat is added to a system that does no external work, how much does the internal energy of the system increase? Example 15-1

31. • Five hundred joules of heat is absorbed by a system that does 200 J of work on its surroundings. What is the change in the internal energy of the system? Example 15-1

15-3 A graph of pressure versus volume helps to describe what happens in a thermodynamic process

32. • Calculate the amount of work done on a gas that undergoes a change of state described by the pV diagram shown in Figure 15-24.

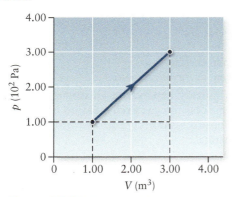

Figure 15-24 Problem 32

33. • Calculate the amount of work done on a gas that undergoes a change of state described by the pV diagram shown in Figure 15-25.

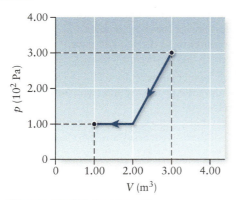

Figure 15-25 Problem 33

34. • A gas is heated and is allowed to expand such that it follows a horizontal line path on a pV diagram from its initial state $(1.0 \times 10^5$ Pa, 1.0 m$^3)$ to its final state $(1.0 \times 10^5$ Pa, 2.0 m$^3)$. Calculate the work done by the gas on its surroundings. Example 15-4

35. • A gas is heated such that it follows a vertical line path on a pV diagram from its initial state $(1.0 \times 10^5$ Pa, 3.0 m$^3)$ to its final state $(2.0 \times 10^5$ Pa, 3.0 m$^3)$. Calculate the work done by the gas on its surroundings. Example 15-4

36. • A sealed cylinder has a piston and contains 8.00×10^3 cm^3 of an ideal gas at a pressure of 8.00 atm. Heat is slowly introduced, and the gas isothermally expands to 1.60×10^4 cm^3. How much work does the gas do on the piston? Example 15-3

37. • An ideal gas expands isothermally, performing 8.80 kJ of work in the process. Calculate the heat absorbed during the expansion. Example 15-3

38. • Heat is added to 8.00 m^3 of helium gas in an expandable chamber that increases its volume by 2.00 m^3. If in the isothermal expansion process 2.00 kJ of work is done by the gas, what was its original pressure? Example 15-3

39. • A cylinder that has a piston contains 2.00 mol of an ideal gas and undergoes a reversible isothermal expansion at 400 K from an initial pressure of 12.0 atm down to 3.00 atm. Determine the amount of work done by the gas. Example 15-3

40. • A gas contained in a cylinder that has a piston is kept at a constant pressure of 2.80×10^5 Pa. The gas expands from 0.500 to 1.50 m^3 when 300 kJ of heat is added to the cylinder. What is the change in internal energy of the gas? Example 15-2

41. • The pressure in an ideal gas is slowly reduced to $\frac{1}{4}$ its initial value, while being kept in a container with rigid walls. In the process 800 kJ of heat leaves the gas. What is the change in internal energy of the gas during this process? Example 15-4

42. • An ideal gas is compressed adiabatically to half its volume. In doing so 1888 J of work is done on the gas. What is the change in internal energy of the gas? Example 15-3

15-4 The concept of molar specific heat helps us understand isobaric, isochoric, and adiabatic processes for ideal gases

43. • Two moles of an ideal monatomic gas expand adiabatically, performing 8.00 kJ of work in the process. What is the change in temperature of the gas during the expansion? Example 15-5

44. • One mole of an ideal monatomic gas ($\gamma = 1.66$), initially at a temperature of 0.00°C, undergoes an adiabatic expansion from a pressure of 10.0 atm to a pressure of 2.00 atm. How much work is done on the gas? Example 15-5

45. • A container holds 32.0 g of oxygen gas at a pressure of 8.00 atm. How much heat is required to increase the temperature by 100°C at constant pressure?

46. • A container holds 32.0 g of oxygen gas at a pressure of 8.00 atm. How much heat is required to increase the temperature by 100°C at constant volume?

47. • The temperature of 4.00 g of helium gas is increased at constant volume by 1.00°C. Using the same amount of heat, the temperature of what mass of oxygen gas will increase at constant volume by 1.00°C?

48. • Heat is added to 1.00 mol of air at constant pressure, resulting in a temperature increase of 100°C. If the same amount of heat is instead added at constant volume, what is the temperature increase? The molar specific heat ratio $\gamma = C_p/C_V$ for the air is 1.4. Example 15-5

49. • The volume of a gas is halved during an adiabatic compression that increases the pressure by a factor of 2.6. What is the molar specific heat ratio $\gamma = C_p/C_V$?

50. •• The volume of a gas is halved during an adiabatic compression that increases the pressure by a factor of 2.5. By what factor does the temperature increase?

51. • What ratio of initial volume to final volume V_i/V_f will raise the temperature of air from 27.0°C to 857°C in an adiabatic process? The molar specific heat ratio $\gamma = C_p/C_V$ for air is 1.4. Example 15-5

52. • A monatomic ideal gas at a pressure of 1.00 atm expands adiabatically from an initial volume of 1.50 m³ to a final volume of 3.00 m³. What is the new pressure? Example 15-5

15-5 The second law of thermodynamics describes why some processes are impossible

53. • An engine doing work takes in 10.0 kJ and exhausts 6.00 kJ. What is the efficiency of the engine? Example 15-6

54. • What is the theoretical maximum efficiency of a heat engine operating between 100°C and 500°C? Example 15-7

55. • A heat engine operating between 473 and 373 K runs at 70.0% of its theoretical maximum efficiency. What is its efficiency? Example 15-7

56. • An engine operates between 10.0°C and 200°C. At the very best how much heat needs to be supplied to output 1.00×10^3 J of work? Example 15-7

57. • A furnace supplies 28.0 kW of thermal power at 300°C to an engine that exhausts waste energy at 20.0°C. At the very best how much work could we expect to get out of the system per second? Example 15-7

58. • A kitchen refrigerator extracts 75.0 kJ per second of energy from a cool chamber while exhausting 1.00×10^2 kJ per second to the room. What is its coefficient of performance? Example 15-8

59. • What is the coefficient of performance of a Carnot refrigerator operating between 0.00°C and 80.0°C? Example 15-8

60. •• An electric refrigerator removes 13.0 MJ of heat from its interior for each kilowatt-hour of electric energy used. What is its coefficient of performance? Example 15-8

61. • A certain refrigerator requires 35.0 J of work to remove 190 J of heat from its interior. (a) What is its coefficient of performance? (b) How much heat is ejected to the surroundings at 22.0°C? (c) If the refrigerator cycle is reversible, what is the temperature inside the refrigerator? Example 15-8

15-6 The entropy of a system is a measure of its disorder

62. • A reservoir at a temperature of 400 K gains 100 J of heat from another reservoir. What is its entropy change? Example 15-9

63. • What is the minimum change of entropy that occurs in 0.200 kg of ice at 273 K when 6.68×10^4 J of heat is added so that it melts to water? Example 15-9

64. • If, in a reversible process, enough heat is added to change a 500-g block of ice to water at a temperature of 273 K, what is the change in the entropy of the ice–water system? Example 15-9

65. • A room is at a constant 295 K maintained by an air conditioner that pumps heat out. How much does the air conditioner change the entropy of the room for each 5.00 kJ of heat it removes? Example 15-9

66. •• One mole of ideal gas expands isothermally from 1.00 to 2.00 m³. What is the entropy change for the gas?

67. •• A 1.80×10^3-kg car traveling at 80.0 km/h crashes into a concrete wall. If the temperature of the air is 27.0°C, what is the entropy change of the universe as a result of the crash? Assume all of the car's kinetic energy is converted into heat. Example 15-9

68. •• A 1.00×10^3-kg rock at 20.0°C falls 1.00×10^2 m into a large lake, also at 20.0°C. Assuming that all of the rock's kinetic energy on entering the lake converts to thermal energy absorbed by the lake, what is the change in entropy of the lake? Example 15-9

69. •• **Astronomy** The surface of the Sun is about 5700 K, and the temperature of Earth's surface is about 293 K. What is the net change in entropy of the Sun–Earth system when 8000 J of heat lost by the Sun is then absorbed by Earth? Example 15-9

70. •• A 0.750-L cup of coffee at 70°C is left outside where the temperature is 4°C. When the coffee reaches thermal equilibrium with the atmosphere, by how much has the entropy of the atmosphere changed? Assume the properties of coffee are identical to those of water. Example 15-9

71. •• A balloon containing 0.50 mol of helium in a 20.0°C chamber undergoes an isothermal expansion as the pressure in the chamber is slowly reduced. If the pressure drops to half its initial value during this process, what is the entropy change of the helium? Example 15-9

General Problems

72. ••• A vertical, insulated cylinder contains an ideal gas. The top of the cylinder is closed off by a piston of mass m that is free to move up and down with no appreciable friction. The piston is a height h above the bottom of the cylinder when the gas alone supports it. Sand is now very slowly poured onto the piston until the weight of the sand is equal to the weight of the piston. Find the new height of the piston (in terms of h) if the ideal gas in the cylinder is (a) oxygen, O_2; (b) helium, He; or (c) hydrogen, H_2. Example 15-5

73. •• A Carnot engine on a ship extracts heat from seawater at 18.0°C and exhausts the heat to evaporating dry ice at −78.0°C. If the ship's engines are to run at 8.00×10^3 horsepower, what is the minimum amount of dry ice the ship must carry for the ship to run for a single day? Example 15-7

74. • A Carnot engine removes 1.20×10^3 J of heat from a high-temperature source and dumps 6.00×10^2 J to the atmosphere at 20.0°C. (a) What is the efficiency of the engine? (b) What is the temperature of the hot reservoir? Example 15-7

75. •• A certain engine operates at 85% of its theoretical maximum efficiency. During each cycle it absorbs 4.80×10^2 J of heat from a reservoir at 300°C and dumps 3.00×10^2 J of heat to a cold temperature reservoir. (a) What is the temperature of the cold reservoir? (b) How much more work could

be done by a Carnot engine working between the same two reservoirs and extracting the same 4.80×10^2 J of heat in each cycle? Example 15-7

76. • A refrigerator is rated at 370 W. Its interior is at 0°C, and its surroundings are at 20°C. If the refrigerator operates at 66% of its theoretical maximum efficiency, how much heat can it remove from its interior in 1 min? Example 15-8

77. •• **Medical** During a high fever a 60.0-kg patient's normal metabolism is increased by 10.0%. This results in an increase of 10.0% in the heat given off by the person. When the person slowly walks up five flights of stairs (20.0 m), she normally releases 1.00×10^5 J of heat. Compare her efficiency when she has a fever to when her temperature is normal. Example 15-6

78. •• A rigid 5.50-L pressure cooker contains steam initially at 100°C under a pressure of 1.00 atm. Consult Table 15-1 as needed and assume that the values given there remain constant. The mass of a water molecule is 2.99×10^{-26} kg. (a) To what temperature (in °C) would you have to heat the steam so that its pressure was 1.25 atm? (b) How much heat would you need in part (a)? (c) Calculate the specific heat of the steam in part (a) in units of J/(kg·K).

79. • Liquid nitrogen, with a latent heat of fusion of $L_F = 25.3 \times 10^3$ J/kg, solidifies at a temperature of 63 K. What is the change in entropy of a 1.5-kg sample of liquid nitrogen as it transitions from a liquid to a solid? Example 15-9

80. •• A certain electric generating plant produces electricity by using steam that enters its turbine at a temperature of 320°C and leaves it at 40°C. Over the course of a year, the plant consumes 4.40×10^{16} J of heat and produces an average electric power output of 600 MW. What is the efficiency of this generating plant compared to a Carnot engine operating between the same temperatures? Example 15-7

81. •• As we drill down into the rocks of Earth's crust, the temperature typically increases by 3.0°C for every 100 m of depth. Oil wells can be drilled to depths of 1830 m. If water is pumped into the shaft of the well, it will be heated by the hot rock at the bottom and the resulting steam can be used as a heat engine. Assume that the surface temperature is 20°C. (a) Using such a 1830-m well as a heat engine, what is the maximum efficiency possible? (b) If a combination of such wells is to produce a 2.5-MW power plant, how much energy will it absorb from the interior of Earth each day? Example 15-7

82. ••• The energy efficiency ratio (or rating)—the EER—for air conditioners, refrigerators, and freezers is defined as the ratio of the input rate of heat ($|Q_C|/t$, in BTU/h) to the output rate of work (W/t, in W): EER $= \dfrac{Q_C/t(\text{BTU/hr})}{W/t(\text{W})}$. (a) Show that the EER can be expressed as $\dfrac{Q_C(\text{BTU})}{W(\text{W}\cdot\text{h})}$ and is therefore nothing more than the coefficient of performance CP expressed in mixed units. (b) Show that the EER is related to the coefficient of performance CP by the equation $CP = \text{EER}/3.412$. (c) Typical home freezers have EER ratings of about 5.1 and operate between an interior freezer temperature of 0.00°F and an outside kitchen temperature of about 70.0°F. What is the coefficient of performance for such a freezer, and how does it compare to the coefficient of performance of the best possible freezer operating between those temperatures? (d) What is the EER of the best possible freezer in part (c)? Example 15-8

83. ••• Your energy-efficient home freezer has an EER of 6.50 (see Problem 15-82). In preparation for a picnic you put 1.50 L of water at 20.0°C into the freezer to make ice at 0.00°C for your ice chest. (See Tables 11-1, 14-3, and 14-4 as needed.) (a) How much electrical energy (which runs the freezer) is required to make the ice? Express your answer in J and kWh. (b) How much heat is ejected into your kitchen, which is at 22.0°C during the process? (c) How much does making the ice change the entropy of your kitchen? Example 15-8

84. •• **Biology** The volume of 20°C air taken in during a typical breath is 0.5 L. The inhaled air is heated to 37°C (the internal body temperature) as it enters the lungs. Because air is about 80% nitrogen N_2, we can model it as an ideal gas. Assume that the pressure does not change during the process. (a) How many joules of heat does it take to warm the air in a single breath? (b) Suppose that you take two breaths every 3.0 s. How many food calories (kcal) are used up per day in heating the air you breathe? Is this a significant amount of typical daily caloric intake?

85. •• A heat engine works in a cycle between reservoirs at 273 and 490 K. In each cycle, the engine absorbs 1250 J of heat from the high-temperature reservoir and does 475 J of work. (a) What is its efficiency? (b) By how much is the entropy of the universe changed when the engine goes through one full cycle? (c) How much energy becomes unavailable for doing work when the engine goes through one full cycle? Example 15-7

86. ••• You have a cabin on the plains of central Saskatchewan. It is built on a 8.50-m by 12.5-m rectangular foundation with walls 3.00 m tall. The wooden walls and flat roof are made of white pine that is 9.00 cm thick. To conserve heat the windows are negligibly small. The floor is well insulated, so you lose negligible heat through it. The cabin is heated by an electrically powered heat pump operating on the Carnot cycle between the inside and outside air. When the outside temperature is a frigid −10.0°F, how much electrical energy does the heat pump consume per second to keep the interior temperature a steady and toasty 70.0°F? Assume that the surfaces of the walls and roof are at the same temperature as the air with which they are in contact and neglect radiation. (Consult Table 14-5 as needed.) Example 15-7

87. • **Sports** In an international diving competition divers fall from a platform 10.0 m above the surface of the water into a very large pool. A diver leaves the platform with negligible initial speed. Assuming that all of a diver's kinetic energy is converted to heat when he hits the water, what is the maximum change in the entropy of the water in the pool at 25.0°C when a 75.0-kg diver executes his dive? Does the pool's entropy increase or decrease? Example 15-9

88. •• A heat engine works in a cycle between reservoirs at 273 and 490 K. In each cycle the engine absorbs 1250 J of heat from the high-temperature reservoir and does 475 J of work. (a) What is its efficiency? (b) What is the change in entropy of the universe when the engine goes through one complete cycle? (c) How much energy becomes unavailable for doing work when the engine goes through one complete cycle? Example 15-9

89. •• **Biology** A 68.0-kg person typically eats about 2250 kcal per day, 20.0% of which goes to mechanical energy and the rest to heat. If she spends most of her time in her apartment at 22.0°C, how much does the entropy of her apartment change

in one day? Does the entropy of the apartment increase or decrease? Example 15-9

90. ••• Consider an engine in which the working substance is 1.23 mol of an ideal gas for which $\gamma = 1.41$. The engine runs reversibly in the cycle shown on the pV diagram (Figure 15-26). The cycle consists of an isobaric (constant-pressure) expansion a at a pressure of 15.0 atm, during which the temperature of the gas increases from 300 to 600 K, followed by an isothermal expansion b until its pressure becomes 3.00 atm. Next is an isobaric compression c at a pressure of 3.00 atm, during which the temperature decreases from 600 to 300 K, followed by an isothermal compression d until its pressure returns to 15 atm. Find the work done by the gas, the heat absorbed by the gas, the internal energy change, and the entropy change of the gas, first for each part of the cycle and then for the complete cycle. Example 15-9

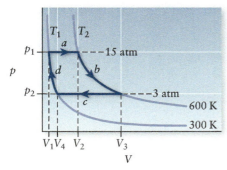

Figure 15-26 Problem 90

91. •• A heat engine with 0.219 mol of a monatomic gas undergoes the cyclic process shown in the pV diagram (Figure 15-27), where $p_1 = 400$ kPa, $p_2 = 640$ kPa, $V_1 = 1.25 \times 10^3$ cm^3, and $V_2 = 2.53 \times 10^3$ cm^3. Between stage 3 and stage 1,

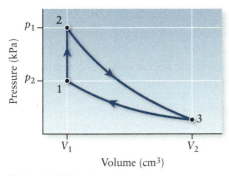

Figure 15-27 Problem 91

the gas is at a constant temperature, and between stage 2 and stage 3, no heat is transferred in or out. The temperature of the gas at stage 2 is 440 K. (a) Identify the process involved in each step of the cycle. (b) What is the temperature between stages 3 and 1? (c) How much work does this heat engine do per cycle? (d) Calculate the efficiency of this heat engine.

92. •• The thermodynamic cycle of a heat engine using ideal helium gas is shown in the pV diagram (Figure 15-28). The ratio of heat capacities γ (also called the ratio of specific heats) is the ratio of the heat capacity at constant pressure C_p to the heat capacity at constant volume C_V. For ideal helium gas $\gamma = 1.67$. A confused physics student determines that the cycle comprises the following processes: $1 \to 2$ adiabatic; $2 \to 3$ isochoric; $3 \to 4$ isothermal; $4 \to 1$ isobaric. Which process did the student identify correctly?

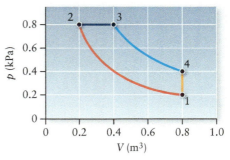

Figure 15-28 Problem 92

93. • A gas is taken through the cyclic process shown in the pV diagram (Figure 15-29). At point 1, the pressure is 2.50×10^3 Pa and the volume 3.50 m^3. At point 2, the volume is 6.00 m^3. At point 3, the pressure is 8.50×10^3 Pa. Over one complete cycle, how much heat is transferred to the system?

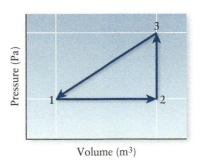

Figure 15-29 Problem 93

16

Electrostatics I: Electric Charge, Forces, and Fields

What do you think?

A shark can detect the electric field produced by a small, electrically charged object. If a charged object 2 m from a shark's nose is moved to be only 1 m from the nose, the magnitude of the electric field that the shark detects will (a) decrease; (b) double; (c) increase by a factor of 4; (d) increase by a factor of 8; (e) increase by a factor of 16.

In this chapter, your goals are to:

- (16-1) Explain why studying electric phenomena is important.
- (16-2) Describe how objects acquire a net electric charge.
- (16-3) Recognize the differences between insulators, conductors, and semiconductors.
- (16-4) Use Coulomb's law to quantitatively describe the force that one charged particle exerts on another.
- (16-5) Explain the relationship between electric force and electric field.
- (16-6) Describe the connection between enclosed charge and electric flux described by Gauss's law.
- (16-7) Apply Gauss's law to symmetric situations and to the distribution of excess charge on conductors.

To master this chapter, you should review:

- (3-2, 3-3, and 3-4) How to add vectors and how to do vector calculations using components.
- (4-2) Newton's second law.

16-1 Electric forces and electric charges are all around you—and within you

You may think of electricity as the shock you get when you walk across a carpet and then touch a metal doorknob, or a commodity that you purchase from the power company. The reality is that electric phenomena are all around you, from the behavior of grains of pollen to the drama of a thunderstorm (Figure 16-1). As you read this sentence, your eye casts an image of the text onto your retina, and this image is transmitted to your brain along the optic nerve as a stream of electrical impulses. Electricity even explains why you don't fall through the floor: The molecules on the surface of the floor exert electric forces that repel the molecules on the underside of your shoes, and these forces are strong enough to balance Earth's downward gravitational force on you.

(a)

Grains of pollen wafting through the air carry a small amount of electric charge. This attracts them to the oppositely charged stigma of a flower and so aids in pollination.

(b)

The motion of air in a thundercloud causes positive and negative charges to separate and build up within the cloud. The release of these charges causes lightning.

Biophoto Associates/Science Source

KingWu/iStockphoto

Figure 16-1 **Electric charges and forces** Electrostatics — the physics that describes how opposite charges attract and like charges repel — plays an important role in (a) biology and (b) the weather.

Electric phenomena are important because *all* ordinary matter is made of electrically charged objects. That's why we'll spend the next several chapters learning about electric charges and their physics. The simplest place to begin our study of electricity is with *electrostatics*, the branch of physics that deals with electric charges at rest and how they interact with each other. In this chapter we'll first learn about the different kinds of electrical charge in nature, and describe the forces that two electrically charged objects exert on each other. We'll see how these forces manifest themselves in two important classes of materials called conductors and insulators. We'll go on to learn about the concept of an *electric field*, a powerful way of understanding how electric charges can exert forces on each other even over a distance. We'll conclude this chapter with a look at *Gauss's law*, an important tool for understanding electric fields.

TAKE-HOME MESSAGE FOR Section 16-1

✔   Electric phenomena are important because all ordinary matter is made of electrically charged objects.

✔   Electrostatics is the study of electric charges at rest and the forces that they exert on each other.

16-2   Matter contains positive and negative electric charge

Figure 16-2 shows some simple experiments that demonstrate the nature of electric charge. If you hold two rubber rods next to each other, they neither attract nor repel each other. The same is true for two glass rods held next to each other, or for a rubber rod and a glass rod (Figure 16-2a). But if you rub the ends of each rubber rod with fur and each glass rod with silk, we find that the ends of the rubber rods repel each other and the ends of the glass rods repel each other, but the ends of the rubber and glass rods attract each other (Figure 16-2b). What's more, the end of the rubber rod attracts the piece of fur used to rub it, and the end of the glass rod attracts the piece of silk used to rub it (Figure 16-2c). In each case where there is a repulsion or an attraction, the repulsive or attractive force is stronger the closer the objects are held to each other.

Here's how we explain these experiments and others like them:

(1) *Matter has electric charge*. In addition to having mass, matter also possesses a property called **electric charge**. This charge comes in two forms, positive and negative. An object that has a net electric charge (more positive than negative, or more negative than positive) is **charged**. Most matter, however, contains equal amounts of positive and negative charge and is electrically **neutral**. In Figure 16-2a the rubber and glass rods are electrically neutral, as are the pieces of fur and silk.

(2) *Electric charge is a property of the constituents of atoms*. All ordinary matter is made up of atoms. Atoms are in turn composed of more fundamental particles called electrons, protons, and (except for the most common type of hydrogen atom) neutrons.

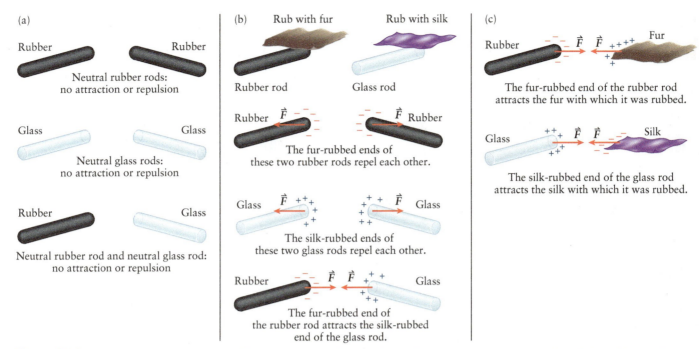

Figure 16-2 Experiments in electrostatics Electric charge can be transferred from one object to another by rubbing. Electrically charged objects exert forces on each other. Plus and minus signs indicate the net charge on each object (+ for positive, − for negative).

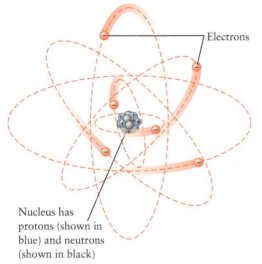

Figure 16-3 A simplified model of an atom Negatively charged electrons orbit the atom's nucleus, which contains most of the atom's mass. The nucleus contains two types of particles, positively charged protons and uncharged neutrons.

Electrons carry a negative charge, while protons carry an equally large positive charge; neutrons are neutral. Protons and neutrons have about the same mass and are found in the dense nucleus at the center of an atom. A negatively charged electron has only about 0.05% the mass of a proton, and the electrons move around the massive, positively charged nucleus (Figure 16-3). A normal atom contains as many electrons as it does protons and so is electrically neutral.

(3) *Charge can be transferred between objects.* When dissimilar materials are rubbed together, negatively charged electrons can be transferred from one material to the other. As an example, consider the rubber rod and fur shown in Figure 16-2b. Initially the rod and fur are electrically neutral, so each has equal amounts of positive and negative charge. But when they are rubbed together, electrons are transferred from the fur to the rubber. This leaves the fur with a net positive charge (more positive charge than negative charge) and the rubber with a net negative charge. By contrast, when the glass rod and silk in Figure 16-2b (both of which are initially neutral) are rubbed together, electrons are transferred from the glass to the silk. So the silk ends up with a net negative charge and the glass rod with a net positive charge.

(4) *Electrically charged objects exert forces on each other.* Objects with electric charges of the same sign (both positive or both negative) repel each other, while objects with electric charges of opposite sign (one positive and the other negative) attract each other (Figure 16-2c). The force between charges is called the **electric force**. The magnitude of the force depends on the amount of charge on each object and on the distance between the objects. The force increases for a greater amount of charge, and it increases if the charged objects are brought closer to each other.

(5) *Charge is never created or destroyed.* In the experiments shown in Figure 16-2, charge is transferred between objects. Although charge moves between the rubber rod and the fur or between the glass rod and the silk, in each case the *total* charge of the two objects remains the same as before they were rubbed together. That is, charge is *conserved*; it is never created or destroyed. No one has ever observed a process in which charge is not conserved, so to the best of our knowledge the conservation of electric charge is an absolute law of nature.

(a)

(b)

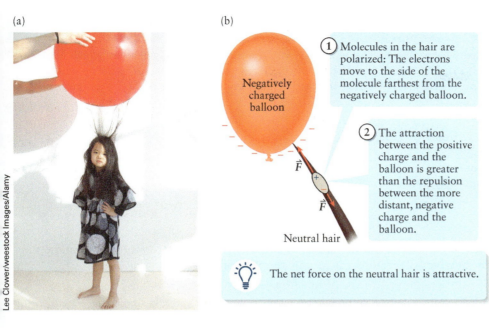

Negatively charged balloon

1. Molecules in the hair are polarized: The electrons move to the side of the molecule farthest from the negatively charged balloon.

$\vec{F}$

$\vec{F}$

Neutral hair

2. The attraction between the positive charge and the balloon is greater than the repulsion between the more distant, negative charge and the balloon.

The net force on the neutral hair is attractive.

Figure 16-4 A hair-raising experiment (a) Rubbing a balloon on your sleeve gives the balloon an electric charge. A charged balloon can attract your hair and make it stand up, even though your hair has no net charge. (b) The nature of the electric force explains why this attraction happens.

The character of the electric force explains the remarkable experiment shown in Figure 16-4a. A balloon is first rubbed against the girl's sleeve, causing electrons to transfer from her sleeve onto the rubber surface of the balloon (just like the rubber rod and fur in Figure 16-2b). When the negatively charged balloon is held next to the girl's head, electric forces attract the hair to the balloon—even though the hair itself is *neutral*. What happens is that the negatively charged electrons on the balloon repel the electrons in the atoms that comprise the hair, but also attract the positively charged nuclei of these atoms. The electrons and nuclei remain within their atoms but end up slightly displaced from each other (Figure 16-4b). We say that the hair becomes *polarized*. As we described above, the electric force is greater the closer charged objects are to each other. Figure 16-4b shows that the nuclei in the polarized hair are pulled slightly closer to the negatively charged balloon than the electrons in the hair. So the attractive force between the hair nuclei and the balloon is slightly greater than the repulsive force between the hair electrons and the balloon. The result is that the hair feels a net attractive force toward the balloon.

The unit of electric charge is the **coulomb** (C), named after the eighteenth-century French physicist Charles-Augustin de Coulomb, who uncovered the fundamental law that governs the interaction of charges. (We'll discuss this law in Section 16-4.) As we mentioned above, electrons and protons have the same magnitude of electric charge; we call this magnitude e. The coulomb is defined so that the value of e is exactly

$$e = 1.602176634 \times 10^{-19} \text{ C}$$

We use q or Q as the symbol for the charge of an object, so the charge on a proton is (to four significant figures) $q_{\text{proton}} = +e = +1.602 \times 10^{-19}$ C, and the charge on an electron is $q_{\text{electron}} = -e = -1.602 \times 10^{-19}$ C. No free particle has ever been detected with a charge smaller in magnitude than e. For that matter, no object has ever been found whose charge was not a multiple of e such as $+2e$, $-3e$, and so on. That is, charge is *quantized*: The charge of an object is always increased or decreased by an amount equal to an integer multiple of the fundamental charge e. It is impossible to add a fraction of a proton or electron to an object. (As we'll see in Chapter 28, there is strong evidence that protons and neutrons are composed of more fundamental particles called *quarks* that have charges of $+2e/3$ and $-e/3$. However, no isolated quarks have ever been observed outside of a larger particle such as a proton or neutron.)

While electrons and protons have opposite signs, it's completely arbitrary whether we choose electrons to have negative charge and protons to have positive charge or the other way around. The choice of sign is due to the American scientist and statesman Benjamin Franklin, who decided in the mid-eighteenth century that the sign of charge on a glass rod rubbed with silk (see Figure 16-2b) is positive. The electron and proton were not discovered until much later (1897 and 1919, respectively).

EXAMPLE 16-1 Electrons in a Raindrop

A water molecule is made up of two hydrogen atoms, each of which has one electron, and an oxygen atom, which has eight electrons. One water molecule has a mass of 2.99×10^{-26} kg. How many electrons are there in a single raindrop with a radius of 1.00 mm = 1.00×10^{-3} m, which has a mass of 4.19×10^{-6} kg? What is the total charge of all of these electrons?

Set Up

Each water molecule has 10 electrons (2 from the hydrogen atoms and 8 from the oxygen atom). So we need to determine the number of water molecules in the raindrop and multiply by 10 to find the number of electrons.

Charge of an electron:

$q_{electron} = -e = -1.602 \times 10^{-19}$ C

Solve

Determine the number of molecules in the raindrop, and from that determine the number of electrons.

The total number of molecules in the raindrop is the mass of the raindrop divided by the mass of a single water molecule:

$$\frac{\text{mass of raindrop}}{\text{mass of one water molecule}} = \frac{4.19 \times 10^{-6} \text{ kg}}{2.99 \times 10^{-26} \text{ kg/molecule}}$$
$$= 1.40 \times 10^{20} \text{ molecules}$$

The number of electrons in the raindrop is

$N = (1.40 \times 10^{20} \text{ molecules})(10 \text{ electrons/molecule})$
$= 1.40 \times 10^{21} \text{ electrons}$

To find the charge of this number of electrons, multiply by the charge per electron.

Total charge of this number of electrons:

$q = Nq_{electron} = (1.40 \times 10^{21} \text{ electrons})(-1.602 \times 10^{-19} \text{ C/electron})$
$= -224 \text{ C}$

Reflect

How large a number is 1.40×10^{21}? The human body contains about 10^{14} cells. Even if you include all of the cells in all of the approximately 9 million (9×10^6) people in New York City, that's only about 9×10^{20} cells—still fewer than the number of electrons in a single raindrop.

The charge of −224 C is quite substantial. But remember that for every electron in a water molecule, there is also one proton (one in each hydrogen atom and eight in each oxygen atom). Since a proton carries exactly as much positive charge as an electron carries negative charge, our raindrop has zero net charge. In Section 16-4 we'll see how difficult it would be to separate the positive and negative charges of this raindrop.

GOT THE CONCEPT? 16-1 Transferring Charge

 Consider the rubber rod and glass rod in Figure 16-2b. Compared to their masses before they are rubbed with fur and silk, respectively, what are their masses after being rubbed? (a) Both rods have more mass. (b) Both rods have less mass. (c) The rubber rod has more mass and the glass rod has less mass. (d) The rubber rod has less mass and the glass rod has more mass. (e) The masses of both rods are unchanged.

TAKE-HOME MESSAGE FOR Section 16-2

✔ Electric charge is quantized. The smallest amount of charge that can be added to or removed from an object is equal to the fundamental charge $e = 1.602 \times 10^{-19}$ C.

✔ All ordinary matter contains positive charge in the form of protons, which have a charge of $+e$, and negative charge in the form of electrons, which have a charge of $-e$. An object is electrically neutral if it contains equal amounts of positive and negative charge. If these amounts are not equal, the object has a net charge.

✔ Charge can be neither created nor destroyed. However, charge can be transferred from one object to another, for example, by moving electrons between objects.

✔ Objects with a net charge exert electric forces on each other. These forces become weaker with increasing distance. If the objects have the same sign of charge (both positive or both negative), the forces are repulsive. If the objects have opposite signs of charge (one positive and one negative), the forces are attractive.

16-3 Charge can flow freely in a conductor but not in an insulator

All substances contain positive and negative charges. But how *mobile* those charges are depends on the specific material. The rubber, glass, silk, and fur shown in Figure 16-2 are all examples of **insulators**, substances in which charges are not able to move freely. All of the electrons in an insulator are bound tightly to the nuclei of atoms, and any excess charge added to an insulator tends to stay wherever it is placed. So when electrons get placed on one end of a rubber rod by rubbing it with fur, the excess electrons cannot redistribute themselves along the rod (Figure 16-5a). Thus there is no excess charge on the opposite end of the rod to attract the positively charged fur. (There is still a very weak attraction between the positively charged fur and the neutral, unrubbed end of the rubber rod. This happens for the same reason that the neutral strands of hair in Figure 16-4 are attracted to a charged balloon.) Most nonmetals are insulators.

By contrast, a metal such as copper is an example of a **conductor**, a substance in which charges can move freely. In a copper atom the outermost, or valence, electron can easily be dislodged. (The valence electron is relatively far from the positively charged nucleus, and the many electrons closer to the nucleus tend to shield the valence electron from the charge of the nucleus.) As a result, electrons—including excess electrons added to the copper—can move between copper atoms relatively freely. This explains what happens when you rub one end of a copper rod with nylon (Figure 16-5b). Electrons are transferred from the nylon to the copper, giving the copper rod a net negative charge and the nylon a net positive charge. But unlike what we saw in Figure 16-5a, after one end of the copper rod has been rubbed, the nylon attracts *both* ends of the copper rod equally. That's because the excess electrons deposited on one end of the copper rod can easily move within the rod. When the positively charged nylon is brought close to *either* end of the copper rod, the excess electrons on the copper rush to that end. As a result, that end has a net negative charge and is attracted to the nylon.

Conductors and insulators are an essential part of all *electric circuits*, which are systems in which there is an ongoing current (a flow of charge) around a closed path. Electric circuits are at the heart of any device that uses a battery (such as a mobile phone, a flashlight, or an electric vehicle) or that you plug into a wall socket (such as a desktop computer, a toaster, or an electric fan). An electric circuit uses moving charges (typically electrons) to transfer energy along a conductor from one point in a circuit to another. In a flashlight, for example, electrons flowing through a copper wire carry energy from the battery (which is a repository of *electric potential energy*) to the light bulb, where the energy is converted into visible light. The electrons then return to the battery through a second copper wire to pick up more energy and repeat the process. Insulators play a crucial role in this process: Each conducting wire is clad in a sheath made of an insulator, which helps ensure that the electrons flow only along the length of the wires. (The visible part of an ordinary extension cord or power cord is actually the insulating sheath. The copper wires are contained within the sheath.)

(a)

The fur-rubbed end of the rubber rod attracts the fur with which it was rubbed...

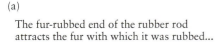

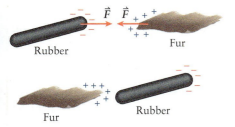

...but the end of the rubber rod that was not rubbed does not feel a strong attraction.

(b)

The nylon-rubbed end of the copper rod attracts the nylon with which it was rubbed...

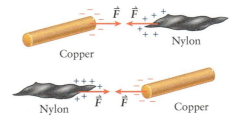

...and the end of the copper rod that was not rubbed also feels a strong attraction.

Figure 16-5 Insulators versus conductors (a) If you place excess charge at one location on an insulator, it remains at that location. (b) If you place excess charge at one location on a conductor, the excess charge can move freely through the conductor.

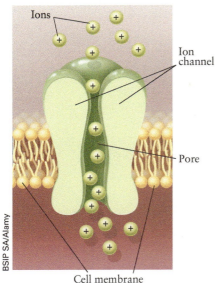

Ions

Ion channel

Pore

Cell membrane

BSIP SA/Alamy

Figure 16-6 Ion flow across cell membranes Many processes in biological cells involve the flow of charged ions across cell membranes. The flow takes place through specialized proteins that form channels through the membrane. While the pore in a typical ion channel is just one or two atoms wide at its narrowest point, ions can pass through the pore at rates of 10^6 per second or higher.

We'll explore the idea of electric potential energy in detail in Chapter 17, and we'll devote Chapters 18 and 21 to exploring various important kinds of electric circuits.

Most metals are good electrical conductors *and* good conductors of heat, with large values of thermal conductivity (see Section 14-7). That's because the flow of electric charge and the flow of heat within a material both require that particles of the material be free to move. In Section 14-3 we learned that the temperature of a material is related to the kinetic energy of that material's particles. For heat to flow from one part of a material to another, the faster-moving particles in the high-temperature part of the material must transfer kinetic energy to the slower-moving particles in the low-temperature part of the object. So the presence of free electrons makes metals good thermal conductors as well as good conductors of electricity. Materials that are electrical insulators, in which the electrons are generally not free to move between atoms, also tend to be thermal insulators with low thermal conductivity.

BioMedical Not all electric conductors are metals. In nearly all biological systems, electric charges are carried by *ions*—atoms with an excess or deficit of electrons. These ions are suspended in water, that is, in aqueous solution. Water is particularly good at holding ions in solution because the water molecule (H_2O), while electrically neutral, has slightly more negative charge at the oxygen atom of the molecule and slightly more positive charge at the hydrogen atoms. So when an ionic compound such as sodium chloride (NaCl) is dissolved in water, the positive ions (Na^+) are attracted to the negative (oxygen) end of (H_2O) molecules while the negative ions (Cl^-) are attracted to the positive (hydrogen) end. Because the dissolved Na^+ and Cl^- ions are free to move in an aqueous solution, they can carry an electrical charge. For this reason, the aqueous solutions which make up living things are conductors (Figure 16-6). It doesn't matter whether charge is carried by an electron or by an ion. As we'll see in Chapter 18, it also doesn't matter whether the moving charges are positive or negative: In either case there is a flow of charge.

GOT THE CONCEPT? 16-2 Sodium in Water

(?) A normal atom of sodium (chemical symbol Na) has 11 electrons and is electrically neutral. But in the compound sodium chloride (NaCl) the sodium atom is actually an ion of charge $+e$. How does a sodium atom acquire this charge? (a) By annihilating one of its electrons; (b) by creating a new electron; (c) by transferring an electron to another atom or molecule; (d) by acquiring an electron from another atom or molecule; (e) by acquiring a proton from another atom or molecule.

There's an important third class of substances called **semiconductors**. These substances have electrical properties that are intermediate between those of insulators and conductors. A common example of a semiconductor is silicon. A silicon atom has four outer electrons, compared to just one for copper, which might suggest that it would be a good conductor. However, in pure silicon each of those electrons forms part of a chemical bond with a neighboring silicon atom, so the outer electrons have limited mobility. As a result, pure silicon conducts electricity far worse than a conductor such as copper, though still far better than an insulator such as rubber.

Semiconductors are of tremendous practical use in electric circuits. That's because it's possible to adjust their electrical properties by *doping*—that is, by adding a small amount of a second substance. Here's an example: If we take pieces of silicon that have been doped in different ways, we can arrange them so that charges will flow through the combination in one direction but will not flow in the opposite direction! Such a combination, called a *diode*, plays the same role in an electric circuit as the valves in the human heart, which allow blood to flow through the heart in one direction only. In Chapter 21 we'll look at some of the applications of semiconductors to modern technologies such as light-emitting diodes (LEDs), solar cells, and integrated circuits.

16-4 Coulomb's law describes the force between charged objects

We learned in Section 16-2 that electrically charged objects, or charges for short, exert forces on each other (see Figure 16-2). Careful measurements reveal that the force between two charges is directly proportional to the magnitude of each charge. The greater the magnitudes of the charges on the objects, the larger the force that acts on each charge. The force also depends on the distance between the charged objects; the closer the charges, the larger the force.

Coulomb's law summarizes the results of these measurements. Specifically, this law tells us about the force between two **point charges**, which are very small charged objects whose size is much smaller than the separation between them (Figure 16-7a). A point charge is an idealization, just like a massless rope or a frictionless incline, but it's a good description in many situations where charged objects interact with each other. Coulomb's law tells us the magnitude of the electric force that two point charges q_1 and q_2 separated by a distance r exert on each other:

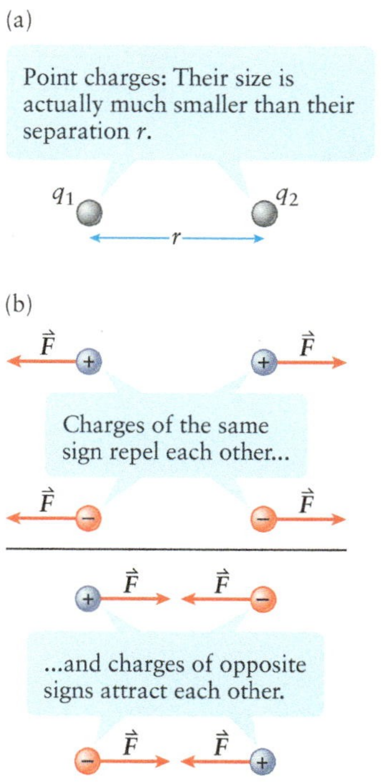

(a)

Point charges: Their size is actually much smaller than their separation r.

q_1 q_2

r

(b)

$\vec{F}$ +   + $\vec{F}$

Charges of the same sign repel each other...

$\vec{F}$ −   − $\vec{F}$

+ $\vec{F}$ $\vec{F}$ −

...and charges of opposite signs attract each other.

− $\vec{F}$ $\vec{F}$ +

Figure 16-7 Coulomb's law (a) Coulomb's law describes the electric force that two point charges (not drawn to scale) exert on each other. (b) The direction of the electric force between point charges depends on the signs (positive or negative) of the charges.

Any two point charges q_1 and q_2 exert equally strong **electric forces** on each other.

Coulomb constant

Absolute values of the **point charges**

Coulomb's law (16-1)

$$F_{q_1 \text{ on } q_2} = F_{q_2 \text{ on } q_1} = \frac{k|q_1||q_2|}{r^2}$$

Distance between the point charges

The value of the **Coulomb constant** k in Equation 16-1 is, to three significant figures,

$$k = 8.99 \times 10^9 \text{ N·m}^2/\text{C}^2$$

Note that Equation 16-1 just tells you the *magnitude* of the electric force between two point charges. As Figure 16-7b shows, the *direction* of the electric force is such that charges of the same sign (both positive or both negative) repel each other, while charges of opposite sign (one positive and one negative) attract each other. Figure 16-8 shows an application of this principle.

You should notice the similarity between Coulomb's law and Newton's law of universal gravitation:

Gravitational constant (same for any two objects)

Masses of the two objects

Any two objects (1 and 2) exert equally strong **gravitational forces** on each other.

Newton's law of universal gravitation (7-3)

$$F_{1 \text{ on } 2} = F_{2 \text{ on } 1} = \frac{Gm_1m_2}{r^2}$$

Center-to-center distance between the two objects

The gravitational forces are attractive: $\vec{F}_{1 \text{ on } 2}$ pulls object 2 toward object 1 and $\vec{F}_{2 \text{ on } 1}$ pulls object 1 toward object 2.

In both Coulomb's law (Equation 16-1) and the law of universal gravitation (Equation 7-3), the force that one object exerts on the other is inversely proportional to the square of the distance between them. In addition, the gravitational force is proportional to the product of the masses of the two particles, while the electric force is

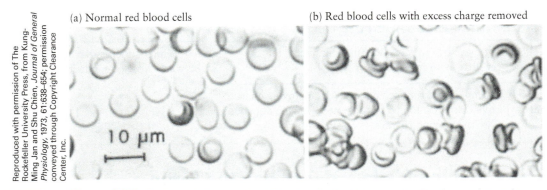

(a) Normal red blood cells

(b) Red blood cells with excess charge removed

10 μm

BioMedical

Figure 16-8 Coulomb's law and red blood cells Red blood cells like those shown in these microscope images carry oxygen from your lungs to other parts of your body through the circulatory system. (a) Each red blood cell has a slight excess of electrons that gives it a net negative charge. Since all cells are negatively charged, they exert repulsive electric forces on each other that help keep the cells apart. (b) The red blood cells in this photo have been treated with an enzyme that removes the excess electrons. Without electric forces to keep them apart, the red blood cells tend to clump. If red blood cells in your body behaved like this, their flow through your circulatory system would be impeded and your body would be starved for oxygen.

proportional to the product of the charges of the two particles. As remarkable as these similarities are, there is one essential difference between the two laws: Newton's law of universal gravitation tells us that two objects with mass always attract each other, but Coulomb's law states that two charged objects can attract or repel each other, depending on the sign of the charges they carry.

The electric force also shares some properties with every other force we've discussed. The electric force is a vector and has a direction. The net electric force on an object is the vector sum of every separate electric force that acts on it. Furthermore, the electric forces that two charged objects exert on each other obey Newton's third law. When point objects with charges q_1 and q_2 interact, the force that object 1 exerts on object 2 is equal in magnitude to the force that object 2 exerts on object 1, but the two forces are in opposite directions (see Figure 16-7b).

EXAMPLE 16-2 **Calculating Electric Force**

(a) What is the electric force (magnitude and direction) between two electrons separated by a distance of 10.0 cm = 0.100 m? (b) Suppose you could remove all the electrons from a drop of water 1.00 mm in radius (see Example 16-1 in Section 16-2) and clump them into a ball 1.00 mm in radius. If this ball of electrons is 10.0 cm from the drop of water from which they were removed, what is the magnitude of the electric force between the drop of water and the ball of electrons?

Set Up

The two electrons repel because both have a negative charge $q = -e = -1.602 \times 10^{-19}$ C. From Example 16-1, the combined charge of all of the electrons in a water drop of this size is −224 C; the water drop was initially neutral, so the charge of the water drop after all of the electrons have been removed is +224 C. Since the ball of electrons and the water drop (with electrons removed) have opposite signs of charge, they attract each other.

We can use Coulomb's law (Equation 16-1) in both parts of this problem. That's because in both cases the charged objects are much smaller than the distance that separates them, so we can treat them as point charges. (Indeed, electrons are very small even compared to the dimensions of an atom.)

Coulomb's law:

$$F_{q_1 \text{ on } q_2} = F_{q_2 \text{ on } q_1} = \frac{k|q_1||q_2|}{r^2}$$

(16-1)

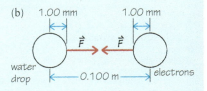

(a)

$\vec{F}$ ← electron   electron → $\vec{F}$

0.100 m

(b) 1.00 mm   1.00 mm

water drop $\vec{F}$ → ← $\vec{F}$ electrons

0.100 m

Solve

(a) Find the magnitude of the force that each electron exerts on the other.

For the two electrons the charges are

$$q_1 = q_2 = -e = -1.602 \times 10^{-19} \text{ C}$$

The distance between the electrons is $r = 0.100$ m. From Equation 16-1 the magnitude of the force that each electron exerts on each other is

$$F = \frac{(8.99 \times 10^9 \text{ N} \cdot \text{m}^2/\text{C}^2) |-1.602 \times 10^{-19} \text{ C}||-1.602 \times 10^{-19} \text{ C}|}{(0.100 \text{ m})^2}$$

$$= 2.31 \times 10^{-26} \text{ N}$$

(b) Find the magnitude of the force between the ball of electrons and the water drop from which they were extracted.

For the ball of electrons and the water drop with electrons removed, the charges are

$$q_1 = -224 \text{ C}, q_2 = +224 \text{ C}$$

The distance between the two objects is $r = 0.100$ m. From Equation 16-1 the two objects exert forces on each other of magnitude

$$F = \frac{(8.99 \times 10^9 \text{ N} \cdot \text{m}^2/\text{C}^2) |-224 \text{ C}||+224 \text{ C}|}{(0.100 \text{ m})^2}$$

$$= 4.51 \times 10^{16} \text{ N}$$

Reflect

The repulsive force between the two electrons is tiny because the electron charge is tiny. By contrast, the attractive force between the ball of electrons and the electron-free water drop is immense. To put this force into perspective, a solid cube of lead with a weight of 4.51×10^{16} N would be 7.4 *kilometers* on a side! This is the magnitude of force that you would have to exert to keep the electrons from flying back into the water drop. There is no known way to produce a force of this magnitude, which is why you'll never see an object with all of its electrons removed. It's relatively easy to remove a small fraction of an object's electrons, as for the fur and the glass rod in Figure 16-2b. But removing *all* of the electrons from a piece of fur or a glass rod is not a practical thing to do.

EXAMPLE 16-3 Three Charges in a Line

A particle with negative charge q is placed halfway between two identical particles, each of which carries the same positive charge: $Q_1 = Q_2 = +Q$. The distance between adjacent charges is d. If each of the three particles experiences a net electric force of zero, what is the magnitude of charge q in terms of Q?

Set Up

We want the net force on each point charge— that is, the *vector* sum of the forces on that charge due to the other two charges—to be equal to zero. We'll use Coulomb's law, Equation 16-1, to solve for the magnitude of the force of one charge on another. We'll also use the idea that charges of the same signs repel while charges of opposite signs attract.

Coulomb's law:

$$F_{q_1 \text{ on } q_2} = F_{q_2 \text{ on } q_1} = \frac{k|q_1||q_2|}{r^2}$$

(16-1)

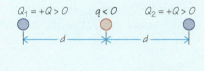

Solve

Let's start by considering the forces on positive charge Q_2. The other positive charge, Q_1, exerts a repulsive force $\vec{F}_{Q_1 \text{ on } Q_2}$ that pushes Q_2 away from Q_1, that is, to the right. The negative charge q exerts an attractive force $\vec{F}_{q \text{ on } Q_2}$ that pulls Q_2 toward q, that is, to the left. For the net force on Q_2 to be zero, these two forces must have the same magnitude.

The net force on charge Q_2 must be zero:

$$\vec{F}_{Q_1 \text{ on } Q_2} + \vec{F}_{q \text{ on } Q_2} = 0$$

For this to be true, $\vec{F}_{Q_1 \text{ on } Q_2}$ and $\vec{F}_{q \text{ on } Q_2}$ must have the same magnitude:

$$F_{Q_1 \text{ on } Q_2} = F_{q \text{ on } Q_2}$$

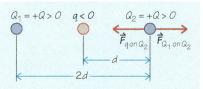

Use Coulomb's law (Equation 16-1) to find the magnitudes $F_{Q_1 \text{ on } Q_2}$ and $F_{q \text{ on } Q_2}$. Set these equal to each other and solve for the magnitude (absolute value) of q. Note that d is the distance between Q_1 and q, and also between Q_2 and q. Notice that Q_1 and Q_2 are separated by distance $2d$. Remember that q is negative, so we need to keep its absolute value.

The distance between Q_1 and Q_2, each of which has a charge of magnitude Q, is $2d$. From Equation 16-1, the force of Q_1 on Q_2 has magnitude

$$F_{Q_1 \text{ on } Q_2} = \frac{k|Q_1||Q_2|}{(2d)^2} = \frac{kQ^2}{4d^2}$$

The distance between q and Q_2 is d, so the magnitude of the force of q on Q_2 is

$$F_{q \text{ on } Q_2} = \frac{k|q||Q_2|}{d^2} = \frac{k|q|Q}{d^2}$$

For the forces to have the same magnitude,

$$\frac{kQ^2}{4d^2} = \frac{k|q|Q}{d^2}$$

Solve for the absolute value of q:

$$|q| = \frac{Q}{4}$$

Reflect

The value $|q| = Q/4$ satisfies the condition that there is zero net force on Q_2. Because Q_1 is twice as far from Q_2 as q, and because the electric force is inversely proportional to the square of the distance, the charge Q_1 must be $(2)^2 = 4$ times greater than the magnitude of q for the forces these charges exert on Q_2 to have the same magnitude.

You can see that since Q_1 and Q_2 have the same charge, the forces on Q_1 are the mirror images of those on Q_2 (a repulsive force to the *left* exerted by Q_2, which is a distance $2d$ from Q_1, and an attractive force to the *right* exerted by q, which is a distance d from Q_1). So the net force on Q_1 will be zero, too.

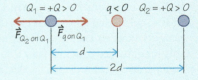

You can see that the net force on the negative charge q is also guaranteed to be zero. This charge is the same distance d from the two equal positive charges Q_1 and Q_2, so the force from Q_1 that pulls q to the left is just as great as the force from Q_2 that pulls q to the right.

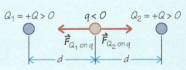

EXAMPLE 16-4 Three Charges in a Plane

Charges Q_1 and Q_3 are both positive and equal to 1.50×10^{-6} C; charge Q_2 is negative and equal to -1.50×10^{-6} C. Charges Q_1 and Q_2 are placed at a fixed position a distance $D = 6.00$ cm apart, and Q_3 is placed at a fixed position a distance $H = 4.00$ cm above the midpoint of the line that connects Q_1 and Q_2. Calculate the magnitude and direction of the force on Q_3 due to the other two charges.

Set Up

The net force on positive charge Q_3 is the vector sum of $\vec{F}_{1 \text{ on } 3}$, the *repulsive* force that the positive charge Q_1 exerts on Q_3, and $\vec{F}_{2 \text{ on } 3}$, the *attractive* force that the negative charge Q_2 exerts on Q_3. We'll use Coulomb's law to find the magnitude of each of these forces. We'll then add the two force vectors using components.

Coulomb's law:

$$F_{q_1 \text{ on } q_2} = F_{q_2 \text{ on } q_1} = \frac{k|q_1||q_2|}{r^2}$$

(16-1)

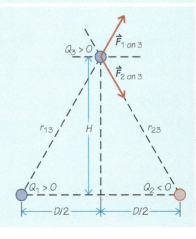

Solve

Use Equation 16-1 to find the magnitudes of the forces $\vec{F}_{1\,on\,3}$ and $\vec{F}_{2\,on\,3}$.

The figure above shows that the distance from Q_1 to Q_3 is the same as the distance from Q_2 to Q_3:

$$r_{13} = r_{23} = \sqrt{\left(\frac{D}{2}\right)^2 + H^2} = \sqrt{\left(\frac{6.00\ cm}{2}\right)^2 + (4.00\ cm)^2}$$

$$= 5.00\ cm = 5.00 \times 10^{-2}\ m$$

All three charges have the same magnitude:

$$|Q_1| = |Q_2| = |Q_3| = 1.50 \times 10^{-6}\ C$$

So from Equation 16-1, $\vec{F}_{1\,on\,3}$ and $\vec{F}_{2\,on\,3}$ have the same magnitude:

$$F_{1\,on\,3} = \frac{k|Q_1||Q_3|}{r_{13}^2}$$

$$= \frac{(8.99 \times 10^9\ N \cdot m^2/C^2)(1.50 \times 10^{-6}\ C)^2}{(5.00 \times 10^{-2}\ m)^2}$$

$$= 8.09\ N$$

$$F_{2\,on\,3} = \frac{k|Q_2||Q_3|}{r_{23}^2} = F_{1\,on\,3} = 8.09\ N$$

Choose the positive x direction to be to the right and the positive y direction to be upward. Then find the x and y components of $\vec{F}_{1\,on\,3}$ and $\vec{F}_{2\,on\,3}$, and use these to calculate the components of the net force on Q_3.

The components of $\vec{F}_{1\,on\,3}$ and $\vec{F}_{2\,on\,3}$ are

$$F_{1\,on\,3,x} = F_{1\,on\,3}\cos\theta$$
$$F_{1\,on\,3,y} = F_{1\,on\,3}\sin\theta$$
$$F_{2\,on\,3,x} = F_{2\,on\,3}\cos\theta$$
$$F_{2\,on\,3,y} = -F_{2\,on\,3}\sin\theta$$

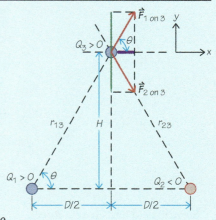

Since the magnitudes $F_{1\,on\,3}$ and $F_{2\,on\,3}$ are equal, the components of the net force on Q_3 are

$$F_{net\,on\,3,x} = F_{1\,on\,3,x} + F_{2\,on\,3,x}$$
$$= F_{1\,on\,3}\cos\theta + F_{2\,on\,3}\cos\theta$$
$$= 2F_{1\,on\,3}\cos\theta$$

$$F_{net\,on\,3,y} = F_{1\,on\,3}\sin\theta + (-F_{2\,on\,3}\sin\theta)$$
$$= F_{1\,on\,3}\sin\theta - F_{1\,on\,3}\sin\theta = 0$$

From the figure,

$$\cos\theta = \frac{(D/2)}{r_{13}} = \frac{3.00\ cm}{5.00\ cm} = 0.600$$

So

$$F_{net\,on\,3,x} = 2(8.09\ N)(0.600) = 9.71\ N$$
$$F_{net\,on\,3,y} = 0$$

The net force on Q_3 is to the right and has magnitude 9.71 N.

Reflect

The two individual forces on Q_3 add to a net force that is neither directly away from Q_1 nor directly toward Q_2.

GOT THE CONCEPT? 16-3 Electric Force

(?) In part (b) of Example 16-2 we imagined removing all of the electrons from a drop of water and moving them to a given distance from the water drop. We then calculated the electric force between the drop and the electrons. Suppose instead we removed only one-half of the electrons from the water drop then moved them to the same distance as in Example 16-2. Compared to the force we calculated in part (b) of Example 16-2, what would be the force between the electrons and the water drop in this case? (a) The same; (b) $1/\sqrt{2}$ as great; (c) 1/2 as great; (d) 1/4 as great; (e) 1/16 as great.

TAKE-HOME MESSAGE FOR Section 16-4

✔ Coulomb's law tells us the magnitude of the electric force that two point charges exert on each other. This magnitude is proportional to the product of the magnitudes of the two charges and inversely proportional to the square of the distance between them.

✔ The electric force between two point charges is repulsive if the two charges have the same sign (both positive or both negative) and attractive if the two charges have opposite signs (one positive and one negative).

16-5 The concept of electric field helps us visualize how charges exert forces at a distance

Most forces in our daily experience arise only when one object is in direct contact with another object, like the normal force that acts on your body when you sit in a chair or when you push directly on an object. Nevertheless, some forces, such as the gravitational force and the Coulomb force, appear to act even between objects separated by a distance. We can describe forces of this kind using the concept of a field. In this view every charged object modifies all of space by producing an electric field, which is strongest closest to the object but extends infinitely far away. A second charged object senses this change in space, interacting with the electric field and experiencing an electric force (Figure 16-9).

If we place a particle carrying charge q in an electric field $\vec{E}$, the force experienced by the particle is

If a particle with charge q is placed at a position where the **electric field** due to other charges is $\vec{E}$...

Electric field and electric force
(16-2)

$$\vec{F} = q\vec{E}$$

...then the **electric force** on the particle is $\vec{F} = q\vec{E}$.

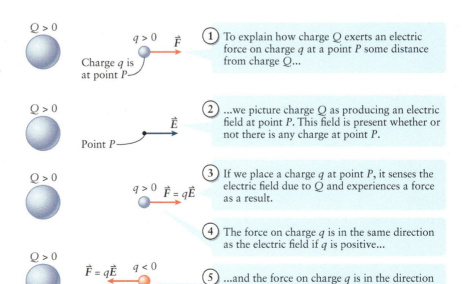

Figure 16-9 Electric field and electric force The electric field concept helps us visualize how a charge Q exerts an electric force on a second charge q.

① To explain how charge Q exerts an electric force on charge q at a point P some distance from charge Q...

② ...we picture charge Q as producing an electric field at point P. This field is present whether or not there is any charge at point P.

③ If we place a charge q at point P, it senses the electric field due to Q and experiences a force as a result.

④ The force on charge q is in the same direction as the electric field if q is positive...

⑤ ...and the force on charge q is in the direction opposite to the electric field if q is negative.

An equivalent way to write Equation 16-2 is

$$\bar{E} = \frac{\bar{F}}{q} \qquad (16\text{-}3)$$

Equation 16-3 tells us that we can interpret the electric field $\bar{E}$ at a certain point as the *electric force per charge* that acts on a charged object placed at that point. If an object with double the charge is placed at that point, it will experience double the force.

WATCH OUT! Note that the direction of *force* and the direction of *field* can be different.

! Although we can think of the electric field as the electric force per unit charge, you must be careful about the direction of the force. In particular, the direction of the electric force that a charge experiences depends on the *sign* of the charge as well as the *direction* of the electric field. Equation 16-2 tells us that a positive charge ($q > 0$) experiences a force in the same direction as the electric field $\bar{E}$, but a negative charge ($q < 0$) experiences a force in the direction opposite to $\bar{E}$ (Figure 16-9). So the electric field at a certain point is in the direction of the electric force that would be exerted on a positive charge placed at that point.

From Equation 16-3 we see that the SI units of electric field are newtons per coulomb, or N/C. If a 1-C charge experiences a 1-N force at a certain point in space, then at that point in space there is an electric field with a magnitude of 1 N/C.

Equations 16-2 and 16-3 are strictly valid only if the particle of charge q is a *point* particle with a very small size. That's because the value of $\bar{E}$ can be different at different places. If the charge q were spread over a large volume (say, a sphere 1 m in diameter), the value of $\bar{E}$ could be different at different points within that volume. In that case it wouldn't be clear which value of $\bar{E}$ to use in Equation 16-2 or 16-3. But if the charge q is within a point particle, the value of $\bar{E}$ to use is the value at the point where that particle is located.

Note that we've actually used the concept of a field before. In Chapter 4 we expressed the gravitational force $\bar{w}$ on an object of mass m as $\bar{w} = m\bar{g}$, where $\bar{g}$ is the acceleration due to gravity. But we can also think of $\bar{g}$ as the *gravitational field* that Earth sets up in the space around it; $\bar{g}$ represents the value of that field vector (magnitude and direction) at a given point. An object of mass m placed at the point then experiences a gravitational force $\bar{w} = m\bar{g}$ (compare to Equation 16-2). Just as the concept of an electric field gives us a way to visualize how two charges can interact over a distance, the concept of a gravitational field helps us visualize how two objects with mass (Earth and the object of mass m) can interact without touching each other.

→ **Go to Interactive Exercise 16-1 for more practice dealing with electric fields.**

EXAMPLE 16-5 Determining Charge-to-Mass Ratio

When released from rest in a uniform electric field of magnitude 1.00×10^4 N/C, a certain charged particle travels 2.00 cm in the direction of the field in 2.88×10^{-7} s. You can ignore any nonelectric forces acting on the particle. (a) What is the *charge-to-mass ratio* of this particle (that is, the ratio of its charge q to its mass m)? (b) Other experiments show that the mass of this particle is 6.64×10^{-27} kg. What is the charge of this particle?

Set Up

The only force that acts on the particle is the electric force given by Equation 16-2. Since the particle accelerates in the direction of the electric field $\bar{E}$, the force on the particle must also be in the direction of $\bar{E}$. So the charge on the particle must be positive. Since $\bar{E}$ is uniform (it has the same value at all points), the force on the particle will be constant. Its acceleration a_x will

Electric field and electric force:

$$\bar{F} = q\bar{E} \qquad (16\text{-}2)$$

Straight-line motion with constant acceleration:

$$x = x_0 + v_{0x}t + \frac{1}{2}a_xt^2 \qquad (2\text{-}9)$$

be constant as well, so we can use one of the constant-acceleration equations from Chapter 2 to determine a_x. We'll use this with Newton's second law and Equation 16-2 to learn what we can about this particle.

Newton's second law:

$$\sum \vec{F}_{ext} = m\vec{a} \qquad (4\text{-}2)$$

Solve

(a) We are given that the particle travels 2.00 cm = 2.00×10^{-2} m in 2.88×10^{-7} s. Use this to determine the particle's constant acceleration.

Take the positive x direction to be the direction in which the particle moves. The particle begins at rest, so $v_{0x} = 0$. If we take the initial position of the particle to be $x_0 = 0$, then Equation 2-9 becomes

$$x = \frac{1}{2} a_x t^2$$

Solve for the acceleration:

$$a_x = \frac{2x}{t^2} = \frac{2(2.00 \times 10^{-2}\ \text{m})}{(2.88 \times 10^{-7}\ \text{s})^2} = 4.82 \times 10^{11}\ \text{m/s}^2$$

Relate the acceleration to the net external (electric) force on the particle and solve for the charge-to-mass ratio.

The net force on the particle of charge q is the electric force in the x direction, which from Newton's second law is equal to the mass m of the particle multiplied by the acceleration a_x:

$$qE_x = ma_x$$

This says that the acceleration of a particle in an electric field depends on the particle's charge-to-mass ratio:

$$a_x = \frac{q}{m} E_x$$

In this example we know both a_x and E_x, so the charge-to-mass ratio is

$$\frac{q}{m} = \frac{a_x}{E_x} = \frac{4.82 \times 10^{11}\ \text{m/s}^2}{1.00 \times 10^4\ \text{N/C}}$$

Since 1 N = 1 kg·m/s², this is

$$\frac{q}{m} = 4.82 \times 10^7\ \text{C/kg}$$

(b) Given the charge-to-mass ratio and the mass of the particle, determine the charge q.

The charge of the particle is

$$q = m\left(\frac{q}{m}\right) = (6.64 \times 10^{-27}\ \text{kg})(4.82 \times 10^7\ \text{C/kg})$$

$$= 3.20 \times 10^{-19}\ \text{C}$$

The charge on a proton is $e = 1.60 \times 10^{-19}$ C; the charge on this particle is $2e$.

Reflect

The particle in this example has about four times the mass of a proton but only double the charge of a proton. For historical reasons it's known as an *alpha particle*; in fact, it's the nucleus of a helium atom, which contains two protons (each with charge e) and two neutrons (each with nearly the same mass as a proton but with zero charge).

An important application of the force produced by an electric field is *electrophoresis*. Chemists use this technique to separate molecules of different kinds according to their charge and mass. In the simplest kind of electrophoresis, a small amount of a sample containing molecules of different kinds is placed on a strip of filter paper, and the paper is soaked with a solution that conducts electricity (Figure 16-10). An electric field of magnitude E is then applied along the length of the strip. Each molecule accelerates in response to the field, and that acceleration (taking into account

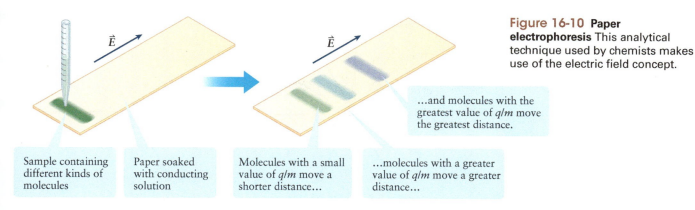

Figure 16-10 Paper electrophoresis This analytical technique used by chemists makes use of the electric field concept.

...and molecules with the greatest value of q/m move the greatest distance.

Sample containing different kinds of molecules

Paper soaked with conducting solution

Molecules with a small value of q/m move a shorter distance...

...molecules with a greater value of q/m move a greater distance...

fluid resistance on the molecules) is proportional to the charge-to-mass ratio q/m of the molecule. As a result, molecules with larger values of q/m move farther along the paper than do those with small values of q/m. The many applications of this specific technique, called *paper electrophoresis*, include analyzing currency to determine whether it is counterfeit (a forger's ink may have a different chemical composition than the ink used in legal currency) and looking for the presence of cancer antibodies or human immunodeficiency virus (HIV) in blood.

A different sort of electrophoresis is used for DNA profiling (also called genetic fingerprinting), which is an essential part of modern forensic science. A sample of human DNA is treated with an enzyme that breaks the long DNA strand into shorter segments. The sizes of these segments are characteristic of the person's genetic code, so measuring the segment sizes is a powerful technique for forensic identification. Unfortunately, the ratio of q/m for a segment of DNA is nearly the same for segments of any size, so paper electrophoresis isn't useful. Instead, a sample containing the DNA segments is placed in a special gel that is permeated by many microscopic pores. When an electric field is applied, all of the segments move in response, but the smaller segments move through the gel pores more easily than large ones do. The result is that the DNA segments are spread out according to their size, allowing a genetic fingerprint to be made. Similar techniques are used in medical research for studying both DNA and proteins.

BioMedical

Electric Field of a Point Charge

Equation 16-2 tells us how a charge q responds to a given electric field $\vec{E}$. It also tells us how to determine the value of $\vec{E}$ at any point. As an example, Figure 16-11 shows how we might determine the electric field around a positive point charge Q. We place a small positive charge q (which we call a *test charge*) at various locations around the charge Q and measure the force $\vec{F}$ on that small charge. The electric field at each location is given by Equation 16-3, $\vec{E} = \vec{F}/q$; since q is positive, the electric field is in the same direction as the force on the test charge. The Coulomb force exerted by Q repels a positive test charge q; thus, the direction of force at each location—and so the direction of the electric field at each location, shown by the blue vectors—is radially away from Q. As the lengths of the vectors show, the electric field magnitude

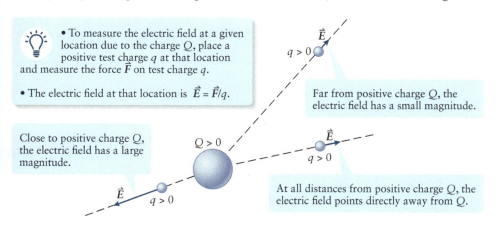

- To measure the electric field at a given location due to the charge Q, place a positive test charge q at that location and measure the force $\vec{F}$ on test charge q.

- The electric field at that location is $\vec{E} = \vec{F}/q$.

$q > 0$

Far from positive charge Q, the electric field has a small magnitude.

Close to positive charge Q, the electric field has a large magnitude.

$Q > 0$

$q > 0$

$q > 0$

At all distances from positive charge Q, the electric field points directly away from Q.

$q > 0$

Figure 16-11 Mapping the electric field We can map out the electric field surrounding a point charge Q—in this case a positive charge—by placing a positive test charge $+q$ at various locations around Q.

(a) Positive point charge: electric field vectors

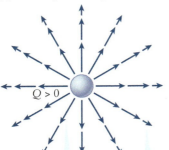

$Q > 0$

(b) Positive point charge: electric field lines

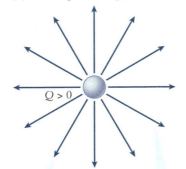

$Q > 0$

(c) Negative point charge: electric field lines

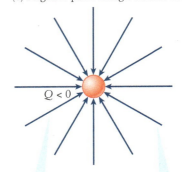

$Q < 0$

| Field due to a *positive* charge points *away* from the charge. | Field strength decreases with increasing distance from the charge. | Field due to a *positive* charge points *away* from the charge. | Farther from the charge, where the field strength is weaker, field lines are farther apart. | Field lines due to a *negative* charge point *toward* the charge. | Farther from the charge, where the field strength is weaker, field lines are farther apart. |

Figure 16-12 Electric field and electric field lines The electric field around a charged object can be represented by (a) electric field vectors or (b), (c) electric field lines.

decreases with increasing distance from Q. That's because the force between Q and q decreases with increasing distance in accordance with Coulomb's law (Section 16-4).

There are two ways we can depict the entire electric field around a positive charge Q. In Figure 16-12a we draw vectors to represent the electric field at a large number of points around Q. Figure 16-12b shows a simpler approach that's easier to draw: We connect adjacent vectors to form lines, called **electric field lines**. The direction of the field line passing through any point represents the direction of the field at that point. The magnitude of the electric field is shown by the density of the field lines—that is, how close they are to each other. Close to Q, where the field lines are close together, the field has a large magnitude. Far from Q, where the field lines are far apart, the magnitude of the electric field is smaller. If the charge Q is negative, a positive test charge q experiences an attractive force $\vec{F}$ toward Q, so the electric field $\vec{E} = \vec{F}/q$ due to charge Q is directed *toward* that charge (Figure 16-12c).

WATCH OUT! Electric fields are three-dimensional.

Figure 16-12 may give you the incorrect impression that the electric field and electric field lines of a point charge Q lie only on the plane of the page. Not so! The electric field completely surrounds the charge; it is three-dimensional. The field lines are arranged around the charge rather like the spines of a sea urchin (Figure 16-13).

Figure 16-13 An electric field analogy The electric field lines of a point charge are arranged radially around the charge in three dimensions, much like the spines on a sea urchin.

We can use Coulomb's law, Equation 16-1, to write an expression for the *magnitude* of the electric field due to a point charge Q. If we place a test charge q a distance r from charge Q, Equation 16-1 tells us that the magnitude of the Coulomb force on q is $F = k|q||Q|/r^2$. From Equation 16-2 we can also write this force magnitude as $F = |q|E$, where E is the magnitude of the electric field due to charge Q at the position of test charge q. Setting these two expressions for F equal to each other, we get

$$|q|E = \frac{k|q||Q|}{r^2}$$

or

Coulomb constant   Absolute value of charge Q

Magnitude of the electric field
due to a point charge Q

$$E = \frac{k|Q|}{r^2}$$

Distance between point charge Q and the location where the field is measured

Magnitude of the electric field
due to a point charge
(16-4)

The *magnitude* of the electric field due to a point charge Q decreases with increasing distance r from the charge. As Figure 16-12 shows, the *direction* of the electric field depends on the sign of the charge Q. The field points radially outward from a positive point charge and radially inward toward a negative point charge.

Electric Field of an Arrangement of Charges

The results in Figure 16-12 and Equation 16-4 describe the electric field due to a single point charge (positive or negative). If several point charges $Q_1, Q_2, Q_3, \ldots$ are present, experiment shows that the net electric force $\vec{F}$ that these charges exert on a test charge q at any location is just the vector sum of the forces $\vec{F}_1, \vec{F}_2, \vec{F}_3, \ldots$ that these charges *individually* exert on q. If we use the symbol $\vec{E}$ for the net electric field produced at a given location by $Q_1, Q_2, Q_3, \ldots$ together, and the symbols $\vec{E}_1, \vec{E}_2, \vec{E}_3, \ldots$ for the electric fields that these charges produce individually, we can use Equation 16-2 to express this experimental result as

$$\vec{F} = \vec{F}_1 + \vec{F}_2 + \vec{F}_3 + \cdots \quad \text{or} \quad q\vec{E} = q\vec{E}_1 + q\vec{E}_2 + q\vec{E}_3 + \cdots$$

If we divide both sides of the second of these equations by q, we get

$$\vec{E} = \vec{E}_1 + \vec{E}_2 + \vec{E}_3 + \cdots \tag{16-5}$$

In other words, *when there are two or more point charges present, the electric field at any point in space is the vector sum of the fields due to each charge separately.* The following examples illustrate how to use this principle.

→ *Go to Picture It 16-1 for more practice dealing with electric fields.*

EXAMPLE 16-6   Where *Is* the Electric Field Zero?

A point charge $Q_1 = +4.00$ nC (1 nC = 1 nanocoulomb = 10^{-9} C) is placed 0.500 m to the left of a point charge $Q_2 = +9.00$ nC. Find the position between the two point charges where the net electric field is zero.

Set Up

At any point between the two charges, the electric field $\vec{E}_1$ due to Q_1 points to the right (away from this positive charge) and the electric field $\vec{E}_2$ due to Q_2 points to the left (away from this positive charge). We want to find the point P where the total field $\vec{E} = \vec{E}_1 + \vec{E}_2$ equals zero. We'll use the symbol D for the 0.500-m distance between the two charges and x for the distance from Q_1 to point P: Then the distance from Q_2 to point P is $D - x$. Our goal is to find the value of x for which $\vec{E} = 0$.

Magnitude of the electric field
due to a point charge Q:

$$E = \frac{k|Q|}{r^2} \tag{16-4}$$

Total electric field:

$$\vec{E} = \vec{E}_1 + \vec{E}_2 \tag{16-5}$$

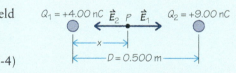

Solve

If the net electric field at P is zero, then $\vec{E}_1$ (to the right) and $\vec{E}_2$ (to the left) must have equal magnitudes so that these two vectors cancel. Use Equation 16-4 to write this statement in equation form.

For the net electric field at P to be zero,

$$E_1 = E_2$$

The distance from Q_1 to P is x, and the distance from Q_2 to P is $D - x$, so from Equation 16-4

$$E_1 = \frac{k|Q_1|}{x^2} \quad \text{and} \quad E_2 = \frac{k|Q_2|}{(D-x)^2}$$

If these are equal to each other,

$$\frac{k|Q_1|}{x^2} = \frac{k|Q_2|}{(D-x)^2}$$

Solve this equation for x.

The factors of k cancel in the above equation, so

$$\frac{|Q_1|}{x^2} = \frac{|Q_2|}{(D-x)^2} \quad \text{or}$$

$$|Q_1|(D-x)^2 = |Q_2|x^2$$

Multiply out the quantity $(D - x)^2$ and rearrange:

$$|Q_1|(D^2 - 2Dx + x^2) = |Q_2|x^2$$

$$(|Q_2| - |Q_1|)x^2 + (2D|Q_1|)x + (-D^2|Q_1|) = 0$$

We can simplify this if we divide through by $|Q_1|$:

$$\left(\frac{|Q_2|}{|Q_1|} - 1\right)x^2 + 2Dx + (-D^2) = 0$$

This is a quadratic equation of the form $ax^2 + bx + c = 0$, with $a = |Q_2|/|Q_1| - 1 = |9.00\ \text{nC}|/|4.00\ \text{nC}| - 1 = 1.25$, $b = 2D = 2(0.500\ \text{m}) = 1.00\ \text{m}$, and $c = -D^2 = -(0.500\ \text{m})^2 = -0.250\ \text{m}^2$. The solutions are

$$x = \frac{-b \pm \sqrt{b^2 - 4ac}}{2a}$$

$$= \frac{-(1.00\ \text{m}) \pm \sqrt{(1.00\ \text{m})^2 - 4(1.25)(-0.250\ \text{m}^2)}}{2(1.25)}$$

$$= \frac{-(1.00\ \text{m}) \pm \sqrt{2.25\ \text{m}^2}}{2.50}$$

$$= \frac{-(1.00\ \text{m}) \pm (1.50\ \text{m})}{2.50}$$

$$= +0.200\ \text{m} \ \text{or} -1.00\ \text{m}$$

We want a positive value of x to correspond to a point to the right of Q_1, so the solution we want is $x = +0.200\ \text{m}$. We conclude that point P is a distance $x = 0.200\ \text{m}$ to the right of charge Q_1 and a distance $D - x = 0.500\ \text{m} - 0.200\ \text{m} = 0.300\ \text{m}$ to the left of charge Q_2.

Reflect

Because charge Q_1 is smaller than Q_2, the location of the point where the net electric field is zero must be closer to Q_1 than to Q_2. That's just what we found. You can check the result $x = 0.200\ \text{m}$ by substituting this value into the above expressions for E_1 and E_2 and confirming that $E_1 = E_2$ for this value of x.

But what's the significance of the second solution, $x = -1.00\ \text{m}$? This refers to a point 1.00 m to the *left* of charge Q_1 and 1.00 m + 0.500 m = 1.50 m to the left of charge Q_2. Our calculation shows that at this point E_1 is equal to E_2. However, at this point $\vec{E}_1$ and $\vec{E}_2$ *both* point to the *left* (both fields point away from the positive charges that produce them). So at this point the electric fields $\vec{E}_1$ and $\vec{E}_2$ do *not* cancel, and the total field is not zero.

EXAMPLE 16-7 Field of an Electric Dipole

A combination of two point charges of the same magnitude but opposite signs is called an **electric dipole.** Figure 16-14 shows an electric dipole made up of a point charge $+q$ and a point charge $-q$ separated by a distance $2d$. (This is a simple model for an ionic molecule like NaCl, which has a charge $+q$ on the Na^+ ion and a charge $-q$ on the Cl^- ion.)

Derive expressions for the magnitude and direction of the net electric field due to these two charges at a point P a distance y along the midline of the dipole.

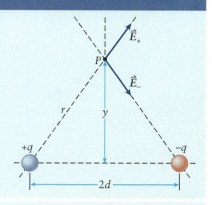

Figure 16-14 An electric dipole The field produced by an electric dipole at any point is the vector sum of the fields $\vec{E}_+$ and $\vec{E}_-$ caused by the positive charge $+q$ and the negative charge $-q$, respectively.

Set Up

The net field is the vector sum of the field $\vec{E}_+$ due to the charge $+q$ (which points away from $+q$) and the field $\vec{E}_-$ due to the charge $-q$ (which points toward $-q$). In Example 16-4 in Section 16-4 we used vector addition to find the net electric *force* exerted by two charges (one positive and one negative) on a third charge; here we use vector addition to find the net electric *field* due to the positive and negative charges. As in Example 16-4, we'll choose the positive x direction to be to the right and the positive y axis to be upward and add the two vectors using components.

Magnitude of the electric field due to a point charge Q:

$$E = \frac{k|Q|}{r^2} \qquad (16\text{-}4)$$

Total electric field:

$$\vec{E} = \vec{E}_+ + \vec{E}_- \qquad (16\text{-}5)$$

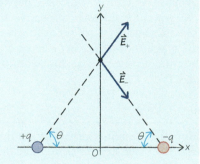

Solve

Use Equation 16-4 to find the magnitudes of the fields $\vec{E}_+$ and $\vec{E}_-$ at P.

The distance from $+q$ to point P is the same as the distance from $-q$ to P. Call this distance r:

$$r = \sqrt{y^2 + d^2}$$

Since $+q$ and $-q$ have the same magnitude (q, which is positive) and are the same distance from P, Equation 16-4 tells us that the fields that the two charges produce at P have the same magnitude:

$$E_+ = E_- = \frac{kq}{r^2} = \frac{kq}{y^2 + d^2}$$

Find the x and y components of $\vec{E}_+$ and $\vec{E}_-$, and use these to calculate the components of the net field at P.

The components of $\vec{E}_+$ and $\vec{E}_-$ are

$$E_{+,x} = E_+ \cos \theta$$
$$E_{+,y} = E_+ \sin \theta$$
$$E_{-,x} = E_- \cos \theta$$
$$E_{-,y} = -E_- \sin \theta$$

Since the magnitudes E_+ and E_- are equal, the components of the net field at P are

$$E_x = E_{+,x} + E_{-,x} = E_+ \cos \theta + E_- \cos \theta$$

$$= \frac{2kq}{y^2 + d^2} \cos \theta$$

$$E_y = E_{+,y} + E_{-,y} = E_+ \sin \theta + (-E_- \sin \theta)$$

$$= \frac{kq}{y^2 + d^2} \sin \theta - \frac{kq}{y^2 + d^2} \sin \theta = 0$$

From the figure

$$\cos \theta = \frac{d}{r} = \frac{d}{\sqrt{y^2 + d^2}}$$

So the components of the net electric field are

$$E_x = \frac{2kq}{(y^2 + d^2)} \frac{d}{\sqrt{y^2 + d^2}} = \frac{2kqd}{(y^2 + d^2)^{3/2}}$$

$$E_y = 0$$

The net electric field at point P is to the right and has magnitude $E = 2kqd/(y^2 + d^2)^{3/2}$.

Reflect

We can check our result by substituting $y = 0$ so that the point P is directly between the two charges and a distance d from each charge. Then $\bar{E}_+$ and $\bar{E}_-$ both point to the right, and the magnitude of the net electric field should be equal to the sum of the magnitudes of $\bar{E}_+$ and $\bar{E}_-$.

At $y = 0$, the net electric field has magnitude

$$E = \frac{2kqd}{(0 + d^2)^{3/2}} = \frac{2kqd}{d^3} = \frac{2kq}{d^2} = 2\left(\frac{kq}{d^2}\right)$$

This is just twice the magnitude of the field due to each individual charge:

$$E_+ = E_- = \frac{kq}{d^2}$$

Note that at points very far from the dipole, so that y is much greater than d, the magnitude of the field is inversely proportional to the *cube* of y: At double the distance, the field of a dipole is $(1/2)^3 = 1/8$ as great. This is a much more rapid decrease with distance than the field of a single point charge, for which E is inversely proportional to the *square* of the distance: At double the distance, the field of a point charge is $(1/2)^2 = 1/4$ as great. The dipole field decreases much more rapidly because the fields of $+q$ and $-q$ partially cancel each other.

If y is much greater than d, $y^2 + d^2$ is approximately equal to y^2. Then the magnitude of the net electric field due to the dipole is approximately

$$E_{net} = \frac{2kqd}{(y^2)^{3/2}} = \frac{2kqd}{y^3}$$

By using techniques like those we employed in Example 16-7, it's possible to calculate and map out the electric field at all points around an electric dipole. Figure 16-15 shows the field lines. Note that as you move away from the dipole along its midline, the field lines become farther apart. This is a graphical way of showing that the magnitude of the field decreases with increasing distance, just as we found in Example 16-7.

GOT THE CONCEPT? 16-4 Electric Field I

 The positive charge in the dipole shown in Figure 16-15 is attracted to the negative charge. To find the force of this attraction, the electric field $\bar{E}$ to use in Equation 16-2 is (a) the field due to the positive charge; (b) the field due to the negative charge; (c) the net field due to both charges; (d) none of these.

GOT THE CONCEPT? 16-5 Electric Field II

 Suppose both of the charges in Figure 16-14 were negative and had the same magnitude. At point P in that figure, the net electric field due to these charges would (a) point to the left; (b) point to the right; (c) point straight up; (d) point straight down; (e) be zero.

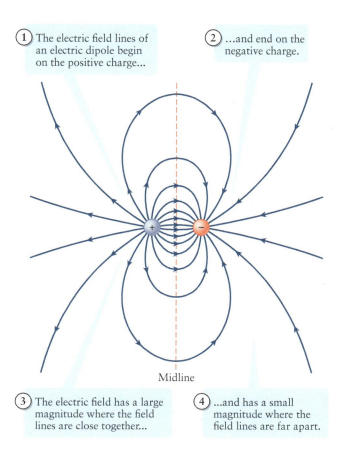

① The electric field lines of an electric dipole begin on the positive charge...

② ...and end on the negative charge.

Midline

③ The electric field has a large magnitude where the field lines are close together...

④ ...and has a small magnitude where the field lines are far apart.

Figure 16-15 **Field lines of an electric dipole** At any point the electric field due to an electric dipole is the vector sum of the field due to the positive charge and the field due to the negative charge.

TAKE-HOME MESSAGE FOR Section 16-5

✔ Any charged object produces an electric field in the space around it. A second charged object responds to this electric field; this is the origin of the electric force that the first object exerts on the second.

✔ The electric field of a positive point charge points directly away from that

charge; the electric field of a negative point charge points directly toward that charge.

✔ The net electric field due to two or more charges is the vector sum of the electric fields due to the individual charges.

16-6 Gauss's law gives us more insight into the electric field

Example 16-7 in the previous section shows how we can find the net electric field due to two point charges. In principle we can extend this approach to calculate the net electric field due to a collection of any number of point charges. If there are many such charges, however, the calculations can become very complex (Figure 16-16).

Happily, there's an alternative and a much easier approach that we can use to find the electric field if the charges are arranged in a very *symmetrical* fashion — for example, uniformly distributed over a spherical volume. This approach uses a principle called *Gauss's law*, which is an alternative way to express Coulomb's law (Equation 16-1). We'll develop Gauss's law in this section and apply it to a variety of physical situations in the following section.

To understand Gauss's law, we first need to define a new quantity called *electric flux*. We'll do this by making an analogy between electric fields and the flow of water.

BioMedical

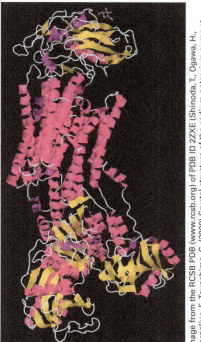

Figure 16-16 **A protein molecule** This protein molecule, a long chain of amino acids, coils on itself because different parts of the chain carry different amounts of electric charge. The electric forces between the parts pull the chain into the complex shape shown here. Understanding the details of protein folding requires knowing the electric field at any point due to the arrangement of charges along the chain. This very complicated problem can only be solved using a computer.

Figure 16-17 Flux of water
The flux of water through a rectangular wire frame depends on the magnitude of the velocity $\vec{v}$ of the water, the area A of the wire frame, and the angle between the direction of $\vec{v}$ and a line perpendicular to the plane of the frame.

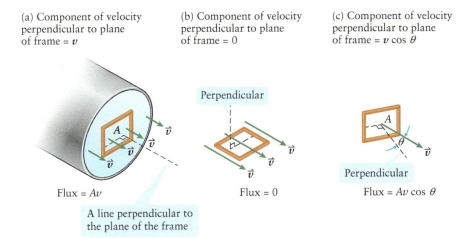

(a) Component of velocity perpendicular to plane of frame = v

Flux = Av

A line perpendicular to the plane of the frame

(b) Component of velocity perpendicular to plane of frame = 0

Perpendicular

Flux = 0

(c) Component of velocity perpendicular to plane of frame = $v \cos \theta$

Perpendicular

Flux = $Av \cos \theta$

Water Flux and Electric Flux

Figure 16-17 shows water flowing through a pipe. The vectors labeled $\vec{v}$ in Figure 16-17a represent the velocity of the water at each point. For simplicity we've assumed that the water velocity is the same everywhere. Now imagine that we place a rectangular wire frame of area A in the flow. We define the flux of water through this wire frame as the product of (i) the area A of the frame and (ii) the component of flow velocity $\vec{v}$ that's perpendicular to the plane of the frame. If we orient the frame so that it's face-on to the water flow as in Figure 16-17a, then $\vec{v}$ is perpendicular to the plane of the frame and the perpendicular component is just v. The flux is then equal to Av. If instead we orient the frame so that it is edge-on to the flow as in Figure 16-17b, the flow velocity $\vec{v}$ has no component perpendicular to the plane of the frame. In this case the flux is zero. If we orient the frame so that a line perpendicular to the frame is at an angle θ to the direction of $\vec{v}$ as in Figure 16-17c, the perpendicular component of $\vec{v}$ is $v \cos \theta$ and the flux is $A(v \cos \theta) = Av \cos \theta$.

Note that we've actually encountered the concept of flux before. In Section 11-9 we learned that if an incompressible fluid is in steady flow through a pipe of varying cross-sectional area A, the product of the area and the flow speed v has the same value at any two points 1 and 2 along the pipe: $A_1 v_1 = A_2 v_2$ (Equation 11-19). In the language we've just introduced, this says that the *flux* of the fluid through the entire pipe maintains the same value even if the cross-sectional area of the pipe changes.

Figure 16-18 shows how we extend the idea of flux to the electric field. Instead of a pipe carrying a fluid, let's look at a region of space where there is an electric field $\vec{E}$. We saw in Section 16-5 that the value of $\vec{E}$ can vary from point to point, so we consider a small enough region that we can treat $\vec{E}$ as having essentially the same value over that region. We then imagine a small rectangular area A that we can orient however we like. By analogy to the flux of water in Figure 16-17, we

Figure 16-18 Electric flux The electric flux through a small rectangular surface depends on the magnitude of the electric field $\vec{E}$, the area A of the surface, and the angle between the direction of $\vec{E}$ and a line perpendicular to the plane of the surface. (Compare Figure 16-17.)

(a) Component of field perpendicular to surface = E

A line perpendicular to the plane of the surface

Electric flux $\Phi = AE$

(b) Component of field perpendicular to surface = 0

Perpendicular

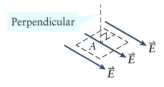

Electric flux $\Phi = 0$

(c) Component of field perpendicular to surface = $E \cos \theta$

Perpendicular

Electric flux $\Phi = AE \cos \theta$

define the **electric flux** Φ (the uppercase Greek letter "phi") through the area A as follows:

| Area of the surface | Magnitude of the electric field |

Electric flux through a surface

$$\Phi = AE_\perp = AE \cos \theta$$

The **component of electric field** perpendicular to the surface

Angle between the electric field and the perpendicular to the surface

Electric flux
(16-6)

In Figure 16-18a we orient the area so that it is face-on to the electric field $\vec{E}$. Then the electric field is perpendicular to the surface and the angle $\theta = 0$. Therefore, $E_\perp$ is equal to the field magnitude E and the electric flux through the surface is $\Phi = AE_\perp = AE$; alternatively, since $\theta = 0$, $\Phi = AE \cos 0 = AE(1) = AE$. In Figure 16-18b the area is edge-on to the electric field, so the electric field has zero component perpendicular to the area A. Then $E_\perp = 0$ and so $\Phi = AE_\perp = 0$. Equivalently, with this orientation $\theta = 90°$ and so $\Phi = AE \cos 90° = AE(0) = 0$. Finally, in Figure 16-18c the area is oriented at an angle θ between 0 and 90°. The component of the electric field perpendicular to the area is $E_\perp = E \cos \theta$, and the electric flux $\Phi = AE \cos \theta$ has a value between AE and zero.

Note that unlike the flow of water in Figure 16-17, there's nothing "flowing" through the area A in Figure 16-18. Unlike velocity $\vec{v}$ the electric field $\vec{E}$ does not represent motion of any kind. But we can still use the analogy between $\vec{v}$ and $\vec{E}$ as expressed by Figures 16-16 and 16-17 to define electric flux Φ in the manner given by Equation 16-6.

WATCH OUT! In calculating electric flux, the area is "imaginary."

(!) In Figure 16-17 we measured the flux of water through an area outlined by a real, physical wire frame. But we really didn't need the frame: It was simply there to help us visualize the area in question. In defining electric flux as in Figure 16-18, we've done away with the wire frame entirely. You can think of the area A in Figure 16-18 and Equation 16-6 as "imaginary" in the sense that there's no physical object outlining the area.

Electric Flux Through a Closed Surface: Gauss's Law

Why is the idea of electric flux through a surface a useful one? To see the answer, let's consider a closed surface—that is, one that encloses a volume. For example, let's find the electric flux through a spherical surface of radius r that is centered on and encloses a positive point charge q (Figure 16-19a). From Equation 16-4 the electric field $\vec{E}$ due to the point charge has the same magnitude $E = kq/r^2$ at every point on the surface because every point is the same distance r from the point charge. However, $\vec{E}$ points in different directions at different points on the spherical surface: straight upward at the top of the surface, to the left at the leftmost point on the surface, and so on. But if we look at a very small rectangular portion of the surface, then $\vec{E}$ points in essentially the same direction at every point on that rectangle. In fact, $\vec{E}$ is perpendicular to the rectangle of area ΔA, so $\theta = 0$ in Equation 16-6 and the flux through that rectangle is $(\Delta A)E = (\Delta A)(kq/r^2)$. If we now imagine that the entire spherical surface is made up of a very large number of such rectangles of area $\Delta A_1, \Delta A_2, \Delta A_3, \ldots, \Delta A_N$, the total electric flux through the spherical surface as a whole is just the sum of the fluxes through the individual rectangles:

$$\Phi = (\Delta A_1)\left(\frac{kq}{r^2}\right) + (\Delta A_2)\left(\frac{kq}{r^2}\right) + (\Delta A_3)\left(\frac{kq}{r^2}\right) + \cdots + (\Delta A_N)\left(\frac{kq}{r^2}\right)$$

$$= (\Delta A_1 + \Delta A_2 + \Delta A_3 + \cdots + \Delta A_N)\left(\frac{kq}{r^2}\right) \qquad (16\text{-}7)$$

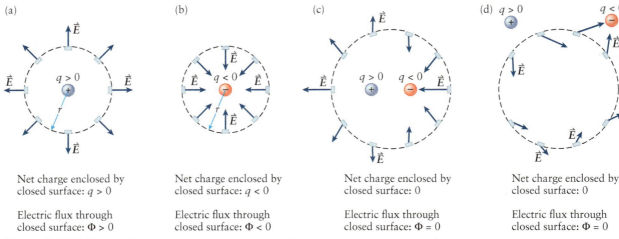

(a) Net charge enclosed by closed surface: $q > 0$

Electric flux through closed surface: $\Phi > 0$

(b) Net charge enclosed by closed surface: $q < 0$

Electric flux through closed surface: $\Phi < 0$

(c) Net charge enclosed by closed surface: 0

Electric flux through closed surface: $\Phi = 0$

(d) Net charge enclosed by closed surface: 0

Electric flux through closed surface: $\Phi = 0$

Figure 16-19 Electric flux through a closed surface The net electric flux through a closed spherical surface is (a) positive if a positive charge is enclosed by the surface, (b) negative if a negative charge is enclosed by the surface, (c) zero if equal amounts of positive and negative charge are enclosed by the surface, and (d) zero if no charge at all is enclosed by the surface.

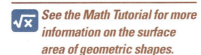 *See the Math Tutorial for more information on the surface area of geometric shapes.*

(16-8)

GOT THE CONCEPT? 16-6
Electric Flux I

(?) When a certain charged particle is placed at the center of a sphere, the net electric flux through the surface of the sphere is Φ_0. If the radius of the sphere is doubled, what would be the new flux through the sphere? (a) $\Phi_0/4$; (b) $\Phi_0/2$; (c) Φ_0; (d) $2\Phi_0$; (e) $4\Phi_0$.

In Equation 16-7 the quantity $\Delta A_1 + \Delta A_2 + \Delta A_3 + \cdots + \Delta A_N$ is the sum of the areas of all of the individual rectangles that make up the spherical surface, and so is equal to the total area A of the surface. The surface area of a sphere of radius r is $A = 4\pi r^2$, so Equation 16-7 becomes

$$\Phi = (4\pi r^2)\left(\frac{kq}{r^2}\right) = 4\pi kq$$

Notice that the radius of the spherical surface surrounding the point charge q cancels out in this equation: The magnitude of the electric field decreases as the square of the sphere's radius r and the surface area increase at the same rate. So their product—the electric flux Φ—is directly proportional to the charge q enclosed within the spherical surface, but does *not* depend on the radius.

Equation 16-8 also holds true if the charge q is negative (Figure 16-19b): The electric flux through the closed surface is *negative* in this case. The interpretation is that the flux is positive if the electric field points out of the closed surface, as in Figure 16-19a, but negative if the electric field points into the closed surface, as in Figure 16-19b.

What happens if we replace the single charge q with an electric dipole like that shown in Figure 16-15, with both a positive charge and a negative charge of equal magnitude? In Figure 16-19c the spherical surface is centered on the dipole and encloses both the positive and negative charges, so the *net* enclosed charge is zero. The figure shows that for each small rectangle on the surface where the electric field $\vec{E}$ points outward, giving a positive contribution to the flux, there is another small rectangle elsewhere on the surface where $\vec{E}$ points inward, giving an equally large negative contribution to the flux. These positive and negative contributions to the flux cancel, so the net electric flux through this surface—which encloses zero net charge—is zero. The same cancellation also happens for the spherical surface in Figure 16-19d, which does not enclose either charge. In this case as well, the net enclosed charge is zero and the net electric flux through the surface is zero.

Here's what we've concluded so far from Figure 16-19:

- Closed surfaces that enclose a point charge q, as in Figure 16-19a and Figure 16-19b, have a nonzero electric flux through them. This flux is proportional to the enclosed charge q, as in Equation 16-8.

- Closed surfaces that enclose zero net charge—either equal amounts of positive and negative charge, as in Figure 16-19c, or no charge at all, as in Figure 16-19d— have zero net electric flux through them. This is consistent with Equation 16-8, but with $q = 0$.

We were able to draw these conclusions because the closed surfaces in Figure 16-19 are spherical in shape and placed symmetrically with respect to the charges inside them. But it can be shown that the same conclusions hold true for a closed surface of *any* shape or placement. We can summarize these conclusions by rewriting Equation 16-8 with q replaced by q_{encl}, which represents the *net* charge enclosed by the closed surface:

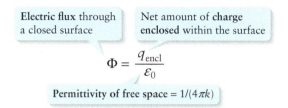

Electric flux through a closed surface

Net amount of **charge enclosed** within the surface

$$\Phi = \frac{q_{encl}}{\varepsilon_0}$$

Gauss's law
(16-9)

Permittivity of free space = $1/(4\pi k)$

This relationship is called **Gauss's law** after Carl Friedrich Gauss, the nineteenth-century German mathematician and physicist who first deduced it. A closed surface used to enclose charge in order to apply Gauss's law is referred to as a **Gaussian surface**. Equation 16-9 holds true for *any* Gaussian surface (Figure 16-20), no matter what its shape or size and no matter where the charges are located inside the surface. As we mentioned previously, such surfaces are imaginary: The surface does not need to be made of any physical substance.

GOT THE CONCEPT? 16-7 Electric Flux II

 When a certain charged particle is placed at the center of a sphere, the net electric flux through the surface of the sphere is Φ_0. If the sphere were elongated into a spheroid (such as a rugby ball or an American football) but still enclosed the same charge, what would be the new electric flux through the surface? (a) Less than Φ_0; (b) Φ_0; (c) more than Φ_0; (d) not enough information given to decide.

In Equation 16-9 we have replaced the combination $4\pi k$ that appears in Equation 16-8 with $1/\varepsilon_0$. Here ε_0, called the **permittivity of free space** for historic reasons, is equal to $1/(4\pi k) = 8.85 \times 10^{-12} \text{ C}^2/(\text{N}\cdot\text{m}^2)$ to three significant figures. So we can state Gauss's law as

The net electric flux through any closed surface equals the net charge enclosed by that surface divided by the permittivity ε_0. Charges outside the surface have no effect on the net electric flux through the surface.

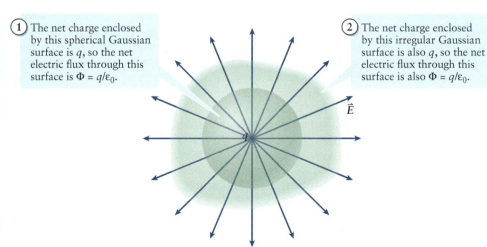

① The net charge enclosed by this spherical Gaussian surface is q, so the net electric flux through this surface is $\Phi = q/\varepsilon_0$.

② The net charge enclosed by this irregular Gaussian surface is also q, so the net electric flux through this surface is also $\Phi = q/\varepsilon_0$.

The net electric flux through a Gaussian surface depends only on the net charge that it encloses, not the shape or size of the surface.

$\vec{E}$

Figure 16-20 Gauss's law Two different Gaussian surfaces that enclose the same net charge.

This law holds true because of the $1/r^2$ character of the electric field due to a point charge (see the discussion following Equation 16-8).

We'll see in the following section how we can use Gauss's law to determine the electric field due to certain charge distributions and what Gauss's law tells us about how charges distribute themselves in a conductor.

GOT THE CONCEPT? 16-8 Electric Flux III

 When a certain charged particle is placed at the center of a sphere, the net electric flux through the surface of the sphere is Φ_0. If a second, identical charged particle is placed outside the sphere, what would be the new electric flux through the surface? (a) Less than Φ_0; (b) Φ_0; (c) more than Φ_0; (d) not enough information given to decide.

TAKE-HOME MESSAGE FOR Section 16-6

✔ The electric flux through a surface is analogous to the flux of water through a wire frame. Electric flux depends on the magnitude of the electric field, the area of the surface, and the relative orientation of the field and the surface.

✔ For a closed surface (one that encloses a volume) an electric field $\vec{E}$ that points out of the closed surface makes a positive contribution to the electric flux. An electric field that points into the closed surface makes a negative contribution to the electric flux.

✔ Gauss's law states that the net electric flux through a closed surface is proportional to the amount of charge enclosed by the surface. If the net enclosed charge is zero, the net electric flux through the surface is zero.

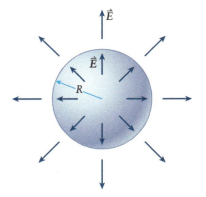

Charge Q is uniformly distributed throughout the volume of this sphere of radius R.

Figure 16-21 **A uniformly charged sphere** The spherical symmetry of this charge distribution tells us that the electric field $\vec{E}$ that it produces must be radial and that the magnitude of $\vec{E}$ can depend only on the distance from the center of the sphere.

16-7 In certain situations Gauss's law helps us calculate the electric field and determine how charge is distributed

Gauss's law tells us the value of the net electric flux through the closed Gaussian surfaces shown in Figure 16-19. But that isn't enough to tell us the value of $\vec{E}$ at any individual point on these surfaces. Yet we can use Gauss's law to determine $\vec{E}$ in cases where the charge that produces the field is distributed in a particularly symmetric way. Let's look at an example.

Electric Field of a Spherical Charge Distribution

In Figure 16-21 charge Q is distributed uniformly throughout a spherical volume of radius R. (This could be a model of how electric charge is distributed over the volume of an atomic nucleus.) This distribution is *spherically symmetric*: You can rotate the sphere through any angle around its center and it looks exactly the same. Hence the electric field $\vec{E}$ caused by the charge distribution must also be spherically symmetric. This implies that the direction of the electric field must be either radially inward or radially outward, like the sea urchin spines shown in Figure 16-13. (If the field pointed in any other direction, the field lines would look different after rotating the sphere through some angle. The spherical symmetry says that's impossible.) So at any point $\vec{E}$ can have only a radial component E_r, which points either directly away from the center of the sphere ($E_r > 0$, as in Figure 16-21) or directly toward the center of the sphere ($E_r < 0$). This must be true for points inside the sphere as well as outside the sphere.

The spherical symmetry of the charge distribution in Figure 16-21 tells us something more: The value of E_r at a given point depends only on the radial distance r from the center of the sphere and not on the point's location around the sphere.

Given what spherical symmetry tells us about the electric field for the situation in Figure 16-21, what more can we learn by using Gauss's law? In Figure 16-22a we've

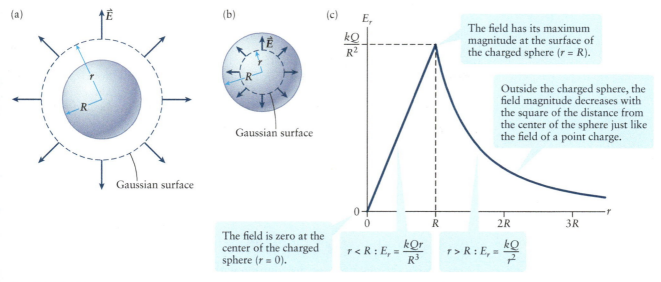

Gaussian surface

(b)

Gaussian surface

(c)

The field has its maximum magnitude at the surface of the charged sphere ($r = R$).

Outside the charged sphere, the field magnitude decreases with the square of the distance from the center of the sphere just like the field of a point charge.

The field is zero at the center of the charged sphere ($r = 0$).

$$r < R : E_r = \frac{kQr}{R^3} \qquad r > R : E_r = \frac{kQ}{r^2}$$

Figure 16-22 **Finding the electric field of a uniformly charged sphere** To find how $\vec{E}$ depends on distance from the center of the sphere of charge Q, we use (a) one Gaussian surface that encloses the entire sphere and (b) a second Gaussian surface that encloses only part of the sphere. (c) The radial electric field E_r as a function of the distance r from the center, graphed for the case $Q > 0$.

drawn a spherical Gaussian surface that's centered on the sphere and that encloses the entire sphere. This Gaussian surface has radius $r > R$ and surface area $A = 4\pi r^2$. Just like the spherical surface in Figure 16-19a or Figure 16-19b, the electric field is perpendicular to the surface at every point and has the same magnitude at every point. Using the same reasoning that we used for Equation 16-8, the net electric flux through this surface is just the area of the surface multiplied by the radial electric field:

$$\Phi = AE_r = 4\pi r^2 E_r \tag{16-10}$$

This flux is positive if the electric field points outward (so $E_r > 0$) and negative if the electric field points inward ($E_r < 0$). Since the entire charged sphere is enclosed within the Gaussian surface, the enclosed charge is $q_{encl} = Q$. Gauss's law, Equation 16-9, tell us that the flux is also equal to q_{encl}/ε_0. Equating Equations 16-9 and 16-10 yields

$$4\pi r^2 E_r = \frac{Q}{\varepsilon_0} \quad \text{and} \quad E_r = \frac{Q}{4\pi\varepsilon_0 r^2} = \frac{kQ}{r^2} \quad (r > R) \tag{16-11}$$

Equation 16-11 says that at points outside the sphere, the electric field is exactly the same as the field due to a point charge Q (Equation 16-4). The field is inversely proportional to the square of the distance r and is proportional to the charge Q on the sphere. The field points radially outward ($E_r > 0$) if the charge Q is positive and points radially inward ($E_r < 0$) if Q is negative. (Compare Figure 16-12.)

We can also use Gauss's law to tell us about the electric field at points inside the sphere. In Figure 16-22b we've drawn another spherical Gaussian surface of radius r and surface area $A = 4\pi r^2$ but with a radius r that's less than the radius R of the charged sphere. Again the net electric flux through this Gaussian surface is given by Equation 16-10. But now the enclosed charge is less than the total charge Q on the sphere. Since the charge is distributed uniformly, the fraction of charge that's enclosed by the Gaussian surface is just equal to the ratio of the volume within the Gaussian surface (a sphere of radius r, with volume $4\pi r^3/3$) to the total volume occupied by the charge Q (a sphere of radius R, with volume $4\pi R^3/3$). So Gauss's law applied to the Gaussian surface in Figure 16-22b tells us

$$4\pi r^2 E_r = \frac{q_{encl}}{\varepsilon_0} = \frac{1}{\varepsilon_0}\left[Q\left(\frac{4\pi r^3/3}{4\pi R^3/3}\right)\right] = \frac{Qr^3}{\varepsilon_0 R^3}$$

If we divide through by $4\pi r^2$, we get

$$E_r = \frac{Qr^3}{4\pi\varepsilon_0 r^2 R^3} = \frac{Qr}{4\pi\varepsilon_0 R^3} = \frac{kQr}{R^3} \quad (r < R) \tag{16-12}$$

Equation 16-12 tells us that the direction of the electric field is the same inside the charged sphere as outside: $E_r > 0$ (an outward field) if Q is positive and $E_r < 0$ (an inward field) if Q is negative. It also says that inside the sphere the magnitude of the field increases with increasing distance r from the center of the sphere. At the surface of the charged sphere, $r = R$, both Equation 16-11 and Equation 16-12 give the same result: $E_r = kQ/R^2$. At the very center of the sphere ($r = 0$), the electric field is zero. Figure 16-22c shows a graph of E_r as a function of r for the case of a positively charged sphere ($Q > 0$).

These conclusions about the electric field of a charged sphere would have been very difficult to obtain without using Gauss's law. (The alternative approach is to divide the charged sphere into a very large number of small segments, treating each segment as an individual point charge, and using vector addition to add the individual electric fields produced by all of the segments. That would take a lot of strenuous mathematics.) The relative ease with which we came to these conclusions shows the power of Gauss's law.

Electric Field of a Large, Flat, Charged Disk

Figure 16-23a shows a charge distribution with a different kind of symmetry: a uniformly charged plate or disk. (We'll see later that charged disks of this kind are found in an important device called a *capacitor*, used in many electric circuits.) This charge distribution has *rotational symmetry*, which means that it looks the same if you rotate it through any angle around an axis that passes vertically through the center of the disk. However, the electric field $\vec{E}$ due to this charge distribution can (and does) vary in a complicated way as you move from the center of the disk toward the edges.

To simplify the problem let's just consider what the electric field is like at points around the disk near the disk center. We imagine that the edges of the disk are so far away that we can regard them as being infinitely distant. Then our problem is that of finding the electric field due to an *infinite sheet of* charge (Figure 16-23b).

We'll use the symbol σ (the lowercase Greek letter "sigma") for the amount of charge per unit area on the sheet, also called the **surface charge density**. The sheet is uniformly charged, so σ has the same value everywhere on the sheet. The units of σ are coulombs per square meter (C/m^2).

Our infinite sheet of charge has *translational* symmetry: No matter which way or how far you move parallel to the disk, the charge distribution looks exactly the same. The same must therefore be true of the electric field produced by the charge distribution. So if the disk lies in the x–y plane, the electric field $\vec{E}$ at any point cannot depend on the x or y coordinate of that point. The field can depend only on the z coordinate, which is the coordinate measured perpendicular to the plane in Figure 16-23.

Translational symmetry also tells us that $\vec{E}$ must be *perpendicular* to the plane of the sheet. This means that a point charge q placed in that field will feel a force $\vec{F} = q\vec{E}$ either directly toward or directly away from the sheet, depending on the sign of q. (If there were a component of $\vec{E}$ in some direction parallel to the plane of the sheet, a positive point charge would be pushed in that direction. This would only be the case if the point charge were repelled or attracted by one part of the sheet. But since the charge distribution is the same everywhere on the sheet, no part attracts or repels the positive charge more than any other part. So there can't be any component of $\vec{E}$ parallel to the sheet.) Thus $\vec{E}$ can depend only on the z coordinate and can have only a z component.

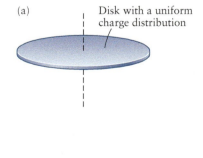

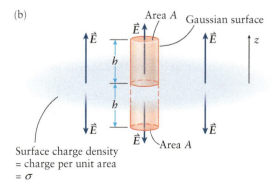

Figure 16-23 Finding the electric field of a uniformly charged disk (a) A uniformly charged disk; (b) to find the electric field of the disk at points close to the center of the disk, we can replace the disk by an infinitely large charged sheet.

The infinite sheet also has *reflection* symmetry: It looks the same if we flip the sheet upside down. The same must be true of the electric field. So if the field points upward above the sheet, it must point downward below the sheet, as we've indicated in Figure 16-23b.

To find the electric field at a certain distance h from the plane of the sheet, we'll use a cylindrical Gaussian surface as in Figure 16-23b. The top and bottom faces of the cylinder each have area A, and each is the same distance h from the sheet. The symmetries we have described tell us that the electric field $\vec{E}$ is perpendicular to the top and bottom faces and has the same value everywhere on each of these faces. Because $\vec{E}$ is parallel to the sides of the cylinder, there is zero electric flux through the sides: $\Phi_{sides} = 0$. The flux through the top and bottom faces of the cylinder, however, is not zero. If E_z is the z component of electric field a distance h above the infinite sheet, the flux through the top face of the cylinder is $\Phi_{top} = AE_z$. Note that this flux is positive if E_z is positive, which means that the electric field above the sheet is upward and points out of the Gaussian surface; this flux is negative if E_z is negative, which means that the electric field above the sheet is downward and points into the Gaussian surface. The reflection symmetry tells us that the flux through the bottom face is the same, so $\Phi_{top} + \Phi_{bottom} = 2\Phi_{top} = 2AE_z$. The net electric flux through the Gaussian surface in Figure 16-23b is therefore

$$\Phi = \Phi_{sides} + \Phi_{top} + \Phi_{bottom} = 0 + 2AE_z = 2AE_z \qquad (16\text{-}13)$$

The area of the sheet enclosed within the Gaussian surface is A, so the amount of charge enclosed within the surface is the charge per unit area σ multiplied by the area A: $q_{encl} = \sigma A$. Using this and Equation 16-13 in Gauss's law, Equation 16-9, gives us an expression for the electric field component E_z a distance h above the sheet:

$$2AE_z = \frac{\sigma A}{\varepsilon_0} \quad \text{or} \quad E_z = \frac{\sigma}{2\varepsilon_0} \qquad (16\text{-}14)$$

Equation 16-14 tells us that E_z has the same sign as the surface charge density σ. So the electric field $\vec{E}$ above the sheet points upward (away from the sheet) if the surface charge density is positive ($\sigma > 0$); $\vec{E}$ points downward (toward the sheet) if the surface charge density is negative ($\sigma < 0$). This agrees with the idea that electric fields point away from positive charges and toward negative charges.

Equation 16-14 also tells us that the value of E_z at a point a distance h above the sheet does *not* depend on h. (Note that h doesn't appear anywhere in this equation.) So for an infinite sheet of charge, the electric field is the same at all distances from the sheet.

These results are only approximate, since there's no such thing as a truly infinite sheet of charge. But they are valid for a charged disk at points that are relatively close to the disk, so the height h above the disk is small compared to the radius of the disk. We'll make use of Equation 16-14 in later chapters.

Excess Charge on Conductors

Gauss's law leads to a remarkable conclusion about a conductor to which we add excess charge so that the conductor has a net nonzero charge. Charges are free to move in a conductor, so if we add excess charges they will move in the conductor until they come to rest in equilibrium so that the net force on each added charge is zero. Because the electric force on a charge q is directly proportional to the electric field $\vec{E}$ (Equation 16-2), $\vec{E}$ inside the conductor must be *zero*. If it were not, excess charges inside the conductor would experience a force and be pushed to some new location. Note that this statement only applies inside the volume of the conductor. Outside the charged block of conductor the electric field need not be zero.

Let's see what Gauss's law tells us in this situation. Imagine a Gaussian surface that lies completely inside the volume of a conductor that carries excess charge (Figure 16-24). Since $\vec{E} = 0$ everywhere inside the conductor, the net electric flux Φ through this surface is zero. From Gauss's law, Equation 16-9, Φ is equal to the net charge q_{encl} enclosed by the surface divided by ε_0. So the net charge inside the Gaussian surface is zero. This holds true for any Gaussian surface, no matter how small, that lies entirely within the conductor. So there can be *no* excess charge within the volume of the conductor. Instead, *all of the excess charge on a conductor in equilibrium must*

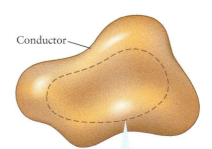

Conductor

There can be no electric field inside the conductor, so the flux through this Gaussian surface must be zero.

Figure 16-24 A charged conductor Gauss's law tells us that any excess charge on a conductor must reside on the surface of the conductor.

reside on the surface. That's why we depicted the charged copper rod in Figure 16-5b with its excess charge spread over the surface of the rod, not its interior.

Our conclusion that $\vec{E} = 0$ inside a conductor holds true *only* if all of the charges are at rest. We will see in later chapters that if an electric current is present inside a conductor, there must be a nonzero electric field to sustain the flow of charge.

EXAMPLE 16-8   Gauss's Law and Surface Charge

A conducting sphere of radius R_1 carries excess charge +7 C. This sphere is enclosed within a concentric spherical conducting shell of inner radius R_2 and outer radius R_3. The conducting shell carries a net charge +2 C. How much charge is on the inner surface of the conducting shell? How much is on the outer surface?

Set Up

In equilibrium the electric field inside the volume of the conducting shell must be zero. So if we imagine a spherical Gaussian surface that lies inside the shell, so that its radius r is intermediate between R_2 and R_3, the net electric flux through that surface will be zero. From Gauss's law, that means that the net charge enclosed by the Gaussian surface must be zero. The +2-C charge on the shell will arrange itself on the surfaces of the shell to make that happen.

Gauss's law:

$$\Phi = \frac{q_{encl}}{\varepsilon_0} \qquad (16\text{-}9)$$

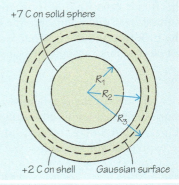

+7 C on solid sphere

R_1
R_2
R_3

+2 C on shell   Gaussian surface

Solve

The charge enclosed within our Gaussian surface includes the charge of the central sphere ($q_{sphere} = +7$ C) and whatever charge q_{inner} is present on the inner surface of the shell. These charges must add to zero.

For the Gaussian surface

$$q_{encl} = q_{sphere} + q_{inner} = 0$$

So the charge on the inner surface of the shell equals the negative of the charge on the central sphere:

$$q_{inner} = -q_{sphere} = -7 \text{ C}$$

The total charge on the shell is $q_{shell} = +2$ C. Because all of the excess charge on a conductor resides on its surfaces, q_{shell} is the sum of the charge on the inner and outer surfaces.

$$q_{shell} = q_{inner} + q_{outer}$$

So

$$\begin{aligned} q_{outer} &= q_{shell} - q_{inner} \\ &= (+2 \text{ C}) - (-7 \text{ C}) \\ &= +9 \text{ C} \end{aligned}$$

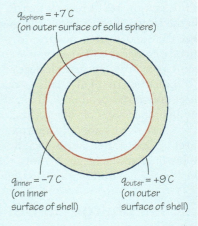

$q_{sphere} = +7$ C
(on outer surface of solid sphere)

$q_{inner} = -7$ C
(on inner surface of shell)

$q_{outer} = +9$ C
(on outer surface of shell)

Reflect

While the shell as a whole carries a *positive* excess charge of +2 C, the inner surface of the shell has a *negative* charge. Here's why: When the positively charged sphere is placed inside the shell, some of the free electrons in the shell are drawn to the shell's inner surface, giving it a negative charge of −7 C. This leaves a deficit of electrons—a positive charge of +7 C—on the outside surface, which adds to the +2 C of excess charge to give the outer surface of the shell a charge of +9 C.

GOT THE CONCEPT? 16-9   Size of a Gaussian Surface

In our derivation of Equation 16-13 for the electric field due to an infinite charged sheet, we used a Gaussian surface in the shape of a cylinder whose top and bottom faces had area A. If we used a larger cylinder with top and bottom faces of area $2A$, how would this change our result for the electric field? (a) E_z would be four times larger. (b) E_z would be twice as large. (c) E_z would be the same. (d) E_z would be one-half as great. (e) E_z would be one-fourth as great.

TAKE-HOME MESSAGE FOR Section 16-7

✔ Gauss's law can be used to calculate the electric field in certain cases where the charge distribution has a high degree of symmetry. These include a spherical distribution of charge and a very large, uniformly charged sheet.

✔ Gauss's law tells us that when excess charge added to a conductor is allowed to come to rest in equilibrium, the excess charge resides only on the surfaces of the conductor.

Key Terms

charged	electric dipole	Gauss's law
conductor	electric field lines	neutral
coulomb	electric flux	permittivity of free space
Coulomb constant	electric force	point charges
Coulomb's law	insulators	semiconductors
electric charge	Gaussian surface	surface charge density

Chapter Summary

Topic	Equation or Figure
Electric charge: Ordinary matter contains equal amounts of positive and negative charge. Charge can be transferred from one object to another, for example, by rubbing. In this and all other processes, charge is conserved: It can be moved from place to place but can be neither created nor destroyed.	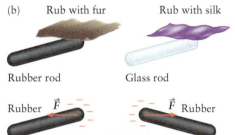 (b) Rub with fur Rub with silk Rubber rod Glass rod The fur-rubbed ends of these two rubber rods repel each other. The silk-rubbed ends of these two glass rods repel each other. The fur-rubbed end of the rubber rod attracts the silk-rubbed end of the glass rod. (Figure 16-2b)
Conductors, insulators, and semiconductors: In a conductor charges are free to move with ease; in an insulator charges can move very little. Semiconductors have properties intermediate between those of conductors and insulators.	(b) The nylon-rubbed end of the copper rod attracts the nylon with which it was rubbed... 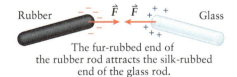Copper Nylon 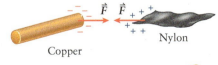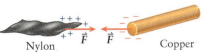Nylon Copper ...and the end of the copper rod that was not rubbed also feels a strong attraction. (Figure 16-5b)

Coulomb's law: The electric forces that two point charges exert on each other are proportional to the magnitudes of the charges and inversely proportional to the square of the distance between the charges. These forces are attractive if the two charges have opposite signs, and repulsive if the two charges have the same sign.

Any two point charges q_1 and q_2 exert equally strong **electric forces** on each other.

Coulomb constant

Absolute values of the **point charges**

$$F_{q_1 \text{ on } q_2} = F_{q_2 \text{ on } q_1} = \frac{k|q_1||q_2|}{r^2}$$

(16-1)

Distance between the point charges

Electric field: We can regard the interaction between charges as a two-step process: One charge sets up an electric field, and the other charge responds to that field. The electric field points away from a positive charge and toward a negative charge. The electric field of a single point charge is given by a simple equation; the field due to a combination of charges is the vector sum of the fields due to the individual charges.

If a particle with charge q is placed at a position where the **electric field** due to other charges is $\vec{E}$...

$$\vec{F} = q\vec{E}$$

(16-2)

...then the **electric force** on the particle is $\vec{F} = q\vec{E}$.

Coulomb constant

Absolute value of charge Q

Magnitude of the electric field due to a point charge Q

$$E = \frac{k|Q|}{r^2}$$

(16-4)

Distance between point charge Q and the location where the field is measured

---

Gauss's law: The electric flux through a surface equals the area of that surface multiplied by the component of electric field perpendicular to the surface. Gauss's law states that the net electric flux through a closed surface (one that encloses a volume) is proportional to the net charge enclosed within that volume. Gauss's law can be used to determine the electric field in situations where the charge distribution is highly symmetric. It also tells us that any excess charge on a conductor resides on the surfaces of the conductor.

Electric flux through a closed surface

Net amount of **charge enclosed** within the surface

$$\Phi = \frac{q_{\text{encl}}}{\varepsilon_0}$$

(16-9)

Permittivity of free space = $1/(4\pi k)$

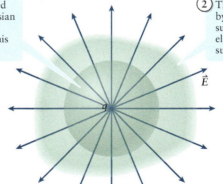

① The net charge enclosed by this spherical Gaussian surface is q, so the net electric flux through this surface is $\Phi = q/\varepsilon_0$.

② The net charge enclosed by this irregular Gaussian surface is also q, so the net electric flux through this surface is also $\Phi = q/\varepsilon_0$.

$\vec{E}$

(Figure 16-20)

 The net electric flux through a Gaussian surface depends only on the net charge that it encloses, not the shape or size of the surface.

Answer to What do you think? Question

(c) The magnitude of the electric field of a charged object measured at a distance r from the object is proportional to $1/r^2$ (see Section 16-5). If r is made one-half as great (decreased from $r = 2$ m to $r = 1$ m), the electric field magnitude changes by a factor of $1/(1/2)^2 = 1/(1/4) = 4$.

Answers to Got the Concept? Questions

16-1 (c) When the rubber rod is rubbed with fur, electrons are transferred to the rod from the fur. The electrons have mass, so the rod gains a little mass and the fur loses an equal amount of mass. By contrast, electrons are transferred from the glass rod to the silk when these two objects are rubbed together. As a result, the glass rod loses a little mass and the silk gains an equal amount of mass. The amount of mass involved is very small but is not zero.

16-2 (c) A sodium atom is neutral because it has 11 protons in its nucleus (each of charge $+e$) and 11 electrons (each of charge $-e$). If a sodium atom loses 1 of its 11 electrons by transferring it to another atom, it is left with 11 protons with a combined charge of $+11e$ and 10 electrons with a combined charge of $-10e$, and so it has a net positive charge of $+e$. In NaCl the electron is transferred to the chlorine atom, which then has a net negative charge of $-e$.

16-3 (d) Equation 16-1 tells us that the magnitude of the electric force between two charged objects is directly proportional to the product of the two charges. If we remove only half of the electrons from the neutral water drop, the ball of electrons has half the negative charge as before and the water drop has half the positive charge as before. So the electric force between the water drop and the ball of electrons will be $\left(\frac{1}{2}\right) \times \left(\frac{1}{2}\right) = \frac{1}{4}$ as great as before.

16-4 (b) Equation 16-2, $\vec{F} = q\vec{E}$, tells us the force $\vec{F}$ that acts on a charge q due to the electric field $\vec{E}$ produced by other charges. So to find the force that acts on the positive charge in Figure 16-15 due to the negative charge, in Equation 16-2 we let q be the positive charge (the charge that experiences the force) and let $\vec{E}$ be the electric field produced at the position of the positive charge by the negative charge (the other charge that exerts the force).

16-5 (d) If both charges were negative, the electric field at point P due to each charge would point directly toward that charge. So the electric field due to the right-hand charge would point down and to the right, while the electric field due to the left-hand charge would point down and to the left. Since point P is the same distance from each charge, and each charge has the same magnitude, the electric fields due to the two charges would have the same magnitude. The horizontal components of the two fields cancel (one points left and the other points right), so what remains is a net electric field that points straight down.

16-6 (c) The flux remains the same. According to Gauss's law, the net electric flux due to an enclosed charge does not depend on the size of the surface that encloses the charge.

16-7 (b) The flux remains the same. According to Gauss's law, the net electric flux due to an enclosed charge does not depend on the shape of the surface that encloses the charge.

16-8 (b) The flux remains the same. According to Gauss's law, the electric flux through a closed surface depends only on the charge enclosed within the surface. A charge outside the surface will cause positive flux on one part of the surface (where its electric field points out of the surface) and negative flux on another part of the surface (where its electric field points into the surface), but will have no net effect on the flux through the surface as a whole.

16-9 (c) The result does not depend on the size of the cylinder we choose. If we double the area, the flux through the top and bottom faces would double, but the amount of enclosed charge would double as well. In the derivation of Equation 16-14 these factors would cancel, so our result for E_z would be the same.

Questions and Problems

In a few problems you are given more data than you actually need; in a few other problems you are required to supply data from your general knowledge, outside sources, or informed estimate.

Interpret as significant all digits in numerical values that have trailing zeros and no decimal points.

For all problems use $g = 9.80$ m/s^2 for the free-fall acceleration due to gravity. Neglect friction and air resistance unless instructed to do otherwise.

- •   Basic, single-concept problem
- ••   Intermediate-level problem; may require synthesis of concepts and multiple steps
- ••• Challenging problem
- Example   See worked example for a similar problem.

Conceptual Questions

1. • How, if at all, would the physical universe be different if the proton were negatively charged and the electron were positively charged?

2. • How, if at all, would the physical universe be different if the proton's charge were very slightly larger in magnitude than the electron's charge?

3. • When an initially electrically neutral object acquires a net positive charge, does its mass increase or decrease? Why?

4. • When you remove socks from a hot dryer, they tend to cling to everything. Two identical socks, however, usually repel. Why?

5. • Describe a set of experiments that might be used to determine if you have discovered a third type of charge other than positive and negative.

6. • How does a person become "charged" as he or she shuffles across a carpet, wearing cloth slippers, on a dry winter day?

7. • After combing your hair with a plastic comb, you find that when you bring the comb near a small bit of paper, the bit of paper moves toward the comb. Then, shortly after the paper touches the comb, it moves away from the comb. Explain these observations.

8. • After combing your hair with a plastic comb, you find that when you bring the comb near an empty aluminum soft-drink can that is lying on its side on a nonconducting table-top, the can rolls toward the comb. After being touched by the comb, the can is still attracted by the comb. Explain these observations.

9. • (a) A positively charged glass rod attracts a lighter object suspended by a thread. Does it follow that the object is negatively charged? (b) If, instead, the rod repels it, does it follow that the suspended object is positively charged?

10. • Some days it can be frustrating to attempt to demonstrate electrostatic phenomena for a physics class. An experiment that works beautifully one day may fail the next day if the weather has changed. Air-conditioning helps a lot while demonstrating the phenomena during the summer. Why?

11. • Discuss the similarities and differences between the gravitational and electric forces.

12. • Why is the gravitational force usually ignored in problems on the scale of particles such as electrons and protons?

13. • (a) What are the advantages of thinking of the force on a charge at a point P as being exerted by an electric field at P, rather than by other charges at other locations? (b) Is the convenience of the field as a calculation device worth inventing a new physical quantity? Or is there more to the field concept than that?

14. • Do electric field lines point along the trajectory of positively charged particles? Why or why not?

15. • An electron and a proton are released in a region of space where the electric field is vertically downward. How do the electric forces acting on the electron and proton compare?

16. • Inside a uniform spherical charge distribution, why is it that as one moves out from the center, the electric field increases as r rather than decreases as $1/r^2$?

17. • Is the electric field $\bar{E}$ in Gauss's law only the electric field due to the charge inside the Gaussian surface, or is it the total electric field due to all charges both inside and outside the surface? Explain your answer.

18. • If the net electric flux out of a closed surface is zero, does that mean the charge density must be zero everywhere inside the surface? Explain your answer.

Multiple-Choice Questions

19. • Electric charges of the opposite sign
   A. exert no force on each other.
   B. attract each other.
   C. repel each other.
   D. repel and attract each other.
   E. repel and attract each other depending on the magnitude of the charges.

20. • If two uncharged objects are rubbed together and one of them acquires a negative charge, then the other one
   A. remains uncharged.
   B. also acquires a negative charge.
   C. acquires a positive charge.
   D. acquires a positive charge equal to twice the negative charge.
   E. acquires a positive charge equal to half the negative charge.

21. • Metal sphere A has a charge of $-Q$. An identical metal sphere B has a charge of $+2Q$. The magnitude of the electric force on B due to A is F. The magnitude of the electric force on A due to B is
   A. $F/4$.
   B. $F/2$.
   C. F.
   D. $2F$.
   E. $4F$.

22. • A balloon can be charged by rubbing it with your sleeve while holding it in your hand. You can conclude from this that the balloon is a(n)
   A. conductor.
   B. insulator.
   C. neutral object.
   D. Gaussian surface.
   E. semiconductor.

23. • A positively charged rod is brought near one end of an uncharged metal bar. The end of the metal bar farthest from the charged rod will be charged
   A. positively.
   B. negatively.
   C. neutral.
   D. twice as much as the end nearest the rod.
   E. none of the above ways.

24. • A free positive charge released in an electric field will
   A. remain at rest.
   B. accelerate in the direction opposite to the electric field.
   C. accelerate in the direction perpendicular to the electric field.
   D. accelerate in the same direction as the electric field.
   E. accelerate in a circular path.

25. • Consider a point charge $+Q$ located outside a closed surface such as a sphere bound by the black circle in **Figure 16-25**. What is the net electric flux through the closed surface?

   A. $\dfrac{+Q}{\varepsilon_0}$

   B. $\dfrac{-Q}{\varepsilon_0}$

   C. 0

   D. $\dfrac{+2Q}{\varepsilon_0}$

   E. $\dfrac{-2Q}{\varepsilon_0}$

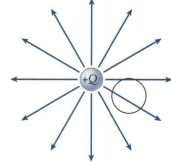

Figure 16-25 Problem 25

26. • If a charge is located at the center of a spherical volume and the electric flux through the surface of the sphere is Φ,

what would be the flux through the surface if the radius of the sphere were tripled?

- A. 3Φ
- B. 9Φ
- C. Φ
- D. $\Phi/3$
- E. $\Phi/9$

27. • A point charge $+Q$ is at the center of a spherical conducting shell of inner radius R_1 and outer radius R_2, as shown in **Figure 16-26**. The charge on the inner surface of the shell is

- A. $+Q$.
- B. $-Q$.
- C. 0.
- D. $+Q/2$.
- E. $-Q/2$.

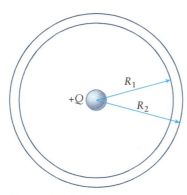

Figure 16-26 Problem 27

Problems

16-1 Electric forces and electric charges are all around you — and within you

16-2 Matter contains positive and negative electric charge

28. • The nucleus of a copper atom has 29 protons and 35 neutrons. What is the total charge of the nucleus? Example 16-1

29. • Five electrons are added to 1.00 C of positive charge. What is the net charge of the system? Example 16-1

30. •• How many coulombs of negative charge are there in 0.500 kg of water? Example 16-1

31. • A particular ion of oxygen is composed of 8 protons, 10 neutrons, and 6 electrons. In terms of the fundamental charge e, what is the total charge of this ion? Example 16-1

32. •• The charge per unit length on a glass rod is 0.00500 C/m. If the rod is 1 mm long, how many electrons have been removed from the glass rod? Example 16-1

33. • How many electrons must be transferred from an object to produce a charge of 1.60 C? Example 16-1

16-3 Charge can flow freely in a conductor but not in an insulator

34. • Suppose 2.00 C of positive charge is distributed evenly throughout a solid sphere of radius 1.27 cm. (a) What is the charge per unit volume for this situation? (b) Is the sphere insulating or conducting? How do you know?

35. • The maximum amount of charge that can be collected on a Van de Graaff generator's conducting sphere (30-cm diameter) is about 30 μC. Calculate the surface charge density, σ, of the sphere in C/m^2.

36. • **Biology** Most workers in nanotechnology are actively monitored for excess static charge buildup. The human body acts like an insulator as one walks across a carpet, collecting -50 nC per step. (a) What charge buildup will a worker in a manufacturing plant accumulate if she walks 25 steps? (b) How many electrons are present in that amount of charge? (c) If a delicate manufacturing process can be damaged by an electrical discharge of greater than 10^{12} electrons, what is the

maximum number of steps that any worker should be allowed to take before touching the components?

16-4 Coulomb's law describes the force between charged objects

37. • Two point charges are separated by a distance of 20.0 cm. The numerical value of one charge is twice that of the other. If each charge exerts a force of magnitude 45.0 N on the other, find the magnitude of the charges. Example 16-2

38. • The mass of an electron is 9.11×10^{-31} kg. How far apart would two electrons have to be for the electric force exerted by each on the other to be equal to the weight of an electron? Example 16-2

39. • Charge A, $+5.00$ μC, is positioned at the origin of a coordinate system. Charge B, -3.00 μC, is fixed on the x axis at $x = 3.00$ m. (a) Determine the magnitude and direction of the force that charge B exerts on charge A. (b) What is the magnitude and direction of the force that charge A exerts on charge B? Example 16-2

40. • Point charge A with charge $q_A = +3.00$ μC is located at the origin. Point charge B with charge $q_B = -4.00$ μC is on the x axis at $x = 3.00$ m. Point charge C with charge $q_C = -2.00$ μC is on the x axis at $x = 6.00$ m. And point charge D with charge $q_D = +6.00$ μC is on the x axis at $x = 8.00$ m. What is the net electric force on point charge A due to the other three charges? Example 16-3

41. • A charge of $+3.00$ μC is located at the origin, and a second charge of -2.00 μC is located on the $x-y$ plane at the point (30.0 cm, 20.0 cm). Determine the electric force (magnitude and direction) exerted by the -2.00 μC charge on the 3.00 μC charge. Example 16-4

42. • A uranium ion and an iron ion are separated by a distance of $R = 40$ nm, as shown in **Figure 16-27**. The uranium atom is singly ionized; the iron atom is doubly ionized. (a) Calculate the distance r from the uranium atom at which an electron will be in equilibrium. Ignore the gravitational attraction between the particles. (b) What is the magnitude of the force on the electron due to the uranium ion? Example 16-3

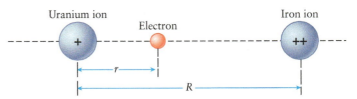

Figure 16-27 Problem 42

43. • Two charges lie on the x axis, a -2.00-μC charge at the origin and a $+3.00$-μC charge at $x = 0.100$ m. At what position along the x axis, if any, should a third $+4.00$-μC charge be placed so that the net force on the third charge is equal to zero? Example 16-3

44. •• Point charge A with a charge of $+3.00$ μC is located at the origin. Point charge B with a charge of $+6.00$ μC is located on the x axis at $x = 7.00$ cm. And point charge C with a charge of $+2.00$ μC is located on the y axis at $y = 6.00$ cm. What is the net force (magnitude and direction) exerted on each charge by the others? Example 16-4

45. • A charge q_1 equal to 0.600 μC is at the origin, and a second charge q_2 equal to 0.800 μC is on the x axis at 5.00 cm. (a) Find the force (magnitude and direction) that each charge exerts on the other. (b) How would your answer change if q_2 were –0.800 μC? Example 16-2

46. •• A charge Q_1 = +7.00 μC is located on the y axis at y = 4.00 cm. A second charge Q_2 = –4.00 μC is located on the x–y plane at the point (–3.00 cm, 4.00 cm). And a third charge Q_3 = +6.00 μC is located on the x axis at x = –3.00 cm. Determine the x and y components of the net electric force that acts on the third charge Q_3. Example 16-4

16-5 The concept of electric field helps us visualize how charges exert forces at a distance

47. • At point P in Figure 16-28 the electric field is zero. What can you conclude about (a) the signs of q_1 and q_2 and (b) the magnitudes of q_1 and q_2? Example 16-6

Figure 16-28 Problem 47

48. • Near the surface of Earth an electric field points radially downward and has a magnitude of approximately 100 N/C. What charge (magnitude and sign) would have to be placed on a penny that has a mass of 3.11 g to cause it to rise into the air with an upward acceleration of 0.190 m/s²? Example 16-5

49. • Two charges are placed on the x axis, +5.00 μC at the origin and –10.0 μC at x = 10.0 cm. (a) Find the electric field on the x axis at x = 6.00 cm. (b) At what point(s) on the x axis is the electric field zero? Example 16-6

50. • In Figure 16-29 the electric field at the origin is zero. If q_1 is 1.00×10^{-7} C, what is q_2? Example 16-6

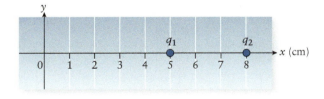

Figure 16-29 Problems 50, 51, and 52

51. • In Figure 16-29 if $q_1 = 1.00 \times 10^{-7}$ C and $q_2 = 2.00 \times 10^{-7}$ C, (a) what is the electric field $\bar{E}$ at the point (x, y) = (0.00 cm, 3.00 cm)? (b) What is the force $\bar{F}$ acting on an electron at that position? Example 16-7

52. • In Figure 16-29 if $q_1 = 1.00 \times 10^{-7}$ C and $q_2 = 2.00 \times 10^{-7}$ C, (a) what is the electric field $\bar{E}$ at the point (x, y) = (6.00 m, 3.00 m)? (b) What is the force $\bar{F}$ acting on a proton at that position? Example 16-7

53. •• In the Bohr model the hydrogen atom consists of an electron in a circular orbit of radius $a_0 = 5.29 \times 10^{-11}$ m around the nucleus. Using this model, and ignoring relativistic effects, what is the speed of the electron? Example 16-5

54. • **Biology** The cockroach *Periplaneta americana* can detect a static electric field of 8.0 kN/C using its long antennae. If we model the excess static charge on a cockroach as a point charge located at the end of each antenna, what magnitude of charge would each antenna possess in order for each antennae to experience a force of 3.0 μN from the external electric field? Example 16-5

16-6 Gauss's law gives us more insight into the electric field

55. • A rectangular area is rotated in a uniform electric field, from a position where the maximum electric flux goes through it to an orientation where only half the maximum flux goes through it. What is the angle of rotation?

56. • A point charge of 4.00×10^{-12} C is located at the center of a cubical Gaussian surface. What is the electric flux through each face of the cube?

57. • The net electric flux through a cubic box with sides that are 20.0 cm long is 4.80×10^3 N·m²/C. What charge is enclosed by the box?

58. •• A 10.0-cm-long uniformly charged plastic rod is sealed inside a plastic bag. The net electric flux through the bag is 7.50×10^5 N·m²/C. What is the linear charge density (charge per unit length) on the rod?

59. •• Figure 16-30 shows a prism-shaped object that is 40.0 cm high, 30.0 cm deep, and 80.0 cm long. The prism is immersed in a uniform electric field of 500 N/C directed parallel to the x axis. (a) Calculate the electric flux out of each of its five faces and (b) the net electric flux out of the entire closed surface. (c) If in addition to the given electric field the prism also enclosed a point charge of –2.00 μC, how would your answers above change qualitatively, if at all?

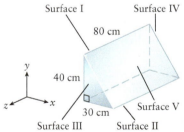

Figure 16-30 Problem 59

16-7 In certain situations Gauss's law helps us calculate the electric field and determine how charge is distributed

60. •• Use Gauss's law to find an expression for the electric field just outside the surface of a sphere carrying a uniform surface charge density σ (charge per unit area). Example 16-8

61. • Determine the charge density for each of the following cases (assume that all densities are uniform): (a) a solid cylinder that has a length L, has a radius R, and carries a charge Q throughout its volume; (b) a flat plate (very thin) that has a width W, has a length L, and carries a charge Q on its surface area; (c) a solid sphere of radius R carrying a charge Q throughout its volume; and (d) a hollow sphere of radius R carrying a charge Q over its surface area.

62. •• An electric field of magnitude 4.00×10^2 N/C and pointing radially outward exists at all points just outside the surface of a 2.00-cm-diameter steel ball bearing. Assuming the ball bearing is in electrostatic equilibrium, (a) what is the total charge on the ball? (b) What is the surface charge density on the ball? Example 16-8

63. •• Consider an infinite plane with a uniform charge distribution σ. (a) Use Gauss's law to find an expression for the electric field due to the plane. (b) What field would be created by two equal but oppositely charged parallel planes? Consider the region between the planes as well as the two regions outside the planes. The simplest way to express your answers is in terms of the surface charge density σ (the charge per unit area) on the plane.

64. •• A –3.20-µC charge sits in static equilibrium in the center of a conducting spherical shell that has an inner radius of 2.50 cm and an outer radius of 3.50 cm. The shell has a net charge of –5.80 µC. Determine the charge on each surface of the shell and the electric field just outside the shell. Example 16-8

General Problems

65. • The magnitude of the repulsive force (F) between two +2.5-µC charges as a function of the distance of separation (r) is listed in the following table. Empirically derive a relationship between F and r by doing a curve fit to find the power n of r in the following formula:

$$F = \frac{kq_1q_2}{r^n}$$

Use a graphing calculator or spreadsheet.

r (m)	F (N)	r (m)	F (N)
0.003	5500	0.080	6
0.004	3000	0.100	5
0.005	2000	0.200	1
0.010	600	0.300	0.5
0.020	175	0.400	0.35
0.040	30	0.500	0.25
0.050	10	0.600	0.15

66. •• A spherical party balloon on Earth's surface that is 25 cm in diameter contains helium at room temperature (20°C) and at a pressure of 1.3 atm. If one electron could be stripped from every helium atom in the balloon and removed to a satellite orbiting Earth 22,000 mi (32,187 km) above the planet, with what force would the balloon and the satellite attract each other when the satellite is directly above the balloon? Example 16-2

67. •• A plutonium-242 atom has a nucleus of 94 protons and 148 neutrons and has 94 electrons. The diameter of its nucleus is approximately 15×10^{-15} m. (a) Make a reasonable physical argument as to why we can treat the nucleus as a point charge for points outside of it. (b) Plutonium decays radioactively by emitting an alpha particle from its nucleus. The mass of the alpha particle is 6.6×10^{-27} kg, and the particle has two protons and two neutrons. If the alpha particle comes from the surface of the ^{242}Pu nucleus, what is its greatest acceleration? Example 16-2

68. • When a test charge of +5.00 nC is placed at a certain point, the force that acts on it has a magnitude of 0.0800 N and is directed northeast. (a) If the test charge were –2.00 nC instead, what force would act on it? (b) What is the electric field at the point in question? Example 16-5

69. •• **Biology** A red blood cell may carry an excess charge of about -2.5×10^{-12} C distributed uniformly over its surface. The cells, modeled as spheres, are approximately 7.5 µm in diameter and have a mass of 9.0×10^{-14} kg. (a) How many excess electrons does a typical red blood cell carry? (b) What is the surface charge density σ on the red blood cell? Express your answer in C/m² and in electrons/m². Example 16-1

70. •• Three point charges are placed on the x–y plane: a +50.0-nC charge at the origin, a –50.0-nC charge on the x axis at 10.0 cm, and a +150-nC charge at the point (10.0 cm, 8.00 cm). (a) Find the total electric force on the +150-nC charge due to the other two. (b) What is the electric field at the location of the +150-nC charge due to the presence of the other two charges? Example 16-7

71. •• Two small spheres each have a mass m of 0.100 g and are suspended as pendulums by light insulating strings from a common point, as shown in Figure 16-31. The spheres are given the same electric charge, and the two come to equilibrium when each string is at an angle of $\theta = 3.00°$ with the vertical. If each string is 1.00 m long, what is the magnitude of the charge on each sphere? Example 16-2

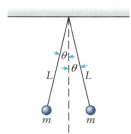

Figure 16-31 Problem 71

72. •• A small 1.00-g plastic ball that has charge $q = 1.00$ C is suspended in a uniform electric field by a string that has length $L = 1.00$ m in a uniform electric field, as shown in Figure 16-32. If the ball is in equilibrium when the string makes a 9.80° angle with the vertical as indicated by θ, what is the electric field strength? Example 16-5

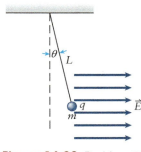

Figure 16-32 Problem 72

73. • **Biology** The 9-inch-long elephant nose fish in the Congo River generates a weak electric field around its body using an organ in its tail. When small prey (or even potential mates) swim within a few feet of the fish, they perturb the electric field. The change in the field is picked up by electric sensor cells in the skin of the elephant nose. These remarkable fish can detect changes in the electric field as small as 3.0 µN/C. (a) How much charge (modeled as a point charge) in the approaching prey would be needed to produce such a change in the electric field at a distance of 75 cm? (b) How many electrons would be required to create the charge? Example 16-6

74. ••• Three charges (q_A, q_B, and q_C) are placed at the vertices of the equilateral triangle that has sides of length s in Figure 16-33. Derive expressions for the electric field at (a) X (at the center of the triangle), (b) Y (at the midpoint of the side between q_B and q_C), and (c) Z (at the midpoint of the side between q_A and q_C). (d) Now use the

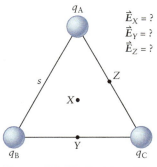

Figure 16-33 Problem 74

following numerical values and calculate the electric field at those same points: $s = 10.0$ cm, $q_A = +20.0$ nC, $q_B = -8.00$ nC, and $q_C = -10.0$ nC. Example 16-7

75. •• Calculate the x and y components of the electric field at the center of the hexagon shown in **Figure 16-34**. Assume the sides of the hexagon are all 5.00 cm long. Example 16-7

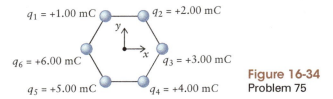

$q_1 = +1.00$ mC

$q_2 = +2.00$ mC

$q_6 = +6.00$ mC

$q_3 = +3.00$ mC

$q_5 = +5.00$ mC

$q_4 = +4.00$ mC

Figure 16-34
Problem 75

76. •• Electric fields up to 2.00×10^5 N/C have been measured inside of clouds during electrical storms. Neglect the drag force due to the air in the cloud and any collisions with air molecules. (a) What acceleration does the maximum electric field produce for protons in the cloud? Express your answer in SI units and as a fraction of g. (b) If the electric field remains constant, how far will the proton have to travel to reach 10% of the speed of light $(3.00 \times 10^8$ m/s) if it started with negligible speed? (c) Can you neglect the effects of gravity? Explain your answer. Example 16-5

77. •• An electron with an initial speed of 5.00×10^5 m/s enters a region in which there is an electric field directed along its direction of motion. If the electron travels 5.00 cm in the field before being stopped, what are the magnitude and direction of the electric field? Example 16-5

78. •• An electron, released in a region where the electric field is uniform, is observed to have an acceleration of 3.00×10^{14} m/s² in the positive x direction. (a) Determine the electric field producing the acceleration. (b) Assuming the electron is released from rest, determine the time required for it to reach a speed of 11,200 m/s, the escape speed from Earth's surface. Example 16-5

79. •• **Chemistry** The iron atom (Fe) has 26 protons, 30 neutrons, and 26 electrons. The diameter of the atom is approximately 1.0×10^{-10} m, while the diameter of its nucleus is about 9.2×10^{-15} m. (You can reasonably model the nucleus as a uniform sphere of charge.) What are the magnitude and direction of the electric field that the nucleus produces (a) just outside the surface of the nucleus and (b) at the distance of the outermost electron? (c) What would be the magnitude and direction of the acceleration of the outermost electron due only to the nucleus, neglecting any force due to the other electrons? Example 16-5

80. ••• An electron with kinetic energy K is traveling along the $+x$ axis, which is along the axis of a cathode-ray tube as shown in **Figure 16-35**. There is an electric field $E = 12.00 \times 10^4$ N/C pointed in the $+y$ direction between the deflection plates, which are 0.0600 m long and are separated by 0.0200 m. Determine the minimum initial kinetic energy the electron can have and still avoid colliding with one of the plates. Example 16-5

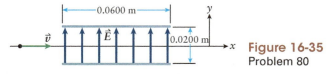

0.0600 m

$\vec{v}$

$\vec{E}$

0.0200 m

Figure 16-35
Problem 80

81. •• In the famous Millikan oil-drop experiment, tiny spherical droplets of oil are sprayed into a uniform vertical electric field. The drops get a very small charge (just a few electrons) due to friction with the atomizer as they are sprayed. The field is adjusted until the drop (which is viewed through a small telescope) is just balanced against gravity and therefore remains stationary. Using the measured value of the electric field, we can calculate the charge on the drop and from this calculate the charge e of the electron. In one apparatus the drops are 1.10 µm in diameter, and the oil has a density of 0.850 g/cm³. (a) If the drops are negatively charged, which way should the electric field point to hold them stationary (up or down)? (b) Why? (c) If a certain drop contains four excess electrons, what magnitude electric field is needed to hold it stationary? (d) You measure a balancing field of 5183 N/C for another drop. How many excess electrons are on this drop? Example 16-5

82. •• **Biology** There is a naturally occurring vertical electric field near Earth's surface, pointing toward the ground. In fair weather conditions and in an open field, the strength of the electric field is 1.00×10^2 N/C. A spherical pollen grain, with a radius of 12 µm, is released from its parent plant by a light breeze, giving it a net charge of -0.80 fC (where 1 fC $= 1 \times 10^{-15}$ C). What is the ratio of the magnitudes of the electric force to the gravitational force, $F_{electric}/F_{gravitational}$, acting on the pollen? Assume the volume mass density of the pollen is the same as water, 1.00×10^3 kg/m³, which is a primary constituent of pollen. Example 16-5

83. •• Two hollow, concentric, spherical shells are covered with charge (**Figure 16-36**). The inner sphere has a radius R_i and a surface charge density of $+\sigma_i$, while the outer sphere has a radius R_o and a surface charge density of $-\sigma_o$. Derive an expression for the electric field in the following three radial regions: (a) $r < R_i$, (b) $R_i < r < R_o$, and (c) $r > R_o$. Example 16-8

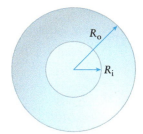

R_o

R_i

Figure 16-36 Problem 83

84. •• A charge $q_1 = +2q$ is at the origin, and a charge $q_2 = -q$ is on the x axis at $x = a$. Find expressions for the total electric field on the x axis in each of the regions (a) $x < 0$; (b) $0 < x < a$; and (c) $a < x$. (d) Determine all points on the x axis where the electric field is zero. (e) Use a graphing calculator or spreadsheet to make a plot of E_x versus x for all points on the x axis, and (f) qualitatively discuss what happens for $-\infty < x < \infty$. Example 16-6

85. • Two point charges are enclosed by a spherical conducting shell that has inner and outer radii of 13.0 and 15.2 cm, respectively. One point charge has a charge of $q_1 = 8.30$ µC, while the second point charge has an unknown charge q_2. The conducting shell is known to have a net electric charge of -2.40 µC, but measurements find that the charge on the outer surface of the shell is $+3.70$ µC. Determine the charge q_2 of the second point charge. Example 16-8

86. • **Biology** The magnitude of the electric field across the membrane of mitochondria is about 3.0×10^7 N/C. What is the magnitude of the resulting electrostatic force on a singly ionized molecule as it transports across the membrane? Example 16-5

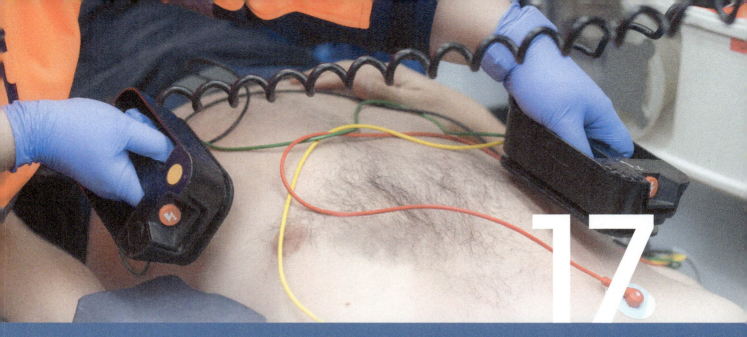

DAVID HERRAEZ/Alamy

Electrostatics II: Electric Potential Energy and Electric Potential

In this chapter, your goals are to:

- (17-1) Explain the significance of energy in electrostatics.
- (17-2) Discuss how the work done on a charged particle by the electric field relates to changes in electric potential energy.
- (17-3) Explain the difference between electric potential and electric potential energy.
- (17-4) Recognize why equipotential surfaces are perpendicular to electric field lines.
- (17-5) Explain what is meant by capacitance and describe how the capacitance of a parallel-plate capacitor depends on the size of the plates and their separation.
- (17-6) Calculate the electric energy stored in a capacitor.
- (17-7) Explain how to treat capacitors attached in series and in parallel as a single equivalent capacitance.
- (17-8) Describe how the capacitance of a capacitor increases when an insulating material other than a vacuum is placed between the plates of the capacitor.

To master this chapter, you should review:

- (6-2)   The work done by a force on a moving object.
- (6-6)   The relationship between a conservative force and the potential energy associated with that force.
- (7-3)   The generalized expression for gravitational potential energy.
- (16-5, 16-7) The electric field due to a point charge and to a sheet of charge.

What do you think?

A cardiac defibrillator works by delivering an electric shock to a malfunctioning heart to jump-start it back into its normal rhythm. Electric energy is stored for this purpose in the defibrillator by separating positive charge $+q$ from an equal amount of negative charge $-q$. To release this energy to the heart, conducting paddles are placed on the patient's chest, and charge $-q$ in the form of electrons is allowed to flow through the paddles and patient until it reaches the stored charge $+q$ within the defibrillator. If the value q of the charge stored in a defibrillator is doubled, by what factor will the stored energy increase? (a) $\sqrt{2}$; (b) 2; (c) 4; (d) 8; (e) 16.

17-1 Electric energy is important in nature, technology, and biological systems

To many, *electricity* implies *energy*. The energy to run your computer, lighting, and television is delivered to your home in the form of electricity. A bolt of lightning releases so much energy that it heats air to a temperature of 30,000°C, causing the air to glow with the characteristic light that we call a lightning flash (see Figure 16-1b).

Figure 17-1 Electric energy
Reservoirs of electric energy in (a) technology and (b) nature.

(a)

An ordinary battery is a storehouse of electric energy. Devices that use multiple batteries have greater energy requirements.

iStockphoto/Thinkstock

(b)

An electric ray (order *Torpediniformes*) is equipped with a large number of organic batteries that work together to deliver intense electric shocks.

A defibrillator saves lives by delivering a sharp punch of electric energy to a malfunctioning heart (see the photo that opens this chapter). When you purchase an ordinary electric battery, you are really paying for the electric energy that the battery can deliver to whatever electric circuit you plug it into (Figure 17-1a). Some specialized species of fish such as electric rays (Figure 17-1b) are equipped with organic batteries; they can deliver an intense burst of electric energy to stun their prey or ward off predators.

In this chapter we'll make clear what is meant by *electric potential energy*. We'll gain insight into this new kind of energy by considering the similarities to and differences from gravitational potential energy. Just as there is a change in gravitational potential energy when a massive object moves up or down in the presence of Earth's gravitational field, there is a change in electric potential energy when a charged object moves along with or opposite to an electric field. We'll go on to introduce the useful concepts of *electric potential*, or electric potential energy per unit charge, and *voltage*, which is the difference in the value of electric potential at two positions. We'll also learn about *equipotential surfaces*, which are surfaces on which the electric potential has the same value at every point.

An important device for storing electric energy is a *capacitor*. In its simplest form a capacitor is just two pieces of metal, called capacitor plates, placed close to each other but not in contact. We'll examine the key properties of capacitors, including how to combine two or more of them for even greater energy storage. We'll conclude the chapter by seeing how capacitors can be made even more effective by inserting an insulating material between the plates. In Chapter 18 we'll learn how batteries like those shown in Figure 17-1a provide electric energy to the components of an electric circuit.

TAKE-HOME MESSAGE FOR Section 17-1

✔ Electric potential energy associated with a point charge is analogous to the gravitational potential energy associated with a massive object.

✔ Electric potential is electric potential energy per charge.

✔ A capacitor is a device for storing electric energy.

17-2 Electric potential energy changes when a charge moves in an electric field

We learned in Section 6-6 that Earth's gravitational force is a *conservative* force: As an object moves from one position to another, the work done on the object by the gravitational force depends only on where the object starts and where it ends up, not on how it gets there. Hence we can express the work done by gravity W_{grav} in terms of a change in the gravitational potential energy U_{grav} of the system of Earth and object:

(6-20)

$$W_{grav} = -\Delta U_{grav}$$

(a) Work done by gravity and change in gravitational potential energy when a baseball undergoes a displacement

(b) Three special cases

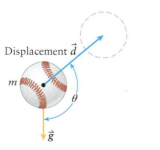

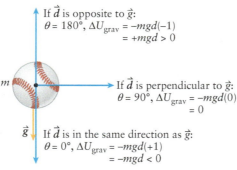

If $\vec{d}$ is opposite to $\vec{g}$:
$\theta = 180°$, $\Delta U_{grav} = -mgd(-1)$
$= +mgd > 0$

If $\vec{d}$ is perpendicular to $\vec{g}$:
$\theta = 90°$, $\Delta U_{grav} = -mgd(0)$
$= 0$

If $\vec{d}$ is in the same direction as $\vec{g}$:
$\theta = 0°$, $\Delta U_{grav} = -mgd(+1)$
$= -mgd < 0$

Displacement $\vec{d}$

m

θ

$\vec{g}$

Work done by gravity:
$W_{grav} = mgd \cos \theta$

Change in gravitational potential energy:
$\Delta U_{grav} = -W_{grav} = -mgd \cos \theta$

Figure 17-2 Change in gravitational potential energy When an object of mass m moves in the presence of gravity, the gravitational force can do work on it and there can be a change in gravitational potential energy.

The minus sign means that if the object moves downward so that gravity does positive work (the gravitational force and the object's displacement are in the same direction), there is a negative change in gravitational potential energy (U_{grav} decreases). If the object moves upward so that gravity does negative work (force and displacement are in opposite directions), there is a positive change in gravitational potential energy (U_{grav} increases).

To be specific, suppose an object of mass m moves near Earth's surface, so the acceleration $\vec{g}$ due to Earth's gravity and the gravitational force $\vec{F} = m\vec{g}$ on the object are the same at all positions. The object's displacement is $\vec{d}$, and that displacement is at an angle θ to the direction of the force $\vec{F} = m\vec{g}$ and so at the same angle θ to the direction of $\vec{g}$ (Figure 17-2a). Then the work done by gravity is $W = Fd \cos \theta$ (Equation 6-2), and the change in gravitational potential energy is

$$\Delta U_{grav} = -W_{grav} = -mgd \cos \theta \qquad (17\text{-}1)$$

Figure 17-2b shows three special cases. If the object moves a distance d straight upward, $\theta = 180°$ and $\cos \theta = \cos 180° = -1$. In this case the downward gravitational force does negative work on the object, and the gravitational potential energy increases: $\Delta U_{grav} = -mgd(-1) = +mgd$. If the object moves a distance d straight downward, in the direction of the gravitational force, $\theta = 0$ and $\cos \theta = \cos 0° = 1$. Then gravity does positive work on the object and the gravitational potential energy decreases: $\Delta U_{grav} = -mgd(+1) = -mgd$. And if the object moves horizontally, gravity does zero work and there is no change in gravitational potential energy: $\theta = 90°$ and $\Delta U_{grav} = -mgd \cos 90° = 0$. These are the same results that we found in Section 6-6.

Like gravity, the electric force due to the interaction between charged objects is a conservative force. You can see this easily for the special case in which an object moves in a region of space where there is a *uniform* electric field $\vec{E}$—that is, where the value of $\vec{E}$ is the same at all positions (Figure 17-3). The electric force on an object of charge q is the same everywhere in this region and equal to $\vec{F} = q\vec{E}$. That's exactly like the situation in Figure 17-2: In a region where $\vec{g}$ is the same at all positions, the gravitational force on an object of mass m is the same everywhere and equal to $\vec{F} = m\vec{g}$. This uniform gravitational force is conservative, so the force $\vec{F} = q\vec{E}$ due to a uniform electric field must be conservative as well. Hence we can express the work done by the electric force in terms of a change in **electric potential energy**.

Electric Potential Energy in a Uniform Field

This analogy between gravitational force and electric force tells us how to write the change in electric potential energy for an object of charge q that undergoes a displacement $\vec{d}$ in the presence of a uniform electric field $\vec{E}$ (Figure 17-3a). Following the same steps that we used to find Equation 17-1 above, you can see that if θ is the angle between the directions of $\vec{d}$ and $\vec{E}$, then the work done on the charge by the electric

√x̄ *See the Math Tutorial for more information on trigonometry.*

 Go to Picture It 17-1 for more practice dealing with electric potential energy.

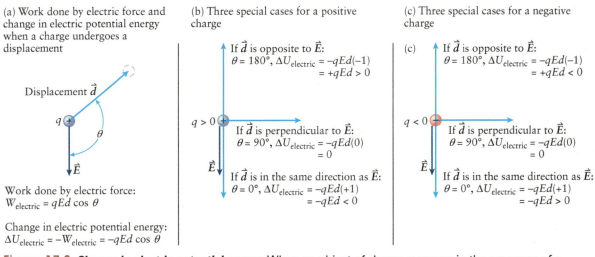

(a) Work done by electric force and change in electric potential energy when a charge undergoes a displacement

Displacement $\vec{d}$

q +

θ

$\vec{E}$

Work done by electric force:
$W_{electric} = qEd \cos \theta$

Change in electric potential energy:
$\Delta U_{electric} = -W_{electric} = -qEd \cos \theta$

(b) Three special cases for a positive charge

$q > 0$ +

$\vec{E}$

If $\vec{d}$ is opposite to $\vec{E}$:
$\theta = 180°$, $\Delta U_{electric} = -qEd(-1)$
$= +qEd > 0$

If $\vec{d}$ is perpendicular to $\vec{E}$:
$\theta = 90°$, $\Delta U_{electric} = -qEd(0)$
$= 0$

If $\vec{d}$ is in the same direction as $\vec{E}$:
$\theta = 0°$, $\Delta U_{electric} = -qEd(+1)$
$= -qEd < 0$

(c) Three special cases for a negative charge

(c)

$q < 0$ −

$\vec{E}$

If $\vec{d}$ is opposite to $\vec{E}$:
$\theta = 180°$, $\Delta U_{electric} = -qEd(-1)$
$= +qEd < 0$

If $\vec{d}$ is perpendicular to $\vec{E}$:
$\theta = 90°$, $\Delta U_{electric} = -qEd(0)$
$= 0$

If $\vec{d}$ is in the same direction as $\vec{E}$:
$\theta = 0°$, $\Delta U_{electric} = -qEd(+1)$
$= -qEd > 0$

Figure 17-3 Change in electric potential energy When an object of charge q moves in the presence of a uniform electric field $\vec{E}$, the electric force can do work on it and there can be a change in electric potential energy (compare Figure 17-2).

force $\vec{F} = q\vec{E}$ is $W_{electric} = qEd \cos \theta$. The change in electric potential energy equals the negative of the electric work done on the charge:

Electric potential energy change for a charge in a uniform electric field

(17-2)

The change in electric potential energy for an object of charge q that moves in a uniform electric field $\vec{E}$...

...equals the negative of the work done on the object by the electric force.

$$\Delta U_{electric} = -W_{electric}$$

$$= -qEd \cos \theta$$

Angle between the displacement and the direction of the electric field

Charge of the object

Magnitude of the electric field

Straight-line displacement of the object

Just as for the gravitational case, Equation 17-2 tells us how much the electric potential energy *changes* when the charge is displaced as shown in Figure 17-3a. Note also that the electric potential energy is a shared property of the charge q and the other charges that produce the electric field $\vec{E}$.

Suppose q is positive, so the force $\vec{F} = q\vec{E}$ on the charged object is in the *same* direction as the uniform electric field $\vec{E}$ (Figure 17-3b). If this positive charge moves a distance d opposite to $\vec{E}$, then $\theta = 180°$ and $\cos \theta = -1$. In this case the electric force does negative work on the charge, $\Delta U_{electric} = (-qEd)(-1) = +qEd$ is positive, and the electric potential energy of the system increases. That's exactly what happens to *gravitational* potential energy when a massive object moves upward from Earth's surface, in the direction opposite to the gravitational force (Figure 17-2b). If the positive charge q instead moves in the same direction as the electric field, and so in the same direction as the electric force, then $\theta = 0$ and $\cos \theta = +1$. Then the electric force does positive work on the object, so $\Delta U_{electric} = (-qEd)(+1) = -qEd$ is negative and the electric potential energy of the system decreases. That's the same thing that happens when a massive object falls toward Earth's surface, in the same direction as the gravitational force on the object: Gravitational potential energy decreases (Figure 17-2b).

For an object with *negative* charge ($q < 0$), the force $\vec{F} = q\vec{E}$ on the object is in the direction *opposite* to the electric field $\vec{E}$ (Figure 17-3c). If such a negative charge moves a distance d opposite to the direction of $\vec{E}$ so that $\theta = 180°$ and $\cos \theta = -1$ in Equation 17-2, the electric force does *positive* work on the charge because the force and displacement are in the same direction. Then, because $q < 0$, $\Delta U_{electric} = +qEd$ is *negative*, and the electric potential energy of the system decreases. In the same way, if a negative charge moves a distance d in the direction of $\vec{E}$ so that $\theta = 0$ and $\cos \theta = +1$ in

Equation 17-2, the electric force does *negative* work on the object (the electric force is opposite to E and so opposite the displacement). The electric potential energy change in this case is $\Delta U_{\text{electric}} = (-qEd)(+1) = -qEd$, which is positive because q is negative. So the electric potential energy of the system increases. These results are exactly opposite to what happens for a positive charge that moves a distance d in the direction of the electric field.

EXAMPLE 17-1 Electric Potential Energy Difference in a Uniform Field

An electron (charge $q = -e = -1.60 \times 10^{-19}$ C, mass $m = 9.11 \times 10^{-31}$ kg) is released from rest in a uniform electric field. The field points in the positive x direction and has magnitude 2.00×10^{2} N/C. Find the speed of the electron after it has moved 0.300 m.

Set Up

The only force acting on the electron of charge q is the conservative electric force. Since only this conservative force does work on the electron, mechanical energy is conserved (Equation 6-27). We'll find the change in electric potential energy using Equation 17-2 and from this find the change in kinetic energy of the electron. The electron's initial kinetic energy is zero because it starts at rest; once we know the final kinetic energy, we can determine the electron's final speed.

The electron has a negative charge, so the force $\vec{F} = q\vec{E}$ is directed opposite to the electric field $\vec{E}$, and the electron will move in the direction opposite to $\vec{E}$ when released from rest. So the angle in Equation 17-2 between electric field and displacement will be $\theta = 180°$.

Conservation of mechanical energy:

$$K_i + U_{\text{electric,i}} = K_f + U_{\text{electric,f}} \quad (6\text{-}27)$$

Electric potential energy change for a charge in a uniform electric field:

$$\Delta U_{\text{electric}} = -qEd \cos \theta \quad (17\text{-}2)$$

Kinetic energy:

$$K = \frac{1}{2}mv^2 \quad (6\text{-}8)$$

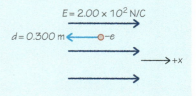

Solve

Use Equation 17-2 to solve for the change in electric potential energy.

The change in electric potential energy equals the difference between the final and initial electric potential energy:

$$\Delta U_{\text{electric}} = U_{\text{electric,f}} - U_{\text{electric,i}}$$

From Equation 17-2,

$$\begin{aligned}\Delta U_{\text{electric}} &= U_{\text{electric,f}} - U_{\text{electric,i}} \\ &= -qEd \cos \theta \\ &= -(-1.60 \times 10^{-19} \text{ C})(2.00 \times 10^{2} \text{ N/C})(0.300 \text{ m}) \cos 180° \\ &= -(-9.60 \times 10^{-18} \text{ N} \cdot \text{m})(-1) \\ &= -9.60 \times 10^{-18} \text{ N} \cdot \text{m} = -9.60 \times 10^{-18} \text{ J}\end{aligned}$$

Find the final kinetic energy and the final speed of the electron.

The initial kinetic energy of the electron is $K_i = 0$. Solve the energy conservation equation for the final kinetic energy of the electron:

$$\begin{aligned}K_f &= K_i + U_{\text{electric,i}} - U_{\text{electric,f}} = 0 - (U_{\text{electric,f}} - U_{\text{electric,i}}) \\ &= 0 - (-9.60 \times 10^{-18} \text{ J}) = 9.60 \times 10^{-18} \text{ J}\end{aligned}$$

Finally, solve for the final speed of the electron:

$$K_f = \frac{1}{2}mv_f^2 \text{ so}$$

$$v_f = \sqrt{\frac{2K_f}{m}} = \sqrt{\frac{2(9.60 \times 10^{-18} \text{ J})}{9.11 \times 10^{-31} \text{ kg}}} = 4.59 \times 10^{6} \text{ m/s}$$

Reflect

The electric potential energy decreases when we release the electron in the electric field, in the same way that the gravitational potential energy decreases when you release a ball in the presence of Earth's gravity. In both situations, the kinetic energy increases by the same amount that the potential energy decreases.

You should verify that the force on the electron is very small, only 3.2×10^{-17} N in magnitude. The electron has only a tiny mass, however, and our results show that it acquires a ferocious speed (about 1.5% of the speed of light) after moving just 0.300 m.

Electric Potential Energy of Point Charges

Electric potential energy is particularly important for the case of two *point charges* interacting with each other. An example from chemistry is the dissociation energy of an ionic compound such as sodium chloride (NaCl)—that is, the energy that must be given to the molecule to break it into its component atoms. The dissociation energy is determined in large part by the change in electric potential energy required to separate the positive sodium ion (Na^+) and the negative chloride ion (Cl^-), both of which behave much like point charges, as well as by the change in electric potential energy required to move an electron from the Cl^- ion to the Na^+ ion to make the atoms neutral.

Just as for a charge moving in a uniform electric field, we can use our knowledge of gravitation to find an expression for the electric potential energy of two point charges q_1 and q_2 separated by a distance r. In Section 7-2 we saw the following expression for the attractive *gravitational* force between two objects with *masses* m_1 and m_2 separated by a distance r:

Newton's law of universal gravitation
(7-3)

Gravitational constant (same for any two objects)   Masses of the two objects

Any two objects (1 and 2) exert equally strong **gravitational forces** on each other.

$$F_{1 \text{ on } 2} = F_{2 \text{ on } 1} = \frac{Gm_1m_2}{r^2}$$

Center-to-center distance between the two objects

The gravitational forces are attractive: $\vec{F}_{1 \text{ on } 2}$ pulls object 2 toward object 1 and $\vec{F}_{2 \text{ on } 1}$ pulls object 1 toward object 2.

The gravitational potential energy of these two objects is given by Equation 7-5 in Section 7-3:

Gravitational potential energy
(7-5)

Gravitational constant (same for any two objects)   Masses of the two objects

Gravitational potential energy of a system of two objects (1 and 2)

$$U_{\text{grav}} = -\frac{Gm_1m_2}{r}$$

Center-to-center distance between the two objects

The gravitational potential energy is zero when the two objects are infinitely far apart. If the objects are brought closer together (so r is made smaller), U_{grav} decreases (it becomes more negative).

Figure 17-4a graphs the gravitational potential energy given by Equation 7-5. We choose the point at which potential energy U_{grav} is zero to be where the two objects are infinitely far apart, so $r \to \infty$. The gravitational potential energy is negative for any finite value of r and increases—that is, becomes less negative—as the objects move farther apart. That's because the change in gravitational potential energy is the negative of the work done by the gravitational force: $\Delta U_{\text{grav}} = -W_{\text{grav}}$ (see Equation 6-20 at the beginning of this section). If we hold object m_1 stationary and move object m_2 farther away, increasing the distance r, the attractive gravitational force does negative work. Then $W_{\text{grav}} < 0$, so $\Delta U_{\text{grav}} > 0$ and the gravitational potential energy increases, as Figure 17-4a shows.

(a) Gravitational potential energy

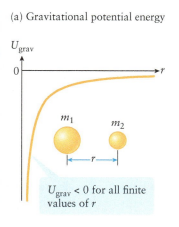

$U_{grav} < 0$ for all finite values of r

(b) Electric potential energy, charges of opposite sign

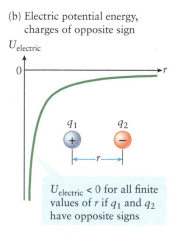

$U_{electric} < 0$ for all finite values of r if q_1 and q_2 have opposite signs

(c) Electric potential energy, charges of the same sign

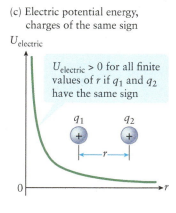

$U_{electric} > 0$ for all finite values of r if q_1 and q_2 have the same sign

Figure 17-4 Potential energy for two masses and for two charges The electric potential energy $U_{electric}$ for two point charges is similar to the gravitational potential energy for two masses. The sign of $U_{electric}$ depends on the signs of the two charges.

Now compare Equation 7-3 to Coulomb's law for the electric force between two point charges, Equation 16-1:

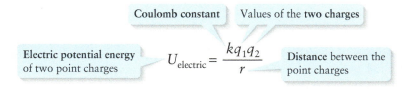

Any two point charges q_1 and q_2 exert equally strong **electric forces** on each other.

Coulomb constant

Absolute values of the **point charges**

$$F_{q_1 \text{ on } q_2} = F_{q_2 \text{ on } q_1} = \frac{k|q_1||q_2|}{r^2}$$

Distance between the point charges

Coulomb's law
(16-1)

If we replace m and G in Equation 7-3 with q and k, we see that Equation 16-1 is identical to Equation 7-3 for the gravitational force, but with an important difference: The electric force is *attractive* (like the gravitational force) if the two charges q_1 and q_2 have opposite signs (one negative and the other positive) but *repulsive* if q_1 and q_2 have the same sign (either both positive or both negative). Here's an expression for the electric potential energy of two point charges that accounts for both of these possibilities:

Coulomb constant | Values of the **two charges**

Electric potential energy of two point charges

$$U_{electric} = \frac{kq_1q_2}{r}$$

Distance between the point charges

Electric potential energy of two point charges
(17-3)

This expression is very similar to Equation 7-5 for gravitational potential energy. Equation 17-3 shows that $U_{electric}$ is inversely proportional to the distance r, so the electric potential energy is zero when the two charges are infinitely far apart ($r \to \infty$). But unlike gravitational potential energy, $U_{electric}$ can be either negative or positive depending on the signs of the two charges. If q_1 and q_2 have different signs (one positive and one negative) so that the two charges attract each other, then $U_{electric} < 0$ for any finite distance r between the charges (Figure 17-4b). This makes sense: If we hold charge q_1 at rest and move charge q_2 farther away, the attractive electric force on q_2 does negative work (the force on q_2 is opposite to the displacement of q_2). In this case the change in electric potential energy is positive, and the electric potential energy increases (becomes less negative) with increasing distance r, just like the gravitational potential energy of two masses. But if q_1 and q_2 have the same sign (both positive or both negative) so that the two charges repel each other, then $U_{electric} > 0$ for any finite distance r between the charges (Figure 17-4c). In this case when q_2 moves away from q_1, the repulsive electric force on q_2 does positive work (the force on q_2 is in the same direction as the displacement of q_2). Then the change in electric potential energy is negative, and the electric potential energy *decreases* (becomes less positive) with increasing distance r.

Here's a useful way to interpret $U_{electric}$ as given by Equation 17-3:

The electric potential energy of a pair of point charges equals the amount of work you would have to do to bring the charges to their current positions from infinitely far away.

If the two charges have opposite signs as in Figure 17-4b, you would have to do *negative* work to oppose the attractive electric force that pulls the two charges together. (If you didn't exert a force to do this work, the electric attraction would make the charges crash into each other instead of stopping a distance r apart.) So in this case the electric potential energy of the two charges is negative, as Figure 17-4b shows. By contrast, if the two charges have the same sign as in Figure 17-4c, you would have to do positive work to push the two charges together to overcome the repulsive electric force that pushes the two charges apart. So the electric potential energy of these two charges is positive, as depicted in Figure 17-4c.

The same idea holds for an assemblage of three or more point charges. To put charges q_1, q_2, and q_3 into proximity to each other, you would first have to do work to bring q_1 and q_2 from infinity to a distance r_{12} from each other. If you then brought in the third charge q_3 while keeping the other two charges stationary, you would have to do additional work against the electric force that q_1 exerts on q_3 *and* against the electric force that q_2 exerts on q_3. If q_3 ends up a distance r_{13} from q_1 and a distance r_{23} from q_2, the electric potential energy of the assemblage is the total amount of work that you did:

(17-4)
$$U_{electric} = \frac{kq_1q_2}{r_{12}} + \frac{kq_1q_3}{r_{13}} + \frac{kq_2q_3}{r_{23}}$$

(electric potential energy of three charges)

The total electric potential energy of a system of three charges is the sum of three terms, one for each pair of charges in the system (q_1 and q_2, q_1 and q_3, and q_2 and q_3). Each such term is the same as Equation 17-3. You can easily extend this idea to an assemblage made up of any number of point charges.

The following examples show how to do calculations using Equations 17-3 and 17-4.

→ **Go to Interactive Exercise 17-1 for more practice dealing with electric potential energy.**

EXAMPLE 17-2   Electric Potential Energy and Nuclear Fission

When a nucleus of uranium-235 (92 protons and 143 neutrons) absorbs an additional neutron, it undergoes a process called *nuclear fission* in which it breaks into two smaller nuclei. One possible fission is for the uranium nucleus to divide into two palladium nuclei, each of which has 46 protons and is 5.9×10^{-15} m in radius. The palladium nuclei then fly apart due to their electric repulsion. If we assume that the two palladium nuclei begin at rest and are just touching each other, what is their combined kinetic energy when they are very far apart?

Set Up

We can treat the two spherical nuclei as though they were point charges of $q = +46e$ located at the centers of the two nuclei. (To motivate this, recall from Section 16-7 that the electric field outside a sphere of charge Q is the same as that of a point charge Q located at the sphere's center.) Equation 17-3 then tells us the electric potential energy of the two palladium nuclei when they begin at rest. We'll then use energy conservation to find the combined kinetic energy when the palladium nuclei are very far apart.

Electric potential energy of two point charges:

$$U_{electric} = \frac{kq_1q_2}{r} \qquad (17\text{-}3)$$

Conservation of mechanical energy:

$$K_i + U_{electric,i} = K_f + U_{electric,f} \qquad (6\text{-}27)$$

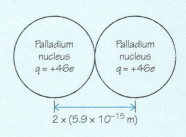

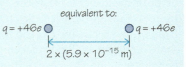

Solve

Use Equation 17-3 to solve for the initial electric potential energy when the two nuclei are just touching.

Each nucleus has charge $+46e$:

$$q_1 = q_2 = +46(1.60 \times 10^{-19} \text{ C}) = +7.36 \times 10^{-18} \text{ C}$$

The separation between the charges is twice the radius of either nucleus:

$$r = 2(5.9 \times 10^{-15} \text{ m})$$

The initial electric potential energy is

$$U_{\text{electric,i}} = \frac{kq_1q_2}{r} = \frac{(8.99 \times 10^9 \text{ N} \cdot \text{m}^2/\text{C}^2)(7.36 \times 10^{-18} \text{ C})^2}{2(5.9 \times 10^{-15} \text{ m})}$$
$$= 4.1 \times 10^{-11} \text{ N} \cdot \text{m} = 4.1 \times 10^{-11} \text{ J}$$

Use energy conservation to find the combined final kinetic energy of the two palladium nuclei.

The palladium nuclei begin at rest, so the initial kinetic energy is zero:

$$K_i = 0$$

The palladium nuclei end up very far apart, so their separation is essentially infinite ($r \to \infty$). The final electric potential energy is therefore zero:

$$U_{\text{electric,f}} = 0$$

From the energy conservation equation, the final kinetic energy is

$$K_f = K_i + U_{\text{electric,i}} - U_{\text{electric,f}}$$
$$= 0 + 4.1 \times 10^{-11} \text{ J} - 0 = 4.1 \times 10^{-11} \text{ J}$$

Reflect

All of the initial electric potential energy is converted into kinetic energy of the palladium nuclei. The energy released by the fission of a single uranium-235 nucleus, 4.1×10^{-11} J, is very small. But imagine that you could get 1.0 kg of uranium-235, which contains about 2.6×10^{24} uranium atoms, to undergo fission at once. (This is the principle of a *fission bomb*.) The released energy would be $(2.6 \times 10^{24}) \times (4.1 \times 10^{-11} \text{ J}) = 1.1 \times 10^{14}$ J, equivalent to the energy given off by exploding 26,000 tons of TNT. This suggests the terrifying amount of energy that can be released by a fission weapon.

EXAMPLE 17-3 Electric Potential Energy of Three Charges

A particle with charge $q_1 = +4.30 \text{ μC}$ is located at $x = 0$, $y = 0$. A second particle with charge $q_2 = -9.80 \text{ μC}$ is located at $x = 0$, $y = 4.00$ cm, and a third particle with charge $q_3 = +5.00 \text{ μC}$ is located at position $x = 3.00$ cm, $y = 0$. (Note that 1 μC = 1 microcoulomb = 10^{-6} C.) What is the total electric potential energy of these three charges?

Set Up

We'll use Equation 17-4 to find the value of U_{electric}. We're given the values of the three charges; we'll use the positions of the charges to find the distances r_{12}, r_{13}, and r_{23} between them.

Electric potential energy of three charges:

$$U_{\text{electric}} = \frac{kq_1q_2}{r_{12}} + \frac{kq_1q_3}{r_{13}} + \frac{kq_2q_3}{r_{23}}$$

$$(17\text{-}4)$$

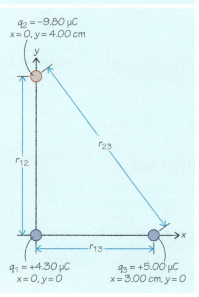

Solve

The figure above shows that charges 1 and 2 are $r_{12} = 4.00$ cm apart, and that charges 1 and 3 are $r_{13} = 3.00$ cm apart. The distance r_{23} between charges 2 and 3 is the hypotenuse of a right triangle of sides r_{12} and r_{13}.	From the figure above, $$r_{12} = 4.00 \text{ cm} = 4.00 \times 10^{-2} \text{ m}$$ $$r_{13} = 3.00 \text{ cm} = 3.00 \times 10^{-2} \text{ m}$$ From the Pythagorean theorem, $$r_{23} = \sqrt{r_{12}^2 + r_{13}^2}$$ $$= \sqrt{(4.00 \times 10^{-2} \text{ m})^2 + (3.00 \times 10^{-2} \text{ m})^2}$$ $$= 5.00 \times 10^{-2} \text{ m}$$

Calculate each term in the expression for electric potential energy, Equation 17-4.	We can write Equation 17-4 as $$U_{\text{electric}} = U_{12} + U_{13} + U_{23}$$ Each term on the right-hand side of this equation represents the contribution to the electric potential energy due to a specific pair of point charges. The contribution due to the interaction between charges q_1 and q_2 is $$U_{12} = \frac{kq_1q_2}{r_{12}}$$ $$= \frac{(8.99 \times 10^9 \text{ N} \cdot \text{m}^2/\text{C}^2)(+4.30 \times 10^{-6} \text{ C})(-9.80 \times 10^{-6} \text{ C})}{4.00 \times 10^{-2} \text{ m}}$$ $$= -9.47 \text{ J}$$ The contribution due to the interaction between charges q_1 and q_3 is $$U_{13} = \frac{kq_1q_3}{r_{13}}$$ $$= \frac{(8.99 \times 10^9 \text{ N} \cdot \text{m}^2/\text{C}^2)(+4.30 \times 10^{-6} \text{ C})(+5.00 \times 10^{-6} \text{ C})}{3.00 \times 10^{-2} \text{ m}}$$ $$= +6.44 \text{ J}$$ The contribution due to the interaction between charges q_2 and q_3 is $$U_{23} = \frac{kq_2q_3}{r_{23}}$$ $$= \frac{(8.99 \times 10^9 \text{ N} \cdot \text{m}^2/\text{C}^2)(-9.80 \times 10^{-6} \text{ C})(+5.00 \times 10^{-6} \text{ C})}{5.00 \times 10^{-2} \text{ m}}$$ $$= -8.81 \text{ J}$$

Finally, calculate the total electric potential energy.	The total electric potential energy is the sum of the three terms calculated above: $$U_{\text{electric}} = U_{12} + U_{13} + U_{23}$$ $$= (-9.47 \text{ J}) + (+6.44 \text{ J}) + (-8.81 \text{ J})$$ $$= -11.84 \text{ J}$$

Reflect

The contributions U_{12} and U_{23} are negative because these pairs of charges (q_1 and q_2 for U_{12}, q_2 and q_3 for U_{23}) have opposite signs. The members of these pairs attract each other, so you must do negative work to move the charges from infinity to the positions shown in the figure. The contribution U_{13}, however, is positive because charges q_1 and q_3 have the same sign (both positive). These two charges repel, so you must do positive work to move these charges from infinity to their positions. The total potential energy is negative, which shows that the total amount of work you would do to move all three charges from infinity is negative.

GOT THE CONCEPT? 17-1 Electric Potential Energy

? Suppose you reversed the signs of the three point charges in Example 17-3 so that $q_1 = -4.30$ μC, $q_2 = +9.80$ μC, and $q_3 = -5.00$ μC. If the positions of the charges remain unchanged, what would be the total electric potential energy of this assemblage? (a) More negative than −11.84 J; (b) −11.84 J; (c) between −11.84 J and +11.84 J; (d) +11.84 J; (e) more positive than +11.84 J.

TAKE-HOME MESSAGE FOR Section 17-2

✔ The electric potential energy associated with a point charge can change if the charge changes position in an electric field.

✔ The electric potential energy of a pair of point charges equals the amount of work you would have to do to move those charges from infinity to their present positions.

✔ The electric potential energy of an assemblage of three or more point charges is the sum of the electric potential energies for each pair of charges in the assemblage.

17-3 Electric potential equals electric potential energy per charge

Our discussion in Section 17-2 shows that if a point charge q changes position, the potential energy change $\Delta U_{electric}$ depends on both the magnitude and the sign (positive or negative) of q (see Figure 17-3). We can simplify things by considering the potential energy *per charge*—that is, the electric potential energy for a charge at a given position divided by the value of that charge. We call this quantity the **electric potential** V:

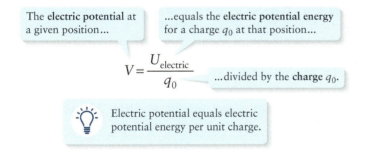

The **electric potential** at a given position...

...equals the **electric potential energy** for a charge q_0 at that position...

$$V = \frac{U_{electric}}{q_0}$$

...divided by the **charge** q_0.

💡 Electric potential equals electric potential energy per unit charge.

Electric potential related to electric potential energy (17-5)

We call the charge q_0 the **test charge**: Its charge has such a small magnitude that it doesn't affect the other charges that create the electric field in which q_0 moves. Because we divide out the value of q_0, the value of the potential V at a given position does *not* depend on the value of the point charge q_0 that we place there. Instead, V is determined by the other charges that produce the electric field at the position where we place the test charge.

Note that electric potential V has the same relationship to electric potential energy $U_{electric}$ as electric field $\vec{E}$ has to electric force $\vec{F}$: V is electric potential energy per charge, just as $\vec{E}$ is electric force per charge. Like electric potential energy, but unlike electric field, potential V is a *scalar* quantity.

For electric potential energy, the value of $U_{electric}$ for a charge at a given position is not as important as the potential energy *difference* $\Delta U_{electric}$ when the charge moves from a point a to a different point b. The same is true for electric potential. From Equation 17-5, the **electric potential difference** ΔV between two points is

Electric potential difference related to electric potential energy difference (17-6)

The **difference in electric potential** between two positions...

...equals the **change in electric potential energy** for a charge q_0 moved between these two positions...

$$\Delta V = \frac{\Delta U_{\text{electric}}}{q_0}$$

...divided by the **charge q_0**.

 Electric potential difference equals electric potential energy difference per unit charge.

The SI unit of electric potential and electric potential difference is the **volt** (V), named after the Italian scientist Alessandro Volta. For example, a common AA or AAA flashlight battery has "1.5 volts" written on its side. This means that the electric potential at the positive terminal of the battery (labeled +) is 1.5 V greater than the electric potential at the negative terminal of the battery (labeled −). In other words, 1.5 V is the electric potential *difference* between the terminals of the battery. In Chapter 18 we'll see how this electric potential difference causes electric charge to flow when a battery is included in an electric circuit. (Note that Volta invented the electric battery in 1800.) Equations 17-5 and 17-6 show that 1 volt is equal to 1 joule per coulomb, and we know that 1 joule is equal to 1 newton multiplied by 1 meter. So

$$1\ \text{V} = 1\ \frac{\text{J}}{\text{C}} = 1\ \frac{\text{N} \cdot \text{m}}{\text{C}}$$

This means that if the electric potential at point b is 1 V higher than at point a, to move a charge $q_0 = +1$ C from a to b means increasing the electric potential energy by 1 J (and so would take 1 J of work). If instead the charge were $q_0 = +2$ C, to move it from a to b would mean a potential energy increase of +2 J and would require 2 J of work. In either case the electric *potential difference* is the same, though: From Equation 17-6, $\Delta V = \Delta U_{\text{electric}}/q_0$ equals (+1 J)/(+1 C) = 1 V in the first case and (+2 J)/(+2 C) = 1 V in the second case.

We saw in Chapter 16 that the SI units of electric field are newtons per coulomb, or N/C. Because 1 V = 1 N·m/C, it follows that 1 V/m = 1 N/C. So an equally good set of units for electric field is V/m (volts per meter).

We'll often abbreviate the terms "electric potential" and "electric potential difference" as simply "potential" and "potential difference," respectively. Since the unit of electric potential is the volt, it's also common to refer to electric potential difference as **voltage**.

WATCH OUT! **Don't confuse the symbol *V* for electric potential and the abbreviation V for volts.**

In this chapter and others you'll see mathematical statements such as "*V* = 10 V." Such statements are perfectly legal because *V* and V are different symbols. *V* (italicized) represents the electric potential at a point in space. V (not italicized) is an abbreviation for the unit of electric potential, the volt. In this textbook you can tell the difference by noticing which font is used and by paying close attention to the context in which *V* and V are used. But it's pretty hard to handwrite something in italics, so make sure you know the distinction between *V* and V in your own work on your homework assignments and exams.

WATCH OUT! **Electric potential and electric potential energy are not the same.**

Remember that electric potential *V* and electric potential energy U_{electric} are related but different scalar quantities. Electric potential energy U_{electric} refers to the energy associated with a particular amount of charge at a given position. Electric potential *V* is the energy associated with a *unit* charge at that position.

Electric Potential in a Uniform Electric Field

We can use Equation 17-6 to find the electric potential difference between two points a and b in a uniform electric field (Figure 17-5). From Equation 17-2 the change or difference in electric potential energy when a charge q_0 is moved from a to b is

$\Delta U_{electric} = -q_0 Ed \cos\theta$. Equation 17-6 tells us that the electric potential difference between a and b is $\Delta U_{electric}$ divided by q_0:

$$\Delta V = \frac{\Delta U_{electric}}{q_0} = \frac{-q_0 Ed \cos\theta}{q_0}$$

$$= -Ed \cos\theta \qquad (17\text{-}7)$$

Equation 17-7 shows that ΔV is positive if $\cos\theta$ is negative, which is the case if the angle θ is greater than 90° as depicted in Figure 17-5. In other words, *if you move in a direction opposite to the electric field, the electric potential increases*. If the angle θ is less than 90°, $\cos\theta$ is positive and ΔV is negative; *if you move in the direction of the electric field, the electric potential decreases*.

We know that the electric force on a positive charge is in the direction of the electric field $\vec{E}$, and the electric force on a negative charge is in the direction opposite to $\vec{E}$. So our observations about how electric potential changes with position tell us that

> *If an object has positive charge, the electric force on that object pushes it toward a region of lower electric potential. If an object has negative charge, the electric force on that object pushes it toward a region of higher electric potential.*

As we'll see below, these observations hold true whether or not the electric field is uniform.

Comparing Equations 17-2 and 17-7 shows that the right-hand side of Equation 17-7 equals the negative of the work done by the electric field ($q_0 Ed \cos\theta$) divided by the charge q_0. So another way to think of the potential difference between two points is as the negative of the work done by the electric field per unit charge when a charged object is moved between those points. Alternatively, the potential difference between two points equals the work that *you* must do per unit charge against the electric force to move a charged object between those points. So if you take a charge of +1 C that is at rest at point a and move it to point b where it is again at rest, and the potential at b is 1 V higher than the potential at a, the electric field does −1 J of work on the charge and you do +1 J of work on that charge.

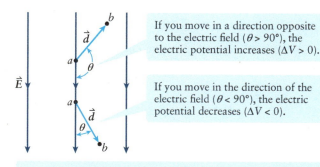

If you move in a direction opposite to the electric field ($\theta > 90°$), the electric potential increases ($\Delta V > 0$).

If you move in the direction of the electric field ($\theta < 90°$), the electric potential decreases ($\Delta V < 0$).

If you move from point a to point b in a uniform electric field $\vec{E}$, the change in electric potential is $\Delta V = -Ed \cos\theta$.

Figure 17-5 Change in electric potential Calculating the difference in electric potential between two points in a uniform electric field $\vec{E}$.

→ *Go to Picture It 17-2 for more practice dealing with electric potential.*

EXAMPLE 17-4 Electric Potential Difference in a Uniform Field I

A uniform electric field points in the positive x direction and has magnitude 2.00×10^2 V/m. Points a and b are both in this field: Point b is a distance 0.300 m from a in the negative x direction. Determine the electric potential difference $V_b - V_a$ between points a and b.

Set Up

We apply Equation 17-7 to the path shown in the figure. Note that the magnitude of the displacement is $d = 0.300$ m, and the angle between the electric field and the displacement is $\theta = 180°$. Note also that the potential difference ΔV equals the potential at the *end* of the displacement $\vec{d}$ (that is, at point b) minus the potential at the *beginning* of the displacement (that is, at point a).

Potential difference between two points in a uniform electric field:

$$\Delta V = V_b - V_a = -Ed \cos\theta \qquad (17\text{-}7)$$

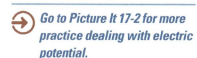
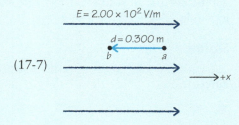

Solve

Use Equation 17-7 to solve for the potential difference.

Calculate the potential difference from the electric field magnitude E, the displacement d, and the angle θ:

$$\Delta V = V_b - V_a = -Ed \cos \theta$$
$$= -(2.00 \times 10^2 \text{ V/m})(0.300 \text{ m}) \cos 180°$$
$$= -(60.0 \text{ V})(-1)$$
$$= +60.0 \text{ V}$$

Reflect

We can check our result by comparing with Example 17-1 in Section 17-2, where we considered the change in electric potential *energy* $\Delta U_{electric}$ for an electron that undergoes the same displacement in this same electric field. Using Equation 17-6 we find the same value of $\Delta U_{electric}$ as in Example 17-1.

The positive value of $\Delta V = V_b - V_a$ means that point b is at a higher potential than point a. This agrees with our observation above that if you travel opposite to the direction of the electric field, the electric potential increases. The value $E = 2.00 \times 10^2$ V/m means that the electric potential increases by 2.00×10^2 V for every meter that you travel opposite to the direction of $\vec{E}$.

If a charge $q_0 = -1.60 \times 10^{-19}$ C travels from a to b, the change in electric potential energy is given by Equation 17-6:

$$\Delta V = \frac{\Delta U_{electric}}{q_0}$$

so

$$\Delta U_{electric} = q_0 \Delta V = (-1.60 \times 10^{-19} \text{ C})(+60.0 \text{ V})$$
$$= -9.60 \times 10^{-18} \text{ V·C} = -9.60 \times 10^{-18} \text{ J}$$

(Recall from above that 1 V = 1 J/C.)

EXAMPLE 17-5 Electric Potential Difference in a Uniform Field II

Determine the electric potential difference $\Delta V = V_c - V_a$ between points a and c in the uniform electric field of Example 17-4. Point c is a distance 0.500 m from point a, and the straight-line path from a to c makes an angle of 126.9° with respect to the electric field.

Set Up

Again we'll use Equation 17-7 to calculate the potential difference between the two points. Since the displacement from point a to point c points generally opposite to the direction of the electric field, we expect that $\Delta V = V_c - V_a$ will be positive, just like $\Delta V = V_b - V_a$ in Example 17-4.

Potential difference between two points in a uniform electric field:

$$\Delta V = V_c - V_a = -Ed \cos \theta \qquad (17\text{-}7)$$

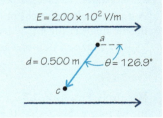

Solve

Use Equation 17-7 to solve for the potential difference.

Calculate the potential difference from the electric field magnitude E, the displacement d, and the angle θ:

$$\Delta V = V_c - V_a = -Ed \cos \theta$$
$$= -(2.00 \times 10^2 \text{ V/m})(0.500 \text{ m}) \cos 126.9°$$
$$= -(2.00 \times 10^2 \text{ V/m})(0.500 \text{ m})(-0.600)$$
$$= +60.0 \text{ V}$$

Reflect

The potential difference between points a and c is the *same* as the potential difference between points a and b in Example 17-4. Equation 17-7 tells us why this should be: The potential difference $\Delta V = -Ed \cos \theta$ involves the magnitude E of the electric field multiplied by $d \cos \theta$, which is the component of the displacement $\vec{d}$ in the direction of $\vec{E}$. In both examples the electric field magnitude has the same value (2.00×10^2 V/m), as does the component of displacement in the direction of the electric field (-0.300 m).

Another way to come to this same conclusion is to recognize that the displacement from a to c can be broken down into two parts: a displacement in the negative x direction from a to b, followed by a displacement from b to c in the y direction. The potential difference for the displacement from a to b is $V_b - V_a = +60.0$ V as we calculated in Example 17-4; the potential difference for the displacement from b to c is $V_c - V_b = -Ed \cos 90° = 0$, since that displacement is perpendicular to the electric field. So the net potential difference between a and c is $V_c - V_a = (V_c - V_b) + (V_b - V_a) = +60.0$ V $+ 0$ V $= +60.0$ V.

EXAMPLE 17-6 Transmission Electron Microscope

A *transmission electron microscope* forms an image by sending a beam of fast-moving electrons rather than a beam of light through a thin sample. As we will see in Chapter 26, such fast-moving electrons behave very much like a light wave. If the electrons have sufficiently high energy, the image that they form can show much finer detail than even the best optical microscope. The electrons are emitted from a heated metal filament and are then accelerated toward a second piece of metal called the *anode* that is at a potential 2.50 kV (1 kV = 1 kilovolt = 10^3 V) higher than that of the filament. If the electrons leave the filament initially at rest, how fast are the electrons traveling when they pass the anode?

Set Up

As the electrons move through the potential difference between the filament and the anode, the electric potential energy will change in accordance with Equation 17-6. Each electron has a negative charge $q_0 = -e$, so an *increase* in electric potential ($\Delta V > 0$) means a *decrease* in electric potential energy ($\Delta U_{electric} < 0$). The total mechanical energy (kinetic energy plus potential energy) is conserved as the electron moves because there are no forces acting on it other than the conservative electric force, so the electron kinetic energy will increase as the potential energy decreases.

Electric potential difference related to electric potential energy difference:

$$\Delta V = \frac{\Delta U_{electric}}{q_0} \qquad (17\text{-}6)$$

Mechanical energy is conserved:

$$K_i + U_{electric,i} = K_f + U_{electric,f} \qquad (6\text{-}27)$$

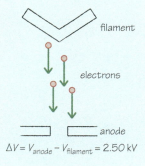

filament

electrons

anode

$\Delta V = V_{anode} - V_{filament} = 2.50$ kV

Solve

Solve Equation 17-6 for the change in electric potential energy as the electron moves from the filament to the anode, which is at a potential 2.50 kV higher than the filament.

From Equation 17-6,

$$\Delta U_{electric} = q_0 \Delta V$$

The electron has change $q_0 = -e$, so

$$\begin{aligned} \Delta U_{electric} &= -e\Delta V \\ &= -(1.60 \times 10^{-19} \text{ C})(+2.50 \times 10^3 \text{ V}) \\ &= -4.00 \times 10^{-16} \text{ J} \end{aligned}$$

(Recall that 1 V = 1 J/C, so 1 J = 1 C·V.)

The conservation of mechanical energy tells us that the change in the kinetic energy of the electron is equal to the negative of the change in the electric potential energy.

Mechanical energy is conserved:

$$E = K + U_{electric} = \text{constant}$$

So the *change* in mechanical energy is zero:

$$\Delta E = \Delta K + \Delta U_{electric} = 0$$

The change in the kinetic energy of the electron is

$$\begin{aligned} \Delta K &= -\Delta U_{electric} = -(-4.00 \times 10^{-16} \text{ J}) \\ &= +4.00 \times 10^{16} \text{ J} \end{aligned}$$

Each electron begins with zero kinetic energy, so the change in its kinetic energy is equal to its final kinetic energy as it reaches the anode.

Since the electron begins with zero kinetic energy at the filament, the change in its kinetic energy is

$$\Delta K = +4.00 \times 10^{-16} \text{ J} = K_{anode} - K_{filament} = K_{anode}$$

Use this to find the speed of the electron at the anode:

$$K_{anode} = \frac{1}{2} m_{electron} v_{anode}^2, \text{ so}$$

$$v_{anode} = \sqrt{\frac{2K_{anode}}{m_{electron}}} = \sqrt{\frac{2(4.00 \times 10^{-16} \text{ J})}{9.11 \times 10^{-31} \text{ kg}}}$$

$$= 2.96 \times 10^7 \text{ m/s}$$

Reflect

The speed of light is $c = 3.00 \times 10^8$ m/s; the electrons are accelerated to nearly one-tenth of that speed.

Example 17-6 is just one of many situations in which an object with a charge of $-e$ (such as an electron) or $+e$ (such as a proton) moves through a potential difference. A common unit for the potential energy change in such situations is the **electron volt** (eV), which is equal to the magnitude e of the charge on the electron multiplied by 1 volt. Since $e = 1.60 \times 10^{-19}$ C, it follows that

$$1 \text{ eV} = (1.60 \times 10^{-19} \text{ C})(1 \text{ V}) = 1.60 \times 10^{-19} \text{ C} \cdot \text{V} = 1.60 \times 10^{-19} \text{ J}$$

In Example 17-6, the electron of charge $-e$ moves through a potential difference of $+2.50 \times 10^3$ V, the change in electric potential energy is -2.50×10^3 eV, and the kinetic energy that the electron acquires is $+2.50 \times 10^3$ eV. We also use the abbreviations 1 keV = 10^3 eV, 1 MeV = 10^6 eV, and 1 GeV = 10^9 eV. (The largest particle accelerator in the world, the Large Hadron Collider at the European Organization for Nuclear Research [CERN] near Geneva, Switzerland, accelerates protons to a kinetic energy of 7×10^{12} eV = 7 TeV. This is equivalent to making the protons pass through a potential difference of 7×10^{12}, or 7 trillion, volts.) In later chapters we'll see that the electron volt is a useful unit for expressing energies on the atomic or nuclear scale.

Electric Potential Due to a Point Charge

Note that Equation 17-7 is useful for calculating potential differences only in the case of a uniform electric field. Another important case is the electric potential due to a *point charge Q*. From Equation 17-3, if we place a test charge q_0 a distance r from a point charge Q, the electric potential energy of the system of two charges is

$$U_{electric} = \frac{kq_0Q}{r}$$

Equation 17-5 tells us that to find the electric potential due to the point charge Q, we must divide $U_{electric}$ by the value of the test charge q_0:

Electric potential due to a
point charge
(17-8)

Electric potential due to a point charge Q Coulomb constant

$$V = \frac{kQ}{r}$$

Value of the point **charge**

Distance from the point charge Q to the location where the potential is measured

Equation 17-8 says that all points that are the same distance r from a point charge have the same electric potential due to that charge. It also says that if $Q > 0$, the electric potential due to the charge is positive and decreases (becomes less positive) as you move farther away from the charge so that r increases (Figure 17-6a). If $Q < 0$, the electric potential due to the charge is negative and decreases (becomes more negative) as you move closer to the charge (Figure 17-6b). For either sign of Q, the electric potential goes to zero at an infinite distance from the point charge.

(a) Electric potential due to a positive point charge; r = distance from charge

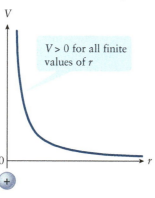

$V > 0$ for all finite values of r

(b) Electric potential due to a negative point charge; r = distance from charge

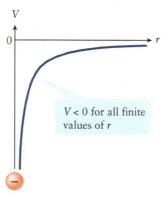

$V < 0$ for all finite values of r

Figure 17-6 **Electric potential due to a point charge** The electric potential V due to a point charge is (a) positive if the charge is positive and (b) negative if the charge is negative.

These observations about Equation 17-8 are consistent with our previous statements about electric potential and electric force. A positive test charge q_0 placed near a positive charge Q feels an electric force that pushes it farther away from Q, toward regions where the potential V due to Q is lower (less positive). If instead that positive test charge is placed near a negative charge Q, the test charge feels an electric force that pulls it toward Q—again toward regions where the potential V is lower (in this case more negative).

WATCH OUT! Don't confuse the formulas for electric potential and electric field due to a point charge.

❗ Be sure that you recognize the differences between Equation 17-8, $V = kQ/r$, and Equation 16-4 for the magnitude of the electric field due to a point charge Q, $E = k|Q|/r^2$. Equation 17-8 says that the potential V due to a point charge is inversely proportional to r and can be positive or negative, depending on the sign of Q. By contrast, Equation 16-4 tells us that the magnitude E of the field due to a point charge is inversely proportional to the *square* of r. Furthermore, because E is the magnitude of a vector, it is always positive (it is proportional to the absolute value of Q).

If there is not a single point charge but an assemblage of charges, the total electric potential at a given position due to these charges is the sum of the individual potentials. For example, for the case of three charges Q_1, Q_2, and Q_3, the potential at a point that is a distance r_1 from the first charge, a distance r_2 from the second charge, and a distance r_3 from the third charge is

$$V = \frac{kQ_1}{r_1} + \frac{kQ_2}{r_2} + \frac{kQ_3}{r_3}$$

(17-9)

(electric potential of three charges)

GOT THE CONCEPT? 17-2 Electric Potential Difference

❓ Four point charges are each moved from one position to another position where the electric potential has a different value. Rank the four charges in order of the electric potential energy change that takes place, from most positive to most negative. (a) A +0.0010 C charge that moves through a potential increase of 5.0 V; (b) a +0.0020 C charge that moves through a potential decrease of 2.0 V; (c) a −0.0015 C charge that moves through a potential decrease of 4.0 V; (d) a −0.0010 C charge that moves through a potential increase of 2.5 V.

TAKE-HOME MESSAGE FOR Section 17-3

✔ The electric potential at a certain position equals the electric potential energy per unit charge for a test charge at that position. Electric potential is a scalar, not a vector, quantity. Electric potential decreases as you move in the direction of the electric field.

✔ The electric potential due to a point charge is positive if the charge is positive and negative if the charge is negative. As

you move farther from an isolated point charge, the electric potential becomes closer to zero.

✔ The electric potential due to a collection of charges is the sum of the potentials due to the individual charges.

✔ The electric potential difference between two points equals the difference in electric potential energy per unit charge between those points.

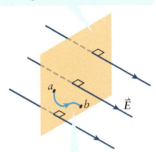

A plane perpendicular to a uniform electric field $\vec{E}$ is an **equipotential surface:** Potential V has the same value at all points on this surface.

Any curve that lies on an equipotential surface, such as this curve from a to b, is an equipotential curve.

Figure 17-7 Equipotential surfaces I In a uniform electric field equipotential surfaces are planes perpendicular to the electric field lines.

17-4 The electric potential has the same value everywhere on an equipotential surface

In Examples 17-4 and 17-5 in the previous section, we looked at the potential differences between two pairs of points in an electric field, a and b in Example 17-4 and a and c in Example 17-5. Although the electric field is the same in both examples, the distance from a to c in Example 17-5 is clearly longer than the distance from a to b in Example 17-4. Nonetheless, the potential differences that we found in the two examples were equal, which tells us that the potential is the *same* at points b and c. You can see why from Equation 17-7, $\Delta V = -Ed \cos \theta$. A displacement $\vec{d}$ from point b to point c is perpendicular to the electric field $\vec{E}$, so the angle in Equation 17-7 is $\theta = 90°$, and $\cos \theta = \cos 90° = 0$. Thus the potential difference ΔV between points b and c must be zero.

In general, the electric potential will be the same at any two points that lie along a curve perpendicular to electric field lines. Such a curve is called an *equipotential curve* or simply an **equipotential**. For the case of a uniform electric field, the electric potential has the same value anywhere on a plane that's perpendicular to the electric field. Such a plane is an example of an **equipotential surface**, one on which the electric potential has the same value at all points (Figure 17-7). No work is required to move a charge from one point to another along any path on an equipotential surface. As we described in Section 17-3, the electric potential decreases as you move in the direction of the electric field $\vec{E}$, so the value of the potential V is lower for equipotential surfaces that are "downstream" in the electric field than on surfaces that are "upstream."

Figure 17-8 shows both electric field lines and equipotentials for the case of a *nonuniform* electric field. The equipotential surfaces are perpendicular at *all* points to the field lines, just as for the case of a uniform field in Figure 17-7. But since the electric field lines are not parallel lines, the equipotential surfaces are not flat planes. In general, *any* surface that is everywhere perpendicular to the electric field is an equipotential surface, and the value of the potential V on an equipotential surface is lower the farther "downstream" in the electric field that surface is.

In Figure 17-8 points e and f are both on the equipotential for which $V = +50$ V, and points s and t are both on the equipotential for which $V = 0$ V. So the potential difference between points e and s is the same as the potential difference between points f and t:

$$V_e - V_s = V_f - V_t = (+50 \text{ V}) - (0 \text{ V}) = +50 \text{ V}$$

However, the distance between points e and s is less than that between points f and t. Equation 17-7 tells us that the potential difference between two points is proportional to the distance d between the points and the electric field magnitude E.

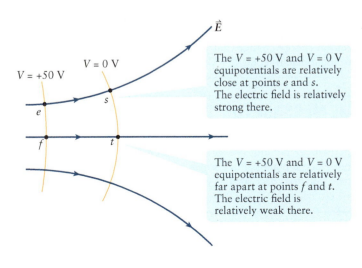

The $V = +50$ V and $V = 0$ V equipotentials are relatively close at points e and s. The electric field is relatively strong there.

The $V = +50$ V and $V = 0$ V equipotentials are relatively far apart at points f and t. The electric field is relatively weak there.

Figure 17-8 Equipotential surfaces II In a nonuniform electric field, equipotential surfaces are curved surfaces (seen here from the side) that are everywhere perpendicular to the electric field lines.

(This equation is strictly valid only for a uniform field, but is still approximately true for a nonuniform field.) So the electric field must be greater between the points e and s that are closer together, and less between the points f and t that are farther apart. This is an example of a general rule:

> Where two adjacent equipotential surfaces are close together, the electric field is relatively strong. Where these surfaces are far apart, the electric field is relatively weak.

Figure 17-9 shows an application of this idea. For a positive point charge (Figure 17-9a) or a negative point charge (Figure 17-9b), the electric field points radially outward or inward. The equipotential surfaces are everywhere perpendicular to the field lines, so they are spheres centered on the point charge. The radial distance from one spherical equipotential surface to the next is the same no matter where you are around the sphere, so the electric field magnitude is the same at all points a given distance from the point charge. This agrees with Equation 16-4 for the field magnitude E due to a point charge, $E = k|Q|/r^2$; the value of E depends only on the distance from the charge, not on where you are around the charge. For an electric dipole, however, the situation is different (Figure 17-9c). The field lines point from the positive charge to the negative one, and the equipotential surfaces are neither spherical nor centered on the point charges. (They are, however, everywhere perpendicular to the field lines.) Adjacent equipotential surfaces are close together between the two charges because the electric field is strong there. To the left of the left-hand charge or to the right of the right-hand charge, the electric field is relatively weak and adjacent equipotential surfaces are farther apart.

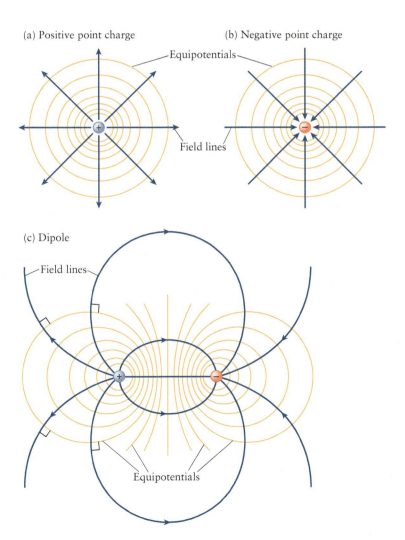

(a) Positive point charge

Equipotentials

(b) Negative point charge

Field lines

(c) Dipole

Field lines

Equipotentials

Figure 17-9 Equipotential surfaces III (a), (b) The equipotential surfaces for a positive or negative point charge are spheres centered on the point charge. (You can see this from Equation 17-8, $V = kQ/r$. This says that the potential V has the same value at all points that are the same distance from a point charge Q—that is, at all points on a sphere of radius r centered on the charge.) (c) For a dipole (point charges $+Q$ and $-Q$) the equipotentials are more complicated.

The equipotential concept is helpful for understanding the electric field around a charged conductor. We learned in Section 16-7 that if we put excess charge on a conductor and allow the individual excess charges to move to their equilibrium positions, all of the excess charge will end up on the conductor's surface and the electric field $\vec{E}$ will be zero everywhere inside the conductor. Since $\vec{E}$ is uniform inside the conductor (it has the same value at all points), we can use Equation 17-7 to calculate the potential difference between two points inside the conductor separated by a distance d: $\Delta V = -Ed \cos \theta = 0$ because $E = 0$. In other words, *the electric potential has the same value everywhere inside a conductor in equilibrium*. We say that a conductor in equilibrium is an **equipotential volume**. That's why we can make statements like "This conductor is at a potential of +20 V" or "The conductor at the positive end of a AA battery is at a potential 1.5 V higher than the conductor at the negative end"—the value of potential is the same *everywhere* throughout the volume of the conductor.

The electric field *outside* a charged conductor, however, is *not* zero. If the excess charge on the conductor is positive, the electric field $\vec{E}$ outside will point away from the surface of the conductor; if the excess charge is negative, $\vec{E}$ outside will point toward the surface. Because the surface of the conductor is part of the equipotential volume, it is itself an equipotential surface. Because field lines and equipotential surfaces are always perpendicular, we conclude that *the electric field just outside a conductor in equilibrium must be perpendicular to the surface of the conductor*. Since the electric field always points toward lower electric potential, we can also conclude that *a positively charged conductor is at a higher electric potential than an adjacent negatively charged conductor*. We'll use these ideas about conductors in the following section to help us understand an important device called a *capacitor*.

GOT THE CONCEPT? 17-3 Equipotentials and Electric Field

(?) The orange vertical lines in Figure 17-10 represent equipotential curves in some region of space. Which statement is correct about the electric field $\vec{E}$ at point A compared to the electric field at point B? (a) At A the field $\vec{E}$ points to the right and has a greater magnitude than at B; (b) at A the field $\vec{E}$ points to the right and has a smaller magnitude than at B; (c) at A the field $\vec{E}$ points to the left and has a greater magnitude than at B; (d) at A the field $\vec{E}$ points to the left and has a smaller magnitude than at B; (e) none of these.

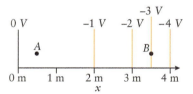

Figure 17-10 What equipotentials tell you about $\vec{E}$ What can you infer from this figure about the electric field at A and at B?

TAKE-HOME MESSAGE FOR Section 17-4

✔ The potential is the same everywhere along an equipotential curve or equipotential surface.

✔ Any curve or surface that is perpendicular to the electric field lines is an equipotential.

✔ Where equipotential surfaces are close, the electric field is relatively strong.

✔ A conductor in equilibrium has the same potential throughout its volume. The electric field at the surface of a conductor is perpendicular to the surface.

17-5 A capacitor stores equal amounts of positive and negative charge

BioMedical

The surface of every cell in your body is a *membrane* composed of a phospholipid bilayer that separates the fluid inside the cell and the fluid outside the cell. Negative charge accumulates on the membrane's interior surface, and this attracts positive charge onto the exterior surface. The result is a potential difference between the inner and outer surfaces of the membrane, and an electric field within the membrane that points from the outside in (Figure 17-11). This field helps drive essential ions through apertures in the membrane. They are also the source of the electrical signal used by the specialized cells called neurons to code, process, and transmit information.

A system or device that can store positive and negative charge like a cell membrane is called a **capacitor**. In technological applications a capacitor uses two pieces of metal called **plates**. One plate holds a positive charge q, and the other carries a negative charge $-q$. (Note that the *net* charge on the capacitor is zero.) It takes work to separate the positive and negative charges against the electric forces that attract them to each other, and this work goes into increasing the electric potential energy of the system of charges. So a capacitor is a device for storing electric potential energy.

We often need to draw capacitors in diagrams that represent electric circuits. The standard symbol for a capacitor in the diagram of a circuit is

The two closely spaced parallel lines represent the two plates of the capacitor, and the straight horizontal lines represent wires that can connect the capacitor to an electric circuit.

The simplest geometry for a capacitor is two large parallel plates, one with charge $+q$ and the other with charge $-q$. Figure 17-12a shows these two plates separated from each other. We saw in Section 16-7 that close to a large charged plate and far from its edges, the electric field due to that plate is uniform, is perpendicular to the plane of the plate, and from Equation 16-4 has magnitude $E = \sigma/2\varepsilon_0$. In this expression σ is the charge per unit area, equal to the charge q divided by the area A of the plate: $\sigma = q/A$. Since each plate has the same magnitude of charge, the field $\vec{E}_+$ due to the charge on the positive plate has the same magnitude as the field $\vec{E}_-$ due to the charge on the negative plate. (Recall from Section 16-6 that the *permittivity of free space* ε_0 is a constant equal to $1/(4\pi k)$, where k is the constant in Coulomb's law, Equation 16-1: $\varepsilon_0 = 8.85 \times 10^{-12}$ C^2/(N·m^2).)

For a capacitor the *net* electric field is the vector sum of $\vec{E}_+$ and $\vec{E}_-$: $\vec{E} = \vec{E}_+ + \vec{E}_-$. Figure 17-12b shows the two plates moved into position to form a **parallel-plate capacitor**. Because the magnitude of each field does not depend on the distance from the plate that generates the field, $\vec{E}_+$ cancels $\vec{E}_-$ in the region above the upper plate and below the lower plate. Between the plates the fields $\vec{E}_+$ and $\vec{E}_-$ have the same magnitude and point in the same direction, so the net field between the plates has twice the magnitude of the field due to either plate by itself:

$$E = 2\left(\frac{\sigma}{2\varepsilon_0}\right) = \frac{\sigma}{\varepsilon_0} = \frac{q}{\varepsilon_0 A} \qquad (17\text{-}10)$$

(electric field in a parallel-plate capacitor)

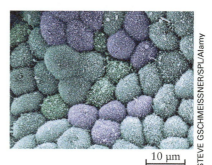

Figure 17-11 Human cells and potential difference Each of the cells in your body, like these cells from the internal lining of the gall bladder, is enclosed by a cell membrane. Typically the electric potential of the membrane's outer surface is about 0.1 V higher than the electric potential of the inner surface.

(a) Two large plates (viewed from the side) carry charge $+q$ and $-q$. Close to the plates the electric fields are nearly constant, so we represent them by straight, parallel field lines.

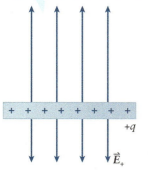

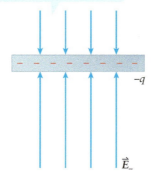

(b) Because the fields are nearly constant, the magnitude of the field due to each plate is the same at any point near the plates...

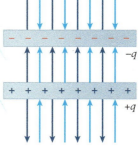

(c) ...so when the plates are placed close together, the fields cancel in the region outside the plates. The fields add in between the plates.

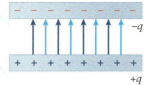

Figure 17-12 Electric field of a parallel-plate capacitor The field due to two oppositely charged plates is the vector sum of the fields due to each plate. Each plate in part (c) has area A, and the distance between the plates is d.

Note that the electric field points from the positive plate to the negative plate, just as we described in our discussion of conductors in Section 17-4.

(Equation 17-10 and the claim that $\vec{E}_+$ and $\vec{E}_-$ cancel outside the capacitor are strictly true only if the plates are large and close together, and when we consider points in space far from the edges of the capacitor. This idealization is physically reasonable for the kinds of problems we will encounter.)

We can substitute Equation 17-10 into Equation 17-7 to calculate the *potential difference* between the plates of a parallel-plate capacitor. The electric field points from the positive plate to the negative plate. If we let d be the distance between the plates, and we travel from the negative plate to the positive plate, the angle between the electric field and the displacement is $\theta = 180°$. The potential difference is therefore

$$V = V_+ - V_- = -Ed \cos \theta = -Ed(-1)$$
$$= Ed = \left(\frac{q}{\varepsilon_0 A} \right) d$$
(17-11)
$$= \frac{qd}{\varepsilon_0 A}$$

WATCH OUT! The symbol *V* sometimes stands for potential, sometimes for potential difference.

Previously in this chapter we've used V to denote the electric potential at a given point and ΔV to denote the difference in electric potential between two points. In Equation 17-11, and throughout our discussion of capacitors, we'll follow common practice and use V as the symbol for the potential difference, or voltage, between the two capacitor plates. Just remember "for capacitors, V means voltage." We'll frequently refer to V as the "voltage *across* the capacitor."

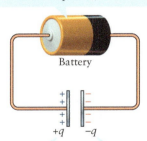

When this battery is connected to the plates of a capacitor, it creates a potential difference V between the plates...

Battery

$+q$ $-q$

...which drives electrons from one plate, through the battery, and onto the other plate. The plate that lost electrons ends up with charge $+q$, and the other plate ends up with charge $-q$.

Figure 17-13 **Charging a capacitor** The charges that appear on the two capacitor plates have the same magnitude but opposite signs.

Capacitance of a parallel-plate capacitor
(17-13)

 Go to Picture It 17-3 for more practice dealing with capacitance.

For a capacitor of plate area A and plate separation d, the potential difference V between the plates is proportional to the magnitude of the charge q on each plate.

We can rewrite Equation 17-11 as an expression for the amount of charge on each plate. To charge an initially uncharged capacitor, we apply a voltage V between its plates as in Figure 17-13. Electrons are driven to one of the plates, giving it a charge $-q$; the plate from which the electrons were taken is left with a charge $+q$. (As we mentioned previously the net charge on the capacitor remains zero.) Equation 17-11 tells us that the magnitude q of the charge that each plate acquires is proportional to the applied voltage V:

(17-12)
$$q = \frac{\varepsilon_0 A}{d} V$$

(charge on a parallel-plate capacitor)

The quantity $\varepsilon_0 A/d$ in Equation 17-12 is called the **capacitance** of the capacitor. We use the symbol C for capacitance:

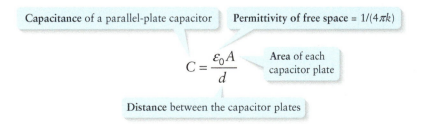

Capacitance of a parallel-plate capacitor   Permittivity of free space = $1/(4\pi k)$

$$C = \frac{\varepsilon_0 A}{d}$$

Area of each capacitor plate

Distance between the capacitor plates

For a parallel-plate capacitor, C depends only on the area A of the plates, the distance d between them, and the material between them. We've assumed that the plates are separated by vacuum; in Section 17-8 we'll explore what happens if the space between the capacitor plates is filled with a different material.

Capacitance tells us the amount of charge that can be stored on a capacitor held at a given voltage. From Equations 17-12 and 17-13,

A capacitor carries a charge $+q$ on its positive plate and a charge $-q$ on its negative plate.

The magnitude of q is directly proportional to V, the **voltage** (potential difference) between the plates.

$$q = CV$$

The constant of proportionality between charge q and voltage V is the **capacitance** C of the capacitor.

Charge, voltage, and capacitance for a capacitor
(17-14)

The greater the capacitance, the more charge is present for a given voltage. Note that while Equations 17-12 and 17-13 are valid for a parallel-plate capacitor only, Equation 17-14 is valid for capacitors of *any* geometry. In the problems at the end of this chapter, you'll analyze some other simple types of capacitor.

Equation 17-14 shows that the unit of capacitance is the coulomb per volt, also known as the **farad** (symbol F): 1 F = 1 C/V. The farad is named for the nineteenth-century English physicist Michael Faraday. The capacitors used in consumer electronics are typically in the range of 10 pF (10×10^{-12} F) to 10,000 μF (10,000 $\times 10^{-6}$ F); 1 F is an extremely large capacitance. Using 1 F = 1 C/V, 1 V = 1 J/C, and 1 J = 1 N·m, we see that 1 F = (1 C)/(1 J/C) = 1 C²/J = 1 C²/(N·m). We can express the value of the constant ε_0 as

$$\varepsilon_0 = 8.85 \times 10^{-12} \, \frac{C^2}{N \cdot m^2} = 8.85 \times 10^{-12} \, \frac{F}{m}$$

These units are useful in calculations of capacitance, as we'll see below.

An everyday application of Equation 17-13 is the touchscreen on a mobile device such as a smartphone or tablet (Figure 17-14). Behind the device's glass screen is a layer of a special transparent conductor called indium tin oxide (ITO), which is actually a solid mixture of indium oxide, In_2O_3, and tin oxide, SnO_2. When you touch your finger to the screen, the conducting ITO layer acts as one plate of a capacitor and your finger—which is also a conductor—acts as the other plate. Sensor circuits in the mobile device detect where on the screen a capacitance appears due to this capacitor, which is how the device "knows" where on the screen it has been touched. (To verify that the object touching the screen has to be a conductor, try using the rubber eraser on a pencil to touch the screen of a mobile device. Because rubber is an insulator, not a conductor, the screen won't respond.) The sensor circuits are adjusted so that they will register only if the capacitance C is above a certain minimum value. That explains why you have to physically touch the screen: According to Equation 17-13, C increases as the distance d between the plates decreases, so the capacitance is largest when your finger is touching the screen and so closest to the ITO layer. Your finger still acts as a capacitor plate if you hold it a slight distance away from the screen, but the capacitance is now too low to trigger the sensor circuits.

iStockphoto/Thinkstock

Figure 17-14 Capacitive touchscreens Mobile devices such as these use the physics of capacitors to determine where your finger touches the screen.

EXAMPLE 17-7 A Parallel-Plate Capacitor

Two square, parallel conducting plates each have dimensions 5.00 cm by 5.00 cm and are placed 0.100 mm apart. Determine the capacitance of this configuration.

Set Up

We'll use Equation 17-13 to determine the value of C for this capacitor.

Capacitance of a parallel-plate capacitor:

$$C = \frac{\varepsilon_0 A}{d} \qquad (17\text{-}13)$$

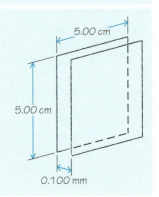

Solve

The area of each plate is the product of the length of the two sides, and the plate separation is given. We need to convert all dimensions into SI units.

Area $A = (5.00 \times 10^{-2} \text{ m})(5.00 \times 10^{-2} \text{ m}) = 2.50 \times 10^{-3} \text{ m}^2$

Plate separation $d = 1.00 \times 10^{-4} \text{ m}$

Using Equation 17-13,

$C = \dfrac{(8.85 \times 10^{-12} \text{ F/m})(2.50 \times 10^{-3} \text{ m}^2)}{1.00 \times 10^{-4} \text{ m}} = 2.21 \times 10^{-10} \text{ F}$

$= 221 \text{ pF}$

(Recall that the prefix "p," or "pico-" represents 10^{-12}.)

Reflect

Capacitors with capacitances of a few hundred picofarads are found in calculators and mobile phones.

To make a 1-F capacitor with the same separation $d = 0.100$ mm between plates, we would have to increase the plate area by a factor of $(1 \text{ F})/(2.21 \times 10^{-10} \text{ F}) = 4.52 \times 10^9$. The length of each side would have to be increased by a factor of the square root of 4.52×10^9, or 6.73×10^4. The sides of the plates would be $(6.73 \times 10^4)(5.00 \times 10^{-2} \text{ m}) = 3.36 \times 10^3$ m, or 3.36 *kilometers* (about 2 miles). One farad is a *very* large capacitance.

BioMedical **EXAMPLE 17-8** **Insulin Release**

The hormone insulin minimizes variations in blood glucose levels. Pancreatic beta cells ("β cells") synthesize insulin and store it in *vesicles*, bubble-like organelles approximately 150 nm in radius within the cytoplasm of the cells. To release insulin, vesicles fuse with the membrane of the β cell. This increases the surface area of the β cell by the surface area of the fused vesicles (Figure 17-15). The thickness of the cell membrane does not change, so the increase in surface area increases the capacitance of the β cell membrane. Experiment shows that the membrane capacitance increases at 1.6×10^{-13} F/s during insulin release. If the membrane capacitance is approximately 1 µF per square centimeter of surface area, estimate the number of vesicles that fuse with the cell membrane per second during insulin release.

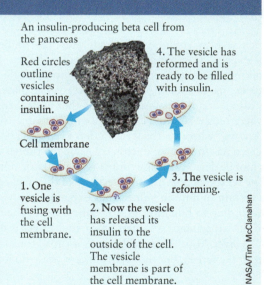

An insulin-producing beta cell from the pancreas

Red circles outline vesicles containing insulin.

Cell membrane

1. One vesicle is fusing with the cell membrane.

2. Now the vesicle has released its insulin to the outside of the cell. The vesicle membrane is part of the cell membrane.

3. The vesicle is reforming.

4. The vesicle has reformed and is ready to be filled with insulin.

NASA/Tim McClanahan

Figure 17-15 Changing membrane, changing capacitance To study this important process in a pancreatic β cell, scientists monitor changes in the capacitance of the cell membrane.

Set Up

Although the cell membrane is not flat like the parallel-plate capacitor shown in Figure 17-12, we can treat it as such because the size of the cell is large compared to the thickness of the cell membrane. (It's like being above Earth's surface at a height that's much smaller than Earth's radius. While Earth is approximately spherical, when seen from such a short distance, it appears to be flat.) We'll use Equation 17-13 to calculate the rate at which the area A of the cell membrane must increase to cause the measured rate of capacitance increase. This increase in area comes from fusing vesicles.

Capacitance of a parallel-plate capacitor:

$C = \dfrac{\varepsilon_0 A}{d}$ (17-13)

Surface area of a sphere of radius r:

$A_{\text{sphere}} = 4\pi r^2$

Solve

Find the rate at which the membrane area must increase to cause this rate of capacitance increase.

Rate of change of the capacitance of the membrane:

$$\frac{\Delta C_m}{\Delta t} = 1.6 \times 10^{-13} \text{ F/s}$$

Equation 17-13 tells us that the capacitance C_m is directly proportional to the membrane surface area A_m. We are told that the capacitance per unit area is approximately $1 \ \mu\text{F/cm}^2$, so the change in capacitance ΔC_m that corresponds to a given change in area ΔA_m is

$$\Delta C_m = \left(1 \frac{\mu\text{F}}{\text{cm}^2}\right) \Delta A_m$$

Since the capacitance increases by 1.6×10^{-13} F in 1 s, the increase in area of the membrane in 1 s must be

$$
\begin{aligned}
\Delta A_m &= \frac{\Delta C_m}{1 \ \mu\text{F/cm}^2} \\
&= (1.6 \times 10^{-13} \text{ F})\left(1 \frac{\text{cm}^2}{\mu\text{F}}\right)\left(\frac{1 \ \mu\text{F}}{10^{-6} \text{ F}}\right)\left(\frac{1 \text{ m}}{100 \text{ cm}}\right)^2 \\
&= 1.6 \times 10^{-11} \text{ m}^2
\end{aligned}
$$

(We used $1 \ \mu\text{F} = 1$ microfarad $= 10^{-6}$ F and 1 m = 100 cm.)

Now we can calculate how many vesicles must fuse with the membrane per second to cause this increase in area.

The surface area of a single vesicle is

$$
\begin{aligned}
A_{\text{vesicle}} &= 4\pi r_{\text{vesicle}}^2 = 4\pi (150 \text{ nm})^2 = 4\pi (150 \times 10^{-9} \text{ m})^2 \\
&= 2.8 \times 10^{-13} \text{ m}^2
\end{aligned}
$$

The number of vesicles that must add their area to the membrane in 1 s to cause an area increase $\Delta A_m = 1.6 \times 10^{-11} \text{ m}^2$ is

$$\frac{\Delta A_m}{A_{\text{vesicle}}} = \frac{1.6 \times 10^{-11} \text{ m}^2}{2.8 \times 10^{-13} \text{ m}^2/\text{vesicle}} = 57 \text{ vesicles}$$

Our final answer should have just one significant figure, since we were given the capacitance per unit area $1 \ \mu\text{F/cm}^2$ to just one significant figure. So our final answer is that about 60 vesicles fuse with the membrane wall per second.

Reflect

The ability to detect small changes in capacitance makes it possible to study the molecular mechanisms of hormone release from single cells and to verify previous biochemical measurements of insulin release from β cells. After insulin release, new vesicles form by pinching off from the cell membrane and are refilled with insulin. By recycling the vesicle membrane, β cells maintain their size over the long term.

GOT THE CONCEPT? 17-4 Capacitors

A parallel-plate capacitor has a potential difference V between its plates. If the potential difference is increased to $2V$, what effect does this have on the capacitance C?

(a) C increases by a factor of 4. (b) C increases by a factor of 2. (c) C becomes $\frac{1}{2}$ as great. (d) C becomes $\frac{1}{4}$ as great. (e) C is unchanged.

TAKE-HOME MESSAGE FOR Section 17-5

✔ A capacitor has two plates, one with charge $+q$ and the other with charge $-q$.

✔ For a given capacitor the potential difference between the plates is proportional to the magnitude of the charge q on each plate. The proportionality constant, called the capacitance, depends on the geometry of the plates and on the material in the space between the plates.

17-6 A capacitor is a storehouse of electric potential energy

In electric circuits, capacitors are most useful because of their ability to store electric potential energy. An applied voltage V such as that supplied by the battery in Figure 17-13 charges a capacitor by effectively pulling negative charges from one of the plates and depositing them on the other. To move the negative charges away from the positive charges requires work, and it is this work that results in electric potential energy being stored in the capacitor. At a later time the potential energy can be transferred by charge leaving the capacitor and passed on to other parts of the circuit. (That's what happens in the electronic flash unit in a camera or mobile phone. The device's battery charges a capacitor, and the energy stored in the charged capacitor is used to produce a short, intense burst of light.)

Let's see how to calculate the amount of electric potential energy stored in a capacitor that has charge $+q$ and $-q$ on its positive and negative plates, respectively. This is equal to the amount of electric potential energy that's added to the capacitor if we start with both plates uncharged (which we can regard as a state of zero potential energy) and move charge $+q$ from the first plate to the second one, leaving charge $-q$ on the first plate. If the potential difference between the plates had a constant value ΔV, Equation 17-6 tells us that the change in electric potential energy $\Delta U_{electric}$ in this process would be

$$\Delta V = \frac{\Delta U_{electric}}{q} \quad \text{so} \quad \Delta U_{electric} = q\Delta V$$

However, the potential difference between the plates (that is, the voltage across the capacitor) does *not* stay constant as we transfer charge from one plate to the other! Equation 17-14 tells us that the potential difference between the plates is proportional to the amount of charge on the positive plate. So as we transfer more charge from one plate to the other, the potential difference across which the charge must move increases from zero (its starting value) to a final value V given by Equation 17-14: $q = CV$, so $V = q/C$. To correctly calculate the amount of potential energy stored in the capacitor when it is charged, we have to replace ΔV in Equation 17-6 by the *average* value of the potential difference during the charging process. Because potential difference increases in direct proportion to the charge, this average value is just the average of the starting potential difference and the final potential difference V:

$$
\begin{aligned}
\textbf{(17-15)} \qquad \Delta U_{electric} &= q\Delta V_{average} \\
&= q\left[\frac{\text{(starting potential difference)} + \text{(final potential difference)}}{2}\right] \\
&= q\left(\frac{0+V}{2}\right) = \frac{1}{2}qV
\end{aligned}
$$

Equation 17-14 tells us that $q = CV$ and $V = q/C$, so we can also write Equation 17-15 as

$$\textbf{(17-16)} \qquad \Delta U_{electric} = \frac{1}{2}(CV)V = \frac{1}{2}CV^2 \quad \text{or} \quad \Delta U_{electric} = \frac{1}{2}q\left(\frac{q}{C}\right) = \frac{q^2}{2C}$$

If we say that the electric potential energy of the initial uncharged capacitor was zero, the final potential energy $U_{electric}$ is just equal to the increase in potential energy $\Delta U_{electric}$ given by Equation 17-15 or Equation 17-16. So we can write the potential energy stored in the capacitor in three ways:

Electric potential energy stored in a capacitor
(17-17)

The **electric potential energy** stored in a charged capacitor...

$$U_{electric} = \frac{1}{2}qV = \frac{1}{2}CV^2 = \frac{q^2}{2C}$$

...can be expressed in three ways in terms of the **charge** q, **potential difference** V, and **capacitance** C.

Equation 17-17 says that the energy stored in a charged capacitor is proportional to the *square* of the charge q or, equivalently, to the square of the potential difference (voltage) V across the capacitor. These results are true for capacitors of all kinds, not just the simple parallel-plate capacitor that we discussed in Section 17-5.

The last of the three expressions for electric potential energy in Equation 17-17, $U_{electric} = q^2/(2C)$, is very similar to the equation for *spring* potential energy that we learned in Section 6-6:

Spring potential energy of a stretched or compressed spring   Spring constant of the spring

**Spring potential energy
(6-23)**

$$U_{spring} = \frac{1}{2}kx^2$$

Extension of the spring ($x > 0$ if spring is stretched, $x < 0$ if spring is compressed)

The only differences between $U_{electric} = q^2/(2C)$ from Equation 17-17 and the expression in Equation 6-23 is that the spring displacement x is replaced by the charge q, and the spring constant k is replaced by the reciprocal of the capacitance, $1/C$. This similarity isn't surprising. To add potential energy to a spring by stretching it, you have to pull against the force of magnitude $F = kx$ that the spring exerts on you. The greater the distance the spring is already stretched, the more force it exerts and the harder it is to stretch it further. In exactly the same way, to add potential energy to a capacitor by increasing the magnitude of charge on the two plates, you have to transfer charge against a voltage $V = (1/C)q$. The greater the charge already on the plates, the greater the voltage and the harder it is to increase q.

BioMedical   EXAMPLE 17-9   A Defibrillator Capacitor

A defibrillator, like the one shown in the photograph that opens this chapter, is essentially a capacitor that is charged by a high-voltage source and then delivers the stored energy to a patient's heart. (a) How much charge does the 80.0-µF capacitor in a certain defibrillator store when it is fully charged by applying 2.50 kV? (b) How much energy can this defibrillator deliver?

Set Up

We're given the capacitance $C = 80.0$ µF (recall 1 µF = 1 micro farad = 10^{-6} F) and the potential difference $V = 2.50$ kV (recall 1 kV = 1 kilovolt = 10^3 V) between the capacitor plates. We'll use Equation 17-14 to determine the magnitude q of the charge on each capacitor plate and Equation 17-17 to find the potential energy $U_{electric}$ stored in the charged capacitor. If we assume that no energy is lost in the process of being transferred to the patient, this is equal to the energy that the defibrillator delivers.

Charge, voltage, and capacitance for a capacitor:
$$q = CV \tag{17-14}$$
Electric potential energy stored in a capacitor:
$$U_{electric} = \frac{1}{2}qV = \frac{1}{2}CV^2 = \frac{q^2}{2C} \tag{17-17}$$

Solve

(a) Substitute the given values of C and V into Equation 17-14 to solve for the charge q.

Charge on the capacitor plates:
$$q = CV = (80.0 \text{ µF})(2.50 \text{ kV})$$
$$= (80.0 \times 10^{-6} \text{ F})(2.50 \times 10^3 \text{ V})$$
$$= 0.200 \text{ F·V} = 0.200 \text{ C}$$
(Recall that 1 F = 1 C/V, so 1 F·V = 1 C.)

(b) Since the values of C and V are given, let's use the second of the three relationships in Equation 17-17 to find the stored electric potential energy.

Energy stored in the capacitor:
$$U_{electric} = \frac{1}{2}CV^2$$
$$= \frac{1}{2}(80.0 \times 10^{-6} \text{ F})(2.50 \times 10^3 \text{ V})^2$$
$$= 2.50 \times 10^2 \text{ F·V}^2$$
$$= 2.50 \times 10^2 \text{ J}$$
(Recall that 1 V = 1 J/C and 1 F = 1 C/V, so
1 F·V^2 = 1 [C/V]V^2 = 1 C·V = 1 C·[J/C] = 1 J.)

Reflect

The American Heart Association recommends that a defibrillator shock should deliver between 40 and 360 J to be effective, so our numerical result is in the recommended range.

We can double-check our answers by using the other two relationships in Equation 17-17. Happily, by using the value of q that we calculated in part (a), we get the same result for $U_{electric}$ in part (b) as we found above.

Remember that in solving any problem, you should *always* take advantage of alternative ways to find the answer in order to check your results.

One alternative way to calculate the energy stored in the capacitor:

$$U_{electric} = \frac{1}{2}qV$$
$$= \frac{1}{2}(0.200 \text{ C})(2.50 \times 10^3 \text{ V})$$
$$= 2.50 \times 10^2 \text{ C} \cdot \text{V}$$
$$= 2.50 \times 10^2 \text{ J}$$

Another alternative way to calculate the energy stored in the capacitor:

$$U_{electric} = \frac{q^2}{2C}$$
$$= \frac{(0.200 \text{ C})^2}{2(80.0 \times 10^{-6} \text{ F})}$$
$$= 2.50 \times 10^2 \frac{\text{C}^2}{\text{F}}$$
$$= 2.50 \times 10^2 \text{ J}$$

(Recall that 1 V = 1 J/C and 1 F = 1 C/V. You should verify that $1 \text{ C}^2/\text{F} = 1 \text{ J}$.)

GOT THE CONCEPT? 17-5 Spreading the Plates of a Capacitor I

Suppose you increase the distance between the plates of a charged parallel-plate capacitor without changing the amount of charge stored on the plates. What will happen to the energy stored in the capacitor? (a) It will decrease. (b) It will remain the same. (c) It will increase. (d) There is not enough information given to decide.

GOT THE CONCEPT? 17-6 Spreading the Plates of a Capacitor II

Suppose you increase the distance between the plates of a parallel-plate capacitor while holding constant the potential difference between the plates. (You could do this by keeping the plates connected to a battery, as in Figure 17-13.) What will happen to the energy stored in the capacitor? (a) It will decrease. (b) It will remain the same. (c) It will increase. (d) There is not enough information given to decide.

TAKE-HOME MESSAGE FOR Section 17-6

✔ A charged capacitor stores electric potential energy. The amount of energy stored is proportional to the square of the magnitude of the charge on each plate and also proportional to the square of the potential difference between the plates.

17-7 Capacitors can be combined in series or in parallel

In both biological systems and electric circuits, it is not uncommon for more than one capacitor to be connected in some way. The net result is that the capacitor combination behaves as though it were a *single* capacitor, with an **equivalent capacitance** that depends on the properties of the individual capacitors present. (An analogy is lifting a heavy conference table to move it across a room. Four people of normal strength could do the job, or it could be done by a single weightlifter. We would say that the four people together have an "equivalent strength" comparable to that of the weightlifter.) Our goal in this section is to find the equivalent capacitance in different situations.

Capacitors in Series

Figure 17-16 shows three initially uncharged capacitors with capacitances C_1, C_2, and C_3 that are connected end to end. The capacitors become charged when the combination is connected to a battery of voltage V as shown in the figure. Note that the negative plate of one capacitor is connected to the positive plate of the next capacitor in the combination. Capacitors connected in this way are said to be in **series**.

What is the equivalent capacitance of this series combination? To answer this question let's first determine the charges q_1, q_2, and q_3 on the individual capacitors and the voltages V_1, V_2, and V_3 across the individual capacitors. For each capacitor the magnitudes of the charge on the positive and negative plates must be equal. So as the battery draws negative charge from the left-hand plate of C_1, whatever positive charge $+q$ that plate acquires must be balanced by charge $-q$ on the right plate. The charging of the right plate of C_1 occurs as negative charge is drawn from the left plate of C_2, and because this whole section is initially uncharged, the left plate of C_2 acquires charge $+q$. If we apply the same reasoning to C_3, we see that all three capacitors acquire the *same* charge:

$$q_1 = q_2 = q_3 = q \tag{17-18}$$
(capacitors in series)

Thus the series combination of three capacitors is equivalent to a single capacitor with charge q. You can think of the charge on the negative plate of C_1 as canceling the charge on the positive plate of C_2 and likewise for the charges on the negative plate of C_2 and the positive plate of C_3.

The charges and voltages for each capacitor are also given by Equation 17-14, $q = CV$. If we substitute this into Equation 17-18, we get

$$q = C_1V_1 = C_2V_2 = C_3V_3$$
(capacitors in series)

which we can rewrite as expressions for the voltages across the individual capacitors:

$$V_1 = \frac{q}{C_1} \quad V_2 = \frac{q}{C_2} \quad V_3 = \frac{q}{C_3} \tag{17-19}$$
(capacitors in series)

Now, the voltage V of the battery equals the voltage across the combination of three capacitors. This is just the sum of the voltages across the individual capacitors:

$$V = V_1 + V_2 + V_3 \tag{17-20}$$
(capacitors in series)

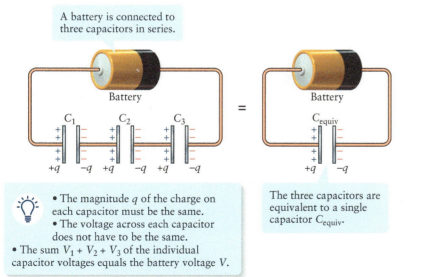

A battery is connected to three capacitors in series.

Battery C_1 C_2 C_3 $=$ Battery C_{equiv}

$+q$ $-q$ $+q$ $-q$ $+q$ $-q$ $+q$ $-q$

- The magnitude q of the charge on each capacitor must be the same.
- The voltage across each capacitor does not have to be the same.
- The sum $V_1 + V_2 + V_3$ of the individual capacitor voltages equals the battery voltage V.

The three capacitors are equivalent to a single capacitor C_{equiv}.

Figure 17-16 Capacitors in series What is the equivalent capacitance of this series combination?

If we substitute Equations 17-19 into Equation 17-20, we get a relationship between the charge q on each capacitor and the voltage V across the combination—that is, between the charge q on the equivalent capacitor and the voltage V across the equivalent capacitor:

(17-21)
$$V = \frac{q}{C_1} + \frac{q}{C_2} + \frac{q}{C_3} = q\left(\frac{1}{C_1} + \frac{1}{C_2} + \frac{1}{C_3}\right)$$

(capacitors in series)

For the equivalent capacitor alone, of capacitance C_{equiv}, Equation 17-14 says that $q = C_{\text{equiv}}V$ or $V = q/C_{\text{equiv}}$. If we compare this to Equation 17-21, we see that

 If capacitors are in series, the reciprocal of the equivalent capacitance is the sum of the reciprocals of the individual capacitances.

Equivalent capacitance of capacitors in series (17-22)

$$\frac{1}{C_{\text{equiv}}} = \frac{1}{C_1} + \frac{1}{C_2} + \frac{1}{C_3}$$

Equivalent capacitance of **capacitors in series**

Capacitances of the individual capacitors

If there are more than three capacitors in series, the same rule given in Equation 17-22 applies. The equivalent capacitance of a series combination is always *less* than the smallest capacitance of any of the individual capacitors.

Capacitors in Parallel

Figure 17-17 shows an alternative way to connect the three initially uncharged capacitors C_1, C_2, and C_3 to a battery with voltage V. Capacitors connected in this way are said to be in **parallel**. Notice the difference in the arrangement of the capacitors in parallel (Figure 17-17) compared to capacitors in series (Figure 17-16). In a series arrangement, the right-hand plate of one capacitor is connected to the left-hand plate of the capacitor to its right. In a parallel arrangement, all of the right-hand plates are connected, and all of the left-hand plates are connected.

To find the equivalent capacitance of capacitors in parallel, first note that all of the right-hand capacitor plates are connected to one terminal of the battery and all

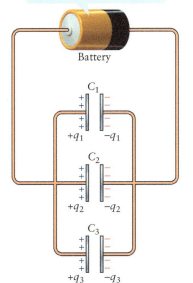

A battery is connected to three capacitors in parallel.

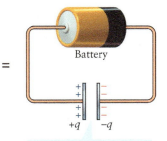

The three capacitors are equivalent to a single capacitor C_{equiv}.

 • The voltage V across each capacitor must be the same.
• The charges on each capacitor do not have to be the same.
• The sum $q_1 + q_2 + q_3$ of the individual capacitor charges equals the charge q on the equivalent capacitor.

Figure 17-17 Capacitors in parallel What is the equivalent capacitance of this parallel combination?

of the left-hand plates are connected to the other terminal. So each of the voltages V_1, V_2, and V_3 across the individual capacitors is equal to the voltage V across the battery:

$$V = V_1 = V_2 = V_3 \qquad \text{(17-23)}$$

(capacitors in parallel)

Equation 17-23 coupled with Equation 17-14, $q = CV$, then tells us the charges q_1, q_2, and q_3 on the individual capacitors:

$$q_1 = C_1V \quad q_2 = C_2V \quad q_3 = C_3V \qquad \text{(17-24)}$$

(capacitors in parallel)

The *total* charge acquired by all three capacitors is the sum of the charges q_1, q_2, and q_3. From Equation 17-24,

$$q = q_1 + q_2 + q_3 = C_1V + C_2V + C_3V$$
$$= (C_1 + C_2 + C_3)V \qquad \text{(17-25)}$$

(capacitors in parallel)

For the equivalent capacitor of capacitance C_{equiv} that corresponds to the three capacitors in parallel, Equation 17-14 says that $q = C_{\text{equiv}}V$. Comparing this to Equation 17-25 shows that

Equivalent capacitance of capacitors in parallel | Capacitances of the individual capacitors

$$C_{\text{equiv}} = C_1 + C_2 + C_3$$

Equivalent capacitance of capacitors in parallel (17-26)

 If capacitors are in parallel, the equivalent capacitance is the sum of the individual capacitances.

If there are more than three capacitors in parallel, the same rule given in Equation 17-26 applies. The equivalent capacitance of a parallel combination is always *greater* than the smallest capacitance of any of the individual capacitors.

Example 17-10 shows how to do calculations with capacitors in series and in parallel. (We'll see a biological application of these calculations in the following section.) Many real networks of capacitors are more complex: They have a mixture of series and parallel combinations. To find the equivalent capacitance of such a network, we identify any small grouping of capacitors that are either entirely in series or entirely in parallel, find the equivalent capacitance of each group, and then combine them in larger and larger groupings, using the series and parallel rules, until we have accounted for all of the capacitors in the network. Example 17-11 illustrates how to do this.

WATCH OUT! Capacitors in series and parallel have different properties.

❗ Here's a summary of the differences between series and parallel combinations of capacitors. In a *series* combination, each capacitor has the same charge, but there are different voltages across capacitors with different capacitances. In a *parallel* combination, there is the same voltage across each capacitor, but there are different charges on capacitors with different capacitances.

EXAMPLE 17-10 Two Capacitors in Series or in Parallel

(a) If two capacitors are connected in series, find their equivalent capacitance for the case when the capacitors have different capacitances $C_1 = 2.00\ \mu\text{F}$ and $C_2 = 4.00\ \mu\text{F}$ and for the case where both have the same capacitance $C_1 = C_2 = 2.00\ \mu\text{F}$. (b) Repeat part (a) for the two capacitors connected in parallel.

Set Up

We'll use Equations 17-22 and 17-26 to find the equivalent capacitance C_{equiv} in each case. Since there are only two capacitors in each combination, we drop the C_3 term from these equations.

Two capacitors in series:

$$\frac{1}{C_{\text{equiv}}} = \frac{1}{C_1} + \frac{1}{C_2} \qquad \text{(17-22)}$$

Two capacitors in parallel:

$$C_{\text{equiv}} = C_1 + C_2 \qquad \text{(17-26)}$$

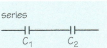

Solve

(a) We first apply Equation 17-22 to the case where $C_1 = 2.00\ \mu F$ and $C_2 = 4.00\ \mu F$.

With capacitors $C_1 = 2.00\ \mu F$ and $C_2 = 4.00\ \mu F$ in series, Equation 17-22 becomes

$$\frac{1}{C_{equiv}} = \frac{1}{2.00\ \mu F} + \frac{1}{4.00\ \mu F}$$
$$= 0.500\ \mu F^{-1} + 0.250\ \mu F^{-1} = 0.750\ \mu F^{-1}$$

To find C_{equiv} take the reciprocal of $1/C_{equiv}$:

$$C_{equiv} = \frac{1}{0.750\ \mu F^{-1}} = 1.33\ \mu F$$

Note that C_{equiv} is less than either C_1 or C_2.

Now repeat the calculation for the case where $C_1 = C_2 = 2.00\ \mu F$.

Follow the same steps but with both capacitances equal to $2.00\ \mu F$:

$$\frac{1}{C_{equiv}} = \frac{1}{2.00\ \mu F} + \frac{1}{2.00\ \mu F}$$
$$= 0.500\ \mu F^{-1} + 0.500\ \mu F^{-1} = 1.00\ \mu F^{-1}$$

$$C_{equiv} = \frac{1}{1.00\ \mu F^{-1}} = 1.00\ \mu F$$

For this case of identical capacitors in series, C_{equiv} is exactly one-half of $C_1 = C_2 = 2.00\ \mu F$.

(b) Apply Equation 17-26 to the case where $C_1 = 2.00\ \mu F$ and $C_2 = 4.00\ \mu F$.

With capacitors $C_1 = 2.00\ \mu F$ and $C_2 = 4.00\ \mu F$ in parallel, Equation 17-26 becomes

$$C_{equiv} = 2.00\ \mu F + 4.00\ \mu F = 6.00\ \mu F$$

Note that C_{equiv} is greater than either C_1 or C_2.

Now repeat the calculation for the case where $C_1 = C_2 = 2.00\ \mu F$.

Follow the same steps but with both capacitances equal to $2.00\ \mu F$:

$$C_{equiv} = 2.00\ \mu F + 2.00\ \mu F = 4.00\ \mu F$$

For this case of identical capacitors in parallel, C_{equiv} is exactly twice as great as $C_1 = C_2 = 2.00\ \mu F$.

Reflect

Our results illustrate the following general results: Connecting capacitors in series reduces the capacitance, while connecting them in parallel increases the capacitance.

EXAMPLE 17-11 **Multiple Capacitors**

Find the equivalent capacitance of the three capacitors shown in Figure 17-18. The individual capacitances are $C_1 = 1.00\ \mu F$, $C_2 = 2.00\ \mu F$, and $C_3 = 6.00\ \mu F$.

Figure 17-18 A capacitor network These three capacitors are neither all in series nor all in parallel.

Set Up

Whenever capacitors are combined in ways other than purely in series or purely in parallel, we look for groupings of capacitors that *are* either in parallel or in series, and then we combine the groups one at a time.

Two capacitors in series:

$$\frac{1}{C_{equiv}} = \frac{1}{C_1} + \frac{1}{C_2} \qquad (17\text{-}22)$$

Replace the parallel capacitors C_1 and C_2 by their equivalent capacitor C_{12}:

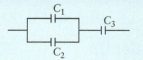

Then find the equivalent capacitance of C_{12} and C_3 in series.

Notice in Figure 17-18 that C_1 and C_2 are in parallel because their two right plates are directly connected, as are their two left plates. We can therefore use Equation 17-26 to find C_{12} (the equivalent capacitance of the combination of C_1 and C_2) by using our relationship for capacitors in parallel.

Capacitor C_3 is in series with C_{12}, so we can find their combined capacitance by using Equation 17-22. This result is C_{123}, the equivalent capacitance of all three capacitors.

Two capacitors in parallel:

$$C_{equiv} = C_1 + C_2 \qquad\qquad (17\text{-}26)$$

Solve

First find the equivalent capacitance of the parallel capacitors C_1 and C_2.

For $C_1 = 1.00~\mu F$ and $C_2 = 2.00~\mu F$ in parallel, the equivalent capacitance C_{12} is given by Equation 17-26:

$$C_{12} = C_1 + C_2 = 1.00~\mu F + 2.00~\mu F = 3.00~\mu F$$

Then find C_{123}, the equivalent capacitance of the series capacitors C_{12} and C_3. This is the equivalent capacitance of the entire network of C_1, C_2, and C_3.

Since $C_{12} = 3.00~\mu F$ and $C_3 = 6.00~\mu F$ are in series, their equivalent capacitance C_{123} is given by Equation 17-22:

$$\frac{1}{C_{123}} = \frac{1}{C_{12}} + \frac{1}{C_3} = \frac{1}{3.00~\mu F} + \frac{1}{6.00~\mu F}$$
$$= 0.333~\mu F^{-1} + 0.167~\mu F^{-1} = 0.500~\mu F^{-1}$$

Take the reciprocal of this to find C_{123}:

$$C_{123} = \frac{1}{0.500~\mu F^{-1}} = 2.00~\mu F$$

Reflect

As we described previously, when capacitors are connected in parallel, the equivalent capacitance is always greater than the greatest individual capacitor. That's why C_{12} is greater than either C_1 or C_2. When capacitors are connected in series, the equivalent capacitance is always less than the least individual capacitor, which is why C_{123} is less than either C_{12} or C_3.

GOT THE CONCEPT? 17-7 Energy in a Capacitor Combination

? Capacitor 1 has capacitance $C_1 = 1.0~\mu F$ and capacitor 2 has capacitance $C_2 = 2.0~\mu F$. You connect the initially uncharged capacitors to each other, then connect the capacitor combination to a battery. Which capacitor stores the greater amount of electric potential energy if the two capacitors are connected in series? If they are connected in parallel? (a) Capacitor 1 for both the series and parallel cases; (b) capacitor 2 for both the series and parallel cases; (c) capacitor 1 for the series case, capacitor 2 for the parallel case; (d) capacitor 2 for the series case, capacitor 1 for the parallel case; (e) not enough information given to decide.

TAKE-HOME MESSAGE FOR Section 17-7

✔ Whenever two or more capacitors are connected, the equivalent capacitance is the capacitance of a single capacitor that is the equivalent of the combined capacitors.

✔ Capacitors are connected in series when they are connected one after another. In a series combination, all capacitors have the same charge, and the equivalent capacitance is less than that of any of the individual capacitors.

✔ Capacitors are connected in parallel when all of their right-hand plates are directly connected to each other and all of their left-hand plates are directly connected to each other. In a parallel combination, the voltage is the same across all capacitors, and the equivalent capacitance is greater than that of any of the individual capacitors.

17-8 Placing a dielectric between the plates of a capacitor increases the capacitance

So far in our discussion of capacitors, we've assumed that there is only vacuum between the capacitor plates. In most situations, however, the two plates are separated by a layer of a **dielectric**, a material that is both an insulator and *polarizable*—that is, in which there is a separation of positive and negative charge within the material when it's exposed to an electric field. Dielectrics used in commercial capacitors include glass, ceramics, and plastics such as polystyrene. These dielectrics not only help to keep the positive and negative plates from touching each other but also increase the capacitance of the capacitor. In this section we'll see how dielectrics make this possible.

Let's consider an isolated parallel-plate capacitor with charges $+q$ and $-q$ on its plates and with vacuum in the space between its plates (Figure 17-19a). The charge creates a uniform electric field $\vec{E}_0$ between the plates. We now insert a dielectric material that fills the space between the plates of this capacitor. What happens then depends on the kind of molecules that make up the dielectric.

If the molecules are *polar*—that is, if one end of the molecule has a positive charge and the other end has a negative charge of the same magnitude—the molecules will orient themselves so that their positive ends are pointed toward the negatively charged plate and their negative ends toward the positively charged plate, as in Figure 17-19b. (The most common molecule in your body, the water molecule H_2O, is a polar molecule. Others include ammonia, NH_3, and sucrose, $C_{12}H_{22}O_{11}$.)

If the molecules are not polar, so there normally is no separation of charge within the molecule, the electric field between the plates of the capacitor will *induce* a slight separation of the positive and negative charges in each molecule. (We described this

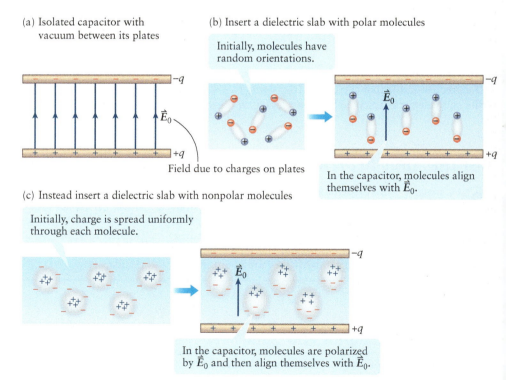

(a) Isolated capacitor with vacuum between its plates

(b) Insert a dielectric slab with polar molecules

Initially, molecules have random orientations.

Field due to charges on plates

In the capacitor, molecules align themselves with $\vec{E}_0$.

(c) Instead insert a dielectric slab with nonpolar molecules

Initially, charge is spread uniformly through each molecule.

In the capacitor, molecules are polarized by $\vec{E}_0$ and then align themselves with $\vec{E}_0$.

(d) With either type of dielectric, the result is layers of positive and negative charge on the dielectric surfaces. These produce a field $\vec{E}_{dielectric}$.

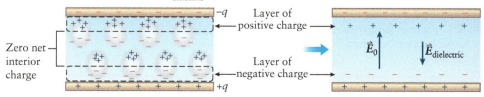

Zero net interior charge

Layer of positive charge

Layer of negative charge

Figure 17-19 A dielectric in a parallel-plate capacitor A dielectric slab inserted between the plates of an isolated capacitor reduces the electric field between the plates.

process in Section 16-2.) The molecules will then orient themselves in the same manner as polar molecules would (Figure 17-19c). In either case we say that the dielectric becomes *polarized* once it has been inserted between the plates of the capacitor.

In the interior of the polarized dielectric in Figures 17-17b or 17-17c, there are as many positive ends of molecules as there are negative ends, so the interior of the dielectric is neutral. But there *is* a net charge at the surfaces of the dielectric: There is a layer of positive charge on the surface next to the negatively charged plate, and there is a layer of negative charge on the surface next to the positively charged plate. (The dielectric as a whole is neutral, so these two layers have the same magnitude of charge.) These two layers of charge create an electric field $\vec{E}_{dielectric}$ that points opposite to the electric field $\vec{E}_0$ created by the charges on the plates of the capacitor (Figure 17-19d). Hence $\vec{E}_{dielectric}$ partially cancels $\vec{E}_0$, and the *net* electric field $\vec{E} = \vec{E}_0 + \vec{E}_{dielectric}$ between the plates is less than $\vec{E}_0$. So the effect of the dielectric is to *reduce* the electric field between the plates. We can express this as

Electric field in an isolated parallel-plate capacitor **with dielectric** between the plates

Electric field in that same capacitor with **no dielectric** between the plates

$$E = \frac{E_0}{\kappa}$$

Dielectric constant; note that $\kappa \geq 1$

Electric field in an isolated parallel-plate capacitor with a dielectric (17-27)

 For an isolated charged capacitor, a dielectric reduces the electric field between the plates.

The quantity κ (the Greek letter "kappa") is the **dielectric constant** of the material. The greater the value of κ, the more the dielectric is polarized when it is placed between the plates of a charged capacitor, so the greater the magnitude of the field $\vec{E}_{dielectric}$ due to the dielectric and the smaller the magnitude E of the net field. Table 17-1 lists values of the dielectric constant for a variety of materials.

What does Equation 17-27 tell us about how the dielectric affects the capacitance of the capacitor? We saw in Section 17-5 that the potential difference V between the positive and negative plates is proportional to the magnitude E of the electric field between the plates (see Equation 17-11, which states that $V = Ed$). Equation 17-27 says that the dielectric reduces the value of the electric field by a factor $1/\kappa$, so V is also reduced by a factor $1/\kappa$. Now, Equation 17-14 tells us that the capacitance C is related to potential difference V and the magnitude q of the charges on the plates by $q = CV$, or $C = q/V$. While the dielectric reduces V by a factor $1/\kappa$, it has no effect on the value of q. (Since we specified that the capacitor is isolated, no charge can leave either plate.) Because the capacitance $C = q/V$ is inversely proportional to the potential difference V, which is reduced by a factor $1/\kappa$, it follows that C *increases* by a factor of $1/(1/\kappa) = \kappa$. If C_0 is the capacitance without the dielectric, the capacitance with the dielectric is

TABLE 17-1 **Dielectric Constants (at 20°C and 1 atm)**	
Material	κ
vacuum	1
air	1.00058
lipid	2.2
paraffin	2.2
paper	2.7
ceramic (porcelain)	5.8
water	80

Capacitance of a parallel-plate capacitor **with dielectric** between the plates

Capacitance of that same capacitor with **no dielectric** between the plates

$$C = \kappa C_0$$

Capacitance of a parallel-plate capacitor with a dielectric (17-28)

 A dielectric increases the capacitance of a capacitor.

Dielectric constant; note that $\kappa \geq 1$

As an example, Table 17-1 tells us that the dielectric constant of porcelain is $\kappa = 5.8$. From Equation 17-28 the capacitance of a capacitor with porcelain filling the space between its plates is 5.8 times greater than an identical capacitor with vacuum between its plates. That's fundamentally because the electric field in the porcelain-filled capacitor is only $1/5.8 = 0.17$ as great as for an identical vacuum capacitor carrying the same charge (Equation 17-27).

Example 17-12 shows how to apply these ideas to the membrane of a cell, which we can regard as a capacitor with multiple layers of dielectric between its plates.

BioMedical EXAMPLE 17-12 Cell Membrane Capacitance

In all cells, it is easier for positive potassium ions (K^+) to flow out of the cell than it is for negative ions. As a result, there is negative charge on the inside of the cell membrane and positive charge on the outside of the membrane, much like a capacitor. Consider a typical cell with a membrane of thickness 7.60 nm $= 7.60 \times 10^{-9}$ m. (a) Find the capacitance of a square patch of membrane 1.00 μm $= 1.00 \times 10^{-6}$ m on a side, assuming that the membrane has dielectric constant $\kappa = 1$. (b) The actual structure of the membrane is a layer of lipid surrounded by layers of a polarized aqueous solution and layers of water (Figure 17-20). Taking account of this structure, find the capacitance of a square patch of membrane 1.00 μm on a side.

Figure 17-20 Membrane capacitance A cell membrane (shown here in highly simplified form) acts as a capacitor filled with a dielectric. What is the capacitance of a membrane that consists of a layer of lipid surrounded by layers of a polarized aqueous solution and water?

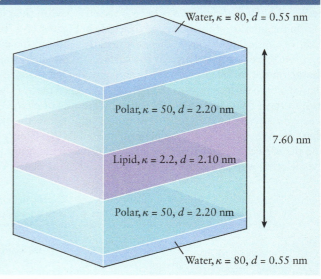

Water, $\kappa = 80$, $d = 0.55$ nm
Polar, $\kappa = 50$, $d = 2.20$ nm
7.60 nm
Lipid, $\kappa = 2.2$, $d = 2.10$ nm
Polar, $\kappa = 50$, $d = 2.20$ nm
Water, $\kappa = 80$, $d = 0.55$ nm

Set Up

A dielectric constant $\kappa = 1$ corresponds to vacuum, so in part (a) we can find the capacitance C_0 of the membrane under the assumption $\kappa = 1$ by using Equation 17-13 from Section 17-5.

In part (b) we must deal with the more complicated situation shown in Figure 17-20. We can't simply use Equation 17-28 to find the capacitance because there's not a single value of κ for the arrangement of layers that make up the membrane. Instead we'll imagine that there's a very thin conducting sheet separating each layer from the next. If there are charges $+q$ and $-q$ on the outer and inner surfaces of the multilayer membrane, there will also be charges $+q$ and $-q$ on the surfaces of these conducting sheets. This is exactly like the situation with capacitors in series (Section 17-7). So we'll treat each of the five layers in Figure 17-20 as an individual capacitor, then use Equation 17-22 to find the capacitance of the combination.

Parallel-plate capacitor with vacuum between the plates:

$$C_0 = \frac{\varepsilon_0 A}{d} \qquad (17\text{-}13)$$

Parallel-plate capacitor with dielectric:

$$C = \kappa C_0 \qquad (17\text{-}28)$$

Five capacitors in series:

$$\frac{1}{C_{equiv}} = \frac{1}{C_1} + \frac{1}{C_2} + \frac{1}{C_3} + \frac{1}{C_4} + \frac{1}{C_5} \qquad (17\text{-}22)$$

layer 1
layer 2
layer 3
layer 4
layer 5

$$= \begin{array}{c} C_1 \\ C_2 \\ C_3 \\ C_4 \\ C_5 \end{array}$$

Solve

(a) First use Equation 17-13 to find the capacitance, assuming that the membrane has $\kappa = 1$ (equivalent to vacuum).

The area A for the capacitor of interest is a square 1.00 μm on a side:
$$A = (1.00 \text{ μm})^2 = (1.00 \times 10^{-6} \text{ m})^2$$
$$= 1.00 \times 10^{-12} \text{ m}^2$$

A parallel-plate capacitor with this area and with $d = 7.60 \times 10^{-9}$ m has capacitance
$$C_0 = \frac{\varepsilon_0 A}{d} = \frac{(8.85 \times 10^{-12} \text{ F/m})(1.00 \times 10^{-12} \text{ m}^2)}{7.60 \times 10^{-9} \text{ m}}$$
$$= 1.16 \times 10^{-15} \text{ F}$$

(b) Calculate the capacitance of each of the five layers separately using Equations 17-13 and 17-28.

The capacitance of the first layer of water is $\kappa_1 = 80$ times the capacitance of a vacuum capacitor of thickness $d_1 = 0.55$ nm:
$$C_1 = \kappa_1 \left(\frac{\varepsilon_0 A}{d_1} \right) = (80) \frac{(8.85 \times 10^{-12} \text{ F/m})(1.00 \times 10^{-12} \text{ m}^2)}{0.55 \times 10^{-9} \text{ m}}$$
$$= 1.3 \times 10^{-12} \text{ F}$$

Similarly, for the second layer of polarized aqueous solution with $\kappa_2 = 50$ and thickness $d_2 = 2.20$ nm, the capacitance is

$$C_2 = \kappa_2 \left(\frac{\varepsilon_0 A}{d_2} \right)$$

$$= (50) \frac{(8.85 \times 10^{-12} \text{ F/m})(1.00 \times 10^{-12} \text{ m}^2)}{2.20 \times 10^{-9} \text{ m}}$$

$$= 2.0 \times 10^{-13} \text{ F}$$

The capacitance of the third lipid layer with $\kappa_3 = 2.2$ and thickness $d_3 = 2.10$ nm is

$$C_3 = \kappa_3 \left(\frac{\varepsilon_0 A}{d_3} \right)$$

$$= (2.2) \frac{(8.85 \times 10^{-12} \text{ F/m})(1.00 \times 10^{-12} \text{ m}^2)}{2.10 \times 10^{-9} \text{ m}}$$

$$= 9.3 \times 10^{-15} \text{ F}$$

The fourth layer of polarized aqueous solution is identical to the second layer, and the fifth water layer is identical to the first layer. So

$$C_4 = C_2 = 2.0 \times 10^{-13} \text{ F}$$
$$C_5 = C_1 = 1.3 \times 10^{-12} \text{ F}$$

Calculate the net capacitance of the five layers in series using Equation 17-22.

The equivalent capacitance C_{equiv} of the five-layer stack is given by

$$\frac{1}{C_{equiv}} = \frac{1}{C_1} + \frac{1}{C_2} + \frac{1}{C_3} + \frac{1}{C_4} + \frac{1}{C_5}$$

Since $C_4 = C_2$ and $C_5 = C_1$, this is

$$\frac{1}{C_{equiv}} = \frac{1}{C_1} + \frac{1}{C_2} + \frac{1}{C_3} + \frac{1}{C_2} + \frac{1}{C_1} = \frac{2}{C_1} + \frac{2}{C_2} + \frac{1}{C_3}$$

$$= \frac{2}{1.3 \times 10^{-12} \text{ F}} + \frac{2}{2.0 \times 10^{-13} \text{ F}} + \frac{1}{9.3 \times 10^{-15} \text{ F}}$$

$$= 1.2 \times 10^{14} \text{ F}^{-1}$$

The reciprocal of this is C_{equiv}:

$$C_{equiv} = \frac{1}{1.2 \times 10^{14} \text{ F}^{-1}} = 8.4 \times 10^{-15} \text{ F}$$

Reflect

The equivalent capacitance of the stack of five layers is less than that of any individual layer. This is just what we saw in Section 17-7 for capacitors in series.

Compared to the capacitance $C_0 = 1.16 \times 10^{-15}$ F calculated assuming $\kappa = 1$, the capacitance $C_{equiv} = 8.4 \times 10^{-15}$ F is greater by a factor $C_{equiv}/C_0 = (8.4 \times 10^{-15} \text{ F})/(1.16 \times 10^{-15} \text{ F}) = 7.2$. So the effective dielectric constant of the membrane is 7.2, intermediate between the largest ($\kappa_1 = 80$) and smallest ($\kappa_3 = 2.2$) values of dielectric constant for the individual layers.

GOT THE CONCEPT? 17-8   Inserting a Dielectric into a Capacitor

(?) An isolated parallel-plate capacitor (one that is not connected to anything else) has a vacuum between its plates and has charges $+q$ and $-q$ on its plates. If you insert a slab of dielectric with dielectric constant κ that fills the space between the plates, the capacitance increases. What happens to the energy stored in the capacitor as you insert the dielectric slab? Will you have to push the slab into the capacitor, or will you feel the capacitor pulling the slab in? (a) Stored energy increases; you will have to push the slab in. (b) Stored energy increases; the slab will be pulled in. (c) Stored energy decreases; you will have to push the slab in. (d) Stored energy decreases; the slab will be pulled in. (e) There is not enough information given to decide.

TAKE-HOME MESSAGE FOR Section 17-8

✔ When a dielectric material is placed in an electric field $\vec{E}_0$, it becomes polarized. This produces an additional electric field that partially cancels $\vec{E}_0$ and so reduces the net field in the dielectric. The dielectric constant is a measure of how much the field is reduced.

✔ If a dielectric material fills the space between the plates of a capacitor, the capacitance is greater than if there is vacuum between the plates.

Key Terms

capacitance
capacitor
dielectric
dielectric constant
electric potential
electric potential
  difference

electric potential energy
electron volt
equipotential
equipotential surface
equipotential volume
equivalent capacitance
farad

parallel (capacitors)
parallel-plate capacitor
plates
series (capacitors)
test charge
volt
voltage

Chapter Summary

Topic	Equation or Figure

Electric potential energy: Like the gravitational force, the electric force is a conservative force and has an associated potential energy. The change in electric potential energy when a charged object moves from one point to another depends on the sign of the object's charge. The sign of the electric potential energy of two point charges depends on whether the two charges have the same sign or different signs.

The **change in electric potential energy** for an object of charge q that moves in a uniform electric field $\vec{E}$...

...equals the **negative of the work done** on the object by the electric force.

$$\Delta U_{\text{electric}} = -W_{\text{electric}}$$
$$= -qEd\cos\theta$$

Charge of the object

Angle between the displacement and the direction of the electric field

Magnitude of the electric field

Straight-line **displacement** of the object

(17-2)

Coulomb constant   Values of the **two charges**

Electric potential energy of two point charges

$$U_{\text{electric}} = \frac{kq_1q_2}{r}$$

Distance between the point charges

(17-3)

Electric potential: The electric potential at a given position equals the electric potential energy for a point charge q_0 at that position, divided by the value of q_0. If you move in a direction opposite to the electric field, the electric potential increases; if you move in the direction of the electric field, the electric potential decreases. The electric potential due to a point charge is inversely proportional to the distance from the charge. Voltage is the difference in electric potential between two locations.

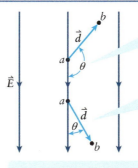

If you move in a direction opposite to the electric field ($\theta > 90°$), the electric potential increases ($\Delta V > 0$).

If you move in the direction of the electric field ($\theta < 90°$), the electric potential decreases ($\Delta V < 0$).

 If you move from point a to point b in a uniform electric field $\vec{E}$, the change in electric potential is $\Delta V = -Ed\cos\theta$.

(Figure 17-5)

Electric potential due to a point charge Q **Coulomb constant**

$$V = \frac{kQ}{r}$$ Value of the point **charge**   (17-8)

Distance from the point charge Q to the location where the potential is measured

---

Equipotentials: The electric potential has the same value everywhere on an equipotential surface. An equipotential surface is everywhere perpendicular to the electric field. The electric potential has the same value throughout the volume of a conductor in equilibrium.

A plane perpendicular to a uniform electric field $\vec{E}$ is an **equipotential surface:** Potential V has the same value at all points on this surface.

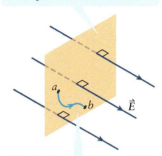

(Figure 17-7)

Any curve that lies on an equipotential surface, such as this curve from a to b, is an equipotential curve.

Capacitors: A capacitor has two conducting plates that store charges $+q$ and $-q$, of the same magnitude but opposite sign. The potential difference V between the plates is proportional to the charge q; the proportionality constant, the capacitance C, depends only on the geometry of the capacitor and the substance between the plates.

A capacitor carries a charge $+q$ on its positive plate and a charge $-q$ on its negative plate.

The magnitude of q is directly proportional to V, the **voltage** (potential difference) between the plates.

$$q = CV$$ (17-14)

The constant of proportionality between charge q and voltage V is the **capacitance** C of the capacitor.

Energy stored in a capacitor: A charged capacitor stores electric potential energy. The stored energy can be expressed by any two of the three quantities: capacitance C, charge q, and potential difference V.

The **electric potential energy** stored in a charged capacitor...

$$U_{\text{electric}} = \frac{1}{2}qV = \frac{1}{2}CV^2 = \frac{q^2}{2C}$$ (17-17)

...can be expressed in three ways in terms of the **charge** q, **potential difference** V, and **capacitance** C.

Capacitors in series and parallel: A collection of capacitors in a circuit behaves as though it were a single capacitor, with an equivalent capacitance C_{equiv}. The equivalent capacitance is different for capacitors in series (with the positive plate of one capacitor connected to the negative plate of another) than for capacitors in parallel (with positive

 If capacitors are in series, the reciprocal of the equivalent capacitance is the sum of the reciprocals of the individual capacitances.

$$\frac{1}{C_{\text{equiv}}} = \frac{1}{C_1} + \frac{1}{C_2} + \frac{1}{C_3}$$ (17-22)

Equivalent capacitance of **capacitors in series**   **Capacitances** of the individual capacitors plates connected to positive plates and negative plates to negative plates).

> **Equivalent capacitance of capacitors in parallel** **Capacitances of the individual capacitors**

$$C_{equiv} = C_1 + C_2 + C_3 \qquad (17\text{-}26)$$

 If capacitors are in parallel, the equivalent capacitance is the sum of the individual capacitances.

Dielectrics: A dielectric is an insulator that becomes polarized when placed in an electric field. When the space between the plates of a capacitor is filled with a dielectric, the electric field between the plates decreases (for a given charge q on the plates) and the capacitance increases.

> **Capacitance of a parallel-plate capacitor with dielectric between the plates** **Capacitance of that same capacitor with no dielectric between the plates**

$$C = \kappa C_0 \qquad (17\text{-}28)$$

 A dielectric increases the capacitance of a capacitor.

Dielectric constant; note that $\kappa \geq 1$

Answer to What do you think? Question

(c) The energy stored by separating charge $+q$ from charge $-q$ in a capacitor is proportional to q^2 (see Section 17-6, especially Equation 17-17). So doubling the value of q increases the stored energy by a factor of $2^2 = 4$.

Answers to Got the Concept? Questions

17-1 (b) Although q_1, q_2, and q_3 all have reversed sign, the signs of the *products* $q_1 q_2$, $q_1 q_3$, and $q_2 q_3$ in Equation 17-4 remain the same. So if you change all of the positive charges in an assemblage to negative and vice versa, there is no effect on the electric potential energy of the assemblage.

17-2 (c), (a), (d), (b). From Equation 17-6, the potential energy change $\Delta U_{electric}$ for a point charge q_0 is related to the potential change ΔV by $\Delta V = \Delta U_{electric}/q_0$, or $\Delta U_{electric} = q_0 \Delta V$. For the four cases we have (a) $\Delta U_{electric} = (+0.0010\ C)(5.0\ V) = +0.0050\ C \cdot V = +0.0050\ J$; (b) $\Delta U_{electric} = (+0.0020\ C)(-2.0\ V) = -0.0040\ J$; (c) $\Delta U_{electric} = (-0.0015\ C)(-4.0\ V) = +0.0060\ J$; (d) $\Delta U_{electric} = (-0.0010\ C)(+2.5\ V) = -0.0025\ J$.

17-3 (b) In the vicinity of A, the potential decreases by 1 V (from $V = 0$ V to $V = -1$ V) in the 2-m distance between $x = 0$ and $x = 2$ m. The electric field $\vec{E}$ points in the direction of decreasing potential, which is to the right (the positive x direction) at both points. In the vicinity of B, the equipotentials are much closer together; the potential decreases by 1 V (from $V = -2$ V to $V = -3$ V) in the 0.5-m distance from $x = 3$ m to $x = 3.5$ m. Where equipotentials are farther apart, as at A, the electric field has a smaller magnitude.

17-4 (e) Equation 17-13 shows that the capacitance of a parallel-plate capacitor depends on its geometry only. It does not depend on the amount of charge on the plates or on the potential difference between the plates.

17-5 (c) The stored energy increases. The positive and negative charges on the two plates attract each other, so to pull the plates apart you must do positive work on the system.

The work you do increases the electric potential energy stored in the capacitor. To verify this conclusion, note from Equation 17-13 that if you increase the spacing d between the plates, the capacitance $C = \varepsilon_0 A/d$ will decrease. Equation 17-17, $U_{electric} = q^2/2C$, then tells us that if the charge q remains constant and the capacitance C decreases, the stored energy $U_{electric}$ will increase.

17-6 (a) The stored energy decreases. Equation 17-13 tells us that as you increase the spacing d between the plates, the capacitance $C = \varepsilon_0 A/d$ decreases. Equation 17-17, $U_{electric} = (1/2)CV^2$, then tells us that if the potential difference V remains constant and C decreases, the stored energy must decrease. Note that in the situation of Got the Concept? 17-5, the stored energy *increases* when the plates are pulled apart. Why is the answer different here? The explanation is that in Got the Concept? 17-5 the charge q remained constant. In the current situation, however, the charge q has to change in accordance with Equation 17-14, $q = CV$: The charge q decreases as the capacitance C decreases. (Electrons flow from the negative plate through the battery and onto the positive plate, so the magnitude of charge on both plates decreases.) Equation 17-17 also tells us that we can express the stored energy as $U_{electric} = (1/2)qV$, so the stored energy will decrease if q decreases while V remains the same.

17-7 (c) Equation 17-17 gives us two useful expressions for the electric potential energy stored in a capacitor with capacitance C: $U_{electric} = (1/2)CV^2$ (which says that for a given voltage V the stored energy is proportional to C) and $U_{electric} = q^2/2C$ (which says that for a given charge q the stored energy is inversely proportional to C). In a series combination both capacitors carry the same charge q, so $U_{electric} = q^2/2C$ tells us that the capacitor with the smaller capacitance ($C_1 = 1.0$ μF) stores the greater amount of energy. In a parallel combination, the voltage V is the same for both capacitors, so $U_{electric} = (1/2)CV^2$ tells us that the capacitor with the greater capacitance ($C_2 = 2.0$ μF) stores the greater amount of energy. We can draw these conclusions without knowing the specific values of q for the series case or V for the parallel case.

17-8 (d) Because the capacitor is isolated, the charge q does not change when the dielectric slab is inserted. So among the expressions for stored energy given in Equation 17-17, the one to use is $U_{electric} = q^2/2C$. Since q remains constant but capacitance C increases as you insert the slab, it follows that $U_{electric}$ decreases. Just as a ball is pulled downward by the gravitational force, in the direction that decreases gravitational potential energy, the slab must be pulled into the capacitor by an electric force because motion in that direction decreases electric potential energy $U_{electric}$.

Questions and Problems

In a few problems you are given more data than you actually need; in a few other problems you are required to supply data from your general knowledge, outside sources, or informed estimate.

Interpret as significant all digits in numerical values that have trailing zeros and no decimal points.

For all problems use $g = 9.80$ m/s² for the free-fall acceleration due to gravity. Neglect friction and air resistance unless instructed to do otherwise.

* •   Basic, single-concept problem
* ••   Intermediate-level problem; may require synthesis of concepts and multiple steps
* •••   Challenging problem
* Example   See worked example for a similar problem

Conceptual Questions

1. • What is the difference between electric potential and electric field?

2. • What is the difference between electric potential and electric potential energy?

3. • Explain why electric potential requires the existence of only one charge, but a finite electric potential energy requires the existence of two charges.

4. • An electron is released from rest in an electric field. Will it accelerate in the direction of increasing or decreasing potential? Why?

5. • Does it make sense to say that the voltage at some point in space is 10.3 V? Explain your answer.

6. •• Explain why an electron will accelerate toward a region of lower electric potential energy but higher electric potential.

7. • (a) If the electric potential throughout some region of space is zero, does it necessarily follow that the electric field is zero? (b) If the electric field throughout a region is zero, does it necessarily follow that the electric potential is zero?

8. • Discuss how a topographical map showing various elevations around a mountain is analogous to the equipotential lines surrounding a charged object.

9. • How much work is required to move a charge from one end of an equipotential path to the other? Explain your answer.

10. • Explain why capacitance depends neither on the stored charge Q nor on the potential difference V between the plates of a capacitor.

11. • Describe three methods by which you might increase the capacitance of a parallel-plate capacitor.

12. • If the voltage across a capacitor is doubled, by how much does the stored energy change?

13. •• You charge a capacitor and then remove it from the battery. The capacitor consists of large movable plates with air between them. You pull the plates a bit farther apart. What happens to the stored energy?

14. •• The capacitance of several capacitors in series is less than any of the individual capacitances. What, then, is the advantage of having several capacitors in series?

15. • What is the advantage to arranging several capacitors in parallel?

16. • Which way of connecting (series or parallel) three identical capacitors to a battery would store more energy?

17. • Qualitatively explain why the equivalent capacitance of a parallel combination of identical capacitors is larger than the individual capacitances.

18. • Does inserting a dielectric into a capacitor increase or decrease the energy stored in the capacitor? Explain your answer.

19. • What are the benefits, if any, of filling a capacitor with a dielectric other than air?

20. •• Capacitors A and B are identical except that the region between the plates of capacitor A is filled with a dielectric. As shown in Figure 17-21, the plates of these capacitors are maintained at the same potential difference by a battery. Is the electric field magnitude in the region between the plates of capacitor A smaller, the same, or larger than the field in the region between the plates of capacitor B? Explain your answer.

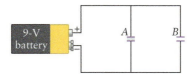

Figure 17-21 Problem 20

Multiple-Choice Questions

21. •• For a positive charge moving in the direction of the electric field, its potential energy

    A. increases and its electric potential increases.
    B. increases and its electric potential decreases.
    C. decreases and its electric potential increases.
    D. decreases and its electric potential decreases.
    E. and its electric potential remain constant.

22. •• If a negative charge is released in a uniform electric field, it will move

    A. in the direction of the electric field.
    B. from high potential to low potential.
    C. from low potential to high potential.

D. in a direction perpendicular to the electric field.

E. in circular motion.

23. • An equipotential surface must be

A. parallel to the electric field at every point.

B. equal to the electric field at every point.

C. perpendicular to the electric field at every point.

D. tangent to the electric field at every point.

E. equal to the inverse of the electric field at every point.

24. • A positive charge is moved from one point to another point along an equipotential surface. The work required to move the charge

A. is positive.

B. is negative.

C. is zero.

D. depends on the sign of the potential.

E. depends on the magnitude of the potential.

25. • The electric potential at a point equidistant from two particles that have charges $+Q$ and $-Q$ is

A. larger than zero.

B. smaller than zero.

C. equal to zero.

D. equal to the average of the two distances times the charges.

E. equal to the net electric field.

26. • Four point charges of equal magnitude but differing signs are arranged at the corners of a square (**Figure 17-22**). The electric field E and the potential V at the center of the square are

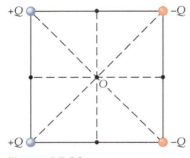

Figure 17-22 Problem 26

A. $E = 0; V \neq 0$.

B. $E = 0; V \neq 0$.

C. $E \neq 0; V \neq 0$.

D. $E \neq 0; V = 0$.

E. $E = 2V^{1/2}$.

27. •• An isolated parallel-plate capacitor carries a charge Q. If the separation between the plates is doubled, the electrical energy stored in the capacitor will be

A. halved.

B. doubled.

C. unchanged.

D. quadrupled.

E. quartered.

28. •• A parallel-plate capacitor is connected to a battery that maintains a constant potential difference across the plates. If the separation between the plates is doubled, the electrical energy stored in the capacitor will be

A. halved.

B. doubled.

C. unchanged.

D. quadrupled.

E. quartered.

29. • Capacitors connected in series have the same

A. charge.

B. voltage.

C. dielectric.

D. surface area.

E. separation.

30. • Capacitors connected in parallel have the same

A. charge.

B. voltage.

C. dielectric.

D. surface area.

E. separation.

Problems

17-1 Electric energy is important in nature, technology, and biological systems
17-2 Electric potential energy changes when a charge moves in an electric field

31. • A point charge q_0 that has a charge of 0.500 μC is at the origin. (a) A second particle that has a charge $q = 1.00$ μC and a mass of 0.0800 g is placed at $x = 0.800$ m. What is the potential energy of this system of charges? (b) If the particle with charge q is released from rest, what will its speed be when it reaches $x = 2.00$ m? Example 17-2

32. • A uniform electric field of 2.00 kN/C points in the $+x$ direction. (a) What is the change in potential energy $U_{electric,b} - U_{electric,a}$ of a +2.00-nC test charge as it is moved from point a at $x = -30.0$ cm to point b at $x = +50.0$ cm? (b) The same test charge is released from rest at point a. What is its kinetic energy when it passes through point b? (c) If a negative charge instead of a positive charge were used in this problem, qualitatively how would your answers change? Example 17-1

33. •• **Biology** Two red blood cells each have a mass of 9.0×10^{-14} kg and carry a negative charge spread uniformly over their surfaces. The repulsion arising from the excess charge prevents the cells from clumping together. One cell carries −2.50 pC of charge and the other −3.10 pC, and each cell can be modeled as a sphere 7.5 μm in diameter. (a) What speed would they need when very far away from each other to get close enough to just touch? Ignore viscous drag from the surrounding liquid. (b) What is the magnitude of the maximum acceleration of each cell in part (a)? Example 17-2

34. • Three charges lie on the x axis. Charge $q_1 = +2.20$ μC is at $x = -30.0$ cm, charge $q_2 = -3.10$ μC is at the origin, and charge $q_3 = +1.70$ μC is at $x = 25.0$ cm. Calculate the potential energy of the system of charges. Example 17-3

35. • Four point charges are arranged in a rectangle of length x and width y as shown in **Figure 17-23**. Two charges are positive and two are negative, but all have the same magnitude Q. Derive an expression for the amount of energy ΔU required to disassemble the configuration of charges so that each charge is far away from all the others. Example 17-3

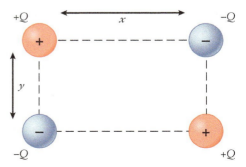

Figure 17-23 Problem 35

17-3 Electric potential equals electric potential energy per charge

36. • A uniform electric field of magnitude 28.0 V/m makes an angle of 30.0° with the x axis. If a charged particle moves along the x axis from the origin to $x = 10.0$ m, what is the potential difference of its final position relative to its initial position? Example 17-5

37. • At a certain point in space, there is a potential of 800 V relative to zero. What is the potential energy of the system when a +1.0-µC charge is placed at that point in space? Example 17-6

38. • How much work is required to move a 2.0-C positive charge from the negative terminal of a 9.0-V battery to the positive terminal? Example 17-6

39. • **Biology** A potential difference exists between the inner and outer surfaces of the membrane of a cell. The inner surface is negative relative to the outer surface. If 1.5×10^{-20} J of work is required to eject a positive sodium ion (Na^+) from the interior of the cell, what is the potential difference between the inner and outer surfaces of the cell? Example 17-6

40. • **Chemistry** What is the electric potential due to the nucleus of hydrogen at a distance of 5.00×10^{-11} m? Assume the potential is equal to zero as $r \to \infty$.

41. • (a) What is the electric potential due to a point charge of +2.00 µC at a distance of 0.500 cm? (b) How will the answer change if the charge has a value of −2.00 µC? Assume the potential is equal to zero as $r \to \infty$.

42. • The electric potential has a value of −200 V at a distance of 1.25 m from a point charge. What is the value of that charge? Assume the potential is equal to zero as $r \to \infty$.

43. • At point P in Figure 17-24 the electric potential is zero. (As usual, we take the potential to be zero at infinite distance.) (a) What can you say about the two charges? (b) Are there any other points of zero potential on the line connecting P and the two charges?

Figure 17-24 Problem 43

44. • Two point charges are placed on the x axis: +0.500 µC at $x = 0$ and −0.200 µC at $x = 10.0$ cm. At what point(s), if any, on the x axis is the electric potential equal to zero?

45. • A charge of +2.00 µC is at the origin and a charge of −3.00 µC is on the y axis at $y = 40.0$ cm. (a) What is the potential at point a, which is on the x axis at $x = 40.0$ cm? (b) What is the potential difference $V_b - V_a$ when point b is at (40.0 cm, 30.0 cm)? (c) How much work is required to move an electron at rest from point a to rest at point b?

46. •• Calculate the electric potential at the origin O due to the point charges in Figure 17-25.

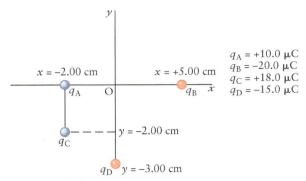

$q_A = +10.0$ µC
$q_B = -20.0$ µC
$q_C = +18.0$ µC
$q_D = -15.0$ µC

Figure 17-25 Problem 46

47. • In the Bohr model of the hydrogen atom, an electron in the lowest energy state moves around the nucleus at a speed of 2.19×10^6 m/s at a distance of 0.529×10^{-10} m from the nucleus. (a) What is the electric potential due to the hydrogen nucleus at this distance? (b) How much energy is required to ionize a hydrogen atom, whose electron is in this lowest energy state? Assume the electric potential goes to zero as $r \to \infty$.

48. • As shown in Figure 17-26, two large parallel plates, which are aligned along the y axis, are separated by a distance $d = 30.0$ cm and are at different electric potentials. The center of each plate has a small opening that lies on the x axis. A proton, traveling on the x axis, passes through the first plate with a speed of 2.50×10^5 m/s, and then leaves through the second plate with a speed of 7.80×10^5 m/s. Calculate the potential difference $V_2 - V_1$ between the two plates. Note that a positive potential difference indicates the second plate is at a higher potential than the first plate. Example 17-4

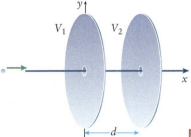

Figure 17-26 Problem 48

17-4 The electric potential has the same value everywhere on an equipotential surface

49. • Electric field lines for a system of two point charges are shown in Figure 17-27. Reproduce the figure and draw on it some equipotential lines for the system.

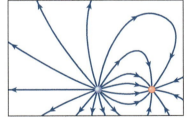

Figure 17-27 Problem 49

50. • In Figure 17-28, equipotential lines are shown at 1-m intervals. What is the electric field at (a) point A and (b) point B?

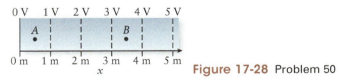

Figure 17-28 Problem 50

51. • Equipotential lines for some region of space are shown in Figure 17-29. What is the approximate electric field at (a) point A and (b) point B?

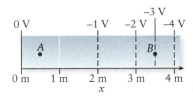

Figure 17-29 Problem 51

52. •• Draw the equipotential lines and electric field lines surrounding (a) two positive charges and (b) two negative charges (**Figure 17-30**).

(a)

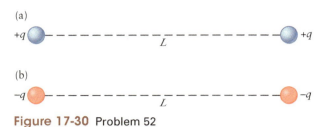

(b)

Figure 17-30 Problem 52

53. •• Draw the equipotential lines and electric field lines surrounding a dipole ($+q$ is a distance L from $-q$) (**Figure 17-31**).

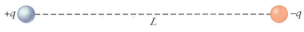

Figure 17-31 Problem 53

54. •• Draw the electric field lines and the electric equipotential lines for the charge distribution in **Figure 17-32**.

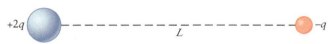

Figure 17-32 Problem 54

17-5 A capacitor stores equal amounts of positive and negative charge

55. • Using a single 10.0-V battery, what capacitance do you need to store 10.0 μC of charge?

56. • A 2.00-μF capacitor is connected to a 12.0-V battery. What is the magnitude of the charge on each plate of the capacitor?

57. • A parallel-plate capacitor has a plate separation of 1.00 mm. If the material between the plates is air, what plate area is required to provide a capacitance of 2.00 pF? Example 17-7

58. • A parallel-plate capacitor has square plates that have edge lengths equal to 1.00×10^2 cm and are separated by 1.00 mm. What is the capacitance of this device? Example 17-7

59. • An air-filled parallel-plate capacitor has plates measuring 10.0 cm × 10.0 cm and a plate separation of 1.00 mm. If you want to construct a parallel-plate capacitor of the same capacitance but with plates measuring 5.00 cm × 5.00 cm, what plate separation do you need? Example 17-7

60. • A parallel-plate capacitor has square plates that have edge length equal to 1.00 m. If the material between the plates is air, what separation distance is required to provide a capacitance of 8850 pF? Example 17-7

17-6 A capacitor is a storehouse of electric potential energy

61. • Using a single 10.0-V battery, what capacitance do you need to store 1.00×10^{-4} J of electric potential energy? Example 17-9

62. •• A parallel-plate capacitor has square plates that have edge length equal to 1.00×10^2 cm and are separated by 1.00 mm. It is connected to a battery and is charged to 12.0 V. How much energy is stored in the capacitor? Example 17-9

63. • You charge a 2.00-μF capacitor to 50.0 V. How much additional energy must you add to charge it to 100 V? Example 17-9

64. • A capacitor has a capacitance of 80.0 μF. If you want to store 160 J of electric energy in this capacitor, what potential difference do you need to apply to the plates? Example 17-9

65. •• (a) You want to store 1.00×10^{-5} C of charge on a capacitor, but you only have a 100-V voltage source with which to charge it. What must be the value of the capacitance? (b) You want to store 1.00×10^{-3} J of energy on a capacitor, and you only have a 100-V voltage source with which to charge it. What must be the value of the capacitance? Example 17-9

66. • **Medical** A defibrillator containing a 20.0-μF capacitor is used to shock the heart of a patient by holding it to the patient's chest. Just prior to discharging, the capacitor has a voltage of 10.0 kV across its plates. How much energy is released into the patient, assuming no energy losses? Example 17-9

17-7 Capacitors can be combined in series or in parallel

67. •• How should four 1.0-pF capacitors be connected to have a total capacitance of 0.75 pF? Example 17-11

68. • Three capacitors have capacitances 10.0 μF, 15.0 μF, and 30.0 μF. What is their effective capacitance if the three are connected (a) in parallel and (b) in series? Example 17-10

69. •• A series circuit consists of a 0.50-μF capacitor, a 0.10-μF capacitor, and a 220-V battery. Determine the charge on each of the capacitors. Example 17-10

70. • Two capacitors provide an equivalent capacitance of 8.00 μF when connected in parallel and 2.00 μF when connected in series. What is the capacitance of each capacitor? Example 17-10

71. •• A 2.00-μF capacitor is first charged by being connected across a 6.00-V battery. It is then disconnected from the battery and connected across an uncharged 4.00-μF capacitor. Calculate the final charge on each of the capacitors. Example 17-10

72. • A 0.0500-μF capacitor and a 0.100-μF capacitor are connected in parallel across a 220-V battery. Determine the charge on each of the capacitors. Example 17-10

73. • What is the equivalent capacitance of the network of three capacitors shown in **Figure 17-33**? Example 17-11

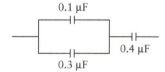

Figure 17-33 Problem 73

74. • Calculate the equivalent capacitance between a and b for the combination of capacitors shown in **Figure 17-34**. Example 17-11

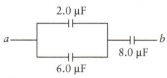

Figure 17-34 Problem 74

75. • A 10.0-μF capacitor, a 40.0-μF capacitor, and a 100.0-μF capacitor are connected in parallel across a 12.0-V battery. (a) What is the equivalent capacitance of the combination? (b) What is the charge on each capacitor? (c) What is the potential difference across each capacitor? Example 17-10

76. • A 10.0-μF capacitor, a 40.0-μF capacitor, and a 100.0-μF capacitor are connected in series across a 12.0-V battery. (a) What is the equivalent capacitance of the combination?

(b) What is the charge on each capacitor? (c) What is the potential difference across each capacitor? Example 17-10

77. •• For the capacitor network shown in Figure 17-35, the potential difference across ab is 75.0 V. How much charge and how much energy are stored in this system? Example 17-11

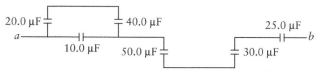

Figure 17-35 Problem 77

17-8 Placing a dielectric between the plates of a capacitor increases the capacitance

78. • What is the dielectric constant of the material that fills the gap in a parallel-plate capacitor with plate area of 20.0 cm² and plate separation of 1.00 mm if the capacitance is measured to be 0.0142 μF? Example 17-12

79. • A parallel-plate capacitor has plates of 1.00 cm by 2.00 cm. The plates are separated by a 1.00-mm-thick piece of paper. What is the capacitance of this capacitor? The dielectric constant for paper is 2.7. Example 17-12

80. •• A 2800-pF air-filled capacitor is connected to a 16-V battery. If you now insert a ceramic dielectric material ($k = 5.8$) that fills the space between the plates, how much charge will flow from the battery? Example 17-12

81. •• A parallel-plate capacitor has square plates 1.00×10^2 cm on a side that are separated by 1.00 mm. It is connected to a battery and charged to 12.0 V. How much energy is stored in the capacitor if a ceramic dielectric material (κ is 5.8) fills the space between the plates? Example 17-12

82. ••• (a) Determine the capacitance of the parallel-plate capacitor shown in Figure 17-36. The dielectric with constant κ_1 fills up one-quarter of the area, but the full separation of the plates. The materials with constants κ_2 and κ_3 fill the other three-quarters of the area, and divide the separation of the plates in half. (b) What happens to the capacitance if the material with constant κ_3 is replaced by air? Example 17-12

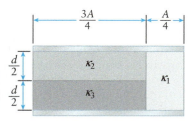

Figure 17-36 Problem 82

83. •• A parallel-plate capacitor that has a plate separation of 0.50 cm is filled halfway with a slab of dielectric material (κ is 5.0) (Figure 17-37). If the plates are 1.25 cm by 1.25 cm in area, what is the capacitance of this capacitor? Example 17-12

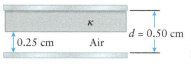

Figure 17-37 Problem 83

General Problems

84. • A pollen grain with a diameter of 55 μm has a maximum electric charge limit it can carry before the electric field it generates exceeds the dielectric limit for air of 3×10^6 N/C. Estimate the maximum voltage at the surface of a pollen grain that is carrying its maximum electric charge, assuming $V = 0$ at infinity.

85. •• A parallel-plate capacitor has a plate separation of 1.5 mm and is charged to 600 V. If an electron leaves the negative plate, starting from rest, how fast is it going when it hits the positive plate? Example 17-6

86. ••• As shown in Figure 17-38, three particles, each with charge q, are at different corners of a rhombus with sides of length a and with one diagonal of length a and the other of length b. (a) What is the electric potential energy of the charge distribution? (b) What is the electric potential at the vacant corner of the rhombus? (c) How much work by an external agent is required to bring a fourth particle, also of charge q, from rest at infinity to rest at the vacant corner? (d) What is the total electric potential energy of the four charges? Example 17-3

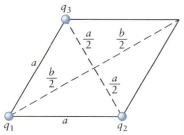

Figure 17-38 Problem 86

87. ••• A lightning bolt transfers 20 C of charge to Earth through an average potential difference of 30 MV. (a) How much energy is dissipated in the bolt? (b) What mass of water at 100°C could this energy turn into steam? Example 17-6

88. • In 2004, physicists at the SLAC National Accelerator Laboratory fired electrons toward each other at very high speeds so that they came within 1.0×10^{-15} m of each other (approximately the diameter of a proton). (a) What was the electric force on each electron at closest approach? (b) Would you be able to feel a force of this magnitude acting on you? (c) To be able to get this close, what kinetic energy must each electron have had when all of them were far apart? Example 17-2

89. • **Biology** Potassium ions (K^+) move across an 8.0-nm-thick cell membrane from the inside to the outside. The potential inside the cell is −70.0 mV, and the potential outside is zero. (a) What is the change in the electrical potential energy of the potassium ions as they move across the membrane? Does their potential energy increase or decrease? Example 17-6

90. • Calculate the equivalent capacitance of the combination in Figure 17-39. Example 17-11

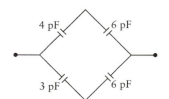

Figure 17-39 Problem 90

91. •• When five capacitors with equal capacitances are connected in series, the equivalent capacitance of the combination is 6.00 mF. The capacitors are then reconnected so that a parallel combination of two capacitors is connected in series with a parallel combination of three capacitors. Determine the equivalent capacitance of this combination in millifarads. Example 17-11

92. •• Determine the equivalent capacitance of the combination in **Figure 17-40**. Example 17-11

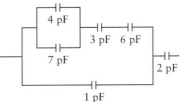

Figure 17-40 Problem 92

93. • The arrangement of four capacitors in **Figure 17-41** has an equivalent capacitance of 8.00 μF. Calculate the value of C_x. Example 17-11

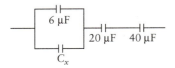

Figure 17-41 Problem 93

94. •• Calculate the charge stored on each capacitor in the circuit shown in **Figure 17-42**. Example 17-10

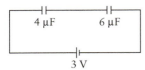

Figure 17-42 Problem 94

95. •• A parallel-plate capacitor is made by sandwiching 0.100-mm sheets of paper (dielectric constant 2.7) between three sheets of aluminum foil (A, B, and C in **Figure 17-43**) and rolling the layers into a cylinder. A capacitor that has an area of 10 m² is fabricated this way. What is the capacitance of this capacitor? Example 17-12

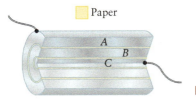

Figure 17-43 Problem 95

96. •• A parallel-plate capacitor with a capacitance of 5.00 μF is fully charged with a 12.0-V battery. The battery is then removed. How much work is required to triple the separation between the plates? Example 17-9

97. •• An air-filled parallel-plate capacitor is connected to a battery with a voltage V. (a) The plates are pulled apart, doubling the gap width, while they remain connected to the battery. By what factor does the potential energy of the capacitor change? (b) If the capacitor is first removed from the battery, what happens to the stored potential energy when the gap width is doubled? Explain your answer. Example 17-9

98. •• An air-filled parallel-plate capacitor is attached to a battery with a voltage V. While attached to the battery, the area of the plates is doubled and the separation of the plates

is halved. During this process, by what factor do (a) the capacitance, (b) the charge on the positive plate, (c) the potential across the plates, and (d) the potential energy stored in the capacitor change? (e) By what factor do all the listed quantities change if, once the capacitor is charged, it is first disconnected from the battery before the area and separation are changed as described? Example 17-7

99. •• A honey bee of mass 130 mg has accumulated a static charge of +1.8 pC. The bee is returning to her hive by following the path shown in **Figure 17-44**. Because Earth has a naturally occurring electric field near ground level of around 100 V/m pointing vertically downward, the bee experiences an electric force as she flies. (a) What is the change in the bee's electric potential energy, $\Delta U_{electric}$, as she flies from point A to point B? (b) Compute the ratio of the bee's change in electric potential energy to her change in gravitational potential energy, $\Delta U_{electric}/\Delta U_{grav}$. Example 17-4

Figure 17-44 Problem 99

100. ••• A parallel-plate capacitor has area A and separation d. (a) What is its new capacitance if a *conducting* slab of thickness $d' < d$ is inserted between, and parallel to, the plates as shown in **Figure 17-45**? (b) Does your answer depend on where the slab is positioned vertically between the plates? Example 17-12

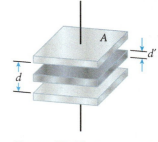

Figure 17-45 Problem 100

101. ••• Three 0.18-μF capacitors are connected in parallel across a 12-V battery (**Figure 17-46**). The battery is then disconnected. Next, one capacitor is carefully disconnected so that it doesn't lose any charge and is reconnected with its positively charged and negatively charged sides reversed. (a) What is the potential difference across the capacitors now? (b) What is the stored energy of the combination of capacitors after they have been rearranged? Example 17-9

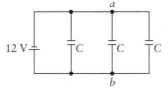

Figure 17-46 Problem 101

102. •• (a) Calculate the charge and energy stored on the 25-μF capacitor after the switch S has been in position A in **Figure 17-47** for a long time. The switch is then thrown to

position B. (b) Repeat the calculations for both the 25-μF and the 20-μF capacitors after the switch has been in position B for a long time. Example 17-10

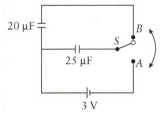

Figure 17-47 Problem 102

103. •• Figure 17-48 shows equipotential curves for 30 V, 10 V, and −10 V. A proton follows the path shown in the figure. If the proton's speed at point A is 80 km/s, what is its speed at point B? Example 17-6

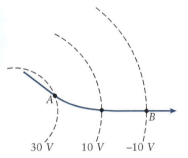

Figure 17-48 Problem 103

104. ••• A parallel-plate, air-filled capacitor has a charge of 20.0 μC and a gap width of 0.100 mm. The potential difference between the plates is 200 V. (a) What is the electric field magnitude between the plates? (b) What is the surface charge density on the positive plate? (c) If the plates are moved closer together while the charge remains constant, how do the electric field magnitude, surface charge density, and potential difference change, if at all? Explain your answers. Example 17-7

105. • Two charges equal in magnitude but opposite in sign are placed on a line. Three points, A, B, and C, are defined, as shown in Figure 17-49. (a) Order the values V_A, V_B, and V_C of the electric potential at points A, B, and C, respectively, from highest to lowest. (b) A single, negative test charge is moved among the three points. Order the potential energies U_A, U_B, and U_C of the system when the test charge is at each point, from highest to lowest.

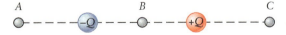

Figure 17-49 Problem 105

106. •• In the region shown in Figure 17-50, there is a uniform electric field of magnitude 40.0 N/C that points in the positive y direction. Points 2, 3, and 4 are all 0.650 m away from

point 1, and the angle $\phi = 35°$. Calculate the following potential differences: (a) $V_2 − V_1$, (b) $V_3 − V_1$, (c) $V_4 − V_1$, and (d) $V_2 − V_4$. Example 17-5

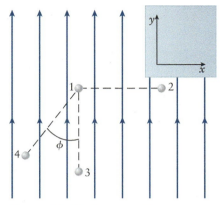

Figure 17-50 Problem 106

107. • Figure 17-51, based on a figure by James Clerk Maxwell, shows the electric field lines and equipotentials of a pair of positive charges. Charge A has four times the magnitude of charge B. Assume that the electric potential is zero infinitely far from the charge configuration. (a) At which point 1 through 8 would a negative charge have the greatest electric potential energy? (b) If a positive charge has a velocity $\bar{v}$ at point 8, moving to which point would result in the greatest kinetic energy loss? (c) A moving charged particle would experience no net change in kinetic energy when moving between which pairs of points?

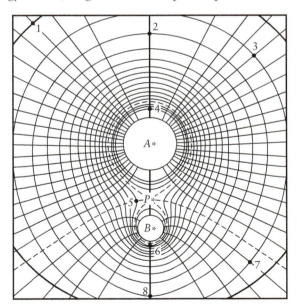

Lines of Force and Equipotential Surfaces.

Figure 17-51 Problem 107

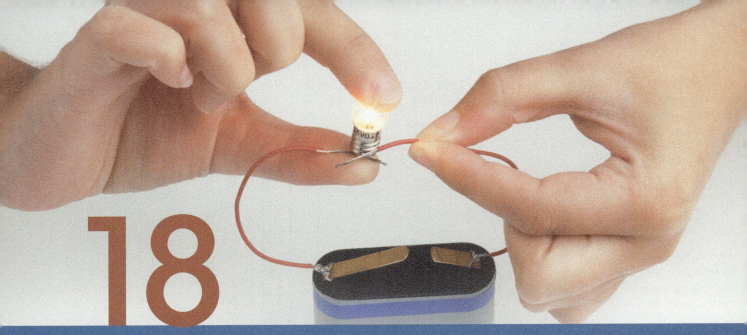

18

DC Circuits: Electric Charges in Motion

What do you think?

The simplest of all electric circuits is a battery connected to a light bulb, such as the one you find inside an ordinary flashlight. The battery causes electrons to move through the circuit. As electrons pass through the light bulb, what is the principal kind of energy that they transfer to the light bulb to make it shine? (a) Kinetic energy; (b) electric potential energy; (c) both kinetic energy and electric potential energy; (d) neither kinetic energy nor electric potential energy; (e) answer depends on the details of how the light bulb is constructed.

In this chapter, your goals are to:

- (18-1) Recognize why moving electric charges are important.
- (18-2) Explain the meaning of current and drift speed, and the difference between direct and alternating current.
- (18-3) Describe the relationships among voltage, current, resistance, and resistivity for charges moving in a wire.
- (18-4) Calculate the resistance of a resistor and the current that a given voltage produces in that resistor.
- (18-5) Discuss Kirchhoff's rules and how to apply them to single-loop and multiloop circuits.
- (18-6) Calculate the power into or out of a circuit element.
- (18-7) Explain what happens when a capacitor in series with a resistor is charged or discharged.

To master this chapter, you should review:

- (5-5)   How fluids exert drag forces on objects moving through them.
- (6-9)   How power is the rate at which energy is transferred.
- (12-3)   How to describe the position of an object oscillating with simple harmonic motion.
- (16-3)   How the loosely bound electrons in metals allow them to conduct electricity.
- (16-7)   Why the electric field outside a uniformly charged sphere is the same as if all the charges were located at the center of the sphere.
- (17-3)   How electric field and potential difference are related, and how potential difference and electric potential energy difference are related.
- (17-5, 17-6, and 17-7)   The relationship among charge, voltage, and capacitance; the energy stored in a capacitor; and the equivalent capacitance of capacitors attached in series and parallel.

18-1 Life on Earth and our technological society are only possible because of charges in motion

In the previous two chapters we've investigated electric force, electric field, electric potential energy, and electric potential. In our investigations we considered electric charges that were fixed in place. But in many important situations in nature and technology, electric charges are in *motion*. You are able to read these words thanks to electric charges that travel along the optic nerve from your eye to your brain, transmitting the image of those words in the form of a coded electrical signal. If you are reading these words after sunset, you are either looking at a printed page illuminated by a light bulb that's powered by moving electric charges or else reading them on a tablet or other electronic device that operates using complex electric circuitry.

(a)
Wires and cables are made of conductors that allow moving charges to flow along their length.

(b)
The current in a mobile device is provided by a battery (a source of emf).

(c)
A camera's electronic flash uses energy stored in a capacitor to produce a burst of light.

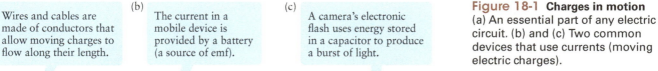

iStockphoto/Thinkstock

Daniel M Ernst/Shutterstock

Henrik Sorensen/Getty Images

Figure 18-1 Charges in motion
(a) An essential part of any electric circuit. (b) and (c) Two common devices that use currents (moving electric charges).

In this chapter we'll look at the basic physics of electric charges in motion. We'll introduce the idea of *current*, which measures the rate at which charges move through a conductor (Figure 18-1a). Ordinary conductors have *resistance* to the flow of charge, so it's necessary to set up a voltage between the ends of a conductor to produce a current. This is the role of a *source of emf*, the most common example of which is an ordinary battery (Figure 18-1b). We'll analyze simple circuits that include a source of emf and one or more *resistors* (circuit elements that have resistance) and see how to treat combinations of resistors.

Fundamentally, an electric circuit is used to transfer energy from one place to another (for instance, from a battery to a light bulb, as in the photograph that opens this chapter). We'll see how to describe the *power*, or rate of energy transfer, associated with any circuit element. Finally, we'll study circuits that include a capacitor, which make it possible to deliver energy in a quick burst (Figure 18-1c).

TAKE-HOME MESSAGE FOR Section 18-1

✔   Technological devices and biological systems depend on electric charges in motion.

✔   An electric circuit is fundamentally a means of transferring energy.

18-2 Electric current equals the rate at which charge flows

Figure 18-2 shows a common situation in which electric charges are in motion. What sets charges into motion is the **battery**, also known as an *electrochemical cell*. Inside the cell are two different substances that, due to their different chemical properties, undergo a chemical reaction so that each substance ends up with an excess or deficit of electrons. In an ordinary alkaline battery, like a AA or D cell, the two substances are zinc and manganese dioxide. The zinc ends up with an electron excess, while the manganese dioxide ends up with an electron deficit.

The two terminals of the battery are each connected to one of these substances. Electrons can flow between each substance and the metal terminal to which it is attached.

Figure 18-2 Current in a circuit Charge flows in a wire loop due to the potential difference (voltage) supplied by a battery.

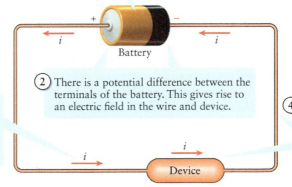

(1) A simple electric circuit is made up of a battery connected to a wire and a device (such as a light bulb), forming a closed circuit.

Battery

(2) There is a potential difference between the terminals of the battery. This gives rise to an electric field in the wire and device.

(3) The electric field causes mobile charges in the wire to flow. The current—the rate at which charge flows past a given point in the wire—is i.

(4) Charge does not pile up at any point in the circuit. So the current has the same value i at all points in the circuit, including inside the battery, in the wire, and inside the device.

Device

As a result, the terminal attached to the substance with an electron excess also has an electron excess and is called the *negative* terminal. The other terminal, attached to the substance with an electron deficit, also has an electron deficit and is called the *positive* terminal. These terminals are marked on the battery (and in Figure 18-2) by a minus sign and a plus sign, respectively. Because of this charge imbalance, the positive terminal is at a higher electric potential than the negative terminal. The potential difference, or voltage, between the terminals depends on the two substances within the battery. For zinc and manganese dioxide, the voltage is 1.5 V, a number that you will see written on the case of any alkaline battery.

We know from Section 17-3 that charges tend to flow between two points when there is a potential difference between those points: The electric force pushes positive charges from high to low potential and negative charges from low to high potential. If a battery is isolated and not connected to anything, there is no way for charge to flow outside the battery from one terminal to another. But charge *can* flow if a metallic wire is used to connect the two terminals to each other as in Figure 18-2. (As we saw in Section 16-3, metals are especially good electrical conductors because one or more of the electrons associated with a metal atom are only weakly bound to that atom. As a result, these loosely bound electrons can move with relative freedom within the metal.) So the battery and wire form a complete loop or **circuit**, and charge flows continuously through the circuit from one terminal of the battery to the other.

Here's another way to see why charges move in the wire. The potential difference between the battery terminals means that there is an electric *field* $\vec{E}$ within the wire. This field points along the length of the wire in the direction from the positive terminal to the negative terminal. A mobile charge with charge q within the wire will feel a force $\vec{F} = q\vec{E}$ from the electric field, and this force pushes the charges through the wire. This force is necessary because mobile charges within the wire collide very frequently with the atoms that make up the wire. The net effect of the collisions is to slow or retard the motion of the mobile charges through the wire, rather like how an algal spore moving through water is retarded by fluid resistance (see Section 5-5). The speed at which charge flows is determined by the balance between the retarding force due to collisions and the forward electric force $\vec{F} = q\vec{E}$.

Later in this chapter we'll see how the flow of charge delivers energy to a device (like a light bulb) that's part of the circuit. For now, however, let's concentrate on the properties of the charge flow itself.

Current

The **current** in a circuit like that in Figure 18-2 equals the rate at which charge flows past any point in the circuit. In particular, if an amount of charge Δq moves past a certain point in a time Δt, the current i is

√x **See the Math Tutorial for more information on direct and inverse proportions.**

Definition of current
(18-1)

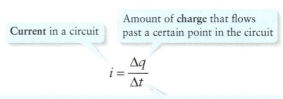

Current in a circuit

Amount of **charge** that flows past a certain point in the circuit

$$i = \frac{\Delta q}{\Delta t}$$

Time required for that amount of charge to flow past that point

The SI unit of current is the **ampere**, named after the French scientist and mathematician André Ampère. One ampere (abbreviated amp or A) is equivalent to one coulomb of charge passing a given point per second: 1 A = 1 C/s.

In much the same way as we treated fluid flow in Chapter 11, we'll begin by limiting our discussion of currents to a *steady* flow of charge. Then the current i has the same value at all *times*. In addition, the current has the same value at all *points* in a simple circuit like that in Figure 18-2 in which charges move around a single loop. The moving charges cannot "pile up" or accumulate at any point in the circuit (if they did, their mutual electrostatic repulsion would make them spread apart again). There is also no way for more moving charges to join the flow or for charges to leave the flow. So the value of the current is the same everywhere in the circuit.

Note that current is *not* a vector quantity. For example, in Figure 18-2 the current is to the left in the upper part of the circuit, downward in the left-hand part of the circuit, to the right in the lower part of the circuit, and upward in the right-hand part of the circuit. So there is no single vector that describes the direction of current in every part of the circuit. Instead, we simply say that the current in Figure 18-2 is counterclockwise around the circuit. If we reversed the battery so that the positive terminal were on the right rather than on the left, the current would be clockwise.

WATCH OUT! The direction of current is chosen to be the direction in which positive charges would flow.

In ordinary wires and in common electric devices such as light bulbs and toasters, the charges that are free to move are negatively charged electrons. In an electric circuit such as that shown in Figure 18-2, electrons flow through the wire from low potential to high potential and so from the negative terminal of the battery toward the positive terminal. But in Figure 18-2 we show the current flowing through the wire from the battery's *positive* terminal to its *negative* terminal. That's because the convention is that current flows in the direction that positively charged objects would move, regardless of whether the moving objects are positively or negatively charged. We'll use this convention throughout this book. (This convention is often attributed to the eighteenth-century American scientist and statesman Benjamin Franklin. The discovery that the moving charges in wires are negatively charged came decades after Franklin's death.) Note that in most biological systems such as neurons and muscle cells, the moving charges actually *are* positive (they are positive ions, atoms that have lost one or more electrons each).

Because an electron has a very small charge ($q = -e = -1.60 \times 10^{-19}$ C), typical currents involve the flow of a very large number of electrons. For example, suppose the current i in the circuit shown in Figure 18-2 is 1.00 A or 1.00 C/s. (That's roughly the current in a large flashlight.) Then the rate at which electrons move past any specific point on the wire is

$$1.00 \text{ A} = 1.00 \frac{\text{C}}{\text{s}} = \left(1.00 \frac{\text{C}}{\text{s}}\right)\left(\frac{1 \text{ electron}}{1.60 \times 10^{-19} \text{ C}}\right) = 6.25 \times 10^{18} \frac{\text{electron}}{\text{s}}$$

(Here we're considering the *magnitude* of the current, so we're ignoring the negative sign on the charge of the electron.) This result says that if we were to pass an imaginary plane through the cross section of the wire, we would find about 6×10^{18} electrons crossing that plane per second.

EXAMPLE 18-1 Charging a Sphere

A large, hollow metal sphere is electrically isolated from its surroundings, except for a wire that can carry a current to charge the sphere. Initially the sphere is uncharged and is at electrical potential zero. If the sphere has a radius of 0.150 m and the current is a steady 5.00 μA = 5.00×10^{-6} A, how long does it take for the sphere to attain a potential of 4.00×10^5 V?

Set Up

The electric potential at the surface of a charged sphere of radius R depends on the amount of excess charge Q on the sphere. In Section 16-7 we learned that the electric *field* outside a uniformly charged sphere is identical to that of a particle with the same charge, located at the center of the sphere. So the electric *potential* at a point just outside the sphere must be given by Equation 17-8, the equation for the potential of a charged particle with total charge Q located at the center of the sphere. We'll rearrange this equation to solve for the amount of charge required to generate the given electric potential. Then we'll find the time required for this charge to reach the sphere.

Electric potential of a charged sphere:

$$V = \frac{kQ}{R} \qquad (17\text{-}8)$$

Definition of current:

$$i = \frac{\Delta q}{\Delta t} \qquad (18\text{-}1)$$

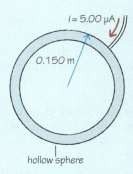

$i = 5.00\ \mu A$

0.150 m

hollow sphere

Solve

Use Equation 17-8 to solve for the final charge on the sphere.

From Equation 17-8,

$$Q = \frac{VR}{k} = \frac{(4.00 \times 10^5\ \text{V})(0.150\ \text{m})}{(8.99 \times 10^9\ \text{N} \cdot \text{m}^2/\text{C}^2)}$$

$$= 6.67 \times 10^{-6}\ \frac{\text{V} \cdot \text{C}^2}{\text{N} \cdot \text{m}}$$

Since $1\ \text{V} = 1\ \text{J/C} = 1\ \text{N} \cdot \text{m/C}$, this becomes

$$Q = \left(6.67 \times 10^{-6}\ \frac{\text{V} \cdot \text{C}^2}{\text{N} \cdot \text{m}}\right)\left(\frac{1\ \text{N} \cdot \text{m/C}}{1\ \text{V}}\right)$$

$$= 6.67 \times 10^{-6}\ \text{C} = 6.67\ \mu\text{C}$$

Our result $Q = 6.67\ \mu\text{C}$ is the amount of charge Δq that must flow onto the sphere in a time Δt. Use Equation 18-1 to solve for Δt. Recall that $1\ \text{A} = 1\ \text{C/s}$.

From Equation 18-1,

$$\Delta t = \frac{\Delta q}{i} = \frac{6.67 \times 10^{-6}\ \text{C}}{5.00 \times 10^{-6}\ \text{A}} = \frac{6.67 \times 10^{-6}\ \text{C}}{5.00 \times 10^{-6}\ \text{C/s}}$$

$$= 1.33\ \text{s}$$

Reflect

A charged sphere like this is used in a Van de Graaff generator, which you may have seen demonstrated in your physics class. If you have, you know that when the generator is turned on to charge the sphere, the sphere can begin to throw off sparks within a second or so. So a result for Δt on the order of 1 s is reasonable.

Drift Speed

The value of the current i tells us what quantity of charge flows past a given point in a circuit per second. Let's see how to relate this to the **drift speed** v_{drift}, which is the average speed at which **mobile charges**—those that are free to move throughout the conducting material in the circuit—move ("drift") through the circuit.

Figure 18-3 shows a wire that has a cross-sectional area A and carries a current i. Charges are moving ("drifting") through the green region at an average speed of v_{drift}. At any time the total moving charge in that region is Δq. Note that Δq is the product of n (the number of mobile charges per volume), the volume $A\Delta x$ of the green region (a cylinder of area A and length Δx), and the amount of charge e on each mobile charge:

$$\Delta q = n(\text{volume})e = n(A\Delta x)e \qquad (18\text{-}2)$$

If the mobile charges are electrons, each has a charge $-e$ rather than e. But since we take the direction of current to be the direction in which positive charges would flow, we'll use the convenient fiction that the charge is positive. The time required for this

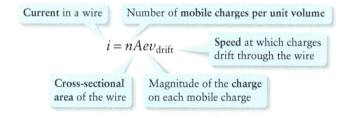

① Current i is present in the wire of cross-sectional area A.

② Within the wire there are n mobile charges per unit volume. Each has a charge e.

③ The mobile charges drift through the wire with speed v_{drift}, and travel a distance Δx in a time $\Delta t = \Delta x/v_{drift}$.

Figure 18-3 Drift speed The current i in a wire is proportional to the speed v_{drift} at which moving charges drift through the wire.

④ A length Δx of the wire (shown in green) has volume $A\Delta x$. This length contains $n(A\Delta x)$ mobile charges, with total charge $\Delta q = n(A\Delta x)e$.

⑤ The current in the wire equals the amount of charge Δq that passes this point divided by the time Δt it takes them to pass this point:

$$i = \frac{\Delta q}{\Delta t} = \frac{n(A\Delta x)e}{\Delta x/v_{drift}} = nAev_{drift}$$

volume of charge to drift the distance Δx along the length of the wire is $\Delta t = \Delta x/v_{drift}$ (time equals distance divided by speed). So from Equations 18-1 and 18-2, the current in the wire is

$$i = \frac{\Delta q}{\Delta t} = \frac{n(A\Delta x)e}{\Delta x/v_{drift}}$$

or, simplifying,

Current in a wire   Number of **mobile charges per unit volume**

$$i = nAev_{drift}$$

Speed at which charges drift through the wire

Cross-sectional area of the wire

Magnitude of the **charge** on each mobile charge

Current and drift speed (18-3)

The number of moving charges per volume (n) is different for different materials. Equation 18-3 tells us that for a wire made of a given material and with a given cross-sectional area A, the current i is directly proportional to the drift speed.

WATCH OUT! The drift speed is not the same as the speed at which individual charges move.

! Even if there is no current through a wire, the mobile charges within that wire are still in motion. However, their motions are in random directions, so there is no *net* motion of charge. It's rather like a swarm of bees flying around a hive: Individual bees move in different directions, but the swarm as a whole stays in the same place. If the swarm moves to a different hive, the motions of individual bees will be different but the swarm as a whole will "drift" together to the new hive. In the same way, if a current is present in the wire in Figure 18-3, the individual charges within the green volume can be moving at different speeds and in different directions, but this "swarm" of charge drifts along the wire at speed v_{drift}.

EXAMPLE 18-2   Electron Drift Speed in a Flashlight

In a large flashlight the distance from the on–off switch to the light bulb is 10.0 cm. How long does it take electrons to drift this distance if the flashlight wires are made of copper, are 0.512 mm in radius, and carry a current of 1.00 A? There are 8.49×10^{28} atoms in 1 m³ of copper, and one electron per copper atom can move freely through the metal.

Set Up

We'll use Equation 18-3 to determine the drift speed from the information given about the number of mobile electrons, the radius of the wire, and the current. From this we'll be able to find the drift time by using the familiar relationship between speed, distance, and time.

Current and drift speed:

$$i = nAev_{drift} \qquad (18\text{-}3)$$

Solve

Calculate the drift speed of the electrons in the wire.

Solve Equation 18-3 for the drift speed:

$$v_{drift} = \frac{i}{nAe}$$

There is one mobile electron per atom and 8.49×10^{28} atoms/m^3, so the value of n is 8.49×10^{28} electrons/m^3. The wire has a circular cross section with radius $r = 0.512$ mm $= 0.512 \times 10^{-3}$ m, so its cross-sectional area is

$$A = \pi r^2 = \pi (0.512 \times 10^{-3} \text{ m})^2 = 8.24 \times 10^{-7} \text{ m}^2$$

The magnitude of the charge per electron is $e = 1.60 \times 10^{-19}$ C/electron, and the current is $i = 1.00$ A $= 1.00$ C/s. So the drift speed is

$$v_{drift} = \frac{i}{nAe}$$

$$= \frac{1.00 \text{ C/s}}{(8.49 \times 10^{28} \text{ electrons/m}^3)(8.24 \times 10^{-7} \text{ m}^2)(1.60 \times 10^{-19} \text{ C/electron})}$$

$$= 8.94 \times 10^{-5} \text{ m/s} = 0.0894 \text{ mm/s}$$

Then calculate how long it takes electrons to drift the distance from switch to light bulb.

At this drift speed the time it takes electrons to travel a distance $d = 10.0$ cm $= 0.100$ m from the flashlight on–off switch to the light bulb is

$$t = \frac{d}{v_{drift}} = \frac{0.100 \text{ m}}{8.94 \times 10^{-5} \text{ m/s}} = 1.12 \times 10^3 \text{ s}$$

$$= (1.12 \times 10^3 \text{ s}) \left(\frac{1 \text{ min}}{60 \text{ s}} \right) = 18.6 \text{ min}$$

Reflect

The phrase "a snail's pace" refers to something that moves very slowly. But an ordinary snail moves at about *twice* the drift speed of electrons in this wire. The drift speed is so slow because electrons in the wire are continually colliding with the copper atoms, which slows their progress tremendously. (More sophisticated physics shows that an electron in copper moves in *random* motion at an average speed of about 10^6 m/s, about 10^{10} times faster than the drift speed. In the analogy we made earlier between electrons and a swarm of bees, you should think of the electrons in this wire as *very* fast-moving bees within a swarm that's drifting very, very slowly.)

At their slower-than-a-snail pace, it takes electrons more than a quarter of an hour to travel from the on–off switch to the light bulb. Why, then, does the flashlight turn on immediately when you move the switch to the on position?

The explanation is that the wire is full of mobile electrons, and these electrons drift in response to the electric field in the wire. With the switch in the off position, there is no electric field and so no drift. An electric field is set up only when the switch is put in the on position, making a complete circuit like that shown in Figure 18-2. Changes in the electric field propagate through the wire at close to the speed of light (3×10^8 m/s), so the field is set up throughout the circuit, and the electrons begin to drift, in a tiny fraction of a second. That's why the light comes on nearly instantaneously.

Direct Current and Alternating Current

In the circuit shown in Figure 18-2 the current always flows around the circuit in the same direction, as shown by the arrows labeled i. Current of this kind is called **direct current** or **dc** for short. You'll find direct current in any device that's powered by a battery, such as a flashlight, a television remote control, or a mobile phone. That's because the potential difference between the two terminals of the battery that powers the circuit always has the same sign: The positive terminal is always at a higher potential than the negative terminal.

Something very different happens in an appliance, such as a toaster or a table lamp, that you plug into a wall socket using a power cord. The potential difference between the two terminals in a wall socket is not constant but instead oscillates or *alternates*. At one instant the left-hand terminal is at a higher potential than the right-hand terminal; a short time later the left-hand terminal is the one at the lower potential, and a short time after that the left-hand terminal is again at the higher potential. As a result, the current in a device plugged into a wall socket alternates direction (Figure 18-4). This is called **alternating current** or **ac** for short. (The third terminal found in most wall sockets, called the *ground*, remains at a constant potential and is not directly involved in producing the current. The ground terminal only comes into play if there is a failure in the wiring in the appliance or its power cord, in which case the resulting current could be dangerously high. In this case the unwanted current is diverted into the ground terminal. As the name suggests, the ground terminal directly connects the circuit to the ground—Earth—which has a nearly infinite capacity to accept excess electrons from such a source.)

The potential difference or voltage between the two terminals of a wall socket varies with time in a sinusoidal way, just as does the position of an object in simple harmonic motion (Section 12-3). We can write this voltage as

$$V(t) = V_0 \sin \omega t = V_0 \sin 2\pi f t \qquad (18\text{-}4)$$

This expression says that the potential difference between the left-hand and right-hand terminals in a wall socket varies between $+V_0$ and $-V_0$. In Equation 18-4 f is the frequency at which the voltage oscillates and $\omega = 2\pi f$ is the corresponding angular frequency. In North America, Central America, and much of South America, the frequency used is $f = 60$ Hz; in most of the rest of the world, $f = 50$ Hz is used. The current in a circuit driven by such a wall socket oscillates with the same frequency. So in an ac circuit electrons just oscillate back and forth around an equilibrium position. That's very different from a dc circuit, in which electrons plod slowly in the same direction around the circuit.

If $f = 60$ Hz, the *period* of oscillation of the current is $T = 1/f = 1/(60$ Hz$) = (1/60)$ s. This means that the current in the circuit shown in Figure 18-4 moves in one direction for $(1/2) \times (1/60)$ s $= (1/120)$ s, then moves in the opposite direction for $(1/120)$ s, and so on. Since the drift speed of electrons in a wire is typically very slow (see Example 18-2), electrons can move only a very short distance (typically around 10^{-6} m or less) in $(1/120)$ s.

In Chapter 21 we'll study alternating current in detail, learn how an alternating voltage is generated, and see why it's used instead of direct current (dc) for wall sockets. For the remainder of this chapter, we'll concentrate exclusively on dc circuits.

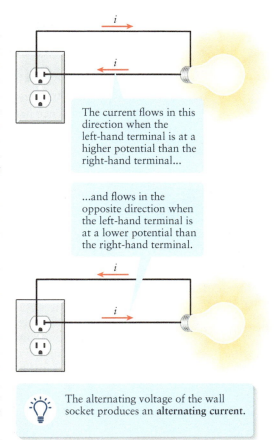

The current flows in this direction when the left-hand terminal is at a higher potential than the right-hand terminal...

...and flows in the opposite direction when the left-hand terminal is at a lower potential than the right-hand terminal.

The alternating voltage of the wall socket produces an **alternating current**.

Figure 18-4 Alternating current The potential difference between the terminals of a wall socket varies sinusoidally. As a result, the current in a circuit that contains this socket alternates direction.

GOT THE CONCEPT? 18-1 Current in a Wire

 Suppose that part of the wire in Example 18-2 were only half the thickness of the rest of the wire, with a radius of 0.256 mm instead of 0.512 mm. Compared to the rest of the wire, would the current in the thinner part be (a) four times greater; (b) twice as great; (c) the same; (d) one-half as great; or (e) one-fourth as great?

GOT THE CONCEPT? 18-2 Drift Speed in a Wire

 Suppose that part of the wire in Example 18-2 were only half the thickness of the rest of the wire, with a radius of 0.256 mm instead of 0.512 mm. Compared to the rest of the wire, would the drift speed in the thinner part be (a) four times faster; (b) twice as fast; (c) the same; (d) one-half as fast; or (e) one-fourth as fast?

TAKE-HOME MESSAGE FOR Section 18-2

✔ An electric field applied in a wire results in an electric current (a net flow of charge in the wire).

✔ The SI unit of current is the ampere (A): 1 A = 1 C/s.

✔ Current has a magnitude and a direction, but it is not a vector.

✔ By convention, the direction assigned to a current is the direction in which positive charge carriers would move. In typical metals like those used in wires, it is the negative electrons that actually move and carry current.

✔ Direct current (dc) always travels the same direction around a circuit. Alternating current (ac) continually changes direction.

18-3 The resistance to current through an object depends on the object's resistivity and dimensions

We saw in Section 18-2 that current exists in a wire only if there is a potential difference between the ends of the wire. This gives rise to an electric field inside the wire, and this field exerts a force that causes mobile charges to move. However, due to collisions between the charges and the atoms of the wire, charges drift through the wire at a relatively slow speed. To better understand current we need to answer three questions:

(1) How is the electric field in a current-carrying wire related to the potential difference between the ends of the wire?

(2) How is the resulting current related to the electric field in the wire?

(3) How is the resulting current related to the potential difference between the ends of the wire?

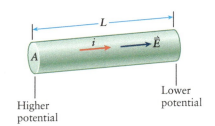

Figure 18-5 Inside a current-carrying wire The electric field in a current-carrying wire points from the higher-potential end toward the lower-potential end. The current is in the same direction.

To answer the first question, consider a straight wire of uniform cross-sectional area A and length L (Figure 18-5). If there is a potential difference between the ends of the wire, the electric field $\vec{E}$ points along the length of the wire from the high-potential end toward the low-potential end. Since the wire is uniform, we expect that the magnitude E of the field will be uniform as well. From Chapter 17 the potential difference V is just equal to the field magnitude multiplied by the length of the wire:

(18-5)
$$V = EL \quad \text{or} \quad E = \frac{V}{L}$$

This is the same as Equation 17-7, $\Delta V = -Ed \cos \theta$. We've changed the symbol for potential difference from ΔV to V, replaced the distance d by L, and used $\theta = 180°$ for the angle between the direction of $\vec{E}$ and the direction that we imagine traveling along the length of the wire to measure the potential difference. With this choice the potential difference V is positive. Remember that potential always increases as we travel opposite to $\vec{E}$ and decreases as we travel in the direction of $\vec{E}$. Remember also that the direction of current, which we choose to be the direction in which positive charges would move, is in the direction of $\vec{E}$ and so from high potential to low potential. For example, if there is a 1.5-V potential difference ($V = 1.5$ V) between the ends of a wire 10 cm in length ($L = 10$ cm $= 0.10$ m), Equation 18-5 tells us that the electric field inside the wire has magnitude $E = V/L = (1.5$ V)$/(0.10$ m$) = 15$ V/m.

For the answer to the second question, we must turn to experiments on current-carrying wires. For many materials (including common conductors such as copper), experiments show that the current i that arises in a wire when an electric field is present inside the wire is directly proportional to both the electric field magnitude E and the cross-sectional area A of the wire. We can write this relationship as

(18-6)
$$i = \frac{EA}{\rho}$$

In Equation 18-6 the quantity ρ (the Greek letter "rho") is called the **resistivity**. The value of ρ depends on the material of which the wire is made and tells us how well or poorly this material inhibits the flow of electric charge. Equation 18-6 says that for a given cross-sectional area A and a given electric field magnitude E, a *greater* value of the resistivity means a *smaller* amount of current i. (It's unfortunate that the same Greek letter that we use as the symbol for density is also used as the symbol for resistivity. We promise never to use density and resistivity in the same equation—and you shouldn't, either!)

The units of resistivity are volt-meters per ampere, or V·m/A. For reasons that we will see below, the unit V/A (volts per ampere) appears very often and is given its own name, the **ohm** (symbol Ω, the uppercase Greek letter "omega"): $1 \Omega = 1$ V/A. In terms of this we can write the units of resistivity as ohm-meters or Ω·m.

For copper, which is a good conductor of electricity, the resistivity is very low: $\rho = 1.725 \times 10^{-8}$ Ω·m at 21°C. For hard rubber, which is a very poor conductor (and a good insulator), $\rho = 10^{13}$ Ω·m at 21°C. The explanation for the huge ratio between these two values of resistivity is that copper has many mobile electrons per cubic meter, while hard rubber has hardly any. The value of resistivity also depends on temperature. In metals atoms move more rapidly and are likely to be less well organized at higher temperatures compared to lower temperatures. As a result, moving charges suffer more collisions with the atoms in a metal when the temperature is higher. Consequently the resistivity of most metals increases as temperature increases. In other materials, such as ceramics, resistivity *decreases* with increasing temperature. Table 18-1 lists the values of resistivity for a variety of substances at 21°C.

In practice, it's more useful to know the answer to the third question: how the current i in a straight wire is related to the potential difference V between the ends of the wire, rather than the electric field E inside the wire. (That's because electrical meters measure potential difference directly, not electric field.) To answer this, substitute the expression $E = V/L$ from Equation 18-5 into Equation 18-6:

$$i = \frac{EA}{\rho} = \left(\frac{V}{L}\right)\left(\frac{A}{\rho}\right) = V\left(\frac{A}{\rho L}\right) \quad \text{or} \quad V = i\left(\frac{\rho L}{A}\right) \qquad \textbf{(18-7)}$$

We define the **resistance** R of a wire as

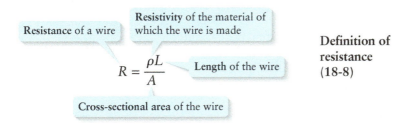

Definition of resistance (18-8)

In terms of this new quantity, we can rewrite Equation 18-7 as

$$V = iR$$

Relationship among potential difference, current, and resistance (18-9)

Equation 18-9 says the potential difference, or voltage, V required to produce a current i in a wire is proportional to i and to the resistance of the wire. Equation 18-8 further tells us that the resistance of a wire depends on the resistivity of the material of which the wire is made, the length of the wire, and the cross-sectional area of the wire. For a given material, the resistance is greater for a wire that is long (large L) and thin (small A) than for a wire that is short (small L) and thick (large A). You can see an example by looking inside the slots of an ordinary kitchen toaster. The wire that makes up the toaster's heating coils is very thin and if uncoiled would be very long. Hence the coils have a high resistance. The resistance is made even higher by making the wire out of nichrome, which has a resistivity about 10^2 times greater than that of copper (see Table 18-1).

Since resistivity ρ has units of ohm-meters (Ω·m), length L has units of meters, and area A has units of meters squared, Equation 18-8 says that the units of resistance R are ohms. This agrees with Equation 18-9, since 1 Ω = 1 V/A: To produce a 1-A current in a wire with a resistance of 1 Ω = 1 V/A, a voltage of 1 V is required.

An ohm is a relatively small resistance. The heating coils of a toaster have a resistance R of about 10 to 20 Ω. A typical **resistor**—a circuit component intended to add resistance to the flow of current—such as you will find for sale in an electronics store will likely have a resistance in the range of 1 kΩ (1 kΩ = 1 kilohm = 10^3 Ω) to

TABLE 18-1 **Resistivity of Some Conductors and Insulators at 21°C**	
Conductor	ρ (Ω·m)
aluminum	2.733×10^{-8}
copper	1.725×10^{-8}
gold	2.271×10^{-8}
iron	9.98×10^{-8}
nichrome	150×10^{-8}
nickel	7.2×10^{-8}
silver	1.629×10^{-8}
titanium	43.1×10^{-8}
tungsten	5.4×10^{-8}
Insulator	ρ (Ω·m)
glass	10^{12}
hard rubber	10^{13}
fused quartz	7.5×10^{17}

10 MΩ (1 MΩ = 1 megohm = 10^6 Ω). The resistance of a human body measured on the skin can be more than 0.5 MΩ. The standard symbol for a resistor is a jagged line:

Equation 18-9 is often referred to as "Ohm's law," after the nineteenth-century Bavarian physicist Georg Ohm, whose pioneering experiments increased our understanding of electric current (and for whom the ohm is named). By itself it suggests that voltage and current are directly proportional to each other. This proportionality holds true if the resistance R remains constant as the voltage is changed, and this is in fact the case for many conducting materials over a wide range of voltages. Materials that have this property are referred to as *ohmic*. However, for many materials (including those used in a variety of electronic devices) the value of the resistance changes as the potential difference across the material changes. We can still use Equation 18-9 for such *nonohmic* materials, provided we keep in mind that R is not a constant.

EXAMPLE 18-3 Stretching a Wire

A 10.0-m-long wire has a radius of 2.00 mm and a resistance of 50.0 Ω. If the wire is stretched to 10.0 times its original length, what will be its new resistance?

Set Up

Equation 18-8 tells us that resistance of the wire depends on its resistivity (which is a property of the material of which the wire is made, and doesn't change if the dimensions change). It also depends on the length and cross-sectional area, both of which change when the wire is stretched. To find the new cross-sectional area, we'll use the idea that the volume of the wire (the product of its length and cross-sectional area) does not change as it's stretched.

Definition of resistance:

$$R = \frac{\rho L}{A} \qquad (18\text{-}8)$$

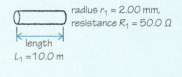

Solve

Find the cross-sectional area of the wire after it has been stretched.

The original cross-sectional area A_1 of the wire of radius $r_1 = 2.00$ mm $= 2.00 \times 10^{-3}$ m is

$$A_1 = \pi r_1^2 = \pi (2.00 \times 10^{-3} \text{ m})^2 = 1.26 \times 10^{-5} \text{ m}^2$$

The volume of the wire is $A_1 L_1$, where $L_1 = 10.0$ m is the wire's initial length. If we stretch the wire to a new length $L_2 = 10.0 L_1 = 1.00 \times 10^2$ m, the cross-sectional area will change to a new value A_2, but the volume will be the same:

$$A_2 L_2 = A_1 L_1$$
$$A_2 = \frac{A_1 L_1}{L_2} = \frac{(1.26 \times 10^{-5} \text{ m}^2)(10.0 \text{ m})}{1.00 \times 10^2 \text{ m}} = 1.26 \times 10^{-6} \text{ m}^2$$

The new cross-sectional area is 1/10.0 of the initial area.

Calculate the new resistance of the wire.

The initial resistance of the wire (before being stretched) is

$$R_1 = \frac{\rho L_1}{A_1} = 50.0 \text{ Ω}$$

The resistance of the wire after being stretched is

$$R_2 = \frac{\rho L_2}{A_2}$$

If we take the ratio of the two resistances, the unknown value of resistivity will cancel out:

$$\frac{R_2}{R_1} = \frac{\rho L_2/A_2}{\rho L_1/A_1} = \frac{L_2 A_1}{L_1 A_2}$$

$$= \left(\frac{1.00 \times 10^2 \ m}{10.0 \ m}\right)\left(\frac{1.26 \times 10^{-5} \ m^2}{1.26 \times 10^{-6} \ m^2}\right)$$

$$= (10.0)(10.0) = 1.00 \times 10^2$$

The length has increased by a factor of 10.0, and the cross-sectional area has decreased by a factor of 10.0, so the new value of resistivity is greater than the old value by a factor of $(10.0)^2 = 1.00 \times 10^2$:

$$R_2 = (1.00 \times 10^2)R_1 = (1.00 \times 10^2)(50.0 \ \Omega)$$

$$= 5.00 \times 10^3 \ \Omega = 5.00 \ k\Omega$$

Reflect

Stretching the wire makes it longer and thinner, and both of these changes make the resistance of the wire increase. Hence the stretched wire in this problem has a much higher resistance than the wire had initially.

EXAMPLE 18-4 Calculating Current

If a 12.0-V potential difference is set up between the ends of each wire in Example 18-3, how much current will flow in each wire?

Set Up

We know the resistance of each wire from Example 18-3, so we can use Equation 18-9 to solve for the current i in each wire.

Relationship among potential difference, current, and resistance:

$$V = iR \qquad (18\text{-}9)$$

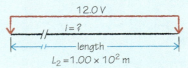

Solve

Rewrite Equation 18-9 as an expression for the current in terms of the voltage and resistance. Then solve for the current in each case.

From Equation 18-9,

$$i = \frac{V}{R}$$

For the wire before it is stretched, $R = R_1 = 50.0 \ \Omega$. The current that results from a 12.0-V potential difference between the ends of this wire is

$$i_1 = \frac{V}{R_1} = \frac{12.0 \ V}{50.0 \ \Omega} = 0.240 \ \frac{V}{\Omega}$$

Recall that $1 \ \Omega = 1 \ V/A$, so $1 \ A = 1 \ V/\Omega$. So the current in the wire before it is stretched is

$$i_1 = 0.240 \ A$$

After the wire is stretched the resistance is $R = R_2 = 5.00 \times 10^3 \ \Omega$. A 12.0-V potential difference between the ends of this wire produces a current

$$i_2 = \frac{V}{R_2} = \frac{12.0 \ V}{5.00 \times 10^3 \ \Omega} = 2.40 \times 10^{-3} \ A = 2.40 \ mA$$

Reflect

The stretched wire has a greater resistance than the original wire, so the same potential difference produces a smaller current. Note that the milliampere (1 mA = 10^{-3} A) is a commonly used unit of current, as are the microampere (1 μA = 10^{-6} A), the nanoampere (1 nA = 10^{-9} A), and the picoampere (1 pA = 10^{-12} A).

GOT THE CONCEPT? 18-3 Is It Ohmic?

(?) A wire is made of a certain conducting material. You apply different potential differences V across the wire and measure the current i that results. Your results are $V = 2.0$ V, $i = 0.15$ A; $V = 4.0$ V, $i = 0.28$ A; $V = 8.0$ V, $i = 0.50$ A. Is the material of which the wire is made (a) ohmic or (b) nonohmic?

TAKE-HOME MESSAGE FOR Section 18-3

✔ Resistivity ρ is a measure of how well or poorly a particular material inhibits an electric current. The SI units of resistivity are volt-meters per ampere or ohm-meters.

✔ The value of resistivity differs from material to material. The resistivity is low for conductors and high for insulators.

✔ The resistance R of an object is a measure of the current through the object for a given potential difference between its ends. The value of R depends on the object's shape and on the resistivity of the material from which it is made. A long, narrow wire has a much higher resistance than a short, thick one.

The circuit elements with colored bands are resistors. The particular colors on each resistor indicate the value of its resistance.

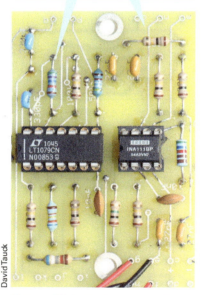

David Tauck

Figure 18-6 Resistors in a circuit A typical electronic circuit is likely to contain many resistors.

18-4 Resistance is important in both technology and physiology

Electrical resistance can be found in every technological device, from the wires in an automobile ignition system to the resistive elements (resistors) in the circuits of a computer or mobile phone. In electronic devices, resistors are often small and cylindrical with colored bands (Figure 18-6). Other resistors look more like tiny, black cubes with the number of ohms etched on one face.

What purpose does a resistor serve? As we have seen, a potential difference V between two points in a conducting material causes a current i. But according to Equation 18-9, $V = iR$, the amount of current that the potential difference produces depends on the resistance of the conducting material. The greater the resistance R, the smaller the current i. Stated another way, the resistance allows us to control the current due to any particular applied voltage.

BioMedical An important application of this idea takes place in every one of the millions of cells in your body. For a cell to live, there must be a higher concentration of positively charged potassium ions (K^+) inside the cell than in the surrounding fluid. This difference in concentration means that K^+ ions tend to leak out of the cell through pathways called *potassium channels* (Figure 18-7). The flow of K^+ ions constitutes a current, and each potassium channel acts like a resistor. For a given potential difference between the interior and exterior of the cell, the amount of current that flows through these channels depends on their resistance. So the flow of potassium through the membranes of your cells is determined by the electrical resistance of the membrane channels.

Why doesn't the flow of K^+ continue until there is equal concentration of these ions inside and outside the cell? Figure 18-7 shows the reason: As positive ions leave the cell, the cell interior is left with a net negative charge and a lower potential than the outside of the cell. This potential difference, called the *membrane potential*, gives rise to an electric field across the membrane that points from the outside to the inside of the cell. This field opposes additional flow of K^+ ions out of the cell. There is still some random flow of K^+ ions in and out of the cell, but the net flow stops when the membrane potential reaches an equilibrium value.

As the following example shows, we can use Equation 18-9 to determine the resistance of the membrane to potassium ion flow.

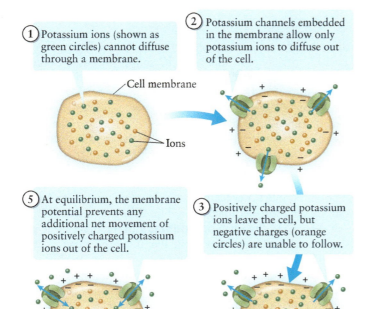

① Potassium ions (shown as green circles) cannot diffuse through a membrane.

② Potassium channels embedded in the membrane allow only potassium ions to diffuse out of the cell.

Cell membrane

Ions

⑤ At equilibrium, the membrane potential prevents any additional net movement of positively charged potassium ions out of the cell.

③ Positively charged potassium ions leave the cell, but negative charges (orange circles) are unable to follow.

④ The result is a potential difference (called the membrane potential) across the membrane.

Figure 18-7 Membrane potential The concentration of positively charged potassium ions (K^+) is always higher inside cells than outside, which causes K^+ ions to leak out. This flow of K^+ ions is a current; the channels that allow K^+ ions to leak out of the cell behave like resistors.

BioMedical EXAMPLE 18-5 Resistance of a Potassium Channel

Using an instrument called a patch clamp, scientists are able to control the electrical potential difference across a tiny patch of cell membrane and measure the current through individual potassium channels. In one experiment a voltage of 0.120 V was applied across a patch of membrane, and as a result K^+ ions carried 6.60 pA of current (1 pA = 1 picoampere = 10^{-12} A). What is the resistance of this K^+ channel?

Set Up

Whether the mobile charges are negative electrons or positive ions, we can use Equation 18-9 to relate potential difference (voltage), current, and resistance.

Relationship among potential difference, current, and resistance:

$$V = iR \qquad (18\text{-}9)$$

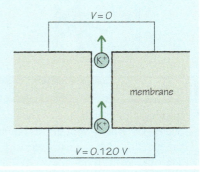

Solve

Rewrite Equation 18-9 to solve for the resistance.

From Equation 18-9,

$$R = \frac{V}{i}$$

The current is $i = 6.60$ pA $= 6.60 \times 10^{-12}$ A, so

$$R = \frac{0.120 \text{ V}}{6.60 \times 10^{-12} \text{ A}} = 1.82 \times 10^{10} \ \Omega$$

Reflect

This is an *immensely* high resistance compared to the values found in circuits like that shown in Figure 18-6. The explanation is that a potassium channel is only about 10^{-9} m wide—so small that K^+ ions must pass through it in single file—and so the resistance to current flow through the channel is very high. There are other types of channels in cell membranes, and all typically have resistances in the range of 10^9 to 10^{11} Ω.

Although the current through a single potassium channel is very small, the electric field required to produce that current is tremendous. The thickness of the cell membrane is only about 7.5 nm $= 7.5 \times 10^{-9}$ m. Can you use Equation 18-5 to show that a potential difference of 0.120 V across this distance corresponds to an electric field magnitude of 1.6×10^7 V/m? (By comparison, the electric field inside the wires of a flashlight is only about 10 V/m.)

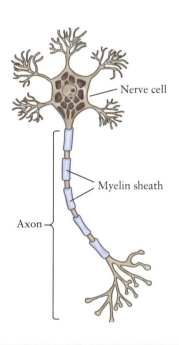

BioMedical

Membrane potential plays a crucial role in the nervous system. When a nerve signal propagates along an axon—a long, cable-like fiber that extends from a nerve cell (Figure 18-8)—it does so in the form of a *variation* in the membrane potential of the axon. This variation, which travels along the length of the axon, is called an *action potential*. At rest the baseline membrane potential is lower on the inside compared to the outside surface of a cell. But if the potential on the interior of the membrane becomes just a little less negative, sodium ion (Na^+) channels in the membrane open. Because the concentration of Na^+ is always higher outside of cells compared to inside, Na^+ ions rush into the axon. This flow is so great that the interior surface of the axon membrane ends up with a positive charge and the exterior with a negative charge, just the reverse of what's shown in Figure 18-7. This inversion of charge is the action potential. The Na^+ channels then start closing as potassium (K^+) channels begin to open, allowing K^+ ions to flow out of the axon and restore the resting polarity of the cell membrane. You can think of an action potential as a momentary voltage "flip" that propagates along the length of the axon like a wave pulse along a stretched string.

The speed with which an action potential can propagate along an axon depends in part on the resistance of the axon to current flowing along its length. As Equation 18-8 tells us, this longitudinal resistance depends on both the length and diameter of the axon. The larger the diameter of the axon, for example, the lower the longitudinal resistance and the faster the electrical signal propagates. (In mammals the signal propagates at about 10 m/s in axons of 3 μm diameter, but at about 50 m/s in axons of 10 μm diameter.) Some axons, such as the giant ones that mediate the escape reflex in squid, have exceptionally large diameters.

Figure 18-8 Structure of a vertebrate nerve cell In a nerve cell at rest, the electric potential inside the membrane that surrounds the axon is lower than the potential outside the membrane. To transmit a nerve signal, a variation in this potential difference (called an action potential) propagates along the length of the axon. In vertebrates many types of axons are sheathed in a layer of a fatty, insulating material called myelin; this helps increase the speed of propagation of the action potential (see Section 18-7).

BioMedical EXAMPLE 18-6 **Giant Axons in Squid**

Running along each side of the back of a squid is a tube-like structure that can be as large as 1.50 mm in diameter in some species. Originally thought to be blood vessels, these structures are actually giant axons that are part of the squid's nervous system. Although the vast majority of axons in the squid range in diameter from about 10.0 to 50.0 μm, as the squid develops, axons from about 30,000 neurons fuse together to form these giant axons. Compare the resistance of a giant axon to current along its length to the resistance of an axon with the same length but a more typical diameter of 15.0 μm.

Set Up

Equation 18-8 tells us the resistance of an axon in terms of its resistivity, length, and cross-sectional area. We'll use this to express the ratio of the resistance of a giant axon to that of an ordinary axon of the same composition (and hence same resistivity) and length but of smaller cross-sectional area.

Definition of resistance:

$$R = \frac{\rho L}{A} \tag{18-8}$$

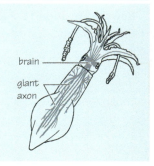

brain

giant axon

Solve

Rewrite Equation 18-8 in terms of the diameter of an axon.	An axon has a circular cross section of radius r, equal to one-half its diameter d, so its cross-sectional area is

$$A = \pi r^2 = \pi \left(\frac{d}{2}\right)^2 = \frac{\pi d^2}{4}$$

If we substitute this into Equation 18-8, we get an expression for the resistance of an axon of length L and diameter d:

$$R = \frac{\rho L}{(\pi d^2/4)} = \frac{4\rho L}{\pi d^2}$$

Write the ratio of the resistance of a giant axon to that of a typical axon.	Let d_{giant} be the diameter of a giant axon and $d_{typical}$ be the diameter of a typical axon. If the two axons have the same length L and contain the same material of resistivity ρ, their resistances are

$$R_{giant} = \frac{4\rho L}{\pi d_{giant}^2} \quad \text{and} \quad R_{typical} = \frac{4\rho L}{\pi d_{typical}^2}$$

The ratio of these two resistances is

$$\frac{R_{giant}}{R_{typical}} = R_{giant} \times \frac{1}{R_{typical}} = \frac{4\rho L}{\pi d_{giant}^2} \times \frac{\pi d_{typical}^2}{4\rho L}$$

The factors of 4, ρ, L, and π cancel, so

$$\frac{R_{giant}}{R_{typical}} = \frac{d_{typical}^2}{d_{giant}^2} = \left(\frac{d_{typical}}{d_{giant}}\right)^2$$

Substitute the numerical values of the diameters of the two axons.	We are given $d_{giant} = 1.50 \text{ mm} = 1.50 \times 10^{-3}$ m and $d_{typical} = 15.0 \text{ }\mu\text{m} = 15.0 \times 10^{-6}$ m $= 1.50 \times 10^{-5}$ m. So the ratio of resistances of the two axons is

$$\frac{R_{giant}}{R_{typical}} = \left(\frac{1.50 \times 10^{-5} \text{ m}}{1.50 \times 10^{-3} \text{ m}}\right)^2 = (1.00 \times 10^{-2})^2$$

$$= 1.00 \times 10^{-4} = \frac{1}{1.00 \times 10^4}$$

The resistance of the giant axon is one ten-thousandth as great as that of a typical axon.

Reflect

When a squid recognizes danger, it sends electrical signals (action potentials) along the giant axons to trigger muscle contraction. Because the longitudinal resistance along an axon determines the speed at which nerve signals propagate, signals get to the muscle as much as 10,000 times more quickly through the giant axon than they would if the squid's nervous system were comprised entirely of ordinary nerve cells with small-diameter axons.

GOT THE CONCEPT? 18-4 Resistance and Diameter

 Two copper wires have the same length, but one has four times the diameter of the other. Compared to the wire that has the smaller diameter, the wire that has the larger diameter has a resistance that is (a) 16 times greater; (b) 4 times greater; (c) the same; (d) 1/4 as great; (e) 1/16 as great.

TAKE-HOME MESSAGE FOR Section 18-4

✔ Electrical resistance is important in technology and in the physiology of cells.

✔ For a given applied voltage, the greater the resistance in an electrical system, the smaller the current.

18-5 Kirchhoff's rules help us to analyze simple electric circuits

Many simple electric circuits are made up of a source of electric potential difference and one or more resistors. An example is a light like that shown in the photograph that opens this chapter: The battery provides the potential difference, and the filament of the light bulb acts as a resistor. We'll refer to batteries and resistors collectively as **circuit elements**. (Later in this chapter we'll consider circuits that also include capacitors as circuit elements.)

In this section we'll see how to analyze circuits made up of a battery and one or more resistors. In particular, we'll see how to determine the voltage across each circuit element as well as the current through each circuit element.

A Single-Loop Circuit and Kirchhoff's Loop Rule

Figure 18-9a shows a battery connected to a single resistor of resistance R. In fact, there are *two* resistors in this circuit; the other one, which we label r, is the **internal resistance** of the battery itself. This reflects the resistance that mobile charges encounter as they pass through the battery. So we can think of the battery as having two components, its **emf** (pronounced "ee-em-eff")—the aspect of the battery that causes charges in the circuit to move—and its internal resistance. Although we draw these as separate entities in Figure 18-9a, they cannot in fact be separated. When we say that a D or AA cell is a "1.5-volt battery" or that the battery in an automobile is "a 12-volt battery," we're actually stating the value of the battery's emf. We use the symbol ε, an uppercase script "e," for emf. So $\varepsilon = 1.5$ V for a D or AA cell and $\varepsilon = 12$ V for a standard automotive battery. (Note that the term "emf" comes from the older term "electromotive force." Although an emf is what pushes mobile charges through the circuit, it has units of volts, not newtons. So it's not accurate to call an emf a "force.") We'll often refer to a battery as a **source of emf**. In later chapters we'll encounter other sources of emf.

The symbol in Figure 18-9a for a source of constant emf is similar to the symbol for a capacitor (see Section 17-5), but with two parallel lines of unequal length:

The longer of the two parallel lines represents the positive terminal of the source. Since we regard current as flowing in the direction that positive charges would move, such a source of emf causes current to flow out of its positive terminal and into its negative terminal (see Figure 18-2).

We call the circuit in Figure 18-9a a **single-loop circuit** because there's only a single path that moving charges can follow around the circuit. Think about what happens to such a charge as it travels through this circuit in the direction of the current i. (We'll continue to use the convenient fiction that these are positive charges.) As the charge passes through the source of emf (the battery) from the negative terminal to the positive terminal, it experiences an *increase* in electric potential (a *voltage rise*) of ε. When it moves through the internal resistance, however, the charge experiences a *decrease* in electric potential (a *voltage drop*) of ir. (Remember that the current moves in the direction of the electric field, and the electric field points in the direction from high potential to low potential. So the potential decreases as you move with the current through a resistor.) The charge experiences an additional decrease in potential (voltage drop) of iR when it moves through the resistor of resistance R.

(a)

A battery has an emf ε...

...and an internal resistance r.

The battery is connected to a resistor of resistance R.

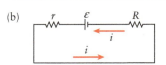

The connecting wires have negligible resistance.

 The sum of the potential changes around the closed loop of this circuit must be zero: $\varepsilon - ir - iR = 0$.

(b)

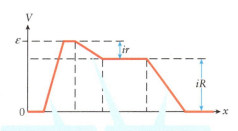

The potential increases by ε from the negative terminal to the positive terminal of the emf.

The potential drops by ir across the internal resistance and drops by iR across the resistor.

Figure 18-9 Kirchhoff's loop rule (a) A circuit made up of a battery (which has an emf ε and an internal resistance r) and a resistor R. (b) The changes in electric potential around the circuit.

There's also resistance in the wires that connect the battery and the resistor. However, the resistance of the connecting wires is generally quite small compared to the values of r and R. So we'll ignore the wire resistance and assume that a moving charge experiences no change in potential as it traverses the wires.

The current in Figure 18-9a does not change with time. So when the charge finishes a trip around the circuit and returns to where it started, the value of the potential it experiences at the starting point must have the same value as when the charge left that point. In other words, the *net change* in potential for a round trip around the loop must be *zero* (Figure 18-9b). This idea was first proposed by the Prussian physicist Gustav Kirchhoff in the mid-nineteenth century and is called **Kirchhoff's loop rule**:

The sum of the changes in potential around a closed loop in a circuit must equal zero.

In equation form, we can write Kirchhoff's loop rule for the circuit in Figure 18-9a as

$$\varepsilon - ir - iR = 0 \qquad\qquad (18\text{-}10)$$

We can rearrange Equation 18-10 to solve for the current in the circuit:

$$\varepsilon = ir + iR = i(r + R)$$
$$i = \frac{\varepsilon}{r + R}$$

To give a specific numerical example, suppose $\varepsilon = 12.0$ V, $r = 1.00\ \Omega$, and $R = 19.0\ \Omega$. Then the current is

$$i = \frac{12.0\ \text{V}}{1.00\ \Omega + 19.0\ \Omega} = \frac{12.0\ \text{V}}{20.0\ \Omega} = 0.600\ \text{A}$$

(Recall that 1 A = 1 V/Ω.) You can see that if the battery had a greater internal resistance r, the current would be smaller. As an example, the internal resistance r of a disposable battery in a flashlight or television remote control increases as the battery is used. Eventually r becomes so great that the current becomes too small to make the device work, which means that it's time to replace the battery. The emf ε of the used battery is almost the same as when it was new; it's the internal resistance that makes the battery no longer useful.

Note that for this numerical example, the voltage drop across the internal resistance is $ir = (0.600\ \text{A})(1.00\ \Omega) = 0.600$ V, and the voltage drop across the 19.0-Ω resistor is $iR = (0.600\ \text{A})(19.0\ \Omega) = 11.4$ V. The sum of the potential changes is zero, just as Kirchhoff's loop rule says it must be:

$$\varepsilon - ir - iR = 12.0\ \text{V} - 0.600\ \text{V} - 11.4\ \text{V} = 0$$

For this single-loop circuit, the *current* $i = 0.600$ A is the same through the internal resistance r as through the resistor R. That's because no charges can appear or disappear at any place in the circuit (see Section 18-2). However, the *voltage drops* are different for r and R because the values of resistance are different.

WATCH OUT! The voltage across a battery in a circuit is less than the emf.

The example we have just given shows that when a battery is in a circuit, the voltage V across the battery (that is, the potential difference between its terminals) is *not* equal to the emf ε. Rather, the voltage across the battery is equal to the emf *minus* the potential drop ir across the internal resistance of the battery. For the battery with emf $\varepsilon = 12.0$ V described above, the voltage is

$V = \varepsilon - ir = 12.0\ \text{V} - (0.600\ \text{A})(1.00\ \Omega) = 12.0\ \text{V} - 0.600\ \text{V} = 11.4$ V. The only time the voltage $V = \varepsilon - ir$ across a battery is equal to the emf is when $i = 0$; that is, when the battery is disconnected from the circuit, there is no current through the battery. So a 1.5-V battery has a 1.5-V potential difference between its terminals only when the battery isn't connected to anything!

(a)

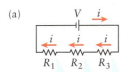

The current through each resistor in series is the same.

(b)

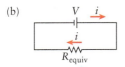

• The current i through R_{equiv} is the same as through each of the resistors in series.
• The voltage drop V across R_{equiv} is the same as across the combination of resistors in series.

Figure 18-10 Resistors in series (a) A circuit contains three resistors connected in series to a source of emf. (b) The three resistors have been replaced by a single, equivalent resistor.

Resistors in Series

An important application of Kirchhoff's loop rule is to resistors in **series**. That is, the resistors are connected end to end. (We used the same nomenclature for capacitors in series in Section 17-7.) Figure 18-10a shows a circuit that has three resistors R_1, R_2, and R_3 in series connected to a source of emf such as a battery with voltage V. This voltage includes the emf and internal resistance of the battery. Imagine we replace the three separate resistors with an **equivalent resistance**—that is, a single resistor of a resistance R_{equiv} that gives the same current as the series combination, as shown in Figure 18-10b. What is the equivalent resistance in terms of R_1, R_2, and R_3?

For R_{equiv} to have the same effect in the circuit as R_1, R_2, and R_3 together, it must give rise to the same current i and the same voltage drop. As we discussed above, the voltage drop across a resistance R that carries current i is iR. The same current is present through each of the three resistors shown in Figure 18-10a, and the total voltage drop across the three resistors in series is the sum of the individual voltage drops: $iR_1 + iR_2 + iR_3$. The voltage drop through the equivalent resistance is iR_{equiv}, so

$$iR_{equiv} = iR_1 + iR_2 + iR_3 = i(R_1 + R_2 + R_3)$$
$$R_{equiv} = R_1 + R_2 + R_3$$

(resistors in series)

For example, if the three resistors have resistances $R_1 = 25.0\ \Omega$, $R_2 = 12.0\ \Omega$, and $R_3 = 36.0\ \Omega$, they are equivalent to a single resistor with resistance $R_{equiv} = 25.0\ \Omega + 12.0\ \Omega + 36.0\ \Omega = 73.0\ \Omega$. This means that if we replace the three resistors by a single 73.0-Ω resistor, the current in the circuit will be exactly the same. The net voltage drop is also the same. If the current in our example is $i = 1.00$ A, the voltage drops across the individual resistors are $V_1 = iR_1 = (1.00\ \text{A})(25.0\ \Omega) = 25.0$ V, $V_2 = iR_2 = (1.00\ \text{A})(12.0\ \Omega) = 12.0$ V, and $V_3 = iR_3 = (1.00\ \text{A})(36.0\ \Omega) = 36.0$ V, and the net voltage drop is 25.0 V $+ 12.0$ V $+ 36.0$ V $= 73.0$ V. The voltage drop across the equivalent resistance is $iR_{equiv} = (1.00\ \text{A})(73.0\ \Omega) = 73.0$ V, the same as for the three resistors in series.

In general, if there are N resistors arranged in series the equivalent resistance is

Equivalent resistance of resistors in series
(18-11)

Equivalent resistance of N resistors in series   Resistances of the individual resistors

$$R_{equiv} = R_1 + R_2 + R_3 + ... + R_N$$

The equivalent resistance of N resistors in series is the sum of the individual resistances.

Equation 18-11 tells us that by combining resistors in *series* we create a circuit with a *higher* equivalent resistance than that of any of the individual resistors. The numerical example above illustrates this: with $R_1 = 25.0\ \Omega$, $R_2 = 12.0\ \Omega$, and $R_3 = 36.0\ \Omega$ in series, $R_{equiv} = 73.0\ \Omega$ is greater than any of R_1, R_2, and R_3.

Our analysis tells us that for resistors in series, the *current* is the same through each resistor, but the *voltage drop* is different for different resistors. As we will see, this is not the case if the resistors are in an arrangement other than series.

Kirchhoff's Junction Rule and Resistors in Parallel

Figure 18-11 shows a circuit with two resistors R_1 and R_2 connected in **parallel** to a battery. (As in Figure 18-10, the potential V includes both the emf and the internal resistance of the battery.) Unlike the single-loop circuits shown in Figures 18-9 and 18-10, this is a **multiloop circuit**: There is more than one pathway that a moving charge can take from the positive terminal of the battery through the circuit to the negative terminal. In particular, the circuit in Figure 18-11 has two **junctions** at A and B where the current either breaks into two currents (as at A) or comes together (as at B). What is the relationship among the current i that passes through the battery, the current i_1 that passes through resistor R_1, and the current i_2 that passes through resistor R_2?

One condition on i, i_1, and i_2 is that charge can be neither created nor destroyed, nor can it pile up anywhere in the circuit. This means that the rate at which charge *arrives* at a junction must be equal to the rate at which charge *leaves* that junction. This is our second rule of electric circuits. Like the loop rule, this was also proposed by Kirchhoff and is called **Kirchhoff's junction rule:**

> *The sum of the currents flowing into a junction equals the sum of the currents flowing out of it.*

Let's apply this rule to the junctions shown in Figure 18-11. Current i flows into junction A, and currents i_1 and i_2 flow out of it. So the junction rule tells us that i (the sole current flowing into junction A) must equal $i_1 + i_2$ (the sum of the currents flowing out of A). At junction B the sum of the currents flowing in is $i_1 + i_2$ and the sole current flowing out is i. So by analyzing either junction we can conclude that

$$i = i_1 + i_2 \qquad (18\text{-}12)$$

The current divides itself into i_1 and i_2 when it reaches junction A, with no extra current being added and no current being lost. These currents rejoin at junction B.

Equation 18-12 by itself doesn't tell us how much of current i takes the branch through resistor R_1 (as current i_1) and how much takes the branch through resistor R_2 (as current i_2). To determine this, let's apply the *loop* rule to two different loops through the circuit in Figure 18-11. First consider a loop that starts at the negative terminal of the battery, passes through the battery to the positive terminal (voltage rise V), then follows the path of current i_1 through resistor R_1 (voltage drop $i_1 R_1$), and returns to the negative terminal of the battery. (As before, we'll ignore the resistance of the wires, so there is no voltage drop as a charge traverses the wires.) From the loop rule, the net change in electric potential for this loop is zero:

$$V - i_1 R_1 = 0 \qquad (18\text{-}13)$$

The second loop we'll consider also starts at the negative terminal of the battery and passes through the battery to the positive terminal (voltage rise V) but then follows the path of current i_2 through resistor R_2 (voltage drop $i_2 R_2$) before returning to the battery's negative terminal. The loop rule says that the net change in electric potential is also zero for this loop:

$$V - i_2 R_2 = 0 \qquad (18\text{-}14)$$

If you compare Equations 18-13 and 18-14, you'll see that these equations can both be true only if

$$i_1 R_1 = i_2 R_2 \qquad (18\text{-}15)$$

In other words, for resistors in parallel the *voltage drop* must be the same for each resistor. However, the *currents* are different for different resistors in parallel: If R_1 is less than R_2, the current will be greater in R_1 (with the smaller resistance) and smaller in R_2 (with the greater resistance). Compare this to resistors in series, for which the current is the same but the voltage drops are different for different resistors.

As an illustration, suppose $V = 12.0$ V, $R_1 = 3.00\ \Omega$, and $R_2 = 6.00\ \Omega$. From Equation 18-13 the current i_1 through resistor R_1 is given by

$$V - i_1 R_1 = 0 \quad \text{so} \quad i_1 R_1 = V \quad \text{and} \quad i_1 = \frac{V}{R_1} = \frac{12.0\ \text{V}}{3.00\ \Omega} = 4.00\ \text{A}$$

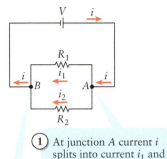

1. At junction A current i splits into current i_1 and current i_2.

2. At junction B current i_1 and current i_2 recombine to current i.

Figure 18-11 Kirchhoff's junction rule At the circuit junctions A and B, the net current into the junction must equal the net current out of that junction.

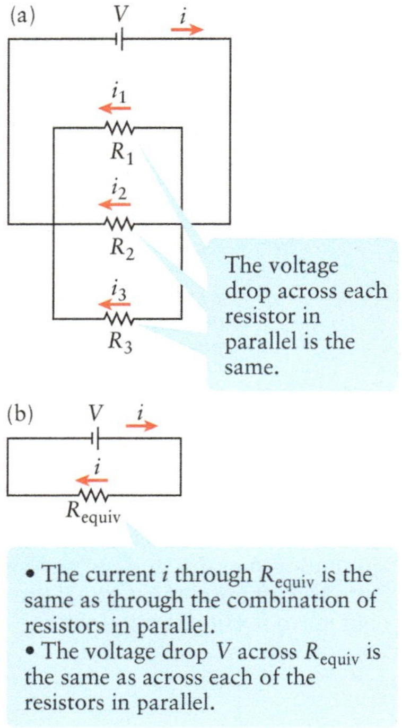

(a)

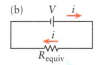

The voltage drop across each resistor in parallel is the same.

(b)

• The current i through R_{equiv} is the same as through the combination of resistors in parallel.
• The voltage drop V across R_{equiv} is the same as across each of the resistors in parallel.

Figure 18-12 Resistors in parallel (a) A circuit contains three resistors connected in parallel to a source of emf. (b) The three resistors have been replaced by a single, equivalent resistor.

 Go to Picture It 18-1 for more practice dealing with resistors in combination.

We can then find the current i_2 using Equation 18-15:

$$i_1 R_1 = i_2 R_2 \quad \text{and} \quad i_2 = \frac{i_1 R_1}{R_2} = \frac{(4.00 \text{ A})(3.00 \text{ }\Omega)}{6.00 \text{ }\Omega} = 2.00 \text{ A}$$

Current $i_1 = 4.00$ A is twice as great as current $i_2 = 2.00$ A because resistance $R_1 = 3.00 \text{ }\Omega$ is half as great as resistance $R_2 = 6.00 \text{ }\Omega$. Note that the total current i that passes through the battery is $i = i_1 + i_2 = 4.00 \text{ A} + 2.00 \text{ A} = 6.00 \text{ A}$.

We can now find the equivalent resistance of a set of resistors in parallel. Figure 18-12a shows three resistors R_1, R_2, and R_3 connected in parallel to a battery with voltage V (including the emf and internal resistance). The currents through these three resistors are i_1, i_2, and i_3, respectively. Figure 18-12b shows the three resistors replaced by an equivalent resistance R_{equiv}. As we did for resistors in series, we'll determine R_{equiv} by demanding that the net current i and the voltage drop be the same for the actual set of resistors in Figure 18-12a and the equivalent resistance in Figure 18-12b. From our discussion above we see that the voltage drop through each of the resistors in Figure 18-12a is equal to V:

$$V = i_1 R_1, \quad V = i_2 R_2, \quad V = i_3 R_3$$

If we divide each of these equations by R_1, R_2, and R_3, respectively, we get expressions for the current through each resistor:

(18-16) $$i_1 = \frac{V}{R_1}, \quad i_2 = \frac{V}{R_2}, \quad i_3 = \frac{V}{R_3}$$

The junction rule tells us that the total current that passes through the battery is the sum of the currents through the individual resistors: $i = i_1 + i_2 + i_3$. From Equations 18-16 we can write this as

(18-17) $$i = \frac{V}{R_1} + \frac{V}{R_2} + \frac{V}{R_3} = V\left(\frac{1}{R_1} + \frac{1}{R_2} + \frac{1}{R_3}\right)$$

The total current must be the same for the circuit in Figure 18-12b with the equivalent resistance. If we apply the loop theorem to this circuit, we get $V - i R_{equiv} = 0$, so $i R_{equiv} = V$ or

(18-18) $$i = \frac{V}{R_{equiv}} = V\left(\frac{1}{R_{equiv}}\right)$$

Since Equations 18-17 and 18-18 are both expressions for the total current i, they can both be valid only if the right-hand sides of these equations are equal to each other:

(18-19) $$\frac{1}{R_{equiv}} = \frac{1}{R_1} + \frac{1}{R_2} + \frac{1}{R_3}$$

(resistors in parallel)

As an illustration, suppose the three resistors in Figure 18-12a are $R_1 = 25.0 \text{ }\Omega$, $R_1 = 12.0 \text{ }\Omega$, and $R_3 = 36.0 \text{ }\Omega$. According to Equation 18-19 they are equivalent to a single resistor with resistance R_{equiv} given by

$$\frac{1}{R_{equiv}} = \frac{1}{25.0 \text{ }\Omega} + \frac{1}{12.0 \text{ }\Omega} + \frac{1}{36.0 \text{ }\Omega} = 0.151 \text{ }\Omega^{-1}$$

$$R_{equiv} = \frac{1}{0.151 \text{ }\Omega^{-1}} = 6.62 \text{ }\Omega$$

If the net current through the battery is 1.00 A, the voltage drop across the equivalent resistance is $i R_{equiv} = (1.00 \text{ A})(6.62 \text{ }\Omega) = 6.62 \text{ V}$. The voltage drop across each of the individual resistors in parallel must be the same, so $6.62 \text{ V} = i_1 R_1 = i_2 R_2 = i_3 R_3$. You can use these relationships to show that $i_1 = 0.265$ A, $i_2 = 0.551$ A, and $i_3 = 0.184$ A (note that the current is greatest through resistor R_2, which has the smallest of the three resistances). The net current through the battery is $i = i_1 + i_2 + i_3 = 1.00$ A, just as for the battery connected to the equivalent resistance.

If we have N resistors in parallel, Equation 18-19 becomes

 The reciprocal of the equivalent resistance of N resistors in parallel is the sum of the reciprocals of the individual resistances.

Equivalent resistance of resistors in parallel
(18-20)

$$\frac{1}{R_{\text{equiv}}} = \frac{1}{R_1} + \frac{1}{R_2} + \frac{1}{R_3} + \cdots + \frac{1}{R_N}$$

Equivalent resistance of N resistors in parallel

Resistances of the individual resistors

Equation 18-20 tells us that by combining resistors in *parallel*, we create a circuit with a *smaller* equivalent resistance than any of the individual resistors. The numerical example above illustrates this: With $R_1 = 25.0 \ \Omega$, $R_2 = 12.0 \ \Omega$, and $R_3 = 36.0 \ \Omega$ in parallel, the equivalent resistance $R_{\text{equiv}} = 6.62 \ \Omega$ is less than any of R_1, R_2, and R_3.

WATCH OUT! Resistors do not combine in the same way as capacitors.

Be careful to distinguish between the rules for equivalent *resistance* that we've developed in this section and the rules for equivalent *capacitance* that we found in Section 17-7. For resistors in series, we find the equivalent resistance by adding the individual resistances ($R_{\text{equiv}} = R_1 + R_2 + \cdots$), and R_{equiv} is greater than any of the individual resistances; for resistors in parallel, we find the reciprocal of the equivalent resistance by adding the reciprocals of the individual resistances ($1/R_{\text{equiv}} = 1/R_1 + 1/R_2 + \cdots$), and R_{equiv} is less than any of the individual resistances. These rules are reversed for capacitors. For capacitors in series, we find the reciprocal of the equivalent capacitance by adding the reciprocals of the individual capacitances ($1/C_{\text{equiv}} = 1/C_1 + 1/C_2 + \cdots$), and C_{equiv} is less than any of the individual capacitances; for capacitors in parallel, we find the equivalent capacitance by adding the individual capacitances ($C_{\text{equiv}} = C_1 + C_2 + \cdots$), and C_{equiv} is greater than any of the individual capacitances.

The following examples illustrate some applications of the relationships we've developed in this section. In the second example, we'll see how to analyze resistors arranged in a combination that is neither purely series nor purely parallel.

BioMedical EXAMPLE 18-7   Giant Axons in Squid Revisited

As we saw in Example 18-6 (Section 18-4), in a squid, axons of approximately 30,000 nerve cells fuse together to form each giant axon. A typical axon has a diameter of 15.0 μm, is 10.0 cm long, and has a resistivity of about 3100 Ω·m. Find the resistance of a giant squid axon by considering it as 30,000 separate axons in parallel.

Set Up

We can find the resistance of a typical axon by using Equation 18-8. Each of these individual axons acts as a separate conducting path that mobile charges can follow between a point of high potential and low potential. That's just like the three resistors in parallel shown in Figure 18-12a, so we can use Equation 18-20 to determine the equivalent resistance of the giant axon as a whole in terms of the resistances of the individual axons.

Definition of resistance:

$$R = \frac{\rho L}{A} \qquad (18\text{-}8)$$

Equivalent resistance of resistors in parallel:

$$\frac{1}{R_{\text{equiv}}} = \frac{1}{R_1} + \frac{1}{R_2} + \frac{1}{R_3} + \cdots + \frac{1}{R_N} \qquad (18\text{-}20)$$

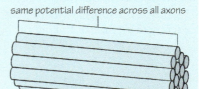
same potential difference across all axons

Solve

Calculate the resistance of an individual axon.

Each individual axon has resistivity $\rho = 3100 \; \Omega \cdot m$, length $L = 10.0 \; cm = 0.100 \; m$, and diameter $15.0 \; \mu m = 15.0 \times 10^{-6} \; m$. The radius r is one-half of the diameter:

$$r = \frac{1}{2}(15.0 \times 10^{-6} \; m) = 7.50 \times 10^{-6} \; m$$

If the axon has a circular cross section, its cross-sectional area is $A = \pi r^2$. So from Equation 18-8 the resistance is

$$R = \frac{\rho L}{A} = \frac{\rho L}{\pi r^2} = \frac{(3100 \; \Omega \cdot m)(0.100 \; m)}{\pi(7.50 \times 10^{-6} \; m)^2} = 1.75 \times 10^{12} \; \Omega$$

This resistance is very large because the axon is long and thin, and because the axon resistivity is much higher than that of metallic conductors (see Table 18-1).

Calculate the resistance of the giant axon, as if it were 30,000 individual axons in parallel.

If N axons are in parallel and each has the same resistance R, there are N identical terms on the right-hand side of Equation 18-20. So

$$\frac{1}{R_{\text{equiv}}} = \frac{1}{R_1} + \frac{1}{R_2} + \frac{1}{R_3} + \cdots + \frac{1}{R_N} = \frac{N}{R}$$

Take the reciprocal of both sides:

$$R_{\text{equiv}} = \frac{R}{N}$$

There are $N = 30,000$ individual axons, each of which has resistance $R = 1.75 \times 10^{12} \; \Omega$, so the equivalent resistance of the giant axon is

$$R_{\text{equiv}} = \frac{1.75 \times 10^{12} \; \Omega}{30,000} = 5.85 \times 10^7 \; \Omega$$

Reflect

In Example 18-6 we found that the resistance of a giant squid axon is about 10,000 times smaller than that of a normal axon. But in this example we've found that the equivalent resistance of a giant axon is 30,000 times smaller than that of an individual axon. Why do our answers differ by a factor of 3?

The explanation is that in this example we made the implicit assumption that the total cross-sectional area of the giant axon is 30,000 times the cross-sectional area of a small axon. However, in an actual squid the cross-sectional area of the giant axon with a diameter of 1.50 mm is smaller than the total cross-sectional area of 30,000 small axons each of diameter 15.0 μm. Although about 30,000 cells contribute to the formation of the giant axon, the process by which the giant axon forms is *not* an actual fusing together of 30,000 fully formed smaller axons. Nevertheless, it's encouraging that our calculations in Example 18-6 and this problem are pretty close.

EXAMPLE 18-8 Resistors in Combination

Figure 18-13 shows two different combinations of three identical resistors, each with resistance R. Find the equivalent resistance of the combination in (a) Figure 18-13a and (b) Figure 18-13b.

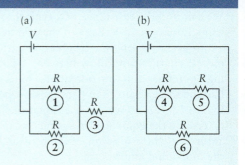

Figure 18-13 **Two combinations of three identical resistors** What is the equivalent resistance of each combination?

Set Up

Neither combination in Figure 18-13 is a simple series or parallel arrangement of resistors. But in Figure 18-13a resistors 1 and 2 are in parallel with each other, and that combination is in series with resistor 3. Similarly, in Figure 18-13b resistors 4 and 5 are in series with each other, and that combination is in parallel with resistor 6. So we can use Equations 18-11 and 18-20 together to find the equivalent resistances of both arrangements of resistors.

Equivalent resistance of resistors in series:

$$R_{equiv} = R_1 + R_2 + R_3 + \cdots + R_N \qquad (18\text{-}11)$$

Equivalent resistance of resistors in parallel:

$$\frac{1}{R_{equiv}} = \frac{1}{R_1} + \frac{1}{R_2} + \frac{1}{R_3} + \cdots + \frac{1}{R_N} \qquad (18\text{-}20)$$

resistors in series

resistors in parallel

Solve

(a) For the arrangement in Figure 18-13a, first find the equivalent resistance of resistors 1 and 2.

Resistors 1 and 2 in Figure 18-13a are in parallel, so their equivalent resistance R_{12} is given by Equation 18-20:

$$\frac{1}{R_{12}} = \frac{1}{R} + \frac{1}{R} = \frac{2}{R}$$

$$R_{12} = \frac{R}{2}$$

The combination of resistors 1 and 2 is in series with resistor 3. This tells us the overall equivalent resistance.

Equivalent resistor R_{12} is in series with resistor 3. The equivalent resistance R_{123} of the entire combination is given by Equation 18-11:

$$R_{123} = R_{12} + R = \frac{R}{2} + R = \frac{3R}{2}$$

(b) For the arrangement in Figure 18-13b, first find the equivalent resistance of resistors 4 and 5.

Resistors 4 and 5 in Figure 18-13b are in series, so their equivalent resistance R_{45} is given by Equation 18-11:

$$R_{45} = R + R = 2R$$

The combination of resistors 4 and 5 is in parallel with resistor 6. This tells us the overall equivalent resistance.

Equivalent resistor R_{45} is in parallel with resistor 6. The equivalent resistance R_{456} of the entire combination is given by Equation 18-20:

$$\frac{1}{R_{456}} = \frac{1}{R_{45}} + \frac{1}{R} = \frac{1}{2R} + \frac{1}{R} = \frac{3}{2R}$$

$$R_{456} = \frac{2R}{3}$$

Reflect

If we had more than three resistors, or if the resistors had different values, we could create a large number of combinations and equivalent resistances.

GOT THE CONCEPT? 18-5 Combinations of Resistors I

(?) Rank the four circuits shown in Figure 18-14 in order of their equivalent resistance, from highest to lowest.

(a)

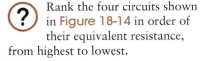

(b)

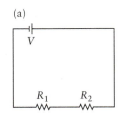

(c)

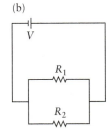

(d)

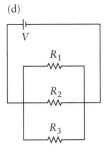

Figure 18-14 Rank the circuits How do these four circuits compare in their equivalent resistances? In the net current through their sources of emf?

GOT THE CONCEPT? 18-6 Combinations of Resistors II

  Rank the four circuits shown in Figure 18-14 in order of the current through the source of emf *V*, from highest to lowest.

TAKE-HOME MESSAGE FOR Section 18-5

✔ In traversing a battery from its negative terminal to its positive terminal, the potential increases (a voltage rise) by an amount that depends on the battery's emf, its internal resistance, and the current *i*. In traversing a resistor of resistance *R* in the direction of the current, the potential drops by *iR* (a voltage drop).

✔ Kirchhoff's loop rule says that the sum of potential changes around a closed loop in a circuit is zero.

✔ Kirchhoff's junction rule says that the net current into a circuit junction equals the net current out of that junction.

✔ If resistors are connected in series, the current is the same in each resistor but the voltage drops are different across different resistors. The equivalent resistance equals the sum of the individual resistances.

✔ If resistors are connected in parallel, the voltage drop is the same for each resistor but the currents are different for different resistors. The reciprocal of the equivalent resistance equals the sum of the reciprocals of the individual resistances.

18-6 The rate at which energy is produced or taken in by a circuit element depends on current and voltage

The aspects of electric circuits that we've concentrated on so far are voltage, current, and resistance. But an electric circuit is fundamentally a way to transfer *energy* from one place to another, such as from a battery to a flashlight bulb (where the energy is converted into visible light) or from a wall socket to a toaster (where the energy is used to heat your morning bread or bagel). In most applications what's of interest is the *rate* at which energy is transferred into or out of a circuit element. For example, for a toaster to be useful it must heat the bread rapidly enough that it becomes toast in a minute or so, not an hour.

As we learned in Section 6-9, **power** is the rate at which energy is transferred into or out of an object. The unit of power is the joule per second, or watt (abbreviated W): 1 W = 1 J/s. You can see the importance of power in electric circuits from the numbers that are used to describe various electric devices: An amplifier for a home audio system is rated by its power output (perhaps 75 to 100 W), and any light bulb is stamped with the power that must be supplied to it for normal operation (say, 13 or 60 W).

In this section we'll see how to calculate the power *output* of a source of emf such as a battery, which is fundamentally a source of electric potential energy. We'll also see how to calculate the power *input* of a resistor, which absorbs electric potential energy and converts it into other forms of energy.

Power in a Circuit Element

The key to understanding energy and power in electric circuits is that there is a potential difference across each circuit element, which means that a change takes place in electric potential energy when a moving charge traverses a circuit element. Remember from Section 17-3 the relationship between electric potential difference and electric potential energy difference:

Electric potential difference related to electric potential energy difference (17-6)

The **difference in electric potential** between two positions...   ...equals the **change in electric potential energy** for a charge q_0 moved between these two positions...

$$\Delta V = \frac{\Delta U_{\text{electric}}}{q_0}$$

...divided by the **charge** q_0.

  Electric potential difference equals electric potential energy difference per unit charge.

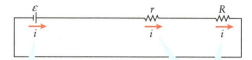

Figure 18-15 Energy and power in a single-loop circuit Energy flows from the emf into the moving charges; energy flows from the moving charges into the internal resistance and the resistor.

Charges moving through the emf undergo a potential change of $+\varepsilon$, so electric potential energy increases; energy is extracted from the emf.

Charges undergo a potential change of $-ir$ moving through the internal resistance and $-iR$ moving through the resistor. Electric potential energy decreases; energy goes into the resistances.

Let's rewrite this equation for the case in which a small quantity Δq of charge moves from one end of a circuit element to the other. If the potential difference between the ends of the element is ΔV, then Equation 17-6 becomes

$$\Delta V = \frac{\Delta U_{\text{electric}}}{\Delta q} \quad \text{or} \quad \Delta U_{\text{electric}} = (\Delta q)\Delta V \qquad (18\text{-}21)$$

We'll continue to use the idea that current is in the direction in which positive charges would flow, so the moving charge is positive: $\Delta q > 0$. Then Equation 18-21 tells us that there is an *increase* in electric potential energy ($\Delta U_{\text{electric}} > 0$) if the charge Δq traverses a circuit element from low potential to high potential, so $\Delta V > 0$. That's the case for a charge that travels through the source of emf in Figure 18-15 from the negative terminal to the positive terminal. Since energy is conserved, it must be that the amount of energy extracted from the source is equal to the increase in electric potential energy.

There is a *decrease* in electric potential energy ($\Delta U_{\text{electric}} < 0$) if the charge Δq traverses a circuit element from high potential to low potential, so $\Delta V < 0$. That's what happens when charge Δq travels through the resistor R in Figure 18-15 in the direction of the current. The lost electric potential energy is deposited into the resistor, which causes an increase in the resistor's temperature. If the resistor is the filament of a conventional flashlight bulb, the increased temperature causes the filament to emit radiation, some of which is in the form of visible light (see Section 14-7).

WATCH OUT! Potential energy may change as charges move around a circuit, but current does not.

It's a common misconception that current is "used up" when it passes through a resistor and that current is "added to" when it passes through a source of emf. In fact, there is *no* change in the current as it passes through either a resistor or a source: Moving charges leave any circuit element at the same rate as they enter the element. All that changes is the electric potential energy associated with the charges.

With a current i in a circuit, the power P for each circuit element is just the rate at which electric potential energy changes in that element. This equals the change in electric potential energy $\Delta U_{\text{electric}}$ for a charge Δq that enters the element, as given by Equation 18-21, divided by the time Δt that it takes each new bit of charge Δq to enter the element:

$$P = \frac{\Delta U_{\text{electric}}}{\Delta t} = \frac{\Delta q \Delta V}{\Delta t} = \left(\frac{\Delta q}{\Delta t}\right)\Delta V = (i)\Delta V \qquad (18\text{-}22)$$

In the last part of Equation 18-22, we've used the idea that the ratio $\Delta q/\Delta t$ equals the rate at which charge enters the circuit element—in other words, this ratio equals the current i through the element (see Equation 18-1).

Equation 18-22 states that the power can be positive or negative depending on the sign of the potential difference ΔV between the ends of the circuit element. Instead of worrying about these signs, we'll replace ΔV in Equation 18-22 with the voltage V across the circuit element, which we'll regard as the absolute value of the potential difference between the ends of the element. Then P is always positive, and we can write

**Power for a circuit element
(18-23)**

Power produced by or transferred into a circuit element

$$P = iV$$

Current through the circuit element

Voltage (absolute value) across the circuit element

 Source of emf: Power flows out of the source and into the moving charges.
Resistor: Power flows out of the charges and into the resistor.

Equation 18-23 says that the power that flows into or out of any circuit element is equal to the product of the current through the element multiplied by the voltage across the element. For a given voltage V, each small amount of charge Δq that traverses the circuit element transfers the same amount of energy into or out of the element; the more charge that traverses the element per unit time and so the greater the current i, the greater the *rate* of energy transfer P.

WATCH OUT! The units of power take time into account.

! The units of power sometimes cause confusion because it seems that watts require an additional time measurement. For example, to find the power in watts that goes into a flashlight bulb, a student might ask, "Find the power for what amount of time?" That's not a sensible question: Time is already included in the units of power, since 1 watt is equal to 1 joule *per second*. When a light bulb is rated at 120 W, that means it requires 120 J of energy every second to operate.

Note that the power company charges its customers based not on how much *power* they use at a given time but on the total amount of *energy* that they use. Since power is energy per time, the units of energy are the units of power multiplied by the units of time. That's why the power company bills in terms of kilowatt-hours (kWh): 1 kWh = 1000 watt-hours, or the amount of energy it takes to run a device that uses 1000 W of power (typical for a microwave oven) for 1 hour. Since 1 h = 3600 s, 1 kWh equals (1000 W)(3600 s) = 3.6×10^6 W·s = 3.6×10^6 J.

If the circuit element is a resistor with resistance R, Equation 18-9 tells us that the voltage V across the resistor is equal to the product of the current and the resistance: $V = iR$, or equivalently $i = V/R$. We can use these two expressions to rewrite Equation 18-23 in two equivalent forms for the special case of a resistor:

$$P = i(iR) \quad \text{or} \quad P = \left(\frac{V}{R}\right)V$$

We can simplify these to

**Power for a resistor
(18-24)**

Power into a **resistor**   **Current** through the resistor

$$P = i^2 R = \frac{V^2}{R}$$

Voltage across the resistor

Resistance of the resistor

The expression $P = i^2 R$ is useful if we know the current through a resistor of known resistance, while $P = V^2/R$ is useful if we know the voltage across that resistor. The following examples show how to use Equations 18-23 and 18-24.

EXAMPLE 18-9 Power in a Single-Loop Circuit

A battery with emf 12.0 V and internal resistance 1.00 Ω is connected to a resistor with resistance 19.0 Ω. Find (a) the rate at which energy is supplied by the emf, (b) the rate at which energy flows into the internal resistance, and (c) the rate at which energy flows into the resistor.

Set Up

We can find the current in the circuit using Kirchhoff's loop rule (the sum of the changes in potential around a closed loop in a circuit must equal zero). Equation 18-9 then tells us the voltage across either resistance. We'll use Equation 18-23 to determine the power for each element of the circuit, and check our results for the two resistances using Equations 18-24.

Relationship among potential difference, current, and resistance:

$$V = iR \qquad (18\text{-}9)$$

Power for a circuit element:

$$P = iV \qquad (18\text{-}23)$$

Power for a resistor:

$$P = i^2R = \frac{V^2}{R} \qquad (18\text{-}24)$$

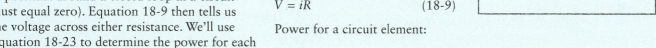

$\varepsilon = 12.0\,V \qquad r = 1.00\,\Omega \qquad R = 19.0\,\Omega$

Solve

(a) First apply Kirchhoff's loop rule to determine the current i. Start at the point a in the circuit and go around in the direction of the current.

There is a voltage rise of $\varepsilon = 12.0$ V going across the emf, a voltage drop of ir going across the internal resistance $r = 1.00\ \Omega$, and a voltage drop of iR going across the resistor with resistance $R = 19.0\ \Omega$. The sum of the potential changes around the circuit is zero:

$$+\varepsilon + (-ir) + (-iR) = 0$$

Rearrange this to solve for the current i:

$$\varepsilon - i(r + R) = 0, \text{ so } i(r + R) = \varepsilon \text{ and}$$

$$i = \frac{\varepsilon}{r + R} = \frac{12.0\ \text{V}}{1.00\ \Omega + 19.0\ \Omega} = \frac{12.0\ \text{V}}{20.0\ \Omega} = 0.600\ \text{A}$$

(Recall that 1 A = 1 V/Ω.)

Use Equation 18-23 to find the power extracted from the emf.

The voltage across the emf itself is $\varepsilon = 12.0$ V. The rate at which energy is extracted from the emf is

$$P_{\text{emf}} = i\varepsilon = (0.600\ \text{A})(12.0\ \text{V}) = 7.20\ \text{A} \cdot \text{V}$$

Since 1 A = 1 C/s, 1 V = 1 J/C, and 1 W = 1 J/s,

$$P_{\text{emf}} = 7.20\ \frac{\text{C}}{\text{s}} \cdot \frac{\text{J}}{\text{C}} = 7.20\ \frac{\text{J}}{\text{s}} = 7.20\ \text{W}$$

(b) First use Equation 18-9 to find the potential difference across the internal resistance $r = 1.00\ \Omega$. Then find the power in the internal resistance using Equation 18-23.

From Equation 18-9 the voltage across the internal resistance is

$$V_r = ir = (0.600\ \text{A})(1.00\ \Omega) = 0.600\ \text{V}$$

From Equation 18-23 the rate at which energy flows into the internal resistance is

$$P_r = iV_r = (0.600\ \text{A})(0.600\ \text{V}) = 0.360\ \text{W}$$

This power into the internal resistance goes into heating the battery.

(c) Repeat part (b) for the resistor of resistance $R = 19.0\ \Omega$.

The voltage across the resistor is

$$V_R = iR = (0.600\ \text{A})(19.0\ \Omega) = 11.4\ \text{V}$$

The rate at which energy flows into the resistor is

$$P_R = iV_R = (0.600\ \text{A})(11.4\ \text{V}) = 6.84\ \text{W}$$

Reflect

Note that the rate at which energy is extracted from the source of emf is equal to the *net* rate at which energy flows into the internal resistance and the resistor. This is equivalent to saying that energy is conserved in the circuit.

The net rate of energy flow into the two resistances is

$$P_r + P_R = 0.360 \text{ W} + 6.84 \text{ W} = 7.20 \text{ W}$$

This is the same as the rate at which energy flows out of the source of emf, $P_{\text{emf}} = 7.20 \text{ W}$.

We can check our result $P_R = 6.84$ W for the power into the resistor by showing that we get the same results using Equations 18-24. Can you use the same approach to check the result $P_r = 0.360$ W for the internal resistance?

From the first of Equations 18-24, the power into the 19.0-Ω resistor is

$$P_R = i^2 R = (0.600 \text{ A})^2 (19.0 \text{ Ω}) = 6.84 \text{ W}$$

(Note that $1 \text{ A}^2 \cdot \text{Ω} = 1 \text{ A} \cdot \text{V} = 1 \text{ W}$.)

To use the second of Equations 18-24, use the voltage $V_R = 11.4$ V across the resistor:

$$P_R = \frac{V_R^2}{R} = \frac{(11.4 \text{ V})^2}{19.0 \text{ Ω}} = 6.84 \text{ W}$$

(Note that $1 \text{ V}^2/\text{Ω} = 1 \text{ V} \cdot (\text{V}/\text{Ω}) = 1 \text{ V} \cdot \text{A} = 1 \text{ W}$.)

This agrees with the result for P_R found above.

EXAMPLE 18-10   Power in Series and Parallel Circuits

Two resistors, one with $R_1 = 2.00$ Ω and one with $R_2 = 3.00$ Ω, are both connected to a battery with emf $\varepsilon = 12.0$ V and negligible internal resistance. Find the power delivered by the battery and the power absorbed by each resistor if the resistors are connected to the battery (a) in series and (b) in parallel.

Set Up

We'll use the same tools as in the previous example, plus Equations 18-11 and 18-20 for the equivalent resistance of resistors in series and in parallel. In each case we'll use the equivalent resistance to find the current through the battery, then use Equation 18-23 to determine the power provided by the battery. If the resistors are in series, the current through each is the same as the current through the battery, so we'll use the first of Equations 18-24 ($P = i^2/R$) to determine the power into each resistor. For resistors in parallel the voltage is the same across each resistor, so in that case we'll find the power into each resistor using the second of Equations 18-24 ($P = V^2/R$).

Relationship among potential difference, current, and resistance:

$$V = iR \qquad (18-9)$$

Equivalent resistance of two resistors in series:

$$R_{\text{equiv}} = R_1 + R_2 \qquad (18-11)$$

Equivalent resistance of two resistors in parallel:

$$\frac{1}{R_{\text{equiv}}} = \frac{1}{R_1} + \frac{1}{R_2} \qquad (18-20)$$

Power for a circuit element:

$$P = iV \qquad (18-23)$$

Power for a resistor:

$$P = i^2 R = \frac{V^2}{R} \qquad (18-24)$$

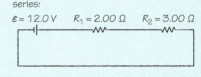

series:

$\varepsilon = 12.0\,\text{V} \qquad R_1 = 2.00\,\text{Ω} \qquad R_2 = 3.00\,\text{Ω}$

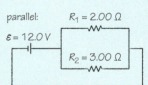

parallel: $R_1 = 2.00\,\text{Ω}$

$\varepsilon = 12.0\,\text{V}$

$R_2 = 3.00\,\text{Ω}$

Solve

(a) The two resistors in series are equivalent to a single resistor R_{equiv}. This is connected directly to the terminals of the battery, across which the voltage is ε (we're told to ignore the internal resistance). So the voltage across R_{equiv} is also equal to ε, which tells us the current through the circuit. Equation 18-23 then tells us the power delivered by the battery.

From Equation 18-11 the equivalent resistance of the two resistors in series is

$$R_{\text{equiv}} = R_1 + R_2 = 2.00 \text{ Ω} + 3.00 \text{ Ω} = 5.00 \text{ Ω}$$

The voltage drop across this equivalent resistance is $V = iR_{\text{equiv}}$ from Equation 18-9, which is also equal to the emf $\varepsilon = 12.0$ V of the battery. So the current through the equivalent resistance is given by

$$\varepsilon = iR_{\text{equiv}} \quad \text{or} \quad i = \frac{\varepsilon}{R_{\text{equiv}}} = \frac{12.0 \text{ V}}{5.00 \text{ Ω}} = 2.40 \text{ A}$$

This is also the current through the battery. Since the voltage across the battery is equal to $\varepsilon = 12.0$ V, the power delivered by the battery is

$$P_{battery} = i\varepsilon = (2.40 \text{ A})(12.0 \text{ V}) = 28.8 \text{ W}$$

The current $i = 2.40$ A is the same through both resistors in this series circuit. Use this to calculate the power that goes into each resistor.

Using the first of Equations 18-24, the power into resistor $R_1 = 2.00 \ \Omega$ is

$$P_1 = i^2 R_1 = (2.40 \text{ A})^2 (2.00 \ \Omega) = 11.5 \text{ W}$$

The power into resistor $R_2 = 3.00 \ \Omega$ is

$$P_2 = i^2 R_2 = (2.40 \text{ A})^2 (3.00 \ \Omega) = 17.3 \text{ W}$$

The net power into the two resistors is equal to the power supplied by the battery, as it should be:

$$P_1 + P_2 = 11.5 \text{ W} + 17.3 \text{ W} = 28.8 \text{ W} = P_{battery}$$

(b) Follow the same steps as in part (a) to find the equivalent resistance of the two resistors in parallel, the current through the battery connected to that parallel arrangement, and the power delivered by the battery in this situation.

From Equation 18-20 the equivalent resistance of the two resistors in parallel is given by

$$\frac{1}{R_{equiv}} = \frac{1}{R_1} + \frac{1}{R_2} = \frac{1}{2.00 \ \Omega} + \frac{1}{3.00 \ \Omega} = 0.833 \ \Omega^{-1}$$

$$R_{equiv} = \frac{1}{0.833 \ \Omega^{-1}} = 1.20 \ \Omega$$

Equation 18-9 tells us that the voltage drop across this equivalent resistance is $V = iR_{equiv}$; since this equivalent resistance is connected to the terminals of the battery, the voltage drop is also equal to the battery emf $\varepsilon = 12.0$ V. So the current through the equivalent resistance and through the battery is given by

$$\varepsilon = iR_{equiv} \text{ or } i = \frac{\varepsilon}{R_{equiv}} = \frac{12.0 \text{ V}}{1.20 \ \Omega} = 10.0 \text{ A}$$

From Equation 18-23 the power delivered by the battery is

$$P_{battery} = i\varepsilon = (10.0 \text{ A})(12.0 \text{ V}) = 1.20 \times 10^2 \text{ W}$$

Note that this is more than four times as much power as the same battery delivers to the same resistors in series.

The voltage is the same across resistors in parallel. Each resistor is effectively connected directly to the terminals of the battery, so the voltage across each resistor is $V = \varepsilon = 12.0$ V. Use this to calculate the power that goes into each resistor.

Using the second of Equations 18-24, we find that the power into resistor $R_1 = 2.00 \ \Omega$ is

$$P_1 = \frac{V^2}{R_1} = \frac{\varepsilon^2}{R_1} = \frac{(12.0 \text{ V})^2}{2.00 \ \Omega} = 72.0 \text{ W}$$

The power into resistor $R_2 = 3.00 \ \Omega$ is

$$P_2 = \frac{V^2}{R_2} = \frac{\varepsilon^2}{R_2} = \frac{(12.0 \text{ V})^2}{3.00 \ \Omega} = 48.0 \text{ W}$$

As for the series case, the net power into the two resistors is equal to the power supplied by the battery:

$$P_1 + P_2 = 72.0 \text{ W} + 48.0 \text{ W} = 1.20 \times 10^2 \text{ W} = P_{battery}$$

Reflect

Although the same battery and same resistors are used in both circuits, the power provided by the battery is *much* different in the two circuits. That's because the current through the battery is different for the two circuits: $i = 2.40$ A for the series circuit, $i = 10.0$ A for the parallel circuit.

What's more, the power into each resistor is very different in the two circuits, because the voltage and current for each resistor are greater for the parallel circuit than for the series circuit. If the resistors are light bulbs that take in power and use it to produce light, the lights will glow brighter in the parallel circuit than in the series circuit.

Use Equation 18-9 to find the voltage across each resistor in the series circuit:

$$V_1 = iR_1 = (2.40 \text{ A})(2.00 \text{ } \Omega) = 4.80 \text{ V}$$
$$V_2 = iR_2 = (2.40 \text{ A})(3.00 \text{ } \Omega) = 7.20 \text{ V}$$

(compared to $V = \varepsilon = 12.0$ V for each resistor in the parallel circuit)

Use Equation 18-9 to find the current through each resistor in the parallel circuit:

$$i_1 = \frac{\varepsilon}{R_1} = \frac{12.0 \text{ V}}{2.00 \text{ } \Omega} = 6.00 \text{ A}$$

$$i_2 = \frac{\varepsilon}{R_2} = \frac{12.0 \text{ V}}{3.00 \text{ } \Omega} = 4.00 \text{ A}$$

(compared to $i = 2.40$ A for each resistor in the series circuit)

Note that the current is the same for both resistors in the series circuit, so more power goes into the resistor with the greater resistance ($P_2 = 17.3$ W into $R_2 = 3.00$ Ω versus $P_1 = 11.5$ W into $R_1 = 2.00$ Ω). That follows from the relationship $P = i^2R$ for resistors. By contrast, the voltage is the same for the two resistors in the parallel circuit, so more power goes into the resistor with the smaller resistance ($P_1 = 72.0$ W into $R_1 = 2.00$ Ω versus $P_2 = 48.0$ W into $R_2 = 3.00$ Ω). That agrees with the relationship $P = V^2/R$ for resistors. So if the resistors are light bulbs, the light bulb with $R_2 = 3.00$ Ω is the brighter one in the series circuit, but the bulb with $R_1 = 2.00$ Ω is the brighter one in the parallel circuit! When comparing the power in resistors, choose wisely among the relationships $P = iV$, $P = i^2R$, and $P = V^2/R$.

We've seen how to apply Equation 18-23, $P = iV$, and Equations 18-24, $P = i^2R = V^2/R$, to dc circuits in which there is a steady current that does not vary with time. But these same equations also apply to circuits in which the current varies with time and is *not* constant. In the following section we'll investigate the time-varying current in a circuit that includes both a resistor and a capacitor.

GOT THE CONCEPT? 18-7 Batteries in Series

? Consider a resistor R connected to a single source of emf ε and negligible internal resistance (Figure 18-16a). If the same resistor is instead connected to two sources of emf ε in series, each with negligible internal resistance (Figure 18-16b), by what factor does the power into the resistor increase? (a) $\sqrt{2}$; (b) 2; (c) 4; (d) 8; (e) 16.

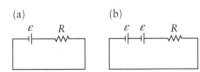

Figure 18-16 **One emf or two** How does adding a second source of emf affect the power delivered to the resistor?

GOT THE CONCEPT? 18-8 Ranking the Power

? Five identical resistors, A, B, C, D, and E, are connected to a source of emf as shown in Figure 18-17. The source has negligible internal resistance. Rank the five resistors in order of the amount of power that goes into each resistor, from greatest to smallest. If the power in two resistors is the same, indicate as such.

Figure 18-17 **Rank the resistors** How do these five identical resistors compare in the power that they absorb?

TAKE-HOME MESSAGE FOR Section 18-6

✔ Power is the rate at which energy is transferred into or out of a system. The unit of power is the watt (1 W = 1 J/s).

✔ The power into or out of any circuit element equals the current through the element multiplied by the voltage across the element.

✔ The power into a resistor is proportional to the square of the current through the resistor or, equivalently, the square of the voltage across the resistor.

18-7 A circuit containing a resistor and capacitor has a current that varies with time

In an ordinary dc circuit the battery provides a steady current and delivers energy to the other circuit elements at a steady rate. In some circuits, however, what's required is a short burst of energy. That's the case for the electronic flash in a camera or mobile phone (Figure 18-1c): Energy has to be delivered to the flash lamp in a very brief time interval to produce a short-duration, high-intensity flash.

The simplest way to deliver a short burst of electric potential energy to a circuit element is by using a *capacitor* as an energy source. (You should review our discussion of capacitors in Sections 17-5 and 17-6.) We'll first examine a circuit in which a battery is used to charge a capacitor, then see what happens when the charged capacitor is discharged and its stored energy is delivered to a resistor.

A Series *RC* Circuit: Charging the Capacitor

Figure 18-18a shows a **series *RC* circuit**, with a resistor of resistance R and a capacitor of capacitance C connected in series to a battery of emf ε. (We will assume that the internal resistance of the battery is so small compared to R that it can be ignored.)

Initially the capacitor in Figure 18-18a is uncharged. If the switch in this circuit is moved to the up position (Figure 18-18b), the circuit is completed and the battery will cause charge to begin to flow. We'll call $t = 0$ the time when the switch is thrown. As time passes, positive charge $+q$ builds up on the upper capacitor plate and an equal amount of negative charge $-q$ builds up on the lower capacitor plate. (The two charges must be of equal magnitude since the current is the same throughout the circuit. Hence positive charge leaves the initially uncharged lower plate at the same rate that it arrives at the initially uncharged upper plate, leaving as much negative charge on the lower plate as there is positive charge on the upper plate.) So the circuit in Figure 18-18b is a *charging* series *RC* circuit.

Let's apply Kirchhoff's loop rule to the circuit shown in Figure 18-18b. In Section 17-5 we found that the magnitude of the charge q on the capacitor plates is proportional to the voltage V across the capacitor:

A **capacitor** carries a **charge** $+q$ on its positive plate and a charge $-q$ on its negative plate.	The magnitude of q is directly proportional to V, the **voltage** (potential difference) between the plates.

$$q = CV$$

The constant of proportionality between charge q and voltage V is the **capacitance** C of the capacitor.

Charge, voltage, and capacitance for a capacitor
(17-14)

We can rewrite Equation 17-14 as $V = q/C$. If we start at point p in Figure 18-18b and move clockwise around the circuit, we encounter a voltage rise of $+\varepsilon$ as we cross the source of emf, a voltage drop $-iR$ as we cross the resistor, and a voltage drop $-q/C$ as we cross the capacitor from the positive plate to the negative plate (which is at lower potential). Kirchhoff's loop rule tells us that the sum of these voltages must be zero:

$$\varepsilon + (-iR) + \left(-\frac{q}{C}\right) = 0 \tag{18-25}$$

If we solve Equation 18-25 for the current i in the circuit, we get

$$i = \frac{\varepsilon}{R} - \frac{q}{RC} \tag{18-26}$$

Equation 18-26 tells us that the current in the circuit of Figure 18-18b *cannot* be constant. As charge builds up on the capacitor, the value of q increases. Hence the right-hand side of Equation 18-26 decreases with time, so the current i must decrease with time as well. What's happening is that the voltage $V = q/C$ across the capacitor increases as the charge increases. This voltage opposes that of the source of emf, so the current decreases. The current is what's causing the charge to increase, so q increases at an ever-slower rate as time goes by.

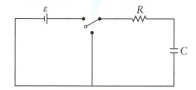

(a) The position of this switch determines whether the source of emf ε is part of the circuit.

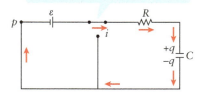

(b) With the switch in the up position, the emf causes charge to flow. This current charges the capacitor, so q increases.

Figure 18-18 Charging a series *RC* circuit (a) In this circuit a resistor and an initially uncharged capacitor are connected in series to a source of emf. (b) The switch is moved to the up position at $t = 0$, beginning the charging process.

We'd like to know the capacitor charge q and the current i as functions of time. Solving Equation 18-26 for these functions is a problem in calculus that's beyond our scope. Instead we'll present the solutions and see that they make sense:

Capacitor charge and current in a charging series *RC* circuit

(18-27)

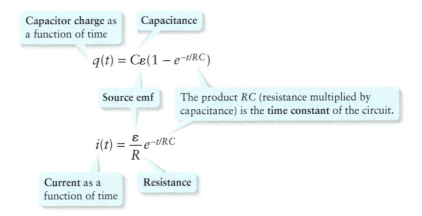

Capacitor charge as a function of time

Capacitance

$$q(t) = C\varepsilon(1 - e^{-t/RC})$$

Source emf

The product RC (resistance multiplied by capacitance) is the **time constant** of the circuit.

$$i(t) = \frac{\varepsilon}{R} e^{-t/RC}$$

Current as a function of time

Resistance

 In these equations, $t = 0$ is when the switch in Figure 18-18b is moved to the up position.

 See the Math Tutorial for more information on exponents and logarithms.

Figure 18-19 graphs the charge q and current i as given by Equations 18-27. These equations involve the **exponential function**, the irrational number $e = 2.71828...$ raised to a power. (Note that e is not the same as the magnitude of the charge on the electron, for which we unfortunately use the same symbol.) The number e has special properties: If a population has N_0 members and grows at a rate r per year (for example, if the population grows by 2% per year, $r = 0.02$), then after t years the population will be $N(t) = N_0 e^{rt}$. In Equations 18-27 the exponential function has a negative exponent, which means that this function decreases with time. In particular, $e^{-t/RC}$ decreases by a factor of $1/e = 0.36787...$ every time the quantity t/RC increases by 1—that is, whenever time t increases by RC. At $t = 0$ (when the switch in Figure 18-18 is closed and charge begins to flow), $e^{-t/RC} = e^0 = 1$; at $t = RC$, $e^{-t/RC} = e^{-1} = 1/e = 0.368$ to three significant figures; at $t = 2RC$, $e^{-t/RC} = e^{-2} = 1/e^2 = 0.135$; and so on.

The quantity RC is called the **time constant** of a series RC circuit. (Note that the product RC has units of ohms times farads. Since $1\ \Omega = 1$ V/A, 1 F $= 1$ C/V, and

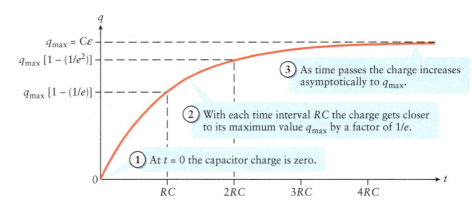

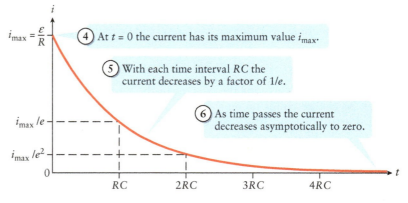

Figure 18-19 Charge and current in a charging series *RC* circuit Capacitor charge q starts at zero and approaches a maximum value; current i starts at a maximum value and approaches zero.

$1\,A = 1\,C/s$, it follows that $1\,\Omega\cdot F = 1\,(V/A)(C/V) = 1\,C/A = 1\,s$. So the quantity RC does indeed have units of time.) The smaller the time constant, the more rapidly the charge q approaches its maximum value q_{max} and the more rapidly the current decreases to its final value of zero.

At $t = 0$ the capacitor is uncharged, so there is zero voltage across the capacitor. As a result, at $t = 0$ the voltage iR across the resistor equals the voltage ε across the source, so $\varepsilon = iR$ and $i = i_{max} = \varepsilon/R$. After a very long time the current has dropped to zero, so the voltage iR across the resistor is zero. Then the voltage q/C across the capacitor equals the voltage ε across the source, so $\varepsilon = q/C$ and $q = q_{max} = C\varepsilon$.

A Series *RC* Circuit: Discharging the Capacitor

Some time after the switch in Figure 18-18b was thrown to the up position, the capacitor is charged to a charge q_{max}. (This may be less than $C\varepsilon$, depending on how long the source has had to charge the capacitor.) We now throw the switch to the down position, as in Figure 18-20, and restart our clock so that $t = 0$ is the time when the switch is moved to the new position. The emf is no longer part of the circuit, so the positive charge on the upper plate of the capacitor is free to move counterclockwise around the circuit to cancel the negative charge on the lower plate. So the value of the capacitor charge q decreases, and the electric potential energy stored in the capacitor is transferred into the resistor. We call this a *discharging* series *RC* circuit.

Kirchhoff's loop rule for the discharging circuit is now the same as Equation 18-25, but with the emf ε removed:

$$(-iR) + \left(-\frac{q}{C}\right) = 0 \qquad (18\text{-}28)$$

Solving Equation 18-28 for the current i in the circuit gives

$$i = -\frac{q}{RC} \qquad (18\text{-}29)$$

The minus sign in Equation 18-29 means that the current flows in the direction opposite to that in the charging *RC* circuit of Figure 18-18b. (You can see this in Figure 18-20.) As the capacitor discharges and the charge q decreases, the current i will decrease in magnitude. The charge and current as functions of time are

Capacitor **charge** as a function of time

$$q(t) = q_{max}\,e^{-t/RC}$$

Capacitor charge at $t = 0$

The product RC (resistance multiplied by capacitance) is the **time constant** of the circuit.

$$i(t) = -\frac{q_{max}}{RC}e^{-t/RC}$$

Current as a function of time

In these equations $t = 0$ is when the switch in Figure 18-20 is moved to the down position.

Capacitor charge and current in a discharging series *RC* circuit
(18-30)

The graphs in Figure 18-21 show the charge q and current i as given by Equations 18-30. Both functions are proportional to $e^{-t/RC}$, so both decrease by a factor of $1/e$ whenever time t increases by one time constant RC. So the value of RC determines how rapidly the capacitor charges *and* how rapidly it discharges.

With the switch in the down position, the voltage across the capacitor causes a current. This current discharges the capacitor, so q decreases.

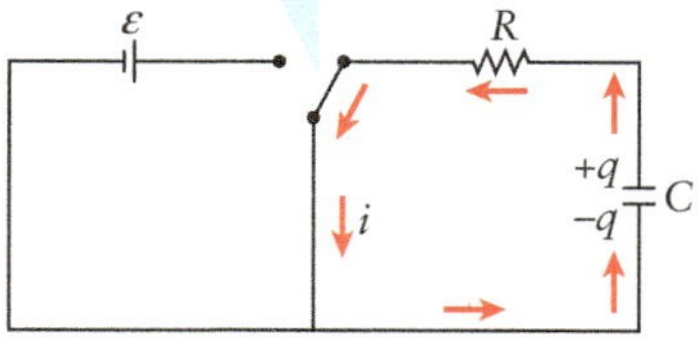

Figure 18-20 Discharging a series *RC* circuit When the switch in the circuit of Figure 18-18 is moved to the down position, the discharging process begins.

 Go to **Interactive Exercises 18-1 and 18-2** for more practice dealing with RC circuits.

 Go to **Interactive Exercise 18-3** for more practice dealing with time constants.

Figure 18-21 Charge and current in a discharging series *RC* circuit Capacitor charge q starts at a maximum value and approaches zero; current i starts at a maximum (negative) value and approaches zero.

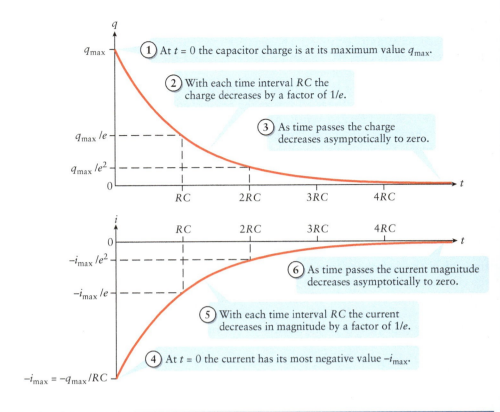

① At $t = 0$ the capacitor charge is at its maximum value q_{max}.

② With each time interval RC the charge decreases by a factor of $1/e$.

③ As time passes the charge decreases asymptotically to zero.

⑥ As time passes the current magnitude decreases asymptotically to zero.

⑤ With each time interval RC the current decreases in magnitude by a factor of $1/e$.

④ At $t = 0$ the current has its most negative value $-i_{max}$.

$-i_{max} = -q_{max}/RC$

EXAMPLE 18-11 A Charging Series *RC* Circuit

A 10.0-MΩ resistor is connected in series with a 5.00-μF capacitor. When a switch is thrown these circuit elements are connected to a 24.0-V battery of negligible internal resistance. The capacitor is initially uncharged. (a) What is the current in the circuit immediately after the switch is moved so that charging begins? (b) What is the charge on the capacitor once it is fully charged? (c) Find the capacitor charge, current, power provided by the battery, power taken in by the resistor, and power taken in by the capacitor at $t = 50.0$ s. (d) When the capacitor is fully charged, find the total energy that has been delivered by the battery and the total energy that has been delivered to the capacitor.

Set Up

We are given $R = 10.0$ MΩ $= 10.0 \times 10^6$ Ω, $C = 5.00$ μF $= 5.00 \times 10^{-6}$ F, and $\varepsilon = 24.0$ V. Equations 18-27 tell us the capacitor charge and current at any time, including at $t = 0$ (when the switch is first closed) and $t \rightarrow \infty$ (long after the switch is closed, so the capacitor is fully charged). We'll use Equations 18-23 and 18-24 to find the power out of the battery and into the resistor and capacitor. (Equation 17-14 will help us in this.) To charge the capacitor to its maximum charge q_{max}, the total charge that must pass through the battery is q_{max}; we'll use this and Equation 17-6 to find the total energy delivered by the battery. Equation 17-17 tells us the total energy that is stored in the charged capacitor.

Capacitor charge and current in a charging series *RC* circuit:

$$q(t) = C\varepsilon(1 - e^{-t/RC})$$

$$i(t) = \frac{\varepsilon}{R}e^{-t/RC} \qquad (18\text{-}27)$$

Power for a circuit element:

$$P = iV \qquad (18\text{-}23)$$

Power for a resistor:

$$P = i^2 R = \frac{V^2}{R} \qquad (18\text{-}24)$$

Charge, voltage, and capacitance for a capacitor:

$$q = CV \qquad (17\text{-}14)$$

Electric potential difference related to electric potential energy difference:

$$\Delta V = \frac{\Delta U_{electric}}{q_0} \qquad (17\text{-}6)$$

Electric potential energy stored in a capacitor:

$$U_{electric} = \frac{1}{2}qV = \frac{1}{2}CV^2 = \frac{q^2}{2C} \qquad (17\text{-}17)$$

$\varepsilon = 24.0$ V $R = 10.0$ MΩ

switch closed at $t = 0$

$C = 5.00$ μF

Solve

(a) Find the current at $t = 0$.

From the second of Equations 18-27, the current when the switch is first closed at $t = 0$ is

$$i(0) = \frac{\varepsilon}{R} e^{-(0)/RC} = \frac{\varepsilon}{R} e^0$$

Since any number raised to the power 0 equals 1, we have $e^0 = 1$ and

$$i(0) = i_{max} = \frac{\varepsilon}{R} = \frac{24.0 \text{ V}}{10.0 \times 10^6 \text{ } \Omega}$$

$$= 2.40 \times 10^{-6} \text{ A} = 2.40 \text{ } \mu\text{A}$$

(b) Find the capacitor charge long after the switch is closed ($t \to \infty$).

The first of Equations 18-27 tells us the capacitor charge $q(t)$. As $t \to \infty$, the exponent $-t/RC \to -\infty$. Any number raised to the power $-\infty$ is zero, so

$$e^{-t/RC} \to 0$$

$$q(t) = C\varepsilon(1 - e^{-t/RC}) \to q_{max} = C\varepsilon(1 - 0) = C\varepsilon$$

$$= (5.00 \times 10^{-6} \text{ F})(24.0 \text{ V})$$

$$= 1.20 \times 10^{-4} \text{ C} = 0.120 \text{ mC}$$

(c) The time constant for this circuit is $RC = 50.0$ s, so we are actually being asked about the behavior of the circuit one time constant after the switch is closed. Use this to find charge q, current i, and the power out of or into each circuit element.

The time constant for this circuit is

$$RC = (10.0 \times 10^6 \text{ } \Omega)(5.00 \times 10^{-6} \text{ F}) = 50.0 \text{ s}$$

so at $t = 50.0$ s, $t/RC = (50.0 \text{ s})/(50.0 \text{ s}) = 1.00$

From Equations 18-27,

$$q = C\varepsilon(1 - e^{-1.00}) = (5.00 \times 10^{-6} \text{ F})(24.0 \text{ V})(1 - 0.368)$$

$$= 7.58 \times 10^{-5} \text{ C} = 0.632 q_{max}$$

$$i = \frac{\varepsilon}{R} e^{-1.00} = \frac{24.0 \text{ V}}{10.0 \times 10^6 \text{ } \Omega}(0.368)$$

$$= 8.83 \times 10^{-7} \text{ A} = 0.368 i_{max}$$

The voltage across the battery is $\varepsilon = 24.0$ V, so from Equation 18-23 the power out of the battery is

$$P_{battery} = i\varepsilon = (8.83 \times 10^{-7} \text{ A})(24.0 \text{ V})$$

$$= 2.12 \times 10^{-5} \text{ W} = 21.2 \text{ } \mu\text{W}$$

From the first of Equations 18-24, the power into the resistor is

$$P_R = i^2 R = (8.83 \times 10^{-7} \text{ A})^2 (10.0 \times 10^6 \text{ } \Omega)$$

$$= 7.80 \times 10^{-6} \text{ W} = 7.80 \text{ } \mu\text{W}$$

Equation 17-14, $q = CV$, tells us that the voltage across the capacitor is $V = q/C$. Combining this with Equation 18-23 gives the power into the capacitor:

$$P_C = i\left(\frac{q}{C}\right) = (8.83 \times 10^{-7} \text{ A})\left(\frac{7.58 \times 10^{-5} \text{ C}}{5.00 \times 10^{-6} \text{ F}}\right)$$

$$= 1.34 \times 10^{-5} \text{ W} = 13.4 \text{ } \mu\text{W}$$

Note that the net power into the resistor and capacitor combined equals the power out of the battery:

$$P_R + P_C = 7.80 \text{ } \mu\text{W} + 13.4 \text{ } \mu\text{W} = 21.2 \text{ } \mu\text{W} = P_{battery}$$

(d) Use the maximum charge stored by the capacitor, which is the total charge moved across the battery, and Equation 17-6 to calculate the change in electric potential energy imparted by the battery. Use Equations 17-7 to calculate the electric potential energy stored in the capacitor.

Long after the switch is closed, the total amount of charge that has passed through the battery and to the positive capacitor plate is $q_{max} = 1.20 \times 10^{-4}$ C. From Equation 17-6 the potential energy change that was imparted by moving this charge across the 24.0-V emf of the battery is

$$\Delta U_{battery} = q_{max}\varepsilon = (1.20 \times 10^{-4} \text{ C})(24.0 \text{ V})$$

$$= 2.88 \times 10^{-3} \text{ J} = 2.88 \text{ mJ}$$

The last of Equations 17-17 tells us the amount of energy that went into the capacitor to store charge q_{max} there:

$$U_C = \frac{q_{max}^2}{2C} = \frac{(1.20 \times 10^{-4} \text{ C})^2}{2(5.00 \times 10^{-6} \text{ F})}$$
$$= 1.44 \times 10^{-3} \text{ J} = 1.44 \text{ mJ}$$

So exactly one-half of the energy taken from the battery goes into the capacitor: $U_C = (1/2)\Delta U_{battery}$.

Reflect

Our results for charge q and current i in part (c) agree with Figure 18-19: After one time constant the capacitor charge has reached $[1 - (1/e)] = 0.632 = 63.2\%$ of its fully charged value q_{max}, and the current has decreased to $1/e = 0.368 = 36.8\%$ of its initial value i_{max}. Can you show that after five time constants ($t = 5RC = 250$ s) the charge will have reached 99.3% of q_{max} and the current will have decreased to just 0.674% of i_{max}?

The power calculations in part (c) show that all of the power extracted from the battery is accounted for: Part of the energy extracted from the battery goes into the resistor, and the rest goes into adding to the electric potential energy stored in the capacitor. We've shown this for a specific instant, but it's true at *all* times during the charging process.

These results from part (c) also help us understand our calculations in part (d): Only one-half of the energy extracted from the battery goes into the capacitor, so the other half must have gone into the resistor. This is a general result for charging *any* series *RC* circuit.

EXAMPLE 18-12 A Discharging Series *RC* Circuit

After the 5.00-µF capacitor in Example 18-11 has attained its full charge of 1.20×10^{-4} C, you disconnect it from the charging circuit and place it in series with a 0.300-kΩ resistor and a switch. When the switch is thrown, the capacitor discharges and produces a current in the 0.300-kΩ resistor. (a) What is the current in the circuit immediately after the switch is moved so that discharging begins? (b) Find the capacitor charge, current, power taken in by the resistor, and power provided by the capacitor at $t = 1.50 \times 10^{-3}$ s $= 1.50$ ms.

Set Up

This situation is similar to Example 18-11 except that the capacitor is now discharging rather than charging and we are using a smaller resistance. We are given $R = 0.300$ kΩ $= 0.300 \times 10^3$ Ω, $C = 5.00$ µF $= 5.00 \times 10^{-6}$ F, and the initial charge on the capacitor $q_{max} = 1.20 \times 10^{-4}$ C. Equations 18-30 tell us the capacitor charge and current at any time, including at $t = 0$ (when the switch is first closed). As in Example 18-11, we'll use Equations 18-23, 18-24, and 17-14 to find the power out of the capacitor and into the resistor.

Capacitor charge and current in a discharging series *RC* circuit:

$$q(t) = q_{max}e^{-t/RC}$$
$$i(t) = -\frac{q_{max}}{RC}e^{-t/RC} \quad (18\text{-}30)$$

Power for a circuit element:

$$P = iV \quad (18\text{-}23)$$

Power for a resistor:

$$P = i^2R = \frac{V^2}{R} \quad (18\text{-}24)$$

Charge, voltage, and capacitance for a capacitor:

$$q = CV \quad (17\text{-}14)$$

$R = 0.300$ kΩ
switch closed at $t = 0$
$C = 5.00$ µF
charge at $t = 0$: 1.20×10^{-4} C

Solve

(a) Find the current at $t = 0$.

The second of Equations 18-30 gives us the current at $t = 0$ when the switch is first closed:

$$i(0) = -\frac{q_{max}}{RC}e^{-(0)/RC} = -\frac{q_{max}}{RC}e^0$$

We saw in Example 18-11 that $e^0 = 1$, so

$$i(0) = -\frac{q_{max}}{RC} = -\frac{1.20 \times 10^{-4} \text{ C}}{(0.300 \times 10^3 \text{ Ω})(5.00 \times 10^{-6} \text{ F})}$$
$$= -8.00 \times 10^{-2} \text{ A} = -0.0800 \text{ A}$$

The minus sign means that the current in this *discharging RC* circuit flows as in Figure 18-20, which is opposite to the direction in a *charging RC* circuit.

(b) The time $t = 1.50$ ms of interest is equal to the time constant for this circuit, $RC = 1.50$ ms. So we are being asked about the properties of the circuit (charge q, current i, and power out of or into each circuit element) one time constant after the switch is closed.

The time constant for this circuit is

$$RC = (0.300 \times 10^3 \ \Omega)(5.00 \times 10^{-6} \ \text{F}) = 1.50 \times 10^{-3} \ \text{s} = 1.50 \ \text{ms}$$

so at $t = 1.50$ ms, $t/RC = (1.50 \ \text{ms})/(1.50 \ \text{ms}) = 1.00$.

From Equations 18-30,

$$q = q_{max}e^{-1.00} = (1.20 \times 10^{-4} \ \text{C})(0.368)$$
$$= 4.41 \times 10^{-5} \ \text{C} = 0.368q_{max}$$

$$i = -\frac{q_{max}}{RC}e^{-1.00} = \frac{1.20 \times 10^{-4} \ \text{C}}{(0.300 \times 10^3 \ \Omega)(5.00 \times 10^{-6} \ \text{F})}(0.368)$$
$$= -2.94 \times 10^{-2} \ \text{A} = 0.368i_{max}$$

The first of Equations 18-24 tells us the power into the resistor:

$$P_R = i^2R = (-2.94 \times 10^{-2} \ \text{A})^2(0.300 \times 10^3 \ \Omega)$$
$$= 0.260 \ \text{W}$$

From Equation 17-14, the voltage across the capacitor is $V = q/C$. Combine this with Equation 18-23 to find the power out of the capacitor:

$$P_C = i\left(\frac{q}{C}\right) = (-2.94 \times 10^{-2} \ \text{A})\left(\frac{4.41 \times 10^{-5} \ \text{C}}{5.00 \times 10^{-6} \ \text{F}}\right)$$
$$= -0.260 \ \text{W}$$

The minus sign means that energy is leaving the capacitor at a rate of $0.260 \ \text{W} = 0.260 \ \text{J/s}$.

Reflect

In agreement with Figure 18-20, after one time constant the current and capacitor charge have both decreased to $1/e = 0.368 = 36.8\%$ of their initial values.

Note that as in Example 18-11, all of the energy in the circuit is accounted for: At $t = 1.50$ ms the power out of the capacitor has the same magnitude (0.260 W) as the power into the resistor. The same will be true at any time after the switch is closed.

If you compare this example with Example 18-11, you'll see that the initial current $i(0)$ is *much* greater in magnitude for the discharging RC circuit (0.0800 A) than for the charging circuit ($2.40 \ \mu\text{A}$). Likewise the power into the resistor after one time constant for the discharging RC circuit (0.260 W) is much greater than for the charging RC circuit (7.80×10^{-6} W). That's because we used a much smaller resistance for the discharging circuit (0.300 kΩ) than for the charging circuit ($10.0 \times 10^6 \ \Omega$), which made the time constant RC much smaller (1.50×10^{-3} s for the discharging circuit compared to 50.0 s for the charging circuit) and so allowed the capacitor to discharge much more rapidly than it charged. Rapid discharge is important in many capacitor applications where a powerful surge of energy is required, such as in an electronic flash unit for photography (see Figure 18-1c) or a cardiac defibrillator (see the photo that opens Chapter 17).

The physics of capacitors in circuits helps explain what happens in the propagation of action potentials along axons, which we discussed in Section 18-4. All biological membranes, including the axonal membrane, act like a capacitor. Under resting conditions the inner surface of the membrane is negatively charged and the outer surface is positively charged (see Figure 18-7). As we described in Section 18-4, as an action potential propagates along an axon, the resting charge distribution reverses: At the peak of an action potential, the inner surface of the membrane is positively charged and the outer surface is negatively charged. When the membrane returns to its initial charge state, it's analogous to a capacitor discharging as in the circuit of Figure 18-20. The value of C is the capacitance of a segment of axon membrane, and the value of R is the resistance of that segment to charge flowing through channels in the membrane. (This is different from the resistance we discussed in Section 18-4, especially Example 18-6. There we considered the *longitudinal* resistance for current flowing along the length of the axon; the value of R we're considering here is the transverse resistance to charges flowing across the axon membrane, perpendicular to the axon's length.)

BioMedical

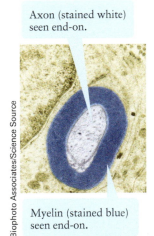

Axon (stained white) seen end-on.

Myelin (stained blue) seen end-on.

Biophoto Associates/Science Source

Figure 18-22 End-on view of a myelinated axon In vertebrates special cells wrap themselves around some axons, forming a thick insulating layer called myelin. This decreases the effective capacitance of the axon and increases the speed at which nerve signals can propagate along the length of the axon. (For a diagram of the entire nerve cell, see Figure 18-8.)

The value of the time constant RC determines the rate of discharge: The smaller the value of RC, the faster the capacitance of the membrane discharges and the more rapidly the signal propagates along the axon. We saw in Section 18-4 that some invertebrates such as squid have large-diameter axons to allow action potentials to propagate quickly from the brain to the muscles that mediate escape reflexes. Another way to make signals propagate more rapidly along an axon is to decrease the capacitance C of the membrane, thus reducing the value of RC. In vertebrates special cells wrap themselves around some axons, forming a thick insulating layer called *myelin* (Figure 18-22; see also Figure 18-8). The myelin acts as a capacitor between the surface of the axon and its environment. This additional capacitance is in series with the capacitance of the axon's cell membrane and so decreases the effective capacitance of the axon. (Recall from Section 17-7 that when capacitors are placed in series, the effective capacitance is less than that of any individual capacitor.)

Only certain sections of the axon are coated with myelin. The action potential in the nonmyelinated gaps, or nodes (see Figure 18-8), produces a signal that propagates rapidly along the myelinated regions of the axon. The signal weakens in strength in the myelinated region where no action potential is generated, but is reinvigorated when it reaches the next node.

GOT THE CONCEPT? 18-9 Discharging an *RC* Circuit

(?) The switch in the circuit of Figure 18-20 is moved to the down position, causing the capacitor to discharge. Suppose that the resistor is a light bulb and that its brightness is proportional to the power into the resistor. Compared to the brightness of the bulb when the switch is moved to the down position at $t = 0$, the brightness of the bulb at $t = RC$ is (a) the same; (b) $1 - (1/e^2) = 0.865$ as great; (c) $1 - 1/e = 0.632$ as great; (d) $1/e = 0.368$ as great; (e) $1/e^2 = 0.135$ as great.

TAKE-HOME MESSAGE FOR Section 18-7

✔ A circuit containing a resistor and a capacitor connected in series is a series *RC* circuit.

✔ Connecting the resistor and an uncharged capacitor to a source of emf causes a current that charges the capacitor. Switching the circuit to remove the source of emf causes the capacitor to discharge through the resistor. In either case an exponential function describes how the current and the capacitor charge vary with time.

✔ The rate of charge or discharge in a series *RC* circuit depends on the time constant, equal to the product RC. In a time interval RC the current decreases by a factor of $1/e$.

Key Terms

ac (alternating current)	emf	parallel (resistors)
alternating current (ac)	equivalent resistance	power
ampere	exponential function	resistance
battery	internal resistance	resistivity
circuit	junction	resistor
circuit element	Kirchhoff's junction rule	series *RC* circuit
current	Kirchhoff's loop rule	series (resistors)
dc (direct current)	mobile charge	single-loop circuit
direct current (dc)	multiloop circuit	source of emf
drift speed	ohm	time constant

Chapter Summary

Topic	Equation or Figure	
Current: Electric charge flows around a circuit in response to an electric potential difference, such as that provided by a battery (a source of emf). Current is the rate of charge flow and is related to the speed at which mobile charges drift through the circuit. In ordinary metals the mobile charges are negatively charged electrons, but it's conventional to take the direction of the current to be the direction in which positive charges would flow.	**Current** in a circuit Amount of **charge** that flows past a certain point in the circuit $$i = \frac{\Delta q}{\Delta t}$$ Time required for that amount of charge to flow past that point	(18-1)
	Current in a wire Number of **mobile charges per unit volume** $$i = nAev_{\text{drift}}$$ **Speed** at which charges drift through the wire **Cross-sectional area** of the wire Magnitude of the **charge** on each mobile charge	(18-3)
Resistance: Resistivity is a measure of how difficult it is for charges to flow through a material. The resistance of a wire or circuit element depends on the dimensions of the object as well as the resistivity of the material of which the object is made. Resistance and resistivity are not constants but vary with temperature. A resistor is a circuit element whose most important property is its resistance.	**Resistance** of a wire **Resistivity** of the material of which the wire is made $$R = \frac{\rho L}{A}$$ **Length** of the wire **Cross-sectional area** of the wire	(18-8)
Voltage, current, and resistance: For current to exist in a conductor with resistance, there must be a potential difference (voltage) between the ends of the conductor. For a given resistance a greater voltage produces a greater current.	**Potential difference** between the ends of a wire $$V = iR$$ **Current** in the wire **Resistance** of the wire	(18-9)
Rules for circuits: Kirchhoff's loop rule states that the sum of the changes in electric potential around a closed loop in a circuit must equal zero. Kirchhoff's junction rule states that the sum of the currents into a junction equals the sum of the currents out of it. These rules allow us to analyze the currents and voltages in electric circuits.	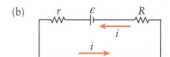 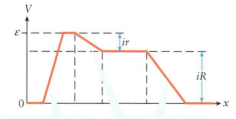 The potential increases by ε from the negative terminal to the positive terminal of the emf. The potential drops by ir across the internal resistance and drops by iR across the resistor.	(Figure 18-9b)

Resistors in series and parallel: A collection of resistors in a circuit behaves as though it were a single resistor, with an equivalent resistance R_{equiv}. The equivalent resistance is different for resistors in series (all of which carry the same current) than for resistors in parallel (all of which have the same voltage).

Equivalent resistance of N resistors in series   Resistances of the individual resistors

$$R_{\text{equiv}} = R_1 + R_2 + R_3 + \ldots + R_N \tag{18-11}$$

 The equivalent resistance of N resistors in series is the sum of the individual resistances.

 The reciprocal of the equivalent resistance of N resistors in parallel is the sum of the reciprocals of the individual resistances.

$$\frac{1}{R_{\text{equiv}}} = \frac{1}{R_1} + \frac{1}{R_2} + \frac{1}{R_3} + \ldots + \frac{1}{R_N} \tag{18-20}$$

Equivalent resistance of N resistors in parallel   Resistances of the individual resistors

Power in circuits: Power is the rate of energy transfer. In an electric circuit, energy flows from a source of emf into the other circuit elements. The power into a resistor can be expressed in terms of the resistance and either the current through the resistor or the voltage across the resistor.

Power produced by or transferred into a circuit element

$$P = iV \tag{18-23}$$

Current through the circuit element   Voltage (absolute value) across the circuit element

 Source of emf: Power flows out of the source and into the moving charges. Resistor: Power flows out of the charges and into the resistor.

Power into a resistor   Current through the resistor

$$P = i^2 R = \frac{V^2}{R} \tag{18-24}$$

Voltage across the resistor

Resistance of the resistor

Series RC circuits: If a resistor R and capacitor C are connected in series to a source of emf, the capacitor charge increases toward a maximum value. The current in the circuit (which carries the charge to the capacitor) begins with a large value and gradually decreases to zero. If the source is taken out of the circuit, the capacitor discharges. The current in the circuit now carries the charge away from the capacitor and again gradually decreases to zero. The rate of charging or discharging depends on the time constant of the circuit, equal to the product RC.

(b) With the switch in the up position, the emf causes charge to flow. This current charges the capacitor, so q increases.

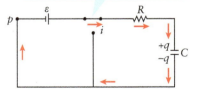

(Figure 18-18b)

With the switch in the down position, the voltage across the capacitor causes a current. This current discharges the capacitor, so q decreases.

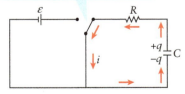

(Figure 18-20)

Answer to What do you think? Question

(b) When electrons pass through the battery, they gain electric potential energy. They lose most of this potential energy as they pass through the filament of the light bulb, and this lost energy heats the filament and causes it to glow. The electrons lose the rest of the potential energy as they pass through the wires that connect the battery and light bulb. If electrons lost *kinetic* energy anywhere in the circuit, they would slow down and create an electron "traffic jam." The electric repulsion between electrons prevents such "jams" from happening, so the electrons must maintain the same average speed and average kinetic energy as they travel around the circuit.

Answers to Got the Concept? Questions

18-1 (c) The current i in a simple circuit (in which the moving charges travel around a single loop) has the same value at all points in the circuit. This is true whether the thickness of the wire varies or is the same at all points.

18-2 (a) As we saw in the answer to the previous Got the Concept? question, the current i has the same value at all points in the circuit. Equation 18-3 shows that for this to be true the product Av_{drift} (the cross-sectional area of the wire multiplied by the drift speed) must also have the same value at all points. (The quantity n, the number of mobile charges per unit volume, depends only on what the wire is made of, not on its dimensions.) The thinner wire has a radius r that is half as great and a cross-sectional area $A = \pi r^2$ that is one-quarter as great as the rest of the wire. So the drift speed of the electrons must be four times greater in the thinner part of the wire.

18-3 (b) For an ohmic material the current and voltage are directly proportional, so doubling the voltage should cause a doubling of current. That is *not* the case for this material: Doubling the voltage from 2.0 to 4.0 V increases the current by a factor of only $(0.28\ \text{A})/(0.15\ \text{A}) = 1.9$, and doubling the voltage again from 4.0 to 8.0 V increases the current by a factor of only $(0.50\ \text{A})/(0.28\ \text{A}) = 1.8$. So this material is nonohmic.

18-4 (e) As in Example 18-6 the resistance of wire of resistivity ρ, length L, and diameter d is $R = 4\rho L/\pi d^2$. This says that the resistance is inversely proportional to the square of the diameter. So if the diameter is increased by a factor of 4 while leaving the resistivity and length the same, the resistance will change by a factor of $1/4^2 = 1/16$.

18-5 (c), (a), (b), (d). Combining resistors in series results in a circuit with larger equivalent resistance; combining them in parallel results in a smaller equivalent resistance. Therefore, circuits (a) and (c) have higher resistance than circuits (b) and (d). For the same reason the equivalent resistance of (c) with three resistors in series is higher than that of circuit (a) with two resistors in series, and the equivalent resistance of circuit (b) with two resistors in parallel is higher than that of circuit (d) with three resistors in parallel. So in order of equivalent resistance from highest to lowest, the circuits are (c), (a), (b), and (d). This conclusion doesn't depend on how the value of R_3 compares to the values of R_1 or R_2: Any additional resistance in series increases the equivalent resistance, and any additional resistance in parallel decreases the equivalent resistance.

18-6 (d), (b), (a), (c). The larger the equivalent resistance the smaller the current for a fixed voltage. So this ranking of the circuits in order of decreasing current is just the opposite of the ranking in order of decreasing equivalent resistance in Got the Concept? 18-5.

18-7 (c) Placing two sources of emf in series is equivalent to increasing the emf from ε to 2ε. This doubles the voltage V across the resistor, which from $V = iR$ (Equation 18-9) means that the current i through the resistor doubles as well. From Equations 18-24, $P = i^2 R = V^2/R$, the power into the resistor is proportional to either the square of the current i or the square of the resistor voltage V. So the power into the resistor increases by a factor of $2^2 = 4$. A device such as a flashlight or television remote control that uses two batteries uses four times as much power as one with a single battery.

18-8 $P_A = P_E > P_D > P_B = P_C$. The full current that passes through the source also passes through resistors A and E. This current divides up to pass through two branches—one with resistors B and C, one with resistor D—so only a fraction of the full current passes through resistors B, C, and D. Since the power into a resistor is given by $P = i^2 R$ (the first of Equations 18-24), it follows that the same power goes into A and E ($P_A = P_E$) and that this is greater than the power into any of the other three resistors. The voltage across the branch with resistors B and C is the same as that across the branch with resistor D (the branches are in parallel). All of this voltage is across resistor D, but only half of this voltage is across resistor B and half across resistor C. From the second of Equations 18-24, $P = V^2/R$, this means that there is more power into D than into B or C but equal amounts into B and C (so $P_D > P_B = P_C$). So $P_A = P_E > P_D > P_B = P_C$. If the resistors are light bulbs, their ranking by brightness will be the same as this ranking by power.

18-9 (e) The second of Equations 18-30 shows that the current in a discharging series RC circuit is proportional to $e^{-t/RC}$, so at $t = RC$ the current is $e^{-1} = 1/e = 0.368$ as great as its value at $t = 0$ (when $e^{-t/RC} = e^0 = 1$). The power $P = i^2 R$ (Equations 18-24) into a resistor is proportional to the square of i, and i is $1/e$ as great as at $t = 0$, so the power is $1/e^2 = (0.368)^2 = 0.135$ as great as at $t = 0$.

Questions and Problems

In a few problems you are given more data than you actually need; in a few other problems you are required to supply data from your general knowledge, outside sources, or informed estimate.

Interpret as significant all digits in numerical values that have trailing zeros and no decimal points.

For all problems use $g = 9.80$ m/s^2 for the free-fall acceleration due to gravity. Neglect friction and air resistance unless instructed to do otherwise.

- • Basic, single-concept problem
- •• Intermediate-level problem, may require synthesis of concepts and multiple steps
- ••• Challenging problem
- Example See worked example for a similar problem.

Conceptual Questions

1. • We distinguish the direction of current in a circuit. Why don't we consider it a vector quantity?

2. • Under ordinary conditions the drift speed of electrons in a metal is around 10^{-4} m/s or less. Why doesn't it take a long time for a light bulb to come on when you flip the wall switch that is several meters away?

3. • We justified a number of electrostatic phenomena by the argument that there can be no electric field in a conductor. Now we say that the current in a conductor is driven by a potential difference and thus there is an electric field in the conductor. Is this statement a contradiction?

4. • Two wires, A and B, have the same physical dimensions but are made of different materials. If A has twice the resistance of B, how do their resistivities compare?

5. • **Biology** When a bird lands with both feet on a high-voltage wire, will the bird be electrocuted? Explain your answer.

6. •• Many ordinary strings of holiday lights contain about 50 bulbs connected in parallel across a 110-V line. Sixty years ago a string of 50 bulbs would be connected in series across the line. What would happen if you could put one of the old-style bulbs into a modern holiday light set? (The light sockets are made differently to prevent this.)

7. • **Biology** Explain how an action potential is generated.

8. • If the only voltage source you have is 36 V, how could you light some 6-V light bulbs without burning them out?

9. • An ammeter measures the current through a particular circuit element. (a) How should it be connected with that element, in parallel or in series? (b) Should an ammeter have a very large or a very small resistance? Why?

10. • For a given source of constant voltage, will more heat develop in a large external resistance connected across it or a small one?

11. • The average drift velocity of electrons in a wire carrying a steady current is constant even though the electric field within the wire is doing work on the electrons. What happens to this energy?

12. • Is current dissipated when it passes through a resistor? Explain your answer.

13. • Give a simple physical explanation for why the charge on a capacitor in an RC circuit can't be changed instantaneously.

14. • (a) Does the time required to fully charge a capacitor through a given resistor with a battery depend on the voltage of the battery? (b) Does it depend on the total amount of charge to be placed on the capacitor? Explain your answers.

Multiple-Choice Questions

15. • If a current-carrying wire has a cross-sectional area that gradually becomes smaller along the length of the wire, the drift velocity
   - A. increases along the length of the wire.
   - B. decreases along the length of the wire.
   - C. remains the same along the length of the wire.
   - D. increases along the length of the wire only if the resistivity increases too.
   - E. decreases along the length of the wire only if the resistivity decreases too.

16. • What causes an electric spark?
   - A. current
   - B. voltage
   - C. both current and voltage
   - D. resistance and current
   - E. resistance and voltage

17. • Two copper wires have the same length, but one has twice the diameter of the other. Compared to the one that has the smaller diameter, the one that has the larger diameter has a resistance that is
   - A. larger by a factor of 2.
   - B. larger by a factor of 4.
   - C. the same.
   - D. smaller by a factor of 1/2.
   - E. smaller by a factor of 1/4.

18. • When a thin wire is connected across a voltage of 1 V, the current is 1 A. If we connect the same wire across a voltage of 2 V, the current is
   - A. (1/4) A.
   - B. (1/2) A.
   - C. 1 A.
   - D. 2 A.
   - E. 4 A.

19. • When a wire that has a large diameter and a length L is connected across the terminals of an automobile battery, the current is 40 A. If we cut the wire to half of its original length and connect one piece that has a length $L/2$ across the terminals of the same battery, the current will be
   - A. 10 A.
   - B. 20 A.
   - C. 40 A.
   - D. 80 A.
   - E. 160 A.

20. • A charge flows from the positive terminal of a 6-V battery, through a light bulb, and through the battery back to the positive terminal. The total voltage change experienced by the charge is
   - A. 1 V.
   - B. 6 V.
   - C. 0 V.
   - D. −1 V.
   - E. −6 V.

21. • When a second light bulb is added in series to a circuit with a single light bulb, the resistance of the circuit
    A. increases.
    B. decreases.
    C. remains the same.
    D. doubles.
    E. triples.

22. • When a light bulb is added in parallel to a circuit with a single light bulb, the resistance of the circuit
    A. increases.
    B. decreases.
    C. remains the same.
    D. doubles.
    E. triples.

23. • If we use a 2-V battery instead of a 1-V battery to charge the capacitor shown in **Figure 18-23**, the time constant will
    A. be four times greater.
    B. double.
    C. remain the same.
    D. be half as much.
    E. be four times less.

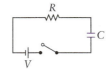

Figure 18-23 Problem 23

Problems

18-1 Life on Earth and our technological society are only possible because of charges in motion

18-2 Electric current equals the rate at which charge flows

24. • A steady current of 35 mA exists in a wire. How many electrons pass any given point in the wire per second? Example 18-1

25. • A light bulb requires a current of 0.50 A to emit a normal amount of light. If the light is left on for 1.0 h, how many electrons pass through the bulb? Example 18-1

26. • A synchrotron radiation facility creates a circular electron beam with a current of 487 mA when the electrons have a speed approximately equal to the speed of light $c = 3 \times 10^8$ m/s. How many electrons pass a given point in the accelerator per hour? Example 18-1

27. •• A copper wire that has a diameter of 2.00 mm carries a current of 10.0 A. Assuming that each copper atom contributes one free electron to the metal, find the drift speed of the electrons in the wire. The molar mass of copper is 63.5 g/mol, and the density of copper is 8.95 g/cm³. Example 18-2

28. • Aluminum wiring can be used as an alternative to copper wiring. Suppose a 40-ft run of 12-gauge aluminum wire (2.05232 mm in diameter) carrying a 7.0-A current is used in a circuit. How long would it take electrons to travel the length of this run and back? There are 6.03×10^{28} atoms in 1 m³ of aluminum, with each atom contributing three free electrons to the metal. Example 18-2

29. •• **Biology** Doubly charged calcium ions, Ca^{2+}, are released into your bloodstream and flow through your capillaries. The diameter of a capillary is about 5 μm, calcium ion concentrations are approximately 1 mol/m³, and typical electric current flow due to the Ca^{2+} ions is around 1.2 nA. Calculate the drift speed of the calcium ions. Example 18-2

30. • **Biology** 2.00×10^{-3} mol of potassium ions pass through a cell membrane in 4.00×10^{-2} s. (a) Calculate the electric current and (b) describe its direction relative to the motion of the potassium ions. Example 18-1

31. • **Biology** Cell membranes contain channels that allow ions to cross the phospholipid bilayer. A particular K^+ channel carries a current of 1.9 pA. How many K^+ ions pass through it in 1.0 ms? Example 18-1

18-3 The resistance to current through an object depends on the object's resistivity and dimensions

32. • How long is a tungsten wire if it has a diameter of 0.150×10^{-3} m and a resistance of 3.00 Ω? The resistivity of tungsten is 5.62×10^{-8} Ω·m. Example 18-3

33. • What is the resistance of a 50.0-m-long aluminum wire that has a diameter of 8.00 mm? The resistivity of aluminum is 2.83×10^{-8} Ω·m. Example 18-3

34. • The resistance ratio of two conductors that have equal cross-sectional areas and equal lengths is 1:3. What is the ratio of the resistivities of the materials from which they are made? Example 18-3

35. • An 8.00-m-long length of wire has a resistance of 4.00 Ω. The wire is uniformly stretched to a length of 16.0 m. Find the resistance of the wire after it has been stretched. Example 18-3

36. • When 120 V is applied to the filament of a 75-W light bulb, the current drawn is 0.63 A. When a potential difference of 3.0 V is applied to the same filament, the current is 0.086 A. Is the filament made of an ohmic material? Explain your answer. Example 18-4

37. • A power transmission line is made of copper that is 1.80 cm in diameter. What is the resistance of 1 mi of the line? Example 18-3

18-4 Resistance is important in both technology and physiology

38. •• There is a current of 112 pA when a certain potential is applied across a certain resistor. When that same potential is applied across a resistor made of the identical material but 25 times longer, the current is 0.044 pA. Calculate the ratio of the diameter of the first resistor to that of the second. Example 18-6

39. • A certain flexible conducting wire changes shape as environmental variables, such as temperature, change. If the diameter of the wire increases by 25% while the length decreases by 12%, by what factor does the resistance of the wire change? Example 18-3

40. • **Biology** Cell membranes contain channels that allow K^+ ions to leak out. Consider a channel that has a diameter of 1.0 nm and a length of 10 nm. If the channel has a resistance of 18 GΩ, what is the resistivity of the solution in the channel? Example 18-6

41. • If a light bulb draws a current of 1.0 A when connected to a 12-V circuit, what is the resistance of its filament? Example 18-5

42. • If a flashlight bulb has a resistance of 12.0 Ω, how much current will the bulb draw when it is connected to a 6.0-V circuit? Example 18-5

43. • A light bulb has a resistance of 8.0 Ω and a current of 0.5 A. At what voltage is it operating? Example 18-5

18-5 Kirchhoff's rules help us to analyze simple electric circuits

44. • Calculate the voltage difference $V_A - V_B$ in each of the situations shown in **Figure 18-24**. Example 18-8

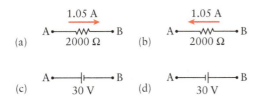

Figure 18-24 Problem 44

45. • An 18.0-Ω resistor and a 6.00-Ω resistor are connected in series. What is their equivalent resistance? Example 18-8

46. • An 18.0-Ω resistor and a 6.00-Ω resistor are connected in parallel. What is their equivalent resistance? Example 18-8

47. • A 9.00-Ω resistor and a 3.00-Ω resistor are connected in series across a 9.00-V battery. Find (a) the current through, and (b) the voltage drop across, each resistor. Example 18-8

48. • A 9.00-Ω resistor and a 3.00-Ω resistor are connected in parallel across a 9.00-V battery. Find (a) the current through, and (b) the voltage drop across, each resistor. Example 18-8

49. • Four wires meet at a junction. In two of the wires, currents $I_1 = 1.25$ A and $I_2 = 2.50$ A enter the junction. In one of the wires, current $I_3 = 5.75$ A leaves the junction. (a) What is the current I_4 in the fourth wire? (b) Does I_4 enter or leave the junction?

50. • A potential difference of 3.6 V is applied between points a and b in **Figure 18-25**. Find (a) the current in each of the resistors and (b) the total current the three resistors draw from the power source. Example 18-8

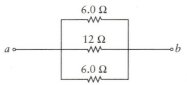

Figure 18-25 Problem 50

51. • The four resistors in **Figure 18-26** have an equivalent resistance of 8 Ω. Calculate the value of R_x. Example 18-8

Figure 18-26 Problem 51

52. • Find the equivalent resistance of the combination of resistors shown in **Figure 18-27**. Example 18-8

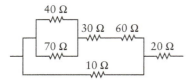

Figure 18-27 Problem 52

53. •• A metal wire of resistance 48 Ω is cut into four equal pieces that are then connected side by side to form a new wire which is one-quarter of the original length. What is the resistance of the new wire? Example 18-7

54. •• Two resistors A and B are connected in series to a 6.0-V battery; the voltage across resistor A is 4.0 V. When A and B are connected in parallel across a 6.0-V battery, the current through B is 2.0 A. What are the resistances of A and B? The batteries have negligible internal resistance. Example 18-8

55. •• A potential difference of 7.50 V is applied between points a and c in **Figure 18-28**. (a) Find the difference in potential between points b and c. (b) Is the current through the 60-Ω resistor larger or smaller than that through the 35-Ω resistor? Why? Example 18-8

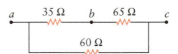

Figure 18-28 Problem 55

56. ••• A circuit is constructed using two batteries and three resistors as shown in **Figure 18-29**. The batteries have voltages $V_A = 3.0$ V and $V_B = 1.5$ V. The resistors have resistances $R_1 = 25$ Ω, $R_2 = 250$ Ω, and $R_3 = 50$ Ω. What is the current through resistor R_2? Define the positive current direction through R_2 to be downward.

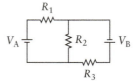

Figure 18-29 Problem 56

18-6 The rate at which energy is produced or taken in by a circuit element depends on current and voltage

57. • A heater is rated at 1500 W. How much current does it draw when it is connected to a 120-V voltage source? Example 18-9

58. • A 4.0-Ω resistor is connected to a 12-V voltage source. What is the power dissipated by the resistor? Example 18-9

59. • How much power is used by a 12-Ω coffeemaker that draws a current of 15 A? Example 18-9

60. •• When connected in parallel across a 120-V source, two light bulbs consume 60 and 120 W, respectively. What powers do the light bulbs consume if instead they are connected in series across the same source? Assume the resistance of each light bulb is constant. Example 18-10

61. • A 40-Ω transmission line carries 1200 A. Calculate the rate at which electrical energy is converted to thermal energy due to resistance. Example 18-9

62. • A stereo speaker has a resistance of 8.0 Ω. If the power output is 40 W, calculate the current passing through the speaker wires. Example 18-9

63. •• Three resistors are connected to a battery with voltage $V_b = 9.00$ V, as shown in **Figure 18-30**. The power dissipated by each resistor is 1.50 W. What is the resistance of each resistor? Example 18-10

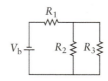

Figure 18-30 Problem 63

64. • A 10.0-V battery is connected in series with two resistors $R_1 = 10.0\ \Omega$ and $R_2 = 40.0\ \Omega$ (Figure 18-31). Calculate the power dissipated in each of the resistors. Example 18-10

Figure 18-31 Problem 64

65. •• A battery with voltage $V_b = 12.0$ V is connected to resistors $R_1 = 8.00\ \Omega$, $R_2 = 10.0\ \Omega$, and $R_3 = 12.0\ \Omega$ as shown in Figure 18-32. Calculate the total power provided by the battery and the power dissipated by each of the three resistors. Example 18-10

Figure 18-32 Problem 65

18-7 A circuit containing a resistor and capacitor has a current that varies with time

66. • A 4.00-MΩ resistor and a 3.00-μF capacitor are connected in series with a power supply. What is the time constant for the circuit? Example 18-11

67. • A capacitor of 20.0 μF and a resistor of $1.00 \times 10^2\ \Omega$ are quickly connected in series to a battery of 6.00 V. What is the charge on the capacitor 0.00100 s after the connection is made? Example 18-11

68. •• A 10.0-μF capacitor has an initial charge of 80.0 μC. If a 25.0-Ω resistor is connected across it, (a) what is the initial current in the resistor? (b) What is the time constant of the circuit? Example 18-12

69. •• A 12.5-μF capacitor is charged to 50.0 V, then discharged through a 75.0-Ω resistor. (a) How long after discharge begins will the capacitor have lost 90.0% of its initial (i) charge and (ii) energy? (b) What is the current through the resistor at both times in part (a)? Example 18-12

General Problems

70. •• The current through a resistor R of 25 Ω is measured at time intervals of 500 μs after closing the switch that connects it to a 5-V battery through a capacitor C with an unknown value (Figure 18-33). The magnitude of the current through the resistor is shown in the table below. Graph the current as a function of time and determine the capacitance.

Time (s)	Current (A)
0	0.200
5.00×10^{-4}	0.109
1.00×10^{-3}	0.060
1.50×10^{-3}	0.032
2.00×10^{-3}	0.018
2.50×10^{-3}	0.010
3.00×10^{-3}	0.005
3.50×10^{-3}	0.003

Figure 18-33 Problem 70

71. • If your local power company charges $0.11 per kW·h, what would it cost to run a 1500-W heater continuously during an 8.0-h night?

72. •• A house is heated by a 24-kW electric furnace. The local power company charges $0.10 per kW·h and the heating bill for January is $218. How long must the furnace have been running on an average January day?

73. •• **Biology** A single ion channel is selectively permeable to K$^+$ and has a resistance of 1.0 GΩ. During an experiment the channel is open for approximately 1.0 ms while the voltage across the channel is maintained at +80.0 mV with a patch clamp. How many K$^+$ ions travel through the channel? Example 18-5

74. •• How much power is dissipated in each resistor shown in Figure 18-34? Example 18-10

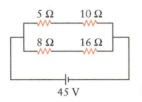

Figure 18-34 Problem 74

75. • Determine the current through each resistor in Figure 18-35. Example 18-8

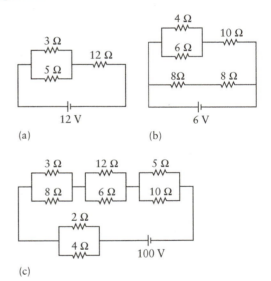

Figure 18-35 Problem 75

76. •• The circuit in Figure 18-36 consists of a battery $V_0 = 6.00$ V and five resistors $R_1 = 450\ \Omega$, $R_2 = 600\ \Omega$, $R_3 = 525\ \Omega$, $R_4 = 375\ \Omega$, and $R_5 = 250\ \Omega$. Determine the current passing through points A, C, and F. Example 18-9

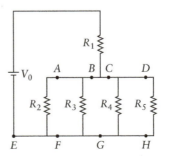

Figure 18-36 Problem 76

77. ••• Lightning bolts can carry as much as 30 C of charge and can travel between a cloud and the ground in around 100 μs. Potential differences have been measured as high as 400 million volts. (a) What is the average current in such a lightning strike? How does it compare with typical household currents? (b) What is the resistance of the air during such a strike? (c) How much energy is transferred during such a strike? (d) What mass of water at 100°C could the lightning bolt evaporate? Example 18-5

78. •• In **Figure 18-37**, a potential difference of 5.00 V is applied between points a and b. Determine (a) the equivalent total resistance, (b) the current in each resistor, and (c) the power dissipated in each resistor. Example 18-10

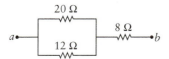

Figure 18-37 Problem 78

79. •• **Biology** Giant electric eels can deliver a voltage shock of 500 V and up to 1.0 A of current for a brief time. A snorkeler in salt water has a body resistance of about 600 Ω. A current of about 500 mA can cause heart fibrillation and death if it lasts too long. (a) What is the maximum power a giant electric eel can deliver to its prey? (b) If the snorkeler is struck by the eel, what current will pass through her body? Is this large enough to be dangerous? (c) What power does the snorkeler receive from the eel? Example 18-10

80. •• **Biology** Most of the resistance of the human body comes from the skin, as the interior of the body contains aqueous solutions that are good electrical conductors. For dry skin, the resistance between a person's hands is typically 500 kΩ. The skin is on average about 2.0 mm thick. We can model the body between the hands as a cylinder 1.6 m long and 14 cm in diameter with the skin wrapped around it. What is the resistivity of the skin? Example 18-6

81. • A capacitor has a value of 160 μF. What is the resistance of the charging circuit if it takes 10.0 s to charge the capacitor to 80.0% of its maximum charge? Example 18-11

82. •• (a) How much time (in terms of the time constant RC) will it take before a discharging capacitor holds only 50.0% of the original charge? (b) What percentage of the original charge will the capacitor hold at a time that is 3 times the time constant? Example 18-12

83. •• A 3.0-μF capacitor is put across a 12-V battery. After it is fully charged the capacitor is disconnected and placed in series through an open switch with a 2.00×10^2-Ω resistor. (a) Determine the charge on the capacitor before it is discharged. (b) What is the initial current through the resistor when the switch is closed? (c) At what time will the current reach 37% of its initial value? Example 18-12

84. •• You need a 75-Ω resistor, but you have only a box of 50-Ω resistors on hand. (a) How can you make a 75-Ω resistor using the resistors you have on hand? (b) How could you use your 50-Ω resistors to make a 60-Ω resistor? Example 18-8

85. •• A probe designed to measure the voltage across a circuit element is connected across that element. The probe should change the resistance or capacitance of the circuit as little as possible. (a) When it is connected across a circuit element, is the probe in series or in parallel with that element? (b) Should a probe connected in this way have (i) a very large or a very small resistance? (ii) a very large or a very small capacitance? Explain your reasoning using the properties of series and parallel connections.

86. •• Batteries release their energy rather slowly and are very damaging environmentally. Capacitors would be much cleaner for the environment and can be quickly recharged. Unfortunately they don't store much energy. (a) A new 1.5-V AAA battery has a "capacity" (*not* capacitance) of 1250 mA·h. What does this "capacity" actually represent? Express it in standard SI units. (b) How many joules of energy can be stored in the AAA battery? (c) At a steady current of 400 mA, how many hours will the AAA battery last? (d) How much energy can be stored in a typical 10-μF capacitor charged to a potential of 1.5 V? How does that compare to the energy stored in the AAA battery?

87. •• An *ultracapacitor* is a very high-capacitance device designed so that the spacing of the plates is around 1000 times smaller than in ordinary capacitors. The plates also contain millions of microscopic nanotubes, which increase their effective area 100,000 times. (a) If the plate separation and effective area of a 10-μF capacitor are changed as described above to make an ultracapacitor, what is its new capacitance? (b) How much energy does the ultracapacitor store if charged to 1.5 V? Compare this to the energy stored in an ordinary 10-μF capacitor at 1.5 V. (c) Compare the energy stored in the ultracapacitor to that of the AAA battery in the previous problem. (d) If the ultracapacitor is to take 1.0 min to decrease to $1/e$ of its initial maximum charge, through what resistance must it discharge? Example 18-12

88. •• **Medical** Early heart pacemakers used an RC circuit to stimulate the heart at regular intervals. A particular pacemaker is designed to operate at 70 beats per minute and sends a pulse to the heart every time the voltage across the capacitor reaches 8.0 volts. If the RC circuit is powered by a 12-V battery and incorporates a 30-μF capacitor, what resistance is required for the RC circuit? Assume that the capacitor discharges completely, and instantly, when it sends the pulse to the heart. Example 18-11

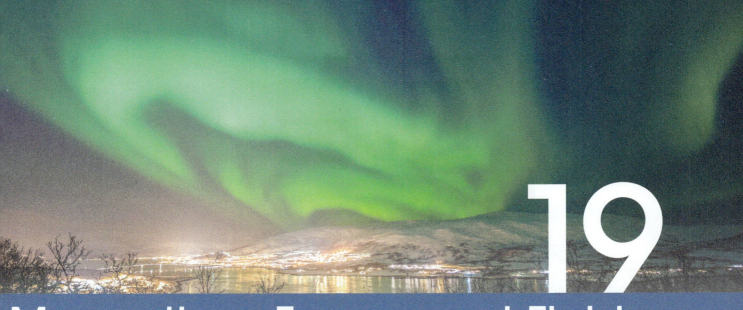

Rowan Romeyn/Alamy

19

Magnetism: Forces and Fields

In this chapter, your goals are to:

- (19-1) Recognize that magnetic forces can act over large distances.
- (19-2) Recognize that magnetism is fundamentally an interaction between moving electric charges.
- (19-3) Calculate the magnitude and direction of the magnetic force on a charged particle.
- (19-4) Describe how a mass spectrometer uses magnetic fields to sort ions according to their mass.
- (19-5) Calculate the magnetic force on a current-carrying wire.
- (19-6) Explain why a current loop in a uniform magnetic field experiences a net torque but zero net force.
- (19-7) Describe the principle of Ampère's law and how to use it.
- (19-8) Calculate the magnetic force that parallel current-carrying wires exert on each other.

To master this chapter, you should review:

- (5-6)   Uniform circular motion.
- (9-3, 9-5) The torque generated by a force and Newton's second law for rotational motion.
- (16-5)   The properties of electric field lines and the nature of electric dipoles.
- (18-2)   The relationship between the current in a wire and the drift speed of charges in the wire.
- (18-6)   Power into a current-carrying resistor.

What do you think?

The aurora borealis ("northern lights") is caused by fast-moving, electrically charged subatomic particles ejected from the Sun. Earth's magnetic field exerts a magnetic force on these particles that steers them toward our planet's north magnetic pole. When these particles enter Earth's upper atmosphere, they collide with the atoms there and cause the atoms to emit an eerie glow. (A similar effect in the southern hemisphere is called the aurora australis.) In what direction are these subatomic particles moving when the magnetic force on them is strongest? (a) In the same direction as the magnetic field; (b) opposite to the magnetic field; (c) perpendicular to the magnetic field; (d) either (a) or (b); (e) The magnetic force doesn't depend on the direction of motion.

19-1 Magnetic forces, like electric forces, act at a distance

If you rub a balloon on your head, then move the balloon a few centimeters away, the hairs on your head will stand up (see Figure 16-4). This is a result of *electric* forces. The rubbing transfers electrons between your hair and the balloon, giving one a net negative charge and the other a net positive charge; these attract each other, even over a distance, and this makes your hair stand up. The effect is rather feeble, however, and only the finest hairs respond noticeably to these electric forces.

Figure 19-1 **Magnetic forces** (a) The force between two bar magnets. (b) A compass needle interacting with Earth's magnetic field. (c) A large electromagnet.

(a)

The north pole (red) of one bar magnet attracts the south pole (blue) of another magnet.

The north poles (red) of the two magnets repel each other.

H.S. Photos/Alamy

(b)

The magnetized compass needle aligns with Earth's magnetic field.

sciencephotos/Alamy

(c)

The magnetic field required to pick up pieces of iron at a scrap yard comes from an electromagnet—a device that generates a magnetic field when there is current through its wires.

Cordelia Molloy/Science Source

Figure 19-1a shows another force that acts at a distance like the electric force but can be *much* stronger than that between the balloon and your hair. Certain objects called **magnets**, made of one of a handful of special materials such as iron, cobalt, and nickel, can exert strong **magnetic forces** on other magnets. (The name "magnet" comes from the region of ancient Greece known as Magnesia, where these objects were discovered more than 2500 years ago.) Just as we use the umbrella term "electricity" to refer to interactions between electrically charged objects, we use the term **magnetism** to describe the interactions between magnets. One application of magnetic interactions is the compass (**Figure 19-1b**). Earth's core acts like a giant magnet and produces a *magnetic field* in the surrounding space. (We'll see that magnetic fields are analogous to electric fields, but with important differences.) The magnetic needle of a compass aligns with Earth's field and points north. Earth's magnetic field is relatively weak; the field produced by the electromagnet in **Figure 19-1c** is thousands of times stronger.

Our goal in this chapter is to understand magnetism. We begin in the next section by investigating the properties of magnets and realizing that they are created by moving charges. From there we develop a full understanding of magnetic forces, which requires us to understand two things: (1) the nature of the magnetic field that moving charges *produce* and (2) how moving charges *respond* to the magnetic fields produced by other moving charges. It turns out to be easiest to look at the second of these first. We'll analyze the magnetic force that acts on a moving charge placed in a magnetic field, as well as the magnetic force on a wire that carries a current (a collection of charges moving in the wire) and is placed in a magnetic field. Later in the chapter we'll see how to calculate the magnetic field produced by a given collection of charges in motion.

TAKE-HOME MESSAGE FOR Section 19-1

✔ Magnetic forces can act over large distances, much like electric forces.

19-2 Magnetism is an interaction between moving charges

Like the electric force, magnetic forces become stronger as the objects are moved closer together. However, the magnetic force is not simply a form of the electric force. The attracting magnets in Figure 19-1a are *not* electrically charged: Their atoms are made of positively charged nuclei and negatively charged electrons, but their net charges are zero. Any electric forces between these magnets are very weak and are not responsible for the strong attraction shown in Figure 19-1a.

As we did for the electric force, we can use the idea of a field to explain how the magnetic force acts at a distance. Just as an electrically charged object sets up an electric field in the space around it, a magnet sets up a **magnetic field** in the space around it (**Figure 19-2a**). A second magnet placed in this field experiences a magnetic force that depends on the magnitude and direction of the field. We use the symbol $\vec{B}$ for magnetic

(a)

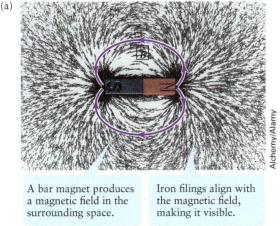

Alchemy/Alamy

A bar magnet produces a magnetic field in the surrounding space.

Iron filings align with the magnetic field, making it visible.

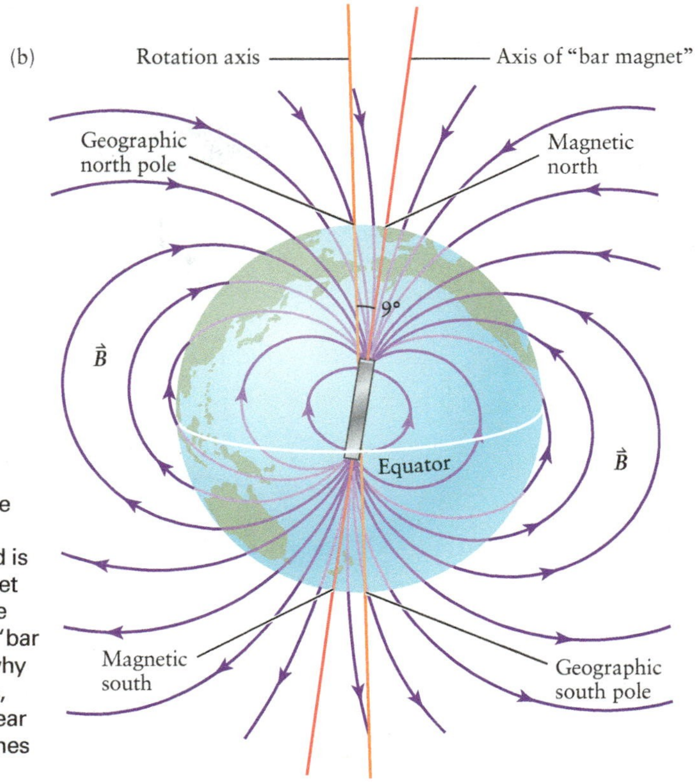

Figure 19-2 **The magnetic fields of a bar magnet and Earth** (a) The magnetic field of a bar magnet points out of its north pole (colored red) and into its south pole (colored black). (b) Earth's magnetic field is much the same, as if there were a giant bar magnet inside our planet (although it is produced in a different way, by electric currents in the liquid portion of our planet's interior). As of this writing (2020), this "bar magnet" is tipped by about 9° from Earth's rotation axis, which is why the magnetic poles are not at the same locations as the geographic, or true, poles. Note that the pole of Earth's internal "bar magnet" near the geographic north pole is actually a *south* pole: Magnetic field lines point into it, and it attracts the north pole of a compass.

(b)

Rotation axis ——————— Axis of "bar magnet"

Geographic north pole

Magnetic north

9°

$\vec{B}$

Equator

$\vec{B}$

Magnetic south

Geographic south pole

field. The magnetic field $\vec{B}$ of a magnet points away from one end of a magnet toward the other end. These two ends are called the **magnetic poles** of the magnet, one of which is called the *north pole* and the other the *south pole*. The meaning of the names is that Earth itself acts like a giant magnet. If allowed to swing freely, a magnet will orient itself so that its north pole points toward a location on Earth called magnetic north (which is near, but not the same as, the geographical north pole) and its south pole points toward a location called magnetic south (Figure 19-2b).

By convention we choose the direction of the magnetic field $\vec{B}$ produced by a magnet to be such that $\vec{B}$ points away from the magnet's north pole and toward the magnet's south pole. This should remind you of an electric dipole (see Section 16-5) made up of a positive charge $+q$ and a negative charge $-q$, for which the electric field $\vec{E}$ points away from the positively charged end and toward the negatively charged end. Indeed, a magnet like that shown in Figure 19-2a is often called a **magnetic dipole**.

The magnetic forces between magnets can be either attractive or repulsive. If the north pole of one magnet is close to the south pole of a second magnet, the force attracts the two magnets toward each other (Figure 19-3a); if the north poles are close together or the south poles are close together, the force makes the two magnets repel each other (Figure 19-3b). This is very different from the behavior of electrically charged objects but analogous to the way in which electric dipoles interact.

WATCH OUT! Magnetic poles are not like electric charges.

! Note that there is an important distinction between electric and magnetic dipoles. You can take an electric dipole apart by separating its component positive and negative charges. But you *cannot* separate the north and south poles of a magnet. If you cut a magnet in half, each half has a north pole and a south pole (Figure 19-3c). The same is true no matter how many small pieces you cut a magnet

into; a very small piece produces only a weak magnetic field, but is still a magnetic dipole with both a north pole and a south pole. (Physicists have speculated about the existence of *magnetic monopoles*, particles that have the properties of an isolated north or south magnetic pole. Many experiments have been performed to look for evidence of a magnetic monopole. No confirmed evidence has yet been found.)

(a) These magnets attract.

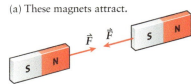

$\vec{F}$ $\vec{F}$

(b) These magnets repel.

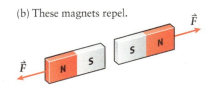

$\vec{F}$

$\vec{F}$

(c) If we cut a magnet in half, each half has a north and south pole.

Figure 19-3 **Bar magnets and magnetic poles** (a) The opposite poles of these magnets attract. (b) Like poles of these magnets repel. (c) Every magnet has both a north and south pole; they cannot be separated.

Figure 19-4 A current-carrying coil acts like a magnet (a) The field of a current-carrying coil is very similar to that of a bar magnet (Figure 19-2a). (b), (c) Such a coil interacts with a bar magnet just as though it were a magnet itself. (d) Two current-carrying coils interact like two bar magnets.

(a) The magnetic field created by current in a coil.

Eli Sidman, Technical Services Group, MIT

(b) This coil and magnet attract.

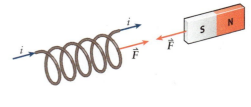

(c) This coil and magnet repel.

(d) These coils attract, just like two magnets.

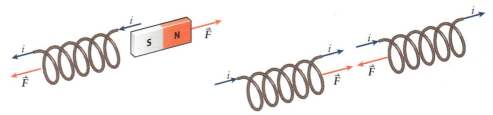

What is it about magnets that causes them to produce and respond to magnetic fields? And why is it impossible to separate their north and south poles? The answers to these questions were revealed by a set of crucial experiments in the nineteenth century. Figure 19-4a shows a version of one of these experiments. A coil of copper wire loops through a flat piece of clear plastic, and iron filings are spread over the plastic. Since copper is not magnetic, the filings lie wherever they were placed. But when there is an electric current through the coil, the filings line up just as they do around the magnet in Figure 19-2. This demonstrates that the moving charges that make up the current in the coil *produce* a magnetic field. The field points out of the coil at one end, which is the "north pole" of the coil, and points into the other end, which is the coil's "south pole."

We can verify that the coil produces a magnetic field by carrying out the experiment shown in Figure 19-4b, in which we've replaced one of the two magnets from Figure 19-3a with a coil. When the current in the coil is turned on, the coil and the magnet are attracted to each other. This can happen only if the current-carrying coil is indeed producing a magnetic field, and the magnet is responding to that field. If we flip the coil over, as shown in Figure 19-4c, the coil and the magnet now repel each other. This is just what would happen if the coil were an iron magnet that we flipped over to interchange the north and south poles (see Figures 19-3a and 19-3b).

For the coil in Figures 19-4b and 19-4c to be attracted to or repelled from the magnet, it must also be true that the moving charges that make up the current in the coil respond to the magnetic field of the magnet. In other words, a current-carrying coil and a magnet are fundamentally the same, in that both objects *produce* as well as *respond to* magnetic fields. Figure 19-4d shows an experiment that verifies this: Two current-carrying coils with the same orientation attract each other, just as do two magnets with the same orientation.

The series of experiments shown in Figure 19-4 suggests that *magnetism is an interaction between charges in motion*. For two objects to exert magnetic forces on each other, there must be moving charges in both objects. We use the umbrella term **electromagnetism** to include both electricity and magnetism, since both involve interactions between charges. The distinction between electricity and magnetism is that two charges exert *electric* forces on each other whether or not the charges are moving, while two charges exert *magnetic* forces on each other only if both charges are in motion. (The objects in the experiments shown in Figure 19-4 are all electrically neutral. So there are no electric forces in those experiments, only magnetic forces.)

You may be wondering if magnetism must always involve charges in motion, since the ordinary magnets shown in Figures 19-1, 19-2, and 19-3 are not attached to sources of emf and so do not carry currents. The answer is that there *are* charges in motion inside an ordinary iron magnet: The moving charges are the electrons within the atoms of the magnet, and their motions within the atom are like that of electrons in the circular coils of Figure 19-4. Each individual iron atom in a magnet produces only a tiny magnetic field. But because a large fraction of the atoms that make up the magnet are oriented so that their electron motions are in the same direction as in the surrounding atoms, their magnetic fields add together to make a substantial total field. This explains why cutting a magnet into pieces doesn't leave you with a separate north pole and south pole: Each piece is simply a smaller version of the original magnet.

GOT THE CONCEPT? 19-1 Electric and Magnetic Forces

(?) Two identical objects each have a positive charge Q. One of the objects is held in place, while the other object is in motion a short distance from the first object. What kinds of forces do the two objects exert on each other? (a) Both electric and magnetic forces; (b) electric forces but not magnetic forces; (c) magnetic forces but not electric forces; (d) neither electric forces nor magnetic forces; (e) answer depends on how the second object is moving.

TAKE-HOME MESSAGE FOR Section 19-2

✔ Magnetism is an interaction between moving charged particles. A moving charged particle produces a magnetic field, and a moving charged particle in a magnetic field experiences a force. A stationary charged particle neither produces nor responds to a magnetic field.

✔ In a magnet, charges are in motion at the atomic level to produce a magnetic field.

19-3 A moving point charge can experience a magnetic force

We'll begin our study of magnetic forces by considering the force on a single charged particle moving in a magnetic field $\vec{B}$. For example, this could be an electron moving in a current-carrying wire in the vicinity of a magnet, or a moving proton in Earth's upper atmosphere that is acted on by our planet's magnetic field. Here's what experiments tell us about the magnetic force on such a moving charged particle:

(1) A charged particle can experience a force when placed in a magnetic field, but *only when it is moving*.

(2) The magnetic force depends on the direction of the charged particle's velocity relative to the direction of the magnetic field. The charged particle does not experience a magnetic force when its velocity is parallel to or opposite to the magnetic field.

(3) The magnitude of the force is proportional to the charge on the particle, to the magnitude of the magnetic field, and to the speed of the particle.

(4) The direction of the force is perpendicular to both the direction of motion of the charged particle and the direction of the magnetic field.

(5) The direction of the force depends on the sign of the charge.

Figure 19-5a shows the magnetic force on a particle with positive charge q moving with velocity $\vec{v}$ in the presence of a magnetic field $\vec{B}$. If θ is the angle between the velocity of the charged particle and the magnetic field, the magnitude of the magnetic force $\vec{F}$ is

Magnitude of the magnetic force on a moving charged particle

Magnitude of the magnetic field

$$F = |q|vB\sin\theta$$

Angle between the direction of the particle's velocity $\vec{v}$ and the direction of the magnetic field $\vec{B}$

Magnitude of the particle's **charge**

Speed of the particle

Magnitude of magnetic force on a moving charged particle
(19-1)

The SI unit of magnetic field is the **tesla** (T). In Equation 19-1, if charge q is in coulombs (C), speed v is in meters per second (m/s), and magnetic field magnitude B is in tesla (T), the force F is in newtons (N). So 1 T is the same as $1\ \text{N}\cdot\text{s}/(\text{C}\cdot\text{m})$ or $1\ \text{N}/(\text{A}\cdot\text{m})$. (Recall that one ampere equals one coulomb per second: $1\ \text{A} = 1\ \text{C/s}$.) The strongest

BioMedical

Figure 19-5 Magnetic force on a moving charged particle The direction of the magnetic force on a particle with charge q depends on whether (a) q is positive or (b) q is negative. The magnitude of the force $F = |q|vB \sin \theta$ goes from (c) a maximum value if $\theta = 90°$ to (d) zero if $\theta = 0$ or $\theta = 180°$.

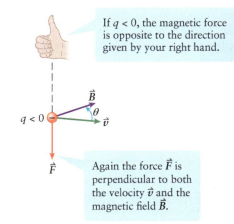

(a) Magnetic force on a positive charge

Your right hand helps determine the direction of the magnetic force.

$q > 0$ $\vec{F}$ $\vec{B}$ θ $\vec{v}$

The force $\vec{F}$ is perpendicular to both the velocity $\vec{v}$ and the magnetic field $\vec{B}$.

(b) Magnetic force on a negative charge

If $q < 0$, the magnetic force is opposite to the direction given by your right hand.

$q < 0$ $\vec{B}$ θ $\vec{v}$ $\vec{F}$

Again the force $\vec{F}$ is perpendicular to both the velocity $\vec{v}$ and the magnetic field $\vec{B}$.

(c) Magnetic force on a positive charge, $\theta = 90°$

For a given speed v, the magnetic force is greatest if the velocity $\vec{v}$ is perpendicular to the magnetic field $\vec{B}$.

$\vec{F}$ $\vec{B}$ $90°$ $\vec{v}$

(d) Magnetic force on a positive charge, $\theta = 0$ or $180°$

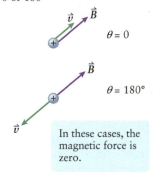

$\vec{v}$ $\vec{B}$ $\theta = 0$

$\vec{B}$ $\theta = 180°$

$\vec{v}$

In these cases, the magnetic force is zero.

magnetic field most of us will experience directly is the roughly 2-T field inside a magnetic resonance imaging (MRI) scanner. Other magnetic fields in our environment are much weaker: The strength of Earth's magnetic field is about 5×10^{-5} T, the electrical impulses that drive the contraction and relaxation of your heart produce magnetic fields of about 5×10^{-11} T, and the magnetic fields in your brain are on the order of 10^{-13} T. An alternative unit for magnetic field is the *gauss* (G): $1 \text{ G} = 10^{-4}$ T.

Here's a **right-hand rule** for determining the direction of the magnetic force on a moving charge. First extend the fingers of your right hand along the direction of the velocity $\vec{v}$ and orient your hand so that your palm is facing in the direction that the magnetic field $\vec{B}$ points. Then extend your right thumb so that it is perpendicular to the four fingers of your right hand and swivel your right hand so that the fingers now point in the direction of $\vec{B}$. If the particle has a positive charge, so $q > 0$, your thumb points in the direction of the force $\vec{F}$. If the particle has a negative charge, so $q < 0$ (Figure 19-5b), the force $\vec{F}$ points in the direction opposite to that given by the right-hand rule. In either case, the force $\vec{F}$ is perpendicular to the velocity $\vec{v}$ *and* perpendicular to the magnetic field $\vec{B}$. If we draw the vectors $\vec{v}$ and $\vec{B}$ with their tails together, they form a plane; the force $\vec{F}$ is perpendicular to that plane, with a direction given by the right-hand rule. (We will encounter a number of other right-hand rules for magnetism in this chapter.)

Note that Equation 19-1 involves the magnitude or absolute value of the charge, $|q|$, which is always positive. So the magnitude F of the magnetic force does *not* depend on whether the charge is positive or negative.

WATCH OUT! The magnetic force on a moving charged particle is never in the same direction as the magnetic field.

From Chapter 16 you're used to the idea that the electric force on a particle with charge q placed in an electric field $\vec{E}$ is $\vec{F}_{\text{electric}} = q\vec{E}$. This force is in the same direction as $\vec{E}$ if q is positive; if q is negative, it is in the direction opposite to $\vec{E}$. Figures 19-5a and 19-5b show that the magnetic force on a moving charged particle is very different: This force is *perpendicular* to the direction of the magnetic field. Electric and magnetic forces are quite different from each other!

Equation 19-1 tells us that the magnitude of the magnetic force on a moving charged particle depends on the angle θ between the directions of $\vec{v}$ and $\vec{B}$. For a given particle speed v, the magnetic force has its greatest magnitude if the velocity $\vec{v}$ is perpendicular to the magnetic field $\vec{B}$; then $\theta = 90°$, $\sin \theta = 1$, and $F = |q|vB$ (Figure 19-5c). If $\theta = 0$ or $\theta = 180°$, so the particle is moving either in the same direction as $\vec{B}$ or in the direction opposite to $\vec{B}$, $\sin \theta = 0$ and the force on the particle is *zero*. Thus a charged particle moving in the direction of the magnetic field, or in the direction opposite to the magnetic field, experiences no magnetic force (Figure 19-5d).

We've shown the vectors $\vec{v}$, $\vec{B}$, and $\vec{F}$ in perspective in Figure 19-5. Drawing in perspective isn't easy to do, so we'll often use a simple convention for drawing vectors that are pointed either into or out of the pages of this book (Figure 19-6). Think of a vector as an arrow, with a sharp point at its head and feathers at its tail. If a vector is directed perpendicular to the plane of the page and pointed toward you—that is, *out of* the page—we'll depict it with the symbol $\odot$. The dot in the center of this symbol represents the sharp point of the arrowhead. If a vector is directed perpendicular to the plane of the page and pointed away from you—that is, *into* the page—we'll depict it with the symbol $\otimes$. The "X" in the center of this symbol represents the feathers on the tail of the arrow.

Go to Interactive Exercises 19-1 and 19-2 for more practice dealing with magnetic forces.

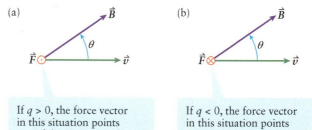

(a)

If $q > 0$, the force vector in this situation points out of the page (as indicated by the circle with a dot).

(b)

If $q < 0$, the force vector in this situation points into the page (as indicated by the circle with an X).

Figure 19-6 Vectors out of the page and into the page We use the symbols $\odot$ and $\otimes$ to denote vectors that point out of the page and into the page, respectively.

EXAMPLE 19-1 Magnetic Forces on a Proton and an Electron

At a location near our planet's equator, the direction of Earth's magnetic field is horizontal (that is, parallel to the ground) and due north, and the magnitude of the field is 2.5×10^{-5} T. Find the direction and magnitude of the magnetic force on a particle moving at 1.0×10^4 m/s if the particle is (a) a proton moving horizontally and due east, (b) an electron moving horizontally and due east, and (c) a proton moving horizontally in a direction 25° east of north. Recall that a proton has charge $e = 1.60 \times 10^{-19}$ C and an electron has charge $-e$.

Set Up

In each case we'll use Equation 19-1 to find the magnitude of the force on the moving proton or electron. The right-hand rule (Figure 19-5) will tell us the direction of the magnetic force.

Magnetic force on a moving charged particle:

$$F = |q|vB \sin \theta \qquad (19\text{-}1)$$

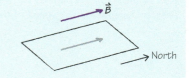

Solve

(a) The velocity vector $\vec{v}$ of the proton is perpendicular to the magnetic field vector $\vec{B}$, so in Equation 19-1 the angle $\theta = 90°$ and $\sin \theta = 1$. The vectors $\vec{v}$ and $\vec{B}$ both lie in a horizontal plane; the magnetic force vector $\vec{F}$ must be perpendicular to this plane, so $\vec{F}$ is vertical. The right-hand rule tells us that the force points vertically upward.

Charge on the proton:

$q = e = 1.60 \times 10^{-19}$ C

The proton velocity vector is perpendicular to the magnetic field vector ($\theta = 90°$), so the magnitude of the magnetic force on the proton is

$F = |q|vB \sin \theta = evB \sin \theta$

$= (1.60 \times 10^{-19} \text{ C})(1.0 \times 10^4 \text{ m/s})(2.5 \times 10^{-5} \text{ T}) \sin 90°$

$= 4.0 \times 10^{-20}$ N

(b) The electron has the same magnitude of charge as the proton and the same velocity, so it experiences the same magnitude of magnetic force. But its charge is negative, so the direction of the magnetic force $\vec{F}$ on the electron is *opposite* to the direction given by the right-hand rule. Hence the force $\vec{F}$ points vertically downward.

Charge on the electron:

$q = -e = -1.60 \times 10^{-19}$ C

The electron velocity vector is perpendicular to the magnetic field vector ($\theta = 90°$), so the magnitude of the magnetic force on the electron is

$F = |q|vB \sin \theta = evB \sin \theta$

$= (1.60 \times 10^{-19} \text{ C})(1.0 \times 10^4 \text{ m/s})(2.5 \times 10^{-5} \text{ T}) \sin 90°$

$= 4.0 \times 10^{-20}$ N

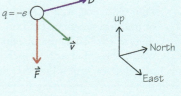

(c) The proton velocity $\vec{v}$ and the magnetic field $\vec{B}$ again both lie in a horizontal plane, but now the angle between these vectors is $\theta = 25°$. So the magnitude of the magnetic force is less than in part (a). The direction of the force is the same as in part (a), however.

The proton velocity vector is at an angle $\theta = 25°$ to the magnetic field vector, so the magnitude of the magnetic force on the proton is

$$F = |q|vB \sin\theta = evB \sin\theta$$
$$= (1.60 \times 10^{-19}\ \text{C})(1.0 \times 10^4\ \text{m/s})$$
$$\times (2.5 \times 10^{-5}\ \text{T}) \sin 25°$$
$$= 1.7 \times 10^{-20}\ \text{N}$$

This is smaller than in part (a) because $\sin 25° = 0.42$ compared to $\sin 90° = 1$.

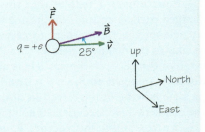

Reflect

This example illustrates how the magnetic force on a moving charged particle depends on the direction in which the particle is moving. Note that the force magnitudes in parts (a), (b), and (c) are very small because a single electron or proton carries very little charge. These particles also have very small mass, however (1.67×10^{-27} kg for the proton and 9.11×10^{-31} kg for the electron), so the resulting *accelerations* are tremendous. Can you use Newton's second law to show that the proton in part (a) and the electron in part (b) have accelerations of 2.4×10^7 m/s^2 and 4.4×10^{10} m/s^2, respectively?

GOT THE CONCEPT? 19-2 A Proton in a Magnetic Field

? A proton is fired into a region of uniform magnetic field pointing into the page (Figure 19-7). The proton's initial velocity is shown by the blue vector, which lies in the plane of the page. In which direction does the proton's trajectory bend? (a) Toward the top of the figure; (b) toward the bottom of the figure; (c) out of the figure; (d) into the figure; (e) answer depends on the speed of the proton.

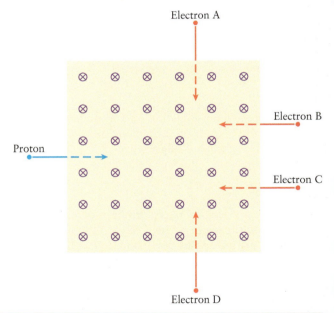

Figure 19-7 Charged particles in a magnetic field How will each particle be deflected when it enters this region of uniform magnetic field?

GOT THE CONCEPT? 19-3 Electrons in a Magnetic Field

? Four electrons, A, B, C, and D, are fired into a region of uniform magnetic field pointing into the page (Figure 19-7). The initial velocities of the electrons are shown by the red vectors, which lie in the plane of the page. Each electron will follow a trajectory that is bent by the magnetic force. For which electron could that trajectory lead to the point on the left-hand side where the proton enters the field region? (There may be more than one correct answer.) (a) Electron A; (b) electron B; (c) electron C; (d) electron D; (e) none of the electrons.

19-4 A mass spectrometer uses magnetic forces to differentiate atoms of different masses

An important application of the magnetic force on a moving charged particle is the **mass spectrometer**, a device used by chemists to determine the masses of individual atoms and molecules. Figure 19-8 shows a simplified version of one type of mass spectrometer. A sample to be analyzed is first vaporized, and its atoms and molecules are ionized by removing one or more electrons so all of the ions have a positive charge. A beam of the moving ions then passes through a *velocity selector*, a region where there is *both* an electric field $\vec{E}$ and a magnetic field $\vec{B}$. The fields $\vec{E}$ and $\vec{B}$ are oriented perpendicular to the ion beam as well as perpendicular to each other. You should be able to show in Figure 19-8 that in this region the electric force on a positive ion acts to the left while the magnetic force acts to the right. An ion will pass through this region without deflection only if these two forces cancel so that the *net* force on the ion is zero. For an ion of positive charge q, the electric force has magnitude $F_{electric} = qE$, and the magnetic force has magnitude $F_{magnetic} = qvB \sin 90° = qvB$ (the ion velocity and the magnetic field are perpendicular, so $\theta = 90°$ in Equation 19-1). So the condition for an ion to continue through this region without being deflected is

$$qE = qvB \quad \text{or} \quad v = \frac{E}{B} \qquad (19\text{-}2)$$

(speed of ions emerging from a velocity selector)

If an ion has a speed different from $v = E/B$, it will be deflected out of the beam. This is why we call this arrangement of fields a velocity selector: After passing through the $\vec{E}$ and $\vec{B}$ fields, the only ions that remain in the beam are those with a speed given by Equation 19-2.

As Figure 19-8 shows, the ion beam then passes into a second region with a magnetic field as before but no electric field. The beam deflects to the right as a result of the magnetic force and continues to deflect. That's because as the direction of the velocity vector changes, the direction of the magnetic force also changes to remain perpendicular to the ion velocity $\vec{v}$. Because $\vec{v}$ and $\vec{B}$ remain perpendicular, the angle θ in Equation 19-1 remains equal to $90°$ as the ions deflect and the magnitude of the magnetic force remains the same, $F_{magnetic} = qvB \sin 90° = qvB$. You probably recognize that what's going on here is exactly the situation that we described in Section 5-6. Because an ion is acted on by a force of constant magnitude that points perpendicular to the ion's velocity, the ion moves at a constant speed in a circular path — that is, in uniform circular motion. We can use Newton's second law to find the radius r of the circular path followed by an ion of charge q and mass m that moves

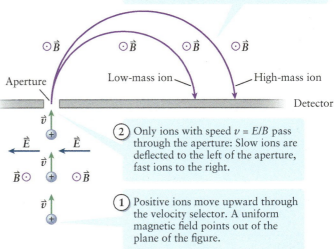

③ In this region ions follow a circular path under the influence of a magnetic field. For ions with the same charge, the trajectory depends on the ion mass.

Aperture   Low-mass ion   High-mass ion

Detector

② Only ions with speed $v = E/B$ pass through the aperture: Slow ions are deflected to the left of the aperture, fast ions to the right.

① Positive ions move upward through the velocity selector. A uniform magnetic field points out of the plane of the figure.

Figure 19-8 A mass spectrometer Electric and magnetic forces are used in this device to separate different ions according to their mass.

with speed v. The acceleration in uniform circular motion has magnitude $a = v^2/r$, and the net force on the ion has magnitude qvB, so

$$(19-3) \qquad qvB = \frac{mv^2}{r} \quad \text{or} \quad r = \frac{mv^2}{qvB} = \frac{mv}{qB}$$

(radius of circular path followed by charged particles in a uniform magnetic field)

Equation 19-3 is the key to understanding how the mass spectrometer works. All of the ions enter the aperture in Figure 19-8 moving at the same speed v (thanks to the velocity selector), and all are exposed to the same magnetic field of the same magnitude B. But ions of different mass m (and the same charge q) will follow paths with different radii and so will land at different locations on the detector in Figure 19-8. By counting how many ions land at each location, we can learn about the composition of the sample being analyzed. Example 19-2 illustrates how this works.

EXAMPLE 19-2   Measuring Isotopes with a Mass Spectrometer

Most oxygen atoms are the isotope ^{16}O ("oxygen-16"), which contains eight electrons, eight protons, and eight neutrons. The mass of this atom is 16.0 u, where 1 u = 1 atomic mass unit = 1.66×10^{-27} kg. The second most common isotope is ^{18}O ("oxygen-18"), which has two additional neutrons and so is more massive than an atom of ^{16}O: Each atom has a mass of 18.0 u. To determine the relative abundances of ^{16}O and ^{18}O, you send a beam of singly ionized oxygen atoms ($^{16}\text{O}^+$ and $^{18}\text{O}^+$, each with a charge $+e = +1.60 \times 10^{-19}$ C) through a mass spectrometer like that shown in Figure 19-8. The magnetic field strength in both parts of the spectrometer is 0.0800 T, and the electric field strength in the velocity selector is 4.00×10^3 V/m. (a) What is the speed of the beam that emerges from the velocity selector? (b) How far apart are the points where the ^{16}O and ^{18}O ions land?

Set Up

We use Equation 19-2 to find the speed v of ions that pass undeflected through the velocity selector and Equation 19-3 to find the radius of the circular path that each isotope follows. The distance from where each ion enters the region of uniform magnetic field to where it strikes the detector is the diameter (twice the radius) of the circular path.

Speed of ions emerging from a velocity selector:

$$v = \frac{E}{B} \qquad (19\text{-}2)$$

Radius of circular path followed by charged particles in a uniform magnetic field:

$$r = \frac{mv}{qB} \qquad (19\text{-}3)$$

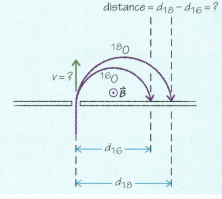

Solve

(a) Calculate the speed of ions that emerge from the velocity selector. Note that this speed does not depend on the charge or mass of the ion, so the $^{16}\text{O}^+$ and $^{18}\text{O}^+$ ions both emerge with this speed.

From Equation 19-2,

$$v = \frac{E}{B} = \frac{4.00 \times 10^3 \text{ V/m}}{0.0800 \text{ T}} = 5.00 \times 10^4 \, \frac{\text{V}}{\text{T} \cdot \text{m}}$$

From above, 1 T = 1 N·s/(C·m), so 1 T·m = 1 N·s/C. We also recall from Chapter 17 that 1 V = 1 J/C = 1 N·m/C. So we can write the speed as

$$v = 5.00 \times 10^4 \, \frac{\text{V}}{\text{T} \cdot \text{m}} = 5.00 \times 10^4 \left(\frac{\text{N} \cdot \text{m}}{\text{C}} \right) \left(\frac{\text{C}}{\text{N} \cdot \text{s}} \right)$$

$$= 5.00 \times 10^4 \text{ m/s}$$

(b) Calculate the radius r and diameter d of the circular path followed by each isotope.

The masses of the two atoms are

For ^{16}O: $m_{16} = (16.0 \text{ u})(1.66 \times 10^{-27} \text{ kg/u}) = 2.66 \times 10^{-26}$ kg

For ^{18}O: $m_{18} = (18.0 \text{ u})(1.66 \times 10^{-27} \text{ kg/u}) = 2.99 \times 10^{-26}$ kg

The mass of each positive ion ($^{16}O^+$ and $^{18}O^+$) is slightly less than the mass of the neutral atom; the difference is the mass of one electron, which is 9.11×10^{-31} kg $= 0.0000911 \times 10^{-26}$ kg. This difference is so small that we can ignore it.

For ^{16}O,

$$r_{16} = \frac{m_{16}v}{qB} = \frac{(2.66 \times 10^{-26} \text{ kg})(5.00 \times 10^4 \text{ m/s})}{(1.60 \times 10^{-19} \text{ C})(0.0800 \text{ T})}$$

$$= 0.104 \, \frac{\text{kg} \cdot \text{m}}{\text{T} \cdot \text{C} \cdot \text{s}}$$

Since $1 \text{ T} = 1 \text{ N} \cdot \text{s}/(\text{C} \cdot \text{m})$ and $1 \text{ N} = 1 \text{ kg} \cdot \text{m/s}^2$, it follows that $1 \text{ T} = 1 \text{ kg}/(\text{C} \cdot \text{s})$ and $1 \text{ T} \cdot \text{C} \cdot \text{s} = 1 \text{ kg}$. So for ^{16}O we have

$$r_{16} = 0.104 \, \frac{\text{kg} \cdot \text{m}}{\text{kg}} = 0.104 \text{ m}$$

$$d_{16} = 2r_{16} = 0.208 \text{ m} = 20.8 \text{ cm}$$

For ^{18}O,

$$r_{18} = \frac{m_{18}v}{qB} = \frac{(2.99 \times 10^{-26} \text{ kg})(5.00 \times 10^4 \text{ m/s})}{(1.60 \times 10^{-19} \text{ C})(0.0800 \text{ T})}$$

$$= 0.117 \text{ m}$$

$$d_{18} = 2r_{18} = 0.234 \text{ m} = 23.4 \text{ cm}$$

The distance between the positions where the ^{16}O and ^{18}O ions land is the difference between the two diameters.

The distance between where the ^{16}O and ^{18}O ions land is

$$d_{18} - d_{16} = 23.4 \text{ cm} - 20.8 \text{ cm} = 2.6 \text{ cm}$$

This is a substantial distance, so the mass spectrometer does a good job of separating the two isotopes.

Reflect

Experiments of this kind show that, on average, 99.8% of oxygen atoms are ^{16}O and 0.2% are ^{18}O. (Making up a small fraction of a percent is a third isotope, ^{17}O.)

Measurements of the ratio of ^{18}O to ^{16}O are important to the science of *paleoclimatology*, the study of Earth's ancient climate. One way to determine the average temperature of the planet in the distant past is to examine ancient ice deposits in Greenland and Antarctica. These deposits endure for hundreds of thousands of years; deposits near the surface are more recent, while deeper deposits are older. The ice comes from ocean water that evaporated closer to the equator and then fell as snow in the far north or far south. Each molecule of water is made up of two hydrogen atoms and one oxygen atom, which could be ^{16}O or ^{18}O. A water molecule can more easily evaporate if it contains lighter ^{16}O than if it contains heavier ^{18}O, so the water that evaporated and fell on Greenland and Antarctica as snow contains an even smaller percentage of ^{18}O than ocean water does. This deficiency becomes even more pronounced in colder climates. It has been shown that a decrease of one part per million of ^{18}O in ice indicates a 1.5°C drop in sea-level air temperature at the time it originally evaporated from the oceans.

Using mass spectrometers to analyze ancient ice from Greenland and Antarctica, paleoclimatologists have been able to determine the variation in Earth's average temperature over the past 160,000 years. They have also analyzed the amount of atmospheric carbon dioxide (CO_2) that was trapped in the ice as it froze. An important result of these studies is that higher levels of atmospheric CO_2 have gone hand-in-hand with elevated temperatures for the past 160,000 years, which is just what we would expect from our discussion of global warming in Section 14-7. In the same way, the tremendous increase in CO_2 levels in the past century due to burning fossil fuels has gone hand-in-hand with recent dramatic increases in our planet's average temperature.

GOT THE CONCEPT? 19-4 Ions in a Mass Spectrometer

? When passed through the mass spectrometer of Example 19-2, which of the following ions would follow nearly the same path as an $^{16}O^+$ ion? Assume that all ions are moving at the speed given by Equation 19-2.

(a) A doubly charged oxygen-16 ion ($^{16}O^{2+}$) of mass 16.0 u;
(b) a singly charged sulfur-32 ion ($^{32}S^+$) of mass 32.0 u;
(c) a doubly charged sulfur-32 ion ($^{32}S^{2+}$) of mass 32.0 u;
(d) more than one of these; (e) none of these.

19-5 Magnetic fields exert forces on current-carrying wires

The magnetic force that we've described may seem to apply only in certain very special circumstances. But in fact, you use magnetic forces whenever you listen to recorded music through earbuds (Figure 19-9a). Within each earbud is a small but powerful magnet adjacent to a flexible plastic cone with a coil of wire attached to it (Figure 19-9b). This coil is connected through the earbud wires to your music player, which sends the musical signal to the coil in the form of a varying electric current. Charges within the coil are thus set into motion, and these moving charges experience magnetic forces exerted by the magnet. These forces pull on the coil that contains the charges and on the plastic cone to which the coil is attached. We learned in Section 19-3 that the direction of magnetic force depends on the direction in which charges move. So the force on the charges, coil, and plastic cone reverses whenever the current in the coil changes direction. The result is that the plastic cone oscillates back and forth in response to the signal coming from your player. This oscillation pushes on the surrounding air, producing a sound wave—the sound of music—that travels to your eardrum.

Let's see how to find the magnitude of the magnetic force on a current-carrying wire, such as a segment of the coil in Figure 19-9b. Figure 19-10 shows a straight wire of length ℓ that carries a current i. A magnetic field $\vec{B}$ points at an angle θ to the direction of the current and has the same value along the entire length of the wire. We learned in Section 18-2 that the current in the wire is related to the speed at which individual charges drift through the wire:

Current and drift speed
(18-3)

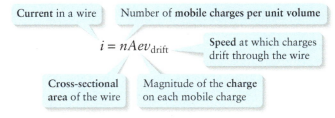

Current in a wire — Number of **mobile charges per unit volume**

$$i = nAev_{\text{drift}}$$

Speed at which charges drift through the wire

Cross-sectional area of the wire — Magnitude of the **charge** on each mobile charge

The force $\vec{F}$ on a current-carrying wire in a magnetic field $\vec{B}$ is perpendicular to both $\vec{B}$ and the length of the wire. The direction is given by a right-hand rule.

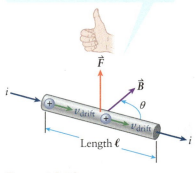

Figure 19-10 Magnetic force on a current-carrying wire The force that a magnetic field exerts on this wire is the sum of the forces on all the moving charges within the wire.

(a)

(b)

Front cover

Transparent plastic cone

Coil (attached to plastic cone)

Magnet (plastic cone is attached to this at its edges)

Jupiterimages/Thinkstock

Christina Micek

Figure 19-9 Earbud physics (a) Earbuds and other speakers use magnetic forces to produce sound. (b) Internal construction of an earbud.

From Equation 19-1, the magnetic force on an individual charge e moving through the wire at speed v_{drift} has magnitude $ev_{drift}B \sin \theta$. Each such moving charge feels a force of the same magnitude and in the same direction. The total number of such moving charges in the wire shown in Figure 19-10 is n (the number of moving charges per unit volume) multiplied by the volume V of the wire, where $V = A\ell$ (the cross-sectional area of the wire multiplied by its length). So the magnitude of the *net* magnetic force on all of the moving charges in the wire is

$$F = (nA\ell)(ev_{drift}B \sin \theta)$$

If we rearrange the terms in this equation, we get

$$F = (nAev_{drift})\ell B \sin \theta \qquad (19\text{-}4)$$

The quantity in parentheses in Equation 19-4 is just the current i as given by Equation 18-3. So the magnitude of the magnetic force on a current-carrying wire is

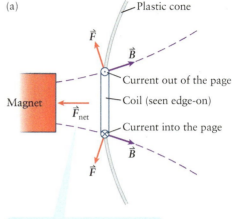

(a)

The net magnetic force on the coil pulls it and the plastic cone toward the magnet.

Magnitude of the magnetic force on a current-carrying wire

Magnitude of the magnetic field (assumed uniform over the length of the wire)

Magnitude of magnetic force on a current-carrying wire (19-5)

$$F = i\ell B \sin \theta$$

Angle between the direction of the current and the direction of the magnetic field $\vec{B}$

Current in the wire   **Length of** the wire

Just as for the magnetic force on a moving charged particle, a right-hand rule tells you the direction of the magnetic force on a current-carrying wire. Extend the fingers of your right hand in the direction of the current and orient your hand so that the palm is facing the direction that the magnetic field vector $\vec{B}$ points. With your right thumb extended, swivel your right hand so that the fingers now point in the direction of $\vec{B}$. Your outstretched thumb points in the direction of the magnetic force exerted on the wire (Figure 19-10).

Equation 19-5 says that there is *no* magnetic force on the wire if the magnetic field $\vec{B}$ points either in the same direction as the current ($\theta = 0$) or in the direction opposite to the current ($\theta = 180°$). In either case, $\sin \theta = 0$ and $F = 0$. For a given magnetic field of magnitude B, the force is greatest if $\vec{B}$ points perpendicular to the current so $\theta = 90°$ and $\sin \theta = \sin 90° = 1$. This idea is used in the design of the earbuds shown in Figure 19-9. The wires in each earbud are curved into a circular coil, not straight as in Figure 19-10. But we can treat the coil as being made up of many short segments, each of which is effectively a straight piece of wire. As Figure 19-11a shows, the magnetic field from the magnet in each earbud is perpendicular to each such segment. The forces on different segments are in different directions, but the vector sum of these forces is toward the magnet. If the current is in the reverse direction, as in Figure 19-11b, all of the individual forces also reverse direction and the vector sum of the forces is away from the magnet. The current from the music player continually reverses direction, so the coil and attached diaphragm oscillate back and forth, producing a sound wave.

In the following section we'll see how to use Equation 19-5 to help us understand how electric motors work. For now, let's see how to use a magnetic field to levitate a current-carrying wire.

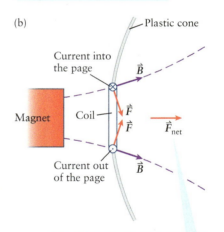

(b)

The net magnetic force on the coil pushes it and the plastic cone away from the magnet.

Figure 19-11 Magnetic forces in an earbud (a) The magnetic field produced by an earbud magnet (see Figure 19-9) is not uniform, so the field exerts a net force on the coil (seen here edge-on) and the plastic cone to which it is attached. (b) Reversing the direction of the current reverses the direction of the net force.

EXAMPLE 19-3 Magnetic Levitation

You set up a uniform horizontal magnetic field that points from south to north and has magnitude 2.00×10^{-2} T. (This is about 400 times stronger than Earth's magnetic field, but easily achievable with common magnets.) You want to place a straight copper wire of diameter 0.812 mm in this field, then run enough current through the wire so that the magnetic force will make the wire "float" in midair. This is called *magnetic levitation*. What minimum current is required to make this happen? The density of copper is 8.96×10^3 kg/m^3.

Set Up

To make the wire "float," there must be an upward magnetic force on the wire that just balances the downward gravitational force. We know from Equation 19-5 that to maximize the magnetic force the current direction should be perpendicular to the magnetic field $\vec{B}$. The right-hand rule then shows that the current should flow from west to east so that the magnetic force is directed upward. We're not given the mass of the wire, but we can express the mass (and hence the gravitational force on the wire) in terms of its density. We're also not given the length of the wire; as we'll see, this will cancel out of the calculation.

Magnetic force on a current-carrying wire:

$$F = i\ell B \sin \theta \qquad (19\text{-}5)$$

Definition of density:

$$\rho = \frac{m}{V} \qquad (11\text{-}1)$$

Volume of a cylindrical wire of cross-sectional area A and length ℓ:

$$V = A\ell$$

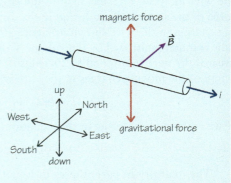

Solve

Write expressions for the two forces (magnetic and gravitational) that act on the wire.

Assume the wire has length ℓ. Since the current flows in a direction perpendicular to the magnetic field, the angle θ in Equation 19-5 is 90°. So the magnitude of the upward magnetic force is

$$F = i\ell B \sin 90° = i\ell B$$

The magnitude of the downward gravitational force on the wire of mass m is

$$w = mg$$

From Equation 11-1, the mass equals the density of copper multiplied by the volume of the wire:

$$m = \rho V = \rho A\ell$$

The cross-sectional area A of the wire is that of a circle of radius r:

$$A = \pi r^2 \quad \text{so} \quad m = \rho A\ell = \rho(\pi r^2)\ell$$

So the magnitude of the gravitational force is

$$w = mg = \rho(\pi r^2)\ell g$$

If the wire is floating in equilibrium, the upward magnetic force must balance the downward gravitational force. Use this to solve for the required current i.

In equilibrium the net vertical force on the wire is zero:

$$F - w = 0, \text{ so } F = w \text{ and } i\ell B = \rho(\pi r^2)\ell g$$

Both the magnetic force and the gravitational force are proportional to the length ℓ of the wire, so ℓ cancels out of the equation:

$$iB = \rho(\pi r^2)g$$

Solve for the current i:

$$i = \frac{\rho(\pi r^2)g}{B}$$

$$= \frac{(8.96 \times 10^3 \text{ kg/m}^3)(\pi)(0.406 \times 10^{-3} \text{ m})^2(9.80 \text{ m/s}^2)}{2.00 \times 10^{-2} \text{ T}}$$

$$= 2.27 \frac{\text{kg}}{\text{T} \cdot \text{s}^2} = 2.27 \text{ A}$$

[Check on units: We know from Section 19-3 that $1 \text{ T} = 1 \text{ N}/(\text{A} \cdot \text{m})$, and we also know that $1 \text{ N} = 1 \text{ kg} \cdot \text{m/s}^2$. Therefore, $1 \text{ T} = 1 \text{ kg}/(\text{A} \cdot \text{s}^2)$, so $1 \text{ kg}/(\text{T} \cdot \text{s}^2) = 1 \text{ A}$.]

Reflect

A current of 2.27 A is relatively small, so this experiment in magnetic levitation is not too difficult to perform. Note that the required current i is inversely proportional to the magnitude B of the magnetic field. You can see that if you tried to make a wire "float" using Earth's magnetic field, which is about 1/400 as strong as the field used here, you would need to use an immense current of 400×2.27 A = 909 A. That's not practical because a current of that magnitude would cause the wire in this example to melt! (Recall from Equation 18-24 in Section 18-6 that the power into a resistor with resistance R that carries current i is $P = i^2 R$. The wire in this example has a small cross-sectional area, so its resistance R will be fairly large. The power delivered to the wire by a 909-A current will quickly increase its temperature to above the melting point of copper, 1085°C.)

A practical application of magnetic levitation is train design. By using magnetic forces to make a train float just above the track, the rolling friction between the wheels and the track can be completely eliminated and very high speeds achieved. (Magnetic forces are also used to propel the train forward.) A train of this type in commercial operation in Shanghai, China, reaches a top speed of 431 km/h (268 mi/h). Such train lines require special magnets and wires capable of sustaining very high currents.

GOT THE CONCEPT? 19-5 Direction of Magnetic Force on a Wire

(?) A long, straight wire carrying a current can be placed in various orientations with respect to a constant magnetic field as shown in Figure 19-12a, b, c, d, e, and f. What is the direction of the force in each case? Give your answers in terms of the positive and negative x, y, and z directions. If the force is zero, say so. (In each case the positive z direction is out of the plane of the figures.)

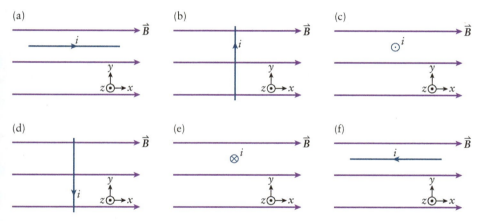

Figure 19-12 Which way is the force? What is the direction of the magnetic force on the wire in each case?

TAKE-HOME MESSAGE FOR Section 19-5

✔ A current-carrying wire in a magnetic field experiences a magnetic force that is perpendicular to both the current direction and the magnetic field direction.

✔ The magnitude of this magnetic force is maximum if the current and magnetic field directions are perpendicular, and zero if the wire axis is along the magnetic field.

19-6 A magnetic field can exert a torque on a current loop

A common everyday application of magnetic forces on current-carrying wires is an electric motor. An electric motor is at the heart of a kitchen blender, a vacuum cleaner, the starter for an internal-combustion automobile, and many other devices (Figure 19-13a). Inside any electric motor you'll find a rotating portion (called the *rotor*) that's wrapped with coils of wire (Figure 19-13b), as well as magnets in the stationary part of the motor. When you turn the motor on, causing a current through the coils, the magnets exert forces on the current-carrying wire of the coils. The net

(a)

iStockphoto/Thinkstock

Figure 19-13 Electric motors and magnetic torque (a) This fan uses an electric motor to rotate the fan blades. (b) The key components of any electric motor are current-carrying coils and a source of magnetic field. The magnetic field exerts forces on the current-carrying wires of the coils, and these produce a torque that makes the rotating part of the motor spin.

WATCH OUT! A current-carrying coil can feel a net magnetic force if the $\vec{B}$ field is not uniform.

! In Figure 19-14a we assumed that the magnetic field is uniform (its magnitude and direction are the same at all points) and found that the net magnetic force on the current loop is zero. But if the magnetic field magnitude were greater at the left-hand side of the loop than at the right-hand side, then the left-hand side would experience a greater magnetic force. This would result in a net force on the loop to the left. In this chapter we'll restrict our discussion to the simple case in which the field is uniform.

(b)

iStockphoto/Thinkstock

effect is that there is a magnetic *torque* that makes the rotor spin. Let's look at a simplified version of this process to see how such a magnetic torque arises.

Figure 19-14a shows a straight wire bent into a single rectangular loop of wire with sides of length L and W. The loop carries a current i provided by a source of emf (not shown), so we call it a **current loop**. This loop is immersed in a uniform magnetic field of magnitude B and is free to rotate around an axis (shown in green). You should apply the right-hand rule for magnetic forces on a current-carrying wire (see Section 19-5) to each side of this loop. You'll see that the left-hand side of the loop experiences a force to the left, the right-hand side feels a force to the right, the top of the loop feels an upward force, and the bottom of the loop feels a downward force, as Figure 19-14a shows. Each segment of the loop carries the same current and is in the same magnetic field. Since the left- and right-hand sides are the same length and at the same angle to the magnetic field, the forces on these two sides have the same magnitude but opposite directions. Thus these forces cancel, and there is zero net force to the left or the right. The forces on the top and bottom segments of the wire also cancel for the same reason. So the net *force* on this rectangular loop is zero. (You should be able to convince yourself that the same would be true if we reversed the direction of the current around the loop or the direction of the magnetic field, or if we changed the angle between the direction of the field and the plane of the loop.) It turns out that there is zero net magnetic force on *any* closed current-carrying loop in a uniform magnetic field, not just loops with the rectangular shape shown in Figure 19-14a.

Although the net magnetic force on the current loop in Figure 19-14a is zero, the net magnetic *torque* around the axis is not. **Figure 19-14b** is a side view of the current

(a) Perspective view

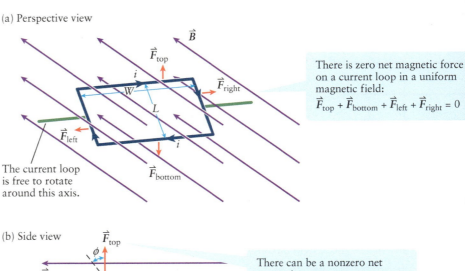

There is zero net magnetic force on a current loop in a uniform magnetic field:
$$\vec{F}_{top} + \vec{F}_{bottom} + \vec{F}_{left} + \vec{F}_{right} = 0$$

The current loop is free to rotate around this axis.

(b) Side view

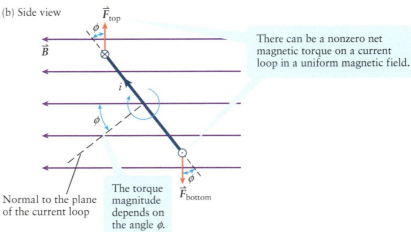

There can be a nonzero net magnetic torque on a current loop in a uniform magnetic field.

Normal to the plane of the current loop

The torque magnitude depends on the angle ϕ.

Figure 19-14 A current loop in a uniform magnetic field (a) The magnetic forces on the sides of the current loop. (b) These forces can give rise to a net magnetic torque.

loop; the axis shown in green in Figure 19-14a is perpendicular to the plane of this figure and passes through the center of the loop. The forces on the near and far sides of the loop, which point out of and into the page, respectively, are directed parallel to this axis and so produce no torque. However, the forces that act on the top and bottom sides of the loop both give rise to a torque. For the situation shown in Figure 19-14b, both of these forces tend to make the loop rotate clockwise around the axis, so the net torque is clockwise.

Let's calculate the magnitude of the magnetic torque on the loop in Figure 19-14b. First note that the magnetic field is perpendicular to the current in both the top side of the loop and the bottom side of the loop. So in Equation 19-5 for the magnetic force on a current-carrying wire, the angle θ equals 90°. Both the top and bottom sides of the loop have length W (see Figure 19-14a), so the magnitude of the magnetic force on each of these sides is

$$F = iWB \sin 90° = iWB$$ (19-6)

To find the torque around the axis that each of these forces produces, we use Equation 9-10:

See the Math Tutorial for more information on trigonometry.

| Magnitude of the torque produced by a force acting on an object | Magnitude of the force |

$$\tau = rF \sin \phi$$

Magnitude of torque
(9-10)

| Distance from the rotation axis of the object to where the force is applied | Angle between the vector $\vec{r}$ (from the rotation axis to where the force is applied) and the force vector $\vec{F}$ |

In Figure 19-14b the distance r is one-half of the length L of the near or far side of the loop, and ϕ is the angle between the direction of the near or far side of the current loop and the force exerted on either the top or bottom of the loop. This angle phi is also equal to the angle between the direction of the magnetic field and an imaginary line that's perpendicular to the plane of the current loop. We call this imaginary line the **normal** to the plane of the loop. From Equations 9-10 and 19-6, the magnitude of the torque due to the force on either the top or bottom side of the loop is

$$\tau_{\text{one side}} = \left(\frac{L}{2}\right)F \sin \phi = \left(\frac{L}{2}\right)(iWB) \sin \phi = \frac{1}{2}i(LW)B \sin \phi$$ (19-7)

The product LW (length times width) in Equation 19-7 is just the area of the current loop: $A = LW$. As we mentioned above, the torque from the top of the loop and the torque from the bottom of the loop both act in the same direction, so the *net* torque on the loop is just double the torque on one side given by Equation 19-7. Using $A = LW$, we have

$$\tau = 2\tau_{\text{one side}} = 2\left(\frac{1}{2}\right)i(LW)B \sin \phi = 2\left(\frac{1}{2}\right)iAB \sin \phi$$

or

| Magnitude of the magnetic torque on a current loop | Magnitude of the magnetic field |

$$\tau = iAB \sin \phi$$

Magnitude of magnetic torque on a current loop
(19-8)

| | Angle between the normal to the plane of the loop and the direction of the magnetic field $\vec{B}$ |
| Current in the loop | Area of the loop |

Although we've assumed a rectangular loop, Equation 19-8 applies to a current loop of area A of *any* shape.

Typically the rotating coil in an electric motor is not just a single loop of wire but has many *turns* (equivalent to many single coils stacked on top of each other). If a coil has N turns, the magnetic torque on it is N times greater than the value given by Equation 19-8.

EXAMPLE 19-4   Angular Acceleration of a Current Loop

A length of copper wire is formed into a square loop with 50 turns. The loop is free to turn about a frictionless axis that lies in the plane of the loop and passes through its center. Each side of the loop is 2.00 cm long, and the moment of inertia of the loop about the axis of rotation is 4.00×10^{-6} kg·m². The loop lies in a region where there is a uniform magnetic field of magnitude 1.50×10^{-2} T that is perpendicular to the rotation axis of the loop. The current in the loop is 0.500 A. Find the angular acceleration of the loop (magnitude and direction) (a) when the loop is released from rest from the orientation shown in Figure 19-15, (b) after the loop has rotated 90°, and (c) after the loop has rotated 180°.

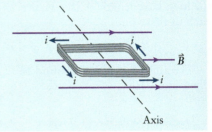

Figure 19-15 A square current loop in a magnetic field When this square current loop is released from rest, how will it begin to rotate?

Set Up

For each orientation we'll use Equation 19-8 to find the magnitude of the magnetic torque on the loop. There is no other torque on the loop (there is no friction, and gravity exerts zero torque since the center of mass of the loop lies on the rotation axis). So the magnetic torque is also the net torque, and we can use Equation 9-16 to find the magnitude of the angular acceleration. We'll find the direction of the angular acceleration by looking at the directions of the magnetic forces on the individual sides of the loop. Figure 19-14 shows that forces on the sides parallel to the rotation axis tend to affect rotation. The forces on the other sides cause no torque (compare Figure 19-14a).

Magnitude of magnetic torque on a current loop:

$$\tau = iAB \sin \phi \qquad (19\text{-}8)$$

Newton's second law for rotational motion:

$$\sum \tau_{\text{ext}, z} = I\alpha_z \qquad (9\text{-}16)$$

Solve

(a) First find the direction of the magnetic torque and hence the direction of the angular acceleration.

The right-hand rule for the magnetic force on a current-carrying wire shows that the forces on the wire tend to cause a clockwise rotation. So the angular acceleration is clockwise, and when the loop is released it will begin to rotate in the clockwise direction.

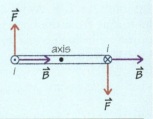

Find the magnitude of the angular acceleration.

The normal to the loop is perpendicular to the magnetic field, so in Equation 19-8 $\sin \phi = \sin 90° = 1$. The loop is square, so the cross-sectional area of the loop is just the square of the length of one side (2.00 cm = 0.0200 m). Since there are 50 turns, the total torque is $N = 50$ times greater than that given by Equation 19-8:

$$\tau = NiAB \sin \phi$$
$$= (50)(0.500 \text{ A})(0.0200 \text{ m})^2(1.50 \times 10^{-2} \text{ T})(1)$$
$$= 1.50 \times 10^{-4} \text{ T·A·m}^2 = 1.50 \times 10^{-4} \text{ N·m}$$

(Recall that 1 T = 1 N/(A·m), so 1 T·A·m² = 1 N·m.)

This is the net torque on the loop, so from Equation 9-16 the magnitude of the loop's angular acceleration is

$$\alpha = \frac{\tau}{I} = \frac{1.50 \times 10^{-4} \text{ N} \cdot \text{m}}{4.00 \times 10^{-6} \text{ kg} \cdot \text{m}^2}$$
$$= 37.5 \text{ rad/s}^2$$

(If the torque is in N·m and the moment of inertia is in kg·m², the angular acceleration in Equation 9-16 is in rad/s².)

When released, the loop will begin to rotate in the clockwise direction, as shown above. Note that as the loop rotates, the angle ϕ and the value of sin ϕ will decrease, so the torque and angular acceleration will decrease: This is *not* a situation with constant angular acceleration.

(b) Find the angular acceleration when the loop has rotated 90° from its initial orientation.

When the loop has rotated 90°, the normal to the loop is in the same direction as the magnetic field, so $\phi = 0$ and sin $\phi = 0$. From Equation 19-8 it follows that there is zero torque on the loop at this point in its rotation. We can also see this using the right-hand rule for the magnetic forces on the wires of the loop: These forces do not exert any torque around the axis. Therefore, this orientation represents an equilibrium position for the loop, and the angular acceleration is zero. Note that the loop will be in motion as it passes through this position (there has been an angular acceleration ever since the loop was released). As a result, the loop doesn't stop at this position but keeps on rotating.

(c) Find the angular acceleration when the loop has rotated by 180° from its initial orientation.

At the 180° position, the normal to the loop is again perpendicular to the direction of the magnetic field, just as in part (a). So again the angle $\phi = 90°$, and again the magnitude of the angular acceleration is $\alpha = 37.5 \text{ rad/s}^2$. However, since the loop has been flipped over relative to its original orientation, the directions of the forces are reversed. So the torque and angular acceleration are now *counterclockwise*. In fact, the angular acceleration has been increasingly counterclockwise ever since the loop moved past the position in (b). Since passing that position, the loop has been rotating in a clockwise direction but has been slowing down due to the counterclockwise angular acceleration.

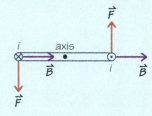

Reflect

The motion of the loop in this example should remind you of the motion of a pendulum (Section 12-5). When displaced from equilibrium and released, the pendulum will swing toward its equilibrium orientation (hanging straight down) but overshoot that equilibrium and swing to the other side of equilibrium. If there is no friction, the pendulum will keep swinging back and forth indefinitely. The same is true for this current loop: If the axis on which it rotates is frictionless, it will oscillate back and forth between the orientation in part (a) and the orientation in part (c).

In an electric motor we want the coil to continue rotating in the same direction, not oscillate back and forth like the coil in Example 19-4. To make this happen, the connection between the coil and the source of emf is arranged so that when the coil is at its equilibrium position (as in part (b) of Example 19-4), the direction of the current *reverses*. As a result, when the coil moves past this equilibrium position the torque is in the same direction as the rotation, and the rotation continues to speed up. The current reverses again after another half-rotation, so the torque is always in the same direction.

With this arrangement the coil would continue to gain rotational speed without limit if there were no other torques acting on it. In practice there are other torques that oppose the rotation, and the rotational speed reaches an upper limit. That's the case for the electric fan in Figure 19-13a: Air resistance on the fan blades increases as the fan turns faster, and the fan speed stabilizes when the torque due to air resistance just balances the torque of the electric motor.

GOT THE CONCEPT? 19-6 Magnetic Torque

? Suppose the number of turns in the current loop of Example 19-4 were increased from 50 to 100. Would this make the maximum angular acceleration of the loop (a) four times greater, (b) twice as great, (c) 1/2 as great, (d) 1/4 as great, or (e) none of these?

TAKE-HOME MESSAGE FOR Section 19-6

✔ There is zero net force on a current loop in a uniform magnetic field.

✔ A uniform magnetic field can exert a torque on a current loop. The torque is maximum if the normal to the plane of the loop is perpendicular to the magnetic field direction.

19-7 Ampère's law describes the magnetic field created by current-carrying wires

So far in our discussion of magnetism, we've looked at the forces and torques that a magnetic field exerts on a current-carrying wire. But current-carrying wires also *produce* magnetic fields (see Figure 19-1c and Figure 19-4). As we described in Section 19-2, all magnetic fields are produced by electric charges in motion, whether it's a current in the coils of the electromagnet shown in Figure 19-1c, electrons in motion within the atoms of an iron bar magnet, or electric currents in the human brain as shown in Figure 19-16. To complete our understanding of magnetic forces, we need to be able to calculate the magnetic field produced by an arrangement of charges in motion. This is much like Chapter 16, where we needed to learn how to calculate the electric field due to an arrangement of charges in order to complete our understanding of electric forces.

We saw in Section 16-5 that the electric field that is due to a charge distribution is just the vector sum of the electric fields due to all of the charges in the distribution. These calculations can be rather challenging unless the charge distribution is very simple. When we look at the analogous problem of finding the *magnetic* field due to a distribution of moving charges, we find that the problem is even more complicated. That's because the magnetic field due to even a single moving charged particle is itself rather complex: The field does not point directly away from or toward the moving charge, but in a direction perpendicular to the velocity vector. Rather than looking at how to do calculations of this kind, we'll look at the result for just one important situation, the magnetic field due to a long, straight, current-carrying wire. We'll then see an alternative approach for calculating a magnetic field using *Ampère's law*. This law is a very powerful one, but like Gauss's law for electric fields (Sections 16-6 and 16-7), it can be used for field calculations only in certain simple situations. We'll conclude the section with a look at the magnetic field produced by a current loop, as well as some of the applications of this field.

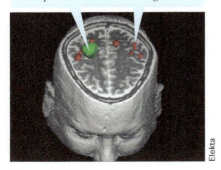

Electric currents in the human brain generate weak magnetic fields. The colors in this magnetoencephalogram represent the strength of the magnetic field produced in different regions.

Figure 19-16 **Currents as sources of magnetic field** Electric currents in biological systems produce magnetic fields.

A Long, Straight Wire and Ampère's Law

Consider a very long, straight wire that carries a constant current i. Experiment and calculation both show that the magnetic field due to this current has the properties shown in Figure 19-17. Note that the wire and the magnetic field that it produces have *cylindrical symmetry*: The wire and the field pattern look

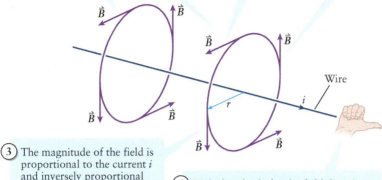

① The magnetic field lines due to a long, straight, current-carrying wire are circles that lie in planes perpendicular to the axis of the wire.

② The magnetic field vectors are tangential to the circular magnetic field lines and perpendicular to the axis of the wire.

Figure 19-17 Magnetic field of a long, straight, current-carrying wire The magnitude of this field is given by Equation 19-9.

③ The magnitude of the field is proportional to the current i and inversely proportional to the distance r from the axis of the wire.

④ Right-hand rule for the field direction: If you point your right thumb in the direction of the current and curl your fingers, the magnetic field curls around the field lines in the direction of the curled fingers of your right hand.

exactly the same if you rotate the wire around its length. The magnitude of the field is given by

Magnitude of the magnetic field due to a long, straight wire

Permeability of free space

$$B = \frac{\mu_0 i}{2\pi r}$$

Current in the wire

Distance from the wire to the location where the field is measured

Magnitude of magnetic field due to a long, straight wire (19-9)

Equation 19-9 and the field properties shown in Figure 19-17 are strictly correct only if the wire is infinitely long. But they are very good approximations if the distance r is small compared to the length of the wire.

The constant μ_0 in Equation 19-9, called the **permeability of free space** for historic reasons, plays a role in magnetism that's comparable to the role of the constant ε_0, the *permittivity* of free space, in electricity (see Section 16-6). Its value determined experimentally is $\mu_0 = 1.25663706 \times 10^{-7}$ T·m/A. Remarkably, this is equal to $4\pi \times 10^{-7}$ T·m/A to nine significant figures. This is not a coincidence, but a result of the way that the ampere was defined prior to 2018. In the old definition of the ampere, at a distance of 1 m from a long, straight wire carrying a current of 1 A, the magnetic field strength was exactly

$$B = \frac{\mu_0 i}{2\pi r} = \frac{(4\pi \times 10^{-7} \text{ T·m/A})(1 \text{ A})}{2\pi(1 \text{ m})} = 2 \times 10^{-7} \text{ T}$$

The current definition of 1 ampere is 1 coulomb per second, with the coulomb defined in terms of the magnitude of charge e on a proton or electron (see Section 16-2). With this definition μ_0 is not exactly equal to $4\pi \times 10^{-7}$ T·m/A, but is close enough that we can safely use $\mu_0 = 4\pi \times 10^{-7}$ T·m/A in calculations.

WATCH OUT! Remember that vectors are straight, never curved.

! When drawing the magnetic field around a long, straight wire, it may be tempting to draw curved arrows to represent how the magnetic field lines curl around the wire. But a vector always denotes a *single* direction and so *cannot* be curved. At any point along a magnetic field line, the direction of the field is always along the *tangent* to the field line, as shown in Figure 19-17.

Figure 19-18 Magnetic field of a solenoid Compare this illustration to the photograph of a solenoid and its field in Figure 19-4a.

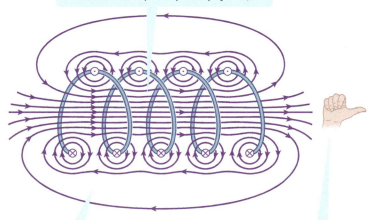

① The magnetic field in the interior of a long, straight solenoid is essentially uniform (the field lines are very nearly evenly spaced).

② The magnetic field outside the solenoid is very weak (the field lines are far apart).

③ Right-hand rule for the field direction: If you curl the fingers of your right hand around the solenoid in the direction of the current, the magnetic field inside the solenoid points in the direction of your right thumb.

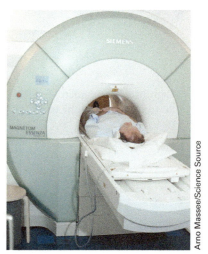

Figure 19-19 **Magnetic resonance imaging (MRI)** The medical imaging technique known as MRI requires that the patient be immersed in a strong, uniform magnetic field. In the MRI device this is done by having the patient lie inside a solenoid.

The magnetic field around current-carrying wires with other geometries is much more complicated. As an example, Figure 19-18 shows some of the magnetic field lines for a straight helical coil of wire. Such a coil is called a **solenoid**. Close to an individual wire, the field lines resemble those around the long, straight wire shown in Figure 19-17. In the space outside the solenoid, the magnetic field is very weak, as you can see from the large spacing between adjacent field lines. (Recall from Section 16-5 that the same is true for electric fields: Where field lines are far apart, the field magnitude is small.) But in the interior of the solenoid, the magnetic field lines are close together and nearly evenly spaced, indicating that the magnetic field there is strong and nearly uniform.

BioMedical This property of solenoids explains why a conventional magnetic resonance imaging (MRI) scanner is in the form of a long tube inside which the patient lies (Figure 19-19). This tube is actually the interior of a solenoid like that shown in Figure 19-18, so the patient is bathed in a strong, uniform magnetic field—which is just what MRI requires. (In Chapter 27 we'll learn more about the physics of MRI.)

Can we use Equation 19-9 to calculate the field inside a solenoid? Not directly, no, because the wires that make up the solenoid are not straight. But we can use Equation 19-9 to illustrate a useful principle about magnetic fields and their sources, and then use that principle to determine the solenoid field. Imagine that we draw a circle of radius r around a long, straight wire as shown in Figure 19-20. Imagine further that we break the circle into a number of segments of length $\Delta\ell$. If $\Delta\ell$ is sufficiently small, we can treat each segment as being straight. Then for each segment take the component $B_\parallel$ of magnetic field parallel to that segment and multiply it by the segment length $\Delta\ell$. If we add up the values of these products for every segment in the circle, the result is a quantity called the **circulation** of the magnetic field around the circle:

$$\text{Circulation} = \sum B_\parallel \Delta\ell \tag{19-10}$$

For the circle shown in Figure 19-20, Equation 19-9 tells us that $B_\parallel$ is equal to $\mu_0 i/(2\pi r)$ at every point around the circle. That's because the magnetic field is everywhere tangent to the circle and so parallel to a short segment of length on the circle. Therefore, the circulation of the magnetic field as defined by Equation 19-10 is

$$\text{Circulation} = \sum \left(\frac{\mu_0 i}{2\pi r}\right)\Delta\ell = \frac{\mu_0 i}{2\pi r}\sum \Delta\ell \tag{19-11}$$

In Equation 19-11 we've taken the quantity $\mu_0 i/(2\pi r)$ outside the sum because it has the same value everywhere around the circle and so in all terms of the sum.

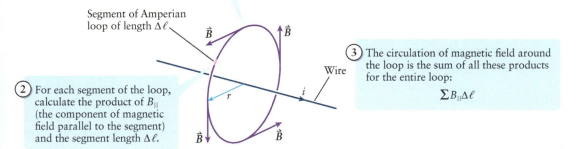

① We draw an Amperian loop that encircles the current-carrying wire. (This particular loop is a circle, and so coincides with a field line.)

Segment of Amperian loop of length $\Delta\ell$

$\vec{B}$

Wire

③ The circulation of magnetic field around the loop is the sum of all these products for the entire loop:

$$\sum B_{\parallel}\Delta\ell$$

② For each segment of the loop, calculate the product of $B_{\parallel}$ (the component of magnetic field parallel to the segment) and the segment length $\Delta\ell$.

Figure 19-20 Ampère's law and circulation Ampère's law states that the circulation of the magnetic field around a closed path (called an *Amperian loop*) is proportional to the current that passes through that path.

The quantity $\sum \Delta\ell$ is the sum of the lengths of all of the segments that make up the circle—that is, the circumference of the circle, which is equal to $2\pi r$. So we can rewrite Equation 19-11 as follows:

$$\text{Circulation} = \frac{\mu_0 i}{2\pi r}(2\pi r) = \mu_0 i \qquad (19\text{-}12)$$

The circulation as given by Equation 19-12 does *not* depend on the radius r of the circle. (If the circle is larger, the magnetic field has a smaller magnitude, but it takes more segments of length $\Delta\ell$ to go all the way around the circle.) Remarkably, the same result holds true even if we draw a noncircular path around the wire: The circulation around *any* path that encloses the wire is equal to $\mu_0 i$. This is an example of **Ampère's law**, which was discovered by the French physicist André-Marie Ampère (pronounced "ahm-pair") in 1826:

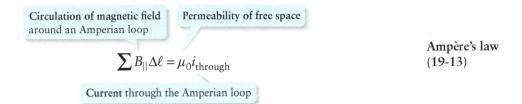

Circulation of magnetic field around an Amperian loop

Permeability of free space

$$\sum B_{\parallel}\Delta\ell = \mu_0 i_{\text{through}}$$

Ampère's law (19-13)

Current through the Amperian loop

An **Amperian loop** is simply a closed path in space; the circle in Figure 19-20 is an example. The subscript "through" reminds us that the right-hand side of Equation 19-13 should include only current that passes through the interior of the Amperian loop.

It can be shown that Ampère's law is true for *any* magnetic field and *any* Amperian loop, no matter what the geometry of the current that produces the magnetic field. (The proof is beyond our scope.) However, it's only in certain very symmetric situations that Ampère's law helps us calculate the value of magnetic field. (This is much like Gauss's law for electric fields. We saw in Sections 16-6 and 16-7 that Gauss's law is always true but is useful for electric field calculations only in certain situations.) Happily, one such situation is the field of a solenoid.

Using Ampère's Law: Magnetic Field of a Solenoid

As Figure 19-18 shows, inside the solenoid the fields from each loop add to create a total field that runs generally parallel to the central axis, particularly at points close to the axis and far from the ends. Outside the solenoid the fields nearly cancel, especially far from the ends. The fields outside the solenoid cancel more completely when the length of the solenoid is large compared to its diameter and when the coils of the solenoid are tightly wound. If there are N turns or *windings* of wire in the solenoid, the *winding density* is $n = N/L$. (The units of n are windings per meter.) Let's see how to use Ampère's law to find the magnetic field of a long, narrow, tightly wound (ideal) solenoid of length L, diameter D, and winding density n carrying a current i. In particular, we'll find the field inside the coil and far from the ends, where we expect the field to be relatively uniform.

To use Ampère's law we must first select an Amperian loop *through* which current flows and *around* which we can calculate the magnetic field. The point at which we want to determine the field—in this case, a point inside the solenoid—must lie on the Amperian loop. Figure 19-21 shows our choice of Amperian loop on a cutaway view of a section of the solenoid. As in Figure 19-18, the current is shown coming out of the page along the top part of the windings and going into the page along the bottom. We picked a rectangle as the Amperian loop and positioned it so that the bottom of the loop is parallel to the uniform field inside the solenoid. The left-hand side of Ampère's law, Equation 19-13, can then be written as

(19-14)
$$\sum B_{\parallel}\Delta\ell = B_{\parallel 1}\ell_1 + B_{\parallel 2}\ell_2 + B_{\parallel 3}\ell_3 + B_{\parallel 4}\ell_4$$

Side 4 is in the region of zero magnetic field outside the solenoid, so $B_{\parallel 4} = 0$ and $B_{\parallel 4}\ell_4 = 0$. Side 1 runs partially through this region of zero field, so $B_{\parallel 1} = 0$ there; inside the solenoid the field is nonzero but is perpendicular to side 1, so $B_{\parallel 1} = 0$ there as well. The same is true for side 3, so in Equation 19-14, $B_{\parallel 1}\ell_1 = B_{\parallel 3}\ell_3 = 0$. The only side of the Amperian loop that makes a nonzero contribution to Equation 19-14 is side 2. Here the magnetic field of magnitude B is parallel to the side of length ℓ_2, so $B_{\parallel 2} = B$ and Equation 19-14 becomes

(19-15)
$$\sum B_{\parallel}\Delta\ell = B\ell_2$$

On the right-hand side of the Ampère's law equation, we need to evaluate i_{through}, the current that passes through the loop. This quantity is *not* just the current in the solenoid i, because every winding of the coil that passes through the loop brings a contribution i to i_{through}. The length of solenoid enclosed by the loop is ℓ_2, and there are n windings per meter, so there are $n\ell_2$ windings that pass through the loop. Each winding carries current i, so the total current through the loop is $n\ell_2$ multiplied by i:

(19-16)
$$i_{\text{through}} = n\ell_2 i$$

We can now substitute Equations 19-15 and 19-16 into the two sides of Equation 19-13 for Ampère's law:

$$B\ell_2 = \mu_0(n\ell_2 i)$$

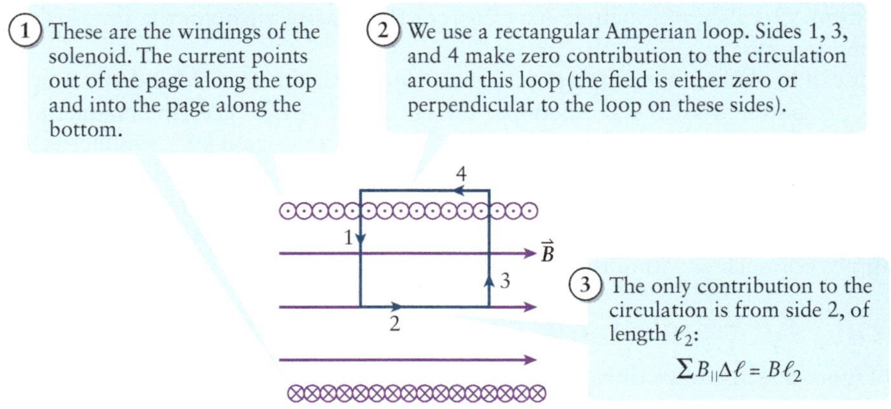

① These are the windings of the solenoid. The current points out of the page along the top and into the page along the bottom.

② We use a rectangular Amperian loop. Sides 1, 3, and 4 make zero contribution to the circulation around this loop (the field is either zero or perpendicular to the loop on these sides).

Figure 19-21 Ampère's law and the field of a solenoid The Amperian loop shown here helps us to calculate the magnetic field inside a long, straight solenoid.

③ The only contribution to the circulation is from side 2, of length ℓ_2:
$$\sum B_{\parallel}\Delta\ell = B\ell_2$$

The length ℓ_2 of the Amperian loop cancels, and we are left with

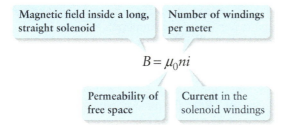

$$B = \mu_0 n i$$

Magnetic field inside a long, straight solenoid | Number of windings per meter

Permeability of free space | Current in the solenoid windings

Magnetic field inside a long, straight solenoid
(19-17)

In deriving Equation 19-17 we've assumed that the field inside the solenoid is perfectly uniform and parallel to the solenoid axis and that the field outside the solenoid is exactly zero. These assumptions are strictly valid only for an infinitely long solenoid. But Equation 19-17 is a good approximation at points near the middle of any solenoid, especially one whose length is large compared to its diameter.

Note that Ampère's law also tells us the *direction* of the magnetic field inside the solenoid. Just as for the case of a long, straight wire (Figure 19-17), point your right thumb in the direction of the current through the Amperian loop in Figure 19-21. This current points out of the page, so you should point your thumb in that direction. If you curl the fingers of your right hand, they will curl in a counterclockwise direction: from left to right below the enclosed windings, and from right to left above the enclosed windings. That's consistent with the direction of the magnetic field inside the solenoid, which is from left to right.

Equation 19-17 tells us that the more windings that can be packed into a length of the solenoid, the greater the field magnitude inside the solenoid. A typical solenoid used for electronic applications is therefore more likely to look like the one in Figure 19-22, with many layers of tightly packed windings, than the one shown in Figure 19-18.

In the following example we use Ampère's law to find the magnetic field due to a rather different distribution of current.

Figure 19-22 A real-life solenoid When the current to this solenoid is turned on, the iron rod that sticks out from its end experiences a strong magnetic force and is pulled into the solenoid. Such a device can be used to unlock a security door, as well as many other applications.

EXAMPLE 19-5 Magnetic Field Due to a Coaxial Cable

A coaxial cable consists of a solid conductor of radius R_1 surrounded by insulation, which in turn is surrounded by a thin conducting shell of radius R_2 made of either fine wire mesh or a thin metallic foil (Figure 19-23). The combination is enclosed in an outer layer of insulation. The inner conductor carries current in one direction, and the outer conductor carries it back in the opposite direction. For a coaxial cable that carries a constant current i, find expressions for the magnetic field (a) inside the inner conductor at a distance $r < R_1$ from its central axis, (b) in the space between the two conductors, and (c) outside the coaxial cable. Assume that the moving charge in the inner conductor is distributed uniformly over the volume of the conductor.

Figure 19-23 A coaxial cable Equal amounts of current flow in opposite directions in the inner and outer conductors of this cable.

Set Up

Both the inner conductor separately and the coaxial cable as a whole have the same cylindrical symmetry as a long, straight wire (Figure 19-17). So we expect that the field lines are circles concentric with the axis of the cable. Just as for the long, straight wire, this means that it's natural to choose circular paths concentric with the cable axis as the Amperian loops. To find the field in the three regions, we'll choose the radius r of the Amperian loop to be less than R_1 in part (a), between R_1 and R_2 in part (b), and greater than R_2 in part (c).

Ampère's law:

$$\sum B_\parallel \Delta\ell = \mu_0 i_{\text{through}} \quad (19\text{-}13)$$

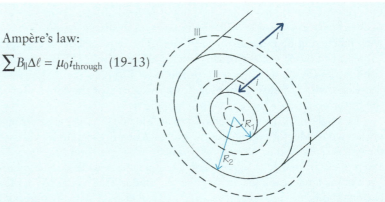

Dashed circles labeled I, II, and III: Amperian loops for parts (a), (b), and (c), respectively

Solve

(a) Find the field inside the inner conductor by using an Amperian loop of radius $r < R_1$.

Inside the inner conductor the magnetic field has magnitude B_{inner} and points tangent to the Amperian loop, so $B_\parallel = B_{inner}$. The left-hand side of the Ampère's law equation is

$$\sum B_\parallel \Delta\ell = B_{inner} \sum \Delta\ell = B_{inner}(2\pi r)$$

The Amperian loop encloses area πr^2, which is less than the cross-sectional area πR_1^2 of the inner conductor. The current through the loop is therefore a fraction $(\pi r^2)/(\pi R_1^2)$ of the total current i in the inner conductor:

$$i_{through} = i\left(\frac{\pi r^2}{\pi R_1^2}\right) = i\frac{r^2}{R_1^2}$$

Insert these into Equation 19-13 and solve for B_{inner}:

$$B_{inner}(2\pi r) = \mu_0 i \frac{r^2}{R_1^2}$$

$$B_{inner} = \mu_0 i \frac{r^2}{2\pi r R_1^2} = \frac{\mu_0 i r}{2\pi R_1^2}$$

(b) Find the field between the conductors by using an Amperian loop of radius r, where $R_1 < r < R_2$.

Between the two conductors the magnetic field of magnitude $B_{between}$ also points tangent to the Amperian loop, so as in part (a) the left-hand side of Equation 19-13 is

$$\sum B_\parallel \Delta\ell = B_{between}(2\pi r)$$

The Amperian loop encloses the entire inner conductor, so $i_{through} = i$. Insert these into Equation 19-13 and solve for $B_{between}$:

$$B_{between}(2\pi r) = \mu_0 i$$

$$B_{between} = \frac{\mu_0 i}{2\pi r}$$

(c) Find the field outside the cable by using an Amperian loop of radius $r > R_2$.

Just as in parts (a) and (b), outside the outer conductor the magnetic field of magnitude B_{outer} points tangent to the Amperian loop, so

$$\sum B_\parallel \Delta\ell = B_{outer}(2\pi r)$$

The Amperian loop encloses both conductors, each of which carries current i. Since the currents flow in opposite directions, the *net* current through the loop is $i_{through} = 0$. So Equation 19-13 tells us

$$B_{outer}(2\pi r) = \mu_0(0)$$
$$B_{outer} = 0$$

Reflect

Our result from (a) says that the magnetic field is zero at the center of the inner conductor ($r = 0$), then increases in direct proportion to r with increasing distance from the center. The field reaches its maximum value at the outer surface of the inner conductor ($r = R_1$). Between the conductors the field is inversely proportional to r, so the magnitude decreases with increasing distance from the center of the cable. Outside the outer conductor there is *zero* magnetic field.

Coaxial cables are often referred to as "shielded" cables. The arrangement of the two conductors eliminates the presence of stray magnetic fields outside the cable. The shielding also serves to isolate the inner conductor from external electromagnetic signals. You'll find a coaxial cable connected to the back of most television sets (it's the "cable" in the term "cable TV"); the signal carried by this cable involves an alternating current and hence a varying magnetic field rather than a steady one, but the shielding principle is the same.

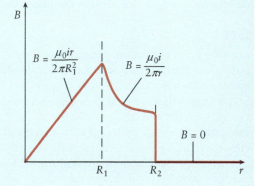

Magnetic Field of a Current Loop

An important special case for which Ampère's law is *not* helpful is the magnetic field produced by a current loop (a current-carrying wire bent into a circle). Figure 19-24 shows some of the magnetic field lines for such a loop.

Unlike the case for a long, straight wire (Figure 19-17) or for a coaxial cable (Example 19-5), the magnetic field does *not* have the same magnitude at all points on a field line: The magnitude B is greater where the lines are closer together. So choosing an Amperian loop that coincides with a field line will not give us a simple equation for the magnitude B, as was the case in Example 19-5. To calculate the magnetic field at a given point in this situation, it's necessary to find the contribution to the field due to each short segment of the loop, then add those contributions using vector arithmetic. Such a calculation is beyond our scope. Here's the result for the magnitude of the magnetic field along the axis of the current loop, labeled y in Figure 19-24:

$$B = \frac{\mu_0 i}{2} \frac{R^2}{(R^2 + y^2)^{3/2}}$$

(19-18)

In Equation 19-18 i is the current in the loop, R is the radius of the loop, and y is the coordinate along the y axis, where $y = 0$ represents the plane of the loop. If we substitute $y = 0$ into Equation 19-18, we get the field magnitude at the very center of the loop. At points far from the loop, so y is much greater than R, we can replace $R^2 + y^2$ with y^2 to good approximation. Equation 19-18 then becomes

$$B = \frac{\mu_0 i}{2} \frac{R^2}{(y^2)^{3/2}} = \frac{\mu_0 i R^2}{2|y|^3}$$

(19-19)

(We've added the absolute value signs because y can be positive or negative, but the field magnitude B must be positive.) So at large distances from a current loop, the magnetic field is inversely proportional to the *cube* of the distance from the loop.

This result is reminiscent of the *electric* field of an electric dipole (a combination of a positive charge q and a negative charge $-q$): We found in Example 16-7

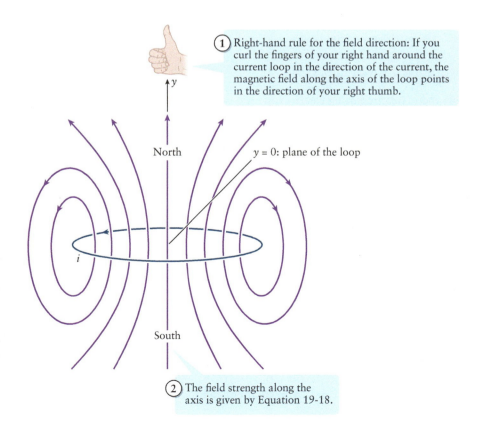

1) Right-hand rule for the field direction: If you curl the fingers of your right hand around the current loop in the direction of the current, the magnetic field along the axis of the loop points in the direction of your right thumb.

y

North $y = 0$: plane of the loop

South

2) The field strength along the axis is given by Equation 19-18.

i

Figure 19-24 Magnetic field due to a current loop This magnetic field pattern is similar to the electric field pattern of an electric dipole (Figure 16-14).

(Section 16-5) that at large distances from an electric dipole, the electric field due to that dipole is inversely proportional to the cube of the distance. The overall magnetic field pattern of a current loop also has some similarities to the electric field pattern of an electric dipole. (Compare Figure 19-24 and Figure 16-15; if you rotate Figure 16-15 clockwise 90°, the similarities will be more evident.)

In light of these similarities, we use the term "magnetic dipole" to refer to a current loop. The two "poles" of a magnetic dipole are the points just above and just below the loop, as Figure 19-24 shows. We call these poles north and south by analogy to the poles of a bar magnet: The magnetic field points away from the current loop at its north pole and points toward the current loop at its south pole, just as for a bar magnet (see Section 19-2).

Note that unlike the two charges that make up an electric dipole, the north and south poles of a current loop can never be separated: The current loop must always have two sides! As we discussed in Section 19-2, a permanent magnet such as a bar magnet acts as a magnetic dipole. That's because a permanent magnet is really just a collection of *atomic* current loops, each the result of electron motions within the atom. Their combined effect is the same as electrons moving around the circular loop of wire in Figure 19-24. The poles of such a magnet can no more be separated than can the two sides of a current loop.

Earth's magnetic field is nearly that of a dipole, with the axis of the field tilted about 9° from Earth's rotation axis (Figure 19-2). The magnetic field is produced because molten material in the outer regions of Earth's core is in a state of continuous motion, and this motion gives rise to electric currents that generate the field. The photo that opens this chapter illustrates one dynamic consequence of our planet having a magnetic field.

BioMedical

Anyone who uses a compass to navigate makes use of Earth's magnetic field, which points generally from south to north. Other living organisms also take advantage of Earth's field to guide them from location to location. Pigeons, honeybees, and sea turtles, among others, rely to some extent on an internal magnetic compass to navigate. Sea turtles, for example, have been observed to travel hundreds of kilometers and still find their way back to their nesting sites along relatively direct paths. Yet when the turtles are transported away from their nests after a magnet has been attached to their heads, they take wildly circuitous routes back to the nesting site. The field of the attached magnet clearly disrupts the turtles' ability to determine their position using Earth's magnetic field.

Magnetic Materials

While there are circulating electrons within every kind of atom, not all materials have the same magnetic properties. In some materials there is a net rotation of electron charge within the atom, so each atom behaves like a current loop. In most cases these atomic current loops are randomly oriented, so their effects cancel out. But if the material is placed in a strong magnetic field, the atomic current loops experience a torque and align themselves with the magnetic field (see Figure 19-14b). As a result, the material behaves like a much larger current loop. If the magnetic field is turned off, random thermal motion will cause the atomic current loops to return to their original, nonaligned orientations. Materials that display this behavior are called **paramagnetic**. Everyday paramagnetic materials include aluminum and sodium. The net magnetic effect in a paramagnetic material is generally quite small; while an empty can made of (paramagnetic) aluminum acts like a current loop when brought next to a magnet, the magnetic force on the aluminum can is so small that a magnet can't pick it up.

In a handful of materials the interactions between adjacent atomic current loops are very strong. As a result, once the material is placed in a magnetic field the atomic current loops not only align with the field but can *remain* aligned after the field is turned off, leaving the material permanently magnetized. Iron is the most common of these materials, which are called **ferromagnetic** (Figure 19-25). ("Ferro" derives from the Latin word for iron.) Any permanent magnet is made of a ferromagnetic material. A permanent magnet can pick up objects made of

Andy Tay

0.5 μm

BioMedical **Figure 19-25**
Ferromagnetic materials in bacteria Several species of bacteria, called *magnetotactic bacteria*, contain tiny pieces of ferromagnetic material (either iron oxide or iron sulfide). These appear as dark dots in this microscope image. Magnetotactic bacteria actively swim along geomagnetic field lines and use these membrane-encased bits of ferromagnetic material to orient themselves in the swampy environments in which they are found.

a ferromagnetic material, such as a steel paper clip. The field of the permanent magnet causes the atomic current loops in the paper clip to align, making the paper clip a magnet itself. The magnetized paper clip is then attracted to the permanent magnet.

WATCH OUT! Earth is not a permanent magnet.

 Our planet's core is made primarily of iron and nickel, both of which are ferromagnetic materials. It's common to conclude from this that the core is magnetized like a permanent magnet and that this gives rise to our planet's magnetic field. However, this cannot be true. Any ferromagnetic material loses its magnetism if it is heated above a certain temperature specific to that material: This critical temperature is 773°C for iron and 354°C for nickel. The temperature in Earth's core is in excess of 4400°C, so the iron and nickel in the core do *not* act like ferromagnetic materials. Instead, Earth's magnetic field is caused by electric currents in the molten material that makes up the outer regions of the core.

Most materials are neither paramagnetic nor ferromagnetic because their atoms have zero net electron current. When placed in a magnetic field, a small amount of atomic current appears, but the current loops end up aligned in the direction *opposite* to what happens for paramagnetic or ferromagnetic materials. (This is a consequence of *electromagnetic induction*, which we'll discuss in Chapter 20.) As a result, these materials, called **diamagnetic**, are slightly repelled by magnets rather than being attracted. In most cases, however, the repulsion is very weak.

GOT THE CONCEPT? 19-7 Ampère's Law

? The field lines in Figure 19-24 are closed curves. If the distance around one such curve is L and the current in the loop is i, what is the average value around the closed curve of the component of magnetic field $B_\parallel$ parallel to the curve? (a) $\mu_0 i$; (b) $\mu_0 i/L$; (c) $2\mu_0 i$; (d) $2\mu_0 i/L$; (e) not enough information given to decide.

TAKE-HOME MESSAGE FOR Section 19-7

✔ Ampère's law relates the current through a wire to the magnetic field it generates.

✔ Current through a long, straight wire produces a magnetic field with circular field lines centered on the wire.

✔ Current through a solenoid (a straight, helical coil of wire) produces a relatively uniform magnetic field along its axis that is proportional to the winding density as well as the current.

✔ A current loop produces a more complicated magnetic field. A magnetic material can be thought of as a collection of atomic current loops.

19-8 Two current-carrying wires exert magnetic forces on each other

We've seen that a current-carrying wire experiences a force when placed in a magnetic field, and also that a current-carrying wire generates a magnetic field. Let's put these ideas together and look at the magnetic interaction between *two* current-carrying wires. (Note that these two wires do not exert *electric* forces on each other. That's because each wire has as much positive charge as negative charge and so is electrically neutral.)

Experiment shows that two parallel, straight wires carrying current in the same direction attract each other, and that two parallel, straight wires carrying current in opposite directions repel. Let's see why this is the case.

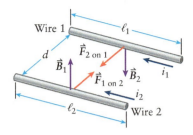

$\vec{B}_1$ = Magnetic field due to wire 1
at the position of wire 2

$\vec{B}_2$ = Magnetic field due to wire 2
at the position of wire 1

Figure 19-26 Magnetic forces between two current-carrying wires These two parallel wires carry current in the same direction and exert attractive magnetic forces on each other. If we reverse the direction of one of the currents, the forces become repulsive.

Figure 19-26 shows the situation. Wires 1 and 2 are long, straight, and parallel to each other and are separated by a distance d. The current i_1 in wire 1 sets up a magnetic field $\vec{B}_1$ at the position of wire 2, which carries current i_2. From Equation 19-9 the magnitude B_1 of this field is

$$(19\text{-}20) \qquad B_1 = \frac{\mu_0 i_1}{2\pi d}$$

The right-hand rule for the field produced by a long, straight wire (Section 19-7) tells us that at the position of wire 2, $\vec{B}_1$ points upward and perpendicular to wire 2. To find the direction of the force $\vec{F}_{1 \text{ on } 2}$ that this field exerts on wire 2, use the right-hand rule for the direction of the magnetic force on a current-carrying wire (Section 19-5): This tells us that $\vec{F}_{1 \text{ on } 2}$ points toward wire 1, so the force attracts wire 2 to wire 1. The magnitude of the force on wire 2, of length ℓ_2, is given by Equation 19-5 with $\theta = 90°$ (since the direction of the current in wire 2 is perpendicular to the direction of $\vec{B}_1$):

$$(19\text{-}21) \qquad F_{1 \text{ on } 2} = i_2 \ell_2 B_1 \sin 90° = i_2 \ell_2 B_1$$

If we substitute B_1 from Equation 19-20 into Equation 19-21, we get

$$(19\text{-}22) \qquad F_{1 \text{ on } 2} = i_2 \ell_2 \left(\frac{\mu_0 i_1}{2\pi d} \right) = \frac{\mu_0 i_1 i_2 \ell_2}{2\pi d}$$

The force per unit length on wire 2 is $F_{1 \text{ on } 2}$ (given by Equation 19-22) divided by the length ℓ_2 of wire 2:

$$(19\text{-}23) \qquad \text{Magnetic force per unit length exerted by wire 1 on wire 2} = F_{1 \text{ on } 2}/\ell_2 = \frac{\mu_0 i_1 i_2}{2\pi d}$$

We can use the same procedure to find the magnetic force per unit length that wire 2 exerts on wire 1. The field $\vec{B}_2$ that the current i_2 in wire 2 produces at the position of wire 1 has magnitude $B_2 = \mu_0 i_2/(2\pi d)$ and points *downward* in Figure 19-26. From the right-hand rule for the force on a current-carrying wire, the force $\vec{F}_{2 \text{ on } 1}$ on wire 1 points toward wire 2 (the force is attractive); its magnitude is $F_{2 \text{ on } 1} = i_1 \ell_1 B_2 = i_1 \ell_1 [\mu_0 i_2/(2\pi d)] = \mu_0 i_1 i_2 \ell_1/(2\pi d)$, where ℓ_1 is the length of wire 1. The force per unit length on wire 1 is then $F_{1 \text{ on } 2}$ divided by ℓ_1:

$$(19\text{-}24) \qquad \text{Magnetic force per unit length exerted by wire 2 on wire 1} = F_{2 \text{ on } 1}/\ell_1 = \frac{\mu_0 i_1 i_2}{2\pi d}$$

The force magnitudes per unit length in Equations 19-23 and 19-24 are equal, and the forces $\vec{F}_{1 \text{ on } 2}$ and $\vec{F}_{2 \text{ on } 1}$ are opposite in direction. That's just what we would expect from Newton's third law.

What changes if we reverse the direction of the current i_1 in wire 1? The force *magnitudes* given by Equations 19-23 and 19-24 won't be affected, but the force *directions* will be. This will reverse the direction of the magnetic field $\vec{B}_1$ that wire 1 produces at the position of wire 2, and so will reverse the direction of the force $\vec{F}_{1 \text{ on } 2}$ on wire 2. So in this case wire 2 will be pushed away from wire 1 (it will be repelled). Reversing the direction of i_1 will also reverse the direction of the force $\vec{F}_{2 \text{ on } 1}$ that wire 2 exerts on wire 1, so this force will push wire 1 away from 2 (again, it will be repelled). So we conclude that

Two parallel current-carrying wires attract each other if they carry current in the same direction, and repel each other if they carry current in opposite directions.

EXAMPLE 19-6 Wires in a Computer

The two long, straight wires that run along the back of a computer case to power the cooling fan carry 0.110 A in opposite directions. The wires are separated by 5.00 mm. (a) Find the force per unit length (magnitude and direction) that these wires exert on each other. (b) The mass per unit length of the wire is 5.00×10^{-3} kg/m. What acceleration does one of the wires experience due to this force?

Set Up

We'll use Equation 19-23 to find the force per unit length that one wire exerts on the other. (As we saw with Equation 19-24, the force per unit length has the same magnitude for either wire.) If we assume that this force equals the net external force on the wire, we can use Newton's second law to calculate the acceleration of the wire.

Magnetic force per unit length exerted by wire 1 on wire 2:

$$F_{1 \text{ on } 2}/\ell_2 = \frac{\mu_0 i_1 i_2}{2\pi d} \qquad (19\text{-}23)$$

Newton's second law:

$$\sum \vec{F}_{\text{ext}} = m\vec{a} \qquad (4\text{-}2)$$

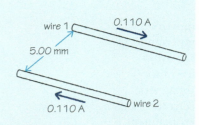

Solve

(a) The currents are in opposite directions, so the force is repulsive (it pushes the two wires apart). Use Equation 19-23 to find the magnitude of the force per unit length.

The two wires are separated by $d = 5.00$ mm $= 5.00 \times 10^{-3}$ m and carry currents with the same magnitude: $i_1 = i_2 = 0.110$ A. The force per unit length on either wire is

$$= \frac{(4\pi \times 10^{-7} \text{ T·m/A})(0.110 \text{ A})(0.110 \text{ A})}{2\pi(5.00 \times 10^{-3} \text{ m})}$$

$$= 4.84 \times 10^{-7} \text{ T·A} = 4.84 \times 10^{-7} \text{ N/m}$$

(Recall that $1 \text{ T} = 1 \text{ N}/(\text{A·m})$.)

(b) Use Newton's second law to find the acceleration of the wire.

Our result for part (a) says that a 1-m length of wire would experience a force of magnitude 4.84×10^{-7} N. Since this wire has mass per unit length 5.00×10^{-3} kg/m, a 1-m length would have mass 5.00×10^{-3} kg. The acceleration is

$$a = |\vec{a}| = \frac{|\sum \vec{F}_{\text{ext}}|}{m} = \frac{4.84 \times 10^{-7} \text{ N}}{5.00 \times 10^{-3} \text{ kg}}$$

$$= 9.68 \times 10^{-5} \text{ m/s}^2$$

Reflect

The force and acceleration are both very gentle, so the effect on these wires will be almost imperceptible. In applications with very large currents, however, the magnetic forces between conductors can be substantial.

GOT THE CONCEPT? 19-8   Forces on a Current-Carrying Coil

(?) A very flexible helical coil is suspended as shown in Figure 19-27. What will happen when a sizable current i is sent through the coil? (a) The coils will be pulled together. (b) The coils will be pushed apart. (c) Some of the coils will be pulled together, while others will be pulled apart. (d) There will be no net effect on the coils.

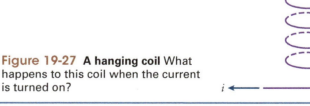

Figure 19-27 A hanging coil What happens to this coil when the current is turned on?

TAKE-HOME MESSAGE FOR   Section 19-8

✔ Two wires attract each other when carrying current in the same direction and repel each other when carrying currents in opposite directions.

✔ The forces per unit length on the wires are equal in magnitude and opposite in direction, exactly as required by Newton's third law.

Key Terms

Ampère's law	magnet	normal
Amperian loop	magnetic dipole	paramagnetic
circulation	magnetic field	permeability of free space
current loop	magnetic force	right-hand rule
diamagnetic	magnetic poles	solenoid
electromagnetism	magnetism	tesla
ferromagnetic	mass spectrometer	

Chapter Summary

Topic	Equation or Figure	
Magnetism and magnetic forces: Magnetic forces are present whenever moving charged objects interact with each other. (By comparison, electric forces are present whenever charged objects interact with each other, whether moving or not.) A magnet is an object that contains charges in continuous motion. A magnet sets up a magnetic field in the space around it; a second magnet responds to that field and can be attracted or repelled, depending on its orientation.	(a) These magnets attract. (b) These magnets repel.	(Figure 19-3 a/b)
Magnetic force on a moving charged particle: A single charged particle can experience a magnetic force when moving in a magnetic field. The magnitude of the force depends on both the speed of the particle and the direction of the particle relative to the magnetic field. The direction of the force is perpendicular to both the velocity $\vec{v}$ and the magnetic field $\vec{B}$ and is given by a right-hand rule.	Magnitude of the magnetic force on a moving charged particle Magnitude of the magnetic field $$F = |q|vB\sin\theta$$ **Angle** between the direction of the particle's velocity $\vec{v}$ and the direction of the magnetic field $\vec{B}$ **Magnitude** of the particle's **charge** **Speed** of the particle	(19-1)
	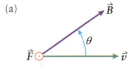 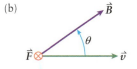 (a) If $q > 0$, the force vector in this situation points out of the page (as indicated by the circle with a dot). (b) If $q < 0$, the force vector in this situation points into the page (as indicated by the circle with an X).	(Figure 19-6)

Particle trajectories in a magnetic field: A charged particle moving in a magnetic field and subject to no other forces can move in a circular trajectory whose radius depends on its speed, mass, and charge as well as the magnetic field magnitude. This is the principle of the mass spectrometer.

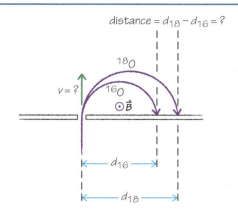

(Example 19-2)

Magnetic forces on current-carrying wires: If a wire carries a current and is placed in a magnetic field, it experiences a magnetic force. This force is the sum of the magnetic forces acting on the individual moving charges within the wire. The force magnitude depends on the amount of current and the orientation of the wire relative to the magnetic field; its direction is given by a right-hand rule.

Magnitude of the magnetic force on a current-carrying wire

Magnitude of the magnetic field (assumed uniform over the length of the wire)

$$F = i\ell B \sin \theta$$

Angle between the direction of the current and the direction of the magnetic field $\vec{B}$

(19-5)

Current in the wire

Length of the wire

The force $\vec{F}$ on a current-carrying wire in a magnetic field $\vec{B}$ is perpendicular to both $\vec{B}$ and the length of the wire. The direction is given by a right-hand rule.

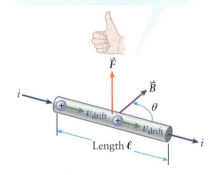

(Figure 19-10)

Magnetic torque on a current loop: A current-carrying loop of wire can experience a torque when placed in a magnetic field. The magnitude and direction of the torque depend on how the current loop is oriented relative to the direction of the magnetic field. Electric motors make use of this principle.

Magnitude of the magnetic torque on a current loop

Magnitude of the magnetic field

$$\tau = iAB \sin \phi$$

Angle between the normal to the plane of the loop and the direction of the magnetic field $\vec{B}$

(19-8)

Current in the loop

Area of the loop

(b) Side view

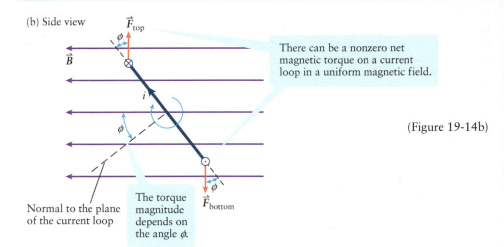

There can be a nonzero net magnetic torque on a current loop in a uniform magnetic field.

(Figure 19-14b)

Normal to the plane of the current loop

The torque magnitude depends on the angle ϕ.

Ampère's law and the field produced by moving electric currents: A long, straight, current-carrying wire produces a relatively simple magnetic field in the space around it. Ampère's law—which relates the circulation of magnetic field around a closed loop to the amount of current through that loop—can be used to find the magnetic field produced by currents with other simple geometries.

Magnitude of the magnetic field due to a long, straight wire

Permeability of free space

$$B = \frac{\mu_0 i}{2\pi r}$$

Current in the wire

Distance from the wire to the location where the field is measured

(19-9)

Circulation of magnetic field around an Amperian loop

Permeability of free space

$$\sum B_{\parallel} \Delta \ell = \mu_0 i_{\text{through}}$$

Current through the Amperian loop

(19-13)

① We draw an Amperian loop that encircles the current-carrying wire. (This particular loop is a circle, and so coincides with a field line.)

(Figure 19-20)

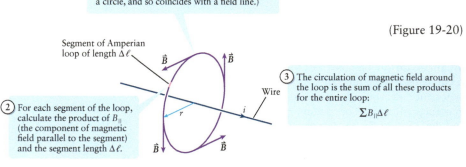

Segment of Amperian loop of length $\Delta \ell$

Wire

③ The circulation of magnetic field around the loop is the sum of all these products for the entire loop:

$$\sum B_{\parallel} \Delta \ell$$

② For each segment of the loop, calculate the product of $B_{\parallel}$ (the component of magnetic field parallel to the segment) and the segment length $\Delta \ell$.

Current loops and magnetic materials: A current loop is called a magnetic dipole because the magnetic field that it produces is similar to the electric field produced by an electric dipole. A permanent magnet is a material in which the atoms behave like individual current loops, many of which are oriented in the same direction so that their individual magnetic fields add to make a strong field.

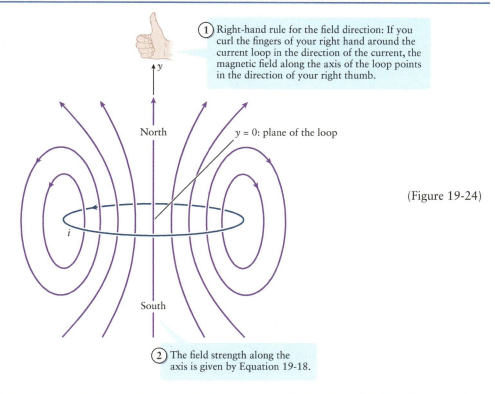

① Right-hand rule for the field direction: If you curl the fingers of your right hand around the current loop in the direction of the current, the magnetic field along the axis of the loop points in the direction of your right thumb.

North

y = 0: plane of the loop

South

(Figure 19-24)

② The field strength along the axis is given by Equation 19-18.

Force between current-carrying wires: Two parallel current-carrying wires exert magnetic forces on each other: The current in one wire produces a magnetic field, and the current in the other wire responds to that field. The two wires attract if the currents are in the same direction and repel if the currents are in opposite directions.

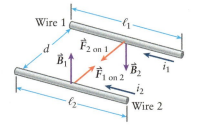

Wire 1

$\vec{F}_{2\text{ on }1}$

d

$\vec{B}_1$

ℓ_1

i_1

$\vec{F}_{1\text{ on }2}$

$\vec{B}_2$

i_2

ℓ_2

Wire 2

(Figure 19-26)

$\vec{B}_1$ = Magnetic field due to wire 1 at the position of wire 2

$\vec{B}_2$ = Magnetic field due to wire 2 at the position of wire 1

Answer to What do you think? Question

(c) Equation 19-1 (Section 19-3) gives the magnitude of the magnetic force on a charged particle moving in a magnetic field. The magnitude is proportional to sin θ, where θ is the angle between the magnetic field and the velocity of the particle. The sine function is greatest when $\theta = 90°$, in which case the particle is moving perpendicular to the field. This is in stark contrast to the *electric* force on a charged particle, which does not depend on the direction or magnitude of the particle's velocity.

Answers to Got the Concept? Questions

19-1 (b) Two charged objects exert *electric* forces on each other whether or not the objects are moving. But for these objects to exert *magnetic* forces on each other, *both* objects must be in motion. (One must be in motion to produce a magnetic field, and the other must be in motion to experience a force due to that field.) Since only one object is in motion, there is no magnetic force between the two objects.

19-2 (a) To apply the right-hand rule to the proton in Figure 19-7, start by pointing the fingers of your right hand in the direction of the proton's initial velocity. Orient your palm so that when you make a "slapping" motion with your hand, your palm points in the direction of the magnetic field (into the page). If you then stick your right thumb out straight, it points toward the top of the page. This is the direction of the force on the proton. This force makes the proton's trajectory bend upward in the figure.

19-3 (a), (c). Electrons have a negative charge, so the direction of the magnetic force each electron experiences is opposite to the direction your thumb points when applying the right-hand rule. For electron A, point the fingers of your right hand in the direction of the electron's initial velocity. Then orient your palm so that when you make a "slapping" motion with your hand, your palm points in the direction of the magnetic field (into the page). If you then stick your right thumb out straight, it points to the right, which means the electron feels a force to the left. So electron A feels a force toward the point at which the proton enters the field. For electron B, the right-hand rule has your right thumb pointing toward the bottom of the page, so the force on the electron is toward the top of the page. It does not bend toward the proton's entry point. The right-hand rule predicts the same direction for the magnetic force on electron C, but in this case a force toward the top of the page does bend the electron's trajectory toward the point where the proton enters the field. Finally, when you apply the right-hand rule to electron D, your right thumb sticks out to the left. For the negatively charged electron the magnetic force is therefore to the right, so the trajectory of electron D does not bend toward the proton's entry point into the field.

19-4 (c) Equation 19-3 shows that for a given speed v, the radius of the circular path followed by an ion in a mass spectrometer is $r = mv/(qB)$. This is directly proportional to the ratio of the ion mass m to its charge q. Compared to the value of the ratio m/q for a $^{16}O^+$ ion ($m = 16$ u, $q = e$), the value for an $^{16}O^{2+}$ ion is 1/2 as great ($m = 16$ u, $q = 2e$), the value for a $^{32}S^+$ ion is twice as great ($m = 32$ u, $q = e$), and the value for a $^{32}S^{2+}$ ion is the same ($m = 32$ u, $q = 2e$).

19-5 (a) force is zero, (b) $-z$ direction, (c) $+y$ direction, (d) $+z$ direction, (e) $-y$ direction, (f) force is zero. The force on a current-carrying wire is zero if the wire axis lies along the direction of the magnetic field, as in cases (a) and (f). In cases (b), (c), (d), and (e), we use the right-hand rule for the magnetic force on a current-carrying wire to find the direction of the force: Swing the extended fingers of your right hand, palm first, from the direction of the current to the direction of $\vec{B}$. Your extended right thumb then points in the direction of the force on the wire.

19-6 (e) Doubling the number of turns of wire will double the magnetic torque on the current loop. But this also doubles the mass and the moment of inertia of the current loop, so the angular acceleration—equal to the torque divided by the moment of inertia—will be unaffected. For a real electric motor, however, the moment of inertia is due partially to the mass of the coil and partially to the mass of what the motor is turning, such as the fan blades in Figure 19-13a. As a result, doubling the number of turns will increase the maximum torque by a factor of two but increase the moment of inertia by less than a factor of two, and the maximum angular acceleration will in fact increase.

19-7 (b) The net current through the closed curve is i, since the current loop passes once through the plane of the curve. From Ampère's law the circulation of the magnetic field around the closed curve is $\sum B_{\parallel}\Delta\ell = \mu_0 i_{\text{through}} = \mu_0 i$. The left-hand side of this equation is the average value of $B_{\parallel}$ multiplied by the total distance around the closed curve, so $(B_{\parallel})_{\text{average}}L = \mu_0 i$ and $(B_{\parallel})_{\text{average}} = \mu_0 i/L$.

19-8 (a) Each segment of the coil is a piece of wire and is attracted to the piece of wire in the coils directly above and below it (in each of which the current flows in the same direction). Each piece of wire is also repelled by the pieces of wire on the opposite side of the coil above it and the opposite side of the coil below it. But these pieces are at a greater distance d, so these repulsive forces are smaller than the attractive forces (see Equation 19-23). The net result is that the coils attract each other and so pull together.

Questions and Problems

In a few problems you are given more data than you actually need; in a few other problems you are required to supply data from your general knowledge, outside sources, or informed estimate.

Interpret as significant all digits in numerical values that have trailing zeros and no decimal points.

For all problems use $g = 9.80$ m/s^2 for the free-fall acceleration due to gravity. Neglect friction and air resistance unless instructed to do otherwise.

- •   Basic, single-concept problem
- ••   Intermediate-level problem; may require synthesis of concepts and multiple steps
- •••   Challenging problem
- Example   See worked example for a similar problem.

Conceptual Questions

1. • You are given three iron rods. Two of them are magnets but the third one is not. How could you use the two magnets to find that the third rod is not magnetized?

2. • In a lightning strike there is a negative charge moving rapidly from a cloud to the ground. In what direction is a lightning strike deflected by Earth's magnetic field?

3. • Physicists refer to crossed electric and magnetic fields as a *velocity selector*. In the same sense, the deflection of charged particles in a strong magnetic field perpendicular to their motion can be thought of as a *momentum selector*. Why?

4. • If a magnetic field exerts a force on moving charged particles, is it capable of doing work on the particles? Explain your answer.

5. •• A velocity selector consists of crossed electric and magnetic fields, with the magnetic field directed toward the top of the page. A beam of positively charged particles passing through the velocity selector from left to right is undeflected by the fields. (a) In what direction is the electric field? (left, right, toward the top of the page, toward the bottom of the page, into the page, out of the page) (b) The direction of the particle beam is reversed so that it travels from right to left. Is it deflected? If so, in what direction? (c) A beam of electrons (negatively charged) moving with the same speed is passed through from left to right. Is it deflected? If so, in what direction?

6. • A current-carrying wire is in a region where there is a magnetic field, but there is no magnetic force acting on the wire. How can this be?

7. • A long, straight current-carrying wire is placed in a cubic region that has a uniform magnetic field as shown in Figure 19-28. Does the force on the wire depend on the width of the magnetic field region? Explain your answer.

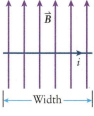

Figure 19-28 Problem 7

8. • How is it possible for an object that experiences no net magnetic force to experience a net magnetic torque?

9. • In telephone lines two wires carrying currents in opposite directions are twisted together. How does this reduce the net magnetic field created by the two wires?

10. • A power cord for an electronic device consists of two parallel straight wires carrying currents in opposite directions. Is there any force between them? Explain your answer.

11. • Parallel wires exert magnetic forces on each other. What about perpendicular wires? Explain your answer.

Multiple-Choice Questions

12. • The magnetic force on a moving charged particle
   A. depends on the sign of the charge on the particle.
   B. depends on the magnetic field at the particle's instantaneous position.
   C. is in the direction which is mutually perpendicular to the direction of motion of the charge and the direction of the magnetic field.
   D. is proportional both to the charge and to the magnitude of the magnetic field.
   E. is described by all of the above options, A through D.

13. • A proton traveling to the right enters a region of uniform magnetic field that points into the page. When the proton enters this region, it will be
   A. deflected out of the plane of page.
   B. deflected into the plane of page.
   C. deflected toward the top of the page.
   D. deflected toward the bottom of the page.
   E. unaffected in its direction of motion.

14. • An electron is moving northward in a magnetic field. The magnetic force on the electron is toward the northeast. What is the direction of the magnetic field?
   A. up
   B. down
   C. west
   D. south
   E. This situation cannot exist because of the orientation of the velocity and force vectors.

15. • A proton with a velocity along the $+x$ axis enters a region where there is a uniform magnetic field $\vec{B}$ in the $+y$ direction. You want to balance the magnetic force with an electric field so that the proton will continue along a straight line. The electric field should be in the
   A. $+x$ direction.
   B. $-x$ direction.
   C. $+z$ direction.
   D. $-z$ direction.
   E. $-y$ direction.

16. • A circular flat coil that has N turns, encloses an area A, and carries a current i has its central axis parallel to a uniform magnetic field $\vec{B}$ in which it is immersed. The net force on the coil is
   A. zero.
   B. $NiAB$.
   C. NiB.
   D. iBA.
   E. NiA.

17. • A circular flat coil that has N turns, encloses an area A, and carries a current i has its central axis parallel to a uniform

magnetic field $\vec{B}$ in which it is immersed. The net torque on the coil is
   A. zero.
   B. $NiAB$.
   C. NiB.
   D. iBA.
   E. NiA.

18. • A very long, straight wire carries a constant current. The magnetic field at distance d from the wire and far from its ends varies with distance d according to
   A. d^{-3}.
   B. d^{-2}.
   C. d^{-1}.
   D. d.
   E. d^2.

19. • A solenoid carries a current. If the radius of the solenoid were doubled and all other quantities remained the same, the magnetic field inside the solenoid would
   A. remain the same.
   B. be twice as strong as initially.
   C. be half as strong as initially.
   D. be one-quarter as strong as initially.
   E. be four times as strong as initially.

20. • Two parallel wires carry currents in opposite directions, as shown in Figure 19-29. Which of the following statements is correct?

Figure 19-29 Problem 20

   A. The force on the i_2 wire is upward, and the force on the i_1 wire is upward.
   B. The force on the i_2 wire is downward, and the force on the i_1 wire is upward.
   C. The force on the i_2 wire is upward, and the force on the i_1 wire is downward.
   D. The force on the i_2 wire is downward, and the force on the i_1 wire is downward.
   E. Neither wire experiences a net force.

21. •• Two current-carrying wires are perpendicular to each other. One wire lies horizontally with the current directed toward the east. The other wire is vertical with the current directed upward. What is the direction of the net magnetic force on the horizontal wire due to the vertical wire?
   A. east
   B. west
   C. south
   D. north
   E. zero force

Problems

19-1 Magnetic forces, like electric forces, act at a distance

19-2 Magnetism is an interaction between moving charges

19-3 A moving point charge can experience a magnetic force

22. • Convert the units for the following expressions for magnetic fields as directed (recall $1\ G = 10^{-4}$ T):
   A. $5.00\ T =$ _____ G
   B. $25{,}000\ G =$ _____ T
   C. $7.43\ mG =$ _____ μT
   D. $1.88\ mT =$ _____ G

23. • Determine the directions of the magnetic forces that act on positive charges moving in the magnetic fields as shown in Figure 19-30. Example 19-1

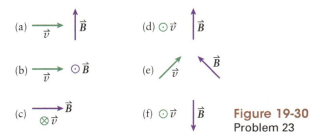

Figure 19-30
Problem 23

24. • Determine the direction of the missing vector, $\vec{v}$, $\vec{B}$, or $\vec{F}$, in the scenarios shown in Figure 19-31. All moving charges are positive. Example 19-1

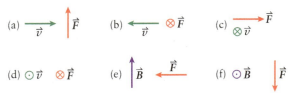

Figure 19-31 Problem 24

25. • A +1 C charge moving at 1 m/s makes an angle of 30° with a uniform, 1-T magnetic field. What is the magnitude of the magnetic force that the charge experiences? Example 19-1

26. • An electron is moving with a speed of 18 m/s in a direction parallel to a uniform magnetic field of 2.0 T. What are the magnitude and direction of the magnetic force on the electron? Example 19-1

27. • A proton P travels with a speed of 18 m/s toward the top of the page through a uniform magnetic field of 2.0 T directed into the page, as shown in Figure 19-32. What are the magnitude and direction of the magnetic force on the proton? Example 19-1

Figure 19-32 Problem 27

28. • A proton is propelled at 2×10^6 m/s perpendicular to a uniform magnetic field. If it experiences a magnetic force of 5.8×10^{-13} N, what is the magnitude of the magnetic field? Example 19-1

29. • An electron moves with a velocity of 1.0×10^7 m/s in the x–y plane at an angle of 30° above the $+x$ axis. There is a magnetic field of 3.0 T in the $+y$ direction. Calculate the magnetic force (magnitude and direction) on the electron. Example 19-1

30. •• There is a uniform magnetic field of magnitude 2.2 T in the $+z$ direction. Find the magnitude of the force on a particle of charge -1.2 nC if its velocity is (a) 1.0 km/s in the y–z plane in a direction that makes an angle of 40° with the z axis and (b) 1.0 km/s in the x–y plane in a direction that makes an angle of 40° with the x axis. Example 19-1

19-4 A mass spectrometer uses magnetic forces to differentiate atoms of different masses

31. •• A beam of protons is directed in a straight line along the $+z$ direction through a region of space in which there are crossed electric and magnetic fields. If the electric field is 500 V/m in the $-y$ direction and the protons move at a constant speed of 10^5 m/s, what must be the magnitude and direction of the magnetic field such that the beam of protons continues along its straight-line trajectory? Example 19-2

32. •• A beam of ions (each ion has a charge $q = -2e$ and a kinetic energy of 4.00×10^{-13} J) is deflected by the magnetic field of a bending magnet as shown in Figure 19-33. The radius of curvature of the beam is 20.0 cm, and the strength of the magnetic field is 1.50 T. (a) What is the mass of one of the ions in the beam? (b) Sketch the path for the given ions in the given magnetic field as a reference path. Then sketch a path for a more massive doubly ionized negative ion and a less massive doubly ionized positive ion, both with the same speed as the given ions, for comparison. Example 19-2

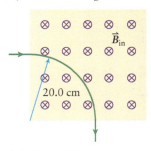

Figure 19-33 Problem 32

33. • An electron is in a region of space containing a uniform 2.00×10^{-5} T magnetic field. The electron's speed is 150 m/s and it travels perpendicularly to the field. Under these conditions, the electron undergoes circular motion. (a) Find the radius of the electron's path, and (b) the frequency of its motion. Example 19-2

19-5 Magnetic fields exert forces on current-carrying wires

34. • A 1.5-m length of straight wire experiences a maximum force of 2.0 N when in a uniform magnetic field that is 1.8 T. What current must be passing through it? Example 19-3

35. • A straight segment of wire 35.0 cm long carrying a current of 1.40 A is in a uniform magnetic field. The segment makes an angle of 53° with the direction of the magnetic field. If the force on the segment is 0.200 N, what is the magnitude of the magnetic field? Example 19-3

36. • A straight wire of length 0.50 m is conducting a current of 2.0 A and makes an angle of 30° with a 3.0-T uniform magnetic field. What is the magnitude of the force exerted on the wire? Example 19-3

37. • A wire of length 0.50 m is conducting a current of 8.0 A in the $+x$ direction through a 4.0-T uniform magnetic field directed parallel to the wire. What are the magnitude and direction of the magnetic force on the wire? Example 19-3

38. • A wire of length 0.50 m is conducting a current of 8.0 A toward the top of the page and through a 4.0-T uniform magnetic field directed into the page as shown in Figure 19-34. What are the magnitude and direction of the magnetic force on the wire? Example 19-3

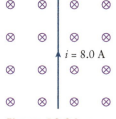

Figure 19-34 Problem 38

39. • A straight wire is positioned in a uniform magnetic field so that the maximum force on it is 4.0 N. If the wire is 80 cm long and carries a current that is 2.0 A, what is the magnitude of the magnetic field? Example 19-3

19-6 A magnetic field can exert a torque on a current loop

40. • A square loop 10.0 cm on a side with 100 turns of wire experiences a minimum torque of zero and a maximum torque of 0.0450 N·m in a uniform magnetic field. If the current in the loop is 2.82 A, calculate the magnetic field magnitude. Example 19-4

41. • What is the torque on a round loop of wire that carries a current of 1.00×10^2 A, has a radius of 10.0 cm, and whose plane makes an angle of 30.0° with a uniform magnetic field of 0.244 T? How does the answer change if the angle decreases to 10.0°? Increases to 50.0°? Example 19-4

42. •• A wire loop with 50 turns is formed into a square with sides of length s. The loop is in the presence of a 1.5-T uniform magnetic field that points in the negative y direction. The plane of the loop is tilted off the x axis by 15° (Figure 19-35). If 2.0 A of current flows through the loop and the loop experiences a torque of magnitude 0.035 N·m, what are the lengths of the sides s of the square loop? Example 19-4

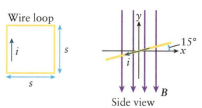

Wire loop

Side view

Figure 19-35 Problem 42

43. •• **Medical** MRI (magnetic resonance imaging) scans are often prohibited for patients with an implanted device such as a pacemaker. There are multiple reasons for this, including electromagnetic interference that confuses the device. Here we consider another potential issue: the physical torque on the device due to the magnetic field. Suppose a patient's pacemaker implant requires 2.0 mA of current to function. The leads of the device travel close together until they reach the heart and then split off to touch the top and bottom of the heart. We can approximate the size of the current loop as half the cross-sectional area of the heart. Treat the heart as a box that is 12 cm long and 8.0 cm wide. Most MRI machines use a magnetic field of 1.5 T. Calculate the maximum torque on the current loop. Example 19-4

19-7 Ampère's law describes the magnetic field created by current-carrying wires

44. • A long, straight wire carries current in the +z direction (out of the page). Determine the direction of the magnetic field due to the current at the points O, P, Q, and R (Figure 19-36). Example 19-5

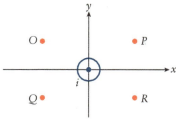

Figure 19-36 Problem 44

45. • A long, straight wire carries current in the +x direction. Determine the direction of the magnetic field due to the current at the points O, P, Q, and R (Figure 19-37). Example 19-5

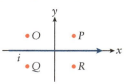

Figure 19-37 Problem 45

46. ••• Using Figure 19-38, derive an expression for the magnetic field at the point C located at the center of the two circular, current-carrying arcs and the connecting radial lines. Assume the radii of the small and large arcs are r_1 and r_2, respectively.

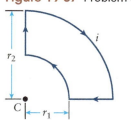

Figure 19-38 Problem 46

47. ••• Derive an expression for the magnetic field at point C at the center of the circular, current-carrying wire segments shown in Figure 19-39.

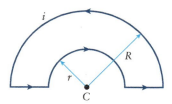

Figure 19-39 Problem 47

48. • Calculate the magnitude of the magnetic field at a perpendicular distance of 2.20 m from a long copper pipe that has a diameter of 2.00 cm and carries a current of 20.0 A. Example 19-5

49. • Jerry wants to predict the magnetic field that a high-voltage line creates in his apartment. A current of 100 A passes through a wire that is 5 m from his window. Calculate the magnitude of the magnetic field. How does the field compare to the magnitude of Earth's magnetic field of about 5×10^{-5} T in New York City? Example 19-5

50. • A solenoid with 25 turns per centimeter carries a current of 25 mA. What is the magnitude of the magnetic field in the interior of the coils?

51. •• You want to wind a solenoid that is 3.5 cm in diameter, is 16 cm long, and will have a magnetic field of 0.0250 T when a current of 3.0 A is in it. What total length of wire do you need?

52. •• A coaxial cable consists of a solid inner conductor of radius R_i, surrounded by a concentric outer conducting shell of radius R_o. Insulating material fills the space between the conductors.

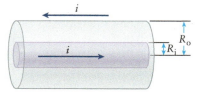

Figure 19-40 Problem 52

The inner conductor carries current to the right, and the outer conductor carries the same current to the left down the outer surface of the cable (Figure 19-40). Using Ampère's law, derive an expression for the magnitude of the magnetic field in three separate regions of space: inside the inner conductor, between the two conductors, and outside of the outer conductor. Example 19-5

19-8 Two current-carrying wires exert magnetic forces on each other

53. • If wire 1 carries 2.00 A of current north, wire 2 carries 3.60 A of current south, and the two wires are separated by 1.40 m, calculate the force (magnitude and direction) acting on a 1.00-cm section of wire 1 due to wire 2. Example 19-6

54. •• What is the net force (magnitude and direction) on the rectangular loop of wire that is 2.00 cm wide, 6.00 cm long, and located 2.00 cm from a long, straight wire that carries $i = 40.0$ A of current as shown in Figure 19-41? Assume a current of 20.0 A in the loop. Example 19-6

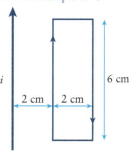

Figure 19-41 Problem 54

55. • The fasteners on overhead power lines are 50.0 cm long. What force must they be able to withstand if two high-voltage lines are 2.00 m apart, each carrying 2500 A in the same direction? Example 19-6

General Problems

56. • Honeybees can acquire a small net charge on the order of 1 pC as they fly through the air and interact with plants. Estimate the magnetic force on a honeybee due to Earth's magnetic field as the bee flies near the ground from east to west. Assume that Earth's magnetic field runs parallel to the surface of the ground from south to north with a magnitude of 5×10^{-5} T.

57. •• The magnetic field due to a current-carrying cylinder of radius 1 cm is measured at various points ($r < 1$ cm and $r > 1$ cm). Graph the magnitude of the magnetic field B as a function of r and use this graph to find the functional relationship between the magnetic field and the radial distance. *Hint:* The magnetic field will be described by two different functions—one for inside the cylinder and one for outside the cylinder. After making an initial graph, you may want to graph the data for the inside and outside separately.

r (m)	B (T)	r (m)	B (T)
0.001	0.00050	0.015	0.00353
0.002	0.00100	0.020	0.00250
0.003	0.00152	0.025	0.00200
0.004	0.00200	0.030	0.00180
0.005	0.00252	0.035	0.00143
0.006	0.00300	0.040	0.00125
0.007	0.00350	0.045	0.00110
0.008	0.00401	0.050	0.00103
0.009	0.00453	0.100	0.000502
0.010	0.00500		

58. • Horizontal electric power lines supported by vertical poles can carry large currents. Assume that Earth's magnetic field runs parallel to the surface of the ground from south to north with a magnitude of 0.50×10^{-4} T and that the supporting poles are 32 m apart. Find the magnitude and direction of the force that Earth's magnetic field exerts on a 32-m segment of power line (a wire) carrying 95 A if the current runs (a) from north to south, (b) from east to west, or (c) toward the northeast making an angle of 30.0° north of east. (d) Are any of the above forces large enough to have an appreciable effect on the power lines? Example 19-3

59. • A levitating train is three cars long (180 m) and has a mass of 100 metric tons (1 metric ton = 1000 kg). The current in the superconducting wires is about 500 kA, and even though the traditional design calls for many small coils of wire, assume for this problem that there is a 180-m-long wire carrying the current. Find the magnitude of the magnetic field needed to levitate the train. Example 19-3

60. •• An electron and a proton have the same kinetic energy upon entering a region of constant magnetic field, and their velocity vectors are perpendicular to the magnetic field. Suppose the magnetic field is strong enough to allow the particles to circle in the field. What is the ratio $r_{\text{proton}}/r_{\text{electron}}$ of the radii of their circular paths? Example 19-2

61. ••• In the mass spectrometer shown in Figure 19-42 a particle with charge $-e = -1.60 \times 10^{-19}$ C enters a region of magnetic field that has a strength of 0.00242 T and points into the page. The velocity of the particle is confirmed with a velocity selector. The electric field is 9.00×10^4 V/m (down) and the magnetic field is 0.00530 T (into page) in the velocity selector. If the radius of curvature of the particle is 4.00 cm, calculate the mass of the particle. Example 19-2

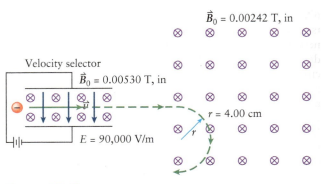

Figure 19-42 Problem 61

62. •• **Biology** The largest controlled magnetic field created has a strength of 1200 T. To see if such a strong magnetic field could pose health risks for nearby workers, calculate the maximum acceleration the field could produce on Na$^+$ ions (of mass 3.8×10^{-26} kg) in blood traveling through the aorta. The speed of blood is highly variable, but 50 cm/s is reasonable in the aorta. Does your result indicate that it would be dangerous to expose workers to such a large magnetic field? Example 19-1

63. •• During electrical storms, a bolt of lightning can transfer 10 C of charge in 2.0 μs (the amount and time can vary considerably). We can model such a bolt as a very long current-carrying wire. (a) What is the magnetic field 1.0 m from such a bolt? What is the field 1.0 km away? How do the fields compare with Earth's magnetic field? (b) Compare the fields in part (a) with the magnetic field produced by a typical household current of 10 A in a very long wire at the same distances from the wire as in (a). (c) How close would you have to get to the wire in part (b) for its magnetic field to be the same as the field produced by the lightning bolt at 1.0 km from the bolt? Example 19-5

64. • A straight wire carries a current of 8.00 A toward the top of the page. What are the magnitude and direction of the magnetic field at point P, which is 8.00 cm to the right of the wire, as shown in Figure 19-43? Example 19-5

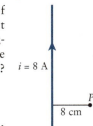

Figure 19-43 Problem 64

65. •• A long, straight wire carries a current as shown in Figure 19-44. A charged particle moving parallel to the wire experiences a force of 0.80 N at point P. Assuming the same charge and same velocity, what would be the magnitude of the magnetic force on the charge at point S? Example 19-5

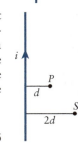

Figure 19-44 Problem 65

66. • Two long, straight wires parallel to the x axis are at $y = \pm 2.5$ cm (Figure 19-45). Each wire carries a current of 16 A in the $+x$ direction. Calculate the magnetic field on the y axis at (a) $y = 0$, (b) $y = 1.0$ cm, and (c) $y = 4.0$ cm. Example 19-5

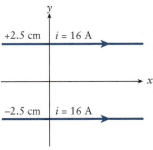

Figure 19-45 Problem 66

67. •• **Medical** Transcranial magnetic stimulation (TMS) is a noninvasive method to stimulate the brain using magnetic fields. It is used in treating strokes, Parkinson's disease, depression, and other physical conditions. In the procedure a circular coil is placed on the side of the forehead to generate a magnetic field inside the brain. Although values can vary, a typical coil would be about 15 cm in diameter and contain 250 thin circular windings. The magnetic field in the cortex (3.0 cm from the coil measured along a line perpendicular to the coil at its center) is typically 0.50 T. (a) What current in the coil is needed to produce the desired magnetic field inside the brain? (b) What is the magnetic field at the center of the coil at the forehead? (c) If the current needed in part (a) seems too large, how could you easily achieve the same magnetic field with a smaller current?

68. •• A wire of mass 40 g slides without friction on two horizontal conducting rails spaced 0.8 m apart. A steady current of 100 A is in the circuit formed by the wire and the rails. A uniform magnetic field of 1.2 T, directed into the plane of the drawing, acts on it. (a) In which direction in **Figure 19-46** will the wire accelerate? (b) What is the magnetic force on the wire? (c) How long must the rails be if the wire, starting from rest, is to reach a speed of 200 m/s? How would your answers differ if the magnetic field were (d) directed out of the page or (e) in the plane of the drawing, directed toward the top of the drawing? Example 19-3

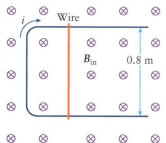

Figure 19-46 Problem 68

69. • A small 20-turn current loop with a 4.00-cm diameter is suspended in a region with a magnetic field of 1.00×10^3 G, with the plane of the loop parallel with the magnetic field direction. (a) What is the current in the loop if the torque exerted by the magnetic field on the loop is 4.00×10^{-5} N·m? (b) Describe the subsequent motion of the loop if it is allowed to rotate. Example 19-4

70. • A long, straight wire carries a current of 1.2 A toward the south. A second, parallel wire carries a current of 3.8 A toward the north and is 2.8 cm from the first wire. What is the magnitude of the magnetic force per unit length each wire exerts on the other? Example 19-6

71. ••• Two straight conducting rods, which are 1.0 m long, exactly parallel, and separated by 0.85 mm, are connected by an external voltage source and a 17-Ω resistance, as shown in **Figure 19-47**. The 0.5-Ω rod "floats" above the 2.5-Ω rod, in equilibrium. If the mass of each rod is 25 g, what must be the potential V of the voltage source? Example 19-6

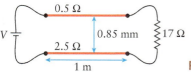

Figure 19-47 Problem 71

72. •• **Medical** When operated on a household 110-V line, typical hair dryers draw about 1650 W of power. We can model the current as a long, straight wire in the handle. During use, the current is about 3.0 cm from the user's head. (a) What is the current in the dryer? (b) What is the resistance of the dryer? (c) What magnetic field does the dryer produce at the user's head? Compare with Earth's magnetic field (5×10^{-5} T) to decide if we should have health concerns about the magnetic field created when using a hair dryer. Example 19-5

73. ••• Three very long, straight wires lie at the corners of a square of side d, as shown in **Figure 19-48**. The magnitudes of the currents in the three wires are the same, but the two diagonally opposite currents are directed into the page while the other one is directed outward. Derive an expression for the magnetic field (magnitude and direction) at the fourth corner of the square. Example 19-5

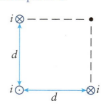

Figure 19-48
Problem 73

74. •• A 2.0-m lamp cord leads from the 110-V outlet to a lamp having a 75-W light bulb. The cord consists of two insulated parallel wires 4.0 mm apart and held together by the insulation. One wire carries the current into the bulb, and the other carries it out. What is the magnitude of the magnetic field the cord produces (a) midway between the two wires and (b) 2.0 mm from one of the wires in the same plane in which the two wires lie? (c) Compare each of the fields in parts (a) and (b) with Earth's magnetic field (5×10^{-5} T). (d) What magnetic force (magnitude and direction) do the two wires exert on one another? Example 19-6

75. •• Some people have raised concerns about the magnetic fields produced by current-carrying high-voltage lines in residential neighborhoods. Currents in such lines can be up to 100 A. Suppose you have such a line near your house. If the wires are supported horizontally 5.0 m above the ground on vertical poles and your living room is 12 m from the base of the poles, what magnetic field strength does the wire produce in your living room if it carries 100 A? Express your answer in teslas and as a multiple of Earth's magnetic field (5×10^{-5} T). Does the magnetic field from such wires seem strong enough to cause health concerns? Example 19-5

76. ••• **Medical magnetoencephalography (MEG)** is a technique for measuring changes in the magnetic field of the brain caused by external stimuli such as touching the body or viewing images of food. Such a change in the field occurs due to electrical activity (current) in the brain. During the process, magnetic sensors are placed on the skin to measure the magnetic field at that location. Typical field strengths are a few femtoteslas (1 femtotesla = 1 fT = 10^{-15} T). An adult brain is about 140 mm wide, divided into two sections (called hemispheres, although the brain is not truly spherical) each about 70 mm wide. We can model the current in one hemisphere as a circular loop, 65 mm in diameter, just inside the brain. The sensor is placed so that it is along the axis of the loop 2.0 cm from the center. A reasonable magnetic field is 5.0 fT at the sensor. According to this model, (a) what is the current in the brain and (b) what is the magnetic field at the center of the hemisphere of the brain?

77. ••• Helmholtz coils are composed of two coils of wire that have their centers on the same axis, separated by a distance that is equal to the radius of the coils (Figure 19-49). The coils have N turns of wire that carry a current i in the same direction. If one coil is centered at the origin, and the other at $x = R$, derive expressions for the net magnetic field due to the coils at the points (a) $x = R/2$ and (b) $x = 2R$.

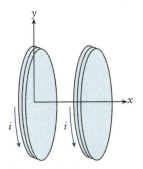

Figure 19-49 Problem 77

78. •• Geophysicists often measure magnetic field in gauss (1 G = 10^{-4} T). Earth's magnetic field at the equator can be taken as 0.7 G directed north. At the center of a flat circular coil that has 10 turns of wire and is 1.4 m in diameter, the coil's magnetic field exactly cancels Earth's field. (a) What must be the current in the coil? (b) How should the coil be oriented?

79. ••• **Biology** Migratory birds use Earth's magnetic field to guide them. Some people are concerned that human-caused magnetic fields could interfere with bird navigation. Suppose that a pair of parallel high-voltage lines, each carrying 100 A, are 3.00 m apart and lie in the same horizontal plane. Find the magnitude and direction of the magnetic field the lines produce at a point 15.0 m above them equidistant from both lines in each of the following cases. (a) The lines run in the north–south direction, and both currents run from north to south. (b) Both lines run in the north–south direction, and the current in the eastern line runs northward while the current in the western line runs southward. (c) The lines run in the east–west direction, and both currents run from west to east. (d) Is it reasonable to think that the fields caused by the wires are likely to interfere with bird migration? Example 19-5

80. ••• A square loop of wire lies on a horizontal table, with one side of the loop constrained to stay on the table, as shown in Figure 19-50. The loop is in the presence of a magnetic field that is parallel to the surface of the table. When a current $i_1 = 0.350$ A flows through the loop, the loop lifts off the table to an angle $\theta_1 = 15.0°$ relative to the surface of the table. When a different current of $i_2 = 1.18$ A flows through the loop, the angle the loop makes relative to the table increases to $\theta_2 = 42.1°$. Calculate the ratio of the magnetic torque on the

loop when current i_1 flows through it to the magnetic torque when current i_2 flows through the loop. Example 19-4

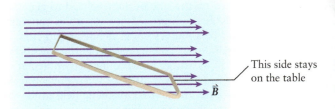

Figure 19-50 Problem 80

81. ••• A square, 35-turn, current-carrying loop is in a 0.75-T uniform magnetic field. The plane of the loop is parallel to the direction of the magnetic field, as shown in

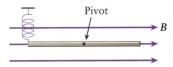

Figure 19-51 Problem 81

Figure 19-51 (an edge-on view of the loop). The loop is kept from rotating by a stretched spring that is attached to one side of the loop. The spring constant is 450 N/m, and the spring is stretched 5.6 cm. (a) If the length of each side of the square loop is 42 cm, what is the magnitude of the current running through the loop? (b) When viewed from above, is the current in the loop clockwise or counterclockwise? Example 19-4

82. ••• A 300-turn wire loop with a 5.0 cm radius is connected to a 0.20-kg pulley by a lightweight rod. The pulley, which has a radius of 3.1 cm, is attached to a 4.2-kg block by a massless rope (Figure 19-52). The loop is in a 0.75-T magnetic field. When

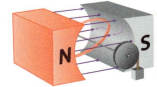

Figure 19-52 Problem 82

no current flows through the wire loop, the plane of the loop is parallel to the magnetic field, and the block attached to the pulley just touches the ground. However, when a 0.76-A current flows through the loop, magnetic forces cause the loop to rotate, which lifts the block off the ground. Calculate the angle the plane of the loop rotates relative to the magnetic field. How high off the ground is the block lifted? Example 19-4

pick-uppath/Getty Images

20

Electromagnetic Induction

In this chapter, your goals are to:

- (20-1) Explain the importance of electromagnetic induction.
- (20-2) Describe what is meant by a motional emf and an induced emf.
- (20-3) Explain what determines the magnitude and direction of an emf in a circuit with a changing magnetic flux.
- (20-4) Define the key properties of an ac generator.

To master this chapter, you should review:

- (6-2)   The work done by a constant force.
- (12-3)   Simple harmonic motion.
- (16-6)   The electric flux through an area.
- (18-6)   The electric power for a resistor.
- (19-3, 19-5) The magnetic force on a moving charged particle and on a current-carrying wire.
- (19-7)   The magnetic field produced by a current-carrying loop.

What do you think?

The stripe on the back of a credit card is magnetized in a pattern that encodes your account information. A credit card reader contains a loop of wire, and when you swipe the card through the reader, the magnetized card's motion generates an electric current in the wire that sends a signal to the credit card company. To make this current flow, what kind of force must act on electrons in the wires of the card reader? (a) A magnetic force; (b) an electric force; (c) a combination of electric and magnetic forces.

20-1 The world runs on electromagnetic induction

In Chapter 18 we discussed *direct-current* electric circuits in which the current always flows in the same direction. In these circuits what makes the current flow is the emf provided by a battery. This electric potential difference between the battery terminals is caused by chemical processes inside the battery (see Section 18-2).

But many of the electric circuits around you are *alternating-current* circuits in which the current constantly changes direction. That includes the current in light fixtures, toasters, electric fans, and other devices plugged into wall sockets. (Alternating current also indirectly powers mobile devices like cell phones and laptop computers. These devices have batteries, but the batteries are recharged by plugging them into a wall socket.) What kind of emf produces an alternating current?

The answer to this question comes from a remarkable discovery made by physicists around 1830: *If the magnetic field in a region of space changes, the change gives rise to an electric field*. This electric field, called an *induced* field, is very different in character from the electric field produced by point charges that we described in Chapter 16: An induced

Figure 20-1 Electromagnetic induction Two examples of the phenomenon of electromagnetic induction, in which electric currents are induced by the presence of a changing magnetic flux.

(a)

An electric generator produces current by electromagnetic induction: Coils of wire move relative to a magnetic field, which generates an emf in the coils. The motion can be powered by the wind, as in these wind turbines.

GregC/iStock/Getty Images

(b)

A changing magnetic field applied to the brain induces an electric field there, causing electric currents. Areas in red are where the currents are strongest.

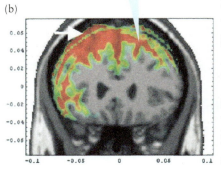

Republished with permission of Elsevier Science and Technology Journals, from Post, A. and Keck, M.E., "Transcranial Magnetic Stimulation as a Therapeutic Tool in Psychiatry: What Do We Know About the Neurobiological Mechanisms?" *Journal of Psychiatric Research 35:4, page 193–215 (2001): permission conveyed through*

electric field does not point away from positive charges and toward negative charges but instead has field lines that form closed loops like magnetic field lines. That's just what's needed for that induced electric field to push charges around a loop of wire and generate an electric current. As we'll see later in this chapter, it's easy to make this induced electric field flip its direction back and forth, which makes a current that flips back and forth—in other words, an alternating current. The vast amount of electric current used by our technological civilization is produced in this way (Figure 20-1a).

We use the term **electromagnetic induction** for the process whereby a changing magnetic field induces an electric field. (The word "electromagnetic" shows that this process involves both electric and magnetic fields.) Electromagnetic induction has many applications beyond producing an alternating current to be delivered to wall sockets. It's how a credit card reader decodes the information on the card's magnetized strip (see the photo that opens this chapter). It's also at the heart of a relatively new medical technique called *transcranial magnetic stimulation* (TMS), which allows physicians to stimulate electrical activity in the brain without sticking electrodes to the scalp or inserting them through the skull. In TMS, a time-varying magnetic field is produced inside the brain by current-carrying coils around the head. This causes an induced electric field, which in turn causes currents to flow within the brain (Figure 20-1b). TMS has been used with some success to treat cases of depression that have not responded to more conventional therapy.

In this chapter we'll begin by describing the relationship between a changing magnetic flux and the electric field that it induces. We'll introduce two important laws that describe electromagnetic induction. The first of these, Faraday's law, will tell us how the emf that appears in a closed loop (such as an electric circuit) due to an induced electric field is related to the rate of change of the magnetic flux through the loop. The second, Lenz's law, will tell us the direction of this induced emf. We'll see how induced emf makes possible the important device called a *generator*, which converts mechanical energy into electric energy and creates an alternating emf. (Each of the wind turbines shown in Figure 20-1 uses its spinning blades to run a generator.) In Chapter 21 we'll see how all of these ideas explain the behavior of alternating-current circuits.

BioMedical

TAKE-HOME MESSAGE FOR Section 20-1

✔ In electromagnetic induction, a time-varying magnetic flux in a certain region gives rise to an electric field in that same region.

✔ Electromagnetic induction is used to produce an alternating current.

20-2 A changing magnetic flux creates an electric field

Figure 20-2 shows an experiment that we can understand with the physics we already know. A loop of wire with an attached ammeter (a device for measuring the current in the loop) is placed near the south pole of a stationary magnet. No current flows if the loop is held stationary. That's not surprising, since there's no source of emf connected to the loop. But a current *does* flow in the loop when it is moved toward the magnet's south pole (Figure 20-2a) and flows in the opposite direction when the loop is moved away from the magnet's south pole (Figure 20-2b). What's happening is that the mobile charges within the loop are moving along with the loop through the magnetic field of the bar magnet and so experience a magnetic force that pushes the charges around the

(a) Moving the loop toward a stationary magnet

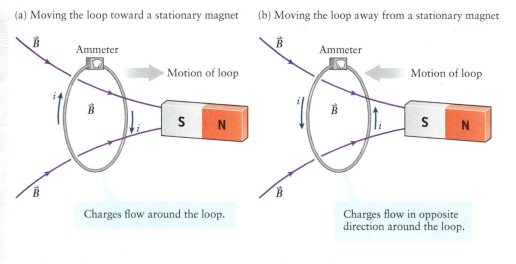

(b) Moving the loop away from a stationary magnet

Charges flow around the loop.

Charges flow in opposite direction around the loop.

Figure 20-2 A loop of wire moving with respect to a magnet If a loop of wire moves toward or away from a magnet, a current flows in the loop. The current is caused by magnetic forces.

(c) Side view of loop moving toward a stationary magnet

(d) Side view of loop moving away from a stationary magnet

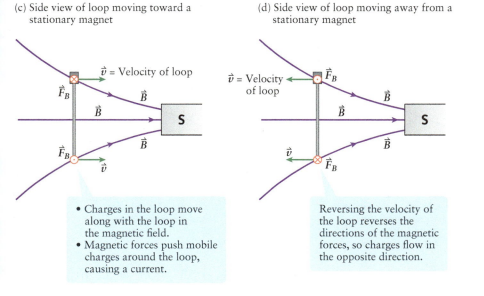

• Charges in the loop move along with the loop in the magnetic field.
• Magnetic forces push mobile charges around the loop, causing a current.

Reversing the velocity of the loop reverses the directions of the magnetic forces, so charges flow in the opposite direction.

loop (Figure 20-2c). Reversing the direction in which the loop and its charges move also reverses the direction of the magnetic force, so the charges are pushed in the opposite direction, and the current direction reverses (Figure 20-2d). In either case we say that the magnetic force on the mobile charges is equivalent to an emf that makes the current flow. There is no magnetic force, and hence no emf, if the loop and its charges are at rest. (Recall from Section 19-2 that magnetic forces act only on *moving* charges.) Because the loop must be in motion for the emf to appear, we call it a **motional emf**.

Figure 20-3 shows an experiment that looks similar but involves entirely different physics. Now we hold the loop stationary and move the south pole of the magnet either toward the loop (Figure 20-3a) or away from the loop (Figure 20-3b). In this case there can be no magnetic force on the mobile charges within the loop because those charges are at rest in the stationary loop. Nonetheless, there is an emf in the loop and a current flows around the loop in response, but only when the magnet is moving relative to the loop. Since there is no magnetic force in this situation, it must be that the emf is due to an *electric* force on the mobile charges (Figures 20-3c and 20-3d).

What's happening is that when the magnet is moving, the magnetic field at the location of the loop is changing: Its magnitude increases when the magnet's south pole moves toward the loop (Figure 20-3a) and decreases when the magnet's south pole moves away from the loop (Figure 20-3b). So this experiment shows that an electric field is *induced* by the changing magnetic field. For this reason we call the emf in the experiment of Figure 20-3 an **induced emf**. We use the term "electromagnetic induction" for situations in which a changing magnetic field causes, or induces, an electric field.

Although the experiments in Figures 20-2 and 20-3 are different, they have the *same* result: Whether the loop moves toward the stationary magnet at 1 m/s, as in Figure 20-2a,

Figure 20-3 **A magnet moving with respect to a loop of wire** If a magnet moves toward or away from a loop of wire, a current flows in the loop. Magnetic forces cannot explain why this happens, so electric forces must be present to produce the current.

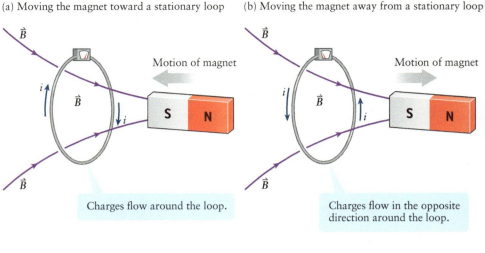

(a) Moving the magnet toward a stationary loop

Charges flow around the loop.

(b) Moving the magnet away from a stationary loop

Charges flow in the opposite direction around the loop.

(c) Side view of magnet moving toward a stationary loop

- Charges in the loop experience electric forces.
- These forces push mobile charges around the loop, causing a current.

(d) Side view of magnet moving away from a stationary loop

Reversing the velocity of the magnet reverses the directions of the electric forces, so charges flow in the opposite direction.

or the magnet moves toward the stationary loop at 1 m/s, as in Figure 20-3a, the same emf appears in the loop. In fact, the same emf appears if the magnet and loop are both moving, as long as the magnet and loop approach each other at a relative speed of 1 m/s. Since the result is the same in each of these cases, we should be able to describe all of these effects in terms of a single equation. But what equation is that?

It turns out that we can describe the emf in any of these situations in terms of the change in *magnetic flux* through the loop in Figures 20-2 and 20-3. We define this in the same way that we defined *electric* flux in Section 16-6: It's the area A of the surface outlined by the loop, multiplied by $B \cos \theta$, the component of the magnetic field that's perpendicular to that surface (see part (a) of Figure 20-4). In equation form the **magnetic flux** Φ_B ("phi-sub-B") through the loop is

Magnetic flux
(20-1)

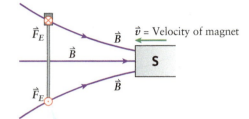

Area of the surface

Angle between the magnetic field and the perpendicular to the surface

Magnetic flux through a surface

$$\Phi_B = AB_\perp = AB \cos \theta$$

The component of the magnetic field perpendicular to the surface

Magnitude of the magnetic field

See the Math Tutorial *for more information on trigonometry.*

(The subscript "B" on the symbol Φ_B in Equation 20-1 reminds us that this is the flux of the magnetic field $\vec{B}$.) As parts (b) and (c) of Figure 20-4 show, the flux Φ_B can be positive or negative. Note that the choice of positive x direction is arbitrary; in Figure 20-4 we chose the positive x direction to be up, so Φ_B is positive for the case shown in Figure 20-4b and negative for the case shown in Figure 20-4c. Had we chosen

(a)

(b)

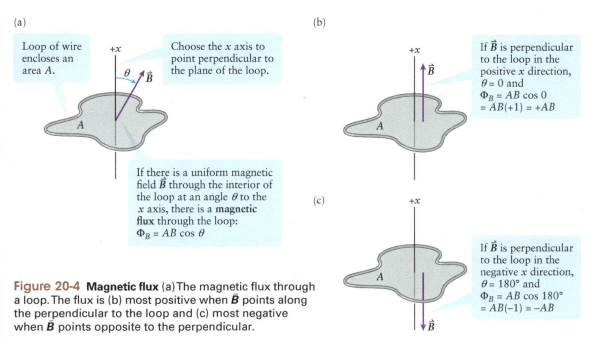

Loop of wire encloses an area A.

Choose the x axis to point perpendicular to the plane of the loop.

If there is a uniform magnetic field $\vec{B}$ through the interior of the loop at an angle θ to the x axis, there is a **magnetic flux** through the loop:
$\Phi_B = AB \cos \theta$

If $\vec{B}$ is perpendicular to the loop in the positive x direction, $\theta = 0$ and
$\Phi_B = AB \cos 0$
$= AB(+1) = +AB$

(c)

If $\vec{B}$ is perpendicular to the loop in the negative x direction, $\theta = 180°$ and
$\Phi_B = AB \cos 180°$
$= AB(-1) = -AB$

Figure 20-4 Magnetic flux (a) The magnetic flux through a loop. The flux is (b) most positive when $\vec{B}$ points along the perpendicular to the loop and (c) most negative when $\vec{B}$ points opposite to the perpendicular.

the positive x direction to be downward, we would have had $\Phi_B < 0$ in Figure 20-4b and $\Phi_B > 0$ in Figure 20-4c. It doesn't matter which one we choose, since the physics will turn out to be the same in either case. (In a similar way, in Chapter 2 we had to make a choice of positive x direction for analyzing motion in a straight line. The actual motion didn't depend on which direction we chose to be positive and which to be negative.)

In Figures 20-2 and 20-3 the magnetic field is not uniform over the area enclosed by the loop, so the perpendicular component $B_\perp = B \cos \theta$ has different values at different points on this area. In such a case $B_\perp$ in Equation 20-1 is the perpendicular component of the magnetic field *averaged* over the area A enclosed by the loop. Note that if the loop is actually a coil with N turns of wire, the net magnetic flux through the coil is N multiplied by the flux through one turn of the coil.

If the magnet and loop in Figures 20-2 and 20-3 are not moving with respect to each other, the magnetic flux through the loop remains the same. In this case there is no emf and no current in the loop. The flux changes, however, when either the loop moves relative to the magnet (Figure 20-2) or the magnet moves relative to the loop (Figure 20-3). In these cases there *is* an emf in the loop and the current. This suggests that *an emf appears in a loop when the magnetic flux through that loop changes*. This observation is known as **Faraday's law of induction**, named for the nineteenth-century English physicist Michael Faraday:

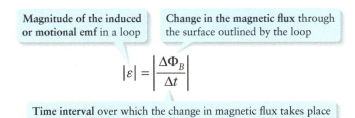

Magnitude of the induced or motional emf in a loop

Change in the magnetic flux through the surface outlined by the loop

$$|\varepsilon| = \left| \frac{\Delta \Phi_B}{\Delta t} \right|$$

Time interval over which the change in magnetic flux takes place

Faraday's law of induction
(20-2)

This law states that the magnitude of the emf that appears in a loop is equal to the magnitude of the *rate of change* of the magnetic flux through the loop. If a large change in flux $\Delta \Phi_B$ happens in a short time interval Δt, the resulting emf has a large magnitude; if the change in flux is relatively small and happens over a long time interval, the resulting emf has a small magnitude.

Note that Equation 20-2 tells us only the *magnitude* of the emf, not its direction. In the following section we'll see how the direction is determined.

WATCH OUT!   It's not the magnetic flux that causes an emf, but the rate at which the flux changes.

 The mere presence of magnetic flux through a loop does not cause an emf to appear in the loop. If a flux is present but does not change, such as what happens when a magnet and loop are held stationary with respect to each other, there is *no* resulting emf. An emf appears only when the flux *changes*, such as when the magnet and loop in Figures 20-2 and 20-3 move either toward or away from each other.

➔ *Go to Interactive Exercise 20-1 for more practice dealing with motional emf.*

In the following example we'll check Faraday's law. We'll do this by considering a situation in which we can use our knowledge of magnetic forces to calculate the emf, then compare this to the emf calculated using Equation 20-2.

EXAMPLE 20-1   Changing Magnetic Flux I: A Sliding Bar in a Magnetic Field

A copper bar of length L slides at a constant speed v along stationary, U-shaped copper rails (Figure 20-5). A uniform magnetic field of magnitude B is directed perpendicular to the plane of the bar and rails. The moving bar and stationary rails form a closed circuit, and an emf is produced in this circuit because the wire is moving in a magnetic field. Determine the emf in the circuit (a) by using the expression for the magnetic force on a charge in the moving wire and (b) by using Faraday's law of induction, Equation 20-2.

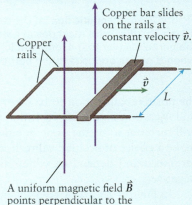

Copper bar slides on the rails at constant velocity $\vec{v}$.

Copper rails

A uniform magnetic field $\vec{B}$ points perpendicular to the plane of the rails.

Figure 20-5 A sliding copper bar What emf is generated in the bar as it slides in the presence of a magnetic field $\vec{B}$?

Set Up

For a battery the magnitude of the emf equals the change in electric potential (potential energy per charge) for a charge that traverses the battery; that is, it's equal to the *work per charge* that the battery does on charges that travel from one terminal to the other. We'll use the same idea in part (a) to calculate the emf in terms of the work done by the magnetic force on a charged particle that travels the length of the moving bar. In part (b) we'll find the emf by instead using Equation 20-2. While the magnetic field doesn't change, the area of the loop outlined by the moving bar and the rails *does* change, and so the magnetic flux through this loop changes.

Magnetic force on a moving charged particle:

$$F = |q|vB \sin \theta \qquad (19\text{-}1)$$

Work done by a constant force that points in the same direction as the straight-line displacement:

$$W = Fd \qquad (6\text{-}1)$$

Magnetic flux:

$$\Phi_B = AB_\perp = AB \cos \theta \qquad (20\text{-}1)$$

Faraday's law of induction:

$$|\varepsilon| = \left| \frac{\Delta \Phi_B}{\Delta t} \right| \qquad (20\text{-}2)$$

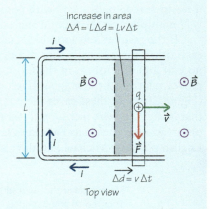

increase in area
$\Delta A = L\Delta d = Lv\Delta t$

Top view

Solve

(a) Find the magnetic force on a charged particle moving along with the copper bar.

For a positive charge q moving with the bar, the velocity $\vec{v}$ is perpendicular to the magnetic field $\vec{B}$. So $\theta = 90°$ in Equation 19-1, and the magnetic force $\vec{F}$ on such a charge has magnitude

$$F = qvB \sin 90° = qvB(1) = qvB$$

The force $\vec{F}$ is perpendicular to both $\vec{v}$ and $\vec{B}$, so it is directed along the length of the moving bar.

Use the magnetic force on a charged particle to find the emf produced in the bar.

The magnetic force $\vec{F}$ on a charge q causes it to move along the length L of the bar. Since $\vec{F}$ is in the same direction as the displacement of the charge, the work done on the charge as it travels this length is

$$W = FL = qvBL$$

The magnitude of the emf in the bar equals the work done per charge:

$$|\varepsilon| = \frac{W}{q} = \frac{qvBL}{q} = vBL$$

(b) Use Faraday's law of induction to find the emf.

The magnetic field $\vec{B}$ points perpendicular to the plane of the loop outlined by the moving copper bar and the copper rails. If the area of this loop is A and we take the positive x direction to point out of the plane of the above figure (in the same direction as $\vec{B}$), then $\theta = 0$ in Equation 20-1. The magnetic flux through the loop is then

$$\Phi_B = AB \cos 0 = AB(1) = AB$$

The magnetic field is constant, but the area A changes with time because the bar moves. The speed v of the bar is just the distance Δd that the bar moves divided by the time Δt that it takes to move that distance, so

$$v = \frac{\Delta d}{\Delta t} \quad \text{and} \quad \Delta d = v\Delta t$$

During time Δt the area A of the loop outlined by the moving bar and rails increases by an amount $\Delta A = L\Delta d = Lv\Delta t$. Therefore, the change in magnetic flux through the loop during this time is

$$\Delta\Phi_B = (\Delta A)B = (Lv\Delta t)B = vBL\Delta t$$

From Equation 20-2 the magnitude of the emf in the loop is

$$|\varepsilon| = \left|\frac{\Delta\Phi_B}{\Delta t}\right| = \left|\frac{vBL\Delta t}{\Delta t}\right| = vBL$$

Reflect

We find the same expression for the emf in both parts (a) and (b), as we must. This gives us added confidence that Equation 20-2 is valid, and a host of experiments back up its validity.

EXAMPLE 20-2   Changing Magnetic Flux II: A Varying Magnetic Field

A uniform magnetic field of magnitude $B = 1.50$ T is directed at an angle of $60.0°$ to the plane of a circular loop of copper wire. The loop is 3.50 cm in diameter. (a) What is the magnetic flux through the loop? What is the induced emf in the loop if the magnetic field decreases to zero (b) in 10.0 s or (c) in 0.100 s?

Set Up

The magnetic flux is given by Equation 20-1. Note that θ in this equation is the angle between the direction of magnetic field $\vec{B}$ and the *perpendicular* to the loop, so $\theta = 90.0° - 60.0° = 30.0°$. The magnetic flux through the loop changes when the field magnitude changes, so an emf will be induced in the loop. We'll use Equation 20-2 to calculate the magnitude of this induced emf.

Magnetic flux:

$$\Phi_B = AB_\perp = AB \cos \theta \qquad (20\text{-}1)$$

Area of a circle of radius r:

$$A = \pi r^2$$

Faraday's law of induction:

$$|\varepsilon| = \left|\frac{\Delta\Phi_B}{\Delta t}\right| \qquad (20\text{-}2)$$

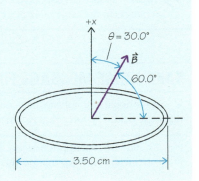

Solve

(a) Find the area of the loop, then use Equation 20-1 to calculate the magnetic flux through the loop.

The radius r of the loop is one-half of the diameter:

$$r = \frac{1}{2}(3.50 \text{ cm}) = 1.75 \text{ cm} = 1.75 \times 10^{-2} \text{ m}$$

The area of the loop is

$$A = \pi r^2 = \pi(1.75 \times 10^{-2} \text{ m})^2 = 9.62 \times 10^{-4} \text{ m}^2$$

From Equation 20-1 the magnetic flux through the loop is

$$\Phi_B = AB \cos\theta = (9.62 \times 10^{-4} \text{ m}^2)(1.50 \text{ T}) \cos 30.0°$$
$$= 1.25 \times 10^{-3} \text{ T·m}^2$$

(b) The change in magnetic flux is the final value (zero) minus the initial value that we found in (a). Equation 20-2 tells us that to find the magnitude of the induced emf, we divide this change by the time $\Delta t = 10.0$ s over which the flux change takes place.

The change in magnetic flux is

$$\Delta\Phi_B = (\text{final flux}) - (\text{initial flux})$$
$$= 0 - 1.25 \times 10^{-3} \text{ T·m}^2 = -1.25 \times 10^{-3} \text{ T·m}^2$$

If the flux decreases to zero in $\Delta t = 10.0$ s, the magnitude of the induced emf is

$$|\varepsilon| = \left|\frac{\Delta\Phi_B}{\Delta t}\right| = \left|\frac{-1.25 \times 10^{-3} \text{ T·m}^2}{10.0 \text{ s}}\right|$$
$$= 1.25 \times 10^{-4} \text{ T·m}^2/\text{s} = 1.25 \times 10^{-4} \text{ V}$$

(c) Repeat part (b) with $\Delta t = 0.100$ s.

If the flux decreases to zero in just $\Delta t = 0.100$ s, the magnitude of the induced emf is

$$|\varepsilon| = \left|\frac{\Delta\Phi_B}{\Delta t}\right| = \left|\frac{-1.25 \times 10^{-3} \text{ T·m}^2}{0.100 \text{ s}}\right|$$
$$= 1.25 \times 10^{-2} \text{ T·m}^2/\text{s} = 1.25 \times 10^{-2} \text{ V}$$

Reflect

The induced emf is 100 times greater in part (c) than in part (b) because the same flux change takes place in 1/100 as much time. The faster the flux change, the greater the induced emf that results. Note that the emf is induced *only* during the time when the magnetic flux is changing. There is zero emf when the magnetic field is at its original value of 1.50 T, and there is zero emf when the magnetic field has stabilized at its final value of zero.

If we replace the loop by a coil of the same diameter with 500 turns of wire, the induced emf is 500 times greater: $500 \times 1.25 \times 10^{-4}$ V = 0.0625 V in part (b), $500 \times 1.25 \times 10^{-2}$ V = 6.25 V in part (c). The key to generating a large induced emf is to have many turns of wire and a rapid change in magnetic flux.

In this example we chose the positive x direction to be upward so that the angle θ between the magnetic field and the perpendicular to the loop was 30.0°. Can you show that we would have found the same results for emf had we chosen the positive x direction to be downward so that $\theta = 150.0°$?

GOT THE CONCEPT? 20-1 A Wooden Loop

 Suppose the loop in Example 20-2 were made out of wood rather than copper wire, but the magnetic field changes in the same manner as in part (b) of Example 20-2. Compared to the emf calculated in part (b) of Example 20-2, the emf induced in the wooden loop would be (a) zero; (b) much smaller but not zero; (c) slightly less; (d) the same; (e) greater.

TAKE-HOME MESSAGE FOR Section 20-2

✔ A motional emf appears in a conductor that moves in a magnetic field. The force that produces the emf is a magnetic one.

✔ An induced emf appears in any loop subjected to a changing magnetic field. The force that produces the emf is an electric one.

✔ Both motional emfs and induced emfs can be described by Faraday's law of induction: The magnitude of the emf in a loop is equal to the absolute value of the change in magnetic flux through the loop divided by the time over which the change takes place.

20-3 Lenz's law describes the direction of the induced emf

Equation 20-2 tells us the *magnitude* of the emf that appears in a loop when there is a change in magnetic flux through that loop: $|\varepsilon| = |\Delta\Phi_B/\Delta t|$. It does not, however, tell us the *direction* in which the emf tends to make current flow around that loop. As we'll see, there's a simple rule for determining this direction that works for both motional emfs (caused by a conductor moving in a magnetic field) and induced emfs (caused by a conductor being exposed to a changing magnetic field).

To learn about this rule let's think again about the loop of wire in Figure 20-2. In addition to the field $\vec{B}_{magnet}$, there's also a magnetic field produced by the current within the loop itself. As we learned in Section 19-7, a current-carrying loop produces a magnetic field of its own. The direction of the field $\vec{B}_{loop}$ due to the loop depends on the direction of the current around the loop and is given by a right-hand rule: Curl the fingers of your right hand around the loop in the direction of the current, and the extended thumb of your right hand will point in the direction of $\vec{B}_{loop}$ in the interior of the loop (see part (a) of Figure 20-6). So whenever an emf appears in a loop—either a motional emf as in Figure 20-2 or an induced emf as in Figure 20-3—the current produced by that emf generates a magnetic field $\vec{B}_{loop}$ whose direction depends on the direction of the current and emf. We call $\vec{B}_{loop}$ an **induced magnetic field**.

The field $\vec{B}_{loop}$ itself produces a magnetic flux through the loop, and it's the sense of this flux that will tell us the direction of the emf in the loop. Let's choose the positive x direction for the loop in Figure 20-6 to point to the right, perpendicular to the plane of the loop. If the loop is close to the south pole of a bar magnet, as in Figure 20-6b, the field of the magnet causes a positive magnetic flux through the loop (the field $\vec{B}_{magnet}$ points generally to the right, in the positive x direction). If the loop moves toward the magnet as in the left-hand side of Figure 20-6b, the field $\vec{B}_{magnet}$ inside the loop increases and the positive flux increases. Experiment shows that in this case the current induced in the loop gives rise to an induced magnetic field $\vec{B}_{loop}$ within the loop, which points in the *opposite* direction to $\vec{B}_{magnet}$. So while the flux due to $\vec{B}_{magnet}$ becomes more positive, $\vec{B}_{loop}$ gives rise to a negative flux that opposes the change in the flux of $\vec{B}_{magnet}$.

If instead the loop moves away from the magnet as in the right-hand side of Figure 20-6b, the field $\vec{B}_{magnet}$ inside the loop decreases and the positive flux decreases. In this case the direction of the induced current in the loop is reversed, as is the direction of the induced magnetic field $\vec{B}_{loop}$: Now $\vec{B}_{loop}$ inside the loop points in the *same* direction as $\vec{B}_{magnet}$. The magnetic flux due to $\vec{B}_{loop}$ is now positive, which opposes the negative change (decrease) in the flux of $\vec{B}_{magnet}$.

In both cases shown in Figure 20-6b, the induced magnetic field is in a direction opposite to the *change* in flux of the external magnetic field (in this case the field due to the bar magnet). Many experiments show that this is always the case, no matter whether the induced magnetic field is due to a motional emf, an induced emf, or a combination of the two. The nineteenth-century Russian physicist Heinrich Lenz summarized these observations in a principle that we now call **Lenz's law**:

The direction of the magnetic field induced within a conducting loop opposes the change in magnetic flux that created it.

It's common to combine Faraday's law and Lenz's law into a single equation:

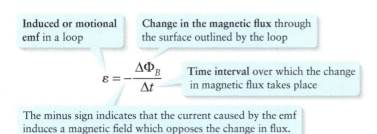

Induced or motional emf in a loop

Change in the magnetic flux through the surface outlined by the loop

$$\varepsilon = -\frac{\Delta\Phi_B}{\Delta t}$$

Time interval over which the change in magnetic flux takes place

The minus sign indicates that the current caused by the emf induces a magnetic field which opposes the change in flux.

Faraday's law and Lenz's law for induction
(20-3)

Figure 20-6 Lenz's law The current induced in a loop by a change in flux always acts to oppose the flux change.

(a)

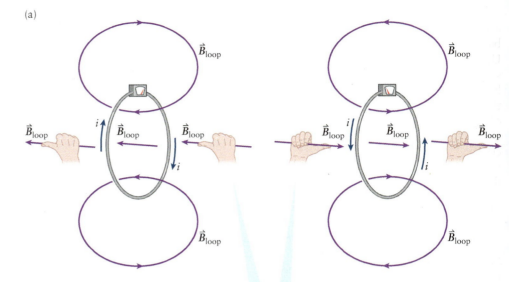

- A current-carrying loop generates a magnetic field $\vec{B}_{\text{loop}}$.
- To find the direction of $\vec{B}_{\text{loop}}$, curl the fingers of your right hand around the loop in the direction of the current i. Your extended right thumb points in the direction of $\vec{B}_{\text{loop}}$ in the interior of the loop.
- $\vec{B}_{\text{loop}}$ itself causes a magnetic flux through the loop.

(b)

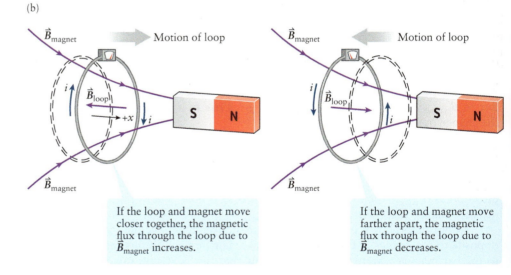

If the loop and magnet move closer together, the magnetic flux through the loop due to $\vec{B}_{\text{magnet}}$ increases.

If the loop and magnet move farther apart, the magnetic flux through the loop due to $\vec{B}_{\text{magnet}}$ decreases.

WATCH OUT! Like Faraday's law, Lenz's law is about the *change* in flux.

Notice that Lenz's law refers to the direction of the *change* in magnetic flux ("Is the flux increasing or decreasing?"), not to the direction of the field that causes the flux. The field $\vec{B}_{\text{magnet}}$ points in the same direction in both of the situations shown in Figure 20-6b, but the flux change is different in the two situations, so the emf is in different directions as well. The minus sign in Equation 20-3 is a reminder about the direction of the induced or motional emf. As we discuss below, this direction will always be such that the resulting current produces a magnetic field which opposes the change in flux.

We can check Lenz's law by revisiting the sliding copper bar from Example 20-1 (Section 20-2). The upward external magnetic field of magnitude B causes an upward magnetic flux through the loop formed by the sliding bar and the rails on which it slides (Figure 20-7). The area enclosed by this loop increases as the bar slides to the right, so the upward flux increases as well. By Lenz's law an induced current will flow in the loop to generate an induced magnetic field that opposes this change in flux. So this induced magnetic field must point downward, and to produce that induced field the current in the loop must be clockwise as seen from above the loop. That's just the direction of current flow that we depicted in the figure that accompanies Example 20-1. So this situation is consistent with Lenz's law.

The sliding copper bar in Figure 20-7 illustrates another aspect of Lenz's law. Once the induced current is flowing in the bar, the external magnetic field exerts a force on that current. Using the right-hand rule for this force (see Section 19-5), we see that this force points opposite to the direction in which the bar is moving. In other words, this force *opposes* the motion that gives rise to the change in flux through the loop made up of the bar and rails. That's always the case when a conductor moves through a magnetic field: A current is induced in the conductor, and the magnetic field exerts a force on the current that opposes the motion of the conductor. We can summarize this in an alternative statement of Lenz's law:

When the magnetic flux through a loop changes, current flows in a direction that opposes that change.

Since a magnetic force opposes the motion of the bar in Figure 20-7, we need to apply an external force to keep the bar in motion. If we make the bar move faster, the magnetic flux through the loop changes more rapidly, the emf and resulting current in the loop are greater, and the magnetic force opposing the motion of the bar is greater. (The magnetic force on the bar is proportional to the speed of the bar, just like the drag force on a microscopic object moving through a fluid; see Section 5-5.) So we must apply a greater force to make the bar slide at a faster speed.

This same effect explains the phenomenon of *magnetic braking*. If you try to make a magnet move past a conductor or a conductor move past a magnet, currents appear in the conductor. (These are called *eddy currents*, since their pattern resembles that of eddies in a body of water. The conductor does *not* need to be in the form of a loop for these currents to appear.) The magnetic force that the magnet exerts on the eddy currents opposes the motion of the conductor relative to the magnet, so by Newton's third law there is a force that opposes the motion of the magnet relative to the conductor. One application of this is to roller coasters. When a roller coaster car enters the part of the ride where it's supposed to slow down, a copper fin on the car passes through powerful permanent magnets mounted on the track. Eddy currents arise in the fin, and the interaction between the eddy currents and the field of the permanent magnets causes a force that smoothly brings the car to a slow speed. The car is then stopped by conventional mechanical braking.

Eddy currents are also used in an *electromagnetic flowmeter*, a device that can measure blood flow in an artery. Blood is an electrical conductor; eddy currents are induced in the blood as it flows past magnets in the flowmeter. The device records the small but measurable magnetic fields due to these currents and uses them to determine the rate of flow. The advantage of an electromagnetic flowmeter is that it is noninvasive: No component of the device need be surgically introduced into the body.

Another application of eddy currents is *magnetic induction tomography*, a relatively new experimental technique for medical imaging. In this technique changing magnetic fields created by coils placed near a part of the body induce eddy currents. Observing the fields produced by these eddy currents is a way to monitor brain swelling. Eddy currents can also be used for the controlled, repeated delivery of medication. A capsule containing the drug is implanted in the body; the capsule is made from a gel that heats up slightly when there are eddy currents, opening pores through which the medication is released. As for an electromagnetic flowmeter, no implanted electronics are required.

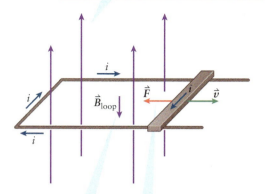

(1) As the bar slides to the right, the upward magnetic flux through the loop increases.

(2) The induced current i produces an induced magnetic field $\vec{B}_{\text{loop}}$ that opposes the change in flux.

(3) The magnetic force on the current in the moving bar opposes the bar's motion.

Figure 20-7 **A sliding copper bar revisited** Lenz's law helps explain the direction in which current flows in this situation.

BioMedical

BioMedical

GOT THE CONCEPT? 20-2 Induced Current I

? In Figure 20-8 a rectangular loop of wire moves to the right into a region of constant, uniform magnetic field. The field points into the plane of the figure, in a direction perpendicular to the plane of the loop. When the loop is entering the field region, as in Figure 20-8a, what is the direction of the current around the loop? (a) Clockwise; (b) counterclockwise; (c) current is zero; (d) not enough information given to decide.

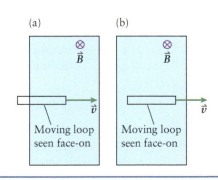

Figure 20-8 **A moving rectangular loop of wire** In each situation what is the direction of the induced current in the moving loop of wire?

GOT THE CONCEPT? 20-3 Induced Current II

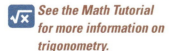

In Figure 20-8 a rectangular loop of wire moves to the right into a region of constant, uniform magnetic field. The field points into the plane of the figure, in a direction perpendicular to the plane of the loop. When the loop is moving and completely inside the field region, as in Figure 20-8b, what is the direction of the current around the loop? (a) Clockwise; (b) counterclockwise; (c) current is zero; (d) not enough information given to decide.

TAKE-HOME MESSAGE FOR Section 20-3

✔ An emf induced by a changing magnetic flux tends to cause a current to flow. This current generates a magnetic field of its own, called the induced magnetic field.

✔ The induced magnetic field is in a direction that opposes the change in flux that created the emf that gave rise to the induced field.

✔ Eddy currents arise whenever a conducting material, even a nonmagnetic one, moves relative to a magnetic field.

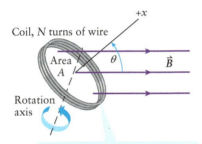

Coil, N turns of wire

+x

Area A

θ

$\vec{B}$

Rotation axis

As the coil rotates with angular speed ω, the angle θ between the magnetic field $\vec{B}$ direction and the perpendicular to the coil changes: θ = ωt + φ.

Figure 20-9 An ac generator As the coil rotates, an oscillating emf is generated in the turns of wire that make up the coil.

√x **See the Math Tutorial for more information on trigonometry.**

20-4 Faraday's law explains how alternating currents are generated

We learned in Section 18-2 about the importance of alternating current in technology. (If you're reading these words in a room lit by electric light, the light bulbs are powered by alternating current. If you're reading on a mobile device such as a tablet, the device's battery was charged by plugging it into a wall socket and using the alternating current delivered by that socket.) We now have the physics we need to understand how alternating current is produced.

Let's look at a coil of wire with N turns, each of which has area A. As Figure 20-9 shows, this coil is free to rotate around an axis that lies along a diameter of the coil. We place the coil in a region of uniform magnetic field $\vec{B}$, then rotate the coil at a constant angular speed ω. As the coil rotates the magnetic flux through each turn of the coil changes, so an emf is generated. As we will see, this is an alternating emf of just the sort required to generate an alternating current. That's why a rotating coil of the sort shown in Figure 20-9 is called an **ac generator**.

We begin by writing an equation for the magnetic flux through the rotating coil. The angle θ between the magnetic field $\vec{B}$ and the perpendicular to the coil changes as the coil rotates:

$$\theta = \omega t + \phi$$ (20-4)

In Equation 20-4 φ is the value of the angle at t = 0. From Equation 20-1 the magnetic flux through the N turns of the coil is N times that through one turn:

$$\Phi_B = NAB \cos \theta = NAB \cos (\omega t + \phi)$$ (20-5)

When θ = ωt + φ = 0 so cos θ = 1, the perpendicular to the coil is in the same direction as $\vec{B}$ and the flux has its most positive value Φ_B = NAB; when θ = ωt + φ = π/2 so cos θ = 0, the coil is edge-on to the magnetic field and the flux is zero; when θ = ωt + φ = π and cos θ = −1, the perpendicular to the coil points opposite to $\vec{B}$ and the flux has its most negative value Φ_B = −NAB; and so on. So the magnetic flux varies with time, and it follows that there will be an emf in the coil.

Faraday's law and Lenz's law (Equation 20-3) tell us that the emf is equal to the negative of the rate of change of Φ_B. We actually know how to find the rate of change of a cosine function like that in Equation 20-5. In Section 12-3 we saw that the position of an object undergoing simple harmonic motion with amplitude A is given by

$$x = A \cos (\omega t + \phi)$$ (12-6)

The rate of change of position x is just the velocity v_x, which we found was equal to

$$v_x = -\omega A \sin (\omega t + \phi)$$ (12-7)

You can see that Equation 20-5 for magnetic flux is identical to Equation 12-6 for position, with amplitude A replaced by NAB (note that A in Equation 20-5 denotes area, not amplitude). Making the same replacement in Equation 12-7 tells us that the rate of change of magnetic flux through the rotating coil is

$$\frac{\Delta \Phi_B}{\Delta t} = -\omega NAB \sin(\omega t + \phi) \tag{20-6}$$

Substituting Equation 20-6 into Equation 20-3 then gives us the emf in the rotating coil:

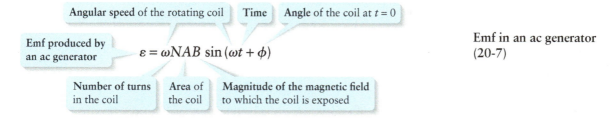

Emf in an ac generator
(20-7)

The emf alternates with angular frequency ω, which is the same as the angular speed of the rotating coil. The maximum value of the emf is $\varepsilon_{max} = \omega NAB$, which shows that we can increase the maximum emf by increasing the angular speed ω, the number of turns N, the coil area A, the magnetic field magnitude B, or a combination of these.

While the emf produced by an ac generator changes from positive to negative, the power delivered by the generator does not. As an example, suppose an ac generator is connected to a circuit device (such as a light bulb or a toaster) that we can represent as a resistor with resistance R. If we ignore the internal resistance of the coil, the emf is equal to the voltage drop across the resistor: $\varepsilon = iR$. The current in the resistor is therefore

$$i = \frac{\varepsilon}{R} = \frac{\omega NAB}{R} \sin(\omega t + \phi) \tag{20-8}$$

The current in the resistor alternates with the same angular frequency ω as the emf. From Section 18-6 the power into such a resistor is

$$P = i^2 R \tag{18-24}$$

If we substitute Equation 20-8 into Equation 18-24, we get

$$P = \left(\frac{\omega NAB}{R} \sin(\omega t + \phi)\right)^2 R = \frac{\omega^2 N^2 A^2 B^2}{R} \sin^2(\omega t + \phi) \tag{20-9}$$

Figure 20-10 shows graphs of the emf ε (Equation 20-7) and resistor power P (Equation 20-9) as functions of time. The power P is *never* negative, which means that

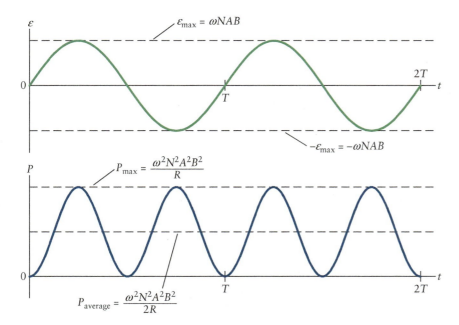

Figure 20-10 An ac generator: Emf and power These graphs show the emf ε generated in the coil shown in Figure 20-9 and the power P that this emf delivers to a resistor R connected to the coil. We assume $\phi = 0$ in Equations 20-7 and 20-9. Note that T is the time it takes for the coil to complete one rotation.

Figure 20-11 **An ac hydroelectric generator** At the Center Hill Dam in Tennessee, water flowing past a turbine turns this rotor, which is equipped with a series of electromagnets (the tall red rectangles around the rotor's rim). These are arranged with alternating orientations: one electromagnet has its north pole at the top, the next one has its north pole at the bottom, the next has its north pole at the top, and so on. (The red rotor shown here has been removed for maintenance.) As it spins, stationary coils around the rotor are exposed to an alternating magnetic field, which produces an alternating emf as depicted in Figure 20-10. Each of the three generators at the Center Hill Dam can produce an average power of 45 megawatts (4.5×10^7 W).

energy always flows from the ac generator into the resistor, never the other way. The average value of the function $\sin^2(\omega t + \phi)$ is $\frac{1}{2}$, so the average power into the resistor is

(20-10)
$$P_{average} = \frac{\omega^2 N^2 A^2 B^2}{2R}$$

It may seem like the power given by Equation 20-10 comes "for free": You let the coil rotate, and an emf is generated that makes power flow into the resistor. Alas, this power comes at a price. As we discussed in Section 20-3, whenever a conductor (such as the coil of an ac generator) moves in the presence of a magnetic field, the conductor experiences a magnetic force that opposes its motion. So left to itself, the coil would quickly slow to a halt. To keep the coil in motion, you must apply a torque that just balances the effects of this magnetic force. At an electric generating station, this torque is applied to the blades of a turbine that is connected to the coil. The blades can be turned by the force of the wind (Figure 20-1a), by the force of flowing water at a hydroelectric plant, or by the force of a fast-moving steam at a coal-fired or nuclear power plant (where heat from burning fossil fuels or radioactive decay is used to boil water and produce steam). Part of the mechanical power used to make the turbine spin goes into the electric power provided by the generator; the rest is lost due to friction in the turbine and generator. Note that in some generators, like the ones shown in Figure 20-11, the coil is stationary and the magnetic field alternates; in either case there is an alternating emf like that shown in Figure 20-10.

The ac generator is one part of a system of power delivery based on alternating current. In Chapter 21 we'll explore in more detail the physics of alternating current in circuits.

→ **Go to Interactive Exercise 20-2** for more practice dealing with coils.

EXAMPLE 20-3 Lighting the Gym with a Bicycle

You attach the coil of an ac generator to an exercise bicycle so that as you work the pedals the coil turns, and an emf is generated. The generator is geared so that it makes 10 rotations for each rotation of the pedals. The coil has an area of 6.40×10^{-3} m², has 2000 turns of wire, and is in a magnetic field of magnitude 0.100 T. How many times a second must you turn the pedals to deliver an average power of 60.0 W to a light bulb with a resistance of 80.0 Ω?

Set Up

The situation is as shown in Figure 20-9. We'll use Equation 20-10 to solve for the angular speed ω of the generator coil. Since the coil makes 10 rotations for every rotation of the pedals, the angular speed of the pedals is equal to ω divided by 10.

Average power delivered by an ac generator to a resistor:

$$P_{average} = \frac{\omega^2 N^2 A^2 B^2}{2R} \tag{20-10}$$

Solve

Rewrite Equation 20-10 to solve for ω.

Find the angular speed of the generator coil from Equation 20-10:

$$\omega^2 = \frac{2RP_{average}}{N^2 A^2 B^2}$$

$$\omega = \frac{\sqrt{2RP_{average}}}{NAB} = \frac{\sqrt{2(80.0 \ \Omega)(60.0 \ \text{W})}}{(2000)(6.40 \times 10^{-3} \ \text{m}^2)(0.100 \ \text{T})}$$

$$= 76.5 \ \text{rad/s}$$

Convert this from radians per second to revolutions per second:

$$\omega = \left(76.5 \ \frac{\text{rad}}{\text{s}}\right)\left(\frac{1 \ \text{rev}}{2\pi \ \text{rad}}\right) = 12.2 \ \text{rev/s}$$

The generator turns 10 times faster than the pedals, so the pedals must turn at a rate of

(12.2 rev/s)/10 = 1.22 rev/s = 73.1 rev/min

Reflect

A cycling cadence of 73.1 rev/min isn't difficult for an amateur cyclist, so you can certainly power a light bulb in this way. Exercise bicycles with generators of this kind are commercially available and are used to return power to the electrical grid in the same manner as residential solar panels.

GOT THE CONCEPT? 20-4   A Flickering Fluorescent Lamp

 Certain types of fluorescent lamps flicker rapidly. That's because these lamps emit a pulse of light every time a burst of electric power is provided to the lamp. If such a lamp is powered by a source of emf that oscillates at 60 Hz, what is the frequency at which the lamp will flicker? (a) 30 Hz; (b) 60 Hz; (c) 120 Hz; (d) 240 Hz; (e) 3600 Hz.

TAKE-HOME MESSAGE FOR Section 20-4

✔ An ac generator consists of a coil that rotates in a magnetic field.

✔ The emf produced by an ac generator oscillates at an angular frequency that equals the angular speed of the rotating coil.

Key Terms

ac generator
electromagnetic induction
Faraday's law of induction

induced emf
induced magnetic field
Lenz's law

magnetic flux
motional emf

Chapter Summary

Topic	Equation or Figure
Motional emf: An emf appears in a loop when that loop moves in the presence of a magnetic field. The emf is a result of magnetic forces on the mobile charges within the loop.	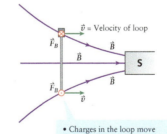 (c) Side view of loop moving toward a stationary magnet • Charges in the loop move along with the loop in the magnetic field. • Magnetic forces push mobile charges around the loop, causing a current. (Figure 20-2c)
Induced emf: An emf also appears in a loop when the magnetic field within the loop changes. Here the forces that create the emf are not magnetic, but electric; an electric field is produced by the changing magnetic field.	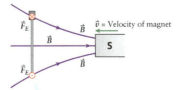 (c) Side view of magnet moving toward a stationary loop • Charges in the loop experience electric forces. • These forces push mobile charges around the loop, causing a current. (Figure 20-3c)

Faraday's law and Lenz's law:
Faraday's law states that if the magnetic flux through a loop changes, an emf is induced around that loop. The magnitude of the induced emf equals the magnitude of the rate of change of the magnetic flux through the loop.

Lenz's law states that the direction of the induced emf is such as to oppose the flux change: The induced current causes its own flux (which helps compensate for the change). A conductor and magnet that move relative to each other experience magnetic forces that oppose this relative motion.

Induced or motional emf in a loop

Change in the magnetic flux through the surface outlined by the loop

$$\varepsilon = -\frac{\Delta\Phi_B}{\Delta t}$$

Time interval over which the change in magnetic flux takes place

(20-3)

The minus sign indicates that the current caused by the emf induces a magnetic field which opposes the change in flux.

Alternating current and an ac generator: To produce an alternating emf (the sort needed to cause an alternating current), rotate a coil in the presence of a constant magnetic field. The changing flux causes an emf that oscillates sinusoidally.

Angular speed of the rotating coil

Time

Angle of the coil at $t = 0$

Emf produced by an ac generator

$$\varepsilon = \omega NAB \sin(\omega t + \phi)$$

(20-7)

Number of turns in the coil

Area of the coil

Magnitude of the magnetic field to which the coil is exposed

Answer to What do you think? Question

(b) The electrons in the wire are initially at rest and so do not experience a magnetic force. (Recall that a magnetic field exerts a force only on a charged object that is in motion.) So it must be an *electric* force that sets the electrons into motion. This is a consequence of electromagnetic induction: As you swipe the credit card through the reader, the wire loop is exposed to a varying magnetic field from the stripe on the back of the card. This causes a changing magnetic flux through the loop, which induces an electric field and an emf. Electrons move in the loop in response to this emf.

Answers to Got the Concept? Questions

20-1 (d) The induced emf does *not* depend on what the loop is made of. (Note that Equation 20-2 makes no reference to the properties of the loop material.) So the emf will be exactly the same whether the loop is made of copper, wood, silver, rubber, or even air. The difference is that because copper is a good conductor while wood is a very poor conductor, a current will be generated in the copper loop but not in the wooden loop.

20-2 (b) As the loop in Figure 20-8a moves into the field region, there is an increasing magnetic flux through the loop due to the magnetic field directed into the plane of the figure. According to Lenz's law, current will flow in the loop to induce a magnetic field $\vec{B}_{loop}$ that will oppose this increase, so $\vec{B}_{loop}$ in the interior of the loop must point out of the plane in Figure 20-8a. The right-hand rule depicted in Figure 20-6a tells us that to induce such a field, current must flow counterclockwise around the loop. (You can confirm this by using the right-hand rule for the magnetic force on a moving charge. A positive charge in the right-hand leg of the moving loop feels an upward magnetic force, and this force drives current in a counterclockwise direction around the loop. There are also magnetic forces on charges in the top and bottom legs of the loop, but these forces have zero component along the length of the wire and so do not induce a current.)

20-3 (c) Although the loop in Figure 20-8b is moving, the magnetic flux through the loop remains constant because it is moving through a region of constant, uniform magnetic field. Since there is no flux change, Faraday's law tells us that no emf is induced and so no current will be generated. (You can confirm this by using the right-hand rule for the magnetic force on a moving charge. A positive charge in the right-hand leg of the moving loop feels an upward magnetic force, and this force by itself would drive current in a counterclockwise direction around the loop. But a positive charge in the left-hand leg of the moving loop also feels an upward magnetic force, which by itself would drive current in a clockwise direction around the loop. There are also magnetic forces on charges in the top and bottom legs of the loop, but these forces have zero component along the length of the wire and so do not induce a current. The net effect is that there is *zero* current in the loop.)

20-4 (c) Figure 20-10 shows that the power delivered by an ac generator goes through two up-and-down cycles during the time T required for the emf to go through a single cycle. So if the emf varies at 60 Hz, the power delivered to the fluorescent lamp varies at 2×60 Hz = 120 Hz.

Questions and Problems

In a few problems you are given more data than you actually need; in a few other problems you are required to supply data from your general knowledge, outside sources, or informed estimate.

Interpret as significant all digits in numerical values that have trailing zeros and no decimal points.

For all problems use $g = 9.80 \text{ m/s}^2$ for the free-fall acceleration due to gravity. Neglect friction and air resistance unless instructed to do otherwise.

•	Basic, single-concept problem
••	Intermediate-level problem; may require synthesis of concepts and multiple steps
•••	Challenging problem
Example	See worked example for a similar problem

Conceptual Questions

1. • Two conducting loops with a common axis are placed near each other, as shown in Figure 20-12. Initially the currents in both loops are zero. If a current is suddenly set up in loop a in the direction shown, is there also a current in loop b? If so, in which direction? What is the direction of the force, if any, that loop a exerts on loop b? Explain your answer.

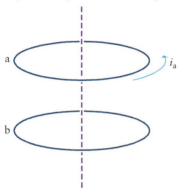

Figure 20-12 Problem 1

2. •• In a popular demonstration of electromagnetic induction, a metal plate is suspended in midair above a large electromagnetic coil, as shown in Figure 20-13. (a) How does this work? (b) If your professor does the demonstration, one thing you'll notice is that the plate gets quite hot. (In fact, you can end the demonstration by frying an egg on the plate!) Why does the plate become hot? (c) Would the trick work if the plate were made of an insulating material?

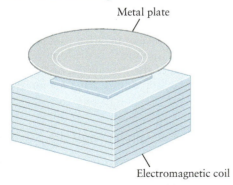

Metal plate

Electromagnetic coil   **Figure 20-13** Problem 2

3. • A conducting rod slides without friction on conducting rails in a magnetic field, as shown in Figure 20-14. The rod is given an initial velocity $\vec{v}$ to the right. Describe its subsequent motion and justify your answer.

Figure 20-14 Problem 3

4. • A common physics demonstration is to drop a small magnet down a long, vertical aluminum pipe. Describe the motion of the magnet and the physical explanation for the motion.

5. • **Medical** In hospitals with magnetic resonance imaging (MRI) facilities and at other locations where large magnetic fields are present, there are usually signs warning people with pacemakers and other electronic medical devices not to enter. Why?

6. •• Figure 20-15 depicts an electron (e^-) between the poles of an electromagnet. Explain how the electron is accelerated if the magnetic field is gradually being increased.

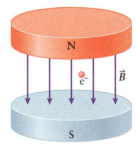

Figure 20-15 Problem 6

Multiple-Choice Questions

7. • Figure 20-16 shows a sequence of sketches depicting a rectangular loop passing from left to right through a region of constant magnetic field. The field points out of the page and perpendicular to the plane of the loop. In which sketch is the magnitude of the magnetic flux through the loop decreasing?
    A. from left to right approaching the magnetic field
    B. entering the magnetic field
    C. inside the magnetic field
    D. leaving the magnetic field
    E. from left to right moving away from the magnetic field

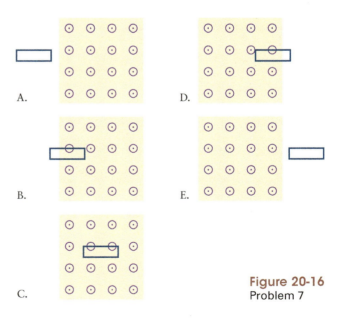

Figure 20-16
Problem 7

8. • On which variable does the magnetic flux depend?
    A. the magnetic field
    B. the area of a region through which the magnetic field passes
    C. the orientation of the field with respect to the region through which it passes
    D. all of the above
    E. none of the above

9. • Two metal rings with a common axis are placed near each other, as shown in Figure 20-17. If current i_a is suddenly set up and is increasing in ring a as shown, the current in ring b is
  A. zero.
  B. parallel to i_a.
  C. antiparallel to i_a.
  D. alternatively parallel and antiparallel to i_a.
  E. perpendicular to i_a.

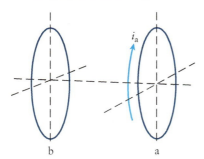

b   a   **Figure 20-17** Problem 9

10. •• Figure 20-18 shows two coils wound around an iron ring, which directs the magnetic field of each coil around the ring. Current appears in the coil on the right
  A. the moment the battery is connected by closing the switch.
  B. the entire time the battery is connected with the switch closed.
  C. the moment the battery is disconnected by opening the switch.
  D. the moment the battery is connected by closing the switch and the moment the battery is disconnected by opening the switch.
  E. the entire time the battery is disconnected with the switch open.

Figure 20-18
Problem 10

11. • The copper ring of radius R in Figure 20-19 lies in a magnetic field pointed into the plane of the figure. The field is uniformly decreasing in magnitude. The induced current in the ring is
  A. clockwise and constant.
  B. clockwise and changing.
  C. zero.
  D. counterclockwise and constant.
  E. counterclockwise and changing.

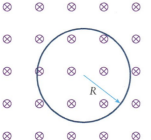

Figure 20-19 Problem 11

12. • A conducting loop moves at a constant speed parallel to a long, straight wire carrying a constant current, as shown in Figure 20-20. The induced current in the loop will be
  A. clockwise.
  B. only parallel to the current i.
  C. counterclockwise.
  D. alternately clockwise and then counterclockwise.
  E. zero.

Figure 20-20 Problem 12

Problems

20-1 The world runs on electromagnetic induction

20-2 A changing magnetic flux creates an electric field

13. • A single-turn circular loop of radius 5.0 cm lies in the plane perpendicular to a uniform magnetic field. During a 0.12-s time interval, the field magnitude increases uniformly from 0.20 to 0.40 T. Determine the magnitude of the emf induced in the loop during the time interval. Example 20-2

14. • A circular coil that has 100 turns and a radius of 10.0 cm lies in a magnetic field that has a magnitude of 0.0650 T directed perpendicular to the coil. (a) What is the magnetic flux through the coil? (b) The magnetic field through the coil is increased steadily to 0.100 T over a time interval of 0.500 s. What is the magnitude of the emf induced in the coil during the time interval? Example 20-2

15. • A 30-turn coil with a diameter of 6.00 cm is placed in a constant, uniform magnetic field of 1.00 T directed perpendicular to the plane of the coil. Beginning at time $t = 0$ s, the field is increased at a uniform rate until it reaches 1.30 T at $t = 10.0$ s. The field remains constant thereafter. What is the magnitude of the induced emf in the coil at (a) $t < 0$ s, (b) $t = 5.00$ s, and (c) $t > 10.0$ s? (d) Plot the magnetic field and the induced emf as functions of time for the range -5.00 s $< t < 15.0$ s. Example 20-2

16. • A student takes a strand of copper wire and forms it into a circular loop of circumference 0.350 m. The student then places the loop in a uniform, constant magnetic field of magnitude 0.00250 T that is oriented perpendicular to the face of the loop. Pulling on the ends of the wire, the student reduces the circumference of the loop to 0.125 m in a time interval of 0.800 s. Assuming that the loop remains circular as it shrinks, what is the magnitude of the average emf induced around the loop during this time interval? Example 20-1

20-3 Lenz's law describes the direction of the induced emf

17. • Determine the direction of the induced current in the loop for each case shown in Figure 20-21. In part (f) the loop spins around its axis but does not change position.

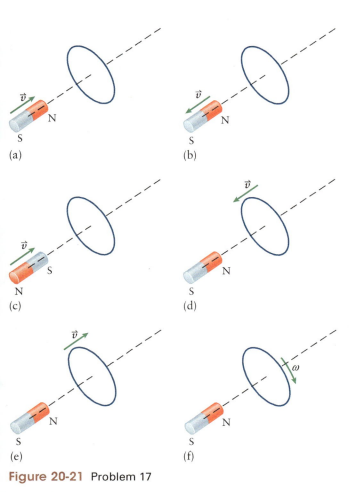

Figure 20-21 Problem 17

18. •• A bar magnet is moved steadily through a wire loop, as shown in Figure 20-22. Make a qualitative sketch of the induced emf in the loop as a function of time (be sure to include the times t_1, t_2, and t_3). Consider the direction of positive emf to be as indicated in the figure.

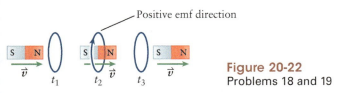

Figure 20-22
Problems 18 and 19

19. •• A bar magnet is moved steadily through a wire loop, as shown in Figure 20-22, except that the leading edge of the magnet is the south pole instead of the north pole. Make a qualitative sketch of the induced voltage in the loop as a function of time (be sure to include the times t_1, t_2, and t_3). Consider the direction of positive emf to be as indicated in the figure.

20. • Each situation shown in Figure 20-23 shows the initial and final magnitude and direction of a changing magnetic

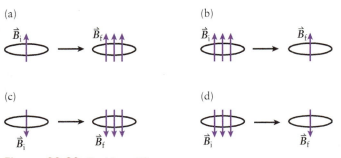

Figure 20-23 Problem 20

field. A conducting wire loop is placed perpendicular to the magnetic field as shown. For each situation, determine if the induced current in the loop is clockwise, counterclockwise, or zero as viewed from above while the magnetic field changes from its initial value to its final value.

21. •• A square, 30-turn coil 10.0 cm on a side with a resistance of 0.820 Ω is placed between the poles of a large electromagnet. The electromagnet produces a constant, uniform magnetic field of 0.600 T directed into the page. As suggested by Figure 20-24, the field drops sharply to zero at the edges of the magnet. The coil moves to the right at a

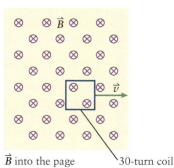

Figure 20-24 Problem 21

constant velocity of 2.00 cm/s. What is the current through the wire coil (a) before the coil reaches the edge of the field, (b) while the coil is leaving the field, and (c) after the coil leaves the field? (d) What is the total charge that flows past a given point in the coil as it leaves the field? (e) Plot the induced current in the loop as a function of the horizontal position of the right side of the current loop. Let the right-hand edge of the magnetic field region be $x = 0$. Your plot should be in the range of -5.00 cm $< x < 20.0$ cm.

22. •• (a) Determine the magnitude and direction of the force on each side of the coil in Problem 21 for situations (a) through (c). (b) As the loop enters the field region from the left, what is the direction of the induced current and the resulting force on each segment of the coil?

20-4 Faraday's law explains how alternating currents are generated

23. • A rectangular coil with sides 0.10 m by 0.25 m has 500 turns of wire. It is rotated about its long axis in a magnetic field of 0.58 T directed perpendicular to the rotation axis. At what frequency must the coil be rotated for it to generate a maximum potential of 110 V?

24. • An electromagnetic generator consists of a coil that has 100 turns of wire, has an area of 400 cm^2, and rotates at 60 rev/s in a magnetic field of 0.25 T directed perpendicular to the rotation axis. What is the magnitude of the emf induced in the coil?

25. •• You decide to build a small generator by rotating a coiled wire inside a static magnetic field of 0.30 T. You construct the apparatus by coiling wire into three loops of radius 0.16 m. If the coil rotates at 3.0 revolutions per second and is connected to a device with 1.0×10^2 Ω resistance, calculate the average power supplied to that device. Example 20-3

26. • Perhaps it has occurred to you that we could tap Earth's magnetic field to generate energy. One way to do this would be to spin a metal loop about an axis perpendicular to Earth's magnetic field. Suppose that the metal loop is a square that is 45.0 cm on each side and that we want to generate an electric potential in the loop of amplitude 120 V at a place where Earth's magnetic field is 5×10^{-5} T. At what angular speed (in rev/s) would we have to spin the coil? Does this appear to be a feasible method to extract energy from Earth's magnetic field?

General Problems

27. •• The induced voltage versus time for a coil that has 100 circular turns of wire with radii 25 cm is given in the table.

Plot $V(t)$ and use this graph to predict the graph of the magnetic field as a function of time $B(t)$ that is passing through the loop (assume the magnetic field is perpendicular to the plane of the loop and that the average value of $B(t)$ is 0).

t (s)	V (V)	t (s)	V (V)
0	0	9	2
1	2	10	4
2	4	11	2
3	2	12	0
4	0	13	-2
5	-2	14	-4
6	-4	15	-2
7	-2	16	0
8	0		

28. ••• A long, rectangular loop of width w, mass m, and resistance R is being pushed into a magnetic field by a constant force $\vec{F}$ (Figure 20-25). Derive an expression for the speed of the loop while it is entering the magnetic field. Example 20-1

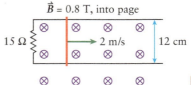

Figure 20-25 Problem 28

29. •• A magnetic field of 0.45 G is directed straight down, perpendicular to the plane of a circular coil of wire that is made up of 250 turns and has a radius of 20 cm. (a) If the coil is stretched, in a time of 15 ms, to a radius of 30 cm, calculate the emf induced in the coil during the process. (b) Assuming the resistance of the coil is a constant 25 Ω, what is the induced current in the coil during the process? (c) What is the direction of the induced current in the coil (clockwise or counterclockwise, as viewed from above)? Example 20-1

30. •• **Astronomy** Activity on the Sun, such as solar flares and coronal mass ejections, hurls large numbers of charged particles into space. When the particles reach Earth, they can interfere with communications and the power grid by causing electromagnetic induction. As one example, a current of millions of amps (known as the *auroral electrojet*) that runs about 100 km above Earth's surface can be perturbed. The change in the current causes a change in the magnetic field it produces at Earth's surface, which induces an emf along Earth's surface and in the power grid (which is grounded). Induced electric fields as high as 6.0 V/km have been measured. We can model the circuit at Earth's surface as a rectangular loop made up of the power lines completed by a path through the ground beneath. We can treat the magnetic field created by the electrojet as being uniform (but not constant). Consider a 1.0-km-long stretch of power line that is 5.0 m above the surface of Earth. If the induced emf in the Earth–power line loop is 6.0 V, at what rate must the magnetic field through the loop be changing? Assume that the plane of the loop is perpendicular to Earth's surface. Example 20-2

31. • **Medical** During transcranial magnetic stimulation (TMS) treatment, a magnetic field typically of magnitude 0.50 T is produced in the brain using external coils. During the treatment the current in the coils (and hence the magnetic field in the brain) rises from zero to its peak in about 75 μs. Assume that the magnetic field is uniform over a circular area of diameter 2.0 cm inside the brain. What is the magnitude of the average induced emf around this area in the brain during the treatment? Example 20-2

32. •• A permanent bar magnet with the north pole pointing downward is dropped into a solenoid. (a) Determine the direction of the induced current that would be measured in the ammeter shown in Figure 20-26. (b) If the magnet is suddenly pulled upward through the solenoid, what is the direction of the induced current that would be measured in the ammeter?

Figure 20-26 Problem 32

33. •• A pair of parallel conducting rails that are 12 cm apart lies at right angles to a uniform magnetic field of 0.8 T directed into the page, as shown in Figure 20-27. A 15-Ω resistor is connected across the rails. A conducting bar is moved to the right at 2 m/s across the rails. (a) What is the current in the resistor? (b) What direction is the current in the bar (up or down)? (c) What is the magnetic force on the bar? Example 20-1

$\vec{B}$ = 0.8 T, into page

15 Ω   2 m/s   12 cm

Figure 20-27 Problem 33

34. ••• A 50-turn square coil with a cross-sectional area of 5.00 cm² has a resistance of 20.0 Ω. The plane of the coil is perpendicular to a uniform magnetic field of 1.00 T. The coil is suddenly rotated about the axis shown in Figure 20-28 through an angle of 60° over a period of 0.200 s. (a) What charge flows past a point in the coil during that time? (b) If the loop is rotated a full 360° around the axis, what is the net charge that passes the point in the loop? Explain your answer.

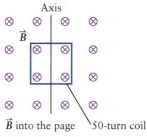

Axis

$\vec{B}$

$\vec{B}$ into the page   50-turn coil

Figure 20-28 Problem 34

35. •• An ordinary gold wedding ring is tossed end over end into a running MRI machine with a 4.0-T field. The ring spends 1.3 s in the field, during which it completes 60 full rotations. If the ring has a resistance of 6.0 Ω and a diameter of 18 mm, what is the average power dissipated by the ring? Example 20-3

36. •• Marisol is designing a system to detect when the door to her room is opened. She has observed that the magnetic field in the vicinity of the door is perpendicular to the door when it is closed and has a magnitude of about $B = 5.5 \times 10^{-5}$ T. Knowing Faraday's law, Marisol uses electromagnetic induction as the basis of her alarm. She wraps a single loop of wire around the perimeter of her door and connects it to a voltmeter that can detect an emf as small as $V_{min} = 1.00 \times 10^{-6}$ V. The voltmeter is configured to take a time-averaged reading and will trigger the alarm if the average emf over a certain period of time exceeds the threshold voltage V_{min}. What is the minimum time interval required for someone to open Marisol's door $\theta = 25.0°$, starting from the closed position, without the average induced emf over that time interval exceeding the threshold? The dimensions of the door are 0.900 m by 2.00 m. Example 20-2

Deyan Georgiev/AGE Fotostock

21 Alternating-Current Circuits

What do you think?

These transformers use Faraday's law to raise and lower the voltage of alternating current. Can they also be used to raise or lower the voltage of *direct* current? (a) Yes; (b) yes, but only to raise the voltage; (c) yes, but only to lower the voltage; (d) no.

In this chapter, your goals are to:

- (21-1) Explain the importance of alternating current.
- (21-2) Describe what is meant by the root mean square value.
- (21-3) Calculate the voltage change produced by a transformer.
- (21-4) Describe why an inductor opposes changes in the current passing through it.
- (21-5) Explain the flow of energy in an *LC* circuit.
- (21-6) Describe what happens in a driven series *LRC* circuit when the driving frequency changes.
- (21-7) Discuss why current can flow in only one direction in a *pn* junction diode.

To master this chapter, you should review:

- (12-3, 12-4) Simple harmonic motion and the energy of a mass–spring system.
- (12-8) How a damped oscillator responds to different driving frequencies.
- (17-5, 17-6) The definition of capacitance and the electric energy stored by capacitors.
- (18-3, 18-6) The current through and voltage across a resistor, and the power into a resistor.
- (18-5) Using Kirchhoff's loop rule to analyze circuits.
- (19-7) The magnetic field created by a solenoid.
- (20-2, 20-3) How Faraday's law and Lenz's law describe the induced emf that opposes a change in magnetic flux through a loop.
- (20-4) How ac generators work and the power they generate.

21-1 Most circuits use alternating current

In Chapter 20 we learned the principles of electromagnetic induction and how they can be used to *generate* an alternating current. In this chapter we'll learn how to *manipulate* and *use* alternating current. Most of the electrical power on our planet is transmitted and used in the form of an alternating current, so by studying this chapter you'll learn an important aspect of how the world around you works.

We'll see how *mutual inductance*—in which a changing emf in one coil makes it possible to induce an emf in another coil—is key for understanding how electric power can be transmitted efficiently over long distances. This same principle explains how the

885

FIGURE 21-1 Alternating current When you (a) use the transformer in your mobile phone charger or (b) tune your television to a different channel, you are using the physics of alternating current.

(a)

(b)

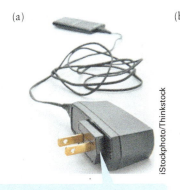

This transformer uses mutual inductance—in which a changing emf in one coil induces an emf in a second coil nearby—to reduce the voltage from a wall socket to a lower value suitable for charging a battery.

When you change the channel on a television, you're telling the TV tuning circuit to change its capacitance. This changes the natural frequency of the circuit so that it matches the carrier frequency of the channel you want to watch.

TAKE-HOME MESSAGE FOR Section 21-1

✔ The same principles of electromagnetic induction that explain how to produce an alternating current (ac) also tell us how to manipulate and use such currents.

✔ Many circuits that include an ac source can be modeled as a combination of resistors, capacitors, and inductors (a third type of circuit element).

---

relatively high voltage of 120 to 240 V available from a wall socket can be used to charge a cell phone that operates at low voltage, typically 5 V or less (Figure 21-1a). We'll also introduce *self-inductance*, an effect in which a changing emf in a coil produces an emf reaction in the coil itself. Self-inductance is at the heart of an important device called an *inductor* that has many applications in circuits.

Many circuits that are connected to an ac source (such as the ac voltage provided by a wall socket) can be modeled as a combination of resistors, capacitors, and inductors. We'll see what happens when any one of these circuit elements by itself is connected to an ac source. We'll then go on to examine an *LRC series circuit* in which all three of these circuit elements are present. Such circuits hold the key to understanding what happens when you tune a radio or television to receive a particular station (Figure 21-1b). Finally, we'll take a brief look at *semiconductors*, a class of material that is essential for the operation of two other important kinds of circuit elements, called *diodes* and *transistors*.

21-2 We need to analyze ac circuits differently than dc circuits

Current arises in a circuit as a result of an applied voltage. The batteries that power your mobile phone or your flashlight are sources of (approximately) constant voltage. A 9-V battery, for example, introduces a roughly constant potential difference of 9 V between the two points at which it connects to a circuit. By convention, we refer to a circuit driven by a fixed voltage source, or one that does not change direction, as a **direct current**, or **dc** circuit. However, when you plug an electrical device into a wall socket, you are accessing an **alternating current**, or **ac** source.

The voltage from an ac source varies with time in a sinusoidal fashion, as we discussed in our description of ac generators in Section 20-4. In that section we wrote the emf delivered by an ac generator as

Emf in an ac generator
(20-7)

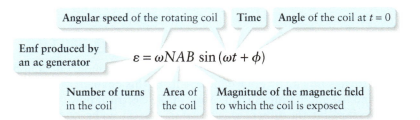

Angular speed of the rotating coil | Time | Angle of the coil at $t = 0$

Emf produced by an ac generator

$$\varepsilon = \omega NAB \sin(\omega t + \phi)$$

Number of turns in the coil | Area of the coil | Magnitude of the magnetic field to which the coil is exposed

Let's choose the time $t = 0$ to be when the angle of the coil is $\phi = 0$, use the symbol V_0 for the combination of factors ωNAB, and use the symbol $V(t)$ for the time-varying emf. (We use V since emf is measured in volts.) If the generator is connected to two terminals, like the two terminals of a wall socket, then $V(t)$ represents the voltage between

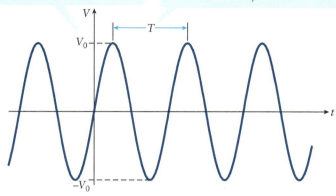

The value V of this ac voltage oscillates between V_0 (the voltage amplitude) and $-V_0$.

The period T of this ac voltage equals the reciprocal of the frequency f:
$$T = \frac{1}{f} = \frac{2\pi}{\omega}$$

FIGURE 21-2 **An ac voltage**
This figure graphs the ac voltage given by Equation 21-1. Note that at $t = 0$ the voltage is zero and increasing.

those terminals. We can then rewrite Equation 20-7 as a general equation that we'll use for ac sources of all kinds:

Time-varying voltage provided by an **ac source**

Angular frequency of the voltage

$$V(t) = V_0 \sin \omega t \quad \text{Time}$$

Voltage amplitude = maximum positive value of $V(t)$

Voltage of an ac source
(21-1)

Figure 21-2 graphs the voltage $V(t)$ as a function of time. The angular frequency ω of the voltage (in rad/s) is related to the frequency f of the voltage (in Hz) by the same relationship we used in Section 12-3 for simple harmonic motion:

$$\omega = 2\pi f \tag{21-2}$$

The period T of the oscillation is equal to $1/f$. For example, in the United States and Canada, ac voltage is applied at a frequency $f = 60$ Hz, so the period of oscillation is $T = 1/f = 1/(60 \text{ Hz}) = 0.017$ s, and the angular frequency is $\omega = 2\pi f = (2\pi \text{ rad})(60 \text{ Hz}) = 3.8 \times 10^2$ rad/s.

If we attach a source of ac voltage described by Equation 21-1 to a resistor of resistance R, we can still use Equation 18-24 to calculate the power that flows into the resistor:

Power into a resistor

Current through the resistor

$$P = i^2 R = \frac{V^2}{R} \quad \text{**Voltage** across the resistor}$$

Resistance of the resistor

Power into a resistor
(18-24)

The only difference between a resistor in a dc circuit and one in an ac circuit is that because the voltage varies with time, the power into the resistor also varies with time. The *instantaneous* power into the resistor at time t is

$$P(t) = \frac{V^2(t)}{R} = \frac{V_0^2 \sin^2(\omega t)}{R} \tag{21-3}$$

The value of $P(t)$ oscillates between zero (when $\sin \omega t = 0$) and V_0^2/R (when $\sin \omega t = 1$ or -1).

Often it's convenient to talk not about the instantaneous power $P(t)$ but the *average* power P_{average}. For example, if an appliance is designed to be powered by an ac voltage, its power rating in watts (such as a 1000-W microwave oven or a 1500-W hair dryer) is always stated in terms of the average power delivered to that appliance when in operation. The voltage amplitude V_0 and the resistance R are constants, so to find $P_{average}$ we have to figure out only the average value of $\sin^2(\omega t)$. As we discussed in Section 20-4, this average value is $\frac{1}{2}$ (see Figure 20-10). So from Equation 21-3 the average power into the resistor is

(21-4)
$$P_{average} = \frac{1}{2}\frac{V_0^2}{R}$$

The average power given by Equation 21-4 is one-half of the maximum value of the instantaneous power given by Equation 21-3.

Equation 21-4 suggests an alternative way to describe the voltage provided by an ac source. Comparing this equation to Equation 21-3, we can write

$$\left(\frac{V^2(t)}{R}\right)_{average} = \frac{1}{2}\frac{V_0^2}{R}$$

If we multiply both sides of this equation by the resistance R, we get

$$(V^2(t))_{average} = \frac{V_0^2}{2}$$

The left-hand side of this equation is called the *mean square value* of the ac voltage: We take the square of the voltage and then calculate the average (or *mean*) of that quantity. The square root of the mean square is called the **root mean square value**, or **rms value**, of the voltage:

The **root mean square (rms) value** of $V(t)$... ...is the square root of the average value of $V^2(t)$.

Root mean square value
(21-5)

$$V_{rms} = \sqrt{(V^2(t))_{average}} = \frac{V_0}{\sqrt{2}}$$

If $V(t)$ is a sinusoidal function, its rms value equals the maximum value of $V(t)$ (the amplitude) divided by $\sqrt{2}$.

WATCH OUT! **The average voltage is zero, but the average power is not.**

The average voltage of an ac source described by Equation 21-1 is zero. But the average power delivered by that source is *not* zero because power is proportional to the square of the voltage, which is always positive.

As an example, in the United States the standard value for the ac voltage supplied by a wall socket is 120 V; this is actually the rms voltage, so $V_{rms} = 120$ V. The voltage amplitude, or peak voltage, is larger by a factor of $\sqrt{2}$, so $V_0 = V_{rms}\sqrt{2} = (120\text{ V})\sqrt{2} = 170$ V.

In terms of the rms voltage, we can write the average power from Equation 21-4 as

(21-6)
$$P_{average} = \frac{1}{2}\frac{V_0^2}{R} = \left(\frac{V_0^2}{2}\right)\frac{1}{R} = \left(\frac{V_0}{\sqrt{2}}\right)^2\frac{1}{R} = \frac{V_{rms}^2}{R}$$

This says that Equation 18-24, $P = V^2/R$, tells us the *average* power if we replace V by the rms value V_{rms} of the voltage.

EXAMPLE 21-1   Traveling Abroad with Your Hair Dryer

Seasoned travelers know that an electric device that works in one country may not work in another. A certain hair dryer made in the United States, where the rms wall voltage is 120 V, is rated at 1.5 kW. Find the power consumed by the hair dryer when connected to a wall outlet in Australia, where the voltage amplitude is 325 V. Treat the resistance of the hair dryer as the same regardless of the wall voltage. (In reality, the resistance depends somewhat on the temperature of the dryer's heating elements, which depends on the operating voltage.)

Set Up

We'll use Equation 21-6 to find the resistance of the hair dryer from the given values of $P_{average}$ and V_{rms} in the United States. We'll then use this same equation to find the average power in Australia. To do this we'll need to know the rms voltage in Australia; we'll find this from the given value of the voltage amplitude V_0 by using Equation 21-5.

Average power into a resistor:

$$P_{average} = \frac{V_{rms}^2}{R} \qquad (21-6)$$

Root mean square (rms) voltage:

$$V_{rms} = \sqrt{(V^2(t))_{average}} = \frac{V_0}{\sqrt{2}} \qquad (21-5)$$

Solve

Use Equation 21-6 to determine the resistance of the hair dryer. We assume this has the same value no matter what the operating voltage.

Solve Equation 21-6 for resistance R:

$$R = \frac{V_{rms}^2}{P_{average}}$$

Substitute the values in the United States, $V_{rms} = 120$ V and $P_{average} = 1.5$ kW $= 1.5 \times 10^3$ W:

$$R = \frac{(120 \text{ V})^2}{1.5 \times 10^3 \text{ W}} = 9.6 \text{ }\Omega$$

(Recall that $1 \text{ W} = 1 \text{ V}^2/\Omega$.)

Find the value of V_{rms} in Australia; then use this to determine the average power into the hair dryer when used in Australia.

In Australia the voltage amplitude is $V_0 = 325$ V, so the rms voltage is

$$V_{rms} = \frac{V_0}{\sqrt{2}} = \frac{325 \text{ V}}{\sqrt{2}} = 230 \text{ V}$$

With this value of V_{rms}, the average power into the hair dryer is

$$P_{average,Australia} = \frac{V_{rms,Australia}^2}{R} = \frac{(230 \text{ V})^2}{9.6 \text{ }\Omega} = 5.5 \times 10^3 \text{ W} = 5.5 \text{ kW}$$

Reflect

Our result of 5.5 kW is *considerably* more power than the hair dryer is designed to take in. You may know how hot the air from a 1500-W hair dryer can be, or perhaps you've felt how hot a 100-W bulb can get after it's been on for a while. So you can probably imagine how hot 5500 W would make the heating element of the hair dryer. To radiate that much power, the required temperature of the heating element might well exceed the melting point of the material! We'll see in the following section how this hair dryer can be used safely even with the higher voltage provided in Australia.

GOT THE CONCEPT? 21-1 Average and Instantaneous Power

A resistor is connected to a source of ac voltage described by Equation 21-1. At which of the following times is the instantaneous power into the resistor equal to the average power into the resistor? (a) $t = 0$; (b) $t = T/8$; (c) $t = T/4$; (d) more than one of these; (e) none of these.

TAKE-HOME MESSAGE FOR Section 21-2

✔ The arithmetic mean of a sinusoidally varying voltage is zero because it is negative as often as it is positive. The average power is not zero because it is proportional to the square of the voltage, which cannot be negative.

✔ It is common to characterize ac voltage by its root mean square (rms) value. The peak voltage is $\sqrt{2}$ times the rms voltage.

21-3 Transformers allow us to change the voltage of an ac power source

If you plug your mobile phone's charger into a wall socket, energy from an ac generator that may be hundreds of kilometers away flows through power lines (Figure 21-3) into your home and your phone. The *voltage* associated with this energy has different values at different places in the circuit, however; the voltage in the power lines is typically hundreds of kilovolts, the voltage provided by a wall socket in the United States or Canada is 120 V, and the voltage supplied to your phone by the charger is 5 V. In this section we'll see the reason different voltages are used in this way, and we'll learn about an important device called a *transformer* that makes it possible to raise or lower an alternating-current voltage.

FIGURE 21-3 Power lines When electric energy is transmitted over long distances in the form of an alternating current, it is most efficient to use very high voltage (typically 110 kV or higher). The power lines shown here are elevated so that people and vehicles cannot touch them, which would present a safety hazard.

To understand why high voltages are used in power lines, recall this relationship from Section 18-6:

Power for a circuit element (18-23)

Power produced by or transferred into a circuit element

$$P = iV$$

Current through the circuit element

Voltage (absolute value) across the circuit element

Source of emf: Power flows out of the source and into the moving charges.
Resistor: Power flows out of the charges and into the resistor.

Equation 18-23 tells us that the same electric power P can be delivered at any voltage V, provided that the product of current and voltage remains the same. This is true for ac circuits as well as dc circuits. For a power line it's best to use high voltage and low current. The reason is that a long power line has a substantial resistance R, so some of the energy being carried by the current goes into heating the power line rather than being transmitted to the end user. The rate at which energy is dissipated in a resistor with resistance R is given by Equation 18-24 (see Section 21-2), which we can write as $P = i^2R$. To minimize this energy loss the current i should be as small as possible, so from Equation 18-23 the voltage V must be as large as possible.

WATCH OUT! In power lines, don't confuse the voltage between adjacent cables with the voltage between the ends of a single cable.

You might think "Equation 18-24 also says that $P = V^2/R$ for a resistor. So why isn't it preferable to have the voltage V be small to minimize energy loss?" The explanation is that in Equation 18-23, $P = iV$, V represents the voltage between the *two* cables that make up a power line. (Figure 21-3 shows that the cables come in pairs.) You need both cables to make a complete circuit, just as the power cord for a household appliance has two wires in it (one wire connects to one prong of the plug at the end of the cord and the other wire to the other prong). This is the voltage that must be large to make the current small. By contrast, in Equation 18-24 V is the potential difference between two ends of a *single* cable and is related to the current in the cable by the relationship $V = iR$ for resistors (Equation 18-9; see Section 18-3). This voltage is low if the current is small, so the power $P = i^2R = V^2/R$ that goes into heating the resistor is small.

BioMedical

In the home, however, the voltage must be kept small. That's because there is a risk that a person could be exposed to that voltage (for instance, by touching an appliance whose internal wiring had failed). If a voltage V is applied between two parts of your body, the resulting current through your body is proportional to V. A current of just 0.1 A can cause ventricular fibrillation, and a current of 0.2 A or more can cause severe burns and stop the heart altogether. To reduce the risk of accidents, the voltage supplied to the home has to be kept at a much lower value (120 to 240 V) than that used in power lines (more than 110 kV = 110,000 V).

A device that can raise or lower an ac voltage to a desired value is called a **transformer** (see Figure 21-1a). The transformer shown in Figure 21-4 consists of two coils formed by winding separate wires around a piece of iron. The **primary coil** on the left is connected to the input voltage. The **secondary coil** on the right is the source of the output voltage, even though there's no direct electric connection between the two coils. Let's see how this works.

When an ac voltage is applied to the primary coil in the transformer, the current in the windings gives rise to a magnetic field. This field oscillates with the same angular frequency ω as the ac voltage and current and so creates a time-varying magnetic flux through the windings of the primary coil. Because iron is a magnetic material, the iron core confines the magnetic field lines so that the same time-varying flux that passes

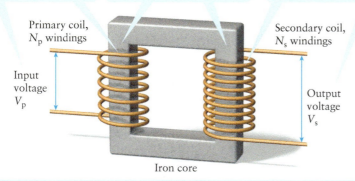

② The alternating current in the primary coil creates an alternating magnetic field of the same angular frequency ω. The field lines are "trapped" by the iron core and all pass through the secondary coil.

③ The alternating magnetic flux through the windings of the secondary coil induces an alternating emf, or output voltage, of the same angular frequency ω but of amplitude V_s.

① An input ac voltage with amplitude V_p and angular frequency ω causes an alternating current in the primary coil.

FIGURE 21-4 A transformer This device uses two coils and a piece of magnetic material such as iron to raise or lower the voltage from an ac source.

Primary coil, N_p windings

Secondary coil, N_s windings

Input voltage V_p

Output voltage V_s

Iron core

The amplitude V_s of the output voltage equals (N_s/N_p) times the amplitude V_p of the input voltage.
- If $N_s > N_p$, then $V_s > V_p$. This is a step-up transformer.
- If $N_s < N_p$, then $V_s < V_p$. This is a step-down transformer.

through each winding of the primary coil also passes through each winding of the secondary coil. Faraday's law (see Sections 20-2 and 20-3) then tells us that a time-varying emf is induced in the secondary coil, which means this coil acts as a voltage source. The output voltage produced by the secondary coil oscillates at the same frequency as the input voltage that feeds into the primary coil. This effect, in which a change in the current in one coil induces an emf and current in a second coil, is called **mutual inductance**.

The *amplitude* of the output voltage in Figure 21-4 will be different from that of the input voltage—that is, the amplitude will be "transformed" by the transformer—if there are different numbers of windings in the primary and secondary coils (N_p and N_s, respectively, in Figure 21-4). To see how this can be, first notice that the time-varying magnetic flux Φ_B in the primary coil induces an emf there. From Equation 20-2 the magnitude of the emf induced in each of the windings of the primary coil is $|\Delta\Phi_B/\Delta t|$, the magnitude of the rate of change of the magnetic flux through that winding. There are N_p windings in the primary coil, so the total emf in that coil has magnitude $N_p|\Delta\Phi_B/\Delta t|$. By Kirchhoff's loop rule (Section 18-5), the sum of the voltage drops around the left-hand circuit that includes the primary coil must be zero, so the emf in the primary coil must have the same magnitude as the input voltage $V_p(t)$. (We're ignoring any voltage drops due to the resistance of the wires in this circuit.) So we can write

$$|\text{input voltage}| = |V_p(t)| = N_p\left|\frac{\Delta\Phi_B}{\Delta t}\right| \tag{21-7}$$

Since the iron core ensures that there is the same magnetic flux Φ_B through each winding of the secondary coil as through each winding of the primary coil, the rate of change of the flux is the same and so the emf per winding is the same as well. There are N_s windings in the secondary coil, so the total emf in the secondary coil has magnitude $N_s|\Delta\Phi_B/\Delta t|$. This emf is the output voltage $V_s(t)$, so the magnitude of the output voltage is

$$|\text{output voltage}| = |V_s(t)| = N_s\left|\frac{\Delta\Phi_B}{\Delta t}\right| \tag{21-8}$$

If we divide Equation 21-8 by Equation 21-7, the factors of $|\Delta\Phi_B/\Delta t|$ cancel:

$$\frac{|\text{output voltage}|}{|\text{input voltage}|} = \frac{|V_s(t)|}{|V_p(t)|} = \frac{N_s}{N_p} \tag{21-9}$$

Equation 21-9 is valid for the *instantaneous* values of the input and output voltages, so it must also be true for the *maximum* values. These are the voltage amplitudes, which we call V_p and V_s for the input (primary) voltage and output (secondary) voltage, respectively. Equation 21-9 is also true for the *rms* values of these voltages. So we are left with the result

Input and output voltages for a transformer (21-10)

Amplitudes of the input (p) and output (s) voltages

$$\frac{V_s}{V_p} = \frac{V_{s,rms}}{V_{p,rms}} = \frac{N_s}{N_p}$$

Number of windings in the primary coil (p) and secondary coil (s)

Rms values of the input (p) and output (s) voltages

(a) Voltage $V_p(t)$ and current $i_p(t)$ in the primary coil

$V_p(t)$ $i_p(t)$

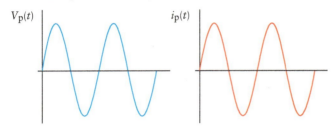

(b) Step-up transformer: Voltage $V_s(t)$ and current $i_s(t)$ in the secondary coil

$V_s(t)$ Secondary voltage > primary voltage $i_s(t)$ Secondary current < primary current

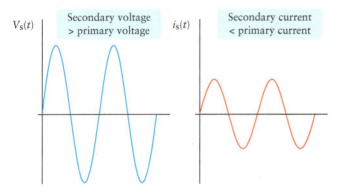

(c) Step-down transformer: Voltage $V_s(t)$ and current $i_s(t)$ in the secondary coil

$V_s(t)$ Secondary voltage < primary voltage $i_s(t)$ Secondary current > primary current

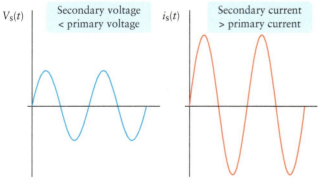

FIGURE 21-5 Current and voltage in a transformer The oscillating voltage and current in (a) the primary coil of a transformer, (b) the secondary coil of a step-up transformer (N_s greater than N_p), and (c) the secondary coil of a step-down transformer (N_s less than N_p).

Go to Interactive Exercise 21-1 for more practice dealing with coils and current.

In other words, in a transformer the ratio of the number of windings in the secondary coil (N_s) to the number of windings in the primary coil (N_p) determines the output voltage relative to the input voltage. In a **step-up transformer** N_s is greater than N_p, so the output voltage is greater than the input voltage. In a **step-down transformer**, N_s is less than N_p, and the output voltage is less than the input voltage. A step-up transformer is used at an electric power generation station to raise the voltage provided by the generator (which is connected to the primary coil) to a much larger value in the transmission lines (which are connected to the secondary coil). Close to where the electric power is to be used in home or industry, step-down transformers are used to reduce the voltage to a safe value. The charger for your mobile phone or other electronic device is also a step-down transformer that lowers the voltage provided by the wall socket to a much lower value suitable for the battery to be recharged.

The *current* in the secondary coil is also different from that in the primary coil (**Figure 21-5**). If we neglect any energy losses in the transformer (for instance, due to heating of the iron core by eddy currents), conservation of energy requires that the rate at which energy is delivered to the primary coil by the input voltage must equal the rate at which energy is transferred from the primary to the secondary coil. The rate of energy delivery is power, and power equals current times voltage (Equation 18-23), so

input power in primary coil = output power into secondary coil

(21-11) $i_p(t)V_p(t) = i_s(t)V_s(t)$

In Equation 21-11, $i_p(t)$ and $V_p(t)$ are the current and voltage in the primary coil at time t, and $i_s(t)$ and $V_s(t)$ are the corresponding quantities in the secondary coil. If we rearrange Equation 21-11 and compare to Equation 21-9, we see that

(21-12) $\dfrac{|\text{output current}|}{|\text{input current}|} = \dfrac{|i_s(t)|}{|i_p(t)|} = \dfrac{|V_p(t)|}{|V_s(t)|} = \dfrac{N_p}{N_s}$

Compare Equation 21-12 with Equation 21-10. For a step-up transformer, for which N_s is greater than N_p, the output voltage is greater than the input voltage, but the output current is smaller than the input current. The reverse is true for a step-down transformer.

EXAMPLE 21-2 A High-Voltage Transformer

Each of the 17 generators employed in the hydroelectric power plant at Hoover Dam in Colorado can generate up to 133 MW of average power. This power is delivered at 8.00 kV rms to a transformer that connects to a long-distance transmission line that operates at 5.00×10^2 kV rms. (a) What is the ratio of the number of windings in the secondary coil of the transformer to the number in the primary coil? (b) What is the rms current in the high-voltage transmission line?

Set Up

In part (a) we'll use Equation 21-10 to relate the ratio N_s/N_p (the number of windings in the secondary coil divided by the number in the primary coil) to the input and output rms voltages. In part (b) we'll assume that the transmission line delivers its power to a resistor, so we can relate the current, voltage, and power using Equations 18-9 and 18-23.

Input and output voltages for a transformer:

$$\frac{V_s}{V_p} = \frac{V_{s,rms}}{V_{p,rms}} = \frac{N_s}{N_p} \quad (21\text{-}10)$$

Relationship among potential difference, current, and resistance:

$$V = iR \quad (18\text{-}9)$$

Power for a circuit element:

$$P = iV \quad (18\text{-}23)$$

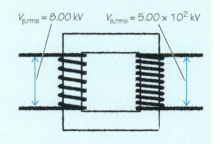

$V_{p,rms} = 8.00$ kV $V_{s,rms} = 5.00 \times 10^2$ kV

Solve

(a) We know both the primary and secondary rms voltages, so we can rearrange Equation 21-10 to solve for the ratio of the number of windings.

From Equation 21-10:

$$\frac{N_s}{N_p} = \frac{V_{s,rms}}{V_{p,rms}} = \frac{5.00 \times 10^2 \text{ kV}}{8.00 \text{ kV}} = 62.5$$

There are 62.5 times as many windings in the secondary coil as in the primary coil (or, equivalently, 125 windings in the secondary for every two windings of the primary).

(b) Since the resistance R is a constant, the current and voltage both have the same sin ωt time dependence. We use this to relate the average power $P_{average} = 133$ MW to the rms values of the current and voltage in the transmission line.

If the voltage is $V(t) = V_0 \sin \omega t$, from Equation 18-9 the current is

$$i(t) = \frac{V(t)}{R} = \frac{V_0 \sin \omega t}{R} = \frac{V_0}{R} \sin \omega t = i_0 \sin \omega t$$

In this expression i_0 is the amplitude of the current. From Equation 18-23 the instantaneous power is

$$P(t) = i(t)V(t) = (V_0 \sin \omega t)(i_0 \sin \omega t) = i_0 V_0 \sin^2 \omega t$$

To find the average power $P_{average}$, replace $\sin^2 \omega t$ by its average value $\frac{1}{2}$:

$$P_{average} = \frac{i_0 V_0}{2} = \left(\frac{i_0}{\sqrt{2}}\right)\left(\frac{V_0}{\sqrt{2}}\right) = i_{rms} V_{rms}$$

The average power is the product of the rms current and the rms voltage. Since $i(t)$ has the same time dependence as $V(t)$, the rms value of current is equal to its maximum value divided by $\sqrt{2}$, just as for the voltage. From this expression the rms current in the transmission line is

$$i_{s,rms} = \frac{P_{average}}{V_{s,rms}} = \frac{133 \text{ MW}}{5.00 \times 10^2 \text{ kV}} = \frac{133 \times 10^6 \text{ W}}{5.00 \times 10^5 \text{ V}} = 266 \text{ A}$$

Reflect

In the process of delivering power to the end user, the voltage must ultimately be stepped down to 120 V. This process is not normally done with a single transformer. One reason is that if only one transformer were used for this stepping-down process, there would need to be more than 4000 windings in the primary coil for every one in the secondary.

Ratio of coil windings to step down 5.00×10^2 kV to 120 V:

$$\frac{N_p}{N_s} = \frac{V_p}{V_s} = \frac{5.00 \times 10^2 \text{ kV}}{120 \text{ V}} = 4170$$

We found a current of 266 A in the transmission line. Although that sounds large, it is only about a factor of 10 more than the current in household wiring. In addition, the power loss in the transmission wire is relatively low at this current. A few hundred kilometers of transmission wire might have a resistance of 25 Ω. The loss in such a line would be about 1.8 MW. Although in absolute terms this is a significant power loss, it represents only 1.4% of the total power transmitted.

The voltage between the two ends of the transmission line is $V_{ends} = iR$, where $R = 25\ \Omega$ is the resistance of the line. Use the formula we derived in part (b) to find the power lost in the transmission line:

$$P_{lost,average} = i_{rms}V_{ends,rms} = i_{rms}(i_{rms}R)$$
$$= i_{rms}^2 R = (266\ \text{A})^2(25\ \Omega) = 1.8 \times 10^6\ \text{W} = 1.8\ \text{MW}$$

Express this as a percentage of the power that is transmitted:

$$\text{percentage} = \frac{\text{power lost}}{\text{power transmitted}} \times 100\%$$
$$= \frac{1.8\ \text{MW}}{133\ \text{MW}} \times 100\% = 1.4\%$$

GOT THE CONCEPT? 21-2 Transforming a Transformer

(?) A certain step-up transformer has 100 windings in its primary coil and 250 windings in its secondary coil. You reverse the connections to the transformer so that the primary coil becomes the secondary and vice versa. If you now apply an ac input voltage of 120 V rms to the reversed transformer, what will be the rms output voltage? (a) 40 V; (b) 48 V; (c) 100 V; (d) 250 V; (e) 300 V.

TAKE-HOME MESSAGE FOR Section 21-3

✔ A transformer uses mutual inductance to raise or lower an ac voltage.

✔ In a step-up transformer, there are more windings in the secondary coil than in the primary coil. The output voltage is greater than the input voltage, and the output current is less than the input current.

✔ In a step-down transformer, there are fewer windings in the secondary coil than in the primary coil. The output voltage is less than the input voltage, and the output current is greater than the input current.

21-4 An inductor is a circuit element that opposes changes in current

We saw in Section 21-3 that a transformer works by the mutual inductance of two coils: A change in current in one coil causes a change in the magnetic flux through the other coil, which induces an emf. The same physics applies equally well to individual windings within a *single* coil. You can think of each turn of the coil as a separate loop; as a changing current passes through any loop, the changing magnetic field that arises induces an emf in the other loops. This induced emf, according to Lenz's law, opposes the change in flux that created it. So if the current is increasing in the coil, the induced emf will oppose an increase in current. If the current is decreasing, the induced emf will be directed so that the current is augmented. This **self-inductance** has the net effect of opposing a change in current in a coil. Let's see how to determine the emf induced in this way.

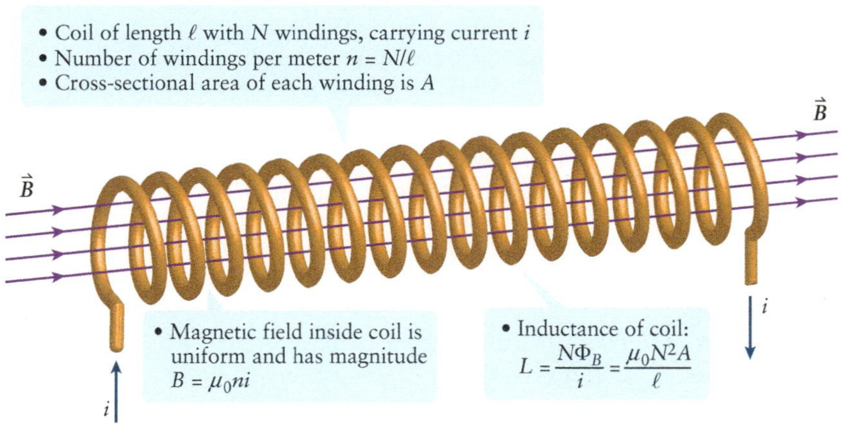

Inductance of a Coil

Consider a coil of N windings in which there is a current i (Figure 21-6). This current causes the coil to produce a magnetic field, and the flux of this field through each winding of the coil is Φ_B. The total flux through

- Coil of length ℓ with N windings, carrying current i
- Number of windings per meter $n = N/\ell$
- Cross-sectional area of each winding is A

$\vec{B}$

$\vec{B}$

- Magnetic field inside coil is uniform and has magnitude $B = \mu_0 ni$

- Inductance of coil:
$$L = \frac{N\Phi_B}{i} = \frac{\mu_0 N^2 A}{\ell}$$

i

i

FIGURE 21-6 A current-carrying coil The inductance L of a coil equals the total magnetic flux through the coil's windings divided by the current in the coil that produces the flux. (This long, straight coil is an idealized version of the solenoid shown in Figure 19-18.)

all N windings is then $N\Phi_B$. We define the **inductance** of the coil (symbol L) as the total flux divided by the current that produces it:

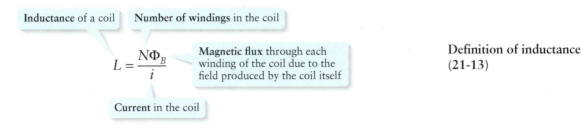

Inductance of a coil   Number of windings in the coil

$$L = \frac{N\Phi_B}{i}$$

Magnetic flux through each winding of the coil due to the field produced by the coil itself

Current in the coil

Definition of inductance
(21-13)

The units of inductance are henrys, abbreviated as H: $1\ \text{H} = 1\ \text{T}\cdot\text{m}^2/\text{A}$.

The magnetic field is directly proportional to the current i that produces it, and the same is true of the flux of that magnetic field. So the numerator in Equation 21-13 is proportional to the current i. Since the denominator in Equation 21-13 is equal to i, the current will cancel out when we calculate the inductance L. The inductance, therefore, does not depend on i but only on geometrical factors such as the dimensions of the coil and physical constants. As an example, let's calculate the inductance of a long, straight coil or solenoid. We learned in Section 19-7 that the magnetic field of an ideal solenoid (one whose length is much greater than its diameter) is uniform and has a magnitude given by Equation 19-17:

Magnetic field inside a long, straight solenoid   Number of windings per meter

$$B = \mu_0 n i$$

Permeability of free space   Current in the solenoid windings

Magnetic field inside a long, straight solenoid
(19-17)

If the solenoid has length ℓ, the quantity n in Equation 19-17 equals the total number of windings N divided by ℓ. We can then write the magnetic flux through the cross-sectional area A of each winding of the solenoid as

$$\Phi_B = BA = (\mu_0 n i)A = \left(\frac{\mu_0 N i}{\ell}\right)A$$

If we substitute this into Equation 21-13, we get the following expression for the inductance of the solenoid:

$$L = \frac{N\Phi_B}{i} = \frac{N}{i}\left(\frac{\mu_0 N i}{\ell}\right)A = \frac{\mu_0 N^2 A}{\ell} \tag{21-14}$$

(inductance of a solenoid)

For the solenoid the inductance depends only on its dimensions and a physical constant (μ_0, the permeability of free space); inductance does not depend on current.

The quantity μ_0, which equals $4\pi \times 10^{-7}\ \text{T}\cdot\text{m/A}$ to nine significant figures, appears in all expressions for the inductance of a coil. Since $1\ \text{H} = 1\ \text{T}\cdot\text{m}^2/\text{A}$, it follows that $1\ \text{T}\cdot\text{m/A} = 1\ \text{H/m}$. So an alternative way to express the permeability of free space is $\mu_0 = 4\pi \times 10^{-7}\ \text{H/m}$. This is useful for problems that involve inductance, as in the following example.

EXAMPLE 21-3 How Large Is One Henry?

You decide to create a solenoid with inductance of 1.00 H by tightly wrapping wire around a plastic pipe. The wire has a diameter of 2.60 mm, and the pipe has an outside radius of 3.00 cm. How long a pipe will you need?

Set Up

Winding the wire tightly around the pipe will create a solenoid. The length of the solenoid may likely be large compared to its diameter, so we'll treat it as ideal. Then we can use Equation 21-14 to calculate the inductance of this solenoid. The pipe has a circular cross section, so to find A we'll use the expression for the area of a circle.

Inductance of a solenoid:

$$L = \frac{\mu_0 N^2 A}{\ell} \qquad (21\text{-}14)$$

Area of a circle of radius r:

$$A = \pi r^2$$

Solve

We don't know how many windings the solenoid will have. However, because the solenoid is tightly wrapped, we can write the number of windings as the total length of the solenoid divided by the diameter of the wire. We substitute this expression for N into the inductance equation and then solve for the length of the pipe.

The number of windings that will fit equals the length ℓ of the pipe divided by the diameter D_W of the wire:

$$N = \frac{\ell}{D_W}$$

Substitute this into Equation 21-14 for the inductance of the solenoid:

$$L = \frac{\mu_0 (\ell/D_W)^2 A}{\ell} = \frac{\mu_0 \ell A}{D_W^2}$$

The length of the pipe is therefore

$$\ell = \frac{L D_W^2}{\mu_0 A}$$

The cross-sectional area of the pipe of radius 3.00 cm is

$$A = \pi r^2 = \pi(3.00 \text{ cm})^2 = \pi(3.00 \times 10^{-2} \text{ m})^2 = 2.83 \times 10^{-3} \text{ m}^2$$

We are given $L = 1.00$ H, $D_W = 2.60 \times 10^{-3}$ m, and $\mu_0 = 4\pi \times 10^{-7}$ H/m, so

$$\ell = \frac{(1.00 \text{ H})(2.60 \times 10^{-3} \text{ m})^2}{(4\pi \times 10^{-7} \text{ H/m})(2.83 \times 10^{-3} \text{ m}^2)}$$

$$= 1.90 \times 10^3 \text{ m} = 1.90 \text{ km}$$

Reflect

We would need a pipe nearly 2 km long to construct a 1.00-H inductor! One henry is a *very* large inductance; practical inductors have much smaller values of inductance. (Note that the length we determined is indeed much greater than the radius of the pipe, so we were justified in treating this as an ideal solenoid.)

The emf of an Inductor

An inductor in a circuit produces an emf if the current through the inductor changes. We can find an expression for this emf if we combine the definition of inductance, Equation 21-13, with Faraday's law and Lenz's law (Section 20-3). From Equation 21-13 the total flux through an inductor is

$$N\Phi_B = Li$$

Faraday's and Lenz's laws tell us that the emf produced by an inductor is equal to the negative of the rate of change of the total flux through the inductor (Equation 20-3). So we can write the induced emf in the inductor as

$$\varepsilon = -\frac{\Delta(N\Phi_B)}{\Delta t} = -\frac{\Delta(Li)}{\Delta t} \qquad (21\text{-}15)$$

Let's ignore the resistance of the wire that makes up the inductor, so this is an *ideal* inductor. Then ε in Equation 21-15 is also equal to the voltage V across the inductor (there is no additional voltage drop due to resistance, since there is zero resistance). Since the inductance L does not change with time, we can write Equation 21-15 as

Voltage across an inductor (21-16)

Voltage across an inductor   Inductance of the inductor

$$V = -L\frac{\Delta i}{\Delta t}$$

Rate of change of the current i in the inductor

The negative sign means that the voltage opposes any change in the current.

(a) An inductor with constant current

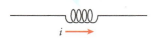

- The symbol for an inductor is a coil.
- An ideal inductor has zero resistance.

$i \longrightarrow$

If the current in the inductor is constant, there is zero voltage across the inductor: $V = 0$. This includes the case where there is no current.

(b) An inductor with increasing current

The voltage *drops* from left to right along the inductor.

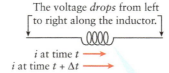

i at time $t \longrightarrow$
i at time $t + \Delta t \longrightarrow$

If current in the inductor is increasing so that $\Delta i / \Delta t > 0$, there is negative voltage across the inductor: $V < 0$. This voltage opposes the increase in the current.

(c) An inductor with decreasing current

The voltage *rises* from left to right along the inductor.

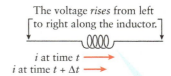

i at time $t \longrightarrow$
i at time $t + \Delta t \longrightarrow$

If current in the inductor is decreasing so $\Delta i / \Delta t < 0$, there is positive voltage across the inductor: $V > 0$. This voltage opposes the decrease in the current.

FIGURE 21-7 Voltage across an inductor The voltage across an ideal inductor (a coil with no resistance) depends on whether the current through the inductor is (a) constant, (b) increasing, or (c) decreasing.

Let's examine Equation 21-16 to see what it means. In Figure 21-7a an inductor (symbolized by a coil) carries a constant current i. The rate of change of this constant current is zero, so from Equation 21-16 there is zero voltage across the inductor: $V = 0$. (Remember, we are ignoring the resistance of the wires that make up the inductor.) In Figure 21-7b the current through the inductor increases, so in Equation 21-16 $\Delta i / \Delta t > 0$ and $V < 0$. The negative value of V means that there is a voltage *drop* from where the current enters the inductor to where it leaves the inductor. That's like the voltage drop for current that passes through a resistor, which means that the inductor is now resisting the current—or, more properly, resisting the increase in current. So in this case the voltage opposes the change in current. In Figure 21-7c the current through the inductor decreases, so in Equation 21-16 $\Delta i / \Delta t < 0$ and $V > 0$. The positive value of V means that there is a voltage *gain* from where the current enters the inductor to where it leaves. That's like the voltage gain for current directed through a battery (a source of emf) from the negative terminal to the positive terminal, so in this case the inductor is assisting the current and so is trying to prevent the current from decreasing. Again the voltage opposes the change in current. Let's summarize these observations:

> *The voltage across an inductor opposes any change in the current through the inductor. If the current is not changing, the voltage is zero.*

This is just what we would expect from Lenz's law, which says that the induced emf in any coil (including an inductor) always acts to oppose any changes. A useful rule is that V in Equation 21-16 represents the voltage *gain* across the inductor when traveling in the direction of the current. If V is negative, there is a voltage drop.

WATCH OUT! Don't confuse inductors, capacitors, and resistors.

It's useful to contrast an inductor with two other devices used in circuits, a capacitor (Section 17-5) and a resistor (Section 18-3). The voltage across a *capacitor* is $V = q/C$, where q is the magnitude of the charge on either plate of the capacitor and C is the capacitance. The greater the amount of charge on each capacitor plate, the greater the capacitor voltage. The voltage across a *resistor* that carries current i is $V = iR$, where R is the resistance of the resistor. This voltage is proportional to the rate at which charge moves through the resistor, that is, the current i. The greater the current, the greater the rate at which charge flows through the resistor and the greater the resistor voltage. By contrast, for an *inductor* Equation 21-16 tells us that the voltage is proportional to the rate of change of current—that is, to the rate of change of the rate at which charge flows through the inductor.

Our observations show that an inductor in a circuit acts to oppose changes in the current within the circuit. The more rapid the change in current—that is, the greater the value of the quantity $\Delta i / \Delta t$ in Equation 21-16—the greater the voltage that will oppose the change. Some computer cables have a built-in inductor (Figure 21-8) that is designed to deal with high-frequency interference from signals radiated from other devices. These high-frequency signals would cause rapidly changing currents in the cables that could compromise the operation of the computer, so it's important to suppress them. If such a signal appears, the inductor will set up a voltage that opposes the signal. The inductor is made of a special material called a *ferrite* that also has a large resistance to high-frequency currents, so the energy in the stray signal is absorbed and dissipated as heat.

As we'll see in the following sections, inductors aren't just for suppressing currents. In the right circumstances they can be used for *sustaining* an alternating current of just the right frequency. This is important in tuning a television or radio receiver.

Ferrite inductor

Todd Ruskell

FIGURE 21-8 Inductor on a cable The ferrite inductor on this monitor cable reduces the interference from external high-frequency electromagnetic signals.

EXAMPLE 21-4 Inductor Voltage

A 0.500-mH inductor is in a dc circuit that carries a current of 0.500 A. If the current increases to 0.900 A in 0.150 ms due to a fault in the circuit, how large is the voltage that appears across the inductor? From the perspective of a positive charge moving through the inductor, is there a voltage rise or drop across the inductor?

Set Up

We'll use Equation 21-16 to calculate the voltage induced across the inductor. To decide whether there's a voltage rise or drop, we'll use the idea that the voltage across the inductor acts to oppose the change in current.

Voltage across an inductor:

$$V = -L\frac{\Delta i}{\Delta t} \qquad (21\text{-}16)$$

$i = 0.500$ A
i increasing

Solve

Find the voltage that appears in the inductor.

The current increases from 0.500 to 0.900 A in 0.150 ms = 0.150×10^{-3} s, so

$$\frac{\Delta i}{\Delta t} = \frac{0.900\ \text{A} - 0.500\ \text{A}}{0.150 \times 10^{-3}\ \text{s}} = 2.67 \times 10^{3}\ \text{A/s}$$

From Equation 21-16 the voltage across the inductor is

$$V = -L\frac{\Delta i}{\Delta t} = -(0.500 \times 10^{-3}\ \text{H})(2.67 \times 10^{3}\ \text{A/s})$$
$$= -1.33\ \text{H} \cdot \text{A/s} = -1.33\ \text{V}$$

The magnitude of the voltage is 1.33 V. The value of V is negative, which means that there is a voltage *drop* of 1.33 V across the inductor. Thus the voltage opposes the flow of current and so opposes the increase in current (compare Figure 21-7b).

Reflect

The inductor voltage is comparable to that produced by a AA or AAA battery. This voltage is present only during the time the current is *changing*, however; when the current is constant, the inductor voltage is zero.

GOT THE CONCEPT? 21-3 Inductors Versus Capacitors Versus Resistors

? For which of the following devices does the voltage across that device depend on the current that flows into that device? (a) An inductor; (b) a capacitor; (c) a resistor; (d) more than one of these; (e) none of these.

TAKE-HOME MESSAGE FOR Section 21-4

✔ The inductance of a coil equals the net magnetic flux through the coil due to the current in the coil, divided by the current.

✔ An inductor is a circuit device that opposes any change in the current through that device. The voltage across an inductor is proportional to its inductance and to the rate of change of current through the inductor.

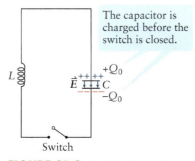

The capacitor is charged before the switch is closed.

L
$\vec{E}$ C
$+Q_0$
$-Q_0$

Switch

FIGURE 21-9 An *LC* circuit An inductor (inductance L) and a capacitor (capacitance C) are connected as shown. Figure 21-10 shows what happens when we close the switch.

21-5 In a circuit with an inductor and capacitor, charge and current oscillate

Figure 21-9 shows a circuit made up of an inductor of inductance L and a capacitor of capacitance C connected by ideal, zero-resistance wires. This is called an **LC circuit**. With the switch open, positive charge $+Q_0$ is placed on the upper plate of the capacitor, and negative charge $-Q_0$ is placed on the lower plate. This can be done by transferring a number of electrons from the upper plate to the lower plate. The switch is then closed, completing the circuit so that charge can flow. What happens? Remarkably, we'll see that the charge in this circuit behaves very much like the block attached to an ideal spring that we studied in Section 12-3: The charge *oscillates* back and forth in simple harmonic motion. Among other applications, this effect is used to generate the oscillating electric field that a microwave oven uses to cook food.

An *LC* Circuit and Simple Harmonic Motion

You might think that the excess electrons in Figure 21-9 would move along the wires from the lower plate to the upper plate until the two plates are electrically neutral, at which point the motion of electrons would stop. Because there's an inductor in the circuit, however, what happens is far more interesting, as Figure 21-10 shows. Once the switch is closed (step 1 in Figure 21-10), a counterclockwise current is established around the circuit. (Remember that by convention the direction of current is that in which positive charges would flow.) The current can't suddenly increase from zero, however, because a voltage appears across the inductor to oppose the increase in current. Instead the current increases gradually and reaches maximum when the capacitor is fully discharged (step 2 in Figure 21-10). The current can't just stop, however; the inductor always opposes any change in the current, so once the current starts to decrease, a voltage will appear across the inductor to slow that decrease. As a result, the motion of charge continues. When the current finally decreases to zero, the capacitor is again charged but with the opposite polarity to its initial configuration (step 3 in Figure 21-10). Now a current is established just as it was between steps 1 and 2, but in the opposite, clockwise direction: This increases gradually until the current is maximum and the capacitor is again fully discharged (step 4 in Figure 21-10). Again the inductor keeps the current from coming to a sudden stop, so the current decreases gradually until the capacitor is again fully charged and has returned to the configuration in step 1. The whole process then starts over!

What's happening here is an *oscillation* of charge and current. These oscillations are another example of *simple harmonic motion*, the special kind of oscillatory motion that we studied in Section 12-3. To see this let's apply Kirchhoff's loop rule to the *LC* circuit in Figure 21-10. This says that if we travel around the circuit and measure the voltage across each element of the circuit, the sum of those voltages must be zero.

We'll begin our trip around the circuit at the upper right-hand corner and move around the circuit clockwise so that we first pass downward through the capacitor. If we let q be the charge on the upper plate of the capacitor at a given instant, the charge on the lower plate at that same instant is $-q$. So if we start at the upper right-hand corner of the circuit and travel downward through the capacitor, we'll encounter a voltage drop

$$V_C = -\frac{q}{C} \quad (LC \text{ circuit})$$

(21-17)

This voltage drop is negative if q is positive, since then we travel from the positive charge to the negative charge and so will encounter a decrease in electric potential

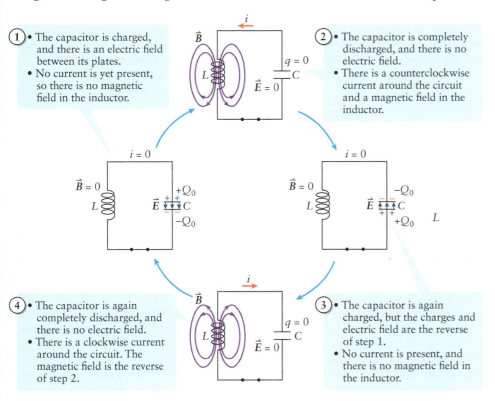

① • The capacitor is charged, and there is an electric field between its plates.
• No current is yet present, so there is no magnetic field in the inductor.

② • The capacitor is completely discharged, and there is no electric field.
• There is a counterclockwise current around the circuit and a magnetic field in the inductor.

④ • The capacitor is again completely discharged, and there is no electric field.
• There is a clockwise current around the circuit. The magnetic field is the reverse of step 2.

③ • The capacitor is again charged, but the charges and electric field are the reverse of step 1.
• No current is present, and there is no magnetic field in the inductor.

FIGURE 21-10 Oscillations of an *LC* circuit When the switch in Figure 21-9 is closed, the capacitor charge oscillates back and forth through the circuit. These electrical oscillations are directly analogous to the mechanical oscillations of a block attached to an ideal spring (Section 12-3).

(that is, a voltage drop). If q is negative, so that there's negative charge on the upper plate and positive charge on the lower plate, we'll encounter a voltage rise.

As we continue around the circuit, we'll encounter zero voltage across the wires (which have zero resistance). We next pass upward through the inductor on the left-hand side of the circuit. We'll take the current i to be positive if it also flows upward through the inductor (that is, clockwise around the circuit). If the current is positive, positive charge will flow onto the upper plate of the capacitor. The rate at which the charge q on the upper plate increases, measured in coulombs per second or amperes, is just equal to the current:

(21-18)
$$i = \frac{\Delta q}{\Delta t} \quad (LC \text{ circuit})$$

From Equation 21-16 the voltage across the inductor is

(21-19)
$$V_L = -L\frac{\Delta i}{\Delta t} \quad (LC \text{ circuit})$$

If the current through the inductor is positive (upward, or clockwise around the circuit) and increasing in magnitude (becoming more positive), then $\Delta i/\Delta t$ is positive and V_L is negative: We'll encounter a voltage drop as we move upward through the inductor. The same is true if the current through the inductor is negative (downward, or counterclockwise around the circuit) and decreasing in magnitude (becoming less negative). If instead the current through the inductor is either positive (clockwise) and decreasing or negative (counterclockwise) and becoming more negative, then $\Delta i/\Delta t$ is negative and V_L is positive: We'll encounter a voltage rise as we move upward through the inductor.

Kirchhoff's loop rule says that the sum of the voltages V_C from Equation 21-17 and V_L from Equation 21-19 must be zero:

$$V_C + V_L = -\frac{q}{C} - L\frac{\Delta i}{\Delta t} = 0$$

If we add $L(\Delta i/\Delta t)$ to both sides of this equation, we get

(21-20)
$$L\frac{\Delta i}{\Delta t} = -\frac{q}{C} \quad (LC \text{ circuit})$$

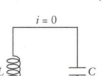

A circuit with an inductor of inductance L and a capacitor of capacitance C...

...has simple harmonic oscillations analogous to those of a block of mass m attached to a spring of spring constant k.

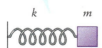

FIGURE 21-11 An *LC* circuit is analogous to a block attached to a spring The oscillations of charge in an *LC* circuit obey the same equations as do the oscillations of a block attached to an ideal spring. The inductor plays the role of the block, and the capacitor plays the role of the spring.

Equations 21-18 and 21-20 look like very difficult equations to solve, since both the capacitor charge q and the current i depend on time. But in fact these are equations that we've solved before, although with different symbols (Figure 21-11). Recall from Section 12-3 that if a block of mass m is attached to an ideal spring of spring constant k and displaced from equilibrium by x, the force that the spring exerts on the block is $F_x = -kx$. (Positive x means that the spring is stretched, and negative x means that the spring is compressed.) If the block is then released, it oscillates. The velocity of the block is the rate of change of the displacement x:

(21-21)
$$v_x = \frac{\Delta x}{\Delta t} \quad (\text{block and ideal spring})$$

Newton's second law says that the force $-kx$ on the block is equal to the mass of the block multiplied by its acceleration a_x. Acceleration is the rate of change of velocity, so we can write Newton's second law for the block as

(21-22)
$$m\frac{\Delta v_x}{\Delta t} = -kx \quad (\text{block and ideal spring})$$

You can see that Equations 21-21 and 21-22 for a block and ideal spring are the *same* as Equations 21-18 and 21-20 for an *LC* circuit, with the names of the variables changed as listed in the first four rows of Table 21-1.

For a block attached to an ideal spring, the spring provides a restoring *force*: Whether the spring is compressed or stretched, this force tries to return the system to equilibrium. The block, however, has *inertia*: Once in motion, it tends to remain in motion. As a result, when the spring reaches the equilibrium position, it does not stop but instead overshoots. As a result, the block oscillates back and forth around its equilibrium position. In a similar way the capacitor in an *LC* circuit provides a restoring *voltage* $-q/C$; whether q is positive or negative, this voltage drives charge around the

TABLE 21-1 Comparing a Block on an Ideal Spring to an *LC* Circuit

Quantity for a block on an ideal spring	Corresponding quantity for an *LC* circuit
spring displacement, x	capacitor charge, q
block velocity, $v_x = \Delta x / \Delta t$	current, $i = \Delta q / \Delta t$
spring constant of the spring, k	reciprocal of the capacitance, $1/C$
mass of the block, m	inductance, L
oscillation amplitude, A	maximum capacitor charge, Q_0
potential energy of the spring, $U_{spring} = \dfrac{1}{2} k x^2$	electric energy in the capacitor, $U_E = \dfrac{q^2}{2C}$
kinetic energy of the block, $K = \dfrac{1}{2} m v_x^2$	magnetic energy in the inductor, $U_B = \dfrac{1}{2} L i^2$

circuit in a direction that tries to neutralize both plates. The inductor opposes changes in current; once a current is in motion, the inductor tends to keep it in motion. As a result, when the capacitor is in equilibrium with both of its plates neutral, the current "overshoots," and the capacitor ends up charged again with reversed polarity. The result is an oscillation that's *exactly* like the simple harmonic motion of a block on an ideal spring. It's like watching the same play but with different actors in the cast: The inductor plays the role of the block, and the capacitor plays the role of the spring.

We can use Table 21-1 to tell us the angular frequency, period, and frequency for the oscillations of an *LC* circuit. From Equation 12-11 these quantities for a block on a spring are

Angular frequency $\omega = \sqrt{\dfrac{k}{m}}$

k = spring constant of the spring

m = mass of the object connected to the spring

Period $T = \dfrac{2\pi}{\omega} = 2\pi \sqrt{\dfrac{m}{k}}$

Angular frequency, period, and frequency for a block attached to an ideal spring (12-11)

Frequency $f = \dfrac{1}{T} = \dfrac{1}{2\pi} \sqrt{\dfrac{k}{m}} = \dfrac{\omega}{2\pi}$

Table 21-1 tells us that k corresponds to $1/C$, and m corresponds to L. So the angular frequency, period, and frequency for the oscillations of an *LC* circuit are

Angular frequency $\omega = \sqrt{\dfrac{1}{LC}}$

C = capacitance

Period $T = \dfrac{2\pi}{\omega} = 2\pi \sqrt{LC}$

L = inductance

Angular frequency, period, and frequency for an *LC* circuit (21-23)

Frequency $f = \dfrac{1}{T} = \dfrac{1}{2\pi} \sqrt{\dfrac{1}{LC}} = \dfrac{\omega}{2\pi}$

The smaller the product LC of inductance and capacitance, the greater the angular frequency and frequency and the shorter the oscillation period.

As an example, imagine an *LC* circuit made with a 2.00-μH inductor and a 10.0-nF capacitor. Then $L = 2.00 \times 10^{-6}$ H and $C = 10.0 \times 10^{-9}$ F, and from the last of Equations 21-23 the oscillation frequency is

$$f = \dfrac{1}{2\pi} \sqrt{\dfrac{1}{LC}} = \dfrac{1}{2\pi} \sqrt{\dfrac{1}{(2.00 \times 10^{-6} \text{ H})(10.0 \times 10^{-9} \text{ F})}}$$
$$= 1.13 \times 10^6 \text{ Hz} = 1.13 \text{ MHz}$$

The charges in this circuit oscillate back and forth 1.13 million times per second and complete one oscillation in a time $T = 1/f = 8.89 \times 10^{-7}$ s $= 0.889$ μs. Such oscillations are much more rapid than those of a block attached to a spring.

The capacitor charge q is analogous to the displacement x for a block on a spring, so the maximum capacitor charge Q_0 is analogous to the amplitude A (the maximum displacement of the block on a spring). We can use this observation to write expressions for the capacitor charge q and current i as functions of time. In Section 12-3 we found that the displacement x and velocity v_x for a block on a spring are given by

(12-6) $$x(t) = A \cos(\omega t + \phi) \quad \text{(block and ideal spring)}$$

(12-7) $$v_x(t) = -\omega A \sin(\omega t + \phi) \quad \text{(block and ideal spring)}$$

Since A is analogous to Q_0, displacement x is analogous to q, and velocity v_x is analogous to i, we can write the following expressions for an LC circuit:

(21-24) $$q(t) = Q_0 \cos(\omega t + \phi) \quad \text{(}LC\text{ circuit)}$$

(21-25) $$i(t) = -\omega Q_0 \sin(\omega t + \phi) \quad \text{(}LC\text{ circuit)}$$

The angular frequency ω is given by the first of Equations 21-23: $\omega = \sqrt{1/LC}$. The cosine is maximum when the sine is zero and vice versa, so the capacitor charge is maximum when the current is zero and the capacitor charge is zero when the current is maximum. That's just what Figure 21-10 shows.

Since $\sin(\omega t + \phi)$ has values between $+1$ and -1, the maximum value of the current in Equation 21-25 is

(21-26) $$i_{max} = \omega Q_0 = \frac{Q_0}{\sqrt{LC}} \quad \text{(}LC\text{ circuit)}$$

As an example, consider the LC circuit we described above with $L = 2.00$ μH $= 2.00 \times 10^{-6}$ H and $C = 10.0$ nF $= 10.0 \times 10^{-9}$ F. If the maximum capacitor charge is $Q_0 = 0.400$ μC $= 4.00 \times 10^{-7}$ C, the maximum current in the LC circuit is

$$i_{max} = \frac{Q_0}{\sqrt{LC}} = \frac{4.00 \times 10^{-7} \text{ C}}{\sqrt{(2.00 \times 10^{-6} \text{ H})(10.0 \times 10^{-9} \text{ F})}} = 2.83 \text{ A}$$

Even though only a tiny amount of charge oscillates back and forth between the capacitor plates, the oscillation is so rapid that the maximum current is appreciable.

Energy in an *LC* Circuit

An electric circuit is first and foremost a means of transferring energy from one part of the circuit to another. The same is true for an LC circuit. To see what kinds of energy are involved, let's first look at the total mechanical energy E in a system made up of a block attached to an ideal spring. From Section 12-4 this is the sum of the kinetic energy K of the moving block and the elastic potential energy U_{spring} in the spring:

$$E = K + U_{spring} = \frac{1}{2}mv_x^2 + \frac{1}{2}kx^2$$

The kinetic and potential energies change, but the total energy E is constant and equal to $(1/2)kA^2$ (A is the amplitude, or maximum displacement).

Table 21-1 tells us how to find the analogous expression for the total energy of an LC circuit: Just replace m with L, k with $1/C$, x with q, and v_x with i. The result is

Magnetic and electric energies in an *LC* circuit

(21-27)

Energy in an LC circuit | Magnetic energy U_B in the inductor | Electric energy U_E in the capacitor

$$E = \frac{1}{2}Li^2 + \frac{q^2}{2C}$$

Charge on the capacitor

Inductance of the inductor | Current in the inductor | Capacitance of the capacitor

We recognize the second term on the right-hand side of Equation 21-27, $q^2/2C$, from Section 17-6: It's just the electric potential energy stored in the capacitor. Since a capacitor plays the same role in an *LC* circuit as the spring does in a block–spring combination, this electric potential energy is analogous to the elastic potential energy of a spring. We use the symbol U_E for this electric energy. You can think of U_E as the energy required to take an uncharged capacitor and move charges $+q$ and $-q$ to the plates of the capacitor, thereby setting up an electric field $\vec{E}$ between the plates.

The first term on the right-hand side of Equation 21-27, $(1/2)Li^2$, is one that we haven't seen before. This term represents energy stored in the inductor as a result of the presence of current. To see how this energy arises, recall from Section 21-4 that an inductor sets up an emf that opposes any change in the current through the inductor. If we want to make current flow through an inductor where none was flowing before, we have to do work against that emf. The quantity $(1/2)Li^2$ is exactly equal to the amount of work we have to do. By building up the current from zero to i, we also create a magnetic field $\vec{B}$, so we can think of $(1/2)Li^2$ as the energy required to set up a magnetic field in and around the inductor. That's why we call this **magnetic energy** and denote it by the symbol U_B (the subscript B reminds us that a $\vec{B}$-field is involved).

In an *LC* circuit the energy oscillates between the electric and magnetic forms, just as the energy in a block–spring combination oscillates between the potential and kinetic forms (see Figure 12-10 in Section 12-4). In Figure 21-10 the energy is purely electric in steps 1 and 3 (where the capacitor charge is maximum and the current is zero) and purely magnetic in steps 2 and 4 (where the capacitor charge is zero and the current is maximum). If we substitute $q(t)$ from Equation 21-24 and $i(t)$ from Equation 21-25 into Equation 21-27, we get an expression for the energy of the *LC* circuit at any time:

$$E = U_B + U_E = \frac{1}{2}L[-\omega Q_0 \sin(\omega t + \phi)]^2 + \frac{[Q_0 \cos(\omega t + \phi)]^2}{2C}$$

$$= \frac{1}{2}L\omega^2 Q_0^2 \sin^2(\omega t + \phi) + \frac{Q_0^2}{2C}\cos^2(\omega t + \phi) \tag{21-28}$$

The first of Equations 21-23 tells us that $\omega = 1/\sqrt{LC}$, so $\omega^2 = 1/LC$ and $L\omega^2 = 1/C$. If we substitute this into the first term on the right-hand side of Equation 21-28 and recall that $\sin^2\theta + \cos^2\theta = 1$, we get

$$E = \frac{Q_0^2}{2C}\sin^2(\omega t + \phi) + \frac{Q_0^2}{2C}\cos^2(\omega t + \phi)$$

$$= \frac{Q_0^2}{2C}[\sin^2(\omega t + \phi) + \cos^2(\omega t + \phi)]$$

$$= \frac{Q_0^2}{2C} \quad (LC \text{ circuit}) \tag{21-29}$$

The electric and magnetic energies in an oscillating *LC* circuit both vary with time, but the total energy $E = Q_0^2/2C$ remains constant.

WATCH OUT! Magnetic energy is not the same as kinetic energy.

! Table 21-1 shows that the magnetic energy in an inductor, $U_B = (1/2)Li^2$, is analogous to the kinetic energy $K = (1/2)mv_x^2$ of the block in an oscillating block–spring system. This magnetic energy is present only if charges are in motion in the inductor, so i is nonzero. However, U_B is *not* the kinetic energy of the moving charges. That kinetic energy is very small because the moving electrons have very little mass. Instead, U_B is the energy stored in the magnetic field of the inductor that is caused by inductor current i.

EXAMPLE 21-5 Analyzing an *LC* Circuit

An oscillating *LC* circuit is made of a 2.00-μH inductor and a 10.0-nF capacitor. The maximum charge on the capacitor is 0.400 μC. (a) Find the total energy in the circuit. (b) At an instant when one-quarter of the total energy is electric energy in the capacitor, find the absolute values of the capacitor charge and the current in the inductor.

Set Up

Equation 21-29 tells us the total energy of the *LC* circuit. This equals the sum of the magnetic and electric energies, which depend on the current i and capacitor charge q as Equation 21-27 tells us. We'll use this to find the absolute values of i and q at the instant in question.

Energy in an *LC* circuit:

$$E = \frac{Q_0^2}{2C} \tag{21-29}$$

Magnetic and electric energies in an *LC* circuit:

$$E = U_B + U_E = \frac{1}{2}Li^2 + \frac{q^2}{2C} \tag{21-27}$$

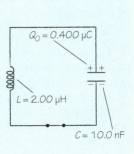

Solve

(a) Use Equation 21-29 to find the total energy.

The maximum capacitor charge is $Q_0 = 0.400 \ \mu C = 0.400 \times 10^{-6}$ C, and the capacitance is $C = 10.0$ nF $= 10.0 \times 10^{-9}$ F. The total energy of the oscillating circuit is

$$E = \frac{Q_0^2}{2C} = \frac{(0.400 \times 10^{-6} \text{ C})^2}{2(10.0 \times 10^{-9} \text{ F})} = 8.00 \times 10^{-6} \text{ J} = 8.00 \ \mu\text{J}$$

(b) The total energy remains constant. So if one-quarter of the energy is in the electric form, the other three-quarters must be in the magnetic form. Use this to determine the absolute values of the current and capacitor charge.

The magnetic energy is $\frac{3}{4}$ of the total energy:

$$U_B = \frac{1}{2}Li^2 = \frac{3}{4}E = \frac{3}{4}(8.00 \times 10^{-6} \text{ J}) = 6.00 \times 10^{-6} \text{ J}$$

The inductance is 2.00 μH $= 2.00 \times 10^{-6}$ H. Solve for the absolute value of the current:

$$i^2 = \frac{2U_B}{L} = \frac{2(6.00 \times 10^{-6} \text{ J})}{2.00 \times 10^{-6} \text{ H}} = 6.00 \text{ A}^2$$

$$|i| = \sqrt{i^2} = \sqrt{6.00 \text{ A}^2} = 2.45 \text{ A}$$

The electric energy is $\frac{1}{4}$ of the total energy:

$$U_E = \frac{q^2}{2C} = \frac{1}{4}E = \frac{1}{4}(8.00 \times 10^{-6} \text{ J}) = 2.00 \times 10^{-6} \text{ J}$$

Solve for the absolute value of the capacitor charge:

$$q^2 = 2CU_E = 2(10.0 \times 10^{-9} \text{ F})(2.00 \times 10^{-6} \text{ J})$$
$$= 4.00 \times 10^{-14} \text{ C}^2$$
$$|q| = \sqrt{q^2} = \sqrt{4.00 \times 10^{-14} \text{ C}^2} = 2.00 \times 10^{-7} \text{ C} = 0.200 \ \mu\text{C}$$

Reflect

We can check our result for $|q|$ by noting that U_E is proportional to q^2. If the electric energy U_E has its maximum value when $q = Q_0 = 0.400 \ \mu$C, then U_E will have $\frac{1}{4}$ of this value when $q = Q_0/2 = 0.200 \ \mu$C.

Note that all that we can determine in this problem are the absolute values of i and q. That's because the magnetic and electric energies are proportional to the squares of i and q, respectively. If the current were negative (counterclockwise in Figure 21-10) so that $i = -2.45$ A, the magnetic energy would be the same as if $i = +2.45$ A (clockwise current in Figure 21-10); if the capacitor charge were negative (so the upper capacitor plate in Figure 21-10 were negatively charged) so $q = -0.400 \ \mu$C, the electric energy would be the same as if $q = +0.400 \ \mu$C (so the upper capacitor plate were positively charged).

GOT THE CONCEPT? 21-4
Doubling the Charge in an *LC* Circuit

(?) A certain *LC* circuit oscillates at frequency f when the maximum capacitor charge is Q_0. If you double the value of Q_0, the new frequency of oscillation is (a) $2f$; (b) $f\sqrt{2}$; (c) $f/\sqrt{2}$; (d) $f/2$; (e) none of these.

You use the physics of *LC* circuits whenever you use a microwave oven. At the heart of any microwave oven is a device called a *cavity resonator*, which is basically a hollow metal tube closed at both ends by metal caps. When equal amounts of positive and negative charges are placed on the caps, the charge flows back and forth between the caps along the inner surfaces of the tube. This gives rise to time-varying electric and magnetic fields inside the tube. The electric field is greatest when the magnetic field is zero and vice versa, just as in Figure 21-10. (The geometry of these fields is more complicated than in Figure 21-10, however.) The frequency at which the fields oscillate—typically 2.45 GHz, or 2.45×10^9 Hz, in a home microwave oven and 0.915 GHz in a large commercial oven—is determined by the size and shape of the cavity resonator. A small hole in the cavity resonator is connected to one end of a metal pipe, whose other end leads into the interior of the oven. The oscillating fields "flow" along the pipe into the oven, where the energy of the electric field is absorbed by water molecules in the food. This sets the molecules into vibration and raises the temperature of the food, cooking it.

There is resistance in a cavity resonator, just as there is in any real circuit. As a result, the oscillations will die away just as do the oscillations of a block on a spring when friction is present. To sustain the oscillations of a circuit, we need to *drive* the circuit with an alternating voltage, in much the same way you sustain the back-and-forth oscillations of a child on a playground swing by pushing on the child once per cycle. We'll analyze such *driven* ac circuits in the following section.

TAKE-HOME MESSAGE FOR Section 21-5

✔ An *LC* circuit, made up of an inductor connected to a capacitor, is the electrical analog of a block attached to a spring. The inductor plays the role of the block, and the capacitor plays the role of the spring.

✔ In an *LC* circuit the capacitor charge and inductor current both oscillate with a frequency determined by the values of the inductance *L* and capacitance *C*.

✔ As the charge and current oscillate, the energy in the circuit oscillates between magnetic energy in the inductor and electric energy in the capacitor. If there is no resistance in the circuit, the energy remains constant.

21-6 When an ac voltage source is attached in series to an inductor, a resistor, and a capacitor, the circuit can display resonance

When you turn on the radio in a car, you expect to listen to only one station at a time. But the car antenna is being simultaneously bombarded by signals from *all* of the local radio stations. How does the radio "know" which signal to play through the car's speakers and which signals to ignore? The explanation is that the radio circuit is essentially a combination of an inductor, resistor, and capacitor in series, and this circuit is driven by an ac voltage coming from the radio signal. As we'll see, such a **series *LRC* circuit** can be designed to give a large current in response to a voltage at one particular frequency, while responding very little to voltages at other frequencies. When you tune a radio, you're adjusting the radio circuit so that the frequency at which it has the greatest response matches the frequency of the station you want to hear.

A driven series *LRC* circuit behaves in much the same way as a damped, driven mechanical oscillator like the one we studied in Section 12-8. (This would be a good time to review that section.) To understand the properties of a driven series *LRC* circuit, it's useful to first look at three simpler ac circuits: one with just an ac source and a resistor, one with just an ac source and a capacitor, and one with just an ac source and an inductor.

An ac Source and a Resistor

In Figure 21-12a an ac source is connected to a resistor of resistance *R*. The voltage provided by the source is

$$V(t) = V_0 \sin \omega t \tag{21-30}$$

What is the current $i(t)$ in the circuit, and how much power does the source deliver to the resistor? To find out, let's apply Kirchhoff's loop rule, which states that the sum of the voltage drops around the circuit is zero. The voltage across the resistor is $i(t)$ multiplied by *R*, so the loop rule says that

$$V(t) - i(t)R = 0$$

and so

$$i(t) = \frac{V(t)}{R} = \frac{V_0}{R} \sin \omega t \quad \text{(resistor and ac source)} \tag{21-31}$$

The graph of $V(t)$ and $i(t)$ versus time in Figure 21-12a shows that the current is in phase with the source voltage. From Equation 18-23, the power that the source delivers to the circuit (in this case to the resistor) is equal to the product of the current through the source multiplied by the voltage across the source: $P(t) = i(t)V(t)$. The graph in Figure 21-12a shows that $i(t)$ and $V(t)$ always have the same sign (both positive or both negative), so the product $i(t)V(t)$ is always positive. Thus power is always being transferred from the ac source into the resistor.

 See the Math Tutorial for more information on trigonometry.

FIGURE 21-12 Three ac circuits
The graphs show the source voltage and currents for three circuits: an ac source connected to (a) a resistor, (b) a capacitor, and (c) an inductor.

(a) An ac source connected to a resistor

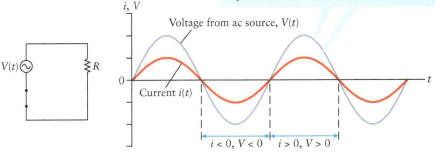

- Current i is in phase with source voltage V.
- Since i and V always have the same sign, the power $P = iV$ delivered by the source is always positive.

(b) An ac source connected to a capacitor

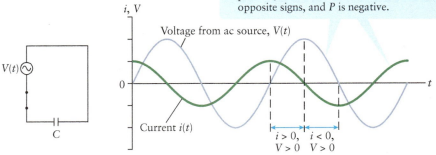

- Current i leads the source voltage V by 1/4 cycle.
- Half of the time i and V have the same sign, so the power $P = iV$ delivered by the source is positive; the other half of the time i and V have opposite signs, and P is negative.

(c) An ac source connected to an inductor

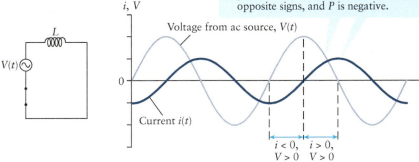

- Current i lags the source voltage V by 1/4 cycle.
- Half of the time i and V have the same sign, so the power $P = iV$ delivered by the source is positive; the other half of the time i and V have opposite signs, and P is negative.

An ac Source and a Capacitor

The circuit shown in Figure 21-12b is the same as in Figure 21-12a, except that we've replaced the resistor by a capacitor. The voltage provided by the source is still given by Equation 21-30, but the voltage across the capacitor is $q(t)/C$, where $q(t)$ is the capacitor charge. The loop rule now says that

$$V(t) - \frac{q(t)}{C} = 0$$

The capacitor charge is therefore

$$q(t) = CV(t) = CV_0 \sin \omega t \quad \text{(capacitor and ac source)}$$

(21-32)

The capacitor charge oscillates between $+CV_0$ and $-CV_0$.

To find the current in the circuit, two trigonometric identities that will be of use are $\sin \theta = \cos (\theta - \pi/2)$ and $\cos \theta = -\sin (\theta - \pi/2)$. Using the first of these, we can rewrite Equation 21-32 as

$$q(t) = CV_0 \cos \left(\omega t - \frac{\pi}{2} \right) \quad \text{(capacitor and ac source)}$$

The current $i(t)$ is the rate of change of the capacitor charge $q(t)$, just as it was for the LC circuit that we discussed in Section 21-5. In that section we saw that if $q(t) = Q_0 \cos (\omega t + \phi)$, then $i(t) = -\omega Q_0 \sin (\omega t + \phi)$ (see Equations 21-24 and 21-25). That's the same situation we have here, with Q_0 replaced by CV_0 and ϕ equal to $-\pi/2$. So the current in the circuit of Figure 21-12b is

$$i(t) = -\omega CV_0 \sin \left(\omega t - \frac{\pi}{2} \right)$$

Using the second trigonometric identity that we stated above, we can write this as

$$i(t) = \omega CV_0 \cos \omega t \quad \text{(capacitor and ac source)} \tag{21-33}$$

From Equation 21-32, the maximum charge on the capacitor is CV_0 no matter what the frequency. If the frequency is low and the period long, there's plenty of time for the charge to build up from zero to CV_0, so the required current will be low. Equation 21-33 shows the same thing: If the frequency is low, the angular frequency ω is likewise low and the current amplitude ωCV_0 is small. So a capacitor connected to an ac source acts to reduce the current amplitude at low frequency.

The graph in Figure 21-12b shows the source voltage and the current versus time. We say that the current *leads* the voltage by 1/4 cycle: The graph of current has its first peak at $t = 0$, but the graph of voltage reaches its first peak 1/4 cycle later. As a result, for half of the time the current and voltage have the same sign, while for the other half of the time the signs of $i(t)$ and $V(t)$ are opposite. Therefore, for half of the time there is positive power $P(t) = i(t)V(t)$ out of the source and into the capacitor, while for the other half of the time the power is negative and energy is flowing out of the capacitor back into the source. What's happening is that the capacitor is alternately charging and discharging; when it's charging the electric energy U_E in the capacitor is increasing, and when it's discharging the electric energy is decreasing. Over one complete cycle, there's *zero* net flow of energy out of the source.

An ac Source and an Inductor

Figure 21-12c shows a third circuit, one that includes only an ac source and an inductor. Again the source voltage is given by Equation 21-30. The voltage drop across the inductor is $L(\Delta i/\Delta t)$, so the loop rule says that

$$V(t) - L \frac{\Delta i(t)}{\Delta t} = 0$$

We can solve this for $\Delta i(t)/\Delta t$, which is the rate of change of the current:

$$\frac{\Delta i(t)}{\Delta t} = \frac{V(t)}{L} = \frac{V_0}{L} \sin \omega t \quad \text{(inductor and ac source)} \tag{21-34}$$

Given Equation 21-34 for the rate of change of current, what is the current itself? As we did for the capacitor circuit shown in Figure 21-12b, let's think for a minute about Equation 21-24, $q(t) = Q_0 \cos (\omega t + \phi)$, and Equation 21-25, $i(t) = -\omega Q_0 \sin (\omega t + \phi)$. In these equations $i(t)$ is the rate of change of $q(t)$. So these equations tell us that if the rate of change of a function is given by a constant multiplied by $\sin (\omega t + \phi)$, the function itself is given by the same constant divided by $-\omega$ and multiplied by $\cos (\omega t + \phi)$. Using this with Equation 21-34, in which the constant is V_0/L and $\phi = 0$, we conclude that the current for the circuit in Figure 21-12c is

$$i(t) = -\frac{V_0}{\omega L} \cos \omega t \quad \text{(inductor and ac source)} \tag{21-35}$$

Equation 21-35 says that the amplitude of the current is $V_0/(\omega L)$. Since ω is in the denominator, the current amplitude is small when the frequency, and hence the angular frequency ω, is high. That agrees with the idea that an inductor opposes changes in current. High frequency corresponds to rapid changes in current, which the inductor suppresses by making the current small. So an inductor connected to an ac source acts to reduce the current amplitude at high frequency. That's in sharp contrast to a capacitor, which acts to reduce the current amplitude at low frequency.

The graph in Figure 21-12c shows $V(t)$ and $i(t)$ versus time for the circuit with an inductor and an ac source. In this circuit the current *lags* behind the voltage by 1/4 cycle; if you look at any peak of the voltage, you'll see that a peak of the current occurs 1/4 cycle later. Just as for the capacitor circuit in Figure 21-12b, for half of the time the current and voltage have the same sign, so power flows out of the source and adds energy to the inductor. In this case it's *magnetic* energy that's being added as the current through the inductor increases in magnitude. For the other half of the time, current and voltage have opposite signs, the source power is negative, and energy flows out of the inductor as the current decreases in magnitude. As for the capacitor, over one cycle there is zero net energy delivered to the inductor.

A Driven Series *LRC* Circuit

Let's now combine all of the devices shown in Figure 21-12 into a single circuit (Figure 21-13). Again the voltage of the ac source is given by Equation 21-30, $V(t) = V_0 \sin \omega t$. We expect that when the switch is closed there will be an oscillating current in the circuit with angular frequency ω. We can also expect that at low frequencies the presence of the capacitor will keep the current amplitude small, while at high frequencies the current amplitude will be small due to the presence of the inductor. At some frequency that's not too low and not too high, the current amplitude will be largest. That's the frequency where the voltage of the ac source produces the greatest response. One of our tasks is to find just what this frequency is.

If we apply Kirchhoff's loop rule to the circuit shown in Figure 21-13, we get the following equation:

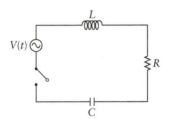

FIGURE 21-13 A driven series *LRC* circuit When the switch is closed, an oscillating current is set up in this circuit. The amplitude and phase of this current depend on the values of the inductance L, resistance R, and capacitance C, as well as the amplitude V_0 and angular frequency ω of the voltage provided by the ac source.

$$(21\text{-}36) \qquad V(t) - L\frac{\Delta i(t)}{\Delta t} - i(t)R - \frac{q(t)}{C} = 0 \quad \text{(driven series } LRC \text{ circuit)}$$

This is a complicated equation, since it involves the capacitor charge $q(t)$; the current $i(t)$, which is the rate of change of the capacitor charge; and the rate of change of the current, $\Delta i(t)/\Delta t$. It is possible to solve Equation 21-36 using some tricks of trigonometry, but it's a rather strenuous exercise. Instead, let's just present the result for the current $i(t)$ and see what it tells us:

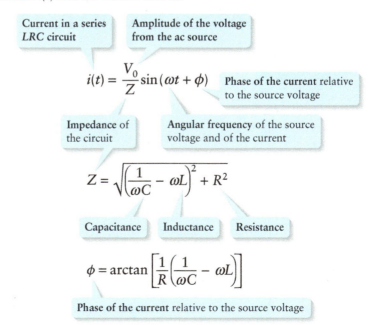

Current in a driven *LRC* circuit
(21-37)

$$i(t) = \frac{V_0}{Z}\sin(\omega t + \phi)$$

Current in a series *LRC* circuit — Amplitude of the voltage from the ac source — Phase of the current relative to the source voltage

$$Z = \sqrt{\left(\frac{1}{\omega C} - \omega L\right)^2 + R^2}$$

Impedance of the circuit — Angular frequency of the source voltage and of the current — Capacitance — Inductance — Resistance

$$\phi = \arctan\left[\frac{1}{R}\left(\frac{1}{\omega C} - \omega L\right)\right]$$

Phase of the current relative to the source voltage

The first of Equations 21-37 shows that the greater the value of the **impedance** Z, the smaller the amplitude V_0/Z of the current. Like resistance, impedance has units of ohms. So you can think of it as similar to an "effective resistance" of the circuit to an ac current. Unlike resistance, however, impedance depends on the angular frequency as given by the second of Equations 21-37. The quantity $[1/(\omega C) - \omega L]^2$ is large at both low and high angular frequencies; that's because the term $1/(\omega C)$ (due to the capacitor) gets very large for low values of ω, while the term ωL (due to the inductor) gets very large for high values of ω. So the impedance Z is large and the current amplitude V_0/Z is small for very low angular frequencies thanks to the capacitor, as well as for very high angular frequencies thanks to the inductor. That's just what we predicted above.

The impedance is smallest, and the current amplitude V_0/Z largest, when the quantity $[1/(\omega C) - \omega L]^2$ equals zero, so

$$\frac{1}{\omega C} - \omega L = 0$$

To solve for the angular frequency at which this happens, multiply this equation by ω/L:

$$\frac{\omega}{L}\left(\frac{1}{\omega C} - \omega L\right) = \frac{1}{LC} - \omega^2 = 0$$

So $\omega^2 = 1/(LC)$, or

$$\omega = \sqrt{\frac{1}{LC}} \tag{21-38}$$

We saw this same angular frequency in Equation 21-23: It's the angular frequency at which an LC circuit oscillates. We'll refer to $\sqrt{1/(LC)}$ as the **natural angular frequency** of the circuit and give it the symbol ω_0. It's the angular frequency at which the current would oscillate if the ac source and resistor weren't in the circuit at all!

In Section 12-8 we saw this same effect, called **resonance**, for an oscillating mechanical system. Equation 21-38 tells us that the current in a driven series LRC circuit is greatest, or in resonance, if the *driving* angular frequency ω equals the *natural* angular frequency $\omega_0 = \sqrt{1/(LC)}$. When $\omega = \omega_0 = \sqrt{1/(LC)}$ the impedance in Equation 21-37 is equal to R and the current amplitude is $V_0/Z = V_0/R$.

Why does resonance happen in a series LRC circuit? The explanation is that at low frequencies the capacitor suppresses the current, and at high frequencies the inductor suppresses the current. But at one special frequency—the natural frequency of the circuit—the effects of the capacitor and inductor cancel exactly so that the current in the circuit has its maximum amplitude. Figure 21-14a shows the current amplitude as a function of the driving frequency ω for three different values of the resistance R. For any value of R the current amplitude is maximum when $\omega/\omega_0 = 1$ and $\omega = \omega_0 = \sqrt{1/(LC)}$. As R decreases, the peak becomes higher (the maximum current amplitude increases).

(a) Amplitude of the current

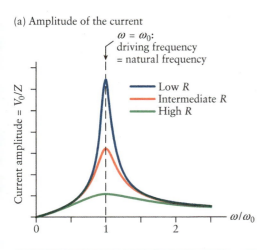

(b) Phase of the current

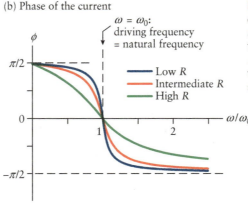

FIGURE 21-14 Resonance in a driven series *LRC* circuit (a) The amplitude and (b) the phase of the current in a driven series *LRC* circuit both depend on the angular frequency ω of the source voltage.

• The current amplitude is maximum at $\omega = \omega_0$ (resonance).
• The smaller the resistance, the greater the current amplitude at resonance.

• At $\omega = \omega_0$ (resonance), $\phi = 0$ and the current is in phase with the source voltage.
• If $\omega < \omega_0$, $\phi > 0$ and the current leads the source voltage. At very low frequencies ϕ approaches $\pi/2$ (current leads voltage by 1/4 cycle).
• If $\omega > \omega_0$, $\phi < 0$ and the current lags behind the source voltage. At very high frequencies ϕ approaches $-\pi/2$ (current lags voltage by 1/4 cycle).

The third of Equations 21-37 describes the phase ϕ of the current (see **Figure 21-14b**). The value of ϕ tells us how much the current leads the source voltage. At resonance, where $\omega/\omega_0 = 1$ and $\omega = \omega_0 = \sqrt{1/(LC)}$, we have $(1/(\omega C)) - \omega L = 0$ and $\phi = \arctan 0 = 0$. So at resonance the current is in phase with the source voltage, just as happens in a circuit with only an ac source and a resistor (Figure 21-12a).

For very low angular frequencies ω, the quantity $1/(\omega C)$ is large while ωL is small, so $(1/(\omega C)) - \omega L$ is large and positive. In this case ϕ is near $\pi/2$ radians, so at low frequencies the current leads the voltage by about 1/4 of a cycle (recall that there are 2π radians in a cycle). That's just like the behavior of a circuit with only an ac source and a capacitor (Figure 21-12b). By contrast, at very high angular frequencies the quantity $1/(\omega C)$ is small while ωL is large. Then $(1/(\omega C)) - \omega L$ is large and negative, and ϕ is near $-\pi/2$ radians or $-1/4$ cycle. The negative value of ϕ means that at high frequencies the current *lags* the source voltage by about 1/4 cycle. The circuit with just an ac source and an inductor in Figure 21-12c behaves in the same way.

As Figure 21-14b shows, changing the resistance doesn't affect the angular frequency at which $\phi = 0$. This always happens when $\omega = \omega_0 = \sqrt{1/(LC)}$. Changing the value of R does change the shape of the graph of ϕ versus ω, however; the smaller the resistance, the closer the driving angular frequency ω must be to the natural angular frequency ω_0 for ϕ to be close to zero.

These ideas show what's involved in tuning a radio. The signal from a broadcast station is at a particular frequency. The radio antenna detects that signal and converts it to an ac voltage with that same frequency. For the circuit of the radio receiver to respond to that frequency, the natural frequency of the receiver circuit must match the driving frequency of the signal from the radio station—that is, the radio receiver must be *tuned* to be in resonance with the station to which you want to listen. Other radio signals from other stations will be present in the circuit as well, but their driving frequencies won't match the natural frequency of the circuit. Hence these undesired signals will produce very little response in the radio (see Figure 21-14a), and you'll hear only the sound of the station to which you've tuned the radio. The following example illustrates this idea.

 Go to Interactive Exercise 21-2 for more practice dealing with LRC circuits.

EXAMPLE 21-6 Tuning an FM Radio

The tuner knob on an FM radio moves the plates of an adjustable capacitor. This capacitor is in series with a 0.130-μH inductor and a net resistance of 755 Ω. The peak current induced in this circuit by a radio wave becomes large when the natural frequency of the circuit matches the carrier frequency of the radio wave. (a) What is the frequency of the FM station that is tuned in when the capacitor in the radio is adjusted to 19.6 pF? (b) If the peak operating voltage in the tuning circuit is 9.00 V, what is the peak current?

Set Up

The tuning circuit resonates when driven by a radio wave with an angular frequency ω that equals the natural angular frequency ω_0 of the circuit. We'll use Equation 21-38 to determine the value of ω_0 for this circuit and convert it to an ordinary frequency in Hz. We'll find the peak current, or current amplitude, using Equations 21-37.

Natural angular frequency of the tuning circuit:

$$\omega_0 = \sqrt{\frac{1}{LC}} \qquad (21\text{-}38)$$

Relationship between frequency and angular frequency:

$$f = \frac{\omega}{2\pi}$$

Amplitude of the oscillating current:

$$i_{0,\text{max}} = \frac{V_0}{Z}$$

$$Z = \sqrt{\left(\frac{1}{\omega C} - \omega L\right)^2 + R^2} \qquad (21\text{-}37)$$

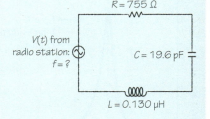

R = 755 Ω

V(t) from radio station: f = ?

C = 19.6 pF

L = 0.130 μH

Solve

(a) Use the given values of capacitance and inductance to determine the natural frequency of the circuit.

We are given $L = 0.130$ µH $= 0.130 \times 10^{-6}$ H and $C = 19.6$ pF $= 19.6 \times 10^{-12}$ F. From Equation 21-38 the natural angular frequency is $\omega_0 = \sqrt{1/(LC)}$, and the natural frequency is ω_0 divided by 2π:

$$f_0 = \frac{\omega_0}{2\pi} = \frac{1}{2\pi}\sqrt{\frac{1}{LC}} = \frac{1}{2\pi}\sqrt{\frac{1}{(0.130 \times 10^{-6}\text{ H})(19.6 \times 10^{-12}\text{ F})}}$$
$$= 99.7 \times 10^6\text{ Hz} = 99.7\text{ MHz}$$

(b) Find the amplitude of the oscillating current.

At resonance the term $(1/(\omega C)) - \omega L$ in the expression for impedance is equal to zero. Then Equation 21-37 tells us that $Z = R = 755\ \Omega$. If the voltage amplitude is 9.00 V, the current amplitude is

$$i_{0,\text{max}} = \frac{V_0}{Z} = \frac{9.00\text{ V}}{755\ \Omega} = 0.0119\text{ A} = 11.9\text{ mA}$$

Reflect

The carrier frequencies of FM radio stations lie in the megahertz (MHz) range. So the frequency we found, 99.7×10^6 Hz, is 99.7 MHz on your FM radio dial. The peak current, about 10 mA, is typical for a portable FM radio.

Note that you tune a television set in the same way. When you use the remote control to change the channel (Figure 21-1b), you're commanding a circuit in the television to change its capacitance. This adjusts the natural frequency of the circuit to match the carrier frequency of the channel you want to watch.

GOT THE CONCEPT? 21-5 Modifying a Series *LRC* Circuit

? We've discussed the behavior of a series *LRC* circuit with $R = 1.00\ \Omega$, $C = 5.00$ µF, $L = 20.0$ µH, and $V_0 = 1.00$ V. If we increased the resistance to $2.00\ \Omega$ but kept all the other values the same, which of the following would be affected?

(a) The natural angular frequency of the circuit; (b) the current amplitude at resonance; (c) the current amplitude when the driving angular frequency is twice the natural angular frequency; (d) two of (a), (b), and (c); (e) all of (a), (b), and (c).

TAKE-HOME MESSAGE FOR Section 21-6

✔ A driven series *LRC* circuit is made up of an inductor, a capacitor, and a resistor connected in series to a source of ac voltage.

✔ The response of a driven series *LRC* circuit depends on the frequency of the ac source. At low frequencies, the capacitor dominates the circuit and the current is small. The current is also small at high frequencies, where the inductor dominates

the circuit. When the frequency of the source equals the natural frequency of the circuit, the current is large and maximum power flows from the source into the resistor. This is resonance.

✔ As the charge and current oscillate, the energy in the circuit oscillates between magnetic energy in the inductor and electric energy in the capacitor. If there is no resistance in the circuit, the energy remains constant.

21-7 Diodes are important parts of many common circuits

In our discussion of electric circuits, we've seen two different types of voltage sources (batteries that provide a constant emf and ac sources that provide a time-varying emf) and three kinds of circuit elements (resistors, capacitors, and inductors). Before leaving the subject of circuits, let's take a look at another important circuit device called a *diode*. This device is possible because of *semiconductors*, a class of materials whose electrical properties are intermediate between those of conductors and insulators.

A common semiconductor is the element silicon (chemical symbol Si). Each silicon atom has four outer or *valence* electrons. In a conducting material such as copper or silver the valence electrons are free to move throughout the material, so the resistivity of the material is low. In silicon, however, the atoms are formed into a crystal structure in which all four of an atom's valence electrons are involved in chemical bonds with adjacent silicon atoms. As a result, the valence electrons are able to move only with great difficulty. That's why pure silicon has about 10^{10} times the resistivity of copper

FIGURE 21-15 Doped semiconductors (a) Silicon (Si) doped with phosphorous (P) is an n-type semiconductor. Because phosphorous has five valence electrons, there is an extra, weakly bound electron that can contribute to electrical conduction. (b) Silicon doped with boron (B) is a p-type semiconductor. Because boron has only three valence electrons, there is a hole in one of its bonds. The hole can move through the semiconductor as though it were a positive charge, contributing to electrical conduction.

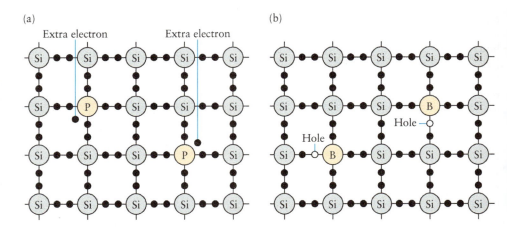

(though only about 10^{-13} the resistivity of an insulating material such as rubber). That can change dramatically by **doping** the silicon—that is, by adding small amounts of a different kind of atom to solid silicon. Two common elements used for this purpose are phosphorous and boron.

Figure 21-15a depicts a silicon crystal doped with a small amount of phosphorous (chemical symbol P), so that the phosphorous atoms replace a few of the silicon atoms. An atom of phosphorous has five valence electrons, four of which form bonds to the adjacent silicon atoms. The fifth electron can move relatively easily through the material, and so the doped silicon has much lower resistivity than does pure silicon. Adding just one part per million of phosphorous can decrease the resistivity by several powers of ten. Since the mobile charges in phosphorous-doped silicon are negatively charged electrons, we call such a material an **n-type semiconductor** (n for negative).

Figure 21-15b shows a different kind of doping in which some of the silicon atoms have been replaced with atoms of boron (chemical symbol B). Boron has only three valence electrons, so for each boron atom there is a "hole" in the electron population. These **holes** are actually free to travel through the boron-doped silicon. (A very rough analogy is to think of these holes as traveling like bubbles of carbon dioxide through a carbonated beverage.) Since a hole represents an absence of negative charge, it's equivalent to a *positive* charge. For this reason boron-doped silicon is called a **p-type semiconductor** (p for positive). Adding holes by doping the silicon with even a small amount of boron also causes a substantial decrease in resistivity.

Now let's take a crystal of silicon and dope one side of the crystal with boron and one side with phosphorous. The side with boron is the p side and the side with phosphorous is the n side (the two sides are p-type and n-type semiconductors, respectively). The region where the two sides meet is called a **pn junction**. Some of the mobile holes from the p side diffuse across the junction to the n side, and some of the mobile electrons from the n side diffuse to the p side. The semiconductor isn't a particularly good conductor, so neither the electrons nor the holes get very far from the junction. The net result is a double layer of charge around the junction, a positive layer of holes on the n side and a negative layer of electrons on the p side (Figure 21-16).

To see what makes a pn junction useful, imagine placing it in a dc circuit with a source of emf ε and a resistor. If the polarity of the source is as shown in Figure 21-17a, a current flows through the junction and around the circuit. What happens is that the source of emf creates an electric field within the pn junction that points from left to right, which helps to drive positively charged holes to the right across the junction and negatively charged electrons to the left across the junction. This enhanced diffusion of electrons and holes gives rise

● – Electrons
○ + Holes

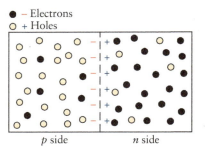

p side   n side

FIGURE 21-16 A pn junction The two sides of this semiconductor are doped with different atoms. Holes diffuse from the p side to the n side, and electrons diffuse from the n side to the p side. The result is a double layer of charge at the junction between the two sides.

FIGURE 21-17 A pn-junction diode This device is like a one-way valve for electric current. (a) Current will flow through the pn junction if the emf is applied as shown here (forward bias). (b) Little or no current will flow through the pn junction if the emf is reversed (reverse bias).

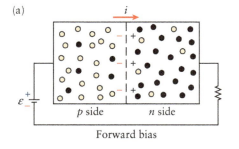

Forward bias

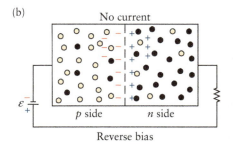

Reverse bias

to a net current from left to right in the semiconductor. In this case the *pn* junction is said to be *forward biased*. If we reverse the polarity of the source as shown in Figure 21-17b, however, the source creates an electric field that points from right to left within the *pn* junction. This tends to push the holes leftward (back into the *p* side from which they came) and the electrons rightward (back into the *n* side from which they came). In this case the *pn* junction is said to be *reverse biased*: The emf suppresses the diffusion of electrons and holes, and there is little or no current. Essentially the junction conducts in only one direction! A single-junction semiconductor device like this is called a **diode**.

Diodes have many uses, one of which is converting alternating current into direct current. If we replace the source of emf in Figure 21-17 with an ac source, the diode will conduct electricity when the alternating voltage is positive (forward bias) but not when the alternating voltage is negative (reverse bias). In this case the current through the resistor in Figure 21-17 won't be constant, but it will be in one direction only.

Another use for *pn* junctions is in **photovoltaic solar cells**. Light shining on the *p* side of the junction can excite electrons and produce new holes in the process. If the electrons happen to migrate to the junction, they will be accelerated into the *n* side by the electric field between the double layer of charge at the junction. The result is an excess negative charge on the *n* side and an excess positive charge on the *p* side. This charge imbalance produces a potential difference between the two regions. If one terminal of a resistor is connected to the *p* side and the other terminal to the *n* side, a current will flow through the resistor due to the potential difference and energy will be transferred to the resistor. The net effect is that the energy of sunlight is converted to electric energy. All solar cells designed for home and industrial use operate on this basic principle.

The basic process used in solar cells can also be run in reverse. In certain kinds of *pn* junctions, a forward bias like that shown in Figure 21-17a produces large concentrations of electrons on the *p* side and holes on the *n* side. When the electrons and holes diffuse across the barrier and recombine, energy is released in the form of light. Such a device is called a **light-emitting diode** or **LED**. LEDs are extremely efficient light sources: They produce much more light for a given power input than do incandescent light bulbs or fluorescent lamps. A television remote control (Figure 21-1b) sends commands to the television in the form of infrared light emitted by an LED.

Many other important semiconductor devices are used in circuits. In a *transistor*, a weak electrical current is used to modulate a stronger current. For example, the alternating current from a music player carries all the musical information about the song you want to hear, but the current is too feeble to drive a speaker or even a pair of earbuds. By using transistors, however, this feeble current can be used to modulate a much stronger current from the music player's battery. This modulated current *is* large enough to drive a speaker or earbuds, and that's what produces the music that you hear.

TAKE-HOME MESSAGE FOR Section 21-7

✔ A semiconductor can be made a better conductor of electricity by doping it with a small amount of a different kind of atom. In an *n*-type semiconductor, there are extra electrons that are able to move through the material and carry a current. In a *p*-type semiconductor, the mobile charges are holes where there is an electron missing. Holes behave like positively charged objects.

✔ In a *pn* junction one side of the semiconductor is *p*-type and the other side is *n*-type. This has the property that it is much easier to establish a current through the junction in one direction compared to the opposite direction.

Key Terms

alternating current (ac)	magnetic energy	root mean square value (rms value)
diode	mutual inductance	secondary coil
direct current (dc)	natural angular frequency	self-inductance
doping	*n*-type semiconductor	series *LRC* circuit
hole	photovoltaic solar cell	step-down transformer
impedance	*pn* junction	step-up transformer
inductance	primary coil	transformer
LC circuit	*p*-type semiconductor	
light-emitting diode (LED)	resonance	

Chapter Summary

Topic	Equation or Figure

Root mean square values for alternating current: To find the root mean square or rms value of a quantity, first find the average (mean) value of the square of that quantity, then take the square root of that average. Alternating voltages are often described in terms of their rms values rather than their amplitudes.

The **root mean square (rms)** value of $V(t)$... ...is the square root of the average value of $V^2(t)$.

$$V_{\text{rms}} = \sqrt{(V^2(t))_{\text{average}}} = \frac{V_0}{\sqrt{2}}$$ (21-5)

If $V(t)$ is a sinusoidal function, its rms value equals the maximum value of $V(t)$ (the amplitude) divided by $\sqrt{2}$.

Transformers: A transformer uses electromagnetic induction to raise or lower the voltage of an alternating current. The ratio of the output voltage to the input voltage depends on the number of windings in the coils of the transformer. Raising the voltage lowers the current and vice versa. In an ideal transformer, no power is lost in this process.

Amplitudes of the input (p) and output (s) voltages

$$\frac{V_s}{V_p} = \frac{V_{s,\text{rms}}}{V_{p,\text{rms}}} = \frac{N_s}{N_p}$$ (21-10)

Number of windings in the primary coil (p) and secondary coil (s)

Rms values of the input (p) and output (s) voltages

Inductors: The simplest inductor is a coil of wire. If the current in the coil changes, there will be a change in the magnetic flux through the coil and an emf will be induced to oppose that change. So an inductor in a circuit always acts to oppose any change in the current in that circuit. The inductance of a coil depends on the coil's geometry.

Inductance of a coil | **Number of windings** in the coil

$$L = \frac{N\Phi_B}{i}$$ (21-13)

Magnetic flux through each winding of the coil due to the field produced by the coil itself

Current in the coil

Voltage across an inductor | Inductance of the inductor

$$V = -L\frac{\Delta i}{\Delta t}$$ (21-16)

Rate of change of the current i in the inductor

The negative sign means that the voltage opposes any change in the current.

LC circuits: A circuit made up of an inductor and capacitor is the electric analog of a block oscillating at the end of an ideal spring. The angular frequency of the oscillation is $\omega = \sqrt{1/(LC)}$. The energy in the LC circuit oscillates between electric energy in the capacitor and magnetic energy in the inductor.

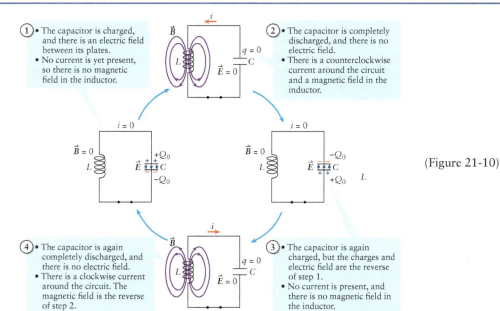

(Figure 21-10)

① • The capacitor is charged, and there is an electric field between its plates.
• No current is yet present, so there is no magnetic field in the inductor.

② • The capacitor is completely discharged, and there is no electric field.
• There is a counterclockwise current around the circuit and a magnetic field in the inductor.

③ • The capacitor is again charged, but the charges and electric field are the reverse of step 1.
• No current is present, and there is no magnetic field in the inductor.

④ • The capacitor is again completely discharged, and there is no electric field.
• There is a clockwise current around the circuit. The magnetic field is the reverse of step 2.

Driven series *LRC* circuits: If an ac source is connected in series to an inductor, a resistor, and a capacitor to make a circuit, the current in the circuit oscillates at the angular frequency of the source. The voltage across the inductor leads the current by 1/4 cycle, the voltage across the resistor is in phase with the current, and the voltage across the capacitor lags the current by 1/4 cycle.

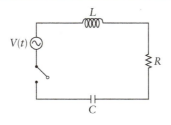

(Figure 21-13)

LRC circuit resonance: The current in a driven series *LRC* circuit has its maximum amplitude when the source (driving) angular frequency equals the natural angular frequency $\omega_0 = \sqrt{1/(LC)}$ of the circuit. When this happens the circuit is in resonance. At resonance the current is in phase with the ac source voltage, and maximum power is delivered to the resistor. In addition, at resonance the voltages across the inductor and capacitor have equal amplitude but are 1/2 cycle out of phase with each other, so their voltages cancel.

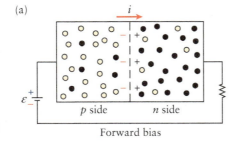

(a) Amplitude of the current

(Figure 21-14a)

Semiconductors and diodes: A semiconductor can be made a better conductor by adding impurities that contribute mobile electrons (an *n*-type semiconductor) or mobile holes (a *p*-type semiconductor). A semiconductor with one side that is *p*-type and one side that is *n*-type can act as a diode that conducts current easily from the *p* side to the *n* side, but conducts poorly in the opposite direction.

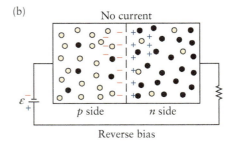

(Figure 21-17)

Answer to What do you think? Question

(d) As we learned in Chapter 20, Faraday's law comes into play only when currents are changing. If the current is constant, as in the case of direct current, Faraday's law has no effect. (See Section 21-3.)

Answers to Got the Concept? Questions

21-1 (b) The voltage provided by the ac source is $V(t) = V_0 \sin \omega t$, where the angular frequency ω is related to the period T by $\omega = 2\pi/T$. The instantaneous power into the resistor is given by Equation 21-3, $P(t) = V^2(t)/R$. At $t = 0$, $V(0) = V_0 \sin 0 = 0$, so $P(0) = 0$. At $t = T/8$, $\omega t = (2\pi/T)(T/8) = \pi/4$, so $V(T/8) = V_0 \sin (\pi/4) = V_0(1/\sqrt{2})$ and $P(T/8) = (V_0/\sqrt{2})^2/R = V_0^2/(2R)$. At $t = T/4$, $\omega t = (2\pi/T)(T/4) = \pi/2$, so $V(T/4) = V_0 \sin (\pi/2) = V_0(1) = V_0$ and $P(T/4) = V_0^2/R$. Equation 21-4 tells us that the average power is $P_{average} = V_0^2/(2R)$, which is equal to the instantaneous power at $t = T/8$ and not at the other times. Note also that at $t = T/8$, the instantaneous voltage V is equal to the rms value $V_{rms} = V_0/\sqrt{2}$.

21-2 (b) With the connections reversed, the number of windings in the primary coil is $N_p = 250$, and the number of windings in the secondary coil is $N_s = 100$. With $V_{p,rms} = 120$ V Equation 21-10 tells us that the output rms voltage is $V_{s,rms} = V_{p,rms}(N_s/N_p) = (120 \text{ V})(100/250) = 48$ V.

21-3 (c) The voltage across an inductor depends on the rate at which the current through the inductor is changing, not on the value of current itself. The voltage across a capacitor depends on the amount of charge on either plate, not on the current. (The current determines how rapidly the capacitor charge is changing and so how rapidly the voltage is changing, but does not affect the present value of the voltage.) The voltage $V = iR$ across a resistor is directly proportional to the current *i*, so this does depend on the value of current.

21-4 (e) As Equations 21-23 show, the oscillation frequency of an *LC* circuit depends on the values of inductance and capacitance alone. It does not depend on the amount of charge placed on the capacitor plates. This is the direct analog of the behavior of a block attached to an ideal spring, which we studied in Section 12-3: The oscillation frequency depends on the spring constant and the mass of the block, but does not depend on the amplitude of the oscillation.

21-5 (d) The natural angular frequency $\sqrt{1/(LC)}$ does not depend on the resistance. The other two quantities, however, do. You can see this from Figure 21-14a: Increasing the resistance decreases the current amplitude at *all* angular frequencies, not just at resonance.

Questions and Problems

In a few problems you are given more data than you actually need; in a few other problems you are required to supply data from your general knowledge, outside sources, or informed estimate.

　Interpret as significant all digits in numerical values that have trailing zeros and no decimal points.

　For all problems use $g = 9.80$ m/s² for the free-fall acceleration due to gravity. Neglect friction and air resistance unless instructed to do otherwise.

•	Basic, single-concept problem
••	Intermediate-level problem; may require synthesis of concepts and multiple steps
•••	Challenging problem
Example	See worked example for a similar problem.

Conceptual Questions

1. • Explain why ac voltage is often described using the root mean square value rather than the average voltage.

2. • Does Equation 18-9 apply to alternating currents? If not, is there any manner in which it can be amended so that it will be valid?

3. • Search the Internet to determine the root mean square value of voltage in a common wall receptacle for five countries that are not mentioned in this chapter.

4. • Give a simple explanation as to why an electric appliance, which is supposed to be operated with a certain root mean square voltage, operated with a slightly increased root mean square voltage will lead to catastrophic results.

5. • Examine the label on the power converter for a laptop computer, digital camera, or cell phone. If you took the converter to a country where the rms voltage is 240 V, would you need a power transformer? Explain your answer.

6. • Why does a transformer whose primary coil has 10 times as many turns as its secondary coil normally deliver about 1/10 of the voltage it receives?

7. • (a) What is a transformer? (b) How does it change the voltage input to some different voltage output? (c) Will a transformer work with a dc input?

8. • **Chemistry** Some of the most toxic substances that have been widely used in the United States (and many other countries in the world) are known as polychlorinated biphenyls or PCBs. (PCBs are a class of organic compounds that have 2 to 10 chlorine atoms attached to biphenyl, a molecule composed of two benzene rings.) Liquid PCBs were often used to fill the interior of transformers. Research PCBs and explain why.

9. • Why is electric power for domestic use in the United States, Canada, and most of the Western Hemisphere transmitted at very high voltages and stepped down to 120 V by a transformer near the point of consumption?

10. • A given length of wire is wound into a solenoid. How will its self-inductance be changed if it is rewound into another coil of (a) twice the length or (b) twice the diameter?

11. • When the switch *S* is opened in the *RL* circuit shown in Figure 21-18, a spark jumps between the switch contacts. Why?

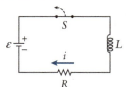

FIGURE 21-18 Problem 11

12. • What change must you make in the current in an inductor to double the energy stored in it?

13. • Discuss the storage of energy in an ideal *LC* circuit with no losses.

14. • Compare the expressions for the energy stored in an inductor and the energy stored in a capacitor and explain the similarities.

15. • Is the current through a resistor in an ac circuit always in phase with the potential applied to the circuit? Why or why not?

16. • What does this statement mean? The voltage drop across an inductor leads the current by 1/4 cycle.

17. • Define the concept of impedance for an *LRC* circuit. What are the units of impedance?

Multiple-Choice Questions

18. • The most important advantage of ac over dc electrical signals is that
　A. electric power can't be delivered by a dc source.
　B. dc could only be used in the early days of electrical power distribution.
　C. dc results in more power loss in wire than ac.
　D. it is relatively straightforward to change the voltage delivered by an ac source.
　E. ac is safer.

19. • In a sinusoidal ac circuit with rms voltage *V*, the peak-to-peak voltage equals
　A. $\sqrt{2}V$.
　B. $2V$.
　C. $V/\sqrt{2}$.
　D. $2\sqrt{2}V$.
　E. $V/2$.

20. • The common electrical receptacle voltage in North America is often referred to as "120 volts ac." One hundred twenty volts is
　A. the arithmetic mean of the voltage as it varies with time.
　B. the root mean square (rms) average of the voltage as it varies with time.
　C. the peak voltage from an ac wall receptacle.
　D. the average voltage over many weeks of time.
　E. one-half the peak voltage.

21. • If a power utility were able to replace an existing 500-kV transmission line with one operating at 1 MV, it would change the amount of heat produced in the transmission line to
A. 1/4 of the previous value.
B. 1/2 of the previous value.
C. 2 times the previous value.
D. 4 times the previous value.
E. 0.

22. • Two solenoids have the same cross-sectional area and length, but the first one has twice as many turns per unit length as the second one. The ratio of the inductance of the second to the first is
A. 1:1.
B. 1:$\sqrt{2}$.
C. 1:1/2.
D. 1:4.
E. 2:1.

23. • For an LC circuit when the charge on the capacitor is 1/2 of the maximum charge, the energy stored in the capacitor is
A. the total energy.
B. 1/2 of the total energy.
C. 1/4 of the total energy.
D. 1/8 of the total energy.
E. twice the total energy.

24. • If the current through an inductor were doubled, the energy stored in the inductor would be
A. halved.
B. the same.
C. doubled.
D. quadrupled.
E. 1/4 as much.

25. • The voltage leads the current
A. in circuits with an ac source and a resistor.
B. in circuits with an ac source and a capacitor.
C. in circuits with an ac source and an inductor.
D. in any ac circuit.
E. in circuits with an ac source and both a capacitor and an inductor.

Problems

21-1 Most circuits use alternating current

21-2 We need to analyze ac circuits differently than dc circuits

26. • A sinusoidally varying voltage is represented by $V(t) = (75.0 \text{ V}) \sin (120\pi t)$. What are its frequency and peak voltage? Example 21-1

27. • Write an expression for the instantaneous voltage delivered by an ac generator supplying 120 V rms at 60 Hz. Example 21-1

28. • What is the rms current provided to a 60-W light bulb that is plugged into a 120-V rms wall receptacle? Example 21-1

29. • The maximum potential difference across the terminals of a 60.0-Hz sinusoidal ac source is +17.0 V at $t = 1/4$ cycle. If at $t = 0$ the potential difference is zero, calculate the potential difference at $t = 2.00$ ms. Example 21-1

30. • An ac voltage is represented by $V(t) = (200 \text{ V}) \sin (120\pi t)$. What is its rms voltage? Example 21-1

31. • The rms current passing through a 50.0-Ω resistor in a sinusoidal ac circuit is 12.0 A. What is the maximum voltage drop across the resistor? Example 21-1

32. • The peak-to-peak current passing through a 150-Ω resistor is 24.0 A. Find the maximum voltage across the resistor and the rms current through the resistor. Example 21-1

33. •• Derive a relationship between the rms current through and the maximum voltage across a resistor R. Apply your result to the following situations: (a) $R = 100 \text{ }\Omega$, $V_{max} = 50.0$ V, $i_{rms} = ?$; (b) $R = 200 \text{ }\Omega$, $i_{rms} = 2.50$ A, $V_{max} = ?$; and (c) $V_{max} = 28.0$ V, $i_{rms} = 127$ mA, $R = ?$ Example 21-1

21-3 Transformers allow us to change the voltage of an ac power source

34. • The primary coil of a transformer makes 240 turns around its core, and the secondary coil of that transformer makes 80 turns. If the primary voltage is 120 V rms, what is the secondary voltage? Example 21-2

35. • The 400-turn primary coil of a step-down transformer is connected to an ac line that is 120 V rms. The secondary coil voltage is 6.50 V rms. Calculate the number of turns in the secondary coil. Example 21-2

36. • The primary coil of a transformer makes 240 turns, and the secondary coil of that transformer makes 80 turns. If an alternating current of 3.00 A (rms) passes through the primary coil of the transformer, what current passes through the secondary coil of that transformer? Assume the transformer is ideal. Example 21-2

37. • An ideal transformer produces an output voltage that is 500% larger than the input voltage. If the input current is 10 A (rms), what is the output current? Example 21-2

38. • A neon sign in a shop operates at around 12 kV rms. A transformer is to step up the 120 V rms line voltage to that value. The secondary coil of the transformer has 20,000 turns. How many turns must be placed into the primary coil? Example 21-2

39. •• The 400-turn primary coil of a step-down transformer is connected to an ac line that is 120 V rms. The secondary coil is to supply 15.0 A at 6.30 V rms. Assuming no power loss in the transformer, calculate (a) the number of turns in the secondary coil and (b) the current in the primary coil. Example 21-2

40. •• The ratio of the number of turns in the primary to the number of turns in the secondary of a step-down transformer is 25:1. If the input voltage across the primary coil is 750 V rms and the current in the output circuit is 25.0 A (rms), how much power is delivered to the secondary? Assume the transformer is 100% efficient. Example 21-2

41. •• **Chemistry** A high-voltage discharge tube is often used to study atomic spectra. The tubes require a large voltage across their terminals to operate. To get the large voltage, a step-up transformer is connected to a line voltage (120 V rms) and is designed to provide 5000 V rms to the discharge tube and to dissipate 75.0 W. (a) What is the ratio of the number of turns in the secondary to the number of turns in the primary? (b) What are the rms currents in the primary and secondary coils of the transformer? (c) What is the effective resistance that the 120-V source is subjected to? Example 21-2

21-4 An inductor is a circuit element that opposes changes in current

42. • How much voltage is produced by an inductor of value 25 μH if the time rate of change of the current is 58 mA/s? Example 21-4

43. • Uncle Leo tunes an old-fashioned radio that has an antenna made from a 3.0-cm-long solenoid with a cross-sectional area of 0.50 cm², composed of 300 turns of fine copper wire. Calculate the inductance of the coil assuming it is air-filled. Example 21-3

44. • A tightly wound solenoid of 1600 turns, cross-sectional area of 6.00 cm², and length of 20.0 cm carries a current of

2.80 A. (a) What is its inductance? (b) If the cross-sectional area is doubled, does anything happen to the value of the inductance? Explain your answer. Example 21-3

21-5 In a circuit with an inductor and capacitor, charge and current oscillate

45. • What energy is stored in a 250-mH inductor with a current of 0.055 A? Example 21-5

46. • Lucinda has a 1.0-mH inductor. What size capacitor should she choose to make an oscillator with a natural frequency of 980 kHz?

47. • An LC circuit is formed with a 15-mH inductor and a 1000-μF capacitor. Calculate the frequency of oscillation f for the circuit.

48. • A 200-pF capacitor is charged to 120 V and then quickly connected to an inductor. Calculate the maximum energy stored in the magnetic field of the inductor as the circuit oscillates. Example 21-5

49. • LC circuits are used for filters in electronics. The selectivity of an LC circuit is defined as the ratio L/C. Calculate the selectivity of the bandwidth for an LC circuit composed of an inductor ($L = 0.250$ H) and a capacitor ($C = 875$ μF).

50. •• If the ratio of the energy stored in a capacitor compared to the total energy stored in an LC circuit is 0.5, calculate the ratio of the charge stored on the capacitor compared to the maximum charge stored on the capacitor in that circuit. Example 21-5

51. •• When the charge on the capacitor in an LC circuit is one-half of the maximum stored charge, calculate the ratio of the energy stored in the capacitor compared to the total energy in both the inductor and the capacitor. Example 21-5

21-6 When an ac voltage source is attached in series to an inductor, a resistor, and a capacitor, the circuit can display resonance

52. • An LRC circuit contains a 1.00-μF capacitor, a 5.00-mH inductor, and a 100-Ω resistor. What is its resonant frequency? Example 21-6

53. • An LRC circuit contains a 500-Ω resistor, a 5.00-H coil, and an unknown capacitor. The circuit resonates at 1000 Hz. What is the value of the capacitance? Example 21-6

54. • The resonant frequency of an LRC circuit is 250 Hz. If the resistance is 200 Ω and the capacitance is 125 nF, what is the value of the inductance? Example 21-6

55. • A sinusoidal voltage of 120 V rms and a frequency of 60 Hz are applied to a 50.0-μF capacitor. Calculate the peak value of the current.

56. • A sinusoidal voltage that is 120 V rms and has a frequency of 60 Hz is applied to a 0.20-H inductor. Calculate the peak value of the current.

57. •• A sinusoidal voltage of 50.0 V (peak) at a frequency of 400 Hz is applied to a capacitor of unknown capacitance. The current in the circuit is 400 mA (rms). (a) What is the capacitance? (b) If the frequency of the voltage is increased, what, if anything, will happen to the rms value of the current in the circuit? Why?

58. • A 100-Ω resistor is connected across a 120-V rms, 60-Hz ac power line. Calculate (a) i_{rms}, (b) i_{max}, and (c) the average power dissipated in the resistor.

59. •• A sinusoidal voltage of 40.0 V rms and a frequency of 100 Hz is applied to (a) a 100-Ω resistor, (b) a 0.200-H

inductor, and (c) a 50.0-μF capacitor. Calculate the peak value of the current and the average power delivered in each case.

60. • A potential of 40.0 V rms and a frequency of 100 Hz is applied to (a) a 0.200-H inductor and (b) a 50.0-μF capacitor. In each case, find the peak value of the current.

61. •• A 35-mH inductor with 0.20-Ω resistance is connected in series to a 200-μF capacitor and a 60-Hz, ac, 45-V source. Calculate the (a) rms current and (b) phase angle for the circuit.

62. •• Assume that the circuit in Figure 21-19 has L equal to 0.60 H, R equal to 250 Ω, and C equal to 3.5 μF. At a frequency of 60 Hz, what are the impedance and the phase angle between the current and voltage?

63. •• Assume that the circuit in Figure 21-19 has L equal to 0.60 H, R equal to 280 Ω, and C equal to 3.5 μF. The amplitude of the driving voltage is 150 V rms. At a frequency of 60 Hz, what is the rms current in the circuit?

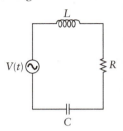

FIGURE 21-19
Problems 62 and 63

64. •• In a driven series LRC circuit, the voltage amplitude and frequency of the source are 100 V and 500 Hz, respectively. The resistance has a value of 500 Ω, the inductance has a value of 0.20 H, and the capacitance has a value of 2.0 μF. (a) What is the impedance of the circuit? (b) What is the amplitude of the current from the source? (c) If the voltage of the source is given by $V(t) = (100$ V$) \sin (1000\pi t)$, how does the current vary with time?

65. • A circuit contains only a 75-nF capacitor and a 120-V rms source. What is the ratio of the rms current generated by the source operating at 60 Hz to that generated by the source operating at 100 Hz?

66. •• The tuner knob on an FM radio moves the plates of an adjustable capacitor that is in series with a 0.400-μH inductor and a net resistance of 1000 Ω. The peak current induced in the circuit by a radio wave of a specific carrier frequency becomes large when the natural frequency of the circuit matches the carrier frequency. What is the frequency of the FM station that is tuned in when the capacitor in the radio is adjusted to 5.80 pF? If the peak operating voltage in the tuning circuit is 9.00 V, what is the peak current? Example 21-6

67. • A series LRC tuning circuit in a TV receiver resonates at 58 MHz. The circuit uses an 18-pF capacitor. What is the inductance of the circuit? Example 21-6

68. • An LRC series circuit contains a 500-Ω resistor, a 5.00-H inductor, and a capacitor. What value of capacitance will cause the circuit to resonate at 1000 Hz? Example 21-6

69. • **Medical** Hearing aids can be tuned to filter out or amplify either high- or low-frequency sounds depending on the frequency range in which a user has suffered hearing loss. If, for instance, a user needed to amplify low-frequency sounds and the hearing aid had a capacitance of 10 μF, what inductance should it have to produce peak signals at 1500 Hz? Example 21-6

General Problems

70. • A technician drives an LRC circuit at various frequencies with an ac voltage supply set to an rms voltage of $V_{rms} = 2.0$ V and records the resulting rms current in a table as shown. She knows that the capacitance in the circuit is $C = 20$ μF and the resistance in the circuit is $R = 5$ Ω. (a) Use the current data to estimate the resonant angular frequency ω_0 and

the inductance L in the circuit. (b) Use the rms current at the resonant frequency and the given rms voltage to verify that the resistance is, in fact, 5 Ω.

ω (krad/sec)	I_{rms} (A)	ω (krad/sec)	I_{rms} (A)
0.5	0.020	5.5	0.374
1.0	0.041	6.0	0.323
1.5	0.065	6.5	0.274
2.0	0.093	7.0	0.236
2.5	0.126	7.5	0.206
3.0	0.170	8.0	0.183
3.5	0.226	8.5	0.164
4.0	0.297	9.0	0.149
4.5	0.368	9.5	0.137
5.0	0.400	10.0	0.126

71. •• A 150-Ω resistor connected across a 60-Hz ac power supply of voltage amplitude 75 V produces heat in the resistor at a certain rate. If you want to replace the ac source with a dc power supply and still produce the same rate of heating, what should be the voltage of the dc source? Example 21-1

72. ••• An LRC series circuit has a 125-Ω resistor, a 12.5-mH inductor, and a parallel plate capacitor having a plate separation of 2.10 mm with a plastic material completely filling the region between the plates. The rectangular capacitor plates each measure 4.25 by 6.20 cm. (a) If you want the maximum rms current through the circuit at an ac frequency of 55.0 Hz, what should be the dielectric constant of the plastic in the capacitor? (b) Consult Table 17-1 to see if it appears feasible to achieve the desired results for the circuit and explain why or why not. Would the use of an ultracapacitor (which can have a capacitance of up to several farads) allow you to achieve the desired results? Example 21-6

73. •• In Europe the standard voltage is 240 V rms at 60 Hz. Suppose you take your 5.00-W electric razor to Rome and plug it into the receptacle (with an adapter to fit the receptacle but not to change the voltage). (a) What power will it draw in Rome? (b) What rms current will run through the razor in the United States and in Rome? Is the razor in danger of being damaged by using it in Rome without a voltage adapter? (c) If you want to use the razor in Rome without damaging it, what type of transformer would you need? Be as quantitative as you can. (d) What is the resistance of your razor? Example 21-1

74. •• Medical A dc current of 60 mA can cause paralysis of the body's respiratory muscles and hence interfere with breathing, but only 15 mA (rms) of ac current will have the same effect. Suppose a person is working with electrical power lines on a warm humid day and therefore has a low body resistance of 1000 Ω. What dc and what ac (amplitude and rms) potentials would it take to cause respiratory paralysis?

75. •• A power cord has a resistance of 0.080 Ω and is used to deliver 1500 W of power. (a) If the power is delivered at 12 V rms, how much power is dissipated in the power cord (assuming the current and voltage are in phase)? (b) If the power is delivered at 120 V rms, how much power is dissipated in the power cord (again assuming the current and voltage are in phase)? (c) Which voltage would you prefer to use to power your electrical device? Why? Example 21-1

76. ••• An ac electrical generator is made by turning a flat coil in a uniform constant magnetic field of 0.225 T. The coil consists of 33 square windings, and each winding is 15.0 cm on each side. It rotates at a steady rate of 745 rpm about an axis perpendicular to the magnetic field passing through the middle of the coil and parallel to two of its opposite sides. An 8.50-Ω light bulb is connected across the generator. (a) Find the voltage and current amplitudes for the light bulb. (b) At what average rate is heat generated in the light bulb? (c) How much energy is consumed by the light bulb every hour? Example 21-1

77. •• A small ac heating coil consisting of 750 windings is 8.50 cm long. It has a diameter of 1.25 cm and a resistance of 2.15 Ω. The coil is connected in series with a 2240-μF ultracapacitor, and the combination is plugged into a household outlet (60.0 Hz and 120 V rms). (a) What is the impedance of the circuit? (b) What is the rms current through the circuit? (c) What is the maximum (or peak) current through the circuit?

78. •• An LRC circuit consists of a 15.0-μF capacitor, a resistor, and an inductor connected in series across an ac power source of variable frequency having a voltage amplitude of 25.0 V. You observe that when the power source frequency is adjusted to 44.5 Hz, the rms current through the circuit has its maximum value of 65.0 mA. What will be the rms current if you change the frequency of the power source to 60.0 Hz? Example 21-6

79. ••• The circuit in Figure 21-20 is known as a low-pass filter because it allows low-frequency signals to pass but attenuates higher frequencies. Suppose its input is an ac signal that is composed of a broad range of frequencies. In this case the voltage across the capacitor is the output that is detected. Show that the ratio of the output voltage to the input voltage is

$$\frac{V_o}{V_i} = \frac{1}{\sqrt{1 + (\omega RC)^2}}.$$

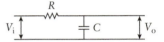

FIGURE 21-20 Problem 79

80. ••• The circuit in Figure 21-21 is known as a high-pass filter because it allows high-frequency signals to pass but attenuates lower frequencies. Suppose its input is an ac signal that is composed of a broad range of frequencies. The voltage across the resistor is the output that is detected. Show that the ratio of the output voltage to the input voltage is

$$\frac{V_o}{V_i} = \frac{1}{\sqrt{1 + 1/(\omega RC)^2}}.$$

FIGURE 21-21 Problem 80

81. •• Medical Resonant wireless charging is an emerging technology where a powered transmitting circuit induces a current in a receiving circuit by the magnetic interaction between the circuits' inductors. Similar to how transformers work, what makes this process unique is that the receiving circuit can be several centimeters to a few meters away from the transmitting circuit, depending on the system. This induction with distance works only when both circuits, which can be treated like effective LRC circuits, operate at the same resonant frequency. One such medical benefit of this technology is the ability to remotely charge an internal battery in a medical implant. Figure 21-22 shows a graph of the induced current in such a medical implant (the receiving LRC circuit) that is operating at its resonant frequency. (a) What is the rms value of the current? (b) What is the implant's resonant frequency ω_0?

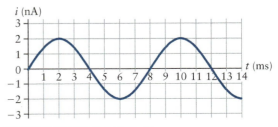

FIGURE 21-22 Problem 81

22

Electromagnetic Waves

Matt Champlin/Moment Open/Getty Images

What do you think?

Our eyes are sensitive to the light from the setting sun, while the receivers in a cell phone tower (at left) are sensitive to radio waves coming from mobile phones. Both visible light and radio waves are kinds of electromagnetic waves. Compared to radio waves, visible light has (a) much higher frequency and much faster speed; (b) much lower frequency and much slower speed; (c) about the same frequency and about the same speed; (d) much higher frequency and about the same speed; (e) much lower frequency and about the same speed.

In this chapter, your goals are to:

- (22-1) Define an electromagnetic wave.
- (22-2) Discuss how speed, frequency, and wavelength are related for electromagnetic waves, and describe the structure of an electromagnetic plane wave.
- (22-3) Explain what Maxwell's equations are and what they tell us about electromagnetic waves.
- (22-4) Calculate the energy density and intensity of an electromagnetic wave, and the energy of a photon.

To master this chapter, you should review:

- (13-2, 13-3, and 13-4) — The properties of mechanical waves
- (16-6) — Electric flux and Gauss's law for the electric field
- (17-5, 17-6) — Capacitors
- (19-7) — Ampère's law
- (20-2, 20-3) — Magnetic flux and Faraday's law
- (21-2) — Root mean square (rms) values
- (21-4, 21-5, and 21-6) — Inductors

22-1 Light is just one example of an electromagnetic wave

What is light? The answer to this question was not discovered until the nineteenth century, when the Scottish physicist James Clerk Maxwell realized that light is an **electromagnetic wave**—a traveling disturbance that, unlike a sound wave, does not require a physical material through which to propagate. Instead, an electromagnetic wave involves oscillating electric and magnetic fields. These can exist even in the vacuum of space, which is why we can see the light from distant stars (Figure 22-1a).

In this chapter we'll study the properties of electromagnetic waves. We'll examine the broad variety of electromagnetic waves, which also includes x rays, microwaves, and radio waves. We'll see how wavelength, frequency, and speed are related for electromagnetic waves that propagate in a vacuum. We'll also look at the inner workings

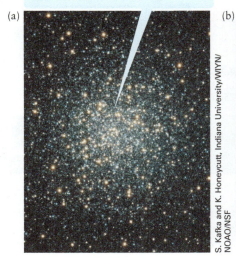

We can see the light from this cluster of stars even though it is separated from us by 100,000 light years (about 10^{18} km) of nearly empty space. This is because electromagnetic waves, including visible light, can propagate in a vacuum.

(a)

S. Kafka and K. Honeycutt, Indiana University/WIYN/NOAO/NSF

The intensity of an electromagnetic wave—including this laser beam used in ophthalmic surgery—depends on the amplitudes of the electric and magnetic fields that make up the wave.

(b)

Will & Deni McIntyre/Science Source

Figure 22-1 Electromagnetic waves Unlike sound waves or water waves, electromagnetic waves do not require the oscillation of any material substance. Instead, what oscillates are electric and magnetic fields. This explains (a) why these waves can propagate in a vacuum and (b) what determines their intensity.

of a particularly simple kind of electromagnetic wave called a *sinusoidal plane wave*. We'll discover how such waves are possible by examining the fundamental equations that govern electric and magnetic fields. We'll then use these equations to help us understand the amount of energy carried by an electromagnetic wave and the wave intensity (Figure 22-1b). We'll find that the energy of an electromagnetic wave comes in small packets called *photons*, whose properties help explain why certain kinds of electromagnetic waves can be harmful while others are not.

22-2 In an electromagnetic plane wave, electric and magnetic fields both oscillate

Experiment shows that in a vacuum all electromagnetic waves—including radio waves, infrared radiation, visible light, ultraviolet radiation, x rays, gamma rays, and others—propagate at the same speed. This is the **speed of light**, to which we give the symbol c:

$$c = 2.99792458 \times 10^8 \text{ m/s } (= 3.00 \times 10^8 \text{ m/s to 3 significant figures})$$ (22-1)

Different kinds of electromagnetic waves have different frequencies and wavelengths. In Section 13-3 we learned that for a mechanical wave the frequency f and wavelength λ are related to the propagation speed of the wave v_p by $v_p = f\lambda$ (Equation 13-2). The same relationship holds for electromagnetic waves in a vacuum with v_p replaced by c:

> Speed of light in a vacuum   Frequency of an electromagnetic wave
>
> $$c = f\lambda$$
>
> Wavelength of the wave in a vacuum

Propagation speed, frequency, and wavelength of an electromagnetic wave
(22-2)

Equation 22-2 tells us that the product of frequency f and wavelength λ has the same value, c, for *all* electromagnetic waves in a vacuum. The longer the wavelength, the lower the frequency; the shorter the wavelength, the higher the frequency.

Electromagnetic waves of any wavelength are possible. Figure 22-2a shows the names given to different wavelength ranges, which we refer to collectively as the **electromagnetic spectrum**. The human eye is sensitive to only a very narrow range of wavelengths known as **visible light** (Figure 22-2b). Visible light encompasses wavelengths from about

Figure 22-2 The electromagnetic spectrum (a) We classify electromagnetic waves according to their wavelength in a vacuum λ (or, equivalently, their frequency f). (b) Visible light makes up a tiny portion of the entire electromagnetic spectrum.

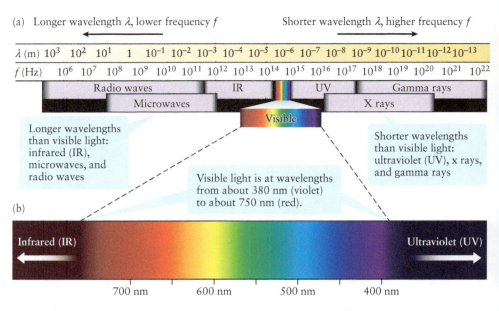

(a) Longer wavelength λ, lower frequency f   Shorter wavelength λ, higher frequency f

(b) Visible light is at wavelengths from about 380 nm (violet) to about 750 nm (red).

Longer wavelengths than visible light: infrared (IR), microwaves, and radio waves

Shorter wavelengths than visible light: ultraviolet (UV), x rays, and gamma rays

Infrared (IR)   Ultraviolet (UV)

700 nm   600 nm   500 nm   400 nm

BioMedical

$\lambda = 380$ nm to about $\lambda = 750$ nm (1 nm = 1 nanometer = 10^{-9} m). We perceive light of different wavelengths as having different colors; the shortest-wavelength light we can see is violet, and the longest-wavelength light we can see is red. At wavelengths shorter than visible light are ultraviolet light (UV), x rays, and gamma rays. At wavelengths longer than visible light are infrared light (IR), microwaves, and radio waves.

Other species can detect wavelengths longer or shorter than those visible to humans. Certain snakes (including pythons and rattlesnakes) have special pit organs on their heads that can sense infrared light. This enables these snakes to detect the radiation that both predators and prey emit due to their body temperature (see Section 14-7). Other species, such as damselfish, are able to sense ultraviolet light (Figure 22-3).

Note that in a medium other than a vacuum, electromagnetic waves propagate at speeds slower than c. For example, visible light travels at about 2.2×10^8 m/s ($0.73c$) in water and at about $0.9998c$ in air. The propagation speed in a medium other than vacuum also depends on the wave frequency; for example, in ordinary glass blue light travels slightly slower than does red light. (We'll see in Chapter 23 that this explains why a prism is able to break white light into colors.) In a vacuum, however, electromagnetic waves of all frequencies propagate at c.

The structure of electromagnetic waves can be quite complex. For example, the waves that make up the beam of laser light shown in Figure 22-1b are strong near the center of the beam, then taper off in strength toward the edge of the beam. A simpler, idealized kind of wave that has all of the key properties of a more general electromagnetic wave is a **sinusoidal plane wave** (Figure 22-4a). As the name suggests, the disturbance in such a wave oscillates in a sinusoidal fashion. There are actually

Figure 22-3 Using ultraviolet light for face recognition The Ambon damselfish (*Pomacentrus amboinensis*, a reef fish native to the western Pacific) has the ability to detect ultraviolet light. Where you see only dark and light bands on a fish's face, an Ambon sees an intricate pattern. This enables the territorial male Ambon damselfish (top) to identify and attack another Ambon to defend its territory but ignore a male lemon damselfish (*Pomacentrus moluccensis*, bottom) with its slightly different facial pattern.

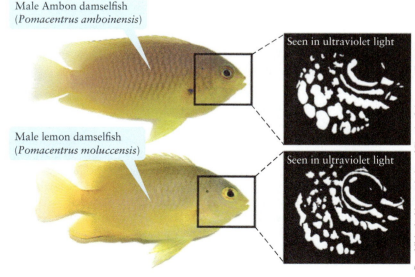

Male Ambon damselfish (*Pomacentrus amboinensis*)

Seen in ultraviolet light

Male lemon damselfish (*Pomacentrus moluccensis*)

Seen in ultraviolet light

Republished with permission of Elsevier Science and Technology Journals, from: Siebeck, U. et. al. "A Species of Reef Fish That Uses Ultraviolet Patterns for Covert Face Recognition" *Current Biology* 20:5, pages 407–410, Fig. 1. (2010); permission conveyed through the Copyright Clearance Center, Inc.

Figure 22-4 **A sinusoidal electromagnetic plane wave**
(a) The characteristics of a simple electromagnetic wave. This illustration shows two complete wavelengths of the wave. (b) How the electric field $\vec{E}$ and magnetic field $\vec{B}$ extend through space.

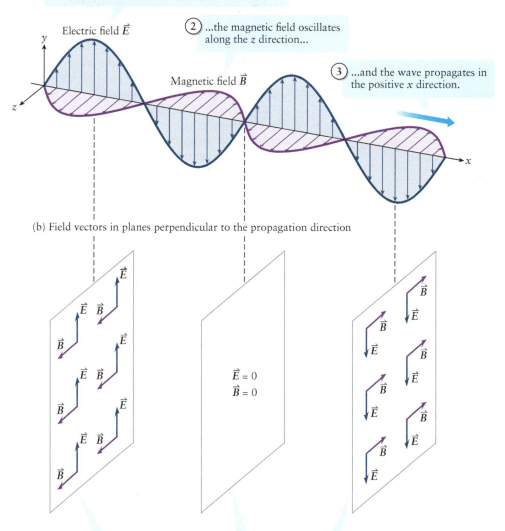

(a) An electromagnetic plane wave propagating in the positive x direction

① In this sinusoidal electromagnetic plane wave, the electric field oscillates along the y direction...

② ...the magnetic field oscillates along the z direction...

③ ...and the wave propagates in the positive x direction.

Electric field $\vec{E}$

Magnetic field $\vec{B}$

(b) Field vectors in planes perpendicular to the propagation direction

$\vec{E} = 0$
$\vec{B} = 0$

④ At any instant the electric field vector has the same value at all points on a plane perpendicular to the direction in which the wave propagates (for this wave, any plane perpendicular to the x axis). The same is true for the magnetic field vector.

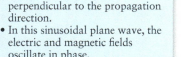

- An electromagnetic wave is transverse: The electric and magnetic fields (the wave disturbances) are perpendicular to the propagation direction.
- In this sinusoidal plane wave, the electric and magnetic fields oscillate in phase.
- This is a plane wave because on any plane perpendicular to the direction in which the wave propagates, $\vec{E}$ has the same value at all points and $\vec{B}$ has the same value at all points.

two disturbances in this wave, an electric field $\vec{E}$ with amplitude E_0 and a magnetic field $\vec{B}$ with amplitude B_0. Both of these fields are *transverse*: They are perpendicular to the direction of propagation, just like the disturbance for waves propagating along a stretched rope or string (Sections 13-3 and 13-4). The fields are also perpendicular to each other. In a snapshot of the wave, as shown in Figure 22-4a, the $\vec{E}$ and $\vec{B}$ fields repeat over the same distance and so have the same wavelength λ. As the wave passes by a given point in space, the $\vec{E}$ and $\vec{B}$ fields both oscillate with the same frequency f. Equation 22-2 then tells us that both fields propagate at the same speed $c = f\lambda$.

WATCH OUT! The fields of an electromagnetic plane wave extend beyond the axis of propagation.

Figure 22-4a may give you the misleading impression that the $\vec{E}$ and $\vec{B}$ fields are present only along the x axis (the axis along which the wave propagates). Such a wave would be like an infinitely narrow laser beam. In fact, in a plane wave the fields have the same value at *all* points on a plane perpendicular to the direction of propagation (Figure 22-4b).

That's the origin of the term "plane wave." A true plane wave would extend infinitely far beyond the x axis in Figure 22-4a, so this is an idealization like a frictionless incline or a massless rope. But the fields shown in Figure 22-4 are a good approximation of the actual fields found near the center of the laser beam in Figure 22-1b.

In Section 13-3 we wrote a *wave function* that describes the disturbance associated with a wave on a rope. In the same fashion we can write wave functions for the electric and magnetic fields depicted in Figure 22-4a:

(22-3)

$$E_y(x,t) = E_0\cos(kx - \omega t + \phi)$$
$$B_z(x,t) = B_0\cos(kx - \omega t + \phi)$$

(sinusoidal electromagnetic plane wave)

In Equations 22-3 we use the same symbols that we used for waves on a rope in Section 13-3: $k = 2\pi/\lambda$ is the angular wave number, $\omega = 2\pi f$ is the angular frequency, and ϕ is the phase angle (which tells us what point in the oscillation cycle corresponds to $x = 0, t = 0$). Note that $E_y(x,t)$ and $B_z(x,t)$ both depend on position x and time t in the same way, which tells us that the electric and magnetic fields oscillate in phase with the same wavelength and frequency. In addition, the electric field amplitude E_0 and the magnetic field amplitude B_0 in a plane wave are directly proportional to each other:

(22-4)

$$B_0 = \frac{E_0}{c}$$

(sinusoidal electromagnetic plane wave)

EXAMPLE 22-1   A Radio Wave

A certain FM radio station broadcasts at a frequency of 98.7 MHz (1 MHz = 10^6 Hz). In the wave that reaches the radio in your car, the electric field amplitude is 6.00×10^{-2} V/m. Calculate the wavelength of the wave and the amplitude of the magnetic field.

Set Up

We are given the wave frequency f and the electric field amplitude E_0. We use Equation 22-2 to find the wavelength λ and Equation 22-4 to find the magnetic field amplitude B_0.

Propagation speed, frequency, and wavelength of an electromagnetic wave:

$$c = f\lambda \qquad (22\text{-}2)$$

Relationship between the electric and magnetic field amplitudes in an electromagnetic wave:

$$B_0 = \frac{E_0}{c} \qquad (22\text{-}4)$$

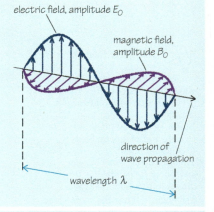

electric field, amplitude E_0

magnetic field, amplitude B_0

direction of wave propagation

wavelength λ

Solve

Use Equation 22-2 and the given frequency $f = 98.7$ MHz $= 98.7 \times 10^6$ Hz to solve for the wavelength.

From Equation 22-2:

$$\lambda = \frac{c}{f} = \frac{3.00 \times 10^8 \text{ m/s}}{98.7 \times 10^6 \text{ Hz}} = 3.04 \frac{\text{m}}{\text{s} \cdot \text{Hz}}$$

Recall that 1 Hz = 1 s^{-1}, so the units of s and Hz cancel:

$$\lambda = 3.04 \text{ m}$$

Use Equation 22-4 and the value $E_0 = 6.00 \times 10^{-2}$ V/m to solve for the magnetic field amplitude.

From Equation 22-4:

$$B_0 = \frac{E_0}{c} = \frac{6.00 \times 10^{-2} \text{ V/m}}{3.00 \times 10^8 \text{ m/s}} = 2.00 \times 10^{-10} \left(\frac{\text{V}}{\text{m}}\right)\left(\frac{\text{s}}{\text{m}}\right)$$

We learned in Section 17-3 that 1 V/m = 1 N/C, and in Section 19-3 we learned that 1 T = 1 (N·s)/(C·m). So

$$B_0 = 2.00 \times 10^{-10} \left(\frac{\text{N}}{\text{C}}\right)\left(\frac{\text{s}}{\text{m}}\right) = 2.00 \times 10^{-10} \frac{\text{N} \cdot \text{s}}{\text{C} \cdot \text{m}}$$

$$= 2.00 \times 10^{-10} \text{ T}$$

Reflect

Because the speed of light c has such a large value in m/s, the magnetic field amplitude B_0 in tesla (T) is much smaller than the electric field amplitude E_0 in volts per meter (V/m). As we'll see later in the chapter, however, the electric and magnetic fields prove to be equally important in an electromagnetic wave in a vacuum.

In this example we've seen how to relate the units of magnetic field (T) to those of electric field (V/m): $1 \text{ T} = 1 \text{ (V/m)} \cdot \text{(s/m)} = 1 \text{ V} \cdot \text{s/m}^2$. We'll make use of this result in later examples.

Why must an electromagnetic wave propagating in a vacuum be transverse? Would it be possible to have a longitudinal component of the wave, with an oscillating electric or magnetic field along the direction of propagation? For that matter, why is it necessary for an electromagnetic wave to include both electric and magnetic fields? Couldn't there be a wave that included an oscillating electric field but no magnetic field, or an oscillating magnetic field but no electric field? To answer these questions we need to look at the fundamental equations that govern electric and magnetic fields.

GOT THE CONCEPT? 22-1 Electromagnetic Wave Speeds

 Four electromagnetic waves in a vacuum have different frequencies. Which of these propagates at the fastest speed? (a) $f = 3.95 \times 10^6$ Hz; (b) $f = 2.44 \times 10^7$ Hz; (c) $f = 1.26 \times 10^{11}$ Hz; (d) $f = 2.26 \times 10^8$ Hz; (e) all have the same speed.

TAKE-HOME MESSAGE FOR Section 22-2

✔ Electromagnetic waves in a vacuum propagate at the speed of light.

✔ Different kinds of electromagnetic waves have different wavelengths and frequencies. The human eye can detect only a very narrow range of wavelengths.

✔ In a sinusoidal electromagnetic plane wave, the electric and magnetic fields both oscillate in phase with the same wavelength and frequency. The field directions are perpendicular to each other and perpendicular to the direction of propagation.

22-3 Maxwell's equations explain why electromagnetic waves are possible

By the middle of the nineteenth century, scientists understood a great deal about electricity and magnetism. The work of Gauss, Ampère, and Faraday had established fundamental relationships that describe electric and magnetic phenomena; for example, it was understood that electric charges give rise to electric fields and that electric currents give rise to magnetic fields. It was the Scottish physicist James Clerk Maxwell who added to these fundamental relationships and forged our understanding of electricity and magnetism into a unified theory. In this section we'll look at four basic equations, known as **Maxwell's equations**, which describe *all* electromagnetic phenomena. We'll then see how these equations help us understand the nature of electromagnetic waves like the sinusoidal plane wave that we discussed in Section 22-2.

Gauss's Laws for Electricity and Magnetism

The first of Maxwell's equations is **Gauss's law for the electric field**, which we first encountered in Section 16-6. It states that the net electric flux through a closed surface (called a *Gaussian surface*) is proportional to the total amount of electric charge enclosed within that surface:

Electric flux through a closed surface   Net amount of **charge enclosed** within the surface

$$\Phi_E = \frac{q_{encl}}{\varepsilon_0}$$

Permittivity of free space $= 1/(4\pi k)$

Gauss's law for the electric field
(16-9)

(a) Gauss's law for the electric field: $\Phi_E = q_{encl}/\varepsilon_0$

Gaussian surface 1 encloses positive charge. Field lines point out of this surface, and there is a net outward (positive) electric flux.

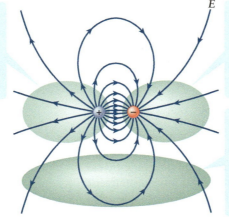

Gaussian surface 2 encloses negative charge. Field lines point into this surface, and there is a net inward (negative) electric flux.

Gaussian surface 3 encloses zero charge. Each field line that points into this surface at one place points out at another place. There is zero net electric flux.

$\vec{E}$

Figure 22-5 Gauss's laws for the electric and magnetic fields (a) The electric flux through a closed surface depends on the charge enclosed by that surface. (b) The magnetic flux through any closed surface is zero.

(b) Gauss's law for the magnetic field: $\Phi_B = 0$

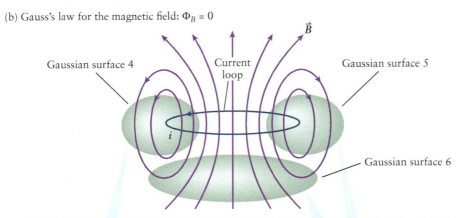

Gaussian surface 4

Current loop

$\vec{B}$

Gaussian surface 5

Gaussian surface 6

- In a region of space where there are no electric charges, there is zero net electric flux through any Gaussian surface in that region.
- In any region of space, there is zero net magnetic flux through any Gaussian surface.

Gaussian surfaces 4 and 5 enclose part of the current loop. **Gaussian surface 6** encloses none of the current loop. For each of these surfaces each field line that points into the surface at one place points out at another place. There is zero net magnetic flux for each surface.

In this equation we've added a subscript E to remind us that the quantity on the left-hand side is the flux of the electric field $\vec{E}$.

Figure 22-5a illustrates Gauss's law for the electric field. If a surface encloses a net positive charge, as for Gaussian surface 1 in Figure 22-5a, there is a net outward (positive) electric flux and electric field lines point out of the surface. If instead there is a net negative charge inside the surface, as for Gaussian surface 2 in Figure 22-5a, the net electric flux is inward (negative) and electric field lines point into the surface. If there is no charge at all inside the surface, as for Gaussian surface 3 in Figure 22-5a, there is zero net electric flux: Each field line that enters the surface at one point exits it at another point.

In addition to Gauss's law for the electric field, there is **Gauss's law for the magnetic field**. This states that for *any* closed Gaussian surface, there is *zero* net flux of the magnetic field $\vec{B}$ through that surface:

Magnetic flux through a closed surface

There is **zero magnetic flux through any closed surface**, no matter what the size or shape of the surface or what it contains.

Gauss's law for the magnetic field (22-5)

$$\Phi_B = 0$$

As an example, Figure 22-5b shows three Gaussian surfaces in the magnetic field of a current loop. For all three surfaces, each field line that enters the surface at one point exits it at another point. This is true whether there is current enclosed within the Gaussian surface (as for surfaces 4 and 5) or if there is no enclosed current (as for surface 6). Just as for the electric flux for Gaussian surface 3 in Figure 22-5a, this implies that there is zero net flux of magnetic field for all three surfaces. (There *would* be a nonzero net magnetic flux through a Gaussian surface if the surface enclosed an isolated north magnetic pole that had no associated south pole, or an isolated south pole that had no associated north pole. These would be the magnetic analogs of an isolated positive or negative electric charge. Physicists have searched for decades for such isolated poles, called *magnetic monopoles*. No confirmed observations have yet been made.)

What Gauss's Laws Tell Us About Electromagnetic Waves

Equations 16-9 and 22-5 are valid not just for static situations in which the electric and magnetic fields are constant, but also in situations where the $\vec{E}$ and $\vec{B}$ fields are changing. So Gauss's laws for the electric and magnetic fields must also hold true for electromagnetic waves. In Figure 22-6a we've drawn a Gaussian surface that encloses part of the sinusoidal plane wave shown in Figure 22-4. Because the wave is transverse, there is no electric or magnetic flux through the front or back face of the Gaussian surface. Although the fields change, at any instant there is as much outward flux of $\vec{E}$ or $\vec{B}$ on one edge of the Gaussian surface as there is inward flux on the opposite edge. So the net electric flux and the net magnetic flux through this surface are both zero, and both of Gauss's laws are obeyed.

Gauss's laws also explain why electromagnetic waves in a vacuum *must* be transverse. Figure 22-6b shows part of an electromagnetic wave with an oscillating *longitudinal* electric field (one that points along or opposite to the direction in which the wave propagates).

Figure 22-6 Gauss's laws applied to a sinusoidal electromagnetic plane wave (a) A transverse electromagnetic wave in a vacuum is compatible with Gauss's laws, but (b) a longitudinal electromagnetic wave in a vacuum is not.

(a) Gauss's laws say that a transverse electromagnetic wave is permitted.

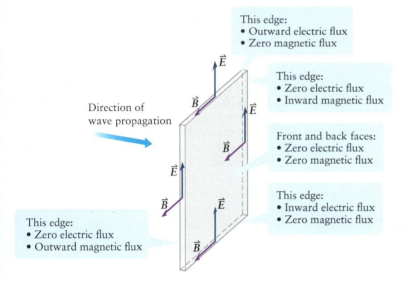

This edge:
• Outward electric flux
• Zero magnetic flux

This edge:
• Zero electric flux
• Inward magnetic flux

Direction of wave propagation

Front and back faces:
• Zero electric flux
• Zero magnetic flux

This edge:
• Inward electric flux
• Zero magnetic flux

This edge:
• Zero electric flux
• Outward magnetic flux

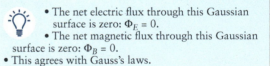

• The net electric flux through this Gaussian surface is zero: $\Phi_E = 0$.
• The net magnetic flux through this Gaussian surface is zero: $\Phi_B = 0$.
• This agrees with Gauss's laws.

(b) Gauss's laws say that a longitudinal electromagnetic wave is not permitted.

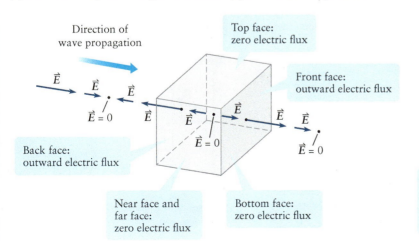

Direction of wave propagation

Top face: zero electric flux

Front face: outward electric flux

$\vec{E} = 0$

Back face: outward electric flux

$\vec{E} = 0$

Near face and far face: zero electric flux

Bottom face: zero electric flux

• The net electric flux through this Gaussian surface is outward, so $\Phi_E \neq 0$.
• This disagrees with Gauss's law for the electric field.

For the Gaussian surface that we've chosen and at the instant of time shown, there is a net nonzero outward flux of electric field. But the wave is in a vacuum, so there can be no charge enclosed by the surface. Equation 16-9 then tells us that there *cannot* be a net electric flux through the surface in Figure 22-6b. We're forced to conclude that the longitudinal wave shown in Figure 22-6b is *impossible* because it contradicts Gauss's law for the electric field. Likewise, it would be impossible for an electromagnetic wave to have a longitudinal magnetic field, since this would violate Gauss's law for the magnetic field as given by Equation 22-5.

Let's now examine the two remaining members of Maxwell's equations and see what they tell us about the nature of electromagnetic waves.

Faraday's Law: A Changing Magnetic Field Generates an Electric Field

The third of Maxwell's equations is **Faraday's law**, which we introduced in Sections 20-2 and 20-3. It states that an emf is induced in a loop if the magnetic flux through that loop changes:

Faraday's law and Lenz's law
for induction
(20-3)

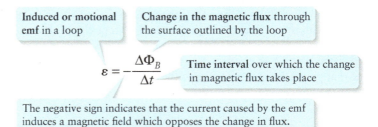

Induced or motional emf in a loop

Change in the magnetic flux through the surface outlined by the loop

$$\varepsilon = -\frac{\Delta \Phi_B}{\Delta t}$$

Time interval over which the change in magnetic flux takes place

The negative sign indicates that the current caused by the emf induces a magnetic field which opposes the change in flux.

As an example, Figure 22-7a shows a wire loop placed between the poles of an electromagnet. As the current in the electromagnet increases, the magnetic field increases and the increasing magnetic flux through the wire loop induces an emf in the loop. It's important to note that the emf is present whether the loop is made of a conductor like copper wire, a semiconductor like silicon, or an insulator like wood; the only difference is the amount of current that's established in response to the emf. The emf is even present if there is no material substance there at all!

Fundamentally, what happens is that the changing magnetic field generates a circulating electric field $\vec{E}$ as Figure 22-7b shows. The emf around a loop is just the *circulation* of the electric field around the loop. We define this in exactly the same way that we defined the circulation of magnetic field in Section 19-7. First imagine breaking the loop into a number of small segments of length $\Delta \ell$; then find the component $E_{\parallel}$ of the electric field that's tangent to each segment; then sum the products $E_{\parallel}\Delta \ell$:

(22-6)

$$\text{circulation of the electric field} = \sum E_{\parallel}\Delta \ell$$

(a) A wire loop in a changing magnetic field

(b) A changing magnetic field produces a circulating electric field

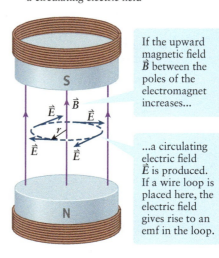

Electromagnet

If the upward magnetic field $\vec{B}$ between the poles of the electromagnet increases...

...an emf is induced in the wire loop because the magnetic flux through the loop changes.

Wire loop

If the upward magnetic field $\vec{B}$ between the poles of the electromagnet increases...

...a circulating electric field $\vec{E}$ is produced. If a wire loop is placed here, the electric field gives rise to an emf in the loop.

Figure 22-7 A changing $\vec{B}$ produces a circulating $\vec{E}$
(a) Faraday's law says that an emf is induced in the loop when the magnetic field changes.
(b) Fundamentally what happens when the magnetic field changes is that a circulating electric field is produced.

(One simple case is the circulating electric field shown in Figure 22-7b. Here $\vec{E}$ is tangential to the circular loop of radius r, so the circulation is just the magnitude E multiplied by the total distance around the loop—that is, its circumference $2\pi r$.)

We can now rewrite Faraday's law by replacing the induced emf ε in Equation 20-3 with the circulation of the electric field, as given by Equation 22-6:

Circulation of electric field around a loop

Change in the magnetic flux through the surface outlined by the loop

$$\sum E_{\parallel}\Delta\ell = -\frac{\Delta\Phi_B}{\Delta t}$$

Time interval over which the change in magnetic flux takes place

Faraday's law in terms of circulation (22-7)

Equation 22-7 tells us that a changing magnetic flux through a loop causes an electric field that circulates around that loop. This is true even if there is no material substance at the location of the loop. The minus sign in Equation 22-7 tells us the direction of the circulating electric field: If a conducting wire is present around the loop, the electric field will cause charges to flow, and the magnetic field due to those moving charges will oppose the change in magnetic field that caused the change in magnetic flux. (This is Lenz's law, which we introduced in Section 20-3.)

EXAMPLE 22-2 An Electric Field Due to a Changing Magnetic Field

The uniform magnetic field shown in Figure 22-7b decreases in magnitude from 1.50 T to zero in a time t, inducing an electric field. What is the magnitude of this electric field around a loop 3.50 cm in diameter if (a) $t = 10.0$ s and (b) $t = 0.100$ s?

Set Up

Equation 20-1 tells us the magnetic flux through the loop of diameter 3.50 cm. The magnetic field $\vec{B}$ is perpendicular to the plane of the loop, so θ (the angle between the direction of $\vec{B}$ and the perpendicular to the loop) is $\theta = 0$.

The electric field in this case has the same magnitude E all the way around the circle and is tangent to the circle, so $E_{\parallel}$ in Equation 22-7 is equal to E.

Magnetic flux:

$$\Phi_B = AB_{\perp} = AB\cos\theta \quad (20\text{-}1)$$

Faraday's law in terms of circulation:

$$\sum E_{\parallel}\Delta\ell = -\frac{\Delta\Phi_B}{\Delta t} \quad (22\text{-}7)$$

Area of a circle of radius r:

$$A = \pi r^2$$

Circumference of a circle of radius r:

$$C = 2\pi r$$

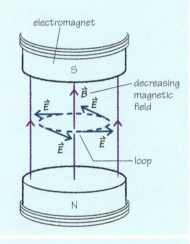

Solve

(a) First determine the initial and final values of magnetic flux and the change in magnetic flux for a circle of diameter 3.50 cm.

The radius of the loop is half the diameter:

$$r = \frac{1}{2}(3.50\text{ cm}) = 1.75\text{ cm} = 1.75\times10^{-2}\text{ m}$$

The area of the loop is

$$A = \pi r^2 = \pi(1.75\times10^{-2}\text{ m})^2 = 9.62\times10^{-4}\text{ m}^2$$

From Equation 20-1 the initial magnetic flux is

$$\Phi_B = AB\cos 0 = (9.62\times10^{-4}\text{ m}^2)(1.50\text{ T})(1)$$
$$= 1.44\times10^{-3}\text{ T}\cdot\text{m}^2$$

The final magnetic field is zero, so the final magnetic flux is zero as well. The change in magnetic flux is

$$\Delta\Phi_B = (\text{final flux}) - (\text{initial flux}) = 0 - 1.44\times10^{-3}\text{ T}\cdot\text{m}^2$$
$$= -1.44\times10^{-3}\text{ T}\cdot\text{m}^2$$

Use Equation 22-7 to calculate the circulation of the electric field in the case where $\Delta t = 10.0$ s. Since the field magnitude E has the same value around the circle and $E_\parallel = E$, we can use this to calculate E.

The circulation of the electric field is equal to $-\Delta\Phi_B/\Delta t$:

$$\text{circulation of the electric field} = \sum E_\parallel \Delta\ell = \frac{\Delta\Phi_B}{\Delta t}$$

$$= -\frac{-1.44 \times 10^{-3}\ \text{T}\cdot\text{m}^2}{10.0\ \text{s}} = 1.44 \times 10^{-4}\ \frac{\text{T}\cdot\text{m}^2}{\text{s}}$$

In Example 22-1 (Section 22-2) we saw that

$1\ \text{T} = 1\ \dfrac{\text{V}\cdot\text{s}}{\text{m}^2}$, so the magnitude of the circulation is

$$1.44 \times 10^{-4}\ \frac{\text{T}\cdot\text{m}^2}{\text{s}} = 1.44 \times 10^{-4}\left(\frac{\text{V}\cdot\text{s}}{\text{m}^2}\right)\left(\frac{\text{m}^2}{\text{s}}\right) = 1.44 \times 10^{-4}\ \text{V}$$

Since $E_\parallel = E$ and E has the same value at all points around the circle, we can write the circulation as

$$\sum E_\parallel \Delta\ell = \sum E\Delta\ell = E\sum \Delta\ell$$

The sum $\sum \Delta\ell$ is the total distance around the loop, equal to the loop circumference $2\pi r$. So

$$E(2\pi r) = 1.44 \times 10^{-4}\ \text{V}$$

$$E = \frac{1.44 \times 10^{-4}\ \text{V}}{2\pi r} = \frac{1.44 \times 10^{-4}\ \text{V}}{2\pi(1.75 \times 10^{-2}\ \text{m})} = 1.31 \times 10^{-3}\ \text{V/m}$$

(b) Repeat the calculation for the case where $\Delta t = 0.100$ s.

Equation 22-7 tells us that the circulation of the electric field, and hence the electric field itself, is inversely proportional to the time Δt over which the magnetic flux changes. If the magnetic field drops to zero in $\Delta t = 0.100$ s rather than $t = 10.0$ s, the elapsed time is smaller by a factor of

$$\frac{0.100\ \text{s}}{10.0\ \text{s}} = 1.00 \times 10^{-2}$$

and so the induced field is larger by a factor of

$$\frac{1}{1.00 \times 10^{-2}} = 1.00 \times 10^{2}$$

Therefore, the electric field in the case where $\Delta t = 0.100$ s is

$$E = (1.00 \times 10^{2})(1.31 \times 10^{-3}\ \text{V/m}) = 0.131\ \text{V/m}$$

Reflect

Our results show that the more rapid the change in magnetic field, the greater the magnitude of the electric field that is induced.

If a circular loop of conducting wire were placed along the circular path of diameter 3.50 cm, a current would be generated so as to produce a magnetic field that would oppose the change in magnetic flux. The upward magnetic field in the figure decreases, so the magnetic field produced in this way would have to be upward. The induced current that produces this magnetic field is in the same direction as the circulating electric field. So the electric field and current must both have the direction shown.

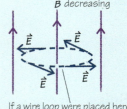

If a wire loop were placed here, current would be induced to produce an upward $\vec{B}$.

The Maxwell–Ampère Law: A Changing Electric Field Generates a Magnetic Field

Equation 22-7 tells us that a circulating electric field is produced by a *magnetic* field that changes over time. In Section 19-7 we learned that a circulating *magnetic* field is produced by electric charges in motion, that is, by a current. The mathematical expression of this statement is *Ampère's law*:

Ampère's law
(19-13)

Circulation of magnetic field around an Amperian loop

Permeability of free space

$$\sum B_\parallel \Delta\ell = \mu_0 i_{\text{through}}$$

Current through the Amperian loop

(a) A current-carrying wire

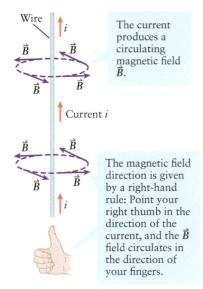

Wire

The current produces a circulating magnetic field $\vec{B}$.

Current i

The magnetic field direction is given by a right-hand rule: Point your right thumb in the direction of the current, and the $\vec{B}$ field circulates in the direction of your fingers.

(b) A current-carrying wire with a capacitor

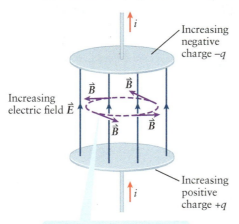

Increasing negative charge $-q$

Increasing electric field $\vec{E}$

Increasing positive charge $+q$

There is no current between the plates, but the changing electric field $\vec{E}$ produces a circulating magnetic field $\vec{B}$.

Figure 22-8 A changing $\vec{E}$ produces a circulating $\vec{B}$
(a) Ampère's law says that a magnetic field circulates around a current-carrying wire. (b) A circulating magnetic field can also be produced by a changing electric field.

Figure 22-8a shows an application of Ampère's law that we introduced in Section 19-7: the magnetic field due to a long, straight, current-carrying wire. The current through each loop in the figure is equal to the current in the wire, so for each loop $i_{\text{through}} = i$. As a result, there is a magnetic field that circulates around each loop, and the circulation $\sum B_{\parallel}\Delta\ell$ of the magnetic field is equal to $\mu_0 i$.

Now suppose we break the wire in Figure 22-8a and insert two metal disks to form a parallel-plate capacitor (**Figure 22-8b**). If the same steady current exists as in Figure 22-8a, positive charge will build up on the lower plate, negative charge will build up on the upper plate, and the electric field between the two plates will increase in magnitude. There is no current from the lower plate to the upper plate, so $i_{\text{through}} = 0$ for the loop between the plates in Figure 22-8b. So Equation 19-13 predicts that there should be no circulating magnetic field between the plates. Yet experiment shows that there *is* a circulating magnetic field between the plates! How can this be?

James Clerk Maxwell's great insight was to propose that if a changing magnetic field could produce a circulating *electric* field, then a changing *electric* field could produce a *magnetic* field. That's what's happening between the capacitor plates shown in Figure 22-8b: The electric field is increasing in magnitude as charge builds on the plates, and a circulating magnetic field results. To explain this effect, Maxwell expanded on Ampère's law as given by Equation 19-13 to include an additional term on the right-hand side. The result is called the **Maxwell–Ampère law:**

Circulation of magnetic field around an Amperian loop	Permittivity of free space	Change in the electric flux through the surface outlined by the loop

$$\sum B_{\parallel}\Delta\ell = \mu_0\left(i_{\text{through}} + \varepsilon_0\frac{\Delta\Phi_E}{\Delta t}\right)$$

Maxwell–Ampère law
(22-8)

Permeability of free space	Current through the Amperian loop	Time interval over which the change in electric flux takes place

The electric flux Φ_E through a loop, calculated using Equation 16-6, is defined in precisely the same manner as the magnetic flux Φ_B through a loop. The quantity $\varepsilon_0(\Delta\Phi_E/\Delta t)$ has units of amperes and is called the **displacement current**. Experiment confirms Equation 22-8: A circulating magnetic field can be produced by an electric current, by a changing electric field, or a combination.

We saw in Section 17-5 that we can write the permittivity of free space as $\varepsilon_0 = 8.85 \times 10^{-12}$ F/m. Since 1 F = 1 C/V and 1 A = 1 C/s, we can write this as

$$\varepsilon_0 = 8.85 \times 10^{-12}\,\frac{\text{F}}{\text{m}} = 8.85 \times 10^{-12}\,\frac{\text{C}}{\text{V} \cdot \text{m}} = 8.85 \times 10^{-12}\,\frac{\text{A} \cdot \text{s}}{\text{V} \cdot \text{m}} \qquad (22\text{-}9)$$

We'll make use of this equation in the following example.

EXAMPLE 22-3 A Magnetic Field Due to a Changing Electric Field

A parallel-plate capacitor like that shown in Figure 22-8b has circular plates 5.00 cm in diameter. The electric field between the plates increases by 8.00×10^5 V/m in 1.00 s, inducing a magnetic field. What is the magnitude of that magnetic field around a loop 3.50 cm in diameter in the space between the capacitor plates?

Set Up

No charge actually moves through the loop in question, so $i_{\text{through}} = 0$ in Equation 22-8. However, the electric flux through the loop changes as the electric field changes, so the term $\Delta\Phi_E/\Delta t$ in Equation 22-8 is not zero, and a circulating magnetic field will result. This example is very similar to Example 22-2, except that now we need to calculate the change in *electric* flux to determine the magnitude of the circulating *magnetic* field.

Equation 16-6 tells us the electric flux through the 3.50-cm-diameter loop; the electric field $\vec{E}$ is perpendicular to the plane of the loop, so $\theta = 0$. The magnetic field has the same magnitude B all the way around the loop and is tangent to the loop, so $B_{\parallel} = B$ in Equation 22-8.

Electric flux:
$$\Phi_E = AE_{\perp} = AE \cos \theta \quad (16\text{-}6)$$
Maxwell–Ampère law:
$$\sum B_{\parallel}\Delta\ell = \mu_0\left(i_{\text{through}} + \varepsilon_0\frac{\Delta\Phi_E}{\Delta t}\right) \quad (22\text{-}8)$$
Permittivity of free space:
$$\varepsilon_0 = 8.85 \times 10^{-12}\frac{\text{A}\cdot\text{s}}{\text{V}\cdot\text{m}} \quad (22\text{-}9)$$
Permeability of free space:
$$\mu_0 = 4\pi \times 10^{-7}\ \text{T}\cdot\text{m/A}$$
Area of a circle of radius r:
$$A = \pi r^2$$
Circumference of a circle of radius r:
$$C = 2\pi r$$

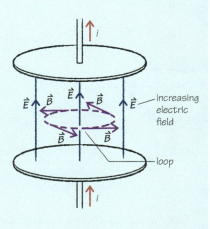

Solve

Find the displacement current $\varepsilon_0(\Delta\Phi_E/\Delta t)$ associated with the change in electric flux through the loop.

The change in electric flux in a time $\Delta t = 1.00$ s is equal to the area of the loop multiplied by the change in electric field in that time (recall that $\theta = 0$). The loop has radius $r = (1/2) \times (3.50\text{ cm}) = 1.75$ cm $= 1.75 \times 10^{-2}$ m, so

$$\Delta\Phi_E = A(\Delta E)\cos 0 = A(\Delta E)(1)$$
$$= (\pi)(1.75 \times 10^{-2}\text{ m})^2(8.00 \times 10^5\text{ V/m})(1)$$
$$= 7.70 \times 10^2\text{ V}\cdot\text{m}$$

The displacement current through the loop is

$$\varepsilon_0\frac{\Delta\Phi_E}{\Delta t} = \left(8.85 \times 10^{-12}\frac{\text{A}\cdot\text{s}}{\text{V}\cdot\text{m}}\right)\left(\frac{7.70 \times 10^2\text{ V}\cdot\text{m}}{1.00\text{ s}}\right)$$
$$= 6.81 \times 10^{-9}\text{ A}$$

Use Equation 22-8 to determine the magnitude of the circulating magnetic field.

Since $i_{\text{through}} = 0$, the circulation of the magnetic field is equal to μ_0 times the displacement current:

circulation of the magnetic field $= \sum B_{\parallel}\Delta\ell = \mu_0\left(\varepsilon_0\dfrac{\Delta\Phi_E}{\Delta t}\right)$

$$= \left(4\pi \times 10^{-7}\frac{\text{T}\cdot\text{m}}{\text{A}}\right)(6.81 \times 10^{-9}\text{ A}) = 8.56 \times 10^{-15}\text{ T}\cdot\text{m}$$

Since $B_{\parallel} = B$ and B has the same value at all points around the circle, we can write the circulation as

$$\sum B_{\parallel}\Delta\ell = \sum B\Delta\ell = B\sum\Delta\ell$$

The sum $\sum\Delta\ell$ is the total distance around the loop and is equal to the loop circumference $2\pi r$. So

$$B(2\pi r) = 8.56 \times 10^{-15}\text{ T}\cdot\text{m}$$
$$B = \frac{8.56 \times 10^{-15}\text{ T}\cdot\text{m}}{2\pi(1.75 \times 10^{-2}\text{ m})} = 7.78 \times 10^{-14}\text{ T}$$

Reflect

The induced magnetic field is very weak (7.78×10^{-14} T) because the displacement current that produces it is very small (6.81×10^{-9} A). If the electric field between the plates were to change more rapidly, the displacement current would be greater and the induced magnetic field stronger.

The displacement current is in the same direction as the current that brings positive charge to the lower plate of the capacitor. The right-hand rule for using Ampère's law (Section 19-7) tells us that the induced magnetic field is in the direction shown.

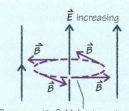

The magnetic field due to an increasing $\vec{E}$ circulates in the same direction as the electric field due to a decreasing $\vec{B}$ (see Example 22-2).

In a vacuum, where no electric currents are present, we can write Faraday's law and the Maxwell–Ampère law as

$$\sum E_\parallel \Delta\ell = -\frac{\Delta\Phi_B}{\Delta t}$$

$$\sum B_\parallel \Delta\ell = +\mu_0\varepsilon_0 \frac{\Delta\Phi_E}{\Delta t}$$

(22-10)

Notice that the first of Equations 22-10 (Faraday's law) has a minus sign, while the second of these equations (the Maxwell–Ampère law) does not. This says that the circulating electric field produced by a *decreasing* magnetic flux (so $\Delta\Phi_B/\Delta t < 0$) is in the same direction as the circulating magnetic field produced by an *increasing* electric flux (so $\Delta\Phi_E/\Delta t > 0$). That's why the electric field that we found in Example 22-2 due to a decreasing magnetic field circulates in the same direction as the magnetic field that we found in Example 22-3 due to an increasing electric field.

What Faraday's Law and the Maxwell–Ampère Law Tell Us About Electromagnetic Waves

Faraday's law says that a varying magnetic field gives rise to a circulating electric field, and the Maxwell–Ampère law says that a varying electric field gives rise to a circulating magnetic field. Taken together, these two laws tell us how electromagnetic waves are possible and why they involve both electric and magnetic fields.

Imagine that you take an electric charge and oscillate it up and down. This will cause the electric field due to the charge to oscillate as well. By the Maxwell–Ampère law, it follows that a magnetic field will be produced which will circulate in one direction when the electric field is increasing and in the other direction when the electric field is decreasing. In other words, the magnetic field will itself oscillate. Faraday's law then tells us that this magnetic field will itself generate an electric field whose direction changes, circulating in one direction when the magnetic field is increasing and in the other direction when the magnetic field is decreasing. These two laws together tell us that a combination of an oscillating electric field and an oscillating magnetic field can sustain each other and will continue to sustain each other even after the original source of the oscillating fields—the oscillating charge—has stopped moving.

Let's look at this in a little more detail for the electromagnetic plane wave depicted in Figure 22-4a. Figure 22-9a shows five snapshots of the wave as it propagates from left to right in the positive x direction. In each snapshot we're looking at the x–y plane, and in each snapshot we've drawn a stationary rectangular loop one-half of a wavelength in width. Because the magnetic field has only a z component, which is perpendicular to the plane of the loop, as the wave propagates there is a varying magnetic flux Φ_B through this loop. When the magnetic flux Φ_B is changing, there is a circulation of electric field around the loop; when the magnetic flux Φ_B has its maximum value either into or out of the x–y plane, the flux is instantaneously not changing, and there is zero circulation of electric field. That's just what Faraday's law tells us must be true.

Figure 22-9b shows five snapshots of the wave at the same instants as in Figure 22-9a, but in these snapshots we're looking at the x–z plane. The rectangular loop in this part of the figure is the same size as those in part (a), but there's a varying *electric* flux Φ_E through this loop because the electric field has only a y component and so is perpendicular to the plane of this loop. When the magnetic flux Φ_E is changing, there is a circulation of magnetic

Figure 22-9 Flux and circulation in a sinusoidal electromagnetic plane wave (a) As an electromagnetic plane wave propagates past the dashed rectangular loop, the electric field circulates around the loop when the magnetic flux through the loop changes (Faraday's law). (b) For this loop, the magnetic field circulates around the loop when the electric flux through the loop changes (the Maxwell–Ampère law).

(a) Viewing the plane of the $\vec{E}$ field

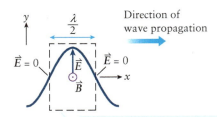

Magnetic flux through loop is maximum outward; zero circulation of $\vec{E}$.

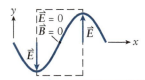

Magnetic flux through loop is changing from outward to inward; counterclockwise circulation of $\vec{E}$.

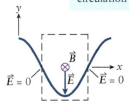

Magnetic flux through loop is maximum inward; zero circulation of $\vec{E}$.

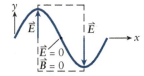

Magnetic flux through loop is changing from inward to outward; clockwise circulation of $\vec{E}$.

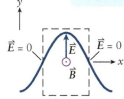

Again magnetic flux through loop is maximum outward; zero circulation of $\vec{E}$.

(b) Viewing the plane of the $\vec{B}$ field

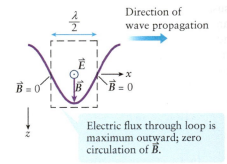

Electric flux through loop is maximum outward; zero circulation of $\vec{B}$.

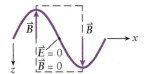

Electric flux through loop is changing from outward to inward; clockwise circulation of $\vec{B}$.

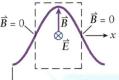

Electric flux through loop is maximum inward; zero circulation of $\vec{B}$.

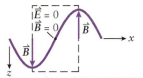

Electric flux through loop is changing from inward to outward; counterclockwise circulation of $\vec{B}$.

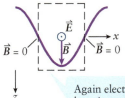

Again electric flux through loop is maximum outward; zero circulation of $\vec{B}$.

field around the loop; when the magnetic flux Φ_E has its maximum value either into or out of the x–z plane, the flux is instantaneously not changing, and there is zero circulation of magnetic field. That's in agreement with the Maxwell–Ampère law. Notice also that the circulation of the electric field in Figure 22-9a is counterclockwise when the magnetic flux Φ_B through the loop is changing from outward to inward, while the circulation of the magnetic field in Figure 22-9b is counterclockwise when the electric flux Φ_E is changing from inward to outward. That's a consequence of there being a minus sign in Faraday's law but no minus sign in the Maxwell–Ampère law (see our discussion above of Equations 22-10).

Faraday's law and the Maxwell–Ampère law show why *both* the electric field and magnetic field are necessary for the propagation of a wave: The varying electric field sustains the magnetic field, and the varying magnetic field sustains the electric field. Once an electromagnetic wave is started, it will continue to propagate through a vacuum even across immense distances (Figure 22-1a). These laws also show why the electric and magnetic fields are naturally perpendicular to each other and why the two fields oscillate together with the same frequency.

Using Faraday's law and the Maxwell–Ampère law, it's possible to calculate the speed c at which such a wave should propagate through a vacuum. It turns out that the speed is related in a simple way to the values of the permittivity of free space ε_0 and the permeability of free space μ_0, both of which appear in Equations 22-10:

$$c = \frac{1}{\sqrt{\mu_0 \varepsilon_0}}$$

Substitute $\varepsilon_0 = 8.85 \times 10^{-12}$ (A·s)/(V·m) and $\mu_0 = 4\pi \times 10^{-7}$ T·m/A:

$$c = \frac{1}{\sqrt{\left(4\pi \times 10^{-7} \, \frac{T \cdot m}{A}\right)\left(8.85 \times 10^{-12} \, \frac{A \cdot s}{V \cdot m}\right)}} = 3.00 \times 10^8 \sqrt{\frac{V}{T \cdot s}}$$

These are very odd units! Note, however, that 1 V = 1 N·m/C and 1 T = 1 N·s/ (C·m), so

$$1\sqrt{\frac{V}{T \cdot s}} = 1\sqrt{\left(\frac{N \cdot m}{C}\right)\left(\frac{C \cdot m}{N \cdot s}\right)\left(\frac{1}{s}\right)} = 1\sqrt{\frac{m^2}{s^2}} = 1\frac{m}{s}$$

The above expression for the speed of an electromagnetic wave then becomes

Speed of light in a vacuum

$$c = \frac{1}{\sqrt{\mu_0 \varepsilon_0}} = 3.00 \times 10^8 \text{ m/s}$$

Permeability of free space   Permittivity of free space

Speed of light in a vacuum (22-11)

This agrees with Equation 22-1 for the speed of light in a vacuum. The same mathematical analysis that leads to Equation 22-11 also relates the amplitudes E_0 and B_0 of the electric and magnetic fields in the wave: The result is that $B_0 = E_0/c$, the same result that we stated in Section 22-2 (Equation 22-4). (Note that historically the value of ε_0 was determined experimentally and the value of μ_0 was determined by how the ampere was defined, as we described in Section 19-7. Since 2018 the value of c has been *defined* to be exactly 2.99792458×10^8 m/s, which is now the basis of how the meter is defined. With this definition the values of both ε_0 and μ_0 are now determined experimentally, consistent with Equation 22-11.)

In the following section we'll use our insight into the nature of electromagnetic waves to help us analyze the energy carried by such waves.

WATCH OUT! The speed of light in a medium other than a vacuum is less than *c*.

! As we noted in Section 22-2, electromagnetic waves propagate at speed c only in a vacuum. We determined the value of c in Equation 22-11 using ε_0, the permittivity of free space, and μ_0, the permeability of free space; "free space" is equivalent to "in a vacuum." Different materials have different values of permittivity and permeability, which is why the speed of light is different in them.

GOT THE CONCEPT? 22-2 Maxwell's Equations

? According to Maxwell's equations, which of the following situations is *possible*? (a) A closed surface that has a net outward magnetic flux through its surface; (b) a closed surface that has a net outward electric flux through its surface; (c) a loop that has zero electric flux through the interior of the loop and has a magnetic field that circulates around the loop; (d) more than one of (a), (b), and (c); (e) none of (a), (b), or (c).

22-4 Electromagnetic waves carry both electric and magnetic energy, and come in packets called photons

Electromagnetic waves carry energy. You can feel the energy delivered to your skin by sunlight (one kind of electromagnetic wave) on a sunny day. A microwave oven is useful for cooking because the water molecules found in food of all kinds absorb the energy in microwaves, which are at wavelengths between infrared and radio (Figure 22-2). And the energy in a laser beam is so tightly concentrated that it can be used as a surgical tool (Figure 22-1b).

The energy in an electromagnetic wave is actually contained within the electric and magnetic fields themselves. To see what this means, let's return to the physics of capacitors (Sections 17-5 and 17-6) and inductors (Sections 21-4 and 21-5). These devices will give us insight into the energy content of electromagnetic waves.

Energy in Electric and Magnetic Fields

Area A Charge $-q$

Volume between the plates: $V = Ad$

d $\vec{E}$

Charge $+q$

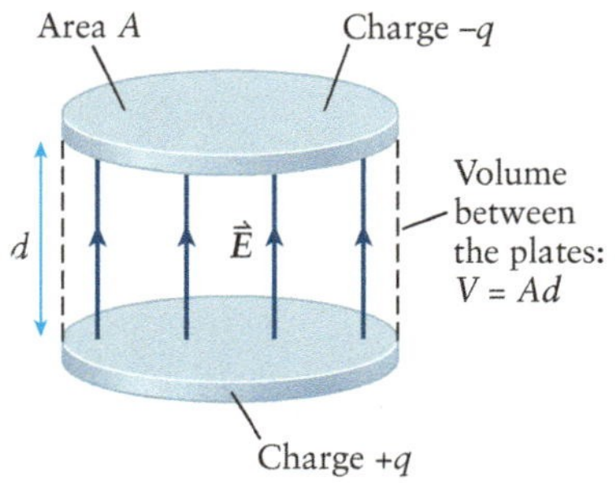

Figure 22-10 Calculating electric energy density The energy stored in this charged capacitor can be thought of as residing in the electric field $\vec{E}$ between the plates.

Figure 22-10 shows a parallel-plate capacitor with plates of area A separated by a distance d. If the two plates are closely spaced, the electric field between the plates is approximately uniform and fills the volume between the plates (that is, there is very little field outside this volume). With a charge $+q$ on one plate and a charge $-q$ on the other plate, the magnitude of this field is

(17-10)
$$E = \frac{q}{\varepsilon_0 A}$$

It takes work to separate the charges $+q$ and $-q$, and this work goes into the electric potential energy U_E stored in the capacitor. The amount of stored energy is

(17-17)
$$U_E = \frac{q^2}{2C}$$

where C, the capacitance of the capacitor, is

(17-13)
$$C = \frac{\varepsilon_0 A}{d}$$

We can think of the electric energy in the capacitor as being stored in the electric field itself. To motivate this idea, let's substitute Equation 17-13 into Equation 17-17 and rearrange:

(22-12)
$$U_E = \frac{q^2}{2}\left(\frac{d}{\varepsilon_0 A}\right) = \frac{1}{2}\varepsilon_0\left(\frac{q^2}{\varepsilon_0^2 A^2}\right)Ad = \frac{1}{2}\varepsilon_0\left(\frac{q}{\varepsilon_0 A}\right)^2 Ad$$

The quantity $q/(\varepsilon_0 A)$ in parentheses on the far right-hand side of Equation 22-12 is just the electric field magnitude E from Equation 17-10. The quantity Ad is the volume V that the electric field occupies. So we can rewrite Equation 22-12 as

(22-13)
$$U_E = \left(\frac{1}{2}\varepsilon_0 E^2\right)V$$

Equation 22-13 tells us that the energy stored in the capacitor is equal to the volume V occupied by the electric field multiplied by a quantity $(1/2)\varepsilon_0 E^2$ that depends on the electric field magnitude E. This quantity has units of energy per volume (J/m^3) and is called the **electric energy density**:

Electric energy density   Electric field magnitude

$$u_E = \frac{1}{2}\varepsilon_0 E^2$$

Permittivity of free space

Electric energy density
(22-14)

Equation 22-14 says that wherever there is an electric field, there is energy. This motivates the idea that the energy stored in a capacitor is stored in the field itself. We derived Equation 22-14 for the special case of a parallel-plate capacitor, but it turns out to be valid in *any* situation where an electric field is present.

We can come to a similar conclusion about the magnetic energy stored in an inductor like the one shown in Figure 22-11. Table 21-1 (Section 21-5) tells us that if an inductor has inductance L and carries a current i, the magnetic energy stored in the inductor is

$$U_B = \frac{1}{2}Li^2 \qquad (22\text{-}15)$$

For an inductor like that in Figure 22-11, which is a long solenoid with N turns of wire, length ℓ, and cross-sectional area A, the inductance is

$$L = \frac{\mu_0 N^2 A}{\ell} \qquad (21\text{-}14)$$

Like the electric field inside the parallel-plate capacitor in Figure 22-10, the magnetic field inside the solenoid in Figure 22-11 is nearly uniform and confined to the volume inside the solenoid. From Equation 19-17 the magnitude of this magnetic field is $B = \mu_0 ni$, where n is the number of turns of wire per meter. This is just the total number of turns N divided by the length ℓ of the solenoid: $n = N/\ell$. So we can write the magnitude of the magnetic field inside the solenoid as

$$B = \frac{\mu_0 Ni}{\ell} \qquad (22\text{-}16)$$

Let's substitute Equation 21-14 for L into Equation 22-15 for the energy in the solenoid, then rearrange:

$$U_B = \frac{1}{2}\left(\frac{\mu_0 N^2 A}{\ell}\right)i^2 = \frac{1}{2\mu_0}\left(\frac{\mu_0^2 N^2 i^2}{\ell^2}\right)A\ell = \frac{1}{2\mu_0}\left(\frac{\mu_0 Ni}{\ell}\right)^2 A\ell \qquad (22\text{-}17)$$

The quantity $\mu_0 Ni/\ell$ in parentheses on the far right of Equation 22-17 is the magnitude B of the magnetic field inside the solenoid. The quantity $A\ell$ is the volume V inside the solenoid, which is also the volume that the magnetic field occupies. So Equation 22-17 becomes

$$U_B = \left(\frac{B^2}{2\mu_0}\right)V \qquad (22\text{-}18)$$

Compare this to Equation 22-13 for the electric energy U_E. We see that according to Equation 22-18, the energy stored in the inductor equals the volume V occupied by the magnetic field multiplied by a quantity $B^2/(2\mu_0)$ that depends on the magnetic field magnitude B and has units J/m^3. We call $B^2/(2\mu_0)$ the **magnetic energy density**:

Magnetic energy density   Magnetic field magnitude

$$u_B = \frac{B^2}{2\mu_0}$$

Permeability of free space

Magnetic energy density
(22-19)

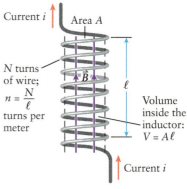

Current i Area A

N turns of wire; $n = \dfrac{N}{\ell}$ turns per meter

$\vec{B}$

ℓ

Volume inside the inductor: $V = A\ell$

Current i

Figure 22-11 Calculating magnetic energy density The energy stored in this current-carrying inductor can be thought of as residing in the magnetic field $\vec{B}$ inside the inductor.

Just as Equation 22-14 tells us that there is electric energy wherever there is an electric field, Equation 22-19 tells us that there is magnetic energy wherever there is a magnetic field. This is true for an inductor, so we can think of the energy stored in a current-carrying inductor as being stored in the magnetic field within the inductor. But Equation 22-19 is valid *wherever* there is a magnetic field.

Energy in an Electromagnetic Plane Wave

Let's apply Equations 22-14 and 22-19 for the electric and magnetic energy densities to the sinusoidal electromagnetic plane wave that we introduced in Section 22-2. From Equation 22-3 the electric field of this plane wave has only a *y* component and the magnetic field has only a *z* component:

$$E_y(x,t) = E_0\cos(kx - \omega t + \phi)$$

(22-3)

$$B_z(x,t) = B_0\cos(kx - \omega t + \phi)$$

Substituting these into Equations 22-14 and 22-19, we find that the electric and magnetic energy densities in the plane wave are

$$u_E = \frac{1}{2}\varepsilon_0 E^2 = \frac{1}{2}\varepsilon_0 E_0^2\cos^2(kx - \omega t + \phi)$$

(22-20)

$$u_B = \frac{B^2}{2\mu_0} = \frac{B_0^2}{2\mu_0}\cos^2(kx - \omega t + \phi)$$

Note that u_E and u_B both depend on position *x* and time *t* in the same way. It may appear from Equation 22-20 that there are different amounts of energy in the electric and magnetic forms, since the coefficients $(1/2)\varepsilon_0 E_0^2$ and $B_0^2/2\mu_0$ are different. But we know from Equations 22-4 and 22-11 that $B_0 = E_0/c$ and $c = 1/\sqrt{\mu_0\varepsilon_0}$, so

$$\frac{B_0^2}{2\mu_0} = \frac{(E_0/c)^2}{2\mu_0} = \frac{(E_0\sqrt{\mu_0\varepsilon_0})^2}{2\mu_0} = \frac{1}{2}\left(\frac{\mu_0\varepsilon_0}{\mu_0}\right)E_0^2 = \frac{1}{2}\varepsilon_0 E_0^2$$

(22-21)

In other words, the coefficients of u_E and u_B in Equation 22-20 are equal. This means that at any position *x* and at any time *t*, an electromagnetic wave in a vacuum has *equal* amounts of electric energy density and magnetic energy density:

(22-22)

$$u_E = u_B = \frac{1}{2}\varepsilon_0 E_0^2\cos^2(kx - \omega t + \phi) = \frac{B_0^2}{2\mu_0}\cos^2(kx - \omega t + \phi)$$

The *total* energy density *u* in the wave is the sum of u_E and u_B. Equation 22-22 tells us that $u_E = u_B$, so *u* is equal to $2u_E$ or $2u_B$:

(22-23)

$$u = u_E + u_B = \varepsilon_0 E_0^2\cos^2(kx - \omega t + \phi) = \frac{B_0^2}{\mu_0}\cos^2(kx - \omega t + \phi)$$

The value of *u* at any position varies with time, so it's often more useful to state its *average* value. The average value of the cosine function squared is 1/2, so

Average energy density in an electromagnetic wave	Electric field magnitude	Electric field rms value	Magnetic field magnitude	Magnetic field rms value

Average energy density in an electromagnetic wave
(22-24)

$$u_{average} = \frac{1}{2}\varepsilon_0 E_0^2 = \varepsilon_0 E_{rms}^2 = \frac{B_0^2}{2\mu_0} = \frac{B_{rms}^2}{\mu_0}$$

Permittivity of free space   Permeability of free space

In Equation 22-24 we've used the root-mean-square (rms) values of the oscillating electric and magnetic fields, $E_{rms} = E_0/\sqrt{2}$ and $B_{rms} = B_0/\sqrt{2}$ (see Section 21-2).

An even more useful way to express the energy carried by an electromagnetic wave is in terms of the wave **intensity**, or average power per unit area. (In Section 13-9 we introduced the idea of intensity in our description of *sound* waves. The same ideas apply to electromagnetic waves.) Figure 22-12 shows a portion of a wave that has cross-sectional area *A* and length ℓ. The energy in this portion of the wave is $u_{average}$ from Equation 22-24 multiplied by the volume $A\ell$ of this portion. This entire portion of the wave moves at speed *c* through the cross-sectional area *A* in a time *t*. The power equals the energy in the wave portion divided by the time $t = \ell/c$ that it takes this portion of the wave to travel at speed c through the cross-sectional area A. The intensity, to which we give the symbol S_{average}, equals the power divided by the area A. So

$$S_{\text{average}} = \frac{\text{energy}}{\text{time}} \times \frac{1}{\text{area}} = \frac{(u_{\text{average}}A\ell)}{\ell/c} \times \frac{1}{A} = u_{\text{average}}c$$

Using Equation 22-24, we can rewrite this as

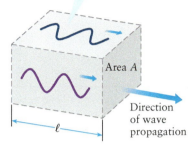

The electromagnetic wave energy in this volume moves with the wave at speed c.

Intensity of an electromagnetic wave	Speed of light	Magnetic field rms value

$$S_{\text{average}} = c\varepsilon_0 E_{\text{rms}}^2 = \frac{cB_{\text{rms}}^2}{\mu_0} = \frac{E_{\text{rms}}B_{\text{rms}}}{\mu_0}$$

Intensity of an electromagnetic wave (22-25)

Permittivity of free space	Electric field rms value	Permeability of free space

Area A

Direction of wave propagation

Figure 22-12 Calculating wave intensity The intensity of an electromagnetic wave equals the amount of wave energy that crosses an area A per unit time, divided by the area A.

(To write the last expression in Equation 22-25, we used the result that since $B_0 = E_0/c$, it follows that $B_{\text{rms}} = E_{\text{rms}}/c$.)

The units of intensity are watts per square meter, or W/m². The intensity of sunlight that reaches Earth is about 1.36×10^3 W/m². As the following example shows, the intensities of other common electromagnetic waves can be much smaller.

EXAMPLE 22-4 Energy Density and Intensity in a Radio Wave

In Example 22-1 (Section 22-2) we considered an FM radio wave of frequency 98.7 MHz with an electric field amplitude 6.00×10^{-2} V/m. We found that the magnetic field has amplitude 2.00×10^{-10} T. Calculate the rms values of the electric and magnetic fields, the average energy density in the wave, and the wave intensity.

Set Up

Each rms value is just equal to the amplitude divided by $\sqrt{2}$. Given the rms values, we'll use Equation 22-24 to calculate the energy density and Equation 22-25 to calculate the intensity.

Root-mean-square values:

$$E_{\text{rms}} = \frac{E_0}{\sqrt{2}}$$

$$B_{\text{rms}} = \frac{B_0}{\sqrt{2}}$$

Average energy density in an electromagnetic wave:

$$u_{\text{average}} = \varepsilon_0 E_{\text{rms}}^2 = \frac{B_{\text{rms}}^2}{\mu_0} \quad (22\text{-}24)$$

Intensity of an electromagnetic wave:

$$S_{\text{average}} = c\varepsilon_0 E_{\text{rms}}^2 = \frac{cB_{\text{rms}}^2}{\mu_0} = \frac{E_{\text{rms}}B_{\text{rms}}}{\mu_0} \quad (22\text{-}25)$$

Permittivity of free space:

$$\varepsilon_0 = 8.85 \times 10^{-12} \frac{\text{C}}{\text{V} \cdot \text{m}} \quad (22\text{-}9)$$

Permeability of free space:
$$\mu_0 = 4\pi \times 10^{-7} \text{ T} \cdot \text{m/A}$$

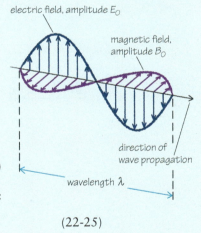

electric field, amplitude E_O

magnetic field, amplitude B_O

direction of wave propagation

wavelength λ

Solve

Calculate the rms values of the electric and magnetic fields.

We are given $E_0 = 6.00 \times 10^{-2}$ V/m and $B_0 = 2.00 \times 10^{-10}$ T. The corresponding rms values are

$$E_{\text{rms}} = \frac{E_0}{\sqrt{2}} = \frac{6.00 \times 10^{-2} \text{ V/m}}{\sqrt{2}} = 4.24 \times 10^{-2} \text{ V/m}$$

$$B_{\text{rms}} = \frac{B_0}{\sqrt{2}} = \frac{2.00 \times 10^{-10} \text{ T}}{\sqrt{2}} = 1.41 \times 10^{-10} \text{ T}$$

Use the value of E_{rms} to calculate the average energy density in the wave.

From Equation 22-24:

$$u_{average} = \varepsilon_0 E_{rms}^2 = \left(8.85 \times 10^{-12} \frac{C}{V \cdot m}\right)\left(4.24 \times 10^{-2} \frac{V}{m}\right)^2$$

$$= 1.59 \times 10^{-14} \frac{C \cdot V}{m^3}$$

One coulomb times one volt is one joule: $1\ C \cdot V = 1$ J. So

$$u_{average} = 1.59 \times 10^{-14}\ J/m^3$$

Find the intensity of the wave.

Comparing Equations 22-24 and 22-25 shows that the wave intensity is c times the average energy density:

$$S_{average} = c\varepsilon_0 E_{rms}^2 = cu_{average}$$

$$= (3.00 \times 10^8\ m/s)(1.59 \times 10^{-14}\ J/m^3)$$

$$= 4.78 \times 10^{-6} \frac{J}{m^2 \cdot s}$$

One joule per second is one watt: 1 J/s $= 1$ W. So

$$S_{average} = 4.78 \times 10^{-6}\ W/m^2$$

Reflect

The energy density and intensity are both very small quantities. It's a testament to the sensitivity of radio receivers that such a wave is quite easy to detect.

We can check our results by using the alternative expressions for $u_{average}$ and $S_{average}$ given in Equations 22-24 and 22-25. As an example, here's a check on the value of $S_{average}$.

From Equation 22-25:

$$S_{average} = \frac{E_{rms}B_{rms}}{\mu_0} = \frac{(4.24 \times 10^{-2}\ V/m)(1.41 \times 10^{-10}\ T)}{4\pi \times 10^{-7}\ T \cdot m/A}$$

$$= 4.78 \times 10^{-6} \frac{V \cdot A}{m^2}$$

One volt times one ampere is one watt ($1\ V \cdot A = 1$ W), so

$$S_{average} = 4.78 \times 10^{-6}\ W/m^2$$

This agrees with our calculation above, as it must.

Photons

BioMedical

Equation 22-25 says that the intensity of an electromagnetic wave depends on the strength of the electric and magnetic fields that make up the wave but not on the wave frequency. (The frequency f doesn't appear anywhere in this equation.) But everyday experience suggests that wave frequency *does* play a role in the energy carried by an electromagnetic wave. As an example, ultraviolet light can trigger a chemical reaction in the skin that causes a suntan or sunburn, but visible light cannot. (That's why sunscreen contains a substance that allows visible light to pass but blocks ultraviolet light.) X rays, with even higher frequency than ultraviolet light, can pierce soft tissue but not bone; as a result, an x-ray image allows a physician to diagnose a broken bone and a dentist to see cavities in your teeth. Gamma rays, with higher frequency than x rays, can damage DNA, cause cancer, and even kill cells. How can we explain these differences?

The explanation is that the energy of an electromagnetic wave propagates as small packets called **photons**. The energy of an individual photon is proportional to the wave frequency, and the proportionality constant h is called **Planck's constant**:

Energy of a photon | Wave frequency

**Energy of a photon
(22-26)**

$$E = hf$$

Planck's constant $= 6.62607015 \times 10^{-34}$ J $\cdot$ s

To three significant figures $h = 6.63 \times 10^{-34}$ J$\cdot$s.

Equation 22-26 explains why ultraviolet light causes a suntan or sunburn, but light in the visible spectrum does not. For tanning or burning to take place in your skin, individual molecules must absorb a certain minimum amount of energy from light to trigger a chemical change. A given molecule must absorb this energy in the form of a single photon, and a visible-light photon lacks sufficient energy to trigger this chemical change. An ultraviolet photon, by contrast, can trigger the change because it has a shorter wavelength, a higher frequency, and more energy per photon. An x-ray photon is more energetic still, which is why it is able to penetrate soft tissue.

Gamma-ray photons, x-ray photons, and short-wavelength, high-frequency ultra-violet photons have enough energy that they can dislodge an electron from an atom. Such **ionizing radiation** breaks apart molecules by pulling electrons from the chemical bonds that hold atoms together. Although ionizing radiation can directly break DNA molecules in living tissue, it's more likely to disrupt some other, more common molecule, such as water, to create highly reactive free radicals that then damage DNA. Depending on the severity of the damage, the cell may be able to recover. However, if the damage cannot be repaired, or if it is repaired incorrectly, the resulting mutations may be lethal. These effects generally go unnoticed until the next time the cell tries to divide. Because cancerous cells divide more frequently than most other cells in the body, they are more susceptible to radiation damage than are most healthy cells.

BioMedical

Planck's constant is very small, so a single photon carries only a miniscule amount of energy. As an example, a photon of red light with wavelength $\lambda = 750$ nm $= 7.50 \times 10^{-7}$ m has frequency $f = c/\lambda = (3.00 \times 10^8$ m/s$)/(7.50 \times 10^{-7}$ m$) = 4.00 \times 10^{14}$ Hz and energy $E = hf = (6.63 \times 10^{-34}$ J·s$)(4.00 \times 10^{14}$ Hz$) = 2.65 \times 10^{-19}$ J. (Recall that 1 Hz $= 1$ s^{-1}.) That's so small that you don't notice individual photons in the light from a lamp, just as you don't notice individual air molecules in a breeze against your face.

It's common to express photon energies in electron volts (eV). We introduced this unit in Section 17-6: 1 eV $= 1.60 \times 10^{-19}$ J. For a red photon of wavelength 750 nm, the energy is $E = (2.65 \times 10^{-19}$ J$)/(1.60 \times 10^{-19}$ J/eV$) = 1.66$ eV. As the following example shows, radio photons have even less energy.

EXAMPLE 22-5 Photons in a Radio Wave

For the radio wave of Examples 22-1 and 22-4, calculate (a) the energy per photon, (b) the number of photons per cubic meter, and (c) the number of photons per second that strike a receiver antenna of area 10.0 cm^2.

Set Up

The radio wave has frequency 98.7 MHz $= 98.7 \times 10^6$ Hz. We'll use this and Equation 22-26 to determine the energy of a single radio photon. From Example 22-4 we know that the energy density (energy per unit volume) of the wave is $u_{average} = 1.59 \times 10^{-14}$ J/m^3 and the intensity (energy per area per time) is $S_{average} = 4.78 \times 10^{-6}$ W/m^2. We'll use these and our calculated value of the photon energy to determine the number of photons per unit volume and the number of photons striking the antenna per time.

Energy of a photon:

$$E = hf \qquad (22\text{-}26)$$

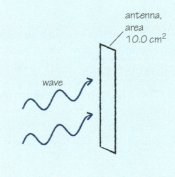

Solve

(a) Calculate the energy of an individual photon of frequency 98.7 MHz.

From Equation 22-26:

$E = hf = (6.63 \times 10^{-34}$ J·s$)(98.7 \times 10^6$ Hz$)$
     $= 6.54 \times 10^{-26}$ J·s·Hz

Since 1 Hz $= 1$ s^{-1}, 1 J·s·Hz $= 1$ J and so

$E = 6.54 \times 10^{-26}$ J or

$$E = \frac{6.54 \times 10^{-26} \text{ J}}{1.60 \times 10^{-19} \text{ J/eV}} = 4.08 \times 10^{-7} \text{ eV}$$

This is much smaller than the energy of a visible-light photon (2.65×10^{-19} J $= 1.66$ eV for red light) because the frequency of the radio wave is much less than the frequency of visible light (4.00×10^{14} Hz for red light).

(b) Calculate the photon density (number of photons per unit volume).

The energy density in the wave is $u_{average} = 1.59 \times 10^{-14}$ J/m³ and the energy per photon is $E = 6.54 \times 10^{-26}$ J/photon. The number of photons per unit volume is

$$\frac{photons}{volume} = \frac{energy}{volume} \times \frac{photon}{energy} = \frac{\left(\dfrac{energy}{volume}\right)}{\left(\dfrac{energy}{photon}\right)}$$

$$= \frac{1.59 \times 10^{-14}\ \text{J/m}^3}{6.54 \times 10^{-26}\ \text{J/photon}} = 2.43 \times 10^{11}\ \text{photons/m}^3$$

Each cubic meter of this wave contains 2.43×10^{11} (243 billion) photons.

(c) Calculate the number of photons that strike an area of 10.0 cm² in 1.00 s.

The intensity (energy per area per time) of the wave is $S_{average} = 4.78 \times 10^{-6}$ W/m², so the rate at which energy arrives at the antenna of area $A = 10.0$ cm² is

$$\frac{energy}{area \cdot time} \times area = S_{average} A$$

$$= (4.78 \times 10^{-6}\ \text{W/m}^2)(10.0\ \text{cm}^2)\left(\frac{1\ \text{m}}{100\ \text{cm}}\right)^2$$

$$= 4.78 \times 10^{-9}\ \text{W} = 4.78 \times 10^{-9}\ \text{J/s}$$

The rate at which photons arrive at the antenna is

$$\frac{photons}{time} = \frac{energy}{time} \times \frac{photon}{energy} = \frac{\left(\dfrac{energy}{time}\right)}{\left(\dfrac{energy}{photon}\right)}$$

$$= \frac{4.78 \times 10^{-9}\ \text{J/s}}{6.54 \times 10^{-26}\ \text{J/photon}} = 7.30 \times 10^{16}\ \text{photons/s}$$

In one second 7.30×10^{16} (73 quadrillion) photons arrive at the antenna.

Reflect

Even this relatively low-intensity wave contains a tremendous number of photons per cubic meter and delivers an astronomical number of photons per second to a receiver.

BioMedical   Our results give us insight into the concerns that some people have expressed about mobile phones causing cancer. The frequencies used by mobile phones are about 7 to 27 times higher (about 700 to 2700 MHz), but even those frequencies correspond to very low photon energies (about 2.9×10^{-6} to 1.1×10^{-5} eV). As we described above, the sort of ionizing radiation that can cause cancer has very high frequency and very high photon energy of several electron volts. The photon energies associated with mobile phones are millions of times smaller, which should be far too low to have any carcinogenic effect.

WATCH OUT! Photons are both particles and waves.

! A common *incorrect* way to think about photons is to visualize them as small particles like miniature marbles and to imagine that a large number of photons acting together behave like a wave. The reality is far different! Each individual photon has aspects of *both* wave and particle, and those particles are very different in character from ordinary objects such as marbles. We'll learn more about the curious properties of photons in Chapter 26.

GOT THE CONCEPT? 22-3 Photons

? Three lasers have equal power output. The first emits a pure violet light, the second emits a pure green light, and the third emits a pure red light. Which laser emits the greater number of photons per second? (a) The violet laser; (b) the green laser; (c) the red laser; (d) all emit the same number of photons per second; (e) answer depends on the value of the power output.

TAKE-HOME MESSAGE FOR Section 22-4

✔ Energy is associated with both electric and magnetic fields. An electromagnetic wave in a vacuum has equal amounts of electric energy and magnetic energy per volume.

✔ The intensity of an electromagnetic wave is the average power per unit area. In a vacuum, the intensity equals the

average energy per volume in the wave multiplied by the speed of light c.

✔ The energy of an electromagnetic wave comes in packets called photons. The energy of a single photon is proportional to the wave frequency.

Key Terms

displacement current
electric energy density
electromagnetic spectrum
electromagnetic wave
Faraday's law
Gauss's law for the electric field

Gauss's law for the magnetic field
intensity
ionizing radiation
magnetic energy density
Maxwell–Ampère law
Maxwell's equations

photon
Planck's constant
sinusoidal plane wave
speed of light
visible light

Chapter Summary

Topic	Equation or Figure	

Speed of electromagnetic waves: In a vacuum all electromagnetic waves propagate at the same speed $c = 3.00 \times 10^8$ m/s. The shorter the wavelength, the higher the frequency of the wave. Our eyes are sensitive to only a narrow band of wavelengths known as the visible spectrum.

Speed of light in a vacuum   Frequency of an electromagnetic wave

$$c = f\lambda \qquad (22\text{-}2)$$

Wavelength of the wave in a vacuum

Electromagnetic plane waves: The simplest electromagnetic wave in a vacuum is a sinusoidal plane wave. The electric and magnetic fields oscillate in phase, are perpendicular to each other, and are transverse (both are perpendicular to the direction of wave propagation). The amplitude B_0 of the magnetic field equals the amplitude E_0 of the electric field divided by the speed of light c.

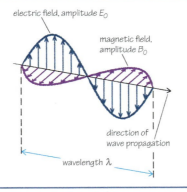

electric field, amplitude E_0
magnetic field, amplitude B_0
direction of wave propagation
wavelength λ

(Example 22-1)

Maxwell's equations: Four equations govern the behavior of electric and magnetic fields in all situations. The two Gauss's laws explain why electromagnetic waves in a vacuum are transverse. Faraday's law and the Maxwell–Ampère law explain how the oscillations of the magnetic field produce the electric field and how the oscillations of the electric field produce the magnetic field. As a result, the wave must have both electric and magnetic aspects, and the electric and magnetic fields are

Electric flux through a closed surface   Net amount of **charge enclosed** within the surface

$$\Phi_E = \frac{q_{\text{encl}}}{\varepsilon_0} \qquad (16\text{-}9)$$

Permittivity of free space $= 1/(4\pi k)$

Magnetic flux through a closed surface   There is **zero magnetic flux through any closed surface**, no matter what the size or shape of the surface or what it contains.

$$\Phi_B = 0 \qquad (22\text{-}5)$$

naturally perpendicular to each other. Maxwell's equations also successfully predict the value of the speed of light in terms of the permittivity of free space ε_0 and the permeability of free space μ_0.

Circulation of electric field around a loop

Change in the magnetic flux through the surface outlined by the loop

$$\sum E_{\parallel}\Delta\ell = -\frac{\Delta\Phi_B}{\Delta t}$$

Time interval over which the change in magnetic flux takes place

(22-7)

Circulation of magnetic field around an Amperian loop

Permittivity of free space

Change in the electric flux through the surface outlined by the loop

$$\sum B_{\parallel}\Delta\ell = \mu_0\left(i_{\text{through}} + \varepsilon_0\frac{\Delta\Phi_E}{\Delta t}\right)$$

(22-8)

Permeability of free space

Current through the Amperian loop

Time interval over which the change in electric flux takes place

Electric and magnetic field energy in electromagnetic waves: The energy density (energy per volume) associated with an electric field is $u_E = (1/2)\varepsilon_0 E^2$, and the energy density associated with a magnetic field is $u_B = B^2/2\mu_0$. In an electromagnetic wave in a vacuum, there are equal amounts of electric energy and magnetic energy. The average energy density in an electromagnetic wave can be expressed in terms of either the field amplitudes or their rms values. The intensity of the wave is the average power per unit area of the wave.

Average energy density in an electromagnetic wave

Electric field magnitude

Electric field rms value

Magnetic field magnitude

Magnetic field rms value

$$u_{\text{average}} = \frac{1}{2}\varepsilon_0 E_0^2 = \varepsilon_0 E_{\text{rms}}^2 = \frac{B_0^2}{2\mu_0} = \frac{B_{\text{rms}}^2}{\mu_0}$$

(22-24)

Permittivity of free space

Permeability of free space

Intensity of an electromagnetic wave

Speed of light

Magnetic field rms value

$$S_{\text{average}} = c\varepsilon_0 E_{\text{rms}}^2 = \frac{cB_{\text{rms}}^2}{\mu_0} = \frac{E_{\text{rms}}B_{\text{rms}}}{\mu_0}$$

(22-25)

Permittivity of free space

Electric field rms value

Permeability of free space

Photons: The energy of an electromagnetic wave comes in packets called photons that have properties of both wave and particle. The higher the wave frequency, the more energy there is per photon.

Energy of a photon

Wave frequency

$$E = hf$$

(22-26)

Planck's constant $= 6.62607015 \times 10^{-34}$ J $\cdot$ s

Answer to What do you think? Question

(d) As we describe in Section 22-2, in a vacuum all varieties of electromagnetic wave propagate at the same speed. Visible light has a much shorter wavelength (λ = about 400 to 700 nm) than radio waves (λ = a few centimeters to several meters) and so has a much higher frequency.

Answers to Got the Concept? Questions

22-1 (e) No matter what the frequency of an electromagnetic wave, its propagation speed in a vacuum is the same: $c = 3.00 \times 10^8$ m/s.

22-2 (d) Situation (a) is impossible: Gauss's law for the magnetic field states that the net magnetic flux through any closed surface must be zero. Situation (b) is possible; according to Gauss's law for the electric field, there will be a net outward electric flux through a closed surface that encloses positive charge. Situation (c) is also possible: One example is the loop around the current-carrying wire in Figure 22-8a. There is zero electric field and hence zero electric flux through the interior of the loop, but the current gives rise to a magnetic field that circulates around the loop.

22-3 (c) All three lasers emit the same amount of electromagnetic wave energy per second. However, the energy per photon is different because each laser emits light of a different wavelength and frequency. Figure 22-2 shows that violet has the shortest wavelength and so the highest frequency and highest photon energy; red has the longest wavelength and so the lowest frequency and lowest photon energy. To emit the same amount of energy per second, the red laser must emit more of its low-energy photons per second than the other lasers do.

Questions and Problems

In a few problems you are given more data than you actually need; in a few other problems you are required to supply data from your general knowledge, outside sources, or informed estimate.

Interpret as significant all digits in numerical values that have trailing zeros and no decimal points.

For all problems use $g = 9.80$ m/s^2 for the free-fall acceleration due to gravity. Neglect friction and air resistance unless instructed to do otherwise.

- •   Basic, single-concept problem
- ••   Intermediate-level problem; may require synthesis of concepts and multiple steps
- •••   Challenging problem
- Example   See worked example for a similar problem.

Conceptual Questions

1. • (a) Rank the following electromagnetic waves from the lowest to the highest wavelength: (i) Microwaves; (ii) red light; (iii) ultraviolet light; (iv) infrared light; (v) gamma rays. (b) Which wavelength has the highest energy per photon? Which has the lowest?

2. • Changing electric fields create changing magnetic fields. These oscillating fields propagate at the speed of light. Describe the directions of the fields and the velocity of the electromagnetic wave.

3. • Describe how the frequency at which the changing electric and magnetic fields oscillate in electromagnetic waves is related to the speed of light.

4. • Does a wire connected to a dc source, such as a battery, emit an electromagnetic wave?

5. • James Clerk Maxwell is credited with compiling the three laws of electricity and magnetism (Gauss's law, Ampère's law, and Faraday's law), adding his own law (also called Gauss's law for magnetism), modifying Ampère's law, and understanding the connections between electricity, magnetism, and optics. Were Maxwell's efforts more important than the individual discoveries of Gauss, Ampère, and Faraday? Explain your answer.

6. • Match the equations that were conceived of by Carl Friedrich Gauss, André Ampère, Michael Faraday, and James Clerk Maxwell with the corresponding written statements:

i. $\sum B_\parallel \Delta \ell = \mu_0 \left(i_{\text{through}} + \varepsilon_0 \dfrac{\Delta \Phi_E}{\Delta t} \right)$

ii. $\sum E_\parallel \Delta \ell = - \dfrac{\Delta \Phi_B}{\Delta t}$

iii. $\Phi_B = 0$

iv. $\Phi_E = \dfrac{q_{\text{encl}}}{\varepsilon_0}$

A. One source of an electric field is an electric charge.
B. There are no magnetic monopoles.
C. Changing magnetic fields induce changing electric fields.
D. Changing electric fields induce changing magnetic fields.

7. • The energy of an ultraviolet light photon is unrelated to the speed of the fundamental electromagnetic waves that make up such radiation. Explain how this is possible.

8. • Name three types of electromagnetic energy that you used today.

9. • Describe how the frequency of an electromagnetic wave is related to the energy of the photons of that wave.

Multiple-Choice Questions

10. • In comparison to x rays in a vacuum, visible light in a vacuum has
  A. a speed that is faster.
  B. wavelengths that are longer.
  C. wavelengths that are equal.
  D. wavelengths that are shorter.
  E. frequencies that are equal.

11. • X rays and gamma rays in a vacuum
  A. have the same frequency.
  B. have the same wavelength.
  C. have the same speed.
  D. have the same "color."
  E. None of the above

12. • In comparison to radio waves in a vacuum, visible light in a vacuum has
  A. a speed that is faster.
  B. wavelengths that are longer.
  C. wavelengths that are equal.
  D. wavelengths that are shorter.
  E. frequencies that are equal.

13. • Which of the following requires a physical medium through which to travel?
  A. radio waves
  B. light
  C. x rays
  D. sound
  E. gamma rays

14. • In an RC circuit the capacitor begins to discharge. In the region of space between the plates of the capacitor, there
  A. is an electric field but no magnetic field.
  B. is a magnetic field but no electric field.
  C. are both electric and magnetic fields.
  D. are no electric and magnetic fields.
  E. is an electric field whose strength is one-half that of the magnetic field.

15. • Maxwell's equations apply
  A. to both electric fields and magnetic fields that are constant over time.
  B. only to electric fields that are time-dependent.
  C. only to magnetic fields that are constant over time.
  D. to both electric fields and magnetic fields that are time-dependent.
  E. to both time-independent and time-dependent electric and magnetic fields.

16. • The phase difference between the electric and magnetic fields in an electromagnetic wave is
    A. 90°.
    B. 180°.
    C. 0°.
    D. alternately 90° and 180°.
    E. alternately 0° and 90°.

Problems

22-1 Light is just one example of an electromagnetic wave

22-2 In an electromagnetic plane wave, electric and magnetic fields both oscillate

17. • Calculate the wavelengths of the electromagnetic waves with the following frequencies and classify the electromagnetic radiation of each (x ray, radio, and so on). Example 22-1
    A. $f = 4.14 \times 10^{15}$ Hz
    B. $f = 7.00 \times 10^{14}$ Hz
    C. $f = 8.00 \times 10^{16}$ Hz
    D. $f = 3.00 \times 10^{13}$ Hz
    E. $f = 9.00 \times 10^{12}$ Hz
    F. $f = 3.44 \times 10^{17}$ Hz
    G. $f = 8.23 \times 10^{15}$ Hz
    H. $f = 6.00 \times 10^{15}$ Hz

18. • Calculate the wavelengths of the electromagnetic waves with the following frequencies. Example 22-1
    A. $f = 7.50 \times 10^{15}$ Hz
    B. $f = 6.00 \times 10^{14}$ Hz
    C. $f = 5.00 \times 10^{14}$ Hz
    D. $f = 4.29 \times 10^{14}$ Hz
    E. $f = 7.50 \times 10^{16}$ Hz
    F. $f = 2.66 \times 10^{16}$ Hz
    G. $f = 8.23 \times 10^{17}$ Hz
    H. $f = 6.00 \times 10^{18}$ Hz

19. • Calculate the frequencies of the electromagnetic waves that have the following wavelengths. Example 22-1
    A. $\lambda = 700$ nm
    B. $\lambda = 600$ nm
    C. $\lambda = 500$ nm
    D. $\lambda = 400$ nm
    E. $\lambda = 100$ nm
    F. $\lambda = 0.0333$ nm
    G. $\lambda = 500$ μm
    H. $\lambda = 63.3$ pm

20. • Calculate the frequencies of the electromagnetic waves that have the following wavelengths. Example 22-1
    A. $\lambda = 800$ nm
    B. $\lambda = 650$ nm
    C. $\lambda = 550$ nm
    D. $\lambda = 450$ nm
    E. $\lambda = 2.22$ nm
    F. $\lambda = 1.10 \times 10^{-8}$ m
    G. $\lambda = 50.0$ μm
    H. $\lambda = 33.4$ mm

21. • The FM radio band is from 88 to 108 MHz. Calculate the corresponding range of wavelengths. Example 22-1

22. • The antenna for an AM radio station is a 75-m-high tower whose height is equivalent to one-quarter wavelength. At what frequency does the station transmit? Example 22-1

23. • How far does light travel in a vacuum in 10 ns?

24. • How long does it take light to travel 300 km in a vacuum?

25. • How long does it take a radio signal from Earth to reach the Moon, which has an orbital radius of approximately 3.84×10^8 m?

26. •• The frequency and wavelength of an electromagnetic wave are related to the speed of light by

$$c = f\lambda$$

Starting with this expression derive a similar relationship between the speed of light, the angular frequency ($\omega = 2\pi f$), and the wave number ($k = 2\pi/\lambda$).

27. •• The magnetic field of an electromagnetic wave is given by

$$B(x,t) = (0.7 \text{ μT})\sin[(8\pi \times 10^6 \text{ m}^{-1})x - (2.40\pi \times 10^{15} \text{ s}^{-1})t]$$

Calculate (a) the amplitude of the electric field, (b) the speed, (c) the frequency, (d) the period, and (e) the wavelength. Example 22-1

28. •• Suppose the electric field associated with the radio transmissions of a medical helicopter is given by

$$E(x,t) = (400 \text{ μN/C})\cos[(40\pi \text{ nm}^{-1})x - (12\pi \text{ s}^{-1})t]$$

Determine (a) the wave number, (b) the angular frequency, (c) the wavelength, and (d) the frequency of the electromagnetic wave associated with the electric field. Example 22-1

22-3 Maxwell's equations explain why electromagnetic waves are possible

29. •• The electric field in a region of space increases from 0 to 3000 N/C in 5.00 s. What is the magnitude of the induced magnetic field around a circular area with a diameter of 1.00 m oriented perpendicular to the electric field? Example 22-3

30. ••• A parallel-plate capacitor has closely spaced plates. Charge is flowing onto the positive plate and off the negative plate at the rate $i = \Delta q/\Delta t = 2.8$ A. What is the displacement current through the capacitor between the plates? Example 22-3

31. •• A parallel-plate capacitor has plates of diameter 8.0 cm separated by 1.0 mm. The electric field between the plates is increasing at the rate 1.0×10^6 V/(m·s). What is the magnetic field strength between the plates at a distance of 5.0 cm from the axis of the capacitor? Example 22-3

32. ••• Charge flows onto the positive plate of a 6.0-cm-diameter parallel-plate capacitor at the rate $i = \Delta q/\Delta t = 1.5$ A. What is the magnetic field between the plates at a distance of 3.0 cm from the axis of the plates? Example 22-3

33. •• A ring is in a region of space that contains a uniform magnetic field directed perpendicular to the plane of the ring. The diameter of the ring is 1.5 cm. Figure 22-13 is a graph of the magnetic field as a function of time. What is the magnitude of the electric field along the perimeter of the ring for the time intervals $t = 0$ to 4.0 ms, 4.0 to 10.0 ms, and 10.0 to 14.0 ms? Example 22-2

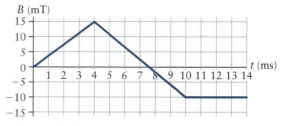

Figure 22-13 Problem 33

34. ••• A 20.0-cm-long solenoid, which has a radius of 1.00 cm and consists of 2500 turns of wire, has a current flowing through it that changes from 13.0 to 86.0 mA over a time of 1.20 s. What is the resulting electric field strength inside the solenoid at a distance of 0.500 cm from the axis of the solenoid? Example 22-2

22-4 Electromagnetic waves carry both electric and magnetic energy, and come in packets called photons

35. •• An electromagnetic plane wave has an intensity $S_{average}$ = 200 W/m². (a) What are the rms values of the electric and magnetic fields? (b) What are the amplitudes of the electric and magnetic fields? Example 22-4

36. •• The amplitude of an electromagnetic wave's electric field is 200 V/m. Calculate (a) the amplitude of the wave's magnetic field and (b) the intensity of the wave. Example 22-4

37. •• The rms value of an electromagnetic wave's magnetic field is 40.0 T. Calculate (a) the amplitude of the wave's electric field and (b) the intensity of the wave. Example 22-4

38. • Calculate the wavelengths and frequencies of photons that have the following energies. Example 22-5
   A. E_{photon} = 2.33×10⁻¹⁹ J
   B. E_{photon} = 4.50×10⁻¹⁹ J
   C. E_{photon} = 3.20×10⁻¹⁹ J
   D. E_{photon} = 8.55×10⁻¹⁹ J
   E. E_{photon} = 63.3 eV
   F. E_{photon} = 8.77 eV
   G. E_{photon} = 1.98 eV
   H. E_{photon} = 4.55 eV

39. • Calculate the wavelengths and frequencies of the photons that have the following energies. Example 22-5
   A. E_{photon} = 3.45×10⁻¹⁹ J
   B. E_{photon} = 4.80×10⁻¹⁹ J
   C. E_{photon} = 1.28×10⁻¹⁸ J
   D. E_{photon} = 4.33×10⁻²⁰ J
   E. E_{photon} = 931 MeV
   F. E_{photon} = 2.88 keV
   G. E_{photon} = 7.88 eV
   H. E_{photon} = 13.6 eV

General Problems

40. • An important news announcement is transmitted by radio waves to people who are 300 km away and sitting next to their radios and also by sound waves to people sitting 3 m from the newscaster in a newsroom. Who receives the news first? Explain your answer.

41. •• (a) Calculate the wave number of an electromagnetic wave that has an angular frequency of 6.28×10¹⁵ rad/s. (b) Calculate the angular frequency, the frequency, and the wavelength of a photon that has a wave number of k = 4π × 10⁶ rad/m. Example 22-1

42. ••• Use a computer-based, graphical program to plot graphs of the electric field (E) versus time (t) and the magnetic field (B) versus time (t) on a three-dimensional graph (see the table). Let t be on the x axis, E be on the y axis, and B be on the z axis. Explain the shape of the graph.

E (N/C)	B (×10⁻⁹ T)	t (s)
100	333	0
70.7	236	π/4
0	0	π/2
−70.7	−236	3π/4
−100	−333	π
−70.7	−236	5π/4
0	0	3π/2
70.7	236	7π/4
100	333	2π
70.7	236	9π/4
0	0	5π/2
−70.7	−236	11π/4
−100	−333	3π
−70.7	−236	13π/4
0	0	7π/2
70.7	236	15π/4
100	333	4π
70.7	236	17π/4
0	0	9π/2
−70.7	−236	19π/4
−100	−333	5π
−70.7	−236	21π/4
0	011	π/2
70.7	236	23π/4
100	333	6π

43. • **Biology** A recent study found that electrons having energies between 3.0 and 20 eV can cause breaks in a DNA molecule even though they do not ionize the molecule. If the energy were to come from light, (a) what range of wavelengths (in nanometers) could cause DNA breaks, and (b) in what part of the electromagnetic spectrum does the light lie? Example 22-5

44. •• **Medical** A dental x ray typically affects 200 g of tissue and delivers about 4.0 μJ of energy using x rays that have wavelengths of 0.025 nm. What is the energy (in electron volts) of such x-ray photons, and how many photons are absorbed during the dental x ray? Assume the body absorbs all of the incident x rays. Example 22-5

45. •• A CO_2 laser produces a cylindrical beam of light with a diameter of 0.750 cm. The energy is pulsed, lasting for 1.50 ns, and each burst contains an energy of 300 mJ. (a) What is the length of each pulse of laser light? (b) What is the average energy per unit volume for each pulse? Example 22-4

46. •• (a) What is the energy of a photon of green light that has a wavelength of 525 nm? Give your answer in joules and electron volts. (b) What is the wave number of the photon? Example 22-5

23

Physical Optics: Wave Properties of Light

What do you think?

Diamonds are renowned for how they reflect light and produce a rainbow of colors. Diamonds have these properties because the speed of light in diamond (a) is faster than in air; (b) is slower than in air; (c) depends on the color of the light; (d) both (a) and (c); (e) both (b) and (c).

In this chapter, your goals are to:

- (23-1) Describe some key properties of light.
- (23-2) Explain Huygens' principle and what it tells us about the laws of reflection and refraction.
- (23-3) Recognize the special circumstances under which total internal reflection can take place.
- (23-4) Explain how a prism is able to break white light into its component colors.
- (23-5) Calculate how the intensity of light is affected by passing through a polarizing filter.
- (23-6) Use the idea of path length difference to calculate what happens in thin-film interference.
- (23-7) Explain two-slit interference in terms of the wave properties of light.
- (23-8) Explain why light spreads out when it passes through a narrow opening.
- (23-9) Calculate how the angular resolution of an optical device is limited by diffraction.

To master this chapter, you should review:

- (13-5) Constructive and destructive interference of waves
- (22-2) Plane waves and the electromagnetic nature of light
- (22-4) Photons

23-1 The wave nature of light explains much about how light behaves

In Chapter 22 we explored how light is an electromagnetic wave, with electric and magnetic fields that oscillate in phase. In this chapter we'll explore several of the consequences of the wave nature of light. For most of this exploration we won't need the details about electric and magnetic fields; what's important is simply that in many cases light can be treated as a wave. As a result, many of the properties of light waves that we'll encounter apply equally well to sound and other types of waves.

Cats, including this sand cat (*Felis margarita*), can see even in very low light levels thanks to reflection by a layer called the *tapetum lucidum* at the back of each eye. Incoming light that isn't absorbed by the retina reflects straight back, and some of the reflected light is detected on the second pass.

The colors of the rainbow are caused by dispersion: The speed of light in water depends on the frequency of the light. As a result, each color of sunlight follows a different path as it enters a raindrop and undergoes refraction, reflects off the back of the raindrop, and undergoes refraction again as it exits the raindrop.

The colors of this soap film are caused by interference. Some light reflects from the front surface of the film, and some enters the film and reflects from the back surface. If the wavelength is just right, the two waves interfere constructively and produce a bright band.

Figure 23-1 Light waves in nature (a) Reflection, (b) dispersion, and (c) interference are among the many phenomena that light waves exhibit in the natural world.

(a)

Malcolm Schuyl/Alamy

(b)

iStockphoto/Thinkstock

(c)
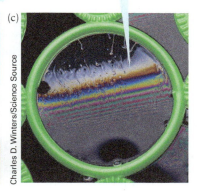
Charles D. Winters/Science Source

We'll begin by introducing *Huygens' principle*, a simplified model that describes how waves propagate through space. We'll use Huygens' principle to explain what happens when light is reflected (Figure 23-1a). Light waves travel at different speeds in different transparent materials, and we'll see how Huygens' principle explains *refraction*—the bending of light when it moves from one transparent material to another. We'll see that in certain circumstances light can be trapped inside a transparent material, just as if that material had mirrored surfaces. This effect, called *total internal reflection*, is essential for the medical technique of endoscopy. We'll see that the speed of light in a transparent material also depends on the wavelength of the light. This phenomenon, called *dispersion*, explains the vivid colors of a rainbow (Figure 23-1b).

One aspect of light that depends on its being a transverse wave is its *polarization*, which describes how the electric field vector of the wave is oriented. We'll see how light can become polarized by scattering or reflection, and we'll examine how polarizing filters work and why they're used in sunglasses.

We'll look at the phenomenon of *interference*, in which two light waves can add together constructively or destructively. Interference explains the colors seen in a thin film of soapy water (Figure 23-1c), as well as why cats and other animals have reflective eyes (see Figure 23-1a). We'll finish with a discussion of *diffraction*, an important effect in which light waves spread out when they pass through a small aperture.

> **TAKE-HOME MESSAGE FOR Section 23-1**
>
> ✔ Many of the key properties of light can be understood simply by using the idea that light can be treated as a wave.

23-2 Huygens' principle explains the reflection and refraction of light

Light travels, or propagates, in a straight line if the material through which the light travels—called the *medium* for the light—is uniform in its properties. But the direction in which light propagates changes when the light strikes a *boundary* between two different media, such as that between air and water (Figure 23-2). In general, some of the light reflects off the boundary, while the remainder travels into the second material at a different angle. The **law of reflection** states that the angle of the reflected light is the same as the angle of the incoming, or **incident**, light: $\theta_1' = \theta_1$. **Refraction** is the change in direction of the light that travels into the second medium: The angle θ_2 for the refracted light is not equal to the angle θ_1 for the incident light.

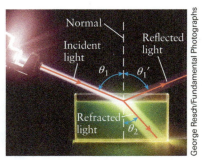

George Resch/Fundamental Photographs

Figure 23-2 Reflection and refraction When a beam of incident light in air strikes the surface of water at an angle θ_1 from the normal, some light is reflected at the same angle ($\theta_1' = \theta_1$) and some light goes into the water at a different angle θ_2.

WATCH OUT! **In reflection and refraction, angles are always measured from the normal to the boundary between two media.**

! Note that in Figure 23-2 the angles θ_1, θ_1', and θ_2 are all measured relative to the **normal** to the boundary, which is a line perpendicular to the boundary at the point where the incident light hits the boundary. A common mistake is to measure these angles relative to the boundary itself. If you make that mistake, you'll end up getting the wrong answer when you use the formulas that we'll derive in this section!

Why is the law of reflection true? And what determines how the direction of the light changes when it travels from one medium into another? We'll answer both of these questions using a model of waves introduced by the seventeenth-century Dutch scientist Christiaan Huygens. (Huygens' model precedes by two centuries Maxwell's complete description of light as an electromagnetic wave, but is consistent with it.) We'll see that what determines the difference between the incident and refracted angles are the *speeds* at which light travels in the two media. Refraction isn't just for visible light but occurs for waves of all kinds: Radio waves, sound waves, and water waves may refract when crossing from one medium to another under the right circumstances.

Huygens' Principle and Reflection

Huygens considered waves that travel in two dimensions (like ripples on the surface of a pond) or in three dimensions (like light waves). He suggested that each point on a wave crest, or **front**, at time t can be treated as a source of tiny **wavelets** that themselves move at the speed of the wave (Figure 23-3). The wave front at a later time $t = t + \Delta t$ is then the superposition of all of these wavelets emitted at time t and is tangent to the leading edges of the wavelets. This idea is called **Huygens' principle**. As Figure 23-3 shows, Huygens' principle helps explain how circular waves in water retain their shape as they propagate outward from a splash in a pond and how light waves spread out in spherical wave fronts from a light source.

To analyze what's happening to the light beams in Figure 23-2, let's apply Huygens' principle to a *plane* wave like the ones we introduced in Section 22-2. A plane wave propagates in a single direction, so it is a good description of the light beams shown in Figure 23-2. For each beam in Figure 23-2, the **ray** is an arrow that points in the direction of light propagation. Note that the ray is always perpendicular to the wave front.

Figure 23-4a shows how to apply Huygens' principle to understand the law of reflection. A plane wave with wave front ABC is directed at an angle toward the boundary between medium 1 (say, air) and medium 2 (say, water or glass). At time t point A on the wave front has just arrived at the boundary. A time Δt later, the wavelets from points A, B, and C have each spread outward by a distance $v_1\Delta t$, where v_1 is the speed at which waves propagate in medium 1. The wavelets from points B and C

② The waves propagate at the same speed v in all directions. So each individual wave crest, or wave front, spreads outward in a circle from its point of origin.

③ In the same way, a light source produces light waves that spread out from the source. In a uniform medium the waves propagate at the same speed v in all directions.

① A drop of water falling into a pond produces waves on the pond's surface.

Light source

Same wave front at time $t + \Delta t$

Wave front at time t

Figure 23-3 **Huygens' principle** This principle provides a simple way to visualize wave propagation in terms of wavelets.

④ We can treat each point (●,●) on a wave front as the source of a wavelet that itself propagates at speed v.

⑤ The superposition of all of the wavelets shows us the shape and size of the wave front at a later time.

propagate forward through medium 1, while the wavelet from point A is reflected at the boundary and so propagates *backward* from the boundary into medium 1. (We haven't drawn the wavelets that propagate into medium 2. We'll return to those a little later to help us understand refraction.)

If we draw a new wave front that's tangent to the leading edges of the wavelets that emanate from points A, B, and C, the result is $A'B'C'$ in Figure 23-4a. The wave front from B' to C' is still propagating toward the boundary; this represents the light still incident on the boundary. But the wave front from A' to B' is propagating away from the boundary and so represents the light reflected from the boundary.

The line BB' in Figure 23-4a is perpendicular to the incident wave front and so points in the direction of the incident ray. Likewise, the line AA' is perpendicular to the reflected wave front and so points in the direction of the reflected ray. To see how these directions are related to each other, notice that triangles ABB' and $AA'B'$ are both right triangles, both have the same hypotenuse of length AB', and both have one side of length $v_1\Delta t$ (Figure 23-4b). These two right triangles are identical, except that triangle $AA'B'$ has been flipped left-to-right compared to triangle ABB'. So the angle θ_1' of the line AA' measured from the vertical (that is, from the normal to the boundary) must be the same as the angle θ_1 of the line BB' measured from the vertical. We conclude that

When light reflects at the boundary between two media, the **angle of the reflected ray** from the normal...

$$\theta_1' = \theta_1$$

...is equal to the **angle of the incident ray** from the normal.

This is just the law of reflection that we mentioned above. We'll use Equation 23-1 extensively in Chapter 24 when we study the properties of mirrors.

(a) Wave front ABC reflects from the boundary between medium 1 and medium 2. Note the right triangles $AA'B'$ and ABB'.

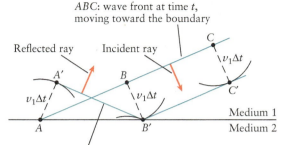

$A'B'C'$: wave front at time $t + \Delta t$, moving away from boundary (reflected) between A' and B', still moving toward the boundary between B' and C'

(b) Comparing the right triangles $AA'B'$ and ABB'.

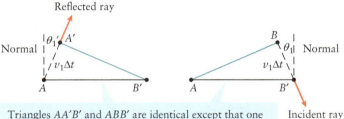

Triangles $AA'B'$ and ABB' are identical except that one is flipped left-to-right compared to the other. So $\theta_1' = \theta_1$.

Figure 23-4 **Huygens' principle and reflection** (a) A wave front reflected at a boundary between two media. (b) Finding the law of reflection, $\theta_1' = \theta_1$.

The law of reflection for light waves at a boundary (23-1)

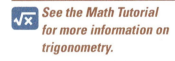
See the Math Tutorial for more information on trigonometry.

Huygens' Principle and Refraction

Let's now use Huygens' principle to determine the direction of the refracted ray in Figure 23-2. Figure 23-5a is similar to Figure 23-4a, except that for point A we've drawn only the wavelet that emanates from that point and propagates into medium 2. We've assumed that the wave speed v_2 in medium 2 is slower than the speed v_1 in medium 1. So in a time Δt the wavelet that propagates into medium 2 travels a short distance $v_2\Delta t$ while the wavelets in medium 1 travel a longer distance $v_1\Delta t$.

If we again draw a new wave front that's tangent to the leading edges of the wavelets from A, B, and C, the result is $A''B'C'$. As in Figure 23-4a, the wave front from B' to C' represents incident light in medium 1 that is still propagating toward the boundary at speed v_1. The wave front from A'' to B' represents *refracted* light that is propagating in medium 2 at speed v_2. The angle of the wave front, and hence the angle of the ray that's perpendicular to the wave front, has changed because the wave speed has changed.

In Figure 23-5b we've redrawn the right triangles ABB' and $AA''B'$ from Figure 23-5a. Both triangles have the same hypotenuse of length AB', but the angles of the two triangles are different. Side BB' of triangle ABB' has length $v_1\Delta t$, points in the direction of the incident ray, and is at an angle θ_1 from the normal to the boundary. Because ABB' is a right triangle, you can see that the angle of side AB from the horizontal is also θ_1, the same as the angle of side BB' from the vertical (the normal to

(a) Wave front ABC propagates from medium 1 into medium 2. Note the right triangles ABB′ and AA″B′.

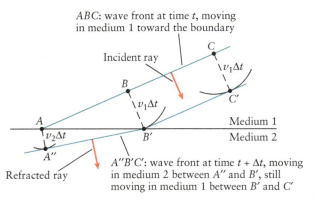

ABC: wave front at time t, moving in medium 1 toward the boundary

Incident ray

Medium 1
Medium 2

A″B′C′: wave front at time $t + \Delta t$, moving in medium 2 between A″ and B′, still moving in medium 1 between B′ and C′

Refracted ray

(b) Comparing the right triangles ABB′ and AA″B′.

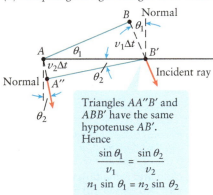

Triangles AA″B′ and ABB′ have the same hypotenuse AB′. Hence

$$\frac{\sin \theta_1}{v_1} = \frac{\sin \theta_2}{v_2}$$
$$n_1 \sin \theta_1 = n_2 \sin \theta_2$$

Figure 23-5 Huygens' principle and refraction
(a) A wave front refracted at a boundary between two media. (b) Finding the law of refraction.

Index of refraction for light
waves in a medium
(23-5)

the boundary). If you look now at triangle AA″B′, you'll see that side AA″ has length $v_2\Delta t$, points in the direction of the refracted ray, and is at a different angle θ_2 from the normal to the boundary. And because AA″B′ is a right triangle, the angle of side A″B′ from the horizontal is θ_2, the same as the angle of side AA″ from the normal.

Recall that the sine of an angle in a right triangle equals the length of the side opposite to that angle divided by the length of the hypotenuse. So the sine of the angle θ_1 in triangle ABB′ is

(23-2)
$$\sin \theta_1 = \frac{\text{length of side } BB'}{\text{length of side } AB'} = \frac{v_1\Delta t}{\text{length of side } AB'}$$

so
$$\frac{\sin \theta_1}{v_1} = \frac{\Delta t}{\text{length of side } AB'}$$

Similarly the sine of the angle θ_2 in triangle AA″B′ is

(23-3)
$$\sin \theta_2 = \frac{\text{length of side } AA''}{\text{length of side } AB'} = \frac{v_2\Delta t}{\text{length of side } AB'}$$

so
$$\frac{\sin \theta_2}{v_2} = \frac{\Delta t}{\text{length of side } AB'}$$

If you compare Equations 23-2 and 23-3, you can see that

$$\frac{\sin \theta_1}{v_1} = \frac{\sin \theta_2}{v_2}$$

Equation 23-4 tells us that the relationship between the angles θ_1 and θ_2 is determined by the speeds v_1 and v_2 of the wave in medium 1 and medium 2, respectively.

It's common to express the speed of light in a given medium in terms of a quantity called the **index of refraction**:

Index of refraction of a medium   Speed of light in a vacuum

$$n = \frac{c}{v}$$

Speed of light in the medium

The index of refraction of vacuum is 1, since light travels at the speed of light c, so $v = c$ and $n = c/c = 1$. In any material medium light travels slower than c, so $v < c$ and $n > 1$. The greater the value of the index of refraction n in a given medium, the slower the speed v at which light propagates in that medium. Table 23-1 lists the index of refraction of some common materials. Note that the index of refraction of air is equal to one to three significant digits, so we'll often take $n_{air} = 1$ in calculations.

Equation 23-5 tells us that $1/v = n/c$, so we can rewrite Equation 23-4 as

$$\left(\frac{n_1}{c}\right)\sin \theta_1 = \left(\frac{n_2}{c}\right)\sin \theta_2$$

If we cancel the factors of c on both sides of this equation, we get

Snell's law of refraction for light
waves at a boundary
(23-6)

Angle of the incident ray
from the normal   Angle of the refracted ray
from the normal

$$n_1 \sin \theta_1 = n_2 \sin \theta_2$$

Index of refraction for the medium
with the incident light   Index of refraction for the medium
with the refracted light

Equation 23-6 is known as **Snell's law of refraction**. (This law is named for the Dutch scientist Willebrord Snellius but was in fact first discovered by the Persian scientist Ibn Sahl in 984, more than 600 years before Snellius.)

Snell's law tells us that when a ray of light crosses from one medium to another, the product of the index of refraction and the sine of the angle the ray makes to the normal remains constant. When light passes into a material of higher index of refraction—for example, from air into glass—so that the speed of light is slower in the second medium and $n_2 > n_1$, the sine of the refracted angle and the angle itself both decrease. In this case $\theta_2 < \theta_1$, and the light bends closer to the normal (Figure 23-6a). When light instead passes into a material of lower index of refraction—for example, from glass into air—so that the speed of light is faster in the second medium and $n_2 < n_1$, the sine of the refracted angle and the angle itself both increase. In this situation $\theta_2 > \theta_1$, and the light bends away from the normal (Figure 23-6b).

BioMedical   The fraction of incident light that is reflected and the fraction that is refracted depend in part on the indices of refraction of the two media. (They also depend on the incident angle and on how the electric field vectors in the light wave are oriented relative to the boundary.) The index of refraction of a medium is often a function of the medium's density; one example is blood plasma, the density and index of refraction of which depend on the concentration of dissolved protein. Veterinarians can use this to estimate protein levels in livestock at the clinic or on the farm: They measure how much light refracts as it passes through a sample of an animal's plasma. Winemakers determine the amount of sugar in their grapes by using the same technique.

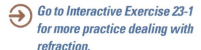

Go to Interactive Exercise 23-1 for more practice dealing with refraction.

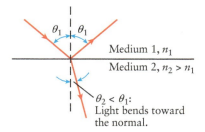

(a) Refraction from one medium into a slower one

θ_1 θ_1

Medium 1, n_1

Medium 2, $n_2 > n_1$

$\theta_2 < \theta_1$: Light bends toward the normal.

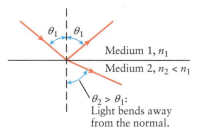

(b) Refraction from one medium into a faster one

θ_1 θ_1

Medium 1, n_1

Medium 2, $n_2 < n_1$

$\theta_2 > \theta_1$: Light bends away from the normal.

Figure 23-6 Refraction toward and away from the normal Which way the refracted ray bends depends on whether the speed of light in the second medium is (a) slower or (b) faster than in the first medium.

TABLE 23-1 Indices of Refraction

Material	Index of refraction
vacuum	1 (exactly)
air at 20°C, 1 atm pressure	1.00029
ice	1.31
water at 20°C	1.33
acetone	1.36
ethyl alcohol	1.36
eye, cornea	1.38
eye, lens	1.41
sugar water (high concentration)	1.49
Plexiglas	1.49
typical crown glass	1.52
sodium chloride	1.54
sapphire	1.77
diamond	2.42

EXAMPLE 23-1   Seeing Under Water

A surveyor (labeled S) looking at an aqueduct is just able to see the underwater edge at point F where the far wall meets the bottom (Figure 23-7). If the aqueduct is 4.2 m wide and her line of sight to the near, top edge at point N is 25° from the horizontal, find the actual depth of the aqueduct.

Figure 23-7   A refracted view If the surveyor just sees the bottom edge of the far wall of the aqueduct, how deep is the aqueduct?

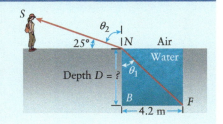

Set Up

A light beam from point F refracts when it reaches the boundary between the water (medium 1) and the air (medium 2) at point N. Figure 23-7 shows that the angle of the refracted ray from the *normal* (shown as a vertical dashed line) is $\theta_2 = 90° - 25° = 65°$. We'll first use Snell's law to determine the angle θ_1 of the incident ray, and then use trigonometry and the given width of the aqueduct (4.2 m) to find the depth D.

Snell's law of refraction:

$$n_1 \sin \theta_1 = n_2 \sin \theta_2 \qquad (23\text{-}6)$$

Solve

Find the angle θ_1 of the incident ray using Snell's law.

Solve Equation 23-6 for the sine of the incident angle θ_1:

$$\sin \theta_1 = \frac{n_2}{n_1} \sin \theta_2$$

Figure 23-7 shows that $\theta_2 = 90° - 25° = 65°$, and from Table 23-1 we see that $n_1 = n_{\text{water}} = 1.33$ and $n_2 = n_{\text{air}} = 1.00$. So

$$\sin \theta_1 = \frac{1.00}{1.33} \sin 65° = 0.68$$
$$\theta_1 = \sin^{-1} 0.68 = 43°$$

Use trigonometry to determine the depth D of the aqueduct.

The incident ray that travels from point F to point N is the hypotenuse of a right triangle NFB with vertical dimension D and horizontal dimension 4.2 m. The side of length 4.2 m is the side opposite the angle θ_1, and the side of length D is the side adjacent to this angle. The tangent of θ_1 equals the opposite side divided by the adjacent side:

$$\tan \theta_1 = \frac{4.2 \text{ m}}{D}$$

Solve for the distance D:

$$D = \frac{4.2 \text{ m}}{\tan \theta_1} = \frac{4.2 \text{ m}}{\tan 43°} = 4.5 \text{ m}$$

The aqueduct is 4.5 m deep.

Reflect

The refracted angle $\theta_2 = 65°$ is larger than the incident angle $\theta_1 = 43°$, just as in Figure 23-6b. This makes sense, since $n_2 < n_1$ (the index of refraction for air is smaller than the index for water).

Our brains are used to the idea that light travels in straight lines. If we extend the refracted ray backward, we see that to the surveyor's eye the light from the far edge of the bottom of the aqueduct appears to be coming from a shallower depth than $D = 4.5$ m. So the *apparent* depth of the pool is less than the *actual* depth D. You can easily see this effect in a swimming pool.

The solid red line represents a ray of light from point F that arrives at your eyes.

Your brain traces the light ray back along the dashed line...

...so that the bottom of the aqueduct appears to you to be at F', at a shallower depth.

The situation shown in Figure 23-8 involves the same effect that makes the aqueduct in Example 23-1 appear shallower than it really is. When light from the submerged part of the chopstick passes from water to air at the boundary between the two, the light rays refract. As a result, it appears to our eyes that the submerged part is in a different position than its true location.

Frequency and Wavelength in Refraction

When a wave travels from one medium to another, the frequency of the wave remains the same. (In a given time interval, as many crests arrive at the boundary as leave the boundary. If this were not true, there would be a "traffic jam" of wave crests at the boundary.) However, since the wave speed is different in the two media, the wavelength must change. This follows from the relationship among the propagation speed v of the wave, the frequency f, and the wavelength λ. From Equation 13-2:

$$v = f\lambda \quad \text{so} \quad \lambda = \frac{v}{f}$$

In a vacuum light waves travel at speed $v = c$, so the wavelength is

$$\lambda_{\text{vacuum}} = \frac{c}{f} \tag{23-7}$$

In a medium with index of refraction n, the wave speed from Equation 23-5 is $v = c/n$. Then we can write the wavelength of light in a medium as

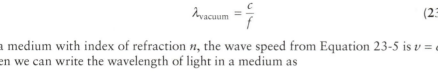

Wavelength of light in a medium
(23-8)

$$\lambda = \frac{c}{nf} = \frac{\lambda_{\text{vacuum}}}{n}$$

Wavelength of light in a medium
Speed of light in a vacuum
Wavelength of the light in a vacuum
Frequency of the light
Index of refraction of the medium

Figure 23-8 A refracted chopstick Due to refraction, the submerged part of the chopstick on the right appears displaced from its actual position.

David Tauck

Equation 23-8 says that the wavelength is shorter in a medium with a higher index of refraction, where the propagation speed is slower. For example, red light that has a wavelength $\lambda_{\text{vacuum}} = 750$ nm in a vacuum has a wavelength in water ($n = 1.33$) equal to $\lambda = \lambda_{\text{vacuum}}/n = (750 \text{ nm})/(1.33) = 564$ nm. The frequency of this light is the same in both media: $f = c/\lambda_{\text{vacuum}} = (3.00 \times 10^8 \text{ m/s})/(750 \times 10^{-9} \text{ m}) = 4.00 \times 10^{14}$ Hz in a vacuum and $f = v/\lambda = c/(n\lambda) = (3.00 \times 10^8 \text{ m/s})/((1.33)(564 \times 10^{-9} \text{ m})) = 4.00 \times 10^{14}$ Hz in water.

We learned in Section 22-4 that the energy of an electromagnetic wave comes in packets called *photons*. A single photon has energy $E = hf$, where h is Planck's constant and f is the frequency (Equation 22-26). Since f is unchanged when a photon passes from one medium to another, the photon energy also remains unchanged. This is one important way that photons are very different from ordinary particles such as marbles, whose kinetic energy changes when their speed changes.

GOT THE CONCEPT? 23-1 **Refraction**

? A beam of light in air travels into a transparent, flat-walled container made of Plexiglas. The incident angle is 5.00°. The light then travels from the Plexiglas into the fresh water inside the container. In each refraction, does the ray bend closer to the normal or farther away from it? (a) Closer in both refractions; (b) farther away in both refractions; (c) closer going from air to Plexiglas, farther away going from Plexiglas to water; (d) farther away going from air to Plexiglas, closer going from Plexiglas to water. (e) In at least one of the refractions, the light does not bend at all.

TAKE-HOME MESSAGE FOR **Section 23-2**

✔ Huygens' principle says that each point on a wave front acts as a source of wavelets. The new wave front is the superposition of the individual wavelets.

✔ When waves encounter a boundary between two media in which the wave speed is different, the waves can bounce back into the first medium (reflect) or pass into the second medium at a different angle (refract).

✔ The angle of the reflected ray is the same as the angle of the incident ray (law of reflection).

✔ Snell's law describes the direction of the refracted ray. If the wave speed is lower in the second medium than in the first, the refracted ray bends toward the normal. If the wave speed is faster in the second medium, the refracted ray bends away from the normal.

✔ The index of refraction is a measure of the speed of light in a medium relative to the speed of light in a vacuum. The slower light travels in a medium, the larger its index of refraction.

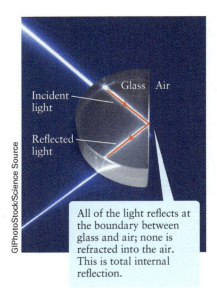

All of the light reflects at the boundary between glass and air; none is refracted into the air. This is total internal reflection.

Figure 23-9 **Total internal reflection** There is nothing unusual about the place where the light beam strikes the glass, yet none of the light escapes into the air on the other side.

23-3 In some cases light undergoes total internal reflection at the boundary between media

In the photograph shown in Figure 23-2, light in air encounters a boundary with water on the other side. Some of the incident light is reflected at the boundary, and some of it is refracted from the first medium (air) into the second medium (water). There are situations, however, in which *none* of the light is refracted into the second medium. Figure 23-9 shows a light beam in glass that encounters a boundary with air on the other side. As the photograph shows, 100% of the light is reflected back into the glass. This effect is called **total internal reflection**.

Figure 23-10 shows how total internal reflection arises. Light travels more slowly in glass than air, so the index of refraction of glass is higher than the index of refraction of air. As light crosses the boundary from glass to air, it is bent away from the normal as in Figure 23-10a. As the incident angle of the light increases (Figure 23-10b), the refracted light gets farther from the normal and decreases in intensity. When the incident angle equals the **critical angle** θ_c, the refracted light lies exactly in the plane of the surface (Figure 23-10c). At this angle the intensity of the refracted light is zero. If the incident angle is greater than the critical angle, as in Figure 23-10d, the light is completely reflected back into the glass. This is total internal reflection.

Total internal reflection is possible only when the first medium (in Figure 23-10, glass) has a higher index of refraction than the second medium (in Figure 23-10, air), so $n_1 > n_2$. Then the refracted angle θ_2 is greater than the incident angle θ_1, and the refracted angle can reach 90°, as in Figure 23-10c. If the first medium has a lower index of refraction than the second medium, so $n_1 < n_2$, the refracted angle θ_2 is less than the incident angle θ_1, and the refracted angle can never reach 90°.

We can calculate the critical angle θ_c using Snell's law of refraction, Equation 23-6. When the incident angle θ_1 equals the critical angle θ_c, the refracted angle θ_2 equals 90°. If we substitute these into Equation 23-6 and recall that sin 90° = 1, we get

$$(23\text{-}9) \qquad n_1 \sin \theta_c = n_2 \sin 90° = n_2 \quad \text{so} \quad \sin \theta_c = \frac{n_2}{n_1}$$

Note that the sine of an angle between 0 and 90° is between 0 and 1. If $n_1 > n_2$, the ratio n_2/n_1 is less than 1, and there will be some angle θ_c for which Equation 23-9 is satisfied. In this case total internal reflection is possible. But if $n_1 < n_2$, the ratio n_2/n_1 is greater than 1 and Equation 23-9 has no solution. This is another way of seeing that total internal reflection is possible only if $n_1 > n_2$.

Figure 23-10 **Approaching total internal reflection** (a) and (b) Light approaches a boundary between glass and air at an incident angle θ_1 less than the critical angle θ_c. (c) When $\theta_1 = \theta_c$, the light ray is refracted along the surface. (d) Total internal reflection occurs for incident angles greater than the critical angle.

(a) Light refracts as it passes from glass into air. The index of refraction of glass is greater than the index of refraction of air, so $\theta_2 > \theta_1$.

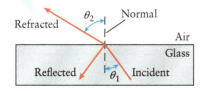

(b) As the angle θ_1 of the incident light increases, so does the angle θ_2 of the refracted light.

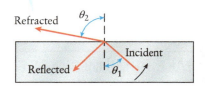

(c) When the angle θ_1 of the incident light equals the critical angle θ_c, the refracted angle is $\theta_2 = 90°$. The intensity of the refracted light becomes zero.

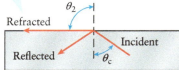

(d) When the angle θ_1 of the incident light is greater than the critical angle θ_c, the light is completely reflected back into the glass. This is total internal reflection.

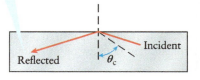

If we solve Equation 23-9 for the critical angle θ_c, we get

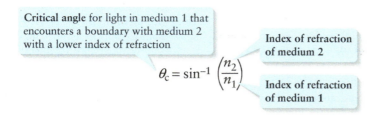

Critical angle for light in medium 1 that encounters a boundary with medium 2 with a lower index of refraction

Index of refraction of medium 2

Index of refraction of medium 1

$$\theta_c = \sin^{-1}\left(\frac{n_2}{n_1}\right)$$

Critical angle for total internal reflection of light waves at a boundary
(23-10)

If the incident angle is greater than the critical angle, there is total internal reflection.

Go to Picture It 23-1 for more practice dealing with refraction and reflection.

Notice that the critical angle depends on the indices of refraction of the media on *both* sides of a boundary.

EXAMPLE 23-2   Critical Angles

(a) A laser is aimed from under the water toward the surface, as in **Figure 23-11**. Find the critical angle of the light incident in the water beyond which total internal reflection occurs. (b) Find the critical angle if the liquid in the tank were replaced by water containing a high concentration of dissolved sugar.

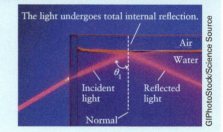

Figure 23-11 Total internal reflection in water Light from a laser aimed from under the water toward the surface is totally internally reflected. What is the minimum incident angle for which total internal reflection will occur?

Set Up

Before hitting the surface (the boundary between water and air), the light is propagating in water, so this is medium 1. In part (a) medium 1 is ordinary water with index of refraction $n_1 = 1.33$; in part (b) medium 1 is sugar water with $n_1 = 1.49$. In both parts, medium 2 on the other side of the surface is air with index of refraction $n_2 = 1.00$. We'll use Equation 23-10 to solve for the critical angle in each case.

Critical angle for total internal reflection:

$$\theta_c = \sin^{-1}\left(\frac{n_2}{n_1}\right) \qquad (23\text{-}10)$$

Solve

(a) Find the critical angle if medium 1 is water.

With $n_1 = 1.33$ and $n_2 = 1.00$, the critical angle is

$$\theta_c = \sin^{-1}\left(\frac{1.00}{1.33}\right) = \sin^{-1} 0.752 = 48.8°$$

If the angle of incidence θ_1 is 48.8° or greater, the light will undergo total internal reflection and no light will go into the air. (Note that θ_1 in Figure 23-11 is approximately 60°, which is indeed greater than 48.8°.)

(b) Find the critical angle if medium 2 is sugar water.

With $n_1 = 1.49$ and $n_2 = 1.00$, the critical angle is

$$\theta_c = \sin^{-1}\left(\frac{1.00}{1.49}\right) = \sin^{-1} 0.671 = 42.2°$$

Reflect

Adding sugar to water decreases the speed of light in the water, which increases the index of refraction. The minimum angle of incidence for which total internal reflection occurs is therefore smaller for sugar water than for pure water. The smaller the critical angle, the larger the range of angles at which light experiences total internal reflection.

These "puddles" are actually a mirage.

Figure 23-12 A mirage What causes these shimmering patches on the road that look like puddles?

Total internal reflection explains the brilliance of a cut diamond (see the image that opens this chapter). If light enters the front surface of a cut diamond, it will undergo total internal reflection at the back surface of the diamond (a boundary with air) if the incident angle is greater than the critical angle. Table 23-1 shows that diamond has a very high index of refraction, 2.42, so from Equation 23-10 the critical angle is very small: $\theta_c = \sin^{-1}(1.00/2.42) = 24.4°$. A talented jeweler cuts a diamond so that two things happen: First, light entering the front of the diamond over a broad range of angles will strike the back at an incident angle greater than $\theta_c = 24.4°$ so that total internal reflection occurs there; and second, this reflected light strikes the *front* surface of the diamond at an incident angle *less* than 24.4° so that this light can escape and be seen by you. The result is a gem that sparkles with brilliant reflections.

Total internal reflection also explains the *mirage* that happens when you look down the highway on a hot day and see what appear to be puddles of water on the road, as in Figure 23-12. The explanation is that air sits in layers above the road, each layer a bit warmer, less dense, and with a lower index of refraction than the one above it. Light from the sky is refracted as it encounters the boundary between one layer of air and the next (Figure 23-13). Because the index of refraction of the layer of air closer to the road is lower, the light is bent farther from the normal. The normal direction in these refractions is vertical, so the light is refracted closer to horizontal. Light that strikes the boundary between layers at a large angle with respect to the normal (a grazing angle as measured from the air layer boundary) can experience total internal reflection and reflect back up into the higher layer. Your eyes trace the light rays back along straight lines, so the light appears to be coming from a point close to the road. What looks like a puddle is actually a refracted image of the blue sky.

Figure 23-13 Explaining a mirage The "puddles" in Figure 23-12 are an illusion caused by total internal reflection.

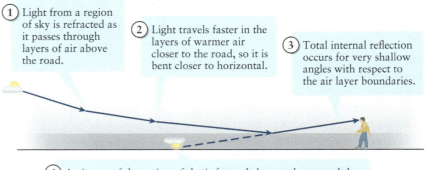

① Light from a region of sky is refracted as it passes through layers of air above the road.

② Light travels faster in the layers of warmer air closer to the road, so it is bent closer to horizontal.

③ Total internal reflection occurs for very shallow angles with respect to the air layer boundaries.

④ An image of the region of sky is formed close to the ground that gives the appearance of shimmery, blue puddles on the road.

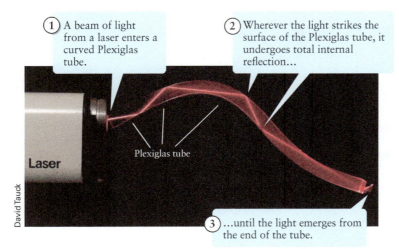

① A beam of light from a laser enters a curved Plexiglas tube.

② Wherever the light strikes the surface of the Plexiglas tube, it undergoes total internal reflection...

Plexiglas tube

Laser

③ ...until the light emerges from the end of the tube.

Figure 23-14 Light trapped by total internal reflection Light propagates through the interior of a curved bar of Plexiglas by a series of total internal reflections.

Bio Medical *Endoscopy* is a medical procedure used to see inside the body. It relies on a light fiber, an optical device used to carry light and sometimes data encoded in pulses of light, from one place to another. A beam of light sent down a light fiber experiences total internal reflection at the surface of the fiber, which results in multiple reflections that keep the beam inside the fiber. The bent bar of Plexiglas in Figure 23-14 carries light in a similar way. Notice in Figure 23-14 that no light leaks out of the bar at the points where the beam hits the surface. This is because the angle of incidence at the surface is greater than the critical angle, so total internal reflection occurs. To maximize total internal reflection, most light fibers are made by surrounding a central core with one or two layers of a material of lower index of refraction than the core. Many endoscopes actually use bundles of several fibers. Light is sent down some fibers to illuminate the subject. Other fibers carry the image back to a camera.

GOT THE CONCEPT? 23-2 Total Internal Reflection

It's possible for a beam of light in Plexiglas to undergo total internal reflection at a boundary if the material on the other side of the boundary is (a) ethyl alcohol; (b) sapphire; (c) diamond; (d) more than one of these; (e) none of these.

TAKE-HOME MESSAGE FOR Section 23-3

✔ When light traveling in a medium with index of refraction n_1 reaches a boundary with a second medium of index of refraction n_2, total internal reflection can happen if $n_1 > n_2$.

✔ Total internal reflection takes place only if the incident angle measured from the normal at the boundary is greater than the critical angle θ_c given by Equation 23-10. In this case none of the light is refracted into the second medium.

23-4 The dispersion of light explains the colors from a prism or a rainbow

White light is a mixture of all the colors of the visible spectrum. You can see these colors by allowing sunlight to pass through a glass prism, as in Figure 23-15: The different colors emerge in different directions. This happens because the speed of light in a medium other than vacuum, such as the glass in a prism, is different for different frequencies of light. (In a vacuum the speed is equal to c for all frequencies.) This is a result of how a light wave interacts with the atoms of the medium. This variation of speed with frequency is called **dispersion**.

Since the speed v of light waves in a medium depends on the frequency, the index of refraction $n = c/v$ (Equation 23-5) depends on frequency as well. In most transparent materials the speed decreases with increasing frequency, from red to yellow to violet (see Figure 22-2): Red light travels fastest, and violet light travels slowest. This means that the index of refraction n increases with increasing frequency. Note that Equation 23-7 tells us that the wavelength in a vacuum is inversely proportional to frequency: $\lambda_{vacuum} = c/f$. So we can also say that the value of n decreases with increasing vacuum wavelength. (We specify *vacuum* wavelength, since the wavelength in the medium depends on the value of n; see Equation 23-8.) Figure 23-16 shows how the index of refraction varies with vacuum wavelength for four different transparent materials. (The indices of refraction given in Table 23-1 are for yellow light, near the middle of the visible spectrum.)

Snell's law of refraction, Equation 23-6, tells us that the angle at which light refracts as it crosses the boundary between two transparent media depends on their indices of refraction. So it follows that different colors of light, with different vacuum wavelengths, refract at different angles. Figure 23-17 shows this for light passing from vacuum into glass. The higher the index of refraction of the second medium, the more the refracted light is bent toward the normal. In common crown glass the index of refraction is about 1.51 for red light and about 1.53 for blue light. Hence the blue light bends more toward the normal than does the red light, and different colors of light are spread out or dispersed. (This is the origin of the term "dispersion.")

The same effect explains the appearance of a rainbow (Figure 23-1b). When raindrops in midair are illuminated by the Sun, sunlight enters each raindrop, is partly reflected off the back of the drop, and then exits out the front of the drop. The index of refraction of water is different for different wavelengths, so each color of light emerges in a slightly different direction to form a rainbow.

The amount of dispersion is different for different transparent materials. In crown glass, for example, the index of refraction varies by 0.02 from red ($n = 1.51$) to blue ($n = 1.53$), while for diamond the index of refraction varies by 0.04 from red ($n = 2.41$) to blue ($n = 2.45$). As a result, the colors of white light are spread out over a wider angle by a cut diamond than by a piece of glass cut to the same shape. This high value of dispersion contributes to the "sparkly" character of a cut diamond.

Figure 23-15 Dispersion Light of different wavelengths in a vacuum, and hence different colors, propagates at different speeds through the glass of which this prism is made. As a result, different colors refract along slightly different paths.

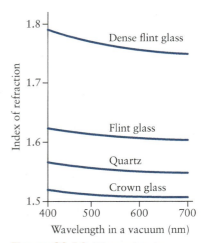

Figure 23-16 Dispersion in different materials The index of refraction in glass varies with the vacuum wavelength of light and with the type of glass.

Figure 23-17 Analyzing dispersion The difference in refracted angle between the red and blue light is greatly exaggerated for clarity.

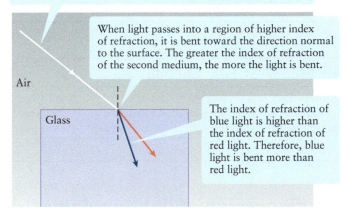

White light is composed of light across all wavelengths (colors) in the visible spectrum. The colors spread apart when light crosses the boundary between two light transmitting media.

When light passes into a region of higher index of refraction, it is bent toward the direction normal to the surface. The greater the index of refraction of the second medium, the more the light is bent.

Air

Glass

The index of refraction of blue light is higher than the index of refraction of red light. Therefore, blue light is bent more than red light.

EXAMPLE 23-3 Dispersion in Dense Flint Glass

A narrow beam of white light enters a rectangular block of dense flint glass at 60.0° from the normal. The block is 0.500 m on a side. How far apart will the red and blue parts of the visible spectrum be when the light leaves the glass? The index of refraction of dense flint glass is 1.75 for red light and 1.79 for blue light.

Set Up

Both colors of light are incident on the block at the same angle: $\theta_1 = 60.0°$. However, the red and blue refract at different angles because the index of refraction is different for the two colors. We'll use Snell's law, Equation 23-6, to calculate the angles θ_{red} and θ_{blue}, and then use trigonometry to find the distances d_{red} and d_{blue} shown in the figure. The difference between these distances tells us how far apart the points are where the two colors emerge from the glass.

Snell's law of refraction:

$$n_1 \sin \theta_1 = n_2 \sin \theta_2 \qquad (23\text{-}6)$$

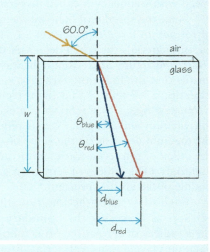

Solve

Find the refracted angles θ_{red} and θ_{blue} for the two colors.

For both colors $n_1 = n_{air} = 1.00$ and $\theta_1 = 60.0°$. For red light $n_2 = 1.75$, so Equation 23-6 becomes

$$1.00 \sin 60.0° = 1.75 \sin \theta_{red}$$

$$\sin \theta_{red} = \frac{1.00 \sin 60.0°}{1.75} = 0.4948$$

$$\theta_{red} = \sin^{-1} 0.4948 = 29.66°$$

(We've kept an extra significant digit in our result; we'll round off at the end of the calculation.) Do the same calculation for blue light, for which $n_2 = 1.79$:

$$1.00 \sin 60.0° = 1.79 \sin \theta_{blue}$$

$$\sin \theta_{blue} = \frac{1.00 \sin 60.0°}{1.79} = 0.4838$$

$$\theta_{blue} = \sin^{-1} 0.484 = 28.93°$$

Find the distances d_{red} and d_{blue} for the two colors of light, and from these find the separation between the two colors as they exit the glass.

For each color of light, the ray that extends from where the light enters the block of glass to where it exits the block forms the hypotenuse of a right triangle. The other two sides are the vertical dimension of the block, $w = 0.500$ m, and the distance d_{red} or d_{blue} that the light is displaced horizontally. In each case the side of length w is adjacent to the angle θ_{red} or θ_{blue}, and the side of length d_{red} or d_{blue} is opposite to that angle. In a right triangle the length of the opposite side divided by the length of the adjacent side equals the tangent of the angle, so

$$\tan \theta_{red} = \frac{d_{red}}{w}$$

$$\tan \theta_{blue} = \frac{d_{blue}}{w}$$

Solve for the distances d_{red} and d_{blue} using the angles that we calculated above:

$$d_{red} = w \tan \theta_{red} = (0.500 \text{ m}) \tan 29.66° = 0.2847 \text{ m}$$

$$d_{blue} = w \tan \theta_{blue} = (0.500 \text{ m}) \tan 28.93° = 0.2764 \text{ m}$$

The distance between where the red light exits the glass and where the blue light exits the glass is

$$d_{red} - d_{blue} = 0.2847 \text{ m} - 0.2764 \text{ m} = 0.0083 \text{ m} = 8.3 \text{ mm}$$

Reflect

The separation between the two colors is fairly substantial, so this block of flint glass does a good job of spreading white light into its constituent colors. We invite you to repeat this calculation for crown glass, the kind of glass commonly used to make windows, for which $n_{red} = 1.51$ and $n_{blue} = 1.53$. Since crown glass has a smaller index of refraction than flint glass, and because the difference between the values of the two indices is smaller for crown glass than for flint glass, you'll find that the distance $d_{red} - d_{blue}$ is smaller for crown glass. Which type of glass would be a better choice for a prism intended to spread apart the different colors of light, as in Figure 23-15?

GOT THE CONCEPT? 23-3 Dispersion

 A flash of white light (containing all of the colors of the visible spectrum) in air shines straight down on the surface of a pond of water. Which color of light from the flash reaches the bottom of the pond first? (a) The blue light; (b) the yellow light; (c) the red light. (d) All reach the bottom at the same time.

TAKE-HOME MESSAGE FOR Section 23-4

✔ The speed of light in a material medium depends on the frequency of the light. This is called dispersion. In most materials the index of refraction for visible light increases from the red end of the spectrum (low frequency, long wavelength) to the blue end of the spectrum (high frequency, short wavelength).

✔ When light crosses a boundary into a medium of different index of refraction, different colors refract by different angles. This causes the colors to spread apart by an amount that depends on how strongly the index of refraction varies with wavelength.

23-5 In a polarized light wave the electric field vector points in a specific direction

When a honeybee finds nectar, it communicates the location to other bees in the hive. In the 1940s Austrian ethologist Karl von Frisch established that if bees can see even a small patch of blue sky, they can use the position of the Sun to describe the path back to the food from the hive (Figure 23-18). How is it that bees can know the position of the Sun even if they can't see it directly?

The explanation is that light is a transverse electromagnetic wave, with an oscillating electric field $\vec{E}$ and magnetic field $\vec{B}$ that are perpendicular to each other and to the

Figure 23-18 Navigating by the light of the sky Honeybees have the ability to detect the polarization of light entering their eyes — that is, the orientation of the electric field $\bar{E}$ in the light wave. The polarization helps them determine the position of the Sun in the sky, even if the Sun isn't directly visible. Karl von Frisch shared the 1973 Nobel Prize in Physiology or Medicine for this and other discoveries about bees and their behavior.

direction of wave propagation (see Figure 22-4). The orientation of the $\bar{E}$ field is called the **polarization** of the light wave. (We don't need to separately state the orientation of $\bar{B}$, since we know that it's perpendicular to both the direction of propagation and the orientation of $\bar{E}$.) Natural light such as that emitted by the Sun or an ordinary light bulb is **unpolarized**: The orientation of the electric field changes randomly from one moment to the next. For example, if the wave is propagating in the positive x direction, at one moment $\bar{E}$ may be oriented along the y axis, a short time later it may be oriented along the z axis, a short time after that it may be oriented at $23.7°$ to the y axis, and so on. The reason for this is that a source of natural light emits light in the form of a stream of photons (see Section 22-4), and the orientation of $\bar{E}$ varies randomly from one photon to another.

BioMedical When sunlight scatters from molecules or small particles in the atmosphere, however, the scattered light that we see has its $\bar{E}$ field oriented predominantly in one direction (Figure 23-19). (Scattering is stronger for short-wavelength light than for long-wavelength light, which is why the color of the sky is dominated by short-wavelength blue light.) Light in which the orientation of the $\bar{E}$ field changes randomly, but is more likely to be in one orientation than in other orientations, is called **partially polarized**. Light for which the $\bar{E}$ field is oriented *completely* along one direction is called **linearly polarized**. For example, the $\bar{E}$ field for the light wave shown in Figure 22-4 has only a y component, so we say this light is linearly polarized along the y axis. Honeybees (Figure 23-18) have the ability to detect the polarization of light coming from the sky, and by using this they can infer the position of the Sun. (Human eyes, by contrast, are only weakly sensitive to polarization.)

Polarizing Light with a Polarizing Filter

A simple way to make polarized light from unpolarized light is by using a **polarizing filter**. This is a transparent sheet which contains long-chain molecules that are all oriented in the same direction. These molecules absorb light whose polarization direction is along the axis of the molecules, but have no effect on light that is polarized perpendicular to that direction. Equivalently, you can think of a polarizing filter as having slots that allow waves to pass if they are polarized along the direction of the slots, called the *polarization direction*. Waves that are polarized perpendicular to the slots are blocked (Figure 23-20). If we send unpolarized light into the filter, at any instant the electric field $\bar{E}$ has a component along the polarization direction of the filter and a component perpendicular to that direction. Only the component of $\bar{E}$ along

Figure 23-19 Polarization by scattering The extent to which sunlight is polarized by scattering depends on the scattering angle.

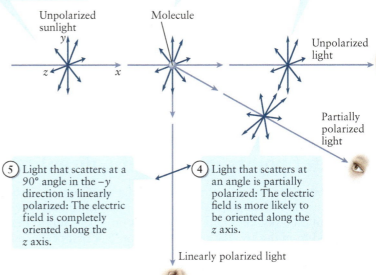

(1) Sunlight propagates in the x direction. The light is unpolarized, so the orientation of the electric field (shown by the blue arrows) changes randomly in the yz plane.

(2) When light strikes a molecule in the atmosphere, the polarization and direction of propagation can both change.

(3) Light that scatters in the forward direction (that is, is not deflected) remains unpolarized.

(5) Light that scatters at a 90° angle in the $-y$ direction is linearly polarized: The electric field is completely oriented along the z axis.

(4) Light that scatters at an angle is partially polarized: The electric field is more likely to be oriented along the z axis.

Unpolarized sunlight

Molecule

Unpolarized light

Partially polarized light

Linearly polarized light

Two waves of different polarizations encounter a polarizing filter.

Only the wave with polarization aligned with the polarization direction of the filter passes through it. The light is now linearly polarized.

These sine waves represent the electric fields of the two waves.

Filter

The green stripes represent the polarization direction of the filter.

Direction of wave propagation

Figure 23-20 A linear polarizing filter I A filter of this kind allows only light with an electric field component aligned in a certain direction, called the polarization direction, to pass through it.

the polarization direction of the filter will pass through, so the light that emerges from the filter will be linearly polarized.

Since a polarizing filter absorbs some of the incident light that falls on it, the light exiting the filter is in general less intense than the incident light. Suppose the incident light has an oscillating electric field with amplitude E_0 that is oriented at an angle θ to the polarization direction of the filter (Figure 23-21). Only the component of this field that is aligned with the polarization direction is allowed to pass through the filter:

$$E_{\text{transmitted}} = E_0 \cos \theta \tag{23-11}$$

The intensity of the light is proportional to the square of the electric field amplitude. Equation 23-11 says that the amplitude $E_{\text{transmitted}}$ of the transmitted light equals $\cos \theta$ times the amplitude E_0 of the incident light. Hence the intensity I of the transmitted light equals $\cos^2 \theta$ times the intensity I_0 of the incident light:

$$I = E_0^2 \cos^2 \theta = I_0 \cos^2 \theta \tag{23-12}$$

If the incident light is unpolarized, the value of the angle θ will vary randomly between 0 and 360°. The average value of $\cos^2 \theta$ over this range is 1/2, so if unpolarized light incident on a polarizing filter has intensity I_0, the intensity of the light that emerges from the filter will be $I = I_0/2$.

Polarizing Light by Reflection

Another way to convert unpolarized light into light that is at least partially polarized is by *reflection*. When light strikes the boundary between two media, a fraction of the

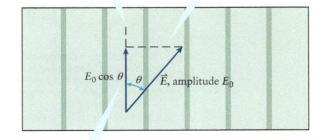

① The polarizing filter allows light polarized along this direction to pass. It blocks light polarized in the perpendicular direction.

② As a result, this component of the electric field of a light wave is blocked by the filter...

- The transmitted light is linearly polarized along the polarization direction of the filter.
- The electric field amplitude of the transmitted light equals the incident amplitude E_0 multiplied by $\cos \theta$.
- The intensity of the transmitted light equals the incident intensity I_0 multiplied by $\cos^2 \theta$.

$E_0 \cos \theta$ θ $\vec{E}$, amplitude E_0

③ ...and only this component is allowed to pass through the filter.

Figure 23-21 A linear polarizing filter II What happens to light that enters a polarizing filter with its electric field $\vec{E}$ at an angle to the polarization direction?

(a) Photographed without a filter

(b) Photographed with a polarizing filter oriented vertically, to block light with horizontal polarization

Figure 23-22 Reducing reflections with a polarizing filter Reflected light is linearly polarized. The polarizing lenses often used for sunglasses can dramatically reduce glare from reflected light.

light is reflected and the remainder is refracted. The fraction that is reflected depends not only on the incident angle of the light but also on the polarization of the incident light. In general, light polarized parallel to the boundary surface is reflected more strongly than light polarized in the perpendicular direction. So if the incident light is unpolarized, the reflected light will be at least partially polarized: It will contain more light in which the electric field is parallel to the boundary surface.

As an example, sunlight that reflects from the windshield and the hood of the car in Figure 23-22a is partially polarized parallel to the reflecting surface, so the electric field of the reflected light has a strong horizontal component and a weak vertical component. In Figure 23-22b we've placed a polarizing filter in front of the camera with its polarizing direction oriented vertically, so light with horizontal polarization is blocked. Hence the reflected light is greatly suppressed. Polarizing sunglasses use this same principle to minimize reflections from the road or the surface of a lake or ocean: Sunlight reflected from these horizontal surfaces is predominantly polarized in the horizontal direction, so to block this reflected light, the filters that make up the sunglass lenses have a vertical polarization direction. If you look at this reflected light through polarizing sunglasses and tilt your head to one side until one eye is directly above the other, you'll see the reflections reappear. That's because the polarization direction of the lenses is now horizontal, the same as the orientation of the electric field of the reflected light.

When unpolarized light is incident on the boundary between two media, there is one particular angle of incidence for which the reflected light is *completely* polarized parallel to the boundary. The angle of incidence θ_1 that results in this special condition is called **Brewster's angle θ_B**. When $\theta_1 = \theta_B$, it turns out that the sum of the incident angle θ_1 and the refracted angle θ_2 equals 90° (Figure 23-23):

$$(23\text{-}13) \qquad \theta_1 + \theta_2 = 90° \quad \text{when } \theta_1 = \theta_B$$

We can use this to solve for the value of Brewster's angle θ_B in terms of the indices of refraction n_1 and n_2 of the two media. Snell's law of refraction, Equation 23-6, gives us this relationship between θ_1 and θ_2:

$$(23\text{-}6) \qquad n_1 \sin \theta_1 = n_2 \sin \theta_2$$

If θ_1 equals Brewster's angle θ_B, Equation 23-13 tells us that

$$\theta_2 = 90° - \theta_1 = 90° - \theta_B$$

If we substitute this into Equation 23-6, we get

$$n_1 \sin \theta_B = n_2 \sin (90° - \theta_B)$$

We know from trigonometry that $\sin (90° - \theta_B) = \cos \theta_B$. So

$$n_1 \sin \theta_B = n_2 \cos \theta_B$$

Figure 23-23 Polarization by reflection Light reflected from a surface is partially polarized and becomes completely polarized when the incident angle equals Brewster's angle.

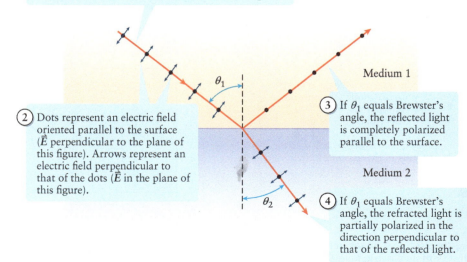

① Unpolarized light propagating in medium 1 strikes the boundary with medium 2 at an incident angle θ_1.

② Dots represent an electric field oriented parallel to the surface ($\vec{E}$ perpendicular to the plane of this figure). Arrows represent an electric field perpendicular to that of the dots ($\vec{E}$ in the plane of this figure).

③ If θ_1 equals Brewster's angle, the reflected light is completely polarized parallel to the surface.

④ If θ_1 equals Brewster's angle, the refracted light is partially polarized in the direction perpendicular to that of the reflected light.

Medium 1

Medium 2

Divide both sides of this equation by n_1, then divide both sides by $\cos \theta_B$. The result is

$$\frac{\sin \theta_B}{\cos \theta_B} = \frac{n_2}{n_1}$$

The sine of an angle divided by its cosine equals the tangent of the angle. So $\tan \theta_B = n_2/n_1$, or

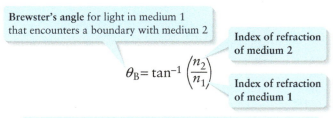

Brewster's angle for light in medium 1 that encounters a boundary with medium 2

Index of refraction of medium 2

$$\theta_B = \tan^{-1}\left(\frac{n_2}{n_1}\right)$$

Index of refraction of medium 1

Brewster's angle for polarization by reflection
(23-14)

 If the incident angle equals Brewster's angle, the reflected light is completely polarized in the direction parallel to the surface of the boundary.

 Go to Interactive Exercise 23-2 for more practice dealing with polarization.

If the angle of incidence for unpolarized light is something other than θ_B, the reflected ray is partly polarized. If unpolarized light strikes the boundary perpendicular to it, so $\theta_1 = 0$, there is no change in polarization of the reflected light.

WATCH OUT! Brewster's angle is not the same as the critical angle.

! Be careful not to confuse Equation 23-14 for Brewster's angle with the similar-looking Equation 23-10 for the critical angle. For a boundary between two transparent media, Brewster's angle (which involves an inverse *tangent* function) is the *one and only* angle of incidence for which the reflected light is completely polarized. This can happen whether the index of refraction n_1 for the first medium is larger or smaller than the index of refraction n_2 for the second medium. The critical angle (which involves an inverse *sine* function) is the *minimum* angle of incidence for which there is total internal reflection (Section 23-3); there is also total internal reflection if the angle of incidence is greater than the critical angle. Total internal reflection is possible only if the index of refraction n_1 for the first medium is greater than the index of refraction n_2 for the second medium.

EXAMPLE 23-4 Brewster's Angle for Air to Water

At what incident angle must light strike the surface of a pond so that the reflected light is completely polarized?

Set Up

This is the situation shown in Figure 23-23, with air as medium 1 and water as medium 2. The reflected light will be completely polarized if the incident angle θ_1 is equal to Brewster's angle θ_B given by Equation 23-14. Table 23-1 tells us the value of the indices of refraction: $n_1 = n_{air} = 1.00$ and $n_2 = n_{water} = 1.33$.

Brewster's angle for polarization by reflection:

$$\theta_B = \tan^{-1}\left(\frac{n_2}{n_1}\right) \qquad (23\text{-}14)$$

Solve

Use Equation 23-14 to find Brewster's angle for this situation.

With $n_1 = 1.00$ and $n_2 = 1.33$, Brewster's angle is

$$\theta_B = \tan^{-1}\left(\frac{1.33}{1.00}\right) = 53.1°$$

The reflected light will be completely polarized if the incident angle θ_1 is equal to $\theta_B = 53.1°$.

Reflect

The boundary between the two media (air and water) is the horizontal surface of the pond, so the normal to this surface is vertical. Our result tells us that if light from the sky strikes the surface at an angle of 53.1° from the vertical, the reflected light will be completely polarized. Light striking close to this angle will be strongly polarized but not 100% polarized.

You should repeat this calculation for light coming from below the water (say, from a diver's flashlight) and striking the surface of the pond from below. Can you show that in this case, the light that reflects back into the water will be completely polarized if the incident light is at an angle of 36.9° to the vertical?

GOT THE CONCEPT? 23-4 Polarizing Filters I

Two polarizing filters, *A* and *B*, are placed one behind the other with their polarization directions perpendicular to each other. A beam of unpolarized light is directed at filter *A*. What fraction of the original light intensity will remain after the light passes through both filters *A* and *B*? (a) 1/2; (b) 1/4; (c) 1/8; (d) zero; (e) none of these.

GOT THE CONCEPT? 23-5 Polarizing Filters II

Two polarizing filters, *A* and *B*, are placed one behind the other with their transmission directions perpendicular to each other. A third polarizing filter, *C*, is placed in between them. The transmission direction of *C* is halfway between those of filters *A* and *B* (that is, at 45° to that of filter *A* and at 45° to that of filter *B*). A beam of unpolarized light is directed at filter *A*. What fraction of the original light intensity will remain after the light passes through filters *A*, *B*, and *C*? (a) 1/2; (b) 1/4; (c) 1/8; (d) zero; (e) none of these.

TAKE-HOME MESSAGE FOR Section 23-5

✔ The orientation of the oscillating electric field in a light wave tells you the polarization of that wave.

✔ In natural light the direction of the electric field changes randomly and is equally likely to be in any direction perpendicular to the propagation direction of the wave. Such light is called unpolarized.

✔ Light that has its electric field oriented in one specific direction is called linearly polarized. Unpolarized light can be polarized by scattering, by passing it through a polarizing filter, or by reflection.

23-6 Light waves reflected from the surfaces of a thin film can interfere with each other, producing dazzling effects

BioMedical

Figure 23-24 Butterfly interference The iridescent colors on the wings of this *Morpho menelaus* butterfly result from the interference of light waves that reflect from the wing surfaces.

The wings of the butterfly *Morpho menelaus* (Figure 23-24) show brilliant colors. Yet the material of which the wings are made is colorless! The explanation for this seeming contradiction is that the colors are produced by the interference of light, a process similar to the interference of sound waves we investigated in Chapter 13. In Section 13-5 we saw that sound waves interfere constructively or destructively depending on how the peaks and troughs of the two waves align. The same is true for light waves.

To begin our investigation of light wave interference, let's consider what happens when light of a single wavelength, and hence a single color—called **monochromatic light**—strikes a thin layer of a transparent material. An example of such a transparent *thin film* is the *tapetum lucidum* ("shining carpet") that lines the back of some animals' eyes behind the retina (see Figure 23-1a). The material behind the *tapetum lucidum* is opaque (light cannot pass through it), so we can think of the *tapetum lucidum* as a **thin film** against an opaque backing.

As Figure 23-25a shows, some of the light waves that strike the front surface of the thin film are reflected. The remaining light enters the thin film (Figure 23-25b), strikes the opaque backing at the back surface of the film, and is reflected back up. (Some light is also absorbed by the backing.) As Figure 23-25c shows, the light waves reflected from the front surface and the light waves reflected from the back surface both end up above the thin film and traveling in the same direction.

The single light wave that was incident on the thin film has now been split into two reflected, outgoing waves that have traveled different paths. These two waves were in phase before they encountered the thin film because they were part of the same incident wave. But in general the two reflected waves are *not* in phase when they recombine above the front surface of the thin film. That's because the light that enters the film and reflects off the back surface travels farther than the light that reflects off the front surface.

Whether the interference that occurs between the two reflected waves is constructive or destructive depends on the number of wave cycles that fit into the extra distance traveled by the light that enters the film. If an integer number of cycles (1, 2, 3, . . . cycles) fit into that extra distance, then the two outgoing waves are in phase and constructively interfere. The surface of the film appears bright. If an odd number of half cycles (1/2, 3/2, 5/2, . . . cycles) fit into the extra distance, however, the two outgoing waves are 180° out of phase and destructively interfere. (We specify an *odd* number of half cycles because an even number of half cycles is the same as an integer number of full cycles; for example, 4/2 = 2. This case gives constructive, not destructive, interference.) When destructive interference occurs, the two outgoing waves cancel each other out and the surface of the film appears dark.

Let's consider the case in which the incident light is normal to the surface of the thin film (that is, the light strikes the film face-on). Then the path length difference Δ_{pl} for the two waves—one that reflects off the front of the film and the other that reflects off the back—is twice the thickness D of the film:

$$\Delta_{pl} = 2D \qquad (23\text{-}15)$$

Constructive interference occurs when an integer number of wavelengths of light fit into this path length difference of $2D$. Destructive interference occurs when an odd number of half-wavelengths fit into the distance $2D$. However, the wavelength of light inside the film is not the same as the wavelength outside the film. Recall from Section 23-2 that the wavelength of light in a medium with index of refraction n is

$$\lambda = \frac{\lambda_{vacuum}}{n} \qquad (23\text{-}8)$$

As in Figure 23-25, let's say that the material outside the film has index of refraction n_1, and the material of which the film is made has index of refraction n_2. Then Equation 23-8 tells us that the wavelengths in the two materials are

$$\lambda_1 = \frac{\lambda_{vacuum}}{n_1} \text{ outside the film}$$

$$\lambda_2 = \frac{\lambda_{vacuum}}{n_2} \text{ inside the film}$$

Comparing these two, we see that

$$\lambda_2 = \frac{n_1 \lambda_1}{n_2} \qquad (23\text{-}16)$$

We now know everything we need to determine how the light reflected from the front surface of the thin film interferes with the light that enters the film and is reflected from the back surface. *Constructive* interference occurs when the path difference equals an integer number of wavelengths λ_2 (the wavelength of light inside the film):

$$2D = m\lambda_2 = m\frac{n_1 \lambda_1}{n_2}, \quad m = 1, 2, 3, \ldots \qquad (23\text{-}17)$$

(constructive interference, thin film with an opaque backing)

If we set $m = 1$ in Equation 23-17 and solve for D, we get the *minimum* thickness that a thin film must have to give constructive interference when light of wavelength λ_1 strikes the front surface face-on:

$$D_{min} = \frac{n_1 \lambda_1}{2n_2} \qquad (23\text{-}18)$$

(minimum thickness for constructive interference, thin film with an opaque backing)

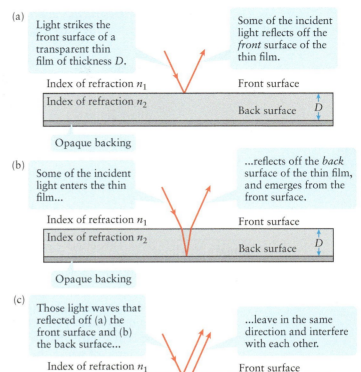

(a) Light strikes the front surface of a transparent thin film of thickness D.

Some of the incident light reflects off the *front* surface of the thin film.

Index of refraction n_1 — Front surface

Index of refraction n_2 — Back surface D

Opaque backing

(b) Some of the incident light enters the thin film...

...reflects off the *back* surface of the thin film, and emerges from the front surface.

Index of refraction n_1 — Front surface

Index of refraction n_2 — Back surface D

Opaque backing

(c) Those light waves that reflected off (a) the front surface and (b) the back surface...

...leave in the same direction and interfere with each other.

Index of refraction n_1 — Front surface

Index of refraction n_2 — Back surface D

Opaque backing

Figure 23-25 **Interference from a thin film I** For clarity in this figure we've drawn light rays that reach the surface of a thin film at an angle. In our calculations we'll assume that the light hits the surface face-on.

Destructive interference occurs when the path difference equals an odd number of one-half the wavelength λ_2 inside the film:

(23-19)
$$2D = (2m-1)\frac{\lambda_2}{2} = (2m-1)\frac{n_1\lambda_1}{2n_2}, \quad m = 1,\, 2,\, 3,\, \ldots$$

(destructive interference, thin film with an opaque backing)

If we set $m = 1$ in Equation 23-19 and solve for D, we get the minimum thickness that a thin film must have to give destructive interference when light of wavelength λ_1 strikes the front surface face-on:

(23-20)
$$D_{\text{min}} = \frac{n_1\lambda_1}{4n_2}$$

(minimum thickness for destructive interference, thin film with an opaque backing)

This effect explains why the *tapetum lucidum* of an animal's eye improves the animal's ability to see in low light levels, and why this ability is greatest at certain wavelengths. The retina detects light by absorbing it, but some light passes through the retina without being absorbed. The *tapetum lucidum* behind the retina reflects some of this "lost" light back into the retina so it can have a second chance to be absorbed and detected. This reflection happens most strongly for certain wavelengths—to be specific, those that satisfy Equation 23-17, where n_2 is the index of refraction of the material of which the *tapetum lucidum* is made. Other colors with other wavelengths do not satisfy that relationship perfectly, so they are not reflected as strongly. Hence the animal's eye is less sensitive to light at these other wavelengths.

Even after a second pass through the retina, some of the light reflected from the *tapetum lucidum* escapes from the retina and exits the eye. This gives rise to the phenomenon of *eyeshine* that you can see in Figure 23-1a. The *tapetum lucidum* for the cat in that photo preferentially reflects green light because an integer number of wavelengths of that color fit into the path difference (twice the thickness of the *tapetum lucidum*). This explains why the cat's eye is most sensitive to green light.

The same effect explains the colors of the *Morpho* butterfly shown in Figure 23-24. The wings of *Morpho* are covered with microscopic scales that act like a thin film. The thickness of these scales is such that there is constructive interference for reflected light at wavelengths in the blue-green part of the spectrum. As a result, *Morpho* appears to glow at those wavelengths.

→ *Go to Interactive Exercise 23-3 for more practice dealing with films.*

EXAMPLE 23-5 A Soapy Film

Monochromatic light that has wavelength 560 nm in air strikes a layer of soapy water, which has an index of refraction of 1.40 and rests on a bathroom tile. (a) If the layer of soapy water is 700 nm thick, does constructive interference, destructive interference, or neither occur when the light strikes the surface close to the normal? (b) What is the minimum thickness of soapy water that would result in no (or minimum) reflection from the surface?

Set Up

The situation is the same as in Figure 23-25. Interference occurs between (i) light that is reflected from the top surface of the soapy layer and (ii) light that enters the soapy layer and is eventually reflected back out. In part (a) we'll compare the given values of wavelength and film thickness to Equations 23-17 and 23-19 to decide whether the interference is constructive, destructive, or something in between. In part (b) we'll use Equation 23-20 to find the minimum thickness for destructive interference. In both parts, Equation 23-16 will help us relate the wavelength of the light in air to its wavelength in the soapy water.

Constructive interference:
$$2D = m\lambda_2 = m\frac{n_1\lambda_1}{n_2}, \quad m = 1, 2, 3, \ldots \tag{23-17}$$

Destructive interference:
$$2D = (2m-1)\frac{\lambda_2}{2} = (2m-1)\frac{n_1\lambda_1}{2n_2}, \quad m = 1, 2, 3, \ldots \tag{23-19}$$

Minimum thickness for destructive interference:
$$D_{\text{min}} = \frac{n_1\lambda_1}{4n_2} \tag{23-20}$$

Wavelength in two different media:
$$\lambda_2 = \frac{n_1\lambda_1}{n_2} \tag{23-16}$$

Solve

(a) In this situation medium 1 is air ($n_1 = 1.00$) and medium 2, of which the film is made, is soapy water ($n_2 = 1.40$). The wavelength in the soapy water is therefore shorter than in air.

Wavelength in air: $\lambda_1 = 560$ nm

Wavelength in soapy water:

$$\lambda_2 = \frac{n_1 \lambda_1}{n_2} = \frac{(1.00)(560 \text{ nm})}{1.40} = 400 \text{ nm}$$

Find how many wavelengths of the wavelength λ_2 in the soapy water fit into the path length difference $\Delta_{pl} = 2D$, where $D = 700$ nm is the film thickness.

The path length difference between the waves that reflect off the top and bottom surfaces of the film is

$$\Delta_{pl} = 2D = 2(700 \text{ nm}) = 1400 \text{ nm}$$

The number of wavelengths that fit into this path length difference equals Δ_{pl} divided by λ_2:

$$\frac{\Delta_{pl}}{\lambda_2} = \frac{1400 \text{ nm}}{400 \text{ nm}} = 3.5 = \frac{7}{2}$$

The path length difference is an odd number of half-wavelengths. So there is destructive interference between the light that reflects from the top surface of the soapy water and the light that reflects from the bottom surface (where the soapy water touches the tile).

(b) The minimum thickness required for destructive interference is such that the path length difference $2D$ is one half-wavelength, so $2D = \lambda_2/2$ and $D = \lambda_2/4$.

From Equation 23-20, the minimum thickness for destructive interference is

$$D_{min} = \frac{n_1 \lambda_1}{4n_2} = \frac{(1.00)(560 \text{ nm})}{4(1.40)} = 100 \text{ nm}$$

Alternatively, since $\lambda_2 = n_1 \lambda_1/n_2$ from Equation 23-16,

$$D_{min} = \frac{\lambda_2}{4} = \frac{400 \text{ nm}}{4} = 100 \text{ nm}$$

Reflect

When light of wavelength 560 nm in air strikes the soapy layer close to the normal, destructive interference occurs both when the layer is 100 nm thick (so twice the thickness is 200 nm, or 1/2 the wavelength in the soapy water) and when the layer is 700 nm thick (so twice the thickness is 1400 nm, or 7/2 the wavelength in the soapy water). For these thicknesses, reflections from the surface would be minimized, and the surface would look dark.

Destructive interference always occurs when twice the thickness of the layer is an odd multiple of one-half of the wavelength. Can you see that this would also occur if the film of soapy water were either 300 or 500 nm thick?

Figure 23-1c shows another example of thin-film interference. This film was made by dipping an open ring into soapy water and then holding the ring vertically. Some of the light that strikes the soap film is reflected from the front surface, while some passes into the film before being reflected at the back surface. When these two light waves recombine, the wavelength of light (the color) that results in constructive interference appears bright. The thickness of the film increases from top to bottom, so the path difference for the two waves is different at different places on the film. That's why the brightest color you see (corresponding to the wavelength for which there is constructive interference) is different at different positions. The thickness of the film is relatively constant *across* the film, however, so the colors appear in bands.

Note that the very top of the soap film in Figure 23-1c appears dark. That may come as a surprise because the top of the film is very thin (far thinner than a wavelength of visible light), so the path difference between light that reflects from the front and back surfaces should be negligible. As a result, there should be constructive interference at the top of the film, and the top should appear bright rather than dark. Why is our prediction incorrect?

To see the explanation, we need to go back to our discussion in Chapter 13 of how waves are reflected from a boundary. In Section 13-6 we saw that a wave pulse on a rope is inverted as it reflects from a fixed boundary. This inversion is equivalent to a phase shift of one-half of a wavelength; the position along the wave that arrived at the boundary as a peak has been reflected as a trough. In general, a wave traveling in one medium is inverted when it reflects (either partially or completely) from a second medium in which the wave speed is lower. (For the rope, the wave speed is zero on the other side of the boundary,

(a) Thin film with an opaque backing

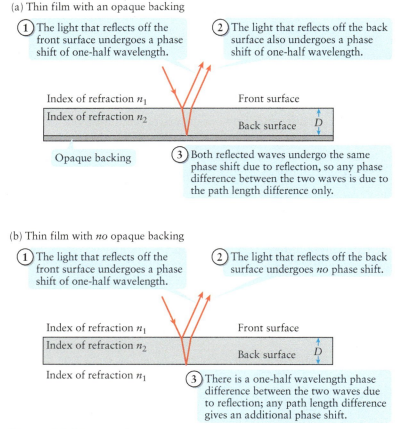

① The light that reflects off the front surface undergoes a phase shift of one-half wavelength.

② The light that reflects off the back surface also undergoes a phase shift of one-half wavelength.

③ Both reflected waves undergo the same phase shift due to reflection, so any phase difference between the two waves is due to the path length difference only.

(b) Thin film with *no* opaque backing

① The light that reflects off the front surface undergoes a phase shift of one-half wavelength.

② The light that reflects off the back surface undergoes *no* phase shift.

③ There is a one-half wavelength phase difference between the two waves due to reflection; any path length difference gives an additional phase shift.

Figure 23-26 **Interference from a thin film II** Phase shifts can occur when light reflects from the surfaces of a thin film.

which is definitely lower than the speed along the rope.) An inversion (a one-half wavelength phase shift) happens to the light waves in **Figure 23-26a** when light waves reflect off the front surface of the thin film because light travels slower in the film of index of refraction $n_2 > n_1$. The same phase shift happens to the light waves that reflect off the opaque background at the back surface of the thin film. Since both waves undergo the same phase shift due to reflection, any phase difference between the two waves is due to the path difference, as we discussed above.

But as shown in **Figure 23-26b**, things are different for the soap film in Figure 23-1c. Again there is a one-half wavelength phase shift for waves that reflect off the front surface of the film. But there is *no* phase shift for waves that reflect off the back surface, since light waves travel *faster* in the medium on the other side of that boundary. So even if the film were extremely thin, so there were no path difference between the light waves that reflect off the front and back surfaces of the film, there would still be a one-half wavelength phase difference between these two waves. This would result in destructive interference. That's just what we see at the top of the soap film in Figure 23-1c, where the film is very thin compared to the wavelength of the light. This explains the dark band across the top of the film.

Thanks to the additional half-wavelength phase shift for a film like that in Figure 23-1c and Figure 23-26b, Equation 23-17 no longer tells us the condition for constructive interference, and Equation 23-19 no longer tells us the conditions for destructive interference. Instead the roles of these equations are reversed: Equation 23-17 is now the condition for *destructive* interference, and Equation 23-19 is now the condition for *constructive* interference.

EXAMPLE 23-6 Reducing the Reflection

The glass in an LCD display is sometimes coated with a transparent thin film to minimize glare. Interference from the front and back surfaces of such an *antireflective coating* minimizes the amount of light reflected from the display, particularly light that strikes normal to the display's surface. One material used in such coatings is zirconium acrylate, which has an index of refraction of 1.54. What is the minimum thickness of zirconium acrylate that will accomplish the desired reduction in reflection for light that has wavelength 560 nm in air? Note that the index of refraction of zirconium acrylate is higher than that of glass.

Set Up

We want the light reflected from the surface of the coating to interfere destructively with the light reflected from the zirconium acrylate-to-glass boundary. This requires that when these two light waves recombine, they are shifted by an odd number of half-wavelengths. One half-wavelength shift occurs because the light in air is inverted when it reflects off the surface of the zirconium acrylate, but the light in zirconium acrylate undergoes no shift when it reflects off the glass. (That's because light travels more slowly in zirconium acrylate than in air, but faster in glass than in zirconium acrylate.) Any additional shift arises from the path difference of $2D$ between the two waves. For the total shift to be equivalent to an odd number of half-wavelengths, the shift due to the path difference must be an *integer* number of wavelengths. The minimum thickness corresponds to $2D$ equal to one wavelength λ_2, the wavelength in the zirconium acrylate as given by Equation 23-16.

Wavelength in two different media:

$$\lambda_2 = \frac{n_1 \lambda_1}{n_2} \quad (23\text{-}16)$$

Solve

First calculate the wavelength of the light in zirconium acrylate.

We are given the wavelength in air ($n_1 = 1.00$): $\lambda_1 = 560$ nm. The wavelength in zirconium acrylate ($n_2 = 1.54$) is given by Equation 23-16:

$$\lambda_2 = \frac{n_1 \lambda_1}{n_2} = \frac{(1.00)(560 \text{ nm})}{1.54} = 364 \text{ nm}$$

The minimum film thickness D_{min} for destructive interference is such that $2D_{min}$ equals λ_2. Use this to solve for D_{min}.

The condition for the minimum thickness that leads to destructive interference is

$$2D_{min} = \lambda_2 = 364 \text{ nm}$$

$$D_{min} = \frac{364 \text{ nm}}{2} = 182 \text{ nm}$$

Reflect

For most people light sensitivity peaks in the range of 555 to 565 nm, which we perceive as yellow. That's why antireflective coatings are usually optimized for yellow light (which includes the 560-nm wavelength we've used here).

Why have we been emphasizing interference due to light reflecting from a *thin* film? The explanation is that in this section we've assumed that a steady, continuous train of light waves is incident on the film. However, the light from ordinary sources such as light bulbs and the Sun is emitted in a sequence of short bursts, each of which is a segment of wave no more than a few micrometers to about a millimeter in length. This is called the *coherence length* of the light. The phase of the wave changes randomly from one burst to the next. If the thickness of the film is small compared to the coherence length, then the two waves that interfere—the light that reflects from the back of the film and the light that reflects from the front of the film—are part of the same burst. In this case the phase relationships we have developed in this section (which assumed a steady train of waves) are valid. But if the thickness of the film is large compared to the coherence length, the two waves are likely to be from different wave bursts and will differ in phase by a random and rapidly changing amount. As a result, the interference between the waves will be neither always constructive nor always destructive, and any interference effects will be wiped out. That's why you won't see interference effects like those we've described in this section from a thick film like an ordinary pane of glass, which is several millimeters deep. (However, you *can* see interference effects for a glass pane if the light source is a laser. A laser produces light in a very different way from an ordinary light bulb, and the coherence length can be several meters.)

GOT THE CONCEPT? 23-6 Inversion on Reflection

(?) Figure 23-27 shows light shining on a thin layer of oil that has an index of refraction of 1.4. The oil layer is atop a piece of glass that has an index of refraction of 1.5. On the underside of the glass is air. At which boundary or boundaries does the reflected light undergo an inversion? (a) The air–oil boundary; (b) the oil–glass boundary; (c) the glass–air boundary; (d) more than one of these; (e) none of these.

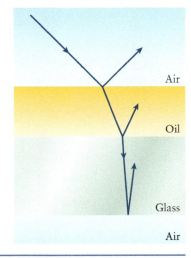

Figure 23-27 Layers of reflection At which boundary or boundaries does the reflected light undergo an inversion?

23-7 Interference can occur when light passes through two narrow, parallel slits

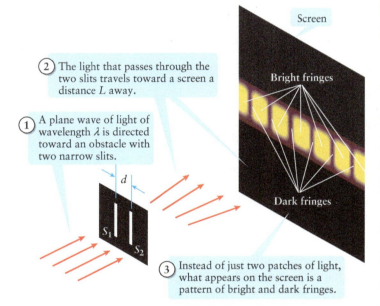

Figure 23-28 The double-slit experiment The light waves that pass through slits S_1 and S_2 produce a pattern of bright and dark fringes.

② The light that passes through the two slits travels toward a screen a distance L away.

① A plane wave of light of wavelength λ is directed toward an obstacle with two narrow slits.

③ Instead of just two patches of light, what appears on the screen is a pattern of bright and dark fringes.

Figure 23-28 shows a remarkable experiment into the properties of light. Light from a laser of a single wavelength λ (that is, monochromatic light) is directed at an opaque obstacle that has two narrow slits, S_1 and S_2, separated by a distance d. The light that emerges from the slits falls on a screen a distance L away.

You might expect that there would be just *two* bright patches on the screen, one produced by the light that passed through slit S_1 and one produced by the light that passed through slit S_2. But in fact what we see is a *series* of bright patches with dark patches between them, collectively called **fringes**. We can explain this curious result using two key ideas about light waves: Huygens' principle and interference.

According to Huygens' principle (Section 23-2), each point on the incident plane wave that strikes the slits in Figure 23-28 acts as a source of wavelets. Since the two slits S_1 and S_2 are very narrow, we can treat each slit as a *single* source of this kind. Each crest of the incident plane wave strikes both slits simultaneously, so the wavelets of light that emanate from each slit start off in phase with each other (Figure 23-29). But because the wavelets travel different distances to reach various locations on the screen, they may or may not arrive at the screen in phase. There are three possibilities:

(1) At certain locations on the screen the two wavelets arrive in phase with each other, so crests from S_1 and S_2 arrive at the same time. So at these locations there is constructive interference, the total wave has maximum amplitude, and the light

Figure 23-29 Huygens' principle and double-slit interference The pattern of bright and dark fringes arises from interference between wavelets emanating from slit S_1 and from slit S_2.

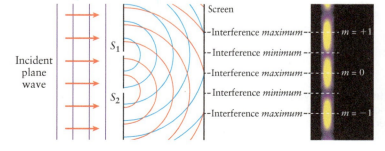

② Wavelets emanate from slit S_1 (shown in blue) and from slit S_2 (shown in red).

① Successive crests of a plane wave of light are incident on the two slits S_1 and S_2.

③ Where wavelets from the two slits arrive at the screen *in phase*, there is an interference *maximum* (a *bright* fringe).

④ Where wavelets from the two slits arrive at the screen *out of phase*, there is an interference *minimum* (a *dark* fringe).

reaching the screen is brightest. These **interference maxima** correspond to the centers of the bright fringes shown in Figures 23-28 and 23-29.

(2) At certain other locations on the screen, the two wavelets arrive out of phase, so a crest from S_1 arrives at the same time as a trough from S_2 and vice versa. At these locations there is destructive interference, the total wave is zero, and no light reaches the screen. The centers of the dark fringes shown in Figure 23-28 and 23-29 correspond to these **interference minima**.

(3) At all other locations on the screen, the interference between the wavelets from S_1 and S_2 is neither completely constructive nor completely destructive. At these locations a crest from S_1 does not arrive at the same time as either a crest or a trough from S_2, so the total wave has neither maximum amplitude nor zero amplitude. These are the locations in Figures 23-28 and 23-29 between the center of a bright fringe and the center of a dark fringe.

Because the wavelets that interfere with each other come from a pair of slits, the experiment shown in Figures 23-28 and 23-29 is called **double-slit interference** and the pattern of bright and dark fringes is called an **interference pattern**. Such interference patterns are conclusive evidence that light is a wave. (In Section 13-5 we discussed the same sort of interference that occurs when there are water waves or sound waves produced by two sources, analogous to the light waves emanating from the two slits in Figures 23-28 and 23-29.)

You may be wondering why it's important to use a laser to demonstrate double-slit interference. The reason is that the waves that enter slit S_1 and the waves that enter slit S_2 must have a constant phase relationship with each other. As we mentioned at the end of Section 23-6, laser light has a high degree of coherence, so the waves entering the two closely spaced slits are guaranteed to have such a constant relationship. (In Figure 23-29 the relationship is that the two waves are in phase.) If you were to use an ordinary light bulb with a color filter to make the light monochromatic before entering the slits, the waves entering slits S_1 and S_2 would be from different short bursts emitted by the light bulb. The phase relationship between these bursts would change from one moment to the next. Hence there would be no definite interference pattern, and on the screen in Figure 23-28 you would see two closely spaced bright spots, one caused by light from slit S_1 and one from slit S_2.

Locating the Interference Maxima and Minima

Let's see how to determine the locations of the double-slit interference maxima and minima on the screen in Figures 23-28 and 23-29. Just as for the thin-film interference that we studied in Section 23-6, what determines whether the interference is completely constructive, completely destructive, or intermediate between those two extremes is the *path difference* $\Delta_{pl} = D_2 - D_1$, the difference between the distance D_1 that a wave from slit S_1 travels to a point P on the screen and the distance D_2 that a wave from slit S_2 travels to that same point (Figure 23-30a). The conditions for constructive and destructive interference are

Constructive interference:

$$\Delta_{pl} = \text{a whole number of wavelengths} = 0, \pm\lambda, \pm2\lambda, \pm3\lambda, \ldots$$

Destructive interference:

$$\Delta_{pl} = \text{an odd number of half-wavelengths} = \pm\lambda/2, \pm3\lambda/2, \pm5\lambda/2, \ldots \qquad \text{(23-21)}$$

Note that the path difference Δ_{pl} is positive for locations in the upper half of the screen. These locations are farther from slit S_2 than from slit S_1, so D_2 is greater than D_1 (the case shown in Figure 23-29). Locations in the lower half of the screen are closer to slit S_2 than to slit S_1, so for these locations D_2 is less than D_1 and the path difference Δ_{pl} is negative.

In typical double-slit experiments the spacing d between the slits is about a millimeter and the distance L from the slits to the screen is a meter or more. Since L is so large compared to d, it's a good approximation to treat the screen as being infinitely far away. Then the straight lines from slit S_1 to the point P on the screen and from slit S_2 to point P are parallel to each other, and are both at an angle θ from the normal to the opaque obstacle with the two slits (Figure 23-30b). The path difference Δ_{pl} is then

Figure 23-30 Calculating the path difference for double-slit interference (a) The path difference Δ_{pl} for a point P on the screen. (b) If the distance to the screen is large compared to the distance d between the slits, Δ_{pl} is related to d and the angle θ to the point P.

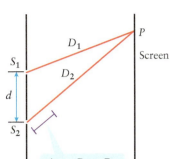

(a) A nearby screen.

$\Delta_{pl} = D_2 - D_1$

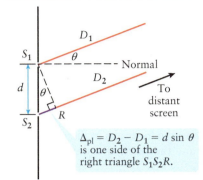

(b) If the screen is very distant, the lines from S_1 to the screen and S_2 to the screen are essentially parallel.

$\Delta_{pl} = D_2 - D_1 = d \sin \theta$ is one side of the right triangle $S_1 S_2 R$.

one side of a right triangle $S_1 S_2 R$ with hypotenuse d. The angle opposite the side of length Δ_{pl} is θ, and the sine of θ equals the length of this opposite side divided by the length d of the hypotenuse:

$$(23\text{-}22) \qquad \sin \theta = \frac{\Delta_{pl}}{d} \quad \text{so} \quad d \sin \theta = \Delta_{pl}$$

If we combine Equations 23-21 and 23-22, we get the following conditions for a bright fringe (constructive interference) and for a dark fringe (destructive interference):

Double-slit experiment: Condition for constructive interference
(23-23)

> Distance between the two slits

> Wavelength of the light illuminating the two slits

Constructive interference (bright fringes): $\quad d \sin \theta = m\lambda$

> Angle between the normal to the two slits and the location of a **bright fringe** on the screen

> Number of the bright fringe: $m = 0, \pm 1, \pm 2, \pm 3,\ldots$

Double-slit experiment: Condition for destructive interference
(23-24)

> Distance between the two slits

> Wavelength of the light illuminating the two slits

Destructive interference (dark fringes): $\quad d \sin \theta = \left(m + \frac{1}{2}\right) \lambda$

> Angle between the normal to the two slits and the location of a **dark fringe** on the screen

> Number of the dark fringe: $m = 0, \pm 1, \pm 2, \pm 3,\ldots$

Equation 23-23 gives the angles θ at which the bright fringes are found. Each of the bright fringes in Figure 23-29 is labeled with the corresponding values of m. For example, for the $m = 0$ bright fringe it follows that $d \sin \theta = 0$, so $\sin \theta = 0$ and $\theta = 0$. Hence the $m = 0$ bright fringe is at the center of the pattern. For the other bright fringes, $\sin \theta = m\lambda/d = +\lambda/d$ (for $m = +1$), $-\lambda/d$ (for $m = -1$), $+2\lambda/d$ (for $m = +2$), $-2\lambda/d$ (for $m = -2$), ...

WATCH OUT! Be careful with numbering the dark fringes.

! Equation 23-24 gives the angles θ at which the dark fringes are found. Note that $m = 0$ in this equation does *not* correspond to the center of the interference pattern (where there is a bright fringe). Instead this is a dark fringe given by $d \sin \theta = (1/2)\lambda$, so $\sin \theta = \lambda/(2d)$. This dark fringe lies between the $m = 0$ bright fringe and the $m = +1$ bright

fringe shown in Figure 23-29. For the $m = +1$ dark fringe, $\sin \theta = (1 + \frac{1}{2})\lambda/d = +3\lambda/(2d)$; this dark fringe lies between the $m = +1$ and $m = +2$ bright fringes. For the $m = -1$ dark fringe, $\sin \theta = (-1 + \frac{1}{2})\lambda/d = -\lambda/(2d)$; this dark fringe lies between the $m = 0$ and $m = -1$ bright fringes.

If we divide both sides of Equation 23-23 and both sides of Equation 23-24 by d, we see that that the value of $\sin\theta$ for either a given bright fringe or a given dark fringe is proportional to λ/d, the ratio of the wavelength to the slit spacing:

Constructive interference (bright fringe): $\sin\theta = m\dfrac{\lambda}{d}$

Destructive interference (dark fringe): $\sin\theta = \left(m + \dfrac{1}{2}\right)\dfrac{\lambda}{d}$ **(23-25)**

Equations 23-25 show that if we increase the wavelength λ, the value of $\sin\theta$ and hence of θ for each fringe increases. This means that the entire interference pattern becomes broader. If instead we increase the slit spacing d, the value of $\sin\theta$ and hence of θ for each fringe decreases. In this case the entire interference pattern becomes narrower.

EXAMPLE 23-7   Measuring Wavelength Using Double-Slit Interference

You send light from a laser through two narrow slits spaced 0.100 mm apart, producing an interference pattern on a screen 5.00 m away. The angle from the central bright fringe to the next bright fringe is 0.300°. (a) What is the wavelength of the light? (b) What is the distance on the screen from the center of the interference pattern to the first dark fringe?

Set Up

We are given the slit spacing $d = 0.100$ mm and the angle $\theta = 0.300°$ for the first bright fringe, which corresponds to $m = 1$ in Equation 23-23. In part (a) we'll use this equation to solve for the wavelength λ. Given the wavelength, in part (b) we'll find the angle of the first dark fringe using Equation 23-24 with $m = 0$. We'll then use trigonometry to find the desired distance.

Bright fringes: $d\sin\theta = m\lambda$ (23-23)

Dark fringes: $d\sin\theta = (m + \frac{1}{2})\lambda$ (23-24)

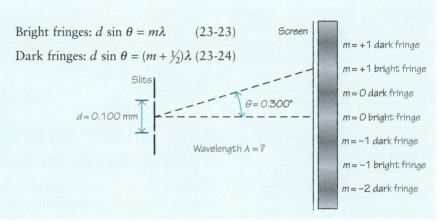

Solve

(a) Solve Equation 23-23 for the wavelength λ.

We have $d = 0.100$ m $= 1.00 \times 10^{-4}$ m for the slit spacing, and $\theta = 0.300°$ for the angle of the first bright fringe away from the central bright fringe. From Equation 23-23 with $m = 1$,

$$\lambda = d\sin\theta = (1.00 \times 10^{-4}\ \text{m})\sin 0.300°$$
$$= (1.00 \times 10^{-4}\ \text{m})(5.24 \times 10^{-3})$$
$$= 5.24 \times 10^{-7}\ \text{m} = 524\ \text{nm}$$

This is in the green part of the visible spectrum.

(b) Use Equation 23-24 to find the angle of the first dark fringe.

The first dark fringe corresponds to $m = 0$ in Equation 23-24. Solve this for θ:

$$d\sin\theta = (0 + \tfrac{1}{2})\lambda = \lambda/2$$

$$\sin\theta = \frac{\lambda}{2d} = \frac{5.24 \times 10^{-7}\ \text{m}}{2(1.00 \times 10^{-4}\ \text{m})}$$
$$= 2.62 \times 10^{-3}$$
$$\theta = \sin^{-1}(2.62 \times 10^{-3}) = 0.150°$$

The point P on the screen where the first dark fringe lies, the center of the interference pattern C, and the point M midway between the two slits define a right triangle. The side of this triangle adjacent to the angle $\theta = 0.150°$ is the distance $L = 5.00$ m from the slits to the screen, and the side opposite this angle is the distance D that we want.

From trigonometry,

$$\tan \theta = \frac{\text{opposite side}}{\text{adjacent side}} = \frac{D}{L}$$

Solve for the distance D from the center of the interference pattern to the first dark fringe:

$$D = L \tan \theta = (5.00 \text{ m}) \tan 0.150°$$
$$= (5.00 \text{ m})(2.62 \times 10^{-3}) = 0.0131 \text{ m} = 1.31 \text{ cm}$$

Reflect

Part (a) shows how a double-slit interference experiment can be used to measure the wavelength of light. Note from part (b) that to three significant digits, the angle of the first dark fringe from the center of the interference pattern is one-half the corresponding angle for the first bright fringe. This agrees with Figure 23-28, which shows that the bright and dark fringes are equally spaced.

GOT THE CONCEPT? 23-7 Modifying a Double-Slit Interference Experiment

(?) In a certain double-slit experiment the $m = 3$ bright fringe is at an angle of 0.500° from the central bright fringe. Which of the following changes would cause the $m = 2$ bright fringe to appear at this angle instead? (There may be more than one correct answer.) (a) Increasing the slit spacing by a factor of 3/2 while leaving the wavelength unchanged; (b) decreasing the slit spacing by a factor of 2/3 while leaving the wavelength unchanged; (c) increasing the wavelength by a factor of 3/2 while leaving the slit spacing unchanged; (d) decreasing the wavelength by a factor of 2/3 while leaving the slit spacing unchanged; (e) doubling the wavelength and simultaneously decreasing the slit spacing by a factor of 1/3.

TAKE-HOME MESSAGE FOR Section 23-7

✔ When light waves strike a pair of narrow slits, the waves that emerge from the two slits give rise to an interference pattern with bright and dark fringes.

✔ The positions of the fringes are determined by the ratio of the slit spacing to the wavelength.

23-8 Diffraction is the spreading of light when it passes through a narrow opening

In Section 23-7 we used Huygens' principle to understand the bright and dark interference fringes formed when light passes through two narrow slits. Huygens' principle will also allow us to understand **diffraction**, in which waves tend to spread out when they pass through a narrow opening or near the sharp edge of an object. We will see that diffraction can also give rise to a pattern of bright and dark fringes.

Consider a wave front of a plane wave that passes through an opening in an obstacle, as in Figure 23-31a. The opening is wide compared to the wavelength of the wave. Many of the wavelets pass through the opening, as in Figure 23-31b. The resulting wave on the other side of the obstacle is mostly a new plane wave, though the ends of the wave front are curved. This means that most of the wave energy continues straight through the opening, with only a small fraction "leaking" to the sides.

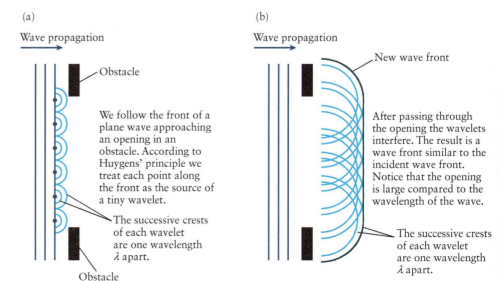

(a)

Wave propagation →

Obstacle

We follow the front of a plane wave approaching an opening in an obstacle. According to Huygens' principle we treat each point along the front as the source of a tiny wavelet.

The successive crests of each wavelet are one wavelength λ apart.

Obstacle

(b)

Wave propagation →

New wave front

After passing through the opening the wavelets interfere. The result is a wave front similar to the incident wave front. Notice that the opening is large compared to the wavelength of the wave.

The successive crests of each wavelet are one wavelength λ apart.

Figure 23-31 Diffraction I: A wide opening Huygens' principle predicts that if the width of the opening is large compared to the wavelength, most of the wave continues straight ahead through the opening. A slight amount diffracts to the sides.

(a)

Wave propagation

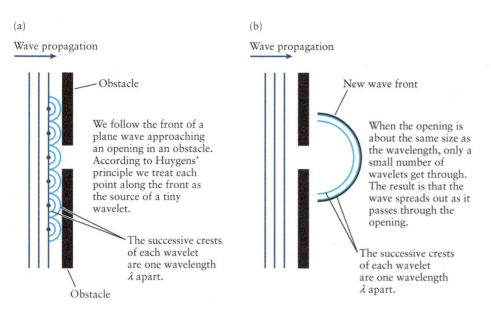

Obstacle

We follow the front of a plane wave approaching an opening in an obstacle. According to Huygens' principle we treat each point along the front as the source of a tiny wavelet.

The successive crests of each wavelet are one wavelength λ apart.

Obstacle

(b)

Wave propagation

New wave front

When the opening is about the same size as the wavelength, only a small number of wavelets get through. The result is that the wave spreads out as it passes through the opening.

The successive crests of each wavelet are one wavelength λ apart.

Figure 23-32 Diffraction II: A narrow opening If the width of the opening is comparable to the wavelength, the wave spreads out substantially after exiting the opening. (Compare Figure 23-31.)

Something rather different happens when the opening is comparable in size to or smaller than the wavelength. That's the situation in Figure 23-32a, in which the opening is much narrower than in Figure 23-31a. Because the opening is narrow, only a very few of the wavelets that comprise each wave front pass through it. And because so few wavelets get through, they cannot reproduce the straight wave front that was incident on the opening. After passing through the narrow opening, the wave spreads out; it has undergone diffraction (Figure 23-32b). Diffraction is also present in Figure 23-31b, as shown by the waves that "leak" to the side of the much wider opening, but to a much smaller greatly reduced extent.

Figure 23-33 shows these effects for water waves passing through an opening in an obstacle. If the width of the opening is large compared to the wavelength, as in Figure 23-33a, there are almost no effects of diffraction. But if the opening is comparable in size to the wavelength as in Figure 23-33b, the wave spreads out. This explains why you hear someone talking on the other side of an open door, even if you're not directly in front of it. The wavelengths of sound used in human speech are in the range of a few meters, comparable in size to the width of a typical door (about one meter).

(a) Water waves pass through a wide opening.

(b) Water waves pass through a narrow opening.

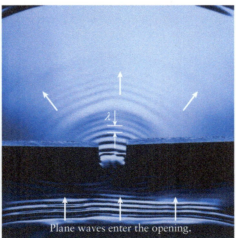

Plane waves enter the opening.

Plane waves enter the opening.

 • The width of the opening is large compared to the wavelength λ.
• Most of the wave energy continues straight ahead, so there is very little diffraction.

 • The width of the opening is comparable in size to the wavelength λ.
• The wave spreads out after passing through the opening; diffraction is important.

Figure 23-33 Water wave diffraction Compare these photographs to Figures 23-31 and 23-32.

Figure 23-34 Diffraction through a single slit A diffraction pattern arises when monochromatic light passes through a narrow slit.

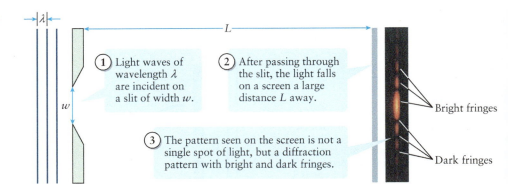

Figure 23-34 Diffraction through a single slit A diffraction pattern arises when monochromatic light passes through a narrow slit.

① Light waves of wavelength λ are incident on a slit of width w.

② After passing through the slit, the light falls on a screen a large distance L away.

③ The pattern seen on the screen is not a single spot of light, but a diffraction pattern with bright and dark fringes.

Bright fringes

Dark fringes

As a result sound waves emerging from a door spread out like the water waves in Figure 23-33b, making it easy to eavesdrop on conversations. However, you can't *see* through an open door if you stand to one side because the wavelengths of visible light (around 550 nm, or 5.5×10^{-7} m) are very small compared to the width of the door. So light waves passing through the door behave like the water waves in Figure 23-33a.

To consider diffraction in more detail, imagine that we send a beam of light of a single wavelength λ through a long, narrow slit of width w. After passing through the slit the light falls on a screen a distance L away (Figure 23-34). You might expect that the pattern on the screen would be a single blob of light. But as the photo in Figure 23-34 shows, what actually appears is a series of bright and dark fringes called a **diffraction pattern**. In Section 23-7 we saw how such fringes arise in the *double*-slit experiment due to interference between Huygens wavelets emanating from the first slit and wavelets emanating from the second slit. In this **single-slit diffraction** experiment, the fringes arise due to interference between wavelets emanating from different parts of the *same* slit. Huygens wavelets emanate from each part of the slit. At some locations, these wavelets interfere destructively; these are the locations of the dark fringes, also called **diffraction minima**. At locations between the dark fringes, the wavelets interfere more or less constructively and we see bright fringes, also called **diffraction maxima**.

Let's see how to determine the positions of the dark fringes. The first dark fringe is found where light waves from the upper half of the slit, of width $w/2$, and the lower half, also of width $w/2$, interfere destructively. This means that the waves from one half of the slit must travel one half-wavelength farther from the slit to the screen than do the waves from the other half, so the path length difference Δ_{pl} equals $\lambda/2$. We can use this to easily determine the position of the first dark fringe if we assume that the distance L to the screen is much greater than the slit width w. With this assumption, Figure 23-35 shows that for a given position on the screen, Δ_{pl} is related to the angle θ between the normal to the slit and a line to that position:

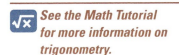

See the Math Tutorial for more information on trigonometry.

$$(23\text{-}26) \qquad \Delta_{pl} = \left(\frac{w}{2}\right)\sin\theta$$

To find the angle for the first dark fringe, substitute $\Delta_{pl} = \lambda/2$ (the condition for destructive interference) into Equation 23-26:

$$\frac{\lambda}{2} = \left(\frac{w}{2}\right)\sin\theta$$

so

$$(23\text{-}27) \qquad \sin\theta = \frac{\lambda}{w}$$

(first dark fringe in single-slit diffraction)

There are two of these first dark fringes, one on either side of the central bright fringe (see Figure 23-35).

To see how the second dark fringe arises, imagine breaking the slit into quarters, each of width $w/4$. Destructive interference occurs when light from the first quarter arrives at the screen out of phase with light from the second quarter, and light from the third quarter arrives at the screen out of phase with light from the fourth quarter. The path length

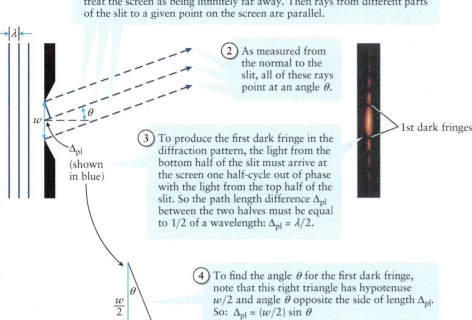

① If the distance L to the screen is much greater than the slit width w, we can treat the screen as being infinitely far away. Then rays from different parts of the slit to a given point on the screen are parallel.

② As measured from the normal to the slit, all of these rays point at an angle θ.

③ To produce the first dark fringe in the diffraction pattern, the light from the bottom half of the slit must arrive at the screen one half-cycle out of phase with the light from the top half of the slit. So the path length difference Δ_{pl} between the two halves must be equal to 1/2 of a wavelength: $\Delta_{pl} = \lambda/2$.

④ To find the angle θ for the first dark fringe, note that this right triangle has hypotenuse $w/2$ and angle θ opposite the side of length Δ_{pl}. So: $\Delta_{pl} = (w/2) \sin \theta$
$$\lambda/2 = (w/2) \sin \theta$$
$$\sin \theta = \lambda/w$$

1st dark fringes

Figure 23-35 Calculating the single-slit diffraction pattern Dark fringes are found where light from each part of the slit interferes destructively with light from another part.

difference Δ_{pl} between light from adjacent quarters is given by Equation 23-26 with $w/2$ replaced by $w/4$. The condition that $\Delta_{pl} = \lambda/2$ then tells us that

$$\frac{\lambda}{2} = \left(\frac{w}{4}\right) \sin \theta$$

If we solve this for $\sin \theta$, we get

$$\sin \theta = \frac{2\lambda}{w} \qquad (23\text{-}28)$$

(second dark fringe in single-slit diffraction)

This gives a larger value for $\sin \theta$, and hence for θ, than for the first dark fringe given by Equation 23-27. For the third dark fringe the factor of 2 in Equation 23-28 is replaced by a 3, for the fourth dark fringe it is replaced by a 4, and so on. In general, the angle of the mth dark fringe is given by

Angle between the normal to the slit and the location of the mth **dark fringe**

Number of the dark fringe: $m = 1, 2, 3,...$

$$\sin \theta = \frac{m\lambda}{w}$$

Wavelength of the light

Width of the slit

Dark fringes in single-slit diffraction
(23-29)

Equations 23-27 and 23-28 are special cases of Equation 23-29, with $m = 1$ and $m = 2$, respectively.

The middle of the central *bright* fringe in Figures 23-34 and 23-35 is where all of the waves from the slit arrive in phase, since they all travel essentially the same distance to this point. This point, which corresponds to $\theta = 0$, is where the intensity is maximum. The next bright fringe, located between the first and second dark fringes, is fainter than the central bright fringe. Roughly speaking, at this bright fringe the light from the upper third of the slit interferes destructively with the light from the middle third of the slit, canceling it out. What remains is the light from the lower third of the slit. This wave has one-third the amplitude and hence $(1/3)^2 = 1/9$ the intensity of the wave that reaches the middle of the central bright fringe. (Recall from Section 22-4 that the intensity of an electromagnetic

Go to Interactive Exercise 23-4 for more practice dealing with diffraction.

wave is proportional to the square of the amplitude of the electric field.) Successive bright fringes are even fainter. Note that the point of greatest intensity in the central bright fringe is at its center; for the other bright fringes the point of greatest intensity is close to (but slightly displaced from) a point halfway between the adjacent dark fringes.

WATCH OUT! Keep in mind the difference between double-slit interference and single-slit diffraction.

! It's common to get confused between double-slit interference and single-slit diffraction, since in both situations there is a pattern of bright and dark fringes. Keep in mind the essential difference between the two situations: In double-slit interference the pattern of fringes is caused by light waves from one slit interfering with light waves from the other slit, while in single-slit diffraction the pattern is caused by light waves from different parts of the same single slit interfering with each other.

EXAMPLE 23-8 Diffraction Through a Slit

A green laser pointer emits light at a wavelength of 532 nm. You aim the beam from this laser at a slit 1.50 μm wide. Find the angles of the first, second, and third dark fringes in the diffraction pattern.

Set Up

We are given the wavelength λ and the slit width ω, and we want to find the value of the angle θ for the $m = 1$, 2, and 3 dark fringes. We'll use Equation 23-29 for this purpose.

Dark fringes in single-slit diffraction:

$$\sin \theta = \frac{m\lambda}{w} \qquad (23\text{-}29)$$

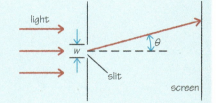

Solve

Find the angle of the first dark fringe ($m = 1$).

We have $\lambda = 532$ nm $= 5.32 \times 10^{-7}$ m and $w = 1.50$ μm $= 1.50 \times 10^{-6}$ m. From Equation 23-29, with $m = 1$,

$$\sin \theta_1 = \frac{\lambda}{w} = \frac{5.32 \times 10^{-7} \text{ m}}{1.50 \times 10^{-6} \text{ m}} = 0.355$$

$$\theta_1 = \sin^{-1} 0.355 = 20.8°$$

Find the angle of the second dark fringe ($m = 2$).

From Equation 23-29, with $m = 2$,

$$\sin \theta_2 = \frac{2\lambda}{w} = \frac{2(5.32 \times 10^{-7} \text{ m})}{1.50 \times 10^{-6} \text{ m}} = 0.709$$

$$\theta_2 = \sin^{-1} 0.709 = 45.2°$$

Find the angle of the third dark fringe ($m = 3$).

From Equation 23-29, with $m = 3$,

$$\sin \theta_3 = \frac{3\lambda}{w} = \frac{3(5.32 \times 10^{-7} \text{ m})}{1.50 \times 10^{-6} \text{ m}} = 1.06$$

This equation has *no* solution! The sine of an angle cannot be greater than 1, so there is no value of θ_3 that satisfies this equation. We are forced to conclude that the diffraction pattern in this situation has a first dark fringe at $\theta_1 = 20.8°$ and a second dark fringe at $\theta_2 = 45.2°$, but there is no third dark fringe before the end of the pattern at $\theta = 90°$.

Reflect

The number of dark fringes present in the diffraction pattern of a slit depends on the relative sizes of the wavelength λ and the slit width w. We explore this further below.

Figure 23-36 shows the intensity in the diffraction pattern of a slit as a function of the angle θ. As the width of the slit is increased from $w = \lambda$ (Figure 23-36a) to $w = 4\lambda$ (Figure 23-36b) to $w = 8\lambda$ (Figure 23-36c), the pattern becomes narrower. Note that if $w = \lambda$, as in Figure 23-36a, the first dark fringe corresponds to $\sin \theta = \lambda/w = 1$, so $\theta = \sin^{-1} 1 = 90°$; there are no dark fringes at smaller angles. The light intensity

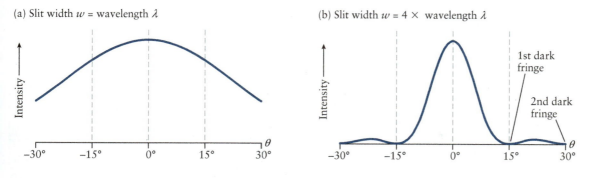

(a) Slit width w = wavelength λ

(b) Slit width $w = 4 \times$ wavelength λ

1st dark fringe

2nd dark fringe

(c) Slit width $w = 8 \times$ wavelength λ

1st dark fringe

3rd dark fringe

2nd dark fringe

4th dark fringe

Figure 23-36 Intensity in single-slit diffraction The intensity in the diffraction pattern from a narrow slit depends on the relative size of the slit width w and the wavelength λ.

is spread out very broadly over all angles, much like what happens to the water waves in Figure 23-33b. If the slit width is large compared to the wavelength, as in Figure 23-36c, the vast majority of the light emerging from the slit goes straight ahead (toward $\theta = 0$) or very nearly so. That's just like what happens to the water waves in Figure 23-33a, for which there is very little diffraction.

The intensity pattern of single-slit diffraction also plays a role in *double-slit* interference, as Figure 23-37 shows. The closely spaced bright fringes in this image are due to interference between light coming from the two slits. However, not all of these fringes are equally bright. The explanation is that light also *diffracts* as it emanates from each of the slits. Each of the slits has a width w that is larger than the wavelength λ, so the effect of diffraction is that the intensity of light emanating from each slit varies with angle θ as shown in Figures 23-36b and 23-36c. Hence the brightness of the bright double-slit interference fringes in Figure 23-37 rises and falls with angle θ just as the intensity curves in Figures 23-36b and 23-36c rise and fall with θ.

In this section we've concentrated on the diffraction that takes place when waves pass through a narrow opening. But diffraction can also happen when waves encounter an obstacle. One example is a sound wave coming from one side of your head. As we mentioned above, sound waves used in speech have wavelengths of a meter or more, which is large compared to the diameter of a human head. As a result, these sound waves are able to diffract around your head, so you can hear the sound with both ears. Because the sound wave must travel a greater distance to one ear than to the other, there will be a phase difference between the waves that the two ears detect. Your brain detects and processes this information about phase and uses this to help determine the direction from which the sound is coming.

① This is the interference pattern in a double-slit experiment.

② The closely spaced bright and dark fringes are due to interference between light waves emanating from the two slits.

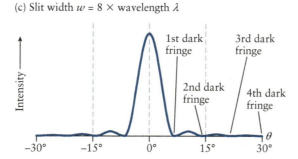

③ The interference pattern disappears at certain locations. This is a result of diffraction of light through each slit (these are the locations of the dark *diffraction* fringes).

④ The intensity of the bright interference fringes decreases with greater angle from the center of the pattern. This is also due to diffraction through the individual slits (see Figure 23-35).

Figure 23-37 Single-slit diffraction affects double-slit interference In a double-slit experiment there is both interference of the light waves emanating from the two slits and diffraction of light waves through each individual slit. The diffraction affects the brightness of the bright interference fringes.

GOT THE CONCEPT? 23-8 Comparing Three Slits

The photographs in Figure 23-38 show the diffraction pattern that is created when red laser light passes through a narrow slit. The wavelength of the light is the same in all three photographs, but the width of the slit is different. Order the photographs from the widest slit to the narrowest one.

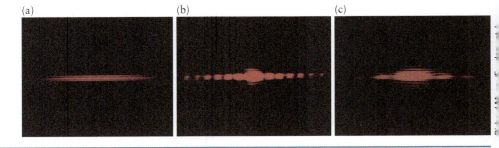

(a)　(b)　(c)

Figure 23-38 Three diffraction patterns Which pattern was produced by the widest slit? By the narrowest slit?

TAKE-HOME MESSAGE FOR Section 23-8

✔ Waves passing through a narrow opening tend to spread out, or diffract. The diffraction is more important the smaller the size of the opening.

✔ The diffraction pattern caused by waves passing through a narrow slit has bright and dark fringes. The positions of these depend on the relative size of the wavelength and the slit width.

23-9 The diffraction of light through a circular aperture is important in optics

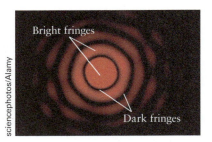

Bright fringes

Dark fringes

Figure 23-39 Diffraction by a circular aperture I Compare the diffraction pattern from a circular aperture with that for a narrow slit (Figure 23-34).

An important real-life application of diffraction is the case of light passing through a *circular* aperture. That's what happens whenever light enters the lens of a microscope, the circular mirror of an astronomical telescope, or the pupil of a human eye. Figure 23-39 shows the diffraction pattern produced by red laser light passing through a circular aperture. Like the diffraction pattern of a slit (Section 23-8), there are bright and dark fringes. But because the aperture is circular, the fringes are circles.

Figure 23-40 shows how we define the angle θ of a point in the diffraction pattern of a circular aperture of diameter D through which passes light of wavelength λ. For a narrow slit we found that the diffraction pattern depends on the relative sizes of the wavelength λ and the slit width w; in the same way, the diffraction pattern of a circular aperture depends on the relative sizes of λ and D. In particular, the location of the center of the first dark fringe is given by

$$\sin \theta = 1.22 \frac{\lambda}{D}$$

(23-30)

(circular aperture, first dark fringe)

This is similar to Equation 23-27 for the first dark fringe formed by light passing through a narrow slit, with the slit width w replaced by the diameter D. The factor of 1.22 results from the different geometry of a circular opening versus a rectangular one.

Figure 23-41 shows the diffraction pattern made by two pointlike objects as the objects are moved closer and closer together. In Figure 23-41a the two objects are so far apart we see only one object and its associated diffraction pattern in the field of view. In Figure 23-41b a second object has been brought close to the first; the diffraction patterns of the two objects overlap but are still distinct. In Figure 23-41c, however, the two objects are very close together. Their diffraction patterns overlap so much that it is barely possible to tell the two objects apart. We have run into the limit on our ability to **resolve**, or optically distinguish, the two objects.

The nineteenth-century English physicist John William Strutt, 3rd Baron Rayleigh, proposed that two pointlike objects observed through a circular aperture can be resolved when the central maximum of one coincides with the center of the first dark fringe of the other. The angle θ_R that separates two point objects that are just barely

sciencephotos/Alamy

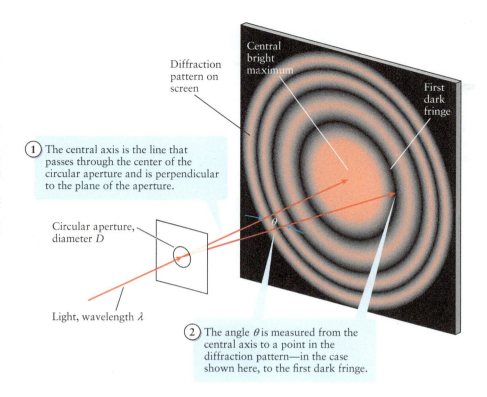

Figure 23-40 Diffraction by a circular aperture II The size of the first dark fringe — and hence the size of the central bright maximum that it surrounds — depends on the ratio of wavelength λ to aperture diameter D.

① The central axis is the line that passes through the center of the circular aperture and is perpendicular to the plane of the aperture.

Circular aperture, diameter D

Light, wavelength λ

② The angle θ is measured from the central axis to a point in the diffraction pattern—in the case shown here, to the first dark fringe.

resolved through a circular aperture, known as the **angular resolution** of the aperture, is then just the angle given by Equation 23-30:

Angle between two pointlike objects that can **barely be resolved** through an optical device

Wavelength of the light

$$\sin\theta_R = 1.22\,\frac{\lambda}{D}$$

Diameter of the circular aperture of the device

Rayleigh's criterion for resolvability
(23-31)

The smaller the value of θ_R, the better the resolution and the smaller the details that can be resolved. Equation 23-31 shows that better angular resolution can be obtained by using light of shorter wavelength λ and by using a larger circular aperture D.

A physician's eye chart is a device for measuring the value of θ_R for each of your eyes. An unaided eye with normal vision can distinguish objects (such as the lines that make up the letter "E" on an eye chart) that are separated by an angle of as small as $1/60$ of a degree. So $\theta_R = (1/60)°$ for a person with normal vision.

BioMedical

(a) (b) (c)

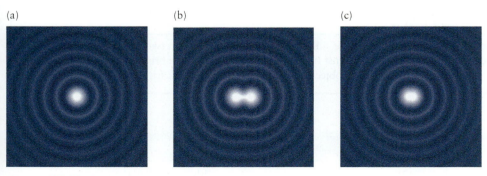

Figure 23-41 Angular resolution (a) Light from a single point source gives rise to a diffraction pattern when it passes through a circular aperture. (b) These diffraction patterns from two distant point sources partially overlap but are distinct. (c) At the limit of our ability to resolve two distant point sources, the diffraction patterns formed as their light passes through a circular aperture overlap and are barely distinguishable.

EXAMPLE 23-9 The Hubble Space Telescope

The Hubble Space Telescope (HST) has a circular aperture 2.4 m in diameter. What is the theoretical angular resolution of the HST for light of wavelength 550 nm?

Set Up

Equation 23-31 gives the angular resolution, the smallest angular separation of two objects that can be resolved as a result of diffraction effects.

Rayleigh's criterion for resolvability:

$$\sin \theta_R = 1.22 \frac{\lambda}{D} \qquad (23\text{-}31)$$

Solve

We are given $\lambda = 550$ nm and $D = 2.4$ m. We calculate θ_R from these using Equation 23-31.

From Equation 23-31:

$$\sin \theta_R = 1.22 \frac{(550 \times 10^{-9} \text{ m})}{(2.4 \text{ m})} = 2.8 \times 10^{-7}$$

$$\theta_R = \sin^{-1}(2.8 \times 10^{-7}) = 1.6 \times 10^{-5} \text{ degree}$$

Reflect

Because the diameter of the HST mirror is so much greater than that of the human eye, θ_R for HST is much smaller than the value $\theta_R = (1/60)° = 0.017°$ for the human eye, which means that HST has far better angular resolution. The photograph shown here is a Hubble Space Telescope (HST) image of the minor planet Pluto and its largest moon, Charon. Pluto and Charon are separated by about 25×10^{-5} degrees in this image. They are easily resolved by HST, for which $\theta_R = 1.6 \times 10^{-5}$ degrees.

Why is it important for a telescope like HST to be in orbit? One key reason is that a telescope of the same diameter on Earth would have a much poorer resolution of about 0.03 degrees and would not be able to resolve Pluto and Charon. That's due to the blurring effects of our atmosphere, which is constantly in motion. By placing HST in orbit high above the atmosphere, this blurring is eliminated. (A second important reason for placing telescopes in orbit is that it makes it possible to see at a variety of wavelengths, including infrared and ultraviolet, that do not penetrate through our atmosphere and so cannot be detected by telescopes on Earth.)

GOT THE CONCEPT? 23-9 Angular Resolution

 Rank the following telescopes from best to worst angular resolution. Assume that diffraction is the only limiting factor. (a) A radio telescope with an effective diameter of 27 km that observes waves at wavelength 21 cm; (b) a telescope in orbit that has a mirror of 0.85 m diameter and that observes infrared light of wavelength 4.5 μm; (c) an inexpensive telescope for amateur astronomers with a lens of 5.0 cm diameter that observes visible light of wavelength 550 nm.

TAKE-HOME MESSAGE FOR Section 23-9

✔ The angular resolution of an optical device is limited by diffraction, which blurs the images of even pointlike objects.

✔ Rayleigh's criterion states that two pointlike objects can just be resolved if the center of the bright maximum for one object coincides with the first dark fringe for the second object.

Key Terms

angular resolution	diffraction pattern	incident
Brewster's angle	dispersion	index of refraction
critical angle	double-slit interference	interference maxima
diffraction	fringe	interference minima
diffraction maxima	front	interference pattern
diffraction minima	Huygens' principle	law of reflection

linearly polarized
monochromatic
normal
partially polarized
polarization

polarizing filter
ray
refraction
resolve
single-slit diffraction

Snell's law of refraction
thin film
total internal reflection
unpolarized
wavelet

Chapter Summary

Topic	Equation or Figure
Speed of light in a medium: In a medium (transparent material) other than vacuum, the speed of light is less than c. This is described in terms of the index of refraction n of the material. The value of n is different for different materials. The wavelength also changes when light enters a medium.	Index of refraction of a medium Speed of light in a vacuum $$n = \frac{c}{v}$$ Speed of light in the medium (23-5)
	Wavelength of light in a medium Speed of light in a vacuum Wavelength of the light in a vacuum $$\lambda = \frac{c}{nf} = \frac{\lambda_{vacuum}}{n}$$ Frequency of the light Index of refraction of the medium (23-8)

| **Huygens' principle, reflection, and refraction:** Each point on a wave front (or wave crest) acts as a source of spherical waves called wavelets. This principle helps us explain the laws of reflection and refraction, which describe what happens when light encounters the boundary between two media. The angle of the reflected light is always equal to the angle θ_1 of the incident light, and the angle θ_2 of the refracted light is given by Snell's law. | (a) Refraction from one medium into a slower one 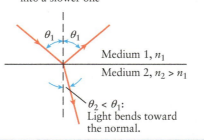 Medium 1, n_1 Medium 2, $n_2 > n_1$ $\theta_2 < \theta_1$: Light bends toward the normal. (Figure 23-6a) |
| | Angle of the incident ray from the normal   Angle of the refracted ray from the normal $$n_1 \sin \theta_1 = n_2 \sin \theta_2$$ Index of refraction for the medium with the incident light   Index of refraction for the medium with the refracted light (23-6) |

| **Total internal reflection:** If light in one medium reaches a boundary with a second medium of lower index of refraction ($n_1 > n_2$), total internal reflection is possible. It will take place only if the incident angle is greater than the critical angle for the two media. | Critical angle for light in medium 1 that encounters a boundary with medium 2 with a lower index of refraction   Index of refraction of medium 2 $$\theta_c = \sin^{-1}\left(\frac{n_2}{n_1}\right)$$ Index of refraction of medium 1 (23-10) |
| | If the incident angle is greater than the critical angle, there is total internal reflection. |

Dispersion: For a given material the index of refraction depends on the frequency of the light (or, equivalently, the wavelength in a vacuum). This is the reason a prism or water droplets can break white light into its constituent colors.

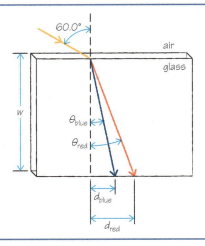

(Example 23-3)

Polarization: The polarization of a light wave is a description of the orientation of its electric field vector. In unpolarized light this orientation changes randomly. In linearly polarized light the orientation is in a fixed direction. Unpolarized light can become polarized by scattering from the atmosphere, by passing through a polarizing filter, or by reflection at Brewster's angle.

Brewster's angle for light in medium 1 that encounters a boundary with medium 2

Index of refraction of medium 2

$$\theta_B = \tan^{-1}\left(\frac{n_2}{n_1}\right)$$

Index of refraction of medium 1

(23-14)

 If the incident angle equals Brewster's angle, the reflected light is completely polarized in the direction parallel to the surface of the boundary.

Thin-film interference: If a thin, transparent film is illuminated with monochromatic light, the light that reflects from the back surface of the film interferes with light that reflects from the front surface. If the two waves emerge in phase, the interference is constructive; if they are out of phase, it is destructive. The details of the interference depend on whether there is an inversion of the wave on each reflection.

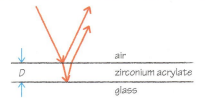

(Example 23-6)

Double-slit interference: When monochromatic light illuminates a pair of closely spaced narrow slits, a pattern of bright and dark fringes results. The bright fringes appear where light waves from the two slits interfere constructively; the dark fringes appear where light waves from the two slits interfere destructively. The closer together the two slits, the broader the interference pattern.

Distance between the two slits

Wavelength of the light illuminating the two slits

Constructive interference (bright fringes): $d \sin \theta = m\lambda$

(23-23)

Angle between the normal to the two slits and the location of a **bright fringe** on the screen

Number of the bright fringe: $m = 0, \pm 1, \pm 2, \pm 3,...$

Distance between the two slits

Wavelength of the light illuminating the two slits

Destructive interference (dark fringes): $d \sin \theta = \left(m + \frac{1}{2}\right)\lambda$

(23-24)

Angle between the normal to the two slits and the location of a **dark fringe** on the screen

Number of the dark fringe: $m = 0, \pm 1, \pm 2, \pm 3,...$

Single-slit diffraction: A pattern of bright and dark fringes also results when monochromatic light illuminates a single narrow slit. When monochromatic light illuminates a narrow slit, a pattern of bright and dark fringes results. The dark fringes appear where light from any one part of the slit interferes destructively with light from some other part of the slit, so that zero net electromagnetic wave reaches the position of the dark fringe. The narrower the width of the slit, the broader the diffraction pattern.

Angle between the normal to the slit and the location of the mth **dark fringe**

Number of the dark fringe: $m = 1, 2, 3,...$

$$\sin \theta = \frac{m\lambda}{w}$$ Wavelength of the light   (23-29)

Width of the slit

Diffraction by a circular aperture: Light also diffracts when it passes through a circular aperture. This sets a fundamental limit on the angular resolution of an optical device: If two objects are closer together than an angle θ_R, it will be impossible to tell with the device whether they are two objects or a single object.

Angle between two pointlike objects that can **barely be resolved** through an optical device

Wavelength of the light

$$\sin \theta_R = 1.22 \frac{\lambda}{D}$$ (23-31)

Diameter of the circular aperture of the device

Answer to What do you think? Question

(e) Light travels more slowly in diamond than in air. So once light enters the front of a diamond, there can be total internal reflection of this light from the cut back surfaces (Section 23-3). This strongly reflected light then emerges from the front of the diamond. The speed of light in the diamond also depends on its color (frequency), so light waves of different colors follow slightly different paths on entering and emerging from the diamond (Section 23-4). Hence you see different colors of light reflected from a cut diamond as you view it at different angles.

Answers to Got the Concept? Questions

23-1 (c) For the first refraction, medium 1 is air and medium 2 is Plexiglas. Table 23-1 shows that the index of refraction of Plexiglas (1.49) is greater than that of air (1.00), so $n_2 > n_1$. Snell's law of refraction (Equation 23-6) tells us that in this situation $\theta_2 < \theta_1$, so the angle θ_2 that the light makes to the normal in Plexiglas is less than the angle θ_1 the light makes in air. So the light bends toward the normal as it crosses the boundary from air into Plexiglas. For the second refraction medium 1 is Plexiglas and medium 2 is water. Since water has a lower index of refraction (1.33) than Plexiglas (1.49), in this case $n_2 < n_1$, and Snell's law tells us that in this situation $\theta_2 > \theta_1$. The angle θ_2 that the light makes to the normal in water is greater than the angle θ_1 the light makes in Plexiglas, so the light bends away from the normal as it crosses the boundary from Plexiglas into water. Can you show that the angle is 3.35° in Plexiglas and 3.76° in water?

23-2 (a) Total internal reflection at a boundary between media is possible only if the second medium has a lower index of refraction than the first medium. Here the first medium is Plexiglas, so from Table 23-1, $n_1 = 1.49$. Total internal reflection is possible for ethyl alcohol, for which $n_2 = 1.36$ is less than n_1. It is not possible for sapphire ($n_2 = 1.77$) or diamond ($n_2 = 2.42$), since both of these have an index of refraction greater than $n_1 = 1.49$.

23-3 (c) For water, like most materials, the index of refraction n is greater, and so the wave speed $v = c/n$ is slower for light of greater frequency and shorter wavelength. This means that the blue light is slowest, and the red light is fastest, so the red light reaches the bottom of the pond slightly before the other colors.

23-4 (d) Once the light has passed through filter A, it will have one-half the intensity of the incident unpolarized light and will be completely polarized in the transmission direction of that filter. Filter B is set so that its transmission direction is perpendicular to that of filter A. So the angle between the polarization direction of the light that reaches filter B and the transmission direction of filter B is $\theta = 90°$. So from Equation 23-12 the intensity of the light that emerges from filter B will be the intensity of light reaching it from filter A multiplied by $\cos^2 90° = 0$. In other words, *no* light emerges from filter B.

23-5 (c) Once the light has passed through filter A, it will have one-half the intensity of the incident unpolarized light and will be completely polarized in the transmission direction of that filter. The transmission direction of filter C is at 45° to that of filter A, so from Equation 23-12 the light that emerges from filter C is reduced in intensity by an additional factor of $\cos^2 45° = (1/\sqrt{2})^2 = 1/2$. This light is polarized in the transmission direction of filter C. This light now enters filter B, which has a transmission direction at 45° to the direction of filter C. So the light that emerges from filter B is polarized in the transmission direction of filter B and has $\cos^2 45° = (1/\sqrt{2})^2 = 1/2$ the intensity of the light that reached it from filter C. So compared to the unpolarized light that was incident on filter A, the polarized light that emerges from filter B has $(1/2) \times (1/2) \times (1/2) = 1/8$ the intensity.

23-6 (d) When light is reflected from a boundary going from a material of higher wave speed to one of lower wave speed, the wave is inverted. No inversion occurs when light is reflected from a boundary where the transition is from a medium of lower speed to one of higher speed. The higher a material's index of refraction, the lower the speed of light is in that material. Another way to state the rule is that a light wave is inverted when it is reflected from a boundary going from a material that has a lower index of refraction to one that has a higher index of refraction. So here the speed of light is highest in air, lower in the oil, and lowest in glass. Light striking the air-to-oil boundary is therefore inverted upon reflection, as is light striking the oil-to-glass boundary. In both cases the boundary separates a material of higher light speed from one of lower light speed. No inversion occurs at the glass-to-air boundary, however, because the speed of light in glass is lower than the speed in air.

23-7 (b), (c). From the first of Equations 23-25, the angle of the $m = 3$ bright fringe is given by $\sin \theta = 3\lambda/d$, and the angle of the $m = 2$ bright fringe is given by $\sin \theta = 2\lambda/d$. To move the $m = 2$ bright fringe to the position of the $m = 3$ bright fringe therefore requires that you increase the ratio λ/d by a factor of $3/2$. You can do this by decreasing d by a factor of $2/3$ while leaving λ alone [choice (b)] or by increasing λ by a factor of

$3/2$ while leaving d unchanged [choice (c)]. You could also increase λ by a factor of 3 and increase d by a factor of 2; note that this is not what choice (e) says, however.

23-8 (b), (c), (a). The narrower the slit, the more the diffracted light is spread out (see Figure 23-36). A good way to compare the three photographs is by looking at the width of the central bright maximum. The central bright maximum is widest in (a), so this is the narrowest slit. Of the other two photos, (c) has the wider central maximum, so the slit width is narrower in photo (c) than in photo (b). Hence the order from the widest to the narrowest slit is (b), (c), (a).

23-9 (b), (a), (c). The smaller the value of θ_R as given by Equation 23-31, the better the diffraction-limited angular resolution. For (a) $\lambda = 21$ cm $= 0.21$ m and $D = 27$ km $= 2.7 \times 10^4$ m, so $\sin \theta_R = 1.22\lambda/D = 9.5 \times 10^{-6}$ and $\theta_R = 5.4 \times 10^{-4}$ degrees. (Note that 27 km is not the diameter of a single telescope. To improve the angular resolution, multiple radio telescopes many kilometers apart are linked together so that they act like a single, gigantic telescope.) For (b) $\lambda = 4.5$ μm $= 4.5 \times 10^{-6}$ m and $D = 0.85$ m, so $\sin \theta_R = 6.5 \times 10^{-6}$ and $\theta_R = 3.7 \times 10^{-4}$ degrees. For (c) $\lambda = 550$ nm $= 550 \times 10^{-9}$ m and $D = 5.0$ cm $= 0.050$ m, so $\sin \theta_R = 1.3 \times 10^{-5}$ and $\theta_R = 7.7 \times 10^{-4}$ degrees. (The actual resolution would be worse due to atmospheric turbulence.) So (b) is the best, (a) is second, and (c) is worst.

Questions and Problems

In a few problems you are given more data than you actually need; in a few other problems you are required to supply data from your general knowledge, outside sources, or informed estimate.

Interpret as significant all digits in numerical values that have trailing zeros and no decimal points.

For all problems use $g = 9.80$ m/s² for the free-fall acceleration due to gravity. Neglect friction and air resistance unless instructed to do otherwise.

- •   Basic, single-concept problem
- ••   Intermediate-level problem; may require synthesis of concepts and multiple steps
- •••   Challenging problem
- Example See worked example for a similar problem

Conceptual Questions

1. • What is Huygens' principle, and why is it necessary to understand Snell's law of refraction?

2. • Does the refraction of light make a swimming pool seem deeper or shallower? Explain your answer.

3. • Does the depth of a pool determine the critical angle that a light ray will have as it travels from the bottom of the pool and heads toward the air above the water? Explain your answer.

4. • Give two common uses of total internal reflection.

5. • In your own words explain why the phenomenon of total internal reflection occurs only when light moves from a medium with a larger index of refraction toward a medium with a smaller index of refraction.

6. • Sunlight striking a diamond throws rainbows of color in every direction. From where do the colors come?

7. • Recently researchers have created materials with a negative index of refraction. Explain what happens to the angle of refraction if light enters such a material from air.

8. • Why do you expect the last color of the sunset to be on the red end of the visible spectrum?

9. • Explain why the Moon appears to change colors during a total lunar eclipse (when Earth's shadow completely blocks the light coming from the Sun). (See problem 8.)

10. • Describe the physical interactions that take place when unpolarized light is passed through a polarizing filter. Be sure to describe the electric field of the light before and after the filter as well as the incident and transmitted intensities of the light source.

11. • Describe how polarized sunglasses work. Why do such sunglasses have *vertically* polarized lenses (as opposed to *horizontally* polarized lenses)?

12. • Linearly polarized light is incident at Brewster's angle on the surface of an optical medium. What can be said about the refracted and reflected beams if the incident beam is polarized (a) parallel to the plane of the surface and (b) perpendicular to the plane of the surface?

13. • Give two or three examples of thin-film interference.

14. • A thin layer of gasoline floating on water appears brightly colored in sunlight. What is the origin of these colors?

15. • In a two-slit interference experiment, why must the incident light striking one slit have the same wavelength and phase as that striking the other slit?

16. • Does the phenomenon of diffraction apply to wave sources other than light? Give an example if it does.

17. • The sound waves used in speech have wavelengths of a few meters. Explain why some of the sound waves emerging from a person's mouth go to the sides so that they can be heard even if the listener is not in front of the person.

18. • A car radio is tuned to receive a signal from a particular radio station. While the car slows to a stop at a traffic signal, the

reception of the radio seems to fade in and out. Use the concept of interference to explain the phenomenon. *Hint*: In broadcast technology, the phenomenon is known as multipathing.

19. •• Suppose you are measuring double-slit interference patterns using an optics kit that contains the following options that you can mix and match: a red laser or a green laser; a slit width of 0.04 or 0.08 mm; a slit separation of 0.25 or 0.50 mm. Which elements would you select to produce an interference pattern with the most widely spaced fringes? Why?

Multiple-Choice Questions

20. • A red laser shines on a pair of slits separated by a distance d, producing an interference pattern on a screen. A green laser of the same intensity is then used to illuminate the same slit pair. Compared to the red laser, the green laser will produce an interference pattern with maxima that are
    A. in the same locations.
    B. brighter.
    C. farther apart.
    D. dimmer.
    E. closer together.

21. • Which kind of wave can refract when crossing from one medium to another with the wave having a different speed in the two media?
    A. electromagnetic waves
    B. sound waves
    C. water waves
    D. electromagnetic, sound, and water waves
    E. only electromagnetic and sound waves

22. • When light enters a piece of glass from air with an angle of θ with respect to the normal to the boundary surface
    A. it bends with an angle larger than θ with respect to the normal to the boundary surface.
    B. it bends with an angle smaller than θ with respect to the normal to the boundary surface.
    C. it does not bend.
    D. it bends with an angle equal to two times θ with respect to the normal to the boundary surface.
    E. it bends with an angle equal to one-half θ with respect to the normal to the boundary surface.

23. • Which phenomenon would cause monochromatic light to enter the prism and follow along the path as shown in Figure 23-42?
    A. reflection
    B. refraction
    C. interference
    D. diffraction
    E. polarization

Figure 23-42 Problem 23

24. • Which color of light, red or blue, travels faster in crown glass?
    A. red
    B. blue
    C. their speeds are the same
    D. depends on the material surrounding the glass
    E. blue if the glass is thin

25. • Two linear polarizing filters are placed one behind the other so that their transmission directions are parallel to one another. A beam of unpolarized light of intensity I_0 is directed at the two filters. What fraction of the light will pass through both filters?
    A. 0
    B. $(1/2)I_0$

C. I_0
D. $(1/4)I_0$
E. $2I_0$

26. • Two linear polarizing filters are placed one behind the other so that their transmission directions form an angle of 45°. A beam of unpolarized light of intensity I_0 is directed at the two filters. What fraction of the light will pass through both?
    A. 0
    B. $(1/2)I_0$
    C. I_0
    D. $(1/4)I_0$
    E. $2I_0$

27. • A monochromatic light passes through a narrow slit and forms a diffraction pattern on a screen behind the slit. As the wavelength of the light decreases, the diffraction pattern
    A. shrinks with all the fringes getting narrower.
    B. spreads out with all the fringes getting wider.
    C. remains unchanged.
    D. spreads out with all the fringes getting alternately wider and then narrower.
    E. becomes dimmer.

28. • A monochromatic light passes through a narrow slit and forms a diffraction pattern on a screen behind the slit. As the slit width increases, the diffraction pattern
    A. shrinks with all the fringes getting narrower.
    B. spreads out with all the fringes getting wider.
    C. remains unchanged.
    D. spreads out with all the fringes getting alternately wider and then narrower.
    E. becomes dimmer.

29. • Figure 23-43 shows two single-slit diffraction patterns created with the same source. The distance between the slit and the viewing screen is the same in both cases. Which of the following is true about the widths, w_a and w_b, of the two slits?

(a)

(b)

Figure 23-43
Problem 29

    A. $w_a > w_b$
    B. $w_a < w_b$
    C. $w_a = w_b$
    D. $w_a = (1/2)w_b$
    E. $w_a = (1/4)w_b$

Problems

23-1 The wave nature of light explains much about how light behaves

23-2 Huygens' principle explains the reflection and refraction of light

30. • The speed of light in a newly developed plastic is 1.97×10^8 m/s. Calculate the index of refraction.

31. • The index of refraction for a vacuum is 1.00000. The index of refraction for air is 1.00029. Determine the ratio of time required for light to travel through 1000 m of air to the time required for light to travel through 1000 m of vacuum.

32. • Calculate the speed of light for each of the following materials:
    A. ice
    B. acetone

C. Plexiglas
D. sodium chloride
E. sapphire
F. diamond
G. water
H. crown glass

33. •• The speed of light in methylene iodide is 1.72×10^8 m/s. The index of refraction of water is 1.33. Through what distance of methylene iodide must light travel such that the time to travel through the methylene iodide is the same as the time required for light to travel through 1000 km of water?

34. •• Light travels from air toward water. If the angle that is formed by the light beam in air with respect to the normal line between the two media is 27°, calculate the angle of refraction of the light in the water. Example 23-1

35. • Determine the unknown angle in each of the situations in Figure 23-44. Example 23-1

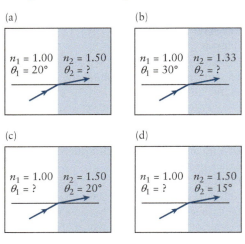

(a)

$n_1 = 1.00$ $n_2 = 1.50$
$\theta_1 = 20°$ $\theta_2 = ?$

(b)

$n_1 = 1.00$ $n_2 = 1.33$
$\theta_1 = 30°$ $\theta_2 = ?$

(c)

$n_1 = 1.00$ $n_2 = 1.50$
$\theta_1 = ?$ $\theta_2 = 20°$

(d)

$n_1 = 1.00$ $n_2 = 1.50$
$\theta_1 = ?$ $\theta_2 = 15°$

Figure 23-44
Problem 35

36. • Determine the unknown index of refraction in each of the situations in Figure 23-45. Example 23-1

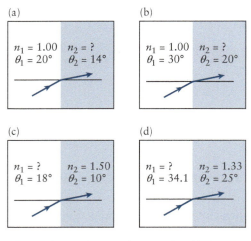

(a)

$n_1 = 1.00$ $n_2 = ?$
$\theta_1 = 20°$ $\theta_2 = 14°$

(b)

$n_1 = 1.00$ $n_2 = ?$
$\theta_1 = 30°$ $\theta_2 = 20°$

(c)

$n_1 = ?$ $n_2 = 1.50$
$\theta_1 = 18°$ $\theta_2 = 10°$

(d)

$n_1 = ?$ $n_2 = 1.33$
$\theta_1 = 34.1$ $\theta_2 = 25°$

Figure 23-45
Problem 36

37. • What is the wavelength of red light (700 nm in air) when it is inside a glass slab with $n = 1.55$?

23-3 In some cases light undergoes total internal reflection at the boundary between media

38. • Calculate the critical angle for the following: Example 23-2
   A. Light travels from plastic ($n = 1.50$) to air ($n = 1.00$).
   B. Light travels from water ($n = 1.33$) to air ($n = 1.00$).
   C. Light travels from glass ($n = 1.56$) to water ($n = 1.33$).
   D. Light travels from air ($n = 1.00$) to glass ($n = 1.55$).

39. • For each of the critical angles given, calculate the index of refraction for the optical materials that light travels out of toward air ($n = 1.00$): Example 23-2
   A. $\theta_c = 48.5°$
   B. $\theta_c = 47.0°$
   C. $\theta_c = 42.6°$
   D. $\theta_c = 35.0°$
   E. $\theta_c = 55.7°$
   F. $\theta_c = 38.5°$
   G. $\theta_c = 22.2°$
   H. $\theta_c = 75.0°$

40. • What is the critical angle for light traveling from sapphire to air? Example 23-2

41. •• What is the largest angle θ_1 that will ensure that light is totally internally reflected in the fiber-optic pipe made of acrylic ($n = 1.50$) shown in Figure 23-46? Example 23-2

Air

$\theta_1 = ?$

Acrylic

Figure 23-46 Problem 41

42. • At what angle with respect to the vertical must a scuba diver look to see her friend standing on the very distant shore? Take the index of refraction of the water to be $n = 1.33$. Example 23-2

43. •• A point source of light is 2.50 m below the surface of a pool. What is the diameter of the circle of light that a person above the water will see? Assume the water has an index of refraction of $n = 1.33$. Example 23-2

44. • A block of glass that has an index of refraction of 1.55 is completely immersed in water ($n = 1.33$). What is the critical angle for light traveling from the glass to the water? Example 23-2

23-4 The dispersion of light explains the colors from a prism or a rainbow

45. •• For a certain optical medium the speed of light varies from a low value of 1.90×10^8 m/s for violet light to a high value of 2.00×10^8 m/s for red light. (a) Calculate the range of the index of refraction of the material for visible light. (b) A white light is incident on the medium from air, making an angle of 30.0° with the normal. Compare the angles of refraction for violet light and red light. (c) Repeat the previous part when the incident angle is 60.0°. Example 23-3

46. ••• A beam of light shines on an equilateral glass prism at an angle of 45° to one face (Figure 23-47a). (a) What is the angle at which the light emerges from the opposite face given that $n_{glass} = 1.57$? (b) Now consider what happens when dispersion is involved (Figure 23-47b). Assume the incident ray

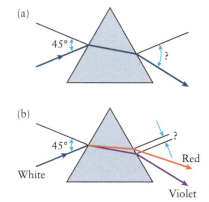

(a)

45°   ?

(b)

45°   ?

White   Red

Violet

Figure 23-47
Problem 46

of light spans the spectrum of visible light between 400 and 700 nm (violet to red, respectively). The index of refraction for violet light in the glass prism is 1.572, and it is 1.568 for red light in the glass prism. Find the distance along the right face of the prism between the points where the red light and violet light emerge back into air. Assume the prism is 10.0 cm on a side and the incident ray hits the midpoint of the left face. Example 23-3

47. • A light beam strikes a piece of glass with an incident angle of 45.0°. The beam contains two wavelengths: 450 nm and an unknown wavelength. The index of refraction for the 450-nm light is 1.482. Determine the index of refraction for the unknown wavelength if the angle between the two refracted rays is 0.275°. Assume the glass is surrounded by air. Example 23-3

48. •• Blue light (500 nm in air) and yellow light (600 nm in air) are incident on a 12-cm-thick slab of glass as shown in Figure 23-48. In the glass the index of refraction for the blue light is 1.545, and for the yellow light it is 1.523. What distance along the glass slab (side AB) separates the points at which the two rays emerge back into air? Example 23-3

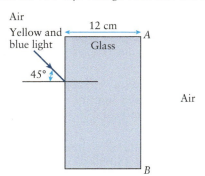

Figure 23-48
Problem 48

23-5 In a polarized light wave the electric field vector points in a specific direction

49. • Unpolarized light is passed through an optical filter that is oriented in the vertical direction. If the incident intensity of the light is 78 W/m², what are the polarization and intensity of the light that emerges from the filter?

50. • What angle(s) does vertically polarized light make relative to a polarizing filter that diminishes the intensity of the light by 25%?

51. •• Light that passes through a series of three polarizing filters emerges from the third filter horizontally polarized with an intensity of 250 W/m². If the polarization angle between the filters increases by 25° from one filter to the next, find the intensity of the incident beam of light, assuming it is initially unpolarized.

52. •• Vertically polarized light that has an intensity of 400 W/m² is incident on two polarizing filters. The first filter is oriented 30.0° from the vertical, while the second filter is oriented 75.0° from the vertical. Predict the intensity and polarization of the light that emerges from the second filter.

53. • What is Brewster's angle when light in water is reflected off a glass surface? Assume $n_{water} = 1.33$ and $n_{glass} = 1.55$. Example 23-4

54. •• The critical angle between two optical media is 60.0°. What is Brewster's angle at the same interface between the two media? Example 23-4

55. • (a) What would be Brewster's angle for reflections off the surface of a pool when the light source is underwater? (b) Calculate the critical angle when light starts in water and reflects off air. Example 23-4

56. •• At what angle θ above the horizontal is the Sun when a person observing its rays reflected off water finds them linearly polarized along the horizontal (Figure 23-49)? Example 23-4

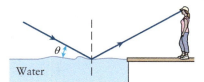

Figure 23-49
Problem 56

23-6 Light waves reflected from the surfaces of a thin film can interfere with each other, producing dazzling effects

57. • A ray of normal-incidence light is reflected from a thin film back into air. If the film is actually a coating on a slab of glass ($n_{film} < n_{glass}$), describe the phase changes that the reflected ray undergoes (a) as it reflects off the front surface of the film and (b) as it reflects off the back surface of the film (the film–glass interface). Example 23-6

58. •• When white light illuminates a thin film with normal incidence, it strongly reflects both indigo light (450 nm in air) and yellow light (600 nm in air) (Figure 23-50). Calculate the minimum thickness of the film if it has an index of refraction of 1.28 and it sits atop a slab of glass that has $n = 1.50$. Example 23-5

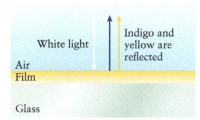

Figure 23-50
Problem 58

59. •• When white light illuminates a thin film normal to the surface, it strongly reflects both blue light (500 nm in air) and red light (700 nm in air) (Figure 23-51). Calculate the minimum thickness of the film if it has an index of refraction of 1.35 and it "floats" on water with $n = 1.33$. Example 23-6

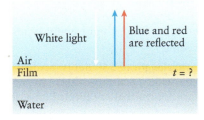

Figure 23-51
Problem 59

60. •• A soap bubble is suspended in air. If the thickness of the soap is 625 nm and both blue light (500 nm in air) and red light (700 nm in air) are *not* observed to reflect from the soap film when viewed normal to the surface of the film, what is the index of refraction of the thin film? Example 23-6

61. •• A thin film of cooking oil ($n = 1.38$) is spread on a puddle of water ($n = 1.33$). What are the minimum and the next three thicknesses of the oil that will strongly reflect blue light having a wavelength in air of 518 nm at normal incidence? Example 23-6

62. •• What is the minimum thickness of a nonreflective coating of magnesium chloride ($n = 1.39$) so that no normal-incident light centered around 550 nm in air will reflect back off a glass lens ($n = 1.56$)? Example 23-5

63. •• Water ($n = 1.33$) in a shallow pan is covered with a thin film of oil that is 450 nm thick and has an index of refraction of 1.45. What visible wavelengths in air will *not* be present in the reflected light when the pan is illuminated with white light and viewed from straight above? Example 23-6

23-7 Interference can occur when light passes through two narrow, parallel slits

64. • Light that has a 650-nm wavelength is incident upon two narrow slits that are separated by 0.500 mm. An interference pattern from the slits is projected onto a screen that is 3.00 m away. (a) What is the separation distance on the screen of the first bright fringe from the central bright fringe? (b) What is the separation distance on the screen of the second dark fringe from the central bright fringe? Example 23-7

65. •• Conducting an experiment with a 532-nm-wavelength green laser, a researcher notices a slight shift in the image generated and suspects the laser is unstable and switching between two closely spaced wavelengths, a phenomenon known as mode-hopping. To determine if this is true, she decides to shine the laser on a double-slit apparatus and look for changes in the pattern. Measuring to the first bright fringe on a screen 0.5 m away, with a slit separation of 80 μm, she measures a distance of 3.325 mm. When the laser shifts, so does the pattern and she then measures the same fringe spacing to be 3.375 mm. What wavelength is the laser "hopping" to? Example 23-7

66. • An interference pattern from a double-slit experiment displays 12 bright fringes per centimeter on a screen that is 8.5 m away. The wavelength of light incident on the slits is 550 nm. What is the distance between the two slits? Example 23-7

67. •• An argon laser that has a wavelength of 455 nm shines on a double-slit apparatus, which produces an interference pattern on a screen that is 10.0 m away from the slits. The slit separation distance is 70.0 μm. (a) How many bright fringes are there on the screen within an angle of ±1° relative to the central axis? (b) How many dark fringes are there on the screen within an angle of ±2° relative to the central axis? Be careful to count *all* the fringes. Example 23-7

23-8 Diffraction is the spreading of light when it passes through a narrow opening

68. • Light that has a wavelength of 550 nm is incident on a single slit that is 10.0 μm wide. Determine the angular location of the first three dark fringes that are formed on a screen behind the slit. Example 23-8

69. • Light that has a wavelength of 475 nm is incident on a single slit that is 800 nm wide. Calculate the angular location of the first three dark fringes that are formed on a screen behind the slit. Example 23-8

70. • What is the highest order dark fringe that is found in the diffraction pattern for light that has a wavelength of 633 nm and is incident on a single slit that is 1500 nm wide? Example 23-8

71. • The highest order dark fringe found in a diffraction pattern is 6. Determine the wavelength of light that is used with the single slit that has a width of 3500 nm. Example 23-8

72. •• When blue light ($\lambda = 500$ nm) is incident on a single slit, the central bright spot has a width of 8.75 cm. If the screen is 3.55 m distant from the slit, calculate the slit width. Example 23-8

73. • A helium–neon laser illuminates a narrow, single slit that is 1850 nm wide. The first dark fringe is found at an angle of 20.0° from the central peak. Determine the wavelength of the light from the laser. Example 23-8

74. •• Yellow light that has a wavelength of 625 nm produces a central maximum peak that is 24.0 cm wide on a screen that is 1.58 m from a single slit. Calculate the width of the slit. Example 23-8

23-9 The diffraction of light through a circular aperture is important in optics

75. •• Light from a helium–neon laser with a wavelength of 633 nm passes through a 0.180-mm-diameter hole and forms a diffraction pattern on a screen 2.0 m behind the hole. Calculate the diameter of the central maximum. Example 23-9

76. • **Biology** The average pupil is 5.0 mm in diameter, and the average normal-sighted human eye is most sensitive at a wavelength of 555 nm. What is the eye's angular resolution in radians? Example 23-9

77. • **Astronomy** The telescope at Mount Palomar has an objective mirror that has a diameter of 508 cm. What is the angular limit of resolution due to diffraction for 560-nm light in degrees and radians? Example 23-9

78. • **Astronomy** The Hubble Space Telescope has a diameter of 2.4 m. What is the angular limit of resolution due to diffraction when a wavelength of 540 nm is viewed? Example 23-9

79. •• The distance from the center of a circular diffraction pattern to the first dark ring is 15,000 wavelengths on a screen that is 0.85 m away. What is the size of the aperture? Example 23-9

80. •• **Biology** Assume your eye has an aperture diameter of 3.00 mm at night when bright headlights are pointed at it. At what distance can you see two headlights separated by 1.50 m as distinct? Assume a wavelength of 550 nm, near the middle of the visible spectrum. Example 23-9

General Problems

81. •• A beam of light travels from medium 1 to medium 2 in the x–y plane. Medium 1 is found in quadrants 2 and 3; medium 2 is in quadrants 1 and 4. The beam touches each of the points in the x–y plane given in the table below. Calculate the ratio of the index of refraction of medium 2 to medium 1. Example 23-1

x (cm)	y (cm)	x (cm)	y (cm)
−4.00	−2.00	+1.00	+0.296
−3.00	−1.52	+2.00	+0.595
−2.00	−1.02	+3.00	+0.901
−1.00	−0.514	+4.00	+1.20
0	0		

82. •• One way of describing the speed of light in an optical material is to specify the ratio of the time that is required for light to travel through a length of vacuum to the time required for light to travel through the same length of the optical material. For example, if light travels through a material in 150% of the time for light to travel through a vacuum, the speed of light in the material would be 2/3 = 1/1.5 that in a vacuum. Complete the table by giving the speed of light

and the index of refraction for each of the following optical materials, listed with the corresponding percentage.

Optical material with percentage of time required for light to pass through compared to an equal length of vacuum	Speed of light	Index of refraction
100%		
125%		
150%		
200%		
500%		
1000%		

83. •• (a) Determine the index of refraction for medium 2 if the distance between points B and C in **Figure 23-52** is 0.75 cm. Assume the index of refraction in medium 1 is 1.00. (b) Suppose $n_2 = 1.55$. Calculate the distance between points B and C. Example 23-1

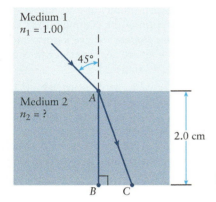

Figure 23-52
Problem 83

84. •• Suppose you are measuring double-slit interference patterns using an optics kit that contains the following options that you can mix and match: a red laser or a green laser; a slit width of 0.04 or 0.08 mm; a slit separation of 0.25 or 0.50 mm. Estimate how far you should place a screen from the double slit to give you an interference pattern on the screen that you can accurately measure using an ordinary ruler. Example 23-7

85. • The closest star to our sun (Proxima Centauri) is 4.0×10^{16} m away. It has a planet (Proxima b) in orbit around it at about 1/20 of the radius of Earth's orbit around our sun. Estimate the minimum diameter of the aperture of a space telescope that would be required to resolve Proxima b from its star, for light in the visible part of the spectrum. Example 23-9

86. ••• Prove that in the case where there are more than two optically different media sandwiched together, with air on the left and air on the right, the angle at which light returns to air is independent of the indices of refraction of the interior media (**Figure 23-53**). In other words, the refraction angle, θ_n, depends *only* on n_1, θ_1, and n_i (not n_2, n_3, n_4, ...). Example 23-1

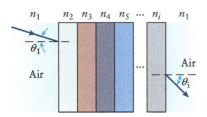

Figure 23-53
Problem 86

87. • One of the world's largest aquaria is the Monterey Bay Aquarium. The viewing wall is made of acrylic and is 0.33 m thick, and the tank holds 1.2 million gallons of water (**Figure 23-54**). If a ray of light is directed into the plastic from air at an angle of 40.0°, calculate the angle that the ray will make (a) when it enters the plastic and (b) when it enters the seawater. The indices of refraction for air, acrylic, and seawater are listed in the figure. Example 23-1

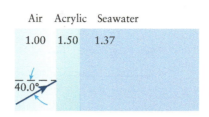

Figure 23-54
Problem 87

88. •• A flat glass surface ($n = 1.54$) has a layer of water ($n = 1.33$) of uniform thickness directly above the glass. At what minimum angle of incidence must light in the glass strike the glass–water interface for the light to be totally internally reflected at the water–air interface? Example 23-2

89. ••• Light rays in air fall normally on the vertical surface of a glass prism ($n = 1.55$), as shown in **Figure 23-55**. (a) What is the largest value of θ such that the ray is totally internally reflected at the slanted face? (b) Repeat the calculation if the prism is immersed in water with $n = 1.33$. Example 23-2

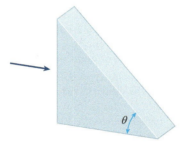

Figure 23-55 Problem 89

90. •• The object in **Figure 23-56** is a depth $d = 0.85$ m below the surface of clear water. (a) How far from the end of the dock, distance D in the figure, must the object be if it cannot be seen from any point on the end of the dock? The index of refraction of water is 1.33. (b) If you could change the index of refraction of the water, how would you change it so that the object could be seen at any distance D beneath the dock? Assume that the dock is 2.0 m long. Example 23-2

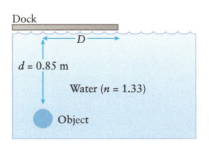

Figure 23-56
Problem 90

91. •• The polarizing angle for light that passes from water ($n = 1.33$) into a certain plastic is 61.4°. What is the critical angle for total internal reflection of the light passing from plastic into air? Example 23-4

92. •• A baseball is hit into a round pool of water that is 4.00 m deep and 17.0 m across (Figure 23-57). It lands right in the center of the pool. A large round raft shaped like a lily pad is floating in the pool, concentrically on top of the location of the ball. What minimum diameter must the raft have to completely obscure the ball from sight? Assume that water has an index of refraction of 1.33. Example 23-2

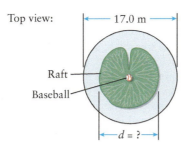

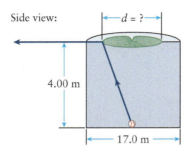

Figure 23-57 Problem 92

93. •• A glass sheet 1.40 μm thick is suspended in air. In reflected light, there are gaps in the visible spectrum at 560 and 640 nm. Calculate the minimum value of the index of refraction of the glass sheet that produces this effect. Example 23-6

94. •• Unpolarized light of intensity 100 W/m² is incident on two ideal polarizing sheets that are placed with their transmission axes perpendicular to each other. An additional polarizing sheet is then placed between the two, with its transmission axis oriented at 30° to that of the first. (a) What is the intensity of the light passing through the stack of polarizing sheets? (b) What orientation of the middle sheet enables the three-sheet combination to transmit the greatest amount of light?

95. •• Unpolarized light that has an intensity of 850 W/m² is incident on a series of polarizing filters as shown in Figure 23-58. If the intensity of the light after the final filter is 75 W/m², what is the orientation of the second filter relative to the x axis? *Hint:* cos (90° − θ) = sin θ and 2 sin θ cos θ = sin 2θ.

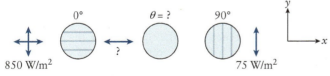

Figure 23-58 Problem 95

96. •• A thin film of soap solution (n = 1.33) has air on either side and is illuminated normally with white light. Interference minima are visible in the reflected light only at wavelengths of 400, 480, and 600 nm in air. What is the minimum thickness of the film? Example 23-6

97. ••• A brass sheet has a thin slit scratched in it. At room temperature (22.0°C) a laser beam illuminates the slit, and you observe that the first dark diffraction spot occurs at ±25.0° on either side of the central maximum. The brass sheet is then immersed in liquid nitrogen at 77.0 K until it reaches the same temperature as the nitrogen. It is removed and the same laser is shined on the slit. At what angle will the first dark spot now occur? Consult Table 14-1 for the thermal properties of brass. Example 23-8

98. •• A wedge-shaped air film is made by placing a small slip of paper between the edges of two thin plates of glass

12.5 cm long. Light of wavelength 600 nm in air is incident normally on the glass plates. If interference fringes with a spacing of 0.200 mm are observed along the plate, how thick is the paper? This form of interferometry is a very practical way of measuring small thicknesses. Example 23-6

99. •• A thin layer of SiO, having an index of refraction of 1.45, is used as a coating on certain solar cells. The refractive index of the cell itself is 3.5. (a) What is the minimum thickness of the coating needed to cancel visible light of wavelength 400 nm in the light reflected from the top of the coating in air? Are any other visible wavelengths also canceled? (b) Suppose that technological limitations require you to make the coating 3.0 times as thick as in part (a). Which, if any, visible wavelengths in the reflected light will be canceled in air and which, if any, will be reinforced? Example 23-6

100. •• You want to coat a pane of glass that has an index of refraction of 1.54 with a 155-nm-thick layer of material that has an index of refraction greater than that of the glass. The purpose of the coating is to cancel light reflected off the top of the film having a wavelength (in air) of 550 nm. (a) What should be the index of refraction of the coating? (b) If, due to technological difficulties, you cannot achieve a uniform coating at the desired thickness, what are the next three thicknesses of the coating you could use? Example 23-6

101. •• Two point sources of light each of which has a wavelength of 500 nm are photographed from a distance of 100 m using a camera with a 50.0-mm focal length lens. The camera aperture is 1.05 cm in diameter. What is the minimum separation of the two sources if they are to be resolved in the photograph, assuming the resolution is limited only by diffraction? Example 23-9

102. •• In 2009 researchers reported on evidence that a giant tsunami had hit the eastern coast of the Mediterranean Sea (present-day Lebanon and Israel) around 1600 B.C.E., causing huge damage to the civilizations located there. It is believed that the tsunami was caused by the eruption of the Thera volcano near the island of Crete. The waves would have passed through the 100-mi-wide opening between Crete and Rhodes, which would cause them to diffract and spread out. Satellite observations of tsunamis show that the waves measure about 250 mi from a crest to the adjacent trough, and the time between successive crests is typically 60 min. (a) How fast do tsunami waves travel? (b) How long after the eruption of Thera would the tsunami reach the eastern shore of the Mediterranean Sea, 600 mi from Thera? (c) For these waves, could we apply the formula $w \sin \theta = m\lambda$ to find the angles at which the waves cancel after passing through the 100-mi "slit" between Crete and Rhodes? Explain why or why not. Example 23-8

103. •• **Biology** The pupil (the opening through which light enters the lens) of a house cat's eye is round under low light, but ciliary muscles narrow it to a thin vertical slit in very bright light. Assume that bright light of wavelength 550 nm in air is entering the eye perpendicular to the lens and that the pupil has narrowed to a slit that is 0.500 mm wide. What are the three smallest angles on either side of the central maximum at which no light will reach the cat's retina (a) if we imagine the eye is filled with air and (b) if we take into consideration that in reality the eye is filled with a fluid having index of refraction of approximately 1.4? Example 23-8

104. •• If you peek through a 0.75-mm-diameter hole at an eye chart, you will notice a decrease in visual acuity.

(a) Calculate the angular limit of resolution if the wavelength is taken as 575 nm. (b) If the horizontal lines on the letter "E" are 2 mm apart, how close would you need to be to resolve them through the hole? Example 23-9

105. •• **Biology** The pupil of the eye is the circular opening through which light enters. Its diameter can vary from about 2.0 to about 8.0 mm to control the intensity of the light reaching the retina. (a) Calculate the angular resolution, θ_R, of the eye for light that has a wavelength of 550 nm in both bright light and dim light. In which light can you see more sharply, dim or bright? (b) You probably have noticed that when you squint, objects that were a bit blurry suddenly become somewhat clearer. In light of your results in part (a), explain why squinting helps you see an object more clearly. Example 23-9

106. •• **Biology** Under bright light the pupil of the eye (the circular opening through which light enters) is typically 2.0 mm in diameter. The diameter of the eye is about 25 mm. Suppose you are viewing something with light of wavelength 500 nm. Ignore the effect of the lens and the vitreous humor in the eye. (a) At what angles (in degrees) will the first three diffraction dark rings occur on either side of the central bright spot on the retina at the back of the eye? (b) Approximately how far (in millimeters) from the central bright spot would the dark rings in part (a) occur? (c) Explain why we do not actually observe such diffraction effects in our vision. Example 23-9

107. •• **Astronomy** Sometime around 2025, astronomers at the European Southern Observatory hope to begin using the ELT (Extremely Large Telescope), which is planned to have a primary mirror 42 m in diameter. Let us assume that the light it focuses has a wavelength of 550 nm. (a) What is the most distant Jupiter-sized planet the telescope could resolve, assuming its resolution is limited only by diffraction? Express your answer in meters and light years. (b) The nearest known exoplanets (planets beyond the solar system) are around 20 light years away. What would have to be the minimum diameter of an optical telescope to resolve a Jupiter-sized planet at that distance using light of wavelength 550 nm? (1 light year = 9.461×10^{15} m.) Example 23-9

108. •• **Astronomy** Under the best atmospheric conditions at the premium site for land-based observing (Mauna Kea, Hawaii, elevation 4.27 km), an optical telescope can resolve celestial objects that are separated by one-fourth of a second of arc (1 second of arc = 1 arcsec = 1/3600 of a degree). The viewing never gets any better than this because of atmospheric turbulence, which makes the images jitter. (a) What minimum diameter aperture is necessary to provide 1/4-arcsec resolution, assuming the resolution is limited only by diffraction? (b) Is there ever any point in building a telescope much bigger than this? Explain your answer. Example 23-9

109. •• **Astronomy** The Herschel infrared telescope, launched into space in 2009, made observations from 2010 to 2013. Its primary mirror is 3.5 m in diameter, and the telescope focuses infrared light in the range of 55 to 672 μm. Because this telescope operated above Earth's atmosphere, its resolution was limited only by diffraction. (a) What

wavelength in its observing range will give the maximum angular resolution? What is that maximum resolution (in radians and seconds, 1° = 60′ and 1′ = 60″)? (b) To achieve the same resolution as in part (a) using visible light of wavelength 550 nm, what should be the mirror diameter of an optical telescope? (c) What is the smallest infrared source that the Herschel infrared telescope can resolve at a distance of 150 light years? (A light year is the distance that light travels in one year—about 9.461×10^{15} m.) Example 23-9

110. •• **Astronomy** The world's largest refracting telescope is at Yerkes Observatory in Williams Bay, Wisconsin. Its objective is 1.02 m in diameter. Suppose you could mount the telescope on a spy satellite 200 km above the ground. (a) Assuming that the resolution is limited only by diffraction, what minimum separation of two objects on the ground could it resolve? Take 550 nm as a representative wavelength for visible light. (b) Because of atmospheric turbulence, objects on the surface of Earth can be distinguished only if their angular separation is at least 1.00 arcsec. (1 arcsec = 1/3600 of a degree.) How far apart would two objects on Earth's surface be if they subtended an angle of 1.00 arcsec as measured from the satellite? Compare this with your answer to part (a). Example 23-9

111. ••• Light from a helium–neon laser that has a wavelength of 633 nm shines on a planar surface with two small slits separated by a distance of 4.00 μm. What is the maximum number of bright fringes that can be seen on a screen 10 m away? Example 23-7

112. ••• In a realistic two-slit experiment, you see two interference effects described in this chapter. There is the large-scale fringe pattern from diffraction that depends on the width of each slit, and within that pattern there is the small-scale interference pattern that depends on the separation distance between the slits. For a slit width of 0.04 mm and a slit separation of 0.50 mm, how many of the bright two-slit interference fringes fit inside the central bright fringe of the diffraction pattern (from the first dark fringe on one side of the center to the first dark fringe on the other side) if the laser has a wavelength of (a) 650 nm or (b) 530 nm? Example 23-7

113. •• An equilateral triangular glass prism rests on a table (Figure 23-59). A beam of white light is incident on one face of the prism, at an angle of 60° below the normal to the surface. If the index of refraction of the glass is 1.51 for red light and 1.53 for blue light, what is the angular separation of the red and blue portions of the spectrum that emerge from the prism? Example 23-3

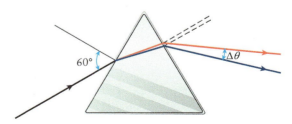

Figure 23-59 Problem 113

moodboard/Alamy

24

Geometrical Optics: Ray Properties of Light

What do you think?

Many older adults can see distant objects clearly but must wear corrective eyeglasses to see nearby objects (for example, to read a book). Are the lenses of these eyeglasses (a) thicker at the middle, (b) thicker at the edges, or (c) of uniform thickness?

In this chapter, your goals are to:

- (24-1) Explain the importance of optical devices.
- (24-2) Describe how a plane mirror forms an image.
- (24-3) Use ray diagrams to explain how the image formed by a concave mirror depends on the position of the object.
- (24-4) Calculate the position and height of an image made by a concave mirror.
- (24-5) Explain the differences between the images made by a convex mirror and a concave mirror.
- (24-6) Calculate the position and height of an image made by a convex mirror.
- (24-7) Describe how the curved surfaces of a lens make light rays converge or diverge.
- (24-8) Calculate the focal length of a lens based on its composition and shape.
- (24-9) Explain how the eye forms images on the retina, and how corrective lenses are used to compensate for deficiencies in vision.
- (24-10) Calculate the angular magnification of a magnifying glass, a microscope, and a telescope.

To master this chapter, you should review:

- (3-7) Calculating the length of a circular arc.
- (23-2) The law of reflection and Snell's law of refraction.
- (23-4) Dispersion.
- (23-9) Diffraction and angular resolution.

24-1 Mirrors or lenses can be used to form images

We saw in Chapter 23 that the direction of a light ray changes when it either reflects from a surface or refracts as it passes from one transparent medium to another. **Geometrical optics** is the branch of physics that uses the laws of reflection and refraction to understand **optical devices**—instruments that change the direction of light rays in a regular way. The simplest optical devices are reflective objects, or *mirrors*, and pieces of transparent material with carefully shaped surfaces, or *lenses*. Mirrors or lenses whose surfaces are curved in just the right way can be used to change the apparent sizes of objects (Figure 24-1a). The lens of your eye is an essential part of vision and needs to be replaced if it becomes clouded with age (Figure 24-1b).

As the name geometrical optics suggests, to study optical devices all we need besides the laws of reflection and refraction is a little bit of geometry. We won't need to refer at all to the wave properties of light. (Indeed, much of the basic physics of mirrors and lenses was deduced before it was understood that light is a wave.)

We'll begin by considering a simple plane mirror. We'll see how the law of reflection explains the kind of image that it forms. We'll then use similar ideas to understand the kind of images formed by a mirror with a surface that's curved either inward or outward. In contrast to mirrors, lenses use the refraction of light to form an image: Light rays can change direction when they enter a lens and again when they exit the lens. Happily, we'll find that many of the same ideas that we'll develop by considering curved mirrors apply equally well to lenses. We'll discuss the human eye, how it forms images, how deficiencies in vision can be corrected, and how it compares to a camera. Finally, we'll look at three optical devices of particular importance: the magnifying glass, the microscope, and the telescope.

This dentist uses a curved mirror to make a magnified image of a patient's tooth. He can also get a magnified view by using the lenses attached to his eyeglasses.

The human eye contains a transparent lens that helps focus images onto the retina. Many people develop cataracts as the clear lens gradually becomes cloudy. The treatment is to surgically remove the lens and replace it with an artificial one.

(a) 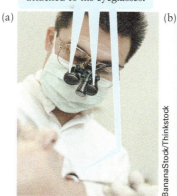 (b)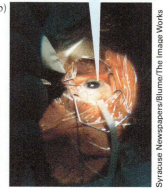

Figure 24-1 Mirrors and lenses Mirrors and lenses are two different devices used to form images.

> ## TAKE-HOME MESSAGE FOR Section 24-1
>
> ✔ A mirror forms images by the reflection of light from the mirror's surface.
>
> ✔ A transparent lens forms images by the refraction of light as it enters and exits the lens.

24-2 A plane mirror produces an image that is reversed back to front

Our visual system can detect only objects that emit or reflect light. (We can't see in the dark!) Although it's easy to find examples of luminous things, such as the Sun or the screen of a mobile device, most of what we see only reflects light. As we learned in Section 23-2, a ray of light that strikes a surface always reflects in such a way that the angle of incidence equals the angle of reflection. (This is the law of reflection.) However, because most objects have uneven surfaces, the light that they reflect goes off in many seemingly random directions. This is called **diffuse reflection** (Figure 24-2a).

If an object has a flat surface, however, light rays that strike that surface are all reflected in the same general direction (Figure 24-2b). Such a flat, reflecting surface is called a **plane mirror**, and reflections from such a surface are called *specular* (from the Latin word for mirror, *speculum*). Your reflection in a bathroom mirror is an example of **specular reflection**. Let's look more closely at how a plane mirror creates an image.

BioMedical   Figure 24-3 shows how your eye and brain interpret light coming from an **object** (a term that refers to anything that acts as a source of light rays for an optical device). Some of the light rays coming from the object go to your eyes. Your brain traces those rays backward to a common origin and interprets that origin as the location of the object. Figure 24-4 shows a similar situation, except that we have added a plane mirror. Some of the light from the object strikes the mirror

(a)

Light that reflects from an object's surface in many random directions is called diffuse. Uneven surfaces produce diffuse light even when the incident rays come from the same direction.

(b)

Light that reflects from a smooth surface is called specular. Light rays coming from the same general direction all reflect in the same general direction.

Figure 24-2 Diffuse and specular reflection Light reflecting from (a) an uneven surface and (b) a smooth surface.

Figure 24-3 Locating an object
When light strikes our eyes, we trace the rays of light back along straight lines to an apparent common source.

(1) Light from a light source strikes an object (in this case, a red ball).

Object

(2) The object reflects light in many directions. Some of the rays go to your eye.

(3) Your brain recognizes that light travels in straight lines, so it traces the rays back along straight lines to a common point. That common point is the location of the object.

and is reflected toward your eye. These rays appear to be coming from a point behind the mirror, so your brain interprets the light as coming from that point. We say that the mirror has formed an **image** of the object, and this image lies behind the mirror.

As you can see in Figure 24-4, no light rays actually pass through the location of the image. An image of this kind is called a **virtual image**. The image formed by a plane mirror is always a virtual image. We will shortly encounter some optical devices that cause light rays to bend toward each other so that the image forms where light rays do actually meet. An image formed by light rays coming together is called a **real image**.

We can determine the position of the image made by a plane mirror by using a **ray diagram** (Figure 24-5). In such a diagram, we draw a few light rays coming from the object and show how they reflect from the mirror. We've drawn the object as a red arrow of height h_O (the **object height**) located a distance d_O (the **object distance**) from

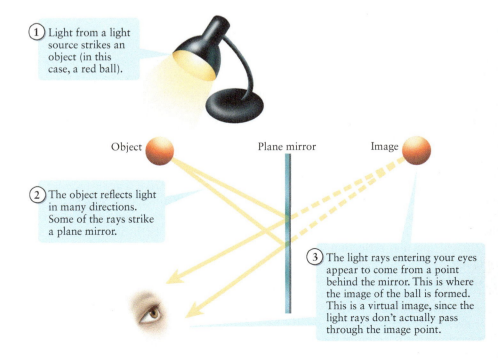

(1) Light from a light source strikes an object (in this case, a red ball).

Object   Plane mirror   Image

(2) The object reflects light in many directions. Some of the rays strike a plane mirror.

(3) The light rays entering your eyes appear to come from a point behind the mirror. This is where the image of the ball is formed. This is a virtual image, since the light rays don't actually pass through the image point.

Figure 24-4 Locating an image
When light from a mirror strikes our eyes, we use the same technique as in Figure 24-3 to determine the position of the image made by the mirror.

the mirror. To determine the image position, we must draw at least two rays that emanate from the tip of the object arrow. (We've drawn three in Figure 24-5.) When each ray strikes the mirror, it obeys the law of reflection: The angle of the reflected ray equals the angle of the incident ray. Note that the horizontal ray strikes the mirror face-on, so it is reflected back horizontally toward the tip of the object arrow.

The reflected rays diverge from each other and never actually meet. But if we trace these rays back to where they would meet, as shown by the dashed lines in Figure 24-5, we find the position of the tip of the image arrow. The geometry of the light rays requires that the image be as far behind the front of the mirror as the object is in front of the mirror. In other words, the **image distance** d_I has the same magnitude as the object distance d_O. We'll use the convention that a point on the reflective side of the mirror (to the left of the mirror in Figure 24-5) is at a positive distance, while a point on the back side of the mirror (to the right of the mirror in Figure 24-5) is at a negative distance. So in Figure 24-5 the object distance d_O is positive, but the image distance d_I is negative. We can write the relationship between d_I and d_O for a plane mirror as

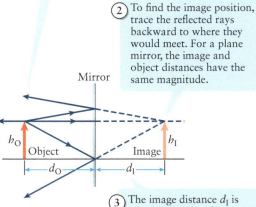

① Light rays from the object reflect from the plane mirror. (The ray that hits the mirror face-on reflects back along its initial path.)

② To find the image position, trace the reflected rays backward to where they would meet. For a plane mirror, the image and object distances have the same magnitude.

③ The image distance d_I is negative because the (virtual) image is behind the mirror: $d_I = -|d_I| = -d_O$

The **image distance** is negative.   The **object distance** is positive.

$$d_I = -|d_I| = -d_O$$

Image distance for a plane mirror
(24-1)

The negative value of d_I indicates that the image is on the opposite side of the mirror from the object.

Figure 24-5 A ray diagram for a plane mirror This diagram helps us determine the position, orientation, and size of the image.

For example, if you stand a distance $d_O = 1.0$ m in front of a plane mirror, your image is at $d_I = -1.0$ m—that is, 1.0 m behind the mirror.

Figure 24-5 also shows that the **image height** h_I is the same as the object height h_O. We define the **lateral magnification** m as the ratio of the image height to the object height. (We will often refer to m simply as the "magnification.")

Lateral magnification   Image height

$$m = \frac{h_I}{h_O}$$

Lateral magnification
(24-2)

Object height

For a plane mirror $h_I = h_O$, so the magnification is $m = 1$. For an optical device that results in magnification greater than 1, the image is larger than the object. Such a device is commonly called a *magnifier*; it forms a magnified image of the object. When m is less than one, the image formed by the optical device is smaller than the object. As we'll see in later sections, a *curved* mirror can form an image that is either larger or smaller than an object placed in front of it.

Go to Picture It 24-1 for more practice dealing with mirrors.

WATCH OUT! The image formed by a plane mirror appears *behind* the mirror.

The image does not form "on" a plane mirror but rather behind it. This is evident from the ray diagrams in Figure 24-5. But if this diagram doesn't convince you, you can prove it to yourself by taping a bit of paper to a plane mirror, then looking at your reflection in the mirror from about 1 m away. You'll find it difficult, likely impossible, to focus your eyes on both your image in the mirror and the piece of paper at the same time. That's because your image in the mirror is twice as far from your eyes as the paper on the surface of the mirror. The image is behind the mirror, not on it.

WATCH OUT! The image formed by a plane mirror is reversed back to front, *not* left to right.

 It's a common misconception that your image in a plane mirror is reversed left to right. A better description is that your image is reversed *back to front*. As an example, in Figure 24-6 a rectangular box *ABCDEFGH* sits in front of a plane mirror, making an image *A'B'C'D'E'F'G'H'*. Note that the face *A'B'C'D'* of the image has the same orientation as the face *ABCD* of the object and is the same distance from the mirror; the same is true of the face *E'F'G'H'* of the image and the corresponding face *EFGH* of the object. You can see that the net result is that the image is identical to the object but flipped from back to front.

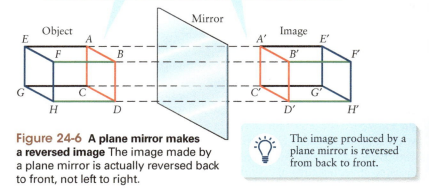

Each point on the image behind the mirror is directly opposite the corresponding point on the object in front of the mirror: *A'* opposite *A*, *B'* opposite *B*, and so on.

Figure 24-6 A plane mirror makes a reversed image The image made by a plane mirror is actually reversed back to front, not left to right.

The image produced by a plane mirror is reversed from back to front.

GOT THE CONCEPT? 24-1 See You, See Me I

 You look into a mirror hanging near the corner of a hallway and see the eyes of someone standing around the corner. Is she able to see you? (a) Yes; (b) no; (c) not enough information given to decide.

GOT THE CONCEPT? 24-2 See You, See Me II

 You look into a mirror hanging near the corner of a hallway. You see the right hand of someone standing around the corner, but not his eyes. Is he able to see you? (a) Yes; (b) no; (c) not enough information given to decide.

TAKE-HOME MESSAGE FOR Section 24-2

✔ When light rays coming from the same general direction hit a plane mirror, they tend to be reflected in the same general direction.

✔ If an object is placed in front of a plane mirror, the image is as far behind the mirror as the object is in front.

The image is the same size as the object, but reversed from back to front.

✔ A plane mirror makes a virtual image; the light rays coming from the image do not actually pass through the image position.

24-3 A concave mirror can produce an image of a different size than the object

Figure 24-7 shows a jalapeño pepper placed in front of two curved mirrors. The mirror in Figure 24-7a is **convex** (its reflective surface is curved outward) and produces an image that is smaller than the object. The mirror in Figure 24-7b is **concave** (its reflective surface is curved inward, or "caved in") and produces an image that is larger than the object. By contrast, the plane mirror that we studied in Section 24-2 always produces an image of the same size as the object. Let's analyze what happens with curved mirrors of this kind.

(a) Convex mirror

(b) Concave mirror

Figure 24-7 Images from curved mirrors A curved mirror can produce an image that is a different size than the object.

David Tauck

Figure 24-8 shows ray diagrams for parallel light rays striking two *spherical* mirrors—that is, mirrors that are shaped like a section cut from a complete sphere. Figure 24-8a shows that parallel light rays converge after they strike a concave mirror, while Figure 24-8b shows that parallel light rays diverge after they strike a convex mirror. By comparison, parallel light rays that strike a plane mirror remain parallel after striking it (see Figure 24-2b). It's important to consider such light rays because light rays that come from distant objects are either parallel or nearly so. The Sun's rays, for example, are essentially parallel when they strike Earth. You can see this from the crisp shadows cast by an object placed in the Sun, such as the hanging frame in Figure 24-9a. In contrast, the shadow cast by a light bulb in Figure 24-9b is fuzzy because the light rays from the light bulb are not parallel.

The concave mirror has many practical uses. For example, a bathroom mirror used for applying makeup or for shaving is concave so that it gives an enlarged image of your face (like the enlarged image of the pepper in Figure 24-7b). Telescopes used by professional and amateur astronomers have a large concave mirror that brings the light from distant objects to a focus, forming an image (we'll discuss this further in Section 24-10). Automobile headlights use the same principle in reverse: A light bulb is placed in front of a concave mirror, and the mirror reflects the light forward to illuminate the road ahead. In this section and the next, we'll look at the concave mirror in detail; we'll return to the convex mirror in Sections 24-5 and 24-6.

Figure 24-10 shows what happens when a series of parallel light rays strike a concave mirror formed from a fairly large section of a sphere. The incoming rays are parallel to each other and also parallel to the **principal axis** of the mirror, the axis that runs through the center of the sphere and also the center of the mirror. Notice that while the reflected rays generally converge along the principal axis of the mirror, they do not converge to the exact same point. This unfortunate characteristic of spherical mirrors is referred to as *spherical aberration*.

To avoid spherical aberration, we limit the reflective surface of the mirrors we consider to a relatively small section of a sphere. Here "relatively small" means that the size of the reflective surface is small compared to the radius of the sphere. There is

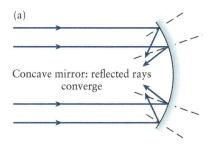

(a)

Concave mirror: reflected rays converge

(b)

Convex mirror: reflected rays diverge

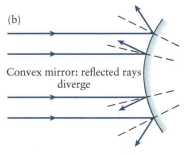

Figure 24-8 Concave and convex mirrors (a) Parallel light rays converge after reflecting from a concave mirror but (b) diverge after reflecting from a convex mirror.

(a)

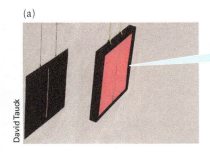

David Tauck

Rays of light from the distant Sun strike the hanging frame. Notice the sharp edges on all of the shadows, including those of the strings holding the frame.

These parallel rays cast a shadow with sharp edges.

(b)

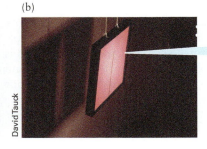

David Tauck

Light from a lamp strikes the frame. Because the light bulb is close to the frame, the central region of the shadow is darker than the outer regions.

These nonparallel rays cast a shadow with fuzzy edges.

Figure 24-9 Parallel and nonparallel light rays These photographs show the difference between light rays that are (a) parallel and (b) nonparallel.

- A concave, spherical mirror causes parallel light rays to nearly converge along the principal axis of the mirror.
- If we consider only rays close to the principal axis, or if the mirror is only a small arc of the complete sphere, the light rays all converge to essentially a single point.

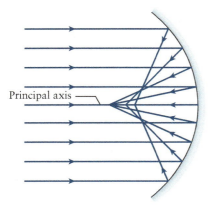

Figure 24-10 Reflection from a spherical mirror How parallel light rays behave when they strike a concave, spherical mirror.

r is the radius of curvature of the full sphere.

f is the focal length.

C is the center of curvature of the full sphere.

F is the focal point of the mirror.

Figure 24-11 Mirror nomenclature This drawing defines many of the variables we use to describe a concave, spherical mirror.

no specific cut-off value; rather, the smaller the mirror compared to the radius, the more tightly focused the reflected rays will be. For the rest of this chapter, we will deal with spherical mirrors small enough that all rays parallel to the principal axis are focused to essentially a single point.

Figure 24-11 defines many of the variables we use to describe a concave, spherical mirror. The **focal point** of the mirror is the point F along the principal axis at which incident rays parallel to the principal axis converge to a common focus when they reflect off the mirror. The distance from the focal point to the center of the mirror is f, the **focal length**. The mirror is a small section of a full sphere, and the point C—the **center of curvature** of the mirror—is the center of that sphere. The distance from C to any point on the mirror is the radius r of the sphere, also called the **radius of curvature** of the mirror.

For a concave mirror small enough that all rays parallel to the principal axis are focused at the focal point, the focal length is *exactly* half of the radius:

Focal length of a spherical mirror (24-3)

Focal length of a spherical mirror   Radius of curvature of the mirror

$$f = \frac{r}{2}$$

Figure 24-11 shows this relationship. Equation 24-3 says that the tighter the curve of a spherical mirror and hence the smaller the radius of curvature r, the shorter the focal length f. (We'll prove the relationship $f = r/2$ in Section 24-4.)

To see how a concave mirror forms an image, let's put our standard arrow at a point far from the mirror, as in Figure 24-12a. "Far" in this case means that the object distance d_O is greater than the radius of the mirror r; in other words, the base of the arrow is farther from the mirror, along the principal axis, than the center of curvature C. Now let's trace two rays of light from the tip of the arrow as they strike and then are reflected from the mirror. The image of the arrow's tip forms where those two rays intersect. Although any two light rays that originate at the tip of the arrow and that strike the mirror will work, we choose two that are particularly convenient. In Figure 24-12b, we trace a ray that starts parallel to the principal axis because all such rays are reflected through the focal point. In Figure 24-12c we add the trace of a ray that strikes the center of the mirror. The normal to the surface at this point lies along the principal axis, so it's easy to apply the law of reflection; the incident and reflected rays are symmetric around the principal axis.

Where does the image of the *base* of the arrow form? The base is on the principal axis, which is normal to the mirror's surface at the center of the mirror. So a light ray coming from the base of the arrow is reflected straight back from the center of the mirror, which means the image of the base of the arrow forms along the principal axis. Figure 24-12d shows the final image of the arrow. This is an **inverted image**: It's flipped upside down compared to the object. The image is also smaller than the object. Finally, the image is real; that is, it forms in front of the mirror where light rays reflected from different parts of the mirror's surface meet.

Figure 24-12d shows both the object distance d_O and the image distance d_I; both distances are positive, since the object and image are both on the reflective side of the mirror. It also shows that if d_O is large, the image is closer to the mirror than the object is, so $d_I < d_O$. Note that the image distance is greater than the focal length (the distance from the mirror to the focal point F). This is how a concave mirror is used in a telescope: The object is very far away, while the image is formed very close to the mirror and is much smaller than the object. (For example, the Moon is 3.84×10^5 km distant and 1738 km in radius, but the Moon's image made by an amateur astronomer's telescope is formed only a meter or so from the mirror and is less than a centimeter in radius.) Figure 24-12e shows the inverted image of one of the authors of this book standing far away from a concave mirror.

Let's see what happens if we move the object closer to the mirror. Figure 24-13a shows the object arrow placed at the center of curvature C, so the object distance equals the radius of curvature r. From Equation 24-3 the focal length $f = r/2$, so for the case shown in Figure 24-13, $d_O = r = 2f$. We trace the same two light rays

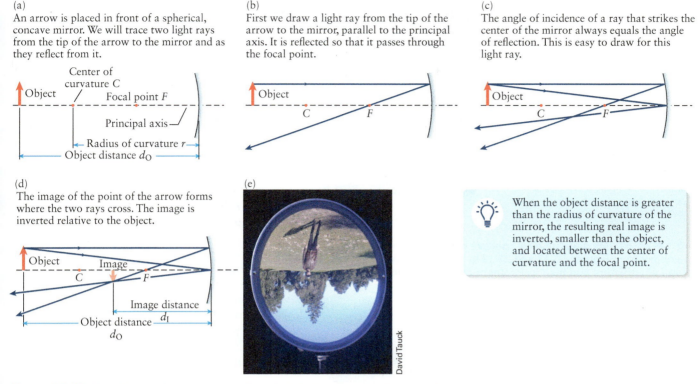

(a)

An arrow is placed in front of a spherical, concave mirror. We will trace two light rays from the tip of the arrow to the mirror and as they reflect from it.

Center of curvature C

Object

Focal point F

Principal axis

Radius of curvature r

Object distance d_O

(b)

First we draw a light ray from the tip of the arrow to the mirror, parallel to the principal axis. It is reflected so that it passes through the focal point.

Object

C

F

(c)

The angle of incidence of a ray that strikes the center of the mirror always equals the angle of reflection. This is easy to draw for this light ray.

Object

C

F

(d)

The image of the point of the arrow forms where the two rays cross. The image is inverted relative to the object.

Object

Image

C

F

Image distance d_I

Object distance d_O

(e)

David Tauck

When the object distance is greater than the radius of curvature of the mirror, the resulting real image is inverted, smaller than the object, and located between the center of curvature and the focal point.

Figure 24-12 Ray diagram for a concave mirror I (a) A distant object in front of a concave mirror. (b), (c), (d) Locating the image of this object. (e) The image of a person standing far from a concave mirror.

as in Figure 24-12; where the two rays meet and the image forms is now farther from the mirror. This image is the same size as the object and, as in the previous case, is both real (the light rays pass through the image) and inverted. With the object placed at the center of curvature, the image forms at the center of curvature, so $d_I = d_O$. In Figure 24-13b one of the authors is standing at the center of curvature of a concave mirror; note that his inverted image is larger than that shown in Figure 24-12e.

In Figure 24-14a we've moved the arrow closer still to the mirror so that it sits between the center of curvature C and the focal point F. In this case the object distance is between $2f$ and f: $2f > d_O > f$. Again we've traced the same two light rays that we considered in the previous two cases. The image is still both real and inverted but has moved farther still from the mirror, so it is now farther from the mirror than the object (so $d_I > d_O$) and larger than the object. Figure 24-14b shows such an image of one of the authors; compare Figures 24-12e and 24-13b.

Comparing Figures 24-12, 24-13, and 24-14 shows that the image gets larger and moves farther from the mirror as we move a distant object closer to the focal point. In Figure 24-15a we place the object *at* the focal point so that the object distance equals the focal length, or $d_O = f$. After reflection, light rays that emanate from the tip of the arrow are parallel. (This is how a concave mirror is used in an automobile headlight: The lamp itself is placed close to the focal point of the curved mirror behind it, and the reflected light forms a beam of nearly parallel light rays.) With the object at the focal point, no image is formed because the parallel reflected rays never meet. The author in Figure 24-15b is standing close to the focal point of the concave mirror; there is no sharp image. If the object is slightly outside the focal point (d_O is slightly larger than f), the image is formed

Figure 24-13 Ray diagram for a concave mirror II (a) We move the object from Figure 24-12 to the center of curvature of the concave mirror. (b) The image of a person standing at the center of curvature of a concave mirror.

(a)

Object

C

F

Image

Image distance d_I
= Object distance d_O

(b)

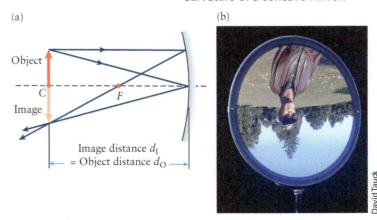

David Tauck

When the object is located at the radius of curvature of the mirror, the resulting real image is inverted, the same size as the object, and located at the radius of curvature.

(a)

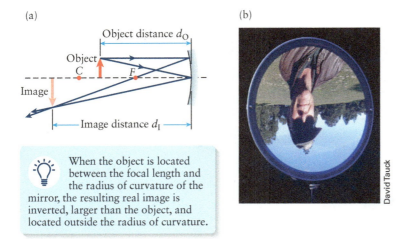

(b)

David Tauck

When the object is located between the focal length and the radius of curvature of the mirror, the resulting real image is inverted, larger than the object, and located outside the radius of curvature.

Figure 24-14 Ray diagram for a concave mirror III (a) We move the object from Figures 24-12 and 24-13 to a point between the center of curvature and the focal point of the concave mirror. (b) The image of a person standing at such a point.

(a)

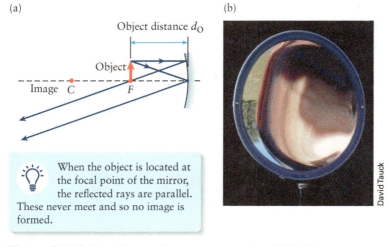

(b)

David Tauck

When the object is located at the focal point of the mirror, the reflected rays are parallel. These never meet and so no image is formed.

Figure 24-15 Ray diagram for a concave mirror IV (a) We move the object from Figures 24-12, 24-13, and 24-14 to the focal point of the concave mirror. (b) The image of a person standing near the focal point.

very far from the mirror and is much larger than the object.

What happens when the object is placed inside the focal point, so the object distance is less than the focal length ($d_O < f$)? In this case the reflected rays never actually meet but rather appear to meet at a point behind the mirror (Figure 24-16a). Hence the image is virtual, like the image formed by the plane mirror of Section 24-2, and the image distance d_I is negative: $d_I = -|d_I| < 0$. The image is larger than the object and is an **upright image** (it has the same orientation as the object). That's the kind of enlarged image you see when you look in a curved bathroom mirror for shaving or applying makeup. Figure 24-16b shows such an enlarged image of one of the authors (compare Figures 24-12e, 24-13b, 24-14b, and 24-15b). The photograph in Figure 24-7b shows another such enlarged image.

So far, our results for image formation by concave mirrors have been qualitative only. In the following section we'll take a more *quantitative* look at how the position of an object placed in front of a concave mirror determines the position and size of the resulting image.

GOT THE CONCEPT? 24-3 A Soup Spoon

A shiny spoon is not so different in shape than a spherical mirror: It's concave on one side and convex on the other. Your reflection from the concave side of the spoon, when held at arm's length, is upside down and appears to float in front of the spoon. When you hold the spoon about 6 cm from your eye, you see only a blur reflected in the spoon, but when you hold it about 4 cm from your eye your reflection is right side up and appears to be behind the spoon. What is the approximate focal length of the spoon? (a) More than 6 cm; (b) 6 cm; (c) between 4 and 6 cm; (d) 4 cm; (e) less than 4 cm.

(a)

When the object is closer to the mirror than the focal point, the resulting virtual image is upright, larger than the object, and located behind the mirror.

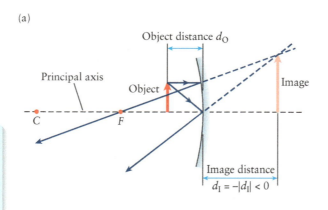

(b)

David Tauck

Figure 24-16 Ray diagram for a concave mirror V (a) We move the object from Figures 24-12, 24-13, 24-14, and 24-15 to a point inside the focal point of the concave mirror. (b) The image of a person standing at such a point.

GOT THE CONCEPT? 24-4 Covering a Concave Mirror

(**?**) You place a light bulb oriented vertically (with its base on the bottom) just outside the focal point of a concave mirror. The resulting real image of the light bulb falls on a wall far from the mirror. This image is inverted and larger than the light bulb. If you were to paint the bottom half of the mirror black so that light rays that strike this half of the mirror would not be reflected, the image would show (a) only the top of the light bulb; (b) only the bottom of the light bulb; (c) only the left side of the light bulb; (d) only the right side of the light bulb; (e) the entire light bulb.

TAKE-HOME MESSAGE FOR Section 24-3

✔ A concave spherical mirror is in the shape of the inner surface of a section of a sphere.

✔ When an object is outside the focal point of a concave mirror, the image is outside the focal point, real and inverted.

✔ As the object is moved closer to the focal point, the image becomes larger and forms farther from the mirror.

✔ When the object is at the focal point, the reflected rays are parallel. Effectively, the image is at infinity and infinitely large.

✔ If the object is placed inside the focal point, the image is virtual and upright.

24-4 Simple equations give the position and magnification of the image made by a concave mirror

Let's return to the concave spherical mirror that we considered in the previous section. We'll see that given the radius of the mirror and the object distance, we can determine exactly how far from the mirror the image forms and how much magnification the mirror provides. All we need are a little geometry and the law of reflection.

The Mirror Equation for a Concave Mirror

To find the mathematical relationship among the object distance d_O, image distance d_I, and the radius of curvature r, we'll imagine that the object is a point located on the principal axis of the mirror as in Figure 24-17. Then the image will also lie along the principal axis. (The explanation is the same one we used in Section 24-3 to show why the image of the base of the arrow in Figure 24-12 must lie on the principal axis.) Figure 24-17 shows an arbitrarily chosen light ray coming from the object at point O and reflecting off the mirror at point P. The image forms where this reflected ray intercepts the principal axis at point I.

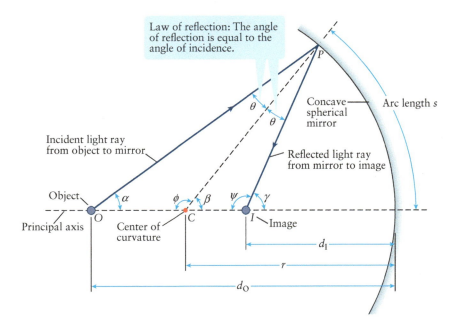

Figure 24-17 Analyzing a concave spherical mirror A ray diagram for a point object on the principal axis of a concave spherical mirror.

WATCH OUT! The path from object to mirror to image isn't just for one special light ray.

! There's nothing special about the light ray we've drawn in Figure 24-17. *Any* ray that emanates from the object point O will reflect off the mirror and pass through the *same* image point I, provided the ray is at a shallow enough angle α to the principal axis of the mirror. So every part of the mirror contributes to forming the image. (Note that in Figure 24-17 we've drawn the angle α as fairly large, even though this angle is actually quite small. We've done this just to make the angles in the figure easier to see.)

$\sqrt{x}$ *See the Math Tutorial for more information on geometry.*

We can find a relationship among d_O, d_I, and r by first finding a relationship among the angles in Figure 24-17 labeled α (the angle between the principal axis and the light ray from O to P), β (the angle between the principal axis and a line from the center of curvature C to P), and γ (the angle between the principal axis and the light ray from P to I). Since we're considering α to be a small angle, then necessarily β and γ are small angles as well. For that reason we can treat each of the three regions formed by one of these angles and the arc length s of the mirror as a sector (a pie-like slice) of a circle. The arc length is common to all three sectors. The length of the arc of a circle of radius r subtended by an angle θ equals $r\theta$ (see Section 3-7), provided the angle θ is in radians. So in Figure 24-17

(24-4)
$$s = d_O\alpha; \quad s = r\beta; \quad s = d_I\gamma$$

Note also that the sum of the angles in any triangle equals 180° or π radians. For the triangle OPC we have

(24-5)
$$\alpha + \theta + \phi = \pi$$

However, the angles ϕ and β in Figure 24-17 must add to 180°, or π radians (together they make up half a circle). So $\phi + \beta = \pi$ and $\phi = \pi - \beta$, and Equation 24-5 becomes

(24-6)
$$\alpha + \theta + \pi - \beta = \pi \quad \text{or} \quad \theta = \beta - \alpha$$

Similarly, for the triangle OPI the sum of the angles is π:

(24-7)
$$\alpha + 2\theta + \psi = \pi$$

Figure 24-17 shows that the angles ψ and γ also make up half a circle, so $\psi + \gamma = \pi$ and $\psi = \pi - \gamma$. Triangle CPI shows that $\beta + \theta + \psi = \pi$. Comparing with Equation 24-7 it follows that $\alpha + 2\theta = \beta + \theta$. Equation 24-7 can then be written as

(24-8)
$$\beta + \theta + \pi - \gamma = \pi \quad \text{or} \quad \theta = \gamma - \beta$$

Equations 24-6 and 24-8 are two different expressions for the angle θ in Figure 24-17. If we set these equal to each other, we get

(24-9)
$$\beta - \alpha = \gamma - \beta \quad \text{or} \quad \alpha + \gamma = 2\beta$$

From Equations 24-4 we have $\alpha = s/d_O$, $\beta = s/r$, and $\gamma = s/d_I$. Substituting these into Equation 24-9 gives

$$\frac{s}{d_O} + \frac{s}{d_I} = 2\frac{s}{r}$$

To simplify we divide through by the arc length s, giving us the relationship we've been seeking among the object distance, image distance, and radius of curvature of the mirror:

(24-10)
$$\frac{1}{d_O} + \frac{1}{d_I} = \frac{2}{r}$$

To help interpret Equation 24-10, recall from Figure 24-11 that if the incident light rays are parallel, they come to a focus at the focal point a distance f from the mirror. So in this case $d_I = f$. The incident rays will be parallel if the object is infinitely far away, so $d_O \to \infty$ and $1/d_O \to 0$. Then Equation 24-10 becomes

$$0 + \frac{1}{f} = \frac{2}{r} \quad \text{or} \quad f = \frac{r}{2}$$

This justifies Equation 24-3: The focal length of a concave spherical mirror is one-half of the radius of curvature.

If we replace $2/r$ with $1/f$ in Equation 24-10, we get the final form of the **mirror equation**:

$$\frac{1}{d_O} + \frac{1}{d_I} = \frac{1}{f}$$

Focal length

Object distance   Image distance

Mirror equation and lens equation relating object distance, image distance, and focal length (24-11)

We also call Equation 24-11 the **lens equation** because, as we shall see in Section 24-8, it's also applicable to the image formed by a lens.

We take the focal length f of a concave mirror to be positive because the center of curvature C of the mirror is in front of the mirror, that is, on its reflective side (to the left in Figure 24-17). Likewise, since the object point O is in front of the mirror, the object distance d_O is positive. The image distance d_I, however, can be positive if the image is real (in front of the mirror) or negative if the image is virtual (behind the mirror). To see when the image distance is positive and when it is negative, let's rewrite Equation 24-11 to solve for d_I:

$$\frac{1}{d_I} = \frac{1}{f} - \frac{1}{d_O} = \frac{d_O}{d_O f} - \frac{f}{d_O f} = \frac{d_O - f}{d_O f} \quad \text{or} \quad d_I = \frac{d_O f}{d_O - f} \tag{24-12}$$

In Equation 24-12, d_O and f are both positive for a concave mirror, so the numerator $d_O f$ is positive. However, the denominator $d_O - f$ can be positive or negative depending on whether d_O is larger or smaller than f.

If the object is outside the focal point so that $d_O > f$, then the denominator $d_O - f$ in Equation 24-12 is positive and the image distance is positive. Therefore, in this case the image made by the mirror is real. As the object moves closer to the focal point, the difference between d_O and f gets smaller and so d_I given by Equation 24-12 gets larger. Hence the image moves farther away from the mirror. If the object is inside the focal point, however, then $d_O < f$ and the denominator $d_O - f$ is negative. Then d_I is negative as well, and the image is virtual. That's exactly the behavior that we deduced in Section 24-3 by analyzing ray diagrams.

As a further check on Equations 24-11 and 24-12, note that if the object is exactly two focal lengths from the mirror so $d_O = 2f$, the image distance is

$$d_I = \frac{d_O f}{d_O - f} = \frac{(2f)f}{2f - f} = \frac{2f^2}{f} = 2f$$

So when the object is a distance $2f$ from the mirror—which, because $r = 2f$, is at the center of curvature—the image is at the same position. That's the same conclusion we came to by using the ray diagram in Figure 24-13.

Magnification for a Concave Mirror

We can also use simple geometry to find an expression for the lateral magnification of the image produced by a concave spherical mirror. In Figure 24-18 we've replaced the point object at O with an upright arrow of height h_O that extends from O (on the principal axis) to O'. Just as in Figure 24-12 we draw two rays coming from the tip of the object arrow at O' to determine the position I' of the tip of the image arrow.

Since the object in Figure 24-18 is outside the focal point F, the image is real and inverted and so the height of the image is negative: $h_I = -|h_I|$. By the law of reflection the ray from O' to the point P' at the center of the mirror makes the same angle θ with the mirror's principal axis (shown as a dashed line in Figure 24-18) as does the reflected ray from P' to I'. If you look at the right triangle $OO'P'$, you'll see that the tangent of θ (the opposite side divided by the adjacent side) is $\tan\theta = h_O/d_O$; if you do the same for the right triangle $II'P'$, you'll see that $\tan\theta = |h_I|/d_I$. Setting these two expressions equal to each other, we see that

$$\frac{h_O}{d_O} = \frac{|h_I|}{d_I} \quad \text{or} \quad \frac{|h_I|}{h_O} = \frac{d_I}{d_O}$$

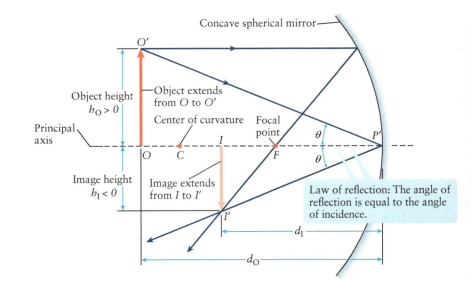

Figure 24-18 Magnification for a concave spherical mirror We replace the point object in Figure 24-17 with an object arrow.

- The right triangles $OO'P'$ and $II'P'$ both include the same angle θ.
- So $\tan \theta = h_O/d_O = |h_I|/d_I$ and $|h_I|/h_O = d_I/d_O$.
- Since h_I is negative, $h_I = -|h_I|$ and $m = h_I/h_O = -|h_I|/h_O = -d_I/d_O$.

Since $h_I = -|h_I|$, we can rewrite this as

(24-13)

$$-\frac{h_I}{h_O} = \frac{d_I}{d_O} \quad \text{or} \quad \frac{h_I}{h_O} = -\frac{d_I}{d_O}$$

From Equation 24-2 the lateral magnification is $m = h_I/h_O$. So Equation 24-13 tells us that

Lateral magnification for a mirror or lens

(24-14)

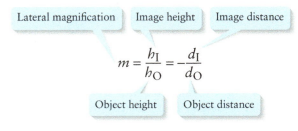

$$m = \frac{h_I}{h_O} = -\frac{d_I}{d_O}$$

 Go to Interactive Exercise 24-1 for more practice dealing with concave mirrors.

A negative value of the magnification m means that the image is inverted, as in Figure 24-18. If m is positive, the image is upright. We've derived Equation 24-14 for the case in which the image is real and in front of the mirror (the same side as the object), but it's also true when the image is virtual and on the back side of the mirror, as in Figure 24-16. (As we'll see in Section 24-8, the same equation also applies to lenses.)

When the object is far from the mirror and the image is close to the focal point, d_I is positive and small compared to d_O, so from Equation 24-14 m is small and negative; the mirror produces a reduced, inverted image (see Figure 24-12). As the object distance decreases, the image distance increases and the ratio $m = -d_I/d_O$ increases in absolute value. As we saw above, when $d_O = 2f$ the image distance d_I is also equal to $2f$; then $m = -1$ and the inverted image is as large as the object (see Figure 24-13). If we move the object even closer to the focal point, but still outside it, the image distance is greater than the object distance and the absolute value of the magnification is greater than 1. Hence the inverted image is larger than the object (see Figure 24-14). If we move the object inside the focal point so that $d_O < f$, the image distance is negative (the image is behind the mirror) and the image is virtual. In this case Equation 24-14 tells us that m is positive, so the virtual image is upright (see Figure 24-16). We see that Equation 24-14 gives us the same results as we deduced from the ray diagrams in Section 24-3.

TABLE 24-1 Sign Conventions for Mirrors

mirror radius of curvature r	• positive for a concave mirror • negative for a convex mirror
focal length f	• positive for a concave mirror • negative for a convex mirror
image distance d_I	• positive if on the reflective side of the mirror (the same side as the object); the image is then a real image • negative if on the nonreflective side of the mirror (the opposite side from the object); the image is then a virtual image
image height h_I	• positive if the image is upright (the same orientation as the object) • negative if the image is inverted (flipped upside down compared to the object)
lateral magnification m	• positive if the image is upright (the same orientation as the object) • negative if the image is inverted (flipped upside down compared to the object)

Table 24-1 summarizes when the radius of curvature r, focal length f, image distance d_I, image height h_I, and magnification m are positive and when they are negative.

EXAMPLE 24-1 Images Made by a Concave Mirror

An object is 1.50 cm tall and is placed in front of a spherical concave mirror that has a radius of curvature equal to 20.0 cm. Find the image distance and height if the object is (a) 14.0 cm from the mirror; (b) 6.00 cm from the mirror. In each case, draw a ray diagram as part of the solution.

Set Up

Our mathematical tools are the expression for the focal length of the mirror, the mirror equation, and the equation for lateral magnification. We're given the radius of curvature $r = 20.0$ cm and the object height $h_O = 1.50$ cm; our goal is to find the values of the image distance d_I and image height h_I for the cases $d_O = 14.0$ cm and $d_O = 6.00$ cm.

Focal length of a spherical mirror:

$$f = \frac{r}{2} \qquad (24\text{-}3)$$

Mirror equation:

$$\frac{1}{d_O} + \frac{1}{d_I} = \frac{1}{f} \qquad (24\text{-}11)$$

Lateral magnification of a mirror

$$m = \frac{h_I}{h_O} = -\frac{d_I}{d_O} \qquad (24\text{-}14)$$

Solve

First calculate the focal length of the mirror.

From Equation 24-3:

$$f = \frac{r}{2} = \frac{20.0 \text{ cm}}{2} = 10.0 \text{ cm}$$

(a) The object distance $d_O = 14.0$ cm is greater than the focal length $f = 10.0$ cm but less than $2f = 20.0$ cm. That is, the object is inside the center of curvature but outside the focal point. We've drawn a ray diagram that's similar to Figure 24-14. This tells us to expect that the image will be real, inverted, farther from the mirror than the object is, and larger than the object.

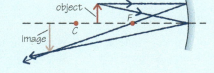

Calculate the image distance using the mirror equation.

From Equation 24-11:

$$\frac{1}{d_I} = \frac{1}{f} - \frac{1}{d_O} = \frac{1}{10.0 \text{ cm}} - \frac{1}{14.0 \text{ cm}}$$
$$= 0.1000 \text{ cm}^{-1} - 0.0714 \text{ cm}^{-1} = 0.0286 \text{ cm}^{-1} \text{ so}$$
$$d_I = \frac{1}{0.0286 \text{ cm}^{-1}} = 35.0 \text{ cm}$$

The image distance is positive, so the image is 35.0 cm in front of the mirror. An image that forms in front of the mirror is a real image.

Calculate the image height using the magnification equation.

From Equation 24-14:

$$m = \frac{h_I}{h_O} = -\frac{d_I}{d_O} = -\frac{35.0 \text{ cm}}{14.0 \text{ cm}} = -2.50$$

The image is 2.50 times larger than the object and, as the minus sign shows, inverted. Solve for the image height h_I:

$$h_I = mh_O = (-2.50)(1.50 \text{ cm}) = -3.75 \text{ cm}$$

The inverted image is 3.75 cm high.

(b) Now the object distance $d_O = 6.00$ cm is less than the focal length $f = 10.0$ cm, so the object is inside the focal point. We've drawn a ray diagram that's similar to Figure 24-16. This tells us to expect that the image will be virtual, upright, farther from the mirror, and larger than the object.

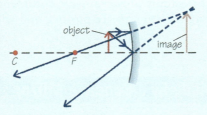

Calculate the image distance using the mirror equation.

From Equation 24-11:

$$\frac{1}{d_I} = \frac{1}{f} - \frac{1}{d_O} = \frac{1}{10.0 \text{ cm}} - \frac{1}{6.00 \text{ cm}}$$

$$= 0.1000 \text{ cm}^{-1} - 0.1667 \text{ cm}^{-1} = -0.0667 \text{ cm}^{-1} \text{ so}$$

$$d_I = \frac{1}{-0.0667 \text{ cm}^{-1}} = -15.0 \text{ cm}$$

The image distance is negative, so the image is 15.0 cm behind the mirror. An image that forms behind the mirror is a virtual image.

Calculate the image height using the magnification equation.

From Equation 24-14:

$$m = \frac{h_I}{h_O} = -\frac{d_I}{d_O} = -\frac{(-15.0 \text{ cm})}{6.00 \text{ cm}} = +2.50$$

Again the image is 2.50 times larger than the object but is now upright as shown by the plus sign. Solve for the image height h_I:

$$h_I = mh_O = (+2.50)(1.50 \text{ cm}) = +3.75 \text{ cm}$$

The upright image is 3.75 cm high.

Reflect

In both parts the position and size of the image are consistent with our ray diagrams. It's always a good idea to draw such diagrams as a check on your calculations using the mirror equation and magnification equation.

GOT THE CONCEPT? 24-5 A Concave Mirror

? If an object is placed 12.0 cm from a concave mirror, the resulting real image is 2.00 times as large as the object. What is the focal length of the mirror? (a) 6.00 cm; (b) 8.00 cm; (c) 16.0 cm; (d) 20.0 cm; (e) 24.0 cm.

TAKE-HOME MESSAGE FOR Section 24-4

✔ The mirror equation relates the focal length of a concave mirror and the positions of the object and image.

✔ A positive value of the image distance d_I indicates that the image is in front of the mirror and is real. A negative value of d_I indicates that the image is behind the mirror and virtual (the light rays never actually go there).

✔ The magnification of the image is positive if the image is upright compared to the object and negative if the image is inverted.

Figure 24-19 Parisian reflections
The Eiffel Tower can be seen reflected from the surface of this person's eye. The eye is relatively spherical, so its outer surface acts like a convex mirror.

24-5 A convex mirror always produces an image that is smaller than the object

Let's now turn our attention to images produced by a *convex* mirror. As Figure 24-7a shows, if an object is held next to a convex mirror, the resulting image is smaller than the object. The same is true if an object is far away from a convex mirror. Figure 24-19 shows the reflection of the Eiffel Tower in the convex surface of a person's eye. Although the Eiffel Tower is 324 m tall, its image is only about a centimeter in height—smaller than the iris of the eye. In this section we'll use ray diagrams to understand the nature of the images formed by a convex mirror.

Recall from Figure 24-8 the key difference between concave and convex mirrors: Parallel light rays that reflect from a concave mirror converge toward a point in front of the mirror, while parallel light rays that reflect from a convex mirror diverge from a point on the back side of the mirror. (If the mirrors are spherical, the rays don't truly converge

Figure 24-20 Ray diagram for a convex mirror How the image is formed for an object in front of a convex mirror.

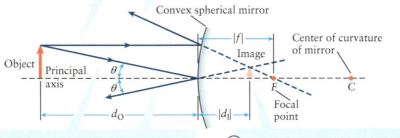

① A light ray that starts off parallel to the principal axis is reflected so that when we trace the ray back behind the mirror, it appears to pass through the virtual focal point F.

- A convex mirror has a negative focal length: $f = -|f|$.
- For any object distance d_O a convex mirror produces a virtual image behind the mirror.
- The image distance d_I is negative: $d_I = -|d_I|$.
- The virtual image is upright and smaller than the object.

② A light ray that strikes the center of the mirror at an angle θ from the principal axis reflects at the same angle.

③ Trace the two reflected rays to where they appear to cross behind the mirror. This is where the virtual image forms.

on or diverge from a single point. But this is a good description of what happens if the rays are all close to the principal axis of the mirror. We'll make this assumption—the same that we made in Sections 24-3 and 24-4 for concave mirrors—throughout this section.)

Because the focal point of a convex mirror is behind the mirror, we say that the focal length f is negative. For a convex mirror, we call the focal point *virtual* because parallel light rays that reflect from the mirror seem to emanate from that point but don't actually pass through it.

Figure 24-20 shows an object placed a distance d_O in front of a convex mirror. The focal point F lies a distance $|f|$ behind the mirror. (Since the focal length f is negative, the distance is the absolute value of f.) The center of curvature C is also behind the mirror, so the radius of curvature r is also negative. It turns out that the focal length f is equal to one-half of the radius of curvature r, just as for a concave mirror:

$$f = \frac{r}{2} \qquad (24\text{-}3)$$

(We'll justify this statement in Section 24-6.) For a concave mirror f and r are both positive; for a convex mirror f and r are both negative.

To find the location of the image in Figure 24-20, we draw two rays that emanate from the tip of the object arrow, just as we did for the ray diagrams in Section 24-3 for a concave mirror. The image of the arrow tip forms where the two reflected rays appear to meet. The image of the base of the arrow must lie along the principal axis. (A light ray coming from the base of the arrow and traveling along the principal axis strikes the mirror normal to the surface and so is reflected straight back. The image of the base of the arrow must therefore form on the principal axis.) As Figure 24-20 shows, the image is virtual because it forms behind the mirror, just like the image formed by a plane mirror (Section 24-2): The reflected rays don't actually cross there. Hence the image distance d_I is negative. We saw in Section 24-3 that a concave mirror produces a virtual image only if the object is inside the focal point; a convex mirror produces a virtual image for *any* position of the object.

Figure 24-20 shows that the image formed by a convex mirror is smaller than the object, no matter what the object distance is. Figure 24-20 also shows that the image formed by the convex mirror will always be upright and will always be closer to the mirror than the virtual focal point. Figure 24-21 shows that as the object distance d_O decreases, the image becomes larger and closer to the convex mirror, but remains virtual and upright. If the object were moved all the way to the surface of the mirror, the image would be exactly the same size as the object.

Because a convex mirror makes objects appear smaller, they provide a wide-angle view. Rearview mirrors on automobiles and trucks often include a convex mirror to allow the driver to see as much of the area behind the vehicle as possible. (Any mirror with the label "Objects in mirror are closer than they appear" is a convex mirror.)

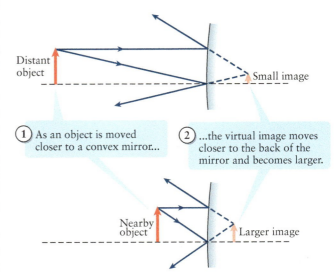

① As an object is moved closer to a convex mirror...

② ...the virtual image moves closer to the back of the mirror and becomes larger.

Figure 24-21 Moving closer to a convex mirror As the object far from a convex mirror is moved closer to the mirror, the image increases in size.

In the following section we'll look at these ideas more quantitatively and see how to calculate the image size and position for a convex mirror.

GOT THE CONCEPT? 24-6 Solar Cooking

 The Sun is so far from Earth that rays of sunlight are effectively parallel. Sunlight reflected from a mirror can be focused onto a pot, raising its temperature enough to pasteurize water in the pot or cook food. What kind of spherical mirror could be used for this purpose? (a) A convex mirror; (b) a concave mirror; (c) either a convex or concave mirror.

TAKE-HOME MESSAGE FOR Section 24-5

✔ Parallel light rays that strike a convex mirror diverge and appear to emanate from a virtual focal point behind the mirror.

✔ If an object is placed anywhere in front of a convex mirror, the resulting image is virtual, upright, closer to the mirror than the object is, and smaller than the object.

24-6 The same equations used for concave mirrors also work for convex mirrors

Just as we did for the concave mirror in Section 24-4, we can use geometry and the law of reflection to find an equation that relates the object and image distances and the radius of curvature for a convex mirror. We'll also find an equation that relates the sizes of the image and object for a convex mirror. Remarkably, we'll see that these equations are exactly the same as those for a concave mirror, provided we're careful with the signs of the radius of curvature, focal length, and image distance.

The Mirror Equation for a Convex Mirror

In Figure 24-22 we've placed a point object on the principal axis of a convex mirror with (negative) radius of curvature r. This is analogous to Figure 24-17, in which we placed a point object on the principal axis of a concave mirror. The object is at position O, a distance d_O from the mirror. We've drawn an arbitrarily chosen light ray that travels away from the object at an angle α from the principal axis and reflects off the mirror at P; if we extend the reflected ray backward, the extension (shown as a dashed blue line) crosses the principal axis at I. A second light ray (not shown) that travels away from the object along the principal axis will be reflected straight back along that axis. If we extend this reflected ray backward, it will meet the extension of the first reflected ray at I. Hence I is the position of the image made by the mirror. The image is behind the mirror, so the image distance is negative. The center of curvature C of the mirror also lies behind the mirror, so the radius of curvature is also negative. Mathematically, $d_I = -|d_I| < 0$ and $r = -|r| < 0$.

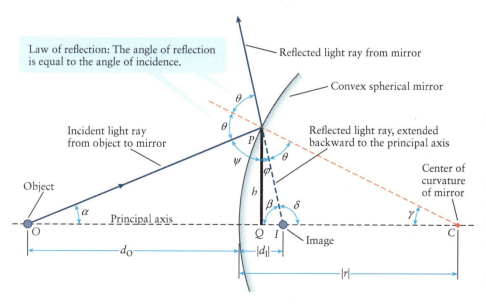

Figure 24-22 **Analyzing a convex spherical mirror** A ray diagram for a point object on the principal axis of a convex spherical mirror.

Note that the red dashed line CP in Figure 24-22 is normal to the mirror at P, so by the law of reflection the incident ray and the reflected ray are both at the same angle θ relative to this normal. It follows that θ is also the angle between the line CP and the backward extension PI of the reflected ray.

To relate the object distance d_O, image distance d_I, and radius of curvature r, we'll first find a relationship among α, the angle between the principal axis and the light ray incident on the mirror; β, the angle between the principal axis and the reflected ray; and γ, the angle between the principal axis and the red dashed line CP that defines the normal at the point P where the light ray is reflected. Notice that each of these three angles is part of a right triangle for which one side is the thick black line PQ in Figure 24-22: triangle OPQ for angle α, triangle IPQ for angle β, and triangle CPQ for angle γ. In each case the line PQ, of length h, is the side opposite the angle. Since we're considering only light rays that are very close to the principal axis, the reflection point P is also very close to the principal axis. Hence point Q (directly below point P) is essentially at the point where the principal axis touches the mirror surface. As a result, we can treat d_O as the base of triangle OPQ, $|d_I|$ as the base of triangle IPQ, and $|r|$ as the base of triangle CPQ. For each triangle the tangent of the angle equals the length h of the side opposite the angle, divided by the length of the triangle's base. So

√x **See the Math Tutorial for more information on trigonometry.**

$$\tan \alpha = \frac{h}{d_O}, \quad \tan \beta = \frac{h}{|d_I|}, \quad \tan \gamma = \frac{h}{|r|} \tag{24-15}$$

If the incident ray OP is at a small angle to the principal axis, then α, β, and γ are all small angles. (We've drawn the angle α fairly large in Figure 24-22 to make the geometry easier to visualize. In reality, it must be quite small to conform to the approximation that the rays are nearly parallel to the principal axis.) The tangent of a small angle is approximately equal to the angle in radians, so Equation 24-15 becomes

$$\alpha = \frac{h}{d_O}, \quad \beta = \frac{h}{|d_I|}, \quad \gamma = \frac{h}{|r|} \tag{24-16}$$

To relate the angles α, β, and γ, we'll do as in Section 24-4 and use the result that the sum of the angles of any triangle must be 180°, or π radians. For the triangle IPC that connects the image point I, reflection point P, and center of curvature C, the three angles are γ, θ, and δ. So

$$\gamma + \theta + \delta = \pi \tag{24-17}$$

Figure 24-22 shows that the angles β and δ must add to π radians (together they make up half a circle around the point I). So $\beta + \delta = \pi$, $\delta = \pi - \beta$, and Equation 24-17 becomes

$$\gamma + \theta + \pi - \beta = \pi \quad \text{or} \quad \theta = \beta - \gamma \tag{24-18}$$

For the triangle OPC that connects the object point O, reflection point P, and center of curvature C, the angles are α, $(\psi + \theta + \phi)$, and γ, so we have

$$\alpha + (\psi + \theta + \phi) + \gamma = \pi \tag{24-19}$$

We can simplify Equation 24-19 by noting from Figure 24-22 that the angles θ, ψ, ϕ, and θ form a half-circle around the point P, so their sum is π radians: $\theta + \psi + \phi + \theta = \pi$, or $\psi + \theta + \phi = \pi - \theta$. If we substitute this expression for $(\psi + \theta + \phi)$ into Equation 24-19, we get

$$\alpha + \pi - \theta + \gamma = \pi \quad \text{or} \quad \theta = \alpha + \gamma \tag{24-20}$$

Equations 24-18 and 24-20 are both expressions for the angle θ. If we set these equal to each other, we get

$$\beta - \gamma = \alpha + \gamma \quad \text{or} \quad \alpha - \beta = -2\gamma \tag{24-21}$$

Equation 24-21 is the relationship among the angles α, β, and γ we've been looking for. We can now get a relationship among the object distance d_O, image distance d_I, and radius of curvature r by substituting the expressions for α, β, and γ from Equation 24-16 into Equation 24-21:

$$\frac{h}{d_O} - \frac{h}{|d_I|} = -\frac{2h}{|r|}$$

If we divide through by the height h of the black line in Figure 24-22 and recall that the image distance and radius of curvature are both negative, so that $d_I = -|d_I|$ and $r = -|r|$, this becomes

(24-22)
$$\frac{1}{d_O} + \frac{1}{d_I} = \frac{2}{r}$$

Equation 24-22 is *exactly* the same as Equation 24-10, which we derived for a *concave* mirror. It's reassuring that we get the same expression for both kinds of mirrors.

Note that if the object is infinitely far away, so that $d_O \to \infty$ and $1/d_O \to 0$, the rays from the object to the mirror will all be parallel to the axis and the virtual image will be formed at the focal point F, a distance $|f|$ behind the mirror. Then $d_I = -|f| = f$ (recall that the focal length f is negative). For this situation, Equation 24-22 becomes

$$\frac{1}{f} = \frac{2}{r} \quad \text{or} \quad f = \frac{r}{2}$$

This result is exactly the same as that for a concave spherical mirror: The focal length f is equal to one-half of the radius of curvature r (Equation 24-3). The only difference is that for a convex mirror f and r are both negative.

If we substitute $1/f = 2/r$ into Equation 24-22, we get the mirror equation for a convex mirror:

(24-11)
$$\frac{1}{d_O} + \frac{1}{d_I} = \frac{1}{f}$$

This is Equation 24-11, the *same* mirror equation that we derived for a *concave* mirror in Section 24-4. This equation works equally well whether the focal length is positive (for a concave mirror) or negative (for a convex mirror).

We can use Equation 24-11 to explore the properties of the image made by a convex mirror. We saw in Section 24-4 that this equation can be rewritten as

(24-12)
$$d_I = \frac{d_O f}{d_O - f}$$

Using $f = -|f|$ for the negative focal length of a convex mirror, Equation 24-12 becomes

(24-23)
$$d_I = \frac{d_O(-|f|)}{d_O - (-|f|)} = -\left(\frac{d_O}{d_O + |f|}\right)|f|$$

The right-hand side of Equation 24-23 is always negative, so the image distance d_I for a convex mirror will always be negative (the image will always be behind the mirror). Note that the fraction in parentheses is always less than or equal to 1, so the image distance is always between 0 and $-f$. That is, the image always forms somewhere between the mirror and the focal point.

Magnification for a Convex Mirror

Figure 24-23 again shows the image made by a convex spherical mirror, but now the object is an upright arrow of height h_O a distance d_O in front of the mirror. Just as for a concave mirror (see Figure 24-18 in Section 24-4), we draw two rays coming from the tip of the object arrow at O'. One ray is parallel to the principal axis of the mirror; after reflection this travels away from the mirror as though it had been emitted from the focal point F. The other ray strikes the center of the mirror at P', and the reflected ray makes the same angle θ with the principal axis as the incident ray from O' to P'. If we extend this reflected ray backward, the extension is at the same angle θ to the principal axis. The extensions of the two reflected rays meet at I', the position of the tip of the image arrow. The base of the image arrow is at I, directly underneath I' on the principal axis. The image is behind the mirror, so the image distance d_I is negative. This image is also upright, so the image height h_I is positive.

To relate the image and object heights, note that the right triangles $OO'P'$ and $II'P'$ both include the same angle θ. For $OO'P'$ the tangent of θ (the opposite side

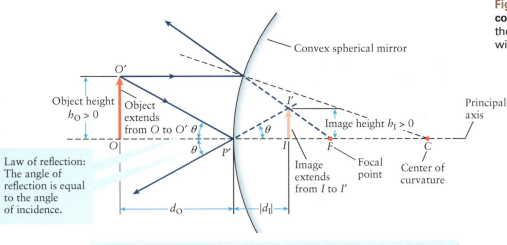

Figure 24-23 Magnification for a convex spherical mirror We replace the point object in Figure 24-22 with an object arrow.

- The right triangles $OO'P'$ and $II'P'$ both include the same angle θ.
- So $\tan \theta = h_O/d_O = h_I/|d_I|$ and $h_I/h_O = |d_I|/d_O$.
- Since d_I is negative, $d_I = -|d_I|$ and $m = h_I/h_O = |d_I|/d_O = -d_I/d_O$.

divided by the adjacent side) equals h_O/d_O; for $II'P'$ the tangent of θ equals $h_I/|d_I|$. These two expressions for $\tan \theta$ must be equal, so

$$\frac{h_O}{d_O} = \frac{h_I}{|d_I|} \quad \text{or} \quad \frac{h_I}{h_O} = \frac{|d_I|}{d_O} = -\frac{d_I}{d_O} \quad (\text{since } d_I = -|d_I|)$$

The lateral magnification m equals h_I/h_O, so it follows that for a convex mirror

$$m = \frac{h_I}{h_O} = -\frac{d_I}{d_O} \tag{24-14}$$

This is the *same* expression for magnification as for a concave mirror. Since the image distance d_I is negative for a convex mirror, Equation 24-14 tells us that the magnification is positive and so the image is upright. In addition, for a convex mirror the image is closer to the mirror than is the object, so $|d_I| < d_O$, and the value of m is less than 1. That is, the image made by a convex mirror is always smaller than the object, just as we saw in Section 24-5.

➔ *Go to Interactive Exercise 24-2 for more practice dealing with concave and convex mirrors.*

WATCH OUT! Positive magnification indicates that an image is upright.

! Remember that the *sign* of the magnification of a mirror doesn't indicate whether the image is larger or smaller than the object, only whether the image is upright or inverted.

An upright image, for example, is always associated with a positive magnification, regardless of whether the image is larger than or smaller than the object.

EXAMPLE 24-2 An Image Made by a Convex Mirror

An object is 1.50 cm high and is placed in front of a spherical convex mirror with a radius of curvature of magnitude 48.0 cm. Find the image distance and height if the object is (a) 68.0 cm from the mirror; (b) 3.00 cm from the mirror.

Set Up

This is similar to Example 24-1 in Section 24-4. The key difference is that the mirror is now convex, so the radius of curvature is negative: $r = -48.0$ cm. We're given the object height $h_O = 1.50$ cm, and want to find the values of the image distance d_I and image height h_I for the cases $d_O = 68.0$ cm and $d_O = 3.00$ cm. Our tools are Equations 24-3, 24-11, and 24-14, which apply to convex mirrors as well as concave ones.

Focal length of a spherical mirror:

$$f = \frac{r}{2} \tag{24-3}$$

Mirror equation:

$$\frac{1}{d_O} + \frac{1}{d_I} = \frac{1}{f} \tag{24-11}$$

Lateral magnification of a mirror

$$m = \frac{h_I}{h_O} = -\frac{d_I}{d_O} \tag{24-14}$$

Solve

Use Equation 24-3 to calculate the focal length of the mirror.

From Equation 24-3:

$$f = \frac{r}{2} = \frac{-48.0 \text{ cm}}{2} = -24.0 \text{ cm}$$

The negative value of focal length means that the focal point is behind the mirror.

In both cases we expect that the image will be virtual (behind the mirror), smaller than the object, and closer to the mirror than the object is. The specific position and size of the image will be different in the two cases, however.

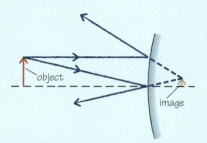

(a) With an object distance $d_O = 68.0$ cm, we use Equation 24-11 to find the image distance and Equation 24-14 to find the image size.

From the mirror equation (Equation 24-11):

$$\frac{1}{d_I} = \frac{1}{f} - \frac{1}{d_O} = \frac{1}{(-24.0 \text{ cm})} - \frac{1}{68.0 \text{ cm}}$$
$$= -0.0417 \text{ cm}^{-1} - 0.0147 \text{ cm}^{-1} = -0.0564 \text{ cm}^{-1}, \text{ so}$$

$$d_I = \frac{1}{(-0.0564 \text{ cm}^{-1})} = -17.7 \text{ cm}$$

The image distance is negative, so the image is 17.7 cm behind the mirror. An image that forms behind the mirror is a virtual image.

The magnification of the image is, from Equation 24-14,

$$m = \frac{h_I}{h_O} = -\frac{d_I}{d_O} = -\frac{(-17.7 \text{ cm})}{68.0 \text{ cm}} = +0.261$$

Since $m > 0$, the image is upright. The image is 0.261 as tall as the object, so its height is

$$h_I = m h_O = (+0.261)(1.50 \text{ cm}) = 0.391 \text{ cm}$$

(b) Repeat the calculations of part (a) with $d_O = 3.00$ cm.

Calculate the image distance:

$$\frac{1}{d_I} = \frac{1}{f} - \frac{1}{d_O} = \frac{1}{(-24.0 \text{ cm})} - \frac{1}{3.00 \text{ cm}}$$
$$= -0.0417 \text{ cm}^{-1} - 0.333 \text{ cm}^{-1} = -0.375 \text{ cm}^{-1}, \text{ so}$$

$$d_I = \frac{1}{(-0.375 \text{ cm}^{-1})} = -2.67 \text{ cm}$$

Again the image distance is negative. Note that with a smaller object distance than in (a), the image distance is also smaller.

The magnification of the image is

$$m = \frac{h_I}{h_O} = -\frac{d_I}{d_O} = -\frac{(-2.67 \text{ cm})}{3.00 \text{ cm}} = +0.889$$

Again $m > 0$, and the image is upright. The image is 0.889 as tall as the object; its height is

$$h_I = m h_O = (+0.889)(1.50 \text{ cm}) = +1.33 \text{ cm}$$

Reflect

As the object is moved closer to the convex mirror, the image moves closer to the mirror and increases in size.

GOT THE CONCEPT? 24-7 A Convex Mirror

 If the image made by a convex mirror of focal length f is 1/2 the height of the object, the distance from the object to the mirror must be equal to (a) $4|f|$; (b) $2|f|$; (c) $|f|$; (d) $|f|/2$; (e) $|f|/4$.

24-7 Convex lenses form images like concave mirrors and vice versa

A curved mirror forms images by reflection. A **lens**—a piece of glass or other transparent material with a curved surface on its front side, back side, or both sides—forms images by the *refraction* of light as the light enters and leaves the lens. Just as for a curved mirror, the images made by a lens can be larger or smaller than the object (Figure 24-24). In this section we'll explore how lenses form images.

This rodent appears larger when viewed through a magnifying glass, a lens that is convex on both sides.

(a)

Monika Graff/The Image Works

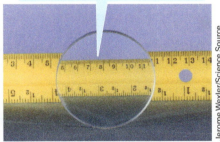

The ruler appears smaller than actual size when viewed through this lens, which is concave on both sides.

(b)

Jerome Wexler/Science Source

Figure 24-24 Lenses can magnify or shrink Different types of lenses can make (a) large or (b) small images.

The key idea that we need to understand lenses is Snell's law of refraction. As we learned in Section 23-2, a light ray changes direction when it moves from one transparent medium to a second medium in which light travels at a different speed. The ray bends toward the normal if the speed of light is slower in the second medium (Figure 24-25a) and bends away from the normal if the speed of light is faster in the second medium (Figure 24-25b).

Figure 24-26 shows how parallel light rays refract when they enter or exit a glass sphere (in which the speed of light is relatively slow) surrounded by air (in which the speed of light is almost as fast as in a vacuum). Both the front and back surfaces of the sphere are convex (they bulge outward). Parallel light rays converge when they enter the glass through the left-hand convex surface, and also converge when they exit the glass through the right-hand convex surface. Figure 24-27 shows what happens when parallel light rays enter or exit a piece of glass with a concave surface (one that bulges inward). In this case, parallel rays diverge as they cross the concave surface into the glass, and diverge as they exit the glass through the concave surface.

The light rays in Figures 24-26 and 24-27 undergo only a single refraction when they enter or exit a glass object with a curved surface. In a lens there are *two* refractions: once when the light enters the front surface of the lens, and once when it exits the back surface. Figure 24-26 shows that the refractions will make the rays converge if each surface is convex, and Figure 24-27 shows that the refractions will make the rays diverge if each surface is concave. So a lens with two convex surfaces—called a *convex* lens—will be a **converging lens** that takes incoming parallel light rays and makes them converge toward the principal axis. The same is true for a lens with one convex surface and one flat surface, called a *plano-convex* lens. The refraction at the flat surface by itself causes neither convergence nor divergence. Similarly, a lens with two concave surfaces—called a *concave* lens—will be a **diverging lens** that takes incoming parallel

(a) Refraction from one medium into a slower one

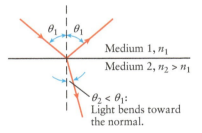

θ_1 θ_1

Medium 1, n_1

Medium 2, $n_2 > n_1$

$\theta_2 < \theta_1$:
Light bends toward the normal.

(b) Refraction from one medium into a faster one

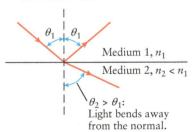

θ_1 θ_1

Medium 1, n_1

Medium 2, $n_2 < n_1$

$\theta_2 > \theta_1$:
Light bends away from the normal.

Figure 24-25 Snell's law of refraction A light ray crossing the boundary between two transparent media can refract either (a) toward or (b) away from the normal, depending on how the speed of light compares in the two media.

Figure 24-26 Refraction by a glass sphere If a glass sphere is surrounded by air, parallel light rays converge whether they (a) enter the sphere or (b) exit the sphere.

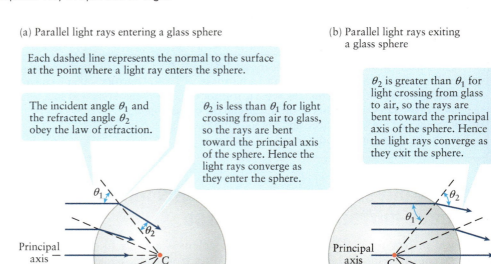

(a) Parallel light rays entering a glass sphere

Each dashed line represents the normal to the surface at the point where a light ray enters the sphere.

The incident angle θ_1 and the refracted angle θ_2 obey the law of refraction.

θ_2 is less than θ_1 for light crossing from air to glass, so the rays are bent toward the principal axis of the sphere. Hence the light rays converge as they enter the sphere.

(b) Parallel light rays exiting a glass sphere

θ_2 is greater than θ_1 for light crossing from glass to air, so the rays are bent toward the principal axis of the sphere. Hence the light rays converge as they exit the sphere.

light rays and makes them diverge away from the principal axis. A lens with one concave surface and one flat surface, called a *plano-concave* lens, will also be a diverging lens (Figure 24-28). Here's a general rule: A lens that is thicker in the middle than at the edges will be a converging lens, while a lens that is thicker at the edges than at the middle will be a diverging lens.

WATCH OUT! The curvature of a lens has the opposite effect to the same curvature in a mirror.

We saw in Sections 24-3 and 24-5 that a concave mirror causes light rays to converge on reflection, while a convex mirror causes light rays to diverge on reflection. By contrast, a concave lens causes light rays to *diverge* as they pass through, and a convex lens causes light rays to *converge* as they pass through. Mirrors and lenses are different!

A lens with spherical surfaces suffers from the same spherical aberration as a spherical mirror (Section 24-3): Parallel light rays do not all focus to the same point for a converging lens or appear to originate from a single point for a diverging lens. However, we can neglect this effect if the rays of light are all close to the principal axis. We can ensure this by using only a small section of a large spherical surface to form each surface of the lens (Figure 24-29). We'll also assume that there is very little thickness of material between the front and back surfaces of the lens. The result is called a **thin lens**.

(a) Parallel light rays entering a piece of glass with a spherical cutout

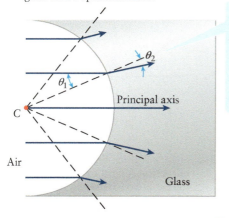

θ_2 is less than θ_1 for light crossing from air to glass, so the rays are bent away from the principal axis of the glass. Hence the light rays diverge as they enter the glass.

(b) Parallel light rays exiting a piece of glass with a spherical cutout

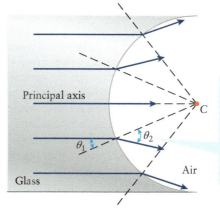

θ_2 is greater than θ_1 for light crossing from glass to air, so the rays are bent away from the principal axis of the glass. Hence the light rays diverge as they exit the glass.

Figure 24-27 Refraction by glass with a spherical cutout If a piece of glass with a spherical cutout is surrounded by air, parallel light rays diverge whether they (a) enter the glass through the cutout or (b) exit the glass through the cutout.

These are converging lenses: Parallel light rays that enter either of these lenses will exit the lens converging toward each other.

These are diverging lenses: Parallel light rays that enter either of these lenses will exit the lens diverging away from each other.

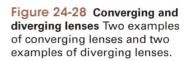

Figure 24-28 Converging and diverging lenses Two examples of converging lenses and two examples of diverging lenses.

Convex   Plano-convex   Concave   Plano-concave

We'll consider only thin lenses for the rest of this chapter so that we can neglect spherical aberration and treat each lens as if parallel rays are focused to a single point. Eyeglasses and contact lenses are everyday examples of thin lenses.

Ray Diagrams for Converging Lenses

We saw earlier that ray diagrams are powerful tools for visualizing how a curved mirror produces an image. Let's see how to draw a ray diagram to help us locate the position of the image made by a converging lens.

Figure 24-30 shows a thin convex lens. Notice that we have marked *two* focal points F. If parallel light rays enter the left-hand side of the lens, they converge at the focal point on the right-hand side; if parallel rays enter the lens on its right-hand side, they converge at the focal point on the left-hand side. For a thin lens, the focal length, the distance from the center of the lens to the focal point, is the same on both sides of the lens even if the two surfaces have a different radius of curvature. That's why we've drawn the two focal points F in Figure 24-30 the same distance from the center of the lens.

In Figure 24-30 we've drawn a red arrow as our object on the left-hand side of the lens and placed this arrow outside the left-hand focal point. We've also drawn three representative light rays emanating from the tip of the object arrow. While each light

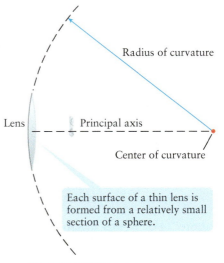

Radius of curvature

Lens   Principal axis

Center of curvature

Each surface of a thin lens is formed from a relatively small section of a sphere.

Figure 24-29 A thin lens Analyzing image formation by a lens is much easier if we assume that the lens is thin and has spherical surfaces.

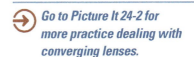

Go to Picture It 24-2 for more practice dealing with converging lenses.

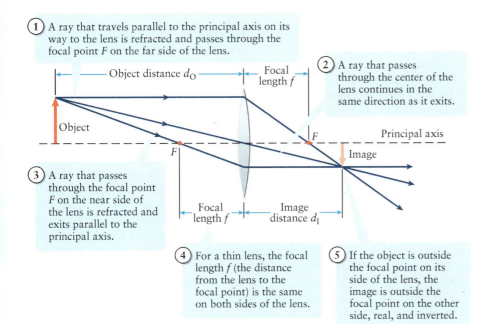

(1) A ray that travels parallel to the principal axis on its way to the lens is refracted and passes through the focal point F on the far side of the lens.

Object distance d_O

Focal length f

(2) A ray that passes through the center of the lens continues in the same direction as it exits.

Object

F

F

Principal axis

Image

(3) A ray that passes through the focal point F on the near side of the lens is refracted and exits parallel to the principal axis.

Focal length f

Image distance d_I

(4) For a thin lens, the focal length f (the distance from the lens to the focal point) is the same on both sides of the lens.

(5) If the object is outside the focal point on its side of the lens, the image is outside the focal point on the other side, real, and inverted.

Figure 24-30 Ray diagram for a converging lens I How the image is formed for an object placed outside the focal point of a converging lens.

ray actually refracts twice, once on entering the lens and once on exiting, for simplicity we've drawn the rays as if they refract only once, along the centerline of the lens (the vertical line that runs through the center of the lens). The three rays are:

(1) A ray that arrives at the lens traveling parallel to the principal axis is refracted so that it passes through the far (right-hand) focal point F.

(2) A ray that enters the lens directly at its center continues in a straight line as it exits. This is only an approximation because it assumes that the ray undergoes no refraction on entering or exiting the lens. But this is a good approximation for a thin lens, which is nearly flat at its very center.

(3) A ray that passes through the near (left-hand) focal point F on its way to the lens exits the lens traveling parallel to the principal axis. To see why this is the case, imagine that we could record a video of light traveling along this path, then run the video backward. Then we would see a light ray coming from the far right in Figure 24-30 and traveling parallel to the principal axis. After this ray passes through the lens from right to left, it would naturally pass through the left-hand focal point F. If we now run that same video forward, we see light from the object following the path shown in Figure 24-30.

Figure 24-31 A real, inverted image made by a converging lens The lens in the person's hand makes a real, inverted, and small image of these pyramids in Sudan.

Marka/Getty Images

The image of the tip of the arrow in Figure 24-30 forms where the rays coming from the tip cross. The image of the base of the arrow must form along the principal axis because a ray that comes from the base and travels along the axis strikes the center of the lens and therefore continues in the same direction. You can see that the image of the arrow is real (the light rays actually cross at the image point) and is inverted.

Note that in Figure 24-30 the object is relatively far from the lens (the object distance d_O is greater than twice the focal length f), and the image is real, inverted, *and* smaller than the object. That's the same result we found for a concave mirror in Section 24-3 (see Figure 24-12). This reinforces the idea that a convex lens behaves similarly to a concave mirror. Figure 24-31 shows the small, inverted image formed when light from a distant object passes through a convex lens. You can also see this in Figure 11-41 (Section 11-12), in which a spherical drop of water acts as a convex lens.

WATCH OUT!   A real image formed by a lens is on the side of the lens opposite to the object.

(!) We've defined a real image as one that forms where reflected or refracted light rays converge. A concave mirror can cause light rays to converge only on the reflective side of the mirror, which is where we would put an object, so this is the only side on which a real image can form. A lens, however, can only cause light rays to converge on the side of the lens *opposite* to the object, as in Figure 24-30. As we'll see below, only a converging lens can produce a real image, and only if the object is outside the focal point as in Figure 24-30.

Recall that for mirrors our convention was that the image distance is positive if the image is real and so is on the same side of the mirror as the object. For a lens we'll also say that the image distance is positive if the image is real. This means that for a lens, a positive image distance d_I implies that the image is on the *opposite* side of the lens from the object. For example, the image distance is positive in Figure 24-30.

In Figure 24-32 we've moved the object to a point between the lens and the focal point, so the object distance is less than the focal length: $d_O < f$. To locate the image, we've drawn just two light rays from the tip of the arrow. One ray travels parallel to the principal axis of the lens and therefore passes through the focal point on the far side after being refracted by the lens. The other ray we have drawn enters the center of the lens and so continues straight. (In Figure 24-30 we drew a third ray through the focal point on the same side of the lens as the object. We don't draw that ray here, since it leads away from the lens rather than toward it.) Notice that these two rays do not meet. But if we extend the rays backward, as shown by the dashed lines, they appear to meet on the same side of the lens as the object. Where they meet is the location of the image. The rays do not actually cross there, so this is a *virtual* image. Because the image is on the same side of the lens as the object, the image distance is negative: $d_I < 0$.

As in Figure 24-30, the base of the image arrow must lie on the principal axis of the lens. It follows that the image is upright. Note that the image is also larger than the object. This is the kind of image shown in Figure 24-24a; when we hold a convex

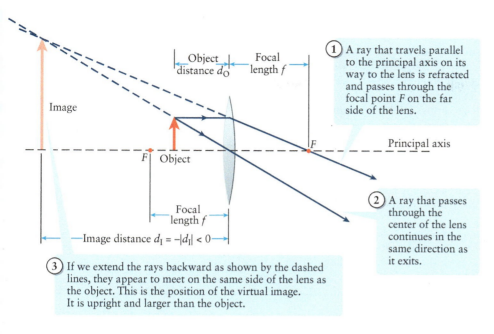

Figure 24-32 Ray diagram for a converging lens II How the image is formed for an object placed inside the focal point of a converging lens.

(1) A ray that travels parallel to the principal axis on its way to the lens is refracted and passes through the focal point F on the far side of the lens.

(2) A ray that passes through the center of the lens continues in the same direction as it exits.

(3) If we extend the rays backward as shown by the dashed lines, they appear to meet on the same side of the lens as the object. This is the position of the virtual image. It is upright and larger than the object.

lens so that the rodent is inside the focal point of the lens, we see an enlarged, upright image of the rodent. This is very similar to the image made by a concave mirror when an object is placed inside the focal point of the mirror (see Figure 24-16).

Ray Diagrams for Diverging Lenses

We can also use ray diagrams to learn about the image made by a diverging lens. In Figure 24-33 we've placed an object arrow in front of a thin lens with two concave surfaces. Since this lens causes parallel light rays to diverge, we say that the focal length is negative: $f = -|f| < 0$. (The focal length of a convex mirror, which also causes parallel light rays to diverge, is also negative.)

We've drawn two light rays to determine where the image forms, and as we did for the thin, convex lens, we've drawn the rays as if they refract only once along the centerline of the lens. The refracted rays don't actually meet, but if we extend these rays backward, we find the location where the extensions meet. The image of the tip of the arrow forms at this point; the image is virtual (since the rays don't actually meet there), upright, and smaller than the object. This is the same behavior we saw in Section 24-5 for convex mirrors. As for a convex mirror, the image is virtual, upright, and smaller no matter what the object distance.

Figure 24-24b shows an image of a ruler made by a concave lens. Just like the image in Figure 24-33, this image is upright and smaller than the object.

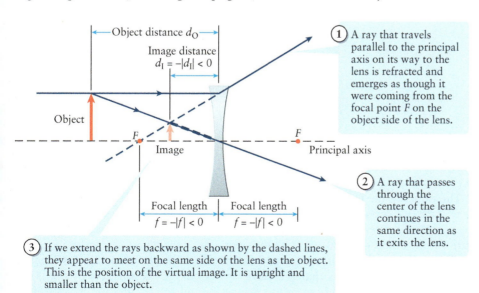

(1) A ray that travels parallel to the principal axis on its way to the lens is refracted and emerges as though it were coming from the focal point F on the object side of the lens.

(2) A ray that passes through the center of the lens continues in the same direction as it exits the lens.

(3) If we extend the rays backward as shown by the dashed lines, they appear to meet on the same side of the lens as the object. This is the position of the virtual image. It is upright and smaller than the object.

Figure 24-33 Ray diagram for a diverging lens How the image is formed for an object placed in front of a diverging lens.

GOT THE CONCEPT? 24-8 A Lens in Sugar Water

(?) You make a lens with two convex surfaces out of ice ($n = 1.31$). You then submerge the lens in a large tank of concentrated sugar water ($n = 1.49$). If you place an object in the tank and in front of the lens, an image is formed at a position inside the tank. What kind of image is this? (a) A real image; (b) a virtual image that is larger than the object; (c) a virtual image that is smaller than the object; (d) either (a) or (b), depending on the distance from the object to the lens; (e) any of (a), (b), or (c), depending on the distance from the object to the lens.

TAKE-HOME MESSAGE FOR Section 24-7

✔ When parallel light rays enter one side of a convex lens, they exit the lens converging toward the focal point on the other side of the lens.

✔ When parallel light rays enter a concave lens, they exit the lens diverging away from the focal point on the same side of the lens that the rays entered.

✔ A convex lens produces a real image if the object is outside the focal point and produces a virtual image if the object is inside the focal point.

✔ A concave lens produces a virtual image no matter what the position of the object.

24-8 The focal length of a lens is determined by its index of refraction and the curvature of its surfaces

As we did for concave mirrors and convex mirrors, we'd like to find equations for the focal length of a lens; for the relationship among the focal length of a lens, the object distance, and the image distance; and for the magnification of an image produced by a lens. As we'll see, the latter two equations turn out to be identical to those for curved mirrors. The expression for the focal length, however, is a little more complicated.

Focal Length of a Thin Lens

The focal length f of a thin lens depends on the index of refraction n of the material of which it is made. The greater the value of n, the more sharply a light ray is bent as it passes either from the surrounding air (for which the index of refraction is essentially 1) into the lens or from the lens into the air. The value of f also depends on how the front and back surfaces of the lens are curved. The mathematical expression of these relationships for a lens in air is called the **lensmaker's equation:**

Lensmaker's equation for the focal length of a thin lens (24-24)

Focal length of a lens surrounded by air | Index of refraction of the lens material

$$\frac{1}{f} = (n - 1)\left(\frac{1}{R_1} - \frac{1}{R_2}\right)$$

Radius of curvature of lens surface 1 (the surface closer to the object) | Radius of curvature of lens surface 2 (the surface farther from the object)

Equation 24-24 can be derived by applying Snell's law of refraction (Equation 23-6) to a ray of light refracted by both surfaces of a lens. The derivation is beyond our scope, however.

The values of the radii of curvature R_1 and R_2 depend on how sharply and in what direction the two surfaces of the lens are curved. As Figure 24-34 shows, we take a radius to be positive if the center of curvature is on the other side of the lens from the object but negative if the center of curvature is on the same side as the object. For example, in Figure 24-34 the radius R_1 of surface 1 is positive, but the radius R_2 of surface 2 is negative.

Equation 24-24 tells us that the more tightly curved the surfaces of a lens are, the *smaller* the magnitudes of the radii R_1 and R_2 and the *smaller* the magnitude of

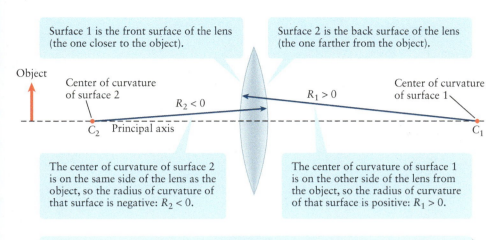

Surface 1 is the front surface of the lens (the one closer to the object).

Surface 2 is the back surface of the lens (the one farther from the object).

Object

Center of curvature of surface 2

$R_2 < 0$

$R_1 > 0$

Center of curvature of surface 1

C_2 Principal axis

C_1

The center of curvature of surface 2 is on the same side of the lens as the object, so the radius of curvature of that surface is negative: $R_2 < 0$.

The center of curvature of surface 1 is on the other side of the lens from the object, so the radius of curvature of that surface is positive: $R_1 > 0$.

Figure 24-34 Interpreting the lensmaker's equation For each surface of a thin lens, the sign of the radius of curvature depends on where the center of curvature is located.

- If the center of curvature of a lens surface is on the other side of the lens from the object, the radius of curvature R of that surface is positive.
- If the center of curvature of a lens surface is on the same side of the lens as the object, the radius of curvature R of that surface is negative.
- A flat surface has an infinite radius of curvature R, so $1/R = 0$ for that surface.

the focal length f. (We saw a similar result for the focal length of a spherical mirror in Section 24-3.) The smaller the magnitude of f, the more sharply light rays are refracted by passing through the lens.

The lensmaker's equation is expressed in terms of the reciprocal of the focal length f. For that reason it's common, especially among optometrists, to characterize lenses (and also mirrors) in terms of the reciprocal of the focal length:

$$P = \frac{1}{f}$$

(24-25)

The quantity P is called the **power** of the lens or mirror. The units of P are m^{-1}; $1\ m^{-1}$ is known as 1 **diopter**. The larger the power of a lens, the greater the amount of refraction it causes (that is, the more "powerfully" it causes light rays to bend). If parallel light rays enter a thin, convex lens of power 2 diopters, they come to a focus $1/2$ m (0.5 m) behind the lens; if instead these parallel light rays enter a 5-diopter converging lens, they are bent more sharply and come to a focus just $1/5$ m (0.2 m) behind the lens. A thin, concave lens has a negative focal length and so has a negative power P: When parallel light rays enter a diverging lens with a power of −5 diopters, they emerge from the lens as though they were emanating from a point $1/5$ m (0.2 m) in front of the lens (see Figure 24-33). In general, the smaller the magnitude of the focal length of a lens, the greater the magnitude of its power.

EXAMPLE 24-3 Calculating Focal Length for a Convex Lens

In Figure 24-34 the front surface (surface 1) of the lens has a radius of curvature of magnitude 15.0 cm, and the back surface (surface 2) has a radius of curvature of magnitude 25.0 cm. The lens is made of crown glass with an index of refraction 1.520. (a) Calculate the focal length of the lens. (b) Now flip the orientation of the lens so that the front surface is now the one with a radius of curvature of magnitude 25.0 cm and the back surface is now the one with a radius of curvature of magnitude 15.0 cm. Calculate the focal length of the lens in this case.

Set Up

For each situation we'll draw the lens and determine on which side of the lens the centers of curvature lie; that will tell us whether R_1 and R_2 are positive or negative. We'll then use Equation 24-24 to calculate the focal length.

Lensmaker's equation for the focal length of a thin lens:

$$\frac{1}{f} = (n - 1)\left(\frac{1}{R_1} - \frac{1}{R_2}\right)$$

(24-24)

Solve

(a) Both surfaces are convex. Hence the center of curvature of each surface is on the other side of the lens from that surface. The center of curvature C_1 of the front surface (on the left in the figure) is on the far side of the lens, so R_1 is positive: $R_1 = +15.0$ cm. The center of curvature C_2 of the back surface (on the right in the figure) is on the near side of the lens, so R_2 is negative: $R_2 = -25.0$ cm.

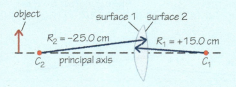

Calculate the focal length.

From Equation 24-24:

$$\frac{1}{f} = (1.520 - 1)\left[\frac{1}{(+15.0 \text{ cm})} - \frac{1}{(-25.0 \text{ cm})}\right]$$

$$= 0.520[(0.0667 \text{ cm}^{-1}) - (-0.0400 \text{ cm}^{-1})]$$

$$= 0.0555 \text{ cm}^{-1}$$

$$f = \frac{1}{0.0555 \text{ cm}^{-1}} = 18.0 \text{ cm} = 0.180 \text{ m}$$

Although R_2 is negative, we subtract rather than add $1/R_2$ in Equation 24-24, so both the $1/R_1$ term and the $1/R_2$ term—that is, both surface 1 and surface 2—contribute to giving $1/f$ a positive value. As a result, the focal length is positive, as we expect for a convex lens.

(b) Because we have flipped the lens around, we have interchanged surfaces 1 and 2. The object is still to the left of the lens, however. As in part (a), R_1 is positive and R_2 is negative, but now $R_1 = +25.0$ cm and $R_2 = -15.0$ cm.

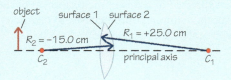

Calculate the focal length.

From Equation 24-24:

$$\frac{1}{f} = (1.520 - 1)\left[\frac{1}{(+25.0 \text{ cm})} - \frac{1}{(-15.0 \text{ cm})}\right]$$

$$= 0.520[(0.0400 \text{ cm}^{-1}) - (-0.0667 \text{ cm}^{-1})]$$

$$= 0.0555 \text{ cm}^{-1}$$

$$f = \frac{1}{0.0555 \text{ cm}^{-1}} = 18.0 \text{ cm} = 0.180 \text{ m}$$

Again both the $1/R_1$ term and the $1/R_2$ term contribute to giving $1/f$ a positive value. The focal length has the same positive value as in part (a).

Reflect

The focal length of this thin lens stays the same after we flip it back to front. As we stated in the previous section, the focal points on either side of the lens are both the same distance (that is, the same focal length) from the lens. Note that the power of this lens is $P = 1/f = 1/(0.180 \text{ m}) = 5.55 \text{ m}^{-1}$, or 5.55 diopters.

EXAMPLE 24-4 Calculating Focal Length for a Concave Lens

A certain lens made of crown glass ($n = 1.520$) has concave front and back surfaces. The front surface has a radius of curvature of magnitude 15.0 cm, and the back surface has a radius of curvature of magnitude 25.0 cm. Calculate the focal length of the lens.

Set Up

As in the previous example, we'll first draw the lens and use our drawing to decide whether R_1 and R_2 are positive or negative. Equation 24-24 will then allow us to calculate the focal length.

Lensmaker's equation for the focal length of a thin lens:

$$\frac{1}{f} = (n-1)\left(\frac{1}{R_1} - \frac{1}{R_2}\right) \tag{24-24}$$

Solve

Both surfaces are concave. Hence the center of curvature of each surface is on the same side of the lens as that surface. The center of curvature C_1 of the front surface (on the left in the figure) is on the near side of the lens, so R_1 is negative: $R_1 = -15.0$ cm. The center of curvature C_2 of the back surface (on the right in the figure) is on the far side of the lens, so R_2 is positive: $R_2 = +25.0$ cm.

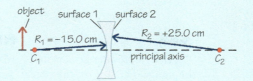

Calculate the focal length.

From Equation 24-24,

$$\frac{1}{f} = (1.520 - 1)\left[\frac{1}{(-15.0 \text{ cm})} - \frac{1}{(+25.0 \text{ cm})}\right]$$

$$= 0.520[(-0.0667 \text{ cm}^{-1}) - (0.0400 \text{ cm}^{-1})]$$

$$= -0.0555 \text{ cm}^{-1}$$

$$f = \frac{1}{-0.0555 \text{ cm}^{-1}} = -18.0 \text{ cm} = -0.180 \text{ m}$$

The first term is negative, and we subtract from it the positive $1/R_2$ term. So both terms contribute to giving $1/f$ a negative value so that the focal length is negative.

Reflect

The focal length is negative, just as we expect for a concave lens. The power of such a lens is negative: $P = 1/f = 1/(-0.180 \text{ m}) = -5.55 \text{ m}^{-1}$, or -5.55 diopters. You should repeat the calculation with the lens reversed back to front, as in part (b) of Example 24-3; you should get the same result for the focal length. Do you?

Image Position and Magnification for a Thin Lens

To see how the image distance is related to the object distance and the focal length for a thin, convex lens, let's look again at a ray diagram like Figure 24-30. We've drawn such a diagram in Figure 24-35. Let's see how to use trigonometry to find relationships between the distances of interest.

The shaded regions in Figure 24-35a are similar triangles. That's because the ray from the tip of the object at O' to the tip of the image at I' passes through the center of the lens at C without deflection, so the angle θ is the same in both triangle $O'C'C$ and triangle ICI'. The ratio of the heights of the two triangles is therefore equal to the ratio of their bases; that is,

$$\frac{|h_I|}{h_O} = \frac{d_I}{d_O} \qquad (24\text{-}26)$$

(The height h_I of the image in Figure 24-35a is negative because the image is inverted. That's why we've used $|h_I|$ in Equation 24-26 for the distance from I to I'.) The two shaded regions in Figure 24-35b — the right triangles $FC'C$ and $FI'I$ — are also similar triangles. That's because the straight ray from C' to I' passes through the focal point, so the angle ϕ is the same on either side of F.

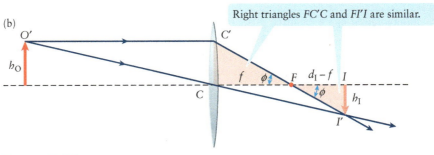

Figure 24-35 Analyzing a converging lens Ray diagrams for an object arrow on the principal axis of a converging lens. The similar triangles in (a) and (b) help us determine the position and magnification of the image.

So for these two triangles as well, the ratio of their heights is equal to the ratio of their bases:

(24-27)

$$\frac{|h_I|}{h_O} = \frac{d_I - f}{f}$$

If we combine Equation 24-26 and Equation 24-27, we find a relationship among the image distance d_I, the object distance d_O, and the focal length f:

$$\frac{d_I}{d_O} = \frac{d_I - f}{f} \quad \text{or} \quad \frac{d_I}{d_O} = \frac{d_I}{f} - 1 \quad \text{or} \quad \frac{d_I}{d_O} + 1 = \frac{d_I}{f}$$

Divide both sides by d_I:

(24-11)

$$\frac{1}{d_O} + \frac{1}{d_I} = \frac{1}{f} \quad \text{(lens equation)}$$

This is the same equation that we deduced for spherical mirrors in Section 24-4. It applies just as well to thin, spherical convex lenses. Now we see why we were justified in calling Equation 24-11 the mirror and lens equation—it works for both.

Equation 24-26 above also tells us the magnification of the image. Since the image height h_I is negative in Figure 24-35a (the image is inverted), h_I is equal to $-|h_I|$ and $|h_I| = -h_I$. If we substitute this into Equation 24-26, we get

$$\frac{(-h_I)}{h_O} = \frac{d_I}{d_O} \quad \text{or} \quad \frac{h_I}{h_O} = -\frac{d_I}{d_O}$$

Lateral magnification is equal to the ratio of image height to object height: $m = h_I/h_O$ (Equation 24-2). So the magnification of the image produced by a thin lens is

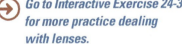
Go to Interactive Exercise 24-3 for more practice dealing with lenses.

(24-14)

$$m = \frac{h_I}{h_O} = -\frac{d_I}{d_O} \quad \text{(lateral magnification)}$$

That's the same Equation 24-14 that we derived in Section 24-4 for a curved mirror.

We've derived Equations 24-11 and 24-14 for a thin convex lens with a positive focal length, but they turn out to be equally valid for a thin concave lens with a negative focal length. As for mirrors, it's important to keep track of the signs of quantities in these equations. Table 24-2 summarizes when the quantities in the lensmaker's equation (Equation 24-24), the lens equation (Equation 24-11), and the magnification equation (Equation 24-14) are positive and when they are negative.

Because the equations for mirrors and thin lenses are effectively identical, many of the same conclusions that we came to for images made by a curved mirror also apply to images made by a thin lens:

- If the lens has a positive focal length (a converging lens) and an object is placed outside the focal point, the image is real and inverted. Depending on the object distance, the image can be smaller, larger, or the same size as the object.

- If the lens has a positive focal length (a converging lens) and an object is placed inside the focal point of a converging lens, the image is virtual, upright, and larger than the object.

- If the lens has a negative focal length (a diverging lens), the image is virtual, upright, and smaller than the object. This is true for any object distance.

One key difference between mirrors and lenses is in the position of the image. For a mirror a real image is on the same side of the mirror as the object, and a virtual

TABLE 24-2 **Sign Conventions for Lenses**	
lens surface radius of curvature R_1 or R_2	• positive if the center of curvature is on the side of the lens opposite from the object • negative if the center of curvature is on the same side of the lens as the object
focal length f	• positive for a converging lens • negative for a diverging lens
image distance d_I	• positive if on the side of the lens opposite from the object; the image is then a real image • negative if on the same side of the lens as the object; the image is then a virtual image
image height h_I	• positive if the image is upright (the same orientation as the object) • negative if the image is inverted (flipped upside down compared to the object)
lateral magnification m	• positive if the image is upright (the same orientation as the object) • negative if the image is inverted (flipped upside down compared to the object)

image is on the other side of the mirror. For a lens a real image is on the opposite side of the lens from the object, and a virtual image is on the same side of the lens as the object.

EXAMPLE 24-5 An Image Made by a Convex Lens

A thin convex lens has a focal length of 15.0 cm. How far from the lens does the image form when an object is placed 9.00 cm from the center of the lens? Is the image virtual or real? Is the image inverted or upright? By what factor is the image magnified relative to the object?

Set Up

We'll begin by drawing a ray diagram to help us visualize the kind of image that will be produced. We'll use Equation 24-11 to calculate the position of this image, and Equation 24-14 to calculate the image height compared to the object height.

Lens equation:

$$\frac{1}{d_O} + \frac{1}{d_I} = \frac{1}{f}$$ (24-11)

Magnification:

$$m = \frac{h_I}{h_O} = -\frac{d_I}{d_O}$$ (24-14)

Solve

We begin by drawing a ray diagram. We've drawn one ray from the object that enters the lens parallel to the principal axis, exits the lens, and passes through the focal point F on the other side of the lens. We've also drawn a ray that passes through the center of the lens. The rays never meet on the other side of the lens, but their extensions do meet on the same side of the lens as the object. So the image will be virtual. As the diagram shows, the image is also upright and larger than the object.

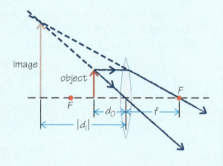

Find the image distance using the lens equation.

We are given $d_O = 9.00$ cm and $f = 15.0$ cm. From Equation 24-11:

$$\frac{1}{d_I} = \frac{1}{f} - \frac{1}{d_O} = \frac{1}{15.0 \text{ cm}} - \frac{1}{9.00 \text{ cm}}$$
$$= 0.06667 \text{ cm}^{-1} - 0.111 \text{ cm}^{-1}$$
$$= -0.0444 \text{ cm}^{-1}$$

$$d_I = \frac{1}{(-0.0444 \text{ cm}^{-1})} = -22.5 \text{ cm}$$

The negative value of d_I means that the image is on the same side of the lens of the object, just as the ray diagram shows. It must therefore be a virtual image.

Find the image height using the magnification equation.

From Equation 24-14:

$$m = \frac{h_I}{h_O} = -\frac{d_I}{d_O} = -\frac{(-22.5 \text{ cm})}{9.00 \text{ cm}} = +2.50$$

The plus sign means the image is upright, in agreement with the ray diagram. The image is 2.50 times as large as the object.

Reflect

Although we didn't try to draw the ray diagram to exact scale, by eye the ratio of the focal length to the object distance looks to be about 2 to 1, which is certainly consistent with the actual values $f = 15.0$ cm and $d_O = 9.00$ cm. So we expect that the image distance in the diagram is about one and a half times the focal length (22.5 cm compared to 15.0 cm)— and it is. Can you verify that the height of the image in the diagram is about 2.5 times the height of the object?

GOT THE CONCEPT? 24-9 **Lateral Magnification**

 You are given a converging lens with focal length f. You place the same object at each of the following object distances. Rank the images that are produced in order of their lateral magnification, from most positive (that is, the largest upright image) to most negative (that is, the largest inverted image). (a) $d_O = 3f$; (b) $d_O = 2f$; (c) $d_O = 3f/2$; (d) $d_O = f/2$; (e) $d_O = f/4$.

TAKE-HOME MESSAGE FOR Section 24-8

✔ The focal length of a thin lens is determined by the index of refraction of the lens material and the radii of curvature of the front and back surfaces of the lens. These radii can be positive or negative.

✔ The more sharply curved the surfaces of a lens, the smaller the magnitudes of their radii of curvature and the smaller the magnitude of the focal length of the lens.

✔ The same equation that relates object distance, image distance, and focal length for a spherical mirror also applies to thin lenses. The magnification equation is also the same for spherical mirrors and thin lenses.

✔ Image distance d_I is positive when the image is real and negative when the image is virtual. The focal length f is positive for a thin, convex lens and negative for a thin, concave lens.

24-9 A camera and the human eye use different methods to focus on objects at various distances

We can now apply the ideas of the last two sections to two important optical devices: a camera and the human eye. Both of these devices use refraction to form a real, inverted image of an object. In a digital camera (including the camera in a smartphone) the image is formed on a light-sensitive sensor and stored electronically; in the human eye the image is formed on the light-sensitive retina and sent via the optic nerve to the brain. As we will see, however, there are essential differences in how a camera and the human eye form images and how they focus on objects at different distances.

The Camera

Figure 24-36a shows a simplified cross section of a typical camera. Unlike the thin lenses we discussed in Sections 24-7 and 24-8, the lens of a camera can be relatively thick and made up of several individual pieces or *elements*. (Figure 24-36a shows a two-element lens; many smartphone cameras have five or more elements, and lenses used by professional photographers may have more than 15 elements.) The elements typically have different shapes and are made of different kinds of glass. The shapes are chosen to minimize spherical aberration (see Section 24-7) and to provide a sharp image at all points on the light-sensitive sensor.

Unfortunately, glass suffers from dispersion (Section 23-4): The index of refraction is slightly different for light of different wavelengths. The focal length of a lens depends on the index of refraction (Equation 24-24), so if all of the elements of a camera lens were made of the same kind of glass, light of different wavelengths would be brought to a focus at different distances behind the lens. As a result, if the yellow colors of an object formed a sharp image on the sensor, the blue and red colors would be slightly blurred (Figure 24-36b). To avoid such *chromatic aberration*, lens designers use different kinds of glass for different lens elements. These kinds of glass are chosen so that the different dispersion of the individual elements largely cancel out. The result is that light of all colors comes to the same focus (Figure 24-36c).

(a) A schematic camera

Parallel light rays from a distant object

Camera body

Electronic sensor at the focal point of the lens

Lens elements

(b) A single-element lens has chromatic aberration

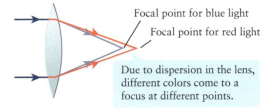

Focal point for blue light
Focal point for red light

Due to dispersion in the lens, different colors come to a focus at different points.

(c) Correcting chromatic aberration

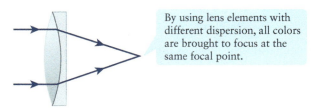

By using lens elements with different dispersion, all colors are brought to focus at the same focal point.

Figure 24-36 A camera and its lens (a) The lens of a camera is composed of two or more elements. (b) A single-element lens produces chromatic aberration. (c) A lens with more than one element of different indices of refraction can largely eliminate chromatic aberration.

The lens equation, Equation 24-11, tells us that for a camera lens of a given focal length f, changing the distance d_O from lens to object will change the distance d_I from lens to image:

$$\frac{1}{d_O} + \frac{1}{d_I} = \frac{1}{f} \qquad (24\text{-}11)$$

If the object is very distant, d_O is very large and $1/d_O$ is nearly zero. Then Equation 24-11 becomes $1/d_I = 1/f$, which tells us that the image distance d_I equals the focal length f. Hence the camera's light-sensitive sensor is placed a distance f behind the lens so that distant objects will be in sharp focus (Figure 24-37a).

As the object approaches the lens, the object distance d_O decreases, so the image distance d_I increases. Hence the sensor must be farther from the lens to record a sharp image. In most professional cameras *focusing* the camera on a nearby object is accomplished by moving the lens farther away from the sensor (Figure 24-37b). On some digital cameras you can actually see the lens move outward from the camera body to focus on a nearby object.

In the camera on a smartphone, a different method is used to focus on nearby objects. The front element of the lens remains fixed, but one or more of the rear elements move to decrease the effective focal length of the elements acting in combination.

The Human Eye

BioMedical   Like a camera lens, the human eye is composed of different elements with different indices of refraction (Figure 24-38). The *cornea* is the front surface of the eye and has index of refraction $n = 1.376$, which is substantially higher than that of the air outside the eye ($n = 1.000$). Behind the cornea is the *aqueous humor*, a fluid that is 98% water and whose index of refraction ($n = 1.336$) is consequently very close to that of pure water ($n = 1.330$). Since the indices of refraction of the cornea and aqueous humor have fairly similar values, most of the refraction caused by the cornea takes place at its front surface rather than its rear surface.

At the rear of the aqueous humor is the *iris*. This is a diaphragm that opens and closes to regulate the amount of light falling on the sensitive retina. After passing through the aperture of the iris, light rays undergo further refraction when they enter and leave the crystalline *lens*. This is a converging lens about 9 mm in diameter and about 4 mm thick. It is made up of transparent cells called lens fibers arranged in concentric layers rather like the layers of an onion.

Unlike the lenses we studied in Sections 24-7 and 24-8, the index of refraction of the lens is *not* the same throughout its volume: Its value varies from about $n = 1.406$ at the center of the lens to about $n = 1.386$ near its outer rim. This actually enhances the ability of the lens to make light rays converge and helps to minimize spherical aberration.

The index of refraction of the lens is only slightly greater than that of the aqueous humor in front of the lens ($n = 1.336$) or that of the *vitreous humor* ($n = 1.337$), the transparent gel that lies behind the lens and fills most of the volume of the eye. Hence the lens causes less convergence of light rays than does the cornea. Roughly 70% of the refraction needed to bend parallel light rays coming from a distant object and focus them on the retina comes from the cornea; the remaining 30% comes from the refraction provided by the lens (Figure 24-39a).

Unlike in a camera lens, the human eye does not focus on nearby objects by moving its elements forward or back: The distance from the cornea to the center of the lens remains the same, as does the distance from the center of the lens to the retina. Instead, the lens (which is made of flexible biological material) changes shape! To focus on a close object, the *ciliary muscles* that hold the lens in place squeeze the lens around its rim. This causes the lens to deform, as shown in

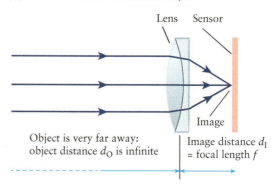

(a) A camera focused on a distant object

Object is very far away: object distance d_O is infinite

Image distance d_I = focal length f

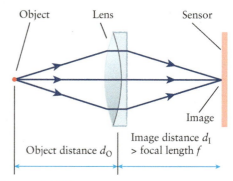

(b) The same camera focused on a nearby object

Image distance d_I > focal length f

Object distance d_O

Figure 24-37   Focusing a camera (a) To produce a sharp image of a distant object on the sensor, the distance from lens to sensor must equal the focal length of the lens. (b) If the object is nearby, the lens must be moved away from the sensor to produce a sharp image.

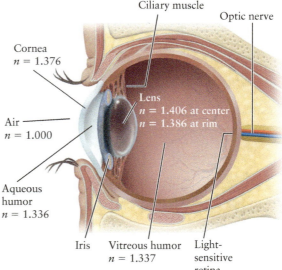

Cornea $n = 1.376$

Air $n = 1.000$

Aqueous humor $n = 1.336$

Ciliary muscle

Optic nerve

Lens $n = 1.406$ at center $n = 1.386$ at rim

Iris   Vitreous humor $n = 1.337$ Light-sensitive retina

Figure 24-38   The human eye The elements of the human eye work together to bring parallel light rays from a distant object to a focus on the light-sensitive retina.

(a) An eye focused on a distant object

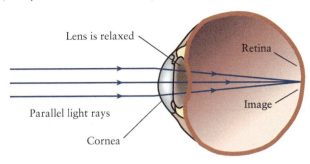

Lens is relaxed

Retina

Parallel light rays

Image

Cornea

(b) The same eye focused on a nearby object

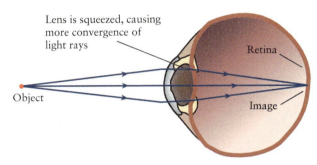

Lens is squeezed, causing more convergence of light rays

Retina

Object

Image

Figure 24-39 Focusing the human eye (a) When the lens is relaxed, parallel light rays from a distant object come to a focus on the light-sensitive retina. (b) To make a sharp image on the retina of a nearby object, the lens must change shape.

(a) Hyperopia (farsightedness): light rays converge too little

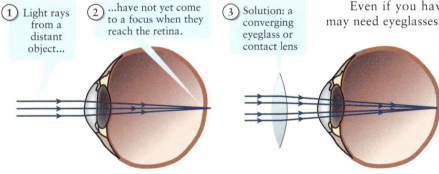

① Light rays from a distant object...

② ...have not yet come to a focus when they reach the retina.

③ Solution: a converging eyeglass or contact lens

(b) Myopia (nearsightedness): light rays converge too much

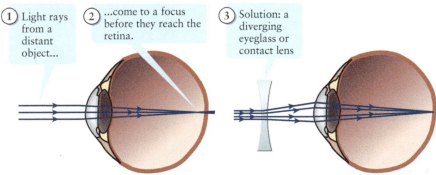

① Light rays from a distant object...

② ...come to a focus before they reach the retina.

③ Solution: a diverging eyeglass or contact lens

Figure 24-40 Correcting deficiencies of vision (a) Hyperopia is corrected with a converging lens; (b) myopia is corrected with a diverging lens.

Figure 24-39b. The front and back surfaces of the lens become more sharply curved, so their radii of curvature are reduced in magnitude. Hence the focal length f of the lens decreases, the power $P = 1/f$ of the lens increases (see Equation 24-25), and light rays are bent more sharply as they traverse the lens. Hence an image of the nearby object can be brought to a focus on the retina.

In many people the focusing mechanism shown in Figure 24-39 does not work perfectly. One common example is the condition known as *hyperopia* or *farsightedness*, in which the cornea and lens provide too little bending of light rays. As a result, parallel light rays that enter the eye have not yet come to a focus when they reach the retina, so objects appear blurred (Figure 24-40a). The lenses of a person with hyperopia must be squeezed, as in Figure 24-39b, to focus both on distant objects and on nearby ones, which causes eye strain and headaches. Even with the lenses squeezed as much as possible, the nearest object that can be seen in sharp focus can still be quite far away. Hyperopia can occur if the eye is too short from front to back, if the cornea has the wrong shape, or if the ciliary muscles are too weak. To compensate for the inadequate convergence of light rays provided by a hyperopic eye, the treatment is to wear converging (convex) eyeglasses or contact lenses.

More common than hyperopia is *myopia* or *nearsightedness*. In this condition the cornea and lens actually do too good of a job of bending light rays and making them converge. As a result, parallel light rays from a distant object come to a focus at a point in front of the retina (Figure 24-40b). Nearby objects do appear in focus, however. Myopia can occur if the eye is too long from front to back, or if the cornea or lens has an incorrect shape. The treatment is to wear diverging (concave) eyeglasses or contact lenses. Then parallel light rays coming from a distant object will be diverging when they enter the cornea, and this divergence compensates for the excess convergence caused by the cornea and lens.

Even if you have eyes without hyperopia or myopia, you may need eyeglasses by the time you reach middle age. The reason is that the lenses in our eyes lose some of their flexibility as we age. As a result, the lens is unable to squeeze as in Figure 24-39b, so light from nearby objects cannot be focused on the retina and these objects appear blurred. This condition, called *presbyopia*, is treated by wearing reading glasses (see the photograph that opens this chapter) with convex lenses. Just as in hyperopia, these eyeglasses compensate for the inadequate converging power of the aging, less flexible lens. Unlike hyperopia, however, a person with presbyopia may not need to use eyeglasses to see distant objects clearly. Many older people have both myopia (which requires concave, diverging eyeglasses to see distant objects) *and* presbyopia (which requires convex, converging eyeglasses to see nearby objects). The solution to this problem is *bifocal* eyeglasses, which have a concave shape in their upper part for distant vision and a convex shape in their lower part for reading.

With age the lenses of the eyes can also become clouded by cataracts. In this case the lenses are surgically removed and replaced by artificial lenses (see Figure 24-1b).

EXAMPLE 24-6 Correcting for Farsightedness and Nearsightedness

(a) Even with the lenses of her eyes squeezed as much as possible, a certain farsighted person has sharp vision only for objects that are 3.00 m or farther away from her corneas. What should be the focal length and power of a contact lens that will allow her to clearly see an object 0.250 m away from her corneas? (b) A certain nearsighted person can only see objects sharply if they are no more than 1.25 m in front of his corneas. What should be the focal length and power of a contact lens that will allow him to see very distant objects clearly?

Set Up

Since a contact lens sits on the cornea and is very thin, the distance d_O from the contact lens to the object in each case is essentially the same as from the cornea to the object [0.250 m in (a), infinity in (b)]. In part (a) the contact lens must make the eye "think" that an object 0.250 m away is actually 3.00 m away. So this contact lens must produce a virtual image of the object that's 3.00 m in front of the contact lens. Similarly, in part (b) the contact lens must make the eye "think" that a distant object is only 1.25 m away. To do this, the contact lens must produce a virtual image of the object 1.25 m in front of the contact lens. In each case we'll use Equation 24-11 to find the required focal length f from the specified object and image distances. Equation 24-25 then tells us the power of the contact lens.

Lens equation:

$$\frac{1}{d_O} + \frac{1}{d_I} = \frac{1}{f} \qquad (24\text{-}11)$$

Lens power:

$$P = \frac{1}{f} \qquad (24\text{-}25)$$

Solve

(a) We draw a ray diagram with a point object a distance $d_O = 0.250$ m from the contact lens. (We draw a point object rather than an object arrow, since we're not concerned with the size of the image.) The light rays emerging from this contact lens must appear to emanate from a point 3.00 m in front of the lens, so the contact lens must make the light rays converge. Hence it must be a converging contact lens with a positive focal length.

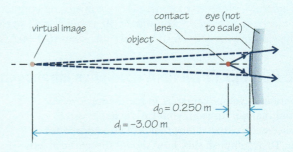

Solve for the focal length of the contact lens using the lens equation.

The distance to the object is $d_O = 0.250$ m, and the distance to the image formed by the contact lens is $d_I = -3.00$ m (negative because the image is on the same side of the contact lens as the object). From Equation 24-11:

$$\frac{1}{f} = \frac{1}{d_O} + \frac{1}{d_I} = \frac{1}{0.250} + \frac{1}{(-3.00 \text{ m})}$$
$$= 4.00 \text{ m}^{-1} - 0.333 \text{ m}^{-1}$$
$$= 3.67 \text{ m}^{-1}$$
$$f = \frac{1}{3.67 \text{ m}^{-1}} = 0.273 \text{ m}$$

The positive focal length means that as predicted, this is a converging contact lens. The power of this contact lens is

$$P = \frac{1}{f} = 3.67 \text{ m}^{-1} = 3.67 \text{ diopters}$$

(b) We again draw a ray diagram, now with parallel light rays from a distant object entering the contact lens. The light rays emerging from the contact lens must appear to emanate from a point 1.25 m in front of the lens. Since the contact lens makes parallel light rays diverge, it must be a diverging contact lens with a negative focal length.

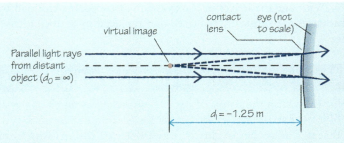

Again use the lens equation to solve for the focal length of the contact lens.

The distance d_O to the object is infinite (so $1/d_O = 0$), and the image formed by the contact lens is at $d_I = -1.25$ m. (This is negative because the image is on the same side of the contact lens as the object.) From Equation 24-11:

$$\frac{1}{f} = \frac{1}{d_O} + \frac{1}{d_I} = 0 + \frac{1}{(-1.25 \text{ m})}$$

$$\frac{1}{f} = \frac{1}{(-1.25 \text{ m})}$$

Take the reciprocal of both sides:

$$f = -1.25 \text{ m}$$

The negative value means that this is a diverging contact lens, as we expect. The power of this contact lens is

$$P = \frac{1}{f} = \frac{1}{(-1.25 \text{ m})} = -0.800 \text{ m}^{-1} = -0.800 \text{ diopter}$$

Reflect

This example reinforces the idea that you should draw a ray diagram for *any* problem that involves image formation by lenses or mirrors. Note that in part (a) the converging contact lens produces a virtual image. This can happen only if the object is placed inside the focal point of the contact lens, so that the object distance is less than the focal length. This is indeed the case in part (a): The object distance was $d_O = 0.250$ m, and we found that the focal length was $f = 0.273$ m.

GOT THE CONCEPT? 24-10 A Corrective Intraocular Lens

(?) A patient develops cataracts in her myopic eye. As an ophthalmic surgeon, your task is to choose a replacement artificial lens that will also correct for her myopia. To accomplish this, how should the artificial lens compare to the patient's natural lens? (There may be more than one correct answer.) (a) Same shape but a higher index of refraction; (b) same shape but a lower index of refraction; (c) same index of refraction but with more sharply curved surfaces; (d) same index of refraction but with more gently curved surfaces.

TAKE-HOME MESSAGE FOR Section 24-9

✔ A camera lens refracts parallel light rays from a distant object so that they come to a focus on a light-sensitive sensor. To focus on nearby objects, either the lens is moved away from the sensor or the focal length of the lens is decreased.

✔ In the human eye, refraction takes place in both the cornea and the lens to focus light rays on the retina. Adjusting the focus from distant to nearby objects is done by reshaping the lens.

✔ Corrective lenses (eyeglasses or contact lenses) are used to either increase or decrease the amount by which light rays are forced to converge within the eye on their way to the retina.

24-10 The concept of angular magnification plays an important role in several optical devices

We've used the idea of *lateral* magnification to compare the *height* of the image made by a mirror or lens to the height of the object. But in many situations the kind of magnification that is most relevant is *angular* magnification. As an example, suppose there's an object you want to inspect that's some distance from your eye (Figure 24-41a). To get a better view of this object, you bring it closer to your eye so that it subtends a larger angle θ (Figure 24-41b). The greater the value of θ, the more detail you can see. However, some objects (like the wings of a fly or human blood cells) are so small that their details cannot be seen by the naked eye. And some objects (like the Moon and planets) are so distant that it's impractical to bring your eye any closer to them. Let's look at three optical devices that address these limitations: the *magnifying glass*, the *microscope*, and the *telescope*.

The Magnifying Glass

The closest that any object can be brought to a normal, relaxed eye and be in focus is about 25 cm. This is called the *near* point of the eye. (You can see an object at a closer distance, but that requires focusing the eye as in Figure 24-39. This can be tiring.) If the height of the object is h_O, Figure 24-42a shows that the angle θ subtended by this object at a 25-cm distance from the eye is given by $\tan \theta = h_O/(25 \text{ cm})$. As we mentioned in Section 24-6, the tangent of a small angle is approximately equal to the angle in radians, so this expression becomes

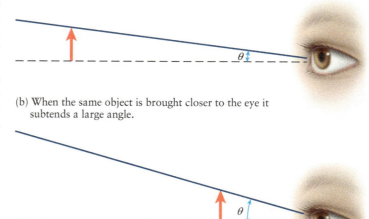

(a) When an object is far from the eye it subtends a small angle θ.

(b) When the same object is brought closer to the eye it subtends a large angle.

Figure 24-41 For a better view, an object should subtend a larger angle Moving an object (a) so that it is closer to the eye (b) means that it subtends a larger angle θ.

$$\theta = \frac{h_O}{25 \text{ cm}} \quad \text{(angle subtended by an object at the near point of the eye)} \quad (24\text{-}27)$$

If the object is small (h_O has a small value), the angle θ is small and it will be difficult to see any detail. The purpose of a **magnifying glass** or *magnifier* is to overcome this limitation. A magnifying glass is simply a converging lens with a focal length f that is shorter than 25 cm. If you place an object at the focal point F on one side of the lens, the light rays emerging from the other side are parallel (Figure 24-42b). What you see is an upright virtual image that is infinitely far away (and so appears in focus to the relaxed eye)

(a) Without a magnifying lens, the closest an object can be to a relaxed eye and be in focus is 25 cm.

(c) The magnified view through a magnifying glass

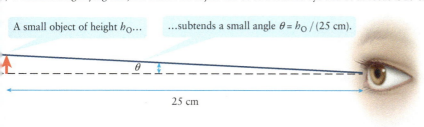

A small object of height h_O... ...subtends a small angle $\theta = h_O / (25 \text{ cm})$.

25 cm

(b) The same object placed at the focal point F of a magnifying glass (a converging lens of short focal length f)

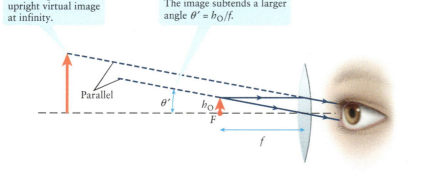

The lens produces an upright virtual image at infinity.

The image subtends a larger angle $\theta' = h_O/f$.

Parallel

θ' h_O F

f

Paul Michael Hughes/Alamy

Figure 24-42 A magnifying glass (a) An object viewed at the near point of a normal relaxed eye. (b) The same object viewed through a magnifying glass. (c) The image of this object seen through a magnifying glass is upright, virtual, and subtends a larger angle.

and that subtends an angle θ', where $\tan \theta' = h_O/f$. Using the same small-angle approximation as in Equation 24-27, we can write the angle subtended by the virtual image as

(24-28)
$$\theta' = \frac{h_O}{f} \quad \text{(angle subtended by the image of an object viewed through a magnifying glass)}$$

The **angular magnification** M of the magnifying glass is the ratio of the angle θ' (with the magnifying glass, Equation 24-28) to the angle θ (without the magnifying glass, Equation 24-27):

(24-29)
$$M = \frac{\theta'}{\theta} = \frac{h_O/f}{h_O/(25 \text{ cm})} = \frac{25 \text{ cm}}{f} \quad \text{(angular magnification of a magnifying glass)}$$

For example, if the magnifying glass has a focal length of 10 cm, the angular magnification is $M = (25 \text{ cm})/(10 \text{ cm}) = 2.5$. As viewed through the magnifying lens, the object will subtend an angle 2.5 times larger than with the naked eye, and so will appear 2.5 times larger (Figure 24-42c).

The smaller the focal length f of the magnifying glass, the greater the angular magnification M given by Equation 24-29. Making f smaller means using more sharply curved lens surfaces, which increases spherical aberration (Section 24-7) and decreases image quality. In practice the largest value of M obtainable with a magnifying glass is about 15. For greater angular magnification, a *microscope* is necessary.

The Microscope

A **microscope** uses *two* lenses of short focal length, an **objective** and an **eyepiece** (Figure 24-43a). Because two lenses are used it's also called a *compound microscope*. The object to be viewed is placed just outside the focal point of the objective lens, so the object distance d_O is just slightly larger than the objective focal length $f_{objective}$. The image distance d_I is therefore much larger than either d_O or $f_{objective}$, as Figure 24-43b shows. (You can also see this from Equation 24-11, the lens equation: $1/d_O + 1/d_I = 1/f_{objective}$, which we can rewrite as $1/d_I = 1/f_{objective} - 1/d_O$. Since d_O is almost the same as $f_{objective}$, the difference $1/f_{objective} - 1/d_O$ is very small. Hence $1/d_I$ is very small and its reciprocal d_I is large.) The image produced by the objective lens is inverted and real, and from Equation 24-14 the *lateral* magnification is

(24-30)
$$m_{objective} = \frac{h_I}{h_O} = -\frac{d_I}{d_O} \approx -\frac{d_I}{f_{objective}} \quad \text{(lateral magnification of a microscope objective lens)}$$

On the far right-hand side of Equation 24-30 we've used the idea that the object distance d_O is almost the same as the objective focal length $f_{objective}$. A typical objective lens for a microscope such as you might use in a biology lab course has a focal length of from 2 to 40 mm.

The second lens, or eyepiece, is just a magnifying glass. The idea is that the real *image* made by the objective lens acts as the *object* for the eyepiece. Just as in Figure 24-42c, we place the eyepiece so that its focal point $F_{eyepiece}$ is at the position of the image formed by the objective. The viewer then sees an enlarged virtual image at infinity. Since the objective produces an inverted image and the eyepiece does not further invert the image, the viewer sees an image that is inverted relative to the object.

The net *angular* magnification of the image delivered to the viewer is the product of the lateral magnification $m_{objective}$ from the objective, Equation 24-30, and the angular magnification $M_{eyepiece}$ from the eyepiece, Equation 24-29. It's customary to always treat angular magnification as positive, so we use the absolute value of $m_{objective}$:

(24-31)
$$M_{microscope} = |m_{objective}|M_{eyepiece} = \left(\frac{d_I}{f_{objective}}\right)\left(\frac{25 \text{ cm}}{f_{eyepiece}}\right) = \frac{d_I(25 \text{ cm})}{f_{objective}f_{eyepiece}}$$

(angular magnification of a microscope)

Equation 24-31 shows that the smaller the focal lengths of the objective and eyepiece lens, the greater the angular magnification. Even relatively inexpensive microscopes can achieve angular magnifications of 2500, called 2500×.

Note that if you want to take a photograph using a microscope, the eyepiece is *not* used. Instead a light sensor like that used in a digital camera is placed at the

(a) A microscope

PhotoAlto/Alamy

(c) A photomicrograph

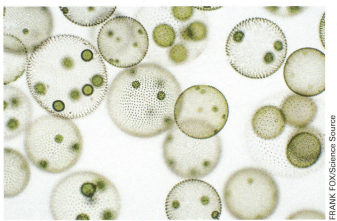

FRANK FOX/Science Source

(b) How a microscope works

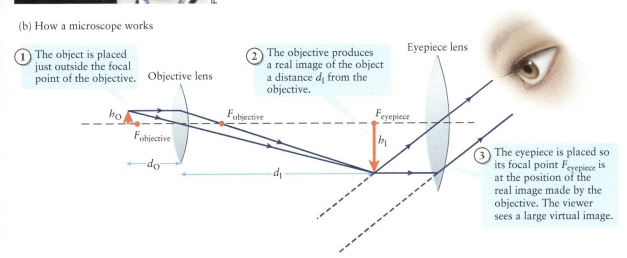

① The object is placed just outside the focal point of the objective.

Objective lens

② The objective produces a real image of the object a distance d_I from the objective.

Eyepiece lens

③ The eyepiece is placed so its focal point $F_{eyepiece}$ is at the position of the real image made by the objective. The viewer sees a large virtual image.

h_O $F_{objective}$ $F_{eyepiece}$

$F_{objective}$

d_O h_I d_I

BioMedical **Figure 24-43** **A microscope** (a) A microscope uses two lenses, an objective and an eyepiece. (b) How a microscope forms an image. (c) To record this microscope image on a light sensor, the eyepiece was not used; instead the sensor was placed at the position of the real image made by the objective shown in (b). The algal colonies shown here (*Volvox*) are about 0.35 mm in diameter and are just visible to the unaided eye; the microscope reveals their detailed structure.

position of the real image made by the object lens, so that this image is focused on the sensor. Figure 24-43c shows an example of a photograph made in this way, called a *photomicrograph*.

The Telescope

Like a microscope, a **telescope** has an objective lens that makes a real image of an object. It also has an eyepiece lens that increases the angular size of that image for the viewer. The difference is that a telescope is used for looking at large, distant objects rather than small objects close at hand.

Figure 24-44 shows a telescope used for astronomy (an *astronomical telescope*). Because astronomical objects are very far away, the angle θ that they subtend is very small. (The full Moon subtends an angle of only about 0.5° as seen from Earth.) The objective lens of focal length $f_{objective}$ makes a real, inverted image of the object. Since the object is essentially infinitely far away, the light rays from the object that enter the objective are parallel and the resulting real image is formed at the focal point F, a distance $f_{objective}$ behind the objective lens. The image has height h_I. As Figure 24-44a shows, the angle θ subtended by the distant astronomical object is the same as the angle subtended by the real image as seen from the objective lens.

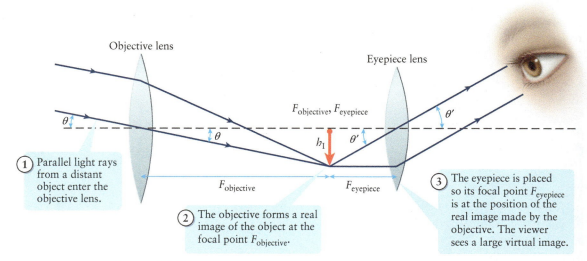

Figure 24-44 **A refracting telescope** In a refracting telescope a converging objective lens makes a real image of a distant object. The eyepiece lens provides a view with an enhanced angular size.

Again using the approximation for small angles that θ in radians is equal to $\tan \theta$, we have $\theta = h_I/f_{objective}$.

To give this image a larger angular size as seen by the viewer, we place the eyepiece lens of focal length $f_{eyepiece}$ so that its focal point is at the same point as the image made by the objective lens. As for the microscope, the image made by the objective lens acts as the object for the eyepiece lens. Figure 24-44 shows that the angle subtended by the final image is $\theta' = h_I/f_{eyepiece}$. The angular magnification of the telescope as a whole is the ratio of θ' (the angular size of the distant object as seen through the eyepiece) to θ (the angular size of the object as seen by the unaided eye), or

(24-32)
$$M_{telescope} = \frac{\theta'}{\theta} = \frac{h_I/f_{eyepiece}}{h_I/f_{objective}} = \frac{f_{objective}}{f_{eyepiece}} \quad \text{(angular magnification of a telescope)}$$

Equation 24-32 shows that you can increase the magnification of the telescope by using an objective lens with a *greater* focal length, an eyepiece lens with a *smaller* focal length, or both. A typical telescope used by beginning amateur astronomers can give angular magnifications from about 35× to about 100×. Note that just as for a microscope, if you want to take a photograph using a telescope you remove the eyepiece and place a light sensor at the point $F_{objective}$ where the objective makes the real image.

WATCH OUT! Magnification is not the most important aspect of an astronomical telescope.

It's a common misconception that the primary purpose of a telescope is to magnify images. But in fact magnification is not the most important aspect of a telescope. The reason is that there are limits to how sharp any astronomical image can be. One limit is due to the blurring caused by Earth's atmosphere: As light rays from space travel through our moving atmosphere, they are randomly refracted as they pass through one air mass to the next. This causes stars to twinkle as seen by the naked eye, and it also causes astronomical images to blur and "wobble" as seen through a telescope. A second limit is due to diffraction, which as we saw in Section 23-9 poses fundamental limits on the angular resolution of an image. Magnifying an image that's blurred by the atmosphere, by diffraction, or a combination may make it look bigger, but will not make it any clearer. Thus, beyond a certain point, there's nothing to be gained by further magnification. Astronomers actually place more priority on the *light-gathering power* of a telescope, which is a measure of how much light the telescope can collect from a distant object. This is important because astronomical objects are generally quite dim. For this reason telescopes have large-diameter objectives to gather as much light as possible. This means brighter images, which makes it easier to see faint details.

The telescope shown in Figure 24-44 is called a **refracting telescope** because it uses an objective *lens*. Almost all telescopes used in astronomy research are **reflecting telescopes** that use a converging *mirror* for an objective instead of a converging lens (Figure 24-45a). One important reason is chromatic aberration, which we discussed in Section 23-9: A simple objective lens has a different focal length for different colors of light. An objective mirror, by contrast, brings all colors to a focus at the same point. Another reason is that because a lens can be supported only around its edges, a large objective lens tends to sag and distort under its own weight as it is pointed to different parts of the sky. This distortion has negative effects on the clarity of the image. By contrast, a large objective mirror can be supported by braces on its back, which minimizes the distortions that can affect a large lens.

One disadvantage of a reflecting telescope is that the objective mirror makes its real image at an inconvenient location in front of the mirror, where it is difficult to access. There are several schemes that astronomers use to get around this, one of which was developed by Isaac Newton. In a *Newtonian telescope* (Figure 24-45b) a small, flat mirror is placed at a 45° angle in front of the focal point. This secondary mirror deflects the light rays and the real image to one side, and the eyepiece is mounted on the side of the telescope. The overall magnification is given by Equation 24-32, the same as for a refracting telescope.

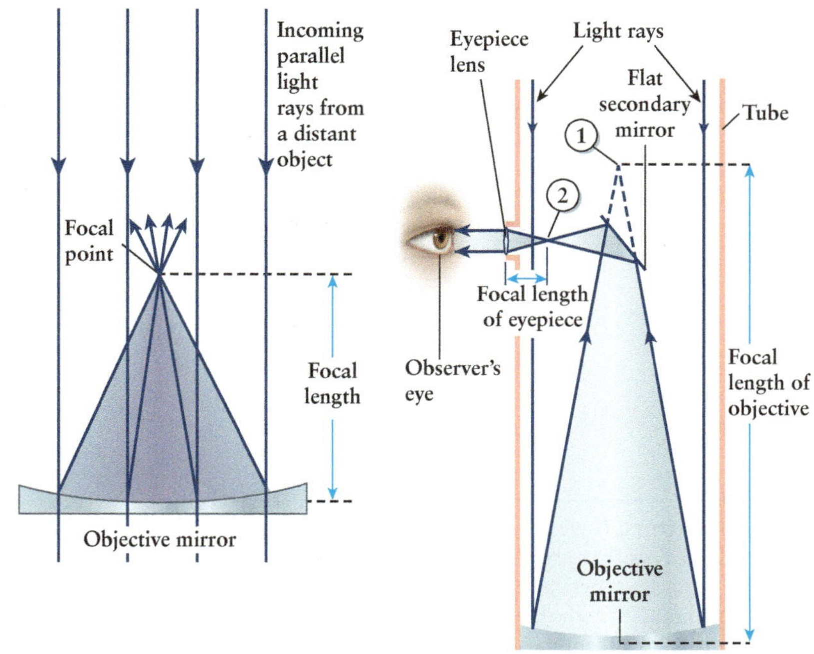

Just as for a microscope, an astronomical telescope produces a final image that is inverted. This is fine for astronomy, but can be distracting if you are using the telescope to look at objects on Earth like a bird in a tree or a distant mountain. A *terrestrial telescope* has an additional lens or prisms between the objective and eyepiece that inverts the image again, giving a final image that is upright. A pair of binoculars is essentially two terrestrial telescopes side by side.

(a) A reflecting telescope

(b) Newton's design

Figure 24-45 **A reflecting telescope** (a) In a reflecting telescope a concave mirror is used as the objective. (b) Newton's seventeenth-century design (still used in telescopes today) uses a flat 45° mirror to move the image made by the objective mirror from point 1 to point 2, where it can be viewed through the eyepiece.

GOT THE CONCEPT? 24-11   Microscope vs. Telescope: Lateral Magnification

 A certain microscope and a certain refracting telescope both advertise an angular magnification of 50×. For which of these two devices does the objective lens produce a real image with the greater absolute value of *lateral* magnification? (a) The microscope; (b) the telescope. (c) The lateral magnification is the same for both devices. (d) Not enough information given to decide.

Key Terms

angular magnification
center of curvature
concave
converging lens
convex
diffuse reflection
diopter
diverging lens
eyepiece
focal length
focal point
geometrical optics
image
image distance

image height
inverted image
lateral magnification
lens
lens equation
lensmaker's equation
magnifying glass
microscope
mirror equation
object
object distance
object height
objective
optical device

plane mirror
power (of a lens or mirror)
principal axis
radius of curvature
ray diagram
real image
reflecting telescope
refracting telescope
specular reflection
telescope
thin lens
upright image
virtual image

Chapter Summary

Topic	Equation or Figure
Plane mirrors: A plane mirror makes an upright, virtual image of any object. The image is the same size as the object, so the lateral magnification is $m = 1$. The image is reversed back to front, not side to side.	The **image distance** is negative.　　The **object distance** is positive. $$d_I = -|d_I| = -d_O$$ (24-1)　　The negative value of d_I indicates that the image is on the opposite side of the mirror from the object.
	Lateral magnification　Image height $$m = \frac{h_I}{h_O}$$ (24-2)　Object height
Spherical mirrors: If we consider only parallel light rays that are close to the principal axis of a spherical concave mirror, the reflected rays all converge at the focal point. The distance from the center of the mirror to the focal point is the focal length. If the mirror is convex rather than concave, parallel light rays diverge rather than converge after reflection. The radius of curvature and the focal length are both negative for a convex mirror.	**Focal length** of a spherical mirror　　**Radius of curvature** of the mirror $$f = \frac{r}{2}$$ (24-3)
Image formation by a concave mirror: We can locate the image made by a concave mirror by drawing a ray diagram. If the object is outside the focal point, the image is real and inverted; its size depends on how far the object is from the mirror. If the object is inside the focal point, the image is virtual, upright, and larger than the object.	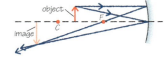 (Example 24-1, figure 1)
	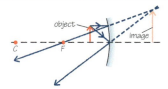 (Example 24-1, figure 2)
Image formation by a convex mirror: A ray diagram also helps us locate the image made by a convex mirror. No matter where the object is placed, the image is virtual, upright, and smaller than the object.	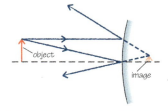 (Example 24-2, figure 1)
The mirror equation and lens equation: The mirror equation and lens equation relate the object distance, image distance, and focal length for either a concave or convex mirror or a converging or diverging lens. The lateral magnification depends on the image and object distances and can be positive (if the image is upright) or negative (if the image is inverted).	$$\frac{1}{d_O} + \frac{1}{d_I} = \frac{1}{f}$$　Focal length (24-11)　Object distance　Image distance
	Lateral magnification　Image height　Image distance $$m = \frac{h_I}{h_O} = -\frac{d_I}{d_O}$$ (24-14)　Object height　Object distance

Lenses: Due to refraction, light rays converge as they enter or exit a converging glass lens and diverge as they enter or exit a diverging glass lens. A converging lens behaves similarly to a concave mirror, and a diverging lens behaves similarly to a convex mirror. The focal length of a lens is given by the lensmaker's equation, which involves the index of refraction of the lens material and the radii of curvature of the lens surfaces. This equation is valid if the lens is thin.

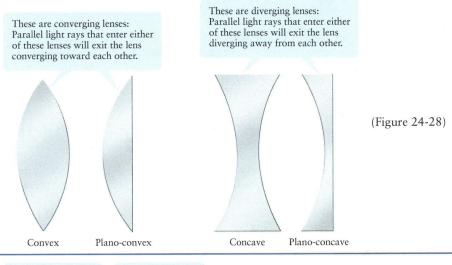

These are converging lenses: Parallel light rays that enter either of these lenses will exit the lens converging toward each other.

These are diverging lenses: Parallel light rays that enter either of these lenses will exit the lens diverging away from each other.

Convex   Plano-convex   Concave   Plano-concave

(Figure 24-28)

Focal length of a lens surrounded by air

Index of refraction of the lens material

$$\frac{1}{f} = (n - 1)\left(\frac{1}{R_1} - \frac{1}{R_2}\right)$$

(24-24)

Radius of curvature of lens surface 1 (the surface closer to the object)

Radius of curvature of lens surface 2 (the surface farther from the object)

Image formation by lenses: A converging lens produces a real, inverted image if the object is outside the focal point and a virtual, upright, enlarged image if the object is inside the focal point. A diverging lens always produces a virtual, upright, reduced image. The same equation that relates object distance d_O, image distance d_I, and focal length f for a spherical mirror also applies to thin lenses, as does the equation for lateral magnification.

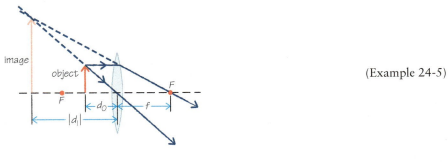

(Example 24-5)

The magnifying glass, microscope, and telescope: A magnifying glass is used for viewing nearby objects: It produces a virtual image that has an angular magnification compared to the object. The objective of a microscope makes a real image of a nearby object, and the objective of a telescope makes a real image of a distant object; in both devices a magnifying lens is used as an eyepiece to provide angular magnification.

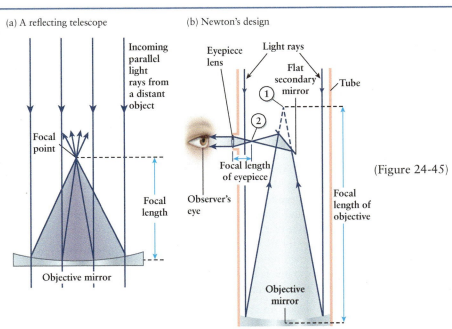

(a) A reflecting telescope

Incoming parallel light rays from a distant object

Focal point

Focal length

Objective mirror

(b) Newton's design

Eyepiece lens   Light rays

Flat secondary mirror   Tube

Focal length of eyepiece

Observer's eye

Focal length of objective

Objective mirror

(Figure 24-45)

Answer to What do you think? Question

(a) These eyeglasses are used by people with presbyopia. Their eyes are able to take essentially parallel light rays from distant objects and make them converge onto the retina to form an image but aren't able to do the same with the diverging light rays coming from a nearby object. The corrective eyeglasses take these diverging light rays and make them nearly parallel, so a converging lens is needed. A converging lens is thicker at the middle than it is at the edges. We discuss the physics of the eye and of corrective lenses in Section 24-9.

Answers to Got the Concept? Questions

24-1 (a) If you see someone's eyes in a mirror, she can see you as well. This must be true because if light rays can reflect off the mirror from her eyes to yours, then light can follow that path in the opposite direction and reflect from your eyes to hers, too.

24-2 (c) From the information given it's not possible to know whether the person around the corner can see you. We've sketched this situation in Figure 24-46. Light rays that reflect off the mirror from his right hand to your eyes, shown in red, enable you to see his hand. You can't see his eyes because the path that light would need to follow for this to happen, shown by the light blue, dashed line, does not intercept the mirror. Whether the other person can see some part of you, however, depends on how far out you are holding your right hand. If it is far enough away from your body, light rays can follow the path drawn in dark blue, from your hand to his eyes.

The light ray drawn in red leaves the right hand of the person around the corner, reflects off the mirror, and arrives at your eyes. You can see his right hand.

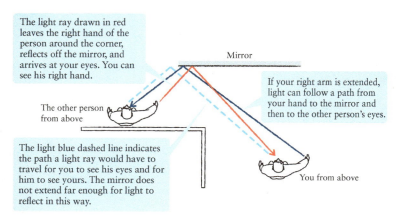

Mirror

The other person from above

If your right arm is extended, light can follow a path from your hand to the mirror and then to the other person's eyes.

The light blue dashed line indicates the path a light ray would have to travel for you to see his eyes and for him to see yours. The mirror does not extend far enough for light to reflect in this way.

You from above

Figure 24-46

24-3 (b) When an object is placed farther from a concave mirror than the focal point, the image is real and inverted. Note that a real image forms in front of a mirror; that is, it appears to be in front of the mirror. So when you hold the spoon at arm's length, the object (you!) must therefore be farther from the reflective surface than the focal point. When you hold the spoon 4 cm from your eye, however, you (your eye) must be closer to the spoon than its focal point. An object placed closer to a concave reflective surface than the focal point forms a virtual, inverted image. Because a virtual image is formed at the point where light rays from any point on an object appear to meet (but don't actually), the image forms behind the mirror and is upright. The focal point of the spoon must be about 6 cm from the spoon. At the focal point no image forms because light rays from any point on your face are reflected parallel to each other.

24-4 (e) A concave mirror makes an image because light rays from a point on the object that strike anywhere on the mirror are all reflected to the same point on the image. If you black out part of the mirror, there are fewer positions on the mirror that contribute to the image, but they still form an image at the same place. This is true for any point on the object, so the image will still show the entire light bulb. The only difference is that the image will be dimmer because only half as much light energy is reflected by the half-painted mirror.

24-5 (b) A real image made by a concave mirror is also an inverted image, so the magnification is $m = -2.00$. Equation 24-14 then tells us the image distance: $m = -d_I/d_O$, so $d_I = -md_O = -(-2.00)(12.0 \text{ cm}) = +24.0 \text{ cm}$. (The positive value of the image distance tells us that the image is in front of the mirror, as must be true for a real image.) The mirror equation, Equation 24-11, then tells us the focal length: $1/f = 1/d_O + 1/d_I = 1/(12.0 \text{ cm}) + 1/(24.0 \text{ cm}) = 0.0833 \text{ cm}^{-1} + 0.0417 \text{ cm}^{-1} = 0.125 \text{ cm}^{-1}$, so $f = 1/(0.125 \text{ cm}^{-1}) = 8.00 \text{ cm}$. Note that the object distance of 12.0 cm is greater than the focal length, as must be the case if the image is to be real.

24-6 (b) A concave mirror can focus parallel light rays; a convex mirror cannot, so it cannot concentrate solar energy onto a pot for cooking. Another way to see this is to think of the mirror as making an image of the Sun that lies at the position of the pot so as to concentrate the Sun's light there. This is possible with a concave mirror, which makes a real image of a distant object. A convex mirror, by contrast, can make only a virtual image whose rays never actually cross. Hence a convex mirror is no good for concentrating solar energy.

24-7 (c) We can solve this problem using the mirror equation, Equation 24-11, and the expression for magnification, Equation 24-14. The image is one-half the height of the object, so the lateral magnification is $m = h_I/h_O = 1/2$. From Equation 24-14 the magnification m is also equal to $-d_I/d_O$, so $-d_I/d_O = 1/2$ and $d_I = -d_O/2$; that is, the image is half as far from the mirror as the object is and is behind the mirror (because d_I is negative). Substitute $d_I = -d_O/2$ into the mirror equation and solve for d_O:

$$\frac{1}{d_O} + \frac{1}{d_I} = \frac{1}{f} \quad \text{so} \quad \frac{1}{d_O} + \left(-\frac{2}{d_O}\right) = -\frac{1}{d_O} = \frac{1}{f}$$
$$\text{and} \quad d_O = -f = |f|$$

(The focal length of a convex mirror is negative, so $f = -|f|$.) The distance from the object to the mirror equals the distance from the mirror to the focal point.

24-8 (c) The lenses we described in Section 24-7 are made of a material in which light travels more slowly than in the surrounding air—that is, a material with a higher index of refraction than the surroundings. In this situation, however, the lens is made of a material with a *lower* index of refraction than the surroundings. Hence light rays refract in the opposite sense as they enter and exit the lens, which means that this convex lens is actually a diverging lens. Like the diverging lens in Figure 24-33, the image made by this lens is virtual and smaller than the object.

24-9 (d), (e), (a), (b), (c). In each case first use the lens equation, Equation 24-11, in the form $1/d_I = 1/f - 1/d_O$ to find the image distance d_I. Then use Equation 24-14, $m = -d_I/d_O$, to find the lateral magnification m. (a) $1/d_I = 1/f - 1/(3f) = 2/(3f)$, so $d_I = 3f/2$ and $m = -(3f/2)/(3f) = -1/2$. The image is real (since d_I is positive), inverted (since m is negative), and 1/2 the size of the object (since the absolute value of m is 1/2). (b) $1/d_I = 1/f - 1/(2f) = 1/(2f)$, so $d_I = 2f$ and $m = -(2f)/(2f) = -1$. The image is real (since d_I is positive), inverted (since m is negative), and the same size as the object (since the absolute value of m is 1). (c) $1/d_I = 1/f - 2/(3f) = 1/(3f)$, so $d_I = 3f$ and $m = -(3f)/(3f/2) = -2$. The image is real (since d_I is positive), inverted (since m is negative), and twice the size of the object (since the absolute value of m is 2). (d) $1/d_I = 1/f - 2/f = -1/f$, so $d_I = -f$ and $m = -(-f)/(f/2) = +2$. The image is virtual (since d_I is negative), upright (since m is positive), and twice the size of the object (since the absolute value of m is 2). (e) $1/d_I = 1/f - 4/f = -3/f$, so $d_I = -f/3$ and $m = -(-f/3)/(f/4) = +4/3$. The image is virtual (since d_I is negative), upright (since m is positive), and 4/3 the size of the object (since the absolute value of m is 4/3).

24-10 (b), (d). In myopia (nearsightedness) the cornea and lens make light rays converge too sharply, so parallel light rays from a distant object come to a focus in front of the retina, as in Figure 24-40b. The replacement lens should therefore cause less convergence of light and have a longer focal length (that is, a lower power). This can be accomplished by reducing the index of refraction of the lens to be closer in value to that of the aqueous and vitreous humors, so that less refraction takes place. It can also be accomplished by giving the lens surfaces a more gentle curve, which means radii of curvature of greater magnitude and hence a longer focal length.

24-11 (a) As Figure 24-43b shows, the real image made by the objective lens of a microscope is *larger* than the object: $|h_I| > |h_O|$. Hence for the microscope the absolute value of the lateral magnification, $|m| = |h_I|/|h_O|$, is greater than 1. By contrast, the real image made by the objective lens of a telescope is *much smaller* than the object: $|h_I| < |h_O|$. (A telescope being used to view the Moon may produce a real image about a centimeter in diameter, while the Moon itself has a diameter of 3476 km.) So for the telescope the absolute value of the lateral magnification, $|m| = |h_I|/|h_O|$, is much less than 1.

Questions and Problems

In a few problems you are given more data than you actually need; in a few other problems you are required to supply data from your general knowledge, outside sources, or informed estimate.

Interpret as significant all digits in numerical values that have trailing zeros and no decimal points.

For all problems use $g = 9.80$ m/s^2 for the free-fall acceleration due to gravity. Neglect friction and air resistance unless instructed to do otherwise.

- •   Basic, single-concept problem
- ••   Intermediate-level problem; may require synthesis of concepts and multiple steps
- •••   Challenging problem
- Example   See worked example for a similar problem.

Conceptual Questions

1. • A plane mirror seems to invert your image left and right but not up and down. Is this what it really does?

2. • Explain why reflected light is usually diffuse.

3. • What is the difference between a real image and a virtual image?

4. • When you view a car's side mirror, you see a smaller image than you would if the mirror were flat. Is the mirror concave or convex? Explain your answer.

5. • Explain the meaning, in terms of physics, of the phrase etched on the right side mirror of most cars: "Objects in mirror are closer than they appear."

6. • What is the radius of curvature of a plane mirror? Explain your answer.

7. • For a certain glass lens in air, both radii of curvature are positive. Is it a converging lens or a diverging lens, or do you need additional information to tell? Explain.

8. • A convex lens made of clear ice ($n = 1.31$) acts as a *converging* lens when it is in air, but as a *diverging* lens when it is surrounded by acetone ($n = 1.36$). Explain why.

9. • A laptop computer is connected to a video projector that projects an image on a screen. If the lens of the projector is half covered, what happens to the image? Explain your answer.

10. • **Biology** The image focused on your retina is actually inverted (sketch a simple ray diagram showing this observation). What does this fact say about our definitions of "right side up" and "upside down"?

11. • **Biology** Experimental subjects who wear inverting lenses (glasses that invert all images) for several days adapt to their new perception of the world so well that they can even ride a bicycle. Several days after the glasses are removed, their perceptions return to normal. Discuss this phenomenon and comment.

12. • **Medical** Explain why converging lenses are used to correct farsightedness (hyperopia) while diverging lenses are used for nearsightedness (myopia).

13. • **Biology** Discuss why nearsightedness is not found in all people, but virtually everyone eventually has difficulty focusing on nearby objects as they age.

14. •• Explain why looking through a small opening often provides visual acuity even to an extremely nearsighted person.

15. • Explain why the lens in a digital camera must move away from the light sensor in order to focus on a nearby object.

Multiple-Choice Questions

16. • Which is true when an object is moved farther from a plane mirror?
   - A. The height of the image decreases, and the image moves farther from the mirror.
   - B. The height of the image stays the same, and the image moves farther from the mirror.
   - C. The height of the image increases, and the image moves farther from the mirror.
   - D. The height of the image stays the same, and the image moves closer to the mirror.
   - E. The height of the image decreases, and the image moves closer to the mirror.

17. • A real image can form in front of
   - A. a plane mirror.
   - B. a concave mirror.
   - C. a convex mirror.

D. any type of mirror.

E. no mirror of any type.

18. • When an object is placed a little farther from a concave mirror than the focal length, the image is

    A. magnified and real.

    B. magnified and virtual.

    C. smaller and real.

    D. smaller and virtual.

    E. smaller and reversed.

19. • If you want to start a fire using sunlight, which kind of mirror would be most efficient?

    A. a plane mirror

    B. a concave mirror

    C. a convex mirror

    D. any type of plane, concave, or convex mirror

    E. It is not possible to start a fire using sunlight and a mirror; you must use a concave lens.

20. • An object is placed at the center of curvature of a concave mirror. The image is

    A. real and upright.

    B. real and inverted.

    C. virtual and upright.

    D. virtual and inverted.

    E. nonexistent; no image is formed.

21. • When an object is placed farther from a convex mirror than the absolute value of the focal length, the image is

    A. larger and real.

    B. larger and virtual.

    C. smaller and real.

    D. smaller and virtual.

    E. smaller and reversed.

22. • **Medical** When a dentist needs a mirror to see an enlarged, upright image of a patient's tooth, what kind of mirror should she use?

    A. a plane mirror

    B. a concave mirror

    C. a convex mirror

    D. either a plane mirror or a concave mirror

    E. either a plane mirror or a convex mirror

23. • A magnifying lens allows one to look at a very near object by forming an image of it farther away. The object appears larger. To create a magnifying lens, one would use a

    A. short focal length ($f < 1$ m) converging lens.

    B. short focal length ($|f| < 1$ m) diverging lens.

    C. long focal length ($f > 1$ m) converging lens.

    D. long focal length ($|f| > 1$ m) diverging lens.

    E. either a converging or a diverging lens.

24. •• A compound microscope is a two-lens system used to look at very small objects. Which of the following statements is correct?

    A. The objective and the eyepiece both have the same focal length, and both serve as magnifying lenses.

    B. The objective is a short focal length, converging lens and the eyepiece functions as a magnifying lens.

    C. The objective is a long focal length, converging lens and the eyepiece functions as a magnifying lens.

    D. The objective is a short focal length, diverging lens and the eyepiece functions as a magnifying lens.

    E. The objective is a long focal length, diverging lens and the eyepiece functions as a magnifying lens.

25. An engineer would like to modify a consumer camera to better capture images in the near infrared, just beyond the visible reds. To save money, the decision is made to keep the same lens for producing the image. The index of refraction of the lens is smaller for infrared radiation than for visible light. Which of the following changes could be made to produce a sharp image in the near infrared?

    A. Relocate the detector closer to the lens.

    B. Relocate the detector farther from the lens.

    C. A sharp image cannot be produced with the same lens system.

    D. There is no need to change anything.

Problems

24-1 Mirrors or lenses can be used to form images

24-2 A plane mirror produces an image that is reversed back to front

26. • If the angle of incidence on a flat mirror is 0°, what is the angle of reflection?

27. • Two flat mirrors are perpendicular to each other. An incoming beam of light makes an angle of $\theta = 30°$ with the first mirror, as shown in **Figure 24-47**. What angle will the outgoing beam make with respect to the normal of the second mirror?

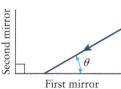

Figure 24-47 Problem 27

28. • A 1.8-m-tall man stands 2.0 m in front of a vertical plane mirror. How tall will the image of the man be?

29. • What must be the minimum height of a plane mirror for a 1.8-m-tall person to see a full image of himself?

30. •• A plane mirror is 10 m away from and parallel to a second plane mirror (**Figure 24-48**). Find the locations of the first five images formed by each mirror when an object is positioned exactly in the middle between the two mirrors.

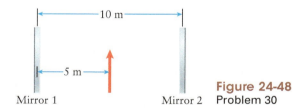

Figure 24-48 Problem 30

31. •• A plane mirror is 10 m away from and parallel to a second plane mirror (**Figure 24-49**). Find the locations of the first five images formed by each mirror when an object is positioned 3 m from one of the mirrors.

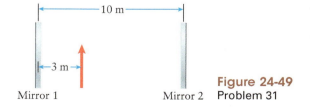

Figure 24-49 Problem 31

32. • At which of the points A through E will the image of the face be visible in the plane mirror of length L (**Figure 24-50**)? The distance x equals $L/2$. Points A through E are collinear and separated by a distance of $3L/4$, and point C lies on a line bisecting the mirror. Assume the face lies directly on point C.

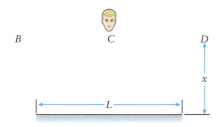

Figure 24-50 Problem 32

33. • Using a ruler, a protractor, and the law of reflection, show the location of the image of your face when you stand a short distance in front of a plane mirror, as shown in Figure 24-51.

Figure 24-51 Problem 33

34. • One person is looking in a plane mirror at the image of a second person (Figure 24-52). Using a ruler, a protractor, and the law of reflection, show the location of the image as seen by the first person.

Figure 24-52 Problem 34

35. • You hold an autofocusing camera 2 m in front of a plane mirror and take a picture of the reflection. You discover that the picture of the reflection is out of focus. An autofocus camera determines the distance to the object by measuring the time it takes an infrared pulse to leave the camera, bounce off the subject, and return to the camera. Knowing this, explain what must have happened.

24-3 A concave mirror can produce an image of a different size than the object

36. • Describe the difference between the images seen in a spherical, concave mirror when the object is "up close" (closer to the image than the focal length) compared to "far away" (outside the focal length).

37. • Are there any situations where a real image is formed in a spherical, concave mirror? Describe the location of the object relative to the focal point for such situations.

38. • Describe the image you see of yourself if your head is at the center of curvature of a spherical concave mirror.

24-4 Simple equations give the position and magnification of the image made by a concave mirror

39. • An object is placed 8.0 cm in front of a concave mirror with a 10.0-cm radius of curvature. Calculate the image distance and the magnification of the image. Determine whether the image is real or virtual and whether it is inverted or upright by using (a) a ray diagram and (b) the mirror equation. Example 24-1

40. • An object 1.0 cm tall is placed 3.0 cm in front of a spherical concave mirror with a radius of curvature equal to 10.0 cm. Calculate the image distance and height by using (a) a ray diagram and (b) the mirror equation. Example 24-1

41. • An object 1.0 cm tall is placed 6.0 cm in front of a spherical concave mirror with a radius of curvature equal to 10.0 cm. Calculate the image distance and height by using (a) a ray diagram and (b) the mirror equation. Example 24-1

42. • The radius of curvature of a spherical concave mirror is 20.0 cm. Describe the image formed when a 10.0-cm-tall object is (a) 5.0 cm from the mirror, (b) 20.0 cm from the mirror, (c) 50.0 cm from the mirror, and (d) 100.0 cm from the mirror. For each case give the image distance, the image height, the type of image (real or virtual), and the orientation of the image (upright or inverted). Example 24-1

43. • The radius of curvature of a spherical concave mirror is 15.0 cm. Describe the image formed when a 20.0-cm-tall object is (a) 10.0 cm from the mirror, (b) 20.0 cm from the mirror, and (c) 100.0 cm from the mirror. For each case give the image distance, the image height, the type of image (real or virtual), and the orientation of the image (upright or inverted). Example 24-1

44. • An object is 24.0 cm from a spherical concave mirror of unknown focal length. The image that is formed is 30.0 cm from the mirror and on the same side of the mirror as the object. (a) Calculate the focal length. (b) Is the image real or virtual? (c) If the object is 10.0 cm tall, determine the height of the image(s). Example 24-1

45. • Construct ray diagrams to locate the images in the following cases. (a) A 10.0-cm-tall object is located 5.0 cm in front of a spherical concave mirror with a radius of curvature of 20.0 cm. (b) A 10.0-cm-tall object is located 10.0 cm in front of a spherical concave mirror with a radius of curvature of 20.0 cm. (c) A 10.0-cm-tall object is located 20.0 cm in front of a spherical concave mirror with a radius of curvature of 20.0 cm. Example 24-1

46. •• Derive a relationship between the radius of curvature of a spherical, concave mirror and the object distance that gives an upright image that is four times as tall as the object. Example 24-1

24-5 A convex mirror always produces an image that is smaller than the object

47. • Describe the difference between the images seen in a spherical, convex mirror when the object is "up close" (a shorter distance from the mirror than the focal distance of the mirror) compared to "far away" (a longer distance from the mirror than the focal distance of the mirror).

48. • Are there any situations where a real image is formed in a spherical, convex mirror? Why or why not?

49. • How can you remember the difference between the shapes of a spherical *concave* mirror and a spherical *convex* mirror?

50. • A 10-cm-tall object is located in front of a spherical, convex mirror with a radius of curvature of 20 cm. Construct ray diagrams to locate the images and estimate the image height in each of the following cases. (a) The object is located 5 cm in front of the mirror. (b) The object is located 10 cm in front

of the mirror. (c) The object is located 20 cm in front of the mirror.

24-6 The same equations used for concave mirrors also work for convex mirrors

51. • The radius of curvature of a spherical convex mirror is 20.0 cm. Describe the image formed when a 10.0-cm-tall object is positioned (a) 20.0 cm from the mirror, (b) 50.0 cm from the mirror, and (c) 100.0 cm from the mirror. For each case, provide the image distance, the image height, the type of image (real or virtual), and the orientation of the image (upright or inverted). Example 24-2

52. • The radius of curvature of a spherical convex mirror is 15.0 cm. Describe the image formed when a 20.0-cm-tall object is positioned (a) 5.0 cm from the mirror, (b) 20.0 cm from the mirror, and (c) 100.0 cm from the mirror. For each case, provide the image distance, the image height, the type of image (real or virtual), and the orientation of the image (upright or inverted). Example 24-2

53. • A car's convex rearview mirror has a radius of curvature equal to 15.0 m. What are the magnification, type, and location of the image that is formed by an object that is 10.0 m from the mirror? Example 24-2

54. • An 18.0-cm-long pencil is placed beside a convex spherical mirror, and its image is 10.5 cm in length. If the radius of curvature of the mirror is 88.4 cm, find the image distance, the object distance, and the magnification of the pencil. Example 24-2

55. • A 1-cm-long horse fly hovers 1.0 cm from a shiny sphere with a radius of 25.0 cm. Calculate the location of the image of the fly, its type (real or virtual), and its length. Example 24-2

56. • A spherical convex mirror is placed at the end of a driveway on a corner with a limited view of oncoming traffic. The mirror has a radius of curvature of 1.85 m. An oncoming car is 12.6 m from the mirror. How far from the mirror does the image of the car appear to be? Example 24-2

57. •• Using the mirror equation, prove that all images in spherical convex mirrors are virtual. Example 24-2

58. • A girl sees her image in a shiny glass sphere tree ornament that has a diameter of 10.0 cm. The image is upright and is located 1.5 cm behind the surface of the ornament. How far from the ornament is the child located? Example 24-2

59. • A shiny sphere, 30.0 cm in diameter, is placed in a garden for aesthetic purposes. Determine the type, location, and height of the image of a 6.00-cm-tall squirrel located 40.0 cm in front of the sphere. Example 24-2

24-7 Convex lenses form images like concave mirrors and vice versa

60. • Under what circumstances will the images formed by converging or diverging lenses be designated as "real"? Indicate the type or types of lenses and the required position of the object.

61. • A real image that is created due to reflection in a spherical mirror appears in front of the mirrored surface. Is this also the case for a real image that is created due to refraction in a lens?

62. • Where does the bending of light physically take place in a typical concave or convex lens? Is this how we draw ray diagrams? Why or why not?

63. • What *minimum* number of rays are required to locate the image that is formed by a lens in a ray diagram? Explain.

24-8 The focal length of a lens is determined by its index of refraction and the curvature of its surfaces

64. • A 10.0-cm-tall object is located in front of a converging lens with a power of 5.00 diopters. Describe the image created (type, location, height) and draw the ray diagrams if the object is located (a) 5.0 cm from the lens, (b) 10.0 cm from the lens, (c) 20.0 cm from the lens, and (d) 50.0 cm from the lens. Example 24-5

65. • A 10.0-cm-tall object is located in front of a diverging lens with a power of −5.00 diopters. Describe the type, location, and height of the image created, and draw the ray diagrams if the object is located (a) 5.0 cm from the lens, (b) 10.0 cm from the lens, (c) 20.0 cm from the lens, and (d) 50.0 cm from the lens. Example 24-5

66. • A 2.00-cm-tall object is located 18.0 cm in front of a converging lens with a focal length of 30.0 cm. (a) Use the lens equation and (b) a ray diagram to describe the type, location, and height of the image that is formed. Example 24-5

67. • A lens is formed from a plastic material that has an index of refraction of 1.55. If the radius of curvature of one surface is 1.25 m and the radius of curvature of the other surface is 1.75 m, use the lensmaker's equation to calculate the focal length and the power of the lens. Example 24-3

68. •• A glass lens ($n = 1.60$) has a focal length of −31.8 cm and a plano-concave shape. (a) Calculate the radius of curvature of the concave surface. (b) If a lens is constructed from the same glass to form a plano-convex shape with the same radius of curvature, what will the focal length be? Example 24-4

69. • A 2.00-cm-tall object is 30.0 cm in front of a converging lens that has a focal length of 18.0 cm. (a) Use the lens equation and (b) a ray diagram to describe the type, location, and height of the image that is formed. Example 24-5

24-9 A camera and the human eye use different methods to focus on objects at various distances

70. • Gbenga needs to get glasses to correct his farsightedness. His eyes currently cannot focus on objects that are within 2 ft (or 61 cm) of his eyes. This is in contrast to people with normal vision who can focus on objects as close as 25 cm in front of them. If the glasses that Gbenga will get will sit 1.6 cm in front of his eyes, what lens focal length and power would correct his vision? That is, what lens focal length and power would allow Gbenga to focus on objects that are 25 cm in front of his eyes? Will they be converging or diverging lenses? Example 24-6

71. • Suppose a given cell phone camera has a single lens and a light sensor that can move to change its distance from the lens. If the camera can focus only on an object 6.5 cm or farther away and the lens has a focal length of 4.3 mm, what are the maximum and minimum distances of the light sensor from the lens? Example 24-5

72. •• At a distance of 7.0 cm, a 51 mm by 89 mm business card fills the screen of a cell phone in camera mode. (a) If the focal length of the camera lens is 4.3 mm, what are the length and height of the camera's light sensor? (b) The light sensor is made up of 16×10^6 individual picture elements, or pixels (so

this is a 16-megapixel sensor). What is the area of each pixel? If the pixels are square, what is the length of each side of a pixel? Example 24-5

73. • Andrea, who is nearsighted, wears glasses with lenses that have a power of –1.60 diopters. If the glasses Andrea wears sit 1.40 cm in front of her eyes, what is the farthest distance that objects can be for Andrea to see them clearly without her glasses? Example 24-6

24-10 The concept of angular magnification plays an important role in several optical devices

74. • To view the craters of the Moon, you construct a refracting telescope using a 95.0-cm focal length as its objective and a 12.5-cm focal length lens as its eyepiece. (a) Determine the angular magnification of your telescope when you look at the Moon. (b) Is the image you see upright or inverted with respect to the object?

75. • You want to use a lens of focal length 9.00 cm that you just happen to have around to examine the hairy details of your favorite pet caterpillar. With the lens close to your eye and the animal at the lens's focal point, what angular magnification do you achieve? Assume that your near point is at 25.0 cm.

76. • A compound microscope has a barrel of length 160 mm. The focal lengths of the objective and the eyepiece are 4.00 and 25.0 mm, respectively. (a) Find the total angular magnification of the microscope for a person whose near point is 25.0 cm away. (b) Is the image seen in the microscope upright or inverted?

77. ••• **Biology** Calculate the overall magnification of a compound microscope that uses an objective with a focal length of 0.50 cm, an eyepiece with a focal length of 2.50 cm, and a distance of 18 cm between the two.

General Problems

Note: In these problems "infinity" means a very large distance compared to the focal length of a lens.

78. • Determine the focal length for an unknown lens with the following object and image distances:

Object distance (cm)	Image distance (cm)	Object distance (cm)	Image distance (cm)
30	98	60	37
35	67	65	35
40	53	70	34
45	47	75	33
50	42	80	32
55	38		

79. • A square plane mirror of side length s hangs on a wall such that its bottom edge is a height *h* above the floor. The wall opposite the mirror is a distance *d* away. Marco, whose eyes are at the exact height of the center of the mirror, stands directly in front of the mirror at the center of its width, as shown in Figure 24-53. Derive an expression for the maximum distance *x* that Marco can stand from the mirror and still see the reflection of the bottom of the wall behind him. Assume that *s* < 2*h*.

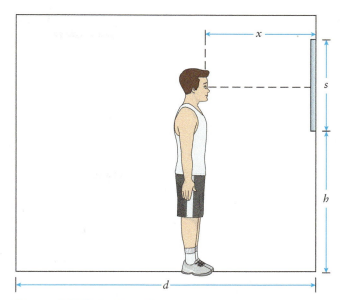

Figure 24-53 Problem 79

80. •• The opposite walls of a barber shop are covered by plane mirrors, so that multiple images arise from multiple reflections, and you see many reflected images of yourself receding to infinity. The width of the shop is 6.50 m, and you are standing 2.00 m from the north wall. (a) How far apart are the first two images of you behind the north wall? (b) What is the separation of the first two images of you behind the south wall? Explain your answer.

81. • An object is 40.0 cm from a concave spherical mirror whose radius of curvature is 32.0 cm. Locate and describe the type and magnification of the image formed by the mirror (a) by calculating the image distance and lateral magnification and (b) by drawing a ray diagram. On the ray diagram draw an eye in a position from which it can view the image. Example 24-1

82. •• **Biology** A typical human eye is nearly spherical and usually about 2.5 cm in diameter. Suppose a person first looks at a coin that is 2.3 cm in diameter, located 30.0 cm from her eye, and then looks up at her friend who is 1.8 m tall and 3.25 m away. (a) Find the approximate size of each image (coin and friend) on her retina. (*Hint*: Just consider rays from the top and bottom of the object that pass through the center of the lens.) (b) Are the images in part (a) upright or inverted, and are they real or virtual?

83. •• **Biology** A typical human lens has an index of refraction of 1.41. The lens has a double convex shape, but its curvature can be varied by the ciliary muscles acting around its rim. At minimum power, the radius of the front of the lens is 10.0 cm, while that of the back is 6.00 mm. At maximum power the radii are 6.00 and 5.50 mm, respectively. (The numbers can vary somewhat.) If the lens were in air, (a) what would be the ranges of its focal length and its power (in diopters)? (b) At maximum power, where would the lens form an image of an object 25 cm in front of the front surface of the lens? (c) Would the image fall on the retina of a human eye? The retina is located approximately 2.5 cm from the lens. Example 24-3

84. •• **Biology** A typical person's eye is 2.5 cm in diameter and has a near point (the closest an object can be and still be seen in focus) of 25 cm and a far point (the farthest an object can

be and still be in focus) of infinity. (a) What is the range of the effective focal lengths of the focusing mechanism (lens plus cornea) of the typical eye? (b) Is the equivalent focusing mechanism of the eye a diverging or a converging lens? Justify your answer without using any mathematics, and then see if your answer is consistent with your result in part (a). Example 24-5

85. ••• A geneticist looks through a microscope to determine the phenotype of a fruit fly. The microscope is set to an overall magnification of 400× with an objective that has a focal length of 0.60 cm. The distance between the eyepiece and objective is 16 cm. Find the focal length of the eyepiece lens assuming a near point of 25 cm (the closest an object can be to the eye and still be seen in focus).

86. •• A thin lens made of glass that has a refractive index equal to 1.60 has surfaces with radii of curvature that have magnitudes equal to 12.0 and 18.0 mm. These surfaces could be either convex or concave. What are the possible values for its focal length? Sketch a cross-sectional view of the lens for each possible combination, making sure to label the radii of curvature of each surface of the lens and the associated focal length of the entire lens. Example 24-3

87. • You are designing lenses that consist of small double convex pieces of plastic having surfaces with radii of curvature of magnitudes 3.50 cm on one side and 4.25 cm on the other side. You want the lenses to have a focal length of 1.65 cm in air. What should be the index of refraction of the plastic to achieve the desired focal length? Example 24-3

88. ••• A lens of focal length +15.0 cm is 10.0 cm to the left of a second lens of focal length −15.0 cm. (a) Where is the final image of an object that is 30.0 cm to the left of the positive lens? (b) Is the image real or virtual? (c) How do the image's size and orientation compare to those of the original object? Explain your answer. (d) Where should your eye be located to see the image?

89. ••• A thin, diverging lens having a focal length of magnitude 45.0 cm has the same principal axis as a concave mirror with a radius of 60.0 cm. The center of the mirror is 20.0 cm from the lens, with the lens in front of the mirror. An object is placed 15.0 cm in front of the lens. (a) Where is the final image due to the lens–mirror combination? (b) Is the final image real or virtual? Upright or inverted? (c) Suppose now that the concave mirror is replaced by a convex mirror of the same radius. Repeat parts (a) and (b) for the new lens–mirror combination.

90. ••• When you place a bright light source 36.0 cm to the left of a lens, you obtain an upright image 14.0 cm from the lens and also a faint inverted image 13.8 cm to the left of the lens that is due to reflection from the front surface of the lens. When the lens is turned around, a faint inverted image is 25.7 cm to the left of the lens. What is the index of refraction of the material? Example 24-4

91. ••• A thin, converging lens having a focal length of magnitude 25.00 cm is placed 1.000 m from a plane mirror that is oriented perpendicular to the principal axis of the lens. A flower, 8.400 cm tall, is 1.450 m from the mirror on the principal axis of the lens. (a) Where is the final image of the flower produced by the lens–mirror combination? Is it real or virtual? Upright or inverted? How tall is the image? (b) If the converging lens is replaced by a diverging lens having a focal length of the same magnitude as the original lens, what will be the answers to part (a)?

92. ••• The eyepiece and objective in a compound microscope are separated by 20.0 cm, and the objective has a focal length of 8.0 mm. (a) If it is to have a magnifying power of 200×, what should be the focal length of the eyepiece? (b) If the final image is viewed at infinity, how far from the objective should the object be placed?

93. •• **Biology** You may have noticed that the eyes of cats appear to glow green in low light. This effect is due to the reflection of light by the *tapetum lucidum*, a highly reflective membrane just behind the retina of the eye (see Figure 23-1a). Light that has passed through the retina without hitting photoreceptors is reflected back to the retina, thus enabling the animal to see much better than humans in low light. The eye of a typical cat is about 1.25 cm in diameter. Assume that the light enters the eye traveling parallel to the principal axis of the lens. (a) If some of the light reflected off the *tapetum lucidum* escapes being absorbed by the retina, where will it be focused? (b) The refractive index of the liquid in the eye is about 1.4. How does this affect the location of the image in part (a)?

94. •• **Medical** A nearsighted eye is corrected by placing a diverging lens in front of the eye. The lens will create a virtual image of a distant object at the far point (the farthest an object can be and still be in focus) of the myopic viewer where it will be clearly seen. In the traditional treatment of myopia, an object at infinity is focused to the far point of the eye. If an individual has a far point of 70 cm, prescribe the correct power of the lens that is needed. Example 24-6

95. •• **Medical** A farsighted eye is corrected by placing a converging lens in front of the eye. The lens will create a virtual image that is located at the near point (the closest an object can be and still be in focus) of the viewer when the object is held at a comfortable distance (usually taken to be 25 cm). If a person has a near point of 75 cm, what power reading glasses should be prescribed to treat this hyperopia? Example 24-6

96. ••• **Medical** (a) Prove that when two thin lenses of focal lengths f_1 and f_2 are pressed next to one another, the effective focal length $f_{combined}$ of the two lenses acting together is given by

$$\frac{1}{f_{combined}} = \frac{1}{f_1} + \frac{1}{f_2}$$

(b) Describe how this relates to a prescription for a contact lens which is placed directly on the eye. (Assume there is no significant separation between the contact lens and the lens of the eye.) (c) Why would an eyeglass prescription that is identical to a contact lens prescription give a very subtle difference in the image seen? Example 24-6

97. •• **Medical** Without glasses, a certain person needs to have his eyes 15.0 cm from a book to read comfortably and can focus clearly only on distant objects up to 2.75 m away, but no farther. A typical normal eye should be able to focus on objects that are between 25.0 cm (the near point) and infinity (the far point) from the eye. (a) What type of correcting lenses does the person need: single focal length or bifocals? Why? (b) What should an optometrist specify as the focal length(s) of the correcting contact lens or lenses? (c) What is the power (in diopters) of the correcting lens or lenses? Example 24-6

98. ••• **Medical** One of the inevitable consequences of aging is a decrease in the flexibility of the lens. This leads to the far-sighted condition called *presbyopia* (elder eye). Almost every aging human will experience it to some extent. However, for the myopic person, at some point, it is possible that far vision will be limited by a subpar far point *and* near vision will be hampered by an expanding near point. One solution is to wear bifocal lenses that are diverging in the upper half to correct the nearsightedness and converging in the lower half to correct the farsightedness.

Suppose one such individual asks for your help. The patient complains that she can't see far enough to safely drive (her far point is 112 cm) and she can't read the font of her smartphone without holding it beyond arm's length (her near point is 83 cm). Prescribe the bifocals that will correct the visual issues for your patient. Example 24-6

99. •• A common zoom lens for a digital camera covers a focal length range of 18 to 200 mm. For the purposes of this problem, treat the lens as a thin lens. If the lens is zoomed out to 200 mm and is focused on a petroglyph that is 15.0 m away and 38 cm wide, (a) how far is the lens from the light sensor of the camera and (b) how wide is the image of the petroglyph on the sensors? (c) If the closest that the lens can get to the sensor at its 18 mm focal length is 5.2 cm, what is the closest object it can focus on at that focal length?

100. •• A macro lens is designed to take very close-range photographs of small objects such as insects and flowers. At its closest focusing distance, a certain macro lens has a focal length of 35.0 mm and forms an image on the light sensor of the camera that is 1.09 times the size of the object. (a) How close must the object be to the lens to achieve this maximum image size? (b) What is the magnification if the object is twice as far from the lens as in part (a)? For this problem, treat the lens as a thin lens.

101. • You desire to observe details of the *Statue of Freedom*, the sculpture by Thomas Crawford that is the crowning feature of the dome of the U.S. Capitol in Washington, D.C. For this purpose, you construct a refracting telescope, using as its objective a lens with a focal length of 90.0 cm. To achieve an angular magnification of magnitude 6.0, what focal length should the eyepiece have?

102. • To heighten your enjoyment of your 27-carat blue diamond, you view it through a lens held close to your right eye at an angular magnification of 5.0. The distance to your right eye's near point is 25 cm. What is the focal length of the lens?

103. • **Astronomy** A refracting astronomical telescope, or refractor, consists of an eyepiece lens at one end of a cylindrical tube and an objective lens at the other end. The objective lens gathers light from a distant object (such as a planet) and focuses it at the focal point of the eyepiece lens. The eyepiece basically acts as a magnifying lens to create a virtual image of the objective's image. The overall magnification, M, is found to be $M = -f_o/f_e$, where f_o is the focal length of the objective and f_e is the focal length of the eyepiece. (a) Calculate the magnification of the 36-in. refractor at the University of California's Lick Observatory on Mount Hamilton near San Jose, California. The focal length of the objective lens is 17.37 m, and the focal length of the eyepiece is 22 mm. (b) What is the significance of the negative sign in the magnification equation?

104. • A compound microscope has a barrel length of 160.0 mm and an objective with a 4.500-mm focal length. The total angular magnification of the microscope is −400.0. Find the angular magnification of the eyepiece.

105. • A certain lens element inside a camera is a converging lens that has symmetric convex sides ($R_1 = 4.300$ cm and $R_2 = -4.300$ cm) and can be modeled as a thin lens. The lens's index of refraction for blue light is $n_{blue} = 1.588$ and for red light is $n_{red} = 1.582$. Calculate the magnitude of the separation distance of the lens's focal lengths for blue and red light, $|f_{blue} - f_{red}|$. Example 24-3

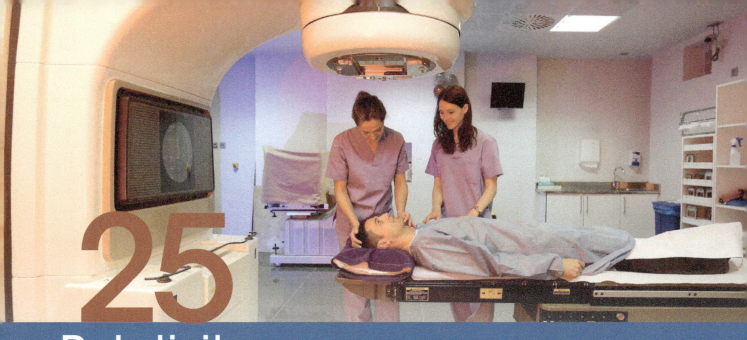

age fotostock/Alamy

25

Relativity

What do you think?

In cancer radiotherapy, a beam of electrons is accelerated to nearly the speed of light. The kinetic energy of the electrons is used to create intense x-ray beams that can be accurately targeted on the location of the cancerous tissue. Which requires more energy: (a) accelerating an electron from rest to 90% of the speed of light, or (b) further accelerating that electron from 90 to 99% of the speed of light?

In this chapter, your goals are to:

- (25-1) Identify some circumstances under which the physics you know breaks down.
- (25-2) Describe how different observers view the same motion in Newtonian physics.
- (25-3) Explain how the Michelson–Morley experiment helped rule out the ether model of the propagation of light.
- (25-4) Describe how the time interval between two events can have different values in different frames of reference.
- (25-5) Calculate how the dimensions of an object change when it is in motion.
- (25-6) Explain why the speed of light in a vacuum is an ultimate speed limit.
- (25-7) Calculate the rest energy of an object with mass.
- (25-8) Explain what the principle of equivalence tells us about the nature of gravity.

To master this chapter, you should review:

- (2-6)   The motion of objects in free fall.
- (3-6)   Projectile motion.
- (4-2, 4-3) Newton's first and second laws.
- (6-3)   Kinetic energy and the work-energy theorem.
- (8-4)   Elastic collisions.
- (8-6)   The relationship between external forces and momentum change.
- (22-3)   The electromagnetic nature of light.

25-1 The concepts of relativity may seem exotic, but they're part of everyday life

Most of our everyday experiences involve objects that move at speeds that are slow compared to the speed of light in a vacuum. In the late nineteenth century, scientists began to realize that the laws of physics that we've developed so far in this book don't properly describe light or objects moving at speeds near the speed of light. Albert Einstein first understood the way to extend physics into this regime of extremely high speed.

(a) Relativity tells us that time flows at a different rate aboard an orbiting GPS satellite than on Earth. A GPS receiver must account for this to accurately determine its position.

(b) Sunlight is a result of reactions deep inside the Sun that convert mass into energy—a process predicted by the special theory of relativity.

(c) At the Large Hadron Collider near Geneva, Switzerland, protons are accelerated to 99.99999999% of the speed of light. The special theory of relativity is needed to explain how particles behave at such speeds.

Figure 25-1 Relativity in your world (a) A GPS-equipped mobile phone and (b) the light from the Sun illustrate applications of the theory of relativity. (c) Subatomic particles are regularly accelerated to just below the speed of light. Their behavior cannot be understood without the theory of relativity.

In this chapter we'll focus on Einstein's *special theory of relativity*. We'll see how discoveries concerning the nature of light helped motivate the central ideas of this theory. We'll also see how a simple postulate—that the speed of light does not depend on the motion of either the emitter or the observer of the light—leads to a radical transformation of our understanding of space and time. We'll discover that the speed of light is an ultimate speed limit, and it's impossible to accelerate an object beyond that speed. We'll also see that mass is simply another form of energy. We'll conclude with a look at Einstein's *general theory of relativity*, which provides new insights into the nature of gravity.

The effects of the special theory of relativity are present in the world around you. Your mobile phone probably has the ability to determine its location using the Global Positioning System, or GPS (Figure 25-1a). A GPS receiver detects signals from a collection of satellites that orbit Earth and calculates its position by timing those signals. However, the satellites move at about 28,000 km/h (18,000 mi/h) relative to Earth, and special relativity tells us that a moving clock or timekeeper runs at a different rate than a stationary one (see Section 25-4). Your mobile phone has to be able to correct for this in order to give you accurate positioning information. Ordinary sunlight is also a consequence of relativity: The Sun shines by converting a fraction of its mass into electromagnetic energy, a direct application of the idea that objects have energy simply as a consequence of having mass (Figure 25-1b). And while even the fastest spacecraft travel at only a tiny fraction of the speed of light, subatomic particles can move much faster (Figure 25-1c)

> **TAKE-HOME MESSAGE FOR Section 25-1**
>
> ✔ The physics we have learned so far must be modified for objects moving at speeds comparable to the speed of light.
>
> ✔ These modifications lead us to new ideas about space, time, and energy.

25-2 Newton's mechanics includes some ideas of relativity

You're on a train, looking out the window at a second train right next to yours. One of the trains is moving (Figure 25-2). Is your train moving and the other one stationary, or vice versa? Perhaps both are moving. How can you tell?

Figure 25-2 Relative motion This is the view of one train as seen from the window of another. Which train is moving?

Figure 25-3 **Two observers**
An observer on the platform in a train station uses coordinates x, y, and z and measures time t. An observer aboard the train uses coordinates x', y', and z' and measures t'.

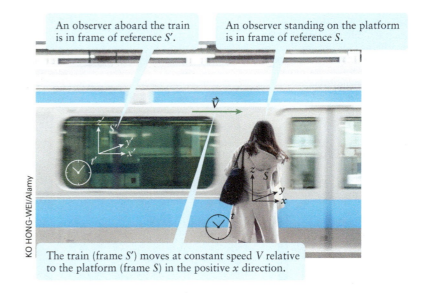

An observer aboard the train is in frame of reference S'.

An observer standing on the platform is in frame of reference S.

KO HONG-WEI/Alamy

The train (frame S') moves at constant speed V relative to the platform (frame S) in the positive x direction.

To address this question let's return to Newton's first law, which we introduced in Section 4-3:

Newton's first law of motion
(4-6)

If the net external force on an object is zero...

...the object does not accelerate...

$$\text{If } \sum \vec{F}_{\text{ext}} = 0, \text{ then } \vec{a} = 0 \text{ and } \vec{v} = \text{constant}$$

...and the velocity of the object remains constant. If the object is at rest, it remains at rest; if it is in motion, it continues in motion in a straight line at a constant speed.

The first law states that an object that experiences no net force could *either* be at rest *or* in motion at a constant velocity. So although we might make a distinction between an object at rest and one in (uniform) motion with respect to us, the laws of physics do not. In the case of the two trains, this tells us something profound: If you have no reference to the ground or the tracks on which the trains move, there is *no* experiment or measurement that would tell you whether the other train is moving at a constant velocity with respect to yours or your train is moving at a constant velocity with respect to the other one.

A useful way to think about this idea is to introduce the concept of a **frame of reference** (also called a **reference frame** or simply *frame*). This is a coordinate system with respect to which we can make observations or measurements. Figure 25-3 shows two different frames of reference. A person standing on the train platform is in frame S and measures the positions of objects using the coordinates x, y, and z. This person has a watch, and the time of a certain **event**—that is, something that happens at a particular location at a particular moment of time—as measured on this watch is t. A person riding on the train is in frame S' and measures the positions of objects using the coordinates x', y', and z'. The time of an event as measured by this person is t'. If neither frame of reference is accelerating, the message of Newton's first law is that *both* frames are equally good for making measurements.

As an example, let's consider a child riding on the train (Figure 25-4). The train is on a straight track and is moving relative to the platform at a constant speed V. As measured by a person on the platform (frame of reference S), the child is moving at a constant velocity with zero acceleration, so the net force on the child must be zero: The upward normal force exerted by the seat exactly balances the downward gravitational force that Earth exerts on the child. As measured by another passenger seated on board the train (frame of reference S'), the child is at rest (the child isn't moving relative to that passenger). So as measured in S', the net force on the child must again be zero. The two observers—one in frame of reference S and one in frame of reference S'—disagree about *how* the child moves. But they agree that the child's motion is in accordance with Newton's first law, Equation 4-6.

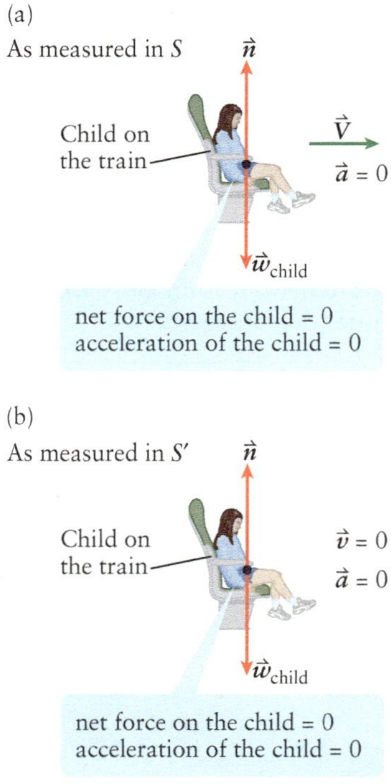

(a)
As measured in S

$\vec{n}$

Child on the train

$\vec{V}$

$\vec{a} = 0$

$\vec{w}_{\text{child}}$

net force on the child = 0
acceleration of the child = 0

(b)
As measured in S'

$\vec{n}$

Child on the train

$\vec{v} = 0$

$\vec{a} = 0$

$\vec{w}_{\text{child}}$

net force on the child = 0
acceleration of the child = 0

Figure 25-4 **Newton's first law in two frames of reference** The free-body diagram for a child on the train as measured in the two frames of reference shown in Figure 25-3.

It's not just Newton's first law that applies in both frames of reference depicted in Figure 25-3. Newton's *second* law also applies in both frames:

If a **net external force** acts on an object...

...the object accelerates. The acceleration is in the same direction as the net force.

$$\sum \vec{F}_{\text{ext}} = m\vec{a}$$

The magnitude of acceleration that the net external force causes depends on the mass m of the object (the quantity of material in the object). The greater the mass, the smaller the acceleration.

Newton's second law of motion
(4-2)

As an illustration, suppose the child riding on the train is tossing a ball up and down (Figure 25-5). A passenger seated on the train (frame of reference S') sees the motion of the ball as purely vertical. By contrast, a person on the platform (frame of reference S) sees the ball following a parabolic path: The ball has the same horizontal component of velocity V as the train, and it maintains that horizontal velocity during its flight. As for the child in Figure 25-4, the observers in the two frames of reference disagree about how the ball moves. Both observers, however, agree that the ball obeys Newton's second law: When the ball is in flight, only the gravitational force acts on it, so the acceleration is downward and has magnitude g. In frame S' the straight up-and-down motion is free fall, as we described in Section 2-6; in frame S the ball is in projectile motion, as we described in Section 3-6. Each description is correct for the frame of reference in which the motion is observed.

We refer to a frame of reference attached to an object that does not accelerate as an inertial frame. An **inertial frame** of reference is one in which Newton's first law is valid: If the net force on an object is zero, it either remains at rest or moves with a constant velocity relative to an observer in that frame of reference. If one frame of reference S is inertial, a second frame of reference S' is also inertial if it moves at a constant velocity relative to S. That's the case for the two frames of reference depicted in Figure 25-3.

By contrast, a frame of reference attached to an accelerated object is a noninertial frame. To an observer in a **noninertial frame**, Newton's first law does not hold true. An example is a frame of reference attached to a car that is accelerating forward. A ball sitting on the floor of this car has zero net force on it (the upward normal force exerted by the floor balances the downward gravitational force), yet the ball accelerates toward the back of the car. A rotating frame of reference, such as a carnival merry-go-round, is also noninertial because an object that follows a circular path is accelerating. Just like a ball in a car that accelerates forward, a ball placed on the merry-go-round floor has zero net force acting on it, yet this ball will tend to roll to the outside of the merry-go-round. In the frame of reference of a person riding on the merry-go-round, Newton's first law does not hold true.

Strictly speaking, an observer at rest on Earth's surface, such as the person in frame S standing on the platform in Figure 25-3, is in a noninertial frame. That's because Earth rotates on its axis like a merry-go-round and also moves along a roughly circular orbit around the Sun. However, the accelerations involved with those motions are so small (each is a small fraction of g) that for many purposes we can ignore them. As a result, we can safely regard the frame of reference S in Figure 25-3 as an effectively inertial one, and likewise for the frame S' attached to the train.

We've seen that Newton's first and second laws of motion work equally well in both inertial frame S and inertial frame S'. The same should be true for any inertial frame of reference. This statement is called the **principle of Newtonian relativity**:

The laws of motion are the same in all inertial frames of reference.

The word "relativity" means that measurements made relative to one inertial frame of reference are just as valid as those made relative to another inertial frame. Since the laws of motion don't distinguish between two inertial frames S and S', it's meaningless to ask which frame is "really" moving and which frame is "really" at rest. (You may think that frame S is the one that's really at rest because it's stationary with respect to the platform. But remember that the platform is on Earth and that our entire

(a)
As measured in S'

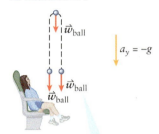

net force on ball = gravitational force
Ball experiences free fall.

(b)
As measured in S

net force on ball = gravitational force
Ball experiences projectile motion.

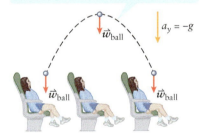

Figure 25-5 A tossed ball in two frames of reference The child on the train from Figure 25-4 tosses a ball straight up and down relative to her reference frame S'. (We draw the ball's motion in S' with a slight horizontal displacement to distinguish the up-and-down motion.)

planet is in motion through the solar system.) So another way to express the principle of Newtonian relativity is

> *There is no way to detect absolute motion. Only motion relative to a selected frame of reference can be detected.*

Let's apply the principle of Newtonian relativity to the ball shown in Figure 25-5. Suppose you are standing on the platform as the train goes by, so you are in reference frame S. You see the ball following a parabolic path, and you make observations of the x, y, and z coordinates of the ball as functions of time t. The child riding in the train is in reference frame S' and sees the ball moving straight up and down relative to her. As the ball moves she measures the x', y', and z' coordinates of the ball as functions of time t'. The two sets of coordinates have the same orientation, as Figure 25-6 shows. To calibrate the two clocks—one in frame S, the other in frame S'—you and the child both set your clocks to read zero at the instant that you pass each other, when the origins of your frames of reference coincide. So at this instant $t = t' = 0$.

Suppose you and the child both measure the same event—that is, the ball being at a certain point in its motion, such as the high point in its path. In your frame S the coordinates of this event in space and time are x, y, z, and t, and the coordinates of this same event as measured in frame S' are x', y', z', and t'. (Note that we are thinking of time as a fourth coordinate of an event.) As Figure 25-6 shows, a simple set of equations relates the coordinates of the same event in the two frames:

Galilean transformation

(25-1)

> Inertial frame of reference S' moves at speed V in the positive x direction relative to inertial frame of reference S.

Coordinates of an event as measured in frame S'

$$x' = x - Vt$$
$$y' = y$$
$$z' = z$$
$$t' = t$$

Coordinates of the same event as measured in frame S

This set of equations is known as the **Galilean transformation**. Note that the relative motion of the two frames of reference along the positive x axis does not affect the y and z coordinates, which are the same in both reference frames.

We can use the Galilean transformation to compare the velocity of the ball as measured in frame S' to its velocity as measured in frame S. To do this we'll look at two events in the motion of the ball separated by a short time interval from t_1 to t_2 as measured in frame S. During this time interval the coordinates of the ball as measured

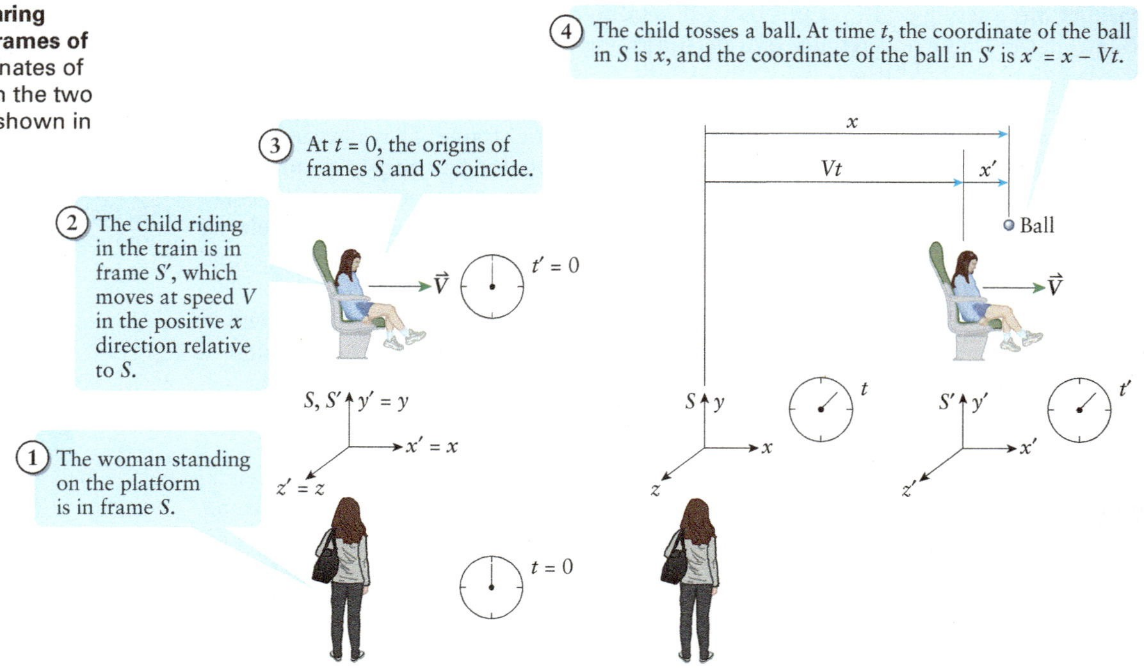

Figure 25-6 Comparing coordinates in two frames of reference The coordinates of a ball as measured in the two frames of reference shown in Figure 25-3.

④ The child tosses a ball. At time t, the coordinate of the ball in S is x, and the coordinate of the ball in S' is $x' = x - Vt$.

③ At $t = 0$, the origins of frames S and S' coincide.

② The child riding in the train is in frame S', which moves at speed V in the positive x direction relative to S.

① The woman standing on the platform is in frame S.

in frame S change from x_1, y_1, and z_1 to x_2, y_2, and z_2. The components of the ball's velocity $\vec{v}$ in frame S are

$$v_x = \frac{\Delta x}{\Delta t} = \frac{x_2 - x_1}{t_2 - t_1}$$

$$v_y = \frac{\Delta y}{\Delta t} = \frac{y_2 - y_1}{t_2 - t_1} \qquad (25\text{-}2)$$

$$v_z = \frac{\Delta z}{\Delta t} = \frac{z_2 - z_1}{t_2 - t_1}$$

For the same two events as measured in frame S', the time interval is from t_1' to t_2', and the coordinates change from x_1', y_1', and z_1' to x_2', y_2', and z_2'. So in frame S' the components of the object's velocity $\vec{v}'$ are

$$v_x' = \frac{\Delta x'}{\Delta t'} = \frac{x_2' - x_1'}{t_2' - t_1'}$$

$$v_y' = \frac{\Delta y'}{\Delta t'} = \frac{y_2' - y_1'}{t_2' - t_1'} \qquad (25\text{-}3)$$

$$v_z' = \frac{\Delta z'}{\Delta t'} = \frac{z_2' - z_1'}{t_2' - t_1'}$$

Now substitute the expressions for x', y', z', and t' from Equations 25-1 into Equations 25-3. We get

$$v_x' = \frac{(x_2 - Vt_2) - (x_1 - Vt_1)}{t_2 - t_1} = \frac{(x_2 - x_1) - V(t_2 - t_1)}{t_2 - t_1} = \frac{(x_2 - x_1)}{t_2 - t_1} - V$$

$$v_y' = \frac{y_2 - y_1}{t_2 - t_1} \qquad (25\text{-}4)$$

$$v_z' = \frac{z_2 - z_1}{t_2 - t_1}$$

If we now compare Equations 25-4 for the velocity components in frame S' to Equations 25-2 for the velocity components in frame S, we see that

Inertial frame of reference S' moves at speed V in the positive x direction relative to inertial frame of reference S.

Velocity components of an object as measured in frame S'

$$v_x' = v_x - V$$
$$v_y' = v_y$$
$$v_z' = v_z$$

Velocity components of the same object as measured in frame S

Galilean velocity transformation
(25-5)

Equations 25-5, which relate the velocity of an object in frame S' to the velocity of the same object in frame S, are called the **Galilean velocity transformation**.

As an example, think again of the ball shown in Figure 25-5. If the ball moves straight up and down as measured in frame S', then in that frame the ball is in free fall with only a y component of velocity. The other two components are zero: $v_x' = v_z' = 0$. As measured in frame S, the velocity of the ball has components

$$v_x = v_x' + V = V$$
$$v_y = v_y'$$
$$v_z = v_z' = 0$$

As measured in frame S, the ball moves up and down along the y direction with the same velocity as measured in frame S': $v_y = v_y'$. At the same time, as measured in frame S, the ball maintains a constant velocity V in the x direction. That's just the behavior we expect for a projectile: Its motion is a combination of up-and-down free fall and constant-velocity horizontal motion (see Section 3-6).

WATCH OUT! There is nothing special about either the S frame or the S' frame.

In Figures 25-5 and 25-6 we've chosen to think of frame S as at rest and frame S' as moving. This selection is purely our choice because the principle of Newtonian relativity says that *all* inertial frames are equivalent. We could just as well say that frame S' is stationary and frame S is moving with speed V in the negative x direction.

You may wonder why we've spent so much time and effort explaining motion as seen from two different inertial frames of reference. As we'll discover in the next few sections, the reason is that something remarkable happens when the relative speed V of the two frames is comparable to c, the speed of light in a vacuum. In that case we'll find that the Galilean transformation equations do *not* hold true: As measured in the two different frames, the time interval between events can be different and objects can have different dimensions. These remarkable observations will radically transform our notions of the nature of time and space themselves.

EXAMPLE 25-1 Two Cars

You observe two racecars approaching you. A red car is in one lane moving at 24 m/s relative to you. In a second lane a blue car is moving at 36 m/s relative to you. (a) What is the velocity of the red car as measured by the driver of the blue car? (b) What is the velocity of the blue car as measured by the driver of the red car?

Set Up

We'll use Equations 25-5 to transform the velocity of a car as measured in one frame of reference to the velocity of the same car as measured in a different frame of reference. Note that these equations assume that frame S' is moving relative to frame S at speed V in the positive x direction. We'll use this to decide which frame of reference corresponds to S and which to S'.

Galilean velocity transformation:

$$v'_x = v_x - V$$
$$v'_y = v_y$$
$$v'_z = v_z \qquad (25\text{-}5)$$

The given speeds of both cars are measured relative to you, that is, in your frame.

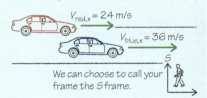

We can choose to call your frame the S frame.

Solve

(a) Let's take the positive x direction to be in the direction both cars are moving relative to you. Then we'll take S to be your frame of reference and S' to be the frame of reference of the driver of the blue car. The relative speed of these two frames is $V = 36$ m/s (the speed of the blue car relative to you). All of the motions in this example are along the x axis, so we don't need the y or z members of Equations 25-5.

Use the x equation from Equations 25-5 to relate the velocity of the red car as measured by you ($v_{red,x} = +24$ m/s) to its velocity as measured by the driver of the blue car ($v'_{red,x}$):

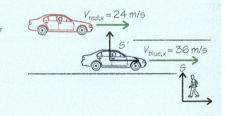

$$v'_{red,x} = v_{red,x} - V$$
$$= +24 \text{ m/s} - 36 \text{ m/s}$$
$$= -12 \text{ m/s}$$

As measured by the driver of the blue car, the red car is moving in the negative x direction—that is, backward—at 12 m/s.

(b) Again we take S to be your frame of reference, but now S' is the frame of reference of the driver of the red car. The relative speed of these two frames is $V = 24$ m/s (the speed of the red car relative to you).

Use the x equation from Equations 25-5 to relate the velocity of the blue car as measured by you ($v_{blue,x} = +36$ m/s) to its velocity as measured by the driver of the red car ($v'_{blue,x}$):

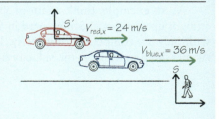

$$v'_{blue,x} = v_{blue,x} - V$$
$$= +36 \text{ m/s} - 24 \text{ m/s}$$
$$= +12 \text{ m/s}$$

As measured by the driver of the red car, the blue car is moving in the positive x direction—that is, forward—at 12 m/s.

Reflect

The driver of the blue car sees the red car falling farther and farther behind her at a rate of 12 m/s, while the driver of the red car sees the blue car moving farther and farther in front of him at a rate of 12 m/s. Note that the *magnitude* of the two answers is the same: Both drivers agree that the other driver is moving at a relative speed of 12 m/s.

GOT THE CONCEPT? 25-1 Groundspeed Versus Airspeed

(?) A typical jet airliner has a cruise airspeed—that is, its speed relative to the air through which it is flying—of 900 km/h. If the wind at the airliner's cruise altitude is blowing at 100 km/h from west to east, what is the speed of the airliner relative to the ground if the airplane is flying from west to east? From east to west? (a) 800 km/h west to east, 1000 km/h east to west; (b) 1000 km/h west to east, 800 km/h east to west; (c) 800 km/h in both directions; (d) 900 km/h in both directions; (e) 1000 km/h in both directions.

TAKE-HOME MESSAGE FOR Section 25-2

✔ Newton's laws of motion treat all nonaccelerating objects identically, whether the objects are in motion or at rest.

✔ The laws of motion are the same in all inertial frames of reference. There is no way to detect absolute motion; only motion relative to a selected frame of reference can be detected.

✔ The Galilean transformations allow you to convert the coordinates and velocity of an object observed in one inertial frame of reference to the coordinates and velocity of the same object observed in a different inertial frame.

25-3 The Michelson–Morley experiment shows that light does not obey Newtonian relativity

We saw in the preceding section that Newton's laws of motion are the same in all inertial frames. Is the same true for the other laws of physics? Physicists asked this very question during the second half of the nineteenth century, specifically about the laws of electromagnetism.

We learned in Section 22-3 that the laws of electromagnetism explain how electromagnetic waves, including visible light, are possible. These laws also predict that the speed of electromagnetic waves in a vacuum is related to the constants ε_0 and μ_0 that appear in the equations of electromagnetism:

Speed of light in a vacuum

$$c = \frac{1}{\sqrt{\mu_0 \varepsilon_0}} = 3.00 \times 10^8 \text{ m/s}$$

Permeability of free space   Permittivity of free space

Speed of light in a vacuum
(22-11)

Our discussion in Section 25-2 tells us that we can measure relative motion but not absolute motion. So we are forced to ask this question: Relative to what inertial frame of reference is the speed of light in a vacuum equal to c?

In the nineteenth century the most common answer to this question was to imagine a substance that fills all space. This substance, which was called the luminiferous ether, was thought to be the medium for electromagnetic waves, just as air is the medium for sound waves in our atmosphere. This substance must be of extraordinarily low density so that its presence is almost undetectable. (The word "luminiferous" comes from the Latin for "light-bearing." "Ether" in this phrase has nothing to do with the organic compounds of the same name.) In this model c is the speed of electromagnetic waves relative to the frame in which the luminiferous ether is at rest.

How can we test whether this model is correct? To see the answer, note that the speed of sound waves in dry air is 343 m/s, but you will measure a different speed if a wind is blowing (that is, the air is moving relative to you). If a sound wave is traveling from west to east and a wind is blowing past you at 10 m/s from west to east, the sound will move relative to you at 343 m/s + 10 m/s = 353 m/s. If the sound wave is traveling from west to east and a 10-m/s wind is blowing from east to west, the speed of the sound wave relative to you will be 343 m/s − 10 m/s = 333 m/s. The same should be true for light waves if the luminiferous ether is moving past you so that there is an "ether wind." If a light wave is traveling from west to east and the ether is moving from west to east relative to you at 10 m/s, you would measure the speed of the wave to be $c + 10$ m/s;

if the ether is instead moving from east to west relative to you at 10 m/s, you would measure the speed of the wave to be $c - 10$ m/s. If we can detect these small changes in the speed of light, that would be evidence that the luminiferous ether really exists.

Nineteenth-century scientists looked to Earth's motion around the Sun as a source of "ether wind." Our planet moves around its orbit at an average speed of 29.8 km/s = 2.98×10^4 m/s, or about $10^{-4} c$. If the luminiferous ether is at rest relative to the solar system as a whole, we should experience an "ether wind" that blows past our moving planet at $10^{-4} c$. Depending on the direction of that "ether wind" relative to the direction of light propagation, we would expect the speed of light to vary between $(1 + 10^{-4})c$ and $(1 - 10^{-4})c$. The challenge is to design an experiment that can detect such small changes in the speed of light.

In 1887, the American scientists Albert Michelson and Edward Morley carried out the first definitive experiment of this kind. Figure 25-7 shows a simplified version of the **Michelson–Morley experiment**. Their apparatus split a beam of light into two, sent the two beams along perpendicular paths, and then allowed them to recombine at a viewing screen. What is seen on the viewing screen is an interference pattern between the waves in the two beams. The nature of this pattern depends on the difference in length between the two paths and also on whether the speeds at which light travels along each path are the same or different. If there is an "ether wind" that is more nearly aligned with one leg of the interferometer than the other, the speed of light should indeed be different along the two legs. Their apparatus was sensitive enough that Michelson and Morley should have been able to measure the effects of an "ether wind" due to Earth's motion around the Sun.

Michelson and Morley found *no* effect due to an "ether wind." They and other scientists refined and repeated the experiment many times, but the results were always the same: There is no evidence for the existence of the luminiferous ether. The conclusion from this and other experiments is that electromagnetic waves do not require the presence of a material medium but can propagate in a complete vacuum.

Under Newtonian relativity, to explain a constant value of the speed of light in a vacuum required that the equations of electromagnetism hold true only in a specific reference frame, one that is at rest relative to the medium for electromagnetic waves. But if there is no luminiferous ether, there is no such medium and hence no such special frame of reference. So physicists had no choice but to conclude that Newtonian relativity does not apply to electromagnetic waves. It was left to Albert Einstein to modify the ideas of Newtonian relativity and find a new way to look at the relationships between measurements made in different inertial frames of reference. We'll explore Einstein's simple yet radical ideas in the following section.

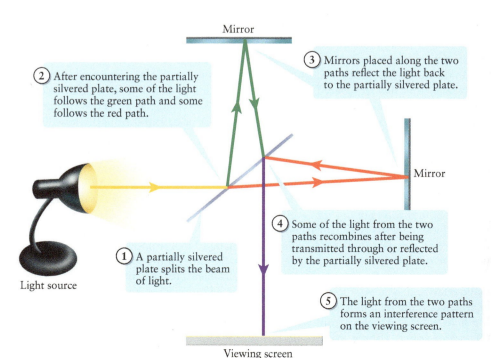

Figure 25-7 The Michelson–Morley experiment simplified If the luminiferous ether exists and is moving relative to this apparatus, its presence will be apparent in the interference pattern on the viewing screen.

Mirror

(2) After encountering the partially silvered plate, some of the light follows the green path and some follows the red path.

(3) Mirrors placed along the two paths reflect the light back to the partially silvered plate.

Mirror

(1) A partially silvered plate splits the beam of light.

Light source

(4) Some of the light from the two paths recombines after being transmitted through or reflected by the partially silvered plate.

(5) The light from the two paths forms an interference pattern on the viewing screen.

Viewing screen

GOT THE CONCEPT? 25-2 Speed of Sound in Water

(?) A bottlenose dolphin (*Tursiops truncatus*) can swim at speeds up to about 10 m/s. These dolphins produce sounds used for social communication and for echolocation (using sound to detect objects around them while swimming in dark or murky waters). If a swimming dolphin travels at top speed and produces a sound wave that propagates forward, how fast does that wave travel relative to the dolphin? The speed of sound in water is 1500 m/s. (a) 10 m/s; (b) 1490 m/s; (c) 1500 m/s; (d) 1510 m/s; (e) answer depends on the frequency of the sound wave.

TAKE-HOME MESSAGE FOR Section 25-3

✔ Experiment shows that electromagnetic waves do not require a material medium; they can easily propagate in a perfect vacuum.

✔ Because electromagnetic waves do not require a material medium, we conclude that electromagnetic waves do not obey Newtonian relativity.

25-4 Einstein's relativity predicts that the time between events depends on the observer

In 1905, German-born theoretical physicist Albert Einstein published his **special theory of relativity**, or *special relativity* for short. Einstein based his theory on two postulates:

- *First postulate*: All laws of physics are the same in all inertial frames.
- *Second postulate*: The speed of light in a vacuum is the same in all inertial frames, independent of both the speed of the source of the light and the speed of the observer.

The adjective "special" means that the theory applies to the special case of inertial frames and constant-velocity motion. (Einstein's *general* theory of relativity, which we'll discuss in Section 25-8, extends these ideas to accelerating, noninertial frames.) The first postulate extends the principle of relativity beyond Newton's laws of motion to include all aspects of physics, including electromagnetic waves. Einstein's second postulate derives from our conclusion in Section 25-3 that there is no special frame of reference in which the speed of light in a vacuum is equal to $c = 3.00 \times 10^8$ m/s.

Perhaps the most astounding consequence of the second postulate of special relativity is that the time interval between two events is *not* an absolute. Rather, this time interval depends on the motion of the frame of reference from which time is measured. To see how this comes about, we'll do a thought experiment.

Imagine a *light clock*, a special clock that uses light to measure intervals of time (Figure 25-8). A laser fires an extremely brief burst, or pulse, of light straight downward. This is reflected by a mirror and arrives at a light-sensitive detector next to the laser. For simplicity we can treat the laser and detector as being at the same position. One tick of the clock is the time interval between the pulse leaving the laser (event 1) and the pulse arriving at the detector (event 2). In the frame of reference of the clock, these two events occur at the same point in space. We use the term **proper time** for the time interval between two events that occur at the same place, and we denote it by the symbol Δt_{proper}. Another way to think of proper time is that it is the time interval as measured in a frame of reference in which the clock is at rest. You can think of the rest frame of the clock as being "attached" to the clock.

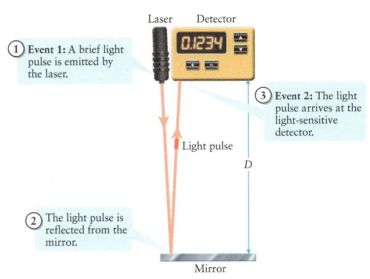

① **Event 1:** A brief light pulse is emitted by the laser.

③ **Event 2:** The light pulse arrives at the light-sensitive detector.

Light pulse

D

② The light pulse is reflected from the mirror.

Laser Detector

Mirror

Figure 25-8 A light clock The duration of each "tick" of this clock is the time between the start of the pulse and the time when the reflected light arrives at the detector.

WATCH OUT! Every object—including a moving object—has a rest frame.

(!) An object is always at rest with respect to itself, which means that an object is not moving in a reference frame that is attached to it. Notice, however, that the statement is a relative one—the object is not moving *relative* to its own rest frame. We can still define any number of other frames with respect to which the object *is* in motion. A person standing on the sidewalk is at rest in her own rest frame but is moving as measured from the frame of reference of a car driving past.

Figure 25-9 A moving light clock and time dilation When the light clock depicted in Figure 25-8 moves, the distance the light pulse travels is longer than twice the distance from the laser to the mirror.

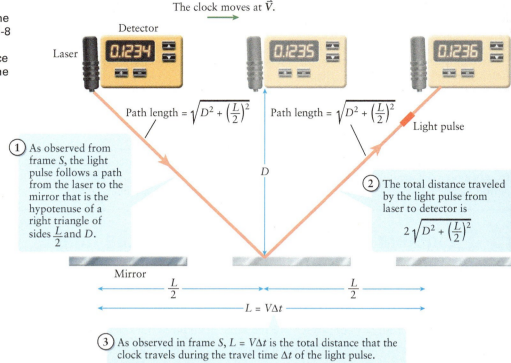

The clock moves at $\vec{V}$.

① As observed from frame S, the light pulse follows a path from the laser to the mirror that is the hypotenuse of a right triangle of sides $\frac{L}{2}$ and D.

Path length = $\sqrt{D^2 + \left(\frac{L}{2}\right)^2}$ Path length = $\sqrt{D^2 + \left(\frac{L}{2}\right)^2}$

Light pulse

② The total distance traveled by the light pulse from laser to detector is

$$2\sqrt{D^2 + \left(\frac{L}{2}\right)^2}$$

Mirror $\frac{L}{2}$ $\frac{L}{2}$

$L = V\Delta t$

③ As observed in frame S, $L = V\Delta t$ is the total distance that the clock travels during the travel time Δt of the light pulse.

During one tick of our light clock, a light pulse travels the distance D to the mirror and then the same distance back to the detector, for a total distance of $2D$. We imagine that the clock is placed in a vacuum so that the speed of the light pulse is c. The time interval for the tick is then

(25-6)
$$\Delta t_{\text{proper}} = \frac{2D}{c}$$

We now let the light clock move at speed V relative to us. In keeping with the way we named frames in discussing Newtonian relativity, we attach a frame S' to the clock and consider ourselves in the S frame. The S' frame therefore moves at speed V relative to the S frame. Figure 25-9 shows the process of a single tick of the clock as observed from our S frame. The entire clock moves during the time that it takes for the light pulse to move from the laser to the mirror and from the mirror to the detector. The total distance L that the clock moves during this time equals the product of the speed V and the time interval of one clock tick. Be careful, however: We *cannot* assume that the time interval for one tick as measured in S is Δt_{proper} because Δt_{proper} is measured in frame S' at rest with respect to the clock. Instead we use the symbol Δt for the time interval of one tick as observed from the S frame. So

(25-7)
$$L = V\Delta t$$

√x̄ **See the Math Tutorial for more information on trigonometry.**

From our vantage point in the S frame, the path followed by the light pulse during time Δt traces out two sides of a triangle (Figure 25-9). The time interval measured in S is therefore equal to the sum of the lengths of these two sides—the distance the light pulse travels—divided by the speed at which the light pulse travels in frame S. The second postulate of special relativity assures us that the speed of the light pulse is c in *all* inertial frames, and so is the same in frame S as in frame S'. Thus

(25-8)
$$\Delta t = \frac{2\sqrt{D^2 + \left(\frac{L}{2}\right)^2}}{c}$$

We now have two expressions for the time interval between the light pulse leaving the laser and the same pulse arriving at the detector. As measured in the clock rest frame S', this time interval is the proper time Δt_{proper} as given by Equation 25-6; as measured in our rest frame S, the time interval is Δt as given by Equation 25-8. To compare these two time intervals, first note that the distance L does not appear in the expression for Δt_{proper}. To eliminate L from the expression for Δt, substitute Equation 25-7, $L = V\Delta t$, into Equation 25-8:

$$\Delta t = \frac{2\sqrt{D^2 + \left(\dfrac{V\Delta t}{2}\right)^2}}{c}$$

Square both sides of this equation and rearrange to bring the two Δt terms together:

$$c^2\Delta t^2 = 4\left(D^2 + \left(\frac{V\Delta t}{2}\right)^2\right) = 4D^2 + V^2\Delta t^2$$

$$c^2\Delta t^2 - V^2\Delta t^2 = 4D^2 \quad \text{or} \quad (c^2 - V^2)\Delta t^2 = 4D^2$$

Rearrange this equation to solve for Δt^2, then take the square root of both sides to solve for Δt:

$$\Delta t^2 = \frac{4D^2}{c^2 - V^2} = \frac{4D^2}{c^2\left(1 - \dfrac{V^2}{c^2}\right)}$$

$$\Delta t = \frac{2D}{c\sqrt{1 - \dfrac{V^2}{c^2}}} \tag{25-9}$$

If we compare Δt (Equation 25-9) to Δt_{proper} (Equation 25-6), we get

Time interval between two events as measured in frame S', in which those events occur at the same place

Time interval between the same two events as measured in frame S

$$\Delta t = \frac{\Delta t_{proper}}{\sqrt{1 - \dfrac{V^2}{c^2}}}$$

Speed of S' relative to S

Speed of light in a vacuum

Time dilation
(25-10)

An observer in S moving relative to the clock measures a time interval Δt for one tick of the light clock. An observer in S' who is at rest with respect to the clock measures the time interval for one tick to be Δt_{proper}. Equation 25-10 shows that Δt and Δt_{proper} are *not* equal: An observer in the S frame sees time running at a *different* rate than an observer in the S' frame!

For any nonzero value of the speed V of one inertial frame relative to another, the denominator in Equation 25-10 is less than 1. It follows that Δt is greater than Δt_{proper}. That is, the time interval as measured in S is longer than as measured in S'. We say that the time interval as measured in S has been expanded or *dilated* (the same term used to refer to an increase in size of the pupil of the eye). That's why this effect of special relativity is known as **time dilation**. If $\Delta t_{proper} = 1$ s, the time interval Δt for one tick as measured in S will be greater than 1 s. Equivalently, if Δt as measured in S equals 1 s, Δt_{proper} as measured in S' will be less than 1 s. So an observer in S says that the light clock, which is moving past her at speed V, is ticking off time slowly: After 1 s has elapsed according to the observer in S, the moving clock has ticked off less than 1 s. So Equation 25-10 says that a *moving clock runs slowly*.

It may seem that Equation 25-10 is valid only for time intervals measured using a light clock. But the first postulate of relativity says that *all* laws of physics are the same in every inertial frame of reference. This implies that time dilation is also valid for time intervals measured using a mechanical clock, such as a grandfather clock that keeps time with an oscillating pendulum or a wristwatch that uses an oscillating piece of quartz (Figure 12-1b). Imagine a pendulum clock on your desk, with its pendulum swinging back and forth once every second. If the clock is moving, then the swing of the pendulum is slower, so the clock ticks more slowly. As we will see, however, the time dilation effect is very small unless V is very large.

BioMedical **WATCH OUT!** Time dilation occurs whether or not a clock is present.

! You don't need to have a clock per se for the effects of time dilation to occur. Imagine, say, that you can place a *Caenorhabditis elegans* worm (Figure 25-10) on a spacecraft that will fly past Earth at high speed. *C. elegans* is popular among geneticists because it grows to adulthood in a series of easily identifiable developmental stages that all together take less than 3 days. Were you to observe a *C. elegans* worm as it moved past you at high speed, the growth to adulthood might take weeks or even years, measured on a clock at rest with respect to you.

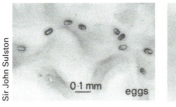   

0.1 mm   eggs   L1   L3   adult

Time ⟶

Figure 25-10 **A worm "clock"** The worm *Caenorhabditis elegans* grows to adulthood in a series of easily identifiable developmental stages, each of which lasts a well-defined period of time.

WATCH OUT! The effect of time dilation arises only when events in one frame are viewed from a second frame in motion relative to the first.

! We imagined, above, a pendulum clock that ticks once every time the pendulum makes one full oscillation. If the clock sits on your desk, you see it swinging back and forth once every second. If the clock moves relative to you, you see the pendulum swing more slowly. But if you and your clock are moving *together* relative to some other specific object or frame, you will see *no* time dilation effect on your clock, regardless of how fast you and the clock are moving relative to that other object or frame. You will still see the pendulum swinging back and forth once every second. There is no time dilation because you and the clock are not moving relative to each other.

We do not observe time dilation in everyday life because most objects travel very slowly compared to the speed of light. As an example, the greatest launch speed ever given to a spacecraft is 16,260 m/s = 58,536 km/h = 36,373 mi/h relative to Earth. Even this tremendous speed is only 5.42×10^{-5} of the speed of light c, and the factor $1/\sqrt{1 - V^2/c^2}$ in Equation 25-10 is larger than 1 by only 1.47×10^{-9}. Due to time dilation, a clock on board the spacecraft does indeed run slowly as measured from Earth, but by only one second every 21.5 years! As the following two examples show, however, time dilation can be substantial if the speeds involved are **relativistic speeds**—that is, an appreciable fraction of the speed of light.

EXAMPLE 25-2 **A Moving Clock**

A clock moves past you. What must be the speed of the clock relative to you so that you see it as running at one-half (0.500) the rate of the clock on your cell phone in your hand?

Set Up

We want a moving clock to be observed as running at 0.500 the rate of a clock at rest with respect to you. This means making Δt (the time interval measured by you) equal to twice Δt_{proper} (the time interval measured in the rest frame of the moving clock). That implies the factor $1/\sqrt{1 - V^2/c^2}$ in Equation 25-10 must be equal to $1/0.500 = 2.00$. We'll use this to solve for the speed V of the moving clock relative to you.

Time dilation:

$$\Delta t = \frac{\Delta t_{proper}}{\sqrt{1 - \frac{V^2}{c^2}}}$$

(25-10)

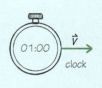

Solve

Determine the value of V that makes $\Delta t = 2.00\Delta t_{\text{proper}}$.

From Equation 25-10 we want

$$\Delta t = \frac{\Delta t_{\text{proper}}}{\sqrt{1 - \dfrac{V^2}{c^2}}} = 2.00\Delta t_{\text{proper}} \quad \text{so} \quad \frac{1}{\sqrt{1 - \dfrac{V^2}{c^2}}} = 2.00$$

Solve for the value of the speed V:

$$\sqrt{1 - \frac{V^2}{c^2}} = \frac{1}{2.00} = 0.500$$

$$1 - \frac{V^2}{c^2} = (0.500)^2 = 0.250$$

$$\frac{V^2}{c^2} = 1 - 0.250 = 0.750$$

$$\frac{V}{c} = \sqrt{0.750} = 0.866$$

$$V = 0.866c$$

$$= 0.866(3.00 \times 10^8 \text{ m/s}) = 2.60 \times 10^8 \text{ m/s}$$

Reflect

For you to observe the moving clock running a factor of 2.00 slower than the clock at rest with respect to you, the relative speed between you and the clock would need to be 86.6% of the speed of light in a vacuum. This is more than 10^4 times faster than the top speed of any craft built by humans.

EXAMPLE 25-3 Muon Decay

Our planet is continually bombarded by fast-moving subatomic particles from space (mostly protons). When these particles collide with the atoms and molecules that make up Earth's upper atmosphere, they can produce a new subatomic particle called the *muon*. These muons travel at high speed, around $0.994c$. Muons naturally decay, however; if you create a number of muons in the lab, on average half of them will decay after 1.56 μs (1 μs = 10^{-6} s). This time is known as the *half-life* of the muon. Of the remaining muons, another half will decay after another 1.56 μs, and so on. If 1.00×10^6 muons are created at an altitude of 15.0 km, (a) how many would you expect to strike Earth if time-dilation effects were ignored? (b) How many would you expect to strike Earth when time dilation is taken into account?

Set Up

The production and decay of the muon occur at the same place in the rest frame of the muon, so the half-life of 1.56 μs is a proper time interval in the muon frame. We'll call this frame S'. This frame moves at speed $V = 0.994c$ relative to our frame on Earth, so we measure a different half-life Δt as given by Equation 25-10. We'll use the idea that the muon population decreases by 1/2 in each half-life to determine how many reach our planet's surface.

Time dilation:

$$\Delta t = \frac{\Delta t_{\text{proper}}}{\sqrt{1 - \dfrac{V^2}{c^2}}} \qquad (25\text{-}10)$$

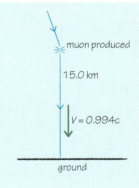

muon produced

15.0 km

$V = 0.994c$

ground

Solve

(a) If there were no time dilation, the half-life of the muon in the Earth frame S would be $\Delta t = \Delta t_{\text{proper}} = 1.56$ μs, the same as in the muon frame S'. Use this to calculate the number of muons that successfully reach Earth's surface.

Time for a muon to travel a distance $d = 15.0$ km at $V = 0.994c$:

$$T = \frac{d}{V} = \frac{15.0 \text{ km}}{0.994(3.00 \times 10^8 \text{ m/s})}\left(\frac{10^3 \text{ m}}{1 \text{ km}}\right)$$

$$= 5.03 \times 10^{-5} \text{ s}\left(\frac{1 \text{ μs}}{10^{-6} \text{ s}}\right) = 50.3 \text{ μs}$$

Express this as a multiple of the half-life:

$$\frac{T}{\Delta t_{\text{proper}}} = \frac{50.3 \text{ μs}}{1.56 \text{ μs}} = 32.2 \text{ half-lives}$$

After one half-life, the number of muons has decreased to 1/2 of its initial value; after two half-lives, to $(1/2) \times (1/2) = 1/2^2$ of its initial value; after three half-lives, to $(1/2) \times (1/2) \times (1/2) = 1/2^3$ of its initial value; and so on. So after 32.2 half-lives, the number of muons remaining would be the original number of 1.00×10^6 multiplied by $1/2^{32.2}$:

$$\text{muons remaining} = (1.00 \times 10^6)\left(\frac{1}{2^{32.2}}\right)$$

$$= (1.00 \times 10^6)(1.97 \times 10^{-10})$$

$$= 1.97 \times 10^{-4}$$

Much less than 1 muon, on average, survives the trip to Earth's surface.

(b) With time dilation, we first calculate the half-life in the Earth frame using Equation 25-10. We then use the same method as in part (a) to calculate the number of muons that reach Earth's surface.

Accounting for time dilation, the half-life as measured in the Earth frame S is the half-life measured in the muon frame S' divided by $\sqrt{1 - V^2/c^2}$:

$$\Delta t = \frac{\Delta t_{\text{proper}}}{\sqrt{1 - \dfrac{V^2}{c^2}}} = \frac{1.56\ \mu s}{\sqrt{1 - \dfrac{(0.994c)^2}{c^2}}} = \frac{1.56\ \mu s}{\sqrt{1 - (0.994)^2}}$$

$$= \frac{1.56\ \mu s}{\sqrt{0.0120}} = \frac{1.56\ \mu s}{0.109}$$

$$= 14.3\ \mu s$$

Due to time dilation the half-life as measured in the Earth frame is about 9 times longer than the half-life as measured in the muon frame.

The effect of time dilation causes time to run more slowly for the muon than we measure in our own frame, so fewer half-lives elapse for a given distance traveled. Because fewer half-lives have elapsed, fewer muons have decayed.

From part (a), the time for a muon to travel to the surface is 50.3 μs. Expressed as a multiple of the time-dilated half-life, this is

$$\frac{T}{\Delta t} = \frac{50.3\ \mu s}{14.3\ \mu s} = 3.53\ \text{half-lives}$$

The number of muons remaining at the surface is the original number of 1.00×10^6 multiplied by $1/2^{3.53}$:

$$\text{muons remaining} = (1.00 \times 10^6)\left(\frac{1}{2^{3.53}}\right)$$

$$= (1.00 \times 10^6)(0.0868)$$

$$= 8.68 \times 10^4$$

About 1 in 11 of the muons produced in the upper atmosphere reaches Earth's surface, far more than would be the case were there no time dilation.

Reflect

One of the earliest confirmations of Einstein's special theory of relativity was a 1941 experiment that compared the number of muons observed in an hour at two elevations in Colorado, 3240 and 1616 m above sea level. Without time dilation, the number of muons observed at the lower elevation would have been about 9% of the number at the higher elevation. The measured number at the lower elevation was more than 80%, in accordance with Einstein's prediction. This experiment provided dramatic evidence that the bizarre phenomenon of time dilation is very real.

The Twin Paradox

Imagine two twins named Bertha and Eartha. Bertha takes a round trip from Earth to another star and back on a fast spaceship that travels at $v = 0.866c$, while her twin Eartha stays at home on Earth. Eartha's clock ticks off 10 years during the time that Bertha is gone. As we saw in Example 25-2, if a clock moves past you at this speed, you will observe it as running at one-half the rate of a clock in your hand. So Eartha expects that when her twin returns, Bertha's clock will have ticked off only 5 years, not 10, and that Bertha will have aged only 5 years. So Eartha predicts that *Bertha* will actually be younger than Eartha after the round trip is done.

But from *Bertha's* frame of reference, *Eartha's* clock is the one that's moving at 0.866c, so Bertha observes Eartha's clock to be running slow. Hence Bertha expects that when she returns to Earth, she will find that Eartha's clock has ticked off half as much time as Bertha's clock on the spaceship. Therefore, Bertha predicts that Eartha will have aged only half as much as Bertha has and that *Eartha* will be the younger twin after the round trip is done.

It seems like the experiences of the two twins are exactly symmetrical: Each twin sees the other one moving, and hence aging more slowly, and predicts that the other one will have aged less. But clearly both twins can't be correct. So how can we resolve this *twin paradox*?

The explanation is that the two observers, Eartha and Bertha, are *not* perfectly symmetrical. Eartha remained in the same inertial reference frame (the Earth) at all times during Bertha's round trip, but Bertha did not: On the outbound leg Bertha was in a frame moving at 0.866c in the direction from Earth toward the other star, while on the return leg Bertha was moving at 0.866c in the *opposite* direction from the star back toward Earth. It turns out that because Bertha changed her reference frame and her velocity—that is, because she *accelerated*—she is the twin who ages less during the round trip. So Eartha is correct, and after the round trip Eartha will have aged 10 years, while Bertha will have aged only 5 years.

This effect has been experimentally verified many times. If an unstable particle like a muon (Example 25-3) is produced in the laboratory and sent at relativistic speed along a circular round-trip path, its half-life is longer than if the particle is produced at rest; in other words, the particle that takes the round trip ages more slowly. The effect has also been tested using very precise clocks, one that remained in the laboratory and one that was flown on a round trip in a fast-moving airplane. The clock that traveled out and back ticked off a fraction of a second less time than did the clock that stayed at rest, just as predicted by special relativity.

GOT THE CONCEPT? 25-3 Proper Time

(?) A clock is placed aboard Starship *Alpha,* which Albert flies past Earth at half the speed of light relative to Earth. Barbara flies Starship *Beta* alongside *Alpha* at the same velocity. George pilots Starship *Gamma* past Earth at half the speed of light relative to Earth, but in the opposite direction. Elena observes from Earth. For which of these observers is the time interval between ticks of the clock equal to the proper time? (a) Albert; (b) Barbara; (c) George; (d) Elena; (e) more than one of these.

GOT THE CONCEPT? 25-4 Comparing Clocks

(?) A clock is placed aboard Starship *Alpha,* which Albert flies past Earth at half the speed of light relative to Earth. Elena observes from Earth, where she has an identical clock. Which pair of words correctly fills in the blanks in this statement: "Elena measures Albert's clock as running _____, and Albert measures Elena's clock as running _____." (a) slow, fast; (b) slow, slow; (c) fast, fast; (d) fast, slow.

TAKE-HOME MESSAGE FOR Section 25-4

✔ Einstein based his special theory of relativity on two postulates: First, that all laws of physics are the same in all inertial frames, and second, that the speed of light in a vacuum is the same in all frames and independent of both the speed of the source of the light and the speed of the observer.

✔ If two events happen at the same place in one frame, the time interval between these events as measured in that frame is called the proper time. As measured from a second frame moving relative to the first one, the time interval between those two events is longer than the proper time. This is called time dilation.

25-5 Einstein's relativity also predicts that the length of an object depends on the observer

We learned in the preceding section that the time interval between two events is not an absolute but depends on the motion of the observer. As we will see in this section, the *length* of an object is also not an absolute: Different observers will measure the same object as having different dimensions. So the nature of space and time is very different from what had been thought previous to the work of Einstein.

We'll conclude this section by looking at the *Lorentz transformation*, a set of equations that allows us to convert the space and time coordinates of an event in one inertial frame of reference to the coordinates in a second inertial frame. This transformation is an extension of the Galilean transformation that we explored in Section 25-2. We'll see that the Galilean transformation is actually just a special case of the Lorentz transformation valid for two frames that are moving with respect to each other at a speed far less than *c*.

Length Contraction

Figure 25-11 shows a thought experiment in which we look at the same motion in two different frames of reference, much as we did in Figures 25-8 and 25-9. In Figure 25-11a a rod is moving to the right at speed *V* relative to frame *S*, which you can think of as our frame of reference. The right-hand end of the rod is at $x = 0$ at time $t = 0$, and the left-hand end of the rod is at $x = 0$ at a later time $t = \Delta t_S$. The length *L* of the rod in our frame is therefore the speed *V* of the rod multiplied by the time Δt_S needed to travel its own length:

$$(25\text{-}11) \qquad\qquad L = V\Delta t_S$$

(length of the rod in frame *S*)

Figure 25-11b shows the same process as observed in the frame of reference *S'* in which the rod is at rest. In this frame the rod has length L_{rest}, and the frame *S* moves to the left at speed *V*. The point $x = 0$ on frame *S* travels from one end of the rod to the other in a time $\Delta t_{S'}$. The length of the rod equals the speed *V* of frame *S* multiplied by the time $\Delta t_{S'}$ that frame *S* needs to travel this length:

$$(25\text{-}12) \qquad\qquad L_{rest} = V\Delta t_{S'}$$

(length of the rod in its own rest frame *S'*)

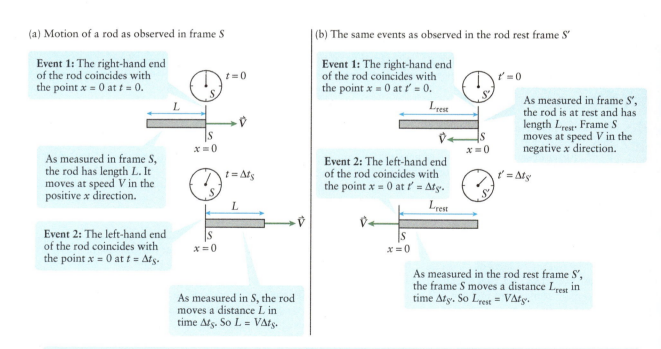

(a) Motion of a rod as observed in frame *S*

Event 1: The right-hand end of the rod coincides with the point $x = 0$ at $t = 0$.

$t = 0$

L

$\vec{V}$

S

$x = 0$

As measured in frame *S*, the rod has length *L*. It moves at speed *V* in the positive *x* direction.

Event 2: The left-hand end of the rod coincides with the point $x = 0$ at $t = \Delta t_S$.

$t = \Delta t_S$

L

$\vec{V}$

S

$x = 0$

As measured in *S*, the rod moves a distance *L* in time Δt_S. So $L = V\Delta t_S$.

(b) The same events as observed in the rod rest frame *S'*

Event 1: The right-hand end of the rod coincides with the point $x = 0$ at $t' = 0$.

$t' = 0$

L_{rest}

$\vec{V}$

S

$x = 0$

As measured in frame *S'*, the rod is at rest and has length L_{rest}. Frame *S* moves at speed *V* in the negative *x* direction.

Event 2: The left-hand end of the rod coincides with the point $x = 0$ at $t' = \Delta t_{S'}$.

$t' = \Delta t_{S'}$

L_{rest}

$\vec{V}$

S

$x = 0$

As measured in the rod rest frame *S'*, the frame *S* moves a distance L_{rest} in time $\Delta t_{S'}$. So $L_{rest} = V\Delta t_{S'}$.

- The two events are (1) the right-hand end of the rod coinciding with $x = 0$ and (2) the left-hand end of the rod coinciding with $x = 0$.
- These two events happen at the same place in frame *S*, so Δt_S is the proper time interval between these events. The time interval $\Delta t_{S'}$ must therefore be greater than Δt_S (time dilation).
- The distance L_{rest} must therefore be greater than the distance *L*. So the length *L* of the moving rod is less than the length L_{rest} of the rod in a frame where it is at rest. This is length contraction.

Figure 25-11 A moving rod and length contraction A rod as observed in (a) a frame in which the rod is moving along its length and (b) a frame in which the rod is at rest.

To see how the lengths as measured in the two frames compare, note that the length measurements in S and S' both involve the same pair of events: the right-hand end of the rod coinciding with the point $x = 0$ in frame S and the left-hand end of the rod coinciding with that same point (events 1 and 2 in Figure 25-11). These two events happen at the same location in frame S but at different locations in frame S'. So Δt_S is the proper time interval between these events. The time interval $\Delta t_{S'}$ measured in frame S' is therefore longer ("dilated") compared to the time interval Δt_S measured in frame S, and the two time intervals are related by Equation 25-10:

$$\Delta t_{S'} = \frac{\Delta t_S}{\sqrt{1 - \dfrac{V^2}{c^2}}} \tag{25-13}$$

Multiply both sides of Equation 25-13 by V, then replace $V\Delta t_S$ by L in accordance with Equation 25-11 and replace $V\Delta t_{S'}$ with L_{rest} according to Equation 25-12:

$$V\Delta t_{S'} = \frac{V\Delta t_S}{\sqrt{1 - \dfrac{V^2}{c^2}}} \quad \text{or} \quad L_{rest} = \frac{L}{\sqrt{1 - \dfrac{V^2}{c^2}}}$$

We can rewrite this as

> Length of an object in frame S', in which it is at rest

$$L = L_{rest}\sqrt{1 - \frac{V^2}{c^2}}$$

> Speed of S' relative to S

> Speed of light in a vacuum

> Length of the same object in frame S, in which the object is moving along its length

Length contraction (25-14)

If the rod is moving, V is greater than zero and the factor $\sqrt{1 - V^2/c^2}$ is less than 1. So the length L of the moving rod is less than the length L_{rest} of the rod at rest. This is called **length contraction**: *A moving object is shortened along the direction in which it is moving.*

WATCH OUT! Length contraction occurs only along the direction of motion.

There is *no* change in length of a moving object in any direction other than the direction of motion. For example, if the rod in Figure 25-11 is moving in the x direction relative to frame S, an observer in S will measure the rod as having a shorter length in the x direction than will an observer in S' at rest with respect to the rod. But both observers will agree about the height and width of the rod (its dimensions in the y and z directions).

Length contraction gives us an alternative way to understand the results of Example 25-3 in the preceding section, in which we used time dilation to explain how short-lived muons produced at an altitude of 15.0 km are able to survive their trip to Earth's surface. Imagine a vertical rod that extends 15.0 km upward from the ground to where the muons are produced. This rod is stationary relative to Earth, so its length as measured by you on the ground is its rest length: $L_{rest} = 15.0$ km $= 1.50 \times 10^4$ m. But as seen from the frame of a descending muon, this rod is moving upward along its length at $V = 0.994c$. As measured by a muon, the length of this rod is contracted in accordance with Equation 25-14:

$$L = L_{rest}\sqrt{1 - \frac{V^2}{c^2}} = (15.0 \text{ km})\sqrt{1 - \frac{(0.994c)^2}{c^2}}$$

$$= (15.0 \text{ km})\sqrt{1 - (0.994)^2}$$

$$= (15.0 \text{ km})\sqrt{0.0120} = (15.0 \text{ km})(0.109)$$

$$= 1.64 \text{ km} = 1.64 \times 10^3 \text{ m}$$

Moving at $V = 0.994c$, this contracted rod travels past the muon in a time

$$\frac{1.64 \times 10^3 \text{ m}}{0.994c} = \frac{1.64 \times 10^3 \text{ m}}{(0.994)(3.00 \times 10^8 \text{ m/s})} = 5.50 \times 10^{-6} \text{ s} = 5.50 \text{ }\mu\text{s}$$

The half-life of the muon in its own rest frame is 1.56 μs, so from the muon's perspective it takes just $(5.50 \text{ }\mu\text{s})/(1.56 \text{ }\mu\text{s}) = 3.53$ half-lives for the contracted rod to move past it—that is, for the muon to move from where it is produced to Earth's surface. That's exactly the result that we found in Example 25-3 using the ideas of time dilation. The ideas of length contraction and time dilation are mutually consistent!

EXAMPLE 25-4 A Flying Meter Stick I

A meter stick (length 1.00 m) hurtles through space at a speed of $0.800c$ relative to you, with its length aligned with the direction of motion. What do you measure as the length of the meter stick?

Set Up

The meter stick's length along its direction of motion is 1.00 m as measured in its own rest frame S', so $L_{rest} = 1.00$ m. Your frame is frame S, and the two frames are moving relative to each other at $V = 0.800c$. We'll use Equation 25-14 to find the length L as measured in your frame.

Length contraction:

$$L = L_{rest}\sqrt{1 - \frac{V^2}{c^2}} \qquad (25\text{-}14)$$

$L_{rest} = 1.00$ m

$V = 0.800c$ = velocity of the meter stick relative to frame S

Solve

Substitute $L_{rest} = 1.00$ m and $V = 0.800c$ into Equation 25-14 and calculate L.

From Equation 25-14:

$$L = L_{rest}\sqrt{1 - \frac{V^2}{c^2}} = (1.00 \text{ m})\sqrt{1 - \frac{(0.800c)^2}{c^2}}$$
$$= (1.00 \text{ m})\sqrt{1 - (0.800)^2} = (1.00 \text{ m})\sqrt{1 - 0.640}$$
$$= (1.00 \text{ m})\sqrt{0.360} = (1.00 \text{ m})(0.600)$$
$$= 0.600 \text{ m} = 60.0 \text{ cm}$$

Reflect

As measured in your frame of reference, the meter stick is only 60.0 cm in length. This is not an optical illusion; the meter stick really is only 60.0% as long in your frame of reference as in the rest frame of the meter stick.

EXAMPLE 25-5 A Flying Meter Stick II

The meter stick from the preceding example again hurtles through space at a speed of $0.800c$ relative to you, but now it is tilted at an angle of $30.0°$ as measured in its rest frame with respect to the direction of motion. Now what do you measure as the length of the meter stick?

Set Up

As measured in the meter stick's rest frame S', the meter stick is inclined at an angle $\theta' = 30.0°$ to its direction of motion relative to frame S. The x dimension of the stick undergoes length contraction given by Equation 25-14, but the y dimension does not. We'll calculate the x and y dimensions of the stick in your frame S, then use the Pythagorean theorem to find the length of the stick in frame S.

Length contraction:

$$L = L_{rest}\sqrt{1 - \frac{V^2}{c^2}} \qquad (25\text{-}14)$$

The meter stick as observed in its rest frame S'

$V = 0.800c$ = velocity of the meter stick relative to frame S

$L_{rest,x}$

$L_{rest} = 1.00$ m

$L_{rest,y}$

$\theta' = 30.0°$

Solve

First calculate the x and y dimensions of the meter stick in its rest frame.

In the rest frame S' of the stick, its dimensions are

$$L_{rest,x} = L_{rest}\cos \theta' = (1.00 \text{ m})\cos 30.0° = 0.866 \text{ m}$$
$$L_{rest,y} = L_{rest}\sin \theta' = (1.00 \text{ m})\sin 30.0° = 0.500 \text{ m}$$

Calculate the x and y dimensions of the meter stick in your frame, in which the stick is moving at $V = 0.800c$ along the x direction.

In your frame S, the x dimension of the stick is contracted:

$$L_x = L_{\text{rest},x}\sqrt{1 - \frac{V^2}{c^2}} = (0.866 \text{ m})\sqrt{1 - \frac{(0.800c)^2}{c^2}}$$

$$= (0.866 \text{ m})\sqrt{1 - 0.640} = (0.866 \text{ m})\sqrt{0.360}$$

$$= 0.520 \text{ m}$$

Length contraction occurs only along the direction of motion, so the y dimension of the stick is *not* contracted:

$$L_y = L_{\text{rest},y} = 0.500 \text{ m}$$

Use the Pythagorean theorem to calculate the length of the meter stick in your frame.

From the Pythagorean theorem, the length of the meter stick in your frame S is

$$L = \sqrt{L_x^2 + L_y^2}$$

$$= \sqrt{(0.520 \text{ m})^2 + (0.500 \text{ m})^2}$$

$$= \sqrt{0.520 \text{ m}^2} = 0.721 \text{ m}$$

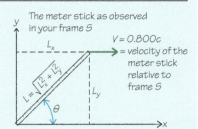

The meter stick as observed in your frame S

$V = 0.800c$ = velocity of the meter stick relative to frame S

Reflect

Because the meter stick is not aligned with the direction of motion, the amount of contraction is not as great as in Example 25-4, in which the meter stick was completely aligned with the direction of motion. Notice also that the angle θ of the meter stick in your frame is different from the angle $\theta' = 30.0°$ in the stick's rest frame. Can you show that $\theta = 43.9°$?

The Lorentz Transformation

The Galilean transformation that we presented in Section 25-2 is not consistent with the postulates of special relativity. A set of transformation equations that *is* consistent with relativity is the **Lorentz transformation**. As in Figure 25-3, we take frame S' to be moving at speed V in the positive x direction relative to frame S. The origins of the two frames coincide at $t = 0$ in frame S and $t' = 0$ in frame S'. If an event takes place at coordinates x, y, z, and t in frame S, the coordinates of that same event in frame S' are

Inertial frame of reference S' moves at speed V in the positive x direction relative to inertial frame of reference S.

$$x' = \frac{x - Vt}{\sqrt{1 - \dfrac{V^2}{c^2}}}$$

Coordinates of an event as measured in frame S'

$$y' = y$$
$$z' = z$$

Coordinates of the same event as measured in frame S

$$t' = \frac{t - \dfrac{V}{c^2} x}{\sqrt{1 - \dfrac{V^2}{c^2}}}$$

Lorentz transformation
(25-15)

Note that at speeds that are far less than the speed of light, the ratio V/c is very small and can be treated as essentially zero. Then $\sqrt{1 - V^2/c^2}$ is essentially equal to 1, and Equations 25-15 become

$$x' = x - Vt, \quad y' = y, \quad z' = z, \quad t' = t$$

These are just the Galilean transformation equations that we presented in Section 25-2. So at speeds far slower than the speed of light, the equations of Einstein's special theory of relativity reduce to the equations of Newtonian relativity.

Equations 25-15 are useful for finding the coordinates of an event in frame S' if we know the event's coordinates in frame S. If we instead want to determine the coordinates of an event in frame S from the coordinates in frame S', it's most convenient to use the *inverse* Lorentz transformation:

> Inertial frame of reference S' moves at speed V in the positive x direction relative to inertial frame of reference S.

**Inverse Lorentz transformation
(25-16)**

$$x = \frac{x' + Vt'}{\sqrt{1 - \dfrac{V^2}{c^2}}}$$

$$y = y'$$

$$z = z'$$

$$t = \frac{t' + \dfrac{V}{c^2}x'}{\sqrt{1 - \dfrac{V^2}{c^2}}}$$

> Coordinates of an event as measured in frame S

> Coordinates of the same event as measured in frame S'

It's possible to derive the equations for time dilation (Equation 25-10) and length contraction (Equation 25-14) from the Lorentz transformation equations. Instead, let's look at another remarkable result of the special theory of relativity that we can deduce from the Lorentz transformation equations.

EXAMPLE 25-6 Simultaneity Is Relative

In your reference frame you have a meter stick that is oriented along the x axis, with one end at $x = 0$ and the other end at $x = 1.00$ m. There is a light bulb at each end of the stick, and you make the two bulbs flash simultaneously (as measured by you) at $t = 0$. A spacecraft flies past you at $V = 0.800c$ in the positive x direction. (a) According to an observer in the spacecraft, are the two light flashes simultaneous? If not, which flash happens first as measured by her? (b) On board the spacecraft is an identical meter stick with a light bulb at each end. The observer on board the spacecraft places the two ends of the stick at $x' = 0$ and $x' = 1.00$ m, and she makes the two bulbs flash simultaneously (as measured by her) at $t' = 0$. According to you are the two light flashes simultaneous? If not, which flash happens first as measured by you?

Set Up

In part (a) we're given the coordinates in frame S of two events, the flash of the left-hand bulb at $x = 0$ and $t = 0$ and the flash of the right-hand bulb at $x = 1.00$ m and $t = 0$. We'll use the t' equation from the Lorentz transformation, Equations 25-15, to determine the times of these two events as measured in the spaceship frame S'. Similarly in part (b) we're given the coordinates in frame S' of two other events, the flash of the left-hand bulb at $x' = 0$ and $t' = 0$ and the flash of the right-hand bulb at $x' = 1.00$ m and $t' = 0$. We'll find the times of these events as measured in your S frame using the t equation from the inverse Lorentz transformation, Equations 25-16.

Time equation from the Lorentz transformation:

$$t' = \frac{t - \dfrac{V}{c^2}x}{\sqrt{1 - \dfrac{V^2}{c^2}}} \qquad (25\text{-}15)$$

Time equation from the inverse Lorentz transformation:

$$t = \frac{t' + \dfrac{V}{c^2}x'}{\sqrt{1 - \dfrac{V^2}{c^2}}} \qquad (25\text{-}16)$$

(a) Meter stick at rest in frame S:
What does an observer in frame S' measure?

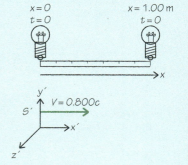

(b) Meter stick at rest in frame S':
What does an observer in frame S measure?

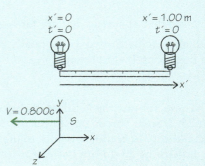

Solve

(a) Use the Lorentz transformation to calculate the times of the simultaneous flashes in S as measured in S'.

The left-hand bulb on the meter stick at rest in frame S flashes at $x = 0, t = 0$. In frame S' this bulb flashes at

$$t'_{left} = \frac{(0) - \dfrac{V}{c^2}(0)}{\sqrt{1 - \dfrac{V^2}{c^2}}} = 0$$

The right-hand bulb on the meter stick at rest in frame S flashes at $x = 1.00$ m, $t = 0$. In frame S' this bulb flashes at

$$t'_{right} = \frac{(0) - \dfrac{V}{c^2}(1.00 \text{ m})}{\sqrt{1 - \dfrac{V^2}{c^2}}} = \frac{-\dfrac{(0.800c)}{c^2}(1.00 \text{ m})}{\sqrt{1 - \dfrac{(0.800c)^2}{c^2}}}$$

$$= \frac{-0.800 \text{ m}}{c\sqrt{1 - (0.800)^2}} = \frac{-0.800 \text{ m}}{(3.00 \times 10^8 \text{ m/s})(0.600)}$$

$$= -4.44 \times 10^{-9} \text{ s}$$

As observed from the spaceship frame S', the two events are *not* simultaneous: The right-hand bulb flashes 4.44×10^{-9} s *before* the left-hand bulb.

(b) Use the inverse Lorentz transformation to calculate the times of the simultaneous flashes in S' as measured in S.

The left-hand bulb on the meter stick at rest in frame S' flashes at $x' = 0, t' = 0$. In frame S this bulb flashes at

$$t_{left} = \frac{(0) + \dfrac{V}{c^2}(0)}{\sqrt{1 - \dfrac{V^2}{c^2}}} = 0$$

The right-hand bulb on the meter stick at rest in frame S' flashes at $x' = 1.00$ m, $t' = 0$. In frame S this bulb flashes at

$$t_{right} = \frac{(0) + \dfrac{V}{c^2}(1.00 \text{ m})}{\sqrt{1 - \dfrac{V^2}{c^2}}} = \frac{+\dfrac{(0.800c)}{c^2}(1.00 \text{ m})}{\sqrt{1 - \dfrac{(0.800c)^2}{c^2}}}$$

$$= \frac{+0.800 \text{ m}}{c\sqrt{1 - (0.800)^2}} = \frac{+0.800 \text{ m}}{(3.00 \times 10^8 \text{ m/s})(0.600)}$$

$$= +4.44 \times 10^{-9} \text{ s}$$

As observed from your frame S, the two events are *not* simultaneous: The right-hand bulb flashes 4.44×10^{-9} s *after* the left-hand bulb.

Reflect

This example illustrates yet another counterintuitive consequence of Einstein's special theory of relativity: Two events that are simultaneous to one observer need not be simultaneous to another observer. So even the simple statement "Two things happen at the same time" has to be qualified by stating in which frame of reference it holds true. Time in relativity is not an absolute!

GOT THE CONCEPT? 25-5 No Contraction?

A rod with a rest length of 1.00 m whizzes past you at $0.995c$. Is it possible that you could measure its length to be equal to its rest length? (a) Yes; (b) no.

25-6 The speed of light is the ultimate speed limit

Suppose you are an outfielder running to catch a batted baseball (Figure 25-12a). The baseball is traveling at 30.0 m/s relative to the ground, and you are running toward the ball at 10.0 m/s relative to the ground. The Galilean velocity transformation that we learned in Section 25-2 says that, relative to you, the baseball travels at 30.0 m/s plus 10.0 m/s, or 40.0 m/s; the velocities simply add. But suppose instead that you are an astronaut flying in your spaceship at 1.00×10^8 m/s, as in Figure 25-12b, and you are moving toward a light beam aimed at you by a stationary astronaut. The same idea that we applied to the baseball predicts that relative to you the light beam travels at $c = 3.00 \times 10^8$ m/s (the speed of the light beam relative to the other astronaut) plus 1.00×10^8 m/s (the speed of your spaceship relative to the other astronaut), or 4.00×10^8 m/s. But that *cannot* be correct: The second postulate of the special theory of relativity says that the speed of light in a vacuum is the same to all inertial observers. So the light beam must also travel at speed $c = 3.00 \times 10^8$ m/s relative to you. Clearly the Galilean transformation for velocities is inadequate. In this section we'll explore the *Lorentz velocity transformation*, which allows us to combine velocities in a way that is consistent with Einsteinian relativity.

It's possible to derive the transformation of velocities from the Lorentz transformation of coordinates that we introduced in Section 25-5 (see Equations 25-15 and 25-16). We'll skip over the derivation and just present the result for the special case in which all motions are along the same line, as in Figure 25-12. Suppose that inertial frame S' is moving at speed V in the positive x direction relative to inertial frame S.

(a)

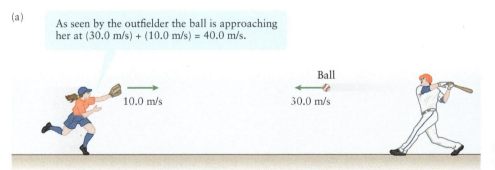

As seen by the outfielder the ball is approaching her at (30.0 m/s) + (10.0 m/s) = 40.0 m/s.

Ball

10.0 m/s   30.0 m/s

(b)

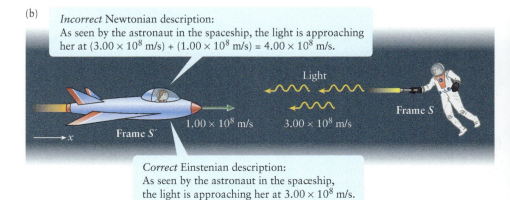

Incorrect Newtonian description:
As seen by the astronaut in the spaceship, the light is approaching her at $(3.00 \times 10^8$ m/s) + $(1.00 \times 10^8$ m/s) = 4.00×10^8 m/s.

Light

Frame S

1.00×10^8 m/s 3.00×10^8 m/s

Frame S'

x

Correct Einsteinian description:
As seen by the astronaut in the spaceship, the light is approaching her at 3.00×10^8 m/s.

Figure 25-12 Velocity addition — Newtonian and Einsteinian (a) An outfielder running toward a batted baseball. (b) An astronaut in her spaceship flying toward a light beam.

An object is moving relative to frame S along the x direction, with an x component of velocity v_x. In frame S' the same object has an x component of velocity given by

Inertial frame of reference S' moves at speed V in the positive x direction relative to inertial frame of reference S.

x component of velocity of an object moving along the x axis as measured in frame S'

$$v_x' = \frac{v_x - V}{1 - \dfrac{V}{c^2}v_x}$$

x component of velocity of the same object as measured in frame S

Lorentz velocity transformation
(25-17)

This is called the **Lorentz velocity transformation**. Notice that if the speed V of one frame relative to the other is small compared to c, the ratio V/c is much less than 1 and the denominator is essentially equal to 1. The same thing happens if the velocity v_x of the object relative to frame S is small compared to c. So if either of the speeds involved is a small fraction of the speed of light, Equation 25-17 reduces to $v_x' = v_x - V$, which is just the Galilean velocity transformation (the first of Equations 25-5). This justifies our use of the Galilean velocity transformation for slow-moving objects.

Equation 25-17 allows us to find the object's velocity relative to frame S' if we know its velocity relative to frame S. If instead we know the object's velocity relative to frame S' and want to calculate its velocity relative to frame S, we use the *inverse* Lorentz velocity transformation:

Inertial frame of reference S' moves at speed V in the positive x direction relative to inertial frame of reference S.

x component of velocity of an object moving along the x axis as measured in frame S

$$v_x = \frac{v_x' + V}{1 + \dfrac{V}{c^2}v_x'}$$

x component of velocity of the same object as measured in frame S'

Inverse Lorentz velocity transformation
(25-18)

Let's see what the Lorentz velocity transformation tells us about the situation shown in Figure 25-12b. The astronaut with the flashlight is in frame S, and you and your spaceship are in frame S'. You are moving to the right (in the positive x direction), which is just how frame S' must move relative to frame S in order to use Equation 25-17 or 25-18. The relative speed of the two frames is V. We know that the light travels relative to the astronaut with the flashlight at the speed of light in the negative x direction, so its velocity relative to S is $v_x = -c$. From Equation 25-17, the velocity v_x' of the light relative to you in frame S' is

$$v_x' = \frac{v_x - V}{1 - \dfrac{V}{c^2}v_x} = \frac{-c - V}{1 - \dfrac{V}{c^2}(-c)} = \frac{-(c+V)}{1 + \dfrac{V}{c}}$$

We can simplify this by factoring c out of the numerator:

$$v_x' = \frac{-c\left(1 + \dfrac{V}{c}\right)}{1 + \dfrac{V}{c}} = -c$$

This says that as measured by you in frame S', the light travels at the speed of light c in the negative x direction just as in frame S. Note that while $V = 1.00 \times 10^8$ m/s in Figure 25-12b, our result doesn't depend on the value of V: No matter what the relative speed of the two frames, each observer will see light propagating in a vacuum at the same speed c.

A direct consequence of this calculation is that *no object can move faster than c in any inertial frame of reference*. If an object (light) is traveling at speed c in one inertial frame, it is traveling at c in all inertial frames; if an object is traveling slower than c in one inertial frame, it is traveling slower than c in all inertial frames. Thus the speed of light in a vacuum represents an ultimate speed limit.

EXAMPLE 25-7 Baseball for Superheroes

Suppose the baseball game in Figure 25-12 is being played by superheroes. The outfielder has super speed and can run at $0.300c$ relative to the ground. The batter has super strength and can bat the ball with such force that the ball ends up traveling horizontally at $0.900c$ relative to the ground. (The bat and ball are made of super materials that can withstand the tremendous forces required.) How fast is the ball moving relative to the outfielder?

Set Up

This is nearly the same situation that we discussed above with the two astronauts and the beam of light, except that the astronauts have been replaced by (super) baseball players and the beam of light has become a baseball. We use Equation 25-17 to calculate the velocity of the ball relative to the outfielder in frame S'.

Lorentz velocity transformation:

$$v'_x = \frac{v_x - V}{1 - \dfrac{V}{c^2} v_x} \qquad (25\text{-}17)$$

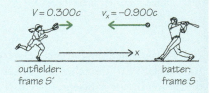

outfielder: frame S' batter: frame S

Solve

Use the velocity of the ball relative to the batter (v_x) and the velocity of the outfielder relative to the batter (V) to find the velocity of the ball relative to the outfielder (v'_x).

The outfielder (frame S') is moving in the positive x direction relative to the batter (frame S) at $V = 0.300c$. The velocity of the ball relative to the batter (frame S) is $v_x = -0.900c$ (negative because the ball is moving in the negative x direction). From Equation 25-17 the velocity of the ball relative to the outfielder is

$$v'_x = \frac{v_x - V}{1 - \dfrac{V}{c^2} v_x} = \frac{-0.900c - 0.300c}{1 - \dfrac{0.300c}{c^2}(-0.900c)}$$

$$= \frac{-1.200c}{1 + (0.300)(0.900)} = \frac{-1.200c}{1.27}$$

$$= -0.945c$$

Relative to the outfielder the ball is moving at $0.945c$ in the negative x direction (to the left).

Reflect

In the Galilean velocity transformation the velocity of the ball relative to the outfielder would have been $v'_x = v_x - V = -0.900c - 0.300c = -1.200c$, which is faster than c. Thanks to the $1 - (V/c^2)v_x$ term in the denominator of Equation 25-17, the actual speed of the ball relative to the outfielder is slower than c. The batter may be super-powered, but the ball cannot exceed the speed of light in a vacuum as measured by any observer.

GOT THE CONCEPT? 25-6 Returning a Super Baseball

 Suppose the super outfielder in Example 25-7 catches the ball and throws it back toward the super batter, who is still standing at home plate. If the outfielder is still running at $0.300c$ and she throws the ball at $0.700c$ relative to her, what is the speed of the ball relative to the batter? (a) $0.331c$; (b) $0.506c$; (c) $0.826c$; (d) $1.00c$; (e) $1.26c$.

TAKE-HOME MESSAGE FOR Section 25-6

✔ At speeds that are an appreciable fraction of the speed of light, we must use the Lorentz velocity transformation to calculate relative velocities. This transformation respects the rule that the speed of light in a vacuum is the same in all inertial frames of reference.

✔ No object can move faster than c, the speed of light in a vacuum, in any inertial frame of reference.

25-7 The equations for kinetic energy and momentum must be modified at very high speeds

We have seen that the speed of light in a vacuum c is an ultimate speed limit: No object can travel faster than c in any inertial frame of reference. As we'll see in this section, this tells us that the expressions we learned earlier in this book for kinetic energy K

(Chapter 6) and momentum $\vec{p}$ (Chapter 8) *cannot* be entirely correct. They fail at speeds that are a reasonable fraction of *c*. We'll see why this must be so and encounter a new kind of energy called *rest energy* that is intrinsic to any object with mass.

In Section 6-3 we introduced kinetic energy through the work-energy theorem:

> **Work done on an object** by the **net force** on that object

$$W_{net} = K_f - K_i$$

> **Kinetic energy** of the object **after** the work is done on it

> **Kinetic energy** of the object **before** the work is done on it

The work-energy theorem
(6-9)

This says that if an object starts at rest so that its initial kinetic energy K_i is zero, the more work the net force on an object does, the greater the final kinetic energy K_f of the object. In principle there is no limit to how much kinetic energy an object can acquire. There is a problem, however: Using Newtonian physics we found that the expression for the kinetic energy *K* of an object of mass *m* moving at speed *v* is

$$K = \frac{1}{2}mv^2$$

Since the speed of light in a vacuum *c* is the maximum speed an object can acquire, this expression says that the maximum kinetic energy that an object can acquire is $(1/2)mc^2$. This contradicts the idea that there should be no limit on an object's kinetic energy. Clearly we need an improved equation for kinetic energy.

There is a similar problem with the Newtonian expression for momentum. We saw in Section 8-6 that the change in momentum of an object is determined by the external forces on the object and the duration of the time interval over which the forces act:

> The **sum of all external forces** acting on an object

> **Duration** of a time interval over which the external forces act

$$\left(\sum \vec{F}_{external\ on\ object} \right) \Delta t = \vec{p}_f - \vec{p}_i = \Delta \vec{p}$$

> **Change in the momentum** $\vec{p}$ of the object during that time interval

External force and momentum
change for an object
(8-23)

Suppose an object starts at rest so that its initial momentum $\vec{p}_i$ is zero and a constant net force acts on it. Then Equation 8-23 says that the longer the time interval Δt that the force acts, the greater the final momentum $\vec{p}_f$ of the object. In principle there is no limit on how long the force can act, so there should be no upper limit on an object's momentum. However, this can't be reconciled with the Newtonian expression for the momentum of an object of mass *m* with velocity $\vec{v}$:

$$\vec{p} = m\vec{v}$$

According to this expression, the maximum magnitude of momentum that an object of mass *m* can have is $p = mc$, which would be attained only when an object is moving at the speed of light. This directly contradicts the notion that there should be no upper limit on momentum. Just as for kinetic energy, we need a new expression for momentum that's consistent with the special theory of relativity.

It's possible to derive the correct expressions for *K* and $\vec{p}$ by analyzing an elastic collision in which both mechanical energy and momentum are conserved (see Section 8-5). The derivation is beyond our scope, so we'll just look at the results. For a particle moving with velocity $\vec{v}$, both the expression for the kinetic energy and the expression for the momentum involve a dimensionless quantity called **relativistic gamma**:

> **Relativistic gamma** for a particle moving at speed *v*

$$\gamma = \frac{1}{\sqrt{1 - \dfrac{v^2}{c^2}}}$$

> Speed of the particle

> Speed of light in a vacuum

Relativistic gamma
(25-19)

Figure 25-13 Relativistic gamma
The quantity $\gamma = 1/\sqrt{1-(v^2/c^2)}$ increases dramatically toward infinity as speed v approaches the speed of light.

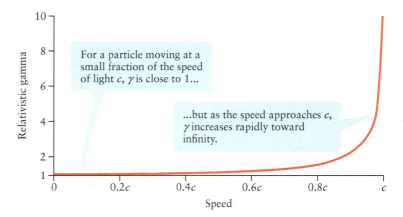

For a particle moving at a small fraction of the speed of light c, γ is close to 1...

...but as the speed approaches c, γ increases rapidly toward infinity.

Relativistic gamma is equal to 1 when $v = 0$ and becomes infinitely large as v approaches c (Figure 25-13). In terms of relativistic gamma, we can write the correct expressions for kinetic energy and momentum as

Kinetic energy of a particle of mass m and velocity $\vec{v}$

Einsteinian expressions for kinetic energy and momentum (25-20)

$$K = (\gamma - 1)mc^2$$

Relativistic gamma for speed v

$$\vec{p} = \gamma m\vec{v}$$

Momentum of a particle of mass m and velocity $\vec{v}$

 At speeds that are a small fraction of the speed of light c, these are approximately equal to the Newtonian expressions

$$K = \frac{1}{2}mv^2 \text{ and } \vec{p} = m\vec{v}$$

Figure 25-14 compares the kinetic energy K and the magnitude of momentum p from Equations 25-20 to the Newtonian expressions for these quantities. If v is a small

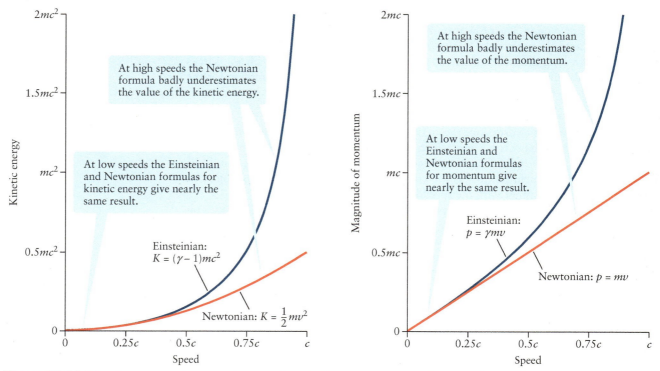

(a) Comparing Newtonian and Einsteinian formulas for kinetic energy

At high speeds the Newtonian formula badly underestimates the value of the kinetic energy.

At low speeds the Einsteinian and Newtonian formulas for kinetic energy give nearly the same result.

Einsteinian: $K = (\gamma - 1)mc^2$

Newtonian: $K = \frac{1}{2}mv^2$

(b) Comparing Newtonian and Einsteinian formulas for momentum

At high speeds the Newtonian formula badly underestimates the value of the momentum.

At low speeds the Einsteinian and Newtonian formulas for momentum give nearly the same result.

Einsteinian: $p = \gamma mv$

Newtonian: $p = mv$

Figure 25-14 Kinetic energy and momentum in relativity These graphs show how the Newtonian and Einsteinian expressions for (a) the kinetic energy of a particle and (b) the momentum of a particle depend on the speed of the particle.

fraction of the speed of light, the Einsteinian and Newtonian expressions give essentially identical results. This justifies our use of these expressions in earlier chapters, in which we considered only relatively slow-moving objects. But as the speed v approaches c, the Einsteinian expressions from Equations 25-20 approach infinity. So even though the speed of light is an upper limit to the speed of an object, there is *no* upper limit on the kinetic energy or momentum of an object.

EXAMPLE 25-8   The Energy and Momentum Costs of High Speed

Electrons can be accelerated to speeds very close to the speed of light. The mass of an electron is 9.11×10^{-31} kg.
(a) How much kinetic energy must be given to an electron to accelerate it from rest to $0.900c$? How much momentum?
(b) How much additional kinetic energy must be given to the electron to accelerate it from $0.900c$ to $0.990c$? How much additional momentum?

Set Up

We use Equations 25-20 to find the kinetic energy K and magnitude of momentum p for each speed. Equation 25-19 tells us the value of relativistic gamma for each speed.

Relativistic gamma:

$$\gamma = \frac{1}{\sqrt{1 - \dfrac{v^2}{c^2}}} \qquad (25\text{-}19)$$

Einsteinian kinetic energy and momentum:

$$K = (\gamma - 1)mc^2$$
$$\bar{p} = \gamma m \bar{v} \qquad (25\text{-}20)$$

Solve

(a) The kinetic energy that must be given to the electron is the difference between its kinetic energy at $v = 0.900c$ and its kinetic energy at $v = 0$, and similarly for the momentum.

At $v = 0$:

$$\gamma = \frac{1}{\sqrt{1 - \dfrac{v^2}{c^2}}} = \frac{1}{\sqrt{1 - 0}} = \frac{1}{1} = 1$$

$$K = (\gamma - 1)\, mc^2 = (1 - 1)\, mc^2 = 0$$
$$p = \gamma m v = (1)\, m\, (0) = 0$$

Just as in Newtonian physics, a particle at rest has zero kinetic energy and zero momentum.

To calculate K and p at nonzero speeds, it's useful to first find the values of mc and mc^2 for an electron:

$$\begin{aligned} mc &= (9.11 \times 10^{-31}\ \text{kg})(3.00 \times 10^8\ \text{m/s}) \\ &= 2.73 \times 10^{-22}\ \text{kg} \cdot \text{m/s} \end{aligned}$$

$$\begin{aligned} mc^2 &= (9.11 \times 10^{-31}\ \text{kg})(3.00 \times 10^8\ \text{m/s})^2 \\ &= 8.20 \times 10^{-14}\ \text{kg} \cdot \text{m}^2/\text{s}^2 \\ &= 8.20 \times 10^{-14}\ \text{J} \end{aligned}$$

At $v = 0.900c$:

$$\gamma = \frac{1}{\sqrt{1 - \dfrac{v^2}{c^2}}} = \frac{1}{\sqrt{1 - \dfrac{(0.900c)^2}{c^2}}} = \frac{1}{\sqrt{1 - (0.900)^2}}$$

$$= \frac{1}{\sqrt{0.190}} = 2.29$$

$$\begin{aligned} K &= (\gamma - 1)\, mc^2 = (2.29 - 1)(8.20 \times 10^{-14}\ \text{J}) \\ &= 1.06 \times 10^{-13}\ \text{J} \end{aligned}$$

$$\begin{aligned} p &= \gamma m v = (2.29)\, m\, (0.900c) = (2.29)(0.900)mc \\ &= (2.29)(0.900)(2.73 \times 10^{-22}\ \text{kg} \cdot \text{m/s}) \\ &= 5.64 \times 10^{-22}\ \text{kg} \cdot \text{m/s} \end{aligned}$$

The electron begins with $K = 0$ and $p = 0$, so it must be given 1.06×10^{-13} J of kinetic energy and 5.64×10^{-22} kg·m/s of momentum to accelerate it from rest to $0.900c$.

(b) Repeat the calculation in part (b) for the additional kinetic energy and momentum that must be given to the electron to accelerate it from $0.900c$ to $0.990c$.

At $v = 0.990c$:

$$\gamma = \frac{1}{\sqrt{1 - \dfrac{v^2}{c^2}}} = \frac{1}{\sqrt{1 - \dfrac{(0.990c)^2}{c^2}}} = \frac{1}{\sqrt{1 - (0.990)^2}}$$

$$= \frac{1}{\sqrt{0.0199}} = 7.09$$

$$K = (\gamma - 1)\, mc^2 = (7.09 - 1)(8.20 \times 10^{-14} \text{ J})$$
$$= 4.99 \times 10^{-13} \text{ J}$$

$$p = \gamma mv = (7.09)\, m\, (0.990c) = (7.09)(0.990)\, mc$$
$$= (7.09)(0.990)(2.73 \times 10^{-22} \text{ kg·m/s})$$
$$= 1.92 \times 10^{-21} \text{ kg·m/s}$$

The difference between these values and the values of K and p at $v = 0.900c$ from part (a) tells us the additional kinetic energy and momentum that must be given to the electron:

$$\Delta K = 4.99 \times 10^{-13} \text{ J} - 1.06 \times 10^{-13} \text{ J}$$
$$= 3.93 \times 10^{-13} \text{ J}$$

$$\Delta p = 1.92 \times 10^{-21} \text{ kg·m/s} - 5.64 \times 10^{-22} \text{ kg·m/s}$$
$$= 1.35 \times 10^{-21} \text{ kg·m/s}$$

Reflect

Compared to accelerating an electron from rest to $0.900c$, accelerating that same electron from $0.900c$ to $0.990c$ requires 3.70 times as much additional kinetic energy and 2.40 times as much additional momentum. As the speed gets closer and closer to c, it requires ever-greater amounts of kinetic energy and momentum to cause an ever-smaller speed increase. You can see that an object can *never* be accelerated from rest to the speed of light. This would require adding infinite amounts of kinetic energy and momentum to the object.

Bio Medical Note that there are objects within your body that are moving at a respectable fraction of the speed of light c, though not as fast as in this example (Figure 25-15). Every one of your red blood cells contains molecules of hemoglobin, and every hemoglobin molecule includes four iron atoms. The two innermost (and fastest-moving) electrons in an iron atom have an average speed of about $0.19c$; at this speed γ is about 1.02, and the Einsteinian kinetic energy of the electrons $K = (\gamma - 1)mc^2$ is about 3% greater than predicted by the Newtonian expression $K = \frac{1}{2}mv^2$.

Image Source/Alamy

Figure 25-15 A sample of Einsteinian particles This sample of a patient's blood includes particles moving at about 19% of the speed of light: the innermost electrons in the iron atoms found in hemoglobin.

We mentioned above that the Einsteinian expressions for kinetic energy and momentum, Equations 25-20, come from an analysis of elastic collisions. In particular, we must demand that if energy and momentum are conserved in one inertial frame of reference, they must be conserved in all inertial frames. That's required if energy and momentum are to be consistent with the first of the postulates of special relativity. It turns out that for this to be the case, we must also include a term mc^2 in the total energy of a particle of mass m. This quantity is called the **rest energy** of a particle, since it is present even when the particle is not in motion.

Rest energy
(25-21)

Rest energy of a particle   Mass of the particle

$$E_0 = mc^2$$

Speed of light in a vacuum

Equation 25-21 is one of the most famous in science, and one of the most misunderstood. Rest energy is not potential energy; potential energy is associated with a force an object experiences. (For example, gravitational potential energy is associated with the gravitational force.) It is not kinetic energy; kinetic energy is associated with motion. Rest energy is energy that is intrinsic to an object because of its mass.

Equation 25-21 tells us that mass is simply one possible manifestation of energy. Stated another way, the mass of an object is a measure of its rest energy content. Figure 25-16 is visual evidence of the equivalence of mass and energy. This image shows the result of a head-on collision between two protons, each of which was moving at just under the speed of light. Dozens of new particles appear after the collision. These are not fragments of the colliding protons but rather new particles that were created in the collision. This is possible because some of the kinetic energy of the colliding protons was converted into the rest energy of these new particles; another portion of this kinetic energy went into the kinetic energies of the new particles, which fly away from the collision site at high speeds.

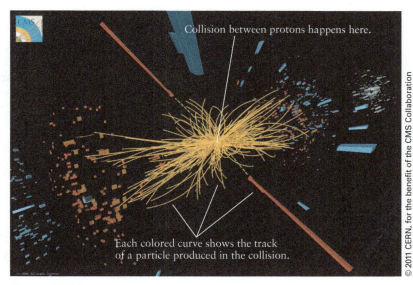

Collision between protons happens here.

Each colored curve shows the track of a particle produced in the collision.

© 2011 CERN, for the benefit of the CMS Collaboration

Figure 25-16 **Converting kinetic energy to new particles** This image from the Large Hadron Collider at CERN in Geneva, Switzerland, shows dozens of new particles produced by a collision between energetic protons.

You can observe the equivalence of mass and energy in action whenever you go outside on a sunny day or look up at the stars on a clear night. The Sun and stars shine thanks to a process occurring in their interiors in which hydrogen nuclei are fused together to form nuclei of helium. The mass of the products of such a reaction is slightly less than the mass of the hydrogen nuclei present before the reaction. The "lost" mass is converted to energy in the form of electromagnetic radiation, and it is that radiation that we see in the form of sunlight and starlight.

EXAMPLE 25-9   Converting Mass to Energy in the Sun

The Sun emits 3.84×10^{26} J of energy every second. (a) At what rate (in kg/s) is the Sun's mass decreasing? (b) How much mass has the Sun lost since it was formed 4.56×10^9 years ago? Assume that it has emitted energy at the same rate over its entire history. Compare to the present-day mass of the Sun, 1.99×10^{30} kg.

Set Up

The emitted energy comes from the rest energy of mass that is "lost" in nuclear reactions in the Sun's interior. We use Equation 25-21 to find the amount of mass equivalent to the energy emitted in one second. The total mass lost over the Sun's lifetime equals this amount of mass multiplied by the number of seconds that have elapsed since the Sun formed.

Rest energy:

$$E_0 = mc^2 \qquad\qquad (25\text{-}21)$$

Solve

(a) Calculate the amount of mass lost by the Sun per second.

Mass equivalent of 3.84×10^{26} J:

$$m = \frac{E_0}{c^2} = \frac{3.84 \times 10^{26} \text{ J}}{(3.00 \times 10^8 \text{ m/s})^2} = 4.27 \times 10^9 \text{ J} \cdot \text{s}^2/\text{m}^2$$

Since 1 J = 1 kg $\cdot$ m^2/s^2, we can write this as

$$m = 4.27 \times 10^9 \text{ kg}$$

The mass of the Sun decreases at a rate of 4.27×10^9 kg/s.

(b) Calculate the total amount of mass lost by the Sun in its history.

The age of the Sun in seconds is

$$(4.56 \times 10^9 \text{ y})\left(\frac{365.25 \text{ d}}{1 \text{ y}}\right)\left(\frac{24 \text{ h}}{1 \text{ d}}\right)\left(\frac{60 \text{ min}}{1 \text{ h}}\right)\left(\frac{60 \text{ s}}{1 \text{ min}}\right) = 1.44 \times 10^{17} \text{ s}$$

In this number of seconds the total mass lost by the Sun is

$$\left(4.27 \times 10^9 \frac{\text{kg}}{\text{s}}\right)(1.44 \times 10^{17} \text{ s}) = 6.14 \times 10^{26} \text{ kg}$$

As a fraction of the Sun's present-day mass, the amount of mass lost is $(6.14 \times 10^{26} \text{ kg})/(1.99 \times 10^{30} \text{ kg}) = 3.09 \times 10^{-4} = 0.0309\%$ of the total mass.

Reflect

In more than 4 billion years of producing energy at a prodigious rate, the Sun has lost only a tiny fraction of its total mass. This is a testament to how much energy can be released by converting even a small amount of mass.

GOT THE CONCEPT? 25-7 Can We Build a Starship?

? The total amount of electric energy produced per year in the United States from all sources is about 1.5×10^{19} J. You propose to use all of this energy to accelerate a spacecraft to $0.990c$ in order to travel to other stars. About how massive could your proposed starship be? (a) About 3×10^6 kg (the mass of an ocean liner); (b) about 3×10^5 kg (the mass of a large airliner); (c) about 3×10^3 kg (the mass of a sport utility vehicle); (d) about 30 kg (the mass of a kayak or canoe); (e) about 3 kg (the mass of a skateboard).

TAKE-HOME MESSAGE FOR Section 25-7

✔ The mathematical expressions for kinetic energy and momentum have to be modified to be consistent with special relativity. Both the kinetic energy and momentum of an object increase without limit as the object's speed approaches c.

✔ Any object with mass has a kind of energy called rest energy.

25-8 Einstein's general theory of relativity describes the fundamental nature of gravity

Sitting in a chair in the patent office in Bern, Switzerland, in 1907, a young Albert Einstein had what he would call "the happiest thought of my life." He imagined a man falling freely from the roof of a house and realized that "at least in his immediate surroundings—there exists no gravitational field." If the man released an object, for example, it would accelerate at the same rate as the man accelerated, and because the man would "not feel his own weight," it would appear to him that neither he nor the object was experiencing a gravitational force.

Einstein's thought experiment led him to postulate a new principle called the **principle of equivalence**. This principle states that

A gravitational field is equivalent to an accelerated frame of reference in the absence of gravity.

The principle of equivalence dictates that it is not possible to distinguish experimentally between a system in an accelerating frame and a system under the influence of gravity. In other words if you were to drop a ball in a windowless elevator car that makes no noise and doesn't shake, you could not tell from the motion of the ball whether the elevator car was sitting stationary on the surface of a planet and experiencing its gravity or accelerating in empty space, far from sources of gravity.

Let's explore physics in this imaginary elevator car further. In Figure 25-17a a ball is thrown horizontally while the elevator car is stationary near Earth's surface. Due to the force of gravity, the ball accelerates downward, following a familiar parabolic arc. The figure shows the position of the ball at five instants, spanning four equal time intervals. What if the stationary elevator car were far from Earth and from any other massive object that could exert a noticeable gravitational force on the ball? In that case the ball would travel along a straight line, as shown in Figure 25-17b.

Now consider the situation shown in Figure 25-18. The elevator car is again far from any object that could exert a noticeable gravitational force on the ball, but now the car is accelerating "upward" (toward the top of the page). Because there is no discernible gravitational force, the ball travels in a straight line, as in Figure 25-17b. However, because the elevator car is accelerating, an observer *in the car* sees the ball trace out a parabolic arc with respect to the walls and floor of the car. Remember that there are no windows in the elevator car, and it makes no noise and does not vibrate as it moves. An observer in the car cannot, therefore, detect its motion. The observer observes the effect of the principle of equivalence: It appears that the ball is falling under the influence of gravity.

Einstein's theory of gravitation is therefore a theory of accelerating frames of reference. Because this is more general than the case of inertial frames, the type that was at the heart of the special theory of relativity, this expanded theory is called the **general theory of relativity**.

Predictions of General Relativity

Einstein's general theory of relativity does more than give us a way to think about accelerating frames of reference. It also makes remarkable predictions about the behavior of light and the nature of time and space, predictions that have been verified by careful experiment and observation. Let's look at a few of these.

Gravity bends light. Imagine that in Figure 25-18 we replace the thrown ball with a beam of light fired horizontally. As seen by an outside observer in an inertial frame of reference, the light beam moves in a horizontal straight line toward the far wall. But because the elevator car is accelerating upward, an observer in the car will see the light beam follow a curved path and hit the far wall below the height from which it started. (The curvature of the beam's path will be very much less than that of the trajectory of the ball in Figure 25-18 because the light travels so much faster. Nevertheless, it *will* curve.)

According to the principle of equivalence, the effects of acceleration are indistinguishable from the effects of gravitation. So we conclude that a light beam fired horizontally on Earth will curve downward, just as the path of a ball thrown horizontally curves. This effect is called the **gravitational bending of light**.

(a)

In a stationary elevator car near Earth's surface...

...a ball thrown horizontally follows a parabolic path.

(b)

In a stationary elevator car far from Earth or any other object...

...a ball thrown horizontally travels in a straight line.

Figure 25-17 Elevator cars on Earth and in space A ball is thrown horizontally in (a) an elevator car on Earth's surface and (b) in an elevator car far from any massive object.

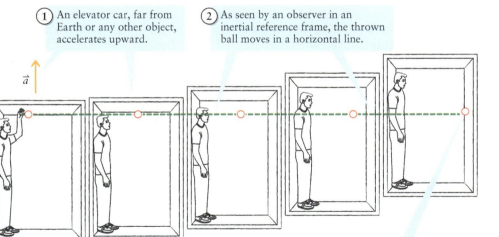

(1) An elevator car, far from Earth or any other object, accelerates upward.

(2) As seen by an observer in an inertial reference frame, the thrown ball moves in a horizontal line.

$\vec{a}$

(3) As seen by an observer in the accelerating elevator car, the ball follows a *curved path*: Where it hits the far wall is lower than where it left the thrower's hand.

The trajectory of the ball in the accelerated frame of reference of the elevator car is identical to that of a ball in a uniform gravitational field. This agrees with Einstein's principle of equivalence.

Figure 25-18 An accelerating elevator car in space When a ball is thrown horizontally in an elevator car, which is both far from any massive object and also accelerating, it follows a straight path. An observer in the elevator, however, would see the ball follow a parabolic path with respect to the floor and walls of the elevator car.

Figure 25-19 Gravity deflects light A massive object positioned between Earth and a distant star bends light from the star so that it can be seen on Earth. This is called gravitational lensing.

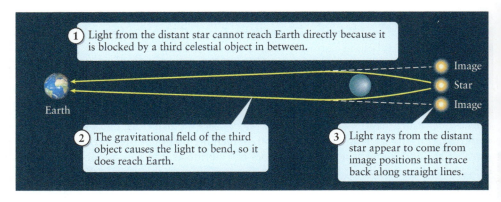

1 Light from the distant star cannot reach Earth directly because it is blocked by a third celestial object in between.

Image
Star
Image

Earth

2 The gravitational field of the third object causes the light to bend, so it does reach Earth.

3 Light rays from the distant star appear to come from image positions that trace back along straight lines.

Because the speed of light is so great, the gravitational bending of light predicted by Einstein is too small to measure on Earth. For example, a light beam traversing the width of a typical elevator would bend downward by less than 10^{-15} m, about the diameter of a proton. But the bending is measurable if a light beam is acted on by a much stronger gravitational field that acts over a much greater distance. The first measurement of gravitational bending of light was made in 1919 during a total solar eclipse. During totality, when the Moon blocked out the Sun's disk, astronomers photographed the stars around the Sun. These stars were shifted from their usual positions by 4.86×10^{-4} of a degree, consistent with Einstein's predictions.

A more stunning phenomenon associated with gravitational bending of light occurs when multiple images of a distant star form as light is bent by a closer, massive celestial object. Figure 25-19 depicts such gravitational lensing, in which a massive object is positioned directly between Earth and a distant star. Light from the star cannot reach Earth directly, but the lensing effect results in light that was initially not propagating toward Earth to be bent back toward us. In this way light from the star can approach Earth from many directions—for example, the two directions shown in the figure. Figure 25-20a presents an example of *gravitational lensing* showing four distinct images. When light is bent around the intervening object and reaches Earth from a full circle around it, the star's light is spread out into a ring around the lensing object, as in Figure 25-20b.

Space is curved, and gravitational waves propagate through space. A way to interpret the gravitational bending of light is that light in fact travels in a straight line in empty space but that *space itself is curved* by the presence of a massive object. Indeed, Einstein envisioned gravity as being caused by a curvature of space. In this picture, a massive object like Earth curves the space around it, and a light beam bends because it follows that curvature. Furthermore, an object like a ball falls toward Earth because it responds to the curvature of space that Earth produces. (A 0.1-kg ball and a 10-kg ball sense the same curvature of space due to Earth, which explains why gravity produces the same acceleration on objects of different mass.) Einstein's full mathematical formulation of the general theory of relativity is a set of equations that describe how the curvature of space and time, collectively referred to as spacetime, are affected by the presence of mass and energy.

Figure 25-20 Gravitational lenses Two examples of a massive, distant galaxy acting as a gravitational lens. (a) This gravitational lens makes four images of an even more distant supernova (an exploding star). (b) A more distant galaxy is located directly along our line of sight to the gravitational lens. The resulting image of the more distant galaxy is a nearly perfect circular ring.

(a)

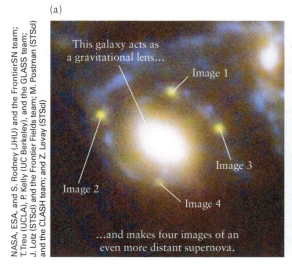

This galaxy acts as a gravitational lens...

Image 1

Image 2

Image 3

Image 4

...and makes four images of an even more distant supernova.

NASA, ESA, and S. Rodney (JHU) and the FrontierSN team; T. Treu (UCLA), P. Kelly (UC Berkeley), and the GLASS team; J. Lotz (STScI) and the Frontier Fields team; M. Postman (STScI) and the CLASH team; and Z. Levay (STScI)

(b)

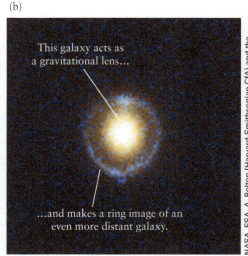

This galaxy acts as a gravitational lens...

...and makes a ring image of an even more distant galaxy.

NASA, ESA, A. Bolton (Harvard-Smithsonian CfA) and the

(a)

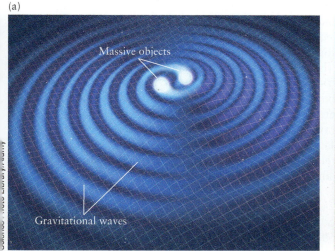

(b)

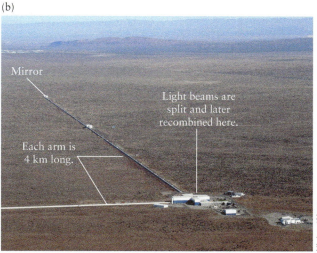

Figure 25-21 Gravitational waves (a) As two massive celestial objects orbit each other, they produce ripples in spacetime called gravitational waves. These gravitational waves carry away energy, causing the objects to spiral inward and emit a strong burst of gravitational waves when they finally collide and merge. (b) This LIGO gravitational wave detector is near Hanford, Washington. (The other is in Livingston, Louisiana). A laser beam is split into two, with each individual beam then sent along one of the 4-km arms and reflected back by a mirror at the end of the arm. The passage of a gravitational wave through the detector changes the lengths of each of the arms, which causes a change in the interference pattern formed when the two light beams are recombined.

How can we test this idea? The general theory of relativity predicts not only that spacetime curves in response to the presence of massive objects, but also that ripples in spacetime called **gravitational waves**—that is, small variations in the curvature of spacetime—should spread away from massive objects that are oscillating in a certain way (Figure 25-21a). (Newton's theory of gravitation makes no such prediction.) Such gravitational waves were detected for the first time in 2015 by the Laser Interferometer Gravitational-Wave Observatory (LIGO), a set of two detectors located 3000 km apart in the U.S. states of Washington and Louisiana (Figure 25-21b).

The operating principle of each LIGO detector is the same as that of the Michelson–Morley apparatus shown schematically in Figure 25-7. A light beam is split in two, sent along two perpendicular arms of the detector, reflected from mirrors at the end of each arm (in the case of the LIGO detectors, 4 km away), and then recombined. If a gravitational wave from space passes through a LIGO detector, the change in the curvature of spacetime will slightly change the length of each arm of the detector and hence the distance that each light beam travels, and this changes the interference pattern formed when the beams recombine. Local disturbances such as an earthquake could also trigger a change in these distances, which would produce a false signal. That's why LIGO has two detectors located 3000 km apart: Only a disturbance coming from space will produce the same signal in both detectors.

As of this writing, LIGO has made 50 confirmed detections of gravitational wave events, most of them caused by a pair of objects many times more massive than the Sun colliding with each other. (These objects are thought to be black holes. In a black hole the material has been so compressed that the gravitational field near the black hole is so strong that nothing, not even light, can escape. *Black holes* are another prediction of general relativity.) Although such a cataclysm emits an immense amount of gravitational wave energy (equivalent to converting several times the mass of the Sun completely into energy), the signal detected by LIGO is miniscule: In each detection each 4-km arm of the detector changed in length by only about 10^{-18} m, about 10^{-3} as large as the radius of a proton. The 2017 Nobel Prize in Physics was awarded to Rainer Weiss, Barry Barish, and Kip Thorne for their work on LIGO.

Gravity affects time. We saw in Sections 25-4 and 25-5 that in the special theory of relativity, both time intervals and distances are affected by motion. Similarly, in the general theory of relativity, a massive object such as Earth affects time as well as curving space. Einstein predicted that clocks on the ground floor of a building should tick slightly more slowly than clocks on the top floor, which are farther from Earth. This **gravitational slowing of time** has been measured even for differences in height as small as 33 cm (1 ft): An experiment in 2010 using extremely precise clocks showed that the lower clock would fall behind by about 9×10^{-8} s over a 79-year human lifetime, in agreement with Einstein's prediction.

An important application of the gravitational slowing of time is to the Global Positioning System, or GPS (Figure 25-1a). A GPS receiver uses signals from orbiting satellites, each of which carries extremely accurate clocks, to triangulate its position on Earth. The effects of gravity slow down time on Earth's surface compared to the satellites' clocks, so the general theory of relativity *must* be taken into account for an accurate GPS result. If the gravitational slowing of time were not taken into account, a GPS receiver would accumulate errors of more than 10 km per day!

The general theory of relativity has never made an incorrect prediction. It now stands as our most accurate and complete description of gravity.

GOT THE CONCEPT? 25-8 Time Passages

 You use your spaceship to take a clock from Earth's surface to the Moon's. Which of these will affect how much time elapses on this clock compared to how much time elapses on an identical clock that you left on Earth? (a) Earth's gravitation; (b) the Moon's gravitation; (c) the speed of your spaceship; (d) both (a) and (b), but not (c); (e) all of (a), (b), and (c).

TAKE-HOME MESSAGE FOR Section 25-8

✔   The principle of equivalence, a central postulate of Einstein's general theory of relativity, states that a gravitational field is equivalent to an accelerated frame of reference in the absence of gravity.

✔   It is not possible to distinguish between a system in an accelerating frame and a system under the influence of gravity.

✔   General relativity predicts the gravitational bending of light, the existence of gravitational waves, and the gravitational slowing of time. All of these have been observed.

Key Terms

event
frame of reference (reference frame)
Galilean transformation
Galilean velocity transformation
general theory of relativity
gravitational bending of light
gravitational slowing of time

gravitational waves
inertial frame
length contraction
Lorentz transformation
Lorentz velocity transformation
Michelson–Morley experiment
noninertial frame
principle of equivalence

principle of Newtonian relativity
proper time
reference frame (frame of reference)
relativistic gamma
relativistic speed
rest energy
special theory of relativity
time dilation

Chapter Summary

Topic	Equation or Figure	
Newtonian relativity: The principle of Newtonian relativity states that the laws of motion are the same in all inertial frames of reference. The Galilean transformation relates the coordinates of an event in one frame to the coordinates of the same event in another frame, and the Galilean velocity transformation relates the velocity of an object relative to one frame to its velocity relative to another frame. These transformations are valid only for speeds that are small compared to the speed of light c.	Inertial frame of reference S' moves at speed V in the positive x direction relative to inertial frame of reference S. Coordinates of an event as measured in frame S' → $\begin{aligned} x' &= x - Vt \\ y' &= y \\ z' &= z \\ t' &= t \end{aligned}$ ← Coordinates of the same event as measured in frame S	(25-1)
	Inertial frame of reference S' moves at speed V in the positive x direction relative to inertial frame of reference S. Velocity components of an object as measured in frame S' → $\begin{aligned} v'_x &= v_x - V \\ v'_y &= v_y \\ v'_z &= v_z \end{aligned}$ ← Velocity components of the same object as measured in frame S	(25-5)

The Michelson–Morley experiment:
In the nineteenth century it was hypothesized that space was filled with a material medium, called the luminiferous ether, that was required for light to propagate through space. The Michelson–Morley experiment provided evidence that the ether does not exist and motivated Einstein's special theory of relativity.

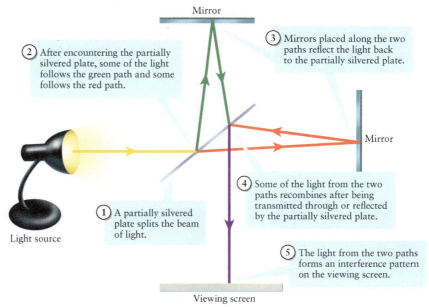

(Figure 25-7)

The special theory of relativity:
Einstein's special theory of relativity is based on the postulates that all laws of physics are the same in all inertial frames, and the speed of light in a vacuum is the same in all inertial frames. Two consequences are that moving clocks run slow, and moving objects are shortened along their direction of motion.

Time interval between two events as measured in frame S', in which those events occur at the same place

Time interval between the same two events as measured in frame S

Speed of S' relative to S

Speed of light in a vacuum

$$\Delta t = \frac{\Delta t_{\text{proper}}}{\sqrt{1 - \dfrac{V^2}{c^2}}}$$

(25-10)

Length of an object in frame S', in which it is at rest

Length of the same object in frame S, in which the object is moving along its length

Speed of S' relative to S

Speed of light in a vacuum

$$L = L_{\text{rest}} \sqrt{1 - \frac{V^2}{c^2}}$$

(25-14)

The Lorentz transformation:
The Lorentz transformation is a generalization of the Galilean transformation that is valid for all speeds. The Lorentz velocity transformation is a generalization of the Galilean velocity transformation: It shows that the speed of light c is an ultimate speed limit and that an object at rest can never be accelerated to c.

Inertial frame of reference S' moves at speed V in the positive x direction relative to inertial frame of reference S.

Coordinates of an event as measured in frame S'

Coordinates of the same event as measured in frame S

$$x' = \frac{x - Vt}{\sqrt{1 - \dfrac{V^2}{c^2}}}$$

$$y' = y$$
$$z' = z$$

$$t' = \frac{t - \dfrac{V}{c^2} x}{\sqrt{1 - \dfrac{V^2}{c^2}}}$$

(25-15)

Inertial frame of reference S' moves at speed V in the positive x direction relative to inertial frame of reference S.

x component of velocity of an object moving along the x axis as measured in frame S'

x component of velocity of the same object as measured in frame S

$$v_x' = \frac{v_x - V}{1 - \dfrac{V}{c^2} v_x}$$

(25-17)

Kinetic energy, momentum, and rest energy: The formulas for kinetic energy and momentum must be modified to account for c being the ultimate speed limit. The relativity postulates also show that an object with mass m has a rest energy $E_0 = mc^2$ even when it is not in motion.

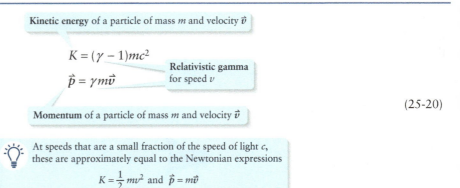

Kinetic energy of a particle of mass m and velocity $\vec{v}$

$$K = (\gamma - 1)mc^2$$

Relativistic gamma for speed v

$$\vec{p} = \gamma m \vec{v}$$

Momentum of a particle of mass m and velocity $\vec{v}$

(25-20)

At speeds that are a small fraction of the speed of light c, these are approximately equal to the Newtonian expressions

$$K = \tfrac{1}{2}mv^2 \text{ and } \vec{p} = m\vec{v}$$

The general theory of relativity: The principle of equivalence states that a gravitational field is equivalent to an accelerated frame of reference in the absence of gravity. This principle shows that light is affected by gravity just as objects with mass are.

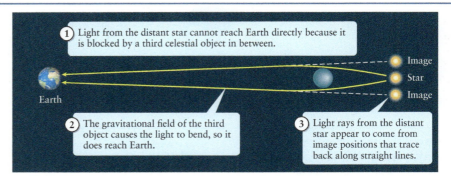

1. Light from the distant star cannot reach Earth directly because it is blocked by a third celestial object in between.

Image
Star
Image

Earth

2. The gravitational field of the third object causes the light to bend, so it does reach Earth.

3. Light rays from the distant star appear to come from image positions that trace back along straight lines.

(Figure 25-19)

Answer to What do you think? Question

(b) As an object approaches the speed of light, it becomes increasingly difficult to further increase its speed. See Example 25-8 in Section 25-7.

Answers to Got the Concept? Questions

25-1 (b) Let S be the frame of reference of an observer on the ground and let S' be the frame of reference of an observer moving with the air (for example, an observer floating in a balloon). Then S' is moving due east at $V = 100$ km/h, so we take the positive x direction to be to the east. First suppose the airliner is flying west to east, so its velocity relative to the air is $v'_x = +900$ km/h. From the first of Equations 25-5, the velocity of the airplane relative to the ground in this situation is $v_x = v'_x + V = +900$ km/h $+ 100$ km/h $= +1000$ km/h; relative to the ground the airliner is moving east at 1000 km/h. If instead the airliner is flying east to west, its velocity relative to the air is $v'_x = -900$ km/h and its velocity relative to the ground is $v_x = v'_x + V = -900$ km/h $+ 100$ km/h $= -800$ km/h; the airliner is moving west at 800 km/h relative to the ground. In North America high-altitude winds typically blow from west to east, as in this example, so a west-to-east airliner trip is typically faster than an east-to-west trip between the same two airports.

25-2 (b) This is similar to Got the Concept? Question 25-1. Let S be the frame of reference of the water and let S' be the frame of reference of the dolphin. The dolphin swims in the positive x direction at speed $V = 10$ m/s. As measured in frame S, the sound wave moves in the positive x direction at velocity $v_x = 1500$ m/s. From the first of Equations 25-5, the velocity of the sound wave relative to S', the frame of reference of

the dolphin, is $v'_x = v_x - V = 1500$ m/s $- 10$ m/s $= 1490$ m/s. Because the dolphin is "chasing" the sound wave, the speed of the wave relative to the dolphin is less than its speed relative to the water. As we will see in Section 25-6, light waves do not behave like this: If you shoot a laser beam forward from a fast-moving vehicle, the speed of the light wave relative to the vehicle is $c = 3.00 \times 10^8$ m/s no matter how fast the vehicle is moving.

25-3 (e) The proper time is measured in any frame that is at rest with respect to the clock. Albert aboard Starship *Alpha* is at rest with respect to the clock, so he measures the clock's proper time. Although Barbara is not on the same craft as the clock, she is also not moving relative to it, so she also measures the clock's proper time. George's speed is the same as the clock's but not in the same direction. So George is moving with respect to the clock and therefore does not observe proper time intervals. The same is true for Elena on Earth, who is moving relative to the clock. We conclude that the time interval between clock ticks is the proper time for both Albert and Barbara, but it is not the proper time for either George or Elena.

25-4 (b) The statement of time dilation is that moving clocks run slow. Albert's clock is moving relative to Elena, and Elena's clock is moving at the same speed relative to Albert. So *both* observers will see the other person's clock running slow.

There is no contradiction, since the two observers are looking at different pairs of events: Elena is looking at the time interval between two clicks of Albert's clock, while Albert is looking at the time interval between two clicks of Elena's clock.

25-5 (a) The length you measure will be the same as the rest length if the rod is oriented so that its length is perpendicular to the direction of motion. There is no length contraction along a perpendicular direction. (The *width* of the rod will be contracted along the direction of motion, but that's not what the question is about.)

25-6 (c) The outfielder's frame is S', so the velocity of the ball relative to her is $v'_x = +0.700c$ (positive, since the ball is now moving in the positive x direction). She is still moving at $V = 0.300c$ relative to the batter. The velocity of the ball relative to the batter is v_x, given by the inverse Lorentz velocity transformation (Equation 25-18):

$$v_x = \frac{v'_x + V}{1 + \frac{V}{c^2}v'_x} = \frac{0.700c + 0.300c}{1 + \frac{(0.300c)}{c^2}(0.700c)}$$

$$= \frac{1.00c}{1 + 0.210} = 0.826c$$

25-7 (d) The kinetic energy of the spacecraft will be $K = (\gamma - 1)mc^2$. From Example 25-8, $\gamma = 7.09$ if $v = 0.990c$, so $K = (7.09 - 1)mc^2 = 6.09mc^2$. If all of the 1.5×10^{19} J goes into the kinetic energy K of the starship, the mass that can be accelerated to $0.990c$ will be

$$m = \frac{K}{6.09c^2} = \frac{1.5 \times 10^{19} \text{ J}}{(6.09)(3.00 \times 10^8 \text{ m/s})^2}$$

$$= 27 \text{ J·s}^2/\text{m}^2 = 27 \text{ kg}$$

That's about the mass of a kayak or canoe, not including any occupants. This calculation shows that it's simply not feasible with current technology to build a spacecraft that can carry humans and travel at speeds approaching the speed of light.

25-8 (e) To account for everything that could cause a difference in time recorded by the two clocks, you would have to include the different gravitational slowing of time on Earth and the Moon (which has weaker gravity than Earth). You would also have to include the effects of time dilation due to the high-speed travel from Earth to Moon.

Questions and Problems

In a few problems you are given more data than you actually need; in a few other problems you are required to supply data from your general knowledge, outside sources, or informed estimate.

Interpret as significant all digits in numerical values that have trailing zeros and no decimal points.

For all problems use $g = 9.80$ m/s^2 for the free-fall acceleration due to gravity. Neglect friction and air resistance unless instructed to do otherwise.

* Basic, single-concept problem
** Intermediate-level problem; may require synthesis of concepts and multiple steps
*** Challenging problem
Example See worked example for a similar problem

Conceptual Questions

1. • How would you change the following Galilean transformation equations in the case where a frame was moving in both the x and the y directions?

$$x' = x - Vt$$
$$y' = y$$
$$z' = z$$
$$t' = t$$

2. • What is a frame of reference? What is an inertial frame of reference?

3. • What kind of reference frame is Earth's surface? Explain your answer.

4. • What does the statement "All motion is relative, there is no absolute motion" mean?

5. • Describe the Michelson and Morley experiment. Why do you think it was repeated so many times?

6. • What are the two postulates of the special theory of relativity?

7. • Explain how the measurement of time enters into the determination of the length of an object.

8. • Is it possible for one observer to find that event A happens after event B and another observer to find that event A happens before event B? Explain your answer.

9. • Is it possible to accelerate an object to the speed of light in a real situation? Explain your answer.

10. • Gene Roddenberry created the popular TV and film series *Star Trek*. Once at a public lecture he was asked if he thought we would ever be able to travel faster than light. He responded, "No, and that's a good thing because it means that we can always do it in science fiction." How have science fiction writers gotten around the cosmic speed limit?

11. • What is the fundamental postulate of the general theory of relativity?

12. • Describe one of the predictions of general relativity.

13. • Why do you think there was so much resistance from the established physics community when Einstein proposed his new theory of relativistic motion in 1905?

Multiple-Choice Questions

14. • Which of the following statements is/are true?
   A. The laws of motion are the same in all inertial frames of reference.
   B. The laws of motion are the same in all reference frames.
   C. There is no way to detect absolute motion.
   D. All of the above statements are true.
   E. Only two of the above statements are true.

15. • If we used radio waves to communicate with an alien spaceship approaching Earth at 10% of the speed of light, we would receive their signals at a speed of
   A. 0.10c.
   B. 0.90c.
   C. 0.99c.
   D. 1.00c.
   E. 1.10c.

16. • Time dilation means that
   A. the slowing of time in a moving frame of reference is only an illusion resulting from motion.
   B. time really does pass more slowly in a frame of reference moving relative to a frame of reference at relative rest.
   C. time really does pass more slowly in a frame of reference at rest relative to a frame of reference that is moving.
   D. time is unchanging regardless of the frame of reference.
   E. no two clocks at rest can ever read the same time.

17. • A meter stick hurtles through space at a speed of $0.95c$ with its length perpendicular to the direction of motion. You measure its length to be equal to
   A. 0 m.
   B. 0.05 m.
   C. 0.95 m.
   D. 1.00 m.
   E. 1.05 m.

18. • A particle has an Einsteinian momentum of p. If its speed doubles, the Einsteinian momentum will be
   A. greater than $2p$.
   B. equal to $2p$.
   C. less than $2p$, but not equal to $2p/c$.
   D. equal to $2p/c$.
   E. equal to p.

19. • Consider two atomic clocks, one at the GPS ground control station near Colorado Springs (elevation 1830 m) and the other one in orbit in a GPS satellite (altitude 20,200 km). According to the general theory of relativity, which atomic clock runs slow?
   A. The clock in Colorado runs slow.
   B. The clock in orbit runs slow.
   C. The clocks keep identical time.
   D. The orbiting clock is 95% slower than the clock in Colorado.
   E. The orbiting clock alternately runs slow and then fast depending on where the Sun is.

20. •• Aamir, Bob, and Cesar are all the same age when they start an interstellar shipping business to a colony several light years from Earth. Aamir takes the first round trip alone while the others stay home on Earth. On the next trip he takes Cesar with him while Bob stays on Earth. Bob and Cesar take the third trip together while Aamir takes some well-deserved rest on Earth. Assuming each trip is identical, what is the order of their ages after the third trip?
   A. Aamir > Bob > Cesar
   B. Aamir > Bob = Cesar
   C. Cesar = Aamir > Bob
   D. Bob > Aamir = Cesar
   E. Bob = Cesar > Aamir

21. • A massive particle speeds up and slows down as it travels along at an appreciable fraction of the speed of light. The relativistic kinetic energy
   A. will always be less than the classical kinetic energy.
   B. will always be the same as the classical kinetic energy.
   C. will always be greater than the classical kinetic energy.
   D. cannot be compared with the classical kinetic energy without more information.
   E. will always be equal to the rest energy.

Problems

25-1 The concepts of relativity may seem exotic, but they're part of everyday life

25-2 Newton's mechanics includes some ideas of relativity

22. • Relative to the ground, a bicyclist rides at 8.00 m/s toward the north. A car is moving at 25.0 m/s, also toward the north, and is initially behind the rider. A truck is moving at 15.0 m/s toward the south, approaching the bicycle and the car. (a) Make a sketch of the three vehicles and label their respective velocity vectors, relative to the ground. (b) Calculate the relative velocities of each vehicle compared to the other two. Example 25-1

23. • A frame of reference, S, is fixed on the surface of Earth with the x axis pointing toward the east, the y axis pointing toward the north, and the z axis pointing up. A second frame of reference, S', is moving at a constant 4.00 m/s toward the east. At time $t = t' = 0$ the origins of both frames of reference coincide. (a) Describe mathematically the relationships between x' and x, y' and y, and z' and z. (b) A picture is taken in the S frame of reference at $t = 4.00$ s at the point (2 m, 1 m, 0 m). Calculate the corresponding values of x', y', and z' for the same event in the S' frame. Example 25-1

24. • A boat sails from the pier at Fisherman's Wharf in San Francisco at 4.00 m/s, directly toward Alcatraz. A kite rider heads directly away from Alcatraz toward the boat at a relative speed of 6.00 m/s, according to the skipper of the sailboat. Calculate the velocity of the kite rider relative to the pier. Example 25-1

25. •• A radio-controlled model car travels at 15.0 m/s, to the right, relative to the parking lot that it is driving on. A girl on her scooter chases after the model car at 4.00 m/s. The parking lot represents reference frame S, the model car is in frame S', and the girl is described by frame S''. Assume that the car and the girl are both at the origin of the parking lot at $t = 0$ s. Write the Galilean transformation equations between (a) S and S', (b) S and S'', and (c) S' and S''. Example 25-1

26. •• Consider an airplane traveling at 50.0 m/s airspeed between two points that are 200 km apart. What is the round-trip time for the plane (a) if there is no wind? (b) if there is a wind blowing at 10.0 m/s along the line joining the two points? Example 25-1

27. • Assume that the origins of S and S' coincide at $t = t' = 0$ s. An observer in inertial frame S measures the space and time coordinates of an event to be $x = 750$ m, $y = 250$ m, $z = 250$ m, and $t = 2.0$ μs. What are the space coordinates of the event in inertial frame S', which is moving in the $+x$ direction at a speed of $0.01c$ relative to S? Example 25-1

28. • At time $t' = 4.00 \times 10^{-3}$ s, as measured in S', a particle is at the point $x' = 10$ m, $y' = 4$ m, and $z' = 6$ m. Compute the corresponding values of x, y, and z, as measured in S, for (a) a relative velocity between S' and S of $+500$ m/s and (b) a relative velocity between S' and S of -500 m/s. Assume that the origins of S and S' coincide at $t = 0$ and that the motion lies along the x and x' axes. Example 25-1

29. • Suppose that at $t = 6.00 \times 10^{-4}$ s, the space coordinates of a particle are $x = 100$ m, $y = 10$ m, and $z = 30$ m according to coordinate system S. Compute the corresponding values as measured in the frame S' if the relative velocity between S' and S is 150,000 m/s along the x and x' axes. The origins of the reference frames coincide at $t = 0$. Example 25-1

30. ••• A floatplane lands on a river. The velocity of the plane relative to the air is 30 m/s, due east; the velocity of the wind is 20 m/s, due north; and the current in the river is 5 m/s, due south. Calculate the velocity of the plane relative to the water just before it lands. Example 25-1

31. • A spaceship moves by Earth at 2.4×10^8 m/s. A satellite moves by Earth in the opposite direction at 1.6×10^8 m/s. Use the Galilean velocity transformation to calculate the speed of the satellite relative to the spaceship and comment on your answer. Example 25-1

25-3 The Michelson–Morley experiment shows that light does not obey Newtonian relativity

32. • The Michelson–Morley experiment was performed hundreds of times in a futile attempt to find the luminiferous ether. Why was the experiment performed on an enormous slab of marble that was floated in a pool of mercury?

33. •• Suppose that the distance in the Michelson–Morley experiment (Figure 25-7) from the partially silvered plate to either mirror is 20,000 m. (a) How much time would light, moving at speed c, need to travel from the partially silvered plate to one mirror and back again? How much *additional* time would be required *if the ether existed*, if the Galilean velocity transformation were valid, and if Earth moved relative to the ether along the direction from the partially silvered plate to this mirror at the following speeds: (b) at $0.01c$, (c) at $0.1c$, (d) at $0.5c$, and (e) at $0.9c$? Example 25-1

25-4 Einstein's relativity predicts that the time between events depends on the observer

34. • An observer in reference frame S observes that a lightning bolt strikes the origin, and 10^{-4} s later a second lightning bolt strikes the same location. What is the time separation between the two lightning bolts determined by a second observer in reference frame S' moving at a speed of $0.8c$ along the collinear x–x' axis? Example 25-2

35. • A radioactive particle travels at $0.80c$ relative to the laboratory observers who are performing research. Calculate the ratio of the half-life of the particle as measured in the laboratory frame compared to the half-life according to the proper frame of the particle. Example 25-3

36. • The elementary particle called a muon is unstable and decays in about 2.20 s, as observed in its rest frame, into an electron, a neutrino, and an antineutrino. What lifetime do you observe for muons approaching you at 0.900 times the speed of light? Example 25-3

37. • When certain unstable subatomic particles traveled through the laboratory at a speed of $0.98c$, scientists measured an average half-life of 11 μs for their decay. If these particles were at rest in the laboratory, what would be their half-life? Example 25-3

38. •• The time dilation effect between two frames of reference is measured in the lab to have a 0.01% difference between the relative time and the proper time $[\Delta t = (t - t_{proper})/t_{proper} \times 100\% = 0.01\%]$. Calculate the relative speed of the two reference frames. Example 25-2

39. •• **Astronomy** The Andromeda galaxy is a spiral galaxy that is a distance of 2.54 million light years from Earth. Is there any possible speed that a spaceship can achieve to deliver a human being to this galaxy? Consider the lifetime of a human to be 80 years. Example 25-2

40. • **Astronomy** The nearest star to our own Sun is Proxima Centauri, 4.25 light years away. If a spaceship travels at $0.75c$, how much time is required for a one-way trip (a) according to an Earth observer and (b) according to the captain of the ship? Example 25-2

41. • A subatomic particle is traveling relative to a nuclear power plant with a relativistic gamma, $\gamma = 1/\sqrt{1 - (v/c)^2}$, equal to 20. The particle is observed to decay 30 ns after it is created in the plant. What is the proper lifetime of this particle? Example 25-3

42. • **Astronomy** A spaceship travels at $0.95c$ toward Alpha Centauri. According to Earthlings, the distance is 4.37 light years. (a) From the perspective of the space travelers, how long does it take to reach this star if the ship starts at Earth? (b) From the perspective of Earthlings, how long does the trip take? Example 25-3

25-5 Einstein's relativity also predicts that the length of an object depends on the observer

43. • A car travels in the positive x direction in the reference frame S. The reference frame S' moves at a speed of $0.80c$, along the x axis. The proper length of the car is 3.20 m. Calculate the length of the car according to observers in the S' frame. Example 25-4

44. • A moving spaceship is measured to have a length that is two-thirds of its proper length. What is the speed of the spaceship relative to the observer? Example 25-4

45. • A stick moves past an observer at a speed of $0.44c$. According to the observer the stick is oriented parallel to the direction of motion and is 0.88 m long. Determine the proper length of the stick. Example 25-4

46. •• A standard tournament domino is 1.5 in. wide and 2.5 in. long. Describe how you might orient a domino so that it will measure 1.5 in. by 1.5 in. as it moves by. What relative speed is required? Example 25-5

47. •• How fast must a positive pion (an unstable particle) be moving to travel 100 m (according to the laboratory frame) before it has a 50% chance of decaying? The half-life, at rest, of a positive pion is 1.80×10^{-8} s. Give your answer in units of meters per second (m/s) and as a fraction of the speed of light (in other words, find both v and v/c). Example 25-5

48. •• You stand on the ground halfway between two mountains, one on your left side and one on your right side. Atop both mountain peaks are bright lamps, which are separated by a distance of 2.00 km. You observe them both turn on, but at slightly different times. As you observe this, a pilot, who is flying a super-advanced rocket plane at a speed of $0.750c$, flies parallel to a line that connects both mountain peaks. According to the pilot, who is flying right-to-left in your reference frame, both lamps turn on at the same time. Which lamp do you observe turning on first, and by how much time does it precede the other lamp? Example 25-6

49. •• A spacecraft passes you traveling at $0.600c$. Your alien friend Gaar on the spaceship measures the length of the ship as 60 m from front to back. He stands in the middle of the ship and fires one photon backward and one photon forward, and each photon strikes a detector at each end of the ship. (a) According to Gaar, what is the travel time of each photon? Verify that he reports the detection events as simultaneous. (b) According to you, which photon registers first? (c) Use the Lorentz transformation, and the fact that the detection events are simultaneous according to Gaar, to find the time difference according to you. Example 25-6

25-6 The speed of light is the ultimate speed limit

50. •• Spaceship A moves at $0.8c$ toward the right, while spaceship B moves in the opposite direction at $0.7c$ (both speeds are measured relative to Earth). (a) Calculate the velocity of Earth relative to spaceship A. (b) Calculate the velocity of Earth relative to spaceship B. (c) Calculate the velocity of spaceship A relative to spaceship B. Example 25-7

51. •• A (very fast) car traveling with a velocity of $+0.35c$ to the right passes an observer sitting on the side of the road. A truck traveling with a velocity of $+0.25c$ passes the same observer. Determine the relative velocity between the car and the truck. Give your answer in terms of "the velocity of the truck relative to the car is . . ." and "the velocity of the car relative to the truck is . . ." Example 25-7

52. •• A spaceship flies by Earth at $0.92c$. It fires a rocket at $0.75c$ in the forward direction, relative to the spaceship. What is the velocity of the rocket relative to Earth? Example 25-7

53. •• Suppose the spaceship in problem 52 continues to fly by Earth at 0.92*c*. This time, however, it fires a rocket at 0.75*c* in the backward direction relative to the spaceship. What is the velocity of the rocket relative to Earth? Example 25-7

54. •• Prove that the relative velocity of a laser fired from a spaceship that is moving at 0.92*c* past Earth will be *c* from the perspective of Earth. Example 25-7

55. •• Galaxy B moves away from galaxy A at 0.550 times the speed of light. Galaxy C moves away from galaxy B in the same direction at 0.750 times the speed of light. How fast does galaxy C recede from galaxy A? Express your answer as a fraction of the speed of light. Example 25-7

56. •• A proton travels at 99.999954% of the speed of light. An antiproton travels in the opposite direction at the same speed. What is the relative velocity of the two particles? Example 25-7

25-7 The equations for kinetic energy and momentum must be modified at very high speeds

57. • A 2.00-kg object moves at 400,000 m/s. (a) Calculate the Newtonian momentum of the object. (b) Calculate the Einsteinian momentum of the object. (c) Which of the answers is correct? What is the percent difference? Example 25-8

58. •• An electron travels at 0.444*c*. Calculate (a) its Einsteinian momentum, (b) its Einsteinian kinetic energy, (c) its rest energy, and (d) the total energy of the electron. Example 25-8

59. • A proton ($m = 1.673 \times 10^{-27}$ kg) is traveling at 0.5*c*. Calculate its Einsteinian momentum and its Einsteinian kinetic energy. Example 25-8

60. • A particle is traveling with respect to an observer such that its Einsteinian energy is twice its rest energy. How fast is it moving with respect to the observer? Example 25-8

61. •• A particle has a rest energy of 5.33×10^{-13} J and a total energy of 9.61×10^{-13} J. Calculate the momentum of the particle. Example 25-8

62. •• A proton has a rest energy of 1.50×10^{-10} J and a momentum of 1.07×10^{-19} kg·m/s. Calculate its speed. Example 25-8

63. • Recent home energy bills indicate that a household used 411 kWh of electrical energy and 201 therms for gas heating and cooking in a period of one month (1.0 therm = 29.3 kWh). How many milligrams of mass would need to be converted directly to energy each month to meet the energy needs for the home? Example 25-9

25-8 Einstein's general theory of relativity describes the fundamental nature of gravity

64. •• What would an observer inside an elevator measure for the free-fall acceleration near the surface of Earth if the elevator accelerates downward at 8.00 m/s²?

65. •• What would an observer inside an elevator measure for the free-fall acceleration near the surface of Earth if the elevator accelerates upward at 18.0 m/s²?

General Problems

66. •• This problem is intended to help you to understand the behavior of relativistic variables such as time, length, and momentum. To begin, construct a table of relativistic gamma γ versus *v*/*c* for the following values of *v*/*c*:

v/*c*	*v*/*c*
0	0.7
0.1	0.8
0.2	0.9
0.3	0.99
0.4	0.999
0.5	0.9999
0.6	

Now plot a graph of γ versus *v*/*c*. As long as γ is close to unity (1.0), the effects of Einstein's special theory of relativity go unnoticed. (a) From your graph (or your table) determine the maximum value of *v*/*c* for which γ is less than 1% larger than unity (1.0). (b) What numerical value does the slope of your graph take at (i) *v*/*c* = 0.1, (ii) *v*/*c* = 0.5, (iii) *v*/*c* = 0.9, and (iv) *v*/*c* = 0.999?

67. •• A super rocket car traverses a straight track 2.40×10^5 m long in 10^{-3} s as measured by an observer next to the track. (a) How much time elapses on a clock in the rocket car during the run? (b) What is the distance traveled in traversing the track as determined by the driver of the rocket car? Example 25-3

68. •• Observers in reference frame S see an explosion located at $x_1 = 580$ m. A second explosion occurs 4.5 μs later at $x_2 = 1500$ m. In reference frame S′, which is moving along the +*x* axis at speed *v*, the explosions occur at the same point in space. What is the separation in time between the two explosions as measured in S′? Example 25-6

69. • Estimate the difference in a 10,000-s time interval as measured by a proper observer and a relative observer traveling on a commercial jetliner. Example 25-2

70. •• A radioactive nucleus traveling at a speed of 0.8*c* in a laboratory decays and emits an electron in the same direction as the nucleus is moving. The electron travels at a speed of 0.6*c* relative to the nucleus. (a) How fast is the electron moving according to an observer in the laboratory? (b) Rocket A travels away from Earth at 0.6*c*, and rocket B travels away from Earth in exactly the opposite direction at 0.8*c*. What is the speed of rocket B as measured by the pilot of rocket A? (c) Why did you get the same answer that you did for part (a)? Example 25-7

71. •• A beam of positive pions (unstable particles) has a speed of 0.88*c*. Their half-life, as measured in the reference frame of the laboratory, is 1.80×10^{-8} s. What is the distance traveled by the laboratory, as measured by the pion, during one half-life? Example 25-3

72. • A car is driving along the freeway. Estimate the ratio of its kinetic energy to its rest energy. Example 25-8

73. ••• Muons have a proper half-life of 1.56×10^{-6} s. Suppose a muon is formed at an altitude of 3000 m and travels at a speed of 0.950*c* straight toward Earth. (a) Does the muon reach Earth's surface within one half-life? Complete the problem from both perspectives: the muon's point of view and Earth's reference frame. (b) Calculate the minimum speed of the muon so that it *just barely* reaches Earth's surface after one half-life. Example 25-3

74. •• Two students, Nora and Allison, are both the same age when Allison hops aboard a flying saucer and blasts off to achieve a cruising speed of 0.800*c* for 20 years (according to Allison). Neglecting the acceleration of the ship during blastoff, landing, and turnarounds, find the difference in age between Nora and Allison when they are reunited. Who is younger? Example 25-2

75. • Flying saucers A and B fly between two planets P_1 and P_2 as shown in Figure 25-22. Saucer A flies toward P_1 with velocity 0.350*c* and saucer B flies toward P_2 with velocity 0.750*c*. The planets are at rest with respect to each other. Flying saucer C is headed toward P_1 with velocity 0.850*c* along a different path, as shown. The longest dimension of each saucer is referred to as its width, and the shortest dimension is called its height. Given that each saucer and planet defines an inertial frame, identify the inertial frame or frames in which observers

measure the proper time for (a) saucer A's trip between the planets, (b) the width of saucer B, (c) the height of saucer B, and (d) the distance between the planets.

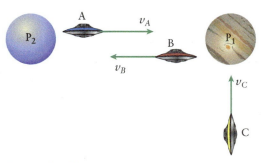

Figure 25-22 Problem 75

76. •• Suppose a jet plane flies at 300 m/s relative to an observer on the ground. Using only special relativity, determine how far the plane must travel, as measured by an observer on the ground, before the clocks aboard the plane are 10 s behind clocks on the ground. Assume the two clocks were originally synchronized to start. Example 25-2

77. •• At the end of the linear accelerator at the SLAC National Accelerator laboratory, electrons have a speed of $0.99999999995c$. (a) Calculate the value of γ for an electron with this speed. (b) What time interval would an observer at rest relative to the accelerator measure for a time interval of 1.66 μs measured from the electron's perspective? Example 25-3

78. •• The half-life of a certain subatomic particle at rest is about 1×10^{-8} s. How fast is a beam of these particles moving if one-half of them decay in 6×10^{-8} s as measured in the laboratory? Example 25-3

79. •• Based on experiments in your lab, you know that a certain radioactive isotope has a half-life of 2.25 μs. As a high-speed spaceship passes your lab, you measure that the same isotope at rest inside the spaceship takes 3.15 s for one-half of it to decay. (a) What is the half-life of the isotope as measured by an astronaut working inside the spaceship? (b) How fast is the spaceship traveling relative to Earth? Example 25-3

80. • We know 1 kg of trinitrotoluene (TNT) yields an energy of 4.2 MJ. The energy released comes from the chemical bonds in the material. How much mass would be required to create an explosion equivalent to 1.8×10^9 kg TNT? Assume all the energy released comes simply from the rest energy of the material. Example 25-9

81. •• **Astronomy** High-speed cosmic rays strike atoms in Earth's upper atmosphere and create secondary showers of particles. Suppose a particle in one of the showers is created 25.0 km above the surface traveling downward at 90.0% the speed of light. Consider the following two events: "A particle is created in the upper atmosphere" and "a particle strikes the ground." We can view the events from two reference frames, one fixed on Earth and one traveling with the created particle. (a) In which of the reference frames are the proper time and the rest length between the two events measured? Explain your reasoning. (b) In the particle's reference frame how long after its creation does it take it to reach the ground? (c) In Earth's reference frame how long after creation does it take the

particle to reach the ground? (d) Show that the times in parts (b) and (c) are consistent with time dilation. Example 25-4

82. ••• A rocket 642 m long is traveling parallel to Earth's surface at $0.5c$ from left to right. At time $t = 0$ a light flashes for an instant at the center of the rocket. Detectors at opposite ends of the rocket record the arrival of the light signal. Call event A the light striking the left detector and event B the light striking the right detector. Observers at rest in the rocket and on Earth record the events. (a) What is the speed of the light signal as measured by the observer (i) at rest in the rocket and (ii) at rest on Earth? (b) At what time after the flash do events A and B occur as measured by (i) the observer in the rocket and (ii) the observer at rest on Earth? Which event occurs first in each case? (c) Show that the results in part (b) are consistent with time dilation. Example 25-6

83. •• **Biology** Twin astronauts, Anselma and Baraka, have identical pulse rates of 70 beats/min on Earth. Anselma remains on Earth, but Baraka is assigned to a space voyage during which he travels at $0.75c$ relative to Earth. What will be Baraka's pulse rate as measured by (a) Anselma on Earth and (b) the physician in Baraka's spacecraft? Example 25-2

84. •• A rocket is traveling at speed v relative to Earth. Inside a lab on the rocket, two laser beams are turned on, one pointing in the forward direction and the other pointing in the backward direction relative to the rocket's velocity. (a) What is the speed of each laser beam relative to the laboratory in the rocket? (b) Use the Lorentz velocity transformation to find the velocity of each laser beam as measured by an observer at rest on Earth. (c) As observed from Earth, how fast are the two laser beams separating from *each other*? Example 25-7

85. ••• A 1000-kg rocket is flying at $0.90c$ relative to your lab. Calculate the kinetic energy of the rocket using the Einsteinian formula and the ordinary Newtonian formula. What is the percent error if we use the Newtonian formula? Does the Newtonian formula overestimate or underestimate the kinetic energy? Example 25-8

86. ••• A spaceship is traveling at $0.50c$ relative to Earth. Inside the ship, a cylindrical piston that is 50.0 cm long and 4.50 cm in diameter contains 1.25 mol of ideal gas at 25.0°C under 2.20 atm of pressure. The cylinder is oriented with its axis parallel to the direction in which the spaceship is flying. What is the particle density (in molecules per cubic meter) of the gas in the cylinder as measured by (a) an astronaut in the rocket ship's lab and (b) a scientist in an Earth lab? Example 25-5

87. ••• **Astronomy** In December 2009 the discovery was announced of a planet that may have a large amount of water and hence would be a good candidate for possible life. The planet, GJ 1214 b, orbits a small star that is 42 light years from Earth. In the future we might decide to send some astronauts to explore the planet. When they arrive there we want them to be young enough to perform tests. Suppose that the captain is 25 years old at launch time. (a) What is the minimum speed the spaceship will need for the captain to be no more than 60 years old at arrival? (b) As soon as the spaceship arrives at the planet, the captain has orders to send a radio signal to Earth to notify Mission Control that the trip was successful. How many years after launch from Earth will it be when the signal arrives at Earth? You can ignore acceleration times and any motion of Earth and GJ 1214 b. Example 25-4

26

Quantum Physics and Atomic Structure

NASA, ESA, and T. Brown (STScI)

What do you think?

Some of the stars in this photo are blue, while others are red. Based on the color alone, you can conclude that the blue stars are (a) hotter than the red stars; (b) cooler than the red stars; (c) made of different materials than the red stars; (d) both (a) and (c); (e) both (b) and (c).

In this chapter, your goals are to:

- (26-1) Recognize the limitations of classical physics for explaining the properties of light and matter.
- (26-2) Describe how the photoelectric effect and blackbody radiation provide evidence for the photon picture of light.
- (26-3) Explain why the wavelength of a photon increases if it scatters from an electron.
- (26-4) Calculate the wavelength of a particle such as an electron.
- (26-5) Describe why atoms absorb light at only certain wavelengths, and why they emit and absorb light at the same wavelengths.
- (26-6) Explain how the Bohr model of the atom explains the spectrum of hydrogen.
- (26-7) Describe the Heisenberg uncertainty principle.

To master this chapter, you should review:

- (3-7)   Uniform circular motion.
- (4-2)   Newton's second law.
- (9-7)   Angular momentum of a moving particle.
- (13-3)   Angular wave number and the wave function of a sinusoidal wave.
- (14-3)   The energy of molecules in an ideal gas.
- (14-7)   Heat transfer by radiation.
- (17-3)   Electric potential.
- (22-2, 22-3, The electromagnetic nature of light.
  and 22-4)
- (25-7)   Relativistic kinetic energy and momentum.

26-1 Experiments that probe the nature of light and matter reveal the limits of classical physics

In Section 22-4 we introduced the idea that light and other electromagnetic waves come in particle-like packets of energy called photons. Is there solid evidence for this? Can a photon strike a particle such as an electron and "bounce" or scatter the way a cue ball

Night vision goggles "amplify light" using the photoelectric effect. Even faint light causes a surface to emit electrons, and the current of these electrons can be amplified in a circuit to generate a brighter image that the wearer of the goggles can see.

The characteristic color of neon lights is due to the structure of the neon atom. When excited by an electric current, neon atoms in the light tube jump to a higher energy level. When they jump down to a lower energy level, they emit photons of a specific frequency, wavelength, and color.

(a)

(b)

Figure 26-1 **Photons, electrons, and atoms** Two examples of the interaction between light and matter. These can only be understood if we use the ideas that light has particle aspects and matter has wave aspects.

does when it strikes a billiard ball? And if waves have a particle-like nature, do particles ever exhibit properties we associate with waves?

In this chapter we'll see that the answer to each of these questions is "yes." The photoelectric effect (Figure 26-1a), in which light striking a surface causes the surface to eject electrons, can be understood only if light comes in the form of photons. The same is true for blackbody radiation, the light emitted by an object as a consequence of its temperature. We'll see direct evidence that photons really do behave like tiny "cue balls" when they scatter from electrons. And we'll learn that particles like electrons also have a wavelike aspect. This wave nature of matter will help us understand the structure of atoms and the manner in which atoms absorb and emit light (Figure 26-1b). We'll find that the energies of atoms are *quantized*—that is, they can have only certain very definite values. The lesson of this chapter is that the microscopic world of atoms and light is very different in character from the macroscopic world of ordinary-sized objects that we see around us.

26-2 The photoelectric effect and blackbody radiation show that light is absorbed and emitted in the form of photons

By the end of the nineteenth century, it was recognized that light was an electromagnetic wave. Maxwell's equations (Section 22-3) describe the properties of these waves in great detail, and so physicists were confident that they had a deep understanding of the nature of light. But experimental studies of two very different phenomena showed that the true nature of light is actually more complex. In one of these phenomena, the *photoelectric effect*, a material absorbs light and the absorbed energy is used to eject electrons; in the other, called *blackbody radiation*, a material emits electromagnetic radiation when it is heated. The discoveries made by studying these two phenomena radically altered how physicists answered the question "What is light?"

The Photoelectric Effect

BioMedical When light strikes certain materials, electrons can be ejected from the surface of those materials (Figure 26-2a). This **photoelectric effect**, discovered in 1886, plays an important role in biological research through a technique called *photoemission electron microscopy* or PEEM. In this technique, a biological sample is illuminated with ultraviolet light or x rays. Different materials in the

---

TAKE-HOME MESSAGE FOR Section 26-1

✔   Light waves have some of the characteristics of particles, and particles such as electrons have some of the characteristics of waves.

✔   The energies of atoms are quantized (restricted to certain specific values).

(a)   Light   Electrons
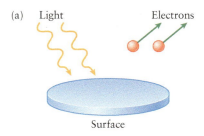
                    Surface

(b)   Fibrinogen (blue dots)
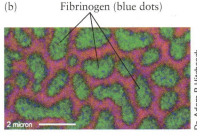
2 micron

Figure 26-2 **The photoelectric effect** (a) In the photoelectric effect electrons escape from a surface when the surface is illuminated with light. (b) An example of photoemission electron microscopy. Protein interactions with polymers are critical to understanding the compatibility of synthetic materials with blood (such as medical implants). This image uses photoemission electron microscopy (PEEM) to map the spatial distribution of fibrinogen, a blood protein, on a sheet of two different types of polymer. This image could be made because fibrinogen (shown as blue dots) and the two polymers (shown in green and red) emit different numbers of electrons when illuminated with ultraviolet light. Data were measured at the Advanced Light Source, Lawrence Berkeley National Laboratory.

surface layer of the sample will emit fewer or greater numbers of electrons in response to this light. By recording the differences in the number of emitted electrons, called **photoelectrons**, it's possible to construct an image of the sample surface that shows where the different materials are located. Figure 26-2b shows an example of an image made in this way.

What makes the photoelectric effect so remarkable is that it does not behave in accordance with the idea that light is an electromagnetic wave. It takes energy to liberate an electron from the surface of a material, and in the photoelectric effect this energy is provided by the electric and magnetic energy in the light absorbed by the surface. We learned in Section 22-4 that according to Maxwell's equations, the intensity of such a wave depends on the rms values of the electric and magnetic fields in that wave, but not on the frequency of the wave:

Intensity of an electromagnetic wave (22-25)

Intensity of an electromagnetic wave	Speed of light	Magnetic field rms value

$$S_{average} = c\varepsilon_0 E_{rms}^2 = \frac{cB_{rms}^2}{\mu_0} = \frac{E_{rms}B_{rms}}{\mu_0}$$

Permittivity of free space	Electric field rms value	Permeability of free space

Equation 22-25 suggests that light of *any* frequency should be able to liberate an electron from the surface of a material, provided the light wave is sufficiently intense. Experiment shows that this is not the case. For example, if the biological sample shown in Figure 26-2 is illuminated with red light, no electrons are ejected no matter how intense the light. But if instead we illuminate the sample with x rays, which have a higher frequency than red light, electrons *are* ejected. (Figure 26-2b was made by using x rays.) This is impossible to understand on the basis of Maxwell's equations.

In 1905, the same year that he published his special theory of relativity, Albert Einstein proposed a simple but radical explanation for the strange behavior of the photoelectric effect. He suggested that light of frequency f comes in small packets, each with an energy E that is directly proportional to the frequency. Today these packets are called **photons**. We first encountered this idea in Section 22-4:

Energy of a photon (22–26)

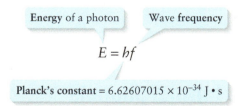

Energy of a photon	Wave frequency

$$E = hf$$

Planck's constant = $6.62607015 \times 10^{-34}$ J • s

We'll see shortly how the value of Planck's constant h is determined from the photoelectric effect. Later in this section we'll see why the constant is named for the German physicist Max Planck.

Let's see how Einstein's idea explains the properties of the photoelectric effect. The minimum amount of energy required to remove a single electron from a material is called the **work function Φ_0** of the material (Φ is the uppercase Greek letter "phi"). The value of Φ_0 varies from one material to another; it is small if electrons are easy to remove and large if electrons are hard to remove. In a given material, some electrons will be more difficult to remove, but Φ_0 represents the energy required to remove the most easily dislodged electron. Einstein proposed that an electron can absorb only a single photon at a time. So for even the most easily dislodged electron to be ejected from the material, it must absorb a photon with an energy equal to or greater than Φ_0. If the energy of the absorbed photon is greater than Φ_0, the energy that remains after the electron is ejected goes into the kinetic energy of the electron as it flies away from the material. Electrons that require more energy to be ejected will emerge from the material with less kinetic energy, but those with *maximum* kinetic energy will be those

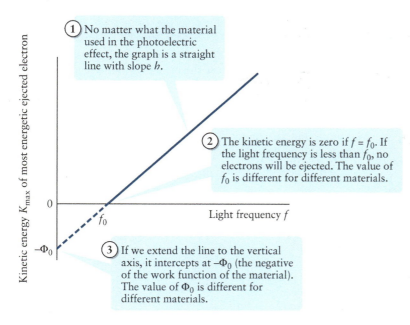

① No matter what the material used in the photoelectric effect, the graph is a straight line with slope h.

② The kinetic energy is zero if $f = f_0$. If the light frequency is less than f_0, no electrons will be ejected. The value of f_0 is different for different materials.

③ If we extend the line to the vertical axis, it intercepts at $-\Phi_0$ (the negative of the work function of the material). The value of Φ_0 is different for different materials.

Figure 26-3 Electron kinetic energy versus frequency in the photoelectric effect The kinetic energy of the most energetic electron ejected in the photoelectric effect depends on the frequency of the light. This cannot be explained using the wave model of light but can be explained by the photon concept.

that were the easiest to dislodge. So in Einstein's picture the most energetic electrons ejected from the material will emerge with kinetic energy K_{max} given by

Maximum kinetic energy of an electron ejected from a material by the **photoelectric effect**

Work function of the material

$$K_{max} = hf - \Phi_0$$

Maximum kinetic energy of an electron in the photoelectric effect (26-1)

Planck's constant

Frequency of the light used to illuminate the material

Equation 26-1 tells us that a graph of K_{max} as a function of the light frequency f should be a straight line of slope h (see **Figure 26-3**). Since kinetic energy can never be negative, Equation 26-1 also tells us that electrons will be emitted only if $hf - \Phi_0 > 0$, or

$$f > f_0 = \frac{\Phi_0}{h}$$

(26-2)

Equation 26-2 says that electrons will be emitted from the surface only if the light frequency is greater than a threshold frequency f_0 equal to the work function Φ_0 divided by Planck's constant h. This agrees with the observation that no electrons can be ejected from a surface by light of too low a frequency, no matter how intense the light.

The graph shown in Figure 26-3 turns out to be an excellent match to experimental measurements of the maximum kinetic energy of ejected electrons for different light frequencies f. The slope of the graph tells us the value of Planck's constant h. Equation 26-1 also shows that if we extend the graph of K_{max} as a function of f to (unphysical) values of K_{max} less than zero, the graph intercepts the vertical axis at $-\Phi_0$. If we repeat the experimental measurements for a second material with a different work function Φ_0, the straight line intercepts the vertical axis at a different point but has the same slope h as for the first material. This reinforces the idea that Planck's constant is a universal constant; its value does not depend on the properties of the material that absorbs the photons.

The remarkable fit of Einstein's theory to experiment is powerful evidence that light is indeed absorbed in the form of photons with energy $E = hf$ as given by Equation 22-26. Einstein was awarded the Nobel Prize in Physics in 1921 for his explanation of the photoelectric effect.

 Go to Interactive Exercise 26-1 for more practice dealing with the photoelectric effect.

WATCH OUT! Some electrons require more energy than Φ_0 to be ejected.

The work function Φ_0 is the smallest amount of energy required to eject an electron from a given material under the most favorable conditions. Other electrons in the material require more energy to eject and will emerge from the material with less kinetic energy than the value K_{max} given by Equation 26-1.

EXAMPLE 26-1 The Photoelectric Effect with Cesium

The work function for a sample of cesium is 3.43×10^{-19} J. (a) What is the minimum frequency of light that will result in electrons being ejected from this sample by the photoelectric effect? (b) What is the maximum wavelength of light that will result in electrons being ejected from this sample by the photoelectric effect?

Set Up

Equation 26-1 tells us that the minimum energy required to eject an electron corresponds to having $K_{max} = 0$, so the electrons just barely make it out of the cesium. We'll use this to find the threshold frequency f_0 that just barely allows an electron to be ejected. We'll find the corresponding wavelength using Equation 22-2.

Maximum kinetic energy of an electron in the photoelectric effect:

$$K_{max} = hf - \Phi_0 \qquad (26\text{-}1)$$

Propagation speed, frequency, and wavelength of an electromagnetic wave:

$$c = f\lambda \qquad (22\text{-}2)$$

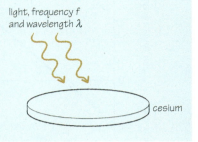
light, frequency f and wavelength λ

cesium

Solve

(a) Use Equation 26-1 to calculate the frequency f that corresponds to $K_{max} = 0$.

From Equation 26-1 with $K_{max} = 0$:

$$0 = hf - \Phi_0$$
$$hf = \Phi_0$$

This says that the photon energy hf is just enough to remove the most easily dislodged electron from the material (which requires energy Φ_0), with nothing left over to give the electron any kinetic energy. So $hf = \Phi_0$ is the minimum photon energy that will eject an electron, and the frequency f of this photon is the minimum (threshold) frequency that will do the job:

$$f_0 = f = \frac{\Phi_0}{h}$$
$$= \frac{3.43 \times 10^{-19} \text{ J}}{6.63 \times 10^{-34} \text{ J} \cdot \text{s}} = 5.17 \times 10^{14} \text{ s}^{-1} = 5.17 \times 10^{14} \text{ Hz}$$

(b) Find the wavelength that corresponds to the frequency that we calculated in part (a).

We can rewrite Equation 22-2 as

$$\lambda = \frac{c}{f}$$

In words, this says that wavelength is inversely proportional to frequency. So the *minimum* frequency of light that will eject an electron corresponds to the *maximum* wavelength that will eject an electron:

$$\lambda_{max} = \frac{c}{f_0} = \frac{3.00 \times 10^8 \text{ m/s}}{5.17 \times 10^{14} \text{ Hz}}$$
$$= 5.80 \times 10^{-7} \text{ m} = 580 \text{ nm}$$

(Recall that 1 nm = 10^{-9} m.)

Reflect

Figure 22-2 shows that a wavelength of 580 nm is in the yellow-green part of the visible spectrum. If we illuminate cesium with light of higher frequency and shorter wavelength than this (for example, blue or violet light), the photons will have more energy than the minimum and electrons will be ejected from the cesium. If instead we illuminate cesium with light of lower frequency and longer wavelength (for example, orange or red light), the photons will have less energy than the required minimum and no electrons will be ejected.

Blackbody Radiation

The photoelectric effect shows that light is *absorbed* in the form of photons. If we are to fully believe the photon concept, however, it must also be true that light is *emitted* in the form of photons. We learned in Section 14-7 that ordinary objects emit electromagnetic radiation as a result of their temperature. If we can find evidence that this emission is in the form of photons, we will have further evidence that the photon description of light is the correct one. Let's take a closer look at radiation of this kind.

Experiment shows that the rate at which an object emits radiation is proportional to its surface area A and to the fourth power of its Kelvin temperature T:

Rate at which an object emits energy in the form of radiation

Emissivity of the object (a number between 0 and 1)

$$P = e\sigma AT^4$$

Temperature of the object on the Kelvin scale

Stefan–Boltzmann constant $= 5.6704 \times 10^{-8} \text{ W} \cdot \text{m}^{-2} \cdot \text{K}^{-4}$

Surface area of the object

Rate of energy flow in radiation (14-22)

The higher the temperature of an object of a given size, the greater the radiated power P and so the more brightly it glows.

Experiment also shows that the *color* of the radiation emitted by an object depends on its temperature T (Figure 26-4). A heated object emits light at all wavelengths, but emits most strongly at a particular frequency called the *frequency of maximum emission*. As the temperature increases, the frequency of maximum emission increases.

Equation 14-22 shows that the radiated power also depends on a quantity e called the *emissivity*, which depends on the properties of the object's surface. This has its greatest value ($e = 1$) for an idealized type of dense object called a **blackbody.** An ideal blackbody does not reflect any light at all but absorbs all radiation falling on it. If a blackbody is in thermal equilibrium with its surroundings, it must emit energy at the same rate that it absorbs it in order for its temperature T to remain constant. So in addition to being a perfect absorber of energy, an ideal blackbody in thermal equilibrium with its surroundings is also a perfect emitter of energy because it emits as much energy as it absorbs.

Ordinary objects, such as tables, textbooks, and people, are not ideal blackbodies; they reflect light, which is why they are visible. (Even a piece of wood darkened with soot or painted a dull black reflects *some* light.) But it is possible to make a nearly ideal blackbody simply by building a box and drilling a small hole in one side (Figure 26-5a). Light that enters the hole will reflect around inside the box, with part of the light energy being absorbed by the walls on each reflection. Eventually all of the light energy will be absorbed, so the interior of the box acts like a perfect absorber and is effectively an ideal blackbody. You can see this effect if you look into another person's eye (Figure 26-5b). The pupil at the center of the iris appears black, even though the tissues that line the interior of the eye are pinkish in color. That's because after multiple reflections, those tissues almost completely absorb light that enters the eye through the pupil.

If the box in Figure 26-5a is in thermal equilibrium at temperature T, the rate at which the walls absorb energy in the form of radiation must be equal to the rate at which the walls

BioMedical

- A hot, dense object emits electromagnetic radiation. The idealized case is called **blackbody radiation.**
- The frequency of maximum emission is directly proportional to the Kelvin temperature T of the object: The higher the temperature T, the greater the frequency of maximum emission.

① This metal bar heated with a flame emits light at all frequencies but glows most strongly at red frequencies.

② As the temperature of the bar increases, it glows most strongly at orange frequencies...

③ ...and at even higher temperatures it glows most strongly at yellow frequencies.

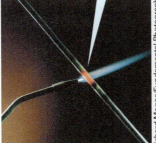
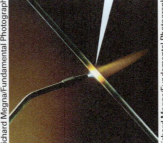

Figure 26-4 Radiation from heated objects The color of the light from a heated object depends on its temperature.

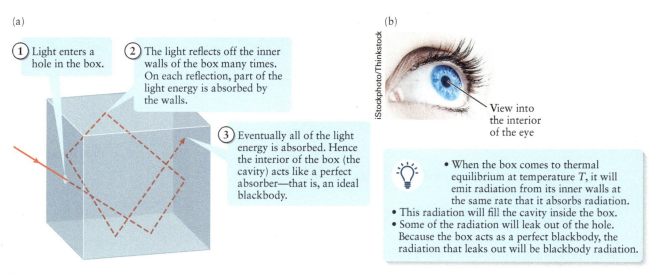

(a)

① Light enters a hole in the box.

② The light reflects off the inner walls of the box many times. On each reflection, part of the light energy is absorbed by the walls.

③ Eventually all of the light energy is absorbed. Hence the interior of the box (the cavity) acts like a perfect absorber—that is, an ideal blackbody.

(b)

iStockphoto/Thinkstock

View into the interior of the eye

- When the box comes to thermal equilibrium at temperature T, it will emit radiation from its inner walls at the same rate that it absorbs radiation.
- This radiation will fill the cavity inside the box.
- Some of the radiation will leak out of the hole. Because the box acts as a perfect blackbody, the radiation that leaks out will be blackbody radiation.

Figure 26-5 A blackbody cavity (a) A box with a small hole in one side is a good approximation to an ideal blackbody. (b) The interior of the eye has properties very similar to the box in (a).

emit energy. The cavity in the interior of the box will be filled with this radiation, which will itself be in thermal equilibrium with the walls of the box. Since the walls act as an ideal blackbody, the light that fills the cavity is effectively **blackbody radiation**—the kind of light that would be emitted by a perfect blackbody of emissivity $e = 1$ at temperature T. We can study this light by examining the small fraction of light that emerges from the hole in Figure 26-5a.

Figure 26-6 shows the experimentally observed *spectrum* of blackbody radiation— that is, the relative amount of light energy present at different frequencies—for two different temperatures. The high-temperature curve lies above the low-temperature curve, which tells us that the higher the temperature of a blackbody, the greater the amount of radiation at all frequencies. Furthermore, as the blackbody temperature increases, the peak of the curve shifts to a higher frequency. This agrees with our observation about how the frequency of maximum emission varies with an object's temperature (Figure 26-4). Figure 26-6 explains why we can't see the radiation from objects at room temperature, about $T = 300$ K. At this relatively low temperature, the frequency of maximum emission is in the infrared, which our eyes cannot see. There is some emission at visible frequencies (in Figure 26-6, to the right of the peak of the curve), but the amount of emission at $T = 300$ K is so low that our eyes can't detect it.

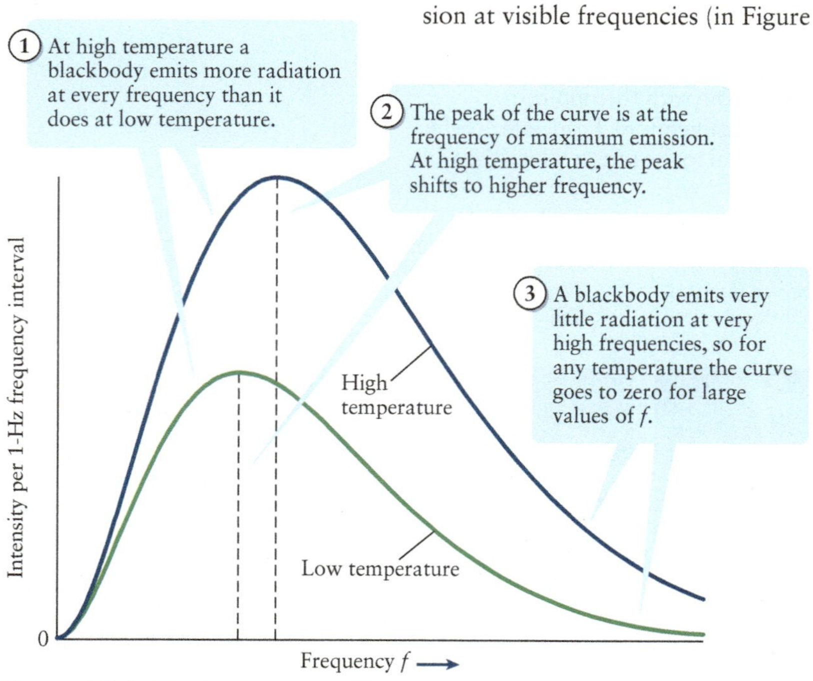

In the late nineteenth century physicists tried to understand the shape of the blackbody spectrum shown in Figure 26-6 using their knowledge of thermodynamics and electromagnetic waves. Their efforts ended in failure. To understand how they failed, we begin by noting that the electromagnetic waves inside the cavity in Figure 26-5a should be in the form of *standing* waves. That's because the waves will bounce back and forth between the walls of the cavity. We learned in Section 13-6 that when waves bounce back and forth between the ends of a string that's tied down at both ends, steady wave patterns arise for waves of certain wavelengths and frequencies. The same is true for electromagnetic waves in a cavity. The standing wave patterns for electromagnetic waves in a cavity are more complex than for waves on a string: Electromagnetic waves are three-dimensional, not just one-dimensional like those along the length of a string, and involve

① At high temperature a blackbody emits more radiation at every frequency than it does at low temperature.

② The peak of the curve is at the frequency of maximum emission. At high temperature, the peak shifts to higher frequency.

③ A blackbody emits very little radiation at very high frequencies, so for any temperature the curve goes to zero for large values of f.

High temperature

Low temperature

Intensity per 1-Hz frequency interval

0

Frequency f ⟶

Figure 26-6 Blackbody spectra The spectrum of light emitted by an ideal blackbody depends on the temperature of the blackbody.

two varying quantities, the electric field and the magnetic field. For our purposes, however, all we need to know is that these standing waves exist.

The shape of the blackbody spectrum in Figure 26-6 indicates how much energy is present in the standing waves in each frequency range: There is relatively little energy at very low frequencies, more energy at frequencies near the frequency of maximum emission, and again relatively little energy at high frequencies. However, nineteenth-century physics suggested that *every* possible standing wave in the cavity should contain on average the same amount of energy. This conclusion came from the equipartition theorem, which we first encountered in Section 14-3. This theorem states that a molecule in a gas at a Kelvin temperature T has, on average, an amount of energy $(1/2)kT$ for each degree of freedom of the molecule, where $k = 1.381 \times 10^{-23}$ J/K is the Boltzmann constant. Arguments from thermodynamics suggest that for the same reason a standing wave inside a cavity in thermal equilibrium at temperature T should also possess an average amount of energy equal to $(1/2)kT$ per degree of freedom. There are two degrees of freedom per standing wave, one for the electric field and one for the magnetic field, so the total average energy per standing wave should be kT. It turns out that the number of standing waves in a given frequency interval increases with increasing frequency. So according to nineteenth-century physics, the total energy per frequency interval (equal to the energy kT per standing wave multiplied by the number of standing waves per frequency interval) should increase with increasing frequency and *never* decrease (Figure 26-7). This is in profound disagreement with the experimentally observed shape of the spectrum.

If we now introduce the photon concept, however, we *can* match the experimentally observed spectrum. We still use the idea that the average energy available for a standing wave of frequency f is kT, but now this energy goes into photons of that frequency which each have energy $E = hf$. This energy fundamentally comes from the walls of the box in Figure 26-5a, since these walls emit the photons. At low frequencies the photon energy hf is small compared to the available energy kT, so there will be many photons present for a low-frequency standing wave. But at high frequencies hf is much larger than kT, which means that the energy required to produce a photon is larger than the average energy available to produce one. Hence the average number of photons present for that standing wave will be very small. (It need not be zero, since kT is only the *average* energy available. From time to time the available energy will be greater than kT, and some photons can be produced.) So even though energy is *available* for a high-frequency standing wave, that energy can't be used to create photons, so the amount of energy *present* is quite small. That's just the effect we need to make the theoretical curve in Figure 26-7 decline at large frequencies.

Using a somewhat different version of this photon argument, in 1900 the German physicist Max Planck was able to make a theoretical prediction for the blackbody spectrum that was in excellent agreement with the experimental spectrum shown in Figure 26-7. Planck's theoretical formula was the first to involve the new quantity h that now bears his name, and the value of h given in Equation 22-26 is the one that gives the best match between this formula and the experimental data. Planck was awarded the 1918 Nobel Prize in Physics for his achievement.

The explanation of blackbody spectra in terms of photons is the evidence we were seeking that light is emitted in the form of photons. In the following section we'll see even more compelling evidence for the photon picture of light.

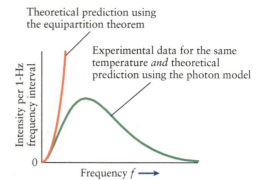

Theoretical prediction using the equipartition theorem

Experimental data for the same temperature *and* theoretical prediction using the photon model

Intensity per 1-Hz frequency interval

Frequency $f \longrightarrow$

- A photon model accurately describes the spectrum of blackbody radiation.
- The model that does not use the photon concept fails to describe the spectrum.

Figure 26-7 The photon model explains blackbody radiation A model for blackbody radiation that does not use the photon concept predicts (incorrectly) that the intensity should increase without limit as the frequency increases.

GOT THE CONCEPT? 26-1   Blackbody Radiation

Two objects of the same size are both perfect blackbodies. One is at a temperature of 3000 K, so its frequency of maximum emission is in the infrared part of the electromagnetic spectrum; the other is at a temperature of 12,000 K, so its frequency of maximum emission is in the ultraviolet part of the spectrum. Compared to the object at 3000 K, the object at 12,000 K (a) emits more infrared light; (b) emits more visible light; (c) emits more ultraviolet light; (d) two of (a), (b), and (c); (e) all of (a), (b), and (c).

26-3 As a result of its photon character, light changes wavelength when it is scattered

Blackbody radiation and the photoelectric effect suggest that photons can be treated like tiny bundles of energy, and so have a particle-like nature. In the photoelectric effect, a photon strikes an electron and is *absorbed*. But if a photon is like a particle, is it possible that, as in the collisions we studied in Chapter 8, a photon could strike an electron and *bounce off*? If so, based on our experience with collisions, linear momentum would be conserved in such an interaction and the momentum of the photon should change as a result. This effect, called *Compton scattering*, is further evidence that light does indeed come in the form of photons.

Let's first see how to express the momentum of a photon. We saw in Section 25-7 that for a particle of mass m, we can write the kinetic energy and momentum as

Einsteinian expressions for kinetic energy and momentum (25-20)

Kinetic energy of a particle of mass m and velocity $\vec{v}$

$$K = (\gamma - 1)mc^2$$

$$\vec{p} = \gamma m \vec{v}$$

Relativistic gamma for speed v

Momentum of a particle of mass m and velocity $\vec{v}$

 At speeds that are a small fraction of the speed of light c, these are approximately equal to the Newtonian expressions:

$$K = \frac{1}{2}mv^2 \text{ and } \vec{p} = m\vec{v}$$

In these expressions the quantity γ (relativistic gamma) is

Relativistic gamma (25-19)

Relativistic gamma for a particle moving at speed v

$$\gamma = \frac{1}{\sqrt{1 - \dfrac{v^2}{c^2}}}$$

Speed of the particle

Speed of light in a vacuum

We also saw that an object of mass m has a rest energy E_0 that is present even when it is not moving:

Rest energy (25-21)

Rest energy of a particle   Mass of the particle

$$E_0 = mc^2$$

Speed of light in a vacuum

If we combine the first of Equations 25-20 with Equations 25-19 and 25-21, we find that the total energy of a particle (kinetic energy plus rest energy) is

$$E = K + E_0 = (\gamma - 1)mc^2 + mc^2 = \gamma mc^2 = \frac{mc^2}{\sqrt{1 - \dfrac{v^2}{c^2}}}$$ (26-3)

From the second of Equations 25-20, the magnitude of the momentum of a particle is

$$p = \gamma mv = \frac{mv}{\sqrt{1 - \dfrac{v^2}{c^2}}}$$ (26-4)

Comparing Equations 26-3 and 26-4, we see that

$$p = \frac{Ev}{c^2} \quad \text{(momentum of a particle of total energy } E\text{)}$$ (26-5)

A photon has zero mass, which is why it can travel at the speed of light. (Any object with nonzero mass would require an infinite amount of kinetic energy to reach $v = c$, as we described in Section 25-7.) As such, we can't apply Equations 26-3 or 26-4 directly to photons. But we can use the combination in Equation 26-5, in which mass does not appear explicitly. Setting $v = c$ in Equation 26-5, we get the following relationship for the momentum of a photon:

$$p = \frac{Ec}{c^2} = \frac{E}{c} \quad \text{(momentum of a photon)}$$ (26-6)

From Equation 22-26 we can write $E = hf$, and we know that for light waves $c = f\lambda$ (Equation 22-2). If we substitute these into Equation 26-6, we get an alternative expression for the momentum of a photon:

$$p = \frac{hf}{f\lambda}$$

or, simplifying,

Magnitude of the **momentum** of a photon   Energy of the photon

$$p = \frac{E}{c} = \frac{h}{\lambda}$$

Planck's constant

Speed of light in a vacuum   Wavelength of the photon

Momentum and energy of a photon
(26-7)

A photon's momentum is directly proportional to the energy that it carries and inversely proportional to its wavelength. Thus a violet photon of wavelength 400 nm has twice the momentum, as well as twice the energy, of an infrared photon of wavelength 800 nm.

In the early 1920s American physicist Arthur Compton showed conclusively that photons have momentum and that the momentum is inversely proportional to the wavelength, as stated in Equation 26-7. In his experiments, an x-ray photon collided with an electron in a carbon atom, a process now called **Compton scattering**. Compton detected both the electron, which is knocked out of the atom, and the scattered photon. Compton could account for the directions and energies of the electron and the scattered photon by requiring that the total momentum of the electron and the photon be conserved in the collision. In other words, he showed that the photon description of light applies not just to the absorption and emission of light but also to what happens to light when it is scattered. For revealing this fundamental aspect of light, Compton was awarded the Nobel Prize in Physics in 1927.

Figure 26-8 shows the situation that Compton studied, the collision of a photon (symbol γ_i) and a stationary electron (symbol e^-). The electron is

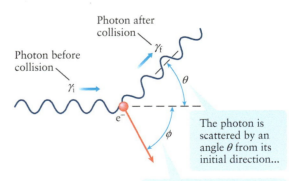

Photon after collision
Photon before collision
The photon is scattered by an angle θ from its initial direction...

...and the scattered electron moves off at an angle ϕ from the initial direction of the photon.

Figure 26-8 Compton scattering A photon undergoes a wavelength shift when it scatters from an electron that is initially at rest.

scattered at angle ϕ relative to the initial direction of the photon, and the photon scatters at angle θ relative to its initial direction. Although we have labeled the photon γ_f after the collision, the scattered photon is the same photon that collided with the electron. The subscripts "i" for "initial" and "f" for "final" instead imply that the energy, wavelength, and other quantities associated with the photon have changed as a result of the collision. For example, because energy is conserved during the collision and because some of the photon's energy is almost always transferred to the electron, the outgoing photon carries less energy than it had initially. From Equation 26-7 the energy of a photon is

$$E = pc = \frac{hc}{\lambda}$$

Because the photon has less energy after the collision than it had initially, E_f is less than E_i, and the final wavelength λ_f is greater than the initial wavelength λ_i. In other words, as the energy of the photon decreases, its wavelength increases. Compton found that the increase in wavelength $\Delta\lambda$ from λ_i to λ_f is a function of the angle θ at which the photon scatters:

**Compton scattering equation
(26-8)**

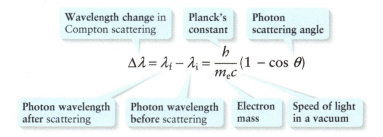

$$\Delta\lambda = \lambda_f - \lambda_i = \frac{h}{m_e c}(1 - \cos\theta)$$

The proportionality constant in Equation 26-8 is known as the *Compton wavelength* λ_C:

(26-9)

$$\lambda_C = \frac{h}{m_e c}$$

 See the Math Tutorial for more information on trigonometry.

➔ **Go to Picture It 26-1 for more practice dealing with Compton scattering.**

The scattering angle θ of the photon ranges from 0° (straight forward) to 180° (straight back). When the photon continues in the same direction after the collision, so that θ equals 0° and $\cos\theta$ equals 1, $\Delta\lambda$ equals zero. In other words, if $\theta = 0°$ there is no change in the photon's wavelength and no change in the photon's energy.

The maximum change in a photon's wavelength and energy when it undergoes Compton scattering occurs when it scatters straight back, in the direction opposite to the one in which it approached the electron. In this case $\theta = 180°$, so $\cos\theta = -1$ and the term in parentheses in Equation 26-8 equals 2. The maximum possible change in the photon's wavelength is therefore $2\lambda_C$, or twice the Compton wavelength. Using the known values of h, m_e, and c, we find

$$\lambda_C = \frac{h}{m_e c} = \frac{6.63 \times 10^{-34} \text{ J·s}}{(9.11 \times 10^{-31} \text{ kg})(3.00 \times 10^8 \text{ m/s})}$$
$$= 2.43 \times 10^{-12} \text{ m} = 2.43 \times 10^{-3} \text{ nm} = 0.00243 \text{ nm}$$

(Recall that 1 nm = 10^{-9} m.) By comparison, the wavelength of visible light ranges between about 380 and 750 nm. So if a photon of wavelength 400 nm is scattered by an electron its wavelength changes by a value on the order of 0.00243 nm, a change of less than one part in 10^5. Such a tiny change is very difficult to measure, so we conclude that a visible-light photon undergoes a negligible shift in wavelength due to Compton scattering. But for an x-ray photon with a wavelength on the order of 10^{-11} m = 0.01 nm, a wavelength shift of 0.00243 nm corresponds to a large percentage of the initial wavelength. That's why Compton first noticed this effect in the scattering of x rays rather than visible light.

If light did not have a photon aspect, we would expect *no* wavelength change on scattering. A light wave of frequency f and wavelength λ encountering an electron would make the electron oscillate at the same frequency f, and the electron would emit radiation with the same frequency f and hence the same wavelength λ as the initial light wave. The change in wavelength that Compton observed is unambiguous evidence that light does indeed have a particle character.

EXAMPLE 26-2 Compton Scattering

A photon carries 2.00×10^{-14} J of energy. It undergoes Compton scattering in a block of carbon. What is the largest fractional change in energy the photon can undergo as a result?

Set Up

Given the initial photon energy $E_i = 2.00 \times 10^{-14}$ J, we can calculate its wavelength λ_i using Equation 26-7. We use Equation 26-8 to calculate the change in wavelength due to scattering; this will be maximum if $\theta = 180°$, so $\cos \theta = -1$. Once we know the final wavelength λ_f, we can use Equation 26-7 again to find the final photon energy. Comparing this to the initial photon energy tells us the fractional change in energy.

Momentum and energy of a photon:

$$p = \frac{E}{c} = \frac{h}{\lambda} \qquad (26\text{-}7)$$

Compton scattering equation:

$$\Delta\lambda = \lambda_f - \lambda_i = \frac{h}{m_e c}(1 - \cos\theta) \qquad (26\text{-}8)$$

Before:

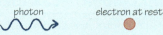
photon   electron at rest

After:

scattered photon   recoiling electron

Solve

First calculate the wavelength of the initial photon using Equation 26-7.

From Equation 26-7 the wavelength of the initial photon is

$$\lambda_i = \frac{hc}{E_i} = \frac{(6.63 \times 10^{-34} \text{ J·s})(3.00 \times 10^8 \text{ m/s})}{2.00 \times 10^{-14} \text{ J}}$$

$$= (9.95 \times 10^{-12} \text{ m})\left(\frac{1 \text{ nm}}{10^{-9} \text{ m}}\right) = 9.95 \times 10^{-3} \text{ nm}$$

Calculate the wavelength shift using Equation 26-8.

The maximum wavelength shift is with $\theta = 180°$ and $\cos \theta = -1$:

$$\Delta\lambda_{max} = \lambda_f - \lambda_i = \frac{h}{m_e c}(1 - \cos 180°)$$

$$= (2.43 \times 10^{-3} \text{ nm})[1 - (-1)] = 4.86 \times 10^{-3} \text{ nm}$$

The wavelength of the final photon equals the wavelength of the initial photon plus the shift $\Delta\lambda$. Use this to find the energy of the final photon.

The final wavelength is

$$\lambda_f = \lambda_i + \Delta\lambda = 9.95 \times 10^{-3} \text{ nm} + 4.86 \times 10^{-3} \text{ nm}$$
$$= 1.481 \times 10^{-2} \text{ nm} = 1.481 \times 10^{-11} \text{ m}$$

The energy of the final photon is

$$E_f = \frac{hc}{\lambda_f} = \frac{(6.63 \times 10^{-34} \text{ J·s})(3.00 \times 10^8 \text{ m/s})}{1.481 \times 10^{-11} \text{ m}}$$
$$= 1.34 \times 10^{-14} \text{ J}$$

This is less than the energy of the initial photon. The lost energy has gone into the kinetic energy of the scattered electron.

Express the energy change as a fraction of the initial photon energy.

The fractional energy change is the energy change $E_f - E_i$ divided by the initial energy E_i:

$$\text{fractional energy change} = \frac{E_f - E_i}{E_i}$$

$$= \frac{1.34 \times 10^{-14} \text{ J} - 2.00 \times 10^{-14} \text{ J}}{2.00 \times 10^{-14} \text{ J}}$$

$$= -0.328 = -32.8\%$$

Reflect

An initial photon of this high energy and short wavelength can lose as much as 32.8% (nearly one-third) of its initial energy when it undergoes Compton scattering.

Example 26-2 suggests why x rays are useful in cancer radiation therapy. If an x-ray photon strikes an electron in a water molecule within a cancerous cell, the photon can scatter and transfer a substantial amount of energy to the electron. This transferred energy is great enough that the electron escapes from the molecule, leaving the water molecule in an ionized state. These ionized water molecules damage the DNA of the cancerous cell and cause cell death.

BioMedical

GOT THE CONCEPT? 26-2 Compton Scattering

Suppose a photon has a wavelength equal to the Compton wavelength λ_C. If this photon collides with an electron, and the photon is scattered through an angle of 90°, what will be the wavelength of the photon after the collision? (a) Zero; (b) $\lambda_C/2$; (c) λ_C; (d) $2\lambda_C$; (e) $3\lambda_C$.

TAKE-HOME MESSAGE FOR Section 26-3

✔ A photon has zero mass but does have momentum. The magnitude of the momentum is proportional to the photon energy and inversely proportional to the wavelength.

✔ In Compton scattering, a photon scatters from an electron. The photon loses energy and momentum, and these are transferred to the electron. The change in wavelength of the photon depends on the angle through which it is scattered.

26-4 Matter, like light, has aspects of both waves and particles

We have seen that light has a dual nature, with attributes of both waves and particles. Light comes in the form of particles (photons), but these particles have a wave aspect: Associated with them is a frequency f, which determines the photon energy $E = hf$, and a wavelength λ, which determines the photon momentum $p = h/\lambda$. Is it possible that ordinary matter, which we know is made of particles such as electrons, protons, and neutrons, also has a dual nature? Could these particles also have a wave aspect?

In 1924 French graduate student Louis de Broglie (pronounced "de broy") proposed precisely that idea. In particular, he suggested that the relationship $p = h/\lambda$ between momentum p and wavelength λ that applies to photons should also apply to particles such as electrons (Figure 26-9). The wavelength of a particle is called its **de Broglie wavelength**:

de Broglie wavelength
(26-10)

A particle has a **de Broglie wavelength**...   ...equal to **Planck's constant**...

$$\lambda = \frac{h}{p}$$

...divided by the **momentum of the particle.** The greater the momentum, the shorter the de Broglie wavelength.

How large should we expect the wavelength of an electron to be? Let's examine the case of an electron of charge $q = -e$ that gains its momentum by moving through a potential difference of V, so the electron starts at position a where the potential is zero and moves to a position b where the potential has a positive value V. (You may want to review the discussion of electric potential in Section 17-3.) The electron potential energy then changes from $U_a = qV_a = (-e)(0)$ to $U_b = qV_b = (-e)V = -eV$. The change in electric potential energy is

$$\Delta U = U_b - U_a = (-eV) - 0 = -eV < 0$$

Particle description	Wave description
Particle, mass m — Speed v_1 Momentum $p_1 = mv_1$	Wavelength $\lambda_1 = \dfrac{h}{p_1}$
Particle, mass m — Speed $v_2 = 2v_1$ Momentum $p_2 = mv_2 = 2p_1$	Wavelength $\lambda_2 = \dfrac{h}{p_2} = \dfrac{h}{2p_1} = \dfrac{\lambda_1}{2}$

Figure 26-9 Wave-particle duality Matter has both particle aspects (speed and momentum) and wave aspects (wavelength).

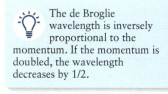 The de Broglie wavelength is inversely proportional to the momentum. If the momentum is doubled, the wavelength decreases by 1/2.

The electric potential energy decreases by an amount eV. Mechanical energy is conserved if the only force acting on the electron is the (conservative) electric force, so the decrease in electric potential energy equals the gain in kinetic energy of the electron:

$$\Delta K = -\Delta U = -(-eV) = +eV > 0 \tag{26-11}$$

If the electron starts at rest, its initial kinetic energy is zero and its final kinetic energy is $K = (1/2)mv^2$, so $\Delta K = (1/2)mv^2 - 0 = (1/2)mv^2$. (We're assuming that the electron is moving at a speed much slower than the speed of light c, so we don't have to use the Einsteinian expression for kinetic energy.) Then, from Equation 26-11, the final kinetic energy of the electron is

$$K = \frac{1}{2}mv^2 = eV$$

Solve for the final speed of the electron:

$$v^2 = \frac{2K}{m} = \frac{2eV}{m} \quad \text{so} \quad v = \sqrt{v^2} = \sqrt{\frac{2eV}{m}} \tag{26-12}$$

The final momentum of the electron is $p = mv$, and its final wavelength is $\lambda = h/p = h/(mv)$ from Equation 26-10. If we substitute v from Equation 26-12 into the formula for the wavelength of the electron, we get

$$\lambda = \frac{h}{mv} = \frac{h}{m}\sqrt{\frac{m}{2eV}} = \frac{h}{\sqrt{2meV}} \tag{26-13}$$

Suppose that the electron is accelerated through a potential difference $V = 50.0$ V. Substituting this into Equation 26-13 along with the values of Planck's constant h, the electron mass m, and the magnitude e of the electron charge, we get

$$\lambda = \frac{6.63 \times 10^{-34} \text{ J} \cdot \text{s}}{\sqrt{2(9.11 \times 10^{-31} \text{ kg})(1.60 \times 10^{-19} \text{ C})(50.0 \text{ V})}}$$
$$= 1.74 \times 10^{-10} \text{ m} = 0.174 \text{ nm}$$

To see how to measure such a short electron wavelength, note that a photon with this wavelength is in the x-ray region of the electromagnetic spectrum (see Figure 22-2). It was known in the 1920s that x rays show interference effects when they reflect from adjacent atoms in a crystal: At certain angles waves that reflect from one atom will interfere constructively (so the reflected intensity is high) with those that reflect from neighboring atoms, while at other angles they interfere destructively (so the reflected intensity is near zero). This can happen because the spacing between adjacent atoms in a crystal is around 0.1 nm, comparable to the wavelength of the x rays. So if electrons have a wave aspect, we expect that a beam of electrons that have been accelerated from rest through 50.0 V should display the same kind of interference effects as a beam of x rays.

In 1927 the American physicists Clinton Davisson and Lester Germer performed precisely this kind of experiment using a beam of electrons directed at a target of crystalline nickel. They found that the intensity of reflected electrons was greater for certain angles, just as for x rays. What's more, the angles at which this maximum intensity occurred were precisely those expected if the wavelength of electrons was given by the de Broglie relation, Equation 26-10. This groundbreaking result was quickly confirmed in experiments carried out by the British physicist G. P. Thomson. These results resoundingly confirmed de Broglie's remarkable hypothesis and showed that matter does indeed have a wave aspect. (The 1927 Nobel Prize in Physics went to de Broglie; the 1937 prize was shared by Davisson and Thomson.) The dual character of *both* light and matter, which have both wave and particle characteristics, is called **wave-particle duality**.

BioMedical Wave-particle duality is both surprising and counterintuitive. It is also of tremendous practical use. One important application that has revolutionized biology is the *electron microscope*. A major limitation of ordinary microscopes is that the smallest detail that can be resolved is about the size of a wavelength of visible light (about 380 to 750 nm). This makes microscopes useless for studying the structure of viruses, for example, which range in size from 5 to 300 nm. But as our above example of the 50.0-V electron shows, the wavelength of electron waves can be a fraction of a nanometer. So images made with an *electron microscope* can reveal details that are forever hidden from an ordinary visible-light microscope. Figure 26-10 is an electron microscope image of an influenza virus, in which details smaller than a nanometer across can be seen.

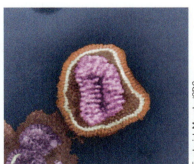

Frederick Murphy/CDC

Figure 26-10 An electron micrograph When a beam of low-energy electrons is shot through a thin slice of a specimen in a transmission electron microscope (TEM), the pattern formed by the diffracted electrons forms an image. A TEM captured this (false-color) image of an influenza virus particle, which is only about 100 nm in diameter. Because the wavelengths of low-energy electrons are so much shorter than those of light, a TEM is capable of significantly better resolution than light microscopes (better than 0.005 nm for a TEM, compared to about 0.2 μm with the most powerful optical microscopes).

If particles have wave aspects, why don't we notice these aspects for objects around us? The answer is that any object large enough to be seen by the naked eye has a relatively large momentum, so its de Broglie wavelength (which is inversely proportional to momentum) is infinitesimal. As an example, a dust mote floating in the air (such as you might see when a shaft of sunlight comes through the window) has a mass of about 8×10^{-10} kg. If it drifts at a speed of 1 mm/s = 10^{-3} m/s, its momentum is

$$p = mv = (8 \times 10^{-10} \text{ kg})(10^{-3} \text{ m/s}) = 8 \times 10^{-13} \text{ kg} \cdot \text{m/s}$$

and its de Broglie wavelength is

$$\lambda = \frac{h}{p} = \frac{6.63 \times 10^{-34} \text{ J} \cdot \text{s}}{8 \times 10^{-13} \text{ kg} \cdot \text{m/s}} = 8 \times 10^{-22} \text{ m}$$

That's about 10^{-6} of the diameter of a proton! It's impossible to see wave effects from a wave with such a tiny wavelength. To see diffraction of such a wave, we would have to create a slit whose width is much smaller than the width of a single proton. Wave effects are even smaller for larger objects (greater mass m) moving faster (greater speed v). So the wave aspect of matter is generally noticeable only on the atomic or subatomic scale. Particles such as electrons exhibit noticeable wave properties; objects such as dust motes, baseballs, and humans do not.

EXAMPLE 26-3 Finding the Wavelength of a Room-Temperature Neutron

A nuclear reactor emits *thermal neutrons*. These are neutrons that behave as though they were particles of an ideal gas at Kelvin temperature T. We learned in Section 14-3 that the average kinetic energy of a particle in an ideal gas at temperature T is $(3/2)kT$, where $k = 1.381 \times 10^{-23}$ J/K is the Boltzmann constant. Calculate the de Broglie wavelength of an average neutron at 293 K (room temperature). The mass of a neutron is 1.67×10^{-27} kg.

Set Up

We'll first calculate the kinetic energy of an average neutron using Equation 14-13, which we learned in Section 14-3. From this we can find the speed and magnitude of momentum of an average neutron. Equation 26-10 will then allow us to calculate the de Broglie wavelength of such a neutron.

de Broglie wavelength:

$$\lambda = \frac{h}{p} \qquad (26\text{-}10)$$

Temperature and average translational kinetic energy of an ideal gas particle:

$$K_{\text{translational, average}} = \frac{1}{2}m(v^2)_{\text{average}} = \frac{3}{2}kT \qquad (14\text{-}13)$$

Solve

Calculate the translational kinetic energy, speed, and momentum of an average neutron at $T = 293$ K.

From Equation 14-13:

$$K_{\text{translational, average}} = \frac{3}{2}kT = \frac{3}{2}(1.381 \times 10^{-23} \text{ J/K})(293 \text{ K})$$
$$= 6.07 \times 10^{-21} \text{ J}$$

Calculate the speed of a neutron with this kinetic energy:

$$K_{\text{translational, average}} = \frac{1}{2}mv^2 \text{ so}$$

$$v = \sqrt{\frac{2K_{\text{translational, average}}}{m}} = \sqrt{\frac{2(6.07 \times 10^{-21} \text{ J})}{1.67 \times 10^{-27} \text{ kg}}}$$
$$= 2.70 \times 10^3 \text{ m/s}$$

The magnitude of momentum of a neutron with this speed is

$$p = mv = (1.67 \times 10^{-27} \text{ kg})(2.70 \times 10^3 \text{ m/s})$$
$$= 4.50 \times 10^{-24} \text{ kg} \cdot \text{m/s}$$

Calculate the de Broglie wavelength of such a neutron.

From Equation 26-10:

$$\lambda = \frac{h}{p} = \frac{6.63 \times 10^{-34} \text{ J} \cdot \text{s}}{4.50 \times 10^{-24} \text{ kg} \cdot \text{m/s}} = 1.47 \times 10^{-10} \text{ m} = 0.147 \text{ nm}$$

(Recall that 1 J = 1 kg·m²/s² and 1 nm = 10⁻⁹ m.)

Reflect

The de Broglie wavelength of a thermal neutron is about 0.147 nm, a distance that is typical of the size of atoms and of the spacing between atoms within a molecule. For this reason thermal neutrons are useful for studying the structure of complex molecules such as proteins. When the neutrons scatter from a protein molecule, they diffract and produce a diffraction pattern that is characteristic of the particular arrangement of atoms in the molecule. X rays can have the same wavelength, but they interact with the charges within atoms and so scatter only weakly from relatively small atoms (with a small amount of internal charge) such as hydrogen, carbon, nitrogen, and oxygen found in proteins. Neutrons, by contrast, are electrically neutral and actually scatter more strongly from smaller atoms than from larger ones. This makes neutrons superior to x rays for studies of protein structure.

GOT THE CONCEPT? 26-3 Ranking de Broglie Wavelengths

 Rank the following objects in order of their de Broglie wavelength, from longest to shortest. (a) A proton moving at 2.00×10^3 m/s; (b) a proton moving at 4.00×10^3 m/s; (c) an electron moving at 2.00×10^3 m/s; (d) an electron moving at 4.00×10^3 m/s.

TAKE-HOME MESSAGE FOR Section 26-4

✔ Particles can exhibit wave properties such as diffraction.

✔ The wavelength of a particle is inversely proportional to its momentum. Hence wave effects are noticeable only for very small particles such as electrons, for which the momentum is very small.

26-5 The spectra of light emitted and absorbed by atoms show that atomic energies are quantized

We have seen that the late nineteenth and early twentieth centuries were years of tremendous change in the study of physics. Studying the photoelectric effect and blackbody radiation led to the revolutionary concept that light has particle aspects, and de Broglie introduced the no less revolutionary idea that matter has wave aspects. During this same time a key set of experiments radically transformed our understanding of the nature of atoms.

The Nuclear Atom

The early Greeks introduced the idea of the atom, a unit of matter so small that it could not be subdivided. (The word "atom" is derived from the Greek term for indivisible.) In 1897 the British physicist J. J. Thomson discovered that atoms are not in fact indivisible but have an internal structure: All atoms contain negatively charged particles (electrons) that can be removed from the atom. It was known that atoms are electrically neutral, so there must also be positively charged material inside an atom. But what form does this positive charge take?

Thomson proposed that most of the mass of the atom is in the form of electrons and that the positively charged material is a low-density sort of jelly in which the electrons are embedded. This model is sometimes called the plum pudding model, since Thomson envisioned electrons scattered throughout the positive charge much like raisins in the traditional English dessert. (If you're not familiar with plum pudding, think of electrons as pieces of fruit embedded in a gelatin dessert or salad.)

A crucial experiment that tested this model was carried out in 1909 at the University of Manchester in England by the New Zealand–born chemist and physicist Ernest Rutherford with his colleagues Hans Geiger and Ernest Marsden. They fired

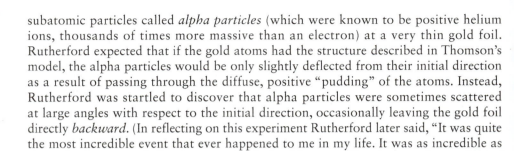

subatomic particles called *alpha particles* (which were known to be positive helium ions, thousands of times more massive than an electron) at a very thin gold foil. Rutherford expected that if the gold atoms had the structure described in Thomson's model, the alpha particles would be only slightly deflected from their initial direction as a result of passing through the diffuse, positive "pudding" of the atoms. Instead, Rutherford was startled to discover that alpha particles were sometimes scattered at large angles with respect to the initial direction, occasionally leaving the gold foil directly *backward*. (In reflecting on this experiment Rutherford later said, "It was quite the most incredible event that ever happened to me in my life. It was as incredible as if you fired a 15-in. shell [a large projectile fired from a military weapon] at a piece of tissue paper and it came back and hit you. On consideration, I realized that this scattering backwards must be the result of a single collision, and when I made calculations I saw that it was impossible to get anything of that order of magnitude unless you took a system in which the mass of the atom was concentrated in a minute nucleus.")

To account for this, Rutherford proposed a model of the atom in which negatively charged electrons orbit a small, positively charged *nucleus* that contains nearly all of the atom's mass. In Rutherford's model most of the volume of each atom is empty, so most alpha particles fired at the gold foil would experience only slight deflections as they passed through. But once in a while, about one time out of every 10,000, an alpha particle would approach a gold nucleus almost head on and be scattered at a large angle, sometimes directly backward.

In Rutherford's model electrons orbit the nucleus like a satellite orbiting Earth. But while we can place an Earth satellite in an orbit of any size we wish, electrons in an atom behave differently: They can move only in very specific orbits around the nucleus. In this section we'll explore how studies of the spectrum of light emitted and absorbed by atoms led physicists to this discovery. In the following section we'll explore the theories that were developed to explain why only certain electron orbits are allowed.

The Discovery of Atomic Spectra

In the early part of the nineteenth century, the English scientist William Hyde Wollaston and the German physicist Joseph von Fraunhofer independently discovered dark lines in the spectrum of visible light coming from the Sun (Figure 26-11). These lines are always in the same locations within the spectrum. Some years later Gustav Kirchhoff, the same physicist we encountered in our study of electric circuits, was able to reproduce these same dark lines in the laboratory. He passed light from a lamp (made to simulate sunlight) through vapors created by heating sodium. The light from the lamp itself had a continuous spectrum like that of a blackbody, but the spectrum of light that had passed through the sodium vapor had two dark lines. Kirchhoff concluded that the dark lines result from certain specific colors of light being absorbed by the sodium vapor. What is more, these lines were at the same position as the closely spaced pair of lines in the yellow-orange region of the Sun's spectrum. (You can easily find these lines in Figure 26-11.) The same mechanism must therefore be happening with sunlight: The light coming from the solar interior has a continuous, blackbody-like spectrum, but certain wavelengths of that light are absorbed by atoms in the Sun's atmosphere. The Sun's atmosphere must contain sodium atoms identical to those in Kirchhoff's laboratory. In light of Kirchhoff's discovery, a spectrum like that shown in Figure 26-11 is called an **absorption spectrum**, and the dark lines are called **absorption lines**.

Scientists soon discovered that each element produces its own characteristic absorption lines when light passes through a vapor containing atoms of that element. Thus an absorption spectrum acts as a "fingerprint" of the chemical composition of the vapor that produced the absorption lines. This is how we know the chemical composition of the Sun's atmosphere. It's also how we determine the chemical composition of the atmospheres of distant stars like those in the photograph that opens this chapter, and how we know that all stars have basically the same chemical makeup (Figure 26-12). What was not understood was *why* atoms should selectively absorb only light of certain wavelengths and why the absorbed wavelengths should be different for atoms of different elements.

Figure 26-12 Cecilia Payne-Gaposchkin and the composition of stars In her 1925 doctoral thesis, British-born astronomer Cecilia Payne (later Payne-Gaposchkin) was the first to use absorption spectra like that in Figure 26-11 to show that all stars are made primarily of hydrogen and helium. She later became professor of astronomy at Harvard University and the first woman at Harvard to chair a department.

(a) The absorption spectrum produced by passing white light through a gas of hydrogen atoms

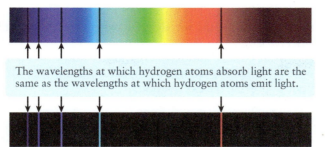

The wavelengths at which hydrogen atoms absorb light are the same as the wavelengths at which hydrogen atoms emit light.

(b) The emission spectrum produced by a heated gas of hydrogen atoms

Figure 26-13 Absorption and emission spectra of atomic hydrogen (a) When light passes through a gas, light of specific wavelengths is absorbed, forming dark lines. (b) When a gas is made to glow by passing an electric current through it, it emits only specific wavelengths of light.

- Atoms of each element absorb and emit light at wavelengths that are characteristic of that element.
- The characteristic wavelengths differ from one element to another.
- Hydrogen has the simplest arrangement of characteristic wavelengths of any element.

An important clue about the mystery of absorption spectra came from studying the light *emitted* by atoms. Physicists of the nineteenth century discovered that light created by heating a vapor gives rise to an **emission spectrum,** a spectrum that consists only of specific emitted wavelengths. What is more, if the vapor contains atoms of a certain element, the wavelengths in the emission spectrum from those atoms (the **emission lines**) are precisely the same as the wavelengths in the absorption spectrum of that same element (Figure 26-13). To explore this, we'll concentrate on the absorption and emission spectra of hydrogen, which has the simplest set of absorption and emission lines of any element. But the underlying physics applies to all elements.

Johann Balmer, a Swiss mathematician, made an analysis of the lines in the absorption and emission spectra of hydrogen. He devised a formula that both reproduced the wavelengths of lines that had been reported and correctly predicted the wavelengths of spectral lines that had not yet been observed. This was later extended by the Swedish physicist Johannes Rydberg. The *Rydberg formula* for the hydrogen spectral lines is

Wavelength of an absorption or emission line in the spectrum of **atomic hydrogen**

$$\frac{1}{\lambda} = R_H \left(\frac{1}{n^2} - \frac{1}{m^2} \right)$$

Rydberg formula for the spectral lines of hydrogen
(26-14)

Rydberg constant = 1.09737×10^7 m^{-1}

n and m are integers: n can be 1, 2, 3, 4,... and m can be any integer greater than n.

The value of the constant R_H in Equation 26-14, called the *Rydberg constant,* is chosen to match the experimental data. To four significant figures $R_H = 1.097 \times 10^7$ m^{-1}. As an example, the hydrogen absorption and emission lines shown in Figure 26-13 all correspond to $n = 2$ in Equation 26-14. The series of wavelengths for which $n = 2$ is called the *Balmer series.* For example, to get the wavelength of the red spectral line in Figure 26-13, set $n = 2$ and $m = 3$ in Equation 26-14 and then take the reciprocal:

$$\frac{1}{\lambda} = R_H \left(\frac{1}{2^2} - \frac{1}{3^2} \right) = (1.097 \times 10^7 \text{ m}^{-1}) \left(\frac{1}{4} - \frac{1}{9} \right) = 1.524 \times 10^6 \text{ m}^{-1}$$

$$\lambda = \frac{1}{1.524 \times 10^6 \text{ m}^{-1}} = 6.563 \times 10^{-7} \text{ m} = 656.3 \text{ nm}$$

The spectral line to the left of this one in Figure 26-13 (in the blue-green part of the spectrum) corresponds to $n = 2$ and $m = 4$; you can show that for this spectral

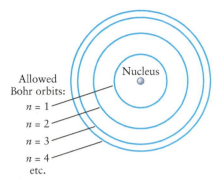

Figure 26-14 Bohr orbits in the hydrogen atom In the model devised by Niels Bohr, electrons in the hydrogen atom are allowed to be in certain orbits only. (The radii of the orbits are not shown to scale.)

line, $\lambda = 486.2$ nm. The wavelengths with $n = 1$ are all in the ultraviolet and are called the *Lyman series*; the wavelengths with $n = 3$ are all in the infrared and are called the *Paschen series*.

What Balmer and Rydberg did not know was *why* the spectral hydrogen lines were given by this relatively simple formula. It was left to the Danish physicist Niels Bohr to provide the explanation.

Energy Quantization

Bohr realized that to fully understand the structure of the hydrogen atom, he had to be able to derive Balmer's formula using the laws of physics. He first made the rather wild assumption that the electron in a hydrogen atom can orbit the nucleus only in certain specific orbits. (This was a significant break with the ideas of Newton, in whose mechanics any orbit should be possible.) Figure 26-14 shows the four smallest of these **Bohr orbits**, labeled by the numbers $n = 1$, $n = 2$, $n = 3$, and so on.

Although confined to one of these allowed orbits while circling the nucleus, an electron can jump from one Bohr orbit to another. For an electron to do this, the hydrogen atom must gain or lose a specific amount of energy. The atom must absorb energy for the electron to go from an inner to an outer orbit; the atom must release energy for the electron to go from an outer to an inner orbit. As an example, Figure 26-15 shows an electron jumping between the $n = 2$ and $n = 3$ orbits of a hydrogen atom as the atom absorbs or emits a photon.

When the electron jumps from one orbit to another, the energy of the photon that is emitted or absorbed equals the difference in energy between these two orbits. This energy difference, and hence the photon energy, is the same whether the jump is from a low orbit to a high orbit (Figure 26-15a) or from the high orbit back to the low one (Figure 26-15b). According to Einstein, if two photons have the same energy E, the relationship $E = hf$ (Equation 22-26) tells us that they must also have the same frequency f and hence the same wavelength $\lambda = c/f$. It follows that if an atom can emit photons of a given energy and wavelength, it can also absorb photons of precisely the same energy and wavelength. Thus Bohr's picture explains Kirchhoff's observation that atoms emit and absorb the same wavelengths of light.

The Bohr picture also helps us visualize what happens to produce an emission spectrum. When a gas is heated its atoms move around rapidly and can collide forcefully with each other. These energetic collisions excite the atoms' electrons into high orbits. The electrons then cascade back down to the innermost possible orbit, emitting photons whose energies are equal to the energy differences between different Bohr orbits. In this fashion a hot gas produces an emission line spectrum with a variety of different wavelengths.

To produce an absorption spectrum, begin with a relatively cool gas so that the electrons in most of the atoms are in inner, low-energy orbits. If a beam of light with a continuous spectrum is shone through the gas, most wavelengths will pass through undisturbed. Only those photons will be absorbed whose energies are just right to excite an electron to an allowed outer orbit. Hence only certain wavelengths will be absorbed, and dark lines will appear in the spectrum at those wavelengths.

As in Figure 26-15, the energy of the photon that is absorbed or emitted in a jump between orbits must be equal to the *difference* between the energy of the atom with the electron in the larger-radius, higher-energy orbit and the energy of the atom with the electron in the smaller-radius, lower-energy

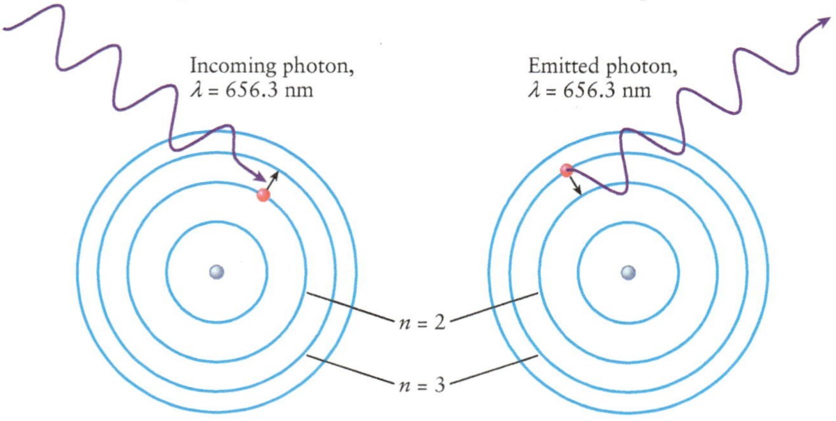

(a) Atom absorbs a 656.3-nm photon; absorbed energy causes electron to jump from the $n = 2$ orbit up to the $n = 3$ orbit

Incoming photon, $\lambda = 656.3$ nm

(b) Electron falls from the $n = 3$ orbit to the $n = 2$ orbit; energy lost by atom goes into emitting a 656.3-nm photon

Emitted photon, $\lambda = 656.3$ nm

$n = 2$

$n = 3$

Figure 26-15 The Bohr model explains absorption and emission spectra When a photon is (a) absorbed or (b) emitted by a hydrogen atom, the electron makes a transition or jump between two allowed orbits. The photon energy equals the difference in energy between the upper and lower electron orbits. (As in Figure 26-14, the radii of the orbits are not shown to scale.)

orbit. Bohr concluded that the numbers n and m in the Rydberg formula correspond to the numbers of the orbits between which an electron jumps as it absorbs or emits a photon. The value of n is the number of the lower orbit, and the value of m is the number of the upper orbit. For example, the jump shown in Figure 26-15 corresponds to $n = 2$ and $m = 3$.

We can better understand Bohr's idea by combining the Rydberg formula, Equation 26-14, with the expressions $E = hf$ and $f = c/\lambda$ for a photon. Together these latter two expressions say that the energy of a photon of wavelength λ is $E = hc/\lambda$. So if we multiply Equation 26-14 by hc, we get an expression for the energy of a photon absorbed or emitted by a hydrogen atom:

$$E_{\text{photon}} = \frac{hc}{\lambda} = hcR_{\text{H}}\left(\frac{1}{n^2} - \frac{1}{m^2}\right) = \frac{hcR_{\text{H}}}{n^2} - \frac{hcR_{\text{H}}}{m^2} = \left(-\frac{hcR_{\text{H}}}{m^2}\right) - \left(-\frac{hcR_{\text{H}}}{n^2}\right) \qquad (26\text{-}15)$$

If we say that a hydrogen atom has energy $E_{\text{atom},n} = -hcR_{\text{H}}/n^2$ when the electron is in the nth orbit and has energy $E_{\text{atom},m} = -hcR_{\text{H}}/m^2$ when the electron is in the mth orbit, where m is greater than n, then we can rewrite Equation 26-15 as

$$E_{\text{photon}} = E_{\text{atom},m} - E_{\text{atom},n} \qquad (26\text{-}16)$$

Equation 26-16 uses the idea that the energy of a hydrogen atom is **quantized**: That is, the energy can only have certain values. These energies are given by

$$E_{\text{atom},n} = -\frac{hcR_{\text{H}}}{n^2} \quad \text{where } n = 1, 2, 3, 4, \ldots \qquad (26\text{-}17)$$

Note that the energy of the atom is negative, and greater values of n correspond to energies that are less negative (that is, closer to zero). This agrees with the idea that the larger the orbit and the greater the value of n for that orbit, the higher the energy.

Each quantized value of the energy is called an **energy level**. Figure 26-16 shows several of the energy levels of the hydrogen atom, along with vertical arrows that show the energy of the photon that must be absorbed or emitted in a jump or transition between levels. The transitions that correspond to the Lyman series involve a photon with a very large amount of energy, so these photons have a high frequency and short wavelength: They are all in the ultraviolet part of the spectrum. By contrast, the transitions that correspond to the Paschen series involve a photon with a very small amount of energy, which is why these photons have a low frequency and long wavelength and are in the infrared part of the spectrum. The Balmer series is intermediate between these two; the wavelengths are in either the visible range (380 to 750 nm) or the ultraviolet range.

Elements other than hydrogen also absorb and emit light at specific wavelengths, although those wavelengths do not follow a simple mathematical pattern like the characteristic wavelengths of hydrogen given by Equation 26-14. The conclusion is that there are quantized energy levels for atoms of other elements, but the arrangement of energy levels is more complex than for hydrogen. In the following section we'll see how Niels Bohr justified the quantization of energy for the relatively simple case of the hydrogen atom, and how he was able to reproduce Equation 26-17 for the energy levels. We'll then use these ideas to gain insight into the structure of other atoms.

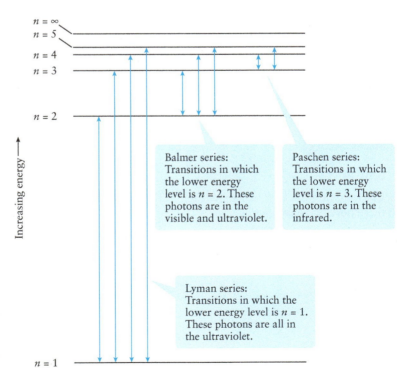

Balmer series: Transitions in which the lower energy level is $n = 2$. These photons are in the visible and ultraviolet.

Paschen series: Transitions in which the lower energy level is $n = 3$. These photons are in the infrared.

Lyman series: Transitions in which the lower energy level is $n = 1$. These photons are all in the ultraviolet.

Figure 26-16 Hydrogen energy levels This figure shows some of the lower-lying energy levels of the hydrogen atom and possible transitions between those levels. Note that the energy difference is greatest between the $n = 1$ and $n = 2$ levels, less between the $n = 2$ and $n = 3$ levels, and even less between the $n = 3$ and $n = 4$ levels.

WATCH OUT! Energy quantization is not just an obscure effect in atomic physics.

! Discrete energy levels play an important role in modern technology. One example is a *laser*, a device that emits an intense beam of light of a very specific wavelength. This is possible because the laser contains a material with two distinct energy levels. Light of the laser's characteristic wavelength is emitted when transitions take place from the upper level to the lower one. What makes the laser unique is that these transitions occur coherently rather than at random. Energy is added to the material to pump its molecules into excited states. The excited molecules naturally want to transition to a lower state, but if left on their own, would do so at random times. However, when a photon of energy equal to the difference between two states is sent into the material, it can *stimulate* this transition and cause a second photon of that same energy to be released. This photon can stimulate further emission, so the number of emitted photons increases. The emitted photons are all in phase and all travel in the same general direction. The net result is an intense beam. If materials did not have quantized energy levels, lasers could not exist.

Fluorescent light bulbs also depend on energy quantization. The material inside the bulb emits ultraviolet photons when an electric current passes through it. These photons are absorbed by the white coating on the inner surface of the bulb. Since ultraviolet photons are very energetic, this excites the material of the coating to very high energy levels. The coating then drops down to its initial energy level in a series of small steps. (It's like taking a big leap to the top of a staircase, then coming carefully down the staircase one step at a time.) Each small step between closely spaced energy levels emits a low-energy, visible-light photon. There are so many such energy levels, with a variety of spacing between them, that the net result is that a mixture of almost all visible colors—that is, white light—is emitted from the bulb.

EXAMPLE 26-4 Photon Possibilities

A collection of hydrogen atoms is excited to the $n = 3$ energy level. What are the possible wavelengths that these atoms could emit as they return to the lowest energy ($n = 1$) level?

Set Up

There are two routes that an atom can take from the $n = 3$ level to the $n = 1$ level. One, it could drop down to the $n = 1$ level in a single step by emitting a single photon whose energy is equal to the difference between the energies of the $n = 3$ and $n = 1$ levels. The wavelength of this photon is given by Equation 26-14 with $n = 1$ and $m = 3$. Two, the atom could first drop to the $n = 2$ level by emitting a photon (with a wavelength given by Equation 26-14 with $n = 2$ and $m = 3$), then emit a second photon as it drops from the $n = 2$ level to the $n = 1$ level (with a wavelength given by Equation 26-14 with $n = 1$ and $m = 2$). So three different wavelengths can be emitted by the excited atoms. We'll calculate each of these in turn.

Rydberg formula for the spectral lines of hydrogen:

$$\frac{1}{\lambda} = R_H \left(\frac{1}{n^2} - \frac{1}{m^2} \right) \quad (26\text{-}14)$$

Solve

Use Equation 26-14 to calculate each of the three possible wavelengths.

For the $n = 3$ to $n = 1$ transition:

$$\frac{1}{\lambda} = R_H \left(\frac{1}{1^2} - \frac{1}{3^2} \right) = (1.097 \times 10^7 \text{ m}^{-1})\left(1 - \frac{1}{9} \right)$$
$$= 9.751 \times 10^6 \text{ m}^{-1}$$
$$\lambda = \frac{1}{9.751 \times 10^6 \text{ m}^{-1}} = 1.026 \times 10^{-7} \text{ m} = 102.6 \text{ nm}$$

This is an ultraviolet wavelength.

For the $n = 3$ to $n = 2$ transition:

$$\frac{1}{\lambda} = R_H \left(\frac{1}{2^2} - \frac{1}{3^2} \right) = (1.097 \times 10^7 \text{ m}^{-1})\left(\frac{1}{4} - \frac{1}{9} \right)$$
$$= 1.524 \times 10^6 \text{ m}^{-1}$$
$$\lambda = \frac{1}{1.524 \times 10^6 \text{ m}^{-1}} = 6.563 \times 10^{-7} \text{ m} = 656.3 \text{ nm}$$

This is a visible wavelength (in the red part of the spectrum).

For the $n = 2$ to $n = 1$ transition:

$$\frac{1}{\lambda} = R_H\left(\frac{1}{1^2} - \frac{1}{2^2}\right) = (1.097 \times 10^7 \text{ m}^{-1})\left(1 - \frac{1}{4}\right)$$

$$= 8.228 \times 10^6 \text{ m}^{-1}$$

$$\lambda = \frac{1}{8.228 \times 10^6 \text{ m}^{-1}} = 1.215 \times 10^{-7} \text{ m} = 121.5 \text{ nm}$$

This is another ultraviolet wavelength.

Reflect

Comparing with Figure 26-16 shows that the 656.3-nm wavelength represents an emission line of the Balmer series, while the 102.6- and 121.5-nm wavelengths represent emission lines of the Lyman series.

GOT THE CONCEPT? 26-4   Ranking Hydrogen Transitions

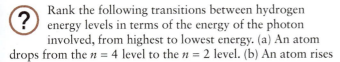

Rank the following transitions between hydrogen energy levels in terms of the energy of the photon involved, from highest to lowest energy. (a) An atom drops from the $n = 4$ level to the $n = 2$ level. (b) An atom rises from the $n = 3$ level to the $n = 5$ level. (c) An atom drops from the $n = 3$ level to the $n = 1$ level. (d) An atom drops from the $n = 4$ to the $n = 3$ level.

TAKE-HOME MESSAGE FOR   Section 26-5

✔ Atoms are composed of electrons orbiting a positively charged nucleus that has most of the mass of the atom.

✔ In the Bohr model an electron's orbits around the nucleus can have only certain well-defined energies. Thus the energy of the atom is quantized and can have only certain values. The atom cannot have energies intermediate between those values.

✔ Electrons that make a transition from one allowed orbit to another either radiate or absorb a photon of a well-defined energy. This gives rise to the lines in emission and absorption spectra.

26-6   Models by Bohr and Schrödinger give insight into the intriguing structure of the atom

The force that pulls the Moon toward Earth is directed toward Earth's center, yet the Moon does not fall into Earth. In the same way, the negatively charged electron in an atom experiences a Coulomb force that points directly toward the positively charged protons at the center of the atom. Why does the electron orbit rather than fall into the proton? You likely have an intuitive answer that both the Moon and the electron *orbit* rather than fall in. (In a real sense each *is* falling in, but it is always missing the target!) Niels Bohr based his model of the hydrogen atom, for which he received the Nobel Prize in Physics in 1922, on the physics of atomic orbits. The **Bohr model** provides a theoretical foundation for the physics of atomic spectra that we encountered in the previous section.

The Bohr Model

Bohr made two fundamental assumptions in describing the orbit of an electron around a positive atomic nucleus. First he modeled the orbit as uniform circular motion, that is, as an electron moving at constant speed in a circular path. This is not exactly correct but more than satisfactory to provide a broad understanding of the atom. Bohr's second assumption was that only specific values are allowed for the angular momentum of an orbiting electron. We will return to this second assumption shortly.

Bohr considered a single electron orbiting a nucleus of charge $+Ze$, where Z, the *atomic number* of an atom, is the number of protons in the nucleus. The orbiting electron experiences only one force, the Coulomb attraction between it and the protons in the atomic nucleus. The magnitude of the Coulomb force on an electron orbiting in a circle of radius r is, from Equation 16-1,

$$F = \frac{k(Ze)(e)}{r^2} = \frac{kZe^2}{r^2}$$

Here $k = 8.99 \times 10^9$ N·m^2/C^2 is the Coulomb constant. Newton's second law (Equation 4-2) requires that this force equal the mass of the electron m_e multiplied by the acceleration it experiences. For an object in uniform circular motion at speed v, the magnitude of the acceleration is, from Equation 3-17,

$$a = \frac{v^2}{r}$$

Combining the above two equations into Newton's second law gives

$$\frac{kZe^2}{r^2} = \frac{m_e v^2}{r}$$

If we multiply both sides of this equation by r, we get

(26-18)
$$\frac{kZe^2}{r} = m_e v^2$$

Let's set Equation 26-18 aside for a moment and examine Bohr's second assumption—a requirement that only specific values are allowed for the angular momentum associated with the electron. Bohr recognized that the dimensions of Planck's constant h are those of angular momentum, so he constrained the electron's angular momentum to be only multiples of h. Specifically, this requirement is

(26-19)
$$L = n\left(\frac{h}{2\pi}\right) = n\hbar$$

(orbital angular momentum in the Bohr model)

where L is the electron's orbital angular momentum and n is any integer starting from 1. The constant $\hbar$ (pronounced "h bar") is defined to be h divided by 2π.

To relate angular momentum to Equation 26-18, we express L in terms of the mass of the electron m_e, its speed v, and its distance from the center of the atom r using Equation 9-29:

Magnitude of the angular momentum of a particle
(9-29)

Magnitude of the **angular momentum of a particle**   Magnitude of the **linear momentum of a particle**

$$L = r_\perp p = rp \sin \phi$$

Angle between the vector from axis to particle and the momentum vector

Perpendicular distance from the axis to the line of the momentum vector

Distance from the axis to the particle

The magnitude of the electron's linear momentum is $p = m_e v$. Since the electron moves in a circle, the vector from the rotation axis to the particle always has the same radius r and is always perpendicular to the momentum vector. So $\phi = 90°$, $\sin \phi = 1$, and

(26-20)
$$L = r m_e v$$

We can now rewrite Equation 26-18 in terms of the angular momentum L by multiplying the right-hand side by 1 in the form of $(r^2/r^2)(m_e/m_e)$:

$$\frac{kZe^2}{r} = \frac{r^2 m_e^2 v^2}{m_e r^2} = \frac{L^2}{m_e r^2}$$

Bohr's requirement that the electron's angular momentum is an integer multiple of $\hbar$ then gives

$$\frac{kZe^2}{r} = \frac{(n\hbar)^2}{m_e r^2}$$

or

$$r_n = \frac{n^2\hbar^2}{m_e k Z e^2} \quad \text{where } n = 1, 2, 3, \ldots$$

(26-21)

(orbital radii in the Bohr model)

We add the subscript "n" to the variable r to indicate that the radius can take on only specific values and that the allowed values of radius depend on n. Because the values of r_n are proportional to the square of an integer, the orbital radii of electrons in an atom are quantized.

Notice that both n and Z in Equation 26-21 are dimensionless. Because r_n is a distance, all of the other terms on the right-hand side, taken as they appear in the equation, must have dimensions of distance as well. This distance, usually written as a_0 and called the *Bohr radius*, is

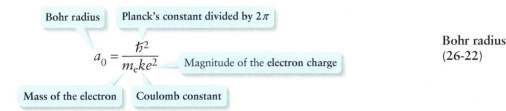

Bohr radius
(26-22)

Using the values for $\hbar$, m_e, k, and e, we find that the value of the Bohr radius a_0 is approximately

$$a_0 = 0.529 \times 10^{-10} \text{ m} = 0.0529 \text{ nm}$$

In terms of a_0 the quantized radii of the electron orbits (Equation 26-21) are

Radius of the nth allowed orbit in a single-electron atom $n = 1, 2, 3,\ldots$

$$r_n = \frac{n^2}{Z} a_0$$

Atomic number of the atom   Bohr radius

Orbital radii in the Bohr model
(26-23)

The integer n identifies the orbit, where the $n = 1$ orbit is the closest to the nucleus. Because every element is distinguished by the number of protons it carries, the atomic number Z specifies a particular element. So, for example, setting Z equal to 1 gives the radii of the electron orbits in a hydrogen atom, and setting Z equal to 1 and n equal to 1 gives the radius of the first electron orbit in hydrogen. This is the normal state, usually referred to as the ground state, of hydrogen. Moreover, notice that the radius of the ground state orbit of hydrogen is the Bohr radius. In other words, the radius of a typical hydrogen atom is about 0.05 nm. We can also conclude from Equation 26-23 together with the value of the Bohr radius that, in general, atoms are no more than a few nanometers in radius. The Bohr model sets the scale for atomic sizes.

What is the energy of an electron orbiting an atomic nucleus according to the Bohr model? This is the sum of its kinetic energy and its electric potential energy. The electric potential energy for two point charges q_1 and q_2 is $U_{\text{electric}} = kq_1q_2/r$ (Equation 17-3), so

$$E = \frac{1}{2}m_e v^2 + \frac{k(-e)(Ze)}{r_n}$$

(26-24)

We can write the kinetic energy term in terms of angular momentum in a way that is similar to our approach when developing the relationship for r_n. Using Equation 26-20 shows that the kinetic energy term becomes

$$\frac{1}{2}m_e v^2 = \frac{1}{2}\frac{L^2}{m_e r_n^2} = \frac{n^2\hbar^2}{2m_e r_n^2}$$

so the total electron energy from Equation 26-24 is

$$E = \frac{n^2\hbar^2}{2m_e r_n^2} - \frac{kZe^2}{r_n}$$

Substituting Equation 26-23 for r_n gives the energy in terms of only the physical constants and the counting integer n:

$$E = \frac{n^2\hbar^2}{2m_e} \left(\frac{m_e kZe^2}{n^2\hbar^2} \right)^2 - kZe^2 \frac{m_e kZe^2}{n^2\hbar^2}$$

Simplifying this is straightforward when you notice that the numerator of both terms is $m_e(kZe^2)^2$ and that both terms have $n^2\hbar^2$ in the denominator:

$$E = \frac{m_e(kZe^2)^2}{2n^2\hbar^2} - \frac{m_e(kZe^2)^2}{n^2\hbar^2}$$

or

(26-25)
$$E_n = -\frac{m_e(kZe^2)^2}{2n^2\hbar^2}, \quad n = 1, 2, 3, \ldots$$

(electron energies in the Bohr model)

We add the subscript "n" to the variable E to indicate that the orbital energy of the electron can take on only specific values that depend on n. Because the values of E_n are proportional to $1/n^2$, the orbital energy of electrons in an atom is quantized. This is in agreement with our conclusion in the previous section that for atomic spectral lines to occur only at specific wavelengths, electrons must orbit the hydrogen nucleus with specific, well-defined energies. Note also that the energy is equal to a negative constant divided by n^2, exactly in accordance with Equation 26-17. (We'll see below that the numerical value of the constant is the same in Equation 26-25 as in Equation 26-17.)

The value of energy of an electron in orbit around an atomic nucleus is negative but closer and closer to zero for increasing values of n. In other words, the lowest energy orbit is the one closest to the nucleus, as we would expect. That the energy is negative emphasizes that the electron is bound to the nucleus, and that energy must be supplied to either move the electron to a higher orbit or to break the electron free from the nucleus altogether.

Because E_n is an energy, and because both n and Z are dimensionless, the combination of other quantities on the right-hand side of Equation 26-25 must have dimensions of energy as well. This energy, usually written as E_0, is called the *Rydberg energy*:

Rydberg energy
(26-26)

Coulomb constant   Magnitude of the **electron charge**

Mass of the electron

Rydberg energy

$$E_0 = \frac{m_e(ke^2)^2}{2\hbar^2} = 2.18 \times 10^{-18}\,\text{J} = 13.6\,\text{eV}$$

Planck's constant divided by 2π

In Equation 26-26 we're given the value of the Rydberg energy in joules and in electron volts (eV). The energy that an electron acquires when it moves through a potential difference of 1 V is 1 eV, or approximately 1.602×10^{-19} J. To give you an idea of the amount of energy 1 eV represents, a photon of visible light carries between about 1.5 and 3 eV.

Using Equation 26-26 we can write Equation 26-25 for the quantized electron orbital energy in terms of the Rydberg energy E_0:

Electron energies in the
Bohr model
(26-27)

Electron energy for the *n*th allowed orbit in a single-electron atom

Atomic number of the atom

$$E_n = -\frac{Z^2}{n^2} E_0$$

$n = 1, 2, 3, \ldots$ Rydberg energy = 13.6 eV

Again the integer n identifies the orbit, and the atomic number Z specifies a particular element. Setting Z equal to 1 and n equal to 1 therefore tells us that the energy of the ground state of hydrogen is −13.6 eV. We can also conclude from Equation 26-27 that, in general, the energy of electrons in orbit around an atomic nucleus is between about −10 eV and, for the largest elements (for which Z is about 100), −10^5 eV. The Bohr model sets the scale for atomic electron energies.

What does the Bohr model predict for the atomic spectral lines of hydrogen? Every line results from a transition between two electron orbits; that is, the energy of the emitted photon is the energy difference ΔE between two allowed orbits (see Equation 26-16). Since the energy of a photon is hf and its frequency $f = c/\lambda$, the wavelength λ of the photon is then

$$\lambda = \frac{hc}{\Delta E}$$

and the reciprocal of the wavelength (which is what appears in the Rydberg formula) is

$$\frac{1}{\lambda} = \frac{\Delta E}{hc} \qquad (26\text{-}28)$$

Let's consider the transition of an electron from a higher orbit m down to a lower orbit n. We can determine the energy difference between these two orbits by applying Equation 26-27. Atomic number Z equals 1 for hydrogen, so

$$\Delta E = -\frac{1}{m^2}E_0 - \left(-\frac{1}{n^2}\right)E_0$$

or

$$\Delta E = \left(\frac{1}{n^2} - \frac{1}{m^2}\right)E_0$$

Substituting this into Equation 26-28 yields

$$\frac{1}{\lambda} = \frac{E_0}{hc}\left(\frac{1}{n^2} - \frac{1}{m^2}\right)$$

Compare this to the Rydberg formula, Equation 26-14. It has exactly the same form! How does the value of E_0/hc compare to the value of R_H, which equals approximately 1.10×10^7 m^{-1}? In SI units

$$\frac{E_0}{hc} = \frac{2.18 \times 10^{-18}\ \text{J}}{(6.63 \times 10^{-34}\ \text{J·s})(3.00 \times 10^8\ \text{m/s})} = 1.10 \times 10^7\ \text{m}^{-1}$$

The ratio E_0/hc equals R_H, so $E_0 = hcR_H$. That's just what we expect if we compare Equation 26-17 (in which we deduced the energies from the Rydberg equation) and Equation 26-27 (in which we derived the energies from the Bohr assumptions). We conclude that the Bohr model is entirely consistent with the observed spectral lines of hydrogen as described by the Rydberg formula.

EXAMPLE 26-5 Lowest Energy Level of a Lithium Ion

Find the (a) radius, (b) energy, and (c) speed of an electron in the lowest energy level of the doubly ionized lithium ion. An atom of lithium ($Z = 3$) has three electrons, so this ion has just one electron.

Set Up

For this ion, $Z = 3$; for the lowest energy level $n = 1$. We'll use Equation 26-23 to calculate the radius of this orbit, Equation 26-27 to calculate the energy, and Equation 26-19 for angular momentum to determine the electron speed.

Orbital radii in the Bohr model:

$$r_n = \frac{n^2 a_0}{Z} \quad \text{where } n = 1, 2, 3, \ldots \qquad (26\text{-}23)$$

Electron energies in the Bohr model:

$$E_n = -\frac{Z^2}{n^2}E_0 \quad \text{where } n = 1, 2, 3, \ldots \qquad (26\text{-}27)$$

Orbital angular momentum in the Bohr model:

$$L = n\left(\frac{h}{2\pi}\right) = n\hbar \qquad (26\text{-}19)$$

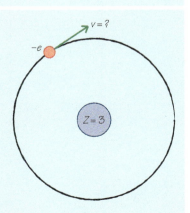

Solve

(a) Calculate the radius of the orbit.	From Equation 26-23 with $Z = 3$ and $n = 1$: $$r_1 = \frac{1^2 a_0}{3} = \frac{a_0}{3} = \frac{0.0529 \text{ nm}}{3} = 0.0176 \text{ nm}$$
(b) Calculate the energy.	From Equation 26-27 with $Z = 3$ and $n = 1$: $$E_n = -\frac{3^2}{1^2} E_0 = -9E_0 = -9(13.6 \text{ eV}) = -122 \text{ eV}$$
(c) To find an expression for the electron speed, combine Equation 26-19 with the equation for the angular momentum of a particle moving in a circular orbit.	Equation 26-20 tells us that the angular momentum of an electron of mass m_e moving in an orbit of radius r at speed v is $L = rm_e v$. Set this equal to the expression for L in Equation 26-19: $rm_e v = n\hbar$ so $$v = \frac{n\hbar}{rm_e}$$ We are interested in the case of $n = 1$ and calculated r_1 in part (a). Substituting the values for $\hbar = h/(2\pi)$ and the mass m_e of the electron gives $$v = \frac{1(6.63 \times 10^{-34} \text{ J·s})}{2\pi (1.76 \times 10^{-11} \text{ m})(9.11 \times 10^{-31} \text{ kg})}$$ $$= 6.58 \times 10^6 \text{ m/s}$$

Reflect

The radius of an electron in the lowest level of the doubly ionized lithium ion ($Z = 3$) is one-third the radius of an electron in the ground state of hydrogen. Because there are three protons in the lithium nucleus, compared to one for hydrogen, the Coulomb force on the electron is greater, so it is reasonable that the orbital radius is smaller. Also, notice that the speed of the electron is about 2% of the speed of light in a vacuum.

Why did we insist that the problem be about a doubly ionized lithium ion, rather than a lithium atom? The difference is that when more than one electron is present, each electron is affected not only by the electric force from the nucleus but also by the electric force from the other electrons. We haven't taken the interaction between electrons into account in any of our calculations: The Bohr model is for *single-electron* atoms and ions only.

Although the Bohr model successfully predicts atomic spectra and other phenomena associated with hydrogen atoms, it nevertheless leaves us with an outstanding question: Why should the electron orbits be quantized in multiples of $\hbar$? For the answer we turn back to Louis de Broglie. Recall from Section 26-4 that de Broglie postulated that particles, such as electrons, have a wavelike nature. For a nonrelativistic electron of mass m_e moving at speed v, the momentum has magnitude $p = m_e v$. From the de Broglie relation, Equation 26-10, the wavelength of such an electron is

$$\lambda = \frac{h}{m_e v}$$

For an electron orbiting an atomic nucleus we express this in terms of angular momentum by making use of Equation 26-20:

$$\lambda = \frac{hr}{m_e vr} = \frac{hr}{L} = \frac{2\pi \hbar r}{L}$$

(We used $\hbar = h/2\pi$, so $h = 2\pi\hbar$.) Now let L be an integer multiple of $\hbar$, as Bohr required. Then

$$\lambda = \frac{2\pi \hbar r}{n\hbar}$$

or

$$n\lambda = 2\pi r$$

We recognize $2\pi r$ as the circumference of the orbital path of the electron. So Bohr's requirement that L be an integer multiple of $\hbar$ is equivalent to demanding that an integer number of electron wavelengths fit into the circumference of the orbit so that the wave joins smoothly onto itself. In a real sense, it is the wavelike nature of particles that results in the quantization of the energy of atomic electrons.

Confirming Energy Quantization

Bohr's model of the atom works relatively well for hydrogen, for a singly ionized helium ion (an atom with Z equal to 2 but only one electron), and for a doubly ionized lithium ion (an atom with Z equal to 3 but only one electron, as in Example 26-5). For atoms with more than one electron, it isn't possible to make calculations of energy levels using Bohr's physics. However, the general picture of the atom it provides, with electrons in quantized energy states, applies to all atoms. This was confirmed experimentally in 1914 by the German physicists James Franck and Gustav Hertz. (Franck and Hertz were awarded the 1925 Nobel Prize in Physics for their work. Gustav Hertz was a nephew of Heinrich Hertz, whom we encountered earlier.)

Franck and Hertz used a device similar to the one shown schematically in Figure 26-17 to measure the effect of bombarding atoms in a gas with electrons. The cathode is heated to give off electrons, which are accelerated by a variable voltage toward the mesh grid. Some electrons pass through the grid and arrive at the anode, where the electron current is detected by the ammeter. Notice that a voltage is also applied between the grid and the anode, which acts against the electrons; only electrons that carry sufficient energy as they pass the grid will make it to the anode. These electrical components sit inside a tube filled with low-pressure mercury vapor, so collisions between an electron and a mercury (Hg) atom can occur. Franck and Hertz used mercury vapor because the spectral lines of low-pressure mercury gas were well studied; one prominent ultraviolet spectral line has a wavelength $\lambda = 253.7$ nm. The energy of this photon is

$$E = hf = \frac{hc}{\lambda} = \frac{(6.63 \times 10^{-34} \text{ J·s})(3.00 \times 10^8 \text{ m/s})}{253.7 \times 10^{-9} \text{ m}}$$

$$= (7.84 \times 10^{-19} \text{ J})\left(\frac{1 \text{ eV}}{1.60 \times 10^{-19} \text{ J}}\right) = 4.90 \text{ eV}$$

This is the energy emitted when an atomic electron falls from an excited energy level down to a lower energy level. Now consider what happens in the collision of an electron and a mercury atom in the Franck–Hertz tube. In general, the more kinetic energy an electron carries as it approaches the grid, the more likely it will make it

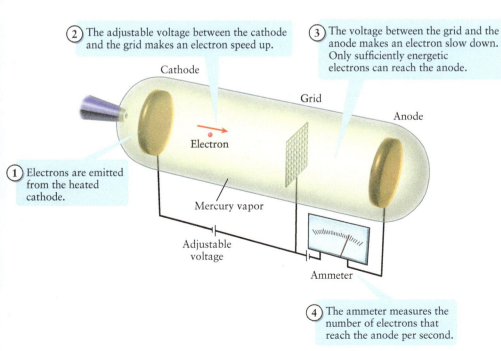

② The adjustable voltage between the cathode and the grid makes an electron speed up.

③ The voltage between the grid and the anode makes an electron slow down. Only sufficiently energetic electrons can reach the anode.

Cathode

Grid

Anode

Electron

① Electrons are emitted from the heated cathode.

Mercury vapor

Adjustable voltage

Ammeter

④ The ammeter measures the number of electrons that reach the anode per second.

Figure 26-17 The Franck–Hertz experiment In this experiment, electrons are accelerated from a cathode toward a mesh grid in a tube filled with mercury vapor. Some electrons pass through the grid, and the current at the anode is measured by the ammeter.

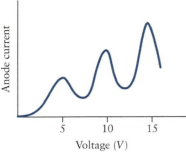

Figure 26-18 Evidence of energy quantization As the voltage between the cathode and the mesh grid in the Franck–Hertz experiment increases, the anode current increases. When the kinetic energy of the accelerated electrons equals the energy required to excite an electron in a mercury atom to a higher energy level, accelerated electrons lose energy. Not as many electrons reach the anode at the corresponding cathode-grid voltages, so the anode current decreases.

through. So the anode current should grow as the voltage between the cathode and the grid is increased. However, if the energy levels of electrons in Hg (and all) atoms are quantized, when the kinetic energy of the incident electron equals the energy difference between two Hg energy levels, the Hg atom can absorb the electron's energy. When this happens it is less likely that the electron will reach the anode, so the anode current should decrease. Figure 26-18 shows a typical curve of current versus voltage from the Franck–Hertz experiment. The general trend, as we would expect, is that the anode current increases as the voltage is increased. But the current drops dramatically at certain values of accelerating voltage, in this case, at 4.9 V, and at 9.8 V and 14.7 V, which are multiples of 4.9 V—the energy of the mercury spectral line.

When the cathode-grid voltage is set to 4.9 V, the kinetic energy of electrons that leave the cathode reaches 4.9 eV just as they approach the mesh grid. (Recall that the unit of energy eV is the amount of energy gained by an electron when it experiences a potential difference of 1 V.) For this reason electrons that collide with a mercury atom in close proximity to the grid give up their energy in the process of causing an electron in the atom to jump to a higher energy level. Because the collision occurs close to the grid, there is no opportunity for the electron to undergo another acceleration; in other words, there is no opportunity for the electron to acquire enough energy to reach the anode. For this reason when the cathode-grid voltage is set to 4.9 V, so that the collisions between electrons and Hg atoms occur near the grid with electron kinetic energy equal to 4.9 eV, the number of electrons reaching the anode decreases. In addition, when the voltage is set to 9.8 V, accelerated electrons that collide with an Hg atom halfway between the cathode and the grid lose their energy and then accelerate up to a kinetic energy of 4.9 eV again by the time they reach the grid. A collision there once again results in the electron transferring its energy to an Hg atom and being unable to reach the anode. A similar phenomenon occurs when the voltage is set to any integer multiple of 4.9 V. This rise and fall in anode current versus cathode-grid voltage is shown dramatically in Figure 26-18. The underlying explanation of this phenomenon is the quantization of electron energy levels in the atom, as predicted by Bohr.

Beyond the Bohr Model: Quantum Mechanics

As powerful as the Bohr model is at providing an understanding of the atom and atomic spectra, it does not tell the whole story. The Bohr atom treats the electrons that orbit atomic nuclei as particles. But no theoretical description of the atom can be complete unless it accounts for the wave nature of the electrons. The theoretical underpinning of our understanding of atoms that includes these wave properties is found in an equation developed by Austrian physicist Erwin Schrödinger. The *Schrödinger equation* relies on matter waves to describe the state of a system as a function of time, much like Newton's laws do while treating physical systems as particles. The Schrödinger equation is thus the fundamental equation of **quantum mechanics**, in which matter is treated as intrinsically wavelike in nature. Schrödinger was awarded the 1933 Nobel Prize in Physics for his work. (The Schrödinger equation itself is too mathematically ornate for the purposes of this book.)

Perhaps the most notable difference between Schrödinger's quantum-mechanical description and Newton's classical description of physics is the ability to specify the position and velocity of objects. In Newton's description at any instant of time we can identify a specific position in space and a specific velocity vector for any particle in a system, for example, an electron orbiting an atomic nucleus. A wave, however, is not localized in space. The result is that while Newton's laws predict the position and velocity of an object, the Schrödinger equation predicts the *probability* of finding a certain value of position or velocity. For this reason, electrons in the quantum model of the atom are described not as tiny marbles orbiting the nucleus at fixed radii but rather as a charge distribution. This distribution, sometimes called a *probability cloud*, gives the probability of finding the electron at any given position; the denser the cloud in some region, the more likely it is that the electron will be found there.

Figure 26-19a shows the probability cloud associated with the ground state of hydrogen. The more dense the color in any region in this figure, the more likely it is that the electron will be found in that region. This probability distribution is spherically symmetric; that is, it varies only as a function of radius from the center of the nucleus. For that reason we can also express the same information in a curve of probability

(a)   (b)

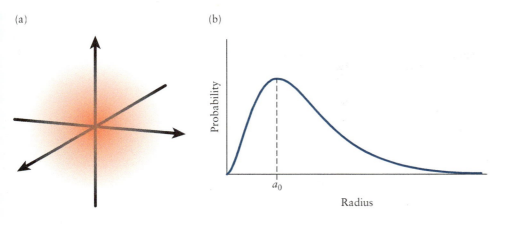

Figure 26-19 **The Schrödinger picture of the hydrogen atom** (a) In the state of lowest energy, an electron in a hydrogen atom can be found anywhere relative to the nucleus. The darker (denser) the color in a region of the probability cloud, the more likely it is that the electron will be found in that region. (b) The probability cloud in part (a) is spherically symmetric, so we can represent the probability of finding the electron as a function of radius from the center of the nucleus. The probability peaks at a distance from the nucleus equal to a_0, the Bohr radius.

versus radius, as in Figure 26-19b. Notice that the most probable radius of an electron in the ground state of a hydrogen atom is a_0, the Bohr radius.

The Bohr model employs a single integer, or **quantum number**, to describe electron states. For the Bohr atom this integer is n, which determines the energy level of the electron. In the fully quantum-mechanical view of the atom, *four* quantum numbers are required. These are n, the principal quantum number; ℓ, the angular momentum (or orbital) quantum number; m_ℓ, the magnetic quantum number; and m_s, the electron spin quantum number. The specific values of each of these four quantum numbers completely describe the state of an electron in an atom.

The principal quantum number n plays a role in the quantum atom similar to that which n plays in the classical Bohr atom; in particular, it specifies the energy level or electron shell. The lowest energy state, or ground state, corresponds to n equal to 1.

The angular momentum quantum number ℓ is a measure of the angular momentum the electron carries. For any value of n, ℓ varies in integer steps from 0 to $n - 1$; each value of ℓ specifies a subshell, or electron orbital, within the energy level specified by n. The shape of each orbital is different. By convention we refer to the orbitals with letters rather than integer numbers; the values of ℓ equal to 0, 1, 2, 3, 4, and 5 correspond to the orbitals s, p, d, f, g, and h. It is also standard to refer to an electron subshell by giving both n and this orbital letter code together. For example, an electron in the p subshell of the $n = 2$ energy level is said to be in the $2p$ subshell. The electron's energy is slightly dependent on the value of ℓ, an effect called *fine structure* not found in the Bohr model.

The magnetic quantum number m_ℓ specifies an orientation of an electron's subshell. It is so called because its value determines the (very small) energy associated with the interaction of a moving charge—the electron—with the magnetic field of the nucleus. The larger the value of ℓ, the more orientations are allowed; m_ℓ can take integer values between $-\ell$ and $+\ell$, including 0. Figure 26-20 shows the shapes of a number of electron orbitals.

The fourth quantum number m_s involves a new feature of the electron that was not discovered until the 1920s. Electrons have an intrinsic characteristic called **spin**, which is akin to the angular momentum of a rotating sphere. Even electrons that do not orbit an atomic nucleus possess spin, which can take on one of two values, often called spin "up" and spin "down." To fully describe an atomic electron, then, we must also specify its spin state. This is described by the electron spin quantum number m_s, which for an electron can be equal to either $+1/2$ or $-1/2$.

These quantum numbers play an important role in multi-electron atoms. Each electron in a multi-electron atom has a specific value of n, ℓ, m_ℓ, and m_s. (The details of the orbitals are affected by the presence of other electrons, but the same four quantum numbers still apply.) What is more, it turns out that there can be only *one* electron with a specific combination of these four quantum numbers. This fundamental restriction on electrons is called the **Pauli exclusion principle**, after the Swiss physicist Wolfgang Pauli, who deduced this principle in 1925. (Pauli received the 1945 Nobel Prize in Physics for his work.) Let's see what the Pauli exclusion principle tells us about the structure of multi-electron atoms.

For the $n = 1$ shell, only one value of ℓ (equal to 0, which is the s orbital), and therefore only one value of m_ℓ, is allowed. Two values of m_s are always possible, so the maximum number of electrons that can occupy the $n = 1$ shell is $1 \times 2 = 2$, the product of the number of possible m_ℓ values and the number of possible m_s values. That is, two electrons can occupy the $1s$ orbital. For $n = 2$, ℓ is allowed to be either 0 or 1. When $\ell = 0$,

Figure 26-20 Probability clouds for different quantum numbers Different atomic orbitals are specified by specific values of the quantum numbers n, ℓ, and m_ℓ. Each orbital has a unique shape.

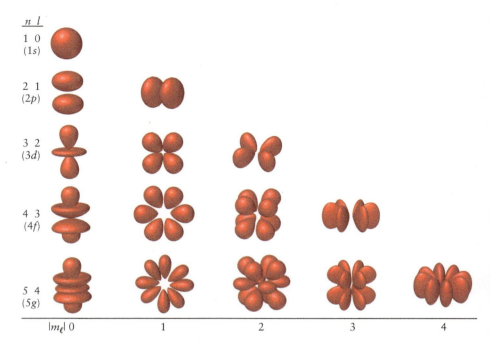

the only possible value of the magnetic quantum number is $m_\ell = 0$. So including the factor of 2 for the electron spin quantum number, one possible value of m_ℓ and two possible values of m_s means two electrons can occupy the 2s orbital. However, when $\ell = 1$ (the p orbital), m_ℓ can be −1, 0, or +1. Including the factor of 2 for the electron spin quantum number, that means that the number of electrons that can occupy the 2p orbital is 3×2, or 6. Because two electrons can occupy the 2s orbital and six electrons can occupy the 2p orbital, an atom can have at most eight electrons in the $n = 2$ energy level.

The pattern above repeats for $n = 3$, up through $\ell = 2$, the d orbital. For this orbital, five values (−2, −1, 0, +1, +2) are possible for m_ℓ, so the maximum number of electrons in this orbital is $5 \times 2 = 10$. Thus 2 electrons can occupy 3s, 6 electrons can occupy 3p, and 10 electrons can occupy 3d. The maximum number of electrons in the $n = 3$ energy shell of an atom is therefore 18.

The arrangement of electrons, according to how many occupy which orbitals, is directly correlated to the chemical properties of the elements. Consider the 18 lightest elements and their electron configurations, listed in Table 26-1. (In the table the number of electrons that occupy a particular orbital is given as a superscript, for example, $2p^4$ indicates that four electrons occupy the 2p orbital.) In hydrogen, lithium, and sodium (as well as potassium, rubidium, cesium, and francium), the outermost, or valence, electron is a single electron in an s orbital. These *alkali metals* share common chemical properties and occupy a single column in the periodic table.

Three of the *noble gases* are listed in Table 26-1. These are helium, neon, and argon; in each the outermost subshell is completely full. As a result, all electrons are relatively tightly bound, so these elements do not easily gain, lose, or share electrons. For that reason, the noble gases are relatively inert. Helium, neon, and argon, as well as the other noble gases, occupy a single column in the periodic table.

WATCH OUT! As atomic number increases, electron orbitals do not fill in a continuous fashion.

The configuration of the 18 electrons of argon, the heaviest element listed in Table 26-1, is $1s^2 2s^2 2p^6 3s^2 3p^6$. The next heaviest element is potassium, for which the configuration of the 19 electrons is $1s^2 2s^2 2p^6 3s^2 3p^6 4s^1$. Notice that the additional electron does not occupy the 3d orbital, although d follows p in our ordering (s, p, d, f, and so on).

Orbitals fill according to the increase in energy required, and for that reason do not fill according to counting up linearly in ℓ (orbital letters) and m_ℓ. Instead, orbitals fill in this order: 1s, 2s, 2p, 3s, 3p, 4s, 3d, 4p, 5s, 4d, 5p, 6s, 4f, 5d, 6p, 7s, 5f, 6d, 7p. Notice that 4s, not 3d, follows the 3p orbital in this sequence, which is why the valence electron in potassium is in a 4s orbital.

Halogens are elements that are highly reactive; that is, they easily form bonds with certain other elements, especially the alkali metals, to form molecules. Two halogens, fluorine and chlorine, are listed in Table 26-1. The outermost shell in both is one electron short of being full, which means that an atom of one of these elements can readily share an electron with another atom. This is particularly true for an atom of an alkali metal, which has one valence electron that is easily shared. So bring a sodium atom near a chlorine atom, and they will readily bond to form NaCl (table salt).

The ideas of quantum mechanics find applications on scales even smaller than that of the atom. In the final two chapters of this book we will see how quantum-mechanical ideas help us understand the nature of the atomic nucleus and of the fundamental particles that are the essential building blocks of all ordinary matter.

Go to Interactive Exercise 26-2 for more practice dealing with atomic energy.

TABLE 26-1 Electron Configurations of Light Elements

Atomic number	Element	Electron configuration
1	hydrogen (H)	$1s^1$
2	helium (He)	$1s^2$
3	lithium (Li)	$1s^2 2s^1$
4	beryllium (Be)	$1s^2 2s^2$
5	boron (B)	$1s^2 2s^2 2p^1$
6	carbon (C)	$1s^2 2s^2 2p^2$
7	nitrogen (N)	$1s^2 2s^2 2p^3$
8	oxygen (O)	$1s^2 2s^2 2p^4$
9	fluorine (F)	$1s^2 2s^2 2p^5$
10	neon (Ne)	$1s^2 2s^2 2p^6$
11	sodium (Na)	$1s^2 2s^2 2p^6 3s^1$
12	magnesium (Mg)	$1s^2 2s^2 2p^6 3s^2$
13	aluminum (Al)	$1s^2 2s^2 2p^6 3s^2 3p^1$
14	silicon (Si)	$1s^2 2s^2 2p^6 3s^2 3p^2$
15	phosphorus (P)	$1s^2 2s^2 2p^6 3s^2 3p^3$
16	sulfur (S)	$1s^2 2s^2 2p^6 3s^2 3p^4$
17	chlorine (Cl)	$1s^2 2s^2 2p^6 3s^2 3p^5$
18	argon (Ar)	$1s^2 2s^2 2p^6 3s^2 3p^6$

GOT THE CONCEPT? 26-5 Ionization Energy

The *ionization energy* of an atom is the energy required to remove an electron from the atom. In terms of the Bohr model, it is equal to the energy difference between the $n = 1$ energy level (the level of lowest energy, in which the electron is closest to the nucleus) and the $n = \infty$ energy level (which has an infinite radius, so the electron has moved infinitely far away). Compared to the ionization energy of hydrogen, the energy required to remove the last electron from doubly ionized lithium is (a) 3 times greater; (b) 9 times greater; (c) 27 times greater; (d) 81 times greater; (e) 243 times greater.

TAKE-HOME MESSAGE FOR Section 26-6

✔ In the Bohr model of the atom, electrons follow Newtonian orbits around the nucleus but with quantized values of orbital angular momentum. As a result, only certain orbital radii and orbital energies are allowed.

✔ The Franck–Hertz experiment confirmed that atomic energies are quantized.

✔ The Schrödinger equation explains the hydrogen atom by describing the electron as a wave. It predicts the probability of finding the electron at a particular location within the atom.

✔ The Pauli exclusion principle allows us to understand the structure of multi-electron atoms.

26-7 In quantum mechanics, it is impossible to know precisely both a particle's position and its momentum

We saw in the previous section that in quantum mechanics we describe the position of an electron in a hydrogen atom in terms of probabilities rather than certainties. We can calculate the probability that we will find the electron in a certain region around the nucleus, but we cannot say with certainty whether or not we will find it there. This fundamental uncertainty is a direct consequence of the wave nature of the electron. To gain more insight into this, let's investigate a fundamental principle of quantum mechanics called the *Heisenberg uncertainty principle*. We can understand this principle using ideas about sinusoidal waves that we introduced in Chapter 13.

In Section 13-3 we wrote the following wave function to describe a sinusoidal wave with wavelength λ and frequency f:

Wave function for a sinusoidal wave propagating in the +x direction (13-6)

| Wave function for a sinusoidal wave propagating in the +x direction | Angular wave number of the wave $= 2\pi/\lambda$ |

$$y(x,t) = A \cos (kx - \omega t + \phi) \quad \text{Phase angle}$$

| Amplitude of the wave | Angular frequency of the wave $= 2\pi f$ |

In Equation 13-6 the angular wave number of the wave is $k = 2\pi/\lambda$ and the angular frequency of the wave is $\omega = 2\pi f$. Note that the function in Equation 13-6 is valid for any value of x from $x = -\infty$ to $x = +\infty$. This means that a wave with a single definite wavelength has an infinite extent (Figure 26-21a). So the answer to the question "Where is this wave?" is "Everywhere!" A more realistic description of a wave is one that has a finite extent in space; for example, if you make a sound with your voice that lasts 1.00 s, the sound wave that emanates from your mouth at a speed of 343 m/s (the speed of sound in dry air at 20°C) will be (343 m/s)(1.00 s) = 343 m in extent from its leading edge to its trailing edge. Mathematically, a wave of a finite spatial extent Δx can be expressed as a sum of waves of infinite extent like the one shown in Figure 26-21a but with a range of angular wave numbers of breadth Δk (Figure 26-21b). To make a wave of shortened spatial extent, we have to add together infinite-extent waves from a broader range of angular wave numbers (Figure 26-21c). We can express the relationship between Δx (the spatial extent of a wave) and Δk (the breadth of angular wave numbers that go into that wave) as

(26-29)
$$\Delta k \Delta x \geq \frac{1}{2}$$

Equation 26-29 says that the product of the wave spatial extent Δx and the angular wave number breadth Δk cannot be less than 1/2. (This number arises from the specific way in which Δx and Δk are defined mathematically.) It says that to minimize the spatial extent of a wave necessarily means increasing the range of angular wave numbers that make up the wave. So the shorter the spatial extent of a wave, the less precisely we can answer the question "What is the angular wave number of the wave?" In other words, a wave of finite spatial extent does not have a single definite angular wave number $k = 2\pi/\lambda$ and so does not have a definite wavelength λ.

Equation 26-29 is true for waves of all kinds, from ocean waves to sound waves to seismic waves. To apply it to quantum-mechanical waves such as those that describe an electron in a hydrogen atom, multiply both sides of the equation by $\hbar$, which is Planck's constant divided by 2π. We get

(26-30)
$$\hbar \Delta k \Delta x \geq \frac{\hbar}{2} \quad \text{or} \quad \Delta(\hbar k)\Delta x \geq \frac{\hbar}{2}$$

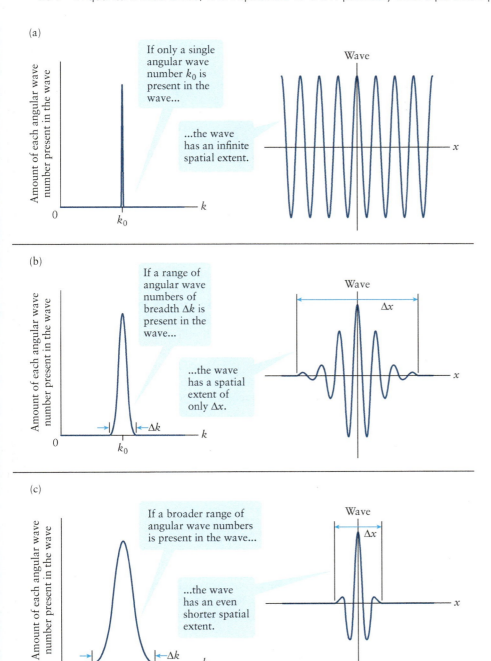

(a)

If only a single angular wave number k_0 is present in the wave...

...the wave has an infinite spatial extent.

Wave

Amount of each angular wave number present in the wave

(b)

If a range of angular wave numbers of breadth Δk is present in the wave...

...the wave has a spatial extent of only Δx.

Wave

Δx

Amount of each angular wave number present in the wave

(c)

If a broader range of angular wave numbers is present in the wave...

...the wave has an even shorter spatial extent.

Wave

Δx

Amount of each angular wave number present in the wave

Figure 26-21 Angular wave number and the spatial extent of a wave (a) A wave of a single, discrete wavelength λ and hence a single, discrete angular wave number $k = 2\pi/\lambda$ has an infinite spatial extent. (b), (c) Combining waves over a breadth of angular wave number yields a total wave that has a finite spatial extent. The shorter the spatial extent of the wave, the greater the range of angular wave numbers present in the wave.

In Equation 26-30 $\Delta(\hbar k)$ is the breadth of values of the quantity $\hbar k$ that must be included in the wave. We know that $\hbar = h/2\pi$ and $k = 2\pi/\lambda$; furthermore, from Equation 26-10 we know that the de Broglie wavelength of a particle of momentum p is $\lambda = h/p$, so $p = h/\lambda$. Putting these together, we find that

$$\hbar k = \left(\frac{h}{2\pi}\right)\left(\frac{2\pi}{\lambda}\right) = \frac{h}{\lambda} = p$$

So the quantity $\hbar k$ is just equal to the *momentum* of the particle that the wave describes. We'll relabel this as p_x since the wave function in Equation 13-6 describes a wave propagating along the x axis, so the particle has only an x component of momentum. So we can interpret $\Delta(\hbar k)$ as Δp_x, the breadth of values of momentum present in the wave, or equivalently the *uncertainty* in the value of p_x for the particle that the wave describes. Similarly we can interpret Δx, the spatial extent of the wave, as the *uncertainty* in the position x of the particle. (We know the particle lies somewhere within the wave's spatial extent, but we cannot say at any instant exactly where it is.) Then we can rewrite Equation 26-29 as

Heisenberg uncertainty principle for momentum and position (26-31)

Uncertainty in the x component of momentum of a particle

$\hbar = \dfrac{h}{2\pi}$ = Planck's constant divided by 2π

$$\Delta p_x \Delta x \geq \dfrac{\hbar}{2}$$

Uncertainty in the x coordinate of the particle

Equation 26-31 is the **Heisenberg uncertainty principle for momentum and position**, which was first expressed by the German physicist Werner Heisenberg in 1927. (Heisenberg was awarded the 1932 Nobel Prize in Physics for his work in the development of quantum mechanics.) It states that for any particle, the product of the uncertainty in momentum Δp_x and the uncertainty in position Δx can never be less than $\hbar/2$. This places a fundamental limitation on how well you can simultaneously know the momentum and position of a particle. For example, suppose you know the position of a particle to within an uncertainty Δx. It follows from Equation 26-31 that the minimum uncertainty in how well you can know the momentum of that particle is $\Delta p_x = \hbar/(2\Delta x)$. In other words, it's fundamentally impossible to measure the momentum of the particle with an uncertainty less than $\hbar/2\Delta x$. In Newtonian physics it's possible in principle to measure exactly both the momentum and position of a particle; the Heisenberg uncertainty principle says this is impossible, no matter how refined your measuring techniques. Note that the Heisenberg uncertainty principle also applies to the uncertainties in momentum and position for the y and z directions:

$$\Delta p_y \Delta y \geq \dfrac{\hbar}{2}$$
$$\Delta p_z \Delta z \geq \dfrac{\hbar}{2}$$

The value of $\hbar/2$ is very small: $\hbar/2 = h/4\pi = 5.27 \times 10^{-35}$ J·s. Hence the Heisenberg uncertainty principle has little consequence for measurements of relatively large objects like baseballs or insects. But on the atomic scale and smaller the uncertainty principle is tremendously important, as the following example shows.

EXAMPLE 26-6 Applying the Heisenberg Uncertainty Principle

If an electron is in the lowest ($n = 1$) energy level of the hydrogen atom, its uncertainty in position is approximately equal to the Bohr radius $a_0 = 0.0529$ nm. Find the minimum uncertainty in its x component of momentum, and find the kinetic energy of an electron with this magnitude of momentum. (b) Repeat these calculations for an electron whose uncertainty in position is one-tenth of a Bohr radius.

Set Up

We use the Heisenberg uncertainty principle for momentum and position, Equation 26-31. For a given position uncertainty Δx, the minimum value momentum uncertainty Δp_x corresponds to using the equals sign in this equation. If we set Δp_x equal to the momentum p, we can find the speed and hence the kinetic energy of the electron.

Heisenberg uncertainty principle for momentum and position:

$$\Delta p_x \Delta x \geq \dfrac{\hbar}{2} \qquad (26\text{-}31)$$

Momentum of a particle of mass m and speed v:

$$p = mv$$

Kinetic energy of a particle of mass m and speed v:

$$K = \dfrac{1}{2}mv^2$$

Solve

(a) Calculate the minimum momentum uncertainty.

Using the equals sign in Equation 26-30:

$$\Delta p_x \Delta x = \frac{\hbar}{2}$$

$$\Delta p_x = \frac{\hbar}{2\Delta x} = \frac{h}{4\pi\Delta x}$$

$$= \frac{6.63 \times 10^{-34} \text{ J·s}}{4\pi(0.0529 \times 10^{-9} \text{ m})}$$

$$= 9.97 \times 10^{-25} \frac{\text{J·s}}{\text{m}} = 9.97 \times 10^{-25} \text{ kg·m/s}$$

Find the speed and kinetic energy of an electron with this magnitude of momentum.

If we let $p = \Delta p_x$ and use $p = mv$, the speed of the electron is

$$v = \frac{p}{m} = \frac{9.97 \times 10^{-25} \text{ kg·m/s}}{9.11 \times 10^{-31} \text{ kg}} = 1.09 \times 10^6 \text{ m/s}$$

The kinetic energy of an electron moving at this speed is

$$K = \frac{1}{2}mv^2 = \frac{1}{2}(9.11 \times 10^{-31} \text{ kg})(1.09 \times 10^6 \text{ m/s})^2$$

$$= 5.46 \times 10^{-19} \text{ J}$$

$$= 5.46 \times 10^{-19} \text{ J}\left(\frac{1 \text{ eV}}{1.602 \times 10^{-19} \text{ J}}\right) = 3.41 \text{ eV}$$

(b) The Heisenberg uncertainty principle states that the product of the minimum uncertainty in momentum Δp_x and the minimum uncertainty in position Δx is $\hbar/2$. So if we make Δx one-tenth as large, Δp_x must become 10 times larger.

If we reduce the value of Δx by one-tenth from a_0 to $a_0/10$, Δp_x will be 10 times greater:

$$\Delta p_x = 10(9.97 \times 10^{-25} \text{ kg·m/s})$$

$$= 9.97 \times 10^{-24} \text{ kg·m/s}$$

If this is the magnitude of momentum, then $p = mv$ is 10 times greater than in part (a) and the speed v is also 10 times greater:

$$v = 10(1.09 \times 10^6 \text{ m/s})$$

$$= 1.09 \times 10^7 \text{ m/s}$$

In this case the kinetic energy is

$$K = \frac{1}{2}mv^2 = \frac{1}{2}(9.11 \times 10^{-31} \text{ kg})(1.09 \times 10^7 \text{ m/s})^2$$

$$= 5.46 \times 10^{-17} \text{ J}$$

$$= 5.46 \times 10^{-17} \text{ J}\left(\frac{1 \text{ eV}}{1.602 \times 10^{-19} \text{ J}}\right) = 341 \text{ eV}$$

This is 100 times greater than the kinetic energy we found in part (a).

Reflect

The Heisenberg uncertainty principle for momentum and position tells us why an electron in an atom has kinetic energy and does not simply remain at rest. Because it is confined to a very small volume and so has a very small uncertainty in position, an electron necessarily has a relatively large uncertainty in momentum and hence in velocity. Consequently we can expect it to have on average a nonzero kinetic energy. For such a rough calculation our result of 3.41 eV is not too different from the actual average kinetic energy of an electron in the $n = 1$ energy level, which is 13.6 eV.

The Heisenberg uncertainty principle also tells us why a negatively charged electron does not simply fall into the positively charged nucleus. By bringing the electron to one-tenth of its original distance from the nucleus, its kinetic energy increased by a factor of 100. Decreasing the distance r from the electron to the nucleus also decreases its (negative) electron potential energy $U_{\text{electric}} = k(Ze)(-e)/r = -kZe^2/r$, where Z is the number of protons in the nucleus (for hydrogen, $Z = 1$). But because U_{electric} is inversely proportional to r, decreasing r from a_0 to $a_0/10$ will only make the value of U_{electric} more negative by a factor of 10. Because the positive kinetic energy increases by a much larger factor, the result is that it would take a tremendous increase in energy to move the electron closer to the nucleus. Hence the electron cannot fall into the nucleus. So the Heisenberg uncertainty principle prevents atoms from collapsing, and so makes it possible for atoms to exist.

We will learn in Chapter 28 that there is a second form of the Heisenberg uncertainty principle that relates the uncertainty in *energy* of a phenomenon to the *duration* of that phenomenon. This will lead to the startling result that it's possible to violate the law of conservation of energy, provided the duration of time during which the law is violated is sufficiently short. This result will help us understand the properties of the fundamental forces of nature.

GOT THE CONCEPT? 26-6 Heisenberg Uncertainty for Different Particles

(?) A neutron has 1839 times the mass of an electron. If a neutron and an electron both have the same uncertainty in position, the minimum uncertainty in momentum of the neutron will be (a) 1839 times the minimum uncertainty in momentum of the electron; (b) 1/1839 the minimum uncertainty of the electron; (c) the same as the minimum uncertainty of the electron.

TAKE-HOME MESSAGE FOR Section 26-7

✔ A wave has a definite wavelength and angular wave number only if it has infinite spatial extent. The shorter the spatial extent of the wave, the greater the breadth of angular wave numbers there must be in the wave.

✔ The Heisenberg uncertainty principle for momentum and position states that the product of the uncertainty in a particle's momentum and the uncertainty in its position must be greater than a certain minimum value. This is true no matter how these quantities are measured.

Key Terms

absorption lines
absorption spectrum
blackbody
blackbody radiation
Bohr model
Bohr orbit
Compton scattering
de Broglie wavelength

emission lines
emission spectrum
energy level
Heisenberg uncertainty principle
    for momentum and position
Pauli exclusion principle
photoelectric effect
photoelectrons

photon
quantized
quantum mechanics
quantum number
spin
wave-particle duality
work function

Chapter Summary

Topic	Equation or Figure

The photoelectric effect: When light shines on a surface, the surface can emit electrons. However, no electrons are emitted if the frequency of the light is below a certain critical value. Einstein showed that this could be explained if light is absorbed in the form of photons whose energy is proportional to their frequency: $E = hf$, where h is Planck's constant.

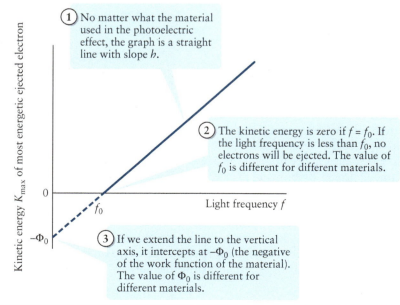

(1) No matter what the material used in the photoelectric effect, the graph is a straight line with slope h.

(2) The kinetic energy is zero if $f = f_0$. If the light frequency is less than f_0, no electrons will be ejected. The value of f_0 is different for different materials.

(3) If we extend the line to the vertical axis, it intercepts at $-\Phi_0$ (the negative of the work function of the material). The value of Φ_0 is different for different materials.

(Figure 26-3)

Blackbody radiation: Objects emit light due to their temperature. An ideal blackbody (one that does a perfect job of absorbing light) is also a perfect emitter of light. The details of the spectrum of a blackbody can be understood only if a blackbody emits light in the form of photons.

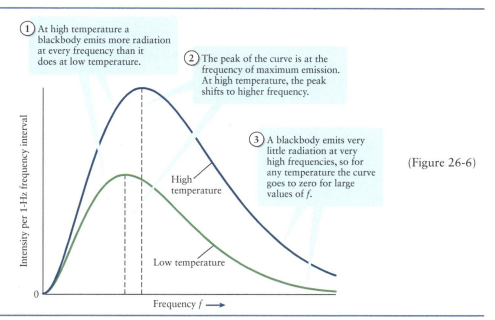

① At high temperature a blackbody emits more radiation at every frequency than it does at low temperature.

② The peak of the curve is at the frequency of maximum emission. At high temperature, the peak shifts to higher frequency.

③ A blackbody emits very little radiation at very high frequencies, so for any temperature the curve goes to zero for large values of f.

High temperature

Low temperature

(Figure 26-6)

Compton scattering: Photons have momentum in inverse proportion to their wavelength. This is demonstrated by Compton scattering, in which a photon undergoes an increase in wavelength when it scatters from an electron.

Magnitude of the **momentum** of a photon | Energy of the photon

$$p = \frac{E}{c} = \frac{h}{\lambda}$$ Planck's constant

(26-7)

Speed of light in a vacuum | Wavelength of the photon

Wavelength change in Compton scattering | Planck's constant | Photon scattering angle

$$\Delta\lambda = \lambda_f - \lambda_i = \frac{h}{m_e c}(1 - \cos\theta)$$

(26-8)

Photon wavelength after scattering | Photon wavelength before scattering | Electron mass | Speed of light in a vacuum

---

Wave-particle duality: Just as photons have particle aspects, matter has wave aspects. The de Broglie wavelength of a particle is inversely proportional to the momentum (the same relationship between wavelength and momentum as for a photon).

A particle has a **de Broglie wavelength**... | ...equal to **Planck's constant**...

$$\lambda = \frac{h}{p}$$

(26-10)

...divided by the **momentum of the particle.** The greater the momentum, the shorter the de Broglie wavelength.

Atomic structure and atomic spectra: Alpha particle scattering shows that the atom is made up of a small positive nucleus surrounded by electrons. The energy of an atom is quantized. It can have only certain definite values. The evidence for this comes from the spectra of atoms, which show that atoms absorb only specific wavelengths of light and that they emit the same wavelengths that they absorb.

(a) The absorption spectrum produced by passing white light through a gas of hydrogen atoms

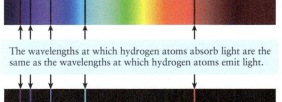

The wavelengths at which hydrogen atoms absorb light are the same as the wavelengths at which hydrogen atoms emit light.

(b) The emission spectrum produced by a heated gas of hydrogen atoms

 • Atoms of each element absorb and emit light at wavelengths that are characteristic of that element.
• The characteristic wavelengths differ from one element to another.
• Hydrogen has the simplest arrangement of characteristic wavelengths of any element.

(Figure 26-13)

The Bohr model of the atom: In the Bohr model of hydrogen, a single electron orbits the nucleus much like a satellite orbiting Earth. The difference is that the orbit can have only certain values of angular momentum, radius, and energy. Transitions between the allowed orbits are the cause of absorption and emission spectra.

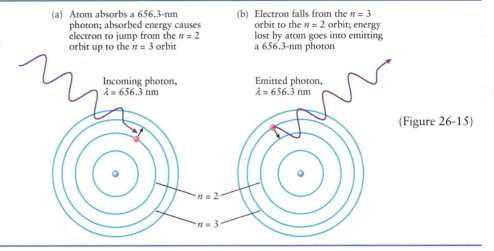

(a) Atom absorbs a 656.3-nm photon; absorbed energy causes electron to jump from the $n = 2$ orbit up to the $n = 3$ orbit

(b) Electron falls from the $n = 3$ orbit to the $n = 2$ orbit; energy lost by atom goes into emitting a 656.3-nm photon

Incoming photon, $\lambda = 656.3$ nm

Emitted photon, $\lambda = 656.3$ nm

(Figure 26-15)

$n = 2$

$n = 3$

The Schrödinger equation and the Pauli exclusion principle: In the more complete Schrödinger description of the atom, the electron is described by a wave. Four quantum numbers describe the state of the electron, and a probability cloud describes the probability of finding an electron at different positions within the atom. The Pauli exclusion principle states that there can be only one electron per state; this helps explain the properties of multi-electron atoms.

(a)

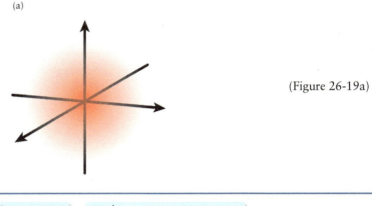

(Figure 26-19a)

The Heisenberg uncertainty principle for momentum and position: Because particles have wave properties, there is a fundamental limit to how well both the momentum and the position of a particle can be known simultaneously. This helps to explain why atoms do not collapse.

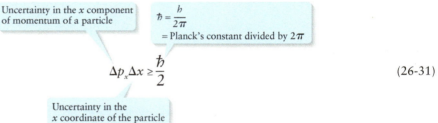

Uncertainty in the x component of momentum of a particle

$\hbar = \dfrac{h}{2\pi}$

= Planck's constant divided by 2π

$$\Delta p_x \Delta x \geq \frac{\hbar}{2}$$

(26-31)

Uncertainty in the x coordinate of the particle

Answer to What do you think? Question

(a) The spectrum of light from a star is very similar to the spectrum of an ideal blackbody (Section 26-2). Blue light has a higher frequency than red light, so the blue stars have a frequency of maximum emission (at which they emit most strongly) that is greater than that of the red stars. The frequency of maximum emission of a blackbody increases with increasing temperature, so the blue stars must be hotter than the red stars.

To determine the chemical composition of the stars, the color of stars isn't enough information: We need to look at the absorption spectrum of the stars (Section 26-5). This can't be inferred from a photograph like the one that opens this chapter. In fact, it turns out that all of the stars in this image have almost the same chemical composition: predominantly hydrogen and helium, with trace amounts of other elements.

Answers to Got the Concept? Questions

26-1 (e) Figure 26-6 shows that the higher the temperature, the more light a blackbody of a given size emits at *all* frequencies. Note that both the 3000-K blackbody and the 12,000-K blackbody emit light at visible frequencies, not just the invisible frequencies at which they emit most strongly. That's why you can see radiation from these objects. The 3000-K blackbody emits more red light than any other visible color, so it

will appear red; the 12,000-K blackbody emits more blue light than any other visible color, so it will appear blue.

26-2 (d) If $\theta = 90°$, then $\cos \theta = \cos 90° = 0$ and the wavelength shift in Compton scattering is $\Delta\lambda = (h/m_e c)(1 - \cos \theta) = (h/m_e c)(1 - 0) = h/m_e c = \lambda_C$. The initial wavelength is λ_C, so the final wavelength is $\lambda_f = \lambda_i + \Delta\lambda = \lambda_C + \lambda_C = 2\lambda_C$. That is, the final photon has twice the wavelength and so half the energy of the initial photon. (The lost energy is transferred to the electron from which the photon scattered.)

26-3 (c), (d), (a), (b). Equation 26-10 tells us that the de Broglie wavelength λ is inversely proportional to the momentum p: $\lambda = h/p$. So a ranking from longest to shortest de Broglie wavelength is the same as a ranking from smallest to largest momentum. A proton has mass 1.67×10^{-27} kg and an electron has mass 9.11×10^{-31} kg, so the momentum in each of the four cases is (a) $(1.67 \times 10^{-27}$ kg$)(2.00 \times 10^3$ m/s$) = 3.34 \times 10^{-24}$ kg $\cdot$ m/s; (b) $(1.67 \times 10^{-27}$ kg$)(4.00 \times 10^3$ m/s$) = 6.68 \times 10^{-24}$ kg $\cdot$ m/s; (c) $(9.11 \times 10^{-31}$ kg$)(2.00 \times 10^3$ m/s$) = 1.82 \times 10^{-27}$ kg $\cdot$ m/s; (d) $(9.11 \times 10^{-31}$ kg$)(4.00 \times 10^3$ m/s$) = 3.64 \times 10^{-27}$ kg $\cdot$ m/s. So the ranking is (c) slow-moving electron, (d) fast-moving electron, (a) slow-moving proton, (b) fast-moving proton.

26-4 (c), (a), (b), (d). The photon energy is equal to the difference between the energies of the upper energy level of the atom and the lower energy level of the atom. To determine the ranking, just look at Figure 26-16 to see the spacing between the energy levels of the atom. The spacing is (c) greatest between the $n = 3$ and $n = 1$ levels, (a) second greatest between the $n = 4$ and $n = 2$ levels, (b) third greatest between the $n = 5$ and $n = 3$ levels, and (d) least between the $n = 4$ and $n = 3$ levels.

26-5 (b) The ionization energy is equal to $E_{\text{ionization}} = E_\infty - E_1$. From Equation 26-27:

$$E_{\text{ionization}} = E_\infty - E_1 = \left(-\frac{Z^2}{\infty^2}E_0\right) - \left(-\frac{Z^2}{1^2}E_0\right)$$
$$= 0 - (-Z^2 E_0) = Z^2 E_0$$

(The reciprocal of infinity is zero.) A doubly ionized lithium ion is like a hydrogen atom—both have a single electron—but with $Z = 3$ instead of $Z = 1$ for hydrogen. Since the ionization energy is proportional to Z^2, the value for doubly ionized lithium is $3^2 = 9$ times greater than for hydrogen. You can see that for hydrogen $E_{\text{ionization}} = E_0 = 13.6$ eV, while for doubly ionized lithium $E_{\text{ionization}} = 9E_0 = 9(13.6$ eV$) = 122$ eV.

26-6 (c) Using the equals sign in Equation 26-30 so we are talking about minimum uncertainties:

$$\Delta p_x \Delta x = \frac{\hbar}{2} \quad \text{so} \quad \Delta p_x = \frac{\hbar}{2\Delta x}$$

For a given position uncertainty Δx, this expression for the minimum momentum uncertainty Δp_x doesn't depend on mass. So the value of Δp_x will be the same for a neutron as for an electron, even though their masses are very different.

Questions and Problems

In a few problems you are given more data than you actually need; in a few other problems, you are required to supply data from your general knowledge, outside sources, or informed estimate. Interpret as significant all digits in numerical values that have trailing zeros and no decimal points.

For all problems use $g = 9.80$ m/s^2 for the free-fall acceleration due to gravity. Neglect friction and air resistance unless instructed to do otherwise.

- •   Basic, single-concept problem
- ••   Intermediate-level problem; may require synthesis of concepts and multiple steps
- •••   Challenging problem
- Example   See worked example for a similar problem

Conceptual Questions

1. • Prior to Einstein's description of the photoelectric effect, light was thought to act like a wave. Explain why the existence of a frequency below which photoelectrons are not emitted favors a description of light as a particle instead.

2. • Describe how the number of photoelectrons emitted from a metal plate in the photoelectric effect would change if (a) the intensity of the incident radiation were increased, (b) the wavelength of the incident radiation were increased, and (c) the work function of the metal were increased.

3. • Is it possible to observe photoelectrons emitted from a metal plate with relativistic speeds?

4. • Consider the photoelectric emission of electrons induced by incident light of a single wavelength. The incoming photons all have the same energy, but the emitted electrons have a range of kinetic energies. Why?

5. • A markedly nonclassical feature of the photoelectric effect is that the energy of the emitted electrons doesn't increase as you increase the intensity of the light striking the metal surface. What change does occur as the intensity is increased?

6. • How does the intensity of light from a blackbody change when its temperature is increased? What changes occur in the body's radiation spectrum?

7. • The Compton effect is practically unobservable for visible light. Why?

8. • Which of the two Compton scattering experiments more clearly demonstrates the particle nature of electromagnetic radiation: a collision of the photon with an electron or a collision with a proton? Explain your answer.

9. • An electron and a proton have the same kinetic energy. Which has the longer wavelength?

10. • Is the wavelength of an electron the same as the wavelength of a photon if both particles have the same total energy?

11. • Does the de Broglie wavelength of a particle increase or decrease as its kinetic energy increases?

12. • Why do we never observe the wave nature of particles for everyday objects such as birds or bumblebees, for example?

13. • According to classical electromagnetic theory, an accelerated charge emits electromagnetic radiation. What would this mean for the electron in the Bohr atom? What would happen to its orbit?

14. • What is the shortest wavelength of electromagnetic radiation that can be emitted by a hydrogen atom?

15. • Why do you think Bohr's model was originally developed for the element hydrogen?

16. • Why do you think that Bohr's model of the hydrogen atom is still taught in undergraduate physics classes?

17. • Are there quantities in classical physics that are quantized?

18. • The Heisenberg uncertainty principle applies not only to very small objects, but also to objects in our everyday world. Why, then, do we not worry about it when simultaneously measuring the position and momentum of a car?

Multiple-Choice Questions

19. • Atoms of an element emit a spectrum that
    A. is the same as all other elements.
    B. is evenly spaced.
    C. is unique to that element.
    D. is evenly spaced and unique to that element.
    E. is indistinguishable from most other elements.

20. • An ideal blackbody is an object that
    A. absorbs most of the energy that strikes it and emits a little of the energy it absorbs.
    B. absorbs a little of the energy that strikes it and emits most of the energy it absorbs.
    C. absorbs half of the energy that strikes it and emits half of the energy it absorbs.
    D. absorbs all the energy that strikes it and emits all the energy it absorbs.
    E. neither absorbs nor emits energy except at ultraviolet ("black light") wavelengths.

21. • The color of light emitted by a solid hot object depends on
    A. the size of the object.
    B. the shape of the object.
    C. the material from which the object is made.
    D. the temperature of the object.
    E. the color of the object when it is cold.

22. • Which photon has more energy?
    A. a photon of ultraviolet radiation
    B. a photon of green light
    C. a photon of yellow light
    D. a photon of red light
    E. a photon of infrared radiation

23. • Light that has a wavelength of 600 nm strikes a metal surface, and a stream of electrons is ejected from the surface. If light of wavelength 500 nm strikes the surface, the maximum kinetic energy of the electrons emitted from the surface will
    A. be greater.
    B. be smaller.
    C. be the same.
    D. be 5/6 smaller.
    E. be unmeasurable.

24. • In the Compton effect experiment, the change in a photon's wavelength depends on
    A. the scattering angle.
    B. the initial wavelength of the photon.
    C. the final wavelength of the photon.
    D. the density of the scattering material.
    E. the atomic number of the scattering material.

25. • The maximum change in a photon's energy when it undergoes Compton scattering occurs when its scattering angle is
    A. 0°.
    B. 45°.
    C. 90°.
    D. 135°.
    E. 180°.

26. • As the scattering angle in the Compton effect increases, the energy of the scattered photon
    A. increases.
    B. stays the same.
    C. decreases.
    D. is proportional to $\sin\theta$.
    E. is proportional to $\cos\theta$.

27. • The de Broglie wavelength depends only on
    A. the particle's mass.
    B. the particle's speed.
    C. the particle's energy.
    D. the particle's momentum.
    E. the particle's charge.

Problems

26-1 Experiments that probe the nature of light and matter reveal the limits of classical physics

26-2 The photoelectric effect and blackbody radiation show that light is absorbed and emitted in the form of photons

28. • Calculate the range of photon frequencies and energies in the visible spectrum of light (approximately 380–750 nm). Example 26-1

29. • What is the energy of a low-frequency 2000-Hz radio photon? Example 26-1

30. • What is the energy of a 0.200-nm x-ray photon? Example 26-1

31. • What are the wavelength and frequency of a 3.97×10^{-19}-J photon? Example 26-1

32. • **Biology** Under most conditions, the human eye will respond to a flash of light if 100 photons hit photoreceptors at the back of the eye. Determine the total energy of such a flash if the wavelength is 550 nm (green light). Example 26-1

33. •• The threshold wavelength for the photoelectric effect for silver is 262 nm. (a) Determine the work function for silver. (b) What is the maximum kinetic energy of an electron emitted from silver if the incident light has a wavelength of 222 nm? Example 26-1

34. •• Light that has a 195-nm wavelength strikes a metal surface, and photoelectrons are produced moving as fast as $0.004c$. (a) What is the work function of the metal? (b) What is the threshold wavelength for the metal above which no photoelectrons will be emitted? Example 26-1

35. • What is the minimum frequency of light required to eject electrons from a metal with a work function of 6.53×10^{-19} J? Example 26-1

36. •• The work functions of aluminum, calcium, potassium, and cesium are 6.54×10^{-19} J, 4.65×10^{-19} J, 3.57×10^{-19} J, and 3.36×10^{-19} J, respectively. For which of the metals will photoelectrons be emitted when irradiated with visible light (wavelengths from 380 to 750 nm)? Example 26-1

37. • Wien's displacement law states that the wavelength of maximum emission for a blackbody is given by the following formula:

$$\lambda_{max} = \frac{0.290 \text{ K} \cdot \text{cm}}{T}$$

The Sun has a wavelength of maximum emission of 475 nm. Using Wien's displacement law, determine the corresponding temperature of the outer layer of the Sun, assuming it is a blackbody.

38. • (a) If the human body acts like a blackbody, use Wien's displacement law (see problem 37) to calculate the body's wavelength of maximum emission. Assume an average skin temperature of 34°C. (b) In what part of the electromagnetic spectrum is the light?

39. •• Suppose a blackbody at 400 K radiates just enough heat in 15 min to boil water for a cup of tea. How long will it take to boil the same water if the temperature of the radiator is 500 K?

26-3 As a result of its photon character, light changes wavelength when it is scattered

40. • The quantity $\lambda_C = h/(mc)$ is called the Compton wavelength. Calculate the numerical value of this quantity for (a) an electron, (b) a proton, and (c) a pi meson (which has a mass of 2.50×10^{-28} kg).

41. • What is the momentum of a photon if its wavelength (a) is 550 nm and (b) is 0.0711 nm? Example 26-2

42. •• X rays that have wavelengths of 0.125 nm are scattered off stationary electrons at an angle of 30.0°. (a) Calculate the wavelength of the scattered electromagnetic radiation. (b) Calculate the fractional wavelength change ($\Delta\lambda/\lambda_i = (\lambda_f - \lambda_i)/\lambda_i$) for the scattered x rays. Example 26-2

43. •• If a photon undergoes Compton scattering from a stationary electron and experiences a fractional wavelength change (see problem 42) of +7.25%, calculate the angle at which the scattered photons are directed if the original photons have a wavelength of 0.00335 nm. Example 26-2

44. • Photons that have a wavelength of 0.00225 nm are Compton scattered off stationary electrons at 45.0°. What is the energy of the scattered photons? Example 26-2

45. • X-ray photons that have a wavelength of 0.140 nm are scattered off carbon atoms (which possess essentially stationary electrons in their valence shells). What are the wavelengths of the Compton-scattered photons and the kinetic energies of the scattered electrons if the photons are scattered at angles of (a) 0.00°, (b) 30.0°, (c) 45.0°, (d) 60.0°, (e) 90.0°, and (f) 180°? Example 26-2

46. • A 0.0750-nm photon Compton scatters off a stationary electron. Determine the maximum speed of the scattered electron. Example 26-2

47. •• A photon Compton scatters off a stationary electron at an angle of 60.0°. The electron moves away with 1.28×10^{-17} J of kinetic energy. Determine the initial wavelength of the photon. Example 26-2

48. •• An x-ray source is incident on a collection of stationary electrons. The electrons are scattered with a speed of 4.50×10^5 m/s, and the photon scatters at an angle of 60.0° from the incident direction of the photons. Determine the wavelength of the x-ray source. Example 26-2

49. • Arthur Compton scattered photons that had a wavelength of 0.0711 nm off a block of carbon during his famous experiment of 1923 at Washington University in St. Louis, Missouri. (a) Calculate the frequency and energy of the photons. (b) What is the wavelength of the photons that are scattered at 90.0°? (c) What is the energy of the photons that are scattered at 90.0°? (d) What is the energy of the electrons that recoil from the Compton scattering with $\theta = 90.0°$? Example 26-2

26-4 Matter, like light, has aspects of both waves and particles

50. • Calculate the de Broglie wavelength of a 0.150-kg ball moving at 40.0 m/s. Comment on the significance of the result. Example 26-3

51. • Calculate the de Broglie wavelength of an electron that has a speed of $0.00730c$. Example 26-3

52. • What is the de Broglie wavelength of a proton ($m = 1.67 \times 10^{-27}$ kg) moving at 4.00×10^5 m/s? Example 26-3

53. •• What is the de Broglie wavelength of an electron that has a kinetic energy of (a) 1.60×10^{-19} J, (b) 1.60×10^{-18} J, (c) 1.60×10^{-17} J, and (d) 1.60×10^{-16} J? Example 26-3

54. •• Calculate the de Broglie wavelength of an alpha particle ($m_\alpha = 6.64 \times 10^{-27}$ kg) that has a kinetic energy of (a) 1.60×10^{-13} J and (b) 8.00×10^{-13} J. Example 26-3

55. •• Calculate the de Broglie wavelength of a thermal neutron that has a kinetic energy of about 6.41×10^{-21} J. Example 26-3

56. •• Write an expression that relates the Newtonian kinetic energy ($K = (1/2)mv^2$) and mass of a nonrelativistic particle to its de Broglie wavelength. (That is, complete the expression $\lambda(K, m) = ?$) Example 26-3

26-5 The spectra of light emitted and absorbed by atoms show that atomic energies are quantized

57. • The Balmer formula can be written as follows:

$$\lambda = (364.56 \text{ nm})\left(\frac{m^2}{m^2 - 4}\right)$$

where m is equal to any integer larger than 2. This represents the wavelengths of visible colors that are emitted from the hydrogen atom. Calculate the first four colors (wavelengths) that are observed in the spectrum of hydrogen due to the Balmer series. Example 26-4

58. •• **Astronomy** Hydrogen atoms in interstellar gas clouds emit electromagnetic radiation at a wavelength of 21 cm when an electron in the ground state of hydrogen switches spin states. Determine the energy difference between the two spin states in this transition. Example 26-4

59. •• Prove that the Balmer formula is a special case of the Rydberg formula with n set equal to 2. Example 26-4

$$\text{Rydberg formula: } \frac{1}{\lambda} = R_H\left(\frac{1}{n^2} - \frac{1}{m^2}\right)$$
$$R_H = 1.09737 \times 10^7 \text{ m}^{-1}$$
$$\text{Balmer formula: } \lambda = b\left(\frac{m^2}{m^2 - 4}\right) \quad b = 364.56 \text{ nm}$$

60. • A hypothetical atom has four unequally spaced energy levels in which a single electron can be found. Suppose a collection of the atoms is excited to the highest of the four levels. (a) What is the maximum number of unique spectral lines that could be measured as the atoms relax and return to the lowest, ground state? (b) Suppose the previous hypothetical atom has 10 energy levels. Now what is the maximum number of unique spectral lines that could be measured in the emission spectrum of the atom? Example 26-4

61. • The Lyman series results from transitions of the electron in hydrogen in which the electron ends in the $n = 1$ energy level. Using the Rydberg formula for the Lyman series, calculate the wavelengths of the photons emitted in the transitions that end in the $n = 1$ level and start in the energy levels that correspond to n equal to 2 through 6, and indicate the initial and final levels of the transition corresponding to each wavelength. State whether each wavelength is visible (380 to 750 nm), ultraviolet (shorter than 380 nm), or infrared (longer than 750 nm). Example 26-4

62. • The Balmer series results from transitions of the electron in hydrogen in which the electron ends in the $n = 2$ energy level. Using the Rydberg formula for the Balmer series, calculate the wavelengths of the photons emitted in the transitions that end in the $n = 2$ level and start in the energy levels that correspond to n equal to 3 through 6 and indicate the initial and final levels of the transition corresponding to each wavelength. State whether each wavelength is visible (380 to 750 nm), ultraviolet (shorter than 380 nm), or infrared (longer than 750 nm). Example 26-4

63. • The Paschen series results from transitions of the electron in hydrogen in which the electron ends in the $n = 3$ energy level. Using the Rydberg formula for the Paschen series, calculate the wavelengths of the photons emitted in the transitions that end in the $n = 3$ level and start in the energy levels that correspond to n equal to 4 through 6 and indicate the initial and final levels of the transition corresponding to each wavelength. State whether each wavelength is visible (380 to 750 nm), ultraviolet (shorter than 380 nm), or infrared (longer than 750 nm). Example 26-4

64. •• Calculate the shortest wavelength (and the highest energy) associated with emitted photons in the (a) Lyman, (b) Balmer, and (c) Paschen series (see problems 61, 62, and 63). Example 26-4

65. •• Express the Balmer formula (see problem 57) in terms of the *frequency* of the photons that are emitted (rather than the wavelength). Extend all numerical values out to five significant figures.

66. •• Express the Rydberg formula in terms of the *frequency* of the photons that are emitted (rather than the wavelength). Extend all numerical values out to five significant figures.

26-6 Models by Bohr and Schrödinger give insight into the intriguing structure of the atom

67. • For an electron in the nth state of the hydrogen atom, write expressions for (a) the angular momentum of the electron, (b) the radius of the electron's orbit, (c) the kinetic energy of the electron, (d) the total energy of the electron, and (e) the speed of the electron. Example 26-5

68. • Set up a chart for the five quantities listed in problem 67 and calculate the values for $n = 1, 2, 3, 4,$ and 5 (4 significant figures, SI units). See the following table. Example 26-5

n	L_n	r_n	K_n	E_n	v_n
1					
2					
3					
4					
5					

69. •• Devise a straightforward method that allows you to calculate (a) the speed and (b) the angular momentum of an electron in the nth Bohr orbit. Describe your method and write the formulas you would use for the calculations.

70. •• Using the formula that you devised in problem 69, calculate the speed of an electron and the angular momentum of an electron in the tenth Bohr orbit.

71. •• **Astronomy** For carbon ($Z = 6$) the frequencies of spectral lines resulting from a single electron transition are

increased over those for hydrogen by the ratio of the Rydberg constants:

$$\frac{R_C}{R_H} = \frac{1 - m_e/m_C}{1 - m_e/m_H}$$

The mass of a hydrogen atom (m_H) is 1837 times greater than the mass of the electron (m_e), and the mass of the carbon atom (m_C) is 12 times greater than the mass of the hydrogen atom. Find the difference in frequency between the spectral lines for the carbon 272α transition and the hydrogen 272α transition. The 272α transition is from $n = 273$ to $n = 272$. Express your answer in megahertz (MHz).

72. • A hydrogen atom that has an electron in the $n = 2$ state absorbs a photon. (a) What wavelength must the photon possess to send the electron to the $n = 4$ state? (b) What possible wavelengths would be detected in the spectral lines that result from the deexcitation of the atom as it returns from $n = 4$ to the ground state? Example 26-5

73. • How much energy is needed to ionize a hydrogen atom that starts in the Bohr orbit represented by $n = 3$? If an atom is ionized, its outer electron is no longer bound to the atom. Example 26-5

74. • Find the energies of the first 10 energy levels (sketch and label an energy-level diagram) for singly ionized helium, He$^+$. Example 26-5

75. • In the Bohr model of hydrogen, there is just one possible state of the electron in the lowest energy level ($n = 1$) and just one possible state of the electron in the next energy level ($n = 2$). In the more accurate Schrödinger picture of the hydrogen atom, in which the electron is described by the four quantum numbers n, ℓ, m_ℓ, and m_s, there are two possible states in the $n = 1$ level and eight possible states in the $n = 2$ level. (a) Give the values of n, ℓ, m_ℓ, and m_s for each of the two possible states in the $n = 1$ energy level. (b) Give the values of n, ℓ, m_ℓ, and m_s for each of the eight possible states in the $n = 2$ energy level.

76. • Which of the following electron configurations for an excited atom of beryllium ($Z = 4$) are possible? Which are impossible? Explain your reasoning.
(i) $1s^2 2s^1 2p^1$
(ii) $1s^1 2s^3$
(iii) $1s^1 2p^3$
(iv) $1s^3 2s^1$

26-7 In quantum mechanics, it is impossible to know precisely both a particle's position and its momentum

77. •• The Heisenberg uncertainty principle for momentum and position plays an important role in the field of microscopy. The simple act of observing a sample involves bombarding the sample with photons. This bombardment results in a change to the physical state of the sample. Consider a 600-nm photon that scatters off an atom located within a crystalline lattice. Using the uncertainty principle, determine the minimum position uncertainty of the atom. Example 26-6

78. • (a) What is the minimum uncertainty in an electron's velocity if the position is known within 1.5 nm? (b) What is the minimum uncertainty in a helium atom's velocity if the position is known within 0.10 nm? The mass of a helium atom is 6.646×10^{-27} kg. Example 26-6

79. Consider a 1350-kg automobile clocked by law-enforcement radar at a speed of 38.0 m/s. (a) If the position of the car is known to within 1.50 m at the time of the measurement, what is the uncertainty in the velocity of the car? (b) If the speed limit is 36.5 m/s, could the driver of the car reasonably evade a speeding ticket by invoking the Heisenberg uncertainty principle? Example 26-6

General Problems

80. Data for the photoelectric effect of silver are given in the table.

Frequency of incident radiation (10^{15} s^{-1})	Maximum kinetic energy (10^{-19} J)
2.00	5.90
2.50	9.21
3.00	12.52
3.50	15.84
4.00	19.15

Using these data, find (a) the experimentally determined value of Planck's constant h and (b) the threshold frequency f_0 for silver.

81. • **Biology** Vitamin D is produced in the skin when 7-dehydrocholesterol reacts with UVB rays (ultraviolet B) having wavelengths between 270 and 300 nm. What is the energy range of the UVB photons? Example 26-1

82. •• A helium-neon laser produces light of wavelength 632.8 nm. The laser beam carries a power of 0.50 mW and strikes a target perpendicular to the beam. (a) How many photons per second strike the target? (b) At what rate does the laser beam deliver linear momentum to the target if the photons are all absorbed by the target? Example 26-1

83. • **Astronomy** In 2009 astronomers detected gamma-ray photons having energy ranging from 700 GeV to around 5 TeV coming from supernovae (exploding giant stars) in the galaxy M82. (1 GeV = 10^9 eV, 1 TeV = 10^{12} eV.) (a) What is the range of wavelengths of the gamma-ray photons detected from M82? (b) Calculate the ratio of the energy of a 5-TeV photon to the energy of a visible light photon having a wavelength of 500 nm. Example 26-1

84. ••• Derive an expression for the change in photon *energy* in Compton scattering (that is, an expression for the difference between the final and initial photon energies E_f and E_i) as a function of the photon scattering angle.

85. •• A photon of frequency 4.81×10^{19} Hz scatters off a free stationary electron. Careful measurements reveal that the photon goes off at an angle of 125° with respect to its original direction. (a) How much energy does the electron gain during the collision? (b) What percent of its original energy does the photon lose during the collision? Example 26-2

86. •• **Chemistry** A laboratory oven that contains hydrogen molecules H_2 and oxygen molecules O_2 is maintained at a constant temperature T. Each oxygen molecule is 16 times as massive as a hydrogen molecule. Find the ratio of the de Broglie wavelength of the hydrogen molecule to that of the oxygen molecule, assuming that each molecule has kinetic energy $(3/2)kT$. Example 26-3

87. •• A hydrogen atom makes a transition from the $n = 5$ state to the ground state and emits a single photon of light in the process. The photon then strikes a piece of silicon, which has a photoelectric work function of 4.8 eV. Is it possible that a photoelectron will be emitted from the silicon? If not, why not? If so, find the maximum possible kinetic energy of the photoelectron. Example 26-4

88. ••• Suppose the electron in the hydrogen atom were bound to the proton by gravitational forces (rather than electrostatic forces). Find (a) the radius and (b) the energy of the first orbit.

89. •• **Biology** The *E. coli* bacterium is about 2.0 μm long. Suppose you want to study it using photons of that wavelength or electrons having that de Broglie wavelength. (a) What is the energy of the photon and the energy of the electron? (b) Which one would be better to use, the photon or the electron? Explain why. Example 26-3

90. ••• In 1913, Niels Bohr proposed a model of the hydrogen atom in which the electron could exist only in specific circular orbits. He did not offer any explanation as to why only those particular orbits would be allowed. In 1924, Louis de Broglie suggested that the electron could only exist in orbits corresponding to certain kinds of wave patterns. Figure 26-22 shows an electron orbit that has been "cut and unwrapped" to better show the electron wave. Classify the wave patterns according to whether or not they are compatible with de Broglie's description of the allowed Bohr orbits of the hydrogen atom.

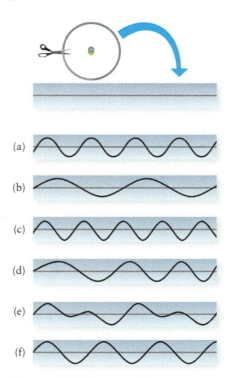

Figure 26-22 Problem 90

91. •• Restate de Broglie's formula for particle waves in the case that the speeds are relativistic. (You will need to use the relativistic relationship between speed and momentum.)

92. •• A relativistic electron has a de Broglie wavelength of 346 fm (1 fm = 10^{-15} m). Determine its speed. (You will

need to use the relativistic relationship between speed and momentum. See problem 91.)

93. • **Astronomy** Radio astronomers use radio frequency waves to identify the elements in distant stars. One of the standard lines that is often studied is designated the 272α line. This spectral line refers to the transition in hydrogen from $n = 273$ to $n = 272$. Calculate the wavelength and frequency of the electromagnetic radiation that is emitted for the 272α transition.

94. • The Lyman series, the Balmer series, and the Paschen series, in which the electron ends up in the $n = 1$, $n = 2$, and $n = 3$ energy levels, respectively, are commonly studied in basic chemistry and physics classes (see problems 61, 62, and 63). The Brackett series (in which the electron ends up in the $n = 4$ energy level) and the Pfund series (in which the electron ends up in the $n = 5$ energy level) are not so well known. (a) Calculate the shortest and longest wavelengths for the spectral lines that are part of the Brackett series. State whether each wavelength is visible (380 to 750 nm), ultraviolet (shorter than 380 nm), or infrared (longer than 750 nm). (b) Calculate the shortest and longest wavelengths for the spectral lines that are part of the Pfund series. State whether each wavelength is visible (380 to 750 nm), ultraviolet (shorter than 380 nm), or infrared (longer than 750 nm). Example 26-4

95. •• When x-ray photons are aimed at a carbon target, a photon can undergo Compton scattering from one of the electrons in a carbon atom. But a photon can undergo Compton scattering from any charged particle, and so can be scattered by a carbon nucleus. (a) Calculate the Compton wavelength $\lambda_C = h/mc$ for a carbon nucleus, which has a mass of 1.99×10^{-26} kg. Express your answer in nanometers (nm). (b) Find the wavelength shift $\Delta\lambda$ for a photon that undergoes a $180°$ scattering from a carbon nucleus. Express your answer in nm. Compare your answer to the wavelength shift of 0.00486 nm for a photon that undergoes a $180°$ scattering from an electron. (c) The x-ray photons that Arthur Compton

used in his 1923 experiment had a wavelength of 0.0711 nm. What fractional change in wavelength $\Delta\lambda/\lambda$ would such a photon undergo due to Compton scattering from an electron? From a carbon nucleus? Explain why Compton was only able to measure the Compton scattering of photons from electrons, not from nuclei. (d) Suppose you want a photon to undergo a 1.00% increase in wavelength (which is more easily measured) as a result of a $180°$ Compton scattering from a carbon nucleus. What would be the wavelength and energy of such a photon before scattering? (This is a gamma-ray photon, and such photons were not available to Compton when he did his experiments.)

96. •• **Biology** You want to use a microscope to study the structure of a mitochondrion about 1 μm in size. To be able to observe small details within the mitochondrion, you want to use a wavelength of 0.0500 nm. (a) If your microscope uses light of this wavelength, what is the momentum p of a photon? What is the energy E of a photon? (b) If instead your microscope uses electrons of this de Broglie wavelength, what is the momentum p of an electron? What is the speed v of an electron? (Recall $p = mv$.) What is the kinetic energy $K = (1/2)mv^2$ of an electron? (c) Comparing your results for the photon momentum and energy in (a) to your results for the electron momentum and kinetic energy in (b), what advantages can you see to using electrons rather than photons?

97. •• (a) When a helium atom jumps down from its first excited energy level (electron configuration $1s^1 2p^1$) to its lowest energy level (electron configuration $1s^2$), it emits an ultraviolet photon of wavelength 58.4 nm. What is the energy of such a photon in eV? (b) Figure 26-18 shows that in a Franck–Hertz experiment using mercury vapor, the anode current drops off for cathode-grid voltages of 4.9 V, 2×4.9 V = 9.8 V, and 3×4.9 V = 14.7 V. If you repeat the Franck–Hertz experiment using helium gas rather than mercury vapor, what are the first three cathode-grid voltages at which you would expect to see a drop in the anode current?

imageBROKER/Alamy

Nuclear Physics

27

In this chapter, your goals are to:

- (27-1) Explain why quantum ideas play an important role in nuclear physics.
- (27-2) Describe how we know that the force that holds the nucleus together is both strong and of short range.
- (27-3) Explain how and why the binding energy per nucleon in a nucleus depends on the size of the nucleus.
- (27-4) Calculate the energy released in nuclear fission.
- (27-5) Describe why very high temperatures are needed for nuclear fusion reactions.
- (27-6) Calculate what happens in the decay of a radioactive substance.

To master this chapter, you should review:

- (3-7) Uniform circular motion.
- (11-2) How to calculate the density of an object.
- (19-6, Current loops and magnetic fields.
  19-7)
- (22-4) The energy of a photon.
- (25-7) The equivalence of mass and energy.
- (26-5) How an atom in an excited energy level emits a photon when the atom decays to a lower energy level.
- (26-6) The spin of an electron.

What do you think?

These stenciled outlines of human hands are found in the Cave of the Hands in the Patagonia region of Argentina. They are known to be 9500 to 13,000 years old, an age determined from the radioactive decay of carbon-14. This type of carbon has 6 protons and 8 neutrons in its nucleus and has a half-life of 5730 years. Of the carbon-14 that was present when these outlines were made, the fraction that remains today is closest to (a) 1/2; (b) 1/4; (c) 1/8; (d) 1/16; (e) 1/32.

27-1 The quantum concepts that help explain atoms are essential for understanding the nucleus

We learned in Chapter 26 that a handful of radical ideas—the notion that energy is quantized, that electromagnetic waves come in packets called photons, and that the laws of quantum mechanics can specify only the probabilities that particles will behave in certain ways—are essential for understanding the nature and behavior of the atom. These same ideas apply on the even smaller scale of the atomic nucleus.

1135

Figure 27-1 Applications of nuclear physics The properties of atomic nuclei explain (a) how magnetic resonance imaging (MRI) works, (b) how the Sun provides energy for life on Earth, and (c) how Earth sustains its internal energy.

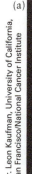

When placed in a strong magnetic field, the nuclei of hydrogen atoms will orient their spins with the field. This effect is at the heart of the diagnostic technique called magnetic resonance imaging.

Photosynthesis in plants depends on energy from sunlight. This energy is released by nuclear reactions that take place in the core of the Sun.

More than 50% of the energy that powers our planet's geological activity, including volcanic eruptions, comes from the radioactive decay of unstable nuclei in Earth's interior.

(a)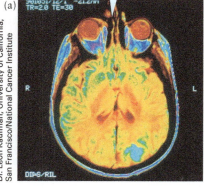

(b)

(c)

Dr. Leon Kaufman, University of California, San Francisco/National Cancer Institute

Brand X Pictures/Thinkstock

iStockphoto/Thinkstock

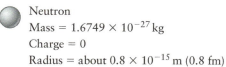

TAKE-HOME MESSAGE FOR Section 27-1

✔ Many of the same concepts that help us understand atomic physics are also important in nuclear physics.

✔ The balance between the strong nuclear force and the electric force determines the stability of nuclei.

We'll see that unlike atoms, in which negatively charged electrons are bound to the positively charged nucleus by the attractive electric force, nuclei are bound by a *strong nuclear force* that keeps the protons and neutrons together. We'll also see that like electrons, nuclei can have an intrinsic angular momentum or *spin*; magnetic resonance imaging, or MRI, makes use of how nuclear spins respond to an external magnetic field (Figure 27-1a). The interplay between the strong nuclear force and the electric force (which makes all of the protons in a nucleus try to repel each other) means that nuclei of different sizes are more tightly or loosely bound. We'll learn that as a result, the largest nuclei are prone to break into smaller fragments through a process called *fission*. We'll also learn that the smallest nuclei can release energy by *fusion*, in which two nuclei join together to form a larger one. Fusion reactions make the Sun shine, and the sunlight that they produce makes life on Earth possible (Figure 27-1b).

Many types of nuclei are **radioactive**: They spontaneously release energy and emit radiation in the form of either particles or photons. Understanding the binding energies of nuclei will also help us understand the three main types of radioactive decay: alpha, beta, and gamma. The energy released by alpha and beta decays in our planet's interior helps power geologic activity such as volcanic eruptions (Figure 27-1c). We'll find that the concept of *half-life* will help us understand the nature of radioactive decays of all kinds.

27-2 The strong nuclear force holds nuclei together

As we learned in Chapter 26, an atom has a small, positively charged nucleus at its center. There are two ways you can see that the nucleus must be positively charged. First, because atoms are neutral, positive charge is required to balance the negative charge of the electrons. Second, the Coulomb attraction between the nucleus and the negatively charged electrons provides the force that holds the atom together.

While the nucleus as a whole is positively charged, not all of its constituents carry a charge. Nuclei contain both protons, each of which has charge $+e = 1.602 \times 10^{-19}$ C, and neutrons, the particle we mentioned briefly in Chapter 16 that has zero charge (Figure 27-2). Protons and neutrons share many similar properties, including similar masses: The masses of the proton and neutron are 1.6726×10^{-27} kg and 1.6749×10^{-27} kg, respectively. Collectively, we refer to protons and neutrons as **nucleons**.

What holds the protons and neutrons together in the nucleus? It cannot be the electric force, because neutrons do not feel the electric force at all and protons (which all have the same positive charge $+e = 1.602 \times 10^{-19}$ C) exert *repulsive* electric forces

⊕ Proton
Mass = 1.6726×10^{-27} kg
Charge = $+e = 1.602 \times 10^{-19}$ C
Radius = about 0.83×10^{-15} m (0.83 fm)

◯ Neutron
Mass = 1.6749×10^{-27} kg
Charge = 0
Radius = about 0.8×10^{-15} m (0.8 fm)

Figure 27-2 Nucleons All atomic nuclei are composed of protons and neutrons, collectively known as nucleons. (The only nucleus that lacks neutrons is the simplest type of hydrogen nucleus, which is made up of a single proton.) The radii of protons and neutrons are very difficult to measure, which is why their values are relatively uncertain.

on each other. These forces are very large because protons are close together inside the nucleus. To see just how large these forces are, note that a typical radius of a nucleus is about 5 fm (1 fm = 1 femtometer = 10^{-15} m). The electric force between two protons separated by a distance $r = 5.00$ fm is

$$F = \frac{k(+e)(+e)}{r^2} = \frac{(8.99 \times 10^9 \ \mathrm{N \cdot m^2/C^2})(1.602 \times 10^{-19} \ \mathrm{C})^2}{(5.00 \times 10^{-15} \ \mathrm{m})^2}$$

$$= 9.21 \ \mathrm{N}$$

This may not seem like a lot of force, but remember that this force is applied to a proton, which has a mass of only 1.6726×10^{-27} kg. If the electric force were the only force acting on the protons in a nucleus, the nucleus would simply fly apart (Figure 27-3).

We conclude that there must be an additional *attractive* force that acts on all nucleons (both protons and neutrons) and that binds them together in the nucleus. This attractive force must be stronger than the repulsive electric force between protons, so we call it the **strong nuclear force**. Over short distances the strong nuclear force is hundreds of times stronger than the electric force.

WATCH OUT! The nucleus is not held together by gravitational forces.

It's true that nucleons (protons and neutrons) attract each other through the gravitational force. But the masses of nucleons are so small that the gravitational forces between them are entirely negligible. For the two protons in Figure 27-3, the repulsive electric forces between them are stronger than the attractive gravitational forces by a factor of 10^{46}! We conclude that gravitational forces play no role on the scale of the atomic nucleus.

If the nuclear force is so strong compared to the electric force, and if protons attract other protons by this force, why are neutrons necessary to help overcome the Coulomb repulsion between protons? The answer lies in the *range* of the strong nuclear force, which is the distance beyond which one nucleon no longer experiences a force due to another. Experiments show that the strong nuclear force between two nucleons has a range of only about 2.0 fm. The radius of a proton or neutron is about 0.8 fm, so two nucleons must almost be touching to experience the strong nuclear force. As a very rough analogy, you can think of protons and neutrons as tiny spheres coated with very strong Velcro, which makes the nucleons stick together if they are brought close enough to each other.

By contrast, the electric repulsion between protons separated by a distance r is proportional to $1/r^2$, so is present even if the distance r is very large. We say that the strong nuclear force is a *short-range* force, whereas the electric force is a *long-range* force.

Because the range of the nuclear force is smaller than the diameter of most nuclei (a few to perhaps 15 fm), each nucleon exerts an attractive nuclear force only on its nearest neighbors. Each proton, however, exerts a repulsive force on *every other* proton in the nucleus. As a result, the nuclear force between neighboring protons cannot overcome the Coulomb repulsion between all of the protons. To prevent a nucleus from spontaneously breaking apart (that is, for it to be *stable*), the nucleus must also contain neutrons. You can think of the neutrons as "spacers" that increase the average distance between protons and so decrease their mutual Coulomb repulsion.

In the smallest stable nuclei, the number of neutrons is about the same as the number of protons; in nuclei with more than about 20 protons, there must be more neutrons than protons for the nucleus to be stable. An unstable nucleus will eventually undergo a spontaneous transformation, termed a *decay*, in which it either splits apart or gives off energy in some other way. Figure 27-4 shows the number of neutrons versus the number of protons in known atomic nuclei. Stable nuclei are shown in black. Notice that as the number of protons increases, more additional neutrons are required for stability.

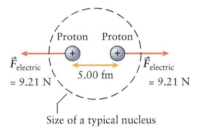

Figure 27-3 Electric repulsion between protons Because protons are so close together inside the nucleus, they exert very substantial electric forces on each other that tend to push them apart. To stabilize the nucleus against these repulsive forces, a much stronger attractive force between nucleons — called the strong nuclear force — must be present to hold the nucleus together.

Nuclides, Isotopes, and Nuclear Sizes

Atoms of each element have a unique number of protons and the same number of electrons: Hydrogen has one, helium two, lithium three, and so on. Many properties of atoms are related to this number, usually designated as the **atomic number** Z. Many properties of nuclei arise from *both* the value of Z and the value of the **neutron number** N, which is the number of neutrons in the nucleus. Nuclei of a given element all have the same number of protons and so have the same value of Z (changing the number of protons changes the

Figure 27-4 **Neutrons versus protons in nuclei** The interplay between the strong nuclear force and the electric force explains the relationship between the numbers of protons and neutrons in nuclei.

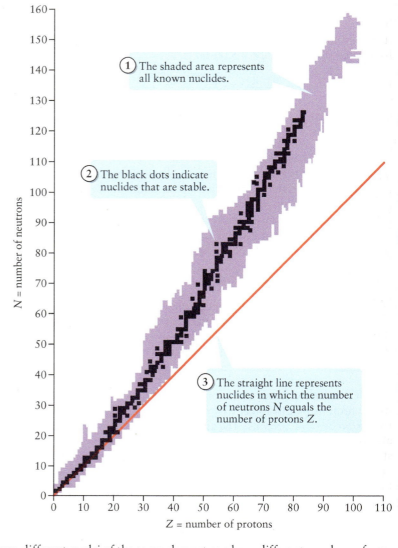

① The shaded area represents all known nuclides.

② The black dots indicate nuclides that are stable.

③ The straight line represents nuclides in which the number of neutrons N equals the number of protons Z.

N = number of neutrons

Z = number of protons

💡 For most nuclides, more neutrons than protons must be present (N must be greater than Z) to prevent the protons from flying apart due to electrostatic repulsion.

element). However, different nuclei of the same element can have different numbers of neutrons and so different values of N. Each combination of Z and N specifies a **nuclide**. For each element, the most common configuration of Z and N corresponds to the most stable nuclide; nuclei of that element with a different number of neutrons are termed **isotopes**. For example, potassium has 19 protons (and 19 electrons). The most stable and most common nuclide of potassium has 20 neutrons, $N = 20$. We denote this nuclide with the symbol ^{39}K, commonly referred to as "potassium-39." The number of protons ($Z = 19$) is understood from the symbol K for potassium, and the total number of protons and neutrons, termed the **mass number** A, is given in the pre-superscript, 39. Note that the mass number of any nucleus equals the atomic number plus the neutron number: $A = Z + N$.

More than 93% of all potassium atoms have a ^{39}K nucleus. About 7% of potassium atoms have a ^{41}K nucleus, however, with 19 protons and 22 neutrons, and 0.012% have a ^{40}K nucleus with 19 protons and 21 neutrons. We say that ^{39}K, ^{40}K, and ^{41}K are three isotopes of potassium.

The *size* of a nucleus (its radius and volume) is related to the mass number A. Experiments show that all nuclei are approximately spherical and have radii proportional to the cube root of A, or $A^{1/3}$:

Radius of a nucleus

(27-1)

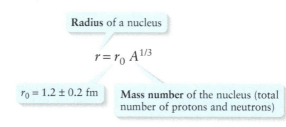

Radius of a nucleus

$$r = r_0 A^{1/3}$$

$r_0 = 1.2 \pm 0.2$ fm

Mass number of the nucleus (total number of protons and neutrons)

This equation states that if we increase the number of nucleons in a nucleus by a factor of 10, say from $A = 20$ to $A = 200$, the radius of the nucleus will increase by a factor of $10^{1/3} = 2.2$ (Figure 27-5).

The volume of a sphere is proportional to r^3. Hence the volume of a nucleus is directly proportional to $(A^{1/3})^3 = A$, which is the mass number of the nucleus and the number of nucleons within that nucleus. This is consistent with our model of nucleons as spheres covered with Velcro. In this model, increasing the number of nucleons by a factor of 10 would result in a ball of nucleons (that is, a nucleus) with 10 times the volume. Because the attractive force is short range, there is no tendency for the nucleons to be compressed together as the size of the nucleus increases.

It's useful to contrast nuclei with planets, which are held together by the *long*-range gravitational force of attraction between all of the parts of the planet. As an example, compare the planets Jupiter and Saturn (which have the same chemical composition): Jupiter has 3.3 times more mass than Saturn, but Jupiter has only 1.7 times greater volume because it is more highly compressed and has a greater density. Nuclei of different sizes do *not* behave like planets of different sizes: All nuclei have basically the same density (see Example 27-2 below). This is further evidence that the strong nuclear force is short-range rather than long-range.

Can we simply add more protons and more neutrons to make larger and larger nuclei? The answer is "no," and for the same reason that neutrons are required for nuclear stability. Figure 27-4 shows that as we work our way up the periodic table to atoms that have more and more protons, more *additional* neutrons are required for stability. Each additional proton exerts a repulsive force on all the others, but neutrons and protons can only attract their nearest neighbors. Those additional neutrons cause the size of the nucleus to grow so that eventually too many neutrons are near the surface of the nucleus and therefore not completely surrounded by neighbors. At that point, the nuclear forces holding the nucleus together are not large enough to overcome the Coulomb repulsion between the protons, and the nucleus cannot be stable. The largest stable nuclide is lead-208 (^{208}Pb), which has 82 protons and 126 neutrons.

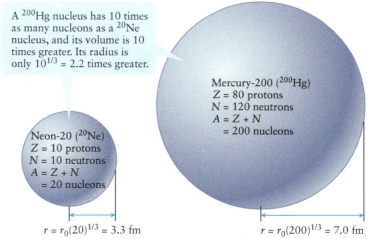

A ^{200}Hg nucleus has 10 times as many nucleons as a ^{20}Ne nucleus, and its volume is 10 times greater. Its radius is only $10^{1/3} = 2.2$ times greater.

Mercury-200 (^{200}Hg)
$Z = 80$ protons
$N = 120$ neutrons
$A = Z + N$
$= 200$ nucleons

Neon-20 (^{20}Ne)
$Z = 10$ protons
$N = 10$ neutrons
$A = Z + N$
$= 20$ nucleons

$r = r_0(20)^{1/3} = 3.3$ fm

$r = r_0(200)^{1/3} = 7.0$ fm

Figure 27-5 Nuclear sizes The volume of an atomic nucleus is proportional to the number of nucleons it contains (the mass number A). The nuclear radius is proportional to the cube root of A.

EXAMPLE 27-1 Nuclear Radii

Estimate the radius of the nucleus of ^{12}C, a relatively small nucleus; ^{118}Sn, a nucleus of medium size; and ^{236}U, a relatively large nucleus. The nuclides ^{12}C and ^{118}Sn are stable; ^{236}U, like all other isotopes of uranium, is unstable.

Set Up

In each case we use Equation 27-1 to calculate the radius of the nucleus. The value of A for each nucleus is given by the pre-superscript: $A = 12$ for carbon (C), $A = 118$ for tin (Sn), and $A = 236$ for uranium (U).

Radius of a nucleus:

$$r = r_0 A^{1/3} \qquad (27\text{-}1)$$

Solve

Apply Equation 27-1 to each nuclide.

We'll use $r_0 = 1.2$ fm in our calculations.
For ^{12}C, which has 6 protons and 6 neutrons,

$r(^{12}\text{C}) = (1.2 \text{ fm})(12)^{1/3} = 2.7$ fm

For ^{118}Sn, which has 50 protons and 68 neutrons,

$r(^{118}\text{Sn}) = (1.2 \text{ fm})(118)^{1/3} = 5.9$ fm

For ^{236}U, which has 92 protons and 144 neutrons,

$r(^{236}\text{U}) = (1.2 \text{ fm})(236)^{1/3} = 7.4$ fm

Reflect

This calculation shows that typical nuclei have radii of just a few femtometers. Although ^{236}U has $236/12 = 19.7$ times as many nucleons as ^{12}C, it is only larger in radius by a factor of $(7.4 \text{ fm})/(2.7 \text{ fm}) = 2.7$. That's a consequence of the $A^{1/3}$ factor in Equation 27-1: Note that $(19.7)^{1/3} = 2.7$.

Notice that in carbon, the smallest of the three nuclei, the number of neutrons equals the number of protons. In tin, which has almost 10 times the mass number of carbon, the number of neutrons required for nuclear stability is 36% greater than the number of protons. And even with nearly 60% more neutrons than protons, the relatively large ^{236}U nucleus is not stable. So we see direct evidence that in larger nuclei more and more additional neutrons, compared to the number of protons, are required for nuclear stability, and that at some size a nucleus is too large for the strong nuclear force to overcome the Coulomb repulsion between the protons.

EXAMPLE 27-2 Nuclear Density

Estimate the density (in kg/m^3) of a nucleus that has mass number A.

Set Up

The density of an object is its mass m divided by its volume V. We'll find the volume of a nucleus from Equation 27-1 and the formula for the volume of a sphere. To estimate the mass of a nucleus, we'll multiply the mass number A (the number of nucleons) by the average mass of a nucleon.

Radius of a nucleus:

$$r = r_0 A^{1/3} \qquad (27\text{-}1)$$

Definition of density:

$$\rho = \frac{m}{V} \qquad (11\text{-}1)$$

Volume of a sphere of radius r:

$$V = \frac{4}{3}\pi r^3$$

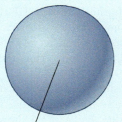

mass $m = A \times$ (average mass of a nucleon)
radius $r = r_0 A^{1/3}$
volume $V = (4/3)\pi r^3$
density $\rho = m/V$

Solve

Find the volume of a nucleus of mass number A.

Substitute Equation 27-1 into the expression for the volume of a sphere of radius r:

$$V = \frac{4}{3}\pi r^3 = \frac{4}{3}\pi(r_0 A^{1/3})^3 = \frac{4}{3}\pi r_0^3 (A^{1/3})^3$$
$$= \frac{4}{3}\pi r_0^3 A$$

Use $r_0 = 1.2 \text{ fm} = 1.2 \times 10^{-15}$ m, as in Example 27-1:

$$V = \frac{4}{3}\pi(1.2 \times 10^{-15} \text{ m})^3 A$$
$$= (7.2 \times 10^{-45} \text{ m}^3)A$$

Take the average mass of a nucleon to be the average of the proton mass m_p and the neutron mass m_n. Use this to write an expression for the mass of a nucleus of mass number A.

The average mass of a nucleon is

$$m_{\text{avg}} = \frac{m_p + m_n}{2} = \frac{(1.6726 \times 10^{-27} \text{ kg}) + (1.6749 \times 10^{-27} \text{ kg})}{2}$$
$$= 1.6738 \times 10^{-27} \text{ kg}$$

A nucleus with A nucleons then has mass

$$m = m_{\text{avg}}A = (1.6738 \times 10^{-27} \text{ kg})A$$

Calculate the density of the nucleus.

The density of the nucleus is

$$\rho = \frac{m}{V} = \frac{(1.6738 \times 10^{-27} \text{ kg})A}{(7.2 \times 10^{-45} \text{ m}^3)A} = \frac{1.6738 \times 10^{-27} \text{ kg}}{7.2 \times 10^{-45} \text{ m}^3}$$
$$= 2.3 \times 10^{17} \text{ kg/m}^3$$

Reflect

Our final expression for the density ρ does not depend on A, the mass number of the nucleus. So the density of *all* nuclei is about the same. This agrees with our statements about the short-range character of the strong nuclear force. Note also that a block of solid iridium, the densest of all stable elements, has a density of $22,650$ kg/m^3 (22.65 times the density of water). Our calculation shows that nuclei are 10^{13} times denser than iridium. Nuclei are *extremely* dense! This makes sense: Most of the mass of an atom is concentrated in its nucleus, which has a far smaller volume than the atom as a whole. So nuclear density (mass divided by volume) must be far greater than what we think of as the "ordinary" density of matter such as water or iridium.

Nuclear Spin and Magnetic Resonance Imaging

We learned in Section 26-6 that electrons have a type of intrinsic angular momentum called *spin*. (This is something of a misnomer because this angular momentum does not correspond directly to a spinning motion of the electrons.) Protons, and neutrons, too, have spin, and like an electron the spin of a proton or neutron can take on one of two values, often referred to as "spin up" and "spin down." In nuclei with an even number of nucleons, typically there are as many "spin up" nucleons as there are "spin down," and most such nuclei have zero net spin. (The orbital angular momentum of nucleons moving inside the nucleus can also contribute to the net spin of the nucleus.) But a nucleus with an odd number of nucleons must have a nonzero net spin. In particular, the nucleus of hydrogen—a single proton—has a net spin.

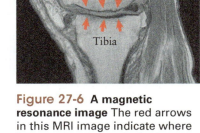

Figure 27-6 A magnetic resonance image The red arrows in this MRI image indicate where bones rub together in the knee of a patient suffering from osteoarthritis.

BioMedical Measurements that make use of the spin of hydrogen nuclei enable us to localize hydrogen in an object or body, as well as to get information about the material in which those hydrogen atoms are embedded. That's the principle of **magnetic resonance imaging** (MRI). In an MRI scanner, spin information is used to form a three-dimensional map of the density of hydrogen atoms in a body. Living organisms are composed largely of water—our bodies, for example, are 60 to 70% water—and each water molecule contains two hydrogen atoms. As such, MRI is an ideal way to probe the internal structures in the body. The MRI scan in Figure 27-6 shows the leg bones rubbing against each other in the knee of a person with osteoarthritis. Figure 27-1a shows an MRI scan of a patient's head.

Protons and neutrons also have associated magnetic fields. This field is a dipole field like that of a current-carrying loop of wire (see Section 19-7). The line connecting the north and south poles of a proton or neutron is along the direction of its spin angular momentum vector. Now, we saw in Section 19-6, a current loop placed in an external magnetic field tends to align with its normal along the direction of the external field. In the same way, when a single proton or neutron is placed in a uniform external magnetic field, the magnetic force tends to align the spin direction either generally parallel to or antiparallel to the external field. The same is true for a nucleus with an odd number A of nucleons. To be precise, the spin direction aligns so that its component along the direction of the external field has a fixed positive or negative value. As such, the spin direction can rotate, or *precess*, around an axis defined by the field direction. Figure 27-7a shows this precession for a nucleus with spin aligned with the external field, and Figure 27-7b shows this precession for a nucleus with spin anti-aligned with (that is, opposed to) the external field.

Consider a large number of hydrogen atoms placed in a uniform magnetic field. For these atoms the nucleus is a single proton. About half of the protons will end up with spin aligned with the field and half with spin anti-aligned. Now, the energy of the spin-aligned orientation of a nucleus in an external magnetic field (Figure 27-7a) is lower than the energy of the anti-aligned orientation (Figure 27-7b) by an amount ΔE that depends on the magnetic field strength. Suppose we now bathe the atoms in an additional alternating magnetic field with frequency f.

(a) A nucleus with its spin aligned with an external magnetic field

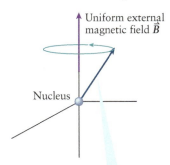

Uniform external magnetic field $\vec{B}$

Nucleus

The spin angular momentum vector of the nucleus traces out a cone, but is generally aligned in the same direction as $\vec{B}$.

(b) A nucleus with its spin anti-aligned with an external magnetic field

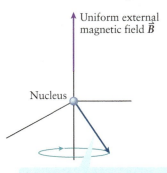

Uniform external magnetic field $\vec{B}$

Nucleus

The spin angular momentum vector of the nucleus traces out a cone, but is generally aligned in the direction opposite to $\vec{B}$.

Figure 27-7 Nuclear spin When a nucleus is placed in an external magnetic field, the component of the spin along the direction of the external field is either (a) aligned with the external field or (b) anti-aligned with the external field.

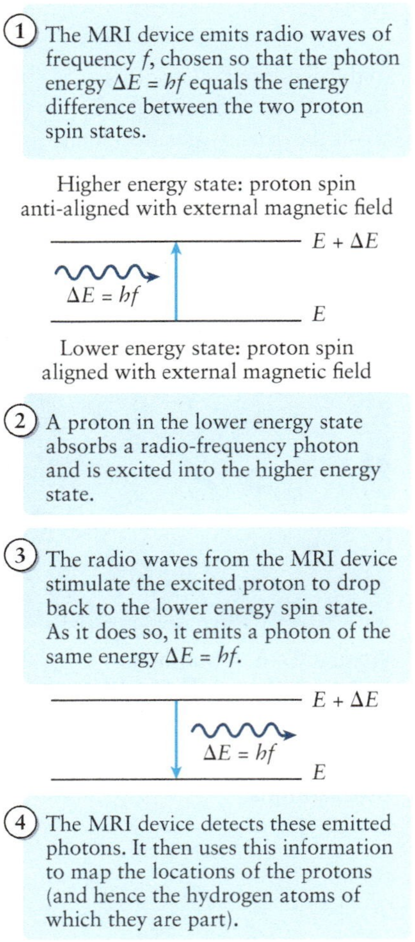

① The MRI device emits radio waves of frequency f, chosen so that the photon energy $\Delta E = hf$ equals the energy difference between the two proton spin states.

Higher energy state: proton spin anti-aligned with external magnetic field

$\Delta E = hf$ ——— $E + \Delta E$

——— E

Lower energy state: proton spin aligned with external magnetic field

② A proton in the lower energy state absorbs a radio-frequency photon and is excited into the higher energy state.

③ The radio waves from the MRI device stimulate the excited proton to drop back to the lower energy spin state. As it does so, it emits a photon of the same energy $\Delta E = hf$.

——— $E + \Delta E$

$\Delta E = hf$

——— E

④ The MRI device detects these emitted photons. It then uses this information to map the locations of the protons (and hence the hydrogen atoms of which they are part).

Figure 27-8 How to make proton spins flip In a magnetic resonance imaging (MRI) scan, the patient is placed in a strong magnetic field. This causes a difference in energy ΔE between the two spin states of the nucleus (a proton) in a hydrogen atom in the patient. Radio waves from the MRI device excite these protons into the higher energy state, flipping their spins; when the proton spin flips back to return to the lower energy state, the proton emits a radio photon that the MRI device detects.

The alternating field is made up of photons of frequency f and energy hf, where h is Planck's constant (see Section 26-2). If we choose the frequency f so that hf is equal to ΔE, the photon energy is just equal to the energy difference between the two spin states. As a result, protons that are initially in the lower energy state, with their spin aligned with the field, can absorb a photon of energy $\Delta E = hf$ and flip their spin to be anti-aligned with the external field. What is more, protons that are initially in the higher energy state, with their spin anti-aligned with the field, can be stimulated by the alternating field to emit a photon of energy $\Delta E = hf$ and so flip their spin to the lower energy, aligned state. (A similar sort of *stimulated emission* also occurs in a laser; see Section 26-5.)

The difference in energy between the two spin states of a hydrogen atom depends on the strength of the external field. With the high field strength required to be able to clearly observe the spin-flipping phenomenon, typically around 3 T, the energy difference ΔE is about 5×10^{-7} eV. (As we have learned before, one eV, or one electron volt, is the amount of energy acquired by an electron when it experiences a potential difference of one volt: 1 eV $= 1.60 \times 10^{-19}$ J. We saw in Section 26-6 that the difference in energy between adjacent energy levels of a hydrogen atom is around 1 to 10 eV. The energy difference ΔE for this spin flip is more than a million times smaller.) From Equation 22-26 the frequency of a photon of this energy is

$$f = \frac{\Delta E}{h} = \frac{5 \times 10^{-7} \text{ eV}}{4.14 \times 10^{-15} \text{ eV} \cdot \text{s}} = 1.2 \times 10^8 \text{ s}^{-1} = 120 \text{ MHz}$$

(We used the value of h in eV·s.) So the frequency of the oscillating field used to induce the hydrogen atoms to flip their spins in a strong magnetic field is around 100 MHz, which is in the radio-frequency part of the electromagnetic spectrum. We use the term "resonance" because the spins flip only if the field frequency has just the right value, similar to the behavior of a forced oscillator (Section 12-8) or sound waves in a pipe (Figure 13-22).

Figure 27-8 shows how we "see" this spin flipping and so locate the hydrogen atoms. Initially there will be more atoms in the lower energy state than the higher energy state (lower energy is more likely for a system than higher energy), so when the hydrogen atoms in an external magnetic field are exposed to radio waves with the right frequency f, there is a net absorption of the electromagnetic energy. The radio-frequency signal from the MRI device then stimulates the excited protons to return to their initial (lower energy) state, and in so doing they generate a radio-frequency field of their own. The MRI device detects that radio-frequency field and uses it to map the density of hydrogen atoms in the body.

As we have mentioned, the difference in energy ΔE between the two hydrogen spin states depends on the strength of the external magnetic field. In an MRI device the magnetic field is made to vary over a body's volume, so the energy absorbed and then re-emitted in the spin–flip process also varies in different parts of the body. The exact frequencies of the radio energy detected by the MRI device therefore provide the information necessary to create images of high spatial resolution. In addition, the time it takes for the spins of the hydrogen nuclei to return to their equilibrium state depends on the particular molecules in the tissue. So timing information in an MRI device provides the means to differentiate one type of tissue from another.

GOT THE CONCEPT? 27-1 Nuclear Radius and Density

 A ^{20}Ne nucleus has 10 protons and 10 neutrons for a total of 20 nucleons, and a ^{160}Dy nucleus has 66 protons and 94 neutrons for a total of 160 nucleons. Compared to a ^{20}Ne nucleus, a ^{160}Dy nucleus has (a) double the radius and a greater density; (b) 8 times the radius and a greater density; (c) double the radius and the same density; (d) 8 times the radius and the same density; (e) double the radius and a lower density.

TAKE-HOME MESSAGE FOR Section 27-2

✔ The strong nuclear force binds nucleons (protons and neutrons) together in the nucleus of an atom.

✔ The volume of a nucleus is proportional to its mass number (the total number of nucleons in the nucleus).

✔ To be stable a light nucleus must have about as many neutrons as protons. More massive nuclei with more than about 20 protons require more neutrons than protons to be stable. Very large nuclei with mass number greater than 208 are always unstable.

✔ Protons and neutrons have spin. Magnetic resonance imaging uses the difference in energy between a state in which a nuclear spin is aligned with an external magnetic field and the state in which the spin is anti-aligned with the field.

27-3 Some nuclei are more tightly bound and more stable than others

Release a ball at the top of a hill, and it rolls down. Pull an object attached to the free end of a spring away from its equilibrium position, and it tends to return to that position. In both cases the systems are finding their way to a more stable configuration. All physical systems do the same; if a more stable configuration exists for a system, it will eventually find itself in that configuration as long as nature provides a mechanism for the transition to take place.

In this context, consider the nucleus of a helium atom, which consists of two protons and two neutrons. This configuration of these four nucleons must be more stable than when they are separate; otherwise the helium nucleus would end up broken apart. This stability results because the attraction of the strong nuclear force between the four nucleons overwhelms the electric repulsive force between the two protons.

Let's be quantitative about *how* stable a given nucleus is. The total mass M_{tot} of the two protons and two neutrons in the nucleus of ^{4}He equals the sum of two proton masses (m_p) and two neutron masses (m_n):

$$M_{tot} = 2m_p + 2m_n$$
$$= 2(1.6726 \times 10^{-27} \text{ kg}) + 2(1.6749 \times 10^{-27} \text{ kg}) = 6.695 \times 10^{-27} \text{ kg}$$

But the actual mass of a helium nucleus is 6.645×10^{-27} kg, which is *less* than the total mass of the two protons and two neutrons (**Figure 27-9**). How is this possible? The answer is the key to nuclear stability; the energy equivalent of the difference in mass is tied up in binding the nucleons together. Recall from Section 25-7 that energy and mass are equivalent and that an object of mass m has a rest energy $E_0 = mc^2$ (Equation 25-21). Since a ^{4}He nucleus has a smaller mass than its constituent nucleons, it also has a smaller rest energy. The difference between the rest energy of the ^{4}He nucleus and the rest energy of its constituent nucleons is the **binding energy** E_B. You can think of this as the energy that is released when the four nucleons come together to form a ^{4}He nucleus. Alternatively, you can think of the binding energy as the energy that would be required to separate a ^{4}He nucleus into two protons and two neutrons. The greater the binding energy *per nucleon* in a nucleus, the more tightly the nucleus is bound and therefore the more stable it is.

Figure 27-10 is a graph of the binding energy per nucleon (E_B/A) for different nuclides as a function of their mass number A. The energy values on the vertical axis are given in MeV, where 1 MeV = 10^6 eV. Figure 27-10 shows that the value of E_B/A for all nuclides is in the range from 1 to 9 MeV, which is why these units are convenient. (By comparison, the binding energy of a hydrogen *atom*—the energy required to separate the single electron in a hydrogen atom from the proton—is 13.6 eV, about 10^{-5} as great as the binding energy of even the most weakly bound nucleus. This indicates how strong the forces are within the nucleus compared with the forces on electrons within atoms.)

Figure 27-10 shows that the nuclide with the smallest binding energy per nucleon is ^{2}H, an isotope of hydrogen with one proton and one neutron. (Just 0.0115% of hydrogen

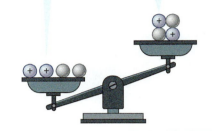

Combined mass of two protons and two neutrons: 6.695×10^{-27} kg

Mass of two protons and two neutrons bound into a ^{4}He nucleus: 6.645×10^{-27} kg

- The mass of a nucleus is less than the sum of the masses of its component nucleons.
- The difference in masses is due to the binding energy of the nucleus.

Figure 27-9 Binding energy Nucleons are bound together in a nucleus by the strong nuclear force. Hence the rest energy of any nucleus made up of two or more nucleons is less than the sum of the rest energies of the constituent nucleons, and the same is true of the masses.

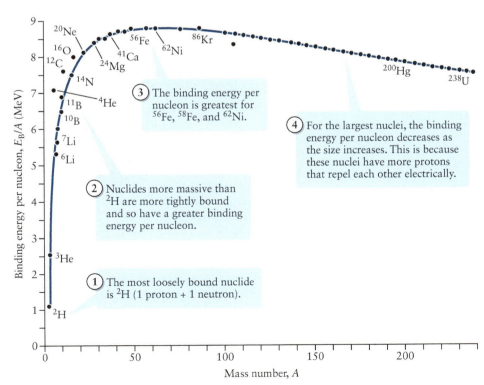

Figure 27-10 The curve of binding energy The binding energy per nucleon in a nucleus depends on the mass number A.

Felix Choo/Alamy

Figure 27-11 **Five cents' worth of the most stable nuclei** A Canadian five-cent coin is made of steel (mostly iron) plated with nickel. Naturally occurring iron is 5.85% ^{54}Fe, 91.75% ^{56}Fe, 2.12% ^{57}Fe, and 0.28% ^{58}Fe; naturally occurring nickel is 68.08% ^{58}Ni, 26.22% ^{60}Ni, 1.14% ^{61}Ni, 3.64% ^{62}Ni, and 0.93% ^{64}Ni. All of these isotopes are stable and have very large values of the binding energy per nucleon E_B/A.

atoms are of this isotope.) As A increases, E_B/A increases rapidly because more nucleons are surrounded by other nucleons to which they are attracted and so are more tightly bound. The value of E_B/A peaks at about 8.8 MeV for A in the range of 56 to 62 and then decreases slowly for higher and higher values of A. This decrease happens because the number of protons Z also increases as A increases, and the electric repulsion between protons destabilizes the nucleus. The most stable nuclei are the nickel isotope ^{62}Ni and the two iron isotopes ^{56}Fe and ^{58}Fe (Figure 27-11). The binding energy per nucleon of these nuclei places them near the peak of the E_B/A curve in Figure 27-10.

Whenever possible nuclei will rearrange themselves to maximize their stability and so maximize their binding energy per nucleon. Figure 27-10 shows that nuclei with relatively *large* values of mass number A (at the far right of the graph) can become more stable by *decreasing* the number of nucleons. One process of this kind is **nuclear fission,** in which nuclei split into smaller pieces. For example, atoms of curium-244 can spontaneously fission into xenon-135 and molybdenum-109. Curium has 96 protons, so ^{244}Cm has 148 neutrons. There are 54 protons in xenon and 42 in molybdenum or 96 total. The total number of nucleons in ^{135}Xe and ^{109}Mo, 244, equals the number of nucleons in ^{244}Cm. In other words, ^{135}Xe and ^{109}Mo contain the same 96 protons and 148 neutrons as in the original ^{244}Cm; the curium atom has split into two fragments. We'll discuss fission in more detail in Section 27-4.

Figure 27-10 also shows that nuclei with relatively *small* values of A (at the far left of the graph) can become more stable by *increasing* the number of nucleons. One way to do this is by **nuclear fusion,** in which two small nuclei join together to make a larger one. For example, the 12 protons and 12 neutrons in two carbon-12 nuclei can fuse to form a magnesium-24 nucleus. More than one nucleus can also be formed in a fusion process. For example, when a helium-3 nucleus fuses with a lithium-6 nucleus, the reaction forms two helium-4 nuclei and one hydrogen-1 nucleus. (Count the nucleons: ^{3}He has two protons and one neutron, and ^{6}Li has three protons and three neutrons, for a total of five protons and four neutrons. The two ^{4}He nuclei have two protons and two neutrons each, leaving the one remaining proton as a ^{1}H nucleus.) In Section 27-5 we'll discuss nuclear fusion more carefully.

Let's see how to calculate the binding energy per nucleon of a nucleus such as ^{4}He. We'll do this by finding the total mass of the protons and neutrons that make up the nucleus and comparing it to the mass of the actual nucleus. We then convert that difference to the equivalent energy.

Instead of measuring mass in kilograms, it's most convenient to use units that take advantage of the equivalence between mass and energy. Since rest energy equals mass multiplied by c^2, we'll measure masses in units of MeV/c^2. Note that 1 MeV/$c^2 = 1.7827 \times 10^{-30}$ kg. In these units, the mass of a proton is 938.27 MeV/c^2, the mass of a neutron is 939.57 MeV/c^2, and the mass of a ^{4}He nucleus is 3727.4 MeV/c^2.

The difference Δ between (i) the mass of the two protons and two neutrons separately and (ii) the mass of the helium nucleus is

$$\Delta = 2m_p + 2m_n - m_{He}$$
$$= 2(938.27 \text{ MeV}/c^2) + 2(939.57 \text{ MeV}/c^2) - 3727.4 \text{ MeV}/c^2$$
$$= 28.3 \text{ MeV}/c^2$$

The energy equivalent of any mass is obtained by multiplying it by c^2, so the energy equivalent of this difference is 28.3 MeV. (Notice how straightforward it is to find the energy equivalent of mass when we write mass in units of MeV/c^2.) In other words, the binding energy of ^{4}He is 28.3 MeV. There are four nucleons in the helium nucleus, so E_B/A is about 7.1 MeV. You can verify this result from the curve in Figure 27-10.

The binding energy per nucleon is much higher for ^{4}He than for other light nuclei. (Figure 27-10 shows that E_B/A is about 2.5 MeV for ^{3}He and about 5.3 MeV for ^{6}Li.) Thanks to its relatively high binding energy per nucleon, ^{4}He is far more stable than other light nuclei. For this reason, when large nuclei break apart to transform to a more stable, more energetically favorable state, in many cases they do so by emitting two protons and two neutrons bound together. In such processes the four bound protons and neutrons are referred to as an *alpha particle* (α particle). **Nuclear radiation** is the emission by a nucleus of either energy (in the form of a photon) or particles, such as an α particle. We'll discuss nuclear radiation in more detail in Section 27-6.

To find the binding energy of ^{4}He, we subtracted the mass of the nucleus from the mass of the two protons and two neutrons separately and then multiplied by c^2 to find

GOT THE CONCEPT? 27-2
Stability

(?) Rank these nuclides in order from most stable to least stable. (a) ^{11}B; (b) ^{20}Ne; (c) ^{86}Kr; (d) ^{200}Hg.

the equivalent energy. In general, for a nucleus consisting of N neutrons and Z protons, E_B is

$$E_B = (Nm_n + Zm_p - m_{nucleus})c^2$$

where m_n is the mass of a neutron, m_p is the mass of a proton, and $m_{nucleus}$ is the mass of the nucleus. In practice it's easier to measure the masses of neutral *atoms* (including their electrons) than the masses of isolated atomic nuclei. In terms of these masses, we can write the binding energy of a nucleus as

Binding energy of a nucleus	Number of neutrons in the nucleus	Number of protons in the nucleus	Mass of a neutral atom (including electrons) containing that nucleus

$$E_B = (Nm_n + Zm_{1_H} - m_{atom})c^2$$

Binding energy of a nucleus
(27-2)

Mass of a neutron	Mass of a neutral hydrogen atom (1 proton + 1 electron)	Speed of light in a vacuum

The terms Zm_{1_H} and m_{atom} each include the mass of Z electrons, so the electron masses cancel. The masses of neutral atoms can be found online, or go to the Table of Atomic Masses in Appendix C found in the online version of this book. Note that values given in units of atomic mass units (u or amu) can be expressed as MeV/c^2, where

$$1\,u = 931.494\ MeV/c^2$$

EXAMPLE 27-3   The Binding Energy of ^{4}He

Previously, we determined the binding energy per nucleon in the ^{4}He nucleus using the mass of the nucleus. Use the following values to determine the binding energy per nucleon (in MeV/c^2) of the ^{4}He nucleus using the *atomic* mass of ^{4}He: neutron mass = 1.008665 u, atomic mass of ^{1}H = 1.007825 u, and atomic mass of ^{4}He = 4.002602 u.

Set Up

We'll use Equation 27-2 and the values given for the neutron mass, the atomic mass of ^{1}H, and the atomic mass of ^{4}He.

Binding energy of nucleus:

$$E_B = (Nm_n + Zm_{1_H} - m_{atom})c^2$$
(27-2)

neutron mass = m_n = 1.008665 u
atomic mass of ^{1}H = m_{1H} = 1.007825 u
atomic mass of ^{4}He = m_{4He} = 4.002602 u

binding energy of ^{4}He = rest energy of two neutrons + rest energy of two ^{1}H atoms − rest energy of a ^{4}He atom

Solve

Calculate the binding energy of the ^{4}He nucleus, which has two neutrons ($N = 2$) and two protons ($Z = 2$).

Substitute the values of m_n, m_{1H}, and m_{4He} into Equation 27-2, with $m_{atom} = m_{4He}$:

$$E_B = (2(1.008665\ u) + 2(1.007825\ u) - 4.002602\ u)c^2$$

$$= 0.030378\ uc^2$$

Since 1 u = 931.494 MeV/c^2,

$$E_B = 0.030378\ uc^2 \left(\frac{931.494\ MeV/c^2}{1\ u} \right)$$

$$= 28.297\ MeV$$

The binding energy per nucleon equals the binding energy of the nucleus divided by the number of nucleons in the nucleus.

The ^{4}He nucleus has four nucleons (two neutrons and two protons), so the binding energy per nucleon is

$$\frac{E_B}{A} = \frac{28.297 \text{ MeV}}{4} = 7.0742 \text{ MeV}$$

Reflect

Previously, we calculated $E_B/A = 7.1$ MeV to two significant figures using the mass of the ^{4}He nucleus; our new calculation is consistent with this.

Why would we do this kind of calculation using atomic masses rather than nuclear masses? The reason is that in general the masses of neutral atoms have been well measured, but precise measurements of the masses of atomic nuclei in isolation are difficult to obtain.

TAKE-HOME MESSAGE FOR Section 27-3

✔ The binding energy of a nucleus is the energy that would be required to separate it into its individual nucleons.

✔ The greater the binding energy per nucleon in a nucleus, the more tightly the nucleus is bound and therefore the more stable it is.

✔ The most stable nuclides are ^{56}Fe, ^{58}Fe, and ^{62}Ni. Smaller and larger nuclei have a lower binding energy per nucleon.

27-4 The largest nuclei can release energy by undergoing fission and splitting apart

As we saw in the previous section, nuclei with higher values of binding energy per nucleon (E_B/A) are more stable than those with lower values. Figure 27-10, a plot of the binding energy E_B per nucleon in nuclei versus mass number A, shows that E_B/A decreases as A increases beyond 60 or so. In other words, large nuclei are less stable than smaller ones for A greater than about 60. As a consequence of this instability, these large nuclei can undergo fragmentation or *fission* into smaller nuclei. Fission of a large nucleus can happen spontaneously, or it can be induced by imparting energy to the nucleus through a collision. In either case the smaller fragments have a higher value of E_B/A and are more stable.

Let's take a look at one of the most important processes of this kind, called **neutron-induced fission**. As an example, the collision of a neutron with a ^{235}U nucleus will cause it to fission. Figure 27-12 shows one possible result. For a brief time the

(1) A uranium nucleus (^{235}U) absorbs a neutron.

(2) The result is a uranium nucleus (^{236}U) in an excited state.

(3) The excited uranium nucleus fissions into two smaller, more tightly bound nuclei...

(4) ...as well as a few neutrons. These can trigger the fission of other ^{235}U nuclei.

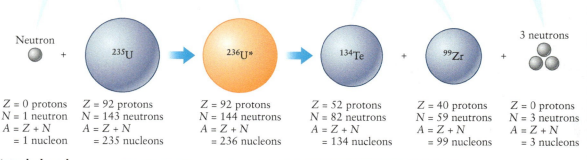

Neutron	^{235}U	^{236}U*	^{134}Te	^{99}Zr	3 neutrons
$Z = 0$ protons	$Z = 92$ protons	$Z = 92$ protons	$Z = 52$ protons	$Z = 40$ protons	$Z = 0$ protons
$N = 1$ neutron	$N = 143$ neutrons	$N = 144$ neutrons	$N = 82$ neutrons	$N = 59$ neutrons	$N = 3$ neutrons
$A = Z + N$	$A = Z + N$	$A = Z + N$	$A = Z + N$	$A = Z + N$	$A = Z + N$
$= 1$ nucleon	$= 235$ nucleons	$= 236$ nucleons	$= 134$ nucleons	$= 99$ nucleons	$= 3$ nucleons

Figure 27-12 Neutron-induced fission When one of the largest nuclei absorbs a slow-moving neutron, it can fission into smaller, more stable fragments.

 Energy is released in this fission reaction: The total kinetic energy of the fission fragments is much greater than the total kinetic energy of the initial neutron and ^{235}U nucleus.

neutron and ^{235}U nucleus remain stuck together as ^{236}U*. (The asterisk indicates that this is an excited and short-lived state of ^{236}U.) This excited nucleus quickly fissions into fragments. In this particular reaction the fragments are an isotope of tellurium (^{134}Te), an isotope of zirconium (^{99}Zr), and three neutrons. Because these fragments are all more stable than the original nucleus, energy is released. The process described by Figure 27-12 occurs even when the colliding neutron is moving very slowly and so has essentially zero kinetic energy. So the released energy is almost entirely due to the change in binding energy between the initial ^{235}U nucleus and the fission products.

We can estimate the energy released in the process shown in Figure 27-12 by comparing the binding energy of the ^{235}U nucleus to the binding energies of the ^{134}Te and ^{99}Zr fragments. (There is no binding energy associated with the initial or final neutrons; they are not bound to any other particle.) If you make measurements on the graph in Figure 27-10, you'll see that the binding energy per nucleon E_B/A is about 7.6 MeV for $A = 235$, about 8.4 MeV for $A = 134$, and about 8.7 MeV for $A = 99$. The binding energy of each nucleus (E_B) equals the binding energy per nucleon (E_B/A) multiplied by the number of nucleons (A). The energy released in the fission reaction equals the difference between the total binding energy of the fragments and the binding energy of the initial ^{235}U nucleus:

(energy released)

$$= \text{(binding energy of } ^{134}\text{Te)} + \text{(binding energy of } ^{99}\text{Zr)}$$
$$- \text{(binding energy of } ^{235}\text{U)}$$
$$= (134)(8.4 \text{ MeV}) + (99)(8.7 \text{ MeV}) - (235)(7.6 \text{ MeV})$$
$$= 200 \text{ MeV}$$

We've given our result to just one significant figure because the values of E_B/A that we measured from Figure 27-10 are just estimates. The actual amount of energy released during this process is about 185 MeV, which is quite close to our estimate. This illustrates the tremendous amount of energy released in fission. By contrast, combustion (a chemical process that involves the electrons in molecules, not the nuclei of atoms) yields only a few eV for every molecule of fuel consumed. The energy release in fission is greater by a factor of several million!

When a heavy nucleus like ^{235}U undergoes fission, a wide variety of fragments can result. Figure 27-12 shows one possible result. Two others are

$$\text{n} + {}^{235}\text{U} \rightarrow {}^{236}\text{U}^* \rightarrow {}^{143}\text{Ba} + {}^{90}\text{Kr} + 3\text{n}$$
$$\text{n} + {}^{235}\text{U} \rightarrow {}^{236}\text{U}^* \rightarrow {}^{140}\text{Xe} + {}^{92}\text{Sr} + 4\text{n}$$

In each case the total number of protons and the total number of neutrons both remain the same. The ^{235}U nucleus has 92 protons and $235 - 92 = 143$ neutrons. In the first of these two processes, ^{143}Ba has 56 protons and 87 neutrons, and ^{90}Kr has 36 protons and 54 neutrons. The total number of protons after the fission has occurred is then $56 + 36 = 92$. The total number of neutrons before the fission is $143 + 1 = 144$ (including the neutron that starts the process). After the fission the number of neutrons is $87 + 54 + 3 = 144$ (including the three neutrons released in the process). You can easily verify that the number of protons and the number of neutrons likewise remain the same in the second process above.

Uranium-235 can also undergo *spontaneous* fission, in which the nucleus fragments without undergoing a collision with a neutron. The following example shows how to calculate the energy released in this process.

EXAMPLE 27-4 Spontaneous Uranium Fission

Determine the energy released when a ^{235}U nucleus spontaneously undergoes fission to ^{140}Xe, ^{92}Sr, and three neutrons. The binding energy per nucleon in the nuclei of ^{235}U, ^{140}Xe, and ^{92}Sr are 7.59 MeV, 8.29 MeV, and 8.65 MeV, respectively.

Set Up

The energy released during the fission process is the difference between the binding energy of ^{140}Xe and ^{92}Sr nuclei and the binding energy of the ^{235}U nucleus. (There is no binding energy associated with the three neutrons.) The binding energy for each nucleus equals the binding energy per nucleon for that nucleus multiplied by the number of nucleons A.

(energy released)

= (total binding energy of fragments)
  – (binding energy of original ^{235}U nucleus)

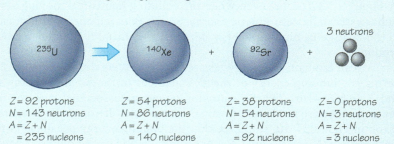

	^{235}U	^{140}Xe	^{92}Sr	3 neutrons
	$Z = 92$ protons	$Z = 54$ protons	$Z = 38$ protons	$Z = 0$ protons
	$N = 143$ neutrons	$N = 86$ neutrons	$N = 54$ neutrons	$N = 3$ neutrons
	$A = Z + N$	$A = Z + N$	$A = Z + N$	$A = Z + N$
	= 235 nucleons	= 140 nucleons	= 92 nucleons	= 3 nucleons

Solve

Calculate the binding energy for each of the nuclei, and then use these values to calculate the energy released.

The binding energies of the individual nuclei are

For ^{235}U: $(235)(7.59 \text{ MeV}) = 1784 \text{ MeV}$

For ^{140}Xe: $(140)(8.29 \text{ MeV}) = 1161 \text{ MeV}$

For ^{92}Sr: $(92)(8.65 \text{ MeV}) = 796 \text{ MeV}$

The energy released in the spontaneous fission is then

(energy released) = (binding energy of ^{140}Xe) + (binding energy of ^{92}Sr)
  – (binding energy of ^{235}U)

= 1161 MeV + 796 MeV – 1784 MeV

= 173 MeV

Reflect

This result is consistent with our earlier claim that a typical amount of energy released in the fission of ^{235}U is around 200 MeV.

Spontaneous fission of ^{235}U is a *very* unlikely process: A given ^{235}U nucleus has a 50% chance of decaying in a period of 7.04×10^8 years, and the probability that it will decay by spontaneous fission is only 7.0×10^{-11}. (Here 7.04×10^8 y is the *half-life* of ^{235}U. We'll discuss this concept more carefully in Section 27-6. In that section we'll see that ^{235}U almost always decays by a different process called *alpha emission*.) By contrast, once a ^{235}U nucleus merges with a neutron to form an excited ^{236}U* nucleus, as in Figure 27-7, it typically undergoes fission within a fraction of a second.

All of the examples of fission that we've described result in the release of neutrons. Imagine what can happen if a large number of ^{235}U atoms are close to each other. Should one nucleus be struck by a neutron and fission, as shown in Figure 27-12, there would then be three neutrons moving among the atoms. Should each of these neutrons strike a ^{235}U nucleus and start a fission process, there would be nine neutrons, so possibly nine more fissions. With a sufficient number of ^{235}U atoms present, this *chain reaction* quickly grows, with an accompanying rapid increase in energy released.

Isotopes that are capable of sustaining a fission chain reaction are used as nuclear fuels. Such isotopes are termed *fissile*; the most common fissile nuclear fuels are ^{233}U, ^{235}U, ^{239}Pu, and ^{241}Pu. In the fission reactions they undergo, typically two or three neutrons are released in addition to larger fragments. These released neutrons don't have to be moving rapidly to trigger additional fission reactions: a chain reaction in a fissile material can be induced by a neutron carrying essentially zero kinetic energy. Indeed, in fuels such as ^{235}U, slower, less energetic neutrons are more efficiently absorbed by the fissile nuclei.

Most nuclear reactors use the energy released in a fission chain reaction to heat water and produce steam to drive an electric generator (see Section 20-4). There are several challenges to producing energy in this way. First, a minimum amount, or *critical mass*, of fissile material must be present to sustain a chain reaction. As it happens, only about 0.7% of the naturally occurring uranium in the world is ^{235}U, and the other three commonly used fissile isotopes do not occur naturally. Most naturally occurring uranium is ^{238}U, which is not fissile. To use uranium as a nuclear fuel, then, it is necessary to separate the ^{235}U atoms from the ^{238}U, a costly and difficult process known as *enrichment*. In addition, the ^{238}U atoms that inevitably remain tend to absorb free neutrons and thereby inhibit a chain reaction. Once the critical mass of ^{235}U has been assembled, controlling the chain reaction is another challenge. If too many of the neutrons produced in the fissions result in a second fission, the energy released increases so rapidly that the fuel and whatever vessel is used to contain it can be damaged or even melt. To control a fission chain reaction in a nuclear reactor, control rods made of a substance that is a good absorber of neutrons are inserted between pieces of fuel.

Operating a fission nuclear reactor safely is a significant challenge. First, the fission fragments are radioactive. Many of these fragments, or the fragments produced when they decay, are long lived and tend to produce dangerous radiation for years, centuries, or even longer. In addition, reactors commonly use water as a *moderator*, a material that tends to slow the free neutrons (to make them more easily absorbed by a ^{235}U nucleus). Should the containment vessel rupture, this hot water can be released into the atmosphere in the form of steam carrying radioactive particles. Perhaps the most significant nuclear accident occurred at the Chernobyl nuclear power plant in Ukraine in 1986, in which an uncontrolled chain reaction caused a catastrophic power increase, leading to a series of explosions and the release of large quantities of radioactive steam, fuel, and smoke into the environment.

GOT THE CONCEPT? 27-3 Fission

Consider the spontaneous fission process ^{20}Ne → ^{10}B + ^{10}B. This process does not occur in nature. Why not? (a) The number of protons does not remain constant. (b) The number of neutrons does not remain constant. (c) Both (a) and (b). (d) This process would absorb energy, not release it. (e) This process would neither release nor absorb energy.

TAKE-HOME MESSAGE FOR Section 27-4

✔ The binding energy per nucleon of nuclides with more than about 60 neutrons and protons decreases with increasing values of *A*. For this reason the fission process, in which a nucleus breaks up into smaller fragments, leads to more stable configurations of the nucleons in bigger nuclei.

✔ Fission can be triggered by allowing a slow-moving neutron to merge with a fissile nucleus.

27-5 The smallest nuclei can release energy if they are forced to fuse together

As we saw in the previous section, Figure 27-10 explains why the largest nuclei can undergo fission: The binding energy per nucleon E_B/A is maximum for mass numbers *A* around 60. By breaking into smaller fragments, nuclei with values of *A* much greater than 60 can therefore increase their binding energy per nucleon E_B/A and so become more stable. The opposite is true for the lightest nuclei, with *A* much less than 60. These nuclei can increase the value of E_B/A and become more stable by becoming *larger*. Processes in which two small nuclei combine to form a larger one are called *fusion* processes.

Figure 27-13 shows how two ^{3}He nuclei (each with two protons and one neutron) fuse together to form a ^{4}He nucleus (with two protons and two neutrons). Two protons are left over, and without more neutrons there is no way for the protons to be bound together in a single nucleus. As a result, these protons fly off separately. Energy is released during this process because the final configuration of the protons and neutrons is more stable than the initial configuration. A photon carries away the energy released.

How much energy is released in the process shown in Figure 27-13? As in fission, the energy released in fusion is the difference between the total binding energy of the final nuclei and the total binding energy of the original nuclei. From Figure 27-10, the binding energy per nucleon for ^{3}He is approximately 2.5 MeV. Each ^{3}He has three nucleons, so the total binding energy of ^{3}He is 3(2.5 MeV) = 7.5 MeV, and the two ^{3}He nuclei together have a combined binding energy equal to 7.5 MeV + 7.5 MeV = 15.0 MeV. In Section 27-3 we found that a ^{4}He nucleus has a binding energy of 28.3 MeV. The two

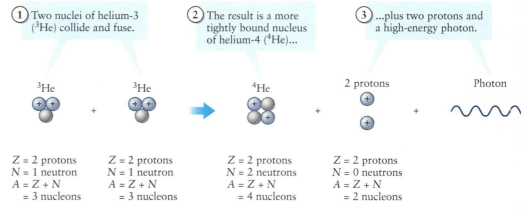

① Two nuclei of helium-3 (³He) collide and fuse.

② The result is a more tightly bound nucleus of helium-4 (⁴He)...

③ ...plus two protons and a high-energy photon.

³He	³He	⁴He	2 protons	Photon

$Z = 2$ protons
$N = 1$ neutron
$A = Z + N$
$= 3$ nucleons

$Z = 2$ protons
$N = 1$ neutron
$A = Z + N$
$= 3$ nucleons

$Z = 2$ protons
$N = 2$ neutrons
$A = Z + N$
$= 4$ nucleons

$Z = 2$ protons
$N = 0$ neutrons
$A = Z + N$
$= 2$ nucleons

Figure 27-13 Nuclear fusion The fusion of two ³He nuclei to make a ⁴He nucleus is one of the energy-releasing reactions that take place in the core of the Sun.

Energy is released in this fusion reaction: The total kinetic energy of the ⁴He nucleus and protons, plus the energy of the photon, is much greater than the total kinetic energy of the initial ³He nuclei.

protons are single particles, so they make no contribution to the binding energy. The energy released in the fusion process is therefore

$$(\text{energy released}) = (\text{binding energy of }^4\text{He nucleus})$$
$$- (\text{binding energy of two }^3\text{He nuclei})$$
$$= (28.3 \text{ MeV}) - (15.0 \text{ MeV}) = 13.3 \text{ MeV}$$

The actual value, found using more accurate values of the binding energies, is closer to 12.86 MeV.

The process shown in Figure 27-13 is the final step in the *proton–proton cycle*, the fusion process that takes place near the center of the Sun and is the source of the Sun's energy (Figure 27-14). The cycle begins with the fusing of two protons (the nuclei of ¹H) to form ²H. This nucleus fuses with another proton, forming ³He, and finally, two ³He nuclei fuse to form ⁴He. We can summarize these three steps as

Step 1: $^1\text{H} + {}^1\text{H} \rightarrow {}^2\text{H} + e^+ + \nu_e$

Step 2: $^2\text{H} + {}^1\text{H} \rightarrow {}^3\text{He} + \gamma$

Step 3: $^3\text{He} + {}^3\text{He} \rightarrow {}^4\text{He} + {}^1\text{H} + {}^1\text{H} + \gamma$

In step 1 the e⁺ particle is a **positron**, a particle with the same mass as an electron but with positive charge $+e$. The particle named ν_e (the Greek letter "nu" with a subscript "e") is a **neutrino**, a nearly massless, neutral particle. Also note that this step involves a proton being converted into a neutron. We'll discuss this conversion, called beta-plus decay, in Section 27-6. Step 1 and the subsequent interaction of the positron with an electron in the Sun (in which the two particles annihilate each other and convert into photons) release 1.44 MeV of energy. The energy released in step 2 is 5.49 MeV. Both of these steps must occur twice before step 3 can occur (because step 3 requires two ³He nuclei). So six protons are used in steps 1 and 2, and in step 3 two of those protons are returned along with a ⁴He nucleus. The net result is therefore that four ¹H nuclei disappear and are replaced by one ⁴He nucleus. The net energy release is 2(1.44 MeV) from step 1 happening twice, plus 2(5.49 MeV) from step 2 happening twice, plus 12.86 MeV from the fusion of two ³He nuclei in step 3. The sum of these is a net energy release of 26.7 MeV as four ¹H nuclei are transformed into a ⁴He nucleus.

Fission processes release more energy than the proton–proton fusion cycle, around 200 MeV compared to 26.7 MeV. Therefore, it might seem that fission is a more effective way to convert fuel to energy. Consider, however, that while 235 nucleons in ²³⁵U are spent to release 200 MeV, in the fusion process described above, the 26.7 MeV released come at the expense of only four nucleons. Comparing energy per nucleon (think miles per gallon or kilometers per liter), fission provides less than 1 MeV per nucleon, while proton–proton fusion gives 26.7 MeV divided by 4, or nearly 7 MeV per nucleon. Fusion processes are *much* more efficient at releasing nuclear binding energy.

Hydrogen makes up about 75% of the Sun's mass of approximately 2×10^{30} kg. If the proton–proton cycle leads to more stability, why doesn't all of that hydrogen

Figure 27-14 The Sun: A nuclear fusion reactor All of the energy released by the Sun is a result of fusion reactions in its core that convert hydrogen nuclei into helium nuclei.

quickly fuse to form ^{4}He, leaving the Sun a gigantic (and cool) ball of helium gas? The answer lies in the same forces at play within nuclei: the Coulomb force that repels protons from each other and the short-ranged strong nuclear force that draws them together. For two protons to fuse to form ^{2}H, they must come within a few femtometers of each other, at which point the strong attraction is able to overcome the Coulomb repulsion. This requires the protons to have considerable kinetic energy, which can come from being at high temperature. A temperature of more than 4×10^6 K is required for the proton–proton cycle to start. The temperature at the core of the Sun is around 15×10^6 K, so the proton–proton cycle can and does occur there. Even at that temperature, however, the probability that two nearby protons will fuse is small. This means that only a small fraction—about 4×10^{-19}—of the hydrogen in the Sun is undergoing fusion at any one time. The Sun won't burn out for a long time.

As a star ages, its core temperature increases and additional fusion reactions become possible. For example, three ^{4}He nuclei can fuse to form a ^{12}C nucleus. This requires a temperature of about 10^8 K because the ^{4}He nuclei are more massive than protons and repel each other more strongly due to their greater charge. At even higher temperatures a ^{4}He nucleus can fuse with a ^{12}C nucleus to form a ^{16}O nucleus, a ^{4}He nucleus can fuse with a ^{16}O nucleus to form a ^{20}Ne nucleus, and so on. So as stars age they manufacture heavier and heavier chemical elements. In the most massive stars, which have the highest core temperatures, so much kinetic energy is available at the very end of the star's evolution that fusion processes can produce even the heaviest nuclei up to uranium. Making these massive nuclei by fusion absorbs rather than releases energy, which is why it can happen only in very special circumstances.

These fusion reactions in stars make life on Earth possible. Here's why: When the universe first originated some 13.8 billion years ago, almost all ordinary matter was in the form of hydrogen or helium. (We'll discuss the origin of the universe in Chapter 28.) All heavier elements had to be manufactured by fusion within stars. After an aging star produces elements heavier than helium and goes through its final stages of evolution, it ejects much of its material into interstellar space (Figure 27-15). This material, which is enriched in heavy elements, can then be incorporated into a later generation of stars. Our Sun is such a "second-generation" star, with an elevated abundance of elements heavier than helium. As part of the process by which the Sun formed, some of these elements went into forming Earth and the Sun's other planets. This means that all of the nuclei of carbon in the organic compounds that make up your body, all of the nuclei in the oxygen that you breathe, and all of the nuclei of iron in the hemoglobin in your blood were manufactured in stars that died billions of years ago. This is one of the great lessons of nuclear physics: You are made of star-stuff.

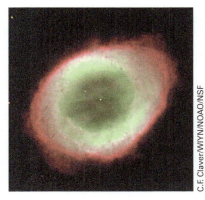

Figure 27-15 Seeding space with fusion products The Ring Nebula is a cloud of gas emitted by an aging star. The cloud includes nitrogen (outermost edge, shown in red) and oxygen (inner region, shown in green) produced by fusion reactions within the star.

BioMedical

GOT THE CONCEPT? 27-4 Fusion

? Ordinary hydrogen gas is in the form of diatomic hydrogen (H_2), and more than 99.9% of the atoms in hydrogen gas have a nucleus that is a single proton (^{1}H). Why don't the two hydrogen nuclei in an H_2 molecule spontaneously undergo fusion as in step 1 of the proton–proton cycle: ^{1}H + ^{1}H $\rightarrow$ ^{2}H + e$^+$ + ν_e? (a) The nuclei are too far apart. (b) The nuclei are moving too slowly relative to each other. (c) The nuclei in an atom are different from those outside an atom. (d) both (a) and (b); (e) all of (a), (b), and (c).

TAKE-HOME MESSAGE FOR Section 27-5

✔ The binding energy per nucleon of nuclides that have fewer than about 60 neutrons and protons is larger for increasing values of mass number A. For this reason, fusion of two small nuclei to form a larger one leads to more stable configuration of the nucleons.

✔ Fusion can only take place if the fusing nuclei come very close to each other. Therefore very high temperatures are required so the nuclei can overcome their mutual electric repulsion.

27-6 Unstable nuclei may emit alpha, beta, or gamma radiation

For many people, phrases such as "radioactivity" and "nuclear radiation" are synonymous with danger. Not all nuclear radiation is dangerous, however, and indeed several types of nuclear radiation have important practical applications. In this section we'll explore a number of aspects of nuclear radiation.

All naturally occurring nuclear processes take place because the final state is more energetically favorable than the initial state. This is true of fission and fusion, and it is true of radiation processes, too. A relatively few nuclides—266 out of more than 3000—are stable. All the rest are radioactive; that is, they decay into another nuclide by radiating away one or more particles. It's also possible for a nucleus in an excited state to radiate energy in the form of a photon as the nucleus transitions to a less excited state.

The three most common kinds of radiation are *alpha*, *beta*, and *gamma* radiation. The terms were coined by Ernest Rutherford, who in his research between 1899 and 1903 classified radiation according to the depth that a radiation particle was able to penetrate other objects. Alpha particles penetrated the least, beta particles more, and gamma particles the most. (Rutherford received the 1908 Nobel Prize in Chemistry for this research, which was the first to show that one element can change into another through radioactive processes.) We now know that alpha particles are ^{4}He nuclei; beta particles are electrons; and gamma particles are photons with energies that can be millions of times greater than the energies of visible-light photons.

Before we consider the properties of these specific types of radiation, let's look at a concept that is common to all of them—the idea of *radioactive half-life*.

Radioactive Decay and Half-Life

Nuclear radiation of all kinds involves physics on the very small scale of the nucleus, so is governed by quantum mechanics. As we learned in Section 26-6, quantum mechanics cannot tell us the position or velocity of a particular object at any time. It can, however, tell us the *probability* that an object will be at a particular place or have a particular velocity at any given instant. In the same way, quantum mechanics cannot tell us when a given radioactive nucleus will decay, but it can predict the probability that this nucleus will decay within a given time interval.

As an example, consider the emission of a beta particle in the decay of ^{137}Cs (cesium), in which a neutron is converted into a proton:

$$^{137}\text{Cs} \rightarrow {}^{137}\text{Ba} + e^- + \bar{\nu}_e$$

The beta particle is the electron (e^-), and $\bar{\nu}_e$ is an **antineutrino**, related to the light, neutral neutrino particle we discussed in the context of fusion in the previous section. Every ^{137}Cs nucleus can, and eventually will, decay radioactively to a barium nucleus (^{137}Ba). Experiment shows that if a sample of ^{137}Cs contains, say, 10,000 atoms, after 30.17 years only about 5000 will be left. However, if we select any individual ^{137}Cs atom of those 10,000, we have no idea when its nucleus will decay. Perhaps it will decay in the next second or perhaps not for thousands of years.

While we can't make a definitive claim about when any particular atomic nucleus will decay, we can quantify the *rate* at which a group of radioactive atoms decays, that is, the number of decays per second. The probability λ that any one nucleus of a given type will decay in the next second is the same for all such nuclei. The quantity λ is called the **decay constant**. It has units of s^{-1}, since it refers to a probability per second. The value of λ is different for different radioactive nuclides: It is greater for nuclides that decay rapidly and smaller for nuclides that decay slowly.

If we have a sample of N such nuclei, the total number of decays that take place in the next second—that is, the *decay rate*—will be equal to the product of the decay constant λ (the probability that any one nucleus decays in the next second) and N (the number of radioactive nuclei present) (Figure 27-16). So the decay rate of a radioactive sample is greater if the sample is larger (the number of nuclei N in the sample is greater) or the nuclei have a higher decay constant λ. The SI unit of decay rate is the **becquerel** (Bq), after the nineteenth-century French physicist Antoine Henri Becquerel, who along with Marie Skłodowska-Curie and Pierre Curie won the 1903 Nobel Prize in Physics for their discovery of radioactivity. The becquerel is equivalent to one radioactive decay per second. Physicists also commonly use units of curies (Ci) for decay rate; 1 Ci = 3.7×10^{10} Bq = 3.7×10^{10} decays/s.

After an elapsed time of Δt seconds, the total number of decays will be equal to $\lambda N \Delta t$. This means that in a time Δt the number of radioactive nuclei decreases by $\lambda N \Delta t$. So the *change* in the number of radioactive nuclei present is

Jose Manuel Revuelta Luna/Alamy

BioMedical **Figure 27-16 Radioactive decays in bananas** Bananas are a good source of potassium. About 0.0117% of the potassium atoms in bananas have the radioactive nucleus ^{40}K, for which the decay constant (the probability per second that a nucleus will decay) is very small: $\lambda = 1.76 \times 10^{-17}$ decays/s = 1.76×10^{-17} Bq. The number N of ^{40}K atoms in a typical 0.15-kg banana is about 9×10^{17} (with a combined mass of about 6×10^{-8} kg), so the number of ^{40}K nuclei that decay per second in a banana is $\lambda N = (1.76 \times 10^{-17}$ decays/s$)(9 \times 10^{17})$, or about 15 decays/s. This is extraordinarily small and poses no health hazard.

(27-3) $$\Delta N = -\lambda N \Delta t$$

The minus sign in Equation 27-3 indicates that the number of nuclei decreases as a result of the decays. If we divide both sides of Equation 27-3 by the elapsed time Δt, we get an expression for the rate of change $\Delta N / \Delta t$ of the number of radioactive nuclei:

In a sample of radioactive material, the **rate of change of the number of radioactive nuclei...**

... is negative because the number of nuclei decreases due to decay...

$$\frac{\Delta N}{\Delta t} = -\lambda N$$

Radioactive decay equation
(27-4)

...is proportional to the **decay constant** (the probability that a given nucleus decays in a one-second interval)...

...and is proportional to the **number of radioactive nuclei that remain.**

Equation 27-4 tells us that as time goes by the decay rate will decrease because the number of radioactive nuclei will decrease. Using the tools of calculus we can solve Equation 27-4 to find the number of nuclei present as a function of time $N(t)$. The result is

$$N(t) = N_0 e^{-\lambda t}$$

(27-5)

In Equation 27-5 N_0 is the number of nuclei present at a specific time that we choose to call $t = 0$. This equation tells us that the number of nuclei present decreases exponentially, as Figure 27-17 shows. The number of decays per second at time t is equal to the decay constant λ (the decay probability per second per nucleus), multiplied by $N(t)$, the number of nuclei remaining at time t. We can write this as

$$R(t) = \lambda N(t) = \lambda N_0 e^{-\lambda t}$$

In this equation λN_0 is equal to the decay rate at $t = 0$, which we call R_0. So the decay rate as a function of time is

$$R(t) = R_0 e^{-\lambda t}$$

(27-6)

Equation 27-6 tells us that the decay rate, too, will decrease exponentially as time goes by. With the passage of time, a radioactive sample will undergo fewer and fewer decays per second.

Figure 27-17 shows that the number of ^{137}Cs nuclei remaining decreases by one-half every 30.17 years due to beta decay. Because the decay rate is proportional to the number of ^{137}Cs nuclei remaining, the decay rate also decreases by one-half every 30.17 years. This time is called the **half-life** of the radioactive decay, to which we give the symbol $\tau_{1/2}$

WATCH OUT! Be careful not to misuse the concept of half-life.

It's important to note that the graph of nuclear decay in Figure 27-17 is for a *large number* of radioactive nuclei. Statistically, half of the nuclei present at any time will have decayed after one half-life. For a *single* nucleus, half-life has a different interpretation: Over one half-life, a given radioactive nucleus has a 50% chance of not decaying and a 50% chance of decaying (and if it does decay, it can happen at any time during that half-life).

Figure 27-17 Nuclear decay and half-life This exponential curve shows the evolution of a sample that originally contains 10,000 cesium-137 (^{137}Cs) nuclei, which decay to barium-137 (^{137}Ba).

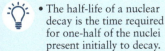

• The half-life of a nuclear decay is the time required for one-half of the nuclei present initially to decay.
• Radioactive decay is a statistical process: It's impossible to predict when any one individual unstable nucleus will decay.

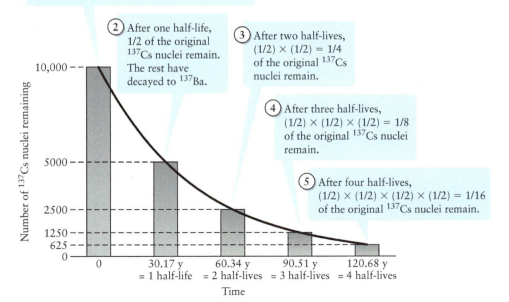

1. Initially we have 10,000 ^{137}Cs nuclei. These nuclei are unstable and decay to ^{137}Ba.

2. After one half-life, 1/2 of the original ^{137}Cs nuclei remain. The rest have decayed to ^{137}Ba.

3. After two half-lives, $(1/2) \times (1/2) = 1/4$ of the original ^{137}Cs nuclei remain.

4. After three half-lives, $(1/2) \times (1/2) \times (1/2) = 1/8$ of the original ^{137}Cs nuclei remain.

5. After four half-lives, $(1/2) \times (1/2) \times (1/2) \times (1/2) = 1/16$ of the original ^{137}Cs nuclei remain.

(the Greek letter "tau"). If there are N_0 radioactive nuclei present at $t = 0$, there will be $N_0/2$ present at $t = \tau_{1/2}$. If we substitute this into Equation 27-5, we get

$$\frac{N_0}{2} = N_0 e^{-\lambda \tau_{1/2}}$$

Divide both sides of this equation by N_0, then take the natural logarithm of both sides:

$$(27\text{-}7) \qquad \ln\left(\frac{1}{2}\right) = \ln(e^{-\lambda \tau_{1/2}})$$

Why do we do this? The reason is that the natural logarithm "undoes" the exponential function: For any x, $\ln(e^x) = x$. Furthermore, $\ln(1/x) = -\ln x$ for any x. If we apply these to Equation 27-7, we get

$$-\ln 2 = -\lambda \tau_{1/2}$$

or

$$(27\text{-}8) \qquad \tau_{1/2} = \frac{\ln 2}{\lambda}$$

Equation 27-8 says that the half-life is inversely proportional to the decay constant λ, the probability that a given nucleus of a certain type will decay in a 1-s interval. The greater the decay constant, the shorter the half-life and the more rapidly a sample of that nucleus will decay.

The half-life of radioactive sources varies widely, from far less than 1 s to billions of years or more. Common radioactive sources include ^{32}P, a beta emitter used in DNA research, which has a half-life of 14.3 days; ^{241}Am, an alpha emitter often found in household smoke detectors, which has a half-life of 432.2 y; and ^{238}U, an alpha emitter with a half-life of 4.47×10^9 y. Radioactive isotopes with half-lives on the order of a second or less aren't terribly useful because they don't stay around long enough.

EXAMPLE 27-5 Technetium

One form of an isotope of technetium, ^{99m}Tc, undergoes gamma decay with a half-life of 6.01 h. (The "m" stands for "metastable." Technetium-99 is widely used as a radioactive tracer for medical purposes because its gamma radiation is easily detected and because technetium doesn't stay in the body for long. Hence the total radiation delivered to the patient is low.) (a) What is the decay constant λ, the probability that a nucleus of ^{99m}Tc will decay per second, in this sample? Does λ change as time goes on? (b) What fraction of the initial number of ^{99m}Tc nuclei will be left after 1.00 day? (c) What fraction will be left after 4.00 days have elapsed?

Set Up

We'll use Equation 27-8 to relate the decay constant λ to the half-life. Equation 27-5 will tell us the number of ^{99m}Tc nuclei remaining after a time t in terms of the initial number of nuclei N_0.

Half-life of a radioactive substance:

$$\tau_{1/2} = \frac{\ln 2}{\lambda} \qquad (27\text{-}8)$$

Number of radioactive nuclei present at time t:

$$N(t) = N_0 e^{-\lambda t} \qquad (27\text{-}5)$$

Solve

(a) Determine the decay constant for ^{99m}Tc.

We can rewrite Equation 27-8 as

$$\lambda = \frac{\ln 2}{\tau_{1/2}}$$

Substitute $\tau_{1/2} = 6.01$ h:

$$\lambda = \frac{\ln 2}{(6.01\ \text{h})}\left(\frac{1\ \text{h}}{60\ \text{min}}\right)\left(\frac{1\ \text{min}}{60\ \text{s}}\right) = 3.20 \times 10^{-5}\ \text{s}^{-1}$$

The probability that a given ^{99m}Tc nucleus will decay in a 1-s interval is 3.20×10^{-5}, corresponding to odds of 1 in $(1/3.20 \times 10^{-5}) = 31,200$. This value depends only on the half-life, so it does not vary with time.

(b) To find the fraction of nuclei remaining after $t = 1.00$ d, substitute this value of t into Equation 27-5.

The fraction of ^{99m}Tc nuclei remaining after a time t equals the number remaining $N(t)$ divided by the number N_0 present initially:

$$\frac{N(t)}{N_0} = \frac{N_0 e^{-\lambda t}}{N_0} = e^{-\lambda t}$$

From part (a) we know the value of λ in s^{-1}, so we need to express t in seconds:

$$t = (1.00\text{ d})\left(\frac{24\text{ h}}{1\text{ d}}\right)\left(\frac{60\text{ min}}{1\text{ h}}\right)\left(\frac{60\text{ s}}{1\text{ min}}\right) = 8.64 \times 10^4\text{ s}$$

The fraction of nuclei remaining is then

$$\frac{N(1.00\text{ d})}{N_0} = e^{-(3.20\times 10^{-5}\text{ s}^{-1})(8.64\times 10^4\text{ s})}$$

$$= e^{-2.77} = 0.0628$$

(c) Repeat part (b) with $t = 4.00$ d.

The elapsed time is now

$$t = (4.00\text{ d})\left(\frac{8.64\times 10^4\text{ s}}{1\text{ d}}\right) = 3.46 \times 10^5\text{ s}$$

and the fraction of nuclei remaining is

$$\frac{N(4.00\text{ d})}{N_0} = e^{-(3.20\times 10^{-5}\text{ s}^{-1})(3.46\times 10^5\text{ s})}$$

$$= e^{-11.1} = 1.55 \times 10^{-5}$$

Reflect

We can check our results by noting that $t = 1.00$ d is almost exactly 4 times the 6.01-h half-life of ^{99m}Tc, and $t = 4.00$ d is almost exactly 16 times the half-life. This check agrees with our calculations, as it should.

After each half-life, the number of ^{99m}Tc nuclei remaining decreases by one-half. So after four half-lives, the fraction of ^{99m}Tc nuclei remaining will be

$$\frac{1}{2} \times \frac{1}{2} \times \frac{1}{2} \times \frac{1}{2} = \frac{1}{2^4} = \frac{1}{16} = 0.0625$$

The fraction actually remaining at $t = 1.00$ d (slightly less than four 6.01-h half-lives) is 0.0628, very close to our estimate.

After 16 half-lives, the fraction of ^{99m}Tc remaining will be

$$\frac{1}{2^{16}} = \frac{1}{65,536} = 1.53 \times 10^{-5}$$

The fraction actually remaining at $t = 4.00$ d (slightly less than 16 times the 6.01-h half-life) is 1.55×10^{-5}, which again is very close to our estimate.

Alpha Radiation

We saw in Section 27-4 that a large nucleus can increase the binding energy per nucleon E_B/A by breaking into smaller fragments. The most likely decay products are those that are more stable, that is, those with larger binding energy per nucleon. As Figure 27-10 shows, ^{4}He has a greater binding energy per nucleon than any other nucleus with a small value of A. So the nucleus of the ^{4}He atom is far more stable than other small nuclei and therefore a far more probable decay product of large nuclei. For this reason, the radioactive emission of a ^{4}He nucleus is the most likely decay process for a large nucleus. Another name for a ^{4}He nucleus is an **alpha particle** (symbol α), and emission of an alpha particle is called **alpha decay**.

The alpha radiation process reduces the number of protons Z of the initial, or parent, nucleus by two and reduces the number of neutrons of the parent by two. The result is a daughter nucleus that has an atomic number $Z - 2$ and a mass number $A - 4$, accompanied by an alpha particle with two protons and two neutrons (Figure 27-18).

The daughter nucleus and the alpha particle both carry away the energy released in alpha decay. However, the kinetic energy of the alpha particle is far greater than the kinetic energy of the daughter. We can confirm this by looking at the ratio of the kinetic energy K_α

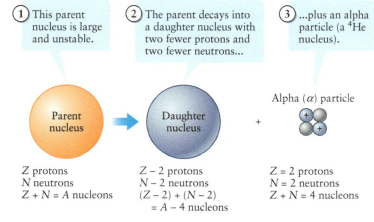

① This parent nucleus is large and unstable.

② The parent decays into a daughter nucleus with two fewer protons and two fewer neutrons...

③ ...plus an alpha particle (a ⁴He nucleus).

Figure 27-18 Alpha decay Large nuclei can increase their stability by emitting an alpha particle and becoming a smaller, more stable daughter nucleus.

Alpha (α) particle

Parent nucleus

Daughter nucleus

+

Z protons
N neutrons
$Z + N = A$ nucleons

$Z - 2$ protons
$N - 2$ neutrons
$(Z - 2) + (N - 2)$
$= A - 4$ nucleons

$Z = 2$ protons
$N = 2$ neutrons
$Z + N = 4$ nucleons

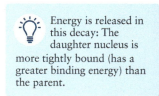

Energy is released in this decay: The daughter nucleus is more tightly bound (has a greater binding energy) than the parent.

of the alpha particle to the kinetic energy K_D of the daughter. The alpha particle has mass m_α and is emitted with speed v_α, and the daughter nucleus has mass m_D and is emitted with speed v_D. The kinetic energies of the alpha particle and the daughter are then

$$K_\alpha = \frac{1}{2} m_\alpha v_\alpha^2$$

$$K_D = \frac{1}{2} m_D v_D^2$$

The ratio of these kinetic energies is

(27-9)
$$\frac{K_\alpha}{K_D} = \frac{(1/2)m_\alpha v_\alpha^2}{(1/2)m_D v_D^2} = \frac{m_\alpha}{m_D} \left(\frac{v_\alpha}{v_D} \right)^2$$

We can relate the speeds v_α and v_D by noting that momentum must be conserved in the alpha decay (because no external forces act on the parent nucleus as it decays). If the parent nucleus is at rest, the total momentum is zero before the decay and so must be zero after the decay. The alpha particle and daughter nucleus must therefore fly off in opposite directions, and each must have the same magnitude of momentum:

$$m_\alpha v_\alpha = m_D v_D \quad \text{so} \quad \frac{v_\alpha}{v_D} = \frac{m_D}{m_\alpha}$$

If we substitute this into Equation 27-9, we find that the ratio of the alpha particle's kinetic energy to that of the daughter nucleus is

$$\frac{K_\alpha}{K_D} = \frac{m_\alpha}{m_D} \left(\frac{v_\alpha}{v_D} \right)^2 = \frac{m_\alpha}{m_D} \left(\frac{m_D}{m_\alpha} \right)^2 = \frac{m_D}{m_\alpha}$$

The mass of the daughter nucleus is much larger than the mass of the alpha particle, so the fraction m_D/m_α is much greater than one and K_α is large compared to K_D.

All nuclei with more than 82 protons are unstable and have some probability of alpha decay. As an example, the element thorium ($Z = 90$ protons) undergoes alpha decay to radium, which contains two fewer protons ($Z = 88$). The α decay of ^{228}Th, for example, is

$$^{228}\text{Th} \rightarrow {}^{224}\text{Ra} + \alpha$$

Just as ^{228}Th decays to ^{224}Ra, ^{224}Ra undergoes alpha decay to ^{220}Rn. This process continues until the daughter nucleus has $Z = 82$ or less. In this case the final alpha decay is to lead, with $Z = 82$:

$$^{228}\text{Th} \rightarrow {}^{224}\text{Ra} + \alpha \quad (\tau_{1/2 \text{ Th}} = 1.91 \text{ y})$$
$$\hookrightarrow {}^{220}\text{Rn} + \alpha \quad (\tau_{1/2 \text{ Ra}} = 3.63 \text{ d})$$
$$\hookrightarrow {}^{216}\text{Po} + \alpha \quad (\tau_{1/2 \text{ Rn}} = 55.6 \text{ s})$$
$$\hookrightarrow {}^{212}\text{Pb} + \alpha \quad (\tau_{1/2 \text{ Po}} = 0.145 \text{ s})$$

Note that in each alpha decay the daughter nucleus has two fewer protons and two fewer neutrons than its parent.

Another nucleus that decays by alpha radiation is ^{235}U, which has a half-life $\tau_{1/2} = 7.04 \times 10^8$ y. (In Example 27-4 in Section 27-4, we looked at the spontaneous fission of ^{235}U. This is a very rare decay mode; ^{235}U undergoes alpha decay rather than spontaneous fission almost 100% of the time.) Substantial amounts of ^{235}U, ^{238}U ($\tau_{1/2} = 4.47 \times 10^9$ y), and ^{232}Th ($\tau_{1/2} = 1.40 \times 10^{10}$ y) are present in Earth's core, and the energy released by the alpha decay of these isotopes helps to sustain our planet's high internal temperatures. All of Earth's geologic activity, including earthquakes, volcanic eruptions (Figure 27-1c), and the drifting of continents, is powered by the motions of our planet's interior. So alpha decay plays an important role in Earth's dynamic geology.

EXAMPLE 27-6 Alpha Decay of ^{238}U

The uranium isotope ^{238}U undergoes alpha decay to ^{234}Th. The binding energy per nucleon is 7.570 MeV in ^{238}U and 7.597 MeV in ^{234}Th. Find the energy released in the process ^{238}U $\rightarrow$ ^{234}Th $+ \alpha$.

Set Up

We'll use the same principle that we used in Example 27-4 (Section 27-4) to find the energy released in fission: The released energy equals the total binding energy of the nuclei present after the decay, minus the binding energy of the original (parent) nucleus. As in that example, we'll find the binding energy for each nucleus by multiplying the binding energy per nucleon times the number of nucleons A.

(energy released)
= (total binding energy of daughter plus alpha particle)
  − (binding energy of original ^{238}U nucleus)

Solve

From Example 27-3 (Section 27-3) the binding energy of a ^{4}He nucleus (an alpha particle) is

$E_B(^4\text{He}) = 28.297$ MeV

The binding energy of a ^{234}Th nucleus ($A = 234$) is

$E_B(^{234}\text{Th}) = (234)(7.597 \text{ MeV}) = 1777.7$ MeV

and the binding energy of a ^{238}U nucleus ($A = 238$) is

$E_B(^{238}\text{U}) = (238)(7.570 \text{ MeV}) = 1801.7$ MeV

The released energy is then

$E_{\text{released}} = E_B(^{234}\text{Th}) + E_B(^4\text{He}) - E_B(^{238}\text{U})$
$= 1777.7 \text{ MeV} + 28.297 \text{ MeV} - 1801.7 \text{ MeV}$
$= 4.3$ MeV

Reflect

The alpha particles emitted by radioactive isotopes with long half-lives, such as ^{238}U, tend to have kinetic energies in the 4 to 5 MeV range.

Beta Radiation

For most possible mass numbers A, there exist a number of nuclides with that same value of A. For example, molybdenum, technetium, ruthenium, and rhodium each have an isotope with 99 nucleons: ^{99}Mo has 42 protons and 57 neutrons, ^{99}Tc has 43 protons and 56 neutrons, ^{99}Ru has 44 protons and 55 neutrons, and ^{99}Rh has 45 protons and 54 neutrons. (It is also possible to create isotopes of other elements with A equal to 99.) The binding energy per nucleon in each is slightly different, however, so only one of them is the most stable: For $A = 99$, the most stable isotope is ^{99}Ru. If there were a process whereby ^{99}Tc could convert one of its neutrons into a proton, or ^{99}Rh could convert one of its protons into a neutron, either of these nuclei could transform into the more stable ^{99}Ru.

The process that makes this possible is called **beta decay**. There are actually two varieties of beta decay. In **beta-minus decay** a neutron (charge zero) changes into a proton (charge $+e$). The net charge cannot change, so an electron (charge $-e$), also known as a beta-minus (β^-) particle, is also produced and escapes from the nucleus.

To account for other conservation requirements a third particle, the neutral and nearly massless antineutrino ($\bar{\nu}_e$), is also created in this process. The full process, then, is

$$n \to p + e^- + \bar{\nu}_e \quad \text{(beta-minus decay)}$$

In **beta-plus decay** a proton (charge $+e$) changes into a neutron (charge zero). To conserve charge a positively charged electron or *positron*, also called a beta-plus (β^+) particle, is also produced and escapes from the nucleus, along with a neutral and nearly massless neutrino (which, for our purposes, is essentially the same particle as an antineutrino). This process is

$$p \to n + e^+ + \nu_e \quad \text{(beta-plus decay)}$$

Figure 27-19a depicts the beta-minus decay of a nucleus with too many neutrons such as ^{99}Tc. We can write this process as

$$^{99}\text{Tc} \to {}^{99}\text{Ru} + e^- + \bar{\nu}_e$$

The number of nucleons ($A = 99$) is the same before and after the decay, but the number of protons Z has increased by 1 (from 43 to 44) and the number of neutrons N has decreased by 1 (from 56 to 55). The decay of ^{137}Cs to ^{137}Ba, which we discussed at the beginning of this section, is another example of beta-minus decay. In this case Z increases from 55 to 56 and N decreases from 82 to 81.

One particularly important example of beta-plus decay is the decay of potassium-40 (^{40}K) to argon-40 (^{40}Ar) (see Figure 27-16). This process has a very long half-life of 1.25×10^9 y. Since potassium is abundant in Earth's interior, the energy released by this decay makes a substantial contribution to keeping our planet's interior in a fluid state and powering its geologic activity.

Figure 27-19b depicts the beta-plus decay of a nucleus with too many protons such as ^{99}Rh. We can write this process as

$$^{99}\text{Rh} \to {}^{99}\text{Ru} + e^+ + \nu_e$$

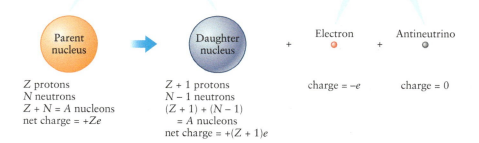

(a) Nuclei with too many neutrons can undergo beta-minus (β^-) decay.

(1) The daughter nucleus has one more proton and one fewer neutron than the parent.　(2) The nucleus also emits an electron (so electric charge is conserved) and an antineutrino.

Parent nucleus → Daughter nucleus + Electron + Antineutrino

Z protons
N neutrons
Z + N = A nucleons
net charge = +Ze

Z + 1 protons
N − 1 neutrons
(Z + 1) + (N − 1)
= A nucleons
net charge = +(Z + 1)e

charge = −e　charge = 0

(b) Nuclei with too many protons can undergo beta-plus (β^+) decay.

(1) The daughter nucleus has one fewer proton and one more neutron than the parent.　(2) The nucleus also emits a positron (so electric charge is conserved) and a neutrino.

Parent nucleus → Daughter nucleus + Positron + Neutrino

Z protons
N neutrons
Z + N = A nucleons
net charge = +Ze

Z − 1 protons
N + 1 neutrons
(Z − 1) + (N + 1)
= A nucleons
net charge = +(Z − 1)e

charge = +e　charge = 0

Figure 27-19 Beta decay Nuclei can increase their stability by (a) converting one neutron into a proton or (b) converting one proton into a neutron.

As for beta-minus decay, the number of nucleons remains the same (in this case $A = 99$). But now the number of protons Z decreases by 1 (from 45 to 44), and the number of neutrons increases by 1 (from 54 to 55).

An important application of beta decay is in carbon-14 dating, a technique used to measure the age of objects that are composed, or partially composed, of organic matter. Almost all carbon atoms in Earth's atmosphere—for example, the carbon in carbon dioxide, CO_2—have a nucleus with a stable isotope of carbon, either ^{12}C (98.9%) or ^{13}C (1.1%). But about 1 in 10^{12} of those carbon atoms has a ^{14}C nucleus. This radioactive isotope of carbon is a β^- emitter with a half-life of 5730 y:

$$^{14}C \rightarrow {}^{14}N + e^- + \bar{v}_e$$

Carbon-14 is constantly produced in the atmosphere by cosmic rays slamming into ^{14}N nuclei. As a result, even though ^{14}C undergoes radioactive decays, the ratio of ^{14}C to ^{12}C in the atmosphere has remained relatively constant for at least tens of thousands of years. The $^{14}C/^{12}C$ ratio is the same in living organisms—for example, plants that breathe in CO_2—as it is in the atmosphere. However, once an organism dies, it no longer replenishes its supply of carbon, so the $^{14}C/^{12}C$ ratio decreases as the ^{14}C decays. As an example, in the Chauvet-Pont-d'Arc Cave in southern France (Figure 27-20) there are smoke stains whose $^{14}C/^{12}C$ ratio has been measured. This makes it possible to determine how much time has elapsed since the firewood that created the smoke was part of a living tree. This carbon-14 dating technique tells us that the smoke stains are 30,000 to 33,000 years old—which must be the age of the cave paintings made by the light of that burning firewood.

Figure 27-20 Carbon-14 dating and human prehistory These cave paintings in France were created by a prehistoric artist. The amount of ^{14}C present in smoke stains on the walls is less than 3% of what would be expected from smoke produced by burning freshly cut firewood. Detailed analysis shows that between 5.2 and 5.8 half-lives of ^{14}C have elapsed since the smoke was produced, indicating an age of the smoke stains (and the paintings) between 30,000 and 33,000 years.

EXAMPLE 27-7 Ötzi the Iceman

In 1991, two German hikers discovered a human corpse in the Ötztal Alps on the border between Austria and Italy. The remains proved to be a well-preserved natural mummy of a man who lived during the last Ice Age. The rate of radioactive decay of ^{14}C in the mummy of "Ötzi the Iceman" was measured to be 0.121 Bq per gram (Bq/g). In a living organism, the rate of radioactive decay of ^{14}C is 0.231 Bq/g. How long ago did Ötzi the Iceman live?

Set Up

Once Ötzi died, his body stopped taking in ^{14}C. After that time the number $N(t)$ of ^{14}C nuclei in his body decreased due to beta-minus decay. The ^{14}C decay rate $R(t)$, which is proportional to $N(t)$, decreased in the same manner. We are given $R(t) = 0.121$ Bq/g for the present-day decay rate and $R_0 = 0.231$ Bq/g (the decay rate for a living organism and hence the decay rate at time $t = 0$, the last date on which Ötzi was still alive). We'll solve Equation 27-6 for the present time t (the elapsed time since Ötzi died). We'll also have to use Equation 27-8 to find the decay constant λ from the known half-life $\tau_{1/2} = 5730$ y of ^{14}C.

Decay rate as a function of time:
$$R(t) = R_0 e^{-\lambda t} \qquad (27\text{-}6)$$
Half-life of a radioactive substance:
$$\tau_{1/2} = \frac{\ln 2}{\lambda} \qquad (27\text{-}8)$$

Solve

Rearrange Equations 27-6 and 27-8 to find an expression for the time t since Ötzi died.

We know the present-day decay rate $R(t)$ and the initial decay rate in a living organism R_0. We want to find the time t since Ötzi died, so we rearrange Equation 27-6. Divide both sides by R_0:
$$\frac{R(t)}{R_0} = e^{-\lambda t}$$
Take the natural logarithm of both sides and recall that $\ln e^x = x$:
$$\ln\left(\frac{R(t)}{R_0}\right) = \ln e^{-\lambda t} = -\lambda t$$

Divide both sides by $-\lambda$:

$$t = -\frac{1}{\lambda} \ln\left(\frac{R(t)}{R_0}\right)$$

To get an expression for $1/\lambda$, divide both sides of Equation 27-8 by $\ln 2$:

$$\frac{1}{\lambda} = \frac{\tau_{1/2}}{\ln 2}$$

Putting everything together, the time t since Ötzi died is

$$t = -\frac{\tau_{1/2}}{\ln 2} \ln\left(\frac{R(t)}{R_0}\right)$$

Substitute the given values into the expression for t.

We are given $\tau_{1/2} = 5730$ y for ^{14}C, $R(t) = 0.121$ Bq/g, and $R_0 = 0.231$ Bq/g:

$$t = -\frac{(5730 \text{ y})}{\ln 2} \ln\left(\frac{0.121 \text{ Bq/g}}{0.231 \text{ Bq/g}}\right) = -\frac{(5730 \text{ y})}{0.693} \ln 0.524$$

$$= -\frac{(5730 \text{ y})}{0.693}(-0.647) = 5350 \text{ y}$$

Reflect

Ötzi died 5350 y ago. His mummy thus gives us a unique look into life in prehistoric Europe.

Carbon-14 dating can be used only on objects less than about 50,000 years old, or about 8 to 10 half-lives of ^{14}C. For older objects the decay rate of ^{14}C has decreased to such a small value that it is hard to measure accurately, so any determination of age with this technique becomes difficult. For much older objects such as rocks, a similar approach is used but with isotopes with much longer half-lives. For example, the age of meteorites that fall to Earth is determined by looking at the ratio of uranium to lead (the endpoint of a series of alpha decays that starts with uranium); the oldest meteorites are more than 4.5×10^9 years old.

Gamma Radiation

A nucleus in an excited state radiates energy in a way analogous to the emission of a photon when an electron in an excited atomic state falls to a state of lower energy (see Section 26-5). Just like atoms, nuclei have excited states of definite energy, and when they transition from an excited state to a less excited one, they will emit a photon of energy equal to the difference in energy between the initial and final states.

The most common way that a nucleus can become excited is following an alpha decay or a beta decay. Although these decay processes result in a more stable configuration of the nucleons, the nucleons that remain in the daughter nucleus may not be, initially, in the most stable arrangement for that particular nuclide. This excited daughter nucleus decays to a more stable configuration, giving off energy in the form of a gamma (γ) ray (Figure 27-21). This process is called **gamma decay**.

The energy carried away when a nucleus in an excited state decays is on the order of 1 MeV. From Equation 22-26 in Section 22-4, $E = hf$, the frequency and wavelength of a photon of energy $E = 1.00$ MeV $= 1.00 \times 10^6$ eV are

$$f = \frac{E}{h} = \frac{1.00 \times 10^6 \text{ eV}}{4.14 \times 10^{-15} \text{ eV} \cdot \text{s}} = 2.42 \times 10^{20} \text{ Hz}$$

$$\lambda = \frac{c}{f} = \frac{3.00 \times 10^8 \text{ m/s}}{2.42 \times 10^{20} \text{ Hz}}$$

$$= 1.24 \times 10^{-12} \text{ m} = 1.24 \times 10^{-3} \text{ nm}$$

Recall that the wavelength of visible light photons is in the range from 380 to 750 nm; a photon emitted by an excited nucleus is in the gamma radiation range, far from the visible part of the spectrum.

Gamma radiation does not change the atomic number Z or the neutron number N of a nucleus; that is, the number of protons and the number of neutrons remain the same after

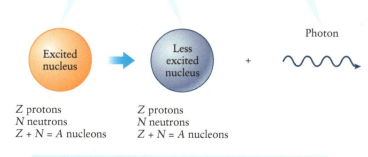

① In gamma (γ) decay a nucleus in an excited state drops into a less excited state.

② The energy lost by the nucleus goes into a high-energy photon (gamma ray).

Photon

Excited nucleus

Less excited nucleus

+

Z protons
N neutrons
Z + N = A nucleons

Z protons
N neutrons
Z + N = A nucleons

 In gamma decay there is no change in the number of protons or the number of neutrons in the nucleus.

Figure 27-21 Gamma decay An excited nucleus can lower its energy by emitting a photon.

a γ ray is emitted. As an example, earlier we considered the beta-minus decay of ^{137}Cs to ^{137}Ba. In 95% of those decays the ^{137}Ba nucleus is formed in an excited state that we denote as ^{137}Ba*:

$$^{137}\text{Cs} \rightarrow {}^{137}\text{Ba*} + \text{e}^- + \bar{\nu}_e$$

The excited barium nucleus then decays to its ground state by emission of a photon of energy 0.662 MeV:

$$^{137}\text{Ba*} \rightarrow {}^{137}\text{Ba} + \gamma$$

The values of $Z = 56$ and $N = 81$ for the barium nucleus do not change in this second step of the radiation process.

Biological Effects of Radiation

We'll finish this section with a discussion of how radiation from nuclei affects living organisms including humans. As they pass through living tissue, energetic particles cause ionization in molecules such as water, creating free radicals (highly reactive molecules and atoms) that can damage DNA and causes potentially lethal mutations to cells. This is the principle of using radiation to destroy cancerous tumors. The cells in these tumors reproduce more rapidly than healthy cells, and so are more quickly destroyed by the mutations produced by ionizing radiation.

BioMedical

Charged particles—such as the α particles produced in alpha decay and the electrons and positrons produced in beta-minus and beta-plus decays—exert electric forces on molecular electrons to cause ionization. High-energy photons such as those produced in gamma decay, by contrast, cause ionization either by colliding with electrons and knocking them out of molecules (Compton scattering, described in Section 26-3) or by having their energy absorbed by molecular electrons (the photoelectric effect, described in Section 26-2).

In each of these cases the amount of damage depends on the amount of energy absorbed by the tissue from the radiation that passes through it. The energy from radiation absorbed per unit mass of tissue is called the *absorbed dose*. The unit of absorbed dose is the *gray* (Gy), equal to 1 joule per kilogram: 1 Gy = 1 J/kg. However, experiment shows that different kinds of radiation produce different amounts of biological effect for the same absorbed dose. This is described by a factor called the **relative biological effectiveness (RBE)**, which has a different value for different types of radiation. The value of the RBE is 1 for x rays, gamma rays, electrons, and positrons, but about 20 for alpha particles. The amount of biological effect of a given amount of radiation, or *equivalent dose*, is the product of the RBE and the absorbed dose. The unit of absorbed dose is the sievert (Sv):

equivalent dose (Sv) = RBE × absorbed dose (Gy)

Since alpha particles have such a large RBE, for a given absorbed dose they produce a much greater equivalent dose and have much greater biological effect than do beta or gamma radiation. However, alpha particles are not very penetrating (a piece of paper is enough to stop the alpha particles from most alpha-emitting nuclei), so they do not pose much radiation hazard if they are coming from outside the body.

Since radioactive materials occur naturally in the environment, we are continually exposed to radiation. In one year a person at sea level has an equivalent dose of about 3 mSv (1 mSv = 10^{-3} Sv), most of which comes from the trace amounts of radon gas found inside homes and other buildings. (Radon, which has no stable isotopes, is a decay product of naturally occurring uranium and thorium. Because it is a gas, radon can seep into buildings, concentrate there, and be inhaled by the occupants. The longest-lived isotope of radon, ^{222}Rn, is an alpha emitter; since alpha particles have a large RBE, even small amounts of inhaled radon have a measurable effect on a person's annual equivalent dose.) At high altitude there is greater exposure to cosmic rays coming from space (which are screened by the atmosphere); the additional equivalent dose from a round-trip airline flight between the west and east coasts of North America is about 0.03 mSv and is about 1.5 mSv per year for a person living at an elevation of 1500 m (5000 ft). A chest x ray adds an additional 0.1 mSv, a mammogram an additional 0.4 mSv, and a CT scan of the abdomen and pelvis an additional 10 mSv.

To put these numbers in perspective, a whole-body equivalent dose up to about 200 mSv (0.2 Sv) causes no immediately detectable effect. To have fatal consequences a short-term whole-body dose would have to be about 5 Sv, equivalent to having 50,000 simultaneous chest x rays.

GOT THE CONCEPT? 27-5 Half-Lives

 A certain radioactive isotope has a half-life of 5 days. You are given a sample containing a number of nuclei of this isotope. About how long would you have to wait before about 1/1000 of the initial number of nuclei remained? (a) 20 days; (b) 40 days; (c) 50 days; (d) 100 days; (e) 1000 days.

TAKE-HOME MESSAGE FOR Section 27-6

✔ Radioactive decay is a statistical process. The number of nuclei and the decay rate both decrease exponentially with time, and both decrease by one-half in a time equal to one half-life.

✔ The three most common modes by which radiation occurs are alpha, beta, and gamma radiation.

✔ Alpha particles are ^{4}He nuclei and are emitted by large nuclei with $Z > 82$. In alpha emission the proton number and neutron number each decrease by 2.

✔ Beta particles are either negatively charged electrons or positively charged positrons. In beta-minus emission a neutron in the nucleus changes into a proton; in beta-plus emission a proton in the nucleus changes into a neutron.

✔ Gamma particles are high-energy photons, typically with energies of about 1 MeV. They are emitted when a nucleus decays from an excited state to a less excited one.

✔ Alpha, beta, and gamma radiation can all damage living tissue by producing ionization.

Key Terms

alpha decay
alpha particle
antineutrino
atomic number
becquerel
beta decay
beta-minus decay
beta-plus decay
binding energy

decay constant
gamma decay
half-life
isotope
magnetic resonance imaging
mass number
neutrino
neutron-induced fission
neutron number

nuclear fission
nuclear fusion
nuclear radiation
nucleon
nuclide
positron
radioactive
relative biological effectiveness (RBE)
strong nuclear force

Chapter Summary

Topic	Equation or Figure
The strong nuclear force and nuclear sizes: The strong nuclear force is a short-range attractive force that acts between nucleons (protons or neutrons). Because this force has a short range, the volume of a nucleus is proportional to the mass number (number of nucleons). In larger nuclei, the number of neutrons must exceed the number of protons to counterbalance the electric repulsion between protons.	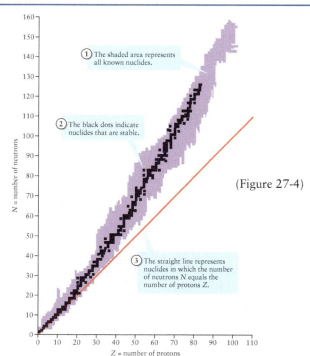 (Figure 27-4)

For most nuclides, more neutrons than protons must be present (N must be greater than Z) to prevent the protons from flying apart due to electrostatic repulsion.

① The shaded area represents all known nuclides.

② The black dots indicate nuclides that are stable.

③ The straight line represents nuclides in which the number of neutrons N equals the number of protons Z.

Nuclear binding energy: The binding energy of a nucleus is the energy required to separate it into its constituent nucleons. The binding energy per nucleon is greatest for nuclei with around 60 nucleons; for larger nuclei the electric repulsion between protons makes nuclei less stable. The binding energy of a particular isotope can be calculated from the mass of a neutral atom containing that isotope.

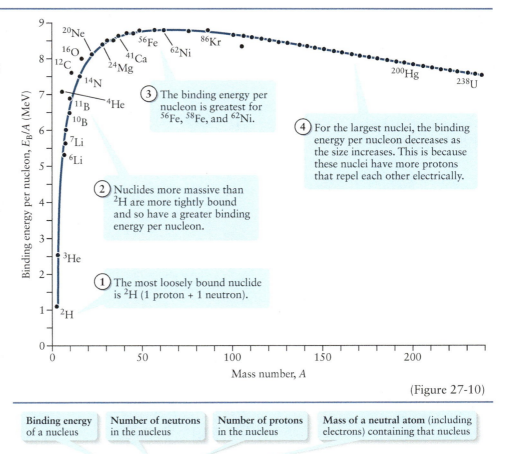

③ The binding energy per nucleon is greatest for ^{56}Fe, ^{58}Fe, and ^{62}Ni.

④ For the largest nuclei, the binding energy per nucleon decreases as the size increases. This is because these nuclei have more protons that repel each other electrically.

② Nuclides more massive than ^{2}H are more tightly bound and so have a greater binding energy per nucleon.

① The most loosely bound nuclide is ^{2}H (1 proton + 1 neutron).

(Figure 27-10)

Binding energy of a nucleus	Number of neutrons in the nucleus	Number of protons in the nucleus	Mass of a neutral atom (including electrons) containing that nucleus

$$E_{\text{B}} = (Nm_{\text{n}} + Zm_{1_{\text{H}}} - m_{\text{atom}})c^2 \qquad (27\text{-}2)$$

Mass of a neutron   Mass of a neutral hydrogen atom (1 proton + 1 electron)   Speed of light in a vacuum

Nuclear fission: When the largest nuclei absorb a neutron, they fragment (fission) into two smaller nuclei plus a few neutrons. The fragments are more tightly bound than the original nucleus, so energy is released in this process. In a sustained fission reaction, the released neutrons trigger other, nearby nuclei to also undergo fission.

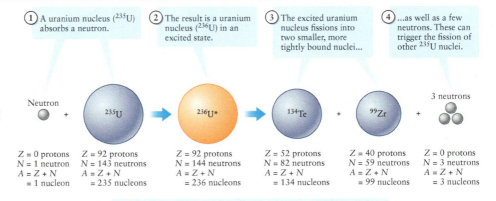

① A uranium nucleus (^{235}U) absorbs a neutron.

② The result is a uranium nucleus (^{236}U) in an excited state.

③ The excited uranium nucleus fissions into two smaller, more tightly bound nuclei...

④ ...as well as a few neutrons. These can trigger the fission of other ^{235}U nuclei.

Neutron ^{235}U   → ^{236}U*   → ^{134}Te   + ^{99}Zr   +   3 neutrons

$Z = 0$ protons	$Z = 92$ protons	$Z = 92$ protons	$Z = 52$ protons	$Z = 40$ protons	$Z = 0$ protons
$N = 1$ neutron	$N = 143$ neutrons	$N = 144$ neutrons	$N = 82$ neutrons	$N = 59$ neutrons	$N = 3$ neutrons
$A = Z + N$	$A = Z + N$	$A = Z + N$	$A = Z + N$	$A = Z + N$	$A = Z + N$
$= 1$ nucleon	$= 235$ nucleons	$= 236$ nucleons	$= 134$ nucleons	$= 99$ nucleons	$= 3$ nucleons

Energy is released in this fission reaction: The total kinetic energy of the fission fragments is much greater than the total kinetic energy of the initial neutron and ^{235}U nucleus.

(Figure 27-12)

Nuclear fusion: The smallest nuclei can merge together to form a larger nucleus, releasing energy in the process. These fusion processes require very high temperatures so that the fusing nuclei have enough kinetic energy to overcome their mutual electric repulsion.

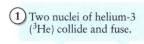

1 Two nuclei of helium-3 (^{3}He) collide and fuse.

2 The result is a more tightly bound nucleus of helium-4 (^{4}He)...

3 ...plus two protons and a high-energy photon.

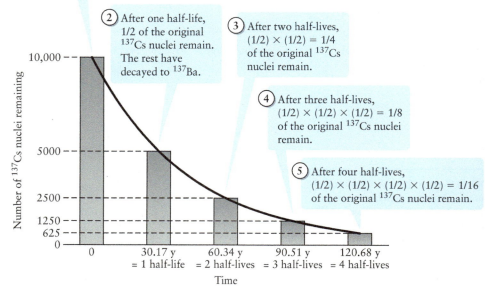

^{3}He ^{3}He ^{4}He   2 protons   Photon

Z = 2 protons
N = 1 neutron
A = Z + N
= 3 nucleons

Z = 2 protons
N = 1 neutron
A = Z + N
= 3 nucleons

Z = 2 protons
N = 2 neutrons
A = Z + N
= 4 nucleons

Z = 2 protons
N = 0 neutrons
A = Z + N
= 2 nucleons

Energy is released in this fusion reaction: The total kinetic energy of the ^{4}He nucleus and protons, plus the energy of the photon, is much greater than the total kinetic energy of the initial ^{3}He nuclei.

(Figure 27-13)

Nuclear decay and half-life: The decay of unstable nuclei is a statistical process. This means that the rate of decay is proportional to the number of unstable nuclei present. As a result the number of nuclei and the decay rate both decline in an exponential manner. The time for the number of nuclei and the decay rate to decrease by one-half is called the half-life.

- The half-life of a nuclear decay is the time required for one-half of the nuclei present initially to decay.
- Radioactive decay is a statistical process: It's impossible to predict when any one individual unstable nucleus will decay.

1 Initially we have 10,000 ^{137}Cs nuclei. These nuclei are unstable and decay to ^{137}Ba.

2 After one half-life, 1/2 of the original ^{137}Cs nuclei remain. The rest have decayed to ^{137}Ba.

3 After two half-lives, (1/2) × (1/2) = 1/4 of the original ^{137}Cs nuclei remain.

4 After three half-lives, (1/2) × (1/2) × (1/2) = 1/8 of the original ^{137}Cs nuclei remain.

5 After four half-lives, (1/2) × (1/2) × (1/2) × (1/2) = 1/16 of the original ^{137}Cs nuclei remain.

Number of ^{137}Cs nuclei remaining

10,000
5000
2500
1250
625
0

0   30.17 y
= 1 half-life
60.34 y
= 2 half-lives
90.51 y
= 3 half-lives
120.68 y
= 4 half-lives

Time

(Figure 27-17)

Alpha, beta, and gamma decays: Large nuclei release energy and become more stable by emitting an alpha particle (a ^{4}He nucleus). Other nuclei with too many neutrons undergo beta-minus decay, in which one neutron changes into a proton; those with too many protons undergo beta-plus decay, in which one proton changes into a neutron. In gamma decay an excited state of a nucleus (indicated by an asterisk) transitions to a less excited state and emits a gamma-ray photon (γ).

An alpha decay:

^{228}Th $\rightarrow$ ^{224}Ra + α

A beta-minus decay:

^{99}Tc $\rightarrow$ ^{99}Ru + e^- + $\bar{\nu}_e$

A beta-plus decay:

^{99}Rh $\rightarrow$ ^{99}Ru + e^+ + ν_e

A gamma decay:

^{137}Ba* $\rightarrow$ ^{137}Ba + γ

Answer to What do you think? Question

(b) In a radioactive material, the amount that remains decreases by 1/2 after one half-life. After two half-lives, the amount that remains is $(1/2) \times (1/2) = (1/2)^2 = 1/4$; after three half-lives, $(1/2) \times (1/2) \times (1/2) = (1/2)^3 = 1/8$; and so on. The age of the cave painting is between $(13{,}000 \text{ y})/(5730 \text{ y}) = 2.3$ half-lives and $(9500 \text{ y})/(5730 \text{ y}) = 1.7$ half-lives, so approximately 2 half-lives have elapsed since the painting was made. Therefore the amount of carbon-14 remaining is about $(1/2) \times (1/2) = (1/2)^2 = 1/4$ of the original.

Answers to Got the Concept? Questions

27-1 (c) Equation 27-1 tells us that the radius of a nucleus is proportional to $A^{1/3}$, the cube root of the number of nucleons (the mass number). Compared to a ^{20}Ne nucleus, a ^{160}Dy nucleus has $160/20 = 8$ times as many nucleons, so its radius is larger by a factor of $8^{1/3} = 2$. As we discussed in Example 27-2, nuclei with any value of A have the same density.

27-2 (c), (b), (d), (a). A ranking in order of stability is a ranking in order of binding energy per nucleon. Figure 27-10 shows that this is greatest for ^{86}Kr (more than 8.5 MeV), second greatest for ^{20}Ne (slightly more than 8 MeV), third greatest for ^{200}Hg (slightly less than 8 MeV), and least for ^{11}B (about 6.9 MeV).

27-3 (d) The number of protons and the number of neutrons both remain the same: A ^{20}Ne nucleus has 10 protons and 10 neutrons, and each ^{10}B nucleus has 5 protons and 5 neutrons. But the binding energy per nucleon for ^{10}B is *less* than for ^{20}Ne (see Figure 27-10). For fission to occur and energy to be released, the fragments must have a greater binding energy per nucleon than the initial nucleus. The process ^{20}Ne $\rightarrow$ ^{10}B + ^{10}B would require a substantial amount of energy to be spontaneously added to the ^{20}Ne nucleus, and there's no way that can happen.

27-4 (d) For two ^{1}H nuclei to fuse, they must come to within a few femtometers (1 fm = 10^{-15} m) of each other. But in a typical molecule, including H_2, the separation between nuclei in adjacent atoms is on the order of 10^{-10} m. So the nuclei are normally too far apart to fuse. If they were moving rapidly with respect to each other and so had a large kinetic energy, the nuclei could overcome their mutual Coulomb repulsion and come close enough to fuse. But atomic nuclei move hardly at all within molecules, so fusion cannot happen this way. Option (c) is incorrect: The properties of a nucleus are affected not at all by the presence or absence of electrons orbiting the nucleus.

27-5 (c) After each 5-day half-life, the number of nuclei decreases by a factor of 1/2. For the number to decrease by a factor of 1/1000 requires about 10 half-lives, since $(1/2)^{10} = 1/2^{10} = 1/1024$.

Questions and Problems

In a few problems you are given more data than you actually need; in a few other problems you are required to supply data from your general knowledge, outside sources, or informed estimate.

Interpret as significant all digits in numerical values that have trailing zeros and no decimal points.

For all problems use $g = 9.80 \text{ m/s}^2$ for the free-fall acceleration due to gravity. Neglect friction and air resistance unless instructed to do otherwise.

- •   Basic, single-concept problem
- ••   Intermediate-level problem; may require synthesis of concepts and multiple steps
- ••• Challenging problem
- Example See worked example for a similar problem

Conceptual Questions

1. • What is an isotope?

2. • What is the difference between atomic number and mass number?

3. • (a) Describe what is meant by the phrase "larger nuclei are neutron rich." (b) Why do most nuclei contain at least as many neutrons as protons?

4. • Some historians would claim that without Einstein's special theory of relativity, nuclear physics would never have developed. Explain why.

5. • A simple idea of nuclear physics can be stated as follows: "The whole nucleus weighs less than the sum of its parts." Explain why.

6. • Describe two characteristics of the binding energy that are comparable to the work function (from the photoelectric effect) and two characteristics that are dissimilar to the concept of the work function.

7. • Describe the basic characteristics of the nuclear force that exists between nucleons. What other competing force between nucleons is present in the nucleus?

8. • What is the difference between fission and fusion?

9. • (a) Which elements in the periodic table are more likely to undergo nuclear fission? (b) Which are more likely to undergo nuclear fusion?

10. • **Astronomy** (a) Describe the nuclear reactions that occur in our Sun. (b) Discuss how the equilibrium state of the Sun is not permanent and discuss the eventual future of our solar system.

11. • Explain how conservation of energy and momentum would be violated if a neutrino were not emitted in beta decay.

12. • The decay constant of a radioactive nucleus is just that, *constant*. It does not depend on the size of the nuclear sample, the temperature, or any external fields (such as gravity, electricity, or magnetism). Define the decay constant and comment on how nuclear radioactivity would change if the quantity were dependent on temperature.

13. • At any given instant a sample of radioactive uranium contains many, many different isotopes of atoms that are *not* uranium. Explain why.

14. • (a) Explain how radioactive ^{14}C is used to determine the age of ancient artifacts. (b) Which types of artifacts can have their age determined in this way and which types cannot?

15. • If atomic masses are used, explain why the mass of the electron is *not* accounted for in the basic beta decay

$$n \rightarrow p + e^- + \bar{v}_e$$

Assume that the mass of the antineutrino ($\bar{v}_e$) is very small and can be neglected.

16. • **Medical** Describe, in broad terms, the health risks associated with the three major forms of radioactivity: alpha, beta, and gamma. Focus on the dangers due to inherent health risks and the ability of each to penetrate shielding material.

17. • A friend says he found a radioactive rock and asks you to carry it to the physics lab in a thin backpack. Is this safe? Explain.

Multiple-Choice Questions

18. • In an atomic nucleus, the nuclear force binds _____ together.
   A. electrons
   B. neutrons
   C. protons
   D. neutrons and protons
   E. neutrons, protons, and electrons

19. • The mass of a nucleus is _____ the sum of the masses of its nucleons.
   A. always less than
   B. sometimes less than
   C. always more than
   D. always equal to
   E. sometimes equal to

20. • Assuming that the final particles in each case are more stable than the initial particles, which of the statements regarding fission and fusion reactions is true?
   A. Fusion absorbs energy and fission releases energy.
   B. Fusion releases energy and fission absorbs energy.
   C. Both fusion and fission absorb energy.
   D. Both fusion and fission release energy.
   E. Both fusion and fission can release or absorb energy.

21. • In fission processes, which of the following statements is true?
   A. Only the total number of nuclei remains the same.
   B. Only the total number of protons remains the same.
   C. Only the total number of neutrons remains the same.
   D. The total number of protons and the total number of nuclei both remain the same.
   E. The total number of protons and the total number of neutrons both remain the same.

22. • In a spontaneous fission reaction, the total mass of the products is _____ the mass of the original element.
   A. greater than
   B. less than
   C. the same as
   D. double
   E. one-half

23. • What is the source of the Sun's energy?
   A. chemical reactions
   B. fission reactions
   C. fusion reactions
   D. gravitational collapse
   E. both fusion reactions and fission reactions

24. • In a fusion reaction the total mass of the products is _____ the mass of the original elements.
   A. greater than
   B. less than
   C. the same as
   D. double
   E. one-half

25. • The decay constant λ depends only on
   A. the number of atoms at the initial time.
   B. the initial decay rate.
   C. the half-life.
   D. the binding energy per nucleon.
   E. whether the decay is alpha, beta, or gamma.

26. • The number of polonium-210 atoms in a radioactive sample of ^{210}Po
   A. decreases linearly with time.
   B. increases linearly with time.
   C. decreases exponentially with time.
   D. increases exponentially with time.
   E. remains constant.

27. • The decay rate for a sample of any isotope
   A. decreases linearly with time.
   B. increases linearly with time.
   C. decreases exponentially with time.
   D. increases exponentially with time.
   E. remains constant.

Problems

Note: In all cases, unless otherwise stated, binding energy, energy released, and the like are to be expressed in MeV. Also, all atomic masses should be calculated to the nearest 10^{-6}.

27-1 The quantum concepts that help explain atoms are essential for understanding the nucleus

27-2 The strong nuclear force holds nuclei together

28. • Provide the elemental abbreviation (for example, ^{16}O for oxygen-16) and give the number of protons, the number of neutrons, and the mass number for each of the following isotopes:
   A. hydrogen-3   E. technetium-100
   B. beryllium-8   F. tungsten-184
   C. aluminum-26   G. osmium-190
   D. gold-197   H. plutonium-239

29. • Calculate the radius of each of the nuclei in problem 28. Example 27-1

30. • Name the element and give the number of protons, the number of neutrons, and the mass number for each of the following nuclei:
   A. ^{2}H   D. ^{12}C   G. ^{131}I
   B. ^{4}He   E. ^{56}Fe   H. ^{235}U
   C. ^{6}Li   F. ^{90}Sr

31. • If our Sun (mass = 1.99×10^{30} kg, radius = 6.96×10^8 m) were to collapse into a neutron star (an object composed of tightly packed neutrons with roughly the same density as a nucleus), what would the new radius of our "neutron-sun" be? Example 27-2

32. •• Given that a nucleus is approximately spherical and has a radius $r = r_0 A^{1/3}$ (where r_0 is about 1.2 fm), determine its approximate mass density. Express your answer in SI units and convert to tons per cubic inch, units that might be used in a news report. Example 27-2

33. •• Calculate (a) the magnitude of the repulsive electric force between two protons separated by a distance of

4.0×10^{-15} m inside a nucleus, and (b) the magnitude of the attractive gravitational force between the same two protons. The proton mass is 1.67×10^{-27} kg. (c) Can gravity hold a nucleus together?

27-3 Some nuclei are more tightly bound and more stable than others

34. • Calculate the atomic mass of each of the isotopes listed below. Give your answer in atomic mass units (u) and in grams (g). The values will include the mass of Z electrons. Example 27-3

  A. ^{1}H   D. ^{12}C   G. ^{131}I
  B. ^{4}He   E. ^{56}Fe   H. ^{238}U
  C. ^{9}Be   F. ^{90}Sr

35. • What is the binding energy of carbon-12? Give your answer in MeV. Example 27-3

36. • What is the binding energy per nucleon for the following isotopes? Example 27-3

  A. ^{2}H   D. ^{12}C   G. ^{129}I
  B. ^{4}He   E. ^{56}Fe   H. ^{235}U
  C. ^{6}Li   F. ^{90}Sr

37. • What minimum energy is needed to remove a neutron from ^{40}Ca and so convert it to ^{39}Ca? The atomic masses of the two isotopes are 39.96259098 and 38.97071972 u, respectively. Example 27-3

38. • What is the binding energy of the last neutron of carbon-13? The atomic mass of carbon-13 is 13.003355 u. Example 27-3

39. • **Medical** Iodine-131 is a radioactive isotope that is used in the treatment of cancer of the thyroid. The natural tendency of the thyroid to take up iodine creates a pathway for which radiation (β^- and γ) that is emitted from this unstable nucleus can be directed onto the cancerous tumor with very little collateral damage to surrounding healthy tissue. Another advantage of the isotope is its relatively short half-life (8 days). Calculate the binding energy of iodine-131 and the binding energy per nucleon. The mass of iodine-131 is 130.906124 u. Example 27-3

27-4 The largest nuclei can release energy by undergoing fission and splitting apart

40. • Calculate the energy released in the following nuclear fission reaction:

$$^{239}\text{Pu} + \text{n} \rightarrow {}^{98}\text{Tc} + {}^{138}\text{Sb} + 4\text{n}$$

The atomic masses are ^{239}Pu = 239.052157 u, ^{98}Tc = 97.907215 u, and ^{138}Sb = 137.940793 u. Example 27-4

41. • Complete the following nuclear fission reaction of thorium-232 and calculate the energy released in the reaction: $^{232}\text{Th} + \text{n} \rightarrow {}^{99}\text{Kr} + {}^{124}\text{Xe} + \_?\_$ The atomic masses are ^{232}Th = 232.038051 u, ^{99}Kr = 98.957606 u, and ^{124}Xe = 123.905894 u. Example 27-4

42. • Complete the following fission reactions: Example 27-4

  A. $^{235}\text{U} + \text{n} \rightarrow {}^{128}\text{Sb} + {}^{101}\text{Nb} + \_?\_$
  B. $^{235}\text{U} + \text{n} \rightarrow \_?\_ + {}^{116}\text{Pd} + 4\text{n}$
  C. $^{238}\text{U} + \text{n} \rightarrow {}^{99}\text{Kr} + \_?\_ + 11\text{n}$
  D. $\_?\_ + \text{n} \rightarrow {}^{101}\text{Rb} + {}^{130}\text{Cs} + 8\text{n}$

43. • Complete the following fission reactions: Example 27-4

  A. $^{242}\text{Am} + \_?\_ \rightarrow {}^{90}\text{Sr} + {}^{149}\text{La} + 4\text{n}$
  B. $^{244}\text{Pa} + \text{n} \rightarrow \_?\_ + {}^{131}\text{Sb} + 12\text{n}$
  C. $\_?\_ + \text{n} \rightarrow {}^{92}\text{Se} + {}^{153}\text{Sm} + 6\text{n}$
  D. $^{262}\text{Fm} + \text{n} \rightarrow {}^{112}\text{Rh} + \_?\_ + 9\text{n}$

44. • Calculate the energy (in MeV) released in the following nuclear fission reaction:

$$^{242}\text{Am} + \_?\_ \rightarrow {}^{90}\text{Sr} + {}^{149}\text{La} + 4\text{n}$$

Start by completing the reaction and use the following nuclear masses: ^{242}Am = 242.059549 u, ^{90}Sr = 89.9077387 u, and ^{149}La = 148.934733 u Example 27-4

45. •• Assuming that in a fission reactor a neutron loses half its energy in each collision with an atom of the moderator, determine how many collisions are required to slow a 200-MeV neutron to an energy of 0.04 eV. Example 27-4

46. • Knowing that the binding energy per nucleon for uranium-235 is about 7.6 MeV/nucleon and the binding energy per nucleon for typical fission fragments is about 8.5 MeV/nucleon, find an average energy release per uranium-235 fission reaction in MeV. Example 27-4

47. •• How many kilograms of uranium-235 must completely fission spontaneously into ^{140}Xe, ^{92}Sr, and three neutrons to produce 1000 MW of power continuously for 1 year, assuming that the fission reactions are 100% efficient? Example 27-4

48. • Repeat problem 47 in the more realistic case where the fission reactions are about 30% efficient in producing 1000 MW of power over 1 year of continuous operation. Example 27-4

49. • Calculate the number of fission reactions per second that take place in a 1000-MW reactor. Assume that 200 MeV of energy is released in each reaction. Example 27-4

27-5 The smallest nuclei can release energy if they are forced to fuse together

50. • Complete the following fusion reactions: Example 27-4

  A. $^2\text{H} + {}^3\text{H} \rightarrow {}^4\text{He} + \_?\_$
  B. $^4\text{He} + {}^4\text{He} \rightarrow {}^7\text{Be} + \_?\_$
  C. $^2\text{H} + {}^2\text{H} \rightarrow {}^3\text{He} + \_?\_$
  D. $^2\text{H} + {}^1\text{H} \rightarrow \gamma + \_?\_$
  E. $^2\text{H} + {}^2\text{H} \rightarrow {}^3\text{H} + \_?\_$

51. •• Calculate the energy released in each of the fusion reactions in problem 50. Give your answers in MeV. Example 27-4

52. •• **Astronomy** Consider the proton–proton cycle that occurs in most stars (including our own Sun):

$$\text{Step 1: } {}^1\text{H} + {}^1\text{H} \rightarrow {}^2\text{H} + e^+ + \nu_e$$
$$\text{Step 2: } {}^2\text{H} + {}^1\text{H} \rightarrow {}^3\text{He} + \gamma$$
$$\text{Step 3: } {}^3\text{He} + {}^3\text{He} \rightarrow {}^4\text{He} + 2\ {}^1\text{H} + \gamma$$

Calculate the net energy released from the three steps. Do *not* ignore the mass of the positron in step 1. (You may ignore the mass of the neutrino.) Example 27-4

53. •• Each fusion reaction of deuterium (^{2}H) and tritium (^{3}H) releases about 20 MeV. What mass of tritium is needed to create 10^{14} J of energy, the same as that released by exploding 25,000 tons of TNT? Assume that an endless supply of deuterium is available. Example 27-4

54. •• How many fusion reactions per second must be sustained to operate a deuterium–tritium fusion power plant that outputs 1000 MW, operating at 33% efficiency? Example 27-4

27-6 Unstable nuclei may emit alpha, beta, or gamma radiation

55. • Complete the following conversions:

  A. 100 μCi = _____ Bq
  B. 1500 decays/min = _____ Bq

C. 16,500 Bq = _____ Ci

D. 7.55×10^{10} Bq = _____ decays/min

56. • The curie unit is defined as 1 Ci = 3.7×10^{10} Bq, which is about the rate at which radiation is emitted by 1.00 g of radium. (a) Calculate the half-life of radium from the definition. (b) What does your calculation tell you about the radiation emission rate of radium? Example 27-5

57. • A certain radioactive isotope has a decay constant of 0.00334 s^{-1}. Find the half-life in seconds and days. Example 27-5

58. •• A radioactive sample is monitored with a radiation detector, which registers 5640 counts per minute. Twelve hours later, the detector reads 1410 counts per minute. Calculate the decay constant and the half-life of the sample. Example 27-5

59. • What fraction of a sample of ^{32}P will be left after 4 months? Its half-life is 14.3 days. Example 27-5

60. • What fraction of a radioactive sample will be left after 6 half-lives? What about 7.5 half-lives? Example 27-5

61. •• **Medical** A patient is injected with 7.88 µCi of radioactive iodine-131 that has a half-life of 8.02 days. Assuming that 90% of the iodine ultimately finds its way to the thyroid, what decay rate do you expect to find in the thyroid after 30 days? Example 27-5

62. •• The ratio of carbon-14 to carbon-12 in living wood is 1.3×10^{-12}. How many decays per second are there in 550 g of wood? Example 27-7

63. •• You take a course in archaeology that includes field work. An ancient wooden totem pole is excavated from your archaeological dig. The beta decay rate is measured at 150 decays/min. If the totem pole contains 225 g of carbon and the ratio of carbon-14 to carbon-12 in living trees is 1.3×10^{-12}, what is the age of the pole? Example 27-7

64. •• Determine the decay rate for 500 g of carbon from a tree limb 12 centuries after it is cut off. Example 27-7

65. •• How many nuclei of radon-222 are present in a sample for which you measure 485 decays/min? Example 27-5

66. •• The ages of rocks that contain fossils can be determined using the isotope ^{87}Rb. This isotope of rubidium undergoes beta decay with a half-life of 4.75×10^{10} y. Ancient samples contain a ratio of ^{87}Sr to ^{87}Rb of 0.0225. Given that ^{87}Sr is a stable product of the beta decay of ^{87}Rb, and there was originally no ^{87}Sr present in the rocks, calculate the age of the rock sample. Assume that the decay rate is constant over the relatively short lifetime of the rock compared to the half-life of ^{87}Rb. Example 27-7

67. • Complete the following alpha decays: Example 27-6

   A. ^{238}U $\rightarrow \alpha + $__?__

   B. ^{234}Th $\rightarrow$ __?__ $+ \, ^{-?}_{-}$Ra

   C. __?__ $\rightarrow \alpha + {}^{236}$U

   D. ^{214}Bi $\rightarrow \alpha + $__?__

68. • Complete the following beta decays:

   A. ^{14}C $\rightarrow e^- + \bar{v}_e + $__?__

   B. ^{239}Np $\rightarrow e^- + \bar{v}_e + $__?__

   C. __?__ $\rightarrow e^- + \bar{v}_e + {}^{60}$Ni

   D. ^{3}H $\rightarrow e^- + \bar{v}_e + $__?__

   E. ^{13}N $\rightarrow e^+ + $__?__ $+ $__?__

69. • Complete the following gamma decays:

   A. ^{131}I* $\rightarrow \gamma + $__?__

   B. ^{145}Pm* $\rightarrow {}^{145}$Pm $+ $__?__

   C. __?__ $\rightarrow \gamma + {}^{24}$Na

70. •• Nickel-64 has an excited state 1.34 MeV above the ground state. The atomic mass of the ground state of this isotope of nickel is 63.927967 u. (a) What is the mass of the atom when the nucleus is in this excited state? (b) What is the wavelength of the gamma ray that is emitted when the nucleus decays to the ground state?

71. •• **Medical** A chest x ray imparts an equivalent radiation dose of about 1×10^{-3} Sv to a person. Approximately how many radon atoms (which emit alpha particles that have an energy of approximately 5 MeV) do you need to inhale before you expose yourself to radiation dose equal to that of 1 chest x ray? Assume the body absorbs x-ray radiation and alpha radiation with the same efficiency, and that all inhaled radon atoms will decay and their emitted radiation be absorbed by the body.

General Problems

72. • (a) Using a spreadsheet or programmable calculator, calculate the binding energy per nucleon for the following isotopes of the five least massive elements. Masses given are atomic masses.

hydrogen-1: 1.007825 u	lithium-8: 8.022486 u	boron-8: 8.024605 u
hydrogen-2: 2.014102 u	lithium-9: 9.026789 u	boron-10: 10.012936 u
hydrogen-3: 3.016049 u	lithium-11: 11.043897 u	boron-11: 11.009305 u
helium-3: 3.016029 u	beryllium-7: 7.016928 u	boron-12: 12.014352 u
helium-4: 4.002602 u	beryllium-9: 9.012174 u	boron-13: 13.017780 u
helium-6: 6.018886 u	beryllium-10: 10.013534	boron-14: 14.025404 u
helium-8: 8.033922 u	beryllium-11: 11.021657 u	boron-15: 15.031100 u
lithium-6: 6.015121 u	beryllium-12: 12.026921 u	
lithium-7: 7.016003 u	beryllium-14: 14.024866 u	

(b) Now calculate the binding energy per nucleon for the following isotopes of the five most massive naturally occurring elements. The masses provided are atomic masses.

radium-221: 221.01391 u	thorium-228: 228.028716 u	uranium-231: 231.036264 u
radium-223: 223.018499 u	thorium-229: 229.031757 u	uranium-232: 232.037131 u
radium-224: 224.020187 u	thorium-230: 230.033127 u	uranium-233: 233.039630 u
radium-226: 226.025402 u	thorium-231: 231.036299 u	uranium-234: 234.040946 u
radium-228: 228.031064 u	thorium-232: 232.038051 u	uranium-235: 235.043924 u
actinium-227: 227.027749 u	thorium-234: 234.043593 u	uranium-236: 236.045562 u
actinium-228: 228.031015 u	protactinium-231: 231.035880 u	uranium-238: 238.050784 u
thorium-227: 227.027701 u	protactinium-234: 234.043300 u	uranium-239: 239.054290 u

(c) Compare your results and comment on any patterns or trends that are obvious.

73. •• (a) What is the approximate radius of the ^{238}U nucleus? (b) What electric force do two protons on opposite ends of the ^{238}U nucleus exert on each other? (c) If the electric force in part (b) were the only force acting on the protons, what would be their acceleration just as they left the nucleus? (d) Why do the protons in part (b) not accelerate apart? Example 27-1

74. • The semi-empirical binding energy formula is given as follows:

$$E_B = (15.8 \text{ MeV})A - (17.8 \text{ MeV})A^{2/3}$$
$$- (0.71 \text{ MeV})\frac{Z(Z-1)}{A^{1/3}} - (23.7 \text{ MeV})\frac{(N-Z)^2}{A}$$

where A is the mass number, N is the number of neutrons, and Z is the number of protons. Using the formula, calculate the binding energy per nucleon for fermium-252. Compare your answer with the standard common expression for the binding energy: $E_B = (Nm_n + Zm_p - m_{whole})c^2$. Example 27-3

75. • The *fissionability parameter* is defined as the atomic number squared divided by the mass number for any given nucleus (Z^2/A). It can be shown that when this parameter is less than 44, a nucleus will be stable against small deformation; essentially, the nucleus will be stable against spontaneous fission. Calculate the value of this parameter for (a) ^{235}U, (b) ^{238}U, (c) ^{239}Pu, (d) ^{240}Pu, (e) ^{246}Cf, and (f) ^{254}Cf.

76. • The stable isotope of sodium is ^{23}Na. What kind of radioactivity would be expected from (a) ^{22}Na and (b) ^{24}Na?

77. • Estimate the number of nuclei that are present in a 50-kg human body.

78. •• In 2010 physicists first created element number 117 (tennessine, Ts) by colliding ^{48}Ca and ^{249}Bk nuclei. The result was two isotopes of the new element, one of which had a half-life of 14 ms and contained 176 neutrons. (a) What is the radius of the nucleus of the new element 117? (b) What percent of the newly created isotope was left 1.0 s after its creation? Example 27-5

79. •• Natural uranium is made up of two isotopes: ^{235}U and ^{238}U. The half-life of ^{235}U is 7.04×10^8 y, and the half-life of ^{238}U is 4.47×10^9 y. Assuming that all uranium isotopes were created simultaneously and in equal amounts at the same time that Earth was formed, estimate the age of Earth. The current percent abundance of ^{235}U is 0.72% and for ^{238}U it is 99.28%. Example 27-7

80. •• **Astronomy** The atom technetium (Tc) has no stable isotopes, yet its spectral lines have been detected in red giant stars (stars at the end of their lifetimes). Tc can be produced artificially on Earth. Its longest-lived isotope, ^{98}Tc, has a half-life of 4.2 million years. (a) If any ^{98}Tc were present when Earth formed 4.5×10^9 y ago, what percentage of it is still present? (Careful! You cannot do this calculation with your calculator. You must use logarithms to express the answer in scientific notation.) (b) What percent of the original ^{98}Tc would be present in a red giant that is 10 billion years old? (Careful again! You'll need to use logarithms.) (c) Explain why the detection of technetium in old stars is strong evidence that stars manufacture the more massive atoms in the universe. Example 27-5

81. • An old wooden bowl unearthed in an archeological dig is found to have one-fourth of the amount of carbon-14 present in a similar sample of fresh wood. Determine the age of the bowl. Example 27-7

82. • In an attempt to determine the age of the cave paintings in Chauvet-Pont-d'Arc Cave in France, scientists used carbon-14 dating to measure the age of bones of bears found in the cave. The bears are depicted in the paintings, so presumably the bones are approximately the same age as the paintings. The results showed that the level of ^{14}C was reduced to 2.35% of its present-day level. How old were the bones (and presumably the paintings)? Example 27-7

83. • In one common type of household smoke detector, the radioactive isotope americium-241 decays by alpha emission. The alpha particles produce a small electrical current because they are charged. If smoke enters the detector, it blocks the alpha particles, which reduces the current and causes the alarm to go off. The half-life of ^{241}Am is 433 y, and its atomic weight is 241 g/mol. Typical decay rates in smoke detectors are 690 Bq. (a) Write the alpha decay reaction of ^{241}Am and identify the daughter nucleus. (b) By how much does the alpha particle current decrease in 1.0 y due to the decay of the americium? How much in 50 y? (c) How many grams of ^{241}Am are there in a typical smoke detector? Example 27-5

84. • In March 2011 a giant tsunami struck the Fukushima nuclear reactor in Japan, resulting in very large radiation leaks, including leaks of cesium-137. The isotope has a 30-y half-life and is a beta-minus emitter. (a) What daughter nucleus is left after cesium-137 decays? (b) How long after the release will it take for the decay rate of the cesium-137 to be reduced by 99%? Example 27-5

85. •• Three isotopes of aluminum are given in the following table:

Isotope	Atomic mass (u)	E_B/nucleon	Decay process
^{26}Al	25.986892		
^{27}Al	26.981538		stable
^{28}Al	27.981910		

Calculate the binding energy per nucleon for each isotope and make a prediction of the decay processes for the unstable isotopes aluminum-26 and aluminum-28. Example 27-3

86. •• **Biology** In February 2010 it was announced that water containing the carcinogen tritium (^{3}H) was leaking from aging pipes at 27 U.S. nuclear reactors. In one well in Vermont, contaminated water registered 70,500 pCi/L; the federal safety limit was 20,000 pCi/L. Tritium is a β^- emitter with a half-life of 12.3 y. (a) How many protons and neutrons does the tritium nucleus contain? (b) Write out the decay equation for tritium and identify the daughter nucleus. (c) If the leak at the Vermont site is stopped, how long will it take for the water in the contaminated well to reach the federal safety level? Example 27-5

87. • **Medical** Iodine-125 is used to treat, among other things, brain tumors and prostate cancer. It decays by gamma decay with a half-life of 59.4 days. Patients who fly soon after receiving ^{125}I implants are given medical statements from the hospital verifying such treatment because their radiation could set off radiation detectors at airports. If the initial decay rate was

525 μCi, (a) what will the rate be at the end of the first year, and (b) how many months after the treatment will the decay rate be reduced by 90%? Example 27-5

88. •• **Medical** Ruthenium-106 is used to treat melanoma in the eye. This isotope decays by β^- emission with a half-life of 373.59 days. One source of the isotope is reprocessed nuclear reactor fuel. (a) How many protons and neutrons does the ^{106}Ru nucleus contain? (b) Could we expect to find significant amounts of ^{106}Ru in ore mined from the ground? Why or why not? (c) Write the decay equation for ^{106}Ru and identify the daughter nucleus. (d) How many years after ^{106}Ru is implanted in the eye does it take for its decay rate to be reduced by 75%? Example 27-5

89. •• **Medical** You are asked to prepare a sample of ruthenium-106 for a radiation treatment. Its half-life is 373.59 days, it is a beta emitter, its atomic weight is 106/g/mol, and its density at room temperature is 12.45 g/cm^3. (a) How many grams will you need to prepare a sample having an activity rate of 125 μCi? (b) If the sample in part (a) is a spherical droplet, what will be its radius? Example 27-5

90. •• Electron capture by an isolated proton is *not* allowed in nature. Explain why. Specifically, describe why the following nuclear reaction does *not* occur (and for good reason!): Example 27-3

$$e^- + p \not\to n + \nu_e$$

91. ••• Taking into account the recoil (kinetic energy) of the daughter nucleus, calculate the kinetic energy of the alpha particle in the following decay of a ^{235}U nucleus at rest: Example 27-6

$$^{235}U \to \alpha + {}^{231}Th$$

92. • A friend suggests that the world's energy problems could be solved if only physicists were to pursue the fusion of *heavy* nuclei rather than the fusion of *light* nuclei. To prove his point he suggests that the following fusion reaction should be considered: $^{157}Nd + {}^{80}Ge \to {}^{235}U + 2n$

Using the insights that you have acquired in this chapter, show that his argument is flawed. The atomic mass of neodymium-157 is 156.939032 u, and the mass of germanium-80 is 79.925373 u.

NASA, ESA, and The Hubble Heritage Team (STScI/AURA). Acknowledgment: P. McCullough (STScI)

28

Particle Physics and Beyond

In this chapter, your goals are to:

- (28-1) Explain why physicists examine both the smallest and largest objects in the universe.
- (28-2) Describe the difference between hadrons and leptons, and explain what hadrons are made of.
- (28-3) Explain how fundamental particles interact with each other, and the differences among the four fundamental forces.
- (28-4) Calculate the distance to a remote galaxy from its recessional velocity.

To master this chapter, you should review:

- (7-4)   How the speed of an object in a gravitational orbit depends on the strength of the gravitational force (distance from the massive central object).
- (13-3)   The wave function of a sinusoidal wave.
- (13-10)   How the Doppler effect describes the shift in frequency of a sound wave coming from a moving object.
- (15-3)   How an ideal gas cools when it undergoes an adiabatic expansion.
- (22-4)   The energy of a photon.
- (25-7)   The equivalence of mass and energy.
- (26-2)   How the frequency spectrum of blackbody radiation depends on temperature.
- (26-5)   Atomic spectra and Rutherford's discovery of the nucleus.
- (26-7)   The Heisenberg uncertainty principle for momentum and position.
- (27-6)   Half-lives and the process of beta decay.

What do you think?

This Hubble Space Telescope image shows numerous stars, a background of excited hydrogen gas glowing red, and dark interstellar clouds containing dust particles. All of these are made of normal matter: atoms and their components: electrons, protons, and neutrons. Of all the matter in the universe, what fraction is normal matter? (a) 100%; (b) between 90 and 99%; (c) between 50 and 89%; (d) between 20 and 49%; (e) less than 20%.

28-1 Studying the ultimate constituents of matter helps reveal the nature of the universe

"What is the world made of?" "Where did I come from?" You probably asked questions like these when you were a small child. Physicists and astronomers ask these questions throughout their professional lives, in search of ever more sophisticated answers about the nature and origin of the physical universe. In this final chapter we'll take a brief look at our present understanding of **particle physics,** the branch of physics that concerns the fundamental constituents of matter and how they interact.

1171

Figure 28-1 Particle physics and cosmology Ordinary objects and common technology connect us to (a) particle physics, the study of fundamental particles and their interactions, and (b) cosmology, the study of the nature and evolution of the universe.

Potatoes are a good source of potassium. A small fraction of the potassium is radioactive ^{40}K, which can decay by emitting a positron—a bit of antimatter. This decay also involves a quark, a neutrino, and an exchange particle called a W$^+$.

(a)

Bombaert Patrick/Alamy

When a television is disconnected from the source, about 1% of the "static" on the screen is from cosmic background radiation— the afterglow of the Big Bang at the beginning of time.

(b)

Evan Sharboneau/Hemera/Thinkstock/Getty Images

TAKE-HOME MESSAGE FOR Section 28-1

✔ Ordinary matter such as atoms is composed of a small variety of different fundamental particles.

✔ To learn about the nature and evolution of our universe, scientists study galaxies and the radiation left over from the early universe.

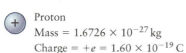

Electron
Mass = 9.1094×10^{-27} kg
Charge = $-e = -1.60 \times 10^{-19}$ C

Proton
Mass = 1.6726×10^{-27} kg
Charge = $+e = 1.60 \times 10^{-19}$ C

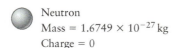

Neutron
Mass = 1.6749×10^{-27} kg
Charge = 0

Figure 28-2 The first three subatomic particles to be discovered All atoms are combinations of electrons (discovered 1897), protons (discovered 1917), and neutrons (discovered 1932). While the electron is a truly fundamental particle, the proton and neutron are not: They are composed of other, more fundamental particles called quarks and gluons.

We'll begin by looking at the fundamental particles that make up all of the ordinary matter that you see around you, including the matter that makes up your body. These include *leptons*, of which the electron is the best-known example, and *quarks*, which have the curious property that their charges are a fraction of the fundamental charge e. We'll see that these fundamental particles interact with each other by exchanging particles back and forth, a process that actually involves violating the law of conservation of energy (but in a way that's nonetheless compatible with the laws of physics). These interactions help explain the strong forces that hold the nucleus together, the electric forces that keep electrons in the atom, and the weak forces that cause radioactive beta decay (Figure 28-1a).

We'll conclude the chapter by redirecting our attention from the smallest particles to *cosmology*, the study of the nature and evolution of our universe as a whole. By studying galaxies, some of the largest structures in the universe, we'll learn that our universe is expanding. We'll also find that the universe is filled with electromagnetic radiation that is left over from the first few hundred thousand years of the history of the universe (Figure 28-1b). These clues are evidence that our universe began in a state of tremendously high temperature and immense density some 13.8 billion years ago. Finally, we'll learn that most of the *matter* in the universe is in a form whose nature is almost a complete mystery and that most of the *energy* in the universe is even more mysterious.

28-2 Most forms of matter can be explained by just a handful of fundamental particles

By the late nineteenth century scientists had concluded that all matter was composed of atoms and that atoms could not be subdivided into more fundamental particles. It was thought that a hydrogen atom was intrinsically different from a carbon atom, which in turn was intrinsically different from an iron atom, and so on.

As we learned in Section 26-5, this conclusion was quite incorrect. In 1897, J. J. Thomson discovered the electron, a particle of charge $-e = -1.602 \times 10^{-19}$ C, which turned out to be a constituent of the atoms of every element. In 1909, Ernest Rutherford discovered the atomic nucleus, and in 1917 he found evidence that all nuclei contain a positively charged particle of charge $+e$ that is identical to a hydrogen nucleus—that is, what we now call a proton. In 1932, the English physicist James Chadwick discovered the neutron, which has zero charge. As we learned in Chapter 27, all nuclei are composed of protons and neutrons (referred to collectively as nucleons), and all atoms are made of nuclei plus electrons. So by 1932 it seemed that these three subatomic particles—electron, proton, and neutron (Figure 28-2)—were the truly fundamental building blocks of matter.

That conclusion, too, turned out to be wildly incorrect. Since 1932 literally hundreds of other subatomic particles have been discovered. Almost all of these are unstable and decay to other particles with a radioactive half-life of a fraction of a second. (In this aspect they resemble the neutron: A free neutron that is not incorporated into a nucleus undergoes beta decay with a half-life of about 15 minutes.) But none of these additional particles can be regarded as simple combinations of protons, neutrons, and electrons. They include the neutrino, which has no charge, interacts hardly at all with other particles, and has a mass so close to zero that it has yet to be accurately measured; the muon, which resembles an electron in almost every way except that it is 207 times more massive; the pion, which like the proton and neutron experiences the strong nuclear force but has only about one-seventh the mass of a proton; and the delta, which resembles a proton or neutron but is about 30% more massive and comes in four varieties, with charges $+2e$, $+e$, 0, and $-e$. The discovery of these additional particles forced physicists to once again ask the question: What *are* the fundamental building blocks of matter?

Hadrons and Quarks

To answer this question, it's useful to distinguish between particles that experience the strong nuclear force, including the proton and neutron, and those that do not, such as the electron. Since protons and neutrons are relatively heavy and electrons are relatively light, we use the term **hadrons** (from the Greek word for stout or thick) for particles that experience the strong force and the term **leptons** (from the Greek word for small or delicate) for those that do not. (The photon is considered to be in a special category of its own, to which we will return later.) The vast majority of new particles discovered since 1932 are hadrons, so we'll look at these first.

In 1964, the American physicists Murray Gell-Mann and George Zweig independently proposed that all hadrons are made of more fundamental entities that Gell-Mann whimsically named **quarks**. The first evidence that quarks really exist came from experiments carried out at the Stanford Linear Accelerator Center, or SLAC, in 1967 (**Figure 28-3a**). These experiments were the same in principle as Rutherford's 1909 experiment that led to the discovery of the atomic nucleus. As we saw in Section 26-5, Rutherford aimed a beam of alpha particles at a target of gold. Had the charge inside the atom been distributed more or less uniformly, the alpha particles would have undergone only gentle deflections as they passed through the gold atoms. Instead, Rutherford found that some alpha particles were scattered by very large angles. This was evidence that charge was highly concentrated into a very small object within the atom—namely, the nucleus. The experiment at SLAC used electrons instead of alpha particles and aimed these electrons at a target of protons (the nuclei of hydrogen atoms inside a tank of liquid hydrogen). The 3.2-km-long accelerator gave each electron a tremendous kinetic energy—some 20,000 MeV, compared to the 7 MeV of Rutherford's alpha particles—and a correspondingly large momentum. This was done so that the electron would have a de Broglie wavelength much smaller than the size of the proton and so would be sensitive to fine details of the proton's internal structure as it passed through the proton.

Much as in the Rutherford experiment six decades before, many physicists expected that the electrons would undergo only small-angle deflections because they thought the charge inside the proton was distributed uniformly over its volume. And just as in the Rutherford experiment, what they found was that a substantial number of electrons were scattered by very large angles (**Figure 28-3b**). The conclusion was that the proton's charge is carried by smaller entities inside the proton, which are the quarks.

These experiments and a host of others confirm that Gell-Mann and Zweig were correct and that quarks are the fundamental building blocks of all hadrons. To explain all of the hundreds of hadron varieties currently known, we need six varieties or *flavors* of quarks. These are known as the up (u), down (d), charm (c), strange (s), top (t), and bottom (b) quarks. **Table 28-1** lists the six quarks, along with their masses and charges. The quarks are divided into three groups, or *generations*, of two quarks: u and d in the first generation, c and s in the second generation, and t and b in the third generation. As Table 28-1 shows, the quarks in each generation are more massive than those in the previous generation. Physicists usually write the quark generations as

$$\begin{pmatrix} u \\ d \end{pmatrix} \begin{pmatrix} c \\ s \end{pmatrix} \begin{pmatrix} t \\ b \end{pmatrix}$$

The quarks along the top row (u, c, and t) all have the same charge, as do the quarks along the bottom row (d, s, and b). However, all six quarks differ not only in

(a)

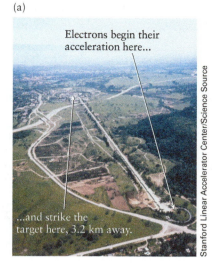

Electrons begin their acceleration here...

...and strike the target here, 3.2 km away.

(b)

When high-energy electrons scatter from protons, a substantial number scatter backward.

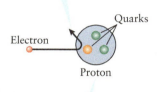

Quarks

Electron

Proton

This is evidence that there are small charged objects (quarks) inside the proton.

Figure 28-3 Discovering quarks (a) In a seminal series of experiments at the Stanford Linear Accelerator Center, electrons were accelerated to a kinetic energy of 20 GeV = 2×10^4 MeV and fired into a target containing protons. (b) Measuring how the electrons scattered from the protons provided evidence of the existence of quarks.

TABLE 28-1 The Six Quarks

Quark	Symbol	Charge	Approximate mass
up	u	$+\frac{2}{3}e$	$2.16\ \mathrm{MeV}/c^2$
down	d	$-\frac{1}{3}e$	$4.67\ \mathrm{MeV}/c^2$
charm	c	$+\frac{2}{3}e$	$1.27\ \mathrm{GeV}/c^2$
strange	s	$-\frac{1}{3}e$	$93\ \mathrm{MeV}/c^2$
top	t	$+\frac{2}{3}e$	$173\ \mathrm{GeV}/c^2$
bottom	b	$-\frac{1}{3}e$	$4.18\ \mathrm{GeV}/c^2$

Note: $1\ \mathrm{GeV}/c^2 = 10^3\ \mathrm{MeV}/c^2 = 10^9\ \mathrm{eV}/c^2$.

(a) Proton (p)

Net charge $= q_u + q_u + q_d$

$= \left(+\frac{2}{3}e\right) + \left(+\frac{2}{3}e\right) + \left(-\frac{1}{3}e\right) = +e$

(b) Neutron (n)

Net charge $= q_d + q_d + q_u$

$= \left(-\frac{1}{3}e\right) + \left(-\frac{1}{3}e\right) + \left(+\frac{2}{3}e\right) = 0$

(c) Delta-plus-plus (Δ^{++})

Net charge $= q_u + q_u + q_u$

$= \left(+\frac{2}{3}e\right) + \left(+\frac{2}{3}e\right) + \left(+\frac{2}{3}e\right) = +2e$

(d) Sigma-minus (Σ^-)

Net charge $= q_s + q_d + q_d$

$= \left(-\frac{1}{3}e\right) + \left(-\frac{1}{3}e\right) + \left(-\frac{1}{3}e\right) = -e$

Figure 28-4 **Baryons** All baryons, including (a) the proton and (b) the neutron, are made of combinations of three quarks. The proton is the only stable baryon; the neutron and all others, like (c) and (d), decay into other particles.

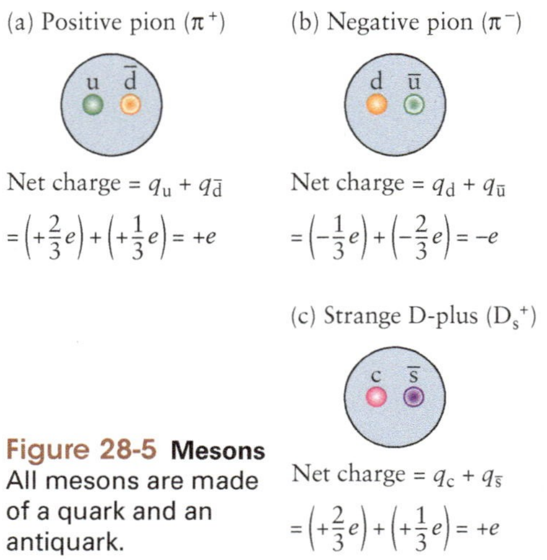

(a) Positive pion (π^+)

Net charge $= q_u + q_{\bar{d}}$

$= \left(+\frac{2}{3}e\right) + \left(+\frac{1}{3}e\right) = +e$

(b) Negative pion (π^-)

Net charge $= q_d + q_{\bar{u}}$

$= \left(-\frac{1}{3}e\right) + \left(-\frac{2}{3}e\right) = -e$

(c) Strange D-plus (D_s^+)

Net charge $= q_c + q_{\bar{s}}$

$= \left(+\frac{2}{3}e\right) + \left(+\frac{1}{3}e\right) = +e$

Figure 28-5 **Mesons** All mesons are made of a quark and an antiquark.

mass but also in other subtle properties. (These properties are beyond our scope in this brief introduction.) All attempts to find evidence of a fourth-generation quark have failed; as best we know, there are only three generations of quarks.

Note that each flavor of quark has a charge that is a *fraction* of e, either $+2e/3$ or $-e/3$. The explanation is that the proton, neutron, and other related particles are actually combinations of three quarks. **Figure 28-4** shows four examples of such combinations. Any hadron that can be made up of three quarks is called a **baryon**. Baryons always have a net charge that is an integer multiple of e. Two examples are the least massive baryons, the proton (**Figure 28-4a**) of net charge $+e$ and the neutron (**Figure 28-4b**) with net charge zero. This picture explains why the neutron, which has zero net charge, nonetheless produces a magnetic field of its own: The u quark and two d quarks within the neutron are each charged, and the motions of these charged particles within the neutron generate a magnetic field.

Note that in Table 28-1 we list only an *approximate* mass for each quark. We know quite precise values of the masses of other particles; for example, the mass of the proton is known to seven significant figures. By contrast, we know only rough values for the masses of the quarks. That's because, unlike protons, neutrons, or electrons, quarks are never found as solitary particles. Instead, they are found only in combinations like those shown in Figure 28-4. This situation is quite unlike that in atoms, in which electrons can be removed by ionization, or in nuclei, in which protons or neutrons can be removed from a nucleus in a sufficiently energetic collision. By contrast, it seems to be impossible to remove an individual quark from a baryon. We'll say more about this phenomenon of **quark confinement** a little later in this section.

Note also that the mass of a baryon is generally much greater than the sum of the masses of its constituent quarks. For example, a proton is composed of two u quarks and a d quark, which from Table 28-1 have a combined mass of approximately $2(2.16\ \mathrm{MeV}/c^2) + 4.67\ \mathrm{MeV}/c^2 = 8.99\ \mathrm{MeV}/c^2$. Yet the mass of the proton is $938.3\ \mathrm{MeV}/c^2$. The difference is associated with the forces that bind the quarks together and keep them confined within the proton. We'll explore these forces in the following section.

Other hadrons are made up of a quark and an **antiquark**, the **antimatter** version of a quark. For every type of matter particle there is an antimatter partner that is identical to it in every way except that it is oppositely charged. We've already encountered one example of antimatter, the positron (e^+) produced in β^+ decay (Section 27-6); the positron and electron have the same mass and are identical except for the sign of their charges ($-e$ for the electron, $+e$ for the positron). Each of the six quarks has an antiquark associated with it. Antiquarks are signified by adding "bar" to the name or placing a bar over the quark's symbol. For example, the antiquark associated with the up quark (u) is the up-bar or $\bar{u}$. The $\bar{u}$ antiquark has charge $-2e/3$, the opposite of the $+2e/3$ charge of the u quark. In the same way, the $\bar{d}$ antiquark carries charge $+e/3$, the opposite of the $-e/3$ charge of the d quark.

Figure 28-5 shows three examples of hadrons made up of a quark and an antiquark. Such quark–antiquark combinations are called **mesons**. The charge of a meson is always an integer multiple of e. Pions are the least massive mesons; the π^+ (**Figure 28-5a**) and the π^- (**Figure 28-5b**) both have a mass of $139.6\ \mathrm{MeV}/c^2$. There is also a neutral pion (π^0), which is made up of a combination of $u\bar{u}$ and $d\bar{d}$ and has a slightly lower mass of $135.0\ \mathrm{MeV}/c^2$. All mesons, including pions, are unstable; they decay with a half-life that is a small fraction of a second.

Just as three quarks make up a baryon, three antiquarks make up an *antibaryon*. For each variety of baryon, there is a corresponding variety of antibaryon. For example, the proton has quark content uud and charge $(+2e/3) + (+2e/3) + (-e/3) = +e$; the corresponding antibaryon is the antiproton, which has quark content $\bar{u}\bar{u}\bar{d}$ and charge $(-2e/3) + (-2e/3) + (+e/3) = -e$.

① What happens if we attempt to remove a quark (in this case, a u quark) from a hadron (in this case, a proton)?

② In our attempt we exert a force over a distance, which means we do work on the u quark and so add energy to it.

③ The force holding the u quark inside the proton does not decrease with distance, so we cannot remove this quark. Instead, the energy we add goes into creating a new quark–antiquark pair (in this case, a d quark and a d̄ antiquark).

④ Rather than removing a quark from the original hadron, the net result is that we end up with *two* hadrons (in this case, a positive pion and a neutron).

Proton (p)

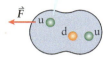

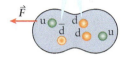

Positive pion (π⁺)   Neutron (n)

Figure 28-6 **Quark confinement** Attempting to remove a quark from a hadron just results in producing additional hadrons.

When a particle and its antimatter partner collide, their total mass can be converted to its equivalent energy. For example, the collision of an electron and a positron results in two or more photons that carry the total energy of the two particles, a process known as *annihilation*. A quark and an antiquark can also annihilate each other; we'll see examples of these processes in the next section. In the reverse process, energy can be converted into a particle and its corresponding antiparticle. This helps us understand what happens when we try to remove a single quark from a hadron such as a proton (Figure 28-6). Experiment shows that unlike the electric force holding an electron inside an atom or the strong nuclear force holding a proton or neutron inside a nucleus, the force that holds a quark inside a hadron does *not* decrease with increasing distance but instead remains constant. So it would take an infinite amount of work (force times distance) to remove a quark from a hadron, which is why we saw that quarks are *confined* inside hadrons. If we do enough work on a quark while trying to remove it, the energy added by this work can produce a new quark–antiquark pair. These combine with the original quarks so that we end up with two hadrons, as Figure 28-6 shows.

In practice the way that physicists experiment with adding energy to quarks is by colliding hadrons with each other; the kinetic energy of the collision goes into producing new hadrons as in Figure 28-6. Recall Figure 25-16 in Section 25-7, which shows the result of a collision between two protons at the Large Hadron Collider at CERN in Switzerland. The kinetic energy in this collision was so great that multiple baryons, antibaryons, and mesons were produced; the paths of these after the collision are shown by the many yellow tracks in Figure 25-16.

Leptons

While all hadrons are composed of quarks or antiquarks, there are other particles that are not composed of quarks at all. The *leptons* are another category of particles that are as fundamental as the quarks; to the best of our knowledge, leptons are not made up of smaller constituents. The electron is a lepton, as is the muon that we encountered in Section 25-4 (see Example 25-3). The electron neutrino that is created in hydrogen fusion (Section 27-5) and in beta decay (Section 27-6) is also a lepton. There are six leptons, listed in Table 28-2, each of which has an antimatter partner.

Notice that in Table 28-2 we have listed the leptons in three groups of two. As is the case for the quarks, leptons form three generations, which we usually write as

$$\begin{pmatrix} e^- \\ \nu_e \end{pmatrix} \begin{pmatrix} \mu^- \\ \nu_\mu \end{pmatrix} \begin{pmatrix} \tau^- \\ \nu_\tau \end{pmatrix}$$

As for the quarks listed in Table 28-1, the leptons in each successive generation are more massive than their counterparts in the preceding generation. In many ways, the muon and tau are more massive versions of the electron, so they share many properties and interact with other particles in similar ways. One way in which electrons, muons, and tau particles *are*

TABLE 28-2 **The Six Leptons**			
Lepton	**Symbol**	**Charge**	**Mass**
electron	e^-	$-e$	$0.5110 \text{ MeV}/c^2$
electron neutrino	ν_e	0	$<1.1 \text{ eV}/c^2$
muon	μ^-	$-e$	$105.7 \text{ MeV}/c^2$
muon neutrino	ν_μ	0	$<0.19 \text{ MeV}/c^2$
tau	τ^-	$-e$	$1.777 \text{ GeV}/c^2$
tau neutrino	ν_τ	0	$<18.2 \text{ MeV}/c^2$

significantly different (in addition to the differences in mass) is that while the electron is stable and does not decay, the other two are unstable: The muon has a half-life of 1.56 μs (see Example 25-3), and the tau has a half-life of about 2.0×10^{-13} s.

Each lepton also has an antimatter particle associated with it. We have already encountered the electron and positron. In the same way, the antimuon (μ^+) and antitau (τ^+) have the same masses as the muon and tau, respectively, but the μ^+ and τ^+ have positive charge $+e$ rather than negative charge $-e$. Like the neutrinos ν_e, ν_μ, and ν_τ, the antineutrinos $\bar{\nu}_e$, $\bar{\nu}_\mu$, and $\bar{\nu}_\tau$ have zero charge; they differ from the neutrinos in another way that we'll describe below.

Notice that in Table 28-2 we've given only upper limits for the masses of the three types of neutrino. That's because neutrinos have zero charge, so the methods physicists normally employ to determine the masses of particles (which involve measuring how a particle responds to electric and magnetic forces) can't be used. The indirect methods used for determining neutrino masses tell us that each of the three neutrinos has a small but nonzero mass; how small is not yet known.

Unlike quarks, which appear in groups of three to form baryons or in quark–antiquark combinations to form mesons, leptons do not group together to form other particles. In the following section we'll see the reason for this. We'll also discover an essential third class of particles in addition to hadrons and leptons, the *exchange particles*.

Conservation Laws for Hadrons and Leptons

In earlier chapters we encountered a number of important conservation laws, including the conservation of energy and the conservation of momentum. There are several additional conservation laws that govern the behavior of hadrons and leptons.

- *Baryon number is conserved.* Every baryon (composed of three quarks) is assigned *baryon number B* equal to +1, and every antibaryon is assigned B equal to −1. Mesons are assigned $B = 0$, as are leptons. Experiments show that in every process that involves baryons, the sum of the values of B for all particles present before the process equals the sum of the values of B for all particles present after the process. Consider, for example, the β^- decay of a neutron:

$$n \rightarrow p + e^- + \bar{\nu}_e$$
$$B = 1 \quad 1 \quad 0 \quad 0$$

(28-1)

The neutron and proton are both baryons, so each has $B = +1$. The electron and the antineutrino are leptons, each with $B = 0$. The total baryon number equals 1 before the decay and equals $1 + 0 + 0 = 1$ after the decay, so baryon number is conserved. Quarks have a fractional baryon number: Every quark has $B = +1/3$, and every antiquark has $B = -1/3$. That's consistent with a baryon with three quarks having $B = 3(+1/3) = +1$ and a meson with a quark and an antiquark having $B = (+1/3) + (-1/3) = 0$.

- *Lepton number is conserved.* Every electron (e^-) and electron neutrino (ν_e) is assigned *electron-lepton number* $L_e = +1$, and every positron (e^+) and electron antineutrino ($\bar{\nu}_e$) is assigned L_e equal to −1. Muons, muon neutrinos, tau particles, and tau neutrinos each have a similarly defined *muon-lepton number* L_μ or a *tau-lepton number* L_τ. A particle that is not a lepton is assigned L_e, L_μ, and L_τ equal to zero. Experiment shows that each of these lepton numbers is separately conserved.

As an example, consider the β^- decay of a neutron from Equation 28-1:

$$n \rightarrow p + e^- + \bar{\nu}_e$$
$$L_e = 0 \quad 0 \quad 1 \quad -1$$

(28-2)

Both the neutron and the proton are baryons, so each has $L_e = 0$; the electron has $L_e = +1$, and the electron antineutrino has $L_e = -1$. The total value of L_e before the decay is zero, and afterward it is $0 + 1 + (-1) = 0$. The electron-lepton number is conserved in β^- decay. A second example is the decay of a tau into an electron, an electron antineutrino, and a tau neutrino:

$$\tau^- \rightarrow e^- + \bar{\nu}_e + \nu_\tau$$
$$L_e = 0 \quad 1 \quad -1 \quad 0$$
$$L_\tau = 1 \quad 0 \quad 0 \quad 1$$

(28-3)

The tau (τ^-) and tau neutrino (ν_τ) each have $L_e = 0$ and $L_\tau = 1$, the electron (e^-) has $L_e = 1$ and $L_\tau = 0$, and the electron antineutrino ($\overline{\nu}_e$) has $L_e = -1$ and $L_\tau = 0$. You can see that the total electron-lepton number L_e equals 0 before and after the decay, and the total tau-lepton number L_τ equals 1 before and after the decay. Both of these lepton numbers are separately conserved in the decay of the tau.

Physicists have searched for processes in which either baryon number or one of the lepton numbers is not conserved. No such process has ever been observed. It appears that these four laws—conservation of baryon number, electron-lepton number, muon-lepton number, and tau-lepton number—are as universal and fundamental as the law of conservation of electric charge (which says that the net electric charge has the same value before and after any process).

GOT THE CONCEPT? 28-1 Particles That May or May Not Exist

Which of the following particles could possibly exist? (a) A meson with charge $-2e$; (b) a baryon with charge $-2e$; (c) an antibaryon with charge $-2e$; (d) more than one of these; (e) none of these.

GOT THE CONCEPT? 28-2 Processes That May or May Not Happen

Electric charge is conserved in each of these processes, but some are never observed to occur. Which *can* occur? (a) $n + p \rightarrow n + p + p + \overline{p}$ ($\overline{p}$ is an antiproton); (b) $n + p \rightarrow n + n + p$; (c) $\pi^- \rightarrow \mu^- + \overline{\nu}_e$; (d) more than one of these; (e) none of these.

TAKE-HOME MESSAGE FOR Section 28-2

✔   Quarks are the constituents of hadrons, the particles that experience the strong nuclear force. There are six varieties of quarks, each of which has a charge that is a fraction of e.

✔   Baryons such as the proton are made up of three quarks. Mesons are made up of a quark and an antiquark. Quarks are never found in isolation but are always confined inside hadrons.

✔   Leptons, which include electrons and neutrinos, have no constituent particles. They do not experience the strong force.

✔   In the interactions of subatomic particles, baryon number and the three lepton numbers are conserved.

28-3   Four fundamental forces describe all interactions between material objects

Since early in our study of physics we have seen the importance of *forces*, the pushes and pulls that one object exerts on another. We've encountered three fundamental kinds of forces: the gravitational force, the electromagnetic force, and the strong nuclear force. *Gravitation* is an attractive force that draws objects closer together. The gravitational force attracts you to Earth and keeps Earth in orbit around the Sun. Electric and magnetic forces, two manifestations of the *electromagnetic force*, cause charged objects to accelerate. As a result, electrons can be bound to atomic nuclei, and atoms can bond together. Note that all contact forces, including the normal force between two objects that touch or the force of friction, are electromagnetic in nature; they arise from the electromagnetic interactions between the atoms of the two surfaces in contact. As we explored in Chapter 27, the *strong nuclear force* binds protons and neutrons together to form atomic nuclei. To this list we will add a force that we have not yet named but the effects of which we encountered in Section 27-6; this **weak force** is at the heart of the interaction that governs beta decay.

Now that we are exploring the fundamental constituents of matter, we are in a position to ask a central question about force: How do objects exert forces on each other? And what is fundamentally different about the four different kinds of forces: gravitational, electromagnetic, strong, and weak? If we can understand how fundamental particles such as quarks and leptons exert forces of different kinds on each other, we will be closer to answering these questions.

As we will see, the manner in which fundamental particles exert forces on each other—that is, how they *interact*—comes from a remarkable aspect of nature: It is possible to *violate* the law of conservation of energy, provided we do it for a sufficiently short time. We can see this by returning to the *Heisenberg uncertainty principle*, which we introduced in Chapter 26.

The Heisenberg Uncertainty Principle Revisited

In Section 26-7 we learned that due to the wave nature of matter, there is a fundamental limitation on how well you can simultaneously know the momentum and position of a particle. This is the *Heisenberg uncertainty principle for momentum and position*:

Heisenberg uncertainty principle for momentum and position (26-31)

Uncertainty in the x component of momentum of a particle

$\hbar = \dfrac{h}{2\pi}$

= Planck's constant divided by 2π

$$\Delta p_x \Delta x \geq \frac{\hbar}{2}$$

Uncertainty in the x coordinate of the particle

We deduced this from the properties of waves. To make a wave that propagates in the x direction and that has a small spatial extent—equivalent to a small uncertainty Δx in the x coordinate of the wave—we have to combine sinusoidal waves with a range of wavelengths λ and hence a range of angular wave numbers $k = 2\pi/\lambda$ (see Figure 26-21). In quantum mechanics the x component of momentum of a particle is given by $p_x = h/\lambda = \hbar k$. So a particle—represented by a wave with a small spatial extent—will necessarily include a range of values of momentum p_x. Because its momentum does not have a single unique value, the particle has a momentum uncertainty Δp_x. Equation 26-31 says that the smaller the spatial extent given by Δx, the greater the range of angular wave numbers and hence the greater the momentum uncertainty Δp_x.

There is a similar relationship between the *duration in time* of a wave and the range of *frequencies* that the wave includes. To see this, recall from Section 13-3 the expression for a sinusoidal wave propagating in the $+x$ direction:

Wave function for a sinusoidal wave propagating in the $+x$ direction (13-6)

Wave function for a sinusoidal wave propagating in the $+x$ direction

Angular wave number of the wave = $2\pi/\lambda$

$$y(x,t) = A \cos (kx - \omega t + \phi)$$

Phase angle

Amplitude of the wave

Angular frequency of the wave = $2\pi f$

In Equation 13-6 the angular wave number of the wave is $k = 2\pi/\lambda$ and the angular frequency of the wave is $\omega = 2\pi f$, where f is the wave frequency. Note that the function in Equation 13-6 is valid for any value of x from $-\infty$ to $+\infty$ *and* for any time t from $-\infty$ to $+\infty$. So a wave with a single definite wavelength and a single frequency has an infinite spatial extent *and* an infinite duration. This means that the wave has always been present and will always be present. A more realistic description of a wave is one that has a finite duration, for example, the wave produced when you turn a source of waves (like a laser pointer) on and then off again.

Note that for waves of all kinds, including the quantum-mechanical waves that describe particles, changing the wavelength also changes the frequency. Note also in Equation 13-6 that k and x appear together in the same way as do ω and t. So if we combine sinusoidal waves of infinite extent and infinite duration with a range of values of k to create a wave with a small *spatial extent* (that is, one that is present over only a limited region of space), that wave will also have a range of values of ω and will have a short *duration* (that is, one that is present for only a limited length of time). Hence a wave of short duration necessarily includes a range of frequencies—that is, its frequency does not have a single definite value, but necessarily has some uncertainty.

To see how the frequency uncertainty of a wave is related to its duration in time, recall from Section 26-7 this relationship between the angular wave number uncertainty Δk and the spatial extent Δx (that is, the uncertainty in position):

$$\Delta k \Delta x \geq \frac{1}{2}$$

(26-29)

Since in Equation 13-6 k and x appear in the same combination as angular frequency ω and time t, it follows that the relationship between the angular frequency uncertainty $\Delta\omega$ and the time duration Δt has the same form as Equation 26-29:

$$\Delta\omega \Delta t \geq \frac{1}{2}$$

(28-4)

Equation 28-4 says that the product of $\Delta\omega$, the range of angular frequencies present in the wave, and the duration Δt of the wave cannot be less than 1/2. (This number arises from the specific way in which $\Delta\omega$ and Δt are defined mathematically.) It says that to minimize the duration of a wave necessarily means increasing the range of frequencies that make up the wave. So the shorter the duration of a wave, the less precisely we can answer the question "What is the frequency of the wave?"

How does Equation 28-4 make it possible to violate energy conservation? To see the answer, let's consider the photon, the particle associated with electromagnetic waves. We learned in Section 22-4 that the energy of a photon is proportional to its frequency:

Energy of a photon   Wave frequency

$$E = hf$$

Planck's constant = $6.62607015 \times 10^{-34}$ J $\cdot$ s

Energy of a photon
(22-26)

We can rewrite this in terms of the angular frequency ω of the photon. Since $\omega = 2\pi f$, it follows that $f = \omega/(2\pi)$ and the energy of a photon is

$$E = h\left(\frac{\omega}{2\pi}\right) = \left(\frac{h}{2\pi}\right)\omega = \hbar\omega$$

(28-5)

(Recall that $\hbar$ is Planck's constant divided by 2π: $\hbar = h/2\pi$.) Let's now take Equation 28-4 and multiply both sides of the equation by $\hbar$. We get

$$\hbar\Delta\omega\Delta t \geq \frac{\hbar}{2} \quad \text{or} \quad \Delta(\hbar\omega)\Delta t \geq \frac{\hbar}{2}$$

(28-6)

Here $\Delta(\hbar\omega)$ is the breadth of values of the quantity $\hbar\omega$ that must be included in the wave—in other words, it is the uncertainty in the value of $\hbar\omega$. But Equation 28-5 tells us that $E = \hbar\omega$ is the energy of a photon associated with the wave. So $\Delta(\hbar\omega)$ is the uncertainty in photon energy ΔE, and Equation 28-6 becomes

$$\Delta E \Delta t \geq \frac{\hbar}{2} \text{ for a photon}$$

(28-7)

This means that just as a wave of finite duration does not have a single definite frequency, a photon of finite duration does not have a single definite energy: It includes energy values that extend over a range of breadth ΔE. We can think of ΔE as the *uncertainty* in the energy of the photon. Equation 28-7 says that if a photon has a duration Δt, the product of Δt and the uncertainty in energy ΔE of the photon cannot be less than $\hbar/2$ and so ΔE cannot be less than $\hbar/(2\Delta t)$. This energy uncertainty is *not* a result of the limitations of an experimental apparatus that we might use to measure the energy of a photon. Rather, it is intrinsic to photons because of their wave nature.

We have used the equations $E = hf$ and $E = \hbar\omega$ to apply to photons only. But this relationship applies to particles of *all* kinds: We can think of anything with an energy E as having an associated angular frequency ω given by $E = \hbar\omega$. Then Equation 28-7 applies to phenomena in general, not just to photons. This implies that any physical phenomenon that has a finite duration Δt will necessarily have an uncertainty ΔE in energy given by Equation 28-7. The minimum value of this uncertainty is found by replacing the ≥ sign (greater than or equal to) in Equation 28-7 with an equals sign. The shorter the duration Δt of the phenomenon, the greater the minimum value of the energy uncertainty ΔE and the more uncertain the energy of the phenomenon. This is known as the **Heisenberg uncertainty principle for energy and time:**

Heisenberg uncertainty principle for energy and time (28-8)

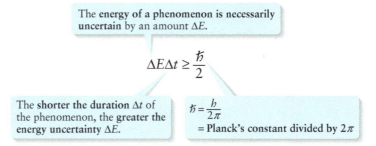

The **energy of a phenomenon is necessarily uncertain** by an amount ΔE.

$$\Delta E \Delta t \geq \frac{\hbar}{2}$$

The **shorter the duration** Δt of the phenomenon, the **greater the** energy uncertainty ΔE.

$\hbar = \dfrac{h}{2\pi}$
= Planck's constant divided by 2π

Another way to interpret Equation 28-8 is to say that it places limits on how precisely the law of conservation of energy must be obeyed. Suppose a system undergoes some kind of process that lasts for a time Δt. Equation 28-8 says that the energy of the system is necessarily uncertain during that process, and the minimum energy uncertainty is given by $\Delta E \Delta t = \hbar/2$ or $\Delta E = \hbar/(2\Delta t)$. So it's fundamentally impossible to measure the energy of the system with an uncertainty less than $\hbar/(2\Delta t)$. This means that during the process the energy E of the system could actually vary by as much as $\hbar/(2\Delta t)$, and there would be no way that we could tell that the energy had changed value—that is, that the energy was not conserved. The shorter the duration Δt of the process, the greater the amount $\hbar/(2\Delta t)$ by which energy conservation can be (temporarily) violated during that duration of time. Stated another way, it's acceptable to violate the law of conservation of energy by an amount ΔE, provided the duration of time during which the law is violated is no more than $\Delta t = \hbar/(2\Delta E)$.

Exchange Particles: The Electromagnetic Force

The Heisenberg uncertainty principle for energy and time helps us understand the following bold statement: *All forces result from the exchange of particles.* To see what we mean by this statement, first consider the electromagnetic force. Up to this point we've used the idea that charged particles exert electromagnetic forces on each other even at a distance, with no physical contact required. But a more sophisticated way to look at this force is to envision that when two charged particles exert an electric or magnetic force on each other, they do so by exchanging a photon: One of the charged particles emits the photon, and the other absorbs it. The exchanged photon has energy, and it violates the law of conservation of energy for this photon to spontaneously appear and be emitted by one of the charged particles. But as we have seen, it's perfectly acceptable to violate the law of conservation of energy by an amount ΔE, provided we do so for a time no longer than $\Delta t = \hbar/(2\Delta E)$. So the uncertainty principle proves that what we've described can take place provided the exchanged photon is absorbed by the second particle (and so disappears) within a time $\hbar/(2\Delta E)$ after the photon of energy ΔE was emitted by the first particle.

In this picture we say that the electromagnetic force is *mediated* by the exchange of a photon and that the exchanged photon is the *mediator* of the force. The particles that mediate forces are called **exchange particles.** You should try to envision a continuous stream of photons going back and forth between charged particles so that the two particles continuously exert an electromagnetic force on each other. Any charged particle can and does emit and absorb photons exchanged in this way. These photons are not the same as the photons emitted by a light bulb or a laser, however; they can exist for only a finite time before they must disappear. We call them **virtual particles.**

We can represent the exchange process by a diagram such as the one shown in Figure 28-7. In this slightly simplified version of a

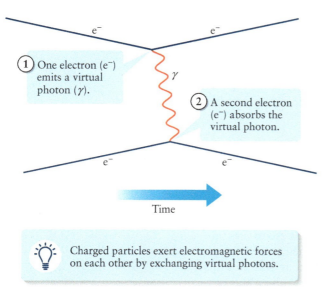

1. One electron (e^-) emits a virtual photon (γ).

2. A second electron (e^-) absorbs the virtual photon.

Time

Charged particles exert electromagnetic forces on each other by exchanging virtual photons.

Figure 28-7 Photon exchange and the electromagnetic interaction All electric and magnetic interactions between charged particles involve the exchange of virtual photons.

Feynman diagram, invented by American physicist Richard Feynman to visualize and analyze processes that involve fundamental particles, time runs from left to right. The lines associated with each particle do not represent actual paths that the particles take through space but only indicate which particles interact with which other particles. In Figure 28-7 two electrons exchange a photon and in so doing each exerts an electromagnetic force on the other. A particle such as an electron is drawn as a solid, straight line in Feynman diagrams. An exchange particle is drawn as either a wavy line (as for the photon) or a spiral line.

This picture helps us understand Coulomb's law, which tells us that the electric force between two charged particles separated by a distance r decreases in proportion to $1/r^2$. In other words, the electric force goes to zero only when the charges are infinitely far apart. This is possible because the photon has zero mass, so its minimum energy $E = hf$ is zero. (If the photon did have a mass m, its minimum energy would be its rest energy mc^2.) If one charged particle violates conservation of energy by creating and emitting a photon of energy ΔE, the uncertainty principle says that this photon can exist no longer than $\Delta t = \hbar/(2\Delta E)$. Even traveling at the speed of light c, this photon can travel no farther than $c\Delta t = c\hbar/(2\Delta E)$, so that distance is the maximum separation at which two charged particles can interact by exchanging photons of energy ΔE. But because the lower limit on the energy of a photon is zero, the amount ΔE by which the particle violates conservation of energy can be as small as we like. So the distance $r = c\Delta t = c\hbar/(2\Delta E)$ can be arbitrarily large, and two charged particles can interact via the electromagnetic force at any distance out to infinity. But because only low-energy photons (with small ΔE) can be exchanged between charged particles separated by great distances r, the force is quite weak if r is large—just as in Coulomb's law.

Exchange Particles: The Strong Force

We use a similar picture to explain the strong interaction between quarks. The Feynman diagram in Figure 28-8 shows two quarks that exert forces on each other by exchanging a particle called the **gluon**. This rather whimsical name expresses the idea that the exchange of gluons provides the force that confines, or glues, quarks inside baryons. Gluon exchange is also how the quark and antiquark inside a meson interact with each other and how the antiquarks inside an antibaryon interact.

Like the photon the gluon has zero mass, so the lower limit on the energy of a gluon is zero. Using the same argument we made above for photons, it follows that quarks or antiquarks separated by any distance, no matter how large, can interact by exchanging gluons. But unlike photons, the kinds of particles that emit and absorb gluons are not simply those with electric charge: They are particles that have a different attribute called *color*. (This is yet another of the light-hearted names associated with quark physics. It has nothing to do with the wavelength of light or the perception of color by the human eye.) Quarks and antiquarks carry color, as do gluons themselves. So unlike photons, gluons can interact with each other by exchanging other gluons. (By contrast, photons have no electric charge and cannot interact directly with other photons.) As a consequence of this curious aspect of gluons, the force that gluons mediate between quarks does *not* get weaker as the quarks move farther apart. This helps to explain why this force makes it impossible to remove an isolated quark from a hadron.

As we mentioned in Section 28-2, the mass of a nucleon (a proton or neutron) is much greater than the sum of the masses of the three quarks that constitute the nucleon. The gluon picture helps us understand why this is: Part of the additional mass is associated (through the relationship $E = mc^2$) with the energy of the virtual gluons that are continuously exchanged between the constituent quarks. The remaining mass is associated with the kinetic energies of the constituent quarks, which are in high-speed motion inside the nucleon.

Since quarks are charged, they also interact electromagnetically by exchanging photons. But these interactions have a relatively small effect compared to the dominant strong interaction mediated by gluons.

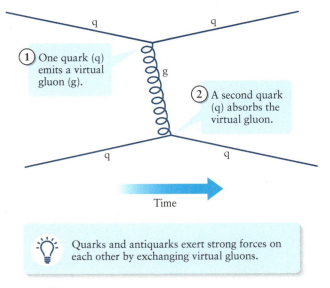

1. One quark (q) emits a virtual gluon (g).

2. A second quark (q) absorbs the virtual gluon.

Time

Quarks and antiquarks exert strong forces on each other by exchanging virtual gluons.

Figure 28-8 Gluon exchange and the strong interaction between quarks The forces that bind quarks together inside baryons and mesons are the result of the exchange of virtual gluons.

(a)

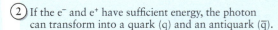

① In electron–positron annihilation, an electron (e^-) and a positron (e^+) collide and are transformed into a virtual photon.

② If the e^- and e^+ have sufficient energy, the photon can transform into a quark (q) and an antiquark ($\bar{q}$).

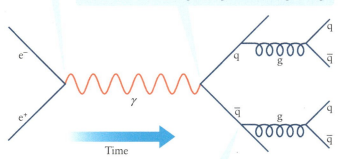

③ The quark and antiquark can emit gluons, which transform into quark–antiquark pairs. The shower of quarks and antiquarks coalesces into hadrons, including baryons (three quarks), antibaryons (three antiquarks), and/or mesons (quark–antiquark pairs).

(b)

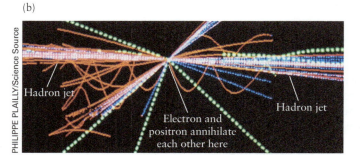

PHILIPPE PLAILLY/Science Source

Figure 28-9 Electron–positron annihilation and hadron production (a) When an electron and a positron collide, the result can be a shower of hadrons produced via virtual photons and gluons. (b) This image shows the result of a high-energy electron–positron collision at CERN in Geneva, Switzerland. Two jets of hadrons are produced along with other particles.

As we mentioned in the previous section, when an electron and a positron (matter and antimatter) collide, they annihilate each other. The Feynman diagram in Figure 28-9a shows one possible outcome of such an annihilation. The electron of charge $-e$ and the positron of charge $+e$ disappear and are replaced by a single photon. This *virtual* photon does not have the proper relationship between energy and momentum that a real photon must have, so it violates the conservation laws that govern the electron–positron collision. Like an exchange photon, this virtual photon can exist for only a finite time before it must disappear. In some cases the virtual photon will transform back into an electron–positron pair. But as Figure 28-9a shows, it's also possible for the virtual photon to become a quark and an antiquark of the same flavor (for instance, a u quark and a $\bar{u}$ antiquark, or an s quark and an $\bar{s}$ antiquark). The quark and antiquark can themselves emit virtual gluons, which can in turn transform into quark–antiquark pairs. Depending on how much energy is available in the original collision between electron and positron, many such quark–antiquark pairs can be produced. These will sort themselves into combinations of quarks (baryons), combinations of antiquarks (antibaryons), and quark–antiquark pairs (mesons). The result will be a shower of hadrons emanating from the site of the electron–positron collision. Figure 28-9b shows "jets" of hadrons emerging from the site of just such a collision. Many varieties of hadrons were first identified in electron–positron collision experiments of this sort.

The notion of gluon exchange also helps us understand the strong force in the context in which we first encountered it in Section 27-2: a force that attracts nucleons (protons or neutrons) to each other. Figure 28-10 shows how this force between nucleons arises. The meson that is produced in this way acts as an exchange particle between the nucleons, and this exchange gives rise to the attractive force between nucleons. The meson exchange particle has a substantial mass m_{exchange}, so the minimum energy that an exchanged meson can have is its rest energy $m_{\text{exchange}}c^2$. Producing such a meson means violating the conservation of energy by an amount of at least $\Delta E = m_{\text{exchange}}c^2$, which means that the meson can exist for a time no longer than $\Delta t = \hbar/(2\Delta E) = \hbar/(2m_{\text{exchange}}c^2)$. Even moving at the speed of light, the maximum distance that this exchange meson can travel during a time $\Delta t = \hbar/(2m_{\text{exchange}}c^2)$ is

$$(28\text{-}9) \qquad c\Delta t = \frac{c\hbar}{2m_{\text{exchange}}c^2} = \frac{\hbar}{2m_{\text{exchange}}c}$$

(range of a force mediated by an exchange particle of mass m_{exchange})

Equation 28-9 tells us the *range* of the force mediated by the exchange meson. There are many types of mesons, but the type whose exchange will have the longest range is the one with the smallest mass. (That's because the range in Equation 28-9 is inversely proportional to the mass of the exchange particle.) The mass of the lightest meson, the neutral pion (π^0), is 135.0 MeV/$c^2 = 135.0 \times 10^6$ eV/c^2, so the maximum range of the strong force between nucleons should be

$$\frac{\hbar}{2m_{\text{neutral pion}}c} = \frac{\hbar c}{2m_{\text{neutral pion}}c^2} = \frac{hc}{4\pi m_{\text{neutral pion}}c^2}$$
$$= \frac{(4.136 \times 10^{-15} \text{ eV}\cdot\text{s})(3.00 \times 10^8 \text{ m/s})}{4\pi(135.0 \times 10^6 \text{ eV}/c^2)c^2}$$
$$= 7.31 \times 10^{-16} \text{ m} = 0.731 \text{ fm}$$

Figure 28-10 The strong force between nucleons The force that binds protons and neutrons together has its origin in the interaction between quarks and gluons.

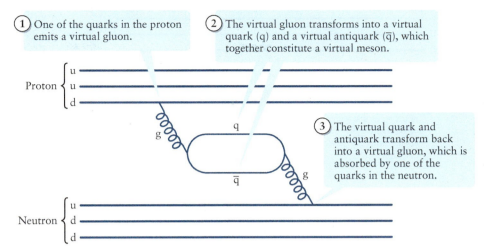

① One of the quarks in the proton emits a virtual gluon.

② The virtual gluon transforms into a virtual quark (q) and a virtual antiquark ($\bar{q}$), which together constitute a virtual meson.

③ The virtual quark and antiquark transform back into a virtual gluon, which is absorbed by one of the quarks in the neutron.

- Nucleons (protons and neutrons) exert the strong force on each other by exchanging virtual mesons.
- This exchange is actually due to interactions between quarks and gluons.

Because this range is so small, the strong force between nucleons is a short-range force, just as we discussed in Section 27-2. Indeed the rough value of 0.731 fm that we calculated here is of the same order of magnitude as the value we gave in Section 27-2 for the range of the strong force between nucleons.

Exchange Particles: The Weak Force and the Gravitational Force

The weak force is mediated by *three* different exchange particles. These are the neutral Z^0 (charge zero) and the positively and negatively charged W particles, W^+ (charge $+e$) and W^- (charge $-e$). These particles are not massless: The Z^0 has mass 91.2 GeV/c^2, about 100 times the mass of a proton, and the W^+ and W^- both have mass 80.4 GeV/c^2. The calculation above shows that the range of the force mediated by an exchange particle of mass $m_{exchange}$ is inversely proportional to the mass. So since the Z^0, W^+, and W^- have roughly 600 times the mass of the neutral pion (135.0 MeV/c^2), the range of the weak force is roughly 1/600 that of the strong force between nucleons, or about 10^{-18} m $= 10^{-3}$ fm. The weak force is not only weak, it is of *extremely* short range!

It turns out that any two particles can exert a weak force on each other. For instance, Figure 28-11 shows two examples of the weak interaction between a neutrino and an electron. Note that even though the weak interaction does not directly involve electric charge, charge is still conserved. In Figure 28-11a the neutrino remains neutral, and the electron retains its charge of $-e$. In Figure 28-11b the neutrino emits a W^+ of charge $+e$ and becomes an electron of charge $-e$, so charge is conserved. The W^+ combines with an electron of charge $-e$ and becomes a neutrino of charge zero, and again charge is conserved.

Figure 28-12 shows how the weak force gives rise to the β^- decay of a neutron (quark content udd) to a proton (quark content uud):

$$n \rightarrow p + e^- + \bar{\nu}_e$$

Compared to a u quark, a d quark is substantially more massive (see Table 28-1) and so has a substantially greater rest energy. So the d quark can lower its energy by becoming a u quark. It does this by emitting a W^- particle, as shown. This is a very short-lived virtual particle: The rest energy of the W^- (80.4 GeV/c^2) is far greater than the rest energy of the original d quark (about 4.67 MeV/c^2), so emitting a W^- means

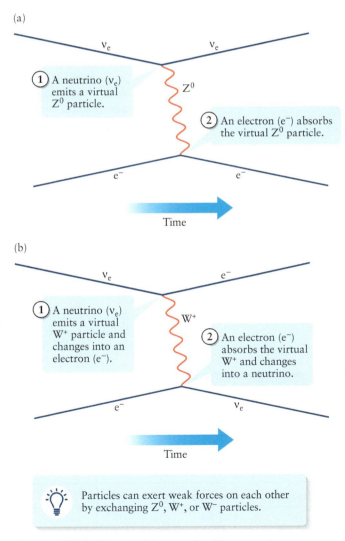

(a)

ν_e ν_e

① A neutrino (ν_e) emits a virtual Z^0 particle.

Z^0

② An electron (e^-) absorbs the virtual Z^0 particle.

e^- e^-

Time

(b)

ν_e e^-

① A neutrino (ν_e) emits a virtual W^+ particle and changes into an electron (e^-).

W^+

② An electron (e^-) absorbs the virtual W^+ and changes into a neutrino.

e^- ν_e

Time

Particles can exert weak forces on each other by exchanging Z^0, W^+, or W^- particles.

Figure 28-11 The weak interaction The weak force between particles involves the exchange of massive virtual particles. These come in both (a) neutral and (b) charged varieties.

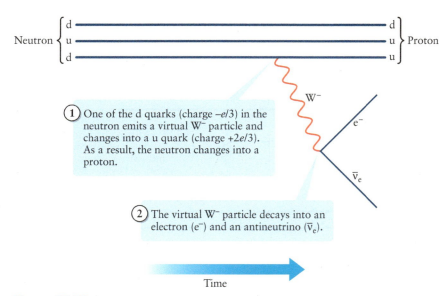

(1) One of the d quarks (charge $-e/3$) in the neutron emits a virtual W⁻ particle and changes into a u quark (charge $+2e/3$). As a result, the neutron changes into a proton.

(2) The virtual W⁻ particle decays into an electron (e⁻) and an antineutrino ($\bar{\nu}_e$).

Time

Figure 28-12 Beta decay of a neutron The weak interaction describes how a free neutron decays into a proton, an electron, and an antineutrino.

violating energy conservation by quite a bit. The W⁻ of charge $-e$ quickly decays into an electron of charge $-e$ and a neutral antineutrino $\bar{\nu}_e$. (It has to be an *anti*neutrino to conserve electron-lepton number L_e: The W⁻ is not a lepton and therefore has $L_e = 0$, so the W⁻ must decay into particles with a net value of L_e equal to zero. The electron, with $L_e = +1$, must therefore be accompanied by a neutral particle with $L_e = -1$, which means an antineutrino rather than a neutrino.)

As we discussed in Section 27-6, in some nuclei it's energetically favorable for β^+ decay to take place, in which a proton changes to a neutron. This is similar to the process shown in Figure 28-12, except that one of the u quarks in a proton changes to a d quark by emitting a W⁺ rather than a W⁻. The W⁺ then decays into a positron (e⁺) and a neutrino (ν_e). One nucleus in which this can happen is the potassium isotope ^{40}K, which is found in foods such as potatoes (see Figure 28-1a). The antimatter produced in this way—the positron—quickly encounters an atomic electron, and the positron and electron annihilate each other. Happily, this happens very infrequently in your food (and releases very little energy when it does happen). So you needn't worry about biting into a bit of antimatter when you eat a potato!

A particle named the *graviton* is thought to mediate the gravitational force. So far no experimental evidence has confirmed the existence of this particle, although no experimental evidence has excluded it either. Because the gravitational force between two masses separated by a distance r is proportional to $1/r^2$, just like the electric force between two charged particles, the graviton is thought to have the same mass as the photon: zero. The graviton also carries no charge.

Table 28-3 summarizes the four fundamental forces and the exchange particles that mediate these forces.

The Standard Model and the Higgs Particle

The picture that Table 28-3 summarizes—in which two classes of fundamental particles, the six quarks and the six leptons, interact by means of six exchange particles, the graviton, the Z^0, the W⁺, the W⁻, the photon, and the gluon—is called the **Standard Model**. What this very simple table does not reflect is the decades of experimental and theoretical effort (and several Nobel Prizes in Physics) that have gone into constructing this model. One long-standing question about the Standard Model that has recently been answered is this: Of the exchange particles listed in Table 28-3, why do the Z^0, W⁺, and W⁻ all have substantial masses (91.2, 80.4, and 80.4 GeV/c^2,

respectively), while the graviton, photon, and gluon each have zero mass? The answer proposed in the 1960s was that there is a field that fills all of space, now called the **Higgs field**, and the interactions of the Z^0, W⁺, and W⁻ particles with this field give these particles their masses. Unlike electric and magnetic fields, which are vectors with a magnitude and direction, the Higgs field is a scalar that has no direction. Because the Higgs field has the same nonzero value everywhere in space, the masses of the Z^0, W⁺, and W⁻ have the same values no matter where in space these particles are found.

This proposal may remind you of a debunked idea that we encountered in Section 25-3. In the nineteenth century, it was thought that there was a substance called the luminiferous ether that filled all space and that acted as a medium for the propagation of light waves.

TABLE 28-3 The Four Fundamental Forces

Force	Range	Mediator(s)	Strength relative to the strong force
gravity	infinite	graviton	10^{-40}
weak	~10^{-3} fm	Z^0, W⁺, W⁻	10^{-6}
electromagnetic	infinite	photon (γ)	10^{-2}
strong*	~1 fm	gluon (g)	1

*Gluons mediate the strong force between quarks; this force has infinite range. As described in the text, the strong force between *nucleons* (protons and neutrons) has a range of only about 1 fm because it involves not just gluons but also virtual meson exchange.

The Michelson–Morley experiment, which we also discussed in Section 25-3, demonstrated that no such luminiferous ether exists. To determine whether the Higgs field exists, twenty-first-century physicists used the idea that for each kind of field there is a corresponding particle. For example, the particle that corresponds to the electric and magnetic fields is the photon. The electric and magnetic forces are mediated by virtual photon exchange, and an electromagnetic wave is made up of photons. To explain the observed masses of the Z^0, W^+, and W^-, the **Higgs particle** that corresponds to the Higgs field would have to have zero charge, have a mass about 50% greater than that of the W^+ or W^-, and be unstable. If such a particle could be produced in a high-energy collision of other particles, it would suggest that the Higgs field does indeed exist.

In 2012, after decades of experimental effort, the Higgs particle was discovered in collisions between high-energy protons at the Large Hadron Collider in Switzerland (Figure 28-13). The observed Higgs particle mass of 125 GeV/c^2 is within the expected range of values, and as predicted the Higgs particle has zero charge and a half-life of about 10^{-22} s. This discovery provided strong experimental support that unlike the luminiferous ether, the Higgs field actually does exist throughout space. This field is now regarded as an essential part of the Standard Model.

The Higgs field is thought to explain not just the masses of the Z^0, W^+, and W^-, but also the masses of the leptons and the quarks. In this picture, the reason the muon has a greater mass than the electron is that muons interact more strongly with the Higgs field than electrons do. In future experiments, physicists will investigate how particles of various kinds interact with the Higgs *field* by measuring how these particles interact with Higgs *particles*. These and other aspects of the Standard Model are the subject of active research by physicists around the globe.

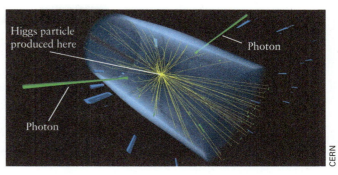

Figure 28-13 Detecting the Higgs particle A Higgs particle was produced in the collision shown here between high-energy protons at the Large Hadron Collider at CERN. The Higgs particle then decayed into two photons. By measuring the energy of these photons, physicists can determine the mass of the Higgs particle.

GOT THE CONCEPT? 28-3 A Quark Decay

A c quark can decay into an s quark via the weak force. In this process, what does the c quark emit? (a) A W^+, which decays into a positron and a neutrino; (b) a W^+, which decays into a positron and an antineutrino; (c) a W^-, which decays into an electron and a neutrino; (d) a W^-, which decays into an electron and an antineutrino.

TAKE-HOME MESSAGE FOR Section 28-3

✔ Particles exert forces on each other through the exchange of virtual particles that are emitted by one particle and absorbed by another.

✔ The photon is the exchange particle that mediates the electromagnetic force; the gluon mediates the strong force; and the Z^0, W^+, and W^- particles mediate the weak force.

The graviton, which has not yet been detected, is thought to mediate the gravitational force.

✓ The Standard Model is our picture of six types of quarks and six types of leptons that interact by means of these various exchange particles. A key part of the Standard Model is the Higgs field, a field that exists throughout all space and explains the large masses of the Z^0, W^+, and W^-.

28-4 We live in an expanding universe, and the nature of most of its contents is a mystery

The Standard Model describes all the normal matter in the world around us. All of the atoms in our bodies and our surroundings are made of leptons and hadrons, the hadrons are all made of quarks, and all of these particles interact by means of exchange particles. But why do we specify that this description applies to "normal" matter only? What other sort of matter could there be, and how much of the contents of the physical universe is so-called normal matter? To answer these questions, let's turn our attention from the smallest fundamental particles to galaxies, some of the largest structures in the universe, and to the nature of the universe itself—the subject of the science called **cosmology**.

WATCH OUT! A galaxy is not a solar system, nor is it the entire universe.

In everyday language, many people use the terms "galaxy," "solar system," and "universe" interchangeably. In fact, a galaxy is a large collection of stars and other matter, while a solar system is a single star and its retinue of planets. Our galaxy contains some 200 billion stars, many of which have planets, and the observable universe contains literally billions of galaxies.

Figure 28-14 A universe of galaxies Each blob of light in this image from the Hubble Space Telescope is a galaxy, a grouping of billions of stars and other matter. The area shown is a very small patch of sky, only about 1/15 the apparent width of the full Moon.

The Hubble Law and the Expansion of the Universe

Figure 28-14 is an image made with the Hubble Space Telescope of a small portion of the night sky. At first glance it may appear that the bright dots in this image are individual stars. But closer inspection shows that each is actually a **galaxy**, a collection of a tremendous number of stars (typically 10^{10} to 10^{12}). Our Sun is one of approximately 2×10^{11} stars in the Milky Way, the local galaxy of which we are part.

In the 1920s the American astronomer Edwin Hubble showed that galaxies are very distant. (He did this by identifying within other galaxies certain types of stars that are found in our own galaxy and whose light output is known. These stars appear very dim when seen in other galaxies, so they must be very far away.) In collaboration with the astronomer Milton Humason, he also measured the spectra of many galaxies. These are really the spectra of the combined light from all of the stars that make up each galaxy, so they have absorption lines (see Section 26-5). Hubble and Humason found that the absorption lines in the spectra of almost all galaxies are shifted to longer wavelengths and lower frequencies than in the spectra of nearby stars in our own galaxy. We call this a **redshift** because in the visible part of the electromagnetic spectrum, red light has the longest wavelength and lowest frequency. What's more, the greater the distance to a galaxy, the greater the wavelength shift. What did this discovery mean?

To answer this question, recall what we learned in Section 13-10 about the *Doppler effect*: If a source of waves is moving away from us, the wave that we receive from that source is shifted to a lower frequency and a longer wavelength. In Section 13-10 we introduced this idea in the context of sound waves, but the same effect also applies to electromagnetic waves. Hubble and Humason concluded that other galaxies are moving away from us, and so the light that we receive from those galaxies is at a longer wavelength. They found that the speed v at which a distant galaxy is receding from us is directly proportional to the distance d to that galaxy (Figure 28-15):

The Hubble law
(28-10)

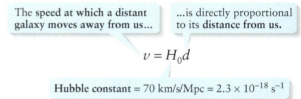

The speed at which a distant galaxy moves away from us... ...is directly proportional to its distance from us.

$$v = H_0 d$$

Hubble constant = 70 km/s/Mpc = 2.3×10^{-18} s^{-1}

This direct proportionality is called the **Hubble law**, and the constant H_0 in Equation 28-10 is called the **Hubble constant**. Astronomers typically measure distances in the cosmos in parsecs (pc), a unit equal to 3.26 light years. One light year (ly) is the distance that light travels in one year, equal to 3.09×10^{16} m. The current best value of H_0 is roughly 70 km/s/Mpc, where 1 Mpc = 1 megaparsec = 10^6 pc = 3.26×10^6 ly.

The interpretation of the Hubble law is that because remote galaxies are getting farther and farther apart as time goes on, the universe is expanding. Figure 28-16 shows how this explains the direct proportionality between the speed v at which a galaxy is moving away from us and the distance d to that galaxy. As the universe expands, the amount of space between widely separated galaxies increases. The expansion of the universe is actually the expansion of space.

The greater the distance to a galaxy, the more rapidly it is moving away from us. This is evidence that we live in an expanding universe.

Figure 28-15 The Hubble law Each black dot in this graph represents the speed at which an individual galaxy is moving away from us, as well as the distance to that galaxy. The straight line is the best fit to that data, corresponding to a value of H_0 = 70 km/s/Mpc in Equation 28-10.

WATCH OUT! The universe is expanding, but the galaxies (and you) are not.

It's important to realize that the expansion of the universe occurs primarily in the vast spaces that separate clusters of galaxies. The galaxies themselves do not expand. Einstein and others have established that an object that is held together by its own gravity, such as a galaxy or a cluster of galaxies, is always contained within a patch of nonexpanding space. A galaxy's gravitational field produces this nonexpanding region. Thus Earth and your body, for example, are not getting any bigger. Only the distance between widely separated galaxies increases with time.

Although distant galaxies are moving away from us, that does *not* mean that we are at the center of the universe. Figure 28-16 shows that in an expanding universe, every galaxy moves away from every other galaxy, so an alien astronomer in a distant galaxy would see the same relationship between the speeds and distances of galaxies—that is, the same Hubble law—as does an Earth astronomer. Since every point in the universe appears to be at the center of the expansion, it follows that our universe has no center at all.

WATCH OUT! "If the universe is expanding, what is it expanding into?"

It's common to think of the expanding universe as being like a balloon that grows in size as it's inflated. Such a balloon expands in three-dimensional space into the surrounding air. But the actual universe includes *all* space; there is nothing "beyond" it because there is no "beyond." Asking "What lies beyond the universe?" is as meaningless as asking "Where on Earth is north of the North Pole?"

The ongoing expansion of space explains why the light from remote galaxies is redshifted. Imagine a photon coming toward us from a distant galaxy. As the photon travels through space, the space is expanding, so the photon's wavelength becomes stretched (Figure 28-17). When the photon reaches our eyes, we see an increased wavelength: The photon has been redshifted. The greater the distance the photon has had to travel and so the longer the amount of time it has had to travel to reach us, the more its wavelength will have been stretched. Thus photons from distant galaxies have larger redshifts than those of photons from nearby galaxies, as expressed by the Hubble law.

Although we've used the idea of a Doppler shift, a redshift caused by the expansion of the universe is properly called a **cosmological redshift**. It is *not* the same as a Doppler shift. Doppler shifts are caused by an object's motion through space, whereas a cosmological redshift is caused by the expansion of space.

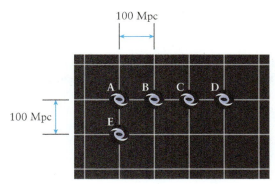

(a) Five galaxies spaced 100 Mpc apart.

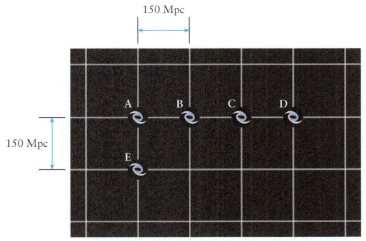

(b) The expansion of the universe spreads the galaxies apart.

Figure 28-16 An expanding universe (a) The distance between adjacent galaxies in this illustration is 100 megaparsecs (Mpc). (b) If the universe expands by 50%, the distance between adjacent galaxies increases to 150 Mpc. Galaxies A and B were 100 Mpc apart and have moved 50 Mpc farther apart; galaxies A and C were 200 Mpc apart and have moved 100 Mpc farther apart. So a galaxy that is at twice the distance *d* from galaxy A has moved twice the additional distance in the same amount of time and therefore is moving away at twice the speed *v*. This agrees with Equation 28-10, which says that *v* is directly proportional to *d*.

	Original distance from A (Mpc)	Later distance from A (Mpc)	Change in distance from A (Mpc)
A↔B	100	150	50
A↔C	200	300	100
A↔D	300	450	150
A↔E	100	150	50

Figure 28-17 Wavelengths expand in an expanding universe If you (a) draw a wave on a rubber band and then (b) stretch the band, the wavelength is also stretched. In the same way, as a photon travels through the space between galaxies, the photon's wavelength is stretched as the space expands.

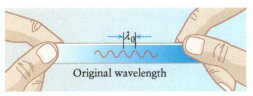

(a) A wave drawn on a rubber band...

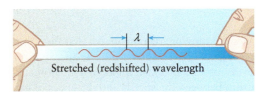

(b) ...increases in wavelength as the rubber band is stretched.

EXAMPLE 28-1 Measuring the Distance to a Galaxy from Its Redshift

Find the distances to the following galaxies: (a) NGC 4889, whose redshifted spectrum shows that it is moving away from us at 6410 km/s (2.14% of the speed of light); (b) 1255-0, which has a larger redshift that shows it to be moving away from us at 0.822c. Express the distances in megaparsecs and in light years.

Set Up

For each galaxy we know the speed v at which it is moving away from us due to the expansion of the universe. So we can use the Hubble law to determine the distance d to that galaxy.

The Hubble law:

$$v = H_0 d \qquad \text{(28-10)}$$

Solve

(a) Use Equation 28-10 and the value $H_0 = 70$ km/s/Mpc to calculate the distance to NGC 4889.

Rewrite Equation 28-10 to solve for the distance d:

$$d = \frac{v}{H_0}$$

We are given $v = 6410$ km/s, so

$$d = \frac{6410 \text{ km/s}}{70 \text{ km/s/Mpc}} = 92 \text{ Mpc}$$

$$= (92 \text{ Mpc})\left(\frac{3.26 \times 10^6 \text{ ly}}{1 \text{ Mpc}}\right) = 3.0 \times 10^8 \text{ ly}$$

(b) Repeat the calculations for the galaxy 1255-0, which is moving away at 0.822c.

For this galaxy:

$$v = 0.822c = 0.822(3.00 \times 10^5 \text{ km/s})$$
$$= 2.47 \times 10^5 \text{ km/s}$$

Use this to calculate the distance to the galaxy as in part (a):

$$d = \frac{2.47 \times 10^5 \text{ km/s}}{70 \text{ km/s/Mpc}} = 3.5 \times 10^3 \text{ Mpc}$$

$$= (3.5 \times 10^3 \text{ Mpc})\left(\frac{3.26 \times 10^6 \text{ ly}}{1 \text{ Mpc}}\right) = 1.1 \times 10^{10} \text{ ly}$$

Reflect

NGC 4889 is some 300 million light years away, so the light we see from NGC 4889 left that galaxy 300 million years ago, before even the first dinosaurs appeared on Earth. Galaxy 1255-0 is far more distant, some 11 *billion* light years away. When the light we receive from 1255-0 left that galaxy some 11 billion years ago, our Earth—which is a mere 4.56×10^9 years old—had not yet formed. When we use telescopes to look at distant astronomical objects, we are not only looking out into space; we are also looking back in time.

The Big Bang, Cosmic Background Radiation, and the Origin of Matter

If the universe is expanding, it must be that in the past the matter in the universe was closer together and therefore denser than it is today. If we look far enough into the very distant past, there must have been a time when the density of matter was almost inconceivably high. This leads us to conclude that some sort of tremendous event caused ultradense matter to begin the expansion that continues to the present day. This event, called the **Big Bang**, marks the beginning of the universe. If we use the observed expansion rate of the universe and work backward, we find that the age of the universe is approximately 13.8 billion (13.8×10^9) years.

We learned in Section 15-3 that the temperature of an ideal gas decreases when it undergoes an adiabatic expansion (one in which there is no heat transfer to its surroundings). The expansion of the universe is much like an adiabatic expansion, so it follows that the average temperature of the universe has decreased over the past 13.8 billion years. (There are places in the universe that are at very high temperature, such as the interiors of stars. But the average temperature of space is very cold.) If we work backward in time, it follows that the early universe must have been at very high temperatures indeed. The hot early universe must therefore have been filled with many high-energy photons of high frequency and short wavelength. The properties of this radiation field depended on its temperature, as described by Planck's blackbody law (Section 26-2).

The universe has expanded so much since those ancient times that all of the short-wavelength, high-energy photons left over from the early universe have had their wavelengths stretched by a factor of about 1100. (Think of stretching the rubber band in Figure 28-17 to 1100 times its original length!) As a result, they have become long-wavelength, low-energy photons. The temperature of this cosmic radiation field is now only a few degrees above absolute zero, and the blackbody spectrum of this radiation has its peak intensity at low frequencies in the microwave part of the electromagnetic spectrum (wavelengths of approximately 1 mm). Hence this radiation field, which fills all of space, is called the **cosmic microwave background** or **cosmic background radiation**. It represents the "afterglow" of the very high-temperature conditions that prevailed when the universe was young.

It's actually possible to detect the cosmic background radiation using an ordinary television (Figure 28-1b). Using much more sensitive detectors, scientists have found that the cosmic background radiation does indeed have a blackbody spectrum with a temperature of 2.725 K, which we can regard as the average temperature of the present-day universe. However, we observe slightly different cosmic background radiation coming from different parts of the sky. Even when the effects of Earth's motion are accounted for, there remain variations in the temperature of the radiation field of about 300 μK (300 microkelvins, or 3×10^{-4} K) above or below the average 2.725 K temperature (Figure 28-18). These tiny temperature variations indicate that matter and radiation were not distributed in a totally uniform way in the early universe. Regions that were slightly denser than average were also slightly cooler than average; less-dense regions were slightly warmer. Over time the denser regions evolved to form the first galaxies, and within them the first stars. By studying these non-uniformities we are really studying our origins.

The map of the sky shown in Figure 28-18 shows the universe as it was some 380,000 years after the Big Bang, when the universe was less than 0.003% of its present age. Prior to that time the universe was so dense as to be opaque. (Think of a hot, dense, luminous fog.) Events prior to that date are forever hidden from our direct view. But by using the laws of particle physics that we described in Sections 28-2 and 28-3, we can infer a great deal about what conditions must have been like in the very early universe.

Our best understanding is that the universe began at extremely high temperatures and that the immense energy gave rise to a sea of quarks, antiquarks, leptons, and antileptons. As the universe expanded and cooled, by 10^{-6} s after the Big Bang quarks and antiquarks had coalesced into baryons, antibaryons, and mesons. But now we have a dilemma. If there had been perfect symmetry between particles and antiparticles, then

WATCH OUT! It's not correct to think of the Big Bang as an explosion.

! When a bomb explodes, pieces of debris fly off into space from a central location. If you could trace all the pieces back to their origin, you could find out exactly where the bomb had been. This process is not possible with the universe, however, because the universe itself always has and always will consist of *all* space. The present-day universe is infinite and was infinite when it first originated in the Big Bang.

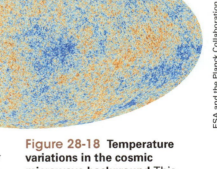

ESA and the Planck Collaboration

Figure 28-18 Temperature variations in the cosmic microwave background This map shows small variations in the temperature of the cosmic background radiation across the entire sky. Lower-temperature regions (shown in blue) show where the early universe was slightly denser than average.

for every proton there should have been an antiproton. For every electron, there should likewise have been a positron. By the time the universe was 1 second old, every particle would have been annihilated by an antiparticle, leaving no matter at all in the universe! Obviously, this did not happen. But why didn't it?

The resolution to this dilemma is that the laws of physics are very slightly asymmetrical between matter and antimatter: For every 10^9 antimatter particles that were created out of the energy of the Big Bang, 10^9 plus one matter particles were created. When those 10^9 antimatter particles encountered 10^9 matter particles, annihilations resulted in no more particles and lots of energy in the form of photons. (These are the photons that gave rise to the cosmic background radiation that we see today.) That one extra matter particle avoided annihilation. All the normal matter that we see today is a result of the very slight imbalance between matter and antimatter in the early universe. Experiments on particle interactions are consistent with this very slight asymmetry between matter and antimatter.

By 15 minutes after the Big Bang, temperatures had dropped enough that protons and neutrons could coalesce into the first atomic nuclei. Only the four lightest elements (hydrogen, helium, lithium, and beryllium) were present in appreciable numbers. The heavier elements would be formed only much later, once stars had formed and nuclear reactions within those stars could manufacture carbon, nitrogen, oxygen, and all the other elements.

Keep in mind that only *nuclei*, not atoms, formed in the first 15 minutes of the history of the universe. It would be another 380,000 years before temperatures became low enough for these nuclei to combine with electrons to form atoms for the first time. Free electrons are very efficient at scattering photons and thereby impeding the free passage of light. As a result, before the formation of atoms the universe remained opaque. But once atoms formed 380,000 years after the Big Bang and there were fewer free electrons present to scatter light, the universe became transparent for the first time. Had you been present in the universe at that time, you would have a seen a bright luminous fog dissipate around you to reveal a transparent sky filled with a thin gas of mostly hydrogen. (There would have been no starlight, however. The first stars did not form until much later, about 200 million years after the Big Bang.) The map of the cosmic background radiation in Figure 28-18 shows the universe at that moment in cosmic history.

Dark Matter and Dark Energy

At this point we might feel that we have understood, at least in broad outline, the nature and origin of matter. But have we in fact accounted for all of the matter in the universe? To answer this question we again look at galaxies, in particular *spiral galaxies* like the ones shown in Figure 28-19. These galaxies have a flattened disk shape, and all of the material in the galaxy rotates around the center. Each part of the galaxy is held in its orbit around the center by the gravitational attraction of the other parts of the galaxy; the greater that gravitational attraction on a given part, the faster that part must move to remain in a circular orbit (see Section 7-4). So if we can measure the speed at which parts of the galaxy orbit around the center, we can calculate the total amount of mass in the galaxy. In a galaxy, most of the *ordinary* matter—by which we mean the same protons, neutrons, and electrons of which all atoms are made—is in its stars. Furthermore, we can estimate the combined mass of all the stars from the brightness of the galaxy. If the total amount of mass of all kinds in the galaxy (measured from orbital motion in the galaxy) is a close match to the combined amount of mass of ordinary matter in its stars, then we can conclude that ordinary matter accounts for most or all of the mass of the galaxy.

We can measure orbital speeds most directly for spiral galaxies that happen to be oriented edge-on to us, like the one shown in Figure 28-19b. We do this by looking at the spectrum of light from the two sides of the galaxy: The spectral lines will be Doppler shifted toward shorter wavelengths on the side where material is rotating toward us, and toward longer wavelengths on the side where material is rotating away from us. (These shifts are in addition to the cosmological redshift of the galaxy's spectrum as a whole, caused by the expansion of the universe.) The amount of Doppler shift tells us the speeds at which material in various parts of the galaxy are moving around the center. The following example illustrates how this information can be used to determine the mass of the galaxy.

(a) We see spiral galaxy NGC 3982 face-on.

NASA, ESA, and the Hubble Heritage Team (STScI/AURA)

All of the matter in the galaxy rotates around the center.

(b) We see spiral galaxy NGC 4013 edge-on.

C. Howk (JHU), B. Savage (U. Wisconsin), N. A. Sharp (NOAO)/ WIYN/NOAO/NSF

Stars at one edge rotate toward us...

...and stars at the other edge rotate away from us.

Figure 28-19 Spiral galaxies
(a) The material in a spiral galaxy is held in orbit around the center by the gravitational attraction of all the other material in the galaxy. (b) For a galaxy that is edge-on to us, we can measure the orbital speeds, and hence the mass of the galaxy, using the Doppler effect.

EXAMPLE 28-2 Finding the Mass of a Galaxy

The large spiral galaxy UGC 2885 contains about 10^{12} stars, about five times more than our Milky Way galaxy. We see it nearly edge-on like the galaxy in Figure 28-19b, so astronomers can use the Doppler shift of light from various parts of UGC 2885 to determine how rapidly material is orbiting the center of this galaxy. The outermost visible part of this galaxy is a distance 72.0 kpc from the center (1 kpc = 1 kiloparsec = 10^3 pc) and orbits the center at 298 km/s. What is the mass of UGC 2885? Give your answer in kilograms and in solar masses (1 solar mass = mass of the Sun = 1.99×10^{30} kg).

Set Up

In Section 7-4 we found an equation for the speed v of a satellite orbiting Earth in a circular orbit of radius r. Since gravitation is universal, this same equation also applies to a star orbiting a galaxy if we replace the mass of Earth in Equation 7-12 with the mass of the galaxy m_{galaxy}. We are given $r = 72.0$ kpc and $v = 298$ km/s for galaxy UGC 2885; our goal is to solve for its mass, m_{galaxy}.

Speed of an Earth satellite in a circular orbit:

$$v = \sqrt{\frac{Gm_{\text{Earth}}}{r}} \qquad (7\text{-}12)$$

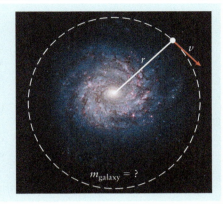

$m_{\text{galaxy}} = ?$

Solve

Replace m_{Earth} with m_{galaxy} in Equation 7-12 and solve for m_{galaxy}.

Rewrite Equation 7-12 for an object orbiting a galaxy:

$$v = \sqrt{\frac{Gm_{\text{galaxy}}}{r}}$$

To solve for m_{galaxy}, square both sides of this equation and rearrange:

$$v^2 = \frac{Gm_{\text{galaxy}}}{r}$$

$$m_{\text{galaxy}} = \frac{rv^2}{G}$$

To use this equation to solve for the mass of UGC 2885, first convert the value of r from pc to m and convert the value of v from km/s to m/s. From Chapter 7 the value of the gravitational constant is $G = 6.67 \times 10^{-11}$ N·m²/kg²; recall that 1 N = 1 kg·m/s².

The distance r is

$$r = 72.0 \text{ kpc}\left(\frac{10^3 \text{ pc}}{1 \text{ kpc}}\right)\left(\frac{3.26 \text{ ly}}{1 \text{ pc}}\right)\left(\frac{3.09 \times 10^{16} \text{ m}}{1 \text{ ly}}\right)$$

$$= 7.25 \times 10^{21} \text{ m}$$

The orbital speed v is

$$r = 298 \text{ km/s}\left(\frac{10^3 \text{ m}}{1 \text{ km}}\right) = 2.98 \times 10^5 \text{ m/s}$$

Substitute these into the expression for the mass of the galaxy:

$$m_{\text{galaxy}} = \frac{rv^2}{G} = \frac{(7.25 \times 10^{21} \text{ m})(2.98 \times 10^5 \text{ m/s})^2}{6.67 \times 10^{-11} \text{ N·m}^2/\text{kg}^2}$$

$$= 9.66 \times 10^{42} \text{ kg}^2\cdot\text{m/N·s}^2\left(\frac{1 \text{ N}}{1 \text{ kg·m/s}^2}\right)$$

$$= 9.66 \times 10^{42} \text{ kg}$$

$$= 9.66 \times 10^{42} \text{ kg}\left(\frac{1 \text{ solar mass}}{1.99 \times 10^{30} \text{ kg}}\right)$$

$$= 4.85 \times 10^{12} \text{ solar masses}$$

Reflect

The Sun is a rather average star. So if most of the mass of a galaxy is in its stars and there are 10^{12} stars in this galaxy, we would expect the mass of galaxy UGC 2885 to be about 10^{12} solar masses. But our result shows that the actual mass of this galaxy is about *five times greater* than what we would expect. These measurements have been made for a wide variety of spiral galaxies, and the results are almost always the same: The total mass of the galaxy deduced from the orbital speeds is about five times greater than the combined mass of all of the stars in the galaxy. (This discovery was made by the American astronomers Vera Rubin and Kent Ford in the 1960s and 1970s; UGC 2885 is named Rubin's Galaxy in her honor.)

The results of Example 28-2 show that ordinary matter, which is concentrated in a galaxy's stars, is only a *small fraction* of the mass of a galaxy. The remaining mass is in a form that does not emit electromagnetic radiation at any wavelength and may not even be composed of the fundamental particles that we described in Sections 28-2 and 28-3. This mysterious matter, which is the dominant form of matter in spiral galaxies and apparently in the universe as a whole, is called **dark matter**. As of this writing, its nature remains a mystery to science.

There is yet another complication to our understanding of the universe. As we look at increasingly distant galaxies, and so look farther back in the history of the universe, we find that as expected these distant galaxies are moving away from us due to the expansion of the universe. But the expansion rate has *not* been constant; observations show that for the past 5 billion years, the expansion of the universe has been *speeding up*! To date the only viable explanation for this increased speed is that, in addition to ordinary matter and dark matter, the universe is suffused with a curious form of energy that causes an accelerated expansion of the universe. We cannot detect this energy from its gravitational effects (the technique astronomers use to detect dark matter), and it does not emit detectable radiation of any kind. We refer to this curious energy as **dark energy**. The nature of dark energy is even more mysterious than that of dark matter.

WATCH OUT! Don't confuse antimatter, dark matter, and dark energy.

! The terms "antimatter," "dark matter," and "dark energy" may all seem equally mysterious, but they describe three very different things. *Antimatter* is the same as normal matter but with the opposite charge: An antihydrogen atom would be made of a negatively charged antiproton and a positively charged positron. *Dark matter* is thought by many scientists to be composed of some kind of fundamental particle that has zero charge and feels neither the strong force nor the electromagnetic force. What this particle is (or whether dark matter is made of something other than fundamental particles) is a topic of intense research. *Dark energy* is not a material substance at all, but appears to be some kind of field that permeates the universe, similar to (but different from) the Higgs field. It, too, is being studied intently by scientists.

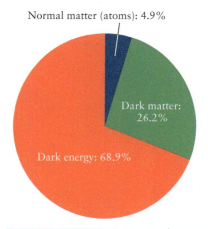

Normal matter (atoms): 4.9%

Dark matter: 26.2%

Dark energy: 68.9%

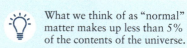

What we think of as "normal" matter makes up less than 5% of the contents of the universe.

Figure 28-20 Recipe for a universe This pie chart shows the three constituents of the universe as a whole. The numbers show what percentage of the total energy content of the universe each constituent represents.

Since matter and energy are equivalent through the Einstein relation $E = mc^2$, we can compare the relative importance of ordinary matter, dark matter, and dark energy by asking what fraction of the total energy in the universe each one represents. The graph in Figure 28-20 shows the results: Dark energy represents 68.9% of the total, dark matter another 26.2%, and "ordinary" normal matter just 4.9%. This is a sobering graph, for it tells us that more than 95% of the energy in the universe is in forms that we do not understand! The quest to understand both dark energy and dark matter is one of the most challenging pursuits in contemporary physics.

GOT THE CONCEPT? 28-4 Dark Matter

? In spiral galaxies like those shown in Figure 28-19, the ratio of dark matter to visible matter is about five to one. The percentage of the light from a spiral galaxy that is emitted by dark matter is closest to (a) 83%; (b) 80%; (c) 20%; (d) 17%; (e) zero.

TAKE-HOME MESSAGE FOR Section 28-4

✔ The redshifts of galaxies reveal that we live in an expanding universe.

✔ The universe began with a Big Bang, with tremendously high temperatures present in the early universe. The cosmic background radiation is the "afterglow" of that early epoch in the history of the universe.

✔ The dominance of matter over antimatter in the universe is the result of a small but important asymmetry in the laws of particle physics.

✔ Most of the matter in the universe is in the form of dark matter, and most of the energy in the universe is in the form of dark energy. The nature of dark matter and the nature of dark energy are unsolved problems in physics.

Key Terms

antimatter	exchange particle	lepton
antiquark	galaxy	meson
baryon	gluon	particle physics
Big Bang	hadron	quark
cosmic background radiation	Heisenberg uncertainty principle	quark confinement
cosmic microwave background	for energy and time	redshift
cosmological redshift	Higgs field	Standard Model
cosmology	Higgs particle	virtual particle
dark energy	Hubble constant	weak force
dark matter	Hubble law	

Chapter Summary

Topic	Equation or Figure	
Hadrons and leptons: Hadrons, including the proton and neutron, are particles that experience the strong force; leptons, including the electron and neutrino, are particles that do not. Hadrons are composed of more fundamental particles called quarks; these are confined to the interior of hadrons and cannot exist in isolation. Hadrons include baryons, which are made of three quarks, and mesons, which are made of a quark and an antiquark. Leptons are themselves fundamental particles; they are not made of anything simpler.	(a) Proton (p) (b) Neutron (n) Net charge $= q_u + q_u + q_d$ Net charge $= q_d + q_d + q_u$ $= \left(+\frac{2}{3}e\right) + \left(+\frac{2}{3}e\right) + \left(-\frac{1}{3}e\right) = +e$ $= \left(-\frac{1}{3}e\right) + \left(-\frac{1}{3}e\right) + \left(+\frac{2}{3}e\right) = 0$	(Figure 28-4)
	(a) Positive pion (π^+) (b) Negative pion (π^-) Net charge $= q_u + q_{\bar{d}}$ Net charge $= q_d + q_{\bar{u}}$ $= \left(+\frac{2}{3}e\right) + \left(+\frac{1}{3}e\right) = +e$ $= \left(-\frac{1}{3}e\right) + \left(-\frac{2}{3}e\right) = -e$	(Figure 28-5)
The four forces and exchange particles: All interactions between particles can be understood in terms of four basic forces. Each of these forces involves the exchange of virtual particles, which is permitted by the Heisenberg uncertainty principle. The strong force between quarks involves the exchange of gluons, and the electromagnetic force involves the exchange of photons; both gluons and photons have zero mass, so these are long-range forces. The weak force responsible for beta decay involves the exchange of massive particles, so this force has a very short range. The gravitational force involves the exchange of gravitons, which have not yet been detected experimentally.	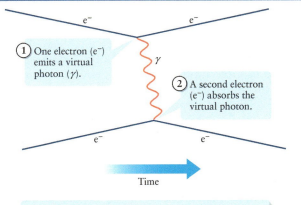 ① One electron (e^-) emits a virtual photon (γ). ② A second electron (e^-) absorbs the virtual photon. Time 💡 Charged particles exert electromagnetic forces on each other by exchanging virtual photons.	(Figure 28-7)
	The **energy of a phenomenon is necessarily uncertain** by an amount ΔE. $$\Delta E \Delta t \geq \frac{\hbar}{2}$$ The **shorter the duration** Δt of the phenomenon, the **greater the energy uncertainty** ΔE. $\hbar = \frac{h}{2\pi}$ = Planck's constant divided by 2π	(28-8)

Cosmology, dark matter, and dark energy: The Hubble law—the more distant a galaxy is from us, the faster it moves away from us—tells us that the universe is expanding and was once very highly compressed and at very high temperature. The cosmic background radiation is a relic of this ancient epoch. Processes in the early universe had a slight preference for matter over antimatter, which is why the present-day universe contains matter but almost no antimatter. Normal matter is actually just a small component of the universe: Mysterious dark matter is about five times as prevalent. Even more important and even more mysterious is dark energy, which causes the expansion of the universe to speed up.

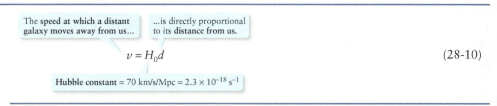

The speed at which a distant galaxy moves away from us... ...is directly proportional to its distance from us.

$$v = H_0 d \qquad (28\text{-}10)$$

Hubble constant = 70 km/s/Mpc = 2.3×10^{-18} s^{-1}

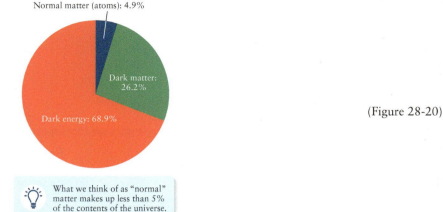

Normal matter (atoms): 4.9%

Dark matter: 26.2%

Dark energy: 68.9%

(Figure 28-20)

What we think of as "normal" matter makes up less than 5% of the contents of the universe.

Answer to What do you think? Question

(e) Most of the matter in the universe is dark matter, a mysterious substance that (as far as we know) does not emit, reflect, or absorb light of any kind. Dark matter interacts with other matter through the gravitational force (and, perhaps, the weak force). Its nature and properties remain a mystery. There is about five times as much dark matter as there is ordinary matter, so ordinary matter makes about one-sixth, or 16%, of the matter in the universe.

Answers to Got the Concept? Questions

28-1 (c) A quark has charge $+2e/3$ or $-e/3$, and an antiquark has charge $-2e/3$ or $+e/3$ (see Table 28-1). Mesons are made up of one quark and one antiquark, and no quark–antiquark combination has a net charge $-2e$. Baryons are made up of three quarks, and no combination of three quarks has a net charge $-2e$. Antibaryons are made up of three antiquarks; if each of the three has charge $-2e/3$, the total charge of the antibaryon will be $(-2e/3) + (-2e/3) + (-2e/3) = -2e$

28-2 (a) The neutron (n) and proton (p) each have baryon number $B = 1$, and the antiproton ($\bar{p}$) has $B = -1$. Baryon number is conserved in process (a), since total B equals $1 + 1 = 2$ before the process and equals $1 + 1 + 1 + (-1) = 2$ after the process. This process can occur if the initial neutron and proton have sufficient kinetic energy so that, when they collide, this energy can be converted into the rest energy of the additional proton and antiproton. Baryon number is *not* conserved in process (b), since total B equals $1 + 1 = 2$ before the process and $1 + 1 + 1 = 3$ after the process. So (b) cannot occur. In process (c) the π^- is a meson with $L_e = 0$ and $L_\mu = 0$, the muon μ^- has $L_e = 0$ and $L_\mu = 1$, and the electron antineutrino $\bar{\nu}_e$ has $L_e = -1$ and $L_\mu = 0$. The total electron-lepton number is 0 before the process and $0 + (-1) = -1$ after the process, and the total muon-lepton number is zero before the process and $1 + 0 = 1$ after the process. So neither electron-lepton number L_e nor muon-lepton number L_μ is conserved in process (c), and process (c) cannot occur.

28-3 (a) Table 28-1 shows that the c quark has charge $+2e/3$ and the s quark has a charge $-e/3$. The quark must therefore emit a W$^+$ with charge $+e$ so that the total charge of the W$^+$ and s quark equals the initial charge of the c quark: $(+e) + (-e/3) = +2e/3$. The W$^+$ must decay into a positron (charge $+e$) to conserve charge. Since the W$^+$ has electron-lepton number $L_e = 0$ (it's not a lepton) and the positron has $L_e = -1$, the positron must be accompanied by a neutral particle with $L_e = +1$. That must be a neutrino, not an antineutrino (which has $L_e = -1$).

28-4 (e) Dark matter is so called because it emits *no* electromagnetic radiation of any kind.

Questions and Problems

In a few problems you are given more data than you actually need; in a few other problems you are required to supply data from your general knowledge, outside sources, or informed estimate.

Interpret as significant all digits in numerical values that have trailing zeros and no decimal points.

•	Basic, single-concept problem
••	Intermediate-level problem; may require synthesis of concepts and multiple steps
•••	Challenging problem
Example	See worked example for a similar problem

Conceptual Questions

1. • Define these terms: (a) baryon, (b) meson, (c) quark, (d) lepton, and (e) antiparticle.

2. • A meson and a baryon come very close to one another. By which of the fundamental forces (gravitational, electromagnetic, weak, or strong) could these particles interact?

3. • What do exchange particles do?

4. • Discuss the similarities and differences between a photon and a gluon.

5. • Write the quark content for the antiparticle of each of the following: (a) n, (b) p, and (c) π^+.

6. • List all of the unique combinations of quarks resulting in a baryon that can be produced from just the up (u), down (d), and strange (s) quarks. (Neglect the antimatter baryons.)

7. • A positron is stable; that is, it does not decay. Why, then, does a positron have only a short existence?

8. • When a positron and an electron annihilate each other at rest, why must more than one photon be created?

9. • Explain how we know the universe is expanding.

10. • Explain the evidence in favor of the existence of (a) dark matter and (b) dark energy.

Multiple-Choice Questions

11. • A particle composed of a quark and an antiquark is classified as a
   A. baryon.   D. lepton.
   B. meson.   E. antiparticle.
   C. photon.

12. • A particle composed of three quarks is classified as a
   A. baryon.   D. lepton.
   B. meson.   E. antiparticle.
   C. photon.

13. • The quark composition of an antiproton is
   A. uud.   D. $\overline{u}\overline{u}\overline{d}$.
   B. uu$\overline{d}$.   E. uuu.
   C. $\overline{u}\overline{u}$d.

14. • The beta decay process is mediated by the
   A. gravitational force.
   B. electromagnetic force.
   C. strong force.
   D. weak force.
   E. more than one of A, B, C, and D.

15. • Through which force does the photon primarily interact with other particles?
   A. strong
   B. electromagnetic
   C. weak
   D. gravitational
   E. More than one of the above are equally important.

16. • Which of these particles interact using the strong force?
   A. neutrinos   D. quarks
   B. leptons   E. gravitons
   C. photons

17. • Recent observations indicate that most of the energy in the universe is in the form of
   A. protons, neutrons, and electrons.
   B. dark matter.
   C. dark energy.
   D. photons.
   E. none of the above.

18. • The accelerating expansion of the universe is thought to be driven by
   A. gravitational potential energy.
   B. electric potential energy.
   C. thermal energy.
   D. dark energy.
   E. kinetic energy.

Problems

28-1 Studying the ultimate constituents of matter helps reveal the nature of the universe

28-2 Most forms of matter can be explained by just a handful of fundamental particles

19. • Is the reaction $n \rightarrow \pi^+ + \pi^- + \mu^+ + \mu^-$ possible? Explain why or why not.

20. • Is the reaction $p \rightarrow e^+ + \gamma$ possible? Explain why or why not.

21. • Is the reaction $e^- + p \rightarrow n + \overline{\nu}_e$ possible? Explain why or why not.

22. • What is the charge of the particle that is composed of (a) the quark combination uds? (b) the quark combination uss? (c) Which particle is likely to be more massive? Explain your answers.

23. • Two protons collide in a particle accelerator and generate new particles from their kinetic energy. Which of the following reactions are possible? Which are not possible? If a reaction is not possible, state which conservation law(s) is/are violated.
   A. $p + p \rightarrow p + p + p + \overline{p}$
   B. $p + p \rightarrow p + p + n + \overline{n}$
   C. $p + p \rightarrow p + K^+$

24. • Which of the following two possibilities for the weak decay of a sigma particle (a baryon) is/are possible? Why?
   A. $\Sigma^- \rightarrow \pi^- + p$
   B. $\Sigma^- \rightarrow \pi^- + n$

25. • Which of the following reactions are possible? If a reaction is not possible, tell which conservation law(s) is/are violated.
   A. $n \rightarrow p + e^- + \overline{\nu}_e$
   B. $\mu^- \rightarrow e^- + \overline{\nu}_e + \nu_\mu$
   C. $\pi^- \rightarrow \mu^- + \overline{\nu}_\mu$

26. •• A neutral η (eta) meson at rest decays into two photons:

$$\eta \rightarrow \gamma + \gamma$$

Calculate the energy, momentum, and wavelength for each of the identical photons. The mass of the eta particle is 547 MeV/c^2.

27. •• Two protons with the same speed collide head-on in a particle accelerator, causing the reaction:

$$p + p \rightarrow p + p + \pi^0$$

Calculate the minimum kinetic energy of each of the incident protons.

28. •• A high-energy photon in the vicinity of a nucleus can create an electron–positron pair:

$$\gamma \rightarrow e^- + e^+$$

(a) What minimum energy must the photon have? (b) Why is the nucleus needed?

29. •• How much energy, in the form of gamma rays, would result due to the annihilation of a positron with kinetic energy 34 MeV and an electron with kinetic energy 16 MeV?

30. •• A proton–antiproton annihilation takes place, leaving two photons with a combined energy of 2.5 GeV. Find the kinetic energy of the proton and antiproton if the proton has

(a) the same kinetic energy as the antiproton and (b) 1.25 times as much kinetic energy as the antiproton.

31. •• The kinetic energy of a neutral pion (π^0) is 860 MeV. This pion decays to two photons, one of which has energy 640 MeV. Calculate the energy of the other photon.

28-3 Four fundamental forces describe all interactions between material objects

32. • Draw the Feynman diagram for beta-plus decay wherein a proton changes into a neutron, a positron, and a neutrino.

33. • Draw the Feynman diagram for the process described in problem 28.

34. • Draw the Feynman diagram for annihilation of a proton and an antiproton into two photons.

35. • Draw the Feynman diagram for the beta decay of the antineutron.

36. • One type of meson that can be exchanged between nucleons is the ρ (rho), with a mass of 775 MeV/c^2. Estimate the range of the interaction due to the exchange of ρ mesons.

37. • Two protons with enough energy to possibly produce the Higgs particle are smashed together in a series of experiments. Each of these experiments produces a unique collection of signals corresponding to the presence of particles produced directly or indirectly as a result of the proton–proton collision. One experiment produces signals indicating the presence of an electron, two muons, two high-energy photons, and a Z^0 boson. Could this collection of particles have been produced by the decay of a Higgs particle? Explain your answer.

28-4 We live in an expanding universe, and the nature of most of its contents is a mystery

38. • A galaxy is observed to recede from Earth with an approximate speed of 0.8c. (a) Approximately how far from Earth is this galaxy? (b) How long ago was the light that we see emitted by the galaxy? Example 28-1

39. • Galaxy NGC 3982 (Figure 28-19a) is 6.8×10^7 ly from Earth. Approximately how fast is it receding from Earth? Example 28-1

40. • Estimate the distance to a galaxy that is moving away from us at one-tenth of the speed of light due to the expansion of the universe. Example 28-1

41. • One proposal about dark matter is that it is made up of particles substantially more massive than a proton that have zero electric charge, so they respond to gravitational forces but not electromagnetic forces. Assume that a dark matter particle has 100 times the mass of a proton, which has mass 1.67×10^{-27} kg. (a) The average density of dark matter in our Milky Way galaxy is thought to be about 5×10^{-23} kg/m^3. How many dark matter particles are there in one cubic meter? (b) There are about 2.7×10^{25} molecules per cubic meter in the air around you, and a total of about 10^{19} grains of sand on all the beaches on Earth. Relatively speaking, which is more likely: finding a dark matter particle in the air around you, or finding one particular grain of sand somewhere on one of Earth's beaches?

General Problems

42. • Suppose that a fifth fundamental force were discovered and that it was mediated by electrons and positrons. Estimate the range of the force.

43. •• The United States used approximately 4.12×10^{12} kWh of electrical energy during 2019. (a) If we could generate all the energy using a matter–antimatter reactor, how many kilograms of fuel would the reactor need, assuming 100% efficiency in the annihilation? (b) Suppose the fuel consisted of iron and anti-iron, each of density 7800 kg/m^3. If the iron and anti-iron were each stored as a cube, what would be the dimensions of each cube?

44. •• Suppose that an electron neutrino and an electron anti-neutrino, both of which are just barely moving, encounter each other in space and completely annihilate each other to form two photons of equal energy. In view of the uncertainty about the mass of the electron neutrino (see Table 28-2), what is the shortest wavelength of light that could be emitted by the annihilation? Would the light be visible to the human eye?

45. •• **Astronomy** To determine what took place during the first 10^{-43} s after the Big Bang, a time interval called the *Planck time* t_P, a theory that describes gravity in terms of quantum mechanics is required. (No such theory yet exists.) Calculate t_P by first using dimensional analysis and finding the proper combination of the fundamental constants: Newton's gravitational constant $G = 6.6743 \times 10^{-11}$ N·m^2/kg^2, the speed of light $c = 2.9979 \times 10^8$ m/s, and Planck's constant divided by 2π, $\hbar = 1.0546 \times 10^{-34}$ J·s. Start your calculation by asking to what power would each of the three constants need to be raised in order to yield units of seconds: $[t_P] = [G]^x[c]^y[\hbar]^z$. Solve for x, y, and z to derive the formula for t_P. Then use the values of the three constants to calculate the precise value of t_P.

46. •• The very massive Higgs particle (mass 125 GeV/c^2) is created when two protons traveling at equally high speeds but in opposite directions collide head-on. The mass of a proton is 938.27 MeV/c^2. To make a Higgs particle when they collide, each proton must have a minimum kinetic energy of (125 GeV)/2 = 62.5 GeV. (a) What is the minimum total energy of each proton? (b) In terms of the speed of light c, what speed is each proton traveling? Calculate the total energy and speed to five significant figures.

47. • A Higgs particle (mass 125 GeV/c^2) can decay into two photons (see Figure 28-13). (a) If a Higgs particle produced in a high-energy collision is at rest, what will be the energy of each photon? (*Hint:* Since the Higgs particle has zero momentum, the two photons must have the same magnitude of momentum but must travel in opposite directions.) Give your answer in GeV and in joules. (b) What will be the wavelength of each photon? (c) In what part of the electromagnetic spectrum does this wavelength lie?

48. •• The photons that make up the cosmic microwave background (Figure 28-18) were emitted about 380,000 years after the Big Bang. Today, 13.8 billion years after the Big Bang, the wavelengths of these photons have been stretched by a factor of about 1100 since they were emitted because lengths in the expanding universe have increased by that same factor of about 1100. (a) Consider a cubical region of empty space in today's universe 1.00 m on a side, with a volume of 1.00 m^3. What was the length of each side of this same cubical region 380,000 years after the Big Bang? What was its volume? (b) Today the average density of ordinary matter in the universe is about 2.4×10^{-27} kg/m^3. What was the average density of ordinary matter at the time that the photons in the cosmic background radiation were emitted? Example 28-1

APPENDIX A
SI Units and Conversion Factors

Base Units*

Time	The *second* (s) is the duration of 9,192,631,770 periods of the radiation corresponding to the transition between the two hyperfine levels of the ground state of the ^{133}Cs atom.
Length	The *meter* (m) is the distance traveled by light in a vacuum in 1/299,792,458 s.
Mass	The *kilogram* (kg) is the mass defined by taking the value of Planck's constant h to be $6.62607015 \times 10^{-34}$ J/s = $6.62607015 \times 10^{-34}$ kg·m^2/s, where the meter and second are defined as above.
Mole	The *mole* (mol) is the amount of substance of a system that contains $6.02214076 \times 10^{23}$ elementary entities.
Current	The *ampere* (A) is an amount of current equal to one coulomb per second (C/s), where the numerical value of the elementary charge e is taken to be $1.602176634 \times 10^{-19}$ C.
Temperature	The *kelvin* (K) is a unit of temperature defined by taking the value of the Boltzmann constant k to be 1.380649×10^{-23} J/K = 1.380649×10^{-23} kg·m^2/(s^2·K).
Luminous intensity	The *candela* (cd) is the luminous intensity in a given direction, of a source that emits monochromatic radiation of frequency 540×10^{12} Hz and that has a radiant intensity, in that direction of 1/683 W/steradian.

*These definitions are found on the Internet at http://physics.nist.gov/cuu/Units/current.html.

Derived Units

Force	newton (N)	1 N = 1 kg·m/s^2
Work, energy	joule (J)	1 J = 1 N·m
Power	watt (W)	1 W = 1 J/s
Frequency	hertz (Hz)	1 Hz = 1 cy/s
Charge	coulomb (C)	1 C = 1 A·s
Potential	volt (V)	1 V = 1 J/C
Resistance	ohm (Ω)	1 Ω = 1 V/A
Capacitance	farad (F)	1 F = 1 C/V
Magnetic field	tesla (T)	1 T = 1 N/(A·m)
Magnetic flux	weber (Wb)	1 Wb = 1 T·m^2
Inductance	henry (H)	1 H = 1 J/A^2

Conversion Factors

Conversion factors are written as equations for simplicity; relations marked with an asterisk are exact.

Length

1 km = 0.6214 mi
1 mi = 1.609 km
1 m = 1.0936 yard = 3.281 ft
    = 39.37 in.
*1 in. = 2.54 cm
*1 ft = 12 in. = 30.48 cm
*1 yard = 3 ft = 91.44 cm
1 light year = 1 $c \cdot$ y = 9.461×10^{15} m
*1 Å = 0.1 nm

Area

*1 m^2 = 10^4 cm^2
1 km^2 = 0.3861 mi^2 = 247.1 acres
*1 $in.^2$ = 6.4516 cm^2
1 ft^2 = 9.29×10^{-2} m^2
1 m^2 = 10.76 ft^2
*1 acre = 43 560 ft^2
1 mi^2 = 640 acres = 2.590 km^2

Volume

*1 m^3 = 106 cm^3
*1 L = 1000 cm^3 = 10^{-3} m^3
1 gal = 3.785 L
1 gal = 4 qt = 8 pt = 128 oz = 231 $in.^3$
1 $in.^3$ = 16.39 cm^3
1 ft^3 = 1728 $in.^3$ = 28.32 L
    = 2.832×10^4 cm^3

Time

*1 h = 60 min = 3.6 ks
*1 d = 24 h = 1440 min = 86.4 ks
1 y = 365.25 day = 3.156×10^7 s

Speed

*1 m/s = 3.6 km/h
1 km/h = 0.2778 m/s = 0.6214 mi/h
1 mi/h = 0.4470 m/s = 1.609 km/h
1 mi/h = 1.467 ft/s

Angle and Angular Speed

*π rad = 180°
1 rad = 57.30°
1° = 1.745×10^{-2} rad
1 rev/min = 0.1047 rad/s
1 rad/s = 9.549 rev/min

Mass

*1 kg = 1000 g
*1 tonne = 1000 kg = 1 Mg
1 u = 1.6605×10^{-27} kg
    = 931.49 MeV/c^2
1 kg = 6.022×10^{26} u
1 slug = 14.59 kg
1 kg = 6.852×10^{-2} slug

Density

*1 g/cm^3 = 1000 kg/m^3 = 1 kg/L
(1 g/cm^3)g = 62.4 lb/ft^3

Force

1 N = 0.2248 lb = 10^5 dyn
*1 lb = 4.448222 N
(1 kg)g = 2.2046 lb

Pressure

*1 Pa = 1 N/m^2
*1 atm = 101.325 kPa = 1.01325 bar
1 atm = 14.7 $lb/in.^2$ = 760 mmHg
    = 29.9 in.Hg = 33.9 ftH_2O
1 $lb/in.^2$ = 6.895 kPa
1 torr = 1 mmHg = 133.32 Pa
1 bar = 100 kPa

Energy

*1 k$\cdot$Wh = 3.6 MJ
*1 cal = 4.186 J
1 ft$\cdot$lb = 1.356 J = 1.286×10^{-3} BTU
*1 L$\cdot$atm = 101.325 J
1 L$\cdot$atm = 24.217 cal
1 BTU = 778 ft$\cdot$lb = 252 cal
    = 1054.35 J
1 eV = 1.602×10^{-19} J
1 u$\cdot c^2$ = 931.49 MeV
*1 erg = 10^{-7} J

Power

1 horsepower = 550 ft$\cdot$lb/s
    = 745.7 W
1 BTU/h = 2.931×10^{-4} kW
1 W = 1.341×10^{-3} horsepower
    = 0.7376 ft$\cdot$lb/s

Magnetic Field

*1 T = 10^4 G

Thermal Conductivity

1 W/(m$\cdot$K) = 6.938 BTU$\cdot$in./(h$\cdot ft^2 \cdot$°F)
1 BTU$\cdot$in./(h$\cdot ft^2 \cdot$°F) = 0.1441 W/(m$\cdot$K)

Viscosity

*1 Pa$\cdot$s = 10 poise

APPENDIX B
Numerical Data

Terrestrial Data

Free-fall acceleration g

Standard value (at sea level at 45° latitude)*	$9.806\ 65\ \text{m/s}^2 = 32.1740\ \text{ft/s}^2$
At equator*	$9.7804\ \text{m/s}^2$
At poles*	$9.8322\ \text{m/s}^2$

Mass of Earth M_E 5.97×10^{24} kg

Radius of Earth R_E, mean 6.37×10^6 m = 3960 mi

Escape speed 1.12×10^4 m/s = 6.96 mi/s

Solar constant[†] $1.37\ \text{kW/m}^2$

Standard temperature and pressure (STP):

Temperature	273.15 K
Pressure	101.3 kPa (1.00 atm)

Molar mass of air   28.97 g/mol

Density of air ρ_{air} (273.15 K, 101.3 kPa) $1.29\ \text{kg/m}^3$

Speed of sound (273.15 K, 101.3 kPa)   331 m/s

Latent heat of fusion of H_2O (0°C, 1 atm)   334 kJ/kg

Latent heat of vaporization of H_2O (100°C, 1 atm)   2.26 MJ/kg

* Measured relative to Earth's surface.
[†] Average power incident normally on 1 m^2 outside Earth's atmosphere at the mean distance from Earth to the Sun.

Astronomical Data*

Earth

Distance to the Moon, mean[†]	3.844×10^8 m = 2.389×10^5 mi
Distance to the Sun, mean[†]	1.496×10^{11} m = 9.32×10^7 mi = 1.00 AU
Orbital speed, mean	2.98×10^4 m/s

Moon

Mass	7.35×10^{22} kg
Radius	1.737×10^6 m
Period	27.32 day
Acceleration of gravity at surface	$1.62\ \text{m/s}^2$

Sun

Mass	1.99×10^{30} kg
Radius	6.96×10^8 m

* Additional solar system data are available from NASA at https://nssdc.gsfc.nasa.gov /planetary/factsheet/.
[†] Center to center.

Physical Constants*

Universal constant of gravitation	G	$6.674\ 30(15) \times 10^{-11}$ N·m²/kg²
Speed of light	c	$2.997\ 924\ 58 \times 10^{8}$ m/s
Fundamental charge	e	$1.602\ 176\ 634 \times 10^{-19}$ C
Avogadro's constant	N_A	$6.022\ 140\ 76 \times 10^{23}$ particles/mol
Gas constant	R	$8.314\ 462\ 618\ldots$ J/(mol·K)
		$1.987\ 204\ 258\ 640\ 83\ldots$ cal/(mol·K)
		$8.205\ 736\ 608\ 095\ 96\ldots \times 10^{-2}$ L·atm/(mol·K)
Boltzmann constant	$k = R/N_A$	$1.380\ 649 \times 10^{-23}$ J/K
		$8.617\ 333\ 262\ldots \times 10^{-5}$ eV/K
Stefan–Boltzmann constant	$\sigma = (\pi^2/60)k^4/(\hbar^3 c^2)$	$5.670\ 374\ 419\ldots \times 10^{-8}$ W/(m²·K⁴)
Atomic mass constant	$m_u = (1/12)m(^{12}\text{C})$	$1.660\ 539\ 066\ 60(50) \times 10^{-27}$ kg = 1 u
Permeability of free space	μ_0	$1.256\ 637\ 062\ 12(19) \times 10^{-6}$ N/A²
Permittivity of free space	$\varepsilon_0 = 1/(\mu_0 c^2)$	$8.854\ 187\ 8128(13) \times 10^{-12}$ C²/(N·m²)
Coulomb constant	$k = 1/(4\pi\varepsilon_0)$	$8.987\ 551\ 7923(14) \times 10^{9}$ N·m²/C²
Planck's constant	h	$6.626\ 070\ 15 \times 10^{-34}$ J·s
		$4.135\ 667\ 696\ldots \times 10^{-15}$ eV·s
	$\hbar = h/(2\pi)$	$1.054\ 571\ 817\ldots \times 10^{-34}$ J·s
		$6.582\ 119\ 569\ldots \times 10^{-16}$ eV·s
Mass of electron	m_e	$9.109\ 383\ 7015(28) \times 10^{-31}$ kg
		$0.510\ 998\ 950\ 00(15)$ MeV/c^2
Mass of proton	m_p	$1.672\ 621\ 923\ 69(51) \times 10^{-27}$ kg
		$938.272\ 088\ 16(29)$ MeV/c^2
Mass of neutron	m_n	$1.674\ 927\ 498\ 04(95) \times 10^{-27}$ kg
		$939.565\ 420\ 52(54)$ MeV/c^2
Bohr magneton	$m_B = e\hbar/(2m_e)$	$9.274\ 010\ 0783(28)) \times 10^{-24}$ J/T
		$5.788\ 381\ 8060(17) \times 10^{-5}$ eV/T
Nuclear magneton	$m_n = e\hbar/(2m_p)$	$5.050\ 783\ 7461(15) \times 10^{-27}$ J/T
		$3.152\ 451\ 258\ 44(96) \times 10^{-8}$ eV/T
Magnetic flux quantum	$\phi_0 = h/(2e)$	$2.067\ 833\ 848\ldots \times 10^{-15}$ T·m²
Quantized Hall resistance	$R_K = h/e^2$	$2.581\ 280\ 745\ldots \times 10^{4}$ Ω
Rydberg constant	R_H	$1.097\ 373\ 156\ 8160(21) \times 10^{7}$ m⁻¹
Josephson frequency–voltage quotient	$K_J = 2e/h$	$4.835\ 978\ 484\ldots \times 10^{14}$ Hz/V
Compton wavelength	$\lambda_C = h/(m_e c)$	$2.426\ 310\ 238\ 67(73) \times 10^{-12}$ m

* Updated values for these and other constants may be found on the Internet at http://physics.nist.gov/cuu/Constants/index.html. The numbers in parentheses represent the uncertainties in the last two digits. (For example, 2.044 43(13) stands for 2.044 43 ± 0.000 13.) Values without uncertainties are exact, including those values with ellipses (such as the value of π, which is exactly 3.1415. . .).

Atomic Numbers and Atomic Weights*

Atomic Number	Name	Symbol	Weight	Atomic Number	Name	Symbol	Weight
1	Hydrogen	H	[1.007 84; 1.008 11]	60	Neodymium	Nd	144.242(3)
2	Helium	He	4.002602(2)	61	Promethium	Pm	
3	Lithium	Li	[6.938; 6.997]	62	Samarium	Sm	150.36(2)
4	Beryllium	Be	9.0121831(5)	63	Europium	Eu	151.964(1)
5	Boron	B	[10.806; 10.821]	64	Gadolinium	Gd	157.25(3)
6	Carbon	C	[12.009 6; 12.011 6]	65	Terbium	Tb	158.92535(2)
7	Nitrogen	N	[14.006 43; 14.007 28]	66	Dysprosium	Dy	162.500(1)
8	Oxygen	O	[15.999 03; 15.999 77]	67	Holmium	Ho	164.93033(2)
9	Fluorine	F	18.998403163(6)	68	Erbium	Er	167.259(3)
10	Neon	Ne	20.1797(6)	69	Thulium	Tm	168.93422(2)
11	Sodium	Na	22.98976928(2)	70	Ytterbium	Yb	173.045(10)
12	Magnesium	Mg	[24.304, 24.307]	71	Lutetium	Lu	174.9668(1)
13	Aluminum	Al	26.9815385(7)	72	Hafnium	Hf	178.49(2)
14	Silicon	Si	[28.084; 28.086]	73	Tantalum	Ta	180.94788(2)
15	Phosphorus	P	30.973761998(5)	74	Tungsten	W	183.84(1)
16	Sulfur	S	[32.059; 32.076]	75	Rhenium	Re	186.207(1)
17	Chlorine	Cl	[35.446; 35.457]	76	Osmium	Os	190.23(3)
18	Argon	Ar	39.948(1)	77	Iridium	Ir	192.217(3)
19	Potassium	K	39.0983(1)	78	Platinum	Pt	195.084(9)
20	Calcium	Ca	40.078(4)	79	Gold	Au	196.966569(5)
21	Scandium	Sc	44.955908(5)	80	Mercury	Hg	200.592(3)
22	Titanium	Ti	47.867(1)	81	Thallium	Tl	[204.382; 204.385]
23	Vanadium	V	50.9415(1)	82	Lead	Pb	207.2(1)
24	Chromium	Cr	51.9961(6)	83	Bismuth	Bi	208.98040(1)
25	Manganese	Mn	54.938044(3)	84	Polonium	Po	
26	Iron	Fe	55.845(2)	85	Astatine	At	
27	Cobalt	Co	58.933194(4)	86	Radon	Rn	
28	Nickel	Ni	58.6934(4)	87	Francium	Fr	
29	Copper	Cu	63.546(3)	88	Radium	Ra	
30	Zinc	Zn	65.38(2)	89	Actinium	Ac	
31	Gallium	Ga	69.723(1)	90	Thorium	Th	232.0377(4)
32	Germanium	Ge	72.630(8)	91	Protactinium	Pa	231.03588(2)
33	Arsenic	As	74.921595(6)	92	Uranium	U	238.02891(3)
34	Selenium	Se	78.971(8)	93	Neptunium	Np	
35	Bromine	Br	[79.901, 79.907]	94	Plutonium	Pu	
36	Krypton	Kr	83.798(2)	95	Americium	Am	
37	Rubidium	Rb	85.4678(3)	96	Curium	Cm	
38	Strontium	Sr	87.62(1)	97	Berkelium	Bk	
39	Yttrium	Y	88.90584(2)	98	Californium	Cf	
40	Zirconium	Zr	91.224(2)	99	Einsteinium	Es	
41	Niobium	Nb	92.90637(2)	100	Fermiun	Fm	
42	Molybdenum	Mo	95.95(1)	101	Mendelevium	Md	
43	Technetium	Tc		102	Nobelium	No	
44	Ruthenium	Ru	101.07(2)	103	Lawrencium	Lr	
45	Rhodium	Rh	102.90550(2)	104	Rutherfordium	Rf	
46	Palladium	Pd	106.42(1)	105	Dubnium	Db	
47	Silver	Ag	107.8682(2)	106	Seaborgium	Sg	
48	Cadmium	Cd	112.414(4)	107	Bohrium	Bh	
49	Indium	In	114.818(1)	108	Hassium	Hs	
50	Tin	Sn	118.710(7)	109	Meitnerium	Mt	
51	Antimony	Sb	121.760(1)	110	Darmstadtium	Ds	
52	Tellurium	Te	127.60(3)	111	Roentgenium	Rg	
53	Iodine	I	126.90447(3)	112	Copernicium	Cn	
54	Xenon	Xe	131.293(6)	113	Nihonium	Nh	
55	Cesium	Cs	132.90545196(6)	114	Flerovium	Fl	
56	Barium	Ba	137.327(7)	115	Moscovium	Mc	
57	Lanthanum	La	138.90547(7)	116	Livermorium	Lv	
58	Cerium	Ce	140.116(1)	117	Tennessine	Ts	
59	Praseodymium	Pr	140.90766(2)	118	Oganesson	Og	

* Some weights are listed as intervals ([*a*; *b*]; *a* ≤ atomic weight ≤ *b*) because these weights are not constant but depend on the physical, chemical, and nuclear histories of the samples used. Atomic weights are not listed for some elements because these elements do not have stable isotopes. Exceptions are thorium, protactinium, and uranium. From http://www.ciaaw.org/atomic-weights.htm.

GLOSSARY

absolute pressure Total pressure at a point in a fluid, equal to the sum of the gauge and the atmospheric pressures.

absolute zero The lowest temperature that is theoretically possible, at which the motion of particles is at a minimum; 0 on the Kelvin scale; −273.15°C or −459.67°F.

absorption lines Dark lines in an otherwise continuous spectrum. These indicate certain wavelengths of light that are absorbed by the atoms of the intervening medium.

absorption spectrum A continuous spectrum, broken by a specific pattern of dark lines or bands, observed when light traverses a particular absorbing medium.

ac (alternating current) An electric current that reverses its direction many times a second at regular intervals.

ac generator Alternating-current generator; as its coil rotates in a magnetic field, an oscillating emf is generated in the turns of wire that make up the coil.

acceleration The rate of change of velocity, due to changes in its direction or magnitude.

acceleration due to gravity Acceleration of a body due to the pull of gravity; an object in free fall near Earth's surface has an acceleration of approximately 9.8 m/s^2.

acceleration vector (*or* instantaneous acceleration vector) The change in a velocity vector per unit over time.

adiabatic process In thermodynamics, a process that occurs without transfer of heat or matter into or out of a system.

alpha decay Type of radioactive decay in which an atomic nucleus emits an alpha particle (helium nucleus) and thereby transforms into a nucleus with a mass number that is reduced by four and an atomic number that is reduced by two.

alpha particle A positively charged particle, indistinguishable from a helium nucleus and consisting of two protons and two neutrons.

alternating current (ac) *See* ac (alternating current).

ampere The SI unit of electric current, equal to a flow of one coulomb per second.

Ampère's law For any closed loop path, the circulation of a magnetic field created by an electric current is equal to the size of that electric current times the permeability of free space; discovered by the French physicist André-Marie Ampère.

Amperian loop An imaginary closed path in space around a current-carrying conductor.

amplitude (of a wave) The maximum displacement from equilibrium that occurs as a wave moves through its medium.

angular acceleration The rate of change of angular velocity.

angular displacement The angle through which an object has been rotated.

angular frequency Frequency of a periodic process (as electric oscillation or sound vibration) expressed in radians per second, equivalent to the frequency in cycles multiplied by 2π.

angular magnification The ratio of the angular size of an object viewed through an optical device (such as a magnifying glass, a microscope, or a telescope) to its angular size viewed with the unaided eye.

angular position On a rotating object, the angle of a line on the object from its original position.

angular resolution The angle that separates two point objects that are just barely resolved through a circular aperture.

angular speed The magnitude or absolute value of angular velocity.

angular velocity The rate of change of angular displacement.

angular wave number The reciprocal of wavelength multiplied by 2π.

antimatter Particles with the same properties as ordinary particles, but with the opposite electric charge.

antineutrino The antiparticle of a neutrino.

antinode Positions along a standing wave at which the oscillation is maximal.

antiquark The antiparticle of a quark.

apparent weight Weight of an object submerged in a fluid. This is less than its true weight due to buoyant forces.

apparent weightlessness The perceived state of an object that is accelerating along with its surroundings.

Archimedes' principle The buoyant force on an object immersed in a fluid is equal to the weight of the fluid that the object displaces.

atmosphere (unit of pressure) The average value of atmospheric pressure at sea level; equal to 1.01325×10^5 Pa or about 14.7 pounds per square inch.

atomic number The number of protons in an atomic nucleus; determines the chemical properties of an element and its place in the periodic table.

average acceleration The change in velocity divided by the elapsed time.

average angular acceleration The change in angular velocity divided by the elapsed time.

average angular velocity The change in angular displacement divided by the elapsed time.

average speed The total distance an object travels divided by the time it takes for the object to travel that distance.

average velocity The total displacement of an object divided by the elapsed time.

baryon Any hadron that can be made up of three quarks.

battery An electrochemical cell that can set charges into motion.

beat frequency Rate of the periodic variations of amplitude when two waves of different frequencies interfere.

beats Periodic variations in amplitude when two waves of different frequency interfere.

becquerel The SI unit of radioactive decay rate, equal to one disintegration per second.

Bernoulli's equation A relationship among pressure, speed, and height in an ideal fluid in motion.

Bernoulli's principle In a moving fluid, the pressure is low where the fluid is moving rapidly.

beta decay A type of radioactive decay in which a beta particle (an electron or a positron) is emitted from an atomic nucleus.

beta-minus decay When a neutron (charge zero) changes into a proton, an electron, and an electron antineutrino.

beta-plus decay When a proton changes into a neutron (charge zero), a positron, and an electron neutrino.

Big Bang The rapid expansion of matter from a state of extremely high density and temperature that marked the origin of the universe.

binding energy The energy required to disassemble an atomic nucleus into its component protons and neutrons.

blackbody An object that does not reflect any light at all but absorbs all radiation falling on it.

blackbody radiation Light emitted by a perfect blackbody of emissivity = 1.

Bohr model Theory of atomic structure in which a small, positively charged nucleus is surrounded by electrons that travel in circular orbits around the nucleus.

Bohr orbit In the Bohr model, one of the orbits in which electrons in an atom travel around the nucleus.

Boltzmann constant Physical constant in the ideal gas law which has the same value for all gases and relates the average kinetic energy of the particles in a gas to the temperature of the gas; equal to 1.38065×10^{-23} J/K.

boundary layer In a fluid, the layer next to a solid surface within which the fluid speed increases from zero at the surface to full speed at the edge of the layer.

Brewster's angle An angle of incidence at which light with a particular polarization is perfectly transmitted through a transparent dielectric surface, with no reflection.

British thermal unit The quantity of heat required to increase the temperature of one pound (1 lb) of pure water from 63°F to 64°F.

bulk modulus A measure of how resistant to compression a substance is, defined as the ratio of the pressure increase to the resulting relative decrease of the volume.

buoyant force The upward force exerted by a fluid on an object placed in it.

calorie The quantity of heat required to increase the temperature of one gram (1 g) of pure water from 14.5°C to 15.5°C.

capacitance The ability of a system to store an electric charge.

capacitor A system or device that can store positive and negative charge, consisting of one or more pairs of conductors that may be separated by an insulator.

Carnot cycle An ideal, reversible thermodynamic cycle consisting of two isothermal processes and two adiabatic processes; the most efficient cycle in a heat engine; first proposed in the early 1800s by Sadi Carnot.

Cavendish experiment Experiment used to determine the value of the gravitational constant.

Celsius scale The most common temperature scale; based on the work of the eighteenth-century Swedish astronomer Anders Celsius. In this scale the freezing point of water is approximately 0°C, and the boiling point is approximately 100°C.

center of curvature (of a mirror) The center of the sphere defined by the surface of a concave mirror.

center of mass A point representing the average position of the matter in a body or system; moves as though all of the body's mass were concentrated at that point and all external forces act on it.

centripetal acceleration The rate of change of tangential velocity of an object in circular motion; points toward the center of the circle defined by the object's trajectory.

centripetal force The force that points toward the inside of an object's curving trajectory and produces the centripetal acceleration.

charged object A body with net electric charge.

circuit A complete loop in which a charge flows through a wire continuously from one terminal of a battery to the other.

circuit element Any component of a circuit, such as a battery, resistor, inductor, or capacitor.

circulation (of a magnetic field) For an Amperian loop, the sum of products (of the component of magnetic field parallel to each loop segment multiplied by the segment length) along the loop.

closed pipe A pipe which is open at one end and blocked at the other.

coefficient of kinetic friction The ratio of the force of friction acting on a sliding object to the object's normal force; depends on the properties of the surfaces in contact.

coefficient of linear expansion The fractional change in length of an object per unit change in temperature.

coefficient of performance (of a heat pump or refrigerator) The ratio of useful heating or cooling provided to work required.

coefficient of rolling friction The ratio of the force of friction acting on the point of a rolling object that is in contact with the surface to the object's normal force; depends on the properties of the surfaces in contact.

coefficient of static friction The ratio of the force of friction acting on an unmoving object to the object's normal force; depends on the properties of the surfaces in contact.

coefficient of volume expansion The fractional change in the volume of an object per unit change in temperature.

completely inelastic collision Encounter in which two objects stick together after they collide and the most mechanical energy is lost.

component (of a vector) The projection of a vector onto a coordinate axis.

compressible fluid A fluid that can be easily compacted by squeezing.

compression Pressure applied to all sides of a body, which may result in a reduction in volume.

compressive strain The length by which an object shrinks expressed as a fraction of its relaxed length.

compressive stress The force applied to an object being squeezed divided by the object's cross-sectional area.

Compton scattering Elastic scattering of a photon by a free charged particle, usually an electron.

concave Curved inward, such as a mirror.

condensation The phase change from gas to liquid.

conduction Transfer of heat by the direct collision of particles, with no net displacement of the particles.

conductor A substance in which charges can move freely.

conservation of angular momentum If there is no net external torque on a system, the angular momentum of the system is conserved.

conservative force A force that can be associated with a potential energy, such as the gravitational force or the force exerted by an ideal spring.

constant acceleration A situation in which velocity changes at a steady rate.

constant velocity Movement at a steady speed in the same direction.

constructive interference The mutual reinforcement of waves such that the amplitude of the total wave is the sum of the amplitudes of the individual waves.

contact force A force that arises only when two objects come in contact with one another.

contact time The amount of time colliding objects are in contact.

convection Energy transfer by the motion of a liquid or gas (such as air) caused by the tendency of hotter, less dense material to rise and colder, denser material to sink under the influence of gravity.

converging lens Lens that takes incoming parallel light rays and brings them to a focus on the principal axis.

convex Curved outward, such as a mirror.

coordinates Quantities indicating the position of an object in reference to the origin on a coordinate system.

coordinate system A system that can be used to denote the position of an object at a given time.

cosmic background radiation (cosmic microwave background) The thermal radiation left over from the Big Bang.

cosmological redshift A redshift caused by the expansion of the universe.

cosmology The science of the origin and evolution of the universe.

coulomb The unit of electric charge, equal to the amount of electricity conveyed in one second by a current of one ampere; named after the eighteenth-century French physicist Charles-Augustin de Coulomb, who uncovered the fundamental law that governs the interaction of charges.

Coulomb's constant A proportionality constant used to determine the magnitude of the electric forces between two point charges; equal to 8.99×10^9 N·m^2/C^2.

Coulomb's law The magnitude of the force of electrostatic attraction or repulsion acting between two electric charges is directly proportional to the product of the charges and inversely proportional to the square of the distance between them.

critical angle The angle of incidence beyond which total internal reflection occurs.

critically damped oscillations Case in which the minimum amount of damping is applied to result in a nonoscillatory response; when displaced from equilibrium, the system returns smoothly to equilibrium with no overshoot and hence no oscillation.

critical point A point on a phase diagram at which both the liquid and gas phases of a substance have the same density and are therefore indistinguishable.

cross product (vector product) The product of two vectors in three dimensions that is itself a vector at right angles to both the original vectors, with a direction given by the right-hand rule and a magnitude equal to the product of the magnitudes of the original vectors and the sine of the angle between their directions.

current The rate at which charge flows past any point in a circuit.

current loop A single loop that carries a current provided by a source of emf.

damped oscillations Oscillations that are diminished by a frictional force.

damping coefficient Proportionality constant related to the physical characteristics of a particular system and determining the degree to which oscillations are diminished.

dark energy A repulsive force that counteracts gravity and causes the universe to expand at an accelerating rate.

dark matter Nonluminous material that is postulated to exist in space and that is the dominant form of matter in the universe.

dc (direct current) An electric current that does not change direction in a circuit.

de Broglie wavelength The wavelength of a particle, given by Planck's constant divided by the momentum of the particle.

decay constant Proportionality between the size of a population of radioactive nuclei and the rate at which the population decreases because of radioactive decay.

decibel The units of sound intensity level.

degree of freedom In thermodynamics, a possible form of motion of an object.

density The mass of a substance divided by the volume that it occupies.

deposition The phase change from gas to solid.

destructive interference When two waves cancel each other out so that the amplitude of the total wave is zero.

diamagnetic Tending to become magnetized in a direction opposite to that of the applied magnetic field.

dielectric A material that is both an insulator and polarizable.

dielectric constant The greater the dielectric constant of a material, the more the material is polarized when it is placed in the electric field between the plates of a charged capacitor.

diffraction The bending of light around an obstacle or aperture.

diffraction maxima Locations in a diffraction pattern where wavelets interfere constructively, producing a bright fringe.

diffraction minima Locations in a diffraction pattern where wavelets interfere destructively, producing a dark fringe.

diffraction pattern The distinctive pattern of bright and dark fringes caused when light is diffracted through a slit or aperture.

diffuse light Light that reflects from an object's surface in many random directions.

dimensional analysis A method of checking the relations of physical quantities by identifying their dimensions.

diode A single-junction semiconductor device.

diopter A unit of power for a lens or mirror that is equal to the reciprocal of the focal length (in meters).

direct current (dc) *See* dc (direct current).

dispersion The separation of light according to wavelength due to differing propagation speeds of different wavelengths of light; a prism separates white light into its component colors because light of each color travels at a different speed through the glass.

displacement The difference between the positions of an object at two separate times.

displacement current A quantity appearing in Maxwell's equations that is defined in terms of the rate of change of the electric displacement field.

displacement vector Vector drawn from the starting point of an object's motion to its endpoint.

diverging lens Lens that takes incoming parallel light rays and causes them to spread away from the principal axis.

doping Adding small amounts of a different kind of atom to a substance.

Doppler effect The change in frequency of a wave caused by an observer moving relative to its source or its source moving relative to the observer.

drag force The force that resists the motion of an object through a liquid or a gas.

drift speed The average speed at which charges move through a conductor.

driving angular frequency The angular frequency of a driving force in an oscillation.

eccentricity Parameter determining the circularity of an ellipse; a perfect circle has zero eccentricity.

efficiency The useful work divided by the amount of energy taken in to do the work.

elapsed time The duration of a time interval.

elastic An elastic object returns to its original shape after being squeezed or stretched.

elastic collision A collision in which the forces between the colliding objects are conservative; both total momentum and total mechanical energy are conserved.

electric charge The physical property of matter that causes it to experience a force near other charged material.

electric dipole A combination of two point charges of the same magnitude but opposite signs.

electric energy density The energy per unit volume stored in an electric field, such as in a capacitor.

electric field lines Lines showing the direction of an electric field.

electric flux The area of a surface multiplied by the component of the electric field that's perpendicular to that surface.

electric force The force between electric charges.

electric potential The electric potential energy for a charge at a given position divided by the value of that charge.

electric potential difference The difference in electric potential between two locations.

electric potential energy Potential energy that results from conservative Coulomb forces; associated with the configuration of a particular set of point charges.

electromagnetic induction The process whereby a changing magnetic field induces an electric field.

electromagnetic spectrum The range of electromagnetic waves according to wavelength.

electromagnetic wave Waves that are propagated by simultaneous periodic variations of electric and magnetic fields. These include radio waves, infrared, visible light, ultraviolet, x rays, and gamma rays.

electromagnetism An umbrella term to cover both electricity and magnetism, since both involve interactions between charges.

electron volt A unit of energy equal to the work done on an electron in accelerating it through a potential difference of one volt.

emf (electromotive force) The voltage developed by any source of electrical energy such as a battery.

emission lines Bright lines in the emission spectrum of a gas.

emission spectrum A spectrum that consists only of specific emitted wavelengths.

emissivity How well or how poorly a surface radiates.

energy The capacity to do work.

energy level Quantized value of energy, for example, of electrons in an atom.

entropy A measure of the amount of disorder present in a system.

equation of continuity In fluid dynamics, $A_1v_1 = A_2v_2$: the product of a pipe's cross-sectional area A and the flow speed v is conserved; it has the same value at point 1 as at point 2.

equation of hydrostatic equilibrium The equation $p = p_0 + pgd$, which must be satisfied for a fluid to remain at rest.

equation of state A relationship among the quantities of pressure, volume, and temperature.

equilibrium A state in which the net external force on an object is zero so it is not accelerating (and, if it is an extended object, the net external force on the object is also zero and the object is not rotating).

equipartition theorem Principle stating that the energy of a molecule is shared equally among each degree of freedom.

equipotential A curve along which the electric potential has the same value at all points.

equipotential surface A surface on which the electric potential has the same value at all points.

equipotential volume Space inside a conductor in which the electric potential has the same value everywhere.

equivalent capacitance Effective capacitance of an arrangement of two or more connected capacitors.

equivalent resistance Effective resistance of an arrangement of two or more connected resistors.

escape speed The minimum speed at which an object must be launched from Earth's surface to escape to infinity.

event Something that happens at a certain point in time.

exchange particle A virtual particle that interacts with ordinary particles to mediate forces, producing the effects of attraction and repulsion.

exponent The superscript following the number 10 that denotes the number of zeros needed to write the long form of a number in scientific notation.

exponential function The irrational number e raised to a power.

external forces Forces exerted on an object by other objects.

eyepiece A magnifying glass used to give an enlarged view of the image produced by the objective lens of a microscope or telescope.

Fahrenheit scale The official temperature scale used in the United States, the Cayman Islands, and Belize; originated by the German scientist Daniel Fahrenheit. On this scale water freezes at 32°F and boils at 212°F.

failure The point at which the structure of a material starts to lose its integrity, which eventually leads to the object breaking apart.

farad The SI unit of electrical capacitance, equal to the capacitance of a capacitor in which one coulomb of charge causes a potential difference of one volt; named after English physicist Michael Faraday.

Faraday's law (of induction) An emf is induced in a loop if the magnetic flux through that loop changes.

ferromagnetic A ferromagnetic material has a high susceptibility to magnetization, the strength of which depends on that of the applied magnetizing field and that may persist after removal of the applied field.

first law of thermodynamics In a thermodynamic process, the change in the internal energy of a system equals the heat that flows into the system during the process minus the work that the system does during the process.

fluid A substance (a gas or a liquid) that can flow because its molecules can move freely with respect to each other.

fluid resistance The resistance experienced by an object as it moves through a fluid.

focal length The distance from the focal point to the center of a mirror or lens.

focal point The point along the principal axis of a mirror or lens at which incident rays parallel to the principal axis converge and come to a common focus.

force A push or a pull.

forced oscillations Case in which a periodic driving force causes a system to oscillate at the frequency of that driving force.

force pair In an interaction between objects A and B, the forces of A on B and of B on A.

frame of reference (*or* reference frame) A coordinate system with respect to which we can make observations or measurements.

free-body diagram A graphical representation of all external forces acting on a body.

free fall The state of an object falling toward Earth without any effect from air resistance.

freezing The phase change from liquid to solid.

frequency The number of cycles of an oscillation per unit of time.

friction force Force resisting the sliding of an object across a surface, acting parallel to the surface and opposite to the motion of an object.

fringe The series of bright and dark patches in an interference or diffraction pattern.

front (of a wave) Wave crest.

fundamental frequency The lowest natural frequency of an oscillating object.

fundamental mode For an oscillating object, the standing wave mode corresponding to the fundamental frequency.

fusion The change from solid to liquid; melting.

galaxy A collection of a tremendous number of stars held together by gravity.

Galilean transformation Set of equations used to translate between the coordinates of two reference frames which differ only by constant relative motion.

Galilean velocity transformation Set of equations used to translate between the velocity of an object in two reference frames which differ only by constant relative motion.

gamma decay A radioactive process in which an atomic nucleus loses energy by emitting a gamma ray without a change in its atomic number or mass number.

gas A fluid that expands to fill whatever volume is available to it.

gauge pressure The amount by which the pressure exceeds atmospheric pressure.

Gaussian surface A closed surface used to enclose charge in order to apply Gauss's law.

Gauss's law (for the electric field) The net electric flux through a closed surface (called a Gaussian surface) equals the net charge enclosed by that surface divided by the permittivity. Charges outside the surface have no effect on the net electric flux through the surface; named after nineteenth-century German mathematician and physicist Carl Friedrich Gauss.

Gauss's law (for the magnetic field) For any closed Gaussian surface, there is zero net flux of the magnetic field through that surface.

general theory of relativity Albert Einstein's theory which provides a unified description of gravity as a geometric property of space and time, or spacetime.

geometrical optics The science of mirrors and lenses.

global warming An increase in the global average surface temperature caused by the greenhouse effect.

gluon A subatomic particle of a class that is thought to bind quarks together.

gravitational bending of light Effect in which a light beam passing near a massive object will curve under the influence of the object's gravity.

gravitational constant The constant involved in the calculation of gravitational force between two objects; equal to 6.67×10^{-11} N·m²/kg².

gravitational force The force of attraction between all masses in the universe; especially the attraction of Earth's mass for bodies near its surface.

gravitational potential energy The ability to do work related to an object's vertical position in the presence of gravity.

gravitational slowing of time Effect in which gravity influences the rate of a ticking clock; clocks on the ground floor of a building tick more slowly than clocks on the top floor, which are farther from Earth's center.

gravitational waves Small variations in the curvature of spacetime that spread away from moving massive objects.

greenhouse effect Warming effect in which the atmosphere prevents some of the radiation emitted by Earth's surface from escaping into space.

greenhouse gas One of several gases in the atmosphere that are transparent to visible light but not to infrared radiation.

hadron A particle that can experience the strong force.

Hagen–Poiseuille equation Relates the pressure difference between the two ends of a pipe to the resulting flow rate of a viscous fluid.

half-life The time taken for the radioactivity of a specified isotope to fall to half its original value.

harmonic property Characteristic of simple harmonic motion in which the angular frequency, period, and frequency of an oscillation are independent of the amplitude if the restoring force obeys Hooke's law.

heat A form of energy arising from the random motion of the molecules of bodies, which may be transferred by conduction, convection, or radiation.

heat engine A system or device that converts heat to work.

Heisenberg uncertainty principle for energy and time In quantum mechanics, the shorter the duration of a phenomenon, the greater the uncertainty in the energy of that phenomenon.

Heisenberg uncertainty principle for momentum and position In quantum mechanics, the smaller the uncertainty in the position of an object, the greater the minimum uncertainty in the momentum of that object.

hertz The SI unit of frequency, equal to one cycle per second.

Higgs field The theoretical field that gives fundamental particles their mass.

Higgs particle Subatomic particle associated with the Higgs field.

hole The lack of an electron at a position where one could exist in an atom.

Hooke's law The force needed to extend a spring by a certain distance is proportional to that distance.

Hubble constant The ratio of the speed of recession of a galaxy (due to the expansion of the universe) to its distance from the observer.

Hubble law The observation that the speed of recession of distant galaxies is proportional to their distance from the observer.

Huygens' principle The wave front at a later time is the superposition of all of the wavelets emitted at the starting time and is tangent to the leading edges of the wavelets.

hydrostatic equilibrium State in which a fluid is at rest.

ideal gas A theoretical gas composed of a set of randomly moving, noninteracting point particles.

ideal gas constant A constant in the ideal gas law, expressed in units of energy per temperature increment per mole; equal to $8.314\ \mathrm{J/(mol \cdot K)}$.

ideal gas law Equation of the state of a hypothetical ideal gas.

image An appearance of an object formed by light rays reflected by a mirror or focused by a lens.

image distance The distance from a mirror to an object's reflected image, or the distance from a lens to the image.

image height The height of an object's image reflected in a mirror or focused by a lens.

impedance The measure of the opposition that a circuit presents to an alternating current when an alternating voltage is applied.

impulse The product of the force acting on an object and the duration of the time interval over which that force acts.

impulse-momentum theorem The impulse on an object over a time interval equals the object's momentum at the end of the time interval minus its momentum at the beginning of the time interval.

incident light Incoming light that strikes a surface.

incompressible fluid A fluid whose volume and density change very little when squeezed.

index of refraction A measure of the speed of light traveling in a medium; the speed of light in a vacuum divided by the speed of light in the medium.

induced emf An emf induced around a loop by a changing magnetic field.

induced magnetic field A magnetic field produced by the current in a loop that is caused by an induced emf in that loop.

inductance For a current-carrying conducting coil, the ratio of the magnetic flux through the coil divided by the current that produces the flux.

inelastic collision A collision in which mechanical energy is not conserved.

inertia The tendency of an object to resist change in motion.

inertial frame of reference A frame of reference attached to an object that does not accelerate.

instantaneous acceleration Acceleration of an object at a specific instant.

instantaneous speed Speed of an object at a specific instant.

instantaneous velocity Velocity of an object at a specific instant.

insulator Substance in which charges are not able to move freely.

intensity Average wave power per unit area.

interference The combination of two or more electromagnetic waves to form a resultant wave in which the displacement is either reinforced or canceled.

interference maxima Locations in an interference pattern where wavelets interfere constructively, producing a bright fringe.

interference minima Locations in an interference pattern where wavelets interfere destructively, producing a dark fringe.

interference pattern Alternating bright and dark fringes produced when two or more light waves combine or cancel each other out.

internal energy The energy within an object due to the kinetic and potential energies associated with the individual molecules that comprise the object.

internal forces Forces exerted by one part of an object or system on another part.

internal resistance The resistance that mobile charges encounter as they pass through a battery.

inverse-square law for waves Intensity is inversely proportional to the square of the distance from the source.

inverted image Image (produced by a mirror or lens) that is flipped upside down relative to the object.

inviscid flow Flow of a fluid that is assumed to have no viscosity.

ionizing radiation Photons with enough energy to dislodge an electron from an atom.

irreversible process A process that cannot return both the system and the surroundings to their original conditions.

irrotational flow Flow in which the speed varies gradually from one part of the fluid to another, with no abrupt jumps.

isobaric process Process in which the pressure of the system remains constant.

isochoric process Process in which the volume of the system remains constant.

isotherm Contours of constant temperature.

isothermal process Process in which the temperature of the system remains constant.

isotope Variants of a particular chemical element that share the same number of protons in the nucleus of each atom but differ in neutron numbers.

joule The SI unit of work or energy, equal to the work done by a force of one newton when its point of application moves one meter in the direction of action of the force.

junction Points in a circuit where either the current breaks into two currents or two currents come together into one.

kelvin The SI base unit of thermodynamic temperature, equal in magnitude to the degree Celsius.

Kelvin scale Temperature scale based on the relationship between pressure and temperature of low-density gases; first proposed by the nineteenth-century Scottish physicist William Thomson (1st Baron Kelvin).

kilogram The SI unit of mass.

kinetic energy The energy that an object possesses by virtue of being in motion.

kinetic friction Force acting on a sliding object that opposes the object's motion.

Kirchhoff's junction rule The sum of the currents flowing into a junction equals the sum of the currents flowing out of it.

Kirchhoff's loop rule The sum of the changes in electric potential around a closed loop in a circuit must equal zero.

laminar flow Smooth fluid flow in which each object follows the object directly in front of it.

latent heat (of fusion or vaporization) The amount of heat per unit mass that must flow into or out of the substance to cause the phase change.

lateral magnification The ratio of the height of an image to the height of the corresponding object.

law of areas A line joining the Sun and a planet sweeps out equal areas in equal intervals of time, regardless of the position of the planet in the orbit.

law of conservation of angular momentum *See* conservation of angular momentum.

law of conservation of energy One kind of energy can transform into another, but the total amount of energy of all forms remains the same.

law of conservation of momentum If the net external force on a system of objects is zero, then the total momentum of the system does not change.

law of orbits The orbit of each planet is an ellipse with the Sun located at one focus of the ellipse.

law of periods The square of the period of a planet's orbit is proportional to the cube of the semimajor axis of the orbit.

law of reflection The angle of the reflected light is the same as the angle of the incident light.

LC circuit A circuit made up of an inductor of inductance L and a capacitor of capacitance C connected by ideal, zero-resistance wires.

length contraction The shortening of an object or distance moving at nearly the speed of light along the direction of motion; predicted by the theory of special relativity.

lens A piece of glass or other transparent material with a curved surface for concentrating or dispersing light rays.

lens equation Expression relating object distance and lens distance to focal length.

lensmaker's equation Expression relating the focal length of a lens in air to its index of refraction and curvature.

Lenz's law The direction of the magnetic field induced within a conducting loop opposes the change in magnetic flux that created it; named after Russian physicist Heinrich Lenz.

lepton A particle that is not affected by the strong force.

lever arm The perpendicular distance from the rotation axis of a body to the line of action of a force applied to that body.

light-emitting diode (LED) A semiconductor device that emits visible light when an electric current passes through it; produces much more light for a given power input than do incandescent light bulbs or fluorescent lamps.

line of action For a force applied to a body, an extension of the force vector through the point of application.

linear mass density Mass per unit length.

linear momentum (momentum) The product of an object's mass and its velocity vector.

linearly polarized light Light for which the electric field is oriented completely along one direction.

liquid A fluid that maintains the same volume regardless of the shape and size of its container.

longitudinal wave (pressure wave) A traveling disturbance or vibration in which the individual parts of the wave medium move in the direction parallel to the direction of wave propagation.

Lorentz transformation Set of equations used to translate between the coordinates of two reference frames moving relative to one another, consistent with special relativity.

Lorentz velocity transformation Set of equations relating the velocity of an object in two reference frames moving relative to one another, consistent with special relativity.

Mach angle For supersonic flow, the angle a shock wave makes with the direction of motion, determined by the velocity of the object and the velocity of shock propagation.

Mach cone The conical pressure wave front produced by a body moving at a speed greater than that of sound.

Mach number The ratio of an object's speed to the speed of sound.

magnet A material or object that produces a magnetic field.

magnetic dipole A pair of equal and opposite magnetic poles separated by a small distance; the magnetic field points away from the magnet's north pole and toward the magnet's south pole.

magnetic energy The energy required to set up a magnetic field in and around an inductor.

magnetic energy density The energy per unit volume stored in a magnetic field, such as in an inductor.

magnetic field A region around a magnetic material or a moving electric charge within which the force of magnetism acts.

magnetic flux The area of a surface multiplied by the component of the magnetic field that's perpendicular to that surface.

magnetic force Force of attraction or repulsion that arises between magnets, or between electrically charged particles because of their motion.

magnetic poles The two ends of a magnetic field.

magnetic resonance imaging (MRI) A form of medical imaging that measures the response of atomic nuclei in body tissues to radio waves when placed in a strong magnetic field, producing detailed images of internal organs.

magnetism The interaction between magnets or between electrically charged particles due to their motion.

magnifying glass A converging lens used to view a nearby object and produce a virtual image with large angular magnification.

magnitude (of a vector) The straight-line distance from the starting point of a vector to its endpoint.

mass The measure of the amount of material in an object.

mass number The total number of protons and neutrons in an atomic nucleus.

mass spectrometer A device used to determine the masses of individual atoms and molecules.

Maxwell–Ampère law Magnetic fields can be generated in two ways: by electrical current and by changing electric fields.

Maxwell's equations Four basic equations that describe all electromagnetic phenomena.

mean free path The average distance that a molecule travels from the time at which it collides with one molecule to when it collides with another molecule.

mechanical wave A propagating oscillation of matter that transfers energy through a medium.

medium The substance through which a mechanical wave propagates.

meson A subatomic particle that is intermediate in mass between an electron and a proton and that transmits the strong interaction that binds nucleons together in the atomic nucleus.

meter The SI unit of measurement for length.

Michelson–Morley experiment An experiment performed in 1887 attempting to detect the velocity of Earth with respect to the hypothetical luminiferous ether; discovered no evidence for its existence.

microscope An optical device used to make enlarged images of small, nearby objects.

millimeters of mercury Unit of measurement for pressure; 760 mmHg equals 1 atm.

mirror equation Expression relating object distance and image distance to the focal length of a mirror.

mobile charge Charged particles that are free to move throughout a conducting material, as in a circuit.

molar specific heat The quantity of heat required to raise the temperature of one mole of the substance by one kelvin.

molar specific heat at constant pressure The quantity of heat required to make one mole of a substance undergo a temperature change of one kelvin if the pressure is held constant.

molar specific heat at constant volume The quantity of heat required to make one mole of a substance undergo a temperature change of one kelvin if the volume is held constant.

mole Unit measuring the quantity of a substance; one mole equals the number of atoms in exactly 12 grams of carbon–12, given by Avogadro's number: 6.022×10^{22}.

moment of inertia A property of a body that defines its resistance to a change in angular velocity about an axis of rotation.

momentum (linear momentum) *See* linear momentum.

monochromatic light Light that has a single definite wavelength.

motion diagram A diagram that visualizes the motion of an object using distance gained between equal time intervals.

motion in a plane Two-dimensional motion.

motion in one dimension Motion in a straight line.

motional emf A changing emf due to the motion of a conductor in a magnetic field.

multiloop circuit A circuit with more than one pathway that a moving charge can take from the positive terminal of the battery through the circuit to the negative terminal.

mutual inductance An effect in a transformer in which a change in the current in one coil induces an emf and current in a second coil.

natural angular frequency The angular frequency at which a system oscillates when not subjected to an external force.

negative displacement The distance, in the negative direction along a defined coordinate axis, between the position of an object at one time and its position at an earlier time.

negative work Work done whenever the angle between the displacement of an object and the force acting on that object is greater than 90°.

net external force (net force) The vector sum of all external forces acting on an object.

neutral matter Matter that contains equal amounts of positive and negative charge.

neutrino A nearly massless, neutral particle.

neutron-induced fission The radioactive decay of an atomic nucleus initiated by the collision of a neutron.

neutron number The number of neutrons in the nucleus of an atom.

newton The SI unit of force; equal to the force that would give a mass of one kilogram an acceleration of one meter per second per second.

Newton's first law An object at rest tends to stay at rest, and an object in uniform motion tends to stay in motion with the same speed and in the same direction, unless acted upon by a net force.

Newton's law of universal gravitation Any two objects exert a gravitational force of attraction on each other in a direction along the line joining the objects, with a magnitude proportional to the product of the masses of the objects and inversely proportional to the square of the distance between them.

Newton's laws of motion Isaac Newton's three fundamental relationships between force and motion.

Newton's second law If a net external force acts on an object, the object accelerates. The net external force is equal to the product of the object's mass and the object's acceleration.

Newton's third law If object A exerts a force on object B, object B exerts a force on object A that has the same magnitude but is in the opposite direction. These two forces act on different objects.

node (of a wave) Any point where the displacement of a wave is always zero.

nonconservative force A dissipative force that does not have a defined potential energy, such as friction.

noninertial frame A frame of reference attached to an accelerated object.

normal A line that is perpendicular to a surface.

normal force The support force exerted upon an object that is in contact with another stable object, acting perpendicular to the surface of contact.

no-slip condition Requirement that the velocity of a fluid be zero next to a solid surface.

***n*-type semiconductor** A type of semiconductor doping that provides extra electrons to the host material, creating an excess of negative electron charge carriers.

nuclear fission A nuclear reaction in which a heavy nucleus splits spontaneously or on impact with another particle, releasing energy.

nuclear fusion A nuclear reaction in which atomic nuclei of low atomic number fuse to form a heavier nucleus, releasing energy.

nuclear radiation The emission by a nucleus of either energy (in the form of a photon) or particles (such as an alpha particle).

nucleon A proton or neutron.

nuclide An atomic species characterized by the specific constitution of its nucleus, that is, by its number of protons and its number of neutrons.

object Anything that acts as a source of light rays for an optical device.

object distance An object's distance from a mirror or lens.

object height The vertical extent of an object that is reflected in a mirror or refracted by a lens.

objective In a microscope, the converging lens used to produce a magnified real image of a small object. In a telescope, the converging lens or mirror used to produce a small real image of a distant object.

ohm The SI unit of electrical resistance; the resistance in a circuit transmitting a current of one ampere when subjected to a potential difference of one volt.

one-dimensional wave A wave that propagates along a single dimension of space.

open pipe A pipe that is unblocked on both ends.

optical device An instrument that changes the direction of light rays in a regular way.

orbital period The time required to complete an orbit.

origin The location from which the points on a coordinate system are measured.

oscillation The regular movement of an object back and forth around a point of equilibrium.

overdamped oscillations Case in which high damping is applied, resulting in a nonoscillatory response; when displaced from equilibrium, the system returns to equilibrium with no overshoot and hence no oscillation.

parabola A particular u-shaped curve; the shape of the path of a projectile under the influence of gravity.

parallel (capacitors) An arrangement of capacitors connected along multiple paths (not in series), resulting in a multiloop circuit; the total capacitance is equal to the sum of all the individual capacitances.

parallel (resistors) An arrangement of resistors connected along multiple paths (not in series), resulting in a multiloop circuit; the reciprocal of the total resistance is equal to the sum of the reciprocals of all the individual resistances.

parallel-axis theorem Expression determining the moment of inertia of an object about a given axis in terms of the moment of inertia about a parallel axis running through its center of mass and the distance between the two axes.

parallel-plate capacitor A capacitor formed using two parallel metal plates.

paramagnetic Tendency to be weakly attracted by the poles of a magnet but not retaining any permanent magnetism; if the magnetic field is turned off, random thermal motion will cause the atomic current loops to return to their original, nonaligned orientations.

partially polarized light Light in which the orientation of the electric field changes randomly but is more likely to be in one orientation than in other orientations.

particle physics The branch of physics that concerns the fundamental constituents of matter and how they interact.

pascal Unit of measurement for pressure; equal to one newton per square meter.

Pascal's principle Pressure applied to a confined, static fluid is transmitted undiminished to every part of the fluid as well as to the walls of the container.

path length difference The difference in the distance traversed by two waves traveling to the same point from different locations.

Pauli exclusion principle The quantum mechanical principle that no two electrons may occupy the same quantum state simultaneously.

pendulum A system that oscillates back and forth due to the restoring force of gravity.

period The time for one complete cycle of an oscillation.

permeability of free space A constant involved in the relationship between an electric current and the magnetic field that it produces in a vacuum.

permittivity of free space A constant involved in the relationship between an electric charge and the electric field that it produces in a vacuum.

phase A physically distinctive form of matter, such as a solid, liquid, gas, or plasma.

phase angle The amount that a wave is shifted, indicating where in the oscillation cycle the object is at $t = 0$.

phase change The transformation from one state of matter to another.

phase diagram A graph of pressure p versus temperature T for a substance, showing the values of p and T for each phase of the substance.

phase difference The mathematical difference between two phase angles, such as the phase angles of two different waves.

photoelectric effect The emission, or ejection, of electrons from the surface of a material in response to incident light.

photoelectrons Electrons emitted through the photoelectric effect.

photon The quantum of electromagnetic energy, regarded as a discrete particle having zero mass, no electric charge, and an indefinitely long lifetime.

photovoltaic solar cell An electrical device that converts the energy of light directly into electricity by the photovoltaic effect.

physical pendulum A pendulum whose mass is distributed throughout its volume.

Planck's constant A physical constant relating the ratio of the energy of a photon to its frequency; equal to 6.626×10^{-34} J·s.

plane mirror A flat, reflecting surface.

plastic The state of an object when the tensile stress exceeds the yield strength and an object deforms permanently.

plates Two pieces of metal used in a capacitor to store charge.

pn junction The region inside a semiconductor where p-type and n-type semiconductors meet.

point charges Very small charged objects whose size is much smaller than the separation between the charges.

polarization The orientation of the electric field in a light wave.

polarizing filter A transparent sheet which contains long-chain molecules that are all oriented in the same direction, used to polarize the light passing through it.

position The location of an object on a coordinate system.

position vector A vector that extends from an origin to a point where an object is located at a specific moment in time.

positive displacement The distance, in the positive direction along a defined coordinate axis, between the position of an object at one time and its position at an earlier time.

positron A subatomic particle with the same mass as an electron and a numerically equal but positive charge.

potential energy An ability to do work based on an object's position.

pound The English unit of force.

power The rate at which work is done or energy is transferred.

power (of a lens or mirror) The reciprocal of the focal length (in meters).

power of ten An alternative name for the exponent given in scientific notation, referring to how many tens must be multiplied together to give the desired number.

pressure The magnitude of the force per unit area on the surface of an object.

pressure amplitude In a sound wave, the maximum pressure variation above or below the pressure of the undisturbed air.

pressure wave (longitudinal wave) *See* longitudinal wave.

primary coil The winding of a transformer connected to the input voltage.

principal axis A line passing through the center of the surface of a lens or spherical mirror and through the centers of curvature of all segments of the lens or mirror.

principle of equivalence A gravitational field is equivalent to an accelerated frame of reference in the absence of gravity.

principle of Newtonian relativity The laws of motion are the same in all inertial frames of reference.

projectile Object undergoing free-fall motion under the influence of gravity.

projectile motion Free-fall motion under the influence of gravity, involving both vertical and horizontal motion.

propagation speed The rate at which a wave travels in a medium.

proper time The time interval between two events in a frame of reference in which the events occur at the same place.

p-type semiconductor A type of semiconductor with an absence of negative charge and hence an abundance of positive charge carriers or holes.

pV diagram A graph that plots the pressure of a system on the vertical axis versus the volume of the system on the horizontal axis.

quantized Restricted to only certain values, as energy.

quantum mechanics Branch of physics that deals with the motions and interactions of atoms and subatomic particles, incorporating the concepts of quantization of energy, wave-particle duality, and the uncertainty principle.

quantum number Number describing the value of a physical quantity in a quantum mechanical system, such as an atom.

quark Any of a number of subatomic particles carrying a fractional electric charge, postulated as building blocks of the hadrons.

quark confinement The phenomenon wherein quarks can never be removed from the hadrons they compose.

radiation Energy transfer by the emission (or absorption) of electromagnetic waves.

radioactive A radioactive nuclide is one that decays into another nuclide by emitting ionizing radiation or particles.

radius of curvature The distance from a curved surface (such as a mirror) to its center of curvature.

ratio of specific heats For a given substance, the ratio of molar specific heat at constant pressure to the molar specific heat at constant volume.

ray An arrow that points in the direction of light propagation.

ray diagram Drawing used to determine the position of an image made by a mirror or lens.

reaction force The force exerted by object A on object B in reaction to having a force exerted by object B on object A.

real image An image formed by light rays coming together.

redshift The displacement of spectral lines toward longer wavelengths (the red end of the spectrum) in radiation from distant galaxies and celestial objects.

reference frame *See* frame of reference.

reflecting telescope A telescope in which the objective is a converging mirror.

refracting telescope A telescope in which the objective is a converging lens.

refraction The change in direction of a beam of light that travels from one medium into another; the angle that the refracted light makes to the normal is not equal to the angle of the incident light to the normal.

refrigerator A device that takes in energy and uses it to transfer heat from an object at low temperature to an object at high temperature.

relative biological effectiveness (RBE) A measure of how much biological effect is produced by a given kind of radiation.

relative velocity The velocity of one object relative to another object.

relativistic gamma A dimensionless quantity that is equal to 1 when an object is at rest and becomes infinitely large as the speed of the object approaches the speed of light.

relativistic speed A speed that is a significant proportion of the speed of light.

reservoir A part of a system large enough either to absorb or supply heat without a change in temperature.

resistance For an electrical conductor, the resistivity of the material of which the conductor is made multiplied by the length of the conductor and divided by its cross-sectional area.

resistivity A measure of how well or poorly a material inhibits the flow of electric charge.

resistor A circuit component intended to add resistance to the flow of current.

resolve To optically distinguish; as in telling two closely spaced objects apart.

resonance The condition in which an object or system is subjected to an oscillating force having a frequency close to its own natural frequency.

rest energy Energy of a particle that is not in motion.

restoring force A force that tends to bring an object back toward equilibrium.

reversible process An ideal process that can return both the system and the surroundings to their original conditions without increasing entropy; throughout the entire reversible process the system is in thermodynamic equilibrium with its surroundings.

Reynolds number The ratio of the forces due to pressure differences acting on a small piece of a fluid to the viscous forces acting on the same piece of fluid.

right-hand rule A rule that uses the right hand to determine the orientation of vector quantities normal to a plane; for example, used to find the direction of the angular momentum vector around an axis of rotation.

rigid object An object with a fixed shape; the distance between any two points in the object remains constant in time regardless of external forces exerted on it.

rolling friction Force acting on the point of contact between a rolling object and the surface opposite to the direction of motion.

rolling without slipping The state in which an object rolls uniformly across a surface without skidding.

root-mean-square speed (rms speed) (of molecules) A measure of how fast gas molecules move; the square root of the average value of the square of individual speeds.

root mean square value (rms value) (of an ac circuit) The square root of the average value of the ac voltage squared.

rotation Motion in which an object spins around an axis.

rotational kinematics The study of rotational motion, including angular velocities and angular acceleration, in the absence of forces.

rotational kinetic energy Energy possessed by an object by virtue of its rotational motion.

scalar A physical quantity that has only magnitude, not direction.

scientific notation A standard shorthand system used by physicists for extremely large or small numbers.

second The SI unit of measurement for time.

second law of thermodynamics The amount of disorder in an isolated system either always increases or, if the system is in equilibrium, stays the same.

secondary coil The winding of a transformer that is the source of the output voltage.

self-inductance The induction of a voltage in a current-carrying coil when the current in the coil itself is changing. The induced emf will oppose any change in the current.

semiconductors Substances with electrical properties that are intermediate between those of insulators and conductors.

semimajor axis Half of the distance of the longest diameter of an ellipse (the major axis).

series (capacitors) An arrangement of capacitors connected along a single path (not in parallel); the reciprocal of the total capacitance is equal to the sum of the reciprocals of the individual capacitances.

series (*LRC* circuit) A circuit containing an inductor, resistor, and capacitor in series.

series (*RC* circuit) A circuit that contains both a resistor and capacitor in series.

series (resistors) An arrangement of resistors connected along a single path (not in parallel); the total resistance is equal to the sum of all the individual resistances.

shear modulus A measure of the rigidity and resistance to deformation of a material.

shear strain The change in an object's shape caused by shear stress.

shear stress Forces applied parallel to the plane in which an object lies, deforming the object without making it expand or contract.

shock wave A sudden pressure increase in a narrow region of a medium (for example, air), such as that caused by a body moving faster than the speed of sound.

significant figures The number of digits in a figure that can be known with some degree of confidence.

simple harmonic motion (SHM) Oscillatory motion under a Hooke's law restoring force (which is proportional to the displacement from the equilibrium position).

simple pendulum A pendulum in which all of the mass is concentrated at a single point.

single-loop circuit A circuit with only a single path that moving charges can follow.

single-slit diffraction Experiment in which a wave passes through a narrow opening, producing a pattern of bright and dark fringes.

sinusoidal function A mathematical curve that describes a smooth repetitive oscillation, such as the sine and cosine functions.

sinusoidal wave A wave in which the wave pattern at any instant is a sinusoidal function.

Snell's law of refraction Expression describing the relationship between the angles of incidence and refraction when referring to light or other waves passing through a boundary between two different media, such as water, glass, and air.

solenoid A straight helical coil of wire.

solid A substance whose individual molecules cannot move freely but remain in essentially fixed positions relative to one another.

sonic boom A loud, explosive noise caused by the shock wave from an aircraft traveling faster than the speed of sound.

sound intensity level The power carried by sound waves per unit area.

sound wave A wave in air consisting of periodic variations in air pressure.

source of emf A device that originates voltage in a circuit, such as a battery.

special theory of relativity Theory developed by Albert Einstein that states: (1) All laws of physics are the same in all inertial frames, and (2) the speed of light in a vacuum is the same in all inertial frames, independent of both the speed of the source of the light and the speed of the observer.

specific gravity The density of a substance divided by the density of 4°C liquid water.

specific heat The amount of heat per unit mass required to raise the temperature by one kelvin.

specular reflection Type of surface reflection in which light rays moving in a single direction reflect from a smooth surface in a single outgoing direction.

speed The rate at which an object is moving; equal to the magnitude of the object's velocity.

speed of light The speed at which all electromagnetic waves—including radio waves, x rays, and others—travel in a vacuum.

spin An intrinsic characteristic of electrons, protons, and neutrons akin to the angular momentum of a rotating sphere.

spring constant Measure of the stiffness of a spring.

spring potential energy The energy stored in a spring, based on whether it is relaxed, stretched, or compressed.

Standard Model A mathematical description of the elementary particles of matter and the electromagnetic, weak, and strong forces by which they interact.

standing wave A wave in which each point in the medium has a constant amplitude, giving it the appearance of being stationary.

standing wave mode The conditions under which a standing wave is possible; for example, a whole number of half-wavelengths must fit onto a string.

state variables Quantities such as volume, temperature, pressure, and internal energy that depend on the state or condition of a substance.

static friction Force acting on a stationary object that opposes the object's sliding motion.

steady flow Type of fluid flow in which the flow pattern does not change with time.

Stefan–Boltzmann constant The constant of proportionality in the Stefan–Boltzmann law: The total energy radiated per unit surface area of a blackbody in unit time is proportional to the fourth power of the thermodynamic temperature; equal to 5.670×10^{-8} W/(m$^2 \cdot$K^4).

step-down transformer A transformer in which the number of windings in the secondary coil is less than the number of windings in the primary coil, so the output voltage is less than the input voltage.

step-up transformer A transformer in which the number of windings in the secondary coil is greater than the number of windings in the primary coil, so the output voltage is greater than the input voltage.

strain The amount of deformation that results from applied stress.

streamlines The paths followed by bits of fluid in laminar flow.

stress Force per area exerted on an object tending to cause the object to change in size or shape.

strong nuclear force An attractive force between protons and neutrons that is stronger than the repulsive electric force between protons.

sublimation The phase change from solid directly to gas.

surface charge density The amount of charge per unit area on a surface.

surface tension The attractive force exerted on molecules at the surface of a liquid by the molecules beneath, causing the liquid to assume the shape having the least surface area.

surface wave A wave that propagates along the interface between two media (e.g., a seismic wave that travels along the surface of the Earth).

Système International (SI) The standard system of units based on the fundamental quantities.

telescope An optical device used to make images of distant objects.

temperature A measure of the kinetic energy associated with molecular motion.

tensile strain The distance an object stretches when pulled, expressed as a fraction of its relaxed length.

tensile stress Stretching force applied to an object divided by the object's cross-sectional area.

tension Stretching force applied at each end of an object; for example, the force exerted by a rope on an object it tows.

tesla The SI units of magnetic field strength.

test charge A point charge with such a small magnitude that it negligibly affects the field in which it is placed.

thermal conductivity A measure of how easily heat passes through a specified material.

thermal contact A state in which two or more systems can exchange thermal energy.

thermal energy Energy associated with the random motion of atoms and molecules.

thermal equilibrium The condition in which two objects in physical contact exchange no heat energy; in thermal equilibrium the objects are said to be at the same temperature.

thermal expansion The tendency of matter to increase in length, area, or volume in response to an increase in temperature.

thermodynamic process Any process that changes the state of a system.

thermodynamics The branch of physics that deals with relationships among properties of substances such as temperature, pressure, and volume, as well as the energy and flow of energy associated with these properties.

thermometer Instrument used to measure temperature changes.

thin film A very fine layer of a substance on a supporting material; for example, the thin lining behind the retina of the eyes of some animals.

thin lens A lens with a thickness that is negligible compared to the radii of curvature of the lens surfaces.

third law of thermodynamics It is possible for the temperature of a system to be arbitrarily close to absolute zero, but it can never reach absolute zero.

tides The tendency of an object (such as Earth's oceans) to elongate due to the gravitational influence of another object (such as the Moon).

time constant The product of resistance multiplied by capacitance.

time dilation A difference of elapsed time between two events as measured by observers either moving relative to each other or located at different distances from large gravitational masses.

time interval A set length of time.

torque A force that causes rotation.

torr A unit measuring pressure; used especially in measuring partial vacuums; equal to 1/760 atm or 133.32 pascals.

total internal reflection The phenomenon occurring when a wave strikes a medium boundary at an angle larger than the critical angle with respect to the normal; 100% of the incident light is reflected back into the first medium.

total mechanical energy The sum of kinetic and potential energy.

total momentum The momentum of a system of objects.

trajectory The path followed by a projectile or an object.

transformer A device that can raise or lower an ac voltage to a desired value.

translation Motion in which an object as a whole moves through space.

translational kinetic energy The energy possessed by an object by virtue of its motion as a whole through space.

transverse wave A traveling disturbance or vibration in which the individual parts of the wave medium move in a direction perpendicular to the direction of wave propagation.

traveling wave A wave of the form $y(x, t) = A \cos(kx - \omega t)$ in which a disturbance propagates from one location to another.

triple point The particular combination of pressure and temperature of a material at which the solid, liquid, and vapor phases all coexist.

turbulent flow Type of fluid motion in which the velocity of the flow at any point is continuously undergoing changes in both magnitude and direction.

two-dimensional motion Vertical and horizontal movement of an object; motion confined to a plane.

ultimate strength The maximum tensile stress that a material can withstand before failure.

underdamped oscillation Case in which a small damping coefficient applied to an oscillating system causes the system to oscillate with ever-decreasing amplitude.

uniform circular motion The motion of an object going around a circular path at a constant speed.

uniform density Constant mass density throughout a volume.

unit conversion The process of changing the value of a quantity from one set of units to another, such as converting a length from meters to kilometers.

units The standard measurement for a specific quantity; for example, the second is a unit of measurement for time.

unpolarized light Natural light such as that emitted by the Sun or an ordinary light bulb, in which the orientation of the electric field changes randomly from one moment to the next.

unsteady flow Fluid motion in which the velocity changes with time.

upright image Image (produced by a lens or mirror) that has the same orientation as the object.

vaporization The phase change from solid to liquid.

vector A physical quantity that has both a magnitude and a direction.

vector addition The combination of two or more displacements to find the vector sum.

vector difference The result of subtracting one vector from another.

vector multiplication by a scalar An operation in which the product of a scalar and a vector pointing in a given direction is a new vector that has a magnitude equal to the product of the absolute value of the scalar and the magnitude of the original vector. The direction of the new vector is either the same as that of the original vector (if the scalar is positive) or opposite to it (if the scalar is negative).

vector product   *See* cross product.

vector subtraction   The process of taking a vector difference; the inverse of vector addition.

vector sum   The result of vector addition.

velocity   The magnitude and direction of the rate of change of an object's position.

velocity vector (*or* **instantaneous velocity vector**)   The rate of change of the position of an object at a given point in time, expressed as a magnitude (speed) and direction.

virtual image   An image from which rays of reflected or refracted light appear to diverge; for example, the image seen in a plane mirror.

virtual particle   A particle whose existence is allowed by the uncertainty principle and that exhibits many of the characteristics of an ordinary particle, but that exists for a limited time.

viscosity   A measure of the resistance to flow of a fluid.

visible light   The range of wavelengths visible to the human eye.

volt   The SI unit of electromotive force or electric potential; the emf required to drive one ampere of current against one ohm resistance.

voltage   Electric potential difference.

volume flow rate   Volume of fluid per unit time passing a given point.

volume strain   The change in volume of an object under stress divided by the original volume.

volume stress   Force applied perpendicularly to all faces of an object; pressure change.

v_x–t graph   Chart depicting an object's velocity along the x axis versus time.

watt   The SI unit of power; equal to one joule per second.

wave   A disturbance or vibration that travels through space.

wave function   A mathematical description of the properties of a wave, expressing the displacement of the wave medium at every position and at every time.

wavelength   The distance between successive crests of a wave.

wavelet   A tiny segment of a larger wave.

wave-particle duality   The dual character of both light and matter, which have both wave and particle characteristics.

weak force   An interaction between elementary particles, often involving neutrinos or antineutrinos, that is responsible for certain kinds of radioactive decay.

weight   The magnitude of the gravitational force that acts on an object.

weighted average   A mean calculated by giving values in a data set more influence according to some attribute of the data, such as how often a given value appears in the data set.

work   The transfer of energy from one object to another.

work-energy theorem   When an object undergoes a displacement, the work done on it by the net force equals the object's kinetic energy at the end of the displacement minus its kinetic energy at the beginning of the displacement.

work function   The minimum amount of energy required to remove a single electron from a material.

x component   The component of a vector parallel to the x axis.

x–t graph   Chart depicting an object's position along the x axis versus time.

x–y plane   The coordinate plane formed by the x axis and the y axis.

y component   The component of a vector parallel to the y axis.

yield strength   The tensile strength at which an object is permanently deformed and can no longer return to its normal strength.

Young's modulus   A measure of the stiffness of a given material.

zeroth law of thermodynamics   If two objects are each in thermal equilibrium with a third object, they are also in thermal equilibrium with each other.

Math Tutorial

In this tutorial, we review some of the basic results of algebra, geometry, trigonometry, and calculus. In many cases, we merely state results without proof. Table M-1 lists some mathematical symbols.

M-1 Significant figures

Many numbers we work with in science are the result of measurement and are therefore known only within a degree of uncertainty. This uncertainty should be reflected in the number of digits used. For example, if you have a 1-meter-long rule with scale spacing of 1 cm, you know that you can measure the height of a box to within a fifth of a centimeter or so. Using this rule, you might find that the box height is 27.0 cm. If there is a scale with a spacing of 1 mm on your rule, you might perhaps measure the box height to be 27.03 cm. However, if there is a scale with a spacing of 1 mm on your rule, you might not be able to measure the height more accurately than 27.03 cm because the height might vary by 0.01 cm or so, depending on where you measure the height of the box. When you write down that the height of the box is 27.03 cm, you are stating that your best estimate of the height is 27.03 cm, but you are not claiming that it is exactly 27.030000... cm high. The four digits in 27.03 cm are called significant figures. Your measured length, 27.03 cm, has four significant digits. Significant figures are also called significant digits.

The number of significant digits in an answer to a calculation will depend on the number of significant digits in the given data. When you work with numbers that have uncertainties, you should be careful not to include more digits than the certainty of measurement warrants. *Approximate* calculations (order-of-magnitude estimates) always result in answers that have only one significant digit or none. When you multiply, divide, add, or subtract numbers, you must consider the accuracy of the results. Listed below are some rules that will help you determine the number of significant digits of your results.

(1) When multiplying or dividing quantities, the number of significant digits in the final answer is no greater than that in the quantity with the fewest significant digits.

(2) When adding or subtracting quantities, the number of decimal places in the answer should match that of the term with the smallest number of decimal places.

(3) Exact values have an unlimited number of significant digits. For example, a value determined by counting, such as 2 tables, has no uncertainty and is an exact value. In addition, the conversion factor 0.0254000... m/in. is an exact value because 1.000... inches is exactly equal to 0.0254000... meters. (The yard is, by definition, equal to exactly 0.9144 m, and 0.9144 divided by 36 is exactly equal to 0.0254.)

(4) Sometimes zeros are significant and sometimes they are not. If a zero is before a leading nonzero digit, then the zero is not significant. For example, the number 0.00890 has three significant digits. The first three zeroes are not significant digits but are merely markers to locate the decimal point. Note that the zero after the nine is significant.

TABLE M-1
Mathematical Symbols

$=$	is equal to
$\neq$	is not equal to
$\approx$	is approximately equal to
$\sim$	is of the order of
$\propto$	is proportional to
$>$	is greater than
$\geq$	is greater than or equal to
$\gg$	is much greater than
$<$	is less than
$\leq$	is less than or equal to
$\ll$	is much less than
Δx	change in x
$\lvert x \rvert$	absolute value of x
$n!$	$n(n-1)(n-2)... 1$
Σ	Sum

(5) Zeros that are between nonzero digits are significant. For example, 5603 has four significant digits.

(6) The number of significant digits in numbers with trailing zeros and no decimal point is ambiguous. For example, 31,000 could have as many as five significant digits or as few as two significant digits. To prevent ambiguity, you should report numbers by using scientific notation or by using a decimal point.

EXAMPLE M-1   Finding the Average of Three Numbers

Find the average of 19.90, −7.524, and −11.8179.

Set Up

You will be adding three numbers and then dividing the result by 3. The first number has four significant digits, the second number has four, and the third number has six.

Solve

Sum the three numbers.

$$19.90 + (-7.524) + (-11.8179) = 0.5581$$

If the problem only asked for the sum of the three numbers, we would round the answer to the least number of decimal places among all the numbers being added—the answer would be 0.56 (0.5581 rounds up to 0.56 to two significant digits). However, we must divide this intermediate result by 3, so we use the intermediate answer with the two extra digits (italicized and red).

$$\frac{0.5581}{3} = 0.1860333...$$

Only two of the digits in the intermediate answer, 0.5581..., are significant digits, so we must round the final number to get our final answer. The number 3 in the denominator is a whole number and has an unlimited number of significant digits. So the final answer has the same number of significant digits as the numerator, which is two.

The final answer is 0.19.

Reflect

The sum in step 1 has two significant digits following the decimal point, the same as the number being summed with the least number of significant digits after the decimal point.

M-2 Equations

An equation is a statement written using numbers and symbols to indicate that two quantities, written on either side of an equal sign (=), are equal. The quantity on either side of the equal sign may consist of a single term, or of a sum or difference of two or more terms. For example, the equation $x = 1 - (ay + b)/(cx - d)$ contains three terms, x, 1, and $(ay + b)/(cx - d)$.

You can perform the following operations on equations:

(1) The same quantity can be added to or subtracted from each side of an equation.
(2) Each side of an equation can be multiplied or divided by the same quantity.
(3) Each side of an equation can be raised to the same power.

These operations are meant to be applied to each *side* of the equation rather than each term in the equation. (Because multiplication is distributive over addition, operation 2— and only operation 2—of the preceding operations also applies term by term.)

Caution: Division by zero is forbidden at any stage in solving an equation; results (if any) would be invalid.

Adding or Subtracting Equal Amounts
To find x when $x - 3 = 7$, add 3 to both sides of the equation: $(x - 3) + 3 = 7 + 3$, so $x = 10$.

Multiplying or Dividing by Equal Amounts
If $3x = 17$, solve for x by dividing both sides of the equation by 3; we get $x = \dfrac{17}{3}$, or 5.7.

EXAMPLE M-2 **Simplifying Reciprocals in an Equation**

Solve the following equation for x.

$$\frac{1}{x} + \frac{1}{4} = \frac{1}{3}$$

Equations containing reciprocals of unknowns occur in many circumstances in physics. Two instances of this are geometric optics and electric circuit analysis.

Set Up

In this equation, the term containing x is on the same side of the equation as a term not containing x. Furthermore, x is found in the denominator of a fraction. We'll start by isolating the $1/x$ term, find common denominators, and then multiply both sides of the equation by appropriate quantities.

Solve

Subtract $\dfrac{1}{4}$ from each side.	$\dfrac{1}{x} = \dfrac{1}{3} - \dfrac{1}{4}$
Simplify the right side of the equation by using the lowest common denominator.	Begin by multiplying both terms on the right-hand side by appropriate forms of 1. $$\frac{1}{x} = \frac{1}{3}\left(\frac{4}{4}\right) - \frac{1}{4}\left(\frac{3}{3}\right) = \frac{4}{12} - \frac{3}{12}$$ $$= \frac{4-3}{12} = \frac{1}{12} \quad \text{so} \quad \frac{1}{x} = \frac{1}{12}$$
Multiply both sides of the equation by $12x$ to determine the value of x.	$$12x\,\frac{1}{x} = 12x\,\frac{1}{12}$$ $$12 = x$$

Reflect

To check our answer, substitute 12 for x in the left side of the original equation.	$\dfrac{1}{x} + \dfrac{1}{4} = \dfrac{1}{12} + \dfrac{3}{12} = \dfrac{4}{12} = \dfrac{1}{3}$

M-3 Direct and inverse proportions

When we say variable quantities x and y are **directly proportional**, we mean that as x and y change, the ratio x/y is constant. To say that two quantities are proportional is to say that they are directly proportional. When we say variable quantities x and y are inversely proportional, we mean that as x and y change, the ratio xy is constant.

Relationships of direct and inverse proportion are common in physics. Objects moving at the same velocity have momenta directly proportional to their masses. The ideal gas law ($PV = nRT$) states that pressure P is directly proportional to (absolute) temperature T, when volume V remains constant, and is inversely proportional to volume, when temperature remains constant. Ohm's law ($V = iR$) states that the voltage V across a resistor is directly proportional to the electric current in the resistor when the resistance remains constant.

Constant of Proportionality

When two quantities are directly proportional, the two quantities are related by a *constant of proportionality*. If you are paid for working at a regular rate R in dollars per day, for example, the money m you earn is directly proportional to the time t you work; the rate R is the constant of proportionality that relates the money earned in dollars to the time worked t in days.

$$\frac{m}{t} = R \quad \text{or} \quad m = Rt$$

If you earn \$400 in 5 days, the value of R is $\$400/(5 \text{ days}) = \$80/\text{day}$. To find the amount you earn in 8 days, you could perform the calculation

$$m = (\$80/\text{day})(8 \text{ days}) = \$640$$

In some proportion problems you don't need to know the value of the constant of proportionality. Because the amount you earn in 8 days is $\frac{8}{5}$ times what you earn in 5 days, this amount is

$$m_{8 \text{ days}} = (8 \text{ days})\left(\frac{\$400}{5 \text{ days}}\right) = \$640$$

EXAMPLE M-3 Painting Cubes

You need 15.4 mL of paint to cover one side of a cube. The area of one side of the cube is 426 cm². What is the relation between the volume of paint needed and the area to be covered? How much paint do you need to paint one side of a cube on which the one side has an area of 503 cm²?

Set Up

To determine the amount of paint for the side whose area is 503 cm² we will set up a proportion.

Solve

The volume V of paint needed increases in proportion to the area A to be covered.	V and A are directly proportional. That is, $\dfrac{V}{A} = k$ or $V = kA$ where k is the proportionality constant.
Determine the value of the proportionality constant using the given values $V_1 = 15.4$ mL and $A_1 = 426$ cm².	$k = \dfrac{V_1}{A_1} = \dfrac{15.4 \text{ mL}}{426 \text{ cm}^2} = 0.0362 \text{ mL/cm}^2$
Determine the volume of paint needed to paint a side of a cube whose area is 503 cm² using the proportionality constant in step 1.	$V_2 = kA_2 = (0.0362 \text{ mL/cm}^2)(503 \text{ cm}^2)$ $\qquad = 18.2 \text{ mL}$

Reflect

Our value for V_2 is greater than the value for V_1, as expected. The amount of paint needed to cover an area equal to 503 cm² should be greater than the amount of paint needed to cover an area of 426 cm² because 503 cm² is larger than 426 cm².

M-4 Linear equations

A **linear equation** is an equation of the form $x + 2y - 4z = 3$. That is, an equation is linear if each term either is constant or is the product of a constant and a variable raised to the first power. Such equations are said to be linear because the plots of these equations form straight lines or planes. The equations of direct proportion between two variables are linear equations.

Graph of a Straight Line

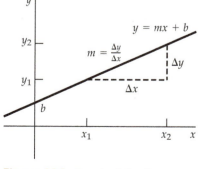

Figure M-1 Graph of the linear equation $y = mx + b$, where b is the y intercept and $m = \Delta y/\Delta x$ is the slope

A linear equation relating y and x can always be put into the standard form

(M-1)
$$y = mx + b$$

where m and b are constants that may be either positive or negative. Figure M-1 shows a graph of the values of x and y that satisfy Equation M-1. The constant b, called the y **intercept**, is the value of y at $x = 0$. The constant m is the slope of the line, which equals the ratio of the change in y to the corresponding change in x. In the figure, we have indicated two points on the line, (x_1, y_1) and (x_2, y_2), and the changes $\Delta x = x_2 - x_1$ and $\Delta y = y_2 - y_1$. The slope m is then

$$m = \frac{y_2 - y_1}{x_2 - x_1} = \frac{\Delta y}{\Delta x}$$

If x and y are both unknown in the equation $y = mx + b$, there are no unique values of x and y that are solutions to the equation. Any pair of values (x_1, y_1) on the line in Figure M-1 will satisfy the equation. If we have two equations, each with the same two unknowns x and y, we can solve the equations simultaneously for the unknowns. Example M-4 shows two methods for simultaneously solving two linear equations.

EXAMPLE M-4 Using Two Equations to Solve for Two Unknowns

Find the values of x and y that simultaneously satisfy

$$3x - 2y = 8 \qquad \text{(M-2)}$$

and

$$y - x = 2 \qquad \text{(M-3)}$$

Set Up

Graph the two equations (Figure M-2). At the point where the lines intersect, the values of x and y satisfy both equations.

  We can solve two simultaneous equations by first solving either equation for one variable in terms of the other variable and then substituting the result into the second equation.

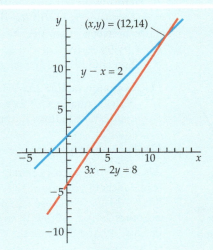

Figure M-2 Graph of Equations M-2 and M-3. At the point where the lines intersect, the values of x and y satisfy both equations.

Solve

Solve Equation M-3 for y.	$y = x + 2$
Substitute this value for y into Equation M-2.	$3x - 2(x + 2) = 8$
Simplify the equation and solve for x.	$3x - 2x - 4 = 8$
	$x - 4 = 8$
	$x = 12$

Use your solution for x and one of the given equations to find the value of y.	Return to Equation M-3 and substitute $x = 12$.
	$y - x = 2$, where $x = 12$
	$y - 12 = 2$
	$y = 2 + 12 = 14$

Reflect

An alternative method is to multiply one equation by a constant such that one of the unknown terms is eliminated when the equations are added or subtracted.	We can multiply both sides of Equation M-3 by 2:
	$2(y - x) = 2(2)$
	$2y - 2x = 4$
Add the result to Equation M-2 and solve for x.	$\cancel{2y} - 2x = 4$
	$3x - \cancel{2y} = 8$
	$\overline{3x - 2x = 12 \text{ so } x = 12}$
Substitute into Equation M-3 and solve for y.	$y - 12 = 2 \text{ so } y = 14$

M-5 Quadratic equations and factoring

A **quadratic equation** is an equation of the form $ax^2 + bxy + cy^2 + ex + fy + g = 0$, where x and y are variables and a, b, c, e, f, and g are constants. In each term of the equation the powers of the variables are integers that sum to 2, 1, or 0. The designation *quadratic equation* usually applies to a much simpler equation of one variable which we can write in the standard form

(M-4)
$$ax^2 + bx + c = 0$$

where a, b, and c are constants. The quadratic equation has two solutions or *roots*—values of x for which the equation is true.

Factoring

We can solve some quadratic equations by **factoring**. Very often we can group or organize the terms of an equation into other terms. When we factor terms, we look for multipliers and multiplicands—which we now call factors—that will yield two or more new terms as a product. For example, we can find the roots of the quadratic equation $x^2 - 3x + 2 = 0$ by factoring the left side to get $(x - 2)(x - 1) = 0$. The roots are $x = 2$ and $x = 1$.

Factoring is useful for simplifying equations and for understanding the relationships between quantities. You should be familiar with the multiplication of the factors $(ax + by)(cx + dy) = acx^2 + (ad + bc)xy + bdy^2$.

You should readily recognize some typical factorable combinations:

(1) Common factor: $2ax + 3ay = a(2x + 3y)$
(2) Perfect square: $x^2 - 2xy + y^2 = (x - y)^2$ (If the expression on the left side of a quadratic equation in standard form is a perfect square, the two roots will be equal.)
(3) Difference of squares: $x^2 - y^2 = (x + y)(x - y)$

Also, look for factors that are prime numbers (2, 5, 7, and so on) because these factors can help you simplify terms quickly. For example, we can simplify the equation $98x^2 - 140 = 0$ because 98 and 140 share the common factor 2. That is, $98x^2 - 140 = 0$ becomes $2(49x^2 - 70) = 0$, so we have $49x^2 - 70 = 0$. We can simplify this result further because 49 and 70 share the common factor 7. So $49x^2 - 70 = 0$ becomes $7(7x^2 - 10) = 0$, so we have $7x^2 - 10 = 0$.

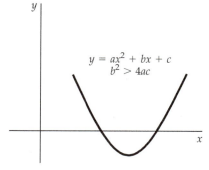

Figure M-3 Graph of y versus x when $y = ax^2 + bx + c$ for the case $b^2 > 4ac$. The two values of x for which $y = 0$ satisfy the quadratic equation (Equation M-4).

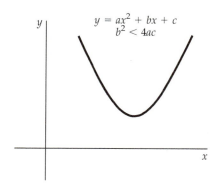

Figure M-4 Graph of y versus x when $y = ax^2 + bx + c$ for the case $b^2 < 4ac$. In this case, there are no real values of x for which $y = 0$.

The Quadratic Formula

Not all quadratic equations can be solved by factoring. However, we can solve *any* quadratic equation in the standard form $ax^2 + bx + c = 0$ by using the quadratic formula,

(M-5)
$$x = \frac{-b \pm \sqrt{b^2 - 4ac}}{2a} = -\frac{b}{2a} \pm \frac{1}{2a}\sqrt{b^2 - 4ac}$$

When b^2 is greater than $4ac$, there are two solutions corresponding to the + and – signs, respectively. Figure M-3 shows a graph of y versus x where $y = ax^2 + bx + c$. The curve, a parabola, crosses the x axis twice. (The simplest representation of a parabola in (x,y) coordinates is an equation of the form $y = ax^2 + bx + c$.) The two roots of this equation are the values for which $y = 0$; that is, they are the x intercepts.

When b^2 is less than $4ac$, the graph of y versus x does not intersect the x axis, as is shown in Figure M-4; there are still two roots, but they are not real numbers. When $b^2 = 4ac$, the graph of y versus x is tangent to the x axis at the point $x = -b/2a$; the two roots are each equal to $-b/2a$.

EXAMPLE M-5 Factoring a Second-Degree Polynomial

Factor the expression $6x^2 + 19xy + 10y^2$.

Set Up

We examine the coefficients of the terms to see whether we can factor the expression without resorting to more advanced methods. We'll compare our expression to the general one $(ax + by)(cx + dy) = acx^2 + (ad + bc)xy + bdy^2$.

Solve

The coefficient of x^2 is 6, which we can factor in two ways.

$ac = 6$

$3 \cdot 2 = 6$ or $6 \cdot 1 = 6$

The coefficient of y^2 is 10, which we can also factor in two ways. | $bd = 10$
 $5 \cdot 2 = 10$ or $10 \cdot 1 = 10$

List the possibilities for a, b, c, and d in a table. Include a column for $ad + bc$.

    If $a = 3$, then $c = 2$, and vice versa. In addition, if $a = 6$, then $c = 1$, and vice versa. For each value of a there are four values for b.

a	b	c	d	$ad + bc$
3	5	2	2	16
3	2	2	5	19
3	10	2	1	23
3	1	2	10	32
2	5	3	2	19
2	2	3	5	16
2	10	3	1	32
2	1	3	10	23
6	5	1	2	17
6	2	1	5	32
6	10	1	1	16
6	1	1	10	61
1	5	6	2	32
1	2	6	5	17
1	10	6	1	61
1	1	6	10	16

Find a combination such that $ad + bc = 19$. The table shows that there are two such combinations.

$ad + bc = 19$
$3 \cdot 5 + 2 \cdot 2 = 19$ and
$2 \cdot 2 + 5 \cdot 3 = 19$

It doesn't matter which combination we choose. To finish this problem, we will use the combination in the second row of the table to factor the expression in question.

$6x^2 + 19xy + 10y^2 = (3x + 2y)(2x + 5y)$

Reflect

As a check, expand $(3x + 2y)(2x + 5y)$ to see if we return to the original equation.

$$(3x + 2y)(2x + 5y) = 6x^2 + 15xy + 4xy + 10y^2$$
$$= 6x^2 + 19xy + 10y^2$$

You should be able to show that the combination in the fifth row is also an acceptable factoring.

M-6 Exponents and logarithms

Exponents

The notation x^n stands for the quantity obtained by multiplying x by itself n times. For example, $x^2 = x \cdot x$ and $x^3 = x \cdot x \cdot x$. The quantity n is called the power, or the exponent, of x (the base). Listed below are some rules that will help you simplify terms that have exponents.

(1) When two powers of x are multiplied, the exponents are added.

$$(x^m)(x^n) = x^{m+n} \tag{M-6}$$

    Example: $x^2 x^3 = x^{2+3} = (x \cdot x)(x \cdot x \cdot x) = x^5$.

(2) Any number (except 0) raised to the 0 power is defined to be 1.

$$x^0 = 1 \tag{M-7}$$

(3) Based on rule 2,

$$x^n x^{-n} = x^0 = 1$$

$$x^{-n} = \frac{1}{x^n} \tag{M-8}$$

(4) When two powers are divided, the exponents are subtracted.

$$\frac{x^n}{x^m} = x^n x^{-m} = x^{n-m} \tag{M-9}$$

(5) When a power is raised to another power, the exponents are multiplied.

(M-10)
$$(x^n)^m = x^{nm}$$

(6) When exponents are written as fractions, they represent the roots of the base. For example,

$$x^{1/2} \cdot x^{1/2} = x$$

so

$$x^{1/2} = \sqrt{x} \quad (x > 0)$$

EXAMPLE M-6   Simplifying a Quantity That Has Exponents

Simplify $\dfrac{x^4 x^7}{x^8}$.

Set Up

According to rule 1, when two powers of x are multiplied, the exponents are added.

$$(x^m)(x^n) = x^{m+n} \qquad \text{(M-6)}$$

Rule 4 states that when two powers are divided, the exponents are subtracted.

$$\frac{x^n}{x^m} = x^n x^{-m} = x^{n-m} \qquad \text{(M-9)}$$

Solve

Simplify the numerator $x^4 x^7$ using rule 1.

$$x^4 x^7 = x^{4+7} = x^{11}$$

Simplify $\dfrac{x^{11}}{x^8}$ using rule 4.

$$\frac{x^{11}}{x^8} = x^{11} x^{-8} = x^{11-8} = x^3$$

Reflect

Use the value $x = 2$ to test our answer.

$$\frac{2^4 2^7}{2^8} = 2^3 = 8$$

$$\frac{2^4 2^7}{2^8} = \frac{(16)(128)}{256} = \frac{2048}{256} = 8$$

Logarithms

We can express any positive number y as some power x of any other positive number a (except the number one). That is, $y = a^x$. The number x is said to be the **logarithm** of y to the **base** a, which we can express as

$$x = \log_a y$$

In other words, logarithms are *exponents*. Hence the rules for working with logarithms correspond to similar rules for exponents. Here are some of these rules that will help you simplify quantities that have logarithms.

(1) If $y_1 = a^n$ and $y_2 = a^m$, then

$$y_1 y_2 = a^n a^m = a^{n+m}$$

Correspondingly,

(M-11)
$$\log_a y_1 y_2 = \log_a a^{n+m} = n + m = \log_a a^n + \log_a a^m = \log_a y_1 + \log_a y_2$$

It then follows that

(M-12)
$$\log_a y^n = n \log_a y$$

(2) Because $a^1 = a$ and $a^0 = 1$,

(M-13)
$$\log_a a = 1$$

and

(M-14)
$$\log_a 1 = 0$$

There are two bases in common use: Logarithms to base 10 are called **common logarithms**, and logarithms to base e (where $e = 2.718...$) are called **natural logarithms**.

In this text, we use the symbol ln for natural logarithms and the symbol log, without a subscript, for common logarithms. So

$$\log_e x = \ln x \quad \text{and} \quad \log_{10} x = \log x \qquad \text{(M-15)}$$

and $y = \ln x$ implies

$$x = e^y \qquad \text{(M-16)}$$

We can easily change logarithms from one base to another. Suppose that

$$z = \log x \qquad \text{(M-17)}$$

Then

$$10^z = 10^{\log x} = x \qquad \text{(M-18)}$$

Taking the natural logarithm of both sides of Equation M-18, we obtain

$$z \ln 10 = \ln x$$

Substituting $\log x$ for z (see Equation M-17) gives the useful result

$$\ln x = (\ln 10)\log x \qquad \text{(M-19)}$$

EXAMPLE M-7   Converting Between Common Logarithms and Natural Logarithms

The steps leading to Equation M-19 show that, in general, $\log_b x = (\log_b a) \log_a x$. So converting logarithms from one base to another requires only multiplication by a constant. Describe the mathematical relation between the constant for converting common logarithms to natural logarithms and the constant for converting natural logarithms to common logarithms.

Set Up

We have a general mathematical formula for converting logarithms from one base to another. We look for the mathematical relation by exchanging a for b and vice versa in the formula.

Solve

We have a formula for converting logarithms from base a to base b.	$\log_b x = (\log_b a) \log_a x$
To convert from base b to base a, exchange all a for b and vice versa.	$\log_a x = (\log_a b) \log_b x$
Divide both sides of the equation in step 1 by $\log_a x$.	$\dfrac{\log_b x}{\log_a x} = \log_b a$
Divide both sides of the equation in step 2 by $(\log_a b)\log_a x$.	$\dfrac{1}{\log_a b} = \dfrac{\log_b x}{\log_a x}$
The results show that the conversion factors $\log_b a$ and $\log_a b$ are reciprocals of one another.	$\dfrac{1}{\log_a b} = \log_b a$

Reflect

For the value of $\log_{10} e$, your calculator will give 0.43429. For ln 10, your calculator will give 2.3026. Multiply 0.43429 by 2.3026; you will get 1.0000.

M-7 Geometry

The properties of the most common **geometric figures**—bounded shapes in two or three dimensions whose lengths, areas, or volumes are governed by specific ratios—are a basic analytical tool in physics. For example, the characteristic ratios within triangles give us the laws of *trigonometry* (see Section M-8), which in turn give us the theory of vectors, essential in analyzing motion in two or more dimensions. Circles and spheres are essential for understanding, among other concepts, angular momentum and the probability densities of quantum mechanics.

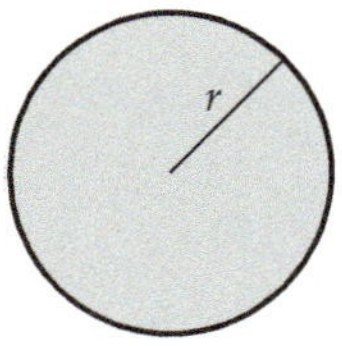

Area of a circle $A = \pi r^2$

Figure M-5 Area of a circle

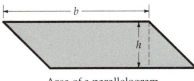

Area of a parallelogram
$A = bh$

Figure M-6 Area of a parallelogram

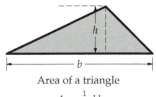

Area of a triangle
$A = \frac{1}{2} bh$

Figure M-7 Area of a triangle

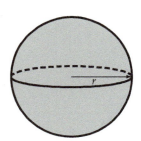

Surface area of a sphere
$A = 4\pi r^2$
Volume of a sphere
$V = \frac{4}{3}\pi r^3$

Figure M-8 Surface area and volume of a sphere

Basic Formulas in Geometry

Circle The ratio of the circumference of a circle to its diameter is a number π, which has the approximate value

$$\pi = 3.14159$$

So the circumference C of a circle is related to its diameter d and its radius r by

(M-20) $\qquad C = \pi d = 2\pi r \quad$ circumference of circle

The area of a circle is (Figure M-5)

(M-21) $\qquad A = \pi r^2 \quad$ area of circle

Parallelogram The area of a parallelogram is the base b multiplied by the height h (Figure M-6).

$$A = bh$$

Triangle The area of a triangle is one-half the base multiplied by the height (Figure M-7).

$$A = \frac{1}{2} bh$$

Sphere A sphere of radius r (Figure M-8) has a surface area given by

(M-22) $\qquad A = 4\pi r^2 \quad$ surface area of sphere

and a volume given by

(M-23) $\qquad V = \frac{4}{3}\pi r^3 \quad$ volume of sphere

Cylinder A cylinder of radius r and length L (Figure M-9) has a surface area (not including the end faces) of

(M-24) $\qquad A = 2\pi r L \quad$ surface of cylinder

and volume of

(M-25) $\qquad V = \pi r^2 L \quad$ volume of cylinder

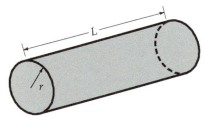

Surface area of the curved sides of a cylinder
$A = 2\pi r L$
Volume of a cylinder
$V = \pi r^2 L$

Figure M-9 Surface area (not including the end faces) and the volume of a cylinder

---

EXAMPLE M-8 Calculating the Volume of a Spherical Shell

An aluminum spherical shell has an outer diameter of 40.0 cm and an inner diameter of 38.0 cm. What is the volume of the aluminum in this shell?

Set Up

The volume of the aluminum in the spherical shell is the volume that remains when we subtract the volume of the inner sphere having $d_i = 2r_i = 38.0$ cm from the volume of the outer sphere having $d_o = 2r_o = 40.0$ cm.

Volume of a sphere:

$$V = \frac{4}{3}\pi r^3 \qquad \text{(M-23)}$$

Solve

Subtract the volume of the sphere of radius r_i from the volume of the sphere of radius r_o.

$$V = V_o - V_i = \frac{4}{3}\pi r_o^3 - \frac{4}{3}\pi r_i^3 = \frac{4}{3}\pi(r_o^3 - r_i^3)$$

Substitute 20.0 cm for r_o and 19.0 cm for r_i.

$$V = \frac{4}{3}\pi[(20.0 \text{ cm})^3 - (19.0 \text{ cm})^3]$$
$$= 4.78 \times 10^3 \text{ cm}^3$$

Reflect

The volume calculated is less than the volume V_o of the outer sphere, as it should be.

$$V_o = \frac{4}{3}\pi r_o^3 = \frac{4}{3}\pi(20.0 \text{ cm})^3$$
$$= 3.35 \times 10^4 \text{ cm}^3$$

M-8 Trigonometry

Trigonometry, which gets its name from Greek roots meaning "triangle" and "measure," is the study of some important mathematical functions, called **trigonometric functions**. These functions are most simply defined as ratios of the sides of right triangles. These right-triangle definitions are of limited use because they are valid only for angles between zero and 90°. However, we can extend the validity of the right-triangle definitions by defining the trigonometric functions in terms of the ratio of the coordinates of points on a circle.

In physics, we first encounter trigonometric functions when we use vectors to analyze motion in two dimensions. Trigonometric functions are also essential in the analysis of any kind of periodic behavior, such as circular motion, oscillatory motion, and wave mechanics.

Angles and Their Measure: Degrees and Radians

The size of an angle formed by two intersecting straight lines is known as its **measure**. The standard way of finding the measure of an angle is to place the angle so that its **vertex**, or point of intersection of the two lines that form the angle, is at the center of a circle located at the origin of a graph that has Cartesian coordinates and one of the lines extends rightward on the positive x axis. The distance traveled *counterclockwise* on the circumference from the positive x axis to reach the intersection of the circumference with the other line defines the measure of the angle. (Traveling clockwise to the second line would simply give us a negative measure; to illustrate basic concepts, we position the angle so that the smaller rotation will be in the counterclockwise direction.)

One of the most familiar units for expressing the measure of an angle is the **degree**, which equals 1/360 of the full distance around the circumference of the circle. For greater precision, or for smaller angles, we either show degrees plus minutes (′) and seconds (″), with $1' = 1°/60$ and $1'' = 1'/60 = 1°/3600$; or show degrees as an ordinary decimal number.

For scientific work, a more useful measure of an angle is the **radian** (rad). Again, place the angle with its vertex at the center of a circle and measure counterclockwise rotation around the circumference. The measure of the angle in radians is then defined as the length of the circular arc from one line to the other divided by the radius of the circle (Figure M-10). If s is the arc length and r is the radius of the circle, the angle θ measured in radians is

$$\theta = \frac{s}{r}$$

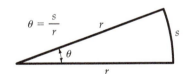

Figure M-10 The angle θ in radians is defined to be the ratio s/r, where s is the arc length intercepted on a circle of radius r.

(M-26)

Because the angle measured in radians is the ratio of two lengths, it is dimensionless. The relation between radians and degrees is

$$360° = 2\pi \text{ rad}$$

or

$$1 \text{ rad} = \frac{360°}{2\pi} = 57.3°$$

Figure M-11 shows some useful relations for angles.

The Trigonometric Functions

Figure M-12 shows a right triangle formed by drawing the line segment BC perpendicular to AC. The lengths of the sides are labeled a, b, and c. The right-triangle

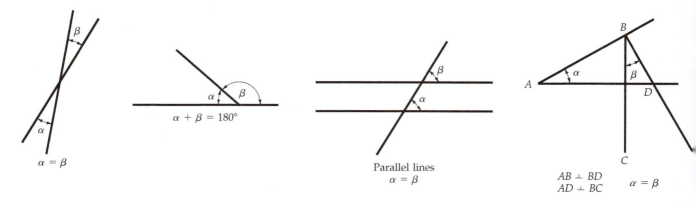

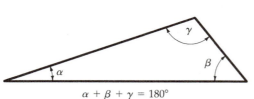
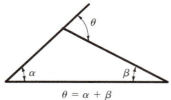

Figure M-11 Some useful relations for angles

definitions of the trigonometric functions $\sin \theta$ (the **sine**), $\cos \theta$ (the **cosine**), and $\tan \theta$ (the **tangent**) for an acute angle θ are

$$\text{(M-27)} \qquad \sin \theta = \frac{a}{c} = \frac{\text{opposite side}}{\text{hypotenuse}}$$

$$\text{(M-28)} \qquad \cos \theta = \frac{b}{c} = \frac{\text{adjacent side}}{\text{hypotenuse}}$$

$$\text{(M-29)} \qquad \tan \theta = \frac{a}{b} = \frac{\text{opposite side}}{\text{adjacent side}} = \frac{\sin \theta}{\cos \theta}$$

(**Acute angles** are angles whose positive rotation around the circumference of a circle measures less than 90° or $\pi/2$.) Three other trigonometric functions—the **secant** (sec), the **cosecant** (csc), and the **cotangent** (cot)—are defined as the reciprocals of these functions.

$$\text{(M-30)} \qquad \csc \theta = \frac{c}{a} = \frac{1}{\sin \theta}$$

$$\text{(M-31)} \qquad \sec \theta = \frac{c}{b} = \frac{1}{\cos \theta}$$

$$\text{(M-32)} \qquad \cot \theta = \frac{b}{a} = \frac{1}{\tan \theta} = \frac{\cos \theta}{\sin \theta}$$

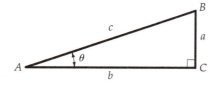

Figure M-12 A right triangle with sides of length a and b and a hypotenuse of length c

The angle θ whose sine is x is called the arcsine of x, and is written $\sin^{-1} x$. That is, if

$$\sin \theta = x$$

then

$$\text{(M-33)} \qquad \theta = \arcsin x = \sin^{-1} x$$

The arcsine is the *inverse* of the sine. We define the inverse of the cosine and tangent similarly. The angle whose cosine is y is the arccosine of y. That is, if

$$\cos \theta = y$$

then

$$\text{(M-34)} \qquad \theta = \arccos y = \cos^{-1} y$$

The angle whose tangent is z is the arctangent of z. That is, if

$$\tan \theta = z$$

then

$$\text{(M-35)} \qquad \theta = \arctan z = \tan^{-1} z$$

Trigonometric Identities

We can derive several useful formulas, called **trigonometric identities**, by examining relationships between the trigonometric functions. One useful tool for discerning such identities is the **Pythagorean theorem**.

$$a^2 + b^2 = c^2 \qquad \text{(M-36)}$$

Simple algebraic manipulation of Equation M-36 gives us three trigonometric identities. First, if we divide each term in Equation M-36 by c^2, we obtain

$$\frac{a^2}{c^2} + \frac{b^2}{c^2} = 1$$

or, from the definitions of $\sin \theta$ (which is a/c) and $\cos \theta$ (which is b/c),

$$\sin^2 \theta + \cos^2 \theta = 1 \qquad \text{(M-37)}$$

Similarly, we can divide each term in Equation M-36 by a^2 or b^2 and obtain

$$1 + \cot^2 \theta = \csc^2 \theta \qquad \text{(M-38)}$$

and

$$1 + \tan^2 \theta = \sec^2 \theta \qquad \text{(M-39)}$$

Table M-2 lists these last three and many more trigonometric identities. Notice that they fall into four categories: functions of sums or differences of angles, sums or differences of squared functions, functions of double angles (2θ), and functions of half angles ($\frac{1}{2}\theta$). Notice that some of the formulas contain paired alternatives, expressed with the signs $\pm$ and $\mp$; in such formulas, remember to always apply the formula with either all the upper alternatives or all the lower ones.

TABLE M-2 **Trigonometric Identities**
$\sin(A \pm B) = \sin A \cos B \pm \cos A \sin B$
$\cos(A \pm B) = \cos A \cos B \mp \sin A \sin B$
$\tan(A \pm B) = \dfrac{\tan A \pm \tan B}{1 \mp \tan A \tan B}$
$\sin A \pm \sin B = 2 \sin\left[\dfrac{1}{2}(A \pm B)\right]\cos\left[\dfrac{1}{2}(A \mp B)\right]$
$\cos A + \cos B = 2 \cos\left[\dfrac{1}{2}(A + B)\right]\cos\left[\dfrac{1}{2}(A - B)\right]$
$\cos A - \cos B = 2 \sin\left[\dfrac{1}{2}(A + B)\right]\sin\left[\dfrac{1}{2}(B - A)\right]$
$\tan A \pm \tan B = \dfrac{\sin(A \pm B)}{\cos A \cos B}$
$\sin^2 \theta + \cos^2 \theta = 1;\ \sec^2 \theta - \tan^2 \theta = 1;$ $\csc^2 \theta - \cot^2 \theta = 1$
$\sin 2\theta = 2 \sin \theta \cos \theta$
$\cos 2\theta = \cos^2 \theta - \sin^2 \theta = 2\cos^2 \theta - 1 = 1 - 2\sin^2 \theta$
$\tan 2\theta = \dfrac{2 \tan \theta}{1 - \tan^2 \theta}$
$\sin\dfrac{1}{2}\theta = \pm\sqrt{\dfrac{1 - \cos \theta}{2}};\ \cos\dfrac{1}{2}\theta = \pm\sqrt{\dfrac{1 + \cos \theta}{2}};$
$\tan\dfrac{1}{2}\theta = \pm\sqrt{\dfrac{1 - \cos \theta}{1 + \cos \theta}}$

Some Important Values of the Functions

Figure M-13 is a diagram of an *isosceles* right triangle (an isosceles triangle is a triangle with two equal sides), from which we can find the sine, cosine, and tangent of 45°. The two acute angles of this triangle are equal. Because the sum of the three angles in a triangle must equal 180° and the right angle is 90°, each acute angle must be 45°. For convenience, let us assume that the equal sides each have a length of 1 unit. The Pythagorean theorem gives us a value for the hypotenuse of

$$c = \sqrt{a^2 + b^2} = \sqrt{1^2 + 1^2} = \sqrt{2} \text{ units}$$

We calculate the values of the functions as follows:

$$\sin 45° = \frac{a}{c} = \frac{1}{\sqrt{2}} = 0.707 \quad \cos 45° = \frac{b}{c} = \frac{1}{\sqrt{2}} = 0.707 \quad \tan 45° = \frac{a}{b} = \frac{1}{1} = 1$$

Another common triangle, a 30°–60°–90° right triangle, is shown in Figure M-14. Because this particular right triangle is in effect half of an *equilateral triangle* (a 60°–60°–60° triangle or a triangle having three equal sides and three equal angles), we can see that the sine of 30° must be exactly 0.5 (Figure M-15). The equilateral triangle must have all sides equal to c, the hypotenuse of the 30°–60°–90° right triangle. It follows that side a is one-half the length of the hypotenuse, and so

$$\sin 30° = \frac{1}{2}$$

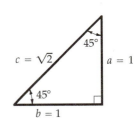

Figure M-13 An isosceles right triangle

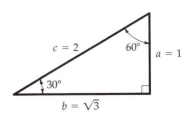

Figure M-14 A 30°–60°–90° right triangle

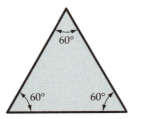 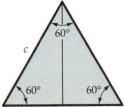

Figure M-15 (a) An equilateral triangle. (b) An equilateral triangle that has been bisected to form two 30°–60°–90° right triangles.

To find the other ratios within the 30°–60°–90° right triangle, let us assign a value of 1 to the side opposite the 30° angle. Then

$$c = \frac{1}{0.5} = 2 \qquad\qquad b = \sqrt{c^2 - a^2} = \sqrt{2^2 - 1^2} = \sqrt{3}$$

$$\cos 30° = \frac{b}{c} = \frac{\sqrt{3}}{2} = 0.866 \qquad \tan 30° = \frac{a}{b} = \frac{1}{\sqrt{3}} = 0.577$$

$$\sin 60° = \frac{b}{c} = \cos 30° = 0.866 \qquad \cos 60° = \frac{a}{c} = \sin 30° = \frac{1}{2}$$

$$\tan 60° = \frac{b}{a} = \frac{\sqrt{3}}{1} = 1.732$$

Small-Angle Approximation

For small angles, the length a is nearly equal to the arc length s, as you can see in Figure M-16. The angle $\theta = s/c$ is therefore nearly equal to $\sin \theta = a/c$.

(M-40)
$$\sin \theta \approx \theta \quad \text{for small values of } \theta$$

Similarly, the lengths c and b are nearly equal, so $\tan \theta = a/b$ is nearly equal to both θ and $\sin \theta$ for small values of θ.

(M-41)
$$\tan \theta \approx \sin \theta \approx \theta \quad \text{for small values of } \theta$$

Equations M-40 and M-41 hold only if θ is measured in radians. Because $\cos \theta = b/c$, and because these lengths are nearly equal for small values of θ, we have

(M-42)
$$\cos \theta \approx 1 \quad \text{for small values of } \theta$$

Figure M-17 shows graphs of θ, $\sin \theta$, and $\tan \theta$ versus θ for small values of θ. If accuracy of a few percent is needed, you can safely use small-angle approximations only for angles of about a quarter of a radian (or about 15°) or less. Below this value, as the angle becomes smaller, the approximation $\theta \approx \sin \theta \approx \tan \theta$ is even more accurate.

Trigonometric Functions as Functions of Real Numbers

So far we have illustrated the trigonometric functions as properties of angles. Figure M-18 shows an *obtuse* angle with its vertex at the origin and one side along the x axis. The trigonometric functions for a "general" angle such as this are defined by

(M-43)
$$\sin \theta = \frac{y}{c}$$

(M-44)
$$\cos \theta = \frac{x}{c}$$

(M-45)
$$\tan \theta = \frac{y}{x}$$

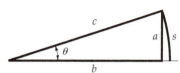

Figure M-16 For small angles, $\sin \theta = a/c$, $\tan \theta = a/b$, and the angle $\theta = s/c$ are all approximately equal.

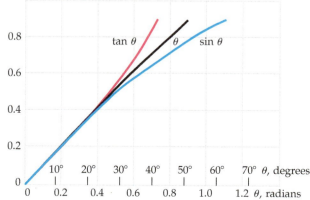

Figure M-17 Graphs of $\tan \theta$, θ, and $\sin \theta$ versus θ for small values of θ.

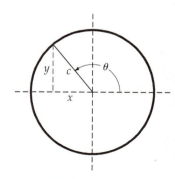

Figure M-18 Diagram for defining the trigonometric functions for an obtuse angle

It is important to remember that values of x to the left of the vertical axis and values of y below the horizontal axis are negative; c in the figure is always regarded as positive. Figure M-19 shows plots of the general sine, cosine, and tangent functions versus θ. The sine and cosine functions have a period of 2π rad. This means that for any value of θ, $\sin(\theta + 2\pi) = \sin \theta$. That is, when an angle changes by 2π rad, the function returns to its original value. The tangent function has a period of π rad: $\tan(\theta + \pi) = \tan \theta$. Some other useful relations are

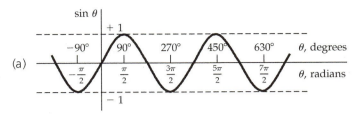

(a)

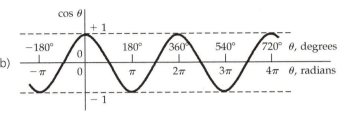

(b)

$$\sin(\pi - \theta) = \sin \theta \qquad \text{(M-46)}$$

$$\cos(\pi - \theta) = -\cos \theta \qquad \text{(M-47)}$$

$$\sin\left(\frac{1}{2}\pi - \theta\right) = \cos \theta \qquad \text{(M-48)}$$

$$\cos\left(\frac{1}{2}\pi - \theta\right) = \sin \theta \qquad \text{(M-49)}$$

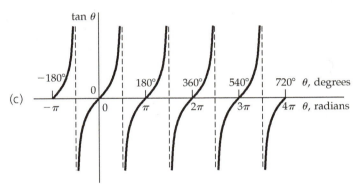

(c)

Because the radian is dimensionless, it is not hard to see from the plots in Figure M-19 that the trigonometric functions are functions of all real numbers.

Figure M-19 The trigonometric functions $\sin \theta$, $\cos \theta$, and $\tan \theta$ versus θ

EXAMPLE M-9 **Cosine of a Sum**

Using the suitable trigonometric identity from Table M-2, find $\cos(135° + 22°)$. Give your answer with four significant figures.

Set Up

As long as all angles are given in degrees, there is no need to convert to radians, because all operations are numerical values of the functions. Be sure, however, that your calculator is in degree mode. The suitable identity is $\cos(A \pm B) = \cos A \cos B \mp \sin A \sin B$, where the upper signs are appropriate.

Solve

Write the trigonometric identity for the cosine of a sum, with $A = 135°$ and $B = 22°$.	$\cos(135° + 22°) = (\cos 135°)(\cos 22°)$ $- (\sin 135°)(\sin 22°)$
Using a calculator, find $\cos 135°$, $\sin 135°$, $\cos 22°$, and $\sin 22°$.	$\cos 135° = -0.7071$ $\cos 22° = 0.9272$ $\sin 135° = 0.7071$ $\sin 22° = 0.3746$
Enter the values in the formula and calculate the answer.	$\cos(135° + 22°) = (-0.7071)(0.9272)$ $- (0.7071)(0.3746)$ $= -0.9205$

Reflect

The calculator shows that $\cos(135° + 22°) = \cos(157°) = -0.9205$.

M-9 **The dot product**

For two vectors $\vec{A}$ and $\vec{B}$ separated by angle θ, as shown in Figure M-20, their dot product C is defined as

$$C = \vec{A} \cdot \vec{B} = AB \cos \theta \qquad \text{(M-50)}$$

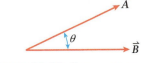

Figure M-20 Two vectors separated by an angle θ

(a)

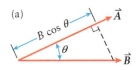

(b)

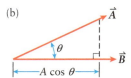

Figure M-21 The dot product is a measure of how parallel two vectors are. (a) $B \cos \theta$ is the component of $\vec{B}$ that is parallel to $\vec{A}$. (b) $A \cos \theta$ is the component of $\vec{A}$ that is parallel to $\vec{B}$.

which you can read as, "C equals A dot B." In Equation M-50, A and B are the magnitudes of vectors $\vec{A}$ and $\vec{B}$, respectively. As a result, the dot product of two vectors is a scalar quantity. This is why $\vec{A} \cdot \vec{B}$ is also called the **scalar product** of $\vec{A}$ and $\vec{B}$.

Physically, the dot product $\vec{A} \cdot \vec{B}$ is a measure of how parallel the two vectors are. We can think of it as the magnitude of vector $\vec{A}$ multiplied by the component of vector $\vec{B}$ that is parallel to $\vec{A}$. Referring to Figure M-21a, we see that $B \cos \theta$ is the component of $\vec{B}$ that is parallel to $\vec{A}$. That is, $B \cos \theta$ tells us how much of $\vec{B}$ points in the direction of $\vec{A}$. Alternatively, you can think of the dot product $\vec{A} \cdot \vec{B}$ as the magnitude of vector $\vec{B}$ multiplied by the component of vector $\vec{A}$ parallel to $\vec{B}$ (Figure M-21b).

The dot product is commutative; the order of the vectors in a dot product does not affect the result. That is,

$$\vec{A} \cdot \vec{B} = \vec{B} \cdot \vec{A}$$

The dot product is also distributive, which means

$$\vec{A} \cdot (\vec{B} + \vec{C}) = \vec{A} \cdot \vec{B} + \vec{A} \cdot \vec{C}$$

Three special cases of the dot product are particularly important in physics. First, the dot product of two vectors $\vec{A}$ and $\vec{B}$ that point in the same direction (so $\theta = 0$ and $\cos \theta = \cos 0 = 1$) equals the product of their magnitudes.

(M-51)
$$\vec{A} \cdot \vec{B} = AB \cos 0 = AB$$
(if $\vec{A}$ and $\vec{B}$ point in the same direction)

(As an example, the dot product of a vector $\vec{A}$ with itself is equal to the square of its magnitude: $\vec{A} \cdot \vec{A} = AA \cos 0 = A^2$.)

Second, the dot product of two perpendicular vectors $\vec{A}$ and $\vec{B}$ (so $\theta = 90°$ and $\cos \theta = \cos 90° = 0$) is zero.

(M-52)
$$\vec{A} \cdot \vec{B} = AB \cos 90° = 0$$
(if $\vec{A}$ and $\vec{B}$ are perpendicular)

Third, if two vectors $\vec{A}$ and $\vec{B}$ point in opposite directions (so $\theta = 180°$ and $\cos \theta = \cos 180° = -1$), their dot product equals the *negative* of the product of their magnitudes.

(M-53)
$$\vec{A} \cdot \vec{B} = AB \cos 180° = -AB$$
(if $\vec{A}$ and $\vec{B}$ point in opposite directions)

Finally, it's useful to know how to calculate the dot product of two vectors $\vec{A}$ and $\vec{B}$ that are expressed in terms of their components A_x, A_y, A_z, and B_x, B_y, and B_z.

(M-54)
$$\vec{A} \cdot \vec{B} = A_x B_x + A_y B_y + A_z B_z$$

You can verify that Equation M-54 is correct by thinking of $\vec{A}$ as the sum of three vectors: $\vec{A}_1$, which has only an x component A_x; $\vec{A}_2$, which has only a y component A_y; and $\vec{A}_3$, which has only a z component A_z. From the definition of the dot product, $\vec{A}_1 \cdot \vec{B}$ is equal to A_x multiplied by the component of $\vec{B}$ in the direction of $\vec{A}_1$, or $\vec{A}_1 \cdot \vec{B} = A_x B_x$. Similarly, $\vec{A}_2 \cdot \vec{B} = A_y B_y$ and $\vec{A}_3 \cdot \vec{B} = A_z B_z$. Since $\vec{A} = \vec{A}_1 + \vec{A}_2 + \vec{A}_3$ and the dot product is distributive, it follows that

$$\vec{A} \cdot \vec{B} = (\vec{A}_1 + \vec{A}_2 + \vec{A}_3) \cdot \vec{B} = \vec{A}_1 \cdot \vec{B} + \vec{A}_2 \cdot \vec{B} + \vec{A}_3 \cdot \vec{B} = A_x B_x + A_y B_y + A_z B_z$$

That's the same as Equation M-54. If the vectors have only x and y components, Equation M-54 simplifies to $\vec{A} \cdot \vec{B} = A_x B_x + A_y B_y$.

EXAMPLE M-10 The Dot Product

(a) Calculate the dot product of vector $\vec{A}$ with magnitude 5.00 that points in a horizontal direction 36.9° north of east and vector $\vec{B}$ of magnitude 1.50 that points in a horizontal direction 53.1° south of west. (b) Calculate the dot product of vector $\vec{C}$ with components $C_x = 4.00$, $C_y = 3.00$ and vector $\vec{D}$ with components $D_x = -0.900$, $D_y = -1.20$.

Set Up

In part (a) we know the magnitude and direction of the vectors, so we'll use Equation M-50. In part (b) the vectors are given in terms of components, so we'll evaluate the dot product using Equation M-54.

$$\vec{A} \cdot \vec{B} = AB \cos \theta \qquad \text{(M-50)}$$

Dot product of two vectors in terms of components:

$$\vec{A} \cdot \vec{B} = A_x B_x + A_y B_y + A_z B_z \qquad \text{(M-54)}$$

Solve

(a) The drawing shows that the angle between $\vec{A}$ and $\vec{B}$ is $\theta = 163.8°$. We use this in Equation M-50 to evaluate the dot product.

$$\vec{A} \cdot \vec{B} = AB \cos \theta$$
$$= (5.00)(1.50) \cos 163.8°$$
$$= (5.00)(1.50)(-0.960)$$
$$= -7.20$$

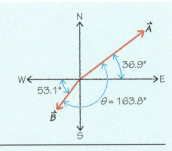

(b) Both $\vec{C}$ and $\vec{D}$ are in the x–y plane and have no z components, so we just need the first two terms in Equation M-54 to calculate their dot product.

$$\vec{C} \cdot \vec{D} = C_x D_x + C_y D_y$$
$$= (4.00)(-0.900) + (3.00)(-1.20)$$
$$= -7.20$$

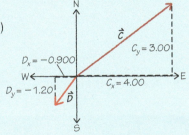

Reflect

It's not a coincidence that we got the same result in part (b) as in part (a): Vectors $\vec{A}$ and $\vec{C}$ are the same, as are vectors $\vec{B}$ and $\vec{D}$. (You can verify this by using the techniques from Chapter 3 to calculate the components of the vectors $\vec{A}$ and $\vec{B}$ in part (a). You'll find that the components are the same as those of $\vec{C}$ and $\vec{D}$ in part (b).) This should give you confidence that the method of calculating the dot product using components gives you the same result as the method that involves the magnitudes and directions of the vectors.

Notice that the angle between vectors $\vec{A}$ and $\vec{B}$ is between 90° and 180°, and the dot product is negative.

M-10 The cross product

The dot product, described in Section M-9, is only one way to multiply two vectors. We can also multiply two vectors $\vec{A}$ and $\vec{B}$ using the **cross product**.

$$\vec{C} = \vec{A} \times \vec{B} \tag{M-55}$$

The symbol × represents the mathematical operation known as the cross product. As you can see from Equation M-55, the result of taking the cross product of two vectors is also a vector. The magnitude of the resulting vector is the product of the magnitudes of the two vectors and the sine of the angle between them. That is, the magnitude of the cross product of $\vec{A}$ and $\vec{B}$ is

$$C = \left| \vec{A} \times \vec{B} \right| = AB \sin \phi \tag{M-56}$$

where according to convention ϕ is defined as the angle that goes from $\vec{A}$ to $\vec{B}$. $\vec{C}$ points in the direction perpendicular to both $\vec{A}$ and $\vec{B}$ as shown in Figure M-22.

We can interpret the magnitude of the cross product $\vec{A} \times \vec{B}$ as the magnitude of vector $\vec{A}$ multiplied by the component of vector $\vec{B}$ perpendicular to $\vec{A}$, or the magnitude of vector $\vec{B}$ multiplied by the component of vector $\vec{A}$ perpendicular to $\vec{B}$.

Note that the order of the two vectors in a cross product makes a difference. The cross product of $\vec{B}$ and $\vec{A}$ is the negative of the cross product of $\vec{A}$ and $\vec{B}$ or

$$\vec{A} \times \vec{B} = -\vec{B} \times \vec{A} \tag{M-57}$$

This results from the definition of the angle ϕ in Equation M-56. Since ϕ is directed from the first vector to the second vector, if you travel the angle from the second vector to the first—in reverse direction—ϕ becomes negative. And the sine of a negative angle is also negative.

In addition, the cross product obeys the distributive law under addition.

$$\vec{A} \times (\vec{B} + \vec{C}) = \vec{A} \times \vec{B} + \vec{A} \times \vec{C} \tag{M-58}$$

To determine the direction of the cross product $\vec{C} = \vec{A} \times \vec{B}$, you can use the right-hand rule. To apply this rule, point the fingers of your right hand in the direction of the first vector of the cross product (in this case $\vec{A}$). Then curl your fingers toward the second vector, $\vec{B}$. If you stick your thumb straight out, it points in the direction of the cross product, vector $\vec{C}$ (Figure M-23a). If you instead want to find the direction of

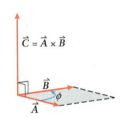

Figure M-22 The cross product is a vector $\vec{C}$ that is perpendicular to both $\vec{A}$ and $\vec{B}$. The magnitude of $\vec{C}$ is $AB \sin \phi$, which equals the area of the parallelogram shown.

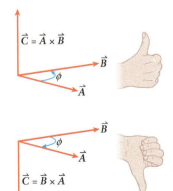

Figure M-23 (a) To find the direction of $\vec{A} \times \vec{B}$, point the fingers of your right hand in the direction of vector $\vec{A}$, then curl them toward vector $\vec{B}$. Your thumb points in the direction of the cross product. (b) The direction of $\vec{B} \times \vec{A}$ points in the opposite direction of $\vec{A} \times \vec{B}$.

the cross product $\vec{B} \times \vec{A}$, begin by pointing the fingers of your right hand in the direction of vector $\vec{B}$. Then curl them toward vector $\vec{A}$. Your thumb again points in the direction of the cross product (Figure M-23b). Note that because you must curl your fingers in the opposite direction as for $\vec{C} = \vec{A} \times \vec{B}$, the cross product of $\vec{B} \times \vec{A}$ points in the opposite direction of $\vec{A} \times \vec{B}$, which is just what we stated in Equation M-58.

There are two special cases of the cross product that are worth pointing out. The first is the cross product for two perpendicular vectors, for which $\phi = 90°$, so $\sin \phi = 1$.

$$\left| \vec{A} \times \vec{B} \right| = AB \sin 90° = AB(1) = AB$$

(magnitude of the cross product of two perpendicular vectors)

The second special case is the cross product of two parallel vectors, for which $\phi = 0$, so $\sin \phi = 0$.

$$\left| \vec{A} \times \vec{B} \right| = AB \sin 0 = AB(0) = 0$$

(magnitude of the cross product for two parallel vectors)

One example of a cross product of two parallel vectors is the cross product of a vector with itself: $\vec{A} \times \vec{A} = 0$.

EXAMPLE M-11 The Cross Product

Evaluate $\vec{A} \times \vec{B}$, in which the components of vector $\vec{A}$ are $A_x = 5$, $A_y = 0$, and the components of vector $\vec{B}$ are $B_x = 9$, $B_y = 7$.

Set Up

We will use the definition of the magnitude of the cross product, Equation M-56, to find the magnitude of the cross product, and the right-hand rule to determine the direction of the cross product.

We will have to use the components of vector $\vec{B}$ to determine its magnitude and the angle it makes with the x axis and vector $\vec{A}$.

$$C = \left| \vec{A} \times \vec{B} \right| = AB \sin \phi \qquad \text{(M-56)}$$

Determine vector magnitude and direction from vector components.

$$A = \sqrt{A_x^2 + A_y^2}$$

$$\tan \theta = \frac{A_y}{A_x} \qquad \text{(3-2)}$$

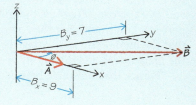

Solve

Begin by determining the magnitude and direction of vector $\vec{B}$ using its components and Equations 3-2.

Determine the magnitude of vector $\vec{B}$ from its components.

$$B = \sqrt{B_x^2 + B_y^2} = \sqrt{9^2 + 7^2} = 11.4$$

Determine the angle $\vec{B}$ makes with the x axis (and vector $\vec{A}$) from its components.

$$\tan \phi = \frac{B_y}{B_x} = \frac{7}{9} = 0.778, \text{ so}$$

$$\phi = \arctan 0.778 = 37.9°$$

Vector $\vec{A}$ has only an x component, and that component is positive. Its magnitude is equal to the absolute value of its x component, and the angle it makes with the x axis is 0.

Determine the magnitude of vector $\vec{A}$ from its components.

$$A = \sqrt{A_x^2 + A_y^2} = \sqrt{5^2 + 0^2} = 5$$

The vector $\vec{A}$ makes an angle of zero degrees with the x axis: It has only an x component, and $A_x = 5$ is positive.

Now that we know both the magnitude and direction of the vectors, we can use Equation M-56 to determine the magnitude of the cross product.

Apply Equation M-56 to the two vectors.

$$\left| \vec{A} \times \vec{B} \right| = AB \sin \phi$$
$$= (5.00)(11.4) \sin 37.9°$$
$$= (5.00)(11.4)(0.614)$$
$$= 35.0$$

Use the right-hand rule to determine the direction of the cross product.

From the figure, if we first point the fingers of our right hand in the direction of $\vec{A}$ (along the x axis), and then curl them toward $\vec{B}$, we see that the thumb points in the positive z direction. So the cross product $\vec{A} \times \vec{B}$ has a magnitude of 35 and points in the +z direction.

Reflect

The vectors $\vec{A}$ and $\vec{B}$ lie in the x–y plane, so the cross product, which must be perpendicular to both vectors, should point along the z axis. That's just what we found.

ANSWERS TO ODD PROBLEMS

Chapter 1

1. The meter (SI unit for length) is defined by the distance light travels in a vacuum in a tiny fraction (1/299,792,458) of a second. The SI unit of time is the second. It is defined as the time required for a cesium atom to emit 9,192,631,770 complete cycles of radio waves due to transitions between two particular energy levels of the cesium atom. The SI unit for mass is the kilogram. It is the mass defined by taking the value of Planck's constant h to be $6.62607015 \times 10^{-34}$ J/s $= 6.62607015 \times 10^{-34}$ kg·m²/s, where the meter and second are defined as above.

3. To be a useful standard of measurement, an object, system, or process should be unchanging, replicable, and possible to measure precisely so that errors in its measurement do not carry over into calibration errors in every other measurement.

5. No. The equation 3 meters = 7 meters has consistent units, but it is false. The same goes for 1 = 2, which consistently has no units.

7. (a) The fewest number of significant figures in 61,000 is two—the "6" and the "1." (b) If the period is acting as a decimal point, then the trailing zeros are significant and the quantity 61,000. would have five significant figures. (c) When numbers are written in scientific notation, all the digits before the power of 10 are significant. Therefore, 6.10×10^4 has three significant figures.

9. B

11. C

13. C

15. D

17. C

19. A. 0.00442
    B. 0.0000709
    C. 828
    D. 6,020,000
    E. 456,000
    F. 0.0224
    G. 0.0000375
    H. 0.000138

21. A. 10^{-12}
    B. 10^{-3}
    C. 10^6
    D. 10^{-6}
    E. 10^{-15}
    F. 10^9
    G. 10^{12}
    H. 10^{-2}

23. A. 72.5 m
    B. 1.96×10^3 cm
    C. 7.87×10^3 cm²
    D. 16.5 L
    E. 4.41×10^5 g
    F. 3.45 m³

G. 3.32 km²
H. 7.24×10^3 cm³

25. A. 127 L
    B. 16.6 gal
    C. 2.66×10^8 L
    D. 6.70 m³
    E. 674 cm³
    F. 0.229 cup
    G. 3.09 acres
    H. 6.80×10^{-4} qt

27. (a) 345 μs; (b) 20.0 pW; (c) 233.7 Mm; (d) 65.4 kg

29. 0.1 N/m

31. A. 4
    B. 1
    C. 6
    D. 1
    E. 5
    F. 2
    G. 3
    H. 4

33. A. 26.2
    B. −32.3
    C. 74.8
    D. 200

35. $L = (L/T) \, T + L = L + L$

37. $[mv^2] = (\text{kg})(\text{m/s})^2 = \text{kg} \cdot \text{m}^2/\text{s}^2$

39. A. correct
    B. not correct
    C. not correct
    D. not correct
    E. correct
    F. not correct

41. (a) 2.5×10^{-3} kg
    (b) 0.15 kg/m³
    (c) 5×10^{-16} m³
    (d) one tablet: 8.1×10^{-5} kg
        one bottle: 8.1×10^{-3} kg
    (e) 2.0×10^{-8} m³/s
    (f) 1400 kg/m³

43. $\text{m}^3/\text{kg} \cdot \text{s}^2$

45. 3.00×10^8 m/s

47. (a) 0.4 g/cm³, 400 kg/m³
    (b) 40% the density of water
    (c) 9×10^{-5} g
    (d) 5×10^{-4}%

Chapter 2

1. It's always a good idea to include the units of every quantity throughout your calculation. Eventually, once they become more comfortable, most people will convert all of their values into SI units and stop including them in the intermediate steps of the calculation. Using SI units in all calculations minimizes

calculation errors and ensures the answer will be in SI units as well.

3. Average velocity is a vector quantity—the *displacement* divided by the time interval. Average speed is a scalar quantity—the *distance* divided by the time interval.

5. The largest speed, 1 m/s, will give the largest displacement in a fixed time.

7. An object will slow down when its acceleration vector points in the direction opposite to its velocity vector. Recall that acceleration is the change in velocity divided by the change in time.

9. (a) displacement versus time: m/s
   (b) velocity versus time: m/s^2
   (c) distance versus time: m/s

11. No, there is no way to tell whether the video is being played in reverse. An object being thrown up in the air will undergo the same acceleration, time of flight, and so on as an object falling from its maximum height.

13. Assuming the initial speed of the ball is the same in both cases, the velocity of the second ball will be the same as the velocity of the first ball. The upward trajectory of the first ball involves the ball going up and reversing itself back toward its initial location. At that point the trajectories of the two balls are identical as the balls hit the ground.

15. D

17. E

19. A. 108 km/h
    B. 22.5 km/h
    C. 2.01×10^5 mi/h
    D. 60.0 mi/h
    E. 40 m/s (rounded to one significant figure)

21. 2.83 m/s

23. $v_{\text{Jack,ave}} = 0.10$ m/s, speed$_{\text{Jack,ave}} = 0.90$ m/s;
    $v_{\text{Jill,ave}} = 0$ m/s, speed$_{\text{Jill,ave}} = 1.0$ m/s

25. 200 mi

27. 17.0 m

29. (a) 18 m/s; (b) 16 m/s; (c) 14 m/s

31. (a) 5 m away
    (b) Its speed is greatest in the interval between 4 and 5 s.
    (c) Its velocity is equal to 0 during this time interval.

33. (a) 2 m/s
    (b) The object changes direction twice.
    (c) −4 m/s^2
    (d) −0.2 m/s^2

35.

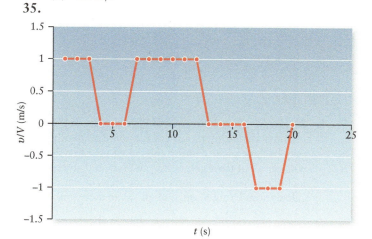

37. The trains will not collide; 62 m.

39. 0.918 s

41. 2.83 m/s^2

43. $\Delta x_{\text{Bugatti}} = 27.9$ m
    $\Delta x_{\text{Saleen}} = 32.7$ m

45. 20.3 m

47. (a) 17.1 m/s; (b) 2.26 s

49. (a) 4.08 m/s; (b) 0.832 s

51. 8.20 m/s; 1.60 m/s; 31.0 m/s; 1.84 s

53. 2.2 m/s

55. (a) 2.87 s; (b) 0.413 m

57. Geoff wins the race by 2.33 s.

59. 2.83 m

61. 125 m

63. (a) 0.494 s; (b) 2.20 m; (c) 0.730 s

65. 10 m

67. -8.6×10^{-3} m/s

69. 60 s

Chapter 3

1. A scalar is a number, whereas a vector has a direction associated with a number. Distance and speed are examples of scalar quantities, whereas displacement and velocity are examples of vectors.

3. If an object is moving at a constant velocity (that is, its speed and direction are constant), then the average velocity and instantaneous velocity are equal. If the object is not moving at a constant velocity (that is, its speed is changing, its direction is changing, or both), then its average velocity can equal its instantaneous velocity only at specific instants.

5. During the motion of a projectile, only v_x, $a_x (= 0)$, and $a_y (= -g$, if up is positive) are constant. The position (x, y) of the projectile is constantly changing while it is moving. The acceleration in the y direction is nonzero, which means v_y is changing.

7. The rock's speed will be greater than the speed with which it was thrown. Gravity is accelerating the rock in the y direction, so the magnitude of the y component of its velocity will increase.

9. No. A long jumper should take off at an angle of 45° to achieve the maximum range.

11. Yes. The force from the vine is accelerating the ape upward; at the bottom of the swing is when the vertical component of the velocity changes from downward to upward, which requires an acceleration in the upward direction.

13. Yes. You are accelerating because the direction of the velocity vector is changing. The acceleration vector points east.

15. Relative motion refers to motion measured relative to some inertial frame. Absolute motion refers to motion measured in the absolute frame, which is defined to be at rest according to all observers in the universe. However, since the laws of motion are the same in all inertial frames, there is no frame that all observers agree is at rest. Therefore, all motion is relative.

17. B

19. D

21. D
23. D
25. B
27. 5.6; 27°
29. A. 3.61; 326°
B. 2.83; 135°
C. 2; 270°
31. $C = 6.40$; 38.7°
33. (a) $C_x = 21$ m/s and $C_y = 61$ m/s
(b) $D_x = 21$ m/s and $D_y = -19$ m/s
(c) $E_x = 42$ m/s and $E_y = 82$ m/s
35. 50 m/s; 53° west of north
37.

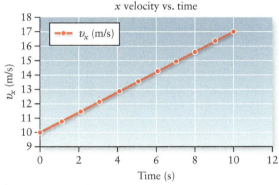

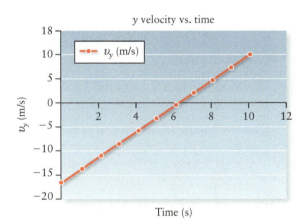

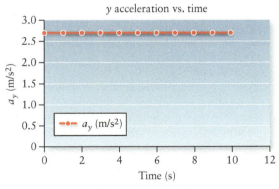

39. (a) $v_x = -4.00$ km/h and $v_y = 2.50$ km/h
(b) The average velocities are the same.
41. $v_{0x} = 29$ m/s and $v_{0y} = 17$ m/s
43. 4 m
45. $v_x = 21.7$ m/s and $v_y = \pm 8.82$ m/s
47. $v_0 = 45$ ft/s = 10 m/s
49. 3.49 s
51. $v = 34.6$ m/s

53. 17 m/s² toward the center of the track
55. (a) 5.9×10^{-3} m/s² toward the center of the Sun
(b) 8.00 m/s² toward the center of the circle
57. $r = 0.07$ m; $v = 700$ m/s
59. (a) 4.7 m/s; (b) 108°
61. −2 m/s
63. 0.25 s
65. 6.7 km east and 13 km south
67. (a) 4.62 s; (b) 6.86 s
69. (a) $t = 9.24$ s; (b) 271 m; (c) 282 m
71. $v_0 = 12.3$ m/s; $v_1 = 12.9$ m/s; $t = 0.391$ s
73. maximum height = 6.05 h; maximum range = 6.05 R
75. Moon: increased by a factor of 6; Mars: increased by a factor of 2.6
77. (a) 68.8 m/s², 7g; (b) 86.6%
79. 0.309 m
81. (a) 17.1 m; (b) 102 m and 6.45 m

Chapter 4

1. Yes, the net force vector will determine the direction of the acceleration vector for the mass in question.
3. $[F] = $ N $=$ kg·m/s²
5. Zero. When you stand on a motionless horizontal bathroom scale, it is not accelerating. Therefore the ground must push up on the scale with the same magnitude of force you push down on the scale.
7. This definition agrees with the one given in the book. Each equal quantity of matter (by the measure of mass) resists acceleration to the same degree, so it has the same inertia.
9. If the motion of an object is along only one axis, then the acceleration is zero along the other axes. The net force along those axes must, therefore, equal zero as well.
11. The forces acting on a water bottle at rest on a desk are gravity acting downward and the normal force from the desk acting upward. The net force is zero, since the water bottle is at rest.
13. When you walk, you push backward on the floor with your shoe. According to Newton's third law, the floor must then push forward on your shoe. It is this forward push that gives you a forward acceleration as you start to walk.
15. You can double up the rope. Each of the two strands pulls only 425 N.
17. The scale always reads the force that it exerts on the chair. As the elevator begins to ascend, everything in it accelerates upward. So the upward force on the chair from the scale must be greater than the downward force of the chair's weight. When moving at constant speed either up or down, the chair does not accelerate. So the scale must exert a force equal in magnitude to the chair's weight. Both when the elevator slows at the top, and when the elevator begins to descend, the acceleration of the chair is downward. So the upward force of the scale on the chair must be less than the scale's weight. As the elevator slows at the bottom, it again accelerates upward, and the force of the scale on the chair is greater than the scale's weight.
19. Seat belts increase the time over which passengers come to rest, which decreases the magnitude of their

accelerations and, correspondingly, the magnitude of the net force acting on them.

21. B
23. D
25. B
27. C
29. D
31. C
33. 4000 N
35. 12.00 m/s²
37. $F_2 = 1.39$ N; $F_3 = 1.63$ N
39. 2450 N
41. 0 N
43.

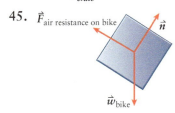

45.

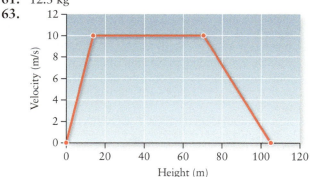

47. (a) 45 N in the vertical, upward direction
    (b) 10 N in the vertical, upward direction
    (c) 0 N
49. 762 N
51. 413 N
53. 10,000 N
55. 482 N
57. 4.20 s
59. 1.50 m/s² to the west
61. 12.3 kg
63.

Velocity (m/s) vs Height (m)

65. 3.50 m/s
67. 157.5 m/s
69. (a) 2.45 m/s²; (b) 147 N
71. (a) –2.80 m/s²; (b) The elevator is moving down.
73. (4)
75. (a) 504 N; (b) 784 N; (c) 1104 N
77. (a) 13 N; (b) 43 N
79. (a) $a = 2.09 \times 10^3$ m/s², time is $t = 1.38$ ms
    (b) $F_n = 25.9$ mN

$F_n = 25.9$ mN

$F_g = 0.12$ mN

(c) $F_{ground} = 25.8$ mN, which is 215 times the weight of the froghopper
81. (a) 644 N; (b) 92.4 kg
83. (a) 466 N; (b) 537 N
85. $T_1 = 2.49$ N and $T_2 = 7.46$ N
87. $a = 3.36$ m/s²; $T = 19.3$ N
89. $a = 3.50$ m/s²; $T_2 = 25.2$ N; $T_1 = 8.40$ N
91. (a) The largest pull force before a string breaks is 73.4 N.
    (b) $T_A = 26.2$ N; $T_B = 45.0$ N

Chapter 5

1. Sometimes. If the coefficient of static friction is less than 1, the normal force will be larger than the static friction force. If the coefficient of static friction is more than 1, the static friction force can exceed the normal force or not, depending on how much friction is required to prevent slipping.
3. When a pickup truck accelerates forward, the contents will also accelerate forward due to the force of static friction between the box and the truck's bed.
5. (a) Weight stays the same; (b) normal force decreases; (c) frictional force is the same; (d) maximum static frictional force decreases
7. An object on a slope (Figure 5-12a); an object in a stack of objects (the lower objects' normal forces are greater than their own weight); an object pressed to a wall (Figure 5-14); an object sitting in an elevator as the elevator starts or stops
9. Although the coefficients of friction between synovial fluid and bone should be the same for everyone, the normal forces involved will be different from person to person. This means the magnitudes of the frictional forces will also be different from person to person.
11. His acceleration decreases because the net force on him decreases. The net force is equal to his weight minus his drag force. The faster an object is moving, the larger the drag force it experiences as a function of its speed. Because drag force increases with increasing speed, net force and acceleration decrease. As the drag force approaches his weight, the acceleration approaches zero.
13. Yes. If he lies flat in relation to the wind and she does not, she will have a smaller profile in relation to the wind. Thus she will experience less drag and be able to catch up with him.
15. If you know the car's instantaneous speed and the radius of curvature, you can use the formula $a_{cent} = v^2/r$.
17. When the bucket is whirled at a high enough speed, all the gravitational force goes into changing the water's direction, leaving none to accelerate the water out of the bucket. At the lowest point on the circle, the normal force on the water is greatest because the gravitational force points opposite the direction of centripetal acceleration; at the highest point, the normal force is least because the gravitational force points entirely in the direction of centripetal acceleration.
19. The frictional force between tires and road provides the centripetal acceleration that keeps the car moving in a circle. If the frictional force is not strong enough to keep

the car moving in a circle, the car continues in a straight line and starts to skid.

21. A
23. B
25. B
27. D
29. C
31. 41.0 N
33. 0.5
35. (a) 245 N; (b) 25.0 N; (c) 0.102
37. 28.9 N
39. 80.2 N
41. 2.20 kg
43. $a_x = 2.21$ m/s^2; $T = 36.1$ N
45. 6.98×10^{-9} N
47. (a) 6.17 m/s; (b) 0.274 kg/m
49. 6.06 m/s
51. (a) 1.75 N; (b) 1.26 N
53. 1.18×10^4 N
55. 9.0×10^{13} m/s^2
57. 6.29×10^3 N
59. 0.885
61. We can plot the force versus time.

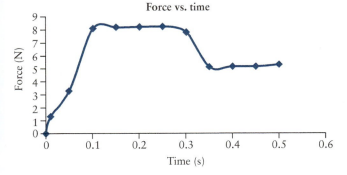

Then we see that the crate is likely stationary up until $t = 0.3$ s. So we use the average of points at $t = 0.1$ s through $t = 0.25$ s to find the coefficient of static friction. We then use the points at $t = 0.35$ s through $t = 0.5$ s for the coefficient of kinetic friction. So $\mu_k = 0.535$.

63. 21.8°
65. 174 N
67. 1.76 m/s^2
69. 0.65
71. (a) 6.84 s; (b) Static friction opposes the relative slipping motion between two surfaces. The bottom of the tire tends to slip backward relative to the wet pavement, so the frictional force opposes the backward motion and hence is forward.
73. 3.77 m/s^2
75. 0.382
77. 2.19×10^4 N; no, the force would not be appreciably different.
79. $T = 2.21$ N; $v = 4.20$ m/s
81. (a) 3.5 m/s^2; (b) 16 N; (c) Yes, headrests would help neck injuries, since the head would pivot through an angle of less than 60° before hitting the headrest. This would affect the strain in the neck muscles but not the contact force necessary in stopping the head from moving. Headrests prevent the head from snapping backward when the car comes to rest.

Chapter 6

1. If the magnitude of the force or the magnitude of the displacement is zero, then the work done will be zero. In addition, if the force and displacement are nonzero but perpendicular, then the work done will also be zero.

3. Zero. Your roommate does positive work on the block as she lifts it (she exerts an upward force in the same direction as the displacement). Assuming she sets the block down gently on the floor, she does an equal amount of negative work in setting the block down (she exerts an upward force, but the block moves downward). Similarly, your roommate must exert a horizontal force in the direction of motion, doing positive work to begin to move it across the room. But she must do an equal amount of negative work to stop the block's horizontal motion. So she does no net work on the block.

5. (a) Yes, she is expending energy. (b) No, she does no work on the boulder. The energy she is expending is being dissipated inside her body.

7. Yes, whenever the kinetic energy decreases from a positive value to a smaller positive value, the change in kinetic energy is negative.

9. The orbital path of the satellite is a circle. Its instantaneous displacement vector is tangent to the circle. The gravitational force points toward the center of the circle along the radius. The radius is perpendicular to the tangent of a circle, meaning that the displacement vector is perpendicular to the force. Therefore, no work can be done.

11. (a) When it reaches the bottom; (b) When it reaches terminal velocity. This will occur when the force of air resistance is equal and opposite to the force of gravity. This may or may not occur before the rock reaches the bottom.

13. The system of you and Earth has more gravitational potential energy when you are at the top of the hill. It is harder to pedal up the hill because you are converting chemical energy in your body to mechanical energy, which is stored as potential energy as you climb the hill.

15. (a) Conservation of mechanical energy would imply that the ball comes back to exactly the same height at which it was released. The small amount of mechanical energy lost as heat through friction implies it can come back no higher than this. (b) You might be hit by the ball if you push it ever so slightly rather than just releasing it.

17. Before each firing of the toy gun, it is "loaded" by a person expending energy to compress its spring, which is held in place by the gun's trigger. The person firing the gun also expends energy in pulling the trigger, which releases the spring.

19. C
21. C
23. E
25. C
27. (a) $W_{crane} = 2.94 \times 10^4$ J
    (b) $W_{gravity} = -2.94 \times 10^4$ J
29. (a) $W_{ext} = 3.50 \times 10^4$ J
    (b) $W_{gravity} = -1.2 \times 10^4$ J
    (c) 0

31. (a) $W_{person} = -5.88 \times 10^3$ J
    (b) $W_{gravity} = 5.88 \times 10^3$ J
33. $W_{net} = 3.13 \times 10^5$ J
35. $W_{net} = -3.6 \times 10^5$ J; a net work of 3.6×10^5 J is required.
37. (a) 4 m/s²; (b) 3.2 N
39. 0.251
41. 4.50 m/s
43. 58.5 N
45. (a) 0.424 m/s; (b) 0.612 m/s; (c) 0.707 m/s
47. 3×10^2 J
49. 5.69×10^5 J
51. -8.79×10^4 J
53. -3.14×10^5 J
55. 9.69×10^{-2} J
57. (a) 103 J; (b) 103 J
59. 227 kJ
61. 85%
63. 15.6 m/s
65. 9.66 m
67. point b: 6 m/s; point c: 8 m/s; point d: 6 m/s; point E: 4 m/s
69. 2.05×10^3 N
71. 88.1% was "lost" to air resistance
73. 13.3°
75. Neil: 14.8 m/s; Gus: 14.2 m/s
77. 2.64×10^5 W
79. 54 s
81. $W_{nonconservative} = -3.63 \times 10^5$ J
83. 0.370 m
85. 7.00 m/s
87. (a) 19.9 m; (b) 9.90 m/s
89. No, there was no friction.
91. 0.882
93. (a) 24 m/s; (b) 710 J/kg is lost to nonconservative forces.
95. $v_i = \sqrt{3gR}$
97. 9.2×10^2 W

Chapter 7

1. Newton reasoned that the Moon, as it orbits, falls toward the center of Earth due to the pull of gravity. Incrementally, it will move tangent to the orbital path; then it will fall toward Earth. The combined motion leads to the familiar circular path. An apple falls straight toward Earth without the tangential motion. The vertical motion is the same for both the apple and the Moon.
3. (a) When the mass of an object is doubled, the force between it and another object doubles. (b) If both masses are halved, the force between two objects is one-fourth the original value.
5. The mass cancels when applying $F = ma$ to get the acceleration.
7. Orbiting the Sun is an incomplete description of the Moon's motion, but the Moon is definitely orbiting the Sun as a part of the Earth–Moon system.
9. The gravitational potential energy is negative because we choose to place the zero point of potential energy at $r = \infty$. With that choice the gravitational potential

energy is negative because the gravitational force is attractive. The two objects will gain kinetic energy as they approach one another.
11. Humans had to evolve to support their body against the force of gravity (that is, to support the weight of a human). The cardiovascular system had to evolve to be able to move blood back from the legs to the heart against the force of gravity. The way humans run has also evolved to the particular gravity of Earth. There are many other examples.
13. The most noticeable effects would be the comparison between gravitation at relatively small distances (such as the Earth–Moon system) and gravitation at extremely large distances (such as galaxies and galaxy clusters). The deviation in the inverse square law would result in a different value for G or m. Also noncircular orbits would not be elliptical, although if the power is close enough to 2, the difference in orbit could be too small to notice without watching for a very long time.
15. From Kepler's law the larger the orbit, the longer the period.
17. B
19. E
21. A
23. B
25. A
27. 4.00×10^{-10} N. The gravitational force exerted by the stump on the boulder points toward the stump.
29. 9.35×10^{-16} N
31. $F_{Earth\ on\ apple} = 1 \times 10^1$ N; $F_{Moon\ on\ apple} = 3 \times 10^{-5}$ N
33. (a) 2.37×10^{20} N; (b) toward the Sun
35. The point between Earth and the Moon where the net gravitational force due to the two is zero is 3.46×10^8 m from Earth and 3.83×10^7 m from the Moon.
37. (a) 2058 N; (b) 340 N; (c) 13.4 N
39. $W_{gravity\ on\ object} = 4.89 \times 10^5$ J
41. (a) $v_f = 1.45 \times 10^3$ m/s; (b) 1.05×10^8 J
43. $T_{Mars} = 1.881$ y
45. $v = 7.73 \times 10^3$ m/s
47. 4.2×10^{11} m
49. 7.20×10^7 m
51. $d = 3.75 \times 10^8$ m
53. (a) 2.21×10^{-6} N
    (b) 4.89×10^{-5} N
    (c) Earth's tidal forces on the Moon are stronger.
55. (a) Europa: 3.551 d, Ganymede: 1.071×10^9 m = 1.071×10^6 km, Callisto: 1.883×10^9 m = 1.070×10^6 km; (b) 1.90×10^{27} kg
57. (a) 5.67×10^5 m = 5.7×10^2 km
    (b) weight in orbit = 6.9×10^2 N, weight on Earth's surface = 8.2×10^2 N
59. (a) 5.3 km; (b) 7.5 km
61. (a) 0.923 au, or 1.38×10^{11} m; (b) 324 d; (c) fastest at the perihelion and slowest at the aphelion; (d) 1.47
63. 1.9×10^{41} kg

Chapter 8

1. Yes, if the tennis ball moves 18 times faster
3. If an object, or a system of objects, experiences no net external force, its momentum is conserved. No external

forces act on the two-sphere system; its momentum is thus conserved. Each individual sphere does experience a net external force during its collision with the other sphere; as a result, individual sphere momentum will not be conserved.

5. Look for "loss" of energy. Did the collision make a sound? Did the objects deform? If you can accurately gauge their velocities, do the objects have less total kinetic energy than before? Any of these conditions indicates an inelastic collision.

7. When two balls are raised and released on a Newton's cradle, the balls collide with the balls at rest on the cradle and transfer their energy and momentum to the two balls on the other side of the collision. The fact that two balls are always moving is required by the combination of conservation of momentum and conservation of energy. If you only applied conservation of momentum, it would be possible for the three stationary balls to move off with a lower speed than the two incoming balls. This does not happen.

9. An impulse, commonly speaking, is a brief push; or, psychologically, a spur-of-the-moment desire. This fits a common use for the physicist's notion of impulse as applied to collisions. The forces in collisions will tend to be brief. Changes in momentum, of course, occur in other situations, and the term makes less sense there.

11. We'll consider the system to be the child and the board, both of which start at rest. (a) The board will move in the direction opposite to that the child is running. (b) The center of mass of the system will remain stationary for all time. (c) The answers to (a) and (b) do not change if the child walks rather than runs.

13. The boat will move in a direction opposite the direction of the dog's motion.

15. E
17. E
19. C
21. C
23. B
25. 2.0×10^5 kg·m/s to the east
27. initial momentum: −6250 kg·m/s
    final momentum: 17,500 kg·m/s
    change in momentum: 2.38×10^4 kg·m/s
29. (a) 8 times larger; (b) 16 times larger
31. (a) 6.40 m/s; (b) She moves in the opposite direction.
33. 0.8 m/s to the right
35. $v_{cue,f} = 1.76$ m/s
37. $v_f = 4.9$ m/s; $\theta_f = 31°$
39. (a) 6.67 m/s to the east; (b) before $K_i = 2.00 \times 10^6$ J, after $K_f = 6.67 \times 10^5$ J
41. (a) 9.82 m/s; (b) before $K_i = 552$ J, after $K_f = 530$ J
43. 5.99 m/s, 45.2° deflected toward the sidelines from the running back's original path
45. The 2.00-kg ball goes 1.00 m/s to the left, and the 4.00-kg ball goes 2.00 m/s to the right.
47. Ball 1's final velocity is 1.72 m/s to the left, and ball 2's final velocity is 4.28 m/s to the right.
49. 2.00 kg
51. 2.3×10^2 m/s
53. (a) 0.363 kg·m/s; (b) 42.6 N
55. x component = −3.6 m; y component = 0.0 m

57. $y_{CM} = -0.085\ R$, $x_{CM} = 0.038\ R$
59. (a) 24 m/s; (b) 10 m/s; (c) −8.5 m/s
61. (a) Remington bullet
    (b) $m_{block} = 2.62$ kg
    (c) $m_{bullet,ix} = 566$ m/s
63. (a) $v_{Sally,fx} = 2.5$ m/s; (b) The system starts at rest and no external forces on the system, so the center of mass will not move. Therefore, the center of mass is located at the point where she threw the boot. (c) $\Delta t = 12$ s
65. $v_{fx} = 19.2$ m/s
67. $v_{fx} = 4.35$ m/s
69. $v_{1f} = 3.10$ m/s
    $\theta = -29°$ relative to the initial direction
71. $v_{fx} = 24.2$ m/s
    $v_{fy} = -16.5$ m/s
73. $y_i = 0.00156$ m = 1.56 mm

Chapter 9

1. One radian is the angle produced by going one radius of length around the edge of a circle. Radians should be included if the resulting quantity is intrinsically angular in nature (for example, one can include it in angular momentum) but not otherwise. For example, rotational kinetic energy is just energy, so the units are joules, not joules multiplied by radians squared.

3. (a) iii; (b) vi; (c) i; (d) vii; (e) iv; (f) ix
   (g)

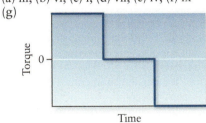

5. The magnitude of the torque exerted on the bolt is equal to $FR \sin \phi$, where F is the magnitude of the force, R is the distance from the rotation axis (the bolt, in this case) to the location of the force, and ϕ is the angle between these two vectors. Since F is the same in all four cases, the torque will be the largest when the force is applied to the end farthest from the bolt *and* perpendicular to the wrench (C). The torques in (B) and (D) are the next largest and are equal to one another. Finally, (A) is the smallest, since R is much smaller in this case compared to the other three. To reiterate: (A) < (B) = (D) < (C).

7. The moment of inertia of an object determines how that object responds to an applied torque.

9. The moment of inertia of a rotating object is calculated relative to the axis of rotation. If we don't explicitly define this axis at the outset of the problem, the calculated moment of inertia will be ambiguous at best and meaningless at worst. Even for axes that pass through the center of mass, axes with different orientations can have different moments.

11. The positive direction is up the ramp.
    (a) Graph (a). The angular acceleration should behave the same as the linear acceleration, which is constant and negative.

(b) Graph (e). The angular velocity should behave the same as the linear velocity. The cylinder initially rolls up the ramp, stops, and then rolls down the ramp.

13. We can use conservation of energy to answer this question. All three objects—a sphere, a cylinder, and a ring—start at rest from the same height and thus have the same initial potential energy. As the objects roll down the inclined plane, potential energy is converted into rotational *and* translational kinetic energy. The smaller the moment of inertia of an object, the more energy that will be converted into translational kinetic energy rather than rotational kinetic energy. The object that reaches the bottom of the ramp has the largest translational kinetic energy, so the sphere

$\left(I_{\text{sphere}} = \dfrac{2}{5}MR^2\right)$ "wins," followed by the cylinder

$\left(I_{\text{cylinder}} = \dfrac{1}{2}MR^2\right)$, and then the hoop $(I_{\text{hoop}} = MR^2)$.

15. An object moving in a straight line can have a nonzero angular momentum relative to a rotation axis as long as its velocity vector cannot be extended through the axis.

17. An external tension force does work on the ball to pull it inward, which means the ball's rotational kinetic energy will not remain constant. However, since this force is directed through the axis of rotation, it applies no torque on the ball, which means the ball's angular momentum will remain constant. Since the ball moves closer to the axis of rotation, its moment of inertia must decrease. As a result, the ball's angular speed must increase to ensure constant angular momentum.

19. (a) The SI unit for rotational kinetic energy is the joule.
(b) The SI unit for moment of inertia is (kilogram)(meter)^2.
(c) The SI unit for angular momentum is (kilogram)(meter)^2/(second).

21. Yes, the cat will fall on its feet, but it will have to undergo a more radical twisting motion of its torso to compensate for its missing tail. The tail, although small, has a large moment arm, which is why it is critical to the cat as it attempts to produce large torques and angular momenta during its fall.

23. D
25. D
27. E
29. E
31. B
33. 1.05 rad/s
35. 4.71 rad/s; 3.49 rad/s; 39.5 rad/s
37. (a) 314.2 s = 3×10^2 s; (b) 157 rev = 2×10^2 rev
39. (a) 0.0438 rad/s^2; (b) 40.5 rad; (c) 3.77 m/s
41. (a) 1.3 rev; (b) 1574 rev
43. (a) 0.262 rad; (b) 6.94×10^{-4} rpm; 7.27×10^{-5} rad/s
45. $\tau = 9.00$ N·m
47. (a) 63.8 N·m; (b) 57.8 N·m; (c) 21.8 N·m; (d) 0
49. $F = 1000$ N
51. (a) 21.5 kg·m^2; (b) 5.63 N·m
53. $I = 90$ kg·m^2
55. $I_{\text{total}} = \dfrac{11}{6}MR^2$
57. $I = \dfrac{3}{2}MR^2$

59. $I = 858$ g·cm^2
61. $I_{\text{total}} = \dfrac{85}{4}MR^2$
63. (a) 65.3 rad/s^2; (b) 3.27 m/s^2; (c) 1.63 N
65. $\tau_{\text{applied},z} = 2.7 \times 10^3$ N·m
67. (a) tension: 0.248 N, angular acceleration: 2.6×10^2 rad/s^2
(b) $I = 2.38 \times 10^{-5}$ kg·m^2
69. $K_{\text{rotational}} = 2.24$ J
71. $\omega = 4.08$ rad/s = 39.0 rev/min
73. 2.57×10^{29} J
75. $K_{\text{rotational}} = 1.05$ J
77. $v_{\text{CM,f}} = 3.98$ m/s
79. $n = 0.13$ N
81. $L_z = 2.95$ kg·m^2/s
83. $L_z = 7.08 \times 10^{33}$ kg·m^2/s
85. $L_z = 5.39 \times 10^3$ kg·m^2/s
87. 54 kg·m^2
89. $\omega_{\text{dough,fz}} = 4.30$ rad/s
91. $F_{\text{right}} = 57.8$ N and $F_{\text{left}} = 124$ N. Using the center of the pole as the rotation axis, the force from the bird produces a clockwise torque, whose vector lies along the $-x$ axis. The left-hand force produces a clockwise torque whose vector lies along the $+x$ axis. The right-hand force produces a counterclockwise torque, whose vector lies along the $-x$ axis.
93. $f_s = 0.515$ N, up the incline
95. 3.00×10^2 N
97. $(0.63)ML^2$
99. (a) 147,000 rad/s^2; (b) 0.292 rotations
101. 2.0×10^2 rpm
103. 5.13 rad/s
105. (a) $R_{\text{ring}} = R\sqrt{\dfrac{2}{5}}$; (b) same angular momentum
107. (a) 14.0 rad/s; (b) 3.0 rad/s; (c) -4.0×10^2 J
109. (a) This relationship is not unique; it will hold as long as $m_1 = 2m_2$. (b) This relationship is unique.
111. (a) 2.57×10^{29} J; (b) 3.9×10^8 y; (c) 2.9×10^8 y
113. (a) decrease; (b) increase
115. (a) 0.03 rad/s^2; (b) 3 rad/s; (c) $9g$; (d) 2×10^1 s
117. 3.09
119. 1.8×10^{-5} kg·m^2

Chapter 10

1. When an object experiences a stress below the yield strength, it should return its original shape unchanged.

3. Yes. Real cables have weight, and that weight can be sufficient to break the cable.

5. The brass wire stretches by $\dfrac{4mgL}{\pi D^2 Y_{\text{brass}}}$, and the steel wire stretches by $\dfrac{4mgL}{\pi D^2 Y_{\text{steel}}}$.

7. The skin loses some of its elasticity as people age and the tension in the skin decreases. Since the skin is not held as tightly, it will wrinkle.

9. When determining the volume of an object by completely immersing it in a liquid, we assume that the object does not absorb any of the liquid. In addition we assume a high bulk modulus for the object, so its volume does not appreciably decrease when subjected

to the increased pressure associated with immersion in the liquid.

11. (a) Young's modulus accounts for the stretching or compressing of an object in one dimension. The bulk modulus describes the expansion or compression of an entire volume. (b) Both of the variables have the same units (Pa, MPa, GPa, and so on).

13. The shear modulus is a measure of how much an object will deform under a given stress. The larger the shear modulus, the less it will deform; it is more rigid. So the term "rigidity" applies because a material will deform less if it has a larger shear modulus.

15. One reason is that a sufficiently tall mountain will collapse under the stress of its own weight. If the same amount of material is distributed in a cone rather than as a vertical cylinder, the weight is distributed over a larger area and the stress is less.

17. Yield strength is the tensile stress at which a material becomes permanently deformed. Ultimate strength is the maximum tensile stress the material can withstand before failure.

19. Leg bones must retain their shape to hold the body together. The small strain of the leg bone means that its length stays nearly constant as weight is first placed on a leg and then taken off the leg. Antlers, on the other hand, need to be able to bend a great deal to survive the strong impacts to which they are subjected.

21. B
23. C
25. A
27. A
29. A
31. $L_0 = 3$ cm
33. $\frac{\Delta L}{L_0} = 0.017 = 1.7\%$
35. $r = 5.89 \times 10^{-4}$ m
37. $\frac{F}{A} = 1.35 \times 10^4$ N/m^2
39. (a) 64.2 mm; (b) 6.42 mm
41. 2.0×10^8 Pa
43. 1×10^7 N
45. 2.0
47. 6.63 m
49. 1.6 kN
51. (a) 5.0; (b) No, they cannot use this rope.
53. (a) yield strength $= 250 \times 10^9$ N/m^2
    (b) ultimate strength $= 400 \times 10^9$ N/m^2
    (c) Young's modulus $= 125{,}000 \times 10^9$ N/m^2
55. 0.02 m = 2 cm
57. 1.3 kN; 2.9×10^2 lb
59. (a) 1.2×10^8 strands; (b) 2.2 cm; yes, this seems reasonable.
61. (a) -0.0023 cm^3; (b) -0.082 cm
63. 5.4×10^{-5} m = 54 µm
65. 1.9×10^5 N/m^2
67. (a) 0.69 cm
    (b) 0.040 cm
    (c) 3.5×10^2 N/m
    (d) A spring constant of 3.5×10^2 N/m is comparable to everyday springs in the lab.
69. (a) 9.8×10^3 N; (b) 0.34 cm^2

71. 3×10^8 N/m^2
73. 2.5×10^4 N/m^2
75. 7 bolts

Chapter 11

1. The dam could be the same thickness.
3. The collection bag should be held below the body so that the blood can flow down into it.
5. While jumping up, the person will have very little blood pumped to the brain, as most of the blood will be in the lower trunk and in the legs due to gravity. This may result in a fainting spell. Once the person is horizontal, the blood will evenly distribute in the body and will reach the brain, and consciousness will be regained.
7. The pressure in a fluid depends only on the depth. Since the height of the water is the same in both vessels, they will have the same pressure at the bottom. The total weight of the fluid is not directly related to the pressure.
9. No. The pieces could be different sizes.
11. The water level will fall. The boat displaces a volume of water equal to the weight of the boat. If the boat is removed from the water, the water will no longer be displaced and the water level will fall.
13. The equation of continuity applies, assuming no water is lost or gained along the way. The speed will increase when the cross-sectional area decreases. Therefore, the speed of the water in the wide valley will be slower than the speed of the water in the narrow channel.
15. Landing airplanes should approach from the east (into the wind). When landing, airplanes maintain a certain speed relative to the air. When flying into the wind, the speed of the airplane relative to the ground is reduced and the airplane can stop in a shorter distance.
17. horizontal
19. D
21. B
23. A
25. A
27. $R = 0.13$ m
29. $L = 0.073$ m = 7.3 cm
31. $\rho_{\text{average}} = 8.9 \times 10^1$ kg/m^3
33. $R_{\text{star}} = 13$ km
35. (a) $p_{1\,\text{foot}} = 7.5 \times 10^4$ Pa; (b) $p_{1\,\text{foot}} = 1.4 \times 10^4$ Pa
37. $p = 1.19 \times 10^5$ Pa
39. $\Delta p = 137$ mmHg
41. The Hoover Dam is approximately 700 ft (213 m) tall and approximately 1200 ft (366 m) long. Pressure increases linearly with water depth, so we can take an average pressure at mid-depth on the water side, which is atmospheric pressure plus roughly 10^6 Pa. Approximating the dam as a flat surface (admittedly a rough estimate) with dimensions 213 m by 366 m, the total hydrostatic force (that is, total force on the water side of the dam) is about 9×10^{10} N.
43. 880 mmHg
45. 1.01 atm
47. 550 kPa
49. 1.0 kN and points outward
51. 3.79 MN toward the inside of the drum

53. $F_2 = 1.4 \times 10^4$ N
55. (a) $F_2 = 1.9 \times 10^3$ N
    (b) $\Delta y_2 = (8.3 \times 10^{-3})\Delta y_1$
    (c) $W_1 = 3.2$ J; $W_2 = 3.2$ J; the work done in slowly pushing the small piston is equal to the work done in raising the large piston.
57. (a) $F_b = 3.53 \times 10^1$ N; (b) $\dfrac{V_{submerged}}{V} = 0.600$;
    (c) $w_{water} = 3.53 \times 10^1$ N
59. (a) $\rho_{fluid}gs^3$; (b) $\rho_{fluid}gs^3 - m_{cube}g$; (c) $\rho_{fluid}s^3 g$
61. The crown is most likely made of gold.
63. (a) 1.02×10^3 kg/m³; (b) 25%
65. 1.0×10^4 m/s
67. $6.25v_1$
69. 1.25 h
71. 2.49×10^5 Pa
73. 6.29 m
75. 1.21×10^3 Pa
77. 0.291 L/s
79. (a) 5.3 kg; (b) 9.6×10^{12}
81. 1.79×10^1 mi/kg
83. (a) 1.90×10^6 Pa; (b) 1.89×10^2 m
85. 3.8×10^{-3} N
87. 18.8 m/s
89. 2

Chapter 12

1. Simple harmonic motion also requires that the direction of the force be opposite to the displacement.
3. Simple harmonic motion is a special case of oscillation for which the period does not depend on the amplitude. For may types of oscillation the period does depend on the amplitude, such as a pendulum swinging with large amplitude.
5. (a) radians/second; (b) radians
7. It is an offset of where in the oscillation cycle we choose to set time $t = 0$. If, for example, the phase angle is π rather than zero, at $t = 0$ the object is at $x = -A$ instead of $x = +A$.
9. In simple harmonic motion, potential energy is at a maximum when kinetic energy is at a minimum. One quarter-cycle later, potential energy is at a minimum, and kinetic energy is at a maximum. The phase angle between kinetic and potential energy is thus 90°, or $\dfrac{\pi}{2}$ radians.
11. A simple pendulum is made from a long, thin string that is tied to a small mass. We can treat it as a point mass attached to the end of a massless string. A physical pendulum has its mass distributed over an extended volume.
13. The force of gravity depends on elevation and determines the period of the pendulum. Measuring the period and length of the pendulum will yield enough information to estimate the strength of the gravitational force.
15. The frequency of the driving force is how often this driving force repeats. The natural frequency of the oscillator is the frequency at which it would oscillate if no damping or driving force were present.

17. B
19. E
21. A
23. D
25. B
27. 80 Hz
29. (a) 0.067 s; (b) 1.8×10^3 cycles
31. (a) $x(t) = A \cos\left(\sqrt{\dfrac{k}{M}}\,t\right)$
    (b) $-\dfrac{A}{2}\sqrt{\dfrac{3k}{M}}$
    (c) 0
33. 0.379 s
35. (a) $\dfrac{d}{2}\cos\left(\dfrac{\pi t}{1800}\right)$
    (b)

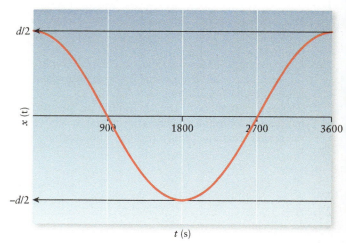

37. $x = \pm\dfrac{A}{\sqrt{2}}$
39. (a) 2.0×10^2 N/m; (b) 4.0×10^1 N
41. $T = 2.24$ s
43. $T = 16.42$ s
45. Mercury: 3.27 s
    Venus: 2.11 s
    Earth: 2.01 s
    Mars: 3.27 s
    Jupiter: 1.26 s
    Saturn: 1.94 s
    Uranus: 2.13 s
    Neptune: 1.88 s
47. (a) 0.122 s; (b) 0.198 s; (c) 1.19 s
49. $T = 0.898$ s
51. $T = 0.76$ s
53. 0.307 kg/s
55. $A = 4.78 \times 10^{-3}$ m = 4.78 mm
57. (a) 2.7×10^{-2} m; (b) 6.0 rad/s
59. 3.2 m/s
61. (a) 0.41 m/s; (b) −0.74 m/s²
63. (a) 1.55 m/s; (b) 7.5 cm or −7.5 cm from equilibrium
65. 30.0% loss per cycle
67. (a) 57 cm; (b) 0.90 m
69. 1.63 s
71. $(3.08$ cm$) \sin((114$ rad/s$)t)$
73. (a) 35 N/m; (b) 17.6 m

75. 0.85 rad; The oscillations are not simple harmonic.
77. (a) 1.45 cm; (b) no
79. (a) (e) only; (b) (a) and (g); (c) (c) and (i)

Chapter 13

1. In longitudinal waves the direction of the wave disturbance is parallel to the direction of propagation; in transverse waves these two directions are parallel. Examples of longitudinal waves: sound waves, waves along a spring. Examples of transverse waves: waves on a string, "the wave" in a stadium.

3.

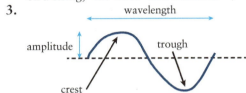

5. The frequency counts the number of oscillations per time. The angular frequency counts the oscillations per time in units of 2π; one complete oscillation is equivalent to 2π rad.

7. (a) The wave speed v depends on the medium, and wave speed increases as the wave passes from air into water. (b) The frequency f is determined by the vibrations of the source, so the frequency does not change. (c) The wavelength is $\lambda = v/f$, so since v increases and f stays the same, λ increases.

9. A *rarefaction* in a sound wave is a point where the pressure variation of the medium is most negative; a *compression* is a point where the pressure variation is most positive (see Figure 13-11). A *phonon* is a miniscule quantum-mechanical disturbance of a solid. A sound wave in a solid, with its rarefactions and compressions, is made up of a large number of phonons.

11. The car receives the signal that is coming directly from the broadcasting tower and from many other reflections (buildings, mountains, low-lying clouds, and so on). When the signals are not directly out of phase, some signal gets through and you will hear the station's broadcast. However, sometimes the path difference between the direct wave and the reflected wave is equal to an odd number of half wavelengths, leading to an "out-of-phase" or destructive interference pattern. As the car comes to a stop, the dead zones are traversed slowly enough to be noticeable.

13. The length L of an organ pipe determines the longest possible wavelength of a standing wave in the pipe (either $4L$ for a closed pipe or $2L$ for an open pipe). The speed of sound waves in the pipe divided by this wavelength equals the fundamental frequency.

15. Greg's string was incorrectly adjusted to Elizabeth's frequency (206 Hz + 8 Hz = 214 Hz). Although both pianos are now out of tune, they are in tune to one another, and the beat frequency is zero.

17. (a) A sonic boom occurs when the speed of the sound source exceeds the speed of sound in the medium. (b) Imagine a stationary listener (so $v_{\text{listener}} = 0$) in front of a source of sound moving at $v_{\text{source}} > v_{\text{sound}}$ (so source and listener are approaching). According to

Equation 13-35, the frequency of sound heard by the listener is *negative*, which is impossible. This means that there is no sound in front of the source, as Figure 13-33a shows.

19. The human ear responds to sound waves. The human voice produces sound waves, as can other parts of the body (such as clapping hands or unwanted sounds from the digestive tract).

21. A
23. C
25. C
27. E
29. C
31. wavelength: 0.92 m = 92 cm; wave number: 6.9 rad/m
33. The wavelength is equal to the speed multiplied by the period. In this case it should be on the order of a few centimeters.

35. $\left[\dfrac{\omega}{k}\right] = \dfrac{\frac{1}{T}}{\frac{1}{L}} = \dfrac{L}{T}$; $[\lambda f] = L \cdot \dfrac{1}{T} = \dfrac{L}{T}$. Therefore, the SI units of radians are dimensionless.

37. (a) 2 Hz; (b) 6.3 m; (c) 1×10^1 m/s
39. $y(0, 4.0) = 0.1$ m
41. $y(x,t) = 0.1 \sin(10\pi x - \pi t)$
43. $v_p = 1.9 \times 10^2$ m/s
45. E string to A string: 0.38 mm
A string to D string: 0.57 mm
D string to G string: 0.86 mm
47. $v_p = 1.5 \times 10^3$ m/s
49. $v_p = 1.3 \times 10^3$ m/s
51. (a) $2A_1$; (b) 0; (c) $2.5A_1$; (d) $\dfrac{A_1}{2}$
53.

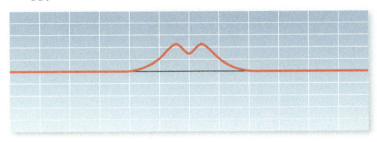

55. 2.9 m
57.

59. (a) $f_1 = 27.0$ Hz; $f_2 = 54.0$ Hz; $f_3 = 81.0$ Hz; $f_4 = 108$ Hz
(b)

61. 11.6 kg
63. 70.5 N
65. open at both ends
67. The length is 3.38 m, and the pipe is open at both ends.
69. 4.8 m. That's definitely on the large size but not outrageous.
71. 8.79 Hz to 44.49 kHz
73. (a) 5 Hz; (b) Try tightening the string just a little tiny bit. If that makes the beat faster, loosen it a bit.

75. 1.04×10^2 N

77. (a) 3.16; (b) 1.73; (c) 1.41; (d) no difference, except that you may be able to get a larger factor decrease by moving to the side rather than directly away from the source

79. 40 J per day. She'd do better by completely replacing the plan with something else. Barring that, she could use a larger "microphone."

81. 102.8 dB

83. (a) 1.26; (b) 100

85. 50 μW

87. 53 Hz

89. 649 Hz

91. (a)

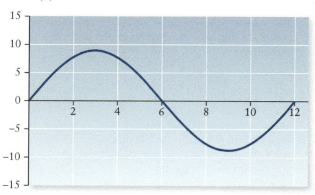

(b) $(9 \text{ m}) \sin\left(\left(\frac{\pi}{6} \text{ m}^{-1}\right)x - (\pi \text{ s}^{-1})t\right)$

93. (a)

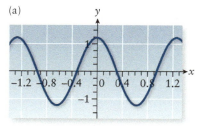

(b)

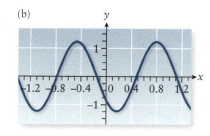

95. (a) 1.69×10^5 N/m²; (b) 1.47 kHz; (c) The parts of the resonator that are in the air all have a frequency increase from the substitution of some helium into the air; the parts of the resonator that are in the body do not. This produces a different mix of frequencies and a different general sound.

97. (a) 1760 Hz; (b) $T/4$

99. 32°C

101. 69.0 dB

103. (a) 1 mm; (b) 4×10^2 J

105. 2.11 m/s

107. 5.1 m

109. (a) 1.26×10^5 W; (b) 1.00×10^5 km

Chapter 14

1. If A is in thermal equilibrium with B, and B is in thermal equilibrium with C, then A is in thermal equilibrium with C. This is used to establish that temperature is actually a property of objects.

3. The temperature at which the pressure of an ideal gas would become zero is −273.15°C. This is the zero point of the Kelvin temperature scale.

5. It is true that a 1-degree Celsius (or Kelvin) change is equal to a 1.8-degree Fahrenheit change, but the length of a thermometer is based on the maximum and minimum values it is designed to read, not the temperature scale it employs.

7. Thermometer A (200–270 K) can be used as a freezer gauge. Thermometer B (230–270 K) can be used as a meteorological gauge (for outdoor temperatures). Thermometer C (300–550 K) can be used as an oven gauge. Thermometer D (300–315 K) can be used as a medical thermometer for humans.

9. Place the thermometer in an ice bath and let it come to equilibrium. The level of the mercury corresponds to 0°C. Then boil some water, place the thermometer in it, and let it reach equilibrium. The level of the mercury in this case corresponds to 100°C. Then we can divide the distance between the 0 mark and the 100 mark by 10 to get the marks for 10°C, 20°C, and so on. Dividing these regions into 10 again will give the individual degree markings.

11. Since the two gases are in thermal contact for a long time, they will eventually reach thermal equilibrium, which means $T_A = T_B$. From the ideal gas law, $pV = NkT$, so $p_A V_A = p_B V_B$. We cannot say anything about the individual values of the pressure and volume, only their product.

13. (a) $pV = NkT$ is most appropriate if one is counting molecules; $pV = nRT$ is most appropriate if one is counting moles. (b) p equals the pressure of the gas, V is the volume the gas occupies, N equals the number of molecules, n is the number of moles of molecules, k is the Boltzmann constant, R is the universal gas constant, and T is the temperature of the gas.

15. The two metals must be chosen to have different thermal expansion coefficients. Because the metals are attached directly, thermal expansion will bend the bimetallic strip, which in turn pushes or pulls on the switch.

17. (a) The 3 appears in the formula because the formula is derived using the sum of the square of the speed in each dimension. In a one-dimensional system, therefore, the 3 should be replaced with 1: $K_{\text{translational,average}} = \frac{1}{2}kT$. (b) $K_{\text{translational,average}} = 5kT$

19. (a) They are the same temperature. (b) The steam can cause a more severe burn because of its latent heat of vaporization.

21. First the temperature of the ice rises to 0°C. (For reference, the specific heat of ice is 2093 J/kg·K.) Second, all the ice melts. (The heat of fusion L_f is 3.34×10^5 J/kg.) Third, the water warms up to 100°C. (The specific heat of water is 4186 J/kg·K.) Fourth, all the water boils. (The heat of vaporization L_V is 2.260×10^6 J/kg.) Finally, the temperature of the steam rises. (The specific heat of steam is 2009 J/kg·K.)

23. (a) Joule, calorie, BTU (British thermal units); (b) Joules are used in physics; calories are used in chemistry and food chemistry; and BTUs are used in engineering.

25. Heated air expands and becomes more buoyant. Cooler, denser air near the ceiling descends, displacing the warmer air and pushing it upward. Thus the warm air rises, and the cooler air falls.

27. C

29. C

31. D

33. E

35. D

37. (a) 82°F; (b) 14°C; (c) −229°F; (d) 290 K; (e) 99°F

39. $T_F = (9/5)T_C + 32$

41. Fahrenheit: $T_{F,low} = 98.1$°F, $T_{F,high} = 98.6$°F
Kelvin: $T_{K,low} = 309.8$ K, $T_{K,high} = 310.2$ K

43. (a) $T_B < T_A$; (b) $T_B > T_A$

45. 6×10^3 J

47. 17.0 L

49. 872 m/s

51. 1.3×10^9 m

53. 70.4 K

55. 1.78×10^{-4} m

57. 8.04 cm

59. (a) length: 20.015 cm, width: 30.022 cm; (b) increases by 0.15%

61. (a) 2.7144 cm; (b) 2.7118 cm

63. 0.004 m

65. 8.79 g/cm³

67. 78.5 kJ

69. 3.85×10^5 J

71. 8.12×10^9 kg

73. 1110 K

75. 35.0°C

77. 19 strokes

79. 78 kJ

81. 248 kJ

83. (a) 35.7 g; (b) 2 cubes

85. 0.244 kg

87. The iron melts 7.86 g of ice, the copper melts 10.1 g of ice, and the water melts 14.6 g of ice. Water melts the most ice.

89. a factor of 18.32 times

91. 65 solar radii or 4.5×10^{10} m

93. 6.12 kW

95. 2.9×10^6 J

97. $c = 0.400$ J/g·K = 400 J/kg·K. Therefore, it is most likely copper or iron/steel.

99. 8.20×10^{-21} J

101. (a) $v_{esc} = 6.18 \times 10^5$ m/s = 618 km/s
(b) $T = 1.55 \times 10^7$ K

103. $2\alpha A_0 \Delta T$

105. Temperature change: 35 K
Expansion:
$\Delta L_{steel} = 0.0091$ m = 9.1 mm
$\Delta L_{Cu} = 0.0058$ m = 5.8 mm

107. (a) 26.7°C; (b) yes; (c) Her body prevents this increase by radiating heat and evaporating sweat.

109. $H = 3.0 \times 10^2$ W

111. (a) $E_{treadmill}/E_{person} = 35.9\%$, remainder of energy goes into heating the body: 64.1%
(b) $m = 0.206$ kg

113. $P = 3.9 \times 10^{26}$ W

115. (a) $m_{water} = 2.4 \times 10^{16}$ kg; (b) $L = 2.9 \times 10^4$ m or 18 mi

Chapter 15

1. Yes, a system can absorb heat without increasing its internal energy. An example is when the heat absorbed by the system is equal to the work done by the system.

3. Yes, this is consistent with the first law of thermodynamics because there is no change in internal energy for an isothermal process. Therefore, the work done by the gas must equal the heat put into the system.

5. The work done on the gas to compress it results in an increase in temperature because there is insufficient time for heat to flow out of the system. Thus the internal energy of the system increases, and the temperature increases.

7. High efficiency is almost always a primary objective for a steam-electric generating plant. In such a plant, the steam is the working substance for the heat engine that drives the electric generator. The temperature of the steam is the temperature of the high-temperature heat source. The temperature of the low-temperature reservoir is usually fixed by circumstances, such as the temperature of a nearby lake. Thus increasing the feed-steam temperature is the only way to increase the Carnot efficiency limit for the generator.

9. No. The term "surroundings" is a boundary we draw at our convenience. If we choose the system as Earth and imagine enclosing it in a perfectly reflecting balloon to prevent heat exchanges with its surroundings, parts of Earth would still be able to exchange heat with each other and undergo other processes that increase entropy within the system.

11. Energy is transferred to the refrigerator's working substance both as heat from the things inside the box and as work done on it by the compressor. The refrigerator then transfers all this energy as heat to the room air. With the refrigerator door kept open, the compressor has to work even harder. This extra energy is transferred by the refrigerator to the room air as heat, causing the room to grow even hotter.

13. A greater mass of water takes longer to freeze. If we use a higher-power refrigerator, then it will take less time to freeze the water. If the ambient air is warmer, then it will take longer to freeze the water.

15. There is just one configuration of the fragments that will reassemble the cup, compared to countless configurations that will not. So there is effectively zero probability that the cup can reassemble.

17. Entropy is a state variable that does not depend on the history of changes to the system. So we can calculate the change in entropy by assuming that instead of a free expansion with no heat flow, the final state of the system was achieved by an isothermal expansion.

But in such an isothermal expansion, heat must have flowed into the system to cause the increase in the system's volume; thus, according to the equation $\Delta S = Q/T$, the entropy of the final state will be greater than the entropy of the initial state. Thus the entropy must also increase for the free expansion.

19. As heat flows from the water to its environment, the temperature of the water decreases and its entropy decreases.

21. B
23. B
25. C
27. D
29. A
31. +300 J
33. 300 J
35. 0 J
37. +8.80 kJ
39. 9.22 kJ
41. −800 kJ
43. −321 kJ
45. 3 kJ
47. 19.0 g
49. 1.4
51. 27.5
53. 40.0%
55. 14.8%
57. 13.7 kJ
59. 3.41
61. (a) 5.43; (b) 225 J; (c) −24°C
63. 245 J/K
65. −16.9 J/K
67. 1.48 kJ/K
69. 25.9 J/K
71. 2.88 J/K
73. 1.83×10^6 kg
75. (a) 47°C; (b) 32.0 J
77. The efficiency drops from 10.5 to 9.66%.
79. −603 J/K
81. (a) 15.8%; (b) 1.37 TJ
83. (a) 329 kJ = 0.0914 kWh; (b) 956 kJ; (c) 3.24 kJ/K
85. (a) 38.0%; (b) 0.288 J/K; (c) 78.6 J
87. The water's entropy increases by 24.7 J/K.
89. 25.5 kJ/K; increases
91. (a) between 1 & 2: isochoric; between 2 & 3: adiabatic; between 3 & 1: isothermal
    (b) 275 K
    (c) 97.6 J
    (d) 0.217
93. −7500 J

Chapter 16

1. There would be no difference. The charges were originally arbitrarily defined as positive and negative.

3. The mass decreases because electrons are removed from the object to make it positively charged.

5. Our current understanding is that like charges repel and opposite charges attract. If a charged insulating object either repels or attracts both charged glass (positive) and charged rubber (negative), then you may have discovered a new kind of charge. However, your insulating object could also be neutral and is attracted due to becoming polarized (see Figure 16-4).

7. When the comb is run through your hair, electrons are transferred to the comb. The paper is polarized by the charged comb. The paper is then attracted to the comb. When they touch, a small amount of charge is transferred to the paper so now the paper and comb are similarly charged and repel each other.

9. (a) No, the object could be attracted after being polarized. (b) Yes, to be repelled, the suspended object should be charged positively.

11. Similarities: The force varies as $\frac{1}{r^2}$; the force is directly proportional to the product of the masses or charges; the force is directed along the line that connects the two particles. Differences: Charges can either attract or repel; masses always attract. The gravitational forces between electrons and protons in an atom are many orders of magnitude weaker than the electric forces.

13. (a) Once you have calculated the electric field at P, you can find the force on any charge placed at P. (b) The field concept helps explain how charges can interact even at a distance.

15. The electric forces are equal in magnitude but opposite in direction. The force on the proton is downward while the force on the electron is upward.

17. The electric field is the total electric field due to all charges both inside and outside the Gaussian surface. However, the charges outside the surface create a net zero electric flux through the surface.

19. B
21. C
23. A
25. C
27. B
29. $q_{net} = 1.00$ C
31. $2e$
33. 9.99×10^{18} electrons
35. $\sigma = 1 \times 10^{-4}$ C/m^2
37. $q_1 = 1.00 \times 10^{-5}$ C; $q_2 = 2.00 \times 10^{-5}$ C
39. (a) $F_{B \text{ on } A} = 0.0150$ N. The force that charge B exerts on charge A is in the $+x$ direction. (b) From Newton's third law, $\vec{F}_{A \text{ on } B} = -\vec{F}_{B \text{ on } A}$. Thus the magnitude of the force that A exerts on B is 0.0150 N and the force that A exerts on B is in the $-x$ direction.
41. Magnitude = 0.415 N, direction = 33.7° from the $+x$ axis
43. $x = -0.445$ m
45. (a) Force of q_1 on q_2: 1.73 N in the $+x$ direction. Force of q_2 on q_1: 1.73 N in the $-x$ direction. (b) The magnitudes of the forces would remain the same, but the charges would attract rather than repel each other.
47. (a) The charges q_1 and q_2 must have opposite signs. (b) Charge q_2 must have a larger magnitude than q_1 to compensate for its greater distance from P.
49. (a) $E_{1x} = 6.87 \times 10^7$ N/C
    (b) $x = -0.241$ m $= -24.1$ cm
51. (a) x component: -4.57×10^5 N/C; y component: $+2.23 \times 10^5$ N/C
    (b) $F_x = +7.33 \times 10^{-14}$ N; $F_y = -3.56 \times 10^{-14}$ N
53. $v = 2.19 \times 10^6$ m/s

55. The rectangle was rotated through an angle of 60°.
57. $q_{encl} = 4.25 \times 10^{-8}$ C
59. (a) $\Phi_I = -1.60 \times 10^2$ N·m^2/C; $\Phi_{II} = 0$; $\Phi_{III} = 0$; $\Phi_{IV} = 0$; $\Phi_V = 1.60 \times 10^2$ N·m^2/C
    (b) $\varphi_{total} = 0$

61. (a) $\dfrac{Q}{\pi R^2 L}$; (b) $\dfrac{Q}{WL}$; (c) $\dfrac{3Q}{4\pi R^3}$; (d) $\dfrac{Q}{4\pi R^2}$

63. (a) $\dfrac{\sigma}{2\varepsilon_0}$ and points outward perpendicularly to the plane.
    (b) In the region between the two plates, the net field is $\dfrac{\sigma}{\varepsilon_0}$ and points from the positive plate toward the negative plate. In all other regions the field is zero.

65. $r^{-2.003}$, therefore, $n = 2$
67. (a) The 94 protons are likely to be concentrated into an approximately uniform sphere of charge, which is equivalent to a point charge for points outside it.
    (b) 1.1×10^{29} m/s^2
69. (a) 1.6×10^7 electrons; (b) $\sigma = 1.4 \times 10^{-2}$ C/m^2 = 8.8×10^{16} electrons/m^2
71. 7.9×10^{-9} C
73. (a) 1.9×10^{-16} C; (b) 1200 electrons
75. $E_x = 1.08 \times 10^{10}$ N/C; $E_y = 1.87 \times 10^{10}$ N/C
77. 14.2 N/C in the direction of travel of the electron
79. (a) 1.8×10^{21} N/C, outward from the center of the nucleus
    (b) 1.5×10^{13} N/C, outward from the center of the nucleus
    (c) 2.6×10^{24} m/s^2, inward toward the center of the nucleus
81. (a) The electric field is pointing downward. (b) Gravity pulls downward on the drop, so the electric force on it must be upward. Since the drop is negative, the electric field must point downward for the force to point upward. (c) 9070 N/C; (d) 7
83. (a) 0; (b) $\dfrac{\sigma_i R_i^2}{\varepsilon_0 r^2}$ pointing radially outward; (c) $\dfrac{\sigma_i R_i^2 - \sigma_o R_o^2}{\varepsilon_0 r^2}$, pointing radially outward if $\sigma_i R_i^2 > \sigma_o R_o^2$ or inward if $\sigma_i R_i^2 < \sigma_o R_o^2$
85. -2.20×10^{-6} C

Chapter 17

1. The electric potential is the electric potential energy per unit charge and is a scalar. The electric field is the electric force per unit charge and is a vector. The electric potential depends on both the electric field and the region over which the field extends.
3. Both electric field and potential result from the existence of a charge. If a charge exists, it will create both of these physical quantities. To have a nonzero potential energy, a force must be able to do work on an object as it is moved. If only one charge exists, then there is no electric force acting on the charge. If there is no force, then that force can do no work, so there is no potential energy.
5. This statement makes sense only if the zero point of the electric potential has been previously defined.
7. (a) Yes, a region of constant potential must have zero electric field. (b) No, if the electric field is zero, the potential need only be constant.

9. Zero. The electric field is perpendicular to an equipotential; therefore, the work done in moving along an equipotential is zero.
11. Increase plate area, decrease separation, and increase the dielectric constant
13. The energy stored in the capacitor increases.
15. The total capacitance is the sum of the individual capacitances. This means that for a given voltage, capacitors in parallel store more charge than any one of the individual capacitors.
17. For a given potential across them, the capacitors in parallel have a greater total area and so can store a greater amount of charge, thus having a larger capacitance.
19. The larger dielectric strength increases the maximum voltage that can be applied across the plates. The larger dielectric constant decreases the voltage required to store a given amount of charge, thus increasing the capacitance. The dielectric material can also be used to maintain the separation of the plates.
21. D
23. C
25. C
27. B
29. A
31. (a) $U_{electric} = 5.62 \times 10^{-3}$ J
    (b) $v_x = 9.18$ m/s
33. (a) $v = 3.2 \times 10^2$ m/s
    (b) $a = 1.4 \times 10^{10}$ m/s^2

35. $2kQ^2 \left(\dfrac{1}{x} + \dfrac{1}{y} + \dfrac{1}{\sqrt{x^2 + y^2}} \right)$

37. $U_{electric} = 8 \times 10^{-4}$ J
39. $\Delta V = 0.094$ V
41. (a) $V = 3.60 \times 10^6$ V
    (b) $V = -3.60 \times 10^6$ V
43. (a) The two charges must be opposite in sign. Charge q_2 must have a greater magnitude than charge q_1. (b) In addition to point P and infinity, there will be a point between the two charges where the potential is zero.
45. (a) $V_a = -2.73 \times 10^3$ V (rounded to three significant figures)
    (b) $V_b - V_a = -2.67 \times 10^4$ V (rounded to three significant figures)
    (c) $W_{required} = 4.28 \times 10^{-15}$ J
47. (a) $V = 27.2$ V
    (b) $E_{ionization} = 2.17 \times 10^{-18}$ J
49.

Equipotential

Electric field line

51. (a) The electric field at point *A* has a magnitude of 0.5 V/m and points to the right.
(b) The electric field at point *B* has a magnitude of 2 V/m and points to the right.

53.

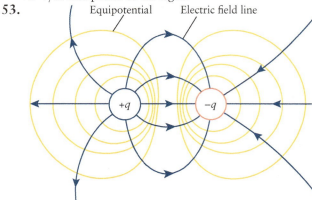

Equipotential   Electric field line

55. $C = 1.00 \times 10^{-6}$ F
57. $A = 2.26 \times 10^{-4}$ m^2
59. $d_2 = 0.250$ mm
61. $C = 2.00 \times 10^{-6}$ F
63. $\Delta U_{electric} = 8 \times 10^{-7}$ J
65. (a) $C = 10^{-7}$ F
(b) $C = 2 \times 10^{-7}$ F
67. $C_{1234} = 0.75$ pF. We know they have to be in series because the total equivalent capacitance is less than the capacitance of any individual capacitor.
69. $q_1 = q_2 = 1.8 \times 10^{-5}$ C
71. $q_{2f} = 8.00 \times 10^{-6}$ C
$q_{1f} = 4.00 \times 10^{-6}$ C
73. $C_{equiv} = 0.2$ μF
75. (a) $C_{equiv} = 150.0$ μF
(b) $q_1 = 1.20 \times 10^{-4}$ C
$q_2 = 4.80 \times 10^{-4}$ C
$q_3 = 1.20 \times 10^{-3}$ C
(c) $v_1 = v_2 = v_3 = 12.0$ V
77. charge: $q_{equiv} = 551 \times 10^{-6}$ C = 551 μC
energy stored: $U_{electric} = 2.07 \times 10^{-2}$ J
79. $C = 4.8 \times 10^{-12}$ F
81. $U_{electric} = 3.7 \times 10^{-6}$ J
83. $C_{equiv} = 4.6 \times 10^{-13}$ F
85. 1.45×10^7 m/s
87. (a) 6×10^8 J; (b) 2.65×10^2 kg
89. 1.12×10^{-20} J; increases
91. 36 mF
93. 14 μF
95. 1×10^{-6} F
97. (a) 1/2; (b) energy doubles, charge remains constant
99. (a) 6.3×10^{-10} J; (b) 1.4×10^{-7}
101. (a) 4.0 V; (b) 1/9
103. 1×10^5 m/s
105. (a) V_A, V_B, V_A; (b) U_A, U_B, U_C
107. (a) 1; (b) 4; (c) 3 and 7, 2 and 8

Chapter 18

1. Current is not a vector because no single vector can describe the direction of current flow along a curved piece of wire, such as in a circuit.

3. There is no contradiction. If the charges on a conductor are at rest, the electric field inside the conductor must be zero. However, the moving charges in a conductor that is carrying a current are not at rest.

5. The bird will not be electrocuted because there is no potential difference between the bird's feet. The bird grabs only one high-voltage wire, and it is not completing a circuit.

7. The cell membrane has the ability to alter its permeability properties. At rest the baseline membrane potential is lower on the inside compared to the outside of a cell. If the membrane potential depolarizes (that is, becomes a little less negative on the inside surface), sodium ion channels open and sodium ions rush into the cell. The inner surface of the membrane becomes positively charged, and the potential difference quickly swings positive. This is the peak of the action potential. Just as suddenly, as the potential difference across the cell membrane changes during the initial phase of an action potential, potassium ion channels begin to open. The membrane quickly becomes much more permeable to potassium ions than to sodium ions. Because the concentration of potassium ions is always higher inside cells, positive charge begins to flow out of the cell. This outward positive current drives the membrane potential to its resting level.

9. (a) Since the same current must flow through the ammeter and the circuit element, the two must be in series. (b) The equivalent resistance of resistors in series is the sum of the individual resistances, so the ammeter should have a very small resistance to have as little effect as possible on the resistance in the circuit.

11. This energy is dissipated as heat in the wire. The rate at which energy dissipated is related to the amount of current running through the wire and the resistance of the wire.

13. It cannot be changed instantaneously because the resistor in the circuit limits the current (the rate of flow of charge). Because the current is finite, not infinite, it requires time for the charge to flow onto and off of the capacitor plates.

15. A
17. E
19. D
21. A
23. C
25. 1.12×10^{22} electrons
27. 2.34×10^{-4} m/s
29. 0.3 mm/s
31. 12,000 K$^+$ ions
33. 0.0282 Ω
35. 16.0 Ω
37. 0.109 Ω
39. 0.56
41. 12 Ω
43. 4.0 V
45. 24.0 Ω
47. (a) 0.750 A in each resistor; (b) $V_9 = 6.75$ V, $V_3 = 2.25$ V
49. (a) 2.0 A; (b) It enters the junction.
51. $R_x = 3$ Ω
53. $R_{equiv} = 3.0$ Ω

55. (a) $V_{R_2} = 4.9$ V; (b) The current through the 60-Ω resistor is larger than the current through the 35-Ω resistor because the potential difference across the two branches is equal and total resistance of the top branch is larger than the resistance of the bottom branch.
57. 12.5 A
59. 2.7 kW
61. 58 MW
63. $R_1 = 6.00\ \Omega$; $R_2 = R_3 = 24\ \Omega$
65. $P_b = 20$ W; $P_1 = 3.56$ W; $P_2 = 4.44$ W; $P_3 = 12$ W
67. 4.72×10^{-5} C
69. (a) (i) 0.00216 s, (ii) 0.00108 s; (b) (i) 0.0667 A, (ii) 0.211 A
71. \$1.32
73. $n = 5 \times 10^5$ ions
75. (a) $R_1 = 3\ \Omega$; $R_2 = 5\ \Omega$; $R_3 = 12\ \Omega$; $V = 12$ V
  $R_1: i_1 = 0.5$ A
  $R_2: i_2 = 0.3$ A
  $R_3: i_3 = 0.9$ A
  (b) $R_1 = 4\ \Omega$; $R_2 = 6\ \Omega$; $R_3 = 10\ \Omega$; $R_4 = 8\ \Omega$; $R_5 = 8\ \Omega$;
  $V = 6$ V
  $R_1: i_1 = 0.3$ A
  $R_2: i_2 = 0.2$ A
  $R_3: i_3 = 0.5$ A
  $R_4: i_4 = 0.4$ A
  $R_5: i_5 = 0.4$ A
  (c) $R_1 = 3\ \Omega$; $R_2 = 8\ \Omega$; $R_3 = 12\ \Omega$; $R_4 = 6\ \Omega$; $R_5 = 5\ \Omega$;
  $R_6 = 10\ \Omega$; $R_7 = 2\ \Omega$; $R_8 = 4\ \Omega$; $V = 100$ V
  $R_1: i_1 = 7$ A
  $R_2: i_2 = 3$ A
  $R_3: i_3 = 3$ A
  $R_4: i_4 = 6$ A
  $R_5: i_5 = 6$ A
  $R_6: i_6 = 3$ A
  $R_7: i_7 = 6$ A
  $R_8: i_8 = 3$ A
77. (a) $i = 3 \times 10^5$ A; household currents are on the order of 10 A, so average lightning currents can be about 30,000 times larger.
  (b) $R = 1 \times 10^3\ \Omega$
  (c) $\Delta E = 1 \times 10^{10}$ J
  (d) $m = 5 \times 10^3$ kg
79. (a) $P_{\text{delivered}} = 500$ W
  (b) $i = 0.8$ A; yes, this is large enough to be harmful.
  (c) $P_{\text{received}} = 400$ W (rounded to two significant figures)
81. $R = 3.9 \times 10^4\ \Omega = 39$ kΩ
83. (a) $q = 3.6 \times 10^{-5}$ C
  (b) $i(0) = 0.06$ A
  (c) $t = 6.0 \times 10^{-4}$ s
85. (a) The probe is in parallel with the circuit element.
  (b) (i) The probe should have a very large resistance because adding a very large resistance in parallel does not appreciably affect the equivalent parallel resistance.
  (ii) The probe should have a very small capacitance because adding a very small capacitance in parallel does not appreciably affect the equivalent parallel capacitance.
87. (a) $C_2 = 1000$ F
  (b) $U_{\text{ultracapacitance}} = 1 \times 10^3$ J;
  $U_{\text{ultracapacitance}}/U_{\text{capacitor}} = 1 \times 10^8$
  (c) $U_{\text{ultracapacitance}}/U_{\text{battery}} = 0.2$
  (d) $R = 0.05\ \Omega$

Chapter 19

1. Bring each end of the third iron rod close to a given end of one of the magnetized rods. If both ends of the third rod experience the same force, the third iron rod is not magnetized.
3. Applying Newton's second law to a charged particle whose velocity is at right angles to a uniform magnetic field B results in the equation $qvB = \dfrac{mv^2}{r}$ or $r = \dfrac{mv}{qB}$.
  As long as the particles have the same charge q, the radius r of their orbits depends on their momentum $p = mv$, not their velocity v alone.
5. (a) The electric field points into the page. (b) The beam is deflected into the page. (c) The electron beam is not deflected.
7. Yes, because the width of the field determines the length of the wire in the magnetic field. The longer the length of wire in the magnetic field, the greater the force exerted on the wire will be.
9. Since the directions of the two currents are opposite, the magnetic field from one wire is opposite to the other. This means the magnetic fields will cancel each other to some extent.
11. There will be no net force on the wires, but there will be a torque. When two wires are perpendicular, each wire will experience a force on one end and another force, equal in magnitude but opposite in direction, on the other end. This results in zero net force. However, the forces will produce a nonzero torque that will make the wires want to align with each other such that the current in each wire is traveling in the same direction as the current in the neighboring wire.
13. C
15. D
17. A
19. A
21. E
23. (a) out of the page; (b) down; (c) down; (d) to the left; (e) out of the page; (f) right
25. 0.5 N
27. 5.8×10^{-18} N to the left
29. 2.4×10^{-12} N
31. 5×10^{-3} T in the $+x$ direction
33. (a) 4.27×10^{-5} m; (b) 5.59×10^5 Hz
35. 0.511 T
37. 0 N
39. 2.5 T
41. 0.664 N·m at 30°
  0.755 N·m at 10°
  0.493 N·m at 50°
43. 1.44×10^{-5} N·m
45. B_O and B_P point in the $+z$ direction (out of the page). B_Q and B_R point in the $-z$ direction (into the page).
47. $B = \dfrac{\mu_0 I}{4}\left(\dfrac{R - r}{rR}\right)$ into the page
49. $B = 4 \times 10^{-6}$ T $= 0.08\ B_{\text{Earth}}$
51. 120 m
53. 1.03×10^{-8} N away from the wire and points to the left.
55. 0.313 N

57.

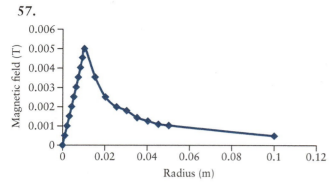

For $r < 0.01$ m: $B(r) = \left(0.501\dfrac{\text{T}}{\text{m}}\right)r$

For $r > 0.01$ m: $B(r) = \dfrac{0.000050 \text{ T}\cdot\text{m}}{r}$

59. 0.0109 T

61. 9.12×10^{-31} kg

63. (a) $B = 1.0$ T at $r = 1.0$ m; $B = 0.001$ T at $r = 1.0$ km. Compared to Earth's magnetic field, the lightning bolt has a magnetic field 20,000 times as large at 1.0 m and 20 times as large at 1.0 km. (b) $B = 2.0 \times 10^{-6}$ T at $r = 1.0$ m; $B = 2.0 \times 10^{-9}$ T at $r = 1.0$ km. In each case the field due to the lightning is 500,000 times greater than the household current. (c) $r = 2.0$ mm

65. 0.40 N

67. (a) 298 A; (b) 0.62 T; (c) The easiest way to achieve the same field with less current is to increase the number of windings, N.

69. (a) 15.9 mA; (b) The loop will initially rotate to point its dipole moment in the direction of the magnetic field. At this instant there is no net torque on the loop, but because of its rotational momentum, it will continue to rotate past this point, slowing down. After it momentarily comes to rest, it will rotate back toward its initial position. If the system is frictionless, this harmonic oscillation will continue indefinitely.

71. 650 V

73. $B = \dfrac{\mu_0 i}{2\sqrt{2}\pi d}$ and makes an angle of 315° with the $+x$ axis.

75. $B = 1.5 \times 10^{-6}$ T $= 0.031\, B_{\text{Earth}}$. Since the field is so much smaller than the Earth's magnetic field, there should be little or no cause for concern.

77. (a) $B = 0.7155\dfrac{\mu_0 Ni}{R}$ in the $+x$ direction

(b) $B = 0.221\dfrac{\mu_0 Ni}{R}$ in the $+x$ direction

79. (a) 2.64×10^{-6} T pointing westward; (b) 2.64×10^{-7} T pointing vertically upward; (c) 2.64×10^{-6} T pointing southward; (d) In (a) and (c), B is about 5% of B_{Earth}. This is fairly small, but it could possibly be enough to interfere with navigation. In (b), B is about 0.5% of B_{Earth}, which seems too small to cause much of a problem.

81. (a) $i = 1.14$ A; (b) clockwise

Chapter 20

1. A current is induced in loop b while the current in loop a is changing. If the current is in the direction shown and is increasing, the flux of its magnetic field through loop b is upward and increasing. In accordance with Lenz's law, the direction of the induced current in loop b is such that the flux of its magnetic field through loop b is downward, opposing the change in flux that produced it. This means that the current in loop b is in the opposite direction as the current in loop a, and the two loops repel. After the current in loop a stops changing, then the current in loop b becomes 0 and there is no force between the loops.

3. The induced emf will create an induced current in the clockwise direction, which will exert a force to the left on the sliding rod. This will gradually slow the rod until it comes to a stop.

5. As someone with a pacemaker or other electronic device walks through regions of spatially varying magnetic fields, the changing magnetic flux through the electronic circuitry will induce an emf in the device. This extra emf could cause unwanted, and possibly fatal, malfunctions of the device.

7. D

9. C

11. A

13. $|\varepsilon| = 1.3 \times 10^{-2}$ V

15. (a) The magnetic field is constant when $t < 0$, so the induced emf in the coil is zero.
(b) $|\varepsilon| = 2.54 \times 10^{-3}$ V $= 2.54$ mV
(c) The magnetic field is constant when $t > 10.0$ s, so the induced emf in the coil is zero.
(d) Magnetic field as a function of time:

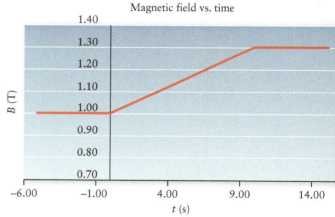

Induced emf as a function of time:

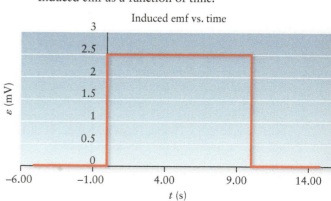

17. (a) counterclockwise; (b) clockwise; (c) clockwise; (d) counterclockwise; (e) clockwise; (f) no induced current

19. *V*

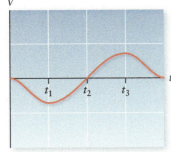

21. (a) $i = 0$
(b) $i = 0.0439$ A
(c) $i = 0$
(d) $\Delta q = 0.220$ C
(e)

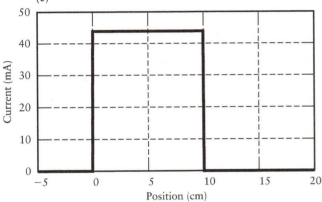

23. $\omega = 3.8$ rad/s
25. $P_{average} = 9.3$ mW
27.

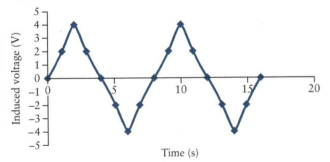

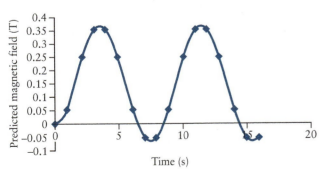

29. (a) 0.12 V; (b) 0.0047 A; (c) counterclockwise
31. 2.1 V
33. (a) 0.01 A; (b) up; (c) 0.001 N
35. 7.3 mW

Chapter 21

1. Many ac voltages are sinusoidal. A sine (or cosine) function has an average value of zero, since equal amounts of the function lie above and below the axis. The root mean square value of the function is not zero, and the larger the rms value, the larger the amplitude of the ac voltage.

3. For example:

Country	V_{rms}
Angola	220
Botswana	231
Ecuador	120–127
Mexico	127
Slovenia	220

5. No. These devices usually work over a range of voltages and will not function properly if a transformer is used. They contain "smart" circuitry that senses if the voltage is 120 or 240 V. (Many people needlessly buy transformers for their laptops. All that may be needed, depending on destination, is an adapter plug.)

7. (a) A transformer is a device that transfers electrical energy from one circuit to another through electro-magnetic induction. (b) Due to the properties of mutual inductance, the voltage output can be larger or smaller than the voltage input depending on the number of turns of wire in the primary and secondary coils. (c) A transformer requires an ac input so that the changing flux in the primary coil will induce a varying electromotive force according to Faraday's law.

9. It saves copper. The primary power loss in transmission equals $i_{rms}^2 R$, where R is the resistance of the transmission lines, and the rate at which energy is transported equals $V_{rms} i_{rms}$. Thus, at high voltage, power can be transmitted with a lower current, and the lower the current, the less the transmission loss. The other way to reduce transmission loss would be to decrease the resistance of the transmission lines. The only practical way to do that is to use thicker wire, which requires more of the metal that the wires are made of. The voltages are stepped down for consumer safety.

11. The current drops suddenly toward zero when the switch is opened. This current drop induces a very large emf in the inductor, which produces a large potential difference across the switch gap. This potential is large enough to cause dielectric breakdown of the air in the gap, which causes the spark. The spark allows current to continue to flow for a brief period.

13. The electrical energy stored in the capacitor is $U_E = \dfrac{q^2}{2C}$, which varies as $\cos^2 \omega t$, and the magnetic energy stored in the inductor is $U_B = \dfrac{Li^2}{2}$, which varies as $\sin^2 \omega t$.

Therefore, when a maximum amount of energy is stored in the capacitor, no energy is stored in the inductor, and vice versa.

15. No. This would be true only if there is negligible inductance and capacitance in the circuit or the circuit is at resonance.

17. Impedance describes the overall opposition to the flow of charge in a circuit driven by a time-varying voltage. The SI units of impedance are ohms (Ω).

19. D

21. A

23. C

25. C

27. $V(t) = (120\sqrt{2}\ \text{V})\sin(120\pi t)$

29. $V(t) = 11.6$ V

31. $V_0 = 849$ V

33. $V_{max} = V_0 = \sqrt{2}i_{rms}R$
   (a) $i_{rms} = 0.4$ A
   (b) $V_{max} = 7 \times 10^2$ V
   (c) $R = 156\ \Omega$

35. $N_s = 22$ turns

37. $i_s = 2$ A

39. (a) $N_s = 21$ turns
   (b) $i_p = 0.79$ A

41. (a) The ratio of the number of coils in the secondary to the number of coils in the primary is 42:1.
   (b) $i_p = 0.625$ A ≈ 0.62 A (rounded to two significant figures)
   $i_s = 0.015$ A ≈ 0.02 A (rounded to one significant figure)
   (c) $R = 190\ \Omega$

43. $L = 1.9 \times 10^{-4}$ H $= 0.19$ mH

45. $U_B = 3.8 \times 10^{-4}$ J

47. $f = 4 \times 10^1$ Hz

49. selectivity $= 286$ H/F

51. $\dfrac{U_E}{U_{E,max}} = \dfrac{1}{4}$

53. $C = 5 \times 10^{-9}$ F $= 5$ nF

55. $i_0 = 3$ A

57. (a) $C = 5\ \mu$F; (b) The root mean square current will increase. We can see this using the equation for the peak current: $i_0 = i_{rms}\sqrt{2} = \omega C V_0 = (2\pi f)C V_0$.

59. (a) $i_{R,0} = 0.6$ A; $P_{R,average} = 16$ W $= 2 \times 10^1$ W (rounded to one significant figure)
   (b) $i_{L,0} = 0.5$ A, $P_{L,average} = 0$
   (c) $i_{C,0} = 2$ A, $P_{C,average} = 0$

61. (a) $i_{rms} = 2 \times 10^2\ \Omega$
   (b) $\varphi = 0.3$ rad

63. $i_{rms} = 0.25$ A

65. The higher the frequency, the greater is the current output from the capacitor.

67. $L = 4.2 \times 10^{-7}$ H

69. $L = 1 \times 10^{-3}$ H $= 1$ mH

71. 53 V

73. (a) 2.0×10^1 W; (b) United States: 0.042 A, Rome: 0.083 A, most likely will be damaged; (c) step-down transformer with twice as many windings in the primary coil as the secondary coil; (d) 2.9×10^3 V

75. (a) 1.2×10^3 W; (b) 12 W; (c) We should use a voltage of 120 V. A much smaller proportion of the total power is dissipated in the cord, which means more is delivered to the device. It is much more efficient than the 12.0-V delivery system.

77. (a) 2.29 V; (b) 52 A; (c) 74 A

79. $\dfrac{V_o}{V_i} = \dfrac{1}{\omega C\sqrt{\left(\dfrac{1}{\omega C}\right)^2 + R^2}} = \dfrac{1}{\sqrt{(\omega C)^2\left(\dfrac{1}{\omega C}\right)^2 + (\omega C)^2(R^2)}}$
   $= \dfrac{1}{\sqrt{1 + (\omega RC)^2}}$

81. (a) $i_{rms} = 1.414$ nA; (b) $\omega_0 = 785.4$ rad/s

Chapter 22

1. (a) (v), (iii), (ii), (iv), (i); (b) Gamma rays have the highest energy per photon. Microwaves have the lowest energy per photon.

3. The speed of light equals the product of the frequency and the wavelength ($c = f\lambda$).

5. Without the seminal work of Faraday, Gauss, Coulomb, Ampère, and others, it would not have been possible for Maxwell to complete his theory. It was Maxwell who really pushed the forefronts of science by understanding the connections among electricity, magnetism, and optics.

7. All photons travel at the same speed ($c = 3.00 \times 10^8$ m/s). The energy of a photon is determined not by its speed but by its frequency.

9. The energy of a photon is proportional to the wave frequency.

11. C

13. D

15. E

17. (a) 7.25×10^{-8} m; ultraviolet light
   (b) 4.29×10^{-7} m; violet visible light
   (c) 3.75×10^{-9} m; x ray
   (d) 1.00×10^{-5} m; infrared light
   (e) 3.33×10^{-5} m; infrared light
   (f) 8.72×10^{-10} m; x ray
   (g) 3.65×10^{-8} m; ultraviolet light
   (h) 5.00×10^{-8} m; ultraviolet light

19. (a) 4.29×10^{14} Hz
   (b) 5.00×10^{14} Hz
   (c) 6.00×10^{14} Hz
   (d) 7.50×10^{14} Hz
   (e) 3.00×10^{15} Hz
   (f) 9.01×10^{18} Hz
   (g) 6.00×10^{11} Hz
   (h) 4.74×10^{18} Hz

21. between 2.8 and 3.4

23. $\Delta x = 3$ m

25. $\Delta t = 1.28$ s

27. (a) $E_0 = 2 \times 10^2$ V/m
   (b) $v = 3.00 \times 10^8$ m/s
   (c) $f = 1.20 \times 10^{15}$ Hz
   (d) $T = 8.33 \times 10^{-16}$ s
   (e) $\lambda = 2.50 \times 10^{-7}$ m

29. $B = 1.67 \times 10^{-15}$ T $\approx 2 \times 10^{-15}$ T

31. $B = 3 \times 10^{-13}$ T

33. $E_{\parallel} = 1.0 \times 10^{-3}$ V/m (perimeter of ring from 0 to 4.0 ms)
   $E_{\parallel} = 15.6 \times 10^{-3}$ V/m (from 4.0 to 10.0 ms)
   $E_{\parallel} = 0$ V/m (from 10.0 to 14.0 ms)

35. (a) $E_{rms} = 274$ V/m ≈ 300 V/m;
$B_{rms} = 9.15 \times 10^{-7}$ T $\approx 9 \times 10^{-7}$ T
(b) $E_0 = 387$ V/m ≈ 400 V/m;
$B_0 = 1.29 \times 10^{-6}$ T $\approx 1 \times 10^{-6}$ T

37. (a) $E_0 = 1.70 \times 10^{11}$ V/m $\approx 2 \times 10^{11}$ V/m
(b) $S_{average} = 3.82 \times 10^{19}$ W/m^2 $\approx 4 \times 10^{19}$ W/m^2

39. A. wavelength = 5.77×10^{-7} m = 577 nm;
frequency = 5.20×10^{14} Hz
B. wavelength = 4.14×10^{-7} m = 414 nm;
frequency = 7.24×10^{14} Hz
C. wavelength = 1.55×10^{-7} m = 155 nm;
frequency = 1.93×10^{15} Hz
D. wavelength = 4.59×10^{-6} m = 4590 nm;
frequency = 6.53×10^{13} Hz
E. wavelength = 1.34×10^{-15} m;
frequency = 2.25×10^{23} Hz
F. wavelength = 4.32×10^{-10} m = 0.432 nm;
frequency = 6.95×10^{17} Hz
G. wavelength = 1.58×10^{-7} m = 158 nm;
frequency = 1.90×10^{15} Hz
H. wavelength = 9.14×10^{-8} m = 91.4 nm;
frequency = 3.28×10^{15} Hz

41. (a) $k = 2.09 \times 10^7$ rad/m
(b) angular frequency: $\omega = 3.77 \times 10^{15}$ rad/s
frequency: $f = 6.00 \times 10^{14}$ Hz
wavelength: $\lambda = 5.00 \times 10^{-7}$ m

43. (a) 62.1 to 414 nm; (b) It extends from violet visible light into the UV.

45. (a) 0.450 m; (b) 1.51×10^4 J/m^3

Chapter 23

1. Christiaan Huygens suggested that every point along the front of a wave be treated as many separate sources of tiny "wavelets" that themselves move at the speed of the wave. This is important to see how light moves from one medium to another, different, medium and allows you to predict the bending that occurs in Snell's law.

3. No, the critical angle depends only on the index of refraction of the water.

5. When light moves from a medium of larger index of refraction toward a medium with a smaller index of refraction, the angle of refraction will increase. If you make the incident angle larger and larger, the refracted angle will ultimately approach 90°. After the incident angle increases past this critical value, all of the incident light is reflected rather than refracted.

7. When light enters a material with a negative index of refraction, the light refracts "back" away from the normal to the surface, as seen in the picture below:

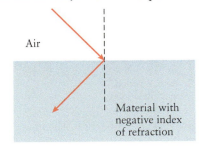

Air

Material with negative index of refraction

9. The light coming from the Sun is refracted as it passes through Earth's atmosphere. Because of the wavelength dependence of the index of refraction, red light bends less than orange than yellow than green than blue than violet. In addition, bluer light is scattered more than red light as it passes through the atmosphere. Because of these effects, more red/orange-colored light can pass into the shadow of the Earth, making the Moon appear red.

11. Polarized sunglasses use polarizing filters to preferentially absorb light that harms your eyes. Since most glare is created by light that is reflected from horizontal surfaces (for example, a lake or a ski slope), it is advantageous to block out the light that is horizontally polarized. As such, the lenses of polarized sunglasses allow only vertically polarized light to enter. Naturally, the intensity of the light that passes through such lenses is decreased in intensity.

13. 1. Soap bubbles reflect different colors.
2. Thin film coatings on photographic lenses ("nonreflective coatings").
3. Oil floating on a puddle of water will show different colors.

15. If the light striking the two slits did not have the same wavelength and a fixed phase relationship, the interference pattern formed on the screen would vary so rapidly that it would not be visible.

17. Waves spread significantly when they pass through an opening that is much smaller than their wavelength. The human mouth is much smaller than the wavelength of a sound wave, so sound waves spread widely when they emerge from the mouth.

19. 10 m

21. D

23. B

25. B

27. A

29. A

31. 1.00029

33. 763 km

35. (a) 13.2°; (b) 22.1°; (c) 30.9°; (d) 22.8°

37. 452 nm

39. A. 1.34
B. 1.37
C. 1.48
D. 1.74
E. 1.21
F. 1.61
G. 2.65
H. 1.04

41. Light will always be totally internally reflected.

43. 5.70 m

45. (a) 1.58 (rounded to three significant figures), 1.50;
(b) 18.5°, 19.5°; (c) 33.3°, 35.3°

47. 1.50 or 1.47

49. vertically polarized and has an intensity of 39 W/m^2

51. 7.4×10^2 W/m^2

53. 49.4°

55. (a) $\theta_B = 36.9°$
(b) $\theta_c = 48.8°$

57. (a) shifted by half a wavelength; (b) shifted by half a wavelength

59. 648 nm

61. $t = 93.8$ nm; $t = 282$ nm; $t = 469$ nm; $t = 657$ nm

63. $\lambda = 652$ nm and $\lambda = 435$ nm are not seen in the film.

65. 540 nm

67. (a) 5; (b) 8

69. $\theta_1 = 36.4°$; additional fringes are not visible.

71. 583 nm

73. 633 nm

75. 17 mm

77. 7.71×10^{-6} degrees $= 1.34 \times 10^{-7}$ radians

79. 6.9×10^{-5} m

81. $n_2/n_1 = 1.56$

83. (a) $n_2 = 2$
(b) $BC = 1.0$ cm

85. 4 m

87. (a) $\theta_2 = 25.4°$
(b) $\theta_3 = 28.0°$

89. (a) $\theta = 49.8°$
(b) $\theta = 30.9°$

91. $\theta_C = 24.2°$

93. $n = 1.60$

95. 29° or 61°

97. $\theta_2 = 25.1°$

99. (a) 70 nm; (b) canceled wavelengths: 400 nm;
(c) reinforced wavelengths: 600 nm

101. Small-angle approximation: 5.81×10^{-3} m ≈ 6 mm
(rounded to one significant figure)

103. (a) $m = 1$: $\theta = 1.1 \times 10^{-3}$ rad
$m = 2$: $\theta = 2.2 \times 10^{-3}$ rad
$m = 3$: $\theta = 3.3 \times 10^{-3}$ rad
(b) $m = 1$: $\theta = 7.9 \times 10^{-4}$ rad
$m = 2$: $\theta = 1.6 \times 10^{-3}$ rad
$m = 3$: $\theta = 2.4 \times 10^{-3}$ rad

105. (a) Bright light: 3.4×10^{-4} rad, dim light: 8.4×10^{-5} rad;
The angular resolution is better in dim light than in bright light because the angular resolution is smaller for dim light. (b) By squinting we partially close our eye, thereby limiting the light entering it. This causes the pupil to enlarge, which improves our angular resolution.

107. (a) $L = 9.0 \times 10^{15}$ m $= 0.95$ ly
(b) $D = 9 \times 10^2$ m

109. (a) Maximum angular resolution:
$\theta_R = 1.9 \times 10^{-5}$ rad $= 3.9''$
(b) $D = 0.035$ m
(c) $x = 2.7 \times 10^{13}$ m

111. 13 bright fringes

113. $\Delta\theta = 1.59°$

Chapter 24

1. In fact your image in a plane mirror is reversed back to front, not left to right (see Figure 24-6).

3. A real image forms where light rays come together. No light rays actually meet where a virtual image forms.

5. Mirrors with this warning are convex mirrors that produce images that are smaller than the object. This gives the driver a wide-angle view, but also gives a misleading impression of the distance to other vehicles.

7. Additional information is needed. In accordance with the lensmaker's equation, if the radius of curvature of the front surface is larger, it is a diverging lens; if the radius of curvature of the front surface is smaller, it is a converging lens.

9. The full image is formed, but it is dimmer than before because it is formed with half as much light.

11. Since the brain can be "trained" to interpret all the nerve inputs that it receives, it is feasible that upright and inverted could be "redefined." It must be very disconcerting, however, for those several days when "up is down and down is up"!

13. Presbyopia (farsightedness) is associated with a weakening of the muscles around the eye and inflexibility in the crystalline lens system as a person ages. This loss in adaptive amplitude leads to the inability to focus on objects that are close up. Myopia (nearsightedness), on the other hand, is a disorder that affects as much as 50% of the world's population, but it is not brought on by the aging process.

15. Since the focal length of the lens is fixed, as the object distance is decreased for a nearby object, the image distance must be increased by moving the lens away from the screen, which is the light detector.

17. B

19. B

21. D

23. A

25. B

27. 30°

29. 0.90 m

31. The first five images to the left are 3, 17, 23, 37, and 43 m to the left of the left mirror. The first five images to the right are 7, 13, 27, 33, and 47 m to the right of the right mirror.

33.

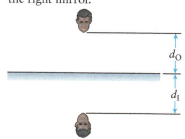

35. The camera will focus the lens on the surface of the mirror (2 m away from the virtual image). It will be out of focus a little bit.

37. As long as $d_0 > f$, the image will be real in a concave, spherical mirror. In this case the focal point is located between the mirror and the object.

39. (a)

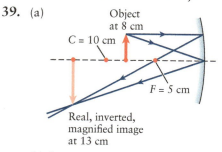

(b) $d_I = 13.3$ cm, $m = -1.66$. The image is real and inverted.

41. (a)

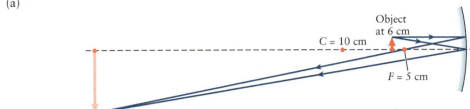

Object
at 6 cm

C = 10 cm

F = 5 cm

Real, inverted,
magnified image
at 30 cm

(b) $d_I = 30$ cm, $h = -5.0$ cm

43. (a) $d_I = 30$ cm, $h = -60$ cm. The image is real and
inverted.
(b) $d_I = 12$ cm, $h = -12$ cm. The image is real and
inverted.
(c) $d_I = 8.11$ cm, $h = -1.62$ cm. The image is real and
inverted.

45.

Virtual, upright,
enlarged image

(a)

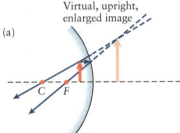

C   F

(b)

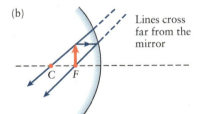

Lines cross
far from the
mirror

C   F

(c)

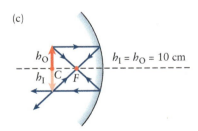

h_O $h_I = h_O = 10$ cm

h_I C   F

47. In both cases, the images are virtual. The image of an
"up close" object is larger than the image of the same
object "far away." In both cases the image is smaller
than the object.

49. A convex mirror is the opposite of a concave mirror.
Use a phrase such as "The cave goes in" to remember
the shape of a concave mirror.

51. (a) $d_I = -6.67$ cm, $h = 3.33$ cm. The image is virtual
and upright.
(b) $d_I = -8.33$ cm, $h = 1.67$ cm. The image is
virtual and upright.
(c) $d_I = -9.09$ cm, $h = 0.909$ cm. The image is virtual
and upright.

53. $d_I = -4.29$ cm, $h = 0.429$ m. The image is virtual and
upright.

55. $d_I = -0.93$ cm, length $= 0.926$ cm. The image is virtual
and upright.

57.
$$\frac{1}{d_0} + \frac{1}{d_I} = \frac{1}{f}$$
$$\frac{1}{d_I} = \frac{1}{f} - \frac{1}{d_0} = \frac{d_0 - f}{f d_0}$$
$$d_I = \frac{f d_0}{d_0 - f}$$

The focal length of a spherical convex mirror is always
negative, and the distance of the object from the mirror,
d_0, is positive for all mirrors, which means the term
$d_0 - f$ will always be positive for a real object.
Therefore,

$$d_I = \frac{f d_0}{d_0 - f} < 0$$

59. $d_I = -6.32$ cm, $h = 0.947$. The image is virtual and
upright.

61. No, a real image in a converging lens occurs on the
opposite side of the lens from the object.

63. Two. Images form where all the refracted rays (or
their backward extensions) intersect. Therefore,
wherever two rays intersect, the rest of the rays also
intersect.

65. (a) $d_I = -4.0$ cm, $h = 8.0$ cm. The image is virtual and
upright.

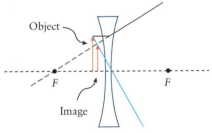

Object

F   F

Image

(b) $d_I = -6.67$ cm, $h = 6.67$ cm. The image is virtual
and upright.

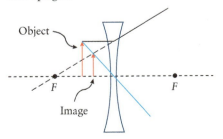

Object

F   F

Image

(c) $d_I = -10.0$ cm, $h = 5.00$ cm. The image is virtual and upright.

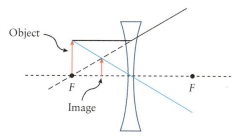

(d) $d_I = -14.3$ cm, $h = 2.86$ m. The image is virtual and upright.

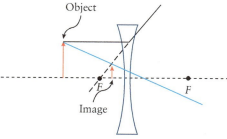

67. 0.7543 m^{-1}; 1.33 m
69. (a) 45.0 cm; real; 3.00 cm tall and inverted
(b)

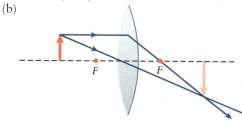

71. minimum = 4.3 mm; maximum = 4.6 mm
73. 63.9 cm
75. 2.78
77. $m_{total} = -360$
79. $x = sd/2h$
81. (a) Image distance: $d_I = 26.7$ cm. The image is real. Magnification: $m = -0.667$. The image is inverted and smaller.
(b)

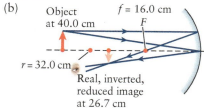

Object at 40.0 cm $\quad f = 16.0$ cm

$r = 32.0$ cm

Real, inverted, reduced image at 26.7 cm

83. (a) minimum power = 109 diopters
focal length at minimum power = 9.15 mm
maximum power = 143 diopters
focal length at maximum power = 7.00 mm
(b) $d_I = 7.20$ mm
(c) The image distance is much shorter than the distance between the lens and the retina, so the image would not fall on the retina.
85. $f_e = -2$ cm (rounded to one significant figure)
87. $n = 2.16$
89. (a) The image formed by the mirror is located 7.5×10^2 cm in front of the mirror.

(b) The image is real because $d_{I,B} > 0$. The image is inverted because $m_{total} < 0$.
(c) The image created by the mirror is virtual because $d_{I,B} < 0$. The image is upright because $m_{total} > 0$.
91. (a) $d_{I,B} = -43.8$ cm. The image formed by the mirror is located 43.8 cm behind the mirror. Since $d_{I,B} < 0$, the image is virtual. The final image is 10.5 cm tall and inverted.
(b) $d_{I,B} = -116$ cm. The image formed by the mirror is located 116 cm behind the mirror. Since $d_{I,B} < 0$, the image is virtual. The final image is 3.00 cm tall and upright.
93. (a) The light will be focused 3.12 mm in front of the retina. (b) Reflection is not affected by the index of refraction, so the answer would be the same as in (a).
95. $P = 2.7$ diopters
97. (a) The person's far point is too close, so he needs a single focal length lens to correct this issue.
(b) $f = -2.75$ m
(c) $P = -0.364$ diopters
99. (a) $d_I = 0.203$ m = 203 mm
(b) $h_I = 0.51$ cm
(c) $d_O = 28$ mm
101. 15 cm
103. (a) $M = -790$; (b) The negative sign tells us that the image is inverted when viewed in the eyepiece.
105. $|f_{blue} - f_{red}| = 0.03770$ mm

Chapter 25

1. If the frame were moving in both x and y directions, the given Galilean transformation equations would be as follows:
$x' = x - V_x t$
$y' = y - V_y t$
$z' = z$
$t' = t$

3. Earth's surface is a noninertial frame because Earth rotates around its axis and orbits around the Sun, both of which involve acceleration. We often approximate Earth's surface as an inertial frame because the accelerations associated with these motions are small compared to the gravitational acceleration g.

5. The Michelson–Morley experiment was performed to determine our motion through the "luminiferous ether" (the hypothesized medium of light). The experiment compared travel times for two beams of light that were moving at right angles to each other. As the beams were turned 90°, it was expected that a noticeable deviation would be perceived due to the "ether wind." No deviation was found, which is consistent with the idea that the ether does not exist.

7. If the object is moving relative to you, you measure its length by finding the difference between the coordinates of its endpoints at the same time.

9. It is impossible to accelerate an object to the speed of light. From the equation $E = \gamma mc^2$ as the speed of an object approaches the speed of light, the amount of energy required to further increase the speed approaches infinity. It would take an infinite amount of work or an infinitely large force to accelerate the object to the speed of light.

11. The equivalence of gravity and acceleration is fundamental to the theory of general relativity.

13. Physics in 1905 was predicated on Newtonian theories regarding motion, gravity, and cosmology. In addition, Maxwell's theory of electromagnetism was firmly in place. Einstein's new ideas were extraordinary, and extraordinary claims require extraordinary evidence. As this experimental evidence was obtained over the years that followed, the theory of relativity became established physics.

15. D
17. D
19. A
21. C
23. (a) $x' = x - 4t$; $y' = y$; $z' = z$
    (b) $x' = -14$ m; $y' = 1$ m; $z' = z$
25. (a) $x' = x - 15t$; $y' = y$; $z' = z$
    (b) $x'' = x - 4t$; $y'' = y$; $z'' = z$
    (c) $x' = x'' - 11t$; $y' = y''$; $z' = z''$
27. $x' = 7.4 \times 10^2$ m; $y' = y = 250$ m; $z' = z = 250$ m; $t' = t = 2.0 \times 10^{-6}$ s
29. $x' = 10$ m; $y' = y = 10$ m; $z' = z = 30$ m; $t' = t = 6.0 \times 10^{-4}$ s
31. $v'_{satellite,x} = -4.0 \times 10^8$ m/s. This is greater than the speed of light, so the Galilean transformation cannot be used to calculate relative velocities for speeds approaching c.
33. (a) 133 μs; (b) 1×10^{-2} μs; (c) 1 μs; (d) 40 μs; (e) 6×10^2 μs
35. 1.67
37. 2.19 μs
39. $0.9999999995c$. Unfortunately, this speed is unattainably high.
41. 1.5 ns
43. 1.92 m
45. 0.98 m
47. $V = 2.9955$ m/s; $V/c = 0.9985$
49. (a) 1×10^{-7} s, for both; (b) The backward photon registers first. (c) 1.5×10^{-7} s
51. The velocity of the truck relative to the car is $-0.110c$. The velocity of the car relative to the truck is $0.110c$.
53. $0.548c$
55. $0.920c$
57. (a) $p_{Newtonian} = 800,000$ kg·m/s
    (b) $p_{Einsteinian} = 800,001$ kg·m/s
    (c) Although they are virtually the same, the Einsteinian version is more accurate. Percent difference: 0.0002%.
59. (a) $p = 3 \times 10^{-19}$ kg·m/s
    (b) $K = 2 \times 10^{-11}$ J
61. $p = 2.67 \times 10^{-21}$ kg·m/s
63. $m = 0.252$ mg
65. 8.20 m/s² in the upward direction
67. (a) 6×10^{-4} s; (b) 1.44×10^5 m
69. $\Delta t - \Delta t_{proper} \approx 1 \times 10^{-10}$ s
71. $\Delta x = 2.3$ m
73. (a) The muon does not make it to the Earth's surface.
    (b) $0.988c$
75. (a) A; (b) B; (c) A, B, P_1, P_2; (d) C, P_1, P_2
77. (a) $\gamma = 100,000$
    (b) $\Delta t = 0.166$ s
79. (a) $\Delta t_{proper} = 2.25 \times 10^{-6}$ s
    (b) $V = 0.700c$
81. (a) The proper time is the time measured by the particle because the two events occur at the same location in the particle's reference frame. The proper length is in Earth's reference frame because the atmosphere and Earth are at rest in that frame.
    (b) $\Delta t_{particle} = 4.04 \times 10^{-5}$ s = 40.4 μs
    (c) $\Delta t_{Earth} = 9.26 \times 10^{-5}$ s = 92.6 μs
    (d) $\Delta t_{Earth} = 92.6$ μs. Yes, the results are consistent.
83. (a) $f = 50$ beats/min
    (b) 70 beats/min
85. (a) $K_{Einsteinian} = 1 \times 10^{20}$ J
    $K_{Newtonian} = 4 \times 10^{19}$ J
    (b) % error = 70%
    (c) The Newtonian formula underestimates the kinetic energy.
87. (a) $0.77c$; (b) 97 y

Chapter 26

1. If light behaved simply as a wave, light of any frequency could produce a photoelectron if the intensity of that light were increased sufficiently to provide the required energy. The observation that light of a frequency lower than the cutoff frequency does not produce a photoelectron is consistent with the idea that light consists of particle-like packets called photons, each of which has an energy that depends on the frequency of light. Each photon can then produce a single photoelectron if the photon's frequency is sufficiently high.

3. Yes, it is possible for photoelectrons to be emitted from a metal at relativistic speeds, as long as the photon energy is sufficiently high.

5. Increasing the intensity means delivering photons to the surface of the metal at a greater rate; correspondingly, the photocurrent—the rate at which the electrons are emitted—increases.

7. Visible light can Compton scatter. However, because the Compton wavelength is so small compared to the wavelength of visible light (0.0024 nm compared to 400 to 750 nm), the change in the wavelength of a visible photon would be negligibly small.

9. For a given kinetic energy, an electron has less momentum than a proton and so has a longer wavelength.

11. As the kinetic energy increases, the momentum increases as well and so the de Broglie wavelength decreases.

13. The electron in the Bohr atom would be constantly accelerating, and so it would be constantly radiating. As the electron radiates, it would lose energy and spiral into the nucleus.

15. Because hydrogen only has one electron, it presents no complications to the orbit of an electron due to the interaction of other electrons.

17. In classical (pre-Bohr) physics the only quantity known to be quantized was electric charge, which was found only in multiples of e (the magnitude of the charge on the electron).

19. C
21. D
23. A
25. E
27. D

29. 1×10^{-30} J

31. 5.01×10^{-7} m = 501 nm; 5.99×10^{14} Hz

33. (a) 7.59×10^{-19} J; (b) 1.37×10^{-19} J

35. 9.85×10^{14} Hz

37. 6110 K

39. 6 min

41. (a) 1.21×10^{-27} kg·m/s = 2.25 eV/c
(b) 9.32×10^{-24} kg·m/s = 17,400 eV/c

43. 25.8°

45. (a) $\lambda = 0.140$ nm; $K = 0$ eV
(b) $\lambda = 0.14033$ nm; $K = 20.8$ eV
(c) $\lambda = 0.14071$ nm; $K = 44.7$ eV
(d) $\lambda = 0.1412$ nm; $K = 76.6$ eV
(e) $\lambda = 0.14243$ nm; $K = 151$ eV
(f) $\lambda = 0.14486$ nm; $K = 297$ eV

47. 0.137 nm

49. (a) $f = 4.22 \times 10^{18}$ Hz, $E = 17,440$ eV; (b) 17,500 eV;
(c) 16,900 eV; (d) 577 eV

51. 3.32×10^{-10} m

53. (a) 1.23×10^{-9} m; (b) 3.88×10^{-10} m; (c) 1.23×10^{-10} m;
(d) 3.88×10^{-11} m

55. 1.43×10^{-10} m

57. 656.21 nm; 486.08 nm; 434.00 nm; 410.13 nm

59. Rydberg formula with $n = 2$:

$$\frac{1}{\lambda} = R_H \left(\frac{1}{2^2} - \frac{1}{m^2} \right) = R_H \left(\frac{1}{4} - \frac{1}{m^2} \right) = R_H \left(\frac{m^2 - 4}{4m^2} \right)$$

$$\lambda = \frac{1}{R_H} \left(\frac{4m^2}{m^2 - 4} \right) = \frac{4}{1.09737 \times 10^7 \text{ m}^{-1}} \left(\frac{m^2}{m^2 - 4} \right)$$

$$= (3.6451 \times 10^{-7} \text{ m}) \left(\frac{m^2}{m^2 - 4} \right)$$

$$= (364.51 \text{ nm}) \left(\frac{m^2}{m^2 - 4} \right) \approx b \left(\frac{m^2}{m^2 - 4} \right)$$

61. 1.215×10^{-7} m = 121.5 nm
1.026×10^{-7} m = 102.6 nm
9.723×10^{-8} m = 97.23 nm
9.496×10^{-8} m = 94.96 nm
9.376×10^{-8} m = 93.76 nm
These are all ultraviolet wavelengths.

63. 1.875×10^{-6} m = 1875 nm
1.282×10^{-6} m = 1282 nm
1.094×10^{-6} m = 1094 nm
These are all infrared wavelengths.

65. $(8.2233 \times 10^{14} \text{ Hz}) \left(\frac{m^2 - 4}{m^2} \right)$

67. (a) $L_n = n\hbar$

(b) $r_n = \frac{n^2 \hbar^2}{m_e k e^2}$

(c) $K_n = \frac{m_e (k e^2)^2}{2n^2 \hbar^2}$

(d) $E_n = -K_n = -\frac{m_e (k e^2)^2}{2n^2 \hbar^2}$

(e) $v_n = \frac{k e^2}{n \hbar}$

69. (a) $L_n = n\hbar$

(b) $L_n = r_n m_e v_n$

$$v_n = \frac{L_n}{r_n m_e} = \frac{n\hbar}{\left(\frac{n^2}{Z} a_0 \right) m_e} = \frac{Z\hbar}{n a_0 m_e}$$

71. $\Delta f = 0.162$ MHz

73. 1.51 eV

75. (a) $n = 1$
State 1: $l = 0$; $m_l = 0$; $m_s = -1/2$
State 2: $l = 0$; $m_l = 0$; $m_s = +1/2$

(b)

n	l	m_l	m_s
2	0	0	$\pm 1/2$
2	1	1	$\pm 1/2$
2	1	0	$\pm 1/2$
2	1	-1	$\pm 1/2$

77. 23.9 nm

79. (a) 2.6×10^{-38} m/s; (b) no

81. 7.37×10^{-19} J; 6.63×10^{-19} J

83. 2×10^{12}

85. (a) 1.21×10^{-14} J; (b) 38.0%

87. 8.3 eV

89. (a) 9.9×10^{-20} J, 6.0×10^{-26} J; (b) electron

91. $\lambda = \frac{h}{mv} \sqrt{1 - \frac{v^2}{c^2}}$

93. (a) 92.2 cm; (b) 325 MHz

95. (a) 1.11×10^{-7} nm
(b) 2.22×10^{-7} nm, ~22,000 times smaller than from an electron
(c) electron: 6.82%; carbon nucleus: 0.0003%; wavelength shift from Compton scattering from carbon nuclei is extremely small and very difficult to measure; (d) 0.022 pm; 56 MeV

97. (a) $E = 21.3$ eV; (b) The first three cathode-grid voltages at which one expects to see a drop in anode current are the first three whole-number multiples of the voltage that would result in the energy gain in (a), thus 21.3 V, 42.6 V, and 63.9 V.

Chapter 27

1. Isotopes are nuclides with the same atomic number Z but different numbers of neutrons N. They have different mass numbers $A = Z + N$.

3. (a) "Neutron rich" means that heavier nuclei (larger than above $A = 40$) have more neutrons than protons. (b) Neutrons are important to help regulate the repulsive Coulomb force between the positive protons. Since the neutrons are neutral, they do not repel one another. However, they do contribute to the attractive nuclear forces between the nucleons.

5. A nucleus has less mass than the sum of its constituent protons and neutrons because some of that mass is converted to energy when the parts are combined. The released energy is the binding energy of the nucleus.

7. The nuclear force is very short ranged. Outside of a few femtometers, the nuclear force quickly goes to zero. Inside about 0.7 fm, the force is repulsive between nucleons, so the radius of the nucleus will never shrink to zero. The major competition inside the nucleus comes from the repulsive Coulomb force between the tightly packed, positively charged protons.

9. (a) Elements with a higher atomic number than iron are more likely to undergo fission. (b) Elements with a lower atomic number than iron are more likely to undergo fusion.

11. Consider the beta decay of a neutron: $n \rightarrow p + \beta^- + \bar{v}_e$. If the electron ($\beta^-$) were the only decay product, application of conservation of energy and momentum to the two-body decay would require that the β particle be ejected with a single unique energy. Instead we observe experimentally that β particles are produced with energies that range from zero to a maximum value.

13. As uranium begins to decay, the daughter nuclei are themselves often radioactive, decaying by various paths. So, at any instant there are isotopes from all the many possible decay modes of all the radioactive progeny of the parent nuclei.

15. There is 1 electron mass included in the atomic mass of the proton. There is no need to add in another one to account for the emitted electron.

17. No. All forms of radiation easily penetrate materials like fabric. For example, to something as tiny as an alpha particle, the weave of backpack fabric consists of vast, open spaces occasionally interrupted by strands of matter. Therefore, none of the radiation from the rock would be shielded.

19. A

21. E

23. C

25. C

27. C

29. (a) 1.7 fm; (b) 2.4 fm; (c) 3.6 fm; (d) 7.0 fm; (e) 5.6 fm; (f) 6.8 fm; (g) 6.9 fm; (h) 7.4 fm

31. 1.3×10^4 m

33. (a) 14 N; (b) 1.2×10^{-35} N; (c) no

35. $E_B = 92.162$ MeV

37. $\Delta E = 15.643$ MeV

39. Binding energy: 1103.33 MeV; binding energy per nucleon: 8.42235 MeV/nucleon

41. The missing product is 10n; $E_{released} = 89.951$ MeV.

43. (a) n; (b) ^{102}Zr (zirconium-102); (c) ^{250}Cm (curium-250); (d) ^{142}Cs (cesium-142)

45. 33 collisions

47. mass of uranium-235 required: 4.16×10^5 g $\approx$ 4.10×10^2 kg

49. 3×10^{19} reactions/s

51. (a) $E_{released} = 17.590$ MeV
(b) $E_{released} = -18.992$ MeV. Therefore, the energy required to initiate this reaction is 18.992 MeV.
(c) $E_{released} = 3.270$ MeV
(d) $E_{released} = 5.494$ MeV
(e) $E_{released} = 4.033$ MeV

53. 156 g $\approx$ 200 g (rounded to one significant figure)

55. (a) 4×10^6 Bq; (b) 25 Bq; (c) 4.5×10^{-7} Ci; (d) 4.53×10^{12} decays/minute

57. $\tau_{1/2} = 208$ s $= 0.00240$ d

59. $0.00298 \approx 0.30\%$ (rounded to two significant figures)

61. $R = 0.5$ μCi

63. age of the pole: 2.6×10^4 y

65. $N = 3.85 \times 10^6$ nuclei

67. (a) ^{234}Th (thorium-234); (b) α, ^{230}Ra (radium-230); (c) ^{240}Pu (plutonium-240); (d) ^{210}Tl (thallium-210)

69. (a) ^{131}I (iodine-131); (b) ^{145}Pm (promethium-145); (c) ^{24}Na* (sodium-24)

71. 6.2×10^7 atoms

73. (a) $r = 7.44$ fm $\approx$ 7.4 fm
(b) $F_{electric} = 1.04$ N $\approx$ 1.0 N
(c) $a = 6.2 \times 10^{26}$ m/s^2

75. (a) 36; (b) 36; (c) 37; (d) 37; (e) 39; (f) 38

77. Assuming the elemental composition of the body is 63% H, 26% O, 10% C, and 1% N, there would be about 2×10^{28} nuclei in a 50-kg body.

79. 5.9×10^9 y

81. 1.15×10^4 y

83. (a) ^{237}Np; (b) 0.26%, 8%; (c) 5.4×10^{-9} g

85. 8.14978 MeV/nucleon; ^{26}Al $\rightarrow$ e$^+$ + v_e + ^{26}Mg; 8.33159 MeV/nucleon; 8.30992 MeV/nucleon; ^{28}Al $\rightarrow$ e$^-$ + $\bar{v}_e$ + ^{28}Si

87. (a) 7.40 μCi; (b) 197 d $\times \frac{1 \text{ mo}}{30 \text{ d}} \approx$ 6.6 mo

89. (a) 3.79×10^{-8} g; (b) 8.99×10^{-4} cm = 8.99 μm

91. 4.599 MeV

Chapter 28

1. (a) Baryons are hadrons (particles that experience the strong interaction) that are made up of three quarks. (b) Mesons are hadrons made up of a quark and an antiquark. (c) Quarks are fractionally charged particles that make up hadrons (baryons and/or mesons). (d) Leptons are particles that with nonzero mass that do not experience the strong interaction. (e) An antiparticle (for example, a positron) has the same mass as the corresponding particle (for example, the electron) but has the opposite charge.

3. When two particles interact through one of the four fundamental forces (strong, weak, electromagnetic, or gravity), the interaction is mediated by an exchange particle that is transferred between the two interacting particles.

5. (a) The antiparticle, $\bar{n}$, is composed of one up-bar antiquark and two down-bar antiquarks: $\bar{u}\bar{d}\bar{d}$. (b) The antiparticle, $\bar{p}$, is composed of two up-bar antiquarks and one down-bar antiquark: $\bar{u}\bar{u}\bar{d}$. (c) The antiparticle, π^-, is composed of one down quark and one up-bar antiquark: $d\bar{u}$.

7. Once formed, a positron quickly meets an electron from the abundant supply of electrons in ordinary matter. The positron and electron then annihilate one another.

9. We know the universe is expanding because the spectra of distant galaxies are redshifted, meaning that they are receding away from us. This is known as the cosmological redshift. Furthermore, the farther away a galaxy is, the faster it is receding away from us. This means that the space between galaxies, and therefore the universe, is expanding.

11. B

13. D

15. A

17. C

19. Charge: A neutron has a charge of 0. The π^+ and μ^+ each have a charge of $+e$, and the π^- and μ^- each have a charge of $-e$. Charge is conserved.
Baryon number: A neutron has a baryon number of +1. Pions and muons each have a baryon number of 0, as

they are not baryons. The baryon number is not conserved.

Lepton number: Neither neutrons nor pions are leptons, so they have a lepton number of 0. A muon has a muon–lepton number equal to +1, and an antimuon has a muon–lepton number of −1. The lepton number is conserved. No, the reaction is not possible because the baryon number is not conserved.

21. Charge: A neutron and an antimatter electron neutrino have a charge of 0, a proton has a charge of +e, and an electron has a charge of −e. Charge is conserved.
Baryon number: A neutron and a proton each have a baryon number of +1. Electrons and neutrinos each have a baryon number of 0, as they are not baryons. The baryon number is conserved.
Electron–lepton number: Neither neutrons nor protons are leptons, so they have an electron–lepton number of 0. An electron has an electron–lepton number equal to +1, and an antimatter electron neutrino has an electron–lepton number of −1. The lepton number is not conserved.
No, the reaction is not possible because the lepton number is not conserved.

23. (a) Protons have a baryon number of +1 and a charge of +e. Antiprotons have a baryon number of −1 and a charge of −e. In this reaction charge and baryon number are both conserved, so this reaction is possible.
(b) Protons have a baryon number of +1 and a charge of +e. Neutrons have a baryon number of +1 and a charge of 0. Antineutrons have a baryon number of −1 and a charge of 0. In this reaction, charge and baryon number are both conserved, so this reaction is possible.
(c) Protons have a baryon number of +1 and a charge of +e. The K^+ particle has a baryon number of 0 and a charge of +e. In this reaction charge is conserved, but baryon number is not, so this reaction is not possible.

25. (a) Charge: A neutron and an electron antineutrino each have a charge of 0. A proton has a charge of +e, and an electron has a charge of −e. Charge is conserved.
Baryon number: A neutron and proton each have a baryon number of +1. An electron and an electron antineutrino each have a baryon number of 0. The baryon number is conserved.
Electron–lepton number: A neutron and a proton each have an electron–lepton number of 0. An electron has an electron–lepton number of +1. An electron antineutrino has an electron–lepton number of −1. The electron–lepton number is conserved.
This reaction is possible.
(b) Charge: An electron antineutrino and a muon neutrino each have a charge of 0. A muon and an electron each have a charge of −e. Charge is conserved.
Baryon number: All of the particles involved have a baryon number of 0. The baryon number is conserved.
Lepton number: An electron has an electron–lepton number of +1. An electron antineutrino has an

electron–lepton number of −1. The electron–lepton number is conserved. A muon and a muon neutrino each have a muon–lepton number of +1. The muon–lepton number is conserved.
This reaction is possible.
(c) Charge: A negatively charged pion and a muon each have a charge of −e. A muon antineutrino has a charge of 0. Charge is conserved.
Baryon number: All of the particles involved have a baryon number of 0. The baryon number is conserved.
Lepton number: A pion has a lepton number of 0. A muon has a muon–lepton number of +1. A muon antineutrino has a muon–lepton number of −1. The muon–lepton number is conserved.
This reaction is possible.

27. K_p = 67.50 MeV

29. E_{total} = 51 MeV

31. $K_{\gamma 2} = 3.6 \times 10^2$ MeV

33.

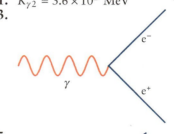

35.

37. No. The Higgs particle has to have zero charge. This collection of particles will have a net negative charge.

39. 1500 km/s

41. (a) 300 particles/m³; (b) The probability of randomly selecting a particle in a cubic meter of air and having it be a (hypothetical) dark matter particle is 1/10,000 as great as the probability of selecting a particular grain of sand randomly from all the grains of sand on all the beaches on Earth.

43. (a) mass necessary to produce this much energy: m = 165 kg
(b) L = 0.219 m = 21.9 cm

45. $t_P = \sqrt{\dfrac{G\hbar}{c^5}}$
$t_P = 5.3912 \times 10^{-44}$ s

47. (a) The energy of each photon is 62.5 GeV. Converting this energy to joules, we obtain 1.00×10^{-8} J. (b) The wavelength of each photon is 1.99×10^{-17} m. (c) These photons are gamma rays.

49. (a) strong; (b) weak; (c) electromagnetic; (d) strong; (e) weak

INDEX

Note: Page numbers preceded by A indicate appendices; those preceded by M indicate the Math Tutorial.

T

FORMULAS FROM GEOMETRY

Triangle:

$h = a \sin \theta$

$\text{Area} = \dfrac{1}{2}bh$

$c^2 = a^2 + b^2 - 2ab \cos \theta$ (Law of Cosines)

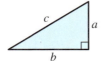

Right Triangle:

Pythagorean Theorem
$c^2 = a^2 + b^2$

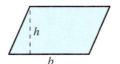

Equilateral Triangle:

$h = \dfrac{\sqrt{3}s}{2}$

$\text{Area} = \dfrac{\sqrt{3}s^2}{4}$

Parallelogram:

$\text{Area} = bh$

Trapezoid:

$\text{Area} = \dfrac{h}{2}(a + b)$

Circle:

$\text{Area} = \pi r^2$

$\text{Circumference} = 2\pi r$

Sector of Circle:

$\text{Area} = \dfrac{\theta r^2}{2}$

$s = r\theta$

θ in radians

Circular Ring:

$\text{Area} = \pi(R^2 - r^2)$

$\quad\quad\quad = 2\pi pw$

$p = $ average radius,

$w = $ width of ring

Sector of Circular Ring:

$\text{Area} = \theta pw$

$p = $ average radius,

$w = $ width of ring,

θ in radians

Ellipse:

$\text{Area} = \pi ab$

$\text{Circumference} \approx 2\pi \sqrt{\dfrac{a^2 + b^2}{2}}$

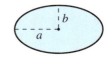

Cone:

$\text{Volume} = \dfrac{Ah}{3}$

$A = $ area of base

Right Circular Cone:

$\text{Volume} = \dfrac{\pi r^2 h}{3}$

$\text{Lateral Surface Area} = \pi r \sqrt{r^2 + h^2}$

Frustum of Right Circular Cone:

$\text{Volume} = \dfrac{\pi(r^2 + rR + R^2)h}{3}$

$\text{Lateral Surface Area} = \pi s(R + r)$

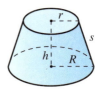

Right Circular Cylinder:

$\text{Volume} = \pi r^2 h$

$\text{Lateral Surface Area} = 2\pi rh$

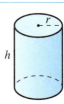

Sphere:

$\text{Volume} = \dfrac{4}{3}\pi r^3$

$\text{Surface Area} = 4\pi r^2$

Wedge:

$A = B \sec \theta$

$A = $ area of upper face,

$B = $ area of base

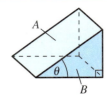

ALGEBRA

Factors and Zeros of Polynomials:

Given the polynomial $p(x) = a_n x^n + a_{n-1} x^{n-1} + \cdots + a_1 x + a_0$. If $p(b) = 0$, then b is a *zero* of the polynomial and a *solution* of the equation $p(x) = 0$. Furthermore, $(x - b)$ is a *factor* of the polynomial.

Fundamental Theorem of Algebra:

If $f(x)$ is a polynomial function of degree n, where $n > 0$, then f has at least one zero in the complex number system.

Quadratic Formula:

If $p(x) = ax^2 + bx + c$, $a \neq 0$ and $b^2 - 4ac \geq 0$, then the real zeros of p are $x = \left(-b \pm \sqrt{b^2 - 4ac}\right)/2a$.

Example

If $p(x) = x^2 + 3x - 1$, then $p(x) = 0$ when

$$x = \frac{-3 \pm \sqrt{13}}{2}.$$

Special Factors:

$x^2 - a^2 = (x - a)(x + a)$

$x^3 - a^3 = (x - a)(x^2 + ax + a^2)$

$x^3 + a^3 = (x + a)(x^2 - ax + a^2)$

$x^4 - a^4 = (x - a)(x + a)(x^2 + a^2)$

$x^4 + a^4 = \left(x^2 + \sqrt{2}ax + a^2\right)\left(x^2 - \sqrt{2}ax + a^2\right)$

$x^n - a^n = (x - a)(x^{n-1} + ax^{n-2} + \cdots + a^{n-1})$

$x^n + a^n = (x + a)(x^{n-1} - ax^{n-2} + \cdots + a^{n-1})$, for n odd

$x^{2n} - a^{2n} = (x^n - a^n)(x^n + a^n)$

Examples

$x^2 - 9 = (x - 3)(x + 3)$

$x^3 - 8 = (x - 2)(x^2 + 2x + 4)$

$x^3 + 4 = \left(x + \sqrt[3]{4}\right)\left(x^2 - \sqrt[3]{4}x + \sqrt[3]{16}\right)$

$x^4 - 4 = \left(x - \sqrt{2}\right)\left(x + \sqrt{2}\right)(x^2 + 2)$

$x^4 + 4 = (x^2 + 2x + 2)(x^2 - 2x + 2)$

$x^5 - 1 = (x - 1)(x^4 + x^3 + x^2 + x + 1)$

$x^7 + 1 = (x + 1)(x^6 - x^5 + x^4 - x^3 + x^2 - x + 1)$

$x^6 - 1 = (x^3 - 1)(x^3 + 1)$

Binomial Theorem:

$(x + a)^2 = x^2 + 2ax + a^2$

$(x - a)^2 = x^2 - 2ax + a^2$

$(x + a)^3 = x^3 + 3ax^2 + 3a^2 x + a^3$

$(x - a)^3 = x^3 - 3ax^2 + 3a^2 x - a^3$

$(x + a)^4 = x^4 + 4ax^3 + 6a^2 x^2 + 4a^3 x + a^4$

$(x - a)^4 = x^4 - 4ax^3 + 6a^2 x^2 - 4a^3 x + a^4$

$(x + a)^n = x^n + nax^{n-1} + \dfrac{n(n-1)}{2!}a^2 x^{n-2} + \cdots + na^{n-1}x + a^n$

$(x - a)^n = x^n - nax^{n-1} + \dfrac{n(n-1)}{2!}a^2 x^{n-2} - \cdots \pm na^{n-1}x \mp a^n$

Examples

$(x + 3)^2 = x^2 + 6x + 9$

$(x^2 - 5)^2 = x^4 - 10x^2 + 25$

$(x + 2)^3 = x^3 + 6x^2 + 12x + 8$

$(x - 1)^3 = x^3 - 3x^2 + 3x - 1$

$\left(x + \sqrt{2}\right)^4 = x^4 + 4\sqrt{2}x^3 + 12x^2 + 8\sqrt{2}x + 4$

$(x - 4)^4 = x^4 - 16x^3 + 96x^2 - 256x + 256$

$(x + 1)^5 = x^5 + 5x^4 + 10x^3 + 10x^2 + 5x + 1$

$(x - 1)^6 = x^6 - 6x^5 + 15x^4 - 20x^3 + 15x^2 - 6x + 1$

Rational Zero Test:

If $p(x) = a_n x^n + a_{n-1} x^{n-1} + \cdots + a_1 x + a_0$ has integer coefficients, then every *rational* zero of $p(x) = 0$ is of the form $x = r/s$, where r is a factor of a_0 and s is a factor of a_n.

Example

If $p(x) = 2x^4 - 7x^3 + 5x^2 - 7x + 3$, then the only possible *rational* zeros are $x = \pm 1, \pm \frac{1}{2}, \pm 3$, and $\pm \frac{3}{2}$. By testing, you find the two rational zeros to be $\frac{1}{2}$ and 3.

Factoring by Grouping:

$acx^3 + adx^2 + bcx + bd = ax^2(cx + d) + b(cx + d)$
$= (ax^2 + b)(cx + d)$

Example

$3x^3 - 2x^2 - 6x + 4 = x^2(3x - 2) - 2(3x - 2)$
$= (x^2 - 2)(3x - 2)$